CAMBRIDGE LIBRARY COLLECTION

Books of enduring scholarly value

Physical Sciences

From ancient times, humans have tried to understand the workings of the world around them. The roots of modern physical science go back to the very earliest mechanical devices such as levers and rollers, the mixing of paints and dyes, and the importance of the heavenly bodies in early religious observance and navigation. The physical sciences as we know them today began to emerge as independent academic subjects during the early modern period, in the work of Newton and other 'natural philosophers', and numerous sub-disciplines developed during the centuries that followed. This part of the Cambridge Library Collection is devoted to landmark publications in this area which will be of interest to historians of science concerned with individual scientists, particular discoveries, and advances in scientific method, or with the establishment and development of scientific institutions around the world.

A Mathematical and Philosophical Dictionary

Born into a Newcastle coal mining family, Charles Hutton (1737–1823) displayed mathematical ability from an early age. He rose to become professor of mathematics at the Royal Military Academy and foreign secretary of the Royal Society. First published in 1795–6, this two-volume illustrated encyclopaedia aimed to supplement the great generalist reference works of the Enlightenment by focusing on philosophical and mathematical subjects; the coverage ranges across mathematics, astronomy, natural philosophy and engineering. Almost a century old, the last comparable reference work in English was John Harris' *Lexicon Technicum*. Hutton's work contains manyhistorical and biographical entries, often with bibliographies, including many for continental analytical mathematicians who would have been relatively unfamiliar to British readers. These features make Hutton's *Dictionary* a particularly valuable record of eighteenth-century science and mathematics. Volume 1 ranges from *abacist* (a user of an abacus) to the English physician and Newtonian scientist James Jurin.

Cambridge University Press has long been a pioneer in the reissuing of out-of-print titles from its own backlist, producing digital reprints of books that are still sought after by scholars and students but could not be reprinted economically using traditional technology. The Cambridge Library Collection extends this activity to a wider range of books which are still of importance to researchers and professionals, either for the source material they contain, or as landmarks in the history of their academic discipline.

Drawing from the world-renowned collections in the Cambridge University Library and other partner libraries, and guided by the advice of experts in each subject area, Cambridge University Press is using state-of-the-art scanning machines in its own Printing House to capture the content of each book selected for inclusion. The files are processed to give a consistently clear, crisp image, and the books finished to the high quality standard for which the Press is recognised around the world. The latest print-on-demand technology ensures that the books will remain available indefinitely, and that orders for single or multiple copies can quickly be supplied.

The Cambridge Library Collection brings back to life books of enduring scholarly value (including out-of-copyright works originally issued by other publishers) across a wide range of disciplines in the humanities and social sciences and in science and technology.

A Mathematical
and
Philosophical Dictionary

*Containing an Explanation of the Terms,
and an Account of the Several Subjects,
Comprized under the Heads Mathematics, Astronomy,
and Philosophy, Both Natural and Experimental*

VOLUME 1

CHARLES HUTTON

CAMBRIDGE
UNIVERSITY PRESS

CAMBRIDGE
UNIVERSITY PRESS

University Printing House, Cambridge, CB2 8BS, United Kingdom

Cambridge University Press is part of the University of Cambridge.

It furthers the University's mission by disseminating knowledge in the pursuit of education, learning and research at the highest international levels of excellence.

www.cambridge.org
Information on this title: www.cambridge.org/9781108077705

© in this compilation Cambridge University Press 2015

This edition first published 1796
This digitally printed version 2015

ISBN 978-1-108-07770-5 Paperback

A

MATHEMATICAL AND PHILOSOPHICAL

DICTIONARY.

VOL. I.

A

MATHEMATICAL and PHILOSOPHICAL
D I C T I O N A R Y:

CONTAINING

AN EXPLANATION OF THE TERMS, AND AN ACCOUNT OF THE SEVERAL SUBJECTS,

COMPRIZED UNDER THE HEADS

MATHEMATICS, ASTRONOMY, AND PHILOSOPHY

BOTH NATURAL AND EXPERIMENTAL:

WITH AN

HISTORICAL ACCOUNT OF THE RISE, PROGRESS, AND PRESENT STATE OF THESE SCIENCES:

ALSO

MEMOIRS OF THE LIVES AND WRITINGS OF THE MOST EMINENT AUTHORS,

BOTH ANCIENT AND MODERN,

WHO BY THEIR DISCOVERIES OR IMPROVEMENTS HAVE CONTRIBUTED TO THE ADVANCEMENT OF THEM.

IN TWO VOLUMES.

WITH MANY CUTS AND COPPER-PLATES.

By CHARLES HUTTON, LL.D.

F. R. SS. OF LONDON AND EDINBURGH, AND OF THE PHILOSOPHICAL SOCIETIES OF HAARLEM AND AMERICA;
AND PROFESSOR OF MATHEMATICS IN THE ROYAL MILITARY ACADEMY, WOOLWICH.

VOL. I.

LONDON:

PRINTED BY J. DAVIS,

FOR J. JOHNSON, IN ST. PAUL'S CHURCH-YARD; AND G. G. AND J. ROBINSON,
IN PATERNOSTER-ROW.

M.DCC.XCVI.

P R E F A C E.

AMONG the Dictionaries of Arts and Sciences which have been published, of late years, in various parts of Europe, it is matter of surprise that Philosophy and Mathematics should have been so far overlooked as not to be thought worthy of a separate Treatise, in this form. These Sciences constitute a large portion of the present stock of human knowledge, and have been usually considered as possessing a degree of importance to which few others are entitled; and yet we have hitherto had no distinct Lexicon, in which their constituent parts and technical terms have been explained, with that amplitude and precision, which the great improvements of the Moderns, as well as the rising dignity of the Subject, seem to demand.

THE only works of this kind in the English language, deserving of notice are Harris's Lexicon Technicum, and Stone's Mathematical Dictionary; the former of which, though a valuable performance at the time it was written, is now become too dry and obsolete to be referred to with pleasure or satisfaction : and the latter, consisting only of one volume in 8vo, must be regarded merely as an unfinished sketch, or brief compendium, extremely limited in its plan, and necessarily deficient in useful information.

IT became, therefore, the only resource of the Reader, in many cases where explanation was wanted, to have recourse to Chambers's Dictionary, in four large Volumes folio, or to the Encyclopædia Britannica, now in eighteen large volumes 4to, or the still more stupendous performance of the French Encyclopedists; and even here his expectations might be frequently disappointed. These great and useful works, aiming at a general comprehension of the whole circle of the Sciences, are sometimes very deficient in their descriptions of particular branches; it being almost impossible, in such

extensive

P R E F A C E.

extenfive undertakings, to appreciate, with exactnefs, the due value of every article: They are, befides, fo voluminous and heterogeneous in their nature, as to render a frequent reference to them extremely inconvenient; and even if this were not the cafe, their high price puts them out of the reach of the generality of readers.

WITH a view to obviate thefe defects, the Public are here prefented with a Dictionary of a moderate fize and price, which is devoted folely to Philofophical and Mathematical fubjects. It is a work for which materials have been collecting through a courfe of many years; and is the refult of great labour and reading. Not only moft of the Encyclopedias already extant, and the various publications of the Learned Societies throughout Europe, have been carefully confulted, but alfo all the original works, of any reputation, which have hitherto appeared upon thefe fubjects, from the earlieft writers down to the prefent times.

FROM the latter of thefe fources, in particular, a confiderable portion of information has been obtained, which the curious reader will find, in many cafes, to be highly interefting and important. The Hiftory of Algebra, for inftance, which is detailed at confiderable length in the Firft Volume, under the head-of that Article, will afford fufficient evidence to fhew in what a fuperficial and partial way the inquiry has been hitherto invefligated, even by profeffed writers on the fubject; the principal of whom are M. Montucla, our countryman the celebrated Dr. Wallis, and the Abbé De Gua, a late French author, who has pretended to correct the Doctor's errors and mifreprefentations.

REGULAR hiftorical details are in like manner given of the origin and progrefs of each of thefe Sciences, as well as of the inventions and improvements by which they have been gradually brought from their firft rude beginnings to their prefent advanced ftate.

IT is alfo to be obferved, that befides the articles common to the generality of Dictionaries of this kind, an interefting Biographical Account is here introduced of the moft celebrated Philofophers and Mathematicians, both ancient and modern; among which will be found the Lives of many eminent characters, who have hitherto been either wholly overlooked, or very imperfectly recorded. Complete lifts of their works are alfo fubjoined to each Article, where they could be procured; which cannot but prove highly acceptable to that clafs of readers, who are defirous of obtaining the moft fatisfactory information upon the fubjects of their particular enquiries and purfuits. On the head of Biography however the Author has ftill to lament the want of many other refpectable names which he was defirous to have added

4

to

to his lift of authors, not having been able to procure any circumftantial accounts of their lives. He could have wifhed to have comprifed in his lift, the lives of all fuch public literary charaƈters as the Univerfity Profeffors of Aftronomy, Philofophy and Mathematics, as well as thofe of the other more refpeƈtable clafs of Authors on thofe Sciences. He will therefore thankfully receive the communication of any fuch memoirs from the hands of gentlemen poffeffed of them ; as well as hints and information on fuch ufeful improvements in the fciences as may have been overlooked in this Dictionary, or any articles that may here have been imperfeƈtly or incorreƈtly treated ; that he may at fome future time, by adding them to this work, render it ftill more complete and deferving the public notice.

As this work is an attempt to feparate the words in the fciences of Aftronomy, Mathematics, and Philofophy, from thofe of other arts or fciences, in feveral of which there are already feparate Diƈtionaries ; as in Chemiftry, Geography, Mufic, Marine and Naval affairs, &c ; words fometimes occurred which it was rather doubtful whether they could be confidered as properly belonging to the prefent work or not ; in which cafe many of fuch words have been here inferted. But fuch as appeared clearly and peculiarly to belong to any of thofe other fubjeƈts, have been either wholly omitted, or elfe have had a very fhort account only given of them. The readers of this work therefore, recolleƈting that it is not a General Diƈtionary of all the Arts and Sciences, will not expeƈt to find all forts of words and fubjeƈts here treated of ; but fuch only as peculiarly appertain to the proper matter of the work. And therefore, although fome few words may inadvertently have been omitted ; yet when the Reader does not immediately find every word which he wifhes to confult, he will not always confider them as omiffions of the Author, but for the moft part as relating to fome other fcience foreign to this Diƈtionary.

In all cafes where it could be conveniently done, the neceffary figures and diagrams are inferted in the fame page with the fubjeƈts which they are defigned to elucidate ; a method which will be found much more commodious than that of putting them in feparate plates at the end of each volume, but, which has added very confiderably to the expence of the undertaking : where the fubjeƈts are of fuch a nature that they could not be otherwife well reprefented, they are engraved on Copperplates.

As the whole of this work was written before it was put to the prefs, the Reader will find it of an equal and uniform nature and conftruƈtion throughout ; in which refpeƈt many publications of this kind are very defeƈtive, from the fubjeƈts being diffufely treated under the firft letters of the alphabet, while articles

of

of equal importance in the latter part are ſo much abridged as to be rendered almoſt uſeleſs, in order that the whole might be compriſed in a limited number of
ſheets, according to propoſals made before the works were compoſed. The preſent Dictionary having been completed without any of theſe unfavourable circumſtances, will be found in moſt caſes equally inſtructive and uſeful, and may be
conſulted with no leſs advantage by the Man of Science than the Student.

A M A.

A

PHILOSOPHICAL and MATHEMATICAL

DICTIONARY.

―――――――――――

A.

ABACIST, an Arithmetician. In this fenfe we find the word ufed by William of Malmefbury, in his Hiftory *de Geftis Anglorum*, written about the year 1150; where he fhews that one Gerbert, a learned monk of France, who was afterwards made pope of Rome in the year 998 or 999, by the name of Silvefter the 2d, was the firft who got from the Saracens the abacus, and that he taught fuch rules concerning it, as the Abacifts themfelves could hardly underftand.

ABACUS, *in Arithmetic*, an ancient inftrument ufed by moft nations for cafting up accounts, or performing arithmetical calculations: it is by fome derived from the Greek αϐαξ, which fignifies a cupboard or beaufet, perhaps from the fimilarity of the form of this inftrument; and by others it is derived from the Phœnician *abak*, which fignifies duft or powder, becaufe it was faid that this inftrument was fometimes made of a fquare board or tablet, which was powdered over with fine fand or duft, in which were traced the figures or characters ufed in making calculations, which could thence be eafily defaced, and the abacus refitted for ufe. But Lucas Paciolus, in the firft part of his fecond diftinction, thinks it is a corruption of Arabicus, by which he meant their Algorifm, or the method of numeral computation received from them.

We find this inftrument for computation in ufe, under fome variations, with moft nations, as the Greeks, Romans, Germans, French, Chinefe, &c.

The Grecian abacus was an oblong frame, over which were ftretched feveral brafs wires, ftrung with little ivory balls, like the beads of a necklace; by the various arrangements of which all kinds of computa-

tions were eafily made. Mahudel, in Hift. Acad. R. Infcr. t. 3. p. 390.

The Roman Abacus was a little varied from the Grecian, having pins fliding in grooves, inftead of ftrings or wires and beads. Philof. Tranf. No. 180.

The Chinefe Abacus, or Shwan-pan, like the Grecian, confifts of feveral feries of beads ftrung on brafs wires, ftretched from the top to the bottom of the inftrument, and divided in the middle by a crofs piece from fide to fide. In the upper fpace every ftring has two beads, which are each counted for 5; and in the lower fpace every ftring has five beads, of different values, the firft being counted as 1, the fecond as 10, the third as 100, and fo on, as with us. See SHWAN-PAN.

The Abacus chiefly ufed in European countries, is nearly upon the fame principles, though the ufe of it is here more limited, becaufe of the arbitrary and unequal divifions of money, weights, and meafures, which, in China, are all divided in a tenfold proportion, like our fcale of common numbers. This is made by drawing any number of parallel lines, like paper ruled for mufic, at fuch a diftance as may be at leaft equal to twice the diameter of a calculus, or counter. Then the value of thefe lines, and of the fpaces between them, increafes, from the loweft to the higheft, in a tenfold proportion. Thus, counters placed upon the firft line, fignify fo many units or ones; on the fecond line 10's, on the third line 100's, on the fourth line 1000's, and fo on: in like manner a counter placed in the firft fpace, between the firft and fecond line, denotes 5, in the fecond fpace 50, in the third fpace 500, in the fourth fpace 5000,

B   and

and fo on. So that there are never more than four counters placed on any line, nor more than one placed in any fpace, this being of the fame value as five counters on the next line below. So the counters on the Abacus, in the figure here below, exprefs the number or fum 47382.

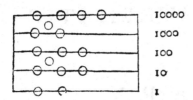

Befides the above inftruments of computation, there have been feveral others invented by different perfons; as *Napier's rods* or *bones*, defcribed in his Rabdologia, which fee under the word NAPIER; alfo the *Abacus Rhabdologicus,* a variation of Napier's, which is defcribed in the firft vol. of *Machines et Inventions approuvées par l'Academie Royale des Sciences.* An ingenious and general one was alfo invented by Mr. Gamaliel Smethurft, and is defcribed in the Philofophical Tranfactions, vol. 46; where the inventor remarks that computations by it are much quicker and eafier than by the pen, are lefs burthenfome to the memory, and can be performed by blind perfons, or in the dark as well as in the light. A very comprehenfive inftrument of this kind was alfo contrived by the late learned Dr. Nicholas Saunderfon, by which he performed very intricate calculations: an account of it is prefixed to the firft volume of his Algebra, and it is there by the editor called *Palpable Arithmetic:* which fee.

ABACUS, *Pythagorean,* fo denominated from its inventor, Pythagoras; a table of numbers, contrived for readily learning the principles of arithmetic; and was probably what we now call the multiplication-table.

ABACUS, or ABACISCUS, in *Architecture,* the upper part or member of the capital of a column; ferving as a crowning both to the capital and to the whole column. Vitruvius informs us that the *Abacus* was originally intended to reprefent a fquare flat tile laid over an urn, or a bafket; and the invention is afcribed to Calimachus, an ingenious ftatuary of Athens, who, it is faid, adopted it on obferving a fmall bafket, covered with a tile, over the root of an Acanthus plant, which grew on the grave of a young lady; the plant fhooting up, encompaffed the bafket all around, till meeting with the tile, it curled back in the form of fcrolls: Calimachus paffing by, took the hint, and immediately executed a capital on this plan; reprefenting the tile by the Abacus, the leaves of the acanthus by the volutes or fcrolls, and the bafket by the vafe or body of the capital. See ACANTHUS.

Abacus is alfo ufed by Scamozzi for a concave moulding in the capital of the Tufcan pedeftal. And the word is ufed by Palladio for other members which he defcribes. Alfo, in the ancient architecture, the fame term is ufed to denote certain compartments in the incruftation or lining of the walls of ftate-rooms, mofaicpavements, and the like. There were *Abaci* of marble,

porphyry, jafper, alabafter, and even glafs; varioufly fhaped, as fquare, triangular, and fuch like.

ABACUS *Logifticus* is a right angled triangle, whofe fides, about the right angle, contain all the numbers from 1 to 60; and its area the products of each two of the oppofite numbers. This is alfo called a *canon of fexagefimals,* and is no other than a multiplication-table carried to 60 both ways.

ABACUS & *Palmulæ,* in the Ancient Mufic, denote the machinery by which the ftrings of the polyplectra, or inftruments of many ftrings, were ftruck, with a plectrum made of quills.

ABACUS *Harmonicus* is ufed by Kircher for the ftructure and difpofition of the keys of a mufical inftrument, either to be touched with the hands or feet.

ABACUS, in *Geometry,* a table or flate upon which fchemes or diagrams are drawn.

ABAS, a weight ufed in Perfia for weighing pearls; and is an eighth part lighter than the European carat.

ABASSI, a filver coin current in Perfia, deriving its name from Schaw Abbas II. King of Perfia, and is worth near eighteen pence Englifh money.

ABATIS, or ABATTIS, from the French *abattre,* to throw down, or beat down, in the Military Art, denotes a kind of retrenchment made by a quantity of whole trees cut down, and laid lengthways befide each other, the clofer the better, having all their branches pointed towards the enemy, which prevents his approach, at the fame time that the trunks ferve as a breaft-work before the men. The Abattis is a very ufeful work on moft occafions, efpecially on fudden emergencies, when trees are near at hand; and has always been practifed with confiderable fuccefs, by the ableft commanders in all ages and nations.

ABBREVIATE; to abbreviate fractions in arithmetic and algebra, is to leffen proportionally their terms, or the numerator and denominator; which is performed by dividing thofe terms by any number or quantity, which will divide them without leaving a remainder. And when the terms cannot be any farther fo divided, the fraction is faid to be in its leaft terms.

So $\frac{16}{24} = \frac{8}{12} = \frac{4}{6} = \frac{2}{3}$,

by dividing the terms continually by 2.

And $\frac{294}{504} = \frac{147}{252} = \frac{49}{84} = \frac{7}{12}$,

by dividing by 2, 3, and 7.

Alfo $\frac{3 \times 8 \times 5}{6} = \frac{8 \times 5}{2} = 4 \times 5 = 20$,

by dividing by 3 and by 2.

And $\frac{12abx^2}{4acx} = \frac{3bx}{c}$, by dividing by 4 ax.

And $\frac{ab^2 + b^2 x}{ax + x^2} = \frac{b^2}{x}$, by dividing by $a + x$.

ABBREVIATION, of fractions, in Arithmetic and Algebra, the reducing them to lower terms.

ABERRATION, in *Aftronomy,* an apparent motion of the celeftial bodies, occafioned by the progreffive motion of light, and the earth's annual motion in her orbit.

This effect may be explained a dnfamiliarized by thn

motion

motion of a line parallel to itſelf, much after the manner that the compoſition and reſolution of forces are explained. If light have a progreſſive motion, let the proportion of its velocity to that of the earth in her orbit, be as the line B C to the line A C; then, by the compoſition of theſe two motions, the particle of light will ſeem to deſcribe the line B A or D C, inſtead of its real courſe B C; and will appear in the direction A B or C D, inſtead of its true direction C B. So that if A B repreſent a tube, carried with a parallel motion by an obſerver along the line A C, in the time that a particle of light would move over the ſpace B C,

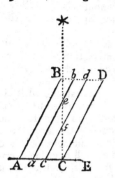

the different places of the tube being A B, *ab*, *cd*, C D; and when the eye, or end of the tube, is at A, let a particle of light enter the other end at B; then when the tube is at *a b*, the particle of light will be at *e*, exactly in the axis of the tube; and when the tube is at *c d*, the particle of light will arrive at *f*, ſtill in the axis of the tube; and laſtly, when the tube arrives at C D, the particle of light will arrive at the eye or point C, and conſequently will appear to come in the direction D C of the tube, inſtead of the true direction B C. And ſo on, one particle ſucceeding another, and forming a continued ſtream or ray of light in the apparent direction D C. So that the apparent angle made by the ray of light with the line A E, is the angle D C E, inſtead of the true angle B C E; and the difference, B C D or A B C, is the quantity of the aberration.

M. de Maupertuis, in his Elements of Geography, gives alſo a familiar and ingenious idea of the aberration, in this manner: " It is thus," ſays he, " concerning the direction in which a gun muſt be pointed to ſtrike a bird in its flight; inſtead of pointing it ſtraight to the bird, the fowler will point a little before it, in the path of its flight, and that ſo much the more as the flight of the bird is more rapid, with reſpect to the flight of the ſhot." In this way of conſidering the matter, the flight of the bird repreſents the motion of the earth, or the line A C, in our ſcheme above, and the flight of the ſhot repreſents the motion of the ray of light, or the line B C.

Mr. Clairaut too, in the Memoires of the Academy of Sciences for the year 1746, illuſtrates this effect in a familiar way, by ſuppoſing drops of rain to fall rapidly and quickly after each other from a cloud, under which a perſon moves with a very narrow tube; in which caſe it is evident that the tube muſt have a certain inclination, in order that a drop which enters at the top, may fall freely through the axis of the tube, without touching the ſides of it; which inclination muſt be more or leſs according to the velocity of the drops in reſpect to that of the tube: then the angle made by the direction of the tube and of the falling drops, is the aberration ariſing from the combination of thoſe two motions.

This diſcovery, which is one of the brighteſt that have been made in the preſent age, we owe to the accuracy and ingenuity of the late Dr. Bradley, Aſtronomer Royal; to which he was occaſionally led by the reſult

of ſome accurate obſervations which he had made with another view, namely, to determine the annual parallax of the fixed ſtars, or that which ariſes from the motion of the earth in its annual orbit about the ſun.

The annual motion of the earth about the ſun had been much doubted, and warmly conteſted. The defenders of that motion, among other proofs of the reality of it, conceived the idea of adducing an inconteſtable one from the annual parallax of the fixed ſtars, if the ſtars ſhould be within ſuch a diſtance, or if inſtruments and obſervations could be made with ſuch accuracy, as to render that parallax ſenſible. And with this view various attempts have been made. Before the obſervations of M. Picard, made in 1672, it was the general opinion, that the ſtars did not change their poſition during the courſe of a year. Tycho Brahe and Ricciolus fancied that they had aſſured themſelves of it from their obſervations; and from thence they concluded that the earth did not move round the ſun, and that there was no annual parallax in the fixed ſtars. M. Picard, in the account of his *Voyage d' Uranibourg*, made in 1672, ſays that the pole ſtar, at different times of the year, has certain variations which he had obſerved for about 10 years, and which amounted to about 40″ a year: from whence ſome who favoured the annual motion of the earth were led to conclude that theſe variations were the effect of the parallax of the earth's orbit. But it was impoſſible to explain it by that parallax; becauſe this motion was in a manner contrary to what ought to follow only from the motion of the earth in her orbit.

In 1674 Dr. Hook publiſhed an account of obſervations which he ſaid he had made in 1669, and by which he had found that the ſtar γ Draconis was 23′ more northerly in July than in October: obſervations which, for the preſent, ſeemed to favour the opinion of the earth's motion, although it be now known that there could not be any truth or accuracy in them.

Flamſteed having obſerved the pole ſtar with his mural quadrant, in 1689 and the following years, found that its declination was 40″ leſs in July than in December; which obſervations, although very juſt, were yet however improper for proving the annual parallax; and he recommended the making of an inſtrument of 15 or 20 feet radius, to be firmly fixed on a ſtrong foundation, for deciding a doubt which was otherwiſe not ſoon likely to be brought to a concluſion.

In this ſtate of uncertainty and doubt, then, Dr. Bradley, in conjunction with Mr. Samuel Molineux, in the year 1725, formed the project of verifying, by a ſeries of new obſervations, thoſe which Dr. Hook had communicated to the public almoſt 50 years before. And as it was his attempt that chiefly gave riſe to this, ſo it was his method in making the obſervations, in ſome meaſure, that they followed; for they made choice of the ſame ſtar, and their inſtrument was conſtructed upon nearly the ſame principles: but had it not greatly exceeded the former in exactneſs, they might ſtill have continued in great uncertainty as to the parallax of the fixed ſtars. And this was chiefly owing to the accuracy of the ingenious Mr. George Graham, to whom the lovers of aſtronomy are alſo indebted for ſeveral other exact and convenient inſtruments.

The succefs then of the intended experiment, evidently depending very much on the accuracy of the inftrument, that leading object was firft to be well fecured. Mr. Molineux's apparatus then having been completed, and fitted for obferving, about the end of November 1725, on the third day of December following, the bright ftar in the head of Draco, marked γ by Bayer, was for the firft time obferved, as it paffed near the zenith, and its fituation carefully taken with the inftrument. The like obfervations were made on the fifth, eleventh, and twelfth days of the fame month; and there appearing no material difference in the place of the ftar, a farther repetition of them, at that feafon, feemed needlefs, it being a time of the year in which no fenfible alteration of parallax, in this ftar, could foon be expected. It was therefore curiofity that chiefly urged Dr. Bradley, being then at Kew, where the inftrument was fixed, to prepare for obferving the ftar again on the 17th of the fame month; when, having adjufted the inftrument as ufual, he perceived that it paffed a little more foutherly this day than it had done before. Not fufpecting any other caufe of this appearance, they firft concluded that it was owing to the uncertainty of the obfervations, and that either this, or the foregoing, was not fo exact as they had before fuppofed. For which reafon they propofed to repeat the obfervation again, to determine from what caufe this difference might proceed: and upon doing it, on the 20th of December, the doctor found that the ftar paffed ftill more foutherly than at the preceding obfervation. This fenfible alteration furprifed them the more, as it was the contrary way from what it would have been, had it proceeded from an annual parallax of the ftar. But being now pretty well fatisfied, that it could not be entirely owing to the want of exactnefs in the obfervations, and having no notion of any thing elfe that could caufe fuch an apparent motion as this in the ftar; they began to fufpect that fome change in the materials, or fabric of the inftrument itfelf, might have occafioned it. Under thefe uncertainties they remained for fome time; but being at length fully convinced, by feveral trials, of the great exactnefs of the inftrument; and finding, by the gradual increafe of the ftar's diftance from the pole, that there muft be fome regular caufe that produced it; they took care to examine very nicely, at the time of each obfervation, how much the variation was; till about the beginning of March 1726, the ftar was found to be 20″ more foutherly than at the time of the firft obfervation: it now indeed feemed to have arrived at its utmoft limit fouthward, as in feveral trials, made about this time, no fenfible difference was obferved in its fituation. By the middle of April it appeared to be returning back again towards the north; and about the beginning of June, it paffed at the fame diftance from the zenith, as it had done in December, when it was firft obferved.

From the quick alteration in the declination of the ftar about this time, increafing about one fecond in three days, it was conjectured that it would now proceed northward, as it had before gone fouthward, of its prefent fituation; and it happened accordingly; for the ftar continued to move northward till September following, when it became ftationary again; being then near 20″ more northerly than in June, and upwards of

39″ more northerly than it had been in March. From September the ftar again returned towards the fouth, till, in December, it arrived at the fame fituation in which it had been obferved twelve months before, allowing for the difference of declination on account of the preceffion of the equinox.

This was a fufficient proof that the inftrument had not been the caufe of this apparent motion of the ftar; and yet it feemed difficult to devife one that fhould be adequate to fuch an unufual effect. A nutation of the earth's axis was one of the firft things that offered itfelf on this occafion; but it was foon found to be infufficient; for though it might have accounted for the change of declination in γ Draconis, yet it would not at the fame time accord with the phenomena obferved in the other ftars, particularly in a fmall one almoft oppofite in right afcenfion to γ Draconis, and at about the fame diftance from the north pole of the equator: for though this ftar feemed to move the fame way, as a nutation of the earth's axis would have made it; yet changing its declination but about half as much as γ Draconis in the fame time, as appeared on comparing the obfervations of both made on the fame days, at different feafons of the year, this plainly proved that the apparent motion of the ftar was not occafioned by a real nutation; fince, had that been the cafe, the alteration in both ftars would have been nearly equal.

The great regularity of the obfervations left no room to doubt, but that there was fome uniform caufe by which this unexpected motion was produced, and which did not depend on the uncertainty or variety of the feafons of the year. Upon comparing the obfervations with each other, it was difcovered that, in both the ftars above mentioned, the apparent difference of declination from the *maxima*, was always nearly proportional to the verfed fine of the fun's diftance from the equinoctial points. This was an inducement to think that the caufe, whatever it was, had fome relation to the fun's fituation with refpect to thofe points. But not being able to frame any hypothefis, fufficient to account for all the phenomena, and being very defirous to fearch a little farther into this matter, Dr. Bradley began to think of erecting an inftrument for himfelf at Wanftead; that, having it always at hand, he might with the more eafe and certainty enquire into the laws of this new motion. The confideration likewife of being able, by another inftrument, to confirm the truth of the obfervations hitherto made with that of Mr. Molineux, was no fmall inducement to the undertaking; but the chief of all was, the opportunity he fhould thereby have of trying in what manner other ftars fhould be affected by the fame caufe, whatever it might be. For Mr. Molineux's inftrument being originally defigned for obferving γ Draconis, to try whether it had any fenfible parallax, it was fo contrived, as to be capable of but little alteration in its direction; not above feven or eight minutes of a degree: and there being but few ftars, within half that diftance from the zenith of Kew, bright enough to be well obferved, he could not, with his inftrument, thoroughly examine how this caufe affected ftars that were differently fituated, with refpect to the equinoctial and folfticial points of the ecliptic.

Thefe confiderations determined him; and by the contrivance and direction of the fame ingenious perfon,

Mr.

Mr. Graham, his instrument was fixed up the 19th of August 1727. As he had no convenient place where he could make use of so long a telescope as Mr. Molineux's, he contented himself with one of but little more than half the length, namely of 12 feet and a half, the other being 24 feet and a half long, judging from the experience he had already had, that this radius would be long enough to adjust the instrument to a sufficient degree of exactness: and he had no reason afterwards to change his opinion; for by all his trials he was very well satisfied, that when it was carefully rectified, its situation might be securely depended on to half a second. As the place where his instrument was hung, in some measure determined its radius; so did it also the length of the arc or limb, on which the divisions were made, to adjust it: for the arc could not conveniently be extended farther, than to reach to about $6\frac{1}{4}$ degrees on each side of his zenith. This however was sufficient, as it gave him an opportunity of making choice of several stars, very different both in magnitude and situation; there being more than two hundred, inserted in the British Catalogue, that might be observed with it. He needed not indeed to have extended the limb so far, but that he was willing to take in *Capella*, the only star of the first magnitude that came so near his zenith.

His instrument being fixed, he immediately began to observe such stars as he judged most proper to give him any light into the cause of the motion already mentioned. There was a sufficient variety of small ones, and not less than twelve that he could observe through all seasons of the year, as they were bright enough to be seen in the day-time, when nearest the sun. He had not been long observing, before he perceived that the notion they had before entertained, that the stars were farthest north and south when the sun was near the equinoxes, was only true of those stars which are near the solsticial colure. And after continuing his observations a few months, he discovered what he then apprehended to be a general law observed by all the stars, namely, that each of them became stationary, or was farthest north or south, when it passed over his zenith at six of the clock, either in the evening or morning. He perceived also that whatever situation the stars were in, with respect to the cardinal points of the ecliptic, the apparent motion of every one of them tended the same way, when they passed his instrument about the same hour of the day or night; for they all moved southward when they passed in the day, and northward when in the night; so that each of them was farthest north, when it came in the evening about six of the clock, and farthest south when it came about six in the morning.

Though he afterwards discovered that the maxima, in most of these stars, do not happen exactly when they pass at those hours; yet, not being able at that time to prove the contrary, and supposing that they did, he endeavoured to find out what proportion the greatest alterations of declination, in different stars, bore to each other; it being very evident that they did not all change their declination equally. It has been before noticed, that it appeared from Mr. Molineux's observations, that γ *Draconis* changed its declination above twice as much as the before-mentioned small star that was nearly op-

posite to it; but examining the matter more nicely, he found that the greatest change in the declination of these stars, was as the sine of the latitude of each star respectively. This led him to suspect that there might be the like proportion between the *maxima* of other stars; but finding that the observations of some of them would not perfectly correspond with such an hypothesis, and not knowing whether the small difference he met with might not be owing to the uncertainty and error of the observations, he deferred the farther examination into the truth of this hypothesis, till he should be farther furnished with a series of observations made in all parts of the year; which would enable him not only to determine what errors the observations might be liable to, or how far they might safely be depended on; but also to judge, whether there had been any sensible change in the parts of the instrument itself.

When the year was completed, he began to examine and compare his observations; and having pretty well satisfied himself as to the general laws of the phenomena, he then endeavoured to find out the cause of them. He was already convinced that the apparent motion of the stars was not owing to a nutation of the earth's axis. The next that occurred to him, was an alteration in the direction of the plumb-line, by which the instrument was constantly adjusted; but this, upon trial, proved insufficient. Then he considered what refraction might do; but here also he met with no satisfaction. At last, through an amazing sagacity, he conjectured that all the phenomena hitherto mentioned, proceeded from the progressive motion of light, and the earth's annual motion in her orbit: for he perceived, that if light were propagated in time, the apparent place of a fixed object would not be the same when the eye is at rest, as when it is moving in any other direction but that of the line passing through the object and the eye; and that when the eye is moving in different directions, the apparent place of the object would be different.

He considered this matter in the following manner. He imagined C A to be a ray of light, falling perpendicularly upon the line B D: then, if the eye be at rest at A, the object must appear in the direction A C, whether light be propagated in time, or in an instant. But if the eye be moving from B towards A, and light be propagated in time, with a velocity that is to the velocity of the eye, as A C to A B; then, light moving from C to A, whilst the eye moves from B to A, that particle of it by which the object will be discerned, when the eye in its motion comes to A, is at C when the eye is at B. Joining the points B, C, he supposed the line B C to be a tube, inclined to the line B D in the angle D B C, and of such a diameter as to admit of but one particle of light: then it was easy to conceive, that the particle of light at C, by which the object must be seen when the eye arrives at A, would pass through the tube B C, so inclined to the line B D, and accompanying the eye in its motion from B to A; and that it would not come

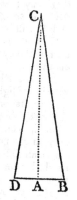

to

to the eye, placed behind such a tube, if it had any other inclination to the line B D. If, instead of supposing B C so small a tube, we conceive it to be the axis of a larger; then, for the same reason, the particle of light at C cannot pass through that axis, unless it be inclined to B D in the same angle D B C.

In the like manner, if the eye move the contrary way, from D towards A, with the same velocity; then the tube must be inclined in the angle B D C. Although therefore the true or real place of an object, be perpendicular to the line in which the eye is moving, yet the visible place will not be so; since that must doubtless be in the direction of the tube. But the difference between the true and apparent place, will be, *cæteris paribus*, greater or less, according to the different proportions between the velocity of light and that of the eye: so that if we could suppose light to be propagated in an instant, then there would be no difference between the real and visible place of an object, although the eye were in motion; for in that case, A C being infinite with respect to A B, the angle A C B, which is the difference between the true and visible place, vanishes. But if light be propagated in time, which was then allowed by most philosophers, then it is evident from the foregoing considerations, that there will always be a difference between the true and visible place of an object, except when the eye is moving either directly towards, or from the object. And in all cases, the sine of the difference between the true and visible place of the object, will be to the sine of the visible inclination of the object to the line in which the eye is moving, as the velocity of the eye, is to the velocity of light.

If light moved only 1000 times faster than the eye, and an object, supposed to be at an infinite distance, were really placed perpendicularly over the plane in which the eye is moving; it follows, from what has been said, that the apparent place of such object will always be inclined to that plane, in an angle of 89° 56'½; so that it will constantly appear 3'½ from its true place, and will seem so much less inclined to the plane, that way towards which the eye tends. That is, if A C be to A B or A D, as 1000 to 1, the angle A B C will be 89° 56'½, and the angle A C B 3'½, and B C D or 2 A C B will be 7', if the direction of the motion of the eye be contrary at one time to what it is at another.

If the earth revolve about the sun annually, and the velocity of light were to the velocity of the earth's motion in its orbit, as 1000 is to 1; then it is easy to conceive, that a star really placed in the pole of the ecliptic, would to an eye carried along with the earth, seem to change its place continually; and, neglecting the small difference on account of the earth's diurnal revolution on its axis, it would seem to describe a circle about that pole, every where distant from it by 3'½. So that its longitude would be varied through all the points of the ecliptic every year, but its latitude would always remain the same. Its right ascension would also change, and its declination, according to the different situation of the sun in respect of the equinoctial points; and its apparent distance from the north pole of the equator, would be 7' less at the autumnal, than at the vernal equinox.

The greatest alteration of the place of a star, in the pole of the ecliptic, or, which in effect amounts to the same, the proportion between the velocity of light and the earth's motion in its orbit, being known, it will not be difficult to find what would be the difference, on this account, between the true and apparent place of any other star at any time; and, on the contrary, the difference between the true and apparent place being given, the proportion between the velocity of light, and the earth's motion in her orbit, may be found.

After the history of this curious discovery, related by the author nearly in the terms above, he gives the results of a multitude of accurate observations, made on a great number of stars, at all seasons of the year. From all which observations, and the theory as related above, he found that every star, in consequence of the earth's motion in her orbit and the progressive motion of light, appears to describe a small ellipse in the heavens, the transverse axis of which is equal to the same quantity for every star, namely 40" nearly; and that the conjugate axis of the ellipse, for different stars, varies in this proportion, namely, as the right sine of the star's latitude; that is, radius is to the sine of the star's latitude, as the transverse axis to the conjugate axis: and consequently a star in the pole of the ecliptic, its latitude being there 90°, whose sine is equal to the radius, will appear to describe a small circle about that pole, as a centre, whose radius is equal to 20". He also gives the following law of the variation of the star's declination: if A denote the angle of position, or the angle at the star made by two great circles drawn from it through the poles of the ecliptic and equator, and B another angle, whose tangent is to the tangent of A, as radius is to the sine of the star's latitude; then B will be equal to the difference of longitude between the sun and the star, when the true and apparent declination of the star are the same. And if the sun's longitude in the ecliptic be reckoned from that point in which it is when this happens; then the difference between the true and apparent declination of the star, will be always as the sine of the sun's longitude from that point. It will also be found that the greatest difference of declination that can be between the true and apparent place of the star, will be to 20", the semitransverse axis of the ellipse, as the sine of A to the sine of B.

The author then shews, by the comparison of a number of observations made on different stars, that they exactly agree with the theory deduced from the progressive motion of light, and that consequently it is highly probable that such motion is the cause of those variations in the situation of the stars. From which he infers, that the parallax of the fixed stars is much smaller, than hath been hitherto supposed by those, who have pretended to deduce it from their observations. He thinks he may venture to say, that in the stars he had observed, the parallax does not amount to 2"; nay, that if it had amounted to 1", he should certainly have perceived it, in the great number of observations that he made, especially of γ Draconis; which agreeing with the hypothesis, without allowing any thing for parallax, nearly as well when the sun was in conjunction with, as in opposition to, this star, it seems very probable

bable that the parallax of it is not so much as one single second; and consequently that it is above 400000 times farther from us than the sun.

From the greatest variation in the place of the stars, namely 40″, Dr. Bradley deduces the ratio of the velocity of light in comparison with that of the earth in her orbit. In the preceding figure, A C is to A B, as the velocity of light to that of the earth in her orbit, the angle A C B being equal to 20″; so that the ratio of those velocities is that of radius to the tangent of 20″, or of radius to 20″, since the tangent has no sensible difference from so small an arc: but the radius of a circle is equal to the arc of 57° $\frac{3}{10}$ nearly, or equal to 206260″; therefore the velocity of light is to the velocity of the earth, as 206260 to 20, or as 10313 to 1.

And hence also the time in which light passes over the space from the sun to the earth, is easily deduced; for this time will be to one year, as A B or 20″ to 360° or the whole circle; that is, 360° : 20″ :: 365¼ days: 8ᵐ 7ˢ, namely, light will pass from the sun to the earth in the time of 8 minutes, 7 seconds; and this will be the same, whatever the distance of the sun is.

Dr. Bradley having annexed to his theory the rules or formulæ for computing the aberration of the fixed stars in declination and right ascension; these rules have been variously demonstrated, and reduced to other practical forms, by Mr. Clairaut in the Memoirs of the Academy of Sciences for 1737; by Mr. Simpson in his Essays in 1740; by M. Fontaine des Crutes in 1744; and several other persons. The results of these rules are as follow: Every star appears to describe in the course of a year, by means of the aberration, a small ellipse, whose greater axis is 40″, and the less axis, perpendicular to the ecliptic, is equal to 40″ multiplied by the sine of the star's latitude, the radius being 1. The eastern extremity of the longer axis, marks the apparent place of the star, the day of the opposition; and the extremity of the less axe, which is farthest from the ecliptic, marks its situation three months after.

The greatest aberration in longitude, is equal to 20″ divided by the cosine of its latitude. And the aberration for any time, is equal to 20″ multiplied by the cosine of the elongation of the star found for the same time, and divided by the cosine of its latitude. This aberration is subtractive in the first and last quadrant of the argument, or of the difference between the longitudes of the sun and star; and additive in the second and third quadrants. The greatest aberration in latitude, is equal to 20″ multiplied by the sine of the star's latitude. And the aberration in latitude for any time, is equal to 20″ multiplied by the sine of the star's latitude, and multiplied also by the sine of the elongation. The aberration is subtractive before the opposition, and additive after it.

The greatest aberration in declination, is equal to 20″ multiplied by the sine of the angle of position A, and divided by the sine of B the difference of longitude between the sun and star when the aberration in declination is nothing. And the aberration in declination at any other time, will be equal to the greatest aberration multiplied by the sine of the difference between the sun's place at the given time and his place when the

aberration is nothing. Also the sine of the latitude of the star is to radius, as the tangent of A the angle of position at the star, is to the tangent of B, the difference of longitude between the sun and star when the aberration in declination is nothing. The greatest aberration in right-ascension, is equal to 20″ multiplied by the cosine of A the angle of position, and divided by the sine of C the difference in longitude between the sun and star when the aberration in right ascension is nothing. And the aberration in right-ascension at any other time, is equal to the greatest aberration multiplied by the sine of the difference between the sun's place at the given time, and his place when the aberration is nothing. Also the sine of the latitude of the star is to radius, as the cotangent of A the angle of position, to the tangent of C.

ABERRATION of the Planets, is equal to the geocentric motion of the planet, the space it appears to move as seen from the earth, during the time that light employs in passing from the planet to the earth. Thus, in the sun, the aberration in longitude is constantly 20″, that being the space moved by the sun, or, which is the same thing, by the earth, in the time of 8ᵐ 7ˢ, which is the time in which light passes from the sun to the earth, as we have seen in the foregoing article. In like manner, knowing the distance of any planet from the earth, by proportion it will be, as the distance of the sun is to the distance of the planet, so is 8ᵐ 7ˢ to the time of light passing from the planet to the earth: then computing the planet's geocentric motion in this time, that will be the aberration of the planet, whether it be in longitude, latitude, right-ascension, or declination.

It is evident that the aberration will be greatest in the longitude, and very small in latitude, because the planets deviate very little from the plane of the ecliptic, or path of the earth; so that the aberration in the latitudes of the planets, is commonly neglected, as insensible; the greatest in Mercury being only 4″⅓, and much less in the other planets. As to the aberrations in declination and right-ascension, they must depend on the situation of the planet in the zodiac. The aberration in longitude, being equal to the geocentric motion, will be more or less according as that motion is; it will therefore be least, or nothing at all, when the planet is stationary; and greatest in the superior planets Mars, Jupiter, Saturn, &c, when they are in opposition to the sun; but in the inferior planets Venus and Mercury, the aberration is greatest at the time of their superior conjunction. These maxima of aberration for the several planets, when their distance from the sun is least, are as below: viz, for

Saturn - - -	27″·0
Jupiter - - -	29 ·8
Mars - - -	37 ·8
Venus - - -	43 ·2
Mercury - - -	59 ·0
The Moon - - -	⅔

And between these numbers and nothing the aberrations of the planets, in longitude, vary according to their situations. But that of the sun varies not, being constantly 20″, as has been before observed. And this may alter his declination by a quantity, which varies from 0 to near 8″; being greatest or 8″ about the equinoxes, and vanishing in the solstices.

The

The methods of computing thefe, and the formulas for all cafes, are given by M. Clairaut in the Memoirs of the Academy of Sciences for the year 1746, and by M. Euler in the Berlin Memoirs, vol. 2, for 1746.

Optic ABERRATION, the deviation or difperfion of the rays of light, when reflected by a fpeculum, or refracted by a lens, by which they are prevented from meeting or uniting in the fame point, called the geometrical focus, but are fpread over a fmall fpace, and produce a confufion of images. Aberration is either lateral or longitudinal: the lateral aberration is meafured by a perpendicular to the axis of the fpeculum or lens, drawn from the focus to meet the refracted or reflected ray: the longitudinal aberration is the diftance, on the axis, between the focus and the point where the ray meets the axis. The aberrations are very amply treated in Smith's Complete Syftem of Optics, in 2 volumes 4to.

There are two fpecies of aberration, diftinguifhed according to their different caufes: the one arifes from the figure of the fpeculum or lens, producing a geometrical difperfion of the rays, when thefe are perfectly equal in all refpects; the other arifes from the unequal refrangibility of the rays of light themfelves; a difcovery that was made by Sir Ifaac Newton, and for this reafon it is often called the Newtonian aberration. As to the former fpecies of aberration, or that arifing from the figure, it is well known that if rays iffue from a point at a given diftance; then they will be reflected into the other focus of an ellipfe having the given luminous point for one focus, or directly from the other focus of an hyperbola; and will be varioufly difperfed by all other figures. But if the luminous point be infinitely diftant, or, which is the fame, the incident rays be parallel, then they will be reflected by a parabola into its focus, and varioufly difperfed by all other figures. But thofe figures are very difficult to make, and therefore curved fpecula are commonly made fpherical, the figure of which is generated by the revolution of a circular arc, which produces an aberration of all rays, whether they are parallel or not, and therefore it has no accurate geometrical focus which is common to all the rays. Let B V F reprefent a concave fpherical fpeculum, whofe centre is C; and let A B, E F be incident rays parallel to the axis CV. Becaufe the angle of incidence is equal to the angle of reflection in

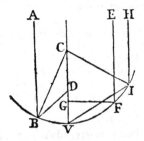

all cafes, therefore if the radii C B, C F be drawn to the points of incidence, and thence B D making the angle C B D equal to the angle C B A, and F G making the angle C F G equal to the angle C F E; then B D, FG will be the reflected rays, and D, G, the points where

they meet the axis. Hence it appears that the point of coincidence with the axis is equally diftant from the point of incidence and the centre: for becaufe the angle C B D is equal to the angle C B A, which is equal to the alternate angle B C D, therefore their oppofite fides C D, D B are equal: and in like manner, in any other, G F is equal to G C. And hence it is evident that when B is indefinitely near the vertex V, then D is in the middle of the radius C V; and the nearer the incident ray is to the axis C V, the nearer will the reflected ray come to the middle point D; and the contrary. So that the aberration D G of any ray E F G, is always more and more, as the incident ray is farther from the axis, or the incident point F from the vertex V; till when the diftance V I is 60 degrees, then the reflected ray falls in the vertex V, making the aberration equal to the whole length D V. And this fhews the reafon why fpecula are made of a very fmall fegment of a fphere, namely, that all their reflected rays may arrive very near the middle point or focus D, to produce an image the moft diftinct, by the leaft aberration of the rays. And in like manner for rays refracted through lenfes.

In fpherical lenfes, Mr. Huygens has demonftrated that the aberration from the figure, in different lenfes, is as follows.

1. In all plano-convex lenfes, having their plane furface expofed to parallel rays, the longitudinal aberration of the extreme ray, or that remoteft from the axis, is equal to $\frac{2}{8}$ of the thicknefs of the lens.

2. In all plano-convex lenfes, having their convex furface expofed to parallel rays, the longitudinal aberration of the extreme ray, is equal to $\frac{7}{6}$ of the thicknefs of the lens. So that in this pofition of the fame planoconvex lens, the aberration is but about one-fourth of that in the former; being to it only as 7 to 27.

3. In all double convex lenfes of equal fpheres, the aberration of the extreme ray, is equal to $\frac{5}{8}$ of the thicknefs of the lens.

4. In a double convex lens, the radii of whofe fpheres are as 1 to 6, if the more convex furface be expofed to parallel rays, the aberration from the figure is lefs than in any other fpherical lens; being no more than $\frac{1.5}{14}$ of its thicknefs.

But the foregoing fpecies of aberration, arifing from the figure, is very fmall, and eafily remedied, in comparifon with the other, arifing from the unequal refrangibility of the rays of light; infomuch that Sir Ifaac Newton fhews in his Optics, pa. 84 of the 8vo. edition, that if the object-glafs of a telefcope be planoconvex, the plane fide being turned towards the object, and the diameter of the fphere, to which the convex fide is ground, be 100 feet, the diameter of the aperture being 4 inches, and the ratio of the fine of incidence out of glafs into air, be to that of refraction, as 20 to 31; then the diameter of the circle of aberrations will in this cafe be only $\frac{96\frac{1}{1}}{72000000}$ parts of an inch: while the diameter of the little circle, through which the fame rays are fcattered by unequal refrangibility, will be about the 55th part of the aperture of the object-glafs, which here is 4 inches. And therefore the error arifing from the fpherical figure of the glafs, is to the error arifing from the different refrangibility of the rays, as $\frac{561}{72000000}$ to $\frac{4}{55}$, that is as 1 to 5449.

So that it may seem strange that objects appear through telescopes so distinct as they do, considering that the error arising from the different refrangibility, is almost incomparably larger than that of the figure. Newton however solves the difficulty by observing that the rays, under their various aberrations, are not scattered uniformly over all the circular space, but collected infinitely more dense in the centre than in any other part of the circle; and that, in the way from the centre to the circumference, they grow more and more rare, so as at the circumference to become infinitely rare; and, by reason of their rarity, they are not strong enough to be visible, unless in the centre, and very near it.

In consequence of the discovery of the unequal refrangibility of light, and the apprehension that equal refractions must produce equal divergencies in every sort of medium, it was supposed that all spherical object-glasses of telescopes would be equally affected by the different refrangibility of light, in proportion to their aperture, of whatever materials they might be constructed: and therefore that the only improvement that could be made in refracting telescopes, was that of increasing their length. So that Sir Isaac Newton, and other persons after him, despairing of success in the use and fabric of lenses, directed their chief attention to the construction of reflecting telescopes.

However, about the year 1747, M. Euler applied himself to the subject of refraction; and pursued a hint suggested by Newton, for the design of making object-glasses with two lenses of glass inclosing water between them; hoping that, by constructing them of different materials, the refractions would balance one another, and so the usual aberration be prevented. Mr. John Dollond, an ingenious optician in London, minutely examined this scheme, and found that Mr. Euler's principles were not satisfactory. M. Clairaut likewise, whose attention had been excited to the same subject, concurred in opinion that Euler's speculations were more ingenious than useful. This controversy, which seemed to be of great importance in the science of optics, engaged also the attention of M. Klingenstierna of Sweden, who was led to make a careful examination of the 8th experiment in the second part of Newton's Optics, with the conclusions there drawn from it. The consequence was, that he found that the rays of light, in the circumstances there mentioned, did not lose their colour, as Sir Isaac had imagined. This hint of the Swedish philosopher led Mr. Dollond to re-examine the same experiment: and after several trials it appeared, that different substances caused the light to diverge very differently, in proportion to their general refractive powers. In the year 1757 therefore he procured wedges of different kinds of glass, and applied them together so that the refractions might be made in contrary directions, that he might discover whether the refraction and divergency of colour would vanish together. The result of his first trials encouraged him to persevere; for he discovered a difference far beyond his hopes in the qualities of different kinds of glass, with respect to their divergency of colours. The Venice glass and English crown glass were found to be nearly allied in this respect: the common English plate glass made the rays diverge more; and the English flint

glass most of all. But without enquiring into the cause of this difference, he proceeded to adapt wedges of crown glass, and of white flint glass, ground to different angles, to each other, so as to refract in different directions; till the refracted light was entirely free from colours. Having measured the refractions of each wedge, he found that the refraction of the white glass was to that of the crown glass, nearly as 2 to 3: and he hence concluded in general, that any two wedges made in this proportion, and applied together so as to refract in contrary directions, would refract the light without any aberration of the rays.

Mr. Dollond's next object was to make similar trials with spherical glasses of different materials, with the view of applying his discovery to the improvement of telescopes: and here he perceived that, to obtain a refraction of light in contrary directions, the one glass must be concave, and the other convex; and the latter, which was to refract the most, that the rays might converge to a real focus, he made of crown glass, the other of white flint glass. And as the refractions of spherical glasses are inversely as their focal distances, it was necessary that the focal distances of the two glasses should be inversely as the ratios of the refractions of the wedges; because that, being thus proportioned, every ray of light that passes through this compound glass, at any distance from its axis, will constantly be refracted, by the difference between two contrary refractions, in the proportion required; and therefore the different refrangibility of the light will be entirely removed.

But in the applications of this ingenious discovery to practice, Mr. Dollond met with many and great difficulties. At length, however, after many repeated trials, by a resolute perseverance, he succeeded so far as to construct refracting telescopes much superior to any that had hitherto been made; representing objects with great distinctness, and in their true colours.

Mr. Clairaut, who had interested himself from the beginning in this discovery, now endeavoured to ascertain the principles of Mr. Dollond's theory, and to lay down rules to facilitate the construction of these new telescopes. With this view he made several experiments, to determine the refractive power of different kinds of glass, and the proportions in which they separated the rays of light: and from these experiments he deduced several theorems of general use. M. D'Alembert made likewise a great variety of calculations to the same purpose; and he shewed how to correct the errors to which these telescopes are subject, sometimes by placing the object-glasses at a small distance from each other, and sometimes by using eye-glasses of different refractive powers. But though foreigners were hereby supplied with the most accurate calculations, they were very defective in practice. And the English telescopes, made, as they imagined, without any precise rule, were greatly superior to the best of their construction.

M. Euler, whose speculations had first given occasion to this important and useful enquiry, was very reluctant in admitting Mr. Dollond's improvements, because they militated against a pre-conceived theory of his own. At last however, after several altercations, being convinced of their reality and importance by M. Clairaut,

aut, he affented ; and he foon after received farther fa tisfaction from the experiments of M. Zeiher, of Pe terfburgh.

M. Zeiher fhewed by experiments that it is the lead, in the compofition of glafs, which gives it this remark able property, namely, that while the refraction of the mean rays is nearly the fame, that of the extreme rays confiderably differs. And, by increafing the lead, he produced a kind of glafs, which occafioned a much greater feparation of the extreme rays than that of the flint glafs ufed by Mr. Dollond, and at the fame time confiderably increafed the mean refraction. M. Zeiher, in the courfe of his experiments, made glafs of minium and lead, with a mixture alfo of alkaline falts ; and he found that this mixture greatly diminifhed the mean refraction, and yet made hardly any change in the dif perfion : and he at length obtained a kind of glafs greatly fuperior to the flint glafs of Mr. Dollond for the conftruction of telefcopes ; as it occafioned three times as great a difperfion of the rays as the common glafs, whilft the mean refraction was only as 1·61 to 1.

Other improvements were alfo made on the new or achromatic telefcopes by the inventor Mr. John Dol lond, and by his fon Peter Dollond ; which may be feen under the proper words. For various differtations on the fubject of the aberration of light, colours, and the figure of the glafs, fee Philof. Tranf. vols. 35, 48, 50, 51, 52, 55, 60 ; Memoirs of the Academy of Sciences of Paris, for the years 1737, 1746, 1752, 1755, 1756, 1757, 1762, 1764, 1765, 1767, 1770 ; the Berlin Ac. 1746, 1762, 1766 ; Swed. Mem. vol. 16 ; Com. Nov. Petripol. 1762 ; M. Euler's Dioptrics ; M. d'Alembert's Opufcules Math. ; M. de Rochon Opufcules ; &c, &c.

ABRIDGING, in *Algebra*, is the reducing a com pound equation, or quantity, to a more fimple form of expreffion. This is done either to fave room, or the trouble of writing a number of fymbols ; or to fimplify the expreffion, either to eafe the memory, or to render the formula more eafy and general.

So the equation $x^3 - ax^2 + abx - abc = 0$, by put ting $p = a$, $q = ab$, and $r = abc$, becomes $x^3 - px^2 + qx - r = 0$.

And the equation $x^2 + (a+b)x - \dfrac{ab}{c} = 0$, by putting $p = a+b$, and $q = \dfrac{ab}{c}$, becomes $x^2 + px - q = 0$.

ABSCISS, ABSCISSE, or ABSCISSA, is a part or fegment cut off a line, terminated at fome certain point, by an ordinate to a curve ; as A P or B P.

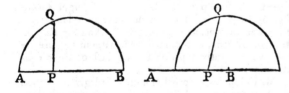

The abfcifs may either commence at the vertex of the curve, or at any other fixed point. And it may be taken either upon the axis or diameter of the curve, or upon any other line drawn in a given pofition.

Hence there are an infinite number of variable abfcif fes, terminated at the fame fixed point at one end, the other end of them being at any point of the given line or diameter.

In the common parabola, each ordinate P Q has but

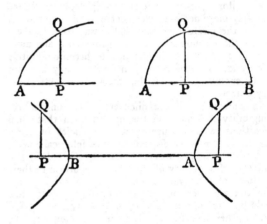

one abfcifs A P ; in the ellipfe or circle, the ordinate has two abfciffes A P, B P lying on the oppofite fides of it ; and in the hyperbola the ordinate P Q has alfo two abfciffes, but they lie both on the fame fide of it. That is, in general, a line of the fecond kind, or a curve of the firft kind, may have two abfciffes to each ordinate. But a line of the third order may have three abfciffes to each ordinate ; a line of the fourth order may have four ; and fo on.

The ufe of the abfciffes is, in conjunction with the ordinates, to exprefs the nature of the curves, either by fome proportion or equation including the abfcifs and its ordinate, with fome other fixed invariable line or lines. Every different curve has its own peculiar equation or property by which it is expreffed, and dif ferent from all others : and that equation or expreffion is the fame for every ordinate and its abfciffes, whatever point of the curve be taken. So, in the circle, the fquare of any ordinate is equal to the rectangle of its two abfciffes, or A P · P B = P Q² ; in the parabola, the fquare of the ordinate is equal to the rectangle of the abfcifs and a certain given line called the parame ter ; in the ellipfe and hyperbola, the fquare of the or dinate is always in a certain conftant proportion to the rectangle of the two abfciffes, namely, as the fquare of the conjugate to the fquare of the tranfverfe, or as the parameter is to the tranfverfe axis ; and fo other proper ties in other curves.

When the natures or properties of curves are ex preffed by algebraic equations, any general abfcifs, as A P, is commonly denoted by the letter x, and the or dinate P Q by the letter y ; the other or conftant lines being reprefented by other letters. Then the equa tions expreffing the nature of thefe curves are as fol low ; namely, for the
circle - - $dx - x^2 = y^2$, where d is the diameter A B ;
parabola - $px = y^2$, where p is the parameter ;
ellipfe - $t^2 : c^2 :: tx - x^2 : y^2$, $\{$ where t is the tranfverfe,
hyperbola $t^2 : c^2 :: tx + x^2 : y^2$, $\{$ & c the conjugate axis.

ABSIS, ABSIDES. See APSIS, APSIDES.

ABSOLUTE

ABSOLUTE EQUATION, in *Aſtronomy*, is the ſum of the optic and excentric equations. The apparent inequality of a planet's motion, ariſing from its not being equally diſtant from the earth at all times, is called its optic equation ; and this would ſubſiſt even if the planet's real motion were uniform. The excentric inequality is cauſed by the planet's motion being not uniform. To illuſtrate this, conceive the ſun to move, or to appear to move, in the circumference of a circle, in whoſe centre the earth is placed. It is manifeſt, that if the ſun move uniformly in this circle, then he muſt appear to move uniformly to a ſpectator at the earth ; and in this caſe there will be no optic nor excentric equation. But ſuppoſe the earth to be placed out of the centre of the circle ; and then, though the ſun's motion ſhould be really uniform, it would not appear to be ſo, being ſeen from the earth ; and in this caſe there would be an optic equation, without an excentric one. Imagine farther, the ſun's orbit to be, not circular, but elliptical, and the earth in its focus : it will be full as evident that the ſun cannot appear to have an uniform motion in ſuch ellipſe ; ſo that his motion will then be ſubject to two equations ; that is, the optic equation, and the excentric equation. *See* EQUATION, and OPTIC INEQUALITY.

ABSOLUTE *Number*, in Algebra, is that term or member of an equation that is completely known, and which is equal to all the other, or unknown terms, taken together ; and is the ſame as what Vieta calls the *homogeneum comparationis*. So, of the equation $x^2 + 16x - 36 = 0$, or $x^2 + 16x = 36$, the abſolute number, or known term, is 36.

ABSOLUTE *Gravity, Motion, Space, Time, &c.* See the reſpective ſubſtantives.

ABSTRACT MATHEMATICS, otherwiſe called pure mathematics, is that which treats of the properties of magnitude, figure, or quantity, abſolutely and generally conſidered, without reſtriction to any ſpecies in particular : ſuch as Arithmetic and Geometry. In this ſenſe, abſtract or pure mathematics, is oppoſed to mixed mathematics, in which ſimple and abſtract properties, and the relations of quantities, primitively conſidered in pure mathematics, are applied to ſenſible objects ; as in aſtronomy, hyd oſtatics, optics, &c.

ABSTRACT *Number*, is a number, or collection of units, conſidered in itſelf, without being applied to denote a collection of any particular and determinate things. So, for example, 3 is an abſtract number, ſo far as it is not applied to ſomething : but when we ſay 3 feet, or 3 perſons, the 3 is no longer an abſtract, but a concrete number.

ABSURD, or ABSURDUM, a term commonly uſed in demonſtrating converſe propoſitions ; a mode of demonſtration, in which the propoſition intended is not proved in a direct manner, by principles before laid down ; but it proves that the contrary is abſurd or impoſſible ; and ſo indirectly as it were proves the propoſition itſelf. The 4th propoſition in the firſt book of Euclid, is the firſt in which he makes uſe of this mode of proof ; where he ſhews that if the extremities of two lines coincide, thoſe lines will coincide in all their parts, otherwiſe they would incloſe a ſpace, which is abſurd or contrary to the 10th axiom. Moſt converſe

propoſitions are proved in this way, which mode of proof is called *reductio ad abſurdum*.

ABUNDANT NUMBER, in *Arithmetic*, is a number whoſe aliquot parts, added all together, make a ſum which is greater than the oumber itſelf. Thus 12 is an abundant number, becauſe its aliquot parts, namely 1, 2, 3, 4, 6, when added together, make 16, which is greater than the number 12 itſelf.

An abundant number is oppoſed to a deficient one, which is leſs than the ſum of its aliquot parts taken together, as the number 14, whoſe aliquot parts 1, 2, 7, make no more than 10 ; and to a perfect number, which is exactly equal to the ſum of all its aliquot parts, as the number 6, which is equal to the ſum of 1, 2, 3, which are its aliquot parts.

ACADEMICIAN, a member of a ſociety called an academy, inſtituted for the promotion of arts, ſciences, or natural knowledge in general.

ACADEMICS, an ancient ſect of philoſophers, who followed the doctrine of Socrates and Plato, as to the uncertainty of knowledge, and the incomprehenſibility of truth.

Academic, in this ſenſe, amounts to much the ſame with Platoniſt ; the difference between them being only in point of time. Thoſe who embraced the ſyſtem of Plato, among the ancients, were called *academici*, academician or academic ; whereas thoſe who did the ſame ſince the reſtoration of learning, have aſſumed the denomination of Platoniſts.

We uſually reckon three ſects of academics ; though ſome make five. The ancient academy was that of which Plato was the chief.

Arceſſilas, one of Plato's ſucceſſors introducing ſome alterations into the philoſophy of this ſect, founded what they call the ſecond academy.

The eſtabliſhment of the third, called alſo the new academy, is attributed to Lacydes, or rather to Carneades.

Some authors add a fourth, founded by Philo ; and a fifth, by Antiochus, called the Antiochan, which tempered the ancient academy with Stoiciſm.

The ancient academy doubted of every thing ; and carried this principle ſo far as to make it a doubt, whether or no they ought to doubt. It was a kind of a principle with them, never to be certain or ſatisfied of any thing ; never to affirm or to deny any thing, either for true or falſe.

The new academy was ſomewhat more reaſonable ; they acknowledged ſeveral things for truths, but without attaching themſelves to any with entire aſſurance. Theſe philoſophers had found that the ordinary commerce of life and ſociety was inconſiſtent with the abſolute and univerſal doubtfulneſs of the ancient academy : and yet it is evident that they looked upon things rather as probable, than as true and certain : by this amendment thinking to ſecure themſelves from thoſe abſurdities into which the ancient academy had fallen.

ACADEMIST, the ſame as Academician.

ACADEMY, ACADEMIA, in *Antiquity*, a fine villa or pleaſure houſe, in one of the ſuburbs of Athens, about a mile from the city ; where Plato, and the wiſe men who followed him, held aſſemblies for diſputes

and

and philosophical conference; which gave the name to the sect of Academics.

The house took its name, *Academy*, from one Academus, or Ecademus, a citizen of Athens, to whom it originally belonged: he lived in the time of Theseus; and here he used to have gymnastic sports or exercises.

The academy was farther improved by Cimon, and adorned with fountains, trees, shady walks, &c, for the convenience of the philosophers and men of learning, who here met to confer and dispute for their mutual improvement. It was surrounded with a wall by Hipparchus, the son of Pisistratus; and it was also used as the burying-place for illustrious persons, who had deserved well of the republic.

It was here that Plato taught his philosophy; and hence it was that all public places, destined for the assemblies of the learned and ingenious, have been since called *Academies*.

Sylla sacrificed the delicious walks and groves of the academy, which had been planted by Cimon, to the ravages of war; and employed those very trees in constructing machines to batter the walls of the city which they had adorned.

Cicero too had a villa, or country retirement, near Puzzuoli, which he called by the same name, *Academia*. Here he used to entertain his philosophical friends; and here it was that he composed his Academical Questions, and his books *De Naturâ Deorum*.

ACADEMY, among the moderns, denotes a regular society or company of learned persons, instituted under the protection of some prince, or other public authority, for the cultivation and improvement of arts or sciences.

Some authors confound Academy with University; but though much the same in Latin, they are very different things in English. An university is properly a body composed of graduates in the several faculties; of professors, who teach in the public schools; of regents or tutors, and students who learn under them, and aspire likewise to degrees. Whereas an academy is not intended to teach, or profess any art or science, but to improve it: it is not for novices to be instructed in, but for those that are more knowing; for persons of learning to confer in, and communicate their lights and discoveries to each other, for their mutual benefit and improvement.

The first modern academy we read of, was established by Charlemagne, by the advice of Alcuin, an English monk: it was composed of the chief geniuses of the court, the emperor himself being a member. In their academical conferences, every person was to give some account of the ancient authors he had read; and each one assumed the name of some ancient author, that pleased him most, or some celebrated person of antiquity. Alcuin, from whose letters we learn these particulars, took that of Flaccus, the surname of Horace; a young lord, named Augilbert, took that of Homer; Adelard, bishop of Corbie, was called Augustin; Recluse, bishop of Mentz, was Dametas; and the king himself, David.

Since the revival of learning in Europe, academies have multiplied greatly, most nations being furnished with several, and from their communications the chief

improvements have been made in the arts and sciences, and in cultivating natural knowledge. There are now academies for almost every art, or species of knowledge; but I shall give a short account only of those institutions of this kind, which regard the cultivation of subjects mathematical or philosophical, which are the proper and peculiar objects of our undertaking.

Italy abounds more in academies than all the world besides; there being enumerated by Jarckius not less than five hundred and fifty in all; and even to the amount of twenty-five in Milan itself. These are however mostly of a private and inferior nature; the consequence of their too great number.

The first academy of a philosophical kind was established at Naples, in the house of Baptista Porta, about the year 1560, under the name of *Academy Secretorum Naturæ*; being formed for the improvement of natural and mathematical knowledge. This was succeeded by the

ACADEMY *of Lyncei*, founded at Rome by prince Frederick Cesi, towards the end of the same century. It was rendered famous by the notable discoveries made by several of its members; among whom was the celebrated Galileo Galilei.

Several other academies contributed also to the advancement of the sciences; but it was by speculations rather than by repeated experiments on the phenomena of nature: such were the academy of Bessarian at Rome, and that of Laurence de Medicis at Florence, in the 15th century; and in the 16th were that of Infiammati at Padua, of Vegna Juoli at Rome, of Ortolani at Placentia, and of Umidi at Florence. The first of these studied fire and pyrotechnia, the second wine and vineyards, the third pot-herbs and gardens, the fourth water and hydraulics. To these may be added that of Venice, called La Veneta, and founded by Frederick Badoara, a noble Venetian; another in the same city, of which Campegio, bishop of Feltro, appears to have been the chief; also that of Cosenza, or La Consentina, of which Bernadin Telesio, Sertorio Quatromanni, Paulus Aquinas, Julio Cavalcanti, and Fabio Cicali, celebrated philosophers, were the chief members. The compositions of all these academies, of the 16th century, were good in their kind; but none of them comparable to those of the Lyncei.

ACADEMY *del Cimento*, that is, of Experiments, arose at Florence, some years after the death of Torricelli, namely in the year 1657, under the protection of prince Leopold of Tuscany, afterwards cardinal de Medicis, and brother to the Grand Duke Ferdinand the Second. Galileo, Toricelli, Aggiunti, and Viviani had prepared the way for it: and some of its chief members were Paul del Buono, who in 1657 invented the instrument for trying the incompressibility of water, namely a thick globular shell of gold, having its cavity filled with water; then the globe being compressed by a strong screw, the water came through the pores of the gold rather than yield to the compression: also, Alphonsus Borelli, well known for his ingenious treatise *De Motu Animalium*, and other works; Candide del Buono, brother of Paul; Alexander Marsili, Vincent Viviani, Francis Rhedi, and the Count Laurence Magalotti, secretary of this academy, who published

lished a volume of their curious experiments in 1667, under the title of *Saggi di Naturali Esperienze*; a copy of which being presented to the Royal Society, it was translated into English by Mr. Waller, and published at London, in 4to, 1684: A curious collection of tracts, containing ingenious experiments on the pressure of the air, on the compressing of water, on cold, heat, ice, magnets, electricity, odours, the motion of sound, projectiles, light, &c, &c. But we have heard little or nothing more of the academy since that time. It may not be improper to observe here, that the Grand Duke Ferdinand, above mentioned, was no mean philosopher and chemist, and that he invented thermometers, of which the construction and use may be seen in the collection of the academy del Cimento.

ACADEMY *degl' Inquieti* at Bologna, incorporated afterwards into that *della traccia* in the same city, followed the example of that *del Cimento*. The members met at the house of the abbot Antonio Sampieri; and here Geminiano Montanari, one of the chief members, made excellent discourses on mathematical and philosophical subjects, some parts of which were published in 1667, under the title of Pensieri Fisico-Mathematici. This academy afterwards met in an apartment of Eustachio Manfredi, and then in that of Jacob Sandri; but it arrived at its chief lustre while its assemblies were held in the palace Marsilli.

ACADEMY *of Rossano*, in the kingdom of Naples, called *La Societa Scientifica Rossanese degl' Incuriosi*, was founded about the year 1540, under the name of *Naviganti*; and was renewed under that of *Spensierati* by Camillo Tuscano, about the year 1600. It was then an academy of belles-lettres, but was afterwards transformed into an academy of sciences, on the solicitation of the learned abbot Don Giacinto Gimma; who, being made president under the title of promoter-general, in 1695, gave it a new set of regulations. He divided the academists into several classes, namely, grammarians, rhetoricians, poets, historians, philosophers, physicians, mathematicians, lawyers, and divines; with a separate class for cardinals and persons of quality. To be admitted a member, it was necessary that the candidate have degrees in some faculty. Members, in the beginning of their books, are not allowed to take the title of *academist* without a written permission from the president, which is not granted till the work has been examined by the censors of the academy. This permission is the highest honour the academy can confer; since they hereby, as it were, adopt the work, and engage to answer for it against any criticisms that may be made upon it. The president himself is not exempt from this law: and it is not permitted that any academist publish any thing against the writings of another, without leave obtained from the society.

There have been several other academies of sciences in Italy, but which have not subsisted long, for want of being supported by the princes. Such were at Naples that of the *Investiganti*, founded about the year 1679, by the marquis d'Arena, Don Andrea Concubletto; and that which, about the year 1698, met in the palace of Don Lewis della Cerda, the duke de Medina, and viceroy of Naples: at Rome, that of *Fisico-Matematici*, which in 1686 met in the house of

Signior Ciampini: at Verona, that of *Aletofili*, founded the same year by Signior Joseph Gazola, and which met in the house of the count Serenghi della Cucca: at Brescia, that of *Filesotici*, founded the same year for the cultivation of philosophy and mathematics, and terminated the year following: that of F. Francisco Lana, a jesuit of great skill in these sciences: and lastly that of Fisico-Critici at Sienna, founded in 1691, by Signior Peter Maria Gabrielli.

Some other academies, still subsisting in Italy, repair with advantage the loss of the former. One of the principal is the academy of Filarmonici at Verona, supported by the marquis Scipio Maffei, one of the most learned men in Italy; the members of which academy, though they cultivate the belles lettres, do not neglect the sciences. The academy of *Ricovrati* at Padua still subsists with reputation; in which, from time to time, learned discourses are held on philosophical subjects. The like may be said of the academy of the *Muti di Reggio*, at Modena. At Bologna is an academy of sciences, in a flourishing condition, known by the name of *The Institute of Bologna*; which was founded in 1712 by count Marsigli, for cultivating physics, mathematics, medicine, chemistry, and natural history. The history of it is written by M. de Limiers, from memoirs furnished by the founder himself. Among the new academies, the first place, after the Institute of Bologna, is given to that of the Countess Donna Clelio Grillo Boromeo, one of the most learned ladies of the age, to whom Signior Gimma dedicates his literary history of Italy. She had lately established an academy of experimental philosophy in her palace at Milan; of which Signior Vallisnieri was nominated president, and had already drawn up the regulations for it, though we do not find it has yet taken place. In the number of these academies may also be ranked the assembly of the learned, who of late years met at Venice in the house of Signior Cristino Martinelli, a noble Venetian, and a great patron of learning.

ACADEMIA *Cosmografica*, or that of the Argonauts, was instituted at Venice, at the instance of F. Coronelli, for the improvement of geography; the design being to procure exact maps, geographical, topographical, hydrographical, and ichnographical, of the celestial as well as terrestrial globe, and their several regions or parts, together with geographical, historical, and astronomical descriptions accommodated to them: to promote which purposes, the several members oblige themselves, by their subscription, to take one copy or more of each piece published under the direction of the academy; and to advance the money, or part of it, to defray the charge of publication. To this end there are three societies settled, namely at Venice, Paris, and Rome; the first under F. Moro, provincial of the Minorites of Hungary; the second under the abbot Laurence au Rue Payenne au Marais; the third under F. Ant. Baldigiani, jesuit, professor of mathematics in the Roman college; to whom those address themselves who are willing to engage in this design. The Argonauts number near 200 members in the different countries of Europe; and their device is the terraqueous globe, with the motto *Plus ultra*. All the globes, maps, and geographical writings of F. Coronelli have been published at the expence of this academy.

2

THE

THE ACADEMY *of Apatifts*, or Impartial Academy, deferves to be mentioned on account of the extent of its plan, including univerfally all arts and fciences. It holds from time to time public meetings at Florence, where any perfon, whether academift or not, may read his works, in whatever form, language, or fubject; the academy receiving all with the greateft impartiality.

In France there are many academies for the improvement of arts and fciences. F. Merfenne, it is faid, gave the firft idea of a philofophical academy in France, about the beginning of the feventeenth century, by the conferences of mathematicians and naturalifts, held occafionally at his lodgings; at which Des Cartes, Gaffendus, Hobbes, Roberval, Pafcal, Blondel, and others, affifted. F. Merfenne propofed to each of them certain problems to examine, or certain experiments to be made. Thefe private affemblies were fucceeded by more public ones, formed by M. Monmort, and M. Thevenot, the celebrated traveller. The French example animated feveral Englifhmen of rank and learning to erect a kind of philofophical academy at Oxford, towards the clofe of Cromwell's adminiftration; which after the reftoration was erected, by public authority, into a Royal Society: an account of which fee under the word. The Englifh example, in its turn, animated the French. In 1666 Louis XIV, affifted by the counfels of M. Colbert, founded an academy of fciences at Paris, called the

ACADEMIE *Royale des Sciences, or Royal Academy of Sciences*, for the improvement of philofophy, mathematics, chemiftry, medicine, belles-lettres, &c. Among the principal members, at the commencement in 1666, were the refpectable names of Carcavi, Huygens, Ro berval, Frenicle, Auzout, Picard, Buot, Du Hamel the Secretary, and Mariotte. There was a perfect equality among all the members, and many of them received falaries fiom the king, as at prefent. By the rules of the academy, every clafs was to meet twice a week; the philofophers and geometricians were to meet, feparately, every Wednefday, and then both together on the Saturday, in a room of the king's library, where the philofophical and mathematical books were kept: the hiftory clafs was to meet on the Monday and Thurfday in the room of the hiftorical books: and the clafs of belles-lettres on the Tuefday and Friday: and on the firft Thurfday of every month all the claffes met together, and by their fecretaries made a mutual report of what had been transacted by each clafs during the preceding month.

In 1699, on the application of the prefident, the abbé Bignon, the academy received, under royal authority and protection, a new form and conftitution; by the articles of which, the academy was to confift of four forts of members, namely honorary, penfionary, affociates, and eleves. The honorary clafs to confift of ten perfons, and the other three claffes of twenty perfons each. The prefident to be chofen annually out of the honorary clafs, and the fecretary and treafurer to be perpetual, and of the penfionary clafs. The meetings to be twice a week, on the Wednefday and Saturday; befides two public meetings in the year.

Of the penfionaries, or thofe who receive falaries, three to be geometricians, three aftronomers, three mechanifts, three anatomifts, three botanifts, and three

chemifts, the other two being the fecretary and treafurer. Of the twenty affociates, of which twelve to be French, and eight might be foreigners, two were to cultivate geometry, two aftronomy, two mechanics, two anatomy, two botany, and two chemiftry. Of the twenty eleves, one to be attached to each penfionary, and to cultivate his peculiar branch of fcience. The penfionaries and their eleves to refide at Paris. No regulars nor religious to be admitted, except into the honorary clafs: nor any perfon to be admitted a penfioner who was not known by fome confiderable work, or fome remarkable difcovery.

In 1716 the Duke of Orleans, then regent of France, by the king's authority made fome alteration in their conftitution. The clafs of eleves was fuppreffed; and inftead of them were inftituted twelve adjuncts, two to each of the fix claffes of penfioners. The honorary members were increafed to twelve: and a clafs of fix free affociates was made, who were not under the obligation of cultivating any particular branch of fcience, and in this clafs only could the regulars or religious be admitted. A prefident and vice-prefident to be appointed annually from the honorary clafs, and a director and fub-director annually from that of the penfioners. And no perfon to be allowed to make ufe of his quality of academician, in the title of any of his books that he publifhed, unlefs fuch book were firft approved by the academy.

The academy has for a device or motto, *Invenit & perficit*. And the meetings, which were formerly held in the king's library, have fince the year 1699 been held in a fine hall of the old Louvre.

Finally, in the year 1785 the king confirmed, by letters patent, dated April 23, the eftablifhment of the academy of fciences, making the following alterations, and adding claffes of agriculture, natural hiftory, mineralogy, and phyfics; incorporating the affociates and adjuncts, and limiting to fix the members of each clafs, namely three penfioners and three affociates; by which the former receive an increafe of falary, and the latter approach nearer to becoming penfioners.

By the articles of this inftrument it is ordained, that the academy fhall confift of eight claffes, namely, that of geometry, 2d aftronomy, 3d mechanics, 4th general phyfics, 5th anatomy, 6th chemiftry and metallurgy, 7th botany and agriculture, and 8th natural hiftory and mineralogy. That each clafs fhall remain irrevocably fixed at fix members; namely, three penfioners and three affociates, independent however of a perpetual fecretary and treafurer, of twelve free-affociates and of eight affociate ftrangers or foreigners, the fame as before, except that the adjunct-geographer for the future be called the affociate-geographer.

The claffes at firft to be filled by the following perfons, namely, that of geometry by Meffieurs de Borda, Jeaurat, Vandermonde, as penfioners; and Meffieurs Coufin, Meufnier, and Charles, as affociates: that of aftronomy by Meffieurs le Monnier, de la Lande, and le Gentil, as penfioners; and Meffieurs Meffier, de Caffini, and Dagelat, as affociates: that of mechanics by Meffieurs l'abbe Boffut, l'abbe Rochon, and de la Place, as penfioners; and Meffieurs Coulomb, le Gendre, and Perrier, as affociates: that of general phyfics by Meffieurs Leroy, Briffon, and

Bailly,

Bailly, as penfioners; and Meffieurs Monge, Mechain, and Quatremere, as affociates: that of anatomy by Meffieurs Daubluton, Tenon, and Portal, as penfioners; and Meffieurs Sabatier, Vicq d'azir, and Brouffonet, as affociates: that of chemiftry and metallurgy by Meffieurs Cadet, Lavoifier, and Beaume, as penfioners; and Meffieurs Cornette, Bertholet, aud Fourcroy, as affociates: that of botany and agriculture by Meffieurs Guettard, Fougeroux, and Adanfon, as penfioners; and Meffieurs de Juffieu, de la Marck, and Desfontaines, as affociates: and that of natural hiftory and mineralogy by Meffieurs Defmaretz, Saye, and l'abbe de Gua, as penfioners; and Meffieurs Darcet, l'abbe Haui, and l'abbe Teffier, as affociates. All names refpectable in the common-wealth of letters; and from whom the world might expect ftill farther improvements in the feveral branches of fcience.

The late M. Rouille de Meflay, counfellor of the parliament of Paris, founded two prizes, the one of 2500 livres, the other of 2000 livres, which the academy diftributed alternately every year: the fubjects of the former prize refpecting phyfical aftronomy, and of the latter, navigation and commerce.

The world is highly indebted to this academy for the many valuable works they have executed, or publifhed, both individually and as a body collectively, efpecially by their memoirs, making upwards of a hundred volumes in 4to, with the machines, indexes, &c. in which may be found moft excellent compofitions in every branch of fcience. They publifh a volume of thefe memoirs every year, with the hiftory of the academy, and eloges of remarkable men lately deceafed: alfo a general index to the volumes every ten years. An alteration was introduced into the volume for 1783, which it feems is to be continued in future, by omitting, in the hiftory, the minutes or extracts from the regifters, containing fome preliminary account of the fubjects of the memoires; but ftill however retaining the eloges of diftinguifhed men, lately deceafed.

M. l'abbe Rozier alfo has publifhed in four 4to volumes, an excellent index of the contents of all the volumes, and the writings of all the members, from the beginning of their publications to the year 1770; with convenient blank fpaces for continuing the articles in writing.

Their hiftory alfo, to the year 1697, was written by M. Du Hamel; and after that time continued from year to year by M. Fontenelle, under the following titles, Du Hamel Hiftoriæ Regiæ Academiæ Scientiarum, Paris, 4to. Hiftoire de l'Academie Royale des Sciences, avec les Memoires de Mathematique & de Phyfique, tirez des Regiftres de l'Academie, Paris, 4to. Hiftoire de l'Academie Royale des Sciences depuis fon etabliffement en 1666, jufqu'en 1699, en 13 tomes, 4to. A new hiftory, from the inftitution of the academy, to the period from whence M. de Fontenelle commences, has been formed; with a feries of the works publifhed under the name of this academy, during the firft interval.

Since the foregoing account was written, it is faid the Academy has been fuppreffed and abolifhed, by the prefent convention of France.

Befides the academies in the capital, there are a great many in other parts of France. The ACADEMIE *Royale*, at Caen, was eftablifhed by letters patent in the year 1705; but it had its rife fifty years earlier in private conferences, held firft in the houfe of M. de Brieux. M. de Segrais retiring to this city, to fpend the reft of his days, reftored and gave new luftre to their meetings. In 1707 M. Foucault, intendant of the generality of Caen, procured the king's letters patent for erecting them into a perpetual academy, of which M. Foucault was to be protector for the time, and the choice afterwards left to the members, the number of whom was fixed to thirty, chofen for this time by M. Foucault. But befides the thirty original members, leave was given to add fix fupernumerary members, from the ecclefiaftical communities in that city.

At Touloufe is the *Academie des jeux floraux*, compofed of forty perfons, the oldeft of the kingdom: befides an academy of fciences and belles-lettres, founded in 1750.

At Montpelier is the royal fociety of fciences, which fince 1708 makes but one body with the royal academy of fciences at Paris.

There are alfo other academies at Bourdeaux, founded in 1703, at Soiffons in 1674, at Marfeilles in 1726, at Lyons in 1700, at Pau in Bearn in 1721, at Montauban in 1744, at Angers in 1685, at Amiens in 1750, at Villefranche in 1679, at Dijon in 1740, at Nimes in 1682, at Befançon in 1752, at Chalons in 1775, at Rochelle in 1734, at Beziers in 1723, at Rouen in 1744, at Metz in 1760, at Arras in 1773, &c. The number of thefe academies is continually augmenting; and even in fuch towns as have no academies, the literati form themfelves into literary focieties, having nearly the fame objects and purfuits.

In Germany and other parts of Europe, there are various academies of fciences, &c. The

ACADEMIE *Royale des Sciences & des Belles-Lettres* of Pruffia, was founded at Berlin, in the year 1700, by Frederic I. king of Pruffia, of which the famous M. Leibnitz was the firft prefident, and its great promoter. The academy received a new form, and a new fet of ftatutes in 1710; by which it was ordained, that the prefident fhall be one of the counfellors of ftate; and that the members be divided into four claffes; the firft to cultivate phyfics, medicine, and chemiftry; the fecond, mathematics, aftronomy, and mechanics; the third, the German language, and the hiftory of the country; and the fourth, oriental learning, particularly as it may concern the propagation of the gofpel among infidels. That each clafs elect a director for themfelves, who fhall hold his poft for life. That they meet in the caftle called the New Marfhal, the claffes to meet in their turns, one each week. And that the members of any of the claffes have free accefs into the affemblies of the reft. Several volumes of their transactions have been publifhed in Latin, from time to time, under the title of Mifcellanea Berolinenfia.

In 1743 the late famous Frederic II. king of Pruffia, made great alterations and improvements in the academy. Inftead of a great lord or minifter of ftate, who had ufually prefided over the academy, he wifely judged that office would be better filled by a man of letters; and he honoured the French academy of fciences by fixing upon one of its members for a prefident, namely M. Maupertuis, a diftinguifhed character in the literary world, and whofe conduct in improving the academy was a proof of the found judgment of the king,

king, who at the same time made new regulations for the academy, and took the title of its Protector. From that time the transactions of the academy have been published, under the title of Histoire de l'Academie Royale des Sciences et Belles Lettres à Berlin, much in the manner of those of the French academy of sciences, and in the French language; and the volumes are now commonly published annually. Besides the ordinary private meetings of the academy, it has two public ones in the year, in January and May, at the latter of which is given a prize gold medal, of the value of 50 ducats, or about 20 guineas. The subject of the prize is successively physics, mathematics, metaphysics, and general literature. For the academy has this peculiar circumstance, that it embraces also metaphysics, logic, and morality; having one class particularly appropriated to these objects, called the class of Speculative Philosophy.

Imperial ACADEMY *of Petersburgh.* This academy was projected by the Czar Peter I, commonly called Peter the Great, who in so many other instances also was instrumental in raising Russia from the state of barbarity in which it had been immerged for so many ages. Having visited France in 1717, and among other things informed himself of the advantages of an academy of arts and sciences, he resolved to establish one in his new capital, whither he had drawn by noble encouragements several learned strangers, and made other preparations, when his death prevented him from fully accomplishing that object, in the beginning of the year 1725. Those preparations and intentions however were carried into execution the same year, by the establishment of the academy, by his consort the czarina Catherine, who succeeded him. And soon after the academy composed the first volume of their works, published in 1728, under the title of Commentarii Academiæ Scientiarum Imperialis Petropolitanæ; which they continued almost annually till 1746, the whole amounting to 14 volumes, which were published in Latin, and the subjects divided and classed under the following heads, namely mathematics, physics, history, and astronomy. Their device a tree bearing fruit not ripe, with the modest motto *paullatim.*

Most part of the strangers who composed this academy being dead, or having retired, it was rather in a languishing state at the beginning of the reign of the empress Elizabeth, when the count Rasomowski was happily appointed president, who was instrumental in recovering its vigour and labours. This empress renewed and altered its constitution, by letters patent dated July 24, 1747, giving it a new form and regulations. It consists of two chief parts, an academy, and a university, having regular professors in the several faculties, who read lectures as in our colleges. The ordinary assemblies are held twice a week, and public or solemn ones thrice in the year; in which an account is given of what has been done in the private ones. The academy has a noble building for their meetings, &c, with a good apparatus of instruments, a fine library, observatory, &c. Their first volume, after this renovation, was published for the years 1747 and 1748, and they have been since continued from year to year, now to the amount of near thirty volumes, under the title of Novi Commentarii Academiæ Scientiarum Imperialis

Petropolitanæ. They are printed in the Latin language, and contain many excellent compositions in all the sciences, especially the mathematical papers of the late excellent M. L. Euler, which always made a considerable portion of every volume. The subjects are classed under heads in the following order, mathematics, physico-mathematics, physics, which include botany, anatomy, &c, and astronomy; the whole prefaced by historical extracts, or minutes, relating to each paper or memoir, after the manner of the volumes of the French academy; but wanting however the eloges of deceased eminent men. Their device is a heap of ripe fruits piled on a table, with the motto *En addit fructus ætate recentes.*

Imperial and Royal ACADEMY *of Sciences and Belles Lettres, at Brussels.* This academy was founded in the year 1773; and several volumes of their memoirs have been published.

Royal ACADEMY *of Sciences,* at Stockholm, was instituted in 1739, and since that time it has published about sixty volumes of transactions, quarterly, in 8vo, in the Swedish language.

For an account of the Royal Society of London, and several other similar institutions, see the words Journal, Society, &c.

American ACADEMY *of Arts and Sciences,* was established in 1780 by the council and house of representatives in the province of Massachuset's Bay, for promoting the knowledge of the antiquities of America, and of the natural history of the country; for determining the uses to which its various natural productions may be applied; for encouraging medicinal discoveries, mathematical disquisitions, philosophical enquiries and experiments, astronomical, meteorological, and geographical observations, and improvements in agriculture, manufactures, and commerce; and, in short, for cultivating every art and science, which may tend to advance the interest, honour, dignity, and happiness, of a free, independent, and virtuous people. The members of this academy are never to be less than forty, nor more than two hundred.

ACADEMY is also used among us for a kind of collegiate school, or seminary; where youth are instructed in the liberal arts and sciences in a private way: now indeed it is used for all kinds of schools.

Frederic I, king of Prussia, established an academy at Berlin in 1703, for educating the young nobility of the court, suitable to their extraction. The expence of the students was very moderate, the king having undertaken to pay the extraordinaries. This illustrious school, which was then called the academy of princes, has now lost much of its first splendour.

The Romans had a kind of military academies established in all the cities of Italy, under the name of *Campi Martis.* Here the youth were admitted to be trained for war at the public expence. And the Greeks, besides academies of this kind, had military professors, called *Tactici,* who taught all the higher offices of war, &c.

We have two royal academies of this kind in England, the expences of which are defrayed by the government; the one at Woolwich, for the artillery and military engineers; and the other at Portsmouth, for the navy. The former was established by his late

majesty

majesty king George II, by warrants dated April the 30th and November the 18th, 1741, for instructing persons belonging to the military part of the ordnance, in the several branches of mathematics, fortification, &c, proper to qualify them for the service of artillery and the office of engineers. This institution is under the direction of the master-general and board of ordnance for the time being: and at first the lectures of the masters in the academy were attended by the practitioner-engineers, with the officers, serjeants, corporals, and private men of the artillery, besides the cadets. At present however none are educated there but the gentlemen cadets, to the number of 90 or 100, where they receive an education perhaps not to be obtained or purchased for money in any part of the world. The master-general of the ordnance is always captain of the cadets' company, and governor of the academy; under him are a lieutenant-governor, and an inspector of studies. The masters have been gradually increased, from two or three at first, now to the number of twelve, namely, a professor of mathematics, and two other mathematical masters, a professor of fortification, and an assistant, two drawing masters, two French masters, with masters for fencing, dancing, and chemistry. This institution is of the greatest consequence to the state, and it is hardly credible that so important an object should be accomplished at so trifling an expence. It is to be lamented however that the academy is fixed in so unhealthy a situation; that the lecture rooms and cadets' barracks are so small as to be insufficient for the purposes of the institution; and that the salaries of the professors and masters should be so inadequate to their labours, and the benefit of their services.

The Royal Naval Academy at Portsmouth was founded by George I, in 1722, for instructing young gentlemen in the sciences useful for navigation, to breed officers for the royal navy. The establishment is under the direction of the board of admiralty, who give salaries to two masters, by one of whom the students are boarded and lodged, the expence of which is defrayed by their own friends, nothing being supplied by the government but their education.

ACANTHUS, *in Architecture*, the leaves of a plant which forms the ornament of the capital of the Corinthian order. Vitruvius ascribes the use of it to the following accident. A young girl dying, her nurse was desirous of consecrating to her manes certain toys which she was fond of in her life-time; which the good woman carried in a little basket, covered with a square tile, and placed it among some green plants which grew on her grave. One of these, which happened to be the acanthus, as it grew up, invironed and in a manner embraced the basket; which Callimachus, a noted Greek sculptor, casting his eyes upon, from thence took the hint of this elegant ornament. See ABACUS.

ACCELERATED *Motion*, is that which receives fresh accessions of velocity; and the acceleration may be either equably or unequably: if the accessions of velocity be always equal in equal times, the motion is said to be equably or uniformly accelerated; but if the

accessions, in equal times, either increase or decrease, then the motion is unequably or variably accelerated.

Acceleration is directly opposite to retardation, which denotes a diminution of velocity.

ACCELERATION comes chiefly under consideration in physics, in the descent of heavy bodies, tending or falling towards the centre of the earth, by the force of gravity.

That bodies are accelerated in their natural descent, is evident both to the sight, and from observing that the greater height they fall from, the greater force they strike with, and the deeper impressions they make in soft substances.

The acceleration of falling bodies has been ascribed to various causes, by different philosophers. Some have attributed it to the pressure of the air downwards: the more a body descends, the longer and heavier, say they, must be the column of atmosphere incumbent upon it; to which they add, that the whole mass of fluid pressing by an infinity of right-lines all ultimately meeting in the earth's centre, such central point must support, as it were, the pressure of the whole mass; and that consequently the nearer a body approaches to it, the more must it receive of the pressure of a multitude of lines tending to unite in the central point.

Mr. Hobbes endeavours to account for this acceleration from a new impression of the cause which makes bodies fall; in which he is so far right. But then he as far mistakes, as to the cause of the fall, which he thinks is the air: at the same time, says he, that one particle of air ascends, another descends; for in consequence of the earth's motion being two-fold, that is circular and progressive, the air must at once both ascend and circulate; whence it follows, that a body falling in this medium, and receiving a new pressure every instant, must have its motion accelerated.

But to both these systems it may be answered, that the air is quite out of the question; for it is very evident that bodies fall, and in falling have their motion accelerated, in vacuo, as in open air, and even more than in the air, in as much as this opposes and somewhat retards their fall.

The Gassendists assign another reason for the acceleration: they pretend that there are continually issuing out of the earth certain attractive corpuscles, directed in an infinite number of rays; those, say they, ascend and then descend, in such sort that the nearer a body approaches to the earth's centre, the more of these attractive rays press upon it, in consequence of which its motion becomes more accelerated.

The peripatetics endeavour to explain the matter thus: the motion of heavy bodies downward, arises, say they, out of an intrinsic principle that causes a tendency in them to the centre, as the place appropriated to their element; where, when they can once arrive, they will be at perfect rest; and therefore, continue they, the nearer bodies approach to it, the more the velocity of their motion is increased: a notion too idle to merit confutation.

The Cartesians account for acceleration, by reiterated impulses of their *materia subtilis*, acting continually

D or

on falling bodies, and propelling them downwards: a conceit equally unintelligible and absurd with the former.

But, leaving all such visionary causes of acceleration, and only admitting the existence of such a force as gravity, so evidently inherent in all bodies, without regard to what may be the cause of it, the whole mystery of acceleration will be cleared up. Consider gravity then, with Galileo, only as a cause or force which acts continually on heavy bodies; and it will be easy to conceive that the principle of gravitation, which determines bodies to descend, must by a necessary consequence accelerate them in falling.

A body then having once begun to descend, through the impulse of gravity; the state of descending is now, by Newton's first law of nature, become as it were natural to it; insomuch that, were it left to itself, it would for ever continue to descend, even though the first cause of its descent should cease. But besides this determination to descend, impressed upon it by the first cause of motion, which would be sufficient to continue to infinity the degree of motion already begun, new impulses are continually superadded by the same cause; which continues to act upon the body already in motion, in the same manner as if it had remained at rest. There being then two causes of motion, acting both in the same direction; it necessarily follows, that the motion which they unitedly produce, must be more considerable than what either could produce separately. And as long as the velocity is thus continued, the same cause still subsisting to increase it more, the descent must of necessity be continually accelerated.

Supposing then that gravity, from whatever principle it arises, acts uniformly upon all bodies at the same distance from the centre of the earth: dividing the time which the heavy body takes up in falling to the earth, into indefinitely small equal parts; gravity will impel the body toward the centre of the earth, in the first indefinitely short instant of the descent. If after this we suppose the action of gravity to cease, the body will continue perpetually to advance uniformly toward the earth's centre, with an indefinitely small velocity, equal to that which resulted from the first impulse.

But then if we suppose that the action of gravity still continues the same after the first impulse; in the second instant, the body will receive a new impulse toward the earth, equal to that which it received in the first instant; and consequently its velocity will be doubled; in the third instant, it will be tripled; in the fourth, quadrupled; in the fifth, quintupled; and so on continually: for the impulse made in any preceding instant, is no ways altered by that which is made in the following one; but they are, on the contrary, always accumulated on each other.

So that the instants of time being supposed indefinitely small, and all equal, the velocity acquired by the falling body, will be, in every instant, proportional to the times from the beginning of the descent; and consequently the velocity will be proportional to the time in which it is produced. So that if a body, by this constant force, acquire a velocity of $16\frac{1}{12}$ feet suppose in one second of time; it will acquire a velocity

I

of $32\frac{1}{6}$ feet in two seconds, $48\frac{1}{4}$ feet in 3 seconds, $64\frac{1}{3}$ in 4 seconds, and so on. Nor ought it to seem strange that all bodies, small or large, acquire, by the force of gravity, the same velocity in the same time. For every equal particle of matter being endued with an equal impelling force, namely its gravity or weight, the sum of all the forces, in any compound mass of matter, will be proportional to the sum of all the weights, or quantities of matter to be moved; consequently, the forces and masses moved, being thus constantly increased in the same proportion, the velocities generated will be the same in all bodies, great or small. That is, a double force moves a double mass of matter, with the same velocity that the single force moves the single mass; and so on. Or otherwise, the whole compound mass falls all together with the same velocity, and in the same manner, as if its particles were not united, but as if each fell by itself, separated all from one another. And thus all being let go at once, they would fall together, just as if they were united into one mass.

The foregoing law of the descent of falling bodies, namely that the velocities are always proportional to the times of descent, as well as the following laws concerning the spaces passed over, &c, were first discovered and taught by the great Galileo, and that nearly in the following manner.

Because the constant velocity with which any body moves, or the space it passes over in a given time, as suppose one second, being multiplied by the time, or number of seconds it is in motion, expresses the space passed over in that time; and the area or space of a rectangular figure being denoted by the length multiplied by the breadth; therefore the space so run over, may be considered as a rectangle compounded of the time and velocity, that is a rectangle of which the time denotes the length, and the velocity the breadth. Suppose then A to be the heavy body which descends, and A B to denote the whole time of any descent; which

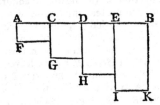

let be divided into a certain number of equal parts, denoting intervals or portions of the given time, as A C, C D, D E, &c. Imagine the body to descend, during the time expressed by the first of the divisions A C, with a certain uniform velocity arising from the force of gravity acting on it, which let be denoted by A F, the breadth of the rectangle C F; then the space run through during the time denoted by A C, with the velocity denoted by A F, will be expressed by the rectangular space C F.

Now the action of gravity having produced, in the first moment, the velocity A F, in the body, before at rest; in the first two moments it will produce the velocity C G, the double of the former; in the third moment, to the velocity C G will be added one degree
more,

more, by which means will be produced the velocity D H, triple of the firſt; and ſo of the reſt; ſo that during the whole time A B, the body will have acquired the velocity B K. Hence, taking the diviſions of the line A B at pleaſure; for example, the diviſions A C, C D, &c, for the times; the ſpaces run through during thoſe times, will be as the areas or rectangles C F, D G, &c; and ſo the ſpace deſcribed by the moving body during the whole time A B, will be equal to all the rectangles, that is, equal to the whole indented ſpace A B K I H G F. And thus it will happen if the increments of velocity be produced, as we may ſay, all at once, at the end of certain portions of finite time; for inſtance at C, at D, &c; ſo that the degree of motion remains the ſame to the inſtant that a new acceleration takes place.

By conceiving the diviſions of time to be ſhorter, for example but half as long as the former, the indentures of the figure will be proportionably more contracted, and it will approach nearer to a triangle; and ſo much the nearer as the diviſions of time are ſhorter: and if theſe be ſuppoſed infinitely ſmall; that is, if increments of the velocity be ſuppoſed to be acquired continually, and at each indiviſible particle of time, which is really the caſe, the rectangles ſo ſucceſſively produced, will form a true triangle, as A B C; the whole time A B conſiſting of minute portions A 1,

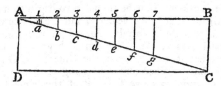

12, 23, &c; and the area of the triangle A B C, of all the minute ſurfaces, or minute trapeziums, which anſwer to the diviſions of the times; the area of the whole triangle A B C, denoting the ſpace run through during the whole time A B; and the area of any ſmaller triangle A 7 g, denoting the ſpace run through during the correſponding time A 7. Bnt the triangles A 1 a, A 7 g, &c, being ſimilar, have their areas to each other as the ſquares of their like ſides A 1, A 7, &c; and conſequently the ſpaces gone through, in any times counted from the beginning, are to each other as the ſquares of the times.

Hence, in any right-angled triangle, as A B C, the one ſide A B repreſents the time, the other ſide B C the velocity acquired in that time, and the area of the triangle the ſpace deſcribed by the falling body.

From the preceding demonſtration is alſo drawn this other general theorem in motions that are uniformly accelerated; namely, that a body deſcending with a uniformly accelerated motion, deſcribes in the whole time of its deſcent, a ſpace, which is exactly the half of that which it would deſcribe uniformly in the ſame time, with the velocity it has acquired at the end of its accelerated fall. For it has been ſhewn that the whole ſpace which the falling body has run through in the time A B, is repreſented by the triangle A B C, the laſt velocity being B C; and the ſpace which the

body would run through uniformly in the ſame time A B, conſtantly with the ſaid greateſt velocity B C, is repreſented by the rectangle A B C D: but it is well known that the rectangle A B C D is double the triangle A B C; and, therefore the latter ſpace run through, is double the former; that is, the ſpace run through by the accelerated motion, is juſt half of that which the body would deſcribe in the ſame time, moving uniformly with the velocity acquired at the end of its accelerated fall.

Hence then, from the foregoing conſiderations are deduced the following general laws of uniformly accelerated motions, namely,

1ſt. That the velocities acquired, are conſtantly proportional to the times; in a double time a double velocity, &c.

2d. That the ſpaces deſcribed in the whole times, each counted from the commencement of the motion, are proportional to the ſquares of the times, or to the ſquares of the velocities; that is, in twice the time, the body will deſcribe 4 times the ſpace; in thrice the time, it will deſcribe 9 times the ſpace; in quadruple the time, 16 times the ſpace; and ſo on. In ſhort, if the times are proportional
to the numbers - 1, 2, 3, 4, 5, &c,
the ſpaces will be as 1, 4, 9, 16, 25, &c,
which are the ſquares of the former. So that if a body, by the natural force of gravity, fall through the ſpace of $16\frac{1}{12}$ feet in the firſt ſecond of time; then in the firſt two ſeconds of time it will fall through four times as much, or $64\frac{1}{3}$ feet; in the firſt three ſeconds it will fall nine times as much, or $144\frac{3}{4}$ feet; and ſo on. And as the ſpaces fallen through are as the ſquares of the times, or of the velocities; therefore the times, or the velocities, are proportional to the ſquare roots of the ſpaces.

3d. The ſpaces deſcribed by falling bodies, in a ſeries of equal inſtants or intervals of time, will be as the odd numbers $\quad$ 1, 3, 5, 7, 9, &c,
which are the differences of $\rbrace$
the ſquares or whole ſpaces $\lbrace$ 1, 4, 9, 16, 25, &c,
that is, the body which has run through $16\frac{1}{12}$ feet in the firſt ſecond, will in the next ſecond run through $48\frac{1}{3}$ feet, in the third ſecond $80\frac{5}{12}$, and ſo on.

4th. If the body fall through any ſpace in any time, it acquires a velocity equal to double that ſpace; that is, in an equal time, with the laſt velocity acquired, if uniformly continued, it would paſs over juſt double the ſpace. So if a body fall through $16\frac{1}{12}$ feet in the firſt ſecond of time, then it has acquired a velocity of $32\frac{1}{3}$ feet in a ſecond; that is, if the body move uniformly for one ſecond, with the velocity acquired, it will paſs over $32\frac{1}{3}$ feet in this one ſecond: and if in any time the body fall through 100 feet; then in another equal time, if it move uniformly with the velocity laſt acquired, it will paſs over 200 feet. And ſo on.

But, as the method of demonſtration uſed by Galileo, by means of infinitely ſmall parts forming a regular triangle, is not approved of by many perſons, the ſame laws may be otherwiſe demonſtrated thus: let the whole time of a body's free deſcent be divided into any number of parts, calling each of theſe parts 1; and let a denote the velocity acquired at the end of the firſt

part

part of time; then will 2 *a*, 3 *a*, 4 *a*, &c, reprefent the velocities at the end of the 2d, 3d, 4th, &c, part of time, becaufe the velocities are as the times; and for the fame reafon ¼*a*, ¾*a*, ⅝*a*, &c, will be the velocities at the middle point of the firft, fecond, third, &c, part of time. But now as the velocities increafe uniformly, the fpace defcribed in any one of thefe parts of time, may be confidered as uniformly defcribed with its middle velocity, or the velocity in the middle of that part of time; and therefore multiplying thofe mean velocities each by their common time 1, we have the fame fractions ½*a*, ¾*a*, ⅝*a*, &c, for the fpaces paffed over in the fucceffive parts of the time; that is, the fpace ½*a* in the firft time, ¾*a* in the fecond, ⅝*a* in the third, and fo on: then add thefe fpaces fucceffively to one another, and we obtain ½*a*, ⁴⁄₂*a*, ⁹⁄₂*a*, ¹⁶⁄₂*a*, &c, for the whole fpaces defcribed from the beginning of the motion to the end of the firft, fecond, third, &c, portion of time; namely ½*a* fpace in one time, ⁴⁄₂*a* in 2 times, ⁹⁄₂*a* in 3 times, and fo on: and it is evident that thefe fpaces are as the numbers 1, 4, 9, 16, &c, which are as the fquares of the times.

And from this mode of demonftration, all the properties above mentioned evidently flow: fuch as that the whole fpaces - - ½*a*, ⁴⁄₂*a*, ⁹⁄₂*a*, &c, are as the fquares of the times - 1, 2, 3, &c, that the feparate fpaces ½*a*, ¾*a*, ⅝*a*, &c, defcribed in the fucceffive times, } 1, 3, 5, &c, are as the odd numbers - - and that the velocity *a* acquired in any time 1, is double the fpace - - ½*a* defcribed in the fame time.

As the laws of acceleration are very important, I fhall here infert the two following propofitions, fent me by my learned friend Mr. Abram Robertfon, of Chrift Church College Oxford, in which thofe laws are demonftrated in a manner fomewhat different.

" Ppoposition 1.

If from the point P in the ftraight line A B, the points M, N begin to move at the fame time, namely, M towards A with a motion uniformly retarded, and N from reft towards B with a motion uniformly accelerated; and if the velocity of M decreafes as much as the velocity of N increafes in the fame time; then the fpace M N is generated by an uniform motion, equal to the velocity with which M begins to move.

A————M————P————N————B

For, by hypothefis, whatever is loft in the velocity of M by retardation, is added to the velocity of N by acceleration: the joint velocities, therefore, of M and N muft always be equal. But it is by the joint velocities of M and N that the fpace M N is generated. Confequently M N is generated by an uniform motion, which is evidently equal to the velocity with which M begins to move.

" Proposition II.

If a point begins to move in the direction of a ftraight line, and continues to move in the fame di-

rection with a velocity uniformly accelerated; the fpace paffed over in any given time, will be equal to half the fpace paffed over in the fame time with the velocity with which the acceleration ends.

Let the point D begin to move from A towards B, along the ftraight line A B, with a motion uniformly accelerated; the fpace A D paffed over, is equal to half the fpace which the point would pafs over, in the fame time with the acquired velocity at D.

A————————D————————B

Let the points M, N begin to move in the ftraight line G H, at the fame time, with equal velocities uniformly accelerated; M beginning to move from G, and N from P; and at the fame time that M comes to the point P, let N come to H. Then as M and N

G————M————P————N————H

move with equal velocities, uniformly accelerated, it is evident that the fpaces, which they pafs over in the fame time, are equal to one another; confequently the fpace G P is equal to the fpace P H. Now as M begins to move from G with a velocity uniformly accelerated, it will arrive at P with an acquired velocity. Hence it is evident, if it be fuppofed to begin to move from P with this acquired velocity, and proceed toward G with a velocity uniformly retarded in the fame degree that it was accelerated when it began to move from G, that it will pafs over the fame fpace G P in the fame time. Wherefore, fuppofing the two points M, N to begin to move from P at the fame time, namely the point M beginning to move with the acquired velocity mentioned above, and proceeding towards G with the velocity uniformly retarded, defcribed above; and the point N as before with the velocity uniformly accelerated: then as the acceleration and retardation are uniform, they will be equal in equal fpaces of time. Again, as M is retarded in the fame degree that it was accelerated when it began to move from G, that is, in the fame degree that N is accelerated, by the former prop. M N is generated by an uniform velocity. But when the point M arrives at G, its velocity becomes equal to 0 or nothing; and at the time that M arrives at G, N arrives at H with the acquired velocity. Wherefore, as the velocities of M and N taken jointly are equal, and confequently uniform, the fpace G H is paffed over with the velocity of N at H, in the fame time that P H is paffed over by N beginning to move from P with a velocity uniformly accelerated to H. But P H is half of G H. " Hence the prop. is manifeft."

And hence the other laws of the fpaces, before mentioned, eafily follow.

Since the fpaces defcended are as the fquares of the times, and the abfciffes of a parabola are as the fquares of the ordinates, therefore the relation of the times and fpaces defcended may be very well reprefented by the ordinates and abfciffes of that figure. Thus if A B be the axis of the parabola A *b d f h*, and A C a tangent at

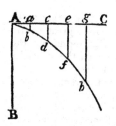

at the vertex perpendicular to the axis, divided into any number of equal parts A a, ac, ce, &c, for the times; and if there be drawn ab, cd, ef, &c, parallel to the axis: hence if ab be the space defcended in the time A a, then cd will be the space defcended in the time A c, and ef the space defcended in the time A e, and fo on continually.

From the properties above-demonftrated, are derived the following practical formulas or theorems for ufe, Namely, if g denote the fpace paffed over in the firft fecond of time, by a body urged by any conftant force, denoted by 1, and t denote the time or number of feconds in which the body paffes over any other fpace s, and v the velocity acquired at the end of that time; then from the foregoing laws we have $v = 2gt$, and $s = gt^2$; and from thefe two equations refult the following general formulas:

$$t = \frac{v}{2g} = \frac{2s}{v} = \sqrt{\frac{s}{g}},$$

$$v = 2gt = 2\sqrt{gs} = \frac{2s}{t},$$

$$s = gt^2 = \frac{vv}{4g} = \frac{tv}{2},$$

$$g = \frac{s}{tt} = \frac{v}{2t} = \frac{vv}{4s}.$$

And here, when the conftant force 1, is the natural force of gravity, then the diftance g defcended in the firft fecond, in the latitude of London, is $16\frac{1}{12}$ feet: but if it be any other conftant force, the value of g will be different, in proportion as the force is more or lefs.

The motion of an afcending body, or of one that is impelled upwards, is diminifhed or retarded by the fame principle of gravity, acting in a contrary direction, after the fame manner that a falling body is accelerated.

A body projected upwards, afcends until it has loft all its motion; which it does in the fame fpace of time, that the body would have taken up in acquiring, by falling, a velocity equal to that with which the falling body began to be projected upwards. And confequently the heights to which bodies afcend, when projected upwards with different velocities, are to each other as the fquares of thofe velocities.

ACCELERATED *Motion of Bodies on Inclined Planes.* The fame general laws obtain here, as in bodies falling freely, or perpendicularly; namely, that the velocities are as the times, and the fpaces defcended down the planes as the fquares of the times, or of the velocities. But thofe velocities are lefs, according to the fine of the plane's inclination; and the fpaces lefs, according to the fquare of the fine. See INCLINED *Plane.*

ACCELERATED *Motion of Pendulums.* See PENDULUM.

ACCELERATED *Motion of Projectiles.* See PROJECTILE.

ACCELERATED *Motion of Compreffed Bodies,* in ex-

panding or reftoring themfelves. See DILATATION, COMPRESSION, and ELASTICITY.

ACCELERATING FORCE, in Phyfics, is the force that accelerates the motion or velocity of bodies; and it is equal to, or expreffed by, the quotient arifing from the motive or abfolute force, divided by the mafs or weight of the body that is moved. In treating of phyfical confiderations refpecting forces, velocities, times, and fpaces gone over, the firft inquiry is the accelerating or accelerative force. This force is greater or lefs in proportion to the velocity it generates in the fame time, and by this velocity it is meafured. All accelerating forces are equal, and generate equal velocities, that have the motive forces directly proportional to the quantities of matter: fo a double motive force will move a double quantity of matter with the fame velocity, as alfo a triple motive force a triple quantity, a quadruple force a quadruple quantity, &c, all with the fame velocity. And this is the reafon why all bodies fall equally fwift by the force of gravity; for the motive force is exactly proportional to their weight or mafs. In general, the accelerating force is in the direct ratio of the motive force, and inverfe ratio of the quantity of matter. When a body is let fall freely, to defcend by the force of its natural gravity, it has been found by experiment that it falls through $16\frac{1}{12}$ feet in one fecond of time, and requires a velocity of $32\frac{1}{6}$ feet in that time: but if the quantity of matter be doubled, and the motive force remain the fame as before, by connecting the falling body to another of equal weight by means of a thread, this other body being laid on a horizontal plane, and the falling body hanging down off the plane, and drawing the other equal body along the plane after it; then the accelerating force will be only half of what it was before, and the fpace fallen in one fecond will be only $8\frac{1}{24}$ feet, and the velocity acquired $16\frac{1}{12}$: and if the quantity of matter be tripled, or the body drawn along the plane doubled; then the accelerating force will be only one-third of what it was at firft, and the fpace defcended in one fecond, and velocity acquired, each one-third of the firft: and fo on.

But accelerating forces are fometimes variable, as well as fometimes conftant; and the variation may be either increafing or decreafing.

The nature of conftant and variable accelerating forces, may be illuftrated in the following manner. Let two weights W, w, be connected by a thread

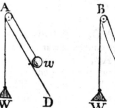
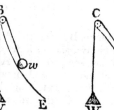

paffing over a pully at A, B, or C; and let the weight W defcend perpendicularly down, while it draws the fmaller weight w up the line A D, or B E, or C F, the

firft

firft being a ftraight inclined plane, and the other two curves, the one convex and the other concave to the perpendicular. Then the fmall weight *w* will always make fome certain refiftance to the free defcent of the large weight W, and that refiftance will be conftantly the fame in every part of the plane A D, the difficulty to draw it up being the fame in every point of it, becaufe every part of it has the fame inclination to the horizon, or to the perpendicular; and confequently the accef- fions to the velocity of the defcending weight W, will be always equal in equal times; that is, in this cafe W defcends by a uniformly accelerating force. But in the two curves B E, C F, the refiftance or oppofition of the fmall weight *w* will be conftantly altering as it is drawn up the curves, becaufe every part of them has a different inclination to the horizon, or to the perpen- dicular : in the former curve, the direction becomes more and more upright, or nearer perpendicular, as the fmall weight *w* afcends, and the oppofition it makes to the defcent of W, becomes more and more; and con- fequently the acceffions to the velocity of W will be always lefs and lefs in equal times; that is, W defcends by a decreafing accelerating force: but in the latter curve C F, as *w* afcends, the direction of the curve be- comes lefs and lefs upright, and the oppofition it makes to the defcent of W, becomes always lefs and lefs; and confequently the acceffions to the velocity of W will be always more and more in equal times; that is, W defcends by an increafing accelerating force. So that although the velocity continually increafes in all thefe cafes, yet whilft it increafes in a conftant ratio to the times of motion, in the plane A D; the velocity increafes in a lefs ratio than the time it afcended up B E, and in a greater ratio than the time increafes in the other curve C F.

Now the relations between the times and velocities in all thefe cafes, may be very well reprefented by the relations between the abfciffes and ordinates of certain lines. Thus let A B and A C be two ftraight lines,

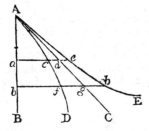

making any angle B A C; and A D, A E two curves, the former concave, and the latter convex towards A B : divide A B into any parts A *a*, A *b*, &c, reprefenting the times of motion; and draw the perpendiculars *a c d e*, *b f g h*, &c, reprefenting the velocities. Then in the right line A C, the ordinates *a d*, *b g*, being as the abfciffes A *a*, A *b*, this reprefents the cafe of uni- formly accelerated motion, in which the velocities are always as the times: but in the curve A D, the ordi- nates *a c*, *b f* increafe in a lefs ratio than the abfciffes A *a*, A *b*; and therefore this reprefents the cafe of decreafing acceleration, in which the velocities increafe

in a lefs ratio than the times: and in the other curve A E, the ordinates *a e*, *b h* increafe in a greater ratio than the abfciffes; and therefore this reprefents the cafe of increafing acceleration, in which the velocities increafe in a greater ratio than the times.

The feveral algebraic formulas or theorems, refpect- ing the time, velocity fpace, for conftant accelerating forces, are delivered above, at the article *Accelerated Mo- tion*, where the value of each circumftance is expreffed in finite determinate quantities. But in the cafes of vari- ably accelerated motions, the formulas will require the help of the method of fluxions to exprefs, not thofe general relations themfelves, but the fluxions of them; and confequently, taking the fluents of thofe ex- preffions, in particular cafes, the relations of time, fpace, velocity, &c, are obtained.

Now if *t* denote the time in motion,

  v the velocity generated by any force,

  s the fpace paffed over,

 and 2 *g* the variable force at any part of the motion, or the velocity the force would generate in one fecond of time, if it fhould continue invariable, like the force of gravity, during that one fecond; and therefore the value of this velocity 2 *g*, will be in proportion to $32\frac{1}{6}$ feet, as that variable force, is to 1 the force of gravity. Then becaufe the force may be fuppofed conftant dur- ing the indefinitely fmall time *t*, and that in uniform motions the fpaces and velocities are proportional to the times, we from thence obtain thefe two general funda- mental porportions,

$$v : \dot{s} :: 1'' : \dot{t}, \text{ or } \dot{s} = \dot{t}\, v\,;$$

$$2\, g : v :: 1'' : \dot{t}, \text{ or } \dot{v} = 2\, g\, \dot{t}.$$

From which are derived the four formulas below, in which the value of each quantity is expreffed in terms of the reft.

$$1. \quad \dot{t} = \frac{\dot{s}}{v} = \frac{\dot{v}}{2\,g}.$$

$$2. \quad \dot{v} = 2\, g\, \dot{t} = \frac{2\, g\, \dot{s}}{v}.$$

$$3. \quad \dot{s} = v\, \dot{t} = \frac{v\, \dot{v}}{2\,g}.$$

$$4. \quad 2\, g = \frac{\dot{v}}{\dot{t}} = \frac{v\, \dot{v}}{\dot{s}}.$$

And thefe theorems equally hold good for the de- ftruction of motion and velocity, by means of retard- ing forces, as for the generation of the fame by means of accelerating forces.

ACCELERATION, in *Mechanics*, the increafe of velo- city in a moving body.

ACCELERATION. *Aftron.* The Diurnal Acceleration of the fixed ftars, is the time which the ftars, in one diur- nal revolution, anticipate the mean diurnal revolution of the fun; which is $3^m\ 55^s\frac{9}{10}$ of mean time, or nearly $3^m\ 56^s$: that is, a ftar rifes, or fets, or paffes the me- ridian, about $3^m\ 56^s$ fooner each day. This accelera- tion of the ftars, which is only apparent in them, arifes from the real retardation of the fun, owing to his apparent

rent motion in his orbit towards the east, which is about $59'\ 8''\frac{2}{10}$ of a degree every day. So that the star which passed the meridian yesterday at the same moment with the sun, is to-day about $59'\ 8''$ past the meridian to the west, when the sun arrives at it; which will take him up about $3^m\ 56^s$ of time to pass over; and therefore the star passes by $3^m\ 56^s$ sooner than the sun each day, or anticipates his motion at that rate. The true quantity of this anticipation, or acceleration, is found by this proportion, $360°\ 59'\ 8''\frac{1}{5} : 59'\ 8''\frac{1}{5} :: 24$ hours : $3^m\ 55^s\frac{9}{10}$, the fourth term of which is the acceleration.

The diurnal acceleration serves to regulate the lengths or vibration of pendulums. If I observe a fixed star set or pass behind a hill, steeple, or such like, when the pendulum marks for instance $8^h\ 10^m$; and the next day, the eye being in the same place as before, the passage be at $8^h\ 6^m\ 4^s$; I thence conclude that the pendulum is well regulated, or truly measures mean time.

ACCELERATION *of a Planet*. A planet is said to be accelerated in its motion, when its real diurnal motion exceeds its mean diurnal motion. And, on the other hand, the planet is said to be retarded in its motion, when the mean exceeds the real diurnal motion. This inequality arises from the change in the distance of the planet from the sun, which is continually varying; the planet moving always quicker in its orbit when nearer the sun, and slower when farther off.

ACCELERATION *of the Moon*, is a term used to express the increase of the moon's mean motion from the sun, compared with the diurnal motion of the earth; by which it appears that, from some uncertain cause, it is now a little quicker than it was formerly. Dr. Halley was led to the discovery, or suspicion, of this acceleration, by comparing the ancient eclipses observed at Babylon, &c, and those observed by Albategnius in the ninth century, with some of his own time; as may be seen in N 218 of the Philosophical Transactions. He could not however ascertain the quantity of the acceleration, because the longitudes of Bagdat, Alexandria, and Aleppo, where the observations were made, had not been accurately determined. But since his time the longitude of Alexandria has been ascertained by Chazelles; and Babylon, according to Ptolemy's account, lies $50'$ east of Alexandria. From these data, Mr. Dunthorne, vol. 46 Philos. Transactions, compared the recorded times of several ancient and modern eclipses, with the calculations of them by his own tables, and thereby verified the suspicion that had been started by Dr. Halley; for he found that the same tables gave the moon's place more backward than her true place in ancient eclipses, and more forward than her true place in later eclipses; and thence he justly inferred that her motion in ancient times was slower, and in later times quicker, than the tables give it.

Not content however with barely ascertaining the fact, he proceeded to determine, as well as the observations would allow, the quantity of the acceleration; and by means of the most authentic eclipse, of which any good account remains, observed at Babylon in the year 721 before Christ, he found that the observed beginning of this eclipse was about an hour and three quarters sooner than the beginning by the tables; and that therefore the moon's true place preceded her place by computation by about $50'$ of a degree at that time.

Then admitting the acceleration to be uniform, and the aggregate of it as the square of the time, it will be at the rate of about $10''$ in 100 years.

Dr. Long, vol. ii. p. 436 of his Astronomy, enumerates the following causes from some one or more of which the acceleration may arise. Either 1st, the annual and diurnal motion of the earth continuing the same, the moon is really carried about the earth with a greater velocity than formerly: or, 2dly, the diurnal motion of the earth, and the periodical revolution of the moon, continuing the same, the annual motion of the earth about the sun is retarded; which makes the sun's apparent motion in the ecliptic a little slower than formerly; and consequently the moon, in passing from any conjunction with the sun, takes up a less time before she again overtakes the sun, and forms a subsequent conjunction: in both these cases, the motion of the moon from the sun is really accelerated, and the synodical month actually shortened: or, 3dly, the annual motion of the earth, and the periodical revolution of the moon, continuing the same, the rotation of the earth upon its axis is a little retarded; in this case, days, hours, minutes, &c, by which all periods of time must be measured, appear of a longer duration; and consequently the synodical month will appear to be shortened, though it really contain the same quantity of absolute time as it always did. If the quantity of matter in the body of the sun be lessened, by the particles of light continually streaming from it, the motion of the earth about the sun may become slower: if the earth increases in bulk, the motion of the moon about the earth may thereby be quickened.

ACCELERATIVE FORCE, &c, the same as ACCELERATING.

ACCESSIBLE, something that may be approached, or to which we can come. In Surveying, it is such a place as will admit of having a distance or length of ground measured from it; or such a height or depth as can be measured by actually applying a proper instrument to it. For the means of doing which, see ALTIMETRY, LONGIMETRY, or HEIGHTS-AND-DISTANCES.

ACCIDENS, ACCIDENT, *Philos.*

Per ACCIDENS is a term often used among philosophers, to denote what does not follow from the nature of a thing, but from some accidental quality of it: in this sense it stands opposed to *per se*, which denotes the nature and essence of a thing. Thus, fire is said to burn *per se*, or considered as fire, and not *per accidens*; but a piece of iron, though red-hot, only burns *per accidens*, by a quality accidental to it, and not considered as iron.

ACCIDENTS, in *Astrology*, denote the most extraordinary occurrences in the course of a person's life, either good or bad: such as a remarkable instance of good fortune, a signal deliverance, a great sickness, &c.

ACCIDENTAL, something that partakes of the nature of an accident; or that is indifferent, or not essential to its subject.—Thus whiteness is accidental to marble, and sensible heat to iron.

ACCIDENTAL *Colours*, so called by M. Buffon, are those which depend on the affections of the eye, in contradistinction to such as belong to light itself.

The impressions made upon the eye, by looking stedfastly on objects of a particular colour, are various according

according to the single colour, or assemblage of colours, in the object; and they continue for some time after the eye is withdrawn, and give a false colouring to other objects that are viewed during their continuance. M. Buffon has endeavoured to trace the connection between these accidental colours, and those that are natural, in a variety of instances. M. d'Arcy contrived a machine for measuring the duration of those impressions on the eye; and from the result of several trials he inferred, that the effect of the action of light on the eye continued about eight thirds of a minute.

The subject has also been considered by M. de la Hire, and M. Aepinus, &c. See Mem. Acad. Paris 1743, and 1765; Nov. Com. Petrop. vol. 10; also Dr. Priestley's Hist. of Discoveries relating to Vision, pa. 631.

ACCIDENTAL *Point*, in *Perspective*, is the point in which a right line drawn from the eye, parallel to another right line, cuts the picture or perspective plane.

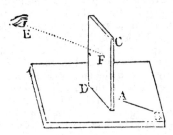

Let A B be the line given to be put into perspective, C F D the picture or perspective plane, and E the eye: draw E F parallel to A B; so shall F be the accidental point of the line A B, and indeed of all lines parallel to it, since only one parallel to them, namely E F, can be drawn from the same point E: and in the accidental point concur or meet the representations of all the parallels to A B, when produced.

It is called the accidental point, to distinguish it from the principal point, or point of view, where a line drawn from the eye perpendicular to the perspective plane, meets this plane, and which is the accidental point to all lines that are perpendicular to the same plane.

ACCIDENTAL *Dignities*, and *Debilities*, in *Astrology*, are certain casual dispositions, and affections, of the planets, by which they are supposed to be either strengthened, or weakened, by being in such a house of the figure.

ACCLIVITY, the slope or steepness of a line or plane inclined to the horizon, taken upwards; in contradistinction to *declivity*, which is taken downwards. So the ascent of a hill, is an *acclivity*: the descent of the same, a *declivity*.

Some writers on fortification use acclivity for *talus*: though more commonly the word talus is used to denote the slope, whether in ascending or descending.

ACCOMPANYMENT, in *Music*, denotes either the different parts of a piece of music for the different instruments, or the instruments themselves which accompany a voice, to sustain it, as well as to make the music more full.

The Accompanyment is used in recitative, as well as in song; on the stage, as well as in the choir, &c.

The ancients had likewise their accompanyments on the theatre; and they had even different kinds of instruments to accompany the chorus, from those which accompanied the actors in the recitation.

The accompanyment among the moderns, is often a different part, or melody, from the song it accompanies. But it is disputed whether it was so among the ancients.

Organists sometimes apply the word to several pipes which they occasionally touch to accompany the treble; as the drone, the flute, &c.

ACCOMPT. See ACCOUNT.

ACCORD, according to the modern French music, is the union of two or more sounds heard at the same time, and forming together a regular harmony.

They divide Accords into *perfect* and *imperfect*; and again into *consonances* and *dissonances*.

Accord is more commonly called CONCORD, which see.

Accord is also spoken of the state of an instrument, when its fixed sounds have among themselves all the justness that they ought to have.

ACCOUNT, or ACCOMPT, in *Arithmetic*, &c, a calculation or computation of the number or order of certain things; as the computation of time, &c.

There are various ways of accounting; as, by enumeration, or telling one by one; or by the rules of arithmetic, addition, subtraction, &c.

ACCOUNT, in *Chronology*, is nearly synonymous with style. Thus, we say the English, the foreign, the Julian, the Gregorian, the Old, or the New account, or style.

We account time by years, months, &c; the Greeks accounted it by olympiads; the Romans, by indictions, lustres, &c.

ACHERNER, or ACHARNER, in *Astronomy*, a star of the first magnitude in the southern extremity of the constellation Eridanus, marked α by Bayer. Its longitude for 1761, ♓ 11° 55′ 1″; and latitude south 59° 22′ 4″.

ACHILLES, a name given by the schools to the principal argument alleged by each sect of philosophers in behalf of their system. In this sense we say this is his Achilles; that is, his master-proof: alluding to the strength and importance of the hero Achilles among the Greeks.

Zeno's argument against motion is peculiarly termed Achilles. That philosopher made a comparison between the swiftness of Achilles, and the slowness of a tortoise, pretending that a very swift animal could never overtake a slow one that was before it, and that therefore there is no such thing as motion: for, said he, if the tortoise were one mile before Achilles, and the motion of Achilles 100 times swifter than that of the tortoise, yet he would never overtake it; and for this reason, namely, that while Achilles runs over the mile, the tortoise will creep over one hundredth part of a mile, and will be so much the foremost; again while Achilles runs over this $\frac{1}{100}$th part, the tortoise will creep over the 100th part of that $\frac{1}{100}$th part, and will still be this last part the foremost; and so on continually, according to an infinite series of 100th parts: from which he concluded that the swifter could never overtake the slower in any finite time, but that they must go on approaching

proaching to infinity. But this sophism lay in their considering as an infinite time, the sum of the infinite series of small times in which Achilles could run over the infinite series of spaces, $1 + \frac{1}{100} + \frac{1}{10000} + \frac{1}{1000000}$ &c, not knowing that the sum of this infinite series is equal to the quantity $1\frac{1}{99}$ of a mile, and that therefore Achilles will overtake the tortoise when the latter has crawled over $\frac{1}{99}$th of a mile.

ACHROMATIC, in *Optics*, without colour; a term which, it seems, was first used by M. de la Lande, in his astronomy, to denote telescopes of a new invention, contrived to remedy aberrations and colours. See *Aberration* and *Telescope*.

ACHROMATIC TELESCOPE, a singular species of refracting telescope, said to be invented by the late Mr. John Dollond, optician to the king, and since improved by his son Mr. Peter Dollond, and others.

Every ray of light passing obliquely from a rarer into a denser medium, changes its direction towards the perpendicular; and every ray passing obliquely from a denser into a rarer medium, changes its direction from the perpendicular. This bending of the ray, caused by the change of its direction, is called its refraction; and the quality of light which subjects it to this refraction, is called its refrangibility. Every ray of light, before it is refracted, is white, though it consists of a number of component rays, each of which is of a different colour. As soon as it is refracted, it is separated into its component rays, which, from that time, proceed diverging from each other, like rays from a centre: and this divergency is caused by the different refrangibility of the component rays, in such sort, that the more the original or component ray is refracted, the more will the compound rays diverge when the light is refracted by one given medium only.

From hence it has been concluded, that any two different mediums that can be made to produce equal refractions, will necessarily produce equal divergencies: whence it should also follow, that equal and contrary refractions should not only destroy each other, but that the divergency of the colours caused by one refraction, should be corrected by the other; and that to produce refraction that would not be affected by the different refrangibility of light, is impossible.

But Mr. Dollond has proved, by many experiments, that these conclusions are not well founded; from which experiments it appeared, that a ray of light, after equal and contrary refractions, was still spread into component rays differently coloured: in other words, that two different mediums may cause equal refraction, but different divergency; and equal divergency, with different refraction. It follows therefore that refraction may be produced, which is not affected by the different refrangibility of light. In other words, that, if the mediums be different, different refractions may be produced, though at the same time the divergency caused by one refraction shall be exactly counteracted by the divergency caused by the other; and so an object may be seen through mediums which, together, cause the rays to converge, without appearing of different colours.

This is the foundation of Mr. Dollond's improvement of refracting telescopes. By subsequent experiments he found, that different sorts of glass differed greatly in their refractive qualities, with respect to the divergency of colours. He found that crown glass causes the least diver-

gency, and white flint the most, when they are wrought into forms that produce equal refractions. He ground a piece of white flint glass into a wedge, whose angle was about 25 degrees; and a piece of crown glass to another, whose angle was about 29 degrees; and these he found refracted nearly alike, but that their divergency of colours was very different.

He then ground several other pieces of crown glass to wedges of different angles, till he got one that was equal, in the divergency it produced, to that of a wedge of flint glass of 25 degrees; so that when they were put together, in such a manner as to refract in contrary directions, the refracted light was perfectly free from colour. Then measuring the fractions of each wedge, he found that that of the white flint glass, was to that of the crown glass, nearly as two to three. And hence any two wedges, made of these two substances, and in this proportion, would, when applied together so as to refract in contrary directions, refract the light without any effect arising from the different refrangibility of the component rays.

Therefore, to make two spherical glasses that refract the light in contrary directions, one must be concave, and the other convex; and as the rays, after passing through both, must meet in a focus, the excess of the refraction must be in the convex one: and as the convex is to refract most, it appears from the experiment that it must be made of crown glass; and as the concave is to refract least, it must be made of white flint.

And farther, as the refractions of spherical glasses are in an inverse ratio of their focal distances, it follows that the focal distances of the two glasses should be in the ratio of the refractions of the wedges; for, being thus proportioned, every ray of light that passes through this combined glass, at whatever distance from its axis, will constantly be refracted by the difference between two contrary refractions, in the proportion required; and therefore the effect of the different refrangibility of light will be prevented.

The removal of this impediment, however, produced another: for the two glasses, which were thus combined, being segments of very deep spheres, the aberrations from the spherical surfaces became so considerable, as greatly to disturb the distinctness of the image. Yet considering that the surfaces of spherical glasses admit of great variations, though the focal distance be limited, and that by these variations their aberration might be made more or less at pleasure; Mr. Dollond plainly saw that it was possible to make the aberrating of any two glasses equal; and that, as in this case the refractions of the two glasses were contrary to each other, and their aberrations being equal, these would destroy each other.

Thus he obtained a perfect theory of making object glasses, to the apertures of which he could hardly perceive any limits: for if the practice could come up to the theory, they must admit of apertures of great extent, and consequently bear great magnifying powers.

The difficulties of the practice are, however, still very considerable. For first, the focal distances, as well as the particular surfaces, must be proportioned with the utmost accuracy to the densities and refracting powers of the glasses, which vary even in the same sort of glass, when made at different times. Secondly, there are four surfaces to be wrought perfectly spherical.

However, Mr. Dollond could construct refracting tele- scopes upon these principles, with such apertures and magnifying powers, under limited lengths, as greatly exceed any that were produced before, in forming the images of objects bright, distinct, and uninfected with colours about the edges, through the whole extent of a very large field or compass of view; of which he has given abundant and undeniable testimony. See TELE- SCOPE.

There has lately appeared in the Gentleman's Maga- zine (1790, pa. 890) a paper on the refracting tele- scope, by an author who signs *Veritus*, in which the invention is ascribed to another person, not heretofore mentioned; in these words: " As the invention has been claimed by M. Euler, M. Klingenstierna, and some other foreigners, we ought, for the honour of England, to assert our right, and give the merit of the discovery to whom it is due; and therefore, without farther preface, I shall observe, that the inventor was Chester More Hall, Esq. of More-hall, in Essex, who, about 1729, as appears by his papers, considering the different humours of the eye, imagined they were placed so as to correct the different refrangibility of light. He then conceived, that if he could find sub- stances having such properties as he supposed these hu- mours might possess, he should be enabled to construct an object glass that would shew objects colourless. After many experiments he had the good fortune to find those properties in two different sorts of glass, and making them disperse the rays of light in different di- rections, he succeeded. About 1733 he completed se- veral achromatic object glasses (though he did not give them this name), that bore an aperture of more than 2½ inches, though the focal length did not exceed 20 inches; one of which is now in the possession of the Rev. Mr. Smith, of Charlotte Street, Rathbone Place.

This glass has been examined by several gentlemen of eminence and scientific abilities, and found to pos- sess the properties of the present achromatic glasses.

Mr. Hall used to employ the working opticians to grind his lenses; at the same time he finished them with the radii of the surfaces, not only to correct the different refrangibility of rays, but also the aberration arising from the spherical figure of the lenses. Old Mr. Bass, who at that time lived in Bridewell precinct, was one of these working opticians, from whom Mr. Hall's invention seems to have been obtained.

In the trial at Westminster-hall about the patent for making achromatic telescopes, Mr. Hall was allowed to be the inventor; but Lord Mansfield observed, that " It was not the person that locked up his invention in his scrutoire that ought to profit by a patent for such an invention, but he who brought it forth for the bene- fit of the public." This, perhaps, might be said with some degree of justice, as Mr. Hall was a gentleman of property, and did not look to any pecuniary advan- tage from his discovery; and, consequently, it is very probable that he might not have an intention to make it generally known at that time.

That Mr. Ayscough, optician on Ludgate Hill, was in possession of one of Mr. Hall's achromatic tele- scopes in 1754, is a fact which at this time will not be disputed."

ACHRONICAL, or *Achronycal*. See ACRONYCHAL.

ACOUSTICS. This term, in physico-mathematical meaning, signifies the doctrine of hearing, and the art of assisting that sense by means of speaking trumpets, hearing trumpets, whispering galleries, and such like. See STENTROPHONIC TUBE.

Sturmius, in his Elements of Universal Mechanics, treating of Acoustics, after examining into the nature of sounds, describes the several parts of the external and internal ear, and their several uses and connexions with each other; and from thence deduces the mecha- nism of hearing: and lastly, he treats of the means of adding an intensity of force to the voice and other sounds; and explains the nature of echoes, otacoustic tubes, and speaking trumpets. See SOUND, EAR, VOICE, and ECHO.

Dr. Hook, in the preface to his Micrography, asserts that the lowest whisper, by certain means, may be heard at the distance of a furlong; and that he knew a way by which it is easy to hear any one speak through a wall of three feet thick; also that by means of an ex- tended wire, sound may be conveyed to a very great distance, almost in an instant.

ACRE, from the Saxon *æcre*, or German *acker*, a *field*, of the Latin *ager*. It is a measure of land, containing, by the ordinance for measuring land, made in the 33d and 34th of Edward I, 160 perches or square poles of land; that is, 16 in length and 10 in breadth, or in that proportion: and as the statute length of a pole is 5½ yards, or 16¼ feet, therefore the acre will contain 4840 square yards, or 43560 square feet. The chain with which land is commonly measured, and which was in- vented by Gunter, is 4 poles or 22 yards in length; and therefore the acre is just 10 square chains; that is, 10 chains in length and one in breadth, or in that pro- portion. Farther, as a mile contains 1760 yards, or 80 chains in length, therefore the square mile contains 640 acres.

The acre, in surveying, is divided into 4 roods, and the rood is 40 perches.

The French acre, *arpent*, is equal to 1¼ English acre; The Strasburg contains about ½ an English acre; The Welch acre contains about 2 English acres; The Irish acre contains 1 ac. 2 r. 19$\frac{7}{21}$ p. English.

Sir William Petty, in his Political Arithmetic, reckons that England contains 39 million acres: but Dr. Greve shews, in the Philos. Transf. Nº 330, that England contains not less than 46 million acres. Whence he infers that England is above 46 times as large as the province of Holland, which it is said con- tains but about one million of acres.

By a statute of the 31st of Elizabeth, it is ordained, that if any man erect a cottage, he shall annex four acres of land to it.

ACRONYCHAL, or ACRONYCAL, in *Astronomy*, is said of a star or planet, when it is opposite to the sun. It is from the Greek ακρωνχος, the point or ex- tremity of night, because the star rose at sun-set, or the beginning of night; and set at sun-rise, on the end of night; and so it shone all the night.

The acronychal is one of the three Greek poetic risings and settings of the stars; and stands distinguish- ed from Cosmical and Heliacal. And by means of which, for want of accurate instruments, and other ob- servations, they might regulate the length of their year.

ACROTERIA,

ACROTERIA, or ACROTERS, in *Architecture*, small pedestals, usually without bases, placed on pediments, and serving to support statues.

Those at the extremities ought to be half the height of the tympanum; and that in the middle, according to Vitruvius, one eighth part more.

ACROTERIA also are sometimes used to signify figures, whether of stone or metal, placed as ornaments or crownings, on the tops of temples, or other buildings.

It is also sometimes used to denote those sharp pinacles or spiry battlements, that stand in ranges about flat buildings, with rails and balustres.

ACTION, in *Mechanics* or *Physics*, a term used to denote, sometimes the effort which some body or power exerts against another body or power, and sometimes it denotes the effects resulting from such effort.

The Cartesians resolve all physical action into metaphysical. Bodies, according to them, do not act on one another; the action comes all immediately from the Deity; the motions of bodies, which seem to be the cause, being only the occasions of it.

It is one of the laws of nature, that action and reaction are always equal, and contrary to each other in their directions.

Action is either instantaneous or continued; that is, either by collision or percussion, or by pressure. These two sorts of action are heterogeneous quantities, and are not comparable, the smallest action by percussion exceeding the greatest action of pressure, as the smallest surface exceeds the longest line, or as the smallest solid exceeds the largest surface: thus, a man by a small blow with a hammer, will drive a wedge below the greatest ship on the stocks, or under any other weight; that is, the smallest percussion overcomes the pressure of the greatest weight. These actions then cannot be measured the one by the other, but each must have a measure of its own kind, like as solids must be measured by solids, and surfaces by surfaces: time being concerned in the one, but not in the other.

If a body be urged at the same time by equal and contrary actions, it will remain at rest. But if one of these actions be greater than its opposite, motion will ensue towards the part least urged.

The actions of bodies upon each other, in a space that is carried uniformly forward, are the same as if the space were at rest; and any powers or forces that act upon all bodies, so as to produce equal velocities in them in the same, or in parallel right lines, have no effect on their mutual actions, or relative motions. Thus the motion of bodies on board of a ship that is carried uniformly forward, are performed in the same manner as if the ship was at rest. And the motion of the earth about its axis has no effect on the actions of bodies and agents at its surface, except in so far as it is not uniform and rectilineal. In general, the actions of bodies upon each other, depend not on their absolute, but relative motion.

Quantity of ACTION, in Mechanics, a name given by M. de Maupertuis, in the Memoirs of the Academy of Sciences of Paris for 1744, and in those of Berlin for 1746, to the continual product of the mass of a body, by the space which it runs through, and by its celerity. He lays it down as a general law, that in the changes made in the state of a body, the quantity of action necessary to produce such change is the least possible. This principle he applies to the investigation of the laws of refraction, and even the laws of rest, as he calls them; that is, of the equilibrium or equipollency of pressures; and even to the modes of acting of the Supreme Being. In this way Maupertuis attempts to connect the metaphysics of final causes with the fundamental truths of mechanics; to shew the dependence of the collision of both elastic and hard bodies, upon one and the same law, which before had always been referred to separate laws; and to reduce the laws of motion, and those of equilibrium, to one and the same principle.

But this quantity of motion, of Maupertuis, which is defined to be the product of the mass, the space passed over, and the celerity, comes to the same thing as the mass multiplied by the square of the velocity, when the space passed over is equal to that by which the velocity is measured; and so the quantity of force will be proportional to the mass multiplied by the square of the velocity; since the space is measured by the velocity continued for a certain time.

In the same year that Maupertuis communicated the idea of his principle, professor Euler, in the supplement to his book, intitled *Methodus inveniendi lineas curvas maximi vel minimi proprietate gaudentes*, demonstrates, that in the trajectories which bodies describe by central forces, the velocity multiplied by what the foreign mathematicians call the element of the curve, always makes a minimum: which Maupertuis considered as an application of his principle to the motion of the planets.

It appears from Maupertuis's Memoir of 1744, that it was his reflections on the laws of refractions, that led him to the theorem above mentioned. The principle which Fermat, and after him Leibnitz, made use of, in accounting for the laws of refraction, is sufficiently known. Those mathematicians pretended, that a particle of light, in its passage from one point to another through two mediums, in each of which it moves with a different velocity, must do it in the shortest time possible: and from this principle they have demonstrated geometrically, that the particle cannot go from the one point to the other in a right line; but being arrived at the surface that separates the two mediums, it must alter its direction in such a manner, that the sine of its incidence shall be to the sine of its refraction, as its velocity in the first medium is to its velocity in the second: whence they deduced the well known law of the constant ratio of those sines.

This explanation, though very ingenious, is liable to this pressing difficulty, namely, that the particle must approach towards the perpendicular, in that medium where its velocity is the least, and which consequently resists it the most: which seems contrary to all the mechanical explanations of the refraction of bodies, that have hitherto been advanced, and of the refraction of light in particular.

Sir Isaac Newton's way of accounting for it, is the most satisfactory of any that has hitherto been offered, and gives a clear reason for the constant ratio of the sines, by ascribing the refraction to the attractive force of the mediums; from which it follows, that the densest mediums, whose attraction is the strongest, should cause the ray to approach the perpendicular; a fact confirmed by experiment. But the attraction of the medium could not cause the ray to approach towards the perpendicular, without increasing its velocity; as may easily be demonstrated. Thus then, according to Newton, the refraction must be towards the perpendicular, when the velocity is increased: contrary to the law of Fermat and Leibnitz.

Maupertuis has attempted to reconcile Newton's explanation with metaphysical principles. Instead of supposing, as the aforesaid gentlemen do, that a particle of light proceeds from one point to another in the shortest time possible; he contends that a particle of light passes from one point to another in such a manner, that the quantity of action shall be the least possible. This quantity of action, says he, is a real expence, in which nature is always frugal. In virtue of this philosophical principle he discovers, that not only the sines are in a constant ratio, but also that they are in the inverse ratio of the velocities, according to Newton's explanation, and not in the direct ratio, as had been pretended by Fermat and Leibnitz.

It is remarkable that, of the many philosophers who have written on refraction, none should have fallen upon so simple a manner of reconciling metaphysics with mechanics; since no more is necessary to that, than making a small alteration in the calculus founded upon Fermat's principle. Now according to that principle, the time, that is, the space divided by the velocity, should be a minimum; so that calling the space run through in the first medium S, with the velocity V, and the space run through in the second medium s, with the velocity v, we shall have $\dfrac{S}{V} + \dfrac{s}{v}$

$= a$ minimum; that is to say, $\dfrac{\dot{S}}{V} + \dfrac{\dot{s}}{v} = 0$, or $\dfrac{\dot{S}}{V} = -$

$\dfrac{\dot{s}}{v}$. Now it is easy to perceive, that the sines of incidence and refraction are to each other, as $\dot{S}$ to $-\dot{s}$; whence it follows, that those sines are in the direct ratio of the velocities V, v; which is exactly what Fermat makes it to be. But in order to have those sines to be in the inverse ratio of the velocities, it is only supposing $V\dot{S} + v\dot{s} = 0$; which gives $SV + svo$ or $S \times V + s \times v$ a minimum: which is Maupertuis's principle.

In the Memoirs of the Academy of Berlin, above cited, may be seen all the other applications which Maupertuis has made of this principle. And whatever may be determined as to his metaphysical basis of it, as also to the idea he has annexed to the quantity of action, it will still hold good, that the product of the space by the velocity is a minimum in some of the most general laws of nature.

ACTIVE, the quality of an agent. or of communicating motion or action to some body. In this sense the word stands opposed to passive: thus we say an active cause, active principle, &c.

Sir Isaac Newton shews that the quantity of motion in the world must be always decreasing, in consequence of the vis inertiæ, &c. So that there is a necessity for certain active principles to recruit it: such he takes the cause of gravity to be, and the cause of fermentation; adding, that we see but little motion in the universe, except what is owing to these active principles.

ACTIVITY, the virtue or faculty of acting. As the activity of an acid, a poison, &c: the activity of fire exceeds all imagination.

According to Sir Isaac Newton, bodies derive their activity from the principle of attraction.

Sphere of ACTIVITY, is the space which surrounds a body, as far as its efficacy or virtue extends to produce any sensible effect. Thus we say, the sphere of activity of a loadstone, of an electric body, &c.

ACUBENE, in *Astronomy*, the Arabic name of a star of the fourth magnitude, in the southern claw of Cancer, marked α by Bayer. Its longitude for 1761, Ω 10° 18′ 9″, south latitude 5° 5′ 56″.

ACUTE, or sharp; a term opposed to obtuse. Thus, ACUTE *Angle*, in *Geometry*, is that which is less than a right angle; and is measured by less than 90°, or by less than a quadrant of a circle. As the angle A B C.

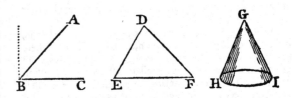

ACUTE *angled Triangle*, is that whose three angles are all acute; and is otherwise called an oxygenous triangle. As the triangle D E F.

ACUTE-*angled Cone*, is that whose opposite sides make an acute angle at the vertex, or whose axis, in a right cone, makes less than half a right angle with the side As the cone G H I.

Pappus, in his Mathematical Collections, says, this name was given to such a cone by Euclid and the ancients, before the time of Apollonius. And they called an

ACUTE-*angled Section of a Cone*, an Ellipsis, which was made by a plane cutting both sides of an acute-angled cone: not knowing that such a section could be generated from any cone whatever, till it was shewn by Apollonius.

ACUTE, in *Music*, is understood of a tone, or sound, which is high, sharp, or shrill, in respect of some other: in which sense the word stands opposed to grave. And both these sounds are independent of loudness or force: so that the tone may be acute or high, without being loud; and loud without being high or acute. For both the affections of acute and grave, depend intirely on the quickness or slowness of the vibrations by which they are produced.

Sounds considered as grave and acute, that is, in the relation of gravity and acuteness, constitute what is called tune, the foundation of all harmony.

ADAGIO, in *Music*, one of the terms used by the Italians to express a degree or distinction of time.

Adagio denotes the slowest time except grave.

Sometimes the word is repeated, as *adagio, adagio*, to denote a still slower time than the former.

Adagio

Adagio alfo fignifies a flow movement, when ufed fubftantively.

ADAMAS, in *Aftrology*, a name given to the moon.

ADAR, in the Hebrew *Chronology*, is the 6th month of their civil year, but the 12th of their ecclefiaftical year. It contains only 29 days; and it anfwers to our February; but fometimes entering into the month of March, according to the courfe of the moon.

ADDITION, the uniting or joining of two or more things together; or the finding of one quantity equal to two or more others taken together.

ADDITION, in *Arithmetic*, is the firft of the four fundamental rules or operations of that fcience; and it confifts in finding a number equal to feveral others taken together, or in finding the moft fimple expreffion of a number according to the eftablifhed notation. The quantity fo found equal to feveral others taken together, is named their fum.

The fign or character of addition is +, and is called *plus*. This character is fet between the quantities to be added, to denote their fum: thus, 3 + 6 = 9, that is, 3 plus 6 are equal to 9; and 2 + 4 + 6 = 12, that is, 2 plus 4 plus 6 are equal to 12.

Simple numbers are either added as above; or elfe by placing them under one another, as in the margin, and adding them together, one after another, beginning at the bottom: thus 2 and 4 make 6, and 6 make 12.

$$\begin{array}{r} 6 \\ 4 \\ 2 \\ \hline 12 \end{array}$$

Compound numbers, or numbers confifting of more figures than one, are added, by firft ranging the numbers in columns under each other, placing always the numbers of the fame denomination under each other, that is, units under units, tens under tens, and fo on; and then adding up each column feparately, beginning at the right hand, fetting down the fum of each column below it, unlefs it amount to ten or fome number of tens, and in that cafe fetting down only the overplus, and carrying one for each ten to the next column. Thus, to add 451 and 326,

$$\left.\begin{array}{r} 451 \\ 326 \end{array}\right\} \text{ that is } \left\{\begin{array}{r} 400 + 50 + 1 \\ 300 + 20 + 6 \end{array}\right.$$

$$\text{Sum } 777 = 700 + 70 + 7$$

Alfo to add the numbers 329 + 1562 + 20347 + 712048; fet them down as in the margin, and beginning at the loweft number on the right hand, fay 8 and 7 make 15, and 2 make 17, and 9 make 26; fet down 6, and carry 2 to the next column, faying 2 and 4 make 6, and 4 make 10, and 6 make 16, and 2 make 18; fet down 8, and carry 1, faying 1 and 3 make 4, and 5 make 9, and 3 make 12; fet down 2, and carry 1, faying 1 and 2 make 3, and 1 make 4, which fet down; then 1 and 2 make 3; and 7 is 7 to fet down: fo the fum of all together is 734286. Or it is the fame as the fums of the columns fet under one another, as in the margin, and then thefe added up in the fame manner.

$$\begin{array}{r} 329 \\ 1562 \\ 20347 \\ 712048 \\ \hline 734286 \end{array}$$

$$\begin{array}{r} 25 \\ 16 \\ 11 \\ 3 \\ 3 \\ 7 \\ \hline 734286 \end{array}$$

When a great number of feparate fums or numbers are to be added, as in long accounts, it is eafier to break or feparate them into two or more parcels, which are added up feverally, and then their fums added together for the total fum. And thus alfo the truth of the addition may be proved, by dividing the numbers into parcels different ways, as the totals muft be the fame in both cafes when the operation is right.

Another method of proving addition was given by Dr. Wallis, in his Arithmetic, publifhed 1657, by cafting out the nines, which method of proof extends alfo to the other rules of arithmetic. The method is this: add the figures of each line of numbers together feverally, cafting out always 9 from the fums as they arife in fo adding, adding the overplus to the next figure, and fetting at the end of each line what is over the nine or nines; then do the fame by the fum-total, as alfo by the former exceffes of 9, fo fhall the laft exceffes be equal when the work is right. So the former example will be proved as below:

$$\begin{array}{r|c|c} 329 & & 5 \\ 156 & \text{Excefs of 9's} & 5 \\ 20347 & & 7 \\ 712048 & & 4 \\ \hline 734286 & & 3 \end{array}$$

When the numbers are of different denominations; as pounds, fhillings, and pence; or yards, feet, and inches; place the numbers of the fame kind under one another, as pence under pence, fhillings under fhillings, &c; then add each column feparately, and carry the overplus as before, from one column to another. As in the following examples:

l.	s.	d.		Yards.	Feet.	Inches.
271	12	3		271	10	3
94	14	7		36	2	7
42	5	10		14	2	5
408	12	8	fums	323	0	3

ADDITION *of Decimals*, is performed in the fame manner as that of whole numbers, placing the numbers of the fame denomination under each other, in which cafe the decimal feparating points will range ftraight in one column; as in this example, to add together thefe numbers 371·0496 + 25·213 + 1·704 + 924·61 + ·0962.

$$\begin{array}{r} 371 \cdot 0496 \\ 25 \cdot 213 \\ 1 \cdot 704 \\ 924 \cdot 61 \\ \cdot 0962 \\ \hline \end{array}$$

The fum 1322·6728

ADDITION *of Vulgar Fractions*, is performed by bringing all the propofed fractions to a common denominator, if they have different ones, which is an indifpenfable preparation; then adding all the numerators together,

together,

together, and placing their fum over the common denominator for the fum total required.

So $\frac{1}{7} + \frac{2}{7} = \frac{3}{7}$. And $\frac{1}{3} + \frac{2}{3} + \frac{3}{3} = \frac{6}{3} = 1\frac{1}{3}$.

And $\frac{1}{4} + \frac{2}{7} = \frac{3}{7} + \frac{4}{7} = \frac{7}{7} = 1\frac{1}{8}$.

And $\frac{5}{6} + \frac{2}{3}$ of $\frac{3}{4} = \frac{5}{6} + \frac{1}{2} = \frac{8}{6} = \frac{4}{3} = 1\frac{1}{3}$.

And $\frac{1}{2}$ of $\frac{3}{4} + \frac{2}{3}$ of $5 = \frac{3}{8} + \frac{10}{3} = \frac{9}{24} + \frac{80}{24} = \frac{89}{24} = 3\frac{17}{24}$.

ADDITION *in Algebra*, or the addition of indeterminate quantities, denoted by letters of the alphabet, is performed by connecting the quantities together by their proper figns, and uniting or reducing fuch as are fufceptible of it; namely fimilar quantities, by adding their co-efficients together if the figns are the fame, but fubtracting them when different. Thus the quantity *a* added to the quantity *b*, makes *a + b*; and *a* joined with —*b*, makes *a — b*; alfo —*a* and —*b* make —*a* — *b*; and 3 *a* and 5 *a* make 3 *a* + 5 *a* or 8 *a*, by uniting the fimilar numbers 3 and 5 to make 8.

Thus alfo $3a + 2bc - 4a^2b + c^3$
added to $2a - 2bc + 3a^2b + 2c^3$
makes - $5a \quad 0 \quad - a^2b + 3c^3$

Alfo $\frac{4a}{3c} + \frac{2a}{3c} = \frac{6a}{3c} = \frac{2a}{c}$;

and $3\sqrt{ac} + 5\sqrt{ac} - 1\frac{1}{2}\sqrt{ac} = 6\frac{1}{2}\sqrt{ac}$;

and $a\sqrt{ac} + b\sqrt{ac} = \overline{a+b}\sqrt{ac}$ or $(a+b)\sqrt{ac}$;

and $5\sqrt{a^2-x^2} + 7\sqrt{a^2-x^2} = 12\sqrt{a^2-x^2}$;

and $6\sqrt{3} + 4\sqrt{3} - 2\sqrt{3} = 8\sqrt{3}$;

and $\frac{3a+7\sqrt{2x}}{a+x} + \frac{7-3a\sqrt{2x}}{a+x} = \frac{14\sqrt{2x}}{a+x}$;

and $\frac{8a\sqrt{ax}}{a+2c} + \frac{16c\sqrt{ax}}{a+2c} = \frac{8a+16c\sqrt{ax}}{a+2c} = 8\sqrt{ax}$;

and $-\frac{3ax}{2c} - \frac{5ax}{2c} = -\frac{8ax}{2c} = -\frac{4ax}{c}$;

and $-a\sqrt{x} - b\sqrt{x} = -(a+b)\sqrt{x}$;

and $+3 - 2 = 1$; and $-3 + 2 = -1$;

and $\frac{7ax}{b} - \frac{4ax}{c} = \frac{7acx}{bc} - \frac{4abx}{bc} = (7c-4b)\frac{ax}{bc}$.

In the addition of furd or irrational quantities, they muft be reduced to the fame denomination, or to the fame radical, if that can be done; then add or unite the rational parts, and fubjoin the common furd. Otherwife connect them with their own figns.

So $\sqrt{8} + \sqrt{18} = \sqrt{4 \times 2} + \sqrt{9 \times 2} = 2\sqrt{2} + 3\sqrt{2} = 5\sqrt{2} = \sqrt{50}$;

and $\sqrt{12} + \sqrt{17} = \sqrt{4 \times 3} + \sqrt{9 \times 3} = 2\sqrt{3} + 3\sqrt{3} = 5\sqrt{3} = \sqrt{75}$;

and $\sqrt{50} + \sqrt{75} = \sqrt{25 \times 2} + \sqrt{25 \times 3} = 5\sqrt{2} + 5\sqrt{3} = (\sqrt{2} + \sqrt{3})5$;

but of $\sqrt{5}$ and $\sqrt{6}$ the fum is fet down $\sqrt{5} + \sqrt{6}$, becaufe the terms are incommenfurable, and not reducible to a common furd.

ADDITION of *Logarithms*. See Logarithms.

ADDITION *of Ratios*, the fame as compofition of ratios; which fee.

ADDITIVE, denotes fomething to be added to another, in contradiftinction to fomething to be taken away or fubtracted. So aftronomers fpeak of additive equations, and geometricians of additive ratios.

ADELARD, or ATHELARD, was a learned monk of Bath, in England, who flourifhed about the year 1130, as appears by fome manufcripts of his in Corpus Chrifti, and Trinity Colleges, Oxford. Voffius fays he was univerfally learned in all the fciences of his time; and that, to acquire all forts of knowledge, he travelled into France, Germany, Italy, Spain, Egypt, and Arabia. He wrote many books himfelf, and tranflated others from different languages: thus, he tranflated, from Arabic into Latin, Euclid's Elements, at a time before any Greek copies had been difcovered; alfo Erichiafarim, upon the feven planets. He wrote a book on the feven liberal arts, another on the aftrolabe, another on the caufes of natural compofitions, befides feveral on phyfics and on medicine.

Although Voffius refers to Oxford for fome of thefe manufcripts, it would yet feem they were not to be found there in Wallis's time; for the Doctor, fpeaking of this author, and other Englifh authors and travellers about the fame age, fays, " A particular account of thefe travels of Shelley and Morley was a while fince to be feen in two prefaces to two manufcript books of theirs in the library of Corpus-Chrifti College in Oxford, but hath lately (by fome unknown hand) been cut out, and carried away; which prefaces (one or both of them) did alfo make mention of the travels of Athelardus Bathonienfis, and are, to that purpofe, cited by Voffius out of that manufcript copy. Whoever hath them, would do a kindnefs (by fome way or other) to reftore them, or at leaft a copy of them. Wallis's Algebra, pa. 6.

ADELM, ALDHELMUS, or ALTHELMUS, a learned Englifhman, who flourifhed about the year 680. He was firft abbot of Malmfbury, and afterward bifhop of Shirburn. He died in the year 709, in the monaftery of Malmfbury.

Adelm was the fon of Kenred or Kenten, who was the brother of Ina, king of the Weft Saxons in England. Befide certain books in theology, he compofed feveral on the mathematical fciences &c; as Arithmetic, and Aftrology, and librum de philofophorum difciplinis. See Bede's Hiftory, lib. 5. cap. 19. He is alfo mentioned by Bale and William of Malmfbury.

ADERAIMIN, or ALDERAIMIN, the Arabic name of a ftar of the third magnitude, in the left fhoulder of Cepheus, marked α by Bayer. Its longitude for 1761, ♈ 9° 30′ 8″ north latitude 68° 56′ 20″.

ADFECTED, fee AFFECTED.

ADHESION, ADHERENCE, in *Phyfics*, is the ftate of two bodies, joined or faftened together, whether by mutual attraction, the interpofition of their own parts, or the impulfe or preffure of external bodies. See COHESION.

Thus two hollow hemifpheres, exhaufted of air, are made to adhere firmly together by the preffure of the atmofphere on their convex or external furfaces; for if they are introduced into an exhaufted receiver, they prefently fall afunder. Alfo two very well polifhed planes

planes adhere firmly together, partly by the external preffure of the atmofphere, and partly by the attraction of their parts.

In No. 389 of the Philof. Tranf. Dr. Defaguliers has given experiments of the adhefion of leaden bullets to each other: the caufe of which he refolves into the principle of attraction.

M. Muffchenbroeck, in his Effai de Phyfique, has given a great many remarks on the adhefion of bodies, and relates various experiments which he had made upon this matter, but chiefly relative to the refiftance made by bodies to fracture, in virtue of the adhefion of their parts; which adhefion he afcribes principally to their mutual attraction. Common experiments prove the mutual adhefion of the parts of water to each other, as well as to the bodies they touch. The fame may be faid of the particles of air, on which M. Petit has a memoir among thofe of the Paris Academy of Sciences for the year 1731.

Some authors however are not willing to admit that the adhefion of the parts of water, or indeed of bodies in general, is to be attributed to the attraction of their parts, and they reafon thus: fuppofe, fay they, that attraction acts at any fmall diftance, as for example to the diftance of one-tenth of an inch, from a particle of water: and about this particle defcribe a circle whofe radius is one-tenth of an inch: then the particle of water will be attracted only by the particles included within the circle; but as thefe particles act in contrary directions, their mutual effects muft deftroy one another, and there can be no attraction of the particle, fince it will have no more tendency one way than another.

ADHIL, in *Aftronomy*, a ftar, of the fixth magnitude, upon the garment of Andromeda, under the laft ftar in her foot.

ADJACENT, whatever lies immediately by the fide of another.

ADJACENT *Angle*, in *Geometry*, is faid of an angle when it is immediately contiguous to another, fo that they have both one common fide. And the term is more particularly ufed when the two angles have not only one common fide, but alfo when the other two fides form one continued right line.

ADJACENT *bodies, in Phyfics,* are underftood of thofe that are near, or next to, fome other body.

ADJUTAGE, or rather AJUTAGE; which fee.

ADSCRIPTS, in *Trigonometry*, is ufed by fome mathematicians, for the tangents of arcs. Vieta calls them alfo profines.

ADVANCE-Fosse, in-Fortification, a ditch thrown round the efplanade or glacis of a place, to prevent its being furprifed by the befiegers.

The ditch fometimes made in that part of the lines or retrenchments neareft the enemy, to prevent him from attacking them, is alfo called the advance-foffe.

The advance-foffe fhould always be full of water, otherwife it will ferve to cover the enemy from the fire of the place, if he fhould become- mafter of the foffe. Beyond the advance-foffe it is ufual to conftruct lunettes, redoüts, &c.

ADVENT, *Adventus,* in the Calendar, the time immediately preceding Chriftmas; and was anciently employed in pious preparation for the *adventus,* or coming on, of the feaft of the Nativity.

Advent includes four Sundays, or weeks; commencing either with the Sunday which falls on St. Andrew's day, namely the 30th day of November, or the neareft Sunday to that day, either before or after.

ÆOLIPILE, *Æolipile,* in Hydraulics, a hollow ball of metal, with a very fmall hole or opening; chiefly ufed to fhew the convertibility of water into elaftic fteam. The beft way of fitting up this inftrument, is with a very flender neck or pipe, to fcrew on and off, for the convenience of introducing the water into the infide; for by unfcrewing the pipe, and immerging the ball in water, it readily fills, the hole being pretty large; and then the pipe is fcrewed on. But if the pipe do not fcrew off, its orifice is too fmall to force its way in againft the included air; and therefore to expel moft of the air, the ball is heated red hot, and fuddenly plunged with its orifice into water, which will then rufh in till the ball is about two-thirds filled with the water. The water having been introduced, the ball is fet upon the fire, which gradually heats the contained water, and converts it into elaftic fteam, which rufhes out by the pipe with great violence and noife; and thus continues till all the water is fo difcharged; though not with a conftant and uniform blaft, but by fits: and the ftronger the fire is, the more elaftic will the fteam be, and the force of the blaft. Care fhould be taken that the ball be not fet upon a violent fire with very little water in it, and that the fmall pipe be not ftopped with any thing; for in fuch cafe, the included elaftic fteam will fuddenly burft the ball with a very dangerous explofion.

This inftrument was known to the ancients, being mentioned by Vitruvius, lib. 1. cap. 6. It is alfo treated of, or mentioned, by feveral modern authors, as Defcartes, in his Meteor. cap. 4; and Father Merfennus, in prop. 29 Phædom. Pneumat. ufes it to weigh the air, by firft weighing the inftrument when red hot, and having no water in it; and afterwards weighing the fame when it becomes cold. But the conclufion gained by this means, cannot be quite accurate, as there is fuppofed to be no air in the ball when it is red hot; whereas it is fhewn by Varenius, in his Geography, cap. 19, fect. 6, prop. 10, that the air is rarefied but about 70 times; and confequently the weight obtained by the above procefs, will be about one-70th too fmall, or more or lefs according to the intenfity of the heat.

In Italy it is faid that the Æolipile is often ufed to cure fmoaky chimneys: for being hung over the fire, the blaft arifing from it carries up the loitering fmoke along with it.

And fome have imagined that the æolipile might be employed as bellows to blow up a fire, having the blaft from the pipe directed into the fire: but experience would foon convince them of their miftake; for it would' rather blow the fire *out* than *up,* as it is not air, but rarefied water, that is thus violently blown through the pipe.

ÆOLUS, in Mechanics, a fmall portable machine, not long fince invented by Mr. Tidd, for refrefhing and changing the air in rooms which are made too clofe.

The machine is adapted to fupply the place of a
square

square of glafs in a fafh-window, where it works with little or no noife, on the principle of the fails of a mill, or a smoke-jack; and thus admitting an agreeable quantity of air, at a convenient part of the room.

ÆOLUS's *Harp*, or *Æolian Harp*, an inftrument fo named, from its producing an agreeable melody, merely by the action of the wind.

Neither the age nor inventor of this inftrument are very well known. It is not mentioned by Merfennus in his Harmonics, where he defcribes moft forts of mufical inftruments: and yet the defcription and ufe of it was given foon after, by Kircher, in his book, Magia Phenotactica & Phonurgia.

The conftruction of this inftrument is thus; let a box be made of as thin deal as poffible, its length anfwering exactly to the width of the window in which it is to be placed; five or fix inches deep, and feven or eight inches wide. Acrofs the top, and near each end, glue on a bit of wainfcot, about half an inch high, and a quarter of an inch thick, to ferve as two bridges for the ftrings to be ftretched over, by means of pins inferted into holes a little behind the bridges, nearer the ends, half the number being at one end, and half at the other end: thefe pins are like thofe of a harpfichord; and for their better fupport in the thin deal, a piece of beech of about an inch fquare, and length equal to the breadth of the box, is glewed on the infide of the lid, immediately under the place of the pins, the holes for receiving them being bored through this piece. It is ftrung with fmall catgut, or blue firft fiddle ftrings, more or lefs at pleafure, on the outfide and lengthways of the lid, fixing one end to one of the fmall pins, and twifting the other end about the oppofite or ftretching pin. A couple of found-holes are cut in the lid; and the thinner this is, the better will be the performance.

When the ftrings are tuned unifon, and the inftrument placed, with the top or ftringed fide outwards, in the window to which it is fitted, the air blowing upon that window, the inftrument will give a found like a diftant choir, increafing or decreafing according to the ftrength of the wind.

ÆRA, in Chronology, is the fame as epoch, or epocha, and means a fixed point of time, from which to begin a computation of the years enfuing.

The word is fometimes alfo written *æra* in ancient authors Its origin is contefted, though it is generally fuppofed that it had its rife in Spain. Some imagine that it is formed from *a. er. a.* the abbreviations of the words, *annus erat Augufti*, or from *a. e. r. a.* the initials of the words *annus erat regni Augufti*, becaufe the Spaniards began their computation from the time that their country came under the dominion of Auguftus. Others derive it from *æs*, *brafs*, the tribute money with which Auguftus taxed the world. It is alfo faid that *æra* originally fignified a number ftamped on money to determine its current value. And that the ancients ufed *æs* or *æra* as an article, as we do the word *item*, to each particular of an account; and hence it came to ftand for a fum or number itfelf.

ÆRA alfo means the way or mode of accounting time. Thus we fay fuch a year of the Chriftian æra, &c.

Spanifh ÆRA, otherwife called the year of Cæfar, was introduced after the fecond divifion of the Roman provinces, between Auguftus, Anthony, and Lepidus, in the 714th year of Rome, the 4676th year of the Julian period, and the 38th year before Chrift. In the 447th year of this æra, the Alani, the Vandals, Suevi, &c, entered Spain. It is frequently mentioned in the Spanifh affairs; their councils, and other public acts, being all dated according to it. Some fay it was abolifhed under Peter IV, king of Arragon, in the year of Chrift 1358, and the Chriftian æra introduced inftead of it. But Mariana obferves that it ceafed in the year of Chrift 1383, under John I, king of Caftile. The like was afterwards done in Portugal.

Chriftian ÆRA. It is generally allowed by Chronologers, that the computation of time from the birth of Chrift, was only introduced in the fixth century in the reign of Juftinian; and it is commonly afcribed to Dionyfius Exiguus. This æra came then into ufe in deeds, and fuch like; before which time either the olympiads, the year of Rome, or that of the reign of the emperors, was ufed for fuch purpofes.

See an account of the other principal æras under the word Epoch.

AERIAL *Perfpective*, is that which reprefents bodies diminifhed and weakened, in proportion to their diftance from the eye.

Aerial Perfpective chiefly refpects the colours of objects, whofe force and luftre it diminifhes more or lefs, to make them appear as if more or lefs emote.

It is founded upon this, that the longer the column of air an object is feen through, the more feebly do the vifual rays emitted from it affect the eye.

AEROGRAPHY, a defcription of the air, or atmofphere, its limits, dimenfions, properties, &c.

AEROLOGY, the doctrine or fcience of the air, and its phænomena, its properties, good and bad qualities, &c. It is much the fame with the foregoing word, Aerography.

AEROMETRY, *Aerometria*, the fcience of meafuring the air, its powers and properties; comprehending not only the quantity of the air itfelf, as a fluid body, but alfo its preffure or weight, its elafticity, rarefaction, condenfation, &c.

The term is not much ufed at prefent; this branch of natural philofophy being ufually called pneumatics, which fee. Wolfius, late profeffor of mathematics at Hall, having reduced feveral properties of the air to geometrical demonftrations, firft publifhed at Leipfic his Elements of Aerometry, in the German language, and afterwards more enlarged in Latin, which have fince been inferted in his *Curfus Mathematicus*, in five volumes in 4to.

AERONAUTICA, the pretended art of failing through the air, or atmofphere, in a veffel, fuftained as a fhip in the fea.

AEROSTATICA, is properly the doctrine of the weight, preffure, and balance of the air and atmofphere.

AEROSTATION, in its proper and primary fenfe denotes the fcience of weights fufpended in the air; but in the modern application of the term, it fignifies the art of navigating or floating in the air, both as to the practice and principles of it. Hence alfo the machines which are employed for this purpofe, are called *aeroftats*,

aeroftats, or *aeroftatic* machines; and which, on account of their round and bell like fhape, are otherwife called *air balloons*. Also *aeronaut* is the name given to the perfon who navigates or floats in the air by means of fuch machines.

Principles of AEROSTATION The fundamental principles of this art have been long and generally known, as well as fpeculations on the theory of it but the fuccefsful application of them to practice feems to be altogether a modern difcovery. Thefe principles chiefly refpect the weight or preffure, and elafticity of the air, with its fpecific gravity, and that of the other bodies to be raifed or floated in it: the particular detail of which principles may be feen under the refpective words in this dictionary. Suffice it therefore in this place to obferve, that any body which is fpecifically, or bulk for bulk, lighter than the atmofl here, or air encompaffing the earth, will be buoyed up by it, and afcend, like as wood, or a cork, or a blown bladder, afcends in water. And thus the body would continue to afcend to the top of the atmofphere, if the air were every where of the fame denfity as at the furface of the earth. But as the air is compreffible and elaftic, its denfity decreafes continually in afcending, on account of the diminifhed preffure of the fuperincumbent air, at the higher elevations above the earth; and therefore the body will afcend only to fuch height where the air is of the fame fpecific gravity with itfelf; where the body will float, and move along with the wind or current of air, which it may meet with at that height. This body then is an aeroftatic machine, of whatever form or nature it may be. And an air-balloon is a body of this kind, the whole mafs of which, including its covering and contents, and the weights annexed to it, is of lefs weight than the fame bulk of air in which it rifes.

We know of no folid bodies however that are light enough thus to afcend and float in the atmofphere; and therefore recourfe muft be had to fome fluid or aeriform fubftance.

Among thefe, that which is called inflammable air, is the moft proper of any that have hitherto been difcovered. It is very elaftic, and from fix to ten or eleven times lighter than common atmofpheric air at the furface of the earth, according to the different methods of preparing it. If therefore a fufficient quantity of this kind of air be inclofed in any thin bag or covering, the weight of the two together will be lefs than the weight of the fame bulk of common air; and confequently this compound mafs will rife in the atmofphere, and continue to afcend till it attain a height at which the atmofphere is of the fame fpecific gravity as itfelf; where it will remain or float with the current of air, as long as the inflammable air does not efcape through the pores of its covering. And this is an inflammable air balloon.

Another way is to make ufe of common air, rendered lighter by warming it, inftead of the inflammable air. Heat, it is well known, rarefies and expands common air, and confequently leffens its fpecific gravity; and the diminution of its weight is proportional to the heat applied. If therefore the air, inclofed in any kind of a bag or covering, be heated, and confequently dilated, to fuch a degree, that the excefs of the weight of an equal bulk of common air, above the

weight of the heated air, be greater than the weight of the covering and its appendages, the whole compound mafs will afcend in the atmofphere, till, by the diminifhed denfity of the furrounding air, the whole become of the fame fpecific gravity with the air in which it floats; where it will remain, till, by the cooling and condenfation of the included air, it fhall gradually contract and defcend again, unlefs the heat is renewed or kept up. And fuch is a heated air-balloon, otherwife called a Montgolfier, from its inventor.

Now it has been difcovered, by various experiments, that one degree of heat, accord to the fcale of Fahrenheit's thermometer, expands the air about one five-hundredth part; and therefore that it will require about 500 degrees, or nearer 484 de rees of heat, to expand the air to juft double its bulk. Which is a degree of heat far above what it is practicable to give it on fuch occafions. And therefore, in this refpect, common air heated, is much inferior to inflammable air in point of levity and ufefulnefs for aeroftatic machines.

Upon fuch principles then depends the construction of the two forts of air-balloons. But before treating of this branch more particularly, it will be proper to give a fhort hiftorical account of this late-difcovered art.

Hiftory of AEROSTATION. Various fchemes for rifing in the air, and paffing through it, have been devifed and attempted, both by the ancients and moderns, and that upon different principles, and with various fuccefs. Of thefe, fome attempts have been upon mechanical principles, or by virtue of the powers of mechanifm: and fuch are conceived to be the inftances related of the flying pigeon made by Archytas, the flying eagle and fly by Regiomontanus, and various others. Again, other projects have been formed for attaching wings to fome part of the body, which were to be moved either by the hands or feet, by the help of mechanical powers; fo that ftriking the air with them, after the manner of the wings of a bird, the perfon might raife himfelf in the air, and tranfport himfelf through it, in imitation of that animal. But of thefe and various other devices of the like nature, a particular account will be given under the article *artificial flying*, as belonging rather to that fpecies or principle of motion, than to our prefent fubject of aeroftation, which is properly the failing or floating in the air by means of a machine rendered fpecifically lighter than that element, in imitation of aqueous navigation, or the failing upon the water in a fhip, on veffel, which is fpecifically lighter than the water.

The firft rational account that we have upon record, for this fort of failing, is perhaps that of our countryman Roger Bacon, who died in the year 1292. He not only affirms that the art is feafible, but affures us that he himfelf knew how to make an engine, in which a man fitting might be able to carry himfelf through the air like a bird; and he farther affirms that there was another perfon who had tried it with fuccefs. And the fecret it feems confifted in a couple of large thin, fhells, or hollow globes, of copper, exhaufted of air; fo that the whole being thus rendered lighter than air, they would fupport a chair, on which a perfon might fit.

Bifhop Wilkins too, who died in 1672, in feveral

of his works, makes mention of similar ideas being entertained by divers persons. "It is a pretty notion to this purpose, says he (in his Discovery of a New World, prop. 14), mentioned by Albertus de Saxonia, and out of him by Francis Mendoza, that the air is in some part of it navigable. And that upon this statick principle, any brass or iron vessel (suppose a kettle), whose substance is much heavier than that of the water; yet being filled with the lighter air, it will swim upon it, and not sink." And again, in his Dedalus, chap. 6, "Scaliger conceives the framing of such volant automata to be very easy. *Volantis columbæ machinulam, cujus autorem Archytam tradunt*, vel facillime *profiteri audeo*. Those ancient motions were thought to be contrived by the force of some included air: So *Gellius, Ita erat scilicet libramentis suspensum, & aura spiritus inclusa, atque occulta consitum, &c.* As if there had been some lamp, or other fire within it, which might produce such a forcible rarefaction, as should give a motion to the whole frame." From which it would seem that Bishop Wilkins had some confused notion of such a thing as a heated air-balloon.

Again F. Francisco Lana, in his Prodroma, printed in 1670, proposes the same method with that of Roger Bacon, as his own thought.

He considered that a hollow vessel, exhausted of air, would weigh less than when filled with that fluid; he also reasoned that, as the capacity of spherical vessels increases much faster than their surface, if there were two spherical vessels, of which the diameter of one is double the diameter of the other; then the capacity of the former will be equal to 8 times the capacity of the latter, but the surface of that only equal to 4 times the surface of this: and the one sphere have its diameter equal to triple the diameter of the other; then the capacity of the greater will be equal to 27 times the capacity of the less, while its surface is only 9 times greater: and so on, the capacities increasing as the cubes of the diameters, while the surfaces increase only as the squares of the same diameters. And from this mathematical principle, father Lana deduces, that it is possible to make a spherical vessel of any given matter, and thickness, and of such a size as, when emptied of air, it will be lighter than an equal bulk of that air, and consequently that it will ascend in that element, together with some additional weight attached to it. After stating these principles, father Lana computes that a round vessel of plate brass, 14 feet in diameter, weighing 3 ounces the square foot, will only weigh 1848 ounces; whereas a quantity of air of the same bulk will weigh 2155½ ounces, allowing only one ounce to the cubic foot; so that the globe will not only be sustained in the air, but will also carry up a weight of 307½ ounces: and by increasing the bulk of the globe, without increasing the thickness of the metal, he adds, a vessel might be made to carry a much greater weight.

Such then were the ingenious speculations of learned men, and the gradual approaches towards this art. But one thing more was yet wanting: although acquainted in some degree with the weight of any quantity of air, considered as a detached substance, it seems they were not aware of its great elasticity, and the universal pressure of the atmosphere; by which pressure, a globe of the dimensions above-described, and exhausted of its

air, would immediately be crushed inwards, for want of the equivalent internal counter pressure, to be sought for in some element, much lighter than common air, and yet nearly of equal pressure or elasticity with it; a property or circumstance attending common air when considerably heated. It is evident then that the schemes of ingenious men hitherto must have terminated in mere speculation; otherwise they could never have recorded schemes, which, on the first attempt to put in practice, must have manifested their own insufficiency, by an immediate failure of success: For instead of exhausting the vessel of air, it must either be filled with common air heated, or with some other equally elastic and lighter air. So that upon the whole it appears, that the art of traversing the air, is an invention of our own time; and the whole history of it is comprehended within a very short period.

The rarefaction and expansion of air by heat, is a property of it that has long been known, not only to philosophers, but even to the vulgar: by this means it is, that the smoke is continually carried up our chimneys; and the effect of heat upon air, is made very sensible by bringing a bladder, only partly full of air, near a fire; when the air presently expands with the heat, and distends the bladder so as almost to burst it: and so well are the common people acquainted with this effect, that it is the common practice of those who kick blown bladders about for foot-balls, to bring them from time to time to the fire, to restore the spring of the air, and distension of the ball, lost by the continual waste of that fluid through the sides of it.

But the great levity of inflammable air, is a very modern discovery. As to the inflammable property of this air itself, it had been long known to miners, and especially in coal mines, by the dreadful effects it sometimes produces by its explosions. Among them it is sometimes vulgarly called sulphur, but more properly the fire damp, or inflammable damp, to distinguish it from the choak damp, and other damps, a species of air sometimes found in deep wells and mines, and which does not explode nor take fire, but presently extinguishes candles, and suffocates the persons who may happen to go into it. But it seems that it was Mr. Cavendish who first discovered with exactness the specific gravity of inflammable air; and his experiments and observations upon it, are published in the 56th volume of the Philosophical Transactions for the year 1766. Soon after this discovery of Mr. Cavendish, it occurred to the ingenious Dr. Black of Edinburgh, that if a bladder, or other vessel, sufficiently light and thin, were filled with this air, it would form altogether a mass lighter than the same bulk of atmospheric air, and consequently that it would ascend in it.

This idea he mentioned in his chemical lectures in the year 1767 or 1768; and he farther proposed to exhibit the experiment, by filling the allantois of a calf with such air. The allantois however was not prepared just at the time when he was at that part of his lectures, and other avocations afterwards prevented his design: so that, considering it only as an amusing experiment, and being fully satisfied of the truth of so evident an effect, he contented himself with barely mentioning the experiment from time to time in his lectures. About the year 1777 or 1778 too it occurred to Mr. Cavallo,

that

that it might be possible to construct a vessel, which, when filled with inflammable air, would ascend in the atmosphere: and there is no doubt but that similar ideas would occur to many other persons, of so evident a consequence of Mr. Cavendish's discovery.

But it seems to have been Mr. Cavallo who first actually attempted the experiment, in which however he succeeded no farther than in being able to raise soap bubbles of two or three inches diameter: a thing which had been done by children for their amusement time immemorial. These experiments Mr. Cavallo made in the beginning of the year 1782, and an account of them was read at a public meeting of the Royal Society on the 20th day of June of that year. From which it appears that he tried bladders and paper of various sorts. But the bladders, however thin they were made by scraping, &c, were still found too heavy to ascend in the atmosphere, when fully inflated with the inflammable air: and in using China paper, he found that this air passed through its pores, like water through a sieve. And having failed of success by blowing the same air into a thick solution of gum, thick varnishes, and oil paint, he was obliged to rest satisfied with soap-balls or bubbles, which, being filled with inflammable air, by dipping the end of a small glass tube, connected with a bladder containing the air, into a thick solution of soap, and gently compressing the bladder, ascended rapidly in the atmosphere, and broke against the ceiling of the room.

Here however it seems the matter might have rested, had it not been for experiments made in France soon after, by the two brothers Stephen and Joseph Montgolfier, upon principles suggested, not by the levity of inflammable air, which probably they had never heard of, but by that of smoke and clouds ascending in the atmosphere. These two brothers it seems were natives of Annonay, a town in the Vivarais, about 36 miles distant from Lyons; and that in their youth, Stephen, the elder, had assiduously studied the mathematics, but the other had applied himself more particularly to natural philosophy and chemistry. They were not intended for any particular way of business, but the death of a brother obliged them to put themselves at the head of a considerable paper manufactory at Annonay. In the intervals of time allowed by their business they amused themselves in several philosophical pursuits, and particularly with the experiments in aerostation, of which we are now to give some account. It would be perhaps impossible to know all the particular steps and ideas which finally produced this discovery: but it has been said that the real principle, upon which the effect of the aerostatic machine depends, was unknown even for a considerable time after its discovery: that M. Montgolfier attributed the effect of the machine, not to the rarefaction of the air, which is the true cause, but to a certain gas, specifically lighter than common air, which was supposed to be developed from burning substances, and which was commonly called Montgolfier's gas. Be this however as it may, it is well known that the two brothers began to think of the experiment of the aerostatic machine about the middle or the latter part of the year 1782. The natural ascension of the smoke and the clouds in the atmosphere, suggested the first idea; and to imitate those bodies,

or to inclose a cloud in a bag, and let the latter be lifted up by the buoyancy of the former, was the first project of those celebrated gentlemen.

Accordingly the first experiment was made at Avignon by Stephen, the elder brother, about the middle of November 1782. Having prepared a bag of fine silk, in the shape of a parallelopipedon, and of about 40 cubic feet in capacity, he applied burning paper to an aperture in the bottom, which rarefied the air, and thus formed a kind of cloud in the bag; and when it became sufficiently expanded, it ascended rapidly to the ceiling.

Soon afterwards the experiment was repeated with the same machine at Annonay, by the two brothers, in the open air; when the bag ascended to the height of about 70 feet. Encouraged by this success, they constructed another machine, of about 650 cubic feet capacity; which, when inflated as before, broke the cords which confined it, and after ascending rapidly to the height of about 600 feet, descended and fell on the adjoining ground. With another larger machine, of 37 feet diameter, they repeated the experiment on the 25th day of April, which answered exceedingly well the machine had such force of ascension, that, breaking abruptly from its confinement of ropes, it rose to the height of more than 1000 feet, and then, being carried by the wind, descended and fell at a place about three quarters of a mile from the place of its ascension. The capacity of this machine was equal to above 23 thousand cubic feet, and, being nearly globular, when inflated, it measured 117 English feet in circumference. The covering was formed of linen, lined with paper; and its aperture at the bottom was fixed to a wooden frame, of about 4 feet square, or 16 feet in surface. When filled with vapour, which it was conjectured might be about half as heavy as common air, it was capable of lifting up about 490 pounds, besides its own weight, which, together with that of the wooden frame, was equal to about 500 pounds. With this same machine the next experiment was publicly performed at Annonay, on the 5th of June 1783, before a great multitude of spectators. The flaccid bag was suspended on a pole 35 feet high; straw and chopped wool were burned under the opening at the bottom; the vapour, or rather smoke, soon inflated the bag, so as to distend it in all its parts; and this enormous mass ascended in the air with such velocity, that in less than ten minutes it reached the height of above 6 thousand feet; when a breeze carried it in an horizontal direction to the distance of 7668 feet, or near a mile and a half, where it descended gently to the ground.

As soon as the news of this experiment reached Paris, the philosophers of that city, conceiving that a new species of gas, of about half the weight of common air, had been discovered by Messrs. Montgolfier; and knowing that the weight of inflammable air was but about the eighth or tenth part of the weight of common air, they justly concluded that inflammable air would answer the purpose of this experiment better than the gas of Montgolfier, and accordingly they resolved to make trial of it.

A subscription was opened by M. Faujas de St Fond, towards defraying the expence of the experiment. A sufficient sum of money having soon been raised,

Messrs.

Meſſrs. Roberts were appointed to conſtruct the machine, and M. Charles, profeſſor of experimental philoſophy, to ſuperintend the work. After a conſiderable time ſpent, and ſurmounting many difficulties in obtaining a ſufficient quantity of inflammable air, and ſearching out a ſubſtance light enough for the covering, they at length conſtructed a globe of the ſilk called luteſtring, which was rendered impervious to the incloſed air by a varniſh of elaſtic gum or caeutchouc, diſſolved in ſome kind of ſpirit or eſſential oil. The diameter of this globe was about 13 feet; and it had only one aperture, like a bladder, to which a ſtop-cock was adapted: and the weight of this covering, when empty, together with that of the ſtop-cock, was 25 pounds

On the 23d of Auguſt 1783, they began to fill the globe with inflammable air; but this, being their firſt attempt, was attended with many obſtructions and diſappointments, which took up two or three days to overcome.

At length however it was prepared for exhibition, and on the 27th it was carried from the Place des Victoires, where it had been prepared, to the Champs de Mars, a ſpacious open ground in the front of the Military School, where, after introducing ſome more inflammable air, and diſengaging it from the cords by which it was held down, it roſe, in leſs than two minutes, to the height of 3123 feet: the ſpecific gravity of the balloon, when it went up, being 35 pounds leſs than that of common air At that height the balloon entered a cloud, but ſoon appeared again; and at laſt it was loſt among other clouds. After floating about in the air for about three quarters of an hour, it fell in a field about 15 miles from the place of its aſcent; where, as we may eaſily imagine, it occaſioned great amazement to the peaſants who found it. Its fall was owing to a rent in the covering, probably occaſioned by the ſuperior elaſticity of the inflammable air, over that of the rare part of the atmoſphere to which it had aſcended.

In conſequence of this brilliant experiment, numberleſs ſmall balloons were made, moſtly of goldbeater's ſkin, from 6 and 9 to 18 or 20 inches diameter; their cheapneſs putting it in the power of almoſt every family to ſatisfy its curioſity relative to the new experiment; and in a few days time balloons were ſeen flying all about Paris, from whence they were ſoon after ſent abroad.

Mr. Joſeph Montgolfier repeted an experiment with a machine of his conſtruction before the commiſſaries of the Academy of Sciences, on the 11th and 12th of September. The machine was about 74 feet high, and 43 feet in diameter; it was made of canvaſs, covered with paper both within and without, and weighed 1000 pounds. It was filled with rarefied air in 9 minutes, and in one trial the weight of eight men was not ſufficient to keep it down. It was not ſuffered to go up, as it had been intended for exhibition before the Royal Family, a few days after. By the violence of the rain, however, which fell about this time, it was ſo much ſpoiled, that he thought proper to conſtruct another for that purpoſe, in which he uſed ſo great diſpatch, that it was completed in the ſhort ſpace of four days time. This machine was conſtructed of cloth made of linnen and cotton

thread, and painted with water colours both within and without. Its height was 60 feet, and diameter 43 feet. Having made the neceſſary preparation for inflating it, the operation was begun about one o'clock on the 19th of the ſame month, before the king and queen, the court, and the inhabitants of the place, as well as all the Pariſians who could procure a conveyance to Verſailles. The balloon was ſoon filled, and in eleven minutes after the commencement of the operation, the ropes being cut, it aſcended, bearing up with it a wicker cage, containing a cock, a duck, and a ſheep, the firſt animals that ever aſcended into the atmoſphere with an aeroſtatic machine. Its power of aſcenſion, or the weight by which it was lighter than an equal bulk of common air, allowing for the animals and their cage, was 696 pounds. The balloon roſe to the height of 1440 feet; and being driven by the wind for the ſpace of eight minutes, it gradually deſcended in conſequence of two large rents made in the covering by the wind, and fell in a wood at the diſtance of 10,200 feet, or about two miles from Verſailles. The animals landed again as ſafe as when they went up, and the ſheep was found feeding.

The ſucceſs of this experiment induced M. Pilatre de Rozier, with a philoſophical intrepidity which will be recorded with applauſe in the hiſtory of aeroſtation, to offer himſelf as the firſt adventurer in this aerial navigation. For this purpoſe M. Montgolfier conſtructed a new machine, of an oval ſhape, in a garden of the fauxbourg St. Antoine; its diameter being about 48 feet, and height 74 feet. To the aperture in the lower part was annexed a wicker gallery about three feet broad, with a balluſtrade of three feet high. From the middle of the aperture an iron grate, or brazier, was ſuſpended by chains, deſcending from the ſides of the machine, in which a fire was lighted for inflating the machine; and towards the aperture port-holes were opened in the gallery, through which any perſon, who might venture to aſcend, might feed the fire on the grate with fuel, and regulate at pleaſure the dilatation of the air incloſed in the machine: the weight of the whole being upwards of 1600 pounds. On the 15th of October 1783, the fire being lighted, and the balloon inflated, M. P. de Rozier placed himſelf in the gallery, and, to the aſtoniſhment of a multitude of ſpectators, aſcended as high as the length of the reſtraining cords would permit, which was about 84 feet from the ground, and there kept the machine afloat about four minutes and a half, by repeatedly throwing ſtraw and wool upon the fire: the machine then deſcended gradually and gently, through a medium of increaſing denſity, to the ground; and the intrepid adventurer aſſured the admiring ſpectators that he had not experienced the leaſt inconvenience in this aerial excurſion. This experiment was repeted on the 17th with nearly the ſame ſucceſs; and again ſeveral times on the 19th, when M. P. de Rozier, by a partial aſcent and deſcent, ſeveral times repeted, evinced to the multitude of obſervers that the machine may be made to aſcend and deſcend at the pleaſure of the aeronaut, by merely increaſing or diminiſhing the fire in the grate. The balloon having been hauled down, by the ropes which always confined it, M. Gironde de Villette placed himſelf in the gallery oppoſite

opposite to I. de Rozier, and the machine being suffered to ascend, it hovered for about 9 minutes over Paris, in the sight of all its inhabitants, at the height of 330 feet. And on their descending, the marquis of Arlandes ascended with M. de Rozier much in the same manner:

In consequence of the report of these experiments, signed by the commissaries of the Academy of Sciences, it was ordered that the annual prize of 600 livres should be given to Messrs. Montgolfier for the year 1783.

In the experiments above-recited, the machine was always secured by long ropes, to prevent its entire escape: but they were soon succeeded by unconfined aerial navigation. For this purpose the same balloon of 74 feet in height was conveyed to La Muette, a royal palace in the Bois de Boulogne: and all things being got ready, on the 21st of November 1783, M. P. de Rozier and the marquis d'Arlandes took their post in opposite sides of the gallery, and at 54 minutes after one the machine was absolutely abandoned to the element, and it ascended calmly and majestically in the atmosphere. On reaching the height of about 280 feet the intrepid aeronauts waved their hats to the astonished multitude: but they soon after rose too high to be distinguished, and it is supposed they rose to more than 3000 feet in height. At first they were driven, by a north-west wind, horizontally over the river Seine and part of Paris, taking care to clear the steeples and high buildings by increasing the fire, and in rising they met with a current of air which carried them southward. Having thus passed the Boulevard, and finally desisting from supplying the fire with fuel, they descended very gently in a field beyond the new Boulevard, about 9000 yards, or a little more than 5 miles distant from the palace de La Muette, having been between 20 and 25 minutes in the air. The weight of the whole apparatus including that of the two travellers, was between 1600 and 1700 pounds.

Notwithstanding the rapid progress of aerostation in France, it is remarkable that we have no authentic account of any experiments of this kind being attempted in other countries. Even in our own island, where all arts and sciences find an indulgent nursery, and many their birth, no aerostatic machine was seen before the month of November 1783. Various speculations have been made on the reasons of this strange neglect of so novel and brilliant an experiment. But none seemed to carry any shew of probability except that it was said to be discouraged by the leader of a philosophical society, expressly instituted for the improvement of natural knowledge, for the reason, as it was said, that it was the discovery of a neighbouring nation. Be this however as it may, it is a fact that the first aerostatic experiment was exhibited in England by a foreigner unconnected and unsupported. This was a count Zambeccari, an ingenious Italian, who happened to be in London about that time. He made a balloon of oiled-silk, 10 feet in diameter, weighing only 11 pounds: it was gilt, both for ornament, and to render it more impermeable to the inflammable air with which it was to be filled. The balloon, after being publicly shewn for several days in London, was

carried to the Artillery Ground, and there being filled about three quarters with inflammable air, and having a direction inclosed in a tin box for any person by whom it should afterwards be found, it was launched about one o'clock on the 25th of November 1783. At half-past three it was taken up near Petworth in Sussex, 48 miles distant from London; so that it travelled at the rate of near 20 miles an hour. Its descent was occasioned by a rent in the silk, which must have been the effect of the rarefaction of the inflammable air when the balloon ascended to a rarer part of the atmosphere.

The French philosophers having executed the first aerial voyage with a balloon inflated by heated air, resolved to attempt a similar voyage with a balloon filled with inflammable air, which seemed to be preferable to dilated air in every respect, the expence of preparing it only excepted. A subscription was opened however to defray that expence, which was estimated at about ten thousand livres; and the balloon was constructed by Messrs. Roberts, of gores of silk, varnished with a solution of elastic gum. Its form was spherical, and it measured 27½ feet in diameter. The upper hemisphere was covered by a net, which was fastened to a hoop encircling its middle, and called its equator. To this equator was suspended by ropes a car or boat, covered with painted linen, and beautifully ornamented, which swung a few feet below the balloon. To prevent the bursting of the machine by the expansion of the inflammable air in a rarer medium, or to cause the balloon to descend, it was furnished with a valve, which might be opened by means of a string descending from it, for discharging a part of the internal air, without admitting the external to enter: And the car was ballasted with bags of sand for the purpose of lightening it occasionally, and causing it to ascend: so that by letting some of the air escape through the valve, they might descend; and by discharging some of their sand ballast, ascend. To this balloon was likewise annexed a long pipe by which it was filled. The apparatus for filling it consisted of several casks placed round a large tub of water, each having a long tin tube, that terminated under a vessel or funnel which was inverted into the water of the tub, and communicated with the long pipe annexed to the lower part of the balloon. Iron filings and diluted vitriolic acid being put into the casks, the inflammable air which was produced from these materials, passed through the tin tubes, thence through the water of the tubs to the inverted funnel, and so through the pipe into the balloon. When inflated, the weight of the common air which was equal in bulk to the balloon, was 771½ pounds; also the power of ascension, or weight just necessary to keep it from ascending, was 20 pounds, and the weight of the balloon, with its car, passengers, and all its appendages, was 604½ pounds, which two together make 624½ pounds: and this taken from 771½ pounds, the weight of common air displaced, leaves 147 pounds for the weight of the inflammable air contained in the balloon, and which is to 771½ pounds, the weight of the same bulk of common air, nearly as 1 to 5¼; that is, the inflammable air used in this experiment was 5¼ times lighter than common air.

The first of December was fixed on for the display of this grand experiment; and every preparation was

made

made for conducting it with advantage. The garden of the Thuilleries at Paris was the scene of operation; which was soon crowded and encompassed with a prodigious multitude of observers. Signals were given, from time to time, by the firing of cannon, waving of flags, &c: and a small montgolfier was launched, for shewing the direction of the wind, and for the amusement of the people previous to the general display. At three quarters after one o'clock, M. Charles and one of the Roberts, having seated themselves in the boat attached to the balloon, and being furnished with proper instruments, cloathing, and provisions, left the ground, and ascended with a moderately accelerated velocity to the height of about 600 yards; the surrounding multitude standing silent with fear and amazement; while the aerial navigators at this height made signals of their safety. When they left the ground, the thermometer, according to Fahrenheit's scale, stood at 59 degrees; and the barometer, at 30·18 inches: and at the utmost height to which they ascended, the barometer fell to 27 inches; from which they deduced their height as above to be 600 yards, or one third part of a mile. During the rest of the voyage the quicksilver in the barometer was generally between 27 and 27·65 inches, rising and falling, as part of the ballast was thrown out, or some of the inflammable air escaped from the balloon. The thermometer generally stood between 53 and 57 degrees. Soon after their ascent, they remained stationary for some time: they then moved horizontally in the direction north north west: and having crossed the Seine, and passed over several towns and villages, to the great amazement of the inhabitants, they descended in a field, about 27 miles distant from Paris, at three quarters past 3 o'clock; so that they had travelled at the rate of near 15 miles an hour, without feeling the least inconvenience.

The balloon still containing a considerable quantity of inflammable air, M. Charles re-ascended alone, and it was computed he went to the height of 3100 yards, or almost 2 miles, the barometer being then at 20 English inches: having amused himself in the air about 33 minutes, he pulled the string of the valve, and descended at 3 miles distance from the place of his ascent. All the inconvenience he experienced in his great elevation, was a dry sharp cold, with a pain in one of his ears and a part of his face, which he ascribed to the dilatation of the internal air: a circumstance that usually happens to persons who suddenly change the density of their atmosphere, either by ascending into a rarer, or descending into a denser one. The small balloon, launched at the beginning by M. Montgolfier, was found to have moved in a direction opposite to that of the aeronauts; from which it is inferred that there were two currents of air at different heights above the earth.

In the month of December this year, several experiments were made at Philadelphia in America with air balloons, by Messrs. Rittenhouse and Hopkins. They constructed and filled a great many small balloons, and connected them together; in which a man went up several times, and was drawn down again; and finally, the ropes being cut, he ascended to the height of 100 feet, and floated to a considerable distance; but,

being afraid, he cut open the balloons with a knife, and so descended.

About the close of this year small balloons were sent up in many places, and were become very common in some parts of France and England. And in the beginning of the year following, their number and magnitude increased considerably; and some of the more remarkable ones were as follow:—On the 19th of January M. Joseph Montgolfier, accompanied by six other persons, ascended from Lyons with a rarefied air balloon, to the height of 1000 yards. This was the largest machine that had been hitherto made, being 131 feet high, and 104 feet in diameter: it was formed of a double covering of linen, with three layers of paper between them; and it weighed, when it went up, 1600 pounds, including the gallery, passengers, &c. It was at first intended for six passengers; but before it went up, it was not judged safe to freight it with more than three: however no authority nor solicitations could prevail upon any of the six to quit their place, nor even to cast lots which three should resign their pretensions: so that the spectators saw them all ascend with terror and anxiety; and to add to their distress, when the ropes were cut, and the machine had ascended a foot or two from the ground, a seventh person suddenly leaped into the gallery, and the fire being increased, the whole ascended together. To add to the terror of the scene, after being in the air about 15 minutes, a large rent of about 50 feet in length was made by the balloon taking fire, in consequence of which it descended very rapidly to the ground, though fortunately without injury to any of the aeronauts.

On the 22d of February an inflammable air balloon was launched from Sandwich in Kent. It was but a small one, being only 5 feet in diameter; but it was rendered remarkable by being the first machine that crossed the sea from England to France. It was found in a field at Warneton, about 9 miles from Lisle in French Flanders, two hours and a half after it left Sandwich, the distance being about 74 miles; so that it floated at the rate of about 30 miles an hour.

The chevalier Paul Andreani, of Milan, was the first aerial traveller in Italy. The chevalier was at the sole expence of this machine, but was assisted in the construction by two brothers of the name of Gerli. They all three ascended together near Milan on the 25th of February, and remained in the atmosphere about 20 minutes, when they descended, all their fuel being exhausted. This machine was a montgolfier, of a spherical shape, and about 68 feet in diameter. From calculations made on the power of this, and other machines of the same sort, it appears that the included air is rarefied commonly but about one-third, or that the included warm air weighs about two-thirds of the same bulk of the external or common air.

The next aerial voyage was performed on the 2d of March 1784, by M. Jean Pierre Blanchard, a man who has since that time made more voyages than any other person, and who has rendered himself famous by being the first who has floated in the air over the channel from England to France. M. Blanchard it seems had for many years been in pursuit of mechanical means for flying through the air; but on hearing of the late invented air balloons, he dropped

his

his former pursuits, and turned his attention to them. He accordingly constructed one of 27 feet diameter, to which a boat was suspended with two wings, and a rudder to steer it by, as also a large parachute spread horizontally between the boat and the balloon, designed to check the fall in case the balloon should burst. The machine being filled with inflammable air, he ascended, from the Champs de Mars at Paris, to the height of near ten thousand feet, or almost 2 miles; and after floating in the air for an hour and a quarter, he descended at Billancourt near Seve, having experienced by turns heat, cold, hunger, and an excessive drowsiness. It appears from his own account, and as might have been expected, that the wings and rudder of his boat had little or no power in turning the balloon from the direction of the wind.

In the course of this year, 1784, aerostatic experiments and aerial voyages became so frequent, that the limits of this article will not allow of any thing farther than mentioning those which were attended with any remarkable circumstances. On the 25th of April, Messrs. de Morveau and Bertrand ascended from Dijon, with an inflammable-air balloon, to the height of thirteen thousand feet, or near 2 miles and a half, where the thermometer marked 25 degrees. They were in the air 1 hour and 25 minutes, in which time they floated 18 miles.

On the 20th of May, four ladies and a gentleman ascended from Paris, in a large montgolfier, above the highest buildings, and remained suspended there a considerable time, the balloon being confined by ropes from flying away.

On the 23d of May, M. Blanchard, with the same balloon as before, ascended from Rouen, to such height that the mercury in the barometer stood at 20·57 inches, which on the earth had been at 30·16. It was observed that in this voyage M. Blanchard's wings or oars could not turn him aside from the direction of the wind.

On the 4th of June M. Fleurant and Madame Thible, the first lady who made an aerial voyage, ascended at Lyons in a machine of 70 feet diameter. They went to the height of 8500 feet, and floated about 2 miles in 45 minutes.

On the 14th of June, M. Coustard de Massi and M. Mouchet ascended at Nantes to a great height, with a balloon of 32½ feet diameter, filled with inflammable air extracted from zink; and they floated to the distance of 27 miles in 58 minutes.

On the 23d of June, the first aerial traveller M. Pilatre de Rozier, accompanied with M. Prouts, ascended at Versailles, in the presence of the royal family and the king of Sweden, with a large montgolfier, whose diameter was 79 feet, and its height 91 feet and a half. They floated to the distance of 36 miles in three-quarters of an hour, when they descended, which is at the rate of 48 miles an hour. In consequence of this experiment the king granted to M. de Rozier a pension of 2000 livres.

On the 15th of July the duke of Chartres, the two brothers Roberts, and a fourth person ascended from the park of St. Cloud, with an inflammable-air machine, of an oblong form, its diameter being 34 feet, and its length, which went in a direction parallel to the horizon, was 55½ feet; and they remained in the atmosphere for 45 minutes in the greatest fear and danger. The machine contained an interior small balloon, filled with common air, by means of which it was proposed to cause the machine to ascend or descend without the loss of any inflammable air or ballast: and the boat was furnished with a helm and oars, which were intended to guide it. Three minutes after ascending, the machine was lost in the clouds, and involved in a dense vapour. A violent agitation of the air, resembling a whirlwind, greatly alarmed the aeronauts, turned the machine three times round in a moment, and gave it such shocks as prevented them from using any of their instruments for managing the machine. After many struggles, with great difficulty they tore about 7 or 8 feet of the lower part of the covering, by which the inflamable air escaped, and they descended to the ground with great rapidity, though without any hurt. At the place of departure the barometer stood at 30·12 inches, and at their greatest elevation it stood at 24·36 inches; so that their ascent was about 5100 feet, or near one mile.

On the 18th of July, M. Blanchard, with a Mr. Boley, made his third voyage with the same balloon as he had before, and rose so high as to sink the barometer from 30·1 to 25·34 inches, answering to a height of about 4600 feet. In 2 hours and a quarter they floated 45 miles, which is at the rate of 20 miles an hour. In this voyage M. Blanchard pretended that he was able to turn the machine with his wings, and to make it ascend and descend at pleasure. After descending, it is said the balloon remained all the night at anchor full of air; and that the next day several ladies amused themselves by ascending successively to the height of 80 feet, the length of the ropes by which it was anchored.

In the course of this summer two persons had nearly lost their lives by ascending with machines of warmed air. The one in Spain, on the 5th of June, by the machine taking fire, was much burnt, and so hurt by the fall that his life was long despaired of. The latter having ascended a few feet, the machinery entangled under the eves of a house, which broke the ropes, and the man fell about twenty feet: the machine presently took fire, and was consumed. Other montgolfiers were also burned about London, by taking fire, through the defects of their construction.

The first aerial voyage performed in England was by one Vincent Lunardi, a native of Italy, who ascended from the Artillery Ground, London, with an inflammable-air balloon on the 15th of September. His machine was made of oiled silk, painted in alternate stripes of blue and red; and its diameter was 33 feet. From a net, which covered about two-thirds of the balloon, 45 cords descended to a hoop hanging below the balloon, to which the gallery was attached. The machine had no valve; and its neck, which terminated in the form of a pear, was the aperture through which the inflammable air was introduced, and through which it might be let out. The balloon was filled with air produced from zink by means of diluted vitriolic acid. And when the aeronaut departed, at 2 o'clock, he took up with him a dog, a cat, and a pigeon. After throwing out some sand to clear the houses, he ascended to a considerable height; and the direction of his motion at first was north-west by west;

but

but as the balloon rose higher, it came into another current of air, which carried it nearly north. In the course of his voyage the thermometer was as low as 29 degrees, and the drops of water which had collected round the balloon were frozen. About half after three he descended very near the ground, and landed the cat, which was almost dead with cold: then rising, he prosecuted his voyage, till at 10 minutes past 4 o'clock he landed near Ware in Hertfordshire. He pretends that he descended by means of his oars or wings; but other circumstances related by him, strongly contradict the fact.

The longest and most interesting voyage performed about this time, was that of Messrs. Roberts and M. Colin Hullin, who ascended at Paris, at noon on the 19th of September, with an aerostat, filled with inflammable air, which was 27³ feet in diameter, and 46¼ feet long, the machine being made to float with its longest part parallel to the horizon, and having a boat of near 17 feet long attached to it. The boat was fitted up with several wings or oars, shaped like an umbrella, and they ascended at 12 o'clock with 450 pounds of sand ballast, and after various manœuvres finally descended, at 40 minutes past 6 o'clock, near Arras in Artois, 150 miles from Paris, having still 200 pounds of ballast remaining in the boat. In one part they found the current of air uniform from 600 to 4200 feet high, which it seems was their greatest height, and the fall of the barometer had been near 5·6 degrees. They found that by means of their oars they could accelerate their course a little in the direction of the wind, when it moved slowly, which may be true; but there is great reason to doubt of the accuracy of their experiments by which they pretended to cause their path to deviate about 22 degrees from the wind, going with a considerable velocity.

The second aerial voyage in England, was performed by Mr. Blanchard, and Mr. Sheldon professor of anatomy to the Royal Academy, being the first Englishman who ascended with an aerostatic machine. They ascended at Chelsea the 16th of October, at 9 minutes past 12 o'clock. Mr. Blanchard having landed Mr. Sheldon at about 14 miles from Chelsea, re-ascended alone, and finally landed near Rumsey in Hampshire, about 75 miles distant from London, having gone nearly at the rate of 20 miles an hour The wings used on this occasion it seems produced no deviation from the direction of the wind. Mr Blanchard said that he ascended so high as to feel a great difficulty of breathing: and that a pigeon, which flew away from the boat, laboured for some time to sustain itself with its wings in the rarefied air, but after wandering a good while, returned, and rested on the side of the boat.

On the 4th of October, Mr. Sadler, an ingenious tradesman at Oxford, ascended at that place with an inflammable-air balloon of his own construction and filling. And again on the 12th of the same month he ascended at Oxford, and floated to the distance of 14 miles in 17 minutes, which is at the rate of near 50 miles an hour.

The 30th of November this year Mr. Blanchard's fifth aerial voyage, still with his old machine, was performed in company with Dr. J Jeffries, a native of America. Their voyage was about 21 miles; and it

does not appear that the greatest action of their oars produced any effect in directing the course of the balloon.

On the 4th of January, 1785, a Mr. Harper ascended at Birmingham with an inflammable air balloon, and went to the distance of 50 miles in an hour and a quarter, and suffered no other inconvenience than a temporary deafness, and what might be expected from the changes of wet and cold. The thermometer descended from 40 to 28 degrees.

On the 7th of January, Mr. Blanchard, accompanied with Dr. Jeffries, performed his sixth aerial voyage, by actually crossing the British channel from Dover to Calais, with the same balloon which had five times before carried him successfully through the air. They ascended with only 30 pounds of sand ballast, besides their provisions, some books, instruments, and other necessaries. The machine parted with the gas very rapidly, and their ballast was soon all exhausted; after which, from time to time they threw out every thing else in the boat, to prevent themselves from dropping into the sea. In this way they disposed of all their provisions, their books and instruments, and finally the most part of their very clothes themselves. This however bringing them near the French coast, they gradually ascended, cleared the cliffs and houses, and landed in the forest of Guiennes. It is remarkable that a bottle, being thrown out when they were in danger of falling into the sea, struck the water with such force, that they heard and felt the shock very sensibly on the car and balloon. In consequence of this voyage the king of France presented M. Blanchard with a gift of 12000 livres, and granted him a pension of 1200 livres a year.

On the 19th of January, Mr. Crosbie ascended at Dublin in Ireland, with an inflammable-air balloon to a great height. He rose so rapidly that he was out of sight in 3 minutes and a half. By suddenly opening the valve he descended just at the edge of the sea, as he was driving towards the channel, being unprovided for properly passing over to England.

On the 23d of March, Count Zambeccari and Admiral Sir Edward Vernon ascended at London, and sailed to Horsham in Sussex, at the distance of 35 miles in less than an hour. The voyage proved very dangerous, owing to some of the machinery about the valve being damaged, which obliged them to cut open some part of the balloon when they were about two miles perpendicular height above the earth, the barometer having fallen from 30·4 to 20·8 inches In descending they passed through a dense cloud, which felt very cold, and covered them with snow. The observations they made were, that the balloon kept perpetually turning round its vertical axis, sometimes so rapidly as to make each revolution in 4 or 5 seconds; that a peculiar noise, like rustling, was heard among the clouds, and that the balloon was greatly agitated in the descent.

On May the 5th, Mr. Sadler, and William Windham, esq. member of parliament for Norwich, ascended at Moulsey-hurst. The machine took a south-east course, and the current of air was so strong that they were in great danger of being driven to sea. They had the good fortune however to descend near the con-

flux

flux of the Thames and Medway, not a mile from the water's edge. By an accident they lost their balloon: for while the aeronauts were busied in securing their instruments, the country people, whom they had employed in holding down the machine, suddenly let go the cords, when the balloon instantly ascended, and was driven many miles out to sea, where it fell, and was taken up by a trading vessel. It was afterwards restored again, and another voyage made with it from Manchester to Pontefract, in which Mr. Sadler was still more unfortunate; for no person being near when it descended, and not being able to confine it by his own strength, he was dragged by it over trees and hedges; and at last was forced to quit it at the utmost peril of his life; after which it rose, and was out of sight in a few minutes. It was afterwards found near Gainsborough.

On the 12th of May Mr. Crosbie ascended, at Dublin, as high as the tops of the houses; but soon descended again with a velocity that alarmed all the spectators for his safety. On his stepping out of the car, in an instant Mr. M'Guire, a college youth, sprung into it, and the balloon ascended with him to the astonishment of the beholders, and presently he was carried with great velocity towards the channel in the direction of Holyhead. This being observed, a crowd of horsemen pursued full speed the course he seemed to take, and could plainly perceive the balloon descending into the sea. Lord H. Fitzgerald, who was among the foremost, instantly dispatched a swift-sailing vessel mounted with oars, and all the boats that could be got, to the relief of the gallant youth; whom they found almost spent with swimming, just time enough to save his life.

The fate of M. Pilatre de Rozier, the first aerial traveller, and his companion M. Romain, has been much lamented. They ascended at Boulogne the 15th of June, with intent to cross the channel to England: for the first 20 minutes they seemed to take the proper direction; when presently the whole apparatus was seen in flames, and the unfortunate adventurers fell to the ground from the height of more than a thousand yards, and were killed on the spot, their bones being broken, and their bodies crushed in a shocking manner. The machine in which they ascended, consisted of a spherical balloon, 37 feet in diameter, filled with inflammable air; and under this balloon was suspended a small montgolfier, or fire balloon, of 10 feet diameter; the gallery which suspended the aeronauts, was attached to the net of the upper balloon by cords, which were fastened to a hoop rather larger than the montgolfier, and descended perpendicularly to the gallery. The montgolfier was intended to promote and prolong the ascension, by rarefying the atmospheric air, and by that means gaining levity. It is not certainly known whether the balloon was actually set on fire by the montgolfier, or, being over-rarefied by the heat beneath, burst, and by that means the inflammable air was set in a blaze.

On the 19th of July, at 20 minutes past 2 o'clock, Mr. Crosbie ascended at Dublin, with intent to cross the channel to Holyhead in England. The usual form of the boat had been changed, for a capacious wicker basket, of a circular form, round the upper

edges of which were fastened a great many bladders, which were intended to render his gallery buoyant, in case of a disaster at sea. About 300 pounds of ballast were put into the basket, but the aeronaut discharged half a hundred on his first rise. At first the current of air carried him due west; but it soon changed his course to nearly north-east, pointing nearly towards Whitehaven. At upwards of 40 miles from the Irish shore, he found himself within clear sight of both lands, and he said it was impossible to give any adequate idea of the unspeakable beauties which the scenery of the sea, bounded by both lands, presented. He rose at one time so high, that by the intense cold his ink was frozen, and the mercury sunk quite into the ball of the thermometer. He was sick, and felt a strong prepulsion on the tympanum of the ears. At his utmost height he thought himself stationary; but on liberating some gas, he descended to a current of air blowing north, and extremely rough. He now entered a thick cloud, and encountered strong blasts of wind, with thunder and lightning, which brought him rapidly towards the surface of the water. Here the balloon made a circuit, but falling lower, the water entered his car, and he left his notes of observation. All his endeavours to throw out ballast were of no avail; the force of the wind plunged him into the ocean; and with much difficulty he put on his cork jacket. The propriety of his idea was now very manifest in the construction of his boat; as by the admission of the water into the lower part of it, and the suspension of his bladders, which were arranged at the top, the water, added to his own weight, became proper ballast; and the balloon maintaining its poise, it became a powerful sail, by means of which, and a snatch block to his car, he went before the wind as regularly as a sailing vessel. In this situation he found himself inclined to eat, and he took a little fowl. At the distance of a league he discovered some vessels crowding after him; but as his progress outstripped all their endeavours, he lengthened the space of the balloon from the car, which gave a check to the rapidity of his sailing, and he was at length overtaken and saved by the Dunleary barge, which took him on board, and steered to Dunleary, towing the balloon after them.

A similar accident happened to Major Money, who ascended at Norwich, on the 22d of July, at 20 minutes past 4 in the afternoon; when meeting with an improper current, and not being able to let himself down, on account of the smallness of the valve, he was driven out to sea, where, after blowing about for near two hours, he dropped into the water. Here the struggles were astonishing which he made to keep the balloon up, which was torn, and hung only like an umbrella over his head. A ship was once within a mile, but, he adds, whether from want of humanity, or by mistaking the balloon for a sea monster, they sheered off, and left him to his fate: but a boat chased him for two hours, till just dark, and then bore away. He now gave up all hopes, and began to wish that providence had given him the fate of Pilatre de Rozier, rather than such a lingering death. Exerting himself however to preserve life as long as possible, by keeping the balloon floating over his head, to keep himself out of the water, into which nevertheless he sunk gradually inch by-inch, as it loft its power, till he was at length breaft deep in water, when he was providentially taken up by a revenue cutter, at half paft eleven at night, but fo weak that he was obliged to be lifted out of the car into the ship.

About the latter end of Auguft, the longeft aerial voyage hitherto made, was performed by Mr. Blanchard, who afcended at Lifle, accompanied by the Chevalier de L'Epinard, and travelled 300 miles in their balloon before it defcended. On this occafion, as on fome former ones, Mr. Blanchard made trial of a parachute, like a large umbrella, invented to break the fall in cafe of an accident happening the balloon : with this machine he dropped a dog from the car foon after his afcenfion, which defcended gently and unhurt.

On September the 8th, Thomas Baldwin, Efq. afcended from the city of Chefter, at 40 minutes paft one o'clock, and defcended at Rixton-Mofs, at 25 miles diftance, after a voyage of 2 hours and a quarter. The greateft perpendicular altitude afcended was about a mile and a half, and the aeronaut computed that in fome parts of the voyage he moved at the rate of 30 miles an hour. Mr. Baldwin publifhed a very circumftantial account of his voyage, with many ingenious philofophical remarks relating to aeroftation, of which fubject his book may be confidered as one of the beft treatifes yet given to the public.

October the 5th, Mr. Lunardi made the firft aerial voyage in Scotland. He afcended at Edinburgh, and after various turnings, landed near Cupar in Fife, having defcribed a track of 40 miles over the fea, and 10 over the land, in an hour and a half. He faid the mercury in the barometer funk as low as 18·3 inches at his greateft elevation.

November the 19th the celebrated Blanchard afcended at Ghent to a great height, and after many dangers defcended at Delft without his car, which he cut away to lighten the machine when he was defcending too rapidly, and flung himfelf by the cords to the balloon, which ferved him then in the nature of a parachute. On his firft afcent, when he was almoft out of fight, he let down a dog, by means of a parachute, which came eafily to the ground.

November the 25th Mr. Lunardi afcended at Glafgow, and in two hours it is faid he defcribed a track of 125 miles. It is further remarkable that, being overcome with drowfinefs, he fays he flept for about 20 minutes in the bottom of the car, during this voyage.

Many other voyages were made in different countries, and with various fuccefs. But fince the year 1785, the rage for balloons has confiderably abated, and we have gradually had lefs and lefs of thefe aerial excurfions, fo that it is now become rather an uncommon thing to hear of one of them performed in any country whatever : which fpeedy decline in this new art is perhaps to be afcribed chiefly to the following caufes ; namely, a lefs degree of eagernefs in people to purfue fuch experiments, from their curiofity having been fatisfied ; fecondly, the trouble, danger, and great expence, attending them ; and laftly, the want of the means of conducting them, and the fmall degree of utility to which they have hitherto been applied. The failure in the many attempts that have been made to direct balloons at pleafure through the air, cannot but be felt as a very difcouraging circumftance : and it is to be feared that it will ever be felt as fuch, notwithftanding the pretenfions of fome perfons on this head ; for they never have caufed, nor is it to be expected they ever can caufe the machine to deviate fenfibly from the courfe of the wind, except only in the cafe when this moves with a very fmall celerity. For when the current blows only at the rate of 10 miles an hour, which is but a very gentle wind, it may be fhewn that a balloon of 50 feet in diameter will require a force equal to the preffure of 72 pounds weight, to caufe it to deviate 30 degrees from the courfe of the wind ; and a force equal to double or triple that weight, when the wind blows with a double or triple velocity, that is, at the rate of 20 or 30 miles an hour ; and fo in proportion To obviate the danger of a fall, arifing from any accident happening to the balloon, fome experiments have been made with a parachute, chiefly by Mr. Blanchard, whofe endeavours and perfeverance it feems have continued longer than in any other perfon : we ftill hear of his excurfions in different parts of Europe, and improvements of the parachute, wings, &c ; and have juft read accounts of two voyages lately performed by him ; with which, being very curious, we fhall conclude our narration of thefe aerial excurfions. They will be beft related in Mr. Blanchard's own words, taken from his letter, dated Leipfick, October the 9th, 1787, to the editors of the Paris Journal. " I did not mention," fays he, " in your interefting paper, my afcenfion at Strafburg on the 26th of laft Auguft : the weather was fo horrible that I mounted only for the fake of contenting the aftonifhing crowd of ftrangers affembled there from all parts of the country. Every body feemed fatisfied at the attempt, but I affure you, gentlemen, that I was far from being pleafed with fo common an experiment. The only remarkable thing that occurred at that time, was the following circumftance : At the height of about 2000 yards, or a mile and half a quarter, I let down a dog tied to the parachute, who, inftead of defcending gently, was forcibly carried, by a whirlwind, above the clouds. I met him foon after, bending his courfe directly downwards, and, as on recollecting his mafter, he began to bark a little, I was going to take hold of the parachute, when another whirlwind lifted him again to a great height. I loft him for the fpace of fix minutes, and perceived him afterwards, with my telefcope, as if fleeping in the cradle or bafket belonging to the machine. Continually agitated, and impetuoufly toffed through every point of the compafs, by the violence of the different currents of air, I determined to end my voyage on the other fide of the Rhine, after having paffed vertically over Zell. I defcended at a fmall village, with an intention to be affifted a little, and about thirty men foon came within reach of the balloon very a-propos, and fixed me to the ground. The wind was fo violent that anchors or ropes would have been of no fervice. I had however added to the large aeroftatic globe a fmaller one, of 60 pounds afcenfional force, which would have contributed to fix me, when once I let it loofe ; but notwithftanding this precaution, the men's affiftance was very neceffary to me. The parachute was ftill wavering in the air, and did not come down till 12 minutes after."

" I per-

Plate I.

AIR-BALLOONS.

Montgolfier's.
Fauxbourg of St. Antoine.

Montgolfier's.
Fauxbourg of St. Germain.

Charles & Robert's.
Champ de Mars.

Versailles.

Blanchard's.

" I performed my 27th afcenfion at Leipfick the 29th of September, in the midft of an incredible number of fpectators, forming one of the moft brilliant affemblies I ever beheld. The fky was as clear and ferene as poffible, and the air fo calm that many of my friends, and multitudes of others, could follow me on horfeback, and even on foot. I was fometimes fo near them that they thought they could reach me, but I could foon find the means of rifing; and once, when they had actually taken hold of the cords, to fee me float with the ftrings in their hands, I fuddenly cut them, and mounted again in the air. All thefe amufing evolutions were in fight of the town and its environs. At length I yielded to the earneft folicitations of the company, and entered the town triumphantly in my car, followed by a concourfe of people tranfported with joy, and amidft the acclamations of thoufands. The next day I emptied the inflammable air into another globe, with which I intended to try fome experiments; and I let it off with a cradle, in which a dog was fixed. The balloon, having reached a confiderable height, made an explofion in its under part, as I had imagined it would, having previoufly difpofed it in a proper manner for that purpofe; by which means the little animal fell gently to the ground."

" The day before yefterday having repeated this experiment, at the town's requeft, I prepared the globe in fuch a manner as to caufe an explofion in its upper part, and added a parachute with two fmall dogs fixed to it. They went fo high that, notwithftanding the ferenity of the fky, the balloon was loft in its immenfe expanfe. Telefcopes of the beft fort became ufelefs, and I began to be apprehenfive for the death of the little animals, on account of the feverity of the cold. They defcended however about two hours after, quite fafe and well, in the town of Delitzfch, three miles from Leipfick. I went yefterday to claim them, and found them again over the town in the air with the parachute. Such experiments had been already tried many times in the courfe of the day, and fome officers had thrown them from the top of a fteeple, in the fight of all the inhabitants of Delitzfch, from whence they defcended fafe."

We have lately heard of Mr. Blanchard's 32d afcenfion at Brunfwick in the month of Auguft 1788, in which he much affifted his afcent by means of his wings.

For feveral figures of balloons, fee plate 1.

Practice of AEROSTATION. The firft confideration in the practice of aeroftation, is the form and the fize of the machine. Various fhapes have been tried and propofed, but the globular, or the egg-like figure, is the moft proper and convenient, for all purpofes; and this form alfo will require lefs cloth or filk than any other fhape of the fame capacity; fo that it will both come cheaper, and have a greater power of afcenfion. The bag or cover of an inflammable-air balloon, is beft made of the filk ftuff called luftring, varnifhed over. But for a montgolfier, or heated-air balloon, on account of its great fize, linen cloth has been ufed, lined within or without with paper, and varnifhed. Small balloons are made either of varnifhed paper, or fimply of paper, unvarnifhed, or of gold-beater's fkin, or fuch-like light fubftances.

With refpect to the form of a balloon, it will be neceffary that the operator remember the common proportions between the diameters, circumferences, furfaces, and folidities of fpheres; for inftance, that of different fpheres, the circumferences are as the diameters; that the furfaces are as the fquares of the diameters; and the folidities as the cubes of the fame diameters: that any diameter is to its circumference as 7 to 22, or as 1 to $3\frac{1}{7}$; and therefore 3 times and $\frac{1}{7}$ of any diameter will be its circumference; fo that if the diameter of a balloon be 35 feet, its circumference will be 110 feet. And if the diameter be multiplied by the circumference, the product will be the furface of the fphere; thus 35 multiplied by 110 gives 3850, which is the furface of the fame fphere in fquare feet: and if this furface be divided by the breadth of the ftuff, in feet, which the balloon is to be made of, the quotient will be the number of feet in length neceffary to conftruct the balloon; fo if the ftuff be 3 feet wide, then 3850 divided by 3, gives $1283\frac{1}{3}$ feet, or 428 yards nearly, the requifite quantity of ftuff of 3 feet or one yard wide, to form the balloon of 35 feet diameter. Hence alfo, by knowing the weight of a given piece of the ftuff, as of a fquare foot, or fquare yard, it is eafy to find the weight of the whole bag, namely by multiplying the furface, in fquare feet or yards, by the weight of a fquare foot or yard: fo if each fquare yard weigh 16 ounces or 1 pound, then the whole bag will weigh 428 pounds. Again, the capacity, or folid contents, of the fphere, will be found by multiplying $\frac{1}{6}$ of the furface by the diameter, or by taking $\frac{11}{21}$ of the cube of the diameter; which gives 22458 cubic feet for the capacity of the faid balloon, that is, it will contain, or difplace, 22458 cubic feet of air. From the content and furface of the balloon, fo found, is to be derived its power or levity, thus: on an average, a cubic foot of common air weighs $1\frac{1}{5}$ ounce, and therefore to the number 22458, which is the content of our balloon, adding its $\frac{1}{5}$th part, we have 26950 ounces, or 1684 pounds, for the weight of the common air difplaced or occupied by the balloon. From this weight muft be deducted the weight of the bag, namely 428 pounds, and then there remains 1256 pounds levity of the balloon, without however confidering the contained air, whether it be heated air, or of the inflammable kind. If inflammable air be ufed, as it is of different weights, from $\frac{1}{4}$ to $\frac{1}{10}$ or $\frac{1}{12}$ the weight of common air, according to the modes of preparing it, let us fuppofe for inftance that it is $\frac{1}{6}$ of the weight of common air; then $\frac{1}{6}$ of 1684 is 261 pounds, which is the weight of the bag full of that air; which being taken from 1256, leaves 995 pounds for the levity of the balloon when fo filled with that inflammable air, or the weight which it will carry up, confifting of the car, the ropes, the paffengers, the neceffaries, and ballaft. But if heated air be ufed; then as it is known from experiment that, by heating, the contained air is diminifhed in denfity about one-third only, therefore from 1684, take $\frac{1}{3}$ of itfelf, and there remains 1123 for the weight of the contained warm air; and this being fubtracted from 1256, leaves only 133 pounds for the levity of the balloon in this cafe; which being too fmall to carry up the car, paffengers, &c, it fhews that for thofe purpofes a larger balloon is neceffary, on

Montgolfier'e

Montgolfier's principles. But if now, from the preceding computation, it be required to find how much the size of the balloon muft be increafed, that its levity, or power of afcenfion, may be equal to any given weight, as fuppofe 1000 pounds; then becaufe the levities are nearly as the cubes of the diameters, therefore the diameters will be nearly as the cube roots of the levities; but the levities 133 and 1000 are nearly as 1 to 8, the cube roots of which are as 1 to 2, and confequently $1 : 2 :: 35 : 70$ feet, the diameter of a montgolfier, made of the fame thicknefs of ftuff as the former, capable of lifting 1000 pounds.

On the fame principles we can eafily find the fize of a balloon that fhall juft float in air when made of ftuff of a given thicknefs or weight, and filled with air of a given denfity; the rule for which is this: from the weight of a cubic foot of common air, fubtract that of a cubic foot of the lighter or contained air; then divide 6 times the weight of a fquare foot of the ftuff, by the remainder; and the quotient will be the diameter, in feet, of the balloon that will juft float at the furface of the earth. Suppofe, for inftance, that the materials are as before, namely, the ftuff 1 pound to the fquare yard, or $\frac{16}{9}$ ounces to the fquare foot, which taken 6 times is $\frac{32}{3}$; then the cubic foot of common air weighing $1\frac{1}{5}$ ounce, and of heated air $\frac{2}{5}$ of the fame, whofe difference is $\frac{2}{5}$; therefore $\frac{32}{3}$ divided by $\frac{2}{5}$, gives $26\frac{2}{3}$ feet, which is the diameter of a montgolfier that will juft float: but if inflammable air be ufed of $\frac{2}{7}$ the weight of common air, the difference between $1\frac{1}{5}$ and $\frac{1}{4}$ of it, is 1; by which dividing $\frac{32}{3}$ or $10\frac{2}{3}$, the quotient is the fame $10\frac{2}{3}$ feet, which therefore is the diameter of an inflammable-air balloon that will juft float. And if the diameter be more than thefe dimenfions, the balloons will rife up into the atmofphere.

The height nearly to which a given balloon will rife in the atmofphere, may be thus found, having given only the diameter of the balloon, and the weight which juft balances it, or that is juft neceffary to keep it from rifing: compute the capacity or content of the globe in cubic feet, and divide its reftraining weight in ounces by that content, and the quotient will be the difference between the denfity or fpecific gravity of the atmofphere at the earth's furface and that at the height to which the balloon will rife; therefore fubtract that difference or quotient from $1\frac{1}{5}$ or $1\cdot2$, the denfity at the earth, and the remainder will be the denfity at that height: then the height anfwering to that denfity will be found fufficiently near in the annexed table. Thus, in the foregoing examples, in which the diameter of the balloon is 35 feet, its capacity 22458, and the levity of the firft one 995 pounds, or 15920 ounces, the quotient of the latter number divided by the former, is ·709, which is the denfity at the utmoft height, and to which in the table anfwers a little more than $2\frac{1}{2}$ miles, or $2\frac{4}{5}$ miles nearly, which therefore is the height to which the balloon will afcend. And when the fame balloon was filled with heated air, its levity was found equal to only 133 pounds, or 2128

Height in miles.	Denfity.
0	1·200
$\frac{1}{4}$	1·141
$\frac{1}{2}$	1·085
$\frac{3}{4}$	1·031
1	0·980
$1\frac{1}{4}$	0·932
$1\frac{1}{2}$	0·886
$1\frac{3}{4}$	0·842
2	0·800
$2\frac{1}{4}$	0·761
$2\frac{1}{2}$	0·723
$2\frac{3}{4}$	0·687
3	0·653

ounces, then dividing this by 22458 the capacity, the quotient 095 taken from 1·200, leaves 1·105 for the denfity; to which in the table correfponds almoft half a mile, or nearer $\frac{3}{4}$ of a mile. And fo high nearly would thefe balloons afcend, if they keep the fame figure, and lofe none of the contained air: or rather, thofe are the heights they would fettle at; for their acquired velocity would firft carry them above that height, fo far as till all their motion fhould be deftroyed; then they would defcend and pafs below that height, but not fo much as they had gone above; after which they would re-afcend, and pafs that height again, but not fo far as they had gone below it; and fo on for many times, vibrating alternately above and below that point, but always lefs and lefs every time. The foregoing rule, for finding the height to which the balloon will afcend, is independent of the different ftates of the thermometer at that higheft point, and at the furface of the earth; but for greater accuracy, including the allowances depending on the different ftates of the thermometer, fee under the word ATMOSPHERE, where the more accurate rules are given at large.

The beft way to make up the whole coating of the balloon, is by different pieces or flips joined lengthways from end to end, like the pieces compofing the furface of a geographical globe, and contained between one meridian and another, or like the flices into which a melon is ufually cut, and fuppofed to be fpread flat out. Now the edges of fuch pieces cannot be exactly defcribed by a pair of compaffes, not being circular, but flatter or lefs round than circular arches; but if the flips are fufficiently narrow, or numerous, they will differ the lefs from circles, and may be defcribed as fuch. But more accurately, the breadths of the flip, at the feveral diftances from the point to the middle, where it is broadeft, are directly as the fines of thofe diftances, radius being the fine of the half length of the flip, or of the diftance of either point from the middle of the flip: that is, If ACBD be one of the flips, AB being half the circumference, or AE a quadrant conceived to be equal to AC or AD; then will CD be to $a b$, as radius or the fine of AC, to the fine of Aa. So that if the quadrant AE or AC be divided into any number of equal parts, as here fuppofe 9, then divide the quadrant or 90 degrees by the number of parts 9, and the quotient 10 is the number of degrees in each part;

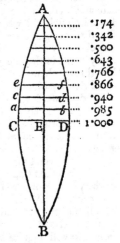

and hence the arcs AC, Aa, Ac, &c, will be refpectively 90°, 80°, 70°, &c; and CD being radius, the feveral breadths ab, cd, ef, &c, will be refpectively the fines of 80°, 70°, 60°, &c, which are here placed oppofite them, the radius being 1. Therefore when it is propofed to cut out flips for a globe of a given diameter; compute the circumference, and make AE or AC a quarter of that circumference, and CD of any breadth, as 3 feet, or 2 feet, or any other

quantity;

quantity; then multiply each of the decimal numbers, set opposite the figure, by that quantity, or breadth of CD, so shall the products be the several breadths *ab*, *cd*, *ef*, &c.

Various schemes have been devised for conducting balloons in any direction, whether vertical or sideways. As to the vertical directions, namely upwards or downwards, the means are obvious, viz. in order to ascend, the aeronaut throws out some ballast; and that he may descend, he opens a valve in the top of his machine by means of a string, to let some of the gas escape; or if it be a montgolfier, he increases or diminishes the fire, as he would ascend or descend. But to direct the machine in a side or horizontal course, is a very difficult operation, and what has hitherto not been accomplished, except in a small degree, and when the current of air is very gentle indeed. The difficulty of moving the balloon sideways, arises from the want of wind blowing upon it; for as it floats along with the current of air, it is relatively in a calm, and the aeronaut feels no more wind than if the machine were at rest in a perfect calm. For this reason, any thing in the nature of sails can be of no use; and all that can be hoped for, is to be attempted by means of oars; and how small the effect of these must be, may easily be conceived from the rarity of the medium against which they must act, and the great magnitude of the machine to be forced through it. We can easily assign what force is necessary to move a given machine in the air with any proposed velocity. From very accurate experiments I have determined, that a globe of $6\frac{3}{4}$ inches in diameter, and moving with a velocity of 20 feet per second of time, suffers a resistance from the air which is just equal to the weight or pressure of one ounce Averdupois; and farther that with different surfaces, and the same velocity, the resistances are directly proportional to the surface nearly, a double surface having a double resistance, a triple surface a triple resistance, and so on; and also that with different velocities, the resistances are proportional to the squares of the velocities nearly, so that a double velocity produced a quadruple resistance, and three times the velocity nine times the resistance, and so on. And hence we can assign the resistance to move a given balloon, with any velocity. Thus, take the balloon as before of 35 feet diameter; then by comparison as above it is found that this globe, if moved with the velocity of 20 feet per second, or almost 14 miles per hour, will suffer a resistance equal to 271 pounds; to move it at the rate of 7 miles an hour, the resistance will be 68 pounds; and to move it $3\frac{1}{2}$ miles an hour, the resistance will be 17 pounds; and so on: and with such force must the aeronauts act on the air in a contrary direction, to communicate such a motion to the machine. And if the balloon move through a rarer part of the atmosphere, than that at the surface of the earth, as $\frac{1}{2}$, or $\frac{1}{3}$, &c, rarer, and consequently the resistance be less in the same proportion; yet the force of the oars will be diminished as much; and therefore the same difficulty still remains. In general, the aeronaut must strike the air, by means of his oars, with a force just equal to the resistance of the air on the balloon, and therefore he must strike that air with a velocity which must be greater as the surface of the oar is less

than the resisted surface of the globe, but not in the same proportion, because the force is as the square of the velocity.

Now suppose the aeronaut act with an oar equal to 100 square feet of surface, to move the balloon above mentioned at the rate of 20 feet per second, or 14 miles an hour; then must he move this oar with the great velocity of 62 feet per second, or near 43 miles per hour: and so in proportion for other velocities of the balloon. From whence it is highly probable, that it will never be in the power of man to guide such machine with any tolerable degree of success, especially when any considerable wind blows, which is almost always the case.

As some aeronauts have thought of using parachutes, made something like umbrellas, to break their fall, in case of any accident happening to the balloon, we shall here consider the principles and power of such a machine. Let us suppose a person wants to know what the size of a parachute must be, that he may descend with it at the uniform rate of 10 feet in a second, which is nearly equal to the velocity he acquires by falling or leaping from the height only of 17 inches, and which it is presumed he may do with safety. Now in order to descend with any uniform velocity, the resistance of the air must be equal to the whole weight that descends: then suppose the weight of the aeronaut to be 150 pounds, and that the parachute is flat, and circular, and made of such materials as that every square foot of its surface weighs 2 ounces, and farther that the weight increases in the same proportion as the surface; then the diameter of the parachute necessary to descend with the moderate velocity of 10 feet per second, must be upwards of 78 feet in diameter: but if the parachute be not a flat surface, but concave on the lower side, its power will be rather the greater, and the diameter may be somewhat less. If it be required to know the power of a flat circular parachute, or what resistance it meets with from air of a mean density, when descending with a given velocity; say as the number 800 is to the square of the velocity in feet, so is the square of the diameter in feet, to a fourth number, which will be the resistance in pounds. And hence, if it be required to know with what velocity a parachute will descend with a given weight; say as the given diameter is to the square root of the weight, so is the number $28\frac{1}{4}$ to a fourth term, which will be the velocity when the descent is in air of a mean density. So if the diameter of a balloon be 50, and its weight together with that of a man be 530 pounds, the square root of which is 23 very nearly; then as $50 : 23 :: 28\frac{1}{4} : 13$, so that the man and parachute will descend with the velocity of 13 feet per second; which it is presumed he may safely do, as he would meet with a shock only equal to that of leaping freely from a height only of 2 feet 2 inches.

The methods of extracting inflammable air from various substances, for filling balloons, and for other purposes, may be seen under the words Air and Gas. And as to the methods of filling and constructing balloons, being matters merely mechanical, they are omitted in this place.

Ample information however on these, and many other particulars, may be met with in several books expressly written on the subject; as in Cavallo's History and

and Practice of Aeroſtation, 8vo, 1785; in Baldwin's Airopaidia, 8vo, 1786; &c.

It has often been diſcuſſed, ſays the former of theſe gentlemen, whether the preference ſhould be given to machines raiſed by inflammable air, or to thoſe raiſed by heated air. Each of them has its peculiar advantages and diſadvantages; a juſt conſideration of which ſeems to decide in favour of thoſe made with inflammable air. The principal comparative advantages of the other ſort are, that they do not require to be made of ſo expenſive materials; that they are filled with little or no expence; and that the combuſtibles neceſſary to fill them are found almoſt every where; ſo that when the ſtock of fuel is exhauſted, the aeronaut may deſcend and recruit it again, in order to proceed on his voyage. But then this ſort of machines muſt be made larger than the other, to take up the ſame weight; and the preſence of a fire is a continual trouble, and a continual danger: in fact, among the many aerial voyages that have been made and attempted with ſuch machines, very few have ſucceeded without an inconvenience, or an accident; and ſome indeed have been attended with dangerous and even fatal conſequences; from which the other ſort is in a great meaſure exempt. But, on the other hand, the inflammable air balloon muſt be made of a ſubſtance impermeable to the ſubtle gas: the gas itſelf cannot be produced without a conſiderable expence; and it is not eaſy to find the materials and apparatus neceſſary for the production of it in every place. However, it has been found that an inflammable-air balloon, of 30 feet in diameter, may be made ſo cloſe as to ſuſtain two perſons, and a conſiderable quantity of ballaſt, in the air for more than 24 hours, when properly managed; and poſſibly one man might be ſupported by the ſame machine for three days: and it is probable that the ſtuff for theſe balloons may be ſo far improved, as to be quite impermeable to the gas, or very nearly ſo; in which caſe, the machine, once filled, would continue to float for a long ſpace of time. At Paris they have already attained to a great degree of perfection in this point; and ſmall balloons have been kept floating in a room for many weeks, without loſing any conſiderable quantity of their levity: but the difficulty lies in the large machines: for in theſe, the weight of the ſtuff itſelf, with the weight and ſtreſs of the ropes and boat, and the folding them up, may eaſily crack and rub off the varniſh, and make them leaky.

In regard to philoſophical obſervations, derived from the new ſubject of aeroſtation, there have been very few made; the novelty of the diſcovery, and of the proſpect enjoyed from the car of an aeroſtatic machine, have commonly diſtracted the attention of the aeronauts; not to remark that many of the adventurers were inadequate to the purpoſe of making improvements in philoſophy, being moſtly influenced either by pecuniary motives, or the vanity of adding their names to the liſt of aerial travellers.—The agreeable ſtillneſs and tranquillity experienced aloft in the atmoſphere, have been matter of general obſervation. Some machines have aſcended to a great height, as far, it has been ſaid, as two miles; and they have commonly paſſed through fogs and clouds, above which they have enjoyed the clear light and heat of the ſun, whilſt the earth beneath

was actually covered by denſe clouds, which poured down abundance of rain. In aſcending very high, the aeronauts have often experienced a pain in their ears, ariſing, it is ſuppoſed, from the internal air being not of the ſame denſity as the air without; but the pain uſually went off in a ſhort time: and it ſeems that this effect is ſimilar to what is experienced by perſons who deſcend by a diving-bell to conſiderable depths in the ſea: I remember often to have heard the late unfortunate Mr. Spalding, the celebrated diver, ſpeak of this effect, with a marked and philoſophical accuracy: after deſcending two or three fathoms below the ſurface, he began to feel a pain in his ears, which gradually increaſed to a very great degree if the deſcent was too quick; his method was therefore to deſcend ſlowly, and to make a ſtop for ſome minutes at the depth of 5 fathom, which is equal nearly to the preſſure of the atmoſphere, and where conſequently the air in his bell was of double the denſity of common air at the ſurface; after reſting here awhile, his ears, as he expreſſed it, gave a crack, and he was ſuddenly relieved of the pain. He then deſcended 5 fathoms more, with the ſame ſymptoms, and the ſame effect: and ſo on continually, from one five fathoms to another, deſcending leiſurely, and ſtopping a little at each ſtage, to give time for his conſtitution to adapt itſelf to the degree of condenſation of the air; after which he felt no more inconvenience, till he came to aſcend again, which was performed with the ſame caution and circumſtances. One experiment is recorded, in which the air of a high region, being brought down, and examined by means of nitrous air, was found to be purer than the air below. The temperature of the upper regions too, it has been found, is much colder than that of the air near the earth; the thermometer, in ſome aeroſtatic machines, having deſcended many degrees below the freezing point of water, while it was conſiderably higher than that degree at the earth's ſurface.

ÆSTIVAL, ſee ESTIVAL.

ÆSTUARY, or ESTUARY, in *Geography*, an arm of the ſea, running up a good way into the land. Such as Briſtol channel, many of the friths in Scotland, and ſuch like.

ÆTHER, ſee ETHER.

AFFECTED, or ADFECTED, EQUATION, in *Algebra*, is an equation in which the unknown quantity riſes to two or more ſeveral powers or degrees. Such, for example, is the equation $x^3 - px^2 + qx = r$, in which there are three different powers of x, namely, x^3, x^2, and x.

The term, affected, is alſo uſed ſometimes in algebra, when ſpeaking of quantities that have co-efficients. Thus in the quantity $2a$, a is ſaid to be affected with the co-efficient 2.

It is alſo ſaid, that an algebraic quantity is affected with the ſign $+$ or $-$, or with a radical ſign; meaning no more than that it has the ſign $+$ or $-$, or that it includes a radical ſign.

The term adfected, or affected, I think, was introduced by Vieta.

AFFECTION, in *Geometry*. a term uſed by ſome ancient writers, ſignifying the ſame as property.

AFFECTION. *Phyſ*. The affections of a body are certain modifications occaſioned or induced by

motion;

motion; in virtue of which the body is difposed after fuch, or fuch a manner.

The affections of bodies, are fometimes divided into *primary* and *fecondary*.

Primary Affections, are thofe which arife either out of the idea of matter, as magnitude, quantity, and figure; or out of the idea of form, as quality and power; or out of both, as motion, place, and time.

Secondary, or *derivative Affections*, are fuch as arife out of primary ones, as divifibility, continuity, contiguity, &c, which arife out of quantity; regularity, irregularity, &c, which arife out of figure, &c.

AFFIRMATIVE QUANTITY, or POSITIVE QUANTITY, one which is to be added, or taken effectively; in contradiftinction to one that is to be fubtracted, or taken defectively.—The term affirmative was introduced by Vieta.

AFFIRMATIVE SIGN, or POSITIVE SIGN, in *Algebra*, the fign of addition, thus marked +, and is called *plus*, or *more*, or *added to*. When fet before any fingle quantity, it ferves to denote that it is an affirmative or a pofitive quantity; when fet between two or more quantities, it denotes their fum, fhewing that the latter are to be added to the former. So + 6, and + *a*, and + AB, are affirmative quantities; alfo + 6 + 8 + 10 denote the fum of 6, 8, and 10, which is 24, and are read thus, 6 plus 8 plus 10. Alfo *a* + *b* + *c* denote the fum of the quantities reprefented by *a*, *b* and *c*, when added together. It feems now not eafy to afcertain with certainty, when, or by whom, this fign was firft introduced; but it was probably by the Germans, as I find it firft ufed by Stifelius in his Arithmetic, printed in 1544.

The early writers on Algebra ufed the word *plus* in Latin, or *piu* in Italian, for addition, and afterwards the initial *p* only, as a contraction; like as they ufed *minus*, or *meno*, or the initial *m* only, for fubtraction: and thus thefe operations were denoted in Italy by Lucas de Burgo, Tartalea, and Cardan, while the figns + and — were employed much about the fame time in Germany by Stifelius, Scheubelius, and others, for the fame operations.

AGE, in *Chronology*, is ufed for a century, being a fyftem or a period of a hundred years.

Chronologifts alfo divide the time fince the creation of the world into three ages: The firft, from Adam till Mofes, which they call the age of nature; the fecond from Mofes to Jefus Chrift, called the age of the law; and the third, or age of grace, from Jefus Chrift till the end of the world.

AGE *of the Moon*, in Aftronomy, is the number of days elapfed fince the laft new moon. To find the moon's age, for any time nearly, for ordinary ufes; add together the epact, the day of the current month, and the number of months from March to the prefent month inclufive; the fum is the moon's age: but if the fum exceed 29, deduct 29 from it in months that have 30 days, or 30 in thofe that have 31; and the remainder will be the age.—At the end of 19 years the moon's age returns upon the fame day of the month, but falls a little fhort of the fame hour of the day.

AGENT, *Agens*, in *Phyfics*, that by which a thing is done or effected; or any thing having a power by which it acts on another, called the patient, or by its action induces fome change in it.

AGGREGATE, the fum or refult of feveral things added together. See *Sum*.

AGITATION, in *Phyfics*, a brifk inteftine motion, excited among the particles of a body. Thus fire agitates the fubtleft particles of bodies. Fermentations, and effervefcences, are produced by a brifk agitation of the particles of the fermenting body.

AGUILON (FRANCIS), or AQUILON, was a jefuit of Bruffels, and profeffor of philofophy at Doway, and of theology at Antwerp. He was one of the firft that introduced mathematical ftudies into Flanders. He wrote a large work on *Optics*, in 6 books, which was publifhed in folio, at Antwerp, in 1613; and a treatife of *Projections of the Sphere*. He promifed alfo to treat upon *Catoptrics* and *Dioptrics*, but this was prevented by his death, which happened at Seville, in the year 1617.

AIR, in *Phyfics*, a thin, fluid, tranfparent, elaftic, compreffible, and dilatable body, which furrounds this terraqueous globe, and covers it to a confiderable height.

Some of the ancients confidered air as an element, namely, one of the four elements, air, earth, water, and fire, of which they conceived all bodies to be compofed; and though it be certain that air, taken in the common acceptation, be far from the fimplicity of an elementary fubftance, yet fome of its parts may properly be fo called. So that air may be diftinguifhed into proper or elementary, and vulgar or heterogeneous.

Elementary AIR, or *Air* properly fo called, is a fubtle, homogeneous, elaftic fluid; being the bafis, or fundamental ingredient, of the whole air of the atmofphere, from which it takes its name. And in this fenfe Dr. Hales, and other modern philofophers, confider it as entering into the compofition of moft, or perhaps all bodies; exifting in them in a folid ftate, devoid of its elafticity, and moft of its diftinguifhing properties, and ferving as their cement; but, by certain proceffes, capable of being difengaged from them, recovering its elafticity, and refembling the air of our atmofphere.

The particular nature of this aerial matter we know but little about: what authors have faid concerning it being chiefly conjectural. There is no way of examining air pure and defecated from the feveral matters with which it is mixed; and confequently we cannot pronounce what are its peculiar properties, abftractedly from other bodies.

Dr. Hook, and fome others, maintain that it is the fame with the *ether*, or that imaginary fine, fluid, active matter, conceived to be diffufed through the whole expanfe of the celeftial regions: which comes to much the fame thing as Newton's fubtle medium, or fpirit. In this fenfe it is fuppofed to be a body *fui generis*, incorruptible, immutable, incapable of being generated, but prefent in all places, and in all bodies.

Other philofophers place its effence in elafticity, making that its diftinctive character. Thefe fuppofe that it may be generated, and that it is nothing elfe but the matter of other bodies, rendered by the

changes

changes it has undergone, susceptible of a permanent elasticity. Mr. Boyle produces a number of experiments, which he made on the production of air, that is, according to him, the extraction of a sensible quantity of air from a body in which there appeared to be little or none at all, by whatever means this may be effected. He observes that among the different methods for this purpose, the chief are fermentation, corrosion, dissolution, decomposition, ebullition of water and other fluids, the reciprocal action of bodies, especially saline ones, upon one another; he adds, that different solid and mineral bodies, in the parts of which no elasticity could be conceived to exist, being plunged into corrosive mediums, which also are quite unelastic, will, by the attenuation of their parts from their mutual collision, produce a considerable quantity of elastic air.

Sir Isaac Newton is of the same opinion, according to whom the particles of a dense compact fixed substance, adhering to each other by a powerful attractive force, cannot be separated but by a violent heat, and perhaps never without fermentation; and these bodies, rarefied by heat and fermentation, are finally transformed into a truly permanent elastic air. On these principles, he adds, gunpowder produces air on explosion. Optics, Qu. 31, &c.

Common, or *heterogeneous* AIR, is an assemblage of corpuscles of various kinds, which together constitute one fluid mass, in which we live and move, and which we constantly breathe; which compound mass altogether, is called the atmosphere.

In this popular and extensive meaning of the term, Mr. Boyle acknowledges that air is the most heterogeneous body in the universe; and Boerhaave proves that it is an universal chaos, a mere jumble of all species of created things. Besides the matter of light or fire, which continually flows into it from the celestial bodies, and perhaps the magnetic effluvia of the earth, whatever fire can volatilize must be found in the air.

Hence, for instance, 1. All sorts of vegetable matter must be contained in the air; being either exhaled from plants growing all over the face of the earth, or rendered volatile by putrefaction, not excepting even the more solid and vascular parts of them.

2. It is no less certain that the air must contain particles of every substance belonging to the animal kingdom. For the copious emanations which are perpetually issuing from the bodies of animals, in the perspiration constantly kept up by the vital heat, are absorbed by the air; and in such quantities too, during the course of an animal life, that, could they be recollected, they would be sufficient to compose a good round number of the like animals. And besides, when a dead animal continues exposed to the air, all its particles evaporate, and are quickly dissipated; so that the substance which composed the animal, is almost wholly incorporated with the air.

3. The whole fossil kingdom must necessarily be found in the atmosphere; for all of that kind, as salts, sulphurs, stones, metals, &c, are convertible into fume, and must consequently take place among aerial substances. Gold itself, the most fixed of all natural bodies, is found among ores, closely adhering to sulphurs in mines, and so is raised along with the mineral.

Of all the emanations which float in the vast ocean of the atmosphere, perhaps the principal are such as consist of saline particles. Many writers suppose that they are of a nitrous kind; but it is probable that they are of all sorts, as vitriol, alum, marine salt, and many others. And Mr. Boyle thinks that there may be great quantities of compound salts, not to be met with on or in the bowels of the earth, formed by the fortuitous concourse and mixture of different saline spirits.

We often find the window-glass of old buildings corroded, as if eaten by worms; though we know of no particular salt that is capable of producing such an effect.

Sulphurs too must make a considerable portion of this compound mass, on account of the many volcanos, grotts, caverns, and mines, dispersed over the face of the globe.

Finally, the various attritions, separations, dissolutions, and other mutual operations of matter of different sorts upon one another, may be regarded as the sources of many other neutral, or anonymous bodies, unknown to us, which rise and float in the air.

Air, taken in this extensive sense, is one of the most general and considerable agents in nature; being concerned in the preservation of animal and vegetable life, and in the production of most of the phenomena that take place in the material world.

Its properties and effects, having been the principal objects of the researches and discoveries of modern philosophers, have been reduced to precise laws and demonstrations, forming no inconsiderable branch of mixed mathematics, under the titles of Pneumatics, Aerometry, &c.

Mechanical Properties and Effects of AIR. Of these the most considerable are its fluidity, its weight, and its elasticity.

1. *Its Fluidity.*—The great fluidity of the air is manifest from the great facility with which bodies traverse it; as in the propagation of, and easy conveyance it affords to, sounds, odours and other effluvia and emanations that escape from bodies: for these effects prove that it is a body whose parts give way to any force, and in yielding are easily moved amongst themselves; which is the definition of a fluid. That the air is a fluid is also proved from this circumstance, that it is found to exert an equal pressure in all directions; an effect which could not take place otherwise than from its extreme fluidity. Neither has it been found that the air can be deprived of this property, whether it be kept for many years together confined in glass vessels, or be exposed to the greatest natural or artificial cold, or condensed by the most powerful pressure; for in none of these circumstances has it ever been reduced to a solid state. It is true indeed that real permanent air may be extracted from solid bodies, and may also be absorbed by them; and we also know that in this case it must be exceedingly condensed, and reduced to a bulk many hundred times less than in its natural state: but in what form it exists in those bodies, or how their particles are combined together, is a mystery which remains hitherto inexplicable.

Those philosophers who, with the Cartesians, make fluidity to consist in a perpetual intestine motion of the parts, think they can prove that this character belongs to air: thus, in a darkened room, where the representations

tations of external objects are introduced by a single ray, the corpuscles with which the air is replete, are seen to be in a continual fluctuation. Some moderns attribute the fluidity of the air, to the fire which is intermixed with it; without which, say they, the whole atmosphere would harden into a solid impenetrable mass: and indeed it must be allowed that the more fire it contains, the greater will its fluidity, mobility, and permeability be; and according as the different positions of the sun augment or diminish the degree of fire, the air always receives a proportional temperature, and is kept in a continual reciprocation.

2. *Its Weight or Gravity.*—The weight or gravity of the air, is a property belonging to it as a body; for gravity is a property essential to matter, or at least a property found in all bodies. But independent of this, we have many direct proofs of its gravity from sense and experiment: thus, the hand laid close upon the end of a vessel, out of which the air is drawn at the other end, soon feels the load of the incumbent atmosphere: thus also, thin glass vessels, exhausted of their air, are easily crushed to pieces by the weight of the external air: and so two hollow segments of a sphere, 4 inches in diameter, exactly fitting each other, being emptied of air, are, by the weight of the ambient air, pressed together with a force which requires the weight of 188 pounds to separate them; and that they are thus forcibly held together by the pressure of the air, is made evident by suspending them in an exhausted receiver, for then they quickly separate of themselves, and fall asunder. Again, if a tube, close at one end, be filled with quicksilver, and the open end be immersed in a bason of the same fluid, and so held upright, the quicksilver in the tube will be kept raised up in it to the height of about 30 inches above the surface of that in the bason, being supported and balanced by the pressure of the external air upon that surface: and that this is the cause of the suspension of the quicksilver in the tube, is made evident by placing the whole apparatus under the receiver of an air-pump; for then the fluid will descend in the tube in proportion as the receiver is exhausted of its air; and then on gradually letting in the air again, the quicksilver reascends to its former height in the tube: and this is what is called, from its inventor, the Terricellian experiment. Nay farther, air can actually be weighed like any other body: for a rigid vessel, full even of common air, by a nice balance is found to weigh more than when the air is exhausted from it; and the effect is proportionally more sensible, if the vessel be weighed full of condensed air, and more still if it be weighed in a receiver void of air.

But although we have innumerable proofs of the gravitating property of the air, yet the full discovery of the laws and circumstances of it are certainly due to the moderns. It cannot indeed be denied, that several of the ancients had some confused notions about this property: thus Aristotle says that all the elements have gravity, and even air itself; and as a proof of it, says that a bladder inflated with air, weighs more than the same when empty; and Plutarch and Stobæus quote him as teaching that the air in its weight is between that of fire and of earth; and farther, he himself, treating of respiration, reports it as the opinion of

Empedocles, that he ascribes the cause of it to the weight of the air, which by its pressure forces itself into the lungs; and much in the same way are the sentiments of Asclepiades expressed by Plutarch, who represents him as saying, among other things, that the external air, by its weight, forcibly opened its way into the breast. But nevertheless it is certain, however unreasonable it may seem, that Aristotle's followers departed in this instance from their master, by asserting the contrary for many ages together. Indeed many of the phenomena arising from this property, have been remarked from the highest antiquity. Many centuries since, it was known that by sucking the air from an open pipe, having its extremity immersed in water, this fluid rises above its level, and occupies the place of the air. In consequence of such observations, sucking pumps were contrived, and various other hydraulic machines; as Heron's syphons, described in his Spiritalia or Pneumatics, and the watering pots known in Aristotle's time under the name of *clepsydræ*, which alternately stop or run as the finger closes or opens their upper orifice. Indeed the reason assigned, by philosophers many ages after, for this phenomenon, was a pretended horror that nature conceives for a vacuum, which, rather then endure it, makes a body ascend contrary to the powerful solicitation of its gravity. Even Galileo, with all his sagacity, could not for some time hit upon any thing more satisfactory; for he only assigned a limit to this dread of vacuity: having observed that sucking pumps would not raise water higher than 16 braffes, or 34 English feet, he limited this abhorring force of nature, to one that was equivalent to the weight of a column of water 34 feet high, on the same base as the void space. Consequently he pointed out a way of making a vacuum, by means of a hollow cylinder, whose piston is charged with a weight sufficient to detach it from the close bottom turned upwards: this effort he called the measure of the force of vacuity, and made use of it for explaining the cohesion of the parts of bodies.

Galileo however was well apprised of the weight of the air as a body: in his Dialogues he shews two ways of demonstrating it, by weighing it in bottles: the transition was easy from one discovery to another: yet still Galileo's knowledge of the matter was imperfect, that is, as to the particular instance of the suspension of a fluid above its level, by the pressure of the external air.

At length Torricelli fell upon the lucky guess, that the counterpoise which keeps fluids above their level, when nothing presses upon their internal surface, is the mass of air resting upon the external one. He discovered it in the following manner: In the year 1643 this disciple of Galileo, on occasion of executing an experiment on the vacuum formed in pumps, above the column of water, when it exceeds 34 feet, thought of using some heavier fluid, such as quicksilver. He conceived that whatever might be the cause by which a column of water of 34 feet high is sustained above its level, the same force would sustain a column of any other fluid, which weighed as much as that column of water, on the same base; whence he concluded that quicksilver, being about 14 times as heavy as water, would not be sustained higher than 29 or 30 inches.

He therefore took a glafs tube of feveral feet in length, fealed it hermetically at one end, and filled it with quick-filver; then inverting it, and holding it upright, by preffing his finger againft the lower or open orifice, he immerfed that end in a veffel of quickfilver; then removing his finger, and fuffering the fluid to run out, the event verified his conjecture; the quickfilver, faithful to the laws of hydroftatics, defcended till the column of it was about 30 inches high above the furface of that in the veffel below. And hence Torricelli concluded that it was no other than the weight of the air incumbent on the furface of the external quickfilver, which counterbalanced the fluid contained in the tube.

By this experiment Torricelli not only proved, what Galileo had done before, that the air had weight, but alfo that it was its weight which kept water and quick-filver raifed in pumps and tubes, and that the weight of the whole column of it was equal to that of a like column of quickfilver of 30 inches high, or of water 34 or 35 feet high; but he did not afcertain the weight of any particular quantity of it, as a gallon, or a cubic foot of it, nor its fpecific gravity to water, which had been done by Galileo, though to be fure with no great accuracy, for he only proved that water was more than 400 times heavier than air.

Torricelli's experiment became famous in a fhort time. Father Merfenne, who kept up a correfpondence with moft of the literati in Italy, was informed of it in 1644, and communicated it to thofe of France, who prefently repeated the experiment: Meffrs. Pafcal and Petit made it firft, and varied it feveral ways; which gave occafion to the ingenious treatife which Pafcal publifhed at 23 years of age, intitled *Experiences Nouvelles touchant la Vuide*. In this treatife indeed he makes ufe of the old principle of *fuga vacui*; but afterwards getting fome notion of the weight of the air, he foon adopted Torricelli's idea, and devifed feveral experiments to confirm it. One of thefe was to procure a vacuum above the refervoir of quickfilver; in which cafe he found the column fink down to the common level: but this appearing to him not fufficiently powerful to diffipate the prejudices of the ancient philofophy, he prevailed on M. Perier, his brother-in-law, to execute the famous experiment of Puy-de-Domme, who found that the height of the quickfilver half-way up the mountain was lefs, by fome inches, than at the foot of it, and ftill lefs at the top: fo that it was now put out of doubt that it was the weight of the atmofphere which counter-poifed the quickfilver.

Des Cartes too had a right notion of this effect of the air, to fuftain fluids above their level, as appears by fome of his letters about this time, and fome years before; and in one of thofe he lays claim to the idea of the Puy-de-Domme experiment: After having defired M. de Carcavi to inform him of the fuccefs of that experiment, which public rumour had advertifed him had been made by M. Pafcal himfelf, he adds, " I had reafon to expect this from him, rather than from you, becaufe I firft propofed it to him two years fince, affuring him at the fame time, that although I had not tried it, yet I could not doubt of the confequence; but as he is a friend of M. Roberval, who profeffes himfelf no friend to me, I fuppofe he is guided by that gentleman's paffions." See more of this hiftory under BAROMETER.

As to the actual weight of any given portion of common air, it feems that Galileo was the firft who determined it experimentally; and he gives two different methods, in his Dialogues, for weighing it in bottles: he did not however perform the experiment very accurately, as he ftated from the refult that the gravity of water was to that of air rather above 400 to 1.

A quantity of air was next weighed by Merfenne in a very ingenious manner. His idea was to weigh a veffel both when full of air, and when emptied of it: to make the vacuum for this purpofe, he knew no better way than by expelling the air out of an eolipile by heating it red hot: by weighing it both when cold and hot, he found a certain difference; which however was not the exact weight of that capacity of air, becaufe the vacuum was not perfect. But by plunging the eolipile, when red hot, into water, juft fo much water entered as was equal in bulk to the air that had been expelled; then he took it out and weighed it with the water, which gave the weight of the fame bulk of water; and on comparing this with the former difference, or weight of air expelled, he found their proportion to be as 1300 to 1. Which is as wide of the truth as Galileo's proportion, namely 400 to 1, but the contrary way. And it is remarkable that the mean between the two, namely 850 to 1, is very near the true proportion as fettled by other more accurate experiments.

Mr. Boyle, by a more accurate experiment, found the proportion to be that of 938 to 1. And Mr. Haukf-bee found it as 850 to 1, proceeding on the fame principles as Merfenne, with a three-gallon glafs bottle, but extracting the air out of it with the air pump, inftead of expelling it by fire; the height of the barometer being at that time 29·7 inches. Alfo by other accurate experiments made before the Royal Society by Mr. Haukfbee, Dr. Halley, Mr. Cotes, and others, the proportion was always between 800 and 900 to 1, but rather nearer the latter, namely, being firft found as 840 to 1, then as 852 to 1, and a third time as 860 to 1; the barometer then ftanding at 29¾ inches, and the weather warm. Mr. Cavendifh determines the ratio 800 to 1, the barometer being 29¾, and the thermometer at 50°; and Sir George Shuckburgh, by a very accurate experiment, finds it 836 to 1, the barometer being at that time at 29·27, and the thermometer at 51°. And the medium of all thefe is about 832 or or 833 to 1, when reduced to the preffure of 30 inches of the barometer, and the mean temperature 55° of the thermometer. Upon the whole therefore it may be fafely concluded that, when the barometer is at 30 inches, and the thermometer at the mean temperature 55°, the denfity or gravity of water is to that of air, as $833\frac{1}{3}$ to 1, that is as $\frac{2500}{3}$ to 1, or as 2500 to 3; and that for any changes in the height of the barometer, the ratio varies proportionally; and alfo that the denfity of the air is altered by the $\frac{1}{440}$th part for every degree of the thermometer above or below temperate.

This number, which is a very good medium among them all, I have chofen with the fraction $\frac{1}{3}$, becaufe it gives exactly $1\frac{1}{4}$ ounce for the mean weight of a cubic foot

foot of air, the weight of the cubic foot of water being juft 1000 ounces averdupois, and that of quickfilver equal to 13600 ounces.

Air, then, having been fhewn to be a heavy fluid fubftance, the laws of its gravitation and preffure muft be the fame as thofe of water and other fluids; and confequently its preffure muft be proportional to its perpendicular altitude. Which is exactly conformable to experiment; for on removing the Terricellian tube to different heights, where the column of air is fhorter, the column of quickfilver which it fuftains is fhorter alfo, and that nearly at the rate of 100 feet for $\frac{1}{10}$ of an inch of quickfilver. And on thefe principles depend the ftructure and ufe of the barometer.

And from the fame principle it likewife follows that air, like other fluids, preffes equally in all directions. And hence it happens that foft bodies endure this preffure without change of figure, and hard or brittle bodies without breaking; being equally preffed on all parts; but if the preffure be taken off, or diminifhed, on one fide, the effect of it is immediately perceived on the other. See ATMOSPHERE, for the total quantity of effects and preffure, and the laws of different altitudes, &c.

From the weight and fluidity of the air, jointly confidered, many effects and ufes of it may eafily be deduced. By the combination of thefe two qualities, it clofely invefts the earth, with all the bodies upon it, conftringing and binding them down with a great force, namely a preffure equal to about 15 pounds upon every fquare inch. Hence, for example, it prevents the arterial veffels of plants and animals from being too much diftended by the impetus of the circulating juices, or by the elaftic force of the air fo copioufly abounding in them. For hence it happens, that on a diminution of the preffure of the air, in the operation of cupping, we fee the parts of the body grow tumid, which caufes an alteration in the circulation of the fluids in the capillary veffels. And the fame caufe hinders the fluids from tranfpiring through the pores of their containing veffels, which would otherwife caufe the greateft debility, and often deftroy the animal. To the fame two qualities of the air, weight and fluidity, is owing the mixture of bodies contiguous to one another, efpecially fluids; for feveral liquids, as oils and falts, which readily mix of themfelves in air, will not mix at all in vacuo. With many other natural phenomena.

3. *Elaflicity.* Another quality of the air, from whence arife a multitude of effects, is its elafticity; a quality by which it yields to the preffion of any other bodies, by contracting its volume; and dilates and expands itfelf again on the removal or diminution of the preffure. This quality is the chief diftinctive property of air, the other two being common to other fluids alfo.

Of this property we have innumerable inftances. Thus, for example, a blown bladder being fqueezed in the hand, we find a fenfible refiftance from the included air; and upon taking off the preffure, the compreffed parts immediately reftore themfelves to their former round figure. And on this property of elafticity depend the ftructure and ufes of the air-pump.

Every particle of air makes a continual effort to dilate itfelf, and fo it acts forcibly againft all the neighbouring particles, which alfo exert the like force in return; but if their refiftance happen to ceafe, or be weakened, the particle immediately expands to an immenfe extent. Hence it is that thin glafs bubbles, or bladders, filled with air, and placed under the receiver of an air-pump, do, upon pumping out the air, burft afunder by the force of the air which they contain. So likewife a clofe flaccid bladder, containing only a fmall quantity of air, being put under the receiver, fwells as the receiver is exhaufted, and at length appears quite full. And the fame thing happens by carrying the flaccid bladder to the top of a very high mountain.

The fame experiment fhews that this elaftic property of the air is very different from the elafticity of folid bodies, and that thefe are dilated after a different manner from the air. For when air ceafes to be compreffed, it not only dilates, but then occupies a far greater fpace, and exifts under a volume immenfely greater than before; whereas folid elaftic bodies only refume the figure they had before they were compreffed.

It is plain that the weight or preffure of the air does not at all depend on its elafticity, and that it is neither more nor lefs heavy than if it were not at all elaftic. But from its being elaftic, it follows that it is fufceptible of a preffure, which reduces it to fuch a fpace, that the force of its elafticity, which re-acts againft the preffing weight, is exactly equal to that weight. Now the law of the elafticity is fuch, that it increafes in proportion to the denfity of the air, and that its denfity increafes in proportion to the forces or weights which comprefs it. But there is a neceffary equality between action and re-action; that is, the gravity of the air, which effects its compreffion, and the elafticity of it, which gives it its tendency to expanfion, are equal.

So that, the elafticity increafing or diminifhing, in the fame proportion as the denfity increafes or diminifhes, that is, as the diftance between its particles decreafe or increafe; it is no matter whether the air be compreffed, and retained in any fpace, by the weight of the atmofphere, or by any other caufe; as in either cafe it muft endeavour to expand with the fame force. And therefore, if fuch air as is near the earth be inclofed in a veffel, fo as to have no communication with the external air, the preffure of fuch inclofed air will be exactly equal to that of the whole external atmofphere. And accordingly we find that quickfilver is fuftained to the fame height, by the elaftic force of air inclofed in a glafs veffel, as by the whole preffure of the atmofphere.—And on this principle of the condenfation and elafticity of the air, depends the ftructure and ufe of the air-gun.

That the denfity of the air is always directly proportional to the force or weight which compreffes it, was proved by Boyle and Mariotte, at leaft as far as their experiments go on this head: and Mr. Mariotte has fhewn that the fame rule takes place in condenfed air. However, this rule is not to be admitted as fcrupuloufly exact; for when air is very forcibly compreffed, fo as to be reduced to $\frac{1}{4}$th of its ordinary bulk, the effect does not anfwer precifely to the rule; for in this cafe the air begins to make a greater refiftance, and requires a ftronger compreffion, than according to the rule. And hence it would feem, that the particles of air cannot, by means of any poffible weight or preffure,

how

how great foever, be brought into perfect contact, or that it cannot thus be reduced to a folid mafs; and confequently that there muft be a limit to which this condenfation of the air can never arrive. And the fame remark is true with regard to the rarefaction of air, namely, that in very high degrees of rarefaction, the elafticity is decreafed rather more than in proportion to the weight or denfity of the air: and hence there muft alfo be a limit to the rarefaction and expanfion of the air, by which it is prevented from expanding to infinity.

We know not however how to affign thofe limits to the elafticity of the air, nor to deftroy or alter it, without changing the very nature of air, which is effected by chemical proceffes. To what degree air is fufceptible of condenfation, by compreffion, is not certainly known. Mr. Boyle condenfed it 13 times more than in its natural ftate, by this means: others have compreffed it into $\frac{1}{60}$th part of its ordinary volume; Dr. Hales made it 38 times more denfe, by means of a prefs; but by freezing water in a hollow caft-iron ball or fhell, he reduced it to 1838 times lefs fpace than it naturally occupies; in which ftate it muft have been of more than twice the denfity or fpecific gravity of water: And as water is not compreffible, except in a very fmall degree, it follows from this experiment, that the particles of air muft be of a nature very different from thofe of water; fince it would otherwife be impoffible to reduce air to a volume above 800 times lefs than in its common ftate; an inference however which militates directly againft an affertion made by Dr. Halley, from fome experiments performed in London, and others at Florence by the Academy del Cimento, namely, that it may be fafely concluded that no force whatever is capable to reduce air into a fpace 800 times lefs than that which it naturally occupies near the furface of the earth.

The elafticity of the air exerts its force equally in all directions; and when it is at liberty, and freed from the caufe which compreffed it, it expands equally in all directions, and in confequence always affumes a fpherical figure in the interftices of the fluids in which it is lodged. This is evident in liquors placed in the receiver of an air pump, by exhaufting the air; at firft there appears a multitude of exceeding fmall bubbles, like grains of fine fand, difperfed through the fluid mafs, and rifing upwards; and as more air is pumped out, they enlarge in fize; but ftill they continue round. Alfo if a plate of metal be immerged in the liquor, on pumping, its furface will be feen covered over with fmall round bubbles, compofed of the air which adhered to it, now expanding itfelf. And for the fame reafon it is that large glafs globes are always blown up of a fpherical fhape, by blowing air through an iron tube into a piece of melted glafs at the end of the pipe.

The expanfion of the air, by virtue of its elaftic property, when only the compreffing force is taken off, or diminifhed, is found to be furprifingly great; and yet we are far from knowing the utmoft dilatation of which it is capable. In feveral experiments made by Mr. Boyle, it expanded firft into 9 times its former fpace; then into 31 times; then into 60, and then into 150 times. Afterwards, it was brought to dilate into

8000 times its firft fpace; then into 10000, and at laft even into 13679 times its fpace; and this folely by its own natural expanfive force, by only removing the preffure, but without the help of fire. And on this principle depends the conftruction and ufe of the MANOMETER.

The elafticity of the air, under one and the fame preffure, is ftill farther increafed by heat, and diminifhed by cold, and that, by fome late accurate experiments made by Sir George Shuckburgh, at the rate of the 440th part of its volume nearly, for each degree of the variation of heat, from that of temperate, in Fahrenheit's thermometer.

Mr. Haukfbee obferved that a portion of air inclofed in a glafs tube, when the temperature was at the freezing point, formed a volume which was to that of the fame quantity of air in the greateft heat of fummer here in England, as 6 to 7. And it has been found by feveral experiments, that air is expanded $\frac{1}{7}$ of its natural bulk by applying the heat of boiling water to it.

Dr. Hales found that the air in a retort, when the bottom of the veffel juft became red hot, was dilated into twice its former fpace; and that in a white, or almoft melting heat, it filled thrice its former fpace: but Mr. Robins found that air was expanded, by means of the white or fufing heat of iron, to 4 times its former bulk.

See feveral ingenious experiments on the elafticity of the air, in the Philof. Tranf. for the year 1777, by Sir George Shuckburgh and Colonel Roy.

This properly explains the common effect obferved on bringing a clofe flaccid bladder near the fire to warm it; when it is prefently found to fwell as if more air were blown into it. And upon this principle depends the ftructure and office of the thermometer; as alfo the air balloons, lately invented by Mr. Montgolfier, for floating in the atmofphere.

M. Amontons firft difcovered that, with the fame degree of heat, air will expand in a degree proportioned to its denfity. And on this foundation the ingenious author has formed a difcourfe, to prove " that the fpring and weight of the air, with a moderate degree of warmth, may enable it to produce even earthquakes, and others of the moft vehement commotions of nature." He computes that at the depth of the 74th part of the earth's radius below the furface, the natural preffure of the air would reduce to the denfity of gold; and thence infers that all matter below that depth, is probably heavier than the heavieft metal that we know of. And hence again, as it is proved that the more the air is compreffed, the more does the fame degree of fire increafe the force of its elafticity; we may infer that a degree of heat, which in our orb can produce only a moderate effect, may have a very violent one in fuch lower orb; and that, as there are many degrees of heat in nature, beyond that of boiling water, it is probable there may be fome whofe violence, thus affifted by the weight of the air, may be fufficiently powerful to tear afunder the folid globe. Mem. de l'Acad. 1703.

Many philofophers have fuppofed that the elaftic property of the air depends on the figure of its corpufcles, which they take to be ramous: fome maintain that

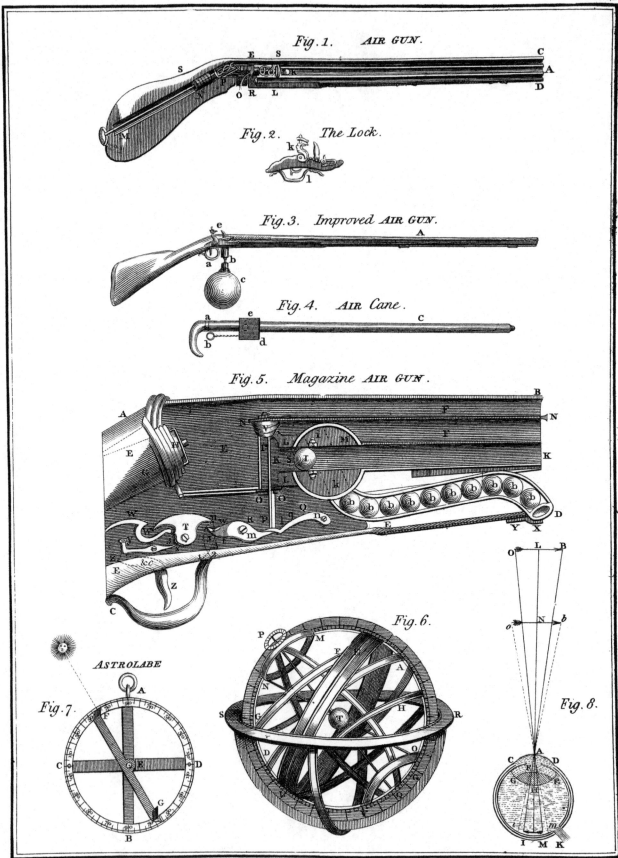

Plate **II**

Fig. 1.   AIR GUN.

Fig. 2.   The Lock.

Fig. 3.   Improved AIR GUN.

Fig. 4.   AIR CANE.

Fig. 5.   Magazine AIR GUN.

ASTROLABE

Fig. 7.

Fig. 6.

Fig. 8.

that they are so many minute *flocculi*, resembling fleeces of wool: others conceive them rolled up like hoops, and curled like wires, or shavings of wood, or coiled like the springs of watches, and endeavouring to expand themselves by virtue of their texture.

But Sir Isaac Newton (Optics, Qu. 31, &c.) explains the matter in a different way; such a contexture of parts he thinks by no means sufficient to account for that amazing power of elasticity observed in air, which is capable of dilating itself into above a million of times more space than it occupied before: but, he observes, as it is known that all bodies have an attractive and a repelling power; and as both these are stronger in bodies, the denser, more compact, and solid they are; hence it follows that when, by heat, or any other powerful agent, the attractive force is overcome, and the particles of the body separated so far as to be out of the sphere of attraction; the repelling power, then commencing, makes them recede from each other with a strong force, proportionable to that with which they before cohered; and thus they become permanent air.

And hence, he says, it is, that as the particles of air are grosser, and rise from denser bodies, than those of transient air, or vapour, true air is more ponderous than vapour, and a moist atmosphere lighter than a dry one.

And M. Amontons makes the elasticity of air to arise from the fire it contains; so that by augmenting the degree of heat, the rarefaction will be increased to a far greater degree than by a mere spontaneous dilatation.

The elastic power of the air becomes the second great source of the remarkable effects of this important fluid. By this property it insinuates itself into the pores of bodies, where, by means of this virtue of expanding, which is so easily excited, it must put the particles of those bodies into perpetual vibrations, and maintain a continual motion of dilatation and contraction in all bodies, by the incessant changes in its gravity and density, and consequently its elasticity and expansion.

This reciprocation is observable in several instances, particularly in plants, in which the tracheæ or air-vessels perform the office of lungs; for as the heat increases or diminishes, the air alternately dilates and contracts, and so by turns compresses the vessels, and eases them again; thus promoting a circulation of their juices. And hence it is found that no vegetation or germination is carried on in vacuo.

It is from the same cause too, that ice is burst by the continual action of the air contained in its bubbles. Thus, too, glasses and other vessels are frequently cracked, when their contained liquors are frozen; and thus also large blocks of stone, and entire columns of marble, sometimes split in the winter season, from some little bubble of included air acquiring an increased elasticity: and for the same reason it is that so few stones will bear to be heated by a fire, without cracking into many pieces, by the increased expansive force of some air confined within their pores. From the same source arise also all putrefaction and fermentation; neither of which can be carried on in vacuo, even in the best disposed subjects. And even respiration, and animal life itself, are supposed, by many authors, to be

conducted, in a great measure, by the same principle of the air. And as we find such great quantities of air generated by the solution of animal and vegetable substances, a good deal must constantly be raised from the dissolution of these elements in the stomach and bowels.

In fact, all natural corruption and alteration seem to depend on air; and even metals, particularly gold, only seem to be durable and incorruptible, in so far as they are impervious to air.

As to the different kinds of air, with its generation, and the effects of different ingredients of it, &c, they are omitted here, as properly belonging to a Chemical Dictionary, or to a General Dictionary of Arts, &c.

For the resistance of the air, see RESISTANCE.

AIR-GUN, in PNEUMATICS, is a machine for propelling bullets with great violence, by the sole means of condensed air.

The first account we meet with of an air-gun, is in the *Elemens d'Artillerie* of David Rivaut, who was preceptor to Louis XIII. of France. He ascribes the invention to one Marin, a burgher of Lisieux, who presented one to Henry IV.

To construct a machine of this kind, it is only necessary to take a strong vessel of any sort, into which the air is to be thrown or condensed by means of a syringe, or otherwise, the more the better; then a valve is suddenly opened, which lets the air escape by a small tube in which a bullet is placed, and which is thus violently forced out before the air.

It is evident then that the effect is produced by virtue of the elastic property of the air; the force of which, as has been shewn in the last article, is directly proportional to its condensation; and therefore the greater quantity that can be forced into the engine, the greater will be the effect. Now this effect will be exactly similar to that of a gun charged with powder, and therefore we can easily form a comparison between them: for inflamed gun-powder is nothing more than very condensed elastic air; so that the two forces are exactly similar. Now it is shewn by Mr. Robins, in his New Principles of Gunnery, that the fluid of inflamed gun-powder has, at the first moment, a force of elasticity equal to about a 1000 times that of common air; and therefore it is necessary that air should be condensed a 1000 times more than in its natural state, to produce the same effect as gun-powder. But then it is to be considered, that the velocities with which equal balls are impelled, are directly proportional to the square roots of the forces; so that if the air in an air-gun be condensed only 10 times, then the velocity it will project a ball with, will be, by that rule, $\frac{1}{10}$th of that arising from gun-powder; and if the air were condensed 20 times, it would communicate a velocity of $\frac{1}{7}$ of that of gun-powder. But in reality the air-gun shoots its ball with a much greater proportion of velocity than as above, and for this reason, namely, that as the reservoir, or magazine of condensed air, is commonly very large in proportion to the tube which contains the ball, its density is very little altered by expanding through that narrow tube, and consequently the ball is urged all the way by nearly the same uniform force as at the first instant; whereas the elastic fluid arising from inflamed gun-powder is but very small in

proportion

proportion to the tube or barrel of the gun, occupying at firſt indeed but a very ſmall portion of it next the but-end : and therefore by dilating into a comparatively large ſpace, as it urges the ball along the barrel, its elaſtic force is proportionally weakened, and it acts always leſs and leſs on the ball in the tube. From which cauſe it happens, that air condenſed into a good large machine only 10 times, will ſhoot its ball with a velocity but little inferior to that given by the gunpowder. And if the valve of communication be ſuddenly ſhut again by a ſpring, after opening it to let ſome air eſcape, then the ſame collection of it may ſerve to impel many balls, one after another.

In all caſes in which a conſiderable force is required, and conſequently a great condenſation of air, it will be requiſite to have the condenſing ſyringe of a ſmall bore, perhaps not more than half an inch in diameter : otherwiſe the force to produce the compreſſion will become ſo great, that the operator cannot work the machine : for, as the preſſure againſt every ſquare inch is about 15 pounds, and againſt every circular inch about 12 pounds, if the ſyringe be one inch in diameter, when one atmoſphere is injected, there will be a reſiſtance of 12 pounds againſt the piſton ; when 2, of 24 pounds ; and when 10 are injected, there will be a force of 120 pounds to overcome ; whereas 10 atmoſpheres act againſt the half-inch piſton, whoſe area is but $\frac{1}{4}$ of the former, with $\frac{1}{4}$ of the force only, namely, 30 pounds ; and 40 atmoſpheres may be injected with ſuch a ſyringe, as well as ten with the larger.

There are air-guns of various conſtructions ; an eaſy and portable one is repreſented in Plate II, fig. 1. which is a ſection lengthways through the axis, to ſhew the inſide. It is made of braſs, and has two barrels ; the inner barrel D A of a ſmall bore, from which the bullets are ſhot ; and a larger barrel E S C D R, on the outſide of it. In the ſtock of the gun there is a ſyringe M N P S, whoſe rod M draws out to take in air ; and by puſhing it in again, the piſton S N drives the air before it, through the valve P E into the cavity between the two barrels. The ball K is put down into its place in the ſmall barrel, with the rammer, as in another gun. There is another valve at S L, which, being opened by the trigger O, permits the air to come behind the ball, ſo as to drive it out with great force. If this valve be opened and ſhut ſuddenly, one charge of condenſed air may make ſeveral diſcharges of bullets ; becauſe only part of the injected air will then go out at a time, and another bullet may be put into the place K : but if the whole air be diſcharged on a ſingle bullet, it will impel it more forcibly. This diſcharge is effected by means of a lock (fig. 2) when fixed to its place as uſual in other guns ; for the trigger being pulled, the cock will go down and drive a lever which opens the valve.

Dr. Macbride (Exper. Eſſ. p. 81) mentions an improvement of the air-gun, made by Dr. Ellis ; in which the chamber for containing the condenſed air is not in the ſtock, which renders the machine heavy and unweildy, but has five or ſix hollow ſpheres belonging to it, of about 3 inches diameter, fitted to a ſcrew on the lock of the gun. Theſe ſpheres are contrived with valves, to confine the air which is forced into their cavities, ſo that a ſervant may carry them ready charged with condenſed air : and thus the gun of this

conſtruction is rendered as light and portable as one of the ſmalleſt fowling-pieces.

Fig. 3 repreſents one made by the late Mr. B. Martin of London, and now by ſeveral of the mathematical inſtrument and gun-makers of the metropolis ; which, for ſimplicity and perfection, perhaps exceeds any other that has been contrived. A is the gun-barrel, of the ſize and weight of a common fowling-piece, with the lock, ſtock, and ramrod. Under the lock, at b, is a round ſteel tube, having a ſmall moveable pin in the inſide, which is puſhed out when the trigger a is pulled, by the ſpringwork within the lock ; to this tube b is ſcrewed a hollow copper ball, perfectly airtight. This copper ball is fully charged with condenſed air by means of a ſyringe, previous to its being applied to the tube b. Hence, if a bullet be rammed down in the barrel, the copper ball ſcrewed faſt at b, and the trigger a be pulled ; then the pin in b will forcibly puſh open a valve within the copper ball, and let out a portion of the condenſed air ; which air will ruſh up through the aperture of the lock, and forcibly act againſt the bullet, driving it to the diſtance of 60 or 70 yards, or farther. If the air be ſtrongly condenſed at every diſcharge, only a portion of the air eſcapes from the ball ; therefore, by re-cocking the piece, another diſcharge may be made ; and this repeated 15 or 16 times. An additional barrel is ſometimes made, and applied for the diſcharge of ſhot, inſtead of the ball above deſcribed.

Sometimes the ſyringe is applied to the end of the barrel C (fig. 4) ; the lock and trigger ſhut up in a braſs caſe d ; and the trigger pulled, or the diſcharge made, by pulling the chain b. In this contrivance there is a round chamber for the condenſed air at the end of the ſpring at e, and it has a valve acting in a ſimilar manner to that of the copper ball. When this inſtrument is not in uſe, the braſs caſe d is made to ſlide off, and the inſtrument then becomes a walking ſtick : from which circumſtance, and the barrel being made of cane, or braſs, &c, it has been called the *Air-cane*. The head of the cane unſcrews and takes off at a, where the extremity of the piſton-rod in the barrel is ſhewn. An iron rod is placed in a ring at the end of this, and the air is condenſed in the barrel in a manner ſimilar to that of the gun as above ; but its force and action is not near ſo ſtrong as in the gun.

Magazine Air-*Gun*. This is an improvement of the common air-gun, made by an ingenious artiſt, called L. Colbe. By his contrivance, ten bullets are ſo lodged in a cavity, near the place of diſcharge, that they may be ſucceſſively drawn into the barrel, and ſhot ſo quickly as to be nearly of the ſame uſe as ſo many different guns ; the only motion required, after the air has been injected, being that of ſhutting and opening the hammer, and cocking and pulling the trigger. Fig. 3 is a longitudinal ſection of this gun, as large in every part as the gun itſelf ; and as much of its length is ſhewn as is peculiar to this conſtruction ; the reſt of it being like the ordinary air-gun. E E is part of the ſtock ; G is the end of the injecting ſyringe, with its valve H, opening into the cavity F F F F between the barrels. K K is the ſmall or ſhooting barrel, which receives the bullets, one at a time, from the magazine D E, being a ſerpentine cavity, in which the bullets b, b, b, &c, are

lodged,

lodged, and clofed at the end D; from whence, by one motion of the hammer, they are brought into the barrel at I, and thence are fhot out by the opening of the valve V, which lets in the condenfed air from the cavity F F F into the channel V K I, and fo along the inner barrel K K K, whence the bullet is difcharged. *s* I *s i* M *k* is the key of a cock, having a hole through it; which hole, in the prefent fituation, makes part of the barrel K K, being juft of the fame bore: fo that the air, which is let in at every opening of the valve V, comes behind this cock, and taking the ball out of it, carries it forward, and fo out of the mouth of the piece.

To bring in another bullet to fucceed I, which is done in an inftant, bring the cylindrical cavity of the key of the cock, which made part of the barrel K K K, into the fituation *i k,* fo that the part 1 may be at K; then turning the gun upfide-down, one bullet next the cock will fall into it out of the magazine, but will go no farther into this cylindrical cavity, than the two little pieces *s s* will permit it; by which means only one bullet at a time will be taken in to the place I, to be difcharged again as before.

A more particular defcription of the feveral parts may be feen in Defaguliers' Exper. Philof. vol. ii. pa. 399 et feq.

AIR-PUMP, in *Pneumatics,* is a machine for exhaufting the air out of a proper veffel, and fo to make what is commonly called a vacuum; though in reality the air in the receiver is only rarefied to a great degree, fo as to take off the ordinary effects of the atmofphere. So that by this machine we learn, in fome meafure, what our earth would be without air; and how much all vital, generative, nutritive, and alterative powers depend upon it.

The principle on which the air-pump is conftructed, is the fpring or elafticity of the air; as that on which the common, or water pump is formed, is the gravity of the fame air: the one gradually exhaufting the air from a veffel by means of a pifton, with a proper valve, working in a cylindrical barrel or tube; and the other exhaufting water in a fimilar manner.

The air-pump has proved one of the principal means of performing philofophical difcoveries, that has been invented by the moderns. The idea of fuch a machine occurred to feveral perfons, nearly about the fame time. But the firft it feems was completed by Otto Guericke, the celebrated conful of Magdeburg, who exhibited his firft public experiments with it, before the emperor and the ftates of Germany, at the breaking up of the imperial diet at Ratifbon, in the year 1654. But it was not till the year 1672 that Guericke publifhed a defcription of the inftrument, with an account of his experiments, in his Experimenta Nova Magdeburgica de Vacuo Spacia: though an account of them had been publifhed by Schottus in 1657, in his Mechanica Hydraulico Pneumatica.

Dr. Hook and M. Duhamel afcribe the invention of the air-pump to Mr. Boyle. But that great man frankly confeffes that Guericke was beforehand with him in the execution. Some attempts, he affures us, he had indeed made upon the fame foundation, before he knew any thing of what had been done abroad: but the information he afterwards received from the account given

by Schottus, enabled him, with the affiftance of Dr. Hook, after two or three unfuccefsful trials, to bring his defign to maturity. The product of their labours was a new air-pump, much more eafy, convenient, and manageable, than the German one. And hence, or rather from the great variety of experiments to which this illuftrious author applied the machine, it was afterwards called *Machina Boyliana,* and the vacuum produced by it, *Vacuum Boylianum.*

Structure of the Air-Pump. Moft of the air-pumps that were firft made, confifted of only one barrel, or hollow cylinder of brafs, with a valve at the bottom, opening inwards; and a moveable embolus or pifton, having likewife a valve opening upwards, and fo exactly fitted to the barrel, that when it is drawn up from the bottom, by means of an indented iron rod or rack, and a handle turning a fmall indented wheel, playing in the teeth of that rod, all the air will be drawn up from the cavity of the barrel: there is alfo a fmall pipe opening into the bottom of the barrel, by means of which it communicates with any proper veffel to be exhaufted of air, which is called a receiver, from its office in receiving the fubjects upon which experiments are to be made in vacuo: the whole being fixed in a convenient frame of wood-work, where the end of the pipe turns up into a horizontal plate, upon which the receiver is placed, juft over that end of the pipe.

The other parts of the machine, being only accidental circumftances, chiefly refpecting conveniency, have been diverfified and improved from time to time, according to the addrefs and feveral views of the makers. That of Otto Guericke was very rude and inconvenient, requiring the labour of two ftrong men, for more than two hours, to extract the air from a glafs, which was alfo placed under water; and yet allowed of no change of fubjects for experiments.

Mr. Boyle, from time to time, removed feveral of thefe inconveniences, and leffened others: but ftill the working of his pump, which had but one barrel, was laborious, by reafon of the preffure of the atmofphere, a great part of which was to be removed at every lift of the pifton, when the exhauftion was nearly completed. Various improvements were fucceffively made in the machine by the philofophers about that time, and foon after, who cultivated this new and important branch of pneumatics; as Papin, Merfenne, Mariotte, and others; but ftill they laboured under a difficulty of working them, from the circumftance of the fingle barrel, till Papin, in his farther improvements of the air-pump, removed that inconvenience, by the ufe of a fecond barrel and pifton, contrived to rife, as the other fell, and to fall as that rofe; by which, and the great improvements made by Mr. Haukfbee, the preffure of the atmofphere on the defcending pifton, always nearly balanced that of the afcending one; fo that the winch, which worked them up and down, was eafily moved by a very gentle force with one hand: and befides, the exhauftion was hereby made in lefs than half the time.

Some of the Germans, and others likewife, made improvements in the air-pump, and contrived it to perform the counter office of a condenfer, in order to examine the properties of the air depending on its condenfation.

8

Mr.

Mr. Boyle contrived a mercurial gauge or index to the air-pump, which is described in his first and second Physico-Mechanical Continuations, for measuring the degrees of the air's rarefaction in the receiver. This gauge is similar to the barometer, being a long glass tube, having its lower end immersed in an open bason of quickfilver, but its other end, which was open also, communicating with the receiver: which being exhausted, this tube is equally exhausted of air at the same time, and the external air presses the quickfilver up into the tube, to a height proportioned to the degree of exhaustion.

Mr. Vream, an ingenious pneumatic operator, made an improvement in Haukfbee's air-pump, by reducing the alternate up-and-down motion of the hand and winch to a circular one. In his method, the winch is turned quite round, and yet the pistons are alternately raised and depressed: by which the trouble of shifting the hand backwards and forwards, as well as the loss of time, and the shaking of the pump, are prevented.

The air-pump, thus improved, is represented in plate III. fig. 1; where *o o* is the receiver to be exhausted, ground truly level at the bottom, set over a hole in the plate, from which descends the bent pipe *h h* to the cistern *d d*, with which the two barrels *a a* communicate, in which the pistons are worked by a toothed wheel, by turning the handle *b b*; by which the racks *c c*, with the pistons, are worked alternately up and down. *l l* is the gauge tube, immersed in a bason of quickfilver *m* at bottom, and communicating with the receiver at top; from which however it may be occasionally disengaged, by turning a cock. And *n* is another cock, by turning of which, the air is again let in to the exhausted receiver; into which it is heard to rush with a considerable hissing noise.

Notwithstanding the great excellency of Mr. Haukfbee's air-pump, it was still subject to inconveniences, from which it was in a great measure relieved by some contrivances of Mr. Smeaton, which are described at large in the Philof. Tranf. for the year 1752. The principal improvements suggested by Mr. Smeaton, relate to the gauge, the valves of the piston, and the piston going closer down to the bottom of the barrel; for his pump has only one. By the last of these, the air was extracted more perfectly at each stroke. By the second, he remedied an inconvenience arising from the valve hole of the piston being too wide properly to support the bladder valve which covered it: instead of the usual circular orifice, Mr. Smeaton perforated the piston with seven small and equal hexagonal holes, one in the centre, and the other six around, forming together the appearance of a transverse section of a honeycomb; the bars or divisions between which, served to support the pressure of the air on the valve. His gage consists of a bulb of glass, of a pear like shape, and capable of holding about half a pound of quickfilver: it is open at the lower end, the other terminating in a tube hermetically sealed; and it has annexed to it a scale, divided into parts of about $\frac{1}{15}$ of an inch, and answering to the 1000th part of the whole capacity. During the exhaustion of the receiver, the gage is suspended in it by a wire; but when the pump has been worked as much as necessary, the gage is pushed down, till the open end be immersed in a bason of quickfilver

placed underneath. The air is then let into the receiver again, and the quickfilver driven by it from the bason, up into the gauge, till the air remaining in it become of the same density as the air without; and as the air always takes the highest place, the tube being uppermost, the expansion will be determined by the number of divisions occupied by the air at the top. This air-pump is made to act also as a condensing engine, as some German machines had done before, by the very simple apparatus of turning a cock.

By means of this gauge, Mr. Smeaton judged that his machine was incomparably better than any former ones, as it seemed to rarefy the air in the receiver 1000, or even 2000 times, while the best of the former construction only rarefied about 140 times: and so the case has since been always understood, an implicit confidence being placed in Mr. Smeaton's accuracy, till the fallacy was accidentally detected in the manner related at large by Mr. Nairne in the Philof. Tranf. for the year 1777. This accurate and ingenious artist wanting to make trial of Mr. Smeaton's pear-gauge, executed an air-pump of his improved construction, in the best manner possible; which, in various experiments made with it, appeared, by the pear-gauge, to rarefy the air to an amazing degree indeed, being at times from 4000 to 10000, or 50000, or even 100000 times rarefied. But upon measuring the same expansion by the usual long and short tube gauges, which both accurately agreed together, he found that these never shewed a rarefaction of more than 600 times: widely different from the same as measured by the pear or internal gauge, by experiments often repeated. ' Finding, says Mr. Nairne, still this disagreement between the pear-gauge and the other gauges, I tried a variety of experiments; but none of them appeared to me satisfactory, till one day in April 1776, shewing an experiment with one of these pumps to the honourable Henry Cavendish, Mr. Smeaton, and several other gentlemen of the Royal Society, when the two gauges differed some thousand times from one another, Mr. Cavendish accounted for it in the following manner. " It appeared, he said, from some experiments of his father's, Lord Cavendish, that water, whenever the pressure of the atmosphere on it is diminished to a certain degree, is immediately turned into vapour, and is as immediately turned back again into water on restoring the pressure. This degree of pressure is different according to the heat of the water: when the heat is 72° of Fahrenheit's scale, it turns into vapour as soon as the pressure is no greater than that of three quarters of an inch of quickfilver, or about 1-40th of the usual pressure of the atmosphere; but when the heat is only 41°, the pressure must be reduced to that of a quarter of an inch of quickfilver before the water turns into vapour. It is true, that water exposed to the open air, will evaporate at any heat, and with any pressure of the atmosphere; but that evaporation is intirely owing to the action of the air upon it; whereas the evaporation here spoken of, is performed without any assistance from the air. Hence it follows, that when the receiver is exhausted to the above-mentioned degree, the moisture adhering to the different parts of the machine will turn into vapour, and supply the place of the air, which is continually drawn away by the working of the

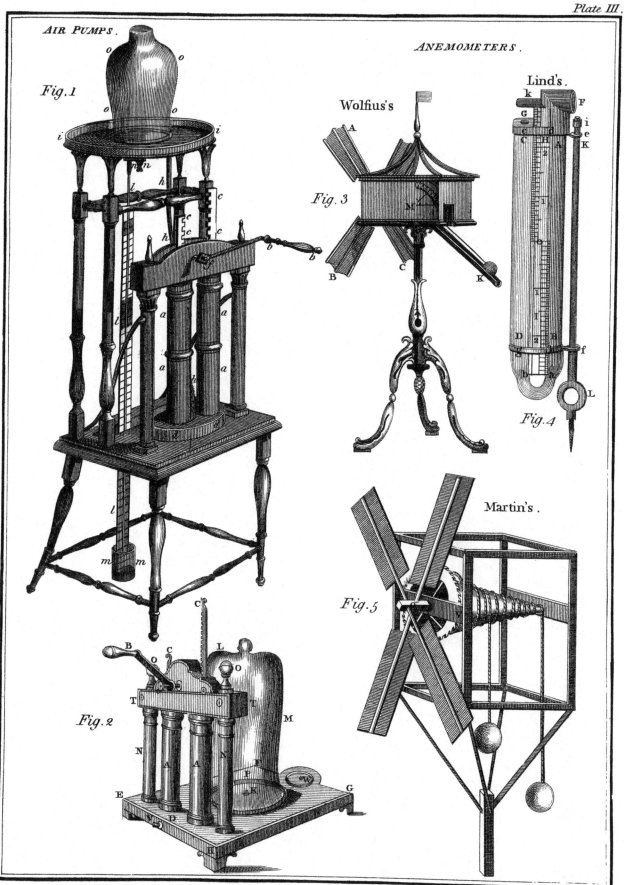

Plate III.

AIR PUMPS.

ANEMOMETERS.

Fig.1

Wolfius's

Lind's.

Fig.3

Fig.4

Fig.5

Martin's.

Fig.2

the pump; so that the fluid in the pear-gauge, as well as that in the receiver, will consist in a good measure of vapour. Now letting the air into the receiver, all the vapour within the pear-gauge will be reduced to water, and only the real air will remain uncondensed; consequently the pear-gauge shews only how much real air is left in the receiver, and not how much the pressure or spring of the included fluid is diminished; whereas the common gauges shew how much the pressure of the included fluid is diminished, and that equally, whether it consist of air or of vapour." Mr. Cavendish having explained so satisfactorily the cause of the disagreement between the two gauges, Mr. Nairne considered that, if he were to avoid moisture as much as possible, the two gauges should nearly agree. And in fact they were found so to do, each shewing a rarefaction of about 600, when all moisture was perfectly cleared away from the pump, and the plate and the edges of the receiver were secured by a cement instead of setting it upon a soaked leather, as in the usual way. But by future experiments, Mr. Nairne found that the same excellent machine would not exhaust more than 50 or 60 times, when the receiver was set upon leather soaked in water, the heat of the room being about 57°. And from the whole, Mr. Nairne concludes that the air-pump of Otto Guericke, and those contrived by Mr. Gratorix, and Dr. Hook, and the improved one by Mr. Papin, both used by Mr. Boyle, as also Hauksbee's, s'Gravesande's, Muschenbroeck's, and those of all who have used water in the barrels of their pumps, could never have exhausted to more than between 40 and 50, if the heat of the place was about 57; and although Mr. Smeaton, with his pump, where no water was in the barrel, but where leather soaked in a mixture of water and spirit of wine was used on the pump-plate, to set the receiver upon, may have exhausted all but a thousandth, or even a ten-thousandth part of the common air, according to the testimony of his pear-gauge; yet so much vapour must have arisen from the wet leather, that the contents of the receiver could never be less than a 70th or 80th part of the density of the atmosphere. But when nothing of moisture is used about this machine, it will, when in its greatest perfection, rarefy its contents of air about 600 times.

It is evident that by means of these two gauges we can ascertain the several quantities of vapour and permanent air which make up the contents of the receiver, after the exhaustion is made as perfect as can be; for the usual external gauge determines the whole contents, made up of the vapour and air, whilst the pear-gauge shews the quantity of real permanent air; consequently the difference is the quantity of vapour.

The principal cause which prevents this pump from exhausting beyond the limit above-mentioned, is the weakened elasticity of the air within the receiver, which, decreasing in proportion as the quantity of the air within is diminished, becomes at last incapable of lifting up the valve of communication between the receiver and the barrel; and consequently no more air can then pass from the former to the latter.

Several ingenious persons have used their endeavours to remove this imperfection in the best air-pumps. Amongst these it seems that one Mr. Haas has succeeded tolerably well; having, by means of a contri-

vance to open the communication valve in the bottom of the barrel, made his machine so perfect, that when every thing is in the greatest perfection, it rarefies the contents of the receiver as far as 1000 times, even when measured by the exterior gauge. The description of this machine, and an account of some experiments performed with it, are given by Mr. Cavallo in the Philos. Transf. for the year 1783.

But the imperfections it seems have more recently been removed by an ingenious contrivance of Mr. Cuthbertson, a mathematical instrument maker at Amsterdam, now of London, whose air-pump has neither cocks nor valves, and is so constructed, that what supplies their place has the advantages of both, without the inconveniences of either. He has also made improvements in the gauges, by means of which he determines the height of the mercury in the tube, by which the degree of exhaustion is indicated, to the hundredth part of an inch. And to obviate the inconvenience of the elastic vapour arising from the wet leather, upon which the receiver is placed, for common experiments, he recommends the use of leather dressed with allum, and soaked in hog's lard, which he found to yield very little of this vapour; but when the utmost degree of exhaustion is required, his advice is, to dry the receiver well, and set it upon the plate without any leather, only smearing its outer edges with hog's lard, or with a mixture of three parts of hog's lard and one of oil. But the use of the leather has long been laid aside by our English instrument-makers, a circumstance which probably had not come to Mr. Cuthbertson's knowledge. An account of this instrument, and of some experiments performed with it, was published at Amsterdam in the year 1787; from which experiments it appears that, by a coincidence of the several gauges, a rarefaction of 1200 times was shewn; but when the atmosphere was very dry, the exhaustion has been so complete, that the gauges have shewn the air in the receiver to be rarefied above 2400 times.

There are made also by different persons, portable, or small air-pumps, of various constructions, to set upon a table, to perform experiments with. In these, the gauge is varied according to the fancy of the maker, but commonly it consists of a bent glass tube, like a syphon, open only at one end. The gauge is placed under a small receiver communicating, by a pipe, with the principal pipe leading from the general receiver to the barrels. The close end of the gauge, of 3 or 4 inches long, before the exhaustion, has the quicksilver forced close up to the top by the pressure of the air on the open end; but when the exhaustion is considerably advanced, it begins to descend, and then the difference of the heights of the quicksilver in the two legs, compared with the height in the barometrical tube, determines the degree of exhaustion: so if the difference between the two be one inch, when the barometer stands at 30, the air is rarefied 30 times; but if the difference be only half an inch, the rarefaction is 60 times, and so on. See Plate III. fig. 2.

The Use of the Air-Pump. In whatever manner or form this machine be made, the use and operation of it are always the same. The handle, which works the piston, is moved up and down in the barrel, by which

means

means a barrel of the contained air is drawn out at every stroke of the piston, in the following manner: by pushing the piston down to the bottom of the barrel, where the air is prevented from escaping downwards, by its elasticity it opens the valve of the piston, and escapes upwards above it into the open air; then raising the piston up, the external atmosphere shuts down its valve, and a vacuum would be made below it, but for the air in the receiver, pipe, &c, which now raises the valve in the bottom of the barrel, and rushes in and fills it again, till the whole air in the receiver and barrel be of one uniform density, but less than it was before the stroke, in proportion as the sum of all the capacities of the receiver, pipe, and barrel together, is to the same sum wanting the barrel. And thus is the air in the receiver diminished at each stroke of the piston, by the quantity of the barrel or cylinder full, and therefore always in the same proportion: so that by thus repeating the operation again and again, the air is rarefied to any proposed degree, or till it has not elasticity enough to open the valve of the piston or of the barrel, after which the exhaustion cannot be any farther carried on: the gauge, in comparison with the barometer, shewing at any time what the degree of exhaustion is, according to the particular nature and construction of it.

But, supposing no vapour from moisture, &c, to rise in the receiver, the degree of exhaustion, after any number of strokes of the piston, may be determined by knowing the respective capacities of the barrel and the receiver, including the pipe, &c. For as we have seen above that every stroke diminishes the density in a constant proportion, namely as much as the whole content exceeds that of the cylinder or barrel; and consequently the sum of as many diminutions as there are strokes of the piston, will shew the whole diminution by all the strokes. So, if the capacity of the barrel be equal to that of the receiver, in which the communication pipe is always to be included; then, the barrel being half the sum of the whole contents, half the air will be drawn out at one stroke; and consequently the remaining half, being dilated through the whole or first capacity, will be of only half the density of the first: in like manner, after the second stroke, the density of the remaining contents will be only half of that after the first stroke, that is only $\frac{1}{4}$ of the original density: continuing this operation, it follows that the density of the remaining air will be $\frac{1}{8}$ after 3 strokes of the piston, $\frac{1}{16}$ after 4 strokes, $\frac{1}{32}$ after 5 strokes, and so on, according to the powers of the ratio $\frac{1}{2}$; that is, such power of the ratio as is denoted by the number of the strokes. In like manner, if the barrel be $\frac{1}{3}$ of the whole contents, that is, the receiver double of the barrel, or $\frac{2}{3}$ of the whole contents; then the ratio of diminution of density being $\frac{2}{3}$, the density of the contents, after any number of strokes of the piston, will be denoted by such power of $\frac{2}{3}$ whose exponent is that number; namely, the density will be $\frac{2}{3}$ after one stroke, $(\frac{2}{3})^2$ or $\frac{4}{9}$ after two strokes, $(\frac{2}{3})^3$ or $\frac{8}{27}$ after 3 strokes, and in general it will be $(\frac{2}{3})^n$ after n strokes: the original density of the air being 1. Hence then, universally, if s denote the sum of the contents of the receiver and barrel, and r that of the receiver only without the barrel, and n any number of strokes of the piston; then, the original density of the air being 1,

the density after n strokes will be $(\frac{r}{s})^n$ o $\frac{r^n}{s^n}$, namely the n power of the ratio $\frac{r}{s}$. So, for example, if the capacity of the receiver be equal to 4 times that of the barrel; then their sum s is 5, and r is 4; and the density of the contents after 30 strokes, will be $(\frac{4}{5})^{30}$, or the 30th power of $\frac{4}{5}$, which is $\frac{1}{808}$ nearly; so that the air in the receiver is rarefied 808 times.

See also the Memoires de l'Acad. Royale des Sciences for the years 1693 and 1705.

From the same formula, namely $(\frac{r}{s})^n = d$ the density, we easily derive a rule for finding the number of strokes of the piston, necessary to rarefy the air any number of times, or to reduce it to a given density d, that of the natural air being 1. For since $(\frac{r}{s})^n = d$, by taking the logarithm of this equation, it is $n \times \log.$ $\frac{r}{s} = \log.$ of d; and hence $n = \dfrac{\log. d}{\log. \frac{r}{s}} = \dfrac{\log. d}{\log. r - \log. s}$ that is, divide the log. of the proposed density by the log. of the ratio of the receiver to the sum of the receiver and barrel together, and the quotient will shew the number of strokes of the piston requisite to produce the degree of exhaustion required. So, for example, if the receiver be equal to 5 times the barrel, and it be proposed to find how many strokes of the piston will rarefy the air 100 times; then $r = 5$, $s = 6$, $d = \frac{1}{100}$, whose log is -2, and $\frac{r}{s} = \frac{5}{6}$, whose log. is $-·07918$; therefore $\dfrac{2}{·07918} = 25\frac{1}{4}$ nearly, which is the number of strokes required.

And, farther, the same formula reduced, would give us the proportion between the receiver and barrel, when the air is rarefied to any degree by an assigned number of strokes of the piston. For since $(\frac{r}{s})^n = d$ the density, therefore, extracting the n root of both sides, it is $\frac{r}{s} = \sqrt[n]{d}$: that is, the n root of the density is equal to the ratio of the receiver to the sum of the receiver and barrel. So, if the density d be $\frac{1}{128}$, and the number of strokes $n = 7$; then the 7th root of $\frac{1}{128}$ is $\frac{1}{2}$; which shews that the receiver is equal to half the receiver and barrel together, or that the capacity of the barrel is just equal to that of the receiver.

Some of the principal effects and phenomena of the air-pump, are the following: That, in the exhausted receiver, heavy and light bodies fall equally swift; so, a guinea and feather fall from the top of a tall receiver to the bottom exactly together. That most animals die in a minute or two: but however, That vipers and frogs, though they swell much, live an hour or two; and after being seemingly quite dead, come to life again in the open air: That snails survive about ten hours; efts, or slow-worms, two or three days; and leeches five or six. That oysters live for 24 hours. That the heart of an eel taken out of the body, continues to
beat

beat for good part of an hour, and that more brifkly than in the air. That warm blood, milk, gall, &c, undergo a confiderable intumefcence and ebullition. That a moufe or other animal may be brought, by degrees, to furvive longer in a rarefied air, than naturally it does. That air may retain its ufual preffure, after it is become unfit for refpiration. That the eggs of filk-worms hatch in vacuo. That vegetation ftops. That fire extinguifhes; the flame of a candle ufually going out in one minute; and a charcoal in about five minutes. That red-hot iron, however, feems not to be affected; and yet fulphur or gun-powder are not lighted by it, but only fufed. That a match, after lying feemingly extinct a long time, revives again on re-admitting the air. That a flint and fteel ftrike fparks of fire as copioufly, and in all directions, as in air. That magnets, and magnetic needles, act the fame as in air. That the fmoke of an extinguifhed luminary gradually fettles to the bottom in a darkifh body, leaving the upper part of the receiver clear and tranfparent; and that on inclining the veffel fometimes to one fide, and fometimes to another, the fume preferves its furface horizontal, after the nature of other fluids. That heat may be produced by attrition. That camphire will not take fire; and that gun-powder, though fome of the grains of a heap of it be kindled by a burning glafs, will not give fire to the contiguous grains. That glow-worms lofe their light in proportion as the air is exhaufted, and at length become totally obfcure; but on re-admitting the air, they prefently recover it all. That a bell, on being ftruck, is not heard to ring, or very faintly. That water freezes. But that a fyphon will not run. That electricity appears like the aurora borealis. With multitudes of other curious and important particulars, to be met with in the numerous writings on this machine, namely, befides the Philof. Tranfactions of moft academies and focieties, in the writings of Torricelli, Pafcal, Merfenne, Guericke, Schottus, Boyle, Hook, Duhamel, Mariotte, Haukfbee, Hales, Mufchenbroeck, Gravefande, Defaguliers, Franklin, Cotes, Helfham, and a great many other authors.

AIR-VESSEL, in *Hydraulics*, is a veffel of air within fome water engines, which being compreffed, by forcing in a confiderable quantity of water, by its uniform fpring, forces it out at the pipe in a conftant uninterrupted ftream, to a great height.

Air-veffel too, in the improved fire engines, is a metallic cylinder, placed between the two forcing pumps, by the action of whofe piftons the water is forced into this veffel, through two pipes, with valves; then the air, previoufly contained in it, is compreffed by the water, in proportion to the quantity admitted, and this air, by its fpring, forces the water through a pipe by a conftant and equal ftream; whereas in the common fquirting engine, the ftream is difcontinued between the feveral ftrokes.

AIRY TRIPLICITY, in *Aftrology*, the figns of Gemini, Libra, and Aquarius.

AJUTAGE, or ADJUTAGE, in *Hydraulics*, part of the apparatus of a *jet d'eau*, or artificial fountain; being a kind of tube fitted to the aperture or mouth of the ciftern, or the pipe; through which the water is to be played in any direction, and in any fhape or figure.

It is chiefly the diverfity in the ajutage, that makes the different kinds of fountains. So that, by having feveral ajutages, to be applied occafionally, one fountain is made to have the effect of many.

Mariotte, Gravefande, and Defaguliers have written pretty fully on the nature of ajutages, or fpouts for jets d'eau, and efpecially the former. He affirms, from experiment, that an even polifhed round hole, made in the thin end of a pipe, gives a higher jet than either a cylindrical or a conical ajutage; but that, of thefe two latter however, the conical is better than the cylindrical figure. See his Traité du Mouvement des Eaux, part 4.

The quantity of water difcharged by ajutages of equal area, but of different figures, is the fame. And for like figures, but of different fizes, the quantity difcharged, is directly proportional to the area of the ajutage, or to the fquare of its diameter, or of any fide or other linear dimenfion: fo, an ajutage of a double diameter, or fide, will difcharge 4 times the quantity of water; of a triple diameter, 9 times the quantity; and fo on; fuppofing them at an equal depth below the furface or head of water. But if the ajutage be at different depths below the head, then the celerity with which the water iffues, and confequently the quantity of it run out in any given time, is directly proportional to the fquare-root of the altitude of the head, or depth of the hole: fo at 4 times the depth, the celerity and quantity is double; at 9 times the depth, triple; and fo on.

It has been found that jets do not rife quite fo high as the head of water; owing chiefly to the refiftance of the air againft it, and the preffure of the upper parts of the jet upon the lower: and for this reafon it is, that if the direction of the ajutage be turned a very little from the perpendicular, it is found to fpout rather higher than when the jet is exactly upright.

It is found by experiment too, that the jet is higher or lower, according to the fize of the ajutage: that a circular hole of about an inch and a quarter in diameter, jets higheft; and that the farther from that fize, the worfe. Experience alfo fhews that the pipe leading to the ajutage, fhould be much larger than it; and if the pipe be a long one, that it fhould be wider the farther it is from the ajutage.

For the other circumftances relating to jets and the iffuing of water under various circumftances, fee EXHAUSTION, FLUX, FOUNTAIN, JET D'EAU, &c, to which they more properly belong.

ALBATEGNI, an Arabic prince of Batan in Mefopotamia, who was a celebrated aftronomer, about the year of Chrift 880, as appears by his obfervations. He is alfo called *Muhammed ben Geber Albatani, Mahomet the fon of Geber*, and *Muhamedes Aractenfis*. He made aftronomical obfervations at Antioch, and at Racah or Aracta, a town of Chaldea, which fome authors call a town of Syria or of Mefopotamia. He is highly fpoken of by Dr. Halley, as a *vir admirandi acuminis, ac in adminiftrandis obfervationibus exercitatiffimus*.

Finding that the tables of Ptolomy were imperfect, he computed new ones, which were long ufed as the beft among the Arabs: thefe were adapted to the meridian of Aracta or Racah. Albategni compofed in

Arabic

Arabic a work under the title of *The Science of the Stars*, comprising all parts of astronomy, according to his own observations and those of Ptolomy. This work, translated into Latin by Plato of Tibur, was published at Nuremberg in 1537, with some additions and demonstrations of Regiomontanus; and the same was reprinted at Bologna in 1645, with this author's notes. Dr. Halley detected many faults in these editions: Philos. Transf. for 1693, Nº 204.

In this work, Albategni gives the motion of the sun's apogee since Ptolomy's time, as well as the motion of the stars, which he makes 1 degree in 70 years. He made the longitude of the first star of Aries to be 18° 2'; and the obliquity of the ecliptic 23° 35'. And upon Albategni's observations were founded the Alphonsine tables of the moon's motions; as is observed by Nic. Muler, in the Tab. Frisicæ, pa. 248.

ALBERTUS MAGNUS, a very learned man in the 13th century, who, among a multitude of books, wrote several upon the various mathematical sciences, as Arithmetic, Geometry, Perspective or Optics, Music, Astrology and Astronomy, particularly under the titles, *de sphæra, de astris, de astronomia, item speculum astronomicum.*

Albertus Magnus was born at Lawingen on the Danube, in Suabia, in 1205, or according to some in 1193; and he died at a great age, at Cologn, November 15, 1280. Vossius and other authors speak of him as a great genius, and deeply skilled in all the learning of the age. His writings were so numerous, that they make 21 volumes in folio, in the Lyons edition of 1615. He has passed also for the author of some writings relating to midwifery, &c, under the title of *De natura rerum*, and *De secretis mulierum*, in which there are many phrases and expressions unavoidable on such a subject, which gave great offence, and raised a clamour against him as the supposed author, and inconsistent with his character, being a Dominican friar, and sometime bishop of Ratisbon; which dignity however he soon resigned, through his love for solitude, to enter again into the monastic life. But the advocates of Albert assert, that he was not the author of either of these two works. It must be acknowledged however, that there are, in his Comment upon the Master of Sentences, some questions concerning the practice of conjugal duty, in which he has used some words rather too gross for chaste and delicate ears: but they allege what he himself used to say in his own vindication, that he came to the knowledge of so many monstrous things at confession, that it was impossible to avoid touching upon such questions. Albert was certainly a man of a most curious and inquisitive turn of mind, which gave rise to other accusations against him; such as, that he laboured to find out the philosopher's stone; that he was a magician; and that he made a machine in the shape of a man, which was an oracle to him, and explained all the difficulties he proposed: the common cant accusations of those times of ignorance and superstition. But having great knowledge in the mathematics and mechanics, by his skill in these sciences he probably formed a head, with springs capable of articulate sounds; like the machines of Boetius and others.

John Matthæus de Luna, in his treatise De Rerum Inventoribus, has attributed the invention of fire-arms

to Albert; but in this he is refuted by Naude, in his Apologie des grands hommes.

ALBUMAZAR, otherwise called ABUASSAR, and JAPHAR, was a celebrated Arabian philosopher and astrologer, of the 9th or 10th century, or according to some authors much earlier. Blancanus, Vossius, &c, speak of him as one of the most learned astronomers of his time, or astrologers, which was then the same thing. He wrote a work *De Magnis Conjunctionibus Annorum Revolutionibus, ac eorum Perfectionibus*, printed at Venice in 1515, at the expence of Melchior Sessa, a work chiefly astrological.

He wrote also *Introductio in Astronomiam*, printed in the year 1489. And it is reported that he observed a comet in his time, above the orb of Venus.

ALCOHOL, in the Arabian Astrology, is when a heavy slow-moving planet receives another lighter one within its orb, so as to come in conjunction with it.

ALDEBARAN, an Arabian name of a fixed star, of the first magnitude, just in the eye of the sign or constellation Taurus, or the bull, and hence it is popularly called the bull's eye. For the beginning of the year 1800, its

Right Ascension is	—	66° 6′ 51″·10
Annual variation in A R	0	0 51 ·31
Declination	—	16 5 52 ·00 N.
And Annual variat. in Decl.	0	0 8 ·30

ALDERAIMIN, a star of the third magnitude in the right shoulder of the constellation Cepheus.

ALDHAFERA, or Aldhaphra, in the Arabian Astronomy, denotes a fixed star of the third magnitude, in the mane of the sign or constellation Leo, the lion.

ALEMBERT (JOHN LE ROND D'), an eminent French mathematician and philosopher, and one of the brightest ornaments of the 18th century. He was perpetual secretary to the French Academy of Sciences, and a member of most of the philosophical academies and societies of Europe.

D'Alembert was born at Paris, the 16th of November 1717. He derived the name of John le Rond from that of the church near which, after his birth, he was exposed as a foundling. But his father, informed of this circumstance, listening to the voice of nature and duty, took measures for the proper education of his child, and for his future subsistence in a state of ease and independence. His mother, it is said, was a lady of of rank, the celebrated Mademoiselle Tencin, sister to cardinal Tencin, archbishop of Lyons.

He received his first education among the Jansenists, in the College of the Four Nations, where he gave early signs of genius and capacity. In the first year of his philosophical studies, he composed a Commentary on the Epistle of St. Paul to the Romans. The Jansenists considered this production as an omen, that portended to the party of Port-Royal a restoration to some part of their former splendor, and hoped to find one day in d'Alembert a second Pascal. To render this resemblance more complete, they engaged their pupil in the study of the mathematics; but they soon perceived that his growing attachment to this science was likely to disappoint the hopes they had formed with respect to his future destination: they therefore endeavoured to divert him from this line; but their endeavours were fruitless.

On

On his quitting the college, finding himself alone, and unconnected in the world, he sought an asylum in the house of his nurse. He hoped that his fortune, though not ample, would enlarge the subsistence, and better the condition of her family, which was the only one that he could consider as his own. It was here therefore that he fixed his residence, resolving to apply himself entirely to the study of geometry.—And here he lived, during the space of 40 years, with the greatest simplicity, discovering the augmentation of his means only by increasing displays of his beneficence, concealing his growing reputation and celebrity from these honest people, and making their plain and uncouth manners the subject of good-natured pleasantry and philosophical observation. His good nurse perceived his ardent activity; heard him mentioned as the writer of many books; but never took it into her head that he was a great man, and rather beheld him with a kind of compassion. " You will never, said she to him one day, be any thing but a philosopher—and what is a philosopher?—a fool, who toils and plagues himself all his life, that people may talk of him when he is dead."

As d'Alembert's fortune did not far exceed the demands of necessity, his friends advised him to think of some profession that might enable him to increase it. He accordingly turned his views to the law, and took his degrees in that faculty: but soon after, abandoning this line, he applied himself to the study of medicine. Geometry however was always drawing him back to his former pursuits; so that after many ineffectual struggles to resist its attractions, he renounced all views of a lucrative profession, and gave himself up entirely to mathematics and poverty.

In the year 1741 he was admitted a member of the Academy of Sciences; for which distinguished literary promotion, at so early an age (24), he had prepared the way by correcting the errors of a celebrated work (The *Analyse Demontrée* of Reyneau), which was esteemed classical in France in the line of analytics. He afterwards set himself to examine, with close attention and assiduity, what must be the motion and path of a body, which passes from one fluid into another denser fluid, in a direction oblique to the surface between the two fluids. Every one knows the phenomenon which happens in this case, and amuses children, under the denomination of *Ducks and Drakes;* but it was d'Alembert who first explained it in a satisfactory and philosophical manner.

Two years after his election to a place in the academy, he published his *Treatise on Dynamics.* The new principle developed in this treatise, consisted in establishing an equality, at each instant, between the changes that the motion of a body has undergone, and the forces or powers which have been employed to produce them: or, to express the same thing otherwise, in separating into *two parts* the action of the moving powers, and considering the *one* as producing alone the motion of the body, in the second instant, and the *other* as employed to destroy that which it had in the first.

So early as the year 1744, d'Alembert had applied this principle to the theory of the equilibrium, and the motion of fluids: and all the problems before resolved

in physics, became in some measure its corollaries. The discovery of this new principle was followed by that of a new calculus, the first essays of which were published in a *Discourse on the General Theory of the Winds,* to which the prize-medal was adjudged by the Academy of Berlin in the year 1746, which proved a new and brilliant addition to the fame of d'Alembert. This new calculus of *Partial Differences* he applied, the year following, to the problem of vibrating chords, the resolution of which, as well as the theory of the oscillations of the air and the propagation of sound, had been but imperfectly given by the mathematicians who preceded him; and these were his masters or his rivals.

In the year 1749 he furnished a method of applying his principle to the motion of any body of a given figure. He also resolved the problem of the precession of the equinoxes; determining its quantity, and explaining the phenomenon of the nutation of the terrestrial axis discovered by Dr. Bradley.

In 1752, d'Alembert published a treatise on the *Resistance of Fluids,* to which he gave the modest title of an *Essay;* though it contains a multitude of original ideas and new observations. About the same time he published, in the Memoirs of the Academy of Berlin, *Researches concerning the Integral Calculus,* which is greatly indebted to him for the rapid progress it has made in the present century.

While the studies of d'Alembert were confined to mere mathematics, he was little known or celebrated in his native country. His connections were limited to a small society of select friends. But his cheerful conversation, his smart and lively sallies, a happy knack at telling a story, a singular mixture of malice of speech with goodness of heart, and of delicacy of wit with simplicity of manners, rendering him a pleasing and interesting companion, his company began to be much sought after in the fashionable circles. His reputation at length made its way to the throne, and rendered him the object of royal attention and beneficence. The consequence was a pension from government, which he owed to the friendship of count d'Argenson.

But the tranquillity of d'Alembert was abated when his fame grew more extensive, and when it was known beyond the circle of his friends, that a fine and enlightened taste for literature and philosophy accompanied his mathematical genius. Our author's eulogist ascribes to envy, detraction, &c, all the opposition and censure that d'Alembert met with on account of the famous Encyclopédie, or Dictionary of Arts and Sciences, in conjunction with Diderot. None surely will refuse the well-deserved tribute of applause to the eminent displays of genius, judgment, and true literary taste, with which d'Alembert has enriched that great work. Among others, the Preliminary Discourse he has prefixed to it, concerning the rise, progress, connections, and affinities of all the branches of human knowledge, is perhaps one of the most capital productions the philosophy of the age can boast of.

Some time after this, d'Alembert published his *Philosophical, Historical,* and *Philological Miscellanies.* These were followed by the *Memoirs of Christina queen of Sweden;* in which d'Alembert shewed that he was acquainted with the natural rights of mankind,

and

and was bold enough to affert them. His *Effay on the Intercourfe of Men of Letters with Perfons high in Rank and Office*, wounded the former to the quick, as it expofed to the eyes of the public the ignominy of thofe fervile chains, which they feared to fhake off, or were proud to wear. A lady of the court hearing one day the author accufed of having exaggerated the defpotifm of the great, and the fubmiffion they require, anfwered flyly, " If he had confulted me, I would have told him ftill more of the matter."

D'Alembert gave elegant fpecimens of his literary abilities in his translations of fome felect pieces of Tacitus. But thefe occupations did not divert him from his mathematical ftudies: for about the fame time he enriched the Encyclopédie with a multitude of excellent articles in that line, and compofed his *Refearches on feveral Important Points of the Syftem of the World*, in which he carried to a higher degree of perfection the folution of the problem concerning the perturbations of the planets, that had feveral years before been prefented to the Academy.

In 1759 he publifhed his *Elements of Philofophy*: a work much extolled as remarkable for its precifion and perfpicuity.

The refentment that was kindled (and the difputes that followed it) by the article *Geneva*, inferted in the Encyclopédie, are well known. D'Alembert did not leave this field of controverfy with flying colours. Voltaire was an auxiliary in the conteft: but as he had no reputation to lofe, in point of candour and decency; and as he weakened the blows of his enemies, by throwing both them and the fpectators into fits of laughter, the iffue of the war gave him little uneafinefs. It fell more heavily on d'Alembert; and expofed him, even at home, to much contradiction and oppofition.

It was on this occafion that the late king of Pruffia offered him an honourable afylum at his court, and the office of prefident of his academy: and the king was not offended at d'Alembert's refufal of thefe diftinctions, but cultivated an intimate friendfhip with him during the reft of his life. He had refufed, fome time before this, a propofal made by the emprefs of Ruffia to entruft him with the education of the Grand Duke; —a propofal accompanied with all the flattering offers that could tempt a man, ambitious of titles, or defirous of making an ample fortune: but the objects of his ambition were tranquillity and ftudy.

In the year 1765, he publifhed his *Differtation on the Deftruction of the Jefuits*. This piece drew upon him a fwarm of adverfaries, who only confirmed the merit and credit of his work by their manner of attacking it.

Befide the works already mentioned, he publifhed nine volumes of memoirs and treatifes, under the title of *Opufcules*; in which he has refolved a multitude of problems relating to aftronomy, mathematics, and natural philofophy; of which his panegyrift, Condorcet, gives a particular account, more efpecially of thofe which exhibit new fubjects, or new methods of inveftigation.

He publifhed alfo *Elements of Mufic*; and rendered, at length, the fyftem of Rameau intelligible: but he did not think the mathematical theory of the fonorous body fufficient to account for the rules of that art.

In the year 1772 he was chofen fecretary to the French Academy of Sciences. He formed, foon after this preferment, the defign of writing the lives of all the deceafed academicians, from 1700 to 1772; and in the fpace of three years he executed this defign, by compofing 70 eulogies.

D'Alembert died on the 29th of October 1783, being nearly 66 years of age. In his moral character there were many amiable lines of candour, modefty, difinterestednefs, and beneficence; which are defcribed, with a diffufive detail, in his eulogium, by Condorcet, in the *Hift. de l'Acad. Royale des Sciences,* 1783.

As it may be curious and ufeful to have in one view an entire lift of d'Alembert's writings, I fhall here infert a catalogue of them, from Rozier's *Nouvelle Table des Articles contenus dans les volumes de l'Academie Royale des Sciences de Paris*, &c, as follows:

Traité de Dynamique, in 4to, Paris, 1743. The 2d ed. in 1758.

Traité de l'Equilibre et du Mouvement des Fluides. Paris, 1744; and the 2d edition in 1770.

Reflexions fur la Caufe Générale des Vents; which gained the prize at Berlin in 1746; and was printed at Paris in 1747, in 4to.

Recherches fur la Précession des Équinoxes, & fur la Nutation de l'Axe de la Terre dans le Syftème Newtonien. Paris, 1749, in 4to.

Effais d'une Nouvelle Théorie du Mouvement des Fluides. Paris, 1752, in 4to.

Recherches fur differens Points importans du Syftème du Monde. Paris, 1754 and 1756, 3 vol. in 4to.

Elemens de Philofophie, 1759.

Opufcules Mathematiques, ou Memoires fur différens Sujets de Géométrie, de Méchaniques, d'Optiques, d'Aftronomie. Paris, 9 vol. in 4to; 1761 to 1773.

Elémens de Mufique, théorique & pratique, fuivant les Principes de M. Rameau, eclairés, développés, & fimplifiés. 1 vol. in 8vo. à Lyon.

De la Deftruction des Jefuites, 1765.

In the Memoirs of the Academy of Paris are the following pieces, by d'Alembert: viz,

Précis de Dynamique, 1743, Hift. 164.

Précis de l'Equilibre & de Mouvement des Fluides, 1744, Hift. 55.

Methode générale pour déterminer les Orbites & les Mouvements de toutes les Planètes, en ayant égard à leur action mutuelle, 1745, p. 365.

Précis des Réflexions fur la Caufe Générale des Vents, 1750, Hift. 41.

Précis des Recherches fur la Précession des Équinoxes, et fur la Nutation de l'Axe de la Terre, 1750, Hift. 134.

Effai d'une Nouvelle Théorie fur la Réfiftance des Fluides, 1752, Hift. 116.

Précis des Effais d'une Nouvelle Théorie de la Réfiftance des Fluides, 1753, Hift. 289.

Précis des Recherches fur les differens Points importans du Syftème du Monde, 1754, Hift. 125.

Recherches fur la Précession des Équinoxes, & fur la Nutation de l'Axe de la Terre, dans l'Hypothefe de la Diffimilitude des Méridiens, 1754, p. 413, Hift. 116.

Reponfe à un Article du Mémoire de M. l'Abbé de
la

la Caille, sur la Théorie du Soleil, 1757, p. 145, Hist. 118.

Addition à ce Mémoire, 1757, p. 567, Hist. 118.

Précis des Opuscules Mathématiques, 1761, Hist. 86.

Précis du troisième volume des Opuscules Mathématiques, 1764, Hist. 92.

Nouvelles Recherches sur les Verres Optiques, pour servir de suite à la théorie qui en à été donnée dans le volume 3e des Opuscules Mathématiques. Premier Mémoire, 1764, p. 75, Hist. 175.

Nouvelles Recherches sur les Verres Optiques, pour servir de suite à la théorie qui en a été donnée dans le troisième volume des Opuscules Mathématiques. Second Mémoire, 1765, p. 53.

Observations sur les Lunettes Achromatiques, 1765, p. 53, Hist. 119.

Suite des Recherches sur les Verres Optiques. Troisième Mémoire, 1767, p. 43, Hist. 153.

Recherches sur le Calcul Intégral, 1767, p. 573.

Accident arrivé par l'Explosion d'une Meule d'Emouleur, 1768, Hist. 31.

Précis des Opuscules de Mathématiques, 4e & 5e volumes. Leur Analyse, 1768, Hist. 83.

Recherches sur les Mouvemens de l'Axe d'une Planète quelconque dans l'hypothese de la Dissimilitude des Méridienes, 1768 p. 1, Hist. 95.

Suite des Recherches sur les Mouvemens, &c, 1768 p. 332, Hist. 95.

Recherches sur le Calcul Intégral, 1769, p. 73.

Mémoire sur les Principes de la Mech. 1769, p. 278.

And in the Memoirs of the Academy of Berlin, are the following pieces, by our author : viz,

Recherches sur le Calcul Intégral, première partie, 1746.

Solution de quelques problèmes d'astronomie, 1747.

Recherches sur la courbe que forme une Corde Tendue, mise en Vibration, 1747.

Suite des recherches sur le Calcul Intégral, 1748.

Lettre à M. de Maupertuis, 1749.

Addition aux recherches sur la courbe que forme une Corde Tendue mise en Vibration, 1750.

Addition aux recherches sur le Calcul Intégral, 1750.

Lettre à M. le professeur Formey, 1755.

Extr. de differ. lettres à M. de la Grange, 1763.

Sur les Tautochrones, 1765.

Extr. de differ. lettres à M. de la Grange, 1769.

Also in the Memoirs of Turin are,

Differentes Lettres à M. de la Grange, en 1764 & 1765, tom. 3 of these Memoirs.

Recherches sur differens sujets de Math. t. 4.

ALFECCA, or Alfeta, a name given to the star commonly called Lucida Coronæ.

ALFRAGAN, ALFERGANI, or FARGANI, a celebrated Arabic astronomer, who flourished about the year 800. He was so called from the place of his nativity, Fergan, in Sogdiana, now called Maràcanda, or Samarcand, anciently a part of Bactria. He is also called Ahmed (or Muhammed) ben-Cothair, or Katir. He wrote the Elements of Astronomy, in 30 chapters or sections. In this work the author chiefly follows Ptolomy, using the same hypotheses, and the same terms, and frequently citing him.

There are three Latin translations of Alfragan's work. The first was made in the 12th century, by Joannes Hispalensis; and was published at Ferrara in 1493, and at Nuremberg in 1537, with a preface by Melancthon. The second was by James Christman, from the Hebrew version of James Antoli, and appeared at Frankfort in 1590. Christman added to the first chapter of the work an ample commentary, in which he compares together the calendars of the Romans, the Egyptians, the Arabians, the Persians, the Syrians, and the Hebrews, and shews the correspondence of their years.

The third and best translation was made by Golius, professor of mathematics and Oriental languages at Leyden: this work, which came out in 1669, after the death of Golius, is accompanied with the Arabic text, and many learned notes upon the first nine chapters; for this author was not spared to carry them farther.

ALGAROTI, commonly called *Count Algaroti*, a celebrated Italian of the present century, well skilled in Architecture and the Newtonian philosophy, &c. Algaroti was born at Padua, but in what year has not been mentioned. Led by curiosity, as well as a desire of improvement, he travelled early into foreign countries; and was very young when he arrived in France in 1736. It was here that he composed his *Newtonian Philosophy for the Ladies*, as Fontenelle had done his Cartesian Astronomy, in the work intitled *The Plurality of Worlds*. He was much noticed by the king of Prussia, who conferred on him many marks of his esteem. He died at Pisa the 23d of May, 1764, and gave orders for his own mausoleum, with this inscription upon it; *Hic jacet Algarotus, sed non omnis*. He was esteemed to be well skilled in painting, sculpture, and architecture. His works, which are numerous, and upon a variety of subjects, abound with vivacity, elegance, and wit : a collection of them has lately been made, and printed at Leghorn; but that for which he is chiefly intitled to a place in this work is his *Newtonian Philosophy for the Ladies*, a sprightly, ingenious, and popular work.

ALGEBRA, a general method of resolving mathematical problems by means of equations. Or, it is a method of performing the calculations of all sorts of quantities by means of general signs or characters. At first, numbers and things were expressed by their names at full length; but afterwards these were abridged, and the initials of the words used instead of them ; and, as the art advanced farther, the letters of the alphabet came to be employed as general representations of all sorts of quantities; and other marks were gradually introduced, to express all sorts of operations and combinations; so as to entitle it to different appellations—universal arithmetic, and literal arithmetic, and the arithmetic of signs.

The etymology of the name, *Algebra*, is given in various ways. It is pretty certain, however, that the word is Arabian, and that from those people we had the name, as well as the art itself, as is testified by Lucas le Burgo, the first European author whose treatise was printed on this art, and who also refers to former authors and masters, from whose writings he had learned it. The Arabic name he gives it, is *Alghebra e Almucabala*, which is explained to signify the art of *restitution*

restitution and comparison, or *opposition and comparison*, or *resolution and equation*, all which agree well enough with the nature of this art. Some however derive it from various other arabic words; as from Geber, a celebrated philosopher, chemist, and mathematician, to whom also they ascribe the invention of this science: some likewise derive it from the word Geber, which with the particle *al*, makes Algeber, which is purely Arabic, and signifies the reduction of broken numbers or fractions to integers.

But Peter Ramus, in the beginning of his Algebra, says " the name Algebra is Syriac, signifying the art and doctrine of an excellent man. For *Geber*, in Syriac, is a name applied to men, and is sometimes a term of honour, as master or doctor among us. That there was a certain learned mathematician, who sent his Algebra, written in the Syriac language, to Alexander the Great, and he named it *Almucabala*, that is, the book of dark or mysterious things, which others would rather call the doctrine of Algebra. And to this day the same book is in great estimation among the learned in the oriental nations, and by the Indians who cultivate this art it is called *Aljabra*, and *Alboret*; though the name of the author himself is not known." But Ramus gives no authority for this singular paragraph. It has however on various occasions been distinguished by other names. Lucas Paciolus, or de Burgo, in Italy, called it *l'Arte Magiore : ditta dal vulgo la Regola de la Cosa over Alghebra e Almucabala*; calling it *l'Arte Magiore*, or the greater art, to distinguish it from common arithmetic, which is called *l'Arte Minore*, or the lesser art. It seems too that it had been long and commonly known in his country by the name *Regola de la Cosa*, or *Rule of the Thing*; from whence came our rule of cofs, cosic numbers, and such like terms. Some of his countrymen followed his denomination of the art; but other Italian and Latin writers called it *Regula rei & census*, the rule of the thing and the product, or the root and the square, as the unknown quantity in their equations commonly ascended no higher than the square or second power. From this Italian word *census*, pronounced *chensus*, came the barbarous word *zenzus*, used by the Germans and others, for quadratics; with the several zenzic or square roots. And hence ♃; 3, ♂, which are derived from the letters *r*, *z*, *c*, the initials of *res*, *zenzus*, *cubus*, or root, square, cube, came to be the signs or characters of these words: like as ℞ and √, derived from the letters *R*, *r*, became the signs of radicality.

Later authors, and other nations, used some the one of those names, and some another. It was also called *Specious Arithmetic* by Vieta, on account of the species, or letters of the alphabet, which he brought into general use; and by Newton it was called *Universal Arithmetic*, from the manner in which it performs all arithmetical operations by general symbols, or indeterminate quantities.

Some authors define algebra to be *the art of resolving mathematical problems*: but this is the idea of analysis, or the analytic art in general, rather than of algebra, which is only one particular species of it.

Indeed algebra properly consists of two parts: first, the method of calculating magnitudes or quantities, as represented by letters or other characters: and secondly the manner of applying these calculations in the solution of problems.

In algebra, as applied to the resolution of problems, the first business is to translate the problem out of the common into the algebraic language, by expressing all the conditions and quantities, both known and unknown, by their proper characters, arranged in an equation, or several equations if necessary, and treating the unknown quantity, whether it be number, or line, or any other thing, in the same way as if it were a known one: this forms the composition. Then the resolution, or analytic part, is the disentangling the unknown quantity from the several others with which it is connected, so as to retain it alone on one side of the equation, while all the other, or known, quantities, are collected on the other side, and so giving the value of the unknown one. And as this disentangling of the quantity sought, is performed by the converse of the operations by which it is connected with the others, taking them always backwards in the contrary order, it hence becomes a species of the analytic art, and is called the modern analysis, in contradistinction to the ancient analysis, which chiefly respected geometry, and its applications.

There have arisen great controversies and sharp disputes among authors, concerning the history of the progress and improvements of Algebra; arising partly from the partiality and prejudices which are natural to all nations, and partly from the want of a closer examination of the works of the older authors on this subject. From these causes it has happened, that the improvements made by the writers of one nation, have been ascribed to those of another; and the discoveries of an earlier author, to some one of much later date. Add to this also, that the peculiar methods of many authors have been described so little in detail, that our information derived from such histories, is but very imperfect, and amounting only to some general and vague ideas of the true state of the arts. To remedy this inconvenience therefore, and to reform this article, I have taken the pains carefully to read over in succession all the older authors on this subject, which I have been able to meet with, and to write down distinctly a particular account and description of their several compositions, as to their contents, notation, improvements, and peculiarities; from the comparison of all which, I have acquired an idea more precise and accurate than it was possible to obtain from other histories, and in a great many instances very different from them. The full detail of these descriptions would employ a volume of itself, and would be far too extensive for this place: I must therefore limit this article to a very brief abridgment of my notes, remarking only the most material circumstances in each author; from which a general idea of the chain of improvements may be perceived, from the first rude beginnings, down to the more perfect state; from which it will appear that the discoveries and improvements made by any one single author, are scarcely ever either very great or numerous; but that, on the contrary, the improvements are almost always very slow and gradual, from former writers, successively made, not by great leaps, and after long intervals of time, but by gradations which, viewed in succession, become almost imperceptible.

As

As to the origin of the analytic art, of which Algebra is a species, it is doubtless as old as any science in the world, being the natural method by which the mind investigates truths, causes, and theories, from their observed effects and properties. Accordingly, traces of it are observable in the works of the earliest philosophers and mathematicians, the subject of whose enquiries most of any require the aid of such an art. And this process constituted their Analytics. Of that part of analytics however which is properly called Algebra, the oldest treatise which has come down to us, is that of Diophantus of Alexandria, who flourished about the year 350 after Christ, and who wrote, in the Greek language, 13 books of Algebra or Arithmetic, as mentioned by himself at the end of his address to Dionysius, though only 6 of them have hitherto been printed; and an imperfect book on multangular numbers, namely in a Latin translation only, by Xilander, in the year 1575, and afterwards in 1621 and 1670 in Greek and Latin by Gaspar Bachet. These books however do not contain a treatise on the elementary parts of Algebra, but only collections of difficult questions relating to square and cube numbers, and other curious properties of numbers, with their solutions. And Diophantus only prefaces the books by an address to one Dionysius, for whose use it was probably written, in which he just mentions certain precognita, as it were to prepare him for the problems themselves. In these remarks he shews the names and generation of the powers, the square, cube, 4th, 5th, 6th, &c, which he calls dynamis, cubus, dynamodinamis, dynamocubus, cubocubus, according to the sum of the indices of the powers; and he marks these powers with the initials thus δ^{ν}, x^{ν}, $\delta\delta^{\nu}$, δx^{ν}, xx^{ν}, &c: the unknown quantity he calls simply ἀριθμος, numerus, the number; and in the solutions he commonly marks it by the final thus ς; also he denotes the monades, or indefinite unit, by μ°. Diophantus there remarks on the multiplication and division of simple species together, shewing what powers or species they produce; declares that minus (λεψις) multiplied by minus produces plus (υπαρξιν); but that minus multiplied by plus, produces minus; and that the mark used for minus is $\uparrow$, namely the ψ inverted and curtailed, but he uses no mark for plus, but a word or conjunction copulative. As to the operations, viz. of addition, subtraction, multiplication, and division of compound species, or those connected by plus and minus, Diophantus does not teach, but supposes his reader to know them. He then remarks on the preparation or simplifying of the equations that are derived from the questions, which we call reduction of equations, by collecting like quantities together, adding quantities that are minus, and subtracting such as are plus, called by the moderns Transposition, so as to bring the equation to simple terms, and then depressing it to a lower degree by equal division when the powers of the unknown quantity are in every term: which preparation, or reduction of the complex equation, being now made, or reduced to what we call a final equation, Diophantus goes no farther, but barely says what the root or res ignota is, without giving any rules for finding it, or for the resolution of equations; thereby intimating that such rules were to be found in some other work, done either by himself or others. Of the body of the

VOL. I.

work, Lib. 1 contains 43 questions, concerning one, two, three, or four unknown numbers, having certain relations to each other, viz. concerning their sums, differences, ratios, products, squares, sums and differences of squares, &c, &c; but none of them concerning either square or cubic numbers. Lib. 2 contains 36 questions. The first five questions are concerning two numbers, though only one condition is given in each question; but he supplies another by assuming the numbers in a given ratio, viz. as 2 to 1. The 6th and 7th contain each two conditions: then in the 8th question he first comes to treat of square numbers, which is this, to divide a given square number into two other squares; and the 9th is the same, but performed in a different way: the rest, to the end, are, almost all, about one, two, or three squares. Lib. 3 contains 24 questions concerning squares, chiefly including three or four numbers. Lib. 4 begins with cubes; the first of which is this, to divide a given number into two cubes whose sides shall have a given sum: here he has occasion to cube the two binomials $5+n$ and $5-n$; the manner of doing which shews that he knew the composition of the cube of a binomial; and many other places manifest the same thing. Only part of the questions in this book are concerning cubes; the rest are relating to squares. Two or three questions in this book have general solutions, and the theorems deduced are general, and for any numbers indefinitely; but all the other questions, in all the four books, find only particular numbers. Lib. 5 is also concerning square and cube numbers, but of a more difficult kind, beginning with some that relate to numbers in geometrical progression. Lib. 6 contains 26 propositions, concerning right-angled triangles; such as to make their sides, areas, perimeters, &c, &c, squares or cubes, or rational, &c. In some parts of this book it appears, that he was acquainted with the composition of the 4th power of the binomial root, as he sets down all the terms of it; and, from his great skill in such matters, it seems probable that he was acquainted with the composition of other higher powers, and with other parts of Algebra, besides what are here treated of. At the end is part of a book, in 10 propositions, concerning arithmetical progressions, and multangular or polygonal numbers. Diophantus once mentions a compound quadratic equation; but the resolution of his questions is by simple equations, and by means of only one unknown letter or character, which he chooses so ingeniously, that all the other unknown quantities in the question are easily expressed by it, and the final equation reduced to the simplest form which it seems the question can admit of. Sometimes he substitutes for a number sought immediately, and then expresses the other numbers or conditions by it: at other times he substitutes for the sum or difference, &c, and thence derives the rest, so as always to obtain the expressions in the simplest form. Thus, if the sum of two numbers be given, he substitutes for their difference; and if the difference be given, he substitutes for their sum: and in both cases he has the two numbers easily expressed by adding and subtracting the half sum and half difference; and so in other cases he uses other similar ingenious notations. In short, the chief excellence in this collection of questions, which seems to be only a set of exercises to some rules which had been given elsewhere, is the neat mode of substitution or notation; which being once made, the reduc-

tion

tion to the final equation is eafy and evident : and there he leaves the folution, only mentioning that the root or αριθμος is fo much. Upon the whole, this work is treated in a very able and mafterly manner, manifefting the utmoft addrefs and knowledge in the folutions, and forcing a perfuafion that the author was deeply fkilled in the fcience of Algebra, to fome of the moft abftrufe parts of which thefe queftions or exercifes relate. However, as he contrives his affumptions and notations fo as to reduce all his conditions to a fimple equation, or at leaft a fimple quadratic, it does not appear what his knowledge was in the refolution of compound or affected equations.

But although Diophantus was the firft author on Algebra that we now know of, it was not from him, but from the Moors or Arabians that we received the knowledge of Algebra in Europe, as well as that of moft other fciences. And it is matter of difpute who were the firft inventors of it ; fome afcribing the invention to the Greeks, while others fay that the Arabians had it from the Perfians, and thefe from the Indians, as well as the arithmetical method of computing by ten characters, or digits ; but the Arabians themfelves fay it was invented amongft them by one *Mahomet ben Mufa*, or fon of Mofes, who it feems flourifhed about the 8th or 9th century. It is more probable, however, that Mahomet was not the inventor, but only a perfon well fkilled in the art ; and it is farther probable, that the Arabians drew their firft knowledge of it from Diophantus or other Greek writers, as they did that of Geometry and other fciences, which they improved and tranflated into their own language ; and from them it was that we received thefe fciences, before the Greek authors were known to us, after the Moors fettled in Spain, and after the Europeans began to hold communications with them, and that our countrymen began to travel amongft them to learn the fciences. And according to the teftimony of Abulpharagius, the Arithmetic of Diophantus was tranflated in Arabic by Mahomet ben-yahya Baziani. But whoever were the inventors and firft cultivators of Algebra, it is certain that the Europeans firft received the knowledge, as well as the name, from the Arabians or Moors, in confequence of the clofe intercourfe which fubfifted between them for feveral centuries. And it appears that the art was pretty generally known, and much cultivated, at leaft in Italy, if not in other parts of Europe alfo, long before the invention of printing, as many writers upon the art are ftill extant in the libraries of manufcripts ; and the firft authors, prefently after the invention of printing, fpeak of many former writers on this fubject, from whom they learned the art.

It was chiefly among the Italians that this art was firft cultivated in Europe. And the firft author whofe works we have in print, was Lucas Paciolus, or Lucas de Burgo, a Cordelier, or Minorite Friar. He wrote feveral treatifes of Arithmetic, Algebra, and Geometry, which were printed in the years 1470, 1476, 1481, 1487, and in 1494 his principal work, intitled *Summa de Arithmetica, Geometria, Proportioni, et Proportionalita*, is a very mafterly and complete treatife on thofe fciences, as they then ftood. In this work he mentions various former writers, as Euclid, St. Auguftine, Sacrobofco or Halifax, Boetius, Prodocimo, Giordano,

Biagio da Parma, and Leonardus Pifanus; from whom he learned thofe fciences. The order of the work is, 1ft Arithmetic, 2d Algebra, and 3d Geometry. Of the Arithmetic the contents, and the order of them, are nearly as follow. Firft, of numbers figurate, odd and even, perfect, prime and compofite, and many others. Then of Common Arithmetic in 7 parts, namely numeration or notation, addition, fubtraction, multiplication, divifion, progreffion, and extraction of roots. Before him, he fays, duplation and mediation, or doubling and halving, were accounted two rules in Arithmetic ; but that he omits them, as being included in multiplication and divifion. He afcribes the prefent notation and method of Arithmetic to the Arabs ; and fays that according to fome the word *Abaco* is a corruption of *Modo Arabico*, but that according to others it was from a Greek word. All thofe primary operations he both performs and demonftrates in various ways, many of which are not in ufe at prefent, proving them not only by what is called cafting out the nines, but alfo by cafting out the fevens, and otherwife. In the extraction of roots he ufes the initial ℞ for a root ; and when the roots can be extracted, he calls them difcrete or rational ; otherwife furd, or indifcrete, or irrational. The fquare root is extracted much the fame way as at prefent, namely, dividing always the laft remainder by double the root found ; and fo he continues the furd roots continually nearer and nearer in vulgar fractions. Thus, for the root of 6, he firfts finds the neareft whole number 2, and the remainder 2 alfo ; then $\frac{2}{4}$ or $\frac{1}{2}$ is the firft correction, and $2\frac{1}{2}$ the fecond root : its fquare is $6\frac{1}{4}$, therefore $\frac{1}{4}$ divided by 5, or $\frac{1}{20}$ is the next correction, and $2\frac{1}{2}$ minus $\frac{1}{20}$, or $2\frac{9}{20}$ is the 3d root : its fquare is $6\frac{1}{400}$, therefore $\frac{1}{400}$ divided by $4\frac{9}{20}$, or $\frac{1}{1780}$ is the 3d correction, which gives $2\frac{881}{1780}$ for the 4th root, whofe fquare exceeds 6 by only $\frac{1}{3847800}$: and fo on continually : and this procefs he calls approximation. He obferves that fractions, which he fets down the fame way as we do at prefent, are extracted, by taking the root of the denominator, and of the denominated, for fo he calls the numerator : and when mixed numbers occur, he directs to reduce the whole to a fraction, and then extract the roots of its two terms as above : as if it be $12\frac{1}{4}$; this he reduces to $\frac{49}{4}$, and then the roots give $\frac{7}{2}$ or $3\frac{1}{2}$: in like manner he finds that $4\frac{1}{2}$ is the root of $20\frac{1}{4}$; $5\frac{1}{2}$ the root of $30\frac{1}{4}$; " and fo on (he adds) in infinitum ;" which fhews that he knew how to form the feries of fquares by addition. He then extracts the cube root, by a rule much the fame as that which is ufed at prefent ; from which it appears that he was well acquainted with the co-efficients of the binomial cubed, namely 1, 3, 3, 1 : and he directs how the operation may be continued " in infinitum" in fractions, like as in the fquare root. After this, he defcribes geometrical methods for extracting the fquare and cube roots inftrumentally : he then treats profeffedly of vulgar fractions, their reductions, addition, fubtraction, and other operations, much the fame as at prefent : then of the rule-of-three, gain-and-lofs, and other rules ufed by merchants.

Paciolus next enters on the algebraical part of this work, which he calls " *L'Arte Magiore ; ditta dal vulgo, la Regola de la Cofa, over Alghebra e Almucabala :*" which laft name he explains by *reftauratio & oppofitio*, and
affigns

affigns as a reafon for the firft name, becaufe it treats of things above the common affairs in bufinefs, which make the *Arte Minore*. Here he afcribes the invention of Algebra to the Arabians, and denominates the feries of powers, with their marks or abbreviations, as *n°.* or *numero*, the abfolute or known number; *co.* or *cofa*, the thing or 1ft power of the unknown quantity; *ce.* or *cenfo*, the product or fquare; *cu.* or *cubo*, the cube, or 3d power; *ce. ce.* or *cenfo de cenfo*, the fquare-fquared, or 4th power; *p°. r°.* or *primo relato*, or 5th power; *ce. cu.* or *cenfo de cubo*, the fquare of the cube, or 6th power; and fo on, compounding the names or indices according to the multiplication of the numbers 2, 3, &c, and not according to their fum or addition, as ufed by Diophantus. He defcribes alfo the other characters made ufe of in this part, which are for the moft part no more than the initials or other abbreviations of the words themfelves; as R̥ for *radici*, the root; R̥. R̥. *radici de radici*, the root of the root; R̥ u. *radici univerfale*, or *radici legata*, or *radici unita*; R̥ cu. *radici cuba*; and q̅3̅ *quantita*, quantity; *p* for *piu* or plus, and *m* for *meno* or minus; and he remarks that the neceffity and ufe of thefe two laft characters are for connecting, by addition or fubtraction, different powers together; as 3 *co.* p. 4 *ce.* m. 5 *cu.* p. 2 *ce. ce.* m. 6 *n*¹. that is, 3 *cofa* piu 4 *cenfa* meno 5 *cubo* piu 2 *cenfa-cenfa* meno 6 *numeri*, or, as we now write the fame thing, $3x + 4x^2 - 5x^3 + 2x^4 - 6$. He firft treats very fully of proportions and proportionalities, both arithmetical and geometrical, accompanied with a large collection of queftions concerning numbers in continued proportion, refolved by a kind of Algebra. He then treats of *el Cataym*, which he fays, according to fome, is an Arabic or Phenician word, and fignifies the Double Rule of Falfe Pofition: but he here treats of both fingle and double pofition, as we do at prefent, dividing the *el Cataym* into fingle and double. He gives alfo a geometrical demonftration of both the cafes of the errors in the double rule, namely when the errors are both plus or both minus, and when the one error is plus and the other minus; and adds a large collection of queftions, as ufual. He then goes through the common operations of Algebra, with all the variety of figns, as to plus and minus; proving that, in multiplication and divifion, like figns give plus, and unlike figns give minus. He next treats of different roots *in infinitum*, and the extraction of roots; giving alfo a copious treatife on radicals or furds, as to their addition, fubtraction, multiplication and divifion, and that both in fquare roots and cube roots, and in the two together, much the fame as at prefent. He makes here a digreffion concerning the 15 lines in the 10th book of Euclid, treating them as furd numbers, and teaching the extraction of the roots of the fame, or of compound furds or binomials, fuch as of 23 *p* R̥ 448, or of R̥ 18 *p* R̥ 10; and gives this rule, among feveral others, namely: Divide the firft term of the binomial into two fuch parts that their product may be ¼ of the number in the fecond term; then the roots of thofe two parts, connected by their proper fign *p* or *m*, is the root of the binomial; as in this 23 *p* R̥ 448, the two parts of 23 are 7 and 16, whofe product, 112, is ¼ of 448, therefore their roots give 4 *p* R̥ 7 for the root R̥ u. 23 *p* R̥ 448. He next treats of equations both fimple and quadratic, or fimple and compound, as he

calls it; and this latter he performs by completing the fquare, and then extracting the root, juft as we do at prefent. He alfo refolves equations of the fimple 4th power, and of the 4th combined with the 2d power, which he treats the fame way as quadratics; expreffing his rules in a kind of bad verfe, and giving geometrical demonftrations of all the cafes. He ufes both the roots or values of the unknown quantity, in that cafe of the quadratics which has two pofitive roots; but he takes no notice of the negative roots in the other two cafes. But as to any other compound equations, fuch as the cube and any other power, or the 4th and 1ft, or 4th and 3d, &c, he gives them up as impoffible, or at leaft fays that no general rule has yet been found for them, any more, he adds, than for the quadrature of the circle. —The remainder of this part is employed on rules in trade and merchandife, fuch as Fellowfhip, Barter, Exchange, Intereft, Compofition or Alligation, with various other cafes in trade. And in the third part of the work, he treats of Geometry, both theoretical and practical.

From this account of Lucas de Burgo's book, we may perceive what was the ftate of Algebra about the year 1500, in Europe; and probably it was much the fame in Africa and Afia, from whence the Europeans had it. It appears that their knowledge extended only to quadratic equations, of which they ufed only the pofitive roots; that they ufed only one unknown quantity; that they had no marks or figns for either quantities or operations, excepting only fome few abbreviations of the words or names themfelves; and that the art was only employed in refolving certain numeral problems. So that either the Africans had not carried Algebra beyond quadratic equations, or elfe the Europeans had not learned the whole of the art, as it was then known to the former. And indeed it is not improbable but this might be the cafe: for whether the art was brought to us by an European, who, travelling in Africa, there learned it; or whether it was brought to us by an African; in either cafe we might receive the art only in an imperfect ftate, and perhaps far fhort of the degree of perfection to which it had been carried by their beft authors. And this fufpicion is rendered rather probable by the circumftance of an Arabic manufcript, faid to be on cubic equations, depofited in the Library of the univerfity of Leyden by the celebrated Warner, bearing a title which in Latin fignifies *Omar Ben Ibrahim al' Ghajamei Algebra cubicarum equationum, five de problematum folidorum refolutione*; and of which book I am in fome hopes of procuring either a copy or a tranflation. by means of my worthy friend Dr. Damen, the learned Profeffor of Mathematics in that univerfity, and by that means to throw fome light on this doubtful fubject.

Since this was written, death has prematurely put an end to the ufeful labours of this ingenious and worthy fucceffor of Gravefande.

After the publication of the books of Lucas de Burgo, the fcience of Algebra became more generally known, and improved, efpecially by many perfons in Italy; and about this time, or foon after, namely about the year 1505, the firft rule was there found out by Scipio Ferreus, for refolving one cafe of a compound cubic equation. But this fcience, as well as other branches of-

Mathe-

Mathematics, was moſt of all cultivated and improved there by Hieronymus Cardan of Bononia, a very learned man, whoſe arithmetical writings were the next that appeared in print, namely in the year 1539, in 9 books, in the Latin language, at Milan, where he practiſed phyſic, and read public lectures on Mathematics; and in the year 1545 came out a 10th book, containing the whole doctrine of cubic equations, which had been in part revealed to him about the time of the publication of his firſt 9 books. And as it is only this 10th book which contains the new diſcoveries in Algebra, I ſhall here confine myſelf to it alone, as it will alſo afford ſufficient occaſion to ſpeak of his manner of treating Algebra in general. This book is divided into 40 chapters, in which the whole ſcience of cubic equations is moſt amply and ably treated. Chap. 1 treats of the nature, number and properties of the roots of equations, and particularly of ſingle equations that have double roots. He begins with a few remarks on the invention and name of the art: calls it *Ars Magna*, or *Coſa*, or *Rules of Algebra*, after Lucas de Burgo and others: ſays it was invented by Mahomet, the ſon of one Moſes an Arabian, as is teſtified by Leonardus Piſanus; and that he left four rules or caſes, which perhaps only included quadratic equations: that afterwards three derivatives were added by an unknown author, though ſome think by Lucas Paciolus; and after that again three other derivatives, for the cube and 6th power, by another unknown author; all which were reſolved like quadratics: that then Scipio Ferreus, Profeſſor of Mathematics at Bononia, about 1505, found out the rule for the caſe *cubum & rerum numero æqualium*, or, as we now write it, $x^3 + bx = c$, which he ſpeaks of as a thing admirable: that the ſame thing was next afterwards found out, in 1535, by Tartalea, who revealed it to him, Cardan, after the moſt earneſt intreaties: that, finally, by himſelf and his quondam pupil Lewis Ferrari, the caſes are greatly augmented and extended, namely, by all that is not here expreſsly aſcribed to others; and that all the demonſtrations of the rules are his own, except only three adopted from Mahomet for the quadratics, and two of Ferrari for cubics.

He then delivers ſome remarks, ſhewing that all ſquare numbers have two roots, the one poſitive, and the other negative, or, as he calls them, *mera & ficta*, true and fictitious or falſe; ſo the *æstimatio rei*, or root, of 9, is either 3 or — 3; of 16 it is 4 or — 4; the 4th root of 81 is 3 or — 3; and ſo on for all even *denominations* or powers. And the ſame is remarked on compound caſes of even powers that are added together; as if $x^4 + 3x^2 = 28$, then the æſtimatio x is = 2 or —2; but that the form $x^4 + 12 = 7x^2$ has four anſwers or roots, in real numbers, two plus and two minus, viz. 2 or — 2, and $\sqrt{3}$ or —$\sqrt{3}$; while the caſe $x^4 + 12 = 6x^2$ has no real roots; and the caſe $x^4 = 2x^2 + 8$ has two, namely 2 and — 2: and in like manner for other even powers. So that he includes both the poſitive and negative roots; but rejects what we now call imaginary ones. I here expreſs the caſes in our modern notation, for brevity ſake, as he commonly expreſſes the terms by words at full length, calling the abſolute or known term the *numero*, the 1ſt power the *res*, the 2d the *quadratum*, the 3d the *cubum*, and ſo on, uſing no mark for the unknown

8

quantity, and only the initials *p* and *m* for plus and minus, and ℞ for radix or root. The *res* he ſometimes calls *poſitio*, and *quantitas ignota*; and in ſtating or ſetting down his equations, he, as well as Lucas de Burgo before him, ſets down the terms on that ſide where they will be plus, and not minus.

On the other hand, he remarks that the odd denominations, or powers, have only one æſtimatio, or root, and that true or poſitive, but none fictitious or negative, and for this reaſon, that no negative number raiſed to an odd power, will give a poſitive number; ſo of $2x = 16$, the root is 8 only; and if $2x^3 = 16$, the root is 2 only: and if there be ever ſo many odd denominations, added together, equal to a number, there will be only one æſtimatio or root; as if $x^3 + 6x = 20$, the only root is 2. But that when the ſigns of ſome of the terms are different as to plus and minus, they may have more roots; and he ſhews certain relations of the co-efficients, when they have two or more roots: ſo the equation $x^3 + 16 = 12x$ has two æſtimatios, the one true or 2, and the other fictitious or — 4, which he obſerves is the ſame as the true æſtimatio of the caſe $x^3 = 12x + 16$, having only the ſign of the abſolute number changed from the former, the 3d root 2 being the ſame as the firſt, which therefore he does not count. He next ſhews what are the relations of the co-efficients when a cubic equation has three roots, of which two are true, and the 3d fictitious, which is always equal to the ſum of the other two, and alſo equal to the true root of the ſame equation with the ſign of the abſolute number changed: thus, in the equation $x^3 + 9 = 12x$, the two true roots are 3 and $\sqrt{5\frac{1}{4}} - 1\frac{1}{2}$, and the fictitious one is — $\sqrt{5\frac{1}{4}} - 1\frac{1}{2}$, which laſt is the ſame as the true root of $x^3 = 12x + 9$, viz. $\sqrt{5\frac{1}{4}} + 1\frac{1}{2}$; and he here infers generally that the fictitious æſtimatio of the caſe $ x^3 + c = bx$, always anſwers to the true root of $ x^3 = bx + c$. Cardan alſo ſhews what the relation of the co-efficients is, when the caſe has no true roots, but only one fictitious root, which is the ſame as the true root of the reciprocal caſe, formed by changing the ſign of the abſolute number. Thus, the caſe $x^3 + 21 = 2x$ has no true root, and only one falſe root, viz. — 3, which is the ſame as the true root of $x^3 = 2x + 21$: and he ſhews in general, that changing the ſign of the abſolute number in ſuch caſes as want the 2d term, or changing the ſigns of the even terms when it is not wanting, changes the ſigns of all the three roots, which he alſo illuſtrates by many examples; thus, the roots of $x^3 + 11x^2 = 72$, are + $\sqrt{40}$ — 4, and — 3, and — $\sqrt{40}$ — 4; and the roots of $x^3 + 72 = 11x^2$, are — $\sqrt{40}$ + 4, and + 3, and + $\sqrt{40}$ + 4.

And he further obſerves, that the ſum of the three roots, or the difference between the true and fictitious roots, is equal to 11, the co-efficient of the 2d term. He alſo ſhews how certain cubic caſes have one, or two, or three roots, according to circumſtances: that the caſe $x^4 + d = bx^2$ has ſometimes four roots, and ſometimes none at all, that is, no real ones: that the caſe $x^3 + bx = ax^2 + c$ may have three true æquatios, or poſitive roots, but no fictitious or negative ones; and for this reaſon, that the odd powers of minus being minus, and the even powers plus, the two terms $x^3 + bx$ would be negative, and equal to a poſitive ſum $ax^2 + c$, which

which is abfurd: and farther, that the cafe $x^3 + ax^2 + bx = c$ has three roots, one plus and two minus, which are the fame, with the figns changed, as the roots of the cafe $x^3 + bx + c = ax^2$. He alfo fhews the relation of the co-efficients when the equation has only one real root, in a variety of cafes: but that the cafe $x^3 + c = ax^2 + bx$ has always one negative root, and either two pofitive roots, or none at all; the number of roots failing by pairs, or the impoffible roots, as we now call them, being always in pairs. Of all thefe circumftances Cardan gives a great many particular examples in numeral co-efficients, and fubjoins geometrical demonftrations of the properties here enumerated; fuch as, that the two correfponding or reciprocal cafes have the fame root or roots, but with different figns or affections; and how many true or pofitive roots each cafe has.

Upon the whole, it appears from this fhort chapter, that Cardan had difcovered moft of the principal properties of the roots of equations, and could point out the number and nature of the roots, partly from the figns of the terms, and partly from the magnitude and relations of the co-efficients. He fhews in effect, that when the cafe has all its roots, or when none are impoffible, the number of its pofitive roots is the fame as the number of changes in the figns of the terms, when they are all brought to one fide: that the co-efficient of the 2d term is equal to the fum of all the roots pofitive and negative collected together, and confequently that when the 2d term is wanting, the pofitive roots are equal to the negative ones: and that the figns of all the roots are changed, by changing only the figns of the even terms: with many other remarks concerning the nature of equations.

In chap. 2, Cardan enumerates all the cafes of compound equations, of the 2d and 3d order, namely, 3 quadratics, and 19 cubics; with 44 derivatives of thefe two, that is, of the fame kind, with higher denominations.

In chap. 3 are treated the roots of fimple cafes, or fimple equations, or at leaft that will reduce to fuch, having only two terms, the one equal to the other. He directs to deprefs the denominations equally, as much as they will, according to the height of the leaft; then divide by the number or co-efficient of the greateft; and laftly extract the root on both fides. So if $20x^3 = 180x^{\frac{1}{5}}$, then $20 = 180x^2$, and $\frac{1}{9} = x^2$, and $x = \frac{1}{3}$.

Chap. 4 treats of both general and particular roots, and contains various definitions and obfervations concerning them. It is here fhewn that the feveral cafes of quadratics and cubics have their roots of the following forms or kinds, namely that the cafe

$x^2 = ax + b$ has its root of this kind $\sqrt{19} + 3$,
$x^2 + ax = b$ has its root of this kind $\sqrt{19} - 3$,
$x^2 + b = ax$ has two roots like $3 + \sqrt{2}$ and $3 - \sqrt{2}$,
$x^3 = bx + c$ has its root of this kind $\sqrt[3]{4} + \sqrt[3]{2}$,
$x^3 + bx = c$ has its root of this kind $\sqrt[3]{4} - \sqrt[3]{2}$,
$x^3 = ax^2 + c$ has this kind of root $\sqrt[3]{16} + 2 + \sqrt[3]{4}$
$x^3 + ax^2 = c$ has this kind of root $\sqrt[3]{16} - 2 + \sqrt[3]{4}$,
where the three parts $\sqrt[3]{16}$, 2, $\sqrt[3]{4}$, are in continual proportion.

Chap. 5 treats of the æftimatio of the loweft degree of compound cafes, that is, affected quadratic equations; giving the rule for each of the three cafes, which con-

fifts in completing the fquare, &c, as at prefent, and which it feems was the method given by the Arabians; and proving them by geometrical demonftrations from Eucl. I. 43, and II. 4 and 5, in which he makes fome improvement of the demonftrations of Mahomet. And hence it appears that the work of this Arabian author was in being, and well known in Cardan's time.

Chap. 6, on the methods of finding new rules, contains fome curious fpeculations concerning the fquares and cubes of binomial and refidual quantities, and the proportions of the terms of which they confift, fhewn from geometrical demonftrations, with many curious remarks and properties, forming a foundation of principles for inveftigating the rules for cubic equations.

Chap. 7 is on the tranfmutation of equations, fhewing how to change them from one form to another, by taking away certain terms out of them; as $x^3 + ax^2 = c$, to $y^3 = by + d$, &c. The rules are demonftrated geometrically; and a table is added, of the forms into which any given cafes will reduce; which transformations are extended to equations of the 4th and 5th order. And hence it appears that Cardan knew how to take away any term out of an equation.

Chap. 8 fhews generally how to find the root of any fuch equation as this $ax^m = x^n + b$, where m and n are any exponents whatever, but n the greater; and the rule is, to feparate or divide the co-efficient a into two fuch parts z and $a - z$, as that the abfolute number b fhall be equal to $\overline{a - z} \cdot z^{\frac{m}{n-m}}$, the product of the one part $a - z$, and the $\frac{m}{n-m}$ power of the other part: then the root x is $= z^{\frac{1}{n-m}}$. The rule is general for quadratics, cubics, and all the higher powers; and could not have been formed without the knowledge of the compofition of the terms from the roots of the equation.

Chap. 9 and 10 contain the refolution of various queftions producing equations not higher than quadratics.

Chap. 11 is of the cafe or form $x^3 + 3bx = 2c$. Cardan now comes to the actual refolution of the firft cafe of cubic equations. He begins with relating a fhort hiftory of the invention of it, obferving that it was firft found out, about 30 years before, by Scipio Ferreus of Bononia, and by him taught to Antonio Maria Florido of Venice, who having a conteft afterwards with Nicolas Tartalea of Brefcia, it gave occafion to Tartalea to find it out himfelf, who after great entreaties taught it to Cardan, but fuppreffed the demonftration. By help of the rule alone, however, Cardan of himfelf difcovered the fource or geometrical inveftigation, which he gives here at large, from Eucl. II. 4. In this procefs he makes ufe of the Greek letters α, β, γ, , &c, to denote certain indefinite numbers or quantities, to render the inveftigation general; which may be confidered as the firft inftance of fuch literal notation in Algebra. He then gives the rule in words at length, which comes to this,

$$x = \sqrt[3]{\sqrt{c^2 + b^3} + c} - \sqrt[3]{\sqrt{c^2 + b^3} - c};$$

illuftrating it in a variety of examples; in the refolution
of

of which, he extracts the cubic roots of such of the binomials as will admit of it, by some rule which he had for that purpose; such as $x^3 + 6x = 20$, which

$$x = \sqrt[3]{\sqrt{108 + 10}} - \sqrt[3]{\sqrt{108 - 10}} - \sqrt{3 - 1} = 2 = \sqrt{3 + 1}.$$

Chap. 12, of the case $x^3 = 3bx + 2c$. This he treats exactly as the last, and finds the rule

$$x = \sqrt[3]{c + \sqrt{c^2 - b^3}} + \sqrt[3]{c - \sqrt{c^2 - b^3}};$$

which he illustrates by many examples, as usual. But when b^3 exceeds c^2, which has since been called the irreducible case, he refers to another following book, called *Aliza*, for other rules of solution, to overcome this difficulty, about which he took infinite pains.

Chap. 13, of the case $x^3 + 2c = 3bx$. This case, by a geometrical process, he reduces to the case in the last chapter: thus, find the æstimatio y of the case $y^3 = 3by + 2c$, having the same co-efficients as the given case $x^3 + 2c = 3bx$; then is $x = \frac{1}{2}y \pm \frac{1}{2}\sqrt{12b - 3y^2}$, giving two roots. He shews also how to find the second root, when the first is known, independent of the foregoing case. From this relation of these two cases he deduces several corollaries, one of which is, that the æstimatio or root of the case $y^3 = 3by + 2c$, is equal to the sum of the roots of the case $x^3 + 2c = 3bx$. As in the example $y^3 = 16y + 21$, whose æstimatio is $\sqrt{9\frac{1}{4}} + 1\frac{1}{2}$, which is equal to the sum of 3 and $\sqrt{9\frac{1}{4}} - 1\frac{1}{2}$, the two roots of the case $x^3 + 21 = 16x$.

In chapters 14, 15, and 16, he treats of the three cases which contain the 2d and 3d powers, but wanting the first power, according to all the varieties of the signs; which he performs by exterminating the 2d term, or that which contains the 2d power of the unknown quantity x, by substituting $y \pm \frac{1}{3}$ the co-efficient of that term for x, and so reducing these cases to one of the former. In these chapters Cardan sometimes also gives other rules; thus, for the case $x^3 + 3ax^2 = 2c$, find first the æstimatio y of the case $y^3 = 3ay + \sqrt{2c}$, then is $x = y^2 - 3a$: also for the case $x^3 + 2c = 3ax^2$, first find the two roots of $y^3 + 2c = 3ay\sqrt[3]{2c}$, then is

$$x = \frac{\sqrt[3]{4c^2}}{y}$$ the two values of x according to the two

values of y. He here also gives another rule, by which a second æstimatio or root is found, when the first is known, namely, if e be the first æstimatio or value of x in the case $x^3 + 2c = 3ax^2$, then is the other value of

$$x = \frac{1}{2}\sqrt{3a - e \cdot 3a + 3e} + \frac{1}{2} \cdot 3a - e.$$

In chapters 17, 18, 19, 20, 21, 22, 23, Cardan treats of the cases in which all the four terms of the equation are present; and this he always effects by taking away the 2d term out of the equation, and so reducing it to one of the foregoing cases which want that term, giving always geometrical investigations, and adding a great many examples of every case of the equations.

Chap. 24, of the 44 derivative cases; which are only higher powers of the forms of quadratics and cubics.

Chap. 25, of imperfect and special cases; containing many particular examples when the co-efficients have certain relations amongst them, with easy rules for

finding the roots; also 8 other rules for the irreducible case $x^3 = bx + c$.

Chap. 26, in like manner, contains easy rules for biquadratics, when the co-efficients have certain special relations.

Then the following chapters, from chap. 27 to chap. 38, contain a great number of questions and applications of various kinds, the titles of which are these: *De transitu capituli specialis in capitulum speciale; De operationibus radicum pronicarum seu mixtarum & Allelarum; De regula modi; De regula Aurea; De regula Magna*, or the method of finding out solutions to certain questions; *De regula æqualis positionis*, being a method of substituting for the half sum and half difference of two quantities, instead of the quantities themselves; *De regula inæqualiter ponendi, seu proportionis; De regula medii; De regula aggregati; De regula liberæ positionis; De regula falsum ponendi*, in which some quantities come out negative; *Quomodo excidant partes & denominationes multiplicando*. Among the foregoing collection of questions, which are chiefly about numbers, there are some geometrical ones, being the application of Algebra to Geometry, such as, In a right-angled triangle, given the sum of each leg and the adjacent segment of the hypotenuse, made by a perpendicular from the right angle, to determine the area &c; with other such geometrical questions, resolved algebraically.

Chap. 39, *De regula qua pluribus positionibus invenimus ignotam quantitatem*; which is employed on biquadratic equations. After some examples of his own, Cardan gives a rule of Lewis Ferrari's, for resolving all biquadratics, namely by means of a cubic equation, which Ferrari investigated at his request, and which Cardan here demonstrates, and applies in all its cases. The method is very general, and consists in forming three squares, thus: first, complete one side of the equation up to a square, by adding or subtracting some multiples or parts of some of its own terms on both sides, which it is always easy to do: 2d, supposing now the three terms of this square to be but one quantity, viz, the first term of another square to which this same side is to be completed, by annexing the square of a new and assumed indeterminate quantity, with double the product of the roots of both; which evidently forms the square of a binomial, consisting of the assumed indeterminate quantity and the root of the first square: 3d, the other side of the equation is then made to become the square of a binomial also, by supposing the product of its 1st and 3d terms to be equal to the square of half its 2d term; for it consists of only three terms, or three different denominations of the original unknown quantity: then this equality will determine the value of the assumed indeterminate quantity, by means of a cubic equation, and from it, that of the original ignota, by the equal roots of the 2d and 3d squares. Here we have a notable example of the use of assuming a new indeterminate quantity to introduce into an equation, long before Des Cartes was born, who made use of a like assumption for a similar purpose. And this method is very general, and is here applied to all forms of biquadratics, either having all their terms, or wanting some of them. To illustrate this rule I shall here set down the process of one of his examples, which is this, $x^4 + 4x + 8 = 10x^2$. Now first subtract

tract $2x^2 + 4x + 7$ from both sides, then the first becomes a square, viz, $x^4 - 2x^2 + 1$ or $\overline{x^2 - 1}^2 = 8x^2 - 4x - 7$. Next assume the indeterminate y, and subtract $2y(x^2 - 1) - y^2$ from both sides, making the first side again a square, viz, $\overline{x - 1}^2 - 2y \cdot x - 1 + y^2$ or $\overline{x^2 - 1 - y}^2 = 8 - 2y \cdot x^2 - 4x + y^2 + 2y - 7$. Of this latter side, make the product of the 1st and 3d terms equal to the square of half the 2d term, that is, $\overline{8 - 2y} \cdot y^2 + 2y - 7 = 2^2$, which reduces to $y^3 + 30 = 2y^2 + 15y$; the positive roots of which are $y = 2$ or $\sqrt{15}$; and hence, using 2 for y, the equation of equal squares becomes $\overline{x^2 - 1 - y}^2$ or $\overline{x^2 - 3}^2 = 4x^2 - 4x + 1$, the roots of which give $x^2 - 3 = 2x$ and 1; and hence $x^2 = 2x + 2$, or $x^2 + 2x = 4$; the two positive roots of which are $\sqrt{3} + 1$ and $\sqrt{5} - 1$, which are two of the values of x in the given equation $x^4 + 4x + 8 = 10x^2$. The other roots he leaves to be tried by the reader.

The 40th, or last, chap. is entitled, Of modes of general supposition relating to this art; with some rules of an unusual kind; and æstimatios or roots of a nature different from the foregoing ones. Some of these are as follow: If $x^3 = ax^2 + c$, and $x - a = y$, and $x : y :: c : d$; then is $y^3 + ay^2 = d$.

Secondly, if $x^3 + ax^2 = c$,
and $y^3 = ay^2 + c$,
then is $x + a : y - a :: y^2 : x^2$.

Thirdly, when $x^3 + c = ax^2$, the square will be taken away, by putting $x = y + \frac{1}{3}a$; and then the equation becomes $y^3 + c - 2(\frac{1}{3}a)^3 = \frac{1}{3}a^2 y$.

Cardan adds some other remarks concerning the solutions of certain cases and questions, all evincing the accuracy of his skill, and the extent of his practice; and then he concludes the book with a remark concerning a certain transformation of equations, which quite astonishes us to find that the same person who, through the whole work, has shewn such a profound and critical skill in the nature of equations, and the solution of problems, should yet be ignorant of one of the most obvious transmutations attending them, namely increasing or diminishing the roots in any proportion. Cardan having observed that the form $x^3 = bx + c$ may be changed into another similar one, viz, $y^3 = \frac{b}{c}y + \sqrt{\frac{1}{c}}$, of which the co-efficient of the term y is the quotient arising from the co-efficient of x divided by the absolute number of the first equation: and that the absolute number of the 2d equation is the root of the quotient of 1 divided by the said absolute number of the first; he then adds, that finding the æstimatio or root of the one equation from that of the other is very difficult, *valdè difficilis*.

It is matter of wonder that Cardan, among so many transmutations, should never think of substituting instead of x in such equations, another positio or root, greater or less than the former in any indefinite proportion, that is, multiplied or divided by a given number; for this would have led him immediately to the same transformation as he makes above, and that by a way which would have shewn the constant proportion be-

tween the two roots. Thus, instead of x in the given form $- - - - - x^3 = bx + c$, substitute dy, and it becomes $- d^3y^3 = bdy + c$; and this divided by d^3 becomes $- - y^3 = \frac{b}{d^2}y + \frac{c}{d^3}$; and here if d be taken $= \sqrt{c}$, it becomes $y^3 = \frac{b}{c}y + \sqrt{\frac{1}{c}}$; which is the transformation in question, and in which it is evident that x is $= y\sqrt{c}$, and $y = \frac{x}{\sqrt{c}}$. Instead of this, Cardan gives the following strange way of finding the one root x from the other y, when this latter is by any means known; viz, Multiply the first given equation by $y^2x + 1$, then add $\frac{x^2}{4y^2}$ to both sides, and lastly extract the roots of both, which can always be done, as they will always be both of them squares; and the roots will give the value of x by a quadratic equation.

Thus, $x^3 = bx + c$ multiplied by $y^2x + 1$ gives $y^2x^4 + x^3 = by^2x^2 + \overline{cy^2 + b} \cdot x + c$; add $\frac{x^2}{4y^2}$, then $y^2x^4 + x^3 + \frac{x^2}{4y^2} = by^2 + \frac{1}{4y^2} \cdot x^2 + \overline{b + cy^2} \cdot x + c$; and the roots are $yx^2 + \frac{x}{2y} = \sqrt{(by^2 + \frac{1}{4y^2})x^2 + (b + cy^2)x + c}$; and this 2d side of the equation he says will always have a root also. It is indeed true that it will have an exact root; but the reason of it is not obvious, which is, because y is the root of the equation $y^3 = \frac{b}{c}y + \sqrt{\frac{1}{c}}$.

Cardan has not shewn the reason why this happens; but I apprehend he made it out in this manner, viz, similar to the way in which he forms the last square in the case of biquadratic equations, namely, by making the product of the 1st and 3d terms equal to the square of half the 2d term: thus, in the present case, it is $4c(by^2 + \frac{1}{4y^2}) = (b + cy^2)^2$, which reduces to $y^3 = \frac{b}{c}y + \sqrt{\frac{1}{c}}$ the equation in question. Therefore taking the root of the equation $y^3 = \frac{b}{c}y + \sqrt{\frac{1}{c}}$, and substituting its value in the quantity $(by^2 + \frac{1}{4y^2})x^2 + (b + cy^2)x + c$, this will become a complete square.

Of Cardan's *Libellus de Aliza Regula*.

Subjoined to the above Treatise on cubic equations, is this *Libellus de Aliza regula*, or the algebraic logistics, in which the author treats of some of the abstruser parts of Arithmetic and Algebra, especially cubic equations, with many more attempts on the irreducible case $x^3 = bx + c$. This book is divided into 60 chapters; but

but I fhall only fet down the titles of fome few of them, whofe contents require more particular notice.

Chap. 4. *De modo redigendi quantitates omnes, quæ dicuntur latera prima ex decimo Euclidis in compendium.* He treats here of all Euclid's irrational lines, as furd numbers, and performs various operations with them.

Chap. 5. *De confideratione binomiorum & recifrum, &c; ubi de æftimatione capitulorum.* Contains various operations of multiplying compound numbers and furds.

Chap. 6. *De operationibus* p: *&* m: (i. e. + and —) *fecundum communem ufum.* Here it is fhewn that, in multiplication and divifion, *plus* always gives the fame figns, and *minus* gives the contrary figns. So alfo in addition, every quantity retains its own fign; but in fubtraction they change the figns. That the √+, or the fquare root of plus, is +; but the √—, or the fquare root of minus, is nothing as to common ufe: (but of this below.) That √—is—; as √—8 is —2. That a refidual, compofed of + and — may have a root alfo compofed of + and —: So √5—√24 is = √3—√2. The rules for the figns in multiplication and divifion are illuftrated by this example; to divide 8 by 2 + √6 or √6 + 2. Take the two corresponding refiduals 2 — √6 and √6 — 2, and by thefe multiply both the divifor and dividend; then the products are + and — refpectively, and the quotients ftill both alike. Thus,

Divid.	Divif.		Divid.	Divif.
8	√6 + 2		8	2 + √6
√6 — 2	√6 — 2		2 — √6	2 — √6
√384 — 16	divide + 4		16 — √384	div. — 2
Quot. √96 — 8.			Quot. √96 — 8.	

And this method of performing divifion of compound furds, was fully taught before him, by Lucas de Burgo, namely, reducing the compound divifor to a fimple quantity, by multiplying by the correfponding quantity, having the fign changed.

In chap. 11 and 18, and elfewhere, Cardan makes a general notation of *a*, *b*, *c*, *d*, *e*, *f*, for any indefinite quantities, and treats of them in a general way.

Cap. 2. *De contemplatione* p: *&* m: (or + and —), *& quod* m: in m: *facit* p: *& de caufis horum juxta veritatem.* Cardan here demonftrates geometrically that, in multiplication and divifion, like figns give plus, and unlike figns give minus. And he illuftrates this numerically, by fquaring the quantity 8, or 6 + 2, or 10 —2, which muft all produce the fame thing, namely 64.

Among many of the chapters which treat of the irreducible cafe $x^3 = bx + c$, there is a peculiar kind of way given in chap. 31, which is entitled *De æftimatione generali* $x^3 = bx + c$ *folida vocata, & operationibus ejus;* in which he fhews how to approximate to the root of that cafe, in a manner fimilar to approximating the fquare root and cube root of a number. The rule he ufes for this purpofe, is the 3d in chap. 25 of the laft book, and it is this: Divide *b* into two parts, fuch that the fum of the products of each, multiplied by the fquare of the other, may be equal to ½*c*; then the fum of the roots of thefe parts is the æftimatio or value of *x* required. So, of this equation $x^3 = 10x + 24$; the two parts are 9 and 1, and their roots 3 and 1, and their fum 4 = *x*, as in the margin.

$$9 \times 1$$
$$3 \quad 1$$
$$3 + 9 \text{ or } 12 = \tfrac{1}{2}c.$$
$$x = 3 + 1 = 4$$

Again, take $x^3 = 6x + 1$. Here he invents a new notation to exprefs the root or *radix*, which he calls *folida*, viz. $x = \sqrt{}$ *folida* 6 in ½, that is, the roots of the two parts of 6, fo that each part multiplied by the root of the other, the two products may be ½ or ½*c*. Then to free this from fractions, and make the operation eafier, multiply that root by fome number as fuppofe 4, that is the fquare part 6 by the fquare of 4, and the folid part ½ by the cube of 4; then $x = \tfrac{1}{4} \sqrt{}$ *folida* 96 in 32. Now, by a few trials, it is found that the parts are nearly $95\tfrac{3}{5}$ and $\tfrac{1}{9}$, which give too much,

or $95\tfrac{9}{10}$ and $\tfrac{1}{15}$, which give too little,

and thereof $95\tfrac{1}{19}$ and $\tfrac{2}{19}$ are ftill nearer. Divide both by 4^2 or 16, then $5\tfrac{151}{152}$ and $\tfrac{1}{152}$ are the quot. And the fum of their roots, or $x = \sqrt{5\tfrac{151}{152}} + \sqrt{\tfrac{1}{152}}$ is nearly the value of the root *x*.

Cap. 42. *De duplici æquatione comparanda in capitulo cubi & numeri æqualium rebus.* Treats of the two pofitive roots of that cafe, neglecting the negative one; and fhewing, not only that that cafe has two fuch roots, but that the fame number may be the common root of innumerable equations.

Cap. 57. *De tractatione æftimationis generalis capituli* $x^3 = bx + c$. Cardan here again refumes the confideration of the irreducible cafe, making ingenious obfervations upon it, but ftill without obtaining the root by a general rule. In this place alfo, as well as elfewhere, he fhews how to form an equation in this cafe, that fhall have a given binomial root, as fuppofe $\sqrt{m} + n$, where the equation will be $x^3 = (m + 3n^2)x + 2n(m - n^2)$, having $\sqrt{m} + n$ for one root, namely the pofitive root. From which it appears that he was well acquainted with the compofition of cubic equations from given roots.

Cap. 59. *De ordine & exemplis in binomiis fecundo & quinto.* Contains a great many numeral forms of the fame irreducible cafe $x^3 = bx + c$, with their roots; from which are derived thefe following cafes, with many curious remarks. When

$$x^3 = (c + 1)x + c, \text{ then } x = \sqrt{c + \tfrac{1}{4}} + \tfrac{1}{2}.$$

$$x^3 = (\tfrac{1}{2}c + 4)x + c, \text{ then } x = \sqrt{\tfrac{1}{2}c + \tfrac{4}{4}} + \tfrac{2}{2}.$$

$$x^3 = (\tfrac{1}{3}c + 9)x + c, \text{ then } x = \sqrt{\tfrac{1}{3}c + \tfrac{9}{4}} + \tfrac{3}{2}.$$

$$x^3 = (\tfrac{1}{4}c + 16)x + c, \text{ then } x = \sqrt{\tfrac{1}{4}c + \tfrac{16}{4}} + \tfrac{4}{2}.$$

$$x^3 = (\tfrac{1}{n}c + n^2)x + c, \text{ then } x = \sqrt{\tfrac{1}{n}c + \tfrac{n^2}{4}} + \tfrac{n}{2}.$$

Cap. 60. *Demonftratio generalis capituli cubi æqualis rebus & numero.* This demonftration of the irreducible cafe is geometrical, like all the reft. Some more ingenious remarks are again added, as if he reluctantly finifhed the book without perfectly overcoming the difficulty of the irreducible cafe. Cardan here alfo ufes the letters *a* and *b* for any two indefinite numbers, in order to fhew the form and manner of the arithmetical operations: thus $\frac{a}{b}$ is the fraction for their quotient, alfo

$\sqrt{\frac{a}{b}}$ or $\frac{\sqrt{a}}{\sqrt{b}}$ the fquare root of that quotient, and

$\sqrt[3]{\frac{a}{b}}$ or $\frac{\sqrt[3]{a}}{\sqrt[3]{b}}$ the cube root of it, &c.

Having

Having confidered the chief contents of Cardan's algebra, it will now be proper to fum them up, and fet down a lift of the improvements made by him, as collected from his writings:

And 1ft, Tartalea having only communicated to him the rules for refolving thefe three cafes of cubic equations, viz,

$$x^3 + bx = c,$$
$$x^3 = bx + c,$$
$$x^3 + c = bx,$$

he from thence raifed a very large and complete work, laying down rules for all forms and varieties of cubic equations, having all their terms, or wanting any of them, and having all poffible varieties of figns; demonftrating all thefe rules geometrically; and treating very fully of almoft all forts of transformations of equations, in a manner heretofore unknown.

2nd, It appears that he was well acquainted with all the roots of equations that are real, both pofitive and negative; or, as he calls them, true and fictitious; and that he made ufe of them both occafionally. He alfo fhewed, that the even roots of pofitive quantities, are either pofitive or negative; that the odd roots of negative quantities, are real and negative; but that the even roots of them are impoffible, or nothing as to common ufe. He was alfo acquainted with,

3d, The number and nature of the roots of an equation, and that partly from the figns of the terms, and partly from the magnitude and relation of the coefficients. He alfo knew,

4th, That the number of pofitive roots is equal to the number of changes of the figns of the terms.

5th, That the coefficient of the fecond term of the equation, is the difference between the pofitive and negative roots.

6th, That when the fecond term is wanting, the fum of the negative roots is equal to the fum of the pofitive roots.

7th, How to compofe equations that fhall have given roots.

8th, That, changing the figns of the even terms, changes the figns of all the roots.

9th, That the number of roots failed in pairs; or what we now call impoffible roots were always in pairs.

10th, To change the equation from one form to another, by taking away any term out of it.

11th, To increafe or diminifh the roots by a given quantity. It appears alfo,

12th, That he had a rule for extracting the cube root of fuch binomials as admit of extraction.

13th, That he often ufed the literal notation a, b, c, d, &c.

14th, That he gave a rule for biquadratic equations, fuiting all their cafes; and that, in the inveftigation of that rule, he made ufe of an affumed indeterminate quantity, and afterwards found its value by the arbitrary affumption of a relation between the terms.

15th, That he applied Algebra to the refolution of geometrical problems. And

16th, That he was well acquainted with the difficulty of what is called the irreducible cafe, viz, $x^3 = bx + c$, upon which he fpent a great deal of time, in attempting to overcome it. And though he did not fully fucceed in this cafe, any more than other perfons have done fince, he neverthelefs made many ingenious obfervations about it, laying down rules for many particular

forms of it, and fhewing how to approximate very nearly to the root in all cafes whatever.

OF TARTALEA.

Nicholas Tartalea, or Tartaglia, of Brefcia, was contemporary with Cardan, and was probably older than he was, but I do not know of any book of Algebra publifhed by him till the year 1546, the year after the date of Cardan's work on Cubic Equations, when he printed his *Quefiti & Inventioni diverfe*, at Venice, where he refided as a public lecturer on mathematics. This work is dedicated to our king Henry the VIIIth of England, and confifts of 9 books, containing anfwers to various queftions which had been propofed to him at different times, concerning mechanics, ftatics, hydroftatics, &c.; but it is only the 9th, or laft, that we fhall have occafion to take notice of in this place, as it contains all thofe queftions which relate to arithmetic and algebra. Thefe are all fet down in chronological order, forming a pretty collection of queftions and folutions on thofe fubjects, with a fhort account of the occafion of each of them. Among thefe, the correfpondence between him and Cardan forms a remarkable part, as we have here the hiftory of the invention of the rules for cubic equations, which he communicated to Cardan, under the promife, and indeed oath, to keep them fecret, on the 25th of March 1539. But, notwithftanding his oath, finding that Cardan publifhed them in 1545, as above related, it feems Tartalea publifhed the correfpondence between them in revenge for his breach of faith; and it elfewhere appears, that many other fharp bickerings paffed between them on the fame account, which only ended with the death of Tartalea, in the year 1557. It feems it was a common practice among the mathematicians, and others, of that time, to fend to each other nice and difficult queftions, as trials of fkill, and to this caufe it is that we owe the principal queftions and difcoveries in this collection, as well as many of the beft difcoveries of other authors. The collection now before us contains queftions and folutions, with their dates, in a regular order, from the year 1521, and ending in 1541, in 42 dialogues, the laft of which is with an Englifh gentleman, namely, Mr. Richard Wentworth, who it feems was no mean mathematician, and who learned fome algebra, &c, of Tartalea, while he refided at Venice. The queftions at firft are moftly very eafy ones in arithmetic, but gradually become more difficult, and exercifing fimple and quadratic equations, with complex calculations of radical quantities: all fhewing that he was well fkilled in the art of Algebra as it then ftood, and that he was very ingenious in applying it to the folutions of queftions. Tartalea made no alteration in the notation or forms of expreffion ufed by Lucas de Burgo, calling the firft power of the unknown quantity, in his language, *cofa*, the fecond power *cenfa*, the third *cubo*, &c, and writing the names of all the operations in words at length, without ufing any contractions, except the initial R for root or radicality. So that the only thing remarkable in this collection, is the difcovery of the rules for cubic equations, with the curious circumftances attending the fame.

The

The first two of these were discovered by Tartalea in the year 1530, namely for the two cases $x^3 + ax^2 = c$, and $x^3 = ax^2 + c$, as appears by Quest. 14 and 25 of this collection, on occasion of a question then proposed to him by one Zuanni de Tonini da Coi or Colle, John Hill, who kept a school at Brescia. And from the 25th letter we learn, that he discovered the rules for the other two cases $x^3 + bx = c$, and $x^3 = bx + c$, on the 12th and 13th of February 1535, at Venice, where he had come to reside the year before. And the occasion of it was this: There was then at Venice one Antonio Maria Fiore or Florido, who, by his own account, had received from his preceptor Scipio Farreo, about thirty years before, a general rule for resolving the case $x^3 + bx = c$. Being a captious man, and presuming on this discovery, Florido used to brave his contemporaries, and by his insults provoked Tartalea to enter into a wager with him, that each should propose to the other two cases thirty different questions; and that he who soonest resolved those of his adversary, should win from him as many treats for himself and friends. These questions were to be proposed on a certain day at some weeks distance; and Tartalea made such good use of his time, that eight days before the time appointed for delivering the propositions, he discovered the rules both for the case $x^3 + bx = c$, and the case $x^3 = bx + c$. He therefore proposed several of his questions so as to fall either on this latter case, or on the cases of the cube and square, expecting that his adversary would propose his in the former. And what he suspected fell out accordingly; the consequence of which was, that on the day of meeting Tartalea resolved all his adversary's questions in the space of two hours, without receiving one answer from Florido in return; to whom, however, Tartalea generously remitted the forfeit of the thirty treats won of him.

Question 31 first brings us acquainted with the correspondence between Tartalea and Cardan. This correspondence is very curious, and would well deserve to be given at full length in their own words, if it were not too long for this place. I may enlarge farther upon it under the article *Cubic Equations*; but must here be content with a brief abstract only. Cardan was then a respectable physician, and lecturer in mathematics at Milan; and having nearly finished the printing of a large work on Arithmetic, Algebra, and Geometry, and having heard of Tartalea's discoveries in cubic equations, he was very desirous of drawing those rules from him, that he might add them to his book before it was finished. For this purpose he first applied to Tartalea, by means of a third person, a bookseller, whom he sent to him, in the beginning of the year 1539, with many flattering compliments, and offers of his services and friendship, &c, accompanied with some critical questions for him to resolve, according to the custom of the times. Tartalea however refused to disclose his rules to any one, as the knowledge of them gained him great reputation among all people, and gave him a great advantage over his competitors for fame, who were commonly afraid of him on account of those very rules. He only sent Cardan therefore, at his request, a copy of the thirty questions which had been proposed to him in the contest with Florido. Not to be rebuffed so easily, Cardan next applied, in the most urgent manner, by letter to Tartalea; which however procured from him only the solution of some other questions proposed by Cardan, with a few of the questions that had been proposed to Florido, but none of their solutions. Finding he could not thus prevail, with all his fair promises, Cardan then fell upon another scheme. There was at Milan a certain Marquis dal Vasto, a great patron of Cardan, and, it was said, of learned men in general. Cardan conceived the idea of making use of the influence of this nobleman to draw Tartalea to Milan, hoping that then, by personal intreaties, he should succeed in drawing the long-concealed rules from him. Accordingly he wrote a second letter to Tartalea, much in the same strain with the former, strongly inviting him to come and spend a few days in his house at Milan, and representing that, having often commended him in the highest terms to the marquis, this nobleman desired much to see him; for which reason Cardan advised him, as a friend, to come to visit them at Milan, as it might be greatly to his interest, the marquis being very liberal and bountiful; and he besides gave Tartalea to understand, that it might be dangerous to offend such a man by refusing to come, who might, in that case, take offence, and do him some injury. This manœuvre had the desired effect: Tartalea on this occasion laments to himself in these words, " By this I am reduced to a great dilemma; for if I go not to Milan, the marquis may take it amiss, and some evil may befal me on that account; I shall therefore go, although very unwillingly." When he arrived at Milan however, the marquis was gone to Vigeveno, and Tartalea was prevailed on to stay three days with Cardan, in expectation of the marquis returning, at the end of which he set out from Milan, with a letter from Cardan, to go to Vigeveno to that nobleman. While Tartalea was at Milan the three days, Cardan plied him by all possible means to draw from him the rules for the cubic equations; and at length, just as Tartalea was about to depart from Milan, on the 25th of March 1539, he was overcome by the most solemn protestations of secrecy that could be made. Cardan says, " I shall swear to you on the holy evangelists, and by the honour of a gentleman, not only never to publish your inventions, if you reveal them to me; but I also promise to you, and pledge my faith as a true christian, to note them down in cyphers, so that after my death no other person may be able to understand them." To this Tartalea replies, " If I refuse to give credit to these assurances, I should deservedly be accounted utterly void of belief. But as I intend to ride to Vigeveno, to see his excellency the marquis, as I have been here now these three days, and am weary of waiting so long; whenever I return therefore, I promise to shew you the whole." Cardan answers, " Since you determine at any rate to go to Vigeveno, to the marquis, I shall give you a letter for his excellency, that he may know who you are. But now before you depart, I intreat you to shew me the rule for the equations, as you have promised." " I am content," says Tartalea: " But you must know, that to be able on all occasions to remember such operations, I have brought the rule into rhyme; for if I had not used that precaution, I should often have forgot it; and although my rhymes are not very good, I do not value that, as it is
sufficient

sufficient that they serve to bring the rule to mind as often as I repeat them. I shall here write the rule with my own hand, that you may be sure I give you the discovery exactly." These rude verses contain, in rather dark and enigmatical language, the rule for these three cases, viz.

$x^3 + bx = c$, ⎫
$x^3 = bx + c$, ⎬
$x^3 + c = bx$, ⎭

which differ however only in the sign of one quantity, and the rule amounts to this: Find two numbers, z and y, such that their difference in the first case, and their sums in the 2d and 3d, may be equal to c the absolute number, and their product equal to the cube of $\frac{1}{3}$ of b the coefficient of the less power; then the difference of their cube roots will be equal to x in the first case, and the sum of their cube roots equal to x in the 2d and 3d cases: that is, taking $z - y = c$ in the 1st case, or $z + y = c$ in the 2d and 3d, and $zy = (\frac{1}{3}b)^3$; then $x = \sqrt[3]{z} - \sqrt[3]{y}$ in the first case, and $x = \sqrt[3]{z} + \sqrt[3]{y}$ in the other two. At parting, T. fails not again to remind C. of his obligation: "Now your excellency will remember not to break your promised faith, for if unhappily you should insert these rules either in the work you are now printing, or in any other, although you should even give them under my name, and as of my invention, I promise and swear that I shall immediately print another work that will not be very pleasing to you." "Doubt not, says C. but that I shall observe what I have promised: Go, and rest secure as to that point: and give this letter of mine to the marquis." It should seem however that T. was much displeased at having suffered himself to be worried as it were out of his rules, for as soon as he quitted Milan, instead of going to wait upon the marquis, he turned his horse's head, and rode straight home to Venice, saying to himself, "By my faith I shall not go to Vigeveno, but shall return to Venice, come of it what will."

After T's departure it seems C. applied himself immediately to resolving some examples in the cubic equations by the new rules, but not succeeding in them, for indeed he had mistaken the words, as it was very easy to do in such bad verses, having mistaken $(\frac{1}{3}b)^3$ for $\frac{1}{3}b^3$, or the cube of $\frac{1}{3}$ of the coefficient, for $\frac{1}{3}$ of the cube of the coefficient; accordingly we find him writing to T. in fourteen days after the above, blaming him much for his abrupt departure without seeing the marquis, who was so liberal a prince he said, and requesting T. to resolve him the example $x^3 + 3x = 10$. This T. did to his satisfaction, rightly guessing at the nature of his mistake; and concludes his answer with these emphatical words, "Remember your promise." On the 12th of May following C. returns him a letter of thanks, together with a copy of his book, saying, "As to my work, just finished, to remove your suspicion, I send you a copy, but unbound, as it is yet too fresh to be beaten. But as to the doubt you express lest I may print your inventions, my faith which I gave you with an oath should satisfy you; for as to the finishing of my book, that could be no security, as I could always add to it whenever I please. But on account of the dignity of the thing, I excuse you for not relying on that which you ought to have done, namely on the faith of a gentleman, instead of the finishing of a book, which might at any time be enlarged by the addition of new chapters; and there are besides a thousand other ways. But the security consists in this, that there is no

greater treachery than to break one's faith, and to aggrieve those who have given us pleasure. And when you shall try me, you will find whether I be your friend or not, and whether I shall make an ungrateful return for your friendship, and the satisfaction you have given me."

It was within less than two months after this, however, that T. received the alarming news of Cardan's shewing some symptoms of breaking the faith he had so lately pledged to him; this was in a letter from a quondam pupil of his, in which he writes, "A friend of mine at Milan has written to me, that Dr. Cardano is composing another algebraical work, concerning some lately-discovered rules; hence I imagine they may be those same rules which you told me you had taught him; so that I fear he will deceive you." To which T. replies, "I am heartily grieved at the news you inform me of, concerning Dr. Cardano of Milan; for if it be true, they can be no other rules but those I gave him; and therefore the proverb truly says, 'That which you wish not to be known, tell to nobody.' Pray endeavour to learn more of this matter, and inform me of it."

Tartalea, after this, kept on the reserve with Cardan, not answering several letters he sent him, till one written on the 4th of August the same year, 1539, complaining greatly of T's neglect of him, and farther requesting his assistance to clear up the difficulty of the irreducible case $x^3 = bx + c$, which C. had thus early been embarrassed with: he says that when $(\frac{1}{3}b)^3$ exceeds $(\frac{1}{2}c)^2$, the rule cannot be applied to the equation in hand, because of the square root of the negative quantities. On this occasion T. turns the tables on C. and plays his own game back upon him; for being aware of the above difficulty, and unable to overcome it himself, he wanted to try if C. could be encouraged to accomplish it, by pretending that the case might be done, though in another way. He says thus to himself, "I have a good mind to give no answer to this letter, no more than to the other two. However I will answer it, if it be but to let him know what I have been told of him. And as I perceive that a suspicion has arisen concerning the difficulty or obstacle in the rule for the case $x^3 = bx + c$, I have a mind to try if he can alter the data in hand, so as to remove the said obstacle, and to change the rule into another form, although I believe indeed that it cannot be done; however there is no harm in trying."—"M. Hieronime, I have received your letter, in which you write that you understand the rule for the case $x^3 = bx + c$, but that when $(\frac{1}{3}b)^3$ exceeds $(\frac{1}{2}c)^2$, you cannot resolve the equation by following the rule, and therefore you request me to give you the solution of this equation $x^3 = 9x + 10$. To which I reply, that you have not used a good method in that case, and that your whole process is intirely false. And as to resolving you the equation you have sent, I must say that I am very sorry that I have already given you so much as I have done, for I have been informed, by a credible person, that you are about to publish another Algebraical work, and that you have been boasting through Milan of having discovered some new rules in algebra. But, take notice, that if you break your faith with me, I shall certainly keep my word with you, nay, I even assure you to do more than I promised." In Cardan's answer to this he says, "You have been misinformed as to my intention to

publish

publish more on Algebra. But I suppose you have heard something about my work *de mysteriis æternitatis*, which you take for some Algebra I intend to publish. As to your repenting of having given me your rules, I am not to be moved from the faith I promised you for any thing you say." To this, and many other things contained in the same letter, T. returned no answer, being still suspicious of Cardan's intentions, and declining any more correspondence with him. This however did not discourage C. for we find him writing again to T. on the 5th of January, 1540, to clear up another difficulty which had occurred in this business, namely to extract the cube root of the binomials, of which the two parts of the rule always consisted, and for which purpose it seems C. had not yet found out a rule. On this occasion he informs T. that his quondam competitor Zuanne Collé had come to Milan, where, in some contests between them, Collè gave Cardan to understand that he had found out the rules for the two cases $x^3 + bx = c$, and $x^3 = bx + c$, and farther that he had discovered a general rule for extracting the cube roots of all such binomials as can be extracted; and that, in particular, the cube root of $\sqrt{108} + 10$ is $\sqrt{3} + 1$, and that of $\sqrt{108} - 10$ is $\sqrt{3} - 1$, and consequently that $\sqrt[3]{\sqrt{108} + 10} \cdot \sqrt[3]{\sqrt{108} - 10}$

is $= \sqrt{3 + 1} - \sqrt{3 - 1} = 2$. He then earnestly entreats T. to try to find out the rule, and the solution of certain other questions which had been proposed to him by Collè. By this letter T. is still more confirmed in his resolution of silence; so that, without returning any answer, he only sets down among his own memorandums some curious remarks on the contents of the letter, and then concludes to himself, "Wherefore I do not choose to answer him again, as I have no more affection for him than for M. Zuanne, and therefore I shall leave the matter between them." Among those remarks he sets down a rule for extracting the cube root of such binomials as can be extracted, and that is done from either member of the binomial alone, thus : Take either term of the binomial, and divide it into two such parts that one of them may be a complete cube, and the other part exactly divisible by 3; then the cube root of the said cubic part will be one term of the required root, and the square root of the quotient arising from the division of $\frac{1}{3}$ of the 2d part by the cube root of the first, will be the other member of the root sought. This rule will be better understood in characters thus : let m be one member of the given binomial, whose cube root is sought, and let it be divided into the two parts a^3 and $3b$, so that $a^3 + 3b$ be $= m$; then is $a + \sqrt{\dfrac{b}{a}}$ the cube root required, if it have one. Thus in the quantity $\sqrt{108} + 10$, taking the term 10 for m, then 10 divides into 1 and 9, where $a^3 = 1$ or $a = 1$, and $3b = 9$ or $b = 3$: therefore $a + \sqrt{\dfrac{b}{a}}$ becomes $1 + \sqrt{3}$ for the cube root of $\sqrt{108} + 10$. And taking the other member $\sqrt{108}$, this divides into the two equal parts $\sqrt{27}$ and $\sqrt{27}$, making $a^3 = \sqrt{27}$, and $3b = \sqrt{27}$; hence $a = \sqrt{3}$, and $b = \sqrt{3}$ also; consequently $a + \sqrt{\dfrac{b}{a}}$ is $= \sqrt{3} + \sqrt{\sqrt{3}}$ or $\sqrt{3} + 1$ for the cube root of the binomial sought, the same as before. "And thus, he adds, we may know whether any proposed binomial

or residual be a cube or a noncube; for if it be a cube, the same two terms for the root must arise from both the given terms separately; and if the two terms of the root cannot thus be brought to agree both ways, such binomial or residual will not be a cube." And thus ends the correspondence between them, at least for this time. But it seems they had still more violent disputes when C. in violation of his faith so often pledged to the contrary, published his work on cubic equations 4 years afterwards, viz, in the year 1545, of which we have before given an account, which disputes, it is said, continued till the death of Tartalea in the year 1557.

. The last article in the volume contains a dialogue on some other forms of the cubic equations, in the year 1541, between T. and a Mr. Richard Wentworth, an English gentleman, who it seems had resided some time at Venice, on some public service from England, as T. in the dedication of the volume to Henry VIII. king of England, makes mention of him as " a gentleman of his sacred majesty." Mr. Wentworth had learned some mathematics of T. and being about to depart for England, requests T. to shew him his newly discovered rules for cubic equations, as a farewell-lesson ; and it is worth while to note a few particulars in this conference, as they shew pretty nicely the limited knowledge of T. at that time, as to the nature and roots of such equations. T. had before, it seems, shewed Mr. W. the rules for the cases of the 3d and 1st powers, and now the latter desires him to do the same as to the three cases in which the 3d and 2d powers only are concerned. On this T. professes great gratitude to Mr. W. for many obligations, but desires to be excused from giving him the rules for these, because he says he intends soon to compose a new work on Arithmetic, Geometry, and Algebra, which he intends to dedicate to him, and in which he means to insert all his new discoveries. On Mr. W. urging him further, T. gives him the roots of some equations of that kind, as for instance :

If $x^3 + 6x^2 = 100$, then

$x = \sqrt[3]{42 + \sqrt{17000}} + \sqrt[3]{42 - \sqrt{17000}} - 2,$
If $x^3 + 9x^2 = 100$, then $x = \sqrt{24} - 2,$
If $x^3 + 3x^2 = 2$, then $x = \sqrt{3} - 1,$
If $x^3 + 4 = 5x^2$, then $x = \sqrt{8} + 2,$
If $x^3 + 6 = 7x^2$, then $x = \sqrt{15} + 3;$
but not the rules for finding them.

In the course of the conversation T. tells him that " all such equations admit of two different answers, and perhaps more ; and hence it follows that they have, or, admit of, two different rules, and perhaps more, the one more difficult than the other." And on Mr. W. expressing his wonder at this circumstance of a plurality of roots, T. replies, " It is however very true, though hardly to be believed, and indeed if experience had not confirmed it, I should scarcely have believed it myself." He then commits a strange blunder in an example which he takes to illustrate this by, namely the equation $x^3 + 3x = 14$, which, he says, it is evident has the number 2 for one of its roots; and yet, he adds, " whoever shall resolve the same equation by my rule, will find the value of x to be $\sqrt[3]{7 + \sqrt{50}} +$ $\sqrt[3]{7 - \sqrt{50}}$, which is proved to be a true root by substituting ftituting it in the equation for x. And therefore, continues he, it is manifeſt that the caſe $x^3 + bx = c$ admits of two rules, namely, one (as in the above example) which ought to give the value of x rational, viz 2, and the other is my rule, which gives the value of x irrational, as appears above; and there is reaſon to think that there may be ſuch a rule as will give the value of $x = 2$, although our anceſtors may not have found it out."——"And theſe two different anſwers will be found not only in every equation of this form $x^3 + bx = c$, when the value of x happens to be rational, as in the example $x^3 + 3x = 14$ above, but the ſame will alſo happen in all the other five forms of cubic equations: and therefore there is reaſon to think that they alſo admit of two different rules; and by certain circumſtances attending ſome of them, I am almoſt certain that they admit of more than two rules, as, God willing, I ſhall ſoon demonſtrate." Now all this diſcourſe ſhews a ſtrange mixture of knowledge and ignorance: it is very probable that he had met with ſome equations which admit of a plurality of roots; indeed it was hardly poſſible for him to avoid it; but it ſeems he had no ſuſpicion what the number of roots might be, nor that his reaſoning in this inſtance was founded on an error of his own, miſtaking the root $x = \sqrt[3]{7 + \sqrt{50}} + \sqrt[3]{7 - \sqrt{50}}$, of the equation $x^3 + 3x = 14$, for a different root from the number or root 2, when in reality it is the very ſame, as he might eaſily have found, if he had extracted the cube roots of the binomials by the rule which he himſelf had juſt given above for that purpoſe: for by that rule he would have found $\sqrt[3]{7 + \sqrt{50}} = 1 + \sqrt{2}$, and $\sqrt[3]{7 - \sqrt{50}} = 1 - \sqrt{2}$, and therefore their ſum is $2 = x$, the ſame root as the other, which T. thought had been different. And beſides this root 2, the equation in hand, $x^3 + 3x = 14$, admits of no other real roots. Nor does any equation of the ſame form, $x^3 + bx = c$, admit of more than one real root.

It ſeems alſo they had not yet diſcovered that all caſes belong to the rules and forms for quadratic equations, which have only two powers in them, in which the exponent of the one is juſt double of the exponent of the other, as $x^{2n} + bx^n = c$; but ſome particular caſes only of this ſort they had as yet ventured to refer to quadratics, as the caſe $x^4 + bx^2 = c$. But in the concluſion of this dialogue T. informs W. of another caſe of this ſort which he had accompliſhed, as a notable diſcovery, in theſe words: "I well remember, ſays he, that in the year 1536, on the night of St. Martin, which was on a Saturday, meditating in bed when I could not ſleep, I diſcovered the general rule for the caſe $x^6 + bx^3 = c$, and alſo for the other two, its accompanying caſes, in the ſame night." And then he directs that they are to be reſolved like quadratics, by completing the ſquare, &c. And in theſe reſolutions it is remarkable that he uſes only the poſitive roots, without taking any notice of the negative ones.

Tartalea alſo publiſhed at Venice, in 1556, &c, a very large work, in folio, on Arithmetic, Geometry, and Algebra. This is a very complete and curious work upon the firſt two branches; but that of Algebra is carried no farther than quadratic equations.

called *book the firſt*, with which the work terminates. It is evidently incomplete, owing to the death of the author, which happened before this latter part of the work was printed, as appears by the dates, and by the prefaces. It appears alſo, from ſeveral parts of this work, that the author had many ſevere conflicts with Cardan and his friend Lewis Ferrari: and particularly, there was a public trial of ſkill between them, in the year 1547; in which it would ſeem that Tartalea had greatly the advantage, his queſtions moſtly remaining unanſwered by his antagoniſts.

OF MICHAEL STIFELIUS.

After the foregoing analyſis of the works of the firſt algebraic writers in Italy, it will now be proper to conſider thoſe of their contemporaries in Germany; where, excepting for the diſcoveries in cubic equations, the art was in a more advanced ſtate, and of a form approaching nearer to that of our modern Algebra; the ſtate and circumſtances indeed being ſo different, that one would almoſt be led to ſuppoſe they had derived their knowledge of it from a different origin.

Here Stifelius and Scheubelius were writers of the ſame time with Cardan and Tartalea, and even before their diſcoveries, or publication, concerning the rules for cubic equations, Stifelius's *Arithmetica Integra* was publiſhed at Norimberg in 1544, being the year before Cardan's work on cubic equations, and is an excellent treatiſe, both on Arithmetic and Algebra. The work is divided into three books, and is prefaced with an Introduction by the famous Melanchthon. The firſt book contains a complete and ample Treatiſe on Arithmetic, the ſecond an Expoſition of the 10th book of Euclid's Elements, and the third a Treatiſe of Algebra, and it is therefore properly the part with which we are at preſent concerned. In the dedication of this part, he aſcribes the invention of Algebra to Geber, an Arabic Aſtronomer; and mentions beſides, the authors Campanus, Chriſt. Rudolph, and Adam Ris, Riſen, or Gigas, whoſe rules and examples he has chiefly given. In other parts of the book he ſpeaks, and makes uſe alſo, of the works of Bretius, Campanus, Cardan (i. e. his Arithmetic publiſhed in 1539, before the work on cubic equations appeared), de Cuſa, Euclid, Jordan, Milichius, Schonerus, and Stapulenſis.

Chap. 1. *Of the Rule of Algebra, and its parts.* Stifelius here deſcribes the notation and marks of powers, or denominations as he calls them, which marks for the ſeveral powers are thus:

1ſt,	2d,	3d,	4th,	5th,	6th,	&c.
♃,	3,	cſ,	33,	fſ,	3 cſ,	&c.

which are formed from the initials of the barbarous way in which the Germans pronounced and wrote the Latin and Italic names of the powers, namely, res or coſa, zenſus, cubo, zenſi-zenſus, ſurſolid, zenſi-cubo, &c. And the coſs or firſt power ♃, he calls the radix or root, which is the firſt time that we meet with this word in the printed authors. He alſo here uſes the ſigns or characters, $+$ and $-$, for addition and ſubtraction, and the firſt of any that I know of: for in Italy they uſed none of theſe characters for a long time after. He has no mark however for equality, but makes uſe of the word itſelf.

Chap.

Chap. 2. Of the Parts of the Rule of Geber or Algebra: teaching the various reductions by addition, subtraction, multiplication, division, involution, and evolution, &c.

Chap. 3. Of the Algorithm of Coffic Numbers: teaching the usual operations of addition, subtraction, multiplication, division, involution, and extraction of roots, much the same as they are at present. Single terms, or powers, he calls simple quantities; but such as 13 + 1 χ a composite or compound, and 2χ — 8 a defective one. In multiplication and division, he proves that like signs give +, and unlike signs —. He shews that the powers 1, 2χ, 3, φ, &c, form a geometrical progression from unity; and that the natural series of numbers 0, 1, 2, 3, &c, from 0, are the exponents of the coffic powers; and he, for the first time, expressly calls them exponents: thus,

Exponents, 0, 1, 2, 3, 4, 5, 6, &c.
Powers, 1, 2χ, 3, φ, 33, $\int s$, 3φ, &c.

And he shews the use of the exponents, in multiplication, division, powers, and roots, as we do at present; viz, adding the exponents in multiplication, and subtracting them in division, &c. And these operations he demonstrates from the nature of arithmetical and geometrical progressions. It is remarkable that these compound denominations of the powers are formed from the simple ones according to the *products* of the exponents, while those of Diophantus are formed according to the *sums* of them; thus the 6th power here is 3 φ or quadrato-cubi, but with Diophantus it is cubo-cubi; and so of others. Which is presumptive evidence that the Europeans had not taken their Algebra immediately from him, independent of other proofs.

Chap. 4. Of the extraction of the roots of coffic numbers. He here treats of quadratic equations, which he resolves by completing the square, from Euclid II. 4 &c. Also quadratics of the higher orders, shewing how to resolve them in all cases, whatever the height may be, provided the exponents be but in arithmetical progression, as

2, 1, 0 } &c; where it is plain that he always
4, 2, 0 } counts 0 for the exponent of the un-
6, 3, 0 } known quantity in the absolute term.
8, 4, 0 }

Chap. 5. Of irrational coffic numbers, and of surd or negative numbers. In this treatise of radicals, or irrationals, he first uses the character $\sqrt{}$ to denote a root, and sets after it the mark of the power whose root is intended; as $\sqrt{3}$ 20 for the square root of 20, and $\sqrt{}$ φ 20 for the cube root of the same, and so on. He treats here also of negative numbers, or what he calls surd or fictitious, or numbers less than 0. On which he takes occasion to observe, that when a geometrical progression is continued downwards below 1, then the exponents of the terms, or the arithmetical progression, will go below 0 into negative numbers, and will yet be the true exponents of the former; as in these,

Expon.	−3	−2	−1	0	1	2	3
Pow.	$\frac{1}{8}$	$\frac{1}{4}$	$\frac{1}{2}$	1	2	4	8

And he gives examples to shew that these negative exponents perform their office the same as the positive ones, in all the operations.

Chap. 6. Of the perfection of the Rule of Algebra, and of Secondary Roots. In the reduction of equations he uses a more general rule than those who had preceded him, who detailed the rule in a multitude of cases; instead of which, he directs to multiply or divide the two sides equally, to transpose the terms with + or —, and lastly to extract such root as may be denoted by the exponent of the highest power.

As to secondary roots, Cardan treated of a 2d *ignota* or unknown, which he called *quantitas*, and denoted it by the initial *q*, to distinguish it from the first. But here Stifelius, for distinction sake, and to prevent one root from being mistaken for others, assigns literal marks to all of them, as A, B, C, D, &c, and then performs all the usual operations with them, joining them together as we do now, except that he subjoins the initial of the power, instead of its numeral exponent:

thus, 3A into 9B makes 27AB,
33 into 4B makes 123B,
2φ into 4A3 makes 8φA3,
1A squared makes 1A3,
6 into 3C makes 18C,
2A3 into 5Aφ makes 10A$\int s$, &c, &c.
8φA3 divided by 4φ makes 2A3, &c.
The square root of 25A3 is 5A, &c.
Also 2A added to 2χ makes 2χ + 2A,
and 2A subtr. from 2χ makes 2χ — 2A.

And he shews how to use the same, in questions concerning several unknown numbers; where he puts a different character for each of them, as 2χ, A, B, C, &c; he then makes out, from the conditions of the question, as many equations as there are characters; from these he finds the value of each letter, in terms of some one of the rest; and so, expelling them all but that one, reduces the whole to a final equation, as we do at present.

The remainder of the book is employed with the solutions of a great number of questions to exercise all the rules and methods; some of which are geometrical ones.

From this account of the state of Algebra in Stifelius, it appears that the improvements made by himself, or other Germans, beyond those of the Italians, as contained in Cardan's book of 1539, were as follow:

1st. He introduced the characters +, —, $\sqrt{}$, for plus, minus, and root, or *radix*, as he calls it.

2d. The initials 2χ, 3, φ, &c. for the powers.

3d. He treated all the higher orders of quadratics by the same general rule.

4th. He introduced the numeral exponents of the powers, —3, —2, —1, 0, 1, 2, 3, &c, both positive and negative, so far as integral numbers, but not fractional ones; calling them by the name *exponens*, exponent: and he taught the general uses of the exponents, in the several operations of powers, as we now use them, or the logarithms.

5th. And lastly, he used the general literal notation A, B, C, D, &c, for so many different unknown or general quantities.

OF SCHEUBELIUS.

John Scheubelius published several books upon Arithmetic and Algebra. The one now before me, is intitled *Algebræ Compendiosa Facilisque Descriptio, quâ depromuntur*

muntur magna Arithmetices miracula. Authore Johanne Scheubelio Mathematicarum Professore in Academia Tubingensi. Parisiis 1552. But at the end of the book it is dated 1551. The work is most beautifully printed, and is a very clear and succinct treatise; and both in the form and matter much resembles a modern printed book. He says that the writers ascribe this art to Diophantus, which is the first time that I find this Greek author mentioned by the modern algebraists: he farther observes, that the Latins call it *Regula Rei & Census,* the rule of the thing and the square (or of the 1st and 2d power); and the Arabs, Algebra. His characters and operations are much the same as those of Stifelius, using the signs and characters $+$, $-$, $\checkmark$, and the powers $\wp$, $\wp$, 3, $\wp$, &c, where the character $\wp$ is used for 1 or unity, or a number, or the o power; prefixing also the numeral coefficients; thus $44\sqrt{3} + 113 + 31\wp - 53\wp$. He uses also the exponents o, 1, 2, 3, &c, of the powers, the same way as Stifelius, before him. He performs the algebraical calculations, first in integers, and then in fractions, much the same as we do at present. Then of equations, which he says may be of infinite degrees, though he treats only of two, namely the first and second orders, or what we call simple and quadratic equations, in the usual way, taking however only the positive roots of these; and adverting to all the higher orders of quadratics, namely, x^4, ax^2, b;

$$x^6, \ ax^3, \ b;$$

$$x^8, \ ax^4, \ b; \&c.$$

Next follows a tract on surds, both simple and compound, quadratic, cubic, binomial, and residual. Here he first marks the notation, observing that the root is either denoted by the initial of the word, or, after some authors, by the mark $\checkmark$:; viz. the sq. root $\checkmark$:, the cube root w$\checkmark$:, and the 4th root, or root of the root thus v$\checkmark$:, which latter method he mostly uses. He then gives the Arithmetic of surds, in multiplication, division, addition, and subtraction. In these last two rules he squares the sum or difference of the surds, and then sets the root to the whole compound, which he calls *radix collecti,* what Cardan calls *radix universalis.* Thus $\sqrt{12} \pm \sqrt{20}$ is ra. col. $32 \pm \sqrt{960}$. But when the terms will reduce to a common surd, he then unites them into one number; as $\sqrt{27} + \sqrt{12}$ is equal $\sqrt{75}$. Also of cubic surds, and 4th roots. In binomial and residual surds, he remarks the different kinds of them which answer to the several irrational lines in the 10th book of Euclid's elements; and then gives this general rule for extracting the root of any binomial or residual $a \pm b$, where one or both parts are surds, and a the greater quantity, namely, that the square root of it is

$$\sqrt{\frac{a + \sqrt{a^2 - b^2}}{2}} \pm \sqrt{\frac{a - \sqrt{a^2 - b^2}}{2}}; \ \text{which he illus-}$$

trates by many examples. This rule will only succeed however, so as to come out in simple terms, in certain cases, namely, either when $a^2 - b^2$ is a square, or when a and $\sqrt{a^2 - b^2}$ will reduce to a common surd, and unite: in all other cases the root is in two compound surds, instead of one. He gives also another rule, which comes however to the same thing as the former, though by the words of them they seem to be different.

Scheubelius wrote much about the time of Cardan and Stifelius. And as he takes no notice of cubic equations, it is probable he had neither seen nor heard any thing about them; which might very well happen, the one living in Italy, and the other in Germany. And, besides, I know not if this be the first edition of Scheubel's book: it is rather likely it is not, as it is printed at Paris, and he himself was professor of mathematics at Tubingen in Germany.

ROBERT RECORDE.

The first part of his Arithmetic was published in 1552; and the second part in 1557, under the title of, "The Whetstone of Witte, which is the seconde parte of Arithmetike: containing the Extraction of Rootes: The Cossike Practise, with the Rule of Equation: and the Workes of Surde Nombers." The work is in dialogue between the master and scholar; and is nearly after the manner of the Germans, Stifelius and Scheubelius, but especially the latter, whom he often quotes, and takes examples from. The chief parts of the work are, 1st. The properties of abstract and figurate numbers. 2nd. The extraction of the square and cube roots, much the same as at present. Here, when the number is not an exact power, but having some remainder over, he either continues the root into decimals as far as he pleases, by adding to the remainders always periods of cyphers; or else makes a vulgar fraction for the remaining part of the root, by taking the remainder for the numerator, and double the root for the denominator, in the square root; but in the cube root he takes for the nominator either the triple square of the root, which is Cardan's rule, or the triple square and triple root, with one more, which is Scheubel's rule. 3d. Of Algebra, or "Cossike Nombers." He uses the notation of powers with their exponents the same as Stifel, with all the operations in simple and compound quantities, or integers and fractions. And he gives also many examples of extracting the roots of compound algebraic quantities, even when the roots are from two to six terms, in imitation of the same process in numbers, just as we do at present; which is the first instance of this kind that I have observed. As of this quantity:

$$253 \wp + 80\sqrt{3} - 263 \jmath - 63 \jmath \ (5 \wp + 83 - 9\wp.$$

4th. "The Rule of Equation, commonly called Algeber's Rule." He here, first of any, introduces the character $=$, for brevity sake. His words are, "And to avoide the tediouse repetition of these woordes: is equalle to: I will sette as I doe often in woorke use, a paire of paralleles, or gemowe lines of one lengthe, thus:$=$, bicause noe 2 thynges can be moare equalle." He gives the rules for simple and quadratic equations, with many examples. He gives also some examples in higher compound equations, with a root for each of them, but gives no rule how to find it. 5th. "Of Surde Nombers." This is a very ample treatise on surds, both simple and compound, and surds of various degrees, as square, cubic, and biquadratic, marking the roots in Scheubel's manner, thus: $\checkmark$, w$\checkmark$, v$\checkmark$. He here uses the names bimedial, binomial, and residual; but says they have been used by others before him, though this is the first place where I have observed the two latter.— Hence it appears that the things which chiefly are new in this author, are these three, viz.

2 1. The

1. The extraction of the roots of compound algebraic quantities.

2. The use of the terms binomial and residual.

3. The use of the sign of equality, or =.

OF PELETARIUS.

The first edition of this author's algebra was printed in 4to at Paris, in 1558, under this title, *Jacobi Peletarii Cenomani, de occulta parte Numerorum, quam Algebram vocant. Lib. duo.*

In the preface he speaks of the supposed authors of Algebra, namely Geber, Mahomet the son of Moses, an Arabian, and Diophantus. But he thinks the art older, and mentions some of his contemporary writers, or a very little before him, as Cardan, Stifel, Scheubel, Chr. Januarius; and a little earlier again, Lucas Paciolus of Florence, and Stephen Villafrancus a Gaul.

Of the two books, into which the work is divided, the first is on rational, and the second on irrational or furd quantities; each being divided into many chapters. It will be sufficient to mention only the principal articles.

He calls the series of powers *numeri creati*, or derived numbers, or also radicals, because they are all raised from one root or *radix*. He names them thus, radix, quadratus cubus, quadrato-quadratus, or biquadratus, supersolidus, quadrato-cubus, &c; and marks them thus $\mathbb{R}$, q, cf, qq, fs, qcf, bfs, &c. Of these he gives the following series in numbers, having the common ratio 2, with their marks set over them, and the exponents set over these again, in an arithmetical series, beginning at 0, thus:

0	1	2	3	4	5	6	7	8	
1	$\mathbb{R}$	q	cf	qq	fs	qcf	bfs	qqq	
1	2	4	8	16	32	64	128	256	&c.

And he shews the use of the exponents, the same as Stifelius and Scheubelius; like whom also he prefixes coefficients to quantities of all kinds, as also the radical $\sqrt{}$. But he does not follow them in the use of the signs + and —, but employs the initials p and m for the same purpose. After the operations of addition, &c, he performs involution and evolution also much the same way as at present: thus, in powers, raise the coefficient to the power required, and multiply the exponent, or sign, as he calls it, by 2, or 3, or 4, &c, for the 2nd, 3d, 4th, &c, power; and the reverse for extraction: and hence he observes, if the number or coefficient will not exactly extract, or the sign do not exactly divide, the quantity is a furd.

After the operations of compound quantities, and fractions, and reduction of equations, namely, simple and quadratic equations, as usual, in chap. 16, *De Inveniendis generatim Radicibus Denominatorum*, he gives a method of finding the roots of equations among the divisors of the absolute number, when the root is rational, whether it be integral or fractional; for then, he observes, the root always lies hid in that number, and is some one of its divisors. This is exemplified in several instances, both of quadratic and cubic equations, and both for integral and fractional roots. And he here observes, that he knows not of any person who has yet given general rules for the solution of cubic equations; which shews that when he wrote this book, either Cardan's last book was not published, or else it had not yet come to his knowledge.

Chap. 17 contains, in a few words, directions for bringing questions to equations, and for reducing these. He here observes, that some authors call the unknown number *res*, and others the *positio*; but that he calls it *radix*, or root, and marks it thus $\mathbb{R}$: hence the term, root of an equation. But it was before called radix by Stifelius.

Chap. 21 &c *seq.* treat of secondary roots, or a plurality of roots, denoted by A, B, C, &c, after Stifelius.

The 2d book contains the like operations in furds, or irrational numbers, and is a very complete work on this subject indeed. He treats first of simple or single furds, then of binomial furds, and lastly of trinomial furds. He gives here the same rule for extracting the root of a binomial and residual as Scheubelius, viz, $\sqrt{a \pm b} =$

$$\sqrt{\frac{a + \sqrt{a^2 - b^2}}{2}} \pm \sqrt{\frac{a - \sqrt{a^2 - b^2}}{2}}.$$ Indivicing by a

binomial or residual, he proceeds as all others before him had done, namely, reducing the divisor to a simple quantity, by multiplying it by the same two terms with the sign of one of them changed, that is by the binomial if it be a residual, or by the same residual if it be a binomial; and multiplying the dividend by the same thing: thus

$$\frac{3.}{\sqrt{5} - 2} = \frac{3}{\sqrt{5} - 2} \times \frac{\sqrt{5} + 2}{\sqrt{5} + 2} = \frac{3\sqrt{5} \mp 6}{5 - 4}$$

$$= 3\sqrt{5} + 6.$$

And, in imitation of this method, in division by trinomial furds, he directs to reduce the trinomial divisor first to a binomial or residual, by multiplying it by the same trinomial with the sign of one term changed, and then to reduce this binomial or residual to a simple nomial as above; observing to multiply the dividend by the same quantities as the divisor. Thus, if the divisor be $4 + \sqrt{2} - \sqrt{3}$; multiplying this by $4 + \sqrt{2} + \sqrt{3}$, the product is $15 + 8\sqrt{2}$; then this binomial multiplied by the residual $15 - 8\sqrt{2}$, gives $225 - 128$ or 97 for the simple divisor: and the dividend, whatever it is, must also be multiplied by the two $4 + \sqrt{2} + \sqrt{3}$ and $15 - 8\sqrt{2}$. Or in general, if the divisor be $a + \sqrt{b} - \sqrt{c}$; multiply it by $a + \sqrt{b} + \sqrt{c}$, which gives $(a + \sqrt{b})^2 - c = a^2 + b - c + 2a\sqrt{b}$; then multiply this by $a^2 + b - c - 2a\sqrt{b}$, and it gives $(a^2 + b - c)^2 - 4a^2 b$, which will be rational, and will all collect into one single term. But Tartalea must have been in possession of some such rule as this, as one of the questions he proposed to Florido was of this nature, namely to find such a quantity as multiplied by a given trinomial furd, shall make it rational: and it appears, from what is done above, that, the given trinomial being $\overline{a + \sqrt{b} - \sqrt{c}}$, the answer will be $\overline{a + \sqrt{b} + \sqrt{c}} \times \overline{a^2 + b - c + 2\sqrt{b}}$.

Chap. 24 shews the composition of the cube of a binomial or residual, and thence remarks on the root of the case or equation $1 cf \, p \, 3 \mathbb{R}$ equal to 10, which he seems to know something about, though he had not Cardan's rules.

Chap. 30, which is the last, treats of certain precepts relating to square and cubic numbers, with a table of such squares and cubes for all numbers to 140; also shewing how to compute them both, by adding always their differences.

He

He then concludes with remarking that there are many curious properties of thefe numbers; one of which is this, that the fum of any number of the cubes, taken from the beginning, always makes a fquare number, the root of which is the fum of the roots of the cubes; fo that the feries of fquares fo formed, have for their roots — 1, 3, 6, 10, 15, 21, &c. whofe diff. are the natural n°s 1, 2, 3, 4, 5, 6, &c. Namely, $1^3 = 1^2$; $1^3 + 2^3 = 3^2$; $1^3 + 2^3 + 3^3 = 6^2$, &c. Or in general, $1^3 + 2^3 + 3^3 - n^3 = (1 + 2 + 3 - n)^2 = \frac{1}{2}n \cdot \overline{n + 1}$.

This work of Peletarius is a very ingenious and mafterly compofition, treating in an able manner of the feveral parts of the fubject then known, excepting the cubic equations. But his real difcoveries, or improvements, may be reduced to thefe three, viz.

1ft. That the root of an eqnation, is one of the divifors of the abfolute term.

2d. He taught how to reduce trinomials to fimple terms, by multiplying them by compound factors.

3d. He taught curious precepts and properties concerning fquare and cube numbers, and the method of conftructing a feries of each by addition only, namely by adding fucceffively their feveral orders of differences.

RAMUS.

Peter Ramus wrote his arithmetic and algebra about the year 1560. His notation of the powers is thus, l, q, c, bq, being the initials of latus, quadratus, cubus, biquadratus. He treats only of fimple and quadratic equations. And the only thing remarkable in his work, is the firft article, on the names and invention of Algebra, which we have noticed at the beginning of this hiftory.

BOMBELLI.

Raphael Bombelli's Algebra was publifhed at Bologna in the year 1579, in the Italian language. It feems however it was written fome time before, as the dedication is dated 1572. In a fhort, but neat, introduction, he firft adverts, in a few words, to the great excellence and ufefulnefs of arithmetic and algebra. He then laments that it had hitherto been treated in fo imperfect and irregular a way ; and declares it is his intention to remedy all defects, and to make the fcience and practice of it as eafy and perfect as may be. And for this purpofe he firft refolved to procure and ftudy all the former authors. He then mentions feveral of thefe, with a fhort hiftory or character of them; as Mahomet the fon of Mofes, an Arabian; Leonard Pifano; Lucas de Burgo, the firft printed author in Europe; Oroncius; Scribelius; Boglione Francefi; Stifelius in Germany; a certain Spaniard, doubtlefs meaning Nunez or Nonius; and laftly Cardan, Ferrari, and Tartalea; with fome others fince, whofe names he omits. He then adds a curious paragraph concerning Diophantus: he fays that fome years fince there had been found, in the Vatican library, a Greek work on this art, compofed by a certain Diophantus, of Alexandria, a Greek author, who lived in the time of Antoninus Pius; which work having been fhewn to him by Mr. Antonio Maria Pazzi Reggiano, public lecturer on mathematics at Rome; and finding it to be a good work, thefe two formed the re-

folution of giving it to the world, and he fays that they had already tranflated five books, of the fix which were then extant, being as yet hindered by other avocations from completing the work. He then adds the following ftrange circumftance, viz. *that they had found that in the faid work the Indian authors are often cited ; by which they learned that this fcience was known among the Indians before the Arabians had it :* a paragraph the more remarkable as I have never underftood that any other perfon could ever find, in Diophantus, any reference to Indian writers: and I have examined his work with fome attention, for that purpofe.

Bombelli's work is divided into three books. In the firft, are laid down the definitions and operations of powers and roots, with various forts of radicals, fimple and compound, binomial, refidual, &c; moftly after the rules and manner of former writers, excepting in fome few inftances, which I fhall here take notice of. And firft of his rule for the cube root of binomials or refiduals, which for the fake of brevity, may be expreffed in modern notation as follows : let $\sqrt{b} + a$ be the binomial, the term $\sqrt{b}$ being greater than a; then the rule for the cube root of $\sqrt{b} + a$ comes to this, $P - Q + \sqrt{\overline{P - Q^2} + \sqrt[3]{b - a^2}}$ where $P = \sqrt[3]{\sqrt{\frac{a^2}{64} + \frac{b - a^2}{64}} + \frac{a}{8}}$, and $Q = \sqrt[3]{\sqrt{\frac{a^2}{64} + \frac{b - a^2}{64}} - \frac{a}{8}}$.

Which is a rule that can be of little or no ufe; for, in the firft place, $\overline{P - Q^2} + \sqrt[3]{b - a^2}$ is the fame as $\overline{P + Q^2}$; and P or $\sqrt[3]{\sqrt{\frac{a^2}{64} + \frac{b - a^2}{64}} + \frac{a}{8}}$ is = $\sqrt[3]{\sqrt{\frac{b}{64}} + \frac{a}{8}} = \sqrt[3]{\frac{\sqrt{b} + a}{8}} = \frac{1}{2}\sqrt[3]{\sqrt{b} + a}$; therefore the whole $P - Q + \sqrt{\overline{P - Q^2} + \sqrt[3]{b - a^2}}$ reduces to $P - Q + P + Q = 2P = 2 \times \frac{1}{2}\sqrt[3]{\sqrt{b} + a} = \sqrt[3]{\sqrt{b} + a}$, the original quantity firft propofed. The next thing remarkable in this 1ft book, is his method for the fquare roots of negative quantities, and his rule for the cube roots of fuch imaginary binomials as arife from the irreducible cafe in cubic equations. His words, tranflated, are thefe: "I have found another fort of cubic root, very different from the former, which arifes from the cafe of the cube equal to the firft power and a number, when the cube of the $\frac{1}{3}$d part of the (coef of the) 1ft power, is greater than the fquare of half the abfolute number, which fort of fquare root hath in its algorifm, names and operations different from the others; for in that cafe, the excefs cannot be called either plus or minus ; I therefore call it *plus of minus* when it is to be added, and *minus of minus* when it is to be fubtracted." He then gives a fet of rules for the figns when fuch roots are multiplied, and illuftrates them by a great many examples. His rule for the cube roots of fuch binomials, viz. fuch as $a + \sqrt{-b}$, is this: Firft find $\sqrt[3]{a^2 + b}$; then, by trials fearch out a number c, and a fq. root $\sqrt{d}$, fuch, that the fum of their fquares $c^2 + d$ may be = $\sqrt[3]{a^2 + b}$

and

and alſo $c^3 - 3cd = a$; then ſhall $c + \sqrt{-d}$ be $= \sqrt[3]{a + \sqrt{-b}}$ ſought. Thus, to extract the cube root of $2 + \sqrt{-121}$: here $\sqrt[3]{a^2 + b} = \sqrt[3]{125} = 5$; then taking $c = 2$, and $d = 1$, it is $c^2 + d = 5 = \sqrt[3]{a^2 + b}$, and $c^3 - 3cd = 8 - 6 = 2 = a$, as it ought; and therefore $2 + \sqrt{-1}$ is $=$ the cube root of $2 + \sqrt{-121}$, as required.

The notation in this book, is the initial R for root, with q or c &c after it, for quadrate or cubic, &c root. Alſo p for *plus*, and m for *minus*.

In the 2d book, Bombelli treats of the algoriſm with unknown quantities, and the reſolution of equations. He firſt gives the definitions and characters of the unknown quantity and its powers, in which he deviates from the former authors, but profeſſes to imitate Diophantus. He calls the unknown quantity *tanto*, and marks it thus   -   - $\underset{\smile}{1}$,
Its ſquare or 2d power *potenza*, $\underset{\smile}{2}$,
Its cube   -   - *cubo*, $\underset{\smile}{3}$,
and the higher names are compounded of theſe, and marked $\underset{\smile}{4}$, $\underset{\smile}{5}$, $\underset{\smile}{6}$, $\underset{\smile}{7}$, &c, ſo that he denotes all the powers by their exponents ſet over the common character $\smile$. And all theſe powers he calls by the general name *dignita*, dignity. He then performs all the algoriſm of theſe powers, by means of their exponents, as we do at preſent, viz, adding them in multiplication, ſubtracting in diviſion, multiplying them by the index in involution, and dividing by the ſame in evolution.

In equations he goes regularly through all the caſes, and varieties of the ſigns and terms; firſt all the ſimple or ſingle powers, and then all the compound caſes; demonſtrating the rules geometrically, and illuſtrating them by many examples.

In compound quadratics, he gives two rules: the firſt is by freeing the *potenza* or ſquare from its coefficient by diviſion, and then completing the ſquare, &c, in the uſual way: and the 2d rule, when the firſt term has its coefficient, may be thus expreſſed; if $ax^2 + bx = c$,

then $x = \dfrac{\sqrt{ac + \frac{1}{4}b^2} - \frac{1}{2}b}{a}$. He takes only the poſitive root or roots; and in the caſe $ax^2 + c = bx$, which has two, he obſerves that the nature of the problem muſt ſhew which of the two is the proper one.

In the cubic equations, he gives the rules and transformations, &c, after the manner of Cardan; remarking that ſome of the caſes have only one root, but others two or three, of which ſome are true, and others falſe or negative. And in one place he ſays that by means of the caſe $x^3 = bx + c$ he *triſects* or *divides an angle into three equal parts*.

When he arrives at biquadratic equations, and particularly to this caſe $x^4 + ax - b$, he ſays, " Since I have ſeen Diophantus's work, I have always been of opinion that his chief intention was to come to this equation, becauſe I obſerve he labours at finding always ſquare numbers, and ſuch, that adding ſome number to them, may make ſquares; and I believe that the ſix books, which are loſt, may treat of this equation, &c." —" But Lewis Ferrari," he adds, " of this city, alſo laboured in this way, and found out a rule for ſuch caſes, which was a very fine invention, and therefore I

ſhall here treat of it the beſt I can." This he accordingly does, in all the caſes of biquadratics, both with reſpect to the number of terms in the equation, and the ſigns of the terms, except I think this moſt general caſe only $rx - qx^2 + px^3 - x^4 = s$; fully applying Ferrari's method in all caſes. Which concludes the 2d book.

The 3d book conſiſts only of the reſolution of near 300 practical queſtions, as exerciſes in all the rules and equations, ſome of which are taken from Diophantus and other authors.

Upon the whole it appears that this is a plain, explicit, and very orderly treatiſe on algebra, in which are very well explained the rules and methods of former writers. But Bombelli does not produce much of improvement or invention of his own, except his notation, which varies from others, and is by means of one general character, with the numeral indices of Stifelius. He alſo firſt remarks that angles are triſected by a cubic equation. But I know not how to account for his aſſertion, that Diophantus often cites the Indian authors; which I think muſt be a miſtake in Bombelli.

CLAVIUS.

Chriſtopher Clavius wrote his Algebra about the year 1580, though it was not publiſhed till 1608, at Orleans. He moſtly follows Stifelius and Scheubelius in his notation and method, &c, having ſcarcely any variations from them; nor does he treat of cubic equations. He mentions the names given to the art, and the opinions about its origin, in which he inclines to aſcribe it to Diophantus, from what Diophantus ſays in his preface to Dyoniſius.

STEVINUS.

The Arithmetic of Simon Stevin of Bruges, was publiſhed in 1585, and his Algebra a little afterwards. They were alſo printed in an edition of his works at Leyden in 1634, with ſome notes and additions of Albert Girard, who it ſeems died the year before, this edition being publiſhed for the benefit of Girrard's widow and children. The Algebra is an ingenious and original work. He denotes the *res*, or unknown quantity, in a way of his own, namely by a ſmall circle $\bigcirc$, within which he places the numeral exponent of the power, as $\textcircled{0}$, $\textcircled{1}$, $\textcircled{2}$, $\textcircled{3}$, &c, which are the 0, 1, 2, 3, &c power of the quantity $\bigcirc$; where $\textcircled{0}$, or the 0 power, is the beginning of quantity, or arithmetical unit. He alſo extends this notation to roots or fractional exponents, and even to radical ones. Thus $\textcircled{\frac{1}{2}}$, $\textcircled{\frac{1}{3}}$, $\textcircled{\frac{1}{4}}$, &c, are the ſq. root, cube root, 4th root, &c;
and $\textcircled{\frac{2}{3}}$ is the cube root of the ſquare;
and $\textcircled{\frac{3}{2}}$ is the ſq. root of the cube. And ſo of others.

The firſt three powers, $\textcircled{1}$, $\textcircled{2}$, $\textcircled{3}$, he alſo calls *coſte* (ſide), *quarre* (ſquare), *cube* (cube); and the firſt of them, $\textcircled{1}$, the prime quantity, which he obſerves is alſo *metaphorically* called the racine or root, (the mark of which is alſo $\sqrt{}$), becauſe it repreſents the root or origin from whence all other quantities ſpring or ariſe, called the *potences* or powers of it. He condemns the terms ſurſolids, and numbers abſurd, irrational, irregular, inexplicable, or ſurd, and ſhews that all numbers are denoted the ſame way, and are all equally proper expreſſions preffions of fome length or magnitude, or fome power of the fame root. He alfo rejects all the compound expreffions of fquare-fquared, cube-fquared, cube-cubed, &c, and fhews that it is beft to name them all by their exponents, as the 1ft, 2d, 3d, 4th, 5th, 6th, &c power or quantity in the feries. And on his extenfion of the new notation he juftly obferves that what was before obfcure, laborious, and tirefome, will by thefe marks be clear, eafy, and pleafant. He alfo makes the notation of algebraic quantities more general in their coefficients, including in them not only integers, as $3\textcircled{1}$, but alfo fractions and radicals, as $\frac{3}{4}\textcircled{2}$, and $\sqrt{2}\textcircled{3}$, &c. He has various other peculiarities in his notations; all fhewing an original and inventive mind. A quantity of feveral terms, *he* calls a multinomial, and alfo binomial, trinomial, &c, according to the number of the terms. He ufes the figns + and —, and fometimes : for equality; alfo × for divifion of fractions, or to multiply crofswife thus, $\frac{3}{4} \times \frac{2}{3} : \frac{1}{1}\frac{8}{2}$.

He teaches the generation of powers by means of the annexed table of numbers, which are the coefficients of all the terms except the firft and laft. And he makes ufe of the fame numbers alfo for extracting all roots whatever: both which things had

$$\begin{array}{c} 2 \\ 3 \cdot 3 \\ 4 \cdot 6 \cdot 4 \\ 5 \cdot 10 \cdot 10 \cdot 5 \\ 6 \cdot 15 \cdot 20 \cdot 15 \cdot 6 \\ \&c. \end{array}$$

firft been done by Stifelius. In extracting the roots of non-quadrate or non-cubic numbers, he has the fame approximations as at prefent, viz, either to continue the extraction indefinitely in decimals, by adding periods of ciphers, or by making a fraction of the remainder in this manner, viz, $\sqrt{N} = n + \dfrac{N - n^2}{2n + 1}$ nearly, and $\sqrt[3]{N} = n + \dfrac{N - n^3}{3n^2 + 3n + 1}$ nearly; where n is the neareft exact root of N; which is Peletarius's rule, and which differs from Tartalea's rule, as this wants the 1 in the denominator. And in like manner he goes on to the roots of higher powers.

He then treats of equations, and their inventors, which according to him are thus:

Mahomet, fon of Mofes, an Arabian, invented thefe -
$$\begin{cases} \textcircled{1} \text{ egale } à \textcircled{0}, \\ \text{its derivatives,} \\ \textcircled{2} \text{ egale } à \textcircled{1}, \textcircled{0}, \end{cases}$$

And fome unknown author, the derivatives of this.

Some unknown author invented thefe -
$$\begin{cases} \textcircled{3} \text{ egale } à \textcircled{1} \textcircled{0}, \\ \textcircled{3} \text{ egale } à \textcircled{2} \textcircled{0}, \end{cases}$$

But afterwards he mentions Ferreus, Tartalea, Cardan, &c, as being alfo concerned in the invention of them.

Lewis Ferrari invented $\textcircled{4}$ egale $à$ $\textcircled{3}$ $\textcircled{2}$ $\textcircled{1}$ $\textcircled{0}$.

He fays alfo that Diophantus once refolves the cafe $\textcircled{2}$ egale $à$ $\textcircled{1}$ $\textcircled{0}$. In his reduction of equations, which is full and mafterly, he always puts the higheft power on one fide alone, equal to all the other terms, fet in their order, on the other fide, whether they be + or —. And he demonftrates all the rules both arithmetically and geometrically. In cubics, he gives up the irreducible cafe, as hopelefs: but fays that Bombelli refolves it by *plus of minus*, and *minus of minus*; thus, if $1\textcircled{3} = 30\textcircled{1} + 36$, then $1\textcircled{1} = \sqrt[3]{18 + \text{of} - 26}$ $+ \sqrt[3]{18 - \text{of} - 26}$, that is, $1\textcircled{1} = \sqrt[3]{18 + 26\sqrt{-1}}$ $+ \sqrt[3]{18 - 26\sqrt{-1}}$. He refolves biquadratics by

means of cubics and quadratics. In quadratics, he takes both the two roots, but looks for no more than two in cubics or biquadratics. He gives alfo a general method of approaching indefinitely near; in decimals, to the root of any equation whatever: but it is very laborious, being little more than trying all numbers, one after another, finding thus the 1ft figure, then the 2d, then the 3d, &c, among thefe ten characters 0, 1, 2, 3, 4, 5, 6, 7, 8, 9. And finally he applies the rules in the refolution of a great many practical queftions.

Although a general air of originality and improvement runs through the whole of Stevinus's work, yet his more remarkable or peculiar inventions, may be reduced to thefe few following: viz,

1ft. He invented not only a new character for the unknown quantity, but greatly improved the notation of powers, by numeral indices, firft given by Stifelius as to integral exponents; which Stevinus extended to fractional and all other forts of exponents, thereby denoting all forts of roots the fame way as powers, by numeral exponents. A circumftance hitherto thought to be of much later invention.

2d. He improved and extended the ufe and notation of coefficients, including in them fractions and radicals, and all forts of numbers in general.

3d. A quantity of feveral terms, he called generally a multinomial; and he denoted all nomials whatever by particular names expreffing the number of their terms, binomial, trinomial, quadrinomial, &c.

4th. A numeral refolution of all equations whatever by one general method.

Befides which, he hints at fome unknown author as the firft inventor of the rules for cubic equations; by whom may probably be intended the author of the Arabic manufcript treatife on cubic equations, given to the library at Leyden by the celebrated Warner.

VIETA.

Moft of Vieta's algebraical works were written about or a little before, the year 1600, but fome of them were not publifhed till after his death, which happened in the year 1603. And his whole mathematical works were collected together by Francis Schooten, and elegantly printed in a folio volume in 1646. Of thefe, the algebraical parts are as follow:

1. Ifagoge in Artem Analyticam.
2. Ad Logifticen Speciofam Notæ priores.
3. Zeteticorum libri quinque.
4. De Æquationum Recognitione, & Emendatione.
5. De Numerofa Poteftatum ad Exegefin Refolutione.

Of all thefe I fhall give a very minute account, efpecially in fuch parts as contain any difcoveries, as we here meet with more improvements and inventions in the nature of equations, than in almoft any former author. And firft of the *Ifagoge* or Introduction to the Analytic Art. In this fhort introduction Vieta lays down certain præcognita in this art, as definitions, axioms, notations, common precepts or operations of addition, fubtraction, multiplication, and divifion, with rules for queftions, &c. From which we find,

1ft. That the names of his powers are latus, quadratum, cubus, quadrato-quadratum, quadrato-cubus, cubo-cubus, &c; in which he follows the method of Diophantus, and not that derived from the Arabians.

2d. That

2d. That he calls powers pure or adfected, and firſt here uſes the terms coefficient, affirmative, negative, ſpecious logiſtics or calculations, homogeneum comparationis, or the abſolute known term of an equation, homogeneum adfectionis, or the 2d or other term which makes the equation adfected, &c. 3d. That he uſes the capital letters to denote the known as well as unknown quantities, to render his rules and calculations general, namely, the vowels A, E, I, O, U, Y for the unknown quantities, and the conſonants B, C, D, &c, for the known ones. 4th. That he uſes the ſign + between two terms for addition; — for ſubtraction, placing the greater before the leſs; and when it is not known which term is the greater, he places = between them for the difference, as we now uſe ∞; thus A = B is the ſame as A ∞ B; that he expreſſes diviſion by placing the terms like a fraction, as at preſent; though he was not firſt in this. But that he uſes no characters for multiplication or equality, but writes the words themſelves, as well as the names of all the powers, as he uſes no exponents, which cauſes much trouble and prolixity in the progreſs of his work; and the numeral coefficients ſet after the literal quantities, have a diſagreeable effect.

II. *Ad Logiſticen Specioſam, Notæ Priores.* Theſe conſiſt of various theorems concerning ſums, differences, products, powers, proportionals, &c, with the geneſis of powers from binomial and reſidual roots, and certain properties of rational right-angled triangles.

III. *Zeteticorum libri quinque.* The zetetics or queſtions in theſe 5 books are chiefly from Diophantus, but reſolved more generally by literal arithmetic. And in theſe queſtions are alſo inveſtigated rules for the reſolution of quadratic and cubic equations. In theſe alſo Vieta firſt uſes a line drawn over compound quantities, as a vinculum.

IV. *De Æquationum Recognitione, & Emendatione.* Theſe two books, which contain Vieta's chief improvements in Algebra, were not publiſhed till the year 1615, by Alexander Anderſon, a learned and ingenious Scotchman, with various corrections and additions. The 1ſt of theſe two books conſiſts of 20 chapters. In the firſt ſix chapters, rules are drawn from the zetetics for the reſolution of quadratic and cubic equations. Theſe rules are by means of certain quantities in continued proportion, but in the reſolution they come to the ſame thing as Cardan's rules. In the cubics, Vieta ſometimes changes the negative roots into affirmative, as Cardan had done, but he finds only the affirmative roots. And he here refers the irreducible caſe to angular ſections for a ſolution, a method which had been mentioned by Bombelli.

Chap. 7 treats of the general method of transforming equations, which is done either by changing the root in various ways, namely by ſubſtituting another inſtead of it which is either increaſed or diminiſhed, or multiplied or divided by ſome known number, or raiſed or depreſſed in ſome known proportion; or by retaining the ſame root, and equally multiplying all the terms. Which ſorts of transformation, it is evident, are intended to make the equation become ſimpler, or more convenient for ſolution. And all or moſt of theſe reductions and transformations were alſo practiſed by Cardan.

Chap. 8 ſhews what purpoſes are anſwered by the foregoing transformations; ſuch as taking away ſome of the terms out of an equation, and particularly the 2d term, which is done by increaſing or diminiſhing the root by the coefficient of the 2d term divided by the index of the firſt: by which means alſo the affected quadratic is reduced to a ſimple one. And various other effects are produced.

Chap. 9 ſhews how to deduce compound quadratic equations from pure ones, which is done by increaſing or diminiſhing the root by a given quantity, being one application of the foregoing reductions.

Chap. 10, the reduction of cubic equations affected with the 1ſt power, to ſuch as are affected with the 2d power; by the ſame means.

In *chap.* 11, by the ſame means alſo, the 2d term is reſtored to ſuch cubic equations as want it.

In *chap.* 12, quadratic and cubic equations are raiſed to higher degrees by ſubſtituting for the root, the ſquare or cube of another root divided by a given quantity.

In *chap.* 13 affected biquadratic equations are deduced from affected quadratics in this manner, when expreſſed in the modern notation:

If $A^2 + BA = Z$, then ſhall $A^4 + \overline{B^2 + 2BZ} \cdot A = Z^2 + B^2Z$.

For ſince $A^2 + BA = Z$, therefore $A^2 = Z - BA$, and its ſquare is
$A^4 = Z^2 - 2BAZ + B^2A^2$: but $B^2A^2 = B^2Z - B^3A$, therefore
$A^4 = Z^2 - 2BAZ + B^2Z - B^3A$, or $A^4 + \overline{B^3 + 2BZ} \cdot A = Z^2 + B^2Z$.

And in like manner for the biquadratic affected with its other terms. And in a ſimilar manner alſo, in chap. 14, affected cubic equations are deduced from the affected quadratics.

In *chap.* 15 it is ſhewn that the quadratic $BA - A^2 = Z$ has two values of the root A, or has ambiguous roots, as he calls them; and alſo that the cubics, biquadratics, &c, which are raiſed or deduced from that quadratic, have alſo double roots.

Having, in the foregoing chapters, ſhewn how the coefficients of equations of the 3d and 4th degree are formed from thoſe of the 2d degree, of the ſame root; and that certain quadratics, and others raiſed from them, have double roots; then in the 16th chap. Vieta ſhews what relation thoſe two roots bear to the coefficients of the two loweſt terms of an equation conſiſting of only three terms. Thus,

$$\text{If } BA - A^2 = Z \left.\right\} \text{ then } B = \frac{A^2 - E^2}{A - E} = A + E,$$
$$\text{and } BE - E^2 = Z \left.\right\} \text{ and } Z = \frac{A^2E - AE^2}{A - E} = AE.$$

$$\text{If } BA - A^3 = Z \left.\right\} \text{ then } B = \frac{A^3 - E^3}{A - E} = A^2 + AE + E^2,$$
$$\text{and } BE - E^3 = Z, \left.\right\} \text{ and } Z = \frac{A^3E - AE^3}{A - E} = A^2E + AE^2.$$

$$\text{If } A^m + BA^n = Z, \left.\right\} \text{ then } B = \frac{E^m - A^m}{E^n + A^n}, \& Z = \frac{A^mE^n + A^nE^m}{E^n + A^n}.$$
$$\text{and } E^m - BE^n = Z, \left.\right\}$$

And ſo on for the ſame terms with their ſigns variouſly changed.

Chap.

Chap. 17 contains several theorems concerning quantities in continued geometrical progression. Which are preparatory to what follows, concerning the double roots of equations, the nature of which he expounds by means of such properties of proportional quantities.

Chap. 18, *Æquationum ancipitum constitutiva* ; treating of the nature of the double roots of equations. Thus, if a, b, c, d, &c. be quantities in continual progression ; then, 1st, of equations affected with the first power,

If $BA - A^2 = Z$; then $B = a+b$, $Z=ab$, & $A=a$ or b.

If $BA - A^3 = Z$; then $B = a^2 + b^2 + c^2$, $Z = a(b^2+c^2)$, & $A = a$ or c.

And in general, if $BA - A^{n+1}$; then $B = a^n + b^n + c^n \ldots k^n$, $Z = a (b^n + c^n + d^n \ldots k^n)$, and $A = a$ or k the first or last term. Where the number of terms a^n, b^n, &c, in B is $n + 1$, and the number of terms in Z is n.

2d, For equations containing only the highest two powers.

If $BA^2 - A^3 = Z$; then $B=a+b+c$, $Z=a(b+c)^2$, or $= c(a+b)^2$, and $A=a+b$, or $b+c$.

If $BA^3 - A^4 = Z$; then $B=a+b+c+d$, $Z=a(b+c+d)^3$ or $=d(a+b+c)^3$, and $A=a+b+c$ or $=b+c+d$.

And, in general, if $BA^n - A^{n+1}=Z$;

then $B=a+b+c \ldots i+k$,

$Z=a(b+c \ldots k)$ or $=k (a+b+ c \ldots i)$,

and $A=a + b + c \ldots i$ or $=b+c+d \ldots k$, the sum of all except the last, or sum of all except the first; where the number of terms in B is $n+1$, and the number of terms in Z is n.

3d. Of equations affected by the intermediate powers.

If $BA^2 - A^4 = Z$; then $B=a^2+b^2$, $Z=a^2b^2$, & $A^2=a^2$ or b^2.

If $BA^3 - A^6 = Z$; then $B=a^3+b^3$, $Z=a^3b^3$, & $A^3=a^3$ or b^3.

If $BA^4 - A^6 = Z$; then $B=a^2+b^2 + c^2$, $Z = a(b^2 + c^2)$, and $A^2 = a + b$ or $b+c$.

4th. Of the remaining cases.

If $BA^2 - A^5 = Z$;

then $B=(a+b)^3 + (b+c)^3 + a(b+c)^2$ or $+ c(a + b)^2$,

and $Z=\overline{B-(a+b)^3} . (a+b)^2$

or $=(b+c)^3 + c(a+b)^2 \cdot (a+b)^2$,

and $A=a + b$ or $b+c$.

If $BA^3 - A^5 = Z$;

then $B=(a+b+c)^2 + (b+c+d)^2 - c(a+b+ c)$.

or $- b(b+c+d$,

and $Z = \overline{B-(b+c+d)^2} . (b+c+d)^3$

or $=\overline{B-(a+b+c)^2} \cdot (a+b+c)^3$;

and $A=a+b+c$ or $b+c+d$.

Chap. 19. *Æqualitatum contradicentium constitutiva.* Of the relation of equations of like terms, but the sign of one term different ; containing these 5 theorems, viz,

1. If $A^2+BA=Z$,} then $B = b-a$, $Z =ab$;
and$E^2 - BE=Z$;{ and $A =a$, $E = b$.

2. If $A^4+BA=Z$, then $B = (b + d)^3 - (a + c)^3$, $Z=a(d^3 - b^3)$ or $=d(c^3 - a^3)$; and$E^4 - BE=Z$; and $A = a$, $E = d$.

3. If $A^6+BA=Z$, then$B=b^5 +d^5 +f^5 - a^5 - c^5 - e^5$, $Z = a(b^5 + d^5 + f^5 - c^5 - e^5)$ or $=f(a^5 + c^5 + e^5 - b^5 - d^5)$; and$E^6 - BE=Z$; and $A = a$, $E =f$.

4. If $A^4+BA^3=Z$, then $B = b + d - a - c$, $Z = a(d-b)^3$ or $= d(c-a)^3$; and$E^4 - BE^3=Z$; and $A = c-a$, $E = d-b$.

5. If $A^6+BA^5=Z$, then $B = b + d + f - a - c - e$, $Z=a(b+d+f - c - e)^5$ or $= f(a + c + e - b - d)^5$; and$E^6 - BE^5=Z$; and $A = a+c+e - b - d$, $E=b+ d + f - c - e$.

Chap. 20. *Æqualitatum inversarum constitutiva.* Containing these six theorems, viz,

1. If $BA - A^3=Z$, then $B=c^2 - a^2$, $Z=a (c^2 - b^2)$ or $=c(b^2 - a^2)$; and $E^3 - BE=Z$; and $A = a$, $E = c$.

2. If $BA - A^5=Z$, then $B=a^4 + c^4 + e^4 - b^4 - d^4$, $Z=a(c^4 + e^4 - b^4 - d^4)$ or $=e(b^4+d^4 - a^4 - c^4)$; and $E^5 - BE=Z$; and $A = a$, $E=e$.

3. If $BA^2+A^3=Z$, then $B=c-a$, $Z = a(c-b)^2$ or $= c(b-a)^2$; and $BE^2 - E^3=Z$; and $A = b-a$, $E=c-b$.

4. If $BA^4+A^5=Z$, then $B=a+c + e - b - d$, $Z=a(c+e - b - d)^4$ or $=e(b + d - a - c)^4$; and $A=b+d - a - c$, $E = c+e - b - d$. and$BE^4 - E^5=Z$;

5. If $BA^3+A^5=Z$, then $B=(a+f) . (d-a)$, $Z=(d - a)^3 B + (d-a)^5$ or $= (e - b)^3 B - (e-b)^5$; and$BE^3 - E^5=Z$; and$A=d-a$, $E = e-b$.

6. If $BA^2+A^5=Z$, then $B=(a + f) . (c - a)^2$, $Z=\overline{B + (c-a)^3} \cdot (c-a)^2$ or$=\overline{B - (d-b)^3} \cdot (d-b)^2$ and$BE^2 - E^5=Z$; and$A=c-a$, $E = d-b$.

Chap. 21. *Alia rursus æqualitatum inversarum constitutiva.* In these two theorems :

1. If $B^zA - A^3=Z$, then $B=a^2+b^2 + c^2$, $Z=(a+c)b^2$ or $=a(b^2+c^2)$ or $= c(a^2+b^2)$; and$E^3 - B^zA=Z$; and $A = a$ or c, $E=a+c$.

2. If $BA^2 - A^3=Z$, then $B=a+b+c$, $Z=(a+b+c)b^2$ or $= a(b + c)^2$ or $=c(a+b)^2$; and$BE^2+E^3=Z$; and $A=a+b$ or $b+c$, $E = b$.

Next follows the 2d of the pieces published by Alexander Anderson, namely,

De Emendatione Æquationum, in 14 chapters.

Chap. 1. Of preparing equations for their resolution in numbers, by taking away the 2d term; by which affected quadraties are reduced to pure ones, and cubic equations affected with the 2d term are reduced to such as are affected with the 3d only. Several examples of both sorts of equations are given. He here too remarks upon the method of taking away any other term out of an equation, when the highest power is combined with that other term only; and this Vieta effects by means of the coefficients, or, as he calls them, the unciæ of the power of a binomial. All which was also performed by Cardan for the same purpose.

Chap. 2. *De transmutatione* Πρῶτον—ἱκατον, *quæ remedium est adversus vitium negationis.* Concerning the transformations by changing the given root A for another root E, which is equal to the homogeneum

compara-

comparationis divided by the firſt root A; by which means negative terms are changed to affirmative, and radicals are taken out of the equation when they are contained in the homogeneum comparationis.

Chap. 3, *De Anaſtrophe,* ſhewing the relation betwĕen the roots of correlate equations; from whence, having given the root of the one equation, that of the other becomes known; and it conſiſts of theſe following 8 theorems, moſtly deduced from the laſt 4 chapters of the foregoing *recognitio æquationum.*

1. If $BA - A^3 = Z$, and $E^3 - BE = Z$; $\}$ then $EA - A^2 = E^2 - B$.

2. If $BA^2 - A^3 = Z$, and $BE^2 + E^3 = Z$, $\}$ then $(E + B)A - A^2 = BD + D^2$.

3. If $BA - A^5 = Z$, and $E^5 - BE = Z$; $\}$ then $E^3 A - E^2 A^2 + EA^3 - A^4 = E^4 - B$.

4. If $BA^4 - A^5 = Z$, and $BE^4 + E^5 = Z$; $\}$ then $(BE^2 + E^3)A - (BE + E^2)A^2 + (B + E)A^3 - A^4 = BE^3 + E^4$.

5. If $A^3 - BA = Z$, and $BE - E^3 = Z$; $\}$ then $A^2 - EA = B - E^2$.

6. If $BA^2 + A^3 = Z$, and $BE^2 - E^3 = Z$, $\}$ then $A^2 + (B - E)A = BE - E^2$.

7. If $BA - A^3 = Z$, and $BE - E^3 = Z$; $\}$ then $A^2 + EA = B - E^2$.

8. If $BA^2 - A^3 = Z$, and $BE^2 - E^3 = Z$; $\}$ then $A^2 + (E - B)A = BE - E^2$.

Chap. 4, *De Iſomæria, adverſus vitium fractionis.* To take away fractions out of an equation. Thus,

if $A^3 + \frac{B}{D} A = Z$. Put $A = \frac{E}{D}$; then $E^3 + BDE = ZD^3$.

Chap. 5, *De Symmetrica Climactiſmo adverſus vitium aſymmetriæ.* To take away radicals or ſurds out of equations, by ſquaring &c the other ſide of the equation.

Chap. 6. To reduce biquadratic equations by means of cubics and quadratics, by methods which are ſmall variations from thoſe of Ferrari and Cardan.

Chap. 7. The reſolution of cubic equations by rules which are the ſame with Cardan's.

Chap. 8. *De Canonica æquationum tranſmutatione, ut coefficientes ſubgraduales ſint quæ præſcribuntur.* To tranſmute the equation ſo that the coefficient of the lower term, or power, may be any given number, he changes the root in the given proportion, thus: Let A be the root of the equation given, E that of the tranſmuted equation, B the given coefficient, and X the required one; then take $A = \frac{BE}{X}$, which ſubſti-tute in the given equation, and it is done.—He commonly changes it ſo, that X may be 1; which he does, that the numeral root of the equation may be the eaſier found; and this he here performs by trials, by taking the neareſt root of the higheſt power alone; and if that does not turn out to be the root of the whole equation, he concludes that it has no rational root.

Chap. 9. To reduce certain peculiar forms of cubics to quadratics, or to ſimpler forms, much the ſame as Cardan had done. Thus,

1. If $A^3 - 2B^2A = B^3$; then is $A^2 - BA = B^2$.
2. If $2B^2A - A^3 = B^3$; then is $A^2 + BA = B^2$.
3. If $A^3 - 3B^2A = 2B^3$; then is $A = 2B$.
4. If $3B^2A - A^3 = 2B^3$; then is $A = B$.
5. If $A^3 - BA^2 + DA = BD$; then is $A = B$.
6. If $A^3 + BA^2 - D^2A = BD^2$; then is $A = D$.
7. If $BA^2 + D^2A - A^3 = BD^2$; then is $A = B$ or $= D$.
8. If $D^2A + BDA - A^3 = B^3$; then is $(D + B)A - A^2 = B^2$
9. If $A^3 - DA^2 + BDA = B^3$; then is $(D - B)A - A^2 = B^2$
10. If $A^3 - 3BA = B\sqrt{2}B$; then is $A = \sqrt{\frac{3}{2}}B - \sqrt{\frac{1}{2}}B$.
11. If $3BA - A^3 = B\sqrt{2}B$; then is $A = \sqrt{\frac{1}{2}}B - \sqrt{\frac{1}{2}}B$.

Chap. 10. *Similium reductionum continuatio.* Being ſome more ſimilar theorems, when the equation is affected with all the powers of the unknown quantity A.

Chap. 11, 12, 13 relate alſo to certain peculiar forms of equations, in which the root is one of the terms of a certain ſeries of continued proportionals.

Chap. 14, which is the laſt in this tract, contains, in four theorems, the general relation between the roots of an equation and the coefficients of its terms, when all its roots are poſitive. Namely,

1. If $\overline{B + D} \cdot A - A^2 = BD$; then is $A = B$ or D.

2. If $A^3 \overline{- B - D - G} \cdot \overline{A^2 + BD + BG + DG} \cdot A = BDG$; then is $A = B$ or D or G.

3. If $\overline{BDG + BDH + BGH + DGH} \cdot A - \overline{BD - BG - BH - DG - DH - GH} . A^2 + \overline{B + D + G + H} \cdot A^3 - A^4 = BDGH$; then is $A = B$ or D or G or H.

4. If $A^5 \overline{- B - D - G - H - K} . A^4 + \overline{BD + BG + BH + BK + DG + DH + DK + GH + GK + HK} . A^3 \overline{- BDG - BDH - BDK - BGH - BGK - BHK - DGH - DGK - DHK - GHK} \cdot A^2 + \overline{BDGH + BDGK + BDHK + BGHK + DGHK} \cdot A = BDGHK$; then is $A = B$ or D or G or H or K.

And from theſe laſt 4 theorems it appears that Vieta was acquainted with the compoſition of theſe equations, that is, when all their roots are poſitive, for he never adverts to negative roots; and from other parts of the work it appears that he was not aware that the ſame properties will obtain in all ſorts of roots whatever. But it is not certain in what manner he obtained theſe theorems, as he has not given any account of the inveſtigations, though that was uſually his way on other occaſions; but he here contents himſelf with barely announcing the theorems as above, and for this ſtrange reaſon, that he might at length bring his work to a concluſion.

To this piece is added, by Alexander Anderſon, an Appendix, containing the conſtruction of the cubic equations by the triſection of an angle, and a demonſtration of the property referred to by Vieta for this purpoſe.

De Numeroſa Poteſtatum Purarum Reſolutione. Vieta here gives ſome examples of extracting the roots of pure powers, in the way that had been long before practiſed, by pointing the number into periods of figures according to the index of the root to be extracted, and then proceeding from one period to another, in the uſual way.

De

De Numerosâ Potestatum adfectarum Resolutione. And here, in close imitation of the above method for the roots of pure powers, Vieta extracts those of adfected ones; or finding the roots of affected equations, placing always the homogeneum comparationis, or absolute term, on one side, and all the terms affected with the unknown quantity, and their proper signs, on the other side. The method is very laborious, and is but little more than what was before done by Stevinus on this subject, depending not a little upon trials. The examples he uses are such as have either one or two roots, and indeed such as are affected commonly with only two powers of the unknown quantity, and which therefore admit only of those two varieties as to the number of roots, namely according as the higher of the two powers is affirmative or negative, the homogeneum comparationis, on the other side of the equation, being always affirmative; and he remarks this general rule, if the higher power be negative, the equation has two roots; otherwise, only one; that is, affirmative roots; for as to negative and imaginary ones, Vieta knew nothing about them, or at least he takes no notice of them. By the foregoing extraction, Vieta finds both the greater and less root of the two that are contained in the equation, and either of them that he pleases; having first, for this purpose, laid down some observations concerning the limits within which the two roots are contained. Also, having found one of the roots, he shews how the other root may be found by means of another equation, which is a degree lower than the given one; though not by depressing the given equation, by dividing it as is now done; but from the nature of proportionals, and the theorems relating to equations, as given in the former tracts, he finds the terms of another equation, different from that last mentioned, from the root &c of which, the 2d root of the original equation may be obtained.

In the course of this work, Vieta makes also some observations on equations that are ambiguous, or have three roots; namely, that the equation $1C - 6Q + 11N = 6$, or as we write it $x^3 - 6x^2 + 11x = 6$ is ambiguous, when the 2d term is negative, and the 3d term affirmative, and when $\frac{1}{3}$ of the square of 6 the coefficient of the 2d term, exceeds 11, the coefficient of the 3d term, and has then three roots. Or in general, if $x^3 - ax^3 + bx = c$, and $\frac{1}{3}a^2 > b$, the equation is ambiguous, and has three roots. He shews also, from the relation of the coefficients, how to find whether the roots are in arithmetical progression or not, and how far the middle root differs from the extremes, by means of a cubic equation of this form $x^3 - ba = c$. In all or most of which remarks he was preceded by Cardan.—Vieta also remarks that the case $x^3 - 9x^2 + 24x = 20$, has three roots by the same rule, viz, 2, 2, 5, but that two of them are equal. And farther, that when $\frac{1}{3}a^2$ is $= b$, then all the three roots are equal, as in the case $x^3 - 6x^2 + 12x = 8$, the three roots of which are 2, 2, 2. But when $\frac{1}{3}a^2$ is less than b, the case is not ambiguous, having but one root. And when $ab = c$, then $a = x$ is one root itself.

Many curious notes are added at the end, with remarks on the method of finding the approximate roots, when they are not rational, which is done in two ways, in imitation of the same thing in the extraction of pure powers, viz, the one by forming a fraction of the remainder after all the figures of the homogeneum comparationis are exhausted; the other by increasing the root of the equation in a 10 fold, or 100 fold, &c, proportion, and then dividing the root which results by 10, or 100, &c: and this is a decimal approximation. And Vieta observes that the roots will be increased 10 or 100 fold, &c, by adding the corresponding number of ciphers to the coefficient of the 2d term, double that number to the 3d, triple the same number to the 4th, and so on. So if the equation were

$$1C + 4Q + 6N = 8,$$

then $1C + 40Q + 600N = 8000$ will have its root 10 fold, and $1C + 400Q + 60000N = 8000000$ will have it 100 fold.

Besides the foregoing algebraical works, Vieta gave various constructions of equations by means of circles and right lines, and angular sections, which may be considered as an algebraical tract, or a method of exhibiting the roots of certain equations having all their roots affirmative, and by means of which he resolved the celebrated equation of 45 powers, proposed to all the world by Adrianus Romanus.

Having now delivered a particular analysis of Vieta's algebraical writings, it will be proper, as with other authors, to collect into one view the particulars of his more remarkable peculiarities, inventions, and improvements.

And first it may be observed, that his writings shew great originality of genius and invention, and that he made alterations and improvements in most parts of algebra; though in other parts and respects his method is inferior to some of his predecessors; as, for instance, where he neglects to avail himself of the negative roots of Cardan; the numeral exponents of Stifelius, instead of which he uses the names of the powers themselves; or the fractional exponents of Stevinus; or the commodious way of prefixing the coefficient before the quantity or factor; and such like circumstances; the want of which gives his Algebra the appearance of an age much earlier than its own. But his real inventions of things before not known, may be reduced to the following particulars.

1st. Vieta introduced the general use of the letters of the alphabet to denote indefinite given quantities; which had only been done on some particular occasions before his time. But the general use of letters for the unknown quantities was before pretty common with Stifelius and his successors. Vieta uses the vowels A, E, I, O, U, Y for the unknown quantities, and the consonants B, C, D, &c, for known ones.

2d. He invented, and introduced many expressions or terms, several of which are in use to this day: such as coefficient, affirmative and negative, pure and adfected or affected, unciæ, homogeneum adfectionis, homogeneum comparationis, the line or vinculum over compound quantities thus $\overline{A + B}$. And his method of setting down his equations, is to place the homogeneum comparationis, or absolute known term, on the right-hand side alone, and on the other side all the terms which contain the unknown quantity, with their proper signs.

3d. In most of the rules and reductions for cubic and other

other equations, he made some improvements, and variations in the modes.

4th. He shewed how to change the root of an equation in a given proportion.

5. He derived or raised the cubic and biquadratic, &c equations, from quadratics; but not by composition in Harriot's way, but by squaring and otherwise multiplying certain parts of the quadratic. And as some quadratic equations have two roots, therefore the cubics and others raised from them, have also the same two roots, and no more. And hence he comes to know what relation these two roots bear to the coefficients of the two lowest terms of cubic and other equations, when they have only 3 terms, namely, by comparing them with similar equations so raised from quadratics. And, on the contrary, what the roots are, in terms of such coefficients.

6. He made some observations on the limits of the two roots of certain equations.

7. He stated the general relation between the roots of certain equations and the coefficients of its terms, when the terms are alternately plus and minus, and none of them are wanting, or the roots all positive.

8. He extracted the roots of affected equations, by a method of approximation similar to that for pure powers.

9. He gave the construction of certain equations, and exhibited their roots by means of angular sections; before adverted to by Bombelli.

OF ALBERT GIRARD.

Albert Girard was an ingenious Dutch or Flemish mathematician, who died about the year 1633. He published an edition of Stevinus's Arithmetic in 1625, augmented with many notes; and the year-after his death was published by his widow, an edition of the whole works of Stevinus, in the same manner, which Girard had left ready for the press. But the work which entitles him to a particular notice in this history, is his " *Invention Nouvelle en l'Algebre, tant pour la solution des equations, que pour recognoistre le nombre des solutions qu'elles reçoivent, avec plusieurs choses qui sont necessaires a la perfection de ceste divine science;*" which was printed at Amsterdam 1629, in small quarto in 63 pages, viz, 49 pages on Arithmetic and Algebra, and the rest on the measure of the superficies of spherical triangles and polygons, by him then lately discovered.

In this work Girard first premises a short tract on Arithmetic; in the notation of which he has something peculiar, viz, dividing the numbers into the ranks of millions, billions, trillions, &c.

He next delivers the common rules of Algebra, both in integers, fractions and radicals; with the notation of the quantities and signs. In this part he uses sometimes the letters A, B, C, &c, after the manner of Vieta, but more commonly the characters of Stevinus, viz, ⓪, ①, ②, ③, &c, for the powers of the unknown quantity, with their roots ⓵⁄₂, ⓵⁄₂, ⓵⁄₄, ⓵⁄₅, ⓵⁄₅, &c, used by Stevinus; and sometimes the more usual marks of the roots as, √ or ³√, ³√, ⁴√, &c; prefixing the coefficients, as 6②, or 3³√32, or 2④. In the signs he follows his predecessors so far as to have + for plus, — or ÷ for minus, = for general or indefinite difference, A + B for the sum, A — B or A = B for

the difference, AB the product, and $\frac{A}{B}$ for the quotient of A and B. He uses the parentheses ( ) for the vinculum or bond of compound quantities, as is now commonly practised on the continent; as A(AB+Bq), or ³√(A cub. — 3AqB); and he introduces the new characters ff for *greater than*, and § for *less than*; but he uses no character for equality, only the word itself.

Girard gives a new rule for extracting the cube root of binomials, which however is in a good measure tentative, and which he explains thus: To extract the cube root of $72 + \sqrt{5120}$.

The squares of the terms $\begin{cases} 5184 \\ 5120 \end{cases}$

their difference 64, and its cube root 4. Which shews that the difference between the squares of the terms required is 4; and the rational part 72 being the greater, the greater term of the root will be rational also; and farther, that the greater terms of the power and root are commensurable, as also the two less terms. Then having made a table as in the margin, where the square of the rational term always exceeds that of the other, by the number 4 above mentioned, one of these binomials must be the cubic root sought, if the given quantity have such a root, and it must

$2 + \sqrt{0}$
$3 + \sqrt{5}$
$4 + \sqrt{12}$
$5 + \sqrt{21}$

be one of these four forms, for it is known to be carried far enough by observing that the cube root of 72 is less than 5, and the cube root of 5120 less than 21; indeed, this being the case, the last binomial is excluded, as evidently too great; and the first is excluded because one of its terms is 0; therefore the root must be either $3 + \sqrt{5}$ or $4 + \sqrt{12}$. And to know whether of these two it must be, try which of them has its two terms exact divisors of the corresponding terms of the given quantity; then it is found that 3 and 4 are both divisors of 72, but that only 5, and not 12, is a divisor of 5120; therefore $3 + \sqrt{5}$ is the root sought, which upon trial is found to answer. It is remarkable here that Girard uses $4 + \sqrt{20}$ instead of $4 + \sqrt{12}$, and $5 + \sqrt{29}$ instead of $5 + \sqrt{20}$, contrary to his own rule.

Girard then gives distinct and plain rules for bringing questions to equations, and for the reduction of those equations to their simplest form, for solution, by the usual modes, and also by the way called by Vieta *Isomeria*, multiplying the terms of the equation by the terms of a geometrical progression, by which means the roots are altered in the proportion of 1 to the ratio of the progression. He then treats of the methods of finding the roots of the several sorts of equations, quadratic, cubic, &c; and adds remarks on the proper number of conditions or equations for limiting questions. The quadratics are resolved by completing the square, and both the positive and negative roots are taken; and he observes that sometimes the equation is impossible, as 2) equ. 6①) — 25, whose roots, he adds, are $3 + \sqrt{-16}$ and $3 - \sqrt{-16}$.

The cubic equations he resolves by Cardan's rule, except the irreducible case, which he the first of any resolves by a table of sines; the other cases also he resolves by tables of sines and tangents; and adds geometrical constructions by means of the hyperbola or the trifection of angles. He next adds a particular mode of refolving all forts of equations, that have rational roots, upon the principle of the roots being divifors of the laft or abfolute term, as before mentioned by Pele-tarius; and then gives the method of approximating to other roots that are not rational, much the fame way as Stevinus.

Having found one root of an equation, by any of the former methods, by means of it he depreffes the equation one degree lower, then finds another root, and fo on till they are all found; for he fhews that every algebraic equation admits of as many folutions or roots, as there are units in the index of the higheft power, which roots may be either pofitive or negative, or imaginary, or, as he calls them, greater than nothing, or lefs than nothing, or involved; fo the roots of the equation 1③ equ. $7 \text{①} - 6$, are 2, 1, and -3; and the roots of the equation 1④ equ. $4 \text{①} - 3$ are

$$1,$$
$$1,$$
$$-1 + \sqrt{-2},$$
$$-1 - \sqrt{-2}.$$

In depreffing an equation to lower degrees, he does not ufe the method of refolution of Harriot, but that which is derived from the general relation of the roots and coefficients of the terms, which he here fully and univerfally ftates, viz, that the coefficient of the 2d term is equal to the fum of all the roots; that of the 3d term equal to the fum of all the products of the roots, taken two by two; that of the 4th term, the fum of the products, taken three by three; and fo on, to the laft or abfolute term, which is the continual product of all the roots; a property which was before ftated by Vieta, as to the equations that have all their roots pofi-tive; and here extended by Girard to all forts of roots whatever: but how either Vieta or he came by this pro-perty, no where appears that I know of. From this general property, among other deductions, Girard fhews how to find the fums of the powers of the roots of an equation; thus, let A, B, C, D, &c, be the 1ft, 2d, 3d, 4th, &c, coefficient, after the firft term, or the fums of the products taken one by one, two by two, three by three, &c; then, in all forts of equations,

$$\left. \begin{array}{l} A \\ Aq - 2B \\ A \text{ cub.} - 3AB + 3C \\ Aqq - 4AqB + 4AC + 2Bq - 4D \end{array} \right\} \text{will be the fum of the} \left\{ \begin{array}{l} \text{roots,} \\ \text{fquares,} \\ \text{cubes,} \\ \text{biquadrates.} \end{array} \right.$$

Girard next explains the ufe of negative roots in Geometry, fhewing that they reprefent lines only drawn in a direction contrary to thofe reprefenting the pofitive roots; and he remarks that this is a thing hitherto unknown. He then terminates the Algebra by fome queftions having two or more unknown quan-tities; and fubjoins to the whole a tract on the men-furation of the furfaces of fpherical triangles and poly-gons, by him lately difcovered.

From the foregoing account it appears that,

1ft, He was the firft perfon who underftood the general doctrine of the formation of the coefficients of the powers, from the fums of their roots, and their products, &c.

2d, He was the firft who underftood the ufe of ne-gative roots in the folution of geometrical problems.

3d, He was the firft who fpoke of the imaginary roots, and underftood that every equation might have as many roots real and imaginary, and no more, as there are

units in the index of the higheft power. And he was the firft who gave the whimfical name of *quantities lefs than nothing* to the negative. And,

4th, He was the firft who difcovered the rules for fumming the powers of the roots of any equation.

OF HARRIOT.

Thomas Harriot, a celebrated aftronomer, philofo-pher, and mathematician, flourifhed about the year 1610, about which time it is probable he wrote his Algebra, as he was then, and had been for many years before, celebrated for his mathematical and aftronomical la-bours. In that year he made obfervations on the fpots in the fun, and on Jupiter's fatellites, the fame year alfo in which Galileo firft obferved them: he left many other curious aftronomical obfervations, and amongft them, fome on the remarkable comets of the years 1607 and 1618. His Algebra was left behind him unpublifhed, as well as thofe other papers, at his death, which hap-pened in the year 1621, being then 60 years of age, and but fix years after the firft publication of the principal parts of Vieta's Algebra by Alexander Anderfon; fo that it is probable that Harriot's Algebra was written before this time, and indeed that he had never feen thefe pieces. Harriot's Algebra was publifhed by his friend Walter Warner, in the year 1631: and it would doubt-lefs be highly grateful to the learned in thefe fciences, if his other curious algebraical and aftronnomical works were publifhed from his original papers in the poffeffion of the Earl of Egremont, to whom they have defcended from Henry Percy, the Earl of Northumberland, that noble Mæcenas of his day. The book is in folio, and intitled *Artis Analyticæ Praxis, ad Æquationes Alg-braicas nova, expedita, & generali methodo, refolvendas;* a work in all parts of it fhewing marks of great genius and originality, and is the firft inftance of the modern form of Algebra in which it has ever fince appeared. It is prefaced by 18 definitions, which are thefe: 1ft, Logiftica Speciofa; 2d, Equation; 3d, Synthefis; 4, Analyfis; 5, Compofition and Refolution; 6, Form-ing an Equation; 7, Reduction of an Equation; 8, Ve-rification; 9, Numerofa & Speciofa; 10, Excogitata; 11, Refolution; 12, Roots; 13 and 14, The kinds and generation of equations by multiplication, from binomial roots or factors, called original equations,

$$\begin{array}{l} \text{as } a + b \mid = aa + ba \\ \quad a - c \mid \quad - ca - bc, \end{array}$$

$$\begin{array}{l} \text{or } a + b \mid = aaa + baa + bca \\ \quad a + c \mid \quad + caa - bda \\ \quad a - d \mid \quad - daa - cda - bcd, \end{array}$$

where he puts a for the unknown quantity, and the fmall confonants, b, c, d, &c, for its literal values or roots; 15, The firft form of canonical equations, which are derived from the above originals, by tranfpofing the homogeneum, or abfolute term,

$$\begin{array}{l} \text{thus } aa + ba \\ \quad - ca = + bc, \text{ &c}; \end{array}$$

16, The fecondary canonicals, formed from the pri-mary by expelling the 2d term,

$$\begin{array}{l} \text{thus } aa = + bb, \\ \text{or } aaa - bba \\ \quad - bca \\ \quad - cca = + bbc \\ \quad \quad + bcc; \end{array}$$

17, That thefe are called canonicals, becaufe they are adapted to canons or rules for finding the numeral roots, &c. 18, Reciprocal equations, in which the homogeneum is the product of the coefficients of the other terms, and the firft term, or higheft power of the root, is equal to the product of the powers in the other terms, as $aaa - caa + bba = + bbc$.

After thefe definitions, the work is divided into two principal parts ; 1ft, of various generations, reductions, and preparations of equations for their refolution in the 2d part. The former is divided into 6 fections as follows.

Sect. 1. *Logiftices Speciofæ*, exemplified in the 4 operations of addition, fubtraction, multiplication, and divifion ; as alfo the reduction of algebraic fractions, and the ordinary reduction of irregular equations to their form proper for the refolution of them, namely, fo that all the unknown terms be on one fide of the equation, and the known term on the other, the powers in the terms ranged in order, the greateft firft, and the firft or higheft power made pofitive, and freed from its coefficient ; as $aa + ba = cd$,
or $aaa + baa - cda = - c^2d$.
In this part he explains fome unufual characters which he introduces, namely

$=$ for equality, as $a = b$.
$>$ for majority, as $a > b$,
$<$ for minority, as $a < b$;

but the firft had been before introduced by Robert Recorde.

Sect. 2. The generation of original equations from binomial factors or roots, and the deducing of canonicals from the originals. He fuppofes that every equation has as many roots as dimenfions in its higheft power ; then fuppofing the values of the unknown letter a in any equation to be b, c, d, f, &c, that is $a=b$, and $a = c$, and $a = d$, &c ; by tranfpofition, or equal fubtraction, thefe become $a - b = 0$, and $a - c = 0$, and $a - d = 0$, &c, or the fame letters with contrary figns, for negative values or roots ; then two of thefe binomial factors multiplied together, gives a quadratic equation, three of them a cubic, four of them a biquadratic, and fo on, with all the terms on one fide of the equation, and o on the other fide, fince, every binomial factor being $= 0$, the continual product of all of them muft alfo be $= 0$. Thus,

$$\begin{array}{c|l} a + b & = aaa + baa + bca \\ a + c & + caa - bda \\ a - d & - daa - cda - bcd = 0 \end{array}$$

an original equation,
and
$$\begin{array}{l} aaa + baa + bca \\ + caa - bda \\ - daa - cda = + bcd \end{array}$$

its canonical, deduced from it. And thefe operations are carried through all the cafes of the 2d, 3d and 4th powers, as to the varieties of the figns + and —, and the proportions of the roots as to equal and unequal, with the reciprocals, &c. From which are made evident, at one glance of the eye, all the relations and properties between the roots of equations, and the coefficients of the terms.

Sect. 3. *Æquationum canonicarum fecundariarum a primariis reductio per gradus alicujus parodici fublationem radice fuppofititia invariata monente.* Containing a great many examples of preparing equations by taking away the 2d, 3d, or any other of the intermediate terms, which is done by making the pofitive coefficients in that term, equal to the negative ones, by which means the whole term vanifhes, or becomes equal to nothing

They are extended as far as equations of the 5th degree ; and at the end are collected, and placed in regular order, all the fecondary canonicals, fo reduced, fo that by the uniform law which is vifible through them all, the feries may be continued to the higher degrees as far as we pleafe.

Sect. 4. *Æquationum canonicarum tam primariarum, quam fecundariarum, radicum defignatio.* A great many literal equations are here fet down, and their roots affigned from the form of the equation, that is all their pofitive roots ; for their negative roots are not noticed here ; and it is every where proved that they cannot have any more pofitive roots than thefe, and confequently the reft are negative. That thofe are roots, he proves by fubftituting them inftead of the unknown letter a in the equation, when they make all the terms on one fide come to the fame thing as the homogeneum on the other fide.

Sect. 5, *In qua æquationum communium per canonicarum æquipollentiam, radicum numerus determinatur.* On the number of the roots of common equations, that is the pofitive roots. This Harriot determines by comparing them with the like cafes found among his canonical forms, which two equations, having the fame number of terms with the fame figns, and the relations of the coefficients and homogeneum correfpondent, he calls equipollents. And whatever was the number of pofitive roots ufed in the compofition of the canonical, the fame, he infers, is the number in the propofed common equation. It is remarkable that in all the examples here ufed, the number of pofitive roots is juft equal to the number of the changes in the figns from + to — and from — to +, which is a circumftance, though not here exprefsly mentioned, that could not efcape the obfervation, or the eye, of any one, much lefs of fo clear and comprehenfive a fight as that of Harriot. In this fection are contained many ingenious difquifitions concerning the limits and magnitudes of quantities, with feveral curious lemmas laid down to demonftrate the propofitions by, which lemmas are themfelves demonftrated in a pure mathematical way, from the magnitudes themfelves, independent of geometrical figures ; fuch as, 1, If a quantity be divided into any two unequal parts, the fquare of half the line will be greater than the product of the two unequal parts. 2, In three continued proportionals, the fum of the extremes is greater than double the mean. 3, In four continued proportionals, the fum of the extremes is greater than the fum of the two means. 4, In any two quantities, one-fourth the fquare of the fum of the cubes, is greater than the cube of the product of the two quantities. 5, Of any two quantities q and r, then $\frac{1}{27}(qq + qr + rr)^3 > \frac{1}{4}(qqr + qrr)^2$. 6, If any quantity be divided into three unequal parts, the fquare of $\frac{1}{3}$ of the whole quantity is greater than $\frac{1}{3}$ of the fum of the three products made of the three unequal parts. 7, Alfo the cube of the $\frac{1}{3}$ part of the whole, is greater than the folid or continual product of the three unequal parts.

Sect. 6. *Æquationum communium reductio per gradus alicujus parodici exclufionem & radicis fuppofititiæ mutationem.* Here are a great many examples of reducing and transforming equations of the 2d, 3d, and 4th degrees; chiefly either by multiplying the roots of equations in any proportion, as was done by Vieta, or increafing or diminifhing the root by a given quantity, after the manner of Cardan. The former of thefe reductions is performed by multiplying the terms of the equation by the correfponding terms of a geometrical progreffion, the 1ft term being 1, and the 2d term the quantity by which the root is to be multiplied. And the other reduction, or transforming to another root, which may be greater or lefs than the given root by a given quantity, is performed commonly by fubftituting $e +$ or $- b$ for the given root a, by which the equation is reduced to a fimpler form. Other modes of fubftitution are alfo ufed; one of which is this, viz, fubftituting $\dfrac{ee \pm bb}{e}$ or $e \pm \dfrac{bb}{e}$ for the root a in the given equation $aaa \mp 3.bba = 2.ccc$, by which it reduces to this quadratic form $e^6 \mp 2c^3 e^3 = - b^6$, from whence Cardan's forms are immediately deduced; namely $e = \surd 3) c^3 - \surd c^6 \pm b^6$, and therefore a or $e \pm \dfrac{bb}{e} = \surd 3) c^3 + \surd c^6 \pm b^6 \pm \surd 3) c^3 - \surd c^6 \pm b^6$; where he denotes the cube or 3d root thus $\surd 3)$, but without any vinculum over the compound quantities.

In this fection, Harriot makes various remarks as they occur: thus he remarks, and demonftrates, that $eee - 3.bbe = -ccc - 2.bbb$ is an impoffible equation, or has no affirmative root. He remarks alfo that the three cafes of the equation $aaa - 3.bba = + 2.ccc$ are fimilar to the three conic fections; namely to the hyperbola when $c > b$, to the parabola when $c = b$, or to the ellipfis when $c < b$, and for which reafon this cafe is not generally refoluble in fpecies.

Having thus fhewn how to fimplify equations, and prepare them for folution, Harriot enters next upon the fecond part of his work, being the

Exegetice Numerofa,

or the numeral refolution of all forts of equations by a general method, which is exemplified in a great number of equations, both fimple and affected as far as the 5th power inclufive; and they are commonly prepared, by the foregoing parts, by freeing them from their 2d term, &c. Thefe extractions are explained and performed in a way different from that of Vieta; and the examples are firft in perfect or terminate roots, and afterwards for irrational or interminate ones, to which Harriot approximates by adding always periods of ciphers to the given number or refolvend, as far as neceffary in decimals, which are continued and fet down as fuch. but with their proper denominator 10, or 100, or 1000, &c.

He then concludes the work with

Canones Directorii,

which form a collection of the cafes or theorems for making the foregoing numeral extractions, ready arranged for ufe, under the various forms of equations, with the factors neceffary to form the feveral refolvends and fubtrahends.

And from a review of the whole work, it appears that Harriot's inventions, peculiarities, and improvements in algebra, may be comprehended in the following particulars.

1ft. He introduced the uniform ufe of the fmall letters a, b, c, d, &c, viz, the vowels a, e, &c for unknown quantities, and the confonants b, c, d, f, &c for the known ones; which he joins together like the letters of a word, to reprefent the multiplication or product of any number of thefe literal quantities, and prefixing the numeral coefficient as we do at prefent, except only feparated by a point, thus $5.bbc$. For a root he fet the index of the root after the mark $\surd$; as $\surd 3)$ for the cube root. He alfo introduced the characters $>$ and $<$ for greater and lefs; and in the reduction of equations, he arranged the operations in feparate fteps or lines, fetting the explanations in the margin on the left hand, for each line. By which, and other means, he may be confidered as the introducer of the modern ftate of Algebra, which quite changed its form under his hands.

2d. He fhewed the univerfal generation of all the compound or affected equations, by the continual multiplication of fo many fimple ones, or binomial roots; thereby plainly exhibiting to the eye the whole circumftances of the nature, myftery and number of the roots of equations; with the compofition and relations of the coefficients of the terms; and from which many of the moft important properties have fince been deduced.

3d. He greatly improved the numeral exegefis, or extraction of the roots of all equations, by clear and explicit rules and methods, drawn from the foregoing generation or compofition of affected equations of all degrees.

OF OUGHTRED'S CLAVIS.

Oughtred was contemporary with Harriot, but lived a long time after him. His Clavis was firft publifhed in 1631, the fame year in which Harriot's Algebra was publifhed by his friend Warner. In this work, Oughtred chiefly follows Vieta, in the notation by the capitals A, B, C, D, &c, in the defignation of products, powers, and roots, though with fome few variations. His work may be comprehended under the following particulars.

1. *Notation.* This extends to both Algebra and Arithmetic, vulgar and decimal. The Algebra chiefly after the manner of Vieta, as abovefaid. And he feparates the decimals from the integers thus, $21|56$, which is the firft time I have obferved fuch a feparation, and the decimals fet down without their denominator.

2. The common rules or operations of Arithmetic and Algebra. In algebraic multiplication, he either joins the letters together like a word, or connects them by the mark $\times$, which is the firft introduction of this character of multiplication: thus $A \times A$ or AA or Aq. But omitting the vinculum over compound factors, ufed by Vieta. He introduces here many neat and ufeful contractions in multiplication and divifion of decimals: as that common one of inverting the multiplier, to have fewer decimals, and abridge the work; that of omitting always one figure at a time, of the divifor, for the fame purpofe; dividing by the component factors of a number inftead of the number itfelf; as 4 and 6 for 24; and many other neat contractions. He

ftates

states his proportions thus 7.9 : : 28.36, and denotes continued proportion thus ÷ ; which is the first time I have observed these characters.

3. Invents and describes various symbolical marks or abbreviations, which are not now used.

4. *The genesis and analysis of powers.* Denotes powers like Vieta, and also roots, thus $\sqrt{q6}$, $\sqrt{c20}$, $\sqrt{qq24}$, &c ; and much in his manner too performs the numeral extraction of roots. He here gives a table of the powers of the binomial A + E as far as the 10th power, with all their terms and coefficients, or unciæ as he calls them, after Vieta.

5. *Equations.* He here gives express and particular directions for the several sorts of reductions, according as the form of the equation may require. And he uses the letter *u* after $\sqrt{}$, for universal, instead of the vinculum of Vieta. And observes that the signs of all the terms of the powers of A + E are positive, but those of A — E are alternately positive and negative.

6. Next follow many properties of triangles and other geometrical figures; and the first instance of applying Algebra to Geometry, so as to investigate new geometrical properties; and after the algebraical resolution of each problem, he commonly deduces and gives a geometrical construction adapted to it. He gives also a good tract on angular sections.

7. The work concludes with the numeral resolution of affected equations, in which he follows the manner of Vieta, but he is more explicit.

OF DESCARTES.

Descartes's Geometry was first published in 1637, being six years after the publication of Harriot's Algebra. That work was rather an application of Algebra to Geometry, than the science either of Algebra or Geometry itself, purely and properly so called. And yet he made improvements in both. We must observe however, that all the properties of equations, &c, which he sets down, are not to be considered as even meant by himself for new inventions or discoveries; but as statements and enumerations of properties, before known and taught by other authors, which he is about to make some use or application of, and for which reason it is that he mentions those properties.

Descartes's Geometry consists of three books. The first of these is, *De Problematibus, quæ construi possunt, adhibendo tantum rectas lineas & circulos.* He here accommodates or performs arithmetical operations by Geometry, supposing some line to represent unity, and then, by means of proportionals, shewing how to multiply, divide, and extract roots by lines. He next describes the notation he uses, but not because it is a new one, for it is the same as had been used by former authors, viz, $a + b$ for the addition of a and b, also $a - b$ for their subtraction, ab multiplication, $\frac{a}{b}$ division, aa or a^2 the square of a, a^3 its cube, &c: also $\sqrt{a^2 + b^2}$ for the square root of $a^2 + b^2$, and $\sqrt{C.a^3 - b^3 + abb}$ for the cube root, &c. He then observes, after Stifelius, that there must be as many equations as there are unknown lines or quantities; and that they must be reduced all to one final equation, by exterminating all the unknown letters except one; when the final equation will appear like these,

$z \infty b$, or

$z^2 \infty - az + b^2$, or

$z^3 \infty + az^2 + b^2z - c^3$, or

$z^4 \infty + az^3 + b^2z^2 - c^3z + d^4$, &c.

Where he uses ∞ for $=$ or equality, setting the highest term or power alone on one side of the equation, and all the other terms on the other side, with their proper signs.

Descartes next defines plane problems, namely, such as can be resolved by right lines and circles, described on a plane superficies; and then the final equation rises only to the 2d power of the unknown letter. He then constructs such equations, viz. quadratics, by the circle, thus finding geometrically the root or roots, that is, the positive ones. But when the lines, by which the roots are determined, neither cut nor touch, he observes that the equation has then no possible root, or that the problem is impossible. He then concludes this book with the algebraical solution of the celebrated problem, before treated of by the antients, namely, to find a point, or the locus of all the points, from whence a line being drawn to meet any number of given lines in given angles, the product of the segments of some of them shall have a given ratio to that of the rest.

Lib. 2. *De Natura Linearum Curvarum.* This is a good algebraical treatise on curve lines in general, and the first of the kind that has been produced by the moderns. Here the nature of the curve is expressed by an equation containing two unknown or variable lines, and others that are known or constant, as $y^2 \infty cy - \frac{cxy}{b}$ $+ ay - ac$. But, not relating to pure Algebra, the particulars will be most properly placed under the article of curve lines, and other terms relating to them. Only one discovery, among many ingenious applications of Algebra to Geometry, may here be particularly noticed, as it may be considered as the first step towards the arithmetic of infinites; and that is the method of tangents, here given, or, which comes to the same thing, of drawing a line perpendicular to a curve at any point, which is an ingenious application of the general form of an equation, generated in Harriot's way, that has two equal roots, to the equation of the curve. Of which a particular account will be given at the article TANGENTS.

Lib. 3. *De Constructione Problematum Solidorum, et Solida excedentium.* Descartes begins this book with remarks on the nature and roots of equations, observing that they have as many roots as dimensions, which he shews, after Harriot, by multiplying a certain number of simple binomial equations together, as $x - 2 \infty 0$, and $x - 3 \infty 0$, and $x - 4 \infty 0$, producing $x^3 - 9xx + 26x - 24 \infty 0$. He here remarks that equations may sometimes have their roots *false*, or what we call negative, which he opposes to those that are positive, or as he calls them *true*, as Cardan had done before. As a natural deduction from the generation or composition of equations, by multiplication, he infers their resolution, or depression, or decomposition, namely, dividing them by the binomial factors which were multiplied to produce the equation: and he observes that by this operation it is known that this divisor is one of the binomial roots, and that there can be no more roots than dimensions, or than those which form with the unknown letter

x, bino-

x, binomials that will exactly divide the equation, as Harriot had shewn before. Descartes adverts to several other properties, mostly known before, which he has occasion to make use of in the progress of his work; such as, that equations may have as many true roots as the terms have changes of the signs + and —, and as many false ones as successions of the same signs: which number and nature of the roots had before been partly shewn by Cardan and Vieta, from the relation of the coefficients, and their signs, and more fully by Harriot in his 5th section. And hence Descartes infers the method of changing the true roots to false, and the false to true, namely by changing the signs of the even terms only, as Cardan had taught before. Descartes then adverts to other reductions and transmutations which had been taught by Cardan, Vieta, and Harriot, such as, To increase or diminish the roots by any quantity; To take away the 2d term : To alter the roots in any proportion, and thence to free the equation from fractions and radicals.

Descartes next remarks that the roots of equations, whether true or false, may be either real or imaginary; as in the equation $x^3 — 6xx + 13x$ $10 \infty 0$, which has only one real root, namely 2. The imaginary roots were first noticed by Albert Girard, as before mentioned. He then treats of the depression of a cubic equation to a quadratic, or plane problem, that it may be constructed by the circle, by dividing it by some one of the binomial factors, which, in Harriot's way, compose the equation. Peletarius having shewn that the simple root is one of the divisors of the known term of the equation, and Harriot that that term is the continual product of all the roots, Descartes therefore tries all the simple divisors of that term, till he finds one of them which, connected with the unknown letter *x*, by + or —, will exactly divide the equation. And the process is the same for higher powers than the cube. But when a divisor cannot be thus found, for depressing a biquadratic equation to a cubic, he gives another rule, which is a new one, for dissolving it into two quadratics, by means of a cubic equation, in this manner :

Let the given biqu. be $+ x^4 * . pxx . qx . r \infty 0$;

Which suppose equal to the product of these two
$$+ xx —yx + \tfrac{1}{2}yy . \tfrac{1}{2}p . . \frac{q}{2y} \infty 0,$$
$$+ xx +yx + \tfrac{1}{2}yy . \tfrac{1}{2}p . \frac{q}{2y} \infty 0;$$

where the sign of $\tfrac{1}{2}p$ in the two quadratics must be the same as the sign of p in the given equation, and in the 1st quadratic the sign of $\frac{q}{2y}$ must be the same as the the sign of q, but in the 2d quadratic the contrary. Then if there be found the root yy of this cubic equation $y^6 . 2py^4 + \genfrac{}{}{0pt}{}{+}{.} \frac{pp}{4r} yy — qq \infty 0,$

where the sign of $2p$ is the same as of p in the given biquadratic, but the sign of $4r$ contrary to that of r in the same : Then the value of y, hence deduced, being substituted for it in the two quadratic equations, and their two pairs of roots taken, they will be the four roots of the proposed biquadratic. And thus also, he hints, may equations of the 6th power be reduced to those of the 5th, and those of the 8th power to those of the 7th,

and so on. Descartes does not give the investigation of this rule; but it has evidently been done, by assuming indeterminate quantities, after the manner of Ferrari and Cardan, as coefficients of the terms of the two quadratic equations, and, after multiplying the two together, determining their values by comparing the resulting terms with those of the proposed biquadratic equation.

After these reductions, which are only mentioned for the sake of the geometrical constructions which follow, by simplifying and depressing the equations as much as they will admit, Descartes then gives the construction of solid and other higher problems, or of cubic and higher equations, by means of parabolas and circles; where he observes that the false roots are denoted by the ordinates to the parabola lying on the contrary side of the axis to the true roots. Finally, these constructions are illustrated by various problems concerning the trisecting of an angle, and the finding of two or four mean proportionals; which concludes this ingenious work.

From the foregoing analysis may easily be collected the real inventions and improvements made in algebra by Descartes. His work, as has been observed before, is not algebra itself, but the application of algebra to geometry, and the algebraical doctrine of curve lines, expressing and explaining their nature by algebraical equations, and on the contrary, constructing and explaining equations by means of the curve lines. What respects the geometrical parts of this tract we shall have occasion to advert to elsewhere; and therefore shall here only enumerate the circumstances which belong more peculiarly to the science of Algebra, which I shall distinguish into the two heads of improvements and inventions. And

1st. Of his improvements. That he might fit equations the better for their application in the construction of problems, Descartes mentions, as it were by-the-bye, many things concerning the nature and reduction of equations, without troubling himself about the first inventors of them, stating them in his own terms and manner, which is commonly more clear and explicit, and often with improvements of his own. And under this head we find that he chiefly followed Cardan, Vieta, and Harriot, but especially the last, and explains some of their rules and discoveries more distinctly, and varies but a little in the notation, putting the first letters of the alphabet for the known, and the latter letters for the unknown quantities; also x^3 for *aaa*, &c; and ∞ for $=$. But Herigone used the numeral exponents in the same manner two years before. Descartes explained or improved most parts of the reductions of equations, in their various transmutations, the number and nature of their roots, true and false, real and what he calls imaginary, called involved by Girard; and the depression of equations to lower degrees.

2d. As to his inventions and discoveries in algebra, they may be comprehended in these particulars, namely, the application of algebra to the geometry of curve lines, the constructing equations of the higher orders, and a rule for resolving biquadratic equations by means of a cubic and two quadratics.

Having now traced the science of Algebra from its origin and rude state, down to its modern and more polished form, in which it has ever since continued, with very little variation; having analysed all or most

of

of the principal authors, in a chronological order, and deduced the inventions and improvements made by each of them; from this time the authors both become too numerous, and their improvements too inconsiderable, to merit a detail in the same minute and circumstantial way: and besides, these will be better explained in a particular manner under the word or article to which each of them severally belongs. It may therefore now suffice to enumerate, or announce only in a cursory manner, the chief improvements and authors on algebra down to the present time.

After the publication of the Geometry of Descartes, a great many other ingenious men followed the same course, applying themselves to algebra and the new geometry, to the mutual improvement of them both; which was done chiefly by reasoning on the nature and forms of equations, as generated and composed by Harriot. Before proceeding upon these however, it is but proper to take notice here of Fermat, a learned and ingenious mathematician, who was contemporary and a competitor of Descartes for his brightest discoveries, which he was in possession of before the geometry of Descartes appeared. Namely, the application of algebra to curve lines, which he expressed by an algebraical equation, and by them constructing equations of the 3d and 4th orders; also a method of tangents, and a method de maximis et minimis, which approach very near to the method of Fluxions or Increments, which they strikingly resemble both in the manner of treating the problems, and in the algebraic notation and process. The particulars of which, see under their proper heads. Besides these, Fermat was deeply learned in the Diophantine problems, and the best edition of Diophantus's Arithmetic, is that which contains the notes of Fermat on that ingenious work.

But to return to the successors of Descartes. His geometry having been published in Holland, several learned and ingenious mathematicians of that country, presently applied themselves to cultivate and improve it; as Schooten, Hudde, Van-Heuraet, De Witte, Slusius, Huygens, &c; besides M. de Beaune, and perhaps some others in France.

Francis Schooten, professor of mathematics in the university of Leyden, was one of the first cultivators of the new geometry. He translated Descartes's Geometry out of French into Latin, and published it in 1649, with his commentary upon it, as also Brief Notes of M. de Beaune; both of them containing many ingenious and useful things. And in 1659 he gave a new edition of the same in two volumes, with the addition of several other ingenious pieces: as two posthumous tracts of de Beaune, the one on the nature and constitution, the other on the limits of equations, shewing how to assign the limits between which are contained the greatest and least roots of equations, extended and completed by Erasmus Bartholine: two letters of M. Hudde on the reduction of equations, and on the maxima and minima of quantities, containing many ingenious rules; among which are some concerning the drawing of tangents, and on the equal roots of equations, which he determines by multiplying the terms of the equation by the terms of any arithmetical progression, o being one of the terms, the equation is commonly depressed one degree lower: also a tract of Van Heuraet on the rectifi-

cation of curve lines; the elements of curves by De Witte; Schooten's principles of universal mathematics, or introduction to Descartes's geometry, which had before been published by itself in 1651; and to the end of the work is added a posthumous piece of Schooten's (for he died while the 2d vol. was printing) intitled *Tractatus de concinnandis demonstrationibus geometricis ex calculo algebraico*. Schooten also published, in 1657, *Exercitationes Mathematicæ*, in which are contained many curious algebraical and analytical pieces, amongst others of a geometrical nature.

An elaborate commentary on Descartes's Geometry was also published by F. Rabuel, a Jesuit; and James Bernoulli, enriched with notes, an edition of the same, printed at Basil in 169—.

The celebrated Huygens also, among his great discoveries, very much cultivated the algebraical analysis: and he is often cited by Schooten, who relates divers inventions of his, while he was his pupil.

Slusius, a canon of Liege, published in 1659, *Mesolabum, seu duæ mediæ propor. per circulum & ellips. vel hyperb. infinitis modis exhibitæ*; by which, any solid problem may be constructed by infinite different ways. And in 1668 he gave a second edition of the same, with the addition of the analysis, and a miscellaneous collection of curious and important problems, relating to spirals, centres of gravity, maxima and minima, points of inflexion, and some Diophantine problems; all shewing him deeply skilled in Algebra and Geometry.

There have been a great number of other writers and improvers of Algebra, of which it may suffice slightly to mention the chief part, as in the following catalogue.

Peter Nonius, or Nunez, a Spaniard, wrote about the time of Cardan, or soon after.

In 1619 several pieces of Van Collen, or Ceulen, were translated out of Dutch into Latin, and published at Leyden by W. Snell; among which are contained a particular treatise on surds, and his proportion of the circumference of a circle, to its diameter.

In 1621 Bachet published, in Greek and Latin, an edition of Diophantus, with many notes. And another edition of the same was published in 1670, with additions by Fermat.

In 1624 Bachet's *Problemes Plaisans et Delectables*, being curious problems in mathematical recreations.

In 1634 Herigone published, at Paris, the first course of mathematics, in 5 vols. 8vo; in the 2d of which is contained a good treatise on Algebra; in which he uses the notation by small letters, introduced by the Algebra of Harriot, which was published three years before, though the rest of it does not resemble that work, and one would suspect that Herigone had not seen it. The whole of this piece bears evident marks of originality and ingenuity. Besides + for *plus*, he uses ∽ for *minus*, and | for *equality*, with several other useful abbreviations and marks of his own. In the notation of powers, he does not repeat the letters like Harriot, but subjoins the numeral exponents, to the letter, as Descartes did two years afterwards. And Herigone uses the same numeral exponents for roots, as $\sqrt{3}$ for the cube root.

In 1635 Cavalerius published his *Indivisibles*; which proved a new æra in analytics, and gave rise to other new modes of computation in analytics.

About

About 1640, et feq. Roberval made feveral notable improvements in analytics, which are publifhed in the early volumes of the Memoirs of the Academy of Sciences; as, 1. A tract on the compofition of motion, and a method of tangents. 2, *De recognitione æquationum.* 3, *De geometrica planarum & cubicarum æquationum refolutione.* 4, A treatife on indivifibles, &c.

In 1643 De Billy publifhed *Nova Geometriæ Clavis Algebra.* And in 1670 *Diophantus Redivivus.* He was an author particularly well fkilled in Diophantine problems.

In 1644 Renaldine publifhed, in 4to, *Opus Algebraicum,* both ancient and modern, with mathematical refolution and compofition. And in 1665, in folio, the fame, greatly enlarged, or rather a new work, which is ery heavy and tedious. In this work Renaldine ufes the parenthefes $(a^2 + b^2)$ as a vinculum, inftead of the line over, as $\overline{a^2 + b^2}$.

In 1655 was publifhed Wallis's *Arithmetica Infinitorum,* being a new method of reafoning on quantities, or a great improvement on the Indivifibles of Cavalerius, and which in a great meafure led the way to infinite feries, the binomial theorem, and the method of fluxions. Wallis here treats ingenioufly of quadratures and many other problems, and gives the firft expreffion for the quadrature of the circle by an infinite feries. Another feries is here added for the fame purpofe, by the Lord Brouncker.

In 1659 was publifhed *Algebra Rhonii Germanice;* which was in 1668, tranflated into Englifh by Mr. Thomas Brancker, with additions and alterations by Dr. John Pell.

In 1661 was publifhed in Dutch, a neat piece of Algebra by Mr. Kinckhuyfen; which Sir I. Newton, while he was profeffor of mathematics at Cambridge, made ufe of and improved, and he meant to republifh it, with the addition of his method of fluxions and infinite feries; but he was prevented by the accidental burning of fome of his papers.

In 1665 or 1666 Sir Ifaac Newton made feveral of his brighteft difcoveries, though they were not publifhed till afterwards: fuch as the binomial theorem; the method of fluxions and infinite feries; the quadrature, rectification, &c of curves; to find the roots of all forts of equations, both numeral and literal, in infinite converging feries; the reverfion of feries, &c. Of each of which a particular account may be feen in their proper articles.

In 1666 M. Frenicle gave feveral curious tracts concerning combinations, magic fquares, triangular numbers, &c; which were printed in the early volumes of Memoirs of the Academy of Sciences.

In 1668 Thomas Brancker publifhed a tranflation of Rhonius's Algebra, with many additions by Dr. John Pell, who ufed a peculiar method of regiftering the fteps in any algebraical procefs, by means of marks and abbreviations in a fmall column drawn down the margin, by which each line, or ftep, is clearly explained, as was before done by Harriot in words at length.

In 1668 Mercator publifhed his *Logarithmotechnia,* or method of conftructing logarithms; in which he gives the quadrature of the hyperbola, by means of an infinite feries of algebraical terms, found by dividing a fimple algebraic quantity by a compound one, and for the firft time that this operation was given to the public, though Newton had before that expanded all forts of compound algebraical quantities into infinite feries.

In the fame year was publifhed James Gregory's *Exercitationes Geometricæ,* containing, among other things, a demonftration of Mercator's quadrature of the hyperbola, by the fame feries.

And in the fame year was publifhed, in the Philofophical Tranfactions, Lord Brouncker's quadrature of the hyperbola by another infinite feries of fimple rational terms, which he had been in poffeffion of fince the year 1657, when it was announced to the public by Dr. Wallis. Lord Brouncker's feries for the quadrature of the circle, had been publifhed by Wallis in his Arithmetic of Infinites.

In 1669 Dr. Ifaac Barrow publifhed his Optical and Geometrical Lectures, abounding with profound refearches on the dimenfions and properties of curve lines; but particularly to be noticed here for his method of tangents, by a mode of calculation fimilar to that of Fluxions, or Increments, from which thefe differ but little, except in the notation.

In 1673 was publifhed, in 2 vols. folio, *Elements of Algebra,* by John Kerfey; a very ample and complete work, in which Diophantus's problems are fully explained.

In 1675 were publifhed Nouveaux Elemens des Mathematiques, par J. Preftet, prétre: a prolix and tedious work, which he prefumptuoufly dedicated to God Almighty.

About 1677 Leibnitz difcovered his *Methodus Differentialis,* or elfe made a variation in Newton's Fluxions, or an extenfion of Barrow's method, for it is not certain which. He gave the firft inftance of it in the Leipfic Acts for the year 1684. He alfo improved infinite feries, and gave a fimple one for the quadrature of the circle, in the fame acts for 1682.

In 1682 Ifmael Bulliald publifhed, in folio, his *Opus Novum ad Arithmeticam Infinitorum,* being a large amplification of Wallis's Arithmetic of Infinites.

In 1683 Tfchirnaufen gave a memoir, in the Leipfic Acts, concerning the extraction of the roots of all equations in a general way; in which he promifed too much, as the method did not fucceed.

In 1684 came out, in Englifh and Latin, 4to, Thomas Baker's *Geometrical Key, or Gate of Equations Unlock'd;* being an improvement of Defcartes's conftruction of all equations under the 5th degree, by means of a circle and only one and the fame parabola for all equations, ufing any diameter inftead of the axis of the parabola.

In 1685 was publifhed, in folio, Wallis's *Treatife of Algebra, both Hiftorical and Practical,* with the addition of feveral other pieces; fhewing the origin, progrefs, and advancement of that fcience, from time to time. It cannot be denied that, in this work, Wallis has fhewn too much partiality to the Algebra of Harriot. Yet, on the other hand, it is as true, that M. de Gua, in his account of it, in the Memoirs of the Academy of Sciences for 1741, has run at leaft as far into the fame extreme on the contrary fide, with refpect to the difcoveries of Vieta; and both thefe I believe from the fame caufe, namely, the want of examining the works of all former writers on Algebra, and fpecifying their feveral difcoveries; as has been done in the courfe of this article.

In

In 1687 Dr. Halley gave, in the Philof. Tranf. the conftruction of cubic and biquadratic equations, by a parabola and circle; with improvements on what had been done by Defcartes, Baker, &c. Alfo, in the fame Tranfaction, a memoir on the number of the roots of equations, with their limits and figns.

In 1690 was publifhed, in 4to, by M. Rolle, *Traité d'Algébre;* in 1699 *Une Methode pour la Refolution des Problemes indeterminés;* and in 1704 *Memoires fur l'inverfe des tangens;* and other pieces.

In 1690 Jofeph Raphfon publifhed *Analyfis Æquationum Univerfalis;* being a general method of approximating to the roots of equations in numbers. And in 1715 he publifhed the *Hiftory of Fluxions,* both in Englifh and Latin.

In 1690 was alfo publifhed, in 4 vols 4to, Dechale's *Curfus feu mundus mathematicus;* in which is a piece of algebra, of a very old-fafhioned fort, confidering the time when it was written.

About 1692, and at different times afterwards, De Lagny publifhed many pieces on the refolution of equations in numbers, with many theorems and rules for that purpofe.

In 1693 was publifhed, in a neat little volume, *Synopfis Algebraica, opus pofthumum Johannis Alexandri.*

In 1694, Dr. Halley gave, in the Philof. Tranf. an ingenious tract on the numeral extraction of all roots, without any previous reduction. And this tract is alfo added to fome editions of Newton's Univerfal Arithmetic.

In 1695 Mr. John Ward, of Chefter, publifhed, in 8vo, *A Compendium of Algebra,* containing plain, eafy, and concife rules, with examples in an eafy and clear way. And in 1706 he publifhed the firft edition of his *Young Mathematician's Guide,* or a plain and eafy introduction to the mathematics: a book which is ftill in great requeft efpecially with beginners, and which has been ever fince the ordinary introduction of the greateft part of the mathematicians of this country.

In 1696 the Marquis de l'Hôpital publifhed his *Analyfe des infiniment petits.* And gave feveral papers to the Leipfic Acts and the Memoires of the Academy of Sciences. He left behind him alfo an ingenious treatife, which was publifhed in 1707, intitled *Traité analytique des Sections Coniques, et de la conftruction des lieux geometriques.*

In 1697, and feveral other years, Mr. Ab. Demoivre gave various papers, in the Philof. Tranf. containing improvements in Algebra: viz. in 1697, A method of raifing an infinite multinomial to any power, or extracting any root of the fame. In 1698, The extraction of the root of an infinite equation. In 1707, Analytical folution of certain equations of the 3d, 5th, 7th, &c degree. In 1722, Of algebraic fractions and recurring feries. In 1738, The reduction of radicals into more fimple forms. Alfo in 1730, he publifhed *Mifcellanea analytica de feriebus & quadraturis,* containing great improvements in feries, &c.

In 169 , Mr. Richard Sault publifhed, in 4to, *A New Treatife of Algebra, apply'd to numeral queftions and geometry. With a converging feries for all manner of Adfected Equations.* The feries here alluded to, is Mr. Raphfon's method of approximation, which had been lately publifhed.

In 1699 Hyac. Chriftopher publifhed at Naples, in 4to, *De conftructione æquationum.*

In 1702 was publifhed Ozanam's Algebra; which is chiefly remarkable for the Diophantine analyfis. He had publifhed his mathematical dictionary in 1691, and in 1693 his courfe of mathematics, in 5 vols 8vo, containing alfo a piece of algebra.

In 1704, Dr. John Harris publifhed his *Lexicon Technicum,* the firft dictionary of arts and fciences: a very plain and ufeful book, efpecially in the mathematical articles. And in 1705 a neat little piece on algebra and fluxions.

In 1705 M. Guifnée publifhed, in 4to, his *Application de l'algebre a la geometrie:* a ufeful book.

In 1706 Mr. William Jones publifhed his *Synopfis Palmariorum Mathefeos,* or a new introduction to the mathematics: a very ufeful compendium in the mathematical fciences. And in 1711 he publifhed, in 4to, a collection of Sir Ifaac Newton's papers, intitled *Analyfis per quantitatum feries, fluxiones, ac differentias: cum enumeratione linearum tertii ordinis.*

In 1707 was publifhed by Mr. Whifton, the firft edition of Sir Ifaac Newton's *Arithmetica Univerfalis: five de compofitione et refolutione arithmetica liber:* and many editions have been publifhed fince. This work was the text book ufed by our great author in his lectures, while he was profeffor of mathematics in the univerfity of Cambridge. And although it was never intended for publication, it contains many and great improvements in analytics; particularly in the nature and tranfmutation of equations; the limits of the roots of equations; the number of impoffible roots; the invention of divifors, both furd and rational; the refolution of problems, arithmetical and geometrical; the linear conftruction of equations; approximating to the roots of all equations, &c. To the later editions of the book is commonly fubjoined Dr. Halley's method of finding the roots of equations. As the principal parts of this work are not adapted to the circumftance of beginners, there have been publifhed commentaries upon it by feveral perfons, as s'Gravefande, Caftilion, Wilder, &c.

In 1708 M. Reyneau publifhed his *Analyfe Demontrée,* in 2 vols 4to. And in 1714 *La Science du Calcul, &c.*

In 1709 was publifhed an Englifh tranflation of Alexander's algebra. With an ingenious appendix by Humphry Ditton.

In 1715 Dr. Brooke Taylor publifhed his *Methodus Incrementorum:* an ingenious and learned work. And in the Philof. Tranf. for 1718, An improvement of the method of approximating to the roots of equations in numbers.

In 1717 M. Nicole gave, in the memoirs of the academy of fciences, a tract on the calculation of finite differences. And in feveral following years, he gave various other tracts on the fame fubject, and on the refolution of equations of the 3d degree, and particularly on the irreducible cafe in cubic equations.

Alfo in 1717 was publifhed a treatife on Algebra by Philip Ronayne.

Alfo in 1717 Mr. James Sterling publifhed *Lineæ tertii Ordinis;* an ingenious work, containing good improvements in analytics. Alfo in 1730 *Methodus Differentialis: five tractatus de fummatione et interpolatione ferierum infinitarum:* with great improvements on infinite feries.

In

In 1726 and 1729 Maclaurin gave, in the Philof. Tranf. tracts on the imaginary roots of equations. And afterwards was publifhed, from his pofthumous papers, his treatife on Algebra, with its application to curve lines.

In 1727 came out s'Gravefande's Algebra, with a fpecimen of a commentary on Newton's univerfal arithmetic.

In 1728 Mr. Campbell gave, in the Philof. Tranf. an ingenious paper on the number of impoffible roots of equations.

In 1732 was publifhed Wolfius's Algebra, in his courfe of mathematics, in 5 vols. 4to.

In 1735 Mr. John Kirkby publifhed his arithmetic and algebra. And in 1748 his doctrine of ultimators.

In 1740 were publifhed Mr. Thomas Simpfon's Effays; in 1743 his Differtations, and in 1757 his Tracts; in all which are contained feveral improvements in feries and other parts of Algebra. As alfo in his algebra, firft printed in 1745, and in his Select Exercifes, in 1752.

Alfo in 1740 was publifhed profeffor Saunderfon's Elements of Algebra, in 2 vols. 4to.

In 1741 M. de la Caille publifhed *leçons de mathematiques; ou elemens d'algebre & de geometrie.*

Alfo in 1741, in the memoirs of the academy of fciences, were given two articles by M. de Gua, on the number of pofitive, negative, and imaginary roots of equations. With an'hiftorical account of the improvements in Algebra; in which he feverely cenfures Wallis for his partiality; a circumftance in which he himfelf is not lefs faulty.

In 1746 M. Clairaut publifhed his Elemens d'algebre, in which are contained feveral improvements, efpecially on the irreducible cafe in cubic equations. He has alfo feveral good papers on different parts of analytics, in the memoirs of the academy of fciences.

In 1747 M. Fontaine gave, in the memoirs ot the academy of fciences, a paper on the refolution of equations. Befides fome analytical papers in the memoirs of other years.

In 1761 M. Caftillion publifhed, in 2 vols 4to, Newton's univerfal arithmetic, with a large commentary.

In 1763 Mr. Emerfon publifhed his Increments. In 1764 his Algebra, &c.

In 1764 Mr. Landen publifhed his Refidual Analyfis. In 1765 his Mathematical Lucubrations. And in 1780 his Mathematical Memoirs. All containing good improvements in infinite feries, &c.

In 1770 was publifhed, in the German language, Elements of Algebra by M. Euler. And in 1774 a French tranflation of the fame. The memoirs of the Berlin and Peterfburgh academies alfo abound with various improvements on feries and other branches of analyfis by this great man.

In 1775 was publifhed at Bologna, in 2 vols 4to, *Compendio d'Analifi di.Girolamo Saladini.*

Befides the foregoing, there have been many other authors who have given treatifes on Algebra, or who have made improvements on feries and other parts of Algebra; as Schonerus, Coignet, Salignac, Laloubere, Hemifchius, Degraave, Mefcher, Henifchius, Roberval, the Bernoullis, Malbranche, Agnefi, Wells, Dodfon, Manfredi, Regnault, Rowning, Maferes, Waring, Lorgna, de la Grange, de la Place, Bertrand, Kuhnius Hales, and many others.

Vol. I.

ALGEBRA, *numeral,* is that which is chiefly concerned in the folution of numeral problems, and in which all the given quantities are expreffed by numbers only. As ufed by the more early authors, Diophantus, Paciolus, Stifelius, &c.

ALGEBRA, *fpecious,* or *literal,* is that commonly ufed by the moderns, in which all the quantities, both known and unknown, are reprefented or expreffed by fpecies or general characters, as the letters of the alphabet, &c; in confequence of which general defignation, all the conclufions become univerfal theorems for performing every operation of the like kind. There are fpecimens of this method from Cardar and others about his time, but it was more generally employed and introduced by Vieta.

ALGEBRAICAL, fomething relating to *algebra.*

Thus we fay *algebraical* folutions, curves, characters or fymbols, &c.

ALGEBRAICAL *Curve,* is a curve in which the general relation between the abfciffes and ordinates can be expreffed by a common algebraical equation.

Thefe are alfo called *geometrical lines* or *curves,* in contradiftinction to *mechanical* or *tranfcendental* ones.

ALGEBRAIST, a perfon fkilled in *algebra.*

ALGENEB, or ALGENIB, a fixed ftar of the fecond magnitude, on the right fide of Perfeus.

ALGOL, or *Medufa's Head,* a fixed ftar of the third magnitude, in the conftellation Perfeus.

ALGORAB, a fixed ftar of the third magnitude, in the right wing of the conftellation Corvus.

ALGORISM, or ALGORITHM, is fimilar to logiftics, fignifying the art of computing in any particular way, or about fome particular fubject; or the common rules of computing in any art. As the *algorithm* of numbers, of algebra, of integers, of fractions, of furds, &c; meaning the common rules for performing the operations of arithmetic, or algebra, or fractions, &c.

ALHAZEN, ALLACEN, or ABDILAZUM, was a learned Arabian, who lived in Spain about the year 1100, according to his editor Rifner, and Weidler. He wrote upon Aftrology; and his work upon Optics was printed, in Latin, at Bafil, in 1572, under the title of *Optica Thefaurus,* by Rifner. Alhazen was the firft who fhewed the importance of refractions in aftronomy, fo little known to the ancients. He is alfo the firft author who has treated on the twilight, upon which he wrote a work, in which he alfo fpeaks of the height of the clouds.

ALIDADE, an Arabic name for the label, index, or ruler, which is moveable about the centre of an aftrolabe, quadrant, &c, and carrying the fights or telefcope, and by which are fhewn the degrees cut off the limb or arch of the inftrument.

ALIQUANT *part,* is that part which will not exactly meafure or divide the whole, without leaving fome remainder. Or the *aliquant* part is fuch, as being taken or repeated any number of times, does not make up the whole exactly, but is either greater or lefs than it. Thus 4 is an aliquant part of 10; for 4 twice taken makes 8 which is lefs than 10, and three times taken makes 12 which is greater than 10.

ALIQUOT

ALIQUOT *part*, is such a part of any whole, as will exactly measure it without any remainder. Or the *aliquot* part is such, as being taken or repeated a certain number of times, exactly makes up, or is equal to the whole. So 1 is an aliquot part of 6, or of any other whole number; 2 is also an aliquot part of 6; being contained just 3 times in 6; and 3 is also an aliquot part of 6, being contained just 2 times: so that all the aliquot parts of 6 are 1, 2, 3.

All the *aliquot* parts of any number may be thus found: Divide the given number by its least divisor; then divide the quotient also by its least divisor; and so on always dividing the last quotient by its least divisor, till the quotient 1 is obtained; and all the divisors, thus taken, are the prime aliquot parts of the given number. Then multiply continually together these prime divisors, viz. every two of them, every three of them, every four of them, &c; and the products will be the other or compound *aliquot* parts of the given number. So if the *aliquot* parts of 60 be required; first divide it by 2, and the quotient is 30: then 30 divided by 2 also, gives 15, and 15 divided by 3 gives 5, and 5 divided by 5 gives 1: so that all the prime divisors or aliquot parts are 1, 2, 2, 3, 5. Then the compound ones, by multiplying every two, are 4, 6, 10, 15; and every three 10, 20, 30. So that all the *aliquot* parts of the given number 60, are 1, 2, 3, 4, 5, 6, 10, 12, 15, 20, 30.—In like manner it will be found that all the *aliquot* parts of 360 are 1, 2, 3, 4, 5, 6, 8, 9, 10, 12, 15, 18, 20, 24, 30, 36, 40, 45, 60, 72, 180.

ALLEN (THOMAS) a celebrated mathematician of the 16th century. He was born at Uttoxeter in Staffordshire, in 1542; was admitted a scholar of Trinity college, Oxford, in 1561; where he took his degree of master of arts in 1567. In 1570 he quitted his college and fellowship, and retired to Glocester hall, where he studied very closely, and became famous for his knowledge in antiquity, philosophy and mathematics. He received an invitation from Henry earl of Northumberland, a great friend and patron of the mathematicians, and he spent some time at the earl's house; where he became acquainted with those celebrated mathematicians Thomas Harriot, John Dee, Walter Warner, and Nathaniel Torporley. Robert earl of Leicester, too, had a great esteem for Allen, and would have conferred a bishopric upon him; but his love for solitude and retirement made him decline the offer. His great skill in the mathematics gave occasion to the ignorant and vulgar to look upon him as a magician or conjurer. Allen was very curious and indefatigable in collecting scattered manuscripts relating to history, antiquity, astronomy, philosophy, and mathematics: which collections have been quoted by several learned authors, and mentioned as in the Bibliotheca Alleniana. He published in Latin the second and third books of Ptolemy, *Concerning the Judgment of the Stars*, or, as it is usually called, of the *quadripartite construction*, with an exposition. He wrote also notes on many of Lilly's books, and some on John Bale's work, *De scriptoribus Maj. Britanniæ*. He died at Glocester hall in 1632, being 90 years of age.

Mr. Burton, the author of his funeral oration, calls him the very soul and sun of all the mathematicians of his age. And Selden mentions him as a person of the most extensive learning and consummate judgment, the bright-

est ornament of the university of Oxford. Also *Camden* says he was skilled in most of the best arts and sciences. A. Wood has also transcribed part of his character from a manuscript in the library of Trinity college, in these words: " He studied polite literature with great application; he was strictly tenacious of academic discipline, always highly esteemed both by foreigners and those of the university, and by all of the highest stations in the church of England, and the university of Oxford. He was a sagacious observer, an agreeable companion, &c."

ALLIGATION, one of the rules in arithmetic, by which are resolved questions which relate to the compounding or mixing together of divers simples or ingredients, being so called from *alligare*, to tie or connect together, probably from certain vincula, or crooked ligatures, commonly used to connect or bind the numbers together.

It is probable that this rule came to us from the Moorish or Arabic writers, as we find it, with all the other rules of arithmetic, in *Lucas de Burgo*, and the other early authors in Europe.

Alligation is of two kinds, *medial* and *alternate*.

ALLIGATION *medial* is the method of finding the rate or quality of the composition, from having given the rates and quantities of the simples or ingredients.

The rule of operation is this: multiply each quantity by its rate, and add all the products together; then divide the sum of the products by the sum of the quantities, or whole compound, and the quotient will be the rate sought.

For example, Suppose it were required to mix together 6 gallons of wine, worth 5s. a gallon; 8 gallons, worth 6s. the gallon; and 4 gallons, worth 8s. the gallon; and to find the worth or value, per gallon, of the whole mixture.

	Gal.		s.		products.
Here	6	mult. by	5	gives	30
	8	- by	6	-	48
	4	- by	8	-	32
whole comp.	18				110 sum of prod.

Then 18) 110 ($6\frac{2}{18}$ or $6\frac{1}{9}$ s. is the rate sought.
108
—
2

ALLIGATION *alternate* is the method of finding the quantities of ingredients or simples, necessary to form a compound of a given rate.

The *rule* of operation is this: 1st, Place the given rates of the simples in a column, under each other; noting which rates are less, and which are greater than the proposed compound. 2d, Connect or link with a crooked line, each rate which is less than the proposed compound rate, with one or any number of those which are greater than the same; and every greater rate with one or any number of the less ones. 3d, Take the difference between the given compound rate and that of each simple rate, and set this difference opposite every rate with which that one is linked. 4th, Then if only one difference stand opposite any rate, it will be the quantity belonging to that rate; but when there are several differences to any one, take their sum for its quantity.

For example, Suppose it were required to mix together gold of various degrees of fineness, viz. of 19, of 21, and of 23 caracts fine, fo that the mixture fhall be of 20 caracts fine. Hence,

	Rates	Diffs.	Sum of Diffs.
Comp. rate 20	$\begin{cases} 21 \\ 19 \\ 23 \end{cases}$	$\begin{cases} \cdots 1 \\ \cdots 1 + 3 \\ \cdots 1 \end{cases}$	$\begin{cases} 1 \text{ of } 21 \text{ caracts fine,} \\ 4 \text{ of } 19 \text{ caracts fine,} \\ 1 \text{ of } 23 \text{ caracts fine.} \end{cases}$

That is, there muft be an equal quantity of 21 and 23 caracts fine, and 4 times as much of 19 caracts fine.

Various limitations, both of the compound and the ingredients, may be conceived; and in fuch cafes, the differences are to be altered proportionally.

Queftions of this fort are however commonly beft and eafieft refolved by common Algebra, of which they form a fpecies of indeterminate problems, as they admit of many, or an indefinite number of anfwers.

There is recorded a remarkable inftance of a difcovery made by Archimedes, both by *alligation* and fpecific gravity at the fame time, namely, concerning the crown of Hiero, king of Syracufe. The king had ordered a crown to be made of pure gold, but when brought to him, a fufpicion arofe that it was mixed with alloy of either filver or copper, and the king recommended it to Archimedes to difcover the cheat without defacing the crown. Archimedes, after long thinking on the matter, without lighting on the means of doing it, being one day in the bath, and obferving how his body raifed the water higher, conceived the idea that different metals of the fame weight would occupy different fpaces, and fo raife or expel different quantities of water. Upon which he procured two other maffes, each of the fame weight with the crown, the one of pure gold, and the other of alloy; then immerfing them all three, feparately, in water, and obferving the fpace each occupied, by the quantity it raifed the water, he from thence computed the quantities of gold and alloy contained in the crown.

ALLIOTH, a ftar in the tail of the great bear. The word in Arabic denotes a horfe; and they gave this name to each of the three ftars, in the tail of the great bear, as they are placed like three horfes, thus arranged for the purpofe of drawing the waggon commonly called Charles's wain, reprefented by the four ftars on the body of the fame conftellation.

ALMACANTAR. See ALMUCANTAR.

ALMAGEST, the name of a celebrated book compofed by Ptolemy; being a collection of a great number of the obfervations and problems of the ancients, relating to geometry and aftronomy, but efpecially the latter. And being the firft work of this kind which has come down to us, and containing a catalogue of the fixed ftars, with their places, befide numerous records and obfervations of eclipfes, the motions of the planets, &c, this work will ever be held dear and valuable to the cultivators of aftronomy.

In the original Greek it is called σνιταξις μεγιςη, the *great compofition* or *collection*. And to the word μεγιςη, *megifte*, the Arabians joined the particle *al*, and thence called it *Almaghefti*, or, as we call it, from them, the Almageft.

Ptolemy was born about the year of Chrift 69, and died in 147, and wrote this work, confifting of 13 books, at Alexandria in Egypt, where the Arabians found it on the capture of that kingdom. It was by them

translated out of Greek, into Arabic, by order of the caliph Almaimon, about the year 827; and firft into Latin about 1230, by favour of the emperor Frederic II. The Greek text however was not known in Europe till about the beginning of the 15th century, when it was brought from Conftantinople, then taken by the Turks, by *George*, a monk of Trabezond, who tranflated it into Latin, which tranflation has feveral times been publifhed.

Riccioli, an Italian jefuit, alfo publifhed, in 1651, a body of Aftronomy, which, in imitation of Ptolemy, he called *Almageftum Novum*, the *New Almageft*; being a large collection of ancient and modern obfervations and difcoveries, in the fcience of Aftronomy.

ALMAMON, caliph of Bagdat, a philofopher and aftronomer in the beginning of the 9th century, he having afcended the throne in the year 814. . He was fon of Harun Al-Rafhid, and grand-fon of Almanfor. His name is otherwife written *Mamon, Almaon, Almamun, Alamoun,* or *Al-Maimon*. Having been educated with great care and with a love for the liberal fciences, he applied himfelf to cultivate and encourage them in his own country. For this purpofe he requefted the Greek emperors to fupply him with fuch books on philofophy as they had among them; and he collected fkilful interpreters to tranflate them into the Arabic language. He alfo encouraged his fubjects to ftudy them; frequenting the meetings of the learned, and affifting at their exercifes and deliberations. He caufed Ptolemy's Almageft to be tranflated in 827, by Ifaac Ben-honain, and Thabet Ben-korah, according to Herbelot, but according to others by Sergius, and Alhazen, the fon of Jofeph. In his reign, and doubtlefs by his encouragement, an aftronomer of Bagdat, named Habafh, compofed three fets of aftronomical tables.

Almamon himfelf made many aftronomical obfervations, and determined the obliquity of the ecliptic to be then 23° 35' (or 23° 33' in fome manufcripts), but Voffius fays 23° 51' or 23° 34. He alfo caufed fkilful obfervers to procure proper inftruments to be made, and to exercife themfelves in aftronomical obfervations; which they did accordingly at Shemafi in the province of Bagdat, and upon Mount Cafius near Damas.

Under the aufpices of Mamon alfo a degree of the meridian was meafured on the plains of Sinjar or Sindgiar (or according to fome Fingar), upon the borders of the Red Sea; by which the degree was found to contain 56⅔ miles, of 4000 coudees each, the coudee being a foot and a half: but it is not known what foot is here meant, whether the Roman, the Alexandrian, or fome other. Riccioli makes this meafure of the degree amount to 81 ancient Roman miles, which value anfwers to 62046 French toifes; a quantity more than the true value of the degree by almoft one-third.

Finally, Mamon revived the fciences in the Eaft to fuch a degree, that many learned men were found, not only in his own time, but after him, in a country where the ftudy of the fciences had been long forgotten. This learned king died near Tarfus in Cilicia, by having eaten too freely of fome dates, on his return from a military expedition, in the year 833.

ALMANAC, a calendar or table, in which are fet down and marked the days and feafts of the year, the common ecclefiaftical notes, the courfe and phafes of

the

the moon, &c, for each month: and anfwers to the *fafti* of the ancient Romans.

The etymology of the word is much controverted among grammarians.—Some derive it from the Arabic, viz, from the particle *al*, and *manah*, to count. While Scaliger, and others, derive it from the fame *al*, and the Greek μιναχος, the courfe of the months. But Golius controverts thefe opinions, and afcribes the word to another origin, though he ftill makes it of Arabic extract, which it more evidently is. He fays that, in the Eaft it is the cuftom for the people, at the beginning of the year, to make prefents to their princes; and that, among the reft, the aftrologers prefent them with their almanacs, or ephemerides, for the year enfuing; whence thefe came to be called *almanha*, that is, new-year's gifts. But this derivation feems rather ftrained and improbable; for, by the fame rule, the gifts or productions of other artifts, or claffes of men, might alfo be called almanacs. There are other gueffes at the etymology; and Verftegan writes the word *almonac*, and makes it of German original. Our anceftors, he obferves, ufed to carve the courfes of the moon, for the whole year, upon a fquare piece of wood, which they called *al-monaght*, which is as much as to fay, in old Englifh or Saxon, *all-moon-heed*.

Almanacs are of various kinds and compofition, fome books, others fheets, &c, fome annual, others perpetual. The effential part is the calendar of months, weeks, and days; the motions, changes, and phafes of the moon; with the rifing and fetting of the fun and moon. To thefe are commonly added various matters, aftronomical, aftrological, chronological, meteorological, and even political, rural, medical, &c; as alfo eclipfes, folar ingreffes, afpects, and configurations of the heavenly bodies, lunations, heliocentric and geocentric motions of the planets, prognoftications of the weather, and predictions of other events, the tides, twilight, equation of time, &c.

Till about the 4th century, almanacs bore the marks of heathenifm only; from thence to the 7th century, they were a mixture of heathenifm and chriftianity; and ever fince they have been altogether chriftian: but at all times, aftrological and other predictions have been confidered as an effential part, and ftill are fo to this day with feveral of them, notwithftanding that moft people *affect* to difbelieve in fuch predictions.

Nautical ALMANAC, and *Aftronomical Ephemeris*, is a kind of national almanac, chiefly for nautical purpofes, which was begun in the year 1767 under the direction of the Board of Longitude, on the recommendation of the prefent worthy Aftronomer Royal, who has the immediate conducting of it. It is ftill publifhed by anticipation for feveral years before hand, for the convenience of fhips going out upon long voyages, for which it is highly ufeful, and was found eminently fo in the courfe of the late voyages round the world for making difcoveries. Befides moft things effential to general ufe, that are to be found in other almanacs, it contains many new and important particulars; more efpecially, the diftances of the moon from the fun and fixed ftars, which are computed for the meridian of the Royal Obfervatory of Greenwich, and fet down to every three hours of time, exprefsly defigned for computing the longitude at fea, by comparing thefe with the like diftances obferved there.

ALMANAR, in the Arabian aftrology, denotes the pre-eminence or prevalence of one planet over another.

ALMUCANTARS, ALMACANTARS, or ALMICANTARS, from the Arabic *almocantharat*, are circles parallel to the horizon, conceived to pafs through every degree of the meridian; ferving to fhew the height of the fun, moon, or ftars, &c; and are the fame as the parallels of altitude.

ALMUCANTAR-*Staff*, was an inftrument formerly ufed at fea to obferve the fun's amplitude at rifing or fetting, and thence to determine the variation of the compafs, &c. The inftrument had an arch of 15 degrees, made of fome fmooth wood.

ALPHONSINE *Tables*, are aftronomical tables compiled by order of Alphonfus, king of Caftile. In the compiling of thefe it is thought that prince himfelf affifted. See *Aftronomical tables*.

ALPHONSUS the 10th, king of Leon and Caftile, who has been furnamed The Wife, on account of his attachment to literature, and is now more celebrated for having been an aftronomer than a king. He was born in 1203; fucceeded his father Ferdinand the 3d, in 1252; and died in 1284, confequently at the age of 81.

The affairs of the reign of Alphonfus were very extraordinary and unfortunate for him. But we fhall here only confider him in that part of his character, on account of which he has a place in this work, namely, as an aftronomer and man of letters. He underftood aftronomy, philofophy, and hiftory, as if he had been only a man of letters; and compofed books upon the motions of the heavens, and on the hiftory of Spain, which are highly commended. "What can be more furprifing," fays Mariana, "than that a prince, educated in a camp, and handling arms from his childhood, fhould have fuch a knowledge of the ftars, of philofophy, and the tranfactions of the world, as men of leifure can fcarcely acquire in their retirements? There are extant fome books of Alphonfus on the motions of the ftars, and the hiftory of Spain, written with great fkill and incredible care." In his aftronomical purfuits he difcovered that the tables of Ptolemy were full of errors; and thence he conceived the firft of any the refolution of correcting them. For this purpofe, about the year 1240, and during the life of his father, he affembled at Toledo the moft fkilful aftronomers of his time, Chriftians, Moors, or Jews, when a plan was formed for conftructing new tables. This tafk was accomplifhed about 1252, the firft year of his reign; the tables being drawn up chiefly by the fkill and pains of Rabbi Ifaac Hazan a learned Jew, and the work called the *Alphonfine Tables*, in honour of the prince, who was at vaft expences concerning them. He fixed the epoch of the tables to the 30th of May 1252, being the day of his acceffion to the throne. They were printed for the firft time in 1483, at Venice, by Radtolt, who excelled in printing at that time; an edition extremely rare: there are others of 1492, 1521, 1545, &c. (Weidler, p. 280).

We muft not omit a memorable faying of Alphonfus, which has been recorded for its boldnefs and pretended impiety; namely, "that if he had been of God's privy council when he made the world, he could have advifed him better." Mariana however fays only in general, that Alphonfus was fo bold as to blame the works of Providence,

Providence, and the conftruction of our bodies; and he fays that this ftory concerning him refted only upon a vulgar tradition. The Jefuit's words are curious: " Emanuel, the uncle of Sanchez (the fon of Alphonfus), in his own name, and in the name of other nobles, deprived Alphonfus of his kingdom by a public fentence: which that prince merited, for daring feverely and boldly to cenfure the works of divine Providence, and the conftruction of the human body, as tradition fays he did. Heaven moft juftly punifhed the folly of his tongue." Though the filence of fuch an hiftorian as Mariana, in regard to Ptolemy's fyftem, ought to be of fome weight, yet we cannot think it improbable, that if Alphonfus did pafs fo bold a cenfure on any part of the univerfe, it was on the celeftial fphere, and meant to glance upon the contrivers and fupporters of that fyftem. For, befides that he ftudied nothing more, it is certain that at that time aftronomers explained the motions of the heavens by intricate and confufed hypothefes, which did no nonour to God, nor anywife anfwered the idea of an able workman. So that, from confidering the multitude of fpheres compofing the fyftem of Ptolemy, and thofe numerous eccentric cycles and epicycles with which it is embarraffed, if we fuppofe Alphonfus to have faid, " That if God had afked his advice when he made the world, he would have given him better council," the boldnefs and impiety of the cenfure will be greatly diminifhed.

ALSTED (John-Henry), a German proteftant divine, and one of the moft indefatigable writers of the 17th century. He was fome time profeffor of philofophy and divinity at Herborn in the county of Naffau: from thence he went into Tranfilvania, to be profeffor at Alba Julia; where he continued till his death, which happened in 1638, being then 50 years of age. He applied himfelf chiefly to compofe methods, and to reduce the feveral branches of arts and fciences into fyftems. His *Encyclopædia* has been much efteemed even by Roman Catholics; it was printed at Lyons, and fold very well throughout all France. Voffius mentions the Encyclopædia in general, but fpeaks of his treatife of *Arithmetic* more particularly, and allows the author to have been a man of great reading and univerfal erudition. His *Thefaurus Chronologicus* is by fome efteemed one of his beft works, and has gone through feveral editions, though others fpeak of it with contempt. In his *Triumphus Biblicus* Alfted endeavours to prove that the materials and principles of all the arts and fciences may be found in the fcriptures; but he gained very few to his opinion. John Himmelius wrote a piece againft his *Theologia Polemica*, which was one of Alfted's beft performances. It feems he was a millenarian, having publifhed, in 1672, a treatife *De Mille Annis*, in which he afferts that the faithful fhall reign with Jefus Chrift upon earth a thoufand years; after which will be the general refurrection, and the laft judgment; and he pretended that this reign would commence in the year 1694.

ALTERNATE *angles*, are the internal angles, A and B, or *a* and *b*, made by a line cutting two parallel lines, and lying on oppofite fides of the cutting line. It is the property of thefe angles to be always equal to each other, namely the angle A = the

angle B, and the angle *a* = the angle *b*. And the exterior alternate angles are alfo equal.

ALTERNATE *Ratio* or *Proportion*, is the ratio of the one antecedent to the other, or of one confequent to the other, in any proportion, in which the quantities are of the fame kind. So if A : B : : C : D, then alternately, or by alternation A : C : : B : D.

ALTERNATION, or *Permutation*, of quantities or things, is the varying or changing the order or pofition of them.

As fuppofe two things *a* and *b*; thefe may be placed either thus *ab* or *ba* that is two ways, or 1×2. If there be three things, *a*, *b*, *c*, then the 3d thing *c*, may be placed three different ways with refpect to each of the two pofitions *ab* and *ba* of the other two things, it may ftand either before them, or between them, or after them both, that is, it may ftand either 1ft, 2d, or 3d; and therefore with three things there will be three times as many changes as with two, that is $1 \times 2 \times 3$ or fix changes with three things. Again, if there be four things *a*, *b*, *c*, *d*; then the fourth thing *d* may be placed in four different ways with refpect to each of the fix pofitions of the other three; for it may be fet either 1ft or 2d or 3d or 4th in the order of each pofition; confequently from four things there will be four times as many alternations as there are from three things; and therefore $1 \times 2 \times 3 \times 4 = 24$ is the number of changes with four things. And fo on, always multiplying the laft found number of alternations by the next number of things; or to find the number of changes for any number of things, as *n*, multiply the feries of natural numbers 1, 2, 3, 4, 5, &c, to *n*, continually together, and the laft product will be the number of alternations fought; fo $1 \times 2 \times 3 \times 4 \times 5 ---- n$ is the number of changes in *n* things.

So if, for example, it were required to find how many changes may be rung on 12 bells; it would be $1 \times 2 \times 3 \times 4 \times 5 \times 6 \times 7 \times 8 \times 9 \times 10 \times 11 \times 12 = 479001600$, the number of changes. Now fuppofing there might be rung 10 changes in one minute, that is 10×12 or 120 ftrokes in a minute, or 2 ftrokes in each fecond of time; then, according to this rate, it would take upwards of 91 years to ring over all thefe changes on the 12 bells only. Alfo, if but two more bells were added, making 14 bells; then, at the fame rate of ringing, it would require about 16575 years to ring all the changes on 14 bells but once over. And if the number of bells were 24, it would require more than 117000000000000000 years to ring all the different changes upon them!

ALTIMETRY, Altimetria, the art of taking or meafuring altitudes or heights, whether acceffible or inacceffible. Or

Altimetria is the part of practical Geometry which refpects the theory and practice of meafuring both heights and depths, and both in refpect of perpendicular and oblique lines.

ALTING (James), was born at Heidelberg in 1618. He travelled into England in 1640, where he was ordained by the learned Dr. Prideaux, bifhop of Worcefter. He afterwards fucceeded Gomarus in the profefforfhip of Groninghen. He died in 1697; and recommended the edition of his works to Menfo Alting (author of Notitia German. Infer. Antiquæ); but they were publifhed in 5 vols folio, with his life, by

Bekker

Bekker of Amſterdam. They contain various analytical, exegetical, practical, problematical, and philoſophical tracts, which ſhew his great induſtry and knowledge.

ALTITUDE, in Geometry is the third dimenſion of body, conſidered with reſpect to its elevation above the ground: and is otherwiſe called its height when meaſured from bottom to top, or its depth when meaſured from top to bottom.

ALTITUDE *of a figure*, is the diſtance of its vertex from the baſe, or the length of a perpendicular let fall from its vertex to the baſe. The altitudes of figures are uſeful in computing their areas or ſolidities.

ALTITUDE, or *Height* of any point of a terreſtrial object, is the perpendicular let fall from that point to the plane of the horizon. *Altitudes* are diſtinguiſhed into *acceſſible* and *inacceſſible*.

Acceſſible ALTITUDE of an object, is that whoſe baſe there is acceſs to, to meaſure the neareſt diſtance to it on the ground, from any place.

Inacceſſible ALTITUDE, of an object, is that whoſe baſe there is not free acceſs to, by which a diſtance may be meaſured to it, by reaſon of ſome impediment, ſuch as water, wood, or the like.

To meaſure or *take* ALTITUDES. If an altitude cannot be meaſured by ſtretching a ſtring from top to bottom, which is the direct and moſt accurate way, then ſome indirect way is uſed, by actually meaſuring ſome other line or diſtance which may ſerve as a baſis, in conjunction with ſome angles, or other proportional lines, either to compute, or geometrically determine, the altitude of the object ſought.

There are various ways of meaſuring altitudes, or depths, by means of different inſtruments, and by ſhadows or reflected images, on optical principles. There are alſo various ways of computing the altitude in numbers, from the meaſurements taken as above, either by geometrical conſtruction, or trigonometrical calculation, or by ſimple numeral computation from the property of parallel lines, &c.

The inſtruments moſtly uſed in meaſuring altitudes, are the quadrant, theodolite, geometrical ſquare, line of ſhadows, &c; the deſcriptions of each of which may be ſeen under their reſpective names.

To meaſure an Acceſſible Altitude Geometrically. Thus, ſuppoſe the height of the acceſſible tower AB be required. Firſt, by means of two rods, the one longer than the other: plant the longer upright at C; then move the ſhorter back from it, till by trials you find ſuch a place, D, that the eye placed at the top of it at E, may ſee the top of the other, F, and the top of the object B ſtraight in a line: next meaſure the diſtances DA or EG and DC or EH, alſo HF the difference between the heights of the rods: then, by ſimilar triangles, as EH: EG:: HF: the 4th proportional GB; to which add AG or DE, and the ſum will be the whole altitude AB ſought.

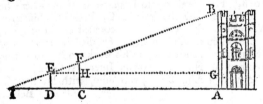

Or, with one rod CF only: plant it at ſuch a place C, that the eye at the ground, or near it, at I, may ſee the tops F and B in a right line: then, having meaſured IC, IA, CF, the 4th proportional to theſe will be the altitude AB ſought.

Or thus, by means of Shadows. Plant a rod *ab* at *a*, and meaſure its ſhadow *ac*, as alſo the ſhadow AC of the object AB; then the 4th proportional to *ac*, *ab*, AC will be the altitude AB ſought.

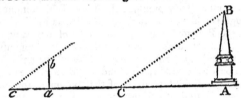

Or thus, by means of Optical Reflection. Place a veſſel of water, or a mirror or other reflecting ſurface, horizontal at C; and move off from it to ſuch a diſtance, D, that the eye E may ſee the image of the top of the object in the mirror at C: then, by ſimilar figures, CD: DE:: CA: AB the altitude ſought.

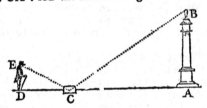

Or thus, by the Geometrical Square. At any place, C, fix the ſtand, and turn the ſquare about the centre of motion, D, till the eye there ſee the top of the object through the ſights or teleſcope on the ſide DE of the quadrant, and note the number of diviſions cut off the other ſide by the plumb line EG: then as EF: FG:: DH: HB; to which add AH or CD, for the whole height AB.

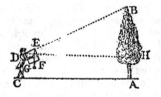

To meaſure an Acceſſible Altitude Trigonometrically. At any convenient ſtation, C, with a quadrant, theodolite, or other graduated inſtrument, obſerve the angle of elevation ACB above the horizontal line AC; and meaſure the diſtance AC. Then, A being a right angle, it will be, as radius is to the tangent of the angle A, ſo is AC to AB ſought.

If AC be not horizontal, but an inclined plane; then the angle above it muſt be obſerved at two ſtations C and D in a right line, and the diſtances AC, CD both meaſured. Then, from the angle C take the angle D, and there remains the angle CBD: hence in the triangle BCD, are given the angles and the ſide DC, to find the ſide CB; and then in the triangle ABC, are given the

two

two fides CA and CB, with the included angle C, to find the third fide AB.

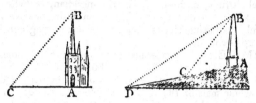

Or thus, meafure only the diftance AC, and the angles A and C: then, in the triangle ABC, are given all the angles and the fide AC, to find the fide AB.

To meafure an Inacceffible Altitude, as a hill, cloud, or other objeć. This is commonly done, by obferving the angle of its altitude at two ftations, and meafuring the diftance between them. Thus, for the height AB of a hill, meafure the diftance CD at the foot of it, and obferve the quantity of the two angles C and D. Then, from the angle C taking the angle D, leaves the angle CBD; hence

As fine ∠CBD : fine ∠D : : CD · CB; and
As rad. : fine ∠ACB : : CB : AB the altitude.

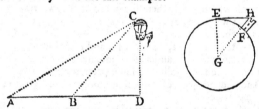

And for a balloon, or cloud, or other moveable objeć C, let two obfervers at A and B, in a plane with C, take at the fame time the angles A and B, and meafure the diftance between them AB; then calculate the altitude CD exaćly as in the laft example.

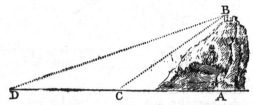

To find the height of an objeć, by knowing the utmoft diftance at which its top can be juft feen in the horizon. As fuppofe the top H of a tower FH can be juft feen from E when the diftance EF is 25 miles, fuppofing the circumference of the earth to be 25000 miles, or the radius 3979 miles or 21009120 feet. Firft as 25000 : 25 : : 360° : 21′ 36″ equal to the angle G; then as radius : fec. ∠G : : EG · GH, which will be found to be 21009536 feet; from which take EG or GF, and there remains 416 feet, for FH the height of the tower fought — Or rather thus, as 10000000 radius : 198 = fec. ∠G — radius : : 21009120 = EG : 416 = FH, as before.

Or the fame may be found eafier thus: The horizon dips nearly 8 inches or ⅔ of a foot at the diftance of 1 mile, and according to the fquare of the diftance for other diftances; therefore as 1² or 1 : 25² or 625 : : ⅔ : ⅔ of 625 or 416 feet, the fame as before.

There is a very eafy method of taking great terreftrial altitudes, fuch as mountains &c, by means of the difference between the heights of the barometer obferved at the bottom and top of the fame. Which fee under the article BAROMETER.

ALTITUDE *of the Eye,* in *Perfpećive,* is a right line let fall from the eye, perpendicular to the geometrical plane.

ALTITUDE, in *Aftronomy,* is the arch of a vertical circle, meafuring the height of the fun, moon, ftar, or other celeftial objeć, above the horizon.

This altitude may be either *true* or *apparent.* The apparent altitude is that which appears by fenfible obfervations made at any place on the furface of the earth. And the true altitude is that which refults by correćing the apparent, on account of refraćion and parallax.

The quantity of the refraćion is different at different altitudes; and the quantity of the parallax is different according to the diftance of the different luminaries: in the fixed ftars this is too fmall to be obferved; in the fun it is but about 8¾ feconds; but in the moon it is about 52 minutes.

Altitudes are obferved by a quadrant, or fextant, or by the fhadow of a gnomon or high pole, and by various other ways, as may be feen in moft books of aftronomy.

Meridian ALTITUDE, is an arch of the meridian intercepted between any point in it and the horizon. So if HO be the horizon, and HEZO the meridian; then the arch HE, or the angle HCE, is the meridian altitude of an objeć in the meridian at the point E.

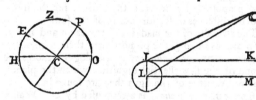

ALTITUDE, or *elevation, of the Pole,* is the angle OCP, or arch OP of the meridian, intercepted between the horizon and pole P.

This is equal to the latitude of the place; and it may be found by obferving the meridian altitude of the pole ftar, when it is both above and below the pole, and taking half the fum, when correćed on account of refraćion. Or the fame may be found by the declination and meridian altitude of the fun.

ALTITUDE, or *elevation,* of the *equator,* is the angle HCE, or arch HE of the meridian, between the horizon and the equator at E; and it is equal to ZP the colatitude of the place.

ALTITUDE of the *Tropics,* the fame as what is otherwife called the *folftitial altitude* of the fun, or his meridian altitude when in the folftitial points.

ALTITUDE, or *height,* of the *horizon,* or of ftars &c feen in it, is the quantity by which it is raifed by refraćion.

Refraćion of ALTITUDE, is an arch of a vertical circle, by which the true altitude of the moon, or a ftar, or other objeć, is increafed by means of the refraćion;

and

and is different at different altitudes, being nothing in the zenith, and greatest at the horizon, where it is about 33.

Parallax of ALTITUDE, is an arch of a vertical circle, by which the true altitude, observed at the centre of the earth, exceeds that which is observed on the surface; or the difference between the angles ☽ LM and ☽ IK of altitude there; and is equal to the angle I ☽ L formed at the moon or other body, and subtended by the radius IL of the earth.

It is evident that this angle is less, as the luminary is farther distant from the earth; and also less, for any one luminary, as it is higher above the horizon; being greatest there, and nothing in the zenith.

ALTITUDE *of the Nonagesimal*, is the altitude of the 90th degree of the ecliptic, counted upon it from where it cuts the horizon, or of the middle or highest point of it which is above the horizon, at any time; and is equal to the angle made by the ecliptic and horizon where they intersect at that time.

ALTITUDE *of the cone of the earth's or moon's shadow*, the height of the shadow of the body, made by the sun, and measured from the centre of the body. To find it, say, As the tangent of the angle of the sun's apparent semidiameter is to radius, so is 1 to a 4th proportional, which will be the height of the shadow, in semidiameters of the body.

So, the greatest height of the earth's shadow, is 217·8 semidiameters of the earth, when the sun is at his greatest distance, or his semidiameter subtends an angle of about 15′ 47″; and the height of the same is 210·7 semidiameters of the earth, when the sun is nearest the earth, or when his semidiameter is about 16′ 19″: And proportionally between these limits for the intermediate distances or semidiameters of the sun.

The altitudes of the shadows of the earth and moon, are nearly as 11 to 3, the proportion of their diameters.

ALTITUDE, or *exaltation*, in astrology, denotes the second of the five essential dignities, which the planets acquire by virtue of the signs they are found in.

ALTITUDE *of motion*, is a term used by Dr. Wallis, for the measure of any motion, estimated in the line of direction of the moving force.

ALTITUDE, in speaking of fluids, is more frequently expressed by the term *depth*. The pressure of fluids, in every direction, is in proportion to their altitude or depth.

ALTITUDE *of the mercury*, in the *barometer* and *thermometer*, is marked by degrees, or equal divisions, placed by the side of the tube of those instruments.

The altitude of the barometer, or of the mercury in its tube, at London, is usually comprised between the limits of 28 and 31 inches; and the mean height, for every day in several years, is nearly 29 87 inches.

ALTITUDE *of the pyramids in Egypt*, was measured so long since as the time of Thales, which he effected by means of their shadow, and that of a pole set upright beside them, making the altitudes of the pole and pyramid proportional to the lengths of their shadows. Plutarch has given an account of the manner of this operation, which is one of the first geometrical observations we have an exact account of.

ALTITUDE, *circles of, parallels of, quadrant of, &c.* See the respective words.

Equal ALTITUDE *Instrument*, is an instrument used to observe a celestial object, when it has the same or an equal altitude, on both sides of the meridian, or before and after it passes the meridian: an instrument very useful in adjusting clocks &c, and for comparing equal and apparent time.

AMBIENT, encompassing round about; as the bodies which are placed about any other body, are called ambient bodies, and sometimes *circum-ambient* bodies; and the whole mass of the air or atmosphere, because it encompasses all things on the face of the earth, is called the *ambient* air.

AMBIGENAL *Hyperbola*, a name given by Newton, in his *Enumeratio linearum tertii ordinis*, to one of the triple hyperbolas EGF of the second order, having one of its infinite legs, as EG, falling within the angle ACD, formed by the asymptotes AC, CD, and the other leg GF falling without that angle.

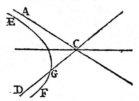

AMBIT, *of a figure*, in Geometry, is the perimeter, or line, or sum of the lines, by which the figure is bounded.

AMBLIGON, or AMBLIGONAL, in Geometry, signifies obtuse-angular, as a triangle which has one of its angles obtuse, or consisting of more than 90 degrees.

AMICABLE *numbers*, denote pairs of numbers, of which each of them is mutually equal to the sum of all the aliquot parts of the other. So the first or least pair of amicable numbers are 220 and 284; all the aliquot parts of which, with their sums, are as follow, viz, of 220, they are 1, 2, 4, 5, 10, 11, 20, 22, 44, 55, 110, their sum — — 284; of 284, they are 1, 2, 4, 71, 142, and their sum is 220.

The 2d pair of amicable numbers are 17296 and 18416, which have also the same property as above.

And the 3d pair of amicable numbers are 9363584 and 9437056.

These three pairs of amicable numbers were found out by F. Schooten, sect. 9 of his *Exercitationes Mathematicæ*, who I believe first gave the name of *amicable* to such numbers, though such properties of numbers it seems had before been treated of by Rudolphus, Descartes, and others.

To find the first pair, Schooten puts $4x$ and $4yz$, or a^2x and a^2yz for the two numbers where $a = 2$; then making each of these equal to the sum of the aliquot parts of the other, gives two equations, from which are found the values of x and z, and consequently, assuming a proper value for y, the two amicable numbers themselves $4x$ and $4yz$.

In like manner for the other pairs of such numbers; in which he finds it necessary to assume $16x$ and $16yz$ or a^4x and a^4yz for the 2d pair, and $128x$ and $128yz$ or a^7x and a^7yz for the 3d pair.

Schooten

Schooten then gives this practical rule, from Descartes, for finding amicable numbers, viz, Assume the number 2, or some power of the number 2, such that if unity or 1 be subtracted from each of these three following quantities, viz;

from 3 times the assumed number,

also from 6 times the assumed number,

and from 18 times the square of the assumed number, the three remainders may be all prime numbers; then the last prime number being multiplied by double the assumed number, the product will be one of the amicable numbers sought, and the sum of its aliquot parts will be the other.

That is, if a be put = the number 2, and n some integer number, such that $3a^n - 1$, and $6a^n - 1$, and $18a^{2n} - 1$ be all three prime numbers; then is $\overline{18a^{2n} - 1}$ $\times 2a^n$ one of the amicable numbers; and the sum of its aliquot parts is the other.

AMONTONS (WILLIAM), an ingenious French experimental philosopher, was born in Normandy the 31st of August 1663. While at the grammar school, he by sickness contracted a deafness that almost excluded him from the conversation of mankind. In this situation he applied himself to the study of geometry and mechanics; with which he was so delighted that it is said he refused to try any remedy for his disorder, either because he deemed it incurable, or because it increased his attention to his studies. Among other objects of his study, were the arts of drawing, of land-surveying, and of building; and shortly after he acquired some knowledge of those more sublime laws by which the universe is regulated. He studied with great care the nature of barometers and thermometers; and wrote his treatise of *Observations and Experiments concerning a new Hour-glass, and concerning Barometers, Thermometers, and Hygroscopes;* as also some pieces in the Journal des Savans. In 1687, he presented a new hygroscope to the Academy of Sciences, which was much approved. He found out a method of conveying intelligence to a great distance in a short space of time: this was by making signals from one person to another, placed at as great distances from each other as they could see the signals by means of telescopes. When the Royal Academy was new regulated in 1699, Amontons was chosen a member of it, as an eleve under the third Astronomer; and he read there his *New Theory of Friction,* in which he happily cleared up an important object in mechanics. In fact he had a particular genius for making experiments: his notions were just and delicate: and he knew how to prevent the inconveniences of his new inventions, and had a wonderful skill in executing them. He died of an inflammation in his bowels, the 11th of October 1705, being only 42 years of age.

The eloge of Amontons may be seen in the volume of the Memoirs of the Academy of Sciences for the year 1705, Hist. pa. 150. And his pieces contained in the different volumes of that work, which are pretty numerous, and upon various subjects, as the air, action of fire, barometers, thermometers, hygrometers, friction, machines, heat, cold, rarefactions, pumps, &c, may be seen in the volumes for the years 1696, 1699, 1702, 1703, 1704, and 1705.

AMPHISCII, or AMPHISCIANS, are the people who inhabit the torrid zone; which are so called, because

they have their shadow at noon turned sometimes one way, and sometimes another, namely, at one time of the year towards the north, and at the other towards the south.

AMPLITUDE, in *gunnery*, the range of the projectile, or the right line upon the ground subtending the curvilinear path in which it moves.

AMPLITUDE, in *astronomy*, is an arch of the horizon, intercepted between the true east or west point, and the centre of the sun or a star at its rising or setting: so that the amplitude is of two kinds; *ortive* or eastern, and *occiduous* or western. Each of these amplitudes is also either northern or southern, according as the point of rising or setting is in the northern or southern part of the horizon: and the complement of the amplitude, or the arch of distance of the point of rising or setting, from the north or south point of the horizon, is the azimuth.

The amplitude is of use in navigation, to find the variation of the compass or magnetic needle. And the rule to find it is this: As the cosine of the latitude is to radius, so is the sine of the sun's or star's declination, to the sine of the amplitude. So in the latitude of London, viz, $51° 31'$, when the sun's declination is $23° 28$ then

cos. $51° 31'$ the lat. - - $- 9.7939907$
sin. $23 28$ the decl. - $+ 9.6001181$

sin. $39 47$ the ampl. 9.8061274

That is, the sun then then rises or sets $39° 47$ from the east or west point, to the north or south according as the declination is north or south.

Magnetical AMPLITUDE, is an arch of the horizon, contained between the sun or star, at the rising or setting, and the magnetical east or west point of the horizon, pointed out by the magnetical compass, or the amplitude or azimuth compass. And the difference between this magnetical amplitude, so observed, and the true amplitude, as computed in the last article, is the variation of the compass.

So if, for instance, the magnetical amplitude be observed, by the compass, to be $61° 47'$, at the time when it is computed to be - $39 47$,

then the difference - - $22 0$ is the variation west.

ANABIBAZON, a name sometimes given to the dragon's tail, or northern node of the moon.

ANACAMPTICS, or the science of the reflections of sounds, frequently used in reference to echoes, which are said to be sounds produced *anacamptically*, or by reflection. And in this sense it was used by the ancients for that part of optics which is otherwise called Catoptrics.

ANACHRONISM, in Chronology, an error in computation of time, by which an event is placed earlier than it really happened. Such is that of Virgil, who makes Dido to reign at Carthage in the time of Æneas, though, in reality, she did not arrive in Africa till 300 years after the taking of Troy.

An error on the other side, by which a fact is placed later, or lower than it should be, is called a *parachronism.* But in common use, this distinction, though proper, is not attended to; and the word *anachronism* is used indifferently for the mistake on both sides.

ANACLASTICS, *or* ANACLATICS, an ancient name for that part of Optics which considers refracted light;

P   being

being the fame as what is more ufually called *dioptrics*. See the Compendium of *Ambrofius Rhodius, lib.* 3. *Optics, pa.* 384 *& feq.*

ANACLASTIC *Curves*, a name given by M. de Mairan to certain apparent curves formed at the bottom of a veffel full of water, to an eye placed in the air ; or the vault of the heavens, feen by refraction through the atmofphere.

M. de Mairan determines thefe curves by a principle not admitted by all authors; but Dr. Barrow, at the end of his Optics, determines the fame curves by other principles.

ANALEMMA, a planifphere, or projection of the fphere, orthographically made on the plane of the meridian, by perpendiculars from every point of that plane, the eye fuppofed to be at an infinite diftance, and in the eaft or weft point of the horizon. In this projection, the folftitial colure, and all its parallels, are projected into concentric circles, equal to the real circles in the fphere; and all circles whofe planes pafs through the eye, as the horizon and its parallels, are projected into right lines equal to their diameters; but all oblique circles are projected into ellipfes, having the diameter of the circle for the tranfverfe axis.

This inftrument, having the furniture drawn on a plate of wood or brafs, with an horizon fitted to it, is ufed for refolving many aftronomical problems; as the time of the fun's rifing and fetting, the length and hour of the day, &c. It is alfo ufeful in dialling, for laying down the figns of the zodiac, with the lengths of days, and other matters of furniture, upon dials.

The oldeft treatife we have on the analemma, was written by Ptolemy, which was printed at Rome in 1562, with a commentary by F. Commandine. Pappus alfo treated of the fame. Since that time, many other authors have treated very well of the analemma; as Aguilonius, Taquet, Dechales, Witty, &c.

ANALOGY, the fame as proportion, or equality, or fimilitude of ratios. Which fee.

ANALYSIS, is, generally, the refolution of any thing into its component parts, to difcover the thing or the compofition. And in mathematics it is properly the method of refolving problems, by reducing them to equations. *Analyfis* may be diftinguifhed into the *ancient* and the *modern*.

The *ancient analyfis*, as defcribed by Pappus, is the method of proceeding from the thing fought as *taken* for granted, through its confequences, to fomething that is *really* granted or known ; in which fenfe it is the reverfe of fynthefis or compofition, in which we lay that down *firft* which was the *laft* ftep of the analyfis, and tracing the fteps of the analyfis back, making that antecedent here which was confequent there, till we arrive at the thing fought, which was taken or affumed as granted in the firft ftep of the analyfis. This chiefly refpected geometrical enquiries.

The principal authors on the ancient analyfis, as recounted by Pappus, in the 7th book of his *Mathematical Collections*, are Euclid in his *Data, Porifmata, & de Locis ad Superficiem* ; Apollonius *de Sectione Rationis, de Sectione Spatii, de Tactionibus, de Inclinationibus, de Locis Planis, & de Sectionibus Conicis* ; Ariftæus, *de Locis Solidis* ; and Eratofthenes, *de Mediis Proportionalibus* : from which

Pappus gives many examples in the fame book. To thefe authors we may add Pappus himfelf. The fame fort of analyfis has alfo been well cultivated by many of the moderns ; as Fermat, Viviani, Getaldus, Snellius, Huygens, Simfon, Stewart, Lawfon, &c, and more efpecially Hugo d'Omerique, in his *Analyfis Geometrica*, in which he has endeavoured to reftore the Analyfis of the ancients. And, on this head, Dr. Pemberton tells us "that Sir Ifaac Newton ufed to cenfure himfelf for not following the ancients more clofely than he did; and fpoke with regret of his miftake, at the beginning of his mathematical ftudies, in applying himfelf to the works of Defcartes, and other algebraical writers, before he had confidered the Elements of Euclid with that attention fo excellent a writer deferves: that he highly approved the laudable attempt of Hugo d'Omerique to reftore the ancient analyfis."

In the application of the ancient analyfis in geometrical problems, every thing cannot be brought within ftrict rules ; nor any invariable directions given, by which we may fucceed in all cafes; but fome previous preparation is neceffary, a kind of mental contrivance and conftruction, to form a connexion between the *data* and *quæfita*, which muft be left to every one's fancy to find out ; being various, according to the various nature of the problems propofed : Right lines muft be drawn in particular directions, or of particular magnitudes ; bifecting perhaps a given angle, or perpendicular to a given line ; or perhaps tangents muft be drawn to a given curve, from a given point ; or circles defcribed from a given centre, with a given radius, or touching given lines, or other given circles ; or fuch-like other operations. Whoever is converfant with the works of Archimedes, Apollonius, or Pappus, well knows that they founded their analyfis upon fome fuch previous operations ; and the great fkill of the analyft confifts in difcovering the moft proper affections on which to found his analyfis : for the fame problem may often be effected in many different ways : of which it may be proper to give here an example or two. Let there be taken, for inftance, this problem, which is the 155th prop. of the 7th book of Pappus.

From the extremities of the bafe A, B, of a given fegment of a circle, it is required to draw two lines AC, BC, meeting at a point C in the circumference, fo that they fhall have a given ratio to each other, fuppofe that of F to G.

The folution of this problem, as given by Pappus, is thus.

ANALYSIS.

Suppofe the thing done, and that the point C is found: then fuppofe CD is drawn a tangent to the circle at C, and meeting the line AB produced in the point D. Now by the hypothefis $AC : BC :: F : G$, and alfo $AC^2 : BC^2 :: DA : DB$, as may be thus proved

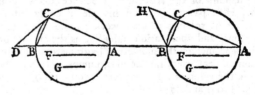

Since DC touches the circle, and BC cuts it, the angle BCD is equal to BAC by Euc. iii. 32 ; alfo the angle

D is

D is common to both the triangles DCA, DCB; thefe are therefore fimilar, and fo, by vi 4. DA : DC : : DC : DB, and hence DA² : DC² : : DA : DB by cor. vi 20. But alfo, by vi 4, DA : AC : : DC : CB, and by permutation DA : DC : : AC : BC, or DA² : DC² : : AC² : BC²; and hence, by equality, AC² : BC² : ; DA : DB.

But the ratio of AC² to BC² is given by prop. LVII of Simfon's edition of the *Data*, becaufe the ratio of AC to BC is given, and confequently that of DA to DB is given. Now fince the ratio of DA to DB is given, therefore alfo, by Data vi, that of DA to AB, and hence, by Data ii, DA is given in magnitude.

And here the analyfis properly ends. For it having been fhewn that DA is given, or that a point D may be found in AB produced, fuch, that a tangent being drawn from it to the circumference, the point of contact will be the point fought; we may now begin the compofition, or fynthetical demonftration; which muft be done by finding the point D, or laying down the line AD, which, it was affirmed, was given, in the laft ftep of the analyfis.

SYNTHESIS.

Conftruction. Make as F¹ : G² : : AD : DB, (which may be done, fince AB is given, by making it as F² — G² : G² : : AB : DB, and then by compofition it will be as F² : G² : : AD : DB); and then from the point D, thus found, draw a tangent to the circle, and from the point of contact C drawing CA and CB, the thing is done.

Demonftration. Since, by the conftr. F² : G² : : AD : DB, and alfo AD : DB : : AC² : BC², which has been already demonftrated in the analyfis, and might be here proved in the fame manner. Therefore F² : G² : : AC² : BC², and confequently F : G : : AC : BC. *Q. E. D.*

Here we fee an inftance of the method of *refolution* and *compofition*, as it was practifed by the ancients, the folution here given being that of Pappus himfelf. But as the method of referring and reducing every thing to the *Data*, and conftantly quoting the fame, may appear now to be tedious and troublefome : and indeed it is unneceffary to thofe who have already made themfelves mafters of the fubftance of that valuable book of Euclid, and have by practice and experience acquired a facility of reafoning in fuch matters : I fhall therefore now fhew how we may abate fomething of the rigour and ftrict form of the ancient method of folution, without diminifhing any part of its admirable elegance and perfpicuity. And this may be done by the inftance of another folution, of the many more which might be given, of the fame problem, as follows.

ANALYSIS.

Let us again fuppofe that the thing is done; viz AC : BC : : F : G, and let there be drawn BH making the angle ABH equal to the angle ACB, and meeting AC produced in H. Then, the angle A being alfo common, the two triangles ABC and ABH are equiangular, and therefore, by vi 4, AC : BC : : AB : BH, in a given ratio; and, AB being given, therefore BH is given in pofition and magnitude.

Conftruction. Draw BH making the angle ABH equal to that which may be contained in the given fegment, and take AB to BH in the given ratio of F to G. Draw ACH, and BC.

Demonftration. The triangles ABC, ABH are equiangular, therefore, vi 4, AC : CB : : AB : BH, which is the given ratio by conftruction.

Modern ANALYSIS, confifts chiefly of algebra, arithmetic of infinites, infinite feries, increments, fluxions, &c; of each of which a particular account may be feen under their refpective articles.

Thefe form a kind of arithmetical and fymbolical analyfis, depending partly on modes of arithmetical computation, partly on rules peculiar to the fymbols made ufe of, and partly on rules drawn from the nature and fpecies of the quantities they reprefent, or from the modes of their exiftence or generation.

The modern analyfis is a general inftrument by which the fineft inventions and the greateft improvements have been made in mathematics and philofophy, for near two centuries paft. It furnifhes the moft perfect examples of the manner in which the art of reafoning fhould be employed; it gives to the mind a wonderful fkill for difcovering things unknown, by means of a fmall number that are given; and by employing fhort and eafy fymbols for expreffing ideas, it prefents to the underftanding things which otherwife would feem to lie above its fphere. By this means geometrical demonftrations may be greatly abridged : a long train of arguments, in which the mind cannot, without the greateft effort of attention, difcover the connection of ideas, is converted into vifible fymbols; and the various operations which they require, are fimply effected by the combination of thofe fymbols. And, what is ftill more extraordinary, by this artifice, a great number of truths are often expreffed in one line only : inftead of which, by following the ordinary way of explanation and demonftration, the fame truths would occupy whole pages or volumes. And thus, by the bare contemplation of one line of calculation, we may underftand in a fhort time whole fciences, which otherwife could hardly be comprehended in feveral years.

It is true that Newton, who beft knew all the advantages of analyfis in geometry and other fciences, laments, in feveral parts of his works, that the ftudy of the ancient geometry is abandoned or neglected. And indeed the method employed by the ancients in their geometrical writings, is commonly regarded as more rigorous, than that of the modern analyfis : and though it be greatly inferior to that of the moderns, in point of difpatch and facility of invention; it is neverthelefs highly ufeful in ftrengthening the mind, improving the reafoning faculties, and in accuftoming the young mathematician to a pure, clear, and accurate mode of inveftigation and demonftration, though by a long and laboured procefs, which he would with difficulty have fubmitted to if his tafte had before been vitiated, as it were, by the more piquant fweets of the modern analyfis. And it is principally on this that the complaints of Newton are founded, who feared left by the too early and frequent ufe of the modern analyfis; the fcience of geometry fhould lofe that rigour and purity which characterife its inveftigations; and the mind become debilitated by the

facility

facility of our analysis. This great man was therefore well founded, in recommending, to a certain extent, the study of the ancient geometricians: for, their demonstrations being more difficult, give more exercife to the mind, accuftom it to a clofer application, give it a greater fcope, and habituate it to patience and refolution, fo neceffary for making difcoveries. But this is the only or principal advantage from it; for if we fhould look no farther than the method of the ancients, it is probable that, even with the beft genius, we fhould have made but few or fmall difcoveries, in comparifon of thofe obtained by means of the modern analyfis. And even with regard to the advantage given to inveftigations made in the manner of the ancients, namely of being more rigorous, it may perhaps be doubted whether this pretenfion be well founded. For to inftance in thofe of Newton himfelf, although his demonftrations be managed in the manner of the ancients; yet at the fame time it is evident that he inveftigates his theorems by a method different from that employed in the demonftrations, which are commonly analytical calculations, difguifed by fubftituting the name of lines for their algebraical value: and though it be true that his demonftrations are rigorous, it is no lefs fo that they would be the fame when tranflated and delivered in algebraic language; and what difference can it make in this refpect, whether we call a line AB, or denote it by the algebraic character *a?* Indeed this laft defignation has this peculiarity, that when all the lines are denoted by algebraic characters, many operations can be performed upon them, without thinking of the lines or the figure. And this circumftance proves of no fmall advantage: the mind is relieved, and fpared as much as poffible, that its whole force may be employed in overcoming the natural difficulty of the problem alone.

Upon the whole therefore the ftate of the comparifon feems to be this; That the method of the ancients is fitteft to begin our ftudies with, to form the mind and to eftablifh proper habits; and that of the moderns to fucceed, by extending our views beyond the prefent limits, and enabling us to make new difcoveries and improvements.

Analyfis is divided, with refpect to its object, into that of *finites,* and that of *infinites.*

Analyfis of finite quantities, is what is otherwife called *algebra,* or *fpecious arithmetic.*

Analyfis of infinites, called alfo the *new analyfis,* is that which is concerned in calculating the relations of quantities which are confidered as infinite, or infinitely little; one of its chief branches being the *method of fluxions,* or the *differential calculus.* And the great advantage of the modern mathematicians over the ancients, arifes chiefly from the ufe of this modern analyfis.

ANALYSIS *of powers,* is the fame as refolving them into their roots, and is otherwife called *evolution.*

ANALYSIS *of curve lines,* fhews their conftitution, nature and properties, their points of inflexion, ftation, retrogradation, variation, &c.

ANALYST, a perfon who analyfes fomething, or makes ufe of the analytical method. In mathematics, it is a perfon fkilled in algebra, or in the mathematical analyfis in general.

ANALYST, the title of an ingenious, though fophiftical book, written by the celebrated Dr. Berkeley, againft the doctrine of fluxions.

ANALYTIC, or ANALYTICAL, fomething belonging to, or partaking of, the nature of analyfis; or performed by the method of analyfis.

Thus we fay *analytical* demonftration, *analytical* enquiry, *analytical* table or fcheme, *analytical* method, &c. The *analytical* ftands oppofed to the *fynthetical,* or that which proceeds by the way of *fynthefis.*

ANALYTICS, the fcience, or doctrine, and ufe of analyfis.

ANAMORPHOSIS, in perfpective and painting, a monftrous projection; or a reprefentation of fome image, either on a plane or curve furface, deformed or diftorted; but which in a certain point of view fhall appear regular, and drawn in juft proportion.

To conftruct an Anamorphofis, or monftrous projection, on a plane.—Draw the fquare ABCD (fig. 1), of any fize at pleafure, and divide it by croffing lines into a number of areolæ or fmaller fquares: and then in this fquare, or reticle, called alfo the *craticular prototype,* draw the regular image which is to be diftorted.—Or, about any image, propofed to be diftorted, draw a reticle of fmall fquares.

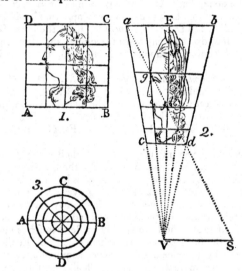

Then draw the line *ab* (fig. 2.) equal to AB, dividing it into the fame number of equal parts, as the fide of the prototype AB; and on its middle point E erect the perpendicular EV, and alfo VS perpendicular to EV, making EV fo much the longer, and VS fo much the fhorter, as it is intended the image fhall be more diftorted. From each of the points of divifion draw right lines to the point V, and draw the right line *a*S. Laftly through the points *c, e, f, g,* &c, draw lines parallel to *ab:* So fhall *abcd* be the fpace upon which the monftrous projection is to be drawn; and is called the *craticular ectype.*

Then, in every areola, or fmall trapezium, of the fpace *abcd,* draw what appears contained in the correfponding areola of the original fpace ABCD: fo fhall there be produced a deformed image in the fpace *abcd,* which yet will appear in juft proportion to an eye diftant

from

3

from it the length of EV, and raifed above it by a height equal to VS.

It will be amufing to contrive it fo, that the deformed image may not reprefent a mere chaos, but fome certain figure: thus, a river with foldiers, waggons, and other objects on the fide of it, have been fo drawn and diftorted, that when viewed by an eye at S, it appeared like the face of a fatyr.

An image may alfo be diftorted mechanically, by perforating through in feveral places with a fine pin; then, placing it againft a candle or lamp, obferve where the rays, which pafs through thefe fmall holes, fall on any plane or curve fuperficies; for they will give the correfpondent points of the image deformed, and by means of which the deformation may be completed.

To draw an Anamorphofis upon the convex furface of a cone. It appears from the conftruction above, that we have only to make a craticular ectype upon the furface of the cone, which may appear equal to the craticular prototype, to an eye placed at a proper height above the vertex of the cone. Hence,

Let the bafe, or circumference, ABCD, of the cone (fig. 3) be divided by radii into any number of equal parts; and let fome one radius be likewife divided into equal parts; then through each point of divifion draw concentric circles: fo fhall the craticular prototype be formed.

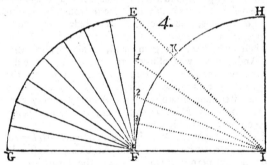

With double the diameter AB, as a radius, defcribe the quadrant EFG (fig. 4) fo as the arch EG be equal to the whole periphery; then this quadrant, being plied or bent round, will form the fuperficies of a cone, whofe bafe is the circle.

Next divide the arch EG into the fame number of equal parts as the craticular prototype is divided into; and draw radii from all the points of divifion. Produce GF to I, fo that FI be equal to FG; and from the centre I, with the radius IF, defcribe the quadrant FKH; and draw the right line IE. Then divide the arch KF into the fame number of equal parts as the radius of the craticular prototype is divided into; and from the centre I draw radii through all the points of divifion, meeting EF in 1, 2, 3, &c. Laftly, from the centre F, with the radii F1, F2, F3, &c, defcribe concentric circles. So will the craticular ectype be formed, whofe areolas will appear equal to each other.

Hence, what is delineated in every areola of the craticular prototype, being transferred into the areolas of the craticular ectype, the images will be diftorted or deformed; and yet they will appear in juft proportion

to an eye elevated above the vertex at a height equal to the height of the cone itfelf.

If the chords of the quadrants be drawn in the craticular prototype, and chords of each of the 4th parts in the craticular ectype, every thing elfe remaining the fame, there will be obtained the craticular ectype in a quadrangular pyramid.

And hence it will be eafy to deform an image, in any other pyramid, whofe bafe is any regular polygon.

Becaufe the illufion is more perfect when the eye, by the contiguous objects, cannot eftimate the diftance of the parts of the deformed image, it is therefore proper to view it through a fmall hole.

Anamorphofes, or monftrous images, may alfo be made to appear in their natural fhape and juft proportions, by means of mirrors of certain fhapes, from which thofe images are reflected again; and then they are faid to be reformed.

For farther particulars, fee Wolfius's *Catoptrics and Dioptrics,* and fome other optical authors.

ANAPHORA, in *Aftrology,* the fecond houfe, or that part of the heavens which is 30 degrees from the horofcope.

The term *anaphora* is alfo fometimes applied promifcuoufly to fome of the fucceeding houfes, as the 5th, the 8th, and the 11th. In this fenfe *anaphora* is the fame as *epanaphora,* and ftands oppofed to *cataphora.*

ANASTROUS *figns,* in *Aftronomy,* a name given to the *duodecatemoria,* or the 12 portions of the ecliptic, which the figns poffeffed anciently, but have fince deferted by the preceffion of the equinox.

ANAXAGORAS, one of the moft celebrated philofophers among the ancients. He was born at Clazomene in Ionia, about the 70th Olympiad. He was a difciple of Anaximenes; and he gave up his patrimony, to be more at leifure for the ftudy of philofophy, giving lectures in that fcience at Athens. Being perfecuted in this place, and at laft banifhed from it, he opened a fchool at Lampfacum, where he was greatly honoured during his life, and ftill more after his death, ftatues having been erected to his memory. It is faid he made fome predictions relative to the phenomena of nature, as earthquakes &c, upon which he wrote fome treatifes. His principal tenets may be reduced to the following:—All things were in the beginning confufedly mixed together, without order and without motion. The principle of things is at the fame time one and multiplex, which had the name of *homœmeries,* or fimilar particles, deprived of life. But there is befide this, from all eternity, another principle, an infinite and incorporeal fpirit, who gave motion to thefe particles; in virtue of which, fuch as are homogeneal united, and fuch as were heterogeneal feparated according to their different kinds. All things being thus put into motion by the fpirit, and every thing being united to fuch as are fimilar, thofe that had a circular motion produced heavenly bodies, the lighter particles afcending, while thofe that were heavier defcended. The rocks of the earth, being drawn up by the whirling force of the air, took fire, and became ftars, beneath which the fun and moon took their ftations. —It was faid he alfo wrote upon the *Quadrature of the Circle;* the treatife upon which, Plutarch fays, he compofed during his imprifonment at Athens.

ANAX.

ANAXIMANDER, a very celebrated Greek philosopher, was born at Miletus in the 42d olympiad; for, according to Apollodorus, he was 64 years of age in the 2d year of the 58th olympiad. He was one of the first who publicly taught philosophy, and wrote upon philosophical subjects. He was the kinsman, companion, and disciple of Thales. He wrote also upon the sphere and geometry, &c. And he carried his researches into nature very far, for the time in which he lived. It is said that he discovered the obliquity of the zodiac; that he first published a geographical table; that he invented the gnomon, and set up the first sun-dial in an open place at Lacedæmon. He taught, that infinity of things was the principal and universal element; that this infinite always preserved its unity, but that its parts underwent changes; that all things came from it; and that all were about to return to it. By this obscure and indeterminate principle he probably meant the chaos of other philosophers. He held that the worlds are infinite; that the stars are composed of air and fire, which are carried about in their spheres, and that these spheres are gods; and that the earth is placed in the midst of the universe, as in a common centre. Farther, that infinite worlds were the produce of infinity; and that corruption proceeded from separation.

ANAXIMENES an eminent Greek philosopher, born at Miletus, the friend, scholar, and successor of Anaximander. He diffused some degree of light upon the obscurity of his master's system. He made the first principle of things to consist in the air, which he considered as infinite or immense, and to which he ascribed a perpetual motion; that this air was the same as spirit or God, since the divine power resided in it, and agitated it. The stars were as fiery nails in the heavens; the sun a flat plate of fire; the earth an extended flat surface, &c.

ANDERSON (ALEXANDER), one of the brightest ornaments of the mathematical world, who flourished about 200 years ago. He was born at Aberdeen in Scotland, it would seem towards the latter part of the 16th century, as he was professor of mathematics at Paris in the early part of the 17th, where he published several ingenious works in geometry and algebra, both of his own, and of his friend Vieta's. Thus he published his "Supplementum Apollonii Redivivi; (of Ghetaldus) sive analysis problematis hactenus desiderati ad Apollonii Pergæi doctrinam περι νευσεων, a Marino Ghetaldo Patritio Ragusino hujusque, non ita pridem restitutam. In qua exhibetur mechanice æqualitatum tertii gradus sive solidarum, in quibus magnitudo omnino data, æquatur homogeneæ sub altero tantum coefficiente ignoto. Huic subnexa est variorum problematum practice." Paris, 1612, in 4to.

" Αιτιολογια: Pro Zetetico Apolloniani problematis a se jam pridem edito in supplemento Apollonii Redivivi. Ad clarissimum & ornatissimum virum Marinum Ghetaldum Patritium Ragusinum. In qua ad ea quae obiter mihi perstrinxit Ghetaldus respondetur, & analytices clarius detegitur." Paris, 1615, in 4to.

He published also,

" Francisci Vietæ Fontenacensis de Aequationum Recognitione & Emendatione Tractatus duo." Paris, 1615, in 4to; with a Dedication, Preface, and an Appendix, by Anderson.

And Vieta's Angulares Sectiones, with the Demonstrations by Anderson.

Alexander was cousin german to a Mr. David Anderson, of Finshaugh, a gentleman who also possessed a singular turn for mathematical and mechanical knowledge. This mathematical genius was hereditary in the family of the Andersons, and from them it seems to have been transmitted to their descendants of the name of Gregory in the same country: the daughter of the said David Anderson having been the mother of the celebrated mathematician James Gregory, and who herself first instructed her son James in the elements of the Mathematics, upon her observing in-him, while yet a child, a strong propensity to those sciences.

The time either of the birth or death of our author Alexander, has not come to my knowledge.

ANDROGYNOUS, an appellation given, by astrologers, to such of the planets as are sometimes hot, and sometimes cold; as mercury, which is accounted hot and dry when near the sun, and cold and moist when near the moon.

ANDROMEDA, in *Astronomy*, a constellation of the northern hemisphere, representing the figure of a woman almost naked, her feet at a distance from each other, and her arms extended and chained; being one of the original 48 asterisms, or figures under which the ancients comprehended the stars, as derived to us from the Greeks, who probably had them from the Egyptians or Indians, and who, it is suspected, altered their names, and accompanied them with fabulous stories of their own. According to them, Cepheus, the father of Andromeda, was obliged to give her up to be devoured by a monster, to preserve his kingdom from the plague; but that she was delivered by Perseus, who slew the monster, and espoused her. And the family were all translated by Minerva to heaven, the mother being the constellation Cassiopeia.

She is sometimes called, in Latin, *Persea, Mulier catenata, Virgo devota*, &c. The Arabians, whose religion did not permit them to draw the figure of the human body on any occasion whatever, have changed this constellation into the figure of a sea-calf. Schickard has changed the name for that of the scripture name *Abigail*. And Schiller has also changed the figure of the constellation, for that of a sepulchre, and calls it the *Holy Sepulchre*.

This constellation contains about 27 stars that are visible to the naked eye; of which the principal are, α Andromeda's head; β in the girdle, and called *mirach* or *mizar*; γ on the south foot, and named *alamak*, and sometimes *alhames*.

The number of stars placed in this constellation by the catalogue of Ptolemy is 23, by that of Tycho Brahe also 23, by that of Hevelius 47, and by that of Flamsteed 66.

ANEMOMETER, an instrument for measuring the force of the wind.

An instrument of this sort, it seems, was first invented by Wolfius in the year 1708, and first published in his *Areometry* in 1709, also in the *Acta Eruditorum* of the same year; afterwards in his *Mathematical Dictionary*, and in his *Elem. Matheseos*. He says he tried the goodness of it, and observes that the internal struc-

ture

ture may be preserved, so as to measure the force of running water, or that of men or horses when they draw or pull. The machine consists of sails, A, B, C, like those of a wind-mill, against which the wind blows, and by turning them about, raises an arm K with a weight L upon it, to different angles of elevation, shewn by the index M, according to the force of the wind. (Plate III. fig. 3)

In the Philos. Transf. another anemometer is described, in which the wind being supposed to blow directly against a flat side, or board, which moves along the graduated arch of a quadrant, the number of degrees it advances shews the comparative force of the wind.

In the same Transactions, for the year 1766, Mr. Alex. Brice describes a method, successfully practised by him, of measuring the velocity of the wind, by means of that of the shadow of clouds passing over a plane upon the earth.

Also in the same Transactions, for the year 1775, Dr. Lind gives a description of a very ingenious portable *Wind-Gauge*, by which the force of the wind is easily measured; a brief description of the principal parts of which here follows. This simple instrument consists of two glass tubes, AB, CD, (Plate III. fig. 4.) which should not be less than 8 or 9 inches long, the bore of each being about $\frac{4}{10}$ of an inch diameter, and connected together by a small bent glass tube *ab*, only of about $\frac{1}{10}$ of an inch diameter, to check the undulations of the water caused by a sudden gust of wind. On the upper end of the leg AB is fitted a thin metal tube, which is bent perpendicularly outwards, and having its mouth open to receive the wind blowing horizontally into it. The two tubes, or rather the two branches of the tube, are connected to a steel spindle KL by slips of brass near the top and bottom, by the sockets of which at *e* and *f* the whole instrument turns easily about the spindle, which is fixed into a block by a screw in its bottom, by the wind blowing in at the orifice at F. When the instrument is used, a quantity of water is poured in, till the tubes are about half full; then exposing the instrument to the wind, by blowing in at the orifice F, it forces the water down lower in the tube AB, and raises it so much higher in the other tube; and the distance between the surfaces of the water in the two tubes, estimated by a scale of inches and parts HI, placed by the sides of the tubes, will be the height of a column of water whose weight is equal to the force or momentum of the wind blowing or striking against an equal base. And as a cubic foot of water weighs 1000 ounces, or $62\frac{1}{2}$ pounds, the 12th part of which is $5\frac{5}{24}$ or $5\frac{1}{4}$ pounds nearly, therefore for every inch the surface of the water is raised, the force of the wind will be equal to so many times $5\frac{1}{4}$ pounds on a square foot. Thus, suppose the water stand 3 inches higher in the one tube, than in the other; then 3 times 5^1 or 15^3 pounds is equal to the pressure or force of the wind on the surface of a foot square.

This instrument of Dr. Lind's, measures only the force or momentum of the wind, but not its velocity. However the velocity of the wind may be deduced from its force so obtained, by help of some experiments performed by me at the Royal Military Academy, in the years 1786, 1787, and 1788; from which experiments it appears that a plane surface of a square foot suffers a resistance of 12 ounces from the wind, when blowing with a velocity of 20 feet per second; and that the force is nearly as the square of the velocity. Hence then, taking the force of $15\frac{3}{4}$ pounds, above found for the force of the wind when it sustains 3 inches of water, and taking the square roots of the forces, it will be, as $\sqrt{12}$: $\sqrt{15\frac{3}{4}}$ :: $20 : 22\frac{3}{4}$ the 4th proportional, that is a velocity of $22\frac{3}{4}$ feet per second, or $15\frac{1}{2}$ miles per hour, is the rate or velocity at which the wind blows, when it raises the water 3 inches higher in the one tube than the other. And farther, as the said height is as the force, and the force as the square of the velocity, we shall have the force and velocity, corresponding to several heights of the water in the one tube, above that in the other, as in the following table.

Table of the corresponding height of water, force on a square foot, and velocity of wind.

Height of water.	Force of wind.	Velocity of wind per hour.
Inches.	Pounds.	Miles.
$0\frac{1}{4}$		4·5
$0\frac{1}{2}$		6·4
1	5·2	9·0
2	10·4	12·7
3	15·6	15·5
4	20·8	18·0
5	26·0	20·1
6	31·25	22·0
7	36·5	23·8
8	41·7	25·4
9	46·9	27·0
10	52·1	28·4
11	57·3	29·8
12	62·5	31·0

In one instance Dr. Lind found that the force of the wind was such as to be equal $34\frac{9}{10}$ pounds, on a square foot; and this by proportion, in the following table, will be found to answer to a velocity of $23\frac{1}{4}$ miles per hour.

Mr. Leutmann improved upon Wolfius's anemometer, by placing the sails horizontal, instead of vertical, which are easier to move, and turn what way soever the wind blows.

Mr. Benjamin Martin also (Plate III. fig. 5) improved upon the same. He made the axis like the fusee of a watch, having a cord winding upon it, with two weights at the ends which make always a balance to the force of the wind on the sails. See his Philos. Britan.

And M. D'Ons-en-Bray invented a new anemometer, which of itself expresses on paper, not only the several winds that have blown during the space of 24 hours, and at what hour each began and ended, but also the different strength or velocity of each. See Mem. Acad. Scienc. an. 1734. See also the article *Wind-Gauge*.

ANEMOSCOPE, is sometimes used to denote a machine invented to foretell the changes of the wind, or weather; and sometimes for an instrument shewing by an index what the present direction of the wind is. Of this latter sort, it seems, was that used by the ancients, and described by Vitruvius; and we have many of them at present in large or public buildings, where an index within side a room or hall, points to the name of the quarter from whence the wind blows without; which is simply effected by connecting an index to the lower end of the spindle of a weather-cock.

It has been obſerved that hygroſcopes made of cat-gut, or ſuch like, prove very good anemoſcopes; ſeldom failing, by the turning of the index, to foretell the ſhifting of the wind. See accounts of two different anemoſcopes; one by Mr. Pickering, vol 43 Philoſ. Tranſ. the other by Mr. B. Martin, vol. 2 of his Philoſ. Britan.

Otto Gueric alſo gave the title anemoſcope to a machine invented by him to foretell the change of the weather, as to rain and fair. It conſiſted of the ſmall wooden figure of a man, which roſe and fell in a glaſs tube, as the atmoſphere was more or leſs heavy. Which was only an application of the common barometer, as ſhewn by M. Couriers in the *Acta Eruditorum* for 1684.

ANGLE, *Angulus*, in *Geometry*, the opening or mu-tual inclination of two lines, or two planes, or three planes, meeting in a point called the vertex or angular point. Such as the angle formed by, or between, the two lines AB and AC, at the vertex or angular point A.—Alſo the two lines AB and AC, are called the *legs* or the *ſides* of the angle.

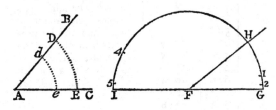

Angles are ſometimes denoted, or named, by the ſingle letter placed at the angular point, as the angle A; and ſometimes by three letters, placing always that of the vertex in the middle. The former method is uſed when only one angle has the ſame vertex; and the latter method it is neceſſary to uſe when ſeveral angles have the ſame vertex, to diſtinguiſh them from one another.

The meaſure of an angle, by which its quantity or magnitude is expreſſed, is an arch, as DE deſcribed from the centre A, with any radius at pleaſure, and contained between its legs AB and AC.—Hence angles are compared and diſtinguiſhed by the ratio of the arcs which ſubtend them, to the whole circum-ference of the circle; or by the number of degrees contained in the arc DE by which they are meaſured, to 360, the number of degrees in the whole circum-ference of the circle. And thus an angle is ſaid to be of ſo many degrees, viz, as are contained in the arc DE.

Hence it matters not, with what radius the arc is deſcribed, by which an angle is meaſured, when great or ſmall, as AD, or A*d,* or any other: for the arcs DE, *de*, being ſimilar, have the ſame ratio as their reſpective radii or circumferences, and therefore they contain the ſame number of degrees.—Hence it follows, that the quantity or magnitude of the angle remains ſtill the ſame, though the legs be ever ſo much in-creaſed or diminiſhed.—And thus, in ſimilar figures, the like or correſponding angles are equal.

The *taking* or *meaſuring* of *angles*, is an operation of great uſe and extent in ſurveying, navigation, geo-graphy, aſtronomy, &c. And the inſtruments chiefly uſed for this purpoſe, are quadrants, ſextants, octants,

theodolites, circumferentors, &c. Mr. Hadley invented an excellent inſtrument for taking the *larger* ſort of angles, where much accuracy is required, or where the motion of the object, or any circumſtance cauſing an unſteadineſs in the common inſtruments, renders the obſervations difficult, or uncertain. And Mr. Dollond contrived an inſtrument for meaſuring *ſmall* angles. See *Hadley's Quadrant,* *Micrometer,* and the Philoſ. Tranſ. Numbers 420, 435, and vol. 48.

To meaſure the Quantity of an Angle.

1. *On paper.* Apply the centre of a protractor to the vertex A of the angle, ſo that the radius may co-incide with one leg, as AB; then the degree on the arch that is cut by the other leg AC, will give the meaſure of the angle required.

Or thus, by a line of chords. Take off the chord of 60 with a pair of compaſſes; and with that radius, from the centre A, deſcribe an arc as DE. Then take this arc DE between the compaſſes, and apply the extent to the ſcale of chords, which will give the degrees in the angle as before.

M. De Lagny gave, in ſeveral memoirs of the Royal Academy of Sciences, a new method of meaſuring angles, which he called *Goniometry.* The method con-ſiſts in meaſuring, with a pair of compaſſes, the arc which ſubtends the propoſed angle, not by applying its extent to a pre-conſtructed ſcale, like chords, but in the following manner: From the angular point as a centre, with a pretty large radius, deſcribe a circle, producing one leg of the angle backwards to cut off a ſemicircle; then ſearch out what part of the ſemicircle the arc is which meaſures the given angle, in this manner; viz, take the extent of this arc with a very fine pair of com-paſſes, and apply it ſeveral times to the arc of the ſemi-circle, to find how often it is contained, with a ſmall part remaining over; in the ſame manner take the ex-tent of this ſmall part, and apply it to the firſt arc, to find how often it is contained in it; and what remains this 2d time, apply in like manner to the firſt remainder; then the 3d remainder apply to the 2d, and ſo on, always counting how often the laſt remainder is con-tained in the next foregoing, till nothing remain, or till the remainder is inſenſible, and too ſmall to be meaſured: Then, beginning at the laſt, and returning backwards, make a ſeries of fractions of which the numerators are always 1, and the denominators are the number of times each remainder is contained in its next remainder, with the fractional part more, as derived from the following remainder; then the laſt fraction, thus obtained, will ſhew what part the given angle is of 180° or the ſemi-circle; and being turned into degrees &c, will be the meaſure of the angle, and nearer, it is aſſerted, than it can be obtained by any other means; whether it be meaſuring, or calculating by trigonometrical tables.— Thus, if it be required to meaſure the angle GFH: With a large radius deſcribe the ſemicircle GHI, meeting the leg FG produced in I; then take the ex-tent of the arc GH in the compaſſes, and applying it from G upon the ſemicircle. ſuppoſe it contains 4 times to the point 4, and the part 4 I over; take 4 I and apply it from H to 1, ſo that HG contains 4 I once, and 1 G over; alſo apply this remainder to the former 4 I, and it contains 5 times, from 4 to 5, and 5 I over; and

and lastly the remainder 51 is just two times contained in the former remainder 1 G or 1 2, without any remainder. Here then, the series of quotients, or numbers of times contained, are 4, 1, 5, 2; therefore, beginning at the last, the first fraction is $\frac{1}{2}$, or the last remainder is half the preceding one; and the 2d fraction is $\frac{1}{5\frac{1}{2}}$ or $\frac{2}{11}$; the 3d is $\frac{1}{1\frac{2}{11}}$ or $\frac{11}{13}$; and the fourth is

$\frac{1}{4\frac{11}{13}}$ or $\frac{13}{63}$; that is, the arc GH is $\frac{13}{63}$ of a semicircle, or the angle GFH is $\frac{13}{63}$ of two right angles, or of 180°, which is equivalent to 37¼ degrees, or 37° 8′ 34″ $\frac{2}{7}$.

2. *On the ground.* Place a surveying instrument with its centre over the angular point to be measured, turning the instrument about till o, the beginning of its arch, fall in the line or direction of one leg of the angle; then turn the index about to the direction of the other leg, and it will cut off from the arch the degrees answering to the given angle.

To plot or lay down any given angle, either on paper or on the ground, may be performed in the same manner; and the method is farther explained under the articles PLOTTING and PROTRACTING, and under the names of the several instruments.

To bisect a given angle, as suppose the angle LKM. From the centre K, with any radius, describe the arc LM; then with the centres L and M, describe two arcs intersecting in N; and draw the line KN, which will bisect the given angle LKM, dividing it into the two equal angles LKN, MKN.

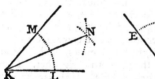 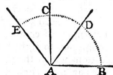

To trisect an angle, see TRISECTION.

Pappus, in his Mathematical Collections, book 4, treats of angular sections, but particularly and more largely, of trisections. He also treats of any section in general, in the 36th and following propositions.

ANGLES are of various kinds and denominations. With regard to the form of their legs, they are divided into *rectilinear*, *curvilinear*, and *mixed*.

Rectilinear, or *right-lined* ANGLE, is that whose legs are both right lines; as the foregoing angle CAB.

Curvilinear ANGLE, is that whose legs are both of them curves.

Mixt, or *mixtilinear* ANGLE, is that of which one leg is a right line, and the other a curve.

With regard to their *magnitude*, angles are again divided into *right* and *oblique*, *acute* and *obtuse*.

Right ANGLE, is that which is formed by one line perpendicular to another; or that which is subtended by a quadrant of a circle. As the angle BAC.—Therefore the measure of a right angle is a quadrant of a circle, or 90°; and consequently all right angles are equal to each other.

Obliqu. ANGLE, is a common name for any angle that is not a right one; and it is either *acute* or *obtuse*.

Acute ANGLE, is that which is less than a right angle, or less than 90 degrees; as the angle BAD. And

Obtuse ANGLE, is greater than a right angle, or whose measure exceeds 90 degrees; as the angle BAE.

With regard to their situation in respect of each other, angles are distinguished into *contiguous*, *adjacent*, *vertical*, *opposite*, and *alternate*.

Contiguous ANGLES, are such as have the same vertex, and one leg common to both. As the angles BAD, CAD, which have AD common.

Adjacent ANGLES, are those of which a leg of the one produced forms a leg of the other: as the angles GFH and IFH, which have the legs IF and FG in a straight line.—Hence adjacent angles are supplements to each other, making together 180 degrees. And therefore if one of these be given, the other will be known by subtracting the given one from 180 degrees. Which property is useful in surveying, to find the quantity of an inaccessible angle; viz, measure its adjacent accessible one, and subtract this from 180 degrees.

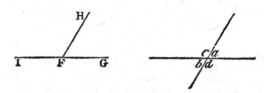

Vertical or *opposite* ANGLES, are such as have their legs mutually continuations of each other; as the two angles *a* and *b*, or *c* and *d*.—The property of these is, that the vertical or opposite angles are always equal to each other, viz, $\angle a = \angle b$, and $\angle c = \angle d$. And hence the quantity of an inaccessible angle of a field, &c, may be found, by measuring its accessible opposite angle.

Alternate ANGLES, are those made on the opposite sides of a line cutting two parallel lines; so, the angles *e* and *f*, or *g* and *h*, are alternates. And these are always equal to each other; viz, the $\angle e = \angle f$, or $\angle g = \angle h$.

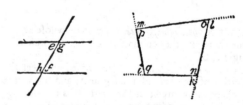

External ANGLES, are the angles of a figure made without it, by producing its sides outwards; as the angles *i*, *k*, *l*, *m*. All the external angles of any right-lined figure, taken together, are equal to 4 right angles; and the external angle of a triangle is equal to both the internal opposite ones taken together; also any external angle of a trapezium inscribed in a circle, is equal to the internal opposite angle.

Internal ANGLES, are the angles within any figure, made by the sides of it; as the angles *n*, *o*, *p*, *q*.—In any right-lined figure, an internal angle as *n*, and its adjacent external angle *k*, together make two right angles, or 180 degrees; and all the internal angles *n*, *o*, *p*, *q*,

taken

taken together, make twice as many right angles, wanting 4 right angles; also any two opposite internal angles of a trapezium inscribed in a circle, taken together, make two right angles, or 180 degrees.

Homologous, or *like* ANGLES, are such angles in two figures, as retain the same order from the first, in both figures.

ANGLE *out of the centre*, as G, is one whose vertex is not in the centre of the circle.—And its measure is half the sum $\frac{a+b}{2}$ of the arcs intercepted by its legs when it is within the circle, or half the difference $\frac{a-b}{2}$ when it is without.

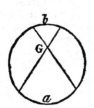 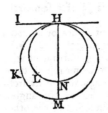

ANGLE *at the centre*, is an angle whose vertex is in the centre; as the angle AFC, formed by two radii AF, FC, and measured by the arc ADC.—An angle at the centre, as AFC, is always double of the angle ABC at the circumference, standing upon the same arc ADC; and all angles at the centre are equal that stand upon the same or equal arcs: also all angles at the centre, are proportional to the arcs they stand upon; and so also are all angles at the circumference.

 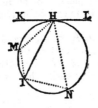

ANGLE *at the circumference*, is an angle whose vertex is somewhere in the circumference of a circle; as the angle ABC.

ANGLE *in a segment*, is an angle whose legs meet the extremities of the base of the segment, and its vertex is anywhere in its arch; as the angle B is in the segment ABC, or standing upon the supplemental segment ADC; and is comprehended between two chords AB and BC.—An angle at the circumference is measured by half the arc ADC upon which it stands; and all the angles ABC, AEC, in the same segment, are equal to each other.

ANGLE *in a semicircle* is an angle at the circumference contained in a semicircle, or standing upon a semicircle, or on a diameter.—An angle in a semicircle, is always a right angle; in a greater segment, the angle is less, and in a less segment the angle is greater than a right angle.

ANGLE *of a segment*, is that made by a chord with a tangent, at the point of contact. So IHK is the angle of the less segment IMH, and IHL, the angle of the greater segment INH.—And the measure of each of these angles, is half the alternate or supplemental segment, or equal to the angle in it; viz, the ∠ IHK = ∠ INH, and the ∠ IHL = ∠ IMH.

ANGLE *of a semicircle*, is the angle which the diameter of a circle makes with the circumference. And Euclid demonstrates that this is less than a right angle, but greater than any acute angle.

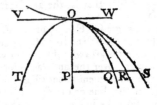

ANGLE *of contact*, is that made by a curve line and a tangent to it, at the point of contact; as the angle IHK. It is proved by Euclid, that the angle of contact between a right line and a circle, is less than any right-lined angle whatever; though it does not therefore follow that it is of no magnitude or quantity. This has been the subject of great disputes amongst geometricians, in which Peletarius, Clavius, Taquet, Wallis, &c, bore a considerable share; Peletarius and Wallis contending that it is no angle at all, against Clavius, who rightly asserts that it is not absolutely nothing in itself, but only of no magnitude in comparison with a right-lined angle, being a quantity of a different kind or nature; like as a line in respect to a surface, or a surface in respect to a solid, &c. And since his time, it has been proved by Sir I. Newton, and others, that angles of contact can be compared to each other, though not to right-lined angles, and what are the proportions which they bear to each other. Thus, the circular angles of contact IHK, IHL, are to each other in the reciprocal subduplicate ratio of the diameters HM, HN. And hence the circular angle of contact may be divided, by describing intermediate circles into any number of parts, and in any proportion. And if, instead of circles, the curves be parabolas, and the point of contact H the common vertex of their axes; the angles of contact would then be reciprocally in the subduplicate ratio of their parameters. But in such elliptical and hyperbolical angles of contact, these will be reciprocally in the subduplicate of the ratio compounded of the ratios of the parameters, and the transverse axes. Moreover, if TOQ be a common parabola, to the axis OP, and tangent VOW, and whose equation is $1 \times x = y^2$, or $x = y^2$, where x is the abscis OP, and y the ordinate PQ, the parameter being 1: and if OR, OS, &c, be other parabolas to the same axis, tangent, and parameter, their ordinate y being PR, or PS, &c, and their equations $x = y^3$, $x = y^4$, $x = y^5$, &c: then the series of angles of contact will be in succession infinitely greater than each other, viz, the angle of contact WOQ infinitely greater than WOR, and this infinitely greater than WOS, and so on infinitely.

And farther, between the angles of contact of any two of this kind, may other angles of contact be found *ad infinitum*, which shall infinitely exceed each other,

and

and yet the greatest of them be infinitely less than the smallest right lined angle. So also $x^2 = y^3$, $x^3 = y^4$, $x^4 = y^5$, &c, denote a series of curves, of which every succeeding one makes an angle with its tangent, infinitely greater than the preceding one; and the least of these, viz, that whose equation is $x^2 = y^3$, or the semicubical parabola, is infinitely greater than any circular angle of contact.

ANGLES are again divided into *plane*, *spherical*, and *solid*.

Plane ANGLES, are all those above treated of; which are defined by the inclination of two lines in a plane, meeting in a point.

Spherical ANGLE, is an angle formed on the surface of a sphere by the intersection of two great circles; or, it is the inclination of the planes of the two great circles.

The measure of a spherical angle, is the arc of a great circle of the sphere, intercepted between the two planes which form the angle, and which cuts the said planes at right angles. For their properties, &c, see SPHERE, SPHERICAL, and SPHERICAL TRIGONOMETRY.

Solid ANGLE, is the mutual inclination of more than two planes, or plane angles, meeting in a point, and not contained in the same plane; like the angles or corners of solid bodies. For their measure, properties, &c, see SOLID *Angle*.

Angles of other less usual kinds and denominations, are also to be found in some books of Geometry. As,

Horned ANGLE, *angulus cornutus*, that which is made by a right line, whether a tangent or secant, with the circumference of a circle.

Lunular ANGLE, *angulus lunularis*, is that which is formed by the intersection of two curve lines, the one concave, and the other convex.

Cissoid ANGLE, *angulus cissoides*, the inner angle made by two spherical convex lines intersecting each other.

Sistroid ANGLE, *angulus sistroides*, is that which is in form of a sistrum.

Pelecoid ANGLE, *angulus pelecoides*, is that in form of a hatchet.

ANGLE, in *Trigonometry*. See TRIANGLE, TRIGONOMETRY, SINE, TANGENT, &c.

ANGLE, in *Mechanics*.—*Angle of Direction*, is that which is comprehended between the lines of direction of two conspiring forces.

ANGLE *of Elevation*, is that which is comprehended between the line of direction and any plane upon which the projection is made, whether horizontal or oblique.

ANGLE *of Incidence*, is that made by the line of direction of an impinging body, at the point of impact. As the angle ABC.

ANGLE *of Reflection*, is that made by the line of direction of the reflected body, at the point of impact. As the angle DBE.

Instead of the angles of incidence and reflection being estimated from the plane on which the body impinges, sometimes the complements of these are understood, viz, as estimated from a perpendicular to the reflecting plane; as the two angles ABF and DBF.

ANGLE in *Optics*.—*Visual* or *Optic Angle*, is the angle included between the two rays drawn from the two extreme points of an object to the centre of the pupil of the eye: as the angle HGI. The apparent magnitude of objects is greater or less, according to the angle under which they appear.—Objects seen under the same or an equal angle, always appear equal— The least *visible* angle, or least angle under which a body can be seen, according to Dr. Hook, is one minute; but Dr. Jurin shews, that at the time of his debate with Hevelius on this subject, the latter could probably discover a single star under so small an angle as 20 But bodies are visible under smaller angles as they are more bright or luminous. Dr. Jurin states the grounds of this controversy, and discusses the question at large, in his Essay upon distinct and indistinct Vision, published in Smith's Optics, pa. 148, & seq.

ANGLE *of the interval*, of two places, is the angle subtended by two lines directed from the eye to those places.

ANGLE of *incidence*, or *reflection*, or *refraction*, &c. See the respective words INCIDENCE, REFLECTION, REFRACTION, &c.

ANGLE in *Astronomy*.—*Angle of Commutation*. See COMMUTATION.

ANGLE *of elongation*, or *Angle at the Earth*. See ELONGATION.

Parallactic ANGLE, or the *parallax*, is the angle made at the centre of a star, the sun, &c, by two lines drawn, the one to the centre of the earth, and the other to its surface. See PARALLACTIC, and PARALLAX.

ANGLE *of the position of the sun, of the sun's apparent semi diameter*, &c. See the respective words.

ANGLE *at the sun*, is the angle under which the distance of a planet from the ecliptic, is seen from the sun.

ANGLE *of the East*. See NONAGESIMAL.

ANGLE *of obliquity*, of the ecliptic, or the angle of inclination of the axis of the earth, to the axis of the ecliptic, is now nearly 23° 28'. See OBLIQUITY, and ECLIPTIC.

ANGLE *of longitude*, is the angle which the circle of a star's longitude makes with the meridian, at the pole of the ecliptic.

ANGLE *of right ascension*, i. the angle which the circle of a star's right ascension makes with the meridian at the pole of the equator.

ANGLE in *Navigation*. ANGLE *of the rhumb*, or *loxodromic angle*. See RHUMB and LOXODROMIC.

ANGLES, in *Fortification*, are understood of those formed by the several lines used in fortifying, or making a place defensible.

These are of two sorts; *real* and *imaginary*.—*Real angles* are those which actually exist and appear in the works. Such as the *flanked angle*, the *angle of the epaule*, *angle of the flank*, and the *re-entering angle of the counterscarp*. Imaginary, or *occult angles*, are those which are only subservient to the construction, and which exist no more after the fortification is drawn. Such as the *angle of the centre, angle of the polygon, flanking angle, saliant angle of the counterscarp*, &c.

ANGLE *of*, or *at, the centre*, is the angle formed at the centre of the polygon, by two radii drawn from the centre to two adjacent angles, and subtended by a side

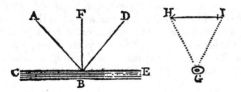

of

of it, as the angle ACB. This is found by dividing 360 degrees by the number of sides in the regular polygon.

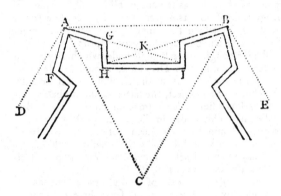

ANGLE *of the Polygon*, is the angle intercepted between two sides of the polygon; as DAB, or ABE. This is the supplement of the angle at the centre, and is therefore found by subtracting the angle C from 180 degrees.

ANGLE *of the Triangle*, is half the angle of the polygon; as CAB or CBA; and is therefore half the supplement of the angle C at the centre.

ANGLE *of the Bastion*, is the angle FAG made by the two faces of the bastion. And is otherwise called the flanked angle.

Diminished ANGLE, is the angle BAG made by the meeting of the exterior side of the polygon with the face AG of the bastion.

ANGLE *of the curtin*, or *of the flank*, is the angle GHI made between the curtin and the flank.

ANGLE *of the epaule*, or *shoulder*, is the angle AGH made by the flank and the face of the bastion.

ANGLE *of the tenaille*, or *exterior flanking angle*, is the angle AKB made by the two rasant lines of defence, or the faces of two bastions produced.

ANGLE *of the counterscarp*, is the angle made by the two sides of the counterscarp, meeting before the middle of the curtin.

ANGLE, *flanking inward*, is the angle made by the flanking line with the curtin.

ANGLE *forming the flank*, is that consisting of one flank and one demigorge.

ANGLE *forming the face*, is that composed of one flank and one face.

ANGLE *of the moat*, is that made before the curtin, where it is intersected.

Re-entering, or *re-entrant* ANGLE, is that whose vertex is turned inwards, towards the place; as H or I.

Saliant, or *sortant* ANGLE, is that turned outwards, advancing its point towards the field; as A or G.

Dead ANGLE, is a re-entering angle, which is not flanked or defended.

ANGLE *of a wall*, in *Architecture*, is the point or corner where the two sides or faces of a wall meet.

ANGLES, in *Astrology*, denote certain houses of a figure, or scheme of the heavens. So the horoscope of the first house, is termed the *angle of the east*.

ANGUINEAL *Hyperbola*, a name given by Sir

I. Newton to four of his curves of the second order, viz, species 33, 34, 35, 36, expressed by the equation $xy^2 + ey = -ax^3 + bx^2 + cx + d$; being hyperbolas of a serpentine figure. See CURVES.

ANGULAR, something relating to, or that hath angles.

At a distance, angular bodies appear round; the angles and small inequalities disappearing at a much less distance than the bulk of the body.

ANGULAR *Motion*, is the motion of a body which moves circularly about a point; or the variation in the angle described by a line, or radius, connecting a body with the centre about which it moves.—Thus, a pendulum has an angular motion about its centre of motion; and the planets have an angular motion about the sun.—Two moveable points M and O, of which the one describes the arc MN, and the other the arc OP, in the same time, have an equal, or the same angular motion, although the real motion of the point O be much greater than that of the point M, viz, as the arc OP is greater than the arc MN. The angular motions of revolving bodies, as of the planets about the sun, are reciprocally proportional to their periodic times. And they are also, as their real or absolute motions directly, and as their radii of motion inversely.

ANGULAR *motion* is also a kind of compound motion composed of a circular and a rectilinear motion; like the wheel of a coach, or other vehicle.

ANIMATED *needle*, a needle touched with a magnet or load-stone.

ANNUAL, in *Astronomy*, something that returns every year, or which terminates with the year.

ANNUAL *motion of the earth*. See EARTH.

ANNUAL *argument of longitude*. See ARGUMENT.

ANNUAL *epacts*. See EPACT.

ANNUAL *equation* of the mean motion of the sun and moon, and of the moon's apogee and nodes. See EQUATION.

ANNUITIES, a term for any periodical income, arising from money lent, or from houses, lands, salaries, pensions, &c; payable from time to time; either annually, or at other intervals of time.

Annuities may be divided into such as are *certain*, and such as depend on some *contingency*, as the continuance of a life, &c.

Annuities are also divided into annuities in *possession*, and annuities in *reversion*; the former meaning such as have commenced; and the latter such as will not commence till some particular event has happened, or till some given period of time has elapsed.

Annuities may be farther considered as payable either *yearly*, or *half yearly*, or *quarterly*, &c.

The *present value* of an annuity, is that sum, which, being improved at interest, will be sufficient to pay the annuity.

The *present value* of an *annuity certain*, payable yearly, is calculated in the following manner.—Let the annuity be 1, and let *r* denote the amount of 1*l.* for a year, or 1*l.* increased by its interest for one year. Then, 1 being the present value of the sum *r*, and having to find the present value of the sum 1, it will be, by proportion

thus

thus, $r : 1 :: 1 : \frac{1}{r}$ the prefent value of $1l$. due a year hence. In like manner $\frac{1}{r^2}$ will be the prefent value of $1l$. due 2 years hence; for $r : 1 :: \frac{1}{r} : \frac{1}{r^2}$. In like manner $\frac{1}{r^3}, \frac{1}{r^4}, \frac{1}{r^5}$, &c, will be the prefent value of $1l$. due at the end of 3, 4, 5, &c, years refpectively; and in general, $\frac{1}{r^n}$ will be the value of $1l$. to be received after the expiration of x years. Confequently the fum of all thefe, or $\frac{1}{r} + \frac{1}{r^2} + \frac{1}{r^3} + \frac{1}{r^4} +$ &c, continued to n terms, will be the prefent value of all the n years annuities. And the value of the perpetuity, is the fum of the feries continued *ad infinitum*.

But this feries, it is evident, is a geometrical progreffion, whofe firft term and common ratio are each $\frac{1}{r}$, and the number of its terms n; and therefore the fum s of all the terms, or the prefent value of all the annual payments, will be $s = \frac{1}{r-1} - \frac{1}{r-1} \times \frac{1}{r^n}$.

When the annuity is a perpetuity, it is plain that the laft term $\frac{1}{r^n}$ vanifhes, and therefore $\frac{1}{r-1} \times \frac{1}{r^n}$ alfo vanifhes; and confequently the expreffion becomes barely $s = \frac{1}{r-1}$; that is, any annuity divided by its intereft for one year, is the value of the perpetuity. So, if the rate of intereft be 5 per cent; then $\frac{100}{5} = 20$ is the value of the perpetuity at 5 per cent. Alfo $\frac{100}{4} = 25$ is the value of the perpetuity at 4 per cent. And $\frac{100}{3} = 33\frac{1}{3}$ is the value of the perpetuity at 3 per cent. intereft. And fo on.

If the annuity is not to be entered on immediately, but after a certain number of years, as m years; then the prefent value of the reverfion is equal to the difference between two prefent values, the one for the firft term of m years, and the other for the end of the laft term n: that is, equal to the difference between

$$\frac{1}{r-1} - \frac{1}{r-1} \times \frac{1}{r^n} \text{ and } \frac{1}{r-1} - \frac{1}{r-1} \times \frac{1}{r^m}, \text{ or } =$$

$$\frac{1}{r-1} \times \overline{\frac{1}{r^m} - \frac{1}{r^n}}$$

Annuities certain differ in value, as they are made payable *yearly*, *half-yearly*, or *quarterly*. And by proceeding as above, ufing the intereft or amount of a half year, or a quarter, as thofe for the whole year were

ufed, the following fet of theorems will arife; where r denotes, as before, the amount of $1l$. and its intereft for a year, and x the number of years, during which, any annuity is to be paid; alfo P denotes the perpetuity $\frac{1}{r-1}$, Y denotes $\frac{1}{r-1} - \frac{1}{r-1} \times \frac{1}{r^n}$ the value of the annuity fuppofed payable yearly, H the value of the fame when it is payable half-yearly, and Q the value when payable quarterly; or univerfally, M the value when it is payable every m part of a year.

THEOR. 1. $Y = P - P \times \left(\frac{1}{r}\right)^n$

THEOR. 2. $H = P - P \times \left(\frac{2}{r+1}\right)^{2n}$.

THEOR. 3. $Q = P - P \times \left(\frac{4}{r+3}\right)^{4n}$

THEOR. 4. $M = P - P \times \left(\frac{m}{r+m-1}\right)^{mn}$

Example 1.

Let the rate of intereft be 4 per cent, and the term 5 years; and confequently $r = 1\cdot04$, $n = 5$, $P = 25$; alfo let $m = 12$, or the intereft payable monthly in theorem 4: then the prefent value of fuch annuity of $1l$. a year, for 5 years, according as it is fuppofed payable $1l$. yearly, or $\frac{1}{2}l$ every half year, or $\frac{1}{4}l$. every quarter, or $\frac{1}{12}l$. every month or $\frac{1}{12}$th part of a year, will be as follows:

$Y = 25 - 25 \times \cdot821928 = 4\cdot4518$
$H = 25 - 25 \times \cdot820348 = 4\cdot4913$
$Q = 25 - 25 \times \cdot819543 = 4\cdot5114$
$M = 25 - 25 \times \cdot818996 = 4\cdot5251$

EXAMPLE 2. Suppofing the annuity to continue 25 years, the rate of intereft and every thing elfe being as before; then the values of the annuities for 25 years will be

$Y = 25 - 25 \times \cdot375118 = 15\cdot6221$
$H = 25 - 25 \times \cdot371527 = 15\cdot7118$
$Q = 25 - 25 \times \cdot369709 = 15\cdot7573$
$M = 25 - 25 \times \cdot368477 = 15\cdot7881$

EXAMPLE 3. And if the term be 50 years, the values will be

$Y = 25 - 25 \times \cdot140713 = 21\cdot4822$
$H = 25 - 25 \times \cdot138032 = 21\cdot5492$
$Q = 25 - 25 \times \cdot136685 = 21\cdot5829$
$M = 25 - 25 \times \cdot135775 = 21\cdot6056$

EXAMPLE 4. Alfo if the term be 100 years, the values will be

$Y = 25 - 25 \times \cdot019800 = 24\cdot5050$
$H = 25 - 25 \times \cdot019053 = 24\cdot5237$
$Q = 25 - 25 \times \cdot018683 = 24\cdot5329$
$M = 25 - 25 \times \cdot018435 = 24\cdot5391$

Hence the difference in the value by making periods of payments fmaller, for any given term of years, is the more as the intervals are fmaller, or the periods more frequent. The fame difference is alfo variable, both as the rate of intereft varies, and alfo as the whole term of years n varies; and, for any given rate of intereft, it is evident that the difference, for any periods m of payments, first increases from nothing as the term n increases, when n is 0, to some certain finite term or value of n, when the difference D is the greatest or a maximum; and that afterwards, as n increases more, that difference will continually decrease to nothing again, and vanish when n is infinite: also the term or value of n, for the maximum of the difference, will be different according to the periods of payment, or value of m. And the general value of n, when the difference is a maximum between the yearly payments and the payments of m times in a year, is expressed by this formula, viz,

$$n = \dfrac{l.\dfrac{ml.\dfrac{m+r-1}{m}}{l.r}}{ml.\dfrac{m+r-1}{m} - l.r},$$ where l. denotes the logarithm of the quantity following it. Hence, taking the different values of m, viz, 2 for half years, 4 for quarters, 12 for monthly payments, &c, and substituting in the general formula, the term or value of n for each case, when the difference in the present worths of the annuities, will be as follows, reckoning interest at 4 per cent, viz,

$$n = \dfrac{l.\dfrac{2l.\dfrac{r+1}{2}}{l.r}}{2l.\dfrac{r+1}{2} - l.r} = 25\cdot3777 \text{ for half-yearly payments,}$$

$$n = \dfrac{l.\dfrac{4l.\dfrac{r+3}{4}}{l.r}}{4l.\dfrac{r+3}{4} - l.r} = 25\cdot3200 \text{ for quarterly payments,}$$

$$n = \dfrac{l.\dfrac{12l.\dfrac{r+11}{12}}{l.r}}{12l.\dfrac{r+11}{12} - l.r} = 25\cdot2643 \text{ for monthly payments.}$$

Annuities may also be considered as in arrears, or as forborn, for any number of years; in which case each payment is to be considered as a sum put out to interest for the remainder of the term after the time it becomes due. And as $1l$. due at the end of 1 year, amounts to r at the end of another year, and to r^2 at the end of the 3d year, and to r^3 at the end of the 4th year, and so on; therefore by adding always the last year's annuity, or 1, to the amounts of all the former years, the sum of all the annuities and ther interests, will be the sum of the following geometrical series, $1 + r + r^2 + r^3 + r^4 \cdots \cdots$ to r^{n-1}, continued till the last term be r^{n-1}, or till the number of terms be n, the number of years the annuity is forborn.

But the sum of this geometrical progression is $\dfrac{r^n - 1}{r - 1}$,

which therefore is the amount of $1l$. annuity forborn for n years. And this quantity being multiplied by any other annuity a, instead of 1, will produce the amount for that other annuity.

But the amounts of annuities, or their present values, are easiest found by the two following tables of numbers for the annuity of $1l$. ready computed from the foregoing principles.

TABLE I.
The Amount of an Annuity of $1l$. at Comp. Interest.

Yrs.	at 3 per cent.	3½ per cent.	4 per cent.	4½ per cent.	5 per cent.	6 per cent.
1	1·00000	1·00000	1·00000	1·00000	1·00000	1·00000
2	2·03000	2·03500	2·04000	2·04500	2·05000	2·06000
3	3·09090	3·10623	3·12160	3·13703	3·15250	3·18360
4	4·18363	4·21494	4·24646	4·27819	4·31013	4·37462
5	5·30914	5·36247	5·41632	5·47071	5·52563	5·63709
6	6·46841	6·55015	6·63298	6·71689	6·80191	6·97532
7	7·66246	7·77941	7·89829	8·01915	8·14201	8·39384
8	8·89234	9·05169	9·21423	9·38001	9·54911	9·89747
9	10·15911	10·36850	10·58280	10·80211	11·02656	11·49132
10	11·46388	11·73139	12·00611	12·28821	12·57789	13·18079
11	12·80780	13·14199	13·48635	13·84118	14·20679	14·97164
12	14·19203	14·60196	15·02581	15·46403	15·91713	16·86994
13	15·61779	16·11303	16·62684	17·15991	17·71298	18·88214
14	17·08632	17·67699	18·29191	18·93211	19·59863	21·01507
15	18·59891	19·29568	20·02359	20·78405	21·57856	23·27597
16	20·15688	20·97103	21·82453	22·71934	23·65749	25·67253
17	21·76159	22·70502	23·69751	24·74171	25·84037	28·21288
18	23·41444	24·49969	25·64541	26·85508	28·13238	30·90565
19	25·11687	26·35718	27·67123	29·06356	30·53900	33·75999
20	26·87037	28·27968	29·77808	31·37142	33·06595	36·78559
21	28·67649	30·26947	31·96920	33·78314	35·71925	39·99273
22	30·53678	32·32890	34·24797	36·30338	38·50521	43·39229
23	32·45288	34·46041	36·61789	38·93703	41·43048	46·99583
24	34·42647	36·66653	39·08260	41·68920	44·50200	50·81558
25	36·45926	38·94986	41·64590	44·56521	47·72710	54·86451
26	38·55304	41·31310	44·31174	47·57064	51·11345	59·15638
27	40·70963	43·75906	47·08421	50·71132	54·66913	63·70577
28	42·93092	46·29063	49·96758	53·99333	58·40258	68·52811
29	45·21885	48·91080	52·96629	57·42303	62·32271	73·63980
30	47·57542	51·62268	56·08494	61·00707	66·43885	79·05819
31	50·00268	54·42947	59·32834	64·75239	70·76079	84·80168
32	52·50276	57·33450	62·70147	68·66625	75·29883	90·38978
33	55·07784	60·34121	66·20953	72·75623	80·06377	97·34316
34	57·73018	63·45315	69·85791	77·03026	85·06696	104·18375
35	60·46208	66·67401	73·65221	81·49662	90·32031	111·43478
36	63·27594	70·00760	77·59831	86·16397	95·83632	119·12087
37	66·17422	73·45787	81·70225	91·04134	101·62814	127·26812
38	69·15945	77·02889	85·97034	96·13820	107·70955	135·90421
39	72·23423	80·72491	90·40915	101·46442	114·09502	145·05846
40	75·40126	84·55028	95·02552	107·03032	120·79977	154·76197
41	78·66330	88·50954	99·82654	112·84669	127·83976	165·04768
42	82·02320	92·60737	104·81960	118·92479	135·23175	175·95054
43	85·48389	96·84863	110·01238	125·27640	142·99334	187·50758
44	89·04841	101·23833	115·41288	131·91384	151·14301	199·75803
45	92·71986	105·78167	121·02939	138·84997	159·70016	212·74351
46	96·50146	110·48403	126·87057	146·09821	168·68516	226·50811
47	100·39650	115·35097	132·94539	153·67263	178·11942	241·09861
48	104·40840	120·38826	139·26321	161·58790	188·02539	256·56453
49	108·54065	125·60185	145·83373	169·85936	198·42666	272·95840
50	112·79687	130·99791	152·66708	178·50303	209·34800	290·33590
51	117·18077	136·58284	159·77377	187·53566	220·81540	308·75606
52	121·69620	142·36324	167·16472	196·97477	232·85617	328·28142
53	126·34708	148·34595	174·85131	206·83863	245·49897	348·97831
54	131·13750	154·53806	182·84536	217·14637	258·77392	370·91701

TAE.

TABLE II.
The present Value of an Annuity of 1l.

Yrs	at 3 per cent.	3½ per cent.	4 per cent.	4½ per cent.	5 per cent.	6 per cent.
1	0·97087	0·96618	0·96154	0·95694	0·95238	0·94340
2	1·91347	1·89969	1·88610	1·87267	1·85941	1·83339
3	2·82861	2·80164	2·77509	2·74896	2·72325	2·67301
4	3·71710	3·67308	3·62990	3·58753	3·54595	3·46511
5	4·57971	4·51505	4·45182	4·38998	4·32948	4·21236
6	5·41719	5·32855	5·24214	5·15787	5·07569	4·91732
7	6·23028	6·11454	6·00205	5·89270	5·78637	5·58238
8	7·01969	6·87396	6·73274	6·59589	6·46321	6·20979
9	7·78611	7·60769	7·43533	7·26879	7·10782	6·80169
10	8·53020	8·31661	8·11090	7·91272	7·72173	7·36009
11	9·25262	9·00155	8·76048	8·52892	8·30541	7·88687
12	9·95400	9·66333	9·38507	9·11858	8·86325	8·38384
13	10·63496	10·30274	9·98565	9·68285	9·39357	8·85268
14	11·29607	10·92052	10·56312	10·22283	9·89864	9·29498
15	11·93794	11·51741	11·11839	10·73955	10·37966	9·71225
16	12·56110	12·09412	11·65230	11·23402	10·83777	10·10590
17	13·16612	12·65132	12·16567	11·70719	11·27407	10·47726
18	13·75351	13·18968	12·65930	12·15999	11·68959	10·82760
19	14·32380	13·70984	13·13394	12·59329	12·08532	11·15812
20	14·87747	14·21240	13·59033	13·00794	12·46221	11·46992
21	15·41502	14·69797	14·02916	13·40472	12·82115	11·76408
22	15·93692	15·16712	14·45112	13·78442	13·16300	12·04158
23	16·44361	15·62041	14·85684	14·14777	13·48857	12·30338
24	16·93554	16·05837	15·24696	14·49548	13·79864	12·55036
25	17·41315	16·48151	15·62208	14·82821	14·09394	12·78336
26	17·87684	16·89035	15·98277	15·14661	14·37519	13·00317
27	18·32703	17·28536	16·32959	15·45130	14·64303	13·21053
28	18·76411	17·66702	16·66306	15·74287	14·89813	13·40616
29	19·18845	18·03577	16·98371	16·02189	15·14107	13·59072
30	19·60044	18·39205	17·29203	16·28888	15·37245	13·76483
31	20·00043	18·73628	17·58849	16·54439	15·59281	13·92909
32	20·38877	19·06887	17·87355	16·78889	15·80268	14·08404
33	20·76579	19·39021	18·14765	17·02286	16·00255	14·23023
34	21·13184	19·70068	18·41120	17·24676	16·19290	14·36814
35	21·48722	20·00066	18·66461	17·46101	16·37419	14·49825
36	21·83225	20·29049	18·90828	17·66604	16·54685	14·62099
37	22·16724	20·57053	19·14258	17·86224	16·71129	14·73678
38	22·49246	20·84109	19·36786	18·04999	16·86789	14·84602
39	22·80822	21·10250	19·58448	18·22966	17·01704	14·94907
40	23·11477	21·35507	19·79277	18·40158	17·15909	15·04630
41	23·41240	21·59910	19·99305	18·56611	17·29437	15·13802
42	23·70136	21·83488	20·18563	18·72355	17·42321	15·22454
43	23·98190	22·06269	20·37079	18·87421	17·54591	15·30617
44	24·25427	22·28279	20·54884	19·01838	17·66277	15·38318
45	24·51871	22·49545	20·72004	19·15635	17·77407	15·45583
46	24·77545	22·70092	20·88465	19·28837	17·88007	15·52437
47	25·02471	22·89944	21·04294	19·41471	17·98102	15·58903
48	25·26671	23·09124	21·19513	19·53561	18·07716	15·65003
49	25·50166	23·27656	21·34147	19·65130	18·16872	15·70757
50	25·72976	23·45562	21·48218	19·76201	18·25593	15·76186
51	25·95123	23·62862	21·61749	19·86795	18·33898	15·81308
52	26·16624	23·79576	21·74758	19·96933	18·41807	15·86139
53	26·37499	23·95726	21·87267	20·06634	18·49340	15·90697
54	26·57766	24·11330	21·99290	20·15918	18·56515	15·94998

THE USE OF TABLE I.

To find the Amount of an annuity forborn any number of years. Take out the amount from the 1st table, for the proposed years and rate of interest; then multiply it by the annuity in question; and the product will be its amount for the fame number of years, and rate of interest.

And the converse to find the rate or time.

Exam. 1. To find how much an annuity of 50l. will amount to in 20 years at 3½ per cent. compound interest.—On the line of 20 years, and in the column of 3½ per cent, stands 28·27968, which is the amount of an annuity of 1l for the 20 years; and therefore 28·27968 multiplied by 50, gives 1413·984l. or 1413l. 19s. 8d. for the answer.

Exam. 2. In what time will an annuity of 20l. amount to 1000l. at 4 per cent. compound interest?—Here the amount of 1000l. divided by 20l. the annuity, gives 50, the amount of 1l. annuity for the fame time and rate. Then, the nearest tabular number in the column of 4 per cent. is 49·96758, which standing on the line of 28, shews that 28 years is the answer.

Exam. 3. If it be required to find at what rate of interest an annuity of 20l. will amount to 1000l. forborn for 28 years.—Here 1000 divided by 20 gives 50 as before. Then looking along the line of 28 years, for the nearest to this number 50, I find 49·96758 in the column of 4 per cent. which is therefore the rate of interest required.

THE USE OF TABLE II.

Exam. 1. To find the present value of an annuity of 50l. which is to continue 20 years, at 3½ per cent.—By the table, the present value of 1l. for the fame rate and time, is 14·21240; therefore 14·2124 × 50 = 710·62l. or 710l. 12s. 4d. is the present value fought.

Exam. 2. To find the present value of an annuity of 20l. to commence 10 years hence, and then to continue for 40 years, or to terminate 50 years hence, at 4 per cent. interest.—In such cases as this, it is plain we have to find the difference between the present values of two equal annuities, for the two given times; which therefore will be effected by subtracting the tabular value of the one term from that of the other, and multiplying by the annuity. Thus,

tabular value for 50 years 21·48218
tabular value for 10 years 8·11090

the difference 13·37128
mult. by   20

gives 267·4256
or 267l. 8s. 6d. the answer.

The foregoing observations, rules, and tables, contain all that is important in the doctrine of *annuities certain.* And for farther information, reference may be had to arithmetical writings, particularly Malcolm's Arithmetic, page 595; Simpson's Algebra, sect. 16; Dodson's Mathematical Repository, page 298, &c; Jones's Synopsis, ch. 10; Philof. Transf. vol. lxvi, page 109.

For what relates to the doctrine of *annuities on lives,* see ASSURANCE, COMPLEMENT, EXPECTATION, LIFE ANNUITIES, REVERSIONS, &c.

ANNULETS, in *Architecture,* are small square members, in the Doric capital, placed under the quarter round.

Annulet is also used for a narrow flat moulding, common to other parts of a column, as well as to the capital; and so called, because it encompasses the column around.

around. In which fenfe annulet is frequently ufed for *baguette*, or little *aftragal*.

ANNULUS, a fpecies of VOLUTA. See alfo RING.

ANOMALISTICAL *Year*, in *Aftronomy*, called alfo *periodical year*, is the fpace of time in which the earth, or a planet, paffes through its orbit. The anomaliftical, or common year, is fomewhat longer than the tropical year; by reafon of the preceffion of the equinox.

And the apfes of all the planets have a like progreffive motion; by which it happens that a longer time is neceffary to arrive at the aphelion, which has advanced a little, than to arrive at the fame fixed ftar. For example, the tropical revolution of the fun, with refpect to the equinox, is 365ᵈᵃ 5ʰ 48ᵐ 45ˢ; but the fidereal, or return to the fame

ftar - 365 6 9 11;

and the anomaliftic revolution is 365 6 15 20, becaufe the fun's apogee advances each year 65″¼ with refpect to the equinoxes, and the fun cannot arrive at the apogee till he has paffed over the 65″¼ more than the revolution of the year anfwering to the equinoxes.

To find the anomaliftic revolution, fay, As the whole fecular motion of a planet *minus* the motion of its aphelion, is to 100 years or 3155760000 feconds, fo is 360°, to the duration of the anomaliftic revolution.

ANOMALOUS, is fomething irregular, or that deviates from the ordinary rule and method of other things of the fame kind.

ANOMALY, in *Aftronomy*, is an irregularity in the motion of a planet, by which it deviates from the aphelion or apogee; or it is the angular diftance of the planet from the aphelion or apogee; that is, the angle formed by the line of the apfes, and another line drawn through the planet.

Kepler diftinguifhes three kinds of anomaly; *mean, eccentric*, and *true*.

Mean or *Simple* ANOMALY, in the ancient aftronomy, is the diftance of a planet's mean place from the apogee. Which Ptolomy calls the angle of the mean motion.

But in the modern aftronomy, in which a planet P is confidered as defcribing an ellipfe APB about the fun S, placed in one focus, it is the time in which the planet moves from its aphelion A, to the mean place or point of its orbit P.

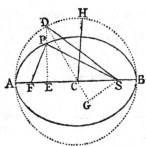

Hence, as the elliptic area ASP is proportional to the time in which the planet defcribes the arc AP, that area may reprefent the mean anomaly.—Or, if PD be drawn perpendicular to the tranfverfe axis AB, and meet the circle in D defcribed on the fame axis; then the mean anomaly may alfo be reprefented by the circular trilineal ASD, which is always proportional to the elliptic one ASP, as is proved in my Menfuration, pr. 8, page 296, fecond edition.—Or, drawing SG perpendicular to the radius DC produced; then the mean anomaly is alfo proportional to SG + the circular arc AD, as is demonftrated by Keil in his *Lect. Aftron.*—Hence, taking DH = SG, the arc AH, or angle ACH will be the mean anomaly in practice, as exprefled in degrees of a circle, the number of thofe degrees being to 360°, as the elliptic trilineal area ASP, is to the whole area of the ellipfe; the degrees of mean anomaly, being thofe in the arc AH, or angle ACH.

Eccentric ANOMALY, or *of the centre*, in the modern aftronomy, is the arc AD of the circle ADB intercepted between the apfis A and the point D determined by the perpendicular DPE to the line of the apfes, drawn through the place P of the planet. Or it is the angle ACD at the centre of the circle.—Hence the eccentric anomaly is to the mean anomaly, as AD to AD + SG, or as AD to AH, or as the angle ACD to the angle ACH.

True or *Equated* ANOMALY, is the angle ASP at the fun, which the planet's diftance AP from the aphelion, appears under; or the angle formed by the radius vector or line SP drawn from the fun to the planet, with the line of the apfes.

The true anomaly being given, it is eafy from thence to find the mean anomaly. For the angle ASP, which is the true anomaly, being given, the point P in the ellipfe is given, and thence the proportion of the area ASP to the whole ellipfe, or of the mean anomaly to 360 degrees. And for this purpofe, the following eafy rules for practice are deduced from the properties of the ellipfe, by M. de la Caille in his Elements of Aftronomy, and M. de la Lande, art. 1240 &c of his aftronomy: 1ft, As the fquare root of SB the perihelion diftance, is to the fquare root of SA the aphelion diftance, fo is the tangent of half the true anomaly ASP, to the tangent of half the eccentric anomaly ACD. 2nd, The difference DH or SG between the eccentric and mean anomaly, is equal to the product of the eccentricity CS, by the fine of SCG the eccentric anomaly juft found. And in this cafe, it is proper to exprefs the eccentricity in feconds of a degree, which will be found by this proportion, as the mean diftance 1 : the eccentricity :: 206264·8 feconds, or 57° 17′ 44″·8, in the arch whofe length is equal to the radius, to the feconds in the arc which is equal to the eccentricity CS; which being multiplied by the fine of the eccentric anomaly, to radius 1, as above, gives the feconds in SG, or in the arc DH, being the difference between the mean and eccentric anomalies. 3d, To find the radius vector SP, or diftance of the planet from the fun, fay either, as the fine of the true anomaly is to the fine of the eccentric anomaly, fo is half the lefs axis of the orbit, to the radius vector SP; or as the fine of half the true anomaly is to the fine of half the eccentric anomaly, fo is the fquare root of the perihelion diftance SB, to the fquare root of the radius vector or planet's diftance SP.

But the mean anomaly being given, it is not fo eafy to find the true anomaly, at leaft by a direct procefs: Kepler, who firft propofed this problem, could not find

a direct

a direct way of resolving it, and therefore made use of an indirect one, by the rule of false position, as may be seen page 695 of Kepler's *Epitom. Astron. Copernic.* See also §628 Wolfius *Elem. Astron.* Now the easiest method of performing this operation, would be to work first for the eccentric anomaly, viz, assume it nearly, and from it so assumed compute what would be its mean anomaly by the rule above given, and find the difference between this result and the mean anomaly given; then assume another eccentric anomaly, and proceed in the same way with it, finding another computed mean anomaly, and its difference from the given one; and treating these differences as in the rule of position for a nearer value of the eccentric anomaly: repeating the operation till the result comes out exact. Then, from the eccentric anomaly, thus found, compute the true anomaly by the 1st rule above laid down.

Of this problem, Dr. Wallis first gave the geometrical solution by means of the protracted cycloid; and Sir Isaac Newton did the same at prop. 31 *lib.* 1 *Princip.* But these methods being unfit for the purpose of the practical astronomer, various series for approximation have been given, viz, several by Sir Isaac Newton in his *Fragmenta Epistolarum,* page 26, as also in the Schol. to the prop. above-mentioned, which is his best, being not only fit for the planets, but also for the comets, whose orbits are very eccentric. Dr. Gregory, in his *Astron.* lib. 3, has also given the solution by a series, as well as M. Reyneau, in his *Analyse Demontrée,* page 713, &c. And a better still for converging is given by Keil in his *Prælect. Astron.* page 375; he says, if the arc AH be the mean anomaly, calling its sine *e,* cosine *f,* the eccentricity *g,* also putting $z = ge$, and $a = 1 + fg$; then the eccentric anomaly AD will be

$$= \frac{rz}{a} \times \left(1 - \frac{z^2}{a^2} \,\&c\right), \text{ supposing } r = 51{\cdot}29578 \text{ de-}$$

grees; of which the first term $\frac{rz}{a}$ is sufficient for all the

planets, even for Mars itself, where the error will not exceed the 200th part of a degree; and in the orbit of the earth, the error is less than the 100000th part of a degree.

Dr. Seth Ward, in his *Astronomia Geometrica,* takes the angle AFP at the other focus, where the sun is not, for the mean anomaly, and thence gives an elegant solution. But this method is not sufficiently accurate when the orbit is very eccentric, as in that of the planet Mars, as is shewn by Bullialdus, in his defence of the *Philolaic. Astron.* against Dr. Ward. However, when Newton's correction is made, as in the Schol. above-mentioned, and the problem resolved according to Ward's hypothesis, Sir Isaac affirms that, even in the orbit of Mars, there will scarce ever be an error of more than one second.

ANSÆ, Anses, in *Astronomy,* those seemingly prominent parts of the ring of the planet Saturn, discovered in its opening, and appearing like handles to the body of the planet; from which appearance the name *ansæ* is taken.

ANSER, in *Astronomy,* a small star, of the 5th or 6th magnitude, in the milky-way, between the eagle and swan, first brought into order by Hevelius.

ANTARCTIC *pole,* denotes the southern pole, or southern end of the earth's axis.—The stars near the antarctic pole never appear above our horizon in these latitudes.

ANTARCTIC *circle,* is a small circle parallel to the equator, at the distance of 23° 28' from the antarctic or south pole.—At one time of the year the sun never rises above the horizon of any part within this circle; and at other times he never sets.

ANTARES, in *Astronomy,* the scorpion's heart; a fixed star of the first magnitude, in the constellation *Scorpio.*

ANTECANIS is used by some astronomers, to denote the constellation otherwise called *canis minor,* or the star *procyon.* It is so called, as preceding, or being the forerunner of the *canis major,* and rising a little before it.

ANTECEDENT, *of a ratio,* denotes the first of the two terms of the ratio, or that term which is compared with the other. Thus, if the ratio be 2 to 3, or *a* to *b;* then 2 or *a* is the antecedent.

ANTECEDENTAL: *Method,* is a branch of general geometrical proportion, or universal comparison, and is derived from an examination of the Antecedents of ratios, having given consequents, and a given standard of comparison, in the various degrees of augmentation and diminution, which they undergo by composition and decomposition. This is a method invented by Mr. James Glenie, and published by him in 1793; a method which he says he always used instead of the fluxional and differential methods, and which is totally unconnected with the ideas of motion and time. See the author's treatise above-mentioned, and also his Doctrine of Universal Comparison, or General Proportion, 1789, upon which it is founded.

ANTECEDENTIA, a term used by astronomers when a planet &c moves westward, or contrary to the order of the signs aries, taurus, &c.—Like as when it moves eastward, or according to the order of the signs aries, taurus, &c, it is then said to move *in consequentia.*

ANTECIANS, or ANTOECI, in *Geography,* the inhabitants of the earth which occupy the same semicircle of the same meridian, but equally distant from the equator, the one north and the other south; as Peloponnesus and the Cape of Good Hope.

These have their noon, or midnight, or any other hour at the same time; but their seasons are contrary, being spring to the one, when it is autumn with the other; and summer with the one, when it is winter with the other; also the length of the day to the one, is equal to the length of night to the other.

ANTES, in *Architecture,* are small pilastres placed at the corners of buildings.

ANTICS, in *Architecture,* figures of men, beasts, &c, placed as ornaments to buildings.

ANTICUM, in *Architecture,* a porch before a door; also that part of a temple, which is called the outer temple, and lies between the body of the temple and the portico.

ANTILOGARITHM, the complement of the logarithm of a sine, tangent, secant, &c, to that of the radius. This is found by beginning at the left hand, subtracting each figure from 9, and the last figure from 10.

ANTINOUS, in *Astronomy*, a part of the conftellation *aquila*, or the eagle.

ANTIOCHIAN &c, or *Academy*, a name given to the fifth academy or branch of academics. It took its name from being founded by Antiochus, a philofopher contemporary with Cicero; and it fucceeded the Philonian academy. Though Antiochus was really a ftoic, and only nominally an academic.

ANTIOCHIAN *epocha*, a method of computing time from the proclamation of liberty granted to the city of Antioch, about the time of the battle of Pharfalia.

ANTIPARALLELS, in *Geometry*, are thofe lines which make equal angles with two other lines, but contrary ways; that is, calling the former pair the firft and 2d lines, and the latter pair the 3d and 4th lines, if the angle made by the 1ft and 3d lines be equal to the angle made by the 2d and 4th, and contrariwife the angle made by the 1ft and 4th equal to the angle made by the 2d and 3d; then each pair of lines are antiparallels with refpect to each other, viz, the firft and 2d, and the 3d and 4th. So, if AB and AC be any two lines, and FC and FE be two others, cutting them fo.

that the angle B is equal to the angle E,
and the angle C is equal to the angle D;

then BC and DE are antiparallels with refpect to AB and AC; alfo thefe latter are antiparallels with regard to the two former.—See alfo SUBCONTRARY.

It is a property of thefe lines, that each pair cuts the other into proportional fegments, taking them alternately,

viz AB : AC :: AE : AD :: DB : EC,
and FE : FC :: FB : FD :: DE : BC.

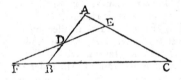

ANTIPODES, in Geography, are the inhabitants of two places on the earth which lie diametrically oppofite to each other, or that walk feet to feet; that is, if a line be continued down from our feet, quite through the centre of the earth, till it arrive at the furface on the other fide, it will fall on the feet of our Antipodes, and *vice verfa*.——Antipodes are 180 degrees diftant from each other every way on the furface of the globe; they have equal latitudes, the one north and the other fouth, but they differ by 180 degrees of longitude: they have therefore the fame climates or degrees of heat and cold, with the fame feafons and length of days and nights; but all of thefe at contrary times, it being day to the one, when it is night to the other, fummer to the one when it is winter to the other, &c: they have alfo the fame horizon, the one being as far diftant on the one fide, as the other on the other fide, and therefore when the fun, &c, rifes to the one, it fets to the other. The Antipodes to London are a part a little fouth of New Zealand.

It has been faid that Plato firft ftarted the notion of Antipodes, and gave them the name; which is likely enough, as he conceived that the earth was of a glo-

bular figure. But there have been many difputes upon this point, and the fathers of the church have greatly oppofed it, efpecially Lactantius and Auguftine, who laughed at it, and were greatly perplexed to think how men and trees fhould hang pendulous in the air with their feet uppermoft, as he thought they muft do, in the other hemifphere.

ANTISCIANS, or ANTISCII, in Geography, are people who dwell in the oppofite hemifpheres of the earth, as to north and fouth, and whofe fhadows at noon fall in contrary directions. This term is more general than *antæci*, with which it is often confounded. The *Antifcians* ftand contradiftinguifhed from *Perifcians*.

Antiscii is alfo ufed fometimes, among *Aftrologers*, for two points of the heavens equally diftant from the tropics. Thus the figns Leo and Taurus are accounted *antifcii* to each other.

ANTŒCI, fee ANTECIANS.

APERTURE, in *Geometry*, is ufed for the fpace left between two lines which mutually incline towards each other, to form an angle.

APERTURE, in Optics, is the hole next the objectglafs of a telefcope or microfcope, through which the light and the image of the object come into the tube, and are thence conveyed to the eye.

Aperture is alfo underftood of that part of the object-glafs itfelf which covers the former, and which is left pervious to the rays.

A great deal depends upon having a juft aperture.— To find it experimentally: apply feveral circles of dark paper, of various fizes, upon the face of the glafs, from the breadth of a ftraw, to fuch as leave only a fmall hole in the glafs; and with each of thefe, feparately, view fome diftinct objects, as the moon, ftars, &c; then that aperture is to be chofen through which they appear the moft diftinctly.

Huygens firft found the ufe of apertures to conduce much to the perfection of telefcopes; and he found by experience (*Dioptr.* prop. 56.) that the beft aperture for an object-glafs, for example of 30 feet, is to be determined by this proportion, as 30 to 3, fo is the fquare root of 30 times the diftance of the focus of any lens, to its proper aperture: and that the focal diftances of the eye-glaffes are proportional to the apertures. And M. Auzout fays he found, by experience, that the apertures of telefcopes ought to be nearly in the fubduplicate ratio of their lengths. It has alfo been found by experience that object glaffes will admit of greater apertures, if the tubes be blacked within fide, and their paffage furnifhed with wooden rings.

It is to be noted, that the greater or lefs aperture of an object-glafs, does not increafe or diminifh the vifible area of the object; all that is effected by this, is the admittance of more or fewer rays, and confequently the more or lefs bright the appearance of the object. But the largenefs of the aperture or focal diftance, caufes the irregularity of its refractions. Hence, in viewing Venus through a telefcope, a much lefs aperture is to be ufed than for the moon, or Jupiter, or Saturn, becaufe her light is fo bright and glaring. And this circumftance fomewhat invalidates and difturbs Azout's proportion, as is fhewn by Dr. Hook, Philof. Tranf. No. 4.

APHELION, or APHELIUM, in Aftronomy, that point

point in the orbit of the earth, or a planet, in which it is at the greatest distance from the sun. Which is the point A (in the fig. to the art. ANOMALY) or extremity of the transverse axis, of the elliptic orbit, farthest from the focus S, where the sun is placed; and diametrically opposite to the perihelion B, or nearer extremity of the same axis. In the Ptolemaic system, or in the supposition that the sun moves about the earth, the aphelion becomes the *apogee*.

The times of the aphelia of the primary planets, may be known by their apparent diameter appearing the smallest, and also by their moving slowest in a given time. Calculations and methods of finding them have been given by many astronomers, as Ricciolli, *Almag. Nov.* lib. 7, sect. 2 and 3; Wolfius, *Elem. Astron.* § 659; Dr. Halley, *Philos. Transf.* No. 128; Sir I. Newton, *Princip.* lib. 3, prop. 14; Dr. Gregory, *Astron.* lib. 3, prop. 14; Keil, *Astron. Lect.*; De la Lande, *Memoires de l'Acad.* 1755, 1757, 1766, and in his *Astron.* liv. 22; also in the writings of MM. Euler, D'Alembert, Clairaut &c, upon attraction.

The aphelia of the planets are not fixed: for their mutual actions upon one another keep those points of their orbits in a continual motion, which is greater or less in the different planets. This motion is made in consequentia, or according to the order of the signs; and Sir I. Newton shews that it is in the sesquiplicate ratio of the distance of the planet from the sun, that is, as the square root of the cube of the distance.

The quantities of this motion, as well as the place of the aphelion for a given time, are variously given by different authors. Kepler states them, for the year 1700 as in the following table.

Planets	Aphelion.			Annual Motion.	
Mercury	♑	8°	25′	30″	1′	45″
Venus	♒	3	24	27	1	18
Mars	♏	0	51	29	1	7
Jupiter	♎	8	10	40	0	47
Saturn	♐	28	3	48	1	10
The Earth	♋	8	25	30		

By De la Hire they are given as follows, for the same year 1700.

Planets.	Aphelion.			Annual Motion.	
Mercury	♑	13°	3′	40″	1′	39″
Venus	♒	6	56	10	1	26
Mars	♏	0	35	25	1	7
Jupiter	♎	10	17	14	1	34
Saturn	♐	29	14	41	1	22

And De la Lande states them as follows, for the year 1750.

Planets.	Aphelion.			Secular Motion.		
Mercury	8ˢ	13°	33′	1°	57′	40″
Venus	10	8	13	4	10	0
Mars	5	1	28	1	51	40
Jupiter	6	10	22	1	43	20
Saturn	8	29	53	2	23	20
The Earth	9	8	38	1	49	10

Of the new planet, Herschel, or Georgium Sidus, the aphelion for 1790 was 11ˢ 23° 29′ 42″, and its annual motion 50″¾. See Connoissance des Temps, 1786 and 1787.

APHRODISIUS, in Chronology, denotes the eleventh month in the Bythinian year, commencing on the 25th of July in ours.

APIAN or APPIAN (PETER), called in German *Bienewitz*, a celebrated astronomer and mathematician, was born at Leisnig or Leipsick in Misnia, 1495, and made professor of mathematics at Ingolstadt, in 1524, where he died in the year 1552, at 57 years of age.

Apian wrote treatises upon many of the mathematical sciences, and greatly improved them; more especially astronomy and astrology, which in that age were much the same thing; also geometry, geography, arithmetic, &c. He particularly enriched astronomy with many instruments, and observations of eclipses, comets, &c. His principal work was the *Astronomicum Cæsareum*, published in folio at Ingolstadt in 1540, and which contains a number of interesting observations, with the descriptions and divisions of instruments. In this work he predicts eclipses, and constructs the figures of them in plano. In the 2d part of the work, or the *Meteoroscopium Planum*, he gives the description of the most accurate astronomical quadrant, and its uses. To it are added observations of five different comets, viz, in the years 1531, 1532, 1533, 1538, and 1539; where he first shews that the tails of comets are always projected in a direction from the sun.

Apian also wrote a treatise on *Cosmography*, or *Geographical Instruction*, with various mathematical instruments. This work Vossius says he published in 1524, and that Gemma Frisius republished it in 1540. But Weidler says he wrote it only in 1530, and that Gemma Frisius published it at Antwerp in 1550 and 1584, with observations of many eclipses. The truth may be, that perhaps all these editions were published.

In 1533 he made, at Norimberg, a curious instrument, which from its figure he called *Folium Populi*; which, by the sun's rays, shewed the hour in all parts of the earth, and even the unequal hours of the Jews.

In 1534 he published his *Inscriptiones Orbis*.

In 1540, his *Instrumentum Sinuum, sive Primi Mobilis*, with 100 problems.

Beside these, Apian was the author of many other works: among which may be mentioned the *Ephemerides* from the year 1534 to 1570: Books upon *Shadows*: *Arithmetical Centiloques*: Books upon *Arithmetic*, with the *Rule of Cofs* (Algebra) demonstrated: Upon *Gauging*: *Almanacs*, with Astrological directions: A book upon *Conjunctions*: Ptolemy, with very correct figures, drawn in a quadrangular form: Ptolemy's works in Greek: Books of Eclipses: the works of Azoph, a very ancient astrologer: the works of Gebre: the Perspective of Vitello: of Critical Days, and of the Rainbow: a new Astronomical and Geometrical Radius, with various uses of Sines and Chords: Universal Astrolabe of Numbers: Maps of the World, and of particular countries: &c, &c.

Apian left a son, who many years afterwards taught mathematics at Ingolstadt, and at Tubinga. Tycho has preserved (Progymn. p. 643) his letter to the Landgrave of Hesse, in which he gives an opinion on the new star in Cassiopeia, of the year 1572.

One

One of the comets obſerved by Apian, viz, that of 1532, had its elements nearly the ſame as of one obſerved 128¼ years after, viz, in 1661, by Hevelius and other Aſtronomers; from hence Dr. Halley judged that they were the ſame comet, and that therefore it might be expected to appear again in the beginning of the year 1789. But it was not found that it returned at this period, although the aſtronomers then looked anxiouſly for it; and it is doubtful whether the diſappointment might be owing to its paſſing unobſerved, or to any errors in the obſervations of Apian, or to its period being diſturbed and greatly altered by the actions of the ſuperior planets, &c.

APIS, *muſca*, the *Bee*, or *Fly*, in Aſtronomy, one of the ſouthern conſtellations, containing 4 ſtars.

APOCATASTASIS, in Aſtronomy, is the period of a planet, or the time employed in returning to the ſame point of the zodiac from whence it ſet out.

APOGEE, *Apogæum*, in Aſtronomy, that point in the orbit of the ſun, moon, &c, which is fartheſt diſtant from the earth. It is at the extremity of the line of the apſides; and the point oppoſite to it is called the *perigee*, where the diſtance from the earth is the leaſt.

The ancient aſtronomers, conſidering the earth as the centre of the ſyſtem, chiefly regarded the apogee and perigee: but the moderns, placing the ſun in the centre, change theſe terms for the *aphelion* and *perihelion*. —The apogee of the ſun, is the ſame thing as the aphelion of the earth; and the perigee of the ſun is the ſame as the perihelion of the earth.

The manner of finding the apogee of the ſun or moon, is ſhewn by Ricciolus, *Almag. Nov.* lib. 3, cap. 24; by Wolfius in *Elem. Aſtr.* § 618; by Caſſini, De la Hire, and many others: ſee alſo *Memoires de l'Academie*, the *Philoſ. Tranſ.* vol. 5, 47, &c.

The quantity of motion in the apogee may be found by comparing two obſervations of it made at a great diſtance of time; converting the difference into minutes, and dividing them by the number of years elapſed between the two obſervations; the quotient gives the annual motion of the apogee. Thus, from an obſervation made by Hipparchus in the year before Chriſt 140, by which the ſun's apogee was found 5° 30′ of ♊; and another made by Ricciolus, in the year of Chriſt 1646, by which it was found 7° 26′ of ♋; the annual motion of the apogee is found to be 1′ 2″. And the annual motion of the moon's apogee is about 1' 10° 39′ 52″

But the moon's apogee moves unequably. When ſhe is in the ſyzygy with the ſun, it moves forwards; but in the quadratures, backwards; and theſe progreſſions and regreſſions are not equable, but it goes forward ſlower when the moon is in the quadratures, or perhaps goes retrograde; and when the moon is in the ſyzygy, it goes forward the faſteſt of all.—See alſo Newton's Theory of the Moon for more upon this ſubject.

APOLLODORUS, a celebrated architect, under Trajan and Adrian, was born at Damaſcus, and flouriſhed about the year of Chriſt 100. He had the direction of the ſtone bridge which Trajan ordered to be built over the Danube in the year 104, which was eſteemed the moſt magnificent of all the works of that emperor. Adrian, one day as Trajan was diſcourſing with this architect upon the buildings he had raiſed at Rome, would needs give his judgment, in which he ſhewed that he knew nothing of the matter. Apollodorus turned upon him bluntly, and ſaid to him, Go paint Citruls, for you are very ignorant of the ſubject we are talking upon. Adrian at this time boaſted of his painting Citruls well. This was the firſt ſtep towards the ruin of Apollodorus; a ſlip which he was ſo far from attempting to retrieve, that he even added a new offence, and that too after Adrian was advanced to the empire, upon the following occaſion: Adrian ſent to him the plan of a temple of Venus; and though he aſked his opinion, yet to ſhew that he had no need of him, and that he did not mean to be directed by it, the temple was already built. Apollodorus wrote his opinion very freely, and remarked ſuch eſſential faults in it, as the emperor could neither deny nor remedy. He ſhewed that it was neither high nor large enough; that the ſtatues in it were diſproportioned to its bulk: for, ſaid he, if the goddeſſes ſhould have a mind to riſe and go out, they could not do it. This put Adrian into a great paſſion, and prompted him to the deſtruction of Apollodorus. He baniſhed him at firſt; then under the pretext of certain ſuppoſed crimes, of which he had him accuſed, he at laſt put him to death.

APOLLONIUS, of Perga, a city in Pamphilia, was a celebrated geometrician who flouriſhed in the reign of Ptolemy Euergetes, about 240 years before Chriſt; being about 60 years after Euclid, and 30 years later than Archimedes. He ſtudied a long time in Alexandria under the diſciples of Euclid; and afterwards he compoſed ſeveral curious and ingenious geometrical works, of which only his books of Conic Sections are now extant, and even theſe not perfect. For it appears from the author's dedicatory epiſtle to Eudemus, a geometrician in Pergamus, that this work conſiſted of 8 books; only 7 of which however have come down to us.

From the Collections of Pappus, and the Commentaries of Eutocius, it appears that Apollonius was the author of various pieces in geometry, on account of which he acquired the title of the Great Geometrician. His *Conics* was the principal of them. Some have thought that Apollonius appropriated the writings and diſcoveries of Archimedes; Heraclius, who wrote the life of Archimedes, affirms it; though Eutocius endeavours to refute him. Although it ſhould be allowed a groundleſs ſuppoſition, that Archimedes was the firſt who wrote upon Conics, notwithſtanding his treatiſe on Conics was greatly eſteemed; yet it is highly probable that Apollonius would avail himſelf of the writings of that author, as well as others who had gone before him; and, upon the whole, he is allowed the honour of explaining a difficult ſubject better than had been done before; having made ſeveral improvements both in Archimedes's problems, and in Euclid. His work upon Conics was doubtleſs the moſt perfect of the kind among the ancients, and in ſome reſpects among the moderns alſo. Before Apollonius, it had been cuſtomary, as we are informed by Eutocius, for the writers on Conics to require three different ſorts of cones to cut the three different ſections from, viz, the parabola from a right angled cone, the ellipſe from an acute, and the

3

the hyperbola from an obtuse cone; becaufe they always fuppofed the fections made by a plane cutting the cones to be perpendicular to the fide of them: but Apollonius cut his fections all from any one cone, by only varying the inclination or pofition of the cutting plane; an improvement that has been followed by all other authors fince his time. But that Archimedes was acquainted with the fame manner, of cutting any cone, is fufficiently proved, againft Eutocius, Pappus, and others, by Guido Ubaldus, in the beginning of his Commentary on the 2d book of Archimedes's Equiponderantes, publifhed at Pifa in 1588.

The firft four books of Apollonius's Conics only have come down to us in their original Greek language; but the next three, the 5th, 6th, and 7th, in an Arabic verfion; and the 8th not at all. Thefe have been commented upon, tranflated, and publifhed by various authors. Pappus, in his Mathematical Collections, has left fome account of his various works, with notes and lemmas upon them, and particularly on the Conics. And Eutocius wrote a regular elaborate commentary on the propofitions of feveral of the books of the Conics.

The firft four books were badly tranflated by Joan. Baptifta Memmius. But a better tranflation of thefe in Latin was made by Commandine, and publifhed at Bononia in 1566.—Voffius mentions an edition of the Conics in 1650; the 5th, 6th, and 7th books being recovered by Golius.—Claude Richard, Profeffor of mathematics in the imperial college of his order at Madrid, in the year 1632, explained, in his public lectures, the firft four books of Apollonius, which were printed at Antwerp in 1655, in folio.—And the Grand Duke Ferdinand the 2d, and his brother Prince Leopold de Medicis, employed a profeffor of the Oriental languages at Rome to tranflate the 5th, 6th, and 7th books into Latin. Thefe were publifhed at Florence in 1661, by Borelli, with his own notes, who alfo maintains that thefe books are the genuine production of Apollonius, by many ftrong authorities, againft Mydorgius and others, who fufpected that thefe three books were not the real production of Apollonius.

As to the 8th book, fome mention is made of it in a book of Golius's, where he had written that it had not been tranflated into Arabic; becaufe it was wanting in the Greek copies, from whence the Arabians tranflated the others. But the learned Merfenne, in the preface to Apollonius's Conics, printed in his Synopfis of the Mathematics, quotes the Arabic philofopher Aben Nedin for a work of his about the year 400 of Mahomet, in which is part of that 8th book, and who afferts that all the books of Apollonius are extant in his language, and even more than are enumerated by Pappus; and Voffius fays he has read the fame; *De Scientiis Mathematicis*, pa. 55.—A neat edition of the firft four books in Latin was publifhed by Dr. Barrow, in 4to, at London in 1675.—A magnificent edition of all the 8 books, was publifhed in folio, by Dr. Halley, at Oxford in 1710; together with the Lemmas of Pappus, and the Commentaries of Eutocius. The firft four in Greek and Latin, but the latter four in Latin only, the 8th book being reftored by himfelf.

The other writings of Apollonius, mentioned by Pappus, are,

1. The Section of a Ratio, or Proportional Sections, two books.
2. The Section of a Space, in two books.
3. Determinate Section, in two books.
4. The Tangencies, in two books.
5. The Inclinations, in two books.
6. The Plane Loci, in two books.

The contents of all thefe are mentioned by Pappus, and many lemmas are delivered relative to them; but none, or very little of thefe books themfelves have defcended down to the moderns. From the account however that has been given of their contents, many reftorations have been made of thefe works, by the modern mathematicians, as follow: viz,

Vieta, Apollonius Gallus. The Tangencies. Paris, 1600, in 4to.

Snellius, Apollonius Batavus. Determinate Section. Lugd. 1601, 4to.

Snellius, Sectio Rationis & Spatii. 1607.

Ghetaldus, Apollonius Redivivus. The Inclinations. Venice, 1607, 4to.

Ghetaldus, Supplement to the Apollonius Redivivus. Tangencies. 1607.

Ghetaldus, Apollonius Redivivus, lib. 2. 1613.

Alex. Anderfon, Supplem. Apol. Redivivi. Inclin. Paris, 1612, 4to.

Alex. Anderfon, Pro Zetetico Apolloniani problematis a fe jam pridem edito in Supplemento Apollonii Redivivi. Paris, 1615, 4to.

Schooten, Loca Plana reftituta. Lug. Bat. 1656.

Fermat, Loca Plana, 2 lib. Tolof. 1679. folio.

Halley, Apol. de Sectione Rationis libri duo ex Arabico MS. Latine verfi duo reftituti. Oxon. 1706, 8vo.

Simfon, Loca Plana, libri duo. Glafg. 1749, 4to.

Simfon, Sectio Determinat. Glafg. 1776, 4to.

Horfley, Apol. Inclinat. libri duo. Oxon. 1770, 4to.

Lawfon, The Tangencies, in two books. Lond. 1771, 4to.

Lawfon, Determinate Section, two books. Lond. 1772, 4to.

Wales, Determinate Section, two books. Lond. 1772, 4to.

Burrow, The Inclinations. Lond. 1779, 4to.

APONO (PETER de), a learned aftronomer and philofopher, was born at Apono near Padua, about the year 1250. He defcribed the *Aftrolabium Planum*, by which were fhewn the equations of the celeftial houfes for any hour and minute, and for any part of the world: it was publifhed at Venice in 1502. He acquired the name of the *Conciliator*, on account of a book of his, in which he reconciles the writings of the ancient philofophers and phyficians: the book was publifhed at Venice in 1483. He refided at Padua, where, from his practifing medicine, and his fkill in aftronomy, he fell under the fufpicion of magic. He died in 1316, at 66 years of age.

APOPHYGE, in Architecture, is a concave part or ring of a column, lying above or below the flat member; and it owes its origin to the ring by which the ends of wooden columns were hooped, to prevent them from fplitting.

APOTOME, the remainder or difference between two lines or quantities which are only commenfurable in power. Such is the difference between 1 and $\sqrt{2}$,

or

or the difference between the side of the square and its diagonal.

The term is used by Euclid; and a pretty full explanation of such quantities is given in the tenth book of his Elements, where he distinguishes six kinds of apotomes, and shews how to find them all geometrically.

APOTOME *Prima*, is when the greater term is rational, and the difference of the squares of the two is a square number; as the difference $3 - \sqrt{5}$.

APOTOME *Secunda*, is when the less number is rational, and the square root of the difference of the squares of the two terms, has to the greater term, a ratio expressible in numbers; such is $\sqrt{18} - 4$, because the difference of the squares 18 and 16 is 2, and $\sqrt{2}$ is to $\sqrt{18}$ as $\sqrt{1}$ to $\sqrt{9}$ or as 1 to 3.

APOTOME *Tertia*, is when both the terms are irrational, and, as in the second, the square root of the difference of their squares, has to the greater term, a rational ratio: as $\sqrt{24} - \sqrt{18}$; for the difference of their squares 24 and 18 is 6, and $\sqrt{6}$ is to $\sqrt{24}$ as $\sqrt{1}$ to $\sqrt{4}$ or as 1 to 2.

APOTOME *Quarta*, is when the greater term is a rational number, and the square root of the difference of the squares of the two terms, has not a rational ratio to it: as $4 - \sqrt{3}$, where the difference of the squares 16 and 3 is 13, and $\sqrt{13}$ has not a ratio in numbers to 4.

APOTOME *Quinta*, is when the less term is a rational number, and the square root of the difference of the squares of the two, has not a rational ratio to the greater: as $\sqrt{6} - 2$, where the difference of the squares 6 and 4 is 2, and $\sqrt{2}$ to $\sqrt{6}$ or $\sqrt{1}$ to $\sqrt{3}$ or 1 to $\sqrt{3}$ is not a rational ratio.

APOTOME *Sexta*, is where both terms are irrational, and the square root of the difference of their squares has not a rational ratio to the greater: as $\sqrt{6} - \sqrt{2}$; where the difference of the squares 6 and 2 is 4, and $\sqrt{4}$ to $\sqrt{6}$ or 2 to $\sqrt{6}$, is not a rational ratio.

The doctrine of apotomes, in lines, as delivered by Euclid in the tenth book, is a very curious subject, and has always been much admired and cultivated by all mathematicians who have rightly understood this part of the elements; and therefore Peter Ramus has greatly exposed his judgment by censuring that book. And the first algebraical writers in Europe commonly employed a considerable portion of their works on an algebraical exposition of that book, which led them to the doctrine of surd quantities; as Lucas de Burgo, Cardan, Tartalea, Stifelius, Peletarius, &c, &c. See also Pappus, lib. 4, prop. 3, and the introduc. to lib. 7. And Dr. Wallis's Algebra, pa. 109.

APOTOME, in *Music*, is the difference between a greater and less semitone, being expressed by the ratio of 128 to 125.

APPARENT, that which is visible, or evident to the eye, or the understanding.

APPARENT *conjunction* of the planets, is when a right line, supposed to be drawn through their centres, passes through the eye of the spectator, and not through the centre of the earth.—And, in general, the apparent conjunction of any objects, is when they appear or are placed in the same right line with the eye.

APPARENT *Altitude, Diameter, Distance, Horizon, Magnitude, Motion, Place, Time*, &c. See the respective substantives, for the quantity and measure of it.

The apparent state of things, is commonly very different from their real state, either as to distance, figure, magnitude, position, &c, &c. Thus,

APPARENT *Diameter*, or *Magnitude*, as for example of the heavenly bodies, is not the real length of the diameter, but the angle which they subtend at the eye, or under which they appear. And hence, the angle, or apparent extent, diminishing with the distance of the object, a very small object, as AB, may have the same apparent diameter as a very large one FG; and indeed the objects have all the same apparent diameter, that are contained in the same angle FEG. And if these are parallel, the real magnitudes are directly proportional to their distances.

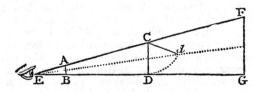

But the apparent magnitude varies not only by the distance, but also by the position of it. So, if the object CD be changed from the direct position to the oblique one C*d*, its apparent magnitude would then be only the angle CE*d*, instead of the angle CED.

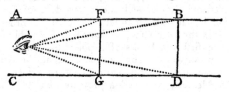

If the eye E be placed between two parallels AB, CD, these parallels will appear to converge or come nearer and nearer to each other the farther they are continued out, and at last they will appear to coincide in that point where the sight terminates, which will happen when the optic angle BED becomes equal to about one minute of a degree, the smallest angle under which an object is visible.—Also the apparent magnitudes of the same object FG or BD, seen at different distances, that is the angles FEG, BED, are in a less ratio than the reciprocal ratio of the distances, or the distance increases in a greater ratio than the angle or apparent magnitude diminishes. But when the object is very remote, or the optic angle is very small, as one degree or thereabouts, the angle then varies nearly as the distance reciprocally.

But although the optic angle be the usual or sensible measure of the apparent magnitude of an object, yet habit, and the frequent experience of looking at distant objects, by which we know that they are larger than they appear, has so far prevailed upon the imagination and judgment, as to cause this too to have some share in our estimation of apparent magnitudes; so that these will be judged to be more than in the ratio of the optic angles.

The apparent magnitude of the same object, at the same distance, is different to different persons, and different animals, and even to the same person, when viewed in different lights, all which may be occasioned

by

by the different magnitudes of the eye, causing the optic angle to differ as that is greater or less: and since, in the same person, the more light there comes from an object, the less is the pupil of the eye, looking at that object; therefore the optic angle will also be less, and consequently the apparent magnitude of the object. Every one must have experienced the truth of this, by looking at another person in a room, and afterwards abroad in the sunshine, when he always appears smaller than in a room where the light is less. So also, objects up in the air, having more light coming from them than when they are upon the ground, or near it, may appear less in the former case than in the latter; like as the ball of the cross on the top of St. Paul's church, which is 6 feet in diameter, appears less than an object of the same diameter seen at the same distance below, near the ground. And this may be the chief reason why the sun and moon appear so much larger when seen in the horizon, where their beams are weak, then when they are raised higher, and their light is more bright and glaring.

Again, if the eye be placed in a rare medium, and view an object through a denser, as glass or water, having plane surfaces; the object will appear larger than it is: and contrariwise, smaller. And hence it is that fishes, and other objects, seen in the water, by an eye in the air, always appear larger than in the air.—In like manner, an object will appear larger when viewed through a globe of glass or water, or any convex spherical segments of these; and, on the contrary, it will appear smaller when viewed through a concave of glass or water.

APPARENT *Distance*, is that distance which we judge an object to be from us, when seen afar off. This is commonly very different from the true distance; because we are apt to think that all very remote objects, whose parts cannot well be distinguished, and which have no other visible objects near them, are at the same distance from us; though perhaps they may be thousands or millions of miles off; as in the case of the sun and moon. The apparent distances of objects are also greatly altered by the refraction of the medium through which they are seen.

APPARENT *Figure*, is the figure or shape which an object appears under when viewed at a distance; and is often very different from the true figure. For a straight line, viewed at a distance, may appear but as a point; a surface, as a line; and a solid, as a surface. Also these may appear of different magnitudes, and the surface and solid of different figures, according to their situation with respect to the eye: thus, the arch of a circle may appear a straight line; a square, a trapezium, or even a triangle; a circle, an ellipsis; angular magnitudes, round; and a sphere, a circle. Also all objects have a tendency to roundness and smoothness, or appear less angular, as their distance is greater: for, as the distance is increased, the smaller angles and asperities first disappear, by subtending a less angle than one minute; after these, the next larger disappear, for the same reason; and so on continually, as the distance is more and more increased; the object seeming still more and more round and smooth. So, a triangle, or square, at a great distance, shews only as a round speck; and the edge of the moon appears round to the eye,

notwithstanding the hills and valleys on her surface. And hence it is also; that near objects, as a range of lamps, or such like, seen at a great distance, appear to be contiguous, and to form one uniform continued magnitude, by the intervals between them disappearing, from the smallness of the angles subtended by them.

APPARENT *Motion*, is either that motion which we perceive in a distant body that moves, the eye at the same time being either in motion or at rest; or that motion which an object at rest seems to have, while the eye itself only is in motion.

The motions of bodies at a great distance, though really moving equally, or passing over equal spaces in equal times, may appear to be very unequal and irregular to the eye, which can only judge of them by the mutation of the angle at the eye. And motions, to be equally visible, or appear equal, must be directly proportional to the distances of the objects moving. Again, very swift motions, as those of the luminaries, may not appear to be any motions at all, but like that of the hour hand of a clock, on account of the great distance of the objects: and this will always happen, when the space actually passed over in one second of time, is less than about the 14000th part of its distance from the eye; for the hour hand of a clock, and the stars about the earth, move at the rate of 15 seconds of a degree in one second of time, which is only the 13751 part of the radius or distance from the eye. On the other hand, it is possible for the motion of a body to be so swift, as not to appear any motion at all; as when through the whole space it describes there constantly appears a continued surface or solid as it were generated by the motion of the object, like as when any thing is whirled very swiftly round, describing a ring, &c.

Also, the more oblique the eye is to the line which a distant body moves in, the more will the apparent motion differ from the true one. So, if a body revolve with an equable motion in the circumference of the circle ABCD &c, and the eye be at E in the plane of the circle; as the body moves from A to B and C, it seems to move slower and slower along the line ALK, till when the body arrives at C, it appears at rest at K; then while it really moves from C by D to F, it appears to move quicker and quicker from K by L to A, where its motion is quickest of all; after this it appears to move slower and slower from A to N while the body moves from F to H: there becoming stationary again, it appears to return from N to A in

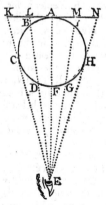

the straight line, while it really moves from H by I to A in the circle. And thus it appears to move in the line KN by a motion continually varying between the least, or nothing, at the extremes K and N, and the greatest of all at the middle point A. Or, if the motion be referred to the concave side of the circle, instead of the line KN, the appearances will be the same.

If an eye move directly forwards from E to O, &c; any remote object at rest at P, will appear to move the
contrary

contrary way, or from P to Q, with the fame velocity. But if the object P move the fame way, and with the fame velocity as the eye; it will feem to ftand ftill. If the object have a lefs velocity than the eye, it will appear to move back towards Q with the difference of the velocities; and if it move fafter than the eye, it will appear to move forwards from Q, with the fame difference of the velocities. And fo likewife when the object P moves contrary to the motion of the eye, it appears to move backwards with the fum of the motions of the two. And the truth of all this is experienced by perfons fitting in a boat moving on a river, or in a wheel-carriage when running faft, and viewing houfes or trees, &c, on the fhore or fide of the road, or other boats or wheel-carriages alfo in motion.

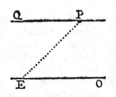

APPARENT PLACE *of an object*, in Optics, is that in which it appears, when feen in or through glafs, water, or other refracting mediums; which is commonly different from the true place. So, if an object be feen in or through glafs, or water, either plane or concave, it will appear nearer to the eye than its true place; but when feen through a convex glafs, it appears more remote from the eye than the real place of it.

APPARENT PLACE *of the Image of an object*, in Catoptrics, is that where the image of an object made by the reflexion of a fpeculum appears to be; and the optical writers, from Euclid downwards, give it as a general rule that this is where the reflected rays meet the perpendicular to the fpeculum drawn from the object: fo that if the fpeculum be a plane, the apparent place of the image will be at the fame diftance behind the fpeculum as the eye is before it; if convex, it will appear behind the glafs nearer to the fame; but if concave, it will appear before the fpeculum. And yet in fome cafes there are fome exceptions to this rule, as is fhewn by Kepler in his *Paralipomena in Vitellionem, prop.* 18. See alfo Wolfius *Catoptr.* § 51, 188, 233, 234.

APPARENT *Place of a Planet*, &c, in Aftronomy, is that point in the furface of the fphere of the world, where the centre of the luminary appears from the furface of the earth.

APPARITION, in Aftronomy, denotes a ftar's or other luminary's becoming vifible, which before was hid. So, the heliacal rifing, is rather an apparition than a proper rifing.

Circle of perpetual APPARITION. See CIRCLE *of perpetual apparition.*

APPEARANCE, in Perfpective, is the reprefentation or projection of a figure, body, or the like object, on the perfpective plane.—The appearance of an objective right line, is always a right line. See PERSPECTIVE.—Having given the appearance of an opake body, and of a luminary, to find the appearance of the fhadow; fee SHADOW.

APPEARANCE *of a ftar or planet.* See APPARITION.

APPEARANCES, in Aftronomy, &c, are more ufually called *phænomena* and *phafes.*—In *Optics,* the term *direct appearance* is ufed for the view or fight of any object by direct rays; without either refraction or reflexion.

APPIAN. See APIAN.

APPLICATE, APPLICATA, *Ordinate* APPLICATE, in Geometry, is a right line drawn to a curve, and bifected by its diameter. This is otherwife called an ORDINATE, which fee.

APPLICATE *Number.* See CONCRETE.

APPLICATION, the act of applying one thing to another, by approaching or bringing them nearer together. So a longer fpace as meafured by the continual *application* of a lefs, as a foot or yard by an inch, &c. And motion is determined by a fucceffive application of any thing to different parts of fpace.

APPLICATION is fometimes ufed, both in Arithmetic and Geometry, for the rule or operation of divifion, or what is fimilar to it in geometry. Thus 20 applied to, or divided by 4, gives 5. And a rectangle ab, applied to a line c, gives the 4th proportional $\dfrac{ab}{c}$, or another line which, with the given line c, will contain another rectangle which fhall be equal to the given rectangle ab. And this is the fenfe in which Euclid ufes the term, *lib.* 6, *pr.* 28.

APPLICATION, in Geometry, is alfo ufed for the act or fuppofition of putting or placing one figure upon another, to find whether they be equal or unequal; which feems to be the primary way in which the mind firft acquires both the idea and proof of equality. And in this way Euclid, and other geometricians, demonftrate fome of the firft or leading properties in geometry. Thus, if two triangles have two fides in the one triangle equal to two fides in the other, and alfo the angle included by the fame fides equal to each other; then are the two triangles equal in all refpects: for by conceiving the one triangle placed on the other, it is proved that they coincide or exactly agree in all their parts. And the fame happens if, of two triangles, one fide and the two adjacent angles of the one triangle, are equal, refpectively, to one fide and the two correfponding angles of the other. Thus alfo it may be proved that the diameter of a circle divides it into two equal parts, as alfo that the diagonal of a fquare or parallelogram bifects or divides it into two equal parts.

APPLICATION of one fcience to another, as of Algebra to Geometry, is faid of the ufe made of the principles and properties of the one for augmenting and perfecting the other. Indeed all arts and fciences mutually receive aid from each other. But the application here meant, is of a more exprefs and immediate nature; as will appear by what follows.

APPLICATION of *Algebra* or of *Analyfis to Geometry.* The firft and principal applications of algebra, were to arithmetical queftions and computations, as being the firft and moft ufeful fcience in all the concerns of human life. Afterwards algebra was applied to geometry and all the other fciences in their turn. The application of algebra to geometry, is of two kinds; that which regards the plane or common geometry, and that which refpects the higher geometry, or the nature of curve lines.

The firft of thefe, or the application of algebra to common geometry, is concerned in the algebraical folution of geometrical problems, and finding out theorems in geometrical figures, by means of algebraical inveftigations or demonftrations. This kind of application

cation

cation has been made from the time of the most early writers on algebra, as Diophantus, Lucas de Burgo, Cardan, Tartalea, &c, &c, down to the present times. Some of the best precepts and exercises of this kind of application, are to be met with in Newton's *Universal Arithmetic*, and in Thomas Simpson's *Algebra* and *Select Exercises*. Geometrical Problems are commonly resolved more directly and easily by algebra, than by the geometrical analysis, especially by young beginners; but then the synthesis, or construction and demonstration, is most elegant as deduced from the latter method. Now it commonly happens that the algebraical solution succeeds best in such problems as respect the sides and other lines in geometrical figures, and on the contrary, those problems in which angles are concerned, are best effected by the geometrical analysis. Newton gives these, among many other remarks on this branch. Having any problem proposed; compare together the quantities concerned in it; and, making no difference between the known and unknown quantities, consider how they depend upon, or are related to, one another; that we may perceive what quantities, if they are assumed, will, by proceeding synthetically, give the rest, and that in the simplest manner. And in this comparison, the geometrical figure is to be feigned and constructed at random, as if all the parts were actually known or given, and any other lines drawn that may appear to conduce to the easier and simpler solution of the problem. Having considered the method of computation, and drawn out the scheme, names are then to be given to the quantities entering into the computation, that is, to some few of them, both known and unknown, from which the rest may most naturally and simply be derived or expressed, by means of the geometrical properties of figures, till an equation be obtained, by which the value of the unknown quantity may be derived by the ordinary methods of reduction of equations, when only one unknown quantity is in the notation; or till as many equations are obtained as there are unknown letters in the notation.

For example, suppose it were required to inscribe a square in a given triangle. Let ABC be the given triangle; and feign DEFG to be the required square; also draw the perpendicular BP of the triangle, which will be given, as well as all the sides of it. Then, considering that the triangles BAC, BEF

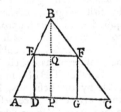

are similar, it will be proper to make the notation as follows, viz, making the base AC = b, the perpendicular BP = p, and the side of the square DE or EF = x. Hence then BQ = BP — ED = $p - x$; consequently, by the proportionality of the parts of those two similar triangles, viz, BP : AC :: BQ : EF, it is $p : b :: p - x : x$; then, multiply extremes and means &c, there arises

$px = bp - bx$, or $bx + px = bp$, and $x = \dfrac{bp}{b + p}$ the

side of the square sought; that is, a fourth proportional to the base and perpendicular, and the sum of the two, taking this sum for the first term, or AC + BP : BP :: AC : EF.

The other branch of the application of algebra to geo-

metry, was introduced by Descartes, in his Geometry, which is the new or higher geometry, and respects the nature and properties of curve lines. In this branch, the nature of the curve is expressed or denoted by an algebraic equation, which is thus derived: A line is conceived to be drawn, as the diameter or some other principal line about the curve; and upon any indefinite points of this line other lines are erected perpendicularly, which are called ordinates, whilst the parts of the first line cut off by them, are called abscisses. Then, calling any abscis x, and its corresponding ordinate y, by means of the known nature, or relations of the other lines in the curve, an equation is derived, involving x and y, with other given quantities in it. Hence, as x and y are common to every point in the primary line, that equation, so derived, will belong to every position or value of the abscis and ordinate, and so is properly considered as expressing the nature of the curve in all points of it; and is commonly called the equation of the curve.

In this way it is found that any curve line has a peculiar form of equation belonging to it, and which is different from that of every other curve, either as to the number of the terms, the powers of the unknown letters x and y, or the signs or coefficients of the terms of the equation. Thus, if the curve line HK be a circle, of which HI is part of the diameter, and IK a perpendicular ordinate: then put HI = x, IK = y, and p = the diameter of the

circle, the equation of the circle will be $px - x^2 = y^2$. But if HK be an ellipse, an hyperbola, or parabola, the equation of the curve will be different, and for all the four curves, will be respectively as follows, viz,

for the circle - - $px - x^2 = y^2$;

for the ellipse - $px - \dfrac{p}{t}x^2 = y^2$,

for the hyperbola - $px + \dfrac{p}{t}x^2 = y^2$,

for the parabola - px - - $= y^2$;

where t is the transverse axis, and p its parameter. And, in like manner for other curves.

This way of expressing the nature of curve lines, by algebraic equations, has given occasion to the greatest improvement and extension of the geometry of curve lines; for thus, all the properties of algebraic equations, and their roots, are transferred and added to the curve lines, whose abscisses and ordinates have similar properties. Indeed the benefit of this sort of application is mutual and reciprocal, the known properties of equations being transferred to the curves they represent; and, on the contrary, the known properties of curves transferred to their representative equations. See CURVES.

APPLICATION *of Geometry to Algebra.* Besides the use and application of the higher geometry, namely, of curve lines, to detecting the nature and roots of equations, and to the finding the values of those roots by the geometrical construction of curve lines, even common geometry may be made subservient to the purposes of algebra. Thus, to take a very plain and simple instance, if it were required to square the binomial $a + b$;

by forming a square, as in the annexed figure, whose side is equal to $a + b$, or the two lines or parts added together denoted by the letters a and b; and then drawing two lines parallel to the sides, from the points where the two parts join, it will be immediately evident that the whole square of the compound quantity $a + b$, is equal to the squares of both the parts, together with two rectangles under the two parts, or a^2 and b^2 and $2ab$, that is the square of $a + b$ is equal to $a^2 + b^2 + 2ab$, as derived from a geometrical figure or construction.

And in this very manner it was, that the Arabians, and the early European writers on algebra, derived and demonstrated the common rule for resolving compound quadratic equations. And thus also, in a similar way, it was, that Tartalea and Cardan derived and demonstrated all the rules for the resolution of cubic equations, using cubes and parallelopipedons instead of squares and rectangles. And many other instances might be given of the use and application of geometry in algebra.

APPLICATION *of Algebra and Geometry to Mechanics.* This is founded on the same principles as the application of algebra to geometry. It consists principally in representing by equations the curves described by bodies in motion, by determining the equation between the spaces which the bodies describe, when actuated by any forces, and the times employed in describing them, &c. A familiar instance also of the application of geometry to mechanics, may be seen at the article ACCELERATION, where the perpendiculars of triangles represent the times, the bases the velocities, and the areas the spaces described by bodies in motion; a method first given by Galileo. In short, as velocities, times, forces, spaces, &c, may be represented by lines and geometrical figures; and as these again may be treated algebraically; it is evident how the principles and properties, of both algebra and geometry, may be applied to mechanics, and indeed to all the other branches of mixt mathematics.

APPLICATION *of Mechanics to Geometry.* This consists chiefly in the use that is sometimes made of the centre of gravity of figures, for determining the contents of solids described by those figures.

APPLICATION *of Geometry and Astronomy to Geography.* This consists chiefly in three articles. 1st, In determining the figure of the globe we inhabit, by means of geometrical and astronomical operations. 2d, In determining the positions of places, by observations of latitudes and longitudes. 3d, In determining, by geometrical operations, the positions of such places as are not far distant from one another.

Geometry and Astronomy are also of great use in Navigation.

APPLICATION *of Geometry and Algebra to Physics or Natural Philosophy.* This application we owe to Newton, whose philosophy may therefore be called the geometrical or mathematical philosophy; and upon this application are founded all the physico-mathematical sciences. Here a single observation or experiment will often produce a whole science: so when we know, as we do by experience, that the rays of light, in reflect-

ing, make the angle of incidence equal to the angle of reflexion; we thence deduce the whole science of catoptrics: for that experiment once admitted, catoptrics become a science purely geometrical, since it is reduced to the comparison of angles and lines given in position. And the same in many other sciences.

APPLICATION *of one thing to another,* in general, is employed to denote the use that is made of the former, to understand or to perfect the latter. Thus, the application of the cycloid to pendulums, means the use made of the cycloidal curve for improving the doctrine and use of pendulums.

APPLY. This term is used two different ways, in geometry.

1st, It signifies to transfer or place a given line, either in a circle or some other figure, so that the extremities of the line shall be in the perimeter of the figure.

2d, It is also used to express division in geometry, or to find one dimension of a rectangle, when the area and the other dimension are given. As the area ab applied to the line c, is $\dfrac{ab}{c}$.

APPROACH, the *curve of equable approach.* It was first proposed by Leibnitz, namely, to find a curve, down which a body descending by the force of gravity, shall make equal approaches to the horizon in equal portions of time. It has been found by Bernoulli and others, that the curve is the second cubical parabola, placed with its vertex uppermost, and which the descending body must enter with a certain determinate velocity.—Varignon rendered the question general for any law of gravity, by which a body may approach towards a given point by equal spaces in equal times. And Maupertuis also resolved the problem in the case of a body descending in a medium which resists as the square of the velocity. *See Hist. de l' Acad. des Sciences* for 1699 and 1730.

Method of APPROACHES, a name given by Dr. Wallis, in his Algebra, to a method of resolving certain problems relating to square numbers, &c. This is done by first assigning certain limits to the quantities required, and then approaching nearer and nearer till a coincidence is obtained.—In this sense, the method of Trial-and-error, or double rule of False Position, may be considered as a method of approaches.

APPROACHES, in *Fortification,* the several works made by the besiegers, for advancing or getting nearer to a fortress or place besieged. Such as the trenches, mines, saps, lodgments, batteries, galleries, epaulments, &c.

APPROACHES, or *Lines of* APPROACH, are particularly used for trenches dug in the ground, and the earth thrown up on the side next the place besieged; under the defence or shelter of which, the besiegers may approach without loss, as near as possible to the place, to raise batteries and plant guns &c, to batter it.—The lines of approach are commonly carried on, in a zig-zag way, parallel to the opposite faces of the besieged work, or nearly so, that they may not be enfiladed by the guns from the enemy's works. And they are also connected by parallels or lines of communication.—The besieged commonly make counter-approaches, to interrupt and defeat the approaches of the besiegers.

The

The ancients made their approaches towards the place besieged, much after the same manner as the moderns. Folard shews, that they had their trenches, their parallels, saps, &c.; which, though usually thought of modern invention, it appears, have been practised long before, by the Greeks, Romans, Asiatics, &c.

APPROXIMATION, a continual approach, still nearer and nearer, to a root or any quantity sought. — Methods of continual approximation for the square roots and cube roots of numbers, have been employed by algebraists and arithmeticians, from Lucas de Burgo down to the present time. And the later writers have given various approximations, not only for the roots of higher powers, or all simple equations, but for the roots of all sorts of compound equations whatever: especially Newton, Wallis, Raphson, Halley, De Lagny, &c, &c; all of them forming a kind of infinite series, either expressed or understood, converging nearer and nearer to the quantity sought, according to the nature of the process.

It is evident that if a number proposed be not a true square, then no exact square root of it can be found, explicable by rational numbers, whether integers or fractions: therefore, in such cases, we must be content with approximations, or coming continually nearer and nearer to the truth. In like manner, for the cube and other roots, when the proposed quantities are not exact cubes, or other powers.

The most easy and general method of approximation, is perhaps by the rule of Double Position, or, what is sometimes called, the Method of Trial-and-error; which method see under its own name. And among all the methods for the roots of pure powers, of which there are many, I believe the best is that which was discovered by myself, and given in the first volume of my Mathematical Tracts, in point of ease, both of execution and for remembering it. The method is this: if N denote any number, out of which is to be extracted the root whose index is denoted by r, and if n be the nearest root first taken; then shall $\dfrac{\overline{r+1}.N + \overline{r-1}.n^r}{\overline{r-1}.N + \overline{r+1}.n^r} \times n$ be the required root of N very nearly; or as $r - 1$ times the given number added to $r + 1$ times the nearest power, is to $r + 1$ times the given number added to $r - 1$ times the nearest power, so is the assumed root n, to the required root, very nearly. Then this last value of the root, so found, if one still nearer is wanted, is to be used for n in the same theorem, to repeat the operation with it. And so on, repeating the operation as often as necessary. Which theorem includes all the rational formulæ of Halley and Dë Lagny.

For example, suppose it were required to double the cube, or to find the cube root of the number 2. Here $r = 3$; consequently $r + 1 = 4$, and $r - 1 = 2$; and therefore the general theorem becomes $\dfrac{4N + 2n^3}{2N + 4n^3} \times n$ or $\dfrac{2N + n^3}{N + 2n^3} \times n$ for the cube root of N, or as $N + 2n^3 : 2N + n^3 :: n :$ the root sought nearly. Now, in this case, $N = 2$, and therefore the nearest root n is 1, and its cube $n^3 = 1$ also: hence $N + 2n^3 = 2 + 2 = 4$, and $2N + n^3 = 4 + 1 = 5$; therefore, as $4 : 5 :: 1 : \frac{5}{4}$ or $1\frac{1}{4} = 1\cdot25$ the first approximation.

Again, taking $r = \dfrac{5}{4}$, and consequently $r^3 = \dfrac{125}{64}$; hence $N + 2n^3 = 2 + \dfrac{250}{64} = \dfrac{378}{64}$, and $2N + n^3 = 4 + \dfrac{125}{64} = \dfrac{381}{64}$; therefore as $378 : 381$, or as $126 : 127 :: \dfrac{5}{4} : \dfrac{635}{504} = 1\cdot259921$, which is the cube root of 2, true in all the figures. And by taking $\dfrac{635}{504}$ for a new value of n, and repeating the process again, a great many more figures may be found.

Of the Roots of Equations by APPROXIMATION.— Stevinus and Vieta gave methods for finding values, always nearer and nearer, of the roots of equations. And Oughtred and others pursued and improved the same. These however were very tedious and imperfect, and required a different process for every degree of equations. But Newton introduced, not only general methods for expressing radical quantities by approximating infinite series, but also for the roots of all sorts of compound equations whatever, which are both easy and expeditious: which will be more particularly described under each respective word or article. His method for approximating of roots, is in substance this: First take a value of the root as near as may be, by trials, either greater or less; then assuming another letter to denote the unknown difference between this and the true value, substitute into the equation the sum or difference of the approximate root and this assumed letter, instead of the unknown letter or root of the equation, which will produce a new equation having only the assumed small difference for its root or unknown letter; and, by any means, find, from this equation, a near value of this small assumed quantity. Assume then another letter for the small difference between this last value and the true one, and substitute the sum or difference of them into the last equation, by which will arise a third equation, involving the second assumed quantity; whose near value is found as before. Proceeding thus as far as we please, all the near values, connected together by their proper signs, will form a series approaching still nearer and nearer to the true value of the root of the first or proposed equation. The approximate values of the several small assumed differences, may be found in different ways: Newton's method is this: As the quantity sought is small, its higher powers decrease more and more, and therefore neglecting them will not lead to any great error, Newton therefore neglects all the terms having in them the 2d and higher powers, leaving only the 1st power and the absolute known term; from which simple equation he always finds the value of the assumed unknown letter nearly, in a very simple and easy manner. Halley's method of doing the same thing, was to neglect all the terms above the square or 2d power, and then to find the root of the remaining quadratic equation; which would indeed be a nearer value of the assumed letter than Newton's was, but then it is much more troublesome to perform.—Raphson has another way, which is a little varied from that of Newton's

ton's

ton's again, which is this: having found a near value of the first assumed small quantity or difference, by this he corrects the first approximation to the root of the proposed equation; and then, assuming another letter for the next, or smaller difference, he introduces it into the original equation in the same way as before. And thus he proceeds, from one correction to another, employing always the first proposed equation to find them, instead of the successive new equations used by Newton.

For example, let it be required to find the root of the equation $x^2 - 5x = 31$, or $x^2 - 5x - 31 = 0$:—Here the root x, it is evident, is nearly $= 8$; for x therefore take $8 + z$, and substitute $8 + z$ for x in the given equation, and the terms will be thus;

$$x^2 = 64 + 16z + z^2$$
$$-5x = -40 - 5z$$
$$-31 = -31$$

the sum is $-7 + 11z + z^2 = 0$.

Then, rejecting z^2, it is $11z - 7 = 0$, and $z = \frac{7}{11} = \cdot6363$, &c, or $= \cdot6$ nearly.

Assume now $z = \cdot6 + y$: then

$$z^2 = \cdot36 + 1\cdot2y + y^2$$
$$11z = 6\cdot6 + 11y$$
$$-7 = -7$$

the sum $-\cdot04 + 12\cdot2y + y^2 = 0$,

where $y = \dfrac{\cdot04}{12\cdot2} = \cdot003278$ nearly.

Assume it $y = \cdot003278 - v$: then

$$y^2 = \cdot000010745284 - \cdot006556v + v^2$$
$$12\cdot2y = \cdot0399916 \qquad - 12.2 \cdot \cdot v$$
$$-\cdot04 = -\cdot04$$

the sum $\cdot000002345284 - 12\cdot206556v + v^2 = 0$,

where $v = \dfrac{\cdot000002345284}{12\cdot206556} = \cdot000000192133$.

Hence, then, collecting all the assumed differences, with their signs, it is found that $x = 8 + z + y - v = 8 + \cdot6 + \cdot003278 - \cdot000000192133 = 8\cdot6003277807867$ the root of the equation required, by Newton's method.

The same by Raphson's way.

First $x = 8 + z$;

then $x^2 = 64 + 16z + z^2$

$$-5x = -40 - 5z$$
$$-31 = -31$$

the sum $-7 + 11z + z^2 = 0$;

hence $z = \frac{7}{11} = \cdot6$ nearly,

and $x = 8 + z = 8\cdot6$ nearly.

Assume it $x = 8\cdot6 + y$;

then $x^2 = 73\cdot96 + 17\cdot2y + y^2$

$$-5x = -43 \qquad - 5y$$
$$-31 = -31$$

the sum $-\cdot04 + 12\cdot2y + y^2 = 0$;

hence $y = \dfrac{\cdot04}{12\cdot2} = \cdot003278$ nearly,

and $x = 8\cdot6 + y = 8\cdot603278$ nearly.

Assume it $x = 8\cdot603278 - v$;

then $x^2 = 74\cdot016392345284 - 17\cdot206556v + v^2$

$$-5x = -43\cdot016390 \qquad +5v$$
$$-31 = -31$$

the sum $\cdot000002345284 - 12\cdot206556v + v^2 = 0$;

hence $v = \cdot000000192133$,

and conseq $x = 8\cdot603277807867$, as before.

EXAMPLE 2. Again, taking the cubic equation $y^3 - 2y - 5 = 0$; Newton proceeds thus:

y is nearly $= 2$; take it therefore $y = 2 + p$;

then $y^3 = 8 + 12p + 6p^2 + p^3$

$$-2y = -4 - 2p$$
$$-5 = -5$$

the sum $-1 + 10p + 6p^2 + p^3 = 0$;

hence $p = \frac{1}{10} = \cdot1$ nearly.

Assume it $p = \cdot1 + q$;

then $p^3 = 0\cdot001 + 0\cdot03q + 0\cdot3q^2 + q^3$

$$+ 6p^2 = 0\cdot06 + 1\cdot2 \qquad + 6$$
$$+ 10p = 1 \qquad + 10$$
$$- 1 = -1$$

the sum $0\cdot061 + 11\cdot23q + 6\cdot3q^2 + q^3 = 0$;

hence $q = -0\cdot0054$ nearly.

Assume it $q = -0\cdot0054 + r$;

then $q^3 = -0\cdot000000157464 + 0\cdot00008748r$, &c.

$$+ 6\cdot3q^2 = +0\cdot000183708 \qquad - 0\cdot06804r,\ \&c.$$
$$+ 11\cdot23q = -0\cdot060642 \qquad + 11\cdot23r$$
$$+ 0\cdot061 = +0\cdot061$$

the sum $+0\cdot000541550536 + 11\cdot16204748r$;

hence $r = -0\cdot000048517$, &c.

Hence, $y = 2 + p + q + r$

$$= 2 + 0\cdot1 - 0\cdot0054 - 0\cdot000048517$$
$$= 2\cdot094551483,$$

the root of the equation $y^3 - 2y = 5$. And in the same manner Newton performs the approximation for the roots of literal equations, that is, equations having literal coefficients; so the root of this equation

$$y^3 + axy + a^2y - x^3 - 2a^3 = 0,\ \text{is}$$

$$y = a - \frac{x}{4} + \frac{x^2}{64a} + \frac{131x^3}{512a^2} + \frac{509x^4}{16384a^3}\ \&c.$$

See also a memoir on this method by the Marquis de Courtivron, in the *Memoires de l'Academie* for 1744.

Other Methods of APPROXIMATION. Besides the foregoing general methods, other particular ways of approximating, for various purposes, have been given by many other persons.—As for example, methods of approximating, by series, to the roots of cubic equations belonging to the irreducible case, by Nicole in the same *Memoirs*, by M. Clairaut in his Algebra, and by myself in the *Philof. Transf.* for 1780. See also several parts of Simpson's works, and my Tracts vol. 1. Also the methods of infinite series by Wallis, Newton, Gregory, Mercator, &c, may be considered as approximations, in quadratures, and other branches of the mathematics, many instances of which may be seen in Wallis's Algebra, and other books:—Likewise the method of exhaustions of the ancients, by which Archimedes and others have approximated to the quadrature and rectification of the circle, &c, which was performed by continually bisecting the sides of polygons, both inscribed in a circle and circumscribed about it; by which means the sum of the sides of the like polygons approach continually nearer and nearer together, and the circumference of the circle is nearly a mean between the two sums. See also EQUATIONS.

APPULSE, in Astronomy, means the actual contact of two luminaries, according to some authors; but others describe it as their near approach to each other, so as to be seen, for instance, within the same telescope. The appulses of the planets to the fixed stars have

always

always been very useful to astronomers, as serving to fix and determine the places of the former. The antients, wanting an easy method of comparing the planets with the ecliptic, which is not visible, had scarce any other way of fixing their situations, but by observing their track among the fixed stars, and marking their appulses to some of those visible points. See Hist. Acad. Scienc. for 1710, pa. 417. And Philos. Transf. No. 369, where Dr. Halley has given a method of determining the places of the planets, by observing their near appulses to the fixed stars. See also Philos. Tranf. No. 76, pa. 361, and Mem. Acad. Scienc. for 1708, where Flamsteed and De la Hire have given observations of the moon's appulses to the Pleiades. See also Flamsteed's Historia Cœlestis, where a multitude of observations of appulses, or small distances, of the moon and planets, from the fixed stars, are recorded. And Dr. Halley has published a map or planisphere of the starry zodiac, in which are accurately laid down all the stars to which the moon's appulse has ever been observed in any part of the world. See Philos. Tranf. No. 369; or Abridg. vol. vi. pa. 170.

APRIL, the 4th month of the year according to the common computation, and the 2d from the vernal equinox.—The word is derived from *Aprilis*, of *aperio*, *I open*; because the earth, in this month, begins to open her bosom for the production of vegetables.—In this month the sun travels through part of the signs Aries and Taurus.

APRON, in Gunnery, a piece of thin or sheet lead, used to cover the vent or touch-hole of a cannon.

APSES, in Astronomy, are the two points in the orbits of planets, where they are at their greatest and least distance, from the sun or the earth. The point at the greatest distance being called the *higher apsis*, and that at the nearest distance the *lower apsis*. And the two apses are also called *auges*. Also the higher apsis is more particularly called the *aphelion*, or the *apogee*; and the lower apsis, the *perihelion*, or the *perigee*. The diameter which joins these two points, is called the *line of the apses* or of the *apsides*; and it passes through the centre of the orbit of the planet, and the centre of the sun or the earth; and in the modern astronomy this line makes the longer or transverse axis of the elliptical orbit of the planet. In this line is counted the excentricity of the orbit; being the distance between the centre of the orbit and the focus, where is placed the sun or the earth.

The foregoing definitions suppose the lines of the greatest and least distances to lie in the same straight line; which is not always precisely the case; as they are sometimes out of a right line, making an angle greater or less than 180 degrees, and the difference from 180 degrees is called the motion of the line of the apses: when the angle is less than 180 degrees, the motion of the apses is said to be contrary to the order of the signs; on the other hand, when the angle exceeds 180 degrees, the motion is according to the order of the signs.

Different means have been employed to determine the motion of the *apses*. Dr. Keil explains, in his Astronomy, the method used by the ancients, who supposed the orbits of the planets to be perfectly circular, and the sun out of the centre. But since it has been

discovered that they describe elliptical orbits, various other methods have been devised for determining it. Halley has given one, which supposes to be known only the time of the planet's revolution, or periodic time. Seth Ward has also given a determination from three different observations of a planet, in any three places of its orbit: but his method being founded on an hypothesis not strictly true, Euler has given one much more exact in vol. 7. of the *Petersburgh Commentaries*. See various ways explained in the Astronomy of Keil and Mounier.

Newton has also given, in the *Principia*, an excellent method of determining the same motion, on the supposition that the orbits of the planets differ but little from circles, which is the case nearly. That great philosopher shews, that if the sun be immoveable, and all the planets gravitate towards him in the inverse ratio of the squares of their distances, then the apses will be fixed, or their motion nothing; that is, the lines of greatest and least distance will form one right line, and the apses will be directly opposite, or at 180 degrees distance from each other. But, because of the mutual tendency of the planets towards each other, their gravitation towards the sun is not precisely in that ratio; and hence it happens, that the apses are not always exactly in a right line with the sun. And Newton has given a very elegant method of determining the motion of the *apses*, on the supposition that we know the force which is thereby added to the gravitation of the planet towards the sun, and that this additional force is always in that direction.

APUS or APOUS, *Avis Indica*, in Astronomy, a constellation of the southern hemisphere, situated near the south pole, between the *triangulum australe* and the chameleon, and supposed to represent the bird of paradise. Also supposed to be one of the birds named *Apodes*, as having no feet.

The number of stars contained in this constellation, are 11 in the British Catalogue, in Bayer's Maps 12, and a still greater number in La Caille's Catalogue; the principal star being but of the 4th or 5th order of magnitude. See *Cælum Australe Stelliferum*, and the *Memoires de l'Acad.* for 1752, pa. 569.

AQUARIUS, in Astronomy, one of the celestial constellations, being the eleventh sign in the zodiac, reckoning from Aries, and is marked by the character ♒, representing part of a stream of water, issuing from the vessel of Aquarius, or the water-pourer. This sign also gives name to the eleventh part of the ecliptic, through which the sun moves in part of the months of January and February.

The poets feign that Aquarius was Ganymede, whom Jupiter ravished under the shape of an eagle, and carried away into Heaven to serve as a cup-bearer, instead of Hebe and Vulcan; whence the name. Others hold, that the sign was thus called, because that when it appears in the horizon, the weather commonly proves rainy.

The stars in the constellation Aquarius, are, in Ptolemy's Catalogue, 45; in Tycho's 41; in Hevelius's 47; and in Flamsteed's 108. See the article CONSTELLATION; also CATALOGUE.

AQUEDUCT, or AQUÆDUCT, as much as to say *ductus aquæ*, a *conduit of water*, is a construction of stone ftone or timber built on uneven ground, to preferve the level of water, and convey it, by a canal, from one place to another.—Some aqueducts are under ground, being conducted through hills, &c; and others are raifed above ground, and fupported on arches, to conduct the water over vallies, &c.

The Romans were very magnificent in their aqueducts; having fome that extended a hundred miles, or more. Frontinus, a man of confular dignity, who had the direction of the aqueducts under the emperor Nerva, fpeaks of nine that emptied themfelves through 13594 pipes, of an inch diameter. And it is obferved by Vigenere, that in the fpace of 24 hours, Rome received from thefe aqueducts not lefs than 500000 hogfheads of water. The chief aqueducts now in being, are thefe: 1ft, that of the Aqua Virginia, repaired by pope Paul IV; 2d, the Aqua Felice, conftructed by pope Sixtus V, and is called from the name he affumed before he was exalted to the papal throne; 3d, the Aqua Paulina, repaired by pope Paul V, in the year 1611; and 4thly, the aqueduct built by Lewis XIV, near Maintenon, to convey the river Bure to Verfailles, which is perhaps the largeft in the world; being 7000 fathoms long, elevated 2560 fathoms in height, and containing 242 arcades. See Philof. Tranf. for 1685, No. 171; or Abridg. vol. 1. pa. 594.

AQUEOUS HUMOUR, or the *watry humour* of the eye, is the firft or outermoft, and the rareft of the three humours of the eye. It is tranfparent and colourlefs, like water; and it fills up the fpace that lies between the cornea tunica, and the cryftalline humour.

AQUILA, *the Eagle*, or the *Vulture* as it is fometimes called, is a conftellation of the northern hemifphere, ufually joined with Antinous. It is one of the 48 old conftellations, according to the divifion of which Hipparchus made his Catalogue of the Fixed Stars, and which are defcribed by Ptolemy. The number of ftars in Aquila, and thofe near it, now in the later-formed conftellation Antinous, amount to 15 in Ptolemy's Catalogue, to 19 in Tycho's, to 42 in that of Hevelius, and to 71 in Flamfteed's. But in Aquila alone, Tycho counts only 12 ftars, and Hevelius 23; the principal ftar being Lucida Aquila, and is between the 1ft and 2d magnitude. The Greeks, as ufual, relate various fables of this conftellation, to make the fcience appear as of their own invention.

ARA, the *Altar*, one of the 48 old conftellations, mentioned by the ancient aftronomers, and is fituated in the fouthern hemifphere; containing only 7 ftars in Ptolemy's Catalogue, and 9 in that of Flamfteed; none of which exceed the 4th magnitude.

ARATUS, celebrated for his Greek poem intitled Φαινόμενα, the *Phenomena*, flourifhed about the 127th Olympiad, or near 300 years before Chrift, while Ptolomy Philadelphus reigned in Egypt. Being educated under Dionyfius Heracleotes, a Stoic philofopher, he efpoufed the principles of that fect, and became phyfician to Antigonus Gonatus, the fon of Demetrius Poliorcetes, King of Macedon. The *Phenomena* of Aratus gives him a title to the character of an aftronomer, as well as a poet. In this work he defcribes the nature and motion of the ftars, and fhews their various difpofitions and relations; he defcribes the figures of

the conftellations, their fituations in the fphere, the origin of the names which they bear in Greece and in Egypt, the fables which have given rife to them, the rifing and fetting of the ftars, and he indicates the manner of knowing the conftellations by their refpective fituations.

The poem of Aratus was commented upon and tranflated by many authors: of whom among the ancients were Cicero, Germanicus Caefar, and Feftus Avienus, who made Latin tranflations of it; a part of the former of which is ftill extant. Aratus muft have been much efteemed by the ancients, fince we find fo great a number of fcholiafts and commentators upon him; among whom are Ariftarchus of Samos, the Aryftylli the geometricians, Apollonius, the Evaeneti, Crates, Numenius the grammarian, Pyrrhus of Magnefia, Thales, Zeno, and many others, as may be feen in Voffius, p. 156. Suidas afcribes feveral other works to Aratus. Virgil, in his Georgics, has tranflated or imitated many paffages from this author: Ovid fpeaks of him with admiration, as well as many others of the poets: And St. Paul has quoted a paffage from him; which is in his fpeech to the Athenians (Acts xvii. 28.) where he tells them that fome of their own poets have faid, *For we are alfo his offspring*, thefe words being the beginning of the 5th line of the Phenomena of Aratus.

His modern editors are as follow: Henry Stephens publifhed his poem at Paris in 1566, in his collection of the poets, in folio. Grotius publifhed an edition of the Phenomena at Leyden in quarto, 1600, in Greek and Latin, with the fragments of Cicero's verfion, and the tranflations of Germanicus and Avienus; all which the editor has illuftrated with curious notes. Alfo a neat and correct edition of Aratus was publifhed at Oxford, 1672, in 8vo. with the Scholia.

ARAEOMETER, fee AREOMETER.

ARC, or ARCH; which fee.

ARCADE, in Architecture, denotes an opening in the wall of a building formed by an arch.

ARC-BOUTANT, is a kind of arched buttrefs, formed of a flat arch, or part of an arch, abutting againft the feet or fides of another arch or vault, to fupport them and prevent them from burfting or giving way.

ARCAS, a name by which fome of the old writers call the ftar Arcturus; a fingle and very bright ftar of the firft magnitude, between the legs of the conftellation Bootes. They fay Arcas, the fon of Califto by Jupiter, when he was about to have killed his mother in the fhape of a bear, was, together with her, fnatched up into Heaven; where fhe was converted into the conftellation of the Great Bear, near the north pole, and the youth into this fingle ftar.

ARCH, ARC, *Arcus*, in Geometry, a part of any curve line; as, of a circle, or ellipfis, or the like.

It is by means of circular arcs, or arches, that all angles are meafured; the arc being defcribed from the angular point as a centre. For this purpofe, every circle is fuppofed to be divided into 360 degrees, or equal parts; and an arch, or the angle it fubtends and meafures, is eftimated according to the number of thofe degrees it contains: thus, an arc, or angle, is faid to be of 30 or 80, or 100 degrees.—Circular arcs are alfo of great ufe in finding of fluents.

Concentric

Concentric ARCS, are such as have the same centre.

Equal ARCS, are such arcs, of the same circle, or of equal circles, as contain the same number of degrees. These have also equal chords, sines, tangents, &c.

Similar ARCS, of unequal circles, &c, are such as contain the same number of degrees, or that are the like part or parts of their respective whole circles. Hence, in concentric circles, any two radii cut off, or intercept, similar arcs MN and OP.—Similar arcs are proportional to the radii LM, LO, or to the whole circumferences.—Similar arcs of other like curves, are also like parts of the wholes, or determined by like parts alike posited.

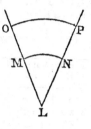

Of the Length of Circular ARCS. The lengths of circular arcs, as found and expressed in various ways, may be seen in my large Treatise on Mensuration, pa. 118, & seq. 2d edition : some of which are as follow. The radius of a circle being 1 ; and of any arc a, if the tangent be t, the sine s, the cosine c, and the versed sine v : then the arc a will be truly expressed by several series, as follow, viz, the arc

$$a = t - \tfrac{1}{3}t^3 + \tfrac{1}{5}t^5 - \tfrac{1}{7}t^7 + \tfrac{1}{9}t^9 \ \&c.$$

$$a = \frac{s}{c} - \frac{1}{3}\cdot\frac{s^3}{c^3} + \frac{1}{5}\cdot\frac{s^5}{c^5} - \frac{1}{7}\cdot\frac{s^7}{c^7} \ \&c.$$

$$a = s + \frac{1}{2\cdot3}s^3 + \frac{1\cdot3}{2\cdot4\cdot5}s^5 + \frac{1\cdot3\cdot5}{2\cdot4\cdot6\cdot7}s^7 \ \&c.$$

$$a = \sqrt{2v} \times (1 + \frac{1}{2\cdot3}\cdot\frac{v}{2} + \frac{1\cdot3}{2\cdot4\cdot5}\cdot\frac{v^2}{2^2} + \frac{1\cdot3\cdot5}{2\cdot4\cdot6\cdot7}\cdot\frac{v^3}{2^3} \ \&c.$$

$$a = \frac{3\cdot14159 \ \&c.}{180} d = \cdot01745329 \ \&c \times d \ ;$$ where d denotes the number of degrees in the given arc. Also $a = \dfrac{8c - C}{3}$ nearly ; where C is the chord of the arc, and c the chord of half the arc ; whatever the radius is.

To investigate the length of the arc of any curve. Put $x =$ the abscifs, $y =$ the ordinate, of the arc z, of any curve whatever. Put $\dot{z} = \sqrt{\dot{x}^2 + \dot{y}^2}$; then, by means of the equation of the curve, find the value of $\dot{x}$ in terms of $\dot{y}$, or of $\dot{y}$ in terms of $\dot{x}$, and substitute that value instead of it in the above expression $\dot{z} = \sqrt{\dot{x}^2 + \dot{y}^2}$; hence, taking the fluents, they will give the length of the arc z, in terms of x or y.

ARCH, in Astronomy. Of this, there are various kinds. Thus, the latitude, elevation of the pole, and the declination, are measured by an arch of the meridian ; and the longitude, by an arch of a parallel circle, &c.

Diurnal ARCH of the sun, is part of a circle parallel to the equator, described by the sun in his course from his rising to the setting. And his *Nocturnal* ARCH is of the same kind ; excepting that it is described from setting to rising.

ARCH *of Progression*, or *Direction*, is an arch of the ecliptic, which a planet seems to pass over, when its motion is direct, or according to the order of the signs.

ARCH of *Retrogradation*, is an arch of the ecliptic, described while a planet is retrograde, or moves contrary to the order of the signs.

ARCH *between the Centres*, in eclipses, is an arch passing from the centre of the earth's shadow, perpendicular to the moon's orbit, meeting her centre at the middle of an eclipse.——If the aggregate of this arch and the apparent semi-diameter of the moon, be equal to the semi-diameter of the shadow, the eclipse will be total for an instant, or without any duration ; and if that sum be less than the radius of the shadow, the eclipse will be total, with some duration ; but if greater, the eclipse will be only partial.

ARCH *of Vision*, is that which measures the sun's depth below the horizon, when a star, before hid by his rays, begins to appear again.—The quantity of this arch is not always the same, but varies with the latitude, declination, right ascension, or descension, and distance, of any planet or star. Ricciol. Almag. v. 1, pa. 42. However, the following numbers will serve nearly for the stars and planets.

TABLE exhibiting the *Arch of Vision* of the PLANETS and FIXED STARS.

PLANETS.			FIXED STARS. Magnitude.	
Mercury	10°	0′	1	12°
Venus	5	0	2	13
Mars	11	30	3	14
Jupiter	10	0	4	15
Saturn	11	0	5	16
			6	17

ARCH, in Architecture, is a concave structure, raised or turned upon a mould, called the centering, in form of the arch of a curve, and serving as the inward support of some superstructure. Sir Henry Wotton says, An arch is nothing but a narrow or contracted vault ; and a vault is a dilated arch.

Arches are used in large intercolumnations of spacious buildings ; in porticoes, both within and without temples ; in public halls, as ceilings, the courts of palaces, cloisters, theatres, and amphitheatres. They are also used to cover the cellars in the foundations of houses, and powder magazines ; also as buttresses and counter-forts, to support large walls laid deep in the earth ; for triumphal arches, gates, windows, &c ; and, above all, for the foundations of bridges and aqueducts. And they are supported by piers, butments, imposts, &c.

Arches are of several kinds, and are commonly denominated from the figure or curve of them ; as circular, elliptical, cycloidal, catenarian, &c, according as their curve is in the form of a circle, ellipse, cycloid, catenary, &c.

There are also other denominations of circular arches, according to the different parts of a circle, or manner of placing them. Thus,

Semicircular ARCHES, which are those that make an exact semicircle, having their centre in the middle of the span or chord of the arch ; called also by the French builders, perfect arches, and *arches en plein centre.* The arches of Westminster Bridge are semicircular.

Scheme ARCHES, or *skene*, are those which are less than semicircles, and are consequently flatter arches ;

containing

containing 120, or 90, or 60, degrees, &c. They are also called *imperfect* and *diminished arches*.

ARCHES *of the third and fourth point*, or *Gothic arches*; or, as the Italians call them, *di terzo* and *quarto acuto*, because they always meet in an acute angle at top. These consist of two excentric circular arches, meeting in an angle above, and are drawn from the division of the chord into three or four or more parts at pleasure. Of this kind are many of the arches in churches and other old Gothic buildings.

Elliptical ARCHES, usually consist of semi-ellipses; and were formerly much used instead of mantle-trees in chimnies; and are now much used, from their bold and beautiful appearance, for many purposes, and particularly for the arches of a bridge, like that at Black-Friars, both for their strength, beauty, convenience, and cheapness.

Straight ARCHES, are those which have their upper and under edges parallel straight lines, instead of curves These are chiefly used over doors and windows; and have their ends and joints all pointing towards one common centre.

ARCH is particularly used for the space between the two piers of a bridge, intended for the passage of the water, boats, &c.

ARCH *of equilibration*, is that which is in equilibrium in all its parts, having no tendency to break in one part more than in another, and which is therefore safer and stronger than any other figure. Every particular figure of the extrados, or upper side of the wall above an arch, requires a peculiar curve for the under side of the arch itself, to form an arch of equilibration, so that the incumbent pressure on every part may be proportional to the strength or resistance there. When the arch is equally thick throughout, a case that can hardly ever happen, then the catenarian curve is the arch of equilibration; but in no other case: and therefore it is a great mistake in some authors to suppose that this curve is the best figure for arches in all cases; when in reality it is commonly the worst. This subject is fully treated in my *Principles of Bridges*, pr. 5, where the proper intrados is investigated for every extrados, so as to form an arch of equilibration in all cases whatever. It there appears that, when the upper side of the wall is a straight horizontal line, as in the annexed figure, the equation

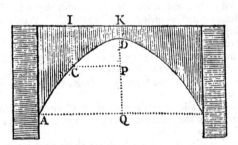

of the curve is thus expressed,

$$y = b \times \frac{\log \text{ of } \dfrac{a + x + \sqrt{2ax + xx}}{a}}{\log \text{ of } \dfrac{a + r + \sqrt{2ar + rr}}{a}};$$

where $x = DP$, $y = PC$, $r = DQ$, $b = AQ$, and $a = DK$. And hence, when a, b, r, are any given numbers, a table is formed for the corresponding values of x and y, by which the curve is constructed for any particular occasion. Thus supposing a or $DK = 6$, b or $AQ = 50$, and r or $DQ = 40$; then the corresponding values of KI and IC, or horizontal and vertical lines, will be as in this table.

Table for constructing the Curve of Equilibration.

Value of KI.	Value of IC.	Value of KI.	Value of IC.	Value of KI.	Value of IC.
0	6·000	21	10·381	36	21·774
2	6·035	22	10·858	37	22·948
4	6·144	23	11·368	38	24·190
6	6·324	24	11·911	39	25·505
8	6·580	25	12·489	40	26·894
10	6·914	26	13·106	41	28·364
12	7·330	27	13·761	42	29·919
13	7·571	28	14·457	43	31·563
14	7·834	29	15·196	44	33·299
15	8·120	30	15·980	45	35·135
16	8·430	31	16·811	46	37·075
17	8·766	32	17·693	47	39·126
18	9·168	33	18·627	48	41·293
19	9·517	34	19·617	49	43·581
20	9·934	35	20·665	50	46·000

The doctrine and use of arches are neatly delivered by Sir Henry Wotton, though he is not always mathematically accurate in the principles. He says; *First*, All matter, unless impeded, tends to the centre of the earth in a perpendicular line. *Secondly*; All solid materials, as bricks, stones, &c, in their ordinary rectangular form, if laid in numbers, one by the side of another, in a level row, and their extreme ones sustained between two supporters; those in the middle will necessarily sink, even by their own gravity, much more if forced down by any superincumbent weight. To make them stand, therefore, either their figure or their position must be altered.—*Thirdly*; Stones, or other materials, being figured *cuneatim*, or wedge-like, broader above than below, and laid in a level row, with their two extremes supported as in the last article, and pointing all to the same centre; none of them can sink, till the supporters or butments give way, because they want room in that situation to descend perpendicularly. But this is a weak structure; because the supporters are subject to too much impulsion, especially where the line is long; for which reason the form of straight arches is seldom used, excepting over doors and windows, where the line is short and the side walls strong. In order to fortify the work, therefore, we must change not only the figure of the materials, but also their position.—*Fourthly*; If the materials be shaped wedge-wise, and be disposed in form of an arch, and pointing to some centre; in this case, neither the pieces of the said arch can sink downwards, for want of room to descend perpendicularly; nor can the supporters or butments suffer much violence, as in the preceding flat form: for the convexity will always make the incumbent rather rest upon the supporters, than thrust or

push

push them outwards. His reasoning, however, afterwards, on the effect of circular and other arches, is not accurate, as he attends only to the side pressure, without considering the effect of different vertical pressures.

The chief properties of arches of different curves, may be seen in the 2d sect. of my *Principles of Bridges*, above quoted. It there appears that none, except the mechanical curve of the arch of equilibration, can admit of a horizontal line at top: that this arch is of a form both graceful and convenient, as it may be made higher or lower at pleasure, with the same span or opening: that all other arches require extrados that are curved, more or less, either upwards or downwards: of these, the elliptical arch approaches the nearest to that of equilibration for equality of strength and convenience; and it is also the best form for most bridges, as it can be made of any height to the same span, its hanches being at the same time sufficiently elevated above the water, even when it is very flat at top: elliptical arches also look bolder and lighter, are more uniformly strong, and much cheaper than most others, as they require less materials and labour. Of the other curves, the cycloidal arch is next in quality to the elliptical one, for those properties, and, lastly, the circle. As to the others, the parabola, hyperbola, and catenary, they are quite inadmissible in bridges that consist of several arches; but may, in some cases, be employed for a bridge of one single arch which may be intended to rise very high, as in such cases as they are not much loaded at the hanches.

ARCH *Mural*. See MURAL *arch*.

ARCHER, or *Sagittarius*, one of the constellations of the northern hemisphere, and one of the twelve signs of the zodiac, placed between the Scorpion and Capricorn. See SAGITTARIUS.

ARCHIMEDES, one of the most celebrated mathematicians among the ancients, who flourished about 250 years before Christ, being about 50 years later than Euclid. He was born at Syracuse in Sicily, and was related to Hiero, who was then king of that city. The mathematical genius of Archimedes set him with such distinguished excellence in the view of the world, as rendered him both the honour of his own age, and the admiration of posterity. He was indeed the prince of the ancient mathematicians, being to them what Newton is to the moderns, to whom in his genius and character he bears a very near resemblance. He was frequently lost in a kind of reverie, so as to appear hardly sensible; he would study for days and nights together, neglecting his food; and Plutarch tells us that he used to be carried to the baths by force. Many particulars of his life, and works, mathematical and mechanical, are recorded by several of the ancients, as Polybius, Livy, Plutarch, Pappus, &c. He was equally skilled in all the sciences, astronomy, geometry, mechanics, hydrostatics, optics, &c, in all of which he excelled, and made many and great inventions.

Archimedes, it is said, made a sphere of glass, of a most surprising contrivance and workmanship, exhibiting the motions of the heavenly bodies in a very pleasing manner. Claudian has an epigram upon this invention, which has been thus translated:

When in a glass's narrow space confin'd,
Jove saw the fabric of th' almighty mind,

He smil'd, and said, Can mortals' art alone,
Our heavenly labours mimic with their own?
The Syracusian's brittle work contains
Th' eternal law, that through all nature reigns.
Fram'd by his art. see stars unnumber'd burn,
And in their courses rolling orbs return:
His sun through various signs describes the year;
And every month his mimic moons appear.
Our rival's laws his little planets bind,
And rule their motions with a human mind.
Salmoneus could our thunder imitate,
But Archimedes can a world create.

Many wonderful stories are told of his discoveries, and of his very powerful and curious machines, &c. Hiero once admiring them, Archimedes replied, these effects are nothing, "But give me, said he, some other place to fix a machine on, and I shall move the earth." He fell upon a curious device for discovering the deceit which had been practiced by a workman, employed by the said king Hiero to make a golden crown. Hiero, having a mind to make an offering to the gods of a golden crown, agreed for one of great value, and weighed out the gold to the artificer. After some time he brought the crown home of the full weight; but it was afterwards discovered or suspected that a part of the gold had been stolen, and the like weight of silver substituted in its stead. Hiero, being angry at this imposition, desired Archimedes to take it into consideration, how such a fraud might be certainly discovered. While engaged in the solution of this difficulty, he happened to go into the bath; where observing that a quantity of water overflowed, equal to the bulk of his body, it presently occurred to him, that Hiero's question might be answered by a like method: upon which he leaped out, and ran homeward, crying out εὑρηκα! εὑρηκα! I have found it! I have found it! He then made two masses, each of the same weight as the crown, one of gold and the other of silver: this done, he filled a vessel to the brim with water, and put the silver mass into it upon which a quantity of water overflowed equal to the bulk of the mass; then taking the mass of silver out he filled up the vessel again, measuring the water exactly, which he put in; this shewed him what measure of water answered to a certain quantity of silver. Then he tried the gold in like manner, and found that it caused a less quantity of water to overflow, the gold being less in bulk than the silver, though of the same weight. He then filled the vessel a third time, and putting in the crown itself, he found that it caused more water to overflow than the golden mass of the same weight, but less than the silver one; so that, finding its bulk between the two masses of gold and silver, and that in certain known proportions, he hence computed the real quantities of gold and silver in the crown, and so manifestly discovered the fraud.

Archimedes also contrived many machines for useful and beneficial purposes: among these, engines for launching large ships; screw pumps, for exhausting the water out of ships, marshes or overflowed lands, as Egypt, &c, which they would do from any depth.

But he became most famous by his curious contrivances, by which the city of Syracuse was so long defended, when besieged by the Roman consul Marcellus; showering upon the enemy sometimes long darts, and stones of vast weight and in great quantities; at other times lifting their ships up into the air, that had come near

T the

the walls, and dashing them to pieces by letting them fall down again; nor could they find their safety in removing out of the reach of his cranes and levers, for there he contrived to fire them with the rays of the sun reflected from burning glasses.

However, notwithstanding all his art, Syracuse was at length taken by storm, and Archimedes was so very intent upon some geometrical problem, that he neither heard the noise, nor minded any thing else, till a soldier that found him tracing of lines, asked him his name, and upon his request to begone, and not disorder his figures, slew him. " What gave Marcellus the greatest concern, says Plutarch, was the unhappy fate of Archimedes, who was at that time in his museum; and his mind, as well as his eyes, so fixed and intent upon some geometrical figures, that he neither heard the noise and hurry of the Romans, nor perceived the city to be taken. In this depth of study and contemplation, a soldier came suddenly upon him, and commanded him to follow him to Marcellus; which he refusing to do, till he had finished his problem, the soldier, in a rage, drew his sword, and ran him through." Livy says he was slain by a soldier, not knowing who he was, while he was drawing schemes in the dust: that Marcellus was grieved at his death, and took care of his funeral; and made his name a protection and honour to those who could claim a relationship to him. His death it seems happened about the 142 or 143 Olympiad, or 210 years before the birth of Christ.

When Cicero was questor for Sicily, he discovered the tomb of Archimedes, all overgrown with bushes and brambles; which he caused to be cleared, and the place set in order. There was a sphere and cylinder cut upon it, with an inscription, but the latter part of the verses quite worn out.

Many of the works of this great man are still extant, though the greatest part of them are lost. The pieces remaining are as follow: 1. Two books on the Sphere and Cylinder.—2. The Dimension of the Circle, or proportion between the diameter and the circumference.—3. Of Spiral lines.—4. Of Conoids and Spheroids.—5. Of Equiponderants, or Centres of Gravity.—6. The Quadrature of the Parabola.—7. Of Bodies floating on Fluids.—8. Lemmata.—9. Of the Number of the Sand.

Among the works of Archimedes which are lost, may be reckoned the descriptions of the following inventions, which may be gathered from himself and other ancient authors. 1. His account of the method which he employed to discover the mixture of gold and silver in the crown, mentioned by Vitruvius.—2. His description of the Cochleon, or engine to draw water out of places where it is stagnated, still in use under the name of Archimedes's Screw. Athenæus, speaking of the prodigious ship built by the order of Hiero, says, that Archimedes invented the cochleon, by means of which the hold, notwithstanding its depth, could be drained by one man And Diodorus Siculus says, that he contrived this machine to drain Egypt, and that by a wonderful mechanism it would exhaust the water from any depth.—3. The Helix, by means of which, Athenæus informs us, he launched Hiero's great ship.—4. The Trispaston, which, according to Tzetzes and Oribasius, could draw the most stupendous weights.—5. The machines, which, according to Polybius, Livy, and Plutarch,

he used in the defence of Syracuse against Marcellus, consisting of Tormenta, Balistæ, Catapults, Sagittarii, Scorpions, Cranes, &c.—6. His Burning Glasses, with which he set fire to the Roman gallies.—7. His Pneumatic and Hydrostatic engines, concerning which subjects he wrote some books, according to Tzetzes, Pappus, and Tertullian.—8. His Sphere, which exhibited the celestial motions. And probably many others.

A whole volume might be written upon the curious methods and inventions of Archimedes, that appear in his mathematical writings now extant only. He was the first who squared a curvilineal space; unless Hypocrates must be excepted on account of his lunes. In his time the conic sections were admitted into geometry, and he applied himself closely to the measuring of them, as well as other figures. Accordingly he determined the relations of spheres, spheroids, and conoids, to cylinders and cones; and the relations of parabolas to rectilineal planes whose quadratures had long before been determined by Euclid. He has left us also his attempts upon the circle: he proved that a circle is equal to a right-angled triangle, whose base is equal to the circumference, and its altitude equal to the radius; and consequently, that its area is equal to the rectangle of half the diameter and half the circumference; thus reducing the quadrature of the circle to the determination of the ratio between the diameter and circumference; which determination however has never yet been done. Being disappointed of the *exact* quadrature of the circle, for want of the rectification of its circumference, which all his methods would not effect, he proceeded to assign an useful approximation to it: this he effected by the numeral calculation of the perimeters of the inscribed and circumscribed polygons: from which calculation it appears that the perimeter of the circumscribed regular polygon of 192 sides, is to the diameter, in a less ratio than that of $3\frac{1}{7}$ or $3\frac{10}{70}$ to 1; and that the perimeter of the inscribed polygon of 96 sides, is to the diameter, in a greater ratio than that of $3\frac{10}{71}$ to 1; and consequently that the ratio of the circumference to the diameter, lies between these two ratios. Now the first ratio, of $3\frac{1}{7}$ to 1, reduced to whole numbers, gives that of 22 to 7, for $3\frac{1}{7} : 1 :: 22 : 7$; which therefore is nearly the ratio of the circumference to the diameter. From this ratio between the circumference and the diameter, Archimedes computed the approximate area of the circle, and he found that it is to the square of the diameter, as 11 is to 14. He determined also the relation between the circle and ellipse, with that of their similar parts. And it is probable that he likewise attempted the hyperbola; but it is not to be expected that he met with any success, since approximations to its area are all that can be given by the various methods that have since been invented.

Beside these figures, he determined the measures of the spiral, described by a point moving uniformly along a right line, the line at the same time revolving with a uniform angular motion; determining the proportion of its area to that of the circumscribed circle, as also the proportion of their sectors.

Throughout the whole works of this great man, we every where perceive the deepest design, and the finest invention. He seems to have been, with Euclid, exceedingly careful of admitting into his demonstrations nothing

nothing but principles perfectly geometrical and unexceptionable : and although his most general method of demonstrating the relations of curved figures to straight ones, be by inscribing polygons in them ; yet to determine those relations, he does not increase the number, and diminish the magnitude, of the sides of the polygon *ad infinitum;* but from this plain fundamental principle, allowed in Euclid's Elements, (viz, that any quantity may be so often multiplied, or added to itself, as that the result shall exceed any proposed finite quantity of the same kind,) he proves that to deny his figures to have the proposed relations, would involve an absurdity. And when he demonstrated many geometrical properties, particularly in the parabola, by means of certain progressions of numbers, whose terms are similar to the inscribed figures; this was still done without considering such series as continued *ad infinitum,* and then collecting or summing up the terms of such infinite series.

There have been various editions of the existing writings of Archimedes. The whole of these works, together with the commentary of Eutocius, were found in their original Greek language, on the taking of Constantinople, from whence they were brought into Italy; and here they were found by that excellent mathematician John Muller, otherwise called Regiomontanus, who brought them into Germany: where they were, with that Commentary, published long afterwards, viz, in 1544, at Basil, being most beautifully printed in folio, both in Greek and Latin, by Hervagius, under the care of Thomas Gechauff Venatorius.—A Latin translation was published at Paris 1557, by Pascalius Hamellius.—Another edition of the whole, in Greek and Latin, was published at Paris 1615, in folio, by David Rivaltus, illustrated with new demonstrations and commentaries: a life of the author is prefixed ; and at the end of the volume is added some account, by way of restoration, of our author's other works, which have been lost ; viz, The Crown of Hiero ; the Cochleon or Water Screw ; the Helicon, a kind of endless screw ; the Trispaston, consisting of a combination of wheels and axles ; the Machines employed in the defence of Syracuse ; the Burning Speculum ; the Machines moved by Air and Water ; and the Material Sphere.—In 1675, Dr. Isaac Barrow published a neat edition of the works, in Latin, at London, in 4to ; illustrated, and succinctly demonstrated in a new method.—But the most complete of any, is the magnificent edition, in folio, lately printed at the Clarendon press, Oxford, 1792. This edition was prepared ready for the press by the learned Joseph Torelli, of Verona, and in that state presented to the University of Oxford. The Latin translation is a new one. Torelli also wrote a preface, a commentary on some of the pieces, and notes on the whole. An account of the life and writings of Torelli is prefixed, by Clemens Sibiliati. And at the end a large appendix is added, in two parts ; the first being a Commentary on Archimedes's paper upon Bodies that float on Fluids, by the Rev. Abram Robertson of Christ Church College ; and the latter is a large collection of various readings in the Manuscript works of Archimedes, found in the library of the late king of France, and of another at Florence, as collated with the Basil edition above mentioned.

There are also extant other editions of certain parts of the works of Archimedes. Thus, Commandine published, in 4to, at Bologna 1565, the two books concerning Bodies that Float upon Fluids, with a Commentary Commandine published also a translation of the Arenarius. And Borelli published, in folio, at Florence 1661, Archimedes's *Liber Assumptorum,* translated into Latin from an Arabic manuscript copy. This is accompanied with the like translation, from the Arabic, of the 5th, 6th, and 7th books of Apollonius's Conics. Mr. G. Anderson published (in 8vo. Lond. 1784) an English translation of the Arenarius of Archimedes, with learned and ingenious notes and illustrations. Dr. Wallis published a translation of the Arenarius. And there may be other editions beside the above, but these are all that I have got, or know of.

ARCHIMEDES's *Screw.* See SCREW *of Archimedes.*

ARCHIMEDES's *Burning-glass.* See BURNING-*glass.*

ARCHITECT, a person skilled in architecture, or the art of building ; who forms plans and designs for edifices, conducts the work, and directs the various artificers employed in it.

The most celebrated architects are, Vitruvius, Palladio, Scamozzi, Serlio, Vignola, Barbaro, Cataneo, Alberti, Viola, Inigo Jones, De Lorme, Perrault, S. Le Clerc, Sir Christopher Wren, and the Earl of Burlington.

ARCHITECTURE, *Architectura,* the art of planning and building or erecting any edifice, so as properly to answer the end proposed, for solidity, conveniency, and beauty ; whether houses, temples, churches, bridges, halls, theatres, &c, &c.—Architecture is divided into civil, military, and naval or marine.

Civil ARCHITECTURE, is the art of designing and erecting edifices of every kind for the uses of civil life in every capacity ; as churches, palaces, private houses, &c ; and it has been divided into five orders or manners of building, under the names of the *Tuscan, Doric, Ionic, Corinthian,* and *Composite.*

There were many authors on architecture among the Greeks and Romans, before Vitruvius ; but he is the first whose work is entire and extant. He lived in the reigns of Julius Cæsar and Augustus, and composed a complete system of architecture, in ten books, which he dedicated to this prince. The principal authors on architecture since Vitruvius, are Philander, Barbarus, Salmasius, Baldus, Alberti, Gauricus, Demoniosius, Perrault, De l'Orme, Rivius, Wotton, Serlio, Palladio, Strada, Vignola, Scamozzi, Dieussart, Catanei, Freard, De Cambray, Blondel, Goldman, Sturmy, Wolfius, De Rosi, Desgodetz, Baratteri, Mayer, Gulielmus, Ware, &c, &c. See also ARCHITECT.

Military ARCHITECTURE, otherwise more usually called *Fortification,* is the art of strengthening and fortifying places, to screen them from the insults or attacks of enemies, and the violence of arms ; by erecting forts, castles, and other fortresses, with ramparts, bastions, &c.—The authors who have chiefly excelled in this art, are Coehorn, Pagan, Vauban, Scheiter, Blondel, and Montalembert.

Naval ARCHITECTURE, or *ship-building,* is the art of constructing ships, galleys, and other vessels proper to float on the water.

ARCHITRAVE, is that part of a column which bears immediately upon the capital. It is the lowest

member

member of the entablature, and is fuppofed to reprefent the principal beam in timber buildings, in which it is fometimes called the *reafon piece*, or *mafter-piece*. Alfo, in chimneys it is called the *mantle-piece;* and the *hyper-thyron* over the jaumbs of doors, or lintels of windows.

ARCHITRAVE *Corniche.* See CORNICHE.

ARCHITRAVE *doors* are thofe which have an archi-trave on the jaumbs, and over the door; upon the cap piece if ftraight; or on the arch, if the top be curved.

ARCHITRAVE *windows*, of timber, are ufually an ogee raifed out of the folid timber, with a lift over it: though fometimes the mouldings are ftruck, and laid on; and fometimes they are cut in brick.

ARCHIVOLT, the contour of an arch; or a band or frame adorned with mouldings, running over the faces of the vouffoirs or arch-ftones, and bearing upon the impofts.

ARCHYTAS, of Tarentum, a celebrated mathe-matician, cofmographer, and Pythagorean philofopher, whom Horace calls

——Maris ac Terræ, numeroque carentis Arenæ Menforem.

He flourifhed about 400 years before Chrift; and was the mafter of Plato, Eudoxus, and Philolaus. He gave a method of finding two mean proportionals between two given lines, and thence the Duplication of the Cube, by means of the conic fections. His fkill in Mechanics was fuch, that he was faid to be the inventor of the crane and the fcrew; and he made a wooden pigeon that could fly about, when it was once fet off, but it could not rife again of itfelf, after it refted. He wrote feveral works, though none of them are now ex-tant, particularly a treatife περὶ τὖ Παντὸς, *de Univerfo*, cited by Simplicius in Ariftot. Categ. It is faid he in-vented the ten categories. He acquired great reputa-tion both in his legiflative and military capacity; having commanded an army feven times without ever being defeated. He was at laft fhipwrecked, and drowned in the Adriatic fea.

ARCTIC *Circle*, is a leffer circle of the fphere, parallel to the equator, and paffing through the north pole of the ecliptic, or diftant from the north or arctic pole, by a quantity equal to the obliquity of the ecliptic, which was formerly eftimated at 23° 30, but its mean quan-tity is now 23° 28' nearly. This, and its oppofite, the antarctic circle, are alfo called the polar circles, where the longeft day and longeft night are 24 hours, and within all the fpace of thefe circles, at one time of the year, the fun never fets, and at the oppofite feafon he never rifes for fome days, more or lefs according as the place is nearer the pole.

ARCTIC *Pole*, the north pole of the world, and fo called from ἀρκτος, *urfa*, the *bear*, from its proximity to the conftellation of that name.

ARCTOPHYLAX, a conftellation otherwife called BOOTES. Which fee.

ARCTURUS, a fixed ftar of the firft magnitude, between the thighs of the conftellation Bootes. So called from ἀρκτος, *bear*, and ὐρα, *tail;* as being near the *bear's tail*.

This ftar is twice mentioned in the book of Job, viz, ix. 9, and xxxviii. 32, by the name Aifh if the tranflation

be right; and by many of the ancients under its Greek name *Arcturus*. The Greeks gave the fabulous hiftory of this ftar, or conftellation, to this purport: Califto, who was afterwards, in form of the great bear, raifed up into a conftellation, they tell us, brought forth a fon to Jupiter, whom they called Arcas. That Lyacon, when Jupiter afterwards came to vifit him, cut the boy in pieces, and ferved him up at table. Jupiter, in re-venge, as well as by way of punifhment, called down lightning to confume the palace, and turned the monarch into a wolf. The limbs of the boy were gathered up, to which the god gave life again, and he was taken and educated by fome of the people. His mother, who was all this time a bear in the woods, fell in his way: he chafed her, ignorant of the fact, and, to avoid him, fhe threw herfelf into the temple of Jupiter: he followed her thither to deftroy her; and this being death by the laws of the country, Jupiter took them both up into heaven, to prevent the punifhment, making her the conftellation of the great bear, and converting the youth into this fingle ftar behind her.

Dr. Hornfby, the Savilian Profeffor of Aftronomy, concludes that Arcturus is the neareft ftar to our fyftem vifible in the northern hemifphere, becaufe the variation of its place, in confequence of a proper motion of its own, is more remarkable than that of any other of the ftars; and by comparing a variety of obfervations re-fpecting both the quantity and direction of the motion of this ftar, he infers, that the obliquity of the ecliptic decreafes at the rate of 58″ in 100 years; a quantity which nearly correfponds to the mean of the compu-tations framed by Euler and De la Lande, upon the principles of attraction. *Philof. Tranf.* v. 63.

ARCTUS, a name given by the Greeks to two conftellations of the northern hemifphere; by the Latins called URSA *major* and *minor*, and by us the *greater* and *leffer* BEAR.

AREA, in general, denotes any plain furface to walk upon; and derived from *arere, to be dry*.

AREA, in *Architecture*, denotes the fpace or fcite of ground on which an edifice ftands. It is alfo ufed for inner courts, and fuch like portions of ground.

AREA, in *Geometry*, denotes the fuperficial content of any figure. The areas of figures are eftimated in fquares and parts of fquares. Thus, fuppofe a rectangle EFGH have its length EH equal to 4 inch-es, or feet, or yards, &c, and its breadth EF equal to 3; its area will then be 3 times 4, or 12 fquares, each fide of which is refpectively one inch, or foot, or yard, &c. The areas of other particular figures may be feen under their refpective names.

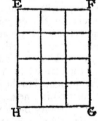

The areas of all fimilar figures, are in the duplicate ratio, or as the fquares of their like fides, or of any like linear dimenfions.——Alfo the law by which the planets move round the fun, is regulated by the areas defcribed by a line connecting the fun and planet; that is, the time in which the planet defcribes, or paffes over, any arc of its elliptic orbit, is proportional to the elliptic area defcribed in that time by the faid line, or the fector contained by the faid arc and two radii

drawn

drawn from its extremities to the focus in which the sun is placed.

AREA, in *Optics*. See FIELD.

ARENARIUS, the name of a book of Archimedes, in which he demonstrated, that not only the sands of the earth, but even a greater quantity of particles than could be contained in the immense sphere of the fixed stars, might be expressed by numbers, in a way by him invented and described. This notation proceeds by certain geometrical progressions; and in denoting and producing certain very distant terms of the progression, he here first of any one makes use of a property similar to that of logarithms, viz, adding the indices of the terms, to find the index of the product of them. See ARCHIMEDES.

AREOMETER, ARÆOMETRUM, an instrument to measure the density or gravity of fluids.

The areometer, or water-poise, is commonly made of glass; consisting of a round hollow ball, which terminates in a long slender neck, hermetically sealed at top; having first as much running mercury put into it, as will serve to balance or keep it swimming in an erect position. The stem, or neck, is divided into degrees or parts which are numbered, to shew, by the depth of its descent into any liquor, the lightness or density of it: for that fluid is heaviest in which it sinks least, and lightest in which it sinks deepest.

Another instrument of this kind is described by Homberg of Paris, in the Memoirs of the Acad. of Sciences for the year 1699; also in the Philos. Transf. Nº 262, where a table of numbers is given, expressing the density of various fluids, as determined by this instrument both in summer and winter. By this table it appears that the density, or specific gravity of quicksilver and distilled water, in the two seasons, were as follow, viz,

in summer as - 13·61 to 1,
in winter as - 13·53 to 1;
and the medium of these two is as 13·57 to 1.

See also the Philos. Transf. vol. 36, or Abridg. vol. 6, for the description and use of another new areometer.

AREOMETRY, the science of measuring the lightness and density of fluids.—See the Philos. Transf. vol. 68, for an essay on areometry, &c.

AREOSTYLE, in Architecture, a sort of intercolumnation in which the columns were placed at a great distance from one another.

ARGENTICOMUS, among Ancient Astrologers, denotes a kind of silver-haired comet, of uncommon lustre, supposed to be the cause of great changes in the planetary system.

ARGETENAR, a star of the fourth magnitude, in the flexure of the constellation Eridanus.

ARGO NAVIS, or *the ship*, is a constellation of fixed stars, in the southern hemisphere, being one of the 48 old constellations. The number of stars in this constellation, are, in Ptolemy's catalogue 45, in Tycho Brahe's 11, in Flamsteed's 64.

The Greeks tell us, that this was the famous ship in which the Argonauts performed that celebrated expedition, which has been so famous in all their history.

ARGUMENT, in Astronomy, is an arch given, by which another arch is found in some proportion to it. Hence,

ARGUMENT *of Inclination*, or ARGUMENT *of Latitude*, of any planet, is an arch of a planet's orbit, intercepted between the ascending node, and the place of the planet from the sun, numbered according to the succession of the signs.

Menstrual ARGUMENT *of Latitude*, is the distance of the moon's true place from the sun's true place.—By this is found the quantity of the real obscuration in eclipses, or how many digits are darkened in any place.

Annual ARGUMENT *of the moon's apogee*, or simply, *Annual Argument*, is the distance of the sun's place from the place of the moon's apogee; that is, the arc of the ecliptic comprised between those two places.

ARIES, or the *Ram*, in *Astronomy*, one of the constellations of the northern hemisphere, and the first of the old twelve signs of the zodiac, and marked ♈ in imitation of a ram's head. It gives name to a twelfth part of the ecliptic, which the sun enters commonly about the 20th of March.—The stars of this constellation in Ptolemy's catalogue are 18, in Tycho Brahe's 21, in Hevelius's 27, and in Flamsteed's 66: but they are mostly very small, only one being of the 2d magnitude, two of the 3d magnitude, and all the rest smaller.

The fabulous account of this constellation, as given by the Greeks, is to this effect. That Nephele gave Phryxus, her son, a ram, which bore a golden fleece, as a guard against the greatest dangers. Juno, the stepmother both of him and Helle, laid designs against their lives; But Phryxus, remembering the admonition of his mother, took his sister with him, and getting upon the back of the ram, they were carried to the sea. The ram plunged in, and the youth was carried over; but Helle dropped off, and was drowned, and so gave name to the Hellespont. When he arrived in Colchis, Æeta, the king, received him kindly; and, sacrificing the ram to Jupiter, dedicated the fleece to the god; which was afterwards carried off by Jason. The animal itself, they say, Jupiter snatched up into the heavens, and made of it the constellation Aries. They have other fables also to account for its origin. But it is most probable that the inventors of this sign, placed it there as the father of those animals which are brought forth about the time the sun approaches to that part of the heavens, and so marking the beginning of spring.

ARIES also denotes a *battering ram*; being a military engine with an iron head, much used by the ancients, to batter and beat down the walls of places besieged. See RAM, and BATTERING *ram*.

ARISTARCHUS, a celebrated Greek philosopher and astronomer, and a native of the city of Samos; but of what date is not exactly known; it must have been however before the time of Archimedes, as some parts of his writings and opinions are cited by that author, viz, in his Arenarius; he probably flourished about 420 years before Christ. He held the opinion of Pythagoras as to the system of the world, but whether before or after him, is uncertain, teaching that the sun and stars were fixed in the heavens, and that the earth moved in a circle about the sun, at the same time that it revolved about its own centre or axis. He taught also, that the annual orbit of the earth, compared with the distance of the fixed stars, is but as a point. On this head Archimedes

chimedes says, "Ariftarchus, the Samian, confuting the notions of aftrologers, laid down certain pofitions, from whence it follows, that the world is much larger than is generally imagined; for he lays it down, that the fixed ftars and the fun are immoveable; and that the earth is carried round the fun in the circumference of a circle." On which account, although he might not fuffer perfecution and imprifonment like Galileo, yet he did not efcape cenfure for his fuppofed impiety; for it is faid Cleanthus was of opinion, that Greece ought to have tried Ariftarchus for irreligion, for endeavouring to preferve the regular appearance of the heavenly bodies, by fuppofing that the heavens themfelves ftood ftill; but that the earth revolved in an oblique circle, and at the fame time turned round its own axis.

Ariftarchus invented a peculiar kind of fun-dials, mentioned by Vitruvius. There is extant of his works only a treatife upon the magnitude and diftance of the fun and moon: this was tranflated into Latin, and commented upon by Commandine, who firft publifhed it with Pappus's Explanations, in 1572. Dr. Wallis afterwards publifhed it in Greek, with Commandine's Latin verfion, in 1688, and which he inferted again in the 3d volume of his Mathematical Works, printed in folio at Oxford, 1699. The piece was animadverted upon by Mr. Fofter, in his Mifcellanies. There is another piece which has gone under the name of Ariftarchus, of the *Mundane Syftem*, its parts, and motions, publifhed in Latin by Robervale, and by Merfenne, in his *Mathematical Synopfis*. But this piece is cenfured, by Menagius (in Diog. Laert.), and Defcartes, in his Epiftles, as a fictitious piece of Robervale's, and not the genuine work of Ariftarchus.

ARISTOTELIAN, fomething that relates to the philofopher *Ariftotle*. Thus we fay, an *Ariftotelian* dogma, the *Ariftotelian* fchool, &c.

ARISTOTELIAN *Philofophy*, the philofophy taught by *Ariftotle*, and maintained by his followers. It is otherwife called the *peripatetic* philofophy, from their practice of teaching while they were walking.—The principles of Ariftotle's philofophy, the learned agree, are chiefly laid down in the four books *de Cælo*. Inftead of the more ancient fyftems, he introduced *matter*, *form*, and *privation*, as the principles of all things; but it does not feem that he derived much benefit from them in natural philofophy. And his doctrines are, for the moft part, fo obfcurely expreffed, that it has not yet been fatisfactorily afcertained, what were his fentiments on fome of the moft important fubjects. He attempted to refute the Pythagorean doctrine concerning the two-fold motion of the earth; and pretended to demonftrate, that the matter of the heavens is ungenerated, incorruptible, and not fubject to any alteration: and he fuppofed that the ftars were carried round the earth in folid orbs.

ARISTOTELIANS, a fect of philofophers, fo called, from their leader *Ariftotle*, and are otherwife called *Peripatetics*.—The Ariftotelians and their dogmata prevailed for a long while, in the fchools and univerfities; even in fpite of all the efforts of the Cartefians, Newtonians, and other corpufcularians. But the fyftems of the latter have at length gained the afcendency; and the Newtonian philofophy in particular is now very generally received.

ARISTOTLE, a Grecian philofopher, the fon of Nicomachus, phyfician to Amyntas king of Macedonia, was born 384 years before Chrift, at Stagira, a town of Macedonia, or, as others fay, of Thrace; from which he is alfo called the Stagirite. Not fucceeding in the profeffion of arms, to which it feems he firft applied himfelf, he turned his views to philofophy, and at 17 years of age entered himfelf a difciple of Plato, and attended in the academy till the death of that philofopher. Ariftotle then retired to Atarna, where the prince Hermias gave him his daughter to wife. Repairing afterwards to the court of king Philip, he became preceptor to his fon, Alexander the Great, whofe education he attended for the fpace of 8 years; and by the magnificent encouragement of this prince he was afterwards enabled to procure all forts of animals, from the infpection of which to write the hiftory of them. On his quitting Macedon, he fettled at Athens, where he eftablifhed his fchool, having the Lyceum affigned him, by the magiftrates, for the place of his inftruction or difputation; where he became the head and founder of the fect called after his name, as alfo Peripatetics, from the circumftance of his giving inftructions while walking. But being here accufed of impiety by Eurymedon, prieft of Ceres, and fearing the fate of Socrates, he retired to Chalcis, where he died at 63 years of age, and 322 years before Chrift. Some fay that he poifoned himfelf, others that he died of a cholic, and others again pretend that he threw himfelf into the fea for grief that he could not difcover the caufe of the flux and reflux of the waters. Laertius, in his life of Ariftotle, eftimates his books at the number of 4000; of which however fcarce 20 have come down to us: thefe may be comprifed under five heads; the firft, relating to poetry and rhetoric; the fecond, to logics; the third, to ethics and politics; the fourth, to phyfics; and the fifth, to metaphyfics. In the fchools, Ariftotle has been called the philofopher, and the prince of philofophers. And fuch was the veneration paid to him, that his opinion was allowed to ftand on a level with reafon itfelf: nor was any appeal from it admitted, the parties, in every difpute, being obliged to fhew, that their conclufions were no lefs conformable to the doctrine of Ariftotle than to truth.

ARITHMETIC, the art and fcience of numbers; or, that part of mathematics which confiders their powers and properties, and teaches how to compute or calculate truly, and with eafe and expedition. It is by fome authors alfo defined the fcience of difcrete quantity. Arithmetic confifts chiefly in the four principal rules or operations of Addition, Subtraction, Multiplication, and Divifion; to which may perhaps be added involution and evolution, or raifing of powers and extraction of roots. But befides thefe, for the facilitating and expediting of computations, mercantile, aftronomical, &c, many other ufeful rules have been contrived, which are applications of the former, fuch as, the rules of proportion, progreffion, alligation, falfe pofition, fellowfhip, intereft, barter, rebate, equation of payments, reduction, tare and tret, &c. Befides the doctrine of the curious and abftract properties of numbers.

Very little is known of the origin and invention of arithmetic. In fact it muft have commenced with mankind, or as foon as they began to hold any fort of commerce

merce together; and muſt have undergone continual improvements, as occaſion was given by the extenſion of commerce, and by the diſcovery and cultivation of other ſciences. It is therefore very probable that the art has been greatly indebted to the Phœnicians or Tyrians; and indeed Proclus, in his commentary on the firſt book of Euclid, ſays, that the Phœnicians, by reaſon of their traffic and commerce, were accounted the firſt inventors of Arithmetic. From Aſia the art paſſed into Egypt, whither it was carried by Abraham, according to the opinion of Joſephus. Here it was greatly cultivated and improved; inſomuch that a conſiderable part of the Egyptian philoſophy and theology ſeems to have turned altogether upon numbers. Hence thoſe wonders related by them about unity, trinity, with the numbers 4, 7, 9, &c. In effect, Kircher, in his Oedip. Ægypt. ſhews, that the Egyptians explained every thing by numbers; Pythagoras himſelf affirming, that the nature of numbers pervades the whole univerſe; and that the knowledge of numbers is the knowledge of the deity.

From Egypt arithmetic was tranſmitted to the Greeks, by means of Pythagoras and other travellers; amongſt whom it was greatly cultivated and improved, as appears by the writings of Euclid, Archimedes, and others: with theſe improvements it paſſed to the Romans, and from them it has deſcended to us.

The nature of the arithmetic however that is now in uſe, is very different from that above alluded to; this art having undergone a total alteration by the introduction of the Arabic notation, about 800 years ſince, into Europe: ſo that nothing now remains of uſe from the Greeks, but the theory and abſtract properties of numbers, which have no dependence on the peculiar nature of any particular ſcale or mode of notation. That uſed by the Hebrews, Greeks, and Romans, was chiefly by means of the letters of their alphabets. The Greeks, particularly, had two different methods; the firſt of theſe was much the ſame with the Roman notation, which is ſufficiently well known, being ſtill in common uſe with us, to denote dates, chapters and ſections of books, &c. Afterwards they had a better method, in which the firſt nine letters of their alphabet repreſented the firſt numbers, from one to nine, and the next nine letters repreſented any number of tens, from one to nine, that is, 10, 20, 30, &c, to 90. Any number of hundreds they expreſſed by other letters, ſupplying what they wanted with ſome other marks or characters: and in this order they went on, uſing the ſame letters again, with ſome different marks, to expreſs thouſands, tens of thouſands, hundreds of thouſands, &c: In which it is evident that they approached very near to the more perfect decuple ſcale of progreſſion uſed by the Arabians, and who acknowledge that they had received it from the Indians. Archimedes alſo invented another peculiar ſcale and notation of his own, which he employed in his Arenarius, to compute the number of the ſands. In the 2d century of chriſtianity lived Cl. Ptolemy, who, it is ſuppoſed, invented the ſexageſimal diviſion of numbers, with its peculiar notation and operations: a mode of computation ſtill uſed in aſtronomy &c, for the ſubdiviſions of the degrees of circles. Thoſe notations however were ill adapted to the practical operations of arithmetic: and hence it is that the art ad-

vanced but very little in this part; for, ſetting aſide Euclid, who has given many plain and uſeful properties of numbers in his Elements, and Archimedes, in his Arenarius, they moſtly conſiſt in dry and tedious diſtinctions and diviſions of numbers; as appears from the treatiſes of Nicomachus, ſuppoſed to be written in the 3d century of Rome, and publiſhed at Paris in 1538; as alſo that of Boethius, written at Rome in the 6th century of Chriſt. A compendium of the ancient arithmetic, written in Greek, by Pſellus, in the 9th century, was publiſhed in Latin by Xylander, in 1556. A ſimilar work was written ſoon after in Greek by Jodocus Willichius; and a more ample work of the ſame kind was written by Jordanus, in the year 1200, and publiſhed with a comment by Faber Stapulenſis in 1480.

Since the introduction of the Indian notation into Europe, about the 10th century, arithmetic has greatly changed its form, the whole algorithm, or practical operations with numbers, being quite altered, as the notation required; and the authors of arithmetic have gradually become more and more numerous. This method was brought into Spain by the Moors or Saracens; whither the learned men from all parts of Europe repaired, to learn the arts and ſciences of them. This, Dr. Wallis proves, began about the year 1000; particularly that a monk, called Gilbert, afterwards pope, by the name of Sylveſter II, who died in the year 1003, brought this art from Spain into France, long before the date of his death: and that it was known in Britain before the year 1150, where it was brought into common uſe before 1250, as appears by the treatiſe of arithmetic of Johannes de Sacro Boſco, or Halifax, who died about 1256. Since that time, the principal writers on this art have been, Barlaam, Lucas de Burgo, Tonſtall, Aventinus, Purbach, Cardan, Scheubelius, Tartalia, Faber, Stifelius, Recorde, Ramus, Maurolyeus, Hemiſchius, Peletarius, Stevinus, Xylander, Kerſey, Snellius, Tacquet, Clavius, Metius, Gemma Friſius, Buteo, Urſinus, Romanus, Napier, Ceulen, Wingate, Kepler, Briggs, Ulacq, Oughtred, Cruger, Van Schooten, Wallis, Dee, Newton, Morland, Moore, Jeake, Ward, Hatton, Malcolm, &c, &c; the particular inventions or excellencies of whom, will be noticed under the articles of the ſeveral ſpecies or kinds of arithmetic here following, which may be included under theſe heads, viz, *theoretical, practical, inſtrumental, logarithmical, numerous, ſpecious, univerſal, common or decadal, fractional, radical or of ſurds, decimal, duodecimal, ſexageſimal, dynamical or binary, tetractycal, political,* &c.

Theoretical ARITHMETIC, is the ſcience of the properties, relations, &c, of numbers, abſtractedly conſidered; with the reaſons and demonſtrations of the ſeveral rules. Such is that contained in the 7th, 8th, and 9th books of Euclid's Elements; the *Logiſtics* of Barlaam the monk, publiſhed in Latin by J. Chambers, in 1600; the *Summa Arithmetica* of Lucas de Burgo, printed 1494, who gives the ſeveral diviſions of numbers from Nicomachus, and their properties from Euclid, with the algorithm, both in integers, fractions, extraction of roots, &c; Malcolm's *New Syſtem of Arithmetic, theoretical and practical,* in 1730, in which the ſubject is very completely treated, in all its branches, &c.

Practical

Practical ARITHMETIC, is the art or practice of numbering or computing; that is, from certain numbers given, to find others which shall have any proposed relation to the former. As, having the two numbers 4 and 6 given; to find their sum, which is 10; or their difference, which is 2; or their product, 24; or their quotient, $1\frac{1}{2}$; or a third proportional to them, which is 9; &c.—Lucas de Burgo's works contain the whole practice of arithmetic, then used, as well as the theory. Tunstall gave a neat practical treatise of Arithmetic in 1526; as did Stifelius, in 1544, both on the practical and other parts. Tartalea gave an entire body of practical arithmetic, which was printed at Venice in 1556, consisting of two parts; the former, the application of arithmetic to civil uses; the latter, the grounds of Algebra. And most of the authors in the list before enumerated, joined the practice of arithmetic with the theory.

Binary or *Dyadic* ARITHMETIC, is that in which only two figures are used, viz 1 and 0. See BINARY. —Leibnitz and De Lagny both invented an arithmetic of this sort, about the same time: and Dangicourt, in the Miscel. Berol. gives a specimen of the use of it in arithmetical progressions; where he shews, that the laws of progression may be more easily discovered by it, than by any other method where more characters are used.

Common or *Vulgar* ARITHMETIC, is that which is concerning integers and vulgar fractions.

Decimal or *Decadal* ARITHMETIC, is that which is performed by a series of ten characters or figures, the progression being ten-fold, or from 1 to 10's, 100's, &c; which includes both integers and decimal fractions, in the common scale of numbers; and the characters used are the ten Arabic or Indian figures 0, 1, 2, 3, 4, 5, 6, 7, 8, 9. This method of arithmetic was not known to the Greeks and Romans; but was borrowed from the Moors while they possessed a great part of Spain, and who acknowledge that it came to them from the Indians. It is probable that this method took its origin from the ten fingers of the hands, which were used in computations before arithmetic was brought into an art. The Eastern missionaries assure us, that to this day the Indians are very expert at computing on their fingers, without any use of pen and ink. And it is asserted, that the Peruvians, who perform all computations by the different arrangements of grains of maize outdo any European, both for certainty and dispatch, with all his rules.

Duodecimal ARITHMETIC, is that which proceeds from 12 to 12, or by a continual subdivision according to 12. This is greatly used by most artificers, in calculating the quantity of their work; as Bricklayers, Carpenters, Painters, Tilers, &c.

Fractional ARITHMETIC, or *of fractions*, is that which treats of fractions, both vulgar and decimal.

Harmonical ARITHMETIC, is so much of the doctrine of numbers, as relates to the making the comparisons, reductions, &c of musical intervals.

ARITHMETIC *of Infinites*, is the method of summing up a series of numbers, of which the numbers of terms is infinite. This method was first invented by Dr. Wallis, as appears by his treatise on that subject; where he shews its uses in geometry, in finding the areas of superficies, the contents of solids, &c. But the method of fluxions, which is a kind of universal arithmetic of infinites, performs all these more easily; as well as a great many other things, which the former will not reach.

Instrumental ARITHMETIC, is that in which the common rules are performed by instruments, or some sort of tangible or palpable substance. Such are the methods of computing by the ten fingers and the grains of maize, by the East Indians and Peruvians, above-mentioned; by the Abacus or Shwanpan of the Chinese; the several sorts of scales and sliding rules; Napier's bones or rods; the arithmetical machine of Pascal, and others; Sir Samuel Morland's instrument, described in 1666; that of Leibnitz, described in the Miscell. Berol.; that of Polenus, published in the Venetian Miscellany, 1709; and that of Dr. Saunderson, of Cambridge, described in the introduction to his algebra.

Integral ARITHMETIC, or *of integers*, is that which respects integers, or whole numbers.

Literal or *Algebra* ARITHMETIC, is that which is performed by letters, which represent any numbers indefinitely.

Logarithmical ARITHMETIC is performed by the tables of logarithms. These were invented by baron Napier; and the best treatise on the subject, is Briggs's Arithmetica Logarithmica, 1624.

Logistical ARITHMETIC. See LOGISTICAL.

Numerous or *Numeral* ARITHMETIC, is that which teaches the calculus of numbers, or of abstract quantities; and is performed by the common numeral or Arabic characters.

Political ARITHMETIC, is the application of arithmetic to political subjects; such as, the strength and revenues of nations, the number of people, births, burials, &c. See POLITICAL Arithmetic. To this head may also be referred the doctrine of Chances, Gaming, &c.

ARITHMETIC of *Radicals*, *Rationals*, and *Irrationals*: See RADICAL, &c.

Sexagesimal or *Sexagenary* ARITHMETIC, is that which proceeds by sixties; or the doctrine of sexagesimal fractions: a method which, it is supposed, was invented by Ptolemy, in the 2d century; at least they were used by him. In this notation, the integral numbers from 1 to 59 were expressed in the common way, by the alphabetical letters: then sixty was called a *sexagena prima*, and marked with a dash to the character, thus I'; twice sixty, or 120, thus II'; and so on to 59 times 60, or 3540, which is LIX'. Again, 60 times 60, or 3600, was called *sexagena secunda*, and marked with two dashes, thus I''; twice 3600, thus II''; and ten times 3600, thus X''; &c. And in this way the notation was continued to any length. But when a number less than sixty was to be joined with any of the sexagesimal integers, their proper expression was annexed without the dash: thus 4 times 60 and 25, is IV'XXV; the sum of twice 60 and 10 times 3600 and 15, is X''II'XV. So near did the inventor of this method approach to the Arabic notation: instead of the sexagesimal progression, he had only to substitute decimal; and to make the signs of numbers, from 1 to 9, simple characters, and to introduce another character, which should signify nothing by itself, but serving only to fill up places. —The *sexagenæ integrorum* were soon laid aside, in ordinary

dinary

dinary calculations, after the introduction of the Arabic notation; but the fexagefimal fractions continued till the invention of decimals, and indeed are still ufed in the fubdivifions of the degrees of circular arcs and angles.

Sam. Reyher has invented a kind of fexagenal rods, in imitation of Napier's bones, by means of which the fexagefimal arithmetic is eafily performed.

Specious ARITHMETIC, is that which gives the calculus of quantities as defigned by the letters of the alphabet: a method which was more generally introduced into algebra, by Vieta; being the fame as literal arithmetic, or algebra.—Dr. Wallis has joined the numeral with the literal calculus; by which means he has demonftrated the rules for fractions, proportions, extraction of roots, &c; of which a compendium is given by himfelf, under the title of Elementa Arithmeticæ, in the year 1698.

Tabular ARITHMETIC, is that in which the operations of multiplication, divifion, &c, are performed by means of tables calculated for that purpofe: fuch as thofe of Herwart, in 1610; and my tables of powers and products, publifhed by order of the Commiffioners of Longitude, in 1781.

Tetractic ARITHMETIC, is that in which only the four characters 0, 1, 2, 3 are ufed. A treatife of this kind of arithmetic is extant, by Erhard or Echard Weigel. But both this, and binary arithmetic, are little better than curiofities, efpecially with regard to practice; as all numbers are much more compendioufly and conveniently expreffed by the common decuple fcale.

Vulgar, or *Common* ARITHMETIC, is that which is converfant about integers and vulgar fractions.

Univerfal ARITHMETIC, is the name given by Newton to the fcience of algebra; of which he left at Cambridge an excellent treatife, being the text-book drawn up for the ufe of his lectures, while he was profeffor of Mathematics in that Univerfity.

ARITHMETICAL, fomething relating to or after the manner of arithmetic.

ARITHMETICAL *Complement*, of a logarithm, is what the logarithm wants of 10·00000 &c; and the eafieft way to find it is, beginning at the left hand, to fubtract every figure from 9, and the laft from 10. So, the arithmetical complement of 8·2501396 is 1·7498604.—It is commonly ufed in trigonometrical calculations, when the firft term of a proportion is not radius; in that cafe, adding all together, the logarithms of the 3d, 2d, and arithmetical complement of the 1ft term.

ARITHMETICAL *Inftruments*, or *Machines*, are inftruments for performing arithmetical computations; fuch as Napier's bones, fcales, fliding rules, Pafcal's machine, &c.

ARITHMETICAL *Mean*, or *Medium*, is the middle term of three quantities in arithmetical progreffion; and is always equal to half the fum of the extremes. So, an arithmetical mean between 3 and 7, is 5; and between a and b, is $\frac{a+b}{2}$, or $\frac{1}{2}a + \frac{1}{2}b$.

ARITHMETICAL *Progreffion*, is a feries of three or more quantities that have all the fame common difference: as 3, 5, 7, &c, which have the common difference 2; and a, $a + d$, $a + 2d$, &c, which have all the fame difference d.

In an arithmetical progreffion, the chief properties are thefe: 1ft, The fum of any two terms, is equal to the fum of every other two that are taken at equal diftances from the two former, and equal to double the middle term when there is one equally diftant between thofe two: fo, in the feries 0, 1, 2, 3, 4, 5, 6, &c, $0 + 6 = 1 + 5 = 2 + 4 =$ twice 3 or 6.—2d, The fum of all the terms of any arithmetical progreffion, is equal to the fum of as many terms of which each is the arithmetical mean between the extremes; or equal to half the fum of the extremes multiplied by the number of terms: fo, the fum of thefe ten terms 0, 1, 2, 3, 4, 5, 6, 7, 8, 9, is $\frac{0+9}{2} \times 10$, or 9×5, which is 45:

and the reafon of this will appear by inverting the terms, fetting them under the former terms, and adding each two together, which will make double the fame feries;

$$
\begin{array}{llllllllll}
\text{thus} & 0, & 1, & 2, & 3, & 4, & 5, & 6, & 7, & 8, & 9, \\
\text{inverted} & 9, & 8, & 7, & 6, & 5, & 4, & 3, & 2, & 1, & 0, \\
\hline
\text{fums} & 9, & 9, & 9, & 9, & 9, & 9, & 9, & 9, & 9 & 9;
\end{array}
$$

where the double feries being the fame number of 9's, or fum of the extremes, the fingle feries muft be the half of that fum.—3d, The laft, or any term, of fuch a feries, is equal to the firft term, with the product added of the common difference multiplied by 1 lefs than the number of terms, when the feries afcends or increafes; or the fame product fubtracted when the feries defcends or decreafes: fo, of the feries 1, 2, 3, 4, &c, whofe common difference is 1, the 50th term is $1 + 1 \times 49$, or $1 + 49$, that is 50; and of the feries 50, 49, 48 &c, the 50th term is $50 - 1 \times 49$, or $50 - 49$, which is 1. Alfo, if a denote the leaft term,

z the greateft term,
d the common difference,
n the number of the terms,
and s the fum of them all;

then the principal properties are expreffed by thefe equations, viz,

$$z = a + d.\overline{n-1},$$
$$a = z - d.\overline{n-1},$$
$$s = \overline{a+z}.\tfrac{1}{2}n,$$
$$s = \overline{z - \tfrac{1}{2}d.n-1}.n,$$
$$s = \overline{a + \tfrac{1}{2}d.n-1}.n.$$

Moreover, when the firft term a, is 0 or nothing; the theorems become $z = d.\overline{n-1}$,

and $s = \tfrac{1}{2}zn$.

ARITHMETICAL *Proportion*, is when the difference between two terms, is equal to the difference between other two terms. So, the four terms, 2, 4, 10, 12, are in arithmetical proportion, becaufe the difference between 2 and 4, which is 2, is equal to the difference between 10 and 12.—The principal property, befides the above, and which indeed depends upon it, is this, that the fum of the firft and laft, is equal to the fum of the two means: fo $2 + 12$, or the fum of 2 and 12, is equal $4 + 10$, or the fum of 4 and 10, which is 14.

ARITHMETICAL *Ratio*, is the same as the difference of any two terms: so, the arithmetical ratio of the series 2, 4, 6, 8, is 2; and the arithmetical ratio of *a* and *b*, is *a* — *b*.

ARITHMETICAL *Scales*, a name given by M. de Buffon, in the *Memoirs of the Acad.* for 1741, to different progressions of numbers, according to which, arithmetical computations might be made. It has already been remarked above, that our common decuple scale of numbers was probably derived from the number of fingers on the two hands, by means of which the earliest and most natural mode of computation was performed; and that other scales of numbers, formed in a similar way, but of a different number of characters, have been devised; such as the binary and tetractic scales of arithmetic. In the memoir above cited, Buffon gives a short and simple method to find, at once, the manner of writing down a number given in any scale of numbers whatever; with remarks on different scales. The general effect of any number of characters, different from ten, is, that by a smaller number of characters, any given number would require more places of figures to express or denote it by, but then arithmetical calculations, by multiplication and division, would be easier, as the small numbers 2, 3, 4, &c, are easier to use than the larger 7, 8, 9; and by employing more than ten characters, although any given number would be expressed by fewer of them, yet the calculations in arithmetic would be more difficult, as by the larger numbers 11, 12, 13, &c. It is therefore concluded, upon the whole, that the ordinary decuple scale is a good convenient medium amongst them all, the numbers expressed being tolerably short and compendious, and no single character representing too large a number. The same might also be said, and perhaps more, of a duodecimal scale, by twelve characters, which would express all numbers in a more compendious way than the decuple one, and yet no single character would represent a number too large to compute by; as is confirmed by the now common practice of extending the multiplication table, in school books, to 12 numbers or dimensions, each way, instead of 10; and every person is taught, with sufficient ease, to multiply and divide by 11 and 12 as easily as by 8 or 9 or 10. Another convenience might be added, namely, that the number 12 admitting of more submultiples than the number 10, there would be fewer expressions of interminate fractions in that way than in decimals. So that on all accounts, it is very probable that the duodecimal would be the best of any scale of numbers whatever.

ARITHMETICAL *Triangle*. See *Arithmetical* TRIANGLE.

ARMED. A magnet or loadstone is said to be armed, when it is capped, cased, or set in iron or steel; to make it take up a greater weight; and also readily to distinguish its poles.

It is surprising, that a little iron fastened to the poles of a magnet, should so greatly improve its power, as to make it even 150 times stronger, or more, than it is naturally, or when unarmed. The effect however, it seems, is not uniform; but that some magnets, by arming, gain much more, and others much less, than one would expect; and that some magnets even lose some of their efficacy by arming. In general, however,

the thickness of the iron armour ought to be nearly proportioned to the natural strength of the magnet; giving thick irons to a strong magnet, and to the weaker ones thinner: so that a magnet may easily be over-loaded.

The usual armour of a load-stone, in form of a right-angled parallelopipedon, consists of two thin pieces of iron or steel, of a square figure, and of a thickness proportioned to the goodness of the stone; the proper thickness being found by trials; always filing it thinner and thinner, till the effect be found to be the greatest possible.—The armour of a spherical load-stone, consists of two steel shells, fastened together by a joint, and covering a good part of the convexity of the stone. This also is to be filed away, till the effect is found to be the greatest.

Kircher, in his book *de Magnete*, says, that the best way to arm a load-stone, is to drill a hole through the stone, from pole to pole, in which is to be placed a steel rod of a moderate length: this rod, he asserts, will take up more weight at the end, than the stone itself when armed in the common way. And Gassendus and Cabæus prescribe the same method of arming. But Muschenbroek found, by repeated trials, that the usual armour already mentioned, is preferable to Kircher's; and he gives the following directions for preparing it. When, by means of steel filings and a small needle, the poles of a magnet have been discovered, he directs that the adjacent parts should be rubbed or ground into parallel planes, without shortening the polar axis; and the magnet may be afterwards shaped into the figure of a cube or parallelopipedon, or any other figure that may be more convenient. Plates of the softest iron are then prepared, of the same length and breadth with the whole polar sides of the magnet: the thickness of which plates, so as that they may admit and convey the greatest quantity of the magnetic virtue, is to be previously determined by experiment, in a manner which he prescribes for the purpose. A thicker piece of iron is to be annexed at right angles to these plates, which is called *pes armaturæ*, the foot or base of the armour: then the plates, nicely smoothed and polished, are to be firmly attached to each of the polar sides, whilst the thicker part or base is brought into close contact with the lower part of the magnet. In this way, he says, almost all the magnetic virtue issuing from the poles, enters into the armour, is directed to the base, and condensed by means of its roundness, so as to sustain the greatest weight of iron. Phys. Exper. and Geom. Dissert. 1729, pa. 131.

ARMILLARY *Sphere*, a name given to the artificial sphere, composed of a number of circles of metal, wood, or paper, which represent the several circles of the system of the world, put together in their natural order. It serves to assist the imagination to conceive the disposition of the heavens, and the motion of the celestial bodies.

This sphere is represented at Plate II, Fig. 6, where P and Q represent the poles of the world, AD the equator, EL the ecliptic and zodiac, PAGD the meridian, or the solstitial colure, T the earth, FG the tropic of cancer, HT the tropic of capricorn, MN the arctic circle, OV the antarctic, N and O the poles of the ecliptic, and RS the horizon.

The Armillary sphere constructed not long since by Dr. Long, in Pembroke-hall, Cambridge, is 18 feet in diameter;

diameter; and will contain more than 30 persons sitting within it, to view, as from a centre, the representation of the celestial spheres. The lower part of the sphere, which is not visible to England, is cut off; and the whole apparatus is so contrived, that it may be turned round with as little labour as is employed to wind up a common jack.

See also Mr. Ferguson's sphere in his Lectures, p. 194.

ARMILLARY *Trigonometer*, an instrument first contrived by Mr. Mungo Murray, and improved by Mr. Ferguson, consisting of five semicircles; viz, meridian, vertical circle, horizon, hour circle, and equator; so adapted to each other by joints and hinges, and so divided and graduated, as to serve for expeditiously resolving many problems in astronomy, dialling, and spherical trigonometry. The drawing, description, and method of using it, may be seen in Ferguson's Tracts, pa. 80, &c.

ARTIFICIAL *Numbers, Sines, Tangents,* &c, are the same as the Logarithms of the natural numbers, sines, tangents, &c.

ARTILLERY, the heavy equipage of war; comprehending all sorts of large fire-arms, with their appurtenances; as cannon, mortars, howitzers, balls, shells, petards, musquets, carbines, &c; being what is otherwise called *Ordnance*. The term is also applied to the larger instruments of war used by the ancients, as the catapult, balista, battering ram, &c.

The term *Artillery*, or *Royal Artillery*, is also applied to the persons employed in that service; and likewise to the art or science itself; and formerly it was used, for what is otherwise called *pyrotechnia*, or the art of fireworks, with the apparatus and instruments belonging to the same.

There have been many authors on the subject of artillery; the principal of which are, Bucherius, Braunius, Tartalea, Collado, Sardi, Ufano, Hanzelet, Digges, Moretti, Simienowitz, Mieth, d'Avelour, Manesson, Mallet, St. Julien; and the later authors, of still more consequence, are Belidor, St. Remy, le Blond, Valiere, Morogue, Puget, Coudray, Robins, Muller, Euler, Antoni, Tignola, Scheele; to which may be added the extensive and accurate experiments published in my 1st vol. of Tracts, and in the Philos. Transf. for 1778.

Park of ARTILLERY, is that place in a camp which is set apart for the Artillery, or large fire arms.

Traile or *Train* of ARTILLERY, a number of pieces of ordnance, mounted on carriages, with all their furniture fit for marching. To this commonly belong mortars, cannon, balls, shells, &c.—There are trains of Artillery in most of the royal magazines; as in the Tower, at Portfmouth, Plymouth, &c, but, above all, at Woolwich, from whence the ships commonly receive their ordnance, and where they are all completely proved before they are received into the public service.

The officers and men of the artillery were formerly called also the Train of Artillery, but are now called the Royal Regiment of Artillery; consisting at present of four battalions, besides a battalion of invalids, and four troops of Horse or Cavalry Artillery.

ASCENDANT, in *Astrology*, denotes the horoscope; or the degree of the ecliptic which rises upon the horizon, at the time of the birth of any one.—This,

it is supposed, has an influence on the person's life and fortune, by giving him a bent and propensity to one thing more than other.—In the science of Astrology, this is called the first house, the Oriental angle, or angle of the East, and the significator of life: and the astrologer says, such a planet ruled in his Ascendant, or Jupiter was in his Ascendant, &c.

ASCENDING, in Astronomy, a term used to denote any star, or degree, or other point of the heavens, rising above the horizon.

ASCENDING *Latitude*, is the latitude of a planet when going towards the north.

ASCENDING *Node*, is that point of a planet's orbit where it crosses the ecliptic, in proceeding northward. It is otherwise called the Northern Node, and is denoted by this character ☊, representing a node, or knot, with the larger part upwards; like as the same character reversed is used to denote the opposite, or Descending Node.

ASCENDING *Signs*, are such as are upon their ascent, or rise, from the nadir or lowest point of the heavens, towards the zenith, or highest point.

ASCENSION-DAY, otherwise called *Holy Thursday*, is a festival of the church, held 10 days before Whitsunday, in memory of our Saviour's Ascension.

ASCENSION, in Astronomy, is either Right or Oblique.

Right ASCENSION of the sun, or of a star, is that degree of the equinoctial, accounted from the beginning of Aries, which rises with them, in a right sphere.—Or, *Right Ascension*, is that point of the equinoctial, counted as before, which comes to the meridian with the sun or star, or other point of the heavens. And the reason of thus referring it to the meridian, is, because this is always at right angles to the equinoctial; whereas the horizon is so only in a right or direct sphere.—The right ascension, stands opposed to the right descension; and is similar to the longitude of places on the earth. All the fixed stars, &c, which have the same right-ascension, that is, which are at the same distance from the first point of Aries, or, which comes to the same thing, which are in the same meridian, rise at the same time in a right sphere, namely to the people who live at the equator. And if they be not in the same meridian, the difference between their times of rising, or of coming to the meridian of any place, is the precise difference of their right ascension.—But, in an oblique sphere, where the horizon cuts all the meridians obliquely, different points of the same meridian never rise or set together: so that two or more stars on the same meridian, or having the same right ascension, never rise or set at the same time in an oblique sphere; and the more oblique the sphere is, the greater is the interval of time between them.

To find the right ascension of the sun, stars, &c, by Trigonometry, say, As radius is to the cosine of the sun's greatest declination, or obliquity of the ecliptic, so is the tangent of the sun's or star's longitude, to the tangent of the right ascension.

RIGHT ASCENSION *of the Mid-heaven*, often used by astronomers, especially in calculating eclipses by means of the nonagesimal degree, is the right ascension of that point of the equator which is in the meridian; and it is equal to the sum of the sun's right ascension and

U 2   the

the horary angle or true time reduced to degrees, or to the sum of the mean longitude and mean time.

Oblique ASCENSION, is an arch of the equator intercepted between the first point of Aries, and that point of the equator which rises together with the star, &c, in an oblique sphere.—The Oblique Ascension is counted from west to east; and is greater or less, according to the various obliquity of the sphere.—To find the Oblique Ascension of the sun, see ASCENSIONAL and GLOBE.

The *Arch of Oblique Ascension*, is an arch of the horizon intercepted between the beginning of Aries, and the point of the equator which rises with a star or planet, in an oblique sphere; and it varies with the latitude of the place.

Refraction of ASCENSION *and Descension.* See REFRACTION.

ASCENSIONAL DIFFERENCE, is the difference between the right and oblique ascension of the same point on the surface of the sphere. Or it is the time the sun rises or sets before or after 6 o'clock.

To find the Ascensional Difference, having given the sun's declination and the latitude of the place, say, As radius is to the tangent of the latitude, so is the tangent of the sun's declination to the sine of the Ascensional Difference sought. The sun's Ascensional Difference converted into time, shews how much he rises before, or sets after, 6 o'clock. When the sun has north declination, the right ascension is greater than the oblique; but the contrary, when the sun has south declination; and the difference, in either case, is the Ascensional Difference.

ASCENT, the motion of a body from below tending upwards; or the continual recess of a body from the earth, or from some other centre of force. And it is opposed to *descent*, or motion downwards.

The Peripatetics attributed the spontaneous ascent of bodies to a principle of levity, inherent in them. But the moderns deny that there is any such thing as spontaneous levity; and they shew, that whatever ascends, does so by virtue of some external impulse or extrusion. Thus it is that smoke, and other rare bodies, ascend in the atmosphere; and oil, light woods, &c, in water: not by any inherent principle of levity; but by the superior gravity, or tendency downwards of the medium in which they ascend and float.

The ascent of light bodies in heavy mediums, is produced after the same manner as the ascent of the lighter scale of a balance. It is not that such scale has an internal principle, by which it immediately tends upwards; but it is impelled upwards by the preponderancy of the other scale; the excess of the weight in the one having the same effect, by augmenting its impetus downwards, as so much real levity in the other: because the tendencies mutually oppose each other, and action and reaction are always equal.—See this farther illustrated under the articles FLUID, and SPECIFIC GRAVITY.

ASCENT *of Bodies on Inclined Planes.* See the doctrine and laws of them under INCLINED PLANE.

ASCENT *of Fluids*, is particularly understood of their rising above their own level, between the surfaces of nearly contiguous bodies, or in slender capillary glass tubes, or in vessels filled with sand, ashes, or the like porous substances. Which is an effect that takes place as well *in vacuo*, as in the open air, and in crooked,

as well as straight tubes. Indeed some fluids ascend swifter than others, as spirit of wine, and oil of turpentine; and some rise after a different manner from others. The phenomenon, with its causes, &c, in the instance of capillary tubes, will be treated more at large under CAPILLARY *Tube.*

As to planes: Two smooth polished plates of glass, metal, stone, or other matter, being placed almost contiguous, have the effect of several capillary tubes, and the fluid rises in them accordingly: the like may be said of a vessel filled with sand, &c; the various small interstices of which form, as it were, a kind of capillary tubes. So that the same principle accounts for the appearance in them all. And to the same cause may probably be ascribed the ascent of the sap in vegetables. And on this subject Sir I. Newton says, "If a large pipe of glass be filled with sifted ashes, well pressed together, and one end dipped into stagnant water, the fluid will ascend slowly in the ashes, so as in the space of a week or fortnight, to reach the height of 30 or 40 inches above the stagnant water. This ascent is wholly owing to the action of those particles of the ashes which are upon the surface of the elevated water; those within the water attracting as much downwards as upwards: it follows, that the action of such particles is very strong; though being less dense and close than those of glass, their action is not equal to that of glass, which keeps quicksilver suspended to the height of 60 or 70 inches, and therefore acts with a force which would keep water suspended to the height of above 60 feet. By the same principle, a spunge sucks in water, and the glands in the bodies of animals, according to their several natures and dispositions, imbibe various juices from the blood." Optics, pa. 367.

Again, if a drop of water, oil, or other fluid, be dropped upon a glass plane, perpendicular to the horizon, so as to stand without breaking, or running off; and another plane touching it at one end, be gradually inclined towards the former, till it touch the drop; then will the drop break and move along towards the touching end of the planes; and it will move the faster in proportion as it proceeds farther, because the distance between the planes is constantly diminishing. And after the same manner, the drop may be brought to any part of the planes, either upward or downward, or sideways, by altering the angle of inclination.

Lastly, if the same perpendicular planes be so placed, as that two of their sides meet, and form a small angle, the other two being only kept apart by the interposition of some thin body; and thus immerged in a fluid, tinged with some colour to render it visible; the fluid will ascend between the planes, and that the highest where the planes are nearest; so as to form a curve line which is found to be a true hyperbola, of which one of the asymptotes is the line of the fluid, the other being a line drawn along the touching sides.

And the physical cause of all these phenomena, is the same power of attraction.

ASCENT *of Vapour.* See CLOUD and VAPOUR.

ASCENT, in Astronomy, &c. See ASCENSION.

ASCII, are those inhabitants of the globe, who, at certain times of the year, have no shadow. Such are the inhabitants of the torrid zone, who twice a year having the sun at noon in their zenith, have then no shadow.

ASELLI,

ASELLI, two fixed stars of the fourth magnitude, in the constellation Cancer.

ASH-Wednesday, the first day of Lent, supposed to have been so called from a custom in the church, of sprinkling ashes that day on the heads of penitents then admitted to penance.

ASPECT, is the situation of the stars and planets in respect of each other. Or, in Astrology, it denotes a certain configuration and mutual relation between the planets, arising from their situations in the zodiac, by which it is supposed that their powers are mutually either increased or diminished, as they happen to agree or disagree in their active or passive qualities. Though such configurations may be varied and combined a thousand ways, yet only a few of them are considered. Hence, Wolfius more accurately defines aspect to be; the meeting of luminous rays emitted from two planets to the earth, either posited in the same right line, or including an angle which is an aliquot part, or some number of aliquot parts, of four right angles, or of 360 degrees.

The doctrine of aspects was introduced by the astrologers, as the foundation of their predictions. And hence Kepler defines aspect to be, an angle formed by the rays of two planets meeting on the earth, capable of exciting some natural power or influence.

The ancients reckoned five aspects, viz, *conjunction*, *sextile, quartile, trine,* and *opposition*.

Conjunction is denoted by this character ☌, and is when the planets are in the same sign and degree, or have the same longitude.

Sextile is denoted by ✱, and is when the planets are distant by the 6th part of a circle, or 2 signs, or 60 degrees.

Quartile is denoted by ☐, and is when the planets are distant ¼ of the circle, or 90 degrees, or 3 signs.

Trine is denoted by △, and is when the planets are distant by ⅓ of the circle, or 4 signs, or 120 degrees. And

Opposition is denoted by ☍, and is when the planets are in opposite points of the circle, or differ by ½ the circle, or 6 signs, or 180 degrees of longitude.

Or their characters and distances are as in this following tablet.

NAME.	CHARACTER.	DISTANCE.
Conjunction - -	☌	0°
Sextile - - -	✱	60
Quartile - -	☐	90
Trine - - -	△	120
Opposition -	☍	180

These intervals are reckoned according to the longitudes of the planets; so that the aspects are the same, whether the planet be in the ecliptic or out of it.

To the five ancient aspects, modern writers have added several more: as *decile*, for the 10th part of a circle; *tridecile*, or $\frac{3}{10}$; and biquintile, or $\frac{2}{5}$ of a circle. And Kepler adds others, from meteorological observations, as he tells us: as the *semisextile*, being $\frac{1}{12}$, or 30°; and *quincunx*, or $\frac{5}{12}$, or 150°. Lastly, to the astrological physicians, &c, we owe *octile*, or $\frac{1}{8}$; *trioctile*, or $\frac{3}{8}$; and *quintile*, $\frac{1}{5}$ of the circle.

The aspects are divided with regard to their supposed influences, into *benign, malign,* and *indifferent.*

The trine and sextile aspects being esteemed benign or friendly; the quartile and opposition, malign or unfriendly; and the conjunction an indifferent aspect.

Aspects are also distinguished into *partile* and *platic*.

Partile Aspect, is when the planets are just so many degrees distant, as are expressed above. And these only are the proper aspects.

Platic Aspect, is when the planets do not regard each other exactly from these very degrees of distance; but the one exceeding as much as the other falls short. So that the one does not cast its rays immediately on the body of the other, but only on its orb or sphere of light.

ASPERITY, signifies the inequality or roughness of the surface of any body; by which some parts of it are more prominent than the rest, so as to hinder the hand, &c, from passing over it with ease and freedom; and thus producing what is called friction.—Asperity, or roughness, stands opposed to smoothness, evenness, politure, &c.

According to the relations of Vermausen, the blind man so celebrated for distinguishing colours by the touch, it seems that every colour has its peculiar degree and kind of asperity. He makes black the roughest, as it is the darkest of colours; but the others are not smoother in proportion as they are lighter; that is, the roughest do not always reflect the least light: for, according to him, yellow is two degrees rougher than blue, and as much smoother than green. See Colours.

ASSIGNABLE *Magnitude*, is used for any finite magnitude that can be expressed or denoted. And,

Assignable *Ratio*, for any expressible ratio.

ASSUMPTION, a feast celebrated in the Romish church, in honour of the miraculous ascent of the Holy Virgin, as they describe it, body and soul, into heaven. It is kept on the 15th of August.

ASSURANCE *on Lives*, a compact by which security is granted for the payment of a certain sum of money on the expiration of the life on which the policy is granted, in consideration of such a previous payment made to the assurer as is accounted a sufficient compensation for the loss and hazard to which he exposes himself.

The sum at which this compensation should be valued, depends principally on these two circumstances, viz, 1st, On the rate of interest given for the use of money; and 2d, On the probability of the duration of the life assured, and the values of annuities. For, 1st, If the interest of money be high, the value of the assurance will be proportionally low, & *è contra*; because the higher the rate of interest, the less will be the present value which amounts to a certain proposed sum in any given time. Also, if the probability of the duration of life be high, the value of the assurance will again be proportionably low, & *è contra*; because the longer the time is, the less will be the principal which will amount to any assigned sum. Thus, if it be required to know the premium or present value, to be given for 100 pounds to be received at the end of any time, as suppose 10 years; then, if the interest of money be at the rate of 5 per cent. the answer, or present premium, would be 61l. 7s. 10d; but at four per cent. it would be 67l. 11s. 1d; and at 3 per cent. it would amount to 74l. 8s. 2d. Again, suppose it were

required

required to affure 100l. on a life, for any time, for inftance 1 year; that is, let 100l. be fuppofed to be payable a year hence, provided a life of a given age fails in that time: here it is evident that, whatever be the rate of intereft, the lefs the probability of the life failing within the year, the lefs the rifk is, and the lefs the premium ought to be. In effect, the rate of intereft being 5 per cent, if it were fure that the life would fail in that year, the value of the affurance would be the fame as the prefent value of 100l. payable at the end of the year, which is 95l. 4s. 9d. But, if it be an equal chance whether the life does or does not fail in the year, in which cafe the probability of fa ling is $\frac{1}{2}$; then the value of the affurance will be but half the former value, or 47l. 12s. 4½d. Or if the odds againft its failing be as 2 to 1, that is, if one perfon out of every 3 die at the age of the propofed life, the probability of dying being only $\frac{1}{3}$, the value of the affurance will be $\frac{1}{3}$ of the firft value, or 31l. 14s. 11d. And if the odds be 19 to 1, or one perfon die out of 20, of that age, the probability of dying will be $\frac{1}{20}$, and the value of the affurance will alfo be $\frac{1}{20}$ of the firft value, or 4l. 15s. 3d. nearly. Laftly, if only one perfon die out of 50 at the given age, the probability of dying will be $\frac{1}{50}$, and the value of the affurance will be accordingly only $\frac{1}{50}$ of the firft fum, or 1l. 18s. 1d: the intereft of money being all along confidered as after the rate of 5 per cent.—Now, according to Dr. Halley's table of obfervations, one perfon dies out of 3, at the age of 87; one in 20 at the age of 64; and one in 50 at the age of 39: It follows, therefore, that the value of the affurance of 100l. for one year, on a life aged 87, is 31l. 14s. 11d; on a life aged 64, it is 4l. 15s. 3d; and on a life aged 39, it is 1l. 18s. 1d: reckoning intereft at 5 per cent. But if intereft were rated at 3 per cent. thefe values would be 32l. 7s. 3d, and 4l. 17s. 1d, and 1l. 18s. 10d.

The affurances moft commonly practifed, are fuch as thefe, on fingle lives, and for fingle years. But many private affurers, and even fome large affuring offices, either from ignorance or impofition, pay no regard to any difference of age, but demand 5l. from all ages indifcriminately, for the affurance of 100l. for one year: a practice very abfurd and inequitable; for it appears that this is more than the value of the affurance of a life of 64 years of age, and even more than double the value of the affurance of a life of 39 years of age; allowing the affured to make 5 per cent. of the money he advances.

When a life is affured for any number of years; the premium or value may be paid, either in one fingle prefent payment; in confequence of which the fum affured will become payable without any farther compenfation, whenever, within the given term, the life fhall happen to drop: or the value may be paid in annual payments, to be continued till the failure of the life, fhould that happen within the term; or, if not, till the determination of the term. And the determination of the value of affurance, in all cafes, is to be made out from the rules for computing annuities on lives; the principal writers on which are Halley, De Moivre, Simpfon, Smart, Kerffeboom, De Parcieux, Price, Morgan, and Maferes. See alfo LIFE ANNUITIES, REVERSION, &c.

Affurances may be made either on *fingle lives;* as above explained; or they may be made on any number of *joint* lives, or on the *longeft* of any lives; that is, an affurer may bind himfelf to pay any fums at the extinction of any *joint* lives, or the *longeft* of any lives, or at the extinction of any one or two of any number of lives. There are further affurances on furvivorfhips; by which is meant an obligation, for the value received, to pay a given fum or annuity, provided a given life fhall furvive any other given life or lives. For which fee SURVIVORSHIP.

The principal offices for making thefe infurances, in England, are the "London and the Royal Exchange Affurance Offices;" "the Amicable Society, incorporated for a perpetual Affurance Office;" "the Society for equitable Affurances on Lives and Survivorfhips;" and "the Weftminfter Society for granting Annuities and infuring Money on Lives."

The firft two of thefe offices, having chiefly in view affurances on fhips and houfes, deal but little in the way of affurances on lives; and all the bufinefs they tranfact in this way, is at 5l. for every 100l. affured on a fingle life for a fingle year, without paying any regard to the ages of the lives affured.

The next, or Amicable Society at Serjeant's Inn, requires an annual payment of 5l. from every member during life, payable quarterly. The whole annual income, hence arifing, is equally divided among the nominees, or heirs, of fuch members as die every year. But this fociety engages that the dividends fhall not be lefs than 150l. to each claimant, though they may be more. No members are admitted whofe ages are greater than 45, or lefs than 12; nor is any difference of contribution allowed on account of difference of age. The fociety has fubfifted ever fince the year 1706, and its credit and ufefulnefs are well eftablifhed.

The Equitable Society for Affurances on Lives and Survivorfhips, which meets at Black-Friars' Bridge, was eftablifhed in the year 1762, in confequence of propofals which had been made, and lectures, recommending fuch a defign, which had been read by Mr. Dodfon, author of the Mathematical Repofitory. It affures any fums or reverfionary annuities, on any life or lives for any number of years, as well as for the whole continuance of the lives; and in any manner that may be beft adapted to the views of the perfons affured: that is, either by making the affured fums payable certainly at the failure of any given lives; or on condition of furvivorfhip; and alfo, either by taking the price of the affurance in one prefent payment, or in annual payments, during any fingle or joint lives, or any terms, lefs than the whole poffible duration of the lives. In fhort, there are no kinds of affurances on lives and furvivorfhips, which this fociety does not make.

In doing this, the Society follows the rules which have been given by the beft mathematical writers on the doctrine of Life Annuities and Reverfions, particularly Mr. Thomas Simpfon, profeffor of Mathematics in the Royal Military Academy. It is to be obferved however that the Society takes the advantage of making its calculations on the fuppofition that the intereft of money is at fo low a rate as 3 per cent, inftead of the ufual intereft of 4 per cent; which confequently raifes the infurance proportionally higher; and it alfo founds its calculations

on

on the tables of the probabilities and values of lives in London; another circumstance which secures a very advantageous profit to the Society, as experience has proved that the deaths are really in a much lower proportion than according to those tables, and even lower than those of Dr. Halley, which are founded on the bills of mortality of Breslaw. By these means the Society finding itself, by experience, well secured against future hazards, and being unwilling to take from the public an extravagant profit, have determined to reduce all the future payments for assurances, one tenth, and also generously to return, to the persons now assured, one tenth of all the payments they have made: and it seems there is reason to expect that this will be only a preparation to farther reduction.

From the foregoing account of this society, it is manifest that its business is such, that none but skilful mathematicians are qualified to conduct it. The interest of the society therefore requires, that it should make the places of those who manage its business sufficiently advantageous, to induce the ablest mathematicians to accept them: and this will render it the more necessary for the society to take care, in filling up any future vacancies, to pay no regard to any other considerations than the ability and integrity of the candidates. The consequence of granting good pay, will be a multitude of solicitations on every vacancy, from persons who, however unqualified, will hope for success from their connexions, and the interest they are able to make. And should the society, in any future time, be led by such causes to trust its business in the hands of persons not possessed of sufficient ability, as mathematicians and calculators, such mistakes may be committed, as may prove, in the highest degree, detrimental and dangerous. There is reason to believe, that at present the society is in no danger of this kind; and one of the great public advantages attending it, is, that it has established an office, where not only the business above described, is transacted with faithfulness and skill; but where also all persons, who want solutions of any questions relating to life annuities and reversions, may apply, and be sure of receiving just answers. The following is a

Table of the rates of assurance on single lives in the Society for Equitable Assurances. The Sum assured 100l.

Age.	For one year.			For seven years at an annual payment of			For the whole life at an annual payment of		
	l.	s.	d.	l.	s.	d.	l.	s.	d.
10	1	9	6	1	10	7	2	2	10
15	1	11	0	1	12	7	2	6	6
20	1	13	11	1	16	0	2	12	10
25	1	17	7	2	0	2	3	0	6
30	2	2	6	2	6	0	3	8	11
35	2	8	7	2	14	2	3	17	9
40	2	19	2	3	5	1	4	7	11
45	3	11	0	3	18	6	5	0	0
50	4	4	8	4	11	2	5	12	11
55	5	0	9	5	11	7	6	9	3
60	5	19	1	6	16	10	7	17	7
65	7	0	11	8	13	0	10	3	9

These rates are 10 per cent. lower than the true values, according to the decrements of life in London, reckoning interest at 3 per cent; but at the same time, it is to be observed that, for all ages under 50, they are near one third higher than all the true values, according to Dr. Halley's table of the decrements of life at Breslaw, and Dr. Price's tables of the decrements of life at Northampton and Norwich. But as the society has lately found that the decrements of life among its members have hitherto been lower than even those given in these last tables, it may reasonably be expected, that they will in time reduce their rates of assurance to the true values, as determined by these tables.

As to the *Westminster Society for granting Annuities, and insuring Money on Lives,* lately established, viz, in the year 1789, from the number and respectability of its members, the equitable terms upon which it proposes to deal, and the known ability and accuracy of the mathematicians and calculators employed in conducting it, there is every reason to expect an honourable and equitable treatment of the public, and a permanent continuance of its usefulness.

ASTERISM, the same with constellation, or a collection of many stars, which are usually represented on globes by some particular image or figure, to distinguish the stars which compose this constellation from those of others.

ASTRÆA, a name given by some to the sign Virgo by others called Erigone, and sometimes Isis. The poets feign that Justice quitted heaven to reside on earth, in the golden age; but, growing weary of the iniquities of mankind, she left the earth, and returned to heaven, placing herself in that part of the zodiac called Virgo, where she became a constellation of stars, and from her orb still looks down on the ways of men. Ovid. Metam. lib. i. ver. 149.

ASTRAGAL, in Architecture, a small round moulding, which encompasses the top of the fust or shaft of a column, like a ring or bracelet. The shaft always terminates at top with an astragal, and at bottom with a fillet, which in this place is called *oxia.*

ASTRAGAL, in Gunnery, is a kind of ring or moulding on a piece of ordnance, at about half a foot distance from the muzzle or mouth; serving as an ornament to the gun, as the former does to a column.

ASTRAL, something belonging to or depending on the stars.

ASTRAL *Year,* or *Sidereal Year.* See YEAR.

ASTRODICTICUM, an astronomical instrument invented by M. Weighel, by means of which many persons shall be able to view the same star at the same time.

ASTROGNOSIA, the art of knowing the fixed stars, their names, ranks, situations in the constellations, and the like.

ASTROLABE, from αϛηρ, star, and λαμβανω, I take; alluding to its use in taking, or observing, the stars. The Arabians call it in their tongue *astharlab;* a word formed by corruption from the common Greek name.

This name was originally used for a system or assemblage of the several circles of the sphere, in their proper order and situation with respect to each other. And the ancient instruments were much the same as our armillary spheres.

The

The firſt, and moſt celebrated of this kind, was that of Hipparchus, which he made at Alexandria, the capital of Egypt, and lodged in a ſecure place, where it ſerved for divers aſtronomical operations. Ptolemy made the ſame uſe of it: but as the inſtrument had ſeveral inconveniences, he contrived to change its figure, though perfectly natural, and agreeable to the doctrine of the ſphere; and to reduce the whole Aſtrolabe to a plane ſurface, to which he gave the name of the *Planiſphere*. Hence,

ATSROLABE is uſed among the moderns for a *Planiſphere*, or a ſtereographic projection of the circles of the ſphere upon the plane of one of the great circles; which is uſually either the plane of the equinoctial, the eye being then placed in the pole of the world; or that of the meridian, the eye being ſuppoſed in the point of interſection of the equinoctial and horizon; or on that of the horizon.

The Aſtrolabe has been treated at large by Stoffler, Gemma Friſius, Clavius, &c. And for a farther account of the nature and kinds of it, ſee the article PLANISPHERE.

ASTROLABE, or SEA ASTROLABE, more particularly denotes an inſtrument chiefly uſed for taking altitudes at ſea; as the altitude of the pole, the ſun, or the ſtars.

The common Aſtrolabe, repreſented Plate II, *fig.* 7, conſiſts of a large braſs ring, about 15 inches in diameter, whoſe limb, or a convenient part of it, is divided into degrees and minutes. It is fitted with a moveable label or index, which turns upon the centre, and carries two ſights; and having a ſmall ring, at A, to hang it by in time of obſervation.

To make uſe of the Aſtrolabe in taking altitudes; ſuſpend it by the ring A, and turn it to the ſun, &c, ſo as that the rays may paſs freely through both the ſights F and G; then will the label cut or point out the altitude on the divided limb. There are many other uſes of the Aſtrolabe; of which Clavius, Henrion, and others have written very largely.

The Aſtrolabe, though now grown into diſuſe, is by many eſteemed equal to any other inſtrument for taking the altitude at ſea; eſpecially between the tropics, where the ſun comes near the zenith.

ASTROLOGER, a perſon profeſſing or practiſing the art of Aſtrology.

ASTROLOGICAL, ſomething relating to Aſtrology.

ASTROLOGY, the art of foretelling future events, from the poſitions, aſpects, and influences of the heavenly bodies.

The word is compounded of αϛηρ, *ſtar*, and λογος, *diſcourſe*; whence, in the literal ſenſe of the term, Aſtrology ſhould ſignify no more than the *doctrine* or *ſcience of the ſtars*; which indeed was its original acceptation, and conſtituted the ancient Aſtrology; which conſiſted formerly of both the branches now called Aſtronomy and Aſtrology, under the name of the latter only; and for the ſake of making judiciary predictions it was that aſtronomical obſervations, properly ſo called, were chiefly made by the ancients. And though the two branches be now perfectly ſeparated, and that of Aſtrology almoſt univerſally rejected by men of real learning, this has but lately been the caſe, as their union ſubſiſted, in

ſome degree, from Ptolemy till Kepler, who had a ſtrong bias towards the ancient aſtrology.

Aſtrology may be divided into two branches, *natural* and *judiciary*.

To NATURAL ASTROLOGY belongs the predicting of natural effects; ſuch as the changes of weather, winds, ſtorms, hurricanes, thunder, floods, earthquakes, &c. But this art properly belongs to *Phyſiology*, or *Natural Philoſophy*; and is only to be deduced, *à poſteriori*, from phenomena and obſervations. And for this ſort of Aſtrology it is that Mr. Boyle makes an apology in his Hiſtory of the Air. Its foundation and merits may be gathered from what is ſaid under the articles AIR, ATMOSPHERE, and WEATHER.

Judicial or *Judiciary* ASTROLOGY, which is what is commonly and properly called *Aſtrology*, is that which profeſſes to foretel moral events, or ſuch as have a dependence on the free will and agency of man; as if they were produced or directed by the ſtars.

The profeſſors of this kind of Aſtrology maintain, "That the heavens are one great volume or book, wherein God has written the hiſtory of the world; and in which every man may read his own fortune, and the tranſactions of his time. The art, ſay they, had its riſe from the ſame hands as Aſtronomy itſelf: while the ancient Aſſyrians, whoſe ſerene unclouded ſky favoured their celeſtial obſervations, were intent on tracing the paths and periods of the heavenly bodies, they diſcovered a conſtant, ſettled relation of analogy, between them and things below; and hence were led to conclude theſe to be the *Parcæ*, the Deſtinies ſo much talked of, which preſide at our births, and diſpoſe of our future fate."

"The laws therefore of this relation being aſcertained by a ſeries of obſervations, and the ſhare each planet has therein; by knowing the preciſe time of any perſon's nativity, they were enabled, from their knowledge in aſtronomy, to erect a ſcheme or horoſcope of the ſituation of the planets, at that point of time; and hence, by conſidering their degrees of power and influence, and how each was either ſtrengthened or tempered by ſome other, to compute what muſt be the reſult thereof."

Judicial Aſtrology, it is commonly ſaid, was invented in Chaldæa, and from thence tranſmitted to the Egyptians, Greeks, and Romans; though ſome inſiſt that it was of Egyptian origin, and aſcribe the invention to Cham. But it is to the Arabs that we owe it. At Rome the people were ſo infatuated with it, that the aſtrologers, or, as they were then called, the mathematicians, maintained their ground in ſpite of all the edicts of the emperors to expel them out of the city. See GENETHLIACI.

Among the Indians, the Bramins, who introduced and practiſed this art in the Eaſt, have hereby made themſelves the arbiters of good and evil hours, which, gives them great authority: they are conſulted as oracles; and they have taken care always to ſell their anſwers at good rates.

The ſame ſuperſtition has prevailed in more modern ages and nations. The French hiſtorians remark, that in the time of queen Catharine de Medicis, Aſtrology was ſo much in vogue, that the moſt inconſiderable thing was not to be done without conſulting the ſtars. And in the reigns of king Henry III. and IV. of France, the predictions

predictions of Aftrologers were the common theme of the court converfation. And this predominant humour in that court was well rallied by Barclay, in his Argenis, *lib.* 2, on occafion of an Aftrologer, who had undertaken to inftruct king Henry in the event of a war which was then threatened by the faction of the Guifes.

ASTROMETEOROLOGIA, the art of foretelling the weather, and its changes, from the afpects and configurations of the moon and planets: a fpecies of Aftrology diftinguifhed by fome under the denomination of *meteorological aftrology.*

ASTRONOMICAL, fomething relating to Aftronomy.

ASTRONOMICAL *Calendar, Characters, Column, Horizon, Hours, Month, Quadrant, Ring-Dial, Sector, Tables, Telefcope, Time, Year.* See the feveral fubftantives.

ASTRONOMICAL *Obfervations.* Of thefe there are records, or mention, in almoft all ages. It is faid that the Chinefe have obfervations for a courfe of many thoufand years. But of thefe, as well as thofe of the Indians, we have never yet had any benefit. But the obfervations of moft of the other ancients, as Babylonians, Greeks, &c, amongft which thofe of Hipparchus make a principal figure, are carefully preferved by Ptolemy, in his Almageft.

About the year 880, Albategni, a Saracen, applied himfelf to the making of obfervations; in which he was followed by others of the fame nation, as well as Perfians and Tartars; among whom were Naffir-Eddin-Ettufi, Arzachel, who alfo conftructed a table of fines, and Ulug Beigh. In 1457 Regiomontanus undertook the province at Norimberg; and his difciples, J. Werner and Ber. Walther, continued the fame from 1475 to 1504. Their obfervations were publifhed together in 1544.—In 1509, Copernicus, and after him the landgrave of Heffe, with his affiftants Rothman and Byrge, obferved; and after them Tycho Brahe, affifted by the celebrated Kepler, from 1582 to 1601.—All the foregoing obfervations, together with Tycho's apparatus of inftruments, are contained in the Hiftoria Cœleftis, publifhed in 1672, by order of the emperor Ferdinand.—In 1651, was publifhed at Bononia, by Ricci-olus, Almageftum Novum, being a complete body of ancient and modern obfervations, which he fo named after the work of the fame nature by Ptolemy.—Soon after, Hevelius, with a magnificent and well-contrived apparatus of inftruments, defcribed in his Machina Cœleftis, began a courfe of obfervations. It has been objected to him, that he only ufed plain fights, and could never be brought to take the advantage of telefcopic ones; which occafioned Dr. Hook to write animadverfions on Hevelius's inftruments, printed in 1674, in which he too rafhly defpifes them, on account of their inaccuracy: but Dr. Halley, who at the inftance of the Royal Society went over to Dantzick in the year 1679, to infpect his inftruments, approved of their juftnefs, as well as of the obfervations made with them. See SIGHTS.—Our two countrymen Jer. Horrox and Will. Crabtree, are celebrated for their obfervations from the year 1635 to 1645, who firft obferved the tranfit of Venus over the fun in the year 1639.—They were followed by Flamfteed, Caffini the father and fon, Halley, de la Hire, Roemer, and Kirchius.—The obfervations of the celebrated Dr. Bradley have not yet been publifhed, though long expected. We have alfo now publifhed,

from time to time, the accurate obfervations of the prefent Britifh Aftronomer Royal: as alfo thofe of the French and other obfervatories, with the obfervations of many ingenious private aftronomers, which are to be found in the Tranfactions and Memoirs of the various Philofophical Societies.—There have been alfo obfervations of many other eminent aftronomers; as, Galileo, Huygens, and our countryman Harriot, whofe very interefting obfervations have lately been brought to light by the earl of Egremont, and count Bruhl, by whofe means they may come to be publifhed. Other publications of celeftial obfervations, are thofe of Caffini, La Caille, Monnier, &c.—See farther under CELESTIAL *Obfervations,* CATALOGUE, OBSERVATORY, &c.

ASTRONOMICAL *Place* of a ftar or planet, is its longitude, or place in the ecliptic, reckoned from the beginning of Aries, *in confequentia,* or according to the order of the figns.

ASTRONOMICALS, a name ufed by fome writers for fexagefimal fractions; on account of their ufe in aftronomical calculations.

ASTRONOMICUS *Radius.* See RADIUS.

ASTRONOMY, the doctrine of the heavens, and their phenomena.

Aftronomy is properly a mixed mathematical fcience, by which we become aequainted with the celeftial bodies, their motions, periods, eclipfes, magnitudes, diftances, and other phenomena. Some, however, underftand the term aftronomy in a more extenfive fenfe, as comprifing in it the theory of the univerfe, with the primary laws of nature: in which fenfe it feems to be rather a branch of phyfics than of mathematics.

Hiftory of Aftronomy.

The invention of aftronomy has been varioufly given, and afcribed to feveral perfons, feveral nations, and feveral ages. Indeed it is probable that mankind never exifted without fome knowledge of aftronomy amongft them. For, befides the motives of mere curiofity, which are fufficient of themfelves to have excited men to a contemplation of the glorious and varying celeftial canopy, it is obvious that fome parts of the fcience anfwer fuch effential purpofes to mankind, as to make the cultivation of it a matter of indifpenfable neceffity. Accordingly we find traces of it, in different degrees of improvement, among all nations.

Adam, in his ftate of innocence, it is fuppofed by fome of the Jewifh rabbins, was endowed with a knowledge of the nature, influence, and ufes of the heavenly bodies; and Jofephus afcribes to Seth and his pofterity a confiderable knowledge of aftronomy: he fpeaks of two pillars, the one of ftone and the other of brick, called the pillars of Seth, upon which they engraved the principles of the fcience; and he fays that the former was ftill entire in his time. But be this as it may, it is evident that the great length of the antediluvian lives would afford fuch excellent opportunities for obferving the heavenly bodies, that we cannot but fuppofe that the fcience of aftronomy was confiderably advanced before the flood. Indeed Jofephus fays that longevity was beftowed upon them for the very purpofe of cultivating the fciences of geometry and aftronomy; obferving that the latter could not be learned in lefs than 600 years; "for that period, he adds, is the *grand year.*"

An expreffion remarkable enough; and by which it may be fuppofed is meant the period in which the fun and moon come again into the fame fituation in which they were at the beginning of it, with regard to the nodes, apogee of the moon, &c. " This period, fays Caffini, of which we find no intimation in any monument of any other nation, is the fineft period that ever was invented: for it brings out the folar year more exactly than that of Hipparchus and Ptolemy; and the lunar month within about one fecond of what is determined by modern aftronomers." If the Antediluvians had fuch a period of 600 years, they muft have known the motions of the fun and moon more exactly than their defcendants knew them fome ages after the flood.

On the building of the Tower of Babel, it is fuppofed that Noah retired with his children, born after the flood, to the north-eaftern part of Afia, where his defcendants peopled the vaft empire of China. And this, fays Dr. Long, "may perhaps account for the Chinefe having fo early cultivated the ftudy of aftronomy, &c." It is faid that the Jefuit miffionaries have found traditional accounts among the Chinefe, of their having been taught this fcience by their firft emperor Fo-hi, who is fuppofed to be the fame with Noah; and Kempfer afferts that Fo hi difcovered the motions of the heavens, divided time into years and months, and invented the 12 figns into which they divide the zodiac, and which they diftinguifh by thefe names following; 1, the moufe; 2, the ox or cow; 3, the tiger; 4, the hare; 5, the dragon; 6, the ferpent; 7, the horfe; 8, the fheep; 9, the monkey; 10, the cock or hen; 11, the dog; and, 12, the boar. They divide the heavens into 28 conftellations, or claffes of ftars, allotting 4 to each of the 7 planets; fo that the year always begins with the fame planet; and their conftellations anfwer to the 28 lunar manfions ufed by the Arabian aftronomers. Thefe conftellations however they do not mark with the figures of animals, like moft other nations, but by connecting the ftars by ftraight lines, and denoting the ftars themfelves by fmall circles: fo, for inftance, the great bear would be marked thus,

The Chinefe themfelves have many records and traditions of the high antiquity of their aftronomy; though not without fufpicion of great miftakes. But, on more certain authority, it is afferted by F. Gaubil, that at leaft 120 years before Chrift, the Chinefe had determined by obfervation the number and extent of their conftellations as they now ftand; the fituation of the fixed ftars with refpect to the equinoctial and folftitial points; and the obliquity of the ecliptic; with the theory of eclipfes: and that they were, long before that, acquainted with the true length of the folar year, the method of obferving meridian altitudes of the fun by the fhadow of a gnomon, and of deducing from thence his declination, and the height of the pole. The fame miffionary alfo fays, that the Chinefe have yet remaining fome books of aftronomy, which were written about 200 years before Chrift; from which it appears,

that the Chinefe had known the daily motion of the fun and moon, and the times of the revolutions of the planets, many years before that period.

Du Halde informs us, that Tcheou cong, the moft fkilful aftronomer that ever China produced, lived more than a thoufand years before Chrift; that he paffed whole nights in obferving the celeftial bodies, and arranging them into conftellations, &c. At prefent however, the ftate of aftronomy is but very low in that country, although it be cultivated at Peking, by public authority, in like manner as in moft of the capital cities of Europe.

The inhabitants of Japan, of Siam, and of the Mogul's empire, have alfo been acquainted with aftronomy from time immemorial; and the celebrated obfervatory at Benares, is a monument both of the ingenuity of the people, and of their fkill in that fcience.

According to Porphyry, aftronomy muft have been of very ancient ftanding in the Eaft. He informs us that, when Babylon was taken by Alexander, there were brought from thence celeftial obfervations for the fpace of 1903 years; which therefore muft have commenced within 115 years after the flood, or within 15 years after the building of Babel.—Epigenes, according to Pliny, affirmed that the Babylonians had obfervations of 720 years engraven on bricks.—Again, Achilles Tatius afcribes the invention of aftronomy to the Egyptians; and adds, that their knowledge of that fcience was engraven on pillars, and by that means tranfmitted to pofterity.

M. Bailly, in his elaborate Hiftory of ancient and modern aftronomy, endeavours to trace the origin of this fcience among the Chaldeans, Egyptians, Perfians, Indians and Chinefe, to a very early period. And thence he maintains, that it was cultivated in Egypt and Chaldea 2800 years before Chrift; in Perfia, 3209; in India, 3101; and in China, 2952 years before that æra. He alfo apprehends, that aftronomy had been ftudied even long before this diftant period, and that we are only to date its revival from thence.

In inveftigating the antiquity and progrefs of aftronomy among the Indians, M. Bailly examines and compares four different fets of aftronomical tables of the Indian philofophers, namely that of the Siamefe, explained by M. Caffini in 1689; that brought from India by M. le Gentil of the Academy of Sciences; and two other manufcript tables, found among the papers of the late M. de Lifle; all of which he found to accord together, and all referring to the meridian of Benares, above-mentioned. It appears that the fundamental epoch of the Indian aftronomy, is a conjunction of the fun and moon, which took place at the amazing diftance of 3102 years before Chrift: and M. Bailly informs us that, by our moft accurate aftronomical tables, fuch a conjunction did really happen at that time. He further obferves that, at prefent, the Indians calculate eclipfes by the mean motions of the fun and moon obferved 5000 years fince; and that their accuracy, with regard to the folar motion, far exceeds that of the beft Grecian aftronomers. They had alfo fettled the lunar motions by computing the fpace through which that luminary had paffed in 1,600,984 days, or a little more than 4383 years. M. Bailly alfo informs us, that they make ufe of the cycle of 19 years, the fame as that

ascribed

afcribed by the Greeks to Meton; that their theory of the planets is much better than Ptolemy's, as they do not fuppofe the earth in the centre of the celeftial motions, and believe that Venus and Mercury move round the fun; and that their aftronomy agrees with the moft modern difcoveries as to the decreafe of the obliquity of the ecliptic, the acceleration of the motion of the equinoctial points, &c.

In the 2d vol. of the Tranfactions of the Royal Society of Edinburgh is alfo a learned and ingenious differtation on the aftronomy of the Brahmins of India, by Mr. Profeffor Playfair; in which the great accuracy and high antiquity of the fcience, among them, is reduced to the greateft probability. It hence appears that their tables and rules of computation, have peculiar reference to an epoch, and to obfervations, 3 or 4 thoufand years before Chrift; and many other inftances are there adduced, of their critical knowledge in the other mathematical fciences, employed in their precepts and calculations.

Aftronomy, it feems, too, was not unknown to the Americans; though in their divifion of time, they made ufe only of the folar, and not of the lunar motions. And that the Mexicans, in particular, had a ftrange predilection for the number 13, by means of which they regulated almoft every thing: their fhorteft periods confifted of 13 days; their cycle of 13 months, each containing 20 days; and their century of 4 periods, of 13 years each: and this exceffive veneration for the number 13, arofe, according to Siguenza, from its being the number of their greater gods. And it is very remarkable, that the Abbé Clavigero afferts it as a fact, that, having difcovered the excefs of a few hours in the folar above the lunar year, they made ufe of intercalary days, to bring them to an equality, as eftablifhed by Julius Cæfar in the Roman Calendar; but with this difference, that, inftead of one day every 4 years, they interpofed 13 days every 52 years, which produces the fame effect.

Moft authors however fix the origin of aftronomy and aftrology, either in Chaldea or in Egypt; and accordingly among the ancients we find the word Chaldean often ufed for aftronomer, or, which was the fame thing, aftrologer. Indeed both of thefe nations pretended to a very high antiquity, and claimed the honour of producing the firft cultivators of this fcience. The Chaldeans boafted of their temple or tower of Belus, and of Zoroafter, whom they placed 5000 years before the deftruction of Troy; while the Egyptians boafted of their colleges of priefts, where aftronomy was taught, and of the monument of Ofymandyas, in which, it is faid, there was a golden circle of 365 cubits in circumference, and one cubit thick, divided into 365 equal parts according to the days of the year, &c.

It is, indeed, evident, that both Chaldea and Egypt were countries very proper for aftronomical obfervations, on account of the extended flatnefs of the country, and the purity and ferenity of the air. The tower of Belus, or of Babel itfelf, of a great height, was probably an aftronomical obfervatory; and the lofty pyramids of Egypt, whatever they were originally defigned for, might perhaps anfwer the fame purpofe; and at leaft they fhew the fkill of this people in practical aftronomy, as they are all placed with their four fronts exactly facing the cardinal points of the compafs. The Chaldeans certainly began to make obfervations foon after the confufion of languages, as appears from the obfervations found there on the taking of Babylon by Alexander; and it is probable they began much earlier. It hence appears that they had determined, with tolerable exactnefs, the length both of a periodical and fynodical month. They had alfo difcovered, that the motion of the moon was not uniform; and they even attempted to affign thofe parts of the orbit in which the motion is quicker or flower. We are alfo affured by Ptolemy that they were not unacquainted with the motion of the moon's apogee and nodes, the latter of which they fuppofed made a complete revolution in 6585¼ days, or a little more than 18 years, and contained 223 complete lunations, which period is called the Chaldean *Saros*. From Hipparchus, the fame author alfo gives us feveral obfervations of lunar eclipfes made at Babylon above 720 years before Chrift. And Ariftotle informs us, that they had many occultations of the planets and fixed ftars by the moon; a circumftance which led them to conceive that eclipfes of the fun were to be attributed to the fame caufe. They had alfo no inconfiderable fhare in arranging the ftars into conftellations. Nor had even thofe eccentric bodies the comets efcaped their obfervation: for both Diodorus Siculus and Appollinus Myndicus, Seneca informs us, accounted thefe to be permanent bodies, having ftated revolutions as well as the planets, but in much more extenfive orbits: although others of them were of opinion, that the comets were only meteors raifed very high in the air, which, blazing for a while, difappear when the matter of which they confift is confumed or difperfed. The branch of dialling was alfo practifed among them long before the Greeks were acquainted with that fcience.

The Egyptians, it appears from various circumftances, were much of the fame ftanding in Aftronomy as the Chaldeans. Herodotus afcribes their knowledge in the fcience to Sefoftris; probably not the fame whom Newton makes contemporary with Solomon, as they were acquainted with aftronomy at leaft many hundred years before that æra. We learn, from the teftimony of fome ancient authors, many particulars relative to the ftate of their knowledge in aftronomy; fuch as, that they believed the figure of the earth was fpherical; that the moon was eclipfed by paffing through the earth's fhadow, though it does not certainly appear that they had any knowledge of the true fyftem of the univerfe; that they attempted to meafure the magnitude of the earth and fun, though their methods of afcertaining the latter were very erroneous; and that they even pretended to foretel the appearance of comets, as well as earthquakes and inundations; and the fame is alfo afcribed to the Chaldeans; though thefe muft probably have been rather a kind of aftrological predictions, than obfervations drawn from aftronomy, properly fo called.

This fcience however fell into great decay with the Egyptians, and in the time of the emperor Auguftus, it was entirely extinct among them.

From Chaldea and Egypt the fcience of aftronomy paffed into Phenicia, which this people applied to the purpofes of navigation, fteering their courfe by the north polar ftar; and hence they became mafters of the fea, and of almoft all the commerce in the world.

The

The Greeks, it is probable, derived their astronomical knowledge chiefly from the Egyptians and Phenicians, by means of several of their countrymen who visited these nations, for the purpose of learning the different sciences. Newton supposes that most of the constellations were invented about the time of the Argonautic expedition; but it is more probable that they were, at least most part of them, of a much older date, and derived from other nations, though cloathed in fables of their own invention or application. Several of the constellations are mentioned by Hesiod and Homer, the two most ancient writers among the Greeks, and who lived about 870 years before Christ. Their knowledge in this science however was greatly improved by Thales the Milesian, and other Greeks, who travelled into Egypt, and brought from thence the chief principles of the science. Thales was born about 640 years before Christ; and he, first of all among the Greeks, observed the stars, the solstices, the eclipses of the sun and moon, and predicted the same. And the same was farther cultivated and extended by his successors Anaximander, Anaximanes, and Anaxagoras; but most especially by Pythagoras, who was born 577 years before Christ, and having resided for several years in Egypt, &c. brought from thence the learning of these people, taught the same in Greece and Italy, and founded the sect of the Pythagoreans. He taught that the sun was in the centre of the universe; that the earth was round, and people had antipodes; that the moon reflected the rays of the sun, and was inhabited like the earth; that comets were a kind of wandering stars, disappearing in the further parts of their orbits; that the white colour of the milky-way was owing to the united brightness of a great multitude of small stars; and he supposed that the distances of the moon and planets from the earth, were in certain harmonic proportions to one another.

Philolaus, a Pythagorean, who flourished about 450 years before Christ, asserted the annual motion of the earth about the sun; and not long after, the diurnal motion of the earth on her own axis, was taught by Hicetas, a Syracusan. About the same time flourished at Athens, Meton and Euctemon, where they observed the summer solstice 432 years before Christ, and observed the risings and settings of the stars, and what seasons they answered to. Meton also invented the cycle of 19 years, which still bears his name.

Eudoxus the Cnidian lived about 370 years before Christ, and was accounted one of the most skilful astronomers and geometricians of antiquity, being accounted the inventor of many of the propositions in Euclid's Elements, and having introduced geometry into the science of astronomy. He travelled into Asia, Africa, Sicily, and Italy, for improvements in astronomy; and we are informed by Pliny, that he determined the annual year to contain 365 days 6 hours, that he determined also the periodical times of the planets, and made other important observations and discoveries.

Calippus flourished soon after Eudoxus, and his celestial sphere is mentioned by Aristotle; but he is better known by a period of 76 which he invented, containing 4 corrected Metonic periods, and which commenced at the summer solstice in the year 330 before Christ. About his time the knowledge of the Pythagorean system was carried into Italy, Gaul, and Egypt, by certain colonies of Greeks.

However, the introduction of Astronomy into Greece is represented by Vitruvius in a manner somewhat different. He maintains, that Berosus, a Babylonian, brought it immediately from Babylon itself, and opened an astronomical school in the isle of Cos. And Pliny says, that in consideration of his wonderful predictions, the Athenians erected him a statue in the gymnasium, with a gilded tongue. But if this Berosus be the same with the author of the Chaldaic histories, he must have lived before Alexander.

After the death of this conqueror, the sciences flourished chiefly in Egypt, under the auspices of Ptolemy Philadelphus and his successors. He founded a school there, which continued to be the grand seminary of learning, till the invasion of the Saracens in the year of Christ 650. From the founding of that school, the science of astronomy advanced considerably. Aristarchus, about 270 years before Christ, strenuously asserted the Pythagorean system, and gave a method of determining the sun's distance by the dichotomy of the moon.——Eratosthenes, who was born at Cyrene in the year 271 before Christ, measured the circumference of the earth by means of a gnomon; and being invited to Alexandria, from Athens, by Ptolemy Euergetes, and made keeper of the royal library there, he set up for that Prince those armillary spheres, which Hipparchus and Ptolemy the astronomer afterwards employed so successfully in observing the heavens. He also determined the distance between the tropics to be $\frac{11}{83}$ of the whole meridian circle, which makes the obliquity of the ecliptic in his time to be 23° 51 $\frac{1}{3}$.——The celebrated Archimedes, too, cultivated astronomy, as well as geometry and mechanics: he determined the distances of the planets from one another, and constructed a kind of planetarium or orrery, to represent the phenomena and motions of the heavenly bodies.

To pass by several others of the ancients, who practised or cultivated astronomy, more or less, we find that Hipparchus, who flourished about 140 years before Christ, was the first who applied himself to the study of every part of astronomy, and, as we are informed by Ptolemy, made great improvements in it: he discovered that the orbits of the planets are eccentric, that the moon moved slower in the apogee than in her perigee, and, that there was a motion of anticipation of the moon's nodes: he constructed tables of the motions of the sun and moon, collected accounts of such eclipses, &c, as had been made by the Egyptians and Chaldeans, and calculated all that were to happen for 600 years to come: he discovered that the fixed stars changed their places, having a slow motion of their own from west to east: he corrected the Calippic period, and pointed out some errors in the method of Eratosthenes for measuring the circumference of the earth: he computed the sun's distance more accurately than any of his predecessors: but his chief work is a catalogue which he made of the fixed stars, to the number of 1022, with their longitudes, latitudes, and apparent magnitudes; which, with most of his other observations, are preserved by Ptolemy in his Almagest.

There was but little progress made in astronomy from the time of Hipparchus to that of Ptolemy, who was born

born at Pelufium in Egypt, in the firft century of chrif-tianity, and who made the greateft part of his obferva-tions at the celebrated fchool of Alexandria in that coun-try. Profiting of thofe of Hipparchus and other an-cient aftronomers, he formed a fyftem of his own, which, though erroneous, was followed for many ages by all na-tions. He compiled a great work, called the Almageft, which contained the obfervations and collections of Hip-parchus and others his predeceffors in aftronomy, on which account it will ever be valuable to the profeffors of that fcience. This work was preferved from the grie-vous conflagration of the Alexandrine library by the Sa-racens, and tranflated out of Greek into Arabic in the year 827, and from thence into Latin in 1230. The Greek original was not known in Europe till the begin-ning of the 15th century, when it was brought from Conftantinople, then taken by the Turks, by George, a monk of Trapezond, by whom it was tranflated into Latin; and various other editions have been fince made.

During the long period from the year 800 till the be-ginning of the 14th century, the weftern parts of Europe were immerfed in grofs ignorance and barbarity, while the Arabians, profiting by the books they had preferved from the wreck of the Alexandrine library, cultivated and improved all the fciences, and particularly that of aftronomy, in which they had many able profef-fors and authors. The caliph Al Manfur firft introduced a tafte for the fciences into his empire. His grandfon Al Mamun, who afcended the throne in 814, was a great encourager and improver of the fciences, and ef-pecially of aftronomy. Having conftructed proper in-ftruments, he made many obfervations; determined the obliquity of the ecliptic to be 23° 35′; and under his aufpices a degree of the circle of the earth was meafured a fecond time in the plain of Singar, on the border of the Red Sea. About the fame time Alferganus wrote elements of aftronomy; and the fcience was from hence greatly cultivated by the Arabians, but principally by Albategnius, who flourifhed about the year 880, and who greatly reformed aftronomy, by comparing his own ob-fervations with thofe of Ptolemy: hence he computed the motion of the fun's apogee from Ptolemy's time to his own; fettled the preceffion of the equinoxes at one degree in 70 years; and fixed the obliquity of the ecliptic at 23° 35′. The tables which he compofed, for the me-ridian of Aracta, were long efteemed by the Arabians. After his time, though the Saracens had many eminent aftronomers, feveral centuries elapfed without producing any very valuable obfervations, excepting thofe of fome eclipfes obferved by Ebn Younis, aftronomer to the caliph of Egypt, by means of which the quantity of the moon's acceleration fince that time may be deter-mined.

Other eminent Arabic aftronomers, were, Arza-chel a Moor of Spain, who obferved the obliquity of the ecliptic: he alfo improved Trigonometry by conftructing tables of fines, inftead of chords of arches, dividing the diameter into 300 equal parts. And Alha-zen, his contemporary, who wrote upon the twilight, the height of the clouds, the phenomenon of the hori-zontal moon; and who firft fhewed the importance of the theory of refractions in aftronomy.

Ulug Beg, grandfon of the celebrated Tartar prince Tamerlane, was a great proficient in practical aftronomy;

he had very large inftruments, particularly a quadrant of about 180 feet high, with which he made good ob-fervations. From thefe he determined the latitude of Samercand, his capital, to be 39° 37′ 23″; and com-pofed aftronomical tables for the meridian of the fame fo exact, that they differ very little from thofe conftruct-ed afterwards by Tycho Brahe; but his principal work was his catalogue of the fixed ftars, made alfo from his own obfervations in the year 1437.

During this period, almoft all Europe was immerfed in grofs ignorance. But the fettlement of the Moors in Spain introduced the fciences into Europe; from which time they have continued to improve, and to be communicated from one people to another, to the prefent time, when aftronomy and all the fciences, have arriv-ed at a very eminent degree of perfection. The em-peror Frederick II, about 1230, firft began to encou-rage learning; reftoring fome decayed univerfities, and founding a new one in Vienna: he alfo caufed the works of Ariftotle, and Ptolemy's Almageft, to be tranflated into Latin; and from the tranflation of this work we may date the revival of aftronomy in Europe. Two years after this, John de Sacro Bofco, that is, of Halifax, compiled, from Ptolemy, Albategnius, Alferganus, and other Arabic aftronomers, his work De Sphæra; which was held in the greateft eftimation for 300 years after, and was honoured with commentaries by Clavius and other learned men. In 1240, Alphonfo, king of Caftile, not only cultivated aftronomy himfelf, but greatly encouraged others; and by the affiftance of fe-veral learned men he corrected the tables of Ptolemy, and compofed thofe which were denominated from him the Alphonfine Tables. About the fame time alfo, Roger Bacon, an Englifh monk, wrote feveral tracts relative to aftronomy, particularly of the lunar afpects, the folar rays, and the places of the fixed ftars. And, about the year 1270, Vitello, a Polander, compofed a treatife on optics, in which he fhewed the ufe of re-fractions in aftronomy.

Little other improvement was made in aftronomy till the time of Purbach, who was born in 1423. He com-pofed new tables of fines for every 10 minutes, making the radius 60, with four ciphers annexed. He con-ftructed fpheres and globes, and wrote feveral aftrono-mical tracts; as, a commentary on Ptolemy's Alma-geft; fome treatifes on Arithmetic and Dialling, with tables for various climates; new tables of the fixed ftars reduced to the middle of that century; and he corrected the tables of the planets, making new equations to them where the Alphonfine tables were erroneous. In his folar tables, he placed the fun's apogee in the beginning of Cancer; but retained the obliquity of the ecliptic 23° 33½′, as determined by the lateft obfervations. He alfo obferved fome eclipfes, made new tables for com-puting them, and had juft finifhed a theory of the pla-nets, when he died in 1462, being only 39 years of age.

Purbach was fucceeded in his aftronomical and mathe-matical labours by his pupil and friend, John Muller, commonly called Regiomontanus, from Monteregio, or Koningfberg, a town of Franconia, where he was born. He completed the epitome of Ptolemy's Alma-geft, which Purbach had begun; and after the death of his friend, was invited to Rome, where he made many

aftronomical

astronomical observations. Being returned to Nuremberg in 1471, by the encouragement of a wealthy citizen named Bernard Walther, he made several instruments for astronomical observations, among which was an armillary astrolabe, like that used at Alexandria by Hipparchus and Ptolemy, with which he made many observations, using also a good clock, which was then but a late invention. He made ephemerides for 30 years to come, shewing the lunations, eclipses, &c; and, the art of printing having then been lately invented, he printed the works of many of the most celebrated ancient astronomers. He wrote the Theory of the Planets and Comets, and a treatise on triangles, still in repute for several good theorems; computing the table of sines for every single minute, to the radius 1000000, and introducing the use of tangents also into trigonometry. After his death, which happened at Rome in 1476, being only 40 years of age, Walther collected his papers, and continued the astronomical observations till his own death also. The observations of both were collected by order of the senate of Nuremberg, and published there in 1544 by John Schoner: they were also afterwards published in 1618 by Snellius, at the end of the observations made by the Landgrave of Hesse; and lastly with those of Tycho Brahe in 1666.

Walther was succeeded, as astronomer at Nuremberg, by John Werner, a clergyman. He observed the motion of the comet in 1500; and wrote several tracts on geometry, astronomy, and geography, in a masterly manner; the most remarkable of which, are those concerning the motion of the 8th sphere, or of the fixed stars; in this tract, by comparing his own observations, made in 1514, with those of Ptolemy, Alphonsus, and others, he shewed that the motion of the fixed stars, since called the precession of the equinoxes, is 1° 10 in 100 years. He made also the first star of Aries 26° distant from the equinoctial point, and the obliquity of the ecliptic only 23° 28'. He constructed a planetarium, representing the celestial motions according to the Ptolemaic hypothesis; and he published a translation of Ptolemy's Geography, with a commentary, in which he first proposed the method of finding the longitude at sea by observing the moon's distance from the fixed stars; now so successfully practised for that purpose. Werner died in 1528, at 60 years of age.

Nicolaus Copernicus was the next who made any considerable figure in astronomy, by whom indeed the old Pythagorean system of the world was restored, which had been till now set aside from the time of Ptolemy. About the year 1507 Copernicus conceived doubts of this system, and entertained notions about the true one, which he gradually improved by a series of astronomical observations, and the contemplation of former authors. By these he formed new tables, and completed his work in the year 1530, containing these, and his renovation of the true system of the universe, in which all the planets are considered as revolving about the sun, placed in the centre. But the work was only printed in 1543, under the care of Schoner and Osiander, by the title of *Revolutiones Orbium Cælestium*; and the author just received a copy of the work a few hours before his death, which happened on the 23d of May 1543, at 70 years of age.

After the death of Copernicus, the science and practice of astronomy were greatly improved by many other persons, as Schoner, Nonius, Appian, Gemma Frisius, Rothman, Byrgius, the Landgrave of Hesse, &c.— Schoner reformed and explained the calendar, improved the methods of making celestial observations, and published a treatise on cosmography; but he died 4 years after Copernicus.—Nonius wrote several works on mathematics, astronomy and navigation, and invented some useful and more accurate instruments than formerly; one of these was the astronomical quadrant, on which he divided the degrees into minutes by a number of concentric circles; the first of which was divided in 90 equal parts or degrees, the second into 89, the third into 88, and so on, to 46; so that, the index of the quadrant always falling upon or near one of the divisions, the minutes would be known by an easy computation.—The chief work of Appian, *The Cæsarean Astronomy*, was published at Ingoldstat in 1540; in which he shews, how to observe the places of the stars and planets by the astrolabe; to resolve astronomical problems by certain instruments; to predict eclipses, and to describe the figures of them; and the method of dividing and using an astronomical quadrant: at the end are added observations of 5 comets, one of which has been supposed the same with that observed by Hevelius, and if so, it ought to have returned again in the year 1789;—but it was not observed then. Gemma Frisius wrote a commentary on Appian's *Cosmography*, accompanied with many observations of eclipses: he also invented the astronomical ring, and several other instruments, useful in taking observations at sea; and he was the first who recommended a timekeeper for determining the longitude at sea.—Rheticus gave up his professorship of mathematics at Wittemberg, that he might attend the astronomical lectures of Copernicus; and, for improving astronomical calculations, he began a very extensive work, being a table of sines, tangents and secants, to a very large radius, and to every 10 seconds, or ⅙ of a minute; which was completed by his pupil Valentine Otho, and published in 1594.

About the year 1561, William IV, Landgrave of Hesse Cassel, applied himself to the study of astronomy, having furnished himself with the best instruments that could then be made: with these he made a great number of observations, which were published by Snellius in 1618, and which were preferred by Hevelius to those of Tycho Brahe. From these observations he formed a catalogue of 400 stars, with their latitudes and longitudes, adapted to the beginning of the year 1593.

Tycho Brahe, a noble Dane, began his observations about the same time with the Landgrave of Hesse, abovementioned, and he observed the great conjunction of Jupiter and Saturn: but finding the usual instruments very inaccurate, he constructed many others much larger and exacter, with which he applied himself diligently to observe the celestial phenomena. In 1571 he discovered a new star in the chair of Cassiopeia; which induced him, like Hipparchus on a similar occasion, to make a new catalogue of the stars; which he composed to the number of 777, and adapted their places to the year 1600. In the year 1576, by favour of the king of Denmark, he built his new observatory, called Uraniburg, on the small island Huenna, opposite

to

to Copenhagen, and which he very amply furnished with many large instruments, some of them so divided as to shew single minutes, and in others the arch might be read off to 10 seconds. One quadrant was divided according to the method invented by Nonius, that is, by 47 concentric circles; but most of them were divided by diagonals; a method of division invented by a Mr. Richard Chanceler, an Englishman. Tycho employed his time at Uraniburg to the best advantage, till the death of the king, when, falling into discredit, he was obliged to remove to Holstein; and he afterwards found means of introducing himself to the Emperor Rodolph, with whom he continued at Prague till the time of his death in 1601.—It is well known that Tycho was the inventor of a system of astronomy, a kind of Semi-Ptolemaic, which he vainly endeavoured to establish instead of the Copernican or true system. His works, however, which are very numerous, shew that he was a man of great abilities; and his discoveries, together with those of Purbach and Regiomontanus, were collected and published together in 1621, by Longomontanus, the favourite disciple of Tycho.

While Tycho resided at Prague with the emperor, he prevailed on Kepler to leave the university of Glatz, and to come to him, which he did with his family and library in 1600: but Tycho dying in 1601, Kepler enjoyed all his life the title of mathematician to the Emperor, who ordered him to finish the tables of Tycho Brahe, which he did accordingly, and published them in 1627, under the title of Rodolphine. He died about the year 1630 at Ratisbon, where he was soliciting the arrears of his pension. From his own observations, and those of Tycho, Kepler discovered several of the true laws of nature, by which the motions of the celestial bodies are regulated. He discovered that all the planets revolved about the sun, not in circular, but in elliptical orbits, having the sun in one of the foci of the ellipse; that their motions were not equable, but varying, quicker or slower as they were near to the sun or farther from him; but that this motion was so regulated, that the areas described by the variable line drawn from the planet to the sun, are equal in equal times, and always proportional to the times of describing them. He also discovered, by trials, that the cubes of the distances of the planets from the sun, were in the same proportion as the squares of their periodical times of revolution. By observations also on comets, he concluded that they are freely carried about among the orbits of the planets, in paths that are nearly rectilinear, but which he could not then determine. Besides many other discoveries, which are to be found in his writings.

In Kepler's time there were many other good proficients in astronomy; as Edward Wright, baron Napier, John Bayer, &c. Wright made several good meridional observations of the sun, with a quadrant of 6 feet radius, in the years 1594, 1595, and 1596; from which he greatly improved the theory of the sun's motion, and computed more accurately his declination, than any person had done before. In 1599 he published also an excellent work entitled "Certain Errors in Navigation discovered and detected," containing a method which has commonly, though erroneously, been ascribed to Mercator.—To Napier we owe some excellent theorems and improvements in spherics, besides the ever memo-

rable invention of logarithms, one of the most useful ever made in the art of numbering, and of the greatest use in all the other mathematical sciences.—Bayer, a German, published his *Uranometria*, being a complete celestial atlas, or the figures of all the constellations visible in Europe, with the stars marked on them, and the stars also accompanied by names, or the letters of the Greek alphabet; a contrivance by which the stars may easily be referred to with distinctness and precision. —About the same time too, astronomy was cultivated by many other persons; namely, abroad by Mercator, Maurolycus, Maginus, Homelius, Schultet, Stevin, Galileo, &c; and in England by Thomas and Leonard Digges, John Dee, Robert Flood, Harriot, &c.

The beginning of the 17th century was particularly distinguished by the invention of telescopes, and the application of them to astronomical observations; an invention to which we owe the most brilliant discoveries, and all the accuracy to which the science is now brought. The more distinguished early observations with the telescope, were made by Galileo, Harriot, Huygens, Hook, Cassini, &c. It is said that from report only, and before he had seen one, Galileo made for himself telescopes, by which he discovered inequalities in the moon's surface, Jupiter's satellites, and the ring of Saturn; also spots on the sun, by which he found out the revolution of that luminary on his axis; and he discovered that the nebulæ and milky way were full of small stars. Harriot also, who has hitherto been known only as an algebraist, made much the same discoveries as Galileo, and as early, if not more so, as appears by his papers not yet printed, in the possession of the earl of Egremont.

Mr. Horrox, a young astronomer of great talents, made considerable discoveries and improvements. In 1633 he found out that the planet Venus would pass over the sun's disc on the 24th of November, 1639; an event which he announced only to his friend Mr. Crabtree; and these two were the only persons in the world that observed this transit, which was also the first time it had ever been seen by mortal eyes. Mr. Horrox made also many other useful observations, and had even formed a new theory of the moon, taken notice of by Newton; but his early death, in the beginning of the year 1640, put a stop to his useful and valuable labours.

About the same time flourished Hevelius, Burgomaster of Dantzic, who furnished an excellent observatory in his own house, where he observed the spots and phases of the moon, from which observations he compiled his *Selenographia;* and an account of his apparatus is contained in his work entitled *Machina Cœlestis,* a book now very scarce, as most of the copies were accidentally burnt, with the whole house and apparatus, in 1679. Hevelius died in 1688, aged 76.

Dr. Hook, a contemporary of Hevelius, invented instruments with telescopic sights, and censured those of the latter; which occasioned a sharp dispute between them, and to settle which the celebrated Dr. Halley was sent over to Hevelius to examine his instruments. The two astronomers made several observations together, very much to their satisfaction, and amongst them was one of an occultation of Jupiter by the moon, when they determined the diameter of the latter to be 30′ 33."

Before

2

Before the middle of the 17th century the construction of telescopes had been greatly improved, particularly by Huygens and Fontana. The former constructed one of 123 feet, with which he long observed the moon and planets, and discovered that Saturn was encompassed with a ring. With telescopes too, of 200 and 300 feet focus, Cassini saw five satellites of Saturn, with his zones or belts, and the shadows of Jupiter's satellites passing over his body. In 1666 Azout applied a micrometer to telescopes, to measure the diameters of the planets, and other small distances in the heavens: but an instrument of this kind had been invented before, by Mr. Gascoigne, though it was but little known abroad.

To obviate the difficulties of the great lengths of refracting telescopes and the aberration of the rays, it is said that Mersennus first started the idea of making telescopes of reflectors, instead of lenses, in a letter to Descartes; and in 1663 James Gregory of Aberdeen shewed how such a telescope might be constructed. After some time spent also by Newton, on the construction of both sorts of telescopes, he found out the great inconvenience which arises to refractors from the different refrangibility of the rays of light, for which he could not then find a remedy; and therefore, pursuing the other kind, in the year 1672 he presented to the Royal Society two reflectors, which were constructed with spherical speculums, as he could not procure other figures. The inconveniences however arising from the different refrangibility of the rays of light, have since been fully obviated by the ingenious Mr. Dollond. Towards the latter part of the 17th, and beginning of the 18th century, practical astronomy it seems rather languished. But at the same time the speculative part was carried to the highest perfection by the immortal Newton in his Principia, and by the Astronomy of David Gregory.

Soon after this however, great improvements of astronomical instruments began to take place, particularly in Britain. Mr. Graham, a celebrated mechanic and watchmaker, not only improved clocks and watch work, but also carried the accuracy of astronomical instruments to a surprising degree. He constructed the old 8 feet mural arch at the Royal Observatory Greenwich, and a small equatorial sector for making observations out of the meridian; but he is chiefly remarkable for contriving the zenith sector of 24 feet radius, and afterwards one of 12½ feet, with which Dr. Bradley discovered the aberration of the fixed stars. The reflecting telescope of Gregory and Newton, was greatly improved by Mr. Hadley, who presented a very powerful instrument of that kind to the Royal Society in 1719. The same gentleman has also immortalized his memory by the invention of the reflecting quadrant or sector, now called by his name, which he presented to the society in 1731, and which is now so universally useful at sea, especially where nice observations are required. It appears however that an instrument similar to this in its principles, had been invented by Newton; and a description with a drawing of it given by him to Dr. Halley, when he was preparing for his voyage in 1701, to discover the variation of the needle: it has also been asserted that a Mr. Godfrey of Philadelphia in America, made the same discovery, and the first instrument of this kind. About the middle of this century, the constructing and dividing of large astronomical instruments were carried to great

perfection by Mr. John Bird; and reflecting telescopes were not less improved by Mr. Short, who also first executed the divided object-glass micrometer, which had been proposed and described by M. Louville and others. Mr. Dollond also brought refracting telescopes to the greatest perfection, by means of his acromatic glasses; and lately the discoveries of Herschel are owing to the amazing powers of reflectors of his own construction.

Thus the astronomical improvements in the present century, have been chiefly owing to the foregoing inventions and improvements in the instruments, and to the establishment of regular observatories in England, France, and other parts of Europe. Roemer, a celebrated Danish Astronomer, first made use of a meridional telescope; and, by observing the eclipses of Jupiter's satellites, he first discovered the progressive motion of light, concerning which he read a dissertation before the Academy of Sciences at Paris in 1675.—Mr. Flamsteed was appointed the first Astronomer Royal at Greenwich in 1675. He observed, for 44 years, all the celestial phenomena, the sun, moon, planets, and fixed stars, of all which he gave an improved theory and tables, viz, a catalogue of 3000 stars with their places, to the year 1689; also new solar tables, and a theory of the moon according to Horrox; likewise, in Sir Jonas Moore's System of Mathematics, he gave a curious tract on the doctrine of the sphere, shewing how, geometrically to construct eclipses of the sun and moon, as well as occultations of the fixed stars by the moon. And it was upon his tables that were constructed, both Halley's tables, and Newton's theory of the moon.—Cassini also, the first French Astronomer Royal, very much distinguished himself, making many observations on the sun, moon, planets and comets, and greatly improved the elements of their motions. He also erected the gnomon, and drew the celebrated meridian line in the church of Petronia at Bologna.

In 1719 Mr. Flamsteed was succeeded by Dr. Halley, as Astronomer Royal at Greenwich. The Doctor had been sent at the early age of 21 to the island of St. Helena, to observe the southern stars, and make a catalogue of them, which was published in 1679. In 1705 he published his *Synopsis Astronomiæ Cometicæ*, in which he ventured to predict the return of a comet in 1758 or 1759. He was the first who discovered the acceleration of the moon, and he gave a very ingenious method for finding her parallax by three observed phases of a solar eclipse. He published, in the Philosophical Transactions, many learned papers, and amongst them some that were concerning the use that might be made of the next transit of Venus in determining the distance of the sun from the earth. He composed tables of the sun, moon, and all the planets, which are still in great repute; with which he compared the observations he made of the moon at Greenwich, amounting to near 1500, and noted the differences. He recommended the method of determining the longitude by the moon's distances from the sun and certain fixed stars; a method which had before been noticed, and which has since been carried into execution, more particularly at the instance of the present Astronomer Royal.

About this time a dispute arose concerning the figure of the earth. Sir Isaac Newton had determined, from a consideration of the laws of gravity, and the diurnal motion of the earth, that the figure of it was an oblate spheroid.

spheroid, and flatted at the poles : but Caffini had determined, from the measures of Picart, that the figure was an oblong spheroid, or lengthened at the poles. To settle this dispute, it was resolved, under Lewis XV, to measure two degrees of the meridian ; one near the equator, and the other as near the pole as possible. For this purpose, the Royal Academy of Sciences sent to Lapland, Mess. Maupertuis, Clairault, Camus, and Le Monier ; being also accompanied by the Abbé Outhier, and by M. Celfus, professor of anatomy at Upsal. And on the southern expedition were sent Mess. Godin, Condamine, and Bouguer, to whom the king of Spain joined Don George Juan and Don Antonio de Ulloa. These set out in 1735, and returned at different times in 1744, 1745, and 1746 ; but the former party, who set out only in 1736, returned the year following ; having both fulfilled their commissions. Picart's measure was also revised by Caffini and De la Caille, which after his errors were corrected, was found to agree very well with the other two ; and the result of the whole served to confirm the determination of the figure before laid down by Newton.—On the southern expedition, it was found that the attraction of the great mountains of Peru had a sensible effect on the plumb-line of one of their largest instruments, deflecting it 7 or 8 seconds from the true perpendicular.

On the death of Dr. Halley, in 1742, he was succeeded by Dr. Bradley, as Astronomer Royal at Greenwich. The accuracy of his observations enabled him to detect the smaller inequalities in the motions of the planets and fixed stars. The consequence of this accuracy was, the discovery of the aberration of light, the nutation of the earth's axis, and a much greater degree of perfection in the lunar tables. He also observed the places, and computed the elements of the comets which appeared in the years 1723, 1736, 1743, and 1757. He made new and accurate tables of the motions of Jupiter's satellites ; and from a multitude of observations of the luminaries, he constructed the most accurate table of refractions yet extant. Also, with a very large transit instrument, and a new mural quadrant of 8 feet radius, constructed by Mr. Bird in 1750, he made an immense number of observations for settling the places of all the stars in the British catalogue, together with near 1500 places of the moon, the greater part of which he compared with Mayer's tables. Dr. Bradley died in 1762.

In the mean time the mathematicians and astronomers elsewhere were assiduous in their endeavours to promote the science of astronomy. The theory of the moon was particularly considered by Messrs Clairault, D'Alembert, Euler, Meyer, Simpson, and Walmisly, and especially Clairault, Euler, and Mayer, who computed complete sets of lunar tables ; those of the last of these authors, for their superior accuracy, were rewarded with a premium of 3000 pounds by the Board of Longitude, who brought them into use in the computation of the Nautical Ephemeris, published by that Board.—The most accurate tables of the satellites of Jupiter were composed, from observations by Mr. Wargentin, an excellent Swedish astronomer.—Among the many French astronomers who contributed to the advancement of the science, it was particularly indebted to M. De la Caille, for an excellent set of solar tables. And in 1750 he went to the Cape of Good Hope to make observations in concert

with the most celebrated astronomers in Europe, for determining the parallax of Mars and the moon, and thence, that of the sun, which it was concluded did not much exceed 10 seconds. Here he re-examined and adjusted, with great accuracy, the stars about the southern pole ; and also measured a degree of the meridian. —In Italy the science was assiduously cultivated by Bianchini, Boscovich, Frisi, Manfredi, Zanotti, and many others ; in Sweden by Wargentin already mentioned, Blingenstern, Mallet, and Planman ; and in Germany by the Eulers, Mayer, Lambert, Grischow, and others.

In the year 1760 all the learned Societies in Europe made preparations for observing the transit of Venus over the sun, which had been predicted by Dr. Halley more than 80 years before, and the use that might be made of it in determining the sun's parallax, and the distances of the planets from the sun. And the same exertions were repeated, to observe the transit in 1769, by sending observers to different parts of the world, for the more convenience in observing. And from the whole, Mr. Short computed that the sun's parallax was nearly $8\frac{2}{3}$ seconds, and consequently the distance of the sun from the earth about 24114 of the earth's diameters, or 96 millions of miles.

Dr. Bradley was succeeded, in 1762, in his office of Astronomer Royal, by Mr. Bliss, Savilian professor of astronomy ; who being in a declining state of health, did not long enjoy it. But, dying in 1765, was succeeded by the learned Nevil Maskelyne, D. D. the present Astronomer Royal, who has discharged the duties of that office with the greatest honour to himself, and benefit to the science. In January 1761 this gentleman was sent by the Royal Society, at a very early age, to the Island of St. Helena, to observe the transit of Venus over the sun, and the parallax of the star Sirius. The first of these objects partly failed, by clouds preventing the sight of the 2d internal contact ; and the 2d also, owing to Mr. Short having suspended the plumbline by a *loop* from the neck of the central pin. However, our astronomer indemnified himself by many other valuable observations : Thus, he observed at St. Helena, the tides ; the horary parallaxes of the moon ; and the going of a clock, to find, by comparison with its previous going which had been observed in England, the difference of gravity at the two places : also, in going out and returning, he practised the method of finding the longitude by the lunar distances taken with a Hadley's Quadrant, making out rules for the use of seamen, and taught the method to the officers on board the ship ; which he afterwards explained in a letter to the Secretary of the Royal Society, inserted in the Philof. Tranf. for the year 1762, and still more fully afterwards, in the British Mariner's Guide, which he published in the year 1763. He returned from St. Helena in the spring of 1762, after a stay there of 10 months ; and in September 1763 sailed for the island of Barbadoes, to settle the longitude of the place, and to compare Mr. Harrison's watch with the time there when this gentleman should bring it out : another object was also to try Mr. Irwin's marine chair, which he did in his way out. While at Barbadoes, he also made many other observations, and amongst them, many relating to the moon's horary parallaxes, not yet published. Returning to England in the latter part of the year 1764, he was appointed in 1765 to fucceed Mr. Blifs as Aftronomer Royal, and immediately recommended to the Board of Longitude the lunar method of finding the longitude, and propofed to them the project of a Nautical Almanac, to be calculated and publifhed to facilitate that method; this they agreed to, and the firft vol. was publifhed for 1767, and it has continued ever fince under his direction, to the great benefit of navigation and univerfal commerce.

A multitude of other ufeful writings by this gentleman are inferted in the volumes of the Philof. Tranf. and particularly in confequence of a propofal, made by him to the Royal Society, the noble project was formed of meafuring very accurately the effect of fome mountain on the plumb line, in deflecting it from the perpendicular; and the mountain of Schehallien, in Scotland, having been found the moft convenient in this ifland for the purpofe, at the requeft of the Society, he went into Scotland to conduct the bufinefs, which he performed in the moft accurate manner, fhewing that the fum of the deflections on the two oppofite fides was about 11¾ feconds of a degree; and proving, to the fatisfaction of the whole world, the univerfal attraction of all matter. From the data refulting from thefe meafures I have computed the mean denfity of the whole matter in the earth, which I have found to be about 4½ times that of common water. Befides many learned and valuable papers in the Philofophical Tranfactions, and the moft affiduous exertions in the duties of the obfervatory, as abundantly appears by the curious and voluminous obfervations which he has publifhed, the world is particularly obliged to his endeavours with the Board of Longitude, for the publication of the Nautical Ephemeris, and the method of obferving the longitude, by the diftances of the moon and ftars, now adopted by all nations, and by which the practice of Navigation has been brought to the greateft perfection.

Finally, the difcoveries of Dr. Herfchel form a new æra in aftronomy. He firft, in 1781, began with obfervations on the periodical ftar *in Collo Ceti*, and a new method of meafuring the lunar mountains, none of which he made more than half a mile in height: and, having conftructed telefcopes vaftly more powerful than any former ones, he proceeded to other obfervations, concerning which he has had feveral papers printed in the Philofophical Tranfactions; as, on the rotation of the planets round their axes; On the parallax of the fixed ftars; Catalogues of double, triple, &c ftars; On the proper motion of the fun and folar fyftem; On the remarkable appearances of the folar regions of the planet Mars, &c. And, above all, his difcovery of a new primary planet, on the 13th of March 1781, which he calls the *Georgian Planet*, but it is named the *Planet Herfchel* by the French and other foreign aftronomers; by which, and its two fatellites, which he has alfo difcovered fince that time, he has greatly enlarged the bounds of the folar fyftem, this new planet being more than twice as far from the fun as the planet Saturn.

Lifts and hiftorical accounts of the principal writings and authors on aftronomy are contained in Weidler's Hiftory of Aftronomy, which is brought down to the year 1737. There is alfo Bailly's Hiftory of aftronomy, ancient and modern. For this purpofe, confult alfo the following authors, viz, Adam, Voffius, Bayle, Chauffepié, Niceron, Perraut, the chronological table of

Riccioli, and that of Sherburn, at the end of his edition of Manilius; alfo the firft volume of De la Lande's aftronomy. The more modern, and popular books on aftronomy, are very numerous, and well known: as thofe of Ferdufon, Long, Emerfon, Vince, De la Lande, Leadbetter, Brent, Keil, Whifton, Wing, Street, &c, &c.

ASTROSCOPE, a kind of aftronomical inftrument, compofed of two cones, on the furface of which are delineated the conftellations, with their ftars, by means of which thefe may eafily be known in the heavens. The aftrofcope was invented by William Shukhard, profeffor of mathematics at Tubingen, upon which he publifhed a treatife in 1698.

ASTROSCOPIA, the art of obferving and examining the ftars, by means of telefcopes, to difcover their nature and properties.

ASTROTHEMATA, the places or pofitions of the ftars, in an aftrological fcheme of the heavens.

ASTROTHESIA, is ufed by fome for a conftellation or collection of ftars in the heavens.

ASTRUM, or ASTRON, a conftellation, or affemblage of ftars: in which fenfe it is diftinguifhed from *After*, which denotes a fingle ftar. Some apply the term, in a more particular fenfe, to the Great Dog, or rather to the large bright ftar in his mouth.

ASYMMETRY, the want of proportion, otherwife called *incommenfurability*, or the relation of two quantities which have no common meafure, as between 1 and $\sqrt{2}$, or the fide and diagonal of a fquare.

ASYMPTOTE, is properly a right line, which approaches continually nearer and nearer to fome curve, whofe afympote it is faid to be, in fuch fort, that when they are both indefinitely produced, they are nearer together than by any affignable finite diftance; or it may be confidered as a tangent to the curve when conceived to be produced to an infinite diftance. Two curves are alfo faid to be afymptotical, when they thus continually approach indefinitely to a coincidence: thus, two parabolas, placed with their axes in the fame right line, are afymptotes to one another.

Of lines of the fecond kind, or curves of the firft kind, that is the conic fections, only the hyperbola has afymptotes, which are two in number. All curves of the fecond kind have at leaft one afymptote; but they may have three. And all curves of the third kind may have four afymptotes. The conchoid, ciffoid, and logarithmic curve, though not reputed geometrical curves, have each one afymptote. And the branch or leg of a curve that has an afymptote, is faid to be of the hyperbolic kind.

The nature of afymptotes will be eafily conceived from the inftance of the afymptote to the conchoid. Thus, if, ABC &c be part of a conchoid, and the line MN be fo drawn that the parts FB, GC, HD, IE, &c, of right lines, drawn from the pole P, be equal to each other; then will the line MN be the afymptote of the curve: becaufe the perpendicular C*c* is fhorter than FB, and D*d* than, C*c*, &c; fo that the two lines continually approach; yet the points E*e* &c can never coincide.

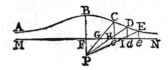

ASYMPTOTES *of the* HYPERBOLA are thus defcribed. Suppofe a right line DE drawn to touch the curve in any point A, and equal to the conjugate *de* of the diamete ACB drawn to that point A, viz, AD or AE equal to the femiconjugate C*d* or C*e* ; then the two lines CDF, CEH, drawn from the centre C, through the points D and E, are the two afymptotes of the curve.

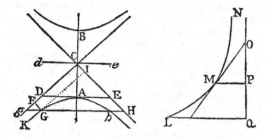

The parts of any right line, lying between the curve of the common hyperbola and its afymptotes, are equal to one another on both fides, that is, *g*G = *h*H. In like manner, in hyperbolas of the fecond kind, if there be drawn any right line cutting both the curve and its three afymptotes in three points, the fum of the two parts of that right line extended in the fame direction from any two of the afymptotes to two points of the curve, is equal to the third part which extends in the contrary direction from the third afymptote to the third point of the curve.

If AGK be an hyperbola of any kind, whofe nature, with regard to the curve and afymptote, is expreffed by this general equation $x^m y^n = a^{m+n}$, where *x* is = CF, and *y* = FG drawn any where parallel to the other afymptote CH ; and the parallelogram CFGI be completed: Then $m−n$ is to n, as this parallelogram CFGI is to the hyperbolic fpace FGK, contained under the determinate line FG, with the afymptote FK and the curve GK, both indefinitely continued towards K. So that, if *m* be greater than *n*, the faid afymptotic fpace is finite and quadrable : but when $m = n$, as in the common or conic hyperbola, then $m − n = o$, the ratio of that fpace to the faid parallelogram, is as *n* to *o*: that is, the hyperbolic fpace is infinitely great, in refpect of the finite parallelogram : and when *m* is lefs than *n*, then, $m − n$ being negative, the afymptotic fpace is to the determinate parallelogram, as a pofitive number is to a negative one; and is what Dr. Wallis calls more than infinite.

ASYMPTOTE *of the* LOGARITHMIC CURVE. If LMN be the logarithmic curve; QON an afymptote, LQ and MP ordinates, MO a tangent, and PO the fubtangent, which in this curve is a conftant quantity. Then the indeterminate fpace LMNQ is equal to LQ × PO, the rectangle under the ordinate LQ and the conftant fubtangent PO ; and the folid generated by the rotation of that curve fpace about the afymptote NQ is equal to half the cylinder, whofe altitude is the faid conftant fubtangent PO, and the radius of its bafe is LQ.

ASYMPTOTES, by fome, are diftinguifhed into various orders. The afymptote is faid to be of the firft order, when it coincides with the bafe of the curvilinear figure : of the 2d order, when it is a right line parallel to the bafe : of the 3d order, when it is a right line ob-

lique to the bafe : of the 4th order, when it is the common parabola, having its axis perpendicular to the bafe : and, in general, of the $n + 2$ order, when it is a parabola whofe ordinate is always as the *n* power of the bafe. The afymptote is oblique to the bafe, when the ratio of the firft fluxion of the ordinate to the fluxion of the bafe, approaches to an affignable ratio, as its limit ; but it is parallel to the bafe, or coincides with it, when this limit is not affignable.

The doctrine and determination of the afymptotes of curves, is a curious part of the higher geometry. Fontenelle has given feveral theorems relating to this fubject, in his *Geometrie de l'Infini*. See alfo Stirling's *Lineæ Tertii Ordinis*, prop. vi, where the fubject of Afymptotes is learnedly treated ; and Cramer's *Introduction à l'analyfe des lignes courbes*, art. 147 *& feq.* for an excellent theory of afymptotes of geometrical curves and their branches. Likewife Maclaurin's Algebra, and his Fluxions, book i, chap. 10, where he has carefully avoided the modern paradoxes concerning infinites and infinitefimals. But the eafieft way of determining afymptotes, it feems, is by confidering them as tangents to the curves at an infinite diftance from the beginning of the abfcifs, that is when the abfcifs *x* is infinite in the equation of the curve, and in the proportion of $\dot{x}$ to $\dot{y}$, or in that of the fubtangent to the ordinate.

The areas bounded by curves and their afymptotes, though indefinitely extended, have fometimes limits to which they may approach indefinitely near : and this happens in hyperbolas of all kinds, except the firft or Apollonian, and in the logarithmic curve ; as was obferved above. But in the common hyperbola, and many other curves, the afymptotical area has no fuch limit, but is infinitely great.—Solids, too, generated by hyperbolic areas, revolving about their afymptotes, have fometimes their limits ; and fometimes they may be produced till they exceed any given folid.—Alfo the furface of fuch folid, when fuppofed to be infinitely produced, is either finite or infinite, according as the area of the generating figure is finite or infinite.

ATLANTIDES, a name given to the Pleiades, or feven ftars, fometimes alfo called Vergiliæ. They were thus called, as being fuppofed by the poets to have been the daughters either of Atlas or his brother Hefperus, who were tranflated to heaven.

ATMOSPHERE, a term ufed to fignify the whole of the fluid mafs, confifting of air, aqueous and other vapours, electric fluids, &c, which furrounds the earth to a confiderable height, and partaking of all its motions, both annual and diurnal.

The compofition of that part of our atmofphere properly called air, was till lately but very little known. Formerly it was fuppofed to be a fimple, homogeneous, and elementary fluid. But the experiments of Dr. Prieftley and others, have difcovered, that even the pureft kind of air, which they call dephlogifticated, is in reality a compound, and might be artificially produced in various ways. This dephlogifticated air, however, is but a fmall part of the compofition of our atmofphere. By accurate experiments, the air we ufually breathe, is compofed of only one-fourth part of this dephlogifticated air, or perhaps lefs ; the other three parts, or more, confifting of what Dr. Prieftley calls *phlogifticated*, and M. Lavoifier *mephitic air*.

Befide

Beside these sorts of air, it is obvious that the whole mass of the atmosphere contains a great deal of water, together with a vast heterogeneous collection of particles raised from all bodies of matter on the surface of the earth, by effluvia, exhalations, &c; so that it may be considered as a chaos of the particles of all sorts of matter confusedly mingled together. And hence the atmosphere has been considered as a large chemical vessel, in which the matter of all kinds of sublunary bodies is copiously floating; and thus exposed to the continual action of that immense surface the sun; from whence proceed innumerable operations, sublimations, separations, compositions, digestions, fermentations, putrefactions, &c.

There is, however, one substance, namely the electrical fluid, which is very distinguishable in the mass of the atmosphere. To measure the absolute quantity of this fluid, either in the atmosphere or any other substance, is perhaps impossible: and all that we know on this subject is, that the electric fluid pervades the atmosphere; that it appears to be more abundant in the superior than the inferior regions; that it seems to be the immediate bond of connection between the atmosphere and the water which is suspended in it; and that by its various operations, the phenomena of hail, rain, snow, lightning, and the other kinds of meteors are occasioned. See those respective articles, and see also Beccaria's Essay on Atmospheric Electricity, annexed to the English translation of his Artificial Electricity.

Uses of the ATMOSPHERE.—The uses of the atmosphere are so many and great, that it seems indeed absolutely necessary, not only to the comfort and convenience of men, but even to the existence of all animal and vegetable life, and to the very constitution of all kinds of matter whatever, and without which they would not be what they are: for by it we live, breathe, and have our being; and by insinuating itself into all the vacuities of bodies, it becomes the great spring of most of the mutations here below; as generation, corruption, dissolution, &c; and without which none of these operations could be carried on. Without the atmosphere, no animal could exist, or indeed be produced; neither any plant, all vegetation ceasing without its aid; there would be neither rain nor dews to moisten the face of the ground; and though we might perceive the sun and stars like bright specks, we should be in utter darkness, having none of what we call day light or even twilight: nor would either fire or heat exist without it. In short, the nature and constitution of all matter would be changed and cease; wanting this universal bond and constituting principle.

By the mechanical force of the atmosphere too, as well as its chemical virtues, many necessary purposes are answered. We employ it as a moving power, in the motion of ships, to turn mills, and for other such uses. And it is one of the great discoveries of the modern philosophers, that the several motions attributed by the ancients to a *fuga vacui*, are really owing to the pressure of the atmosphere. Galileo, having observed that there was a certain standard altitude, beyond which no water could be elevated by pumping, took an occasion from thence to call in question the doctrine of the schools, which ascribed the ascent of water in pumps, to the

fuga vacui, and instead of it he happily substituted the hypothesis of the weight and pressure of the air. It was with him, indeed, little better than an hypothesis, since it had not then those confirmations from experiment, afterwards found out by his pupil Torricelli, and other succeeding philosophers, particularly Mr Boyle.

Nor have the attempts to fly or float through the air been altogether without success. F. De Lana thought he had contrived an aeronautic machine for navigating the atmosphere: and Sturmius, who examined it, asserted that it was not impracticable: though Dr Hook was of a different opinion, and detected the fallacy of the contrivance. Roger Bacon, long before, proposed something of the same kind. The great secret of this art is to contrive a machine so much lighter than the air, that it will rise up and float in the atmosphere, and together with itself, buoy up and carry men along with it. The principle on which it is to be effected is, by means of an air pump, to exhaust the air from a very thin and light, yet firm, metalline vessel. But the hopes of success in such an enterprize will appear very small if it be considered, that if a globe were formed of brass of the thickness only of $\frac{1}{72}$ of an inch, such a globe would require to be about 277 feet in diameter to float in the air: and if, as De Lana supposes, the diameter of the globe were but 25 feet, the thickness of the metal could not exceed $\frac{1}{13}$ of an inch. See Herman's Phoronomia pa. 158. However, what is not to be expected from metalline globes or shells, has now been successfully accomplished by the balloons of cloth, silk, or skin, of Montgolfier and others. See the article *Aerostation*, &c.

Salubrity of the ATMOSPHERE. On the tops of mountains the air is generally more salubrious than in pits or very deep places. Indeed dense air is always more proper for respiration, as to the mere quality of density only, than that which is rarer. But then the air on mountains, though rarer, is freer from phlogistic vapours than that of pits; and hence it has been found that people can live very well on the tops of mountains, even when the air is but about half the density of that below. But it would seem that at some intermediate height between the two extremes, the air is the most salubrious and proper for animal life; and this height, according to M. de Saussure, is about 500 or 600 yards above the level of the sea.

Besides the difference arising from the mere difference of altitude, the salubrity of the atmosphere is greatly affected by many other circumstances. The air, when confined or stagnant, is commonly more impure than when agitated and shifted: thus, all close places are unhealthy, and even the air in a bed chamber is less salubrious in a morning, after it has been slept in, than in the evening. Dr White, in vol. 68 Philos. Trans. gives an account of experiments on this quality of the air, and remarks one instance when the air was particularly impure, viz September 13, 1777; when the barometer stood at 30·30, the thermometer at 69°; the air being then dry and sultry, and no rain having fallen for more than two weeks. A slight shock of an earthquake was perceived that day. In vol. 70 of the same Transactions Dr Ingenhousz gives an account of some experiments on this head, made in various places and situations:

fituations: he finds," That the air at fea, and clofe to it, is in general purer, and fitter for animal life, than the air on the land:" but the Doctor did not find much difference between the air of the towns and of the country, nor between one town and another. The Abbé Fontana, made nearly the fame conclufions, from accurate experiments, afferting, " that the difference between the air of one country and that of another, at different times, is much lefs than what is commonly believed; and yet that this difference in the purity of the air at different times, is much greater than the difference between the air of the different places obferved by him." Finally M. Fontana concludes, that " Nature is not fo partial as we commonly believe. She has not only given us an air almoft equally good every where at every time, but has allowed us a certain latitude, or a power of living and being in health in qualities of air which differ to a certain degree. By this I do not mean to deny the exiftence of certain kinds of noxious air in fome particular places; but only fay, that in general the air is good every where, and that the fmall differences are not to be feared fo much as fome people would make us believe. Nor do I mean to fpeak here of thofe vapours and other bodies which are accidentally joined to the common air in particular places, but do not change its nature and intrinfical property. This ftate of the air cannot be known by the teft of nitrous air; and thofe vapours are to be confidered in the fame manner as we fhould confider fo many particles of arfenic fwimming in the atmofphere. In this cafe it is the arfenic, and not the degenerated air, that would kill the animals who ventured to breathe it."

Figure of the ATMOSPHERE.—As the atmofphere envelops all parts of the furface of our globe, if they both continued at reft, and were not endowed with a diurnal motion about their common axis, then the atmofphere would be exactly globular, according to the laws of gravity; for all the parts of the furface of a fluid in a ftate of reft, muft be equally removed from its centre. But as the earth and the ambient parts of the atmofphere revolve uniformly together about their axis, the different parts of both have a centrifugal force, the tendency of which is more confiderable, and that of the centripetal lefs, as the parts are more remote from the axis; and hence the figure of the atmofphere muft become an oblate fpheroid; fince the parts that correfpond to the equator are father removed from the axis, than the parts which correfpond to the poles. Befides, the figure of the atmofphere muft, on another account, reprefent a flattened fpheroid, namely becaufe the fun ftrikes more directly the air which encompaffes the equator, and is comprehended between the two tropics, than that which pertains to the polar regions: for, from hence it follows, that the mafs of air, or part of the atmofphere, adjoining to the poles, being lefs heated, cannot expand fo much, nor reach fo high. And yet, notwithftanding, as the fame force which contributes to elevate the air, diminifhes its gravity and preffure on the furface of the earth, higher columns of it about the equatorial parts, all other circumftances being the fame, may not be heavier than thofe about the poles.

In the Tranfactions of the Royal Irifh Academy for 1788 Mr Kirwin has an ingenious differtation on the figure, height, weight, &c, of the atmofphere. He obferves that, in the natural ftate of the atmofphere, that is, when the barometer would every where, at the level of the fea, ftand at 30 inches, the weight of the atmofphere, at the furface of the fea, muft be equal all over the globe; and in order to produce this equality, as the weight proceeds from its denfity and height, it muft be loweft where the denfity is greateft, and higheft where the denfity is leaft; that is, higheft at the equator and loweft at the poles, with feveral intermediate gradations.

Though the equatorial air however be lefs denfe to a certain height than the polar, yet at fome greater heights it muft be more denfe: for fince an equatorial and polar column are equal in total weight or mafs, the lower part of the equatorial column, being more expanded by heat &c than that of the polar, muft have lefs mafs, and therefore a proportionally greater part of its mafs muft be found in its fuperior fection; fo that the lower extremity of the fuperior fection of the equatorial column is more compreffed, and confequently denfer, than the correfponding part of the polar column. The fame thing is to be underftood alfo of the extra-tropical columns with refpect to each other, where differences of heat prevail.

Hence, in the higheft regions of the atmofphere, the denfer equatorial air, not being fupported by the collateral extra-tropical columns, gradually flows over, and rolls down to the north and fouth.

Thefe fuperior tides confift chiefly of inflammable air, as it is much lighter than any other, and is generated in great plenty between the tropics; it furnifhes the matter of the auroræ borealis and auftralis, by whofe combuftion it is deftroyed, elfe its quantity would in time become too great, and the weight of the atmofphere annually increafed; but its combuftion is the primary fource of the greateft perturbations of the atmofphere.

Weight or Preffure of the ATMOSPHERE.—It is evident that the mafs of the atmofphere, in common with all other matter, muft be endowed with weight and preffure; and this principle was afferted by almoft all philofophers, both ancient and modern. But it was only by means of the experiments made with pumps and the barometrical tube, by Galileo and Torricelli, that we came to the proof, not only that the atmofphere is endued with a preffure, but alfo what the meafure and quantity of that preffure is. Thus, it is found that the preffure of the atmofphere fuftains a column of quickfilver, in the tube of the barometer, of about 30 inches in height; it therefore follows, that the whole preffure of the atmofphere is equal to the weight of a column of quickfilver, of an equal bafe, and 30 inches height: and becaufe a cubical inch of quickfilver is found to weigh nearly half a pound averdupois, therefore the whole 30 inches, or the weight of the atmofphere on every fquare inch of furface, is equal to 15 pounds. Again, it has been found that the preffure of the atmofphere balances, in the cafe of pumps &c, a column of water of about $34\frac{1}{2}$ feet high; and, the cubical foot of water weighing juft 1000 ounces, or $62\frac{1}{2}$ pounds, $34\frac{1}{2}$ times $62\frac{1}{2}$, or 2158lb, will be the weight of the column of water, or of the atmofphere on a bafe of a fquare foot; and confequently the 144th part of this, or 15lb, is the weight of the atmofphere on a fquare inch;

inch; the same as before. Hence Mr Cotes computed that the pressure of this ambient fluid on the whole surface of the earth, is equivalent to that of a globe of lead of 60 miles in diameter. And hence also it appears, that the pressure upon the human body must be very considerable; for as every square inch of surface sustains a pressure of 15 pounds, every square foot will sustain 144 times as much, or 2160 pounds; then, if the whole surface of a man's body be supposed to contain 15 square feet, which is pretty near the truth, he must sustain 15 times 2160, or 32400 pounds, that is nearly $14\frac{1}{2}$ tons weight, for his ordinary load. By this enormous pressure we should undoubtedly be crushed in a moment, if all parts of our bodies were not filled either with air or some other elastic fluid, the spring of which is just sufficient to counterbalance the weight of the atmosphere. But whatever this fluid may be, it is certain that it is just able to counteract the weight of the atmosphere, and no more: for, if any considerable pressure be superadded to that of the air, as by going into deep water, or the like, it is always severely felt let it be ever so equable, at least when the change is made suddenly; and if, on the other hand, the pressure of the atmosphere be taken off from any part of the human body, as the hand for instance, when put over an open receiver, from whence the air is afterwards extracted, the weight of the external atmosphere then prevails, and we imagine the hand strongly sucked down into the glass.

The difference in the weight of the air which our bodies sustain at one time more than another, is also very considerable, from the natural changes in the state of the atmosphere. This change takes place chiefly in countries at some distance from the equator; and as the barometer varies at times from 28 to 31 inches, or about one tenth of the whole quantity, it follows that this difference amounts to about a ton and a half on the whole body of a man, which he therefore sustains at one time more than at another. On the increase of this natural weight, the weather is commonly fine, and we feel ourselves what we call braced and more alert and active; but, on the contrary, when the weight of the air diminishes, the weather is bad, and people feel a listlessness and inactivity about them. And hence it is no wonder that persons suffer very much in their health, from such changes in the atmosphere, especially when they take place very suddenly, for it is to this circumstance chiefly that a sensation of uneasiness and indisposition is to be attributed; thus, when the variations of the barometer and atmosphere are sudden and great, we feel the alteration and effect on our bodies and spirits very much; but when the change takes place by very slow degrees, and by a long continuance, we are scarcely sensible of it, owing, undoubtedly, to the power with which the body is naturally endowed, of accommodating itself to this change in the state of the air, as well as to the change of many other circumstances of life, the body requiring a certain interval of time to effect the alteration in its state, proper to that of the air &c. Thus, in going up to the tops of mountains, where the pressure of the atmosphere is diminished two or three times more than on the plain below, little or no inconvenience is felt from the rarity of the air, if it is not mixed with other noxious vapours &c; because

that, in the ascent the body has had sufficient time to accommodate itself gradually to the slow variation in the state of the atmosphere: but, when a person ascends with a balloon, very rapidly to a great height in the atmosphere, he feels a difficulty in breathing and an uneasiness of body; and the same is soon felt by an animal when inclosed in a receiver, and the air suddenly drawn or pumped out of it. So also, on the condensation of the air, we feel little or no alteration in ourselves, except when the change happens suddenly, as in very rapid changes in the weather, and in descending to great depths in a diving bell, &c. I have often heard the late unfortunate Mr. Spalding speak of his experience on this point: he always found it absolutely necessary to descend with the bell very slowly, and that only from one depth to another, resting a while at each depth before he began to descend farther: he first descended slowly for about 5 or 6 fathom, and then stopped a while; he felt an uneasiness in his head and ears, which increased more and more as he descended, till he was obliged to stop at the depth above mentioned, where the density of the air was nearly doubled; having remained there a while, he felt his ears give a sudden crack, and after that he was soon relieved from any uneasiness in that part, and it seemed as if the density of the air was not altered. He then descended other 5 fathoms or 30 feet more, with the same precaution and the same sensations as before, being again relieved, in the same manner, after remaining awhile stationary at the next stage of his descent, where the density of the air was tripled. And thus he continued proceeding to a great depth, always with the same circumstances, repeated at every 5 or 6 fathoms, and adding the pressure of one more atmosphere at every period of the progress.

It is not easy to assign the true reason for the variations that happen in the gravity of the atmosphere in the same place. One cause of it however, either immediate or otherwise, it seems, is the heat of the sun; for where this is uniform, the changes are small and regular; thus between the tropics it seems the change depends on the heat of the sun, as the barometer constantly sinks about half an inch every day, and rises again to its former station in the night time. But in the temperate zones the barometer ranges from 28 to near 31 inches, shewing, by its various altitudes, the changes that are about to take place in the weather. If we could know therefore, the causes by which the weather is influenced; we should also know those by which the gravity of the atmosphere is affected. These may perhaps be reduced to immediate ones, viz, an emission of latent heat from the vapour contained in the atmosphere, or of electric fluid from the same, or from the earth; as it is observed that they both produce the same effect with the solar heat in the tropical climates, viz, to rarefy the air, by mixing with it, or setting loose a lighter fluid, which did not before act in such large proportion in any particular place.

With regard to the alteration of heat and cold in the atmosphere, many reasons and hypotheses have been given, and many experiments made; as may be seen by consulting the authors upon this subject, viz, M. Bouguer's observations in Peru, Lambert, De Luc, Saussure's journeys on the Alps, Sex's and Darwin's experiments

experiments in vol. 78 Philof. Tranf. This laft gentleman hence infers, " There is good reafon to conclude that in all circumftances where air is mechanically expanded, it becomes capable of attracting the fluid matter of heat from other bodies in contact with it. Now, as the vaft region of air which furrounds our globe is perpetually moving along its furface, climbing up the fides of mountains, and defcending into the valleys ; as it paffes along it muft be perpetually varying the degree of heat according to the elevation of the country it traverfes : for, in rifing to the fummits of mountains, it becomes expanded, having fo much of the preffure of the fuperincumbent atmofphere taken away ; and when thus expanded, it attracts or abforbs heat from the mountains in contiguity with it ; and, when it defcends into the valleys and is compreffed into lefs compafs, it again gives out the heat it has acquired to the bodies it comes in contact with. The fame thing muft happen to the higher regions of the atmofphere, which are regions of perpetual froft, as has lately been difcovered by the aerial navigators. When large diftricts of air, from the lower parts of the atmofphere, are raifed two or three miles high, they become fo much expanded by the great diminution of the preffure over them, and thence become fo cold, that hail or fnow is produced by the precipitation of the vapour : and as there is, in thefe high regions of the atmofphere, nothing elfe for the expanded air to acquire heat from after it has parted with its vapour, the fame degree of cold continues till the air, on defcending to the earth, acquires its former ftate of condenfation and of warmth. The Andes, almoft under the line, refts its bafe on burning fands : about its middle height is a moft pleafant and temperate climate covering an extenfive plain, on which is built the city of Quito ; while its forehead is encircled with eternal fnow, perhaps coeval with the mountain. Yet, according to the accounts of Don Ulloa, thefe three difcordant climates feldom encroach much on each other's territories. The hot winds below, if they afcend, become cooled by their expanfion ; and hence they cannot affect the fnow upon the fummit ; and the cold winds that fweep the fummit, become condenfed as they defcend, and of temperate warmth before they reach the fertile plains of Quito."

Height and Denfity of the ATMOSPHERE. Various attempts have been made to afcertain the height to which the atmofphere is extended all round the earth. Thefe commenced foon after it was difcovered by means of the Torricellian tube, that air is endued with weight and preffure, And had not the air an elaftic power, but were it every where of the fame denfity, from the furface of the earth to the extreme limit of the atmofphere, like water, which is equally denfe at all depths it would be a very eafy matter to determine its height from its denfity and the column of mercury which it would counterbalance in the barometer tube : for, it having been obferved that the weight of the atmofphere is equivalent to a column of 30 inches or $2\frac{1}{2}$ feet of quickfilver, and the denfity of the former to that of the latter, as 1 to 11040 ; therefore the height of the uniform atmofphere would be 11040 times $2\frac{1}{2}$ feet, that is 27600 feet, or little more than 5 miles and a quarter. But the air, by its elaftic quality, expands and contracts ; and it being found by repeated experiments in

6

moft nations of Europe, that the fpaces it occupies, when compreffed by different weights, are reciprocally -proportional to thofe weights themfelves ; or, that the more the air is preffed, fo much the lefs fpace it takes up ; it follows that the air in the upper regions of the atmofphere muft grow continually more and more rare, as it afcends higher ; and indeed that, according to that law, it muft neceffarily be extended to an indefinite height. Now, if we fuppofe the height of the whole divided into innumerable equal parts ; the quantity of each part will be as its denfity ; and the weight of the whole incumbent atmofphere being alfo as its denfity ; it follows, that the weight of the incumbent air, is every where as the quantity contained in the fubjacent part ; which caufes a difference between the weights of each two contiguous parts of air. But, by a theorem in arithmetic, when a magnitude is continually diminifhed by the like part of itfelf, and the remainders the fame, thefe will be a feries of continued quantities decreafing in geometrical progreffion : therefore if, according to the fuppofition, the altitude of the air, by the addition of new parts into which it is divided, do continually increafe in arithmetical progreffion, its denfity will be diminifhed, or, which is the fame thing, its gravity decreafed, in continued geometrical proportion. And hence, again, it appears that, according to the hypothefis of the denfity being always proportional to the compreffing force, the height of the atmofphere muft neceffarily be extended indefinitely. And, farther, as an arithmetical feries adapted to a geometrical one, is analogous to the logarithms of the faid geometrical one ; it follows therefore that the altitudes are proportional to the logarithms of the denfities, or weights of air ; and that any height taken from the earth's furface, which is the difference of two altitudes to the top of the atmofphere, is proportional to the difference of the logarithms of the two denfities there, or to the logarithm of the ratio of thofe denfities, or their correfponding compreffing forces, as meafured by the two heights of the barometer there. This law was firft obferved and demonftrated by Dr. Halley, from the nature of the hyperbola ; and afterwards by Dr. Gregory, by means of the logarithmic curve. See Philof. Tranf. Nº. 181, or Abridg. vol. 2, p. 13, and Greg. Aftron. lib. v, prop. 3.

It is now eafy, from the foregoing property, and two or three experiments, or barometrical obfervations, made at known altitudes, to deduce a general rule to determine the abfolute height anfwering to any denfity ; or the denfity anfwering to any given altitude above the earth. And accordingly, calculations were made upon this plan by many philofophers, particularly by the French ; but it having been found that the barometrical obfervations did not correfpond with the altitudes as meafured in a geometrical manner, it was fufpected that the upper parts of the atmofpherical regions were not fubject to the fame laws with the lower ones, in regard to the denfity and elafticity. And indeed, when it is confidered that the atmofphere is a heterogeneous mafs of particles of all forts of matter, fome elaftic, and others not, it is not improbable but this may be the cafe, at leaft in the regions very high in the atmofphere, which it is likely may more copioufly abound with the electrical fluid. Be this however as it may, it has lately been difcovered that the law

above

above given, holds very well for all such altitudes as are within our reach, or as far as to the tops of the highest mountains on the earth, when a correction is made for the difference of the heat or temperature of the air only, as was fully evinced by M. De Luc, in a long series of observations, in which he determined the altitudes of hills both by the barometer, and by geometrical measurement, from which he deduced a practical rule to allow for the difference of temperature. See his Treatise on the Modifications of the Atmosphere. Similar rules have also been deduced from accurate experiments, by Sir George Shuckburgh and General Roy, both concurring to shew, that such a rule for the altitudes and densities, holds true for all heights that are accessible to us, when the elasticity of the air is corrected on account of its density: and the result of their experiments shewed, that the difference of the logarithms of the heights of the mercury in the barometer, at two stations, when multiplied by 10000, is equal to the altitude in English fathoms, of the one place above the other; that is, when the temperature of the air is about 31 or 32 degrees of Fahrenheit's thermometer; and a certain quantity more or less, according as the actual temperature is different from that degree.

But it may here be shewn, that the same rule may be deduced independent of such a train of experiments as those above, merely by the density of the air at the surface of the earth alone. Thus, let D denote the density of the air at one place, and d the density at the other; both measured by the column of mercury in the barometrical tube: then the difference of altitude between the two places, will be proportional to the log. of D — the log. of d, or to the log. of $\frac{D}{d}$. But as this formula expresses only the relation between different altitudes, and not the absolute quantity of them, assume some indeterminate, but constant quantity h, which multiplying the expression log. $\frac{D}{d}$, may be equal to the real difference of altitude a, that is, $a = h \times$ log. of $\frac{D}{d}$. Then, to determine the value of the general quantity h, let us take a case in which we know the altitude a which corresponds to a known density d; as for instance, taking $a = 1$ foot, or 1 inch, or some such small altitude: then because the density D may be measured by the pressure of the whole atmosphere, or the uniform column of 27600 feet, when the temperature is 55°; therefore 27600 feet will denote the density D at the lower place, and 27599 the less density d at 1 foot above it; consequently $1 = h \times$ log. of $\frac{27600}{27599}$, which, by the nature of logarithms, is nearly

$= h \times \frac{\cdot 43429448}{27600}$ or $\frac{1}{63551}$ nearly; and hence we find $h = 63551$ feet; which gives us this formula for any altitude a in general, viz, $a = 63551 \times$ log. of $\frac{D}{d}$,

or $a = 63551 \times$ log. of $\frac{M}{m}$ feet, or $10592 \times$ log. of $\frac{M}{m}$ fathoms; where M denotes the column of mercury

in the tube at the lower place, and m that at the upper. This formula is adapted to the mean temperature of the air 55°: but it has been found, by the experiments of Sir Geo. Shuckburgh and General Roy, that for every degree of the thermometer, different from 55°, the altitude a will vary by its 435th part; hence, if we would change the factor h from 10592 to 10000, because the difference 592 is the 18th part of the whole factor 10592, and because 18 is the 24th part of 435; therefore the change of temperature, answering to the change of the factor h, is 24°, which reduces the 55° to 31°. So that, $a = 10000 \times$ log. of $\frac{M}{m}$ fathoms, is the easiest expression for the altitude, and answers to the temperature of 31°, or very nearly the freezing point: and for every degree above that, the result must be increased by so many times its 435th part, and diminished when below it.

From this theorem it follows, that, at the height of $3\frac{1}{2}$ miles, the density of the atmosphere is nearly 2 times rarer than it is at the surface of the earth; at the height of 7 miles, 4 times rarer; and so on, according to the following table:

Height in miles.	Number of times rarer.
$3\frac{1}{2}$	2
7	4
14	16
21	64
28	256
35	1024
42	4096
49	16384
56	65536
63	262144
70	1048576

And, by pursuing the calculations in this table, it might be easily shewn, that a cubic inch of the air we breathe would be so much rarefied at the height of 500 miles, that it would fill a sphere equal in diameter to the orbit of Saturn.

Hence we may perceive how very soon the air becomes so extremely rare and light, as to be utterly imperceptible to all experience; and that hence, if all the planets have such atmospheres as our earth, they will, at the distances of the planets from one another, be so extremely attenuated, as to give no sensible resistance to the planets in their motion round the sun for many, perhaps hundreds or thousands of ages to come. Even at the height of about 50 miles, it is so rare as to have no sensible effect on the rays of light: for it was found by Kepler, and De la Hire after him, who computed the height of the sensible atmosphere from the duration of twilight, and from the magnitude of the terrestrial shadow in lunar eclipses, that the effect of the atmosphere to reflect and intercept the light of the sun, is only sensible to the altitude of between 40 and 50 miles: and at that altitude we may collect, from what has been already said, that the air is above 10000 times rarer than at the surface of the earth. It is well known that the twilight begins and ends when the centre of the sun is about 18 degrees below the horizon, or only 17° 27′, by subtracting 33′ for refraction,

refraction, which raifes the fun fo much higher than he would be. And a ray coming from the fun in that pofition, and entering the earth's atmofphere, is refracted and bent into a curve line in paffing through it to the eye. M. de la Hire took great pains to demonftrate, that, fuppofing the denfity of the atmofphere proportional to its weight, this curve is a cycloid: he alfo fays, that if the ray be a tangent to the atmofphere, the diameter of its generating circle will be the height of the atmofphere; and that this diameter increafes, till at laft, when the rays are perpendicular, it becomes infinite, or the circle degenerates into a right line. This reafoning fuppofes that the refracting furface of the atmofphere is a plane; but fince it is in reality a curve, he obferves that thefe cycloids become in fact epicycloids. But Herman detected the error of M. de la Hire, and fhewed that this curve is infinitely extended, and has an afymptote. And it is obferved by Dr. Brook Taylor, in his Methodus Increm. pa. 168, &c, that this curve is one of the moft intricate and perplexed that can well be propofed. The fame ingenious author computes, that the refractive power of the air is to the force of gravity at the furface of the earth, as 320 millions to 1.

Confidering the extreme rarity of the atmofphere at only 40 or 50 miles in height, it feems to be furprizing that fome meteors fhould be enflamed at fuch great heights as they have been obferved at. A very remarkable one of this kind was obferved by Dr. Halley in the month of March 1719, the altitude of which he computed at between 69 and 73½ Englifh miles; its diameter 2800 yards, or more than a mile and a half; and its velocity about 350 miles in a minute. Others, apparently of the fame kind, but whofe altitude and velocity were ftill greater, have been obferved; particularly that very remarkable one, of Auguft 18th 1783, whofe diftance from the earth could not be lefs than 90 miles, its diameter at leaft as large as the former, while its velocity was certainly not lefs than 1000 miles in a minute. Now, from analogy of reafoning, it feems very probable, that the meteors which appear at fuch great heights in the air, are not effentially different from thofe which are feen on or near the furface of the earth. The difficulty with regard to the former is, that at the great heights above-mentioned, the atmofphere ought not to have any denfity fufficient to fupport flame, or to propagate found; and yet fuch meteors are commonly fucceeded by one explofion or more, and it is faid are even fometimes accompanied with a hiffing noife as they pafs over our heads. The meteor of 1719 was not only very bright, feeming for a fhort time to turn night into day, but was attended with an explofion heard over all the ifland of Britain, caufing a violent concuffion in the atmofphere, and feeming to fhake the earth itfelf: And yet, in the regions in which this meteor moved, the air ought to have been 300 thoufand times rarer than the air we breathe, or 1000 times rarer than the vacuum commonly made by a good air-pump. Dr. Halley offers a conjecture, indeed, that the vaft magnitude of fuch bodies might compenfate for the thinnefs of the medium in which they moved. But

appearances of this kind are, by fome others, attributed to electricity; though the circumftances of them cannot be reconciled to that caufe; for the meteors move with all different degrees of velocity; and though the electrical fire eafily pervades the vacuum of an air-pump, yet it does not in that cafe appear in bright well defined fparks, as in the open air, but rather in long ftreams refembling the aurora borealis; and from fome late experiments it has been concluded that the electric fluid cannot even penetrate a perfect vacuum.

Of the Refractive and Reflective Power of the ATMOSPHERE. It has been obferved above, that the atmofphere has a refractive power, by which the rays of light are bent from the right lined direction, as in the cafe of the twilight; and many other experiments manifeft the fame virtue, which is the caufe of many phenomena. Alhazen, the Arabian, who lived about the year 1100, it feems was more inquifitive into the nature of refraction than former writers. But neither Alhazen, nor his follower Vitello, knew any thing of its juft quantity, which was not known, to any tolerable degree of exactnefs, till Tycho Brahe, with great diligence, fettled it. But neither did Tycho nor Kepler difcover in what manner the rays of light were refracted by the atmofphere. Tycho thought the refraction was chiefly caufed by denfe vapours, very near the earth's furface: while Kepler placed the caufe wholly at the top of the atmofphere, which he thought was uniformly denfe; and thence he determined its altitude to be little more than that of the higheft mountains. But the true conftitution of the denfity of the atmofphere, deduced afterwards from the Torricellian experiment, afforded a jufter idea of thefe refractions, efpecially after it appeared, by a repetition of Mr. Lowthorp's experiment, that the refractive power of the air is proportional to its denfity. By this variation in the denfity and refractive power of the air, a ray of light, in paffing through the atmofphere, is continually refracted at every point, and thereby made to defcribe a curve, and not a ftraight line, as it would have done were there no atmofphere, or were its denfity uniform.

The atmofphere, or air, has alfo a reflective power; and this power is the means by which objects are enlightened fo uniformly on all fides. The want of this power would occafion a ftrange alteration in the appearance of things; the fhadows of which would be fo very dark, and their fides enlightened by the fun fo very bright, that probably we could fee no more of them than their bright halves; fo that for a view of the other halves, we muft turn them half round, or if immoveable, muft wait till the fun could come round upon them. Such a pellucid unreflective atmofphere would indeed have been very commodious for aftronomical obfervations on the courfe of the fun and planets among the fixed ftars, vifible by day as well as by night; but then fuch a fudden tranfition from darknefs to light, and from light to darknefs, immediately upon the rifing and fetting of the fun, without any twilight, and even upon turning to or from the fun at noon day, would have been very inconvenient and offenfive to our eyes. However, though the atmofphere be greatly affiftant in the illumination of objects, yet it muft alfo be obferved that it ftops a great deal of light. By M. Bouguer's experiments, it feems that the light of the moon is often 2000 times weaker in the horizon, than at the altitude of 66 degrees; and that the proportion of her light at the altitudes of 66 and 19 degrees, is about 3 to 2; and the lights of the fun muft bear the fame proportion to each other at thofe heights; which Bouguer made choice of, as being the meridian heights of the fun, at the fummer and winter folftices, in the latitude of Croific in France. Smith's optics. Rem. 95.

For the Atmofphere of the fun, moon, and planets, fee the refpective articles.

ATMOSPHERE *of Solid or Confiftent Bodies*, is a kind of fphere formed by the effluvia, or minute corpufcles, emitted from them. Mr. Boyle endeavours to fhew, that all bodies, even the hardeft and moft coherent, as gems, &c, have their atmofpheres.

ATMOSPHERE, *in Electricity*, denotes that medium which is conceived to be diffufed over the furface of electrified bodies, and to fome diftance around them, and confifting of effluvia iffuing from them; by which, other bodies immerged in it become endued with an electricity contrary to that of the body to which the atmofphere belongs. This was firft noticed at a very early period in the hiftory of this fcience by Otto Guericke, and afterwards by the academicians *del Cimento*, who contrived to render the electric atmofphere vifible, by means of fmoke attracted by a piece of amber, and gently rifing from it, but vanifhing as the amber cooled. Dr. Franklin exhibited this electric atmofphere with greater advantage; by dropping rofin on hot iron plates held under electrified bodies, from which the fmoke arofe and encompaffed the bodies, giving them a very beautiful appearance. But the theory of electric atmofpheres was not well explained and underftood for a confiderable time; and the inveftigation led to many curious experiments and obfervations. The experiments of Mr. Canton and Dr. Franklin prepared the way for the conclufion that was afterwards drawn from them by Meff. Wilcke and Epinus, though they retained the common opinion of electric atmofpheres, and endeavoured to explain the phenomena by it. The conclufion was, that the electric fluid, when there is a redundancy of it in any body, repels the electric fluid in any other body, when they are brought within the fphere of each other's influence, and drives it into the remote parts of the body, or quite out of it, if there be any outlet for that purpofe.

By Atmofphere, M. Epinus fays, no more is to be underftood than the fphere of action belonging to any body, or the neighbouring air electrified by it. Sig. Beccaria agrees in the fame opinion, that electrified bodies have no other atmofphere than the electricity communicated to the neighbouring air, and which goes with the air, and not with the electrified bodies. Mr. Canton alfo, having relinquifhed the opinion that electrical atmofpheres were compofed of effluvia from excited or electrified bodies, maintained that they only refult from an alteration in the ftate of the electric fluid contained in it, or belonging to the air furrounding thefe bodies to a certain diftance; for inftance, that

excited glafs repels the electric fluid from it, and confequently beyond that diftance makes it more denfe; whereas excited wax attracts the electric fluid exifting in the air nearer to it, making it rarer than it was before. In the courfe of experiments that were performed on this occafion, Meff. Wilcke and Epinus fucceeded in charging a plate of air, by fufpending large boards of wood covered with tin, with the flat fides parallel to one another, and at fome inches afunder: for they found, upon electrifying one of the boards pofitively, that the other was always negative; and a fhock was produced by forming a communication between the upper and lower plates. Beccaria has largely confidered the fubject of electric atmofpheres, in his Artificial Electricity, pa. 179 &c, Eng. edit. See alfo Dr Prieftley's Hift. of Electricity, vol. ii. fect. 5. and Cavallo's Electricity, pa. 241.

ATMOSPHERE, *Magnetic, &c*, is underftood of the fphere within which the virtue of the magnet, &c, acts.

ATOM, a particle of matter indivifible on account of its folidity, hardnefs, and impenetrability; which preclude all divifion, and leave no vacancy for the admiffion of any foreign force to feparate or difunite its parts. As atoms are the firft matter, it is neceffary they fhould be indiffolvable, that they may be incorruptible. Newton adds, it is alfo required that they be immutable, that the world may continue in the fame ftate, and bodies be of the fame nature now as formerly.

ATOMICAL PHILOSOPHY, or the doctrine of atoms, a fyftem which accounted for the origin and formation of things, from the hypothefis that atoms are endued with weight and motion. This philofophy was firft taught by Mofchus, fome time before the Trojan war: but it was moft cultivated by Epicurus; whence it is called the Epicurean philofophy.

ATTRACTION, or ATTRACTIVE POWER, a general term ufed to denote the caufe, power, or principle, by which all bodies mutually tend towards each other, and cohere, till feparated by fome other power. The laws, phenomena, &c, of attraction, form the chief fubject of Newton's philofophy, being the principal agent of nature, in almoft all her wonderful operations.

The principle of attraction, in the Newtonian fenfe of it, it feems was firft furmifed by Copernicus. "As for gravity, fays he, I confider it as nothing more than a certain natural appetence *(appetentia)* that the Creator has impreffed upon all the parts of matter, in order to their uniting or coalefcing into a globular form, for their better prefervation; and it is probable that the fame power is alfo inherent in the fun and moon, and planets, that thofe bodies may conftantly retain that round form in which we fee them. *De Revol. Orb. Cæleft.* lib. i, cap. 9. Kepler calls gravity a corporeal and mutual affection between fimilar bodies, in order to their union. *Aft. Nov. in Introd.* And he pronounced more pofitively that no bodies whatever were abfolutely light, but only relatively fo; and confequently that all matter was fubjected to the power and law of gravitation. *Ibid.*

The firft in this country who adopted the notion of attraction, was Dr. Gilbert, in his book *De Magnete*; and the next was the celebrated Lord Bacon, in his *Nov. Organ.* lib. ii, aphor. 36, 45, 48. *Sylv.* cent. i,
expᵃ

exp. 33; also in his treatise *De Motu*, particularly under the articles of the 9th and the 13th sorts of Motion. In France it was received by Fermat and Roberval; and in Italy by Galileo and Borelli. But till Newton appeared, this principle was very imperfectly defined and applied.

It must be observed, that though this great author makes use of the word attraction, in common with the school philosophers, yet he very studiously distinguishes between the ideas. The ancient attraction was conceived to be a kind of quality inherent in certain bodies themselves, and arising from their particular or specific forms. But the Newtonian attraction is a more indefinite principle; denoting not any particular kind or mode of action; nor the physical cause of such action; but only a general tendency, a *conatus accedendi*, to whatever cause, physical or metaphysical, such effect be owing; whether to a power inherent in the bodies themselves, or to the impulse of an external agent. Accordingly, that author remarks, in his *Philos. Nat. Prin. Math.* "that he uses the words *attraction, impulse*, and *propension* to the centre, indifferently; and cautions the reader not to imagine that by attraction he expresses the modus of the action, or its efficient cause, as if there were any proper powers in the centres, which in reality are only mathematical points; or as if centres could attract." Lib. 1, pa. 5. So, he "considers centripetal powers as attractions, though, physically speaking, it were perhaps more just to call them impulses. Ib. pa. 147. He adds, " that what he calls attraction may possibly be effected by impulse, though not a common or corporeal impulse, or after some other manner unknown to us. " *Optic.* p. 322.

Attraction, if considered as a quality arising from the specific forms of bodies, ought, together with sympathy, antipathy, and the whole tribe of occult qualities, to be exploded. But when these are set aside, there will remain innumerable phenomena of nature, and particularly the gravity or weight of bodies, or their tendency to a centre, that argue a principle of action seemingly distinct from impulse; where, at least, there is no sensible impulsion concerned. Nay, what is more, this action, in some respects, differs from all impulsion we know of; impulse being always found to act in proportion to the surfaces of bodies; whereas gravity acts according to their solid content, and consequently it must arise from some cause that penetrates or pervades the whole substance of it. This unknown principle, unknown we mean in respect of its cause, for its phenomena and effects are most obvious, with all its species and modifications, is called attraction; being a general name, under which may be ranged all mutual tendencies, where no physical impulse appears, and which consequently cannot be accounted for upon any known laws of nature.

And hence arise divers particular kinds of attraction; as *Gravity, Magnetism, Electricity, &c*, which are so many different principles, acting by different laws; and only agreeing in this, that we do not perceive any physical causes of them: but that, as to our senses, they may really arise from some power or efficacy in such bodies, by which they are enabled to act even upon distant bodies; though our reason absolutely disallows of any such action.

Attraction may be divided, with respect to the law it observes, into two kinds.

1. That which extends to a sensible distance. As the attraction of gravity, which is found in all bodies; and the attractions of magnetism and electricity, found only in particular bodies. The several laws and phenomena of each, see under their respective articles.

The attraction of gravity, called also among mathematicians the *centripetal force*, is one of the greatest and most universal principles of all nature. We see and feel it operate on bodies near the earth, and find by observation that the same power (i. e. a power which acts in the same manner, and by the same rules, viz, always proportionally to the quantities of matter, and inversely as the squares of the distances) does also obtain in the moon, and the other planets, both primary and secondary, as well as in the comets; and even that this is the very power by which they are all retained in their orbits, &c. And hence, as gravity is found in all the bodies which come under our observation, it is easily inferred, by one of the established rules of philosophizing, that it obtains in all others. And since it is found to be proportional to the quantity of matter in any body, it must exist in every particle of it: and hence it is proved that every particle in nature attracts every other particle.

From this attraction arises all the motion, and consequently all the mutation, in the great world. By this heavy bodies descend, and light ones are made to ascend: by this projectiles are directed, vapours and exhalations rise, and rains &c fall: by this rivers glide, the ocean swells, the air presses, &c. In short, the motions and forces arising from this principle, constitute the subject of that extensive branch of mathematics, called *mechanics* or *statics*, with the parts or appendages of it, as hydrostatics, pneumatics, hydraulics, &c.

2. That which does not extend to sensible distances. Such is found to obtain in the minute particles of which bodies are composed, attracting each other at or extremely near the point of contact, with forces often much superior to that of gravity, but which at any distance decrease much faster than the power of gravity. This power a late ingenious author calls the *attraction of cohesion*, as being that by which the atoms or insensible particles of bodies are united into sensible masses.

This kind of attraction owns Sir Isaac Newton for its discoverer; as the former does for its improver. The laws of motion, percussion, &c, in sensible bodies, under various circumstances, as falling, projected, &c, ascertained by the later philosophers, do not reach those more recluse, intestine motions in the component particles of the same bodies, on which depend the changes in the texture, colour, properties, &c, of bodies. So that our philosophy, if it were only founded on the principle of gravitation, and even carried as far as this would lead us, would still be very deficient.

But besides the common laws of sensible masses, the minute parts they are composed of are found subject to some others, which have but lately been noticed, and are even yet imperfectly known. Newton himself, to whose happy penetration we owe the hint, limits himself with establishing that there are such motions in the

minima

minima naturæ, and that they flow from certain powers or forces, not reducible to any of those in the great world. He shews that, by virtue of these powers, " the small particles act on one another even at a distance; and that many of the phenomena of nature result from it. Sensible bodies, we have already observed, act on one another divers ways; and as we thus perceive the tenor and course of nature, it appears highly probable that there may be other powers of the like kind; nature being very uniform and consistent with herself. Those just mentioned, reach to sensible distances, and so have been observed by vulgar eyes; but there may be others which reach to such small distances as have hitherto escaped observation; and it is probable electricity may reach to such distances, even without being excited by friction."

The great author just mentioned proceeds to confirm the reality of these suspicions from a great number of phenomena and experiments, which plainly argue such powers and actions between the particles, for example of salts and water, oil of vitriol and water, aquafortis and iron, spirit of vitriol and saltpetre. He also shews, that these powers, &c, are unequally strong between different bodies; stronger, for instance, between the particles of salt of tartar and those of aquafortis than those of silver, between aquafortis and lapis calaminaris than iron, between iron than copper, and copper than silver or mercury. So spirit of vitriol acts on water, but more on iron or copper, &c. And the other experiments are innumerable which countenance the existence of such principle of attraction in the particles of matter.

These actions, by virtue of which the particles of the bodies above-mentioned tend towards each other, the author calls by a general indefinite name, *attraction*; a name equally applicable to all actions by which bodies tend towards one another, whether by impulse, or by any other more latent power: and from hence he accounts for an infinity of phenomena, otherwise inexplicable, to which the principle of gravity is inadequate.

Thus, adds our author, " will nature be found very conformable to herself, and very simple; performing all the great motions of the heavenly bodies by the attraction of gravity, which intercedes those bodies, and almost all the small ones of their parts, by some other attractive power diffused through their particles. Without such principles, there never would have been any motion in the world; and without the continuance of it, motion would soon perish, there being otherwise a great decrease or diminution of it, which is only supplied by these active principles."

It need not be said how unjust it is in the generality of foreign philosophers to declare against a principle which furnishes so beautiful a view, for no other reason but because they cannot conceive how one body should act on another at a distance. It is indeed true, that philosophy allows of no action but what is by immediate contact and impulsion; for how can a body exert any active power where it does not exist? yet we see effects, without perceiving any such impulse; and where effects are observed, there must exist causes whether we see them or not. But we may contemplate such effects, without entering into the consideration of the causes, as indeed it seems the business of a philosopher to do: for to exclude a number of phenomena which we do see, would be to leave a great chasm in the history of nature; and to argue about actions which we do not see, would be to raise castles in the air. It follows therefore, that the phenomena of attraction are matter of physical consideration, and as such intitled to a share in the system of physics; but that their causes will only become so when they become sensible, that is when they appear to be the effect of some other higher causes; for a cause is no otherwise seen than as itself is an effect, so that the first cause must needs be always invisible: we are therefore at liberty to suppose the causes of attractions what we please, without any injury to the effects. The illustrious author himself seems to be a little indetermined as to the causes; inclining sometimes to attribute gravity to the action of an immaterial cause (*Optics*, pa. 343 &c), and sometimes to that of a material one, *Ib*. pa. 325.

In his philosophy, the research into causes is the last thing, and never comes under consideration till the laws and phenomena of the effect be settled; it being to these phenomena that the cause is to be accommodated. The cause even of any, the grossest and most sensible action, is not adequately known. How impulse or percussion itself produces its effects, that is how motion is communicated from body to body, confounds the deepest philosophers; yet is impulse received not only into philosophy, but into mathematics: and accordingly the laws and phenomena of its effects make the chief part of common mechanics.

The other species of attraction, therefore, in which no impulse is observable, when their phenomena are sufficiently ascertained, have the same title to be promoted from physical to mathematical consideration; and this without any previous inquiry into their causes, to which our conceptions may not be proportionate.

Our great philosopher, then, far from adulterating science with any thing foreign or metaphysical, as many have reproached him with doing, has the glory of having thrown every thing of this kind out of his system, and of having opened a new source of sublimer mechanics, which, duly cultivated, might be of far greater extent than all the mechanics yet known. Hence it is alone that we must expect to learn the manner of the changes, productions, generations, corruptions, &c, of natural things; with all that scene of wonders opened to us by the operations of chemistry.

Some of our own countrymen have prosecuted the discovery with laudable zeal. Dr. Keil particularly has endeavoured to deduce some of the laws of this new action, and applied them in resolving several of the more general phenomena of matter, as cohesion, fluidity, elasticity, softness, fermentation, coagulation, &c: and Dr. Freind, seconding his endeavours, has made a farther application of the same principles, at once to account for almost all the phenomena that chemistry presents. So that some philosophers are inclined to think that the new mechanics should seem already raised to a complete science, and that nothing now can occur but what we have an immediate solution of, from the principles of attractive forces.

But

But this feems a little too precipitate: a principle fo fertile fhould have been further explained; its particular laws, limits, &c, more induftrioufly detected and laid down, before we had proceeded to the application. Attraction in the grofs is fo complex a thing, that it may folve a thoufand different phenomena alike. The notion is but one degree more fimple and precife than action itfelf; and, till its properties are more fully afcertained, it were better to apply it lefs, and ftudy it more. It may be added, that fome of Newton's followers have been charged with falling into that error which he induftrioufly avoided, viz, of confidering attraction as a caufe or active-property in bodies, not merely as a phenomenon or effect.

For the laws, properties, &c, of the different forts of Attraction, fee their particular articles COHESION, GRAVITY, MAGNETISM, &c.

ATTRACTION, *Centre of*, See CENTRE *of Attraction*.
ATTRACTION *of Mountains*, See MOUNTAINS.

ATTRITION, the ftriking or rubbing of bodies againft one another, fo as to throw off fome of their fuperficial particles: fuch as the grinding and polifhing of bodies. Or fimply the act of rubbing: as when amber and other electric bodies are rubbed, to make them attract, or emit their electric force.

AVANT-FOSS, or Ditch of the Counterfcarp, in Fortification, is a wet ditch furrounding the counterfcarp, on the outer fide, next to the country, at the foot of the glacis. It would not be proper to have fuch a ditch if it could be laid dry, as it would then ferve as a lodgment for the enemy.

AVERROES, or ABEN-ROES, a very fubtile Arabian philofopher, who flourifhed about the end of the 11th century, when the Moors had poffeffion of part of Spain. He was the fon of the high prieft and chief judge of Corduba or Cordova in Spain: but he was educated in the univerfity of Morocco, where he was profeffor, and where he died in 1206, having there ftudied natural philofophy, medicine, mathematics, law, and divinity. After the death of his father, he enjoyed his pofts in Spain, to which was afterward added that of judge of Morocco and-Mauritania, where having fettled deputies, he returned to his duty in Spain. Notwithftanding he was very rich, and had a very great income, his liberality to men of letters in neceffity, whether they were his friends or his enemies, kept him always in debt. He was afterwards ftripped of all his pofts, and thrown into prifon, for herefy, by the inftigations of bad men, his enemies; but the oppreffions of the judge who fucceeded him, caufed him to be reftored to his former employments.

He was exceffively fat, though he eat but once a day, and fpent moft part of the night in the ftudy of philofophy, when he was fatigued, amufing himfelf with reading poetry or hiftory. He was never feen to play at any game, or to partake in any diverfion. He was extremely fond of Ariftotle's works, and wrote commentaries upon them; whence he was ftyled *the Commentator*, by way of eminence. He wrote many other pieces; among them a work on the Whole Art of Phyfic; an Epitome of Ptolomy's Almageft, which Voffius dates about the year 1149; alfo a Treatife of

Aftrology, which was tranflated into Hebrew by R. Jacob Ben Samfon, and faid to be extant in the French king's library. He wrote alfo feveral poems, and many amorous verfes, but thefe laft he threw into the fire when he grew old. His other poems are loft, except a fmall piece, in which he fays, "that when he was young, he acted againft his reafon; but that when he was in years, he followed its dictates;" upon which he utters this wifh, "Would to God I had been born old, and that in my youth I had been in a ftate of perfection!" As to religion, his opinions were, that Chriftianity is abfurd; Judaifm, the religion of children; Mahometanifm, the religion of fwine.

AVICENA, AVICENNE, or AVICENES, has been accounted the prince of Arabian philofophers and phyficians. He was born at Affena, near Bokhara, in 978; and died at Hamadan in 1036, being 58 years of age.

The firft years of Avicena were employed on the ftudy of the Belles Lettres, and the Koran, and at 10 years of age he was perfect mafter of the hidden fenfes of that book. Then applying to the ftudy of logic, philofophy and mathematics, he quickly made a rapid progrefs. After ftudying under a mafter the firft principles of logic, and the firft 5 or 6 propofitions of Euclid's elements, he became difgufted with the flow manner of the fchools, applied himfelf alone, and foon accomplifhed all the reft by the help of the commentators only.

Poffeffed with an extreme avidity to be acquainted with all the fciences, he ftudied medicine alfo. Perfuaded that this art confifts as much in practice as in theory, he fought all opportunities of feeing the fick; and afterwards confeffed that he had learned more from fuch experience than from all the books he had read. Being now in his 16th year, and already celebrated for being the light of his age, he determined to refume his ftudies of philofophy, which medicine, &c, had made him for fome time neglect: and he fpent a year and a half in this painful labour, without ever fleeping all this time a whole night together. At the age of 21, he conceived the bold defign of incorporating, in one work, all the objects of human knowledge; and he carried it into execution in an Encyclopedia of 20 volumes, to which he gave the title of the *Utility of Utilities*.

Many wonderful ftories are related of his fkill in medicine, and the cures which he performed. Several princes had been taken dangeroufly ill, and Avicenes was the only one that could know their ailments, and cure them. His reputation increafed daily, and all the princes of the eaft defired to retain him in their families, and in fact he paffed through feveral of them. But the irregularities of his conduct fometimes loft him their favour, and threw him into great diftreffes. His exceffes in pleafures, and his infirmities, made a poet fay, who wrote his epitaph, that the profound ftudy of philofophy had not taught him good morals, nor that of medicine the art of preferving his own health.

After his death however, he enjoyed fo great a reputation, that till the 12th century he was preferred for the ftudy of philofophy and medicine to all his predeceffors. Even in Europe his works were the only writings in vogue in the fchools. They were very numerous, and various, the titles of which are as follow:

1. Of

. Of the Utility and Advantage of the Sciences, in 20 books.—2. Of Innocence and Criminality, 2 books. —3. Of Health and Remedies, 18 books.—4. On the Means of preserving Health, 3 books.—5. Canons of Physic, 14 books.—6. On Astronomical Observations, 1 book.—7. On Mathematical Sciences.—8. Of Theorems, or Mathematical and Theological Demonstrations, 1 book.—9. On the Arabic Language, and its Properties, 10 books.—10. On the Last Judgment.—11. On the Origin of the Soul, and the Resurrection of Bodies.—12. On the end we should propose to ourselves in Harangues and Philosophical Argumentations. —13. Demonstration of the Collateral Lines in the Sphere.—14. Abridgment of Euclid.—15. On Finity and Infinity.—16. On Physics and Metaphysics.— 17. On Animals and Vegetables, &c.—18. Encyclopedie, 20 volumes.

AUGUST, the 8th month of the year, containing 31 days. In the antient Roman calendar this was called *sextilis*, as being the 6th month from March, with which their year began; but changed to its present name by the emperor Augustus, calling it after his own name on account of his having obtained many victories and honours in that month.

AVOIRDUPOIS *Weight*, a weight used in England for weighing all the larger and coarser sorts of goods; as groceries, cheese, butter, flesh, wool, salt, hops, &c, and all metals except gold and silver. Avoirdupois weight is thus divided, viz.

16 dr. or drams make	1 ounce, marked *oz.*
16 oz. - - - -	1 pound, - - *lb.*
112 lb. - - - -	1 hundred weight, *cwt.*
20 cwt. - - - -	1 ton - - *ton.*

The Avoirdupois ounce is less than the Troy ounce, in the proportion of 700 to 768, but the Avoirdupois pound greater than the Troy pound in the proportion of 700 to 576;

for 1lb Avoird. is = 7000 grains Troy,
but 1lb Troy is = 5760 grains Troy,
also 1 oz Avoird. is = $437\frac{1}{2}$ grains Troy,
and 1 oz Troy is = 480 grains Troy.

AURIGA, the *Waggoner*, a constellation in the northern hemisphere, consisting of 14 stars in Ptolemy's catalogue; but in Tycho's, 27; in Hevelius's, 40; and in the Britannic catalogue, 66.

This is one of the 48 old asterisms, mentioned by all the most ancient astronomers. It is represented by the figure of an old man, in a posture somewhat like sitting, with a goat and her kids in his left hand, and a bridle in his right.

The Greeks probably received this and all their other constellations from the Egyptians; but, wanting to appear the inventors of them themselves, and not understanding the meaning of the figures, they have cloathed them with some of their own fabulous dresses, to favour the deceit. They accordingly tell us that this figure of a waggoner was an honourable character, and Erichthonius, the inventor of coaches. Vulcan, say they, once fell in love with Minerva, and when he could not prevail with her to marry him, he would have obtained her upon less honourable terms. There was a struggle between them, and some way or other Erichthonius was begotten, though it does not seem that Minerva had much share in it: she took care of the offspring however. Some have supposed it was only a serpent; but the graver authors say, Erichthonius was a man with legs only like the body of a serpent, and that to hide this monstrous part of his figure he invented coaches to carry him about. They add, that Jupiter, doing him honour for an invention that was, in some degree, imitating the sun's carriage on the earth, raised him up among the stars.

But others, ill satisfied with a story which so badly agreed with the figure, have said that it belonged to Myrtillus, a son of Mercury and Clytie, and charioteer to Aenomanus; they say, that at his death, his father Mercury, by permission of his superiors, raised him up into the skies. All this however does not at all account for the goat and her two kids in the hands of Auriga. To set this right, they afterward made Auriga to be Olenus, a son of Vulcan, and the father of Aega and Helice, two of the Cretan nymphs that nursed the infant Jupiter. They talk of a goat that was used for giving milk to the young deity, and they suppose that this creature, and its two young ones, were placed in the hands of the father of the virgins, to commemorate the creature they took into their service on that occasion.

Besides the Hœdi, this constellation contains also another of those stars which the ancients honoured with peculiar names, the goat Capra, and Amalthæa Capra: this is the bright one near the shoulder, and supposed to be the mother of the Hœdi, and the nurse of Jupiter.

Although the whole constellation of Auriga is not mentioned among those from which the ancients formed presages of the succeeding weather, the two stars in his arm were of the foremost in that rank. It is these they called by the name Hœdi, and dreaded so extremely on account of the storms and tempests that succeeded their rising, that it is said they shut up the navigation of the sea for their season. And the day of their influence being over, we find, was celebrated as a festival with sports and games, under the name of Natalis Navigationis. Germanicus calls them unfriendly stars to mariners; and Virgil couples them with Arcturus, mentioning their setting and its rising as things of the most important presage. Horace also puts them together as the most formidable of all the stars to those who follow the traffic of the sea. And to the same purpose speak all the ancient writers, thus making a part of the constellation Auriga, if not the whole constellation, a thing to be observed with the utmost attention, and to be feared as much as the blazing Arcturus.

AURORA, the morning twilight; or that faint light which appears in the morning when the sun is within 18 degrees of the horizon.

AURORA BOREALIS, NORTHERN LIGHT, or *Streamers;* a kind of meteor appearing in the northern part of the heavens, mostly in the winter season, and in frosty weather. It is usually of a reddish colour, inclining to yellow, and sends out frequent coruscations of pale light, which seem to rise from the horizon in a pyramidal undulating form, and shooting with great velocity up to the zenith. It appears often in form of an arch, which

which is partly bright, and partly dark, but generally transparent. And the matter of it is not found to have any effect on the rays of light, which pass freely through it. Dr. Hamilton observes, that he could plainly discern the smallest speck in the Pleiades through the density of those clouds which formed part of the Aurora borealis in 1763, without the least diminution of its splendour, or increase of twinkling. Philof. Essays, pa. 106.

Sometimes it produces an Iris. Hence M. Godin judges, that most of the extraordinary meteors and phenomena in the skies, related as prodigies by historians, as battles, and the like, may probably enough be reduced to the class of Auroræ boreales. Hist. Acad. R. Scienc. for 1762, pa. 405.

This kind of meteor never appears near the equator; but, it seems, is frequent enough towards the south pole, like as towards the north, having been observed there by voyagers. See Philof. Transf. N° 461, and vol. 54; also Forster's account of his voyage round the world with Captain Cook, where he describes their appearance as observed for several nights together, in sharp frosty weather, which was much the same as those observed in the north, excepting that they were of a lighter colour.

It seems that meteors of this kind have appeared sometimes more frequently than others. They were so rare in England, or else so little regarded, that none are recorded in our annals since that remarkable one of Nov. 14, 1574, till the surprising Aurora borealis of March 6, 1716, which appeared for three nights successively, but by far more strongly on the first: except that five small ones were observed in the year 1707 and 1708. Hence it would seem, that the air, or earth, or both, are not at all times disposed to produce this phenomenon.

The extent of these appearances is also amazingly great. That in March 1716 was visible from the west of Ireland, to the confines of Russia and the east of Poland; extending at least near 30 degrees of longitude, and from about the 50th degree in latitude, over almost all the north of Europe; and in all places, at the same time, it exhibited the like wondrous appearances. Father Boscovich has determined the height of an aurora borealis, which was observed by the Marquis of Polini the 16th of December, 1737, and found it was 825 miles high; and Mr. Bergman, from a mean of 30 computations, makes the average height of the aurora borealis amount to 70 Swedish, or 460 English miles. But Euler supposes the height to be several thousands of miles; and Mairan also assigns to them a very elevated region.

Many attempts have been made to determine the cause of this phenomenon. Dr. Halley imagines that the watery vapours, or effluvia, exceedingly rarefied by subterraneous fire, and tinged with sulphureous streams, which many naturalists have supposed to be the cause of earthquakes, may also be the cause of this appearance: or that it is produced by a kind of subtile matter, freely pervading the pores of the earth, and which, entering into it nearer the southern pole, passes out again with some force into the æther, at the same distance from the northern. This subtile matter, by becoming more dense, or having its velocity increased, may perhaps be capable of producing a small degree of light, after the manner of effluvia from electric bodies, which, by a strong and quick friction, emit light in the dark; to which sort of light this seems to have a great affinity. Philof. Transf. N° 347. See also Mr. Cotes's description of this phenomenon, and his method of explaining it, by streams emitted from the heterogeneous and fermenting vapours of the atmosphere, in Smith's Optics, pa. 69; or Philof. Transf. abr. vol. 6, part 2.

The celebrated M. de Mairan, in an express treatise on the Aurora Borealis, published in 1731, supposes its cause to be the zodiacal light, which, according to him, is no other than the sun's atmosphere: this light happening, on some occasions, to meet the upper parts of our atmosphere about the limits where universal gravity begins to act more forcibly towards the earth than towards the sun, falls into our air to a greater or less depth, as its specific gravity is greater or less, compared with the air through which it passes. See Tract. Phyf. et Hist. de l'Aurore Boreale. Suite des Memoires de l'Acad. R. des Scien. 1731. Also Philof. Transf. N° 433, or Abridg. vol. 8, pa. 540.

However M. Euler thinks the cause of the aurora borealis not owing to the zodiacal light, as M. de Mairan supposes; but to particles of our atmosphere, driven beyond its limits by the impulse of the solar light. And on this supposition he endeavours to account for the phenomena observed concerning this light. He supposes the zodiacal light, and the tails of comets, to be owing to a similar cause.

But ever since the identity of lightning and the electric matter has been determined, philosophers have been naturally led to seek for the explication of aerial meteors in the principles of electricity; and there is now no doubt but most of them, and especially the aurora borealis, are electrical phenomena. Besides the more obvious and known appearances which constitute a resemblance between this meteor and the electric matter by which lightning is produced, it has been observed that the aurora occasions a very sensible fluctuation in the magnetic needle; and that when it has extended lower than usual in the atmosphere, the flashes have been attended with various sounds of rumbling and hissing, especially in Russia and the other more northern parts of Europe; as noticed by Sig. Beccaria and M. Messier. Mr. Canton, soon after he had obtained electricity from the clouds, offered a conjecture, that the aurora is occasioned by the dashing of electric fire positive towards negative clouds at a great distance, through the upper part of the atmosphere, where the resistance is least: and he supposes that the aurora which happens at the time when the magnetic needle is disturbed by the heat of the earth, is the electricity of the heated air above it: and this appears chiefly in the northern regions, as the alteration in the heat of the air in those parts is the greatest. Nor is this hypothesis improbable, when it is considered, that electricity is the cause of thunder and lightning; that it has been extracted from the air at the time of the aurora borealis; that the inhabitants of the northern countries observe it remarkably strong when a sudden thaw succeeds very cold severe weather; and that the tourmalin is known to emit and absorb the electric fluid only by the increase or diminution of its heat.

heat. Positive and negative electricity in the air, with a proper quantity of moisture to serve as a conductor, will account for this and other meteors, sometimes seen in a serene sky. Mr. Canton has since contrived to exhibit this meteor by means of the Torricellian vacuum, in a glass tube about 3 feet long, and sealed hermetically. When one end of the tube is held in the hand, and the other applied to the conductor, the whole tube will be illuminated from end to end, and will continue luminous without interruption for a considerable time after it has been removed from the conductor. If, after this, it be drawn through the hand either way, the light will be remarkably intense through the whole length of the tube. And though a great part of the electricity be discharged by this operation, it will still flash at intervals, when held only at one extremity, and kept quite still; but if, at the same time, it be grasped by the other hand in a different place, strong flashes of light will dart from one end to the other; and these will continue 24 hours or more, without a fresh excitation. Sig. Beccaria conjectures that there is a constant and regular circulation of the electric fluid from north to south; and he thinks that the aurora borealis may be this electric matter performing its circulation in such a state of the atmosphere as renders it visible, or approaching nearer than usual to the earth. Though probably this is not the mode of its operation, as the meteor is observed in the southern hemisphere, with the same appearances as in the northern. Dr. Franklin supposes, that the electric fire discharged into the polar regions, from many leagues of vaporised air raised from the ocean between the tropics, accounts for the aurora borealis; and that it appears first, where it is first in motion, namely in the most northern part; and the appearance proceeds southward, though the fire really moves northward. Franklin's Exper. and Obs. 1769, pa. 49. Philof. Tranf. vol. 48, pa. 358, 784; *Ib.* vol. 51, pa. 403; Lettere dell' Ellettricifmo, pa. 269; or Prieftley's Hift. of Electricity. See also an ingenious solution of this phenomenon, on the same principles, by Dr. Hamilton, in his Philof. Effays. Mr. Kirwan (in the Transactions of the Royal Irifh Academy, ann. 1788) has some ingenious remarks on the *auroræ borealis & auftralis.* He gives his reasons for suppofing the rarefaction of the atmofphere in the polar regions to proceed from them, and these from a combustion of inflammable air caufed by electricity. He obferves, that after an aurora borealis the barometer commonly falls, and high winds from the fouth generally follow.

AURUM FULMINANS, a preparation from gold, which being thrown into the fire, it explodes with a violent noise, like thunder. The matter is produced by diffolving gold in aqua regia, and precipitating the folution by oil of tartar *per deliquium,* or volatile fpirit of fal ammoniac. The powder being wafhed in warm water, and dried to the confiftence of a pafte, is afterwards formed into fmall grains of the fize of hempfeed.

It is inflammable, not only by fire, but alfo by a gentle warmth; and gives a report much louder than that of gunpowder. A fingle grain laid on the point of a knife, and lighted at a candle, explodes with a greater report than a mufquet: and a fcruple of this powder, it is faid, acts more loudly than half a pound of gunpowder; and yet it is faid that, by mixture, it does not increafe the elaftic force of fired gunpowder. Dr. Black attributes the increafe of weight, and alfo the explofive property of this powder, to adhering fixable air.—This is a very dangerous preparation, and fhould be ufed with great caution.

AUSTRAL, the fame with *fouthern.* Thus, Auftral figns, are the laft 6 figns of the zodiac; and are fo called becaufe they are on the fouth fide of the equinoctial.

AUSTRALIS *Corona*; fee CORONA *Auftralis.*

AUSTRALIS PISCIS, the *Southern Fifh,* is a conftellation of the fouthern hemifphere. See PISCIS *Auftralis.*

AUTOMATON, a feemingly felf-moving machine; or one fo conftructed, by means of weights, levers, pullies, fprings, &c, as to move for a confiderable time, as if it were endued with animal life. And according to this defcription, clocks, watches, and all machines of that kind, are automata.

It is faid, that Archytas of Tarentum, 400 years before Chrift, made a wooden pigeon that could fly; that Archimedes alfo made fuch-like automatons; that Regiomontanus made a wooden eagle that flew forth from the city, met the emperor, faluted him, and returned; alfo that he made an iron fly, which flew out of his hand at a feaft, and returned again after flying about the room; that Dr. Hook made the model of a flying chariot, capable of fupporting itfelf in the air. Many other furprizing automatons we have been eye-witneffes of, in the prefent age: thus, we have feen figures that could write, and perform many other actions in imitation of animals: M. Vaucanfon made a figure that played on the flute; the fame gentleman alfo made a duck, which was capable of eating, drinking, and imitating exactly the voice of a natural one; and, what is ftill more furprizing, the food it fwallowed was evacuated in a digefted ftate, or confiderably altered on the principles of folution; alfo the wings, vifcera, and bones were formed fo as ftrongly to refemble thofe of a living duck; and the actions of eating and drinking fhewed the ftrongeft refemblance, even to the muddling the water with its bill. M. Le Droz of la Chaux de Fonds, in the province of Neufchatel, has alfo executed fome very curious pieces of mechanifm: one was a clock, prefented to the king of Spain; which had, among other curiofities, a fheep that imitated the bleating of a natural one, and a dog watching a bafket of fruit, that barked and fnarled when any one offered to take it away; befides a variety of human figures, exhibiting motions truly furprifing. But all thefe feem to be inferior to M. Kempell's chefs-player, which may truly be confidered as the greateft-mafter-piece in mechanics that ever appeared in the world. See alfo Baptifta Porta's *Magia Nat.* c. 19, and Scaliger's *Subtil.* 326.

AUTUMN, the third feafon, when the harveft and fruits are gathered in. This begins at the defcending equinox, which, in the northern hemifphere,

is

is when the sun enters the sign Libra, or about the 22d day of August; and it ends, when winter commences, about the same day in December.

AUTUMNAL, something belonging to autumn. Thus,

AUTUMNAL *Equinox*, the time when the sun enters the descending point of the ecliptic, where it crosses the equinoctial; and is so called, because the nights and days are then equal.

AUTUMNAL *Point*, the point of the ecliptic answering to the autumnal equinox.

AUTUMNAL *Signs*, are the signs Libra, Scorpio, Sagittary, through which the sun passes during the autumn.

AXIOM, a self-evident truth, or a proposition assented to by every person at first sight. Such as, that the whole is greater than its part; that a thing cannot both be and not be at the same time; and that from nothing, nothing can arise.

Some Axioms are in effect, strictly speaking, no other than identical propositions. Thus, to say that all right angles are equal to each other, is as much as to say, all right angles are right angles; such equality being implied in the very definition, or the very name or term itself.

AXIOM is also an established principle in some art or science. Thus, it is an axiom in physics, that nature does nothing in vain; that effects are proportional to their causes; &c. It is an axiom in geometry, that two things equal to the same third thing, are also equal to each other; that if to equal things equals be added, the sums will be equal. And it is an axiom in optics, that the angle of incidence is equal to the angle of reflection. In this sense also the general laws of motion are called axioms; as, that all motion is rectilinear, that action and reaction are equal, &c.

AXE or AXIS, in *Geometry*, the straight line in a plane figure, about which it revolves, to produce or generate a solid. Thus, if a semicircle be moved round its diameter at rest, it will generate a sphere, whose axis is that diameter. And if a right-angled triangle be turned about its perpendicular at rest, it will describe a cone, whose axis is that perpendicular.

AXIS is yet more generally used for a right line conceived to be drawn from the vertex of a figure to the middle of the base. So the

AXIS *of a circle or sphere*, is any line drawn through the centre, and terminated at the circumference, on both sides.

AXIS *of a cone*, is the line from the vertex to the centre of the base.

AXIS *of a cylinder*, is the line from the centre of the one end to that of the other.

AXIS *of a conic section*, is the line from the principal vertex, or vertices, perpendicular to the tangent at that point. The ellipse and hyperbola have each two axes, which are finite and perpendicular to each other; but the parabola has only one, and that infinite in length.

Transverse AXIS, in the Ellipse and Hyperbola, is the diameter passing through the two foci, and the two principal vertices of the figure. In the hyperbola it is the shortest diameter, but in the ellipse it is the longest.

Conjugate AXIS, or *Second Axis*, in the Ellipse and Hyperbola, is the diameter passing through the centre, and perpendicular to the transverse axis; and is the shortest of all the conjugate diameters.

AXIS, of a curve line, is still more generally used for that diameter which has its ordinates at right angles to it, when that is possible. For, like as in the conic sections, any diameter bisects all its parallel ordinates, making the two parts of them on both sides of it equal; and that diameter which has such ordinates perpendicular to it, is an Axis: So, in curves of the second order, if any two parallel lines each meeting the curve in three points; the right line which cuts these two parallels so, that the sum of the two parts on one side of the cutting line, between it and the curve, is equal to the third part terminated by the curve on the other side, then the said line will in like manner cut all other parallels to the former two lines, viz. so that, of every one of them, the sum of the two parts, or ordinates, on one side, will be equal to the third part or ordinate on the other side. Such cutting line then is a diameter; and that diameter whose parallel ordinates are at right angles to it, when possible, is an Axis. And the same for other curves of still higher orders. Newton, Enumeratio Linearum Tertii Ordinis, sect. 2, art. 1.

AXIS, in *Astronomy*. As, the AXIS *of the world*, is an imaginary right line conceived to pass through the centre of the earth, and terminating at each end in the surface of the mundane sphere. About this line, as an axis, the sphere, in the Ptolomaic system, is supposed daily to revolve.

AXIS *of the Earth*, is the line connecting its two poles, and about which the earth performs its diurnal rotation, from west to east. This is a part of the axis of the world, and always remains parallel to itself during the motion of the earth in its orbit about the sun, and perpendicular to the plane of the equator.

AXIS *of a Planet*, is the line passing through its centre, and about which the planet revolves.—The Sun, Earth, Moon, Jupiter, Mars, and Venus, it is known from observation, move about their several axes; and the like motion is easily inferred of the other three, Mercury, Saturn, and Georgian planet.

AXIS of the *Horizon, Equator, Ecliptic, Zodiac,* &c, are right lines passing through the centres of those circles, perpendicular to their planes.

AXIS *of a Magnet*, or *magnetical* AXIS, is a line passing through the middle of a magnet, lengthwise; in such manner, that however the magnet be divided, provided the division be made according to a plane passing through that line, the magnet will then be cut into two loadstones. And the extremities of such lines are called the poles of the stone.

AXIS *in Mechanics*.—The axis of a balance, is the line upon which it moves or turns.

AXIS *of Oscillation*, is a line parallel to the horizon, passing through the centre about which a pendulum vibrates, and perpendicular to the plane in which it oscillates.

AXIS *in Peritrochio*, or *wheel and axle*, is one of the five mechanical powers, or simple machines; contrived chiefly for the raising of weights to a considerable height, as water from a well, &c. This machine

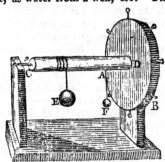

confifts of a circle AB, concentric with the bafe of a cylinder, and moveable together with it about its axis CD. This cylinder is called the *axis*; and the circle, the *peritrochium*; and the radii, or fpokes, which are fometimes fitted immediately into the cylinder, without any circle, the *fcytalæ*. About the axis winds a rope, or chain, by means of which great weights are raifed by turning the wheel.—The axis in peritrochio takes place in the motion of every machine, in which a circle may be conceived as defcribed about a fixed axis, concentric with the plane of a cylinder about which it is placed; as in Crane wheels, Mill wheels, Capftans, &c.

The chief properties of the Axe-in-peritrochio, are as follow:

1. If the power F applied in the direction AF a tangent to the circumference, or perpendicular to the fpoke, be to a weight E, as the radius of the axis C*e* is to the radius of the wheel AD, or the length of the fpoke; the power will juft fuftain the weight; that is, the power and the weight will be in equilibrio, when they are in the reciprocal proportion of their diftances from the centre.

2. When the wheel moves, with the power and weight; the velocities of their motion, and the fpaces paffed over by them, will be both in the fame proportion as above, namely, directly proportional to their diftances from the centre, and reciprocally proportional to their own weights when they are in equilibrio.

3. A power, and a weight, being given to conftruct an axis-in-peritrochio, by which it fhall be fuftained and raifed. Let the axis be taken large enough to fupport the weight and power without breaking: then, as the weight is to the power, fo make the radius of the wheel to the radius of the axis. Hence, if the power be very fmall in refpect of the weight, the radius of the wheel will be vaftly great. For example, fuppofe the weight 4050, and the power only 50; then the radius of the wheel will be 81 times that of the axis; which would be a very inconvenient fize. But this inconvenience is provided againft by increafing the number of the wheels and axes; making one to turn another, by means of teeth or pinions. And to find the effect of a number of wheels and axes, thus turning one another, multiply together, all the radii of the axes, and all the radii of the wheels, and then it will be, as the product of the former is to the product of the latter, fo is the power to the weight. So, if there be 4 wheels and axes, the radius of each axis being 1 foot, and the

radius of each wheel 3 feet; then the continual product of all the wheels is $3 \times 3 \times 3 \times 3$ or 81 feet, and that of the axis only 1; therefore the effect is as 81 to 1, or the weight is 81 times the power. And, on the contrary, if it be required to find the diameter of each of four equal wheels, by which a weight of 4050lb fhall be balanced by a power of 50lb, the diameter of each axis being 1 foot: dividing 4050 by 50, the quotient is 81; extract the 4th root of 81, or twice the fquare root, and it will give 3, for the diameter of the four wheels fought.

AXIS *of a veffel*, is that quiefcent right line paffing through the middle of it, perpendicular to its bafe, and equally diftant from its fides.

AXIS *in Optics.—Optic Axis*, or *vifual axis*, is a ray paffing through the centre of the eye, or falling perpendicularly on the eye.

AXIS *of a lens*, or *glafs*, is the axis of the folid of which the lens is a fegment. Or the axis of a glafs, is the line joining the two vertices or middle points of the two oppofite furfaces of the glafs.

AXIS *of Incidence*, in Dioptrics, is the line paffing through the point of incidence, perpendicularly to the refracting furface.

AXIS *of Refraction*, is the line continued from the point of incidence or refraction, perpendicularly to the refracting furface, along the further medium.

AZIMUTH, of the fun, or ftar, &c, is an arch of the horizon, intercepted between the meridian of the place, and the azimuth or vertical circle paffing through the fun or ftar; and is equal to the angle at the zenith formed by the faid meridian and vertical circle. Or it is the complement to the eaftern or weftern amplitude.—The azimuth is thus found by trigonometry;

As radius is to the tangent of the latitude,
So is the tangent of the altitude of the fun or ftar,
To the cofine of the azimuth from the fouth, at
   the time of the equinox.

AZIMUTH, *magnetical*, an arch of the horizon contained between the magnetical meridian, and the azimuth or vertical circle of the object; or its apparent diftance from the north or fouth point of the compafs. This is found by obferving the fun, or ftar, &c, with an azimuth compafs, when it is 10 or 15 degrees high, either before or after noon.

AZIMUTH COMPASS, an inftrument for finding either the magnetical azimuth or amplitude of a celeftial object. The defcription and ufe of this inftrument, fee under the article COMPASS.

AZIMUTH DIAL, a dial whofe ftile or gnomon is perpendicular to the plane of the horizon.

AZIMUTHS, or *Vertical Circles*, are great circles of the fphere interfecting each other in the zenith and nadir, and cutting the horizon at right angles.—Thefe azimuths are reprefented by the rhumbs on common fea charts; and on the globe by the quadrant of altitude, when fcrewed in the zenith. On thefe azimuths is counted the height of the fun or ftars, &c, when out of the meridian.

B.

BACK-STAFF, an inſtrument formerly uſed for taking the ſun's altitude at ſea; being ſo called becauſe the back of the obſerver is turned towards the ſun when he makes the obſervation. It was ſometimes called Davis's quadrant, from its inventor captain John Davis, a Welchman, and a celebrated navigator, who produced it about the year 1590.

This inſtrument conſiſts of two concentric arches of box-wood, and three vanes: the arch of the longer radius is of 30 degrees, and the other 60 degrees, making between them 90 degrees, or a quadrant: alſo the vane A at the centre is called the *horizon-vane*, that on the arch of 60° at B the *ſhade-vane*, and that on the other arch at C the *ſight-vane*.

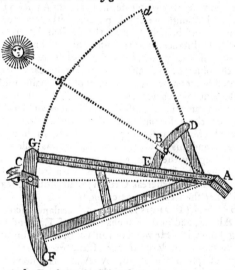

To uſe the Back-Staff. The ſhade-vane is to be ſet upon the 60 arch, at an even degree of ſome latitude, leſs by 10 or 15 degrees than you judge the complement of the ſun's altitude will be; alſo the horizon-vane being put on at A, and the ſight-vane on the 30 arch F G, the obſerver turns his back to the ſun, lifts up the inſtrument, and looks through the ſight-vane, raiſing or falling the quadrant, till the ſhadow of the upper edge of the ſhade-vane fall on the upper edge of the ſlit in the horizon-vane; and then if he can ſee the horizon though the ſaid ſlit, the obſervation is exact, and the vanes are right ſet: But if the ſea appear inſtead of the horizon, the ſight-vane muſt be moved downward towards F; or if the ſky appear, it muſt be moved upward towards G; thus trying till it comes right: the obſerver then examines how many degrees and minutes are cut by that edge of the ſight-vane that anſwers to the ſight hole, and to them he adds the degrees cut by the upper edge of the ſhade-vane; then the ſum is the ſun's diſtance from the zenith, or the

complement of the altitude; that is, of his upper limb when the upper end of the ſhade-vane is uſed in the obſervation, or of his lower limb when the lower part of that vane is uſed; therefore in the former caſe add 16 minutes, the ſun's ſemidiameter, and ſubtract 16 minutes in the latter caſe, to give the zenith diſtance or co-altitude of the ſun's centre.

Mr. Flamſteed contrived a glaſs lens, or double convex, to be placed in the middle of the ſhade-vane, which throws a ſmall bright ſpot on the ſlit of the horizon-vane, inſtead of the ſhade; which is a great improvement, if the glaſs be truly made; for by this means, the inſtrument may be uſed in hazy weather, and a much more accurate obſervation made at all times.

BACON (ROGER), an Engliſh monk of the Franciſcan order, an amazing inſtance of genius and learning, was born near Ilcheſter in Somerſetſhire, in the year 1214. He commenced his ſtudies at Oxford; from whence he removed to the univerſity of Paris, which at that time was eſteemed the centre of literature; and where it ſeems he made ſuch progreſs in the ſciences, that he was eſteemed the glory of that univerſity, and was there greatly careſſed by ſeveral of his countrymen, particularly by Robert Groothead or Grouthead, afterwards biſhop of Lincoln, his great friend and patron. Having taken the degree of doctor, he returned to England in 1240, and took the habit of the Franciſcan order, being but about 26 years of age; but according to ſome he became a monk before he left France. He now purſued his favourite ſtudy of experimental philoſophy with unremitting ardour and aſſiduity. In this purſuit, in experiments, inſtruments, and in ſcarce books, he informs us he ſpent, in the courſe of 20 years, no leſs than 2000*l*, an amazing ſum in thoſe days, and which ſum it ſeems was generouſly furniſhed to him by ſome of the heads of the univerſity, to enable him the better to purſue his noble reſearches. By ſuch extraordinary talents, and amazing progreſs in the ſciences, which in that ignorant age were ſo little known to the reſt of mankind, while they raiſed the admiration of the more intelligent few, could not fail to excite the envy of his illiterate fraternity, whoſe malice he farther drew upon him by the freedom with which he treated the clergy in his writings, in which he ſpared neither their ignorance nor their want of morals: theſe therefore found no difficulty in poſſeſſing the vulgar with the notion of Bacon's dealing with the devil. Under this pretence he was reſtrained from reading lectures; his writings were confined to his convent; and at length, in 1278, he himſelf was impriſoned in his cell, at 64 years of age. However, being allowed the uſe of his books, he ſtill proceeded in the rational purſuit of knowledge, correcting his former labours, and writing ſeveral curious pieces.

When

When Bacon had been 10 years in confinement, Jerom de Afcoli, general of his order, who had condemned his doctrine, was chosen pope by the name of Nicholas IV; and being reputed a person of great abilities, and one who had turned his thoughts to philosophical studies, Bacon resolved to apply to him for his discharge; and to shew both the innocence and the usefulness of his studies, addressed to him a treatise *On the means of avoiding the infirmities of old age.* What effect this had on the pope does not appear; it did not at least produce an immediate discharge: however, towards the latter end of his reign, by the interposition of some noblemen, Bacon obtained his liberty; after which he spent the remainder of his life in the college of his order, where he died in the year 1294, at 80 years of age, and was buried in the Franciscan church. Such are the few particulars which the most diligent researches have been able to discover concerning the life of this very extraordinary man.

Bacon's printed works are, 1. *Epistola Fratris Rogeri Baconis de Secretis Operibus Artis et Naturæ, et de Nullitate Magiæ:* Paris, 1542, in 4to. Basil, 1593, in 8vo. 2. *Opus Majus:* London, 1733, in fol. published by Dr. Jebb. 3. *Thesaurus Chemicus:* Francf. 1603 and 1620. These printed works of Bacon contain a considerable number of essays, which have been considered as distinct books in the catalogue of his writings by Bale, Pitts, &c; but there remain also in different libraries several manuscripts not yet published.

By an attentive perusal of his works, the reader is astonished to find that this great luminary of the 13th century was deeply skilled in all the arts and sciences, and in many of them made the most important inventions and discoveries. He was, says Dr. Peter Shaw, beyond all comparison the greatest man of his time, and he might perhaps stand in competition with the greatest that have appeared since. It is wonderful, considering the ignorant age in which he lived, how he came by such a depth of knowledge on all subjects. His writings are composed with that elegance, conciseness and strength, and adorned with such just and exquisite observations on nature, that, among all the chemists, we do not know his equal. In his chemical writings, he attempts to shew how imperfect metals may be ripened into perfect ones; making, with Geber, mercury the common basis of all metals, and sulphur the cement.

His other physical writings shew no less genius and force of mind. In his treatise *Of the Secret Works of Art and Nature,* he shews that a person perfectly acquainted with the manner observed by nature in her operations, would be able to rival, and even to surpass her. In another piece, *Of the Nullity of Magic,* he shews with great sagacity and penetration, whence the notion of it sprung, and how weak all pretences to it are. From a perusal of his works, adds the same author, we find Bacon was no stranger to many of the capital discoveries of the present and past ages. Gunpowder he certainly knew: thunder and lightning, he tells us, may be produced by art; for that sulphur, nitre and charcoal, which when separate have no sensible effect, yet when mixed together in due proportion, and closely confined, and fired, they yield a loud report. A more precise description of gunpowder cannot be

given in words. He also mentions a sort of unextinguishable fire prepared by art: which shews he was not unacquainted with phosphorus: and that he had a notion of the rarefaction of the air, and the structure of an air-pump, is past contradiction. He was the miracle, says Dr. Freind, of the age he lived in, and the greatest genius, perhaps, for mechanical knowledge, that ever appeared in the world since Archimedes. He appears likewise to have been master of the whole science of optics: he has accurately described the uses of reading-glasses, and shewn the way of making them. Dr. Freind adds, that he also describes the camera obscura, and all sorts of glasses, which magnify or diminish any object, or bring it nearer to the eye, or remove it farther off. Bacon says himself, that he had great numbers of burning-glasses: and that there were none ever in use among the Latins, till his friend Peter de Mahara Curia applied himself to the making of them. That the telescope was not unknown to him, appears from a passage where he says, that he was able to form glasses in such a manner, with respect to our sight and the objects, that the rays shall be refracted and reflected wherever we please, so that we may see a thing under what angle we think proper, either near or at a distance, and be able to read the smallest letters at an incredible distance, and to count the dust and sand, on account of the greatness of the angle under which we see the objects; and also that we shall scarce see the greatest bodies near us, on account of the smallness of the angle under which we view them. His skill in astronomy was amazing: he discovered that error which occasioned the reformation of the calendar; one of the greatest efforts, according to Dr. Jebb, of human industry: and his plan for correcting it was followed by pope Gregory the 13th, with this variation, that Bacon would have had the correction to begin from the birth of our Saviour, whereas Gregory's amendment reaches no higher than the Nicene council.

BACON (FRANCIS), baron of Verulam, viscount of St. Albans, and lord high chancellor of England under king James I. He was born in 1560, being son of Sir Nicholas Bacon lord keeper of the great seal in the reign of queen Elizabeth, by Anne daughter of Sir Anthony Cook, eminent for her skill in the Latin and Greek languages. He gave even in his infancy tokens of what he would one day become; and queen Elizabeth had many times occasion to admire his wit and talents, and used to call him her young lord keeper. He studied the philosophy of Aristotle at Cambridge; where he made such progress in his studies, that at 16 years of age he had run through the whole circle of the liberal arts as they were then taught, and even began to perceive those imperfections in the reigning philosophy, which he afterwards so effectually exposed, and thence not only overturned that tyranny which prevented the progress of true knowledge, but laid the foundation of that free and useful philosophy which has since opened a way to so many glorious discoveries. On his leaving the university, his father sent him to France; where, before he was 19 years of age, he wrote a general view of the state of Europe: but his father dying, he was obliged suddenly to return to England; where he applied himself to the study of the common law, at Gray's-inn. His merit at length raised him to the
highest

highest dignities in his profession, attorney-general, and lord high chancellor. But being of an easy and liberal disposition, his servants took advantage of that temper, and their situation under him, by accepting presents in the line of his profession. Being abandoned by the king, he was tried by the house of lords, for bribery and corruption, and by them sentenced to pay a fine of 40,000*l.* and to remain prisoner in the Tower during the king's pleasure. The king however soon after remitted the fine and punishment: but his misfortunes had given him a distaste for public affairs, and he afterwards mostly lived a retired life, closely pursuing his philosophical studies and amusements, in which time he composed the greatest part of his English and Latin works. Though even in the midst of his honours and employments he forgot not his philosophy, but in 1620 published his great work *Novum Organum*. After some years spent in his philosophical retirement, he died in 1626, being 66 years of age.

The chancellor Bacon is one of those extraordinary geniuses who have contributed the most to the advancement of the sciences. He clearly perceived the imperfection of the school philosophy, and he pointed out the only means of reforming it, by proceeding in the opposite way, from experiments to the discovery of the laws of nature. Addison has said of him, That he had the sound, distinct, comprehensive knowledge of Aristotle, with all the beautiful light graces and embellishments of Cicero. Mr. Walpole calls him the *Prophet of Arts*, which Newton was afterwards to reveal; and adds, that his genius and his works will be universally admired as long as science exists. He did not yet, said another great man, understand nature, but he knew and pointed out all the ways that lead to her. He very early despised all that the universities called philosophy; and he did every thing in his power that they should not disgrace her by their quiddities, their horrors of a vacuum, their substantial forms, and such like impertinencies.

He composed two works for perfecting the sciences. The former *On the Dignity and Augmentation of the Sciences*. He here shews the state in which they then were, and points out what remains to be discovered for perfecting them; condemning the unnatural way of Aristotle, in reversing the natural order of things. He here also proposes his celebrated division of the sciences.

To remedy the faults of the common logic, Bacon composed his second work, the *New Organ of Sciences*, above-mentioned. He here teaches a new logic, the chief end of which is to shew how to make a good inference, as that of Aristotle's is to make a syllogism. Bacon was 18 years in composing this work, and he always esteemed it as the chief of his compositions.

The pains which Bacon bestowed upon all the sciences in general, prevented him from making any considerable applications to any one in particular: and as he knew that natural philosophy is the foundation of all the other sciences, he chiefly endeavoured to give perfection to it. He therefore proposed to establish a new system of physics, rejecting the doubtful principles of the ancients. For this purpose he took the resolution of composing every month a treatise on some

branch of physics; he accordingly began with that of the winds; then he gave that of heat; next that of motion; and lastly that of life and death. But as it was impossible that one man alone could so compose the whole circle of sciences with the same precision, after having given these patterns, to serve as a model to those who might choose to labour upon his principles, he contented himself with tracing in a few words the design of four other tracts, and with furnishing the materials, in his Silva Silvarum, where he has amassed a vast number of experiments, to serve as a foundation for his new physics. In fact, no one before Bacon understood any thing of the experimental philosophy; and of all the physical experiments which have been made since his time, there is scarcely one that is not pointed out in his works.

This great precursor of philosophy was also an elegant writer, an historian, and a wit. His moral essays are valuable, but are formed more to instruct than to please. There are excellent things too in his work *On the Wisdom of the Ancients*, in which he has moralized the fables which formed the theology of the Greeks and Romans. He wrote also *The History of Henry the VIIth king of England*, by which it appears that he was not less a great politician than a great philosopher.

Bacon had also some other writings, published at different times; the whole of which were collected together, and published at Frankfort, in the year 1665, in a large folio volume, with an introduction concerning his life and writings. Another edition of his works was published at London in 1740; the enumeration of which is as below:

1. De Dignitate et Augmentis Scientiarum.
2. Novum Organum Scientiarum, sive Judicia vera de Interpretatione Naturæ; cum Parasceve ad Historiam Naturalem & Experimentalem.
3. Phænomena Universi, sive Historia Naturalis & Experimentalis de Ventis; Historia Densi & Rari; Historia Gravis & Levis; Historia Sympathiæ & Antipathiæ Rerum; Historia Sulphuris, Mercurii, & Salis; Historia Vitæ & Mortis; Historia Naturalis & Experimentalis de Forma Calidi; De Motus, sive Virtutis activæ variis speciebus; Ratio inveniendi causas Fluxus & Refluxus Maris; &c, &c.
4. Silva Silvarum, sive Historia Naturalis.
5. Novus Atlas.
6. Historia Regni Henrici vii Angliæ Regis.
7. Sermones fideles, Ethici, Politici, Oeconomici.
8. De Sapientia Veterum.

BACULE, in *Fortification*, a kind of portcullis, or gate, made like a pit-fall, with a counterpoise, and supported with two great stakes. It is usually made before the *corps de garde*, not far from the gate of a place.

BACULOMETRY, the art of measuring either accessible or inaccessible lines, by the help of *baculi*, staves, or rods. Schwenter has explained this art in his *Geometria Practica*; and the rules of it are delivered by Wolfius, in his Elements: Ozanam also gives an illustration of the principles of Baculometry.

BAILLY (JEAN SYLVAIN), a celebrated French astronomer, historiographer, and politician, was born at Paris the 15th of September 1736, and has figured as

one

one of the greateft men of the age, being a member of feveral academies, and an excellent fcholar and writer. He enjoyed for feveral years the office of keeper of the king's pictures at Paris. He publifhed, in 1766, a volume in 4to, *An Effay on the Theory of Jupiter's Satellites*, preceded by a Hiftory of the Aftronomy of thefe Satellites. In the Journal Encyclopedique for May and July 1773, he addreffed a letter to M. Bernoulli, Aftronomer Royal at Berlin, upon fome difcoveries relative to thefe fatellites, which he had difputed. In 1768 he publifhed the Eulogy of Leibnitz, which obtained the prize at the Academy of Berlin, where it was printed. In 1770 he printed at Paris, in 8vo, the Eulogies of Charles the Vth, of de la Caille, of Leibnitz, and of Corneille. This laft had the fecond prize at the Academy of Rouen, and that of Moliere had the fame honour at the French Academy.

M. Bailly was admitted into the Academy as Adjunct the 29th of January 1763, and as Affociate the 14th of July 1770.—In 1775 came out at Paris, in 4to, his *Hiftory of the Ancient Aftronomy*, in 1 volume: In 1779 the *Hiftory of Modern Aftronomy* in 2 volumes: and in 1787 the *Hiftory of the Indian and Oriental Aftronomy*, being the 2d vol. of the Ancient Aftronomy.

M. Bailly's memoirs publifhed in the volumes of the Academy, are as follow:

Memoir upon the Theory of the Comet of 1759.

Memoir upon the Epoques of the Moon's motions at the end of the laft century.

Firft, fecond, and third Memoirs on the Theory of Jupiter's Satellites, 1763.

Memoir on the Comet of 1762: vol. for 1763.

Aftronomical Obfervations, made at Noflon: 1764.

On the Sun's Eclipfe of the 1ft of April, 1764.

On the Longitude of Polling; 1764.

Obfervations made at the Louvre from 1760 to 1764: 1765.

On the caufe of the Variation of the Inclination of the Orbit of Jupiter's fecond Satellite; 1765.

On the Motion of the Nodes, and on the Variation of the Inclination of Jupiter's Satellites; 1766.

On the Theory of Jupiter's Satellites, publifhed by M. Bailly, and according to the Tables of their Motions and of thofe of Jupiter, publifhed by M. Jeaurat; 1766.

Obfervations on the Oppofition of the Sun and Jupiter; 1768.

On the Equation of Jupiter's Centre, and on fome other Elements of the Theory of that Planet; 1768.

On the Tranfit of Venus over the Sun, the 3d of June 1769; and on the Solar Eclipfe the 4th of June the fame year; 1769.

In the beginning of the revolution in France, in 1789, M. Bailly took an active part in that bufinefs, and was fo popular and generally efteemed, that he was chofen the firft prefident of the States General, and of the National Affembly, and was afterwards for two years together the Mayor of Paris; in both which offices he conducted himfelf with great fpirit, and gave general fatisfaction.

He foon afterward however experienced a fad reverfe of fortune; being accufed by the ruling party of favouring the king, he was condemned for incivifm and wanting to overturn the Republic, and died by the Guillotine at Paris on the eleventh day of November, 1793, at 57 years of age.

BAINBRIDGE (John), an eminent phyfician, aftronomer, and mathematician. He was born in 1582, at Afhby de la Zouch, Leicefterfhire. He ftudied at Cambridge, where having taken his degrees of Bachelor and Mafter of Arts, he returned to Leicefterfhire, where for fome years he kept a grammar-fchool, and at the fame time practifed phyfic; employing his leifure hours in ftudying mathematics, efpecially aftronomy, which had been his favourite fcience from his earlieft years. By the advice of his friends, he removed to London, to better his condition, and improve himfelf with the converfation of learned men there; and here he was admitted a fellow of the college of phyficians. His defcription of the comet, which appeared in 1618, greatly raifed his character, and procured him the acquaintance of Sir Henry Savile, who, in 1619, appointed him his firft profeffor of aftronomy at Oxford. On his removal to this univerfity, he entered a mafter commoner of Merton college; the mafter and fellows of which appointed him junior reader of Linacer's lecture in 1631, and fuperior reader in 1635. As he refolved to publifh correct editions of the ancient aftronomers, agreeably to the ftatutes of the founder of his profefforfhip, that he might acquaint himfelf with the difcoveries of the Arabian aftronomers, he began the ftudy of the Arabic language when he was above 40 years of age. Before completing that work however he died, in the year 1643, at 61 years of age.

Dr. Bainbridge wrote many works, but moft of them have never been publifhed; thofe that were publifhed, were the three following, viz:

1. An Aftronomical Defcription of the late Comet, from the 18th of November 1618, to the 16th of December following; 4to, London, 1619.—This piece was only a fpecimen of a larger work, which the author intended to publifh in Latin, under the title of Cometographia.

2. Procli Sphæra, Ptolomæi de Hypothefibus Planetarum liber fingularis. To which he added Ptolomy's Canon Regnorum. He collated thefe pieces with ancient manufcripts, and gave a Latin verfion of them, illuftrated with figures: printed in 4to, 1620.

3. Canicularia. A treatife concerning the Dog-ftar, and the Canicular Days: publifhed at Oxford in 1648, by Mr. Greaves, together with a demonftration of the heliacal rifing of Sirius, the dog-ftar, for the parallel of Lower Egypt. Dr. Bainbridge undertook this work at the requeft of archbifhop Ufher, but he left it imperfect; being prevented by the breaking out of the civil war, or by death.

There were alfo feveral differtations of his prepared for and committed to the prefs the year after his death, but the edition of them was never completed. The titles of them are as follow:

1ft, Antiprognofticon, in quo Μανλικῆς Aftrologicæ, Cœleftium Domorum, et Triplicitatum Commentis, magnifque Saturni et Jovis (cujufmodi anno 1623, et 1643, contigerunt, et vicefimo fere quoque deinceps anno, ratis naturæ legibus, recurrent) Conjunctionibus innixæ, Vanitas breviter detegitur.

2nd, De Meridionorum five Longitudinum Differentiis inveniendis Differtatio.

3d, De

3d, De Stella Veneris Diatriba.

Beside the foregoing, there were several other tracts, never printed, but left by his will to archbishop Usher; among whose manuscripts they are preserved in the library of the college of Dublin. Among which are the following: 1. A Theory of the Sun. 2. A Theory of the Moon. 3. A Discourse concerning the Quantity of the Year. 4. Two volumes of Astronomical, Observations. 5. Nine or ten volumes of Miscellaneous Papers relating to Mathematical subjects.

BAKER (Thomas), a mathematician of some eminence, was born at Ilton in Somersetshire, in 1625. He entered upon his studies at Oxford in 1640, where he remained seven years. He was afterwards appointed vicar of Bishop's-Nymmet in Devonshire, where he lived a studious and retired life for many years, chiefly pursuing the mathematical sciences; of which he gave a proof of his critical knowledge, in the book he published, concerning the general construction of bi-quadratic equations, by a parabola and a circle; the title of which book at full length is, " The Geometrical Key; or the Gate of Equations unlocked: or a new discovery of the Construction of all Equations, howsoever affected, not exceeding the 4th degree, viz, of Linears, Quadratics, Cubics, Biquadratics, and the finding of all their Roots, as well False as True, without the use of Mesolabe, Trisection of Angles, without Reduction, Depression, or any other previous preparations of equations by a circle, and any (and that one only) Parabole, &c" : 1684, 4to, in English and Latin.

There is some account of this work in the Philos. Transf. an. 1684. And a little before his death, the Royal Society sent him some mathematical queries; to which he returned such satisfactory answers, as procured the present of a medal, with an inscription full of honour and respect.—Mr. Baker died at Bishop's-Nymmet, 1690, in the 65th year of his age.

BAKER (Henry), an ingenious and diligent naturalist, was born in London about the beginning of the 18th century. He was brought up under an eminent bookseller, but being of a philosophical turn of mind, he quitted that line of business soon after the expiration of his apprenticeship, and took to the employment of teaching deaf and dumb persons to speak and write &c, in which occupation, in the course of his life he acquired a handsome fortune. For his amusement he cultivated various natural and philosophical sciences, particularly botany, natural history, and microscopical subjects, in which he especially excelled, having, in the year 1744, obtained the Royal Society's gold medal, for his microscopical experiments on the crystallizations and configurations of saline particles. He had various papers published in the Philos. Transf. of the Royal Society, of which he was a worthy member, as well as of the Societies of Antiquaries, and of Arts. He was author of many pieces, on various subjects, the principal of which were, his Treatise on the Water Polype, and two Treatises on the Microscope, viz, *The Microscope made easy*, and *Employment for the Microscope*, which have gone through several editions.

Mr. Baker married Sophia, youngest daughter of the celebrated Daniel Defoe, by whom he had two sons, who both died before him. He terminated an honourable and useful life, at his apartments in the Strand on the 25th of November 1774, being then upwards of 70 years of age.

BAKER's Central Rule, *for the Construction of Equations,* is a method of constructing all equations, not exceeding the 4th degree, by means of a given parabola and a circle, without any previous reduction of them, or first taking away their second term. See Central *Rule.*

BALANCE, one of the six simple powers in mechanics, chiefly used in determining the equality or difference of weight in heavy bodies, and consequently their masses or quantities of matter.

The balance is of two kinds, the *ancient* and *modern.* The ancient or Roman, called also *Statera Romana,* or *Steelyard,* consists of a lever or beam, moveable on a centre, and suspended near one of its extremities. The bodies to be weighed are suspended from the shorter end, and their weight is shewn by the division marked on the beam, where the power or constant weight, which is moveable along the lever, keeps the steelyard in equilibrio. This balance is still in common use for weighing heavy bodies.

The modern balance, now commonly used, consists of a lever or beam suspended exactly in the middle, and having scales suspended from the two extremities, to receive the weights to be weighed.

In either case the lever is called the *jugum* or the *beam,* and its two halves on each side the axis, the *brachia* or *arms ;* also the line on which the beam turns, or which divides it in two, is called the *axis;* and when considered with regard to the length of the brachia, is esteemed only a point, and called the *centre of the balance,* or *centre of motion:* the extremities where the weights are applied, are the *points of application* or *suspension;* the handle by which the balance is held, or by which the whole apparatus is suspended, is called *trutina;* and the slender part perpendicular to the beam, by which is determined either the equilibrium or preponderancy of bodies, is called the *tongue of the balance.*

From these descriptions we easily gather the characteristic distinction between the Roman balance and the common one, viz, that in the Roman balance, there is one constant weight used as a counterpoise, the point where it is suspended being varied; but, on the contrary, in the common balance or scales, the points of suspension remain the same, and the counterpoise is varied. The principle of both of them may be easily understood from the general properties of the lever, and the following observations.

The beam ABC, the principal part of the balance, is a lever of the first kind; but instead of resting on a fulcrum, it is suspended by a handle, &c, fastened to its centre of motion B: and hence the mechanism of the balance depends on the same theorems as that of the lever. Consequently as the distance between the centre of motion and the place of the unknown weight, is to the distance between the same centre and the place of the known weight, so is the latter weight, to the former. So that the unknown weight is discovered by means

of

of the known one, and their diſtances from the common centre of motion ; viz, if the diſtances from the centre be equal, then the two weights will be equal alſo, as in the common balance ; but if the diſtances be unequal, then the weights will alſo be unequal, and in the very ſame proportion, alternately, the leſs weight having ſo much the greater diſtance, as in the ſteelyard.

The *Common Balance or Scales.*

The two brachia AB, BC, ſhould be exactly equal in length, and in weight alſo when their ſcales D and E are fixed on their ends ; the beam ſhould hang exactly level or horizontal in the caſe of an equipoiſe ; and for this purpoſe the centre of gravity of the whole ſhould fall a little below the centre of motion, and but a little, that the balance be ſufficiently ſenſible to the leaſt variation of weight : the friction on the centre ſhould alſo be as ſmall as poſſible.

The *Steel Yard.*

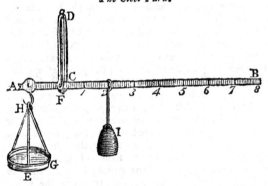

Having made a proper bar of ſteel AB, tapering at the longer end, and very ſtrong at the other, ſuſpend it by a centre C near the ſhorter or thicker end, ſo that it may exactly balance itſelf in equilibrio, and prepare a conſtant weight I to weigh with : then hang on any weight, as one pound for inſtance, at the ſhorter arm, and ſlide the conſtant weight backwards and forwards upon the longer arm, till it be juſt in equilibrio with the former ; and there make a notch and number 1, for the place of 1 pound : take off the 1lb, and hang a two pound weight in its ſtead at the ſhorter arm ; then ſlide the conſtant weight back on the longer arm, till the whole come again into equilibrio, making a notch at the place of the conſtant weight and the number 2, for the place of 2lb. Proceed in the ſame manner for all other weights 3, 4, 5, &c ; as alſo for the intermediate halves and quarters, &c, if it be neceſſary ; always ſuſ-

pending the variable weights at the end of the ſhorter arm, ſhifting the conſtant weight ſo as to balance them, and marking and numbering the places on the longer arm where the conſtant weight always makes a counterpoiſe. The uſe of the Steelyard is hence very evident : the thing whoſe weight is required being ſuſpended by a hook at the ſhort end, move the conſtant weight backwards and forwards on the longer arm, till the beam is balanced horizontally : then look what notch the conſtant weight is placed at, and its number will ſhew the weight of the body that was required. DC is the handle and tongue ; F the centre of motion ; EG a ſcale ſometimes hung on at the end by the hook H.

The *Bent-Lever Balance.*

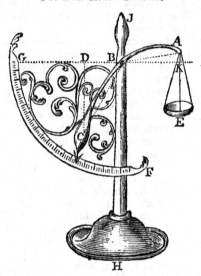

This inſtrument operates by a fixed weight, C, increaſing in power as it aſcends along the arc FG of a circle, and pointing by an index to the number or diviſion of the arc which denotes the weight of any body put into the ſcale at E. And thus one conſtant weight ſerves to weigh all others, by only varying the poſition of the arms of the balance, inſtead of varying the places or points of ſuſpenſion in the arms themſelves.

The Deceitful Balance. This operates in the ſame manner as the ſteelyard, and cheats or deceives by having one arm a little longer than the other ; though the deception is not perceived, becauſe the ſhorter arm is made ſomewhat heavier, ſo as to compenſate for its ſhortneſs, by which means the beam of the balance, when no weights are in the ſcales, hangs horizontal in equilibrio. The conſequence of this conſtruction is, that any commodity put in the ſcale of the longer arm, requires a greater weight in the other ſcale to balance it ; and ſo the body is fallaciouſly accounted heavier than it really is. But the trick will eaſily be detected by making the body and the weight change places, removing them to the oppoſite ſcales, when the weight will immediately be ſeen to preponderate.

Aſſay-Balance. This is a very nice balance, uſed in determining the exact weights of very ſmall bodies. Its ſtructure is but little different from the common ſort ; except

Plate V

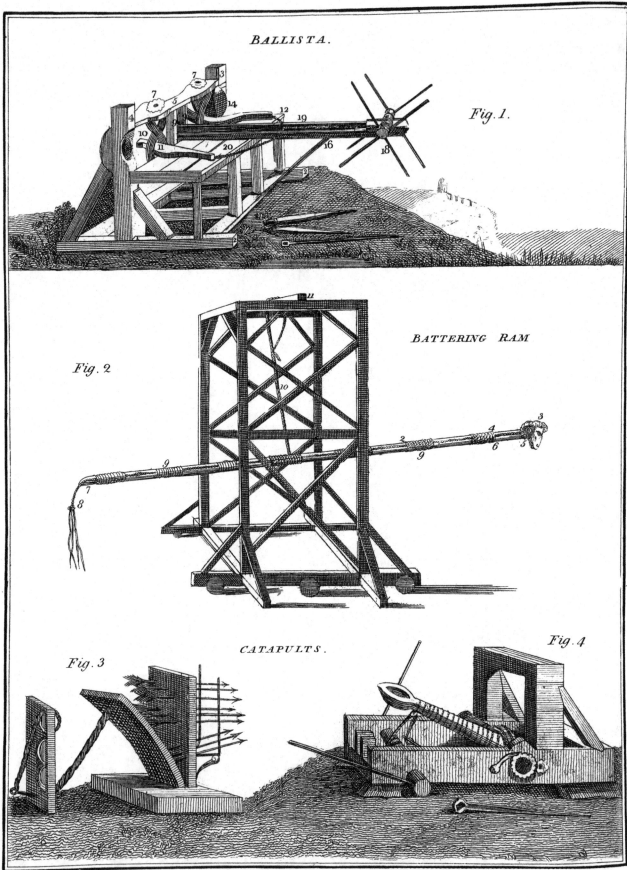

BALLISTA.

Fig. 1.

BATTERING RAM

Fig. 2

CATAPULTS.

Fig. 3

Fig. 4

except that it is made of the beft and hardeft fteel, and made to turn with the fmalleft weight.

Hydroftatical Balance. This is an inftrument for determining the fpecific gravity of bodies. See HYDRO-STATICAL, and SPECIFIC *Gravity.*

BALANCE, in Aftronomy, the fame as *Libra.*

BALANCE *of a Clock or Watch,* is that part which, by its motion, regulates and determines the beats. The circular part of it is called the *rim,* and its fpindle the *verge;* there belong to it alfo two pallets or nuts, that play in the fangs of the crown-wheel: in pocket watches, that ftrong ftud in which the lower pivot of the verge plays, and in the middle of which one pivot of the crown-wheel runs, is called the *potence:* the wrought piece which covers the balance, and in which the upper pivot of the balance plays, is the *cock;* and the fmall fpring in the new pocket watches, is called the *regulator.*

BALCONY, a projecture in the front of a houfe, or other building, commonly fupported by pillars or confoles, and encompaffed by a balluftrade.

BALL, any fpherical, globular, or round body.

BALL, in the military art, fignifies all forts of bullets for fire arms, from the piftol up to the largeft cannon. Cannon balls are made of caft iron; but the mufket and piftol balls of lead, as thefe are both heavier under the fame bulk, and do not furrow the barrels of the pieces.

BALL *of a Pendulum,* is the weight at the bottom of it; and is fometimes, efpecially in fhorter pendulums, called the *bob.*

BALLS *of Fire,* in Meteorology. See FIRE *balls.*

BALLS, in Electricity, invented by Mr. Canton, are two pieces of cork, or pith of elder tree, nicely turned in a lathe to the fize of a fmall pea, and fufpended by fine linen threads. They are ufed as electrometers, and are of excellent ufe to difcover fmall degrees of electricity; and to obferve its changes from pofitive to negative, or the reverfe; as alfo to eftimate the force of a fhock before the difcharge, fo that the operator fhall always be able to tell very nearly before hand what the explofion will be, by knowing how high he has charged his jars.

BALLISTA, a military engine much ufed by the ancients for throwing ftones, darts, and javelins; and fomewhat refembling our crofs-bows, but much larger and ftronger. The word is Latin, fignifying a crofs-bow; but is derived from the Greek βαλλω, to *fhoot,* or *throw.*

Vegetius informs us, that the ballifta difcharged darts with fuch violence and rapidity, that nothing could refift their force: and Athenæus adds, that Agiftratus made one of little more than 2 feet in length, that fhot darts 500 paces. Authors have often confounded the ballifta with the catapulta, attributing to the one what belongs to the other. According to Vitruvius, the ballifta was made after divers manners, though all were ufed to the fame purpofe: one fort was framed with levers and bars; another with pullies; fome with a crane; and others again with a toothed wheel. Marcellinus defcribes the ballifta thus; A round iron cylinder is faftened between two planks, from which reaches a hollow fquare beam placed crofs-wife, faftened with cords, to which are added fcrews. At one end of this ftands the engineer, who puts a wooden fhaft, or arrow, with a large head, into the cavity of the beam; this done, two men bend the engine, by drawing fome

VOL. I.

wheels; when the top of the head is drawn to the utmoft end of the cords, the fhaft is driven out of the ballifta, &c.

The ballifta is ranked by the ancients in the fling kind, and its ftructure and effect reduced to the principles of that inftrument; whence it is called by Hero and others, *funda,* and *fundibulus.* Gunther calls it *Balearica machina,* as a fling peculiar to the Balearic iflands.—Perrault, in his notes on Vitruvius, gives a new contrivance of a like engine for throwing bombs without powder.

Fig. 1, Plate v, reprefents the ballifta ufed in fieges, according to Folard: where 2, 2, denote the bafe of the ballifta; 3, 4, upright beams; 5, 6, tranfverfe beams; 7, 7, the two capitals in the upper tranfverfe beam, (the lower tranfverfe beam has alfo two fimilar capitals, which cannot be feen in this tranfverfe figure); 9, 9, two pofts or fupports for ftrengthening the tranfverfe beams; 10, 10, two fkains of cords faftened to the capitals; 11, 11, two arms inferted between the two ftrands, or parts of the fkains; 12, a cord faftened to the two arms; 13, darts which are fhot by the ballifta; 14, 14, curves in the upright beams, and in the concavity of which cufhions are faftened, in order to break the force of the arms, which ftrike againft them with great force when the dart is difcharged; 16, the arbor of the machine, in which a ftraight groove or canal is formed to receive the darts, in order to their being fhot by the ballifta; 17, the nut of the trigger; 18, the roll or windlafs, about which the cord is wound; 19, a hook, by which the cord is drawn towards the centre, and the ballifta cocked; 20, a ftage or table on which the arbor is in part fuftained.

BALLISTA, in Practical Geometry, the fame as the geometrical crofs, called alfo *Jacob's Staff.* See CROSS-*Staff.*

BALLISTIC PENDULUM, an ingenious machine invented by Benjamin Robins, for afcertaining the velocity of military projectiles, and confequently the force of fired gun-powder. It confifts of a large block of wood, annexed to the end of a ftrong iron ftem, having a crofs fteel axis at the other end, placed horizontally, about which the whole vibrates together like the pendulum of a clock. The machine being at reft, a piece of ordnance is pointed ftraight towards the wooden block, or ball of this pendulum, and then difcharged: the confequence is this; the ball difcharged from the gun ftrikes and enters the block, and caufes the pendulum to vibrate more or lefs according to the velocity of the projectile, or the force of the blow; and by obferving the extent of the vibration, the force of that blow becomes known, or the greateft velocity with which the block is moved out of its place, and confequently the velocity of the projectile itfelf which ftruck the blow and urged the pendulum.

The more minute and particular defcription may be feen in my Tracts, vol. 1, where are given all the rules for ufing it, and for computing the velocities, with a multitude of accurate experiments performed with cannon balls, by means of which the moft ufeful and important conclufions have been deduced in military projectiles and the nature of phyfics. I have alfo fince that publication, made many other experiments of the fame kind, by difcharging cannon balls at various diftances from the block; from which have refulted the

B b   difcovery

discovery of a complete series of the resistances of the air to balls passing through it with all degrees of velocity, from 0 up to 2000 feet in a second of time.

Other writers on this subject are Euler, Antoni, Le Roy, Darcy, &c. See also Robins's Mathematical Tracts.

BALLISTICS, is used by some for projectiles, or the art of throwing heavy bodies. Mersennus has published a treatise on the projection of bodies, under this title.

BALLOON, or BALLON, in a general sense, signifies any spherical hollow body. Thus, with chemists, it denotes a round short-necked vessel, used to receive what is distilled by means of fire: in architecture, a ball or globe on the top of a pillar, &c: and among engineers, a kind of bomb made of pasteboard, and played off in fireworks, in imitation of a real iron bomb-shell.

*Air-*BALLOON. See AEROSTATION and AIR-*Balloon*.

BALLUSTER, a small kind of column or pillar, used for ballustrades.

BALLUSTRADE, a series or row of ballusters, joined by a rail; serving for a rest to the arms, or as a fence or inclosure to balconies, altars, staircases, &c.

BAND, in Architecture, denotes any flat low member, or moulding, that is broad, but not very deep. The word *lace* sometimes means the same thing.

BANQUET, or BANQUETTE, in Fortification, a little foot-bank, or elevation of earth, forming a path along the inside of a parapet, for the soldiers to stand upon to discover the counterscarp, or to fire on the enemy, in the moat, or in the covert way. It is commonly about 3 feet wide, and a foot and a half high.

BARLOWE (WILLIAM), an eminent mathematician and divine, in the 16th century. He was born in Pembrokeshire, his father (William Barlowe) being then bishop of St. David's. In 1560 he was entered commoner of Baliol college in Oxford; and in 1564, having taken a degree in arts, he left the university, and went to sea; but in what capacity is uncertain: however he thence acquired considerable knowledge in the art of navigation, as his writings afterwards shewed. About the year 1573, he entered into orders, and became prebendary of Winchester, and rector of Easton, near that city. In 1588 he was made prebendary of Litchfield, which he exchanged for the office of treasurer of that church. He afterwards was appointed chaplain to prince Henry, eldest son of king James the first; and, in 1614, archdeacon of Salisbury. Barlowe was remarkable, especially for having been the first writer on the nature and properties of the loadstone, 20 years before Gilbert published his book on that subject. He was the first who made the inclinatory instrument transparent, and to be used with a glass on both sides. It was he also who suspended it in a compass-box, where, with 2 ounces weight, it was made fit for use at sea. He also found out the difference between iron and steel, and their tempers for magnetical uses. He likewise discovered the proper way of touching magnetical needles; and of piecing and cementing of loadstones; and also why a loadstone, being double-capped, must take up so great a weight.

Barlowe died in the year 1625.—His works are as follow:

•1. *The Navigator's Supply, containing many things of principal importance belonging to Navigation, and use of diverse Instruments framed chiefly for that purpose.* Lond. 1597, 4to; dedicated to Robert earl of Essex.

2. *Magnetical Advertisement, or Diverse Pertinent Observations and improved Experiments concerning the nature and properties of the Loadstone.* Lond. 1616, 4to.

3. *A Brief Discovery of the idle Animadversions of Mark Ridley, M. D. upon a treatise entitled Magnetical Advertisements.* Lond. 1618, 4to.

In the first of these pieces, Barlowe gave a demonstration of Wright's or Mercator's division of the meridian line, as communicated by a friend; observing that " This manner of carde has been publiquely extant in print these thirtie yeares at least [he should have said 28 only], but a cloude (as it were) and thicke miste of ignorance doth keepe it hitherto concealed: And so much the more, because some who were reckoned for men of good knowledge, have by glauncing speeches (but never by any one reason of moment) gone about what they could to disgrace it."

This work of Barlowe's contains descriptions of several instruments for the use of Navigation, the principal of which is an Azimuth Compass, with two upright sights; and as the author was very curious in making experiments on the loadstone, he treats well and fully upon the Sea-Compass. And he treated still farther on the same instrument in his second work, the Magnetical Advertisement.

BAROMETER, an instrument for measuring the weight or pressure of the atmosphere; and by that means the variations in the state of the air, foretelling the changes in the weather, and measuring heights or depths, &c.

This instrument is founded on what is called the Torricellian experiment, related below, and commonly consists of a glass tube, open at one end; which being first filled with quicksilver, and then inverted with the open end downwards into a bason of the same, the mercury descends in the tube till it remains at about the height of 29 or 30 inches, according to the weight or pressure of the atmosphere at the time, which is just equal to the weight of that column of the quicksilver. Hence it follows that, if by any means the pressure of the air be altered, it will be indicated by the rising or falling of the mercury in the tube; or if the barometer be carried to a higher station, the quicksilver will descend lower in the tube, but when carried to a lower place, it will rise higher in the tube, according to the difference in elevation between the two places.

History of the Barometer —About the beginning of the last century, when the doctrine of a plenum was in vogue, it was a common opinion among philosophers, that the ascent of water in pumps was owing to what they called nature's abhorrence of a vacuum; and that thus fluids might be raised by suction to any height whatever. But an accident having just discovered, that water could not be raised in a pump unless the sucker reached to within 33 feet of the water in the well, it was conjectured by Galileo, who flourished about that time, that there might be some other cause of the ascent of water in pumps, or at least that this abhorrence was limited to the finite height of 33 feet. Being unable to satisfy himself on this head, he recommended the consideration of the difficulty to Torricelli, who had been his disciple. After some time Torricelli fell upon the suspicion that the pressure of the atmosphere was the cause of the ascent of water in pumps; that a column

of

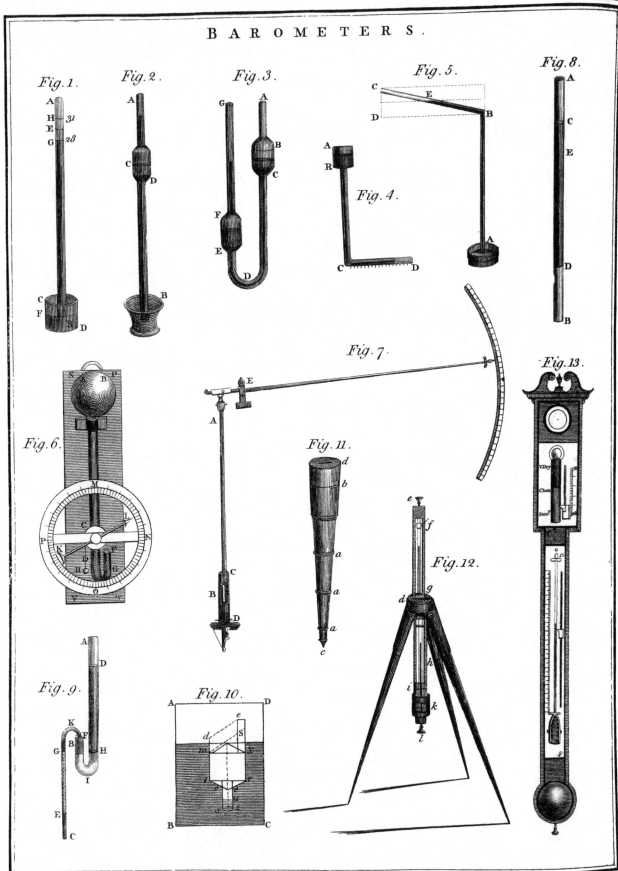

BAROMETERS.

Plate IV

Fig. 1. *Fig. 2.* *Fig. 3.* *Fig. 5.* *Fig. 8.*

Fig. 4.

Fig. 7.

Fig. 6.

Fig. 11.

Fig. 13.

Fig. 12.

Fig. 9.

Fig. 10.

of water 33 feet high was a juft counterpoife to a column of air, of the fame bafe, and which extended up to the top of the atmofphere; and that this was the true reafon why the water did not follow the fucker any farther. And this fufpicion was foon after confirmed by various experiments. Torricelli confidered, that if a column of water 33 feet high were a counterpoife to a whole column of the atmofphere, then a column of mercury of about 2 feet and a half high would alfo be a counterpoife to it, fince quickfilver is near 14 times heavier than water, and fo the 14th part of the height, or near 2 feet and a half, would be as heavy as the column of water. This reafoning was foon verified; for having filled a glafs tube with quickfilver, and inverted it into a bafon of the fame, the mercury prefently defcended till its height, above that in the bafon, was about two feet and a half, juft as he expected. And this is what has, from him, been called the Torricellian experiment.

The new opinion, with this confirmation of it, was readily acquiefced in by moft of the philofophers, who repeated the experiment in various ways. Others however ftill adhered to the old doctrine, and raifed feveral pretended objections againft the new one; fuch as that there was a film or imperceptible *rope of mercury*, extended through the upper part of the tube, which fufpended the column of mercury, and kept it from falling into that in the bafon. This and other objections were however foon overcome by additional confirmations of the true doctrine, particularly by varying the elevation of the place. It was hinted by Defcartes and Pafcal, that if the mercury be fuftained in the tube by the preffure of the atmofphere, by carrying it to a higher fituation, it would defcend lower in the tube, having a fhorter column of the atmofphere to fuftain it, and vice verfa. And Pafcal engaged his brother-in-law, M. Perier, to try that experiment for him, being more conveniently fituated for that purpofe than he was at Paris. This he accordingly executed, by obferving the height of the quickfilver in the tube, firft at the bottom of a mountain in Auvergne, and then at feveral ftations, or different altitudes, in afcending, by which it was found that the mercury fell lower and lower all the way to the top of the mountain; and fo confirming the truth of the doctrine relating to the univerfal preffure of the atmofphere, and the confequent fufpenfion of the mercury in the tube of the barometer. Thus, by the united endeavours of Torricelli, Defcartes, Pafcal, Merfenne, Huygens, and others, the caufe of the fufpenfion of the quickfilver in the tube of the barometer, became pretty generally eftablifhed.

It was fome time however after this general confent, before it was known that the preffure of the air was various at different times, in the fame place. This could not however remain long unknown, as the frequent meafuring of the column of mercury, muft foon fhew its variations in altitude; and experience and obfervation would prefently fhew that thofe variations in the mercurial column, were always fucceeded by certain changes in the weather, as to rain, wind, frofts, &c. Hence this inftrument foon came into ufe as the means of foretelling the changes of the weather; and on this account it obtained the name of the *weather-glafs*, as it did that of *barometer* from its being the meafure of the weight or preffure of the air. We may now proceed to take a view of its various forms and ufes.

The Common Barometer. This is reprefented at fig. 1, plate iv, fuch as it was invented by Torricelli. AB is a glafs tube, of $\frac{1}{4}$, or $\frac{1}{3}$, or $\frac{1}{2}$ inch wide, the more the better, and about 34 inches long, being clofe at the top A, and the open end B immerfed in a bafon of quickfilver CD, which is the better the wider it is. To fill this, or any other barometer; take a clean new glafs tube, of the dimenfions as above, and pour into it well purified quickfilver, with a fmall funnel either of glafs or paper, in a fine continued ftream, till it wants about half an inch or an inch of being full; then ftopping it clofe with the finger, invert it flowly, and the air in the empty part will afcend gradually to the other end, collecting into itfelf fuch other fmall air bubbles as unavoidably get into the tube among the mercury, in filling it with the funnel: and thus continue to invert it feveral times, turning the two ends alternately upwards, till all the air bubbles are collected, and brought up to the open end of the tube, and when the part filled fhall appear, without fpeck, like a fine polifhed fteel rod. This done, pour in a little more quickfilver, to fill the empty part quite full, and fo exclude all air from the tube: then, ftopping the orifice again with the finger, invert the tube, and immerfe the finger and end, thus ftopped, into a bafon of like purified quickfilver; in this pofition withdraw the finger, fo fhall the mercury defcend in the tube to fome place as E, between 28 and 31 inches above that in the bafon at F, as thefe are the limits between which it always ftands in this country on the common furface of the earth. Then meafure, from the furface of the quickfilver in the bafon at F, 28 inches to G, and 31 inches to H, dividing the fpace between them into inches and tenths, which are marked on a fcale placed againft the fide of the tube; and the tenths are fubdivided into hundredth parts of an inch by a fliding index carrying a vernier or nonius. Thefe 3 inches, between 28 and 31, fo divided, will anfwer for all the ordinary purpofes of a ftationary or chamber barometer; but for experiments on altitudes and depths, it is proper to have the divifions carried on a little higher up, and a great deal lower down. In the proper filling and otherwife fitting up of the barometer, feveral circumftances are to be carefully noted; as, that the bore of the tube be pretty wide, to allow the freer motion of the quickfilver, without being impeded by an adhefion to the fides; that the bafon below it be alfo pretty large, in order that the furface of the mercury at F may not fenfibly rife or fall with that in the tube; that the bottom of the tube be cut off rather obliquely, that when it refts on the bottom of the bafon there may be a free paffage for the quickfilver: and that, to have the quickfilver very pure, it is beft to boil it in the tube, which will expel all the air from it. This barometer is commonly fitted up in a neat mahogany cafe, together with a thermometer and hygrometer, as reprefented in plate 4, fig. 13.

As the fcale of variation is but fmall, being included within 3 inches in the common barometer, feveral contrivances have been devifed to enlarge the fcale, or to render the motion of the quickfilver more fenfible.

Defcartes firft fuggefted a method of increafing the fenfibility, which was executed by Huygens. This

was

was effected by making the barometrical tube end in a pretty large cylindrical veffel at top, into which was inferted alfo the lower or open end of a much finer tube than the former, which was partly filled with water, to give little obftruction by its weight to the motion of the mercury, while it moved through a pretty long fpace of the very fine tube by a fmall variation of the mercury below it, and fo rendered the fmall changes in the ftate of the air very fenfible. But the inconvenience was this, that the air contained in the water gradually difengaged itfelf, and efcaped through into the vacuum in the top of the fmall tube, till it was collected in a body there, and by its elafticity preventing the free rife of the fluids in the tubes, fpoiled the inftrument as a barometer. And this, it may be obferved by-the-bye, is the reafon why a water barometer cannot fucceed. This barometer is here reprefented in fig. 2, where CD is the veffel, in which are united the upper or fmall water tube AC, with the lower or mercurial one CB.

To remedy this inconvenience, Huygens thought of placing the mercury at top, and the water at bottom, which he thus contrived. ADG (fig. 3) is a bent tube hermetically fealed at A, but open at G, of about one line in diameter, and paffing through the two equal cylindrical veffels BC, EF, which are about 20 inches apart, and of 15 lines diameter, their length being 10. The mercury being put into the tube, will ftand between the middle of the veffels EF and BC, the remaining fpace to A being void both of air and mercury. Laftly, common water, tinged with a 6th part of aqua regis, to prevent its freezing, is poured into the tube FG, till it rifes a foot above the mercury in DF. To prevent the water from evaporating, a drop of oil of fweet almonds floats on the top of it. But the column of water will be fenfibly affected by heat and cold, which fpoils the accuracy of the inftrument. For which reafon other contrivances have been made, as below.

The Horizontal or Rectangular Barometer, fig. 4, was invented by J. Bernoulli and Caffini; where AB is a pretty wide cylindrical part at the top of the tube, which tube is bent at right angles at C, the lower part of it CD being turned into the horizontal direction, and clofe above at A, but open at the lower end D, where however the mercury cannot run out, being there oppofed by the preffure of the atmofphere. This and the foregoing contrivance of Huygens are founded on the theorem in hydroftatics, that fluids of the fame bafe prefs according to their perpendicular altitude, not according to the quantity of their matter; fo that the fame preffure of the atmofphere fuftains the quickfilver that fills the tube ACD, and the ciftern B, as would fupport the mercury in the tube alone. Hence, having fixed upon the fize of the fcale, as fuppofe the extent of 12 inches, inftead of the 3, in the common barometer from 28 to 31, that is 4 times as long; then the area of a fection of the cylinder AB muft be 4 times that of the tube, and confequently its diameter double, fince the areas of circles are as the fquares of their diameters: then for every natural variation of an inch in the cylinder AB, there will be a variation of four inches in the tube CD.—But on account of the attrition of the mercury againft the fides of the glafs, and the great momentum from the quick motion in CD, the quickfilver is apt to break, and the rife and fall is no longer equa-

ble; and befides, the mercury is apt to be thrown out of the orifice at D by fudden motions of the machine.

The Diagonal Barometer of Sir Samuel Moreland, fig. 5, is another method of enlarging the natural fcale of three inches perpendicular, or CD, by extending it to any length BC in an oblique direction. This is liable in fome degree to the fame inconvenience, from friction and breaking, as the horizontal one; and hence it is found that the diagonal part BC cannot properly be bent from the perpendicular more than in an angle of 45°, which only increafes the fcale nearly in the proportion of 7 to 5.

Doctor Hook's Wheel Barometer, fig. 6. This was invented about 1668, and is meant to render the alterations in the air more fenfible. Here the barometer tube has a large ball AB at top, and is bent up at the lower or open end, where an iron ball G floats on the top of the mercury in the tube, to which is connected another ball H by a cord, hanging freely over a pulley, turning an index KL about its centre. When the mercury rifes in the part FG, it raifes the ball, and the other ball defcends and turns the pulley with the index round a graduated circle from N towards M and P; and the contrary way when the quickfilver and the ball fink in the bent part of the tube. Hence the fcale is eafily enlarged 10 or 12 fold, being increafed in proportion of the axis of the pulley to the length of the index KL. But then the friction of the pulley and axis is fome obftruction to the free motion of the quickfilver. Contrivances to leffen the friction &c, may alfo be feen in the *Philof. Tranf.* vol. 52, art. 29, and vol. 60, art. 10.

The Steelyard Barometer, for fo that may be called which is reprefented by fig. 7, which enlarges the fcale in the proportion of the fhorter to the longer arm of a fteelyard. AB is the barometer tube, clofe at A and open at B, immerfed in a cylindrical glafs ciftern CD, which is but very little wider than the tube AB is. The barometer tube is fufpended to the fhorter arm of an index like a fteelyard, moving on the fulcrum E, and the extremity of its longer arm pointing to the divifions of a graduated arch, with which index the tube is nearly in equilibrio. When the preffure of the atmofphere is leffened, the mercury defcends out of the tube into the ciftern which raifes the tube and the fhorter arm of the index, and confequently the extremity of the longer index, moves downwards, and paffes over a part of the graduated arch. And on the contrary this moves upwards when the preffure of the atmofphere increafes.

The Pendant Barometer, fig. 8, was invented by M. Amontons, in 1695. It confifts of a fingle conical tube AB, hung up by a thread, the larger or open end downwards, and having no veffel or ciftern, becaufe the conical figure fupplies that, and the column of mercury fuftained is always equal to that in the common barometer tube; which is effected thus; when the preffure of the air is lefs, the mercury finks down to a lower and wider part of the tube, and confequently the altitude of its column will be lefs; and on the contrary, by a greater preffure of the atmofphere the mercury is forced up to a higher and narrower part, till the length of the column CD be equal to that in the tube of the common barometer.—The inconvenience of this barometer is, that as the bore muft be made very fmall, to prevent the mercury from falling out by an accidental fhake, the friction

friction and adhesion to the sides of the tube prevent the free motion of the mercury.

Mr. Rowning's Compound Barometers. This gentleman has several contrivances for enlarging the scale, and that in any proportion whatever. One of these is described in the *Philof. Tranf.* N° 427, and also in his *Nat. Philof.* part 2; and another in the same part, which is here represented at fig. 9. ABC is a compound tube, hermetically sealed at A, and open at C; empty from A to D, filled with mercury from thence to B, and from hence to E with water. Hence by varying the proportions of the two tubes AF and FC, the scale of variation may be changed in any degree.

The Marine Barometer. This was first invented by Dr. Hook, to be used on board of ship, being contrived so as not to be affected or injured by the motion of the ship. His contrivance consisted of a double thermometer, or a couple of tubes half filled with spirit of wine; the one sealed at both ends, with a quantity of air included; the other sealed at one end only. The former of these is affected only by the warmth of the air; but the other is affected both by the external warmth and by the variable pressure of the atmosphere. Hence, considering the spirit thermometer as a standard, the excess of the rise or fall of the other above it will shew the increase or decrease of the pressure of the atmosphere. This instrument is described by Dr. Halley, in the *Philof. Tranf.* N° 269, where he says of it, " I had one of these barometers with me in my late southern voyage, and it never failed to prognosticate and give early notice of all the bad weather we had, so that I depended thereon, and made provision accordingly; and from my own experience I conclude, that a more useful contrivance hath not for this long time been offered for the benefit of navigation."

Mr. Nairne, an ingenious artist in London, has lately invented a new kind of *Marine Barometer*; which differs from the common barometer by having the lower part of the tube, for about 2 feet long, made very small, to check the vibrations of the mercury, which would otherwise arise from the motions of the ship. This is also assisted by being hung in gimbals, by a part which subjects it to be the least affected by such motions.

Another sort of Marine Barometer has also been invented by M. Passemente, an ingenious artist at Paris. This contrivance consists only in twisting the middle of the tube into a spiral of two revolutions; by which contrivance the impulses which the mercury receives from the motions of the ship are destroyed, by being transmitted in contrary directions.

The Statical Barofcope, or *Barometer,* of Mr. Boyle, &c. This consists of a large glass bubble, blown very thin, and then balanced by a small brass weight Hence these two bodies being of unequal bulk, the larger will be very much affected by a change of the density of the medium, but the less not at all as to sense: So that, when the atmosphere becomes denser, the ball loses more of its weight, and the brass weight preponderates; and contrariwise when the air grows lighter.

Mr. Cafwell's Barofcope, or *Barometer.* This is described in the *Philof. Tranf.* vol. 24, and seems to be the most sensible and exact of any. It is thus described: Suppose ABCD, (fig. 10) is a bucket of water, in

which is the baroscope *xrezyofm,* which consists of a body *xrfm,* and a tube *ezyo,* which are both concave cylinders, made of tin, or rather glass, and communicating with each other. The bottom of the tube *zy* has a leaden weight to sink it, so that the top of the body may just swim even with the surface of the water by the addition of some grain weights on the top. When the instrument is forced with its mouth downwards, the water ascends into the tube to the height *yu.* To the top is added a small concave cylinder, or pipe, to keep the instrument from sinking down to the bottom: *md* is a wire: and *mS, de* are two threads oblique to the surface of the water, which perform the office of diagonals: for while the instrument sinks more or less by an alteration in the gravity of the air, where the surface of the water cuts the thread is formed a small bubble, which ascends up the thread while the mercury of the common baroscope ascends, and vice versa.

It appears from a calculation which the author makes, that this instrument shews the alterations in the air 1200 times more accurately than the common barometer. He observes, that the bubble is seldom known to stand still even for a minute; that a small blast of wind, which cannot be heard in a chamber, will sensibly make it sink; and that a cloud passing over it always makes it descend, &c.

While some have been increasing the sensibility of the barometer by enlarging the variations, others have endeavoured to make it more convenient by reducing the length of the tube. *M. Amontons,* in 1688, first proposed this alteration in the structure of barometers, by joining several tubes to one another, alternately filled with mercury and with air, or some other fluid; and the number of these tubes may be increased at pleasure: but the contrivance is perhaps more ingenious than useful.

M. Mairan's reduced Barometer, which is only 3 inches long, serves the purpose of a manometer, in shewing the dilatations of the air in the receiver of an air-pump; and instruments of this kind are now commonly applied to this use.

The Portable Barometer, is so contrived that it may be carried from one place to another without being disordered. The end of the tube is tied up in a leathern bag not quite full of mercury; which being pressed by the air, forces the mercury into the tube, and keeps it suspended at its proper height. This bag is usually inclosed in a box, through the bottom of which passes a screw, by means of which the mercury may be forced up to the top of the tube, and prevented from breaking it by dashing against the top when the instrument is removed from one station to another. It seems Mr. Patrick first made a contrivance of this kind: but the portable barometer has received various improvements since; and the most complete of this kind has been described by M. De Luc, in his *Recherches,* vol. 2, pa. 5 &c, together with the apparatus belonging to it, the method of construction and use, and the advantages attending it. Improvements have also been suggested by Sir George Shuckburgh, and Col. Roy, which have been carried into execution, with farther improvements also, by Mr. Ramsden, and other ingenious artists in London.

3 Fig.

Fig. 11 reprefents this inftrument, as inclofed in its mahogany cafe by means of three metallic rings *a a a*. This cafe is a hollow cone, fo fhaped within as to contain fteadily the body of the barometer, and is divided into three branches from *b* to *c*, forming three legs or fupports for the inftrument when obfervations are making, and fuftaining it at the part *d* of the cafe, as it appears in Fig. 12, by an improved kind of gimbals, in which its own weight renders it fufficiently fteady at any time. In the part of the frame *f g* where the barometer tube appears, is made a long flit or opening, that the column of mercury may be feen againft the light, and the vernier piece *f* brought down to coincide very nicely with the edge of the mercury. When the inftrument is fixed in its ftand, the fcrew *l* is to be turned to let the mercury down to its proper ftation, and a peg at *i* muft be loofened, to admit the external air to act upon the mercury contained in the box *k*. The proper adjuftment, or mode of obferving what is called the *zero* or *o* divifion of the column of mercury, is by obferving it in the tranfparent part of the box *k*, which has a glafs tube or refervoir for the quickfilver, and an edged piece of metal attached to the external part of it; with the edge of which the mercury is to be brought into contact by turning the fcrew *l* to the right or left as occafion requires. The vernier piece at *f*, which determines the altitude of the mercurial column, is firft brought down by the hand to a near contact, and then accurately adjufted by turning the fcrew *e* at the top. The divifions annexed to the tube of this inftrument may be of any fort, or of any degree of fmallnefs, according to the purpofes it is intended to ferve. To accommodate it to the ufe of foreigners as well as natives, there are commonly added fcales of both French and Englifh inches, with their fubdivifions to any extent required. It is ufual to place the French fcale of inches on the right fide at *f g*, from 19 to 31 inches, meafured from the zero or furface of the mercury in the box *k* below; each inch being divided into lines or 12th parts, and each line fubdivided by the vernier into 10th parts, or 120th parts of inches; by means of which therefore the length of the mercurial column may be determined to the 120th part of a French inch. The other fcale, which is placed on the left fide of the inftrument, is divided into Englifh inches, and each inch into 20th parts, which by a vernier are fubdivided into 25th parts, or 500th parts of inches; by this means fhewing the height of the mercury to the 500th part of an Englifh inch. But this vernier is figured double or each divifion is accounted 2, which reduces the meafures to 1000ths of an inch for the conveniency of calculation, in meafuring altitudes of hills &c.

A thermometer is always attached to the inftrument, as a neceffary appendage to it, being faftened to the body at *h*, and funk into the furface of the frame, to preferve it from injury: the degrees of this thermometer are marked on two fcales, one on each fide of it, viz, the fcale of Fahrenheit, and that of Reaumur; the freezing point of the former being at 32, and of the latter at 0. Alfo on the right hand fide of thefe two fcales there is a third, called a fcale of *correction*,

placed oppofitely to that of Fahrenheit, with the word *add* and *fubtract* marked; which fhews the neceffary correction of the obferved altitude of the mercury at any given temperature of the air, indicated by the thermometer.

There are feveral other pieces of mechanifm about the inftrument, which will be evident by infpection; and the manner of making the obfervations, with the neceffary calculations, are fully explained in M. de Luc's *Recherches fur les Modifications de l'Atmofphere*, and the *Philof. Tranf.* vol. 67 and 68, before cited.

The Common Chamber Weatherglafs, is alfo ufually fitted up in a neat mahogany frame, and other embellifhments, to make it an ornamental piece of furniture. It confifts of the common tube barometer, with a thermometer by the fide of it, and an hygrometer at the top, as exhibited in fig. 13.

To the foregoing may be added a new fort of *Barometer*, or *Weather Inftrument by the Sound of a Wire*. This is mentioned by M. Lazowfki in his Tour through Switzerland: it is as yet but in an imperfect ftate, and was lately difcovered there by accident. It feems that a clergyman, though near-fighted, often amufed himfelf with firing at a mark, and contrived to ftretch a wire fo as to draw the mark to him to fee how he had aimed. He obferved that the wire fometimes founded as if it vibrated like a mufical cord; and that after fuch foundings, a change always enfued in the ftate of the atmofphere; from whence he came to predict rain or fine weather. On making farther experiments, it was found that the founds were moft diftinct when extended in the plane of the meridian. And according to the weather which was to follow, it was found that the founds were more or lefs foft, or more or lefs continued; alfo fine weather, it is faid, was announced by the tones of counter-tenor, and rain by thofe of bafs. It has been faid that M. Volta mounted 15 chords in this way at Pavia, to bring this method to fome precifion, but no accounts have yet appeared of the fuccefs of his obfervations.

The Phænomena and Obfervations of the Barometer [*]
The phænomena of the barometer are various; but authors are not yet agreed upon the caufes of them; nor is the ufe of it, as a weather-glafs, yet perfectly afcertained, though daily obfervations and experience lead us ftill nearer to precifion. Mr. Boyle obferves that the phænomena of the barometer are fo precarious, that it is exceedingly difficult to form any certain general rules concerning the rife and fall of the mercury. Even that rule fails which feems to hold the moft generally, viz, that the mercury is low in high winds. The beft rules however that have been deduced by feveral authors are as follow.

Dr Halley's Rules for judging of the Weather.

1. In calm weather, when the air is inclined to rain, the mercury is commonly low.

2. In ferene, good, and fettled weather, the mercury is generally high.

3. Upon very great winds, though they be not accompanied with rain, the mercury finks loweft of all, according to the point of the compafs the wind blows from.

---

[*] An ingenious author obferves that, by means of barometers we may regain the knowledge that ftill refides in brutes, and which we forfeited by not continuing in the open air, as they moftly do; and, by our intemperance, corrupting the *crafis* of our organs of fenfe.

4. The

4. The greateſt heights of the mercury are found upon eaſterly or north-eaſterly winds, other circumſtances alike.

5. In calm froſty weather, the mercury commonly ſtands high.

6. After very great ſtorms of wind, when the mercury has been very low, it generally riſes again very faſt.

7. The more northerly places have greater alterations of the barometer than the more ſoutherly, near the equator.

8. Within the tropics, and near them, there is little or no variation of the barometer, in all weathers. For inſtance, at St. Helena it is little or nothing, at Jamaica 3-10ths of an inch, and at Naples the variation hardly ever exceeds an inch; whereas in England it amounts to 2 inches and a half, and at Peterſburgh to 3½ nearly.

Dr. Beal, who followed the opinion of M. Paſcal, obſerves that, *cæteris paribus*, the mercury is higher in cold weather than in warm: and in the morning and evening uſually higher than at mid-day.—That in ſettled and fair weather, the mercury is higher than either a little before or after, or in the rain; and that it commonly deſcends lower after rain than it was before it. And he aſcribes theſe effects to the vapours with which the air is charged in the former caſe, and which are diſperſed by the falling rain in the latter. If it chance to riſe higher after rain, it is uſually followed by a ſettled ſerenity. And that there are often great changes in the air, without any perceptible alteration in the barometer.

Mr Patrick's Rules for judging of the Weather. Theſe are eſteemed the beſt of any general rules hitherto made:

1. The riſing of the mercury preſages, in general, fair weather; and its falling, foul weather, as rain, ſnow, high winds, and ſtorms.

2. In very hot weather, the falling of the mercury indicates thunder.

3. In winter, the riſing preſages froſt: and in froſty weather, if the mercury falls 3 or 4 diviſions, there will certainly follow a thaw. But in a continued froſt, if the mercury riſes, it will certainly ſnow.

4. When foul weather happens ſoon after the falling of the mercury, expect but little of it; and on the contrary, expect but little fair weather when it proves fair ſhortly after the mercury has riſen.

5. In foul weather, when the mercury riſes much and high, and ſo continues for 2 or 3 days before the foul weather is quite over, then expect a continuance of fair weather to follow.

6. In fair weather, when the mercury falls much and low, and thus continues for 2 or 3 days before the rain comes; then expect a great deal of wet, and probably high winds.

7. The unſettled motion of the mercury, denotes uncertain and changeable weather.

8. You are not ſo ſtrictly to obſerve the words engraved on the plates, as the mercury's riſing and falling; though in general it will agree with them. For if it ſtands at *much rain*, and then riſes up to *changeable*, it preſages fair weather; though not to continue ſo long as if the mercury had riſen higher. And ſo, on

the contrary, if the mercury ſtood at *fair*, and falls to *changeable*, it preſages foul weather; though not ſo much of it as if it had ſunk lower.

Upon theſe rules of Mr Patrick, the following *Remarks* are made by *Mr Rowning*. That it is not ſo much the abſolute height of the mercury in the tube that indicates the weather, as its motion up and down: wherefore, to paſs a right judgment of what weather is to be expected, we ought to know whether the mercury is actually riſing or falling; to which end the following rules are of uſe.

1. If the ſurface of the mercury is convex, ſtanding higher in the middle of the tube than at the ſides, it is a ſign that the mercury is then riſing.

2. But if the ſurface be concave, or hollow in the middle, it is then ſinking. And,

3. If it be plain, or rather a very little convex, the mercury is ſtationary: for mercury being put into a glaſs tube, eſpecially a ſmall one, naturally has its ſurface a little convex, becauſe the particles of mercury attract one another more forcibly than they are attracted by glaſs. Farther,

4. If the glaſs be ſmall, ſhake the tube; then if the air be grown heavier, the mercury will riſe about half a 10th of an inch higher than it ſtood before; but if it be grown lighter, it will ſink as much. And, it may added, in the wheel or circular barometer, tap the inſtrument gently with the finger, and the index will viſibly ſtart forwards or backwards according to the tendency to riſe or fall at that time. This proceeds from the mercury's ſticking to the ſides of the tube, which prevents the free motion of it till it be diſengaged by the ſhock: and therefore when an obſervation is to be made with ſuch a tube, it ought to be firſt ſhaken; for ſometimes the mercury will not vary of its own accord, till the weather is preſent which it ought to have indicated.

And to the foregoing may be added the following additional rules, more accurately drawn from later and more cloſe obſervation of the motions of the barometer, and the conſequent changes in the air in this country.

1. In winter, ſpring, and autumn, the ſudden falling of the mercury, and that for a large ſpace, denotes high winds and ſtorms; but in ſummer it denotes heavy ſhowers, and often thunder: and it always ſinks loweſt of all for great winds, though not accompanied with rain; though it falls more for wind and rain together than for either of them alone. Alſo, if, after rain, the wind change into any part of the north, with a clear and dry ſky, and the mercury riſe, it is a certain ſign of fair weather.

2. After very great ſtorms of wind, when the mercury has been low, it commonly riſes again very faſt. In ſettled fair and dry weather, except the barometer ſink much, expect but little rain; for its ſmall ſinking then, is only for a little wind, or a few drops of rain; and the mercury ſoon riſes again to its former ſtation. In a wet ſeaſon, ſuppoſe in hay-time and harveſt, the ſmalleſt ſinking of the mercury muſt be minded; for when the conſtitution of the air is much inclined to ſhowers, a little ſinking in the barometer then denotes more rain, as it never then ſtands very high. And, if

in such a season, it rise suddenly, very fast, and high, expect not fair weather more than a day or two, but rather that the mercury will fall again very soon, and rain immediately to follow: the flow gradual rising, and keeping on for 2 or 3 days, being most to be depended on for a week's fair weather. And the unset tled state of the quickfilver always denoting uncertain and changeable weather, especially when the mercury stands any where about the word *changeable* on the scale.

3. The greatest heights of the mercury, in this country, are found upon easterly and north-easterly winds; and it may often rain or snow, the wind being in these points, and the barometer sink little or none, or it may even be in a rising state, the effect of those winds counteracting. But the mercury sinks for wind, as well as rain, in all the other points of the compass; but rises as the wind shifts about to the north or east, or between those points: but if the barometer should sink with the wind in that quarter, expect it soon to change from thence; or else, should the fall of the mercury be much, a heavy rain is then likely to ensue, as it sometimes happens.

Cause of the Phænomena of the Barometer.

To account for the foregoing phænomena of the barometer, many hypotheses have been framed, which may be reduced to two general heads, viz, *mechanical* and *chemical*. The chief writers upon these causes, are Pascal, Beal, Wallis, Garcin, Garden, Lister, Halley, Garsten, De la Hire, Mariotte, Le Cat, Woodward, Leibnitz, De Mairan, Hamberger, D. Bernoulli, Muschenbroek, Chambers, De Luc, Black, &c; and an account of most of their hypotheses may be seen at large in M. De Luc's *Recherches sur les Modifications de l'Atmosphere*, vol. 1. chap. 3; see also the *Philos. Transf.* and various other works on this subject. It may suffice to notice here slightly a few of the principal of them.

Dr. Lister accounts for the changes of the barometer from the alterations by heat and cold in the mercury itself; contracting by cold, and expanding by heat. But this, it is now well known, is quite insufficient to account for the whole of the effect.

The changes in the weight or pressure of the atmosphere must therefore be regarded as the principal cause of those in the barometer. But then, the difficulty will be to assign the cause of that cause, or whence arise those alterations that take place in the atmosphere, which are sometimes so great as to alter its pressure by the 10th part of the whole quantity. It is probable that the winds, as driven about in different directions, have a great share in them; vapours and exhalations, rising from the earth, may also have some share; and some perhaps the flux and reflux occasioned in the air by the moon; as well as some chemical causes operating between the different particles of matter.

Dr. Halley thinks the winds and exhalations sufficient; and on this principle gives a theory, the substance of which may be comprised in what follows:

1st, That the winds must alter the weight of the air in any particular country; and this, either by bringing together a greater quantity of air, and so load-ing the atmosphere of any place; which will be the case as often as two winds blow from opposite parts, at the same time, towards the same point: or by sweeping away some part of the air, and giving room for the atmosphere to expand itself; which will happen when two winds blow opposite ways from the same point at the same time: or lastly by cutting off the perpendicular pressure of the air; which is the case when a single wind blows briskly any way; it being found by experience, that a strong blast of wind, even made by art, will render the atmosphere lighter; and hence the mercury in a tube below it, as well as in others more distant, will considerably subside. See *Philos. Transf.* Nº 292.

2dly, That the cold nitrous particles, and even the air itself condensed in the northern regions, and driven elsewhere, must load the atmosphere, and increase its pressure.

3dly, That heavy dry exhalations from the earth must increase the weight of the atmosphere, as well as its elastic force; as we find the specific gravity of menstruums increased by dissolved salts and metals.

4thly, That the air being rendered heavier by these and the like causes, is thence better able to support the vapours; which being likewise intimately mixed with it, make the weather serene and fair. Again, the air being made lighter from the contrary causes, it becomes unable to support the vapours with which it is replete; these therefore precipitating, are collected into clouds, the particles of which in their progress unite into drops of rain.

Hence he infers, it is evident that the same causes which increase the weight of the air, and render it more able to support the mercury in the barometer, do likewise produce a serene sky, and a dry season; and that the same causes which render the air lighter, and less able to support the mercury, do likewise generate clouds and rain.

But these principles, though well adapted to many of the particular cases of the barometer, seem however to fall short of some of the principal and most obvious ones, besides being liable to several objections.

Leibnitz accounted for the fall of the mercury before rain by another principle, viz, That as a body specifically lighter than a fluid, while it is sustained by it, adds more weight to that fluid than when, by being reduced in bulk, it becomes specifically heavier, and descends; so the vapour, after it is reduced into the form of clouds, and descends, adds less weight to the air than it did before; and hence the mercury sinks in the tube.—But here, granting that the drops of rain formed from the vapours always increasing in size as they fall lower, were continually accelerated also in their motion, and so the air suffer a continued loss of their weight as they descend; it may however be objected, that by the descent of the mercury the rain is foretold a much longer time before it comes, than the vapour can be supposed to take up in falling: that many times, and in different places, there falls a great deal of rain, without any sinking of the mercury at all; as also that there often happens a fall of the mercury without any rain ensuing: and that sometimes the mercury will suddenly sink, in a short space of of time, half an inch or more, which answers to 7 inches of rain, or about one third of the whole quantity falling in the whole year.

Mr. *De Luc* supposes that the changes observed in the pressure of the atmosphere, are chiefly produced by the greater or less quantity of vapours floating in it: as others have attributed them to the same cause, but have given a different explanation of it. His opinion is, that vapours diminish the specific gravity, and consequently the absolute weight, of those columns of the atmosphere into which they are received, and which, notwithstanding this admixture, still remain of the same height with adjoining columns that consist of pure or dry air. He afterwards vindicates and more fully explains this theory, and applies it to the solution of the principal phenomena of the barometer, as depending on the varying density and weight of the atmosphere.

Dr. *James Hutton*, in his *Theory of Rain*, printed in the Transactions of the Royal Society of Edinburgh, vol. 1, gives ingenious and plausible reasons for thinking that the lessening the weight of the atmosphere by the fall of rain, is not the cause of the fall of the barometer; but that the principal, if not the only cause, arises from the commotions in the atmosphere, which are chiefly produced by sudden changes of heat and cold in the air. "The barometer, says he, is an instrument necessarily connected with motions in the atmosphere; but it is not equally affected with every motion in that fluid body. The barometer is chiefly affected by those motions by which there are produced accumulations and abstractions of this fluid, in places or regions of sufficient extent to affect the pressure of the atmosphere upon the surface of the earth. But as every commotion in the atmosphere may, under proper conditions, be a cause for rain, and as the want of commotion in the atmosphere is naturally a cause of fair weather, this instrument may be made of great importance for the purpose of meteorological observations, although not in the certain and more simple manner in which it has been, with the increase of science, so successfully applied to the measuring of heights." See RAIN.

In the *Encyclopædia Britannica* there is another theory of the changes in the barometer, as depending on the *heat* in the atmosphere, not as producing commotions there, but as altering the specific gravity of the air by the changes of heat and cold. The principles of this theory are, 1st, That vapour is formed by an intimate union between the elements of fire and water, by which the fire or heat is so totally enveloped, and its action so perfectly suspended by the aqueous particles, that it not only loses its properties of burning and of giving light, but becomes incapable of affecting the most sensible thermometer, in which case it is said to be in a latent state: and 2d, That if the atmosphere be affected by any unusual degree of heat, it thence becomes incapable of supporting so long a column of mercury as before; for which reason it is that the barometer sinks.

From these axioms it would follow, that as vapour is formed by an union of fire with water, whether by attraction or a solution of the water in the fire, the vapour cannot be condensed till this union, attraction,

or solution, is at an end. Hence the beginning of the condensation of the vapour, or the first signs of approaching rain, must be the separation of the fire which is latent in the vapour. In the beginning, this may be either slow and partial, or it may be sudden and violent: in the first case, the rain will come on slowly, and after a considerable time; but in the other, it will come very quickly, and in a great quantity. But Dr. Black has proved, that when fire quits its latent state, however long it may have lain dormant and insensible, it always reassumes its proper qualities, and affects the thermometer just the same as if it had never been absorbed. The consequence of this is, that in proportion as the latent heat is discharged from the vapour, those parts of the atmosphere into which it is discharged must be sensibly affected by it; and in proportion to the heat communicated to those parts, they will become specifically lighter, and the mercury will sink of course.

In the *Memoirs of the Literary Society of Manchester*, vol. 4, is also a curious paper on this subject, viz. *Meteorological Observations made on different Parts of the Western Coast of Great Britain: arranged by T. Garnett, M. D.* This paper is composed of materials furnished by several observers; those of Mr. Copland, surgeon at Dumfries, are of special importance. This gentleman is of opinion that the changes of the barometer indicate approaching hot and cold weather, with much more certainty than dry and wet. "Every remarkable elevation of the barometer, says he, where it is of any duration, is followed by very warm or dry weather, and moderate as to wind, or by all of them; but heat seems to have most influence and connexion; and when it is deficient, the continuance of the other two will be longer and more remarkable; therefore the calculation must be in a compound ratio of the excess and deficiency of the heat, and of the dryness of the weather in comparison of the medium of the season; and with regard to the want of strong wind, it appears to be intimately connected with the last, as they shew that no precipitation is going on in any of the neighbouring regions."

In his 14th and 15th remarks, he had said,

'14th, That the barometer being lower, and continuing so longer than what can be accounted for by immediate falls, or stormy weather, indicates the approach of very cold weather for the season; and also, cold weather, though dry, is always accompanied by a low barometer, till near its termination.'

'15th, That warm weather is always preceded and mostly accompanied by a high barometer; and the rising of the barometer in the time of broken or cold weather, is a sign of the approach of warmer weather: and also if the wind is in any of the cold points, a sudden rise of the barometer indicates the approach of a southerly wind, which in winter generally brings rain with it.'

In the two following remarks, Mr. Copland had explained certain phenomena from a principle similar to that on which Dr. Darwin has so much insisted: (Botanic Garden, I. notes p. 79, &c.)

'That the falling of the barometer may proceed from a decomposition of the atmosphere occurring around or near that part of the globe where we are placed, which will occasion the electricity of the atmosphere to be repelled upwards in fine lambent portions; or driven downwards or upwards in more compacted balls of fire; or lastly, to be carried along with the rain, &c, in an imperceptible manner to the surface of the earth: the precipitation of the watery parts generally very soon takes place, which diminishes the real gravity of the atmosphere, and also by the decomposition of some of the more active parts, the air loses part of that elastic and repulsive power which it so eminently possessed, and will therefore press with less force on the mercury of the barometer than before, by which means a fall ensues.

'That the cause of the currents of air, or winds, may also be this way accounted for: and in very severe storms, where great decompositions of the atmosphere take place, this is particularly evident, such as generally occur in one or more of the West India islands at one time, a great loss of real gravity, together with a considerable diminution of the spring of the air immediately ensues; hence a current commences, first in that direction whence the air has most gravity, or is most disposed to undergo such a change; but it being soon relieved of its superior weight or spring on that side, by the decomposition going on as fast as the wind arrives on the island, it immediately veers to another point, which then rushes in mostly with an increase of force; thus it goes on till it has blown more than half way round the points of the compass during the continuation of the hurricane. For in this manner the West India phenomena, as well as the alteration of the wind during heavy rains in this country, can only be properly accounted for.' See remark No. 4.

Mr. C.'s 4th aphorism is, 'That the heaviest rains, when of long continuance, generally begin with the wind blowing easterly, when it gradually veers round to the south; and that the rain does not then begin to cease till the wind has got to the west, or rather a little to the northward of it, when, it may be added, it commonly blows with some violence.'

Many other observations on the barometer, the weather, &c. may be seen in various parts of the Philos. Trans. And for other curious papers on the same, and other subjects connected with the barometer, see the Gentleman's Magazine for 1789, p. 317; also Greu's Journal of Nat. Philos. printed at Leipzig 1792, for the influence of the sun and moon upon the barometer.

The Barometer applied to the measuring of Altitudes.

The secondary character of the barometer, namely as an instrument for measuring accessible heights or depths, was first proposed by Pascal, and Descartes, as has been before observed; and succeeding philosophers have been at great pains to ascertain the proportion between the fall of the barometer and the height to which it is carried; as Halley, Mariotte, Maraldi, Scheuchzer, J. Cassini, D. Bernoulli, Horrebow, Bouguer, Shuckburgh, Roy, and more especially by De Luc, who has given a critical and historical detail of most of the attempts that have at different times been made for applying the motion of the mercury in the barometer to the measurement of accessible heights.

And for this purpose serves the portable barometer, before described, (fig. 11 and 12, plate 4,) which should be made with all the accuracy possible. Various rules have been given by the writers on this subject, for computing the height ascended from the given fall of the mercury in the tube of the barometer, the most accurate of which was that of Dr. Halley, till it was rendered much more accurate by the indefatigable researches of De Luc, by introducing into it the corrections of the columns of mercury and air, on account of heat. And other corrections and modifications of the same may be seen inserted under the article Atmosphere, where the most correct rule is deduced from one single experiment only. This rule is as follows:

The Rule for Computing Altitudes, is this,

Viz, $10000 \times$ log. of $\frac{M}{m}$ is the altitude in fathoms, in the mean temperature of $31°$; and for every degree of the thermometer above that, the result must be increased by so many times its 435th part, and diminished when below it: in which theorem M denotes the length of the column of mercury in the barometer tube at the bottom, and m that at the top of the hill, or other eminence; which lengths may be expressed in any one and the same sort of measures, whether feet, or inches, or tenths, &c, and either English, or French, or of any other nation; but the result is always in fathoms, of 6 English feet each.

And the *Precepts,* in words, for the practice of measurements by the barometer, are these following:

1st, Observe the height of the barometer at the bottom of any height or depth, proposed to be measured; together with the temperature of the mercury by means of the thermometer attached to the barometer, and also the temperature of the air in the shade by another thermometer which is detached from the barometer.

2dly, Let the same thing be done also at the top of the said height or depth, and as near to the same time with the former as may be. And let those altitudes of mercury be reduced to the same temperature, if it be thought necessary, by correcting either the one or the other, viz, augmenting the height of the mercury in the colder temperature, or diminishing that in the warmer, by its 9600th part for every degree of difference between the two; and the altitudes of mercury so corrected, are what are denoted by M and m, in the algebraic formula above.

3dly, Take out the common logarithms of the two heights of mercury, so corrected, and subtract the less from the greater, cutting off from the right hand side of the remainder three places for decimals; so shall those on the left be fathoms in whole numbers, the tables of logarithms being understood to be such as have 7 places of decimals.

4thly, Correct the number last found, for the difference of the temperature of the air, as follows: viz, Take half the sum of the two temperatures of the air, shewn by the detached thermometers, for the mean one; and for every degree which this differs from the standard temperature of $31°$, take so many times the 435th part of the fathoms above found, and add them if the mean temperature

3

temperature be more than 31°, but subtract them if it be below 31°; so shall the sum or difference be the true altitude in fathoms, or being multiplied by 6, it will give the true altitude in English feet.

Example 1. Let the state of the barometers and thermometers be as follows, to find the altitude: viz,

Thermometers.		Barometers.
detached.	attached.	
57	57	29·68 lower
42	43	25·28 upper
mean 49½	dif. 14	

As 9600 : 14 : : 29·68 : ·04
              cor. ·04 logs.

mean 49½   M = 29·64 - 4718782
stand. 31   m = 25·28 - 4027771
dif. 18½

As 435 : 18½ : : 691·01 : 29·388
                29·388

the altitude { 720·399 fath.
sought is { or 4322·394 feet.

Example 2. To find the altitude of a hill, when the state of the barometer and thermometer, as observed at the bottom and top of it, is as follows; viz,

Thermometers.		Barometers.
detached.	attached.	
35	41	29·45
31	38	26·82
mean 33	dif. 3	

As 9600 : 3 : : 29·45 : ·01
              ·01 logs.

mean 33   M = 29·44 - 4689378
stand. 31   m = 26·82 - 4284588
dif. 2

As 435 : 2 : : 404·790 : 1·86
                1·86

the altitude { 406·05 fathoms.
sought is { or 2430·90 feet.

See this rule investigated under the article PNEUMATICS, at the end.

N. B. The mean height of the barometer in London, upon an average of two observations in every day of the year, kept at the house of the Royal Society, for many years past, is 29·88; the medium temperature, or height of the thermometer, according to the same, being 58°. But the medium height at the surface of the sea, according to Sir Geo. Shuckburgh (Philos. Transf. 1777, p. 586) is 30·04 inches, the heat of the barometer being 55°, and of the air 62°.

BAROSCOPE, a machine for shewing the alterations in the weight or pressure of the atmosphere. See BAROMETER.

BARREL, an English vessel or cask, containing 36 gallons of beer measure, or 32 gallons of ale measure. The barrel of beer, vinegar, or of liquor preparing for vinegar, ought to contain 34 gallons, according to the standard of the ale quart.

BARREL, in Clock-work, is the cylinder about which the spring is wrapped.

BARRICADE, or BARRICADO, a military term for a fence, or retrenchment, hastily made with vessels, or baskets of earth, carts, trees, stakes, or the like, to preserve an army from the shot or assault of an enemy.

BARRIER, a kind of fence made at a passage, retrenchment, gate, or such like, to stop it up against an enemy.

BARROW (ISAAC), a very eminent mathematician and divine of the 17th century, was born at London in October, 1630, being the son of Thomas Barrow, then a linen-draper of that city, but descended from an ancient family in Suffolk. He was at first placed at the Charter-house school for two or three years; where his behaviour afforded but little hopes of success in the profession of a scholar, being fond of fighting, and promoting it among his school-fellows: but being removed to Felsted in Essex, his disposition took a different turn; and having soon made a great progress in learning, he was first admitted a pensioner of Peter House in Cambridge; but when he came to join the university, in Feb. 1645, he was entered at Trinity college. He now applied himself with great diligence to the study of all parts of literature, especially natural philosophy. He afterward turned his attention to the profession of physic, and made a considerable progress in anatomy, botany, and chemistry: he next studied divinity; then chronology, astronomy, geometry, and the other branches of the mathematics; with what success, his writings afterwards most eminently shewed.

When Dr. Duport resigned the chair of Greek professor, he recommended his pupil Mr. Barrow for his successor, who, in his probation exercise, shewed himself equal to the character that had been given him by this gentleman; but being suspected of favouring Arminianism, he was not preferred. This disappointment it seems determined him to quit the college, and visit foreign countries; but his finances were so low, that he was obliged to dispose of his books, to enable him to execute that design.

He left England in June 1655, and visited France, Italy, Turkey, &c. At several places, in the course of this tour, he met with kindness and liberal assistance from the English ambassadors, &c, which enabled him to benefit the more from it, by protracting his stay, and prolonging his journey. He spent more than a year in Turkey, and returned to England by way of Venice, Germany, and Holland, in 1659. At Constantinople he read over the works of St. Chrysostom, once bishop of that see, whom he preferred to all the other fathers.

On his return home Barrow was episcopally ordained by bishop Brownrig; and in 1660, he was chosen to the Greek professorship at Cambridge. In July 1662, he was elected professor of geometry in Gresham college: in which station, he not only discharged his own duty, but supplied likewise the absence of Dr. Pope the astronomy professor. Among his lectures, some were upon the projection of the sphere and perspective, which are lost; but his Latin oration, previous to his lectures, is still extant. About this time Mr. Barrow was offered a good living; but the condition annexed, of teaching the patron's son, made him refuse it, as thinking it too

like

like a fimonial contract. Upon the 20th of May 1663 he was elected a fellow of the Royal Society, in the firft choice made by the council after their charter. The fame year the executors of Mr. Lucas having, according to his appointment, founded a mathematical lecture at Cambridge, they felected Mr. Barrow for the firft profeffor; and though his two profefforfhips were not incompatible with each other, he chofe to refign that of Grefham-college, which he did May the 20th, 1664. In 1669 he refigned the mathematical chair to his learned friend Mr. Ifaac Newton, being now determined to quit the ftudy of mathematics for that of divinity. On quitting his profefforfhip, he had only his fellowfhip of Trinity-college, till his uncle gave him a fmall finecure in Wales, and Dr. Seth Ward bifhop of Salifbury conferred upon him a prebend in his church. In the year 1670 he was created doctor in divinity by mandate; and, upon the promotion of Dr. Pearfon mafter of Trinity college to the fee of Chefter, he was appointed to fucceed him by the king's patent bearing date the 13th of February 1672: upon which occafion the king was pleafed to fay, "he had given it to the beft fcholar in England." In this, his majefty did not fpeak from report, but from his own knowledge; the doctor being then his chaplain, he ufed often to converfe with him, and, in his humourous way, to call him an "unfair preacher," becaufe he exhaufted every fubject, and left no room for others to come after him. In 1675 he was chofen vice-chancellor of the univerfity; and he omitted no endeavours for the good of that fociety, nor in the line of his profeffion as a divine, for the promotion of piety and virtue; but his ufeful labours were abruptly terminated by a fever on the 4th of May 1677, in the 47th year of his age. He was interred in Weftminfter abbey, where a monument, adorned with his buft, was foon after erected, by the contribution of his friends.

Dr. Barrow's works are very numerous, and indeed various, mathematical, theological, poetical, &c, and fuch as do honour to the Englifh nation. They are principally as follow:

1. Euclidis Elementa. Cantab. 1655, in 8vo.

2. Euclidis Data. Cantab. 1657, in 8vo.

3. Lectiones Opticæ xviii, Lond. 1669, 4to.

4. Lectiones Geometricæ xiii, Lond. 1670, 4to.

5. Archimedis Opera, Apollonii Conicorum libri iv, Theodofii Sphericorum lib. iii; nova methodo illuftrata, et fuccincte demonftrata. Lond. 1675, in 4to.

The following were publifhed after his deceafe, viz:

6. Lectio, in qua theoremata Archimedis de fphæra et cylindro per methodum indivifibilium inveftigata, ac breviter inveftigata, exhibentur. Lond. 1678, 12mo.

7. Mathematicæ Lectiones habitæ in fcholis publicis academiæ Cantabrigienfis, an. 1664, 5, 6, &c. Lond. 1683.

8. All his Englifh works in 3 volumes, Lond. 1683, folio.—Thefe are all theological, and were publifhed by Dr. John Tillotfon.

9. Ifaaci Barrow Opufcula, viz, Determinationes, Conciones ad Clerum, Orationes, Poemata, &c. volumen quartum. Lond. 1687, folio.

Dr. Barrow left alfo feveral curious papers on mathematical fubjects, written in his own hand, which were communicated by Mr. Jones to the author of "The Lives of the Grefham Profeffors," a particular account of which may be feen in that book, in the Life of Barrow.

Several of his works have been tranflated into Englifh, and publifhed; as the Elements and Data of Euclid; the Geometrical Lectures, the Mathematical Lectures. And accounts of fome of them were alfo given in feveral volumes of the Philof. Tranf.

Dr. Barrow muft ever be efteemed, in all the fubjects which exercifed his pen, a perfon of the cleareft perception, the fineft fancy, the foundeft judgment, the profoundeft thought, and the clofeft and moft nervous reafoning. "The name of Dr. Barrow (fays the learned Mr. Granger) will ever be illuftrious for a ftrength of mind and a compafs of knowledge that did honour to his country. He was unrivalled in mathematical learning, and efpecially in the fublime geometry; in which he has been excelled only by his fucceffor Newton. The fame genius that feemed to be born only to bring hidden truths to light, and to rife to the heights or defcend to the depths of fcience, would fometimes amufe itfelf in the flowery paths of poetry, and he compofed verfes both in Greek and Latin. He at length gave himfelf up entirely to divinity; and particularly to the moft ufeful part of it, that which has a tendency to make men wifer and better."

Several good anecdotes are told of Barrow, as well of his great integrity, as of his wit, and bold intrepid fpirit and ftrength of body. His early attachment to fighting when a boy is fome indication of the latter; to which may be added the two following anecdotes: In his voyage between Leghorn and Smyrna the fhip was attacked by an Algerine pirate, which after a ftout refiftance they compelled to fheer off, Barrow keeping his poft at the gun affigned him to the laft. And when Dr. Pope in their converfation afked him, "Why he did not go down into the hold, and leave the defence of the fhip to thofe, to whom it did belong? He replied, It concerned no man more than myfelf: I would rather have loft my life, than to have fallen into the hands of thofe mercilefs infidels."

There is another anecdote told of him, which fhewed not only his intrepidity, but an uncommon goodnefs of difpofition, in circumftances where an ordinary fhare of it would have been probably extinguifhed. Being once on a vifit at a gentleman's houfe in the country, where the neceffary was at the end of a long garden, and confequently at a great diftance from the room where he lodged; as he was going to it before day, for he was a very early rifer, a fierce maftiff, that ufed to be chained up all day, and let loofe at night for the fecurity of the houfe, perceiving a ftrange perfon in the garden at that unufual time, fet upon him with great fury. The doctor caught him by the throat, grappled with him, and, throwing him down, lay upon him: once he had a mind to kill him; but he altered his refolution, on recollecting that this would be unjuft, fince the dog did only his duty, and he himfelf was in fault for rambling out of his room before it was light. At length he called out fo loud, that he was heard by fome of the family, who came prefently out, and freed the doctor and the dog from the danger they both had been in.

Among

Among other inſtances of his wit and vivacity, they relate the following rencontre between him and that wicked wit lord Rocheſter. Theſe two meeting one day at the court, while the doctor was king's chaplain in ordinary, Rocheſter, thinking to banter him, with a flippant air, and a low formal bow, accoſted him with, "Doctor, I am yours to my ſhoe-tie:" Barrow perceiving his drift, and determined upon defending himſelf, returned the ſalute, with, " My lord, I am yours to the ground." Rocheſter, on this, improving his blow, quickly returned it, with, " Doctor, I am yours to the centre ;" which was as ſmartly followed up by Barrow, with, " My lord, I am yours to the antipodes." Upon which, Rocheſter, diſdaining to be foiled by a muſty old piece of divinity, as he uſed to call him, exclaimed, "Doctor, I am yours to the loweſt pit of hell ;" upon which Barrow, turning upon his heel, with a ſarcaſtic ſmile, archly replied, " There, my lord, I leave you."

BARS, in Muſic, are the ſpaces quite through any compoſition, ſeparated by upright lines drawn acroſs the five horizontal lines, each of which either contains the ſame number of notes of the ſame kind, or ſo many other notes as will make up a like interval of time ; for all the bars, in any piece, muſt be of the ſame length, and played in the ſame time.

BARTER, or TRUCK, is the exchanging of one commodity for another ; and forms a rule in the commercial part of arithmetic, by which the commodities are properly calculated and equalled, by computing firſt the value of the commodity which is given, and then the quantity of the other which will amount to the ſame ſum.

BASE, BASIS, in Architecture, denotes the lower part of a column or pedeſtal.

BASE, in Geometry, the loweſt ſide of any figure. Any ſide of a figure may be conſidered as its baſe, according to the poſition in which it may be conceived as ſtanding ; but commonly it is underſtood of the loweſt ſide : as the baſe of a triangle, of a cone, cylinder, &c.

BASE LINE, in Perſpective, denotes the common ſection of the picture and the geometrical plane.

BASE RING, of a Cannon, is the great ring next behind the vent or touch hole.

BASE, alternate. See ALTERNATE.

BASEMENT, in Architecture, a continued baſe, extended a conſiderable length, as about a houſe, a room, or other piece of building:

BASILIC, in the ancient Architecture, was a large hall, or court of judicature, where the magiſtrates ſat to adminiſter juſtice.

BASILICA, or BASILICUS, the ſame as *Regulus*, or *Cor Leonis*, being a fixed ſtar of the firſt magnitude in the conſtellation *Leo*.

BASILISK, in the older Artillery, was a large piece of ordnance ſo called from its reſemblance to the ſuppoſed ſerpent of that name. It threw an iron ball of 200 pounds weight ; and was in great repute in the time of Solyman emperor of the Turks, in the wars of Hungary ; but it is now grown out of uſe in moſt parts of Europe. Paulus Jovius relates the terrible ſlaughter made in a Spaniſh ſhip by a ſingle ball from one of theſe baſiliſks ; after paſſing through the beams and planks in the ſhip's head, it killed upwards of 30 men. And Maffeus ſpeaks of baſiliſks made of braſs, each of which

required 100 yoke of oxen to draw them.—More modern writers alſo give the name baſiliſk to a much ſmaller and ſizeable piece of ordnance, made of 15 feet long by the Dutch, but of only 10 by the French, and carrying a ball of 48 pounds. The largeſt ſize of cannon now uſed by the Engliſh, are the 32 pounders.

BASIS, in Geometry, the ſame as BASE.

BASS, the loweſt in the four parts of muſic ; by ſome eſteemed the baſis and principal part of all, and by others as ſcarcely neceſſary in ſome tunes.

BASSANTIN (JAMES), a Scotch aſtronomer of the 16th century, born in the reign of James the 4th of Scotland. He was a ſon of the Laird of Baſſantin in the Merſe. After finiſhing his education at the univerſity of Glaſgow, he travelled through Germany and Italy, and then ſettled in the univerſity of Paris, where he taught mathematics with great applauſe. Having acquired ſome property in this employment, he returned to Scotland in 1562, where he died 6 years after.

From his writings it appears he was no inconſiderable aſtronomer, for the age he lived in ; but, according to the faſhion of the times, he was not a little addicted to judicial aſtrology. It was doubtleſs to our author that Sir James Melvil alludes in his Memoirs, when he ſays that his brother Sir Robert, when he was uſing his endeavours to reconcile the two queens Elizabeth and Mary, met with one Baſſantin a man learned in the high ſciences, who told him, " that all his travel would be in vain ; for, ſaid he, they will never meet together ; and next, there will never be any thing but diſſembling and ſecret hatred for a while, and at length captivity and utter wreck to our queen from England." He added, " that the kingdom of England at length ſhall fall, of right, to the crown of Scotland : but it ſhall coſt many bloody battles ; and the Spaniards ſhall be helpers, and take a part to themſelves for their labour." A prediction in which Baſſantin partly gueſſed right, which it is likely he was enabled to do from a judicious conſideration of probable circumſtances and appearances.

Baſſantin's works are,

1. *Aſtronomia Jacobi Baſſantini Scoti, opus abſolutiſſimum, &c*; ter. edit. *Latine et Gallice*. Genev. 1599, fol. This is the title given it by Tornœſius, who tranſlated it into Latin from the French, in which language it was firſt publiſhed.

2. *Paraphraſe de l'Aſtrolabe, avec vne amplification de l'uſage de l'Aſtrolabe.* Lyons 1555. Paris 1617 8vo.

3. *Mathematica Genethliaca.*

4. *Arithmetica.*

5. *Muſica ſecundum Platonem.*

6. *De Matheſi in Genere.*

BASSOON, a muſical inſtrument of the wind kind, ſerving for a baſs to the haut-boy. It is blown with a reed, and furniſhed with eleven holes.

BASS-VIOL, a baſs to the viol.

BASTION, in the modern fortification, a large maſs of earth at the angles of a work, connecting the curtains to each other ; and anſwers to the bulwark of the ancients. It is formed by two faces, two flanks, and two demigorges. The two faces form the ſaliant angle, or angle of the baſtion ; the two flanks form with the faces, the *epaule* or ſhoulders ; and the union of the

the other two ends of the flanks with the curtains forms the two angles of the flanks.

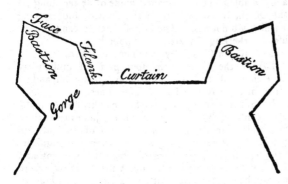

Solid BASTION, are those that are entirely filled up with earth to the height of the rampart, without any void space towards the centre.

Void or *Hollow* BASTION, has the rampart and parapet ranging only round the flanks and spaces, so that a void space is left within towards the centre, where the ground is so low that if the rampart be taken, no retrenchment can be made in the centre, but what will lie under the fire of the besieged.

Regular BASTION, is that which has its due proportion of faces, flanks, and gorges.

Deformed or *Irregular* BASTION, is when the irregularity of the lines and angles throws the bastion out of shape: as when it wants one of the demigorges, one side of the interior polygon being too short, &c.

Demi BASTION, or *Half bastion*, also otherwise called an *Epaulment*, has but one face and flank.

Double BASTION, is when one bastion is raised within, and upon the plane of another bastion.

Flat BASTION, is one built in the middle of the curtain, when it is too long to be defended by the usual bastions at the extremities.

Composed BASTION, is when the two sides of the interior polygon are very unequal, which makes the gorges also unequal.

Cut BASTION, is that which has a re-entering angle at the point, and is sometimes called a BASTION *with a Tenaille*, whose point is cut off, making an angle inwards, and two points outwards. This is used when the saliant angle would be too sharp, or when water or some other impediment prevents it from being carried out to its full extent.

BASTON, or BATOON, in Architecture, a moulding in the base of a column, called also a *Tore* or *Torus*.

BATTEN, a name given by workmen to a scantling or piece of wooden stuff, about an inch thick, and from to 4 inches broad; of a considerable but indeterminate length.

BATTERING, the attacking a place, work, or the like, with heavy artillery.

BATTERING-RAM, a military engine used for beating down walls, before the invention of gunpowder and the modern artillery. It was no other than a long heavy beam of timber, armed with an iron head, something like the head of a ram. This being pushed violently with constant successive blows against a wall, gradually shakes it with a vibratory motion, till the stones are disjointed and the wall falls down. There were several kinds of the battering-ram, the first rude and plain, which the soldiers carried in their arms by main force, and so struck the head of it against the wall. The second was slung by a rope about the middle to another beam lying across upon a couple of posts; which was the kind described by Josephus as used at the siege of Jerusalem. A third sort was covered over with a shell or screen of boards, to defend the men from the stones and darts of the besieged upon the walls, and thence called *testudo arietaria*. And Felibien describes a fourth sort of battering-ram, which ran upon wheels; and was the most perfect and effectual of any.

Vitruvius affirms, that the battering-ram was first invented by the Carthaginians, while they laid siege to Cadiz: yet Pliny assures us, that the ram was invented or used at the siege of Troy; and that it was this that gave occasion to the fable of the wooden horse. In fact there can be no doubt but that the use of some sort of a battering-ram is as old as the art of war itself. And it has even been suspected that the walls of Jericho, mentioned in the book of Joshua, were beaten down by this instrument, the rams horns there mentioned, by means of which they were overthrown, being no other than the horns of the battering-rams. Pephasmenos, a Tyrian, afterwards contrived to suspend it with ropes; and lastly, Polydus, the Thessalian, mounted it on wheels, at the siege of Byzantium, under Philip of Macedon.

Plutarch relates, that Marc Anthony, in the Parthian war, made use of a ram 80 feet long: and Vitruvius affirms that they were sometimes 106, and even 120 feet long; which must have given an amazing force to this engine. The ram required 100 soldiers to work and manage it at one time; who being exhausted, another century relieved them; by which means in was kept playing continually without intermission. See fig. 2, plate V, which represents the battering-ram suspended in its open frame; in which 3 denotes the form of the head, fastened to the enormous beam 2, by three or four bands (4) of iron, of about four feet in breadth. At the extremity of each of these bands was an iron chain (5), one end of which was fastened to a hook (6), and to the last link at the other extremity was firmly bound a cable, which ran the whole length of the beam to the end of the ram 7, where these cables were bound all together as fast as possible with small ropes. To the end of these cables was fastened another, that consisted of several strong cords platted together to a certain length, and then running single (8), at each of which were placed several men, to balance and work the machine. 10 Is the chain or cable by which the ram was hung to the cross beam (11), fixed on the top of the frame; and 12 is the base of the machine.

The unsuspended ram differed from this only in the manner of working it; as it moved on small wheels upon another large beam, instead of being slung by a chain or cable.

BATTERY, in the Military Art, a place raised to plant cannon upon, to play with more advantage upon the enemy. It consists of an epaulment or a breastwork, of about 8 feet high, and 18 or 20 feet thick. In all batteries, the open spaces through which the muzzles

muzzles of the cannon are pointed, are called *Embrafures*, and the diftances between the embrafures, *merlons*. The guns are placed upon a platform of planks &c, afcending a little from the parapet, to check the recoil, and that the gun may be the eafier brought back again to the parapet: they are placed from 12 to 16 feet diftant from one another, that the parapet may be ftrong, and the gunners have room to work.

Mortar BATTERIES differ from the others, in that the flope of the parapet is inwards, and it is without embrafures, the fhells being fired quite over the parapet, commonly at an angle of 45 degrees elevation.

Open BATTERY, is nothing more than a number of cannon, commonly field-pieces, ranged in a row abreaft of one another, perhaps on fome fmall natural elevation of the ground, or an artificial bank a little raifed for the purpofe.

Covered or *Mafked* BATTERY, is when the cannon and gunners are covered by a bank or breaft-work, commonly made of brufh-wood, faggots, and earth, called a fafcine battery.

Sunk or *Buried* BATTERY, is when its platform is funk, or let down into the ground, fo that trenches muft be cut in the earth oppofite the muzzles of the guns, to ferve as embrafures to fire through. This is moftly ufed on the firft making of approaches in befieging and battering a place.

Crofs BATTERIES, are two batteries playing athwart each other upon the fame object, forming an angle there, and battering to more effect, becaufe what one battery fhakes, the other beats down.

BATTERY *d'Enfilade*, is one that fcours or fweeps the whole length of a ftraight line.

BATTERY *en Echarpe*, is one that plays obliquely.

BATTERY *de Reverfe*, or *Murdering Battery*, is one that plays upon the enemy's back.

Camerade or *Joint* BATTERY, is when feveral guns play upon one place at the fame time.

BATTERY, *in Electricity*, is a combination of coated furfaces of glafs, commonly jars, fo connected together that they may be charged at once, and difcharged by a common conductor. Mr. Gralath, a German electrician, firft contrived to increafe the fhock by charging feveral phials at the fame time.—Dr Franklin, having analyfed the Leyden phial, and found that it loft at one furface the electric fire received at the other, conftructed a battery of eleven large panes of fafh window glafs, coated on both fides, and fo connected that the whole might be charged together, and with the fame labour as one fingle pane; then by bringing all the *giving* fides into contact with one wire, and all the *receiving* fides with another, he contrived to unite the force of all the plates, and to difcharge them at once.—Dr. Prieftley defcribes a ftill more complete battery. This confifts of 64 jars, each 10 inches long, and 2½ inches in diameter, all coated within an inch and a half of the top, forming in the whole about 32 fquare feet of coated furface. A piece of very fine wire is twifted about the lower end of the wire of each jar, to touch the infide coating in feveral places; and it is put through a pretty large piece of cork, within the jar, to prevent any part of it from touching the fide, by which a fpontaneous difcharge might be made. Each wire is turned round fo as to make a loop at the

upper end; and through thefe loops paffes a pretty thick brafs rod with knobs, each rod ferving for one row of the jars: and thefe rods are made to communicate together by a thick chain laid over them, or as many of them as may be wanted. The jars ftand in a box, the bottom of which is covered with a tin plate; and a bent wire touching the plate paffes through the box, and appears on the outfide. To this wire is faftened any conductor defigned to communicate with the outfide of the battery; and the difcharge is made by bringing the brafs knob to any of the knobs of the battery. When a very great force is required, the fize or number of the jars may be increafed, or two or more batteries may be ufed.—But the largeft and moft powerful battery of all, is that employed by Dr. Van Marum, to the amazing large electrical machine, lately conftructed for Teyler's mufeum at Haarlem. This grand battery confifts of a great number of jars coated as above, to the amount of about 130 fquare feet; and the effects of it, which are truly aftonifhing, are related by Dr. Van Marum in his defcription of this machine, and of the experiments made with it, at Haarlem 1785, &c. See alfo Franklin's Exper. and Obferv. and Prieftley's Hiftory of Electricity.

BATTLEMENTS, in Architecture, are notches or indentures in the top of a wall or other building, like embrafures, to look through.

BAY, in Geography, denotes a fmall gulph, or an arm of the fea ftretching up into the land; being larger in the middle within, than at its entrance, which is called the mouth of the bay.

BAYER (JOHN), a German lawyer and aftronomer of the latter part of the 16th and beginning of the 17th century, but in what particular year or place he was born, is not certainly known: however, his name will be ever memorable in the annals of aftronomy, on account of that great and excellent work which he firft publifhed in the year 1603, under the title of *Uranometria*, being a complete celeftial atlas, or large folio charts of all the conftellations, with a nomenclature collected from all the tables of aftronomy, ancient and modern, with the ufeful invention of denoting the ftars in every conftellation by the letters of the Greek alphabet, in their order, and according to the order of magnitude of the ftars in each conftellation. By means of thefe marks, the ftars of the heavens may, with as great facility, be diftinguifhed and referred to, as the feveral places of the earth are by means of geographical tables; and as a proof of the ufefulnefs of this method, our celeftial globes and atlaffes have ever fince retained it; and hence it is become of general ufe through all the literary world; aftronomers, in fpeaking of any ftar in the conftellation, denoting it by faying it is marked by Bayer, α, or β, or γ, &c.

Bayer lived many years after the firft publication of this work, which he greatly improved and augmented by his conftant attention to the ftudy of the ftars. At length, in the year 1627, it was republifhed under a new title, viz, *Coelum Stellatum Chriftianum*, that is, the *Chriftian Stellated Heaven*, or the *Starry Heavens Chriftianized*: for in this work, the heathen names and characters, or figures of the conftellations, were rejected, and others, taken from the fcriptures, were inferted

ferted in their ftead, to circumfcribe the refpective conftellations. This was the project of one Julius Schiller, a civilian of the fame piace. But this attempt was too great an innovation, to find fuccefs, or a general reception, which might occafion great confufion. And, we even find in the later editions of this work, that the ancient figures and names were reftored again; at leaft fo I find them in two editions, of the years 1654, and 1661, which are now before me.

BEAD, in Architecture, is a round moulding, carved in fhort embofiments, like beads in necklaces: and fometimes an aftragal is thus carved. There is alfo a fort of plain bead often fet on the edge of each facia of an architrave; as alfo fometimes on the lining board of a door cafe, the upper edge of fkirting boards, &c.

BEAD, in affaying, the fmall ball or mafs of pure metal feparated from the *fcoria*, and feen diftinct and pure in the middle of the coppel while in the fire.

BEAM, in Architecture, a large timber laid acrofs a building, into which the principal rafters are framed. Several ingenious authors have confidered the force or ftrength of beams, as fupporting their own weight and any other additional weight; particularly Varignon, and Parent in the Memoir Acad. R. Scien. an. 1708, and Mr. Emerfon, on the Strength and Strefs of Timber, in his Mechanics. Mr. Parent makes the proportion of the depth to the breadth of a beam to be as 7 to 5 when it is ftrongeft.

BEAMS *of a fhip*, are the large, main, crofs timbers, ftretched from fide to fide, to fupport the decks, and keep the fides of the fhip from falling together.

BEAM *of a balance*, is the horizontal piece of wood or iron fupported on a pivot in the middle, and at the extremities of which the two fcales are fufpended, for weighing any thing.

BEAM-*Compafs*, an inftrument confifting of a wooden or brafs fquare beam, having fliding fockets carrying fteel or pencil points; and are ufed for defcribing large circles, the radii of which are beyond the extent of the common compaffes.

BEAR, in Aftronomy, a name given to two conftellations, called the *greater* and the *leffer bear*, or URSA *major* and *minor*. The pole ftar is in the tail of the little bear, and is within lefs than 2 degrees of the north pole. See URSA, *major* and *minor*.

BEARD, *of a Comet*, the rays which it emits in the direction in which it moves; as diftinguifhed from the tail, or the rays emitted or left behind it as it moves along, being always in a direction from the fun.

BEARER, in Architecture, a poft or brick wall, trimmed up between the two ends of a piece of timber, to fhorten its bearing, or to prevent its bearing with the whole weight at the ends only.

BEARING, in Geography and Navigation, the fituation of one place from another, with regard to the points of the compafs; or an arch of the horizon between the meridian of a place and a line drawn through this and another place, or the angle formed by a line drawn through the two places and their meridians.— The bearings of places on the ground are ufually determined by the magnetic needle.

BEATS, in a Clock or Watch, are the ftrokes made by the fangs or pallets of the fpindle of the balance; or of the pads in a royal pendulum. For the number and ufe of the beats, fee Derham's Artificial Clock Maker, pa. 14 and feq.

BED, *of a Great Gun*, a plank laid between the cheeks of the carriage, on the middle tranfum, for the breech of the gun to reft upon.

BED, or *Stool*, of a mortar, a thick and ftrong planking on which a mortar is placed, hollowed a little to receive the breech and trunions.

BED-MOULDING, in Architecture, a term ufed by workmen for thofe members in a cornice which are placed below the coronet, or crown. It ufually confifts of thefe four members, an ogee, a lift, a large boultine, and another lift under the coronet.

BELIDOR (BERNARD FOREST DE), an engineer in the fervice of France, but born in Catalonia in 1698. He was profeffor in the new fchool of artillery at la Fere, where he publifhed his courfe of mathematics for the ufe of the artillery and engineers. He was the firft who ferioufly confidered the quantity of gunpowder proper for charges, and reduced it to 2-3ds the quantity. He was named Affociate in the Academy of Sciences in 1751; and died Sept. 8, 1761, at 63 years of age.

His works that have been publifhed, are:

1. Sommaire d'un Cours d'Architecture militaire, civile & hydraulique, in 12mo, 1720.

2. Nouveau Cours de Mathematiques, &c. in 4to, 1725.

3. La Science des Ingénieurs, in 4to, 1729.

4. Le Bombardier Francois, in 4to, 1734.

5. Architecture Hydraulique, 4 vols. in 4to, 1737.

6. Dictionnaire portatif de l'Ingénieur, in 8vo.

7. Traite des Fortifications, 4 vols. in 4to.

Befides feveral pieces inferted in the volumes of the Memoirs of the Academy of Sciences, for the years 1737, 1750, 1753, and 1756.

BELLATRIX, in Aftronomy, a ruddy, glittering ftar of the 2d magnitude, in the left fhoulder of Orion. Its name is from the Latin *bellum*, as being anciently fuppofed to have great influence in kindling wars, and forming warriors.

BELTS, *Fafciæ*, in Aftronomy, two zones or girdles furrounding the planet Jupiter, brighter than the reft of his body, and terminated by parallel lines. They are obferved however to be fometimes broader and fometimes narrower, and not always occupying exactly the fame part of the difc. Jupiter's belts were firft obferved and defcribed by Huygens, in his Syft. Saturn. Dark fpots have often been obferved on the belts of Jupiter; and M. Caffini obferved a permanent one on the northern fide of the moft fouthern belt, by which he determined the length of Jupiter's days, or the time in which the planet revolves upon its axis, which is 9h. 56m. Some aftronomers fuppofe that thefe belts are feas, which alternately cover and leave bare large tracts of the planet's furface: and that the fpots are gulphs in thofe feas, which are fometimes dry, and fometimes full. But Azout conceived that the fpots are protuberances of the belts; and others

again

again are of opinion that the transparent and moveable spots are the shadows of Jupiter's satellites.

Cassini also speaks of the belts of Saturn; being three dark, straight, parallel bands, or *fasciæ*; on the disc of that planet. But it does not appear that Saturn's belts adhere to his body, as those of Jupiter do; but rather that they are large dark rings surrounding the planet at a distance. Some imagine that they are clouds in the atmosphere of Saturn, though it would seem that the middlemost is the shadow of his ring.

BENDING, the reducing a body to a curved or crooked form. The bending of boards, planks, &c, is effected by means of heat, whether by boiling or otherwise, by which their fibres are so relaxed that they may be bent into any figure. Bernoulli has a discourse on the bending of springs, or elastic bodies. And Amontons gives several experiments concerning the bending of ropes. The friction of a rope bent or wound about an immoveable cylinder, is sufficient, with a very small power, to sustain very great weights.

BERENICE's Hair; see COMA *Berenices.*

BERKELEY (GEORGE), the virtuous and learned bishop of Cloyne in Ireland, was born in that kingdom, at Kilcrin, the 12th of March 1684. After receiving the first part of his education at Kilkenny school, he was admitted a pensioner of Trinity College, Dublin, at 15 years old; and chosen fellow of that college in 1707.

The first public proof he gave of his literary abilities was, *Arithmetica absque Algebra aut Euclide demonstrata;* which from the preface it appears he wrote before he was 20 years old, though he did not publish it till 1707. It is followed by a Mathematical Miscellany, containing observations and theorems inscribed to his pupil Samuel Molineux.

In 1709 came out the *Theory of Vision;* which of all his works it seems does the greatest honour to his sagacity; being, it has been observed, the first attempt that ever was made to distinguish the immediate and natural objects of sight, from the conclusions we have been accustomed from infancy to draw from them. The boundary is here traced out between the ideas of sight and touch; and it is shewn, that though habit hath so connected these two classes of ideas in the mind, that they are not without a strong effort to be separated from each other, yet originally they have no such connection; insomuch, that a person born blind, and suddenly made to see, would at first be utterly unable to tell how any object that affected his sight would affect his touch; and particularly would not from sight receive any idea of distance, or external space, but would imagine all objects to be in his eye, or rather in his mind.

In 1710 appeared *The Principles of Human Knowledge;* and in 1713 *Dialogues between Hylas and Philonous:* the object of both which pieces is, to prove that the commonly received notion of the existence of matter, is false; that sensible material objects, as they are called, are not external to the mind, but exist in it, and are nothing more than impressions made upon it by the immediate act of God, according to certain rules termed laws of nature.

Acuteness of parts and beauty of imagination were so conspicuous in Berkeley's writings, that his reputation was now established, and his company courted; men of opposite parties concurred in recommending him. For Steele he wrote several papers in the Guardian, and at his house became acquainted with Pope, with whom he always lived in friendship. Swift recommended him to the celebrated earl of Peterborough, who being appointed ambassador to the king of Sicily and the Italian States, took Berkeley with him as chaplain and secretary in 1713, with whom he returned to England the year following.

His hopes of preferment expiring with the fall of queen Anne's ministry, he some time after embraced an offer made him by Ashe, bishop of Clogher, of accompanying his son in a tour through Europe. In this he employed four years; and besides those places which fall within the grand tour, he visited some that are less frequented, and with great industry collected materials for a natural history of those parts; but which were unfortunately lost in the passage to Naples. He arrived at London in 1721; and being much affected with the miseries of the nation, occasioned by the South-sea scheme in 1720, he published the same year *An Essay towards preventing the ruin of Great Britain:* reprinted in his *Miscellaneous Tracts.*

His way was now open into the very first company. Pope introduced him to lord Burlington, by whom he was recommended to the duke of Grafton, then appointed lord-lieutenant of Ireland, who took Berkeley over as one of his chaplains in 1721. The latter part of this year he accumulated the degrees of bachelor and doctor in divinity: and the year following he had a very unexpected increase of fortune from the death of Mrs. Vanhomrigh, the celebrated Vanessa, to whom he had been introduced by Swift: this lady had intended Swift for her heir; but perceiving herself to be slighted by him, she left her fortune, of 8000l. between her two executors, of whom Berkeley was one. In 1724 he was promoted to the deanery of Derry, worth 1100l. a year.

In 1725 he published, and it has since been reprinted in his Miscellaneous Tracts, *A Proposal for converting the savage Americans to Christianity, by a college to be erected in the Summer Isles, otherwise called the Isles of Bermuda.* The proposal was well received, at least by the king; and he obtained a charter for founding the college, with a parliamentary grant of 20,000l. toward carrying it into execution: but he could never get the money, it being otherwise employed by the minister; so that after two years stay in America on this business, he was obliged to return, and the scheme dropped.

In 1732 he published *The Minute Philosopher,* in 2 volumes 8vo, against Freethinkers. In 1733 he was made bishop of Cloyne; and might have been removed in 1745, by lord Chesterfield, to Clogher; but declined it. He resided constantly at Cloyne, where he faithfully discharged all the offices of a good bishop, yet continued his studies with unabated attention.

About this time he engaged in a controversy with the mathematicians, which made a good deal of noise in the literary world; and the occasion of it was this:

Addison

Addison had given the bishop an account of the behaviour of their common friend Dr. Garth in his last illness, which was equally unpleasing to both these advocates of revealed religion. For when Addison went to see the doctor, and began to discourse with him seriously about another world, " Surely, Addison, replied he, I have good reason not to believe those trifles, since my friend Dr. Halley, who has dealt so much in demonstration, has assured me, that the doctrines of christianity are incomprehensible, and the religion itself an imposture." The bishop therefore took up arms against Halley, and addressed to him, as to an Infidel Mathematician, a discourse called *The Analyst*; with a view of shewing that mysteries in faith were unjustly objected to by mathematicians, who he thought admitted much greater mysteries, and even falshoods in science, of which he endeavoured to prove that the doctrine of Fluxions furnished a clear example. This attack gave occasion to *Robins's Discourse concerning the Method of Fluxions*, to *Maclaurin's Fluxions*, and to other smaller works upon the same subject; but the direct answers to *The Analyst* were made by a person under the name of Philalethes Cantabrigiensis, but commonly supposed to be Dr. Jurin, whose first piece was, *Geometry no Friend to Infidelity*, 1734. To this the bishop replied in *A Defence of Freethinking in Mathematics; with an Appendix concerning Mr. Walton's Vindication*, 1735; which drew a second answer the same year from Philalethes, under the title of *The Minute Mathematician, or the Freethinker no just Thinker*, 1735. Other writings in this controversy, beside those before mentioned, were

1. A Vindication of Newton's Principles of Fluxions against the objections contained in the Analyst, by J. Walton, Dublin, 1735.

2. The Catechism of the Author of the Minute Philosopher fully answered, by J. Walton, Dublin, 1735.

3. Reasons for not replying to Mr. Walton's Full Answer, in a letter to P. T. P. by the author of the Minute Philosopher, Dublin, 1735.

4. An Introduction to the Doctrine of Fluxions, and Defence of the Mathematicians against the objections of the author of the Analyst, &c. Lond. 1736.

5. A new Treatise of Fluxions; with answers to the principal objections in the Analyst, by James Smith, A. M. Lond. 1737.

6. Mr. Robins's Discourse of Newton's Methods of Fluxions, and of Prime and Ultimate Ratios, 1735.

7. Mr. Robins's Account of the preceding Discourse, in the Repub. of Letters, for October 1735.

8. Philalethes's Considerations upon some passages contained in two letters to the author of the Analyst &c, in Repub. of Letters, Novemb. 1735.

9. Mr. Robins's Review of some of the principal objections that have been made to the doctrine of Fluxions &c. Repub. of Letters for Decem. 1735.

10. Philalethes's Reply to ditto, in the Repub. of Letters, Jan. 1736.

11. Mr. Robins's Dissertation, shewing that the account of the doctrines of Fluxions &c, is agreeable to the real sense and meaning of their great Inventor, &c, Repub. of Letters, April 1736.

12. Philalethes's Considerations upon ditto, in Repub. of Letters, July 1736.

13. Mr. Robins's Remarks on ditto, in Repub. of Letters, Aug. 1736.

14. Mr. Robins's Remainder of ditto, in an Appendix to the Repub. of Letters, Sept. 1736.

15. Philalethes's Observations upon ditto, in an Appendix to the Repub. of Letters, Nov. 1736.

16. Mr. Robins's Advertisement in Repub. of Letters, Decemb. 1736.

17. Philalethes's Reply to ditto, in an Appendix to the Repub. of Letters for Decem. 1736.

18. Some Observations on the Appendix to the Repub. of Letters for Decem. 1736, by Dr. Pemberton, in the Works of the Learned for Feb. 1737. With some smaller pieces in the same.

In 1736 bishop Berkeley published *The Querist*, " a discourse addressed to magistrates, occasioned by the enormous licence and irreligion of the times;" and many other things afterward of a smaller kind. In 1744 came out his celebrated and curious book, " *Siris*; a Chain of Philosophical Reflections and Inquiries concerning the virtues of Tar-water." It had a second impression, with additions and emendations, in 1747; and was followed by *Farther Thoughts on Tar-water*, in 1752. In July the same year he removed, with his lady and family, to Oxford, partly to superintend the education of a son, but chiefly to indulge the passion for learned retirement, which had always strongly possessed him. He would have resigned his bishoprick for a canonry or headship at Oxford; but it was not permitted him. Here he lived highly respected, and collected and printed the same year all his smaller pieces in 8vo. But this happiness did not long continue, being suddenly cut off by a palsy of the heart Jan. 14, 1753, in the 69th year of his age, while listening to a sermon that his lady was reading to him.

The excellence of Berkeley's moral character is conspicuous in his writings: he was certainly a very amiable as well as a very great man; and it is thought that Pope scarcely said too much, when he ascribed

" To Berkeley every virtue under heaven."

BERME, in Fortification, a small space of ground, 4 or 5 feet wide, left without the rampart, between it and the side of the moat, to receive the earth that rolls down from the rampart, and prevent its falling into the ditch and filling it up.—Sometimes, for greater security, the berme is pallisadoed.

BERNARD (Dr. EDWARD), a learned astronomer, critic and linguist, was born at Perry St. Paul, near Towcester, the 2d of May 1638, and educated at Merchant-Taylor's school, and at St. John's college Oxford. Having laid in a good fund of classical learning at school, in the Greek and Latin languages, he applied himself very diligently at the university to the study of history, the eastern languages, and mathematics under the celebrated Dr. Wallis. In 1668 he went to Leyden to consult some Oriental manuscripts left to that university by Joseph Scaliger and Levin Warner, and especially the 5th, 6th, and 7th books of Apollonius's Conics, the Greek text of which is lost, and this Arabic version having been brought from the east by the celebrated

Golius,

Golius, a transcript of which was thence taken by Bernard, and brought with him to Oxford, with intent to publish it there with a Latin translation; but he was obliged to drop that design for want of encouragement. This however was afterwards carried into effect by Dr. Halley in 1710, with the addition of the 8th book, which he supplied by his own ingenuity and industry.

At his return to Oxford, Bernard examined and collated the most valuable manuscripts in the Bodleian library. In 1669, the celebrated Christopher Wren, Savilian professor of astronomy at Oxford, having been appointed surveyor-general of his majesty's works, and being much detained at London by this employment, obtained leave to name a deputy at Oxford, and pitched upon Mr. Bernard, which engaged the latter in a more particular application to the study of astronomy. But in 1673 he was appointed to the professorship himself, which Wren was obliged to resign, as, by the statutes of the founder, Sir Henry Saville, the professors are not allowed to hold any other office either ecclesiastical or civil.

About this time a scheme was set on foot at Oxford, of collecting and publishing the ancient mathematicians. Mr. Bernard, who had first formed the project, collected all the old books published on that subject since the invention of printing, and all the manuscripts he could discover in the Bodleian and Savilian libraries, which he arranged in order of time, and according to the matter they contained; of this he drew up a synopsis or view; and as a specimen he published a few sheets of Euclid, containing the Greek text, and a Latin version, with Proclus's commentary in Greek and Latin, and learned scholia and corollaries. The synopsis itself was published by Dr. Smith, at the end of his life of our author, under the title of *Veterum Mathematicorum Græcorum, Latinorum, et Arabum, Synopsis.* And at the end of it there is a catalogue of some Greek writers, whose works are supposed to be lost in their own language, but are preserved in the Syriac or Arabic translations of them.

Mr. Bernard undertook also an edition of the *Parva Syntaxis Alexandrina;* in which, besides Euclid, are contained the small treatises of Theodosius, Menelaus, Aristarchus, and Hipsicles; but it never was published.

In 1676 he was sent to France, as tutor to the dukes of Grafton and Northumberland, sons to king Charles the 2d by the dutchess of Cleveland, who then lived with their mother at Paris: but the simplicity of his manners not suiting the gaiety of the dutchess's family, he returned about a year after to Oxford, and pursued his studies; in which he made great proficiency, as appears by his many learned and critical works. In 1691, being presented to the rectory of Brightwell in Berkshire, he quitted his professorship at Oxford, in which he was succeeded by David Gregory, professor of mathematics at Edinburgh.

Toward the latter end of his life he was much afflicted with the stone; yet notwithstanding this, and other infirmities, he undertook a voyage to Holland, to attend the sale of Golius's manuscripts, as he had once before done at the sale of Heinsius's library. On his return to England, he fell into a languishing consumption, which put an end to his life the 12th of January 1696, in the 58th year of his age.

Beside the works of his before mentioned, he was author of many other compositions. He composed tables of the longitudes, latitudes, right ascensions, &c. of the fixed stars: he wrote Observations on the Obliquity of the Ecliptic; and other pieces inserted in the Philosophical Transactions. He wrote also,

1. *A Treatise of the Ancient Weights and Measures.*
2. *Chronologiæ Samaritanæ Synopsis,* in two tables.
3. *Testimonies of the Ancients concerning the Greek Version of the Old Testament by the Seventy.*

And several other learned works. Besides a great number of valuable manuscripts left at his death.

BERNARD (Dr. JAMES), professor of philosophy and mathematics, and minister of the Walloon church at Leyden, was born September the 1st 1658, at Nions in Dauphine. Having studied at Geneva, he returned to France in 1679, and was chosen minister of Venterol, a village in Dauphine; but some time after he was removed to the church of Vinsobres in the same province. To avoid the persecutions against the protestants in France, he went into Holland, where he was appointed one of the pensionary ministers of Ganda. He here published several political and historical works. And in 1699 he began the *Nouvelles de la Republique des Lettres,* which continued till December 1710. In 1705 he was chosen minister of the Walloon church at Leyden; and about the same time, Mr. de Volder, professor of philosophy and mathematics at Leyden, having resigned, Mr. Bernard was appointed his successor; upon which occasion the university also presented him with the degrees of doctor of philosophy and master of arts. In 1716 he published a supplement to Moreri's dictionary in 2 volumes folio. The same year he resumed his *Nouvelles de la Republique des Lettres;* which he continued till his death, which happened the 27th of April 1718, in the 60th year of his age.

BERNOULLI (JAMES), a celebrated mathematician, born at Basil the 27th of December 1654. Having taken his degrees in that university, he applied himself to divinity at the entreaties of his father, but against his own inclination, which led him to astronomy and mathematics. He gave very early proofs of his genius for these sciences, and soon became a geometrician, without a preceptor, and almost without books; for if one by chance fell into his hands, he was obliged to conceal it, to avoid the displeasure of his father, who designed him for other studies. This situation induced him to choose for his device, Phaeton driving the chariot of the sun, with these words, *Invito patre sidera verso,* "I traverse the stars against my father's will:" alluding particularly to astronomy, to which he then chiefly applied himself.

In 1676 he began his travels. When he was at Geneva, he fell upon a method to teach a young girl to write who had been blind from two months old. At Bourdeaux he composed universal gnomonic tables; but they were never published. He returned from France to his own country in 1680. About this time there appeared a comet, the return of which he foretold; and wrote a small treatise upon it. Soon after this he went into Holland, where he applied himself to the study of the new philosophy. Having visited Flanders and Brabant, he passed over to England; where he formed an acquaintance with the most eminent men in the sciences, and was frequent at their philosophical meetings. He

returned

returned to his native country in 1682; and exhibited at Bafil a courfe of experiments in natural philofophy and mechanics, which confifted of a variety of new difcoveries. The fame year he publifhed his Effay on a new Syftem of Comets; and the year following, his Differtation on the weight of the air. About this time Leibnitz having publifhed, in the Acta Eruditorum at Leipfic, fome effays on his new *Calculus Differentialis*, but concealing the art and method of it, Mr. Bernoulli and his brother John difcovered, by the little which they faw, the beauty and extent of it: this induced them to endeavour to unravel the fecret; which they did with fuch fuccefs, that Leibnitz declared that the invention belonged to them as much as to himfelf.

In 1687 James Bernoulli fucceeded to the profefforfhip of mathematics at Bafil; a truft which he difcharged with great applaufe; and his reputation drew a great number of foreigners from all parts to attend his lectures. In 1699 he was admitted a foreign member of the Academy of Sciences of Paris; and in 1701 the fame honour was conferred upon him by the Academy of Berlin: in both of which he publifhed feveral ingenious compofitions, about the years 1702, 3, and 4. He wrote alfo feveral pieces in the *Acta Eruditorum* of Leipfic, and in the *Journal des Scavans*. His intenfe application to ftudy brought upon him the gout, and by degrees a flow fever, which put a period to his life the 16th of Auguft 1705, in the 51ft year of his age — Archimedes having found out the proportion of a fphere and its circumfcribing cylinder, ordered them to be engraven on his monument: In imitation of him, Bernoulli appointed that a logarithmic fpiral curve fhould be infcribed on his tomb, with thefe words, *Eadem mutata refurgo*; in allufion to the hopes of the refurrection, which are in fome meafure reprefented by the properties of that curve, which he had the honour of difcovering.

James Bernoulli had an excellent genius for invention and elegant fimplicity, as well as a clofe application. He was eminently fkilled in all the branches of the mathematics, and contributed much to the promoting the new analyfis, infinite feries, &c. He carried to a great height the theory of the quadrature of the parabola; the geometry of curve lines, of fpirals, of cycloids and epicycloids.

His works, that had been publifhed, were collected, and printed in 2 volumes 4to, at Geneva in 1744. At the time of his death he was occupied on a great work entitled *De Arte Conjectandi*, which was publifhed in 4to, in 1713. It contains one of the beft and moft elegant introductions to Infinite Series, &c. This pofthumous work is omitted in the collection of his works above mentioned.

BERNOULLI (JOHN), the brother of James, laft mentioned, and a celebrated mathematician, was born at Bafil the 7th of Auguft 1667. His father intended him for trade; but his own inclination was at firft for the Belles-Lettres, which however, like his brother, he left for mathematics. He laboured with his brother to difcover the method ufed by Leibnitz, in his effays on the Differential Calculus, and gave the firft principles of the Integral Calculus. Our author, with Meffieurs Huygens and Leibnitz, was the firft who gave the folution of the problem propofed by James

Bernoulli, concerning the catenary, or curve formed by a chain fufpended by its two extremities.

John Bernoulli had the degree of doctor of phyfic at Bafil, and two years afterward was named proffeffor of mathematics in the univerfity of Groningen. It was here that he difcovered the mercurial phofphorus or luminous barometer; and where he refolved the problem propofed by his brother concerning Ifoperimetricals.

On the death of his brother James, the profeffor at Bafil, our author returned to his native country, againft the preffing invitations of the magiftrates of Utrecht to come to that city, and of the univerfity of Groningen, who wifhed to retain him. The Academic Senate of Bafil foon appointed him to fucceed his brother, without affembling competitors, and contrary to the eftablifhed practice: an appointment which he held during his whole life.

In 1714 was publifhed his treatife on the management of fhips; and in 1730, his memoir on the elliptical figure of the planets gained the prize of the Academy of Sciences. The fame academy alfo divided the prize, for their queftion concerning the inclination of the planetary orbits, between our author and his fon Daniel.

John Bernoulli was a member of moft of the academies of Europe, and received as a foreign affociate of that of Paris in 1699. After a long life fpent in conftant ftudy and improvement of all the branches of the mathematics, he died full of honours the 1ft of January 1748, in the 81ft year of his age. Of five fons which he had, three purfued the fame fciences with himfelf. One of thefe died before him; the two others, Nicolas and Daniel, he lived to fee become eminent and much refpected in the fame fciences.

The writings of this great man were difperfed through the periodical memoirs of feveral academies, as well as in many feparate treatifes. And the whole of them were carefully collected and publifhed at Laufanne and Geneva, 1742, in 4 volumes, 4to.

BERNOULLI (DANIEL), a celebrated phyfician and philofopher, and fon of John Bernoulli laft mentioned, was born at Groningen Feb. the 9th, 1700, where his father was then profeffor of mathematics. He was intended by his father for trade, but his genius led him to other purfuits. He paffed fome time in Italy; and at 24 years of age he declined the honour offered him of becoming prefident of an academy intended to have been eftablifhed at Genoa. He fpent feveral years with great credit at Peterfburgh; and in 1733 returned to Bafil, where his father was then profeffor of mathematics; and here our author fucceffively filled the chair of phyfic, of natural and of fpeculative philofophy.

In his work *Exercitationes Mathematicae*, 1724, he took the only title he then had, viz, "Son of John Bernoulli," and never would fuffer any other to be added to it. This work was publifhed in Italy, while he was there on his travels; and it claffed him in the rank of inventors. In his work, *Hydrodynamica*, publifhed in 4to at Argentoratum or Strafbourg, in 1738, to the fame title was alfo added that of *Med. Prof. Bafil*.

Daniel Bernoulli wrote a multitude of other pieces, which have been publifhed in the Mem. Acad. of Sciences at Paris, and in thofe of other Académies. He gained and divided ten prizes from the Academy of

Sciences, which were contended for by the moft illuftrious mathematicians in Europe. The only perfon who has had fimilar fuccefs of the fame kind, is Euler, his countryman, difciple, rival, and friend. His firft prize he gained at 24 years of age. In 1734 he divided one with his father; which hurt the family union; for the father confidered the conteft itfelf as a want of refpect; and the fon did not fufficiently conceal that he thought (what was really the cafe) his own piece better than his father's. And befides, he declared for Newton, againft whom his father had contended all his life. In 1740 our author divided the prize, "On the Tides of the Sea," with Euler and Maclaurin. The Academy at the fame time crowned a fourth piece, whofe chief merit was that of being Cartefian; but this was the laft public act of adoration paid by the Academy to the authority of the author of the Vortices, which it had obeyed but too long. In 1748 Daniel Bernoulli fucceeded his father John in the Academy of Sciences, who had fucceeded his brother James; this place, fince its firft erection in 1699, having never been without a Bernoulli to fill it.

Our author was extremely refpected at Bafil; and to bow to Daniel Bernoulli, when they met him in the ftreets, was one of the firft leffons which every father gave every child. He was a man of great fimplicity and modefty of manners. He ufed to tell two little adventures, which he faid had given him more pleafure than all the other honours he had received. Travelling with a learned ftranger, who, being pleafed with his converfation, afked his name; "I am Daniel Bernoulli," anfwered he with great modefty; "And I," faid the ftranger (who thought he meant to laugh at him), "am Ifaac Newton." Another time having to dinner with him the celebrated Koenig the mathematician, who boafted, with fome degree of felf-complacency, of a difficult problem he had refolved with much trouble, Bernoulli went on doing the honours of his table, and when they went to drink coffee he prefented Koenig with a folution of the problem more elegant than his own.

After a long, ufeful, and honourable life, Daniel Bernoulli died the 17th of March 1782, in the 83d year of his age.

BETELGEUSE, a fixed ftar of the firft magnitude in the right fhoulder of Orion.

BEZOUT (STEPHEN), a celebrated French mathematician, Member of the Academies of Sciences and the Marine, and Examiner of the Guards of the Marine and of the Eleves of Artillery, was born at Nemours the 31ft of March 1730. In the courfe of his ftudies he met with fome books of geometry, which gave him a tafte for that fcience; and the Eloges of Fontenelle, which fhewed him the honours attendant on talents and the love of the fciences. His father in vain oppofed the ftrong attachment of young Bezout to the mathematical fciences. April 8, 1758, he was named adjoint-mechanician in the French academy of fciences; having before that fent them two ingenious memoirs on the integral calculus, and given other proofs of his proficiency in the fciences. In 1763, he was named to the new office of Examiner to the Marine, and appointed to compofe a Courfe of Mathematics for their ufe; and in 1768, on the death of

M. Camus, he fucceeded as Examiner of the Artillery Eleves.

Bezout fixed his attention more particularly to the refolution of algebraic equations; and he firft found out the folution of a particular clafs of equations of all degrees. This method, different from all former ones, was general for the cubic and biquadratic equations, and juft became particular only at thofe of the 5th degree. Upon this work of finding the roots of equations, our author laboured from 1762 till 1779, when he publifhed it. He compofed two courfes of mathematics; the one for the Marine, the other for the Artillery: The foundation of thefe two works was the fame; the applications only being different, according to the two different objects: thefe courfes have every where been held in great eftimation. In his office of examiner he difcharged the duties with great attention, care, and tendernefs; a trait of his juftice and zeal is remarkable in the following inftance: During an examination, which he held at Toulon, he was told that two of the pupils could not be prefent, being confined by the fmall-pox: he himfelf had never had that difeafe, and he was greatly afraid of it; but as he knew that if he did not fee thefe two young men, it would much impede their improvement; he ventured therefore to their bed-fides, to examine them, and was happy to find them fo deferving of the hazard he put himfelf into for their benefit.

Mr. Bezout lived thus feveral years beloved of his family and friends, and refpected by all, enjoying the fruits and the credit of his labours. But the trouble and fatigues of his offices, with fome perfonal chagrines, had reduced his ftrength and conftitution; he was attacked by a malignant fever, of which he died Sept. 27, 1783, in the 54th year of his age, regretted by his family, his friends, the young ftudents, and by all his acquaintance in general.

The books publifhed by him, were:

1. Courfe of Mathematics for the ufe of the Marine, with a Treatife on Navigation; 6 vols. in 8vo, Paris, 1764.

2. Courfe of Mathematics for the Corps of Artillery; 4 vols. in 8vo, 1770.

3. General Theory of Algebraic Equations; 1779.

His papers printed in the volumes of the Memoirs of the Academy of Sciences, are:

1. On curves whofe rectification depends on a given quantity; in the vol. for 1758.

2. On feveral claffes of equations that admit of an algebraic folution; 1762.

3. Firft vol. of a courfe of mathematics, 1764.

4. On certain equations, &c; 1764.

5. General refolution of all equations; 1765.

6. Second vol. of a courfe of mathematics; 1765.

7. Third vol. of the fame; 1766.

8. Fourth vol. of the fame; 1767.

9. Integration of differentials, &c. vol. 3, Sav. Etr.

10. Experiments on cold; 1777.

BIANCHINI (FRANCIS), a very learned Italian philofopher and mathematician of the 17th century, was born at Verona the 13th of December 1662. He was much efteemed by the learned, and was a member of feveral academies; and was even the founder of that at Verona, called the Academy of Aletofili, or

Lovers

Lovers of Truth. He went to Rome in 1684; and was made librarian to cardinal Ottoboni, who was afterwards Pope by the name of Alexander the 8th. He entered into the church, and became canon of St. Mary de la Rotondo, and afterward of St. Lawrence in Damaso.

Bianchini was author of several learned and ingenious differtations. In 1697 he published *La Iftoria univer-fale provata con monumenti, & figurata con fimboli de gli antichi*. In 1701 pope Clement the 11th named him fecretary of the conferences for the reformation of the calendar; and he published in 1703, *De Calendario & Cyclo Cæfaris, ac de Canone Pafchali fancti Hyppoliti, Martyris, Differtationes duæ*. Bianchini was employed on the construction of the large gnomon in the church of the Chartreux at Rome, upon which he published an ample differtation intitled, *De Nummo & Gnomone Clementino*. The research concerning the parallax and the spots of Venus occupied him a long time; but his most remarkable discovery is that of the parallelism of the axis of Venus in her orbit. He propofed to trace a meridian line through the whole extent of Italy. He was admitted a foreign Affociate in the Paris Academy of Sciences, in 1706; and he had many astronomical differtations inferted in their Memoirs, particularly in those of the years 1702, 1703, 1704, 1706, 1707, 1708, 1713, and 1718.—Bianchini died the 2d of March 1729, in the 67th year of his age.

BIMEDIAL *Line*, is the fum of two Medials. Euclid reckons two of these bimedials, in pr. 38 and 39 lib. x; the first is when the rectangle is rational, which is contained by the two medial lines whose fum makes the bimedial; and the second when that rectangle is a medial, or contained under two lines that are commenfurable only in power.

BINARY Number, that which is compofed of two units.

BINARY *Arithmetic*, that in which two figures or characters, viz, 1 and 0, only are ufed; the cipher multiplying every thing by 2, as in the common arithmetic by ten: thus, 1 is one, 10 is 2, 11 is 3, 100 is 4, 101 is 5, 110 is 6, 111 is 7, 1000 is 8, 1001 is 9, 1010 is ten; being founded on the fame principles as common arithmetic.—This fort of arithmetic was invented by Leibnitz, who pretended that it is better adapted than the common arithmetic, for difcovering certain properties of numbers, and for conftructing tables; but he does not venture to recommend it, for ordinary ufe, on account of the great number of places of figures requifite to exprefs all numbers, even very fmall ones. Jof. Pelican of Prague has more largely explained the principles and practice of the binary arithmetic, in a book entitled *Arithmeticus Perfectus, qui tria numerare nefcit*; 1712. And De Lagni propofed a new fyftem of logarithms, on the plan of the binary arithmetic; which he finds fhorter, and more eafy and natural than the common ones.

BINOCLE, or BINOCULAR TELESCOPE, is one by which an object is viewed with both eyes at the fame time. It confifts of two tubes, each furnifhed with glaffes of the fame power, by which means it has been faid to fhew objects larger and more clearly than a mo-

nocular or fingle telefcope; though this is probably only an illufion, occafioned by the ftronger impreffion which two equal images, alike illuminated, make upon the eyes; and they have been found more embarraffing than ufeful in practice. This telefcope has been chiefly treated of by the fathers Reita and Cherubin of Orleans.—There are alfo microfcopes of the fame kind, though but little ufed; being fubject to the fame inconveniences as the telefcopes.

BINOMIAL, a quantity confifting of two terms or members connected by the fign + or —, viz, plus or minus; as $a + b$, or $3a - 2c$, or $a^2 + b$, or $x^2 - 2\sqrt{c}$, &c; denoting the fum or the difference of the two terms. But the difference is alfo fometimes named a *refidual*, and by Euclid an *apotome*. The term binomial was first introduced by Robert Recorde; fee his algebra, pa. 46².

BINOMIAL *Line*, or *Surd*, is that in which at leaft one of the parts is a furd. Euclid enumerates fix kinds of binomial lines or furds, in the 10th book of his Elements, which are exactly fimilar to the fix refiduals or apotomes there treated of alfo, and of which an account is given under the art. APOTOME, which fee. Thofe apotomes become binomials by only changing the fign of the latter term from minus to plus, which therefore are as below.

Euclid's 6 Binomial Lines.

First binomial $3 + \sqrt{5}$,
2d binomial $\sqrt{18} + 4$,
3d binomial $\sqrt{24} + \sqrt{18}$,
4th binomial $4 + \sqrt{3}$,
5th binomial $\sqrt{6} + 2$,
6th binomial $\sqrt{6} + \sqrt{2}$.

To extract the Square Root of a Binomial, as of $a + \sqrt{b}$, or $\sqrt{c} + \sqrt{b}$. Various rules have been given for this purpofe. The first is that of Lucas De Burgo, in his *Summa de Arith.* &c, which is this: When one part, as a, is rational, divide it into two parts fuch, that their product may be equal to ¼th of the number under the radical b; then fhall the fum of the roots of thofe parts be the root of the binomial fought: or their difference is the root when the quantity is refidual. That is, if $c + e = a$, and $ce = \frac{1}{4}b$; then is $\sqrt{c} + \sqrt{e} = \sqrt{a + \sqrt{b}}$ the root fought. As if the binomial be $23 + \sqrt{448}$; then the parts of 23 are 16 and 7, and their product is 112, which is ¼th of 448; therefore the fum of their roots $4 + \sqrt{7}$ is the root fought of $23 + \sqrt{448}$.

De Burgo gives alfo another rule for the fame extractions, which is this: The given binomial being, for example, $\sqrt{c} + \sqrt{b}$, its root will be $\sqrt{\frac{1}{2}\sqrt{c} + \frac{1}{2}\sqrt{c - b}} + \sqrt{\frac{1}{2}\sqrt{c} - \frac{1}{2}\sqrt{c - b}}$.—So in the foregoing example, $23 + \sqrt{448}$, here $\sqrt{c} = 23$, and $\sqrt{b} = \sqrt{448}$; hence $\frac{1}{2}\sqrt{c} = 11\frac{1}{2}$, and $\frac{1}{2}\sqrt{c - b} = \frac{1}{2}\sqrt{23^2 - 448} = \frac{1}{2}\sqrt{81} = 4\frac{1}{2}$;

theref. $\sqrt{\frac{1}{2}\sqrt{c} + \frac{1}{2}\sqrt{c - b}} = \sqrt{11\frac{1}{2} + 4\frac{1}{2}} = \sqrt{16} = 4$,

and $\sqrt{\frac{1}{2}\sqrt{c} - \frac{1}{2}\sqrt{c - b}} = \sqrt{11\frac{1}{2} - 4\frac{1}{2}} = \sqrt{7}$; confeq. $4 + \sqrt{7}$ is the root fought, as before. Again, if the binomial be $\sqrt{18} + \sqrt{10}$; here $c = 18$, and

and $b = 10$; theref. $\frac{1}{2}\sqrt{c} = \frac{1}{2}\sqrt{18} = \frac{3}{2}\sqrt{2}$, and $\frac{1}{2}\sqrt{c-b} = \frac{1}{2}\sqrt{8} = \sqrt{2}$; hence,

$$\sqrt{\tfrac{1}{2}\sqrt{c}+\tfrac{1}{2}\sqrt{c-b}} = \sqrt{\tfrac{3}{2}\sqrt{2} + \sqrt{2}} = \sqrt{\tfrac{5}{2}\sqrt{2}} = \sqrt[4]{\tfrac{25}{2}}, \text{ and}$$

$$\sqrt{\tfrac{1}{2}\sqrt{c}-\tfrac{1}{2}\sqrt{c-b}} = \sqrt{\tfrac{3}{2}\sqrt{2} - \sqrt{2}} = \sqrt{\tfrac{1}{2}\sqrt{2}} = \sqrt[4]{\tfrac{1}{2}};$$

conseq. $\sqrt[4]{\frac{25}{2}} + \sqrt[4]{\frac{1}{2}}$ or $\frac{\sqrt{5}+1}{\sqrt[4]{2}}$ is the root of $\sqrt{18} + \sqrt{10}$ sought. And this latter rule has been used by all other authors, down to the present time.

To extract the Cubic and other higher Roots of a Binomial. This is useful in resolving cubic and higher equations, and was introduced with the resolution of those equations by Tartalea and Cardan. The rules for such extractions are in great measure tentative; and some of the principal ones are the following.

Tartalea's Rule for the Cube Root of a Binomial p+q. This rule is given in his 9th book of Miscellaneous Questions, quest. 40; and it is made out from either of the terms, p or q, of the binomial, taken singly, in this manner: Separate either term, as p, into two such parts that the one of them may be a cubic number, and the other part divisible by 3 without a remainder; then the cube root of the said cubic part will be one term of the root, and the other term will be the square root of the quotient arising from dividing the aforesaid third part by the first term just found. So if p be divided into $r^3 + 3s$, then the root is $r + \sqrt{\frac{s}{r}}$. For example, to extract the cube root of $\sqrt{108} + 10$. Suppose the part 10 to be taken: this separates into the parts 1 and 9, the former of which is a cube, and the latter divisible by 3; that is $r^3 = 1$, and $3s = 9$; hence $r = 1$, and $s = 3$; consequently $r + \sqrt{\frac{s}{r}} = 1 + \sqrt{3}$ is the cubic root of $\sqrt{108} + 10$ sought. Again, to use the other term $\sqrt{108}$: this divides into $\sqrt{27} + \sqrt{27}$, of which the former is a cube, and the latter divisible by 3; that is, $r^3 = \sqrt{27}$, and $3s = \sqrt{27}$; therefore $r = \sqrt{3}$, and $s = \frac{1}{3}\sqrt{27} = \sqrt{3}$ also: hence $r + \sqrt{\frac{s}{r}} = \sqrt{3} + \sqrt{\frac{\sqrt{3}}{\sqrt{3}}} = \sqrt{3} + 1$ the cube root, the same as before.

Bombelli's Rule for the Cubic Root of the Binomial $a + \sqrt{-b}$. First find $\sqrt[3]{a^2 + b}$; then, by trials, search out a number c, and a square root $\sqrt{d}$, such that the sum of their squares $c^2 + d$ be $= \sqrt[3]{a^2 + b}$, and also $c^3 - 3cd$ be $= a$; then shall $c + \sqrt{-d}$ be the cube root of $a + \sqrt{-b}$ sought. For example, to find the cube root of $2 + \sqrt{-121}$: here $\sqrt[3]{a^2 + b} = \sqrt[3]{125} = 5$; then taking $c = 2$, and $d = 1$, it is $c^2 + d = 5 = \sqrt[3]{a^2+b}$, and $c^3 - 3cd = 8 - 6 = 2 = a$, as it ought; therefore $2 + \sqrt{-1}$ is the cube root of $2 + \sqrt{-121}$ sought.—Bombelli gave also a rule for the cube root of the binomial $a + \sqrt{b}$, but it is good for nothing.

Albert Girard's Rule for the Cube Root of a Binomial. This is given in his *Invention Nouvelle en l'Algebre*, and is explained by him thus: Let $72 + \sqrt{5120}$ be the given binomial whose cubic root is sought.

The square of 72 the greater term is 5184
and of the less term is 5120
their difference 64
its cube root 4,

which 4 must be the difference between the squares of the two terms of the root sought; and as the rational part 72 of the given binomial is the greater term, therefore the rational part of the required root will be the greater part also; consequently the cubic root sought must be one of the binomials here set in the margin, where the difference of the squares of the terms is always 4, as required; and to find out

$2 + \sqrt{0}$
$3 + \sqrt{5}$
$4 + \sqrt{12}$
$5 + \sqrt{20}$

which of them it must be, proceed thus: The first, $2 + \sqrt{0}$ must be rejected, because one term of it is 0 or nothing; also because 5 exceeds the cube root of 72, or $\sqrt{20}$ exceeds the cube root of $\sqrt{5120}$, therefore $5 + \sqrt{20}$, and all after it must be rejected too; so that the root must be either $3 + \sqrt{5}$ or $4 + \sqrt{12}$, if the given quantity has a binomial root: to know which of these is to be taken, it must be considered that the rational term of the root must measure the rational term given; and also the irrational term of the root must measure the irrational term given; then, on examination it is found that both 3 and 4 measure or divide the 72 without a remainder, but that only the $\sqrt{5}$, and not $\sqrt{12}$, measures $\sqrt{5120}$; consequently none but $3 + \sqrt{5}$ can be the cube root of the given quantity $72 + \sqrt{5120}$; which is found to answer, by cubing the said root $3 + \sqrt{5}$.

Dr. Wallis's Rule for the Cube Root of a Binomial $a \pm m\sqrt{b}$ or $a \pm m\sqrt{-b}$. In these forms the greatest rational part m is extracted out of the radical part, leaving only b the least radical part possible under the radical sign. He then observes that if the given quantity have a binomial root, it must be of this form $c \pm n\sqrt{b}$, with the same radical b. Then to find the value of c and n, he raises this root to the 3d power, which gives $c^3 + 3cn^2b \pm \overline{3c^2n + n^3b} \cdot \sqrt{b}$, which must be $= a \pm m\sqrt{b}$ the given quantity; hence putting the rational part of the one quantity equal to that of the other, and also the radical part of the one equal to that of the other, gives $c^3 + 3cn^2b = a$, and $3c^2n + n^3b = m$. Then assuming several values of n, from the last equation he finds the value of c; hence if these values of c and n, substituted in the first equation, make it just, they are right; but if not, another value of n is assumed, and so on, till the first equation hold true. And it is to be noted that n is always an integer or else the half of an integer. For example, if the cube root of $135 \pm \sqrt{1825}$ be required, or $135 \pm 78\sqrt{3}$; here $a = 135$, $m = 78$, and $b = 3$; hence $3c^2n + n^3b = m$ is $3c^2n + 3n^3 = 78$, or $c^2n + n^3 = 26$; then assuming $n = 1$, this last equation becomes $c^2 + 1 = 26$, from whence c is found $= 5$; which values of c and n being substituted in the first equation $c^3 + 3cn^2b = a$, makes $5^3 + 3 \cdot 5 \cdot 3 = 170$, but ought to be 135, shewing that c is too great, and consequently n taken too little. Let n therefore be assumed $= 2$, so shall $2c^2 + 8 = 26$, and $c = 3$; and the first equation becomes $3^3 + 3 \cdot 3 \cdot 2^2 \cdot 3 = 27 \cdot 5 = 135 = a$ as it ought, which shews that the true value of n is 2, and that of c is 3; hence then the cube root of $135 \pm 78\sqrt{3}$ or $c \pm n\sqrt{b}$ is

5   3 ±

$3 \pm 2\sqrt{3}$ or $3 \pm \sqrt{12}$. And in like manner is the process instituted when the number in the radical is negative, as the cube root of $81 \pm 30\sqrt{-3}$, which is $\frac{3}{2} \pm \frac{1}{2}\sqrt{-3}$.

Another rule for extracting the cube root of an imaginary binomial was also given by Demoivre, at the end of Saunderson's Algebra, by means of the trisection of an arc or angle.

Sir I. Newton's Rule for any Root of a Binomial $a \pm b$. In his Universal Arith. is given a rule for the square root of a binomial, which is the same as the 2d of Lucas de Burgo, before given; and also a general rule for any root of a binomial, which I have not met with elsewhere; and it is this: Of the given quantity $a \pm b$, let a be the greater term, and c the index of the root to be extracted. Seek the least number n whose power n^c can be divided by $aa - bb$ without a remainder, and let the quotient be q; Compute $\sqrt[c]{a + b} . \sqrt{q}$ in the nearest integer number, which call r; divide $a\sqrt{q}$ by its greatest rational divisor, calling the quotient s; and let the nearest integer number above $\frac{r + \frac{n}{s}}{2 s}$ be t: so shall $\frac{t s \pm \sqrt{t^2 s^2 - n}}{2 \sqrt[c]{q}}$ be the root sought, if the root can be extracted. And this rule is demonstrated by *s'Gravesande* in his commentary on Newton's Arithmetic. And many numeral examples, illustrating this rule, are given in s'Gravesande's Algebra, abovementioned, pa. 160, as also in Newton's Univerf. Arith. pa. 53 2d edit. and in Maclaurin's Algebra pa. 118. Other rules may be found in Schooten's Commentary on the Geometry of Descartes, and elsewhere.

Impossible or *Imaginary* BINOMIAL, is a binomial which has one of its terms an impossible or an imaginary quantity; as $a + \sqrt{-b}$.

BINOMIAL *Curve*, is a curve whose ordinate is expressed by a binomial quantity; as the curve whose ordinate is $x^1 \times \overline{b + dx}^c$. Stirling, Method. Diff. pa. 58.

BINOMIAL *Theorem*, is used to denote the celebrated theorem given by Sir I. Newton for raising a binomial to any power, or for extracting any root of it by an approximating infinite series. It was known by Stifelius, and others, about the beginning of the 16th century, how to raise the integral powers, not barely by a continued multiplication of the binomial given, but Stifelius formed also a table of numbers by a continued addition, which shewed by inspection the coefficients of the terms of any power of the binomial, contained within the limits of the table; but still they could not independent of a table, and of any of the lower powers, raise any power of a binomial at once, by determining its terms one from another only, viz, the 2d term from the 1st, the 3d from the 2d, and so on as far as we please, by a general rule; and much less could they extract general algebraic roots in infinite series by any rule whatever.

For although the nature and construction of that table, which is a table of figurate numbers, was so early known, and employed in the raising of powers, and the extracting the roots of pure numbers; yet it was only by raising the numbers one from another by continual additions, and then taking them from the table

when wanted, till Mr. Briggs first pointed out the way of raising any line in the table by itself, without any of the preceding lines; and thus teaching to raise the terms of any power of a binomial, independent of any of the other powers; and so gave the substance of the binomial theorem in words, wanting only the algebraic notation in symbols; as is shewn at large at pa. 75 of the historical introduction to my Mathematical Tables. Whatever was known however of this matter, related only to pure or integral powers, no one before Newton having thought of extracting roots by infinite series. He happily discovered that, by considering powers and roots in a continued series, roots being as powers having fractional exponents, the same binomial series would equally serve for them all, whether the index should be fractional or integral, or whether the series be finite or infinite. The truth of this method however was long known only by trial in particular cases, and by induction from analogy; nor does it appear that even Newton himself ever attempted any direct proof of it: however, various demonstrations of the theorem have since been given by the more modern mathematicians, some of which are by means of the doctrine of fluxions, and others, more legally, from the pure principles of algebra only: for a full account of which, see pa. 71 &c, of my Mathematical Tracts, vol. 1.

This theorem was first discovered by Sir I. Newton in 1669, and sent in a letter of June 13, 1676, to Mr. Oldenburgh, Secretary to the Royal Society, to be by him communicated to Mr. Leibnitz; and it was in this form: $\overline{p + pq}^{\frac{m}{n}} = p^{\frac{m}{n}} + \frac{m}{n} aq + \frac{m - n}{2n} bq + \frac{m - 2n}{3n} cq + \frac{m - 3n}{4n} dq + $ &c: where $p + pq$ signifies the quantity whose root, or power, or root of any power, is to be found; p being the first term of that quantity; q the quotient of all the rest of the terms divided by that first term; and $\frac{m}{n}$ the numeral index of the power or root of the quantity $p + pq$, whether it be integral or fractional, positive or negative; and lastly $a, b, c, d,$ &c, are assumed to denote the several terms in their order as they are found, viz, $a = $ the first term $p^{\frac{m}{n}}$, $b = $ the 2d term $\frac{m}{n} aq$, $c = $ the 3d term $\frac{m - n}{2n} bq$, and so on. As Newton's general notation of indices was not commonly known, he takes this occasion to explain it; and then he gives many examples of the application of this theorem, one of which is the following.

Ex. 1. To find the value of $\sqrt{c^2 + x^2}$ or $\overline{c^2 + x^2}^{\frac{1}{2}}$, that is, to extract the square root of $c^2 + x^2$ in an infinite series. Here $p = c^2$, $q = \frac{x^2}{c^2}$, $m = 1$, and $n = 2$; therefore $a = p^{\frac{m}{n}} = \overline{c^2}^{\frac{1}{2}} = c$, $b = \frac{m}{n} aq = \frac{x^2}{2c}$, $c = \frac{m - n}{2n} bq = \frac{-x^4}{8c^3}$, &c; and therefore the root sought is $c + \frac{x^2}{2c} - \frac{x^4}{8c^3} + \frac{x^6}{16c^5} - \frac{5x^8}{128c^7} + $ &c.

A variety

A variety of other examples are also given in the same place, by which it is shewn that the theorem is of universal application to all sorts of quantities whatever.

This theorem is sometimes represented in other forms, as

$$p^{\frac{m}{n}} \times : 1 + \frac{m}{n} q + \frac{m}{n} \cdot \frac{m-n}{2n} q^2 + \frac{m}{n} \cdot \frac{m-n}{2n} \cdot \frac{m-2n}{3n} q^3 + \&c;$$

which comes to the same thing. Or also thus

$$p^{\frac{m}{n}} + \frac{m}{n} p^{\frac{m-n}{n}} r + \frac{m}{n} \cdot \frac{m-n}{2n} p^{\frac{m-2n}{n}} r^2 \&c;$$

where the binomial is $\overline{p+r}^{\frac{m}{n}}$

In another letter to Mr. Oldenburgh, of Oct. 24, 1676, Newton explains the train of reasoning by which he obtained the said theorem, as follows: "In the beginning of my mathematical studies, when I was perusing the works of the celebrated Dr. Wallis (see his Arith. of Infinites, prop. 118, and 121, also his Algebra chap. 82), and considering the series by the interpolation of which he exhibits the area of the circle and hyperbola; for instance, in this series of curves, whose common base or axis is x, and the ordinates respectively $\overline{1-xx}^{\frac{0}{2}}, \overline{1-xx}^{\frac{1}{2}}, \overline{1-xx}^{\frac{2}{2}}, \overline{1-xx}^{\frac{3}{2}}, \overline{1-xx}^{\frac{4}{2}}$, &c; I perceived that if the areas of the alternate curves, which are x,

$x - \frac{1}{3}x^3$,
$x - \frac{2}{3}x^3 + \frac{1}{5}x^5$,
$x - \frac{3}{3}x^3 + \frac{3}{5}x^5 - \frac{1}{7}x^7$,

&c; could be interpolated, we should obtain the areas of the intermediate ones; the first of which, or $\overline{1-xx}^{\frac{1}{2}}$, is the area of the circle; now in order to this, it appeared that in all the series the first term was x; that the 2d terms $\frac{0}{3}x^3, \frac{1}{3}x^3, \frac{2}{3}x^3, \frac{3}{3}x^3$, &c, were in arithmetical progression; and consequently that the first two terms of all the series to be interpolated would be

$$x - \frac{\frac{1}{2}x^3}{3} \quad x - \frac{\frac{3}{2}x^3}{3} \quad x - \frac{\frac{5}{2}x^3}{3}, \&c.$$

"Now for the interpolation of the rest, I considered that the denominators 1, 3, 5, 7, &c, were in arithmetical progression; and that therefore only the numeral coefficients of the numerators were to be investigated. But these in the alternate areas, which are given, were the same with the figures of which the several powers of 11 consist, viz, of 11^0, 11^1, 11^2, 11^3 &c; that is, the first   1,
the second   1, 1
the third   1, 2, 1
the fourth   1, 3, 3, 1
the fifth   1, 4, 6, 4, 1
&c.

"I enquired therefore how, in these series, the rest of the terms may be derived from the first two being given; and I found that by putting m for the 2d figure or term,

the rest would be produced by the continued multiplication of the terms of this series,

$$\frac{m-0}{1} \times \frac{m-1}{2} \times \frac{m-2}{3} \times \frac{m-3}{4} \times \frac{m-4}{5} \&c.$$

"For instance, if the 2d term $m = 4$; then shall $4 \times \frac{m-1}{2}$, or 6, be the 3d term; and $6 \times \frac{m-2}{3}$, or 4, the 4th term; and $4 \times \frac{m-3}{4}$, or 1, the 5th term; and $1 \times \frac{m-4}{5}$, or 0, the 6th; which shews that in this case the series terminates.

"This rule therefore I applied to the series to be interpolated. And since, in the series for the circle, the 2d term was $\frac{\frac{1}{2}x^3}{3}$ I put $m = \frac{1}{2}$, which produced the terms $\frac{1}{2} \times \frac{\frac{1}{2}-1}{2}$ or $-\frac{1}{8}$; $-\frac{1}{8} \times \frac{\frac{1}{2}-2}{3}$ or $+\frac{1}{16}$; $+\frac{1}{16} \times \frac{\frac{1}{2}-3}{4}$ or $-\frac{5}{128}$; and so on *ad infinitum*. And hence I found that the required area of the circular segment is

$$x - \frac{\frac{1}{2}x^3}{3} - \frac{\frac{1}{8}x^5}{5} - \frac{\frac{1}{16}x^7}{7} - \frac{\frac{5}{128}x^9}{9} - \&c.$$

"And in the same manner might be produced the interpolated areas of the other curves: as also the area of the hyperbola and the other alternates in this series $\overline{1+xx}^{\frac{0}{2}}, \overline{1+xx}^{\frac{1}{2}}, \overline{1+xx}^{\frac{2}{2}}, \overline{1+xx}^{\frac{3}{2}}$, &c. And in the same way also may other series be interpolated, and that too if they should be taken at the distance of two or more terms.

"This was the way then in which I first entered upon these speculations; which I should not have remembered, but that in turning over my papers a few weeks since, I chanced to cast my eyes on those relating to this matter.

"Having proceeded so far, I considered that the terms $\overline{1-xx}^{\frac{0}{2}}, \overline{1-xx}^{\frac{2}{2}}, \overline{1-xx}^{\frac{4}{2}}, \overline{1-xx}^{\frac{6}{2}}$, &c, that is, 1
$1 - x^2$
$1 - 2x^2 + x^4$
$1 - 3x^2 + 3x^4 - x^6$, &c, might be interpolated in the same manner as the areas generated by them: and for this, nothing more was required but to omit the denominators 1, 3, 5, 7, &c, in the terms expressing the areas; that is, the coefficients of the terms of the quantity to be interpolated, $\overline{1-xx}^{\frac{1}{2}}$, or $\overline{1-xx}^{\frac{3}{2}}$, or generally $\overline{1-xx}^{m}$, will be produced by the continued multiplication of the terms of this series

$$m \times \frac{m-1}{2} \times \frac{m-2}{3} \times \frac{m-3}{4} \&c.$$

"Thus, for example, there would be found

$$\overline{1-xx}^{\frac{1}{2}} = 1 - \frac{1}{2}x^2 - \frac{1}{8}x^4 - \frac{1}{16}x^6 \&c.$$

$$\overline{1-xx}^{\frac{3}{2}} = 1 - \frac{3}{2}x^2 + \frac{3}{8}x^4 + \frac{1}{16}x^6 \&c]$$

$$\overline{1-xx}^{\frac{1}{3}} = 1 - \frac{1}{3}x^2 - \frac{1}{9}x^4 - \frac{5}{81}x^6 \&c.$$

     Thes

"Thus then I difcovered a general method of re-ducing radical quantities into infinite feries, by the theorem which I fent in the beginning of the former letter, before I knew the fame by the extraction of roots.

"But having difcovered that way, this other could not long remain unknown : for, to prove the truth of thofe operations, I multiplied $1 - \frac{1}{2}x^2 - \frac{1}{8}x^4 - \frac{1}{16}x^6$ &c, by itfelf, and the product is $1 - x^2$, all the reft of the terms vanifhing after thefe, *in infinitum.* In like manner, $1 - \frac{1}{3}x^2 - \frac{1}{9}x^4 - \frac{5}{81}x^6$ &c, twice multiplied by itfelf, produced $1 - x^2$. But as this was a certain proof of thofe conclufions, fo I was naturally led to try converfely whether thefe feries, which were thus known to be the roots of the quantity $1 - x^2$, could not be extracted out of it after the manner of arithmetic ; and upon trial I found it to fucceed. The procefs for the fquare root is here fet down

$$
\begin{array}{l}
1 - x^2\ (1 - \tfrac{1}{2}x^2 - \tfrac{1}{8}x^4 - \tfrac{1}{16}x^6\ \&c \\
\underline{1} \\
0 - x^2 \\
\quad \underline{-x^2 + \tfrac{1}{4}x^4} \\
\qquad -\tfrac{1}{4}x^4 \\
\qquad \underline{-\tfrac{1}{4}x^4 + \tfrac{1}{8}x^6 + \tfrac{1}{64}x^8} \\
\qquad\qquad -\tfrac{1}{8}x^6 - \tfrac{1}{64}x^8
\end{array}
$$

"Thefe methods being found, I laid afide the other way by interpolation of feries, and ufed thefe operations only as a more genuine foundation. Neither was I ignorant of the reduction by divifion, which is fo much eafier." See Collins's *Commercium Epiftolicum.*

And this is all the account that Newton gives of the invention of this theorem, which is engraved on his monument in Weftminfter Abbey, as one of his greateft difcoveries.

BIPARTIENT, is a number that divides another into two equal parts without a remainder. So 2 is a bipartient to 4, and 5 a bipartient to 10.

BIPARTITION, is a divifion into two equal parts.

BIQUADRATE, or BIQUADRATIC *Power*, is the fquared fquare, or 4th power of any number or quantity. Thus 16 is the biquadrate or 4th power of 2, or it is the fquare of 4 which is the 2d power of 2.

BIQUADRATIC *Root*, of any quantity, is the fquare root of the fquare root, or the 4th root of that quantity. So the biquadratic root of 16 is 2, and the biquadratic root of 81 is 3.

BIQUADRATIC *Equation*, is that which rifes to 4 dimenfions, or in which the unknown quantity rifes to the 4th power ; as $x^4 + ax^3 + bx^2 + cx + d = 0$.

Any biquadratic equation may be conceived to be generated or produced from the continual multiplication of four fimple equations,

as $\overline{x - p} \times \overline{x - q} \times \overline{x - r} \times \overline{x - s} = 0$;

or from that of two quadratic equations,

as $\overline{x^2 + px + q} \times \overline{x^2 + rx + s} = 0$;

or, laftly, from that of a cubic and a fimple equation,

as $\overline{x - p} \times \overline{x^3 + qx^2 + rx + s} = 0$: which was the invention of Harriot. And, on the contrary, a biquadratic equation may be refolved into four fimple equations, or into two quadratics, or into a cubic and a fimple equation, having all the fame roots with it.

1. Ferrari's Method for Biquadratic Equations :

The firft refolution of a biquadratic equation was given in Cardan's Algebra, chap. 39, being the invention of his pupil and friend Lewis Ferrari, about the year 1540. This is effected by means of a cubic equation, and is indeed a method of depreffing the biquadratic equation to a cubic, which Cardan demonftrates, and applies in a great variety of examples. The principle is very general, and confifts in completing one fide of the equation up to a fquare, by the help of fome multiples or parts of its own terms and an affumed unknown quantity ; which it is always eafy to do ; and then the other fide is made to be a fquare alfo, by affuming the product of its 1ft and 3d terms equal to the fquare of half the 2d term ; for it confifts only of three terms, or three denominations of the original letter ; then this equality will determine the value of the affumed quantity by a cubic equation: other circumftances depend on the artift's judgment. But the method will be farther explained by the following examples, extracted from Cardan's book.

Ex. 1. Given $x^4 + 6x^2 + 36 = 60x$, to be refolved. Add $6x^2$ to both fides of the equation, fo fhall

$x^4 + 12x^2 + 36$ or $\overline{x^2 + 6}|^2 = 6x^2 + 60x$.

Affume y, and add $\overline{x^2 + 6}.2y + y^2$ to both fides, then is $\overline{x^2 + 6}|^2 + \overline{x^2 + 6}.2y + y^2 = 6x^2 + 60x + \overline{x^2 + 6}.2y + y^2$, or $\overline{x^2 + 6 + y}|^2 = \overline{6 + 2y}.x^2 + 60x + 12y + y^2$.

Make now the 1ft $\times$ 3d term $= \frac{1}{4}$ fq. 2d, this gives $\overline{6 + 2y}.\overline{12y + y^2}.x^2 = 900x^2$, or $y^3 + 15y^2 + 36y = 450$; and hence

$y = \sqrt[3]{287\frac{1}{2} + \sqrt{80449\frac{1}{4}}} + \sqrt[3]{287\frac{1}{2} - \sqrt{80449\frac{1}{4}}} - 5.$ From which x may be found by a quadratic equation.

Ex. 2. Given $x^4 = x + 2$.

Before applying Ferrari's method to this example, Cardan refolves it by another way as follows: fubtract 1, then is $x^4 - 1 = x + 1$; divide by $x + 1$, then is $x^3 - x^2 + x - 1 = 1$, or $x^3 + x = x^2 + 2$; and hence

$x = \sqrt[3]{\sqrt{\frac{2241}{2916}} + \frac{47}{54}} - \sqrt[3]{\sqrt{\frac{2241}{2916}} - \frac{47}{54}} + \frac{1}{3}.$

But to refolve it by Ferrari's rule :

Becaufe $x^4 = x + 2$. therefore

$x^4 + 2yx^2 + y^2$ or $\overline{x^2 + y}|^2 = 2yx^2 + x + \overline{2 + y^2}$; hence $2y^3 + 4y = \frac{1}{4}$, or $y^3 + 2y = \frac{1}{8}$; and the root is

$y = \sqrt[3]{\sqrt{\frac{2075}{6912}} + \frac{1}{16}} - \sqrt[3]{\sqrt{\frac{2075}{6912}} - \frac{1}{16}} :$

by means of which x is found by a quadratic equation.

Ex. 3. Given $x^4 + 32x^2 + 16 = 48x$.—Add 240, then $x^4 + 32x^2 + 256$ or $\overline{x^2 + 16}|^2 = 48x + 240$; complete fquare again, then

$\overline{x^2 + 16}|^2 + 2y.\overline{x^2 + 16} + y^2 = 2yx^2 + 48x + y^2 + 32y + 240$;

make the laft fide a fq. by the rule, which gives $y^3 + 32y^2 + 240y = 24 \times 12 = 288$.

Put now $z = y + 10\frac{2}{3}$, and the laft transforms to $z^3 = 101\frac{1}{3}z + 420\frac{10}{27}$; then the value of z found from this, gives the value of y, and hence the value of x, as before.

2. *Def-*

2. Descartes's Rule for Biquadratic Equations.

Another solution was given of biquadratic equations by Descartes, in the 3d book of his Geometry. In this solution he resolved the given biquadratic equation into two quadratics, by means of a cubic equation, in this manner: First, let the 2d term or 3d power be taken away out of the equation, after which it will stand thus,

$$x^4 + px^2 + qx + r = 0.$$ Find y in this

cubic equation $y^6 + 2py^4 + \left.{+p^2 \atop -4r}\right\} y^2 - q^2 = 0$; and then

the values of x in these two quadratics.
$$\begin{cases} x^2 - yx + \tfrac{1}{2}y^2 + \tfrac{1}{2}p + \dfrac{q}{2y} = 0 \\ x^2 + yx + \tfrac{1}{2}y^2 + \tfrac{1}{2}p - \dfrac{q}{2y} = 0, \end{cases}$$

and these values of x will be the roots of the given biquadratic equation.

Ex. Let the equ. be $x^4 - 17x^2 - 20x - 6 = 0$. Hence $p = -17$, $q = -20$, & $r = -6$; and the cubic equ. is $x^6 - 34y^4 + 313y^2 - 400 = 0$, the root of which is $y^2 = 16$, or $y = 4$; hence the two quadratics are $\begin{cases} x^2 - 4x - 3 = 0 \\ x^2 + 4x + 2 = 0, \end{cases}$ the four roots of which are $2 \pm \sqrt{7}$ and $-2 \pm \sqrt{2}$.

3. Euler's Method for Biquadratic Equations.

The celebrated Leonard Euler gave, in the 6th volume of the Petersburgh Ancient Commentaries, for the year 1738, an ingenious and general method of resolving equations of all degrees, by means of the equation of the next lower degree, and among them of the biquadratic equation by means of the cubic; and this last was also given more at large in his treatise of Algebra, translated from the German into French in 1774, in 2 volumes 8vo. The method is this: Let $x^4 - ax^2 - bx - c = 0$, be the given biquadratic equation, wanting the 2d term. Take $f = \tfrac{1}{2}a$, $g = \tfrac{1}{16}aa + \tfrac{1}{4}c$, and $h = \tfrac{1}{64}bb$; with which values of f, g, h, form the cubic equation $z^3 - fz^2 + gz - h = 0$. Find the three roots of this cubic equation, and let them be called p, q, r. Then shall the four roots of the proposed biquadratic be these following, viz,

When $\tfrac{1}{8}b$ is positive:	When $\tfrac{1}{8}b$ is negative:
1st. $\sqrt{p} + \sqrt{q} + \sqrt{r}$	$\sqrt{p} + \sqrt{q} - \sqrt{r}$,
2d. $\sqrt{p} + \sqrt{q} - \sqrt{r}$	$\sqrt{p} - \sqrt{q} + \sqrt{r}$,
3d. $\sqrt{p} - \sqrt{q} + \sqrt{r}$	$-\sqrt{p} + \sqrt{q} + \sqrt{r}$,
4th. $\sqrt{p} - \sqrt{q} - \sqrt{r}$	$-\sqrt{p} - \sqrt{q} - \sqrt{r}$.

Ex. Let the eq. be $x^4 - 25x^2 + 60x - 36 = 0$. Here $a = 25$, $b = -60$, and $c = 36$; theref. $f = \dfrac{25}{2}$, $g = \dfrac{625}{16} + 9 = \dfrac{769}{16}$, and $h = \dfrac{225}{4}$. Conseq. the cubic equation will be

$$z^3 - \frac{25}{2}z^2 + \frac{769}{16}z - \frac{225}{4} = 0.$$

The three roots of which are

$$z = \frac{9}{4} = p, \text{ and } z = 4 = q, \text{ and } z = \frac{25}{4} = r;$$

the roots of which are $\sqrt{p} = \tfrac{3}{2}$, $\sqrt{q} = 2$ or $\tfrac{4}{2}$, $\sqrt{r} = \tfrac{5}{2}$. Hence, as the value of $\tfrac{1}{8}b$ is negative, the four roots are

1st. $x = \tfrac{3}{2} + \tfrac{4}{2} - \tfrac{5}{2} = 1$,
2d. $x = \tfrac{3}{2} - \tfrac{4}{2} + \tfrac{5}{2} = 2$,
3d. $x = -\tfrac{3}{2} + \tfrac{4}{2} + \tfrac{5}{2} = 3$,
4th. $x = -\tfrac{3}{2} - \tfrac{4}{2} - \tfrac{5}{2} = -6$.

4. Simpson's Rule for Biquadratic Equations.

Mr. Simpson gave also a general rule for the solution of biquadratic equations, in the 2d edit. of his Algebra, pa. 150, in which the given equation is also resolved by means of a cubic equation, as well as the two former ways; and it is investigated on the principle, that the given equation is equal to the difference between two squares; being indeed a kind of generalization of Ferrari's method.

Thus, he supposes the given equation, viz,
$$x^4 + px^3 + qx^2 + rx + s = \overline{x^2 + \tfrac{1}{2}px + A}^2 - \overline{Bx + C}^2;$$ then from a comparison of the like terms, the values of the assumed letters are found, and the final equation becomes
$$A^3 - \tfrac{1}{2}qA^2 + kA - \tfrac{1}{2}l = 0,$$
where $k = \tfrac{1}{4}pr - s$, and $l = \tfrac{1}{4}r^2 + s.\overline{\tfrac{1}{4}p^2 - q}$.
The value of A being found in this cubic equation, from it will be had the values of B and C, which have these general values, viz, $B = \sqrt{2A + \tfrac{1}{4}p^2 - q}$, and $C = \dfrac{pA - r}{2B}$. Hence, finally, the root x will be obtained from the assumed equation $\overline{x^2 + \tfrac{1}{2}px + A}^2 - \overline{Bx + C}^2 = 0$, or $x^2 + \tfrac{1}{2}px + A = \pm Bx \pm C$, in four several values.

Ex. Given the equ. $x^4 - 6x^3 - 58x^2 - 114x - 11 = 0$.
Here $p = -6$, $q = -58$, $r = -114$, and $s = -11$, whence
$$k \text{ or } \tfrac{1}{4}pr - s = 182, \quad l \text{ or } \tfrac{1}{4}r^2 + s.\overline{\tfrac{1}{4}p^2 - q} = 2512;$$
and therefore the cubic equation becomes
$$A^3 + 29A^2 + 182A - 1256 = 0,$$
the root of which is $A = 4$.
Hence then B or $\sqrt{2A + \tfrac{1}{4}p^2 - q} = \sqrt{75}$, and C or $\dfrac{pA - r}{2B} = \sqrt{27}$:
and the quadratic equation becomes
$$x^2 - 3x + 4 = \pm x\sqrt{75} \pm \sqrt{27},$$ the four roots of which are $x = 1\tfrac{1}{2} \pm 2\tfrac{1}{2}\sqrt{3} + \sqrt{17 \pm 11\tfrac{1}{2}\sqrt{3}}$,
and $x = 1\tfrac{1}{2} \pm 2\tfrac{1}{2}\sqrt{3} - \sqrt{17 \pm 11\tfrac{1}{2}\sqrt{3}}$.

Mr. Simpson here subjoins an observation which it has since been found is erroneous, viz, that "The value of A, in this equation, will be *commensurate* and *rational* (and therefore the easier to be discovered), not only when all the roots of the given equation are *commensurate*, but when they are *irrational* and even *impossible*; as will appear from the examples subjoined." This is a strange reason for Simpson to give for the proof of a proposition; and it is wonderful that he fell upon no examples that disprove it, as the instances in which it holds true, are very few indeed, in comparison with the number of those in which it fails.

Note. In any biquadratic equation having all its terms, if $\tfrac{3}{8}$ of the square of the coefficient of the 2d term be greater than the product of the coefficients of the 1st and 3d terms, or $\tfrac{3}{8}$ of the square of the coefficient

of

of the 4th term be greater than the product of the coefficients of the 3d and 5th terms, or $\frac{4}{9}$ of the square of the coefficient of the 3d term greater than the product of the coefficients of the 2d and 4th terms; then all the roots of that equation will be real and unequal; but if either of the said parts of those squares be less than either of those products, the equation will have imaginary roots.

For the construction of biquadratic equations, see *Construction*. See also *Descartes's Geometry*, with the *Commentaries of Schooten* and others; *Baker's Geometrical Key*; *Slusius's Mesolabium*; *l'Hospital's Conic Sections*; *Wolfius's Elementa Matheseos*; &c.

BIQUADRATIC *Parabola*, a curve of the 3d order, having two infinite legs, and expressed by one of these three equations, viz,

$a^3 x = y^4$, as in fig. 1,

$a^3 x = y^4 - a^2 y^2$, as in fig. 2,

$a^3 x = y^4 - \overline{a + b} \cdot y^3 - aby^2$, as in fig. 3;

where $x = $ AP the abscifs, $y = $ PQ the ordinate, $b = $ AB, $c = $ AC, and $a = $ a certain given quantity.

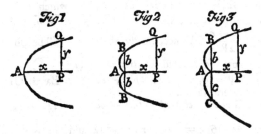

But the most general equation of this curve is the following, which belongs to fig. 4, viz,

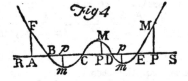

$ay = x^4 + bx^3 + cx^2 + dx + e;$

where $x = $ Ap or AP the abscifs, and $-y$ or $+y$ is the ordinate pm or PM, also a, b, c, d, e, are constant quantities; the beginning of the abscifs being at any point A in the indefinite line AP.

But if the beginning of the abscifs A be where this line interfects the curve, as in fig. 5, then the nature of the curve will be defined by this equation $a \times pm = $ A$p \times$ B$p \times$ C$p \times$ Dp, wherever the point p is taken in the infinite line RS.

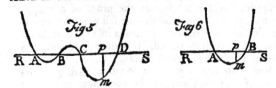

When the curve has no serpentine part, as fig. 6, the equation is more simple, being in this case barely

$a \times pm = $ A$p^2 \times p$B^2. See CURVE *Lines*, and GEOMETRICAL *Lines*.

BIQUINTILE, an aspect of the planets when the distance between them is 144 degrees, or twice the 5th part of 360 degrees.

BISECTION, or BISSECTION, the division of a quantity into two equal parts, otherwise called *bipartition*.

BISSEXTILE, or *Leap-year*, a year confisting of 366 days, and happening every 4th year, by the addition of a day in the month of February, which that year confifts of 29 days. And this is done to recover the 6 hours which the fun takes up nearly in his courfe, more than the 365 days commonly allowed for it in other years.

The day thus added was by Julius Cæsar appointed to be the day before the 24th of February, which among the Romans was the 6th of the calends, and which on this occasion was reckoned twice; whence it was called the *bissextile*.

By the statute *De anno bissextile*, 21 Hen. III, to prevent misunderstandings, the intercalary day and that next before it are to be accounted as one day.

To find what year of the period any given year is; divide the given year by 4; then if o remains, it is leap year; but if any thing remain, the given year is so many after leap year.

But the astronomers concerned in reforming the calendar in 1582, by order of pope Gregory XIII, observing that in 4 years the bissextile added 44 minutes more than the fun spent in returning to the same point of the ecliptic; and computing that in 133 years these supernumerary minutes would form a day; to prevent any changes being thus insensibly introduced into the seafons, directed, that in the courfe of 400 years there should be three sextiles retrenched; so that every centesimal year, which is a leap year according to the Julian account, is a common year in the Gregorian account, unless the number of centuries can be divided by 4 without a remainder. So 1600 and 2000 are bissextile; but 1700, 1800, and 1900 are common years.

The Gregorian computation has been received in most foreign countries ever since the reformation of the calendar in 1582; excepting some northern countries, as Ruffia, &c. And by act of parliament, paffed in 1751, it commenced in all the dominions under the crown of Great Britain in the year following; it being ordered by that act that the natural day next following the 2d of September, should be accounted the 14th; omitting the intermediate 11 days of the common calendar. The supernumerary day, in leap years, being added at the end of the month February, and called the 29th of that month.

BLACK, a colour so called, or rather a privation of all colour. This, it seems, arises from such a peculiar texture and situation of the superficial parts of a black body, that they abforb all or most of the rays of light, reflecting little or none to the eye: and hence it happens that black bodies, thus imbibing the rays, are always found to be hotter than those of a lighter colour. Dr. Franklin obferves that black cloaths heat more, and dry fooner in the fun than white cloaths; that therefore black is a bad colour for cloaths in hot climates; but a fit colour for the linings of ladies' summer hats;

2 and

and that a chimney painted black, when exposed to the sun, will draw more strongly. Franklin's Experim. &c.—Dr. Watson, the present bishop of Landaff, covered the bulb of a thermometer with a black coating of Indian ink, and the consequence was that the mercury rose 10 degrees higher. Philos. Transf. vol. 63, pa. 40.—And a virtuoso of unsuspected credit assured Mr. Boyle, that in a hot climate by blacking the shells of eggs, and exposing them to the sun, he had seen them thus well roasted in a short time.

BLACKNESS, the quality of a black body, as to colour; arising from its stifling or absorbing the rays of light, and reflecting little or none. In which sense it stands directly opposed to whiteness; which consists in such a texture of parts, as indifferently reflects all the rays thrown upon it, of whatever colour they may be.

Descartes, it seems, first rightly distinguished these causes of black and white, though he might be mistaken with respect to the general nature of light and colours. —Sir Isaac Newton shews, in his Optics, that to produce black colours, the corpuscles must be smaller than for exhibiting the other colours; because, where the sizes of the component particles of a body are greater, the light reflected is too much for constituting this colour: but when they are a little smaller than is requisite to reflect the white, and very faint blue of the first order, they will reflect so little light, as to appear intensely black; and yet they may perhaps reflect it variously to and fro within them so long, till it be stifled and lost.

And hence, it appears, why fire, or putrefaction, by dividing the particles of substances, turn them black: why small quantities of black substances impart their colours very freely, and intensely, to other substances, to which they are applied; the minute particles of these, on account of their very great number, easily overspreading the gross particles of others. Hence it also appears, why glass ground very elaborately on a copper plate with sand, till it be well polished, makes the sand, with what is rubbed off from the copper and glass, become very black: also why blacks commonly incline a little towards a blueish colour; as may be seen by illuminating white paper with light reflected from black substances, when the paper usually appears of a blueish white; the reason of which is, that black borders on the obscure blue of the first order of colours, and therefore reflects more rays of that colour than of any other: and lastly why black substances do sooner than others become hot in the sun's light, and burn; an effect which may proceed partly from the multitude of refractions in a little space, and partly from the easy commotion among such minute particles.

BLAGRAVE (John), an eminent mathematician, who flourished in the 16th and 17th centuries. He was the second son of John Blagrave, of Bulmarsh-court near Sunning in Berkshire, descended from an ancient family in that country. From a grammar school at Reading he was sent to St. John's college in Oxford, where he applied himself chiefly to the study of mathematics. From hence he retired to his patrimonial seat of Southcote-lodge near Reading, where he spent the rest of his life, in a retired manner, without marrying, that he might have more leisure to pursue his favourite studies; which he did with great

application and success. After a life thus spent in study, and in acts of benevolence to all around him, he died in the year 1611; and was buried at Reading in the church of St. Lawrence, where a sumptuous monument was erected to his memory.

He left the bulk of his fortune to the posterity of his three brothers, which were very numerous. There have been mentioned various acts of his beneficence in private life, for the encouragement of learning, the reward of merit, and the relief of distress. Some of these were the result of a quaint, humorous disposition, discovered chiefly in his legacies: One of these was 10 pounds left to be annually disposed of in the following manner: On Good-friday, the church-wardens of each of the three parishes of Reading send to the town hall "one virtuous maid who has lived five years with her master:" there, in the presence of the magistrates these three virtuous maids throw dice for the ten pounds. The year following the two losers are returned with a fresh one, and again the third year, till each has had three chances. He also left an annuity to 80 poor widows, who should attend annually on Good-friday also, and hear a sermon, for the preaching of which he left ten shillings to the minister. He took care also for the maintenance of his servants, rewarding their diligence and fidelity, and providing amply for their support. Thus it appears he was not more remarkable for his scientific knowledge, than for his generosity and philanthropy. His works are,

1. A Mathematical Jewel. Lond. 1585, folio.

2. Of the Making and Use of the Familiar Staff. Lond. 1590, 4to.

3. Astrolabium Uranicum Generale. Lond. 1596, 4to.

4. The Art of Dyalling. Lond. 1609, 4to.

BLAIR (John), an eminent chronologist, was educated at Edinburgh. Afterward, coming to London, he was for some time usher of a school in Hedge-lane. In 1754 he first published "The Chronology and History of the World, from the Creation to the year of Christ 1753; illustrated in 56 tables. In 1755 he was elected a fellow of the Royal Society, and in 1761 of the Society of Antiquaries. In 1756 he published a 2d edition of his Chronological Tables; and in 1768 an improved edition of the same with the addition of 14 maps of Ancient and Modern Geography, for illustrating the Tables of Chronology and History; to which is prefixed a Dissertation on the Progress of Geography. In 1757 he was appointed chaplain to the Princess Dowager of Wales, and Mathematical Tutor to the Duke of York; whom he attended in 1763 in a tour to the continent, from which they returned the year after. Dr. Blair had successively several good church livings: as, a prebendal stall at Westminster, the vicarage of Hinckley, and the rectory of Burton Coggles in Lincolnshire, all in 1761; the vicarage of St. Bride's in London, in 1771, in exchange for that of Hinckley; the rectory of St. John the Evangelist in Westminster, in 1776, in exchange for the vicarage of St. Bride's; in the same year the rectory of Horton near Colebrooke, Bucks. Dr. Blair died the 24th of June, 1782.

BLIND, an epithet applied to an animal deprived of the use of eyes; or one from whom light, colours, and

and all the glorious objects of the visible creation, are intercepted by some natural or accidental cause.

BLINDNESS, a privation of the sense of sight. The ordinary causes of blindness are, some external violence, vicious conformation, growth of a cataract, gutta serena, small-pox, &c; or a decay of the optic nerve; an instance of which we have in the Academy of Sciences, where, upon opening the eye of a person long blind, it was found that the optic nerve was extremely shrunk and decayed, and without any medulla in it. The more extraordinary causes of blindness are malignant stenches, poisonous juices dropped into the eye, baneful vermin, long confinement in the dark, or the like.

We find various recompenses for blindness, or substitutes for the use of eyes, in the wonderful sagacity of many blind persons, related by Zahnius in his *Oculus Artificialis*, and others. In some, the defect has been supplied by a most excellent gift of remembering what they had seen before; others by a delicate nose or the sense of smelling; others by a very nice ear; and others again by an exquisite touch, or sense of feeling, which they have had in such perfection, that as it has been said of some, they learned to hear with their eyes, it may be said of these that they taught themselves to see with their hands.

Some have been able to perform all sorts of curious works in the nicest and most dexterous manner. Aldrovandus speaks of a sculptor, who had become blind at 20 years of age, and yet 10 years afterwards he made a perfect marble statue of Cosmo II de Medicis, and another of clay like pope Urban VIII. Bartholin speaks of a blind sculptor in Denmark, who, by mere touch, distinguished perfectly well all sorts of wood, and even colours; and father Grimaldi relates an instance of the same kind; besides the blind organist, lately living in Paris, who it was said did the same thing. What seems more extraordinary still, we are told, by authors of good report, of a blind guide, who used to conduct the merchants through the sands and desarts of Arabia: and a not less marvellous instance is now existing in this country, in one John Metcalf near Manchester, who became quite blind at a very early age; and yet passed many years of his life as a waggoner, and occasionally, as a guide in different roads during the night, or when the paths were covered with snow; and, what is stranger still, his present occupation is that of surveyor and projector of highways in difficult and mountainous parts, particularly about Buxton, and the Peak in Derbyshire.

There are also many instances of blind men who have been highly distinguished for their mental and literary talents, not to speak of the poets Homer, Milton, Ossian, &c; of which we have a remarkable instance in the late Dr. Sanderson, professor of mathematics in the university of Cambridge, and in the present Dr. Henry Moyes, public lecturer in philosophy, who both of them lost their sight by the small pox at an age before they had any recollection; these men were well skilled in all branches of the mathematics, philosophy, and optics, &c, which they taught with the greatest reputation; besides the monument of fame which the former has left behind him in his mathematical and philosophical works.

The effects of a sudden recovery of sight in such as have been born blind, are also very remarkable: devoid of the experience of distance and figure arising from sight, they are liable to the greatest mistakes in this respect, in so much it has been said that they could not distinguish by the mere sight which was a cube and which a globe, without first touching them. Mr. Boyle mentions a gentleman of this sort, who having been restored to sight at eighteen years of age, was near going distracted with the joy: see Boyle's works abridg vol. 1. pa. 4. See also a remarkable case of this kind in the Tatler, Nº 55, vol. 1. And the gentleman couched by Mr. Cheselden had no ideas of colour, shape, or distance: though he knew the colours asunder in a good light during his blind state; yet when he saw them after he had been couched, the faint ideas he had of them before, were not sufficient for him to know them by afterwards: as to distance, his ideas were so deficient, that he thought all the objects he saw touched his eyes, as what he felt did his skin; and it was a considerable time before he could remember which was the dog and which the cat, though often informed, without feeling them.

BLINDS, or BLINDES, in Fortification, a kind of defence usually made of oziers or branches interwoven, and laid across between two rows of stakes, about a man's height, and 4 or 5 feet asunder. They are used particularly at the heads of trenches, when these are extended in front towards the glacis; serving to defend the workmen, and prevent the enemy from overlooking them.

BLOCKADE, is the blocking up a place, by posting troops all about it at the avenues, to prevent supplies of men and provisions from getting into it; and thus starving it out, without forming any regular siege or attacks.

BLONDEL (FRANCIS), a celebrated French mathematician and military engineer. He was born at Ribemond in Picardy in 1617. While he was yet but young, he was chosen Regius Professor of Mathematics and Architecture at Paris. Not long after he was appointed governor to Lewis-Henry de Lomenix, Count de Brienne, whom he accompanied in his travels from 1652 to 1655, of which he published an account. He enjoyed many honourable employments, both in the navy and army; and was entrusted with the management of several negociations with foreign princes. He arrived at the dignity of marshal de camp, and counsellor of state, and had the honour to be appointed mathematical preceptor to the Dauphin. He was a member of the Royal Academy of Sciences, director of the Academy of Architecture, and lecturer to the Royal College: in all which he supported his character with dignity and applause. Blondel was no less versed in the knowledge of the Belles Lettres than in the mathematical sciences, as appears by the comparison he published between Pindar and Horace. He died at Paris the 22d of February 1686, in the 69th year of his age. His chief mathematical works were,

1. Cours d'Architecture. Paris, 1675, in folio.

2. Resolution des quatre principaux problemes d'Architecture. Paris, 1676, in folio.

3. Histoire du Calendrier Romain. Paris, 1682, in 4to.

4. Cours

4. *Cours de Mathematiques.* Paris, 1683, in 4to.

5. *L'Art de jetter des Bombes.* La Haye, 1685, in 4to.

Besides a New Method of fortifying Places, and other works.

Blondel had also many ingenious pieces inserted in the Memoirs of the French Academy of Sciences, particularly in the year 1666.

BLOW, in a general sense, denotes a stroke given either with the hand, a weapon, or an instrument. The effect of a blow is estimated like the force of percussion, and so is expressed by the velocity of the body multiplied by its weight.

BLOWING *of a Fire-arm,* is when the vent or touch-hole is run or gullied, and becomes wide, so that the powder will flame out

BLUE, one of the seven primitive colours of the rays of light, into which they are divided when refracted through a glass prism. See Newton's Optics &c. See also CHROMATICS.

BLUENESS, that quality of a body, as to colour, from whence it is called blue; depending on such a size and texture of the parts that compose the surface of a body, as disposes them to reflect only the blue or azure rays of light to the eye.

The blueness of the sky is thus accounted for by De la Hire, after Da Vinci; viz, that a black body viewed through a thin white one, gives the sensation of blue, like the immense expanse viewed through the air illuminated and whitened by the sun. For the same reason he says it is, that soot mixed with white, makes a blue; for that white bodies, being always a little transparent, when mixing with a black behind, give the perception of blue. From the same principle too he accounts for the blueness of the veins on the surface of the skin, though the blood they are filled with be a deep red.

In the same manner was the blueness of the sky accounted for by many other of the earlier writers, as Fromondus, Funceius, Otto Guericke, and many others, together with several of the more modern writers, as Wolfius, Muschenbroek, &c. But in the explication of this phenomenon, Newton observes that all the vapours, when they begin to condense and coalesce into natural particles, become first of such a magnitude as to reflect the azure rays, before they can constitute clouds of any other colour. This being therefore the first colour they begin to reflect, must be that of the finest and most transparent skies, in which the vapours are not yet arrived at a grossness sufficient to reflect other colours.

Bouguer however ascribes this blueness of the sky to the constitution of the air itself, being of such a nature that these fainter-coloured rays are incapable of making their way through any considerable tract of it. And as to the blue shadows which were first observed by Buffon in the year 1742, he accounts for them by the aerial colour of the atmosphere, which enlightens these shadows, and in which the blue rays prevail; whilst the red rays are not reflected so soon, but pass on to the remoter regions of the atmosphere. And the Abbé Mazeas accounts for the phenomenon of blue shadows by the diminution of light; observing that,

of two shadows which were cast upon a white wall from an opaque body, illuminated by the moon and by a candle at the same time, that from the candle was reddish, while the other from the moon was blue. See Newton's Optics pa. 228, Bouguer Traite d'Optique pa. 368, Edinb. Ess. vol. 2 pa. 75, or Priestley's Hist. of Vision &c, pa. 436.

BOB *of a Pendulum,* the same as the ball, which see.

BODY, in Geometry, is a figure conceived to be extended in all directions, or what is usually said to consist of length, breadth, and thickness; being otherwise called a *Solid.* A body is conceived to be formed or generated by the motion of a surface; like as a surface by the motion of a line, and a line by the motion of a point.—Similar bodies, or solids, are in proportion to each other, as the cubes of their like sides, or linear dimensions.

BODY, in Physics, or Natural Philosophy, a solid, extended, palpable substance; of itself merely passive, being indifferent either to motion or rest, and capable of any sort of motion or figure.

Body is composed, according to the Peripatetics, of matter, form, and privation; according to the Epicureans and Corpuscularians, of an assemblage of hooked, heavy atoms; according to the Cartesians, of a certain extension; and according to the Newtonians, of a system or association of solid, massy, hard, impenetrable, moveable particles, ranged or disposed in this or that manner; from which arise bodies of this or that form, and distinguished by this or that name. These elementary or component particles of bodies, they assert, must be perfectly hard, so as never to wear or break in pieces; which, Newton observes, is necessary, in order to the world's persisting in the same state, and bodies continuing of the same nature and texture in several ages.

BODY *of a piece of Ordnance,* the part contained between the centre of the trunnions and the cascabel. This should always be more fortified or stronger than the rest. See CANNON.

BODY *of the Place,* in Fortification, denotes either the buildings inclosed, or more generally the inclosure itself. Thus, to construct the body of the place, is to fortify or inclose the place with bastions and curtains.

BODY *of a Pump,* the thickest part of the barrel or pipe of a pump, within which the piston moves.

BODIES, *Regular* or *Platonic,* are those which have all their sides, angles, and planes, similar and equal. Of these there are only 5; viz,

the *tetraedron,* contained by 4 equilateral triangles;
the *hexaedron* or *cube,* by 6 squares;
the *octaedron,* by 8 triangles;
the *dodecaedron,* by 12 pentagons; and
the *icosaedron,* by 20 triangles.

To form the five Regular Bodies.

Let the annexed figures be exactly drawn on pasteboard, or stiff paper, and cut out from it by the extreme or bounding lines: then cut the others, or internal lines, only half through, so that the parts may be turned up by them, and then glued or otherwise fastened together with paste, sealing-wax, &c; so shall they form the respective body marked with the corresponding number;

number; viz, N° 1 the tetraedron, N° 2 the hexaedron or cube, N° 3 the octaedron, N° 4 the dodecaedron, and N° 5 the icosaedron.

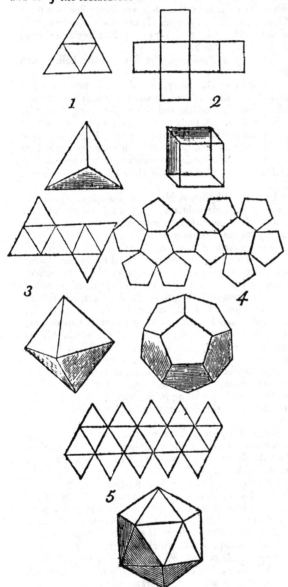

To find the Superficies or Solidity of the Regular Bodies.

1. Multiply the proper tabular area (taken from the following table) by the square of the linear edge of the solid, for the superficies.

2. Multiply the tabular solidity by the cube of the linear edge, for the solid content.

No. of Faces	Names	Surfaces	Solidities
4	Tetraedron	1·73205	0·11785
6	Hexaedron	6·00000	1·00000
8	Octaedron	3·46410	0·47140
12	Dodecaedron	20·64573	7·66312
20	Icosaedron	8·66025	2·18169

Table of the Surfaces and Solidities of the five Regular Bodies, the linear edge being 1.

For more particular properties, see each respective word. See also my large *Mensuration*, pa. 248, edit. 2.

These bodies were called *platonic*, because they were said to have been invented, or first treated of, by *Plato*, who conceived certain mysteries annexed to them.

BOFFRAND (Germain), a celebrated French architect and engineer, was born at Nantes in Bretagne in 1667. He was brought up under Harduin Mansarad, who trusted him with conducting his greatest works. Boffrand was admitted into the French Academy of Architecture in 1709. Many German princes chose him for their architect, and raised considerable edifices on his plans. His manner of building approached that of Palladio; and there was much of grandeur in all his designs. As engineer and inspector-general of bridges and highways, he directed and constructed a number of canals, sluices, bridges, and other mechanical works. He published a curious and useful book, containing the general principles of his art; with an account of the plans, profiles, and elevations of the principal works, which he executed in France and other countries. Boffrand died at Paris in 1755, dean of the Academy of Architecture, first engineer and inspector-general of the bridges and highways, architect and administrator of the general hospital.

BOILING, or EBULLITION, the bubbling up of any fluid, by the application of heat. This is, in general, occasioned by the discharge of an elastic vapour through the fluid that boils; whether that be common air, fixed air, or steam, &c. It is proved by Dr. Hamilton of Dublin, in his Essay on the ascent of vapour, that the boiling of water is occasioned by the lowermost particles of it being heated and rarefied into vapour, or steam; in consequence of this diminution of their specific gravity, they ascend through the surrounding heavier fluid with great velocity, lacerating and throwing up the body of water in the ascent, and so giving it the tumultuous motion called *boiling*.

That this is occasioned by elastic steam, and not by particles of fire or air, as some have imagined, is easily proved by the following simple experiment: Take a common drinking glass filled with hot water, and invert it into a vessel of the same: then, as soon as the water in the vessel begins to boil, large bubbles will be seen to ascend in the glass, by which the water in it will be displaced, and there will soon be a continued bubbling from under its edge: but if the glass be then drawn up, so that its mouth may just touch the water, and a cloth wetted in cold water be applied to the outside, the elastic steam within it will be instantly condensed, upon which the water will ascend so as nearly to fill it again. Some small parts of air &c, that may happen to be lodged in the fluid, may also perhaps be expelled, as well as the rarefied steam. And this is particularly recommended as a method of purifying quicksilver, for making more accurately barometers and thermometers.

We commonly annex the idea of a certain very great degree of heat to the boiling of liquids, though often without reason; for different liquids boil with different degrees of heat; and any one given liquid also, under different pressures of the atmosphere. Thus, a vessel of tar being set over the fire till it boils, it is said a person may then put his hand into it without injury:

injury : and by putting water under the receiver of an air-pump, and applying the flame of a candle or lamp under it, by gradually exhausting, the water is made to boil with always less and less degrees of heat ; and without applying any heat at all, the water, or even the moisture about the bottom or edges of the receiver, will rise in an elastic vapour up into it, when the exhaustion is near completed.

Spirit of wine boils still sooner in vacuo than water. And Dr. Freind gives a table of the different times required to make several fluids boil by the same heat. See also Philos. Transf. N° 122.

BOMB, in Artillery, a shell, or hollow ball of cast-iron, having a large vent, by which it is filled with gun-powder, and which is fitted with a fuze or hollow plug to give fire by, when thrown out of a mortar, &c : about the time when the shell arrives at the intended place, the composition in the pipe of the fuze sets fire to the powder in the shell, which blows it all in pieces, to the great annoyance of the enemy, by killing the people, or firing the houses, &c. They are now commonly called *shells* simply, in the English artillery.

These shells, or bombs, are of various sizes, from that of 17 or 18 inches diameter downwards. The very large ones are not used by the English, that of 13 inches diameter being the highest size now employed by them ; the weight, dimensions, and other circumstances of them, and the others downwards, are as in the following table.

Diameter of the Shell.	Weight of the Shell.	Powder to fill them.		Powder to burst them into most pieces.	
	lbs.	lb.	oz.	lb.	oz.
13 inch	195	9	4½	7	8
10	89	4	14½	3	4
8	46	2	3½	2	0
5⅖ Royal	14½	1	1½	0	14
4⅗ Cohorn	7½	0	8	0	7

Mr. Muller gives the following proportion for all shells. Dividing the diameter of the mortar into 30 equal parts, then the other dimensions, in 30ths of that diameter, will be thus :

Diameter of the bore, or mortar - -   30
Diameter of the shell, - - -   29½
Diameter of the hollow sphere   -   21
Thickness of metal at the fuze hole -   3¼
Thickness at the opposite part - -   5
Diameter of the fuze hole - -   4
Weight of shell empty - - $\frac{10}{177}d$
Weight of powder to fill it - $\frac{2}{73}d$

where d denotes the cube of the diameter of the bore in inches.—But shells have also lately been made with the metal all of the same thickness quite around.

In general, the windage, or difference between the diameter of the shell and mortar, is $\frac{1}{20}$ of the latter ; also the diameter of the hollow part of the shell is $\frac{7}{10}$ of the same.

Bombs are thrown out of mortars, or howitzers ; but they may also be thrown out of cannon ; and a very small sort are thrown by the hand, which are called granados : and the Venetians at the siege of Candia,

when the Turks had possessed themselves of the ditch used large bombs without any piece of ordnance, but barely rolled them down upon the enemy along a plank set aslope, with ledges on the sides to keep the bomb right forwards.

Mr. Blondel, in his *Art de jetter des Bombes*, says the first bombs were those thrown into the city of Watchtendonch in Guelderland, in 1588; and they are described by our countryman Lucar, in his book on Artillery published this same year 1588; though it is pretended by others that they were in use near a century before, namely at the siege of Naples in 1495. They only came into common use, however, in 1634, and then only in the Dutch and Spanish armies. It is said that one Malthus, an English engineer, was sent for from Holland by Lewis the 14th, who used them for him with much success, particularly at the siege of Cohoure in 1642.

The art of throwing bombs, or shells, forms a principal branch of Gunnery, founded on the theory of projectiles, and the quantities and laws of force of gunpowder. And the principal writers on this art are Mess. Blondel, Guisnee, De Ressons, De La Hire, &c.

BOMB-CHEST, is a kind of chest usually filled with bombs, and sometimes only with gunpowder, placed under ground, to blow it up into the air with those who stand upon it ; being set on fire by means of a saucisse fastened at one end. But they are now much out of use.

BOMBARD, an ancient piece of ordnance, now out of use. It was very short and thick, with a large mouth ; some of which it is said threw balls of 300 pounds weight, requiring the use of cranes to load them. The Bombard is by some called *basilisk*, and by the Dutch *donderbus*.

To BOMBARD, is to attack by throwing of bombs, or shells.

BOMBARDIER, a person employed about throwing bombs or shells. He adjusts the fuze, and loads and fires the mortar.

BONES, *Napier's*. See NAPIER.

BONING, in Surveying and Levelling, &c, is the placing three or more rods or poles, all of the same length, in or upon the ground, in such a manner that the tops of them be all in one continued straight line, whether it be horizontal or inclined, so that the eye can look along the tops of them all, from one end of the line to the other.

BONNET, in Fortification, a small work of two faces, having only a parapet, with two rows of palisadoes at about 10 or 12 feet distance. It is commonly placed before the saliant angle of the counterscarp, and having a communication with the covered way, by means of a trench cut through the glacis, and palisadoes on each side.

BONNET *a Pretre*, or *Priest's Cap*, is an outwork, having three saliant angles at the head, besides two inwards. It differs from the double tenaille only in this, that its sides, instead of being parallel, grow narrower, or closer, at the gorge, and opening at the front ; from whence it is called *queue d'aronde*, or swallow's tail.

BOOTES, a constellation of the northern hemisphere, and one of the 48 old ones ; having 23 stars in

Ptolemy's catalogue, 28 in Tycho's, 34 in Bayer's, 52 in Hevilus's, and 54 in Flamsteed's; of which one, in the skirt of his coat, is of the first magnitude, and called *Arcturus.*

Bootes is represented as a man in the posture of walking; his right hand grasping a club, and his left extended upwards, and holding the cord of the two dogs which seem barking at the Great Bear.

The Greeks, contrary to their usual custom, do not give any certain account of the origin of this constellation. Those who in very early days made the stars which were afterwards formed into the great bear represent a waggon drawn by oxen, made this Bootes the driver of them, from whence he was called the waggoner: others continued the office when the waggon was destroyed, and made a celestial bearward of Bootes, making it his office to drive the two bears round about the pole: and some, when the greater waggon was turned into the greater bear, were still for preserving the form of that machine in those stars which constitute Bootes.

This constellation is called by various other names; as *Arcas, Arctophylax, Arcturus-Minor, Bubulcus, Bubulus, Canis-Latrans, Clamator, Icarus, Lycaon, Philometus, Plaustri-Custos, Plorans, Thegnis,* and *Vociferator;* by Hesychius it is called *Orion,* and by the Arabs *Aramech,* or *Archamech.* Schiller, instead of Bootes, makes the figure of St. Sylvester; Schickhard, that of Nimrod; and Weigelius, the three Swedish crowns. See Wolf. Lex Math. p. 266.

BORE, of a gun, or other piece of ordnance, is the chase, cylinder, or hollow part of the piece.

BOREAL Signs, are the first six signs of the Zodiac, or those on the northern side of the equinoctial; viz, the signs ♈ aries, ♉ taurus, ♊ gemini, ♋ cancer, ♌ leo, ♍ virgo.

BOREALIS, Aurora. See Aurora *Borealis.*

BORELLI (John Alphonso), a celebrated philosopher and mathematician, born at Naples the 28th of January 1608. He was professor of philosophy and mathematics in some of the most celebrated universities of Italy, particularly at Florence and Pisa, where he became highly in favour with the princes of the house of Medicis. But having been concerned in the revolt of Messina, he was obliged to retire to Rome, where he spent the remainder of his life under the protection of Christina queen of Sweden, who honoured him with her friendship, and by her liberality towards him softened the rigour of his hard fortune. He continued two years in the convent of the regular clergy of St. Pantaleon, called the *Pious Schools,* where he instructed the youth in mathematical studies. And this study he prosecuted with great diligence for many years afterward, as appears by his correspondence with several ingenious mathematicians of his time, and the frequent mention that has been made of him by others, who have endeavoured to do justice to his memory. He wrote a letter to Mr. John Collins, in which he discovers his great desire and endeavours to promote the improvement of those sciences: he also speaks of his correspondence with, and great affection for, Mr. Henry Oldenburgh, Secretary of the Royal Society; of Dr. Wallis; of the then late learned Mr. Boyle, and lamented the loss sustained by his

death to the commonwealth of learning. Mr. Baxter, in his *Enquiry into the Nature of the Human Soul,* makes frequent use of our author's book *De Motu Animalium,* and tells us, that he was the first who discovered that the force exerted within the body prodigiously exceeds the weight to be moved without, or that nature employs an immense power to move a small weight. But he acknowledges that Dr. James Keil had shewn that Borelli was mistaken in his calculation of the force of the muscle of the heart; but that he nevertheless ranks him with the most authentic writers, and says he is seldom mistaken: and, having remarked that it is so far from being true, that great things are brought about by small powers, that, on the contrary, a stupendous power is manifest in the most ordinary operations of nature, he observes that the ingenious Borelli first observed this in animal motion; and that Dr. Stephen Hales, by a course of experiments in his *Vegetable Statics,* had shewn the same in the force of the ascending sap in vegetables.

After a course of unceasing labours, Borelli died at Pantaleon of a pleurisy, the 31st of December 1679, at 72 years of age.

Beside several books on physical subjects, Borelli published the following mathematical ones: viz.

1. Apollonii Pergæi Conicorum Lib. 5, 6, & 7. Floren. 1661, fol.

2. Theoriæ Medicorum Planetarum ex causis physicis deductæ. Flor. 1666, 4to.

3. De Vi Percussionis. Bologna, 1667, 4to.—This piece was reprinted, with his celebrated treatise De Motu Animalium, and that other De Motionibus Naturalibus, in 1686.

4. Euclides Restitutus, &c. Pisa, 1668, 4to.

5. Osservatione intorno alla vistu ineguali degli Occi. —This piece was inserted in the Journal of Rome, for the year 1669.

6. De Motionibus Naturalibus de Gravitate pendentibus. Regio Julio, 1670, 4to.

7. Meteorologia Aetnea, &c. Regio Julio, 1670, 4to.

8. Osservatione dell' Ecclissi Lunare, 11 Gennaro 1675.—Inserted in the Journal of Rome 1675, p. 34.

9. Elementa Conica Appollonii Pergæi, & Archimedis Opera, nova & breviori methodo demonstrata.— Printed at Rome in 1679, in 12mo, at the end of the 3d edition of his Euclides Restitutus.

10. De Motu Animalium. Pars prima in 1680, and Pars altera in 1681, 4to.—These were reprinted at Leyden 1685, revised and purged from many errors; with the addition of John Bernoulli's Mathematical Meditations concerning the Motion of the Muscles.

11. At Leyden, 1686, in 4to, a more correct and accurate edition, revised by J. Broen, M. D. of Leyden, of his two pieces, De Vi Percussionis, & De Motionibus de Gravitate pendentibus.

BOUGUER (Peter), a celebrated French mathematician, was born at Croisic, in Lower Bretagne, the 10th of February 1698. He was the son of John Bouguer, Professor Royal of Hydrography, a tolerably good mathematician, and author of A complete Treatise on Navigation. Young Bouguer was accustomed to learn mathematics from his father, from the

time

time he was able to speak, and thus became a proficient in those sciences while he was yet a child. He was sent very early to the Jesuits' college at Vannes, where he had the honour to instruct his regent in the mathematics, at eleven years of age.

Two years after this he had a public contest with a professor of mathematics, upon a proposition which the latter had advanced erroneously; and he triumphed over him; upon which the professor, unable to bear the disgrace, left the country.

Two years after this, when young Bouguer had not yet finished his studies, he lost his father; whom he was appointed to succeed in his office of hydrographer, after a public examination of his qualifications; being then only 15 years of age; an occupation which he discharged with great respect and dignity at that early age.

In 1727, at the age of 29, he obtained the prize proposed by the Academy of Sciences, for the best way of masting of ships. This first success of Bouguer was soon after followed by two others of the same kind; he successively gained the prizes of 1729 and 1731; the former, for the best manner of observing at sea the height of the stars, and the latter, for the most advantageous way of observing the declination of the magnetic needle, or the variation of the compass.

In 1729, he gave an *Optical Essay upon the Gradation of Light;* a subject quite new, in which he examined the intensity of light, and determined its degrees of diminution in passing through different pellucid mediums, and particularly that of the sun in traversing the earth's atmosphere. Mairan gave an extract of this first essay in the Journal des Savans, in 1730.

In this same year, 1730, he was removed from the port of Croisic to that of Havre, which brought him into a nearer connection with the Academy of Sciences, in which he obtained, in 1731, the place of associate geometrician, vacant by the promotion of Maupertuis to that of pensioner; and in 1735 he was promoted to the office of pensioner-astronomer. The same year he was sent on the commission to South America, along with Messieurs Godin, Condamine, and Jeuffieu, to determine the measure of the degrees of the meridian, and the figure of the earth. In this painful and troublesome business, of 10 years duration, chiefly among the lofty Cordelier mountains, our author determined many other new circumstances, beside the main object of the voyage; such as the expansion and contraction of metals and other substances, by the sudden and alternate changes of heat and cold among those mountains; observations on the refraction of the atmosphere from the tops of the same, with the singular phenomenon of the sudden increase of the refraction, when the star can be observed below the line of the level; the laws of the density of the air at different heights, from observations made at different points of these enormous mountains; a determination that the mountains have an effect upon a plummet, though he did not assign the exact quantity of it; a method of estimating the errors committed by navigators in determining their route; a new construction of the log for measuring a ship's way; with several other useful improvements.

Other inventions of Bouguer, made upon different occasions, were as follow: The heliometer, being a telescope with two object glasses, affording a good method of measuring the diameters of the larger planets with ease and exactness: his researches on the figure in which two lines or two long ranges of parallel trees appear: his experiments on the famous reciprocation of the pendulum: and those upon the manner of measuring the force of the light: &c, &c.

The close application which Bouguer gave to study, undermined his health, and terminated his life the 15th of August 1758, at 60 years of age.—His chief works, that have been published, are,

1. The Figure of the Earth, determined by the observations made in South America; 1749, in 4to.

2. Treatise on Navigation and Pilotage; Paris, 1752, in 4to. This work has been abridged by M. La Caille, in 1 volume, 8vo, 1768.

3. Treatise on Ships, their Construction and Motions; in 4to, 1756.

4. Optical Treatise on the Gradation of Light; first in 1729; then a new edition in 1760, in 4to.

His papers that were inserted in the Memoirs of the Academy, are very numerous and important: as, in the Memoirs for 1726, Comparison of the force of the solar and lunar light with that of candles.—1731, Observations on the curvilinear motion of bodies in mediums.—1732, Upon the new curves called the *lines of pursuit.*—1733, To determine the species of conoid, to be constructed upon a given base which is exposed to the shock of a fluid, so that the impulse may be the least possible.—Determination of the orbit of comets.—1734, Comparison of the two laws which the earth and the other planets must observe in the figure which gravity causes them to take.—On the curve lines proper to form the arches in domes.—1735, Observations on the equinoxes.—On the length of the pendulum.—1736, On the length of the pendulum in the torrid zone.—On the manner of determining the figure of the earth by the measure of the degrees of latitude and longitude.—1739, On the astronomical refractions in the torrid zone.—Observations on the lunar eclipse, of the 8th of September 1737, made at Quito. —1744, Short account of the voyage to Peru, by the members of the Royal Academy of Sciences, to measure the degrees of the meridian near the equator, and from thence to determine the figure of the earth. —1745, Experiments made at Quito and divers other places in the torrid zone, on the expansion and contraction of metals by heat and cold.—On the problem of the masting of ships.—1746, Treatise on ships, their structure and motions.—On the impulse of fluids upon the fore parts of pyramidoids having their base a trapezium.—Continuation of the short account given in 1744, of the voyage to Peru for measuring the earth.—1747, On a new construction of the log, and other instruments for measuring the run of a ship.— 1748, Of the diameters of the larger planets. The new instrument called a *heliometer*, proper for determining them; with observations of the sun.—Observation of the eclipse of the moon the 8th of August 1748.—1749, Second memoir on astronomical refractions, observed in the torrid zone, with remarks on the manner of constructing the tables of them.—

Figure

Figure of the earth determined by MM. Bouguer and Condamine, with an abridgment of the expedition to Peru.—1750, Observation of the lunar eclipse of the 13th of December 1750.—1751, On the form of bodies most proper to turn about themselves, when they are pushed by one of their extremities, or any other point.—On the moon's parallax, with the estimation of the changes caused in the parallaxes by the figure of the earth.—Observation of the lunar eclipse, the 2d of December 1751.—1752, On the operations made by seamen, called *Corrections.*—1753, Observation of the passage of Mercury over the sun, the 6th of May 1753.—On the dilatations of the air in the atmosphere.—New treatise of navigation, containing the theory and practice of pilotage, or working of ships.—1754, Operations, &c, for distinguishing, among the different determinations of the degree of the meridian near Paris, that which ought to be preferred.—On the direction which the string of a plummet takes.—Solution of the chief problems in the working of ships.—1755, On the apparent magnitude of objects.—Second memoir on the chief problems in the working of ships.—1757, Account of the treatise on the working of ships.—On the means of measuring the light.—1758, His Eulogy.

In the volumes of the prizes given by the academy, are the following pieces by Bouguer:

In vol. 1, on the masting of ships.—Vol. 2, On the method of exactly observing at sea the height of the stars; and the variation of the compass. Also on the cause of the inclination of the planets' orbits.

BOUL'TINE, in Architecture, a convex moulding, of a quarter of a circle, and placed next below the plinth in the Tuscan and Dorick capital.

BOW, an offensive weapon made of wood, horn, steel, or other elastic matter, by which arrows are thrown with great force. This instrument was of very ancient and general use, and is still found among all savage nations who have not the use of fire arms, by which it has been superseded among us. There are two species of the Bow, the *Long*, and the *Cross* Bow.

The *Long Bow* is simply a bow, or a rod, with a string fastened to each end of it, to the middle of which the end of an arrow being applied, and then drawn by the hand, on suddenly quitting the hold, the bow returns by means of its elasticity, and impels the arrow from the string with great violence. The old English archers were famous for the long bow, by means of which they gained many victories in France and elsewhere.

The *Cross Bow*, called also *arbalest* or *arbalet*, is a bow strung and set in a shaft of wood, and furnished with a trigger; serving to throw bullets, darts, and large arrows, &c. The ancients had large machines for throwing many arrows at once, called *arbalets*, or *balistæ*.

The force of a bow may be calculated on this principle, that its spring, *i. e.* the power by which it restores itself to its natural position, is always proportional to the space or distance it is bent or removed from it.

Bow, a mathematical instrument formerly used at sea for taking the sun's altitude. It consisted of a large arch divided into 90 degrees, fixed on a staff, and furnished with three vanes, viz, a side vane, a sight vane, and a horizon vane.

Bow-*Compass*, an instrument for drawing arches of very large circles, for which the common compasses are too small. It consists of a beam of wood or brass, with three long screws that govern or bend a lath of wood or steel to any arch.

BOX AND NEEDLE, the small compass of a theodolite, circumferentor, or plain-table.

BOYAU, in Fortification, a ditch covered by a parapet, and serving as a communication between two trenches. It runs parallel to the works of the body of the place; and serves as a line of contravallation, both to hinder the sallies of the besieged; and to secure the miners. When it is a particular cut running from the trenches, to cover some spot of ground, it is drawn so as not to be enfiladed or scoured by the enemy's shot.

BOYLE (ROBERT), one of the greatest philosophers, as well as best men, that any country has ever produced, was the 7th son and the 14th child of Richard earl of Cork, and was born at Lismore in the province of Munster in Ireland, the 25th of January, 1626-7; the very year of the death of the learned Lord Bacon, whose plans of experimental philosophy our author afterwards so ably seconded. While very young, he was instructed in his father's house to read and write, and to speak French and Latin. In 1635, when only 8 years old, he was sent over to England, to be educated at Eton school. Here he soon discovered an extraordinary force of understanding, with a disposition to cultivate and improve it to the utmost.

After remaining at Eton between 3 and 4 years, his father sent our author and his brother Francis, in 1638, on their travels upon the continent. They passed through France to Geneva, where they settled for some time to pursue their studies: here our author resumed his acquaintance with the elements of the mathematics, which he had commenced at Eton when 10 years old, on occasion of an illness which prevented his other usual studies.

In the autumn of 1641, he quitted Geneva, and travelled through Switzerland and Italy to Venice, from whence he returned again to Florence, where he spent the winter, studying the Italian language and history, and the works of the celebrated astronomer Galileo, who died in a village near this city during Mr. Boyle's residence here.

About the end of March 1642, he set out from Florence, visited Rome and other places in Italy, then returned to the south of France. At Marseilles, in May 1642, Mr. Boyle received letters from his father, which informed him that the rebellion had broken out in Ireland, and with how much difficulty he had procured 250l. then remitted to help him and his brother home. This remittance however never reached them, and they were obliged to return to Geneva with their governor Mr. Marcombes, who contrived on his own credit, and by selling some jewels, to raise money enough to send them to England, where they arrived in 1644. On their arrival they found that their father was dead, and had left our author the manor of Stalbridge in England, with some other considerable estates in Ireland.

From

From this time Mr. Boyle's chief refidence, for fome years at leaft, was at his manor of Stalbridge, from whence he made occafional excurfions to Oxford, London, &c; applying himfelf with great induftry to various kinds of ftudies, but efpecially to philofophy and chemiftry; and feizing every opportunity of cultivating the acquaintance of the moft learned men of his time. He was one of the members of that fmall but learned body of men who, when all academical ftudies were interrupted by the civil wars, fecreted themfelves about the year 1645; and held private meetings, firft in London, afterwards at Oxford, to cultivate fubjects of natural knowledge upon that plan of experiment which Lord Bacon had delineated. They ftyled themfelves then *The Philofophic College*; but after the reftoration, when they were incorporated, and diftinguifhed openly, they took the name of the *Royal Society*.

In the fummer of 1654 he retired to fettle at Oxford, the Philofophical Society being removed from London to that place, that he might enjoy the converfation of the other learned members, his friends, who had retired thither, fuch as Wilkins, Wallis, Ward, Willis, Wren, &c. It was during his refidence here that he improved that admirable engine the air-pump; and by numerous experiments was enabled to difcover feveral qualities of the air, fo as to lay a foundation for a complete theory. He declared againft the philofophy of Ariftotle, as having in it more of words than things; promifing much, and performing little; and giving the inventions of men for indubitable proofs, inftead of building upon obfervation and experiment. He was fo zealous for this true method of learning by experiment, and fo careful about it, that though the Cartefian philofophy then made a great noife in the world, yet he could never be perfuaded to read the works of Defcartes, for fear he fhould be amufed and led away by plaufible accounts of things founded on conjecture, and merely hypothetical. But philofophy, and enquiries into nature, though they engaged his attention deeply, did not occupy him entirely; as he ftill continued to purfue critical and theological ftudies. He had offers of preferment to enter into holy orders, by the government, after the reftoration. But he declined the offer, choofing rather to purfue his ftudies as a layman, in fuch a manner as might be moft effectual for the fupport of religion; and began to communicate to the world the fruits of thefe ftudies. Thefe fruits were very numerous and important, as well as various: the principal of which, as well as of fome other memorable occurrences of his life, were nearly in the following order.

In 1660 came out, 1. New experiments, phyfico-mechanical, touching the fpring of the air and its effects.—2. Seraphic love; or fome motives and incentives to the love of God, pathetically difcourfed of in a letter to a friend. A work which it has been faid was owing to his courtfhip of a lady, the daughter of Cary earl of Monmouth; though our author was never married.—3. Certain phyfiological effays and other tracts, in 1661.—4. Sceptical chemift, 1662; reprinted about the year 1679, with the addition of divers experiments and notes on the produciblenefs of chemical principles.

In the year 1663, the Royal Society being incorporated by king Charles the 2d, Mr. Boyle was named one of the council; and as he might juftly be reckoned among the founders of that learned body, fo he continued one of the moft ufeful and, induftrious of its members during the whole courfe of his life. His next publications were, 5. Confiderations touching the ufefulnefs of experimental natural philofophy, 1663.—6. Experiments and confiderations upon colours; to which was added a letter, containing Obfervations on a diamond that fhines in the dark, 1663. This treatife is full of curious and ufeful remarks on the hitherto unexplained doctrine of light and colours; in which he fhews great judgment, accuracy, and penetration; and which may be faid to have led the way to Newton, who made fuch great difcoveries in that branch of phyfics.—7. Confiderations on the ftyle of the holy fcriptures, 1663. This was an extract from a larger work, intitled An effay on fcripture; which was afterwards publifhed by Sir Peter Pett, a friend of Mr. Boyle's.

In 1664 he was elected into the company of the royal mines; and was all this year occupied in profecuting various good defigns, which was probably the reafon that he did not publifh any works in this year. Soon after came out, 8. Occafional reflections upon feveral fubjects, 1665. This piece expofed our author to the cenfure of the celebrated Dean Swift, who, to ridicule thefe difcourfes, wrote *A pious meditation upon a broomftick, in the ftyle of the honourable Mr. Boyle.*—9. New experiments and obfervations upon cold, 1665.—10. Hydroftatical paradoxes made out by new experiments, for the moft part phyfical and eafy, 1666.—11. The origin of forms and qualities, according to the corpufcular philofophy, 1666.—Both in this and the former year, our author communicated to his friend Mr. Oldenburgh, then fecretary to the Royal Society, feveral curious and excellent fhort pieces of his own, upon a great variety of fubjects, and others tranfmitted to him by his learned friends, which are printed in the Philof. Tranf.

In the year 1668 Mr. Boyle refolved to fettle in London for life; and for that purpofe he removed to the houfe of his fifter, the lady Ranelagh, in Pall-Mall. This removal was to the great benefit of the learned in general, and particularly of the Royal Society, to whom he gave great and continual affiftance, as abundantly appears by the feveral pieces communicated to them from time to time, and printed in their Tranfactions. To avoid improper wafte of time, he had fet hours in the day appointed for receiving fuch perfons as wanted to confult him, either for their own affiftance, or to communicate new difcoveries to him: And he befides kept up an extenfive correfpondence with the moft learned men in Europe; fo that it is wonderful how he could bring out fo many new works as he did. His next publications were, 12. A continuation of new experiments touching the weight and fpring of the air; to which is added, A difcourfe of the atmofphere of confiftent bodies, 1669.—13. Tracts about the cofmical qualities of things; cofmical fufpicions; the temperature of the fubterraneous regions; the bottom of the fea; to which is prefixed an introduction to the hiftory of particular qualities, 1669.—14. Confiderations on the ufefulnefs of experimental and natural philofophy,

lofophy, the 2d part, 1671.—15. A collection of tracts upon feveral ufeful and important points of practical philofophy, 1671.—16. An effay upon the origin and virtues of gems, 1672.—17. A collection of tracts upon the relation between flame and air; and feveral other ufeful and curious fubjects, 1672. Befides furnifhing, in this and the former year, a number of fhort differtations upon a great variety of topics, addreffed to the Royal Society, and inferted in their Tranfactions.—18. Effays on the ftrange fubtilty, great efficacy, and determinate nature, of effluvia; with a variety of experiments on other fubjects, 1673.—19. The excellency of theology compared with philofophy, 1673. This difcourfe was written in the year 1665, while our author, to avoid the great plague which then raged in London, was forced to go from place to place in the country, having little or no opportunity of confulting his books.—20. A collection of tracts upon the faltnefs of the fea, the moifture of the air, the natural and preternatural ftate of bodies; to which is prefixed a dialogue concerning cold, 1674.—21. A collection of tracts containing fufpicions about hidden qualities of the air; with an appendix touching celeftial magnets; animadverfions upon Mr. Hobbes's problem about a vacuum; a difcourfe of the caufe of attraction and fuction, 1674.—22. Some confiderations about the reafonablenefs of reafon and religion; by T. E. (the final letters of his names). To which is annexed a difcourfe about the poffibility of the refurrection; by Mr. Boyle, 1675. The fame year feveral papers communicated to the Royal Society, among which were two upon quickfilver growing hot with gold.—23. Experiments and notes about the mechanical origin or production of particular qualities, in feveral difcourfes on a great variety of fubjects, and among the reft on electricity, 1676.—He then communicated to Mr. Hook a fhort memorial of fome obfervations made upon an artificial fubftance that fhines without any preceding illuftration; publifhed by Hook in his *Lectiones Cutlerianæ.*—24. Hiftorical account of a degradation of gold made by an anti-elixir.—25. Aerial noctiluca; or fome new phænomena, and a procefs of a factitious felf-fhining fubftance, 1680. This year the Royal Society, as a proof of the juft fenfe of his great worth, and of the conftant and particular fervices which through a courfe of many years he had done them, made choice of him for their prefident; but he being extremely, and, as he fays, peculiarly tender in point of oaths, declined that honour.—26. Difcourfe of things above reafon; inquiring, whether a philofopher fhould admit any fuch, 1681.—27. New experiments and obfervations upon the icy noctiluca; to which is added a chemical paradox, grounded upon new experiments, making it probable that chemical principles are tranfmutable, fo that out of one of them others may be produced, 1682.—28. A continuation of new experiments, phyfico-mechanical, touching the fpring and weight of the air, and their effects, 1682.—29. A fhort letter to Dr. Beale, in relation to the making of frefh water out of falt, 1683.—30. Memoirs for the natural hiftory of human blood, efpecially the fpirit of that liquor, 1684.—31. Experiments and confiderations about the porofity of bodies, 1684.—32. Short memoirs for the natural experimental hiftory of mineral waters, &c, 1685.—

33. An effay on the great effects of even languid and unheeded motion, &c, 1685.—34. Of the reconcileablenefs of fpecific medicines to the corpufcular philofophy, &c, 1685.—35. Of the high veneration man's intellect owes to God, peculiarly for his wifdom and power, 1685.—36. Free inquiry into the vulgarly received notion of nature, 1686.—37. The martyrdom of Theodora and Didymia, 1687. A work he had drawn up in his youth.—38. A difquifition about the final caufes of natural things, and about vitiated light, 1688.

Mr. Boyle's health declining very much, he abridged greatly his time given to converfations and communications with other perfons, to have more time to prepare for the prefs fome others of his papers, before his death, which were as follow:—39. *Medicina Hydroftatica,* &c, 1690.—40. The Chriftian virtuofo, &c, 1690. 41. *Experimenta et Obfervationes Phyficæ,* &c, 1691; which is the laft work that he publifhed.

Mr. Boyle died on the laft day of December of the fame year 1691, in the 65th year of his age, and was buried in St. Martin's church in the Fields, Weftminfter; his funeral fermon being preached by Dr. Gilbert Burnet bifhop of Salifbury; in which he difplayed the excellent qualities of our author, with many circumftances of his life, &c. But as the limits of this work will not allow us to follow the bifhop in the copious and eloquent account he has given of this great man's abilities, we muft content ourfelves with adding the following fhort eulogium by the celebrated phyfician, philofopher, and chemift, Dr. Boerhaave; who, after having declared lord Bacon to be the father of experimental philofophy, afferts, that " Mr. Boyle, the ornament of his age and country, fucceeded to the genius and inquiries of the great chancellor Verulam. Which, fays he, of Mr. Boyle's writings fhall I recommend? All of them. To him we owe the fecrets of fire, air, water, animals, vegetables, foffils: fo that from his works may be deduced the whole fyftem of natural knowledge."

Mr. Boyle left alfo feveral papers behind him, which have been publifhed fince his death. Beautiful editions of all his works have been printed at London, in 5 volumes folio, and 6 volumes 4to. Dr. Shaw alfo publifhed in 3 volumes 4to, the fame works " abridged, methodized, and difpofed under the general heads of Phyfics, Statics, Pneumatics, Natural Hiftory, Chymiftry, and Medicine;" to which he has prefixed a fhort catalogue of the philofophical writings, according to the order of time when they were firft publifhed, &c, as follows:

Phyfico-mechanical experiments on the fpring and weight of the air - -	1661
The Sceptical Chymift - -	1661
Phyfiological Effays - -	1662
Hiftory of Colours - -	1663
Ufefulnefs of Experimental Philofophy	1663
Hiftory of Cold - - -	1665
Hiftorical Paradoxes - -	1666
Origin of Forms and Qualities -	1666
Cofmical Qualities - -	1670
The admirable Rarefaction of the air	1670
The Origin and Virtues of Gems -	1672
The Relation betwixt Flame and Air -	1672

2

Efflu-

Effluviums - - -	1673
Saltnefs of the Sea - -	1674
Hidden Qualities in the Air -	1674
The Excellence &c of the Mechanical Hypothefis -	1674
Confiderations on the Refurrection -	1675
Particular Qualities -	1676
Aerial Noctiluca - -	1680
Icy Noctiluca - -	1680
Things above Reafon -	1681
Natural Hiftory of Human Blood -	1684
Porofity of Bodies -	1684
Natural Hiftory of Mineral Waters	1684
Specific Medicines -	1685
The High Veneration due to God -	1685
Languid Motion -	1685
The Notion of Nature -	1685
Final Caufes -	1688
Medicina Hydroftatica -	1690
The Chriftian Virtuofo -	1690
Experimenta & Obfervationes Phyficæ	1691
Natural Hiftory of the Air -	1692
Medicinal Experiments -	1718

BRADLEY (Dr. James), a celebrated Englifh aftronomer, the third fon of William Bradley, was born at Sherborne in Gloucefterfhire in the year 1692. He was fitted for the univerfity at Northleach in the fame county, at the boarding-fchool of Mr. Egles and Mr. Brice. From thence he was fent to Oxford, and admitted a commoner of Baliol college March 15, 1710; where he took the degree of bachelor the 14th of October 1714, and of mafter of arts the 21ft of January 1716. His friends intending him for the church, his ftudies were regulated with that view; and as foon as he was of a proper age to receive holy orders, the bifhop of Hereford, who had conceived a great efteem for him, gave him the living of Bridftow, and foon after he was inducted to that of Landewy Welfry in Pembrokefhire.

He was nephew to Mr. Pound, a gentleman well known in the learned world, by many excellent aftronomical and other obfervations, and who would have enriched it much more, if the journals of his voyages had not been burnt at Pulo Condor, when the place was fet on fire, and the Englifh who were fettled there cruelly maffacred, Mr. Pound himfelf very narrowly efcaping with his life. With this gentleman, at Wanftead, Mr. Bradley paffed all the time that he could fpare from the duties of his function; being then fufficiently acquainted with the mathematics to improve by Mr. Pound's converfation. It may eafily be imagined that the example and converfation of this gentleman did not render Bradley more fond of his profeffion, to which he had before no great attachment: he continued however as yet to fulfil the duties of it, though at this time he had made fuch obfervations as laid the foundation of thofe difcoveries which afterward diftinguifhed him as one of the greateft aftronomers of his age. Thefe obfervations gained him the notice and friendfhip of the lord chancellor Macclesfield, Mr. Newton afterward Sir Ifaac, Mr. Halley, and many other members of the Royal Society, into which he was foon after elected a member.

Soon after, the chair of Savilian profeffor of aftro-

nomy at Oxford became vacant, by the death of the celebrated Dr. John Keil; and Mr. Bradley was elected to fucceed him on the 31ft of October 1721, at 29 years of age: his colleague being Mr. Halley, who was profeffor of geometry on the fame foundation. Upon this appointment, Mr. Bradley refigned his church livings, and applied himfelf wholly to the ftudy of his favourite fcience. In the courfe of his obfervations, which were innumerable, he difcovered and fettled the laws of the alterations of the fixed ftars, from the progreffive motion of light, combined with the earth's annual motion about the fun, and the nutation of the earth's axis, arifing from the unequal attraction of the fun and moon on the different parts of the earth. The former of thefe effects is called the *aberration* of the fixed ftars, the theory of which he publifhed in 1727; and the latter the *nutation* of the earth's axis, the theory of which appeared in 1737: fo that in the fpace of about 10 years, he communicated to the world two of the fineft difcoveries in modern aftronomy; which will for ever make a memorable epoch in the hiftory of that fcience. See ABERRATION and NUTATION.

In 1730 our author fucceeded Mr. Whitefide, as lecturer in aftronomy and experimental philofophy in the Mufeum at Oxford: which was a confiderable emolument to him, and which he held till within a year or two of his death; when the ill ftate of his health induced him to refign it.

Our author always preferved the efteem and friendfhip of Dr. Halley; who, being worn out by age and infirmities, thought he could not do better for the fervice of aftronomy, than procure for Mr. Bradley the place of regius profeffor of aftronomy at Greenwich, which he himfelf had many years poffeffed with the greateft reputation. With this view he wrote many letters, defiring Mr. Bradley's permiffion to apply for a grant of the reverfion of it to him, and even offered to refign it in his favour, if it fhould be thought neceffary: but Dr. Halley died before he could accomplifh this kind object. Our author however obtained the place, by the intereft of lord Macclesfield, who was afterward prefident of the Royal Society; and upon this appointment the univerfity of Oxford fent him a diploma of doctor of divinity.

The appointment of aftronomer royal at Greenwich, which was dated the 3d of February 1741-2, placed our author in his proper element; and he purfued his obfervations with unwearied diligence. However numerous the collection of aftronomical inftruments at that obfervatory, it was impoffible that fuch an obferver as Dr. Bradley fhould not defire to increafe them, as well to anfwer thofe particular views, as in general to make obfervations with greater exactnefs. In the year 1748 therefore, he took the opportunity of the vifit of the Royal Society to the obfervatory, annually made to examine the inftruments and receive the profeffor's obfervations for the year, to reprefent fo ftrongly the neceffity of repairing the old inftruments, and providing new ones, that the fociety thought proper to make application to the king, who was pleafed to order 1000 pounds for that purpofe. This fum was laid out under the direction of our author, who, with the affiftance of the late celebrated Mr. Graham and Mr. Bird, furnifhed the obfervatory with as complete a col-

loction.

lection of aftronomical inftruments, as the moft fkilful and diligent obferver could defire. Dr. Bradley, thus furnifhed with fuch affiftance, purfued his obfervations with great affiduity during the reft of his life; an immenfe number of which was found after his death, in 13 folio volumes, and were prefented to the univerfity of Oxford in the year 1776, on condition of their printing and publifhing them; but which however, unfortunately for the improvement of aftronomy, now after a lapfe of almoft 20 years, has never yet been done.

During Dr. Bradley's refidence at the Royal Obfervatory, the living of the church at Greenwich became vacant, and was offered to him: upon his refufing to accept it, from a confcientious fcruple, "that the duty of a paftor was incompatible with his other ftudies and neceffary engagements," the king was pleafed to grant him a penfion of 250l. over and above the aftronomer's original falary from the Board of Ordnance, "in confideration (as the fign manual, dated the 15th Feb. 1752, expreffes it) of his great fkill and knowledge in the feveral branches of aftronomy and other parts of the mathematics, which have proved fo ufeful to the trade and navigation of this kingdom." A penfion which has been regularly continued to the aftronomers royal ever fince.

About 1748, our author became entitled to bifhop Crew's benefaction of 30l. a year, to the lecture reader in experimental philofophy at Oxford. He was elected a member of the Academy of Sciences at Berlin, in 1747; of that at Paris, in 1748; of that at Peterfburgh, in 1754; and of that at Bologna, in 1757. He was married in the year 1744, but never had more than one child, a daughter.

By too clofe application to ftudy and obfervations, Dr. Bradley became afflicted, for near two years before his death, with a grievous oppreffion on his fpirits; which interrupted his ufeful labours. This diftrefs arofe chiefly from an apprehenfion that he fhould outlive his rational faculties: but this fo much dreaded evil never came upon him. In June 1762 he was feized with a fuppreffion of urine, occafioned by an inflammation in the reins, which terminated his exiftence the 13th of July following. His death happened at Chalfont in Gloucefterfhire, in the 70th year of his age; and he was interred at Minchinhampton in the fame county.

As to his character, Dr. Bradley was remarkable for a placid and gentle modefty, very uncommon in perfons of an active temper and robuft conftitution. Although he was a good speaker, and poffeffed the rare but happy art of expreffing his ideas with the utmoft precifion and clearnefs, yet no man was a greater lover of filence, for he never fpoke but when he thought it abfolutely neceffary. Nor was he more inclined to write than to fpeak, as he has publifhed very little: he had a natural diffidence, which made him always afraid that his works might injure his character; fo that he fuppreffed many which might have been worthy of publication.

His papers, which have been inferted in the Philof. Tranf. are,

1. Obfervations on the comet of 1703. Vol. 33, p. 41.

2. The longitude of Lifbon and of the fort of New York from Wanfted and London determined by the eclipfe of the firft fatellite of Jupiter. Vol. 34, p. 85.

3. An account of a new difcovered Motion of the Fixed Stars. Vol. 35, p. 637.

4. On the Going of Clocks with Ifochronal Pendulums. Vol. 38, p. 302.

5. Obfervations on the Comet of 1736-7. Vol. 40, p. 111.

6. On the apparent Motion of the fixed Stars. Vol. 45, p. 1.

7. On the Occultation of Venus by the Moon, the 15th of April 1751. Vol. 46, p. 201.

8. On the Comet of 1757. Vol. 50, p. 408.

9. Directions for ufing the Common Micrometer. Vol. 62, p. 46.

BRADWARDIN (Thomas), archbifhop of Canterbury, was born at Hartfield in Suffex, about the clofe of the 13th century. He was educated at Merton College, Oxford, where he took the degree of doctor of divinity; and acquired the reputation of a profound fcholar, a fkilful mathematician, and confummate divine. It has been faid he was profeffor of divinity at Oxford; that he was chancellor of the diocefe of London, and confeffor to Edward the 3d, whom he conftantly attended during his war with France. After his return from the war, he was made prebendary of Lincoln, and afterward archbifhop of Canterbury. He died at Lambeth in the year 1349, forty days after his confecration. His works are, 1. *De Caufa Dei*, printed London 1618, publifhed by J. H. Savil.—2. *De Geometria fpeculativa, &c.* Paris, 1495, 1512, 1530.— 3. *De Arithmetica practica*, Paris, 1502, 1512.—4. *De Proportionibus*, Paris, 1495. Venice, 1505, folio.— 5. *De Quadratura Circuli*, Paris, 1495, folio.

BRAHE (Tycho), a celebrated aftronomer, defcended from a noble family originally of Sweden but fettled in Denmark, was born the 14th of December 1546, at Knudftorp in the county of Schonen, near Helfimbourg. He was taught Latin when 7 years old, and ftudied 5 years under private tutors. His father dying while our author was very young, his uncle, George Brahe, having no children, adopted him, and fent him, in 1559, to ftudy philofophy and rhetoric at Copenhagen. The great eclipfe of the fun, on the 21ft of Auguft 1560, happening at the precife time the aftronomers had foretold, he began to confider aftronomy as fomething divine; and purchafing the tables of Stadius, he gained fome notion of the theory of the planets. In 1562 he was fent by his uncle to Leipfic to ftudy the law, where his acquirements gave manifeft indications of extraordinary abilities. His natural inclination however was to the ftudy of the heavens, to which he applied himfelf fo affiduoufly, that, notwithftanding the care of his tutor to keep him clofe to the ftudy of the law, he made ufe of every means in his power for improving his knowledge of aftronomy; he purchafed with his pocket money whatever books he could meet with on the fubject, and read them with great attention, procuring affiftance in difficult cafes from Bartholomew Scultens his private tutor; and having procured a fmall celeftial globe, he took opportunities, when his tutor was in bed, and when the weather was clear, to examine the conftellations in the heavens, to learn their names

names from the globe, and their motions from observation.

After a course of 3 years study at Leipsic, his uncle dying, he returned home in 1565. In this year a difference arising between Brahe and a Danish nobleman, they fought, and our author had part of his nose cut off by a blow; a defect which he so artfully supplied with one made of gold and silver, that it was not perceivable. About this time he began to apply himself to chemistry, proposing nothing less than to obtain the philosopher's stone. But becoming greatly disgusted to see the liberal arts despised, and finding his own relations and friends uneasy that he applied himself to astronomy, as thinking it a study unsuitable to a person of his quality, he went to Wirtemberg in 1566, from whence the breaking out of the plague soon occasioned his removal to Rostock, and in 1569 to Augsburg, where he was visited by Peter Ramus, then professor of astronomy at Paris, and who greatly admired his uncommon skill in this science.

In 1571 he returned to Denmark; and was favoured by his maternal uncle, Steno Billes, a lover of learning, with a convenient place at his castle of Herritzvad near Knudstorp, for making his observations, and building a laboratory. And here it was he discovered, in 1573, a new star in the constellation Cassiopeia. But soon after, his marrying a country girl, beneath his rank, occasioned so violent a quarrel between him and his relations, that the king was obliged to interpose to reconcile them.

In 1574, by the king's command, he read lectures at Copenhagen on the theory of the planets. The year following he began his travels through Germany, and proceeded as far as Venice. He then resolved to remove his family, and settle at Basil; but Frederic the 2d, king of Denmark, being informed of his design, and unwilling to lose a man who was capable of doing so much honour to his country, he promised to enable him to pursue his studies, and bestowed upon him for life the island of Huen in the Sound, and promised that an observatory and laboratory should be built for him, with a supply of money for carrying on his designs: and accordingly the first stone of the observatory was laid the 8th of August 1576, under the name of Uranibourg: The king also gave him a pension of 2000 crowns out of his treasury, a fee in Norway, and a canonry of Roshild, which brought him in 1000 more. This situation he enjoyed for the space of about 20 years, pursuing his observations and studies with great industry: here he kept always in his house ten or twelve young men, who assisted him in his observations, and whom he instructed in astronomy and mathematics. Here it was that he received a visit from James the 6th, king of Scotland, afterward James the 1st of England, having come to Denmark to espouse Anne, daughter of Frederick the 2d: James made our author some noble presents, and wrote a copy of Latin verses in his praise.

Brahe's tranquillity however in this happy situation was at length fatally interrupted. Soon after the death of king Frederick, by the aspersions of envious and malevolent ministers, he was deprived of his pension, fee, and canonry, in 1596. Being thus rendered incapable of supporting the expences of his establish-

ment, he quitted his favourite Uranibourg, and withdrew to Copenhagen, with some of his instruments, and continued his astronomical observations and chemical experiments in that city, till the same malevolence procured from the new king, Charles the 4th, an order for him to discontinue them. This induced him to fall upon means of being introduced to the emperor Rodolphus, who was fond of mechanism and chemical experiments: and to smooth the way to an interview, Tycho now published his book, *Astronomia instaurata Mechanica*, adorned with figures, and dedicated it to the emperor. That prince received him at Prague with great civility and respect; gave him a magnificent house, till he could procure one for him more fit for astronomical observations; he also assigned him a pension of 3000 crowns; and promised him a fee for himself and his descendants. Here then he settled in the latter part of 1598, with his sons and scholars, and among them the celebrated Kepler, who had joined him. But he did not long enjoy this happy situation; for, about 3 years after, he died, on the 24th of October 1601, of a retention of urine, in the 55th year of his age, and was interred in a very magnificent manner in the principal church at Prague, where a noble monument was erected to him; leaving, beside his wife, two sons and four daughters. On the approach of death, he enjoined his sons to take care that none of his works should be lost; exhorted the students to attend closely to their exercises; and recommended to Kepler the finishing of the Rudolphine tables he had constructed for regulating the motion of the planets.

Brahe's skill in astronomy is universally known; and he is famed for being the inventor of a new system of the planets, which he endeavoured, though without success, to establish on the ruins of that of Copernicus. He was very credulous with regard to judicial astrology and presages: If he met an old woman when he went out of doors, or a hare upon the road on a journey, he would turn back immediately, being persuaded that it was a bad omen: Also, when he lived at Uranibourg, he kept at his house a madman, whom he placed at his feet at table, and fed himself; for as he imagined that every thing spoken by mad persons presaged something, he carefully observed all that this man said; and because it sometimes proved true, he fancied it might always be depended on. He was of a very irritable disposition: a mere trifle put him in a passion; and against persons of the first rank, whom he thought his enemies, he openly discovered his resentment. He was very apt to rally others, but highly provoked when the same liberty was taken with himself.—The principal part of his writings, according to Gassendus, are,

1. An account of the New Star, which appeared Nov. 11th 1572, in Cassiopeia; Copenh. 1573, in 4to. —2. An Oration concerning the Mathematical Sciences, pronounced in the university of Copenhagen, in the year 1574: published by Conrad Aslac, of Bergen in Norway.—3. A treatise on the Comet of the year 1577, immediately after it disappeared. Nine years afterward, he revised it, and added a 10th chapter. Printed at Uranibourg, 1589.—4. Another treatise on the New Phenomena of the heavens. In the first part of which he treats of the Restitution, as he calls it, of the sun, and of the fixed stars. And in the 2d part, of

a New Star, which then had made its appearance.—
5. A collection of Astronomical Epistles: printed in
4to, at Uranibourg in 1596; Nuremberg in 1602; and
at Franckfort in 1610. It was dedicated to Maurice
landgrave of Hesse; because there are in it a considerable number of letters of the landgrave William his
father, and of Christopher Rothmann, the mathematician of that prince, to Tycho, and of Tycho to them.
—6. The Mechanical Principles of Astronomy restored: Wandesburg, 1598, in folio.—7. An Answer to
the Letter of a certain Scotchman, concerning the
comet, in the year 1577.—8. On the composition of an
Elixir for the Plague; addressed to the emperor Rodolphus.—9. An elegy upon his Exile: Rostock, 1614,
4to.—10. The Rudolphine Tables; which he had not
finished when he died; but were revised, and published
by Kepler, as Tycho had desired.—11. An accurate
Enumeration of the Fixed Stars: addressed to the emperor Rodolphus.—12. A complete Catalogue of 1000
of the Fixed Stars; which Kepler has inserted in the
Rudolphine Tables.—13. *Historia Cœlestis;* or a History of the Heavens; in two parts: The 1st contains
the Observations he had made at Uranibourg, in 16
books: The latter contains the Observations made at
Wandesburg, Wittenberg, Prague, &c; in 4 books.—
14. Is an Epistle to Caster Pucer; printed at Copenhagen 1668.

BRANCKER, or BRANKER (THOMAS), an eminent mathematician of the 17th century, son of Thomas
Brancker some time bachelor of arts in Exeter College,
Oxford, was born in Devonshire in 1636, and was admitted butler of the said college Nov. 8, 1652, in the
17th year of his age. In 1655, June the 15th, he took
the degree of bachelor of arts, and was elected probationary fellow the 30th of the same month. In 1658,
April the 22d, he took the degree of master of arts,
and became a preacher; but after the restoration, refusing to conform to the ceremonies of the church of
England, he quitted his fellowship in 1662, and retired
to Chester: But not long after, he became reconciled to
the service of the church, took orders from a bishop,
and was made a minister of Whitegate. He had however, for some time, enjoyed great opportunity and leisure for pursuing the bent of his genius in the mathematical sciences; and his skill both in the mathematics
and chemistry procured him the favour of lord Brereton, who gave him the rectory of Tilston. He was
afterward chosen master of the well-endowed school at
Macclesfield, in that county, where he spent the remaining years of his life, which was terminated by a
short illness in 1676, at 40 years of age; and he was
interred in the church at Macclesfield.

Brancker wrote a piece on the Doctrine of the
Sphere, in Latin, which was published at Oxford in
1662; and in 1668, he published at London, in 4to, a
translation of Rhonius's Algebra, with the title of *An
Introduction to Algebra;* which treatise having communicated to Dr. John Pell, he received from him some
assistance towards improving it; which he generously
acknowledges in a letter to Mr. John Collins; with
whom, and some other gentlemen, proficients in this
science, he continued a correspondence during his life.

BREACH, in Fortification, a gap or opening made
in any part of the works of a town, by the cannon or
mines of the besiegers, with intent to storm or attack
the place.

BREAKING *Ground*, in Military Affairs, is beginning the works for carrying on the siege of a place;
more especially the beginning to dig trenches, or approaches.

BREECH *of a Gun*, the hinder part, from the cascabel to the lower part of the bore.

BREREWOOD (EDWARD), a learned mathematician and antiquary, was the son of Robert Brerewood, a reputable tradesman, who was three times
mayor of Chester. Our author was born in that city
in 1565, where he was educated in grammar learning
at the free school; and was afterward admitted, in
1581, of Brazen-nose College, Oxford; where he soon
acquired the character of a hard student; as he has
shewn by the commentaries he wrote upon Aristotle's
Ethics, which were written by him about the age
of 21.

In the year 1596 he was chosen the first Professor
of Astronomy in Gresham College, being one of the
two who, at the desire of the electors, were recommended to them by the university of Oxford. He
loved retirement, and wholly devoted himself to the
pursuit of knowledge. And though he never published
any thing himself, yet he was very communicative, and
ready to impart what he knew, to others, either in
conversation or in writing. His retired situation at
Gresham College being agreeable, it did not appear
that he had any other views, but continued there the
remainder of his life, which was terminated by a fever
the 4th of November 1613, at 48 years of age, in the
midst of his pursuits, and before he had taken proper
care to collect and digest his learned labours; which
however were not lost; being reduced to order, and
published after his death. These were little or nothing mathematical, being of a miscellaneous nature,
upon the several subjects of Weights, Money, Languages, Religion, Logic, the Sabbath, Meteors, the
Eye, Ethics, &c.

BRIDGE, a work of carpentry or masonry, built
over a river, canal, or the like, for the convenience of
passing from one side to the other; and may be considered as a road over water, supported by one or more
arches, and these again supported by proper piers or
buttments. Besides these essential parts, may be added
the paving at top, the banquet, or raised footway, on
each side, leaving a sufficient breadth in the middle for
horses and carriages, also the parapet wall either with
or without a balustrade, or other ornamental and useful
parts. The breadth of a bridge for a great city
should be such, as to allow an easy passage for three
carriages and two horsemen abreast in the middle way,
and for 3 foot passengers in the same manner on each
banquet: but for other smaller bridges, a less breadth.

Bridges are commonly very difficult to execute, on
account of the inconvenience of laying foundations and
walling under water. The earliest rules and instructions for building of bridges are given by Alberti, in
his *Archit.* 1, 8. Other rules were afterwards laid
down by Palladio, Serlio, and Scamozzi, which are
collected by Blondel, in his *Cours d'Archit.* pa. 629
&c. The best of these rules were also given by Goldman, Baukhurst, and in Hawkesmoor's History of
London

London Bridge. M. Gautier has a considerable volume expressly on bridges, antient and modern. See also Riou's Short Principles for the Architecture of Stone Bridges; as also Emerson, Muller, Labelye, and my own Principles of Bridges.

The conditions required in a bridge are, That it be well designed, commodious, durable, and suitably decorated. It should be of such a height as to be quite convenient for the passage over it, and yet easily admitting through its arches the vessels that navigate upon it, and all the water, even at high tides and floods: the neglect of this precept has been the ruin of many bridges. Bridges are commonly continued in a straight direction perpendicular to the stream; though some think they should be made convex towards the stream, the better to resist floods, &c. And bridges of this sort have been executed in some places, as Pont St. Esprit near Lyons. Again, a bridge should not be made in too narrow a part of a navigable river, or one subject to tides or floods: because the breadth being still more contracted by the piers, this will increase the depth, velocity, and fall of the water under the arches, and endanger the whole bridge and navigation. There ought to be an uneven number of arches, or an even number of piers; both that the middle of the stream or chief current may flow freely without the interruption of a pier; and that the two halves of the bridge, by gradually rising from the ends to the middle, may there meet in the highest and largest arch; and also, that by being open in the middle, the eye in viewing it may look directly through there. When the middle and ends are of different heights, their difference however ought not to be great in proportion to the length, that the ascent and descent may be easy; and in that case also it is more beautiful to make the top in one continued curve, than two straight lines forming an angle in the middle. Bridges should rather be of few and large arches, than of many smaller ones, if the height and situation will possibly allow of it; for this will leave more free passage for the water and navigation, and be a great saving in materials and labour, as there will be fewer piers and centres, and the arches &c will require less materials; a remarkable instance of which appears in the difference between the bridges of Westminster and Blackfriars, the expence of the former being more than double the latter.

For the proper execution of a bridge, and making an estimate of the expence, &c, it is necessary to have three plans, three sections, and an elevation. The three plans are so many horizontal sections, viz, first a plan of the foundation under the piers, with the particular circumstances attending it, whether of gratings, planks, piles, &c; the 2d is the plan of the piers and arches; and the 3d is the plan of the superstructure, with the paved road and banquet. The three sections are vertical ones; the first of them a longitudinal section from end to end of the bridge, and through the middle of the breadth; the 2d a transverse one, or across it, and through the summit of an arch; and the 3d also across, but taken upon a pier. The elevation is an orthographic projection of one side or face of the bridge, or its appearance as viewed at a distance, shewing the exterior aspect of the materials, with the manner in which they are disposed &c.

For the figure of the arches, some prefer the semi-circle, though perhaps without knowing any good reason why; others the elliptical form, as having many advantages over the semi-circular; and some talk of the catenarian arch, though its pretended advantages are only chimerical; but the arch of equilibration is the only perfect one, so as to be equally strong in every part: see my Principles of Bridges. The piers are of diverse thickness, according to the figure, span, and height of the arches; as may be seen in the work above mentioned.

With the Romans, the repairing and building of bridges were committed to the priests, thence named *pontifices;* next to the censors, or curators of the roads; but at last the emperors took the care of the bridges into their own hands. Thus, the Pons Janiculensis was built of marble by Antoninus Pius; the Pons Cestius was restored by Gordian; and Arian built a new one which was called after his own name. In the middle age, bridge-building was counted among the acts of religion; and, toward the end of the 12th century, St. Benezet founded a regular order of hospitallers, under the name of *pontifices,* or bridge-builders, whose office was to assist travellers, by making bridges, settling ferries, and receiving strangers into hospitals, or houses, built on the banks of rivers. We read of an hospital of this kind at Avignon, where the hospitallers resided under the direction of their first superior St. Benezet: and the Jesuit Raynaldus has a treatise on St. John the bridge-builder.

Among the bridges of antiquity, that built by Trajan over the Danube, it is allowed, is the most magnificent. It was demolished by his next successor Adrian, and the ruins are still to be seen in the middle of the Danube, near the city Warhel in Hungary. It had 20 piers, of square stone, each of which was 150 feet high above the foundation, 60 feet in breadth, and 170 feet distant from one another, which is the span or width of the arches; so that the whole length of the bridge was more than 1530 yards, or one mile nearly.

In France, the Pont de Garde is a very bold structure; the piers being only 13 feet thick, yet serving to support an immense weight of a triplicate arcade, and joining two mountains. It consists of three bridges, one over another; the uppermost of which is an aqueduct.

The bridge of Avignon, which was finished in the year 1188, consists of 18 arches, and measures 1340 paces, or about 1000 yards in length.

The famous bridge at Venice, called the Rialto, passes for a master piece of art, consisting of only one very flat and bold arch, near 100 feet span, and only 23 feet high above the water: it was built in 1591.— Poulet also mentions a bridge of a single arch, in the city of Munster in Bothnia, much bolder than that of the Rialto at Venice. Yet these are nothing to a bridge in China, built from one mountain to another, consisting only of a single arch, 400 cubits long, and 500 cubits high, whence it is called the flying bridge; and a figure of it is given in the Philos. Transf. Kircher also speaks of a bridge in the same country 360 perches long without any arch, but supported by 300 pillars.

There are many bridges of considerable note in our

own country. The triangular bridge at Crowland in Lincolnshire, it is said, is the most ancient Gothic structure remaining intire in the kingdom; and was erected about the year 860.

London bridge is on the old Gothick structure, with 20 small locks or arches, each of only 20 feet wide; but there are now only 18 open, two having lately been thrown into one in the centre, and another next one side is concealed or covered up. It is 900 feet long, 60 high, and 74 wide; the piers are from 25 to 34 feet broad, with starlings projecting at the ends; so that the greatest water-way, when the tide is above the starlings, was 450 feet, scarce half the breadth of the river; and below the starlings, the water-way was reduced to 194 feet, before the late opening of the centre. London bridge was first built with timber between the years 993 and 1016; and it was repaired, or rather new built with timber, 1163. The stone bridge was begun in 1176, and finished in 1209. It is probable there were no houses on this bridge for upwards of 200 years; since we read of a tilt and tournament held on it in 1395. Houses it seems were erected on it afterwards; but being found of great inconvenience and nuisance, they were removed in 1758, and the avenues to it enlarged and the whole made more commodious; the two middle arches were then thrown into one, by removing the pier from between them; the whole repairs amounting to above 80,000l.

There are still some more bridges in England built in the old manner of London bridge; as the bridge at Rochester, and some others; also the late bridge at Newcastle upon Tyne, which was broken down by a great flood in the year 1771, for want of a sufficient quantity of water-way through the arches.

The longest bridge in England is that over the Trent at Burton, built in the 12th century, of squared free-stone, and is strong and lofty; it contains 34 arches, and the whole length is 1545 feet. But this falls far short of the wooden bridge over the Drave, which according to Dr. Brown is at least 5 miles long.

But one of the most singular bridges in Europe is that built over the Taaf in Glamorganshire, by William Edward, a poor country mason, in the year 1756. This remarkable bridge consists of only one stupendous arch, which, though only 8 feet broad, and 35 feet high, is no less than 140 span, being part of a circle of 175 feet diameter.

There is also a remarkable bridge of one arch, built at Colebrook Dale in 1779, of cast iron: and another still larger of the same metal, is now raising over the river Wear, at Sunderland, the arch being of 240 feet span.

Of modern bridges, perhaps the two finest in Europe, are the Westminster and Blackfriars bridges over the river Thames at London. The former is 1220 feet long, and 44 feet wide, having a commodious broad foot path on each side for passengers. It consists of 13 large and two small arches, all semicircular, with 14 intermediate piers. The arches all spring from about 2 feet above low-water mark; the middle arch is 76 feet wide, and the others on each side decrease always by 4 feet at a time. The two middle piers are each 17 feet thick at the springing of the arches; and the others decrease equally on each side by one foot at

a time; every pier terminating with a saliant right angle against either stream. This bridge is built of the best materials, and in a neat and elegant taste, but the arches are too small for the quantity of masonry contained in it. This bridge was begun in 1738, and opened in 1750; and the whole sum of money granted and paid for the erection of this bridge, with the purchase of houses to take down, and widening the avenues, &c, amounted to 589,500l.

Blackfriars bridge, nearly opposite the centre of the city of London, was begun in 1760, and was completed in 10 years and three quarters; and is an exceeding light and elegant structure, but the materials unfortunately do not seem to be of the best, as many of the arch stones are decaying. It consists of 9 large, elegant, elliptical arches; the centre arch being 100 feet wide, and those on each side decreasing in a regular gradation, to the smallest, at each extremity, which is 70 feet wide. The breadth of the bridge is 42 feet, and the length from wharf to wharf 995. The upper surface is a portion of a very large circle, which forms an elegant figure, and is of convenient passage over it. The whole expence was 150,840l.

There are various sorts of bridges, of stone, wood, or metal, of boats or floats, pendant or hanging bridges, draw bridges, flying bridges, &c, &c, and even natural bridges, or such as are found formed by nature, of which kind a most wonderful one is described by Mr. Jefferson, in his *State of Virginia*; and another, but smaller, is described by Don Ulloa, in the province of Angaraez in South America.

BRIGGS (HENRY), one of the greatest mathematicians in the 16th and 17th centuries, was born at Warleywood, near Halifax, in Yorkshire, in 1556. From a grammar school in that country he was sent to St. John's College, Cambridge, 1579; where after taking both the degrees in arts, he was chosen fellow of his college in 1588. He applied himself chiefly to the study of the mathematics, in which he greatly excelled; in consequence in 1592 he was made examiner and lecturer in that faculty; and soon after, reader of the physic lecture, founded by Dr. Linacer.

Upon the settlement of Gresham College, in London, he was chosen the first professor of geometry there, in 1596. Soon after this, he constructed a table, for finding the latitude, from the variation of the magnetic needle being given. In the year 1609 he contracted an acquaintance with the learned Mr. James Usher, afterwards archbishop of Armagh, which continued many years after by letters, two of Mr. Briggs being still extant in the collection of Usher's letters that were published: in the former of these, dated August 1610, he writes among other things, that he was engaged in the subject of eclipses; and in the latter, dated the 10th of March 1615, that he was wholly taken up and employed about the noble invention of logarithms, which had come out the year before, and in the improvement of which he had afterwards so great a concern. For Briggs immediately set himself to the study and improvement of them; expounding them also to his auditors in his lectures at Gresham college. In these lectures he proposed the alteration of the scale of logarithms, from the hyperbolic form which Napier

had

had given them, to that in which I should be the logarithm of the ratio of 10 to 1; and soon after he wrote to Napier to make the same proposal to himself. In the year 1616 Briggs made a visit to Napier at Edinburgh, to confer with him upon this change; and the next year he did the same also. In these conferences, the alteration was agreed upon accordingly, and upon Briggs's return from his second visit, in 1617, he published the first chiliad, or 1000 of his logarithms. See the Introduction to my Logarithms.

In 1619 he was made the first Savilian professor of geometry; and resigned the professorship of Gresham college the 25th of July 1620. At Oxford he settled himself at Merton college, where he continued a most laborious and studious life, employed partly in the duties of his office as geometry lecturer, and partly in the computation of the logarithms, and in other useful works. In the year 1622 he published a small tract on the "North-west passage to the South Seas, through the continent of Virginia and Hudson's Bay;" the reason of which was probably, that he was then a member of the company trading to Virginia. His next performance was his great and elaborate work, the *Arithmetica Logarithmica* in folio, printed at London in 1624; a stupendous work for so short a time! containing the logarithms of 30 thousand natural numbers, to 14 places of figures beside the index. Briggs lived also to complete a table of logarithmic sines and tangents for the 100th part of every degree, to 14 places of figures beside the index; with a table of natural sines for the same 100th parts to 15 places, and the tangents and secants for the same to ten places; with the construction of the whole. These tables were printed at Gouda in 1631, under the care of Adrian Vlacq, and published in 1633, with the title of *Trigonometria Britannica*. In the construction of these two works, on the logarithms of numbers, and of sines and tangents, our author, beside extreme labour and application, manifests the highest powers of genius and invention; as we here for the first time meet with several of the most important discoveries in the mathematics, and what have hitherto been considered as of much later invention; such as the Binomial Theorem; the Differential Method and Construction of Tables by Differences; the Interpolation by Differences; with Angular Sections, and several other ingenious compositions: a particular account of which may be seen in the Introduction to my Mathematical Tables.

This truly great man terminated his useful life the 26 of January 1630, and was buried in the choir of the chapel of Merton College. As to his character, he was not less esteemed for his great probity and other eminent virtues, than for his excellent skill in mathematics. Doctor Smith gives him the character of a man of great probity; easy of access to all; free from arrogance, moroseness, envy, ambition and avarice; a contemner of riches, and contented in his own situation; preferring a studious retirement to all the splendid circumstances of life. The learned Mr. Thomas Gataker, who attended his lectures when he was reader of mathematics at Cambridge, represents him as highly esteemed by all persons skilled in mathematics, both at home and abroad; and says, that desiring him once to give his judgment concerning judicial astrology, his

answer was, "that he conceived it to be a mere system of groundless conceits." Oughtred calls him the mirror of the age, for his excellent skill in geometry. And one of his successors at Gresham college, the learned Dr. Isaac Barrow, in his oration there upon his admission, has drawn his character more fully; celebrating his great abilities, skill, and industry, particularly in perfecting the invention of logarithms, which, without his care and pains, might have continued an imperfect and useless design.

His writings were more important than numerous: some of them were published by other persons: the list of the principal part of them as follows.

1. *A Table to find the Height of the Pole; the Magnetical Declination being given.* This was published in Mr. Thomas Blundevile's Theoriques of the Seven Planets: London 1602, 4to.

2. *Tables for the improvement of Navigation.* These consist of, A table of declination of every minute of the ecliptic, in degrees, minutes and seconds: A table of the sun's prosthaphaereses: A table of equations of the sun's ephemerides: A table of the sun's declination: Tables to find the height of the pole in any latitude, from the height of the pole star. These tables are printed in the 2d edition of Edward Wright's treatise, intitled, Certain Errors in Navigation detected and corrected; London 1610, 4to.

3. *A description of an Instrumental Table to find the Part Proportional, devised by Mr. Edward Wright.* This is subjoined to Napier's table of logarithms, translated into English by Mr. Wright, and after his death published by Briggs, with a preface of his own: Lond. 1616 and 1618, 12mo.

4. *Logarithmorum chilias prima.* Lond. 1617, 8vo.

5. *Lucubrationes & Annotationes in opera posthuma J. Neperi:* Edinb. 1619, 4to.

6. *Euclidis Elementorum vi libri priores &c.* Lond. 1620, folio. This was printed without his name to it.

7. *A treatise of the North-west passage to the South Sea &c.* By H. B. Lond. 1622, 4to. This was reprinted in Purchas's Pilgrims, vol. 3, p. 852.

8. *Arithmetica Logarithmica, &c.* Lond. 1624, folio.

9. *Trigonometria Britannica, &c.* Goudæ 1633, folio.

10. *Two letters to archbishop Usher.*

11. *Mathematica ab antiquis minus cognita.*—This is a summary account of the most observable inventions of modern mathematicians, communicated by Mr. Briggs to Dr. George Hakewill, and published by him in his *Apologie;* Lond. folio.

Beside these publications, Briggs wrote some other pieces, that have not been printed; as,

(1). *Commentaries on the Geometry of Peter Ramus.*

(2). *Duæ Epistolæ ad celeberrimum virum, Chr. Sever. Longomontanum.* One of these letters contained some remarks on a treatise of Longomontanus, about squaring the circle; and the other a defence of arithmetical geometry.

(3). *Animadversiones Geometricæ:* 4to.

(4). *De eodem Argumento:* 4to.—These two were in the possession of the late Mr. Jones. They both contain a great variety of geometrical propositions, concerning the properties of many figures, with several arithme-

arithmetical computations, relating to the circle, angular sections, &c.—The two following were also in possession of Mr. Jones.

(5). *A treatise of Common Arithmetic;* folio.

(6). *A letter to Mr. Clarke of Gravesend,* dated 25 Feb. 1606; with which he sends him the description of a ruler, called Bedwell's ruler, with directions how to use it.

BRIGGS (WILLIAM), an eminent physician in the latter part of the 17th century, was born at Norwich, for which town his father was four times member of parliament. He studied at the university of Cambridge. He afterwards travelled into France, where he attended the lectures of the famous anatomist Vieussens, at Montpelier. Upon his return he published his *Ophthalmographia,* in 1676. The year following he was made doctor of medicine at Cambridge, and soon after fellow of the college of physicians at London. In 1682 he resigned his fellowship to his brother; and the same year his *Theory of Vision* was published by Hook. The ensuing year he sent to the Royal Society a continuation of that discourse, which was published in their Transactions; and the same year he was appointed physician to St. Thomas's hospital. In 1684 he communicated to the Royal Society two remarkable cases relating to vision, which were likewise printed in their Transactions; and in 1685 he published a Latin version of his *Theory of Vision,* at the desire of Mr. Newton, afterwards Sir Isaac, then professor of mathematics at Cambridge, with a recommendatory epistle from him prefixed to it. He was afterwards made physician in ordinary to king William, and continued in great esteem for his skill in his profession till he died the 4th of September 1704.

BRIGGS's *Logarithms,* that species of them in which 1 is the logarithm of the ratio of 10 to 1, or the logarithm of 10. See LOGARITHMS.

BROKEN *Number,* the same as *Fraction;* which see.

BROKEN *Ray,* or *Ray of Refraction,* in Dioptrics, is the line into which an incident ray is refracted or broken, in crossing the second medium.

BROUNCKER, or BROUNKER, (WILLIAM), lord viscount of Castle Lyons in Ireland, son of Sir William Brounker, afterwards made viscount in 1645, was born about the year 1620. He very early discovered a genius for mathematics, in which he afterwards became very eminent. He was made doctor of physic at Oxford June 23, 1646. In 1657 and 1658, he was engaged in a correspondence by letters on mathematical subjects with Dr. John Wallis, who published them in his *Commercium Epistolicum,* printed 1658, at Oxford. He was one of the nobility and gentry who signed the remarkable declaration concerning king Charles the 2d, published in April 1660.

After the restoration, lord Brounker was made chancellor and keeper of the great seal to the queen consort, one of the commissioners of the navy, and master of St. Katherine's hospital near the tower of London. He was one of those great men who first formed the Royal Society, of which he was by the charter appointed the first president in 1662: which office he held, with great advantage to the Society, and honour to himself, till the anniversary election, Nov. 30, 1677. He died at his house in St. James's street, Westmin-

ster, the 5th of April 1684; and was succeeded in his title by his younger brother Harry, who died in Jan. 1687.

Lord Brounker had several papers inserted in the Philosophical Transactions, the chief of which were, 1. Experiments concerning the Recoiling of Guns.— 2. A Series for the Quadrature of the Hyperbola; which was the first series of the kind upon that subject. —3. Several of his letters to archbishop Usher were also printed in Usher's letters; as well as some to Dr. Wallis, in his Commercium Epistolicum, above mentioned.

BROWN (Sir WILLIAM), a noted physician and miscellaneous writer, of the 18th century. He was settled originally at Lynn in Norfolk, where he published a translation of Dr. Gregory's Elements of Catoptrics and Dioptrics; to which he added, 1. A Method for finding the Foci of all Specula and Lenses universally; as also Magnifying or Lessening a given object by a given Speculum or lens, in any assigned proportion.—2. A Solution of those Problems which Dr. Gregory has left undemonstrated.—3. A particular account of Microscopes and Telescopes, from Mr. Huygens; with the discoveries made by Catoptrics and Dioptrics.

Having acquired a competence by his profession, he removed to Queen's Square, Ormond Street, London, where he resided till his death, in 1774, at 82 years of age; leaving by his will two prize-medals to be annually contended for by the Cambridge poets.

Sir William Brown was a very facetious man; and a great number of his lively essays, both in prose and verse, were printed and circulated among his friends. The active part taken by him in the contest with the licentiates, in 1768, occasioned his being introduced by Mr. Foote in his *Devil upon Two Sticks.*— Upon Foote's exact representation of him with his identical wig and coat, tall figure, and glass stiffly applied to his eye, he sent him a card complimenting him on having so happily represented him; but as he had forgot his muff, he had sent him his own.— This good-natured way of resenting disarmed Foote.— He used to frequent the annual ball at the ladies boarding-school, Queen Square, merely as a neighbour, a good-natured man, and fond of the company of sprightly young folks. A dignitary of the church being there one day to see his daughter dance, and finding this upright figure stationed there, told him he believed he was Hermippus redivivus who lived *anhelitu puellarum.* —When he lived at Lynn, a pamphlet was written against him; which he nailed up against his house-door. —At the age of 80, on St. Luke's day 1771, he came to Batson's coffee-house in his laced coat and band, and fringed white gloves, to shew himself to Mr. Crosby, then lord mayor. A gentleman present observing that he looked very well, he replied, he had neither wife nor debts.

BULLIALD (ISMAEL), an eminent astronomer and mathematician, was born at Laon in the Isle of France in 1605. He travelled in his youth, for the sake of improvement, and gave very early proofs of his astronomical genius; and his riper years rendered him beloved and admired. Riccioli styled him, *Astronomus profundæ indaginis.* He first published his dissertation

tation intitled, *Philolaus, five de vero Syftemate Mundi*; or his true fyftem of the world, according to Philolaus, an ancient philofopher and aftronomer. Afterward, in the year 1645, he fet forth his *Aftronomia Philolaica*, grounded upon the hypothefis of the earth's motion, and the elliptical orbit defcribed by the planet's motion about a cone. To which he added tables intitled, *Tabulæ Philolaicæ*: a work which Riccioli fays ought to be attentively read by all ftudents of aftronomy.—He confidered the hypothefis, or approximation of bifhop Ward, and found it not to agree with the planet Mars; and fhewed in his defence of the Philolaic aftronomy againft the bifhop, that from four obfervations made by Tycho on the planet Mars, that planet in the firft and third quarters of the mean anomaly, was more forward than it ought to be according to Ward's hypothefis; but in the 2d and 4th quadrant of the fame, the planet was not fo far advanced as that hypothefis required. He therefore fet about a correction of the bifhop's hypothefis, and made it to anfwer more exactly to the orbits of the planets, which were moft eccentric, and introduced what is called, by Street in his *Caroline Tables*, the Variation: for thefe tables were calculated from this correction of Bulliald's, and exceeded all in exactnefs that went before. This correction is, in the judgment of Dr. Gregory, a very happy one, if it be not fet above its due place; and be accounted no more than a correction of an approximation to the true fyftem: For by this means we are enabled to gather the coequate anomaly *a priori* and directly from the mean, and the obfervations are well enough anfwered at the fame time; which, in Mercator's opinion, no one had effected before.—It is remarkable that the ellipfis which he has chofen for a planet's motion, is fuch a one as, if cut out of a cone, will have the axis of the cone paffing through one of its foci, viz, that next the aphelion.

In 1657 was publifhed his treatife *De Lineis Spiralibus, Exerc. Geom. & Aftron.* Paris, 4to.—In 1682 came out at Paris, in folio, his large work intitled, *Opus novum ad Arithmeticam Infinitorum*: A work which is a diffufe amplification of Dr. Wallis's Arithmetic of Infinites, and which Wallis treats of particularly in the 80th chapter of his hiftorical treatife of Algebra.—He wrote alfo two Admonitions to Aftronomers. The firft, concerning a new ftar in the neck of the Whale, appearing at fome times, and difappearing at others. The 2d, concerning a nebulous ftar in the northern part of Andromeda's girdle, not difcovered by any of the ancients. This ftar alfo appeared and difappeared by turns. And as thefe phenomena appeared new and furprizing, he ftrongly recommended the obferving them to all that might be curious in aftronomy.

BURNING, the action of fire on fome pabulum or fuel, by which its minute parts are put into a violent motion, and fome of them, affuming the nature of fire, fly off *in orbem*, while the reft are diffipated in vapour or reduced to afhes.

BURNING-*Glafs*, or *Burning-Mirror*, a machine by which the fun's rays are collected into a point; and by that means their force and effect are extremely heightened, fo as to burn objects placed in it.

Burning glaffes are of two kinds, *convex* and *concave*. The convex ones are lenfes, which acting according to the laws of refraction, incline the rays of light towards the axis, and unite them in a point or focus. The concave ones are mirrors or reflectors, whether made of polifhed metal or filvered glafs, and which acting by the laws of reflection, throw the rays back into a point or focus before the glafs.

The ufe of burning glaffes it appears is very ancient, many of the old authors relating fome effects of them. Diodorus Siculus, Lucian, Dion, Zonaras, Galen, Anthemius, Euftatius, Tzetzes, and others, relate that by means of them Archimedes fet fire to the Roman fleet at the fiege of Syracufe. Tzetzes is fo particular in his account of this matter, that his defcription fuggefted to Kircher the method by which it was probably accomplifhed. That author fays that " Archimedes fet fire to Marcellus's navy by means of a burning glafs compofed of fmall fquare mirrors, moving every way upon hinges; which when placed in the fun's rays, directed them upon the Roman fleet, fo as to reduce it to afhes at the diftance of a bow-fhot." And the burning power of reflectors is mentioned in Euclid's Optics, theor. 31. Again, Ariftophanes, in his comedy of The Clouds, introduces Socrates as examining Strepfiades about a method he had difcovered of getting clear of his debts. He replies, that " he thought of making ufe of a burning-glafs which he had hitherto ufed in kindling his fire; for fhould they bring a writ againft me, I'll immediately place my glafs in the fun at fome little diftance from it, and fet it on fire." Pliny and Lactantius have alfo fpoken of glaffes that burn by refraction. The former calls them *balls* or *globes* of *cryftal* or *glafs*, which being expofed to the fun, tranfmit a heat fufficient to fet fire to cloth, or corrode the dead flefh of thofe patients who ftand in need of cauftics; and the latter, after Clemens Alexandrinus, obferves that fire may be kinkled by interpofing glaffes filled with water between the fun and the object, fo as to tranfmit the rays to it.

Among the ancients the moft celebrated burning mirrors were thofe of Archimedes and Proclus; by the former was burnt the fleet of Marcellus, as above mentioned; and by the latter, the navy of Vitellius, befieging Byzantium, according to Zonaras was burnt to afhes.

Among the moderns, the moft remarkable burning-glaffes, are thofe of Magine of 20 inches diameter: of Sepatala of Milan, near 42 inches diameter, and which burnt at the diftance of 15 feet; of Settala of Villette, of Tfchirnhaufen, of Buffon, of Trudaine, and of Parker.

Villette, a French artift at Lyons, made a large mirror, which was bought by Tavernier, and prefented to the king of Pruffia; a fecond, bought by the king of Denmark; a third, prefented to the Royal Academy by the king of France; and a 4th came to England, and was publicly fhewn. This mirror is 47 inches wide, being a fegment of a fphere of 76 inches radius; fo that its focus is about 38 inches from the vertex; and its fubftance is a compofition of tin, copper, and tin-glafs. Some of its effects were as follow:

A filver

		sec.
A silver sixpence melted in	- -	7½
A George the 1st's halfpenny in	-	16
and runs with a hole in	-	34
Tin melts in	- - -	3
Cast iron in	- - - -	16
Slate in	- - -	3
A fossil shell calcines in	- -	7
Piece of Pompey's pillar vitrifies, the black part in		50
the white part in		54
Copper ore in	- - -	8
Bone calcines in 4, and vitrifies in	-	33

An emerald melts into a substance like a torquois stone; a diamond weighing 4 grains loses ⅞ of its weight: the asbestos vitrifies; as all other bodies will do if kept long enough in the focus; but when once vitrified, the mirror can go no farther with them. Philos. Transf. vol. iv. pa. 198.

Tschirnhausen's reflecting mirrors produced equally surprizing effects; as they may be seen described in the Acta Erudit. for 1687, pa. 52. And other persons have made very good ones of wood, straw, paper, ice, and other substances capable of taking a proper form and polish.

Every lens, whether convex, plano-convex, or convexo-convex, collects the sun's rays, dispersed over its convexity, into a point by refraction; and it is therefore a burning-glass. The most considerable of this kind is that made by Tschirnhausen, and described in the same Acta Erudit. The diameters of his lenses are from 3 to 4 feet, having the focus at the distance of 12 feet, and its diameter an inch and a half. To make the focus more vivid, the rays are collected a second time, by a second lens parallel to the first, and placed at such a distance that the diameter of the cone of rays formed by the first lens is equal to the diameter of the second; so that it receives them all; and the focus is reduced from an inch and a half to half the quantity, and consequently its force is quadrupled. This glass vitrifies tiles, slates, pumice-stones &c. in a moment. It melts sulphur, pitch, and all rosins, under water; the ashes of vegetables, woods and other matters, are transmuted into glass; and every thing applied to its focus is either melted, changed into a calx, or into fumes. The author observes that it succeeds best when the matter applied is laid on a hard charcoal well burnt. —But though the force of the solar rays be thus found so surprizing, yet the rays of the full moon, collected by the same burning-glass, do not shew the least increase of heat.

Sir Isaac Newton presented a burning-glass to the Royal Society, consisting of 7 concave glasses, so placed that all their foci join in one physical point. Each glass is about 11½ inches diameter: six of them are placed contiguous to, and round the seventh, forming a kind of spherical segment, whose subtense is about 34½ inches: the common focus is about 22½ inches distant, and about an inch in diameter. This glass vitrifies brick or tile in 1 second, and melts gold in 30 seconds.

M. Buffon also made a variety of very powerful burning-glasses, both as mirrors and as lenses; but at length concluded with one which is probably of the same nature with that of Archimedes, and consisted of 400 mirrors reflecting their rays all to one point, and with which he could melt lead and tin at the distance of 140 feet; and with others he consumed substances at the distance of 210 feet. See Philos. Transf. vol. 44; or Buffon's Histoire Naturelle, Suppl. vol. 1; or Montucla's Histoire des Math. vol. i. pa. 246.

It would seem there is no substance capable of resisting the efficacy of modern burning-glasses; though water &c. are not affected by them at all. Thus, Messrs Macquer and Baumé have succeeded in melting small portions of platina by means of a concave glass, 22 inches diameter, and 28 inches focus; though this metal is not fusible by the strongest fires that can be excited in furnaces, or sustained by any chemical apparatus. Yet it was long since observed, by the Academicians del Cimento, that spirit of wine could not be fired by any burning-glass which they used; and notwithstanding the great improvements these instruments have since received, M. Nollet has not been able, by the most powerful burning mirrors, to set fire to any inflammable liquors whatever.

However, a large burning lens, for fusing and vitrifying such substances as resist the fires of furnaces, and especially for the application of heat in vacuo, and in certain other circumstances in which heat cannot be applied by other means, has long been a desideratum with persons concerned in philosophical experiments: and this it appears is now in a great measure accomplished by Mr. Parker, an ingenious glass manufacturer in Fleet-street, London. His lens is made of flint glass, and is 3 feet in diameter, but when fixed in its frame exposes a surface of 32 inches in the clear; the length of the focus is 6 feet 8 inches, and its diameter one inch. The rays from this large lens are received and transmitted through a smaller, of 13 inches diameter in the clear within the frame, its focal length 29 inches, and diameter of its focus 3-8ths of an inch: so that this second lens increases the power of the former more than 7 times, or as the square of 8 to the square of 3.

From a great number of experiments made with this lens, the following are selected to serve as specimens of its powers:

Substances fused; with their weight, and time of fusion.	Time in sec.	Weight in grs.
Scoria of wrought iron	- -	2	12
Common slate - - -	2	10
Silver, pure - - -	3	20
Platina, pure - - -	3	10
Nickell - - -	3	16
Cast Iron, a cube - -	3	10
Kearsh - - -	3	10
Gold, pure - - -	4	20
Crystal pebble - - -	6	7
Cauk, or terra ponderosa - -	7	10
Lava - - -	7	10
Asbestos - - -	10	10
Bar Iron, a cube - -	12	10
Steel, a cube - - -	12	10
Garnet - - -	17	10
Copper, pure - - -	20	33
Onyx - - -	20	10

Zeolite

SUBSTANCES FUSED, &c.		Time in sec.	Wgt. in grs.
Zeolites	- - - -	23	10
Pumice Stone	- - -	24	10
Oriental Emerald	- - -	25	2
Jasper	- - - -	25	10
White Agate	- - -	30	10
Flint, oriental	- - -	30	10
Topaz, or chrysolite	- - -	45	3
Common Limestone	- - -	55	10
White rhomboidal Spar	- -	60	10
Volcanic Clay	- - -	60	10
Cornish Moorstone	- - -	60	10
Rough Cornelian	- - -	75	10
Rotten Stone	- - -	80	10

BURNING *Zone*, or *Torrid Zone*, the space within 23¼ degrees of the equator, both north and south.

BUSHEL, a measure of capacity for dry goods; as grain, pulse, fruits, &c; containing 4 pecks, or 8 gallons, or ⅛ of a quarter. By act of Parliament, made in 1697, it was ordained that "Every round bushel with a plain and even bottom, being made 18½ inches wide throughout, and 8 inches deep, shall be esteemed a Legal Winchester Bushel, according to the standard in his majesty's exchequer." Now a bushel being thus made will contain 2150·42 cubic inches, and consequently the corn gallon contains only 268·4 cubic inches.

BUTMENTS, are those supporters, or props, by which the feet of arches, or the extremities of bridges are supported; and should be made very strong and firm.

BUTTRESS, is an arch, or a mass of masonry, serving to support the sides of a building, wall, or the like, on the outside. See ARCH, and ARCH-BOUTANT.

C.

CAILLE (NICHOLAS LEWIS DE LA), an eminent French mathematician and astronomer, was born at Rumigny in the diocese of Rheims in 1713. His father having quitted the army, in which he had served, amused himself in his retirement with studying mathematics and mechanics, in which he proved the happy author of several inventions of considerable use to the public. From this example of his father, our author almost in his infancy took a fancy to mechanics, which proved of signal service to him in his maturer years. At school he discovered early tokens of genius. He next came to Paris in 1729; where he studied the classics, philosophy and mathematics. He afterwards studied divinity in the college de Navarre, with the view of embracing the ecclesiastical life: however he never entered into priest's orders, apprehending that his astronomical studies, to which he had become much devoted, might too much interfere with his religious duties. His turn for astronomy soon connected him with the celebrated Cassini, who procured him an apartment in the observatory; where, assisted by the counsels of this master, he soon acquired a name among the astronomers. In 1739 he was joined with M. Cassini de Thury, son to M. Cassini, in verifying the meridian through the whole extent of France: and in the same year he was named professor of mathematics in the college of Mazarine. In 1741 our author was admitted into the Academy of Sciences as an adjoint member for astronomy; and had many excellent papers inserted in their memoirs; beside which he published several useful treatises, viz, Elements of Geometry, Astronomy, Mechanics, and Optics. He also carefully computed all the eclipses of the sun and moon that had happened

since the christian era, which were printed in the work entitled *l'Art de verifier les dates*, &c, Paris, 1750, in 4to. He also compiled a volume of astronomical ephemerides for the years 1745 to 1755; another for the years 1755 to 1765; and a third for the years 1765 to 1775: as also the most correct solar tables of any; and an excellent work entitled *Astronomiæ fundamenta novissimis solis & stellarum observationibus stabilita*.

Having gone through a seven years series of astronomical observations in his own observatory in the Mazarine college, he formed the project of going to observe the southern stars at the Cape of Good Hope: being countenanced by the court, he set out upon this expedition in 1750, and in the space of two years he observed there the places of about 10 thousand stars in the southern hemisphere that are not visible in our latitudes, as well as many other important elements, viz, the parallaxes of the sun, moon, and some of the planets, the obliquity of the ecliptic, the refractions, &c. Having thus executed the purpose of his voyage, and no present opportunity offering for his return, he thought of employing the vacant time in another arduous attempt; no less than that of taking the measure of the earth, as he had already done that of the heavens. This indeed had been done before by different sets of learned men both in Europe and America; some determining the quantity of a degree at the equator, and others at the arctic circle: but it had not as yet been decided whether in the southern parallels of latitude the same dimensions obtained as in the northern. His labours were rewarded with the satisfaction he wished for; having determined a distance of 410814 feet from a place called *Klip-Fontyn* to the

Cape, by means of a bafe of 38802 feet, three times actually meafured : whence he difcovered a new fecret of nature, namely, that the radii of the parallels in fouth latitude are not the fame length as thofe of the correfponding parallels in north latitude. About the 23d degree of fouth latitude he found a degree on the meridian to contain 342222 Paris feet. The court of Verfailles alfo fent him an order to go and fix the fituation of the Ifles of France and of Bourbon. While at the Cape too he obferved a wonderful effect of the atmofphere in fome ftates of it : Although the fky at the Cape be generally pure and ferene, yet when the foutheaft wind blows, which is pretty often, it is attended with fome ftrange and even terrible effects : the ftars look larger, and feem to dance ; the moon has an undulating tremor ; and the planets have a fort of beard like comets.

M. de la Caille returned to France in the autumn of 1754, after an abfence of about 4 years ; loaded, not with the fpoils of the eaft, but with thofe of the fouthern heavens, before then almoft unknown to aftronomers. Upon his return, he firft drew up a reply to fome ftrictures which the celebrated Euler had publifhed relative to the meridian : after which he fettled the refults of the comparifon of his obfervations for the parallaxes, with thofe of other aftronomers : that of the fun he fixed at 9½″ ; of the moon at 56′56″ ; of Mars in his oppofition, 36″ ; of Venus 38″. He alfo fettled the laws by which aftronomical refractions are varied by the different denfity or rarity of the air, by heat or cold, and by drynefs or moifture. And laftly he fhewed an eafy and practicable method of finding the longitude at fea by means of the moon. His fame being now celebrated every where, M. de la Caille was foon elected a member of moft of the academies and focieties of Europe, as London, Bologna, Peterfburgh, Berlin, Stockholm, and Gottingen.

In 1760 our author was attacked with a fevere fit of the gout ; which however did not interrupt the courfe of his ftudies ; for he then planned out a new and large work, no lefs than a hiftory of aftronomy through all ages, with a comparifon of the ancient and modern obfervations, and the conftruction and ufe of the inftruments employed in making them. Towards the latter part of 1761, his conftitution became greatly reduced ; though his mind remained unaffected, and he refolutely perfifted in his ftudies to the laft ; death only putting an end to his labours the 21ft of March 1762, at 49 years of age ; after having committed his manufcripts to the care and difcretion of his efteemed friend M. Maraldi.

Befide the publications before mentioned, and perhaps fome others alfo, he had a vaft number inferted in the volumes of the Memoirs of the French Academy of Sciences, much too numerous indeed, though very important, to be here all mentioned particularly ; fuffice it therefore juft to diftinguifh the years of thofe volumes in which his pieces are to be found, by the following lift of them, viz, 1741, 1742, 1743, 1744, 1745, 1746, 1747, 1748, 1749, 1750, 1751, 1752, 1753, 1754, 1755, 1756, 1757, 1758, 1759, 1760, 1761, 1763 ; in all or moft of which years there are two or three or more of his papers.

CAISSON, in Architecture, a kind of cheft or flat-bottomed boat, in which the pier of a bridge is built, then funk to the bottom of the water, and the fides loofened and taken off from the bottom, by a contrivance for that purpofe ; the bottom of the caiffon being left under the pier as a foundation to it. The caiffon is kept afloat till the pier is built to above the height of low-water mark ; and for that purpofe, its fides are either made of more than that height at firft, or elfe gradually raifed to it as it finks by the weight of the work, fo as always to keep its top above water. Mr. Labelye tells us, that the caiffons in which he built fome of the piers of Weftminfter bridge, contained above 150 load of fir timber, of 40 cubic feet each, and that it was of more tonnage or capacity than a 40 gun fhip of war.

CAISSON, in Military Affairs, is fometimes ufed for a cheft ; and in particular for a bomb or fhell cheft, and is ufed as a fuperficial mine, or fourneau. This is done by filling a cheft either with gun powder and loaded fhells, or elfe with fhells alone, and burying it in a fpot where an enemy, befieging a place, is expected to come, and then firing it by a train to blow the men up.

CALCULATION, the act of computing feveral fums, by adding, fubtracting, multiplying, dividing, &c. From *calculus*, in allufion to the practice of the ancients, who ufed *calculi*, or little ftones, in making computations, in taking fuffrages, and in keeping accounts, &c ; as we now ufe counters, figures, &c. Calculation is more particularly ufed to fignify the computations in aftronomy, trigonometry, &c, for making tables of aftronomy, of logarithms, ephemerides, finding the times of eclipfes, and fuch like.

CALCULATOR, a perfon who makes or performs calculations.—It is alfo the name given by Mr. Fergufon to a machine in the fhape of an orrery, which he conftructed for exhibiting the motions of the earth and moon, and refolving a variety of aftronomical problems. See his Aftron. 4to pa. 265, or 8vo pa. 393.

CALCULATORES, were anciently accountants who reckoned their fums by *calculi*, or little ftones, or counters.—In ancient canons too we find a fort of diviners or enchanters, cenfured under the denomination of *calculatores*.

CALCULUS denotes primarily a fmall ftone, pebble, or counter, ufed by the ancients in making calculations or computations, taking of fuffrages, playing at tables, and the like.

CALCULUS denotes now a certain way of performing mathematical inveftigations and refolutions. Thus, we fay the Arithmetical or Numeral Calculus, the Algebraical Calculus, the Differential Calculus, the Exponential Calculus, the Fluxional Calculus, the Integral Calculus, the Literal or Symbolical Calculus, &c ; for which, fee each refpective word.

Arithmetical or *Numeral* CALCULUS, is the method of performing arithmetical computations by numbers. See ARITHMETIC, and NUMBER.

Algebraical, *Literal*, or *Symbolical* CALCULUS, is the method of performing algebraical calculations by letters or other fymbols. See ALGEBRA.

Differential CALCULUS, is the arithmetic of the indefinitely fmall differences of variable quantities ; a mode of computation much ufed by foreign mathematicians, and introduced by Leibnitz, as fimilar to the method

method of Fluxions of Newton. See DIFFERENTIAL &c.

Exponential CALCULUS, is the applying the fluxional or differential methods to exponential quantities; such as a^x, or x^x, or ay^x, &c. See EXPONENTIAL.

Fluxional CALCULUS, is the method of fluxions, invented by Newton. See FLUXIONS.

Integral CALCULUS, or *Summatorius*, is a method of integrating, or summing up differential quentities; and is similar to the finding of fluents. See INTEGRAL and FLUENT.

CALCULUS *Literalis*, or *Literal Calculus*, is the same with algebra, or specious arithmetic, so called from its using the letters of the alphabet; in contradistinction to numeral arithmetic, in which figures are used.—

CALENDAR, or KALENDAR, a distribution of time as accommodated to the uses of life; or an Almanac, or table, containing the order of days, weeks, months, feasts, &c, occurring in the course of the year: being so called from the word *Calendæ*, which among the Romans denoted the first days of every month, and anciently was written in large characters at the head of each month. See ALMANAC, CALENDS, MONTH, TIME, YEAR, &c.

In Calendars the days were originally divided into octoades, or eights; but afterwards, in imitation of the Jews, they were divided into hebdomades, or sevens, for what we now call a week: which custom, Scaliger observes, was not in use among the Romans till after the time of Theodosius.

Divers calendars are established in different countries, according to the different forms of the year, and distributions of time: As the Persian, the Roman, the Jewish, the Julian, the Gregorian, &c, calendars. — The ancient Roman Calendar is given by Ricciolus, Struvius, Danet, and others; in which we perceive the order and number of the Roman holy-days and work-days.—The Jewish calendar was fixed by Rabbi Hillel, about the year 360; from which time the days of their year may be reduced to those of the Julian calendar.—The three Christian calendars are given by Wolfius in his Elements of Chronology; as also the Jewish and Mohamedan calendars. Other writers on the calendars are Vieta, Clavius, Scaliger, Blondel, &c.

The Roman CALENDAR was first formed by Romulus, who distributed time into several periods for the use of his followers and people. He divided the year into 10 months, of 304 days; beginning on the first of March, and ending with December.

Numa reformed the calendar of Romulus. He added the months of January and February, making it to commence on the first of January, and to consist of 355 days. But as this was evidently deficient of the true year, he ordered an intercalation of 45 days to be made every 4 years, in this manner, viz, Every 2 years an additional month of 22 days, between February and March; and at the end of each two years more, another month of 23 days; the month thus interposed, being called Marcedonius, or the intercalary February.

Julius Cæsar, with the aid of Sosigenes, a celebrated astronomer of those times, farther reformed the Roman calendar, from whence arose the Julian calendar, and the Julian or old style. Finding that the sun performed his annual course in 365 days and a quarter nearly, he divided the year into 365 days, but every 4th year 366 days, adding a day that year before the 24th of February, which being the 6th of the calends, and being thus reckoned twice, gave occasion to the name *bissextile*, or what we also call leap-year.

This calendar was farther reformed by order of the pope Gregory XIII, from whence arose the term Gregorian calendar and style, or what we also call the new style, which is now observed by almost all European nations. The year of Julius was too long by nearly 11 minutes, which amounts to about 3 days in 400 years; the pope therefore, by the advice of Clavius and Ciaconius, ordained that there should be omitted a day in every 3 centuries out of 4; so that every century, which would otherwise be a bissextile year, is made to be only a common year, excepting only such centuries as are exactly divisible by 4, which happens once in 4 centuries. See BISSEXTILE. This reformation of the calendar, or the new style, as we call it, commenced in the countries under the popish influence, on the 4th of October 1582, when 10 days were omitted at once, which had been over-run since the time of the council of Nice, in the year 325, by the surplus of 11 minutes each year. But in England it only commenced in 1752, when 11 days were omitted at once, the 3d of September being accounted the 14th that year; as the surplus minutes had then amounted to 11 days.

Julian Christian CALENDAR, is that in which the days of the week are determined by the letters A, B, C, D, E, F, G, by means of the solar cycle; and the new and full moons, particularly the paschal full moon, with the feast of Easter, and the other moveable feasts depending upon it, by means of golden numbers, or lunar cycles, rightly disposed through the Julian year. See CYCLE, and GOLDEN NUMBER.

In this calendar, it is supposed that the vernal equinox is fixed to the 21st day of March; and that the golden numbers, or cycles of 19 years, constantly indicate the places of the new and full moons; though both are erroneous; and from hence arose a great irregularity in the time of Easter.

Gregorian CALENDAR, is that which, by means of Epacts, rightly disposed through the several months, determines the new and full moons, with the time of Easter, and the moveable feasts depending upon it, in the Gregorian year. This differs therefore from the Julian calendar, both in the form of the year, and in as much as epacts are substituted instead of golden numbers. See EPACT.

Though the Gregorian calendar be more accurate than the Julian, yet it is not without imperfections, as Scaliger and Calvisius have fully shewn; nor is it perhaps possible to devise any one that shall be quite perfect. Yet the Reformed Calendar, and that which is ordered to be observed in England, by act of Parliament made the 24th of George II, come very near to the point of accuracy: For, by that act it is ordered that " Easter-day, on which the rest depend, is always the first Sunday after the full moon, which happens upon, or next after the 21st day of March; and if the full-moon happens upon a Sunday, Easter-day is the Sunday after."

Reformed, or *Corrected*, CALENDAR is that which, rejecting

rejecting all the apparatus of golden numbers, epacts, and dominical letters, determines the equinox, and the paschal full-moon, with the moveable feasts depending upon it, by computation from astronomical tables. This calendar was introduced among the protestant states of Germany in the year 1700, when 11 days were omitted in the month of February, to make the corrected style agree with the Gregorian. This alteration in the form of the year, they admitted for a time: in expectation that, the true quantity of the tropical year being at length more accurately determined by observation, the Romanists would agree with them on some more convenient intercalation.

French New CALENDAR, is a quite new form of calendar that commenced in France on the 22d of September 1792. At the time of printing this (viz, in July 1794), it does not certainly appear whether this new calendar will be made permanent or not; but merely as a curiosity in the science of chronology, a very brief notice of it may here be added, as follows.

The year, in this calendar, commences at midnight the beginning of that day in which falls the true autumnal equinox for the observatory of Paris. The year is divided into 12 equal months, of 30 days each; after which 5 supplementary days are added, to complete the 365 days of the ordinary year: these 5 days do not belong to any month. Each month is divided into three decades of 10 days each; distinguished by 1st, 2d, and 3d decade. All these are named according to the order of the natural numbers, viz, the 1st, 2d, 3d, &c, month, or day of the decade, or of the supplementary days. The years which receive an intercalary day, when the position of the equinox requires it, which we call embolismic or bissextile, they call olimpic; and the period of four years, ending with an olimpic year, is called an olimpiade; the intercalary day being placed after the ordinary five supplementary days, and making the last day of the olimpic year Each day, from midnight to midnight, is divided into 10 parts, each part into 10 others, and so on to the last measurable portion of time.

In this calendar too the months and days of them have new names. The first three months of the year, of which the autumn is composed, take their etymology, the first from the vintage which takes place from September to October, and is called *vendemaire;* the second, *brumaire,* from the mists and low fogs, which shew as it were the transudation of nature from October to November: the third, *frimaire,* from the cold, sometimes dry and sometimes moist, which is felt from November to December. The three winter months take their etymology, the first, *nivose,* from the snow which whitens the earth from December to January; the second, *pluviose,* from the rains which usually fall in greater abundance from January to February; the third, *ventose,* from the wind which dries the earth from February to March. The three spring months take their etymology, the first, *germinal,* from the fermentation and development of the sap from March to April; the second, *floreal,* from the blowing of the flowers from April to May: the third, *prairial,* from the smiling fecundity of the meadow crops from May to June. Lastly, the three summer months take their

etymology, the first, *messidor,* from the appearance of the waving ears of corn and the golden harvests which cover the fields from June to July; the second, *thermidor,* from the heat, at once solar and terrestrial, which inflames the air from July to August; the third, *fructidor,* from the fruits gilt and ripened by the sun from August to September. Thus, the whole 12 months are,

AUTUMN.	SPRING.
Vendemaire	Germinal
Brumaire	Floreal
Frimaire.	Prairial.
WINTER.	SUMMER.
Nivose	Messidor
Pluviose	Thermidor
Ventose.	Fructidor.

From these denominations it follows, that by the mere pronunciation of the name of the month, every one readily perceives three things and all their relations, viz, the kind of season, the temperature, and the state of vegetation: for instance, in the word *germinal,* his imagination will easily conceive, by the termination of the word, that the spring commences; by the construction of the word, that the elementary agents are busied; and by the signification of the word, that the buds unfold themselves.

As to the names of the days of the week, or decade of 10 days each, which they have adopted instead of seven, as these bear the stamp of judicial astrology and heathen mythology, they are simply called from the first ten numbers; thus,

Primdi	Sextidi
Duodi	Septidi
Tridi	Octidi
Quartidi	Nonidi
Quintidi	Decadi.

In the almanac, or annual calendar, instead of the multitude of saints, one for each day of the year, as in the popish calendars, they annex to every day the name of some animal, or utensil, or work, or fruit, or flower, or vegetable, &c, appropriate and most proper to the times.

Astronomical CALENDAR, an instrument engraven upon copper-plates, printed on paper, and pasted on board, with a brass slider which carries a hair, and shews by inspection, the sun's meridian, altitude, right ascension, declination, rising, setting, amplitude, &c, to a greater exactness than can be shewn by the common globes.

CALENDS, *Calendæ,* in the Roman Chronology, denoted the first days of each month; being so named from καλεω, *calo, I call,* or *proclaim;* because that, before the publication of the Roman *Fasti,* and counting their months by the motion of the moon, a priest was appointed to observe the first appearance of the new moon; who, having seen her, gave notice to the president of the sacrifices to offer one; and calling the people together, he proclaimed unto them how they should reckon the days until the nones; pronouncing the word *Caleo* 5 times if the nones should happen on the 5th day, or seven times if they happened on the 7th day of the month.

The calends were reckoned backwards, or in a retrograde

trograde order: thus, for example, the first of May being the calends of May; the last or 30th day of April, was the *pridie calendarum*, or 2d of the calends of May; the 29th of April, the 3d of the calends, or before the calends: and so back to the 13th, where the ides commence; which are likewise numbered backwards to the 5th, where the nones begin; which are also reckoned after the same manner to the first day of the month, which is the calends of April.

Hence comes this rule to find the day of the calends answering to any day of the month, viz, Consider how many days of the month are yet remaining after the day proposed, and to that number add 2, for the number of or from the calends. For example, suppose it were the 23d day of April, it would then be the 9th of the caledns of May: for April containing 30 days, from which 23 being taken, there remains 7; to which 2 being added, makes the sum 9. And the reason for this addition of the constant number 2, is because the last day of the month is called the 2d of the calends of the month following.

CALIBER, or CALIPER, is the thickness or diameter of a round body, particularly the bore or width of a piece of ordnance, or that of its ball.

CALIBER *Compasses*, or CALIPER-*Compasses*, or simple CALIPERS, a sort of compasses made with bowed or arched legs, the better to take the diameter of any round body; as the diameters of balls, or the bores of guns; or the diameter, and even the length of casks, and such like. The best sort of calipers usually contain the following articles, viz, 1st, the measure of convex diameters in inches &c; 2d, of concave diameters: 3d, the weight of iron shot of given diameters; 4th, the weight of iron shot for given gun bores; 5th, the degrees of a semicircle; 6th, the proportion of troy and averdupois weight; 7th, the proportion of English and French feet and pounds weight; 8th, factors used in circular and spherical figures; 9th, tables of the specific gravities and weights of bodies; 10th, tables of the quantity of powder necessary for the proof and service of brass and iron guns; 11th, rules for computing the number of shot or shells in a complete pile; 12th, rules for the fall or descent of heavy bodies; 13th, rules for the raising of water; 14th, rules for firing artillery and mortars; 15th, a line of inches; 16th, logarithmic scales of numbers, sines, versed sines, and tangents; 17th, a sectoral line of equal parts, or the line of lines; 18th, a sectoral line of planes and superficies; and 19th, a sectoral line of solids.

CALIPPIC *Period*, in Chronology, a period of 76 years, continually recurring; at every repetition of which, it was supposed, by its inventor Calippus, an Athenian astronomer, that the mean new and full moons would always return to the same day and hour.

About a century before, the golden number, or cycle of 19 years, had been invented by Meton, which Calippus finding to contain 19 of Nabonassar's year, 4 days and $\frac{1}{76}$, to avoid fractions he quadrupled it, and so produced his period of 76 years, or 4 times 19; after which he supposed all the lunations &c would regularly return to the same hour. But neither is this exact, as it brings them too late by a whole day in 225 years.

CAMBER-BEAM, a piece of timber cut arch-wise, or with an obtuse angle in the middle. They are commonly used in platforms, as for church-roofs, and other occasions where long timbers are wanted to lie at a small slope. A camber-beam is much stronger than another of the same dimensions; for being laid with the hollow side downwards, and having good butments at the ends, they serve for a kind of arch.

CAMELEON, one of the constellations of the southern hemisphere, near the south pole, and invisible in our latitude. There are 10 stars marked in this constellation in Sharp's catalogue.

CAMELOPARDALUS, a new constellation of the northern hemisphere, formed by Hevelius, consisting of 32 stars first observed by him. It is situated between Cepheus, Cassiopeia, Perseus, the Two Bears, and Draco; and it contains 58 stars in the British catalogue.

CAMERA *Æolia*, a name given by Kircher to a contrivance for blowing the fire, for the fusion of ores, without bellows. This is effected by means of water falling through a funnel into a close vessel, which sends from it so much air or vapour, as continually blows the fire. See Hook's Philos. Coll. n° 3, pa. 80.

CAMERA *Lucida*, a contrivance of Dr. Hook to make the image of any thing appear on a wall in a light room, either by day or night. See Philos. Trans. n° 38, pa. 741.

CAMERA *Obscura*, or *Dark Chamber*, an optical machine or apparatus, representing an artificial eye, by which the images of external objects, received through a double convex glass, are shewn distinctly, and in their native colours, on a white ground placed within the machine, in the focus of the glass. The first invention of the camera obscura is ascribed to John Baptista Porta. See his *Magia Naturalis*, *lib*. 17, *cap*. 6, where he largely describes the effects of it. See also the end of s'Gravesande's Perspective, and other optical writers, for the construction and uses of various sorts of camera obscuras.

This machine serves for many useful and entertaining purposes. For example, it is very useful in explaining the nature of vision, representing a kind of artificial eye: it exhibits very diverting sights or spectacles; shewing images perfectly like their objects, clothed in their natural colours, but more intense and vivid, and at the same time accompanied with all their motions; an advantage which no art can imitate: and by this instrument, a person unacquainted with painting, or drawing, may delineate objects with the greatest accuracy of drawing and colouring.

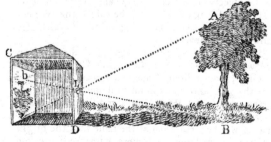

Theory of the Camera Obscura. The theory and

3   principle

principle of this instrument may be thus explained. If any object AB radiate through a small aperture L, upon a white ground opposite to it, within a darkened room, or box, &c; the image of the object will be painted on that ground in an inverted situation. For, by the smallness of the aperture, the rays from the object will cross each other there, the image of the point A being at a, and that of B at b; so that the whole object AB will appear inverted, as at ab. And as the corresponding rays make equal angles on both sides of the aperture, if the ground be parallel to the object, their heights will be to each other directly as their distances from the aperture.

Construction of a Camera Obscura, by which the images of external objects shall be represented distinctly, and in their genuine colours. 1st, Darken a chamber that has one of its windows looking towards a place containing various objects to be viewed; leaving only a small aperture open in one shutter. 2d, In this aperture fit a proper lens, either plano-convex, or convex on both sides; the convexity forming a small portion of a large sphere. But note, that if the aperture be made very small, as of the size of a pea, the objects will be represented even without any lens at all. 3d, At a proper distance to be determined by trials, stretch a per or white cloth, unless there be a white wall at that distance, to receive the images of the objects: or the best way is to have some plaister of Paris cast on a convex mould, so as to form a concave smooth surface, and of a curvature and size adapted to the lens, to be placed occasionally at the proper distance. 4th, If it be rather desired to have the objects appear erect, instead of inverted, this may be done either by placing a concave lens between the centre and the focus of the first lens; or by reflecting the image from a plane speculum inclined to the horizon in an angle of 45 degrees; or by having two lenses included in a draw-tube, instead of one.

That the images be clear and distinct, it is necessary that the objects be illuminated by the sun's light shining upon them from the opposite quarter: so that, in a western prospect the images will be best seen in a forenoon, an eastern prospect the afternoon, and a northern prospect about noon; a southern aspect is the least eligible of any. But the best way of any is, if the lens be fixed in a proper frame, on the top of a building, and made to move easily round in all directions, by a handle extended to the person who manages the instrument; the images being then thrown down into a dark room immediately below it, upon a horizontal round plaister of Paris ground: for thus a view of all the objects quite around may easily be taken in the space of a few minutes; as is the case of the excellent camera obscura placed on the top of the Royal Observatory at Greenwich.

The objects will be seen brighter, if the spectator first wait a few minutes in the dark. Care should also be taken, that no light escape through any chinks; and that the ground be not too much illuminated. It may further be observed, that the greater distance there is between the aperture and the ground, the larger the images will be; but then at the same time the brightness is weakened more and more with the increase of distance.

To construct a Portable Camera Obscura. 1st, Provide a small box or chest of dry wood, and of about 10 inches broad, and 2 feet long or more, according to the size of the lenses. 2d, In one side of it, as BD, fit a sliding tube EF with two lenses; or, to have the image at a less distance from the tube, with three lenses, convex on both sides; the diameter of the two outer ones to be about 7 inches, but that of the inner to be less, as 4¾ or 5 inches. 3d, At a proper distance, within the box, set up perpendicularly an oiled paper GH, so that images thrown upon it may be seen through. 4th, In the opposite side, at I, make a round hole, for a person to look conveniently through with both eyes. Then if the tube be turned towards the objects, and the lenses be placed by trials at the proper distance, by sliding the tube in and out, the objects will be seen delineated on the paper, erect as before.

The machine may be better accommodated for drawing, by placing a mirror to pass from G to C; for this will reflect the image upon a rough glass plane, or an oiled paper, placed horizontally at AB; and a copy of it may there be sketched out with a black-lead pencil.

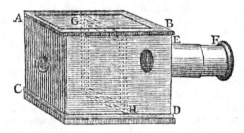

Another Portable Camera Obscura is thus made. 1st. On the top of a box or chest raise a little turret HI, open towards the object AB. 2d, Behind the aperture, incline a small mirror ab at an angle of 45 degrees, to reflect the rays Aa and Bb upon a lens G convex on both sides, and included in a tube GL. Or the lens may be fixed in the aperture. 3d, At the dis-

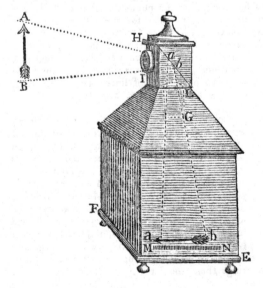

tance of the focus place a table, or board EF, covered with

with a white paper, to receive the image *ab*. Laftly, In MN make an oblong aperture to look through ; and an opening may alfo be made in the fide of the box, for the convenience of drawing.

This fort of camera is eafily changed into a fhow-box, for viewing prints, &c : placing the print at the bottom of the box, with its upper part inwards, where it is enlightened through the front, left open for this purpofe, either by day or candle-light ; and the print may be viewed through the aperture in HI.—A variety of contrivances for this purpofe may be feen defcribed in Harris's Optics, b. ii. fect. 4.— Mr. Storer has alfo procured a patent for an inftrument of this fort, which he calls a delineator ; being formed of two double convex lenfes and a plane mirror, fitted into a proper box. One lens is placed clofe to the mirror, making with it an angle of 45 degrees ; the other being placed at right angles to the former, and fixed in a moveable tube. If the moveable lens be directed towards the object, which is to be viewed or copied, and moved nearer to or farther from the mirror, till the image is diftinctly formed on a greyed glafs, laid upon that furface of the upper lens which is next the eye, it will be found more fharp and vivid than thofe formed in the common inftruments ; becaufe the image is taken up fo near the upper lens. And by increafing the diameter and curvature of the lenfes, the effect will be much heightened.

CAMUS, *(Charles-Stephen-Lewis)*, a celebrated French mathematician, Examiner of the royal Schools of Artillery and Engineers, Secretary and Profeffor of the Royal Academy of Architecture, Honorary member of that of the Marine, and fellow of the Royal Society of London, was born at Creffy en Brie, the 25th of Auguft 1699. His early ingenuity in mechanics and his own intreaties induced his parents to fend him to ftudy at a college in Paris, at 10 years of age ; where in the fpace of two years his progrefs was fo great, that he was able to give leffons in mathematics, and thus to defray his own expences at the college, without any farther charge to his parents. By the affiftance of the celebrated Varignon, young Camus foon ran through the courfe of the higher mathematics, and acquired a name among the learned. He made himfelf more particularly known to the Academy of Sciences in 1727, by his memoir upon the fubject of the prize which they had propofed for that year, viz, ' To determine the moft advantageous way of mafting fhips ;' in confequence of which he was named that year Adjoint-Mechanician to the Academy ; and in 1730 he was appointed profeffor of Architecture. In lefs than three years after, he was honoured with the fecretaryfhip of the fame ; and the 18th of April 1733, he obtained the degree of Affociate in the Academy, where he diftinguifhed himfelf greatly by his memoirs upon living forces, or bodies in motion acted upon by forces, on the figure of the teeth of wheels and pinions, on pump work, and feveral other ingenious memoirs.

In 1736 he was fent, in company with Meffieurs Clairaut, Maupertuis, and Monnier, upon the celebrated expedition to meafure a degree at the north polar circle ; in which he rendered himfelf highly ufeful, not only as a mathematician, but alfo as a mechanician and an artift, branches for which he had a remarkable talent.

In 1741 Camus had the honour to be appointed Penfioner-Geometrician in the Academy ; and the fame year he invented a gauging-rod and fliding-rule proper at once to gauge all forts of cafks, and to calculate their contents. About the year 1747 he was named Examiner of the Schools of Artillery and Engineers : and, in 1756, one of the eight mathematicians appointed to examine by a new meafurement, the bafe which had formerly been meafured by Picard, between Villejuifve and Juvifi ; an operation in which his ingenuity and exactnefs were of great utility. In 1765 M. Camus was elected a fellow of the Royal Society of London ; and died the 4th of May 1768, in the 69th year of his age ; being fucceeded by the celebrated d'Alembert in his office of Geometrician in the French Academy ; and leaving behind him a great number of manufcript treatifes on various branches of the mathematics.

The works publifhed by M. Camus, are :

1. Courfe of Mathematics for the ufe of the Engineers, 4 vols. in 8vo.

2. Elements of Mechanics.

3. Elements of Arithmetic.

And his memoirs printed in the volumes of the Academy, are :

1. Of Accelerated Motions by living forces: vol. for 1728.

2. Solution of a Geometrical Problem of M. Cramer : 1732.

3. On the figure of the teeth and pinions in clocks : 1733.

4. On the action of a Mufket ball, piercing a pretty thick piece of wood, without communicating any confiderable velocity to it : 1738.

5. On the beft manner of employing buckets for raifing water : 1739.

6. A Problem in Statics : 1740.

7. On an Inftrument for gauging of veffels : 1741.

8. On the Standard of the Ell Meafure : 1746.

9. On the Tangents of Points common to feveral branches of the fame curve : 1747.

10. On the Operations in meafuring the diftance between the centres of the pyramids of Villejuive and Juvify, to difcover the beft meafure of the degree about Paris : 1754.

11. On the Mafting of Ships : Prize Tom. 2.

12. The Manner of working Oars : Mach. tom. 2.

13. A Machine for moving many Colters at once : Mach. tom. 2.

CANCER, the *Crab*, one of the twelve figns of the zodiac, ufually drawn on the globe in the form of a crab, and in books of Aftronomy denoted by a character refembling the number fixty-nine, turned fideways, thus ♋.

This is one of the 48 old conftellations ; and, from the hieroglyphic mode of writing among the Egyptians &c, it is probable that they gave the name and figure to this conftellation from the following circumftance, viz, that as the crab is an animal that goes fideling backwards, fo the fun, in his annual courfe through the zodiac, when he arrives at this part of the ecliptic, having reached his utmoft limit northwards, begins there to return back again towards the fouth. But the Greeks, who adapted fome fable of their own to every thing of this kind, pretend that when Hercules was fighting with the Lernæan hydra, there was a crab upon the marsh which seized his foot. The hero crushed the reptile to pieces under his heel; but Juno, in gratitude for the offered service, little as it was, raised the creature into the heavens.

The number of stars in the sign cancer, Ptolemy makes 13, Tycho 15, Bayer and Hevelius 29, and Flamsteed 83.

Tropic of CANCER, a little circle of the sphere parallel to the equinoctial, and passing through the beginning of the sign cancer.

CANDLEMAS, or the *Purification*, a feast of the church, held on the 2d of February, in memory of the purification of the Virgin; taking its name of Candlemas, either from the number of lighted candles used by the Romish church, in the processions of this day, or because that the church then consecrated candles for the whole year.

CANES *Venatici*, the *Hounds*, or the *Greyhounds*, one of the new constellations of the northern hemisphere, which Hevelius has formed out of the unformed stars of the old catalogues. These two dogs are farther distinguished by the names of *asterion* and *chara*. They contain 23 stars according to Hevelius, but 25 in the British catalogue.

CANICULA, a name given by many of the earlier astronomers to the constellation which we call the Lesser Dog, and Canis Minor, but some Procyon and Antecanis. See CANIS MINOR.

It is also used for one of the stars of the constellation Canis Major; called also simply the Dog-star; and by the Greeks Σειριος, Sirius. It is situated in the mouth of the constellation, and is the largest and brightest of all the stars in the heavens. From the heliacal rising of this star, that is, its emersion from the sun's rays, which now happens with us about the 11th of August, the ancients reckoned their *dies caniculares*, or dog-days.

The Egyptians and Ethiopians began their year at the heliacal rising of Canicula; reckoning to its rise again the next year, which is called the *Annus Canarius*.

CANICULAR *Days*, or *Dog-days*, denote a certain number of days, before and after the heliacal rising of canicula, or the dog-star, in the morning. The ancients imagined that this star, so rising, occasioned the sultry weather usually felt in the latter part of the summer, or dog-days; with all the distempers of that sickly season: Homer's Il. lib 5, v. 10, and Virgil's Æn. lib. 10, v. 270. Some authors say, from Hippocrates and Pliny, that the day this star first rises in the morning, the sea boils, wine turns sour, dogs begin to grow mad, the bile increases and irritates, and all animals grow languid; also that the diseases it usually occasions in men, are burning fevers, dysenteries, and phrensies. The Romans too sacrificed a brown dog every year to Canicula at his first rising, to appease its rage. All this however arose from a groundless idea that the dog-star, so rising, was the occasion of the extreme heat and the diseases of that season; for the star not only varies in its rising, in any one year, as the latitude varies, but it is always later and later every year in all latitudes; so that in time the star may, by the same rule, come to be charged with bringing frost and snow, when he comes to rise in winter.

The dog-days were commonly counted for about 40 days, viz, 20 days before and 20 days after the heliacal rising; and almanac-makers have usually set down the dog-days in their almanacs to the changing time of the star's rising, by which means they had at length fallen considerably after the hottest time of the year, till of late we have observed an alteration of them in the almanacs, and very properly, from July 3 to August 11. For, by the dog-days, the ancients meant to express the hottest time of the year, which is commonly during the month of July, about which month the dog-star rose heliacally in the time of the most ancient astronomers that we know of: but the precession of the equinoxes has carried this heliacal rising into a much later and cooler part of the year; and because Hesiod tells us that the hot time of the year ends on the 50th day after the summer solstice, which brings us to about August 10 or 11, therefore the above alteration seems to be very proper.

CANICULAR *Year*, denotes the Egyptian natural year, which was computed from one heliacal rising of canicula, to the next. This year was also called *annus canarius*, and *annus cynicus*; and by the Egyptians themselves the *Sethic year*, from *Seth*, by which name they called Sirius. Some call it also the *heliacal year*. This year consisted ordinarily of 365 days, and every 4th year of 366; by which means it was accommodated to the civil year, like the Julian account. And the reason why they chose this star, in preference to others, to compute their time by, was not only the superior brightness of that star, but because that in Egypt its heliacal rising was a time of very singular note, as coinciding with the greatest augmentation of the Nile, the reputed father of Egypt. Ephestion adds, that from the aspect of canicula, its colour &c, the Egyptians drew prognostics concerning the rise of the Nile; and, according to Florus, predicted the future state of the year. So that it is no wonder the first rising of this star was observed with great attention. Bainbrigge, *Canicul.* cap. 4. p. 26.

CANIS *Major*, the *Great Dog*, a constellation of the southern hemisphere, below the feet of Orion, and one of the old 48 constellations. The Greeks, as usual, have many fables of their own about the exaltation of the dog into the skies; but the origin of this constellation, as well as its other name Sirius, lies more probably among the Egyptians, who carefully watched the rising of this star, and by it judged of the swelling of the Nile, calling the star the sentinel and watch of the year; and hence, according to their manner of hieroglyphic writing represented it under the figure of a dog. They also called the Nile *Siris*; and hence their *Osiris*.

The stars in this constellation, Ptolemy makes 29; Tycho however observed only 13, and Hevelius 21; but in Flamsteed's catalogue they are 31.

CANIS *Minor*, a constellation of the northern hemisphere, just below Gemini, and is one of the 48 old constellations. The Greeks fabled that this is one of Orion's hounds; but the Egyptians were most probably the inventors of this constellation, and they may have given it this figure to express a little dog, or watchful creature, going before as leading in the larger, or rising before it: and hence the Latins have called it Antecanis, the star before the dog.

The

The stars in this conftellation are, in Ptolomy's catalogue 2, the principal of which is the ftar Procyon; in Tycho's 5, in Hevelius's 13; and in Flamfteed's 14.

CANNON, in Military Affairs, a long round hollow engine, made of iron or brafs, &c, for throwing balls, &c, by means of gunpowder. The length is diftinguifhed into three parts; the firft reinforce, the fecond reinforce, and the chafe: the infide hollow where the charge is lodged, being alfo called the chafe, or bore. But for the feveral parts and members of a cannon, fee ASTRAGAL, BASE-RING, BORE, BREECH, CASCABEL, CHASE, MUZZLE, OGEE, REINFORCE-RING, TRUNNIONS, &c. See alfo GUN, and GUNNERY.

Cannon were firft made of feveral bars of iron adapted to each other lengthways, and hooped together with ftrong iron rings. They were employed in throwing ftones and metal of feveral hundred weight. Others were made of thin fheets of iron rolled up, and hooped: and on emergencies they have been even made of leather, with plates of iron or copper. They are now made of caft iron or brafs; being caft folid, and the tube bored out of the middle of the folid metal.

Larrey makes brafs cannon the invention of J. Owen; and afferts that the firft known in England, were in 1535; and farther that iron cannon were firft caft here in 1547. He acknowledged that cannon were known before; and remarks that at the battle of Creffi, in 1346, there were 5 pieces of cannon in the Englifh army, which were the firft ever feen in France. Mezeray alfo obferves that king Edward ftruck terror into the French army, by 5 or 6 pieces of cannon; it being the firft time they had met fuch thundering machines.

In the lift of aids raifed for the redemption of king John of France, in 1368, mention is made of an officer in the French army called *mafter of the king's cannon*, and of his providing 4 large cannon for the garrifon of Harfleur. But father Daniel, in his life of Philip of Valois, produces a proof from the records of the chamber of accounts at Paris, that cannon and gunpowder were ufed in the year 1338. And Du-Cange even finds mention of the fame engines in Froiffart, and other French hiftorians, fome time earlier.

The Germans carry the invention of cannon farther back, and afcribe it to Albertus Magnus, a Dominican monk, about the year 1250. But Ifaac Voffius finds cannon in China upwards of 1700 years ago; being ufed by the emperor Kitey, in the year of Chrift 85. The ancients too, of Europe and Afia, had their fiery tubes, or *cannæ*, which being loaden with pitch, ftones, and iron balls, were exploded with a vehement noife, fmoke, and great effect.

Cannon were formerly made of a very great length, which rendered them exceedingly heavy, and their ufe very troublefome and confined. But it has lately been found by experiment that there is very little added to the force of the ball by a great length of the cannon, and therefore they have very properly been much reduced both in their length and weight, and rendered eafily manageable upon all occafions. They were formerly diftinguifhed by many hard and terrible names, but are now only named from the weight of their ball; as a 6 pounder, a 12 pounder, a 24 pounder, or a 42 pounder, which is the largeft fize now ufed by the Englifh for battering.

CANON, in Algebra, Arithmetic, Geometry, &c,

is a general rule for refolving all cafes of a like nature with the prefent enquiry. Thus the laft ftep of every equation is fuch a canon, and if turned into words, becomes a rule to refolve all cafes or queftions of the fame kind with that propofed.

Tables of fines, tangents, &c, whether natural or artificial, are alfo called canons.

CANON, in Ancient Mufic, is a rule or way of determining the intervals of mufical notes. Ptolomy, rejecting the Ariftoxenian way of meafuring the intervals in mufic by the magnitude of a tone, formed by the difference between a diapente and a diateffaron, thought that they fhould be diftinguifhed by the ratios which the founds terminating thofe intervals bear to one another, when confidered according to their degree of acutenefs or gravity; which, before Ariftoxenus, was the old Pythagorean way. He therefore made the diapafon confift in a double ratio; the diapente confift in a fefquialterate; the diateffaron, in a fefquitertian; and the tone itfelf, in a fefquioctave; and all the other intervals, according to the proportion of the founds that terminate them: wherefore, taking the canon, as it is called, for a determinate line of any length, he fhews how this is to be cut, that it may reprefent the respective intervals: and this method anfwers exactly to experiments in the different lengths of mufical chords. From this canon, Ptolomy and his followers have been called *Canonici*; as thofe of Ariftoxenus were called *Mufici*.

Pafcal CANON, a table of the moveable feafts, fhewing the day of Eafter, and the other feafts depending upon it, for a cycle or period of 19 years. It is faid that the Pafcal Canon was the calculation of Eufebius of Cæfarea, and that it was made by order of the council of Nice.

CANOPUS, a name given by fome of the old aftronomers to a ftar under the 2d bend of Eridanus. Thefe writers fay that the river in the heavens is not the Eridanus, but the Nile, and that this ftar commemorates an ifland made by that river, which was called by the fame name.

CANOPUS is alfo the name of a bright ftar of the firft magnitude in the rudder of the fhip Argo, one of the fouthern conftellations. Its fituation, as given by feveral authors, at different times, is as follows:

Authors	Dates	Longit.	Lat.	Rt. Afcen.	Declin. So.
F. Thomas	Jan. 1682	8° ♋ 52'	75° 15'	93° 32' 20"	52° 31' 33"
F. Noel	1697			93 54	52 29
Dr. Halley	170	10 52	72 49		
F. Feuille	Mar. 1709				52 30 4

CANTALIVERS, in Architecture, are the fame with modillions, except that the former are plain, and the latter carved. They are both a kind of cartoufes, fet at equal diftances under the corona of the cornice of a building.

CANTON (JOHN), an ingenious natural philofopher, was born at Stroud, in Gloucefterfhire, in 1718; and was placed, when young, under the care of Mr Davis, an able mathematician of that place, with whom he had learned both vulgar and decimal arithmetic before he was quite 9 years of age. He next proceeded to higher parts of the mathematics, and particularly to algebra and aftronomy, in which he had made a confiderable progrefs when his father took him from fchool, and fet him to learn his own bufinefs, which was that of a broad cloth weaver. This circumstance was not able to damp his zeal for acquiring knowledge. All his leisure time was devoted to the assiduous cultivation of astronomical science; by which he was soon able to calculate lunar eclipses and other phenomena, and to construct various kinds of sun-dials, even at times when he ought to have slept, being done without the knowledge and consent of his father, who feared that such studies might injure his health. It was during this prohibition, and at these hours, that he computed, and cut upon stone, with no better an instrument than a common knife, the lines of a large upright sun-dial, on which, beside the hour of the day, were shewn the sun's rising, his place in the ecliptic, and some other particulars. When this was finished and made known to his father, he permitted it to be placed against the front of his house, where it excited the admiration of several neighbouring gentlemen, and introduced young Canton to their acquaintance, which was followed by the offer of the use of their libraries. In the library of one of these gentlemen he found Martin's Philosophical Grammar, which was the first book that gave him a taste for natural philosophy. In the possession of another gentleman he first saw a pair of globes; a circumstance that afforded him great pleasure, from the great ease with which he could resolve those problems he had hitherto been accustomed to compute.

Among other persons with whom he became acquainted in early life, was Dr. Henry Miles of Tooting; who perceiving that young Canton possessed abilities too promising to be confined within the narrow limits of a country town, prevailed on his father to permit him to come up to London. Accordingly he arrived at the metropolis the 4th of March 1737, and resided with Dr. Miles at Tooting till the 6th of May following; when he articled himself, for the term of 5 years, as a clerk to Mr. Samuel Watkins, master of the academy in Spital Square. In this situation his ingenuity, diligence, and prudence, were so distinguished, that on the expiration of his clerkship in May 1742, he was taken into partnership with Mr. Watkins for 3 years; which gentleman he afterward succeeded in the school, and there continued during his whole life.

Towards the end of 1745, electricity received a great improvement by the discovery of the famous Leyden phial. This event turned the thoughts of most of the philosophers of Europe to that branch of natural philosophy; and our author, who was one of the first to repeat and to pursue the experiment, found his endeavours rewarded by many notable discoveries.—Towards the end of 1749, he was engaged with his friend, the late ingenious Benjamin Robins, in making experiments to determine the height to which rockets may be made to ascend, and at what distance their light may be seen. —In 1750 was read at the Royal Society, Mr. Canton's "Method of making Artificial Magnets, without the use of, and yet far superior to, any natural ones." This paper procured him the honour of being elected a member of the Society, and the present of their gold medal. The same year he was complimented with the degree of M. A. by the university of Aberdeen. And in 1751 he was chosen one of the council of the Royal Society; an honour which was twice repeated afterwards.

In 1752, our philosopher was so fortunate as to be the first person in England who, by attracting the electric fire from the clouds during a thunder-storm, verified Dr. Franklin's hypothesis of the similarity of lightning and electricity. Next year his paper intitled " Electrical Experiments, with an attempt to account for their several phenomena," was read at the Royal Society. In the same paper Mr. Canton mentioned his having discovered, by many experiments, that some clouds were in a positive, and some in a negative state of electricity: a discovery which was also made by Dr. Franklin in America much about the same time. This circumstance, together with our author's constant defence of the doctor's hypothesis, induced that excellent philosopher, on his arrival in England, to pay Mr. Canton a visit, and gave rise to a friendship which ever after continued between them.—In the Ladies' Diary for 1756, our author answered the prize query that had been proposed in the preceding year, concerning the meteor called shooting stars. The solution, though only signed A. M. was so satisfactory to his friend, the excellent mathematician Mr. Thomas Simpson, who then conducted that ingenious and useful little work, that he sent Mr. Canton the prize, accompanied with a note, in which he said he was sure that he was not mistaken in the author of it, as no one besides, that he knew of, could have given that answer.— Our philosopher's next communication to the public, was a letter in the Gentleman's Magazine for September 1759, on the electrical properties of the tourmalin, in which the laws of that wonderful stone are laid down in a very concise and elegant manner. On the 13th of December in the same year was read at the Royal Society, " An attempt to account for the Regular Diurnal Variation of the Horizontal Magnetic Needle; and also for its Irregular Variation at the time of an Aurora Borealis." A complete year's observations of the diurnal variations of the needle are annexed to the paper. —Nov. 5, 1761, our author communicated to the Royal Society an account of the Transit of Venus of the 6th of June that year, observed in Spital Square. His next communication to the Society, was a Letter, read the 4th of Feb. 1762, containing some remarks on Mr. Delaval's electrical experiments. On the 16th of Dec. the same year, another curious addition was made by him to philosophical knowledge, in a paper, intitled, " Experiments to prove that Water is not Incompressible." And on Nov. 8, the year following, were read before the Society, his farther " Experiments and Observations on the Compressibility of Water, and some other fluids." These experiments are a complete refutation of the famous Florentine experiment, which so many philosophers have mentioned as a proof of the incompressibility of water. For this communication he had a second time the Society's prize gold medal.

Another communication was made by our author to the Society, on Dec. 22, 1768, being " An easy method of making a phosphorus that will imbibe and emit light like the Bolognian Stone; with experiments and observations." When he first shewed to Dr. Franklin the instantaneous light acquired by some of this phosphorus from the near discharge of an electrified bottle, the doctor immediately exclaimed, " And God said let there be light, and there was light."

The

The Dean and Chapter of St. Paul's having, in a letter, dated March 6, 1769, requested the opinion of the Royal Society relative to the best method of fixing electrical conductors to preserve that cathedral from damage by lightning, Mr. Canton was one of the committee appointed to take the letter into consideration, and to report their opinion upon it. The gentlemen joined with him in this business were, Mr. Delaval, Dr. Franklin, Dr. Watson, and Mr. Wilson. Their report was made on the 8th of June following: and the mode recommended by them has been carried into execution. —Our author's last communication to the Royal Society, was a paper read Dec. 21, 1769, containing "Experiments to prove that the Luminousness of the Sea arises from the Putrefaction of its animal Substances."

Besides the papers above mentioned, Mr. Canton wrote a number of others, both in the earlier and the later parts of his life, which appeared in several publications, and particularly in the Gentleman's Magazine.— He died of a dropsy, the 22d of March 1772, in the 54th year of his age.

CAPACITY, is the solid content of any body. Also our hollow measures for corn, beer, wine, &c, are called measures of capacity.

CAPE, or *Promontory*, is any high land, running out with a point into the sea ; as Cape Verde, Cape Horn, the Cape of Good Hope, &c.

CAPELLA, a bright star of the first magnitude, in the left shoulder of Auriga.

CAPILLARY *Tubes*, in Physics, are very small pipes, whose canals are exceedingly narrow; being so called from their resemblance to a hair in smallness. Their usual diameter may be from $\frac{1}{20}$ to $\frac{1}{30}$ of an inch: though Dr. Hook assures us that he drew tubes in the flame of a lamp much smaller, and resembling a spider's thread.

The *Ascent of Water &c*, in capillary tubes, is a noted phenomenon in philosophy. Take several small glass tubes, of different diameters, and open at both ends; immerse them a little way into water, and the fluid will be seen to stand higher in the tubes than the surface of the water without, and higher as the tube is smaller, almost in the reciprocal ratio of the diameter of the tube; and that both in open air, and in vacuo. The greatest height to which Dr. Hook ever observed the water to stand, in the smallest tubes, was 21 inches above the surface in the vessel.

This does not however happen uniformly the same in all fluids ; some standing higher than others; and in quicksilver the contrary takes place, as that fluid stands lower within the tube than its surface in the vessel, and the lower as the tube is smaller. See Philos. Transf. N° 355, or Abr. vol. 4, pa. 423, &c, or Cotes's Hydr. and Pneum. Lect. pa. 265.

Another phenomenon of these tubes is, that such of them as would only naturally discharge water by drops, when electrified, yield a continued and accelerated stream ; and the acceleration is proportional to the smallness of the tube: indeed the effect of electricity is so considerable, that it produces a continued stream from a very small tube, out of which the water had not before been able to drop. Priestley's Hist. Electr. 8vo. vol. 1, pa. 171, ed. 3d.

This ascent and suspension of the water in the tube, is by Dr. Jurin, Mr. Haukfbee, and other philosophers, ascribed to the attraction of the periphery of the concave surface of the tube, to which the upper surface of the water is contiguous and adheres.

CAPITAL, in Architecture, the uppermost part of a column or pilaster, serving as a head or crowning to it ; being placed immediately over the shaft, and under the entablature. It is made differently in the different orders, and is that indeed which chiefly distinguishes the orders themselves.

CAPITAL of a *Bastion*, is an imaginary line dividing any work into two equal and similar parts ; or a line drawn from the angle of the polygon to the point of the bastion, or from the point of the bastion to the middle of the gorge.

CAPONIERE, or CAPONNIERE, in Fortification, is a passage made from one work to another, of 10 or 12 feet wide, and about 5 feet deep, covered on each side by a parapet, terminating in a glacis or slope. Sometimes it is covered with planks and earth.

CAPRA, or the *She-goat*, a name given to the star Capella, on the left shoulder of Auriga ; and sometimes to the constellation Capricorn. Some again represent Capra as a constellation in the northern hemisphere, consisting of 3 stars, comprised between the 45th and 55th degree of latitude.

The poets fable her to be Amalthea's goat, which suckled Jupiter in his infancy.

CAPRICORN, the *Goat*, a southern constellation, and the 10th sign of the zodiac, as also one of the 48 original constellations received by the Greeks from the Egyptians. The figure of this sign is drawn as having the fore part of a goat, but the hinder part of a fish ; and sometimes simply under the form of a goat. In writing, it is denoted by a character representing the crooked horns of a goat's head, thus ♑.

As to the figure of this constellation, the Greeks pretend that Pan, to avoid the terrible giant Typhon, threw himself into the Nile, and was changed into the figure here drawn ; in commemoration of which exploit, Jupiter took it up to heaven. But it is probable, as Macrobius observes, that the Egyptians marked the point of the ecliptic appropriated to this sign, where the sun begins again to ascend up towards the north, with the figure of a goat, an animal which is always climbing the sides of mountains.

The stars in this constellation, in Ptolomy's and Tycho's catalogue, are 28; in that of Hevelius 29 ; though it is to be remarked that one of those in the tail, of the 6th magnitude, marked the 27th in Tycho's book, was lost in Hevelius's time. Flamsteed gives 51 stars to this sign.

Tropic of CAPRICORN, a little circle of the sphere, parallel to the equator, passing through the beginning of Capricorn, or the winter solstice, or the point of the sun's greatest south declination.

CAPSTAN, a large massy column shaped like a truncated cone ; being set upright on the deck of a ship, and turned by levers or bars, passing through holes in its upper extremity. The capstan is a kind of perpetual lever, or an axis-in-peritrochio, which, by means of a strong rope or cable passed round, serves to raise very great weights ; such as to hoist sails, to weigh

the

the anchors, to draw the veffels on fhore, and hoift them up to be refitted, &c.

CAPUT Draconis, or dragon's head, a name given by fome to a fixed ftar of the firft magnitude, in th; head of the conftellation Draco.

CARACT, or Carat, a name given to the weight which expreffes the degree of goodnefs or finenefs of gold. The whole quantity of metal is confidered as confifting of 24 parts, which are the carats, fo that the carat is the 24th part of the whole; this carat is divided into 4 equal parts, called grains of a carat, and the grain into halves and quarters.

When gold is purified to the utmoft degree poffible, fo that it lofes no more by farther trials, it is confidered as quite pure, and faid to be 24 carats fine; if it lofe 1 carat, or 1—24th in purifying, it was of 23 carats fine; and if it lofe 2 carats, it was 22 carats fine; and fo on.

CARCASS, is a hollow cafe formed of ribs of iron, and covered over with pitched cloth &c, about the fize of bomb-fhells; or fometimes made all of iron except two or three holes for the fire to blaze through. Thefe are filled with various matters and combuftibles, to fire houfes, when thrown out of mortar pieces into befieged places.

CARCAVI (Peter de), was born at Lyons, but in what year is not known. He was Counfellor to the Parliament of Touloufe, afterward Counfellor to the Grand Council, and Keeper of the King's Library. He was appointed Geometrician to the French Academy of Sciences in 1666; and died at Paris in 1684. There are extant fome letters of his, printed among thofe of Defcartes.

CARDAN (Hieronymus, or Jerom), one of the moft extraordinary geniufes of his age, was born at Pavia, in Italy, Sept. 24, 1501. At 4 years old he was carried to Milan, his father being an advocate and phyfician in that city: at the age of 20 he went to ftudy in the univerfity of the fame city; and two years afterward he explained Euclid. In 1524, he went to Padua: the fame year he was admitted to the degree of mafter of arts; and the year following, that of doctor of phyfic. He married about the year 1531; and became profeffor of mathematics, and practifed medicine at Milan about 1533. In 1539 he was admitted a member of the college of phyficians at Milan: in 1543 he read public lectures in medicine there; and the fame at Pavia the year following; but he difcontinued them becaufe he could not get payment of his falary, and returned to Milan.

In 1552 he went into Scotland, having been fent for by the archbifhop of St. Andrews, to cure him of a grievous diforder, after trying the phyficians of the king of France and of the emperor of Germany, without benefit. He began to recover from the day that Cardan prefcribed for him: our author took his leave of him at the end of fix weeks and three days, leaving him prefcriptions which in two years wrought a complete cure. Upon this vifit Cardan paffed through London, and calculated king Edward's nativity; for he was famous for his knowledge in aftrology, as well as thofe of mathematics and medicine. Returning to Milan, after four months abfence, he remained there till the beginning of Oct. 1552; and then went to

Pavia, from whence he was invited to Bologna in 1562. He taught in this laft city till the year 1570; at which time he was thrown into prifon; but fome months after he was fent home to his own houfe. He quitted Bologna in 1571; and went to Rome, where he lived for fome time without any public employment. He was however admitted a member of the college of phyficians, and received a penfion from the Pope, till the time of his death, which happened at Rome on the 21ft of September 1575.

Cardan, at the fame time that he was one of the greateft geniufes and moft learned men of his age, in all the fciences, was one of the moft eccentric and fickle in conduct of all men that ever lived: defpifing all good principles and opinions, and without one friend in the world. The fame capricioufnefs that was remarkable in his outward conduct, is alfo obfervable in the compofition of his numerous and elaborate works. In many of his treatifes the reader is ftopped almoft every moment by the obfcurity of his text, or by digreffions from the point in hand. In his arithmetical writings there are feveral difcourfes on the motions of the planets, on the creation, on the Tower of Babel, and fuch like. And the apology which he made for thefe frequent digreffions is, that he might by that means enlarge and fill up his book, his bargain with the bookfeller being at fo much per fheet; and that he worked as much for his daily fupport as for fame. The Lyons edition of his works, printed in 1663, confifts of no lefs than 10 volumes in folio.

In fact, when we confider the tranfcendent qualities of Cardan's mind, it cannot be denied that he cultivated it with every fpecies of knowledge, and made a greater progrefs in philofophy, in the medical art, in aftronomy, in mathematics, and the other fciences, than the moft part of his contemporaries who had applied themfelves but to one only of thofe fciences. In particular, he was perhaps the very beft algebraift of his time, a fcience in which he made great improvements; and his labours in cubic equations efpecially have rendered his name immortal, the rules for refolving them having ever fince borne his name, and are likely to do fo as long as the fcience fhall exift, although he received the firft knowledge of them from another perfon; the account of which, and his difputes with Tartalea, have been given at large under the article Algebra.

Scaliger affirms, that Cardan, having by aftrology predicted and fixed the time of his death, abftained from all food, that his prediction might be fulfilled, and that his continuance to live might not difcredit his art. It is farther remarkable, that Cardan's father alfo died in this manner, in the year 1524, having abftained from fuftenance for nine days. Our author too informs us that his father had white eyes, and could fee in the night-time.

CARDINAL Points, in Geography, are the eaft, weft, north, and fouth points of the horizon.

Cardinal Points of the Heavens, or of a Nativity, are the rifing and fetting of the fun, the zenith and nadir.

Cardinal Signs, are thofe at the four quarters, or the equinoxes and folftices, viz, the figns Aries, Libra, Cancer, and Capricorn.

Cardinal

CARDINAL *Winds*, are thofe that blow from the four cardinal points, viz, the eaft, weft, north, and fouth winds.

CARDIOIDE, the name of a curve fo called by Caftilliani.—But it was firft treated of by Koerfma, and by Carré. See Philof. Tranf. 1741, and Memoires de l'Acad. 1705.

The Cardioide is thus generated. APB is a circle, and AB its diameter. Through one extremity A of the diameter draw a number of lines APQ, cutting the circle in P; upon thefe fet off always PQ equal to the diameter AB; fo fhall the points Q be always in the curve of the cardioide.

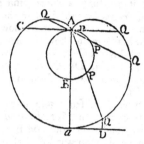

From this generation of the curve, its chief properties are evident, viz, that,

everywhere PQ = AB,
CQ, or QQ is = Aa or 2 AB,
AQ = AB ∓ AP,
P always bifects QQ.

The cardioide is an algebraical curve, and the equation expreffing its nature is thus:

Put a = AB the diameter,
x = aD perp. AB,
y = DQ perp. AD; then is

$$y^4 - 6ay^3 + 2x^2y^2 - 6ax^2y + x^4 \atop + 12a^2y^2 - 8a^3y + 3a^2x^2 \Big\} = 0,$$

which is the equation of the curve.

Many properties of the cardioide may be feen in the places above cited.

CARRÉ (LEWIS), was born in the year 1663, in the province of Brie in France. His father, a fubftantial farmer, intended him for the church. But young Carré, after going through the ufual courfe of education for that purpofe, having an utter averfion to it, he refufed to enter upon that function; by which he incurred his father's difpleafure. His refources being thus cut off, he was obliged to quit the univerfity, and look out into the world for fome employment. In this exigency he had the good fortune to be engaged as an amanuenfis by the celebrated father Malebranche; by which he found himfelf tranfported all at once from the mazes of fcholaftic darknefs, to the fource of the moft brilliant and enlightened philofophy. Under this great mafter he ftudied mathematics and the moft fublime metaphyfics. After feven years fpent in this excellent fchool, M. Carré found it neceffary, in order to procure himfelf fome lefs precarious eftablifhment, to teach mathematics and philofophy in Paris; but efpecially that philofophy which, on account of its tendency to improve our morals, he valued more than all the mathematics in the world. And accordingly his greateft care was to make geome-

try ferve as an introduction to his well beloved metaphyfics.

Moft of M. Carré's pupils were of the fair fex. The firft of thefe, who foon perceived that his language was rather the reverfe of elegant and correct, told him pleafantly that, as an acknowledgement for the pains he took to teach her philofophy, fhe would teach him French; and he ever after owned that her leffons were of great fervice to him. In general he feemed to fet more value upon the genius of women than that of men.

M. Carré, although he gave the preference to metaphyfics, did not neglect mathematics; and while he taught both, he took care to make himfelf acquainted with all the new difcoveries in the latter. This was all that his conftant attendance on his pupils would allow him to do, till the year 1697, when M. Varignon, fo remarkable for his extreme fcrupuloufnefs in the choice of his eleves, took M. Carré to him in that ftation. Soon after, viz. in the year 1700, our author thinking himfelf bound to do fomething that might render him worthy of that title, publifhed the firft complete work on the Integral Calculus, under the title of " A method of meafuring Surfaces and Solids, and finding their Centres of Gravity, Percuffion, and Ofcillation." He afterwards difcovered fome errors in the work, and was candid enough to own and correct them in a fubfequent edition.

In a little time M. Carré became Affociate, and at length one of the Penfioners of the Academy. And as this was a fufficient eftablifhment for one, who knew fo well how to keep his defires within juft bounds, he gave himfelf up entirely to ftudy; and as he enjoyed the appointment of Mechanician, he applied himfelf more particularly to mechanics. He took alfo a furvey of every branch relating to mufic; fuch as the doctrine of founds, the defcription of mufical inftruments; though he defpifed the practice of mufic, as a mere fenfual pleafure. Some fketches of his ingenuity and induftry in this way may be feen in the Memoirs of the French Academy of Sciences. M. Carré alfo compofed fome treatifes on other branches of natural philofophy, and fome on mathematical fubjects; all which he bequeathed to that illuftrious body; though it does not appear that any of them have yet been publifhed. It is not unlikely that he was hindered from putting the laft hand to them by a train of diforders proceeding from a bad digeftion, which, after haraffing him during the fpace of five or fix years, at length brought him to the grave in 1711, at 48 years of age.

His memoirs printed in the volumes of the Academy, with the years of the volumes, are as below.

1. The Rectification of Curve Lines by Tangents: 1701.

2. Solution of a problem propofed to Geometricians, &c. 1701.

3. Reflections on the Table of Equations: 1701.

4. On the Caufe of the Refraction of Light: 1702.

5. Why the Tides are always augmenting from Breft to St. Malo, and diminifhing along the coafts of Normandy: 1702.

6. The Number and the Names of Mufical Inftruments: 1702.

7. On

7. On the Vinegar which caufes fmall ftones to roll upon an inclined plane : 1703.

8. On the Rectification &c. of the Cauftics by reflection : 1703.

9. Method for the Rectification of Curves : 1704.

10. Obfervations on the Production of Sound : 1704.

11. On a Curve formed from a Circle : 1705.

12. On the Refraction of Mufket-balls in water, and on the Refiftance of that fluid : 1705.

13. Experiments on Capillary Tubes : 1705.

14. On the Proportion of Pipes to have a determinate quantity of water : 1705.

15. On the Laws of Motion : 1706.

16. On the Properties of Pendulums ; with fome new properties of the Parabola : 1707.

17. On the Proportion of Cylinders that their founds may form the mufical chords : 1709.

18. On the Elafticity of the Air : 1710.

19. On Catoptrics : 1710.

20. On the Monochord : in the Machines, tom. 1. with fome other pieces, not mathematical.

CARRIAGE, *of a Cannon*, is the machine upon which it is mounted ; ferving to point or direct it for fhooting, and to convey it from place to place.

Wheel CARRIAGE, one that is mounted and moved about upon wheels. Horfes draw in general, to moft advantage, when the direction of their draft is parallel to the ground, or rather a little upwards. A carriage alfo goes eafieft when the centre of gravity is placed very high ; fince then, when once put in motion, it continues it with very little labour to the horfes.

CARTES (RENE DES), one of the moft eminent philofophers and mathematicians of the 17th century, or indeed of any age whatever. He was defcended of an ancient noble family in Touraine in France, being a younger fon of a counfellor in the parliament of Rennes, and was born March 31, 1596. His father gave him a liberal education, and the more fo as he obferved in him the appearance of a promifing genius, ufing to call him the philofopher, on account of his infatiable curiofity in afking the reafons of every thing that he did not underftand.

Des Cartes was fent to the Jefuits college at La Fleche in 1604, and put under the tuition of Father Charlet. Here he made a great progrefs in the learned languages and polite literature ; but having paffed through his courfe of philofophy without any great fatisfaction to himfelf, he left the college in 1612, and began to learn military arts, to ride and fence, and other fuch like exercifes. But notwithftanding his inclination to military achievements, the weaknefs of his conftitution not permitting him early to expofe himfelf to the fatigues of war, he was fent to Paris in 1613. Here he formed an acquaintance with feveral learned perfons, who helped to reclaim him from his intention of declining his ftudies, particularly Father Merfenne, whofe converfation revived in him a love for truth, and induced him to retire from the world to purfue his ftudies without interruption ; which he did for two years : but in May 1616, at the repeated folicitations of his relations, he fet out for Holland, and entered as a volunteer under the Prince of Orange.

Whilft he lay in garrifon at Breda, during the truce between the Spanifh and Dutch, an unknown perfon caufed a problem in mathematics, in the Dutch language, to be fixed up in the ftreets : when Des Cartes, feeing a concourfe of people ftop to read it, defired one who ftood near him to explain it to him in Latin or French. The perfon promifed to fatisfy him, upon condition that he would engage to refolve the problem ; and Des Cartes agreed to the condition with fuch an air, that though he little expected fuch a thing from a young military cadet, he gave him his addrefs, defiring he would bring him the folution. Des Cartes next day vifited Beekman, principal of the college of Dort, who was the perfon that had tranflated the problem to him. Beekman was furprifed at his having refolved it in fuch a fhort time ; but his wonder was much increafed, to find, in the courfe of converfation, that the young man's knowledge was much fuperior to his own in thofe fciences, in which he had employed his whole time for feveral years. During his ftay at Breda, Des Cartes wrote in Latin, a treatife on Mufic, and laid the foundation of feveral of his other works.

In 1619, he entered himfelf a volunteer in the army of the duke of Bavaria. In 1621, he made the campaign in Hungary, under the count de Bucquoy ; but the lofs of his general, who was killed at a fiege that year, determined him to quit the army. He foon after began his travels into the north, and vifited Silefia, Poland, Pomerania, the coafts of the Baltic, Brandenburgh, Holftein, Eaft Friefland, Weft Friefland ; in his paffage to which laft place he was in danger of being murdered. The failors fancied he was a merchant, who had a large fum of money about him ; and perceiving that he was a foreigner who had little acquaintance in the country, and a man of a mild difpofition, they refolved to kill him, and throw his body into the fea. They even difcourfed of their defign before his face, thinking he underftood no language but French, in which he always fpoke to his fervant. Des Cartes fuddenly ftarted up ; and drawing his fword, fpoke to them in their own language, in fuch a tone as ftruck terror into them : upon which they behaved very civilly. The year following he went to Paris, where he cleared himfelf from the imputation of having been received among the Roficrufians, whom he confidered as a company of vifionaries and impoftors.

Dropping the ftudy of mathematics, he now applied himfelf again to ethics and natural philofophy. The fame year he took a journey through Switzerland to Italy. Upon his return he fettled at Paris ; but his ftudies being interrupted by frequent vifits, he went in 1628 to the fiege of Rochelle. He returned to Paris in November ; but in the following fpring he repaired to Amfterdam ; and from thence to a place near Franeker in Friefland, where he began his Metaphyfical Meditations, and fpent fome time in Dioptrics ; about this time too he wrote his thoughts upon Meteors. After about fix months he returned to Amfterdam.

Des Cartes imagined that nothing could more promote the temporal felicity of mankind, than the union of natural philofophy with mathematics. But before he fhould fet himfelf to relieve men's labours, or multiply the conveniencies of life by mechanics, he thought it neceffary to difcover fome means of fecuring the human

man body from difeafe and debility: this led him to the ftudy of anatomy and chemiftry, in which he employed the winter at Amfterdam.

He now, viz, about 1630 or 1631, took a fhort tour to England, and made fome obfervations near London on the variation of the compafs. In the fpring of 1533 he removed to Deventer, where he completed feveral works that were left unfinifhed the year before, and refumed his ftudies in aftronomy. In the fummer he put the laft hand to his " Treatife of the World." The next year he returned to Amfterdam; but foon after took a journey into Denmark, and the lower parts of Germany. In autumn 1635 he went to Lewarden in Friefland, where he remained till 1637, and wrote his " Treatife of Mechanics." The fame year he publifhed his four treatifes concerning Method, Dioptrics, Meteors, and Geometry. About this time he received an invitation to fettle in England from Sir Charles Cavendifh, brother to the earl of Newcaftle, with which he did not appear backward to comply, efpecially upon being affured that the king was a catholic in his heart: but the breaking out of the civil wars in this country prevented his journey.

At the end of 1641, Lewis the 13th, of France, invited him to his court, upon very honourable terms; but he could not be perfuaded to quit his retirement. This year he publifhed his Meditations concerning the Exiftence of God, and the Immortality of the Soul. In 1645 he again applied to anatomy; but was a little diverted from this ftudy, by the queftion concerning the Quadrature of the Circle, which was at that time agitated. During the winter of the fame year he compofed a fmall tract againft Gaffendus's Inftitutes; and another on the Nature of the Paffions. About this time he carried on an epiftolary correfpondence with the princefs Elizabeth, daughter to Frederick the 5th, elector palatine, and king of Bohemia, who had been his pupil in Holland.

A difpute arifing between Chriftina, queen of Sweden, and M. Chanut, the refident of France, concerning the following queftion; When a man carries love or hatred to excefs, which of thefe two irregularities is the worft? The refident fent the queftion to Des Cartes, who upon that occafion drew up the differtation upon Love, that is publifhed in the firft volume of his letters, which proved highly fatisfactory to the queen. In June 1647 he took a journey to France, where the king fettled on him a penfion of 3000 livres; but he returned to Holland about the end of September. In November he received a letter from M. Chanut, in queen Chriftina's name, defiring his opinion of the fovereign good; which he accordingly fent her, with fome letters upon the fame fubject formerly written to the princefs Elizabeth, and his treatife on the Paffions. The queen was fo highly pleafed with them, that fhe wrote him a letter of thanks with her own hand, and invited him to come to Sweden. He arrived at Stockholm in Oct. 1648. The queen engaged him to attend her every morning at five o'clock, to inftruct her in his philofophy: and defired him to revife and digeft all his unpublifhed writings, and to draw up from them a complete body of philofophy. She purpofed alfo to fix him in Sweden, by allowing him a revenue of 3000 crowns a year, with an eftate which fhould

defcend to his heirs and affigns for ever; and to eftablifh an academy, of which he was to be the director. But thefe defigns were fruftrated by his death, which happened the 11th of Feb. 1650, in the 54th year of his age. His body was interred at Stockholm: but 17 years after it was removed to Paris, where a magnificent monument was erected to him in the church of Genevieve du Mont.

As to the character of our author:

Dr. Barrow in his *Opufcula* tells us, that Des Cartes was doubtlefs a very ingenious man, and a real philofopher, and one who feems to have brought thofe affiftances to that part of philofophy relating to matter and motion, which perhaps no one had done before: namely, a great fkill in mathematics; a mind habituated, both by nature and cuftom, to profound meditation; a judgment exempt from all prejudices and popular errors, and furnifhed with a good number of certain and felect experiments; a great deal of leifure; an entire difengagement, by his own choice, from the reading of ufelefs books, and the avocations of life; with an incomparable acutenefs of wit, and an excellent talent of thinking clearly and diftinctly, and of expreffing his thoughts with the utmoft perfpicuity.

Dr. Halley, in a paper concerning Optics, communicated to Mr. Wotton, and publifhed by the latter in his " Reflections upon Ancient and Modern Learning," writes as follows: As to dioptrics, though fome of the ancients mention refraction, as a natural effect of tranfparent media; yet Des Cartes was the firft, who in this age has difcovered the laws of refraction, and brought dioptrics to a fcience.

Dr. Keil, in the introduction to his " Examination of Dr. Burnet's Theory of the Earth," tells us, that Des Cartes was fo far from applying geometry and obfervations to natural philofophy, that his whole fyftem is but one continued blunder on account of his negligence in that point; which he could eafily prove, by fhewing that his theory of the vortices, upon which his fyftem is founded, is abfolutely falfe, for that Newton has fhewn that the periodical times of all bodies, that fwim in vortices, muft be directly as the fquares of their diftances from the centre of them: but it is evident from obfervations, that the planets, in moving round the fun, obferve a law quite different from this; for the fquares of their periodical times are always as the cubes of their diftances: and therefore, fince they do not obferve that law, which of necefity they muft if they fwim in a vortex, it is a demonftration that there are no vortices in which the planets are carried round the fun.

" Nature, fays Voltaire, had favoured Des Cartes with a ftrong and clear imagination, whence he became a very fingular perfon, both in private life, and in his manner of reafoning. This imagination could not be concealed even in his philofophical writings, which are every where adorned with very brilliant ingenious metaphors. Nature had almoft made him a poet; and indeed he wrote a piece of poetry for the entertainment of Chriftina queen of Sweden, which however was fuppreffed in honour of his memory. He extended the limits of geometry as far beyond the place where he found them, as Newton did after him; and firft taught the method of expreffing curves by equations. He applied

plied

plied this geometrical and inventive genius to dioptrics, which when treated by him became a new art; and if he was mistaken in some things, the reason is, that a man who discovers a new tract of land, cannot at once know all the properties of the soil. Those who come after him, and make these lands fruitful, are at least obliged to him for the discovery." Voltaire acknowledges, that there are innumerable errors in the rest of Des Cartes' works; but adds, that geometry was a guide which he himself had in some measure formed, and which would have safely conducted him through the several paths of natural philosophy: nevertheless he had at last abandoned this guide, and gave entirely into the humour of framing hypotheses; and then philosophy was no more than an ingenious romance, fit only to amuse the ignorant.

It has been pretty generally acknowledged, that he borrowed his improvements in Algebra from Harriot's *Artis Analyticæ Praxis*; which is highly probable, as he was in England about the time when Harriot's book was published, and as he follows the manner of Harriot, except in the method of noting the powers. Upon this head the following anecdote is told by Dr. Pell, in Wallis's Algebra, pa. 198. Sir Charles Cavendish, then resident at Paris, discoursing there with M. Roberval, concerning Des Cartes's Geometry, then lately published: I admire, said Roberval, that method in Des Cartes, of placing all the terms of the equation on one side, making the whole equal to nothing, and how he lighted upon it. The reason why you admire it, said Sir Charles, is because you are a Frenchman; for if you were an Englishman, you would not admire it. Why so? asked Roberval. Because, replied Sir Charles, we in England know whence he had it; namely from Harriot's Algebra. What book is that? says Roberval, I never saw it. Next time you come to my chamber, saith Sir Charles, I will shew it to you. Which a while after he did; and upon perusal of it, Roberval exclaimed with admiration, *Il l'a vu! il l'a vu! He had seen it! he had seen it!* finding all that in Harriot which he had before admired in Des Cartes, and not doubting but that Des Cartes had it from thence. See also Montucla's History of Mathematics.

The real improvements of Des Cartes in Algebra and Geometry, I have particularly treated of under the article ALGEBRA; and his philosophical doctrines are displayed in the article CARTESIAN *Philosophy*, here following. He was never married, but had one natural daughter, who died when she was but five years old. There have been several editions of his works, and commentaries upon them; particularly those of Schooten on his Geometry.

CARTESIAN *Philosophy*, or *Cartesianism*, the system of philosophy advanced by Des Cartes, and maintained by his followers, the Cartesians.

The Cartesian philosophy is founded on two great principles, the one metaphysical, the other physical. The metaphysical one is this: *I think, therefore I am, or I exist*: the physical principle is, that *nothing exists but substances.* Substance he makes of two kinds; the one a substance that thinks, the other a substance extended: so that actual thought and actual extension make the essence of substance.

The essence of matter being thus fixed in extension,

Des Cartes concludes that there is no vacuum, nor any possibility of it in nature; but that the universe is absolutely full: by this principle, mere space is quite excluded; for extension being implied in the idea of space, matter is so too.

Des Cartes defines motion to be the translation of a body from the neighbourhood of others that are in contact with it, and considered as at rest, to the neighbourhood of other bodies: by which he destroys the distinction between motion that is absolute or real, and that which is relative or apparent. He maintains that the same quantity of motion is always preserved in the universe, because God must be supposed to act in the most constant and immutable manner. And hence also he deduces his three laws of motion. See MOTION.

Upon these principles Des Cartes explains mechanically how the world was formed, and how the present phenomena of nature came to arise. He supposes that God created matter of an indefinite extension, which he separated into small square portions or masses, full of angles: that he impressed two motions on this matter; the one, by which each part revolved about its own centre; and another, by which an assemblage, or system of them, turned round a common centre. From whence arose as many different vortices, or eddies, as there were different masses of matter, thus moving about common centres.

The consequence of these motions in each vortex, according to Des Cartes, is as follows: The parts of matter could not thus move and revolve amongst one another, without having their angles gradually broken; and this continual friction of parts and angles must produce three elements: the first of these, an infinitely fine dust, formed of the angles broken off; the second, the spheres remaining, after all the angular parts are thus removed; and those particles not yet rendered smooth and spherical, but still retaining some of their angles, and hamous parts, from the third element.

Now the first or subtilest element, according to the laws of motion, must occupy the centre of each system, or vortex, by reason of the smallness of its parts: and this is the matter which constitutes the sun, and the fixed stars above, and the fire below. The second element, made up of spheres, forms the atmosphere, and all the matter between the earth and the fixed stars; in such sort, that the largest spheres are always next the circumference of the vortex, and the smallest next its centre. The third element, formed of the irregular particles, is the matter that composes the earth, and all terrestrial bodies, together with comets, spots in the sun, &c.

He accounts for the gravity of terrestrial bodies from the centrifugal force of the ether revolving round the earth: and upon the same general principles he pretends to explain the phenomena of the magnet, and to account for all the other operations in nature.

CARY (ROBERT), a learned English chronologer and divine, was born at Cockington, in the county of Devon, about the year 1615. He took his degrees in arts, and LL.D. in Oxford. After returning from his travels he was presented to the rectory of Portlemouth, near Kingsbridge in Devonshire: but not long after he was drawn over by the presbyterian ministers to their party, and chosen moderator of that part of the second division of the county of Devon, which was appointed

pointed to meet at Kingsbridge. And yet, upon the restoration of Charles the 2d, he was one of the first to congratulate that prince upon his return, and soon after was preferred to the archdeaconry of Exeter; but from which he was however some time afterward ejected. He spent the rest of his days at his rectory at Portlemouth, and died in 1688, at 73 years of age.— He published *Palælogia Chronica*, a chronological account of ancient time, in three parts. 1, Didactical; 2, Apodidactical; 3, Canonical: in 1677.

CASATI (PAUL), a learned Jesuit, born at Placentia in 1617. He entered early among the Jesuits; and after having taught mathematics and divinity at Rome, he was sent into Sweden to queen Christina, whom he prevailed on to embrace the popish religion. His writings are as follow:

1. *Vacuum Proscriptum.*—2. *Terra Machinis mota.*— 3. *Mechanicorum, libri octo.*—4. *De Igne Dissertationes.* —5. *De Angelis Disputatio Theolog.*—6. *Hydrostaticæ Dissertationes.*—7. *Opticæ Disputationes.* It is remarkable that he wrote this treatise on optics at 88 years of age, and after he was blind. He was also author of several books in the Italian language.

CASCABEL, the knob or button of metal behind the breech of a cannon, as a kind of handle by which to elevate and direct the piece; to which some add the fillet and ogees as far as the base-ring.

CASEMATE, or CAZEMATE, in Fortification, a kind of vault or arch, of stone-work, in that part of the flank of a bastion next the curtain; serving as a battery, to defend the face of the opposite bastion, and the moat or ditch.

The casemate sometimes consists of three platforms, one above another; the highest being on the rampart; though it is common to withdraw this within the bastion.

The casemate is also called the low place, and low flank, as being at the bottom of the wall next the ditch; and sometimes the retired flank, as being the part of the flank nearest the curtain, and the centre of the bastion. It was formerly covered by an epaulement, or a massive body, either round or square, which prevented the enemy from seeing within the batteries; whence it was also called *covered flank*.

It is now seldom used, because the batteries of the enemy are apt to bury the artillery of the casemate in the ruins of the vault: beside, the great smoke made by the discharge of the cannon, renders it intolerable to the men. So that, instead of the ancient covered casemates, later engineers have contrived open ones, only guarded by a parapet, &c.

CASEMATE is also used for a well with several subterraneous branches, dug in the passage of the bastion, till the miner is heard at work, and air given to the mine.

CASERNS, or CAZERNS, in Fortification, small rooms, or huts, erected between the ramparts and the houses of fortified towns, or even on the ramparts themselves; to serve as lodgings for the soldiers on immediate duty, to ease the garrison.

CASE-SHOT, or CANNISTER-SHOT, are a number of small balls put into a round tin cannister, and so shot out of great guns. These have superseded, and been substituted instead of the grape-shot, which have been laid aside.

CASSINI (JOHN DOMINIC), an eminent astrono-

mer, was born of noble parents, at a town in Piedmont in Italy, June 8, 1625. After laying a proper foundation in his studies at home, he was sent to continue them in a college of Jesuits at Genoa. He had an uncommon turn for Latin poetry, which he exercised so very early, that some of his poems were published when he was but 11 years old. At length he met with books of astronomy, which he read, with great eagerness. Pursuing the bent of his inclinations in this way, in a short time he made so amazing a progress, that in 1650 the senate of Bologna invited him to be their public mathematical professor. Cassini was but 25 years of age when he went to Bologna, where he taught mathematics; and made observations upon the heavens, with great care and assiduity. In 1652 a comet appeared, which he observed with great accuracy; and he discovered that comets were not bodies accidentally generated in the atmosphere, as had been supposed, but of the same nature, and probably governed by the same laws, as the planets. The same year he resolved an astronomical problem, which Kepler and Bulliald had given up as insolvable; viz, to determine geometrically the apogee and eccentricity of a planet, from its true and mean place.—In 1653, when a church in Bologna was repaired and enlarged, he obtained leave of the senate to correct and settle a meridian line, which had been drawn by an astronomer in 1575.—In 1657 he attended, as an assistant, a nobleman, who was sent to Rome to compose some differences, which had arisen between Bologna and Ferrara, from the inundations of the Po; and he shewed so much skill and judgment in the management of the affair, that in 1663 the pope's brother appointed him inspector general of the fortifications of the castle of Urbino: and he had afterward committed to him the care of all the rivers in the ecclesiastical state.

Mean while he did not neglect his astronomical studies, but cultivated them with great care. He made several discoveries relating to the planets Mars and Venus, particularly the revolution of Mars upon his own axis: but the point he had chiefly in view, was to settle an accurate theory of Jupiter's satellites; which, after much labour and observation, he happily effected, and published it at Rome, among other astronomical pieces, in 1666.

Picard, the French astronomer, getting Cassini's tables of Jupiter's satellites, found them so very exact, that he conceived the highest opinion of his skill; and from that time his fame increased so fast in France, that the government desired to have him a member of the academy. Cassini however could not leave his station without leave of his superiors; and therefore the king, Lewis the 14th, requested of the pope and the senate of Bologna, that Cassini might be permitted to come into France. Leave was granted for 6 years; and he came to Paris in the beginning of 1669, where he was immediately made the king's astronomer. When this term of 6 years was near expiring, the pope and the senate of Bologna insisted upon his return, on pain of forfeiting his revenues and emoluments, which had hitherto been remitted to him: but the minister Colbert prevailed on him to stay, and he was naturalized in 1673; the same year also in which he was married.

The Royal Observatory of Paris had been finished

some time. The occasion of its being built was this: In 1638, the celebrated Merfenne was the chief institutor and promoter of a society, where several ingenious and learned men met together to talk upon physical and aftronomical subjects; among whom were Gaffend, Defcartes, Monmort, Thevenot, Bulliald, our countryman Hobbes, &c: and this society was kept up by a succession of learned men for many years. At length the government confidering that a number of such men, acting in a body, would fucceed much better in the promotion of fcience, than if they acted feparately, each in his particular art or province, eftablished under the direction of Colbert, in 1666, the Royal Academy of Sciences: and for the advancement of aftronomy in particular, erected the Royal Obfervatory at Paris, and furnifhed it with all kinds of inftruments that were neceffary to make obfervations. The foundation of this noble pile was laid in 1667, and the building completed in 1670. Of this obfervatory, Caffini was appointed to be the firft inhabiter; which he took poffeffion of in Sept. 1671, when he fet himfelf with frefh alacrity to attend the duties of his profeffion. In 1672 he endeavoured to determine the parallax of Mars and the fun: and in 1677 he proved that the diurnal rotation of Jupiter round his axis was performed in 9 hours 58 minutes, from the motion of a fpot in one of his larger belts: alfo in 1684 he difcovered four fatellites of Saturn, befides that which Huygens had found out. In 1693 he publifhed a new edition of his " Tables of Jupiter's Satellites," corrected by later obfervations. In 1695 he took a journey to Bologna, to examine the meridian line, which he had fixed there in 1655; and he fhewed, in the prefence of eminent mathematicians, that it had not varied in the leaft, during that 40 years. In 1700 he continued the meridian line through France, which Picard had begun, to the very fouthern limits of that country.

After our author had refided at the royal obfervatory for more than 40 years, making many excellent and ufeful difcoveries, which he publifhed from time to time, he died September the 14th, 1712, at 87 years of age; and was fucceeded by his only fon James Caffini. His publications were very numerous, far too much fo, even to be enumerated in this place.

CASSINI (JAMES), a celebrated French aftronomer, and member of the feveral Academies of Sciences of France, England, Pruffia, and Bologna, was born at Paris Feb. 18, 1677, being the younger fon of John-Dominic Caffini, above mentioned, whom he fucceeded as aftronomer at the royal obfervatory, the elder fon having loft his life at the battle of La Hogue.

After his firft ftudies in his father's houfe, in which it is not to be fuppofed that mathematics and aftronomy were neglected, he was fent to ftudy philofophy at the Mazarine college, where the celebrated Varignon was then profeffor of mathematics; from whofe affiftance young Caffini profited fo well, that at 15 years of age he fupported a mathematical thefis with great honour. At the age of 17 he was admitted a member of the Academy of Sciences; and the fame year he accompanied his father in his journey to Italy, where he affifted him in the verification of the meridian at Bologna, and other meafurements. On his return he made other fimilar operations in a journey into Holland, where he

difcovered fome errors in the meafure of the earth by Snell, the refult of which was communicated to the Academy in 1702. He made alfo a vifit to England in 1696, where he was made a member of the Royal Society.—In 1712 he fucceeded his father as aftronomer royal at the obfervatory.—In 1717 he gave to the Academy his refearches on the diftance of the fixed ftars, in which he fhewed that the whole annual orbit, of near 200 million of miles diameter, is but as a point in comparifon of that diftance. The fame year he communicated alfo his difcoveries concerning the inclination of the orbits of the fatellites in general, and efpecially of thofe of Saturn's fatellites and ring.—In 1725 he undertook to determine the caufe of the moon's libration, by which fhe fhews fometimes a little towards one fide, and fometimes a little on the other, of that half which is commonly behind or hid from our view.

In 1732 an important queftion in aftronomy exercifed the ingenuity of our author. His father had determined, by his obfervations, that the planet Venus revolved about her axis in the fpace of 23 hours: and M. Bianchini had publifhed a work in 1729, in which he fettled the period of the fame revolution at 24 days 8 hours. From an examination of Bianchini's obfervations, which were upon the fpots in Venus, he difcovered that he had intermitted his obfervations for the fpace of 3 hours, from which caufe he had probably miftaken new fpots for the old ones, and fo had been led into the miftake. He foon afterwards determined the nature and quantity of the acceleration of the motion of Jupiter, at half a fecond per year, and of that of the retardation of Saturn at two minutes per year; that thefe quantities would go on increafing for 2000 years, and then would decreafe again.—In 1740 he publifhed his Aftronomical Tables; and his Elements of Aftronomy; very extenfive and accurate works.

Although aftronomy was the principal object of our author's confideration, he did not confine himfelf abfolutely to that branch, but made occafional excurfions into other fields. We owe alfo to him, for example, Experiments on Electricity, or the light produced by bodies by friction. Experiments on the recoil of fire arms; Refearches on the rife of the mercury in the barometer at different heights above the level of the fea; Reflections on the perfecting of burning-glaffes; and other memoirs.

The French Academy had properly judged that one of its moft important objects, was the meafurement of the earth. In 1669 Picard meafured a little more than a degree of latitude to the north of Paris; but as that extent appeared too fmall from which to conclude the whole circumference with fufficient accuracy, it was refolved to continue that meafurement on the meridian of Paris to the north and the fouth, through the whole extent of the country. Accordingly, in 1683, the late M. de la Hire continued that on the north fide of Paris, and the older Caffini that on the fouth fide. The latter was affifted in 1700 in the continuation of this operation by his fon our author. The fame work was farther continued by the fame Academicians; and finally the part left unfinifhed by de la Hire in the north, was finifhed in 1718 by our author, with the late Maraldi, and de la Hire the younger.

Thefe operations produced a confiderable degree of precifion.

precifion. It appeared alfo, from this meafured extent of 6 degrees, that the degrees were of different lengths in different parts of the meridian; and in fuch fort that our author concluded, in the volume publifhed for 1718, that they decreafed more and more towards the pole, and that therefore the figure of the earth was that of an oblong fpheroid, or having its axe longer than the equatorial diameter. He alfo meafured the perpendicular to the fame meridian, and compared the meafured diftance with the differences of longitude as before determined by the eclipfes of Jupiter's fatellites; from whence he concluded that the length of the degrees of longitude was fmaller than it would be on a fphere, and that therefore again the figure of the earth was an oblong fpheroid; contrary to the determination of Newton by the theory of gravity. In confequence of thefe affertions of our author, the French government fent two different fets of meafurers, the one to meafure a degree at the equator, the other at the polar circle; and the comparifon of the whole determined the figure to be an oblate fpheroid, contrary to Caffini's determination.

After a long and laborious life, our author James Caffini loft his life by a fall in April 1756, in the 80th year of his age, and was fucceeded in the Academy and Obfervatory by his fecond fon Cefar-François de Thury. He publifhed, A Treatife on the Magnitude and Figure of the Earth; as alfo The Elements or Theory of the Planets, with Tables; befide an infinite number of papers in the Memoirs of the Academy, from the year 1699 to 1755.

CASSINI de Thury (Cesar-Francois), a celebrated French aftronomer, director of the obfervatory, penfioner aftronomer, and member of moft of the learned focieties of Europe, was born at Paris June 17, 1714, being the fecond fon of James Caffini, whofe occupations and talents our author inherited and fupported, with great honour. He received his firft leffons in aftronomy and mathematics from MM. Maraldi and Camus. He was hardly 10 years of age when he calculated the phafes of the total eclipfe of the fun of 1727. At the age of 18 he accompanied his father in his two journies undertaken for drawing the perpendicular to the obfervatory meridian from Strafbourg to Breft. From that time a general chart of France was devifed; for which purpofe it was neceffary to traverfe the country by feveral lines parallel and perpendicular to the meridian of Paris, and our author was charged with the conduct of this bufinefs. He did not content himfelf with the meafure of a degree by Picard: fufpecting even that the meafures which had been taken by his father and grandfather were not exempt from fome errors, which the imperfections of their inftruments at leaft would be liable to, he again undertook to meafure the meridian of Paris, by means of a new feries of triangles, of a fmaller number, and more advantageoufly difpofed. This great work was publifhed in 1740, with a chart fhewing the new meridian of Paris, by two different feries of triangles, paffing along the fea coafts, to Bayonne, traverfing the frontiers of Spain to the Mediterranean and Antibes, and thence along the eaftern limits of France to Dunkirk, with parallel and perpendicular lines defcribed at the diftance of 6000 toifes from one another, from fide to fide of the country.—In

1735, he had been received into the academy as adjoint fupernumerary at 21 years of age.

A tour which our author made in Flanders, in company with the king, about 1741, gave rife to the particular chart of France, at the inftance of the king. Caffini publifhed different works relative to thefe charts, and a great number of the fheets of the charts themfelves.

In 1761, Caffini undertook an expedition into Germany; for the purpofe of continuing to Vienna the perpendicular of the Paris meridian; to unite the triangles of the chart of France with the points taken in Germany; to prepare the means of extending into this country the fame plan as in France; and thus to eftablifh fucceffively for all Europe a moft ufeful uniformity. Our author was at Vienna the 6th of June 1761, the day of the tranfit of the planet Venus over the fun, of which he obferved as much as the ftate of the weather would permit him to do, and publifhed the account of it in his *Voyage en Allemagne.*

Finally, M. Caffini, always meditating the perfection of his grand defign, profited of the late peace to propofe the joining of certain points taken upon the Englifh coaft with thofe which had been determined on the coaft of France, and thus to connect the general chart of the latter with that of the Britifh ifles, like as he had before united it with thofe of Flanders and Germany. The propofal was favourably received by the Englifh government, and prefently carried into effect, under the direction of the Royal Society, the execution being committed to the late General Roy; after whofe death the bufinefs was for fome time fufpended; but it has lately been revived under the aufpices of the duke of Richmond, Mafter General of the Ordnance, and the execution committed to the care of Col. Edward Williams and Capt. William Mudge, both refpectable officers of the Artillery, and Mr. Ifaac Dalby, who had before accompanied and affifted General Roy; from whofe united fkill and zeal the happieft profecution of this bufinefs may be expected.

M. Caffini publifhed in the volumes of Memoirs of the French Academy a prodigious number of pieces, chiefly aftronomical, too numerous to particularize in this place, between the years 1735 and 1770; confifting of aftronomical obfervations and queftions; among which are obfervable, Refearches concerning the Parallax of the fun, the moon, Mars, and Venus; On aftronomical refractions, and the effect caufed in their quantity and laws by the weather; Numerous obfervations on the obliquity of the ecliptic, and on the law of its variations. In fhort, he cultivated aftronomy for 50 years, of the moft important for that fcience that ever elapfed, for the magnitude and variety of objects, in which he commonly fuftained a principal fhare.

M. Caffini was of a very ftrong and vigorous conftitution, which carried him through the many laborious operations in geography and aftronomy which he conducted. An habitual retention of urine however rendered the laft 12 years of his life very painful and diftreffing, till it was at length terminated by the fmall-pox the 4th of September 1784, in the 71ft year of his age; being fucceeded in the academy, and as director of the obfervatory, by his only fon the prefent count John-Dominic Caffini; who is the 4th in order by direct defcent in that honourable ftation.

CASSIOPEIA, one of the 48 old constellations, placed near Cepheus, not far from the north pole. The Greeks probably received this figure, as they did the rest, from the Egyptians, and in their fables added it to the family in the neighbouring part of the heavens, making her the wife of Cepheus, and mother of Andromeda. They pretend she was placed in this situation, to behold the destruction of her favourite daughter Andromeda, who is chained just by her on the shore, to be devoured; and that as a punishment for her pride and vanity in presuming to stand the comparison of beauty with the Nereids.

In the year 1572 there burst out all at once in this constellation a new star, which at first surpassed Jupiter himself in magnitude and brightness; but it diminished by degrees, till it quite disappeared at the end of 18 months. This star alarmed all the astronomers of that age, many of whom wrote dissertations upon it; among the rest Tycho Brahe, Kepler, Maurolycus, Lycetus, Gramineus, and others. Beza, the Landgrave of Hesse, Rosa, and others, wrote to prove it a comet, and the same that appeared to the Magi at the birth of Christ, and that it came to declare his second coming: these were answered by Tycho.

The stars in the constellation Cassiopeia, are in Ptolomy's catalogue 13, in Hevelius's 37, in Tycho's 46, and in Flamsteed's 55.

CASTOR, a moiety of the constellation Gemini; called also Apollo. Also a star in this constellation, whose latitude, for the year 1700, according to Hevelius, was 10° 4' 20" north; and its longitude ♊ 16° 4' 14"

CASTOR and POLLUX. See GEMINI.

CASTOR and POLLUX, in Meteorology, is a fiery meteor, which at sea appears sometimes adhering to a part of the ship, in the form of a ball, or even several balls. When one is seen alone, it is properly called Helena; but two are called Castor and Pollux, and sometimes Tyndaridæ.

By the Spaniards, Castor and Pollux are called San Elmo; by the French, St. Elme, St. Nicholas, St. Clare, St. Helene; by the Italians, Hermo; and by the Dutch, Vree Vuuren.

The meteor Castor and Pollux, it is commonly thought, denotes a cessation of the storm, and a future calm; as it is rarely seen till the tempest is nigh spent. But Helena alone portends ill weather, and denotes the severest part of the storm yet behind.

When the meteor adheres to the masts, yards, &c, it is concluded, from the air not having motion enough to dissipate this flame, that a profound calm is at hand; but if it flutter about, that it denotes a storm.

CASTRAMETATION, the art, or act, of encamping an army.

CATACAUSTICS, or *Catacaustic Curves*, in the Higher Geometry, are the species of caustic curves formed by reflection.

These curves are generated after the following manner: If there be an infinite number of rays A B, A C, A D, &c, proceeding from the radiating point A, and reflected at any given curve B C D H, so that the angles of incidence be still equal to the angles of reflection; then the curve B E G, to which the reflect-

ed rays B I, C E, D F, &c, are always tangents, as at the points I, E, F, &c, is the catacaustic, or caustic-by-reflection. Or it is the same thing as to say, that a caustic curve is that formed by joining the points of concurrence of the several reflected rays.

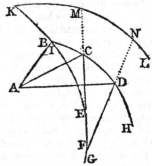

Some properties of these curves are as follow. If the reflected ray I B be produced to K, so that A B = B K, and the curve K L be the evolute of the caustic B E G, beginning at the point K; then the portion of the caustic B E is $= \overline{AC - AB} + \overline{CE - BI}$, that is, the difference of the two incident rays added to the difference of the two reflected rays.

When the given curve B C D is a geometrical one, the caustic will be so too, and will always be rectifiable. The caustic of the circle, is a cycloid, or epicycloid, formed by the revolution of a circle upon a circle.

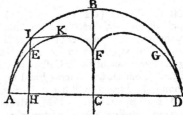

Thus, A B D being a semicircle exposed to parallel rays; then those rays which fall near the axis C B will be reflected to F, the middle point of B C; and those which fall at A, as they touch the curve only, will not be reflected at all; but any intermediate ray H I will be reflected to a point K, somewhere between A and F. And since every different incident ray will have a different focal point, therefore those various focal points will form a curve line A E F in one quadrant, and F G D in the other, being the cycloid above-mentioned. And this figure may be beautifully exhibited experimentally by exposing the inside of a smooth bowl, or glass, to the sun beams, or strong candle light; for then this curve A E F G D will appear plainly delineated on any white surface placed horizontally within the same, or on the surface of milk contained in the bowl.

The caustic of the common cycloid, when the rays are parallel to its axis, is also a common cycloid, described by the revolution of a circle upon the same base. The caustic of the logarithmic spiral, is the same curve.

The principal writers on the caustics, are l'Hôpital, Carré, &c. See Memoires de l'Acad. an. 1666 & 1703.

CATA-

CATACOUSTICS, or *Cataphonics*, is the science of reflected sounds; or that part of acoustics which treats of the properties of echoes.

CATADIOPTRICAL *Telescope*, the same as Reflecting telescope; which see.

CATALOGUE *of the Stars*, is a list of the fixed stars, disposed according to some order; in their several constellations; with the longitudes, latitudes, right-ascensions, &c, of each.

Catalogues of the stars have usually been disposed, either as collected into certain figures called constellations, or according to their right ascensions, that is the order of their passing over the meridian. All the catalogues, from the most ancient down to Flamsteed's inclusively, were of the first of these forms, or in constellations: but most of the others since that have been of the latter form, as being much more convenient for most purposes. Indeed one has lately been disposed in classes according to zones or degrees of polar distance.

Hipparchus of Rhodes first undertook to make a catalogue of the stars, about 128 years before Christ; in which he made use of the observations of Timocharis and Aristyllus, for about 140 years before him. Ptolomy retained Hipparchus's catalogue, containing 1026 fixed stars in 48 constellations, though he himself made abundance of observations, with a view to a new catalogue, an. dom. 140. Albategni, a Syrian, brought the same down to his own time, viz. about the year of Christ 880. Anno 1437, Ulugh Beigh, or Beg, king of Parthia and India, made a new catalogue of 1022 fixed stars, or according to some 1016; since translated out of Persian into Latin by Dr. Hyde, in 1665. The third person who made a catalogue of stars from his own observations was Tycho Brahe, who determined the places of 777 stars for the end of the year 1600; which Kepler, from other observations of Tycho, afterwards increased to the number of 1000 in the Rudolphine tables; adding those of Ptolomy and other authors, omitted by Tycho; so that his catalogue amounts to above 1160. About the same time, William, landgrave of Hesse, with his mathematicians Byrgius and Rothman, determined the places of 400 stars from new observations, rectifying them for the year 1593; which Hevelius prefers to those of Tycho. Ricciolus, in his Astronomia Reformata, determined the places of 101 stars for the year 1700, from his own observations: for the rest he followed Tycho's catalogue; altering it where he thought fit. Anno 1667 Dr. Halley, in the island of St. Helena, observed 350 of the southern stars, not visible in our horizon. The same labour was repeated by Father Noel in 1710, who published a new catalogue of the same stars constructed for the year 1687. Also De la Caille, at the Cape of Good Hope, made accurate observations of about 10 thousand stars near the south pole, in the years 1751 and 1752; the catalogue of which was published in the Memoirs of the French Academy of Sciences for the year 1752, and in some of his own works, as more particularly noticed below.

Bayer, in his Uranometria, published in 1603 a catalogue of 1160 stars, compiled chiefly from Ptolomy and Tycho, in which every star is marked with some letter of the Greek alphabet; the brightest or principal star in any constellation being denoted by the first letter of the alphabet, the next star in order by the 2d letter, and so on; and when the number of stars exceeds the Greek alphabet, the remaining stars are marked by the letters of the Roman alphabet; which letters are preserved by Flamsteed in his catalogue, and by Senex on his globes, and indeed by most astronomers since that time.

In 1673, the celebrated John Hevelius, of Dantzick, published, in his *Machina Cœlestis*, a catalogue of 1888 stars, of which 1553 were observed by himself; and their places set down for the end of the year 1660. But this catalogue, as it stands in Flamsteed's *Historia Cœlestis* of 1725, contains only 1520 stars.

The most complete catalogue ever given from the labours of one man, was the Britannic catalogue, compiled from the observations of the accurate and indefatigable Mr. Flamsteed, the first Royal Astronomer at Greenwich; who for a long series of years devoted himself wholly to that business. As there was nothing wanting either in the observer or apparatus, his may be considered as a perfect work, so far as it goes. It is however to be regretted that the edition had not passed through his own hands: that now extant was published by authority, but without the author's consent, and contains 2734 stars. Another edition was published in 1725, pursuant to his testament, containing 3000 stars, with their places adapted to the beginning of the year 1689; to which is added Mr. Sharp's catalogue of the southern stars not visible in our hemisphere, set down for the year 1726. See vol. 3 of his *Historia Cœlestis*, in which are printed the catalogues of Ptolomy, Ulugh Beigh, Tycho, the Prince of Hesse, and Hevelius; with an account of each of them in the Prolegomena.

The first catalogue we believe that was printed in the new or second form, according to the order of the right ascensions, is that of De la Caille, given in his Ephemerides for the 10 years between 1755 and 1765, and printed in 1755. It contains the right ascensions and declinations of 307 stars, adapted to the beginning of the year 1750.—In 1757, De la Caille published his *Astronomiæ Fundamenta*, containing a catalogue of the right ascensions and declination of 398 stars, likewise adapted to the beginning of 1750.—And in 1763, the year after his death, was published the *Cœlum Australe Stelliferum* of the same author; containing a catalogue of the places of 1942 stars, all situated to the southward of the tropic of Capricorn, and observed by him while he was at the Cape of Good Hope, in 1751 and 1752; their places being also adapted to the beginning of 1750.—In the same year was published his Ephemerides for the 10 years between 1765 and 1775; in the introduction to which are given the places of 515 zodiacal stars, all deduced from the observations of the same author; the places adapted to the beginning of the year 1765.

In the Nautical Almanac for 1773, is given a catalogue of 387 stars, in right ascension, declination, longitude, and latitude, derived from the observations of the late celebrated Dr. Bradley, and adjusted to the beginning of the year 1760. This small catalogue, and the results of about 1200 observations of the moon, are all that the public have yet seen of the multiplied labours

of

of this moft accurate and indefatigable obferver, although he has now (1794) been dead upwards of 32 years.

In 1775 was publifhed a thin volume entitled *Opera Inedita*, containing feveral papers of the late Tobias Mayer, and among them a catalogue of the right afcenfions and declinations of 998 ftars, which may be occulted by the moon and planets; the places being adapted to the beginning of the year 1756.

At the end of the firft volume of 'Aftronomical Obfervations made at the Royal Obfervatory at Greenwich,' publifhed in 1776, Dr. Mafkelyne, the prefent Aftronomer Royal, has given a catalogue of the places of 34 principal ftars, in right-afcenfion and north-polar diftance, adapted to the beginning of the year 1770. Thefe, being the refult of feveral years' repeated obfervations, made with the utmoft care, and the beft inftruments, it may be prefumed are exceedingly accurate.

In 1782, M. Bode, of Berlin, publifhed a very extenfive catalogue of 5058 of the fixed ftars, collected from the obfervations of Flamfteed, Bradley, Hevelius, Mayer, De la Caille, Meffier, Monnier, D'Arquier, and other aftronomers; all adapted to the beginning of the year 1780; and accompanied with a Celeftial Atlas, or fet of maps of the conftellations, engraved in a moft delicate and beautiful manner.

To thefe may be added, Dr. Herfchel's catalogue of double ftars, printed in the Philof. Tranf. for 1782 and 1783; Meffier's nebulæ and clufters of ftars, publifhed in the Connoiffance des Temps for 1784; and Herfchel's catalogue of the fame kind, given in the Philof. Tranf. for 1786.

In 1789 Mr. Francis Wollafton publifhed ' A Specimen of a General Aftronomical Catalogue, in Zones of North-polar Diftance, and adapted to Jan. 1, 1790.' Thefe ftars are collected from all the catalogues beforementioned, from that of Hevelius downwards. This work contains five diftinct catalogues; viz, Dr. Mafkelyne's new catalogue of 36 principal ftars; a general catalogue of all the ftars, in zones of north-polar diftance; an index to the general catalogue; a catalogue of all the ftars, in the order in which they pafs the meridian; and a catalogue of zodiacal ftars, in longitude and latitude.

Finally, in 1792, Dr. Zach publifhed at Gotha, *Tabulæ Motuum Solis*, to which is annexed a new catalogue of the principal fixed ftars, from his own obfervations made in the years 1787, 1788, 1789, 1790. This catalogue contains the right afcenfions and declinations of 381 principal ftars, adapted to the beginning of the year 1800.

CATAPULT, *Catapulta*, a military engine, much ufed by the ancients for throwing huge ftones, and fometimes large darts and javelins, 12, 15, or even 18 feet long, on the enemy. It is fometimes confounded with the Balliftta, which is more peculiarly adapted for throwing ftones; fome authors making them the fame, and others different.

The catapulta, which it is faid was invented by the Syrians, confifted of two huge timbers, like mafts of fhips, placed againft each other, and bent by an engine for the purpofe; thefe being fuddenly unbent again by the ftroke of a hammer, threw the javelins with prodigious force. Its ftructure and the manner of

working it are defcribed by Vitruvius; and a figure of it is alfo given by Perrault. M. Folard afferts that the catapult made infinitely more diforder in the ranks than our cannon charged with iron balls. See Vitruv. Archit. lib. x. cap. 15 and 18; and Perr. notes on the fame; alfo Rivius, pa. 597.

Jofephus takes notice of the furprifing effects of thefe engines, and fays, that the ftones thrown out of them, of a hundred weight or more, beat down the battlements, knocked the angles off the towers, and would level a whole file of men from one end to the other, were the phalanx ever fo deep.

See plate V, fig. 3 and 4, for two forms of the catapult, the one for throwing darts and javelins, the other for ftones.

CATENARY, a curve line which a chain, cord, or fuch like, forms itfelf into, by hanging freely from two points of fufpenfion, whether thefe be in the fame horizontal line or not; as the curve ACB, formed by a heavy flexible line fufpended by any two points A and B.

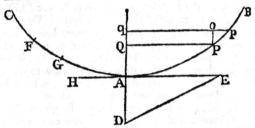

The nature of this curve was fought after by Galileo, who thought it was the fame with the parabola; but though Jungius detected this miftake, its true nature was not difcovered till the year 1691, in confequence of M. John Bernoulli having publifhed it as a problem in the Acta Eruditorum, to the mathematicians of Europe. In 1697 Dr. David Gregory publifhed an inveftigation of the properties before difcovered by Bernoulli and Leibnitz; in which he pretends that an inverted catenary is the beft figure for an arch of a bridge &c. See Philof. Tranf. abr. vol. 1. pa. 39; alfo Bernoulli Opera, vol. 1. pa. 48, and vol. 3. pa. 491; and Cotes's Harmon. Menfur. pa. 108.

The catenary is a curve of the mechanical kind, and cannot be expreffed by a finite algebraical equation, in fimple terms of its abfcifs and ordinate; but is eafily expreffed by means of fluxions; thus if AQ be its axis perpendicular to the horizon, and PQ an ordinate parallel to the fame, or perp. to AQ; alfo pq another ordinate indefinitely near the former, and po parallel to AQ; then, a being fome given or conftant quantity, the fundamental property of the curve is this, viz, Po : op :: AP : a, or $\dot{x} : \dot{y} :: z : a$, that is, the fluxion of the axis, is to the fluxion of the ordinate, as the length of the curve is to the given quantity a; where $x = $ AQ, $y = $ PQ, and $z = $ AP. This, and the other properties of the curve, will eafily appear from the following confiderations: Firft, fuppofing the curve hung up by its two points B and C againft a perpendicular or upright wall: then, every lower part of the curve being kept in its pofition by the tenfion of that which is immediately above it; the lower parts of the curve will retain the fame pofition unvaried; by whatever points it is fufpended

pended above; thus, if it were fixed to the wall by the point F, or G, &c, the whole curve CAB will remain just as it was; for the tensions at F and G have the same effect upon the other parts of the curve as when it is fixed by those points: and hence it follows that the tension of the curve at the point A, in the horizontal direction, is a constant quantity, whether the two legs or branches of the curve, on both sides of it, be longer or shorter: which constant tension at A let be denoted by the quantity a.

Now because any portion of the curve, as AP, is kept in its position by three forces, viz, the tensions at its extremities A and P, and its own weight, of which the tension at A acts in the direction AH or po, and the tension at P acts in the direction Pp, and the wt. of the line acts in the perpendicular direction oP; that is, the three forces which retain the curve AP in its position, act in the directions of the sides of the elementary triangle opP; but, by the principles of mechanics, any three forces, keeping a body in equilibrio, are proportional to the three sides of a triangle drawn in the directions in which those forces act; therefore it follows that the forces keeping AP in its position, viz, the tension at A, the tension at P, and the wt. of AP, are respectively as op, pP, and oP,

that is, as $\dot{y}$, $\dot{z}$, and $\dot{x}$.

But the tension at A is the constant quantity a, and the wt. of the uniform curve AP may be expounded by its length z; therefore it follows that $\dot{x} : \dot{y} :: z : a$; which was to be proved.

Also from this last proportion, by proper analogy, or similar combinations of the terms, there arises this other property, $\dot{x} : \sqrt{\dot{x}\dot{x} + \dot{y}\dot{y}}$ or $\dot{z} :: z : \sqrt{aa + zz}$, or $z\dot{z} = \dot{x}\sqrt{aa + zz}$, or $\dot{x} = \dfrac{z\dot{z}}{\sqrt{a^2 + z^2}}$; and the fluents of these give $x = \sqrt{a^2 + z^2}$. But at the vertex of the curve, where $x = 0$, and $z = 0$, this becomes $0 = \sqrt{a^2 + 0} = a$; and therefore by correction the true equation of the fluents is $x = \sqrt{a^2 + z^2} - a$, or $a + x = \sqrt{a^2 + z^2}$: and hence also $z = \sqrt{\overline{a + x}^2 - a^2} = \sqrt{2ax + x^2}$ and $a = \dfrac{z^2 - x^2}{2x}$.

Any of which is the equation of the curve in terms of the arch and its abscifs; in which it appears that $a + x$ is the hypothenuse of a right-angled triangle whose two legs are a and z. So that, if in QA and HA produced, there be taken AD $= a$, and AE $=$ the curve z or AP; then will the hypothenuse DE be $= a + x$ or DQ. And hence, any two of these three, a, x, z, being given, the third is given also.

Again, from the first simple property, viz, $x : y ::$ $z : a$, or $ax = zy$, by substituting the value of z above found, it becomes $a\dot{x} = \dot{y}\sqrt{2ax + x^2}$, or $y = \dfrac{a\dot{x}}{\sqrt{2ax + x^2}}$; and the fluent of this equation is $y = 2a \times$ hyp. log. of $\sqrt{x} + \sqrt{2a + x}$. But at the vertex of the curve, where $x = 0$ and $y = 0$, this becomes $0 = 2a \times$ hyp. log. of $\sqrt{2a}$; therefore the correct equation of the fluents is $y = 2a \times$ hyp. log. of

$\dfrac{\sqrt{x} + \sqrt{2a + x}}{\sqrt{2a}}$; an equation to the curve also, in terms of x and y, but not in simple algebraic terms. This last equation however may be brought to much simpler terms in different ways; as first by squaring the logarithmic quantity and dividing its coef. by 2, then

$y = a \times$ hyp. log. of $\dfrac{a + x + \sqrt{2ax + x^2}}{a} = a \times$

hyp. log. $\dfrac{a + x + z}{a}$; and 2d by multiplying both numerator and denominator by $\sqrt{2a + x} - \sqrt{x}$, then squaring the product, and dividing the coef. by 2, which gives $y = a \times$ hyp. log. $\dfrac{z + x}{\sqrt{z^2 - x^2}} = a \times$

hyp. log. $\dfrac{\overline{z + x}^2}{z^2 - x^2} = a \times$ hyp. log. $\dfrac{z + x}{z - x}$.

CATHETUS, in Geometry, a name by which the perpendicular leg of a right-angled triangle is sometimes called. Or it is in general any line or radius falling perpendicularly on another line, or surface.

Cathetus of *Incidence*; in Catoptrics, is a right line drawn from a radiant point, or point of incidence, perpendicular to the reflecting line, or plane of the speculum.

Cathetus of *Reflection*, or *of the Eye*, a right line drawn from the eye, perpendicular to the plane of reflection.

Cathetus, in Architecture, denotes the axis of a column &c. In the Ionic Capital, it denotes a line passing perpendicularly through the eye or centre of the volute.

CATOPTRICS, the science of reflex vision, or that part of optics which explains the laws and properties of light reflected from mirrors, or specula.

The first treatise extant on catoptrics, is that which was composed by Euclid: this was published in Latin in 1604 by John Pena; it is also contained in Herigon's Course of Mathematics, and in Gregory's edition of the works of Euclid: though it is suspected by some that this piece was not the work of that great geometrician, notwithstanding that it is ascribed to him by Proclus in lib. 2, and by Marinus in his Preface to Euclid's Data. Alhazen, an Arabian author, composed a large volume of optics about the year 1100, in which he treats pretty fully of catoptrics: and after him Vitello, a Polish writer, composed another about the year 1270. Tacquet, in his Optics, has very well demonstrated the chief propositions of plane and spherical speculums. And the same is very ably done by Dr. Barrow in his Optical Lectures. There are also Trabe's Catoptrics; David Gregory's Elements of Catoptrics; Wolfius's Elements of Catoptrics; and those of Dr. Smith, contained in his learned and very elaborate Treatise on Optics; and many others of less note.

As this subject is treated under the general term Optics, the less need be said of it here. The whole doctrine of Catoptrics depends upon this simple principle, that the angle of incidence is equal to the angle of reflection, that is, that the angle in which a ray of light falls upon any surface, called the angle of incidence,

is

is equal to the angle in which it quits it when reflected from it, called the angle of reflection; though it is sometimes defined that the angles of incidence and reflection, are those which the incident and reflected rays make, not with the reflecting surface itself, but with a perpendicular to that surface, at the point of contact, which are the complements to the others: but it matters not by what name these angles are called, as to the truth and principles of the science; since, if the angles are equal, their complements are also equal. This principle of the equality between the angles of incidence and reflection, is mere matter of experience, being a phenomenon that has always been observed to take place, in every case that has fallen under observation, as near at least as mechanical measurements can ascertain; and hence it is inferred that it is a universal law of nature, and to be considered as matter of fact in all cases. Thus, let AC be an incident ray falling upon the reflecting surface DE, and CB the reflected ray, also CO perpendicular to DE; then is the angle ACD = BCE, or the angle ACO = BCO.

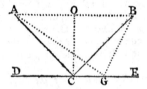

Of this law in nature, viz, the equality between the angles of incidence and reflection, it is remarkable, that in this way, the length or rout AC + CB, in a ray passing from any point A to another given point B, by being reflected from any surface DE, is the shortest possible, namely AC + CB is shorter than the sum AG + GB of any other two lines inflected at the line DE; and hence also the passage of the ray from A to B is performed in a shorter time than if it had passed by any other way.

From this simple principle, and the common properties of lineal geometry, the chief phenomena of catoptrics are easily deduced, and are as here following.

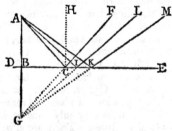

1. Rays of light reflected from a plane surface, have the same inclination to each other after reflection as they had before it. Thus, the rays AC, AI, AK, issuing from the radiant point A, and reflected by the surface DE into the lines CF, IL, KM; these latter lines will have the same inclination to each other as the former AC, AI, AK have. For draw ABG perpendicular to DE, and produce FC, LI, MK backwards to meet this perpendicular, so shall they all meet in the same point G, and AB will in every case be equal to BD: for the incident ∠ ACB is equal to the re-

flected ∠ FCE, which is equal to the opposite angle BCG; so that the two triangles ABC, GBC, have the angles at C equal, as also the right angles at B equal, and consequently the 3d angles at A and G equal; and having also the side BC common, they are equal in all respects, and so AB = BG. And the same for the other rays. Consequently the angles BGC, BGI, BGK are respectively equal to the angles BAC, BAI, BAK; that is, the reflected rays have the same inclinations as the incident ones have.

2. Hence it is that the image of an object, seen by reflection from a plain mirror, seems to proceed from a place G as far beyond, or on the other side of the reflecting plane DE, as the object A itself is before the plane. This is when the incident rays diverge from some point as A.

But if the case be reversed, and FC, LI, MK be considered as incident rays, issuing from points F, L, M, and converging to some point G beyond the reflecting plane; then CA, IA, KA will become the reflected rays, and they will converge to the point A as far before the plane, as the point G is beyond it.

So that universally, when the incident rays diverge from a point A, the reflected rays will also diverge from a point G; and when the incident rays converge towards a point G, the reflected ones will also converge to a point A; and in both cases these two points are at equal distances on the opposite sides of the reflecting plane DE.

3. Parallel rays reflected from a concave spherical surface, converge after reflection. For, let AF, CD, EB be three parallel rays falling upon the concave surface FB, whose centre is C. To the centre draw the perpendiculars FC and BC; also draw FM making the reflected angle CFM equal to the incident angle CFA; and in like manner BM to make the angle CBM = the angle CBE; so shall the rays AF and EB be reflected into the converging rays FM and BM. As to the ray CD, being perpendicular to the surface, it is reflected back again in the same line DC.

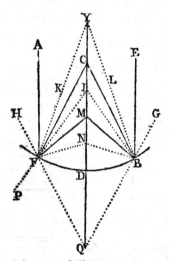

4. Converging rays falling upon the concave surface are made to converge more. Thus, let GB and HF be the incident rays: then because the incident angle HFC is larger than the angle AFC, therefore the equal reflected

reflected angle NFC is greater than the reflected angle MFC, and so the point N is below the point M, or the line FN below the line FM; and in like manner BN is below BM; that is, the reflected rays FN and BN are more converging in this case, than FM and BM in the other.

5. The focus to which all parallel rays, falling near the vertex D, are reflected, is in the middle of the radius M. For, because the ∠ MFC = ∠ AFC which is = the alternate ∠ FCM, therefore the sides opposite these angles are also equal, namely the side FM = CM; consequently when the point F is very near the vertex D, then the sum CM + MF is nearly = CD, and so CM nearly = MD, or the focus of the parallel rays is nearly in the middle of the radius.—But the focus of other reflected rays is either above or below that of the parallel rays; namely, below when the incident rays are converging, and above when they are diverging; as is evident by inspection; thus, N the reflected focus of the converging rays GB and HF, is below M; I that of the diverging rays YB and YF, is above M.

6. Incident and reflected rays are reciprocal, or so that if the reflected rays be returned back, or considered as incident ones, they will be reflected back into what were before their incident rays. And hence it follows that diverging rays, after reflection from a concave spherical surface, become either parallel or less diverging than before. Thus the incident rays MF and MB are reflected into the parallel rays FA and BE, and the rays NF and NB are reflected into FH and BG, which are less diverging; also the rays IF and IB are reflected into FK and BL, which converge.—And hence all the phenomena of concave mirrors will be evident.

7. Rays reflected from a convex speculum, become quite contrary to those reflected from a concave one; so that the parallel rays become diverging, and the diverging rays become still more diverging; also converging rays will become either diverging, or parallel, or else less converging. Thus BDF being a spherical surface, whose centre is C, produce the radii CBV and CFT which are perpendicular to the surface; then it is evident that the parallel rays AF and EB will be reflected into the diverging ones FK and BL; and the diverging rays YB and YF become BO and FP which are more diverging; also the converging rays HF and GB become FR and BS which diverge, or else KF and LB become FA and BE which are parallel, or else lastly PF and OB become FY and BY which are converging.

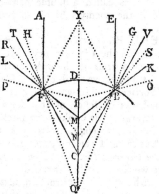

8. Hence, as in the concave speculum, so also in the convex one, of parallel incident rays AF and EB, the imaginary focus M of their reflected rays FL and BK, is in the middle of the radius when the speculum is a small segment of a sphere: but the reflected imaginary focus of other rays is either above or below the middle point M, viz N being that of the converging rays GB and HF, below M; but I, that of the diverging rays YB and YF, above M.

9. When the speculum is the small segment of a sphere, either convex or concave, and the incident rays either converging or diverging, the distances of the foci, or points of concurrence, of the incident rays, and of the reflected rays, from the vertex of the speculum, are directly proportional to the distances of the same from the centre of it;
that is YD : ID :: YC : IC,
and QD : ND :: QC : NC.
For because the radius CF, or the same produced, bisects the angle YFI in the concave speculum, or the external angle YFP in the convex one, therefore YF : IF :: YC : IC; but when F is very near to D, then YF and IF become nearly YD and ID; consequently YD : ID :: YC : IC.

In like manner, because CF bisects the angle QFN in the convex, or its external angle NFH in the concave speculum, therefore QF : FN :: QC : NC; but when F is very near to D, then QF and FN become nearly QD and ND; and therefore QD : ND :: QC : NC.

For example, suppose it were required to find the focal distance of diverging rays incident upon a convex surface, the radius of the sphere being 5 inches, and the distance of the radiant point from the surface 20 inches. Here then are given YD = 20, and CD = 5, to find ID : then
the theorem $\qquad$ YD : ID :: YC : IC,
in numbers is $\qquad$ 20 : ID :: 25 : 5 — ID,
or by permutation 20 : 25 :: ID : 5 — ID,
and by composition 45 : 20 :: 5 : ID $= \frac{100}{45} = \frac{20}{9} =$ $2\frac{2}{9}$ the focal distance sought.

And if it should happen in any case that the value of ID in the calculation should come out a negative quantity, the focal distance must then be taken on the contrary side of the surface.

From the foregoing principles may be deduced and collected the following practical maxims, for plane and spherical mirrors, viz;

I. In a Plane Mirror,

(1). The image will appear as far behind the mirror, as the object is before it.

(2). The image will appear of the same size, and in the same position as the object.

(3). Any plain mirror will reflect the image of an object of twice its own length and breadth.

II. In a Spherical Convex Mirror,

(1). The image will always appear behind the mirror, or within the sphere.

(2). The image will be in the same position, but less than the object.

(3). The image will be curved, but not spherical, like the mirror.

(4). Parallel rays falling on this mirror, will have the

the image at half the distance of the centre from the mirror.

(5). In converging rays, the distance of the object must be equal to half the distance of the centre, to make the image appear behind the mirror.

(6). Diverging rays will have their image at less than half the distance of the centre.

III. *In a Spherical Concave Mirror,*

(1). Parallel rays have their focus, or the image, at half the distance of the centre.

(2). In the centre of the sphere the image appears of the same dimensions as the object.

(3). Converging rays form an image before the mirror.

(4). In diverging rays, if the object be at less than half the distance of the centre, the image will be behind the mirror, erect, curved, and magnified; but if the distance of the object be greater, the image will be before the mirror, inverted and diminished.

(5). The solar rays, being parallel, will be collected in a focus at half the distance of its centre, where their heat will be augmented in proportion as the surface of the mirror exceeds that of the focal spot.

(6). If a luminous body be placed in the focus of a concave mirror, its rays, being reflected in parallel lines, will strongly enlighten a space of the same dimensions with the mirror, at a great distance. If the luminous object be placed nearer than the focus, its rays will diverge, and so enlighten a larger space, but not so strongly. And upon this principle it is that reverberators are constructed.

CATOPTRIC *Dial*, a dial that exhibits objects by reflected rays. See *Reflecting* DIAL.

CATOPTRIC *Telescope*, a telescope that exhibits objects by reflection. See *Reflecting* TELESCOPE.

CATOPTRIC *Cistula*, a machine, or apparatus, by which small bodies are represented extremely large, and near ones extremely wide, and diffused through a vast space; with other very pleasing phenomena, by means of mirrors, disposed by the laws of catoptrics, in the concavity of a kind of chest.

There are various kinds of these machines, accommodated to the various intentions of the artificer: some multiply the objects, some magnify, some deform them, &c. The structure of one or two of them will suffice to shew how many more may be made.

To make a Catoptric Cistula to represent several different scenes of objects, when viewed by different holes.

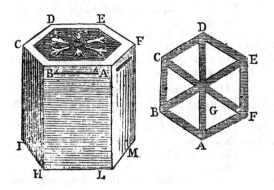

Provide a polygonal cistula, or box, like the multangular prism ABCDEF, and divide its cavity by diagonal planes AD, BE, CF, intersecting in the centre, into as many triangular cells as the chest has sides. Line those diagonal partitions with plain mirrors; and in the sides of the box make round holes, through which the eye may peep within the cells of it. These holes are to be covered with plain glasses, ground within-side, but not polished, to prevent the objects in the cells from appearing too distinctly. In each cell are to be placed the different objects whose images are to be exhibited; then covering up the top of the chest with a thin transparent membrane, or parchment, to admit the light, the machine is complete.

For, from the laws of reflection, it follows, that the images of objects, placed within the angles of mirrors, are multiplied, and appear some more remote than others; by which the objects in one cell will appear to take up more room than is contained in the whole box. Therefore by looking through one hole only, the objects in one cell will be seen, but those multiplied, and diffused through a space much larger than the whole box. Thus every hole will afford a new scene; and according to the different angles the mirrors make with each other, the representations will be different: if they be at an angle greater than a right one, the images will be monstrous, &c.

To make a Catoptric Cistula to represent the objects within it prodigiously multiplied, and diffused through a vast space.

Make a polygonous cistula or box, as before, but without dividing the inner cavity into any apartments, or cells; line the insides CBHI, BHLA, ALMF, &c, with plane mirrors, and at the holes pare off the tin and quicksilver, to look through; place any object in the bottom MI, as a bird in a cage, &c.

Now by looking through the aperture *hi*, each object placed at the bottom will be seen vastly multiplied, and the images removed at equal distances from one another, like a great multitude of birds, or a large aviary.

CAVALIER, in Fortification, a mount of earth raised in a fortress higher than the other works, on which to place cannon &c for scouring the field, or opposing a commanding work. Cavaliers are of different shapes; and are bordered with a parapet, to cover the cannon mounted upon them; their situation is also various, either in the curtain, bastion, or gorge. The cavalier is sometimes called a double bastion, and its use is to overlook the enemy's batteries, and to scour their trenches.

CAVALIERI (BONAVENTURA), an eminent Italian mathematician in the 17th century. He was a native of Milan, and a friar of the order of the Jesuati of St. Jerome. Cavalieri was a disciple of Galileo, and the friend of Torricelli. He was a very eminent mathematician, and was professor of that science at Bologna; where several of his books were published, and where he died in the year 1647. His works that have been published, as far as I can find, are as follow:

1. *Directorium Generale Uranometricum;* 4to, *Bononiæ,* 1632.—In this work the author treats of Trigonometry; and Logarithms, their construction, uses, and applications. The work includes also tables of logarithms

3   of

of common numbers; with trigonometrical tables, of natural fines, and logarithmic fines, tangents, fecants and verfed fines.

2. *Lo Specchio Uftorio overo Trattato delle Settioni Coniche : 4to, Bologna,* 1632.—An ingenious treatife of conic fections.

3. *Geometria Indivifibilibus continuorum nova quadam ratione promota : 4to, Bononiæ,* 1635; and a 2d edition in 1653.—This is a curious original work in geometry, in which the author conceives the geometrical figures as refolved into their very fmall elements, or as made up of an infinite number of infinitely fmall parts, and on account of which he paffes in Italy for the inventor of the infinitefimal calculus.

4. *Trigonometria Plana & Spharica, Linearis, & Logarithmica : 4to, Bononiæ,* 1643.—A very neat and ingenious treatife on Trigonometry; with the tables of fines, tangents, and fecants, both natural and logarithmical.

5. *Exercitationes Geometriæ Sex : 4to, Bononiæ,* 1647. This work contains Exercifes on the method of Indivifibles; Anfwers to the objections of Guldini; The ufe of Indivifibles in coffic powers or algebra, and in confiderations about gravity; with a mifcellaneous collection of problems.

CAUDA *Capricorni,* a fixed ftar of the 4th magnitude, in the tail of Capricorn; called alfo by the Arabs, Dineb Algedi; and marked γ by Bayer.

CAUDA *Ceti,* a fixed ftar of the 3d magnitude; called alfo by the Arabs, Dineb Kaetos; marked β by Bayer.

CAUDA *Cygni,* a fixed ftar of the 2d magnitude in the Swan's tail; called by the Arabs, Dineb Adigege, or Eldegiagich; and marked α by Bayer.

CAUDA *Delphini,* a fixed ftar of the 3d magnitude, in the tail of the Dolphin; marked ε by Bayer.

CAUDA *Draconis,* or Dragon's tail, the moon's fouthern or defcending node.

CAUDA *Leonis,* a fixed ftar of the firft magnitude in the Lion's tail; called alfo by the Arabs, Dineb Eleced; and marked β by Bayer. It is called alfo Lucida Cauda.

CAUDA *Urfæ Majoris,* a fixed ftar of the 3d magnitude, in the tip of the Great Bear's tail; called alfo by the Arabs, Alalioth, and Benenath; and marked η by Bayer.

CAUDA *Urfæ Minoris,* a fixed ftar of the 3d magnitude, at the end of the Leffer Bear's tail; called alfo the Pole Star, and by the Arabs, Alrukabah; and marked α by Bayer.

CAVETTO, a hollow member or moulding, containing a quadrant of a circle, and having an effect juft contrary to that of a quarter round. It is ufed as an ornament in cornices.

CAUSTIC Curves. See *Catacauftics,* and *Diacauftics.*

CAZEMATE. See CASEMATE.

CAZERN. See CASERN.

CEGINUS, a fixed ftar of the 3d magnitude, in the left fhoulder of Bootes; and marked γ by Bayer.

CELERITY, is the velocity or fwiftnefs of a body in motion; or that affection of a body in motion by which it can pafs over a certain fpace in a certain time.

CELESTIAL Globe, &c. See Globe, &c.

CELLARIUS (CHRISTOPHER), a learned geographer and hiftoriographer of the 17th century. He was born in 1638, at Smalcalde in Franconia, where his father was minifter. Our author was fucceffively rector of the colleges at Weymar, Zeits, and Merfbourg, and profeffor of eloquence and hiftory in the univerfity founded by the king of Pruffia at Hall in 1693, where he compofed the greateft part of his works.

His great application to ftudy haftened the infirmities of old age; for it has been faid, he would fpend whole days and nights together over his books, without any attention to his health, or even the calls of nature. He died in 1707, at 69 years of age.

Cellar was author of an amazing number of books, upon various fubjects; but thofe on account of which he has a place here, are his geographical works, which are as follow:

1. *Notitia Orbis Antiqui,* 2 vols in 4to; and is efteemed the beft work extant on the ancient geography.

2. *Atlas Cæleftis;* in folio.

3. *Hiftoria Antiqua,* 2 vols in 12mo; being an abridgement of univerfal hiftory.

CENTAURUS, the *Centaur,* one of the 48 old conftellations, being a fouthern one, and is in form half man and half horfe. It is fabled by the Greeks that it was Chiron the Centaur, who was the tutor of Achilles and Efculapius. The ftars of this conftellation are, in Ptolomy's catalogue 37, in Tycho's 4, and in the Britannic catalogue, with Sharp's appendix, 35.

CENTER. See CENTRE.

CENTESM, the 100th part of any thing.

CENTRAL, fomething relating to a centre. Thus we fay central eclipfe, central forces, central rule, &c.

CENTRAL *Eclipfe,* is when the centres of the luminaries exactly coincide, and come in a line with the eye.

CENTRAL *Forces,* are forces having a tendency directly towards or from fome point or centre; or forces which caufe a moving body to tend towards, or recede from, the centre of motion. And accordingly they are divided into two kinds, in refpect to their different relations to the centre, and hence are called centripetal, and centrifugal.

The doctrine of central forces makes a confiderable branch of the Newtonian philofophy, and has been greatly cultivated by mathematicians, on account of its extenfive ufe in the theory of gravity, and other phyfico-mathematical fciences.

In this doctrine, it is fuppofed that matter is equally indifferent to motion or reft; or that a body at reft never moves itfelf, and that a body in motion never of itfelf changes either the velocity or the direction of its motion; but that every motion would continue uniformly, and its direction rectilinear, unlefs fome external force or refiftance fhould affect it, or act upon it. Hence, when a body at reft always tends to move, or when the velocity of any rectilinear motion is continually accelerated or retarded, or when the direction of a motion is continually changed, and a curve line is thereby defcribed, it is fuppofed that thefe circumftances proceed from the influence of fome power that acts inceffantly; which power may be meafured, in the firft cafe, by the preffure of the quiefcent body againft the obftacle which prevents it from moving, or by the velocity gained or loft in the fecond cafe, or by the flexure of the curve defcribed in the 3d cafe : due regard being had to the time in which thefe effects are produced, and other circumftances, according to the principles of mechanics. Now the power or force of gravity produces effects of each of thefe kinds, which fall under our conftant obfervation near the furface of the earth ; for the fame power which renders bodies heavy, while they are at reft, accelerates their motion when they defcend perpendicularly ; and bends the track of the motion into a curve line, when they are projected in a direction oblique to that of their gravity. But we can judge of the forces or powers that act on the celeftial bodies by effects of the laft kind only. And hence it is, that the doctrine of central forces is of fo much ufe in the theory of the planetary motions.

Sir I. Newton has treated of central forces in lib. 1 fec. 2 of his Principia, and has demonftrated this fundamental theorem of central forces, viz, that the areas which revolving bodies defcribe by radii drawn to an immoveable centre, lie in the fame immoveable planes, and are proportional to the times in which they are defcribed. Prop. 1.

It is remarked by a late eminent mathematician, that this law, which was originally obferved by Kepler, is the only general principle in the doctrine of centripetal forces ; but fince this law, as Newton himfelf has proved, cannot hold in cafes where a body has a tendency to any other than one and the fame point, there feems to be wanting fome law that may ferve to explain the motions of the moon and fatellites which gravitate towards two different centres : the law he lays down for this purpofe is, That when a body is urged by two forces tending conftantly to two fixed points, it will defcribe, by lines drawn from the two fixed points, equal folids in equal times, about the line joining thofe fixed points. See Machin, on the Laws of the Moon's Motion, in the Poftfcript. See alfo a demonftration of this law by Mr. William Jones, in the Philof. Tranf. vol. 59. Very learned tracts have alfo been fince given, when the motion refpects, not two only, but feveral centres, by many ingenious authors, and practical rules deduced from them for computing the places &c of planets and fatellites ; as by La Grange, De la Place, Waring, &c, &c. See Berlin Memoirs : thofe of the Academy of Sciences at Paris ; and the Philof. Tranf. of London.

M. De Moivre gave elegant general theorems relating to central forces, in the Philof. Tranf. and in his Mifcel. Analyt. pa. 231.—Let MPQ be any given curve, in which a body moves : let P be the place of the body at any time ; S the centre of force, or the point to which the central force acting on the body is always directed ; PG the radius of curvature at the point P ; and ST perpendicular to the tangent PT ; then will the centripetal force be everywhere proportional to the quantity $\frac{SP}{GP. \times ST^2}$. Vid. ut fupra.

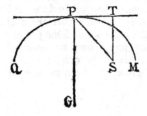

M. Varignon has alfo given two general theorems on this fubject in the Memoirs of the Acad. an. 1700, 1701 ; and has fhewn their application to the motions of the planets. See alfo the fame Memoirs, an. 1706, 1710.

Mr. MacLaurin has alfo treated the fubject of central forces very ably and fully, in his Treatife on Fluxions, art. 416 to 493 ; where he gives a great variety of expreffions for thefe forces, and feveral elegant methods of inveftigating them.

Laws of CENTRAL FORCES.

1. The following is a very clear and comprehenfive rule, for which we are obliged to the marquis de l'Hôpital : Suppofe a body of any determinate weight to revolve uniformly about a centre, with any given velocity ; find from what height it muft have fallen, by the force of gravity, to acquire that velocity ; then, as the radius of the circle it defcribes is to double that height, fo is its weight to its centrifugal force. So that, if b be the body, or its weight or quantity of matter, v its velocity, and r the radius of the circle defcribed, alfo $g = 16\frac{1}{12}$ feet ; then, firft $4g^2 : v^2 :: g : \frac{v^2}{4g}$ the height due to the velocity v ; and as

$r : \frac{v^2}{2g} :: b : \frac{v^2 b}{2gr} = f$ the centrifugal force. And hence, if the centrifugal force be equal to the gravity, the velocity is equal to that acquired by falling through half the radius.

2. The central force of a body moving in the periphery of a circle, is as the verfed fine AM of the indefinitely fmall arc AE ; or it is as the fquare of that arc AE directly, and as the diameter AB inverfely. For AM is the fpace through which the body is drawn from the tangent in the given time, and 2AM is the proper meafure of the central force. But, AE being very fmall, and therefore nearly equal to its chord, by the nature of the circle

$$AB : AE :: AE : AM = \frac{AE^2}{AB}$$

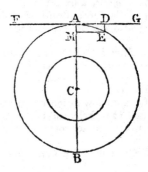

3. If two bodies revolve uniformly in different circles ; their central forces are in the duplicate ratio of their velocities directly, and the diameters or radii of the circles inverfely ;

that is $F : f :: \frac{V^2}{D} : \frac{v^2}{d} :: \frac{V^2}{R} : \frac{v^2}{r}$

For the force, by the laft article, is as

$\frac{AE^2}{AB}$ or $\frac{AE^2}{D}$; and the velocity v is as the space AE uniformly described.

4. And hence, if the radii or diameters be reciprocally in the duplicate ratio of the velocities, the central forces will be reciprocally in the duplicate ratio of the radii, or directly as the 4th power of the velocities; that is, if $V^2 : v^2 :: r : R$, then $F : f :: r^2 : R^2 :: V^4 : v^4$.

5. The central forces are as the diameters of the circles directly, and squares of the periodic times inversely. For if c be the circumference described in the time t, with the velocity v; then the space $c = tv$, or $v = \frac{c}{t}$; hence, using this value of v in the 3d rule, it becomes

$$F : f :: \frac{C^2}{DF^2} : \frac{c^2}{dt^2} :: \frac{D}{T^2} : \frac{d}{t^2} :: \frac{R}{T^2} : \frac{r}{t^2};$$ since the diameter is as the circumference.

6. If two bodies, revolving in different circles, be acted on by the same central force; the periodic times are in the subduplicate ratio of the diameters or radii of the circles; for when $F = f$, then $\frac{D}{T^2} = \frac{d}{t^2}$, and $D : d :: T^2 : t^2$, or $T : t :: \sqrt{D} : \sqrt{d} :: \sqrt{R} : \sqrt{r}$.

7. If the velocities be reciprocally as the distances from the centre, the central forces will be reciprocally as the cubes of the same distances, or directly as the cubes of the velocities. That is, if $V : v :: r : R$, then is $F : f :: r^3 : R^3 :: V^3 : v^3$.

8. If the velocities be reciprocally in the subduplicate ratio of the central distances, the squares of the times will be as the cubes of the distances: for if $V^2 : v^2 :: r : R$, then is $T^2 : t^2 :: R^3 : r^3$.

9. Wherefore, if the forces be reciprocally as the squares of the central distances, the squares of the periodic times will be as the cubes of the distances; or when $F : f :: r^2 : R^2$, then is $T^2 : t^2 :: R^3 : r^3$.

Exam. From this, and some of the foregoing theorems, may be deduced the velocity and periodic time of a body revolving in a circle, at any given distance from the earth's centre, by means of its own gravity. Put $g = 16\frac{1}{12}$ feet, the space described by gravity at the surface, in the first second of time, viz $=$ AM in the foregoing fig. and by rule 2; then, putting $r =$ the radius AC; it is $AE = \sqrt{AB \times AM} = \sqrt{2gr}$ the velocity in a circle at its surface, in one second of time; and hence, putting $c = 3.14159$ &c, the circumference of the earth being $2cr = 25,000$ miles, or 132,000,000 feet, it will be $\sqrt{2gr} : 2cr :: 1'' : c\sqrt{\frac{2r}{g}} = 5078$ seconds nearly, or $1^h 24^m 38^s$, the periodic time at the circumference: Also the velocity there, or $\sqrt{2gr}$ is $= 26000$ feet per second nearly. Then, since the force of gravity varies in the inverse duplicate ratio of the distance, by rules 8 and 9, it is $\sqrt{R} : \sqrt{r} :: v$ or $26000 : 26000\sqrt{\frac{r}{R}} = V$ the velocity of a body revolving about the earth at the distance R; and $\sqrt{r^3} : \sqrt{R^3} :: t$ or $5078'' : 5078\sqrt{\frac{R^3}{r^3}} = T$ the time of revolution in the same. So if, for instance, it be the moon revolving about the earth at the distance of 60 semidiameters; then $R = 60r$, and the above expressions become $V = 26000\sqrt{\frac{1}{60}} = 3357$ feet per second, or $38\frac{1}{7}$ miles per minute, for the velocity of the moon in her orbit; and $T = 5078\sqrt{\frac{R^3}{r^3}} = 2360051$ seconds or $27\frac{3}{10}$ days nearly, for the periodic time of the moon in her orbit at that distance.

Thus also the ratio of the forces of gravitation of the moon towards the sun and earth may be estimated. For, 1 year or $365\frac{1}{4}$ days being the periodic time of the earth and moon about the sun, and $27\frac{3}{10}$ days the periodic time of the moon about the earth, also 60 being the distance of the moon from the earth, and 23920 the distance from the sun, in semidiameters of the earth, by art. 5 it is

$$\frac{60}{27\cdot3^2} : \frac{23920}{365\cdot25^2} :: f \text{ or } 1 : \frac{23920}{60} \times \frac{27\cdot3^2}{365\cdot25^2} = 2\frac{2}{9};$$

that is, the proportion of the moon's gravitation towards the sun, is to that towards the earth, as $2\frac{2}{9}$ to 1 nearly.

Again, we may hence compute the centrifugal force of a body at the equator, arising from the earth's rotation. For, the periodic time when the centrifugal force is equal to the force of gravity, it has been shewn above, is 5078 seconds, and 23 hours, 56 minutes, or 86160 seconds, is the period of the earth's rotation on its axis; therefore, by art. 5, as $86160^2 : 5078^2 :: 1 : \frac{1}{289}$, the centrifugal force required, which therefore is the 289th part of gravity at the earth's surface. Simpson's Flux. pa. 240, &c.

Also for another example, suppose A to be a ball of 1 ounce, which is whirled about the centre C, so as to describe the circle ABE, each revolution being made in half a second; and the length of the cord AC equal to 2 feet. Here then $t = \frac{1}{2}$, $r = 2$, and it having been found above that $c\sqrt{\frac{2R}{g}} = T$ is the periodic time at the circumference of the earth when the centrifugal force is equal to gravity; hence then by art. 5, as $\frac{R}{T^2} : \frac{r}{t^2} :: F$ or $1 : f$, which proportion becomes

$$\frac{g}{2c^2} : \frac{r}{t^2} :: 1 : \frac{2c^2r}{gt^2} = \frac{16c^2}{g} = \frac{16 \times 3\cdot1416^2}{16\frac{1}{12}} = 9\cdot819$$

$=$ the centrifugal force, or that by which the string is stretched, viz, nearly 10 ounces, or 10 times the weight of the ball.

Lastly, suppose the string and ball be suspended from a point D, and describes in its motion a conical surface ADB; then putting DC $= a$, AC $= r$, and AD $= b$; and putting F $=$ 1 the force of gravity as before; then will the body A be affected by three forces, viz, gravity acting parallel to DC, a centrifugal force in the direction CA, and the tension of the string, or force by which it is stretched, in the direction DA; hence these three powers will be as the three sides of the triangle ADC respectively, and therefore as CD or a : AD or b :: 1 : $\frac{b}{a}$ the tension of the string as compared with the weight of the body. Also AC or a : AC or r :: 1 : $\frac{2c^2r}{gt}$ the general expression for the centrifugal force above found; hence

hence, $g t^2 = 2ac^2$ and so $t = c \sqrt{\dfrac{2a}{g}} = 1.108 \sqrt{a} =$ the periodic time. And

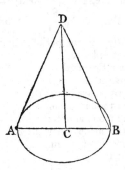

10. When the force by which a body is urged towards a point is not always the same, but is either increased or decreased as some power of the distance; several curves will thence arise according to that power. If the force decrease as the squares of the distances increase, the body will describe an ellipsis, and the force is directed towards one of its foci; so that in every revolution the body once approaches towards it, and once recedes from it: also the eccentricity of the ellipse is greater or less, according to the projectile force; and the curve may sometimes become a circle, when the eccentricity is nothing; the body may also describe the other two conic sections, the parabola and hyperbola, which do not return into themselves, by supposing the velocity greater in certain proportions. Also if the force increase in the simple ratio as the distance increases, the body will still describe an ellipse; but the force will in this case be directed to the centre of the ellipse; and the body, in each revolution, will twice approach towards it, and again twice recede from that point.

CENTRAL RULE, is a rule or method discovered by Mr. Thomas Baker, rector of Nympton in Devonshire, which he published in his Geometrical Key, in the year 1684, for determining the centre of a circle which shall cut a given parabola in as many points as a given equation, to be constructed, has real roots; which he has applied with good success in the construction of all equations as far as the 4th power inclusive.

The *Central Rule* is chiefly founded on this property of the parabola; that if a line be inscribed in the curve perpendicular to any diameter, the rectangle of the segments of this line, is equal to the rectangle of the intercepted part of the diameter and the parameter of the axis.

The Central Rule has the advantage over the methods of constructing equations by Des Cartes and De Latteres, which are liable to the trouble of preparing the equations by taking away the second term; whereas Baker's method effects the same thing without any previous preparation whatever. See also Philos. Transf. N° 157.

CENTRE, or CENTER, in a general sense, signifies a point equally remote from the extremes of a line, plane, or solid; or a middle point dividing them so that some certain effects are equal on all sides of it.

CENTRE *of Attraction,* or *Gravitation,* is the point to which bodies tend by gravity; or that point to which

a revolving planet or comet is impelled or attracted, by the force or impetus of gravity.

CENTRE *of a Bastion,* is a point in the middle of the gorge, where the capital line commences, and which is usually at the angle of the inner polygon of the figure. Or it is the point where the two adjacent curtains produced intersect each other.

CENTRE *of a Circle,* is the point in the middle of a circle, or circular figure, from which all lines drawn to the circumference are equal.

CENTRE *of a Conic Section,* is the middle point of any diameter, or the point in which all the diameters intersect and bisect one another.

In the ellipse the centre is within the figure; but in the hyperbola it is without, or between the conjugate hyperbolas; and in the parabola it is at an infinite distance from the vertex.

CENTRE *of Conversion,* in Mechanics, a term first used by M. Parent, and may be thus conceived: Suppose a stick laid on stagnant water, and then drawn by a thread fastened to it, so that the thread always makes the same angle with the stick, either a right angle or any other; then it will be found that the stick will turn about one point of it, which will be immoveable; and this point is termed the centre of conversion.

This effect arises from the resistance of the fluid to the stick partly immersed in it. And if, instead of the body thus floating on a fluid, the same be conceived to be laid on the surface of another body; then the resistance of this plane to the stick will always have the same effect, and will determine the same centre of conversion. And this resistance is precisely what is called friction, so prejudicial to the effects of machines.

M. Parent has determined this centre in some certain cases, with much laborious calculation. When the thread is fastened to the extremity of the stick, he found that the distance of the centre from this extremity would be nearly $\frac{1}{2}\frac{2}{8}$ of the whole length. But when it is a surface or a solid, there will be some change in the place of this centre, according to the nature of the figure. See Mem. of the Acad. of Sciences, vol. 1, pa. 191.

CENTRE *of a Curve,* of the higher kind, is the point where two diameters meet.—When all the diameters meet in the same point, it is called, by Sir Isaac Newton, the *general centre.*

CENTRE *of a Dial,* is the point where its gnomon or stile, which is placed parallel to the axis of the earth, meets the plane of the dial; and from hence all the hour-lines are drawn, in such dials as have centres, viz, all except that whose plane is parallel to the axis of the world; all the hour-lines of which are parallel to the stile, and to one another, the centre being as it were at an infinite distance.

CENTRE *of an Ellipse,* is the middle of any diameter, or the point where all the diameters intersect.

CENTRE *of the Equant,* in the Old Astronomy, is a point in the line of the aphelion; being as far distant from the centre of the eccentric, towards the aphelion, as the sun is from the same centre of the eccentric towards the perihelion.

CENTRE *of Equilibrium,* is the same with respect to bodies immersed in a fluid, as the centre of gravity is to bodies in free space; being a certain point, upon **which**

which if the body or bodies be fuspended, they will reft in any pofition. To determine this centre, fee Emmerfon's Mechanics, prop. 92, pa. 134

CENTRE *of Friction*, is that point in the bafe of a body on which it revolves, into which if the whole furface of the bafe, and the mafs of the body were collected, and made to revolve about the centre of the bafe of the given body, the angular velocity deftroyed by its friction would be equal to the angular velocity deftroyed in the given body by its friction in the fame time.—See Vince on the Motion of Bodies affected by friction, in the Philof. Tranf 1785.

CENTRE *of Gravity*, is that point about which all the parts of a body do in any fituation exactly balance each other. Hence, by means of this property, if the body be fupported or fufpended by this point, the body will reft in any pofition into which it is put; as alfo that if a plane pafs through the fame point, the fegments on each fide will equiponderate, neither of them being able to move the other.

The whole gravity, or the whole matter, of a body may be conceived united in the centre of gravity; and in demonftrations it is ufual to conceive all the matter as really collected in that point.

Through the centre of gravity paffes a right line, called the *diameter of gravity*; and therefore the interfection of two fuch diameters determines the centre. Alfo the plane upon which the centre of gravity is placed, is called the *plane of gravity*; fo that the common interfection of two fuch planes determines the diameter of gravity.

In homogeneal bodies, which may be divided lengthways into fimilar and equal parts, the centre of gravity is the fame with the centre of magnitude. Hence therefore the centre of gravity of a line is in the middle point of it, or that point which bifects the line. Alfo the centre of gravity of a parallelogram, or cylinder, or any prifm whatever is in the middle point of the axis. And the centre of gravity of a circle or any regular figure, is the fame as the centre of magnitude.

Alfo, if a line can be fo drawn as to divide a plane into equal and fimilar parts, that line will be a diameter of gravity, or will pafs through the centre of gravity; and it is the fame as the axis of the plane. Thus the line drawn from the vertex and perpendicular to the bafe of the ifofceles triangle, is a diameter of gravity; and thus alfo the axis of an ellipfe, or a parabola, &c, is a diameter of gravity. The centre of gravity of a fegment or arc of a circle, is in the radius or line perpendicularly bifecting its chord or bafe.

Likewife, if a plane divide a folid in the fame manner, making the parts on both fides of it perfectly equal and fimilar in all refpects, it will be a plane of gravity, or will pafs through the centre of gravity. Thus, as the interfection of two fuch planes determines the diameter of gravity, the centre of gravity of a right cone, or fpherical fegment, or conoid, &c, will be in the axis of the fame.

Common Centre of Gravity of two or more bodies, or the different parts of the fame body, is fuch a point as that, if it be fufpended or fupported, the fyftem of bodies will equiponderate, and reft in any pofition. Thus, the point of fufpenfion in a common balance beam, or fteelyard, is the centre of gravity of the fame.

Laws and Determination of the Centre of Gravity.

1. In two equal bodies, or maffes, the centre of gravity is equally diftant from their two refpective centres. For thefe are as two equal weights fufpended at equal diftances from the point of fufpenfion; in which cafe they will equiponderate, and reft in any pofition.

2. If the centres of gravity of two bodies A and B be connected by the right line AB, the diftances AC and BC from the common centre of gravity C, are reciprocally as the weights or bodies A and B; that is, AC : BC : : B : A.

See this demonftrated under the article BALANCE.

Hence, if the weights of the bodies A and B be equal, their common centre of gravity C will be in the middle of the right line AB, as in the foregoing article. Alfo fince A : B : : BC : AC, therefore A × AC = B × BC; whence it appears that the powers of equiponderating bodies are to be eftimated by the product of the mafs multiplied by the diftance from the centre of gravity; which product is ufually called the *momentum* of the weights.

Further, from the foregoing proportion, by compofition it will be A + B : A : : AB : BC, or A + B : B : : AB : AC. So that the common centre of gravity C of two bodies will be found, if the product of one weight by the whole diftance between the two, be divided by the fum of the two weights. Suppofe, for example, that A = 12 pounds, B = 4lb, and AB = 36 inches; then 16 : 12 : : 36 : 27 = BC, and confequently AC = 9, the two diftances from the common centre of gravity.

3. *The Common Centre of Gravity of three or more given bodies or points* A, B, C, D, &c, will be thus determined.—If the given bodies lie all in the fame ftraight line AD; by the laft article, find P the centre of gravity of the two A and B, and Q the centre of gravity of C and D; then, confidering P as the place of a body equal to the fum of A and B, and Q as the place of another body equal to both C and D, find S the common centre of gravity of thefe two fums, viz A + B collected in P, and C + D united in Q; fo fhall S be the common centre of gravity of all the four bodies A, B, C, D. And the fame for any other number of bodies, always confidering the fum of any number of them as united or placed in their common centre of gravity, when found.

Otherwife, thus. Take the diftances of the given bodies from fome fixed point as V, calling the diftance VA = *a*, VB = *b*, VC = *c*, VD = *d*, and the diftance of the centre of gravity VS = *x*; then SA = *x* − *a*, SB = *x* − *b*, SC = *c* − *x*, SD = *d* − *x*, and by the nature of the lever A · $\overline{x-a}$ + B · $\overline{x-b}$ = C·$\overline{c-x}$ + D·$\overline{d-x}$; hence A*x* + B*x* + C*x* + D*x* = A*a* + B*b* + C*c* + D*d*, and $x = \dfrac{Aa + Bb + Cc + Dd}{A + B + C + D}$ = VS the diftance fought; which therefore is equal to the

fum

sum of all the momenta, divided by the sum of all the weights or bodies.

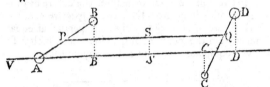

Or thus. When the bodies are not in the same straight line, connect them with the lines AB, CD; then, as before, find P the common centre of A and B, and Q the common centre of C and D; then, conceiving A and B united in P, and C and D united in Q, find S the common centre of P and Q, which will again be the centre of gravity of the whole.

Or the bodies may be all reduced to any line V*AB* &c, drawn in any direction whatever, by perpendiculars B*B*, C*C*, &c, and then the common centre S in this line, found as before, will be at the same distance from V as the true centre S is; and consequently the perpendicular from S will pass through S the real centre.

4. From the foregoing general expression, viz,

$$x = \frac{Aa + Bb + Cc + \&c}{A + B + C + \&c},$$ for the centre of

gravity of any system of bodies, may be derived a general method for finding that centre; for A, B, C, &c, may be considered as the elementary parts of any body, whose sum or mass is M = A + B + C &c, and A*a*, B*b*, C*c*, &c, are the several momenta of all these parts, viz, the product of each part multiplied by its distance from the fixed point V. Hence then, in any body, find a general expression for the sum of the momenta, and divide it by the content of the body, so shall the quotient be the distance of the centre of gravity from the vertex, or from any other fixed point, from which the momenta are estimated.

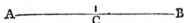

5. *Thus, in a right line* AB, all the particles which compose it may be considered as so many very small weights, each equal to $\dot{x}$, which is therefore the fluxion of the weights, or of the line denoted by *x*. So that the small weight $\dot{x}$ multiplied by its distance from A, viz *x*, is $x\dot{x}$ the momentum of that weight *x*; that is, $x\dot{x}$ is the fluxion of all the momenta in the line AB or *x*; and therefore its fluent $\frac{1}{2}x^2$ is the sum of all those momenta; which being divided by *x* the sum of all the weights, gives $\frac{1}{2}x$ or $\frac{1}{2}$AB for the distance of the centre of gravity C from the point A; that is, the centre is in the middle of the line.

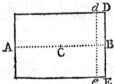

6. *Also in the parallelogram,* whose axis or length AB = *x*, and its breadth DE = *b*; drawing *de* parallel and indefinitely near DE, the areola *d*DE*e* = *b*$\dot{x}$

will be the fluxion of all the weights, which multiplied by its distance *x* from the point A, gives *bx*$\dot{x}$ for the fluxion of all the momenta, and consequently the fluent $\frac{1}{2}bx^2$ is the sum of all those momenta themselves; which being divided by *bx* the sum of all the weights, gives $\frac{1}{2}x = \frac{1}{2}$AB for the distance of the centre C from the extremity at A, and is therefore in the middle of the axis, as is known from other principles.

And the process and conclusion will be exactly the same for a cylinder, or any prism whatever, making *b* to denote the area of the end or of a transverse section of the body.

7. *In a Triangle* ABC; the line AD drawn from one angle to bisect the opposite side, will be a diameter of gravity, or will pass through the centre of gravity; for if that line be supported, or conceived to be laid upon the edge of something, the two halves of the triangle on both sides of that line will just balance one another, since all the parallels EF &c to the base will be bisected, as well as the base itself, and so the two halves of each line will just balance each other. Therefore, putting the base BC = *b*, and the axis or bisecting line AD = *a*, the variable part AS = *x*; then, by similar triangles AD : BC :: AS : EF, that is *a* : *b* :: *x* : $\frac{bx}{a}$ = EF; which, as a weight, multiplied by $\dot{x}$, gives $\frac{bx\dot{x}}{a}$ for the fluxion of the weights; and this again multiplied by *x* = AS, the distance from A, gives $\frac{bx^2\dot{x}}{a}$ for the fluxion of the momenta; the fluent of which, or $\frac{bx^3}{3a}$ divided by $\frac{bx^2}{a}$ the fluent for the weights, gives $\frac{1}{3}x = \frac{1}{3}$AS for the distance of the centre of gravity from the vertex A in the triangle AEF; and when *x* = AD, then $\frac{2}{3}$AD is the distance of the centre of gravity of the triangle ABC.

The Same Otherwise, without Fluxions.—Since a line drawn from any angle to the middle of the opposite side passes through the centre of gravity, therefore the intersection of any two of such lines, will be that centre: thus then the centre of gravity is in the line AD; and it is also in the line CG bisecting AB; it is therefore in their intersection S. Now to determine the distance of S from any angle, as A, produce CG to meet BH parallel to AS in H; then the two triangles AGS, BGH are mutually equal and similar; for the opposite angles at G are equal, as are the alternate angles at H and S, and at A and B, also the side AG = BG; therefore the other sides BH, AS are equal. But the triangles CDS, CBH are similar, and the side CB = 2CD, therefore BH or its equal AS = 2DS, that is AS = $\frac{2}{3}$AD, the same as was found before. And in like manner CS = $\frac{2}{3}$CG.

8. In

8. *In a Trapezium.* Divide the figure into two triangles by the diagonal AC, and find the centres of gravity E and F of these triangles; join EF, and find the common centre G of these two by this proportion, ABC : ADC :: FG : EG, or ABCD : ADC :: EF : EG.

In like manner, for any other figure, whatever be the number of sides, divide it into several triangles, and find the centre of gravity of each; then connect two centres together, and find their common centre as above; then connect this and the centre of a third, and find the common centre of these; and so on, always connecting the last found common centre to another centre, till the whole are included in this process; so shall the last common centre be that which is required.

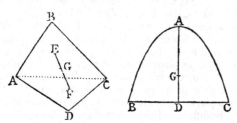

9. *In the Parabola* BAC. Put AD $= x$, BD $= y$, and the parameter $= p$. Then, by the nature of the figure, $px = y^2$, and $2y = 2\sqrt{px}$; hence $2\dot{x}\sqrt{px}$ is the fluxion of the weights, and $2x\dot{x}\sqrt{px}$ is the fluxion of the momenta; then the fluent of the latter divided by that of the former, or $\frac{4}{5}x^{\frac{5}{2}}\sqrt{p}$ divided by $\frac{4}{3}x^{\frac{3}{2}}\sqrt{p}$, gives $\frac{3}{5}x = \frac{3}{5}$AD, for AG, the distance of the centre of gravity G from the vertex A of the parabola.

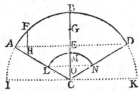

10. *In the Circular Arc* ABD, considered as a physical line having gravity. It is manifest that the centre of gravity G of the arc, will be somewhere in the axis, or middle radius BC, C being the centre of the circle, which is considered as the point of suspension. Suppose F indefinitely near to A, and FH parallel to BC. Put the radius BC or AC $= r$, the semiarc AB $= z$, and the semichord AE $= x$; then is AH $= \dot{x}$, and AF $= \dot{z}$ the fluxion of the weights, and therefore CE $\times \dot{z}$ is the fluxion of the momenta. But, by similar triangles, AC or r : CE :: AF or $\dot{z}$: AH or $\dot{x}$, therefore $r\dot{x} =$ CE $\times \dot{z}$, and so $r\dot{x}$ is also the fluxion of the momenta; the fluent of which is rx, and this divided by z the weight, gives $\frac{rx}{z} = \frac{\text{AC} \times \text{AE}}{\text{AB}} = \frac{\text{AC} \times \text{AED}}{\text{ABD}} =$ CG the distance of the centre of gravity from the centre C of the circle; being a 4th proportional to the given arc, its chord, and the radius of the circle.

Hence, when the arc becomes the semicircle ABK, the above expression becomes $\frac{\text{IC}^2}{\text{IB}}$ or $\frac{r^2}{1\cdot5708r} = \frac{r}{1\cdot5708}$ $= \cdot6366r$, viz a third proportional to a quadrant and the radius.

11. *In the Circular Sector* ABDC. Here also the centre of gravity will be in the axis or middle radius BC. Now with any smaller radius describe the concentric arc LMN, and put the radius AC or BC $= r$, the arc ABD $= a$, its chord AED $= c$, and the variable radius CL or CM $= y$; then as $r : y :: a : \frac{ay}{r} =$ the arc LMN, and $r : y :: c : \frac{cy}{r} =$ the chord LON; also, by the last article, the distance of the centre of gravity of the arc LMN is $\frac{\text{CM} \times \text{LON}}{\text{LMN}} = \frac{\text{CM} \times \text{AED}}{\text{ABD}} = \frac{cy}{a}$; hence the arc LMN or $\frac{ay}{r}$ multiplied by $\dot{y}$ gives $\frac{ay\dot{y}}{r}$ the fluxion of the weights, and this multiplied by $\frac{cy}{a}$ the distance of the common centre of gravity, gives $\frac{cy^2\dot{y}}{r}$ the fluxion of the momenta; the fluent of which, viz $\frac{cy^3}{3r}$, divided by $\frac{ay^2}{2r}$, the fluent of the weights, gives $\frac{2cy}{3a}$ for the distance of the centre of gravity of the sector CLEN from the centre C; and when $y = r$, it becomes $\frac{2cr}{3a} =$ CG for that of the sector CABD proposed; being $\frac{2}{3}$ of a 4th proportional to the arc of the sector, its chord, and the radius of the circle.

Hence, when the sector becomes a semicircle, the last expression becomes $\frac{4r^2}{3a} = \frac{2\text{IC}^2}{3\text{IB}}$ or $\frac{2}{3}$ of a 3d proportional to a quadrantal arc and the radius. Or it is equal to $\frac{4r}{3p} = \cdot4244r$ from the centre C; where $p = 3\cdot1416$.

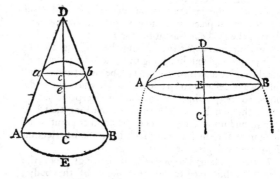

12. *In the Cone* ADB. Putting $a =$ DC, $b =$ area of the base AEB, and $x =$ Dc any variable altitude; then as $a^2 : x^2 :: b : \frac{bx^2}{a^2} =$ area aeb; hence the fluxion of the weights is $\frac{bx^2\dot{x}}{a^2}$, whose fluent, or the solid, is $\frac{bx^3}{3a^2}$; and the fluxion of the momenta is $\frac{bx^3\dot{x}}{a^2}$, whose fluent is $\frac{bx^4}{4a^2}$; then this fluent divided by the former fluent gives $\frac{3}{4}x$ or $\frac{3}{4}Dc$ for the distance of the centre of gravity of the cone Dab, or $\frac{3}{4}DC$ for that of the cone DAB below the vertex D.

And the same is the distance in any other pyramid. So that all pyramids of the same altitude, have the same centre of gravity.

13. In like manner are we to proceed for the centre of gravity in other bodies. Thus, the altitude of the segment of a sphere, or spheroid, or conoid, being x, a being the whole of that axis itself; then the distance of the centre of gravity in each of these bodies, from the vertex, will be as follows, viz,

$\frac{4a - 3x}{6a - 4x}x$ in the sphere or spheroid,

$\frac{3}{4}x$ in the semisphere or semispheroid,

$\frac{2}{3}x$ in the parabolic conoid,

$\frac{4a + 3x}{6a + 3x}x$ in the hyperbolic conoid.

14. *To determine the Centre of Gravity in any Body Mechanically.* Lay the body on the edge of any thing, as a triangular prism, or such like, moving it backward and forward till the parts on both sides are in equilibrio; then is that line just in, or under the centre of gravity. Balance it again in another position to find another line passing through the centre of gravity; then the intersection of these two lines will give the place of that centre itself.

The same may be done by laying the body on an horizontal table, as near the edge as possible without its falling, and that in two positions, as lengthwise and breadthwise: then the common intersection of the two lines contiguous to the edge, will be its centre of gravity. Or it may be done by placing the body on the point of a style, &c, till it rest in equilibrio. It was by this method that Borelli found that the centre of gravity in a human body, is between the nates and pubis; so that the whole gravity of the body is collected into the place of the genitals; an instance of the wisdom of the creator, in placing the membrum virile in the part, which is the most convenient for copulation.

The same otherwise thus. Hang the body up by any point; then a plumb-line hung over the same point, will pass through the centre of gravity; because that centre will always descend to the lowest point when the body comes to rest, which it cannot do except when it falls in the plumb line. Therefore, marking that line upon it, and suspending the body by another point, with the plummet, to find another such line, the intersection of the two will give the centre of gravity.

Or thus. Hang the body by two strings from the same tack, but fixed to different points of the body; then a plummet, hung by the same tack, will fall on the centre of gravity.

In the 4th volume of the New Acts of the Academy of Petersburgh, is the demonstration of a very general theorem concerning centres of gravity, by M. Lhuilier;

a particular example only of the general proposition, will be as follows: Let A, B, C, be the centres of gravity of three bodies; a, b, c their respective masses, and Q their common centre of gravity. Let right lines QA, QB, QC, be drawn from the common centre to that of each body, and the latter be connected by right lines AB, AC, and BC; then

$$QA^2 \times a + QB^2 \times b + QC^2 \times c =$$
$$AB^2 \times \frac{ab}{a+b+c} + AC^2 \times \frac{ce}{a+b+c} + BC^2 \times \frac{bc}{a+b+c}.$$

Uses of the Centre of Gravity. This point is of the greatest use in mechanics, and many important concerns in life, because the place of that centre is to be considered as the place of the body itself in computing mechanical effects; as in the oblique pressures of bodies, banks of earth, arches of bridges, and such like.

The same centre is even useful in finding the superficial and solid contents of bodies; for it is a general rule, that the superficies or solid generated by the rotation of a line or plane about any axis, is always equal to the product of the said line or plane drawn into the circumference or path described by the centre of gravity. For example, it was found above at art 11, that in a semicircle, the distance of the centre of gravity from the centre of the circle, is $\frac{4r}{3p}$; and therefore the path of that centre, or circumference described by it whilst the semicircle revolves about its diameter, is $\frac{8}{3}r$; also the area of the semicircle is $\frac{1}{2}pr^2$; hence the product of the two is $\frac{4}{3}pr^3$, which, it is well known, is equal to the solidity of the sphere generated by the revolution of the semicircle.

And hence also is obtained another method of finding mathematically the centre of gravity of a line or plane, from the contents of the superficies or solid generated by it. For if the generated superficies or solid be divided by the generating line or plane, the quotient will be the circumference described by the centre of gravity; and consequently this divided by $2p$ gives the radius, or distance of that centre from the axis of rotation. So, in the semicircle, whose area is $\frac{1}{2}pr^2$, and the content of the sphere generated by it $\frac{4}{3}pr^3$; here the latter divided by the former is $\frac{8}{3}r$, and this divided by $2p$ gives $\frac{4r}{3p}$ for the distance of the centre of gravity from the axis, or from the centre of the semicircle. The property last mentioned, relative to the relation between the centre of gravity and the figure generated by the revolution of any line or plane, is mentioned by Pappus, in the preface to his 7th book; and father Guldin has more fully demonstrated it in his 2d and 3d books on the Centre of Gravity.

The principal writers on the centre of gravity are Archimedes, Pappus, Guldini, Wallis, Casatus, Carré, Hays, Wolfius, &c.

CENTRE *of Gyration*, is that point in which if the whole mass be collected, the same angular velocity will be generated in the same time, by a given force acting at any place, as in the body or system itself. This point differs from the centre of oscillation, in as much as in this latter case the motion of the body is produced by the gravity of its own particles, but in the case of the
centre

centre of gyration the body is put in motion by some other force acting at one place only.

To determine the Centre of Gyration, in any body, or system of bodies composed of the parts A, B, C, &c, moving about the point S, when urged by a force f acting at any point P. Let R be that centre: then, by mechanics, the angular velocity generated in the system by the force f, is as $\dfrac{f \cdot SP}{A \cdot SA^2 + B \cdot SB^2 + C \cdot SC^2 \&c}$, and, by the same, the angular velocity of the matter placed all in the point R, is $\dfrac{f \cdot AP}{\overline{A + B + C \&c} \times SR^2}$; then since these two are to be equal, their equation will give $SR = \sqrt{\dfrac{A \cdot SA^2 + B \cdot SB^2 + C \cdot SC^2 \&c}{A + B + C \&c}}$, for the distance of the centre of gyration sought, below the axis of motion.

Now because the quantity $A \cdot SA^2 + B \cdot SB^2 + \&c$ is $= SG \cdot SO \cdot b$, where G is the centre of gravity, O the centre of oscillation, and b the whole body or sum of A, B, C, &c; therefore it follows that $SR^2 = SG \cdot SO$; that is, the distance of the centre of gyration, is a mean proportional between those of gravity and oscillation.

And hence also, if p denote any particle of a body, placed at the distance d from the axis of motion; then is $SR = \dfrac{\text{sum of all the } pd^2}{\text{body } b}$; from whence the point R may be determined in bodies by means of Fluxions.

Centre *of an Hyperbola*, is the middle of the axis, or of any other diameter, being the point without the figure in which all the diameters intersect one another; and it is common to all the four conjugate hyperbolas.

Centre *of Magnitude*, is the point which is equally distant from all the similar external parts of a body. This is the same as the centre of gravity in homogeneal bodies that can be cut into like and equal parts according to their length, as in a cylinder or any other prism.

Centre *of Motion*, is the point about which any body, or system of bodies, moves, in a revolving motion.

Centre *of Oscillation*, is that point in the axis or line of suspension of a vibrating body, or system of bodies, in which if the whole matter or weight be collected, the vibrations will still be performed in the same time, and with the same angular velocity, as before. Hence, in a compound pendulum, its distance from the point of suspension is equal to the length of a simple pendulum whose oscillations are isochronal with those of the compound one.

Mr. Huygens, in his Horologium Oscillatorium, first shewed how to find the centre of oscillation. At the beginning of his discourse on this subject, he says, that Mersennus first proposed the problem to him while he was yet very young, requiring him to resolve it in the cases of sectors of circles suspended by their angles, and by the middle of their bases, both when they oscillate sideways and flatways; as also for triangles and the segments of circles, either suspended from their vertex or the middle of their bases. But, says he, not

having immediately discovered any thing that would open a passage into this business, I was repulsed at first setting out, and stopped from a further prosecution of the thing; till being farther incited to it by adjusting the motion of the pendulums of my clock, I surmounted all difficulties. going far beyond Descartes, Fabry, and others, who had done the thing in a few of the most easy cases only, without any sufficient demonstration; and solving not only the problems proposed by Mersennus, but many others that were much more difficult, and shewing a general way of determining this centre, in lines, superficies, and solids.

In the Leipsic Acts for 1691 and 1714, this doctrine is handled by the two Bernoullis: and the same is also done by Herman, in his treatise De Motu Corporum Solidorum et Fluidorum.

It may also be seen in treatises on the Inverse Method of Fluxions, where it is introduced as one of the examples of that method. See Hayes, Carré, Wolfius, &c.

To determine the Centre of Oscillation, in any Compound Mass or Body MN, or of any System of Bodies A, B, C, &c.

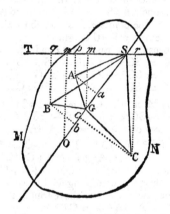

Let MN be the plane of vibration, to which plane conceive all the matter to be reduced by letting fall perpendiculars to this plane from every particle in the body; a supposition which will not alter the vibration of the body, because the particles are still at the same distance from the axis of motion. Let O be the centre of oscillation, and G the centre of gravity; through the axis S draw SGO, and the horizontal line ST; then from every particle A, B, C, &c, let fall perpendiculars Aa and Ap, Bb and Bq, Cc and Cr, &c, to these two lines; and join SA, SB, SC; also draw Gm and On perpendicular to ST.

Now the forces of the weights A, B, C, to turn the body about the axis, are $A \cdot Sp$, $B \cdot Sq$, $-C \cdot Sr$; and, by mechanics, the forces opposing that motion are $A \cdot SA^2$, $B \cdot SB^2$, $C \cdot SC^2$; therefore the angular motion generated in the system is $\dfrac{A \cdot Sp + B \cdot Sq - C \cdot Sr}{A \cdot SA^2 + B \cdot SB^2 + C \cdot SC^2}$. In like manner, the angular velocity which any body or particle p, situated in O, generates in the system, by its weight, is $\dfrac{p \cdot Sn}{p \cdot SO^2}$, or $\dfrac{Sn}{SO^2}$, or $\dfrac{Sm}{SG \cdot SO}$ because of the

the similar triangles SG*m*, SO*n*. But, by the conditions of the problem, the vibrations are performed alike in both these cases; therefore these two expressions must be equal to each other, that is $\dfrac{Sm}{SG \cdot SO} =$

$\dfrac{A \cdot Sp + B \cdot Sq - C \cdot Sr}{A \cdot SA^2 + B \cdot SB^2 + C \cdot SC^2}$, and consequently SO

$= \dfrac{Sm}{SG} \times \dfrac{A \cdot SA^2 + B \cdot SB^2 + C \cdot SC^2}{A \cdot Sp + B \cdot Sq - C \cdot Sr}$. But, by mechanics

again, the sum of the forces $A \cdot Sp + B \cdot Sq - C \cdot Sr$ is equal $\overline{A + B + C} \cdot Sm$ the force of the same matter collected all into its centre of gravity G; and therefore

$SO = \dfrac{A \cdot SA^2 + B \cdot SB^2 + C \cdot SC^2}{(A + B + C) \cdot SG}$; which is the

distance of the centre of oscillation O below the axis of suspension.

Farther, because it was found under the article *Centre of Gravity*, that $(A+B+C) \cdot SG = A \cdot Sa + B \cdot Sb + C \cdot Sc$, therefore $SO = \dfrac{A \cdot SA^2 + B \cdot SB^2 + C \cdot SC^2}{A \cdot Sa + B \cdot Sb + C \cdot Sc}$

is the same distance of the centre of oscillation; where any of the products $A \cdot Sa$, $B \cdot Sb$, &c are to be taken negatively when the points *a*, *b*, &c lie above the point S, or where the axis passes through.

Again, because, by Eucl. II 12 and 13,
it is $SA^2 = SG^2 + GA^2 - 2SG \cdot Ga$,
and $SB^2 = SG^2 + GB^2 + 2SG \cdot Gb$,
and $SC^2 = SG^2 + GC^2 + 2SG \cdot Gc$, &c;

and because by Mechanics, the sum of the last terms is nothing, namely $-2SG \cdot Ga + 2SG \cdot Gb + 2SG \cdot Gc$ &c $= 0$; therefore the sum of the others, or $A \cdot SA^2 + B \cdot SB^2$ &c

$= \overline{A + B + \&c} \cdot SG^2 + A \cdot GA^2 + B \cdot GB^2 + C \cdot GC^2$ &c, or
$= b \cdot SG^2 + A \cdot GA^2 + B \cdot GB^2 + C \cdot GC^2$ &c;

where *b* denotes the body, or sum $A + B + C$ &c of all the parts: this value then being substituted in the numerator of the 2d value of SO above-found, it becomes

$SO = \dfrac{b \cdot SG^2 + A \cdot GA^2 + B \cdot GB^2 \&c}{b \cdot SG}$, or

$SO = SG + \dfrac{A \cdot GA^2 + B \cdot GB^2 \&c}{b \cdot SG}$.

From which it appears that the centre of oscillation is always below the centre of gravity, and that the difference or distance between them is

$GO = \dfrac{A \cdot GA^2 + B \cdot GB^2 \&c}{b \cdot SG}$.

It farther follows from hence, that $SG \cdot GO = \dfrac{A \cdot GA^2 + B \cdot GB^2 \&c}{\text{the body } b}$; that is, the rectangle SG·GO is

always the same constant quantity, wherever the point of suspension S is placed, since the point G and the bodies A, B, &c, are constant. Or GO is always reciprocally as SG, that is GO is less as SG is greater; and the points G and O coincide when SG is infinite; but when S coincides with G, then GO is infinite, or O is at an infinite distance.

To find the Centre of Oscillation by means of Fluxions. From the premises is derived this general method for the centre of oscillation, viz, let *x* be the abscissa of an oscillating body, and *y* its corresponding ordinate or section; then will the distance SO of the centre of oscil-

lation below the axis of suspension S, be equal to the fluent of $yx^2\dot{x}$ divided by the fluent of $yx\dot{x}$. So that, if from the nature or equation of any given figure, the value of *y* be expressed in terms of *x*, or otherwise, and substituted in these two fluxions; then the fluents being duly found, and the one divided by the other, the quotient will be the distance to the centre of oscillation in terms of the abscissa *x*.

But when the body is suspended by a very fine thread of a given length *a*, then the fluent of $\overline{a + x}|^2 \cdot y\dot{x}$ divided by the fluent of $\overline{a + x} \; y\dot{x}$ gives the distance of the same centre of oscillation below the point of suspension.

Ex. For example, in a right line, or rectangle or cylinder or any other prism, whose constant section is *y*, or the constant quantity *a*; then $yx^2\dot{x}$ is $ax^2\dot{x}$, whose fluent is $\frac{1}{3}ax^3$; also $yx\dot{x}$ is $ax\dot{x}$, whose fluent is $\frac{1}{2}ax^2$; and the quotient of the former $\frac{1}{3}ax^3$ divided by the latter $\frac{1}{2}ax^2$, is $\frac{2}{3}x$ for the distance of the centre of oscillation below the vertex in any such figure, namely having every where the same breadth or section, that is, at two-thirds of its length.

In like manner the centre of oscillation is found for various figures, vibrating flatways, and are as they are. expressed below, viz,

Nature of the Figure.	When suspended by Vertex.
Isosceles triangle	$\frac{3}{4}$ of its altitude
Common Parabola	$\frac{5}{7}$ of its altitude
Any Parabola	$\dfrac{2m+1}{3m+1} \times$ its altitude.

As to figures moved laterally or sideways, or edgeways, that is about an axis perpendicular to the plane of the figure, the finding the centre of oscillation is somewhat difficult; because all the parts of the weight in the same horizontal plane, on account of their unequal distances from the point of suspension, do not move with the same velocity; as is shewn by Huygens, in his Horol. Oscil. He found, in this case, the distance of the centre of oscillation below the axis, viz,

In a circle, - - $\frac{3}{4}$ of the diameter:
In a rectangle, susp. by one angle, $\frac{2}{3}$ of the diagonal:
In a parabola susp by its vertex, $\frac{5}{7}$ axis $+ \frac{1}{4}$ param.
The same susp. by mid. of base, $\frac{4}{7}$ axis $+ \frac{1}{2}$ param.

In a sector of a circle - $\dfrac{3 \text{ arc} \times \text{radius}}{4 \text{ chord}}$:

In a cone - $\frac{4}{5}$ axis $+ \dfrac{\text{radius base}^2}{5 \text{ axis}}$:

In a sphere - $g + \dfrac{2r^2}{5g}$, where *r* is

the radius, and $g = a + r$ the rad. added to the length of the thread.

See also Simpson's Fluxions, art. 183 &c.

To find the Centre of Oscillation Mechanically or Experimentally. Make the body oscillate about its point of suspension; and hang up also a simple pendulum of such a length that it may vibrate or just keep time with the other body: then the length of the simple pendulum is equal to the distance of the centre of oscillation of the body below the point of suspension.

Or it will be still better found thus: Suspend the body very freely by the given point, and make it vibrate

in

in small arcs, counting the vibrations it makes in any portion of time, as a minute, by a good stop watch; and let that number of oscillations made in a minute be called n: then shall the distance of the centre of oscillation be SO $= \dfrac{140850}{nn}$ inches. For, the length of the pendulum vibrating seconds, or 60 times in a minute, being $39\frac{1}{8}$ inches, and the lengths of pendulums being reciprocally as the square of the number of vibrations made in the same time, therefore $n^2 : 60^2 :: 39\frac{1}{8} : \dfrac{140850}{nn}$ the length of the pendulum which vibrates n times in a minute, or the distance of the centre of oscillation below the axis of motion.

CENTRE *of Percussion*, in a moving body, is that point where the percussion or stroke is the greatest, in which the whole percutient force of the body is supposed to be collected; or about which the impetus of the parts is balanced on every side, so that it may be stopt by an immoveable obstacle at this point, and rest on it, without acting on the centre of suspension.

1. When the percutient body revolves about a fixed point, the centre of percussion is the same with the centre of oscillation; and is determined in the same manner, viz, by considering the impetus of the parts as so many weights applied to an inflexible right line void of gravity; namely, by dividing the sum of the products of the forces of the parts multiplied by their distances from the point of suspension, by the sum of the forces. And therefore what has been above shewn of the centre of oscillation, will hold also of the centre of percussion when the body revolves about a fixed point. For instance, that the centre of percussion in a cylinder is at $\frac{2}{3}$ of its length from the point of suspension, or that a stick of a cylindrical figure, supposing the centre of motion at the hand, will strike the greatest blow at a point about two-thirds of its length from the hand.

2. But when the body moves with a parallel motion, or all its parts with the same celerity, then the centre of percussion is the same as the centre of gravity. For the momenta are the products of the weights and celerities; and to multiply equiponderating bodies by the same velocity, is the same thing as to take equimultiples; but the equimultiples of equiponderating bodies do also equiponderate; therefore equivalent momenta are disposed about the centre of gravity, and consequently in this case the two centres coincide, and what is shewn of the one will hold in the other.

Centre of Percussion in a fluid, is the same as out of it.

CENTRE *of a Parallelogram*, the point in which its diagonals intersect.

CENTRE *of Pressure*, of a fluid against a plane, is that point against which a force being applied equal and contrary to the whole pressure, it will just sustain it, so as that the body pressed on will not incline to either side.—This is the same as the centre of percussion, supposing the axis of motion to be at the intersection of this plane with the surface of the fluid; and the centre of pressure upon a plane parallel to the horizon, or upon any plane where the pressure is uniform, is the same as the centre of gravity of that plane. Emerson's Mechanics, prop. 91.

CENTRE *of a Regular Polygon*, or *Regular Body*, is the same as that of the inscribed, or circumscribed circle or sphere.

CENTRE *of a Sphere*, is the same as that of its generating semicircle, or the middle point of the sphere, from whence all right lines drawn to the superficies, are equal.

CENTRING *of an Optic Glass*, the grinding it so as that the thickest part be exactly in the middle.

Cassini the younger has a discourse expressly on the necessity of well centring the object glass of a large telescope, that is, of grinding it so as that the centre may fall exactly in the axis of the telescope. Mem. Acad. 1710.

Indeed one of the greatest difficulties in grinding large optic glasses is, that in figures so little convex, the least difference will throw the centre two or three inches out of the middle. And yet Dr. Hook remarks, that though it were better the thickest part of a long object glass were exactly in the middle, yet it may be a very good one when it is an inch or two out of it. Philos. Transf. N° 4.

CENTRIFUGAL FORCE, is that by which a body revolving about a centre, or about another body, endeavours to recede from it. And

CENTRIPETAL FORCE, is that by which a moving body is perpetually urged towards a centre, and made to revolve in a curve, instead of a right line.

Hence, when a body revolves in a circle, these two forces, viz, the centrifugal and centripetal, are equal and contrary to each other, since neither of them gains upon the other, the body being in a manner equally balanced by them. But when, in revolving, the body recedes farther from the centre, then the centrifugal exceeds the centripetal force; as in a body revolving from the lower to the higher apsis, in an ellipse, and respecting the focus as the centre. And when the revolving body approaches nearer to the centre, the centrifugal is less than the centripetal force; as while the body moves from the farther to the nearer extremity of the transverse axis of the ellipse: the two forces being equal to each other only at the very extremities of that axis.

It is one of the established laws of nature, that all motion is of itself rectilinear, and that the moving body never recedes from its first right line, till some new impulse be superadded in a different direction: after that new impulse the motion becomes compounded, but it is still rectilinear, though not in the same line or direction as before. To move in a curve, it must receive a new impulse in a different direction every moment; a curve not being reducible to any number of finite right lines. If then a body, continually drawn towards a centre, be projected in a line that does not pass through that centre, it will describe a curve; in each point of which, as A, it will endeavour to recede from the curve, and proceed in the tangent AD; and if nothing hindered, it would actually proceed in it; so as in the same time in which it describes the arch AE, it would recede the length of the line DE, perpendicular to AD, by its centrifugal force: Or being projected in the direction AD, but being continually drawn out of its direction into a curve by a centripetal force,

force, so as to fall below the line of direction by the perpendicular space DE: Then the centrifugal or centripetal force is as this line of deviation DE; supposing the arch AE indefinitely small.

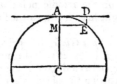

The doctrine of centrifugal forces was first mentioned by Huygens, at the end of his Horologium Oscillatorium, published in 1673, and demonstrated in the volume of his Posthumous Works, as also by Guido Grando; where he has given a few easy cases in bodies revolving in the circumference of circles. But Newton, in his Principia, was the first who fully handled this doctrine; at least as far as regards the conic sections. After him there have been several other writers upon this subject; as Leibnitz, Varignon in the Mem. de l'Acad. Keil in the Philos. Transf. and in his Physics, Bernoulli, Herman, Cotes in his Harmonia Mensurarum, Maclaurin in his Geometrica Organica, and in his Fluxions, and Euler in his book de Motu, where he considers the curves described by a body acted on by centripetal forces tending to several fixed points.

See also the art. *Central Forces*, where this doctrine is more fully explained.

CENTROBARICO, the same as centre of gravity.

CENTROBARIC-*Method*, is a method of determining the quantity of a surface or solid, by means of the generating line or plane, and its centre of gravity. The doctrine is chiefly comprized in this theorem:

Every figure, whether superficial or solid, generated by the motion of a line or plane, is equal to the product of the generating magnitude and the path of its centre of gravity, or the line which its centre of gravity describes.

See more of this subject under the article *Centre of Gravity*.

CENTRUM, in Geometry, Mechanics, &c. See CENTRE.

CENTRUM *Phonicum*, in Acoustics, is the place where the speaker stands in polysyllabical and articulate echoes.

CENTRUM *Phonocampticum*, is the place or object that returns the voice in an echo.

CEPHEUS, a constellation of the northern hemisphere, being one of the 48 old asterisms. The Greeks fable that Cepheus was a king of Ethiopia, and the father of Andromeda, the princess who was delivered up to be devoured by a sea monster, from which she was rescued by Perseus.

The stars of this constellation, in Ptolomy's catalogue are 13, in Tycho's 11, in Hevelius's 51, and in the Britannic catalogue 35.

CERBERUS, one of the new constellations formed by Hevelius out of the unformed stars, and added to the 48 old asterisms. It contains only 4 stars, which are enumerated under Hercules in the Britannic catalogue.

CETUS, *the Whale*, a southern constellation, and one of the 48 old asterisms. The Greeks pretend that it was the sea-monster sent by Neptune to devour Andromeda, but was killed by Perseus.

In the neck of the whale is a remarkable star, Collo Ceti, which appears and disappears periodically, or rather grows brighter and fainter by turns, owing it is supposed to the alternate turning of its bright and dark sides towards us, as it revolves upon its axis, or else owing to the star having a flattish form. The period of its changes is about 312 days. Bullialdus in Phil. Transf. vol. 2, Hevelius ibid. vol. 6, Herschel ibid. vol. 70, Marald. in Mem. Acad. 1719.

The stars in the constellation Cetus, in Ptolomy's catalogue, are 22, in Tycho's 21, in Hevelius's 45, and in the Britannic catalogue 97.

CHAIN, in Surveying, is a lineal measure, consisting of a certain number of iron links, usually 100: serving to take the dimensions of fields &c.

At every 10th link is usually fastened a small brass plate, with a figure engraven upon it, or else cut into different shapes, to shew how many links it is from one end of the chain.

Chains are of various kinds and lengths; as

1. A chain of 100 feet long, each link one foot, for measuring of large distances only, when regard is not proposed to be had to acres &c, in the superficial content.

2. A chain of one pole or 16 feet and a half in length; especially useful in measuring and laying out gardens and orchards, or the like, by the pole or rod measure.

3. A chain of 4 poles, or 66 feet, or 22 yards, in length, called Gunter's chain, and is peculiarly adapted to the business of Surveying or Land-measuring, because that 10 square chains just make an English acre of land; so that the dimensions being taken in these chains, and thence the contents computed in square chains, they are readily turned into acres by dividing by 10, or barely cutting off the last figure from the square chains. But it is still better in practice to proceed thus, viz, count the dimensions, not in chains, but all in links; then the contents are in square links; and five figures being cut off for decimals, the rest are acres; that is four figures to bring the square links to square chains, and one more to bring the square chains to acres.

In this chain, the links are each 7 inches and $\frac{92}{100}$, or 7·92 inches in length, which is very nearly $\frac{2}{3}$ of a foot. And hence any number of chains or links are easily brought to feet or inches, or the contrary: the best way of doing which is this: multiply the number of links by 66, then cut off two figures for decimals, and the rest are feet: or multiply links by 22 for yards, cutting off two figures.

CHALDRON, of Coals, an English dry measure of capacity consisting of 36 bushels heaped up.

The chaldron of coals is accounted to weigh about 2000 pounds.—On ship board, 21 chaldrons of coals are allowed to the score.

CHAMBER of a *Mortar*, or *some cannon*, is a cell or cavity at the bottom of the bore, to receive the charge of powder.

It is not found by experience that chambers have any

any senfible effect on the velocity of the fhct, unlefs in the largeft ordnance, as mortars or very large cannon. Neither is it found that the form of them is very material; a fmall cylinder is as good as any; though mathematical fpeculations may fhew a preference of one form over another. But in practice, the chief point to be obferved, is to have the chamber of a fize juft to contain the charge of powder, and no more, that the ball may lie clofe to the charge; and that its entra ce may point exactly to the centre of the ball.

CHAMBERS (Ephraim), author of the dictionary of fciences called the *Cyclopædia*. He was born at Milton in the county of Weftmoreland, where he received the common education for qualifying a youth for trade and commerce. When he became of a proper age, he was put apprentice to Mr. Senex the globemaker, a bufinefs which is connected with literature, efpecially with geography and aftronomy. It was during Mr. Chambers's refidence with this fkilful artift, that he acquired that tafte for literature which accompanied him through life, and directed all his purfuits. It was even at this time that he formed the defign of his grand work, the Cyclopædia; fome of the firft articles of which were written behind the counter. To have leifure to purfue this work, he quitted Mr. Senex, and took chambers at Gray's-Inn, where he chiefly refided during the reft of his life. The firft edition of the Cyclopædia, which was the refult of many years intenfe application, appeared in 1728, in 2 vols. folio. The reputation that Mr. Chambers acquired by the execution of this work, procured him the honour of being elected F. R. S. Nov. 6, 1729. In lefs than ten years time, a fecond edition became neceffary; which accordingly was printed, with corrections and additions, in 1738; and this was followed by a third edition the very next year.

Although the Cyclopædia was the chief bufinefs of Mr. Chambers's life, and may be regarded as almoft the fole foundation of his fame, his attention was not wholly confined to this undertaking. He was concerned in a periodical publication, called, *The Literary Magazine*, which was begun in 1735. In this work he wrote a variety of articles; particularly a review of Morgan's *Moral Philofopher*. He was alfo concerned with Mr. John Martyn, profeffor of botany at Cambridge, in preparing for the prefs a tranflation and abridgment of the *Philofophical Hiftory and Memoirs of the R. Acad. of Sciences at Paris*; which work was not publifhed till 1742, fome time after our author's deceafe, in 5 volumes 8vo. Mr. Chambers was alfo author of the tranflation of the *Jefuit's Perfpective*, from the French, in 4to; which has gone through feveral editions.

Mr. Chambers's clofe and unremitting attention to his ftudies at length impaired his health, and obliged him occafionally to take a country lodging, but without much benefit; he afterwards vifited the fouth of France, but ftill with little effect; he therefore returned to England, where he foon after died, at Iflington, May 15, 1740, and was buried at Weftminfter Abbey.

After the author's death, two more editions of his Cyclopædia were publifhed. The proprietors afterwards procured a fupplement to be compiled, by Mr. Scott and Dr. Hill, but chiefly by the latter, which extended to two volumes more; and the whole has fince been reduced into one alphabet in 4 volumes, by Dr. Rees, forming a very valuable body of the fciences.

CHAMBRANLE, the border, frame, or ornament of ftone or wood, furrounding the three fides of doors, windows, and chimneys. This is different in the different orders: when it is plain, and without mouldings, it is called fimply and properly, *band*, *cafe*, or *frame*. In an ordinary door, it is moftly called *door-cafe*; in a window, the *window-frame*.

The Chambranle confifts of three parts; the two fides, called afcendants; and the top, called the traverfe or fupercilium.

CHAMFER, or Chamferet, an ornament, in architecture, confifting of half a fcotia; being a kind of fmall furrow or gutter on a column.

CHAMFERING, is ufed for cutting the edge or the end of any thing bevel, or aflope.

CHANCE, *the Doctrine and Laws of*, are the fame as thofe of expectation, or probability, &c; which fee. Chances, in play, confift of the number of ways by which events may happen. Thus, if a halfpenny, or other piece of money, be toffed up, there are two events, or chances, or fides that may turn up, namely one chance for turning up a head, and one for the contrary; that is, it is an equal chance to throw a head or not. And in throwing a common die, which has 6 faces, there are in all 6 chances, that is one chance for throwing an ace or any other fingle point, and 5 chances againft it; or it is 5 to 1 that fuch affigned point does not come up.

Upon this fubject, fee De Moivre, Simpfon, &c.

CHANDELIERS, in Fortification, a kind of wooden parapet, confifting of upright timbers fupporting others laid acrofs the tops of them, 6 feet high, and fortified with fafcines &c. They are ufed to cover the workmen in approaches, galleries, and mines. And they differ from blinds only in this, that the former ferve to cover the men before, and the latter over head.

CHANGES, the permutations or variations of any number of things, with regard to their pofition, order, &c; as how many changes may be rung on a number of bells, or how many different ways any number of perfons may be placed, or how many feveral variations may be made of any number of letters, or any other things propofed to be varied.

To find out fuch number of changes, multiply continually together all the terms in a feries of arithmetical progreffion, whofe firft term and common difference are each unity or 1, and the laft term the number of things propofed to be varied, thus $1 \times 2 \times 3 \times 4 \times 5$ &c. till the laft number be the propofed number of things. For,

If there be only two things, as a and b, they admit of a double order or pofition only; for they may be placed either thus ab or thus ba, viz, $1 \times 2 = 2$ ways. If there be three things, a, b, and c, they will admit of 6 variations $= 1 \times 2$ fince as in the margin, and no more; three each of the three may be combined there different ways with each of the other two.

a	b	c
a	c	b
b	a	c
b	c	a
c	a	b
c	b	a

And

And if there be 4 things, each of them may be combined 4 ways with each order of the other three, that is 4 times 6 ways, or 1 × 2 × 3 × 4 = 24 ways.

In like manner, the combinations

of 5 things are 1 × 2 × 3 × 4 × 5 = 120
of 6 things are 1 × 2 × 3 × 4 × 5 × 6 = 720
&c.

So that if it be proposed to assign how many different ways a company of 6 persons may be placed, at table for instance, the answer will be 720 ways. Also the number of changes that can be rung on 7 bells, are 1 × 2 × 3 × 4 × 5 × 6 × 7 or 720 × 7 = 5040 changes.

CHAPITERS, the crowns or upper parts of a pillar or column.

CHAPPE (Jean d'Auteroche), a French astronomer, was born at Mauriac, in Auvergne, March 2, 1728. A taste for drawing and mathematics appeared in him at a very tender age; and he owed to Dom Germain a knowledge of the first elements of mathematics and astronomy. M. Cassini, after assuring himself of the genius of this young man, undertook to improve it. He employed him upon the map of France, and the translation of Halley's tables, to which he made considerable additions. The king charged him in 1753 with drawing the plan of the county of Bitche, in Lorraine, all the elements of which he determined geographically. He occupied himself greatly with the two comets of 1760; and the fruit of his labour was his Elementary treatise on the theory of those comets, enriched with observations on the zodiacal light, and on the aurora borealis. He soon after went to Tobolsk, in Siberia, to observe the transit of Venus over the sun; a journey which greatly impaired his health. After two years absence he returned to France in 1762, where he occupied himself for some time in putting in order the great quantity of observations he had made. M. Chappe also went to observe the next transit of Venus, viz that of 1769, at California, on the west side of North America, where he died of a dangerous epidemic disease, the 1st of August 1769. He had been named Adjunct Astronomer to the Academy the 17th of January 1759.

The published works of M. Chappe, are,

1. The Astronomical Tables of Dr. Halley; with observations and additions: in 8vo, 1754.

2. Voyage to California to observe the transit of Venus over the sun, the 3d of June 1769: in 4to, 1772.

3. He had a considerable number of papers inserted in the Memoirs of the Academy, for the years 1760, 1761, 1764, 1765, 1766, 1767, and 1768; chiefly relating to astronomical matters.

CHAPTREL, the same with Impost.

CHARACTERISTIC, *of a Logarithm*, the same as Index, or Exponent. This term was first used by Briggs in the 4th section of his Arithmetica Logarithmica, where he treats particularly of it; meaning by it, the integral or first part of a logarithm towards the left hand, which expresses 1 less than the integer places or figures in the number answering to that logarithm, or how far the first figure of this number is removed from the place of units; namely, that 0 is the characteristic of all numbers from 1 to 10; and 1 the characteristic of all those from 10 to 100; and 2 the characteristic of all those from 100 to 1000; and so on.

CHARACTERS, are certain marks used by Astronomers, Mathematicians, &c, to denote certain things whether for the sake of brevity, or perspicuity, in their operations.

1. Astronomical Characters.

Planets &c.	*The twelve Signs or Constellations of the Zodiac.*
☉ The Sun	♈ Aries, the Ram
☾ The Moon	♉ Taurus, the Bull
⊕ The Earth	♊ Gemini, the Twins
☿ Mercury	♋ Cancer, the Crab
♀ Venus	♌ Leo, the Lion
♂ Mars	♍ Virgo, the Maid
♃ Jupiter	♎ Libra, the Balance
♄ Saturn	♏ Scorpio, the Scorpion
♅ Herschel, or the Georgian Planet	♐ Sagittary, the Archer
	♑ Capricorn, the Goat
☊ Ascending Node	♒ Aquarius, the Water-bearer
☋ Descending Node	♓ Pisces, the Fishes.

The Aspects, Time, Motion, &c.

☌ Conjunction	° Degrees
☍ Opposition	′ Minutes or Primes
⚹ Sextile	″ Seconds, &c.
□ Quartile	A. M. Ante merid. or m morn.
△ Trine	P. M. Post merid. or a aftern.
	h, m, s, Hours, min. sec.

2. Mathematical &c. Characters.

Numeral Characters used by different Nations.

The most common numeral characters, are those called Arabic or Indian, viz. 1, 2, 3, 4, 5, 6, 7, 8, 9, with 0 or cipher for nothing.

The Roman numeral characters are seven, viz, I one, V five, X ten, L fifty, C a hundred, D or IↃ five hundred, M or DⅭ or CIↃ a thousand. Other combinations are as in the following synopsis of the Roman Notation.

1 = I
2 = II: As often as any character is repeated,
3 = III so many times its value is repeated.
4 = IIII or IV: A less character before a
5 = V greater diminishes its value.
6 = VI: A less character after a greater in-
7 = VII creases its value.
8 = VIII
9 = IX
10 = X
50 = L
100 = C
500 = D or IↃ: For every Ↄ added, this becomes 10 times as many.
1000 = M or CIↃ: For every C and Ↄ, set one at
2000 = MM each end, it becomes 10 times as much.
5000 = IↃↃ or $\overline{V}$: A line over any number in-
6000 = $\overline{VI}$ creases it 1000 fold.
10000 = $\overline{X}$ or CCIↃↃ
50000 = IↃↃↃ
60000 = $\overline{LX}$
100000 = $\overline{C}$ or CCCIↃↃↃ
1000000 = $\overline{M}$ or CCCCIↃↃↃↃ
2000000 = $\overline{MM}$, &c.

Greek

Greek Numerals.

The Greeks had three ways of expressing numbers. First, The moſt ſimple was, for every ſingle letter, according to its place in the alphabet, to denote a number from α 1 to ω 24; in which manner the books of Homer's Ilias are diſtinguiſhed. Secondly, Another way was by dividing the alphabet into *(firſt)* 8 units, α 1, β 2, &c; *(2nd)* 8 tens, ι 10, κ 20, &c; *(3d)* 8 hundreds, ρ 100, σ 200, &c: And thouſands they expreſſed by a point or accent under a letter, as α 1000, β 2000, &c. Thirdly, A third way was by ſix capital letters, thus, Ι (ια for μια) 1, Π (πεντε) 5, Δ (δεκα) 10, Η (Ηεκατον) 100, Χ (χιλια) 1000, Μ (μυρια) 10000: and when the letter Π incloſed any of theſe, except Ι, it ſhewed that the incloſed letter was five times its own value, as Δ̄ 50, H̄ 500, X̄ 5000, M̄ 50000.

Hebrew Numerals.

The Hebrew alphabet was divided into, Nine Units, as א 1, ב 2, &c; Nine Tens, as י 10, כ 20, &c; Nine Hundreds, as ק 100, ר 200, &c, ך 500, ם 600, ן 700, ף 800, ץ 900. Thouſands were ſometimes expreſſed by the units prefixed to hundreds, as אלדר 1534, &c; and even to tens, as אע 1070, &c. But more commonly thouſands were expreſſed by the word אלף 1000, אלפים 2000; and א אלפים with the other numerals prefixed to ſignify the number of thouſands, as נאלפים 3000, &c.

Characters uſed in Arithmetic and Algebra.

The firſt letters of the alphabet, *a, b, c,* &c, denote given quantities; and the laſt letters *z, y, x,* &c, denote ſuch as are unknown or ſought. Stifelius firſt uſed the capitals A, B, C, &c, for the unknown or required quantities. After that, Vieta uſed the capital vowels A, E, I, O, U, Y for the unknown or required quantities, and the conſonants B, C, D, &c, for known or given numbers. Harriot changed Vieta's capitals into the ſmall letters, viz *a, e, i, o, u,* for unknown, and *b, c, d,* &c, for known quantities. And Deſcartes changed Harriot's vowels for the latter letters *z, y, x,* &c, and the conſonants for the leading letters *a, b, c, d,* &c.

Newton denotes the ſeveral orders of the fluxions of variable quantities by as many points over the latter letters;

as $\dot{x}, \dot{y}, \dot{z}$ are the 1ſt fluxions,
$\ddot{x}, \ddot{y}, \ddot{z}$ are the 2d fluxions,
$\dddot{x}, \dddot{y}, \dddot{z}$ are the 3d fluxions,

&c, of *x, y, z.* And Leibnitz denotes the differentials of the ſame quantities by prefixing *d* to each of them, thus *dx, dy, dz.*

Powers of quantities are denoted by placing the index or exponent after them, towards the upper part; thus a^2 is the 2d power, a^3 the third power, and a^n the *n* power of *a.* Diophantus marked the powers by their initials, thus δͮ, κͮ, δδͮ, δκͮ, κκ, &c, for dynamis, cubus, dynamodynamis, &c, or the 2d, 3d, 4th, &c powers; and the ſame method has been uſed by ſeveral of the early writers, ſince the introduction of Algebra into Europe: but the firſt of them, as Paciolus, Cardan, &c, uſed no mark for powers, but the words

themſelves. Stifel, and others about his time, uſed the initials or abbreviations, ꝝ, 3, ſſ, 3 3, &c, of res or coſs, zenzus, cubus, zenzizenzus, &c, barbarous corruptions of the Italian coſa, cenſus, cubo, cenſi cenſus, &c. But he uſed alſo numeral exponents, both poſitive and negative, to the general characters or roots A, B, C, &c. Bombelli uſed a half circle thus ⌣ as a general character for the unknown or quantity required to be found in any queſtion, and the ſeveral powers of it he denoted by figures ſet above it; thus ⌣̇, ⌣̈, ⌣̇̈, are the 1ſt, 2d, 3d powers of ⌣; which powers he called dignities. Stevinus uſed a whole circle for the ſame unknown quantity, with the numeral index within it, and that both integral and fractional; thus ⓪, ①, ②, ③, are the 0, 1, 2, 3 powers of the general quantity ○; alſo ②̸, ③̸, ⑥̸, he uſes as the ſquare root, cubic root, 4th root of the ſame;

and ②̸/③, the cube root of the ſquare,

and ③̸/②, the ſquare root of the cube, and ſo on.
And theſe fractional exponents were adopted and farther uſed by his commentator Albert Girard. So that Stevinus ought to be eſteemed the firſt perſon who rendered general the notation of all powers and roots in the ſame way, the former by integral, and the latter by fractional exponents. Harriot denoted his powers by a repetition of the letters; thus *a, aa, aaa,* &c. And Deſcartes, inſtead of this, ſet the numeral index at the upper part of the letters, as at preſent thus *a, a^2, a^3,* &c. Though, I am informed, by ſuch as have ſeen Harriot's poſthumous papers, that he alſo there makes uſe of exponents.

The character √ is the ſign of radicality, or of a root, being derived from the initial R or *r,* which was uſed at firſt by Paciolus, Cardan, &c. This character √ I firſt find uſed by Stifel, in 1544, and by Robert Recorde in 1557. The character √ alone denotes the ſquare root only; but at firſt they uſed the initial of the name after it, to denote the ſeveral roots: as √q the quadrate or ſquare root, and √c the cubic root. But the numeral indices of the root were prefixed by Albert Girard, exactly the ſame as they are uſed at preſent, viz $\sqrt[3]{}, \sqrt[3]{}, \sqrt{}$, the 2d, 3d, or 4th root.

The character + denotes addition, and a poſitive quantity. At firſt the word itſelf was uſed, plus, piu, or the initial *p.* by Paciolus, Cardan, Tarfalea, &c. And the character + for addition occurs in Stifelius.

The character — denotes ſubtraction, and a negative quantity; which alſo firſt occurs in the ſame author Stifelius. Before that, the word minus, mene, or the initial *m.* was uſed. Other characters have alſo been ſometimes uſed by other authors, for addition and ſubtraction; but they are now no longer in uſe.

× denotes multiplication, and was introduced by Oughtred.

÷ denoting diviſion, was introduced by Dr. Pell. Diviſion is alſo denoted like a fraction,

thus $\frac{a}{b}$ or $\frac{6}{3} = 2$.

= denotes equality, and was uſed by Robert Recorde. Deſcartes uſes ∞ for the ſame purpoſe.

The character :: for proportionality, or equality of ratios, was introduced by Oughtred; as was alſo the mark ÷ for continued proportion.

> for

$>$ for greater, and $<$ for lefs, were ufed by Harriot.

And $\sqsupset$ and $\sqsupset$ were ufed by Oughtred for the fame purpofes.

Dr. Pell ufed $\odot$ for involution, and $\nu\nu$ for evolution. ∞ denotes a general difference between any two quantities, and was introduced by Dr. Wallis.

The Parenthefis ( ), as a vinculum, was invented by Albert Girard, and ufed in fuch expreffions as thefe, $\sqrt[3]{(72 + \sqrt{5120})}$, and B $(Bq + Cq)$, both for univerfal roots, and multiplication, &c. _____

The ftraight-lined vinculum, _____ , was ufed by Vieta for the fame purpofe; thus $\overline{A - B}$ in $\overline{B + C}$.

Characters in Geometry and Trigonometry.

□	A Square	∠	An Angle
△	A Triangle	∟	A Rightangle
▭	A Rectangle	⊥	Perpendicular
O or ⊙	A Circle	‖	Parallel.

CHARGE, in Electricity, in a ftrict fenfe, imports the accumulation of the electric matter on one furface of an electric, as the Leyden phial, a pane of glafs, &c, whilft an equal quantity paffes off from the oppofite furface. Or, more generally, electrics are faid to be charged, when the equilibrium of the electric matter on the oppofite furface is deftroyed, by communicating one kind of electricity to one fide, and the contrary kind to the oppofite fide: nor can the equilibrium be reftored till a communication be made by means of conducting fubftances between the two oppofite furfaces: and when this is done, the electric is faid to be difcharged. The charge properly refers to one fide, in contradiftinction from the other; fince the whole quantity in the electric is the fame before and after the operation of charging; and the operation cannot fucceed, unlefs what is gained on one fide is loft by the other, by means of conductors applied to it, and communicating either with the earth, or with a fufficient number of non-electrics. To facilitate the communication of electricity to an electric plate &c, the oppofite furfaces are coated with fome conducting fubftance, ufually with tin-foil, within fome diftance of the edge; in confequence of which the electricity communicated to one part of the coating, is readily diffufed through all parts of the furface of the electric in contact with it; and a difcharge is eafily made by forming a communication with any conductor from one coating to the other. If the oppofite coatings approach too near each other, the electric matter forces a paffage from one furface to the other before the charge is complete. And fome kinds of glafs have the property of conducting the electricity over the furface, fo that they are altogether unfit for the operation of charging and difcharging. If indeed the charge be too high, and the glafs plate or phial too thin, the attraction between the two oppofite electricities forces a paffage through the glafs, making a fpontaneous difcharge, and the glafs becomes unfit for farther ufe. See *Conductors, Electrics, Leyden Phial*, &c.

CHARGE, in Gunnery, the load of a piece of ordnance, or the quantity of powder and ball, or fhot, with which it is prepared for execution.

The charge of powder, for proving guns, is equal to the weight of the ball; but for fervice, the charge is $\frac{1}{2}$ or $\frac{1}{3}$

the weight of the ball, or ftill lefs; and indeed in moft cafes of fervice, the quantity of powder ufed is too great for the intended execution. In the Britifh navy, the allowance for 32 pounders is but $\frac{1}{7}$ of the weight of the ball. But it is probable that, if the powder in all fhip guns was reduced to $\frac{1}{4}$ the weight of the ball, or even lefs, it would be a confiderable advantage, not only by faving ammunition, but by keeping the guns cooler and quieter, and at the fame time more effectually injuring the veffels of the enemy. With the prefent allowance of powder, the guns are heated, and their tackle and furniture ftrained, and all this only to render the ball lefs efficacious: for a ball which can but juft pafs through a piece of timber, and in the paffage lofes almoft all its motion, is found to rend and fracture it much more, than when it paffes through with a much greater velocity. See Robins's Tracts, vol. 1. pa. 290, 291.

Again, the fame author obferves, that the charge is not to be determined by the greateft velocity that may be produced; but that it fhould be fuch a quantity of powder as will produce the leaft velocity neceffary for the purpofe in view; and if the windage be moderate, no field-piece fhould ever be loaded with more than $\frac{1}{4}$, or at the utmoft $\frac{1}{3}$ of the weight of its ball in powder; nor fhould the charge of any battering piece exceed $\frac{1}{3}$ of the weight of its bullet. Ib. pa. 266.

Different charges of powder, with the fame weight of ball, produce different velocities in the ball, which are in the fubduplicate ratio of the weights of powder; and when the weight of powder is the fame, and the ball varied, the velocity produced is in the reciprocal fubduplicate ratio of the weight of the ball: which is agreeable both to theory and practice. See my paper on Gunpowder in the Philof. Tranf. 1778, pa. 50; and my Tracts, vol. 1. pa. 266.

But this is on a fuppofition that the gun is of an indefinite length; whereas, on account of the limited length of guns, there is fome variation from this law in practice, as well as in theory; in confequence of which it appears that the velocity of the ball increafes with the charge only to a certain point, which is peculiar to each gun, where the velocity is the greateft; and that by farther increafing the charge, the velocity gradually diminifhes, till the bore is quite full of powder. By an eafy fluxionary procefs it appears that, calling the length of the bore of the gun b, the length of the charge producing the greateft velocity, ought to be $\dfrac{b}{2\cdot718281828}$,

or about $\frac{1}{3}$ of the length of the bore; where 2·718281828 is the number whofe hyp. log. is 1. But, for feveral reafons, in practice the length of the charge producing the greateft velocity, falls fhort of that above mentioned, and the more fo as the gun is longer. From many experiments I have found the length of the charge producing the greateft velocity, in guns of various lengths of bore, from 15 to 40 calibres, as follows.

Length of Bore in Calibres.	Length of Charge for greateft Veloc.
15	$\frac{3}{10}$
20	$\frac{3}{12}$
30	$\frac{3}{18}$
40	$\frac{3}{20}$.

CHARLES's

CHARLES's WAIN, a name by which some of the astronomical writers, in our own language, have called Ursa Major, or the great bear; though some writers say the lesser bear. Indeed both of the two bears have been called waggons or wains, and by the Latins, who have followed the Arabians, two biers, Feretrum majus & minus.

CHART, or Sea-Chart, a hydrographical or sea-map, for the use of navigators; being a projection of some part of the sea in plano, shewing the sea coasts, rocks, sands, bearings, &c. Fournier ascribes the invention of sea-charts to Henry son of John king of Portugal. These charts are of various kinds, the Plain chart, Mercator's or Wright's chart, the Globular chart, &c.

In the construction of charts, great care should be taken that the several parts of them preserve their position to one another, in the same order as on the earth; and it is probable that the finding out of proper methods to do this, gave rise to the various modes of projection.

There are many ways of constructing maps and charts; but they depend chiefly on two principles. First, by considering the earth as a large extended flat surface; and the charts made on this supposition are usually called Plain Charts. Secondly, by considering the earth as a sphere; and the charts made on this principle are sometimes called Globular Charts, or Mercator's Charts, or Reduced Charts, or Projected Charts.

Plain Charts have the meridians, as well as the parallels of latitude, drawn parallel to each other, and the degrees of longitude and latitude everywhere equal to those at the equator. And therefore such charts must be deficient in several respects. For, 1st, since in reality all the meridians meet in the poles, it is absurd to represent them, especially in large charts, by parallel right lines. 2dly, As plain charts shew the degrees of the several parallels as equal to those of the equator, therefore the distances of places lying east and west, must be represented much larger than they really are. And 3dly, In a plain chart, while the same rhumb is kept, the vessel appears to sail on a great circle, which is not really the case. Yet plain charts made for a small extent, as a few degrees in length and breadth, may be tolerably exact, especially for any part within the torrid zone; and even a plain chart made for the whole of this zone will differ but little from the truth.

Mercator's Chart, like the plain charts, has the meridians represented by parallel right lines, and the degrees of the parallels, or longitude, everywhere equal to those at the equator, so that they are increased more and more, above their natural size, as they approach towards the pole; but then the degrees of the meridians, or of latitude, are increased in the same proportion at the same part; so that the same proportion is preserved between them as on the globe itself. This chart has its name from that of the author, Girard Mercator, who first proposed it for use in the year 1556, and made the first charts of this kind; though they were not altogether on true or exact principles, nor does it appear that he perfectly understood them. Neither indeed was the thought originally his own, viz. of lengthening the degrees of the meridian in some proportions for this was hinted by Ptolemy near two thousand years ago. It was not perfected however till Mr. Wright

first demonstrated it about the year 1590, and shewed a ready way of constructing it, by enlarging the meridian line by the continual addition of the secants. See his Correction of Errors in Navigation, published in 1599.

Globular Chart, is a projection so called from the conformity it bears to the globe itself; and was proposed by Messrs Senex, Wilson, and Harris. This is a meridional projection, in which the parallels are equidistant circles, having the pole for their common centre, and the meridians curvilinear and inclined, so as all to meet in the pole, or common centre of the parallels. By which means the several parts of the earth have their proper proportion of magnitude, distance, and situation, nearly the same as on the globe itself; which renders it a good method for geographical maps.

Hydrographical Charts, are sheets of large paper, on which several parts of the land and sea are described, with their respective coasts, harbours, sounds, flats, rocks, shelves, sands, &c, also the points of the compass, and the latitudes and longitudes of the places.

Selenographic Charts, are particular descriptions of the appearances, spots and maculæ of the moon.

Topographic Charts, are draughts of some small parts only of the earth, or of some particular place, without regard to its relative situation, as London, York, &c.

For the Construction of Charts, see Geography, Maps, Projection, &c.

CHASE, of a Gun, is its bore or cylinder.

CHAULNES (The Duke De), a peer of France, but more honourable and remarkable as an astronomer and mathematician. He was born at Paris Dec. 30, 1714. He soon discovered a singular taste and genius for the sciences; and in the tumults of armies and camps, he cultivated mathematics, astronomy, mechanics, &c. He was named honorary-academician the 27th of February 1743, and few members were more punctual in attending the meetings of that body; where he often brought different constructions and corrections of instruments of astronomy, of dioptrics, and achromatic telescopes. These researches were followed with a new parallactic machine, more solid and convenient than those that were in use; as also with many reflections on the manner of applying the micrometer to those telescopes, and of measuring exactly the value of the parts of that instrument. The duke of Chaulnes proposed many other works of the same kind, when death surprised him the 23d Sept. 1769.

He had several papers published in the volumes of Memoirs of the Academy of Sciences, as follow:

1. Observations on some Experiments in the 4th part of the 2d book of Newton's Optics: an. 1755.

2. Observations on the Platform for dividing mathematical instruments: 1765.

3. Determination of the distance of Arcturus from the Sun's limb, at the summer solstice: 1765.

4. On some means of perfecting astronomical instruments: 1765.

5. Of some experiments relative to dioptrics: 1767.

6. The art of dividing mathematical instruments: 1768.

7. Observations of the Transit of Venus, June 3, 1769: 1769.

8. New method of dividing mathematical and astronomical instruments.

CHAUSE.

CHAUSE TRAPPES, or *Calirops*, or *Crowsfeet*, are iron inftruments of fpikes about 4 inches long, made like a ftar, in fuch a manner that whichever way they fall, one point ftands always upwards, like a nail. They are ufually thrown and fcattered into moats and breaches, to gall the horfes feet, and ftop the hafty approach of the enemy.

CHAZELLES (John Matthew), a French mathematician and engineer, was born at Lyons in 1657, and educated there in the college of Jefuits, from whence he removed to Paris in 1675. He firft became acquainted with Du Hamel, fecretary to the Academy of Sciences, and through him with Caffini, who employed him with himfelf at the Obfervatory, where Chazelles greatly improved himfelf, and alfo affifted Caffini in the meafurement of the fouthern part of the meridian of France. Having, in 1684, inftructed the duke of Montemar in the mathematical fciences, this nobleman procured him the appointment of hydrography-profeffor to the galleys of Marfeilles. In difcharging the duties of this department, he made numerous geometrical and aftronomical obfervations, from which he drew a new map of the coaft of Provence.— He alfo performed many other fervices in that department, and as an engineer, along with the armies and naval expeditions. To make obfervations in Geography and Aftronomy, he undertook alfo a voyage to the Levant, and among other things he meafured the pyramids of Egypt, and found the four fides of the largeft of them exactly to face the four cardinal points of the compafs. He made a report of his voyage, on his return, to the Academy of Sciences, upon which he was named a member of their body in 1695, and had many papers inferted in the volumes of their Memoirs, from 1693 to 1708. Chazelles died at Marfeilles the 16th of January 1710.

CHEMIN *des Ronds*, in Fortification, the way of the rounds, or a fpace between the rampart and the low parapet under it, for the rounds to go about it.

CHEMISE, a wall that lines a baftion, or ditch, or the like, for its greater fupport and ftrength.

CHERSONESUS, a peninfula, or part of the land almoft encompaffed round with the fea, only joining to the main land by a narrow neck or ifthmus. Varenius enumerates 14 of thefe.

CHEVAL *de Frife*, pl. *Chevaux de Frife*, or Frifeland horfe, fo called becaufe it was firft ufed in that country. It confifts of a joift or piece of timber, about a foot in diameter, and 10 or 12 long, pierced and tranfverfed with a great number of wooden fpikes of 5 or 6 feet long, and armed or pointed with iron. It is fometimes alfo called *turnpike*, or *tourniquet*. It is chiefly ufed to ftop a breach, defend a paffage, or make a retrenchment to ftop the cavalry.

CHEVRETTE, in Artillery, an engine to raife guns or mortars into their carriage. It is formed of two pieces of wood of about 4 feet long, ftanding upright upon a third, which is fquare. The uprights are about a foot afunder, and pierced with holes exactly oppofite to one another, to receive a bolt of iron, which is put in either higher or lower at pleafure, to ferve as a fupport to a handfpike by which the gun is raifed up.

CHEYNE (George), a Britifh phyfician, and mathematician, was born in Scotland, 1671, and educated at Edinburgh under Dr. Pitcairn. He paffed his youth in clofe ftudy and great abftemioufnefs; but coming to London, when about 30 years of age, he fuddenly changed his whole manner of living; which had fuch an effect upon his conftitution, that his body grew to a moft enormous bulk, weighing it is faid about 448 pounds. From this load of oppreffion however he was afterward in a great meafure relieved, by means of a milk and vegetable diet, which reduced his weight to about one-third of what it had been, and reftored him to a good ftate of health; by which his life was prolonged to the 72d year of his age.

Dr. Cheyne was author of various medical and other tracts, and of a treatife on the Inverfe method of Fluxions, under the title of *Fluxionum Methodus Inverfa; five quantitatum fluentium leges generaliores*: in 4to, 1703. Upon this book De Moivre wrote fome animadverfions in an 8vo vol. 1704; which were replied to by Cheyne in 1705.

CHILIAD, an affemblage of feveral things ranged by thoufands. It was particularly applied to tables of logarithms, becaufe they were at firft divided into thoufands. Thus, in the year 1624, Mr. Briggs publifhed a table of logarithms for 20 chiliads of abfolute numbers; afterward, he publifhed 10 chiliads more; and laftly, one more; making in all 31 chiliads.

CHILIAGON, a regular plane figure of a thoufand fides and angles.

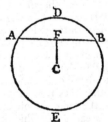

CHORD, a right line connecting the two extremes of an arch; fo called from its refemblance to the chord or ftring of a bow; as AB, which is common to the two parts or arches ADB, AEB that make up the whole circle. The chords have feveral properties:

1. The Chord is bifected by a perpendicular CF drawn to it from the centre.

2. Chords of equal arcs, in the fame or equal circles, are themfelves equal.

3. Unequal Chords have to one another a lefs ratio than that of their arcs.

4. The chord of an arc, is a mean proportional between the diameter and the verfed fine of that arc.

Scale or Line of Chords. See *Plane Scale.*

Chord, or Cord, in Mufic, denotes the ftring or line by whofe vibrations the fenfation of found is excited; and by whofe divifions the feveral degrees of tune are determined.

To divide a Chord AB in the moft fimple manner, fo as to exhibit all the original concords.

A   C E D   B

Divide

Divide the given line into two equal parts at C; then subdivide the part CB equally in two at D, and again the part CD into two equal parts at E. Here AC to AB is an octave; AC to AD a fifth; AD to AB a fourth; AC to AE a greater third, and AE to AD a less third; AE to EB a greater sixth, and AE to AB a less sixth. Malcolm's Treatise of Music, ch. 6. sec. 3. See MONOCHORD.

To find the number of Vibrations made by a Musical Chord or String in a given time; having given its weight, length, and tension. Let l be the length of the chord in feet, 1 its weight, or rather a small weight fixed to the middle and equal to that of the whole chord, and w the tension, or a weight by which the chord is stretched. Then shall the time of one vibration be expressed by $\frac{1}{7}\sqrt{\frac{l}{32\frac{1}{6}w}}$, and consequently the number of vibrations per second is equal to $\frac{7}{1}\sqrt{\frac{32\frac{1}{6}w}{l}}$.

For example, suppose $w = 28800$, or the tension equal to 28800 times the weight of the chord, and the length of it 3 feet; then the last theorem gives 354 nearly for the number of vibrations made in each second of time.

But if w were 14400, there would be made but 250 vibrations per second; and if w were only 288, there would be no more than $35\frac{4}{5}$ vibrations per second. See my Select Exerc. prob. 21. pag. 200

CHORD, in Music, is used for the union of two or more sounds uttered at the same time, and forming together a complete harmony.

Chords are divided into perfect and imperfect. The perfect chord is composed of the fundamental sound below, of its third, its fifth, and its octave: they are likewise subdivided into major and minor, according as the thirds which enter into their composition are flat or sharp. Imperfect chords are those in which the sixth, instead of the fifth, prevails; and in general all those whose lowest are not their fundamental sounds.

Chords are again divided into consonances and dissonances. The consonances are the perfect chord, and its derivatives. Every other chord is a dissonance. A table of both, according to the system of M. Rameau, may be seen in Rousseau's Musical Dictionary, vol. 1, pa. 27.

CHOROGRAPHY, the art of delineating or describing some particular country or province.

This differs from geography as the description of a particular country differs from that of the whole earth: And from topography, as the description of a country differs from that of a town or a district.

CHROMATIC, a species of music which proceeds by semitones and minor thirds. The word is derived from the Greek χρωμα, which signifies colour, and perhaps the shade or intermediate shades of colour, which mingle and connect colours, like as the small intervals in this scale easily slide or run into each other.

Boethius and Zarlin ascribe the invention of the chromatic genus to Timotheus, a Milesian, in the time of Alexander the Great. The Spartans banished it their city on account of its softness. The character of this genus, according to Aristides Quintillianus, was sweetness and pathos.

CHROMATICS, is that part of optics which explains the several properties of the colours of light, and of natural bodies.

Before the time of Sir I. Newton, the notions concerning colour were very vague and wild. The Pythagoreans called colour the superficies of bodies: Plato said that it was a flame issuing from them: According to Zeno, it is the first configuration of matter: And Aristotle said it was that which made bodies actually transparent. Descartes accounted colour a modification of light, and he imagined that the difference of colour proceeds from the prevalence of the direct or rotatory motion of the particles of light. Grimaldi, Dechales, and many others, imagined that the differences of colour depended upon the quick or slow vibrations of a certain elastic medium with which the universe is filled. Rohault conceived, that the different colours were made by the rays of light entering the eye at different angles with respect to the optic axis. And Dr. Hooke imagined that colour is caused by the sensation of the oblique or uneven pulse of light; which being capable of no more than two varieties, he concluded there could be no more than two primary colours.

Sir I. Newton, in the year 1666, began to investigate this subject; when finding that the coloured image of the sun, formed by a glass prism, was of an oblong, and not of a circular form, as, according to the laws of equal refraction, it ought to be, he conjectured that light is not homogeneal; but that it consists of rays of different colours, and endued with divers degrees of refrangibility. And, from a farther prosecution of his experiments, he concluded that the different colours of bodies arise from their reflecting this or that kind of rays most copiously. This method of accounting for the different colours of bodies soon became generally adopted, and still continues to be the most prevailing opinion. It is hence agreed that the light of the sun, which to us seems white and perfectly homogeneal, is composed of no fewer than seven different colours, viz red, orange, yellow, green, blue, purple, and violet or indigo: that a body which appears of a red colour, has the property of reflecting the red rays more plentifully than the rest; and so of the other colours, the orange, yellow, green, &c: also that a body which appears black, instead of reflecting, absorbs all or the most part of the rays that fall upon it; while, on the contrary, a body which appears white, reflects the greatest part of all the rays indiscriminately, without separating them one from another.

The foundation of a rational theory of colours being thus laid, the next inquiry was, by what peculiar mechanism, in the structure of each particular body, it was fitted to reflect one kind of rays more than another; and this is attributed, by Sir I. Newton, to the density of these bodies. Dr. Hooke had remarked, that thin transparent substances, particularly soap-water blown into bubbles, exhibited various colours, according to their thinness; and yet, when they have a considerable degree of thickness, they appear colourless. And Sir Isaac himself had observed, that as he was compressing two prisms hard together, in order to make their sides (which happened to be a little convex) to touch one another, in the place of contact they were

both

both perfectly tranfparent, as if they had been but one continued piece of glafs: but round the point of contact, where the glaffes were a little feparated from each other, rings of different colours appeared. And when he afterwards, farther to elucidate this matter, employed two convex glaffes of telefcopes, preffing their convex fides upon one another, he obferved feveral feries of circles or rings of fuch colours, different, and of various intenfities, according to their diftance from the common central pellucid point of contact.

As the colours were thus found to vary according to the different diftances between the glafs plates, Sir Ifaac conceived that they proceeded from the different thicknefs of the plate of air intercepted between the glaffes: this plate of air being, by the mere circumftance of thinnefs or thicknefs, difpofed to reflect or tranfmit the rays of this or that particular colour. Hence therefore he concluded, that the colours of all natural bodies depend on their denfity, or the magnitude of their component particles: and hence alfo he conftructed a table, in which the thicknefs of a plate neceffary to reflect any particular colour, was expreffed in millionth parts of an inch.

From a great variety of fuch experiments, and obfervations upon them, our author deduced his theory of colours. And hence it feems that every fubftance in nature is tranfparent, provided it be made fufficiently thin; as gold, the denfeft fubftance we know of, when reduced into thin leaves, tranfmits a bluifh-green light through it. If we fuppofe any body therefore, as gold for inftance, to be divided into a vaft number of plates, fo thin as to be almoft perfectly tranfparent; it is evident that all, or the greateft part of the rays, will pafs through the upper plates, and when they lofe their force will be reflected from the under ones. They will then have the fame number of plates to pafs through which they had penetrated before; and thus, according to the number of thofe plates through which they are obliged to pafs, the object appears of this or that colour, juft as the rings of colours appeared different in the experiment of the two plates, according to their diftance from one another, or the thicknefs of the plate of air between them.

This theory of the colours has been illuftrated and confirmed by various experiments, made by other phylofophers. Mr. E. H. Delaval produced fimilar effects by the infufions of flowers of different colours, and by the intimate mixture of the metals with the fubftance of glafs, when they are reduced to very fine parts; the more denfe metals imparting to the glafs the lefs refrangible colours, and the lighter ones thofe colours that are more eafily refrangible. Dr. Prieftley and Mr. Canton alfo, by laying very thin leaves or flips of the metals upon glafs, ivory, wood, or metal, and paffing an electrical ftroke through them, found that the fame effect was produced, viz, that the fubftrated was tinged with different colours, according to the diftance from the point of explofion.

However, the Abbe Mazeas and M. du Tour contended, that the colours between the glaffes are not to be afcribed to the thin ftratum of air, fince they equally produced them by rubbing and preffing together two flat glaffes, which cohered fo clofely that it required

the greateft force to move or flide them over one another. See Prieftley's Hiftory of Vifion.

Of Newton's 8th Exper. in the 2d Book of Optics.

The event of this experiment, which has been contradicted by repetitions of the fame by other philofophers, having been the occafion of much controverfy; and relating to a material part of the doctrine of chromatics, it will not be improper here to give an account of what has paffed concerning it. Newton found, he fays, that when light, by contrary refractions through different mediums, is fo corrected, that it emerges in lines parallel to the incident rays, it continues ever after to be white. But that if the emergent rays be inclined to the incident ones, the whitenefs of the emerging light will, by degrees, in paffing on from the place of emergence, become tinged at its edges with colours. And thefe laws he inferred from experiments made by refracting light with prifms of glafs, placed within a prifmatic veffel of water.

By theorems deduced from this experiment he infers, that the refraction of the rays of every fort, made out of any medium into air, are known by having the refraction of the rays of any one fort: and alfo, that the refraction out of one medium into another is found, whenever we have the refractions out of them both, into any third medium.

Now the fame experiment, when fince performed by other perfons, turning out contrary to what is ftated above, fome rather free reflections have been thrown upon Newton concerning it; but which however have been very fatisfactorily obviated by Mr. Peter Dollond, in a late pamphlet on this fubject; as we fhall fhew below.

In the firft place then, M. Klingenftierna, a Swedifh philofopher, having in the year 1755 confidered the controverfy between Euler and Mr. John Dollond, relative to the refraction of light, formed a theorem of his own, from geometrical reafoning, by which he was induced to believe that the refult of Newton's experiment could not be as he had related it; except when the angles of the refracting mediums are fmall. See the paper on this matter by Klingenftierna in the pamphlet above cited by Mr. Peter Dollond.

This paper of Klingenftierna being communicated to Mr. John Dollond by Mr. Mallet, to whom it was fent for that purpofe, made Dollond entertain doubts concerning Newton's report of the refult of his experiment, and determined him to have recourfe to experiments of his own, which he did in the year 1757, as follows.

He cemented two glafs planes together by their edges, fo as to form a prifmatic veffel when clofed at the ends or bafes; and the edge being turned downward, he placed it in a glafs prifm with one of its edges upward, filling up the vacancy with clear water; fo that the refraction of the prifm was contrived to be contrary to that of the water, in order that a ray of light, tranfmitted through both thefe refracting mediums, might be affected by the difference only between the two refractions. As he found the water to refract more or lefs than the glafs prifm, he diminifhed or augmented the angle between the glafs plates, till the two contrary refractions became equal, which he difcovered by viewing

ing

ing an object through this double prism. For when it appeared neither raised nor depressed, he was satisfied that the refractions were equal, and that the emergent rays were parallel to the incident ones.

Now, according to the prevailing opinion, he observes, that the object ought to have appeared through this double prism in its natural colour; for if the difference of refrangibility had been in all respects equal, in the two equal refractions, they would have rectified each other. But this experiment fully proved the fallacy of the received opinion, by shewing that the divergency of the light by the glass prism, was almost double of that by the water; for the image of the object, though not at all refracted, was yet as much infected with prismatic colours, as if it had been seen through a glass wedge only, having its angle of near 30 degrees.

This experiment is the very same with that of Sir Isaac Newton above-mentioned, notwithstanding the result was so remarkably different. Mr. Dollond plainly saw however, that if the refracting angle of the water-vessel could have admitted of a sufficient increase, the divergency of the coloured rays would have been greatly diminished, or entirely rectified; and that there would have been a very great refraction without colour, as he had already produced a great discolouring without refraction: but the inconveniency of so large an angle as that of the prismatic vessel must have been, to bring the light to an equal divergency with that of the glass prism, whose angle was about 60 degrees, made it necessary to try some experiments of the same kind with smaller angles.

Accordingly he procured a wedge of plate-glass, whose angle was only 9 degrees; and, using it in the same circumstances, he increased the angle of the water-wedge, in which it was placed, till the divergency of the light by the water was equal to that by the glass; that is, till the image of the object, though considerably refracted by the excess of the refraction of the water, appeared nevertheless quite free from any colours proceeding from the different refrangibility of the light.

Many conjectures were made as to the cause of so striking a difference in the results of the same experiment; but none that gave any great satisfaction, till lately that it has been shewn to be probably owing to the nature of the glass then used by Newton. This conjecture is made by Mr. Peter Dollond, son of John, the inventor of the achromatic telescope, in a pamphlet by him lately published in defence of his father's invention, against the misrepresentations of some persons who have unjustly attempted to give the invention to other philosophers, who themselves never imagined that they had any right to it. After a full and satisfactory vindication of his father, Mr. P. Dollond then adds,

" I now come to a more agreeable part of this paper, which is, to endeavour to reconcile the different results of the 8th experiment of the 2d part of the 1st book of Newton's Optics, as related by himself, and as it was found by Dollond, when he tried the same experiment, in the year 1757. Newton says, that light, as often as by contrary refractions it is so corrected, that it emergeth in lines parallel to the incident, continues ever after to be white. Now Dollond says, when he tried the same

experiment, and made the mean refraction of the water equal to that of the glass prism, so that the light emerged in lines parallel to the incident, he found the divergency of the light by the glass prism to be nearly double to what it was by the water prism. The light appeared to be so evidently coloured, that it was directly said by some persons, that if Newton had actually tried the experiment, he must have perceived it to have been so. Yet who could for a moment doubt the veracity of such a character? Therefore different conjectures were made by different persons. Mr. Murdoch in particular gave a paper to the Royal Society in defence of Newton; but it was such as very little tended to clear up the matter. Philos. Transf. vol. 53. pa. 192.—Some have supposed that Newton made use of water strongly impregnated with saccharum saturni, because he mentions sometimes using such water, to increase the refraction, when he used water prisms instead of glass prisms. Newton's Opt. pa. 62.—And others have supposed, that he tried the experiment with so strong a persuasion in his own mind that the divergency of the colours was always in the same proportion to the mean refraction, in all sorts of refracting mediums, that he did not attend so much to that experiment as he ought to have done, or as he usually did. None of these suppositions having appeared at all satisfactory, I have therefore endeavoured to find out the true cause of the difference, and thereby shew, how the experiment may be made to agree with Newton's description of it, and to get rid of those doubts, which have hitherto remained to be cleared up.

" It is well known, that in Newton's time the English were not the most famous for making optical instruments: Telescopes, opera-glasses, &c, were imported from Italy in great numbers, and particularly from Venice; where they manufactured a kind of glass which was much more proper for optical purposes than any made in England at that time. The glass made at Venice was nearly of the same refractive quality as our own crown glass, but of a much better colour, being sufficiently clear and transparent for the purpose of prisms. It is probable that Newton's prisms were made with this kind of glass; and it appears to be the more so, because he mentions the specific gravity of common glass to be to water as 2·58 to 1, Newton's Opt. pa. 247, which nearly answers to the specific gravity we find the Venetian glass generally to have. Having a very thick plate of this kind of glass, which was presented to me about 25 years ago by the late professor Allemand, of Leyden, and which he then informed me had been made many years, I cut a piece from this plate of glass to form a prism, which I conceived would be similar to those made use of by Newton himself. I have tried the Newtonian experiment with this prism, and find it answers so nearly to what Newton relates, that the difference which remains may very easily be supposed to arise from any little difference which may and does often happen in the same kind of glass made at the same place at different times. Now the glass prism made use of by Dollond to try the same experiment, was made of English flint-glass, the specific gravity of which I have never known to be less than 3·22. This difference in the densities of the prisms, used

used by Newton and Dollond, was sufficient to cause all the difference which appeared to the two experimenters in trying the same experiment.

"From this it appears, that Newton was accurate in this experiment as in all others, and that his not having discovered that, which was discovered by Dollond so many years afterwards, was owing entirely to accident; for if his prism had been made of glass of a greater or less deasity, he would certainly have then made the discovery, and refracting telescopes would not have remained so long in their original imperfect state." See *Achromatic*, and *Telescope*.

Mr. Delaval's experiments on the colours of opaque bodies.—Beside the experiments of this gentleman, before-mentioned, on the colours of transparent bodies, he has lately published an account of some made upon the permanent colours of opaque substances, the discovery of which must be of the utmost consequence in the arts of colour-making and dyeing.

The changes of colour in permanently coloured bodies, our author observes, are produced by the same laws that take place in transparent colourless substances; and the experiments by which they are investigated consist chiefly of various methods of uniting the colouring particles into larger masses, or dividing them into smaller ones. Sir Isaac Newton made his experiments chiefly on transparent substances; and in the few places where he treats of others, he acknowledges his want of experiments. He makes the following remark however on those bodies which reflect one kind of light and transmit another, viz. that if these glasses or liquors were so thick and massy that no light could get through them, he questioned whether they would not, like other opaque bodies, appear of one and the same colour in all positions of the eye; though he could not yet affirm it from experience. Indeed it was the opinion of this great philosopher, that all coloured matter reflects the rays of light, some reflecting the more refrangible rays most copiously, and others those that are less so; and that this is at once the true and only reason of these colours. He was likewise of opinion that opaque bodies reflect the light from their anterior surface, by some power of the body evenly diffused over and external to it. With respect to transparent coloured bodies he thus expresses himself: " A transparent body which looks of any colour by transmitted light, may also look of the same colour by reflected light; the light of that colour being reflected by the farther surface of that body, or by the air beyond it: and then the reflected colour will be diminished, and perhaps cease, by making the body very thick, and pitching it on the back-side to diminish the reflection of its farther surface, so that the light reflected from the tinging particles may predominate. In such cases the colour of the reflected light will be apt to vary from that of the light transmitted."

To search out the truth of these opinions, Mr. Delaval entered upon a course of experiments with transparent coloured liquors and glasses, as well as with opaque and semitransparent bodies. And from these experiments he discovered several remarkable properties of the colouring matter; particularly, that in transparent coloured substances it does not reflect any light;

and when, by intercepting the light which was transmitted, it is hindered from passing through such substances, they do not vary from their former colour to any other, but become entirely black.

This incapacity of the colouring particles of transparent bodies to reflect light, being deduced from very numerous experiments, may therefore be taken as a general law. It will appear the more extensive, if it be considered that, for the most part, the tinging particles of liquors, or other transparent substances, are extracted from opaque bodies; that the opaque bodies owe their colours to those particles, in like manner as the transparent substances do; and that by the loss of them they are deprived of their colours.

Notwithstanding these and many other experiments, the theory of colours seems not yet determined with certainty; and it must be acknowledged that very strong objections might be brought against every hypothesis on this subject that has been invented. The discoveries of Sir Isaac Newton however are sufficient to justify the following Aphorisms.

Aphorism 1. All the colours in nature arise from the rays of light.

2. There are seven primary colours, namely red, orange, yellow, green, blue, indigo, and violet.

3. Every ray of light may be separated into these seven primary colours.

4. The rays of light, in passing through the same medium, have different degrees of refrangibility.

5. The difference in the colours of light arises from its different refrangibility: that which is the least refrangible producing red; and that which is the most refrangible, violet.

6. By compounding any two of the primary, as red and yellow, or yellow and blue, the intermediate colour, orange or green, may be produced.

7. The colours of bodies arise from their dispositions to reflect one sort of rays, and to absorb the others: those that reflect the least refrangible rays appearing red; and those that reflect the most refrangible, violet.

8. Such bodies as reflect two or more sorts of rays, appear of various colours.

9. The whiteness of bodies arises from their disposition to reflect all the rays of light promiscuously.

10. The blackness of bodies proceeds from their incapacity to reflect any of the rays of light.—And from their thus absorbing all the rays of light that are thrown upon them, it arises, that black bodies, when exposed to the sun, become hot sooner than all others.

Of the Diatonic Scale of Colours.—Sir Isaac Newton, in the course of his investigations of the properties of light, discovered that the lengths of the spaces occupied in the spectrum by the seven primary colours, exactly correspond to the lengths of chords that sound the seven notes in the diatonic scale of music: which is made evident by the following experiment.

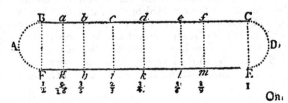

On

On a paper, in a dark chamber, let a ray of light be largely refracted into the spectrum ABCDEF, marking upon it the precise boundaries of the several colours, as *a*, *b*, *c*, &c; and across the spectrum draw the perpendicular lines *ag*, *bh*, &c. Then it will be found that the spaces, by which the several colours are bounded, viz, B*ag*F containing the red, *abhg* containing the orange, *bcih* containing the yellow, &c, will be in exact proportion to the divisions of a musical chord for the notes of an octave; that is, as the intervals of these numbers 1, $\frac{8}{9}$, $\frac{5}{6}$, $\frac{3}{4}$, $\frac{2}{3}$, $\frac{3}{5}$, $\frac{9}{16}$, $\frac{1}{2}$.

CHRONOLOGICAL, belonging to chronology.

Chronological Characters, are characters by which times are distinguished. Of these, some are natural, or astronomical; others are artificial or historical. The natural characters are such as depend on the motions of the stars or luminaries; as eclipses, solstices, equinoxes the different aspects of planets, &c. And the artificial characters are those that have been invented and established by men; as the solar cycle, the lunar cycle, &c. Historical Chronological Characters are those supported by the testimonies of historians, when they fix the dates of certain events to certain periods.

CHRONOLOGY, the art of measuring and distinguishing time; with the doctrine of dates, epochs, eras, &c.

The measurement of time in the most early periods, was by means of the seasons, or the revolutions of the sun and moon. The succession of Juno's priestesses at Argos served Hellanicus for the regulation of his narrative; while Ephorus reckoned his matters by generations. Even in the histories of Herodotus and Thucydides, there are no regular dates for the events recorded; nor were there any endeavours to establish a fixed era until the time of Ptolomy Philadelphus, who attempted it by comparing and correcting the dates of the olympiads, the kings of Sparta, and the succession of the priestesses of Juno at Argos. Eratosthenes and Apollodorus digested the events related by them, according to the succession of the olympiads and of the Spartan kings.

The chronology of the Latins is still more uncertain. The records of the Romans were destroyed by the Gauls; and Fabius Pictor, the most ancient of their historians, was obliged to borrow the chief part of his information from the Greeks. In other European nations the chronology is still more imperfect, and of a later date: and even in modern times a considerable degree of confusion and inaccuracy has arisen, from the want of attention in the historians to ascertain the dates and epochs with precision.

Hence it is evident, how necessary a proper system of chronology must be for the right understanding of history, and also how difficult it must be to establish such a system. For this purpose, however, several learned men have spent much time, particularly Julius Africanus, Eusebius of Caesarea, George Cyncelle, John of Antioch, Dennis, Petau, Clavius, Calvisius, Scaliger, Vieta, Newton, Usher, Simson, Marsham, Helvicus, Vossius, Strauchius, Blair, and Playfair.

Such a system is founded, 1st, On astronomical observations, especially of the eclipses of the sun and moon, combined with calculations of the years and eras of different nations. 2d, The testimonies of credible authors. 3d, Such epochs in history as are so well attested and determined, that they have never been controverted. 4th, Ancient medals, coins, monuments, and inscriptions.

The most obvious division of time, as has been observed, is derived from the apparent or real revolutions of the luminaries, the sun and moon. Thus, the apparent revolution of the sun, or the real rotation of the earth on her axis causing the sun to appear to rise and set, constitutes the vicissitudes of day and night, which must be evident to the most barbarous and ignorant nations. The moon, by her revolution about the earth, and her changes, as naturally and obviously forms months; while the great annual course of the sun through the several constellations of the zodiac, points out the larger division of the year.

The Day is divided into hours, minutes, &c; while the month is divided into weeks, and the year into months, having particular names, and a certain number of days.—See a particular account of each of these under the respective words.

Beside the natural divisions of time arising immediately from the revolutions of the heavenly bodies, there are others which are formed from some of the less obvious consequences of these revolutions, and are called *cycles*, or circles. The most remarkable of these are, 1, The *Solar Cycle*, or cycle of the sun, a period or revolution of 28 years, in which time the days of the months return to the same days of the week, the sun's place to the same signs and degrees of the ecliptic on the same months and days, and the leap-years begin the same course over again with respect to the days of the week on which the days of the months fall. 2, The *Lunar Cycle*, or cycle of the moon, commonly called the Golden Number, is a revolution of 19 years; in which time the conjunctions, oppositions, and other aspects of the moon, are on the same days of the months as they were 19 years before, and within an hour and a half of the same time of the day.

The *Indiction*, or Roman Indiction, is a period of 15 years, used only by the Romans for indicating the times of certain payments made by the subjects to the republic.

The *Cycle of Easter*, called also the *Dionysian Period*, is a revolution of 532 years, and is produced by multiplying the solar cycle 28, by the lunar cycle 19.

The *Julian Period*, is a revolution of 7980 years, and is produced from the continual multiplication of the three numbers 28, 19, 15, of the three former cycles, viz, the solar, lunar, and indiction.

As there are certain fixed points in the heavens, from which astronomers begin their computations, so there are certain points of time, from which historians begin to reckon; and these points or roots of time are called *eras* or *epochs*. The most remarkable of these are, those of the Creation, the Greek Olympiads, the building of Rome, the era of Nabonnassar, the death of Alexander, the birth of Christ, the Arabian Hegira, or flight of Mahomet, and the Persian Jesdegird. All which, with some others of less note, have their beginnings fixed by chronologers to the years of the Julian period, to the age of the world, and to the years before and after the birth of Christ.

The testimony of authors is the second principal part

part of historical chronology. Though no man has a right to be considered as infallible, it would however be making a very unfair judgment of mankind, to treat them all as dupes or impostors; and it would be an injury offered to public integrity, to doubt the veracity of authors universally esteemed, and facts that are truly worthy of belief. When the historian is allowed to be completely able to judge of an event, and to have no intent of deceiving by his relation, his testimony cannot be refused.

The *Epochs* form the 3d principal part of chronology; being those fixed points in history that have never been contested, and of which there cannot reasonably be any doubt. Notwithstanding that chronologers fix upon the events which are to serve as epochs, in a manner quite arbitrary; yet this is of little consequence, provided the dates of these epochs agree, and that there is no contradiction in the facts themselves.

Medals, Monuments, and Inscriptions, form the last of the four principal parts of chronology; and this study is but of very modern date, scarce more than 150 years having elapsed since close application has been made to the study of these. To the celebrated Spanheim we owe the greatest obligations, for the progress that is made in this method; and it is by the aid of medals that M. Vaillant has composed his judicious history of the kings of Syria, from the time of Alexander the Great to that of Pompey. Nor have they been of less service in elucidating all ancient history, especially that of the Romans; and even sometimes that of the middle ages.

Besides the foregoing general account, there are some few systems of chronology which may deserve some more particular notice, as follows.

Sacred Chronology. There have been various systems relating to sacred chronology; which is not to be wondered at, as the three chief copies of the Bible give a very different account of the first ages of the world. For while the Hebrew text reckons about 4000 years from the creation to the birth of Christ, and to the flood 1656 years; the Samaritan makes the former much longer, though it counts from the creation to the flood only 1307 years; and the Septuagint removes the creation of the world to 6000 years before Christ, and 2250 years before the flood. Many attempts have been made to reconcile these differences; though none of them are quite satisfactory. Walton and Vossius give the preference to the account of the Septuagint; while others have defended the Hebrew text. See an abstract of the different opinions of learned men on this subject, in Strauchius's Brev. Chron. translated by Sault, p. 166 and 176.

The Chinese Chronology. No nation has boasted more of its antiquity than the Chinese: but though they be allowed to trace their origin as far back as the deluge, they have few or no authentic records of their history for so long a period as 500 years before the Christian era. This indeed may be owing to the general destruction of ancient remains by the tyrant Tsin-chi-hoang, in the year 213, or some say 246, before Christ. From a chronology of the Chinese history (for which we are obliged to an illustrious Tartar who was viceroy of Canton in the year 1724,

4

and of which a Latin translation was published at Rome in 1730), we learn that the most remote epoch of the Chinese chronology does not surpass the first year of Guei-sie-wang, or 424 years before our vulgar era. And this opinion is confirmed by the practice of two of the most approved historians of China, who admit nothing into their histories previous to this period.

The Chinese, in their computation, make use of a cycle of 60 years, called kia-tse, from tne name given to the first year of it, which serves as the basis of their whole chronology. Every year of this cycle is marked with two letters which distinguish it from the others; and all the years of the emperors, for upwards of 2000 years, have names in history common to them with the corresponding years of the cycle. Philos. Transf. Abridg. vol. 8, part 4, pag. 13.

According to M. Freret, in his Essays, the Chinese date the epocha of Yao, one of their first emperors, about the year 2145, or 2057, before Christ; and they reckon that their first astronomical observations, and the composition of their calendar, preceded Yao 150 years: from whence it is inferred that the era of their astronomical observations coincides with that of the Chaldeans. But later authors date the rise and progress of the sciences in China from the grand dynasty of Tcheou, about 1200 years before the Christian era, and shew that all historical relations of events prior to the reign of Yao are fabulous. Mem. de l'Histoire des Sciences &c. Chinois. vol. 1, Paris 1776.

Babylonian, Egyptian, and Chaldean Annals. These, M. Gibert has attempted to reduce to our chronology, in a letter published at Amsterdam in 1743. He begins with shewing, by the authorities of Macrobius, Eudoxus, Varro, Diodorus Siculus, Pliny, Plutarch, St. Augustin, &c, that by a year, the ancients meant the revolution of any planet in the heavens; so that it might consist sometimes of only one day. Thus, according to him, the solar day was the astronomical year of the Chaldeans; and so the boasted period of 473,000 years, assigned to their observations, is reduced to 1297 years 9 months; the number of years which, according to Eusebius, elapsed from the first discoveries of Atlas in astronomy, in the 384th year of Abraham, to the march of Alexander into Asia in the year 1682 of the same era. And the 17,000 years added by Berosus to the observations of the Chaldeans, reduced in the same manner, will give 46 years and 6 or 7 months; being the exact interval between Alexander's march, and the first year of the 123d Olympiad, or the time to which Berosus carried his history.

Epigenius ascribes 720,000 years to the observations preserved at Babylon; but these, according to M. Gibert's system, amount only to 1971 years 3 months; which differ from Callisthenes's period of 1903 years, allotted to the same observations, only by 68 years, the period elapsed from the taking of Babylon by Alexander, which terminated the latter account, and to the time of Ptolomy Philadelphus, to which Epigenius extended his account.

The Newtonian Principles of Chronology.—Sir Isaac Newton has shewn, that the chronology of ancient kingdoms is involved in the greatest uncertainty: that the

the Europeans in particular had no chronology before the Perfian empire, which commenced 536 years before the birth of Chrift, when Cyrus conquered Darius the Mede: that the antiquities of the Greeks are full of fables, because their writings were in verfe only, till the conqueft of Afia by Cyrus the Perfian; about which time profe was introduced by Pherecides Syrius and Cadmus Milefius. After this time feveral of the Greek hiftorians introduced the computation by generations. The chronology of the Latins was ftill more uncertain: their old records having been burnt by the Gauls 120 years after the expulfion of their kings, or 64 years before the death of Alexander the Great, anfwering to 388 before the birth of Chrift. The chronologers of Gaul, Spain, Germany, Scythia, Sweden, Britain, and Ireland, are of a ftill later date. For Scythia, beyond the Danube, had no letters till Ulphilas, their bifhop, formed them, about the year 276. Germany had none till it received them from the weftern empire of the Latins, about the year 400. The Huns had none in the days of Procopius, about the year 526. And Sweden and Norway received them ftill later.

Sir Ifaac Newton, after a general account of the obfcurity and defects of the ancient chronology, obferves that, though many of the ancients computed by fucceffions and generations, yet the Egyptians, Greeks, and Latins, reckoned the reigns of kings equal to generations of men, and three of them to a hundred, and fometimes to 120 years; and this was the foundation of their technical chronology. He then proceeds, from the ordinary courfe of nature, and a detail of hiftorical facts, to fhew the difference between reigns and generations; and that, though a generation from father to fon may at an average be reckoned about 33 years, or three of them equal to 100 years, yet, when they are taken by the eldeft fons, three of them cannot be eftimated at more than about 75 or 80 years; and the reigns of kings are ftill fhorter; fo that 18 or 20 years may be allowed a juft medium. Sir Ifaac then fixes on four remarkable periods, viz, the return of the Heraclidæ into the Peloponnefus, the taking of Troy, the Argonautic expedition, and the return of Sefoftris into Egypt, after his wars in Thrace; and he fettles the epoch of each by the true value of a generation. To inftance only his eftimate of that of the Argonautic expedition: Having fixed the return of the Heraclidæ to about the 159th year after the death of Solomon, and the deftruction of Troy to about the 76th year after the fame period, he obferves, that Hercules the Argonaut was the father of Hyllus, the father of Clerdius, the father of Ariftomachus, the father of Ariftodemus, who conducted the Heraclidæ into Peloponnefus; fo that, reckoning by the chief of the family, their return was four generations later than the Argonautic expedition, which therefore happened about 43 years after the death of Solomon. This is farther confirmed by another argument: Æfculapius and Hercules were Argonauts: Hippocrates was the 18th inclufively from the former by the father's fide, and the 19th from the latter by the mother's fide: now, allowing 28 or 30 years to each of them, the 17 intervals by the father, and the 18 intervals by the mo-

ther, will on a medium give 507 years; and thefe, reckoning back from the commencement of the Peloponnefian war, or the 431ft year before Chrift, when Hippocrates began to flourifh, will place the Argonautic expedition in the 43d year after the death of Solomon, or 937 years before Chrift.

The other kind of reafoning by which Newton endeavours to eftablifh this epoch, is purely aftronomical. The fphere was formed by Chiron and Mufæus for the ufe of the Argonautic expedition, as is plainly fhewn by feveral of the afterifms referring to that event: and at the time of the expedition the cardinal points of the equinoxes and folftices were placed in the middle of the conftellations Aries, Cancer, Chelæ, and Capricorn. This point is eftablifhed by Newton from the confideration of the ancient Greek calendar, which confifted of 12 lunar months, and each month of 30 days, which required an intercalary month. Of courfe this lunifolar year, with the intercalary month, began fometimes a week or two before or after the equinox or folftice; and hence the firft aftronomers were led to the before-mentioned difpofition of the equinoxes and folftices: and that this was really the cafe, is confirmed by the teftimonies of Eudoxus, Aratus, and Hipparchus. Upon thefe principles Sir Ifaac proceeds to argue in the following manner. The equinoctial colure in the end of the year 1689 cut the ecliptic in ♉ 6° 44′; and by this reckoning the equinox had then gone back 36° 44′ fince the time of the Argonautic expedition. But it recedes 50′ in a year, or 1° in 72′ years, and confequently 36° 44′ in 2645 years; and this, counted backwards from the beginning of 1690, will place this expedition about 25 years after the death of Solomon. But as there is no neceffity for allowing that the middle of the conftellations, according to the general account of the ancients, fhould be precifely the middle between the prima Arietis and ultima Caudæ, our author proceeds to "examine what were thofe ftars through which Eudoxus made the colures to pafs in the primitive fphere, and in this way to fix the pofition of the cardinal points." Now from the mean of five places he finds, that the great circle, which in the primitive fphere, defcribed by Eudoxus, or which at the time of the Argonautic expedition was the equinoctial colure, did in the end of 1689 cut the ecliptic in ♉ 6° 29′ 15″. In the fame manner our author determines that the mean place of the folftitial colure is ♌ 6° 28′ 46″, and as it is at right angles with the other, he concludes that it is rightly drawn. And hence he infers that the cardinal points, in the interval between that expedition and the year 1689, have receded from thofe colures 1ˢ 6° 29′; which, allowing 72 years to a degree, amounts to 2627 years; and thefe counted backwards, as above, will place the Argonautic expedition 43 years after the death of Solomon. Our author has, by other methods alfo of a fimilar nature, eftablifhed this epoch, and reduced the age of the world 500 years.

This elaborate fyftem has not efcaped cenfure; Meff. Freret and Souciet having both attacked it, and on much the fame ground: but the former has confounded reigns and generations, which are carefully diftinguifhed in this fyftem. The aftronomical objections of both

have

have been anfwered by Sir Ifaac Newton himfelf, and by Dr. Halley. Philof. Tranf. abr. vol. 8, part 4, pa. 4. Newton's Chronol. ch. 1.

CHRONOMETER, is any inftrument or machine ufed in meafuring time; fuch as dials, clocks, watches, &c.

The term is however more particularly ufed for a kind of clock, fo contrived as to meafure a fmall portion of time, even to the 16th, or the 40th part of a fecond; one of this latter kind I have feen, made by an ingenious artift; but it could not be ftopped to the 10th part of the propofed degree of accuracy. There is a defcription of one alfo in Defaguliers's Experimental Philofophy, invented by the late ingenious Mr. George Graham; which might be of great ufe for meafuring fmall portions of time in aftronomical obfervations, the time of the fall of bodies, the velocity of running waters, &c. But long intervals of time cannot be meafured by it with fufficient exactnefs, unlefs its pendulum be made to vibrate in a cycloid; for otherwife it is liable to err confiderably, as is the cafe of all clocks with fhort pendulums that fwing in large arches of the circle.

Various other contrivances, befides clocks, have been ufed for meafuring time for fome particular purpofes. See a mufical chronometer defcribed in Malcolm's Treatife of Mufic, pa. 407.

CHRONOSCOPE, a word fometimes ufed for a pendulum, or machine, to meafure time.

CHRYSTALLINE. See CRYSTALLINE.

CIMA, or SIMA, in Architecture, a member, or moulding, called alfo ogee, and cimatium.

CINCTURE, in Architecture, is a ring or lift around the fhaft of a column, at its top and bottom.

CINTRE, in Building, the mould on which an arch is turned or built; popularly called centre, and fometimes a cradle.

CIPHER, or Cypher, one of the numeral characters, or figures, thus formed o. The word comes from the Hebrew *faphar*, to number.

The cipher of itfelf fignifies nothing, or implies a privation of value; but when combined with other numeral characters, it alters their value in a tenfold proportion, for every cipher fo annexed; viz, when fet after a figure in common integral arithmetic, it increafes its value in that proportion, though it has no effect when fet before or to the left hand fide of figures; but on the contrary, in decimal arithmetic, it decreafes their value in that proportion when fet before the figures, but has no effect when fet after them.

Thus, 5   is five,
  but 50 is fifty,
  and 500 is five hundred;
whereas 05, or 005, &c, is ftill but 5 or five.

  Alfo ·5   is five tenths,
  but ·05 is five hundredths,
  and ·005 is five thoufandths;
whereas ·50, or ·500, &c, is ftill but ·5 or five tenths.

The invention and ufe of the cipher, as in the common arithmetic and notation of numbers, is one of the happieft devices that can be imagined; and is afcribed to the Indians, by the Arabians, through whom it came into Europe, along with the revival of literature.

CIRCLE, a plane figure, bounded by a curve line which returns into itfelf, called its circumference, and which is every where equally diftant from a point within, called its centre.

The circumference or periphery itfelf is called the circle, though improperly, as that name denotes the fpace contained within the circumference.

A circle is defcribed with a pair of compaffes, fixing one foot in the centre, and turning the other round to trace out the circumference.

The circumference of every circle is fuppofed to be divided into 360 equal parts, called degrees, and marked °; each degree into 60 minutes or primes, marked ′; each minute into 60 feconds, marked ″; and fo on. So 24° 12′ 15″ 20‴, is 24 degrees 12 minutes 15 feconds and 20 thirds.

Circles have many curious properties, fome of the moft important of which are thefe:

1. The circle is the moft capacious of all plain figures, or contains the greateft area within the fame perimeter, or has the leaft perimeter about the fame area; being the limit and laft of all regular polygons, having the number of its fides infinite.

2. The area of a circle is always lefs than the area of any regular polygon circumfcribed about it, and its circumference always lefs than the perimeter of the polygon. But on the other hand, its area is always greater than that of its infcribed polygon, and its circumference greater than the perimeter of the faid infcribed polygon. However, the area and perimeter of the circle approach always nearer and nearer to thofe of the two polygons, as the number of the fides of thefe is greater; the circle being always limited between the two polygons.

3. The area of a circle is equal to that of a triangle whofe bafe is equal to the circumference, and perpendicular equal to the radius. And therefore the area of the circle is found by drawing half the circumference into half the diameter, or the whole circumference into the whole diameter, and taking the 4th part of the product. Demonftrated by Euclid.

4. Circles, like other fimilar plane figures, are to one another, as the fquares of their diameters. And the area of the circle is to the fquare of the diameter, as 11 to 14 nearly, as proved by Archimedes; or as ·7854 to 1 more nearly; or ftill more nearly as

·7853981633,9744830961,5660845819,8757210492, 9234984377,6455243736,1480769541,0157155224, 9657008706,3355292669,9553702162,8318076661, 7734611 + to 1;

as it has been found by modern mathematicians.

In Wallis's Arithmetic of Infinites are contained the firft infinite feries for expreffing the ratio of a circle to the fquare of its diameter: viz,

1ft, The circle is to the fquare of its diameter,

$$\text{as 1 to } \frac{3\times3\times5\times5\times7\times7\&c}{2\times4\times4\times6\times6\times8\&c}$$

$$\text{or 1 to } \frac{9}{8}\times\frac{25}{24}\times\frac{49}{48}\ \&c,$$

found out by Wallis himfelf;

$$\text{or as 1 to } 1+\cfrac{1}{2+\cfrac{9}{2+\cfrac{25}{2+\cfrac{49}{2+\&c}}}},\ \text{by Ld. Brounker;}$$

or as $1 - \dfrac{1}{2\times3} - \dfrac{1}{2\times4\times5} - \dfrac{1\times3}{2\times4\times6\times7}$

$- \dfrac{1\times3\times5}{2\times4\times6\times8\times9}$ &c to 1, by Sir I. Newton;

or as $1 - \dfrac{1}{3} + \dfrac{1}{5} - \dfrac{1}{7} + \dfrac{1}{9} - \dfrac{1}{11}$ &c to 1, by Gregory and Leibnitz; and a great many other forms of series have been invented by different authors, to express the same ratio between the circle and circumscribed square.

5. The circumferences of circles are to one another, as their diameters, or radii. And as the areas of circles are proportional to the rectangles of their radii and circumferences; therefore the quadrature of the circle will be effected by the rectification of its circumference; that is, if the true length of the circumference could be found, the true area could be found also. But whilst several mathematicians have endeavoured to determine the true area and circumference, others have even doubted of the possibility of the same. Of this latter opinion is Dr. Isaac Barrow: towards the end of his 15th Mathematical Lecture he says, he is of opinion, that the radius and circumference of a circle, are lines of such a nature, as to be not only incommensurable in length and square, but even in length, square, cube, biquadrate, and all other powers to infinity: for, continues he, the side of the inscribed square is incommensurable to the radius, and the square of the side of the inscribed octagon is incommensurable to the square of the radius; and consequently the square of the octagonal perimeter is incommensurable to the square of the radius: and thus the ambits of all regular polygons, inscribed in a circle, may have their superior powers incommensurate with the co-ordinate powers of the radius; from whence the last polygon, that is, the circle itself, seems to have its periphery incommensurate with the radius. Which, if true, will put a final stop to the quadrature of the circle, since the ratio of the circumference to the radius is altogether inexplicable from the nature of the thing, and consequently the problem requiring the explication of such a ratio is impossible to be solved, or rather it requires that for its solution which is impossible to be apprehended. But, concludes he, this great mystery cannot be explained in a few words: But if time and opportunity had permitted, I would have endeavoured to produce many things for the explication and confirmation of this conjecture. Sir Isaac Newton too, in book 1 of his Principia, has attempted to demonstrate the impossibility of the general quadrature of oval figures, by the description of a spiral, and the impossibility of determining, by a finite equation, the intersections of that oval and spiral, which must be the case, if the oval be quadrable. And several other authors have attempted to demonstrate the impossibility of the general quadrature of the circle by any means whatever. On the other hand, many authors not only believe in the possibility of the quadrature of the circle, but some have even pretended to have discovered the same, and have published to the world their pretended discoveries: of which no one has rendered himself more remarkable than our countryman Mr. Hobbes, though a great scholar, and of excellent understanding in other matters. See QUADRATURE.

The approximate quadrature of the circle however, or the determination of the ratio between the diameter and the circumference, is what the mathematicians of all ages have successfully attempted, and with different degrees of accuracy, according to the improved state of the science. Archimedes, in his book *de Dimensione Circuli*, first gave a near value of that ratio in small numbers, being that of 7 to 22, which are still used as very convenient numbers for this purpose in common measurements. Other and nearer ratios have since been successively assigned, but in larger numbers,

as 106 to 333,
or 113 to 355,
or 1702 to 5347,
or 1815 to 5702, &c,

which are each more accurate than the foregoing. Vieta, in his *Universalium Inspectionum ad Canonum Mathematicum*, published 1579, by means of the inscribed and circumscribed polygons of 393216 sides, has carried the ratio to ten places of figures, shewing that if the diameter of a circle be 1000 &c, the circumference will be greater than 314,159,265,35,
        but less than 314,159,265,37.
And Van Colen, in his book *de Circulo & Adscriptis*, has, by the same means, carried that ratio to 36 places of figures; which were also recomputed and confirmed by Willebrord Snell. After these, that indefatigable computer, Mr. Abraham Sharp, extended the ratio to 72 places of figures, in a sheet of paper, published about the year 1706, by means of the series of Dr. Halley, from the tangent of an arc of 30 degrees. And the ingenious Mr. Machin carried the same to a hundred places, by other series, depending on the differences of arcs whose tangents have certain relations to one another. See this method explained in my Mensuration, pa. 120 second edit. And, finally, M. De Lagny, in the Memoirs de l'Acad. 1719, by means of the tangent of the arc of 30 degrees, has extended the same ratio to the amazing length of 128 places of figures; finding, that, if the diameter be 1000 &c, the circumference will be

3.1415,92653,58979,32384,62643,38327,95028,
84197,16939,93751,05820,97494,45923,07816,
40628,62089,98628,03482,53421,17067,98214,
80865,13272,30664,70938,446 + or 447 −

From such methods as the foregoing, a variety of series have been discovered for the length of the circumference of a circle, such as the following, viz, If the diameter be 1, the circumference c will be variously expressed thus,

$$c = 4 \times \left(1 - \frac{1}{3} + \frac{1}{5} - \frac{1}{7} + \frac{1}{9} - \frac{1}{11} + \frac{1}{13} - \frac{1}{15} \&c\right),$$

$$c = \sqrt{8} \times \left(1 + \frac{1}{3} - \frac{1}{5} - \frac{1}{7} + \frac{1}{9} + \frac{1}{11} - \frac{1}{13} - \frac{1}{15} \&c\right),$$

$$c = \sqrt{12} \times \left(1 - \frac{1}{3\cdot3} + \frac{1}{5\cdot3^2} - \frac{1}{7\cdot3^3} + \frac{1}{9\cdot3^4} \&c\right),$$

$$c = 8 \times \left(\frac{1}{1\cdot1\cdot3} + \frac{1}{1\cdot3\cdot5} - \frac{1}{3\cdot5\cdot7} + \frac{1}{5\cdot7\cdot9} - \frac{1}{7\cdot9\cdot11} \&c\right),$$

$$c = 8 \times \left(\frac{2}{3} - \frac{1}{5} - \frac{1}{4\cdot7} - \frac{1\cdot3}{4\cdot6\cdot9} - \frac{1\cdot3\cdot5}{4\cdot6\cdot8\cdot11} \&c\right),$$

$$c = 4\sqrt{2} \times \left(\frac{2}{3} - \frac{1}{5\cdot2} - \frac{1}{4\cdot7\cdot2^2} - \frac{1\cdot3}{4\cdot6\cdot9\cdot2^3} - \frac{1\cdot3\cdot5}{4\cdot6\cdot8\cdot11\cdot2^4} \&c\right),$$

$$c = 4 \times \left(1 - \frac{1}{2\cdot3} - \frac{1}{2\cdot4\cdot5} - \frac{1\cdot3}{2\cdot4\cdot6\cdot7} - \frac{1\cdot3\cdot5}{2\cdot4\cdot6\cdot8\cdot9} \&c\right).$$

And

And many other feries might here be added. See my Menfuration in feveral places; alfo my paper on fuch feries in the Philof. Tranf. 1776; Euler's *Introductio in Analyfin Infinitorum*; and many other authors.

6. Some of the more remarkable properties of the circle, are as follow.

If two lines AB, CD cut the circle, and interfect within it, the angle of interfection E is meafured by half the *fum* of the intercepted arcs AC, DB.

But if the lines interfect without the circle, the angle E is meafured by half the *difference* of the intercepted arcs AC, DB.

7. The angle at the centre of a circle is double the angle at the circumference, ftanding on the fame arc; and all angles in the fame fegment are equal. Alfo the angle at the centre is meafured by the arc it ftands upon, and the angle at the circumference by half the fame arc.

8. If the chords FG, HI crofs at right angles, the fums of the oppofite arcs are equal; viz FH + GI = FI + GH.

9. If one fide NO of a trapezium infcribed in a circle be continued out, the external angle, LOP will be equal to the oppofite internal angle M.

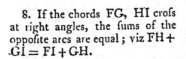

10. An angle, as RQS, formed by a tangent QR and chord QS, is meafured by half the arc of the chord QS, and is equal to any angle T formed in the oppofite arc QTS.

11. If VW be a diameter, and. XYZ a chord perpendicular to it; then is XZ or ZY a mean proportional between the fegments YZ, ZW. So that if d denote the diameter VW, x the abfcifs VZ, and y the ordinate ZX; then is $y^2 = \overline{d-x} \times x$ or $= dx - x^2$; which is called the equation to the circle.

The chord VX is a mean proportional between the diameter VW and the verfed fine VZ; and the chord WX is a mean proportional between the diameter and the verfed fine WZ; alfo each verfed fine is proportional to the fquare of the corresponding chord; viz VZ : WZ :: VX² : WX².

12. When two lines cut the circle, whether they interfect within the circle, or without it, as in the two figures to article 6, the fegments between the common interfection and the two points where each line cuts the curve, are reciprocally proportional, and their rectan-

gles are equal; viz, EA : EC :: ED : EB,
or EA × EB = EC × ED.

13. In a trapezium infcribed in a circle, the rectangle of the two diagonals is equal to the fum of the two rectangles of the two pairs of oppofite fides; viz, AC × BD = AB × DC + AD × BC.

14. If any chords EF, EG, drawn from the fame point E in the circumference, be cut by any other line HI, the rectangles will be all equal which are made of each chord and the part intercepted by this line, viz, EF × EI = EG × EH = EK². :

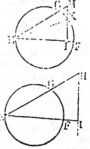

15. In a circle whofe centre is N and radius NO, if two points M, P, in the radius produced, be fo placed that the three NM, NO, MP, be in continued proportion; then if from the points M and P lines be drawn to any, or every point in the circumference, as Q; thefe lines will be always in the given ratio of MO to PO; viz, MQ : PQ :: MO : PO.

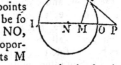

16. If VW, be two points in the diameter, equidiftant from the centre T; and if two lines be drawn from thefe to any point X in the circumference; the fum of their fquares will be equal to the fum of the fquares of the fegments of the diameter made by either point; viz, VX² + WX² = RV² + VS², or = RW² + WS², or = 2RT² + 2VT².

17. If a line FE perpendicular to the diameter AB, meet any other chord CD in the point E; then is AF × FB = CE × ED + EF².

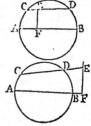

18. If upon the diameter GH of a circle there be formed a rectangle GHKI, whofe breadth GI or HK is equal to GL or HL, the chord of a quadrant, or fide of the infcribed fquare; then if from I and K lines be drawn to any point M in the circle GMH, they will cut the diameter GH in fuch a manner that GO² + HN² = GH².

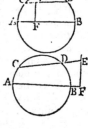

19. If the arcs PQ, QR, RS, &c, be equal, and there be drawn the chords PQ, PR, PS, PT, &c, then it
will

will be PQ : PR :: PR : PQ + PS :: PS : PR + PT :: PT : PS + PV, &c.

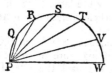

20. The centre of a circle being O, and P a point in the radius, or in the radius produced; if the circumference be divided into as many equal parts AB, BC, CD, &c, as there are units in 2n, and lines be drawn from P to all the points of division; then shall the continual product of all the alternate lines viz PA × PC × PE &c be $= r^n - x^n$ when P is within the circle, or $= x^n - r^n$ when P is without the circle; and the product of the rest of the lines, viz PB × PD + PF &c $= r^n + x^n$: where $r = $ AO the radius, and $x = $ OP the distance of P from the centre.

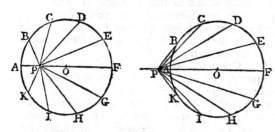

21. A circle may thus be divided into any number of parts that shall be equal to one another both in area and perimeter. Divide the diameter QR into the same number of equal parts at the points S, T, V, &c; then, on one side of the diameter describe semicircles on the diameters QS, QT, QV, and on the other side of it describe semicircles on RV, RT, RS; so shall the parts 1 7, 3 5, 5 3, 7 1 be all equal, both in area and perimeter. See my Tracts, pa. 93.

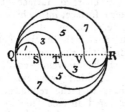

22. *To describe a Circle either about or within a given Regular Polygon.* Bisect two of its angles, or two of its sides, with perpendiculars, and the intersection of the bisecting lines will, in either case, be the centre of the circles.

Parallel, or *Concentric* CIRCLES, are such as are equally distant from each other in every point of their peripheries; or that have the same centre. As, on the other hand, those are called the *eccentric* circles, that have not the same point for their centres.

The Quadrature of the CIRCLE, is the manner of describing, or assigning, a square, whose surface shall be perfectly equal to that of a circle. This problem has exercised the geometricians of all ages, but it is now generally given up as a problem impossible to be effected,

by most persons that have any just claim to that rank. Des Cartes insists on the impossibility of it, for this reason, that a right line and a circle being of different natures, there can be no strict proportion between them. Dr. Barrow shews the strongest probability of the same thing; and not only that the diameter and circumference themselves, but that all powers of them to infinity, are incommensurate.

The Emperor Charles V offered a reward of 100,000 crowns to any person who should resolve this celebrated problem: and the States of Holland also proposed a reward for the same thing. See *Quadrature.*

CIRCLES *of the Higher Orders,* are curves in which
$$WY^m : YZ^m :: YZ : YX,$$
$$\text{or } WY^m : YZ^m :: YZ_n : YX_n.$$

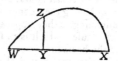

When m and n are each equal to 1, then WY : YZ :: YZ : YX, which is the property of the common circle.

Put WY $= x$, YZ $= y$, WX $= a$; then is YX $- a - x$, and the above proportions become
$$x^m : y^m :: y : \overline{a-x}, \text{ or } y^{m+1} = x^m . \overline{a-x}, \text{ and}$$
$$x^m : y^m :: y^n : \overline{a-x}^n, \text{ or } y^{m+n} = x^m . \overline{a-x}^n, \text{ the}$$
equations to curves of this kind.

Curves defined by this equation will be ovals when m is an odd number. Thus suppose $m = 1$, then the equation becomes $y^2 = x . \overline{a-x}$ or $ax - x^2$, the equation to the common circle. And if $m = 3$, it becomes $y^4 = x^3 . \overline{a-x}$ or $ax^3 - x^4$, which denotes a curve of this form AB.

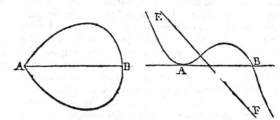

But when m denotes an even number, the curve will have two infinite legs. So if $m = 2$, the equation becomes $y^3 = x^2 . \overline{a-x}$ or $ax^2 - x^3$, for a circle of the 2d order, and which defines one of Newton's defective hyperbolas, being his 37th species of curves, whose asymptote is the right line EF, making an angle of 40 degrees with the absciss AB.

CIRCLE *of Curvature,* or circle of equi-curvature, is that circle which has the same curvature with a given curve at a certain point; or that circle whose radius is equal to the radius of curvature of the given curve at that point.

CIRCLES *of the Sphere,* are such as cut the mundane sphere, and have their circumference in its surface.

These circles are either fixed or moveable.

The latter are those whose peripheries are in the
moveable

moveable or revolving surface; and which therefore move or turn with it; as the meridians, &c. . The former, having their periphery in the immoveable surface, do not revolve; as the ecliptic, equator, and its parallels.

The circles of the sphere are either great or little.

A Great Circle of the Sphere, is that which divides it into two equal parts or hemispheres, having the same centre and diameter with it. As the horizon, meridian, equator, ecliptic, the colures, and the azimuths.

A Little, or *Lesser Circle* of the Sphere, divides the sphere into two unequal parts, having neither the same centre nor diameter with the sphere; its diameter being only some chord of the sphere less than its axis. Such as the parallels of latitude, &c.

CIRCLES *of Altitude*, or *Almucantars*, are little circles parallel to the horizon, having their common pole in the zenith, and still diminishing as they approach it. They are so called from their use, which is to shew the altitude of a star above the horizon.

CIRCLES *of Declination*, are great circles intersecting each other in the poles of the world.

CIRCLE *of Dissipation*, in Optics. See the article DISSIPATION.

Diurnal CIRCLES, are parallels to the equinoctial, supposed to be described by the several stars, and other points of the heavens, in their apparent diurnal rotation about the earth.

CIRCLE *Equant*, in the Ptolomaic Astronomy, is a circle described on the centre of the equant. Its chief use is, to find the variation of the first inequality.

CIRCLES *of Excursion*, are little circles parallel to the ecliptic, and at such a distance from it, as that the excursions of the planets towards the poles of the ecliptic, may be included within them; being usually fixed at about 10 degrees.

It may here be observed, that all the circles of the sphere, described above, are transferred from the heavens to the earth; and so come to have a place in geography as well as in astronomy: all the points of each circle being conceived as let fall perpendicularly on the surface of the terrestrial globe, and thus tracing out circles perfectly similar to them. So, the terrestrial equator is a circle conceived precisely under the equinoctial line, which is in the heavens: and so of the rest.

Horary CIRCLES, in Dialling, are the lines which shew the hours on dials. These are straight lines on the dials, but called circles as being the projections of the meridians.

Horary CIRCLE, or *Hour Circle*, on the globe, is a small brazen circle fixed to the north pole, divided into 24 hours, and furnished with an index to point them out, thereby shewing the difference of meridians in time, and serving for the solution of many problems, on the artificial globes.

CIRCLE *of Illumination*, is that imaginary circle on the surface of the earth, which separates the illuminated side or hemisphere of the earth from the dark side: and all lines passing from the sun to the earth, being physically parallel, are perpendicular to the plane of this circle.

CIRCLES *of Latitude*, or *Secondaries of the Ecliptic*, are great circles perpendicular to the plane of the ecliptic,

intersecting one another in its poles, and passing through every star and planet, &c.—These are so called, because they serve to measure the latitude of the stars, which is an arch of one of these circles, intersected between the star and the ecliptic.

CIRCLES *of Longitude*, are lesser circles parallel to the ecliptic, diminishing more and more as they recede from it, or as they approach the pole of that circle

They are so called, because the longitudes of the stars are counted upon them.

CIRCLE *of Perpetual Apparition*, one of the lesser circles parallel to the equator, described by the most northern point of the horizon, as the sphere revolves round by its diurnal motion.—All the stars included within this circle, are continually above the horizon, and so never set.

CIRCLE *of Perpetual Occultation*, is another lesser circle at a like distance from the equator, but on the other side of it, being described by the most southern point of the horizon, and contains all those stars which never appear in our hemisphere, or which never rise.

All other stars, being contained between these two circles, do alternately rise and set, at certain moments of the diurnal rotation.

Polar CIRCLES, are immoveable circles, parallel to the equator, and at such a distance from the pole as is equal to the greatest declination of the ecliptic, which now is 23° 28′. That next the northern pole is called the arctic, and that next the southern one the antarctic.

CIRCLES *of Position*, are circles passing through the common intersections of the horizon and meridian, and through any degree of the ecliptic, or the centre of any star, or other point in the heavens; and are used for finding out the situation or position of any star. These are usually six in number, cutting the equinoctial into 12 equal parts, which the astrologers call the Celestial Houses, and hence they are sometimes called Circles of the Celestial Houses.

CIRCUIT, *Electrical*, denotes the course of the electric fluid from the charged surface of an electric body, to the opposite surface, into which the discharge is made. Some electricians at first apprehended, that the same particles of the electric fluid that were thrown on one side of the charged glass, actually made the whole circuit of the intervening conductors, and arrived at the opposite side: whereas Dr. Franklin's theory only requires, that the redundancy of electric matter on the charged surface should pass into those bodies which form that part of the circuit which is contiguous to it, driving forward that part of the fluid which they naturally possess; and that the deficiency of the exhausted surface should be supplied by the neighbouring conductors, which form the last part of the circuit. On this supposition, a vibrating motion is successively communicated through the whole length of the circuit.

Many attempts were made, both in France and England, at an early period in the practice of electricity, to ascertain the distance to which the electric shock might be carried, and the velocity of its motion. The French philosophers, at different times, caused it to pass through circuits of 900 and even 2000 toises, or about 2 English miles and a half; and they discharged the Leyden phial through a bason of water, whose surface

was

was equal to about one acre. M. Monier found that, in paffing through an iron wire of 950 toifes in length, it did not fpend a quarter of a fecond; and that its motion was inftantaneous through a wire of 1319 feet. In 1747, Dr. Watfon, and other Englifh philofophers, after many experiments of a fimilar kind, conveyed the electric matter through a circuit of 4 miles; and, from two feveral trials, they concluded that its paffage is inftantaneous. By all which doubtlefs is meant, that its motion is too rapid to be meafured. Prieftley's Hift. of Elect. vol. 1, pa. 128, 8vo edit

CIRCULAR, appertaining to a circle; as a circular form, circular motion, &c.

CIRCULAR *Lines*, a name given by fome authors to fuch ftraight lines as are divided by means of the divifions made in the arch of a circle. Such as the Sines, Tangents, Secants, &c.

CIRCULAR *Numbers*, are fuch as have their powers ending in the roots themfelves. As the number 5, whofe fquare is 25, and its cube 125, &c.

Napier's CIRCULAR *Parts*, are five parts of a right-angled or a quadrantal fpherical triangle; they are the two legs, the complement of the hypothenufe, and the complements of the two oblique angles.

Concerning thefe circular parts, Napier gave a general rule in his *Logarithmorum Canonis Defcriptio*, which is this; "*The rectangle under the radius and the fine of the middle part, is equal to the rectangle under the tangents of the adjacent parts, and to the rectangle under the cofines of the oppofite parts.* The right angle or quadrantal fide being neglected, the two fides and the complements of the other three natural parts are called the circular parts; as they follow each other as it were in a circular order. Of thefe, any one being fixed upon as the middle part, thofe next it are the adjacent, and thofe fartheft from it the oppofite parts." Lord Buchan's Life of Napier, pa. 98.

This rule contains within itfelf all the particular rules for the folution of right-angled fpherical triangles, and they were thus brought into one general comprehenfive theorem, for the fake of the memory; as thus, by charging the memory with this one rule alone: All the cafes of right-angled fpherical triangles may be refolved, and thofe of oblique ones alfo, by letting fall a perpendicular, excepting the two cafes in which there are given either the three fides, or the three angles.—And for thefe a fimilar expedient has been devifed by Lord Buchan and Dr. Minto. "M. Pingre, in the Memoires de Mathematique et de l'hyfique for the year 1756, reduces the folution of all the cafes of fpherical triangles to four analogies. Thefe four analogies are in fact, under another form, Napier's rule of the circular parts, and his fecond or fundamental theorem, with its application to the fupplemental triangle. Although it would be no difficult matter to get by heart the four analogies of M Pingre, yet there are few perfons bleffed with a memory capable of retaining them for any confiderable time. For this reafon, the rule for the circular parts ought to be kept under its prefent form. If the reader attends to the circumftance of the fecond letters of the words *tangents* and *cofines* being the fame with the firft of the words *adjacent* and *oppofite*, he will find it almoft impoffible to forget the rule. And

the rule for the folution of the two cafes of fpherical triangles, for which the former of itfelf is infufficient, may be thus expreffed: *Of the circular parts of an oblique fpherical triangle, the rectangle under the tangents of half the fum and half the difference of the fegments at the middle part* (formed by a perpendicular drawn from an angle to the oppofite fide), *is equal to the rectangle under the tangents of half the fum and half the difference of the oppofite parts.* By the circular parts of an oblique fpherical triangle are meant its three fides and the fupplements of its three angles. Any of thefe fix being affumed as a middle part, the oppofite parts are thofe two of the fame denomination with it, that is, if the middle part is one of the fides, the oppofite parts are the other two, and, if the middle part is the fupplement of one of the angles, the oppofite parts are the fupplements of the other two.—Since every plane triangle may be confidered as defcribed on the furface of a fphere of an infinite radius, thefe two rules may be applied to plane triangles, provided the middle part be reftricted to a fide.

"Thus it appears that two fimple rules fuffice for the folution of all the poffible cafes of plane and fpherical triangles. Thefe rules, from their neatnefs and the manner in which they are expreffed, cannot fail of engraving themfelves deeply on the memory of every one who is a little verfed in trigonometry. It is a circumftance worthy of notice, that a perfon of a very weak memory may carry the whole art of trigonometry in his head." Napier's Life, pa. 102.

CIRCULAR *Sailing*, is that performed in the arch of a great circle.—It is chiefly on account of the fhorteft diftance that this method of failing has been propofed; and for the moft part it is advantageous for a fhip to reach her port by the fhorteft courfe.

As the folutions of the cafes in Mercator's failing are performed by plane triangles; fo the cafes in great-circle failing are refolved by the folution of fpherical triangles. But, after all, the feveral cafes in this kind of failing ferve rather for exercifes in the folution of fpherical triangles, than for any real ufe towards the navigating of a fhip.

CIRCULAR *Spots* are made on pieces of metal by large electrical explofions. See Philof. Tranf. vol. 58, pa. 68; alfo Prieftley's Hift. of Electricity, vol. 2, fect. 9, edit. 8vo.

Thefe beautiful fpots, produced by the moderate charge of a large battery, difcharged between two fmooth furfaces of metals, or femi-metals, lying at a fmall diftance from each other, confift of one central fpot, and feveral concentric circles, which are more or lefs diftinct, and more or fewer in number, as the metal upon which they are marked is more eafy or difficult of fufion, and as a greater or lefs force is employed. They are compofed of dots or cavities, which indicate a real fufion. If the explofion of a battery, iffuing from a pointed body, be repeatedly taken on the plain furface of a piece of metal near the point, or be received from the furface on a point, the metal will be marked with a fpot, confifting of all the prifmatic colours difpofed in circles, and formed by fcales of the metal feparated by the force of the explofion.

CIRCULAR *Velocity*, a term in aftronomy fignifying that

that velocity of a planet, or revolving body, which is measured by the arch of a circle.

CIRCULATING *Decimals*, called also recurring or repeating decimals, are those in which a figure or several figures are continually repeated. They are distinguished into *single* and *multiple*, and these again into *pure* and *mixed*.

A *pure single* circulate, is that in which one figure only is repeated; as ·222 &c, and is marked thus ·2.

A *pure multiple* circulate, is that in which several figures are continually repeated; as ·232323 &c, marked ·23; and ·524524 &c, marked ·524.

A *mixed single* circulate, is that which consists of a terminate part, and a single repeating figure; as 4·222 &c, or 4·2. And

A *mixed multiple* circulate is that which contains a terminate part with several repeating figures; as 45·524.

That part of the circulate which repeats, is called the *repetend*: and the whole repetend, supposed infinitely continued, is equal to a vulgar fraction, whose numerator is the repeating number, or figures, and its denominator the same number of nines: so ·2 is = ²⁄₉; and 23 is = ²³⁄₉₉; and ·524 is = ⁵²⁴⁄₉₉₉.

It seems it was Dr. Wallis who first distinctly considered, or treated of infinite circulating decimals, as he himself informs us in his Treatise of Infinites. Since his time many other authors have treated on this part of arithmetic; the principal of these however, to whom the art is mostly indebted, are Messrs. Brown, Cunn, Martin, Emerson, Malcolm, Donn, and Henry Clarke, in whose writings the nature and practice of this art may be fully seen, especially in the last mentioned ingenious author.

CIRCUMFERENCE, in a general sense denotes the line or lines bounding any figure. But it is commonly used in a more limited sense, to denote the curve line which bounds a circle, and which is otherwise called the *periphery*; the boundary of a right-lined figure being expressed by the term *perimeter*.

The circumference of a circle is every where equidistant from the centre. And the circumferences of different circles are to one another as their radii or diameters, or the ratio of the diameter to the circumference is a constant ratio, in every circle, which is nearly as 7 to 22, as it was found by Archimedes, or, more nearly, as 1 to 3·1416. Under the article *Circle* may be seen various other approximations to that ratio, one of which is carried to 128 places of figures, viz by M. De Lagny.

The Circumference of every circle is supposed to be divided into 360 equal parts, called *degrees*.—Any part of a circumference is called an *arc* or *arch*; and a right line drawn from one end of an arc to the other, is called its *chord*.—The *angle* at the circumference is equal to half the angle at the centre, standing on the same arc; and therefore it is measured by the half of that arc.

CIRCUMFERENTOR, a particular instrument used by surveyors for taking angles. It consists of a brass circle and index all of a piece; the diameter of the circle is commonly about 7 inches; the index about 14 inches long, and an inch and a half broad. On the circle is a card or compass, divided into 360 degrees; the

meridian line of which answers to the middle of the breadth of the index. On the limb or circumference of the circle is soldered a brass ring; which, with another fitted with a glass, forms a kind of box for the needle, which is suspended on a pivot in the centre of the circle. There are also two sights to screw on, and slide up and down the index, as also a spangle and socket screwed on the under side of the circle, to receive the head of the three-legged staff.

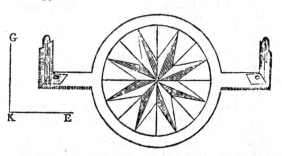

To take, or observe the Quantity of an Angle by the Circumferentor. The angle proposed being EKG; place the instrument at K, with the flower-de-luce of the card towards you; then direct the sights to E, and observe what degrees are cut by the south end of the needle, which let be 295; then, turning the instrument about on its stand, direct the sights to G, noting again what degrees are cut by the south end of the needle; which suppose are 213. This done, subtract the less number from the greater, viz, 213 from 295, and the remainder, or 82 degrees, is the quantity of the angle EKG sought.

CIRCUMGYRATION, is the whirling motion of any body about a centre; as of the planets about the sun, &c.

CIRCUM POLAR *Stars*, are those stars which, by reason of their vicinity to the pole, move round it, without setting.

CIRCUMSCRIBED *Figure*, is a figure which is drawn about another, so that all its sides or planes touch the latter or inscribed figure.

The area and perimeter of every polygon that can be circumscribed about a circle, are greater than those of the circle; and the area and perimeter of every inscribed polygon, are less than those of the circle; but they approach always nearer to equality as the number of sides is more. And on these principles Archimedes, and some other authors since his time, attempted the quadrature of the circle; which is nothing else, in effect, but the measuring the area or capacity of a circle.

CIRCUMSCRIBED *Hyperbola*, is one of Newton's hyperbolas of the 2d order, that cuts its asymptotes, and contains the parts cut off within its own space.

CIRCUMVALLATION, or *Line of Circumvallation*, in the Art of War, is a trench, bordered with a parapet, thrown up around the besieger's camp, as a security against any army that may attempt to relieve the place, as well as to prevent desertion.

CIRCUMVOLUTION, in Architecture, the torus of the spiral line of the Ionic Order.

CISSOID, is a curve line of the second order, invented

vented by Diocles for the purpose of finding two continued mean proportionals between two other given lines. The generation or description of this curve is as follows:

On the extremity B of the diameter AB of the circle AOB, erect the indefinite perpendicular CBD, to which from the other extremity A draw several lines, cutting the circle in I,O,N, &c; and upon these lines set off the corresponding equal distances, viz, HM = AI, and FO = AO, and CL = AN, &c; then the curve line drawn through all the points M, O, L, &c, is the cissoid of Diocles, who was an ancient Greek geometrician.

This curve is, by Newton, reckoned among the defective hyperbolas, being the 42d species in his *Enumeratio Linearum tertii ordinis*. And in his appendix *de Æquationum Constructione Lineari*, at the end of his *Arithmetica Universalis*, he gives another elegant method of describing this curve by the continual motion of a square ruler. Other methods have also been devised by different authors for the same thing.

The Properties of the Cissoid are the following:

1. The curve has two infinite legs AMOL, A*mol* meeting in a cusp A, and tending continually towards the indefinite line CBD, which is their common asymptote.

2. The curve passes through O and *o*, points in the circle equally distant from A and B; or it bisects each semicircle.

3. Letting fall perpendiculars MP, IK from any corresponding points I, M; then is AP = BK, and AM = HI, because AI = MH.

4. AP : PB :: MP² : AP². So that, if the diameter AB be = a, the abscis AP = x, and the ordinate. PM = y; then is $x : a - x :: y^2 : x^2$, or $x^3 = a - x \times y^2$; which is the equation of the curve.

5. Sir Isaac Newton, in his last letter to M. Leibnitz, has shewn how to find a right line equal to one of the legs of this curve, by means of the hyperbola; but he suppressed the investigation, which however may be seen in his Fluxions.

6. The whole infinitely long cissoidal space, contained between the infinite asymptote BCD, and the curves LOA*ol* &c, of the cissoid, is equal to triple the generating circle AOB*o*A.

See more of this curve in Dr. Wallis, vol. 1, pa. 545.

CIVIL *Day*. See *Day*.

CIVIL *Month*. See *Month*.

CIVIL *Year*, is the legal year, or annual account of time, which every government appoints to be used within its own dominions.

It is so called in contradistinction to the natural year,

which is measured exactly by the revolution of the heavenly bodies.

CLAIRAULT (ALEXIS-CLAUDE), a celebrated French mathematician and academician, was born at Paris the 13th of May 1713, and died the 17th of May 1765, at 52 years of age. His father, a teacher of mathematics at Paris, was his sole instructor, teaching him even the letters of the alphabet on the figures of Euclid's Elements, by which he was able to read and write at 4 years of age. By a similar stratagem it was that calculations were rendered familiar to him. At 9 years of age he put into his hands Guisnée's Application of Algebra to Geometry; at 10 he studied l'Hopital's Conic Sections; and between 12 and 13 he read a memoir to the Academy of Sciences concerning four new Geometrical curves of his own invention. About the same time he laid the first foundation of his work upon curves that have a double curvature, which he finished in 1729, at 16 years of age. He was named Adjoint-Mechanician to the Academy in 1731, at the age of 18, Associate in 1733, and Pensioner in 1738; during his connection with the Academy, he had a great multitude of learned and ingenious communications inserted in their Memoirs, beside several other works which he published separately; the list of which is as follows:

1. On Curves of a Double Curvature; in 1730, 4to.
2. Elements of Geometry; 1741, 8vo.
3. Theory of the Figure of the Earth; 1743, 8vo.
4. Elements of Algebra; 1746, 8vo.
5. Tables of the Moon; 1754, 8vo.

His papers inserted in the Memoirs of the Academy are too numerous to be particularised here; but they may be found from the year 1727, for almost every year till 1762; being upon a variety of subjects, astronomical, mathematical, optical, &c.

CLAVIUS (CHRISTOPHER), a German Jesuit, was born at Bamberg in Germany, in 1537. He became a very studious mathematician, and elaborate writer; his works, when collected, and closely printed, making 5 large folio volumes; being a complete body or course of the mathematics. They are mostly elementary, and commentaries on Euclid and others; having very little of invention of his own. His talents and writings have been variously spoken of, and it must be acknowledged that they are heavy and elaborate. He was sent for to Rome, to assist, with other learned men, in the reformation of the calendar, by pope Gregory; which he afterward undertook a defence of, against Scaliger, Vieta, and others, who attacked it. He died at Rome, the 6th of February, 1612, at 75 years of age, after more than 50 years close application to the mathematical sciences.

CLEFF. See *Cliff*.

CLERC (JOHN LE), a celebrated writer and universal scholar, was born at Geneva in 1657. After passing through the usual course of study at Geneva, he went to France in 1678; but returning the year after, he took holy orders. In 1682 Le Clerc visited England, to learn the language: but the smoky air of London not agreeing with his lungs, he soon returned to Holland, where he settled; and was appointed professor of philosophy, polite literature, and the Hebrew tongue, in the school at Amsterdam. Here he long continued to read lectures; for which purpose he drew up and published his Logic, Ontology, Pneumatology, and Natural Philosophy. He published also *Ars Critica*; a Commentary on the Old Testament; a Compendium of Universal History; an Ecclesiastical History of the two first centuries; a French translation of the New Testament, and other works. In 1686, he began, jointly with M. de la Crose, his *Bibliotheque Univerfelle et Historique*, in imitation of other literary journals; which was continued to the year 1693, making 26 volumes. In 1703 he began his *Bibliotheque Choisie*, and continued it to 1714, when he commenced another work on the same plan, called *Bibliotheque Ancienne et Moderne*, which he continued to the year 1728; all of them justly esteemed excellent stores of useful knowledge. He published also, in 1713, a neat little treatise on Practical Geometry, in 2 vols. small 8vo, called, *Pratique de la Geometrie, sur le papier et sur le terrain*. In 1728 he was seized with a palsy and fever; and, after spending the last six years of his life with little or no understanding, he died in 1736, at 79 years of age.

CLEPSYDRA, a kind of water clock, or an hourglass serving to measure time by the fall of a certain quantity, commonly out of one vessel into another.—There have been also clepsydræ made with quicksilver; and the term is also used for hour-glasses of sand.

By this instrument the Egyptians measured their time and the course of the sun. Also Tycho Brahe, in modern times, made use of it to measure the motion of the stars, &c; and Dudley used the same contrivance in making all his maritime observations.

The use of Clepsydræ is very ancient. They were probably invented in Egypt under the Ptolemys; though some authors ascribe the invention of them to the Greeks, and others to the Romans. Pliny informs us, that Scipio Nafica, about 150 years before Christ, gave the first hint for the construction of them: and Pancirollus has particularly described them. According to his account, the clepsydra was a vessel made of glass, with a small hole in the bottom, edged with gold: in the upper part of this vessel a line was drawn, and marked with the 12 hours: the vessel was filled with water, and a cork with a pin fixed in it floated on the surface, pointing to the first hour; and as the water sunk in the vessel by issuing out of the small hole, the pin indicated the other hours as it descended.

Clepsydræ were chiefly used in the winter; as sundials served for the summer. They had however two great defects; the one, that the water ran out more or less easily, as the air was more or less dense; the other, that the water flowed more rapidly at the beginning, than towards the conclusion when its quantity and pressure were much decreased. Amontons has invented a clepsydra which, it is said, is free from both these inconveniences; and the same effect is produced by one described by Mr. Hamilton, in the Philof. Transf. vol. 44, pa. 171, or Abridg. vol. 10, pa. 248. Varignon too, in the *Memoires de l'Acad.* 1699, delivers a general geometrical method of making clepsydræ, or water-clocks, with any kind of vessels, and with any given orifices for the water to run through.

Vitruvius, in lib. 9 of his Achitecture, treats of these instruments; and Pliny in chap. 60, lib. 7, says that Scipio Nafica was the first who measured time at Rome

by clepsydræ, or water-clocks. Gesner, in his Pandects pa. 91, gives several contrivances for these instruments. Solomon de Caus also treats on this subject in his Reasons of Moving Forces &c. So also does Ozanam, in his Mathematical Recreations, in which is contained a Treatise on Elementary Clocks, translated from the Italian of Dominique Martinelli. There is likewise a treatise on Hour-Glasses by Arcangelo Maria Radi, called *Nova Sceinza de Horologi Polvere*. See also the *Technica Curiosa* of Gasper Schottus; and Amonton's Remarques & Experiences Physiques sur la Construction d'une nouvelle Clepsydre, exempte des défauts des autres.

CLIFF, or CLEFF, a term in Music, for a certain mark, from the position of which the proper places of all other notes in a piece of music are known.

CLIMACTERIC, a critical year in a person's life.

According to some, this is every 7th year: but others allow it only to those years produced by multiplying 7 by the odd numbers 3, 5, 7, 9. These years, say they, bring with them some remarkable change with respect to health, life, or fortune: the grand climacteric is the 63d year; but some add also the 81st to it: the other remarkable climacterics are the 7th, 21st, 35th, 49th, and 56th.

CLIMATE, or *Clime*, in Geography, a part of the surface of the earth, bounded by two lesser circles parallel to the equator; and of such a breadth, as that the longest day in the parallel nearer the pole exceeds the longest day in that next the equator, by some certain space, as half an hour, or an hour, or a month.

The beginning of a climate, is a parallel circle in which the day is the shortest; and the end of the climate, is that in which the day is the longest. The climates therefore are reckoned from the equator to the pole; and are so many zones or bands, terminated by lines parallel to the equator: though, in strictness, there are several climates, or different degrees of light or temperature, in the breadth of one zone. Each climate only differs from its contiguous ones, in that the longest day in summer is longer or shorter, by half an hour, for instance, in the one place than in the other.

As the climates commence at the equator, at the beginning of the first climate, that is at the equator, the day is just 12 hours long; but at the end of it, or at the beginning of the 2d climate, the longest day is 12 hours and a half long; and at the end of the 2d, or beginning of the 3d climate, the longest day is 13 hours long; and so of the rest, as far as the polar circles, where the hour climates terminate, and month climates commence. And as an hour climate is a space comprised between two parallels of the equator, in the first of which the longest day exceeds that in the latter by half an hour; so the month climate is a space contained between two circles parallel to the polar circles, and having its longest day longer or shorter than that of its contiguous one, by a month, or 30 days. But some authors, as Ricciolus, make the longest day of the contiguous climates to differ by half hours, to about the latitude of 45 degrees; then to differ by an hour, or sometimes 2 hours, to the polar circle; and after that by a month each. See tables of climates in Varenius, chap. 25, prop. 13.

The ancients, who confined the climates to what they thought the habitable parts of the earth, reckoned only

seven,

feven, the middles of which they made to pafs through fome remarkable places; as the 1ft through Meroe, the 2d through Sienna, the 3d through Alexandria, the 4th through Rhodes, the 5th through Rome, the 6th through Pontus, and the 7th through the mouth of the Boryfthenes. But the moderns, who have failed farther toward the poles, make 30 climates on each fide.

Vulgarly the term Climate is beftowed on any country or region differing from another either in refpect of the feafons, the quality of the foil, or even the manners of the inhabitants; without any regard to the length of the longeft day. Abulfeda, an Arabic author, diftinguifhes the firft kind of climates by the term *real climates*, and the latter by that of *apparent climates*.

CLOCK, a machine now conftructed in fuch a manner, and fo regulated by the uniform motion of a pendulum, as to meafure time, and all its fubdivifions, with great exactnefs. Before the invention of the pendulum, a balance, not unlike the fly of a kitchen-jack, was ufed inftead of it.—They were at firft called nocturnal dials to diftinguifh them from fun-dials, which fhewed the hour by the fhadow of the fun.

The invention of clocks with wheels is afcribed to Pacificus, archdeacon of Verona, in the 9th century, on the credit of an epitaph quoted by Ughelli, and borrowed by him from Panvinius. Others attribute the invention to Boethius, about the year 510.

Mr. Derham, however, makes clock-work of a much older date; ranking Archimedes's fphere, mentioned' by Claudian, and that of Pofidonius, mentioned by Cicero, among machines of this kind: not that either their form or ufe were the fame with thofe of ours; but that they had their motion from fome hidden weights, or fprings, with wheels or pulleys, or fome fuch clock-work principle.

In the *Difquifitiones Monafticæ* of Benedictus Haëften, publifhed in the year 1644, he fays, that clocks were invented by Silvefter the 4th, a monk of his order, about the year 998, as Dithmarus and Bozius have fhewn; for before that time, they had nothing but fundials and clepfydræ to fhew the hour.—Conrade Gefner, in his Epitome, pa. 604, fays, that Richard Wallingford, an Englifh abbot of St. Albans, who flourifhed in the year 1326, made a wonderful clock by a moft excellent art, the like of which could not be produced by all Europe.—Moreri, under the word *Horologe* du Palais, fays, that Charles the Fifth, called the wife king of France, ordered at Paris the firft large clock to be made by Henry de Vie, whom he fent for from Germany, and fet it upon the tower of his palace, in the year 1372.—John Froiffart, in his *Hiftoire & Chronique*, vol. 2, ch. 28, fays, the duke of Bourgogne had a clock, which founded the hour, taken away from the city of Courtray, in the year 1382: and the fame thing is faid by Wm. Paradin, in his *Annales de Bourgogne*.

Clock-makers were firft introduced into England in 1368, when Edward the 3d granted a licence for three artifts to come over from Delft in Holland, and practife their occupation in this country.

The water-clocks, or clepfydræ, and fun-dials, have both a much better claim to antiquity. The French annals mention one of the former kind, fent by Aaron, king of Perfia, to Charlemagne, about the year 807, which it would feem bore fome refemblance to the

modern clocks: it was of brafs, and fhewed the hours by 12 little balls of the fame metal, which at the end of each hour fell upon a bell, and made a found. There were alfo figures of 12 cavaliers, which at the end of each hour came out through certain apertures, or windows, in the fide of the clock, and fhut them again, &c.

The invention of pendulum clocks is owing to the happy induftry of the laft age; and the honour of that difcovery is difputed between Galileo and Huygens. The latter, who wrote an excellent volume on the fubject, declares it was firft put in practice in the year 1657, and the defcription of it printed in 1658. Becher, *De Nova Temporis dimetiendi Theoria*, anno 1680, contends for Galileo; and relates, though at fecond-hand, the whole hiftory of the invention; adding that one Trefler, clock-maker to the father of the then grand duke of Tufcany, made the firft pendulum clock at Florence, under the direction of Galileo Galilei, a pattern of which was brought to Holland. And the Academy del Cimento fay exprefsly, that the application of the pendulum to the movement of a clock was firft propofed by Galileo, and put in practice by his fon Vincenzo Galilei, in 1649. But whoever may have been the inventor, it is certain that the invention never flourifhed till it came into the hands of Huygens, who infifts on it, that if ever Galileo thought of fuch a thing, he never brought it to any degree of perfection. The firft pendulum clock made in England was in the year 1662, by one Fromantil, a Dutchman.

Among the modern clocks, thofe of Strafburg and Lyons are very eminent for the richnefs of their furniture, and the variety of their motions and figures. In the former, a cock claps his wings, and proclaims the hour: the angel opens a door, and falutes the Virgin; and the holy fpirit defcends on her, &c. In the latter, two horfemen encounter, and beat the hour upon each other: a door opens, and there appears on the theatre the Virgin, with Jefus Chrift in her arms; the Magi, with their retinue, marching in order, and prefenting their gifts; two trumpeters founding all the while to proclaim the proceffion.

Thefe, however, are far excelled by two that have lately been made by Englifh artifts, as a prefent from the Eaft-India company to the emperor of China. Thefe two clocks are in the form of chariots, in each of which a lady is placed, in a fine attitude, leaning her right hand upon a part of the chariot, under which appears a clock of curious workmanfhip, little larger than a fhilling, that ftrikes and repeats, and goes for eight days. Upon the lady's finger fits a bird, finely modelled, and fet with diamonds and rubies, with its wings expanded in a flying pofture, and actually flutters for a confiderable time, on touching a diamond button below it; the body of the bird, in which are contained part of the wheels that animate it as it were, is lefs than the 16th part of an inch. The lady holds in her left hand a golden tube little thicker than a large pin, on the top of which is a fmall round box, to which is fixed a circular ornament not larger than a fixpence, fet with diamonds, which goes round in near three hours in a conftant regular motion. Over the lady's head is a double umbrella, fupported by a fmall fluted pillar not thicker than a quill, and under the larger of which a bell is fixed at a confiderable diftance from the

clock,

clock, with which it seems not to have any connection; but from which a communication is secretly conveyed to a hammer, that regularly strikes the hour, and repeats the same at pleasure, by touching a diamond button fixed to the clock below. At the feet of the lady is a golden dog; before which, from the point of the chariot, are two birds fixed on spiral springs, the wings and feathers of which are set with stones of various colours, and they appear as if flying away with the chariot, which, from another secret motion, is contrived to run in any direction, either straight or circular, &c; whilst a boy, that lays hold of the chariot behind, seems also to push it forwards. Above the umbrella are flowers and ornaments of precious stones; and it terminates with a flying dragon set in the same manner. The whole is of gold, most curiously executed, and embellished with rubies and pearls.

The ingenious Dr. Franklin contrived a clock to shew the hours, minutes, and seconds, with only three wheels and two pinions in the whole movement. The dial-plate has the hours engraven upon it in spiral spaces along two diameters of a circle, containing four times 60 minutes. The index goes round in four hours, and counts the minutes from any hour by which it has passed to the next following hour. The small hand, in an arch at top, goes round once in a minute, and shews the seconds. The clock is wound up by a line going over a pulley, on the axis of the great wheel, like a common 30 hour clock. Many of these very simple machines have since been constructed, that measure time exceedingly well. This clock is subject, however, to the inconvenience of requiring frequent winding, by drawing up the weight; as also to some uncertainty as to the particular hour shewn by the index.

Mr. Ferguson has proposed to remedy these inconveniences by another construction, which is described in his Select Exercises, pa. 4. This clock will go a week without winding, and always shews the precise hour; but, as Mr. Ferguson acknowledges, it has two disadvantages which do not belong to Dr. Franklin's clock: when the minute hand is adjusted, the hour plate must also be set right, by means of a pin; and the smallness of the teeth in the pendulum wheel will cause the pendulum ball to describe but small arcs in its vibrations; and therefore the momentum of the ball will be less, and the times of the vibrations will be more affected by any unequal impulse of the pendulum wheel on the pallets. Besides, the weight of the flat ring, on which the seconds are engraven, will load the pivots of the axis of the pendulum wheel with a great deal of friction, which ought by all possible means to be avoided. To remedy this inconvenience, the seconds plate might be omitted.

Mr. Ferguson also contrived a clock, shewing the apparent diurnal motions of the sun and moon, the age and phases of the moon, with the time of her coming to the meridian, and the times of high and low water; and all this by having only two wheels and a pinion added to the common movement. See his Select Exercises before mentioned. In this clock the figure of the sun serves as an hour index, by going round the dial-plate in 24 hours; and a figure of the moon goes round in 24 h. 50½ min. the time of her going round in the heavens from any meridian to the same meridian again.

A clock of this kind was adapted by Mr. Ferguson to the movement of an old watch. See also a description and drawing of an astronomical clock, shewing the apparent daily motions of the sun, moon, and stars, with the times of their rising, southing, and setting; the places of the sun and moon in the ecliptic, and the age and phases of the moon for every day of the year, in the same book, pa. 19.

There have been several treatises upon clocks; the principal of which are the following. Hieronymus Cardan, de Varietate Rerum libri 17.—Conrade Dasypodius, Descriptio Horologii Astronomici Argentinensis in summa Templi erecti.—Guido Pancirollus, Antiqua deperdita & nova reperta.—L'Usage du Cadran, ou de l'Horloge Physique Universelle, par Galilée, Mathematicien du Duc de Florence.—Oughtred's Opuscula Mathematica.—Huygens's Horologium Oscillatorium.—Pendule perpetuelle, par l'Abbe de Hautefeuille.—J. J. Becheri Theoria & Experientia de nova Temporis dimetiendi Ratione & Horologiorum Constructione.——Clark's Oughtredus explicatus, ubi de Constructione Horologiorum.—Horological Disquisitions.—Huygens's Posthumous Works.—Sully's Regle Artificielle du Temps, &c.—Serviere's Recueil d'Ouvrages Curieux.—Derham's Artificial Clock-maker.—Camus's Traités des Forces Mouvantes.—Alexandre's Traité Général des Horologies.—Also Treatises and Principles of Clock-making, by Hatton, Cuming, &c. &c.

CLOUD, a collection of vapours suspended in the atmosphere, and rendered visible.

Although it be generally allowed that the clouds are formed from the aqueous vapours, which before were so closely united with the atmosphere as to be invisible: it is, however, not easy to account for the long continuance of some very opaque clouds without dissolving; or to assign the reason why the vapours, when they have once begun to condense, do not continue to do so till they at last fall to the ground in the form of rain or snow, &c. It is now known that a separation of the latent heat from the water of which vapour is composed is attended with a condensation of that vapour in some degree; in such case, it will first appear as a smoke, mist, or fog; which, if interposed between the sun and earth, will form a cloud; and the same causes continuing to operate, the cloud will produce rain or snow. It is however abundantly evident that some other cause besides mere heat or cold is concerned in the formation of clouds, and the condensation of atmospherical vapours. This cause is esteemed in a great measure the electrical fluid; indeed electricity is now so generally admitted as an agent in all the great operations of nature, that it is no wonder to find the formation of clouds attributed to it; and this has accordingly been given by Beccaria as the cause of the formation of all clouds whatsoever, whether of thunder, rain, hail, or snow.

But whether the clouds are produced, that is, the atmospheric vapours rendered visible, by means of electricity or not, it is certain that they do often contain the electric fluid in prodigious quantities, and many terrible and destructive accidents have been occasioned by clouds very highly electrified. The most extraordinary instance of this kind perhaps on record happened

in

in the island of Java, in the East-Indies, in August, 1772. On the 11th of that month, at midnight, a bright cloud was observed covering a mountain in the district called Cheribou, and several reports like those of a gun were heard at the same time. The people who dwelt upon the upper parts of the mountain not being able to fly fast enough, a great part of the cloud, eight or nine miles in circumference, detached itself under them, and was seen at a distance, rising and falling like the waves of the sea, and emitting globes of fire so luminous, that the night became as clear as day. The effects of it were astonishing ; every thing was destroyed for 20 miles round ; the houses were demolished ; plantations were buried in the earth ; and 2140 people lost their lives, besides 1500 head of cattle, and a vast number of horses, goats, &c. Another remarkable instance of the dreadful effects of electric clouds, which happened at Malta the 29th of October 1757, is related in Brydone's Tour through Malta.

The height of the clouds is not usually great : the summits of high mountains being commonly quite free from them, as many travellers have experienced in passing these mountains. It is found that the most highly electrified clouds descend lowest, their height being often not more than 7 or 800 yards above the ground ; and sometimes thunder-clouds appear actually to touch the ground with one of their edges : but the generality of clouds are suspended at the height of a mile, or little more, above the earth.

The motions of the clouds, though often directed by the wind, are not always so, especially when thunder is about to ensue. In this case they are seen to move very slowly, or even to appear quite stationary for some time. The reason of this probably is, that they are impelled by two opposite streams of air nearly of equal strength ; and in such cases it seems that both the aerial currents ascend to a considerable height ; for Mess. Charles and Robert, when endeavouring to avoid a thunder cloud, in one of their aerial voyages with a balloon, could find no alteration in the course of the current, though they ascended to the height of 4000 feet above the earth. In some cases the motions of the clouds evidently depend on their electricity, independent of any current of air whatever. Thus, in a calm and warm day, small clouds are often seen meeting each other in opposite directions, and setting out from such short distances, that it cannot be supposed that any opposite winds are the cause. Such clouds, when they meet, instead of forming a larger one, become much smaller, and sometimes quite vanish ; a circumstance most probably owing to the discharge of opposite electricities into each other. And this serves also to throw some light on the true cause of the formation of clouds ; for if two clouds, the one electrified positively, and the other negatively, destroy each other on contact, it follows that any quantity of vapour suspended in the atmosphere, while it retains its natural quantity of electricity, remains invisible, but becomes a cloud when electrified either plus or minus.

The shapes of the clouds are also probably owing to their electricity ; for in those seasons in which a great commotion has been excited in the atmospherical electricity, the clouds are seen assuming strange and whimsical shapes, that are continually varying. This, as

3

well as the meeting of small clouds in the air, and vanishing upon contact, is a sure sign of thunder.

The uses of the clouds are evident, as from them proceeds the rain that refreshes the earth, and without which, according to the present state of nature, the whole surface of the earth must become a mere desert. They are likewise useful as a screen interposed between the earth and the scorching rays of the sun, which are often so powerful as to destroy the grass and other tender vegetables. In the more secret operations of nature too, where the electric fluid is concerned, the clouds bear a principal share ; and chiefly serve as a medium for conveying that fluid from the atmosphere into the earth, and from the earth into the atmosphere : in doing which, when electrified to a great degree, they sometimes produce very terrible effects ; an instance of which is related above.

CLOUTS, in Artillery, are thin plates of iron nailed on that part of the axle-tree of a gun carriage which comes through the nave, and through which the linspin goes.

CLUVIER, or CLUVERIUS, (PHILIP), a celebrated geographer, was born at Dantzic in 1580. After an education at home, he travelled into Poland, Germany, and the Netherlands, to improve himself in the knowledge of the law. But, when at Leyden, Joseph Scaliger persuaded him to give way to his genius for geography. In pursuance of this advice, Cluvier visited the greatest part of the European states. He was well skilled in many languages, speaking half a score with facility, viz, Greek, Latin, German, French, English, Dutch, Italian, Hungarian, Polish, and Bohemian. On his return to Leyden, he taught there with great applause ; and died in 1623, being only 43 years of age, justly esteemed the first geographer who had put his researches in order, and reduced them to certain principles. He was author of several ingenious works in geography, viz :

1. *De Tribus Rheni Alveis.*
2. *Germania Antiqua.*
3. *Italia Antiqua, Sicilia, Sardinia, & Corsica.*
4. *Introductio in Universum Geographiam.*

COASTING, is that part of Navigation in which the places are not far asunder, so that a ship may sail in sight of land, or within soundings between them.

COCHLEA, one of the five Mechanical powers, otherwise called the Screw ; being so named from the resemblance a screw bears to the spiral shell of a snail, which the Latins call Cochlea. See SCREW, and MECHANICAL *Powers.*

COCK *of a Dial*, the pin, style, or gnomon.

COEFFICIENTS, in Algebra, are numbers, or given quantities, usually prefixed to letters, or unknown quantities, by which it is supposed they are multiplied ; and so, with such letters, or quantities, making a product, or *coefficient* production ; whence the name.

Thus, in $3a$ the coefficient is 3, in bx it is b, and in cx^2 it is c. If a quantity have no number prefixed, unity or 1 is understood ; as a is the same as $1a$, and bc the same as $1bc$. The name *coefficient* was first given by Vieta.

In any equation so reduced as that its highest power or term has 1 for its coefficient ; then the coefficient of the 2d term is equal to the sum of all the roots, both

positive

pofitive and negative; fo that if the 2d term is wanting in an equation, the fum of the pofitive roots of that equation is equal to the fum of the negative roots, as they mutually balance and cancel each other. Alfo the coefficient of the 3d term of an equation is equal to the fum of all the rectangles arifing by the multiplication of every two of the roots, how many ways foever they can be combined by twos; as once in the quadratic, 3 times in the cubic, 6 times in the biquadratic equation, &c. And the coefficient of the 4th term of an equation, is the fum of all the folids made by the continual multiplication of every three of the roots, how often foever fuch a ternary can be had; as once in a cubic, 4 times in a biquadratic, 10 times in an equation of 5 dimenfions, &c. And thus it will go on infinitely.

COEFFICIENTS *of the fame Order*, is a term fometimes ufed for the coefficients prefixed to the fame unknown quantities, in different equations.

Thus in the equations $\begin{cases} ax + by + cz = m, \\ dx + ey + fz = n, \\ gx + hy + lz = p, \end{cases}$

the coefficients a, d, g, are of the fame order, being the coefficients of the fame letter x; alfo b, e, h are of the fame order, being the coefficients of y; and fo on.

Oppofite COEFFICIENTS, fuch as are taken each from a different equation, and from a different order of coefficients. Thus, in the foregoing equations, a, e, k, or a, h, f, or d, b, k, &c, are oppofite coefficients.

COELESTIAL. See CELESTIAL.

COFFER, in Architecture, a fquare depreffure or finking, in each interval between the modillions of the Corinthian cornice; ufually filled up with a rofe; fometimes with a pomegranate, or other enrichment.

COFFER, in Fortification, denotes a hollow lodgment, athwart a dry moat, 6 or 7 feet deep, and 16 or 18 broad. The upper part of it is made of pieces of timber, raifed 2 feet above the level of the moat; the elevation having hurdles laden with earth for its covering, and ferving as a parapet with embrazures.

The coffer is nearly the fame with the caponiere, excepting that this laft is fometimes made beyond the counterfcarp on the glacis, and the coffer always in the moat, taking up its whole breadth, which the caponiere does not.

It differs from the traverfe and gallery, in that thefe are made by the befiegers, and the coffer by the befieged.

The befieged commonly make ufe of coffers to repulfe the befiegers, when they endeavour to pafs the ditch. And, on the other hand, the befiegers, to fave themfelves from the fire of thefe coffers, throw up the earth on that fide towards the coffer.

COFFER-*Dams*, or *Batardeaux*, in Bridge-building, are inclofures formed for laying the foundation of piers, and for other works in water, to exclude the furrounding water, and fo prevent it from interrupting the workmen.

Thefe inclofures are fometimes fingle, and fometimes double, with clay rammed between them; fometimes they are made with piles driven clofe by one another, and fometimes the piles are notched or dove-tailed into one another; but the moft ufual method is to drive piles with grooves in them, at the diftance of five or fix feet from each other, and then boards are let down

between them, after which the water is pumped out.

COGGESHALL's *Sliding-Rule*, an inftrument ufed in Gauging, and fo called from its inventor. See the defcription and ufe under SLIDING-*Rule*.

COHESION, one of the four fpecies of attraction, denoting that force by which the parts of bodies adhere or ftick together.

This power was firft confidered by Newton as one of the properties effential to all matter, and the caufe of all that variety obferved in the texture of different terreftrial bodies. He did not, however, abfolutely determine that the power of cohefion was an immaterial one; but that it might poffibly arife, as well as that of gravitation, from the action of another. His doctrine of cohefion Newton delivers in thefe words: " The particles of all hard homogeneous bodies, which touch one another, cohere with a great force; to account for which, fome philofophers have recourfe to a kind of hooked atoms, which in effect is nothing elfe but to beg the thing in queftion. Others imagine that the particles of bodies are connected by reft, i. e. in effect by nothing at all; and others by confpiring motions, i. e. by a relative reft among themfelves. For myfelf, it rather appears to me, that the particles of bodies cohere by an attractive force; whereby they tend mutually toward each other: which force, in the very point of contact, is very great; at little diftances is lefs, and at a little farther diftance is quite infenfible."

It is uncertain in what proportion this force decreafes as the diftance increafes. Defaguliers conjectures, from fome phenomena, that it decreafes as the biquadratic or 4th power of the diftance, fo that at twice the diftance it acts 16 times more weakly, &c.

" Now if compound bodies be fo hard, as by experience we find fome of them to be, and yet have a great many hidden pores within them, and confift of parts only laid together; no doubt thofe fimple particles which have no pores within them, and which were never divided into parts, muft be vaftly harder. For fuch hard particles, gathered into a mafs, cannot poffibly touch in more than a few points: and therefore much lefs force is required to fever them, than to break a folid particle, whofe parts touch throughout all their furfaces, without any intermediate pores or interftices. But how fuch hard particles, only laid together, and touching only in a few points, fhould come to cohere fo firmly, as in fact we find they do, is inconceivable; unlefs there be fome caufe, whereby they are attracted and preffed together. Now the fmalleft particles of matter may cohere by the ftrongeft attractions, and conftitute larger, whofe attracting force is feebler: and again, many of thefe larger particles cohering, may conftitute others ftill larger, whofe attractive force is ftill weaker; and fo on for feveral fucceffions, till the progreffion end in the biggeft particle, on which the operations in chemiftry, and the colours of natural bodies, do depend; and which by cohering compofe bodies of a fenfible magnitude.''

Again, the opinion maintained by many is that which is fo ftrongly defended by J. Bernoulli, *De Gravitate Ætheris*; who attributes the cohefion of the parts of matter to the uniform preffure of the atmofphere; confirming this opinion by the known experiment of two polifhed marble planes, which cohere very ftrongly in the open air, but eafily drop afunder in an exhaufted receiver. However, if two plates of this kind be fmeared with oil, to fill up the pores in their furfaces, and prevent the lodgment of air, and one of them be gently rubbed upon the other, they will adhere fo ftrongly, even when fufpended in an exhaufted receiver, that the weight of the lower plate will not be able to feparate it from the upper one. But although this theory might ferve tolerably well to explain the cohefion of compofitions, or greater collections of matter; yet it falls far fhort of accounting for that firft cohefion of the atoms, or primitive corpufcles, of which the particles of hard bodies are compofed.

Again, fome philofophers have pofitively afferted, that the powers, or means, are immaterial, by which matter coheres; and, in confequence of this fuppofition, they have fo refined upon attractions and repulfions, that their fyftems feem but little fhort of fcepticifm, or denying the exiftence of matter altogether. A fyftem of this kind is adopted by Dr. Prieftley, from Meffrs. Bofcovich and Michell, to folve fome difficulties concerning the Newtonian doctrine of light. See his Hiftory of Vifion, vol. 1. pa. 392. " The eafieft method," fays he, " of folving all difficulties, is to adopt the hypothefis of Mr. Bofcovich, who fuppofes that matter is not impenetrable, as has been perhaps univerfally taken for granted; but that it confifts of phyfical points only, endued with powers of attraction and repulfion in the fame manner as folid matter is generally fuppofed to be: provided therefore that any body move with a fufficient degree of velocity, or have a fufficient momentum to overcome any powers of repulfion that it may meet with, it will find no difficulty in making its way through any body whatever; for nothing elfe will penetrate one another but powers, fuch as we know do in fact exift in the fame place, and counterbalance or over-rule one another. The moft obvious difficulty, and indeed almoft the only one, that attends this hypothefis, as it fuppofes the mutual penetrability of matter, arifes from the idea of the nature of matter, and the difficulty we meet with in attempting to force two bodies into the fame place. But it is demonftrable, that the firft obftruction arifes from no actual contact of matter, but from mere powers of repulfion. This difficulty we can overcome; and having got within one fphere of repulfion, we fancy that we are now impeded by the folid matter itfelf. But the very fame is the opinion of the generality of mankind with refpect to the firft obftruction. Why, therefore, may not the next be only another fphere of repulfion, which may only require a greater force than we can apply to overcome it, without diforder ing the arrangement of the conftituent particles; but which may be overcome by a body moving with the amazing velocity of light?"

Other philofophers have fuppofed that the powers both of gravitation and cohefion are material; and that they are only different actions of the etherial fluid, or elementary fire. In proof of this doctrine, they allege the experiment with the Magdeburg hemifpheres, as they are called. The preffure of the atmofphere we fee is, in this cafe, capable of producing a very ftrong cohefion; and if there be in nature any fluid more penetrating, as well as more powerful in its effects, than the air we breath, it is poffible that what is called

the attraction of cohefion may in fome meafure be an effect of the action of that fluid. Such a fluid as this is the element of fire. Its activity is fuch as to penetrate all bodies whatever and in the ftate in which it is commonly called fire, it acts according to the quantity of folid matter contained in the body. In this ftate, it is capable of diffolving the ftrongeft cohefions obferved in nature. Fire, therefore, being able to diffolve cohefions, muft alfo be capable of caufing them, provided its power be exerted for that purpofe, which poffibly it may be, when we confider its various modes or appearances, viz, as fire or heat, in which ftate it confumes, deftroys, and diffolves; or as light, when it feems deprived of that deftructive power; and as the electric fluid, when it attracts, repels and moves bodies, in a great variety of ways. In the Philof. Tranf. for 1777 this hypothefis is noticed, and in fome meafure adopted by Mr. Henly. " Some gentlemen (fays he) have fuppofed that the electric matter is the caufe of the cohefion of the particles of bodies. If the electric matter be, as I fufpect, a real elementary fire inherent in all bodies, that opinion may probably be well founded; and perhaps the foldering of metals, and the cementation of iron, by fire, may be confidered as ftrong proofs of the truth of their hypothefis."

But whatever the caufe of cohefion may be, its effects are evident and certain. The different degrees of it conftitute bodies of different forms and properties. Thus, Newton obferves, the particles of fluids, which do not cohere too ftrongly, and are fmall enough to render them fufceptible of thofe agitations which keep liquors in a fluor, are moft eafily feparated and rarefied into vapour, and make what the chemifts call *volatile bodies*; being rarefied with an eafy heat, and again condenfed with a moderate cold. Thofe that have groffer particles, and fo are lefs fufceptible of agitation, or cohere by a ftronger attraction, are not feparable without a greater degree of heat; and fome of them not without fermentation: and thefe make what the chemifts call *fixt bodies*.

Air, in its fixed ftate, poffeffes the interftices of folid fubftances, and probably ferves as a bond of union to their conftituent parts; for when thefe parts are feparated, the air is difcharged, and recovers its elafticity. And this kind of attraction is evinced by a variety of familiar experiments; as, by the union of two contiguous drops of mercury; by the mutual approach of two pieces of cork, floating near each other in a bafon of water; by the adhefion of two leaden balls, whofe furfaces are fcraped and joined together with a gentle twift, which is fo confiderable, that, if the furfaces are about a quarter of an inch in diameter, they will not be feparated by a weight of 100 lb; by the afcent of oil or water between two glafs planes, fo as to form the hyperbolic curve, when they are made to touch on one fide, and kept feparate at a fmall diftance on the other; by the depreffion of mercury, and by the rife of water in capillary tubes, and on the fides of glafs veffels; alfo in fugar, fponge, and all porous fubftances. And where this cohefive attraction ends, a power of repulfion begins.

To determine the force of cohefion, in a variety of different fubftances, many experiments have been made, and particularly by profeffor Mufchenbroek. The ad-

hefion

hefion of polifhed planes, about two inches in diameter, heated in boiling water, and fmeared with greafe, required the following weights to feparate them:

	Cold greafe	Hot greafe
Planes of Glafs	130 lb	300 lb
Brafs	150	800
Copper	200	850
Marble	225	600
Silver	150	250
Iron	300	950

But when the Brafs planes were made to adhere by other forts of matter, the refults were as in the following table:

With Water	12 oz
Oil	18
Venice Turpentine	24
Tallow Candle	800
Rofin	850
Pitch	1400

In eftimating the *Abfolute Cohefion* of folid pieces of bodies, he applied weights to feparate them according to their length: his pieces of wood were long fquare parallelopipedons, each fide of which was ·26 of an inch, and they were drawn afunder by the following weights:

Fir	600 lb
Elm	950
Alder	1000
Linden tree	1000
Oak	1150
Beech	1250
Afh	1250

He tried alfo wires of metal, 1-10th of a Rhinland inch in diameter: the metals and weights were as follow:

Of Lead	$29\frac{1}{4}$ lb
Tin	$40\frac{1}{4}$
Copper	$299\frac{1}{4}$
Yellow Brafs	360
Silver	370
Iron	450
Gold	500

He then tried the *Relative Cohefion*, or the force with which bodies refift an action applied to them in a direction perpendicular to their length. For this purpofe he fixed pieces of wood by one end into a fquare hole in a metal plate, and hung weights towards the other end, till they broke at the hole: the weights and diftances from the hole are exhibited in the following table.

	Diftance	Weight
Pine	$9\frac{1}{2}$ inc	$36\frac{1}{2}$ oz
Fir	9	40
Beech	7	$56\frac{1}{2}$
Elm	9	44
Oak	$8\frac{1}{2}$	48
Alder	$9\frac{1}{4}$	48

See his Elem. Nat. Philof. cap. 19.

COLD, the privation of heat, or the oppofite to it.

As it is fuppofed that heat confifts in a particular motion of the parts of the hot body, hence the nature of cold, which is its oppofite, is deduced; for it is found that cold extinguifhes, or rather abates heat;

3

hence it would feem to follow, that thofe bodies are cold, which check and reftrain the motion of the particles in which heat confifts.

In general, cold contracts moft bodies, and heat expands them: though there are fome inftances to the contrary, efpecially in the extreme cafes or ftates of thefe qualities of bodies. Thus, though iron, in common with other bodies, expand with heat, yet, when melted, it is always found to expand in cooling again. So alfo, though water always is found to expand gradually as it is heated, and to contract as it cools, yet in the act of freezing, it fuddenly expands again, and that with a moft enormous force, capable of rending rocks, or burfting the very thick fhells of metal, &c. A computation of the force of freezing water has been made by the Florentine Academicians, from the burfting of a very ftrong brafs globe or fhell, by freezing water in it; when, from the known thicknefs and tenacity of the metal, it was found that the expanfive power of a fpherule of water only one inch in diameter, was fufficient to overcome a refiftance of more than 27,000 pounds, or 13 tons and a half. See alfo experiments on burfting thick iron bomb-fhells by freezing water in them by Major Edward Williams of the Royal Artillery, in the Edinb. Philof. Tranf. vol. 2.

Such a prodigious power of expanfion, almoft double that of the moft powerful fteam engines, and exerted in fo fmall a mafs, feemingly by the force of cold, was thought a very powerful argument in favour of thofe who fuppofed that cold, like heat, is a pofitive fubftance. Dr. Black's difcovery of latent heat, however, has now afforded a very eafy and natural explication of this phenomenon. He has fhewn, that, in the act of congelation, water is not cooled more than it was before, but rather grows warmer: that as much heat is difcharged, and paffes from a latent to a fenfible ftate, as, had it been applied to water in its fluid ftate, would have heated it to 135°. In this procefs, the expanfion is occafioned by a great number of minute bubbles fuddenly produced. Formerly thefe were fuppofed to be cold in the abftract; and to be fo fubtle, that, infinuating themfelves into the fubftances of the fluid, they augmented its bulk, at the fame time that, by impeding the motion of its particles upon each other, they changed it from a fluid to a folid. But Dr. Black fhews that thefe are only air extricated during the congelation; and to the extrication of this air he afcribes the prodigious expanfive force exerted by freezing water. The only queftion therefore now remaining, is, By what means this air comes to be extricated, and to take up more room than it naturally does in the fluid. To this it may be anfwered, that perhaps part of the heat which is difcharged from the freezing water, combines with the air in its unelaftic ftate, and, by reftoring its elafticity, gives it that extraordinary force, as is feen alfo in the cafe of air fuddenly extricated in the explofion of gun-powder.

Cold alfo ufually tends to make bodies electric, which are not fo naturally, and to increafe the electric properties of fuch as are fo. And it is farther found that all fubftances do not tranfmit cold equally well; but that the beft conductors of electricity, viz metals, are likewife the beft conductors of cold. It may farther be added, that when the cold has been carried to such an extremity as to render any body an electric, it then ceases to conduct the cold so well as before. This is exemplified in the practice of the Laplanders and Siberians; where, to exclude the extreme cold of the winters from their habitations the more effectually, and yet to admit a little light, they cut pieces of ice, which in the winter time must always be electric in those countries, and put them into their windows; which they find to be much more effectual in keeping out the cold than any other substance.

Cold is the destroyer of all vegetable life, when increased to an excessive degree. It is found that many garden plants and flowers, which seem to be very stout and hardy, go off at a little increase of cold beyond the ordinary standard. And in severe winters, nature has provided the best natural defence for the corn fields and gardens, namely, a covering of snow, which preserves such parts green and healthy as are under it, while such as are uncovered by it are either killed or greatly injured.

Dr. Clarke is of opinion, that cold is owing to certain nitrous, and other saline particles, endued with particular figures proper to produce such effects. Hence, sal-ammoniac, saltpetre, or salt of urine, and many other volatile and alkalizate salts, mixed with water, very much increase its degree of cold. In the Philos. Transf. number 274, M. Geoffroy relates some remarkable experiments with regard to the production of cold. Four ounces of sal-ammoniac dissolved in a pint of water, made his thermometer descend 2 inches and $\frac{3}{4}$ in less than 15 minutes. An ounce of the same salt put into 4 or 5 ounces of distilled water, made the thermometer descend 2 inches and $\frac{1}{4}$. Half an ounce of sal-ammoniac mixed with 3 ounces of spirit of nitre, made the thermometer descend 2 inches and $\frac{5}{12}$; but, on using spirit of vitriol instead of nitre, it sunk 2 inches and $\frac{1}{2}$. In this last experiment it was remarked, that the vapours raised from the mixture had a considerable degree of heat, though the liquid itself was so extremely cold. Four ounces of saltpetre mixed with a pint of water, sunk the thermometer an inch and $\frac{1}{4}$; but a like quantity of sea salt sunk it only $\frac{1}{2}$ of an inch. Acids always produced heat, even common salt with its own spirit. Volatile alkaline salts produced cold in proportion to their purity, but fixed alkalies heat.

But the greatest degree of cold produced by the mixture of salts and aqueous fluids, was that shewn by Homberg; who gives the following receipt for making the experiment: Take a pound of corrosive sublimate, and as much sal-ammoniac; powder them separately, and mix the powders well; put the mixture into a vial, pouring upon it a pint and a half of distilled vinegar, shaking all well together. This composition grows so cold, that it can scarce be held in the hand in summer; and it happened, as M. Homberg was making the experiment, that the matter froze. The same thing once happened to M. Geoffroy, in making an experiment with sal-ammoniac and water, but it never was in his power to make it succeed a second time.

If, instead of making these experiments with fluid water, it be taken in its congealed state of ice, or rather snow degrees of cold will be produced greatly superior to any that have yet been mentioned. A mixture of snow and common salt sinks Fahrenheit's thermometer to 0; pot ashes and pounded ice sunk it 8 degrees far-

ther; two affusions of spirit of salt on pounded ice sunk it $14\frac{1}{2}$ below 0; and by repeated affusions of spirit of nitre M. Fahrenheit sunk it to 40° below 0. This is the ultimate degree of cold which the mercurial thermometer will measure; for the mercury itself begins then to congeal; and therefore recourse must afterwards be had to spirit of wine, naptha, or some other fluid that will not congeal. The greatest degree of cold hitherto produced by artificial means, has been 80° below 0; which was done at Hudson's Bay by means of snow and vitriolic acid, the thermometer standing naturally at 20° below 0. Indeed greater degrees of cold than this have been supposed: Mr. Martin, in his Treatise on Heat, relates, that at Kirenga in Siberia, the mercurial thermometer sunk to 118° below 0; and professor Brown at Petersburg, when he made the first experiment of congealing quicksilver, fixed the point of congelation at 350° below 0; but from later experiments it has been more accurately determined, that 40° below 0 is the freezing point of quicksilver.

The most remarkable experiment however was made by Mr. Walker of Oxford, with spirit of nitre poured on Glauber's salt, the effect of which was found to be similar to that of the same spirit poured on ice or snow; and the addition of sal-ammoniac rendered the cold still more intense. The proportions of these ingredients recommended by Mr. Walker, are concentrated nitrous acid two parts by weight, water one part; of this mixture, cooled to the temperature of the atmosphere, 18 ounces; of Glauber's salt, a pound and a half avoirdupois; and of sal-ammoniac, 12 ounces. On adding the Glauber's salt to the nitrous acid, the thermometer fell 52°, viz from 50 to −2; and on the addition of the sal-ammoniac, it fell to − 9°. Thus Mr. Walker was able to freeze quicksilver without either ice or snow, when the thermometer stood at 45°; viz, by putting the ingredients in 4 different pans, and inclosing these within each other.

Excessive degrees of cold occur naturally in many parts of the globe in the winter season.

Although the thermometer in this country hardly ever descends so low as 0, yet in the winter of 1780, Mr. Wilson of Glasgow observed, that a thermometer laid on the snow sunk to 25° below 0; and Mr. Derham, in the year 1708, observed in England, that the mercury stood within one-tenth of an inch of its station when plunged into a mixture of snow and salt. At Petersburg, in 1732, the thermometer stood at 28° below 0; and when the French academicians wintered near the polar circle, the thermometer sunk to 33° below 0; and in the Asiatic and American continents, still greater degrees of cold are often observed.

The effects of these extreme degrees of cold are very surprising. Trees are burst, rocks rent, and rivers and lakes frozen several feet deep: metallic substances blister the skin like red hot iron: the air, when drawn in by breathing, hurts the lungs, and excites a cough: even the effects of fire in a great measure seem to cease; and it is observed, that though metals are kept for a considerable time before a strong fire, they will still freeze water when thrown upon them. When the French mathematicians wintered at Tornea in Lapland, the external air, when suddenly admitted into their rooms, converted the moisture of the air into whirls of snow; their breasts seemed to be rent when they breathed

it,

it, and the contact of it was intolerable to their bodies ; and the spirit of wine, which had not been highly rectified, burst some of their thermometers by the congelation of the aqueous part.

Extreme cold too often proves fatal to animals in those countries where the winters are very severe ; thus 7000 Swedes perished at once in attempting to pass the mountains which divide Norway from Sweden. But it is not necessary that the cold, in order to prove fatal to human life, should be so very intense as has just been mentioned ; it is only requisite to be a little below 32° of Fahrenheit, or the freezing point, accompanied with snow or hail, from which shelter cannot be obtained. The snow which falls upon the clothes, or the uncovered parts of the body, then melts, and by a continual evaporation carries off the animal heat to such a degree, that a sufficient quantity is not left for the support of life. In such cases, the person first feels himself extremely chill and uneasy ; he turns listless, unwilling to walk or use exercise to keep himself warm, and at last turns drowsy, sits down to refresh himself with sleep, but wakes no more.

COLLIMATION, *Line of*, in a telescope, is a line passing through the intersection of those wires that are fixed in the focus of the object-glass, and the centre of the same glass.

COLLINS (JOHN), an eminent accountant and mathematician, was born at Wood Eaton near Oxford, March 5, 1624. At 16 years of age he was put apprentice to a bookseller at Oxford ; but his genius appeared so remarkable for the study of the mechanical and mathematical sciences, that he was taken under the tuition of Mr. Marr, who drew several curious dials, which were placed in different positions in the king's gardens ; under whom Mr. Collins made no small progress in the mathematics. In the course of the civil wars, he travelled abroad, to prosecute his favourite study ; and on his return he took upon him the profession of an accountant, and published, in the year 1652, a large work entitled, *An Introduction to Merchants Accompts ;* which was followed by several other publications on different branches of accounts. In 1658, he published a treatise called *The Sector on a Quadrant ;* containing the description and use of four several quadrants, each accommodated to the making of sun-dials, &c ; to which he afterward added an appendix concerning reflected dialling, from a glass placed reclining.—In 1659, he published his *Geometrical Dialling ;* and the same year also his *Mariner's Plain Scale new plained.*—Collins now became a fellow of the Royal Society in London, to which he made various communications ; particularly some ingenious chronological rules for the calendar, printed in the Philos. Trans. number 46, for April 1669 : also a curious dissertation concerning the resolution of equations in numbers, in number 69, for March 1671 : an elegant construction of the curious problem, having given the mutual distances of three objects in a plane, with the angles made by them at a fourth place in that plane, to find the distance of this place from each of the three former, vol. 6. pa. 2093 : and thoughts about some defects in algebra, vol. 14. pa. 375.

Collins wrote also several commercial tracts, highly acceptable to the public ; viz, A Plea for bringing over Irish cattle, and keeping out the fish caught by foreigners :—For the promotion of the English fishery :—For the working the Tin-mines :—A Discourse of Salt and Fishery. He was frequently consulted in nice and critical cases of accounts, of commerce, and engineering. On one of these occasions, being appointed to inspect the ground for cutting a canal or river between the Isis and the Avon, he contracted a disorder by drinking cyder when he was too warm, which ended in his death, the 10th of November 1683, at 59 years of age.

Mr. Collins was a very useful man to the sciences, keeping up a constant correspondence with the most learned men, both at home and abroad, and promoting the publication of many valuable works, which, but for him, would never have been seen by the public ; particularly Dr. Barrow's optical and geometrical lectures ; his abridgment of the works of Archimedes, Apollonius, and Theodosius ; Branker's translation of Rhonius's algebra, with Dr. Pell's additions, &c ; which were procured by his frequent solicitations.

It was a considerable time after, that his papers were all delivered into the hands of the learned and ingenious Mr. William Jones, F. R. S. among which were found manuscripts, upon mathematical subjects, of Briggs, Oughtred, Barrow, Newton, Pell, and many others. From a variety of letters from these, and many other celebrated mathematicians, it appears that Collins spared neither pains nor cost to procure what tended to promote real science : and even many of the late discoveries in physical knowledge owe their improvement to him ; for while he excited some to make known every new and useful invention, he employed others to improve them. Sometimes he was peculiarly useful, by shewing where the defect was in any useful branch of science, pointing out the difficulties attending the enquiry, and at other times setting forth the advantages, and keeping up a spirit and warm desire for improvement. Mr. Collins was also as it were the register of all the new improvements made in the mathematical sciences ; the magazine to which the curious had frequent recourse : in so much that he acquired the appellation of the English Mersennus. If some of his correspondents had not obliged him to conceal their communications, there could have been no dispute about the priority of the invention of a method of analysis, the honour of which evidently belongs to Newton ; as appears undeniably from the papers printed in the *Commercium Epistolicum D. Joannis Collins & aliorum de Analysi promota ; jussu Societatis Regiae in lucem editum,* 1712 ; a work that was made out from the letters in the possession of our author.

COLLINS's *Quadrant.* See QUADRANT.

COLLISION, is the friction, percussion, or striking of bodies against one another.

Striking bodies are considered either as elastic, or non-elastic. They may also be either both in motion, or one of them in motion, and the other at rest.

When non-elastic bodies strike, they unite together as one mass ; which, after collision, either remains at rest, or moves forward as one body. But when elastic bodies strike, they always separate after the stroke.

The principal theorems relating to the collision of bodies, are the following :

1. If any body impinge or act obliquely on a plane surface ; the force or energy of the stroke, or action, is as the sine of the angle of incidence. Or the force

upon

upon the furface, is to the fame when acting perpendicularly, as the fine of incidence is to radius.

2. If one body act on another, in any direction, and by any kind of force ; the action of that force on the fecond body is made only in a direction perpendicular to the furface on which it acts.

3. If the plane, acted on, be not absolutely fixed, it will move, after the stroke, in the direction perpendicular to its furface.

4. If a body A strike another body B, which is either at reft, or elfe in motion, either towards A or from it ; then the momenta, or quantities of motion, of the two bodies, estimated in any one direction, will be the very fame after the stroke that they were before it.

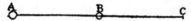

Thus, first, if A with a momentum of 10, strike B at reft, and communicate to it a momentum of 4, in the direction AB. Then there will remain in A only a momentum of 6 in that direction : which together with the momentum of B, viz 4, makes up ftill the fame momentum between them as before.—But if B were in motion before the stroke, with a momentum of 5, in the fame direction, and receive from A an additional momentum of 2 : then the motion of A after the stroke will be 8, and that of B, 7 ; which between them make up 15, the fame as 10 and 5, the motions before the stroke.—Lastly, if the bodies move in oppofite directions, and meet one another, namely A with a motion of 10, and B, of 5 ; and A communicate to B a motion of 6 in the direction AB of its motion : then, before the stroke, the whole motion from both, in the direction AB, is 10—5, or 5 : but after the stroke the motion of A is 4 in the direction AB, and the motion of B is 6 — 5, or 1 in the fame direction AB ; therefore the fum 4 + 1, or 5, is ftill the fame motion from both as it was before.

5. If a hard and fixed plane be struck either by a foft or a hard unelastic body ; the body will adhere to it. But if the plane be struck by a perfectly elastic body, it will rebound from it with the fame velocity with which it struck the plane.

6. The effect of the blow of the elastic body, upon the plane, is double to that of the non-elastic one ; the velocity and mafs being the fame in both.

7. Hence, non-elastic bodies lofe, by their collifion, only half the motion that is loft by elastic bodies ; the maffes and velocities being equal.

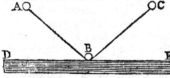

8. If an elastic body A impinge upon a firm plane DE at the point B, it will rebound from it in an angle equal to that in which it struck it ; or the angle of incidence will be equal to the angle of reflection : namely, the angle ABD = CBE.

9. If the non-elastic body B, moving with the velocity V in the direction Bb, and the body b with the velocity v, strike each other, the direction of the mo-

tion being in the line BC ; then they will move after the stroke with a common velocity, which will be more or lefs according as, before the stroke, b moved towards B, or from B, or was at reft ; and that common velocity, in each of thefe cafes, will be as follows : viz, it will be

$\dfrac{BV + bv}{B + b}$ when b moved from B,

$\dfrac{BV - bv}{B + b}$ when b moved towards B,

$\dfrac{BV}{B + b}$ when b was at reft.

For example, if the bodies or weights, B and b, be 5lb and 3lb ; and their velocities V and v, 60 feet and 40 feet per fecond ; then 300 and 120 will be their momenta BV and bv, and 18 = B + b the fum of the weights. Confequently the common velocity after the stroke, in the three cafes above mentioned, will be thus, viz,

$\dfrac{300 + 120}{18} = \dfrac{420}{18}$ or $23\frac{1}{4}$ in the first cafe,

$\dfrac{300 - 120}{18} = \dfrac{180}{18}$ or 10 in the fecond cafe,

$\dfrac{300}{18}$ - - or $16\frac{2}{3}$ in the third cafe.

10. If two perfectly elastic bodies impinge on each other ; their relative velocity is the fame both before and after the impulfe ; that is, they will recede from each other with the fame velocity with which they approached and met.

It is not meant however by this theorem, that each body will have the very fame velocity after the impulfe as it had before ; but that the velocity of the one, after the stroke, will be as much increafed, as that of the other is decreafed, in one and the fame direction. So, if the elastic body B move with the velocity V, and overtake the elastic body b, moving the fame way, with the velocity v ; then their relative velocity, or that with which they strike, is only V — v ; and it is with this fame velocity that they feparate from each other after the stroke : but if they meet each other, or the body b move contrary to the body B ; then they meet and strike with the velocity V + v, and it is with the fame velocity that they feparate again, and recede from each other after the stroke : in like manner, they would feparate with the velocity V of B, if b were at reft before the stroke. Alfo the fum of the velocities of the one body, is equal to the fum of the others. But whether they move forwards or backwards after the impulfe, and with what particular velocities, are circumstances that depend on the various maffes and velocities of the bodies before the stroke, and are as fpecified in the next theorem.

11. If the two elastic bodies B and b move directly towards each other, or directly from each other, the former with the velocity V, and the latter with the velocity v ; then, after their meeting and impulfe, the refpective velocities of B and b in the direction BC, in the three cafes of motion, will be as follow : viz,

$\dfrac{2bv + \overline{B - b} \cdot V}{B + b}$ the velocity of B,

$\dfrac{2BV - \overline{B - b} \cdot v}{B + b}$ the velocity of b,

when the bodies both moved towards C before the stroke ; and

$-2bv$

$$\frac{-2bv + \overline{B - b} \cdot V}{B + b} \text{ the velocity of B,}$$

$$\frac{2BV + \overline{B - b} \cdot v}{B + b} \text{ the velocity of } b,$$

when B moved towards C, and b towards B before the stroke;

$$\frac{B - b}{B + b} \times V \text{ the velocity of B,}$$

$$\frac{2B}{B + b} \times V \text{ the velocity of } b,$$

when b was at rest before the stroke.

12 The motions of bodies after impact, that strike each other obliquely, are thus determined.

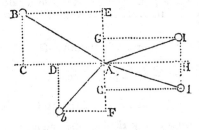

Let the two bodies B, b, move in the oblique directions BA, bA, and strike each other at A with velocities which are in proportion to the lines BA, bA. Let CAH represent the plane in which the bodies touch in the point of concourse; to which draw the perpendiculars BC, bD, and complete the rectangles CE, DF. Now the motion in BA is resolved into the two BC, CA; and the motion in bA is resolved into the two bD, DA; of which the antecedents BC, bD are the velocities with which they irectly meet, and the consequents CA, DA are parallel, and therefore by these the bodies do not impinge on each other, and consequently the motions according to these directions will not be changed by the impulse; so that the velocities with which the bodies meet, are as BC or EA, and bD or FA. The motions therefore of the bodies B, b, directly striking each other with the celerities EA, FA, will be determined by art. 11 or 9, according as the bodies are elastic or non-elastic; which being done, let AG be the velocity, so determined, of one of them, as A; and since there remains also in the same body a force of moving in the direction parallel to BE, with a velocity as BE, make AH equal to BE, and complete the rectangle GH: then the two motions in AH and AG, or HI, are compounded into the diagonal AI, which therefore will be the path and celerity of the body B after the stroke. And after the same manner is the motion of the other body b determined after the impact.

13. The state of the common centre of gravity of bodies is not affected by the collision or other actions of those bodies on one another. That is, if it were at rest before their collision, so will it be also at rest after collision: and if it were moving in any direction, and with any velocity, before collision; it will do the very same after it.

See more upon this subject under the article Per-cussion.

COLONNADE, a Peristyle of a circular figure; or a series of columns disposed in a circle, and insulated within-side.

COLOUR, a property inherent in light, by which, according to the various sizes of its parts, or from some other cause, it excites different vibrations in the optic nerve; which, propagated to the sensorium, affect the mind with different sensations. See the doctrine of colours fully explained under *Chromatics*. See also *Optics*, *Achromatic*, and *Telescope*.

COLUMBA *Noachi*, Noah's Dove, a small constellation in the southern hemisphere, consisting of 10 stars.

COLUMN, in Architecture, a round pillar, made to support or adorn a building.

The column is the principal part of an architectonical order, and is composed of three parts, the *base*, the *shaft*, and the *capital*; each of which is subdivided into a number of lesser parts, called members, or mouldings.

Columns are different according to the different orders they are used in; and also according to their matter, construction, form, disposition, and use. The proportion of the length of each to its diameter, and the diminution of the diameter upwards, are diversly stated by different authors. The medium of them is nearly as follows:

The *Tuscan* is the simplest and shortest of all; its height $3\frac{1}{2}$ diameters, or 7 modules; and it diminishes $\frac{1}{4}$ part of its diameter.

The *Doric* is more delicate, and adorned with flutings; its height $7\frac{1}{2}$ or 8 diameters.

The *Ionic* is more delicate still, being 9 diameters long. It is distinguished from the rest by the volutes, or curled scrolls in its capital, and by its base which is peculiar to it.

The *Corinthian* is the richest and most delicate of all the columns, being 10 diameters in length, and adorned with two rows of leaves, and stalks or stems, from whence spring out small volutes.

The *Composite* Column is also 10 diameters long, its capital adorned with rows of leaves like the Corinthian, and with angular volutes like the Ionic.

COLURES, are two great circles imagined to intersect at right angles in the poles of the world, and to pass, the one through the equinoctial points Aries and Libra, and the other through the solstitial points Cancer and Capricorn; from whence they are called the Equinoctial and Solstitial Colures. By thus dividing the ecliptic into four equal parts, they mark the four seasons, or quarters of the year.

It is disputed over what part of the back of Aries the equinoctial colure passed in the time of Hipparchus. Newton, in his Chronology, takes it to have been over the middle of the constellation. Father Souciet insists that it passed over the dodecatemorion of Aries, or midway between the rump and first of the tail. There are some observations in the Philos. Transf. number 466, concerning the position of this colure in the ancient sphere, from a draught of the constellation Aries, in the Aratæa published at Leyden and Amsterdam in 1652, which seem to confirm Newton's opinion; but the antiquity and authority of the original draught may still remain in question.

COMA BERENICES, *Berenice's Hair*, a modern constellation

ftellation of the northern hemifphere; compofed of un-formed ftars between the Lion's tail and Bootes.

It is faid that this conftellation was formed by Conon, an aftronomer, to confole the queen of Pto-lomy Euergetes, for the lofs of a lock of her hair, which was ftolen out of the temple of Venus, where fhe had dedicated it on account of a victory obtained by her hufband.

The ftars in this conftellation are, in Tycho's cata-logue 14, in Hevelius's 21, and in the Britannic ca-talogue 43.

COMBINATIONS, denote the alternations or va-riations of any number of quantities, letters, founds, or the like, in all poffible ways.

Father Merfenne gives the combinations of all the notes and founds in mufic, as far as 64; the fum of which amounts to a number expreffed by 90 places of figures. And the number of poffible combinations of the 24 letters of the alphabet, taken firft two by two, then three by three, and fo on, according to Preftet's calculation, amounts to

1391724288887252999425128493402200.

Father Truchet, in Mem. de l'Acad. fhews, that two fquare pieces, each divided diagonally into two colours, may be arranged and combined 64 different ways, fo as to form fo many different kinds of chequer-work: a thing that may be of ufe to mafons, paviours, &c.

Doctrine of COMBINATIONS.

I. *Having given any number of things, with the num-ber in each combination; to find the number of combinations.*

1. *When only two are combined together.*

One thing admits of no combination.

Two, *a* and *b*, admit of one only, viz *ab*.

Three, *a*, *b*, *c*, admit of three, viz *ab*, *ac*, *bc*.

Four admit of fix, viz, *ab*, *ac*, *ad*, *bc*, *bd*, *cd*.

Five admit of 10, viz, *ab*, *ac*, *ad*, *ae*, *bc*, *bd*, *be*, *cd*, *ce*, *de*.

Whence it appears that the numbers of combina-tions, of two and two only, proceed according to the triangular numbers 1, 3, 6, 10, 15, 21, &c, which are produced by the continual addition of the ordinal feries 0, 1, 2, 3, 4, 5, &c. And if *n* be the number of things, then the general formula for expreffing the fum of all their combinations by twos, will be $\frac{n \cdot n - 1}{1 \cdot 2}$.

Thus, if *n* = 2; this becomes $\frac{2 \cdot 1}{2} = 1$.

If *n* = 3; it is $\frac{3 \cdot 2}{2}$ -   -   = 3.

If *n* = 4; it is $\frac{4 \cdot 3}{2}$ -   -   = 6. &c.

2. *When three are combined together; then*

Three things admit of one order, *abc*.

Four admit of 4; viz *abc*, *abd*, *acd*, *bcd*.

Five admit of 10; viz *abc*, *abd*, *abe*, *acd*, *ace*, *ade*, *bcd*, *bce*, *bde*, *cde*. And fo on according to the firft py-ramidal numbers 1, 4, 10, 20, &c, which are formed by the continual addition of the former, or triangular numbers 1, 3, 6, 10, &c. And the general formula for any number *n* of combinations, taken by threes, is $\frac{n \cdot n - 1 \cdot n - 2}{1 \cdot 2 \cdot 3}$.

So, if *n* = 3; it is $\frac{3 \cdot 2 \cdot 1}{1 \cdot 2 \cdot 3} = 1$.

If *n* = 4; it is $\frac{4 \cdot 3 \cdot 2}{6} = 4$.

If *n* = 5; it is $\frac{5 \cdot 4 \cdot 3}{2} = 6$. &c.

Proceeding thus, it is found that a general formula for any number *n* of things, combined by *m* at each time, is $s = \frac{n \cdot n-1 \cdot n-2 \cdot n-3 \, \&c}{1 \cdot 2 \cdot 3 \cdot 4 \, \&c}$, continued to *m* factors, or terms, or till the laft factor in the deno-minator be *m*.

So, in 6 things, combined by 4's, the number of combinations is $\frac{6 \cdot 5 \cdot 4 \cdot 3}{1 \cdot 2 \cdot 3 \cdot 4} = 15$.

3. By adding all thefe feries together, their fum will be the whole number of poffible combinations of *n* things combined both by twos, by threes, by fours, &c. And as the faid feries are evidently the coeffi-cients of the power *n* of a binomial, wanting only the firft two 1 and *n*; therefore the faid fum, or whole number of all fuch combinations, will be $1 + 1^1 - n - 1$, or $2^n - n - 1$. Thus if the num-ber of things be 5; then $2^5 - 5 - 1 = 32 - 6 = 26$.

II. *To find the number of Changes and Alterations which any number of quantities can undergo, when com-bined in all poffible varieties of ways, with themfelves and each other, both as to the things themfelves, and the Order or Pofition of them.*

One thing admits but of one order or pofition.

Two things may be varied four ways; thus, *aa*, *ab*, *ba*, *bb*.

Three quantities, taken by twos, may be varied nine ways; thus *aa*, *ab*, *ac*, *ba*, *ca*, *bb*, *bc*, *cb*, *cc*.

In like manner four things, taken by twos, may be varied 4^2 or 16 ways; and 5 things, by twos, 5^2 or 25 ways; and, in general, *n* things, taken by twos, may be changed or varied n^2 different ways.

For the fame reafon, when taken by threes, the changes will be n^3; and when taken by fours, they will be n^4; and fo generally, when taken by *n*'s, the changes will be n^n.

Hence, then, adding all thefe together, the whole number of changes, or combinations in *n* things, taken both by 2's, by 3's, by 4's, &c, to *n*'s, will be the fum of the geometrical feries $n + n^2 + n^3 + n^4 \cdots n^n$, which fum is $= \frac{n^n - 1}{n - 1} \times n$.

For example, if the number of things *n* be 4; this gives $\frac{4^4 - 1}{4 - 1} \times 4 = \frac{255}{3} \times 4 = 340$.

And if *n* be 24, the number of letters in the alpha-bet; the theorem gives

$$\frac{24^{24} - 1}{24 - 1} \times 24 = \overline{24^{24} - 1} \times \frac{24}{23} =$$

1391724288887252999425128493402200. In fo ma-ny different ways, therefore, may the 24 letters of the alphabet be varied or combined among themfelves, or fo many different words may be made out of them.

COMBUST, or COMBUSTION, is faid of a planet

when

when it is in conjunction with the sun, or not distant from it above half their disc.

But according to Argol, a planet is Combust, or in Combustion, when it is within eight degrees and a half of the sun.

COMET, a heavenly body in the planetary region, appearing suddenly, and again disappearing; and during the time of its appearance moving in a proper, though very eccentric orbit, like a planet.

Comets are vulgarly called *Blazing Stars*, and have this to distinguish them from other stars, that they are usually attended with a long train of light, tending always opposite to the sun, and being of a fainter lustre the farther it is from the body of the comet. And hence arises a popular division of comets, into three kinds; viz. *bearded*, *tailed*, and *hairy* comets; though in reality, this division rather relates to the several circumstances of the same comet, than to the phenomena of several. Thus, when the comet is eastward of the sun, and moves from him, it is said to be *bearded*, because the light precedes it in the manner of a beard: When the comet is westward of the sun, and sets after him, it is said to be *tailed*, because the train of light follows it in the manner of a tail: And lastly, when the sun and comet are diametrically opposite, the earth being between them, the train is hid behind the body of the comet, excepting the extremities, which, being broader than the body of the comet, appear as it were around it, like a border of hair, or *coma*, from which it is called *hairy*, and a comet.

But there have been comets whose disc was as clear, round, and well defined, as that of Jupiter, without either tail, beard or coma.

Of the Nature of Comets.—Philosophers and Astronomers, of all ages, have been much divided in their opinions as to the nature of comets. Their strange appearance has in all ages been matter of terror to the vulgar, who have uniformly considered them as evil omens, and forerunners of war, pestilence, &c. Diodorus Siculus and Appollinus Myndius, in Seneca, inform us, that many of the Chaldeans held them to be lasting bodies, having stated revolutions as well as the planets, but in orbits vastly more extensive; on which account they are only visible while near the earth, but disappear again when they go into the higher regions. Others of them were of opinion, that the comets were only meteors raised very high in the air, which blaze for a while, and disappear when the matter of which they consist is consumed or dispersed.

Some of the Greeks, before Aristotle, supposed that a comet was a vast heap or assemblage of very small stars meeting together, by reason of the inequality of their motions, and so uniting into a visible mass, by the union of all their small lights; which must again disappear, as those stars separated, and each proceeded in its course. Pythagoras, however, accounted them a kind of planets or wandering stars, disappearing in the superior parts of their orbits, and becoming visible only in the lower parts of them.

But Aristotle held, that comets were only a kind of transient fires, or meteors, consisting of exhalations raised to the upper region of the air, and there set on fire; far below the course of the moon.

Seneca, who lived in the first century, and who had seen two or three comets himself, plainly intimates that he thought them above the moon; and argues strongly against those who supposed them to be meteors, or who held other absurd opinions concerning them; declaring his belief that they were not fires suddenly kindled, but the eternal productions of nature. He points out also the only way to come at a certainty on this subject, viz. by collecting a number of observations concerning their appearance, in order to discover whether they return periodically or not. " For this purpose, says he, one age is not sufficient; but the time will come when the nature of comets and their magnitudes will be demonstrated, and the routes they take, so different from the planets, explained. Posterity will then wonder, that the preceding ages should be ignorant of matters so plain and easy to be known."

For a long time this prediction of Seneca seemed not likely to be fulfilled; and Tycho Brahe was the first among the moderns, who restored the comets to their true rank in the creation; for after diligently observing the comet of 1577, and finding that it had no sensible diurnal parallax, he assigned it its true place in the planetary regions. See his book De Cometa, anni 1577.

Before this however, there were various opinions concerning them. In the dark and superstitious ages, comets were held to be forerunners of every kind of calamity, and it was supposed they had different degrees of malignity, according to the shape they assumed; from whence also they were differently denominated. Thus, it was said that some were bearded, some hairy; that some represented a beam, sword, or spear; others a target, &c; whereas modern astronomers acknowledge only one species of comets, and account for their different appearances from their different situations with respect to the sun and earth.

Kepler, in other respects a very great genius, indulged the most extravagant conjectures, not only concerning comets, but the whole system of nature in general. The planets he imagined were huge animals swimming round the sun; and the comets monstrous and uncommon animals generated in the celestial spaces.

A still more ridiculous opinion, if possible, was that of John Bodin, a learned Frenchman in the 16th century; who maintained that comets " are spirits, which having lived on the earth innumerable ages, and being at last arrived on the confines of death, celebrate their last triumph, or are recalled to the firmament like shining stars! This is followed by famine, plague, &c, because the cities and people destroy the governors and chiefs who appease the wrath of God."—Others again have denied even the existence of comets, and maintained that they were only false appearances, occasioned by the refraction or reflection of light.

Hevelius, from a great number of observations, proposed it as his opinion, that the comets, like the solar maculæ or spots, are formed or condensed out of the grosser exhalations of his body; in which he differs but little from the opinion of Kepler.

James Bernoulli, in his Systema Cometarum, imagined that comets were no other than the satellites of

some

fome very diftant planet, which was itfelf invifible to us on account of its diftance, as were alfo the fatellites unlefs when in a certain part of their orbits.

Des Cartes advances another opinion: He conjectures that comets are only ftars, formerly fixed, like the reft, in the heavens; but which becoming gradually covered with maculæ or fpots, and at length wholly deprived of their light, cannot keep their places, but are carried off by the vortices of the circumjacent ftars; and in proportion to their magnitude and folidity, moved in fuch a manner, as to be brought nearer the orb of Saturn; and thus coming within reach of the fun's light, rendered vifible.

But the vanity of all thefe hypothefes now abundantly appears from the obferved phenomena of comets, and from the doctrine of Newton, which is as follows:

The comets, he fays, are compact, folid, fixed, and durable bodies; in fact a kind of planets, which move in very oblique and eccentric orbits, every way with the greateft freedom; perfevering in their motions, even againft the courfe and direction of the planets: and their tail is a very thin and flender vapour, emitted by the head or nucleus of the comet, ignited or heated by the fun. This theory of the comets at once folves their principal phenomena, which are as below.

The Principal Phenomena of the Comets.

1. Firft then, thofe comets which move according to the order of the figns, do all, a little before they difappear, either advance flower than ufual, or elfe go retrograde, if the earth be between them and the fun; but more fwiftly, if the earth be placed in a contrary part. On the other hand, thofe which proceed contrary to the order of the figns, move more fwiftly than ufual, if the earth be between them and the fun; and more flowly, or elfe retrograde, when the earth is in a contrary part.—For fince this courfe is not among the fixed ftars, but among the planets; as the motion of the earth either confpires with them, or goes againft them; their appearance, with refpect to the earth, muft be changed; and, like the planets, they muft fometimes appear to move fwifter, fometimes flower, and fometimes retrograde.

2. So long as their velocity is increafed, they nearly move in great circles; but towards the end of their courfe, they deviate from thofe circles; and when the earth proceeds one way, they go the contrary way. Becaufe, in the end of their courfe, when they recede almoft directly from the fun, that part of the apparent motion which arifes from the parallax, muft bear a greater proportion to the whole apparent motion.

3. The comets move in ellipfes, having one of their foci in the centre of the fun; and by radii drawn to the fun, defcribe areas proportional to the times. Becaufe they do not wander precarioufly from one fictitious vortex to another; but, making a part of the folar fyftem, return perpetually, and run a conftant round. Hence, their elliptic orbits being very long and eccentric, they become invifible when in that part which is moft remote from the fun. And from the curvity of the paths of comets, Newton concludes, that when they difappear, they are much beyond the orbit of Jupiter; and that in their perihelion they frequently defcend within the orbits of Mars and the inferior planets.

4. The light of their nuclei, or bodies, increafes as they recede from the earth toward the fun; and on the contrary, it decreafes as they recede from the fun. Becaufe, as they are in the regions of the planets, their accefs towards the fun bears a confiderable proportion to their whole diftance.

5. Their tails appear the largeft and brighteft, immediately after their tranfit through the region of the fun, or after their perihelion. Becaufe then, their heads being the moft heated, will emit the moft vapours.—From the light of the nucleus we infer their vicinity to the earth, and that they are by no means in the region of the fixed ftars, as fome have imagined; fince, in that cafe, their heads would be no more illuminated by the fun, than the planets are by the fixed ftars.

6. The tails always incline from a juft oppofition to the fun towards thofe parts which the nuclei or bodies pafs over, in their progrefs through their orbits. Becaufe all fmoke, or vapour, emitted from a body in motion, tends upwards obliquely, ftill receding from that part towards which the fmoking body proceeds.

7. This declination, cæteris paribus, is the fmalleft when the nuclei approach neareft the fun: and it is alfo lefs near the nucleus, or head, than towards the extremity of the tail. Becaufe the vapour afcends more fwiftly near the head of the comet, than in the higher extremity of its tail; and alfo when the comet is nearer the fun, than when it is farther off.

8. The tails are fomewhat brighter, and more diftinctly defined in their convex, than in their concave part. Becaufe the vapour in the convex part, which goes firft, being fomewhat nearer and denfer, reflects the light more copioufly.

9. The tails always appear broader at their upper extremity, than near the centre of the comet. Becaufe the vapour in a free fpace continually rarefies and dilates.

10. The tails are always tranfparent, and the fmalleft ftars appear through them. Becaufe they confift of infinitely thin vapour.

The Phafes of Comets.—The nuclei, which are alfo called the heads, and bodies, of comets, viewed through a telefcope, fhew a face very different from thofe of the fixed ftars or planets. They are liable to apparent changes, which Newton afcribes to changes in the atmofphere of comets: and this opinion was confirmed by obfervations of the comet in 1744. Hift. Acad. Scienc. 1744. Sturmius fays that, obferving the comet of 1680 with a telefcope, it appeared like a coal dimly glowing, or a rude mafs of matter illuminated with a dufky fumid light, lefs fenfible to the extremes than in the middle; whereas a ftar appears with a round difc, and a vivid light.

Of the comet of 1661, Hevelius obferves, that its body was of a yellowifh colour, very bright and confpicuous, but without any glittering light: in the middle was a denfe ruddy nucleus, almoft equal to Jupiter, encompaffed by a much fainter, thinner matter. February 5th, its head was fomewhat larger and brighter, and of a gold colour; but its light more dufky than the ftars: and here the nucleus appeared divided into feveral parts. Feb. 6th, the difc was leffened; the parts of the nucleus ftill exifted, though less than before: one of them, on the lower part of the disc, on the left, much denser and brighter than the rest; its body round, and representing a very lucid little star: the nuclei still encompassed with another kind of matter. Feb. 10th, the head somewhat more obscure, and the nuclei more confused, but brighter at top than bottom. Feb. 13th, the head diminished much both in size and splendor. March 2d, its roundness a little impaired, and its edges lacerated, &c. March 28th, very pale, and exceeding thin; its matter much dispersed; and no distinct nucleus at all appearing.

Weigelius, who saw the comet of 1664, as also the moon, and a small cloud in the horizon illuminated by the sun at the same time, observed, that through the telescope the moon appeared of a continued luminous surface: but the comet very different; being exactly like the little cloud. And from these observations it was that Hevelius formed his opinion, that comets are like maculæ or spots formed out of the solar exhalations.

Of the Magnitude of Comets.—The estimates that have been given of the magnitude of comets by Tycho Brahe, Hevelius, and some others, are not very accurate; as it does not appear that they distinguished between the nucleus and the surrounding atmosphere. Thus Tycho computes that the true diameter of the comet in 1577 was in proportion to the diameter of the earth, as 3 is to 14; and Hevelius made the diameter of the comet of 1652 to that of the earth, as 52 to 100. But the diameter of the atmosphere is often 10 or 15 times as great as that of the nucleus: the former, in the comet of 1682, was measured by Flamsteed, and found to be 2′, when the diameter of the nucleus alone was only 11 or 12″. Though some comets, estimated by a comparison of their distance and apparent magnitude, have been judged much larger than the moon, and even equal to some of the primary planets. The diameter of that of 1744, when at the distance of the sun from us, measured about 1′, which makes its diameter about three times that of the earth: at another time the diameter of its nucleus was nearly equal to that of the planet Jupiter.

Of the Tails of Comets.—There have been various conjectures about the nature of the tails of comets, the principal of which are those of Newton, and the others that follow. Newton shews that the atmospheres of comets will furnish vapour sufficient to form their tails. This he argues from that wonderful rarefaction in our air at a distance from the earth; which is such, that a cubic inch of common air, expanded to the rarity of that at the distance of half the earth's diameter, or 4000 miles, would fill a space larger than the whole region of the stars. Since then the coma, or atmosphere of a comet, is 10 times higher than the surface of the nucleus, from the centre; the tail, ascending still much higher, must necessarily be immensely rare: so that it is no wonder the stars are visible through it.

Now the ascent of vapours into the tail of the comet, he supposes occasioned by the rarefaction of the matter of the atmosphere at the time of the perihelion. Smoke, it is observed, ascends the chimney by the impulse of the air in which it floats; and air, rarefied by heat, ascends by the diminution of its specific gravity carrying up the smoke along with it: in the same manner then it may be supposed that the tail of a comet is raised by the sun.

The tails therefore thus produced in the perihelion of comets, will go off along with their head into remote regions; and either return from thence, together with the comets, after a long series of years; or rather be there lost, and vanish by little and little, and the comet be left bare; till at its return, descending towards the sun, some short tails are again gradually produced from the head; which afterwards, in the perihelion, descending down into the sun's atmosphere, will be immensely increased.

Newton farther observes, that the vapours, when thus dilated, rarefied, and diffused through all the celestial regions, may probably, by means of their own gravity, be gradually attracted down to the planets, and become intermingled with their atmospheres. He adds that this intermixture may be useful and necessary for the conservation of the water and moisture of the planets, dried up or consumed in various ways. And I suspect, adds our author, that the spirit, which makes the finest, subtilest, and best part of our air, and which is absolutely requisite for the life and being of all things, comes principally from the comets.—On this principle there may seem to be some foundation for the popular opinion of presages from comets; since the tail of a comet thus intermingled with our atmosphere, may produce changes very sensible in animal and vegetable bodies.

It may here be added that another use which Newton conjectures comets may be designed to serve, is that of recruiting the sun with fresh fuel, and repairing the consumption of his light by the streams continually sent forth in every direction from that luminary. In support of this conjecture he observes, that comets in their perihelion may suffer a diminution of their projectile force, by the resistance of the solar atmosphere; so that by degrees their gravitation towards the sun may be so far increased, as to precipitate their fall into his body.

Other opinions on the tails of comets, are the following.

Apian, Tycho Brahe, and some others, think they were produced by the sun's rays transmitted through the nucleus of the comet, which they supposed was transparent, and there refracted as in a glass lens, so as to form a beam of light behind the comet. Des Cartes accounted for the phenomenon of the tail by the refraction of light from the head of the comet to the spectator's eye. Mairan supposes that the tails are formed out of the luminous matter composing the sun's atmosphere: and M. De la Lande combines this hypothesis with that of Newton recited above. But Mr. Rowning, not satisfied with Newton's opinion, accounts for the tails of comets in the following manner: It is well known, says he, that when the sun's light passes through the atmosphere of any body, as the earth, that which passes on one side, is by the refraction made to converge towards that which passes on the opposite side; and this convergency is not wholly effected either at the entrance of the light into the atmosphere, or at its exit on going out; but beginning at

2   its

its entrance, it increases in every point of its progress: It is also agreed that the atmospheres of the comets are very large and dense: he therefore supposes that by such time as the light of the sun has passed through a considerable part of the atmosphere of the comet, the rays are so far refracted towards each other, that they then begin sensibly to illuminate it, or rather the vapours floating in it; and so render that part they have yet to pass through, visible to us: and that this portion of the atmosphere of a comet thus illuminated, appears to us in form of a beam of the sun's light, and passes under the denomination of a comet's tail. Rowning's Nat. Philos. part 4. chap. 11.

M. Euler, Mem. Berlin tom. 2. pa. 117, thinks there is a great affinity between the tails of comets, the zodiacal light, and the aurora borealis, and that the common cause of all of them, is the action of the sun's light on the atmospheres of the comets, of the sun, and of the earth. He supposes that the impulse of the rays of light on the atmosphere of comets, may drive some of the finer particles of that atmosphere far beyond its limits; and that this force of impulse combined with that of gravity towards the comet, would produce a tail, which would always be in opposition to the sun, if the comet did not move. But the motion of the comet in its orbit, and about an axis, must vary the position and figure of the tail, giving it a curvature, and deviation from a line joining the centres of the sun and comet; and that this deviation will be greater, as the orbit of the comet has the greater curvature, and as the motion of the comet is more rapid. It may even happen, that the velocity of the comet, in its perihelion, may be so great, that the force of the sun's rays may produce a new tail, before the old one can follow; in which case the comet might have two or more tails. The possibility of this is confirmed by the comet of 1744, which was observed to have several tails while it was in its perihelion.

Dr. Hamilton urges several objections against the Newtonian hypothesis; and concludes that the tail of a comet is formed of matter which has not the power of refracting or reflecting the rays of light; but that it is a lucid or self-shining substance: and from its similarity to the Aurora borealis, that it is produced by the same cause, and is properly an electrical phenomenon. Dr. Halley too seemed inclined to this hypothesis, when he said, that the streams of light in an Aurora borealis so much resembled the long tails of comets, that at first sight they might well be taken for such: and that this light seems to have a greater affinity to that which the effluvia of electric bodies emit in the dark. Philos. Transf. N° 347. Hamilton's Philos. Essays, pa. 91.

The Motion of Comets.—If it be supposed that the paths of comets are perfectly parabolical, as some have imagined, it will follow that, being impelled towards the sun by a centripetal force, they descend as from spaces infinitely distant; and that by their falls they acquire such a velocity as will carry them off again into the remotest regions, never more to return. But the frequency of their appearance, and their degree of velocity, which does not exceed what they might acquire by their gravity towards the sun, seem to put it past doubt that they move like the planets, in elliptic orbits,

though exceedingly eccentric; and so return again after very long periods.

The apparent velocity of the comet of 1472, as observed by Regiomontanus, was such as to carry it through 40° of a great circle in 24 hours: and it was observed that the comet of 1770 moved through more than 45° in the last 25 hours.

About the return of comets there have been different opinions. Newton, Flamsteed, Halley, and other English astronomers, seem satisfied of the return of comet: Cassini and some of the French think it highly probable; but De la Hire and others oppose it. Those on the affirmative side suppose that the comets describe orbits prodigiously eccentric, insomuch that we can see them only in a very small part of their revolution: out of this, they are lost in the immensity of space; hid not only from our eyes, but our telescopes: that little part of their orbit next us passing sometimes within those of all the inferior planets.

M. Cassini gives the following reasons in favour of the return of comets. 1. It is found that they move a considerable time in the arch of a great circle, when referred to the fixed stars, that is a circle whose plane passes through the centre of the earth; deviating but a little from it chiefly towards the end of their appearance; a deviation however common to them with the planets.—2. Comets, as well as planets, appear to move so much the faster as they are nearer the earth; and when they are at equal distances from their perigee, their velocities are nearly the same. By subtracting from their motion the apparent inequality of velocity occasioned by their different distance from the earth, their equal motion might be found: but we should not still be certain that this is their true motion; because they might have considerable inequalities, not distinguishable in that small part of their orbit visible to us. It is rather probable that their real motion, as well as that of the planets, is unequal in itself; and hence we have a reason why the observations made during the appearance of a comet, cannot give the just period of their revolution.—3. There are no two different planets whose orbits cut the ecliptic in the same angle, whose nodes are in the same points of the ecliptic, and having the same apparent velocity in their perigee: consequently, two comets seen at different times, yet agreeing in all those three circumstances, can only be one and the same comet. Not that this exact agreement, in these circumstances, is absolutely necessary to determine their identity: for the moon herself is irregular in all of them, so that it seems there may be cases in which the same comet, at different periods of revolution, may disagree in these points.

As to the objections against the return of comets, the principal is that of the rarity of their appearance, with regard to the number of revolutions assigned to them. In 1702 there was a comet, or rather the tail of one, seen at Rome, which M. Cassini takes to be the same with that observed by Aristotle, and again lately in the year 1668; which would imply a period of 34 years: Now, it may seem strange that a star which has so short a revolution, and of consequence such frequent returns, should be so seldom seen. Again, in April of the same year 1702, a comet was observed by Messrs. Bianchini and Maraldi, which the latter supposed

posed

posed was the same with that of 1664, both on account of its motion, velocity, and direction. M. de la Hire thought it had some relation to another he had observed in 1698, which Cassini refers to that of 1652; which would make it a period of 43 months, and the number of revolutions between 1652 and 1692, 14: now, it is hard to suppose, that in this age, when the heavens are so narrowly watched, a star should make 14 revolutions unperceived; especially such a star as this, which might appear above a month together; and consequently be often disengaged from the crepuscula. For this reason M. Cassini is very reserved in maintaining the hypothesis of the return of comets, and only proposes those for planets where the motions are easy and simple, and are solved without straining, or allowing any irregularities.

M. de la Hire proposes one general difficulty against the whole system of the return of comets, which would seem to prevent any comet from returning as a planet: which is this; that by the disposition necessarily given to their courses, they ought to appear as small at first as at last; and always increase till they arrive at their nearest proximity to the earth; or if they should chance not to be observed, as soon as they are capable of being seen, it is yet hardly possible but they must often shew themselves before they have arrived at their full magnitude and brightness: but, adds he, none were ever yet observed till they had arrived at it. However, the appearance of a comet in the month of October 1723, while at a great distance, so as to be too small and dim to be viewed without a telescope, as well as the observations of several others since, may serve to remove this obstacle, and set the comets still on the same footing with the planets.

It is a conjecture of Newton, that as those planets which are nearest to the sun, and revolve in the least orbits, are the smallest; so among the comets, such as in their perihelion come nearest the sun, are the smallest, and revolve in the least orbits.

Of the Writings and Lists of Comets.

There have been many writings upon the subject of comets, beside the notices of historians as to the appearance of certain particular ones.

Regiomontanus first shewed how to find the magnitudes of comets, their distance from the earth, and their true place in the heavens. His 16 problems De Co-

metæ Magnitudine, Longitudine, ac Loto, are to be found in a book published in the year 1544, with the title of Scripta Joannis Regiomontani.

Peter Apian observed and wrote upon the comets of 1631, 1632, &c. Other writers are Tycho Brahe, in his Progymnasmata Astronomiæ Instauratæ.—Kepler, of the comet in the year 1607, and de Cometis Libelli tres.—Ricciolus, in his Almagestum Novum, published 1651, enumerates 154 comets cited by historians down to the year 1618.—Hevelius's Prodromus Cometicus, containing the history of the comet of the year 1664. Also his Cometographia.—Lubienietz, in a large folio work expresly on this subject, published 1667, extracts, with immense labour, from the passages of all historians, an account of 415 comets, ending with that of 1665.—Dr. Hook, in his Posthumous works.—M. Cassini's little Tract of Comets.—Sturmius's Dissertatio de Cometarum Natura.—Newton, in his Principia, lib. 3; who first assigned their proper orbits, and by calculations compared the observations of the great comet of 1680 with his theory.—Dr. Halley, his Synopsis Cometica, in the Philos. Transf. number 218, &c; who computed the elements and orbits of 24 comets, and who first ventured to predict the return of one in 1759, which happened accordingly.—De la Lande, Théorie des Comètes, 1759; also, in his Astronomie, vol. 3.—Clairaut, Théorie du mouvement des Comètes, 1760—D'Alembert Opuscules Mathématiques, vol. 2 pa. 97.—M. Albert Euler, 1762.—Séjour, Essai sur les Comètes, 1775.—Besides Boscovich, De la Grange, De la Place, Frisi, Lexel, Barker, Hancocks, Cole, with many others.—And M. Pingré's Cométographie, in 2 vols. 4to, 1784; in which is contained the most ample list of such comets as have been well observed, and their elements computed, to the number of 67. And accounts of a very few more that have been observed since that time, may be seen in the Mem. de l'Acad. and in the Philos. Transf.—And while this work is printing, there has just come out a very ingenious and ample work upon comets, by Sir Henry Englefield, entitled, " On the Determination of the Orbits of Comets."

The whole list of comets that have been noticed, on record, amount to upwards of 500, but the following is a complete list of all that have been properly observed, and their elements computed, the mean distance of the earth from the sun being 100000.

TABLE OF THE ELEMENTS OF COMETS.

Year	Ascending Node	Inclin. of Orbit	Perihelion	Perihel. Dist.	Time of Perihelion	Motion	Calculated by
	s o ′	o ′	s o ′		d h		
837	6 26 33	11 abt.	9 19 3	58000	March 1 0	Ret.	Pingré
1231	0 13 30	6 5	4 14 48	94776	Jan. 30 7	Dir.	Pingré
1264	5 19 0	36 30	9 21 0	44500	July 6 8	Dir.	Duntherne
	5 28 45	30 25	9 5 45	41081	17 6	Dir.	Pingré
1299	3 17 8	68 57	0 3 20	31793	March 31 8	Ret.	Pingré
1301	0 15 abou	70 abt.	9 about	45700	Oct. 22 about	Ret.	Pingré
1337	2 24 21	32 11	1 7 59	40666	June 2 7	Ret.	Halley
	2 6 22	32 11	0 20 0	64450	1 1	Ret.	Pingré
1456	1 18 30	17 56	10 1 0	58550	June 8 22	Ret.	Pingré
1472	9 11 46	5 20	1 15 34	54273	Feb. 28 23	Ret.	Halley

Year	Ascending Node	Inclin. of Orbit	Perihelion	Perihel. Dist.	Time of Perihelion	Motion	Calculated by
	s o '	o '	s o '		d h		
1531	1 19 25	17 56	10 1 39	56700	Aug. 24 21	Ret.	Halley
1532	2 20 27	32 36	3 21 7	50910	Oct. 19 22	Dir.	Halley
1533	4 5 44	35 49	4 27 16	20280	June 16 20	Ret.	Douwes
1556	5 25 42	32 7	9 8 50	66390	April 21 20	Dir.	Halley
1577	0 25 52	74 33	4 9 22	18342	Oct. 26 19	Ret.	Halley
1580	0 18 57	64 40	3 19 6	59628	Nov. 28 15	Dir.	Halley
	0 19 8	64 52	3 19 12	59553	28 14	Dir.	Pingré
1582	7 21 7	61 28	8 5 23	22570	May 6 16	Ret.	Pingré
	7 4 43	59 29	9 11 27	4006	7 9	Ret.	Pingré
				NEW STYLE			
1585	1 7 43	6 4	0 8 51	109358	Oct. 7 20	Dir.	Halley
1590	5 15 31	29 41	7 6 55	57661	Feb. 8 4	Ret.	Halley
1593	5 14 15	87 58	5 26 19	8911	July 18 14	Dir.	La Caille
1596	10 12 13	55 12	7 18 16	51293	Aug. 10 20	Ret.	Halley
	10 15 37	52 10	7 28 31	54942	8 16	Ret.	Pingré
1607	1 20 21	17 2	10 2 16	58680	Oct. 16 4	Ret.	Halley
1618	9 23 25	21 28	10 18 20	51298	Aug. 17 3	Dir.	Pingré
1618	2 16 1	37 34	0 2 14	37975	Nov. 8 13	Dir.	Halley
652	2 28 10	79 28	0 28 19	84750	Nov. 12 16	Dir.	Halley
1661	2 22 31	32 36	3 25 59	44851	Jan. 26 24	Dir.	Halley
1664	2. 21 14	21 19	4 10 41	102575	Dec. 4 12	Ret.	Halley
1665	7 18 2	76 5	2 11 55	10649	April 24 5	Ret.	Halley
1672	9 27 31	83 22	1 17 0	69739	March 1 9	Dir.	Halley
1677	7 26 49	79 3	4 17 37	28059	May 6 1	Ret.	Halley
1678	5 11 40	3 4	10 27 46	123801	Aug. 26 14	Dir.	Douwes
1680	9 2 2	60 56	8 22 40	612½	Dec. 18 0	Dir.	Halley
	9 2 2	61 7	8 22 44	617	17 23	. . .	Halley
	9 2 59	58 40	8 23 27	656	17 21	. . .	Euler
	9 1 53	61 20	8 23 43	592	18 0	. . .	Newton
	9 1 57	61 23	8 22 40	603	18 0	. . .	Pingre
1682	1 21 17	17 56	10 2 53	58328	Sept. 14 8	Ret.	Halley
	1 20 48	17 42	10 1 36	58250	14 22	. . .	Halley
1683	5 23 23	83 11	2 25 30	56020	July 13 3	Ret.	Halley
1684	8 28 15	65 49	7 28 52	96015	June 8 10.	Dir.	Halley
1686	11 20 35	31 22	2 17 1	32500	Sept. 16 15	Dir.	Halley
1689	10 23 45	69 17	8 23 45	1689	Dec. 1 15	Ret.	Pingré
1698	8 27 44	11 46	9 0 51	69129	Oct. 18 17	Ret.	Halley
1699	10 21 46	69 20	7 2 31	74400	Jan. 13 9	Ret.	La Caille
1702	6 9 25	4 30	4 18 41	64590	March 13 14	Dir.	La Caille
1706	0 13 12	55 14	2 12 29	42581	Jan. 30 .5	Dir.	La Caille
	0 13 11	55 14	2 12 36	42686	30 5	. . .	Struyck
1707	1 22 8	88 50	2 17 4	86350		Dir.	Houtteryn
	1 22 47	88 36	2 19 55	85974	Dec. 12 0	. . .	La Caille
	1 22 50	88 38	2 19 58	85904	12 0	. . .	Struyck
1718	4 8 43	30 20	4 1 30	102655	Jan. 15 0	Ret.	La Caille
	4 7 55	31 13	4 1 27	102565	15 1		Douwes
1723	0 14 16	49 59	1 12 52	99865	Sept. 27 16	Ret.	Bradley
1729	10 10 35	77 2	10 22 17	406980	June 23 7	Dir.	Douwes
	10 10 33	76 58	10 22 40	426140	25 11	. . .	La Caille
	10 10 17	76 53	10 27 22	416927	23 0	. . .	Maraldi
	10 10 52	77 19	10 16 27	394927	May 22 11	. . .	Kies
	10 10 33	77 1	10 22 37	408165	June 25 9	. . .	De l'Isle
1737	7 16 22	18 21	10 25 55	22282	Jan. 30 9	Dir.	Bradley
1739	6 27 18	55 53	3 12 34	67160	June 17 11	Ret.	Zanotti
	6 27 25	55 43	3 12 39	67358	17 10	. . .	La Caille
1742	6 5 35	67 4	7 7 33	76555	Feb. 8 5	Ret.	Struyck
	6 5 33		7 7 32	76550	8 4	. . .	Le Monnier
	6 5 38	66 59	7 7 35	76568	8 5	. . .	La Caille
	6 5 43	66 52	7 7 39	76530	8 8	. . .	Zanotti
	6 16 9	56 35	7 16 42	73766	1 22	. . .	Euler

Year	Ascending Node			Inclin. of Orbit		Perihelion			Perihel. Dist.	Time of Perihelion			Motion	Calculated by
	s	°	′	°	′	s	°	′			d	h		
1742	6	9	32	61	44	7	10	49	73668	Feb.	7	4	Ret.	Euler
	6	5	47	68	14	7	7	33	76890		7	22	..	Wrigt
	6	5	29	67	11	7	7	26	76620		8	5	...	Klinkenberg
	6	5	42	66	51	7	7	38	76545		8	7		Houtteryn
1743	2	8	11	2	16	3	2	58	83811	Jan.	10	21	Dir.	Struyck
	2	8	21	2	20	3	2	42	83501		10	21	...	La Caille
1743	0	5	16	45	48	8	6	34	52057	Sept.	20	21	Ret.	Klinkenberg
1744	1	15	45	47	9	6	17	13	22206	March	1	8	Dir.	Bets and Blifs
	1	15	47	47	4	6	17	6	22322		1	8	...	Maraldi
1744	1	15	46	47	5	6	17	10	22250	March	1	8	Dir.	La Caille
	1	15	51	47	18	6	17	18	22156		1	8	...	Zanotti
	1	16	5	47	50	6	17	19	22192		1	9	...	Chéfeaux
	1	15	46	47	11	6	17	12	22222		1	8	...	Euler
	1	15	48	47	8	6	17	13	22223		1	8	...	Pingré
	1	15	49	47	18	6	17	15	22200		1	8	...	Klinkenberg
	1	15	49	47	14	6	17	16	22176		1	9	...	Hiorter
1747	4	26	58	77	57	9	10	6	229388	Feb.	28	12	Ret.	Chefeaux
	4	27	19	79	7	9	7	2	219859	March	3	10	...	Maraldi
	4	27	19	79	6	9	7	2	219851		3	7	...	La Caille
1748	7	22	52	85	27	7	5	1	84066	April	28	20	Ret.	Maraldi
	7	22	46	85	35	7	4	39	84150		29	1	...	Le Monnier
	7	22	52	85	28	7	5	23	84040		28	19	...	La Caille
1748	1	4	40	56	59	9	6	9	65525	June	18	2	Dir.	Struyck
1757	7	4	13	12	50	4	2	58	33754	Oct.	21	8	Dir.	Bradley
	7	4	6	12	39	4	2	39	33907		21	10	...	La Caille
	7	4	4	12	48	4	2	49	33797		21	10	...	Pingré
	7	4	7	12	41	4	2	36	33932		21	9	...	De Ratte
1758	7	20	50	68	19	8	27	38	21535	June	11	3	Dir.	Pingré
1759	1	23	48	17	38	10	3	14	58255	March	12	14	Ret.	Meffier
	1	23	46	17	40	10	3	8	58490		12	14	...	De la Lande
	1	23	49	17	35	10	3	16	58360		12	13	...	Maraldi
	1	23	49	17	38	10	3	16	58380		12	13$\frac{1}{2}$	...	La Caille
	1	23	49	17	39	10	3	16	58350		12	13$\frac{2}{3}$	...	La Caille
	1	23	46	17	40	10	3	19	58298		12	13	...	Klinkenberg
	1	24	7	17	29	10	1	0	59708		13	10	...	Klinkenberg
	1	23	45	17	41	10	3	23	58234		12	13	...	Bailly
1759	4	19	40	79	7	1	23	34	80139	Nov.	27	0	Dir.	Pingré
	4	19	39	78	59	1	23	24	79851		27	2	...	La Caille
	4	19	40	79	3	1	23	38	80208		27	1	...	Chappe
1759	2	19	51	4	52	4	18	25	96599	Dec.	16	21	Ret.	La Caille
1759	2	19	20	4	42	4	19	4	96180	Dec.	16	13	Ret.	Chappe
1762	11	18	56	85	22	3	15	22	101415	May	29	0	Dir.	Maraldi
	11	19	20	84	45	3	15	15	101249		28	15	...	De la Lande
	11	18	58	85	12	3	14	24	101065		29	2	...	Bailly
	11	18	35	85	40	3	13	43	100686		28	2	...	Klinkenberg
	11	19	2	85	3	3	14	30	100986		28	7		Struyck
1763	11	26	23	72	41	2	24	52	49876	Nov.	1	20	Dir.	Pingré
1764	4	0	5	52	54	0	15	15	55522	Feb.	12	14	Ret.	Pingré
1766	8	4	11	40	50	4	23	15	50532	Feb.	17	9	Ret.	Pingré
1766	2	14	23	11	8	8	2	18	33275	April	22	21	Dir.	Pingré
1769	5	25	1	40	38	4	24	6	12376	Oct.	7	13	Dir.	De la Lande
	5	25	2	40	43	4	24	14	12287		7	12	...	Wallot
	5	25	3	40	47	4	24	11	12258		7	13	...	Caffin, jun.
	5	25	7	40	49	4	24	11	12272		7	14	...	Profperin
	5	25	3	40	41	4	24	9	12289		7	14	...	Audiffrédi
	5	25	11	41	1	4	24	33	12100		7	12	...	Slop
	5	19	41	29	41	4	13	15	15880		16	10	...	Zanotti
	5	25	5	40	41	4	24	7	12308		7	14	...	Afclépi
	5	24	42	41	28	4	25	46	11640		7	11	...	Lambert
	5	25	14	40	43	4	24	22	12280		7	18	...	Widder

Year	Ascending Node	Inclin. of Orbit	Perihelion	Perihel. Dist.	Time of Perihelion		Motion	Calculated by
	s o '	o '	s o '			d h		
1769	5 25 3	40 50	4 24 16	12264	Oct.	7 15	Dir.	Euler
	5 25 5	40 50	4 24 11	12269		7 16	. . .	Lexell
	5 25 6	40 47	4 24 16	12274		7 16	. . .	Pingré
1770	4 16 39	1 44	11 26 7	62959	Aug.	9 0	Dir.	Pingré
	4 13 39	1 41	11 25 5	65800		10 22	. . .	Pingré
	4 15 29	1 47	11 26 7	62955		9 0	. . .	Prosperin
	4 15 4	1 45	11 22 51	64456		8 9	. . .	Prosperin
	4 14 30	1 23	0 7 14	71717		25 2	. . .	Prosperin
	4 12 56	1 46	11 29 45	64946		12 21	. . .	Widder
1770	4 12 0	1 34	11 26 16	67438	Aug.	13 13	Dir.	Lexell
	4 12 17	1 35	11 26 26	67689		14 0	. . .	Pingré
	4 16 14	1 45	11 26 13	62872		9 1	. . .	Slop
	4 12 0	1 55	11 25 57	63100		9 4	. . .	Lambert
	4 14 22	1 49	11 26 19	62758		8 19	. . .	Rittenhouse
1770	3 18 42	31 26	6 28 23	52824	Nov.	22 6	Ret.	Pingré
1771	0 27 51	11 15	3 13 28	90576	April	18 22	Dir.	Pingré
	0 27 50	11 17	3 13 48	90188		19 1	. . .	Prosperin
1772	8 12 43	19 0	3 18 6	101814	Feb.	18 21	Dir.	La Lande
1773	4 1 16	61 25	2 15 36	113390	Sept.	5 11	Dir.	Pingré
	4 3 15	62 33	2 21 40	123800		2 12	. . .	Lambert
	4 3 35	62 36	2 20 43	121550		2 19	. . .	Schultz
	4 1 12	61 21	2 15 16	113010		5 6	. . .	Lexell
	4 1 5	61 13	2 14 58	112530		5 9	. . .	Pingré
1774	6 0 57	82 48	10 16 28	142525	Aug.	14 4	Dir.	De Saron
	6 0 50	82 49	10 16 48	142525		14 18	. . .	De Saron
	6 1 22	82 21	10 17 26	142600		15 5	. . .	Boscowich
	6 0 50	83 0	10 17 22	142860		15 11	. . .	Mechain
1779	0 25 3	32 26	2 27 14	71322	Jan.	4 3	Dir.	De Saron
	0 25 6	32 24	2 27 13	71313		4 2	. . .	Mechain
	0 25 4	32 26	2 27 14	71319		4 2	. . .	D'Angos
1780	4 4 0	53 56	8 6 30	9781	Sept.	30 20	Ret.	Lexell
	4 4 30	53 15	8 6 19	10047		30 16	. . .	Lexell
	4 4 9	53 48	8 6 21	9926		30 18	. . .	Mechain
1781	2 25 13	5 16	5 22 17	1027558	March	13 0	Dir.	Boscowich
			5 28 13	944040	Jan.	27 6	. . .	La Place
1781	2 23 1	81 43	7 29 11	77586	July	7 5	Dir.	Mechain
1781	2 17 23	27 13	0 16 3	96101	Nov.	29 13	Ret.	Mechain

M. Facio has suggested, that some of the comets have their nodes so very near the annual orbit of the earth, that if the earth should happen to be found in that part next the node at the time of a comet's passing by; as the apparent motion of the comet will be immensely swift, so its parallax will become very sensible; and its proportion to that of the sun will be given: whence, such transits of comets will afford the best means of determining the distance between the earth and sun.

The comet of 1472, for instance, had a parallax above 20 times greater than the sun's: and if that of 1618 had come down in the beginning of March to its descending node, it would have been much nearer the earth, and its parallax much more notable. But hitherto none has threatened the earth with a nearer appulse than that of 1680: for, Dr. Halley finds, by calculation, that Nov. 11th, at 1 h. 6 min. afternoon, that comet was not more than one semidiameter of the earth to the northward of the earth's path; at which time had the earth been in that part of its orbit, the comet would have had a parallax equal to that of the moon.—

What might have been the consequence of so near an appulse, a contact, or lastly, a shock of these bodies? Mr. Whilton says, a deluge!

To determine the Place and Course of a Comet.—Observe the distance of the comet from two fixed stars, whose longitudes and latitudes are known: then from the distances thus known, calculate the place of the comet by spherical trigonometry.

Longomontanus shews an easy method of finding and tracing out the places of a comet mechanically, which is, to find two stars in the same line with the comet, by stretching a thread before the eye over all the three; then do the same by two other stars and the comet: this done, take a celestial globe, or a planisphere, and draw a line upon it first through the former two stars, and then through the latter two; so shall the intersection of the two lines be the place of the comet at that time. If this be repeated from time to time, and all the points of intersection connected, it will shew the path of the comet in the heavens.

COMETARIUM, a machine adapted to give a representation of the revolution of a comet about the sun.

It

It is so contrived as, by elliptical wheels, to shew the unequal motion of a comet in every part of its orbit. The comet is represented by a small brass ball, carried by a radius vector, or wire, in an elliptical groove about the sun in one of its foci; and the years of its period are shewn by an index moving with an equable motion over a graduated silvered circle. See a particular description, with a figure of it, in Ferguson's Astron. 8vo. pa. 400.

COMMANDINE (FREDERICK), a celebrated mathematician and linguist, was born at Urbino in Italy, in 1509; and died in 1575; consequently at 66 years of age. He was famous for his learning and knowledge in the sciences. To a great depth, and just taste in the mathematics, he joined a critical skill in the Greek language; a happy conjunction which made him very well qualified for translating and expounding the writings of the Greek mathematicians. And accordingly, with a most laudable zeal and industry, he translated and published several of their works, to which no former writer had done that good office. On which account, Francis Moria, duke of Urbino, who was very conversant in those sciences, proved a very affectionate patron to him. He is greatly applauded by Bianchanus, and other writers; and he justly deserved their encomiums.

Of his own works Commandine published the following:

1. Commentarius in Planisphærium Ptolomæi: 1558, in 4to.

2. De Centro Gravitatis Solidorum: Bonon. 1565, in 4to.

3. Horologiorum Descriptio: Pom. 1562, in 4to.

He translated and illustrated with notes the following works, most of them beautifully printed, in 4to. by the celebrated printer Aldus:

1. Archimedis Circuli Dimensio; de Lineis Spiralibus; Quadratura Parabolæ; de Conoidibus & Sphæroidibus; de Arenæ Numero: 1558.

2. Ptolomæi Planisphærium; & Planisphærium Jordani: 1558.

3. Ptolomæi Analemma: 1562.

4. Archimedis de iis quæ vehuntur in aqua: 1565.

5. Apollonii Pergæi Conicorum libri quatuor, una cum Pappi Alexandrini Lemmatibus, & Commentariis Eutocii Ascalonitæ, &c: 1566.

6. Machometes Bagdadinus de Superficierum Divisionibus: 1570.

7. Elementa Euclidis: 1572.

8. Aristarchus de Magnitudinibus & Distantiis Solis & Lunæ: 1572.

9. Heronis Alexandrini Spiritualium liber: 1583.

10. Pappi Alexandrini Collectiones Mathematicæ: 1588.

COMMANDING Ground, in Fortification, an eminence, or rising ground, overlooking any post or strong place. This is of three sorts. 1st, A Front Commanding Ground, or a height opposite to the face of the post, which plays upon its front. 2dly, A Reverse Commanding Ground, or an eminence that can play upon the rear or back of the post. 3dly, An Enfilade Commanding Ground, or an eminence in flank which can, with its shot, scour all the length of a straight line.

COMMENSURABLE Quantities, or Magnitudes, are such as have some common aliquot part, or which may be measured or divided, without a remainder, by one and the same measure or divisor, called their common measure. Thus, a foot and a yard are commensurable, because there is a third quantity that can measure each, viz an inch; which is 12 times contained in the foot, and 36 times in the yard.—Commensurables are to each other, as one rational whole number is to another; but incommensurables are not so: And therefore the ratio of commensurables is rational; but that of incommensurables, irrational: hence also the exponent of the ratio of commensurables, is a rational number.

COMMENSURABLE Numbers, whether integers, or fractions, or surds, are such as have some other number, which will measure or divide them exactly, or without a remainder. Thus, 6 and 8 are commensurable, because 2 measures or divides them both. And $\frac{2}{3}$ and $\frac{3}{4}$, or $\frac{8}{12}$ and $\frac{9}{12}$ are commensurable fractions, because the fraction $\frac{1}{2}$, or $\frac{1}{12}$, &c, will measure them both: and in this sense, all fractions may be said to be commensurable. Also, the surds $2\sqrt{2}$ and $3\sqrt{2}$ are commensurable, being measured by $\sqrt{2}$, or being to each other as 2 to 3.

COMMENSURABLE in Power. Euclid says, right lines are commensurable in power, when their squares are measured by one and the same space or superficies.

COMMON, is applied to an angle, line, measure, or the like, that belongs to two or more figures, or other things. As, a common angle, a common side, a common base, a common measure, &c.

COMMON MEASURE, or divisor, is that which measures two or more things without a remainder. So of 8 and 12, a common measure is 2, and so is 4.

The greatest COMMON Measure, is the greatest number that can measure two other numbers. So, of 8 and 12, the greatest common measure is 4.

To find the greatest common measure of two numbers. Divide the greater term by the less; then divide the divisor by the remainder, if there be any; and so on continually, always dividing the last divisor by the last remainder, till nothing remains; and then is the last divisor the greatest common measure sought.

Thus, to find the greatest common measure of 816 and 1488.

$$816)1488(1$$
$$816$$

$$672)816(1$$
$$672$$

$$144)672(4$$
$$576$$

$$96)144(1$$
$$96$$

the common meas. $48)96(2$
$$96$$

$$0$$

Therefore

Plate VI

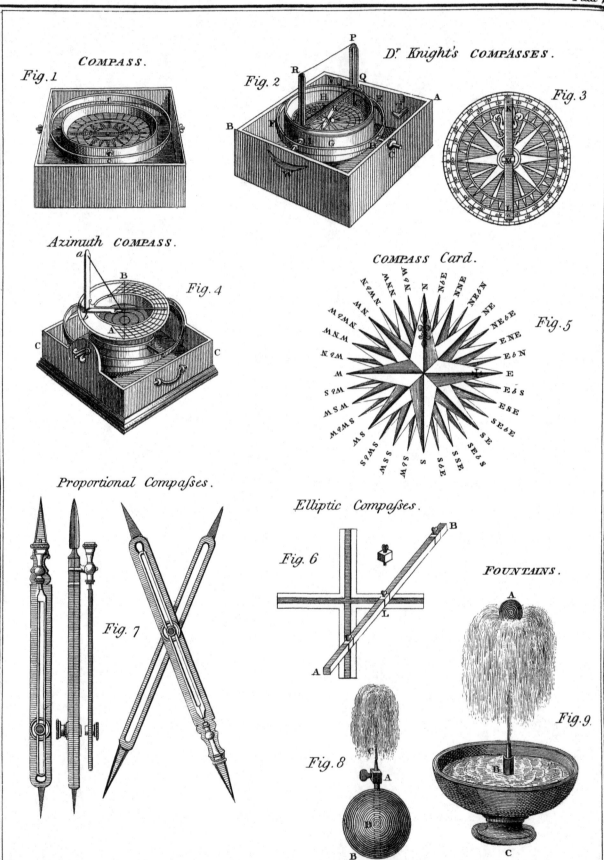

COMPASS.

Fig. 1

Dr. Knight's COMPASSES.

Fig. 2

Fig. 3

Azimuth COMPASS.

Fig. 4

COMPASS Card.

Fig. 5

Proportional Compasses.

Fig. 7

Elliptic Compasses.

Fig. 6

FOUNTAINS.

Fig. 8

Fig. 9

Therefore 48 is the greateſt common meaſure of 816 and 1488, thus:

$$48) 816 (17 \qquad 48) 1488 (31$$
$$\quad \underline{48} \qquad\qquad\qquad \underline{144}$$
$$\quad 336 \qquad\qquad\qquad 48$$
$$\quad \underline{336} \qquad\qquad\qquad \underline{48}$$

The common meaſure is uſeful in fractions, to reduce a fraction to its leaſt terms, by dividing thoſe that are given by their greateſt common meaſure. So $\frac{816}{1488}$ reduces to $\frac{17}{31}$, by dividing 816 and 1488 both by their greateſt common meaſure 48.

COMMUNICATION *of Motion*, that act of a moving body, by which it gives motion, or transfers its motion to another body.

Father Mallebranche conſiders the communication of motion, as ſomething metaphyſical; that is, as not neceſſarily ariſing from any phyſical principles, or any properties of bodies, but flowing from the immediate agency of God.

The communication of motion reſults from, and is an evidence of the impenetrability and inertia of matter, as ſuch; unleſs we admit the hypotheſis of the penetrability of matter, advanced by Boſcovich and Michell, and aſcribe to the power of repulſion thoſe effects which have been uſually aſcribed to its ſolidity and actual reſiſtance.

Newton ſhews that action and reaction are equal and oppoſite; ſo that one body ſtriking or acting againſt another and thence cauſing a change in its motion, does itſelf undergo the very ſame change in its own motion, the contrary way. And hence, a moving body ſtriking directly another at reſt, it loſes juſt as much of its motion as it communicates to the other. For the laws and quantity of motion ſo communicated, either in elaſtic or noneſaſtic bodies, ſee COLLISION.

COMMUTATION, *Angle of*, is the diſtance between the ſun's true place ſeen from the earth, and the place of a planet reduced to the ecliptic: which therefore is found by taking the difference between the ſun's longitude and the heliocentric longitude of the planet.

COMPANY, *Rule of*, or *Rule of Fellowſhip*, in Arithmetic, is a rule by which are determined the true ſhares of profit or loſs, due to the ſeveral partners, or aſſociates, in any enterprize, or trade, in due proportion to the ſtock contributed by each, and the time it was employed. To do which properly, ſee the *Rule of Fellowſhip*.

COMPARTMENT, a deſign compoſed of ſeveral different figures, diſpoſed with ſymmetry; to adorn a parterre, a cieling, pannel of joinery, or the like.

COMPARTITION, the uſeful and graceful diſtribution of the whole ground-plot of an edifice, into rooms of office, and of reception, or entertainment.

COMPASS, or *Mariner's Compaſs*, is an inſtrument uſed at ſea by mariners to direct and aſcertain the courſe of their ſhips. It conſiſts of a circular braſs box, which contains a paper card with the 32 points of the compaſs, or winds, fixed on a magnetic needle that always turns to the north, excepting a ſmall deviation, which is variable at different places, and at the ſame

place at different times. See VARIATION *of the Compaſs*.

The needle with the card turns on an upright pin fixed in the centre of the box. To the middle of the needle is fixed a braſs conical ſocket or cap, by which the card hanging on the pin turns freely round the centre.

The top of the box is covered with a glaſs, to prevent the wind from diſturbing the motion of the card. The whole is incloſed in another box of wood, where it is ſuſpended by braſs hoops or gimbals, to keep the card in a horizontal poſition during the motions of the ſhip. The whole is to be ſo placed in the ſhip, that the middle ſection of the box, parallel to its ſides, may be parallel to the middle ſection of the ſhip along its keel.

The invention of the compaſs is uſually aſcribed to Flavio Gioia, or Flavio of Malphi, about the year 1302; and hence it is that the territory of Principato, the part of the kingdom of Naples where he was born, has a compaſs for its arms. He divided his compaſs only into 8 points. Others aſcribe the invention to the Chineſe; and Gilbert, in his book *de Magnete*, affirms that Marcus Paulus, a Venetian, making a journey to China, brought back the invention with him in 1260. What ſtrengthens this conjecture is, that at firſt they uſed the compaſs, in the ſame manner as the Chineſe ſtill do, viz, letting it float on a ſmall piece of cork, inſtead of ſuſpending it on a pivot. It is added, that their emperor Chiningus, a celebrated aſtrologer, had a knowledge of it 1120 years before Chriſt. The Chineſe divide their compaſs only into 24 points. But Ludi Vertomanus affirms, that when he was in the Eaſt-Indies, about the year 1500, he ſaw a pilot of a ſhip direct his courſe by a compaſs, faſtened and framed as thoſe now commonly uſed. And Barlow, in his book called The Navigator's Supply, anno 1597, ſays, that in a perſonal conference with two Eaſt-Indians, they affirmed, that inſtead of our compaſs, they uſe a magnetical needle of 6 inches, and longer, upon a pin in a diſh of white China earth, filled with water; in the bottom of which they have two croſs lines for the 4 principal winds, the reſt of the diviſions being left to the ſkill of their pilots. Alſo in the ſame book he ſays that the Portugueſe, in their firſt diſcovery of the Eaſt-Indies, got a pilot of Mahinde, who brought them from thence in 33 days, within ſight of Calicut.

But Fauchette relates ſome verſes of Guoyot de Provence, who lived in France about the year 1200, which ſeem to make mention of the compaſs under the name of *marinette*, or *mariner's ſtone*; which ſhew it was uſed in France near 100 years before either the Malfite or Venetian one. The French even lay claim to the invention, from the fleur de lys with which moſt people diſtinguiſh the north point of the card. With as much reaſon Dr. Wallis aſcribes it to the Engliſh, from its name *compaſs*, by which name moſt nations call it, and which he obſerves is uſed in many parts of England to ſignify a circle.

The mariner's compaſs was long very rude and imperfect, but at length received great improvement from the invention and experiments of Dr. Knight, who diſcovered the uſeful practice of making artificial magnets; and the farther emendations of Mr. Smeaton, and

Mr.

Mr. M'Culloch, by which the needles are larger and stronger than formerly, and instead of swinging in gimbals, the compass is supported in its very centre upon a prop, and the centres of motion, gravity, and magnetism are brought almost all to the same point.

After the discovery of that most useful property of the magnet, or loadstone, viz, its giving a polarity to hardened iron or steel, the compass was many years in use before it was known in anywise to deviate from the poles of the world. About the middle of the 16th century, so confident were some persons that the needle invariably pointed due north, that they treated with contempt the notion of the variation, which about that time began to be suspected. However, careful observations soon discovered that in England, and its neighbourhood, the needle pointed to the eastward of the true north line; and the quantity of this deviation being known, mariners became as well satisfied as if the compass had none; because the true course could be obtained by making allowance for the true variation.

From succeeding observations it was afterwards found, that the deviation of the needle from the north was not a constant quantity, but that it gradually diminished, and at last, namely, about the year 1657, it was found that the needle pointed due north at London, and has ever since been going to the westward, till now the variation is upwards of two points of the compass: indeed it was 22° 41 about the middle of the year 1781, as appears by the Philos. Transf. pa. 225, for that year, and is probably now somewhat more, which it would be of consequence to know; but why such useful observations and experiments, as those of the variation and dip of the magnetic needle, have been so long discontinued, to the prejudice of science, is best known to the learned President of that Society, and his Council. So that in any one place it may be suspected the variation has a kind of libratory motion, traversing through the north, to unknown limits eastward and westward. But the settling of this point must be left to time and future experiments. See *Variation*, also *Inclination*, and *Dip*. Also for a farther description of different compasses and their uses, see that useful book, Robertson's Navigation, vol. 2. p. 231.

The Azimuth COMPASS differs from the common sea compass in this; that the circumference of the card or box is divided into degrees; and there is fitted to the box an index with two sights, which are upright pieces of brass, placed diametrically opposite to each other, having a slit down the middle of them, through which the sun or star is to be viewed at the time of observation.

The Use of the Azimuth Compass, is to take the bearing of any celestial object, when it is in, or above the horizon, that from the magnetical azimuth or amplitude, the variation of the needle may be known. See *Azimuth*, and *Amplitude*.

The figure of the compass card, with the names of the 32 points or winds, are as in fig. 5, plate vi; where other compasses are also exhibited.

As there are 32 whole points quite around the circle, which contains 360 degrees, therefore each point of the compass contains the 32d part of 360, that is 11¼ degrees, or 11° 15'; consequently the half point is 5° 37' 30", and the quarter point 2° 48' 45".

The points of the compass are otherwise called Rhumbs; and the numbers of degrees, minutes and seconds made by every quarter point with the meridian, are exhibited in the following table.

A TABLE of Rhumbs, shewing the Degrees, Minutes, and Seconds, that every Point and Quarter-point of the Compass makes with the Meridian.

North		Pts.qr.	°	'	"	Pts.qr.	South	
		0 1	2	48	45	0 1		
		0 2	5	37	30	0 2		
		0 3	8	26	15	0 3		
N b E	N b W	1 0	11	15	0	1 0	S b E	S b W
		1 1	14	3	45	1 1		
		1 2	16	52	30	1 2		
		1 3	19	41	15	1 3		
N N E	N N W	2 0	22	30	0	2 0	S S E	S S W
		2 1	25	18	45	2 1		
		2 2	28	7	30	2 2		
		2 3	30	56	15	2 3		
NE b N	NW b N	3 0	33	45	0	3 0	SE b S	SW b S
		3 1	36	33	45	3 1		
		3 2	39	22	30	3 2		
		3 3	42	11	15	3 3		
N E	N W	4 0	45	0	0	4 0	S E	S W
		4 1	47	48	45	4 1		
		4 2	50	37	30	4 2		
		4 3	53	26	15	4 3		
NE b E	NW b W	5 0	56	15	0	5 0	SE b E	SW b W
		5 1	59	3	45	5 1		
		5 2	61	52	30	5 2		
		5 3	64	41	15	5 3		
E N E	W N W	6 0	67	30	0	6 0	E S E	W S W
		6 1	70	18	45	6 1		
		6 2	73	7	30	6 2		
		6 3	75	56	15	6 3		
E b N	W b N	7 0	78	45	0	7 0	E b S	W b S
		7 1	81	33	45	7 1		
		7 2	84	22	30	7 2		
		7 3	87	11	15	7 3		
East	West	8 0	90	0	0	8 0	East	West

COMPASS *Dials*, are small dials, fitted in boxes, for the pocket, to shew the hour of the day by the direction of the needle that indicates how to place them right, by turning the dial about till the cock or style stand directly over the needle. But these can never be very exact, because of the variation of the needle itself; unless that variation be allowed for, in making and placing the instrument.

COMPASSES, or *Pair of* COMPASSES, a mathematical instrument for describing circles, measuring and dividing lines, or figures, &c.

The common compasses consist of two sharp-pointed branches or legs, of iron, steel, brass, or other metal, joined together at the top by a rivet, about which they move as on a centre. Those compasses are of the best sort in which the pin or axle, on which the joint turns, is made of steel, and also half the joint itself, as the opposite metals wear more equally: the points should also be made of hard steel, well polished; and the joint should open and shut with a smooth, easy, and uniform motion. In some compasses, the points are both fixed; but in others, one is made to take out occasionally, and a drawing-pen, or pencil, put in its place.

There are in use compasses of various kinds and contrivances, adapted to the various purposes they are intended for; as,

COMPASSES *of three Legs*, or Triangular Compasses; the construction of which is like that of the common compasses, with the addition of a third leg or point, which

which has a motion every way. Their use is to take three points at once, and so to form triangles, and lay down three positions of a map to be copied at once.

Beam COMPASSES consist of a long straight beam or bar, carrying two brass cursors; one of these being fixed at one end, the other sliding along the beam, with a screw to fasten it on occasionally. To the cursors may be screwed points of any kind, whether steel, pencils, or the like. To the fixed cursor is sometimes applied an adjusting or micrometer screw, by which an extent is obtained to very great nicety. The beam compasses are used to draw large circles, to take great extents, or the like.

Bow COMPASSES, or *Bows*, are a small sort of compasses, that shut up in a hoop, which serves for a handle. Their use is to describe arcs or circumferences with a very small radius.

Caliber COMPASSES. See CALIBER.

Clockmakers COMPASSES are jointed like the common compasses, with a quadrant or bow, like the spring compasses; only of different use, serving here to keep the instrument firm at any opening. They are made very strong, with the points of their legs of well-tempered steel, as being used to draw or cut lines in pasteboard, or copper, &c.

Cylindrical and Spherical COMPASSES, consist of four branches, joined in a centre, two of them being circular and two flat, a little bent at the ends. The use of them is to take the diameter, thickness, or caliber of round or cylindrical bodies; as cannons, balls, pipes, &c.

There are also spherical compasses, differing in nothing from the common ones, but that their legs are arched; serving to take the diameters of round bodies.

There is also another sort of compasses lately invented, for measuring the diameter of round bodies, as balls, &c. which consist of two flat pieces of metal set at right angles on a straight bar or beam of the same; the one piece being fixed, and the other sliding along it, so far as just to receive the round body between them; and then its diameter, or distance between the two pieces, is shewn by the divisions marked on the beam.

Elliptical COMPASSES, are used to draw ellipses or ovals of any kind. The instrument consists of a beam AB (Plate vi. fig. 6.) about a foot long, bearing three cursors; to one of which may be screwed points of any kind; and to the bottom of the other two are rivetted two sliding dove-tails, adjusted in grooves made in the cross branches of the beam. The dove-tails having a motion every way, by turning about the long branch, they go backward and forward along the cross; so that when the beam has gone half way round, one of these will have moved the whole length of one of the branches; and when the beam has gone quite round, the same dove-tail has gone back the whole length of the branch. Understand the same of the other dove-tail.

Note, the distance between the two sliding dove-tails, is the distance between the two foci of the ellipse; so that by changing that distance, the ellipse will be rounder or flatter. Under the ends of the branches of the cross, are placed four steel points to keep it fast.

The use of this compass is easy: by turning round the long branch, the pen, pencil, or other points will draw the ellipse required.

Its figure shews both its use and construction.

German COMPASSES have their legs a little bent outwards, near the top; so that when shut, the points only meet.

Hair COMPASSES are so contrived within side by a small adjusting screw to one of the legs, as to take an extent to a hair's breadth, or great exactness

Proportional COMPASSES are those whose joint lies, not at the end of the legs, but between the points terminating each leg. These are either simple, or compound. In the former sort the centre, or place of the joint is fixed; so that one pair of these serves only for one proportion.

Compound Proportional COMPASSES have the joint or centre moveable. They consist of two parts or sides of brass, which lie upon each other so nicely as to seem but one when they are shut. These sides easily open, and move about the centre, which is itself moveable in a hollow canal cut through the greatest part of their length. To this centre on each side is fixed a sliding piece, of a small length, with a fine line drawn on it serving as an index, to be set against other lines or divisions placed upon the compasses on both sides. These lines are, 1, A line of lines; 2, a line of superficies, areas, or planes, the numbers on which answer to the squares of those on the line of lines; 3, a line of solids, the numbers on which answer to the cubes of those on the line of lines; 4, a line of circles, or rather of polygons to be inscribed in circles. These lines are all unequally divided, the first three from 1 to 20, and the last from 6 to 20. The use of the first is to divide a line into any number of equal parts; by the 2d and 3d are found the sides of like planes or solids in any given proportion; and by the 4th, circles are divided into any number of equal parts, or any polygons inscribed in them. See Plate vi. fig. 7.

Spring COMPASSES, or *Dividers*, are made of hardened steel, with an arched head, which by its spring opens the legs; the opening being directed by a circular screw fastened to one of the legs, let through the other, and worked with a nut.

Trisecting COMPASSES, for the trisecting of angles geometrically, for which purpose they were invented by M. Tarragon.

The instrument consists of two central rules, and an arch of a circle of 120 degrees, immoveable, with its radius: the radius is fastened with one of the central rules, like the two legs of a sector, that the central rule may be carried through all the points of the circumference of the arch. The radius and rule should be as thin as possible; and the rule fastened to the radius should be hammered cold, to be more elastic; and the breadth of the other central rule must be triple the breadth of the radius: in this rule also is a groove, with a dove-tail fastened on it, for its motion; there must also be a hole in the centre of each rule.

Turn-up COMPASSES, a late contrivance to save the trouble of changing the points: the body is like the common compasses; and towards the bottom of the legs without side, are added two other points, besides the usual ones; the one carrying a drawing pen point, and

the

the other a port-crayon; both being adjufted to turn up, to be ufed or not, as occafion may require.

COMPLEMENT in general, is what is wanting, or neceffary, to complete fome certain quantity or thing. As, the

COMPLEMENT *of an arch or angle*, as of 90° or a quadrant, is what any given arch or angle wants of it; fo the complement of 50° is 40°, and the complement of 100 degrees is —10°, a negative quantity.—The complement to 180° is ufually called the fupplement, to diftinguifh it from the complement to 90°, properly fo called.—The fine of the complement of an arc, is contracted into the word cofine; the tangent of the complement, into cotangent; &c.

Arithmetical COMPLEMENT, is what a number or logarithm wants of unity or 1 with fome number of ciphers. It is beft found, by beginning at the left-hand fide, and fubtracting every figure from 9, except the laft, or right-hand figure, which muft be fubtracted from 10. So, the arithmetical comp. of the log. 9·5329714, by fubtracting from 9's, &c,   is 0·4670286.

The arithmetical complements are much ufed in operations by logarithms, to change fubtractions into additions, which are more conveniently performed, efpecially when there are more than one of them in the operation.

COMPLEMENT, in Aftronomy, is ufed for the diftance of a ftar from the zenith; or the arc contained between the zenith and the place of a ftar which is above the horizon. It is the fame as the complement of the altitude, or co-altitude, or the zenith diftance.

COMPLEMENT *of the Courfe*, in Navigation, is the quantity which the courfe wants of 90°, or 8 points, viz, a quarter of the compafs.

COMPLEMENT *of the Curtain*, in Fortification, is that part of the anterior fide of the curtain, which makes the demigorge.

COMPLEMENT *of the Line of Defence*, is the remainder of that line, after the angle of the flank is taken away.

COMPLEMENTS *of a Parallelogram*, or *in a Parallelogram*, are the two leffer parallelograms, made by drawing two right lines parallel to each fide of the given parallelogram, through the fame point in the diagonal. So P and Q are the complements in the parallelogram ABCD.

In every cafe, thefe complements are always equal, viz, the parallelogram P = Q.

COMPLEMENT *of Life*, a term much ufed, in the doctrine of Life Annuities, by De Moivre, and, according to him, it denotes the number of years which a given life wants of 86, this being the age which he confidered as the utmoft probable extent of life. So 56 is the complement of 30, and 30 is the complement of 56.

That author fuppofed an equal annual decrement of life through all its ftages, till the age of 86. Thus, if there be 56 perfons living at 30 years of age, it is fuppofed that one will die every year, till they be all dead in 56 years. This hypothefis in many cafes is very near the truth; and it agrees fo nearly with Halley's

table, formed from his obfervations of the mortuary bills of Breflaw, that the value of lives deduced either from the hypothefis, or the table, need not be diftinguifhed; hence it very much eafes the labour of calculating them. See *Life* ANNUITIES, alfo De Moivre's Treatife on Annuities, pa. 83, and Price on Reverfionary Payments, pa. 2.

COMPOSITE *Number*, is one that is compounded of, or made up by the multiplication of two other numbers, greater than 1, or which can be meafured by fome other number greater than 1. As 12, which is compofed, or compounded of 2 and 6, or 3 and 4, viz by multiplying together 2 and 6, or 3 and 4, both products making the fame number 12; which therefore is a compofite number.

Compofites are oppofed to prime numbers, or primes, which cannot be exactly meafured by any other number, and cannot be produced by multiplying together two other factors.

Compofite Numbers between themfelves, are the fame with commenfurable numbers, or fuch as have a common meafure or factor; as 15 and 12, which have the common term 3.

The doctrine of Prime and Compofite numbers is pretty fully treated in the 7th and 8th books of Euclid's Elements.

COMPOSITE *Order*, in Architecture, is the laft of the five orders of columns; and is fo called becaufe its capital is compofed out of thofe of the other orders. Thus, it borrows a quarter-round from the Tufcan and Doric; a double row of leaves from the Corinthian; and volutes from the Ionic. Its cornice has fingle modillions, or dentils; and its column is 10 diameters in height.

This order is alfo called the Roman order, and Italic order, as having been invented by the Romans, like as the other orders are denominated from the people among whom they had their rife.

COMPOSITION, is a fpecies of reafoning by which we proceed from things that are known and given, ftep by ftep, till we arrive at fuch as were before unknown or required; viz, proceeding upon principles felf-evident, on definitions, poftulates, and axioms, with a previoufly demonftrated feries of propofitions, ftep by ftep, till it gives a clear knowledge of the thing to be known or demonftrated. Compofition, otherwife called the fynthetical method, is oppofed to Refolution, or the analytical method, and is chiefly ufed by the ancients, Euclid, Apollonius, &c. See Pappus; alfo the term *Analyfis*.

COMPOSITION *of forces*, or *of motion*, is the union or affemblage of feveral forces or motions that are oblique to one another, into an equivalent one in another direction.

1. When feveral forces or motions are united, that act in the fame line of direction, the combined force or motion will be in the fame line of direction ftill. But when oblique forces are united, the compounded force takes a new direction, different from both, and is either a right line or a curve, according to the nature of the forces compounded.

2. If two compounding motions be both equable ones, whether equal to each other or not, the line of the compound

compound motion will be a ſtraight line. Thus, if the one equable in the direction AB be ſufficient to carry a body over the ſpace AB in any time, and the other motion ſufficient to paſs over AC in the ſame time; then by the compound motion, or both acting on the body together, it would in the ſame time paſs over the diagonal AD of the parallelogram ABDC. For becauſe the motions are uniform, any

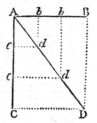

ſpaces A*b*, A*c* paſſed over in the ſame time, are proportional to the velocities, or to AB and AC; and conſequently all the points A, *d*, *d*, D, of the path are in the ſame right line.

3. And though the compounding motions be not equable, but variable, either accelerated or retarded, provided they do but vary in a ſimilar manner, the compounded motion will ſtill be in a ſtraight line. Thus, ſuppoſe, for inſtance, that the motions both vary in ſuch a manner, as that the ſpaces paſſed over in the ſame time, whether they be equal to each other or not, are both as the ſame power *n* of the time; then AB^n : AC^n : : Ab^n : Ac^n, and hence $AB : AC : : Ab : Ac$, and therefore, as before, A*dd*D is ſtill a right line.

4. But if the compounding motions be not ſimilar to each other, as when the one is equable and the other variable, or when they are varied in a diſſimilar manner; then the compounded motion is in ſome curve line. So if the motion in the one direction EF be in a leſs proportion, with reſpect to the time, than that in the direction EG is, then the path will be a curve line E*i*H concave towards EF; but if the motion in EF be in a greater proportion than that in EG, then the path of the compound motion will be a curve E*h*H convex towards EF: that is, in general, the curvilineal path is convex towards that direction in which the motion is in the leſs proportion to the time. Hence, for a

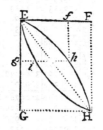

particular inſtance, if the motion in the direction EF be a motion of projection, which is an equable motion, and the motion in the direction EG that ariſing from gravity, which is a uniformly accelerated motion, or in proportion to the ſquares of the times; then is EG as GH^2, and E*g* as gh^2, that is EG : E*g* : : GH^2 : gh^2, which is the property of the parabola; and therefore the path E*h*H of any body projected, is the common parabola.

5. If there be three forces united, or acting againſt the ſame point A at the ſame time, viz, the force or weight B in the direction AB, and the forces or tenſions in the directions AC, AD; and if theſe three forces mutually balance each other, ſo as to keep the common point A in equilibrio; then are theſe forces directly proportional to the reſpective ſides of a triangle formed by drawing lines parallel to the directions of theſe forces; or indeed perpendicular to thoſe directions, or making any one and the ſame angle with them. So that, if BE be drawn parallel, for inſtance, to AD, and meet CA produced in E, forming the triangle

ABE; then are the three forces in the directions AB, AC, AD, reſpectively proportional to the ſides AB, AE, BE.

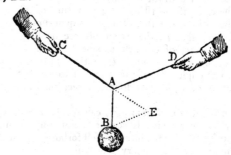

And this theorem, with its corollaries, Dr. Keil obſerves, is the foundation of all the new mechanics of M. Varignon: by help of which may the force of the muſcles be computed, and moſt of the mechanic theorems in Borelli, De Motu Animalium, may immediately be deduced.

See more of the Compoſition of Forces under the article COLLISION.

COMPOSITION *of Numbers and Quantities.* See COMBINATION.

COMPOSITION *of Proportion,* according to the 15th definition of the 5th book of Euclid's Elements, is when, of four proportionals, the ſum of the 1ſt and 2d is to the 2d, as the ſum of the 3d and 4th is to the 4th:

as if it be $a \; : b :: \; c \; : d,$

then by compoſition $a+b : b :: c+d : d.$

Or, in numbers, if   2   : 4 ::   9   : 18,

then by compoſition   6   : 4 ::   27   : 18.

COMPOSITION *of Ratios,* is the adding of ratios together: which is performed by multiplying together their correſponding terms, viz, the antecedents together, and the conſequents together, for the antecedent and conſequent of the compounded ratio; like as the addition of logarithms is the ſame thing as the multiplication of their correſponding numbers. Or, if the terms of the ratios be placed fraction-wiſe, then the addition or compoſition of the ratios, is performed by multiplying the fractions together.

Thus, the ratio of $a : b$, or of 2 : 4,

added to the ratio of $c : d$, or of 6 : 8,

makes the ratio of $ac : bd$, or of 12 : 32; and ſo the ratio of ac to bd is ſaid to be compounded of the ratios of a to b, and c to d. So likewiſe, if it were required to compound together the three ratios, viz, of a to b, c to d, and e to f; then $\frac{a}{b} \times \frac{c}{d} \times \frac{e}{f} = \frac{ace}{bdf}$ are the terms of the compound ratio; or the ratio of ace to bdf is compounded, or made up of the ratios of a to b, c to d, and e to f.

Hence, if the conſequent of each ratio be the ſame as the antecedent of the preceding ratio, then is the ratio of the firſt term to the laſt, compounded, or made up of all the other ratios, viz, the ratio of a to e, equal to the ſum of all the ratios of a to b, of b to c, of c to d, and of d to e; for $\frac{a}{b} \times \frac{b}{c} \times \frac{c}{d} \times \frac{d}{e} = \frac{a}{e}$ the terms or exponents of the compounded ratio.

Hence

Hence also, in a series of continual proportionals, the ratio of the first term to the third is double of the ratio of the first to the second,

and the ratio of the 1st to the 4th is triple of it, and the ratio of the 1st to the 5th is quadruple of it, and so on ; that is, the exponents are double, triple, quadruple, &c, of the first exponent : as in the series 1, a, a^2, a^3, a^4, &c ; where the ratio 1 to a^2 is double, of 1 to a^3 triple, &c, of the ratio of 1 to a ; or the exponent of a^2, a^3, a^4, &c, double, triple, quadruple, &c, of a.

COMPOUND INTEREST, called also *Interest upon Interest*, is that which is reckoned not only upon the principal, but upon the interest itself forborn, which thus becomes a sort of secondary principal.

If r be the amount of 1 pound for 1 year, that is the sum of the principal and interest together for one year ; then is r^2 the amount for 2 years,

and r^3 the amount for 3 years, and in general r^t the amount for t years ; that is r^t is the sum or total amount of all the principals and interests together of 1l. for the whole time or number of years t ; consequently, if p be any other principal sum, forborn for t years, then its amount in that time at compound interest, is $a = pr^t$.

The Rule therefore in words is this, to one pound add its interest for one year, or half year, or for the first time at which the interest is reckoned ; raise the sum r to the power denoted by the time or number of terms ; then this power multiplied by the principal, or first sum lent, will produce the whole amount.

For example, To find how much 50l. will amount to in 5 years at 5 per cent. per annum, compound interest.——Here the interest of 1l. for 1 year is ·05, and therefore $r = 1·05$; hence the 5th power of it for 5 years, is $r^5 = 1·27628$ &c ;

multiply this by p or - 50

gives the amount pr^t or 63·8141. or 63l. 16s. 3¼d. for the amount sought.

But Compound Interest is best computed by means of such a table as the following, being the amounts of 1 pound for any number of years, and at several rates of compound interest.

As an example of the use of this table, suppose it were required to find the amount of 250l. for 35 years at 4 per cent. compound interest.

In the column of 4 per cent, and line of 35 years, is - - - 3·94609, which multiplied by the principal - 250

gives - - - - 986·52250 or - - - - 986l. 10s. 5¼d, which is the amount sought.

Note, By a bare inspection of this table, it appears how many years are required for any sum of money to double itself, at any rate of compound interest ; viz, by looking in the columns when the amount becomes the number 2. So it is found that at the several rates the respective times requisite for doubling any sum, are nearly thus : viz,

Rate   3   3½   4   4½   5   6
Years 23½ 20¼ 17¾ 15¾ 14¼ 12

TABLE of the Amount of 1l. at Compound Interest for many Years and several Rates of Interest.

Yrs.	at 3 per cent	at 3½ per cent	at 4 per cent	at 4½ per cent	at 5 per cent	at 6 per cent
1	1·03000	1·03500	1·04000	1·04500	1·05000	1·06000
2	1·06090	1·07123	1·08160	1·09203	1·10250	1·12360
3	1·09273	1·10872	1·12486	1·14117	1·15763	1·19102
4	1·12551	1·14752	1·16986	1·19252	1·21551	1·26248
5	1·15927	1·18769	1·21665	1·24618	1·27628	1·33823
6	1·19405	1·22926	1·26532	1·30226	1·34010	1·41852
7	1·22987	1·27228	1·31593	1·36086	1·40710	1·50363
8	1·26677	1·31681	1·36857	1·42210	1·47746	1·59385
9	1·30477	1·36290	1·42331	1·48610	1·55133	1·68948
10	1·34392	1·41060	1·48024	1·55297	1·62890	1·79085
11	1·38423	1·45997	1·53945	1·62285	1·71034	1·89830
12	1·42576	1·51107	1·60103	1·69588	1·79586	2·01220
13	1·46853	1·56396	1·66507	1·77220	1·88565	2·13293
14	1·51259	1·61869	1·73168	1·85194	1·97993	2·26090
15	1·55797	1·67535	1·80094	1·93528	2·07893	2·39656
16	1·60471	1·73399	1·87298	2·02237	2·18287	2·54035
17	1·65285	1·79468	1·94790	2·11338	2·29202	2·69277
18	1·70243	1·85749	2·02582	2·20848	2·40662	2·85434
19	1·75351	1·92250	2·10685	2·30786	2·52695	3·02560
20	1·80611	1·98979	2·19112	2·41171	2·65330	3·20714
21	1·86029	2·05943	2·27877	2·52024	2·78596	3·39956
22	1·91610	2·13151	2·36992	2·63365	2·92526	3·60354
23	1·97359	2·20611	2·46472	2·75217	3·07152	3·81975
24	2·03279	2·28333	2·56330	2·87601	3·22510	4·04893
25	2·09378	2·36324	2·66584	3·00543	3·38635	4·29187
26	2·15659	2·44596	2·77247	3·14068	3·55567	4·54938
27	2·22129	2·53157	2·88337	3·28201	3·73346	4·82235
28	2·28793	2·62017	2·99870	3·42970	3·92013	5·11169
29	2·35657	2·71188	3·11865	3·58404	4·11614	5·41839
30	2·42726	2·80679	3·24340	3·74532	4·32194	5·74349
31	2·50008	2·90503	3·37313	3·91386	4·53804	6·08810
32	2·57508	3·00671	3·50806	4·08998	4·76494	6·45339
33	2·65234	3·11194	3·64838	4·27403	5·00319	6·84059
34	2·73191	3·22086	3·79432	4·46636	5·25335	7·25103
35	2·81386	3·33359	3·94609	4·66735	5·51602	7·68609
36	2·89828	3·45027	4·10393	4·87738	5·79182	8·14725
37	2·98523	3·57103	4·26809	5·09686	6·08141	8·63609
38	3·07478	3·69601	4·43881	5·32622	6·38548	9·15425
39	3·16703	3·82537	4·61637	5·56590	6·70475	9·70351
40	3·26204	3·95926	4·80102	5·81636	7·03999	10·28572
41	3·35990	4·09783	4·99306	6·07810	7·39199	10·90286
42	3·46070	4·24126	5·19278	6·35162	7·76159	11·55703
43	3·56452	4·38970	5·40050	6·63744	8·14967	12·25045
44	3·67145	4·54333	5·61652	6·93612	8·55715	12·98548
45	3·78160	4·70236	5·84118	7·24825	8·98501	13·76461
46	3·89504	4·86694	6·07482	7·57442	9·43426	14·59049
47	4·01190	5·03728	6·31781	7·91527	9·90597	15·46592
48	4·13225	5·21359	6·57053	8·27146	10·40127	16·39387
49	4·25622	5·39606	6·83335	8·64367	10·92133	17·37750
50	4·38391	5·58493	7·10668	9·03264	11·46740	18·42015

COMPOUND *Motion*, that motion which is the effect of several conspiring powers or forces, viz, such forces as are not directly opposite to each other : as when the radius of a circle is considered as revolving about a centre, and at the same time a point as moving straight along it ; which produces a kind of a spiral for the path of the point. And hence it is easily perceived, that all curvilinear motion is compound, or the effect of two or more forces ; although every compound motion is not curvilinear.

It is a popular theorem in Mechanics, that in uniform compound motions, the velocity produced by the conspiring powers or forces, is to that of either of the two compounding powers separately, as the diagonal of of a parallelogram, according to the direction of whofe fides they act feparately, is to either of the fides. See COMPOSITION *of Motion*, and COLLISION.

COMPOUND *Numbers*, thofe compofed of the multiplication of two or more numbers; as 12, compofed of 3 times 4. See COMPOSITE.

COMPOUND *Pendulum*, that which confifts of feveral weights conftantly keeping the fame diftance, both from each other, and from the centre about which they ofcillate.

COMPOUND *Quantities*, are fuch as are connected together by the figns + or —. Thus, $a + b$, or $a - c + d$, or $aa - 2a$, are compound quantities.

Compound quantities are diftinguifhed into binomials, trinomials, quadrinomials, &c, according to the number of terms in them; viz, the binomial having two terms; the trinomial, three ; the quadrinomial, 4; &c. Alfo, thofe that have more than two terms, are called by the general name of multinomials, as alfo polynomials.

COMPOUND *Ratio*, is that which is made by adding two or more ratios together ; viz, by multiplying all their antecedents together for the antecedent, and all the confequents together for the confequent of the compound ratio. So 6 to 72 is a ratio compounded of the ratios of 2 to 6, and 3 to 12; becaufe $\frac{6}{72} = \frac{2}{6} \times \frac{3}{12}$: alfo ab to cd is a ratio compounded of the ratio of a to c, and b to d; for $\frac{ab}{cd} = \frac{a}{c} \times \frac{b}{d}$. See COMPOSITION *of Ratios*.

COMPRESSION, the act of preffing, or fqueezing fomething, fo as to bring its parts nearer together, and make it occupy lefs fpace.

Compreffion differs from condenfation as the caufe from the effect, compreffion being the action of any force on a body, without regarding its effects; whereas condenfation denotes the ftate of a body that is actually reduced into a lefs bulk, and is an effect of compreffion, though it may be effected alfo by other means. Neverthelefs, compreffion and condenfation are often confounded.

Pumps, which the ancients imagined to act by fuction, do in reality act by compreffion ; the pifton, in working in the narrow pipe, compreffes the inclofed air, fo as to enable it, by the force of its increafed elafticity, to raife the valve, and make its efcape; upon which, the balance being deftroyed, the preffure of the atmofphere on the ftagnant furface, forces up the water in the pipe, thus evacuated of its air.

It was long thought that water was not compreffible into lefs bulk : and it was believed, till lately, that after the air had been purged out of it, no art or violence was able to prefs it into lefs fpace. In an experiment made by the Academy del Cimento, water, when violently fqueezed, made its way through the fine pores of a globe of gold, rather than yield to the compreffion.

But the ingenious Mr. Canton, attentively confidering this experiment, found that it was not fufficiently accurate to juftify the conclufion which had always been drawn from it ; fince the Florentine philofophers had no method of determining that the alteration of figure in their globe of gold, occafioned

fuch a diminution of its internal capacity, as was exactly equal to the quantity of water forced into its pores. To bring this matter therefore to a more accurate and decifive trial, he procured a fmall glafs tube of about two feet long, with a ball at one end, of an inch and a quarter in diameter. Having filled the ball and part of the tube with mercury, and brought it exactly to the heat of 50° of Fahrenheit's thermometer, he marked the place where the mercury ftood in the tube, which was about fix inches and a half above the ball; he then raifed the mercury by heat to the top of the tube, and there fealed the tube hermetically ; then upon reducing the mercury to the fame degree of heat as before, it ftood in the tube $\frac{32}{100}$ of an inch higher than the mark. The fame experiment was repeated with water exhaufted of air, inftead of mercury, and the water ftood in the tube $\frac{43}{100}$ of an inch above the mark. Since the weight of the atmofphere on the outfide of the ball, without any counterbalance from within, will comprefs the ball, and equally raife both the mercury and water, it appears that the water expands $\frac{11}{100}$ of an inch more than the mercury by removing the weight of the atmofphere. Having thus determined that water is really compreffible, he proceeded to eftimate the degree of compreffion correfponding to any given weight. For this purpofe he prepared another ball, with a tube joined to it ; and finding that the mercury in $\frac{1}{175}$ of an inch of the tube was the hundred thoufandth part of that contained in the ball, he divided the tube accordingly. He then filled the ball and part of the tube with water exhaufted of air ; and leaving the tube open, placed this apparatus under the receiver of an air-pump, and obferved the degree of expanfion of the water anfwering to any degree of rarefaction of the air: and again by putting it into the glafs receiver of a condenfing engine, he noted the degree of compreffion of the water correfponding to any degree of condenfation of the air. He thus found, by repeated trials, that, in a temperature of 50°, and when the mercury has been at its mean height in the barometer, the water expands one part in 21740; and is as much compreffed by the weight of an additional atmofphere ; or the compreffion of water by twice the weight of the atmofphere, is one part in 10870 of its whole bulk. Should it be objected, that the compreffibility of the water was owing to any air which it might be fuppofed to contain, he anfwers, that more air would make it more compreffible ; he therefore let into the ball a bubble of air, and found that the water was not more compreffed by the fame weight than before.

In fome farther experiments of the fame kind, Mr. Canton found that water is more compreffible in winter than in fummer ; but he obferved the contrary in fpirit. of wine, and oil of olives.

The following table was formed, when the barometer was at 29 inches and a half, and the thermometer at 50 degrees.

Compreffion of	Millionth parts.	Spec. grav.
Spirit of wine	66	846
Oil of olives	48	918
Rain water	46	1000
Sea water	40	1028
Mercury	3	13595

He

He infers that these fluids are not only compressible, but elastic; and that the compressions of them, by the same weight, are not in the inverse ratio of their densities, or specific gravities, as might be supposed. Phil. Transf. vol. lii. 1762. art. 103. and vol. liv. 1764. art. 47.

The compression of the air, by its own weight, is surprisingly great: but the air may be still further compressed by art. See *Elasticity of* AIR.

This immense compression and dilatation, Newton observes, cannot be accounted for in any other way, but by a repelling force, with which the particles of air are endued; by virtue of which, when at liberty, they mutually fly each other.

This repelling power, he adds, is stronger and more sensible in air, than in other bodies; because air is generated out of very fixed bodies, but not without great difficulty, and by the help of fermentation: now those particles always recede from each other with the greatest violence, and are compressed with the greatest difficulty, which, when contiguous, cohere the most strongly. See AIR, ATTRACTION, COHESION, DILATATION, and REPULSION.

COMPUTATION, the manner of accounting and estimating time, weights, measures, money, &c. See CALCULATION, which it is also used for.

CONCAVE, an appellation used in speaking of the inner surface of hollow bodies, more especially of spherical or circular ones.

CONCAVE glasses, lenses, and mirrors, have either one side or both sides concave.

The property of all concave lenses is, that the rays of light, in passing through them, are deflected, or made to recede from one another; as in convex lenses they are inflected towards each other; and that the more as the concavity or convexity has a smaller radius. Hence parallel rays, as those of the sun, by passing through a concave lens, become diverging; diverging rays are made to diverge more; and converging rays are made either to converge less, or to become parallel, or go out diverging. And hence it is, that objects viewed through concave lenses, appear diminished; and the more so, as they are portions of less spheres. See LENS.

Concave mirrors have the contrary effect to lenses: they reflect the rays which fall on them, so as to make them approach more to, or recede from each other, than before, according to the situation of the object; and that the more as the concavity is greater, or as the radius of concavity is less. Hence it is that concave mirrors magnifying objects that are presented to them; and that in a greater proportion, as they are portions of greater spheres. And hence also concave mirrors have the effect of burning glasses. See MIRROR, and BURNING GLASS.

CONCAVITY, that side of a figure or body which is hollow.

An arch of a curve has its concavity turned all one way, when the right lines that join any two of its points are all on the same side of the arch.

Archimedes, intending to include in his definition such lines as have rectilinear parts, says, a line has its concavity turned one way, when the right lines that join any two of its points, are either all upon one side

of it, or while some fall upon the line itself, none fall upon the opposite side. Archim. de Sphær. et Cyl. Def. 2, and Maclaurin's Fluxions, art. 180.

When two lines that have their concavity turned the same way, have the same extremes, and the one includes the other, or has its concavity towards it, the perimeter of that which includes, is greater than that which is included. Archim. ib. ax. 2.

CONCENTRIC, having the same centre. It is opposed to excentric, or having different centres.

The word is chiefly used in speaking of round bodies and figures, such as circular, and elliptic ones; but it may likewise be used for polygons that are drawn parallel to each other, from the same centre.

Nonnius's method of graduating instruments consists in describing with the same quadrant 45 concentric arches, dividing the outermost into 90 equal parts, the next into 89, and so on.

CONCHOID, or CONCHILES, the name of a curve invented by Nicomedes. It was much used by the ancients in the construction of solid problems, as appears by what Pappus says.

It is thus constructed: AP and BD being two lines intersecting at right angles; from P draw a number of other lines PFDE, &c, on which take always DE = DF = AB or BC; so shall the curve line drawn through all the points E, E, E, be the first conchoid, or that of Nicomedes; and the curve drawn through all the other points F, F, F, is called the second conchoid; though in reality, they are both but parts of the same curve, having the same pole P, and four infinite legs, to which the line DBD is a common asymptote.

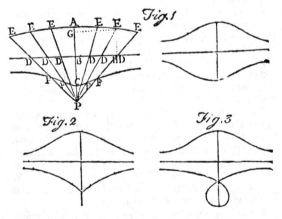

The inventor, Nicomedes, contrived an instrument for describing his conchoid by a mechanical motion: thus, in the rule AD is a channel or groove cut, so that a smooth nail, firmly fixed in the moveable rule CB, in the point F, may slide freely within it: into the rule EG is fixed another nail at K, for the moveable rule CB to slide upon. If therefore the rule BC be so moved, as that the nail F passes along the canal AD; the style, or point in C, will describe the first conchoid.

To determine the equation of the curve: put AB = BC = DE = DF = *a*, PB = *b*, BG = EH = *x*, and GE = BH = *y*; then the equation to the first conchoid

choid will be $x^2 \times \overline{b+x}|^2 + x^2 y^2 = a^2 \times \overline{b+x}|^2$, or $x^4 + 2bx^3 + b^2 x^2 + x^2 y^2 = a^2 b^2 + 2a^2 bx + a^2 x^2$; and, changing only the sign of x, as being negative in the other curve, the equation to the 2d conchoid will be $x^2 \times \overline{b-x}|^2 + x^2 y^2 = a^2 \times \overline{b-x}|^2$, or $x^4 - 2bx^3 + b^2 x^2 + x^2 y^2 = a^2 b^2 - 2a^2 bx + a^2 x^2$.

Of the whole conchoid, expreſſed by theſe two equations, or rather one equation only, with different ſigns, there are three caſes or ſpecies; as firſt,

when BC is leſs than BP, the conchoid will be as in fig. 1;
when BC is equal to BP, the conchoid will be as in fig. 2;
and when BC is greater than BP, the conchoid will be as in fig. 3.

Newton approves of the uſe of the conchoid for triſecting angles, or finding two mean proportionals, or for conſtructing other ſolid problems. Thus, in the Linear Conſtruction of equations, towards the end of his Univerſal Arithmetic, he ſays, " The antients at firſt endeavoured in vain at the triſection of an angle, and the finding of two mean proportionals by a right line and a circle. Afterwards they began to conſider ſeveral other lines, as the conchoid, the ciſſoid, and the conic ſections, and by ſome of theſe to ſolve theſe problems." Again, " Either therefore the trochoid is not to be admitted at all into geometry, or elſe, in the conſtruction of problems, it is to be preferred to all lines of a more difficult deſcription : and there is the ſame reaſon for other curves; for which reaſon we approve of the triſections of an angle by a conchoid, which Archimedes in his Lemmas, and Pappus in his Collections, have preferred to the inventions of all others in this caſe; becauſe we ought either to exclude all lines, beſides the circle and right line, out of geometry, or admit them according to the ſimplicity of their deſcriptions, in which caſe the conchoid yields to none, except the circle." Laſtly, " That is *arithmetically* more ſimple which is determined by the more ſimple equations, but that is *geometrically* more ſimple which is determined by the more ſimple drawing of lines; and in geometry, that ought to be reckoned beſt which is geometrically moſt ſimple : wherefore I ought not to be blamed, if, with that prince of mathematicians, Archimedes, and other antients, I make uſe of the conchoid for the conſtruction of ſolid problems."

CONCRETE *Numbers* are thoſe that are applied to expreſs or denote any particular ſubject; as 3 men, 2 pounds, &c. Whereas, if nothing be connected with a number, it is taken abſtractedly or univerſally : thus, 4 ſignifies only an aggregate of 4 units, without any regard to a particular ſubject, whether men or pounds, or any thing elſe.

CONCURRING, or CONGRUENT *Figures*, in Geometry, are ſuch as, being laid upon one another, do exactly correſpond to, and cover one another, and conſequently muſt be equal among themſelves. Thus, triangles having two ſides and the contained angle equal, each to each, are equal to each other in all reſpects.

CONDENSATION, is the compreſſing or reducing of a body into a leſs bulk or ſpace; by which means it is rendered more denſe and compact.

Wolfius, and ſome other writers, reſtrain the uſe of the word *condenſation* to the action of cold : that which is done by external application, they call compreſſion.

Condenſation however, in general, conſiſts in bring-

ing the parts cloſer to each other, and increaſing their contact, whatever be the means by which it is effected : in oppoſition to rarefaction, which renders the body lighter and looſer, by ſetting the parts farther aſunder, and diminiſhing their contact, and of conſequence their coheſion.

Air eaſily condenſes, either by cold, or by preſſure, but much more by the latter; but moſt of all by chemical proceſs. Water condenſes alſo both by cold and by preſſure; but it ſuddenly expands by congelation : indeed almoſt all matter, both ſolids and fluids, has the ſame property of condenſation by thoſe means. See COMPRESSION. So alſo vapour is condenſed, or converted into water, by diſtillation, or naturally in the clouds. The way in which vapour commonly condenſes, is by the application of ſome cold ſubſtance. On touching it, the vapour parts with its heat which it had before abſorbed : and on doing ſo, it immediately loſes the proper characteriſtics of vapour, and becomes water. But though this be the moſt common and uſual way in which we obſerve vapour to be condenſed, nature certainly proceeds after another manner; ſince we often obſerve the vapours moſt plentifully condenſed when the weather is really warmer than at other times.

CONDENSER, a pneumatic engine, or ſyringe, by which an extraordinary quantity of air may be crowded or puſhed into a given ſpace; ſo that frequently ten atmoſpheres, or ten times as much air as the ſpace naturally contains, without the engine, may be thrown in by means of it, and its egreſs prevented by valves properly diſpoſed.

The condenſer is made either of metal, or glaſs, and either in a cylindrical or globular form, into which the air is thrown with an injecting ſyringe.

The receiver, or veſſel containing the condenſed air, ſhould be made very ſtrong, to bear the force of the air's elaſticity thus increaſed; for which reaſon it is commonly made of braſs. When glaſs is uſed, it will not bear ſo great a condenſation of air; but then the experiment will be more entertaining, as the effect may be viewed of the condenſed air upon any ſubject put within it.

CONDUCTOR, in Electricity, a term firſt introduced in this ſcience by Dr. Deſaguliers, and uſed to denote thoſe ſubſtances which are capable of receiving and tranſmitting electricity; in oppoſition to electrics, in which the matter or virtue of electricity may be excited and accumulated, or retained. The former are alſo called *non-electrics*, and the latter *non-conductors*. And all bodies are ranked under one or other of theſe two claſſes, though none of them are *perfect electrics*, nor *perfect conductors*, ſo as wholly to retain, or freely and without reſiſtance to tranſmit the electric fluid.

To the claſs of conductors, belong all metals and ſemi-metals, ores, and all fluids (except air and oils), together with the ſubſtances containing them, the effluvia of flaming bodies, ice (unleſs very hard frozen), and ſnow, moſt ſaline and ſtony ſubſtances, charcoals, of which the beſt are thoſe that have been expoſed to the greateſt heat, ſmoke, and the vapour of hot water.

It ſeems probable that the electric fluid paſſes through the ſubſtance, and not merely over the ſurfaces of metallic conductors; becauſe, if a wire of any kind of metal be covered with ſome electric ſubſtance, as reſin, fealing-wax, &c, and a jar be difcharged through it, the charge will be conducted as well as without the electric coating.

It has alfo been alleged, that electricity will pervade a vacuum, and be tranfmitted through it almoft as freely as through the fubftance of the beft conductor: but Mr. Walfh found, that the electric fpark or fhock would no more pafs through a perfect vacuum, than through a ftick of folid glafs. In other inftances however, when the vacuum has been made with all poffible eare, the experiment has not fucceeded.

It may alfo be obferved, that many of the forementioned fubftances are capable of being electrified, and that their conducting power may be deftroyed and recovered by different proceffes: for example, green wood is a conductor; but baked, it becomes a non-conductor: again its conducting power is reftored by charring it; and laftly it is deftroyed by reducing this to afhes.

Again, many electric fubftances, as glafs, refin, air, &c, become conductors by being made very hot: however, air heated by glafs muft be excepted.

See, on this fubject, Prieftley's Hift. of Electricity, vol. 1; Franklin's Letters &c, pa. 96 and 262 edit. 1769; Cavallo's Complete Treat. of Electr. chap. 2; Henley's Exper. and Obfer. in Electr. alfo Philof. Tranf. vol. 67 pa. 122; and elfewhere in the different volumes of the Tranfactions.

Prime CONDUCTOR, is an infulated conductor, fo connected with the electrical machine, as to receive the electricity immediately from the excited electric.

Mr. Grey firft employed metallic conductors in this way, in 1734; and thefe were feveral pieces of metal fufpended on filken ftrings, which he charged with electricity. Mr. Du Fay faftened to the end of an iron bar, which he ufed as his prime conductor, a bundle of linen threads, to which he applied the excited tube: but thefe were afterwards changed for fmall wires fufpended from a common gun-barrel, or other metallic rod.

In the prefent advanced ftate of the fcience, this part of the electrical apparatus has been confiderably improved. The prime conductor is made of hollow brafs, and ufually of a cylindrical form. Care fhould be taken, that it be perfectly fmooth and round, without points and fharp edges. The ends of the conductor are fpherical; and it is neceffary, that the part moft remote from the electric fhould be made round and much larger than the reft, the better to prevent the electric matter from efcaping, which it always endeavours moft to do at the greateft diftance from the electric: and the other end fhould be furnifhed with feveral pointed wires or needles, either fufpended from, or fixed to an open metallic ring, and pointing to the globe or cylinder, or plate, to collect the fire. It is beft fupported by pillars of folid glafs, covered with fealing-wax or good varnifh. Prime conductors of a large fize are ufually made of pafte-board, covered with tin-foil or gilt paper; thefe being ufeful for throwing off a longer and denfer fpark than thofe of a fmaller fize: they fhould terminate in a fmaller knob or obtufe edge, at which the fparks fhould be folicited. Mr. Nairne prepared a conductor 6 feet in length, and 1 foot in diameter, from which he drew electrical fparks at the diftance of 16, 17, or 18 inches; and Dr. Van Marum

ftill far exceeded this, with a conductor of 8 inches diameter, and upwards of 20 feet long, formed of different pieces, and applied to the large electrical machine in Teyler's Mufeum at Harlem, the moft powerful machine of the kind ever yet conftructed. But the fize of the conductor is always limited by that of the electrio, there being a maximum which the fize of the former fhould not exceed; for it may be fo large, that the diffipation of the electricity from its furface may be greater than that which the electric is capable of fupplying.

Dr. Prieftley recommends a prime conductor of polifhed copper, in the form of a pear, fupported by a pillar and a firm bafis of baked wood: this receives its fire by a long arched wire of foft brafs, which may be eafily bent, and raifed or lowered to the globe: it is terminated by an open ring, in which fome fharp-pointed wires are hung. In the body of this conductor are holes for the infertion of metalline rods. This, he fays, collects the fire perfectly well, and retains it equally everywhere. Philof. Tranf. vol. 64, art. 7. Hift. Elect. vol. 2, § 2.

Mr. Henly has contrived a new kind of prime conductor, which, from its ufe, is called the *luminous* conductor. It confifts of a glafs tube 18 inches long, and 2 inches diameter. The tube is furnifhed at both ends with brafs caps and ferules about 2 inches long, cemented and made air-tight, and terminated by brafs balls. In one of thefe caps is drilled a fmall hole, which is covered by a ftrong valve, and ferves for exhaufting the tube of its air. Within the tube at each end there is a knobbed wire, projecting to the diftance of 2 inches and a half from the brafs caps. To one of the balls is annexed a fine-pointed wire for receiving and collecting the electricity, and to the other a wire with a knob or ball for difcharging it. The conductor, thus prepared, is fupported on pillars of fealing-wax or glafs. Befide the common purpofes of a prime conductor to an electrical machine, this apparatus ferves to exhibit and afcertain the direction of the electric matter in its paffage through it. See a figure of this conductor in the Philof. Tranf. with a defcription of experiments, &c. with it, vol. 64, pa. 403.

CONDUCTORS *of Lightning*, are pointed metallic rods fixed to the upper parts of buildings, to fecure them from ftrokes of lightning. Thefe were invented and propofed by Dr. Franklin for this purpofe, foon after the identity of electricity and lightning was afcertained; and they exhibit a very important and ufeful application of modern difcoveries in this fcience. This ingenious philofopher, having found that pointed bodies are better fitted for receiving and throwing off the electric fire, than fuch as are terminated by blunt ends or flat furfaces, and that metals are the readieft and beft conductors, foon difcovered that lightning and electricity refembled each other in this and other diftinguifhing properties: he therefore recommended a pointed metalline rod, to be raifed fome feet above the higheft part of a building, and to be continued down into the ground, or the neareft water. The lightning, fhould it ever come within a certain diftance of this rod or wire, would be attracted by it, and pafs through it preferably to any other part of the building, and be conveyed into the earth or water, and there diffipated, without doing any damage to the building. Many facts have occurred
red

red to evince the utility of this simple and seemingly trifling apparatus. And yet some electricians, of whom Mr. Wilson was the chief, have objected to the pointed termination of this conductor; preferring rather a blunt end: because, they pretend, a point invites the electricity from the clouds, and attracts it at a greater distance than a blunt conductor. Philof. Transf. vol. 54, pa. 234; vol. 63, pa. 49; and vol. 68, pa. 232.

This subject has indeed been very accurately examined and discussed; and pointed conductors are almost universally, and for the best reasons, recommended as the most proper and eligible. A sharp-pointed conductor, as it attracts the electric fire of a cloud at a greater distance than the other, draws it off gradually: and by conveying it away gently, and in a continued stream, prevents an accumulation and a stroke; whereas a conductor with a blunt termination receives the whole discharge of a cloud at once, and is much more likely to be exploded, whenever a cloud comes within a striking distance. To this may be added experience; for buildings guarded by either natural or artificial conductors te minating in a point, have very seldom been struck by lightning; but others, having flat or blunt terminations, have often been struck and damaged by it.

The best conductor for this purpose, is a rod of iron, or rather of copper, as being a better conductor of electricity, and less liable to rust, about 3 quarters of an inch thick, which is either to be fastened to the walls of a building by wooden cramps, or supported by wooden posts, at the distance of a foot or two from the wall; though less may do: the upper end of it should terminate in a pyramidal form, with a sharp point and edges; and, when made of iron, gilt or painted near the top, or else pointed with copper; and be elevated 5 or 6 feet above the highest part of the building, or chimneys, to which it may be fastened. The lower end should be driven 5 or 6 feet into the ground, and directed away from the foundations of the building, or continued till it communicates with the nearest water: and if this part be made of lead, it will be less apt to decay. When the conductor is formed of different pieces of metal, care should be taken that they are well joined: and it is farther recommended, that a communication be made from the conductor by plates of lead, 8 or 10 inches broad, with the lead on the ridges and gutters, and with the pipes that carry down the rain water, which should be continued to the bottom of the building, and be made to communicate either with water or moist earth, or with the main pipe which serves the house with water. If the building be large, two, three, or more conductors should be applied to different parts of it, in proportion to its extent. Philof. Transf. vol. 64, pa. 403.

Chains have been used as conductors for preserving ships; but as the electric matter does not pass readily through the links of it, copper wires, a little thicker than a goose quill, have been preferred, and are now generally used. They should reach 2 or 3 feet above the highest mast, and be continued down in any convenient direction, so as always to touch the sea water. Philof. Transf. vol. 52, pa. 633. See also Franklin's Letters &c 1769, pa. 65, 124, 479, &c; and Cavallo's Electr. chap. 9.

For the *Construction* and management of *Electrical Kites*, and *Conductors* or *Machines* for drawing electricity from the clouds, see Priestley's Hift. of Electr. vol. 2, pa. 103 edit. 1775.

CONE, a kind of round pyramid, or a solid body having a circle for its base, and its sides formed by right lines drawn from the circumference of the base to a point at top, being the vertex or apex of the cone.

Euclid defines a cone to be a solid figure, whose base is a circle, and is produced by the entire revolution of a right-angled triangle about its perpendicular leg, called the axis of the cone. If this leg, or axis, be greater than the base of the triangle, or radius of the circular base of the cone, then the cone is *acute-angled*, that is, the angle at its vertex is an acute angle; but if the axis be less than the radius of the base, it is an *obtuse-angled* cone; and if they are equal, it is a *right-angled* cone.

But Euclid's definition only extends to a *right cone*, that is, to a cone whose axis is perpendicular or at right-angles to its base; and not to oblique ones, in which the axis is oblique to the base, the general definition, or description of which may be this: If a line VA continually pass through the point V, turning upon that point as a joint, and the lower part of it be carried round the circumference ABC of a circle; then the space inclosed between that circle and the path of the line, is a cone. The circle ABC is the base of the cone; V is its vertex; and the line VD, from the vertex to the centre of the base, is the axis of the cone. Also the other part of the revolving line, produced above V, will describe another cone V*acb*, called the opposite cone, and having the same common axis produced DV*d*, and vertex V.

Properties of the CONE.—1. The area or surface of every right cone, exclusive of its base, is equal to a triangle whose base is the periphery, and its height the slant side of the cone. Or, the curve superficies of a right cone, is to the area of its circular base, as the slant side is to the radius of the base. And therefore, the same curve surface of the cone is equal to the sector of a circle whose radius is the slant side, and its arch equal to the circumference of the base of the cone.

2. Every cone, whether right or oblique, is equal to one-third part of a cylinder of equal base and altitude; and therefore the solid content is found by multiplying the base by the altitude, and taking ⅓ of the product; and hence also all cones of the same or equal base and altitude, are equal.

3. Although the solidity of an oblique cone be obtained in the same manner with that of a right one, it is otherwise with regard to the surface, since this cannot be reduced to the measure of a sector of a circle, because all the lines drawn from the vertex to the base are not equal. See a Memoir on this subject, by M. Euler, in the Nouv. Mem. de Peterfburgh vol. 1. Dr. Barrow has demonstrated, in his Lectiones Geometricæ, that the solidity of a cone with an elliptic base, forming part of a right cone, is equal to the product of its surface by a third part of one of the perpendiculars

draws

drawn from the point in which the axis of the right cone intersects the ellipse; and that it is also equal to ¼ of the height of the cone multiplied by the elliptic base: consequently that the perpendicular is to the height of the cone, as the elliptic base is to the curve surface. For the curve surface of all the oblique parts of a cone, see my Mensur. pa. 234 &c.

4. *To find the Curve Surface of the Frustum of a Cone.* Multiply half the sum of the circumferences of the two ends, by the slant side, or distance between these circumferences.

5. *For the Solidity of the Frustum of a Cone,* add into one sum the areas of the two ends and the mean proportional between them, multiply that sum by the perpendicular height, and ⅓ of the product will be the solidity. See also my Mensuration, pa. 189.

6. The Centre of Gravity of a cone is ¾ of the axis distant from the vertex.

CONES *of the Higher Kinds,* are those whose bases are circles of the higher kinds; and are generated, like the common cone, by conceiving a line turning on a point or vertex on high, and revolving round the circle of the higher kind.

CONE *of Rays,* in Optics, includes all the several rays which fall from any point of a radiant object, on the surface of a glass.

Double CONE, or *Spindle,* in Mechanics, is a solid formed of two equal cones joined at their bases. If this be laid on the lower part of two rulers, making an angle with each other, and elevated in a certain degree above the horizontal plane, the cones will roll upwards the raised ends, and seem to ascend, though in reality its centre of gravity descends perpendicularly lower.

CONFIGURATION, the exterior surface or shape that bounds bodies, and gives them their particular figure.

CONFIGURATION *of the planets,* in Astrology, is a certain distance or situation of the planets in the zodiac, by which it is supposed that they assist or oppose each other.

CONFUSED *Vision.* See VISION.

CONGELATION, or FREEZING, the act of fixing the fluidity of any liquid, by cold, or the application of cold bodies: in which it differs from coagulation, which is produced by other causes. See FREEZING, FROST, and ICE.

CONGRUITY, in Geometry, is applied to lines and figures, which exactly correspond when laid over one another; as having the same terms, or bounds. It is assumed, as an axiom, that those things are equal and similar, between which there is a congruity. Euclid, and most geometricians after him, demonstrate great part of their elements from the principle of congruity: though Leibnitz and Wolfius substitute the notion of Similitude instead of that of congruity.

CONIC SECTIONS, are the figures made by cutting a cone by a plane.

2. According to the different positions of the cutting plane, there arise five different figures or sections, viz, a *triangle,* a *circle,* an *ellipse,* a *parabola,* and an *hyperbola:* the last three of which only are peculiarly called conic sections.

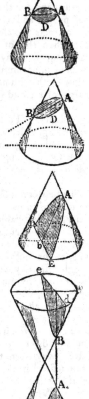

3. If the cutting plane pass through the vertex of the cone, and any part of the base, the section will evidently be a *triangle;* as VAB.

4. If the plane cut the cone parallel to the base, or make no angle with it, the section will be a *circle,* as ABD.

5. The section DAB is an *ellipse,* when the cone is cut obliquely through both sides, or when the plane is inclined to the base in a less angle than the side of the cone is.

6. The section is a *parabola,* when the cone is cut by a plane parallel to the side, or when the cutting plane and the side of the cone make equal angles with the base.

7. The section is an *hyperbola,* when the cutting plane makes a greater angle with the base than the side of the cone makes. And if the plane be continued to cut the opposite cone, this latter section is called the opposite hyperbola to the former; as *dBe.*

8. The *vertices* of any section, are the points where the cutting plane meets the opposite sides of the cone, or the sides of the vertical triangular section; as A and B. —Hence, the ellipse and the opposite hyperbolas have each two vertices; but the parabola only one; unless we consider the other as at an infinite distance.

9. The *axis,* or *transverse diameter* of a conic section, is the line or distance AB between the vertices.— Hence the axis of a parabola is infinite in length.

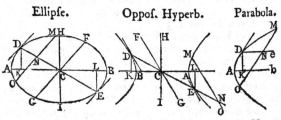

Ellipse.　　　Oppos. Hyperb.　　Parabola.

10. The *centre* C is the middle of the axis.—Hence the centre of a parabola is infinitely distant from the vertex. And of an ellipse, the axis and centre lie within the curve; but of an hyperbola, without.

11. A *Diameter* is any right line, as AB or DE, drawn through the centre, and terminated on each side by the curve: and the extremities of the diameter, or its interfections with the curve, are its *vertices.*—Hence all the diameters of a parabola are parallel to the axis, and infinite in length; because drawn through the centre, a point at an infinite diftance. And hence alfo every diameter of the ellipfe and hyperbola have two vertices; but of the parabola, only one; unlefs we confider the other as at an infinite diftance.

12. The *conjugate* to any diameter, is the line drawn through the centre, and parallel to the tangent of the curve at the vertex of the diameter. So FG, parallel to the tangent at D, is the conjugate to DE; and HI, parallel to the tangent at A, is the conjugate to AB. —Hence the conjugate HI, of the axis AB, is perpendicular to it; but the conjugates of other diameters are oblique to them.

13. An *ordinate* to any diameter, is a line parallel to its conjugate, or to the tangent at its vertex, and terminated by the diameter and curve. So DK and EL are ordinates to the axis AB; and MN and NO ordinates to the diameter DE.—Hence the ordinates of the axis are perpendicular to it; but of other diameters, the ordinates are oblique to them.

14. An *abfcifs* is a part of any diameter, contained between its vertex and an ordinate to it; as AK or BK, and DN or EN.—Hence, in the ellipfe and hyperbola, every ordinate has two abfciffes; but in the parabola, only one; the other vertex of the diameter being infinitely diftant.

15. The *parameter* of any diameter, is a third proportional to that diameter and its conjugate.

16. The *focus* is the point in the axis where the ordinate is equal to half the parameter: as K and L, where DK or EL is equal to the femiparameter.—— Hence, the ellipfe and hyperbola have each two foci; but the parabola only one.——The foci, or burning points, were fo called, becaufe all rays are united or reflected into one of them, which proceed from the other focus, and are reflected from the curve.

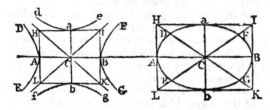

17. If DAE, FBG be two oppofite hyperbolas, having AB for their firft or tranfverfe axis, and *ab* for their fecond or conjugate axis; and if *dae, fbg* be two other oppofite hyperbolas, having the fame axis, but in the contrary order, viz, *ab* their firft axis, and AB their fecond; then thefe two latter curves *dae, fbg,* are called the *conjugate hyperbolas* to the two former DAE, FBG; and each pair of oppofite curves mutually conjugate to the other.

18. And if tangents be drawn to the four vertices of the curves, or extremities of the axis, forming the infcribed rectangle HIKL; the diagonals HCK and ICL, of this rectangle, are called the *afymptotes* of the curves.

19. *Scholium.* The rectangle infcribed between the

four conjugate hyperbolas, is fimilar to a rectangle circumfcribed about an ellipfe, by drawing tangents, in like manner, to the four extremities of the two axes; alfo the afymptotes or diagonals in the hyperbola, are analogous to thofe in the ellipfe. cutting this curve in fimilar points, and making the pair of equal conjugate diameters. Moreover, the whole figure, formed by the four hyperbolas, is, as it were, an ellipfe turned infide out, cut open at the extremities D, E, F, G, of the faid equal conjugate diameters, and thofe four points drawn out to an infinite diftance, the curvature being turned the contrary way, but the axes, and the rectangle paffing through their extremities, remaining fixed, or unaltered.

From the foregoing definitions are eafily derived the following general corollaries to the fections.

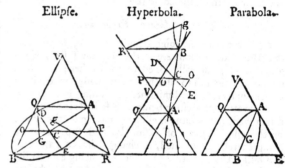

Ellipfe.   Hyperbola.   Parabola.

20. *Corol.* 1. In the ellipfe, the femiconjugate axis, CD or CE, is a mean proportional between CO and CP, the parts of the diameter OP of a circular fection of the cone, drawn through the centre C of the ellipfe, and parallel to the bafe of the cone. For DE is a double ordinate in this circle, being perpendicular to OP as well as to AB.

21. In like manner, in the hyperbola, the length of the femiconjugate axis, CD or CE, is a mean proportional between CO and CP; drawn parallel to the bafe, and meeting the fides of the cone in O and P. Or, if AO' be drawn parallel to the fide VB, and meet PC produced in O', making CO' = CO; and on this diameter O'P a circle be drawn parallel to the bafe: then the femiconjugate CD or CE will be an ordinate of this circle, being perpendicular to O'P as well as to AB.

Or, in both figures, the whole conjugate axis DE is a mean proportional between QA and BR, parallel to the bafe of the cone. See my Conic Sections, pa. 6.

In the parabola, both the tranfverfe and conjugate are infinite; for AB and BR are both infinite.

22. *Corol.* 2. In all the fections, AG will be equal to the parameter of the axis, if QG be drawn making the angle AQG equal to the angle BAR. In like manner Bg will be equal to the fame parameter, if Rg be drawn to make the angle BRg = the angle ABQ.

23. *Corol.* 3. Hence the upper hyperbolic fection, or fection of the oppofite cone, is equal and fimilar to the lower one. For the two fections have the fame tranfverfe or firft axis AB, and the fame conjugate or fecond axis DE, which is the mean proportional between AQ and RB; and they have alfo equal parame-

ters

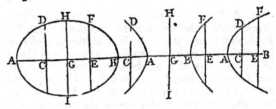

ters AG, B*g*. So that the two opposite sections make, as it were, but the two opposite ends of one entire section or hyperbola, the two being every where mutually equal and similar. Like the two halves of an ellipse, with their ends turned the contrary way.

24. *Corol.* 4. And hence, although both the transverse and conjugate axis in the parabola be infinite, yet the former is infinitely greater than the latter, or has an infinite ratio to it. For the transverse has the same ratio to the conjugate, as the conjugate has to the parameter, that is, as an infinite to a finite quantity, which is an infinite ratio.

The peculiar properties of each particular curve, will be best referred to the particular words ELLIPSE, HYPERBOLA, PARABOLA; and therefore it will only be proper here to lay down a few of the properties that are common to all the conic sections.

Some other General Properties.

25. From the foregoing definitions, &c, it appears, that the conic sections are in themselves a system of regular curves, naturally allied to each other; and that one is changed into another perpetually, when it is either increased, or diminished, in infinitum. Thus, the curvature of a circle being ever so little increased or diminished, passes into an ellipse; and again, the centre of the ellipse going off infinitely, and the curvature being thereby diminished, is changed into a parabola; and lastly, the curvature of a parabola being ever so little changed, there ariseth the first of the hyperbolas; the innumerable species of which will all of them arise orderly by a gradual diminution of the curvature; till this quite vanishing, the last hyperbola ends in a right line. From whence it is manifest, that every regular curvature, like that of a circle, from the circle itself to a right line, is a conical curvature, and is distinguished with its peculiar name, according to the divers degrees of that curvature.

26. That all diameters in a circle and ellipse intersect one another in the centre of the figure within the section: that in the parabola they are all parallel among themselves, and to the axis: but in the hyperbola, they intersect one another, without the figure, in the common centre of the opposite and conjugate sections.

27. In the circle, the *latus rectum*, or parameter, is double the distance from the vertex to the focus, which is also the centre. But in ellipses, the parameters are in all proportions to that distance, between the double and quadruple, according to their different species. While, in the parabola, the parameter is just quadruple that distance. And, lastly in hyperbolas, the parameters are in all proportions beyond the quadruple, according to their various kinds.

28. The first general property of the conic sections, with regard to the abscisses and ordinates of any diameter, is, that the rectangles of the abscisses are to each other, as the squares of their corresponding ordinates. Or, which is the same thing, that the square of any diameter is to the square of its conjugate, as the rectangle of two abscisses of that diameter, to the square of the ordinate which divides them. That is, in all the figures,

the rect. AC . CB : rect. AE . EB :: CD² : EF² :

But as, in the parabola the infinites CB and EB are in a ratio of equality, for this curve the same property becomes AC : AE :: CD² : EF², that is, in the parabola, the abscisses are as the squares of their ordinates.

Or, when one of the ordinates is the semiconjugate GH, dividing the diameter equally in the centre, the same general property becomes,

AG . GB or AG² : AC . CB :: GH² : CD²,
or   AB² : HI² :: AC . CB : CD².

29. From hence is derived the equation of the curves of the conic sections; thus, putting the diameter AB $= d$, its conjugate HI $= c$, abscis AC $= x$, and its ordinate CD $= y$; then is the other abscis CB $= d - x$ in the ellipse, or $d + x$ in the hyperbola, or d in the parabola; and hence the last analogy above, becomes

$d^2 : c^2 :: x . \overline{d \mp x}$ or $dx \mp x^2 : y^2$,

or $dy^2y^2 = c^2 . \overline{dx \mp x^2}$ is the general equation for all the conic sections; and, in particular, it is

$d^2y^2 = c^2 . \overline{dx - x^2}$ in the ellipse,
$d^2y^2 = c^2 . \overline{dx + x^2}$ in the hyperbola, and
$d^2y^2 = c^2dx$, or $dy^2 = c^2x$ in the parabola: Or the three equations may be otherwise expressed thus:

$y^2 = \dfrac{c^2}{d^2} . \overline{dx - x^2}$ in the ellipse,

$y^2 = \dfrac{c^2}{d^2} . \overline{dx + x^2}$ in the hyperbola, and

$y^2 = \dfrac{c^2}{d} x$ or $= px$ in the parabola, where the para-

meter $p = \dfrac{c^2}{d}$ the third proportional to the diameter and its conjugate, by the definition of it.

And from this one general proposition alone, which is easily derived from the section in the solid cone itself, together with the definitions only, as laid down above, all the other properties of all the sections may easily be derived, without any farther reference to the cone, and without mechanical descriptions of the curves in plano; as is done in my Treatise on Conic Sections, for the use of the Royal Mil. Acad.; in which also all the similar propositions in the ellipse and hyperbola are carried on word for word in them both.

The more ancient mathematicians, before the time of Apollonius Pergæus, admitted only the right cone into their geometry, and they supposed the section of it to be made by a plane perpendicular to one of its sides; and as the vertical angle of a right cone may be either right, acute, or obtuse, the same method of cutting these several cones, viz, by a plane perpendicular to one side, produced all the three conic sections. The parabola was called the section of a right-angled cone; the ellipse, the section of the acute-angled cone; and the hyperbola, the section of the obtuse-angled cone. But Apollonius, who, on account of his writings on this subject,

subject, obtained the appellation of *Magnus Geometra*, the Great Geometrician, observed, that these three sections might be obtained in every cone, both oblique and right, and that they depended on the different inclinations of the plane of the section to the cone itself. Apollon. Con. Halley's edit. lib. 1, p. 9.

Instead of considering these curves as sections cut from the solid cone, which is the true genuine way of all the ancients, and of the most elegant writers among the moderns, Descartes, and some others of the moderns, have given arbitrary constructions of curves on a plane, from which constructions they have demonstrated the properties of these, and have afterwards proved that some principal property of them belongs to such curves or sections as are cut from a cone; and hence it is inferred by them that those curves, so described on a plane, are the same with the conic sections.

The doctrine of the conic sections is of great use in physical and geometrical astronomy, as well as in the physico-mathematical sciences. The doctrine has been much cultivated by both ancient and modern geometricians, who have left many good treatises on the subject. The most ancient of these is that of Apollonius Pergæus, containing 8 books, the first 4 of which have often been published; but Dr. Halley's edition has all the eight. Pappus, in his Collect. Mathem. lib. 7, says that the first four of these were written by Euclid, though perfected by Apollonius, who added the other 4 to them. Among the moderns, the chief writers are Mydorgius de Sectionibus Conicis; Gregory St. Vincent's Quadratura Circuli & Sectionum Coni; De la Hire de Sectionibus Conicis; Trevigar Elem. Section. Con.; De Witt's Elementa Curvarum; Dr. Wallis's Conic Sections; De l'Hospital's Anal. Treat. of Conic Sections; Dr. Simson's Section. Con.; Milne's Elementa Section. Conicarum; Muller's Conic Sections; Steel's Conic Sections; Dr. Hamilton's elegant treatise; my own treatise, above cited; and at the writing of this, my friend Mr. Abram Robertson of Oxford is preparing a curious work on this subject, containing at the same time a treatise on the science, and a history of the writings relating to it.

CONICS, that part of the higher geometry, or geometry of curves, which considers the cone, and the several curve lines arising from the sections of it.

CONJUGATE *Axis*, or *Diameter*, in the Conic Sections, is the axis, or a diameter parallel to a tangent to the curve at the vertex of another axis, or diameter, to which that is a conjugate. Indeed the two are mutually conjugates to each other, and each is parallel to the tangent at the vertex of the other.

CONJUGATE HYPERBOLAS, also called *Adjacent Hyperbolas*, are such as have the same axes, but in the contrary order, the first or principal axis of the one being the 2d axis of the other, and the 2d axis of the former, the 1st axis of the latter. See art. 17 of CONIC SECTIONS.

CONJUNCTION, in Astronomy, is the meeting of the stars and planets in the same point or place in the heavens; and is either true or apparent.

True CONJUNCTION is when the line drawn through the centres of the two stars passes also through the centre of the earth. And *Apparent* CONJUNCTION is when that line does not pass through the earth's centre.

CONOID, is a figure resembling a cone, except that the slant sides from the base to the vertex are not straight lines as in the cone, but curved. It is generated by the revolution of a conic section about its axis; and it is therefore threefold, answering to the three sections of the cone, viz, the *Elliptical Conoid*, or spheroid, the *Hyperbolic Conoid*, and the *Parabolic Conoid*.

If a conoid be cut by a plane in any position, the section will be of the figure of some one of the conic sections; and all parallel sections, of the same conoid, are like and similar figures. When the section of the solid returns into itself, it is an ellipse; which is always the case in the sections of the spheroid, except when it is perpendicular to the axis; which position is also to be excepted in the other solids, the section being always a circle in that position. In the parabolic conoid, the section is always an ellipse, except when it is parallel to the axis. And in the hyperbolic conoid, the section is an ellipse, when its axis makes with the axis of the solid, an angle greater than that made by the said axe of the solid and the asymptote of the generating hyperbola; the section being an hyperbola in all other cases, but when those angles are equal, and then it is a parabola.

But when the section is parallel to the fixed axis, it is of the same kind with, and similar to the generating plane itself; that is, the section parallel to the axis, in the spheroid, is an ellipse similar to the generating ellipse; in the parabolic conoid it is a parabola, similar to the generating one; and in the hyperbolic conoid, it is an hyperbola similar to the generating one.

The section through the axis, which is the generating plane, is, in the spheroid the greatest of the parallel sections, but in the hyperboloid it is the least, and in the paraboloid those parallel sections are all equal.

The analogy of the sections of the hyperboloid to those of the cone, are very remarkable, all the three conic sections being formed by cutting an hyperboloid in the same positions as the cone is cut. Thus, let an hyperbola and its asymptote be revolved together about the transverse axis, the former describing an hyperboloid, and the latter a cone circumscribing it: then let it be supposed that they are both cut by one plane in any position; so shall the two sections be like, similar, and concentric figures: that is, if the plane cut both the sides of each, the sections will be concentric and similar ellipses; but if the cutting plane be parallel to the asymptote, or to the side of the cone, the sections will be parabolas; and in all other positions, the sections will be similar and concentric hyperbolas.

And this analogy of the sections will not seem strange, when it is considered that a cone is a species of the hyperboloid; or a triangle a species of the hyperbola, the axes being infinitely small. See my Mensuration, prop. 1, part 3, sect. 4, pag. 265 edit. 8vo.

CONON (*of* SAMOS), a respectable mathematician and philosopher, who flourished about the 130th olympiad, being a contemporary and friend of Archimedes, to whom Conon communicated his writings, and sent him some problems, which Archimedes received with approbation, saying they ought to be published while Conon was living, for he comprehends them with ease, and can give a proper demonstration of them.

At another time he laments the loss of Conon, thus admiring

admiring his genius. " How many theorems in geometry, says he, which at first seemed impossible, would in time have been brought to perfection ! Alas! Conon, though he invented many, with which he enriched geometry, had not time to perfect them, but left many n the dark, being prevented by death." He had an uncommon skill in mathematics, joined to an extraordinary patience and application. This is farther confirmed by a letter sent to Archimedes by a friend of Conon's. " Having heard of Conon's death, with whose friendship I was honoured, and with whom you kept an intimate correspondence ; as he was thoroughly versed in geometry, I greatly lament the loss of a sincere friend, and a person of surprising knowledge in mathematics. I then determined to send to you, as I had before done to him, a theorem in geometry, hitherto observed by no one."

Conon had some disputes with Nicoteles, who wrote against him, and treated him with too much contempt. Apollonius confesses it ; though he acknowledges that Conon was not fortunate in his demonstrations.

Conon invented a kind of volute, or spiral, different from that of Dynostratus ; but because Archimedes explained the properties of it more clearly, the name of the inventor was forgotten, and it was hence called Archimedes's volute or spiral.

As to Conon's astrological or astronomical knowledge, it may in some measure be gathered from the poem of Catullus, who describes it in the beginning of his verses on the hair of Berenice, the sister and wife of Ptolomy Euergetes, upon the occasion of Conon having given out that it was changed into a constellation among the stars, to console the queen for the loss, when it was stolen out of the temple, where she had consecrated it to the gods.

CONSECTARY, or *Corollary*, a consequence deduced from some foregoing principles.

CONSEQUENT, is the latter of the two terms of a ratio ; or that to which the antecedent is referred and compared. Thus, in the ratio *a : b,* or *a* to *b,* the latter term *b* is the consequent, and *a* is the antecedent.

CONSISTENT *Bodies*, is a term much used by Mr. Boyle, for such as are usually called *firm,* or *fixed bodies ;* in opposition to *fluid* ones.

CONSOLE, in Architecture, is an ornament cut upon the key of an arch, having a projecture or jetting, and occasionally serving to support small cornices, busts, and bases.

CONSONANCE, in Music, is commonly used in the same sense with *concord,* viz, for the union or agreement of two sounds produced at the same time, the one grave, the other acute, which is compounded together by such a proportion of each, as proves agreeable to the ear.

An unison is the first consonance, an eighth is the 2d, a fifth is the 3d ; and then follow the fourth, with the third and sixths, major and minor.

CONSTANT *Quantities* are such as remain invariably the same, while others increase or decrease. Thus, the diameter of a circle is a constant quantity ; for it remains the same while the abscisses and ordinates, or the sines, tangents, &c, are variable.

These are sometimes called *given,* or *invariable* or *per-*

5

manent quantities ; and in algebra it is now usual to represent them by the leading letters of the alphabet, *a, b, c,* &c ; while the variable ones are denoted by the last latters, *z, y, x,* &c.

CONSTELLATIONS, certain imaginary figures of birds, beasts, fishes, and other things in the heavens, within which are arranged certain stars. These assemblages are also sometimes called asterisms.

The ancients portioned out the firmament into several parts, or constellations ; reducing a certain number of stars under the representation of certain images, to assist the imagination and memory, to conceive or retain their number, order, and disposition, or even to distinguish the virtues they attributed to them.

The division of the heavens into constellations is very ancient ; being known to the most early authors, whether sacred or profane. In the book of Job the names of some of them are mentioned ; witness that sublime expostulation, *Canst thou restrain the sweet influence of the* Pleiades, *or loosen the bands of* Orion ? And the same may be observed of the oldest among the heathen writers, Hesiod and Homer.

The division of the ancients took in only the visible firmament, or so much as came under their notice, as visible to the naked eye. The first or earliest of these, is contained in the catalogue of Ptolomy, given in the 7th book of his Almagest, prepared, as he assures us, from his own observations, compared with those of Hipparchus, and the other ancient astronomers. In this catalogue Ptolomy has formed 48 constellations. Of these, 12 are about the ecliptic, commonly called the 12 signs ; 21 to the north of it ; and 15 to the south. The northern constellations are, the Little Bear, the Great Bear, the Dragon, Cepheus, Bootes, the Northern Crown, Hercules, the Harp, the Swan, Cassiopeia, Perseus, Auriga, Ophiucus or Serpentary, the Serpent, the Arrow, the Eagle, the Dolphin, the Horse, Pegasus, Andromeda, and the Triangle.

The constellations about the ecliptic are Aries, Taurus, Gemini, Cancer, Leo, Virgo, Libra, Scorpio, Sagittarius, Capricorn, Aquarius, and Pisces : or according to the English names, the Ram, the Bull, the Twins, the Crab, the Lion, the Virgin, the Balance, the Scorpion, the Archer, the Goat, the Water bearer, and the Fishes.

The Southern constellations are, the Whale, Orion, the Eridanus, the Hare, the Great Dog, the Little Dog, the Ship, the Hydra, the Cup, the Raven, the Centaur, the Wolf, the Altar, the Southern Crown, and the Southern Fish.

The other stars not comprehended under these constellations, yet visible to the naked eye, the ancients called *informes,* or *sporades,* some of which the modern astronomers have since reduced into new figures, or constellations. Ptolomy has set down the longitude and latitude of all these stars to about the year of Christ 137, amounting to the number of 1022, viz,

in the northern constellations - -	360
in the zodiacal constellations - -	346
in the southern constellations - -	316
in all of Ptolomy's catalogue - -	1022

Among the modern astronomers, Tycho Brahe is the first

firſt who determined, with exactneſs, and in conſequence of his own obſervations, the long. and lat. of the fixed ſtars, out of which he formed 45 conſtellations; of theſe, 43 were of the old ones deſcribed by Ptolomy, to which Tycho added the Coma Berenices, and Antinous; but he omits 5 of the old ſouthern conſtellations, viz, the Centaur, the Wolf, the Altar, the Southern Crown, and Southern Fiſh; which he could not obſerve, becauſe of the high northern latitude of Uranibourg.

After Tycho, Bayer gave the figures of 60 conſtellations, very exactly repreſented, and with tables annexed, having added, to the 48 old ones of Ptolomy, the following 12 about the ſouth pole, viz, the Peacock, the Toucan, the Crane, the Phœnix, the Dorado, the Flying Fiſh, the Hydra, the Chameleon, the Bee, the Bird of Paradiſe, the Triangle, and the Indian. Beſides accurately diſtinguiſhing the relative ſize and the ſituation of every ſtar, Bayer marks the ſtars in each conſtellation with the letters of the Greek and Roman alphabets, ſetting the firſt letter α to the firſt or principal ſtar in each conſtellation, β to the 2d in order, γ to the 3d, and ſo on; a very uſeful method of noting and deſcribing the ſtars, which has been uſed by all aſtronomers ſince, and who have farther enlarged this method, by adding the ordinal numbers 1, 2, 3, &c, to the other ſtars diſcovered ſince his time, when any conſtellation contains more than can be marked by the two alphabets. The number and order of the ſtars, as mentioned by Bayer, are,

of the 1ſt magnitude	17
of the 2d magnitude	63
of the 3d magnitude	196
of the 4th magnitude	415
of the 5th magnitude	348
of the 6th magnitude	341
of the unformed ſtars	326
in all	1706

After Bayer, a catalogue, with new conſtellations, was publiſhed by Schiller, in 1627, in a work called Coelum Stellatum Chriſtianum, the Chriſtian Starry Heaven, in which he ſubſtitutes, very improperly, other figures of the conſtellations, and names, taken from the ſacred ſcriptures, inſtead of the old ones.

In the year 1665, Riccioli publiſhed his Aſtronomy Reformed, containing a catalogue of the ſtars in 62 conſtellations, viz, the 60 of Bayer, with the Coma Berenices and Antinous of Tycho. He diſtributes the ſtars in all the conſtellations into four claſſes. In the firſt of theſe claſſes are contained thoſe ſtars determined by his own obſervations, and thoſe of Grimaldi. In the ſecond are thoſe ſtars which had been aſcertained by Tycho Brahe and Kepler. In the 3d are the ſtars determined by Hipparchus and Ptolomy. And the 4th claſs conſiſts of thoſe of the ſouthern hemiſphere diſcovered by Navigators, who have aſcertained their places in a more or leſs accurate manner; in which he has marked the longitudes and latitudes for the year 1700, the period to which he has reduced all his obſervations. This catalogue was followed by a number of celeſtial ſchemes and maps of the heavens, publiſhed in 1673 by Pardies, who has repreſented very carefully all the conſtellations, with the ſtars they contain. After this, Vi-

talis publiſhed a catalogue of the fixed ſtars in his Tables of the Primum Mobile, in which their longitudes and latitudes, with the right-aſcenſions and declinations are ſet down for the year 1675.

Some time after this, Royer publiſhed maps of the heavens, reduced into 4 tables, with a catalogue of the fixed ſtars for the year 1700. To the ſtars marked by Bayer, he adds a number of ſtars not before ſeen, with others taken from the tables of Riccioli, and not mentioned by Bayer: he alſo forms, out of the unformed ſtars, eleven other conſtellations. Five of theſe are to the north, and are called the Giraffe, the River Jordan, the River Tigris, the Sceptre, and the Flower-de-luce; with 6 on the ſouth part, which are the Dove, the Unicorn, the Croſs, the Great Cloud, the Little Cloud, and the Rhomboide. To this work Royer has joined the catalogue of the ſouthern ſtars obſerved by Dr. Halley at the iſland of St. Helena.

Hevelius has alſo improved upon the labours of thoſe who went before him, and collected together ſeveral ſtars of the before unformed claſs into ſome new conſtellations. Theſe are, the Unicorn, the Cameloparda-lis, deſcribed by Bartſchius, the Sextant of Urania, the Dogs, the Little Lion, the Lynx, the Fox and Gooſe, the Sobieſki's Crown, the Lizard, the Little Triangle, and the Cerberus; to which Gregory has added the Ring and the Armilla. Some of theſe new conſtellations however anſwer to thoſe of Royer, as the Camelopardal to the Giraffe, the Dogs to the River Jordan, and the Fox to the River Tigris. The latitudes and longitudes are added for the year 1700.

Finally, Flamſteed has given a catalogue of the fixed ſtars, not only much more correct, but much larger than thoſe of all that went before him. He has ſet down the longitude, latitude, right aſcenſion, and polar diſtance of 2934 ſtars, as they were at the beginning of 1690, all determined from his own obſervations. He diſtinguiſhes all the ſtars into ſeven claſſes, or orders of magnitude, diſtinguiſhing thoſe of Bayer by his letters, and marking their variation in right aſcenſion, for ſhewing their ſituation in the ſucceeding years. See the term CATALOGUE.

This catalogue was followed by an Atlas Cœleſtis, publiſhed at London in the year 1729, deſcribing, in ſeveral ſchemes, the figures of the conſtellations ſeen in our hemiſphere, with the exact poſition of the fixed ſtars, with reſpect to the circles of the ſphere, as reſulting from the laſt catalogue corrected by Flamſteed. And ſtill later obſervations, made with farther improved teleſcopes, have greatly enlarged the number and accuracy of the ſtars; but the number of the conſtellations remains the ſame as above deſcribed, except that an attempt has lately been made by Dr. Hill to add to the liſt 14 new ones, formed out of more of the cluſters of unformed ſtars.

Beſide the literal marks of the ſtars introduced by Bayer, it is uſual alſo to diſtinguiſh them by that part of the conſtellation in which they are placed; and many of them again have their peculiar names; as Arcturus, between the knees of Bootes; Gemina, or Lucida, in the Corona Septentrionalis, or Northern Crown; Palilitium, or Aldebaran, in the Bull's eye, Pleiades in his neck, and Hyades in his forehead; Caſtor and Pollux in the heads of Gemini; Capella, with the Hœdi in the

ſhoulder

shoulder of Auriga; Regulus, or Cor Leonis, the Lion's Heart; Spica Virginis in the hand, and Vindemiatrix in the shoulder of Virgo; Antares or Cor Scorpionis, the Scorpion's Heart; Fomalhaut, in the mouth of Piscis Australis, or Southern Fish; Regel, in the foot of Orion; Sirius, in the mouth of Canis Major, the Great Dog; Procyon, in the back of Canis Minor, the Little Dog; and the Pole Star, the last in the tail of Ursa Minor, the Little Bear.

The Greek and Roman poets, from the ancient theology, give wild and romantic fables about the origin of the constellations, probably derived from the hieroglyphics of the Egyptians, and transmitted, with some alterations, from them to the Greeks, who probably obscured them greatly with their own fables. See Hyginus's Poeticon Astron.; Riccioli Almagest. lib. 6. cap. 3, 4, 5; Shelburne's Notes upon Manilius; Bailly's Antient Astronomy; and Gebelin's Monde Primitif, vol. 4: from the whole of which it is made probable, that the invention of the signs of the zodiac, and probably of most of the other constellations of the sphere, is to be ascribed to some very ancient nation, inhabiting the northern temperate zone, probably what is now called Tartary, or the parts to the northward of Persia and China; and from thence transmitted through China, India, Babylon, Arabia, Egypt, Greece, &c.

It is a very probable conjecture, that the figures of the signs in the zodiac, are descriptive of the seasons of the year, or months, in the sun's path: thus, the first sign Aries, denotes, that about the time when the sun enters that part of the ecliptic, the lambs begin to follow the sheep; that on the sun's approach to the 2d constellation, Taurus, the Bull, is about the time of the cows bringing forth their young. The third sign, now Gemini, was originally two kids, and signified the time of the goats bringing forth their young, which are usually two at a birth, while the former, the sheep and cow, commonly produce only one. The 4th sign, Cancer, the Crab, an animal that goes side-ways and backwards, was placed at the northern solstice, the point where the sun begins to return back again from the north to the southward. The 5th sign, Leo, the Lion, as being a very furious animal, was thought to denote the heat and fury of the burning sun, when he has left Cancer, and entered the next sign Leo. The succeeding constellation, the 6th in order, received the sun at the time of ripening corn and approaching harvest; which was aptly expressed by one of the female reapers, with an ear of corn in her hand; viz, Virgo the maid. The ancients gave to the next sign Scorpio, two of the 12 divisions of the zodiac: Autumn, which affords fruits in great abundance, affords the means and causes of diseases, and the succeeding time is the most unhealthy of the year; expressed by this venemous animal, here spreading out his long claws into the one sign, as threatening mischief, and in the other brandishing his tail to denote the completion of it. The fall of the leaf was the season of the ancient hunting; for which reason the stars which marked the sun's place at this season, into the constellation Sagittary, a huntsman with his arrows and his club, the weapons of destruction for the large creatures he pursued. The reason of the Wild Goat's being chosen to mark the southern sol-

stice, when the sun has attained his extreme limit that way, and begins to return and mount again to the northward, is obvious enough; the character of that animal being, that it is mostly climbing, and ascending some mountain as it browzes. There yet remain two of the signs of the zodiac to be considered with regard to their origin, viz, Aquarius and Pisces. As to the former, it is to be considered that the winter is a wet and uncomfortable season; this therefore was expressed by Aquarius, the figure of a man pouring out water from an urn. The last of the zodiacal constellations was Pisces, a couple of fishes, tied together, that had been caught: The lesson was, the severe season is over, your flocks do not yet yield their store; but the seas and rivers are open, and there you may take fish in abundance.

Through a vain and blind zeal, rather than through any love for the science, some persons have been moved to alter either the figures of the constellations, or their names. Thus, venerable Bede, instead of the profane names and figures of the twelve zodiacal constellations, substituted those of the 12 apostles; which example was followed by Schiller, who completed the reformation, and gave scripture names to all the constellations in the heavens. Thus, Aries, or the Ram, was changed into Peter; Taurus, or the Bull, into St. Andrew; Andromeda, into the Sepulchre of Christ; Lyra, into the Manger of Christ; Hercules, into the Magi coming from the East; the Great Dog, into David; and so on. And Weigelius, professor of Mathematics in the uni versity of Jena, made a new order of constellations; changing the firmament into a Cœlum Heraldicum; and introducing the arms of all the princes in Europe, by way of constellations. Thus Ursa major, the Great Bear, he transformed into the elephant of the kingdom of Denmark; the Swan, into the Ruta with swords of the House of Saxony; Ophiuchus, into the Cross of Cologne; the Triangle, into Compasses, which he calls the Symbol of Artificers; and the Pleiades into the Abacus Pythagoricus, which he calls that of merchants; &c.

But the more judicious among astronomers never approved of such innovations; as they only tend to introduce confusion into astronomy. The old constellations are therefore still retained; both because better could not be substituted, and likewise to keep up the greater correspondence and uniformity between the old astronomy and the new. See CATALOGUE.

CONSTRUCTION, in Geometry, the art or manner of drawing or describing a figure, scheme, the lines of a problem, or such like.

CONSTRUCTION of Equations, in Algebra, is the finding the roots or unknown quantities of an equation, by geometrical construction of right lines or curves; or the reducing given equations into geometrical figures. And this is effected by lines or curves according to the order or rank of the equation.

The roots of any equation may be determined, that is, the equation may be constructed, by the intersections of a straight line with another line or curve of the same dimensions as the equation to be constructed: for the roots of the equation are the ordinates of the curve at the points of intersection with the right line; and it is well known that a curve may be cut by a right line in as many points as its dimensions amount to. Thus, then,

then, a fimple equation will be conftructed by the interfection of one right line with another: a quadratic equation, or an affected equation of the 2d rank, by the interfections of a right line with a circle, or any of the conic fections, which are all lines of the 2d order; and which may be cut, by the right line, in two points, thereby giving the two roots of the quadratic equation. A cubic equation may be conftructed by the interfection of the right line with a line of the 3d order: and fo on.

But if, inftead of the right line, fome other line of a higher order be ufed; then the 2d line, whofe interfections with the former are to determine the roots of the equation, may be taken as many dimenfions lower, as the former is taken higher. And, in general, an equation of any height will be conftructed by the interfections of two lines whofe dimenfions, multiplied together, produce the dimenfion of the given equation. Thus, the interfections of a circle with the conic fections, or of thefe with each other, will conftruct the biquadratic equations, or thofe of the 4th power, becaufe $2 \times 2 = 4$; and the interfections of the circle or conic fections with a line of the 3d order, will conftruct the equations of the 5th and 6th power; and fo on.—For example,

To conftruct a Simple Equation. This is done by refolving the given fimple equation into a proportion, or finding a third or 4th proportional, &c. Thus, 1. If the equation be $ax = bc$; then $a : b :: c : x = \dfrac{bc}{a}$, the fourth proportional to a, b, c.

2. If $ax = b^2$; then $a : b :: b : x = \dfrac{b^2}{a}$, a third proportional to a and b.

3. If $ax = b^2 - c^2$; then, fince $b^2 - c^2 = \overline{b + c} \times \overline{b - c}$, it will be $a : b + c :: b - c : x = \dfrac{\overline{b + c} \times \overline{b - c}}{a}$, a fourth proportional to a, $b + c$ and $b - c$.

4. If $ax = b^2 + c^2$; then conftruct the right-angled triangle ABC, whofe bafe is b, and perpendicular is c, fo fhall the fquare of the hypothenufe be $b^2 + c^2$, which call h^2; then the equation is $ax = h^2$, and $x = \dfrac{h^2}{a}$ a third proportional to a and h.

To conftruct a Quadratic Equation.

1. If it be a fimple quadratic, it may be reduced to this form $x^2 = ab$; and hence $a : x :: x : b$, or $x = \sqrt{ab}$ a mean proportional between a and b. Therefore upon a ftraight line take AB $= a$, and BC $= b$; then upon the diameter AC defcribe a femicircle, and raife the perpendicular BD to meet it in D; fo fhall BD be $= x$ the mean proportional fought between AB and BC, or between a and b.

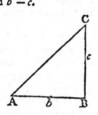

2. If the quadratic be affected, let it firft be $x^2 + 2ax = b^2$; then form the right-angled triangle whofe bafe AB is a, and perpendicular BC is b; and with the centre A and radius AC defcribe the femi-

circle DCE; fo fhall DB and BE be the two roots of the given quadratic equation $x^2 + 2ax = b^2$.

3. If the quadratic be $x^2 - 2nx = b^2$, then the conftruction will be the very fame as of the preceding one $x^2 + 2ax = b^2$.

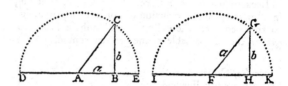

4. But if the form be $2ax - x^2 = b^2$: form a right-angled triangle whofe hypothenufe FG is a, and perpendicular GH is b; then with the radius FG and centre F defcribe a femi-circle IGK; fo fhall IH and HK be the two roots of the given equation $2ax - x^2 = b^2$, or $x^2 - 2ax = - b^2$. See Maclaurin's Algebra, part 3, cap. 2, and Simpfon's Algebra, pa. 267.

To conftruct Cubic and Biquadratic Equations.— Thefe are conftructed by the interfections of two conic fections; for the equation will rife to 4 dimenfions, by which are determined the ordinates from the 4 points in which thefe conic fections may cut one another; and the conic fections may be affumed in fuch a manner, as to make this equation coincide with any propofed biquadratic: fo that the ordinates from thefe 4 interfections will be equal to the roots of the propofed biquadratic. When one of the interfections of the conic fection falls upon the axis, then one of the ordinates vanifhes, and the equation, by which thefe ordinates are determined, will then be of 3 dimenfions only, or a cubic; to which any propofed cubic equation may be accommodated. So that the three remaining ordinates will be the roots of that propofed cubic. The conic fections for this purpofe fhould be fuch as are moft eafily defcribed; the circle may be one, and the parabola is ufually affumed for the other.

Vieta, in his Canonica Recenfione Effectionum Geometricarum, and Ghetaldus, in his Opus Pofthumum de Refolutione & Compofitione Mathematica, as alfo Des Cartes, in his Geometria, have fhewn how to conftruct fimple and quadratic Equations. Des Cartes has alfo fhewn how to conftruct cubic and biquadratic equations, by the interfection of a circle and a parabola: And the fame has been done more generally by Baker in his Clavis Geometrica, or Geometrical Key. But the genuine foundation of all thefe conftructions was firft laid and explained by Slufius in his Mefolabium, part 2. This doctrine is alfo pretty well handled by De la Hire, in a fmall treatife, called La Conftruction des Equations Analytiques, annexed to his Conic Sections. Newton, at the end of his Algebra, has given the conftruction of cubic and biquadratic equations mechanically; as alfo by the conchoid and ciffoid, as well as the conic fections. See alfo Dr. Halley's Conftruction of Cubic and Biquadratic Equations; Colfon's, in the Philof. Tranf.; the Marquis de l'Hofpital's Traite Analytique des Sections Coniques; Maclaurin's Algebra, part 3, c. 3 &c.

CONTACT, the relative ftate of two things that touch each other, but without cutting or entering; or whofe furfaces join to each other without any interftice.

The

The contact of curve lines or surfaces, with either straight or curved ones, is only in points; and yet these points have different proportions to one another, as is shewn by Mr. Robartes, in the Philos. Transf. vol. 27 pa. 470; or Abr. vol. 4. pa. 1.

Because few or no surfaces are capable of touching in all points, and the cohesion of bodies is in proportion to their contact, those bodies will adhere fastest together, that are capable of the greatest contact.

Angle of CONTACT is the opening between a curve line and a tangent to it, particularly the circle and its tangent; as the angle formed at A between BA and AC, at the point of contact A.

It is demonstrated by Euclid, that the line CA standing perpendicular to the radius DA, touches the circle only in one point: and that no other right line can be drawn between the tangent and the circle.

Hence, the angle of contact is less than any rectilinear angle; and the angle of the semi-circle between the radius DA and the arch AB, is greater than any rectilinear acute angle.

This seeming paradox of Euclid has exercised the wits of mathematicians: it was the subject of a long controversy between Peletarius and Clavius; the former of whom maintained that the angle of contact is heterogeneous to a rectilinear one; as a line is to a surface; the latter maintained the contrary.

Dr. Wallis has a formal treatise on the angle of contact, and of the semi-circle; where, with other great mathematicians, he approves of the opinion of Peletarius.

CONTENT, a term often used for the measurement of bodies and surfaces, whether solid or superficial; or the capacity of a vessel and the area of a space; being the quantity either of matter or space included within certain bounds or limits.

CONTIGUITY, the relation of bodies touching one another.

CONTIGUOUS, a relative term, understood of things disposed so near each other, that they join their surfaces, or touch.

CONTIGUOUS *Angles*, are such as have one leg or side common to each angle; and are otherwise called *adjoining angles*; in contradistinction to those made by continuing their legs through the point of contact, which are called *opposite* or *vertical angles*.

CONTINENT, a terra firma, main-land, or a large extent of country, not interrupted by seas: so called, in opposition to island, peninsula, &c.

The world is usually divided into two grand continents, the old and the new: the old continent comprehends Europe, Asia, and Africa; the new, North and South America. Since the discovery of New Holland and New South Wales, it is a doubt with many whether to call that vast country an island or a continent.

CONTINGENT *Line*, the same with tangent line in Dialling, being the intersection of the planes of the dial and equinoctial, and at right angles to the substilar line.

CONTINUAL PROPORTIONALS, are a series of three or more quantities compared together, so that the ratio is the same between every two adjacent terms, viz between the 1st and 2d, the 2d and 3d, the 3d and 4th, &c. As 1, 2, 4, 8, 16, &c, where the terms continually increase in a double ratio; or 12, 4, $\frac{4}{3}$, $\frac{4}{9}$, where the terms decrease in a triple ratio.

A series of continual or continued proportionals, is otherwise called a *progression*.

CONTINUED *Quantity*, or *Body*, is that whose parts are joined and united together.

CONTINUED *Proportion*, is that in which the consequent of the first ratio is the same with the antecedent of the second; as in these, 3 : 6 :: 6 : 12. See CONTINUAL *Proportion*.

On the contrary, if the consequent of the first ratio be different from the antecedent of the second, the proportion is called *discrete*: as 3 : 6 :: 4 : 8.

CONTRA-HARMONICAL *Proportion*, that relation of three terms, in which the difference of the first and second is to the difference of the 2d and 3d, as the 3d is to the first. Thus, for instance, 3, 5, and 6, are numbers contra-harmonically proportional; for 2 · 1 :: 6 : 3.

CONTRA-MURE, in Fortification, is a little wall built before another partition wall, to strengthen it, so that it may receive no damage from the adjacent buildings.

CONTRATE-WHEEL, is that wheel in watches which is next to the crown, whose teeth and hoop lie contrary to those of the other wheels; from whence comes its name.

CONTRAVALLATION, *Line of*, in Fortification, is a trench, guarded with a parapet; being made by the besiegers, between them and the place besieged, to secure themselves on that side, and stop the sallies of the garrison. It is made without musket-shot of the town; sometimes going quite around it, and sometimes not, as occasion may require. The besiegers lie between the lines of circumvallation and contravallation: but it is now seldom used.

CONVERGING *Curves*. See CURVE.

CONVERGING, or CONVERGENT *Lines*, in Geometry, are those that continually approximate, or whose distance becomes continually less and less the farther they are continued, till they meet: in opposition to *divergent* lines, whose distance becomes continually greater.

Lines that converge the one way, diverge the other.

CONVERGING *Rays*, in Optics, are such as incline towards one another in their passage, and in Dioptrics, are those rays which, in their passage out of one medium into another of a different density, are refracted towards one another; so that, if far enough continued, they will meet in a point or focus.

CONVERGING *Series*, a series of terms or quantities, that always decrease the farther they proceed, or which tend to a certain magnitude or limit: in opposition to diverging series, or such as become larger and larger continually. See SERIES.

CONVERSE. A proposition is said to be the converse of another, when, after drawing a conclusion from something first supposed, we return again, and, making a supposition of what had before been concluded, draw from thence as a conclusion what before was made the supposition.

Plate XXXI.

The COPERNICAN or SOLAR SYSTEM

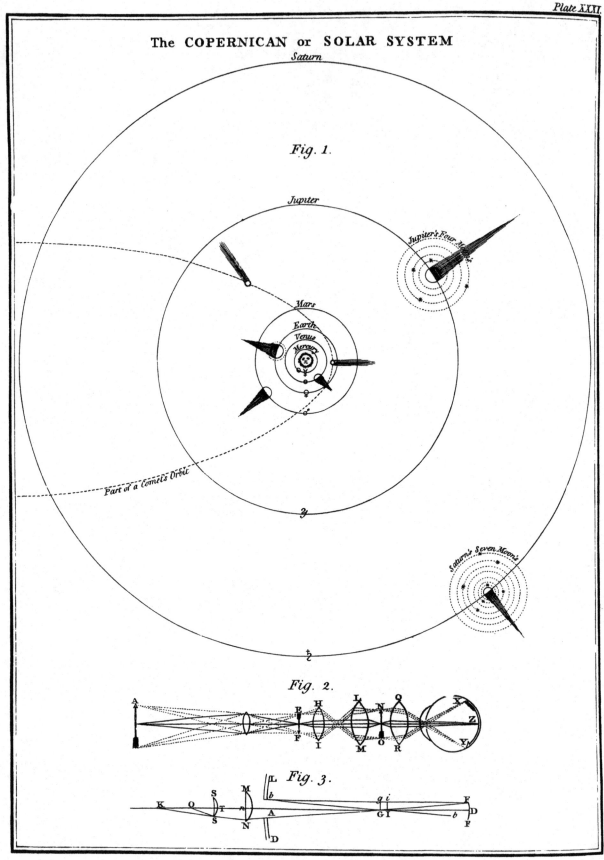

Saturn

Fig. 1.

Jupiter

Jupiter's Four Moons

Mars
Earth
Venus
Mercury

Part of a Comet's Orbit

♃

Saturn's Seven Moons

♄

Fig. 2.

Fig. 3.

suppofition. Thus, when it is fuppofed that the two fides of a triangle are equal, and thence demonftrate or conclude that the two angles oppofite to thofe fides are equal alfo; then the converfe is to fuppofe that the two angles of a triangle are equal, and thence to prove or conclude that the fides oppofite to thofe angles are alfo equal.

CONVERSE *Direction*, in Aftrology, is ufed in oppofition to *direct* direction; that is, by the latter the promoter is carried to the fignificator, according to the order of the figns: whereas by the other it is carried from eaft to weft, contrary to the order of the figns.

CONVERSION, or CONVERTENDO, is when there are four proportionals, and it is inferred, that the firft is to its excefs above the 2d, as the third to its excefs above the 4th: according to Euclid, lib. 5, def. 17.

Thus, if it be - - - $8 : 6 :: 4 : 3$,
then convertendo, or by converfion, $8 : 2 :: 4 : 1$.
Or if there be $a : b :: c : d$,
then convertendo, or by converfion, $a : a - b :: c : c - d$.

CONVEX, round or curved and protuberant outwards, as the outfide of a globular body.

CONVEX *Lens, Mirror*; &c. See LENS, MIRROR, &c.

CONVEXITY, the exterior or outward furface of a convex or round body.

COPERNICAN, fomething relating to Copernicus. As, the

COPERNICAN *Sphere*. See SPHERE.

COPERNICAN *Syftem*, is that fyftem of the world, in which it is fuppofed that the fun is at reft in the centre, and the earth and planets all moving around him in their own orbits.

Here it is fuppofed, that the heavens and ftars are at reft; and the diurnal motion which they appear to have, from eaft to weft, is imputed to the earth's diurnal motion from weft to eaft.

This fyftem was maintained by many of the ancients; particularly Ecphantus, Seleucus, Ariftarchus, Philolaus, Cleanthes Samius, Nicetas, Heraclides Ponticus, Plato, and Pythagoras; from the laft of whom it was anciently called the Pythagoric, or Pythagorean Syftem.

This fyftem was alfo held by Archimedes, in his book of the number of the Grains of Sand; but after him it became neglected, and even forgotten, for many ages; till about 300 years fince, when Copernicus revived it; from whom it took the new name of the Copernican Syftem. See the next article.

COPERNICUS (NICHOLAS), an eminent aftronomer, was born at Thorn in Pruffia, January 19, 1473. He was inftructed in the Latin and Greek languages at home; and afterward fent to Cracow, where he ftudied philofophy, mathematics, and medicine: though his genius was naturally turned to mathematics, which he chiefly ftudied, and purfued through all its various branches.

He fet out for Italy at 23 years of age; ftopping at Bologna, that he might converfe with the celebrated aftronomer of that place, Dominic Maria, whom he affifted for fome time in making his obfervations. From hence he paffed to Rome, where he was prefently confidered as not inferior to the famous Regiomontanus. Here he foon acquired fo great a reputation, that he

was chofen profeffor of mathematics, which he taught there for a long time with the greateft applaufe: and here alfo he made fome aftronomical obfervations about the year 1500.

Afterward, returning to his own country, he began to apply his fund of obfervations and mathematical knowledge, to correcting the fyftem of aftronomy which then prevailed. He fet about collecting all the books that had been written by philofophers and aftronomers, and to examine all the various hypothefes they had invented for the folution of the celeftial phenomena; to try if a more fymmetrical order and conftitution of the parts of the world could not be difcovered, and a more juft and exquifite harmony in its motions eftablifhed, than what the aftronomers of thofe times fo eafily admitted. But of all their hypothefes, none pleafed him fo well as the Pythagorean, which made the fun to be the centre of the fyftem, and fuppofed the earth to move both round the fun, and alfo round its own axis. He thought he difcerned much beautiful order and proportion in this; and that all the embarraffment and perplexity, from epicycles and excentrics, which attended the Ptolemaic hypothefes, would here be entirely removed.

This fyftem he began to confider, and to write upon when he was about 35 years of age. He carefully contemplated the phenomena; made mathematical calculations; examined the obfervations of the antients, and made new ones of his own; till, after more than 20 years chiefly fpent in this manner, he brought his fcheme to perfection, eftablifhing that fyftem of the world which goes by his name, and is now univerfally received by all philofophers.

This fyftem however was at firft looked upon as a moft dangerous herefy, and his work had long been finifhed and perfected, before he could be prevailed upon to give it to the world, being ftrongly urged to it by his friends. At length yielding to their intreaties, it was printed, and he had but juft received a perfect copy, when he died the 24th of May 1543, at 70 years of age; by which it is probable he was happily relieved from the violent fanatical perfecutions of the church, which were but too likely to follow the publication of his aftronomical opinions; and which indeed was afterward the fate of Galileo, for adopting and defending them.

The above work of Copernicus, firft printed at Norimberg in folio, 1543, and of which there have been other editions fince, is intitled *De Revolutionibus Orbium Cæleftium*, being a large body of aftronomy, in 6 books.

When Rheticus, the difciple of our author, returned out of Pruffia, he brought with him a tract of Copernicus, on plane and fpherical trigonometry, which he had printed at Norimberg, and which contained a table of fines. It was afterward printed at the end of the firft book of the Revolutions. An edition of our author's great work was alfo publifhed in 4to at Amfterdam in 1617, under the title of *Aftronomia Inftaurata*, illuftrated with notes by Nicolas Muler of Groningen.

COPERNICUS, the name of an aftronomical inftrument, invented by Whifton, to fhew the motion and phenomena of the planets, both primary and fecondary. It is founded upon the Copernican fyftem, and therefore called by his name.

COR CAROLI, Charles's Heart, an extra-conftellated ftar of the 2d magnitude in the northern hemifphere, between the Coma Berenices and Urfa Major; fo called by Sir Charles Scarborough, in honour of king Charles I.

COR HYDRÆ, the Hydra's Heart, a ftar of the 2d magnitude, in the Heart of the conftellation Hydra.

COR LEONIS, Lion's Heart, or Regulus, a ftar of the firft magnitude in the conftellation Leo.

COR SCORPII. See ANTARES.

CORBEILS, in Fortification, are little bafkets about a foot and a half high, 8 inches broad at the bottom, and 12 at the top; which being filled with earth, are fet againft one another on the parapet, or elfewhere, leaving certain port-holes, from whence to fire under cover upon the enemy.

CORBEL, in Architecture, the reprefentation of a bafket, fometimes feen on the heads of caryatides.

CORBEL, or CORBIL, is alfo ufed, in Building, for a fhort piece of timber placed in a wail, with its end projecting out 6 or 8 inches, as occafion ferves, in the manner of a fhouldering-piece.

CORBET, the fame as CORBEL.

CORDON, in Fortification, a row of ftones jutting out between the rampart and the bafis of the parapet, like the tore of a column. The cordon ranges round the whole fortrefs, and ferves to join the rampart, which is aflope, and the parapet, which is perpendicular, more agreeably together.

In fortifications raifed of earth, this fpace is filled up with pointed ftakes, inftead of a cordon.

CORDS, in Mufic, are the founds produced by an inftrument or the voice.

CORIDOR, or CORRIDOR, in Fortification, is the covert-way lying round about the whole compafs of the works of a place, between the outfide of the moat and the pallifadoes, being about 20 yards broad.

CORIDOR is alfo ufed, in Architecture, for a gallery or long aile, around a building, leading to feveral chambers at a diftance from each other, fometimes wholly inclofed, and fometimes open on one fide.

CORINTHIAN Order, of Architecture, is the 4th in order, or the 5th and laft according to Scamozzi and Le Clerc.

This order was invented by an Athenian Architect, and is the richeft and moft delicate of them all; its capital being adorned with rows of leaves, and of 8 volutas, which fupport the abacus. The height of its column is 10 diameters, and its cornice is fupported by modillions.

CORNEA Tunica, the fecond coat of the eye; fo called from its fubftance refembling the horn of a lantern. This is fituated in the fore-part; and is furrounded by the fclerotica. It has a greater convexity than the reft of the eye, and is a portion of a fmall fphere, or rather fpheroid, and confolidates the whole eye.

CORNICE, CORNICHE, or CORNISH, the third and uppermoft part of the entablature of a column, or the uppermoft ornament of any wainfcotting, &c.

COROLLARY, or CONSECTARY, a confequence drawn from fome propofition or principles already advanced or demonftrated, and without the aid of any other propofition: as if from this theorem, *That a triangle which has two equal fides, has alfo two equal*

angles, this confequence fhould be drawn, *that a triangle which hath the three fides equal, has alfo its three angles equal.*

CORONA, *Crown* or *Crowning,* in Architecture, the flat and moft advanced part of the cornice; fo called becaufe it crowns the cornice and entablature: by the workmen it is called the *drip,* as ferving by its projecture to fcreen the reft of the building from the rain.

CORONA, in Optics, a luminous circle, ufually coloured, round the fun, moon, or largeft planets. See HALO.

CORONA *Borealis,* or *Septentrionalis,* the *Northern Crown* or *Garland,* a conftellation of the northern hemifphere, being one of the 48 old ones. It contains 8 ftars according to the catalogue of Ptolomy, Tycho, and Hevelius; but according to the Britannic Catalogue, 21.

CORONA *Auftralis,* or *Meridianalis,* the *Southern Crown,* a conftellation of the fouthern hemifphere, whofe ftars in Ptolomy's catalogue are 13; in the Britifh Catalogue, 12.

CORPUSCLE the diminutive of corpus, ufed to exprefs the minute parts, or particles, that conftitute natural bodies; meaning much the fame as *atoms.*

Newton fhews a method of determining the fizes of the corpufcles of bodies, from their colours.

CORPUSCULAR *Philofophy,* that fcheme or fyftem of phyfics, in which the phenomena of bodies are accounted for, from the motion, reft, pofition, &c, of the corpufcles or atoms of which bodies confift.

The Corpufcular philofophy, which now flourifhes under the name of the mechanical philofophy, is very ancient. Leucippus and Democritus taught it in Greece; from them Epicurus received it, and improved it; and from him it was called the *Epicurean Philofophy.*

Leucippus, it is faid, received it from one Mochus, a Phenician phifiologift, before the time of the Trojan war, and the firft who philofophized about atoms: which Mochus is, according to the opinion of fome, the Mofes of the Scriptures.

After Epicurus, the corpufcular philofophy gave way to the peripatetic, which became the popular fyftem. Thus, inftead of atoms, were introduced fpecific and fubftantial forms, qualities, fympathies, &c, which amufed the world, till Gaffendus, Charleton, Defcartes, Boyle, Newton, and others, retrieved the corpufcularian hypothefes; which is now become the bafis of the mechanical and experimental philofophy.

Boyle reduces the principles of the corpufcular philofophy to the 4 following heads.

1. That there is but one univerfal kind of matter, which is an extended, impenetrable, and divifible fubftance, common to all bodies, and capable of all forms. —On this head, Newton finely remarks thus: " All things confidered, it appears probable to me, that God in the beginning created matter in folid, hard, impenetrable, moveable particles; of fuch fizes and figures, and with fuch other properties, as moft conduced to the end for which he formed them: and that thefe primitive particles, being folids, are incomparably harder than any of the fenfible porous bodies compounded of them; even fo hard as never to wear, or break in pieces: no other power being able to divide what God made one in the firft creation. While thefe corpufcles remain entire, they may compofe bodies of one

3   and

and the fame nature and texture in all ages: but should they wear away, or break in pieces, the nature of things depending on them would be changed: water and earth, composed of old worn particles, of fragments of particles, would not be of the fame nature and texture now, with water and earth composed of entire particles at the beginning. And therefore, that nature may be lasting, the changes of corporeal things are to be placed only in the various feparations, and new affociations, of thefe permanent corpufcles."

2. That this matter, in order to form the vast variety of natural bodies, must have motion in fome, or all its affignable parts; and that this motion was given to matter by God, the creator of all things; and has all manner of directions and tendencies.— " Thefe corpufcles, fays Newton, have not only a vis inertiæ, accompanied with fuch paffive laws of motion as naturally refult from that force; but also are moved by certain active principles; fuch as that of gravity, and that which caufes fermentation, and the cohefion of bodies."

3. That matter must also be actually divided into parts; and each of thefe primitive particles, fragments, or atoms of matter, must have its proper magnitude, figure, and fhape.

4. That thefe differently fized and fhaped particles, have different orders, pofitions, fituations, and poftures, from whence all the variety of compound bodies arifes.

CORRIDOR. See CORIDOR.

CORVUS, the *Raven*, a fouthern conftellation, fabled by the Greeks, as taken up to heaven by Apollo, to whom it tattled that the beautiful maid Coronis, the daughter of Phlegeos, and mother of Efculapius by Apollo, played the deity falfe with Ifchys, under a tree upon which the animal happened to be perched.

The stars in this conftellation, in Ptolomy's and Tycho's catalogues are 7; but in the Britannic catalogue, 9.

COSECANT, COSINE, COTANGENT, CO-VERSED SINE, are the fecant, fine, tangent, and verfed fine of the complement of an arch or angle; *Co* being, in this cafe, a contraction of the word complement, and was first introduced by Gunter.

COSMICAL ASPECT, among aftrologers, is the afpect of a planet with refpect to the earth.

COSMICAL *Rifing*, or *Setting*, is faid of a star when it rifes or fets at the fame time when the fun rifes.

But, according to Kepler, to rife or fet cofmically, is only fimply to rife or fet, that is, to afcend above, or defcend below, the horizon; as much as to fay, to rife or fet to the world.

COSMOGONY, the fcience of the formation of the univerfe; as diftinguifhed from cofmography, which is the fcience of the parts of the univerfe, fuppofing it formed, and in the ftate as we behold it; and from cofmology, which reafons on the actual and permanent ftate of the world as it now is; whereas cofmogony reafons on the variable ftate of the world at the time of its formation.

COSMOGRAPHY, the defcription of the world; or the art that teaches the conftruction, figure, difpofition, and relation of all the parts of the world, with the manner of reprefenting them on a plane. It confifts chiefly of two parts; viz, *Aftronomy*, which fhews the ftructure of the heavens, with the difpofition of the stars; and *Geography*, which fhews thofe of the earth.

COSMOLOGY, the fcience of the world in general.

COSS, *Rule of*, meant the fame as Algebra, by which name it was for fome time called, when first introduced into Europe through the Italians, who named it *Regola de Cofa, the Rule of the thing*; the unknown quantity, or that which was required in any queftion, being called *cofa*, the *thing*; from whence we have Cofs, and Coffic numbers, &c.

COTES (ROGER), a very eminent mathematician, philofopher, and aftronomer, was born July 10, 1682, at Burbach in Leicefterfhire, where his father Robert was rector. He was first placed at Leicefter fchool; where, at 12 years of age, he difcovered a strong inclination to the mathematics. This being obferved by his uncle, the Rev. Mr. John Smith, he gave him all the encouragement he could; and prevailed on his father to fend him for fome time to his houfe in Lincolnfhire, that he might affift him in thofe studies: and here he laid the foundation of that deep and extenfive knowledge in that fcience, for which he was afterward fo defervedly famous. He was hence removed to St. Paul's fchool, London, where he made a great progrefs in claffical learning; and yet he found fo much leifure as to fupport a conftant correfpondence with his uncle, not only in mathematics, but also in metaphyfics, philofophy, and divinity. His next remove was to Trinity College Cambridge, where he took his degrees, and became fellow.

Jan. 1706, he was appointed profeffor of aftronomy and experimental philofophy, upon the foundation of Dr Thomas Plume, archdeacon of Rochefter; being the first that enjoyed that office, to which he was unanimoufly chofen, on account of his high reputation and merits. He entered into orders in 1713; and the fame year, at the defire of Dr. Bentley, he publifhed at Cambridge the fecond edition of Newton's Mathematica Principia; inferting all the improvements which the author had made to that time. To this edition he prefixed a moft admirable preface, in which he pointed out the true method of philofophifing, fhewing the foundation on which the Newtonian philofophy was raifed, and refuting the objections of the Cartefians and all other philofophers againft it.

The publication of this edition of Newton's Principia added greatly to his reputation; nor was the high opinion the public now conceived of him in the leaft diminifhed, but rather much increafed, by feveral productions of his own, which afterward appeared. He gave in the Philof. Tranfactions, two papers, viz, 1, Logometria, in vol. 29; and a Defcription of the great fiery meteor that was feen March 6, 1716, in vol. 31.

This extraordinary genius in the mathematics died, to the great regret of the univerfity, and all the lovers of the fciences, June 5, 1716, in the very prime of his life, being not quite 34 years of age.

Mr. Cotes left behind him fome very ingenious, and indeed admirable tracts, part of which, with the Logometria

gometria above mentioned, were publifhed, in 1722, by Dr. Robert Smith, his coufin and fucceffor in his profefforfhip, afterward mafter of Trinity-College, under the title of *Harmonia Menfurarum*, which contains a number of very ingenious and learned works: fee the Introduction to my Logarithms. He wrote alfo a *Compendium of Arithmetic*; of the *Refolution of Equations*; of *Dioptrics*; and of the *Nature of Curves*. Befide thefe pieces, he drew up, in the time of his lectures, a courfe of *Hydroftatical and Pneumatical Lectures*, in Englifh, which were publifhed alfo by Dr. Smith in 8vo, 1737, and are held in great eftimation.

So high an opinion had Sir Ifaac Newton of our author's genius, that he ufed to fay, " If Cotes had lived, we had known fomething."

COTESIAN *theorem*, in Geometry, an appellation ufed for an elegant property of the circle difcovered by Mr. Cotes. The theorem is this:

If the factors of the binomial $a^c + x^c$ be required, the index c being an integer number. With the centre O, and radius AO $= a$, defcribe a circle, and di-

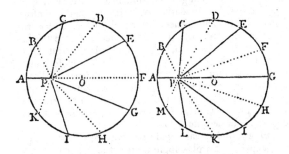

vide its circumferance into as many equal parts as there are units in $2c$, at the points A, B, C, D, &c; then in the radius, produced if neceffary, take OP $= x$, and from the point P, to all the points of divifion in the circumference, draw the lines PA, PB, PC, &c; fo fhall thefe lines taken alternately be the factors fought; viz,

PB $\times$ PD $\times$ PF &c $= a^c + x^c$,

and PA $\times$ PC $\times$ PE &c $= a^c \backsim x^c$, viz, $a^c - x^c$ or $x^c - a^c$, according as the point P is within or without the circle.

For inftance, if $c = 5$, divide the circumference into 10 equal parts, and the point P being within the circle, then will OA^5 + OP^5 = BP $\times$ DP $\times$ FP $\times$ HP $\times$ KP, and OA^5 $-$ OP^5 = AP $\times$ CP $\times$ EP $\times$ GP $\times$ IP.

In like manner, if $c = 6$, having divided the circumference into 12 equal parts, then will OA^6 + OP^6 = BP $\times$ DP $\times$ FP $\times$ HP $\times$ KP $\times$ MP, OA^6 $-$ OP^6 = AP $\times$ CP $\times$ EP $\times$ GP $\times$ IP $\times$ LP.

The demonftration of this theorem may be feen in Dr. Pemberton's Epift. de Cotefii inventis. See alfo Dr. Smith's Theoremata Logometrica and Trigonometrica, added to Cotes's Harm. Menf. pa. 114; De Moivre Mifcel. Analyt. pa. 17; and Waring's Letter to Dr. Powell, pa. 39.

By means of this theorem, the acute and elegant author was enabled to make a farther progrefs in the inverfe method of Fluxions, than had been done before. But in the application of his difcovery there

still remained a limitation, which was removed by Mr. De Moivre. Vide ut fupra.

COVERT-WAY, in fortification, a fpace of ground level with the adjoining country, on the outer edge of the ditch, ranging quite round all the works. This is otherwife called the *corridor*, and has a parapet with its banquette and glacis, which form the height of the parapet. It is fometimes alfo called the counter-fcarp, becaufe it is on the edge of the fcarp.

One of the greateft difficulties in a fiege, is to make a lodgment on the covert-way; becaufe it is ufual for the befieged to palifade it along the middle, and undermine it on all fides.

COVING, in Building, si when houfes are built projecting over the ground plot, and the turned projecture formed into an arch.

COVING *Cornice*, is one that has a large cafemate or hollow in it.

COUNT-WHEEL, is a wheel in the ftriking part of a clock, moving round once in 12 or 24 hours. It is fometimes called the *locking-wheel*, becaufe it has ufually 11 notches in it at unequal diftances from one another, to make the clock ftrike.

COUNTER-APPROACHES, in Fortification, lines or trenches made by the befieged, where they come out to attack the lines of the befiegers in form.

COUNTER-BATTERY, a battery raifed to play on another, to difmount the guns, &c.

COUNTER-BREAST-WORK, the fame as *Fauffe-Braye*.

COUNTER-FORTS, *Buttreffes*, or *Spurs*, are pillars of mafonry ferving to prop or fuftain walls, or terraces, fubject to bulge, or be thrown down.

COUNTER-FUGUE, in Mufic, is when fugues proceed contrary to one another.

COUNTER-GUARD, in Fortification, a work commonly ferving to cover a baftion. It is compofed of two faces, forming a falient angle before the flanked angle of a baftion.

COUNTER-HARMONICAL. See CONTRA-HARMONICAL.

COUNTER-MINE, a fubterraneous paffage, made by the befieged, in fearch of the enemy's mine, to give air to it, to take away the powder; or by any other means to fruftrate the effect of it.

COUNTER-PART, a term in Mufic, only denoting that one part is oppofite to another, fo, the bafs and treble are counterparts to each other.

COUNTER-POINT, in Mufic, the art of compofing harmony; or difpofing and concerting feveral parts fo together, as that they may make an agreeable whole.

COUNTER-POISE, any thing ferving to weigh againft another; particularly a piece of metal, ufually of brafs or iron, making an appendage to the Roman *ftatera*, or fteel-yard. It is contrived to flide along the beam; and from the divifion at which it keeps the balance in equilibrio, the weight of the body is determined. It is fometimes called the *pear*, on account of its figure and *mafs*, by reafon of its weight.

Rope-dancers make ufe of a pole by way of counter-poife, to keep their bodies in equilibrio.

COUNTERSCARP, is that fide of the ditch that is next the country; or properly the talus that fupports the earth of the covert-way: though by this word is often underftood the whole covert-way, with

its

its parapet and glacis. And so it muft be underftood when it is faid, The enemy lodged themfelves on the counterfcarp.

COUNTER-SWALLOWS-TAIL, is an outwork in Fortification, in form of a fingle tenaille, wider towards the place, or at the gorge, than at the head, or next the country.

COUNTER-TENOR, one of the mean or middle parts of mufic : fo called, as being oppofite to the tenor.

COURSE, in Navigation, the point of the compafs, or horizon, which a fhip fteers on ; or the angle which the rhumb line on which it fails makes with the meridian ; being fometimes reckoned in degrees, and fometimes in points of the compafs.

When a fhip fails either due north or fouth, fhe fails on a meridian, makes no departure, and her diftance and difference of latitude are the fame.

When fhe fails due eaft or weft, her courfe makes right-angles with the meridian, and fhe fails either upon the equator, or a parallel to it ; in which cafe fhe makes no difference of latitude, but her diftance and departure are the fame.

But when the fhip fails between the cardinal points, on a courfe making always the fame oblique angle with the meridians, her path is then the loxodromic curve, being a fpiral cutting all the meridians in the fame angle, and terminating in the pole.

COURTAIN. See CURTIN.

CRAB, in Mechanics, an engine ufed for mounting guns on their carriages. See GIN.

CRAB, on fhip-board, is a wooden pillar, whofe lower end is let down through the fhip's decks, and refts upon a focket like the capftan : in its upper end are three or four holes, at different heights, through the middle of it, above one another, to receive long bars, againft which men act by pufhing or thrufting.— It is employed to wind in the cable, and for other purpofes requiring a great mechanical power.

The Crab with three claws is ufed to launch fhips, and to heave them into the dock, or off the key.

CRANE, a machine ufed in building, and in commerce, for raifing large ftones and other weights.

M. Perrault, in his notes on Vitruvius, makes the crane the fame with the corvus, or raven, of the ancients.

The modern crane confifts of feveral members or pieces, the principal of which is a ftrong upright beam, or arbor, firmly fixed in the ground; and fuftained by eight arms, coming from the extremities of four pieces of wood laid acrofs, through the middle of which the foot of the beam paffes. About the middle of the arbor the arms meet, and are mortifed into it : its top ends in an iron pivot, on which is borne a tranfverfe piece, advancing out to a good diftance like a crane's neck ; whence the name. The middle and extremities of this are again fuftained by arms from the middle of the arbor: and over it comes a rope, or cable, to one end of which the weight is fixed ; the other is wound about the fpindle of a wheel, which, turned, draws the rope, and that heaves up the weight ; to be afterwards applied to any fide or quarter, by the mobility of the tranfverfe piece on the pivot.

Several improvements of this ufeful machine are mentioned in Defaguliers's Exper. Philof. pa. 178 &

VOL. I.

seq. particularly how to prevent the inconveniences arifing from fudden jerks, as well as to increafe its force by ufing a double axis in peritrochio, and two handles.

The Crane is of two kinds ; in the firft kind, called the rat-tailed crane, the whole machine, with the load, turns upon a ftrong axis : in the fecond kind, the gibbet alone moves on its axis. See Defaguliers, as above, for a particular account of the different cranes, and of the gradual improvements they have received. See alfo the Supplement to Fergufon's Lectures, pa. 3, &c, or Philof. Tranf. vol. 54, pa. 24, for a defcription of a new and fafe crane, with four different powers adapted to different weights.

CRANE is the name of a fouthern conftellation. See GRUS.

CRANE is alfo a popular name for a fyphon.

CRANK, a contrivance in machines, in manner of an elbow, only of a fquare form ; projecting out from an axis, or fpindle ; and ferving, by its rotation, to raife and fall the piftons of engines for raifing water, or the like.

CRATER, the Cup, a conftellation in the fouthern hemifphere ; whofe ftars, in Ptolomy's catalogue, are 7 ; in Tycho's, 8 ; in Hevelius's, 10 ; and in the Britannic catalogue, 31.

CREEK, a part of a haven, where any thing is landed from the fea. It is alfo faid to be a fhore or bank on which the water beats, running in a fmall channel from any part of the fea.

CREPUSCULUM, *Twilight*; the time from the firft dawn or appearance of the morning, to the rifing of the fun ; and again, between the fetting of the fun, and the laft remains of day.

The Crepufculum, or twilight, it is fuppofed, ufually begins and ends when the fun is about 18 degrees below the horizon ; for then the ftars of the 6th magnitude difappear in the morning, and appear in the evening. It is of longer duration in the folftices than in the equinoxes, and longer in an oblique fphere, than in a right one ; becaufe, in thofe cafes the fun, by the obliquity of his path, is longer in afcending through 18 degrees of altitude.

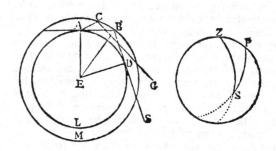

Twilight is occafioned by the fun's rays refracted in our atmofphere, and reflected from the particles of it to the eye. For let A be the place of an obferver on the earth ADL, AB the fenfible horizon, meeting in B the circle CBM bounding that part of the atmofphere which is capable of refracting and reflecting light to the eye. It is plain that when the fun is under this horizon, no direct rays can come to the eye at A: but the sun being in the refracted line CG, the particle C will be illuminated by the direct rays of the sun; and that particle may reflect those rays to A, where they enter the eye of the spectator. And thus the sun's light illuminating an innumerable multitude of particles, may be all reflected to the spectator at A.—From B draw BD touching the circle ADL in D; and let the sun be in the line BD at S: Then the ray SB will be reflected into BA, and will enter the eye, because the angle of incidence DBE is equal to the angle of reflection ABE: And that will be the first ray that reaches the eye in the morning, when the dawning begins; or the last that falls upon the eye at night, when the twilight ends: for when the sun goes lower down, the particles at B can be no longer illuminated.

Kepler indeed assigns another cause of the crepusculum, viz, the luminous matter or atmosphere about the sun; which, arising near the horizon, in a circular figure, exhibits the crepusculum; in no wise, he thinks, owing to the refraction of the atmosphere. The sun's luminous atmosphere indeed, though neither the sole nor principal cause of twilight, may lengthen its duration, by illuminating our air, when the sun is too low to reach it with his own light. Gregor. Astr. lib. 2, prop. 8.

The depth of the sun below the horizon, at the beginning of the morning, or end of the evening twilight, is determined in the same manner as the arch of vision; viz, by observing the moment when the air first begins to shine in the morning, or ceases to shine in the evening; then finding the sun's place for that moment, and thence the time till his rising in the horizon, or from his setting in it in the evening. It is now generally agreed that this depth is about 18 degrees upon an average.—Alhazen found it to be 19°; Tycho, 17°; Rothmann, 24°; Stevenius, 18°; Cassini, 15°; Riccioli, in the equinox in the morning 16°, in the evening 20° 30; in the summer solstice in the morning 21° 25', in the winter solstice in the morning 17° 25'.

Nor is this difference among the determinations of astronomers to be wondered at; the cause of the crepusculum being inconstant: for, if the exhalations in the atmosphere be either more copious, or higher, than ordinary; the morning twilight will begin sooner, and the evening hold longer than ordinary: for the more copious the exhalations are, the more rays will they reflect, consequently the more will they shine; and the higher they are, the sooner will they be illuminated by the sun. On this account too, the evening twilight is longer than the morning, at the same time of the year in the same place. To this it may be added, that in a denser air, the refraction is greater; and that not only the brightness of the atmosphere is variable, but also its height from the earth: and therefore the twilight is longer in hot weather than in cold, in summer than in winter, and also in hot countries than in cold, other circumstances being the same. But the chief differences are owing to the different situations of places upon the earth, or to the difference of the sun's place in the heavens. Thus, the twilight is longest in a parallel sphere, and shortest in a right sphere, and longer to places in an oblique sphere in proportion as they

are nearer to one of the poles; a circumstance which affords relief to the inhabitants of the more northern countries, in their long winter nights. And the twilights are longest in all places of north latitude, when the sun is in the tropic of cancer; and to those in south latitude, when he is in the tropic of capricorn. The time of the shortest twilight is also different in different latitudes; in England, it is about the beginning of October and of March, when the sun is in the signs ♎ and ♓. For the method of determining it by trigonometry, see Gregor. Astron. lib. 2, prob. 41. See also Robertson's Navigation, book 5, prob. 12.—Hence, when the difference between the sun's declination and the depth of the equator is less than 18°, so that the sun does not descend more than 18° below the horizon; the crepusculum will continue the whole night, as is the case in England from about the 22d of May to the 22d of July.

Given the latitude of the place, and the sun's declination; to find the beginning of the morning, and end of the evening twilight.—In the oblique-angled spherical triangle ZPS, are given ZP the colatitude, PS the codeclination, and ZS = 108°, being the sum of 90° the quadrant and 18° the depression at the extremity of the twilight. Then, by spherical trigonometry, calculate the angle ZPS the hour-angle from noon; which changed into time, at the rate of 15° to the hour, gives the time from noon at the beginning or end of twilight. See Robertson, ubi supra.

Of the Height of the sensible Atmosphere, as determined from the duration of twilight, see Keil's Astron. Lect. lect. 20, pa. 235, ed. 1721; or Long's Astron. vol. 1, pa. 260; where it is determined that the height where the atmosphere is dense enough to reflect the rays of light, is about 42 miles.

CRESCENT, the new moon, which, as it begins to recede from the sun, shews a small rim of light, terminating in horns or points, which are still increasing, till it becomes full, and round in the opposition.

The term is sometimes also used for the same figure of the moon in her wane, or decrease, but improperly; both because the horns are then turned towards the west, and because the figure is on the decrease; the *crescent* properly signifying increase, from *cresco, I grow*.

CRONICAL. See ACRONICAL.

CRONOS, a name given to Saturn by some of the old astronomical writers.

CROSIER, four stars, in form of a cross; by the help of which, those that sail in the southern hemisphere find the antarctic pole.

CROSS, in Surveying, is an instrument consisting of a brass circle, divided into 4 equal parts, by two lines crossing each other in the centre. At each extremity of these lines is fixed a perpendicular sight, with small holes below each slit, for the better discovering of distant objects. The cross is mounted on a staff, or stand, to fix it in the ground, and is very useful for measuring small pieces of land, and taking offsets, &c.

Ex. Suppose it be required to survey the field ABCDE with the Cross. Measure along the diagonal line AC, and observe, with the Cross, when you are perpendicularly opposite to the corners, as at F, G, H, and from thence measure the perpendiculars EE, GB, HD.

5

HD. When you think you are nearly opposite a corner, set up the cross, with one of the bars or cross lines in the direction AC; then look through the sights of the other cross bar for the corner, as B; if it be seen through them, the cross is fixed in the right place; if not, take it up and move it backward or forward in the line AC, till the point B be seen through those sights; and then you have the true place of the perpendicular.

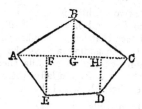

Invention of the Cross, *Inventio Crucis*, an ancient feast, which is still retained in our calendar, and solemnized on the 3d of May, in memory of the finding of the true Cross of Christ, deep in the ground, on Mount Calvary, by St. Helena, the mother of Constantine; where she erected a church for the preservation of part of it: the rest being brought to Rome, and deposited in the church of the Holy Cross of Jerusalem.

Exaltation of the Cross, an ancient feast, held on the 14th of September, in memory of this, that Heraclitus restored to Mount Calvary the true cross, in 642, which had been carried off, 14 years before, by Cosroes king of Persia, upon his taking Jerusalem from the emperor Phocas. This feast is still retained in our calendar, on Sept. 14, under the denomination of *Holy Rood*, or *Holy Cross*.

Cross-Multiplication, a method used chiefly by artificers in multiplying feet and inches by feet and inches, or the like; so called, because the factors are multiplied cross-wise, thus:

f.	i.
9	10
6	8
59	0
6	6 8
65	6 8

Cross-Staff, or *Fore-Staff*, is a mathematical instrument of box, or pear tree, consisting of a square staff, of about 3 feet long, having each of its faces divided like a line of tangents, and having 4 cross pieces of unequal lengths to fit on to the staff, the halves of these being as the radii to the tangent lines on the faces of the staff.—The instrument was used in taking the altitudes of the celestial bodies at sea.

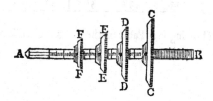

CROUSAZ (John Peter de), a learned philosopher and mathematician, was born at Lausanne in Switzerland, April 13, 1663; where he died in 1748, at 85 years of age. Having made great progress in mathematics and the philosophy of Des Cartes, he travelled into Geneva, Holland, and France. He was successively professor in several universities; and at length was chosen governor to Prince Frederick of Hesse-Cassel, nephew to the king of Sweden.

Crousaz was author of many works, in various branches; belles-lettres, logic, philosophy, divinity, &c, &c; but the most esteemed of them are, 1. His Logic; the best edition of which is that of 1741, in 6 vols. 8vo.—2. A Treatise on Beauty.—3. A Treatise on Education, 2 vols, 12mo.—4. A Treatise on the Human Understanding.—5. Several Treatises on Philosophy and Mathematics; as a Treatise on Motion, &c. with several papers inserted in the Memoirs of the French Academy of Sciences.

CROW, in Mechanics, an iron lever, made with a sharp point at one end, and two claws at the other; being used in heaving and purchasing great weights, &c.

CROWN, in Astronomy, a name given to two constellations, the southern and the northern.

Crown, in Geometry, a plane ring included between two parallel or concentric peripheries, of unequal circles.

The area of this is had, by multiplying its breadth by the length of a middle periphery, which is an arithmetical mean between the two peripheries that bound it; or by multiplying half the sum of the circumferences by half the difference of the diameters; or lastly by multiplying the sum of the diameters by the difference of the diameters, and this last product by .7854. See my Mensuration, pa. 148, 2d ed.

Crown-*Post*, is a post in some buildings standing upright in the middle, between two principal rafters; and from which proceed struts or braces to the middle of each rafter. It is otherwise called a *king-post*, or *king's-piece*, or *joggle-piece*.

Crown-*Wheel*, of a Watch, is the upper wheel next the balance, or that which drives the balance.

Crown-*Work*, in Fortification, is an out-work running into the field; designed to keep off the enemy, gain some hill, or advantageous post, and cover the other works of the place. It consists of two demi-bastions at the extremities, and an entire bastion in the middle, with curtains.

Crowned *Horn-work*, is a Horn-work with a crown-work before it.

CRYSTALLINE *Humour*, is a thick compact humour of the eye, in form of a flattish convex lens, placed in the middle of the eye, and serving to make that refraction of the rays of light which is necessary to have them meet in the retina, and form an image there, by which vision may be performed.

Crystalline *Heavens*, in the Old Astronomy, two orbs imagined between the primum mobile and the firmament, in the Ptolomaic system, which supposed the heavens solid, and only susceptible of a single motion.

King Alphonsus of Arragon, it is said, introduced the Crystallines, to explain what they called the *motion of trepidation*, or *titubation*.

The first Crystalline, according to Regiomontanus,

X x 2   &c,

&c, serves to account for the flow motion of the fixed stars; by which they advance a degree in about 70 years, according to the order of the signs, or from west to east; which occasions a precession of the equinox. The 2d serves to account for the motion of libration, or trepidation; by which the celestial sphere librates from one pole towards the other, causing a difference in the sun's greatest declination.

CUBATURE, or CUBATION, of a solid, is the measuring the space contained in it, or finding the solid content of it, or finding a cube equal to it.

The cubature regards the content of a body, as the quadrature does the superficies or area of a figure.

CUBE, a regular or solid body, consisting of six equal sides or faces, which are squares.—A die is a small cube.

It is also called a *hexaedron*, because of its six sides, and is the 2d of the five Platonic or Regular bodies.

The cube is supposed to be generated by the motion of a square plane, along a line equal and perpendicular to one of its sides.

To describe a Rete, or Net, for forming a cube, or with which it may be covered.—Describe six squares as in the annexed figure, upon card paper, paste-board, or the like, of the size of the faces of the proposed cube; and cut it half through by the lines AB, CD, EF, AC, BD; then fold up the several squares till their edges meet, and so form the cube, or a covering over one, as in the figure annexed.

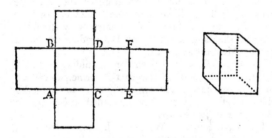

To determine the Surface and Solidity of a Cube.—Multiply one side by itself, which will give one square or face; then this multiplied by 6, the number of faces, will give the whole surface. Also multiply one side twice by itself, that is, cube it, and that will be the solid content.

Duplication of a CUBE. See DUPLICATION.

CUBES, or CUBIC *Numbers*, are formed by multiplying any numbers twice by themselves. So the cubes of 1, 2, 3, 4, 5, 6, &c, are 1, 8, 27, 64, 125, 216, &c.

The third differences of the cubes of the natural numbers are all equal to each other, being the constant number 6. For, let m^3, n^3, p^3 be any three adjacent cubes in the natural series as above, that is, whose roots m, n, p have the common difference 1; then because $n = m + 1$, theref. $n^3 = m^3 + 3m^2 + 3m + 1$, $p = n + 1$, theref. $p^3 = n^3 + 3n^2 + 3n + 1$; so that

the difference between the 1st and 2d, and between the 2d and 3d cubes, are
$$n^3 - m^3 = 3m^2 + 3m + 1, \Big\} \text{ the 1st differences;}$$
$$p^3 - n^3 = 3n^2 + 3n + 1, \Big\}$$
and the dif. of these differences, is
$$3 . \overline{n^2 - m^2} + 3 . \overline{n - m} = 3 . \overline{n + m + 1} = 6 . \overline{m + 1},$$
the 2d difference.

In like manner the next 2d dif. is $6 . \overline{n + 1}$: hence the dif. of these two 2d diffs. is $6 . \overline{n - m} = 6$, which is therefore the constant 3d difference of all the series of cubes. And hence that series of cubes will be formed by addition only, viz, adding always the 3d dif. 6 to find the column or series of 2d diffs, then these added always for the 1st diffs, and lastly these always added for the cubes themselves, as below:

3d Difs.	2d Difs.	1st Difs.	Cubes.
6	6	1	0
6	12	7	1
6	18	19	8
6	24	37	27
6	30	61	64
6	36	91	125
6	42	127	216
6	48	169	343

Peletarius, among various speculations concerning square and cubic numbers, shews that the continual sums of the cubic numbers, whose roots are 1, 2, 3, &c, form the series of squares whose roots are 1, 3, 6, 10, 15, 21, &c.

Thus,
$$1 = 1 = 1^2,$$
$$1 + 8 = 9 = 3^2,$$
$$1 + 8 + 27 = 36 = 6^2,$$
$$1 + 8 + 27 + 64 = 100 = 10^2, \&c.$$
Or, in general,
$$1^3 + 2^3 + 3^3 + 4^3 \&c. \text{ to } n^3 = \overline{1 + 2 + 3 + 4 - - n}|^2 = \tfrac{1}{2} n . \overline{n + 1}.$$

It is also a pretty property, that any number, and the cube of it, being divided by 6, leave the same remainder; the series of remainders being 0, 1, 2, 3, 4, 5, continually repeated. Or that the differences between the numbers and their cubes, divided by 6, leave always 0 remaining; and the quotients, with their successive differences, form the several orders of figurate numbers. Thus,

Num.	Cubes.	Difs.	Quot.	1 Dif.	2 Dif.
1	1	0	0	0	0
2	8	6	1	1	1
3	27	24	4	3	2
4	64	60	10	6	3
5	125	120	20	10	4
6	216	210	35	15	5
7	343	336	56	21	6

The following is a Table of the first 1000 cubic numbers.

TABLE

TABLE OF CUBES.

Num.	Cubes	Num.	Cubes	Num.	Cubes	Num.	Cubes	Num.	Cubes	Num.	Cubes
1	1	60	216000	119	1685159	178	5639752	237	13312053	296	25934336
2	8	61	226981	120	1728000	179	5735339	238	13481272	297	26198073
3	27	62	238328	121	1771561	180	5832000	239	13651919	298	26463592
4	64	63	250047	122	1815848	181	5929741	240	13824000	299	26730899
5	125	64	262144	123	1860867	182	6028568	241	13997521	300	27000000
6	216	65	274625	124	1906624	183	6128487	242	14172488	301	27270901
7	343	66	287496	125	1953125	184	6229504	243	14348907	302	27543608
8	512	67	300763	126	2000376	185	6331625	244	14526784	303	27818127
9	729	68	314432	127	2048383	186	6434856	245	14706125	304	28094464
10	1000	69	328509	128	2097152	187	6539203	246	14886936	305	28372625
11	1331	70	343000	129	2146689	188	6644672	247	15069223	306	28652616
12	1728	71	357911	130	2197000	189	6751269	248	15252992	307	28934443
13	2197	72	373248	131	2248091	190	6859000	249	15438249	308	29218112
14	2744	73	389017	132	2299968	191	6967871	250	15625000	309	29503629
15	3375	74	405224	133	2352637	192	7077888	251	15813251	310	29791000
16	4096	75	421875	134	2406104	193	7189057	252	16003008	311	30080231
17	4913	76	438976	135	2460375	194	7301384	253	16194277	312	30371328
18	5832	77	456533	136	2515456	195	7414875	254	16387064	313	30664297
19	6859	78	474552	137	2571353	196	7529536	255	16581375	314	30959144
20	8000	79	493039	138	2628072	197	7645373	256	16777216	315	31255875
21	9261	80	512000	139	2685619	198	7762392	257	16974593	316	31554496
22	10648	81	531441	140	2744000	199	7880599	258	17173512	317	31855013
23	12167	82	551368	141	2803221	200	8000000	259	17373979	318	32157432
24	13824	83	571787	142	2863288	201	8120601	260	17576000	319	32461759
25	15625	84	592704	143	2924207	202	8242408	261	17779581	320	32768000
26	17576	85	614125	144	2985984	203	8365427	262	17984728	321	33076161
27	19683	86	636056	145	3048625	204	8489664	263	18191447	322	33386248
28	21952	87	658503	146	3112136	205	8615125	264	18399744	323	33698267
29	24389	88	681472	147	3176523	206	8741816	265	18609625	324	34012224
30	27000	89	704969	148	3241792	207	8869743	266	18821096	325	34328125
31	29791	90	729000	149	3307949	208	8998912	267	19034163	326	34645976
32	32768	91	753571	150	3375000	209	9123329	268	19248832	327	34965783
33	35937	92	778688	151	3442951	210	9261000	269	19465109	328	35287552
34	39304	93	804357	152	3511808	211	9393931	270	19683000	329	35611289
35	42875	94	830584	153	3581577	212	9528128	271	19902511	330	35937000
36	46656	95	857375	154	3652264	213	9663597	272	20123648	331	36264691
37	50653	96	884736	155	3723875	214	9800344	273	20346417	332	36594368
38	54872	97	912673	156	3796416	215	9938375	274	20570824	333	36926037
39	59319	98	941192	157	3869893	216	10077696	275	20796875	334	37259704
40	64000	99	970299	158	3944312	217	10218313	276	21024576	335	37595375
41	68921	100	1000000	159	4019679	218	10360232	277	21253933	336	37933056
42	74088	101	1030301	160	4096000	219	10503459	278	21484952	337	38272753
43	79507	102	1061208	161	4173281	220	10648000	279	21717639	338	38614472
44	85184	103	1092727	162	4251528	221	10793861	280	21952000	339	38958219
45	91125	104	1124864	163	4330747	222	10941048	281	22188041	340	39304000
46	97336	105	1157625	164	4410944	223	11089567	282	22425768	341	39651821
47	103823	106	1191016	165	4492125	224	11239424	283	22665187	342	40001688
48	110592	107	1225043	166	4574296	225	11390625	284	22906304	343	40353607
49	117649	108	1259712	167	4657463	226	11543176	285	23149125	344	40707584
50	125000	109	1295029	168	4741632	227	11697083	286	23393656	345	41063625
51	132651	110	1331000	169	4826809	228	11852352	287	23639903	346	41421736
52	140608	111	1367631	170	4913000	229	12008989	288	23887872	347	41781923
53	148877	112	1404928	171	5000211	230	12167000	289	24137569	348	42144192
54	157464	113	1442897	172	5088448	231	12326391	290	24389000	349	42508549
55	166375	114	1481544	173	5177717	232	12487168	291	24642171	350	42875000
56	175616	115	1520875	174	5268024	233	12649337	292	24897088	351	43243551
57	185193	116	1560896	175	5359375	234	12812904	293	25153757	352	43614208
58	195112	117	1601613	176	5451776	235	12977875	294	25412184	353	43986977
59	205379	118	1643032	177	5545233	236	13144256	295	25672375	354	44361864

Num.	Cubes	Num.	Cubes	Num.	Cubes	Num	Cubes	Num.	Cubes	Num.	Cubes	Num.	Cubes
355	44738875	417	72511713	479	109902239	541	158340421	603	219256227	665	294079625
356	45118016	418	73034632	480	110592000	542	159220088	604	220348864	666	295408296
357	45499293	419	73560059	481	111284641	543	160103007	605	221445125	667	296740963
358	45882712	420	74088000	482	111980168	544	160989184	606	222545016	668	298077632
359	46268279	421	74618461	483	112678587	545	161878625	607	223648543	669	299418309
560	46656000	422	75151448	484	113379904	546	162771336	608	224755712	670	300763000
361	47045881	423	75686967	485	114084125	547	163667323	609	225866529	671	302111711
362	47437928	424	76225024	486	114791256	548	164566592	610	226981000	672	303464448
363	47832147	425	76765625	487	115501303	549	165469149	611	228099131	673	304821217
364	48228544	426	77308776	488	116214272	550	166375000	612	229220928	674	306182024
365	48627125	427	77854483	489	116930169	551	167284151	613	230346397	675	307546875
366	49027896	428	78402752	490	117649000	552	168196608	614	231475544	676	308915776
367	49430863	429	78953589	491	118370771	553	169112377	615	232608375	677	310288733
368	49836032	430	79507000	492	119095488	554	170031464	616	233744896	678	311665752
369	50243409	431	80062991	493	119823157	555	170953875	617	234885113	679	313046839
370	50653000	432	80621568	494	120553784	556	171879616	618	236029032	680	314432000
371	51064811	433	81182737	495	121287375	557	172808693	619	237176659	681	315821241
372	51478848	434	81746504	496	122023936	558	173741112	620	238328000	682	317214568
373	51895117	435	82312875	497	122763473	559	174676879	621	239483061	683	318611987
374	52313624	436	82881856	498	123505992	560	175616000	622	240641848	684	320013504
375	52734375	437	83453453	499	124251499	561	176558481	623	241804367	685	321419125
376	53157376	438	84027672	500	125000000	562	177504328	624	242970624	686	322828856
377	53582633	439	84604519	501	125751501	563	178453547	625	244140625	687	324242703
378	54010152	440	85184000	502	126506008	564	179406144	626	245314376	688	325660672
379	54439939	441	85766121	503	127263527	565	180362125	627	246491883	689	327082769
380	54872000	442	86350888	504	128024064	566	181321496	628	247673152	690	328509000
381	55306341	443	86938307	505	128787625	567	182284263	629	248858189	691	329939371
382	55742968	444	87528384	506	129554216	568	183250432	630	250047000	692	331373888
383	56181887	445	88121125	507	130323843	569	184220009	631	251239591	693	332812557
384	56623104	446	88716536	508	131096512	570	185193000	632	252435968	694	334255384
385	57066625	447	89314623	509	131872229	571	186169411	633	253636137	695	335702375
386	57512456	448	89915392	510	132651000	572	187149248	634	254840104	696	337153536
387	57960603	449	90518849	511	133432831	573	188132517	635	256047875	697	338608873
388	58411072	450	91125000	512	134217728	574	189119224	636	257259456	698	340068392
389	58863869	451	91733851	513	135005697	575	190109375	637	258474853	699	341532099
390	59319000	452	92345408	514	135796744	576	191102976	638	259694072	700	343000000
391	59776471	453	92959677	515	136590875	577	192100033	639	260917119	701	344472101
392	60236288	454	93576664	516	137388096	578	193100552	640	262144000	702	345948008
393	60698457	455	94196375	517	138188413	579	194104539	641	263374721	703	347428927
394	61162984	456	94818816	518	138991832	580	195112000	642	264609288	704	348913664
395	61629875	457	95443993	519	139798359	581	196122941	643	265847707	705	350402625
396	62099136	458	96071912	520	140608000	582	197137368	644	267089984	706	351895816
397	62570773	459	96702579	521	141420761	583	198155287	645	268336125	707	353393243
398	63044792	460	97336000	522	142236648	584	199176704	646	269586136	708	354894912
399	63521199	461	97972181	523	143055667	585	200201625	647	270840023	709	356400829
400	64000000	462	98611128	524	143877824	586	201230056	648	272097792	710	357911000
401	64481201	463	99252847	525	144703125	587	202262003	649	273359449	711	359425431
402	64964808	464	99897344	526	145531576	588	203297472	650	274625000	712	360944128
403	65450827	465	100544625	527	146363183	589	204336469	651	275894451	713	362467097
404	65939264	466	101194696	528	147197952	590	205379000	652	277167808	714	363994344
405	66430125	467	101847563	529	148035889	591	206425071	653	278445077	715	365525875
406	66923416	468	102503232	530	148877000	592	207474688	654	279726264	716	367061696
407	67419143	469	103161709	531	149721291	593	208527857	655	281011375	717	368601813
408	67911312	470	103823000	532	150568768	594	209584584	656	282300416	718	370146232
409	68417929	47	104487111	533	151419437	595	210644875	657	283593393	719	371694959
410	68921000	472	105154048	534	152273304	596	211708736	658	284890312	720	373248000
411	69426531	473	105823817	535	153130375	597	212776173	659	286191175	721	374805361
412	69934528	474	106496424	536	153990656	598	213847192	660	287496000	722	376367048
413	70444997	475	107171875	537	154854153	599	214921799	661	288804781	723	377933067
414	70957944	476	107850176	538	155720872	600	216000000	662	290117528	724	379503424
415	71473375	477	108531333	539	156590819	601	217081801	663	291434247	725	381078125
416	71991296	478	109215352	540	157464000	602	218167208	664	292754944	726	382657176

Num	Cubes.	Num	Cubes.	Num.	Cubes.	Num.	Cubes.	Num.	Cubes.
727	384240583	782	478211768	837	586376253	892	709732288	947	849278123
728	385828352	783	480048687	838	588480472	893	712121957	948	851971392
729	387420489	784	481890304	839	590589719	894	714516984	949	854670349
730	389017000	785	483736625	840	592704000	895	716917375	950	857375000
731	390617891	786	485587656	841	594823321	896	719323136	951	860085351
732	392223168	787	487443403	842	596947688	897	721734273	952	862801408
733	393832837	788	489303872	843	599077107	898	724150792	953	865523177
734	395446904	789	491169069	844	601211584	899	726572699	954	868250664
735	397065375	790	493039000	845	603351125	900	729000000	955	870983875
736	398688256	791	494913671	846	605495736	901	731432701	956	873722816
737	400315553	792	496793088	847	607645423	902	733870808	957	876467493
738	401947272	793	498677257	848	609800192	903	736314327	958	879217912
739	403583419	794	500566184	849	611960049	904	738763264	959	881974079
740	405224000	795	502459875	850	614125000	905	741217625	960	884736000
741	406869021	796	504358336	851	616295051	906	743677416	961	887503681
742	408518488	797	506261573	852	618470208	907	746142643	962	890277128
743	410172407	798	508169592	853	620650477	908	748613312	963	893056347
744	411830784	799	510082399	854	622835864	909	751089429	964	895841344
745	413493625	800	512000000	855	625026375	910	753571000	965	898632125
746	415160936	801	513922401	856	627222016	911	756058031	966	901428696
747	416832723	802	515849608	857	629422793	912	758550528	967	904231063
748	418508992	803	517781627	858	631628712	913	761048497	968	907039232
749	420189749	804	519718464	859	633839779	914	763551944	969	909853209
750	421875000	805	521660125	860	636056000	915	766060875	970	912673000
751	423564751	806	523606616	861	638277381	916	768575296	971	915498611
752	425259008	807	525557943	862	640503928	917	771095213	972	918330048
753	426957777	808	527514112	863	642735647	918	773620632	973	921167317
754	428661064	809	529475129	864	644972544	919	776151559	974	924010424
755	430368875	810	531441000	865	647214625	920	778688000	975	926859375
756	432081216	811	533411731	866	649461896	921	781229961	976	929714176
757	433798093	812	535387328	867	651714363	922	783777448	977	932574833
758	435519512	813	537366797	868	653972032	923	786330467	978	935441352
759	437245479	814	539353144	869	656234909	924	788889024	979	938313739
760	438976000	815	541343375	870	658503000	925	791453125	980	941192000
761	440711081	816	543338496	871	660776311	926	794022776	981	944076141
762	442450728	817	545338513	872	663054848	927	796597983	982	946966168
763	444194947	818	547343432	873	665338617	928	799178752	983	949862087
764	445943744	819	549353259	874	667627624	929	801765089	984	952763904
765	447697125	820	551368000	875	669921875	930	804357000	985	955671625
766	449455096	821	553387661	876	672221376	931	806954491	986	958585256
767	451217663	822	555412248	877	674526133	932	809557568	987	961504803
768	452984832	823	557441767	878	676836152	933	812166237	988	964430272
769	454756609	824	559476224	879	679151439	934	814780504	989	967361669
770	456533000	825	561515625	880	681472000	935	817400375	990	970299000
771	458314011	826	563559976	881	683797841	936	820025856	991	973242271
772	460099648	827	565609283	882	686128968	937	822656953	992	976191488
773	461889917	828	567663552	883	688465387	938	825293672	993	979146657
774	463684824	829	569722789	884	690807104	939	827936019	994	982107784
775	465484375	830	571787000	885	693154125	940	830584000	995	985074875
776	467288576	831	573856191	886	695506456	941	833237621	996	988047936
777	469097433	832	575930368	887	697864103	942	835896888	997	991026973
778	470910952	833	578009537	888	700227072	943	838561807	998	994011992
779	472729139	834	580093704	889	702595369	944	841232384	999	997002999
780	474552000	835	582182875	890	704969000	945	843908625	1000	1000000000
781	476379541	836	584277056	891	707347971	946	846590536		

The Cube of a Binomial, is equal to the cubes of the two parts or members, together with triple of the two parallelopipedons under each part and the square of the other; viz. $\overline{a+b}|^3 = a^3 + 3a^2b + 3ab^2 + b^3$. And hence the common method of extracting the cube root.

CUBIC *Equations*, are those in which the unknown quantities rise to three dimensions; as $x^3 = a$, or $x^3 + ax^2 = b$, or $x^3 + ax^2 + bx = c$, &c.

All cubic equations may be reduced to this form, $x^3 + px = q$; viz. by taking away the 2d term.

All cubic equations have three roots; which are either all real, or else one only is real, and the other two imaginary; for all roots become imaginary by pairs.

But the nature of the roots as to real and imaginary, is known partly from the sign of the co-efficient p, and partly from the relation between p and q: for the equation has always two imaginary roots when p is positive; it has also two imaginary roots when p is negative, provided $(\frac{1}{3}p)^3$ is less $\frac{1}{2}q)^2$, or $4p^3$ less than $27q^2$; otherwise the roots are all real, namely, whenever p is negative, and $4p^3$ either equal to, or greater than $27q^2$.

Every cubic equation of the above form, viz. wanting the 2d term, has both positive and negative roots, and the greatest root is always equal to the sum of the two less roots; viz. either one positive root equal to the sum of the two negative ones, or else one negative root equal to the sum of two smaller and positive ones. And the sign of the greatest, or single root, is positive or negative, according as q is positive or negative when it stands on the right-hand side of the equation, thus $x^3 + px = q$; and the two smaller roots have always the contrary sign to q.

So that, in general, the sign of p determines the nature of the roots, as to real and imaginary; and the sign of q determines the affection of the roots, as to positive and negative. See my Tract on Cubic Equations in the Philos. Transf. for 1780.

To find the Values of the Roots of Cubic Equations. Having reduced the equation to this form $x^3 + px = q$, its root may be found in various ways; the first of these, is that which is called Cardan's Rule, by whom it was first published, but invented by Ferreus and Tartalea. See ALGEBRA. The rule is this: Put $a = \frac{1}{3}p$, and $b = \frac{1}{2}q$; then is Cardan's root

$x = \sqrt[3]{b + \sqrt{b^2 + a^3}} + \sqrt[3]{b - \sqrt{b^2 + a^3}}$; or if there be put $s = \sqrt[3]{b + \sqrt{b^2 + a^3}}$, and $d = \sqrt[3]{b - \sqrt{b^2 + a^3}}$; then $s + d = x$, the 1st or Cardan's root,

also $-\dfrac{s+d}{2} + \dfrac{s-d}{2}\sqrt{-3}$ is the 2d root,

and $-\dfrac{s+d}{2} - \dfrac{s-d}{2}\sqrt{-3}$ is the 3d root.

Now the first of these, or Cardan's root, is always a real root, though it is not always the greatest root, as it has been commonly mistaken for. And yet this rule always exhibits the root in the form of an imaginary quantity when the equation has no imaginary roots at all; but in the form of a real quantity when the equation has two imaginary roots. See the reason of this explained in my Tract above cited, pa. 407. As to

6

the other two roots, viz. $-\dfrac{s+d}{2} \pm \dfrac{s-d}{2}\sqrt{-3}$, though, in their general form, they have an imaginary appearance; yet, by substituting certain particular numbers, they come out in a real form in all such cases as they ought to be so.

But, after the first root is found, by Cardan's rule, the other two roots may be found, or exhibited, in several other different ways; some of which are as follow:

Let r denote the 1st, or Cardan's root,
and v and w the other two roots:
then is $v + w = -r$, and $vwr = q$;
and the resolution of these two equations will give the other two roots v and w.

Or resolve the quadratic equation $x^2 + rx + r^2 + p = 0$, and its two roots will be those sought.
Or the same two roots will be

either $-\frac{1}{2}r + \frac{1}{2}\sqrt{-p - \dfrac{3q}{r}}$ and $-\frac{1}{2}r - \frac{1}{2}\sqrt{-p - \dfrac{3q}{r}}$

or $-\frac{1}{2}r + \frac{1}{2}\sqrt{r^2 - \dfrac{4q}{r}}$ and $-\frac{1}{2}r - \frac{1}{2}\sqrt{r^2 - \dfrac{4q}{r}}$,

or $-\frac{1}{2}r + \frac{1}{2}\sqrt{1 - \dfrac{4q}{q - pr}}$ and $-\frac{1}{2}r - \frac{1}{2}r\sqrt{1 - \dfrac{4q}{q - pr}}$.

Ex. 1. If the equation be $x^3 - 15x = 4$: here $p = -15$, $q = 4$, $a = -5$, $b = 2$; hence $\sqrt{b^2 + a^3} = \sqrt{4 - 125} = \sqrt{-121} = 11\sqrt{-1}$, $s = \sqrt[3]{2 + \sqrt{-121}} = 2 + \sqrt{-1}$, and $d = \sqrt[3]{2 - \sqrt{-121}} = 2 - \sqrt{-1}$: therefore $r = c + d = 4$, the 1st root; and $-\dfrac{s+d}{2} \pm \dfrac{s-d}{2}\sqrt{-3} = -2 \mp \sqrt{3}$, the other two roots.

Ex. 2. If $x^3 - 6x = 4$: here $a = -2$, and $b = 2$; therefore $\sqrt{b^2 + a^3} = \sqrt{4 - 8} = \sqrt{-4} = 2\sqrt{-1}$, $s = \sqrt[3]{2 + 2\sqrt{-1}} = -1 + \sqrt{-1}$, & $d = \sqrt[3]{2 - 2\sqrt{-1}} = -1 - \sqrt{-1}$: hence then $r = s + d = -2$, the first root; and $1 \pm \sqrt{3}$ the other two roots.

Ex. 3. If $x^3 + 18x = 6$: here $a = 6$, and $b = 3$; then $\sqrt{b^2 + a^3} = \sqrt{9 + 216} = \sqrt{225} = 15$, $s = \sqrt[3]{3 + 15} = \sqrt[3]{18}$, and $d = \sqrt[3]{3 - 15} = \sqrt[3]{-12} = -\sqrt[3]{12}$: therefore $r = s + d = \sqrt[3]{18} - \sqrt[3]{12} = \cdot 331313$, the 1st root, and $-\dfrac{\sqrt[3]{18} - \sqrt[3]{12}}{2} \pm \dfrac{\sqrt[3]{18} + \sqrt[3]{12}}{2}\sqrt{-3}$ are the two other roots.

2. Another method for the roots of the equation $x^3 + px = q$, is by means of infinite series, as shewn at pa. 415 and seq. of my Tract above cited; whence it appears that the roots are exhibited in various forms of series as follow: viz.

$2\sqrt[3]{b} \times \quad : 1 - \dfrac{2c^2}{3 \cdot 6b^2} - \dfrac{2 \cdot 5 \cdot 8c^4}{3 \cdot 6 \cdot 9 \cdot 12b^4}$ &c

for the 1st root, and

$-\sqrt[3]{b} \times \quad : 1 - \dfrac{2c^2}{3 \cdot 6b^2} - \dfrac{2 \cdot 5 \cdot 8c^4}{3 \cdot 6 \cdot 9 \cdot 12b^4}$ &c

$\pm \dfrac{c\sqrt{-3}}{\sqrt[3]{b^2}} \times : \frac{1}{3} + \dfrac{2 \cdot 5c^2}{3 \cdot 6 \cdot 9b^2} + \dfrac{2 \cdot 5 \cdot 8 \cdot 11c^4}{3 \cdot 6 \cdot 9 \cdot 12 \cdot 15b^4}$ &c

for the two other roots: where $a = \frac{1}{3}p$, $b = \frac{1}{2}q$, and $c = \sqrt{b^2 + a^3}$.

And

And various other series for the same purpose may also be seen in my Tract, so often before cited.

3. A third method for the roots of cubic equations, is by angular sections, and the table of sines. It was first hinted by Bombelli, in his Algebra, that angles are trisected by the resolution of cubic equations. Afterwards, Vieta gave the resolution of cubics, and the higher equations, by angular sections. Next, Albert Girard, in his Invention Nouvelle en l'Algebre, shews how to resolve the irreducible case in cubics by a table of sines: and he also constructs the same, or finds the roots, by the intersection of the hyperbola and circle. Halley and De Moivre also gave rules and examples of the same sort of resolutions by a table of sines. And, lastly, Mr. Anthony Thacker invented, and Mr. William Brown computed, a large set of similar tables, for resolving affected quadratic and cubic equations, with their application to the resolution of biquadratic ones.

4. Lastly, the several methods of approach, or approximation, for the roots of all affected equations, which have been used in various ways by Stevin, Vieta, Newton, Halley, Raphson, and others.

To these may be added the method of Trial-and-error, or of Double Position, one of the easiest and best of any. Of this method, let there be taken the last example, viz, $x^3 + 18x = 6$, in which it is evident that x is very nearly equal to $\frac{1}{3}$, but a little less; take it therefore $x = \cdot 33$; then $x^3 + 18x = 5 \cdot 975937$, but should be 6, and therefore the error is $\cdot 024063$ in defect.

Again suppose $x = \cdot 34$; then $x^3 + 18x = 6 \cdot 159304$, which is $\cdot 159304$ in excess.

Therefore $\dfrac{\cdot 34 \times \cdot 024063 + \cdot 33 \times \cdot 159304}{\cdot 024063 + \cdot 159304} = \dfrac{60752}{183367} =$ $\cdot 3313$, the root as before very nearly.

For the construction of cubic equations, see CONSTRUCTION.

CUBIC *Foot*, of any thing, is so much of it as is contained in a cube whose side is one foot.

CUBIC *Hyperbola*, is a figure expressed by the equation $xy^2 = a$, having two asymptotes, and consisting of two hyperbolas, lying in the adjoining angles of the asymptotes, and not in the opposite angles, like the Apollonian hyperbola; being otherwise called by Newton, in his Enumeratio Linearum Tertii Ordinis, an hyperbolismus of a parabola; and is the 65th species of those lines according to him.

CUBIC *Numbers.* See CUBES.

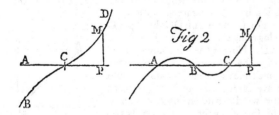

Fig 2

CUBIC *Parabola*, a curve, as BCD, of the 2d order, having two infinite legs CB, CD, tending contrary

VOL. I

ways. And if the abscifs, AP or x, touch the curve in C, the relation between the abscifs and ordinate, viz, $AP = x$, and $PM = y$, is expressed by the equation $y = ax^3 + bx^2 + cx + d$; or when A coincides with C, by the equation $y = ax^3$, which is the simplest form of the equation of this curve.

If the right line AP (fig. 2) cut the cubical parabola in three points A, B, C; and from any point P there be drawn the right line or ordinate PM, cutting the curve in one point M only: then will PM be always as the solid $AP \times BP \times CP$; which is an essential property of this curve.

And hence it is easy to construct a cubic equation, as $x^3 + a^2x = b^3$, by the intersection of this curve and a right line. See the Construction of a cubic equation by means of the cubic parabola and a right line, by Dr. Wallis, in his Algebra: As also the Construction of equations of 6 dimensions, by means of the cubic parabola and a circle, by Dr. Halley, in a lecture formerly read at Oxford.

The curve of this parabola cannot be rectified, not even by means of the conic sections. But a circle may be found equal to the *Curve Surface*, generated by the rotation of the curve AM about the tangent AP to the principal vertex A. Let MN be an ordinate, and MT a tangent at the point M; and let PM be parallel to AN. Divide MN in the point O, so that MO be to ON as TM to MN. Then a mean proportional between $TM + ON$ and $\frac{1}{3}$ of AN will be the radius of a circle, whose area is equal to the superficies described by that rotation, viz, of AM about AP.

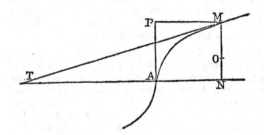

The Area of a Cubic Parabola is $\frac{3}{4}$ of its circumscribing parallelogram.

CUBIC *Root*, of any number, or quantity, is such a quantity as being cubed, or twice multiplied by itself, shall produce that which was given. So, the cubic root of 8 is 2, because 2^3 or $2 \times 2 \times 2$ is equal to 8.

The common method of extracting the cube root, founded on the property given above, viz, $\overline{a + b}|^3 = a^3 + 3a^2b + 3ab^2 + b^3$, is found in every book of common arithmetic, and is as old at least as Lucas de Burgo, where it is first met with in print. Other methods for the cube root may be seen under the article EXTRACTION *of Roots*, particularly this one, viz, the cube root of n, or $\sqrt[3]{n} = \dfrac{2n + a^3}{n + 2a^3} \times a$, very nearly, or $2a^3 + n$: $2n + a^3 :: a : \sqrt[3]{n}$ the cube root of n nearly; where n

is

is any number given whose cube root is fought, and a^3 is the nearest complete cube to n, whether greater or less.

For example, suppose it were proposed to double the cube, or, which comes to the same thing, to extract the cube root of the number 2. Here the nearest cube is 1, whose cube root is 1 also, that is, $a^3 = 1$, and $a = 1$, also $n = 2$; therefore
$2a^3 + n$ $2n + a^3$ a

4 : 5 :: 1 : $\frac{5}{4}$ = 1·25 = $\sqrt[3]{2}$ nearly.

But, for a nearer value, assume now $a = \frac{5}{4}$, or $a^3 = \frac{125}{64}$; then is $2a^3 + n = \frac{250}{64} + 2 = \frac{378}{64}$, and $2n + a^3 = 4 + \frac{125}{64} = \frac{381}{64}$; hence $378 : 381 :: \frac{5}{4} : \frac{635}{504}$ or 1·259921 = $\sqrt[3]{2}$, or the cube root of 2, which is true in the last place of decimals.

And this is the simplest and easiest method for the cube root of any number. See its investigation in my Tracts, vol. 1, pa. 49.

Every number or quantity has three cubic roots, one that is real, and two imaginary: So, the cube root of 1 is either 1, or $-\frac{1 + \sqrt{-3}}{2}$, or $-\frac{1 - \sqrt{-3}}{2}$; and if r be the real root of any cube r^3, the two imaginary cubic roots of it will be
$-\frac{1 + \sqrt{-3}}{2}r$, and $-\frac{1 - \sqrt{-3}}{2}r$: for any one of these being cubed, gives the same cube r^3.

CUBING *of a Solid.* See CUBATURE.

CUBO-CUBE, the 6th power.

CUBO-CUBO-CUBE, the 9th power.

CULMINATION, the passage of a star or planet over the meridian, or that point of its orbit which it is in at its greatest altitude. Hence a star is said to culminate, when it passes the meridian.

To find the time of a Star's culminating. Estimate the time nearly; and find the right ascension both of the sun and star corrected for this estimated time; then the difference between these right ascensions, converted into solar time, at the rate of 15 degrees to the hour, gives the time of southing. See an example of this calculation every year in White's Ephemeris, pa. 45.

CULVERIN, was the name of a piece of ordnance; but is now disused.

CUNETTE. See CUVETTE.

CUNEUS. See WEDGE.

CUNITIA (MARIA), a lady of Silesia, who was famous for her extensive knowledge in many branches of learning, but more particularly in mathematics and astronomy, upon which she wrote several ingenious treatises; particularly one, entitled *Urania Propitia,* printed in 1650, in Latin and German, and dedicated to Ferdinand the third, emperor of Germany. In this work are contained astronomical tables of great ease and accuracy, founded upon Kepler's hypothesis. But notwithstanding her merit shines with such peculiar lustre as to reflect honour on her sex, history does not inform us of the time of her birth or death.

CURRENT, a stream or flux of water in any direction. The *setting* of the current, is that point of the compass towards which the waters run; and the *drift* of a current is the rate it runs an hour.

Currents in the sea, are either natural and general, as arising from the diurnal rotation of the earth on its axis, or the tides, &c; or accidental and particular, caused by the waters being driven against promontories, or into gulphs and streights; from whence they are forced back, and thus disturb the natural flux of the sea.

The currents are so violent under the equator, where the motion of the earth is greatest, that they hurry vessels very speedily from Africa to America; but absolutely prevent their return the same way: so that ships are forced to run as far as the 40th or 45th degree of north latitude, to fall into the return of the current again, to bring them home to Europe. It is shewn by Governor Pownal, that this current performs a continual circulation, setting out from the Guinea coast in Africa, for example, from thence crossing straight over the Atlantic ocean, and so setting into the Gulph of Mexico by the south side of it; then sweeping round by the bottom of the Gulph, it issues out by the north side of it, and thence takes a direction north-easterly along the coast of North America, till it arrives near Newfoundland, where it is turned by a rounding motion backward across the Atlantic again, upon the coasts of Europe, and from thence southward again to the coast of Africa, from whence it set out.

In the streights of Gibraltar, the currents set in by the south side, sweep along the coast of Africa to Egypt, by Palestine, and so return by the northern side, or European coasts, and issue out again by the northern side of the streights. In St. George's Channel too they usually set eastward. The great violence and danger of the sea in the Streights of Magellan, is attributed to two contrary currents setting in, one from the south, and the other from the north sea.

Currents are of some consideration in the art of navigation, as a ship is by them either accelerated or retarded in her course, according as the set is with or against the ship's motion. If a ship sail along the direction of a current, it is evident that the velocity of the current must be added to that of the vessel: but if her course be directly against the current, it must be subtracted: and if she sail athwart it, her motion will be compounded with that of the current; and her velocity augmented or retarded, according to the angle of her direction with that of the current; that is, she will proceed in the diagonal of the parallelogram formed according to the two lines of direction, and will describe or pass over that diagonal in the same time in which she would have described either of the sides, by the separate velocities.

For suppose ABDC be a parallelogram, the diagonal of which is AD. Now if the wind alone would drive the ship from A to B, in the same time as the current alone would drive it from A to C: Then, as the wind neither helps nor hinders the ship from coming towards the line CD, the current will bring it there in the same time as if the wind did not act: And as the current neither helps nor hinders the ship from

coming

coming towards the line BD, the wind will bring it there in the same time as if the current did not act: Therefore the ship muft, at the end of that time, be found in both thofe lines, that is, in their meeting D: Confequently the ship muft have paffed from A to D in the diagonal AD. See COMPOSITION *of Forces*.

See the Sailing in Currents largely exemplified in Robertfon's Navigation, vol. 2, book 7, feét. 8.

CURSOR, a fmall piece of brafs &c that flides; as, the piece in an equinoctial ring dial that flides to the day of the month; or the little ruler or label of brafs fliding in a groove along the middle of another label, reprefenting the horizon in the analemma; or the point that flides along the beam compafs; &c.

CURTATE *Diftance*, is the diftance of a planet's place from the fun, reduced to the ecliptic; or, the interval between the fun and that point where a perpendicular, let fall from the planet, meets the ecliptic.

CURTATION, the interval between a planet's diftance from the fun, and the curtate diftance.

From the foregoing article it is eafy to find the curtate diftance; whence the manner of conftructing tables of curtation is obvious; the quantity of inclination, reduction, and curtation of a planet, depending on the argument of latitude. Kepler, in his Rodolphine Tables, reduces the tables of them all into one, under the title of *Tabulæ Latitudinariæ*.

CURTIN, CURTAIN, or COURTINE, in Fortification, that part of a wall or rampart that joins two baftions, or lying between the flank of one and that of another.—The curtain is ufually bordered with a parapet 5 feet high; behind which the foldiers ftand to fire upon the covert-way, and into the moat.

CURVATURE *of a Line*, is its bending, or flexure; by which it becomes a curve, of any peculiar form and properties. Thus, the nature of the curvature of a circle is fuch, as that every point in the periphery is equally diftant from a point within, called the centre; and fo the curvature of the fame circle is every where the fame; but the curvature in all other curves is continually varying.—The curvature of a circle is fo much the more, as its radius is lefs, being always reciprocally as the radius; and the curvature of other curves is meafured by the reciprocal of the radius of a circle having the fame degree of curvature as any curve has, at fome certain point.

Every curve is bent from its tangent by its curvature, the meafure of which is the fame as that of the angle of contact formed by the curve and tangent. Now the fame tangent AB is common to an infinite number of circles, or other curves, all touching it and each other in the fame point of contact C. So that any curve DCE may be touched by an infinite number of different circles at the fame point C; and fome of thefe circles fall wholly within it, being more curved, or having a greater curvature than that curve; while others fall without it near the point of contact, or between the curve and tangent at that point, and fo, being lefs deflected from the tangent than the curve is, they have a lefs degree of curvature there. Confequently there is one, of this infinite number of circles, which neither falls below it nor above it, but, being equally deflected from the tangent, coincides with it moft intimately of all the circles; and the radius of this circle is called the

radius of curvature of the curve; alfo the circle itfelf is called the *circle of curvature*, or the *ofculatory* circle of that curve, becaufe it touches it fo clofely that no other circle can be drawn between it and the curve.

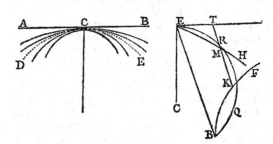

As a curve is feparated from its tangent by its flexure or curvature, fo it is feparated from the ofculatory circle by the increafe or decreafe of its curvature; and as its curvature is greater or lefs, according as it is more or lefs deflected from the tangent, fo the variation of curvature is greater or lefs, according as it is more or lefs feparated from the circle of curvature.

It appears, however, from the demonftration of geometricians, that circles may touch curve lines in fuch a manner, that there may be indefinite degrees of more or lefs intimate contact between the curve and its ofculatory circle; and that a conic feétion may be defcribed that fhall have the fame curvature with a given line at a given point, and the fame variation of curvature, or a contact of the fame kind with the circle of curvature.

If we conceive the tangent of any propofed curve to be a bafe, and that a new line or curve be defcribed, whofe ordinate, upon the fame bafe or abfcifs, is a 3d proportional to the ordinate and bafe of the firft; this new curve will determine the chord of the circle of curvature, by its interfeétion with the ordinate at the point of contact; and it will alfo meafure the variation of curvature, by means of the tangent of the angle in which it cuts that circle: the lefs this angle is, the clofer is the contact of the curve and circle of curvature; and of this contact there may be indefinite degrees.

For example, let EMH be any curve, to which ET is a tangent at the point E; then let there be always taken MT : ET :: ET : TK, and through all the points K draw the curve BKF; then from the point of contact E draw EB parallel to the ordinate TK, meeting the laft curve in B; and finally, defcribe a circle ERQB through the point B and touching ET in E; and it fhall be the ofculatory circle to the given curve EMH. And the contact of EM and ER is always the clofer, the lefs the angle KBQ is. See Maclaurin's Fluxions, art. 366.

Hence it follows, that the contact of the curve EMH and its ofculatory circle is clofeft, when the curve BK touches the arch BQ in B, the angle BET being given; and it is fartheft from this, or moft open, when BK touches the right line BE in B.

Hence alfo there may be indefinite degrees of more and more intimate contact between a circle and a curve. The firft degree is when the fame right line touches

touches them both in the same point; and a contact of this fort may take place between any circle and any arch of a curve. The 2d is when the curve EMH and circle ERB have the fame curvature, and the tangents of the curve BKF and circle BQE interfect each other at B in any affignable angle. The contact of the curve EM and circle of curvature ER at E, is ot the 3d degree, or order, and their osculation is of the 2d, when the curve BKF touches the circle BQE at B, but fo as not to have the fame curvature with it. The contact is of the 4th degree, or order, and their osculation of the 3d, when the curve BKF has the fame curvature with the circle BQE at B; but fo as that their contact is only of the 2d degree. And this gradation of more and more intimate contact, or of approximation towards coincidence, may be continued indefinitely, the contact of EM and ER at E being always of an order two degrees closer than that of BK and BQ at B. There is also an indefinite variety comprehended under each order: thus, when EM and ER have the fame curvature, the angle formed by the tangents of BK and BQ admits of indefinite variety, and the contact of EM and ER is the closer the lefs that angle is. And when that angle is of the fame magnitude, the contact of EM and ER is the closer, the greater the circle of curvature is. When BK and BQ touch at B, they may touch on the fame or on different fides of their common tangent; and the angle of contact KBQ may admit of the fame variety with the angle of contact MER; but as there is feldom occasion for confidering thofe higher degrees of more intimate contact of the curve EMH and circle of curvature ERB, Mr. Maclaurin calls the contact or osculation of the fame kind, when, the chord EB and angle BET being given, the angle contained by the tangents of BK and BQ is of the fame magnitude.

When the curvature of EMH increafes from E towards H, and confequently corresponds to that of a circle gradually lefs and lefs, the arch EM falls within ER; the arch of the osculatory circle, and BK is within BQ. The contrary happens when the curvature of EM decreafes from E towards H, and confequently corresponds to that of a circle which is gradually greater and greater, the arch EM falls without ER, and BK is without BQ. And according as the curvature of EM varies more or lefs, it is more or lefs unlike to the uniform curvature of a circle; the arch of the curve EMH feparates more or lefs from the arch of the osculatory circle ERB, and the angle contained by the tangents of BKF and BQE at B, is greater or lefs. Thus the quality of curvature, as it is called by Newton, depends on the angle contained by the tangents of BK and BQ at B; and the measure of the inequability or variation of curvature, is as the tangent of this angle, the radius being given, and the angle BET being a right one.

The radii of curvature of fimilar arcs in fimilar figures, are in the fame ratio as any homologous lines of thefe figures; and the variation of curvature is the fame. See Maclaurin, art. 370.

When the propofed curve EMH is a conic fection, the new line BKF is alfo a conic fection; and it is a right line when EMH is a parabola, to the axis of which the ordinates TK are parallel. BKF is alfo a right line when EMH is an hyperbola, to one afymptote of which the ordinate TK is parallel.

When the ordinate EB, at the point of contact E inftead of meeting the new curve BK, is an afymptote to it, the curvature of EM will be lefs than in any circle; and this is the cafe in which it is faid to be infinitely little, or that the radius of curvature is infinitely great. And of this kind is the curvature at the points of contrary flexure in lines of the 3d order.

When the curve BK paffes through the point of contact E, the curvature is greater than in any circle, or the radius of curvature vanifhes; and in this cafe the curvature is faid to be infinitely great. Of this kind is the curvature at the cufps of the lines of the 3d order.

As to the degree of curvature in lines of the 3d and higher orders, fee Maclaurin, art. 379; alfo art. 380, when the propofed curve is mechanical.

As curves which pafs through the fame point have the fame tangent when the firft fluxions of the ordinates are equal, fo they have the fame curvature when the 2d fluxions of the ordinate are likewife equal; and half the chord of the osculatory circle that is intercepted between the points where it interfects the ordinate, is a 3d proportional to the right lines that meafure the 2d fluxion of the ordinate and firft fluxion of the curve, the bafe being suppofed to flow uniformly. When a ray revolving about a given point, and terminated by the curve, becomes perpendicular to it, the firft fluxion of the radius vanifhes; and if its 2d fluxion vanifh at the fame time, that point muft be the centre of curvature. The fame may be faid, when the angular motion of the ray about that point is equal to the angular motion of the tangent of the curve; as the angular motion of the radius of a circle about its centre is always equal to the angular motion of the tangent of the circle. Hence the various properties of the circle may fuggeft feveral theorems for determining the centre of curvature.

See art. 396 of the faid book, alfo the following, concerning the curvature of lines that are defcribed by means of right lines revolving about given poles, or of angles that either revolve about fuch poles, or are carried along fixed lines.

It is to be obferved that, as when a right line interfects an arc of a curve in two points, if by varying the pofition of that line the two interfections unite in one point, it then becomes the tangent of the arc; fo when a circle touches a curve in one point, and interfects it in another, if, by varying the centre, this interfection joins the point of contact, the circle has then the clofeft contact with the arc, and becomes the circle of curvature; but it ftill continues to interfect the curve at the fame point where it touches it, that is, where the fame right line is their common tangent, unlefs another interfection join that point at the fame time. In general, the circle of curvature interfects the curve at the point of osculation, only when the number of the fucceffive orders of fluxions of the radius of curvature, that vanifhes when this radius comes to the point of osculation, is an even number.

It has been fuppofed by fome, that two points of contact, or four interfections of the curve and circle of curvature, muft join to form an osculation. But Mr.

James

James Bernoulli juftly infifted, that the coalition of one point of contact and one interfection, or of three interfections, was fufficient. In which cafe, and in general, when an odd number of interfections only join each other, the point where they coincide continues to be an interfection of the curve and circle of curvature, as well as a point of their mutual contact and ofculation. See Maclaurin's Flux. art. 493.

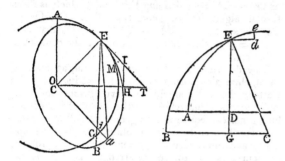

From thefe principles may be determined the circle of curvature at any point of a conic fection. Thus, fuppofe AEMHG be any conic fection, to the point E of which the circle or radius of curvature is to be found. Let ET be a tangent at E, and draw EGB and the tangent HI parallel to the chords of the circle of curvature; then take EB to EG as EI² to HI²; or, when the fection has a centre O, as in the ellipfe and hyperbola, as the fquare of the femi-diameter Oa parallel to ET, is to the fquare of the femi-diameter OA parallel to EB; and a circle EB defcribed upon the chord EB that touches ET, will be the circle of curvature fought.

When BET is a right angle, or EB is the diameter of the circle of curvature, EG will be the axis of the conic fection, and EB will be the parameter of this axis; alfo when the point G, where the conic fection cuts EB, and the point B, are on the fame fide of E, then EMG will be an ellipfis, and EG the greater or lefs axis, according as EG is greater or lefs than EB.

The propofitions relating to the curvature of the conic fections, commonly given by authors, follow with. out much difficulty from this conftruction.

1. When the chord of curvature, thus found, paffes through the centre of the conic fection, it will then be equal to the parameter of the diameter that paffes through the point of contact.

2. The fquare of the femi-diameter Oa, is to the rectangle of half the tranfverfe and half the conjugate axis, as the radius of curvature CE is to Oa. And therefore the cube of the femi-diameter Oa, parallel to the tangent ET, is equal to the folid contained by the radius of curvature CE, and the rectangle of the two axes.

3. The perpendicular to either axis bifects the angle made by the chord of curvature, and the common tangent of the conic fection and circle of curvature.

4. The chord of the ofculatory circle that paffes through the focus, the diameter conjugate to that which paffes through the point of contact, and the tranfverfe axis of the figure, are in continued proportion.

5. When the fection is an ellipfe, if the circle of curvature at E meet Oa in d, the fquare of Ed will be equal to twice the fquare of Oa. Hence Ed : Oa :: $\sqrt{2}$: 1. Which gives an eafy method of determining the circle of curvature to any point E, when the femidiameter Oa is given in magnitude and pofition.

Several other properties of the circle of curvature, and methods of determining it when the fection is given; or vice verfa, of determining the fection when the circle of curvature is given, may be feen in Maclaurin's Flux. art. 375. See alfo the Appendix to Maclaurin's Algebra, fect. 1.

To determine the Radius and Circle of Curvature by the Method of Fluxions. Let AE*e* be any curve, concave towards its axis AD; draw an ordinate DE to the point E where the curve is required to be found; and fuppofe EC perpendicular to the curve, and equal to the radius of the circle BE*e* of curvature fought; laftly, draw E*d* parallel to AD, and *de* parallel and indefinitely near to DE; thereby making E*d* the fluxion or increment of the abfcifs AD, alfo *de* the fluxion of the ordinate DE, and E*e* that of the curve AE. Now put $x = $ AD, $y = $ DE, $z = $ AE, and $r = $ CE the radius of curvature; then is ED $= \dot{x}$, $de = \dot{y}$, and E*e* $= \dot{z}$.

Now, by fim. tri. the 3 lines - E*d* , *de* , E*e* ,
or - $\dot{x}$, $\dot{y}$, $\dot{z}$,
are refpectively as the three - GE , GC , CE ;
therefore - GC . $\dot{x} = $ GE $\dot{y}$;
and the flux. of this equa. is

$$GC . \ddot{x} + \dot{GC} . \dot{x} = GE . \ddot{y} + \dot{GE} \, \dot{y},$$

or becaufe $\dot{GC} = - \dot{BG}$, it is

$$GC . \ddot{x} - \dot{BG} . \dot{x} = GE . \ddot{y} + \dot{GE} \, \dot{y}$$

But, fince the two curves AE and BE have the fame curvature at the point E, their abfciffes and ordinates have the fame fluxions at that point, that is E*d* or $\dot{x}$ is the fluxion both of AD and BG, and *de* or $\dot{y}$ is the fluxion both of DE and GE. In the above equation therefore fubftitute $\dot{x}$ for $\dot{BG}$, and $\dot{y}$ for $\dot{GE}$, and it becomes $GC . \ddot{x} - \dot{x}\dot{x} = GE . \ddot{y} + \dot{y}\dot{y}$,
or $GC . \ddot{x} - GE . \ddot{y} = \dot{x}^2 + \dot{y}^2$ or $\dot{z}^2$.
Now mult. the three terms of this equation refpectively by thefe three quantities, $\dfrac{\dot{y}}{GC}, \dfrac{\dot{x}}{GE}, \dfrac{\dot{z}}{CE}$, which are

equal, and it becomes $\dot{y}\ddot{x} - \dot{x}\ddot{y} = \dfrac{\dot{z}^3}{CE}$, or $\dfrac{\dot{z}^3}{r}$;

and hence $r = \dfrac{\dot{z}^3}{\dot{y}\ddot{x} - \dot{x}\ddot{y}}$, which is the general value of the radius of curvature for all curves whatever, in terms of the fluxions of the abfcifs and ordinate.

Farther, as in any cafe either x or y may be fuppofed to flow equably, that is, either $\dot{x}$ or $\dot{y}$ conftant quantities, or $\ddot{x}$ or $\ddot{y} = $ to nothing, by this fuppofition either of the terms in the denominator of the value of r may be made to vanifh. So that when $\dot{x}$ is conftant, the value of r is $\dfrac{\dot{z}^3}{-\dot{x}\ddot{y}}$, but r is $= \dfrac{\dot{z}^3}{\dot{y}\ddot{x}}$ when y is conftant.

For example, fuppofe it were required to find the radius or circle of curvature to any point of a parabola, its vertex being A, and axis AD.—Now the equation

tion of the curve is $ax = y^2$; hence $a\dot{x} = 2y\dot{y}$, and $a\ddot{x} = 2\dot{y}^2$, supposing $\dot{y}$ constant, also $a^2\ddot{x}^2 = 4\dot{y}^2\dot{y}^2$; hence r or

$$\frac{\dot{z}^3}{\dot{y}\ddot{x}} \text{ or } \frac{\overline{x^2 + \dot{y}^2}^{\frac{3}{2}}}{\dot{y}\ddot{x}} \text{ is } \frac{\overline{a^2 + 4y^2}^{\frac{3}{2}}}{2a^2} \text{ or } \frac{\overline{a + 4x}^{\frac{3}{2}}}{2\sqrt{a}},$$

the general value of the radius of curvature for any point E, the ordinate to which cuts offthe abfcifs AD $= x$.

Hence, when x or the abfcifs is nothing, the laft expreffion becomes barely $\dfrac{a^{\frac{3}{2}}}{2\sqrt{a}} = \frac{1}{2}a = r$ for the radius of curvature at the vertex of the parabola; that is, the diameter of the circle of curvature at the vertex of a parabola, is equal to a the parameter of axis.

Variation of CURVATURE. See VARIATION.

Double CURVATURE, is ufed for the curvature of a line, which twifts fo that all the parts of it do not lie in the fame plane.

CURVE, a line whofe feveral parts proceed bowing, or tend different ways; in oppofition to a ftraight line, all whofe parts have the fame courfe or direction.

The doctrine of curves, and of the figures and folids generated from them, conftitute what is called the higher geometry.

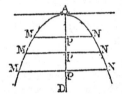 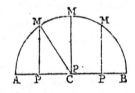

In a curve, the line AD, which bifects all the parallel lines MN, is called a *diameter;* and the point A, where the diameter meets the curve, is called the *vertex:* if AD bifect all the parallels at right angles, it is called the *axis.* The parallel lines MN are called *ordinates,* or *applicates;* and their halves, PM, or PN, *femi-ordinates.* The portion of the diameter AP, between the vertex, or any other fixed point, and an ordinate, is called the abfcifs; alfo the concourfe of all the diameters, if they meet all in one point, is the *centre.* This definition of the diameter, as bifecting the parallel ordinates, refpects only the conic fections, or fuch curves as are cut only in two points by the ordinates; but in the lines of the 3d order, which may be cut in three points by the ordinates, then the diameter is that line which cuts the ordinates fo, that the fum of the two parts that lie on one fide of it, fhall be equal to the part on the other fide: and fo on for curves of higher orders, the fum of the parts of the ordinates on one fide of the diameter, being always equal to the fum of the parts on the other fide of it.

Curve lines are diftinguifhed into *algebraical* or *geometrical,* and *tranfcendental* or *mechanical.*

Algebraical or *Geometrical* CURVES, are thofe in which the relation of the abfciffes AP, to the ordinates PM, can be expreffed by a common algebraic equation

And *Tranfcendental* or *Mechanical Curves,* are fuch as cannot be fo defined or expreffed by an algebraical equation. See TRANSCENDENTAL *Curve.*

Thus, fuppofe, for inftance, the curve be the circle; and that the radius AC $= r$, the abfcifs AP $= x$, and the ordinate PM $= y$; then, becaufe the nature of the circle is fuch, that the rectangle AP $\times$ PB is always $=$ PM², therefore the equation is $x . \overline{2r - x} = y^2$, or $2rx - x^2 = y^2$, defining this curve, which is therefore an algebraical or geometrical line. Or, fuppofe CP $= x$; then is CM² $-$ CP² $=$ PM², that is $r^2 - x^2 = y^2$; which is another form of the equation of the curve.

The doctrine of curve lines in general, as expreffed by algebraical equations, was firft introduced by Des Cartes, who called algebraical curves geometrical ones; as admitting none elfe into the conftruction of problems, nor confequently into geometry. But Newton, and after him Leibnitz and Wolfius, are of another opinion; and think, that in the conftruction of a problem, one curve is not to be preferred to another for its being defined by a more fimple equation, but for its being more eafily defcribed.

Algebraical or geometrical lines are beft diftinguifhed into orders according to the number of dimenfions of the equation expreffing the relation between its ordinates and abfciffes, or, which is the fame thing, according to the number of points in which they may be cut by a right line. And curves of the fame kind or order, are thofe whofe equations rife to the fame dimenfion. Hence, of the firft order, there is the right line only; of the 2d order of lines, or the firft order of curves, are the circle and conic fections, being 4 fpecies only, viz, $dx - x^2 = y^2$ the circle, $\dfrac{c^2}{d^2} \cdot \overline{dx - x^2} = y^2$ the ellipfe, $\dfrac{c^2}{d^2} \cdot dx + x^2 = y^2$ the hyperbola, and $dx = y^2$ the parabola: the lines of the 3d order, or curves of the 2d order, are expreffed by an equation of the 3d degree, having three roots; and fo on. Of thefe lines of the 3d order, Newton wrote an exprefs treatife, under the title of Enumeratio Linearum Tertii Ordinis, fhewing their diftinctive characters and properties, to the number of 72 different fpecies of curves: but Mr. Stirling afterwards added four more to that number; and Mr. Nic. Bernoulli and Mr. Stone added two more.

Curves of the 2d and other higher kinds, Newton obferves, have parts and properties fimilar to thofe of the 1ft kind: Thus, as the conic fections have diameters and axes; the lines bifected by thefe are ordinates; and the interfection of the curve and diameter, the vertex: fo, in curves of the 2d kind, any two parallel right lines being drawn to meet the curve in 3 points; a right line cutting thefe parallels fo, as that the fum of the two parts between the fecant and the curve on one fide, is equal to the 3d part terminated by the curve on the other fide, will cut, in the fame manner, all other right lines parallel to thefe, and that meet the curve in three points, that is, fo as that the fum of the two parts on one fide, will ftill be equal to the 3d part on the other fide. Thefe three parts therefore thus equal, may be called *ordinates,* or *applicates;* the cutting line, the *diameter;*

meter ; and where it cuts the ordinates at right angles, the *axis ;* the interfection of the diameter and the curve, the *vertex ;* and the concourfe of two diameters, the *centre ;* alfo the concourfe of all the diameters, the *common* or *general centre.*

Again, as an hyperbola of the firft kind has two afymptotes ; that of the 2d has 3 ; that of the 3d has 4 ; &c : and as the parts of any right line between the conic hyperbola and its two afymptotes, are equal on either fide ; fo, in hyperbolas of the 2d kind, any right line cutting the curve and its three afymptotes in three points ; the fum of the two parts of that right line, extended from any two afymptotes, the fame way, to two points of the curve, will be equal to the 3d part extended from the 3d afymptote, the contrary way, to the 3d point of the curve.

Again, as in the conic fections that are not parabolical, the fquare of an ordinate, i. e. the rectangle of the ordinates drawn on the contrary fides of the diameter, is to the rectangle of the parts of the diameter terminated at the vertices of an ellipfe or hyperbola, in the fame proportion as a given line called the *latus rectum,* is to that part of the diameter which lies between the vertices, and called the *latus transverfum:* fo, in curves of the 2d kind, not parabolical, the parallelopiped under three ordinates, is to the parallelopiped under the parts of the diameter cut off at the ordinates and the three vertices of the figure, in a given ratio : in which, if there be taken three right lines fituate at the three parts of the diameter between the vertices of the figure, each to each ; then thefe three right lines may be called the *latera recta* of the figure ; and the parts of the diameter between the vertices, the *latera transverfa.*

And, as in a conic parabola, which has only one vertex to one and the fame diameter, the rectangle under the ordinates is equal to the rectangle under the part of the diameter cut off at the ordinates and vertex, and a given right line called the latus rectum : fo, in curves of the 2d kind, which have only two vertices to the fame diameter, the parallelopiped under three ordinates, is equal to the parallelopiped under two parts of the diameter cut off at the ordinates and the two vertices, and a given right line, which may therefore be called the *latus transverfum.*

Further, as in the conic fections, where two parallels, terminated on each fide by a curve, are cut by two other parallels terminated on each fide by a curve, the 1ft by the 3d, and the 2d by the 4th ; the rectangle of the parts of the 1ft is to the rectangle of the parts of the 3d, as that of the 2d is to that of the 4th : fo, when four fuch right lines occur in a curve of the 2d kind, each in three points ; the parallelopiped of the parts of the 1ft, will be to that of the parts of the 3d, as that of the 2d to that of the 4th.

Laftly, the legs of curves, both of the 1ft, 2d, and higher kinds, are either of the parabolic or hyperbolic kind : an hyperbolic leg being that which approaches infinitely towards fome afymptote ; and a parabolic one, that which has no afymptote.

Thefe legs are beft diftinguifhed by their tangents ; for, if the point of contact go off to an infinite diftance, the tangent of the hyperbolic leg will coincide with the afymptote ; and that of the parabolic leg, recede infinitely, and vanifh. Therefore the afymptote of any leg is found, by feeking the tangent of that leg to a point infinitely diftant ; and the direction of an infinite leg is found, by feeking the pofition of a right line parallel to the tangent, where the point of contact is infinitely remote, for this line tends that way towards which the infinite leg is directed.

Reduction of CURVES of the 2d kind.

Newton reduces all curves of the 2d kind to four cafes of equations, expreffing the relation between the ordinate and abfcifs, viz,

in the 1ft cafe, $xy^2 + ey = ax^3 + bx^2 + cx + d$;

in the 2d, $\quad xy = ax^3 + bx^2 + cx + d$;

in the 3d, $\quad y^2 = ax^3 + bx^2 + cx + d$;

in the 4th, $\quad y = ax^3 + bx^2 + cx + d.$

See Newton's Enumeratio, fect. 3 ; and Stirling's Lineæ, &c, pa. 83.

Enumeration of the CURVES of the 2d kind.

Under thefe four cafes, the author brings a great number of different forms of curves, to which he gives different names. An hyperbola lying wholly within the angle of the afymptotes, like a conic hyperbola, he calls an *infcribed hyperbola ;* that which cuts the afymptotes, and contains the parts cut off within its own periphery, a *circumfcribed hyperbola ;* that which has one of its infinite legs infcribed and the other circumfcribed, he calls *ambigenal ;* that whofe legs look towards each other, and are directed the fame way, *converging ;* that where they look contrary ways, *diverging ;* that where they are convex different ways, *crofslegged ;* that applied to its afymptote with a concave vertex, and diverging legs, *conchoidal ;* that which cuts its afymptote with contrary flexures, and is produced each way into contrary legs, *anguineous,* or *fnake-like ;* that which cuts its conjugate acrofs, *cruciform ;* that which returning around cuts itfelf, *nodated ;* that whofe parts concur in the angle of contact, and there terminate, *cufpidated ;* that whofe conjugate is oval, and infinitely fmall, i. e. a point, *pointed ;* that which, from the impoffibility of two roots, is without either oval, node, cufp, or point, *pure.* And in the fame manner he denominates a parabola *converging, diverging, cruciform,* &c. Alfo when the number of hyperbolic legs exceeds that of the conic hyperbola, that is more than two, he calls the hyperbola *redundant.*

Under thofe 4 cafes the author enumerates 72 different curves : of thefe, 9 are redundant hyperbolas, without diameters, having three afymptotes including a triangle ; the firft confifting of three hyperbolas, one infcribed, another circumfcribed, and the third ambigenal, with an oval ; the 2d, nodated ; the 3d, cufpidated ; the 4th, pointed ; the 5th and 6th, pure ; the 7th and 8th, cruciform ; the 9th or laft, anguineal. There are 12 redundant hyperbolas, having only one diameter : the 1ft, oval ; the 2d, nodated ; the 3d, cufpidated ; the 4th, pointed ; 5th, 6th, 7th, and 8th, pure ; the 9th and 10th, cruciform ; the 11th and 12th, conchoidal. And to this clafs Stirling adds 2 more. There are 2 redundant hyperbolas, with three diameters. There are 9 redundant hyperbolas, with three afymptotes converging to a common point ; the 1ft being formed of the 5th and 6th redundant parabolas, whofe whose asymptotes include a triangle; the 2d formed of the 7th and 8th; the 3d and 4th; of the 9th; the 5th is formed of the 5th and 7th of the redundant hyperbolas, with one diameter; the 6th, of the 6th and 7th; the 7th, of the 8th and 9th; the 8th, of the 10th and 11th; the 9th, of the 12th and 13th: all which conversions are effected, by diminishing the triangle comprehended between the asymptotes, till it vanish into a point.

Six are defective parabolas, having no diameters: the 1st, oval; the 2d, nodated; the 3d, cuspidated; the 4th, pointed; the 5th, pure; &c.

Seven are defective hyperbolas, having diameters: the 1st and 2d, conchoidal, with an oval; the 3d, nodated; the 4th, cuspidated, which is the cissoid of the ancients; the 5th and 6th, pointed; the 7th, pure.

Seven are parabolic hyperbolas, having diameters: the 1st, oval; the 2d, nodated; the 3d, cuspidated; the 4th, pointed; the 5th, pure; the 6th, cruciform; the 7th, anguineous.

Four are parabolic hyperbolas: four are hyperbolisms of the hyperbola: three, hyperbolas of the ellipsis: two, hyperbolisms of the parabola.

Six are diverging parabolas; one, a trident; the 2d, oval; the 3d, nodated; the 4th, pointed; the 5th, cuspidated (which is Neil's parabola, usually called the semi-cubical parabola); the 6th, pure.

Lastly, one, commonly called the cubical parabola.

Mr. Stirling and Mr. Stone have shewn that this enumeration is imperfect, the former having added four new species of curves to the number, and the latter two, or rather these two were first noticed by Mr. Nic. Bernoulli. Also Mr. Murdoch and Mr. Geo. Sanderson have found some new species; though some persons dispute the reality of them. See the Genesis Curvarum per umbras, and the Ladies' Diary 1788 and 1789, the prize question.

Organical Description of CURVES.—Sir Isaac Newton shews that curves may be generated by shadows. He says, if upon an infinite plane, illuminated from a lucid point, the shadows of figures be projected; the shadows of the conic sections will always be conic sections; those of the curves of the 2d kind, will always be curves of the 2d kind; those of the curves of the 3d kind, will always be curves of the 3d kind; and so on ad infinitum

And, like as the projected shadow of a circle generates all the conic sections, so the 5 diverging parabolas, by their shadows, will generate and exhibit all the rest of the curves of the 2d kind: and thus some of the most simple curves of the other kinds may be found, which will form, by their shadows upon a plane, projected from a lucid point, all the other curves of that same kind. And in the French Memoirs may be seen a demonstration of this projection, with a specimen of a few of the curves of the 2d order, which may be generated by a plane cutting a solid formed from the motion of an infinite right line along a diverging parabola, having an oval, always passing through a given or fixed point above the plane of that parabola. The above method of Newton has also been pursued and illustrated with great elegance by Mr. Murdoch, in his treatise entitled *Newtoni Genesis Curvarum per umbras, seu Perspectiva Universalis Elementa.*

Mr. Maclaurin, in his *Geometria Organica*, shews how to describe several of the species of curves of the 2d order, especially those having a double point, by the motion of right lines and angles: but a good commodious description by a continued motion of those curves which have no double point, is ranked by Newton among the most difficult problems. Newton gives also other methods of description, by lines or angles revolving above given poles; and Mr. Brackenridge has given a general method of describing curves, by the intersection of right lines moving about points in a given plane. See Philos. Transf. No. 437, or Abr vol. 8, pa. 58; and some particular cases are demonstrated in his *Exerc. Geometrica de Curvarum Descriptione.*

CURVES *above the 2d Order.* The number of species in the higher orders of curves increase amazingly, those of the 3d order only it is thought amounting to some thousands, all comprehended under the following ten particular equations,

$$\text{viz. 1. } y^4 + fx^2y^2 + gxy^3 + hx^3y + iy^2 + hxy + ly,$$
$$\text{o. 2. } y^4 + fxy^3 + gx^2y + hxy^2 + ixy + ky,$$
$$\text{or 3. } x^2y^2 + fy^3 + gx^2y + hy^3 + ky,$$
$$\text{or 4. } x^2y^2 + fy^3 + gy^2 + hxy + iy,$$
$$\text{or 5. } y^3 + fxy^2 + gx^2y + hy,$$
$$\text{or 6. } y^3 + fxy^2 + gxy + hy$$
$$= ax^4 + bx^3 + cx^2 + dx + e;$$
$$\text{or 7. } y^4 + ex^3y + fxy^3 + gxy^2 + hy^2 + ixy + ky,$$
$$\text{or 8. } x^3y + exy^3 + fx^2y + gy^2 + hxy + iy,$$
$$\text{or 9. } x^3y + ey^3 + fxy^2 + gxy + hy,$$
$$\text{or 10. } x^3y + ey^3 + fy^2 + gxy + hy,$$
$$= ax^3 + bx^2 + cx + d.$$

Those who wish to see how far this doctrine has been advanced, with regard to curves of the higher orders, as well as those of the 1st and 2d orders, may consult Mr. Maclaurin's Geometria Organica, and Brackenridge's Exerc. Geom.

All geometrical lines of the odd orders, viz, the 3d, 5th, 7th, &c, have at least one leg running on infinitely; because all equations of the odd dimensions have at least one real root. But vast numbers of the lines of the even orders are only ovals; among which there are several having very pretty figures, some being like single hearts, some double ones, some resembling fiddles, and others again single knots, double knots, &c.

Two geometrical lines of any order will cut one another in as many points, as are denoted by the product of the two numbers expressing those orders.

The theory of curves forms a considerable branch of the mathematical sciences. Those who are curious of advancing beyond the knowledge of the circle and the conic sections, and to consider geometrical curves of a higher nature, and in a general view, will do well to study Cramer's Introduction à l'Analyse des Lignes Courbes Algebraiques, which the learned and ingenious author composed for the use of beginners. There is an excellent posthumous piece too of Maclaurin's, printed as an Appendix to his Algebra, and entitled De Linearum Geometricarum Proprietatibus Generalibus. The same author, at a very early age, gave a remarkable specimen of his genius and knowledge in his Geometria Organica; and he carried these speculations farther afterwards, as may be seen in the theorems he

has

has given in the Philof. Tranf. See Abr. vol. 8, pa. 62. Other writings on this fubject, befide the Treatifes on the Conic Sections, are Archimedes de Spiralibus; Des Cartes Geometria; Dr. Barrow's Lectiones Geometricæ; Newton's Enumeratio Linearum Tertii Ordinis; Stirling's Illuftratio Tractatûs Newtoni de Lineis Tertii Ordinis; Maclaurin's Geometria Organica; Brackenridge's Defcriptio Linearum Curvarum; M. De Gua's Ufages de l'Analyfe de Des Cartes; befide many other Tracts on Curves in the Memoirs of feveral Academies &c.

Ufe of Curves *in the Conftruction of Equations.* One great ufe of curves in Geometry is, by means of their interfections, to give the folution of problems. See Construction.

Suppofe, *ex. gr.* it were required to conftruct the following equation of 9 dimenfions,

$$x^9 + bx^7 + cx^6 + dx^5 + ex^4 + \overline{m + f}.x^3 + gx^2 + hx + k = 0:$$

affume the equation to a cubic parabola $x^3 = y$; then, by writing y for x^3, the given equation will become $y^3 + bxy^2 + ey^2 + dx^2y + exy + my + fx^3 + gx^2 + hx + k = 0$; an equation to another curve of the 2d kind, where m or f may be affumed $= 0$ or any thing elfe: and by the defcriptions and interfections of thefe curves will be given the roots of the equation to be conftructed. It is fufficient to defcribe the cubic parabola once. When the equation to be conftructed, by omitting the two laft terms hx and k, is reduced to 7 dimenfions; the other curve, by expunging m, will have the double point in the beginning of the abfcifs, and may be eafily defcribed as above: If it be reduced to 6 dimenfions, by omitting the laft three terms, $gx^2 + hx + k$; the other curve, by expunging f, will become a conic fection. And if, by omitting the laft three terms, the equation be reduced to three dimenfions, we fhall fall upon Wallis's conftruction by the cubic parabola and right line.

Rectification, Inflection, Quadrature, &c of Curves. See the refpective terms.

Curve *of a Double Curvature,* is fuch a curve as has not all its parts in the fame plane.

M. Clairaut has publifhed an ingenious treatife on curves of a double curvature. See his Recherches fur les Courbes à Double Courbure. Mr. Euler has alfo treated this fubject in the Appendix to his Analyfis Infinitorum, vol. 2, pa. 323.

Family of Curves, is an affemblage of feveral curves of different kinds, all defined by the fame equation of an indeterminate degree; but differently, according to the diverfity of their kind. For example, Suppofe an equation of an indeterminate degree, $a^{m-1}x = y^m$: if $m = 2$, then will $ax = y^2$; if $m = 3$, then will $a^2x = y^3$; if $m = 4$, then is $a^3x = y^4$; &c: all which curves are faid to be of the fame family or tribe.

The equations by which the families of curves are defined, are not to be confounded with the tranfcendental ones: for though with regard to the whole family, they be of an indeterminate degree; yet with refpect to each feveral curve of the family, they are determinate; whereas tranfcendental equations are of an indefinite degree with refpect to the fame curve.

All Algebraical curves therefore compofe a certain family, confifting of innumerable others, each of which comprehends infinite kinds. For the equations by

which curves are defined involve only products, either of powers of the abfciffes and ordinates by conftant coefficients; or of powers of the abfciffes by powers of the ordinates; or of conftant, pure, and fimple quantities by one another. Moreover, every equation to a curve may have o for one member or fide of it; for example, $ax = y^2$, by tranfpofition becomes $ax - y^2 = 0$. Therefore the equation for all algebraic curves will be

$$\left. \begin{array}{c} ay^m + bxy^{m-1} + nx^2y^{m-2} \, \&c \, - \, - \, - fy^m \\ + fy^{m-1} + kxy^{m-2} \\ + qy^{m-2} \end{array} \right\} = 0.$$

Catacauftic, and Diacauftic Curves. See Catacaustic, and Diacaustic.

Exponential Curve, is that which is defined by an exponential equation, as $ax^z = y$, &c.

Curves *by the Light,* or Courbes *a la Lumiere,* a name given to certain curves by M. Kurdwanowfki, a Polifh gentleman. He obferved that any line, ftraight or curved, expofed to the action of a luminous point, received the light differently in its different parts, according to their diftance from the light. Thefe different effects of the light upon each point of the line, may be reprefented by the ordinates of fome curve, which will vary precifely with thefe effects. Prieftley's Hift. of Vifion, pa. 752.

Logarithmic Curve. See Logarithmic *Curve.*

Curve *Reflectoire,* fo called becaufe it is the appearance of the plane bottom of a bafon covered with water, to an eye perpendicularly over it. In this pofition, the bottom of the bafon will appear to rife upwards, from the centre outwards; but the curvature will be lefs and lefs, and at laft the furface of the water will be an afymptote to it. M. Mairan, who firft conceived this idea from the phenomena of light, found alfo feveral kinds of thefe curves; and he gives a geometrical deduction of their properties, fhewing their analogy to cauftics by refraction. Mem. Ac. 1740; Prieftley's Hift. of Vifion, pa. 752.

Radical Curves, a name given by fome authors to curves of the fpiral kind, whofe ordinates, if they may be fo called, do all terminate in the centre of the including circle, and appear like fo many radii of that circle: whence the name.

Regular Curves, are fuch as have their curvature turning regularly and continually the fame way; in oppofition to fuch as bend contrary ways, by having points of contrary flexure, which are called irregular curves.

Characteriftic Triangle of a Curve, is the differential or elementary right-angled triangle whofe three fides are, the fluxions of the abfcifs, ordinate, and curve; the fluxion of the curve being the hypothenufe. So, if pq be parallel to, and indefinitely near to the ordinate PQ, and Qr parallel to the abfcifs AP; then Qr is the fluxion of the abfcifs AP, and qr the fluxion of the ordinate PQ, and Qq the fluxion of the curve AQ; hence the elementary triangle Qqr is the characteriftic triangle of the curve AQ; and the three fides are $\dot{x}$, $\dot{y}$, $\dot{z}$; in which $\dot{x}^2 + \dot{y}^2 = \dot{z}^2$.

CURVILINEAR *Angle, Figure, Superficies, &c,*
are

are such as are formed or bounded by curves; in oppofition to rectilinear ones, which are formed by straight lines or planes.

CUSP, in Aftronomy, is ufed to exprefs the points or horns of the moon, or other luminary.

Cusp, in Aftrology, is ufed for the 1ft point of each of the twelve houfes, in a figure or fcheme of the heavens. See House.

Cusp, in the Higher Geometry, is ufed for the point or corner formed by two parts of a curve meeting and terminating there. See Curve.

CUSPIDATED *Hyperbola*, &c. See Curve.

CUT-Bastion. See Bastion.

CUVETTE, or Cunette, in Fortification, is a kind of ditch within a ditch, being a pretty deep trench, about four fathoms broad, funk and running along the middle of the great dry ditch, to hold water; ferving both to keep off the enemy, and prevent him from mining.

CYCLE, a certain period or feries of numbers proceeding orderly from firft to laft, then returning again to the firft, and fo circulating perpetually.

Cycles have chiefly arifen from the incommenfurability of the revolutions of the earth and celeftial bodies to one another. The apparent revolution of the fun about the earth, has been arbitrarily divided into 24 hours, which is the bafis or foundation of all our menfuration of time, whether days, years, &c. But neither the annual motion of the fun, nor that of the other heavenly bodies, can be meafured exactly, and without any remainder, by hours, or their multiples. That of the fun, for example, is 365 days 5 hours 49 minutes nearly; that of the moon, 29 days 12 hours 44 minutes nearly.

Hence, to fwallow up thefe fractions in whole numbers, and yet in numbers which only exprefs days and years, cycles have been invented; which, comprehending feveral revolutions of the fame body, replace it, after a certain number of years, in the fame points of the heaven whence it firft departed; or, which is the fame thing, in the fame place of the civil calendar.

There are various cycles; as, the cycle of Indiction, the cycle of the moon, the cycle of the fun, &c.

Cycle of *Indiction*, is a feries of 15 years, returning conftantly around like the other cycles; and commenced from the third year before Chrift; whence it happens that if 3 be added to any given year of Chrift, and the fum be divided by 15, what remains is the year of the indiction. See Indiction.

Cycle of *the Moon*, or the *Lunar Cycle*, is a period of 19 years; in which time the new and full moons return to the fame day of the Julian year. See Calippic.

This cycle is alfo called the *Metonic period* or *cycle*, from its inventor Meton, the Athenian; and alfo the *Golden Number*, from its excellent ufe in the calendar: though, properly fpeaking, the golden number is rather the particular number which fhews the year of the lunar cycle, which any given year is in. This cycle of the moon only holds true for $310\frac{7}{10}$ years: for, though the new moons do return to the fame day after 19 years; yet not to the fame time of the day, but near an hour and a half fooner; an error which in $310\frac{7}{10}$

years amounts to an entire day. Yet thofe employed in reforming the calendar went on a fuppofition that the lunations return precifely from 19 years to 19 years, for ever.

The ufe of this cycle, in the ancient calendar, is to fhew the new moon of each year, and the time of Eafter. In the new one, it only ferves to find the Epacts; which fhew, in either calendar, that the new moon falls 11 days too late.

As the Orientals began the ufe of this cycle at the time of the Council of Nice in 325, they affumed, that the firft year of the cycle the pafchal new moon fell on the 13th of March: on which account the lunar cycle 3 fell on the 1ft of January in the third year.

The Occidentals, on the contrary, placed the number 1 to the 1ft of January, which occafioned a confiderable difference in the time of Eafter. Hence, Dionyfius Exiguus, on framing a new calendar, perfuaded the Chriftians of the weft to falve the difference, and come into the practice of the church of Alexandria.

To find the Year of the Lunar Cycle, is to find the golden number. See Golden-*Number*.

Cycle of the Sun, or *Solar Cycle*, is a period or revolution of 28 years; beginning with 1, and ending with 28; which elapfed, the Dominical or Sunday-letters, and thofe that exprefs the other feafts, &c, return into their former place, and proceed in the fame order as before. The days of the month return again to the fame days of the week; the fun's place to the fame figns and degrees of the ecliptic on the fame months and days, fo as not to differ one degree in a hundred years; and the leap years begin the fame courfe with refpect to the days of the week on which the days of the month fall.

This is called the cycle of the *fun*, or the *folar cycle*, not from any regard to the fun's courfe, which has no concern in it; but from *Sunday*, anciently called *dies folis*, the *fun's day*; as the dominical or funday letter is chiefly fought for from this revolution.

The reformation of the calendar under pope Gregory the 13th, occafioned a confiderable alteration of this cycle: In the Gregorian calendar, the folar cycle is not conftant and perpetual; becaufe every 4th fecular year is common; whereas, in the Julian, it is biffextile. The epoch, or beginning of the folar cycle, both Julian and Gregorian, is the 9th year before Chrift. And therefore,

To find the Cycle of the Sun for any given year: add 9 to the number given, and divide the fum by 28; the remainder will be the number of the cycle, and the quotient the number of revolutions fince Chrift. If there be no remainder, it will be the 28th or laft year of the cycle.

CYCLE of the Sun, with the correfpondent Sunday letters, in Julian Years.						
1 GF	5 BA	9 DC	13 FE	17 AG	21 CB	25 ED
2 E	6 G	10 B	14 D	18 F	22 A	26 C
3 D	7 F	11 A	15 C	19 E	23 G	27 B
4 C	8 E	12 G	16 B	20 D	24 F	28 A

CYCLE of the Sun, and Sunday Letters, from the Gregorian year 1700 to the year 1800.						
1 DC	5 FE	9 AG	13 CB	17 ED	21 GF	25 BA
2 B	6 D	10 F	14 A	18 C'	22 E	26 G
3 A	7 C	11 E	15 G	19 B	23 D	27 F
4 G	8 B	12 D	16 F	20 A	24 C	28 E

Great Pafcal CYCLE, is another name for the Victorian or Dionyfian Period. Which fee.

CYCLOID, or TROCHOID, a mechanical or tranfcendental curve, which is thus generated: Suppofe a wheel, or a circle, AE, to roll along a ftraight line AB, beginning at the point A, and ending at B, where it has completed juft one revolution, thereby meafuring out a right line AB exactly equal to the circumference of the generating circle AE, whilft a nail or point A in the circumference of the wheel, or circle, traces out or defcribes a curvilineal path ADB; then this curve ADB is the cycloid, or trochoid.

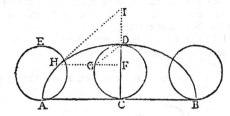

Schooten, in his Commentary on Des Cartes, fays that Des Cartes firft conceived the notion of this elegant curve, and after him it was firft publifhed by Father Merfenne, in the year 1615. But Torricelli, in the Appendix de Dimenfione Cycloidis, at the end of his treatife De Dimenfione Parabolæ, publifhed 1644, fays that this curve was confidered and named a cycloid, by his predeceffors, and particularly by Galileo about 45 years before, i. e. about 1599. And Dr. Wallis fhews that it is of a much older ftanding, having been known to Bovilli about the year 1500, and even confidered by cardinal Cufanus much earlier, viz, before the year 1451. Philof. Tranf. Abr. vol. 1, pa. 116. It would feem however that Torricelli's was the firft regular treatife on the Cycloid; though feveral particular properties of it might be known prior to his work. He firft fhewed, that the cycloidal fpace is equal to triple the generating circle, (though Pafcal contends that Roberval fhewed this): alfo that the folid generated by the rotation of that fpace about its bafe, is to the circumfcribing cylinder, as 5 to 8: about the tangent parallel to the bafe, as 7 to 8: about the tangent parallel to the axis, as 3 to 4: &c.

Honoratus Fabri, in his Synopfis Geom. has a fhort treatife on the cycloid, containing demonftrations of the above, and many other theorems concerning the centres of gravity of the cycloidal fpace, &c; which he fays he found out before the year 1658.

From the preface to Dr. Wallis's treatife on the cycloid we learn, that, in the year 1658, M. Pafcal publicly propofed at Paris, under the name of D'Ettonville, the two following problems as a challenge, to

be folved by the mathematicians of Europe, with a reward of 20 piftoles for the folution : viz, to find the area of any fegment of the cycloid, cut off by a right line parallel to the bafe; alfo the content of the folid generated by the rotation of the fame about the axis, and about the bafe of that fegment. This challenge fet the Doctor upon writing that treatife upon the cycloid, which is a much better and compleater piece than had been given before upon this curve. He here gives the curve furfaces of the folids generated by the rotation of the cycloidal fpace about its axis, and about its bafe, with determinations of the centres of gravity, &c. He here afferts too, that Sir Chriftopher Wren, in 1658, was the firft who found out a right line equal to the curve of the cycloid; and Mr. Huygens, in his Herolog. Ofcillat. fays that he himfelf was the firft inventor of the fegment of a cycloidal fpace, cut off by a right line parallel to the bafe at the diftance of $\frac{1}{4}$ the axis of the curve from the centre, being equal to a rectilinear fpace, viz, to a regular hexagon infcribed in the generating circle; the demonftration of which may be feen in Wallis's treatife.

Several other authors have fpoken or treated of the cycloid : as Pafcal, in his treatife, under the name of D'Ettonville : Schooten in his Commentary on Des Cartes's Geometry, near the end of the 2d book; M. Reinau, in his Analyfe Démontrée, tom. 2, pa. 595: alfo Newton, Leibnitz, de la Loubere, Roberval, Des Cartes, Wren, Fabri, the Bernoulli's, De la Hire, Cotes, &c, &c.

Properties of the CYCLOID.—The circle AE, by whofe revolution the cycloid is traced out, is called the *generating circle;* the line AB, which is equal to the circumference of the circle, is the *bafe* of the cycloid; and the perpendicular DC on the middle of the bafe, is its *axis.* The properties of the cycloid are among the moft beautiful and ufeful of all curve lines: fome of the moft remarkable of which are as follow :

1. The circular arc DG = the line GH parallel to AB.
2. The femicircumf. DGC = the femibafe AC.
3. The arc DH = double the chord DG.
4. The arc DA = double the diam. DC.
5. The tang. HI is parallel to the chord DG.
6. The fpace ADBA = triple the circle AE or CGD &c.
7. The fpace ADGCA = the fame circle AE, &c.

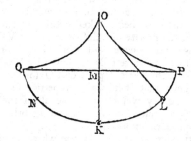

8. A body falls through any arc KL of a cycloid reverfed, in the fame time, whether that arc be great or fmall; that is, from any point L, to the loweft point K, which is the vertex reverfed: and that time

is

is to the time of falling perpendicularly through the axis MK, as the femicircumference of a circle is to its diameter, or as 3·1416 to 2. And hence it follows that, if a pendulum be made to vibrate in the arc LKN of a cycloid, all the vibrations will be performed in the fame time.

9. The evolute of a cycloid, is another equal cycloid. So that if two equal femicycloids OP, OQ, be joined at O, fo that OM be = MK the diameter of the generating circle, and the ftring of a pendulum hung up at O, having its length = OK or = the curve OP; then, by plying the ftring round the curve OP, to which it is equal, and then the ball let go, it will defcribe, and vibrate in the other cycloid PKQ.

10. The cycloid is the curve of fwifteft defcent: or a heavy body will fall from one given point to another, by the way of the arc of a cycloid paffing through thofe two points, in a lefs time, than by any other rout. See the Works of James and John Bernoulli for many other curious properties concerning the defcents in cycloids, &c.

CYCLOIDS are alfo either *curtate* or *prolate*.

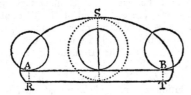

CYCLOID, *Curtate*, or *contraƈted*, is the path defcribed by fome point without the circle, while the circumference rolls along a ftraight line; and a

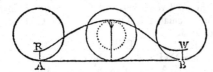

CYCLOID, *Prolate*, or *Infleƈted*, is in like manner the path of fome point taken within the generating circle.

Thus, if, while the circle rolls along the line AB, the point R be taken without the circle, it will defcribe or trace out the curtate or contraƈted cycloid RST; but the point being taken within the circle, it will defcribe the prolate or infleƈted cycloid RVW.

Thefe two curves were both noticed by Torricelli and Schooten, and more fully treated of by Wallis, in his Treatife on the Cycloid, printed at Oxford in 1659; where he fhews that thefe have properties fimilar to the firft or primary cycloid; only the laft of thefe is a curve having a point of infleƈtion, and the other croffing itfelf, and forming a node.

By continuing the motion of the wheel, or circle, fo as to defcribe a right line equal to the generating circumference feveral times repeated, there will be produced as many repetitions of the cycloids, which fo united together will appear as in thefe figures following:

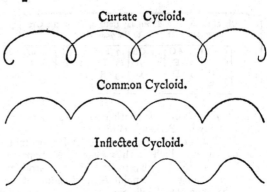

Curtate Cycloid.

Common Cycloid.

Infleƈted Cycloid.

CYGNUS, the *Swan*, a conftellation of the northern hemifphere, being one of the 48 old ones, and fabled by the Greeks to be the fwan, under the form of which Jupiter deceived Leda or Nemefis, from which embrace fprung the beauteous Helen.

The ftars in the conftellation Cygnus, in Ptolomy's catalogue are 19, in Tycho's 18, in Hevelius's 47, and in the Britannic catalogue 81.

CYLINDER, a folid having two equal circular ends, and every plane feƈtion parallel to the ends a circle equal to them alfo.

The cylinder may be conceived to be thus generated:

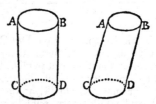

Suppofe two parallel circles AB and CD, and a right line carried continually round them, always parallel to itfelf; this line will defcribe the curve furface of a cylinder, ABDC, of which the two parallel circles AB and CD form the two ends. When the line, or fides is perpendicular to the ends, the cylinder is a right or perpendicular one; otherwife it is oblique.

Or the right cylinder may be conceived to be generated by the rotation of a reƈtangle about one of its fides. The *axis* of the cylinder is the line conneƈting the centres of its two parallel circular ends; and is equal to the altitude of the cylinder when this is a right one, but exceeds the altitude in the oblique cylinder, in the proportion of radius to the fine of the angle of its inclination to the bafe.

The convex furface of a cylinder is equal to the produƈt of the axis multiplied by the circumference of its bafe.

The folidity of a cylinder is equal to the area of its bafe multiplied by its perpendicular altitude.

Cylinders of equal bafes and altitudes, are equal.

Cylinders are to each other, as the produƈt of their bafes and altitudes. And equal cylinders have their bafes reciprocally as their altitudes.

A cylinder is to its infcribed fphere, or fpheroid, as 3 to 2: and to its infcribed cone as 3 to 1.

The

The oblique plane fections of a cylinder, are ellipfes; but all the fections parallel to the ends, are circles.

For the furfaces and folidities of the ungulas, or oblique flices, of a cylinder, fee my Menfuration, pa. 218, 2d edition.

CYLINDRICAL, pertaining to a cylinder.

CYLINDROID, a folid refembling the figure of a cylinder; but differing from it as having ellipfes for its ends or bafes, inftead of circles, in the cylinder.

In the cylindroid, the folidity and curve fuperficies are found the fame way as thofe of the cylinder; viz,

by multiplying the circumference of the bafe by the length or axis, for the furface; and the area of the bafe by the altitude, for the folidity.

CYMATIUM, CIMATIUM, or CIMA, in Architecture, a member, or moulding of the cornice, whofe profile is waved; i. e. concave at top, and convex at bottom.

CYNOSURA, a name given by the Greeks, to urfa minor, or little bear, otherwife called Charles's wain; the ftar in the extremity of the tail being called the pole ftar.

CYPHER, or nought. See Cipher.

D.

DAR

DACTYLONOMY, the art of counting or numbering by the fingers.—The rule is this: the left thumb is reckoned 1, the index or fore-finger 2, and fo on to the right thumb, which is the tenth or laft, and confequently is denoted by the cipher o.

DADO, that part in the middle of the pedeftal of a column &c, between its bafe and cornice.

DAILY, in Aftronomy. See DIURNAL.

DARCY (COUNT), an ingenious philofopher and mathematician, was born in Ireland in 1725; but his friends being attached to the Stuart family, he was fent to France, at 14 years of age, where he fpent the reft of his life. Being put under the care of the celebrated Clairaut, he improved fo rapidly in the mathematics, that at 17 years of age he gave a new folution of the problem concerning the curve of equal preffure in a refifting medium. This was followed the year after by a determination of the curve defcribed by a heavy body, fliding by its own weight along a moveable plane, at the fame time that the preffure of the body caufes a horizontal motion in the plane. Darcy ferved in the war of 1744, and was taken prifoner by the Englifh: and yet, during the courfe of the war he gave two memoirs to the academy; the firft of thefe contained a general principle in mechanics, that of *the prefervation of the rotatory motion;* a principle which he again brought forward in 1750, by the name of *the principle of the prefervation of action.*

In 1760, Darcy publifhed *An Effay on Artillery,* containing fome curious experiments on the charges of gunpowder, &c, &c, and improvements on thofe of the ingenious Robins; a kind of experiments which our author carried on occafionally to the end of his life.

In 1765, he publifhed his *Memoir on the Duration of the Senfation of Sight,* the moft ingenious of his works: the refult of thefe refearches was, that a body may fometimes pafs by our eyes without being feen, or marking its prefence, otherwife than by weakening the brightnefs of the object it covers.

All Darcy's works bear the character which refults

DAT

from the union of genius and philofophy; but as he meafured every thing upon the largeft fcale, and required extreme accuracy in experiment, neither his time, fortune, nor avocations, allowed him to execute more than a very fmall part of what he projected. In his difpofition, he was amiable, fpirited, lively, and a lover of independence, a paffion to which he nobly facrificed, even in the midft of literary fociety.—He died of a cholera morbus in 1779, at 54 years of age.

Darcy was admitted of the French academy in 1749, and was made penfioner-geometrician in 1770.—His effays, printed in the Memoirs of the Academy of Sciences, are various and very ingenious, and are contained in the volumes for the years 1742, 1747, 1749, 1750, 1751, 1752, 1753, 1754, 1758, 1759, 1760, 1765, and in tom. 1, of the Savans Etrangers.

DARK *Chamber.* See CAMERA *Obfcura.*

DARK *Tent,* a portable camera obfcura, made fomewhat like a defk, and fitted with optic glaffes, to take profpects of landfcapes, buildings, &c.

DATA, in General Mathematics, denote certain things or quantities, fuppofed given or known, from which other quantities are difcovered that were unknown, or fought. A problem or queftion ufually confifts of two parts, *data* and *quæfita.*

Euclid has an exprefs and excellent treatife of *Data;* in which he ufes the word for fuch fpaces, lines, angles, &c, as are given; or to which others can be found equal.

Euclid's *Data* is the firft in order of the books that have been written by the ancient geometricians, to facilitate and promote the method of refolution or analyfis. In general a thing is faid to be given which is either actually exhibited, or can be found out, that is, which is either known by hypothefis, or that can be demonftrated to be known: and the propofitions in the book of Euclid's Data fhew what things can be found out or known, from thofe that by hypothefis are already known: fo that in the analyfis or inveftigation of a problem, from the things that are laid down as given

or

or known, by the help of these propositions, it is demonstrated that other things are given, and from these last that others again are given, and so on, till it is demonstrated that that which was proposed to be found out in the problem is given; and when this is done, the problem is solved, and its composition is made and derived from the compositions of the Data which were employed in the analysis. And thus the Data of Euclid are of the most general and necessary use in the solution of problems of every kind.

Marinus, at the end of his Preface to the Data, is mistaken in asserting that Euclid has not used the synthetical, but the analytical method in delivering them: for, though in the analysis of a theorem, the thing to be demonstrated is assumed in the analysis; yet, in the demonstrations of the Data, the thing to be demonstrated, which is, that something is given, is never once assumed in the demonstration; from which it is manifest that every one of them is demonstrated synthetically: though indeed if a proposition of the Data be turned into a problem, the demonstration of the proposition becomes the analysis of the problem. See Simson's edition of Euclid's Data, which is esteemed the best.

DAVIS's *Quadrant*, the common sea quadrant, or back-staff. See BACK-STAFF. See also Robertson's Navigation, book 9, sect. 7.

DAY, a division of time arising from the appearance and disappearance of the sun.

DAY is either *natural* or *artificial*.

Artificial DAY is that which is primarily meant by the word Day, and is the time of its being light, or the time while the sun is above the horizon. Though sometimes the twilight is included in the term daylight; in opposition to night or darkness, being the time from the end of twilight to the beginning of daylight.

Natural DAY is the portion of time in which the sun performs one revolution round the earth; or rather the time in which the earth makes a rotation on its axis. And this is either astronomical or civil.

Astronomical DAY begins at noon, or when the sun's centre is on the meridian, and is counted 24 hours to the following noon.

Civil DAY is the time allotted for day in civil purposes, and begins differently in different nations, but still including one whole rotation of the earth on its axis; beginning either at sun-rise, sun-set, noon, or midnight.

1st, At sun-rising, among the ancient Babylonians, Persians, Syrians, and most other eastern nations, with the present inhabitants of the Balearic islands, the Greeks, &c. 2dly, At sun-setting, among the ancient Athenians and Jews, with the Austrians, Bohemians, Marcomanni, Silesians, modern Italians, and Chinese. 3dly, At noon, with astronomers, and the ancient Umbri and Arabians. And 4thly, at midnight, among the ancient Egyptians and Romans, with the modern English, French, Dutch, Germans, Spaniards, and Portuguese.

The day is divided into hours; and a certain number of days makes a week, a month, or a year.

The different length of the natural day in different climates, has been matter of controversy, viz, whether the natural days be all equally long throughout the year; and if not, what their difference is? A professor of mathematics at Seville, in the Philos. Transf. vol. 10, pa. 425, asserts, from a continued series of observations for three years, that they are all equal. But Mr. Flamsteed, in the same Transf. pa. 429, refutes the opinion; and shews that one day, when the sun is in the equinoctial, is shorter than when he is in the tropics, by 40 seconds; and that 14 tropical days are longer than so many equinoctial ones, by 10 minutes. This inequality of the days flows from two several principles: the one, the eccentricity of the earth's orbit; the other, the obliquity of the ecliptic with regard to the equator, which is the measure of time. As these two causes happen to be differently combined, the length of the day is varied. See EQUATION *of time*.

DAY's-*Work*, in Navigation, denotes the reckoning or account of the ship's course, during 24 hours, or between noon and noon.

DECAGON, a plane geometrical figure of ten sides and ten angles. When all the sides and angles are equal, it is a *regular* decagon, and may be inscribed in a circle; otherwise, not.

If the radius of a circle, or the side of the inscribed hexagon, be divided in extreme and mean proportion, the greater segment will be the side of a decagon inscribed in the same circle. And therefore, as the side of the decagon is to the radius, so is the radius to the sum of the two. Whence, if the radius of the circle be r, the side of the inscribed decagon will be $\frac{\sqrt{5}-1}{2} \times r$.

If the side of a regular decagon be 1, its area will be $\frac{5}{2}\sqrt{5+2\sqrt{5}} = 7.6942088$; therefore as 1 is to 7.6942088, so is the square of the side of any regular decagon, to the area of the same: so that, if s be the side of such a decagon, its area will be equal to $7.6942088 s^2$. See REGULAR *Figure*.

To inscribe a decagon in a circle geometrically. See my Mensuration, prob. 35, pa. 25, 2d edit.

DECEMBER, the last month of the year; in which the sun enters the tropic of Capricorn, making the winter solstice.

In the time of Romulus, December was the 10th month; whence the name, viz, from decem, ten; for the Romans began their year in March, from which December is the 10th month.

The month of December was under the protection of Vesta. Romulus assigned it 30 days; Numa reduced it to 29; which Julius Cæsar increased to 31.

At the latter part of this month they had the *Juveniles Ludi*, and the country people kept the feast of the goddess Vacuna in the fields, having then gathered in their fruits, and sown their corn; whence it seems is derived our popular festival called Harvest-home.

DECHALES (CLAUD-FRANCIS-MILLIET), an excellent mathematician, mechanist, and astronomer, was born at Chambery, the capital of Savoy, in 1611. He chiefly excelled in a just knowledge of the mathematical and mechanical sciences: not that he was bent upon new discoveries, or happy in making them; as his talent rather lay in explaining those sciences with ease and accuracy; which perhaps rendered him equally useful and deserving of esteem. Indeed it was generally allowed that he made the best use of the productions of other men, and that he drew the several parts of the mathematical sciences together with great judgment

and

and perspicuity. It is also said of him, that his probity was not inferior to his learning; and that both these qualities made him generally admired and beloved at Paris, where for four years together he read public mathematical lectures in the college of Clermont.——From hence he removed to Marseilles, where he taught the art of navigation and the practical mathematical sciences.—He afterward became professor of mathematics in the university of Turin, where he died March 28, 1678, at 67 years of age.

Among other works which do honour to his memory, are,

1. An edition of Euclid's Elements; in which he has omitted the less important propositions, and explained the uses of those he has retained.

2. A Discourse on Fortification; and another on Navigation.

3. These performances, with some others, were collected in 3 volumes folio, under the title of *Mundus Mathematicus*, being indeed a complete course of mathematics. And the same was afterward much enlarged, and published at Lyons, 1690, in 4 large volumes, folio.

DECIL, *Decilis*, an aspect or position of two planets, when they are distant from each other a 10th part of the zodiac, or 36 degrees; and is one of the new aspects invented by Kepler.

DECIMALS, any thing proceeding by tens; as Decimal arithmetic, Decimal fractions, Decimal scales, &c.

DECIMAL *Arithmetic*, in a general sense, may be considered as the common arithmetical computation in use, in which the decimal scale of numbers is used, or in which the places of the figures change their value in a tenfold proportion, being 10 times as much for every place more towards the left hand, or 10 times less for every place more towards the right hand; the places being supposed indefinitely continued, both to the right and left. In this sense, the word includes both the arithmetic of integers, and decimal fractions. In a more restrained sense however, it means only

DECIMAL *Fractions*, which are fractions whose denominator is always a 1 with some number of ciphers annexed, more or fewer according to the value of the fraction, the numerator of which may be any number whatever; as $\frac{1}{10}$, $\frac{3}{100}$, $\frac{7}{1000}$, &c.

As the denominator of a decimal is always one of the numbers 10, 100, 1000, &c, the inconvenience of writing these denominators down may be saved, by placing a proper distinction before the figures of the numerator only, to distinguish them from integers, for the value of each place of figures will be known in decimals, as well as in integers, by their distance from the 1st or unit's place of integers, having similar names at equal distances, as appears by the following scale of places, both in decimals and integers:

&c	8	8	8	8	8	8	·	8	8	8	8	8	8	&c
	millions	hund. of thousands	tens of thousands	thousands	hundreds	tens	units	tenths	hundredths	thousandths	ten thousandths	hund. thousandths	millionths	

The mark of distinction for decimals, called the *separatrix*, has been various at different times, according to the fancy of different authors; sometimes a semiparenthesis, or a semicrotchet, or a perpendicular bar, or the same with a line drawn under the figures, or simply this line itself, &c; but it is usual now to write either a comma or a full point near the bottom of the figures; I place the point near the upper part of the figures, as was done also by Newton; a method which prevents the separatrix from being confounded with mere marks of punctuation.

In setting down a decimal fraction without its denominator, the numerator must consist of as many places as there are ciphers in the denominator; and if it has not so many figures, the defect must be supplied by setting before them as many ciphers as will make them up so many: thus $\frac{3}{10}$ is ·3; and $\frac{14}{100}$ is ·14; and $\frac{14}{1000}$ is ·014; and $\frac{3}{1000}$ is ·003; &c.

So that, as ciphers on the right-hand side of integers *increase* their value decimally, or in a tenfold proportion, as 2, 20, 200, &c; so, when set on the left-hand of decimal fractions, they *decrease* the value decimally, or in a tenfold proportion, as ·2, ·02, ·002, &c. But ciphers set on the other sides of these numbers, make no alteration in their value, neither of increase nor decrease, viz, on the left-hand of integers, or on the right-hand of decimals; so 2, or 02, or 002, &c, are all the same; as are also ·2, or ·20, or ·200, &c.

Decimal fractions may be considered as having been introduced by Regiomontanus, about the year 1464, viz, when he transformed the tables of sines from a sexagesimal to a decimal scale. They were also used by Ramus, in his Arithmetic, written in 1550; and before his time by our countrymen Buckley and Recorde. But it was Stevinus who first wrote an express treatise on decimals, viz, about the year 1582, in *La Pratique d'Arithmetique*; since which time, this has commonly made a part in most treatises on arithmetic.

To reduce any Vulgar fraction, or parts of any thing, as suppose $\frac{3}{8}$, to a decimal fraction of the same value; add ciphers at pleasure to the numerator, and divide by the denominator: thus,

$$8)3·000$$

$$·375 = \tfrac{3}{8};$$

and therefore ·375 or $\frac{375}{1000}$ is a decimal of the same value with the proposed vulgar fraction $\frac{3}{8}$.

Some vulgar fractions can never be reduced into decimals without defect; as $\frac{1}{3}$; which by division is ·33333 &c infinitely.

Such numbers are very properly called circulating decimals, and repetends, because of the continual return of the same figures. See REPETENDS and CIRCULATES.

The common arithmetical operations are performed the same way in decimals, as they are in integers; regard being had only to the particular notation, to distinguish the fractional from the integral part of a sum. Thus,

In Addition and Subtraction, all figures of the same place or denomination are set straight under each other, the separatrix, or decimal points, forming a straight column.

In Multiplication, set down the numbers, and multiply
tiply

tiply them as integers; and point off from the product as many places of decimals as there are in both factors; prefixing ciphers if there be any defect of figures.

In Division, set down the numbers and divide also as in integers; making as many decimals in the quotient, as those in the dividend are more than those in the divisor.

Examples are numerous and common in most books of arithmetic.

DECIMAL *Scales*, are any scales divided decimally, or by tens.

DECLINATION, in Astronomy, is the distance of the sun, star, planet, &c, from the equinoctial, either northward or southward; being the same with latitude in geography, or distance from the equator.

Declination is either *real* or *apparent*, according as the real or apparent place of the point or object is considered.

The declination of any point S is an arch of the meridian SE, contained between the given point and the equinoctial EQ. The declination of a star &c, is found by knowing or observing the latitude of the place, i. e. the height of the pole, and then the meridian altitude of the star, &c; hence the difference between the co-latitude and the altitude of the star &c, is the declination, viz, the difference between EH the co-latitude, and SH the altitude, is ES the declination. For ex. Tycho found at Uranibourg the meridian

altitude of Cauda Leonis, viz, HS = 50° 59′ 00″
the co-latitude is   HE = 34 5 45

rem. declin. north   ES = 16 53 15

To find the Sun's Declination at any time; having given his place in the ecliptic; the rule is, as radius is to the sine of the sun's longitude, so is the sine of the greatest declination, or obliquity of the ecliptic, to his present declination sought.

In constructing tables of declination of the sun, planets, and stars, regard should be had to refraction, aberration, nutation, and parallax.

By comparing ancient observations with the modern, it appears that the declination of the fixed stars is variable; and that differently in different stars; for in some it increases, and in others decreases, and that in different quantities.

Circles of DECLINATION, are great circles of the sphere passing through the poles of the world, on which the declination is measured; and consequently are the same as meridians in geography.

Parallax of DECLINATION, is an arch of the circle of declination, by which the parallax in altitude increases or diminishes the declination of a star.

Parallels of DECLINATION, are lesser circles parallel to the equinoctial. The tropic of Cancer is a parallel of declination at 23° 28′ distance from the equinoctial northward; and the tropic of Capricorn is the parallel of declination as far distant southward.

Refraction of the DECLINATION, an arch of the circle of declination, by which the declination of a star is increased or diminished by means of the refraction.

DECLINATION *of the Compass,* or *Needle,* is its deviation from the true meridian. See VARIATION.

DECLINATION *of a Vertical Plane,* or *Wall,* in Dialling, is an arch of the horizon, comprehended either between the plane and the prime vertical circle, when it is counted from the east or west; or between the plane and the meridian, if it be accounted from the north or south.

DECLINATOR, or DECLINATORY, an instrument in dialling, by which the declination, inclination, and reclination of planes are determined.

DECLINERS, or DECLINING *Dials,* are those which cut obliquely, either the plane of the prime vertical circle, or the plane of the horizon.

The use of declining vertical dials is very frequent; because the erect walls of houses, on which dials are commonly drawn, mostly decline from the cardinal points. But incliners and recliners are very rare.

DECLIVITY, a sloping or oblique descent.

DECREMENT, *Equal,* of *Life.* See COMPLEMENT *of life.*

DECREMENTS are the small parts by which a variable and decreasing quantity becomes less and less. The indefinitely small decrements are proportional to the fluxions, which in this case are negative. See FLUXIONS, also INCREMENTS.

DECUPLE, a term of relation or proportion in arithmetic, implying a tenfold change or scale of variation, or one thing 10 times as much as another.

DECUSSATION, a term in geometry and optics, signifying the crossing of any two lines or rays &c: or the action itself of crossing.

The rays of light decussate in the chrystalline, before they reach the retina.

Many of the lines of the 3d order decussate themselves. See Newton's Enumeratio, &c.

DEE (JOHN), a famous mathematician and astrologer, was born at London 1527. In 1542 he was sent to St. John's College, Cambridge. After five years close application to study, chiefly in the mathematical and astronomical sciences, he went over to Holland, to visit some mathematicians on the Continent; whence, after a year's absence, he returned to Cambridge, and was there elected one of the Fellows of Trinity College, then first erected by King Henry the 8th. In 1548 he left England a second time, his stay at home being rendered uneasy to him, by the suspicions that were entertained of his being a conjurer, arising chiefly from his application to astronomy, and from some mechanical inventions of his.

He now visited the university of Louvain; where he was much caressed, and visited by several persons of high rank. After two years he went into France, and read lectures, in the college of Rheims, upon Euclid's Elements. In 1551, he returned to England, and was introduced to King Edward, who assigned him a pension of 100 crowns, which he afterward relinquished for the rectory of Upton upon Severn. But soon after the accession of Queen Mary, having some correspondence with her sister Elizabeth, he was accused of practising against the queen's life by enchantment: on which account he suffered a tedious confinement, and was several times examined; till, in the year 1555, he obtained his liberty by an order of council.

When

2

When Queen Elizabeth afcended the throne, Dee was confulted concerning a propitious day for the coronation: on which occafion he was introduced to the queen, who made him great promifes, which were but ill performed. In 1564, he made another voyage to the continent, to prefent a book which he had dedicated to the Emperor Maximilian. He returned to England the fame year; but in 1571 we find him in Lorrain; where, being dangeroufly ill, the queen fent over two phyficians to his relief. Having once more returned to his native country, he fettled at Mortlake in Surry, where he continued his ftudies with much ardour, and collected a great library of printed books and manufcripts, with a number of inftruments; moft of which were afterward deftroyed by the mob, as belonging to one who dealt with the devil.

In 1578, the queen being much indifpofed, Mr. Dee was fent abroad to confult with German phyficians and philofophers (aftrologers no doubt) on the occafion; though fome have faid fhe employed him as a fpy; probably he acted in a double capacity. We next find him again in England, where he was foon after employed in a more rational fervice. The queen, defirous to be informed concerning her title to thofe countries which had been difcovered by Englifhmen, ordered Dee to confult the ancient records, and to furnifh her with proper geographical defcriptions. Accordingly, in a fhort time, he prefented to the queen, at Richmond, two large rolls, in which the difcovered countries were geographically defcribed and hiftorically illuftrated. His next employment was the reformation of the calendar, on which fubject he wrote a rational and learned treatife, preferved in the Afhmolean library at Oxford.

Hitherto the extravangancies of our eccentrical philofopher feem to have been tempered with a tolerable proportion of reafon and fcience; but henceforward he is to be confidered as a mere necromancer and credulous alchymift. In the year 1581 he became acquainted with one Edward Kelly, by whofe affiftance he performed divers incantations, and maintained a frequent imaginary intercourfe with fpirits and angels; one of whom made him a prefent of a black fpeculum (a polifhed piece of cannel-coal), in which thefe appeared to him as often as he had occafion for them, anfwering his queftions, &c. Hence Butler fays,

> Kelly did all his feats upon
> The devil's looking-glafs, a ftone.
>
> HUDIBRAS.

In 1583 they became acquainted with a certain Polifh nobleman, then in England, named Albert Lafki, a perfon equally addicted to the fame ridiculous purfuits: he was fo charmed with Dee and Kelly, that he perfuaded them to accompany him to his native country; by whofe means they were introduced to Rodolph king of Bohemia; who, though a credulous man, was foon difgufted with their nonfenfe. They were afterward introduced to the king of Poland, but with no better fuccefs. Soon after this they were entertained at the caftle of a rich Bohemian nobleman, where they lived for fome time in great affluence; owing, as they afferted, to their art of tranfmutation by means of a certain powder in the poffeffion of Kelly.

VOL. I.

Dee, now quarrelling with his companion, quitted Bohemia, and returned to England, where he was once more gracioufly received by the queen; who, in 1595, made him warden of Manchefter college, in which town he refided feveral years. In 1604 he returned to his houfe at Mortlake, where he died in 1608, at 81 years of age; leaving a large family and many works behind him.

The books that were printed and publifhed by Dee, are, 1. *Propædumata Aphoriftica, &c.* in 1558, in 12mo. —2. *Monas Hieroglyphica ad Regem Romanorum Maximilianum;* 1564.—3. *Epiftola ad eximium ducis Urbini mathematicum, Fredericum Commandinum, prefixa libello Machometi Bagdadini de Superficierum Divifionibus &c;* 1570.—4. *The Britifh Monarchy, otherwife called, The Petty Navy Royal;* 1576.—5. *Preface Mathematical to the Englifh Euclid,* publifhed by Henry Billingfley, 1570: certainly a very curious and elaborate compofition, and where he fays, many more arts are wholly invented by name, definition, property, and ufe, than either the Grecian or Roman mathematicians have left to our knowledge.—6. *Divers and many annotations and inventions difperfed and added after the 10th book of Englifh Euclid;* 1570.—7. *Epiftola prefixa Ephemeridibus Joannis Feldi à 1557, cui rationem declaraverat Ephemerides confcribendi.*—8. *Parallaticæ Commentationis Praxeofque Nucleus quidam;* 1573.

This catalogue of Dee's printed and publifhed works is to be found in his *Compendious Rehearfal &c,* as well as in his letter to Abp. Whitgift: and from the fame places might be tranfcribed more than 40 titles of books unpublifhed, that were written by him.

DEFENCE, in *Sieges,* is ufed for any thing that ferves to preferve or fcreen the foldiers, or the place. So the parapets, flanks, cafemates, ravelins, and outworks, that cover the place, are called the defences, or covers of the place: and when the cannon have beaten down or ruined thefe works, fo that the men cannot fight under cover, the defences of the place are faid to be demolifhed.

Line of DEFENCE, is that which flanks a baftion, being drawn from the flank oppofite to it.

The line of defence fhould not exceed a mufket fhot, i. e. 120 fathoms: indeed Melder allows 130, Scheiter 140, Vauban and Pagan 150.

Line of DEFENCE, *greater,* or *fichant,* is a line drawn from the point of the baftion to the concourfe of the oppofite flank and curtin.

Line of DEFENCE, *leffer,* or *rafant,* or *flanquant,* is the face of the baftion continued to the curtin.

DEFERENT, or DEFERENS, in the ancient aftronomy, an imaginary circle, which, as it were, carries about the body of a planet, and is the fame with the eccentric; being invented to account for the eccentricity, perigee, and apogee of the planets.

DEFICIENT *Hyperbola,* is a curve having only one afymptote, though two hyperbolic legs running out infinitely by the fide of the afymptote, but contrary ways. See CURVE.

This name was given to the curves by Newton, in his *Enumeratio Linearum tertii Ordinis.* There are 6 different fpecies of them, which have no diameters, expreffed by the equation $xyy + ey = -ax^3 + bx^2 + cx + d$, the term ax^3 being negative. When the equation ax^4

$= bx^3 + cx^2 + dx + \frac{1}{4}ee$ has all its roots real and unequal, the curve has an oval joined to it. When the two middle roots are equal, the oval joins to the legs, which then cut one another in shape of a noose. When three roots are equal, the nodus is changed into a very acute cusp or point. When, of three roots with the same sign, the two greatest are equal, the oval vanishes into a point. When any two roots are imaginary, there is only a pure serpentine hyperbola, without any oval, decussation, cusp, or conjugate point; and when the terms b and d are wanting, it is of the 6th species.

There are also 7 different species of these curves, having each one diameter, expressed by the above equation when the term ey is wanting: according to the various conditions of the roots of the equation $ax^3 = bx^2 + cx + d$, as to their reality, equality, their having the same signs, or two of them being imaginary.

DEFICIENT *Numbers*, are those whose aliquot parts added together, make a sum less than the whole number: as 8, whose parts 1, 2, 4, make only 7; or the number 16, whose parts 1, 2, 4, 8 make only 15.

DEFILE, *in Fortification*, a narrow line or passage through which troops can pass only in file, making a small front, so that the enemy may easily stop their march, and charge them with the more advantage, as the front and rear cannot come to the relief of one another.

DEFINITION, an enumeration, or specification of the chief simple ideas of which a compound idea consists, in order to ascertain or explain its nature and character.

Definitions are of two kinds; the one *nominal*, or of the *name*; the other *real*, or of the *thing*.

Nominal DEFINITION, is an enumeration of such known characters as are sufficient for distinguishing any proposed thing from others; as is that of a square, when it is said that it is a quadrilateral, equilateral, rectangular figure.

Real DEFINITION, a distinct notion, explaining the genesis of a thing; that is, how the thing is made or done: as is this definition of a circle, viz, that it is a figure described by the motion of a right line about a fixed point.

DEFLECTION, the turning any thing aside from its former course, by some adventitious or external cause.

The word is often applied to the tendency of a ship from her true course, by reason of currents, &c, which turn her out of her right way.

DEFLECTION *of the Rays of Light*, is a property which Dr. Hook observed in 1675. He found it different both from reflection and refraction; and that it was made perpendicularly towards the surface of the opacous body.

This is the same property which Newton calls *inflection*. And by others it is called *diffraction*.

DEGREE, *in Algebra*, is used in speaking of equations, when they are said to be of such a degree according to the highest power of the unknown quantity. If the index of that power be 2, the equation is of the 2d degree; if 3, it is of the 3d degree, and so on.

DEGREE, in Geometry or Trigonometry, is the 360th part of the circumference of any circle; for every circle is considered as divided into 360 parts, called degrees; which are marked by a small ° near the top of the figure; thus 45° is 45 degrees.

The degree is subdivided into 60 smaller parts, called minutes, meaning first minutes; the minute into 60 others, called seconds; the second into 60 thirds; &c. Thus $45°\ 12'\ 20''$ are 45 degrees, 12 minutes, 20 seconds.

The magnitude or quantity of angles is accounted in degrees; for because of the uniform curvature of a circle in all its parts, equal angles at the centre are subtended by equal arcs, and by similar arcs in peripheries of different diameters; and an angle is said to be of so many degrees, as are contained in the arc of any circle comprehended between the legs of the angle, and having the angular point for its centre. Thus we say an angle of 90°, or of 45° 24', or of 12° 20' 30''. It is also usual to say, such a star is mounted so many degrees above the horizon, or declines so many degrees from the equator; or such a town is situate in so many degrees of latitude or longitude.—A sign of the ecliptic, or zodiac, contains 30 degrees.

The division of the circle into 360 degrees is usually ascribed to the Egyptians, probably from the circle of the sun's annual course, or according to their number of days in the year, allotting a degree to each day. It is a convenient number too, as admitting of a great many aliquot parts, as 2, 3, 4, 5, 6, 8, 9, &c. The sexagesimal subdivision, however, has often been condemned as improper, by many eminent mathematicians, as Stevinus, Oughtred, Wallis, Briggs, Gellibrand, Newton, &c; who advise a decimal division instead of it, or else that of centesms; as the degree into 100 parts, and each of these into 100 parts again, and so on. Stevinus even holds, that this division of the circle which he contends for, obtained in the wise age, *in sæculo sapienti*. Stev. Cosmog. lib. 1, def. 6. And several large tables of sines &c have been constructed according to that plan, and published, by Briggs, Newton, and others. And I myself have carried the idea still much farther, in a memoir published in the Philos. Transf. of 1783, containing a proposal for a new division of the quadrant, viz, into equal decimal parts of the radius; by which means the degrees or divisions of the arch would be the real lengths of the arcs, in terms of the radius: and I have since computed those lengths of the arcs, with their sines, &c, to a great extent and accuracy.

DEGREE *of Latitude*, is the space or distance on the meridian through which an observer must move, to vary his latitude by one degree, or to increase or diminish the distance of a star from the zenith by one degree; and which, on the supposition of the perfect sphericity of the earth, is the 360th part of the meridian.

The quantity of a degree of a meridian, or other great circle, on the surface of the earth, is variously determined by different observers: and the methods made use of are also various.

Eratosthenes, 250 years before Christ, first determined the magnitude of a degree of the meridian, between Alexandria and Syene on the borders of Ethiopia, by measuring the distance between those places, and comparing it with the difference of a star's zenith distances at those places; and found it $694\frac{4}{9}$ stadia.

Posidonius,

Pofidonius, in the time of Pompey the Great, by means of the different altitudes of a ftar near the horizon, taken at different places under the fame meridian, compared in like manner with the diftance between thofe places, determined the length of a degree only 600 ftadia.

Ptolomy fixes the degree at 68⅔ Arabic miles, counting 7½ ftadia to a mile. The Arabs themfelves, who made a computation of the diameter of the earth, by meafuring the diftance of two places under the fame meridian, in the plains of Sennar, by order of Almamon, make it only 56 miles. Kepler, determining the diameter of the earth by the diftance of two mountains, makes a degree 13 German miles; but his method is far from being accurate. Snell, feeking the diameter of the earth from the diftance between two parallels of the equator, finds the quantity of a degree,

by one method 57064 Paris toifes, or 342384 feet;
by anothe. meth. 57057 - - toifes, or 342342 feet.

The mean between which two numbers, M. Picard found by menfuration, in 1669, from Amiens to Malvoifin, the moft certain, and he makes the quantity of a degree 57060 toifes, or 342360 feet. However, M. Caffini, at the king's command, in the year 1700, repeated the fame labour, and meafuring the space of 6° 18', from the obfervatory at Paris, along the meridian, to the city of Collioure in Rouffillon, that the greatnefs of the interval might diminifh the error, found the length of the degree equal to 57292 toifes, or 343742 Paris feet, amounting to 365184 Englifh feet.

And with this account nearly agrees that of our countryman Norwood, who, about the year 1635, meafured the diftance between London and York, and found that diftance 905751 Englifh feet; the difference of latitude being 2° 28', hence, he determined the quantity of one degree at 367196 Englifh feet, or 57300 Paris toifes, or 69 miles, 288 yards. See Newt. Princ. Phil. prop. 19; and Hift. Acad. Scienc. anno 1700, pa. 153.

M. Caffini, the fon, completed the work of meafuring the whole arc of the meridian through France, in 1718. For this purpofe he divided the meridian of France into two arcs, which he meafured feparately. The one from Paris to Collioure gave him 57097 toifes; the other from Paris to Dunkirk - - 56960; and the whole arc from Dunkirk to Collioure 57060, the fame as M. Picard's.

M. Mufchenbroek, in 1700, refolving to correct the errors of Snell, found by particular obfervations, that the degree between Alcmaer and Bergen-op-zoom contained 57033 toifes.

Meffieurs Maupertuis, Clairaut, Camus, Monnier, and Outheir of France, were fent on a northern expedition, and began their operations, affifted by M. Celfus, an eminent aftronomer of Sweden, in Swedifh Lapland, in July 1736, and finifhed them by the end of May following. They obtained the meafure of that degree, whofe middle point was in lat. 66° 20' north, and found it 57439 toifes, when reduced to the level of the fea. About the fame time another company of philofophers was fent to South America, viz, Meffieurs Godin, Bouguer, and Condamine of France, to whom were joined Don Jorge Juan, and Don An-

tonio de Ulloa of Spain. They left Europe in 1735, and began their operations in the province of Quito in Peru, about October 1736, and finifhed them, after many interruptions, about 8 years after. The Spanifh gentlemen publifhed a feparate account, and affign for the meafure of a degree of the meridian at the equator 56768 toifes. M. Bouguer makes it 56753 toifes, when reduced to the level of the fea; and M. Condamine ftates it at 56749 toifes.

M. Caille, being at the Cape of Good Hope in 1752, found the length of a degree of the meridian in lat. 33° 18' 30" fouth, to be 57037 toifes. In 1755, father Bofcovich found the length of a degree in lat. 43° north to be 56972 toifes, as meafured between Rome and Rimini in Italy. In the year 1740, Meffrs Caffini and La Caille again examined the former meafures in France, and, after making all the neceffary corrections, found the meafure of a degree, whofe middle point is in lat. 49° 22' north, to be 57074 toifes; and in the lat. of 45°, it was 57050 toifes.

In 1764, F. Beccaria completed the meafurement of a portion of the meridian near Turin; from which it is deduced that the length of a degree, whofe middle lat. is 44° 44' north, is 57024 Paris toifes.

At Vienna, 3 degrees of the meridian were meafured; and the medium, for the latitude of 47° 40' north may be taken at 57091 Paris toifes. See an account of this meafurement, by father Jofeph Liefganig, in the Philof. Tranf. 1768, pa. 15.

Finally, in the fame vol. too is an account of the meafurement of a part of the meridian in Maryland and Penfilvania, North America, 1766, by Meffrs Mafon and Dixon; from which it follows that the length of a degree whofe middle point is 39° 12' north, was 363763 Englifh feet, or 56904½ Paris toifes.

Hence, from the whole we may collect the following table of the principal meafures of a degree in different parts of the earth, as meafured by different perfons, viz,

Mean Latitude.	Length of a Degree in Paris toifes.	Names of the Meafurers.	Years of Meafurement.
66° 20' N	57422	Maupertuis &c	1736 & 1737
49 23 N	57074	Maupertuis &c and Caffini	1739 & 1740
47 40 N	57091	Liefganig	1766
45 , 0 N	57028	Caffini	1739 & 1740
44 44 N	57069	Beccaria	1760 to 1764
43 0 N	56979	Bofcovich and Le Maire	1752
39 12 N	56888	Mafon & Dixon	1764 to 1768
0 0	56750	Bouguer and Condamine	1736 to 1744
33 18 S	57037	La Caille	1752

The method of obtaining the length of a degree of the terreftrial meridian, is to meafure a certain diftance upon it by a feries of triangles, whofe angles may be found by actual obfervation, connected with a bafe, whofe length may be taken by an actual furvey, or otherwife; and then to obferve the different altitudes

of

of some star at the two extremities of that distance, which gives the difference of latitude between them: then, by proportion, as this difference of latitude is to one degree, so is the measured length to the length of one degree of the meridian sought. This method was first practised by Eratosthenes, in Egypt. See GEOGRAPHY, and the beginning of this article.

DEGREE *of longitude*, is the space between two meridians that make an angle of 1° with each other at the poles; the quantity or length of which is variable, according to the latitude, being every where as the cosine of the latitude; viz, as the cosine of one lat. is to the cosine of another, so is the length of a degree in the former lat. to that in the latter; and from this theorem is computed the following Table of the length of a degree of long. in different latitudes, supposing the earth to be a globe.

Degr. lat.	English miles.	Degr. lat.	English miles.	Degr. lat.	English miles.
0	69·07	31	59·13	61	33·45
1	69·06	32	58·51	62	32·40
2	69·03	33	57·87	63	31·33
3	68·97	34	57·20	64	30·24
4	68·90	35	56·51	65	29·15
5	68·81	36	55·81	66	28·06
6	68·62	37	55·10	67	26·96
7	68·48	38	54·37	68	25·85
8	68·31	39	53·62	69	24·73
9	68·15	40	52·85	70	23·60
10	67·95	41	52·07	71	22·47
11	67·73	42	51·27	72	21·32
12	67·48	43	50·46	73	20·17
13	67·21	44	49·63	74	19·02
14	66·95	45	48·78	75	17·86
15	66·65	46	47·93	76	16·70
16	66·31	47	47·06	77	15·52
17	65·98	48	46·16	78	14·35
18	65·62	49	45·26	79	13·17
19	65·24	50	44·35	80	11·98
20	64·84	51	43·42	81	10·79
21	64·42	52	42·48	82	9·59
22	63·97	53	41·53	83	8·41
23	63·51	54	40·56	84	7·21
24	63·03	55	39·58	85	6·00
25	62·53	56	38·58	86	4·81
26	62·02	57	37·58	87	3·61
27	61·48	58	36·57	88	2·41
28	60·93	59	35·54	89	1·21
29	60·35	60	34·50	90	0·00
30	59·75				

Note, This table is computed on the supposition that the length of the degrees of the equator are equal to those of the meridian at the medium latitude of 45°, which length is 69$\frac{1}{15}$ English miles.

The expressions *Latitude* and *Longitude*, are borrowed from the ancients, who happened to be acquainted with a much larger extent of the earth in the direction east and west, than in that of north and south; the former of which therefore passed, with them, for the length of the earth, or longitude, and the latter for the breadth or shorter dimension, viz, the latitude.

DEJECTION, in Astrology, is applied to the planets when in their detriment, as astrologers speak, i. e. when they have lost their force, or influence, as is pretended, by reason of their being in opposition to some others, which check and counteract them.

Or, it is used when a planet is in a sign opposite to that in which it has its greatest effect, or influence, which is called its exaltation. Thus, the sign Aries being the exaltation of the Sun, the opposite sign Libra is its dejection.

DEINCLINERS, or DEINCLINING *Dials*, are such as both decline and incline, or recline, at the same time. Suppose, for instance, a plane cutting the prime vertical circle at an angle of 30 degrees, and the horizontal plane at an angle of 24 degrees, the latitude of the place being 52 degrees; a dial drawn on this plane, is called a deincliner.

DELIACAL *Problem*, a celebrated problem among the ancients, concerning the duplication of the cube.

DELPHINUS, the *Dolphin*, a constellation of the northern hemisphere; whose stars, in Ptolomy's catalogue, are 10; in Tycho's the same; in Hevelius's 14; and in Flamsteed's 18.

DEMETRIUS, a Cynic philosopher, and disciple of Apollonius Thyaneus, in the age of Caligula. That emperor wishing to gain the philosopher to his interest by a large present, he refused it with indignation, saying, If Caligula wishes to bribe me, let him send me his crown. Vespasian was displeased with his insolence, and banished him to an island. The cynic derided the punishment, and bitterly inveighed against the emperor.

Demetrius lived to a very great age. And Seneca observes, that " nature had brought him forth to shew mankind that an exalted genius can live securely without being corrupted by the vice of the surrounding world."

DEMI-*Bastion*, in Fortification, one that has only one face and one flank.

DEMI-*Cannon*, and DEMI-*Culverin*, names of certain species of cannon, now no longer used.

DEMI-*Cross*, an instrument used by the Dutch to take the altitude of the sun or a star at sea; instead of which we use the cross staff, or forestaff.

DEMI-*Gorge*, is half the gorge or entrance into the bastion; not taken directly from angle to angle, where the bastion joins to the curtin, but from the angle of the flank to the centre of the bastion; or the angle the two curtins would make, were they thus protracted to meet in the bastion.

DEMI-*Lune*, or *Half-moon*, an outwork consisting of two faces, and two little flanks. It is often built before the angle of a bastion, and sometimes also before the curtin; though now it is very seldom used.

DEMOCRITUS, one of the greatest philosophers of antiquity, was born at Abdera, a town of Thrace, about the 80th olympiad, or about 400 years before Christ. His father, says Valerius Maximus, was able to entertain the army of Xerxes; and Diogenes Laertius adds, upon the testimony of Herodotus, that the king, in requital, presented him with some Magi and Chaldeans. From these, it seems, Democritus received the first part of his education; and from them, whilst yet a boy, he learned theology and astronomy. He next applied to Leucippus, from whom he learned the

system

fyſtem of atoms and a vacuum. His father dying, he and his two brothers divided his effects. Democritus made choice of that part which confifted in money, as being, though the leaſt ſhare, the moſt convenient for travelling; and it is ſaid that his portion amounted to more than 100 talents, which is near 20 thouſand pounds ſterling. His extraordinary inclination for knowledge and the ſciences, induced him to travel into all parts of the world where he might find learned men. He went to viſit the prieſts of Egypt, from whom he learned geometry: He conſulted the Chaldean and Perſian philoſophers: and it is ſaid that he penetrated even into India and Ethiopia, to confer with the Gymnoſophiſts. In theſe travels he waſted his ſubſtance; after which, at his return he was obliged for ſome time to be maintained by his brother. Settling himſelf at Abdera, he there governed in the moſt abſolute manner, by virtue of his conſummate wiſdom. The magiſtrates of that city made him a preſent of 500 talents, and erected ſtatues to him, even in his lifetime: but being naturally more inclined to contemplation than delighted with public honours and employments, he withdrew into ſolitude and retirement.

Democritus always laughed at human life, as a continued farce, which made the people think he was mad; on which they ſent for Hippocrates to cure him: but that celebrated phyſician having diſcourſed with the philoſopher, told the people that he had a great veneration for Democritus; and that, in his opinion, thoſe who eſteemed themſelves the moſt healthy, were the moſt diſtempered.

It is ſaid, though with little probability, that Democritus put out his own eyes, that he might meditate more profoundly on philoſophical ſubjects. He died, according to Diogenes Laertius, in the 361ſt year before the Chriſtian era, at 109 years of age. He was the author of many books, which are loſt; from which Epicurus borrowed his philoſophy.

DEMOIVRE (ABRAHAM), a celebrated mathematician, of French original, but who ſpent moſt of his life in England. He was born at Vitri in Champagne 1667. The revocation of the edict of Nantz, in 1685, determined him, with many others, to take ſhelter in England; where he perfected his mathematical ſtudies, the foundation of which he had laid in his own country. A mediocrity of fortune obliged him to employ his talent in this way in giving leſſons, and reading public lectures, for his better ſupport: in the latter part of his life too, he chiefly ſubſiſted by giving anſwers to queſtions in chances, play, annuities, &c, and it is ſaid moſt of theſe reſponſes were delivered at a Coffee-houſe in St. Martin's-lane, where he ſpent moſt of his time. The Principia Mathematica of Newton, which chance is ſaid to have thrown in his way, ſoon convinced Demoivre how little he had advanced in the ſcience he profeſſed. This induced him to redouble his application; which was attended by a conſiderable degree of ſucceſs; and he ſoon became connected with, and celebrated among, the firſt rate mathematicians. His eminence and abilities in this line, opened him an entrance into the Royal Society of London, and into the academies of Berlin and Paris. By the former his merit was ſo well known and eſteemed, that they judged

him a fit perſon to decide the famous conteſt between Newton and Leibnitz, concerning the invention of Fluxions.

The collection of the Academy of Paris contains no memoir of this author, who died at London Nov. 1754, at 87 years of age, ſoon after his admiſſion into it. But the Philoſophical Tranſactions of London have ſeveral, and all of them intereſting, viz, in the volumes 19, 20, 22, 23, 25, 27, 29, 30, 32, 40, 41, 43.

He publiſhed alſo ſome very reſpectable works, viz,

1. *Miſcellanea Analytica, de Seriebus & Quadraturis &c*; 1730, in 4to. But perhaps he has been more generally known by his

2. *Doctrine of Chances; or, Method of Calculating the Probabilities of Events at Play.* This work was firſt printed, 1718, in 4to, and dedicated to Sir Iſaac Newton: it was reprinted in 1738, with great alterations and improvements; and a third edition was afterwards printed.

3. *Annuities on Lives;* firſt printed 1724, in 8vo.— In 1742 the ingenious Thomas Simpſon (then only 33 years of age) publiſhed his *Doctrine of Annuities and Reverſions;* in which he paid ſome handſome compliments to our author. Notwithſtanding which, Demoivre preſently brought out a ſecond edition of his Annuities, in the preface to which he paſſed ſome harſh reflections upon Simpſon. To theſe the latter gave a handſome and effectual anſwer, 1743, in *An Appendix, containing ſome Remarks on a late book on the ſame ſubject, with anſwers to ſome perſonal and malignant miſrepreſentations, in the preface thereof.* At the end of this anſwer, Mr. Simpſon concludes, " Laſtly, I appeal to all mankind, whether, in his treatment of me, he has not diſcovered an air of ſelf-ſufficiency, ill-nature and inveteracy, unbecoming a gentleman." Here it would ſeem the controverſy dropped: Mr. Demoivre publiſhed the 3d edition of his book in 1750, without any farther notice of Simpſon, but omitted the offenſive reflections that had been in the preface.

DEMONSTRATION, a certain or convincing proof of ſome propoſition: ſuch as the demonſtrations of the propoſitions in Euclid's Elements.

The method of demonſtrating in mathematics, is the ſame with that of drawing concluſions from principles in logic. Indeed, the demonſtrations of mathematicians are no other than ſeries of enthymemes; every thing is concluded by force of ſyllogiſm, only omitting the premiſes, which either occur of their own accord, or are recollected by means of quotations.

DENDROMETER, an inſtrument lately invented by Meſſrs Duncombe and Whittel; ſo called, from its uſe in meaſuring trees.

DENEB, an Arabic term, ſignifying tail; uſed by aſtronomers as a name to ſome of the fixed ſtars, but eſpecially for the bright ſtar in the Lion's tail.

DENOMINATOR, *of a fraction,* is the number or quantity placed below the line, which ſhews the whole integer, or into how many parts the integer is ſuppoſed to be divided by the fraction; as that which gives denomination or name to the parts of the fraction. Thus, in the fraction $\frac{5}{12}$, five-twelfths, the number 12 is the denominator, and ſhews that the integer is here divided into 12 parts; or that it conſiſts of 12 of thoſe

parts

parts of which the numerator contains 5. Also *b* is the denominator of the fraction $\frac{a}{b}$.

DENOMINATOR *of a ratio*, is the quotient arising from the division of the antecedent by the consequent. Thus, 6 is the denominator of the ratio 30 to 5, because 30 divided by 5 gives 6. It is otherwise called the *exponent* of the ratio.

DENSITY, that property of bodies, by which they contain a certain quantity of matter, under a certain bulk or magnitude. Accordingly a body that contains more matter than another, under the same bulk, is said to be denser than the other, and that in proportion to the quantity of matter; or if the quantity of matter be the same, but under a less bulk, it is said to be denser, and so much the more so as the bulk is less. So that, in general, the density is directly proportional to the mass or quantity of matter, and reciprocally or inversely proportional to the bulk or magnitude under which it is contained.

The quantities of matter in bodies, or at least the proportions of them, are known by their gravity or weight; every equal particle of matter being endowed with an equal gravity, it is inferred that equal masses or quantities of matter have an equal weight or gravity; and unequal masses have proportionally unequal weights. So that, when body, or mass, or quantity of matter is spoken of, we are to understand their weight or gravity.

From the foregoing general proportion of the density of bodies, viz, that it is as the mass directly, and as the bulk inversely, may be inferred the proportion of the masses, or of the magnitudes; viz, that the mass or quantity of matter, is in the compound ratio of the bulk and density; and that the bulk or magnitude, is as the mass directly, and the density inversely. Hence, if B, *b* be two bodies, or masses, or weights;

and D, *d* their respective densities;

also M, *m* their magnitudes, or bulks:

Then the theorems above are thus expressed,

viz, $D \propto \frac{B}{M}$, and $B \propto DM$, and $M \propto \frac{B}{D}$;

or $D : d :: \frac{B}{M} : \frac{b}{m}$, and $B : b :: DM : dm$, &c;

or $\frac{D}{d} = \frac{Bm}{bM}$, and $\frac{B}{b} = \frac{DM}{dm}$, and $\frac{M}{m} = \frac{Bd}{bD}$.

No body is absolutely or perfectly dense; or no space is perfectly full of matter, so as to have no vacuity or interstices; on the contrary, it is the opinion of Newton, that even the densest bodies, as gold &c, contain but a small portion of matter, and a very great portion of vacuity; or that it contains a great deal more of pores or empty space, than of real substance.

It has been observed above, that the relative density of bodies may be known by their weight or gravity; and hence the most general way of knowing those densities, is by actually weighing an equal bulk or magnitude of the bodies, whether solid or fluid; if solid, by shaping them to the same figure and dimensions; if fluid, by filling the same vessel with them, and weighing it.

For fluids, there are also other methods of finding their density: as 1st, by making an equilibrium between them in tubes that communicate; for, the diameters of the tubes being equal, and the weights or quantities of matter also equal, the densities will be inversely as the altitudes of the liquids in them, that is inversely as the bulk.

2dly, The densities of fluids are also compared together by immerging a solid in them; for if the solid be lighter than the liquids, the part immerged by its own weight, will be inversely as the density of the fluid; or if it be heavier, and sink in the liquids, by weighing it in them; then the weights lost by the body will be directly proportional to the densities of the fluids.

DENSITY *of the Air*, is a property that has much employed the later philosophers, since the discovery of the Torricellian experiment, and the air-pump. By means of the barometer it is demonstrated that the air is of the same density at all places at the same distance from the level of the sea; provided the temperature, or degree of heat, be the same. Also the density of the air always increases in proportion to the compression, or the compressing forces. And hence the lower parts of the atmosphere are always denser than the upper: yet the density of the lower air is not exactly proportional to the weight of the atmosphere, by reason of heat and cold, which make considerable alterations as to rarity and density; so that the barometer measures the elasticity of the air, rather than its density. If the height of the barometer be considered as the measure both of the density and elasticity of the air, when the thermometer is at 31°, and *b* be any other height of the barometer, when the thermometer is at *t* degrees; then in this case, *b* is the measure of the elasticity,

and $\frac{466 - t}{435} b$ is the measure of the density of the air.

DENSITY *of the Planets*: In homogeneous, unequal, spherical bodies, the gravities on their surfaces, are as their diameters when the densities are equal, or the gravities are as the densities when the bulks are equal; therefore, in spheres of unequal magnitude and density, the gravity is in the compound ratio of the diameters and densities, or the densities are as the gravities divided by the diameters. Knowing therefore the diameters of the planets by observation and comparison, and the gravities at their surface by means of the revolution of the satellites, the relation of their densities becomes known. And as I have found the mean density of the earth to be about 4¼ times that of water, Philos. Transf. 1778; hence the densities of the planets, with respect to water, become known, and are as below:

	Densities.
Water	1
The Sun	$1\frac{2}{13}$
Mercury	$9\frac{1}{4}$
Venus	$5\frac{1}{3}\frac{1}{5}$
The Earth	$4\frac{1}{2}$
Mars	$3\frac{2}{7}$
The Moon	$3\frac{1}{14}$
Jupiter	$1\frac{1}{24}$
Saturn	$0\frac{1}{2}\frac{3}{4}$
Georgian Planet	$0\frac{99}{100}$

As it is not likely that any of these bodies are homogeneal,

geneal, the denfities here determined are fuppofed to be the mean denfities, or fuch as the bodies would have if they were homogeneal, and of the fame mafs of matter and magnitude.

DENTICLES, or DENTILS, are ornaments in a cornice, cut after the manner of teeth. Thefe are moftly affected in the Ionic and Corinthian orders; and of late alfo in the Doric. The fquare member on which they are cut, is called the Denticule.

DEPARTURE, in Navigation, is the eafting or wefting of a fhip, with regard to the meridian fhe *departed* or failed from. Or, it is the difference in longitude, either eaft or weft, between the prefent meridian the fhip is under, and that where the laft reckoning or obfervation was made. This departure, any where but under the equator, muft be accounted according to the number of miles in a degree proper to the parallel the fhip is in.

The Departure, in Plane and Mercator's Sailing, is always reprefented by the bafe of a right-angled plane triangle, where the courfe is the angle oppofite to it, and the diftance failed is the hypothenufe, the perpendicular or other leg being the difference of latitude. And then the theorem for finding it, is always this: As radius is to the fine of the courfe, fo is the diftance failed, to the departure fought.

DEPRESSION *of the Pole.* So many degrees &c as you fail or travel from the pole towards the equator, fo many it is faid you deprefs the pole, becaufe it becomes fo much lower, or nearer the horizon.

DEPRESSION *of a Star,* or *of the Sun,* is its diftance below the horizon; and is meafured by an arc of a vertical circle, intercepted between the horizon and the place of the ftar.

DEPRESSION *of the Vifible Horizon,* or *Dip of the Horizon,* denotes its finking or dipping below the true horizontal plane, by the obferver's eye being raifed above the furface of the fea; in confequence of which, the obferved altitude of an object is by fo much too great.

Thus, the eye being at E, the height AE above the furface of the earth, whofe centre is C; then EH is the real horizon, and E*h* the vifible one, below the former by the angle HE*h*, by reafon of the elevation AE of the eye.

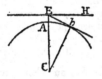

To compute the Depreffion or Dip of the Horizon.

In the right-angled triangle CE*h*, are given C*h* the earth's radius = 21000000 feet, and the hypothenufe CE = the radius increafed by the height AE of the eye; to find the angle C which is = the angle HE*h*, or depreffion fought;

viz, as C*h* : CE :: radius : fec. ∠ C,
or as CE : C*h* :: radius : cofin. ∠ C.

By either of thefe theorems are computed the numbers in the following table, which fhews the depreffion or dip of the horizon of the fea for different heights of the eye, from 1 foot to 100 feet.

Height of the eye	Dip of the horizon	Height of the eye	Dip of the horizon	Height of the eye	Dip of the horizon
feet	′ ″	feet	″	feet	″
1	0 57	13	3 26	26	4 52
2	1 21	14	3 34	28	5 3
3	1 39	15	3 42	30	5 14
4	1 55	16	3 49	35	5 39
5	2 8	17	3 56	40	6 2
6	2 20	18	4 3	45	6 24
7	2 31	19	4 10	50	6 44
8	2 42	20	4 16	60	7 23
9	2 52	21	4 22	70	7 59
10	3 1	22	4 28	80	8 32
11	3 10	23	4 34	90	9 3
12	3 18	24	4 40	100	9 33

See Robertfon's Navigation, book 9 appendix; and Tables requifite to be ufed with the Nautical Ephemeris, pa. 1. See alfo LEVELLING.

DEPTH, the oppofite of Height, and one of the dimenfions of bodies, or of fpace. See HEIGHT, ALTITUDE, ELEVATION, &c.

DERHAM (Doctor WILLIAM), an eminent Englifh philofopher and divine, was born at Stowton, near Worcefter, 1657, and educated in Trinity College, Oxford. In 1682, he was prefented to the vicarage of Wargrave in Berkfhire; and, in 1689, to the valuable rectory of Upminfter in Effex; which, lying at a convenient diftance from London, afforded him an opportunity of converfing and correfponding with the principal literary geniufes of the nation. Applying himfelf there with great eagernefs to the purfuit of his ftudies in natural and experimental philofophy, he foon became a diftinguifhed and ufeful member of the Royal Society, whofe Philofophical Tranfactions contain a great variety of curious and valuable pieces, the fruits of his laudable induftry, in all or moft of the volumes, from the 20th to the 39th, both inclufive; the principal of which are:

1. Experiments on the Motion of Pendulums in vacuo.
2. A Defcription of an inftrument for finding the Meridian.
3. Experiments and Obfervations on the Motion of Sound.
4. On the Migration of Birds.
5. Hiftory of the Spots in the Sun, from 1703 to 1711.
6. Obfervations on the Northern Lights, Oct. 8, 1726, and Oct. 13, 1728.
7. Tables of the Eclipfes of Jupiter's Satellites.
8. The difference of Time in the meridian of different places.
9. Of the meteor called Ignis Fatuus.
10. The Hiftory of the Death Watch.
11. Meteorological Diaries for feveral years.

In his younger days he publifhed his Artificial Clockmaker, a very ufeful little work, that has gone through feveral editions. In 1711, 1712, 1714, he preached thofe fermons at Boyle's lecture, which he afterward digefted under the well-known titles of *Phyfico-Theology* and *Aftro-Theology,* or Demonftrations of the being and attributes

attributes of God, from his works of creation, and a survey of the heavens.

In 1716 he was made a canon of Windsor, being at that time chaplain to the Prince of Wales; and in 1730 received, from the university of Oxford, the degree of Doctor of Divinity. He revised the *Miscellanea Curiosa*, in 3 vols. 8vo, containing many curious papers of Dr. Halley and several other ingenious philosophers. To him also the world is indebted for the publication of the Philosophical Experiments of the late eminent Dr. Hooke, and other ingenious men of his time; as well as notes and illustrations of several other works.

Dr. Derham was very well skilled in medical as well as in physical knowledge; and was constantly a physician to the bodies as well as the souls of his parishioners. This great and good man, after spending his life in the most agreeable and improving study of nature, and the diligent and pious discharge of his duty, died at Upminster in 1735, at 78 years of age.

DESAGULIERS (John Theophilus), an eminent experimental philosopher, was the son of the Rev. John Desaguliers, a French Protestant refugee, and born at Rochelle in 1683. His father brought him to England an infant; and having taught him the classics himself, he sent him at a proper age to Christ church College, Oxford; where in 1702 he succeeded Dr. Keil in reading lectures on experimental philosophy at Hart Hall. In 1712 he married, and settled in London, when he first of any introduced the reading of lectures in experimental philosophy in the metropolis, which he continued during the rest of his life with the greatest applause, having several times the honour of reading his lectures before the king and royal family. In 1714 he was elected F. R. S. and proved a very useful member, as appears from the great number of his papers that are printed in their Philos. Transf. on the subjects of optics, mechanics, and meteorology. The magnificent duke of Chandos made Dr. Desaguliers his chaplain, and presented him to the living of Edgware, near his seat at Cannons; and he became afterward chaplain to Frederick prince of Wales. In the latter part of his life, he removed to lodgings over the Great Piazza in Covent Garden, where he carried on his lectures with great success till the time of his death in 1749, at 66 years of age.

He was a member of several foreign academies, and corresponding member of the Royal Academy of Sciences at Paris; from which academy he obtained the prize, proposed by them for the best account of electricity. He communicated a multitude of curious and valuable papers to the Royal Society, for the year 1714 to 1743, or from vol. 29 to vol. 42.

Beside those numerous communications, he published a valuable *Course of Experimental Philosophy*, 1734, in 2 large vols. 4to; and gave an edition of *Gregory's Elements of Catoptrics and Dioptrics*, with an Appendix on Reflecting Telescopes, 8vo, 1735. This appendix contains some Original Letters that passed between Sir Isaac Newton and Mr. James Gregory, relating to those telescopes.

DESCENDING, a going or moving from above, downwards.

There are ascending and descending stars, and ascending and descending degrees, &c.

DESCENDING *Latitude*, is the latitude of a planet in its return from the nodes to the equator.

DESCENSION, in Astronomy, is either *right*, or *oblique*.

Right DESCENSION is a point, or arch, of the equator, which descends with a star, or sign, below the horizon, in a *right* sphere, and

Oblique DESCENSION is a point, or arch, of the equator, which descends at the same time with a star, or sign, below the horizon, in an oblique sphere.

Descensions, both right and oblique, are counted from the first point of Aries, or the vernal intersection, according to the order of the signs, i. e. from west to east. And, as they are unequal, when it happens that they answer to equal arcs of the ecliptic, as for example to the 12 signs of the zodiac, it follows, that sometimes a greater part of the equator rises or descends with a sign, in which case the sign is said to ascend or descend rightly: and sometimes again, a less part of the equator rises or sets with the same sign, in which case it is said to ascend or descend obliquely See ASCENSION.

Refraction of the DESCENSION. See REFRACTION.

DESCENSIONAL *Difference*, is the difference between the right and oblique descension of the same star, or point of the heavens.

DESCENT, or *Fall*, in mechanics, &c, is the motion, or tendency, of a body towards the centre of the earth, either directly or obliquely.

The descent of bodies may be considered either as freely, like as in a vacuum, or as clogged or resisted by some external force, as an opposing body, or a fluid medium, &c.

1st, If the body b descend freely, and perpendicularly, by the force of gravity; then the motive force urging it downwards, is equal to its whole weight b; and the quantity of matter being b also, the accelerative force will be $\dfrac{b}{b}$ or 1.

2dly, If the body b descending, be opposed by some mechanical power, suppose a wedge or inclined plane, that is, instead of pursuing the perpendicular line of gravity, it is made to descend in a sloping direction down the inclined plane: then if the sine of the angle the plane makes with the horizon be s, to the radius 1, the motive force urging the body down the plane will be bs; and therefore the accelerative force is $\dfrac{bs}{b}$ or s; which is less than in the former case in the proportion of s to 1.

3dly, In a medium, a body suspended loses as much of its weight, as is the weight of a like bulk of the medium; and when descending, it loses the same, beside the obstruction arising from the cohesion of the parts of the medium, and the opposing force of the particles struck, which last produces a greater or less resistance, according to the velocity of the motion. But, the weight of the body being b, and that of a like bulk of the fluid medium m, the motive force urging the body to descend, is only $b - m$; that is, the body only falls by the excess of its weight above that of an equal bulk of the medium.

Hence, the power that sustains a body in a medium,

is equal to the excefs of the abfolute weight of the body above an equal bulk of the medium. Thus, a piece of copper weighing 47⅓ lb, lofes 5⅓ lb of its weight in water: and therefore a power of 42 lb will fuftain it in the water.

4thly, If two bodies have the fame fpecific gravity, the lefs the bulk of the defcending body is, the more of its gravity does it lofe, and the flower does it defcend, in the fame medium. For, though the proportion of the fpecific gravity of the body to that of the fluid be ftill the fame, whether the bulk be greater or lefs, yet the fmaller the body, the more the furface is, in proportion to the mafs; and the more the furface, the more the refiftance of the parts of the fluid, in proportion.

5thly, If the fpecific gravities of two bodies be different; that which has the greateft fpecific gravity will defcend with greater velocity in the air, or refifting medium, than the other body. Thus, a ball of lead defcends fwifter than wood or cork, becaufe it lofes lefs of its weight, though in a vacuum they both fall equally fwift.

The caufe of this defcent, or tendency downwards, has been greatly controverted. Two oppofite hypothefes have been advanced; the one, that it proceeds from an internal principle, and the other from an external one: the firft is maintained by the Peripatetics, Epicureans, and the Newtonians; and the latter by the Cartefians and Gaffendifts. See alfo ACCELERATION.

Laws of DESCENT of Bodies.

1ft, Heavy bodies, in an unrefifting medium, fall with an uniformly accelerated motion. For, it is the nature of all conftant and uniform forces, fuch as that of gravity at the fame diftance from the centre of the earth, to generate or produce equal additions of velocity in equal times. So that, if in one fecond of time there be produced 1 degree of velocity, in 2 feconds there will be 2 degrees of velocity, in 3 feconds 3 degrees, and fo on, the degree or quantity of velocity being always proportional to the length of the time.

2nd, The fpace defcended by an uniform gravity, in any time, is juft the half of the fpace that might be uniformly defcribed in the fame time by the laft velocity acquired at the end of that time, if uniformly continued. For, as the velocity increafes uniformly in an arithmetic progreffion, the whole fpace defcended by the variable velocity, will be equal to the fpace that would be defcribed with the middle velocity uniformly continued for the fame time; and this again will be only half the fpace that would be defcribed with the laft velocity, alfo uniformly continued for the fame time, becaufe the laft velocity is double of the middle velocity, being produced in a double time.

3d, The fpaces defcended by an uniform gravity, in different times, are proportional to the fquares of the times, or to the fquares of the velocities. For the whole fpace defcended in any number of particles of time, confifts of the fums of all the particular fpaces, or velocities, which are in arithmetical progreffion; but the fum of fuch an arithmetical progreffion, beginning at 0, and having the laft term and the number of terms the fame quantity, is equal to half the fquare of the laft term, or of the number of terms; therefore

the whole fums are as the fquares of the times, or of the velocities.

This theory of the defcents by gravity was firft difcovered and taught by Galileo, who afterwards confirmed the fame by experiments; which have often been repeated in various ways by many other perfons fince his time, as Grimaldi, Riccioli, Huygens, Newton, and many others, all confirming the fame laws.

The experiments of Grimaldi and Riccioli were made by dropping a number of balls, of half a pound weight, from the tops of feveral towers, and meafuring the times of falling by a pendulum. Ricciol. Almag. Nov. tom. 1 lib. 2, cap. 21, prop. 4. An abftract of their experiments is exhibited here below:

Vibrations of the pendulum	The time		Space at the end of the time	Space defcended each time
	''	'''	Rom. feet	Rom. feet
5	0	50	10	10
10	1	40	40	30
15	2	30	90	50
20	3	20	160	70
25	4	10	250	90
6	1	0	15	15
12	2	0	60	45
18	3	0	135	75
24	4	0	240	105

The fpace defcended by a heavy body in any given time, being determined by experiment, is fufficient, in connection with the preceding theorems, for determining every inquiry concerning the times, velocities, and fpaces defcended, depending on an uniform force of gravity. From many accurate experiments made in England, it has been found that a heavy body defcends freely through 16 feet 1 inch, or $16\frac{1}{12}$ feet, in the firft fecond of time; and confequently, by theorem 2, the velocity gained at the end of 1 fecond, is $32\frac{1}{6}$ feet per fecond. Hence, by the fame, and theorem 3, the velocity gained in any other time t is $32\frac{1}{6}t$, and the fpace defcended is $16\frac{1}{12}t^2$. So that, if v denote the velocity, and s the fpace due to the time t, and there be put $g = 16\frac{1}{12}$; then

$$\text{is } v = 2gt = 2\sqrt{gs} = \frac{2s}{t}.$$

$$s = gt^2 = \frac{v^2}{4g} = \frac{1}{2}tv.$$

$$t = \frac{v}{2g} = \sqrt{\frac{s}{g}} = \frac{2s}{v}.$$

The experiments with pendulums give alfo the fame fpace for the defcent of a heavy body in a fecond of time. Thus, in the latitude of London, it is found by experiment, that the length of a pendulum vibrating feconds is juft $39\frac{1}{8}$ inches; and it being known that the circumference of a circle is to its diameter, as the time of one vibration of any pendulum, is to the time in which a heavy body will fall through half the length of the pendulum; therefore as $3\cdot1416 : 1 :: 1 : \dfrac{1}{3\cdot1416}$ which is the time of defcending through $19\frac{9}{16}$ inches, or

half

half the length of the pendulum; then, spaces being as the squares of the times, as $\dfrac{1}{3\cdot1416^2} : 1^2 :: 19\frac{2}{7} :$ 193 inches, or 16 feet 1 inch, which therefore is the space a heavy body will descend through in one second; the very same as before.

4th, For any other constant force, instead of the perpendicular free descent by gravity, find by experiment, or otherwise, the space descended in one second by that force, and substitute that instead of $16\frac{1}{12}$ for the value of g in these formulæ: or, if the proportion of the force to the force of gravity be known, let the value of g be altered in the same proportion, and the same formulæ will still hold good. So, if the descent be on an inclined plane, making, for instance, an angle of 30° with the horizon; then, the force of descent upon the plane being always as the sine of the angle it makes with the horizon, in the present case it will be as the sine of 30°, that is, as ½ the radius; therefore in this case the value of g will be but half the former, $8\frac{1}{24}$, in all the foregoing formulæ.

Or, if one body descending perpendicularly draw another after it, by means of a cord sliding over a pulley; then it will be, as the sum of the two bodies is to the descending body, so is $16\frac{1}{12}$ to the value of g in this case; which value of it being used in the said formulæ, they will still hold good. And in like manner for any other constant forces whatever.

5th, The time of the oblique descent down any chord of a circle, drawn either from the uppermost point or lowermost point of the circle, is equal to the perpendicular descent through the diameter of the circle.

6th. The descent, or vibration, through all arcs of the same cycloid are equal, whether great or small.

7th. But the descent, or vibration, through unequal arcs of a circle, are unequal; the times being greater in the greater arcs, and less in the less.

8th, For Descents by Forces that are variable, see FORCES, &c. See also INCLINED PLANE, CYCLOID, PENDULUM, &c.

Line of Swiftest DESCENT, is that which a body, falling by the action of gravity, describes in the shortest time possible, from one given point to another. And this line, it is proved by philosophers, is the arc of a cycloid, when the one point is not perpendicularly over the other. See CYCLOID.

DESCRIBENT, a term in Geometry, signifying a line or superficies, by the motion of which a superficies or solid is described.

DETENTS, in a clock, are those stops which, by being lifted up or let down, lock and unlock the clock in striking,

DETENT-*Wheel*, or HOOP-*Wheel*, that wheel in a clock which has a hoop almost round it, in which there is a vacancy, where the clock locks.

DETERMINATE *Number*. See NUMBER.

DETERMINATE *Problem*, is that which has but one solution, or a certain limited number of solutions; in contradistinction to an indeterminate problem, which admits of infinite solutions.

Such, for instance, is the problem, To form an isosceles triangle on a given line, so that each of the angles at the base shall be double of that at the ver-

tex; which has only one solution: or this, To find an isosceles triangle whose area and perimeter are given; which admits of two solutions.

DETERMINATE *Section*, the name of a Tract, or General Problem, written by the ancient geometrician Apollonius. None of this work has come down to us, excepting some extracts and an account of it by Pappus, in the preface to the 7th book of his Mathematical Collections. He there says that the general problem was, " To cut an infinite right line in one point so, that, of the segments contained between the point of section sought, and given points in the said line, either the square on one of them, or the rectangle contained by two of them, may have a given ratio, either to the rectangle contained by one of them and a given line, or to the rectangle contained by two of them "

Pappus farther informs us, that this Tract of Apollonius was divided into two books; that the first book contained 6 problems, and the second 3; that the 6 problems of the first book contained 16 epitagmas, or cases, respecting the dispositions of the points; and the second book 9. Farther, that of the epitagmas of the 6 problems of the first book, 4 were maxima, and one a minimum: that the maxima are at the 2d epitagma of the 2d problem, at the 3d of the 4th, the 3d of the 5th, and the 3d of the 6th; but that the minimum was at the 3d epitagma of the 3d problem. Also, that the second book contained three determinations; of which the 3d epitagma of the 1st problem, and the 3d of the 2d were minima, and the 3d of the 3d a maximum. Moreover, that the first book had 27 lemmas, and the second book 24; and lastly, that both books contained 83 theorems.

From such account of the contents of this Tract, and the lemmas also given by Pappus, several persons have attempted to restore, or recompose what they thought might be nearly the form of Apollonius's tract, or the subject of each problem, case, determination, &c; among whom are, Snellius, an eminent Dutch mathematician of the last century; a translation of whose work was published in English by Mr. John Lawson, in 1772, together with a new restoration of the whole work by his friend Mr. William Wales.

DEW, a thin light insensible mist, or rain, ascending with a slow motion, and falling while the sun is below the horizon.

To us it appears to differ from rain, as less from more: Its origin and matter are doubtless from the vapours and exhalations that rise from the earth and water. See EXHALATION. Some define it a vapour liquefied, and let fall in drops. M. Huet, in one of his letters, shews that dew does not fall, but rises; and others have adopted the same opinion.

M. du Fay made several experiments, first with glasses, then with pieces of cloth stretched horizontally at different heights; and he found that the lower bodies, with their under surfaces, were wetted before those that were placed higher, or their upper surfaces. And Du Fay and Muschenbroek both found, that different substances, and even different colours, receive the dew differently, and some little or not at all.

From the principles laid down under the article EVAPORATION, the several phenomena of dews are easily accounted for. Such as, for instance, that dews are *more*

more copious in the spring, than in the other seasons of the year; there being then a greater stock of vapour in readiness, than at other times, by reason of the small expence of it in the winter's cold and frost. Hence it is too, that Egypt, and some other hot countries, abound with dews throughout all the heats of summer; for the air there being too hot to constipate the vapours in the day-time, they never gather into clouds; and hence they have no rain: but in climates that are excessively hot, the nights are remarkably cold; so that the vapours raised after sun-set, are readily condensed into dews.

It is natural to conclude, from the different substances which are combined with dew, that it must be either salutary or injurious, both to plants and animals.

It is not easy to ascertain the quantity of dew that rises every night, or in the whole year, because of the winds which disperse it, the rains which carry it down, and other inconveniences: but it is known that it rises in greater abundance after rain than after dry weather, and in warm countries than in cold ones. There are some places in which dew is observed only to ascend, and not to fall; and others again in which it is carried upwards in greater plenty than downwards, being dispersed by the winds.

Dr. Hales made some experiments, to determine the quantity of dew that falls in the night. For this purpose, on the 15th of August, at 7 in the evening, he filled two glazed earthen pans with moist earth; the dimensions of the pans being, 3 inches deep, and 12 inches diameter: and he observes, that the moister the earth, the more dew falls on it in a night; and that more than a double quantity of dew falls on a surface of water, than on an equal surface of moist earth. These pans increased in weight by the night's dew, 180 grains; and decreased in weight by the evaporation of the day, 1 oz 282 grs: so that 540 grains more are evaporated from the earth every 24 hours in summer, than the dew that falls in the night; i. e. in 21 days near 26 ounces from a circular area of a foot diameter. Now if 180 grains of dew, falling in one night on such an area, which is equal to 113 square inches, be equally spread on the surface, its depth will be the 159th part of an inch. He likewise found that the depth of dew in a winter's night was the 90th part of an inch. If therefore we allow 159 nights for the extent of the summer's dew, it will in that time amount to one inch in depth; and reckoning the remaining 206 nights for the extent of the winter's dew, it will produce 2·28 inches depth; and the dew of the whole year will amount to 3·28 inches depth. But the quantity which evaporated in a fair summer's day from the same surface, being 1 oz and 282 grs, gives the 40th part of an inch deep for evaporation, which is 4 times as much as fell at night. Dr. Hales observes that the evaporation of a winter's day is nearly the same as in a summer's day; the earth's greater moisture in winter compensating for the sun's greater heat in summer. Hales's Vegetable Statics, vol. 1, pa. 52 of 4th edit. See EVAPORATION.

Signor Beccaria made several experiments to demonstrate the existence of the electricity that is produced by dew. He observes in general, that such electricity took place in clear and dry weather, during which no strong wind prevailed; and that it depends on the quantity of the dew, as the electricity of the rain depends on the quantity of the rain. He sometimes found that it began before sun-set; at other times not till 11 o'clock at night. *Artificial Electricity*, Appendix, letter 3.

DE WIT (JOHN), the famous Dutch pensionary, was born at Dort, in 1625; where he prosecuted his studies so, diligently, that at 23 years of age, he published *Elementa Curvarum Linearum*, one of the deepest books in mathematics at that time. After taking his degrees, and travelling, he, in 1650, became pensionary of Dort, and distinguished himself very early in the management of public affairs, which soon after raised him to the rank of pensionary of Holland. After rendering the greatest benefits to his country in many important instances, and serving it in several high capacities, with the greatest ability, diligence, and integrity, by some intrigues of the court, it is said, he and his brother were thrown into prison, from whence they were dragged by the mob, and butchered with the most cruel and savage barbarity.

DIACAUSTIC *Curve*, or the *Caustic by Refraction*, is a species of caustic curves, the genesis of which is in the following manner. Imagine an infinite number of rays BA, BM, BD, &c, issuing from the same luminous point B, refracted to or from the perpendicular MC, by the given curve AMD; and so, that CE the sines of the angles of incidence CME, be always to CG the sines of the refracted angles CMG, in a given ratio: then the curve HFN that touches all the refracted rays AH, MF, DN, &c, is called the Diacaustic, or Caustic by Refraction.

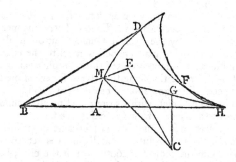

DIACOUSTICS, or DIAPHONICS, the consideration of the properties of sound refracted in passing through different mediums; that is, out of a denser into a more subtile, or out of a more subtile into a denser medium. See SOUND.

DIADROME, a term sometimes used for the vibration, motion, or swing of a pendulum.

DIAGONAL, is a right line drawn across a figure, from one angle to another; and is sometimes called a diameter. It is used chiefly in quadrilateral figures, viz, in parallelograms and trapeziums.

1. Every diagonal, as AC, divides a parallelogram into two equal parts or triangles ABC, ADC.

2. Two diagonals, AC, BD, drawn in a parallelogram, do mutually bisect each other; as in the point E,

3. Any

3. Any line, as FG, drawn through the middle of the diagonal of a parallelogram, is bisected by it at the point E; and it divides the parallelogram into two equal parts, BFGC and AFGD.

4. The diagonal of a square is incommensurable with its side.

5. In any parallelogram, the sum of the squares of the four sides is equal to the sum of the squares of the two diagonals.

6. In any trapezium, the sum of the squares of the four sides is equal to the sum of the squares of the two diagonals together with 4 times the square of the distance between the middle points of the diagonals.

7. In any trapezium, the sum of the squares of the two diagonals is double the sum of the squares of two lines bisecting the two pairs of opposite sides.

8. In any quadrilateral inscribed in a circle, the rectangle of the two diagonals is equal to the sum of the two rectangles under the two pairs of opposite sides.

DIAGONAL *Scale*. See SCALES.

DIAGRAM, is a scheme for the explanation or demonstration of any figure, or of its properties.

DIAL, or SUN-DIAL, an instrument for measuring time by means of the sun's shadow. Or, it is a draught or description of certain lines on the surface of a body, so that the shadow of a style, or ray of the sun through a hole, should touch certain marks at certain hours.

Sun-Dials are doubtless of great antiquity. But the first upon record is, it seems, the dial of Ahaz, who began to reign 400 years before Alexander, and within 12 years of the Building of Rome: it is mentioned in Isaiah, chap. 38, ver. 8.

Several of the antients are spoken of, as makers of dials; as Anaximenes Milesius, Thales. Vitruvius mentions one made by the ancient Chaldee historian Berosus, on a reclining plane, almost parallel to the equator. Aristarchus Samius invented the hemispherical dial. And there were at the same time some spherical ones, with a needle for a gnomon. The discus of Aristarchus was an horizontal dial, with its rim raised up all around, to prevent the shadow from stretching too far.

It was late before the Romans became acquainted with dials. The first sun-dial at Rome was set up by Papyrius Cursor, about the 460th year of the city; before which time, Pliny says there is no mention of any account of time but by the sun's rising and setting: the first dial was set up near the temple of Quirinus; but being inaccurate, about 30 years after, another was brought out of Sicily by the consul M. Valerius Messala, which he placed on a pillar near the Rostrum; but neither did this shew time truly, because not made for that latitude; and, after using it 99 years, Martius Philippus set up another more exact.

The diversity of sun-dials arises from the different situation of the planes, and from the different figure of the surfaces upon which they are described; whence they become denominated *equinoctial, horizontal, vertical, polar, direct, erect, declining, inclining, reclining, cylindrical,* &c. For the general principles of their construction, see DIALLING.

Dials are sometimes distinguished into *primary* and *secondary*.

Primary DIALS are such as are drawn either on the plane of the horizon, and thence called *horizontal* dials; or perpendicular to it, and called *vertical* dials; or else drawn on the polar and equinoctial planes, though neither horizontal nor vertical. And

Secondary DIALS are all those that are drawn on the planes of other circles, beside those last mentioned; or those which either decline, incline, recline, or deincline.

Each of these again is divided into several others, as follow:

Equinoctial DIAL, is that which is described on an equinoctial plane, or one parallel to it.

Horizontal DIAL, is described on an horizontal plane, or a plane parallel to the horizon.—This dial shews the hours from sun-rise to sun-set.

South DIAL, or an *Erect, direct South Dial*, is that described on the surface of the prime vertical circle looking towards the south.—This dial shews the time from 6 in the morning till 6 at night.

North DIAL, or an *Erect, direct North Dial*, is that which is described on the surface of the prime vertical looking northward. This dial only shews the hours before 6 in the morning, and after 6 in the evening.

East DIAL, or *Erect, direct East Dial*, is that drawn on the plane of the meridian, looking to the east.—This can only shew the hours till 12 o'clock.

West Dial, or *Erect, direct West Dial*, is that described on the western side of the meridian.—This can only shew the hours after noon. Consequently this, and the last preceding one, will shew all the hours of the day between them.

Polar DIAL, is that which is described on a plane passing through the poles of the world, and the east and west points of the horizon. It is of two kinds; the first looking up towards the zenith, and called the *upper*; the latter, down towards the nadir, called the *lower*. The polar dial therefore is inclined to the horizon in an angle equal to the elevation of the pole.—The upper polar dial shews the hours from 6 in the morning till 6 at night, and the lower one shews the hours before 6 in the morning, and after 6 in the evening, viz, from sun-rise and till sun-set.

Declining DIALS, are erect or vertical dials which decline from any of the cardinal points; or they are such as cut either the plane of the prime vertical, or of the horizon, at oblique angles.

Declining dials are of very frequent use; as the walls of houses, on which dials are mostly drawn, commonly deviate from the cardinal points.

Of declining dials there are several kinds, which are denominated from the cardinal points which they are nearest to; as decliners from the south, and from the north, and even from the zenith.

Inclined DIALS, are such as are drawn on planes not erect, but inclining, or leaning forward towards the south, or southern side of the horizon, in an angle, either greater or less than the equinoctial plane.

Reclining DIALS, are those drawn on planes not erect, but reclined, or leaning backwards from the zenith towards the north, in an angle greater or less than the polar plane.

Deinclined DIALS, are such as both decline and incline, or recline.—These last three sorts of dials are very rare.

DIALS

DIALS *without Centres*, are those whose hour lines converge so slowly, that the centre, or point of their concourse, cannot be expressed on the given plane.

Quadrantal DIAL. See *Horodictical* QUADRANT.

Reflecting DIAL. See REFLECTING *Dial*.

Cylindrical DIAL, is one drawn on the curve surface of a cylinder. This may first be drawn on a paper plane, and then pasted round a cylinder of wood, &c. It will shew the time of the day, the sun's place in the ecliptic, and his altitude at any time of observation.

There are also *Portable* DIALS, or *on a Card*, and *Universal* DIALS on a *Plain Cross*, &c.

Refracted DIALS, are such as shew the hour by means of some refracting transparent fluid.

Ring DIAL, is a small portable dial, consisting of a brass ring or rim, about 2 inches in diameter, and one-third of an inch in breadth. In a point of this rim there is a hole, through which the sun beams pass, and form a bright speck in the concavity of the opposite semi-circle, which gives the hour of the day in the divisions marked within it.

When the hole is fixed, the dial only shews true about the time of the equinox. But to have it perform throughout the whole year, the hole is made moveable, the signs of the zodiac, or the days of the month, being marked on the convex side of the ring; hence, in using it, the moveable hole is set to the day of the month, or the degree of the zodiac the sun is in; then suspending the dial by the little ring, turn it towards the sun, and his rays through the hole will shew the hour on the divisions within side.

Universal, or *Astronomical Ring* DIAL, is a ring dial which shews the hour of the day in any part of the earth; whereas the former is confined to a certain latitude. Its figure see represented below.

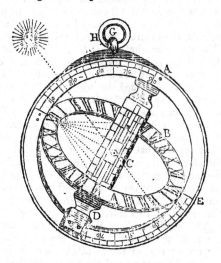

It consists of two rings or flat circles, from 2 to 6 inches in diameter, and of a proportionable breadth &c. The outward ring A represents the meridian of any place you are at, and contains two divisions of 90 degrees each, diametrically opposite to one another, the one serving from the equator to the north pole, the

other to the south pole. The inner ring represents the equator, and turns exactly within the outer, by means of two pivots in each ring at the hour of 12.

Across the two circles goes a thin reglet or bridge, with a cursor C, sliding along the middle of the bridge, and having a small hole for the sun to shine through. The middle of this bridge is conceived as the axis of the world, and the extremities as the poles: on the one side are drawn the signs of the zodiac, and on the other the days of the month. On the edge of the meridian slides a piece, to which is fitted a small ring to suspend the instrument by.

In this dial, the divisions on the axis are the tangents of the angles of the sun's declination, adapted to the semi-diameter of the equator as radius, and placed on either side of the centre: but instead of laying them down from a line of tangents, a scale of equal parts may be made, of which 1000 shall answer exactly to the length of the semi-axis, from the centre to the inside of the equinoctial ring; and then 434 of these parts may be laid down from the centre towards each end, which will limit all divisions on the axis, because 434 is the natural tangent of 23° 28′. And thus, by a nonius fixed to the sliding piece, and taking the sun's declination from an ephemeris, and the tangent of that declination from the table of natural tangents, the slider might be always set true within 2 minutes of a degree. This scale of 434 equal parts might be placed right against the 23° 28′ of the sun's declination, on the axis, instead of the sun's place, which is there of little use. For then the slider might be set in the usual way, to the day of the month, for common use; or to the natural tangent of the declination, when great accuracy is required.

To use this Dial: Place the line *a* (on the middle of the sliding piece) over the degree of latitude of the place, as for instance 51½ degrees for London: put the line which crosses the hole of the cursor to the degree of the sign, or day of the month. Open the instrument so as that the two rings be at right angles to each other, and suspend it by the ring H, that the axis of the dial, represented by the middle of the bridge, may be parallel to the axis of the world. Then turn the flat side of the bridge towards the sun, so that his rays, striking through the small hole in the middle of the cursor, may fall exactly on a line drawn round the middle of the concave surface of the inner ring; in which case the bright spot shews the hour of the day in the said concave surface of the ring.

Nocturnal or *Night*-DIAL, is that which shews the hour of the night, by the light, or shadow projected from the moon or stars.

Lunar or Moon Dials may be either purposely described and adapted to the moon's motion; or the hour may be found on a sun-dial by the moon shining upon it, thus: Observe the hour which the shadow of the index points at by moon light; find the days of the moon's age in the calendar, and take 3-4ths of that number, for the hours to be added to the hour shewn by the shadow, to give the hour of the night. The reason of which is, that the moon comes to the same horary circle later than the sun by about three quarters of an hour every day; and at the time of new moon the solar and lunar hour coincide.

DIAL

DIAL *Planes*, are the plane superficies upon which the hour lines of dials are drawn.

Tide DIAL. See TIDE *Dial*.

DIALLING, the art of drawing sun, moon, and star-dials on any sort of surface, whether plane or curved.

Dialling is wholly founded on the first motion of the heavenly bodies, and chiefly the sun; or rather on the diurnal rotation of the earth: so that the elements of spherics, and spherical trigonometry, should be understood, before a person advances to the doctrine of dialling.

The principles of dialling may be aptly deduced from, and illustrated by, the phenomena of a hollow or transparent sphere, as of glass. Thus, suppose *aPcp* to re-

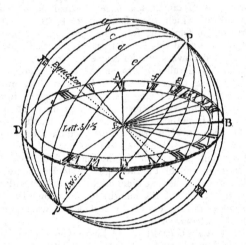

present the earth as transparent; and its equator as divided into 24 equal parts by so many meridian semicircles *a, b, c, d, e,* &c, one of which is the geographical meridian of any given place, as London, which it is supposed is at the point *a*; and if the hour of 12 were marked at the equator, both upon that meridian and the opposite one, and all the rest of the hours in order on the other meridians, those meridians would be the hour circles of London: because, as the sun appears to move round the earth, which is in the centre of the visible heavens, in 24 hours, he will pass from one meridian to another in an hour. Then, if the sphere had an opake axis, as PE*p*, terminating in the poles P and *p*, the shadow of the axis, which is in the same plane with the sun and with each meridian, would fall upon every particular meridian and hour, when the sun came to the plane of the opposite meridian, and would consequently shew the time at London, and at all other places on the same meridian. If this sphere were cut through the middle by a solid plane ABCD in the rational horizon of London, one half of the axis EP would be above the plane, and the other half below it; and if straight lines were drawn from the centre of the plane to those points where its circumference is cut by the hour circles of the sphere, those lines would be the hour lines of an horizontal dial for London; for the shadow of the axis would fall upon each particular hour line of the dial, when it fell upon the like hour circle of the sphere.

If the plane which cuts the sphere be upright, as AFCG, touching the given place, for ex. London, at F,

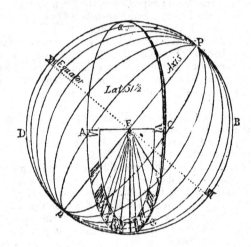

and directly facing the meridian of London, it will then become the plane of an erect direct south dial; and if right lines be drawn from its centre E, to those points of its circumference where the hour circles of the sphere cut it, these will be the hour lines of a vertical or direct south dial for London, to which the hours are to be set in the figure, contrary to those on an horizontal dial; and the lower half E*p* of the axis will cast a shadow on the hour of the day in this dial, at the same time that it would fall upon the like hour circle of the sphere, if the dial plane was not in the way.

If the plane, still facing the meridian, be made to incline, or recline, any number of degrees, the hour circles of the sphere will still cut the edge of the plane in those points to which the hour lines must be drawn straight from the centre; and the axis of the sphere will cast a shadow on these lines at the respective hours. The like will still hold, if the plane be made to decline by any number of degrees from the meridian towards the east or west; provided the declination be less than 90 degrees, or the reclination be less than the co-latitude of the place; and the axis of the sphere will be the gnomon: otherwise, the axis will have no elevation above the plane of the dial, and cannot be a gnomon.

Thus it appears that the plane of every dial represents the plane of some great circle on the earth, and the gnomon the earth's axis; the vertex of a right gnomon the centre of the earth or visible heavens; and the plane of the dial is just as far from this centre as from the vertex of this stile. The earth itself, compared with its distance from the sun, is considered as a point; and therefore, if a small sphere of glass be placed upon any part of the earth's surface, so that its axis be parallel to the axis of the earth, and the sphere have such lines upon it, and such planes within it, as above described; it will shew the hours of the day as truly as if it were placed at the earth's centre, and the shell of the earth were as transparent as glass. Ferguson, lect. 10.

The principal writers on Dials, and Dialling, are the following: Vitruvius, in his Architecture, cap. 4

and

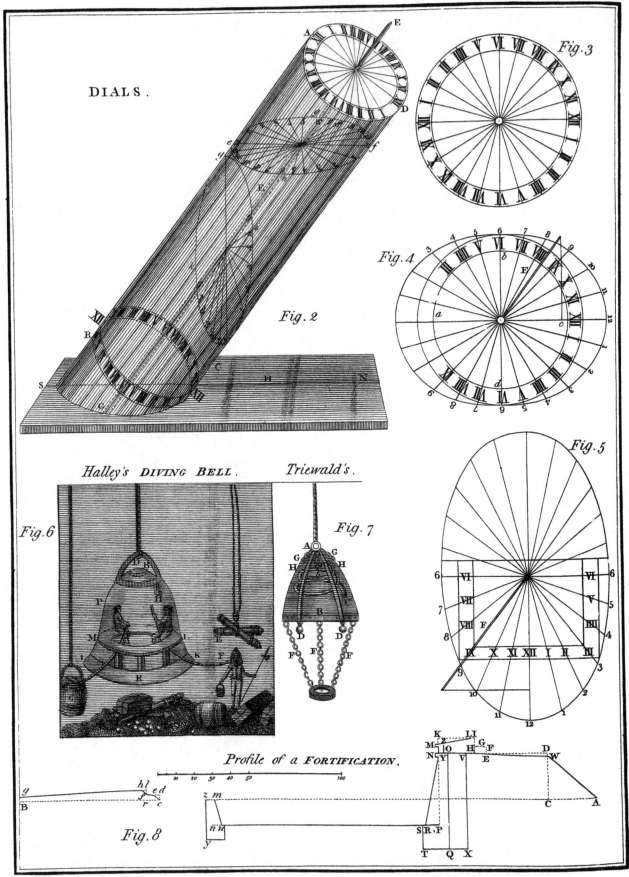

Plate VII.

DIALS.

Fig. 3

Fig. 4

Fig. 2

Fig. 5

Halley's DIVING BELL.   Triewald's.

Fig. 6   Fig. 7

Profile of a FORTIFICATION.

Fig. 8

and 7, lib. 9: Sebaſtian Munſter, his Horolographia: John Dryander de Horologiorum varia Compoſitione: Conrade Geſner's Pandectæ: Andrew Schoner's Gnomonicæ: Fred. Commandine de Horologiorum Deſcriptione: Joan. Bapt. Benedictus de Gnomonum Umbrarumque Solarium Uſu: Joannes Georgius Schomberg, Exegeſis Fundamentorum Gnomonicorum: Solomon de Caus, Traité des Horologes Solaires: Joan. Bapt. Trolta, Praxis Horologiorum: Deſargues, Maniere Univerſelle pour poſer l'Eſſieu & placer les Heures & autres Choſes aux Cadrans Solaires: Ath. Kircher, Ars magna Lucis & Umbræ: Hallum, Explicatio Horologii in Horto Regio Londini: Tractatus Horologiorum Joannis Mark: Clavius, Gnomonices de Horologiis; in which he demonſtrates both the theory and the operations after the rigid manner of the ancient mathematicians: Dechales, Ozanam, and Schottus, gave much eaſier treatiſes on this ſubject; as did alſo Wolfius in his Elementa: M. Picard gave a new method of making large dials, by calculating the hour lines; and M. De la Hire, in his Dialling, printed in 1683, gave a geometrical method of drawing hour lines from certain points, determined by obſervation. Everhard Walper, in 1625, publiſhed his Dialling, in which he lays down a method of drawing the primary dials on a very eaſy foundation; and the ſame foundation is alſo deſcribed at length by Sebaſtian Munſter, in his Rudimenta Mathematica, publiſhed in 1651. In 1672, Sturmius publiſhed a new edition of Walper's Dialling, with the addition of a whole ſecond part, concerning inclining and declining dials, &c. In 1708, the ſame work, with Sturmius's additions, was re-publiſhed, with the addition of a 4th part, containing Picard's and De la Hire's methods of drawing large dials, which makes much the beſt and fulleſt book on the ſubject. Peterſon, Michael, and Muller, have each written on Dialling, in the German language: Coetſius, in his Horologiographia Plana, printed in 1689: Gauppen, in his Gnomonica Mechanica: Leybourn, in his Dialling: Bion, in his Uſe of Mathematical Inſtruments: Wells, in his Art of Shadows. There is alſo a treatiſe by M. Deparceux, 1740. Mr. Ferguſon has alſo written on the ſame ſubject in his Lectures on Mechanics; beſides Emerſon, in his Dialling; and Mr. W. Jones, in his Inſtrumental Dialling.

Univerſal DIALLING *Cylinder*, is repreſented in fig. 2, plate vii, where ABCD is a glaſs cylindrical tube, cloſed at both ends with braſs plates, in the centres of which a wire or axis EFG is fixed. The tube is either fixed to an horizontal board H, ſo that its axis may make an angle with the board equal to that which the earth's axis makes with the horizon of any given place, and be parallel to the axis of the world; or it may be made to move on a joint, and elevated for any particular latitude. The twenty-four hour lines are drawn with a diamond on the outſide of the glaſs, equidiſtant from each other, and parallel to the axis. The XII next B ſtands for midnight, and the XII next the board H for noon. When the axis of this inſtrument is elevated according to the latitude, and the board ſet level, with the line HN in the plane of the meridian, and the end towards the north; the axis EFG will ſerve as a gnomon or ſtile, and caſt a ſhadow on the hour of the day among the pa-

rallel hour lines, when the ſun ſhines on the inſtrument. As the plate AD at the top is parallel to the equator, and the axis EFG perpendicular to it, right lines drawn from the centre to the extremities of the parallels will be the hour lines of an equinoctial dial, and the axis will be the ſtile. An horizontal plate *ef* put down into the tube, with lines drawn from the centre to the ſeveral parallels, cutting its edge, will be an horizontal dial for the given latitude; and a vertical plate *gc*, fronting the meridian, and touching the tube with its edge, with lines drawn from its centre to the parallels, will be a vertical ſouth dial: the axis of the inſtrument ſerving in both caſes for the ſtile of the dial: and if a plate be placed within the tube, ſo as to decline, incline, or recline, by any given number of degrees, and lines be drawn, as above, a declining, inclining, or reclining dial will be formed for the given latitude. If the axis with the ſeveral plates fixed to it be drawn out of the tube, and ſet up in ſunſhine in the ſame poſition as they were in the tube, AD will be an equinoctial dial, *ef* an horizontal dial, and *gc* a vertical ſouth dial; and the time of the day will be ſhewn by the axis EFG. If the cylinder were wood, inſtead of glaſs, and the parallel lines drawn upon it in the ſame manner, it would ſerve to facilitate the operation of making theſe ſeveral dials. The upper plate with lines drawn to the ſeveral interſections of the parallels, which appears obliquely in fig. 2, would be an equinoctial dial as in fig. 3, and the axis perpendicular to it be its ſtile. An horizontal dial for the latitude of the elevation of the axis might be made, by drawing out the axis and cutting the cylinder, as at *efgh*, parallel to the horizontal board H; the ſection would be elliptic as in fig. 4. A circle might be deſcribed on the centre, and lines drawn to the diviſions of the ellipſe would be the hour lines; and the wire put in its place again, as E, would be the ſtile. If this cylinder were cut by a plane perpendicular to the horizontal board H, or to the line SHN, beginning at *g*, the plane of the ſection would be elliptical as in fig. 5, and lines drawn to the points of interſection of the parallels on its edge would be the hour lines of a vertical direct ſouth dial, which might be made of any ſhape, either circular or ſquare, and F the axis of the cylinder would be its ſtile. Thus alſo inclining, declining, or reclining dials might be eaſily conſtructed, for any given latitude. Ferguſon, ubi ſupra.

DIALLING *Globe*, is an inſtrument made of braſs, or wood, with a plane fitted to the horizon, and an index; particularly contrived to draw all ſorts of dials, and to give a clear exhibition of the principles of that art.

DIALLING *Lines*, or *Scales*, are graduated lines, placed on rules or the edges of quadrants, and other inſtruments, to expedite the conſtruction of dials. The principal of theſe lines are, 1. A ſcale of ſix hours, which is only a double tangent, or two lines of tangents each of 45 degrees, joined together in the middle, and equal to the whole line of ſines, with the declination ſet againſt the meridian altitudes in the latitude of London, ſuppoſe, or any place for which it is made: the radius of which line of ſines is equal to the dialling ſcale of ſix hours. 2. A line of latitudes, which is fitted to the hour ſcale, and is made by this

canon.

caron: as the radius is to the chord of 90 degrees; so are the tangents of each respective degree of the line of latitudes, to the tangents of other arches: and then the natural sines of those arches are the numbers, which, taken from a diagonal scale of equal parts, will graduate the divisions of the line of latitude to any radius. The line of hours and latitudes is generally for pricking down all dials with centres. For the method of constructing these scales, see SCALE.

DIALLING *Sphere*, is an instrument made of brass, with several semicircles sliding over one another, on a moving horizon, to demonstrate the nature of the doctrine of spherical triangles, and to give a true idea of the drawing of dials on all manner of planes.

DIAMETER, *of a circle*, is a right line passing through the centre, and terminated at the circumference on both sides.

The diameter divides the circumference, and the area of the circle, into two equal parts. And half the diameter, or the semi-diameter, is called the radius.

For the proportion between the diameter and the circumference of a circle, see CIRCLE and CIRCUMFERENCE.

DIAMETER *of a Conic Section*, or *Transverse Diameter*, is a right-line passing through the centre of the section, or the middle of the axis.——The diameter bisects all ordinates, or lines drawn parallel to the tangent at its vertex. See CONIC *Sections*.

Conjugate DIAMETER, is a diameter, in Conic Sections, parallel to the ordinates of another diameter, called the transverse; or parallel to the tangent at the vertex of this other.

DIAMETER, *of any Curve*, is a right line which divides two other parallel right lines, in such manner that, in each of them, all the segments or ordinates on one side, between the diameter and different points of the curve, are equal to all those on the other side. This is Newton's sense of a Diameter.

But, according to some, a diameter is that line, whether right or curved, which bisects all the parallels drawn from one point to another of a curve. So that in this way every curve will have a diameter; and hence the curves of the 2d order, have, all of them, either a right-lined diameter, or else the curves of some one of the conic sections for diameters. And many geometrical curves of the higher orders, may also have for diameters, curves of more inferior orders.

DIAMETER *of Gravity*, is a right line passing through the centre of gravity.

DIAMETER *in Astronomy*. The diameters of the heavenly bodies are either *apparent*, i. e. such as they appear to the eye; or *real*, i. e. such as they are in themselves.

The apparent diameters are best measured with a micrometer, and are estimated by the measure of the angle they subtend at the eye. These are different in different circumstances and parts of the orbits, or according to the various distances of the luminary; being in the inverse ratio of the distance.

The sun's vertical diameter is found by taking the height of the upper and lower edge of his disk, when he is in the meridian, or near it; correcting the altitude of each edge on account of refraction and parallax; then the difference between the true altitudes of the two, is the true apparent diameter sought. Or the apparent diameter may be determined by observing, with a good clock, the time which the sun's disc takes in passing over the meridian: and here, when the sun is in or near the equator, the following proportion may be used; viz, as the time between the sun's leaving the meridian and returning to it again, is to 360 degrees, so is the time of the sun's passing over the meridian, to the number of minutes and seconds of a degree contained in his apparent diameter: but when the sun is in a parallel at some distance from the equator, his diameter measures a greater number of minutes and seconds in that parallel than it would do in a great circle, and takes up proportionally more time in passing over the meridian; in which case say, as radius is to the cosine of the sun's declination, so is the time of the sun's passing the meridian reduced to minutes and seconds of a degree, to the arc of a great circle which measures the sun's apparent horizontal diameter. See TRANSIT.

The sun's apparent diameter may also be taken by the projection of his image in a dark room.

There are several ways of finding the apparent diameters of the planets: but the most certain method is that with the micrometer.

The following is a table of the apparent diameters of the sun and planets, in different circumstances, and as determined by different astronomers.

TABLE OF APPARENT DIAMETERS.								
1. *Of the Sun, according to*	Greatest			Mean			Least	
	′	″	‴	′	″	‴	′	″	‴
Aristarchus and Archimedes	30	0	0	30	0	0	30	0	0
Ptolomy	33	20	0	32	18	0	31	20	0
Albategnius	33	40	0	32	28	0	31	20	0
Regiomontanus	34	0	0	32	27	0	31	0	0
Copernicus	33	54	0	32	44	0	31	49	0
Tycho	32	0	0	31	0	0	30	0	0
Kepler	31	4	0	30	30	0	30	0	0
Riccioli	32	8	0	31	40	0	31	0	0
J. D. Cassini	32	46	0	32	13	0	31	40	0
Gascoigne	32	50	0	-	-	-	31	40	0
Flamsteed	32	48	0	-	-	-	31	30	0
Mouton	32	32	0	-	-	-	30	29	0
De la Hire	32	44	0	32	11	0	31	38	0
Louville	32	37	7	32	4	36	31	32	50
M. Cassini, jun.	32	37	30	32	5	0	31	32	30
Monnier	-	-	-	32	5	0	-	-	-
Short	32	33	0	-	-	-	31	28	0
2. *Of the Moon.*								
Ptolomy	35	30	0	-	-	-	31	20	0
Tycho, in Conjunc.	28	48	0	-	-	-	25	36	0
Tycho, in Opposit.	36	0	0	-	-	-	32	0	0
Kepler	32	44	0	-	-	-	30	0	0
De La Hire	33	30	0	-	-	-	29	30	0
Newton, in Syzygy	-	-	-	31	30	0	-	-	-
Newton, in Quadrat.	-	-	-	31	3	0	-	-	-
Mouton, Full, in Perigee	33	29	0	-	-	-	-	-	-
Monnier, in Syzygy	-	-	-	31	30	0	-	-	-
Monnier, in Quadrat	-	-	-	31	0	0	-	-	-

7

3. Of Mercury.	Greatest ' " '''	Mean ' " '''	Least ' " '''
Albategnius	- - -	2 5 10	- - -
Alfraganus	- - -	1 15 12	- - -
Tycho	3 57 0	2 10 0	1 29 0
Hevelius	0 11 0	0 6 0	0 4 0
Hortensius	0 28 0	0 19 0	0 10 0
Riccioli	0 25 12	0 13 48	0 9 20
Bradley	0 10 45	- - -	- - -
Monnier	- - -	- - -	0 10 0
4. Of Venus.			
Albategnius	3 8 0	- - -	- - -
Alfraganus	1 34 0	- - -	- - -
Tycho	4 40 0	3 15 0	1 52 0
Hevelius	1 5 0	0 16 0	0 9 0
Hortensius	1 40 0	0 53 0	0 15 20
Kepler	7 6 0	- - -	- - -
Riccioli	4 8 0	1 4 12	0 33 30
Huygens	- - -	- - -	1 25 0
Flamsteed	1 12 0	- - -	- - -
Horrox	1 18 30	- - -	- - -
Crabtree	1 9 0	- - -	- - -
Monnier	- - -	1 17 0	- - -
Transit of 1761	- - -	0 58 0	- - -
Transit of 1769	- - -	0 59 0	- - -
5. Of Mars.			
Albateg. and Alfrag.	- - -	1 34 0	- - -
Tycho	2 46 0	1 40 0	0 57 0
Hevelius	0 20 0	0 5 0	0 2 0
Hortensius	1 4 0	0 36 0	0 9 0
Kepler	6 30 0	- - -	- - -
Riccioli	1 32 0	0 22 0	0 10 6
Huygens	- - -	- - -	0 30 0
Flamsteed	0 33 0	- - -	- - -
Monnier	- - -	- - -	0 26 0
Herschel, Polar Diam.	- - -	- - -	0 21 29
Herschel, Equat. Diam.	- - -	- - -	0 22 25
6. Of Jupiter.			
Albateg. and Alfrag.	- - -	2 36 40	- - -
Tycho	3 59 0	2 45 0	2 14 0
Hevelius	0 24 0	0 18 0	0 14 0
Hortensius	1 1 40	0 50 0	0 38 30
Kepler	0 50 0	- - -	- - -
Riccioli	1 8 46	0 49 46	0 38 18
Huygens	- - -	- - -	1 4 0
Flamsteed	0 54 0	- - -	- - -
Newton, from Pound's Obf.	- - -	0 37 15	- - -
Monnier	- - -	0 37 0	- - -
7. Of Saturn.			
Albateg and Alfrag.	- - -	1 44 28	- - -
Tycho	2 12 0	1 50 0	1 34 0
Hevelius	0 19 0	0 16 0	0 14 0
Hortensius	0 42 40	0 37 0	0 31 0
Kepler	0 30 0	- - -	- - -

Of Saturn	Greatest ' "	Mean ' " '''	Least ' " '''
Riccioli	1 12 0	0 57 0	0 46 0
Huygens	- - -	- - -	0 30 0
Flamsteed	0 25 0	- - -	- - -
Newton, from Pound's Obferv.	- - -	0 16 0	- - -
Monnier	- - -	0 16 0	- - -
Huygens, ♄'s Ring	- - -	1 4 0	1 8 0
Newton, from Pound's Obf.	- - -	0 40 0	- - -
Monnier	- - -	0 42 0	- - -
8. New Planet.			
Herschel	- - -	0 3 54	- - -

The Mean Apparent diameters of the planets, as seen from the fun, are as follow :

Mercury,	Venus,	Earth,	Moon,	Mars,	Jup.	Sat.	Herfch
20"	30"	17"	6"	11"	37"	16"	4"

For the true diameters of the fun and planets, and their proportions to each other, fee Planets, Semidiameter, and Solar System.

DIAMETER *of a Column*, is its thicknefs juft above the bafe. From this the module is taken, which meafures all the other parts of the column.

DIAMETER *of the Diminution*, is that taken at the top of the fhaft.

DIAMETER *of the Swelling*, is that taken at the height of one third from the bafe.

DIAPASON, a mufical interval, otherwife called an octave, or eighth: fo called becaufe it contains all the poffible diverfities of found.

If the lengths of two ftrings be to each other as 1 to 2, the tenfions being equal, their tones will produce an octave.

DIAPENTE, is the perfect fifth or fecond of the concords, making an octave with the Diateffaron. The lengths of the chords are as 3 to 2.

DIAPHONOUS Body or Medium, one that is tranflucent, or through which the rays of light eafily pafs ; as water, air, glafs, talc, fine porcelain, &c. See Transparency.

DIAPHONICS, is fometimes ufed for the fcience of refracted found, as it paffes through different mediums.

DIASTYLE, a fort of edifice, in which the pillars ftand at fuch a diftance from one another, that three diameters of their thicknefs are allowed for the intercolumnation.

DIATESSARON, is the perfect fourth ; or a mufical interval, confifting of one greater tone, one leffer, and one greater femitone. The lengths of ftrings to found the diateffaron, are as 3 to 4.

DIATONIC, a term fignifying the ordinary fort of mufic, which proceeds by tones or degrees, both afcending and defcending. It contains or admits only the greater and leffer tone, and the greater femitone.

DIESIS, a divifion of a tone, lefs than a femitone ; or an interval confifting of a leffer, or imperfect femitone. The Diefis is the fmalleft and fofteft change, or

inflexion,

inflexion, of the voice imaginable. It is also called a *feint*, and is expressed by a St. Andrew's Cross, or saltier,

DIFFERENCE, is the excess by which one magnitude or quantity exceeds another. When a less quantity is subtracted from a greater, the remainder is otherwise called difference.

Ascensional DIFFERENCE. See ASCENSIONAL.

DIFFERENCE *of Longitude*, of two places, is an arch of the equator contained between the meridians of those two places, or the measure of the angle formed by their meridians.

DIFFERENTIAL, an indefinitely small quantity, part, or difference. By some, the Differential is considered as infinitely small, or less than any assignable quantity; and also as of the same import as fluxion.

It is called a Differential, or Differential Quantity, because often considered as the difference between two quantities; and as such it is the foundation of the Differential Calculus. Newton used the term *moment* in a like sense, as being the momentary increase or decrease of a variable quantity. M. Leibnitz and others call it also an infinitesimal.

DIFFERENTIAL *of the 1st, 2d, 3d, &c degree.* See *Differentio-*DIFFERENTIAL.

DIFFERENTIAL *Calculus*, or *Method*, is a method of differencing quantities. See DIFFERENTIAL *Method*, CALCULUS, and FLUXIONS.

DIFFERENTIO-DIFFERENTIAL *Calculus*, is a method of differencing differential quantities.

As the sign of a differential is the letter *d* prefixed to the quantity, as *dx* the differential of *x*; so that of a differential of *dx* is *ddx*, and the differential of *ddx* is *dddx*, &c; similar to the fluxions $\dot{x}, \ddot{x}, \dddot{x}$, &c.

Thus we have degrees of differentials. The differential of an ordinary quantity, is a differential of the first order or degree, as *dx*; that of the 2d degree is *ddx*; that of the 3d degree, *dddx*, &c. The rules for differentials, are the very same as those for fluxions. See FLUXIONS.

DIFFERENTIAL, in Logarithms. Kepler calls the logarithms of tangents, *differentiales;* which we usually call *artificial tangents*.

DIFFERENTIAL *Equation*, is an equation involving or containing differential quantities; as the equation $3x^2dx - 2axdx + aydx + axdy = 0$. Some mathematicians, as Stirling, &c, have also applied the term differential equation in another sense, to certain equations defining the nature of series.

DIFFERENTIAL *Method*, a method of finding quantities by means of their successive differences.

This method is of very general use and application, but especially in the construction of tables, and the summation of series, &c. This method was first used, and the rules of it laid down by Briggs, in his Construction of Logarithms and other Numbers, much the same as they were afterwards taught by Cotes, in his Constructio Tabularum per Differentias; as I have shewn in the Introduction to my Logs. pa. 69 & seq. See Briggs's Arithmetica Logarithmica, cap. 12 and 13, and his Trigonometria Britannica.

The method was next treated in another form by Newton in the 5th Lemma of the 3d book of his Principia, and in his Methodus Differentialis, published

by Jones in 1711, with the other tracts of Newton. This author here treats it as a method of describing a curve of the parabolic kind, through any given number of points. He distinguishes two cases of this problem; the first, when the ordinates drawn from the given points to any line given in position, are at equal distances from one another; and the 2d, when these ordinates are not at equal distances. He has given a solution of both cases, at first without demonstration, which was afterwards supplied by himself and others; see his Methodus Differentialis above mentioned; and Stirling's Explanation of the Newtonian Differential Method, in the Philos. Transf. Nº 362; Cotes, De Methodo Differentiali Newtoniana, published with his Harmonia Mensurarum; Herman's Phoronomia; and Le Seur & Jacquier, in their Commentary on Newton's Principia. It may be observed, that the methods there demonstrated by some of these authors extend to the description of any algebraic curve through a given number of points, which Newton, writing to Leibnitz, mentions as a problem of the greatest use.

By this method, some terms of a series being given, and conceived as placed at given intervals, any intermediate term may be found nearly; which therefore gives a method for interpolations. Briggs's Arith. Log. ubi supra; Newton Meth. Differ. prop. 5; Stirling, Methodus Differentialis.

Thus also may any curvilinear figure be squared nearly, having some few of its ordinates. Newton, ibid. prop 6; Cotes De Method. Differ.; Simpson's Mathematical Differt. pa. 115. And thus may mathematical tables be constructed by interpolation: Briggs, ibid. Cotes Canonotechnia.

The successive differences of the ordinates of parabolic curves, becoming ultimately equal, and the intermediate ordinate required, being determined by these differences of the ordinates, is the reason for the name Differential Method.

To be a little more particular.—The first case of Newton's problem amounts to this: A series of numbers, placed at equal intervals, being given, to find any intermediate number of that series, when its interval or distance from the first term of the series is given.—— Subtract each term of the series from the next following term, and call the remainders first differences, then subtract in like manner each of these differences from the next following one, calling these remainders 2d differences; again, subtract each 2d difference from the next following, for the 3d differences; and so on: then if A be the 1st term of the series,

d' the first of the 1st differences,
d'' the first of the 2d differences,
d''' the first of the 3d differences, &c;

and if *x* be the interval or distance between the first term of the series and any term sought, T, that is, let the number of terms from A to T, both included, be $= x + 1$; then will the term sought, T, be $=$

$$A + \frac{x}{1}d' + \frac{x}{1} \cdot \frac{x-1}{2}d'' + \frac{x}{1} \cdot \frac{x-1}{2} \cdot \frac{x-2}{3}d''' \&c.$$

Hence, if the differences of any order become equal, that is, if any of the diffs. d'', d''', &c, become $= 0$, the above series will give a finite expression for T the term sought; it being evident, that the series must terminate when any of the diffs. d'', d''', &c, become $= 0$.

It

It is also evident that the co-efficients $\frac{x}{1}, \frac{x}{1} \cdot \frac{x-1}{2}$, &c. of the differences, are the same as to the terms of the binomial theorem.

For ex. Suppose it were required to find the log. tangent of $5' 1'' 12''' 24''''$, or $5' 1'' \frac{62}{100}$, or $5' 1'' \cdot 2066$ &c.

Take out the log. tangents to several minutes and seconds, and take their first and second differences, as below:

	Tang.	d'	d''
$5'$ $0''$ - - - -	$7 \cdot 1626964$		
5 1 - - - -	$7 \cdot 1641417$	14453	
5 2 - - - -	$7 \cdot 1655821$	14404	-49
5 3 - - - -	$7 \cdot 1670178$	14357	-47

$\}-48$

Here $A = 7 \cdot 1641417$; $x = \frac{62}{100}$; $d' = 14404$; and the mean 2d difference $d'' = -48$. Hence

$$A - - - - - - 7 \cdot 1641417$$
$$xd' - - - - - 2977$$
$$\frac{x}{1} \cdot \frac{x-1}{1} d'' \qquad\qquad 4$$

Theref. the tang. of $5' 1'' 12''' 24''''$ is $7 \cdot 1644398$

Hence may be deduced a method of finding the sums of the terms of such a series, calling its terms A, B, C, D, &c. For, conceive a new series having its 1st term $= 0$, its 2d $= A$, its 3d $= A+B$, its 4th $= A+B+C$, its 5th $= A+B+C+D$, and so on; then it is plain that assigning one term of this series, is finding the sum of all the terms A, B, C, D, &c. Now since these terms are the differences of the sums, 0, A, A+B, A+B+C, &c; and as some of the differences of A, B, C, &c, are $= 0$ by supposition; it follows that some of the differences of the sums will be

$= 0$; and since in the series $A + \frac{x}{1} d' + \frac{x}{1} \cdot \frac{x-1}{2} d''$ &c, by which a term was assigned, A represented the 1st term; d' the 1st of the 1st differences, and x the interval between the first term and the last; we are to write 0 instead of A, A instead of d', d' instead of d'', d'' instead of d''', &c, also $x+1$ instead of x; which being done, the series expressing the sums will be

$0 + \frac{x+1}{1} A + \frac{x+1}{1} \cdot \frac{x}{2} d' + \frac{x+1}{1} \cdot \frac{x}{2} \cdot \frac{x-1}{3} d''$, &c.

Or, if the real number of terms of the lines be called z, that is, if $z = x+1$, or $x = z-1$, the sum of the series will be $Az + \frac{z}{1} \cdot \frac{z-1}{2} d' + \frac{z}{1} \cdot \frac{z-1}{2} \cdot \frac{z-2}{3} d''$ &c. See De Moivre's Doct. of Chances, pa. 59, 60; or his Miscel. Analyt. pa. 153; or Simpson's Essays, pa. 95.

For ex. to find the sum of six terms of the series of squares $1 + 4 + 9 + 16 + 25 + 36$, of the natural numbers.

Terms	d'	d''	d'''
1			
4	3		
9	5	2	0
16	7	2	0
25	9	2	

Here $A = 1$, $d' = 3$, $d'' = 2$, d''' &c $= 0$, and $z = 6$; therefore the sum is $6 + \frac{6}{1} \cdot \frac{5}{2} \cdot 3 + \frac{6}{1} \cdot \frac{5}{2} \cdot \frac{4}{3} \cdot 2$ $= 6 + 45 + 40 = 91$ the sum required, viz. of $1 + 4 + 9 + 16 + 25 + 36$.

A variety of examples may be seen in the places above cited, or in Stirling's Methodus Differentialis, &c.

As to the Differential method, it may be observed, that though Newton and some others have treated it as a method of describing an algebraic curve, at least of the parabolic kind, through any number of given points; yet the consideration of curves is not at all essential to it, though it may help the imagination. The description of a parabolic curve through given points, is the same problem as the finding of quantities from their given differences, which may always be done by Algebra, by the resolution of simple equations. See Stirling's Method. Differ. pa. 97. This ingenious author has treated very fully of the differential method, and shewn its use in the solution of some very difficult problems. See also SERIES.

DIFFERENTIAL *Scale*, in Algebra, is used for the scale of relation subtracted from unity. See *Recurring* SERIES.

DIFFRACTION, a term first used by Grimaldi, to denote that property of the rays of light, which others have called Inflection; the discovery of which is attributed by some to Grimaldi, and by others to Dr. Hook.

DIGBY (*Sir* KENELM), a famous English philosopher, was born at Gothurst in Buckinghamshire, 1603. He was descended of an ancient family: his great grandfather, with six of his brothers, fought valiantly at Bosworth-field on the side of Henry the 7th, against Richard the 3d. His father, Everard, engaged in the gunpowder plot against James the 1st, for which he was beheaded. His son, however, was restored to his estate; and had afterwards several appointments under king Charles the 1st. He granted him letters of reprisal against the Venetians, from whom he took several prizes with a small fleet which he commanded. He fought the Venetians near the port of Scanderoon, and bravely made his way through them with his booty.

In the beginning of the civil wars, he exerted himself greatly in the king's cause. He was afterwards imprisoned, by order of the parliament; but was set at liberty in 1643. He afterward compounded for his estate; but being banished from England, he retired to France, and was sent on two embassies to Pope Innocent the 10th, from the queen, widow to Charles the 1st, whose chancellor he then was. On the restoration of Charles the 2d, he returned to London; where he died in 1665, at 62 years of age.

Digby was a great lover of learning, and translated several authors into English, as well as published several works of his own; as, 1. *Observations upon Dr. Brown's Religio Medici*, 1643.—2. *Observations on part of Spenser's Fairy Queen*, 1644.—3. *A Treatise of the Nature of Bodies*, 1644.—4. *A Treatise declaring the Operations and Nature of Man's Soul, out of which the Immortality of reasonable souls is evinced*: works that discover great penetration and extensive knowledge.

He applied much to chemistry; and found out several useful medicines, which he distributed with a liberal hand. He particularly distinguished himself by his sympathetic powder for the cure of wounds at a distance; his discourse concerning which made great noise for a while. He held several conferences with Des Cartes, about the nature of the soul, and the principles of things. At the beginning of the Royal Society, he became a distinguished member, being one of the first council. And he had at his own house regular levees or meetings of learned men, to improve themselves in knowledge, by conversing with one another.

This eminent person was, for the early pregnancy of his talents, and his great proficiency in learning, compared to the celebrated Picus de Mirandola, who was one of the wonders of human nature. Yet his knowledge, though various and extensive, probably appeared greater than it really was; as he had all the powers of elocution and address to recommend it. He knew how to shine in a circle, either of ladies or philosophers; and was as much attended to when he spoke on the most trivial subjects, as when he spoke on the most important. It has been said that one of the princes of Italy, who had no child, was desirous that his princess should bring him a son by Sir Kenelm, whom he esteemed a just model of perfection.

DIGGES (LEONARD), a considerable mathematician in the 16th century, was descended from an ancient family, and born at Digges-court in the parish of Barham in Kent; but in what year is not known; and died about the year 1574. He was educated for sometime at Oxford, where he laid a good foundation of learning. Retiring from thence, he prosecuted his studies, and became an excellent mathematician, a skilful architect, and an expert surveyor of land, &c. He composed several books: as, 1. *Tectonicum: briefly shewing the exact Measuring, and speedy Reckoning of all manner of Lands, Squares, Timber, Stones, Steeples, &c;* 1556, 4to. Augmented and published again by his son Thomas Digges, in 1592; and also reprinted in 1647. —2. *A Geometrical Practical Treatise, named Pantometria, in three books.* This he left in manuscript; but after his death, his son supplied such parts of it as were obscure and imperfect, and published it in 1591, folio; subjoining, " A Discourse Geometrical of the five regular and Platonic bodies, containing sundry theoretical and practical propositions, arising by mutual conference of these solids, Inscription, Circumscription, and Transformation."—3. *Prognostication Everlasting of right good effect: or Choice Rules to judge the Weather by the Sun, Moon, and Stars, &c;* in 4to, 1555, 1556, and 1564: corrected and augmented by his son, with divers general tables, and many compendious rules, in 4to, 1592.

DIGGES (THOMAS), only son of Leonard Digges, after a liberal education from his tenderest years, went and studied for some time at Oxford; and by the improvements he made there, and the subsequent instructions of his learned father, became one of the best mathematicians of his age. When queen Elizabeth sent some forces to assist the oppressed inhabitants of the Netherlands, Mr. Digges was appointed muster-master general of them; by which he became well skilled in military affairs; as his writings afterward shewed. He died in 1595.

Mr. Digges, beside revising, correcting, and enlarg-

ing some pieces of his father's, already mentioned, wrote and published the following learned works himself: viz, 1. *Alæ sive Scalæ Mathematicæ : or Mathematical Wings or Ladders,* 1573, 4to. A book which contains several demonstrations for finding the parallaxes of any comet, or other celestial body, with a correction of the errors in the use of the radius astronomicus.—2. *An Arithmetical Military Treatise, containing so much of Arithmetic as is necessary towards military discipline,* 1579, 4to.—3. *A Geometrical Treatise, named Stratioticos, requisite for the perfection of Soldiers,* 1579, 4to. This was begun by his father, but finished by himself. They were both reprinted together in 1590, with several additions and amendments, under this title: " An Arithmetical Warlike Treatise, named Stratioticos, compendiously teaching the science of Numbers, as well in Fractions as Integers, and so much of the Rules and Equations Algebraical, and art of Numbers Cossical, as are requisite for the profession of a souldier. Together with the Moderne militaire-discipline, offices, lawes, and orders in every well-governed campe and armie, inviolably to be observed." At the end of this work there are two pieces; the first, " A briefe and true report of the proceedings of the Earle of Leycester, for the reliefe of the towne of Sluce, from his arrival at Vlishing, about the end of June 1587, untill the surrendrie thereof 26 Julii next ensuing. Whereby it shall plainelie appear, his excellencie was not in anie fault for the losse of that towne:" the second, " A briefe discourse what orders were best for repulsing of foraine forces, if at any time they should invade us by sea in Kent, or elsewhere."—4. *A perfect Description of the Celestial Orbs, according to the most ancient doctrine of the Pythagoreans, &c.* This was placed at the end of his father's " Prognostication Everlasting, &c." printed in 1592, 4to.—5. *A humble Motive for Association to maintain the religion established,* 1601, 8vo. To which is added, *his Letter to the same purpose to the archbishops and bishops of England.*—6. *England's Defence: or, A Treatise concerning Invasion.* This is a tract of the same nature with that printed at the end of his Stratioticos, and called, A briefe Discourse, &c. It was written in 1599, but not published till 1686.—7. *A Letter printed before Dr. John Dee's Parallaticæ Commentationis praxeosque nucleus quidam,* 1573, 4to.—Beside these, and his *Nova Corpora,* he left several mathematical treatises ready for the press; which, by reason of lawsuits and other avocations, he was hindered from publishing.

If our author was great in himself, he was not less so in his son, Sir Dudley Digges, so celebrated as a politician and elegant writer.

DIGIT, in Arithmetic, one of the ten characters 0, 1, 2, 3, 4, 5, 6, 7, 8, 9, by means of which all numbers are expressed.

DIGIT, in Astronomy, is the measure by which the part of the luminaries in eclipses is estimated, being the 12th part of the diameter of the luminary. Thus, an eclipse is said to be of 10 digits, when 10 parts out of 12 of the diameter are in the eclipsed part; when the whole of the luminary is just all covered, the digits eclipsed are just 12; and when the luminary is more than covered, as often happens in lunar eclipses, then more than 12 digits are said to be eclipsed: Thus, if the diameter or breadth of the earth's shadow, where the

the moon paffes through, be equal to one diameter and a half of the moon, then 18° or digits are faid to be eclipfed.

Thefe digits are by Wolfius, and fome others, called *digiti ecliptici.*

DIGIT is alfo a meafure taken from the breadth of the finger; being eftimated at 3-4ths of an inch, and equal to 4 grains of barley, laid breadth-ways, fo as to touch each other.

DILATATION, a motion of the parts of a body by which it expands, or opens itfelf, fo as to occupy a greater fpace.

Many authors confound dilatation with rarefaction; but the more accurate writers diftinguifh between them; defining dilatation as the expanfion of a body into a greater bulk, by its own elaftic power; and rarefaction, the like expanfion produced by means of heat.

The moderns have obferved, that bodies which, after being compreffed, and again left at liberty, reftore themfelves perfectly, do endeavour to dilate themfelves with the fame force by which they are compreffed; and accordingly they fuftain a force, and raife a weight equal to that with which they are compreffed.

Again, bodies, in dilating by their elaftic power, exert a greater force at the beginning of their dilatation, than towards the end; as being at firft more compreffed; and the greater the compreffion, the greater the elaftic power and endeavour to dilate. So that thefe three, the compreffing power, the compreffion, and the elaftic power, are always equal.

Finally, the motion by which compreffed bodies reftore themfelves, is ufually accelerated: thus, when compreffed air begins to reftore itfelf, and dilate into a greater fpace, it is ftill compreffed; and confequently a new impetus is ftill impreffed upon it, from the dilatative caufe; and the former remaining, with the increafe of the caufe, the effect, that is the motion and velocity, muft be increafed likewife. Indeed it may happen, that where the compreffion is only partial, the motion of dilatation fhall not be accelerated, but retarded; as is evident in the compreffion of a fpunge, foft bread, gauze, &c.

DILUTE. To dilute a body, is to render it liquid; or, if it were liquid before, to render it more fo, by the addition of a thinner to it.

DIMENSION, the extenfion of a body, confidered as meafurable. Hence, as we conceive a body extended, and meafurable in length, breadth, and depth, dimenfion is confidered as threefold, viz. length, breadth, and thicknefs. So a line has one dimenfion only, viz length; a fuperficies two, length and breadth; and a body or folid has three, viz. length, breadth, and thicknefs.

DIMENSION is alfo particularly ufed with regard to the powers of quantities in equations. Thus, in a fimple equation, $x = a + b$, the unknown quantity is only of one dimenfion; in a quadratic equation, $x^2 = a^2 + b^2$, it is of two dimenfions; in a cubic, $x^3 = a^3 + b^3$, it is of three dimenfions; and fo on.

DIMETIENT, has fometimes been ufed for diameter.

DIMINISHED *Angle*, a term in Fortification. See ANGLE.

DIMINUTION, in Mufic, is the abating fomething of the full value or quantity of any note.

DIMINUTION, in Architecture, is a contraction of the upper part of a column, by which its diameter is made lefs than that of the lower part.

DINOCRATES, a celebrated ancient architect of Macedonia, of whom feveral extraordinary things are related. He was taken, by Alexander the Great, into Egypt, where he employed him in marking out and building the city of Alexandria. He formed a defign, in which Mount Athos was to be laid out into the form of a man, in whofe left hand were defigned the walls of a great city, and all the rivers of the mount flowing into his right, and from thence into the fea. Another memorable inftance of Dinocrates's architectonic fkill, is his reftoring and building, in a more auguft and magnificent manner than before, the celebrated temple of Diana at Ephefus, after Heroftratus, for the fake of immortalizing his name, had deftroyed it by fire. A third inftance, more extraordinary and wonderful than either of the former, is related by Pliny in his Natural Hiftory; who fays he had formed a fcheme, by building the dome of the temple of Arfinoë at Alexandria of loadftone, to make her image all of iron to hang in the middle of it, as if it were in the air; but the king's death, and his own, prevented the execution or attempt of this project.

DIONYSIUS, the *Periegetit*, an ancient geographer and poet. Pliny fays, he was a native of the Perfian Alexandria, afterwards called Antioch, and at laft Charrax; that he was fent by Auguftus, to furvey the eaftern part of the world. Dionyfius wrote a great number of pieces, enumerated by Suidas and his commentator Euftathius: but his *Periegefis*, or *Survey of the World*, is the only one now extant; which may be well efteemed one of the moft exact fyftems of ancient geography, fince Pliny himfelf propofed it as his pattern.

DIONYSIAN, or *Victorian Period*. See PERIOD.

DIOPHANTUS, a celebrated mathematician of Alexandria, who has been reputed to be the inventor of Algebra; at leaft his is the earlieft work extant on that fcience. It is not certain when Diophantus lived. Some have placed him before Chrift, and fome after, in the reigns of Nero and the Antonines; but all with equal uncertainty. It feems he is the fame Diophantus who wrote the *Canon Aftronomicus*, which Suidas fays was commented on by the celebrated Hypatia, daughter of Theon of Alexandria. His reputation muft have been very high among the ancients, fince they ranked him with Pythagoras and Euclid in mathematical learning. Bachet, in his notes upon the 5th book *De Arithmeticis*, has collected, from Diophantus's epitaph in the Anthologia. the following circumftances of his life; namely, that he was married when he was 33 years old, and had a fon born 5 years after; that this fon died when he was 42 years of age, and that his father did not furvive him above 4 years; from which it appears, that Diophantus was 84 years old when he died.

DIOPHANTUS wrote 13 books of Arithmetic, or Algebra, which Regiomontanus in his preface to Alfraganus, tells us, are ftill preferved in manufcript in the Vatican library. Indeed Diophantus himfelf tells us

that

that his work confisted of 13 books, viz, at the end of his addrefs to Dionyfius, placed at the beginning of the work; and from hence Regiomontanus might be led to fay the 13 books were in that library. No more than 6 whole books, with part of a feventh, have ever been publifhed; and I am of opinion there are no more in being; indeed Bombelli, in the preface to his Algebra, written 1572, fays there were but 6 of the books then in the library, and that he and another were about a tranflation of them.

Thofe 6 books, with the imperfect 7th, were firft publifhed at Bafil by Xylander in 1575, but in a Latin verfion only, with the Greek fcholia of Maximus Planudes upon the two firft books, and obfervations of his own. The fame books were afterwards publifhed in Greek and Latin at Paris in 1621, by Bachet, an ingenious and learned Frenchman, who made a new Latin verfion of the work, and enriched it with very learned commentaries. Bachet did not entirely neglect the notes of Xylander in his edition, but he treated the fcholiaft Planudes with the utmoft contempt. He feems to intimate, in what he fays upon the 28th queftion of the 2d book, that the 6 books which we have of Diophantus, may be nothing more than a collection made by fome novice, of fuch propofitions as he judged proper, out of the whole 13: but Fabricius thinks there is no juft ground for fuch a fuppofition.

DIOPHANTINE *Problems*, are certain queftions relating to fquare and cubic numbers, and to right-angled triangles, &c; the nature of which were firft and chiefly treated of by Diophantes, in his Arithmetic, or rather Algebra.

In thefe queftions, it is chiefly intended to find commenfurable numbers to anfwer indeterminate problems; which often bring out an infinite number of incommenfurable quantities. For example, let it be propofed to find a right-angled triangle, whofe three fides x, y, z are exprefled by rational numbers; from the nature of the figure it is known that $x^2 + y^2 = z^2$, where z denotes the hypothenufe. Now it is plain that x and y may alfo be fo taken, that z fhall be irrational; for if $x = 1$, and $y = 2$, then is $z = \sqrt{5}$.

Now the art of refolving fuch problems, confifts in ordering the unknown quantity or quantities, in fuch a manner, that the fquare or higher power may vanifh out of the equation, and then by means of the unknown quantity in its firft dimenfion, the equation may be refolved without having recourfe to incommenfurables. For ex. in the equation above, $x^2 + y^2 = z^2$, fuppofe $z = x + u$, then is $x^2 + y^2 = x^2 + 2xu + u^2$, out of which equation x^2 vanifhes, and then it is $y^2 = 2xu + u^2$, which gives $x = \dfrac{y^2 - u^2}{2u}$. Hence, affuming y and u equal to any numbers at pleafure, the three fides of the triangle will be $y, \dfrac{y^2 - u^2}{2u}$, and $\dfrac{y^2 + u^2}{2u}$, which are all rational whenever y and u are rational. For ex. if $y = 3$, and $u = 1$, then $\dfrac{y^2 - u^2}{2u} = 4$, and $x + u$ or $\dfrac{y^2 + u^2}{2u} = 5$. It is evident that this problem admits of infinite numbers of folutions, as y or u may be affumed infinitely various. See ALGEBRA, and DIOPHANTUS.

Abundant information on this fort of problems may be found in the writings of a great many authors, particularly Fermat, Bachet, Ozanam, Kerfey, Saunderfon, Euler, &c.

DIOPTER, or DIOPTRA, the fame with the index or alhidade of an aftrolabe, or other fuch inftrument.

DIOPTRA was an inftrument invented by Hipparchus, which ferved for feveral ufes; as, to level water-courfes; to take the height of towers, or places at a diftance; to determine the places, magnitudes, and diftances of the planets, &c.

DIOPTRICS, called alfo *anaclaftics*, is the doctrine of refracted vifion; or that part of Optics which explains the effects of light as refracted by paffing through different mediums, as air, water, glafs, &c, and especially lenfes.

Dioptrics is one of the moft ufeful and pleafant of all the human fciences; bringing the remoteft objects near hand, enlarging the fmalleft objects fo as to fhew their minute parts, and even giving fight to the blind; and all this by the fimple means of the attractive power in glafs and water, caufing the rays of light in their paffage through them to alter their courfe according to the different fubftances of the medium; whence it happens, that the object feen through them, do, in appearance, alter their magnitude, diftance, and fituation.

The ancients have treated of direct and reflected vifion; but what we have of refracted vifion, is very imperfect. J. Baptifta Porta wrote a treatife on refraction, in 9 books, but without any great improvement. Kepler was the firft who fucceeded in any great degree, on this fubject; having demonftrated the properties of fpherical lenfes very accurately, in a treatife firft publifhed anno 1611. After Kepler, Galileo gave fomewhat of this doctrine in his Letters; as alfo an Examination of the Preface of Johannes Pena upon Euclid's Optics, concerning the ufe of optics in aftronomy. Des Cartes alfo wrote a treatife on Dioptrics, commonly annexed to his Principles of Philofophy, which is one of his beft works; in which the true manner of vifion is more diftinctly explained than by any former writer, and in which is contained the true law of refraction, which was found out by Snell, though the name of the inventer is fupprefled: here are alfo laid down the properties of elliptical and hyperbolical lenfes, with the practice of grinding glafles. Dr. Barrow has treated on Dioptrics in a very elegant manner, though rather too briefly, in his Optical Lectures, read at Cambridge. There are alfo Huygens's Dioptrics, an excellent work of its kind. Molyneux's Dioptrics, a work rather heavy and dull. Hartfoeker's Effai de Dioptrique. Cherubin's Dioptrique Oculaire, et La Vifion Parfaite. David Gregory's Elements of Dioptrics. Traber's Nervus Opticus. Zahn's Oculus Artificialis Teledioptricus. Dr. Smith's Optics, a complete work of its kind. Wolfius's Dioptrics, contained in his Elementa Mathefeos Univerfalis. But over all, the Treatife on Optics, and the Optical Lectures of Newton, in whofe experiments are contained far more difcoveries than in all the former writers. Laftly, this fcience was perfected by Dollond's difcovery of the acromatic glafles, by which the colours are obviated in refracting telefcopes.

The laws of Dioptrics fee delivered under the article REFRACTION, LENS, &c; and the application of it in the conftruction of telefcopes, mifcrofcopes, and other **dioptrical**

dioptrical inftruments, under the articles Telescope, and Microscope.

DIP *of the Horizon.* See Depression.

Dipping-*Needle,* or *Inclinatory Needle,* a magnetical needle, fo hung, as that, inftead of playing horizontally, and pointing out north and fouth, one end dips, or inclines to the horizon, and the other points to a certain degree of elevation above it.

The inventor of the Dipping-needle was one Robert Norman, a compafs-maker at Ratcliffe, about the year 1580; this is not only teftified by his own account, in his New Attractive, but alfo by Dr. Gilbert, Mr. William Burrowes, Mr. Henry Bond, and other writers of that time, or foon after it. The occafion of the difcovery he himfelf relates, viz, that it being his cuftom to finifh, and hang the needles of his compaffes, before he touched them, he always found that, immediately after the touch, the north point would dip or decline downward, pointing in a direction under the horizon; fo that, to balance the needle again, he was always forced to put a piece of wax on the fouth end, as a counterpoife. The conftancy of this effect led him, at length, to obferve the precife quantity of the dip, or to meafure the greateft angle which the needle would make with the horizon. This, in the year 1576, he found at London was 71° 50'. It is not quite certain whether the dip varies, as well as the horizontal direction, in the fame place. Mr. Graham made a great many experiments with the dipping-needle in 1723, and found the dip between 74 and 75 degrees. Mr. Nairne, in 1772, found it fomewhat above 72°. And by many obfervations made fince that time at the Royal Society, the medium quantity is 72°½. The trifling difference between the firft obfervations of Norman, and the laft of Mr. Nairne and the Royal Society, lead to the opinion that the dip is unalterable; and yet it may be difficult to account for the great difference between thefe and Mr. Graham's numbers, confidering the well-known accuracy of that ingenious gentleman. Philof. Tranf. vol. 45, pa. 279, vol. 62, pa. 476. vol. 69, 70, 71.

It is certain however, from many experiments and obfervations, that the dip is different in different latitudes, and that it increafes in going northward. It appears from a table of obfervations made with a marine dipping needle of Mr. Nairne's, in a voyage towards the north pole, in 1773, that

in latitude 60° 18' the Dip was 75° 0,
in latitude 70 45 the Dip was 77 52,
in latitude 80 12 the Dip was 81 52, and
in latitude 80 27 the Dip was 82 2½.

See Phipps's Voyage, pa. 122. See alfo the Obfervations of Mr. Hutchins, made in Hudfon's Bay and Straits, Philof. Tranf. vol. 65, pa. 129.

Burrowes, Gilbert, Ridley, Bond, &c, endeavoured to apply this difcovery of the dip to the finding of the latitude; and Bond, going ftill farther, firft of any propofed finding the longitude by it; but for want of obfervations and experiments, he could not go any length. Mr. Whifton, being furnifhed with the farther obfervations of colonel Windham, Dr. Halley, Mr Pound, Mr. Cunningham, M. Noel, M. Feuille, and his own, made great improvements in the doctrine and ufe of the dipping-needle, brought it to more certain rules, and endeavoured in good earneft to find the longitude by it. For

this purpofe, he obferves, 1ft, That the true tendency of the north or fouth end of every magnetic needle, is not to that point of the horizon, to which the horizontal needle points, but towards another, directly under it, in the fame vertical, and in different degrees under it, in different ages, and at different places. 2dly, That the power by which the horizontal needle is governed, and all our navigation ufually directed, it is proved is only one quarter of the power by which the dipping-needle is moved; which fhould render the latter far the more effectual and accurate inftrument. 3dly, That a dipping-needle of a foot long will plainly fhew an alteration of the angle of inclination, in thefe parts of the world, in half a quarter of a degree, or 7½ geographical miles; and a needle of 4 feet, in 2 or 3 miles; i. e. fuppofing thefe diftances taken along, or near a meridian. 4thly, A dipping-needle, 4 feet long, in thefe parts of the world, will fhew an equal alteration along a parallel, as another of a foot long will fhew along a meridian; i. e. that will, with equal exactnefs, fhew the longitude, as this the latitude.

This depends on the pofition of the lines of equal dip, in thefe parts of the world, which it is found do lie about 14 or 15 degrees from the parallels. Hence he argues, that as we can have needles of 5, 6, 7, 8, or more feet long, which will move with ftrength fufficient for exact obfervation; and fince microfcopes may be applied for viewing the fmalleft divifions of degrees on the limb of the inftrument, it is evident that the longitude at land may thus be found to lefs than 4 miles.

And as there have been many obfervations made at fea with the fame inftrument by Noel, Feuille, &c, which have determined the dip ufually within a degree, fometimes within ½ or ⅓ of a degree, and this with fmall needles, of 5 or 6, or at the moft 9 inches long; it is inferred, that the longitude may be found, even at fea, to lefs than half a quarter of a degree. This premifed, the obfervation itfelf follows.

To find the Longitude or Latitude by the Dipping-Needle.—If the lines of equal dip, below the horizon, be drawn on maps, or fea-charts, from good obfervations, it will be eafy, from the longitude known, to find the latitude; and from the latitude known, to find the longitude either at fea or land.

Suppofe, for example, a perfon travelling or failing along the meridian of London, fhould find that the angle of dip, with a needle of one foot, was 75°; the chart will fhew that this meridian, and the line of dip, meet in the latitude of 53° 11'; which is therefore the latitude fought.

Or if he be travelling or failing along the parallel of London, i. e. in 51° 31' north latitude, and find the angle of dip 74°; then this parallel, and the line of this dip, will meet on the map in 1° 46' of eaft longitude from London; which therefore is the longitude fought.

DIRECT, in Arithmetic, is when the proportion of any terms, or quantities is in the natural or direct order in which they ftand; being the oppofite to *inverfe,* which confiders the proportion in the inverted order of the terms. So, 3 : 4 :: 6 : 8 *directly;* or 3 : 4 :: 8 : 6 inverfely.

Rule of Three Direct, is when both pairs of terms are in *direct* proportion.

Direct, in Aftronomy. A planet is faid to be di-

rect

rect, or its motion direct, when it goes forward by its proper motion in the zodiac, according to the succession or order of the signs; or when it appears so to do, to an observer standing upon the earth. Whereas it is said to be, or to move *retrograde*, when it appears to go the contrary way, or backward; and to be *stationary*, when it seems not to move either way.

DIRECT *Dials*. See DIAL and DIALLING.

DIRECT, in Optics. *Direct* vision is that performed by direct rays; in contradistinction to vision by refracted, or reflected rays. Direct vision is the subject of Optics, which prescribes the laws and rules of it.

DIRECT *Rays*, are those which pass on in right lines from the object to the eye, without being turned out of their rectilinear direction by any intervening body, either opaque or pellucid; or without being either reflected or refracted.

DIRECT *Sphere*. See RIGHT *Sphere*.

DIRECTION, in Astronomy, the motion and other phenomena of a planet, when direct.

DIRECTION, in Astrology, is a kind of calculus, by which they pretend to find the time in which any notable accident shall befall the person whose horoscope is drawn.

For instance, having established the sun, moon, or ascendant, as masters or significators of life; and Mars or Saturn as promisers or portenders of death; the direction is a calculation of the time in which the significator shall meet the portender.

The significator they likewise call *apheta*, or giver of life; and the promiser, *anereta, promissor*, or giver of death.

They work the directions of all the principal points of the heavens and stars, as the ascendant, mid-heaven, sun, moon, and part of fortune. The like is done for the planets and fixed stars; but all differently, according to the different authors.

Line of DIRECTION, in Gunnery, is the direct line in which a piece is pointed. Sometimes a line of direction is marked on the upper side of the gun, by a small notch or slit, or knob, in the base and muzzle rings: but, unless the two wheels of the carriage stand equally high, this line will be fallacious; for which reason the gunners commonly find a new line of direction every time, by means of a plummet.

Line of DIRECTION, in Mechanics, denotes the line in which a body moves, or endeavours to proceed.

Angle of DIRECTION, is that comprehended between the lines of direction of two conspiring powers.

Quantity of DIRECTION, is used for the product arising from multiplying the velocity of the common centre of gravity, in a system of bodies, by the sum of their masses. In the collision of bodies the quantity of direction is the same both before and after the impulse.

DIRECTION *of the Load Stone*, that property by which the magnet, or a needle touched by it, always presents one of its ends toward one of the poles of the world, and the opposite end to the other pole. This is also called the *polarity* of the magnet or needle.

The attractive property of the magnet was known long before its directive; and the directive long before the inclinatory.

Number of DIRECTION, is the number of days that Septuagesima Sunday falls after the 17th of January. See NUMBER.

DIRECTLY, in Geometry: we say two lines lie directly against each other, when they are parts of the same right line. Also quantities are said to be directly proportional, when the proportion is according to the order of the terms; in contradistinction to *inversely*, or *reciprocally* proportional, which is taking the proportion contrary to the order of the terms.

In Mechanics, a body is said to strike or impinge *directly* against another body, when the stroke is in a direction perpendicular to the surface at the point of impact.

And a sphere in particular strikes directly against another, when the line of direction passes through both their centres.

DIRECTRIX. See DIRIGENT.

DIRECTRIX, in a Parabola, a line perpendicular to the axis produced, at the distance of the focus without the vertex.

DIRIGENT, a term expressing the line of motion, along which a describent line, or surface, is carried in the genesis of any plane or solid figure.

Thus, if the line AB move parallel to itself, and along the line AD, so that the point A always keeps in the line AD, and the point B in the line BC; a parallelogram ABCD will be formed; of which the line AB is the describent, and the line AD the dirigent. So also, if

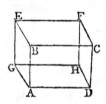

the surface ABEG be supposed carried along the line AD, in a position always parallel to itself at its first situation, the solid AF will be formed; where the surface AE is the describent, and the line AD is the dirigent.

DISC, or DISK, the body or face of the sun or moon; such as it appears to us; for though they be really spherical bodies, they are apparently circular planes.

The diameter of the disc is considered as divided into 12 equal parts, called digits; by means of which it is, that the magnitude of an eclipse is measured, or estimated.—In a total eclipse of either of those luminaries, the whole disc is obscured, or darkened; in a partial eclipse, only part of them.

Illuminated DISC of the Earth. See CIRCLE *of Illumination*.

DISC, in Optics, the magnitude of a telescope glass, or the width of its aperture, whatever its figure be, whether a plane, convex, meniscus, or the like.

DISCHARGER, or DISCHARGING *Rod*, in Electricity, consists of a handle of glass or baked wood, A,

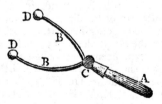

and two bent metal rods BB, terminating in points, and capable of being screwed into the knobs DD, which

move

move by a joint C, fixed to the handle A. Thus it may be used either with the points or balls, as occasion requires; and by being made moveable on a joint, it may be applied to larger or smaller jars at pleasure. By bringing one of these knobs or points to one coated side of a charged electric, and the other to the other side, or to any conductor connected with it, the communication is completed between the two sides, and the electric is discharged.

For the description and use of an Universal Discharger, by Mr. Henly, with which many curious experiments may be performed, see Cavallo's Electricity, pa. 164.

DISCORD, the relation of two sounds which are always, and of themselves, disagreeable, whether applied in succession or consonance.

DISCOUNT, or *Rebate*, is used for an allowance made on a bill, or any other debt not yet become due, in consideration of making present payment of the bill or debt.

Among merchants and traders, it is usual to allow a sum for discount that is equal to the interest of the debt, calculated for the time till it becomes due: but this is not just; for as the true value of the discount is equal to the difference between the debt and its present worth, it is equal only to the interest of that present worth, instead of the interest on the whole debt. And therefore the rule for finding the true discount is this:

As the amount of 100l, for the given rate and time :
Is to the given sum or debt : :
So is the interest of 100l, for the given rate and time :
To the discount of the debt.

So that, if p be the principal or debt, r the rate of interest per cent. and t the time;

then as $100+rt : p :: rt : \frac{prt}{100+rt}$, which is the true

discount. Hence also $p - \frac{prt}{100+rt} = \frac{100p}{100+rt}$ is the present worth, or sum to be received.

For ex. Suppose it be required to find the discount of 250l, for five months, at the rate of 5 per cent. per annum interest. Here $p=250$, $r=5$, and $t = \frac{5}{12}$ or 5 months; then

$$\frac{prt}{100+rt} = \frac{250\times5\times\frac{5}{12}}{100+5\times\frac{5}{12}} = \frac{250\times25}{1200+25} = \frac{250}{49} =$$

$5l. 2s.\frac{2}{49}$ the discount sought.

A Table of Discounts may be seen in Smart's Tables of Interest, the use of which makes calculations of discount very easy.

Discount is also the name of a rule in books of Arithmetic, by which calculations of Discount are made.

DISCRETE, or DISJUNCT, *Proportion*, is that in which the ratio between two or more pairs of numbers is the same, and yet the proportion is not continued, so as that the ratio may be the same between the consequent of one pair and the antecedent of the next pair.

Thus, if the numbers or proportion 6 : 8 :: 3 : 4 be considered: the ratio of 6 to 8 is the same as that of 3 to 4, and therefore these four numbers are proportional: but it is only *discretely* or *disjunctly*, and not continued; for 8 to 3 is not the same ratio as the former; that is, the proportion is broken off between 8 and 3, and not continued all along, as it is in these following four numbers, which are called *continual proportionals*, viz; 3 : 6 :: 12 : 24.

DISCRETE *Quantity*, is such as is not continued and joined together. Such for instance is any number; for its parts, being distinct units, cannot be united into one *continuum*; for in a *continuum* there are no actual determinate parts before division, but they are potentially infinite: so that it is usually and truly said that continued quantity is divisible *in infinitum*.

DISDIAPASON, in Music, a compound concord, in the quadruple ratio of 4 to 1, being a fifteenth or double eighth, and is produced when the voice goes from the first tone to the fifteenth.

DISJUNCT *Proportion*. See DISCRETE *Proportion*.

DISK. See DISC.

DISPART, a term in Gunnery, used for a mark set upon the muzzle-ring of a piece of ordnance, of such height, that a sight-line taken from the top of the base ring near the vent or touch hole to the top of the dispart near the muzzle, may be parallel to the axis of the concave cylinder; for which reason it is evident that the height of the dispart is equal to the difference between the radii of the piece at the base and muzzle-rings, or to half the difference of the diameters there. Hence comes the common method of disparting the gun, which is this: Take, with the calipers, the two diameters, viz, of the base ring and the place where the dispart is to stand, subtract the less from the greater, and take half the difference, which will be the length of the dispart, which is commonly cut to that length from a small bit of wood, and so fixed upright in its place with a bit of wax or pitch.

DISPERSION, in Dioptrics, is the divergency of refracted rays of light.

Point of DISPERSION, is a point from which refracted rays begin to diverge, when their refraction renders them divergent.—It is called *point of dispersion* in opposition to the *point of concourse*, or point in which converging rays concur, after refraction. But it is more usual to call the latter the *focus*, and the former the *virtual focus*.

DISPERSION *of Light*, occasioned by the refrangibility of the rays, or the nature of the refracting medium. See ABERRATION, and INFLECTION.

DISSIPATION, in Physics, a gradual, slow, insensible loss or consumption of the minute parts of a body; or, more properly, the flux by which they fly off and are lost. See EFFLUVIA.

Circle of DISSIPATION, or ABERRATION, in Optics, denotes that circular space upon the retina of the eye, which is occupied by the rays of each pencil in indistinct vision: thus, if the distance of the object, or the constitution of the eye, be such, that the image falls beyond the retina, as when objects are too near; or before the retina, when the rays have not a sufficient divergency, the rays of a pencil, instead of being collected into a central point, will be dissipated over this circular space: and, all other circumstances being alike, this circle will be greater or less, according to the distances from the retina of the foci of refracted rays. But this circle causes no perceptible difference in the distinctness of vision, unless it exceed a certain magnitude; as soon as that is the case, we begin to perceive an indistinct-

ness,

nefs, which increafes as that circle increafes, till at length the object is loft in confufion. This circle is alfo greater or lefs, according to the greater or lefs magnitude of the vifible object: and though it be not eafy to affign the diameter of the faid circle, it feems very probable that vifion continues diftinct for all fuch diftances, or fo long as thefe circles, or the pencils of light from them, do not touch one another upon the retina; and the indiftinctnefs begins when the faid circles begin to interfere. It has been often obferved, that a precife union of the refpective rays upon the retina, is not neceffary to diftinct vifion; but the firft author who afcertained the fact beyond all doubt, was Dr. Jurin. See a variety of obfervations and experiments on this fubject, in his Effay on Diftinct and Indiftinct Vifion, in Smith's Optics, Appendix. In the Philof. Tranf. for 1789, pa. 256, is an excellent paper on this fubject by Dr. Mafkelyne; in which he computes the diameter of the circle of diffipation at ·002667 of an inch, making it anfwer to an external angle of 15°, which he fhews is very compatible with diftinct vifion. See alfo MOON, and VISION.

Radius of DISSIPATION, is the radius of the circle of diffipation.

DISSOLVENT, fomething that diffolves; i. e. divides, and reduces a body into its fmalleft parts.

DISSOLUTION, is a feparation of the ftructure of a body, into fmall or minute parts.

According to Newton, and others, this is effected by certain powerful attractions.

DISSONANCE, or DISCORD, is a falfe confonance or concord; being produced by the mixture or meeting of two founds which are difagreeable to the ear.

DISTANCE, properly fpeaking, denotes the fhorteft line between two points, objects, &c.

DISTANCE, in Aftronomy, as of the fun, planets, comets, &c.

The Real Diftances are found from the parallaxes of the planets, &c. See PARALLAX, and PLANET, and TRANSIT. The diftance of the earth from the fun has been determined at 95 millions of miles, by the late tranfits of Venus; and from this one real diftance, and the feveral relative diftances, by analogy are found all the other real diftances, as in the table below.

The Proportional or Relative Diftances of the planets are very well deduced from the theory of gravity: for Kepler has long fince difcovered, and Newton has demonftrated, that the fquares of their periodical times are proportional to the cubes of their diftances. Kepler's Epit. Aftron. lib. 4; Newton's Principia, lib. 3, phæn. 4; and Gregory's Aftron. book 1, prop. 40. If therefore the mean diftance of the earth from the fun be affumed, or fuppofed 10000, we fhall then, from the foregoing analogy, and the known periodical times, obtain the relative diftances of the other planets: thus,

The Periodical Revolutions in Days and Parts.

Mercury.	Venus.	Earth.	Mars.	Jupiter.	Saturn.	Herfchel.
$87\frac{23}{24}$	$224\frac{17}{24}$	$365\frac{1}{4}$	$686\frac{23}{24}$	$4332\frac{1}{2}$	$10759\frac{7}{24}$	30445

Relative Mean Diftances from the Sun.

3871	7233	10000	15237	52010	95401	190818

Real Diftance in Millions of Miles.

37	66	95	145	$493\frac{1}{2}$	$903\frac{1}{2}$	1813

For the diftances of the fecondary planets from the centres of their refpective primaries, fee SATELLITES.

As to that of the fixed ftars, as having no fenfible parallax, we can do little more than guefs at.

DISTANCE *of the Sun from the Moon's node,* or *apogee,* is an arch of the ecliptic, intercepted between the fun's true place and the moon's node, or apogee. See NODE.

Curtate DISTANCE. See CURTATE.

DISTANCE *of the Baftions,* in Fortification, is the fide of the exterior polygon.

Acceffible DISTANCES, in Geometry, are meafured with the chain, decempeda or ten-foot rod, or the like.

Inacceffible DISTANCES, are found by taking bearings to them, from the two extremities of a line whofe length is given. Various ways of performing this may be feen in my Treatife on Menfuration, fect. 3, on Heights and Diftances.

DISTANCE, in Geography, is the arch of a great circle intercepted between two places.

To find the diftance of two places, A and B, far remote from each other. Affume two ftations, C and D, from which both the places A and B may be feen; and there, with a theodolite, obferve the quantity of the angles ACD, BCD, ADC, BDC, and meafure any diftance as AC.

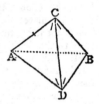

Then, in the triangle ACD, there are given the angles ACD, ADC, and the fide AC; to find the fide CD.—Next, in the triangle BCD, there are given the angles BCD, BDC, and the fide CD; to find the fide BC.—Laftly, in the triangle ABC, there are given the angle ACB, and the fides AC, CB; to find the fide AB, which is the diftance fought.

And in thefe operations, the triangles may be computed either as plane triangles, or as fpherical ones, as the cafe may require, or according to the magnitude of the diftances.

The DISTANCE *of a remote object may alfo be found from its height.* This admits of feveral cafes, according as the diftances are large or fmall, &c. 1ft, Suppofe that from the top of a tower at A, whofe height AB is 120 feet, there be taken the angle BAC = 33°, and the angle BAD = 64°$\frac{1}{2}$, to two trees, or other objects, C, D; to find the diftance between them CD, and the diftance of each from the bottom of the tower at B.

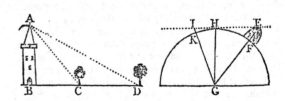

Firft, rad. : tang. ∠ BAD :: AB : BD = 251·585,
next, rad. : tang. ∠ BAC :: AB : BC = 77·929,

their difference is the dift.   CD = 173·656.

2d, Suppofe

2d, Suppose it be required to find the distance to which an object can be seen, by knowing its altitude; *ex. gr.* the Pike of Teneriffe, whose height is accounted 3 miles above the level of the sea, supposing the circumference of the earth 25,000 miles, or the diameter 7958 miles. Let FG be the radius = 3979, EF = 3 the height of the mountain, and EI a tangent to the earth at the point H, which is the farthest distant point to which the top of the mountain E canbe seen. Here in the right angled triangle EGH, are given the hypothenuse EG = 3982, and the leg FG = 3979; to find the other leg HE = 154½ miles = the distance sought nearly. Or, rather, as EG : GH :: rad. : cosin. ∠ G = 2° 13½′; then as 360° : 2° 13½′ :: 25,000 : 154½ miles = the arch of dist. HF sought, the same as before.

3d, If the eye, instead of being in the horizon at H, were elevated above it at I, any known height, as suppose 264 feet, or ₂₀¹th of a mile, as on the top of a ship's mast, &c; then the mountain can be seen much farther off along the line IE, and the distance will be the two tangents IH and HE, or rather the two arcs KH and HF. Hence, as above, as IG : GH :: rad. cosin. ∠ IGH = 17′⅔; then as 360° : 17′⅔ :: 25,000 : 20 miles = the arc KH : this added to the former arc HF = 154½, makes the whole arc KF = 174½ miles, for the whole distance to which the top of the mountain can be seen in this case.

Apparent DISTANCE, in Optics, that distance which we judge an object is placed at when seen afar off, being usually very different from the true distance; because we are apt to think that all very remote objects, whose parts cannot well be distinguished, and which have no other object in view near them, are at the same distance from us, though perhaps the one is thousands of miles nearer than the other, as is the case with regard to the sun and moon.

M. De la Hire enumerates five circumstances, which assist us in judging of the distance of objects; viz, their apparent magnitude, the strength of the colouring, the direction of the two eyes, the parallax of the objects, and the distinctness of their small parts. On the contrary, Dr. Smith maintains, that we judge of distance principally, or solely, by the apparent magnitude of objects; and concludes universally, that the apparent distance of an object seen in a glass, is to its apparent distance seen by the naked eye, as the apparent magnitude to the naked eye is to its apparent magnitude in the glass: But it was long since observed by Alhazen, that we do not judge of distance merely by the angle under which objects are seen; and Mr. Robins clearly shews that Dr. Smith's hypothesis is contrary to fact, in the most common and simple cases. Thus, if a double convex glass be held upright before some luminous object, as a candle, there will be seen two images, one erect, and the other inverted; the first is made simply by reflexion from the nearest surface; the second by reflexion from the farther surface, the rays undergoing a refraction from the first surface both before and after the reflexion. If this glass has not too short a focal distance, when it is held near the object, the inverted image will appear larger than the other, and also nearer; but if the glass be carried off from the object, though the eye remain as near to it as before,

the inverted image will be diminished so much faster than the other, that at length it will appear much less than it, but still nearer. Here, says Mr. Robins, two images of the same object are seen under one view, and their apparent distances immediately compared; and it is evident that those distances have no necessary connexion with the apparent magnitude. This experiment may be made still more convincing, by sticking a piece of paper on the middle of the lens, and viewing it through a short tube. He observes farther, that the apparent magnitude of very distant objects is neither determined by the magnitude of the angle only under which they are seen, nor is the exact proportion of that angle compared with their true distance, but is compounded also with a deception concerning that distance; so that if we had no idea of difference in the distance of objects, each would appear in magnitude proportional to the angle under which it was seen; and if our apprehension of the distance were always just, our idea of their magnitude would be unvaried, in all distances; but in proportion as we err in our conception of their distance, the greater angle suggests a greater magnitude. By not attending to this compound effect, Mr. Robins apprehends that Dr. Smith was led into his mistake.

Dr. Porterfield has made several remarks on the five methods of judging concerning the distance of objects above recited from M. De la Hire; and he has also added to them one more, viz, the conformation of each eye. See *Circle of* DISSIPATION. This, he says, can be of no use to us, with respect to objects that are placed without the limits of distinct vision. But the greater or less confusion with which the object appears, as it is more or less removed from those limits, will assist the mind in judging of its distance: the more confused it appears, the farther will it be thought distant. However, this confusion has its limits; for when an object is placed at a certain distance from the eye, to which the breadth of the pupil bears no sensible proportion, the rays proceeding from a point in the object may be considered as parallel; in which case, the picture on the retina will not be sensibly more confused, though the object be removed to a much greater distance. The most universal, and often the most sure means of judging of the distance of objects, he says, is the angle made by the optic axes: our two eyes are like two different stations, by the assistance of which, distances are taken; and this is the reason why those persons who have lost the sight of one eye, so frequently miss their mark in pouring liquor into a glass, snuffing a candle, and such other actions as require that the distance be exactly distinguished. With respect to the method of judging by the apparent magnitude of objects, he observes that this can only serve when we are otherwise acquainted with their real magnitude. Thus he accounts for the deception to which we are liable in estimating distances, by any extraordinary magnitudes that terminate them; as, in travelling towards a large city, castle, or cathedral, we fancy they are nearer than they really are. Hence also, animals and small objects seen in a valley contiguous to large mountains, or on the top of a mountain or high building, appear exceedingly small. Dr. Jurin accounts for the last recited phenomenon, by observing that we have no distinct idea of

2

distance

distance in that oblique direction, and therefore judge of them merely by their pictures on the eye.

Dr. Porterfield observes, with respect to the strength of colouring, that if we are assured they are of a similar colour, and one appears more bright and lively than the other, we judge that the brighter object is the nearer. When the small parts of objects appear confused, or do not appear at all, we judge that they are at a great distance, and *vice versa ;* because the image of any object, or part of an object, diminishes as the distance of it increases. Finally, we judge of the distance of objects by the number of intervening bodies, by which it is divided into separate and distinct parts ; and the more this is the case, the greater will the distance appear. Thus distances upon uneven surfaces appear less than upon a plane, because the inequalities do not appear, and the whole apparent distance is diminished by the parts that do not appear in it : and thus the banks of a river appear contiguous to a distant eye, when the river is low and not seen. Accidens de la Vue, pa. 358. Smith's Optics, vol. 1, pa. 52, and Rem. pa. 51. Robins's Tracts, vol. 2, pa. 230, 247, 251. Porterfield on the Eye, vol. 1, pa. 105, vol. 2, pa. 387. See Priestley's Hist. of Vision, pa. 205, and pa. 689.

DISTANCE, in Navigation, is the number of miles or leagues that a ship has sailed from any point or place. See SAILING.

Line of DISTANCE, in Perspective. is a right line drawn from the eye to the principal point : as the line OF, drawn between the eye at O, and the principal point F. As this is perpendicular to the plane, or

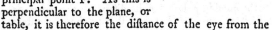

table, it is therefore the distance of the eye from the table.

Point of DISTANCE, in Perspective, is a point in the horizontal line at the same distance from the principal point, as the eye is from the same. Such are the points P and Q, in the horizontal line PQ, whose distance from the principal point F, is equal to that of the eye from the same F.

DISTINCT *Base,* in Optics, is that distance from the pole of a convex glass, at which objects, beheld through it, appear distinct and well defined : so that the Distinct base is the same with what is otherwise called the focus.

The Distinct base is caused by the collection of the rays proceeding from a single point in the object, into a single point in the representation : and therefore concave glasses, which do not unite, but scatter and dissipate the rays, can have no real Distinct base.

DISTINCT *Vision.* See VISION.

DITCH, in Fortification, called also *Foss,* and *Moat,* is a trench dug round the rampart, or wall of a fortified place, between the scarp and counterscarp.

Ditches are either *dry,* or *wet,* that is having water in them ; both of which have their particular advantages. The earth dug out of the ditch serves to raise the rampart.

The ditch in front should be of such breadth as that tall trees may not reach over it, being from 12 to 24 fathoms wide, and 7 or 8 feet deep. The ditches on

the sides are made smaller. But the most general rule is perhaps, that the dimensions of the ditch be such as that the earth dug out may be sufficient to build the rampart of a proper magnitude. The space sometimes left between the rampart and ditch, being about 6 or 8 feet, is called the *berm,* or *list,* serving to pass and repass, and to prevent the earth from rolling into the ditch.

DITONE, in Music, an interval comprehending two tones, a greater and a less. The ratio of the sounds that form the Ditone, is of 4 to 5 ; and that of the semi-ditone, of 5 to 6.

DITTON (HUMPHREY), an eminent mathematician, was born at Salisbury, May 29, 1675. Being an only son, and his father observing in him an extraordinary good capacity, determined to cultivate it with a good education. For this purpose he placed him in a reputable private academy ; upon quitting of which, he, at the desire of his father; though against his own inclination, engaged in the profession of divinity, and began to exercise his function at Tunbridge in the county of Kent, where he continued to preach some years ; during which time he married a lady of that place.

But a weak constitution, and the death of his father, induced Mr. Ditton to quit that profession. And at the persuasion of Dr. Harris and Mr. Whiston, both eminent mathematicians, he engaged in the study of mathematics, a science to which he had always a strong inclination. In the prosecution of this science, he was much encouraged by the success and applause he received : being greatly esteemed by the chief professors of it, and particularly by Sir Isaac Newton, by whose interest and recommendation he was elected master of the new Mathematical School in Christ's Hospital ; where he continued till his death, which happened in 1715, in the 40th year of his age, much regretted by the philosophical world, who expected many useful and ingenious discoveries from his assiduity, learning, and penetrating genius.

Mr. Ditton published several mathematical and other tracts, as below.—1. *Of the Tangents of Curves,* &c. Philos. Transf. vol. 23.

2. A Treatise on *Spherical Catoptrics,* published in the Philos. Transf. for 1705 ; from whence it was copied and reprinted in the Acta Eruditorum 1707, and also in the Memoirs of the Academy of Sciences at Paris.

3. *General Laws of Nature and Motion ;* 8vo, 1705. Wolfius mentions this work, and says, that it illustrates and renders easy the writings of Galileo, Huygens, and the Principia of Newton. It is also noticed by La Roche, in the Memoires de Literature, vol. 8, pa. 46.

4. *An Institution of Fluxions, containing the first Principles, Operations, and Applications, of that admirable Method, as invented by Sir Isaac Newton ;* 8vo, 1706. This work, with additions and alterations, was again published by Mr. John Clarke, in the year 1726.

5. In 1709 he published the *Synopsis Algebraica* of John Alexander, with many additions and corrections.

6. His *Treatise on Perspective* was published in 1712. In this work he explained the principles of that art mathematically ; and besides teaching the methods then generally practised, gave the first hints of the new method

method afterward enlarged upon and improved by Dr. Brook Taylor; and which was published in the year 1715.

7. In 1714, Mr. Ditton published several pieces, both theological and mathematical; particularly his *Difcourfe on the Refurrection of Jefus Chrift; and The New Law of Fluids, or a Difcourfe concerning the Afcent of Liquids, in exact Geometrical Figures, between two nearly contiguous Surfaces.* To this was annexed a tract, to demonftrate the impoffibility of thinking or perception being the refult of any combination of the parts of matter and motion: a fubject much agitated about that time. To this work alfo was added an advertifement from him and Mr. Whifton, concerning a method for difcovering the longitude, which it feems they had publifhed about half a year before. This attempt probably coft our author his life; for although it was approved and countenanced by Sir Ifaac Newton, before it was prefented to the Board of Longitude, and the method has been fuccefsfully put in practice, in finding the longitude between Paris and Vienna, yet that Board then determined againft it: fo that the difappointment, together with fome public ridicule (particularly in a poem written by Dean Swift), affected his health, fo that he died the enfuing year, 1715.

In an account of Mr. Ditton, prefixed to the German tranflation of his Difcourfe on the Refurrection, it is faid that he had publifhed, in his own name only, another method for finding the longitude; but which Mr. Whifton denied. However, Raphael Levi, a learned Jew, who had ftudied under Leibnitz, informed the German editor, that he well knew that Ditton and Leibnitz had correfponded upon the fubject; and that Ditton had fent to Leibnitz a delineation of a machine he had invented for that purpofe; which was a piece of mechanifm conftructed with many wheels like a clock, and which Leibnitz highly approved of for land ufe; but doubted whether it would anfwer on fhipboard, on account of the motion of the fhip.

DIVERGENT *Point.* See *Virtual* Focus.

DIVERGENT, or DIVERGING *Lines,* in Geometry, are thofe whofe diftance is continually increafing.—Lines which diverge one way, converge the other way.

DIVERGENT, or DIVERGING, in Optics, is particularly applied to rays which, iffuing from a radiant point, or having, in their paffage, undergone a refraction, or reflexion, do continually recede farther from each other.

In this fenfe the word is oppofed to convergent, which implies that the rays approach each other, or that they tend to a centre, where they interfect, and, being continued, go on diverging. Indeed all interfecting rays, or lines, diverge both ways from the centre, or point of interfection.

Concave glaffes render the rays diverging; and convex ones, converging.—Concave mirrors make the rays converge; and convex ones, diverge.—It is demonftrated in Optics, that as the diameter of a pretty large pupil does not exceed $\frac{1}{5}$ of a digit; diverging rays, flowing from a radiant point, will enter the pupil as parallel, to all intents and purpofes, if the diftance of the radiant from the eye amount to 40,000 feet. See Focus, Light, and Vision.

DIVERGING HYPERBOLA, is one whofe legs turn their convexities toward each other, and run out quite contrary ways. See HYPERBOLA.

DIVERGING *Parabola.* See *Diverging* PARABOLA.

DIVIDEND, in Arithmetic, is the number given to be divided by fome other number, called the *divifor.* Or it is the number given to be divided, or feparated, into a certain number of equal parts, viz. as many as the divifor contains units; and the number of fuch equal parts is called the *quotient.* Or, more generally, the dividend contains the divifor, as many times as the quotient contains unity.

The Dividend is the numerator of a fraction, whofe denominator is the divifor, and the quotient is the value of the fraction. Thus, $\frac{8}{2} = 4$, and $\frac{3}{4} = .75$.

DIVIDUAL. By this name fome authors diftinguifh the feveral parts of a dividend, from which each feparate figure of the quotient is found.

DIVING, the art, or act of defcending under water, to confiderable depths, and remaining there a competent time.

The ufes of Diving are very confiderable, particularly in the fifhing for pearls, corals, fponges, &c.

Various methods have been propofed, and engines contrived, to render the bufinefs of diving more fafe and eafy. The great point in all thefe, is to furnifh the diver with frefh air, without which he muft either make but a fhort ftay, or perifh.

Thofe who dive for fponges in the Mediterranean, help themfelves by carrying down fponges dipt in oil in their mouths. But confidering the fmall quantity of air that can be contained in the pores of a fponge, and how much that little will be contracted by the preffure of the incumbent water, fuch a fupply cannot long fubfift the diver. For it is found by experiment, that a gallon of air included in a bladder, and by a pipe reciprocally infpired and expired by the lungs, becomes unfit for refpiration in little more than one minute of time. For though its elafticity be but little altered in paffing the lungs, yet it lofes its vivifying fpirit, and is rendered effete. In effect, a naked diver, Dr. Halley affures us, without a fponge, cannot remain above two minutes inclofed in water; nor much longer with one, without fuffocating; nor without long practice, near fo long; ordinary perfons beginning to be fuffocated in about half a minute. Befides, if the depth be confiderable, the preffure of the water on the veffels makes the eyes blood fhotten, and frequently occafions a fpitting of blood. Hence, where there has been occafion to continue long at the bottom, fome have contrived double flexible pipes, to circulate air down into a cavity inclofing the diver, as with armour, both to furnifh air, and to bear off the preffure of the water, and give leave to his breaft to dilate upon infpiration; the frefh air being forced down one of the pipes with bellows, and returning by the other, not unlike to an artery, and vein.

But this method is impracticable when the depth exceeds three fathoms; the water embracing the bare limbs fo clofely, as to obftruct the circulation of the blood in them; and withal preffing fo ftrongly on all the junctures where the armour is made tight with leather; that if there be the leaft defect in any of them, the water rufhes in, and inftantly fills the whole engine, to the great danger of the diver's life.

DIVING.

DIVING-*Bell*, is a machine contrived to remedy all these inconveniencies. In this the diver is safely conveyed to any reasonable depth, and may stay more or less time under the water, as the bell is greater or less. It is most conveniently made in form of a truncated cone, the smallest base being closed, and the larger open. It is to be poised with lead, and so suspended, that it may sink full of air, with its open basis downward, and as near as may be in a situation parallel to the horizon, so as to close with the surface of the water all at once.

Under this covercle the diver sitting, sinks down with the included air to the depth desired; and if the cavity of the vessel can contain a ton of water, a single man may remain a full hour, without much inconvenience, at five or six fathoms deep. But the lower he goes, still the more the included air contracts itself, according to the weight of the water that compresses it; so that at thirty-three feet deep, the bell becomes half full of water; the pressure of the incumbent water being then equal to that of the atmosphere; and at all other depths, the space occupied by the compressed air in the upper part of its capacity, is to the space filled with water, as thirty-three feet to the depth of the surface of the water in the bell below the common surface of it. And this condensed air, being taken in with the breath, soon insinuates itself into all the cavities of the body, and has no ill effect, provided the bell be permitted to descend so slowly as to allow time for that purpose.

One inconvenience that attends it, is found in the ears, within which there are cavities which open only outwards, and that by pores so small, as not to give admission even to the air itself, unless they be dilated and distended by a considerable force. Hence, on the first descent of the bell, a pressure begins to be felt on each ear, which, by degrees, grows painful, till the force overcoming the obstacle, what constringes these pores, yields to the pressure, and letting some condensed air slip in, presently ease ensues. The bell descending lower, the pain is renewed, and afterwards it is again eased in the same manner. But the greatest inconvenience of this engine is, that the water entering it, contracts the bulk of air into so small a compass, that it soon heats, and becomes unfit for respiration: so that there is a necessity for its being drawn up to recruit it; besides the uncomfortable abiding of the diver, who is almost covered with water.

To obviate the difficulties of the diving-bell, Dr. Halley, to whom we owe the preceding account, contrived some further apparatus, by which not only to recruit and refresh the air from time to time, but also to keep the water wholly out of it at any depth; which he effected after the following manner:

His diving-bell (plate vii, fig. 6.) was of wood, three feet wide at top, five feet at bottom, and eight feet high, containing about sixty-three cubic feet in its concavity, coated externally with lead so heavy, that it would sink empty; a particular weight being distributed about its bottom R, to make it descend perpendicularly, and no otherwise. In the top was fixed a meniscus glass D, concave downwards, like a window, to let in light from above; with a cock, as at B, to let out the hot air; and a circular seat, as at LM, for the divers to sit on: and, below, about a yard under the bell, was a stage suspended from it by three ropes, each charged with a hundred weight, to keep it steady, and for the divers to stand upon to do their business. The machine was suspended from the mast of a ship by a sprit, which was secured by stays to the mast-head, and was directed by braces to carry it overboard clear of the side of the ship, and to bring it in again.

To supply air to this bell when under water, he had a couple of barrels, as C, holding thirty-six gallons each, cased with lead, so as to sink empty, each having a bung-hole at bottom, to let in the water as they descended, and let it out again as they were drawn up. In the top of the barrels was another hole, to which was fixed a leathern pipe, or hose, well prepared with bees wax and oil, long enough to hang below the bung-hole; being kept down by a weight appended. So that the air driven to the upper part of the barrel by the encroachment of the water, in the descent, could not escape up this pipe, unless the lower end were lifted up.

These air-barrels were fitted with tackle, to make them rise and fall alternately, like two buckets; being directed in their descent by lines fastened to the under edge of the bell: so that they came readily to the hand of a man placed on the stage, to receive them; and who taking up the ends of the pipes, as soon as they came above the surface of the water in the barrels, all the air included in the upper part of it was blown forcibly into the bell; the water taking its place.

One barrel thus received, and emptied; upon a signal given, it was drawn up, and at the same time the other let down; by which alternate succession, fresh air was furnished so plentifully, that the learned Doctor himself was one of five, who were all together in nine or ten fathoms deep of water for above an hour and a half, without the least inconvenience; the whole cavity of the bell being perfectly dry.

All the precaution he observed, was, to be let down gradually about twelve feet at a time, and then to stop, and drive out the water that had entered, by taking in three or four barrels of fresh air, before he descended farther. And, being arrived at the depth intended, he let out as much of the hot air that had been breathed, as each barrel would replace with cold, by means of the cock B, at the top of the bell, through whose aperture, though very small, the air would rush with so much violence, as to make the surface of the sea boil.

Thus, he found, any thing could be done that was required to be done underneath. And by taking off the stage, he could, for a space as wide as the circuit of the bell, lay the bottom of the sea so far dry as not to be over shoes in water. Besides, by the glass window so much light was transmitted, that, when the sea was clear, and especially when the sun shone, he could see perfectly well to write or read, much more to fasten, or lay hold of any thing under him that was to be taken up. And by the return of the air barrel he often sent up orders written with an iron pen on a plate of lead, directing how he would be moved from place to place.

At other times, when the water was troubled and thick, it would be as dark as night below; but in such cases he was able to keep a candle burning in the bell.

Dr. Halley observes, that they were subject to one inconvenience in this bell; they felt at first a small pain

in

in their ears, as if the end of a tobacco pipe were thruſt into them; but after a little while there was a ſmall puff of air, with a little noiſe, and they were eaſy.

This he ſuppoſes to be occaſioned by the condenſed air ſhutting up a valve leading from ſome cavity in the ear, full of common air; but when the condenſed air preſſed harder, it forced the valve to yield, and filled every cavity. One of the divers, in order to prevent this preſſure, ſtopped his ear with a pledget of paper; which was puſhed in ſo far, that a ſurgeon could not extract it without great difficulty.

The ſame author intimates, that by an additional contrivance he has found it practicable for a diver to go out of the bell to a good diſtance from it; the air being conveyed to him in a continued ſtream by ſmall flexible pipes, which ſerve him as a clue to direct him back again to the bell. For this purpoſe, one end of theſe pipes, kept open againſt the preſſure of the ſea, by a ſmall ſpiral wire, and made tight without by painted leather, and ſheep's guts drawn over it, being open, was faſtened in the bell, as at P, to receive air, and the other end was fixed to a leaden cap on the man's head, reaching down below his ſhoulders, open at bottom, to ſerve him as a little bell, full of air, for him to breathe at his work, which would keep out the water from him, when at the level of the great bell, becauſe of the ſame denſity as the air in the great bell. But when he ſtooped down lower than the level of the great bell, he ſhut the cock F, to cut off the communication between the two bells. Phil. Tranſ. abr. vol. iv. part ii. p. 188, &c. vol. vi. p. 550, &c.

The air in this bell would ſerve him for a minute or two; and he might inſtantly change it, by raiſing himſelf above the great bell, and opening the cock F. The diver was furniſhed with a girdle of large leaden weights, and clogs of lead for the feet, which, with the weight of the leaden cap, kept him firm on the ground; he was alſo well clothed with thick flannels, which being firſt made wet, and then warmed in the bell by the heat of his body, kept off the chill of the cold water for a conſiderable time, when he was out of the bell.

Mr. Martin Triewald, F. R. S. and military architect to the king of Sweden, contrived to conſtruct a diving-bell on a ſmaller ſcale, and leſs expence, than that of Dr. Halley, and yet capable of anſwering the ſame intents and purpoſes. This bell, AB (fig. 7.) ſinks with leaden weights DD, ſuſpended from the bottom of it. It is made of copper, and tinned all over on the inſide; three ſtrong convex lenſes GGG, defended by the copper lids HHH, illuminate this bell. The iron plate E ſerves the diver to ſtand upon, when he is at work; this is ſuſpended by chains FFF, at ſuch a diſtance from the bottom of the bell, that when he ſtands upright, his head is juſt above the water in the bell, where he has the advantage of air fitter for reſpiration, than when he is much higher up; but as there is occaſion for the diver to be wholly in the bell, and conſequently his head in the upper part of it, Mr. Triewald has contrived, that, even there, after he has breathed the hot air as long as he well can, by means of a ſpiral copper tube placed cloſe to the inſide of the bell, he may draw the cooler and freſher air from the lowermoſt parts; for which purpoſe a flexible leather

pipe, about two feet long, is fixed to the upper end of the tube at b; and to the other end of the pipe is faſtened an ivory mouth-piece, for the diver to hold in his mouth, by which to reſpire the air from below. We ſhall only remark, that as air rendered effete by reſpiration is ſomewhat heavier than common air, it muſt naturally ſubſide in the bell; but it may probably be reſtored by the agitation of the ſea water, and thus become fitter for reſpiration. See Fixed Air. Phil. Tranſ. abr. vol. viii. p. 634. Or Deſaguliers's Exper. Phil. vol. ii. p. 220, &c.

The famous Corn. Drebell had an expedient in ſome reſpects ſuperior even to the diving bell, if what is related of it be true. He contrived not only a veſſel to be rowed under water, but alſo a liquor to be carried in the veſſel, which ſupplied the place of freſh air.

The veſſel was made for king James I. carrying twelve rowers, beſides the paſſengers. It was tried in the river Thames; and one of the perſons in that ſubmarine navigation, then living, told it one, from whom Mr. Boyle had the relation.

As to the liquor, Mr. Boyle aſſures us, he diſcovered by a phyſician, who married Drebell's daughter, that it was uſed from time to time, when the air in that ſubmarine boat was clogged by the breath of the company, and rendered unfit for reſpiration: at which time, by unſtopping the veſſel full of this liquor, he could ſpeedily reſtore to the troubled air ſuch a proportion of vital parts, as would make it ſerve again a good while. The ſecret of this liquor Drebell would never diſcloſe to above one perſon, who himſelf aſſured Mr. Boyle what it was. Boyle's Exp. Phyſ. Mech. of the Spring of the Air.

We have had many projects of diving machines, and diving ſhips of various kinds, which have proved abortive.

Diving-Bladder, a term uſed by Borelli for a machine which he contrived for diving under the water to great depths, with great facility, which he prefers to the common diving-bell. The veſica, or bladder, as it is uſually called, is to be of braſs or copper, and about two feet in diameter. This is to contain the diver's head, and is to be fixed to a goat's ſkin habit, exactly fitted to the ſhape of the body of the perſon. Within this veſica there are pipes, by means of which a circulation of air is contrived; and the perſon carries an air pump by his ſide, by means of which he may make himſelf heavier or lighter, as the fiſhes do, by contracting or dilating their air bladder: by this means, the objections all other diving machines are liable to are obviated, and particularly that of the air; the moiſture by which it is clogged in reſpiration, and by which it is rendered unfit for the ſame uſe again, being here taken from it by its circulation through the pipes, to the ſides of which it adheres, and leaves the air as free as before. Borelli Opera Poſthuma.

DIVISIBILITY, a property in quantity, body, or extenſion, by which it becomes ſeparable into parts; either actually, or at leaſt mentally.

Such diviſibility is infinite, if not actually, at leaſt potentially; for no part can be conceived ſo ſmall, but another may be conceived ſtill ſmaller; for every part of matter muſt have ſome finite extenſion, and that extenſion may be biſected, or otherwiſe divided; for the ſame

same reason, these parts may be divided again, and so on without end.

We are not here contending for the possibility of an actual division *in infinitum :* it is only asserted that however small a body is, it may be still farther divided; which it is presumed may be called a division *in infinitum,* because what has no limits, is called *infinite.*

The infinite, or indefinite divisibility of mathematical quantity is thus proved, and illustrated by mathematicians : Suppose a line AD perpendicular to BF; and another as GH also perpendicular to the same BF; with the centres C, C, C, &c, and distances CA, CA, &c, describe circles cutting the line GH in the points *e, e,* &c. Now, the greater the radius AC is, the less is the part *e*G; but the radius may be augmented in infinitum, and therefore the part *e*G may be diminished in the same manner; and yet it can never be reduced to nothing, because the circle can never coincide with the right line BF. Consequently the parts of any magnitude may be diminished in infinitum.

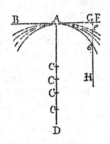

All that is supposed, in strict geometry, concerning the divisibility of magnitude, amounts to no more, than that a given magnitude may be conceived as divided into a number of parts, equal to any given or proposed number.

It is true that there are no such things as parts infinitely small; yet the subtilty of the particles of several bodies is such, that they far surpass our conception; and there are innumerable instances in nature of such parts actually separated from one another.

Several instances of this are given by Mr. Boyle. He speaks of a silken thread 300 yards long, that weighed but two grains and a half. He measured leaf-gold, and found by weighing it, that 50 square inches weighed but one grain : if the length of an inch be divided into 200 parts, the eye may distinguish them all; therefore in one square inch there are 40,000 visible parts; and in one grain of it there are two millions of such parts; which visible parts no one will deny are still farther divisible.

Again, an ounce weight of silver may be gilt over with 8 grains of gold, which may be afterwards drawn into a wire 13,000 feet long, and still be all covered with the same gilding.

In odoriferous bodies a still greater subtilty of parts is perceived, and even such as are actually separated from one another : several bodies scarce lose any sensible part of their weight in a long time, and yet continually fill a very large space with odoriferous particles. Dr. Keil, in his Vera Physica, Lect. 5, has calculated the magnitude of a particle of Assafœtida, which will be the $\dfrac{57}{1000000000000000000}$th part of a cubic inch. And in the same Lecture he shews that the particles of the blood in animalculæ, observed in fluids by means of microscopes, must be less than that part of a cubic inch which is expressed by a fraction whose numerator is 8, and denominator unity with 30 ciphers after it.

The particles of light, if light consist of real particles, furnish another surprising instance of the minuteness of some parts of matter. A small lighted candle placed on a plain, will be visible two miles, and consequently its light fills a sphere of 4 miles diameter, before it has lost any sensible part of its weight. Now, as the force of any body is directly in proportion to its quantity of matter multiplied by its velocity; and since it is demonstrated that the velocity of the particles of light is at least a million of times greater than the velocity of a cannon-ball, it is plain, that if a million of these particles were round, and of the size of a small grain of sand, we durst no more open our eyes to the light, than expose them to sand shot point-blank from a cannon.

By help of microscopes, such objects as would otherwise escape our sight, appear very large : there are some small animals scarce visible with the best microscopes; and yet these have all the parts necessary for life, as blood, and other fluids. How wonderful then must the subtilty of the parts be, which make up such fluids !

Whence is deducible the following theorem :

Any particle of matter, how small soever, and any finite space, how large soever, being given; it is possible for that small sand, or particle of matter, to be diffused through all that great space, and to fill it in such manner, as that there shall be no pore in it, whose diameter shall exceed any given line; as is demonstrated by Dr. Keil. Introduct. ad Ver. Phys.

DIVISIBLE, the faculty or quality of being capable of being divided.

DIVISION, is one of the four principal Rules of Arithmetic, being that by which we find out how often one quantity is contained in another, so that Division is in reality only a compendious method of Subtraction; its effect being to take one number from another as often as possible; that is, as often as it is contained in it. There are therefore three numbers concerned in Division : 1st, That which is given to be divided, called the *dividend;* 2d, That by which the dividend is to be divided, called the *divisor;* 3d, That which expresses how often the divisor is contained in the dividend; or the number resulting from the division of the dividend by the divisor, called the *quotient.*

There are various ways of performing Division, one called the English, another the Flemish, another the Italian, another the Spanish, another the German, and another the Indian way, all equally just, as finding the quotient with the same certainty, and only differing in the manner of arranging and disposing the numbers.

There is also division in integers, division in fractions, and division in species, or algebra.

Division is performed by seeking how often the divisor is contained in the dividend; and when the latter consists of a greater number of figures than the former, the dividend must be taken into parts, beginning on the left, and proceeding to the right, and seeking how often the divisor is found in each of those parts.

For ex. If it be required to divide 6758 by 3. First seek how often 3 is contained in 6, which is 2 times; then how often in 7, which is likewise 2 times, with 1 remaining; which joined to the next figure 5 makes 15, then the 3's in 15 are 5 times; and lastly the

the 3's in 8 are 2 times, and 2 remaining. All the numbers expressing how often 3 is contained in each of those parts, are to be written down according to the order of the parts of the dividend, or from left to right, and separated from the dividend itself by a crooked line, thus:

Divisor Dividend Quotient
$$3 \;)\; 6758 \;(\; 2252\tfrac{2}{3}$$

It appears therefore, that 3 is contained 2252 times in 6758, with 2 remaining over; or that 6758 being divided into 3 parts, each part will be $2252\tfrac{2}{3}$, viz, the figures of the quotient before found, together with the fraction $\tfrac{2}{3}$ formed of the remainder and the divisor.

When the divisor is a single digit, or even as large as the number 12, the division is easily performed by setting down only the quotient as above. But when the divisor is a larger number, it is necessary to set down the several remainders and products &c. This process may be seen at large in most books of arithmetic, as well as various contractions adapted to particular cases: such as, 1st, when the divisor has any number of ciphers at the end of it, they are cut off, as well as the same number of figures from the end of the dividend, and then the work is performed without them both, annexing only the figures last cut off, to the last remainder; 2d, when the divisor is equal to the product of several single digits, it is easier to divide successively by those digits, instead of the divisor at once; 3d, when it is required to continue a quotient to a great many places of figures, as in decimals, a very expeditious method of performing it, is as follows: Suppose it were required to divide 1 by 29, to a great many places of decimals. Adding ciphers to the 1, first divide 10000 by 29 in the common way, till the remainder become a single figure, and annex the fractional supplement to complete the quotient, which gives $\tfrac{1}{29} = 0.03448\tfrac{8}{29}$: next multiply each of these by the numerator 8, so shall $\tfrac{8}{29} = 0.27584\tfrac{64}{29}$ or rather $0.27586\tfrac{6}{29}$; which figures substituted instead of the fraction $\tfrac{8}{29}$ in the first value of $\tfrac{1}{29}$, it becomes $\tfrac{1}{29} = 0.0344827586\tfrac{6}{29}$: again, multiply both of these by the last numerator 6, and it will be $\tfrac{6}{29} = 0.2068965517\tfrac{7}{29}$; which figures substituted for $\tfrac{6}{29}$ in the last-found value of $\tfrac{1}{29}$, it becomes $\tfrac{1}{29} = 0.03448275862068965517\tfrac{7}{29}$: and again, multiplying these by the numerator 7, gives $\tfrac{7}{29} = 0.2413793103448275862O\tfrac{10}{29}$; which figures substituted instead of $\tfrac{7}{29}$ in the last-found value of $\tfrac{1}{29}$, it becomes $\tfrac{1}{29} =$ $0.0344827586206896551724137931034482758620\tfrac{10}{29}$ and so on; where every operation will at least double the number of figures before found by the last one.

Proof of DIVISION. In every example of division, unity is always in the same proportion to the divisor, as the quotient is to the dividend; and therefore the product of the divisor and quotient is equal to the product of 1 and the dividend, that is, the dividend itself. Hence, to prove division, multiply the divisor by the quotient, to the product add the remainder, and the sum will be equal to the dividend when the work is right; if not, there is a mistake.

DIVISION, *in Decimal Fractions*, is performed the same way as in integers, regard being had to the number of decimals, viz, making as many in the quotient as those of the dividend exceed those in the divisor.

DIVISION, *in Vulgar Fractions*, is performed by dividing the numerators by each other, and the denominators by each other, if they will exactly divide; but if not, then the dividend is multiplied by the reciprocal of the divisor, that is, having its terms inverted; for, taking the reciprocal of any quantity, converts it from a divisor to a multiplier, and from a multiplier to a divisor. For ex. $\tfrac{15}{16} \div$ by $\tfrac{5}{8}$ gives $\tfrac{3}{2}$, by dividing the numerators and denominators; but $\tfrac{15}{16} \div$ by $\tfrac{4}{7}$ is the same as $\tfrac{15}{16} \times \tfrac{7}{4}$, which is $= \tfrac{2}{4}\tfrac{3}{4}$. Where $\times$ is the sign of multiplication, and the character $\div$ is the mark of division. Or division is also denoted like a vulgar fraction; so 3 divided by 2, is $\tfrac{3}{2}$.

DIVISION, in *Algebra*, or *Species*, is performed like that of common numbers, either making a fraction of the dividend and divisor, and cancelling or dividing by the terms or parts that are common to both; or else dividing after the manner of long division, when the quantities are compound ones. Thus,

ab divided by a, gives b for the quotient:
and $12ab$ divided by $4b$, gives $3a$ for the quotient:
and $16abc^2$ divided by $8ac$, gives $2bc$:

and a divided by $3b$, gives $\dfrac{a}{3b}$:

and $15abc^3$ divided by $12bc^2$, gives $\dfrac{15abc^3}{12bc^2} = \dfrac{5ac}{4}$

and $a^2 - b^2$ by $a+b$ gives $a-b$; thus

$$a+b \;)\; a^2 - b^2 \;(\; a-b$$
$$\underline{a^2 + ab}$$
$$-ab - b^2$$
$$\underline{-ab - b^2}$$

again, to divide $x^6 - 8x^4 - 124x^2 - 64$ by $x^2 - 16$:

$$x^2 - 16 \;)\; x^6 - 8x^4 - 124x^2 - 64 \;(\; x^4 + 8x^2 + 4$$
$$\underline{x^6 - 16x^4}$$
$$8x^4 - 124x^2$$
$$\underline{8x^4 - 128x^2}$$
$$4x^2 - 64$$
$$\underline{4x^2 - 64}$$

In some cases, the quotient will run out to an infinite series; and then, after continuing it to any certain number of terms, it is usual to annex, by way of a fraction, the remainder with the divisor set under it.

It is to be noted that, in dividing any terms by one another, if the signs be both alike, either both plus, or both minus, the sign of the quotient will be plus; but when the signs are different, the one plus and the other minus, the sign of the quotient will be minus.

DIVISION *by Logarithms.* See LOGARITHMS.

DIVISION *of Mathematical Instruments.* See GRADUATION, and MURAL *Arc* or *Quadrant*.

DIVISION in Music, is the dividing the interval of an octave into a number of lesser intervals.

DIVISION *by Napier's Bones.* See NAPIER'S *Bones.*

DIVISION *of Powers*, is performed by subtracting their exponents. Thus, $a^6 \div a^4$ is $= a^2$; and $4a^{\frac{3}{4}}b^2 \div 2a^{\frac{1}{2}}b^{\frac{4}{3}}$ is $= 2a^{\frac{1}{4}}b^{\frac{3}{3}}$.

DIVISION

DIVISION *of Proportion*, is comparing the difference between the antecedent and confequent, with either of them. Thus,

if
$$\begin{cases} a : b :: c : d, \\ 12 : 4 :: 6 : 2, \end{cases}$$

then by divifion
$$\begin{cases} a-b : a :: c-d : c, \\ 8 : 12 :: 4 : 6, \end{cases}$$

or
$$\begin{cases} a-b : b :: c-d : d, \\ 8 : 4 :: 4 : 2. \end{cases}$$

DIVISOR, is the dividing number; or that which fhews how many parts the dividend is to be divided into.

Common DIVISOR. See COMMON *Divifor*.

DIURNAL, fomething relating to the day; in oppofition to nocturnal, relating to the night.

DIURNAL *Arch*, is the arch defcribed by the fun, moon, or ftars, between their rifing and fetting.

DIURNAL *Circle*, is the apparent circle defcribed by the fun, moon, or ftars, in confequence of the rotation of the earth.

DIURNAL *Motion of a Planet*, is fo many degrees and minutes &c as any planet moves in 24 hours.

The DIURNAL *Motion of the Earth*, is its rotation round its axis, the duration of which conftitutes the natural day.—The reality of the diurnal rotation of the earth is now paft all difpute.

DIURNAL *Parallax*. See PARALLAX.

DIURNAL is alfo ufed in fpeaking of what belongs to the nycthemeron, or natural day of 24 hours: in which fenfe it is oppofed to annual, menftrual, &c.

The diurnal phenomena of the heavenly bodies are folved from the diurnal revolution of the earth; that is, from the rotation of the earth round its own axis in 24 hours. This rotation is equable, and from weft to eaft, about an axis whofe inclination to the ecliptic is now 66° 32′. Since the earth is an opaque body, that fmall part of its furface which comes at the fame time under the confined view of the fpectator, though really fpherical, will feem to be extended like a plane : and the eye, taking a view of the heavens all around, defines a concave fpherical fuperficies, concentric with the earth, or rather with the eye, which the faid plane of the earth's fuperficies will divide into two equal parts, the one of which is vifible, but the other, becaufe of the earth's opacity, hid from the view.

And as the earth revolves about its axis, the fpectator, ftanding upon it, together with the faid plane he ftands upon, called his horizon, dividing the vifible from the invifible hemifphere of the heavens, is carried round the fame way, viz, towards the eaft. From hence it is, that the fun and ftars, placed towards the eaft, being before hid, now become vifible, the horizon as it were finking below them; and the ftars &c towards the weft are covered or hid, and become invifible, the horizon being elevated above them. So that the former ftars, to the fpectator, who reckons the place he ftands on as immoveable, will appear to afcend above the horizon, or rife; and the latter to defcend below the horizon, or fet.

Since the earth, with the horizon of the fpectator fixed to it, continues to move always towards the fame parts, and about the fame axis equally; all bodies, and all phenomena, that do not partake of the faid motion, (that is, all fuch things as are entirely feparate from the earth) will feem to move in the fame time uniformly, but towards the oppofite parts, or from eaft to weft : and every one of thefe objects, according to fenfe, will defcribe the circumference of a circle, whofe plane is perpendicular to the axis of the earth. And becaufe all thefe circles, together with the vifible objects defcribing them, appear to be in the concave fpherical fuperficies of the heavens, every vifible object will feem to defcribe a greater or lefs circle, according to its greater or lefs diftance from the poles, or extremities of the earth's axis produced; the middle circle between thefe poles, called the equator, is confequently the greateft.

It may farther be obferved, that whereas, by the diurnal revolution of the earth, all the feveral luminaries feem to move in the heavens from eaft to weft, hence this feeming diurnal motion of the celeftial lights is called their common motion, as being common to all of them. Befides which, all the luminaries, except the fun, have a proper motion; from which arife their proper phenomena: as for the proper phenomena of the fun, they likewife feem to arife from the proper motion of the fun; though they are really produced by another motion, which the earth has, and by which it moves round the fun once every year, and thence called the annual motion of the earth.

DODECAGON, a regular polygon of 12 equal fides and angles.

If the fide of a Dodecagon be 1, its area will be equal to 3 times the tang. of $75° = 3 \times 2 + \sqrt{3} = 11\cdot1961524$ nearly; and, the areas of plane figures being as the fquares of their fides, therefore $11\cdot1961524$ multiplied by the fquare of the fide of any Dodecagon, will give its area. See my Menfuration, pa. 114, 2d ed.

To infcribe a Dodecagon in a given Circle. Carry the radius 6 times round the circumference, which will divide it into 6 equal parts, or will make a hexagon; then bifect each of thofe parts, which will divide the whole into 12 parts, for the Dodecagon. See alfo other methods of defcribing the fame figure in my Menfur. pa. 26, &c. See POLYGON.

DODECAHEDRON, one of the Platonic bodies, or five regular folids, being contained under a furface compofed of twelve equal and regular pentagons.

To form a Dodecahedron. See *Regular* BODY.

If the fide, or linear edge, of a Dodecahedron be s, its furface will be

$$15s^2 \sqrt{1 + \tfrac{2}{3}\sqrt{5}} = 20\cdot6457788 s^2$$

and its folidity $\quad 5s^3 \sqrt{\dfrac{47 + 21\sqrt{5}}{40}} = 7\cdot66311896 s^3$

If the radius of the fphere that circumfcribes a Dodecahedron be r, then is

its fide or linear edge $= \dfrac{\sqrt{15} - \sqrt{3}}{3} r,$

its fuperficies $\quad = 10r^2 \sqrt{2 - \tfrac{2}{3}\sqrt{5}},$

and its folidity $\quad = \dfrac{20r^3}{3} \sqrt{\dfrac{3 + \sqrt{5}}{30}}.$

The fide of a Dodecahedron infcribed in a fphere, is equal

equal to the greater part of the side of a cube inscribed in the same sphere, and cut according to extreme and mean proportion.

If a line be cut according to extreme and mean proportion, and the lesser segment be taken for the side of a Dodecahedron, the greater segment will be the side of a cube inscribed in the same sphere.

The side of the cube is equal to the right line which subtends the angle of a pentagon of the Dodecahedron, inscribed in the same sphere. See POLYHEDRON; also my Mensur. pa. 253, &c.

DODECATEMORY, the 12 houses or parts of the zodiac of the primum mobile. Also the 12 signs of the zodiac are sometimes so called, because they contain each the 12th part of the zodiac.

DOG, a name common to two constellations, called the Great and Little Dog; but more usually Canis Major, and Canis Minor.

DOME, is a round, vaulted, or arched roof, of a church, hall, pavilion, vestibule, stair-case, &c, by way of crowning, or acroter.

DOMINICAL *Letter*, otherwise called the SUNDAY *Letter*, is one of these first seven letters of the alphabet ABCDEFG, used in almanacs &c, to mark or denote the Sundays throughout the year.

The reason for using seven letters, is because that is the number of days in a week; and the method of using them is this: the first letter A is set opposite the 1st day of the year, the 2d letter B opposite the 2d day of the year, the 3d letter C opposite the 3d day of the year, and so on through the seven letters; after which they are repeated over and over again, all the way to the end of the year, the letter A denoting the 8th day, the letter B the 9th, &c. Then whichever of the letters so placed, falls opposite the first Sunday in the year, the same letter, it is evident, will fall opposite every future Sunday throughout the year, because the number of the letters is the same as the number of days in the week, being both 7 in number; that is, in common years; for as to leap years, an interruption of the order takes place in them. For, on account of the intercalary day, either the letters must be thrust out of their places for the whole year afterwards, so as that the letter, for ex. which answers to the 1st of March, shall likewise answer to the 2d &c; or else the intercalary day must be denoted by the same letter as the preceding one. This latter expedient was judged the better, and accordingly all the Sundays in the year after the intercalary day have another Dominical letter.

The Dominical letters were introduced into the calendar by the primitive christians, instead of the nundinal letters in the Roman calendar; and in this manner were those seven letters set opposite the days of the year, to denote the days in the week, in most of our common almanacs, till the year 1771, when the initial letters of the days of the week were generally introduced instead of them, excepting the Sunday letter itself, which is still retained.

From the foregoing account it follows that,

1st, As the common year consists of 365 days, or 52 weeks, and one day over; the letters go one day backwards every common year: so that in such a year, if the beginning or first day fall on a Sunday, the next year it will fall on Saturday, the next on Friday, and

so on. Consequently, if G be the Dominical letter for the present year, F will be that for the next year; and so on, in a retrograde order.

2d, As the bissextile or leap year consists of 366 days, or 52 weeks, and 2 days over, the beginning of the year next after bissextile goes back 2 days. Whence, if in the beginning of, the bissextile year, the Dominical letter were G, that of the following year will be E.

3d, Since in leap-years the intercalary day falls on the 24th of February, in which case the 24th and 25th days are considered as one day, and denoted by the same letter; after the 24th day of February the Dominical letter goes back one place: thus, if in the beginning of the year the Dominical letter be G, it will afterwards change to the letter F for the remaining Sundays of the year. With us however, this day is now added at the end of February, and from thence it is that the change takes place.

4th, As every fourth year is bissextile, or leap-year, and as the number of letters is 7; the same order of Dominical letters only returns in 4 times 7, or 28 years; which without the interruption of bissextiles, would return in 7 years.

5th, Hence the invention of the solar cycle of 28 years; upon the expiration of which the Dominical letters are restored successively to the same days of the month, or the same order of the letters returns.

To find the DOMINICAL *Letter of any given year.* Find the cycle of the sun for that year, as directed under cycle; and the Dominical letter is found corresponding to it. When there are two letters, the proposed year is bissextile; the former of them serving till the end of February, and the latter for the rest of the year.

The Dominical letter for any year of the present century may be found by this canon:

Divide the odd years, their fourth and 4, by 7,
What is left take from 7, the letter is given.

Thus, for the year 1794, the odd years 94
their 4th 23
and - 4
divided by 7) 121 (17
remains 2
from 7
leaves 5

which answers to E the 5th letter in the alphabet.

The Dominical letter may be found universally, for any year of any century, thus:

Divide the centuries by 4; and twice what does remain
Take from 6; and then add to the number you gain
Their odd years and their 4th; which dividing by 7,
What is left take from 7, the letter is given.

Thus, for the year 1878 the letter is F.

For the centuries 18 divided by 4, leave 2; the double of which taken from 6 leaves 2 again; to which add the odd years 78, and their 4th part 19, the sum 99 divided by 7 leaves 1; which taken from 7, leaves 6 answering to F the 6th letter in the alphabet.

By the reformation of the calendar under pope Gregory the 13th, the order of the Dominical letters was again disturbed in the Gregorian year; for the year

1582,

1582, which had G for its Dominical letter at the beginning; by retrenching 10 days after the 4th of October, came to have C for its Dominical letter: by which means the Dominical letter of the ancient Julian calendar is 4 places before that of the Gregorian, the letter A in the former anfwering to D in the latter.

DONJON, in Fortification, ufually denotes a large ftrong tower, or redoubt, of a fortrefs, where the garrifon may retreat in cafe of neceffity, and capitulate with greater advantage. See DUNGEON.

DORADO, a fouthern conftellation, not vifible in our latitude; called alfo *Xiphias*, or the *Sword-fifh*. The ftars of this conftellation, in Sharp's catalogue, are fix.

DORIC *Order*, of Architecture, is the fecond of the five orders, being placed between the Tufcan and the Ionic. The Doric feems the moft natural, and beft proportioned, of all the orders; all its parts being founded on the natural pofition of folid bodies; for which reafon it is moft proper to be ufed in great and maffy buildings, as the outfide of churches and public places.

The Doric order has no ornaments on its bafe, nor its capital. Its column is 8 diameters high, and its freeze is divided between triglyphs and metopes.

DORMER, or DORMANT, in Architecture, denotes a window made in the roof of a building, or above the entablature; being raifed upon the rafters.

DOUBLE *Afpect, Baftion, Concave, Convex, Cone, Defcant, Eccentricity, Pofition, Ratio, Tenaille, &c.* See the refpective words.

DOUBLE *Horizontal Dial*, one with a double gnomon, the one pointing out the hour on the outer circle, the other the hour on the ftereographic projection drawn upon it. This dial finds the meridian, the hour, the fun's place, rifing, fetting, &c, and many other problems of the globe.

DOUBLE *Point*, in the Higher Geometry, is a point which is common to two parts or legs or branches of fome curve of the 2d or higher order: fuch as, an infinitely fmall oval, or a cufp, or the cruciform interfection, &c, of fuch curves. See Newton's Enumeratio Linearum &c, de Curvarum Punctis Duplicibus.

DOUBLING *a Cape*, or *Point of Land*, in Navigation, fignifies the coming up with it, paffing by it, and leaving it behind the fhip.

The Portuguefe pretend that they firft doubled the Cape of Good Hope, under their admiral Vafquez de Gama; but there are accounts in hiftory, particularly in Herodotus, that the Egyptians, Carthaginians, &c, had done the fame long before them.

DOUCINE, in Architecture, is an ornament on the higheft part of the cornice, or a moulding cut in the figure of a wave, half convex, and half concave.

DOVE-*tail*, in Carpentry, is a method of faftening boards or timbers together, by letting or indenting one piece into another, with a Dove-tail joint, or a joint in form of a Dove's tail.

DRACHM, or DRAM, is the name of a fmall weight ufed with us, and is of two kinds, viz, the 8th part of an ounce in Apothecaries weight, and the 16th part of an ounce in Avoirdupois weight.

DRACO, the DRAGON, a conftellation of the northern hemifphere, whofe origin is varioufly fabled by the Greeks; fome of them reprefenting it as the Dragon which guarded the Hefperian fruit, or golden apples, but being killed by Hercules, Juno, as a reward for its faithful fervices, took it up to heaven, and fo formed this conftellation; while others fay, that in the war of the giants, this Dragon was brought into the combat, and oppofed to Minerva, when the goddefs taking the Dragon in her hand, threw him, twifted as he was, up to the fkies, and fixed him to the axis of the heavens, before he had time to unwind his contortions.

The ftars in this conftellation are, according to Ptolomy, 31; according to Tycho, 32; according to Hevelius, 40; according to Bayer, 33; and according to Flamfteed, 80.

DRAGON, in Aftronomy. See DRACO.

DRAGON'S *Head*, and *Tail*, are the nodes of the planets, but more particularly of the moon, being the points in which the ecliptic is interfected by her orbit, in an angle of about 5° 18'.

One of thefe points looks northward, the moon beginning then to have north latitude; and the other fouthward, where fhe commences fouth latitude; the former point being reprefented by the knot ☊ for the head, and the other by the fame reverfed, or ☋ for the tail. And near thefe points it is that all eclipfes of the fun and moon happen.

This deviation of the path of the moon from the ecliptic feems, according to the fancy of fome, to make a figure like that of a dragon, whofe belly is at the part where fhe has the greateft latitude; the interfections reprefenting the head and tail, from which refemblance the denomination arifes.

Thefe interfections are not always in the fame two points of the ecliptic, but fhift by a retrograde motion, at the rate of 3′ 11″ per day, and completing their circle in 18 years 225 days.

DRAGON-*Beams*, in Architecture, are two ftrong braces or ftruts, ftanding under a breaft-fummer, and meeting in an angle on the fhoulder of the king-piece.

DRAM. See DRACHM.

DRAUGHT-*Compaffes*, thofe provided with feveral moveable points, to draw fine draughts in architecture &c. See COMPASS.

DRAUGHT-*Hooks*, are large hooks of iron, fixed on the cheeks of a gun-carriage, two on each fide, one near the trunnion hole, and the other at the train.

DRAW-*Bridge*, a bridge made after the manner of a floor, to be drawn up, or let down, as occafion requires, before the gate of a town or caftle.

DRIFT, in Navigation, denotes the angle which the line of a fhip's motion makes with the neareft meridian, when fhe drives with her fide to the wind and waves, and is not governed by the power of the helm; and alfo the diftance which the fhip drives on that line, fo called only in a ftorm.

DRIP, in Architecture. See LARMIER.

DRY-*Moat*. See MOAT.

DUCTILITY, a property of certain bodies, by which they are capable of being beaten, preffed, drawn, or ftretched forth, without breaking; or by which they are capable of great alterations in their figure and dimenfions, and of gaining in one way as they lofe in another.

Such are metals, which, being urged by the hammer, gain in length and breadth what they lofe in thicknefs; or, being drawn into wire through holes in iron, grow

longer

longer as they become more slender. Such also are gums, glues, refins, and some other bodies; which, though not malleable, may yet be denominated ductile, in as much as, when softened by water, fire, or some other menstruum, they may be drawn into threads.

Some bodies are ductile both when they are hot and cold, and in all circumstances: such are metals, and especially gold and silver; other bodies are ductile only when they have a certain degree of heat; such as glass, and wax, and such like substances: others again are ductile only when cold, and brittle when hot; as some kinds of iron, viz, those called by workmen redshort, as also brass, and some metallic alloys.

The cause of ductility is very obscure; as depending much on hardness, a quality whose nature we know very little about. It is true, it is usual to account for hardness from the force of attraction between the particles of the hard body; and for ductility, from the particles of the ductile body being, as it were jointed, and entangled with each other. But without dwelling on any fanciful hypotheses about ductility, we may amuse ourselves with some truly amazing circumstances and phenomena of it, in the instances of gold, glass, and spider's-webs. Observing however that the ductility of metals decreases in the following order: gold, silver, copper, iron, tin, lead.

DUCTILITY of Gold. One of the properties of gold is, to be the most ductile of all bodies; of which the gold beaters and gold wire-drawers, furnish us with abundant proof.

Fa. Merfenne, M. Rohault, Dr. Halley, &c, have made computations of it: but they trusted to the reports of the workmen. M. Reaumur, in the Memoires de l'Academie Royale des Sciences, an. 1713, took a surer way; he made the experiment himself. A single grain of gold, he found, even in the common gold leaf, used in most of our gildings, is extended into 36 and a half square inches; and an ounce of gold, which, in form of a cube, is not half an inch either high, broad, or long, is beat under the hammer into a surface of 146 and a half square feet; an extent almost double to what could be done in former times. In Fa. Merfenne's time, it was deemed prodigious, that an ounce of gold should form 1600 leaves; which, together, only made a surface of 105 square feet.

But the diftension of gold under the hammer (how considerable foever) is nothing to that which it undergoes in the drawing-iron. There are gold leaves, in some parts scarce the $\frac{1}{372850}$ part of an inch thick; but $\frac{1}{320000}$ part of an inch is a considerable thickness, in comparison of that of the gold spun on silk in our gold thread.

To conceive this prodigious ductility, it is necessary to have some idea of the manner in which the wire drawers proceed. The wire, and thread we commonly call gold thread, &c, (which is only silver wire gilt, or covered over with gold), is drawn from a large ingot of silver, usually about thirty pounds weight. This they round into a cylinder, or roll, about an inch and a half in diameter, and twenty-two inches long, and cover it with the leaves prepared by the gold beater, laying one over another, till the cover is a good deal thicker than that in our ordinary gilding; and yet, even then, it is

2

very thin; as will be easily conceived from the quantity of gold that goes to gild the thirty pounds of silver: two ounces ordinarily do the business; and, frequently, little more than one.

In effect, the full thickness of the gold on the ingot rarely exceeds $\frac{1}{200}$ or $\frac{1}{160}$ part; and, sometimes not $\frac{1}{1080}$ part of an inch.

But this thin coat of gold must be yet vastly thinner: the ingot is successively drawn through the holes of several irons, each smaller than the other, till it be as fine as, or finer than a hair. Every new hole lessens its diameter; but it gains in length what it loses in thickness; and, of consequence, increases in surface: yet the gold still covers it; it follows the silver in all its extension, and never leaves the minutest part bare, not even to the microscope. Yet, how inconceivably must it be attenuated while the ingot is drawn into a thread, whose diameter is 9000 times less than that of the ingot.

M. Reaumur, by exact weighing, and rigorous calculation, found, that one ounce of the thread was 3232 feet long; and the whole ingot 1163520 feet, Paris measure, or 96 French leagues; equal to 1264400 English feet, or 240 miles English; an extent which far surpasses what Fa. Merfenne, Furetiere, Dr. Halley, &c, ever dreamt of.

Merfenne says, that half an ounce of the thread is 100 toises, or fathoms long; on which footing, an ounce would only be 1200 feet: whereas, M. Reaumur finds it 3232. Dr. Halley makes 6 feet of the wire one grain in weight, and one grain of the gold 98 yards; and, consequently, the ten thousandth part of a grain, above one third of an inch. The diameter of the wire he found one-186th part of an inch; and the thickness of the gold one-154500th part of an inch. But this, too, comes short of M. Reaumur; for, on this principle, the ounce of wire would only be 2680 feet.

But the ingot is not yet extended to its full length. The greatest part of our gold thread is spun, or wound on silk; and, before it is spun, they flat it, by passing it between two rolls, or wheels of exceedingly well polished steel; which wheels, in flatting it, lengthen it by above one seventh. So that our 240 miles are now got to 274.

The breadth, now, of these laminæ, or plates, M. Reaumur finds, is only one-8th of a line, or one-96th of an inch; and their thickness one-3072d. The ounce of gold, then, is here extended to a surface of 1190 square feet; whereas, the utmost the gold beaters can do, we have observed, is to extend it to 146 square feet. But the gold, thus exceedingly extended, how thin must it be! From M. Reaumur's calculus, it is found to be one-175000th of a line, or one-2100000th of an inch; which is scarce one-13th of the thickness of Dr. Halley's gold.

But he adds, that this supposes the thickness of the gold every where equal, which is no ways probable; for in beating the gold leaves, whatever care they can bestow, it is impossible to extend them equally. This we easily find, by the greater opacity of some parts than others; for, where the leaf is thickest, it will gild the wire the thickest.

M. Reaumur,

M. Reaumur, computing what the thickness of the gold must be where thinnest, finds it only one-3150000th part of an inch. But what is the one-3150000th part of an inch? Yet this is not the utmost ductility of gold. for, instead of two ounces of gold to the ingot, which we have here computed upon, a single one might have been used; and, then, the thickness of the gold, in the thinnest places, would only be the 6300000th part of an inch.

And yet, as thin as the plates are, they might be made twice as thin, yet still be gilt; by only pressing them more between the flatter's wheels, they are extended to double the breadth and proportionally in length. So that their thickness, at last, will be reduced to one thirteen or fourteen millionth part of an inch.

Yet, with this amazing thinness of the gold, it is still a perfect cover for the silver: the best eye, or even the best microscope, cannot discover the least chasm, or discontinuity. There is not an aperture to admit alcohol of wine, the subtilest fluid in nature, or even light itself, unless it be owing to cracks occasioned by repeated strokes of the hammer. Add, that if a piece of this gold thread, or gold plate be laid to dissolve in aquafortis, the silver will be all excavated, or eat out, and the gold left entire, in little tubules.

It should be observed, that gold, when it has been struck for some time by a hammer, or violently compressed, as by gold wire drawers, becomes more hard, elastic and stiff, and less ductile, so that it is apt to be cracked or torn: the same thing happens to the other metals by percussion and compression. But ductility and tractability may be restored to metals in that state, by annealing them, or making them red hot. Gold seems to be more affected by percussion and annealing, than any other metals.

As to the DUCTILITY *of soft bodies*, it is not yet carried to that pitch. The reader, however, must not be surprised that, among the ductile bodies of this class, we give the first place to the most brittle of all other, glass.

DUCTILITY *of Glass*. We all know, that, when well penetrated with the heat of the fire, the workmen can figure and manage glass like soft wax; but what is most remarkable, it may be drawn, or spun out into threads exceedingly fine and long.

Our ordinary spinners do not form their threads of silk, flax, or the like, with half the ease, and expedition, as the glass spinners do threads of this brittle matter. We have some of them used in plumes for children's heads, and divers other works, much finer than any hair, and which bend and wave like it with every wind.

Nothing is more simple and easy than the method of making them: there are two workmen employed; the first holds one end of a piece of glass over the flame of a lamp; and, when the heat has softened it, a second operator applies a glass hook to the metal thus in fusion; and, withdrawing the hook again, it brings with it a thread of glass, which still adheres to the mass: then, fitting his hook on the circumference of a wheel about two feet and a half in diameter, he turns the wheel as fast as he pleases; which, drawing out the thread winds it on its rim; till, after a certain number of resolutions, it is covered with a skain of glass thread.

The mass in fusion over the lamp diminishes insensibly: being wound out, as it were, like a pelotoon, or clue of silk, upon the wheel; and the parts, as they recede from the flame, cooling, become more coherent to those next to them; and this by degrees: the parts nearest the fire are always the least coherent, and, of consequence, must give way to the effort the rest make to draw them towards the wheel.

The circumference of these threads is usually a flat oval, being three or four times as broad as thick: some of them seem scarce bigger than the thread of a silk worm, and are surprisingly flexible. If the two ends of such threads be knotted together, they may be drawn and bent, till the aperture, or space in the middle of the knot, doth not exceed one-4th of a line, or one-48th of an inch diameter.

Hence M. Reaumur advances, that the flexibility of glass increases in proportion to the fineness of the threads; and that, probably, had' we but the art of drawing threads as fine as a spider's web, we might weave stuffs and cloths of them for wear. Accordingly, he made some experiments this way: and found he could make threads fine enough, viz, as fine, in his judgment, as spider's thread, but he could never make them long enough to do any thing with them.

DUCTILITY *of Spider's-webs*. See WEB.

DUNGEON, DONJON, in Fortification, the highest part of a castle built after the ancient mode; serving as a watch-tower, or place of observation; and also for the retreat of a garrison, in case of necessity, so that they may capitulate with greater advantage.

DUPLE, or DOUBLE *Ratio*, is that in which the antecedent term is double the consequent; or, where the exponent of the ratio is 2. Thus, 6 to 3 is in a duple ratio.

Sub-DUPLE *Ratio*, is that in which the consequent is double the antecedent; or, in which the exponent of the ratio is ½. As in 3 to 6, which is in subduple ratio.

DUPLICATE *Ratio*, is the square of a ratio, or the ratio of the squares of two quantities. Thus, the duplicate ratio of *a* to *b*, is the ratio of a^2 to b^2, or of the square of *a* to the square of *b*.—In a series of geometrical proportionals, the 1st term is to the 3d, in a duplicate ratio of the 1st to the 2d, or as the square of the first to the square of the 2d: Thus, in the geometricals 2, 4, 8, 16, &c, the ratio of 2 to 8, is the duplicate of that of 2 to 4, or as the square of 2 to the square of 4, that is, as 4 to 16. So that duplicate ratio is the ratio of the squares, as triplicate ratio is the ratio of the cubes, &c.

DUPLICATION, is the doubling of a quantity, or multiplying it by 2, or adding it to itself.

DUPLICATION *of a Cube*, is finding out the side of a cube that shall be double in solidity to a given cube: which is a celebrated problem, much cultivated by the ancient geometricians, about 2000 years ago.

It was first proposed by the oracle of Apollo at Delphos; which, being consulted about the manner of stopping a plague then raging at Athens, returned for answer, that the plague should cease when Apollo's altar, which was cubical, should be doubled. Upon this, they applied themselves in good earnest, to seek the

the duplicature of the cube, which from thence was called the *Delian problem*.

This problem cannot be effected geometrically, as it requires the solution of a cubic equation, or requires the finding of two mean proportionals, viz, between the fide of the given cube and the double of the fame, the first of which two mean proportionals is the fide of the double cube, as was first obferved by Hippocrates of Chios. For, let a be the fide of the given cube, and z the fide of the double cube fought; then it is $z^3 =$ $2a^3$, or $a^2 : z^2 :: z : 2a$; fo that, if a and z be the first and 2d terms of a fet of continued proportionals, then $a^2 : z^2$ is the ratio of the fquare of the 1ft to the fquare of the 2d, which, it is known, is the fame. as the ratio of the 1ft term to the 3d, or of the 2d to the 4th, that is of z to $2a$; therefore z being the 2d term, $2a$ will be the 4th. So that z, the fide of the cube fought, is the 2d of four terms in continued proportion, the 1ft and 4th being a and $2a$, that is, the fide of the double cube is the firft of two mean proportionals between a and $2a$.

Eutocius, in his Commentaries on Archimedes, gives feveral ways of performing this by the mefolabe. In Pappus too are found three different ways; the firft according to Archimedes, the fecond according to Hero, and the 3d by an inftrument invented by Pappus, which gives all the proportions required. The fieur de Comiers has likewife publifhed a demonftration of the fame problem, by means of a compafs with three legs. But all thefe methods are only mechanical. See Valerius Maximus, lib. 8; alfo Eutocius's Com. on lib. 2. Archimedes de Sphæra & Cylindro; and Pappus, lib. 3, prop. 5, & lib. 4, prop. 22.

DURER (ALBERT), defcended of an Hungarian family, but born at Nuremberg, in 1471, was one of the beft engravers, painters, and practical geometricians of his age. He was at the fame time a man of letters and a philofopher; and he was an intimate friend of Erafmus, who revifed fome of the pieces which he publifhed. He was alfo a man of bufinefs, and for many years the leading magiftrate of Nuremberg.

Though not the inventor, he was one of the firft and greateft improvers of the art of engraving. He was however the inventor of cutting in wood, which he devifed and practifed in great perfection, ufing this way for expedition, as he had a multitude of defigns to execute; and as his work was ufually done in the moft exquifite manner, his pieces took him up much time. For in many of thofe prints which he executed on copper, the engraving is elegant to a great degree. His Hell-Scene in particular, engraven in the year 1513, is as highly finifhed a print as ever was engraved, and as hap-

pily executed. In his wooden prints too it is furprifing to fee fo much meaning in fo early a mafter. In fact, Durer was a man of a very extenfive genius. His pictures were excellent; as well as his prints, which were very numerous. They were much admired, from the firft, and eagerly bought up; which put his wife, who was another Xantippe, upon urging him to fpend more time upon engraving than he was inclined to do: for he was rich; and chofe rather to practife his art as an amufement, than as a bufinefs. He died at Nuremberg, in 1528, at 57 years of age.

Albert Durer wrote feveral books, in the German language, which were tranflated into Latin by other perfons, and publifhed after his death. viz,

1. His book upon the rules of painting, intitled, *De Symmetria Partium in rectis formis Humanorum Corporum*, is one of them: printed in folio, at Nuremberg, in 1532, and at Paris in 1557. An Italian verfion alfo was publifhed at Venice, in 1591.

2. *Inflitutiones Geometricæ*; Paris 1532.

3. *De Urbibus, Arcibus, Caftellifque condendis & muniendis*; Paris 1531.

4. *De Varietate Figurarum, et Flexuris Partium, et Geftibus Imaginum*; Nuremberg 1534.

The figures in thefe books, which are from wooden plates, are very numerous, and moft admirably well executed, indeed far beyond any thing of the kind done in our own days. Some of them alfo are of a very large fize, as much as 16 inches in length, and of a proportional breadth, which being exquifitely worked, muft have coft great labour. His geometry is chiefly of the practical kind, confifting of the moft curious defcriptions, infcriptions, and circumfcriptions of geometrical lines, planes, and folids. We here meet, for the firft time, with the plane figures, which folded up make the five regular or platonic bodies, as well as that curious conftruction of a pentagon, being the laft method in prob. 23 of my Menfuration.

DYE, in Architecture, the trunk of the pedeftal, or that part between the bafe and the cornice, being fo called, becaufe it is often made in the form of a dye or cube.

DYMANICS, is the fcience of moving powers; more particularly of the motion of bodies that mutually act on one another. See MECHANICS, MOTION, COMMUNICATION *of Motion*, OSCILLATION, PERCUSSION, &c.

DYPTERE, or DIPTERE, was a kind of temple, encompaffed round with a double row of columns; and the pfeudo-diptere, or falfe diptere, was the fame, only this was encompaffed with a fingle row of columns, inftead of a double row.

EAGLE,

E.

EAGLE, *Aquila*, is a conſtellation of the northern hemiſphere, having its right wing contiguous to the equinoctial. For the ſtars in this conſtellation, ſee AQUILA.

EARTH, *Terra*, in Natural Philoſophy, one of the four vulgar, or Peripatetical elements; defined a ſimple, dry, and cold ſubſtance; and, as ſuch, an ingredient in the compoſition of all natural bodies.

The EARTH, in Geography, this terraqueous globe or ball, which we inhabit, conſiſting of land and ſea.

Figure of the EARTH. The ancients had various opinions as to the figure of the earth: ſome, as Anaximander and Leucippus, held it cylindrical, or in form of a drum: but the principal opinion was, that it was flat; that the viſible horizon was the bounds of the earth, and the ocean the bounds of the horizon; that the heavens and earth above this ocean were the whole viſible univerſe; and that all beneath the ocean was Hades: and of this ſame opinion were alſo ſome of the Chriſtian fathers, as Lactantius, St. Auguſtine, &c. See Lactan. lib. 3, cap. 24; St. Aug. lib. 16, de Civitate Dei; Ariſtotle de Cœlo, lib. 2, cap. 13.

Such of the ancients however as underſtood any thing of aſtronomy, and eſpecially the doctrine of eclipſes, muſt have been acquainted with the round figure of the earth; as the ancient Babylonian aſtronomers, who had calculated eclipſes long before the time of Alexander, and Thales the Grecian, who predicted an eclipſe of the ſun. It is now indeed agreed on all hands, unleſs perhaps by the moſt vulgar and ignorant, that the form of the terraqueous globe is globular, or very nearly ſo.

That the exterior of the earth is round, or rotund, is manifeſt to the moſt common perception, in the caſe of a ſhip ſailing either from the land, or towards it; for when a perſon ſtands upon the ſhore, and ſees a ſhip ſail from the land, out to ſea; at firſt he loſes ſight of the hull and lower parts of the ſhip, next the rigging and middle parts, and laſtly of the tops of the maſts themſelves, in every caſe the rotundity of the ſea between the ſhip and the eye being very viſible: the contrary happens when a ſhip ſails towards us; we firſt ſee the tops of the maſts appear juſt over the rotundity of the ſea; next we perceive the rigging, and laſtly the hull of the ſhip itſelf: all which is well illuſtrated by the following figure,

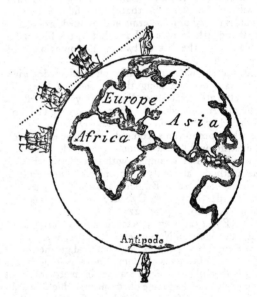

The round figure of the earth is alſo evident from the eclipſes of the ſun and moon; for in all eclipſes of the moon, which are cauſed by the moon paſſing through the earth's ſhadow, that ſhadow always appears circular upon the face of the moon, what way ſoever it be projected, whether eaſt, weſt, north, or ſouth, and howſoever its diameter vary, according to the greater or leſs diſtance from the earth. Hence it follows, that the ſhadow of the earth, in all ſituations, is really conical; and conſequently the body that projects it, i. e. the earth, is at leaſt nearly ſpherical.

The ſpherical figure of the earth is alſo evinced from the riſing and ſetting of the ſun, moon, and ſtars; all which happen ſooner to thoſe who live to the eaſt, and later to thoſe living weſtwardly; and that more or leſs ſo, according to the diſtance and roundneſs of the earth.

So alſo, going or ſailing to the northward, the north pole and northern ſtars become more elevated, and the ſouth pole and ſouthern ſtars more depreſſed; the elevation northerly increaſing equally with the depreſſion ſoutherly; and either of them proportionably to the diſtance gone. The ſame thing happens in going to the ſouthward. Beſides, the oblique aſcenſions, deſcenſions, fcenfions, emerfions, and amplitudes of the rifing and fetting of the fun and ftars in every latitude, are agreeable to the fuppofition of the earth's being of a fpherical form: all which could not happen if it was of any other figure.

Moreover, the roundnefs of the earth is farther confirmed by its having been often failed round: the firft time was in the year 1519, when Ferd. Magellan made the tour of the whole globe in 1124 days. In the year 1557 Francis Drake performed the fame in 1056 days: in the year 1586, Sir Tho. Cavendifh made the fame voyage in 777 days; Simon Cordes, of Rotterdam, in the year 1590; in the year 1598, Oliver Noort, a Hollander, in 1077 days; Van Schouten, in the year 1615, in 749 days; Jac. Heremites and Joh. Huygens, in the year 1623, in 802 days: and many others have fince performed the fame navigation, particularly Anfon, Bougainville, and Cook. Sometimes failing round by the eaftward, fometimes to the weftward; till at length they arrived again in Europe, from whence they fet out; and in the courfe of their voyage, obferved that all the phenomena, both of the heavens and earth, correfpond to, and evince this fpherical figure.

The fame globular figure is likewife inferred from the operation of Levelling, in which it is found neceffary to make an allowance for the difference between the apparent and the true level.

The natural caufe of this fphericity of the globe is, according to Sir Ifaac Newton, the great principle of attraction, which the Creator has ftamped on all the matter in the univerfe; and by which all bodies, and all the parts of bodies, mutually attract one another.— And the fame is the caufe of the fphericity of the drops of rain, quickfilver, &c.

What the earth lofes of its fphericity by mountains and valleys, is nothing confiderable; the higheft eminence being fcarce equivalent to the minuteft protuberance on the furface of an orange. Its difference from a perfect fphere however is more confiderable in another refpect, by which it approaches nearly to the fhape of an orange, or to an oblate fpheroid, being a little flatted at the poles, and raifed about the equatorial parts, fo that the axis from pole to pole is lefs than the equatorial diameter. What gave the firft occafion to the difcovery of this figure of the earth, was the obfervations of fome French and Englifh philofophers in the Eaft-Indies, and other parts, who found that pendulums, the nearer they came to the equator, performed their vibrations flower: from whence it follows, that the velocity of the defcent of bodies by gravity, is lefs in countries nearer to the equator; and confequently that thofe parts are farther removed from the centre of the earth, or from the common centre of gravity. See the Hiftory of the Royal Academy of Sciences, by Du Hamel, p. 110, 156, 206; and l'Hift. de l'Acad. Roy. 1700 and 1701.—This circumftance put Huygens and Newton upon finding out the caufe, which they attributed to the revolution of the earth about its axis. If the earth were in a fluid ftate, its rotation round its axis would neceffarily make it put on fuch a figure, becaufe the centrifugal force being greateft towards the equator, the fluid would there rife and fwell moft; and that its figure really fhould be fo now, feems neceffary, to keep the fea in the equinoctial re-

gions from overflowing the earth about thofe parts. See this curious fubject well handled by Huygens, in his difcourfe De Caufa Gravitatis, pa. 154, where he ftates the ratio of the polar diameter to that of the equator, as 577 to 578. And Newton, in his Principia, firft publifhed in 1686, demonftrates, from the theory of gravity, that the figure of the earth muft be that of an oblate fpheroid generated by the rotation of an ellipfe about its fhorteft diameter, provided all the parts of the earth were of an uniform denfity throughout, and that the proportion of the polar to the equatorial diameter of the earth, would be that of 689 to 692, or nearly that of 229 to 230, or as ·9956522 to 1.

This proportion of the two diameters was calculated by Newton in the following manner. Having found that the centrifugal force at the equator is $\frac{1}{289}$th of gravity, he affumes, as an hypothefis, that the axis of the earth is to the diameter of the equator as 100 to 101, and thence determines what muft be the centrifugal force at the equator to give the earth fuch a form, and finds it to be $\frac{4}{585}$ths of gravity: then, by the rule of proportion, if a centrifugal force equal to $\frac{4}{585}$ths of gravity would make the earth higher at the equator than at the poles by $\frac{1}{100}$th of the whole height at the poles, a centrifugal force that is the $\frac{1}{289}$th of gravity will make it higher by a proportional excefs, which by calculation is $\frac{1}{229}$th of the height at the poles; and thus he difcovered that the diameter at the equator is to the diameter at the poles, or the axis, as 230 to 229. But this computation fuppofes the earth to be every where of an uniform denfity; whereas if the earth is more denfe near the centre, then bodies at the poles will be more attracted by this additional matter being nearer; and therefore the excefs of the femi diameter of the equator above the femi-axis, will be different. According to this proportion between the two diameters, Newton farther computes, from the different meafures of a degree, that the equatorial diameter will exceed the polar, by 34 miles and $\frac{1}{4}$.

Neverthelefs, Meffrs Caffini, both father and fon, the one in 1701, and the other in 1713, attempted to prove, in the Memoirs of the Royal Academy of Sciences, that the earth was an oblong fpheroid; and in 1718, M. Caffini again undertook, from obfervations, to fhew that, on the contrary, the longeft diameter paffes through the poles; which gave occafion for Mr. John Bernoulli, in his Effai d'une Nouvelle Phyfique Celefte, printed at Paris in 1735, to triumph over the Britifh philofopher, apprehending that thefe obfervations would invalidate what Newton had demonftrated. And in 1720, M. De Mairan advanced arguments, fuppofed to be ftrengthened by geometrical demonftrations, farther to confirm the affertions of Caffini. But in 1735 two companies of mathematicians were employed, one for a northern, and another for a fouthern expedition, the refult of whofe obfervations and meafurement plainly proved that the earth was flatted at the poles. See DEGREE.

The proportion of the equatorial diameter to the polar, as ftated by the gentlemen employed on the northern expedition for meafuring a degree of the meridian, is as 1 to 0·9891; by the Spanifh mathematicians as 266 to 265, or as 1 to 0·99624; by M. Bouguer as 179 to 178, or as 1 to 0·99441.

As to all conclufions however deduced from the

length of pendulums in different places, it is to be observed that they proceed upon the suppofition of the uniform denfity of the earth, which is a very improbable circumftance; as juftly obferved by Dr. Horfley in his letter to Capt. Phipps. "You finifh your article, he concludes, relating to the pendulum with faying, 'that thefe obfervations give a figure of the earth nearer to Sir Ifaac Newton's computation, than any others that have hitherto been made;' and then you ftate the feveral figures given, as you imagine, by former obfervations, and by your own. Now it is very true, that *if* the meridians be ellipfes, or *if* the figure of the earth be that of a fpheroid generated by the revolution of an ellipfis, turning on its fhorter axis, the particular figure, or the ellipticity of the generating ellipfis, which your obfervations give, is nearer to what Sir Ifaac Newton faith it fhould be, if the globe were homogeneous, than any that can be derived from former obfervations. But yet it is not what you imagine. Taking the gain of the pendulum in latitude 79° 50' exactly as you ftate it, the difference between the equatorial and the polar diameter, is about as much lefs than the Newtonian computation makes it, and the hypothefis of homogeneity would require, as you reckon it to be greater. The proportion of 212 to 211 fhould indeed, according to your obfervations, be the proportion of the force that acts upon the pendulum at the poles, to the force acting upon it at the equator. But this is by no means the fame with the proportion of the equatorial diameter to the polar. If the globe were homogeneous, the equatorial diameter would exceed the polar by $\frac{1}{230}$ of the length of the latter: and the polar force would alfo exceed the equatorial by the like part. But if the difference between the polar and equatorial force be greater than $\frac{1}{230}$ (which may be the cafe in an heterogeneous globe, and feems to be the cafe in ours,) then the difference of the diameters fhould, according to theory, be lefs than $\frac{1}{230}$, and vice verfa.

"I confefs this is by no means obvious at firft fight; fo far otherwife, that the miftake, which you have fallen into, was once very general. Many of the beft mathematicians were mifled by too implicit a reliance upon the authority of Newton, who had certainly confined his inveftigations to the homogeneous fpheroid, and had thought about the heterogeneous only in a loofe and general way. The late Mr. Clairault was the firft who fet the matter right, in his elegant and fubtle treatife on the figure of the earth. That work hath now been many years in the hands of mathematicians, among whom I imagine there are none, who have confidered the fubject attentively, that do not acquiefce in the author's conclufions.

"In the 2d part of that treatife, it is proved, that putting P for the polar force, Π for the equatorial, δ for the true ellipticity of the earth's figure, and ε for the ellipticity of the homogeneous fpheroid,

$$\frac{P - \Pi}{\Pi} = 2\varepsilon - \delta: \text{ therefore } \delta = 2\varepsilon - \frac{P - \Pi}{\Pi};$$

and therefore, according to your obfervation, $\delta = \frac{1}{231}$. This is the juft conclufion from your obfervations of the pendulum, taking it for granted, that the meridians are ellipfes: which is an hypothefis, upon which all the reafonings of theory have hitherto proceeded. But plaufible as it may feem, I muft fay, that there is much

reafon from experiment to call it in queftion. If it were true, the increment of the force which actuates the pendulum, as we approach the poles, fhould be as the fquare of the fine of the latitude: or, which is the fame thing, the decrement, as we approach the equator, fhould be as the fquare of the cofine of the latitude. But whoever takes the pains to compare together fuch of the obfervations of the pendulum in different latitudes, as feem to have been made with the greateft care, will find that the increments and decrements do by no means follow thefe proportions; and in thofe which I have examined, I find a regularity in the deviation which little refembles the mere error of obfervation. The unavoidable conclufion is, that the true figure of the meridians is not elliptical. If the meridians are not ellipfes, the difference of the diameters may indeed, or it may not, be proportional to the difference between the polar and the equatorial force; but it is quite an uncertainty, what relation fubfifts between the one quantity and the other; our whole theory, except fo far as it relates to the homogeneous fpheroid, is built upon falfe affumptions, and there is no faying what figure of the earth any obfervations of the pendulum give."

He then lays down the following table, which fhews the different refults of obfervations made in different latitudes; in which the firft three columns contain the names of the feveral obfervers, the places of obfervation, and the latitude of each; the 4th column fhews the quantity of P—Π in fuch parts as Π is 100000, as deduced from comparing the length of the pendulum at each place of obfervation, with the length of the equatorial pendulum as determined by M. Bouguer, upon the fuppofition that the increments and decrements of force, as the latitude is increafed or lowered, obferve the proportion which theory affigns. Only the 2d and the laft value of P—Π are concluded from comparifons with the pendulum at Greenwich and at London, not at the equator. The 5th column fhews the value of δ corresponding to every value of P—Π, according to Clairault's theorem:

Obfervers.	Places.	Latitudes.		P—Π	δ
Bouguer	Equator	0°	0'		
Bouguer	Porto Bello	9	34	741·8	$\frac{1}{178}$
Green	Otaheitee	17	29	563·2	$\frac{1}{328}$
Bouguer	San Domingo	18	27	591·0	$\frac{1}{358}$
Abbé de la Caille	Cape of Good Hope	33	55	731·5	
- - - -	Paris	48	50	585·1	$\frac{1}{133}$
The Academicians	Pello	66	48	565·9	$\frac{1}{323}$
Capt. Phipps	- - - -	79	50	471·2	$\frac{1}{231}$

"By this table it appears, that the obfervations in the middle parts of the globe, fetting afide the fingle one at the Cape, are as confiftent as could reafonably be expected; and they reprefent the ellipticity of the earth as about $\frac{1}{320}$. But when we come within 10 degrees of the equator, it fhould feem that the force of gravity fuddenly becomes much lefs, and within the like diftance

4

distance of the poles much greater than it could be in such a spheroid."

The following problem, communicated by Dr Leatherland to Dr. Pemberton, and publifhed by Mr. Robertfon, ferves for finding the proportion between the axis and the equatorial diameter, from meafures taken of a degree of the meridian in two different latitudes, fuppofing the earth an oblate fpheroid.

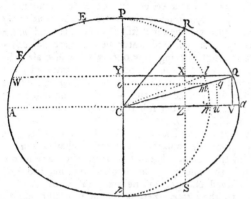

Let AP*ap* be an ellipfe reprefenting a fection of the earth through the axis P*p*; the equatorial diameter, or the greater axis of the ellipfe, being A*a*; let E and F be two places where the meafure of a degree has been taken; thefe meafures are proportional to the radii of curvature in the ellipfe at thofe places; and if CQ, CR be conjugates to the diameters whofe vertices are E and F, CQ will be to CR in the fubtriplicate ratio of the radius of curvature at E to that at F, by Cor. 1, prop. 4, part 6 of Milnes's Conic Sections, and therefore in a given ratio to one another; alfo the angles QCP, RCP are the latitudes of E and F; fo that, drawing QV parallel to P*p*, and QXYW to A*a*, thefe angles being given, as well as the ratio of CQ to CR, the rectilinear figure CVQXRY is given in fpecies; and the ratio of VC² − ZC² (= QX × XW) to RZ² − QV² (= RX × XS) is given, which is the ratio of CA² to CP²; therefore the ratio of CA to CP is given.

Hence, if the fine and cofine of the greater latitude be each augmented in the fubtriplicate ratio of the meafure of the degree in the greater latitude to that in the lefier, then the difference of the fquares of the augmented fine, and the fine of the lefier latitude, will be to the difference of the fquares of the cofine of the lefier latitude and the augmented cofine, in the duplicate ratio of the equatorial to the polar diameter. For, C*q* being taken in CQ equal to CR, and *qv* drawn parallel to QV, C*v* and *vq*, CZ and ZR will be the figns and cofines of the refpective latitudes to the fame radius; and CV, VQ will be the augmentations of C*v* and C*q* in the ratio named.

Hence, to find the ratio between the two axes of the earth, let E denote the greater, and F the lefier of the two latitudes, M and N the refpective meafures taken in each; and let P denote $\sqrt[3]{\dfrac{M}{N}}$: then

$$\sqrt{\frac{\text{cof.}^2\,F - P^2 \times \text{cof.}^2\,E}{P^2 \times \text{fin.}^2\,E - \text{fin.}^2\,F}} \text{ is} = \frac{\text{lefier axis}}{\text{greater axis}}.$$

It alfo appears by the above problem, that when one of the degrees meafured, is at the equator, the cofine of the latitude of the other being augmented in the fubtriplicate ratio of the degrees, the tangent of the latitude will be to the tangent anfwering to the augmented cofine, in the ratio of the greater axis to the lefs. For fuppofing E the place out of the equator; then if the femi-circle P*lmnp* be defcribed, and *l*C joined, and *mo* drawn parallel to *a*C: C*o* is the cofine of the latitude to the radius CP, and CY that cofine augmented in the ratio before-named; YQ being to Y*l*, that is C*a* to C*n* or CP, as the tangent of the angle YCQ, the latitude of the point E, to the tangent of the angle YC*l*, belonging to the augmented cofine. Thus, if M reprefent the meafure in a latitude denoted by E, and N the meafure at the equator, let A denote an angle whofe meafure is

cof. E × $\sqrt[3]{\dfrac{M}{N}}$. Then $\dfrac{\tan. A}{\tan. E}$ is $= \dfrac{\text{lefier axis}}{\text{greater axis}}$.

But M, or the length of a degree, obtained by actual menfuration in different latitudes, is known from the following table:

Names.	Latit.	Value of M.
	° ′	toifes
Maupertuis and Affoc.	66 20	M = 57438
Caffini and {	49 22	M = 57074
La Caille {	45 00	M = 57050
Bofcovich	43 00	M = 56972
De la Caille	33 18	M = 57037
Juan and Ulloa	00 00	M = 56768 } at the
Bouguer	00 00	M = 56753 } equa-
Condamine	00 00	M = 56749 } tor.

Now, by comparing the 1ft with each of the following ones; the 2d with each of the following; and in like manner the 3d, 4th, and 5th, with each of the following; there will be obtained 25 refults, each fhewing the relation of the axes or diameters; the arithmetical means of all of which will give that ratio as 1 to 0·9951989.

If the meafures of the latitudes of 49° 22′, and of 45°, which fall within the meridian line drawn through France, and which have been re-examined and corrected fince the northern and fouthern expedition, be compared with thofe of Maupertuis and his affociates in the north, and that of Bouguer at the equator, there will refult 6 different values of the ratio of the two axes, the arithmetical mean of all which is that of 1 to 0·9953467, which may be confidered as the ratio of the greater axis to the lefs; which is as 230 to 228·92974, or 215 to 214, or very near the ratio as affigned by Newton.

Now, the magnitude as well as the figure of the earth, that is the polar and equatorial diameters, may be deduced from the foregoing problem. For, as half the latus rectum of the greater axis A*a* is the radius of curvature at A, it is given in magnitude from the degree meafured there, and thence the axes themfelves are given. Thus, the circular arc whofe length is equal to the radius being 57·29578 degrees, if this number be multiplied by 56750 toifes, the meafure of a degree at the

the equator, as Bouguer has stated it, the product will be the radius of curvature there, or half the latus rectum of the greater axis ; and this is to half the lesser axis in the ratio of the less axis to the greater, that is, as 0·9953467 to 1 ; whence the two axes are 6533820 and 6564366 toises, or 7913 and 7550 English miles ; and the difference between the two axes about 37 miles. See Robertson's Navigation, vol. 2, pa. 206 &c. See also Suite des Mem. de l'Acad. 1718, pa. 247, and Maclaurin's Fluxions, vol. 2, book 1, ch. 14.

And very nearly the same ratio is deduced from the lengths of pendulums vibrating in the same time, in different latitudes ; provided it be again allowed that the meridians are real ellipses, or the earth a true spheroid, which however can only take place in the case of an uniform gravity in all parts of the earth.

Thus, in the new Petersburgh Acts, for the years 1788 and 1789, are accounts and calculations of experiments relative to this subject, by M. Krafft. These experiments were made at different times and in various parts of the Russian empire. This gentleman has collected and compared them, and drawn the proper conclusions from them : thus he infers that the length x of a pendulum that swings seconds in any given latitude λ, and in a temperature of 10 degrees of Reaumur's thermometer, may be determined by this equation :

$x = 439\cdot178 + 2\cdot321$ sine $^2\lambda$, lines of a French foot,
or $x = 39\cdot0045 + 0\cdot206$ sine $^2\lambda$, in English inches,
in the temperature of 53 of Fahrenheit's thermometer.

This expression nearly agrees, not only with all the experiments made on the pendulum in Russia, but also with those of Mr. Graham in England, and those of Mr. Lyons in 79° 50′ north latitude, where he found its length to be 431·38 lines. It also shews the augmentation of gravity from the equator to the parallel of a given latitude λ : for, putting g for the gravity under the equator, G for that under the pole, and y for that under the latitude λ, M. Krafft finds
$y = (1 + 0\cdot0052848$ sine $^2\lambda) g$; and theref. G = 1·0052848 g.

From this proportion of gravity under different latitudes, the same author infers that, in case the earth is a homogeneous ellipsoid, its oblateness must be $\frac{1}{577}$; instead of $\frac{1}{230}$, which ought to be the result of this hypothesis : but on the supposition that the earth is a heterogeneous ellipsoid, he finds its oblateness, as deduced from these experiments, to be $\frac{1}{337}$; which agrees with that resulting from the measurement of some of the degrees of the meridian. This confirms an observation of M. De la Place, that, if the hypothesis of the earth's homogeneity be given up, then theory, the measurement of degrees of latitude, and experiments with the pendulum, all agree in their result with respect to the oblateness of the earth. See Memoires de l' Acad. 1783, pa. 17.

In the Philos. Transf. for 1791, pa. 236, Mr. Dalby has given some calculations on measured degrees of the meridian, from whence he infers, that those degrees measured in middle latitudes, will answer nearly to an ellipsoid whose axes are in the ratio assigned by Newton, viz, that of 230 to 229. And as to the deviations of some of the others, viz, towards the poles

and equator, he thinks they are caused by the errors in the observed celestial arcs.

Tacquet draws some pretty little inferences, in the form of paradoxes, from the round figure of the earth ; as, 1st, That if any part of the surface of the earth were quite plane, a man could no more walk upright upon it, than on the side of a mountain. 2d, That the traveller's head goes a greater space than his feet ; and a horseman than a footman ; as moving in a greater circle. 3d, That a vessel, full of water, being raised perpendicularly, some of the water will be continually flowing out, yet the vessel still remain full ; and on the contrary, if a vessel of water be let perpendicularly down, though nothing flow out, yet it will cease to be full : consequently there is more water contained in the same vessel at the foot of a mountain, than on the top ; because the surface of the water is compressed into a segment of a smaller sphere below than above. Tacq. Astron. lib. 1, cap. 2.

Changes of the EARTH. Mr. Boyle suspects that there are great, though slow internal changes, in the mass of the earth. He argues from the varieties observed in the change of the magnetic needle, and from the observed changes in the temperature of climates. But as to the latter, there is reason to doubt that he could not have diaries of the weather sufficient to direct his judgment. Boyle's Works abr. vol. 1, pa. 292, &c.

Magnetism of the EARTH. The notion of the magnetism of the earth was started by Gilbert ; and Boyle supposes magnetic effluvia moving from one pole to the other. See his Works abr. vol. 1, pa. 285, 290.

Dr. Knight also thinks that the earth may be considered as a great loadstone, whose magnetical parts are disposed in a very irregular manner ; and that the south pole of the earth is analogous to the north pole in magnets, that is, the pole by which the magnetical stream enters. See MAGNET.

He observes, that all the phenomena attending the direction of the needle, in different parts of the earth, in great measure correspond with what happens to a needle, when placed upon a large terrella ; if we make allowances for the different dispositions of the magnetical parts, with respect to each other, and consider the south pole of the earth as a north pole with regard to magnetism. The earth might become magnetical by the iron ores it contains, for all iron ores are capable of magnetism. It is true, the globe might notwithstanding have remained unmagnetical, unless some cause had existed capable of making that repellent matter producing magnetism move in a stream through the earth.

Now the doctor thinks that such a cause does exist. For if the earth revolves round the sun in an ellipsis, and the south pole of the earth is directed towards the sun, at the time of its descent towards it, a stream of repellent matter will thence be made to enter at the south pole, and issue out at the north. And he suggests, that the earth's being in its perihelion in winter may be one reason why magnetism is stronger in this season than in summer.

This cause here assigned for the earth's magnetism must continue, and perhaps improve it, from year to year. Hence the doctor thinks it probable, that the earth's magnetism has been improving ever since the creation, and that this may be one reason why the use

of

of the compass was not discovered sooner. See Dr. Knight's Attempt to demonstrate, that all the phenomena in nature may be explained by Attraction and Repulsion, prop. 87.

Magnitude and Constitution of the EARTH. This has been variously determined by different authors, both ancient and modern. The usual way has been, to measure the length of one degree of the meridian, and multiply it by 360, for the whole circumference. See DEGREE. Diogenes Laertius informs us, that Anaximander, a scholar of Thales, who lived about 550 years before the birth of Christ, was the first who gave an account of the circumference of the sea and land; and it seems his measure was used by the succeeding mathematicians, till the time of Eratosthenes. Aristotle, at the end of lib. 2 De Cœlo, says, the mathematicians who have attempted to measure the circuit of the earth, make it 40000 stadiums: which it is thought is the number determined by Anaximander.

Eratosthenes, who lived about 200 years before Christ, was the next who undertook this business; which, as Cleomedes relates, he performed by taking the sun's zenith distances, and measuring the distance between two places under the same meridian; by which he deduced for the whole circuit about 250000 stadiums, which Pliny states at 31500 Roman miles, reckoning each at 1000 paces. But this measure was accounted false by many of the ancient mathematicians, and particularly by Hipparchus, who lived 100 years afterwards, and who added 25000 stadiums to the circuit of Eratosthenes.

Possidonius, in the time of Cicero and Pompey the Great, next measured the earth, viz, by means of the altitudes of a star, and measuring a part of a meridian; and he concluded the circumference at 240000 stadiums, according to Cleomedes, but only at 180000 according to Strabo.

Ptolomy, in his Geography, says that Marinus, a celebrated geographer, attempted something of the same kind; and, in lib. 1, cap. 3, he mentions that he himself had tried to perform the business in a way different from any other before him, which was by means of places under different meridians: but he does not say how much he made the number; for he still made use of the 180000, which had been found out before him.

Snellius relates, from the Arabian Geographer Abelfedea, who lived about the 1300th year of Christ, that about the 800th year of Christ, Almaimon, an Arabian king, having collected together some skilful mathematicians, commanded them to find out the circumference of the earth. Accordingly these made choice of the fields of Mesopotamia, where they measured under the same meridian from north to south, till the pole was depressed one degree lower: which measure they found equal to 56 miles, or 56½: so that according to them the circuit of the earth is 20160 or 20340 miles.

It was a long time after this before any more attempts were made in this business. At length however, the same Snell, above mentioned, professor of mathematics at Leyden, about the year 1620, with great skill and labour, by measuring large distances between two parallels, found one degree equal to 28500 perches, each of which is 12 Rhinland feet, amounting to 19

Dutch miles, and so the whole periphery 6840 miles; a mile being, according to him, 1500 perches, or 18000 Rhinland feet. See his treatise called *Eratosthenes Batavus.*

The next that undertook this measurement, was Richard Norwood, who in the year 1635, by measuring the distance from London to York with a chain, and taking the sun's meridian altitude, June 11th old style, with a sextant of about 5 feet radius, found a degree contained 367200 feet, or 69 miles and a half and 14 poles; and thence the circumference of a great circle of the earth is a little more than 25036 miles, and the diameter a little more than 7966 miles. See the particulars of this measurement in his *Seaman's Practice.*

The measurement of the earth by Snell, though very ingenious and troublesome, and much more accurate than any of the ancients, being still thought by some French mathematicians, as liable to certain small errors, the business was renewed, after Snell's manner, by Picard and other mathematicians, by the king's command; using a quadrant of 3⅛ French feet radius; by which they found a degree contained 342360 French feet. See Picard's treatise, *La Mesure de la Terre.*

M. Cassini the younger, in the year 1700, by the king's command also, renewed the business with a quadrant of 10 feet radius, for taking the latitude, and another of 3¼ feet for taking the angles of the triangles; and found a degree, from his calculation, contained 57292 toises, or almost 69¼ English miles.

See the results of many other measurements under the article DEGREE. From the mean of all which, the following dimensions may be taken as near the truth:

the circumference	25000 miles,
the diameter	7957¾ miles,
the superficies	198944206 square miles,
the solidity	263930000000 cubic miles.

Also the seas and unknown parts of the earth, by a measurement of the best maps, contain 160522026 square miles, the inhabited parts 38922180; of which Europe contains 4456065; Asia, 10768823; Africa, 9654807; and America, 14110874.

It is now generally granted that the terraqueous globe has two motions, besides that on which the precession of the equinoxes depends; the one diurnal around its own axis in the space of 24 hours, which constitutes the natural day or nycthemeron; the other annual, about the sun, in an elliptical orbit or track, in 365 days 6 hours, constituting the year. From the former arise the diversities of night and day; and from the latter, the vicissitudes of seasons, spring, summer, autumn, winter.

The terraqueous globe is distinguished into three parts or regions, viz, 1st, The external part or crust, being that from which vegetables spring and animals are nursed. 2d, The middle, or intermediate part, which is possessed by fossils, extending farther than human labour ever yet penetrated. 3d, The internal or central part, which is utterly unknown to us, though by many authors supposed of a magnetic nature; by others, a mass or sphere of fire; by others, an abyss or collection of waters, surrounded by the strata of earth; and by others, a hollow, empty space, inhabited by animals, who have their sun, moon, planets, and

other

other conveniences within the fame. But others divide the body of the globe into two parts, viz, the external part, called the cortex, including the internal, which they call the nucleus, being of a different nature from the former, and poffeffed by fire, water, or more probably by a confiderable portion of metals, as it has been found, by calculation, that the mean denfity of the whole earth is near double the denfity of common ftone. See my determination of it, Philof. Tranf. 1778, pa. 781.

The external part of the globe either exhibits inequalities, as mountains and valleys; or it is plane and level; or dug in channels, fiffures, beds, &c, for rivers, lakes, feas, &c. Thefe inequalities in the face of the earth moft naturalifts fuppofe have arifen from a rupture or fubverfion of the earth, by the force either of the fubterraneous fires or waters. The earth, in its natural and original ftate, it has been fuppofed by Des Cartes, and after him, Burnet, Steno, Woodward, Whifton, and others, was perfectly round, fmooth, and equable; and they account for its prefent rude and irregular form, principally from the great deluge. See DELUGE.

In the external, or cortical part of the earth, there appear various ftrata, fuppofed the fediments of feveral floods; the waters of which, being replete with matters of divers kinds, as they dried up, or oozed through, depofited thefe different matters, which in time hardened into ftrata of ftone, fand, coal, clay, &c.

Dr. Woodward has confidered the circumftances of thefe ftrata with great attention, viz, their order, number, fituation with refpect to the horizon, depth, interfections, fiffures, colour, confiftence, &c. He afcribes the origin and formation of them all, to the great flood or cataclyfmus. At that terrible revolution he fuppofed that all forts of terreftrial bodies had been diffolved and mixed with the waters, forming all together a chaos or confufed mafs. This mafs of terreftrial particles, intermixed with water, he fuppofes was at length precipitated to the bottom; and that generally according to the order of gravity, the heavieft finking firft, and the lighteft afterwards. By fuch means were the ftrata formed of which the earth confifts; which, attaining their folidity and hardnefs by degrees, have continued fo ever fince. Thefe fediments, he farther concludes, were at firft all parallel and concentrical; and the furface of the earth formed of them, perfectly fmooth and regular; but that in courfe of time, divers changes happening, from earthquakes, volcanos, &c, the order and regularity of the ftrata was difturbed and broken, and the furface of the earth by fuch means brought to the irregular form in which it now appears.

M. De Buffon furmifes that the earth, as well as the other planets, are parts ftruck off from the body of the fun by the collifion of comets; and that when the earth affumed its form, it was in a ftate of liquefaction by fire. But that could not be the method of producing the planets; for if they were ftruck off from the body of the fun, they would move in orbits that would always pafs through the fun, inftead of having the fun for their focus, or centre, as they are now found; fo that having been ftruck off they would fall down into the fun again, terminating their career as it were after one revolution only.

EARTH, in *Aftronomy*, is one of the primary planets, according to the fyftem of Copernicus, or Pythagoras; its aftronomical character or mark being ⊖: but according to the Ptolomaic hypothefis, the earth is the centre of the fyftem. For, whether the earth move or remain at reft, that is, whether it be fixed in the centre, having the fun, the heavens and ftars moving round it from eaft to weft; or, thefe being at reft, whether the earth only moves from weft to eaft, is the great article that diftinguifhes the Ptolomaic fyftem from the Copernican.

Motion of the EARTH. It is now univerfally agreed that, befides the fmall motion of the earth which caufes the preceffion of the equinoxes, the earth has two great and independent motions; viz, the one by which it turns round its own axis, in the fpace of 24 hours nearly, and caufing the continual fucceffion of day and night; and the other an abfolute motion of its whole mafs in a large orbit about the fun, having that luminary for its centre, in fuch manner that its axis keeps always parallel to itfelf, inclined in the fame angle to its path, and by that means caufing the viciffitudes of feafons, fpring, fummer, autumn, winter.

It is indeed true that, as to fenfe, the earth appears to be fixed in the centre, with the fun, ftars and heavens moving round it every day; and fuch doubtlefs would be confidered as the true nature of the motions in the rude ages of mankind, as they are ftill by the rude and unlearned. But to a thinking and learned mind, the contrary will foon appear.

Indeed there are traces of the knowledge of thefe motions in the earlieft age of the fciences. Cicero, in his Tufc. Quæft. fays that Nicetas of Syracufe firft difcovered that the earth had a diurnal motion, by which it revolved round its axis every 24 hours; and Plutarch, de Placit. Philofoph. informs, that Philolaus difcovered its annual motion round the fun; and Ariftarchus, about 100 years after Philolaus, propofed the motion of the earth in ftronger and clearer terms, as we are affured by Archimedes, in his Arenarius. And the fame, we are farther affured, was the opinion and doctrine of Pythagoras.

But the religious opinions of the heathen world prevented this doctrine from being more cultivated. For, Ariftarchus being accufed of facrilege by Cleanthes for moving Vefta and the tutelar deities of the univerfe out of their places, the philofophers were obliged to diffemble, and feem to relinquifh fo perilous a pofition.

Many ages afterwards, Nic. Cufanus revived the ancient fyftem, in his Doct. de Pignorant. and afferted the motion of the earth: but the doctrine gained very little ground till the time of Copernicus, who fhewed its great ufe and advantages in aftronomy; and who had immediately all the philofophers and aftronomers on his fide, who dared to differ from the crowd, and were not afraid of ecclefiaftical cenfure, which was not lefs dangerous under the chriftian difpenfation, than it had been under that of the heathen. For, becaufe certain parts of fcripture make mention of the ftability

of

of the earth, and of the motion of the sun, as the rifing and fetting, &c, the fathers of the church thought their religion required that they fhould defend, with all its power, what they conceived to be its doctrines, and to cenfure and punifh every attempt at innovation on fuch points. They have now however been pretty generally convinced that in fuch inftances the expreffions are only to be confidered as accommodated to appearances, and the vulgar notions of things.

By the diurnal rotation of the earth on its axis, the fame phenomena will take place as if it had no fuch motion, and as if the fun and ftars moved round it. For, turning round from weft to eaft, caufes the fun and all the vifible heavens to feem to move the contrary way, or from eaft to weft, as we daily fee them do. So, when in its rotation it has brought the fun or a ftar to appear juft in the horizon in the eaft, they are then faid to be rifing; and as the earth continues to revolve more and more towards the eaft, other ftars feem to rife and advance weftwards, paffing the meridian of the obferver, when they are due fouth from him, and at their greateft altitude above his horizon; after which, by a continuance of the fame motions, viz, of the earth's rotation eaftwards, and the luminaries apparent counter motion weftwards, thefe decline from the meridian, or fouth point, towards the weft, where being arrived, they are faid to fet and defcend below it; and fo on continually from day to day; thus making it day while the fun is above the horizon, and night while he is below it.

While the earth is thus turning on its axis, it is at the fame time carried by its proper motion in its orbit round the fun, as one of the planets, namely, between the orbits of Venus and Mars, having the orbits of Venus and Mercury within its own, or between it and the fun, in the centre, and thofe of Mars, Jupiter, Saturn, &c, without or above it; which are therefore called fuperior planets, and the others the inferior ones. This is called the annual motion of the earth, becaufe it is performed in a year, or 365 days 6 hours nearly; or rather 365 days 5^h 49^m, from any equinox or folftice to the fame again, making the tropical year; but from any fixed ftar to the fame again, as feen from the fun, in 365 days 6 hours 9 minutes, which is called the fidereal year. The figure of this orbit is elliptical, having the fun in one focus, the mean diftance being about 95 millions of miles, which is upon the fuppofition that the fun's parallax is about 8″⅓, or the angle under which the earth's femi-diameter would appear to an obferver placed in the fun: and the eccentricity of the orbit, or diftance of the fun, in the focus, from the centre of this elliptic orbit, is about $\frac{1}{60}$th of the mean diftance.

Now this annual motion is performed in fuch a manner, that the earth's axis is every where parallel, or in the fame direction in every part of the orbit; by which means it happens, that at one time of the year the fun enlightens more of the north polar parts, and at the oppofite feafon of the year more of the fouthern parts, thus fhewing all the varieties of feafons, fpring, fummer, autumn and winter; which may be illuftrated in the following manner: Let the candle I (fig. 1, plate viii) reprefent the fun, about which the

earth E, or F, &c, is moved in its elliptical orbit ABCD, or ecliptic, and cutting the equator *abcd* in the nodes E and G: then, fufpending the terrella by its north pole, and moving it, fo fufpended, round the ecliptic, its axis will always be parallel to its firft pofition, and the various feafons will be reprefented at the different parts of the path. Thus, when the earth is at ♋ or F, the enlightened half of it includes the fouth pole, and leaves the north pole in darknefs, making our winter; at G it is fpring, and the two poles are equally illuminated, and the days are every where of the fame length; at H or ♑ it is our fummer, the north polar parts being in the illuminated hemifphere, and the fouthern in the dark one; laftly at E it is autumn, the poles being equally illuminated again, and the days of equal length every where.

EARTH, *its Quantity of Matter, Denfity, and Attractive Power*. Although the relative denfities of the earth and moft of the other planets have been known a confiderable time, it is but very lately that we have come to the knowledge of the abfolute gravity or denfity of the whole mafs of the earth. This I have calculated and deduced from the obfervations made by Dr. Mafkelyne, Aftronomer Royal, at the mountain Schehallien, in the years 1774, 5, and 6. The attraction of that mountain on a plummet, being obferved on both fides of it, and its mafs being computed from a number of fections in all directions, and confifting of ftone; thefe data being then compared with the known attraction and magnitude of the earth, gave by proportion its mean denfity, which is to that of water as 9 to 2, and to common ftone as 9 to 5: from which very confiderable mean denfity, it may be prefumed that the internal parts contain fome great quantities of metals.

From the denfity, now found, its quantity of matter becomes known, being equal to the product of its denfity by its magnitude. From various experiments too, we know that its attractive force, at the furface, is fuch, that bodies fall there through a fpace of 16$\frac{1}{12}$ feet in the firft fecond of time: from whence the force at any other place, either within or without it, becomes known; for the force at any part within it, is directly as its diftance from the centre; but the force of any part without it, reciprocally as the fquare of its diftance from the centre.

EAST, one of the cardinal points of the horizon, or of the compafs, being the middle point of it between north and fouth, on that fide where the fun rifes, or the point in which it is interfected on that fide by the prime vertical.

EASTER, a feaft of the church, held in memory of our Saviour's refurrection. This feaft has been annually celebrated ever fince the time of the apoftles, and is one of the moft confiderable feftivals in the chriftian calendar; being that which regulates and determines the times of all the other moveable feafts.

The rule for the celebration of Eafter, fixed by the council of Nice, in the year 325, is, that it be held on the Sunday which falls upon, or next after, the full moon which happens next after the 21ft of March; that is, the Sunday which falls upon, or next after the firft full moon after the vernal equinox. The reafon of which

which decree was, that the chriſtians might avoid cele-brating their Eaſter at the ſame time with the Jewiſh Paſſover, which, according to the inſtitution of Moſes, was held the very day of the full moon.

To find EASTER *according to the Old, or Julian Style.*

In the annexed table, find the golden number, with the day of the paſchal full moon, and the Sunday let-ter annexed; compare this letter with the dominical letter of the given year, that it may appear how many days are to be added to the day of the paſchal full moon, to give Eaſter-day.

For ex. In the year 1715, the dominical letter is B, and the golden number is 6, op-ſite to which ſtands April 10 for the day of the paſ-chal full moon; oppoſite to which is the Sunday letter B, which happening to be the ſame with that of the year given, that day is a Sunday; and therefore Eaſter will fall 7 days after, viz, on the 17th of April.

But in this computation, the vernal equinox is ſup-poſed fixed to the 21ſt of March; and the cycle of 19 years, or golden num-bers, is ſuppoſed to point out the places of the new and full moons exactly; both which ſuppoſitions are

Gold. Num.	Paſchal full moons.		Dom. letter.
16	March	21	C
5		22	D
		23	E
13		24	F
2		25	G
		26	A
10		27	B
		28	C
18		29	D
7		30	E
		31	F
15	April	1	G
4		2	A
		3	B
12		4	C
1		5	D
		6	E
9		7	F
		8	G
17		9	A
6		10	B
		11	C
14		12	D
3		13	E
		14	F
11		15	G
		16	A
19		17	B
8		18	C

erroneous: ſo that the Julian Eaſter never happens at its due time, unleſs by accident. For inſtance, in the above example the vernal equinox falls on the 10th of March, eleven days before the rule ſuppoſes it; and the paſchal full moon on the 7th of April, or 3 days earlier than was ſuppoſed: and therefore Eaſter-day ſhould be held on the 10th of April, inſtead of the 17th.

This error had grown to ſuch a height, that pope Gregory the 13th thought it neceſſary to correct it; and accordingly, in the year 1582, by the advice of Aloyſius Lilius and others, he ordered 10 days to be thrown out of October, to bring the vernal equinox back again to the 21ſt of March: and hence ariſe the terms Gregorian calendar, Gregorian year, &c.

This correction however did not entirely remove the error; for the equinoxes and ſolſtices ſtill anticipate 28′ 20″ in every 100 Gregorian years; but the dif-ference is ſo inconſiderable as not to amount to a whole day, or 24 hours, in leſs than 5082 Gregorian years.

The Gregorian, or New Style, was not introduced into England till the year 1752, when eleven days were thrown out, viz, between the 3d and 14th of September, the error amounting then to that quantity.

To find EASTER *according to the New or Gregorian Style, till the year* 1900 *excluſive.* Look for the golden num-ber of the year in the firſt column of the table, againſt which ſtands the day of the paſchal full moon; then look in the 3d column for the Sunday letter, next after the day of the full moon, and the day of the month ſtand-ing againſt that Sunday let-ter is Eaſter-day. When the full moon happens on a Sunday, then the next Sun-day after is Eaſter-day.

For Ex. For the year 1790, the golden number is 5; againſt which ſtands March the 30th, and the next Sunday letter, which is C, below that, ſtands oppoſite April 4, which is therefore the Eaſter day for the year 1790.

Though the Gregorian calendar be much prefera-ble to the Julian, it is yet not without its defects. It cannot, for inſtance, keep the equinox fixed on the 21ſt of March, but it will ſometimes fall on the 19th, and ſometimes on the 23d. Add, that the full moon happening on the 20th of March, might ſometimes be paſchal; yet it is not al-lowed as ſuch in the Gregorian computation; as on the contrary, the full moon of the 22d of March may be allowed for paſchal, which it is not. Scaliger and Cal-viſius have alſo pointed out other inaccuracies in this ca-lendar. An excellent paper on this ſubject by the earl of Macclesfield, may be ſeen in the Philoſ. Tranſ. vol. 40, pa. 417.

EASTER *Term.* See TERM.

EAVES, the margin or lower edge of the roof of a houſe, being the loweſt courſe of tiles, ſlates, or the like, which hang over the walls to throw off the water to a diſtance from the wall.

EAVES-*Board*, or EAVES-*Lath*, a thick feather-edged board, uſually nailed round the eaves of a houſe for the lowermoſt tiles, ſlate, or ſhingles, to reſt upon.

EBBING *and* FLOWING *of the Sea.* See TIDES.

ECCENTRIC. See EXCENTRIC.

ECCENTRICITY. See EXCENTRICITY.

ECHO, or ECCHO, a ſound reflected, or reverbe-rated from ſome body, and thence returned or repeated to the ear.

For an echo to be heard, the ear muſt be in the line of reflection; that the perſon who made the ſound, may hear the echo, it is neceſſary he ſhould be in a perpendicular to the place which reflects it; and for a multiple

Gold. Num.	Paſchal full moon.		Sund. letter.
14	March	21	C
3		22	D
11		23	E
		24	F
		25	G
19		26	A
8		27	B
		28	C
16		29	D
5		30	E
		31	F
13	April	1	G
2		2	A
		3	B
10		4	C
		5	D
18		6	E
7		7	F
		8	G
15		9	A
4		10	B
		11	C
12		12	D
1		13	E
		14	F
9		15	G
		16	A
17		17	B
6		18	C

I

multiple or tautological echo, it is neceffary there be a number of walls and vaults, rocks, and cavities, either placed behind each other, or fronting each other. Thofe murmurs in the air, that are occafioned by the difcharge of great guns, &c, are a kind of indefinite echoes, and are produced from the vaporous particles fufpended in the atmofphere, which refift the undulations of found, and reverberate them to the ear.

There can be no echo, unlefs the direct and reflex founds follow one another at a fufficient diftance of time; for if the reflex found arrive at the ear before the impreffion of the direct found ceafes, the found will not be doubled, but only rendered more intenfe. Now if we allow that 9 or 10 fyllables can be pronounced in a fecond, in order to preferve the founds articulate and diftinct, there fhould be about the 9th part of a fecond between the times of their appulfe to the ear; or, as found flies about 1142 feet in a fecond, the faid difference fhould be $\frac{1}{9}$ of 1142, or 127 feet; and therefore every fyllable will be reflected to the ear at the diftance of about 70 feet from the reflecting body; but as, in the ordinary way of fpeaking, 3 or 4 fyllables only are uttered in a fecond, the fpeaker, that he may have the echo returned as foon as they are expreffed, fhould ftand about 500 feet from the reflecting body; and fo in proportion for any other number of fyllables. Merfenne allows for a monofyllable the diftance of 69 feet; Morton, 90 feet; for a diffyllable 105 feet, a trifyllable 160 feet, a tetrafyllable 182 feet, and a pentafyllable 204 feet. Nat. Hift. Northampton, cap. 5, pa. 358.

From what has been faid, it follows that echoes may be applied for meafuring inacceffible diftances. Thus, Mr. Derham, ftanding upon the banks of the Thames, oppofite to Woolwich, obferved that the echo of a fingle found was reflected back from the houfes in 3 feconds; confequently the fum of the direct and reflex rays muft have been $1142 \times 3 = 3426$ feet, and the half of it, 1713 feet, the breadth of the river in that place.

It alfo follows that the echoing body being removed farther off, it reflects more of the found than when nearer; which is the reafon why fome echoes repeat but one fyllable, or one word, and fome many. Of thefe, fome are tonical, which only return a voice when modulated into fome particular mufical tone; and others polyfyllabical. That fine echo in Woodftock park, Dr. Plot affures us, in the day-time will return very diftinctly 17 fyllables, and in the night 20. Nat. Hift. Oxf. cap. 1, pa. 7.

Echoing bodies may be fo contrived, and placed, as that reflecting the found from one to the other, a multiple echo, or many echoes, fhall arife.—At Rofneath, near Glafgow, in Scotland, there is an echo that repeats a tune played with a trumpet three times completely and diftinctly.—At the fepulchre of Metella, wife of Craffus, there was an echo, which repeated what a man faid five times.—Authors mention a tower at Cyzicus, where the echo repeated feven times.—There is an echo at Bruffels, that anfwers 15 times.

One of the fineft echoes we read of, is that mentioned by Barthius, in his notes on Statius's Thebais, lib. 6, ver. 30, which repeated the words a man uttered 17 times. This was on the banks of the Naha, between Coblentz and Bingen. And whereas, in common echoes, the repetition is not heard till fome time after hearing the words

spoken, or the notes fung; in this, the perfon who fpeaks, or fings, is fcarce heard at all; but the repetition very clearly, and always in furprifing varieties; the echo feeming fometimes to approach nearer, and fometimes farther off; fometimes the voice is heard very diftinctly, and fometimes fcarce at all: one perfon hears only one voice, and another feveral; one hears the echo on the right, and the other on the left, &c.

Addifon, and other travellers in Italy, mention an echo at Simonetta palace, near Milan, ftill more extraordinary, returning the found of a piftol 56 times. The echo is heard behind the houfe, which has two wings; the piftol is difcharged from a window in one of thefe wings, the found is returned from a dead wall in the other wing, and heard from a window in the back-front. See Addif. Travels, pa. 32; Miffon; Voyag. d'Ital. tom. 2, pa. 196; Philof. Tranf. N° 480, pa. 220.

Farther, a multiple echo may be made, by fo placing the echoing bodies, at unequal diftances, as that they may reflect all one way, and not one on the other; by which means, a manifold fucceffive found will be heard; one clap of the hands like many; one ha like a laughter; one fingle word like many of the fame tone and accent; and fo one mufical inftrument like many of the fame kind, imitating each other.

Laftly, echoing bodies may be fo ordered, that from any one found given, they fhall produce many echoes, different both as to tone and intenfion. By which means a mufical room may be fo contrived, that not only one inftrument playing in it fhall feem many of the fame fort and fize, but even a concert of different ones; this may be contrived by placing certain echoing bodies fo, as that any note played, fhall be returned by them in 3ds, 5ths, and 8ths.

Echo is alfo ufed for the place where the repetition of the found is produced, or heard. This is either natural or artificial.

In echoes, the place where the fpeaker ftands, is called the *centrum phonicum*; and the object or place that returns the voice, the *centrum phonocampticum*.

Echo, in Architecture, is applied to certain vaults and arches, moftly of elliptical or parabolical figures; ufed to redouble founds, and produce artificial echoes. —The method of making them is taught by F. Blancani, in his Echometria, at the end of his book on the Sphere.

Vitruvius tells us, that in divers parts of Greece and Italy there were brazen veffels, artfully ranged under the feats of the theatres, to render the found of the actors' voices more clear, and make a kind of echo; by which means, every one of the prodigious multitude of perfons, prefent at thofe fpectacles, might hear with eafe and pleafure.

ECLIPSAREON, an inftrument invented by Mr. Ferguson for fhewing the phenomena of eclipfes; as their time, quantity, duration, progrefs, &c. Ferguson's Aftron., or Philof. Tranf. vol. 48, pa. 520.

ECLIPSE, a privation of the light of one of the luminaries, by the interpofition of fome opaque body, either between it and the eye, or between it and the fun.

The ancients had terrible ideas of eclipfes: fuppofing them prefages of fome dreadful events. Plutarch affures us, that at Rome it was not allowed to talk publicly of any natural caufes of eclipfes; the popular opinion running fo ftrongly in favour of their fupernatural production, at leaft thofe of the moon; for as to thofe of the fun, they had fome idea that they were caufed by the interpofition of the moon between us and the fun; but were at a lofs for a body to interpofe between us and the moon, which they thought muft be the way, if the eclipfes of the moon were produced by natural caufes. They therefore made a great noife with brazen inftruments, and fet up loud fhouts, during eclipfes of the moon; thinking by that means to eafe her in labour: whence Juvenal, fpeaking of a talkative woman, fays, *Una laboranti poterit fuccurrere lunæ*. Others attributed the eclipfe of the moon to the arts of magicians, who, by their inchantments, plucked her out of heaven, and made her fkim over the grafs.

The natives of Mexico keep faft during eclipfes; and particularly their women, who beat and abufe themfelves, drawing blood from their arms, &c; imagining the moon has been wounded by the fun, in fome quarrel between them.

The Chinefe fancy that eclipfes are occafioned by great dragons, who are ready to devour the fun and moon; and therefore when they perceive an eclipfe, they rattle drums and brafs kettles, till they think the monfter, terrified by the noife, lets go his prey.

The fuperftitious notions entertained of eclipfes, were once of confiderable advantage to Chriftopher Columbus, the difcoverer of America, who being driven on the ifland of Jamaica in the year 1493, and diftreffed for want of provifions, was refufed relief by the natives; but having threatened them with a plague, and foretelling an eclipfe as a token of it, which happened according to his prediction, the barbarians were fo terrified, that they ftrove who fhould be the firft in bringing him fupplies, throwing them at his feet, and imploring forgivenefs.

Duration of an Eclipse, is the time of its continuance, or between the immerfion and emerfion.

Immerfion, or Incidence, of an Eclipse, is the moment when the eclipfe begins, or when part of the fun, moon, or planet firft begins to be obfcured.

Emerfion, or Expurgation, of an Eclipse, is the time when the eclipfed luminary begins to re-appear, or emerge out of the fhadow.

Quantity of an Eclipse, is the part of the luminary eclipfed. To determine the quantity eclipfed, it is ufual to divide the diameter of the luminary into 12 equal parts, called digits; whence the eclipfe is faid to be of fo many digits according to the number of them contained in that part of the diameter which is eclipfed or obfcured.

Eclipfes are divided, with refpect to the luminary eclipfed, into *Eclipfes of the fun, of the moon*, and *of the fatellites*; and with refpect to the circumftances, into *total, partial, annular, central*, &c.

Annular Eclipse, is when the whole is eclipfed, except a ring, or annulus, which appears round the border or edge.

Central Eclipse, is one in which the centres of the two luminaries and the earth come into the fame ftraight line.

Partial Eclipse, is when only a part of the luminary is eclipfed. And a

Total Eclipse, is that in which the whole luminary is darkened.

Eclipse *of the Moon*, is a privation of the light of the moon, occafioned by an interpofition of the body of the earth directly between the fun and moon, and fo intercepting the fun's rays that they cannot arrive at the moon, to illuminate her. Or, the obfcuration of the moon may be confidered as a fection of the earth's conical fhadow, by the moon paffing through fome part of it.

The manner of this eclipfe is reprefented in this figure, where S is the fun, E the earth, and M or M the moon.

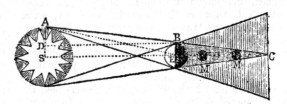

Lunar Eclipfes only happen at the time of full moon; becaufe it is only then the earth is between the fun and moon: nor do they happen every full moon, becaufe of the obliquity of the moon's path with refpect to the fun's; but only in fuch full moons as happen either at the interfection of thofe two paths, called the moon's nodes, or very near them; viz, when the moon's latitude, or diftance between the centres of the earth and moon, is lefs than the fum of the apparent femidiameters of the moon and the earth's fhadow.

The chief Circumftances in Lunar Eclipfes, are the following:—1. All lunar Eclipfes are univerfal, or vifible in all parts of the earth which have the moon above their horizon; and are every where of the fame magnitude, with the fame beginning and end.—2. In all lunar eclipfes, the eaftern fide is what firft immerges and emerges again, i. e. the left-hand fide of the moon as we look towards her, from the north; for the proper motion of the moon being fwifter than that of the earth's fhadow, the moon approaches it from the weft, overtakes and paffes through it with the moon's eaft fide foremoft, leaving the fhadow behind, or to the weftward.—3. Total eclipfes, and thofe of the longeft duration, happen in the very nodes of the ecliptic; becaufe the fection of the earth's fhadow, then falling on the moon, is confiderably larger than her difc. There may however be total eclipfes within a fmall diftance of the nodes; but their duration is the lefs as they are farther from it; till they become only partial ones, and at laft, none at all.—4. The moon, even in the middle of an eclipfe, has ufually a faint appearance of light, refembling tarnifhed copper; which Gaffendus, Ricciolus, Kepler, &c, attribute to the light of the fun, refracted by the earth's atmofphere, and fo tranfmitted thither. —Laftly, fhe grows fenfibly paler and dimmer, before entering into the real fhadow; owing to a penumbra which furrounds that fhadow to fome diftance.

Aftronomy of Lunar Eclipses, *or the method of calculating their times, places, magnitudes, and other phenomena*.

The

The 1st preliminary is to find the length of the earth's conical shadow. This may be found either from the distance between the earth and sun, and the proportion of their diameters, or from the angle of the sun's apparent magnitude at the time. Thus, suppose the semiaxis of the earth's orbit 95,000000 miles, and the eccentricity of the orbit 1,377000 miles, making the greatest distance 96,377000 miles, or 24194 semidiameters of the earth; and the sun's semidiameter being to the earth's, as 112 to 1; then as AD : BE : : DB : EC, that is, 111 : 1 : : 24194 : 218 semidiameters of the earth = EC the length of the earth's shadow. Otherwise, suppose the angle AES, or the sun's apparent semidiameter be 15′ 56″, and the angle BAE, or the sun's parallax 8·6″, which is their difference, or the angle ACE = 15′ 47·4″; hence, as tang. 15′ 47·4″ : radius : : BE or 1 : 218 nearly = CE, the same distance as before. Hence, as the moon's least distance from the earth is scarce 56 semidiameters, and the greatest not more than 64, the moon, when in opposition to the sun, in or near the nodes, will fall into the earth's shadow, and will be eclipsed, as the length of the shadow is almost 4 times the moon's distance.

2. *To find the apparent semidiameter of the earth's shadow*, in the place where the moon passes through it, at any given time. Add together the sun and moon's parallaxes, and from the sum subtract the apparent semidiameter of the sun; so shall the remainder be the apparent semidiameter of the shadow at the place of the moon's passage. For ex. the 28th of April 1790, at midnight, the moon's parallax is 61′9″, to which add 8·6″, or ″9″, for the sun's parallax, from the sum 61′ 18″ take 15′ 56″, the sun's apparent semidiameter, and the remainder 45′ 22″ is the semidiameter of the shadow at the place where the moon passes through at that time. N. B. Some omit the sun's parallax, as of no consequence; but increase the apparent semidiameter of the shadow by one whole minute, for the shadow of the atmosphere; which would give the semidiameter of the shadow, in the case above, 46′ 13″.

3. There must also be had, the true distance of the moon from the node, at the mean opposition; also the true time of the opposition, with the true place of the sun and moon, reduced to the ecliptic; likewise the moon's true latitude at the time of the true opposition; the angle of the moon's way with the ecliptic, and the true horary motions of the sun and moon: from which all the circumstances of her eclipse may be computed by common arithmetic and trigonometry.

To Construct an Eclipse of the Moon.

Let EW be a part of the ecliptic, and C the centre of the earth's shadow, through which draw perpendicular to EW, the line CN towards the north, if the moon have north latitude at the time of the eclipse, or CS southward, if she have south latitude. Make the angle NCD equal to the angle of the moon's way with the ecliptic, which may be always taken at 5° 35′, on an average, without any sensible error; and bisect this angle by the right line CF; in which line it is that the true equal time of opposition of the sun and moon falls, as given by the tables.

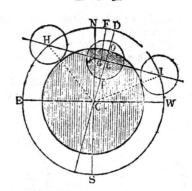

From a convenient scale of equal parts, representing minutes of a degree, take the moon's latitude at the true time of full moon, and set it from C to G, on the line CF; and through the point G, at right-angles to CD, draw the right line HKGLI for the path of the moon's centre. Then is L the point in the earth's shadow, where the moon's centre is at the middle of the eclipse; G the point where her centre is at the tabular time of her being full; and K the point where her centre is at the instant of her ecliptic opposition: also I the moon's centre at the moment of immersion, and H her centre at the end of the eclipse.

With the moon's semidiameter as a radius, and the points I, L, H, as centres, describe circles for the moon at the beginning, middle, and end of the eclipse.

Finally, the length of the line of path IH, measured on the same scale, will serve to determine the duration of the eclipse, viz, by saying, As the moon's horary motion from the sun is to IH : : 1 hour or 60 min. to the whole duration of the eclipse.

To Compute a Lunar Eclipse. This will be very easy from the foregoing construction. For, 1st, in the triangle CGL, right-angled at L, there are given the hypothenuse CG = the moon's latitude at the time of full moon, and the angle GCL = the half of 5° 35′; to find the legs CL and LG.—2d, In the right-angled triangle CHL or CIL, are given the leg CL, and CH or CI, the sum of the semidiameters of the moon and the earth's shadow; to find LH or LI, half the difference of the sun's and moon's motions during the time of the eclipse.—3d, As the difference of the horary motions of the luminaries is to one hour, or 60 min. : : HL to the semiduration of the eclipse, and : : GL to the difference between the opposition and middle of the eclipse; this last therefore taken from the time of full moon, gives the time of the middle of the eclipse; from which subtracting the time in LI, or semiduration before found, gives the beginning of the eclipse; or add the same, and it gives the end of it.—Lastly, from CO the semidiameter of the shadow, take CL, leaves LO; to which add LP, the moon's semidiameter, when necessary, gives OP the quantity eclipsed.

Note, When the moon's distance from the node exceeds 12°, there can be no eclipse of the moon; or, more accurately, the limit is from 10½ and 12$\frac{1}{36}$ degrees, according to the distances of the sun, earth, and moon.

Eclipse *of the Sun*, is an occultation of the sun's body,

body, occafioned by an interpofition of the moon between the earth and fun. On which account it is by fome confidered as an eclipfe of the earth, fince the light of the fun is hid from the earth by the moon, whofe fhadow involves a part of the earth.

The manner of a folar eclipfe is reprefented in this figure; where S is the fun, *m* the moon, and CD the

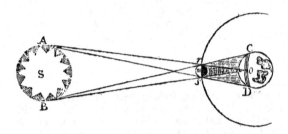

earth, *rmso* the moon's conical fhadow, travelling over a part of the earth CoD, and making a complete eclipfe to all the inhabitants refiding in that track, but no where elfe; excepting that for a large fpace around it there is a fainter fhade, included within all the fpace *r*CD*s*, which is called the *Penumbra*.

Hence, Solar Eclipfes happen when the moon is in conjunction with the fun, or at the new moon, and alfo in the nodes or near them, the limit being about 17 degrees on each fide of it; and fuch eclipfes only happen when the latitude of the moon, viewed from the earth, is lefs than the fum of the apparent femidiameters of the fun and moon; becaufe the moon's way is oblique to the ecliptic, or fun's path, making an angle of nearly 5° 35′ with it.

In the nodes, when the moon has no vifible latitude, the occultation is total; and with fome continuance, when the difc of the moon in perigee appears greater than that of the fun in apogee, and its fhadow is extended beyond the furface of the earth; and without continuance at moderate diftances when the cufp, or point of the moon's fhadow, barely touches the earth. Laftly, out of the nodes, but near them, the eclipfes are partial.

The other circumftances of folar eclipfes are, 1. That none of them are univerfal; that is, none of them are feen throughout the whole hemifphere which the fun is then above; the moon's difc being much too little, and much too near the earth, to hide the fun from the whole difc of the earth. Commonly the moon's dark fhadow covers only a fpot on the earth's furface, about 180 miles broad when the fun's diftance is greateft, and the moon's leaft. But her partial fhadow, or penumbra, may then cover a circular fpace of 4900 miles in diameter, within which the fun is more or lefs eclipfed, as the places are nearer to or farther from the centre of the penumbra. In this cafe the axis of the fhade paffes through the centre of the earth, or the new moon happens exactly in the node, and then it is evident that the fection of the fhadow is circular; but in every other cafe the conical fhadow is cut obliquely by the furface of the earth, and the fection will be an oval, and very nearly a true ellipfis.

2. Nor does the Eclipfe appear the fame in all parts

8

of the earth, where it is feen; but when in one place it is total, in another it is only partial. Farther, when the moon appears much lefs than the fun, as is chiefly the cafe when fhe is in apogee and he in perigee, the vertex of the lunar fhadow is then too fhort to reach the earth, and though fhe be in a central conjunction with the fun, is yet not large enough to cover his whole difc, but lets his limb appear like a lucid ring or bracelet, and fo caufes an *Annular Eclipfe*.

3. A Solar Eclipfe does not happen at the fame time, in all places where it is feen; but appears more early to the weftern parts, and later to the eaftern; as the motion of the moon, and confequently of her fhadow, is from weft to eaft.

4. In moft Solar Eclipfes the moon's difc is covered with a faint light; which is attributed to the reflexion of the light from the illuminated part of the earth.

Laftly, in total Eclipfes of the fun, the moon's limb is feen furrounded by a pale circle of light; which fome aftronomers confider as an indication of a lunar atmofphere; but others as the atmofphere of the fun, becaufe it has been obferved to move equally with the fun, and not with the moon; and befides, it is generally believed that the moon is without any atmofphere, unlefs it be one that is very fmall, and very rare.

To determine the Bounds of a Solar ECLIPSE.

If the moon's parallax were infenfible, the bounds of a folar eclipfe would be determined after the fame manner as thofe of a lunar; but becaufe here is a fenfible parallax, the method is a little altered, viz.

1. Add together the apparent femidiameters of the luminaries, both in apogee and perigee; which gives 33′ 6″ for the greateft fum of them, and 30′ 31″ for the leaft fum.

2. Since the parallax diminifhes the northern latitude and augments the fouthern, therefore let the greateft parallax in latitude be added to the former fums, and alfo fubtracted from them: Thus in each cafe there will be had the true latitude, beyond which there can be no eclipfe. This latitude being given, the moon's diftance from the nodes, beyond which eclipfes cannot happen, is found as for a lunar eclipfe. This limit is nearly between 16¼ and 18⅓ degrees diftance from the nodes.

To find the Digits eclipfed. Add the apparent femidiameters of the luminaries into one fum; from which fubtract the moon's apparent latitude; the remainder is the fcruples, or parts of the diameter eclipfed. Then fay, As the femidiameter of the fun is to the fcruples eclipfed; fo are 6 digits reduced into fcruples, viz 360 fcruples or minutes, to the digits &c eclipfed.

To determine the Duration of a Solar ECLIPSE. Find the horary motion of the moon from the fun, for one hour before the conjunction, and another hour after: then fay, As the former horary motion is to the feconds in an hour, fo are the fcruples of half duration (found as in a lunar eclipfe) to the time of immerfion; and as the latter horary motion is to the fame feconds, fo are the fame fcruples of half duration to the time of emerfion. Laftly, adding the times of immerfion and emerfion together, the aggregate is the total duration.

The moon's apparent diameter when largeft, exceeds

eeeds the fun's when leaft, only 2′ of a degree; and at the greateft folar eclipfe that can happen at any time and place, the total darknefs cannot continue any longer than whilft the moon is moving through this 2′ from the fun in her orbit, which is about 4 minutes of time: for the motion of the fhadow on the earth's difc is equal to the moon's motion from the fun, which on account of the earth's rotation on its axis towards the fame way, or eaftward, is about 30½ minutes of a degree every hour, at a mean rate; but fo much of the moon's orbit is equal to 30½ degrees of a great circle on the earth, becaufe the circumference of the moon's orbit is about 60 times that of the earth; and therefore the moon's fhadow goes 30½ degrees, or 1830 geographical miles in an hour, or 30½ miles in a minute.

To determine the Beginning, Middle, and End, of a Solar Eclipfe. From the moon's latitude, for the time of conjunction, find the arch GL (laft fig. but one), or the diftance of the greateft obfcurity. Then fay, as the horary motion of the moon from the fun, before the conjunction, is to 1 hour; fo is the diftance of the greateft darknefs, to the interval of time between the greateft darknefs and the conjunction. Subtract this interval, in the 1ft and 3d quarter of the anomaly, from the time of the conjunction; and in the other quarters, add it to the fame; the refult is the time of the greateft darknefs. Laftly, from the time of the greateft darknefs subtract the time of incidence, and add it to the time of emerfion; the difference in the firft cafe will be the beginning; and the fum, in the latter cafe, the end of the eclipfe.

To Calculate ECLIPSES *of the Sun.* Firft, find the mean new moon, and thence the true one; with the place of the luminaries for the apparent time of the true one.—2. For the apparent time of the true new moon, compute the apparent time of the new moon obferved.—3. For the apparent time of the new moon feen, compute the latitude feen.—4. Thence determine the digits eclipfed.—5. Find the times of the greateft darknefs, immerfion, and emerfion.—6. Thence determine the beginning, and ending of the eclipfe.

From the foregoing problems, it is evident that all the trouble and fatigue of the calculus arifes from the parallaxes of longitude and latitude, without which, the calculation of folar eclipfes would be the fame with that of lunar ones.

See the Conftruction and Calculation of Eclipfes by Flamfteed in Sir Jonas Moor's Syftem of Mathematics, and in Ferguson's Aftronomy, &c.

In the Philof. Tranf. N° 461 is a contrivance to reprefent folar eclipfes, by means of the terreftrial globe, by M. Seguer, profeffor of Mathematics at Gottingen. And Mr. Ferguson has fitted a terreftrial globe, fo as to fhew the time, quantity, duration, and progrefs of folar eclipfes, at any place of the earth where they are vifible; which he calls the Eclipfareon. He has alfo given a large catalogue of ancient and modern eclipfes, including thofe recorded in hiftory, from 721 years before Chrift, to A. D. 1485; alfo computed eclipfes from 1485 to 1700, and all the eclipfes vifible in Europe from 1700 to 1800. See his Aftron.

The Number of ECLIPSES, of both luminaries, in any year, cannot be lefs than two, nor more than feven; the moft ufual number is 4, and it is rare to have more than 6. The reafon is obvious; becaufe the fun paffes by both the nodes but once in a year, unlefs he pafs by one of them in the beginning of the year; in which cafe he will pafs by the fame again a little before the end of the year; becaufe the nodes move backwards 19½ degrees every year, and therefore the fun will come to either of them 173 days after the other. And if either node be within 17° of the fun at the time of new moon, the fun will be eclipfed; and at the fubfequent oppofition, the moon will be eclipfed in the other node, and come round to the next conjunction before the former node be 17° beyond the fun, and eclipfe him again. When three eclipfes happen about either node, the like number commonly happens about the oppofite one; as the fun comes to it in 173 days afterward, and 6 lunations contain only 4 days more. Thus there may be two eclipfes of the fun, and one of the moon, about each of the nodes. But when the moon changes in either of the nodes, fhe cannot be near enough the other node at the next full, to be eclipfed; and in 6 lunar months afterward fhe will change near the other node; in which cafe there cannot be more than two eclipfes in a year, both of the fun.

Period of ECLIPSES, is the period of time in which the fame eclipfes return again; and as the nodes move backwards 19½ degrees every year, they would fhift through every point of the ecliptic in 18 years and 225 days; and this would be the regular period of their return, if any complete number of lunations were finifhed without a fraction; but this is not the cafe. However, in 223 mean lunations, after the fun, moon, and nodes have been once in a line of conjunction, they return fo nearly to the fame ftate again, as that the fame node which was in conjunction with the fun and moon at the beginning of the firft of thefe lunations, will be within 28′ 12″ of the line of conjunction with the fun and moon again, when the laft of thefe lunations is completed; and in this period there will be a regular return of eclipfes for many ages. To the mean time of any folar or lunar eclipfe, by adding this period, or 18 Julian years 11 days 7 hours 43 minutes 20 feconds, when the laft day of February in leap years is 4 times included, or a day lefs when it occurs 5 times, we fhall have the mean time of the return of the fame eclipfe. In an interval of 6890 mean lunations, containing 557 years 21 days 18 hours 30 minutes 11 feconds, the fun and node meet fo nearly, as to be diftant only 11 feconds.

The Ufe of ECLIPSES. In Aftronomy, eclipfes of the moon determine the fpherical figure of the earth; they alfo fhew that the fun is larger than the earth, and the earth than the moon. Eclipfes alfo, that are fimilar in all circumftances, and that happen at confiderable intervals of time, ferve to afcertain the period of the moon's motion. In Geography, eclipfes difcover the longitude of different places; for which purpofe thofe of the moon are the more ufeful, becaufe they are more often vifible, and the fame lunar eclipfe is of equal magnitude and duration at all places where it is feen. In Chronology, both folar and lunar eclipfes ferve to determine exactly the time of any paft event.

ECLIPSES *of the Satellites.* See SATELLITES *of Jupiter.*

The chief circumftances here obferved, are, 1. That the fatellites of Jupiter undergo two or three kinds of eclipfes;

eclipses; the first of which are proper, being such as happen when Jupiter's body is directly interposed between them and the sun: and these happen almost every day. Various authors have given tables for computing eclipses of the satellites of Jupiter; as Flamsteed, Cassini, &c, but the latest and best of all, are those of professor Wargentin of Upsal.

The second sort are occultations, rather than observations; when the satellites, coming too near the body of Jupiter, are lost in his light; which Riccioli calls *occidere zeusiace, setting jovially*. In which case, the nearest or first satellite exhibits a third kind of eclipse; being observed like a round macula, or dark spot, transiting the disc of Jupiter, with a motion contrary to that of the satellite; like as the moon's shadow projected on the earth, will appear to do, to the lunar inhabitants.

The eclipses of Jupiter's satellites furnish very good means of finding the longitude at sea. Those especially of the first satellite are much surer than the eclipses of the moon, and they also happen much oftener: the manner of applying them is also very easy. See LONGITUDE.

ECLIPTIC, in Astronomy, a great circle of the sphere conceived to pass through the middle of the zodiac. It is sometimes called the *via solis*, or *sun's path*, being the track which he appears to describe among the fixed stars; though more properly it is the apparent path of the earth, as viewed from the sun, and thence called the heliocentric circle of the earth. It is called the ecliptic, because all the eclipses of the sun or moon happen when the moon crosses it, or is nearly in one of those two parts of her orbit where it crosses the ecliptic, which points are called the moon's nodes.

Upon the ecliptic are marked and counted the 12 celestial signs, Aries, Taurus, Gemini, &c; and upon it is counted the longitude of the planets and stars. It is placed obliquely with respect to the equator, which it cuts in two opposite points, viz, the beginning of Aries and Libra, which are directly opposite to each other, and called the equinoxes, making the one half of the ecliptic to the north, and the other half on the south side of the equator; the two extreme points of it, to the north and south, which are opposite to each other, and at a quadrant distance from the equinoctial points both ways, are called the solstices, or solstitial points, or also the two tropics, which are at the beginning of Cancer and Capricorn, and which are at the farthest distance of any points of it from the equator, which distance is the measure of the sun's greatest declination, which is the same with the obliquity of the ecliptic, or the angle it makes with the equator.

This obliquity of the ecliptic is not permanent, but is continually diminishing. by the ecliptic approaching nearer and nearer to a parallelism with the equator, at the rate of half a second in a year nearly, or from 50″ to 55″ in 100 years, as is deduced from ancient and modern observations compared together; and as the mean obliquity of the ecliptic was 23° 28′ about the end of the year 1788, or beginning of 1789, by adding half a second for each preceding year, or subtracting the same for each following year, the mean obliquity will be found nearly for any year either before or since that period. The quantity however of this change is variously stated by different authors, from 50″ to 60″ or

70″ for each century or 100 years. Hipparchus, almost two thousand years since, observed the obliquity of the ecliptic, and found it about 23° 51′; and all succeeding astronomers, to the present time, having observed the same, have found it always less and less; being now rather under 23° 28′; a difference of about 23′ in 1950 years; which gives a medium of 70″ in 100 years. There is great reason however to think that the diminution is variable.

This diminution of the obliquity of the ecliptic to the equator, according to Mr. Long and some others, is chiefly owing to the unequal attraction of the sun and moon on the protuberant matter about the earth's equator. For if it be considered, say they, that the earth is not a perfect sphere, but an oblate spheroid, having its axis shorter than its equatorial diameter; and that the sun and moon are constantly acting obliquely upon the greater quantity of matter about the equator, drawing it, as it were, towards a nearer and nearer coincidence with the ecliptic; it will not appear strange that these actions should gradually diminish the angle between the planes of those two circles. Nor is it less probable that the mutual attractions of all the planets should have a tendency to bring their orbits to a coincidence: though this change is too small to become sensible in many ages.

It is now however well known that this change in the obliquity of the ecliptic, is wholly owing to the actions of the planets upon the earth, and especially the planets Venus and Jupiter, but chiefly the former. See La Grange's excellent paper upon this subject in the Memoirs of the French Academy for 1774; Cassini's in 1778; and La Lande's Astron. vol. 3, art. 2737. According to La Grange, who proceeds upon theory, the annual change of obliquity is variable, and has its limits: about 2000 years ago, he thinks it was after the rate of about 38″ in 100 years; that it is now, and will be for 400 years to come, 56″ per century; but 2000 years hence, 49″ per century. According to Cassini, who computes from observations of the obliquity between the years 1739 and 1778, the annual change at present is 60″ or 1′ in 100 years. But according to La Lande, the diminution is at the rate of 88″ per century; while Dr. Maskelyne makes it only 50″ in the same time.

Beside the regular diminution of the obliquity of the ecliptic, at the rate of near 50 seconds in a century, or half a second a year, which arises from a change of the ecliptic itself, it is subject to two periodical inequalities, the one produced by the unequal force of the sun in causing the precession of the equinoxes, and the other depending on the nutation of the earth's axis. See the Explanation and Use of Dr. Maskelyne's Tables and Observations, pa. vi, where we are shewn how to calculate those inequalities, and where he shews that, from his own observations, the mean obliquity of the ecliptic to the beginning of the year 1769, was 23° 28′ 9″·7.

To find the Obliquity of the Ecliptic, or the greatest declination of the sun: about the time of the summer solstice observe very carefully the sun's zenith distance for several days together; then the difference between this distance and the latitude of the place, will be the obliquity sought, when the sun and equator are both on one side of the place of observation; but their sum will

will be the obliquity when they are on different sides of it. Or, it may be found by observing the meridian altitude, or zenith distance, of the sun's centre, on the days of the summer and winter solstice; then the difference of the two will be the distance between the tropics, the half of which will be the obliquity sought.

By the same method too, the declination of the sun from the equator for any other day may be found; and thus a table of his declination for every day in the year might be constructed. Thus also the declination of the stars might be found.

Authors' Names	Years before Christ	Obliquity		
		o	'	''
Pytheas - - -	324	23	49	23
Eratosthenes and Hipparchus	230 & 140	23	51	20
	after Christ			
Ptolomy - - -	140	23	48	45
Almahmon - -	832	23	35	
Albategnius - -	880	23	35	
Thebat - -	911	23	33	30
Abul Wasi and Hamed	999	23	35	
Persian Tables in Chrysococea	1004	23	35	
Albatrunius - -	1007	23	35	
Arzachel - -	1104	23	33	30
Almæon - -	1140	23	33	30
Choja Naffir Oddin -	1290	23	30	
Prophatius the Jew -	1300	23	32	
Eon Shattir -	1363	23	31	
Purbach and Regiomontanus	1460	23	30	
Ulugh Beigh - -	1463	23	30	17
Walther - -	1476	23	30	
Do. corrected by refraction &c	———	23	29	8
Werner -	1510	23	28	30
Copernicus -	1525	23	28	24
Egnatio Danti -	1570	23	29	
Prince of Hesse -	1570	23	31.	
Rothmann and Byrge	1570	23	30	20
Tycho Brahe -	1584	23	31	30
Ditto corrected		23	29	
Wright - -	1594	23	30	
Kepler - -	1627	23	30	30
Gaffendus - -	1630	23	31	
Ricciolus - -	1646	23	30	20
Ditto corrected -	1655	23	29	
Heyelius - -	1653	23	30	20
Ditto corrected -	1661	23	28	52
Caffini -	1655	23	29	15
Montons, corrected, &c	1660	23	29	3
Richer corrected	1672	23	28	52
De la Hire -	1686	23	29	
Ditto corrected -		23	29	28
Flamsteed - -	1690	23	29	
Bianchini -	1703	23	28	25
Roemer -	1706	23	28	41
Louville -	1715	23	28	24
Godin -	1730	23	28	20
Bradley -	1750	23	28	18
Mayer -	1756	23	28	16
Maskelyne	1769	23	28	10
Hornsby	1772	23	28	8

The observations of astronomers of all ages, on the obliquity of the ecliptic, have been collected together; and although some of them may not be quite accurate, yet they sufficiently shew the gradual and continual decrease of the obliquity from the times of the earliest observations down to the present time. The chief of those observations may be seen in the foregoing table; where the first column contains the name of the observer, or author, the 2d the year before or after Christ, and the 3d the obliquity of the ecliptic for that time.

See Ptolom. Alm. lib. 1, cap. 10; Riccioli Alm. vol. 1, lib. 3, cap. 27; Flamsteed Proleg. Hist. Cœl. vol. 3; Philof. Transf. number 163; ib. vol. 63, pt. 1; Long's Astron. vol. 1, cap. 16; Memoirs of the Acad. an. 1716, 1734, 1762, 1767, 1774, 1778; Acta Erud. Lipsiæ 1719; Naut. Alm. 1779; Maskelyne's Observ. Explan. pa. vi; &c.

According to an ancient tradition of the Egyptians, mentioned by Herodotus, the ecliptic had formerly been perpendicular to the equator: they were led into this notion by observing, for a long series of years, that the obliquity was continually diminishing; or, which amounts to the same thing, that the ecliptic was continually approaching to the equator. From thence they took occasion to suspect that those two circles, in the beginning, had been as far off each other as possible, that is, perpendicular to each other. Diodorus Siculus relates, that the Chaldeans reckoned 403,000 years from their first observations to the time of Alexander's entering Babylon. This enormous account may have some foundation, on the supposition that the Chaldeans built on the diminution of the obliquity of the ecliptic at the rate of a minute in 100 years. M. de Louville, taking the obliquity such as it must have been at the time of Alexander's entrance into Babylon, and going back to the time when the ecliptic, at that rate, must have been perpendicular to the equator, actually finds 402,942 Egyptian, or Chaldean years; which is only 58 years short of that epocha. Indeed there is no way of accounting for the fabulous antiquity of the Egyptians, Chaldeans, &c, so probable, as from the supposition of long periods of very slow celestial motions, a small part of which they had observed, and from which they calculated the beginning of the period, making the world and their own nation to commence together. Or perhaps they sometimes counted months or days for years.

Should the diminution always continue at the rate it has lately done, viz at 50'' or 56'' a century, it would take 96,960 years, from the year 1788, to bring the ecliptic exactly to coincide with the equator.

ECLIPTIC, *in Geography*, a great circle on the terrestrial globe, in the plane of, or directly under, the celestial ecliptic.

ECLIPTIC, *Eclipticus*, something belonging to the ecliptic, or to eclipses; as ecliptic conjunction, opposition, &c.

ECLIPTIC *Bounds*, or *Limits*, are the greatest distances from the nodes at which the sun or moon can be eclipsed; namely, near 18 degrees for the sun, and 12 degrees for the moon.

ECLIPTIC *Digits, digiti ecliptici*. See DIGITS.

Poles of the ECLIPTIC, are the two opposite points

of

of the sphere which are each everywhere equally distant from the ecliptic quite around, or 90° distant from it. The distance of the poles of the ecliptic from the poles of the equator, or of the world, is always equal to the varying distance of the obliquity of the ecliptic, and at the beginning of the year 1789 it was just 23° 28'.

Reduction to the ECLIPTIC. See REDUCTION.

EFFECT, the result or consequence of the application of a cause, or agent, on some subject. It is one of the great axioms in philosophy, that full effects are always proportional to the powers of their adequate causes.

EFFECTION, denotes the geometrical construction of a proposition. The term is also used in reference to problems and practices; which, when they are deducible from, or founded upon some general propositions, are called the *geometrical effection* of them.

EFFERVESCENCE, is popularly used for a light ebullition, or a brisk intestine motion, produced in a liquor by the first action of heat, with any remarkable separation of its parts.

EFFICIENT *Cause*, is that which produces an effect. See CAUSE and EFFECT.

EFFICIENTS, in Arithmetic, are the numbers given for an operation of multiplication, and are otherwise called the factors. Hence the term coefficients in Algebra, which are the numbers prefixed to, or that multiply the letters or algebraic quantities.

EFFLUVIUM, a flux or exhalation of minute particles from any body; or an emanation of subtile corpuscles from a mixed, sensible body, by a kind of motion of transpiration.

ELASTIC, an appellation given to all bodies endowed with the property of elasticity or springiness.

ELASTIC *Body*, is that which changes its figure, and yields to any impulse or pressure, but endeavours by its own nature and force to restore the same again; or, it is a springy body, which, when compressed or condensed, or the like, makes an effort to set itself at liberty, and to repel the body that constrained it. Such, for instance, as a bow, or a sword blade, &c, which are easily bent, but presently return to their former figure and extension. All bodies partake of this property in some degree, though perhaps none are perfectly elastic, as none are found to restore themselves with a force equal to that with which they are compressed.

The principal phenomena observable in Elastic bodies, are 1, That an elastic body (i. e. a body perfectly elastic, if any such there be) endeavours to restore itself with the same force with which it is pressed or bent. 2, An elastic body exerts its force equally towards all sides; though the effect is chiefly found on that side where the resistance is weakest; as is evident in the case of a gun exploding a ball, a bow shooting out an arrow, &c.—3, Elastic bodies, in what manner soever struck, or impelled, are inflected and rebound after the same manner: thus a bell yields the same musical sound, in what manner, or on what side soever it be struck; the same of a tense or musical chord; and a body rebounds from a plane in the same angle in which it meets or strikes it, making the angle of incidence equal to the angle of reflection, whether the intensity of the stroke be greater or less.—4, A body perfectly fluid, if any such

there be, cannot be elastic, if it be allowed that its parts cannot be compressed.—5, A body perfectly solid, if any such there be, cannot be elastic; because, having no pores, it is incapable of being compressed.—6, The elastic properties of bodies seem to differ, according to their greater or less density or compactness, though not in an equal degree: thus, metals are rendered more compact and elastic by being hammered: tempered steel is much more elastic than soft steel; and the density of the former is to that of the latter as 7809 to 7738: cold condenses solid bodies, and renders them more elastic; whilst heat, that relaxes them, has the opposite effect: but, on the contrary, air, and other elastic fluids, are expanded by heat, and rendered more elastic.—For the laws of Motion and Percussion in Elastic bodies, see MOTION, and PERCUSSION.

ELASTIC *Curve*. See CATENARIA.

ELASTIC *Fluids*. See AIR, ELECTRICITY, GAS, ELASTIC VAPOURS, &c.

ELASTIC *Gum*. The same as CAOUTCHOUC, or *Indian Rubber*.

ELASTIC *Vapours*, or *Fluids*, are such as may be compressed mechanically into a less space, and which resume their former state when the compressing force is withdrawn. Such as atmospherical air, and all the aerial fluids, with all kinds of fumes raised by means of heat, whether from solid or fluid bodies.

Of these, some remain elastic only while a considerable degree of heat is applied to them, or to the substance which produces them; while others continue elastic in every degree of cold that has yet been observed. Of the former kind, are the vapours of water, spirit of wine, mercury, sal-ammoniac, and all kinds of sublimable salts: of the latter, those of spirit of salt, mixtures of vitriolic acid and iron, nitrous acid, and various other metals, and in short the several species of aerial fluids indiscriminately.

The elastic force with which any one of these fluids is endowed, has not yet been calculated, as being ultimately greater than any obstacle we can put in its way. Thus, on compressing the atmospherical air, we find that for some little time at first it easily yields to any force applied; but at every succeeding moment the resistance becomes always the stronger, and a greater and greater force must be applied, to compress it farther. As the compression goes on, the vessel containing the air becomes hot; but no power whatever has yet been able in any degree to destroy the elasticity of the contained fluid; for, upon removing the pressure, it is always found to occupy the very same space that it did before. The case is the same with the steam of water, to which a sufficient heat is applied to keep it from condensing into water.

ELASTICITY, or ELASTIC *Force*, that property of bodies by which they restore themselves to their former figure, after any external pressure.

The cause or principle of this important property, elasticity, is variously accounted for. The Cartesians ascribe it to their subtile matter making an effort to pass through pores that are too narrow for it. Thus, say they, in bending or compressing a hard elastic body, as a bow, for instance, its parts recede from each other on the convex side, and approach on the concave one:

conse-

consequently the pores are contracted or straitened on the concave side; and, if they were before round, are now perhaps oval: so that the materia subtilis, or matter of the second element, endeavouring to pass out of the pores thus straitened, must make an effort, at the same time, to restore the body to the state it was in when the pores were rounder, i. e. before the bow was bent: and in this consists its Elasticity.

Other later philosophers account for Elasticity much after the same manner as the Cartesians; with this only difference, that instead of the subtile matter of the Cartesians, these substitute *Ether*, or a fine ethereal medium that pervades all bodies.

Others, setting aside the precarious notion of a materia subtilis, account for Elasticity from the great law of nature, Attraction, or the cause of the cohesion of the parts of solid and firm bodies. Thus, say they, when a hard body is struck or bent, so that the component parts are moved a little from each other, but not quite disjointed or broken off, or separated so far as to be out of the power of that attracting force by which they cohere; they must, on removing the external violence, spring back to their former natural state.

Others again resolve Elasticity into the pressure of the atmosphere: for a violent tension, or compression, though not so great as to separate the constituent particles of bodies far enough to let in any foreign matter, must yet occasion many little vacuola between the separated surfaces; so that on the removal of the force they will close again by the pressure of the aerial fluid upon the external parts.

Lastly, others attribute the Elasticity of all hard bodies to the power of resilition in the air included within them; and so make the elastic force of the air the principle of Elasticity in all other bodies. See Desaguliers's Exper. Philos. vol. 2, pa. 38, &c.

The ELASTICITY *of Fluids* is accounted for from their particles being all endowed with a centrifugal force; whence Sir Isaac Newton demonstrates, prop. 23, lib. 2, that particles, which naturally avoid or fly off from one another by such forces as are reciprocally proportional to the distances of their centres, will compose an elastic fluid, whose density shall be proportional to its compression; and vice versa, if any fluid be composed of particles that fly off or avoid one another, and have its density proportional to its compression, then the centrifugal forces of those particles will be reciprocally proportional to the distances of their centres.

ELASTICITY *of the Air* is the force with which that element endeavours to expand, and with which it does actually dilate itself, on removing the force that compressed it. See AIR, and ATMOSPHERE.

The Elasticity or spring of the air was first discovered by lord Bacon, and farther established by Galileo. Its existence is proved by this experiment of that philosopher: An extraordinary quantity of air being intruded, by means of a syringe, into a hollow ball or shell of glass or metal, till such time as the ball, with this accession of air, weigh considerably more in the balance than it did before; then, opening the mouth of the ball, the air rushes out, till the ball sink to its former weight. From hence we infer, that there is just as much air gone out, as compressed air had been crowded

in. Air therefore returns to its former degree of expansion, upon removing the force that compressed or resisted its expansion; and consequently it is endowed with an elastic force. It may be added, that as the air is found to rush out in every situation or direction of the orifice, the elastic force acts every way, or in every direction alike.

The cause of Elasticity in air hath been usually ascribed to a repulsion between its particles; but what is the cause of that repulsion? The term repulsion, like that of attraction, requires to be defined; and probably it will be found in most cases to be the effect of the action of some other fluid. Thus, it is found that the Elasticity of the atmosphere is very considerably affected by heat. Supposing a quantity of air heated to such a degree as to raise Fahrenheit's thermometer to 212, it will then occupy a considerable space; but if it be cooled again to such a degree, as to sink the thermometer to o, it will shrink up to less than half the former bulk. The quantity of repulsive power therefore acquired by the air, while passing from one of these states to the other, is evidently owing to the heat added to it, or taken away from it. Nor does there seem to be any reason to suppose, that the quantity of Elasticity or repulsive power it still possesses, is owing to any other cause than the fire contained in it. The supposition that repulsion is a primary cause, independent of all others, has given rise to many erroneous theories, and very much embarrassed philosophers in accounting for the phenomena of Elasticity.

The Elasticity of the air is not only proportional to its density, but is always equal to the force which compresses it, because these two exactly balance each other. This Elasticity, in the atmospheric air, is measured by the height of the barometer at any time, allowing for its heat or temperature, after this rate, viz. the 434th part for each degree of Fahrenheit's thermometer, above or below some mean temperature, as 55°; for by that part of the whole it is that air expands or contracts, or else increases or decreases in its Elasticity, for each degree of the thermometer. Sir Geo. Shuckburgh, in the Philos. Transf. for 1777, pa. 561.

ELECTIONS, or *Choice*, signify the several different ways of taking any number of things proposed, either separately, or as combined in pairs, in threes, in fours, &c; not as to the order, but only as to the number and variety of them. Thus, of the things a, b, c, d, e, &c, the elections of

one thing are $(a,)$ $1 = 2^1 - 1$,
two things are (a, b, ab) $3 = 2^2 - 1$,
three things are $(a, b, c, ab, ac, bc, abc)$ $7 = 2^3 - 1$,
&c; and of any number, n, all the elections are $2^n - 1$; that is, one less than the power of 2 whose exponent is n, the number of single things to be chosen, either separately or in combination.

ELECTRIC, in Physics, is a term applied to those substances, in which the electric fluid can be excited, and accumulated, without transmitting it; and which are therefore called *non-conductors*. They are also called *original Electrics*, and *Electrics per se*.

The word is derived from ηλεκτρον, amber, one of the most observable non-conductors. To this class also belong glass, and all vitrifications, even of metals; all precious stones, of which the most transparent are the beſt; all reſins, and reſinous compoſitions; alſo ſulphur, baked wood, all bituminous ſubſtances, wax, ſilk, cotton, all dry animal ſubſtances, as feathers, wool, hair, &c; alſo paper, white ſugar, and ſugarcandy; likewiſe air, oils, chocolate, calces of metals and ſemi-metals, the aſhes of animal and vegetable ſubſtances, the ruſt of metals, all dry vegetable ſubſtances, and ſtones, of which the hardeſt are the beſt.

Subſtances of this kind may be excited, ſo as to exhibit the Electric appearances of attracting and repelling light bodies, emitting a ſpark of light, attended with a ſnapping noiſe, and yielding a current of air, the ſenſation of which reſembles that of a ſpider's web drawn over the face, &c, and a ſmell like that of phoſphorus; and this exciting may be either by friction, or by heating and cooling, or by melting, and pouring one melted ſubſtance into another.

The term is peculiarly applied to the electric, viz. the globe, or cylinder, &c, uſed in electrical machines, to collect the electrical matter by rubbing it.

ELECTRICAL *Air Thermometer*, an inſtrument contrived by Mr. Kinnerſley of Philadelphia, and uſed in determining the effects of the electrical exploſion upon air. The deſcription may be ſeen in Franklin's Letters, &c, pa. 389, 4to, 1769.

ELECTRICAL *Apparatus*, conſiſts of glaſs tubes, about 3 feet long, and an inch and a half in diameter, one of which ſhould be cloſed at one end, and furniſhed at the other end with a braſs cap and ſtop-cock, to rarefy or condenſe the incloſed air; ſticks of ſealing wax, or tubes of rough glaſs, or glaſs tubes covered with ſealing-wax, or cylinders of baked wood for producing the negative electricity; with proper rubbers, as black oiled ſilk, with amalgam upon it for the former, and ſoft new flannel, or hare ſkins, or cat ſkins, tanned with the hair on, for the latter; coated jars, or plates of glaſs, either ſingle, or combined in a battery, for accumulating electricity; metal rods, as diſchargers; an electrical machine; electrometers, and inſulated ſtools, ſupported by pillars of glaſs, covered with ſealing-wax, or baked wood, varniſhed or boiled in linſeed oil.

ELECTRICAL *Atmoſphere*, is a ſtream or maſs of the Electrical fluid which ſurrounds an excited or electrified body, to ſome diſtance.

ELECTRICAL *Balls*. See BALLS and ELECTROMETER.

ELECTRICAL *Battery*, conſiſts of a large quantity of coated jars, placed near each other in a convenient manner. Theſe being charged, or electrified, and connected with each other, are then ſuddenly exploded or diſcharged, with a prodigious effect.

ELECTRICAL *Fluid*, is a fine rare fluid which iſſues from, and ſurrounds electrified bodies.

ELECTRICAL *Kite*, was contrived by Dr. Franklin, to verify his hypotheſis of the identity of electricity and lightning. It conſiſted of a large thin ſilk handkerchief, extended and faſtened at the four corners to two ſlender ſtrips of cedar, and accommodated with a tail, loop, and ſtring, ſo as to riſe in the air like a paper kite. To the top of the upright ſtick of the croſs was fixed a very ſharp pointed wire, riſing a foot or more above the wood; and to the end of the twine, next the hand, a ſilk ribband was tied. From a key

ſuſpended at the junction of the twine and ſilk, when the kite is raiſed during a thunder-ſtorm, a phial may be charged, and electric fire collected, as is uſually done by means of a rubbed glaſs tube or globe. Philoſ. Tranſ. vol. 47, pa. 565, or Franklin's Letters, pa. 111 and 112.

Kites made of paper, covered with varniſh, or with well boiled linſeed oil, to preſerve them from the rain, with a ſtick and cane bow, like the common ones uſed by boys, will anſwer the purpoſe extremely well, and are very uſeful in determining the electricity of the atmoſphere. See CONDUCTOR.

ELECTRICAL *Machine*, is a part of the Electrical apparatus, contrived for collecting a great quantity of electricity, and exhibiting its effects in a very ſenſible manner. It conſiſts of the electric, the moving engine, the rubber, and the prime conductor. In the early ſtate of this ſcience, for the electric, was uſed ſealing-wax, ſulphur, or rough glaſs; but, ſince the method of inſulating the rubber, and ſo producing negative electricity, was introduced, ſmooth glaſs has been uſed. The form is commonly either that of a globe, or of a cylinder. Each figure has its advantages, and its inconveniences. Dr. Van Marum, a late German writer, has conſtructed a machine, in which gumlac, in the form of a diſc, is uſed as an electric inſtead of glaſs; which has the effect of depending very little on the temperature of the air; deſcribed in his Verhandeling over het Electrizeeren, &c, or a Treatiſe concerning the method of electrifying. Groningen, 1776. But he has ſince procured ſome others to be made by Mr. Cuthbertſon, a very ingenious artiſt, of large diſcs, or round plates of glaſs: one of theſe is now placed in Teyler's Muſeum at Harlem, having two of theſe glaſs plates, of 65 inches diameter, excited, on both ſides of them, by rubbers of waxed taffaty; with which, effects are produced that are truly aſtoniſhing and tremendous. See his Deſcription of this machine, and its effects, publiſhed in 4to, at Harlem, 1785, &c.

There have been various contrivances for giving motion to the electric of a machine. The common method is by a wheel turned by a winch or handle; a cord going round a groove in the periphery of the wheel, and over a pulley in the neck of the globe or cylinder. Others have uſed multiplying wheels, which are eaſily turned by a winch; and others again make uſe of a wheel and pinion, or a wheel and endleſs ſcrew. But Van Marum's machine it ſeems has the completeſt movement, its operation being very uniform, and eaſily worked; it is kept in motion by a weight, which, after being wound up to the height of 12 feet, will continue the motion uniformly for 6 hours; yielding alſo a negative power, as well as the poſitive; and the conductors annexed to it ſerving eaſily to convey the electrical power wherever it is required, without the addition of any chain, or wires, &c.

The Rubber is the next material part of a machine. Theſe were formerly made of red baſil ſkins, ſtuffed with hair, wool, flax, or bran: Dr. Nooth introduced ſilk cuſhions ſtuffed with hair, over which is laid a piece of leather, rubbed with amalgam, which are better than the others. The rubber may be inſulated in any way that beſt ſuits the conſtruction of the machine:

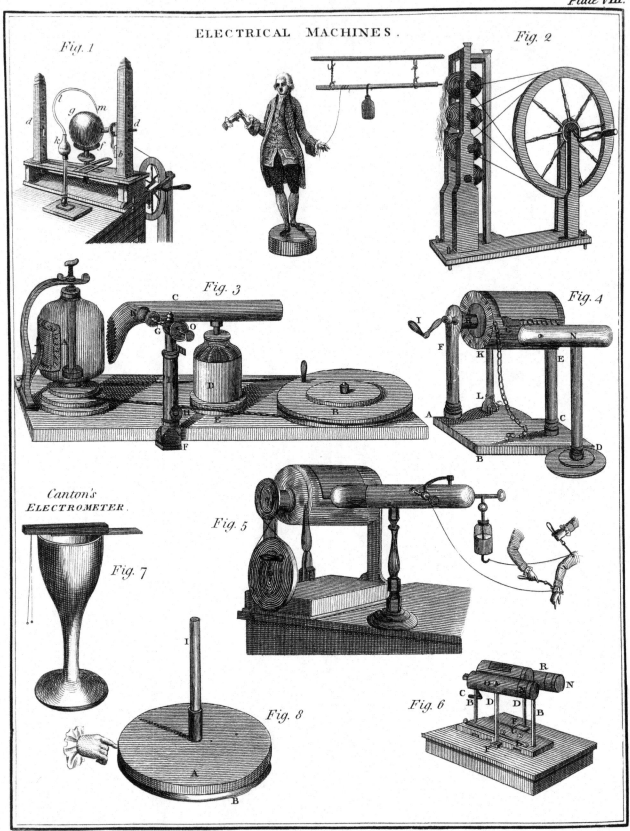

Plate VIII.

ELECTRICAL MACHINES.

Fig. 1

Fig. 2

Fig. 3

Fig. 4

Canton's
ELECTROMETER.

Fig. 7

Fig. 5

Fig. 8

Fig. 6

chine: and a chain or wire may eafily be fufpended from it, to communicate with the floor, whenever the infulation is not neceffary; and thus pofitive and negative electricity may be produced at pleafure. Van Marum ufes mercury to his rubbers.

The Prime Conductor is another neceffary appendage to the Electrical Machine: its ufe is to receive the electricity from the electric, as it is produced, and accumulate it as in a magazine, ready to be drawn off and employed on all occafions. See *Prime* CONDUCTOR.

Defcription of the moft ufeful Electrical Machines.

Fig. 1, plate ix, reprefents Dr. Prieftley's Machine, a very extenfively ufeful one, defcribed in his Hiftory of Electricity; in which g is the globe, or electric; f the rubber; in the two pillars d, d, of baked wood, are feveral holes to receive the fpindles of different globes or cylinders, feveral of which may be put on together, to increafe the electricity: klm is the prime conductor, being a copper tube, fupported on a ftand of glafs or baked wood.

Fig. 2 is Dr. Watfon's Machine, for ufing feveral globes at once, to accumulate a great quantity of electricity.

Fig. 3 reprefents a very portable Electrical Machine invented by Mr. Read, and improved by Mr. Lane. A is the glafs cylinder, moved vertically by means of the pulley at the lower end of the axis, the pulley being turned by the large wheel B parallel to the table; there are feveral pulleys, of different fizes, either of which may be ufed, according as the motion is required to be quicker or flower. The conductor C is furnifhed with points to collect the fluid, and is fcrewed to the wire of a coated jar D. The figure fhews alfo the manner of applying Mr. Lane's electrometer to this machine.

Electrical Machines have of late years undergone fome very effential alterations and improvements; both from the fuggeftions of private electricians, and the inventions of Meffrs. Adams, Nairne, and Jones, inftrument makers in London; fome of which are as follow:

Fig. 4 reprefents a very convenient machine for practice. The frame of this machine confifts of the bottom board ABCD; which, when the machine muft be ufed, is faftened to the table by two metal cramps. EF are two round pillars, of baked wood, which fupport the cylinder G by the axles of the brafs or wooden caps H, turned fometimes by a fimple winch I, and fometimes by a pulley and wheel, as in the next fig. The rubber is fixed to a glafs pillar K, which is faftened to a wooden bafis L at the bottom. The conductor N is ufually made of brafs or tin japanned, and is infulated by a glafs pillar, fcrewed into a wooden bafis or foot, which is moft conveniently placed parallel to the cylinder.

Fig. 5 reprefents an Electrical Machine, with a conductor in the fhape of a T; and an improved medical apparatus, where it is neceffary to give the fhock in the arms.

Fig. 6 fhews Mr. Nairne's patent machine for medical purpofes. Its glafs cylinder is about 7 inches in diameter, and 12 long, with two conductors parallel to it. The rubber is faftened to the conductor R; and confifts of a cufhion of leather ftuffed, having a piece of filk glewed to its under part. The conductors are of tin covered with black lacker, each of them containing a large coated glafs jar, and likewife a fmaller one, or a coated tube, which are vifible when the caps NN are removed. To each conductor is fixed a knob O, for the occafional fufpenfion of a chain to produce pofitive or negative electricity. That part of the winch C which acts as a lever in turning the cylinder, is of glafs. Thus every part of the machine is infulated, the cylinder itfelf and its brafs caps not excepted; by which means very little of the electricity is diffipated, and hence of courfe the effects are likely to be the more powerful. And to this the inventor has adapted fome flexible conducting joints, a difcharging electrometer, &c, for the practice of medical electricity.

The large Electrical Machine placed in Teyler' Mufeum at Harlem, has been partly defcribed above. It was conftructed by Mr. John Cuthbertfon, an Englifh inftrument maker; and it has, for the electric, two glafs plates of 65 inches diameter, made of French glafs, as this is found to produce the moft electricity next to Englifh flint glafs, which could not be made of a fufficient fize: thefe plates are fet on the fame horizontal axis, at the diftance of $7\frac{1}{2}$ inches, and are excited by 8 rubbers, each $15\frac{1}{4}$ inches long; and both fides of the plates are covered with a refinous fubftance to the diftance of $16\frac{1}{4}$ inches from the centre, both to ftrengthen the plates, and to prevent any electricity from being carried off by the axis. Its battery of jars contains 225 fquare feet of coated furface, and its effects are aftonifhingly great.

ELECTRICAL *Phial*. See LEYDEN *Phial*.

ELECTRICAL *Rubber*. See ELECTRICAL *Apparatus*, and ELECTRICAL *Machine*.

ELECTRICAL *Shock*, is the fudden explofion between the oppofite fides of a charged electric; fo called becaufe if the difcharge be made through the body of an animal, it occafions a fudden motion by the contraction of the mufcles through which it paffes, accompanied with a difagreeable fenfation. The force of this fhock is proportioned to the quantity of coated furface, the thinnefs of the glafs, and the power of the machine by which it is charged. Its velocity is almoft inftantaneous, and it has not been found to take up the leaft fenfible time in paffing to the greateft diftances.

It has been obferved that the Electrical Shock is weakened by being communicated through feveral perfons in contact with one another. Indeed it is obftructed in its paffage, even through the beft conductors, as it will prefer a fhort paffage through the air to a long one through the moft perfect conductors; and if the circuit be interrupted, either by electrics, or very imperfect conductors of a moderate thicknefs, the fhock will rend them in its paffage, difperfe them in every direction, and exhibit the appearance of a fudden expanfion of the air about the centre of the fhock. A ftrong fhock made to pafs through or over the belly of a mufcle, forces it to contract; and fent through a fmall animal body, deprives it inftantly of life, and haftens putrefaction. It gives polarity to magnetic needles, reverfes their poles, and produces effects precifely fimilar, though inferior in degree, to thofe of lightning.

ELECTRICAL *Star*. See STAR.

ELECTRICITY,

ELECTRICITY, or ELECTRICAL *Force*, is that power or property, which was firft obferved in amber, the lyneurium, or tourmalin, and which fealing-wax, glafs, and a variety of other fubftances, called electrics, are now known to poffefs, of attracting light bodies, when excited by heat or friction; and which is also capable of being communicated in particular circumftances to other bodies.

ELECTRICITY also denotes the fcience, or that part of natural philofophy, which propofes to inveftigate the nature and effects of this power. From ηλεκτρον, the Greek name for amber, is derived the term Electricity, which is now very extenfively applied, not only to the power of attracting light bodies inherent in amber, but to other fimilar powers, and their various effects, in whatever bodies they refide, or to whatever bodies they may be communicated.

Mufchenbroek and Æpinus have obferved a confiderable analogy, in a variety of particulars, between the powers of Electricity and Magnetifm; and they have also pointed out many inftances in which they differ.

Hiftory of ELECTRICITY.——The property which amber poffeffes of attracting light bodies, was very anciently obferved. Thales of Miletus, 600 years before Chrift, concluded from hence that it was animated. But the firft perfon who exprefsly mentioned this fubftance, was Theophraftus, about 300 years before Chrift. The attractive property of amber is also occafionally noticed by Pliny, and other later naturalifts, particularly Gaffendus, Kenelm Digby, and Sir Thomas Brown. But it was generally apprehended that this quality was peculiar to amber and jet, and perhaps agate, till W. Gilbert, a native of Colchefter, and a phyfician in London, publifhed his treatife *De Magnete*, in the year 1600. Dr. Gilbert made many confiderable experiments and difcoveries, confidering the then infant ftate of the fcience. He enlarged the lift both of electrics, and of the bodies on which they act: he remarked, that a dry air was moft favourable to electrical appearances, whilft a moift air almoft annihilates the electric virtue: he also obferved the conical figure affumed by electrified drops of water: he confidered electrical attraction feparately from repulfion, which he thought had no place in Electricity, as a phenomenon fimilar to the attraction of cohefion, and he imagined, that electrics were brought into contact with the bodies on which they act by their effluvia, excited by friction.

The ingenious Mr. Boyle added to the catalogue of electric fubftances; but he thought that glafs poffeffed this power in a very low degree: he found, that the Electricity of all bodies, in which it might be excited, was increafed by wiping and warming them before they were rubbed; that an excited electric was acted upon by other bodies as ftrongly as it acted upon them; that diamonds rubbed againft any kind of ftuff, emitted light in the dark; and that feathers would cling to the fingers, and to other fubftances, after they had been attracted by electrics. He accounted for electrical attraction, by fuppofing a glutinous effluvia emitted from electrics, which laid hold of fmall bodies, in its way, and carried them back to the body from which it proceeded.

7

Otto Guericke, the celebrated inventor of the airpump, lived about the fame time. This ingenious philofopher difcovered, by means of a globe of fulphur, that a body once attracted by an electric, was next repelled, and continued in this ftate of repulfion till it fhould be touched by fome other body: he also obferved the found and light produced by the excitation of his globe; and that bodies immerged in electrical atmofpheres are themfelves electrified with an electricity oppofite to that of the atmofphere.

The light emitted by electrical bodies was, not long after, obferved to much greater advantage by Dr. Wall, who afcribes to light the electrical property which they poffefs; and he fuggefts a fimilarity between the effects of electricity and lightning.

Sir Ifaac Newton was not inattentive to this fubject: he obferved that excited glafs attracts light bodies on the fide oppofite to that on which it is rubbed; and he afcribes the action of electric bodies to an elaftic fluid, which freely penetrates glafs, and the emiffion of it to the vibratory motions of the parts of excited bodies.

Mr. Hawkfbee wrote on this fubject in the year 1709, when a new æra commenced in the hiftory of this fcience. He firft took notice of the great electrical power of glafs, and the light proceeding from it; though others had before obferved the light proceeding from other electrified fubftances: he also noted the noife occafioned by it, with a variety of phenomena relating to electrical attraction and repulfion. He firft introduced a glafs globe into the electrical apparatus, to which circumftance it was that many of his important difcoveries were owing.

After his time there was an interval of near 20 years in the progrefs of this fcience, till Mr. Stephen Grey eftablifhed a new æra in the hiftory of Electricity. To him we owe the capital difcovery of communicating the power of native electrics to other bodies, in which it cannot be excited, by fupporting them on filken lines, hair lines, cakes of refin or glafs; and a more accurate diftinction than had hitherto obtained between electrics and non-electrics: he also fhewed the effect of electricity on water much more obvioufly than Gilbert had done in the infancy of this fcience.

The experiments of Mr. Grey were repeated by M. du Fay, member of the Academy of Sciences at Paris, to which he added many new experiments and difcoveries of his own. He obferved, that electrical operations are obftructed by great heat, as well as by a moift air; that all bodies, both folid and fluid, would receive electricity, when placed on warm or dry glafs, or fealing-wax; that thofe bodies which are naturally the leaft electric, have the greateft degree of electricity communicated to them by the approach of the excited tube. He tranfmitted the electric virtue through a diftance of 1256 feet; and firft obferved the electric fpark from a living body, fufpended on filken lines, and noted feveral circumftances attending it. M. du Fay also eftablifhed a principle, firft fuggefted by Otto Guericke, that electric bodies attract all thofe that are not fo, and repel them as foon as they are become electric, by the vicinity or contact of the electric body. He likewife inferred from other experiments, that

that there were two kinds of electricity; one of which he called the *vitreous*, belonging to glass, rock crystal, &c; and the other *resinous*, as that of amber, gumlac, &c, distinguished by their repelling those of the same kind, and attracting each other. He farther observed, that communicated electricity had the same property as the excited; and that electric substances attract the dew more than conductors.

Mr. Grey, resuming his experiments in 1734, suspended several pieces of metal on silken lines, and found that by electrifying them they gave sparks; which was the origin of metallic conductors: and on this occasion he discovered a cone or pencil of electric light, such as is now known to issue from an electrified point. From other experiments he concludes, that the electric power seems to be of the same nature with that of thunder and lightning.

Dr. Defaguliers succeeded Mr. Grey in the prosecution of this science. The account of his first experiments is dated in 1739. To him we owe those technical terms of *conductors* or *non-electrics*, and *electrics per se*; and he first ranked pure air among the electrics per se, and supposed its Electricity to be of the vitreous kind.

After the year 1742, in which Dr. Defaguliers concluded his experiments, the subject was taken up and pursued in Germany: the globe was substituted for the tube, which had been used ever since the time of Hawksbee, and a cushion was soon after used as a rubber, instead of the hand. About this time too, some used cylinders instead of the globes; and some of the German electricians made use of more globes than one at the same time. By thus increasing the electrical power, they were the first who succeeded in setting fire to inflammable substances: this was first done by Dr. Ludolf, in the beginning of the year 1744, who, with sparks excited by the friction of a glass tube, kindled the ethereal spirit of Frobenius. Winkler did the same by a spark from his own finger, by which he kindled French brandy, and other spirits, after previously heating them. Mr. Gralath fired the smoke of a candle just blown out, and so lighted it again; and Mr. Boze fired gun-powder, by means of its inflammable vapour. About this time Ludolf the younger demonstrated, that the luminous barometer was made perfectly electrical by the motion of the quicksilver. The electrical star and electrical bells were also of German invention.

In England Dr. Watson made a distinguished figure from this period in the history of Electricity: he fired a variety of substances by the electrical spark, and first discovered that they are capable of being fired by the repulsive power of Electricity. In the year 1745, the accumulation of the electrical power in glass, by means of the Leyden phial, was first discovered. See LEYDEN *phial:* and for the method practised about this time, of measuring the distance to which the electrical shock may be conveyed, see *Electrical* CIRCUIT. Dr. Watson discovered that the glass tubes and globes do not contain the electric matter in themselves, but only serve as *first-movers* or *determiners*, as he expresses it, of that power; which was also confirmed towards the end of 1746, by Mr. Benjamin Wilson, who made the same discovery, that the electric fluid does not come

from the globe, but from the earth, and other non-electric bodies about the apparatus. Dr. Watson also discovered what Dr. Franklin observed about the same time in America, and called the *plus* and *minus* in Electricity. He likewise shewed that the electric matter passed through the substance of the metal of communication, and not merely over the surface. The history of medical Electricity commenced in the year 1747. We must omit other experiments, and conclusions drawn from them, by Mr Wilson, Mr. Smeaton, and Dr. Miles in England, and by the Abbé Nollet, with regard to the effect of Electricity on the evaporation of fluids, on solids, and on animal and other organized bodies, in France.

Whilst the philosophers of Europe were busily employed in electrical experiments and pursuits, those of America, and Dr. Franklin in particular, were equally industrious, and no less successful. His discoveries and observations in Electricity were communicated in several letters to a friend; the first of which is dated in 1747, and the last in 1754; and the particulars of his system may be seen under the articles, *Theory of* ELECTRICITY, LEYDEN *Phial*, POINTS, CHARGING, CONDUCTORS, ELECTRICS, &c.

The similarity between Electricity and Lightning had been suggested by several writers: Dr. Franklin first proposed a method of bringing the matter to the test of experiment, by raising an electrical kite; and he succeeded in collecting electrical fire by this means from the clouds, in 1752, one month after the same theory had been verified in France, and without knowing what had been done there: and to him we owe the practical application of this discovery, in securing buildings from the damage of lightning, by erecting metallic conductors. See CONDUCTORS, and LIGHTNING.

In the subsequent period of the history of this science, Mr. Canton in England, and Signior Beccaria in Italy, acquired distinguished reputation. They both discovered, independently of each other, that air is capable of receiving Electricity by communication, and of retaining it when received. Mr. Canton also, towards the latter end of the year 1753, pursued a series of experiments, which prove that the appearances of positive and negative Electricity, which had hitherto been deemed essential and unchangeable properties of different substances, as of glass and sealing-wax for instance, depend upon the surface of the electrics, and that of the rubber.

This hypothesis, verified by numerous experiments, occasioned a controversy between Mr. Canton, and Mr. Delaval, who still maintained that these different powers depended entirely on the substances themselves. About this time too, some curious experiments were performed by four of the principal electricians of that period, viz. Dr. Franklin, and Messrs Canton, Wilcke, and Æpinus, to ascertain the nature of electric atmospheres; the result of which see under that article.

The theory of two electric fluids, always co-existent and counteracting each other, though not absolutely independent, was maintained by a course of experiments on silk stockings of different colours, communicated to the Royal Society by Mr. Symmer, in the year 1759, which were farther pursued by Mr. Cigna

of.

of Turin, who published an account of them in the Memoirs of the Academy at Turin for the year 1765.

Many inftances occur in the hiftory of the fcience about this period, of the aftonishing force of the electric fhock, in melting wires, and producing other fimilar effects: but the moft remarkable is an experiment of S. Beccaria, in which he thus revivified metals. Several experiments were alfo made by Dr. Watfon, Mr. Smeaton, Mr. Canton, and others, on the paffage of the electric fluid through a vacuum, and its luminous appearance, and on the power poffeffed by certain fubftances of retaining the light communicated to them by an electric explofion. Mr. Canton, S. Beccaria, and others, made many experiments to identify Electricity and lightning, to afcertain the ftate of the atmofphere at different times, and to explain the various phenomena of the Aurora Borealis, Water-Spouts, Hurricanes, &c, on the principles of this fcience.

Thofe who are defirous of farther information with refpect to the hiftory of electrical experiments and difcoveries, may confult Dr. Prieftley's Hiftory and Prefent State of Electricity. This author however is not merely an hiftorian: his work contains many original experiments and difcoveries made by himfelf. He afcertained the conducting power of charcoal, and of hot glafs, the Electricity of fixed and inflammable air, and of oil; the difference between new and old glafs, with refpect to the diffufion of Electricity over its furface; the lateral explofion in electrical difcharges; a new method of fixing circular-coloured fpots on the furfaces of metals, and the moft probable difference between electrics and conductors, &c. The fcience is alfo greatly indebted to many other perfons, either for their experiments and improvements of it, or for treatifes and other writings upon it; as Mr. Henley, to whom we owe feveral curious experiments and obfervations on the electrical and conducting quality of different fubftances, as chocolate, vapour, &c, with the reafon of the difference between them; the fufion of platina; the nature of the electric fluid, and its courfe in a difcharge; the method of eftimating the quantity of it in electrical bodies by an electrometer; the influence of points; &c, &c. Alfo Meffrs Van Marum, Van Swinden, Fergufon, Cavallo, Lord Mahon, Nairne, &c, &c, for their feveral treatifes on the fubject of Electricity, any of which may be confulted with advantage for the experiments and principles of the fcience.

Medical ELECTRICITY. It is natural to imagine that a power of fuch efficacy as that of electricity would be applied to medical purpofes; efpecially, fince it has been found invariably to increafe the fenfible perfpiration, to quicken the circulation of the blood, and to promote the glandular fecretion: accordingly, many inftances occur in the latter period of the hiftory of this fcience, in which it has been applied with confiderable advantage and fuccefs. And among the variety of cafes in which it has been tried, there are none in which it has been found prejudicial except thofe of pregnancy and the venereal difeafe. In moft diforders, in which it has been ufed with perfeverance, it has given at leaft a temporary and partial relief, and in many it has effected a total cure. Of which numerous inftances may be feen in the Philof. Tranf. and the writ-

ings on this fcience by Meffrs Lovet, Weftley, Fergufon, Cavallo, &c. &c.

Theory of ELECTRICITY. It is hardly neceffary to recite the ancient hypothefes on this fubject; fuch as that of the fympathetic powder of the Peripatetics; that of unctuous effluvia emitted by excited bodies, and returning to them again, adopted by Gilbert, Gaffendus, Sir Kenelm Digby, &c; or that of the Cartefians, who afcribed electricity to the globules of the firft elements, difcharged through the pores of the rubbed fubftance, and in their return carrying with them thofe light bodies, in whofe pores they were entangled: thefe hypothefes were framed in the infancy of the fcience, and of philofophy in general, and have long fince been exploded. In the more advanced ftate of electricity there have been two principal theories, each of which has had its advocates. The one, is that of two diftinct electric fluids, repulfive with refpect to themfelves, and attractive of one another, adopted by M. du Fay, on difcovering the two oppofite fpecies of electricity, viz, the vitreous and refinous, and fince new-modelled by Mr. Symmer. It is fuppofed that thefe two fluids are equally attracted by all bodies, and exift in intimate union in their pores; and that in this ftate they exhibit no mark of their exiftence. But that the friction of an electric by a rubber feparates thefe fluids, and caufes the vitreous electricity of the rubber to pafs to the electric, and then to the prime conductor of a machine, while the refinous electricity of the conductor and electric is conveyed to the rubber: and thus the quality of the electric fluid, poffeffed by the conductor and the rubber, is changed, while the quantity remains the fame in each. In this ftate of feparation, the two electric fluids will exert their refpective powers; and any number of bodies charged with either of them will repel each other, attract thofe bodies that have lefs of each particular fluid than themfelves, and be ftill more attracted by bodies that are wholly deftitute of it, or that are loaded with the contrary. According to this theory, the electric fpark makes a double current; one fluid paffing to an electrified conductor from any fubftance prefented to it, whilft the fame quantity of the other fluid paffes from it; and when each body receives its natural quantity of both fluids, the balance of the two powers is reftored, and both bodies are unelectrified. For a further account of the explication of fome of the principal phenomena of electricity by this theory, fee Dr. Prieftley's Hiftory, vol. 2, § 3.

The other theory is commonly diftinguifhed under the denomination of *pofitive and negative electricity*, being firft fuggefted by Dr. Watfon, but digefted, illuftrated, and confirmed by Dr. Franklin; and fince that it has been known by the appellation of the Franklinian hypothefis. It is here fuppofed that all the phenomena of electricity depend on one fluid, *fui generis*, extremely fubtile and elaftic, difperfed through the pores of all bodies, by which the particles of it are as ftrongly attracted as they are repelled by one another. When bodies poffefs their natural fhare of this fluid, or fuch a quantity as they can retain by their non-attraction, it is then faid they are in an unelectrified ftate; but when the equilibrium is difturbed, and they either acquire an additional quantity *from* other bodies, or lofe part of their own natural fhare by communication *to* other bodies, they

they exhibit electrical appearances. In the former case it is said they are electrified positively, or *plus ;* and in the other negatively, or *minus.* This electric fluid, it is supposed, moves with great ease in those bodies that are called conductors, but with extreme difficulty and slowness in the pores of electrics ; whence it comes to pass, that all electrics are impermeable to it. It is farther supposed that electrics contain always an equal quantity of this fluid, so that there can be no surcharge or increase on one side without a proportionable decrease or loss on the other, and vice versa ; and as the electric does not admit the passage of the fluid through its pores, there will be an accumulation on one side, and a corresponding deficiency on the other. Then when both sides are connected together by proper conductors, the equilibrium will be restored by the rushing of the redundant fluid from the overcharged surface to the exhausted one. Thus also, if an electric be rubbed by a conducting substance, the electricity is only conveyed from one to the other, the one giving what the other receives ; and if one be electrified positively, the other will be electrified negatively, unless the loss be supplied by other bodies connected with it, as in the case of the electric and insulated rubber of a machine. This theory serves likewise to illustrate the other phenomena and operations in the science of electricity. Thus, bodies differently electrified will naturally attract each other, till they mutually give and receive an equal quantity of the electric fluid, and the equilibrium is restored between them. Beccaria supposes, that this effect is produced by the electric matter making a vacuum in its passage, and the contiguous air afterwards collapsing, and so pushing the bodies together.

The influence of points, in drawing or throwing off the electric fluid, depends on the less resistance it finds to enter or pass off through fewer particles than through a greater number, whose resistance is united in flat or round surfaces. The electric light is supposed to be part of the electric fluid, which appears when it is properly agitated ; and the sound of an explosion is produced by vibrations, occasioned by the air's being displaced by the electric fluid, and again suddenly collapsing.

As to the nature of the electric fluid, philosophers have entertained very different sentiments : some, and among them Mr. Wilson, have supposed that it is the same with the ether of Sir Isaac Newton, to which the phenomena of attraction and repulsion are ascribed ; whilst the light, smell, and other sensible qualities of the electric fluid, are referred to the grosser particles of bodies, driven from them by the forcible action of this ether ; and other appearances are explained by means of a subtile medium diffused over the surfaces of all bodies, and resisting the entrance and exit of the ether ; which medium, it is supposed, is the same with the electric fluid, and is more rare on the surfaces of conductors, and more dense and resisting on those of electrics : but Dr. Priestley remarks that, though they may possess some common properties, they have others essentially distinct ; the ether is repelled by all other matter, whereas the electric fluid is strongly attracted by it. Others have had recourse to the element of fire ; and from the supposed identity of fire and the electric fluid, as well as from the similarity of some of their effects, the

latter has been usually called the electric fire : but most electricians have supposed that it is a fluid *sui generis.* Mr. Cavendish has published an attempt to deduce and explain some of the principal phenomena of electricity in a mathematical and systematic manner, from the nature of this fluid, considered as composed of particles that repel each other, and attract the particles of all other matter, with a force inversely as some less power of the distance than the cube, whilst the particles of all other matter repel each other, and attract those of the electric fluid, according to the same law. Philos. Transf. vol. 61, pa. 584—677. And a similar hypothesis and method of reasoning was also proposed by M. Æpinus, in his Tentamen Theoriæ Electricitatis & Magnetismi.

Dr. Priestley concludes, from experiments, that the electric matter either is phlogiston, or contains it, since he found that both produced similar effects. Mr. Henley also apprehends, that the electric fluid is a modification of that element, which, in its quiescent state, is called phlogiston ; in its first active state, electricity ; and when violently agitated, fire. Perhaps we may be allowed to enlarge our views, and consider the sun as the fountain of the electric fluid, and the zodiacal light, the tails of comets, the aurora borealis, lightning, and artificial electricity, as its various and not very dissimilar modifications. On this subject, see Priestley's Hist. of Electr. vol. 2, part 3, § 1, 2, 3 ; Wilson's Essay towards an Explication of the Phenomena of Electricity, &c ; Wilson and Hoadley's Obs. &c, pa. 55, 1759 ; Freke's Essay on the Cause of Electricity, 1746 ; Priestley on Air, vol. 1, pa. 186, 274, &c ; Philos. Transf. vol. 67, pa. 129 ; and Mr. Eeles's Letters, on the same subject.

ELECTROMETER, is an instrument that measures the quantity, and determines the quality of electricity, in any electrified body. Previous to the invention of instruments of this kind, Mr. Canton estimated the quantity of electricity in a charged phial, by presenting the phial with one hand to an insulated conductor, and giving it a spark, which he took off with the other ; proceeding in this manner till the phial was discharged, when he determined the height of the charge by the number of sparks. Electrometers are of 4 kinds : 1, the single thread ; 2, the cork or pith balls ; 3, the quadrant ; and 4, the discharging Electrometer.

The 1st, or most simple Electrometer, is a linen thread, called by Dr. Desaguliers, the *thread of trial ;* which, being brought near an electrified body, is attracted by it : but this does little more than determine whether the body is in any degree electrified or not ; without determining with any precision its quantity, much less the quality of it. The Abbé Nollet used two threads, shewing the degree of electricity by the angle of their divergency exhibited in their shadow on a board placed behind them.

Mr. Canton's Electrometer consisted of two balls of cork, or pith of elder, about the size of a small pea, suspended by fine linen threads, about 6 inches long, which may be wetted in a weak solution of salt. See fig. 7. If the box containing these balls be insulated, by placing it on a drinking glass, &c, and an excited smooth glass tube be brought near them, they will first be attracted by it, and then be repelled both from the glafs, and from each other; but on the approach of excited wax, they will gradually approach and come together; and vice verfa. This apparatus will alfo ferve to determine the electricity of the clouds and air, by holding them at a fufficient diftance from buildings, trees, &c; for if the electricity of the clouds or air be pofitive, their mutual repulfion will increafe by the approach of excited glafs, or decreafe by the approach of amber or fealing-wax; on the contrary, if it be negative, their repulfion will be diminifhed by the former, and increafed by the latter. See Philof. Tranf. vol. 48, part 1 and 2, for an account of Mr. Canton's curious experiments with this apparatus.

If two balls of this kind be annexed to a prime conductor, they will ferve to determine both the degree and quality of its electrification, by their mutual repulfion and divergency.

The Difcharging Electrometer, fig. 3, plate ix, was invented by Mr. Lane. It confifts of brafs work G, the lower part of which is inclofed in the pillar F, made of baked wood, and boiled in linfeed oil, and bored cylindrically about two thirds of its length; the brafs work is fixed to the pillar by the fcrew H, moveable in the groove I; and through the fame is made to pafs a fteel fcrew L, to the end of which, and oppofite to K, a polifhed hemifpherical piece of brafs, attached to the prime conductor, is fixed a ball of brafs M well polifhed. To this fcrew is annexed a circular plate O, divided into 12 equal parts. The ufe of this Electrometer is to difcharge a jar D, or any battery connected with the conductor, without a difcharging rod, and to give fhocks fucceffively of the fame degree of ftrength; on which account it is very fit for medical purpofes. Then, if a perfon holds a wire faftened to the fcrew H in one hand, and another wire fixed to E, a loop of brafs wire paffing from the frame of the machine to a tin plate, on which the phial D ftands, he will perceive no fhock, when K and M are in contact; and the degree of the explofion, as well as the quantity of electricity accumulated in the phial, will be regulated by the diftance between K and M. Philof. Tranf. vol. 57, pa. 451.— Mr. Henley much improved Mr. Lane's Electrometer, by taking away the fcrew, the double milled nut, and the fharp-edged graduated plate, and adding other contrivances in their ftead. Mr. Henley's difcharger of this kind has two tubes, one fliding within the other, to lengthen and accommodate it to larger apparatus.

The Quadrant Electrometer of Mr. Henley, confifts of a ftem, terminating at its lower end with a brafs ferrule and fcrew, for faftening it upon any occafion; and its upper part ends in a ball. Near the top is fixed a graduated femicircle of ivory, on the centre of which the index, being a very light rod with a cork ball at its extremity, reaching to the brafs ferrule of the ftem, is made to turn on a pin in the brafs piece, fo as to keep near the graduated limb of the femicircle. When the Electrometer is not electrified, the index hangs parallel to the ftem; but as foon as it begins to be electrified, the index, repelled by the ftem, will begin to move along the graduated edge of the femicircle, and fo mark the degree to which the conductor is electrified, or the height to which the charge of any jar or battery is advanced.

Mr. Cavallo has alfo contrived feveral ingenious Elec-

trometers, for different ufes; as may be feen in his Treatife on Electricity, pa. 370, &c, and in the Philof. Tranf. vol. 67, pa. 48 and 399.

ELECTROPHOR, or ELECTROPHORUS, an inftrument for fhewing perpetual electricity; which was invented by Mr. Volta, of Como, near Milan, in Italy. The machine confifts of two plates, fig. 8, one of which B is a circular plate of glafs, covered on one fide with fome refinous electric, and the other A is a plate of brafs, or a circular board, coated with tinfoil, and furnifhed with a glafs handle I, which may be fcrewed into its centre by means of a focket. If the plate B be excited by rubbing it with new white flannel, and the plate A be applied to its coated fide, a finger, or any other conductor, will receive a fpark on touching this plate; and if the plate A be then feparated, by means of the handle I, it will be found ftrongly electrified, with an electricity contrary to that of the plate B. By replacing the plate A, touching it with the finger, and feparating it again, it will be found electrified as before, and give a fpark to any conductor, attended with a fnapping noife; and by this means a coated phial may be charged. The fame phenomena may repeatedly be exhibited, without any renewed excitation of the electric plate B; the electric power of B having continued for feveral days, and even weeks, after excitation; though there is no reafon to imagine that it is perpetual.

Mr. Cavallo prepares this machine by coating the glafs plate with fealing-wax; and Mr. Adams, philofophical inftrument maker, prepares them with plates formed from a compofition of two parts of fhell-lac, and one of Venice turpentine, without any glafs plate.

The action of this plate depends on a principle difcovered and illuftrated by the experiments of Franklin, Canton, Wilcke, and Æpinus, viz, that an excited electric repels the electricity of another body, brought within its fphere of action, and gives it a contrary electricity. Thus the plate A, touched by a conductor, whilft in contact with the plate B, electrified negatively, will acquire an additional quantity of the electric fluid from the conductor; but if it were in contact with a plate electrified pofitively, it would part with its electricity to the conductor connected with it. See an account of feveral curious experiments with this machine, by Mr. Henley, Mr. Cavallo, and Dr. Ingenhoufz, in the Philof. Tranf. vol. 66, pa. 513; vol. 67, pa. 116 and 389; and vol. 68, pa. 1027 and 1049.

ELEMENTARY, fomething that relates to the principles or elements of bodies, or fciences; as Elementary Air, Fire, Geometry, Mufic, &c.

ELEMENTS, the firft principles, of which all bodies and things are compofed. Thefe are fuppofed few in number, unchangeable, and by their combinations producing that extenfive variety of objects to be met with in the works of nature.

Democritus ftands at the head of the Elementary Philofophers, in which he is followed by Epicurus, and many others after them, of the Epicurean and corpufcular philofophers.

Among thofe who hold the Elements corruptible, fome will have only one, and fome feveral. Of the former, the principal are Heraclitus, who held fire; Anaximenes, air; Thales Milefius, water; and Hefiod, earth;

earth; as the only Element. Hesiod is followed by Bernardin, Telesius; and Thales by many of the chemists.

Among those who admit several corruptible Elements, the principal are the Peripatetics; who, after their leader Aristotle, contend for four Elements, viz, fire, air, water, and earth. Aristotle took the notion from Hippocrates; Hippocrates from Pythagoras; and Pythagoras from Ocellus Lucanus, who it seems was the first author of it.

The Cartesians admit only three Elements, fire, air, and earth. See CARTESIAN *Philosophy.*

Newton observes, that it seems probable that God, in the beginning, formed matter in solid, massive, hard, impenetrable, moveable particles, of such sizes and figures, &c, as most conduced to the end for which he formed them; and that these primitive particles, being solids, are incomparably harder than any porous body compounded of them; even so hard as never to wear out; no ordinary power being able to divide what God made one in the first creation. While the particles remain entire, they may compose bodies of one and the same nature and texture in all ages; but should they wear away, or break in pieces, the nature of things, depending on them, would be changed; water and earth, composed of old worn particles, and fragments of particles, would not be of the same nature and texture now, with water and earth composed of entire particles in the beginning. And therefore, that things may be lasting, the changes of corporeal things are to be placed only in the various separations, and new associations and motions of those permanent particles; compound bodies being apt to break, not in the midst of solid particles, but where those particles are laid together, and only touch in a few points. It seems to him likewise, that these particles have not only a vis inertiæ, with the passive laws of motion thence resulting, but are also moved by certain active principles; such as gravity, and the cause of fermentation, and the cohesion of bodies.

ELEMENTS, a term also used for the first grounds and principles of arts and sciences; as the Elements of geometry, Elements of mathematics, &c. So Euclid's Elements, or simply the Elements, as they were anciently and peculiarly named, denotes the treatise on the chief properties of geometrical figures by that author.

The ELEMENTS of Mathematics have been delivered by several authors in their courses, systems, &c. The first work of this kind is that of Herigon, in Latin and French, and published in 1664, in 10 tomes; which contains Euclid's Elements and Data, Apollonius, Theodosius, &c; with the modern Elements of arithmetic, algebra, trigonometry, architecture, geography, navigation, optics, spherics, astronomy, music, perspective, &c. The work is remarkable for this, that a kind of real and universal characters are used throughout; so that the demonstrations may be understood by such as only remember the characters, without any dependence on language or words at all.

Since Herigon, the Elements of the several parts of mathematics have been also delivered by others; particularly the Jesuit Schottus, in his Cursus Mathematicus, in 1674; De Chales, in his Cursus, 1674; Sir Jonas Moore, in his New System of Mathematics, in

1681; Ozanam, in his Cours de Mathematique, in 1699; Jones, in his Synopsis Palmariorum Matheseos, in 1706; and many others, but above all, Christ. Wolfius, or Wolf, in his Elementa Matheseos Universæ, in 2 vols 4to, the 1st published in 1713, and the 2d in 1715; a very excellent work of the kind. Another edition of the work was published at Geneva, in 5 vols 4to, of the several dates 1732, 1733, 1735, 1738, and 1741.

The ELEMENTS of Euclid, as they were the first, so they continue still the best system of geometry, are in 15 books. There have been numerous editions and commentaries of this work. Proclus wrote a commentary on it. Orontius Fineus first gave a printed edition of the first 6 books, in 1530, with notes, to explain Euclid's sense. Peletarius did the same in 1557. Nic. Tartaglia, about the same time, made a comment on all the 15 books, with the addition of many things of his own. And the same was also done by Billingsley in 1570; and by Flussates Candalla, a noble Frenchman, in the year 1578, with considerable additions as to the comparison and inscriptions of solid bodies; which work was afterwards republished with a prolix commentary, by Clavius. Commandine gave also a good edition of it. In 1703, Dr. Gregory published an edition of the whole works of Euclid, in Greek and Latin, including his Elements. But it would be endless to relate all the other editions of these Elements, either the whole, or in part, that have been given; some of the best of which are those of De Chales, Tacquet, Ozanam, Whiston, Stone, and most especially that of Dr. Rob. Simson, of Glasgow.

Other writers on the Elements of Geometry are almost out of number, in all nations.

ELEMENTS, in the Higher or Sublime Geometry, are the infinitely small parts, or differentials, of a right line, curve, surface, or solid.

ELEMENTS, in Astronomy, are those principles deduced from astronomical observations and calculations, and those fundamental numbers which are employed in the construction of tables of the planetary motions. Thus, the Elements of the theory of the sun, or rather of the earth, are his mean motion and eccentricity, with the motion of the aphelia. And the Elements of the theory of the moon, are her mean motion, that of the node and apogee, the eccentricity, the inclination of her orbit to the plane of the ecliptic; &c.

ELEVATION, the height or altitude of any thing.

ELEVATION, in Architecture, denotes a draught or description of the principal face or side of a building; called also its upright or orthography.

ELEVATION, in Astronomy and Geography, is various; as Elevation of the equator, of the pole, of a star, &c.

ELEVATION *of the Equator*, is the height of the equator above the horizon; or an arc of the meridian intercepted between the equator and the horizon of the place.—The Elevation of the equator and of the pole together always make up a quadrant; the one being the complement of the other. Therefore, the Elevation of the pole being found, and subtracted from 90°, leaves the elevation of the equator.

ELEVATION *of the Pole*, is its height above the

horizon; or an arc of the meridian comprehended between the equator and the horizon of the place.

The elevation of the pole is always equal to the latitude of the place; that is, the arc of the meridian intercepted between the pole and the horizon, is every where equal to the arc of the same meridian intercepted between the equator and the zenith. Thus the north pole is elevated 51° 31 above the horizon of London; and the distance, or number of degrees, is the same between London and the equator; so that London is also in 51° 31 of north latitude.

ELEVATION *of a Star*, or of any other point in the sphere, is the angular height above the horizon; or an arc of the vertical circle intercepted between the star and the horizon. The meridian altitude of any such point, or its altitude when in the meridian, is the greatest of all.

ELEVATION *of a Cannon*, or *Mortar*, is the angle which the bore or the axis of the piece makes with the horizontal plane.

Angle of ELEVATION, is the angle which any line of direction makes above a horizontal line.

ELEVATION is also used by some writers on Perspective, for the scenography, or perspective representation of the whole body or building.

ELLIPSE, or ELLIPSIS, is one of the conic sections, popularly called an oval; being called an Ellipse or Ellipsis by Apollonius, the first and principal author on the conic sections, because in this figure the squares of the ordinates are *less* than, or *defective* of, the rectangles under the parameters and abscisses.

This figure is differently defined by different authors; either from some of its properties, or from mechanical construction, or from the section of a cone, which is the best and most natural way. Thus;

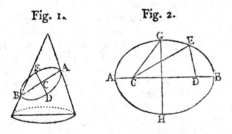

Fig. 1.   Fig. 2.

1. An Ellipse is a plane figure made by cutting a cone by a plane obliquely through the opposite sides of it; or so as that the plane makes a less angle with the base than the side of the cone makes with it; as ABD fig. 1.

The line AB connecting the uppermost and lowest points of the section, is the transverse axis; the middle of it C, is the centre; and the perpendicular to it DCE, through the centre, is the conjugate axis. The parameter, or latus rectum, is a 3d proportional to the transverse and conjugate axes; and the foci are two points in the transverse axis, at such equal distances from the centre, that the double ordinates passing through those points, and perpendicular to the transverse, are equal to the parameter.

2. The Ellipse is also variously described from some of its properties. As first, That it is a figure of such a

nature that if two lines be drawn from two certain points C and D in the axis, fig. 2, to any point E in the circumference, the sum of those two lines CE and DE will be every where equal to the same constant quantity, viz, the axis AB. Or, secondly, that it is

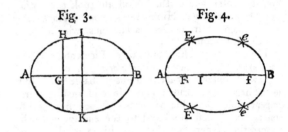

Fig. 3.   Fig. 4.

a figure of such a nature, that the rectangle AG × GB (fig. 3) of the abscisses, tending contrary ways, is to GH² the square of the ordinate, as AB² to IK², the square of the transverse axis to the square of the conjugate; or, which is the same thing, as the transverse axis is to the parameter. And so of other properties.

3. Or the Ellipse is also variously described from its mechanical constructions, which also depend on some of its chief properties. Thus; 1st, If in the axis AB, there be taken any point I (fig. 4); and if with the radii AI, BI, and centres F and *f*, the two foci, arcs be described, these arcs will intersect in certain points E, E, *e*, *e*, which will be in the curve or circumference of the figure: and thus several points I being taken in the axis AB, as many more points E, *e*, &c, will be found; then the curve line drawn through all these points E, *e*, will be an Ellipse. Or, thus; if there be taken a thread of the exact length of the transverse axis AB, and the ends of the thread be fixed by pins in the two foci F and *f*; (fig. 5) then moving a pen or pencil within the thread, so as to keep it always stretched out, it will describe the curve called an Ellipse.

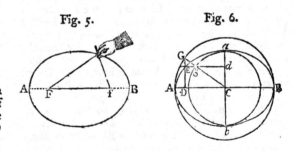

Fig. 5.   Fig. 6.

To Construct an ELLIPSE. There are many other ways of describing or constructing an Ellipse, besides those just now given: as

1st. If upon the given transverse axis there be described a circle AGB (fig. 6), to which draw any ordinate DG, and DE a 4th proportional to the transverse, the conjugate, and the ordinate DG; then E is a point in the curve. Or if the circle *agb* be described on the conjugate axis *ab*, to which any ordinate *dg* is drawn, in which taking *dE* in like manner a 4th proportional to the conjugate, the transverse, and ordinate *dg*, then shall

shall E be in the curve. Or, having described the two circles, and drawn the common radius CgG cutting them in G and g; then dgE drawn parallel to the transverse, and DGE parallel to the conjugate, the intersection E of these two lines will be in the curve of the ellipse. And thus several points E being found, the curve may be drawn through them all with a steady hand.

Fig. 7.   Fig. 8.

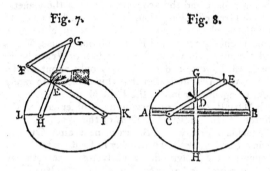

2. If there be provided three rulers, of which the two GH and FI (fig. 7) are of the length of the transverse axis LK, and the third FG equal to HI the distance between the foci; then connecting these rulers so as to be moveable about the foci H and I, and about the points F and G, their intersection E will always be in the curve of the ellipse; so that by moving the rulers about the joints, with a pencil passed through the slits made in them, it will trace out the Ellipse.

Fig. 9.   Fig. 10.

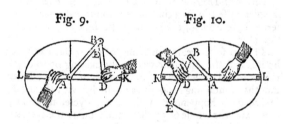

3. If one end A of any two equal rulers AB, DB, (fig. 9 and 10) which are moveable about the point B, like a carpenter's joint-rule, be fastened to the ruler LK, so as to be moveable about the point A; and if the end D of the ruler DB be drawn along the side of the ruler LK; then any point E, taken in the side of the ruler DB, will describe an ellipse, whose centre is A, conjugate axis $= 2DE$, and transverse $= 2AB + 2BE$.

Another method of description is by the Elliptical Compass. See that article, below.

Some of the more Remarkable Properties of the Ellipse.
—1. The rectangles under the abscisses are proportional to the squares of their ordinates; or as the square of any axis, or any diameter, is to the square of its conjugate, so is the rectangle under two abscisses of the former, to the square of their ordinate parallel to the latter; or again, as any diameter is to its parameter, so is the said rectangle under two abscisses of that diameter, to the square of their ordinate. So that if d be

any diameter, c its conjugate, p its parameter $= \frac{c^2}{d}$, x the one abscifs, $d-x$ the other, and y the ordinate; then,

as $d^2 : c^2 :: x . \overline{d-x} : y^2$, or $d^2y^2 = c^2x . \overline{d-x}$; or as $d : p :: x . \overline{d-x} : y^2$, or $dy^2 = px . \overline{d-x}$.
From either of which equations, called the equation of the curve, any one of the quantities may be found, when the other three are given.

2. The sum of two lines drawn from the foci to meet in any point of the curve, is always equal to the transverse axis; that is, $CE + DE = AB$, in the 2d fig. Consequently the line CG drawn from the focus to the end of the conjugate axis, is equal to AI the semitransverse.

Fig. 11.

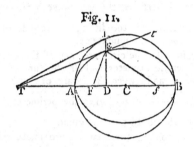

3. If from any point of the curve, there be an ordinate to either axis, and also a tangent meeting the axis produced; then half that axis will be a mean proportional between the distances from the centre to the two points of intersection; viz, CA a mean proportional between CD and CT. And consequently all the tangents TE, TE, meet in the same point of the axis produced, which are drawn from the extremities E, E, of the common ordinates DE, DE, of all Ellipses described on the same axis AB.

4. Two lines drawn from the foci to any point of the curve, make equal angles with the tangent at that point: that is, the $\angle FET = \angle fEt$.

5. All the parallelograms are equal to each other, that are circumscribed about an Ellipsis; and every such parallelogram is equal to the rectangle of the two axes.

6. The sum of the squares of every pair of conjugate diameters, is equal to the same constant quantity, viz, the sum of the squares of the two axes.

7. If a circle be described upon either axis, and from any point in that axis an ordinate be drawn both to the circle and ellipsis; then shall the ordinate of the circle be to the ordinate of the Ellipse, as that axis is to the other axis: viz,
AB : ab :: DG : DE,
and ab : AB :: dg : dE. (in the 6th fig.)
And in the same proportion is the area of the circle to the area of the Ellipse, or any corresponding segments ADG, ADE. Also the area of the Ellipse is a mean proportional between the areas of the inscribed and circumscribed circles. Hence therefore,

8. *To find the Area of an Ellipse.* Multiply the two axes together, and that product by ·7854, for the area. Or

9. *To find the Area of any Segment* ADE. Find the

area

area of the corresponding segment ADG of a circle on the same diameter AB; then say, as the axis AB: its conj. ab :: circ. seg. ADG : elliptic seg. ADE.

10. *To find the length of the whole circumference of the Ellipse.* Multiply the circumference of the circumscribing circle by the sum of the series

$$1 - \frac{d}{2^2} - \frac{3d^2}{2^2.4^2} - \frac{3^2.5d^3}{2^2.4^2.6^2} - \frac{3^2.5^2.7d^4}{2^2.4^2.6^2.8^2} \ \&c,$$

for the area: where d is the difference between an unit and the square of the lefs axis divided by the square of the greater.

Or, for a near approximation, take the circumference of the circle whose diameter is an arithmetical mean between the two axes, or half their sum: that is,

$$\frac{t + c}{2} \times 3\cdot1416 = \text{ the perimeter nearly; being}$$

about the 200th part too little; where t denotes the transverse, and c the conjugate axis.

Or, again, take the circumference of the circle, the square of whose diameter is half the sum of, or an arithmetical mean between the squares of the two axes: that is, $\sqrt{\dfrac{t^2 + c^2}{2}} \times 3\cdot1416 =$ the perimeter nearly; being about the 200th part too great.

Hence combining thefe two approximate rules together, the periphery of the Ellipse will be very nearly equal to half their fum, or equal to

$$\frac{3\cdot1416}{2} \times \left(\frac{t+c}{2} + \sqrt{\frac{t^2+c^2}{2}}\right), \text{ within about the}$$

30000th part of the truth.

For the length of any particular arc, and many other parts about the Ellipfe, fee my Menfuration, pa. 283, &c, 2d edit. See also my Conic Sections, for many other properties of the Ellipfe, efpecially fuch as are common to the hyperbola alfo, or to the conic fections in general.

Infinite ELLIPSES. See ELLIPTOIDE.

ELLIPSOID, is an elliptical fpheroid, being the folid generated by the revolution of an ellipfe about either axis. See SPHEROID.

ELLIPTIC or ELLIPTICAL, fomething relating to an ellipfe.

ELLIPTIC *Compaffes*, or ELLIPTICAL *Compafs*, is an inftrument for defcribing ellipfes at one revolution of the index. It confifts of a crofs ABGH (fig. 8.) with grooves in it, and an index CE, fliding in dovetail grooves; by which motion the end E defcribes the curve of an ellipfe.

ELLIPTICAL *Conoid*, is fometimes ufed for the fpheroid.

ELLIPTICAL *Dial*, an inftrument ufually made of brafs, with a joint to fold together, and the gnomon to fall flat, for the convenience of the pocket. By this inftrument are found the meridian, the hour of the day, the rifing and fetting of the fun, &c.

ELLIPTOIDE, an infinite or indefinite Ellipfis, defined by the indefinite equation $ay^{m+n} = bx^m.\overline{a-x^n}$ when m or n are greater than 1: for when they are each $= 1$, it denotes the common ellipfe.

There are feveral kinds or degrees of Elliptoides, denominated from the exponent $m + n$ of the ordinate y.

7

As the cubical Elliptoide, expreffed by $ay^3 = bx^2.\overline{a-x}$; the biquadratic, or furfolid $ay^4 = bx^2.\overline{a-x^2}$; &c.

ELONGATION, in Aftronomy, the diftance of a planet from the fun, with refpect to the earth; or the angle formed by two lines drawn from the earth, the one to the fun, and the other to the planet; or the arc meafuring that angle: Or it is the difference between the fun's place and the geocentric place of the planet.

The Greateft Elongation, is the greateft diftance to which the planets recede from the fun, on either fide. This is chiefly confidered in the inferior planets, Venus and Mercury; the Greateft Elongation of Venus being about 48 degrees, and of Mercury only about 28 degrees; which is the reafon that this planet is fo rarely feen, being ufually loft in the light of the fun.

EMBER-*Days*, are certain days obferved by the church at four different feafons of the year; viz, the Wednefday, Friday, and Saturday next after Quadragefima Sunday, or the 1ft Sunday in Lent; after Whitfunday; after Holyrood, or Holycrofs, the 14th day of September; and after St. Lucy, the 13th day of December. The name, it feems, is derived from Embers, or afhes, which it is fuppofed were ftrewed on the head, on thefe folemn fafts.

EMBER-*Weeks*, are thofe weeks in which the Ember-days fall. Thefe Ember-weeks are now chiefly noticed on account of the ordination of priefts and deacons; becaufe the canon appoints the Sundays next after the Ember-weeks for the folemn times of ordination; though the bifhops, if they pleafe, may ordain on any Sunday or holiday.

EMBOLIMÆAN, and EMBOLISMIC, *Intercalary*, is chiefly ufed in fpeaking of the additional months inferted by chronologifts to form the lunar cycle of 19 years.

The 19 folar years confifting of 6939 days and 18 hours, and the 19 lunar years only making 6726 days, it was found neceffary to intercalate or infert 7 lunar months, containing 209 days; which, with the 4 bif fextile days happening in the lunar cycle, make 213 days, and the whole 6939 days, the fame as the 19 folar years, which make the lunar cycle.

In the courfe of 19 years there are 228 common moons, and 7 Embolifmic moons, which are diftributed in this manner, viz, the 3d, 6th, 9th, 11th, 14th, 17th, and 19th years, are Embolifmic, and fo contain 384 days each. And this was the method of computing time among the Greeks: though they did not keep regularly to it, as it feems the Jews did. And the method of the Greeks was followed by the Romans till the time of Julius Cæfar.

The Embolifmic months, like other lunar months, are fometimes of 30 days, and fometimes only 29 days.

The *Embolifmic Epacts* are thofe between 19 and 29; which are fo called, becaufe, with the addition of the epact 11, they exceed the number 30: or rather, becaufe the years which have thefe epacts, are Embolifmic; having 13 moons each, the 13th being the Embolifmic.

EMBOLISMUS, in Chronology, fignifies intercalation. As the Greeks ufed the lunar year, which contains only 354 days, that they might bring it to the

solar

folar year, of 365 days, they had an Embolifm every two or three years, when they added a 13th lunar month.

EMBOLUS, the moveable part of a pump or fyringe; called alfo the pifton, and popularly the fucker. The pipe or barrel of a fyringe, &c, being clofe fhut, the embolus cannot be drawn up without a very confiderable force; which force being withdrawn, the embolus returns again with violence; owing to the greater preffure of air above than below it.

EMBRASURE, in Architecture, an enlargement of the aperture or opening of a door, or window, within fide the wall, floping back inwards, to give the greater play for the opening of the door, cafement, &c, or to take in the more light.

EMBRASURES, in Fortification, are the apertures or holes through which the cannon are pointed, whether in cafemates, batteries, or in the parapets of walls. In the navy, thefe are called port-holes. The Embrafures are placed 12 or 15 feet apart from each other; being made floping or opening outwards, from 6 to 9 feet wide on the outfide of the wall, and from 2 to 3 within, to allow the gun to traverfe from fide to fide. Their bafe is about 2½ or 3 feet above the platform on the infide of the wall, but floping down outwards, fo as to be only about 1½ above it on the outfide; in order that the muzzle on occafion may be depreffed, and fo the gun fhoot low, or downwards.

EMERGENT *Year*, in Chronology, is the epoch, or date, from whence any people begin to compute their time or dates. So, our Emergent year is fometimes the year of the creation, but more ufually the year of the birth of Chrift. The Jews ufed that of the Deluge, or the Exodus, &c. The Emergent year of the Greeks, was the beginning of the Olympic games; while that of the Romans was the date of the building of their city.

EMERSION, in Aftronomy, is the re-appearance of the fun, moon, or other planet, after having been eclipfed, or hid by the interpofition of the moon, earth, or other body.

The Emerfions and immerfions of Jupiter's firft fatellite, are particularly ufeful for finding the longitudes of places; the immerfions being obferved from Jupiter's conjunction with the fun, till his oppofition; and the Emerfions from the oppofition till the conjunction. But within 15 days of the conjunction, both before and after it, they cannot be obferved, becaufe the planet and his fatellites are then loft in the fun's light.

EMERSION is alfo ufed when a ftar, after being hid by the fun, begins to re-appear, and to get out of his rays.

Minutes or Scruples of EMERSION, an arc of the moon's orbit, which her centre paffes over, from the time fhe begins to emerge out of the earth's fhadow, to the end of the eclipfe.

EMERSION, in Phyfics, the rifing of any folid above the furface of a fluid that is fpecifically heavier than the folid, into which it had been violently immerged, or pufhed.

It is one of the known laws of hydroftatics, that a lighter folid, being forced down into a heavier fluid, immediately endeavours to emerge; and that with a force equal to the excefs of the weight of a quantity of

the fluid above that of an equal bulk of the folid. Thus, if the body be immerged in a fluid of double its fpecific gravity, it will emerge again till half its bulk be above the furface of the fluid.

EMERSON (WILLIAM), a late eminent mathematician, was born in June 1701, at Hurworth, a village about three miles fouth of Darlington, on the borders of the county of Durham; at leaft it is certain that he refided here from his childhood. His father Dudley Emerfon taught a fchool, and was a tolerable proficient in the mathematics; and without his books and inftructions, perhaps his fon's genius, though eminently fitted for mathematical ftudies, might never have been unfolded. Befide his father's inftructions, our author was affifted in the learned languages by a young clergyman, then curate of Hurworth, who was boarded at his father's houfe. In the early part of his life he attempted to teach a few fcholars: but whether from his concife method (for he was not happy in explaining his ideas), or the warmth of his natural temper, he made no progrefs in his fchool; he therefore foon left it off; and fatisfied with a moderate competence left him by his parents, he devoted himfelf to a ftudious retirement, which he thus clofely purfued, in the fame place, through the courfe of a long life, being moftly very healthy, till towards the latter part of his days, when he was much afflicted with the ftone. Toward the clofe of the year 1781, being fenfible of his approaching diffolution, he difpofed of the whole of his mathematical library to a bookfeller at York; and on May the 20th, 1782, his lingering and painful diforder put an end to his life at his native village, being near 81 years of age.

Mr. Emerfon, in his perfon, was rather fhort, but ftrong and well made, with an open countenance and ruddy complexion, being of a healthy and hardy difpofition. He was very fingular in his behaviour, drefs, and converfation. His manner and appearance were that of a rude and rather boorifh country man; he was of very plain converfation, and indeed feemingly rude, commonly mixing oaths in his fentences, though without any ill intention. He had ftrong good natural mental parts, and could difcourfe fenfibly on any fubject, but was always pofitive and impatient of any contradiction. He fpent his whole life in clofe ftudy, and writing books, from the profits of which, he redeemed his little patrimony from fome original incumbrance. In his drefs he was as fingular as in every thing elfe. He poffeffed commonly but one fuit of clothes at a time, and thofe very old in their appearance. He feldom ufed a waiftcoat; and his coat he wore open before, except the lower button; and his fhirt quite the reverfe of one in common ufe, the hind-fide turned foremoft, to cover his breaft, and buttoned clofe at the collar behind. He wore a kind of rufty coloured wig, without a crooked hair in it, which probably had never been tortured with a comb from the time of its being made. A hat he would make to laft him the beft part of a life time; gradually leffening the flaps, bit by bit, as it loft its elafticity and hung down, till little or nothing but the crown remained.

He often walked up to London when he had any book to be publifhed, revifing fheet by fheet himfelf:— trufting no eye but his own was always a favourite maxim.

maxim with him. In mechanical fubjects, he always tried the propositions practically, making all the different parts himfelf on a fmall fcale; fo that his houfe was filled with all kinds of mechanical inftruments, together or disjointed. He would frequently ftand up to his middle in water while fifhing; a diverfion he was remarkably fond of. He ufed to ftudy inceffantly for fome time, and then for relaxation take a ramble to any pot-alehoufe where he could get any body to drink with and talk to. The late Mr. Montagu was very kind to Mr. Emerfon, and often vifited him, being pleafed with his converfation, and ufed often to come to him in the fields where he was working, and accompany him home, but could never perfuade him to get into a carriage: on thefe occafions he would fometimes exclaim, " Damn your whim-wham! I had rather walk." He was a married man, and his wife ufed to fpin on an old-fafhioned wheel, of his own making, a drawing of which is given in his Mechanics.

Mr. Emerfon, from his ftrong vigorous mind and clofe application, had acquired a deep knowledge of all the branches of mathematics and phyfics, upon all parts of which he wrote good treatifes, though in a rough and unpolifhed ftyle and manner. He was not remarkable however for genius or difcoveries of his own, as his works fhew hardly any traces of original invention. He was well fkilled in the fcience of mufic, the theory of founds, and the various fcales both ancient and modern; but he was a very poor performer, though he could make and repair fome inftruments, and fometimes went about the country tuning harpfichords.

The following is a lift of Mr. Emerfon's works; all of them printed in 8vo, excepting his Mechanics and his Increments in 4to, and his Navigation in 12mo. 1. The Doctrine of Fluxions.—2. The Projection of the Sphere, orthographic, ftereographic, and gnomonical. —3. The Elements of Trigonometry.—4. The Principles of Mechanics.—5. A Treatife of Navigation on the Sea.—6. A Treatife on Arithmetic.—7. A Treatife on Geometry.—8 A Treatife of Algebra, in 2 books.—9. The Method of Increments.—10. Arithmetic of Infinites, and the Conic Sections, with other Curve Lines.—11. Elements of Optics and Perfpective.—12. Aftronomy.—13. Mechanics, with Centripetal and Centrifugal Forces.—14. Mathematical Principles of Geography, Navigation, and Dialling.—15. Commentary on the Principia, with the Defence of Newton.—16. Tracts.—17. Mifcellanies.

EMINENTIAL *Equation*, a term ufed by fome algebraifts, in the inveftigation of the areas of curvilineal figures, for a kind of affumed equation that contains another equation Eminently, the latter being a particular cafe of the former. Hayes's Flux. pa. 97.

ENCEINTE, a French term, in Fortification, fignifying the whole inclofure, circumference, or compafs of a fortified place, whether built with ftone or brick, or only made of earth, and whether with or without baftions, &c.

ENCYCLOPÆDIA, the circle or chain of arts and fciences; fometimes alfo written Cyclopædia.

ENDECAGON, a plane geometrical figure of eleven fides and angles, otherwife called Undecagon. If each fide of this figure be 1, its area will be 9 3656399

$= \frac{11}{4}$ of the tang. of $73\frac{7}{11}$ degrees, to the radius 1. See my Menfuration, pa. 114, &c, 2d edit. See alfo REGULAR *Figure*.

ENFILADE, a French term, applied to thofe trenches, and other lines, that are ranged in a right line, and fo may be fcoured or fwept by the cannon lengthways, or in the direction of the line.

To ENFILADE, is to fweep lengthways by the firing of cannon, &c.

*A Battery d'*ENFILADE, is that where the cannon fweep a right line.

A Poft, or *Command d'*ENFILADE, is a height from whence a whole line may be fwept at once.

ENGINE, in Mechanics, a compound machine, confifting of feveral fimple ones, as wheels, fcrews, levers, or the like, combined together, in order to lift, caft, or fuftain a weight, or, produce fome other confiderable effect, fo as to fave either force or time.

There are numberlefs kinds of engines; of which fome are for war, as the Balifta, Catapulta, Scorpio, Aries or Ram, &c; others for the arts of peace, as Mills, Cranes, Preffes, Clocks, Watches, &c, &c.

ENGINEER, or INGINEER, is applied to a contriver or maker of any kind of ufeful engines or machines; or who is particularly fkilled or employed in them. And he is denominated either a civil or military Engineer, according as the objects of his profeffion refpect civil or military purpofes.

A military Engineer fhould be an expert mathematician and draughtfman, and particularly verfed in fortification and gunnery, being the perfon officially employed to direct the operations both for attacking and defending works. When at a fiege the Engineers have narrowly furveyed the place, they are to make their report to the general, or commander, by acquainting him which part they judge the weakeft, and where approaches may be made with moft fuccefs. It is their bufinefs alfo to draw the lines of circumvallation and contravallation; alfo to mark out the trenches, places of arms, batteries, and lodgments, and in general to direct the workmen in all fuch operations.

ENGONASIS, in Aftronomy, the fame as Hercules, one of the northern conftellations; which fee.

ENGYSCOPE, the fame as MICROSCOPE.

ENHARMONIC, the laft of the three kinds of mufic. It abounds in diefes, or the leaft fenfible divifions of a tone. See Philof. Tranf. number 481; alfo Wallis Appendix ad Ptolom. pa. 165, 166.

ENNEADECATERIS, in Chronology, a cycle or period of 19 folar years, being the fame as the Golden Number and Lunar Cycle, or Cycle of the Moon; which fee; as alfo EMBOLISMIC.

ENNEAGON, a plane geometrical figure of 9 fides and angles; and is otherwife called a Nonagon. If each fide of this figure be 1, its area will be 6·1818242 $= \frac{9}{4}$ of the tang. of 70 degrees, to the radius 1. See my Menfuration, pa. 114, 2d edit. See alfo the article REGULAR *Figure*.

ENTABLATURE, in Architecture, is that part of an order of column which is over the capital, comprehending the Architrave, Frize, and Corniche.

ENTABLATURE, or ENTABLEMENT, is fometimes alfo ufed for the laft row of ftones on the top of the wall of a building, on which the timber and covering ĩng reſt; ſometimes alſo called the Drip, becauſe it projects a little, to throw the water off.

ENVELOPE, in Fortification, is a mound of earth, ſometimes raiſed in the ditch of a place, and ſometimes beyond it, being either in form of a ſingle parapet, or of a ſmall parapet bordered with a parapet. Theſe Envelopes are made only to cover weak parts with ſingle lines, without advancing towards the field, which cannot be done without works that require a great deal of room, ſuch as horn-works, half-moons, &c. Envelopes are ſometimes called Sillons, Contregards, Conſerves, Lunettes, &c.

ENUMERATION, a numbering or counting. Sir Iſaac Newton wrote an ingenious treatiſe, being an Enumeration of the lines of the 3d order.

EOLIPILE. See ÆOLIPILE.

EPACT, in Chronology, the exceſs of the ſolar month above the lunar ſynodical month; or of the ſolar year above the lunar year of 12 ſynodical months; or of ſeveral ſolar months above as many ſynodical months; or of ſeveral ſolar years above as many dozen of ſynodical months.

The Epacts then are either Annual or Menſtrual.

Menſtrual EPACTS, are the exceſſes of the civil calendar month above the lunar month. Suppoſe, for example, it were new moon on the firſt day of January: then ſince the month of January contains 31 days,

and the lunar month	-	29^{ds}	12^h	44^m 3^s;
the menſtrual Epact is	-	1	11	15 57

Annual EPACTS, are the exceſſes of the ſolar year above the lunar. Hence,

as the Julian ſolar year is	365^{ds}	6^h	0^m	0^s,
and the Julian lunar year is	354	8	48	38,
the annual Epact will be	10	21	11	22,

that is, almoſt 11 days. Conſequently the Epact of 2 years, is 22 days; of 3 years, 33 days; or rather 3, ſince 30 days make an emboliſmic, or intercalary month. Then, adding ſtill 11, the Epact of 4 years is 14 days; and ſo of the reſt as in the following table, where they do not become 30, or 0 again, till the 19th year; ſo that at the 20th year the Epact is 11 again; and hence the cycle of Epacts expires with the Golden Number, or Lunar Cycle of 19 years, and begins with the ſame again.

TABLE of Julian Epacts.

Golden Numb.	Epacts	Golden Numb.	Epacts	Golden Numb.	Epacts
I	11	VIII	28	XV	15
II	22	IX	9	XVI	26
III	3	X	20	XVII	8
IV	14	XI	1	XVIII	19
V	25	XII	12	XIX	30 or 0
VI	6	XIII	23		
VII	17	XIV	4		

Again, as the new moons are the ſame, or fall on the ſame day, every 19 years, ſo the difference between the ſolar and lunar years is the ſame every 19 years. And becauſe the ſaid difference is always to be added to the lunar year, to adjuſt or make it equal to the ſolar

year; hence the ſaid difference reſpectively belonging to each year of the moon's cycle, is called the *Epact of the ſaid year*, that is, the number to be added to the ſaid year, to mak it equal to the ſolar year. Upon this mutual reſpect between the cycle of the moon and the cycle of the Epacts, is founded this

Rule for finding the Julian Epact, belonging to any year of the Moon's Cycle.

Multiply the Golden Number, or the given year of the Moon's Cycle, by 11, and the product will be the Epact if it be leſs than 30; but if it exceed 30, then throw out as many 30's as the product contains, and the remainder will be the Epact.

Rule to find the Gregorian Epact.

1ſt, The difference between the Julian and Gregorian years being equal to the difference between the ſolar and lunar year, or 11 days, therefore the Gregorian Epact for any year is the ſame with the Julian Epact for the preceding year; and hence the Gregorian Epact will be found, by ſubtracting 1 from the golden number, multiplying the remainder by 11, and rejecting the 30's. This rule will ſerve till the year 1900; but after that year, the Gregorian Epact will be found by this rule: Divide the centuries of the given year by 4; multiply the remainder by 17; then to this product add 43 times the quotient, and alſo the number 86, and divide the whole ſum by 25, reſerving the quotient: next multiply the golden number by 11, and from the product ſubtract the reſerved quotient, ſo ſhall the remainder, after rejecting all the 30's contained in it, be the Epact ſought.

The following table contains the Golden Numbers, with their correſponding Epacts, till the year 1900.

TABLE of Gregorian Epacts.

Golden Numb	Epacts	Golden Numb.	Epacts	Golden Numb.	Epacts
I	0	VIII	17	XV	4
II	11	IX	28	XVI	15
III	22	X	9	XVII	26
IV	3	XI	20	XVIII	7
V	14	XII	1	XIX	18
VI	25	XIII	12	I	0
VII	6	XIV	23		

On the ſubject of Epacts, ſee Wolfius's Elementa Chronologiæ, apud Opera, tom. 4; pa. 133; alſo Philoſ. Tranſ. vol. 46. pa. 417, or numb. 495, art. 5.

EPAULE, or ESPAULE, in Fortification, the ſhoulder of the baſtion, or the angle made by the face and flank, otherwiſe called the Angle of the Epaule.

EPAULEMENT, in Fortification, a ſide-work haſtily thrown up, to cover the cannon or the men; and is made either of earth thrown up, or bags filled with earth or ſand, or of gabions, or faſcines, &c, with earth: of which latter ſort are commonly the Epaulements of the places of arms for the cavalry behind the trenches.

EPAULEMENT, is alſo uſed for a demi-baſtion, conſiſting

fifting of a face and flank, placed at the point of a horn-work or crown-work. Also for a little flank added to the fides of horn-works, to defend them when they are too long. Also for the redoubts made on a right line, to fortify it. And laftly, for an orillon, or mafs of earth almoft fquare, faced and lined with a wall, and defigned to cover the cannon of a cafemate.

EPHEMERIS, EPHEMERIDES, tables calculated by aftronomers, fhewing the prefent ftate of the heavens for every day at noon; that is, the places in which all the planets are found at that time; differing but little from an Aftronomical Almanac. It is from fuch tables as thefe that the eclipfes, conjunctions, and afpects of the planets are made out; as alfo horofcopes, or celeftial fchemes, conftructed, &c.

There have been Ephemerides of Origan, Kepler, Argoli, Heckerus, Mezzarachis, Wing, Gadbury, Parker, De la Hire, &c.

In France the Academy of Sciences have publifhed annually, from the beginning of the prefent century, a kind of Ephemeris, under the title of Connoiffance des Temps, which is ftill continued, and is in great efteem; as are alfo the Ephemerides, publifhed there every ten years, by M. Defplaces, and De la Lande.

There are now publifhed fuch Ephemerides by the Academies of feveral other nations; but that which is in moft efteem for its accuracy and ufe in finding the longitude, is the Nautical Almanac, or Aftronomical Ephemeris, publifhed in England by the Board of Longitude, under the direction of the Rev. Dr. Mafkelyne, Aftronomer Royal, which commenced with the year 1767.

EPICHARMUS, an ancient poet and philofopher, born in Sicily, was a fcholar of Pythagoras, and flourifhed in the time of Hiero, in whofe reign it is faid he introduced comedy at Syracufe. He wrote alfo treatifes concerning philofophy and medicine; but none of his works have been preferved. He died at 90 years of age, according to Laertius, who has preferved four verfes infcribed on his ftatue.

EPICURUS, a celebrated ancient philofopher, was born at Gargettium in Attica, in the 109th Olympiad, or about 340 years before Chrift. He fettled at Athens in a fine garden he had bought; where he lived with his friends in much tranquillity, and educated a great number of difciples; who lived all in common with their mafter. His fchool was never divided, but his doctrine was followed as an oracle; and the refpect which his followers paid to his memory is admirable; his birth day being ftill kept in Pliny's time, and even the very month he was born in obferved as a continued feftival, and his picture placed every where. He wrote a great many books, and valued himfelf upon making no quotations. He raifed the atomical fyftem to great reputation, though he was not the inventor of it, but only made fome change in that of Democritus. As to his doctrine concerning the fupreme good or happinefs, it was very liable to be mifreprefented, and fome ill effects proceeded from thence, which difcredited his fect, though undefervedly. He was charged with perverting the worfhip of the gods, and inciting men to debauchery. But he did not forget himfelf on this occafion: he publifhed his opinions to the whole world; wrote fome books of devotion; recommended the ve-

neration of the gods, fobriety, and chaftity, living in an exemplary manner, and conformably to the rules of philofophical wifdom and frugality. He died of a fuppreffion of urine, at 72 years of age.—Gaffendus has given us all he could collect from the ancients concerning the perfon and doctrine of this philofopher.

EPICUREAN Philofophy, the doctrine, or fyftem of philofophy maintained by Epicurus and his followers. This confifted of three parts; canonical, phyfical, and ethical. The firft refpected the canons or rules of judging; in which foundnefs and fimplicity of fenfe, affifted by fome natural reflections, chiefly formed his art. His fearch after truth proceeded only by the fenfes; to the evidence of which he gave fo great a certainty, that he confidered them as an infallible rule of truth, and termed them the Firft natural light of mankind.

In the 2d part of his philofophy he laid down atoms, fpace, and gravity, as the firft principles of all things. He afferted the exiftence of God, whom he accounted a bleffed immortal being, but who did not concern himfelf with human affairs.

As to his ethics, he made the fupreme good of man to confift in pleafure, and confequently fupreme evil in pain. Nature itfelf, fays he, teaches us this truth; and prompts us from our birth to procure whatever gives us pleafure, and avoid what gives us pain. To this end he propofes a remedy againft the fharpnefs of pain, which was to divert the mind from it, by turning our whole attention upon the pleafures we have formerly enjoyed. He held that the wife man muft be happy, as long as he is wife: the pain, not depriving him of his wifdom, cannot deprive him of his happinefs: from which it would feem that his pleafure confifted rather in intellectual than in fenfual enjoyments: though this is a point ftrongly contefted.

EPICUREANS, the fect of philofophers holding or following the principles and doctrine of Epicurus. As the nature of the pleafure, in which the chief happinefs is fuppofed to be feated, is a great problem in the morals of Epicurus, there hence arife two kinds of Epicureans, the rigid and the remifs: the firft were thofe who underftood Epicurus's notion of pleafure in the beft fenfe, and placed all their happinefs in the pure pleafures of the mind, arifing from the practice of virtue: while the loofe or remifs Epicureans, taking the words of that philofopher in a grofs fenfe, placed all their happinefs in bodily pleafures or debauchery.

EPICYCLE, in the ancient aftronomy, a little circle having its centre in the circumference of a greater one: or a fmall orb or fphere, which being fixed in the deferent of a planet, is carried along with it, and yet, by its own peculiar motion, carries the planet faftened to it round its proper centre.

It was by means of Epicycles that Ptolomy and his followers folved the various phenomena of the planets, but more efpecially their ftations and retrogradations.

EPICYCLOID, is a curve generated by the revolution of a point of the periphery of a circle, which rolls along or upon the circumference of another circle, either on the convex or concave fide of it.

When a circle rolls along a ftraight line, a point in its circumference defcribes the curve called a cycloid. But

But if, instead of the right line, the circle roll along the circumference of another circle, either equal to the former or not, then the curve described by any point in its circumference is what is called the Epicycloid.

If the generating circle roll along the convexity of the circumference, the curve is called an *Upper*, or *Exterior* Epicycloid; but if along the concavity, it is called a *Lower*, or *Interior* Epicycloid. Also the circle that revolves is called the *Generant;* and the arc of the other circle along which it revolves, is called the *Base* of the Epicycloid. Thus, ABC or BLV is the Generant; DPVE the Exterior Epicycloid, its axis BV; DPUE the Interior Epicycloid; and DBE their common Base.

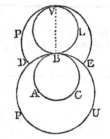

For the Length of the Curve.

The length of any part of the curve of an Epicycloid, which any given point in the revolving circle has described, from the position where it touched the circle upon which it revolved, is to double the versed side of half the arc which all the time of revolving touched the quiescent circle, as the sum of the diameters of the circles, is to the semidiameter of the quiescent circle in the Exterior Cycloid; or as the difference of the diameters is to that semidiameter, for the Interior one.

For the Area of the Epicycloid.

Dr. Halley has given a general proposition for the measuring of all cycloids and Epicycloids: thus, the area of a cycloid, or Epicycloid, either primary, or contracted, or prolate, is to the area of the generating circle; and also the areas of the parts generated in those curves, to the areas of analogous segments of the circle; as the sum of double the velocity of the centre and the velocity of the circular motion, is to this velocity of the circular motion. See the Demonstr. in the Philos. Transf. number 218.

Spherical EPICYCLOIDS are formed by a point of the revolving circle, when its plane makes a constant angle with the plane of the circle on which it revolves. Messrs. Bernoulli, Maupertuis, Nicole, and Clairaut, have demonstrated several properties of these Epicycloids, in Hist. Acad. Sci. for 1732.

Parabolic, Elliptic, &c. EPICYCLOIDS.

If a parabola roll upon another equal to it; its focus will describe a right line perpendicular to the axis of the quiescent parabola: also the vertex of the rolling parabola will describe the cissoid of Diocles; and any other point of it will describe some one of Newton's defective hyperbolas, having a double point in the like point of the quiescent parabola.

In like manner, if an ellipse revolve upon another ellipse, equal and similar to it, its focus will describe a circle, whose centre is in the other focus, and consequently the radius is equal to the axis of the ellipsis; and any other point in the plane of the ellipse will describe a line of the 4th order.

The same may be said also of an hyperbola, revolving upon another, equal and similar to it; for one of the foci will describe a circle, having its centre in the other focus, and the radius will be the principal axis of the hyperbola; and any other point of the hyperbola will describe a line of the 4th order.

Concerning these lines, see Newton's Principia, lib. 1; also De la Hire's Memoires de Mathematique &c, where he shews the nature of this line, and its use in Mechanics; see also Maclaurin's Geometria Organica.

EPIPHANY, a christian festival, otherwise called the Manifestation of Christ to the Gentiles, observed on the 6th of January, in honour of the appearance of our Saviour to the three magi or wise men, who came to adore him and bring him presents.

EPISTYLE, in the ancient Architecture, a term used by the Greeks for what we call Architrave, viz a massive stone, or a piece of wood, laid immediately over the capital of a column.

EPOCHA, or EPOCH, a term or fixed point of time, from whence the succeeding years are numbered or reckoned.

Different nations make use of different Epochs. The christians chiefly use the Epoch of the nativity or incarnation of Jesus Christ; the Mahometans, that of the Hegira; the Jews, that of the creation of the world, or that of the Deluge; the ancient Greeks, that of the Olympiads; the Romans, that of the building of their city; the ancient Persians and Assyrians, that of Nabonassar; &c.

The doctrine and use of Epochs is of very great extent in chronology. To reduce the years of one Epoch to those of another, i. e. to find what year of one corresponds to a given year of another; a period of years has been invented, which, commencing before all the known Epochs, is, as it were, a common receptacle of them all, called the Julian Period. To this period all the Epochs are reduced; i. e. the year of this period when each Epoch commences, is determined. So that, adding the given year of one Epoch to the year of the period corresponding with its rise, and from the sum subtracting the year of the same period corresponding to the other Epoch, the remainder is the year of that other Epoch.

EPOCH *of Christ*, is the common Epoch throughout Europe, commencing at the supposed time of our Saviour's nativity, December 25; or rather, according to the usual account, from his circumcision, or the 1st of January. The author of this Epoch was an Abbot of Rome, one Dionysius Exiguus, a Scythian, about the year 507 or 527. Dionysius began his account from the conception or incarnation, usually called the Annunciation, or Lady Day; which method obtained in the dominions of Great Britain till the year 1752, before which time the Dionysian was the same as the English Epoch: but in that year the Gregorian calendar having been admitted by act of parliament, they now reckon from the first of January, as in the other parts of Europe, except in the court of Rome, where the Epoch of the Incarnation still obtains for the date of their bulls.

A TABLE of the Years of the most remarkable Epochs or Eras and Events.

N. B. The years before Christ, are those before the reputed year of his birth, and not reckoned back from the first year of his age, as is generally done in such tables.	Julian Period	Year of the World.	Years before Christ.
The Creation of the World	706	0	4007
The Deluge, or Noah's flood	2362	1656	2351
Assyrian monarchy founded by Nimrod	2537	1831	2176
The birth of Abraham	2714	2008	1999
Kingdom of Athens founded by Cecrops	3157	2451	1556
Entrance of the Israelites into Canaan	3262	2556	1451
The destruction of Troy	3529	2823	1184
Solomon's temple founded	3701	2995	1012
The Argonautic expedition	3776	3070	937
Lycurgus formed his laws	3829	3103	884
Arbaces, 1st king of the Medes	3838	3132	875
Olympiads of the Greeks began	3938	3232	775
Rome built, or Roman Era	3961	3255	752
Era of Nabonassar	3967	3261	746
First Babylonish captivity, by Nebuchadnezzar	4107	3401	606
The 2d ditto, and birth of Cyrus	4114	3408	599
Solomon's temple destroyed	4125	3419	588
Cyrus began to reign in Babylon	4177	3471	536
Peloponnesian war began	4282	3576	431
Alexander the great died	4390	3684	323
Captivity of 100,000 Jews by Ptolomy	4393	3687	320
Archimedes killed at Syracuse	4506	3800	207
Julius Cæsar invaded Britain	4659	3953	54
He corrected the calendar	4667	3961	46
The true year of Christ's birth	4709	4003	4

The Christian Era begins here.

	Julian Period	Year of the World.	Years since Christ
Dionysian, or vulgar era of Christ's birth	4717	4007	0
Christ crucified, Friday April 3d	4746	4040	33
Jerusalem destroyed	4783	4077	70
Adrian's wall built in Britain	4833	4127	120
Dioclesian Epoch, or that of Martyrs	4997	4291	284
The council of Nice	5038	4332	325
Constantine the great died	5050	4344	337
The Saxons invited into Britain	5158	4452	445
Hegira, or flight of Mohammed	5335	4629	622
Death of Mohammed	5343	4637	630
The Persian Yesdegird	5344	4638	631
Sun, Moon, and Planets ♄, ♃, ♂, ♀, ☿ in ♎, seen from the earth	5899	5193	1186
Art of printing discovered	6153	5447	1440
The reformation begun by Martin Luther	6230	5524	1517
The Calendar corrected by pope Gregory	6295	5589	1582
Oliver Cromwell died	6371	5665	1658
Sir Isaac Newton born, Dec. 25	6355	5649	1642
Made President of the Royal Society	6416	5710	1703
Died, March 20th	6440	5734	1727
New Planet discovered by Herschel	6494	5788	1781

EQUABLE *Motion*, Celerity, Velocity, &c, is that which is uniform, or without alteration, or by which equal fpaces are paffed over in equal times. Hence, the fpaces, paffed over in Equable motions, are proportional to the times. So that if a body pafs over 20 feet in 1 fecond of time, it will pafs over 40 feet in 2 feconds, and fo on.

EQUABLY *Accelerated* or *Retarded*, &c, is when the motion or change is increafed or decreafed by equal quantities or degrees in equal times.

EQUAL, a term of relation between different things, but of the fame kind, magnitude, quantity, or quality.——Wolfius defines Equals to be thofe things that may be fubftituted for each other, without any alteration of their quantity.—It is an axiom in mathematics &c, that two things which are equal to the fame third, are alfo equal to each other. And if Equals be equally altered, by Equal addition, fubtraction, multiplication, divifion, &c, the refults will be alfo Equal.

EQUAL *Circles*, are thofe whofe diameters are equal.

EQUAL *Angles*, are thofe whofe fides are equally inclined, or which are meafured by fimilar arcs of circles.

EQUAL *Lines*, are lines of the fame length.

EQUAL *Plane Figures*, are thofe whofe areas are equal; whether the figures be of the fame form or not.

EQUAL *Solids*, are fuch as are of the fame fpace, capacity, or folid content; whether they be of the fame kind or not.

EQUAL *Curvatures*, are fuch as have the fame or equal radii of curvature.

EQUAL *Ratios*, are thofe whofe terms are in the fame proportion.

EQUAL, in Optics, is faid of things that are feen under equal angles.

EQUALITY, the exact agreement of two things in refpect of quantity. Thofe figures are equal which may occupy the fame fpace, or may be conceived to poffefs the fame fpace, by the flexion or tranfpofition of their parts. See a learned difcourfe upon this, by Dr. Barrow, in the 11th and 12th of his Mathematical Lectures.

EQUALITY, in Algebra, the relation or comparifon between two quantities that are really or effectually equal. See EQUATION.

Equality, in Algebra, is ufually denoted by two equal parallel lines, as = : thus $2 + 3 = 5$, i. e. 2 plus 3, are equal to 5. This character =, was firft introduced by Robert Recorde. Des Cartes, and fome others after him, ufe the mark $\propto$ inftead of it : as $2 + 3 \propto 5$.

EQUALITY, in Aftronomy. *Circle of* EQUALITY, or the EQUANT. See CIRCLE and EQUANT.

Ratio or *Proportion* of EQUALITY, is that between two equal numbers or quantities.

Proportion of EQUALITY *evenly ranged*, or *ex æquo ordinata*, is that in which two terms, in a rank or feries, are proportional to as many terms in another feries, compared to each other in the fame order, i. e. the firft of one rank to the firft of another, the 2d to the 2d, &c.

Proportion of EQUALITY *evenly difturbed*, called alfo

ex æquo perturbata, is that in which more than two terms of one rank, are proportional to as many terms of another, compared to each other in a different and interrupted order ; viz, the 1ft of one rank to the 2d of another, the 2d to the 3d, &c.

EQUANT, or ÆQUANT, in Aftronomy, a circle formerly conceived by aftronomers, in the plane of the deferent, or eccentric ; for regulating and adjufting certain motions of the planets, and reducing them more eafily to a calculus.

EQUATED *Anomaly*. See ANOMALY.

EQUATED *Bodies*. On Gunter's Sector there are fometimes placed two lines, anfwering to one another, and called the Lines of Equated Bodies. They are fituated between the lines of fuperficies and folids, and are marked with the letters D, I, C, S, O, T, to fignify dodecahedron, icofahedron, cube, fphere, octahedron, and tetrahedron.

The ufes of thefe lines are, 1ft, When the diameter of the fphere is given, to find the fides of the five regular bodies, each equal to that fphere ; 2d, From the fide of any one of thofe bodies being given, to find the diameter of the fphere, and the fides of the other bodies, which fhall be each equal to the firft given body. So, when the fphere is given, take its diameter, and apply it over on the fector in the points S, S ; but when one of the other five bodies is given, apply its fide over in its proper points ; then the parallels taken from between the points of the other bodies, or fphere, fhall be the fides or diameter, equal feverally to the fphere or body firft given.

EQUATION, in Algebra, an expreffion of equality between two different quantities ; or two quantities, whether fimple or compound, with the mark of equality between them : as $2 + 3 = 7 - 2$; or $2 \times 3 = 6$; or $5 \times 3 = \frac{30}{2}$; or $a + b = c$; or $x^2 + ax = b$; &c. When the things, or two fides of the equation, are the fame, the expreffion becomes an Identity, as $5 = 5$, or $a = a$, &c. Sometimes the quantities are placed all on one fide, and made equal to 0 or nothing on the other fide ; as $5 - 5 = 0$, or $a - b = 0$: which is no more than fetting down the difference of two equal quantities equal to nothing.

The character or fign ufually employed to denote an Equation, is =, which is placed between the two equal quantities, called the two fides of the Equation.

The *Terms* of an EQUATION, are the feveral quantities or parts of which it is compofed. Thus, of the Equation $a + b = c$, the terms are a, b, and c : and the tenor or import of the expreffion is, that fome quantity reprefented by c, is equal to two others reprefented by a and b.

Equations are either Simple or Affected.

A *Simple* EQUATION is that which has only one power of the unknown quantity : as $a + x = 3b$, or $ax^2 = bc$, $a^3 + 2x^3 = 5b$, &c ; where x denotes the unknown quantity, and the other letters known ones. But

An *Affected*, or *Adfected* EQUATION contains two or more different powers : as $x^2 + ax = b$, or $3x - 4x^2 + x^3 = 25$, &c.

3 K 2   Again,

Again, Equations are denominated from the higheſt power contained in them; as quadratic, cubic, biquadratic, &c. Thus,

A *Quadratic* EQUATION, is that in which the unknown quantity riſes to two dimenſions, or to the ſquare or 2d power: as $x^2 + 20x = 200$, or $ax - x^2 = b$.

A *Cubic* EQUATION, is that in which the unknown quantity is of three dimenſions, or riſes to the cube or 3d power: as $x^3 = 25$, or $x^3 - 2x^2 = 27$, or $x^3 - ax^2 + bx = c$.

A *Biquadratic* EQUATION, is that in which the unknown quantity is of 4 dimenſions, or riſes to the 4th or biquadratic power: as $x^4 = 25$, or $x^4 - 20x = 10$, or $x^4 - ax^3 + bx^2 - cx = d$.

And ſo for other higher orders of Equations.

The *Root* of an EQUATION, is the value of the unknown letter or quantity contained in it. And this value being ſubſtituted in the terms of the Equation inſtead of that letter or quantity, will cauſe both ſides to vaniſh, or will make the one ſide exactly equal to the other. So the root of the Equation $\frac{3x}{5} + 18 = 24$, is 10; becauſe that uſing 10 for x, it becomes $\frac{3 \times 10}{5} + 18 = 6 + 18 = 24$.

Every Equation has as many roots as it has dimenſions, or as it contains units in the index of the higheſt power, when the powers are all reduced to integral exponents. So the ſimple Equation of the 1ſt power, has only one root; but the quadratic has 2, the cubic 3, the biquadratic 4, &c. Thus the two roots of this equation $x^2 - 4x = -3$ are 1 and 3; for either of theſe ſubſtituted for x makes $x^2 - 4x$ come out equal to -3. Alſo the three roots of $x^3 - 4x^2 - 11x = -30$, or $x^3 - 4x^2 - 11x + 30 = 0$, are 2, 5, and -3; as will appear by ſubſtituting each of theſe inſtead of x in the equation, which will make all the terms on one ſide equal to the other ſide. And ſo of others.

The Relation between the Roots of Equations, and the Coefficients of their Terms.——In every Equation, when the terms are ranged in order according to the order of the powers, the greater before the leſs; the firſt term or higheſt power freed from its coefficient, by dividing all the terms by it, and all brought to one ſide, and made equal to nothing on the other ſide, when it will appear in this form,

$x^n + ax^{n-1} + bx^{n-2} + cx^{n-3} \dots = 0$;

then the relations between the roots and coefficients, are as follow:

1ſt, The coefficient a of the 2d term, is equal to the ſum of all the roots.

2d, The coefficient b of the 3d term, is equal to the ſum of all the products of the roots that can be made by multiplying every two of them together. In like manner,

3d, The coefficients c, d, e, &c, of the following terms, are reſpectively equal to the ſum of the products of the roots made by multiplying every three together, or every four together, or every five together, &c, the ſigns of all the roots being changed. All which will appear below, in the Generation of Equations.

The Roots of Equations are Poſitive or Negative,

and Real or Imaginary. Thus, the two roots of the Equation $x^2 - 4x = -3$, are 1 and 3, real and both poſitive; but the roots of the Equation $x^3 - 4x^2 - 11x = -30$, are 2, 5, & -3, are real, two poſitive and one negative; and the roots of the equation $x^3 + 9x = 10$, are 1 and $-\frac{1}{2} \pm \frac{1}{2} \sqrt{-39}$, one real and two imaginary.

The *Generation* of EQUATIONS, is the multiplying of certain aſſumed ſimple equations together, to produce compound ones, with intent to ſhew the nature of theſe; a method invented by Harriot, which is this: Suppoſe x to denote the unknown quantity of any equation, and let the roots of that equation, or the values of x, be, a, b, c, d, &c; that is $x = a$, and $x = b$, and $x = c$, &c; or $x - a = 0$, and $x - b = 0$, and $x - c = 0$, &c; then multiply theſe laſt equations together, thus,

$x - a = 0$
$x - b = 0$

$x^2 - ax$
$\quad - bx + ab$

$\left. \begin{matrix} x^2 - a \\ - b \end{matrix} \right\} . x + ab = 0$, the quadratic Equation.
$x - c \ \text{-----} = 0$

$x^3 - \overline{a + b} . x^2 + abx$
$\qquad - c . x^2 + \overline{ac + bc} . x - abc$

$\begin{matrix} x^3 - \\ - \\ - \end{matrix} \begin{matrix} a \\ b \\ c \end{matrix} . x^2 + \begin{matrix} ab \\ ac \\ bc \end{matrix} \quad x - abc = 0$, the cubic Equ.
$x - d \qquad\qquad\qquad = 0$

$\begin{matrix} x^4 - \\ - \\ - \\ - \end{matrix} \begin{matrix} a \\ b \\ c \\ d \end{matrix} \ x^3 + \begin{matrix} ab \\ ac \\ bc \\ ad \\ bd \\ cd \end{matrix} . x^2 \begin{matrix} - abc \\ - abd \\ - acd \\ - bcd \end{matrix} . x + abcd = 0$, the biquadratic Equ.

Now the roots of theſe equations are a, b, c, d, &c; and it is obvious that the ſum of all the roots is the coefficient of the 2d term, the ſum of all the products of every two is the coefficient of the 3d term, the ſum of all the products of every three that of the 4th term, and ſo on, to the laſt term, which is the continual product of all the roots.

Reduction of EQUATIONS, is the transforming or changing them to their ſimpleſt and moſt commodious form, to prepare them for finding or extracting their roots. The moſt convenient form is, that the terms be ranged according to the powers of the unknown letter, the higheſt power foremoſt next the left hand, and that term to have only $+ 1$ for its coefficient; alſo all the terms containing the unknown letter to be on one ſide of the equation, and the abſolute known term only on the other ſide.

Now this reduction chiefly reſpects the firſt term, or that which contains the higheſt power of the unknown quantity; and the general rule for reducing it is, to conſider in what manner it is involved or connected with other quantities, and then perform the counter or op-

poſite

pofite relation or operation; for every operation is undone or counteracted by the reverfe of it; as addition by fubtraction, multiplication by divifion, involution by evolution, &c: then bring all the unknown terms to one fide, and the known term to the other fide, changing the figns, from + to —, or from — to +, of thofe terms which are changed from one fide to the other; and laftly divide by the coefficient of the firft term, with its fign.

Thus, for ex. if $5x - 1z = 3x + 4$:
then is $5x - 3x = 1z + 4$, or $2x = 16$;
and fo $x = \frac{16}{2} = 8$.

Again, if $ax - b = cx + d$:
then $ax - cx = b + d$,
and $x = \dfrac{b + d}{a - c}$

And, if $6x^2 - 20 = 12x + 2x^2$:
then $6x^2 - 2x^2 = 12x + 20$,
or $4x^2 - 12x = 20$,
and $x^2 - 3x = 5$.

Alfo, if $\dfrac{a}{a + x} + \dfrac{b}{x} = 3$:
then $ax + ab + bx = 3ax + 3x^2$,
and $-3x^2 + ax + bx - 3ax = -ab$,
or $-3x^2 + bx - 2ax = -ab$,
and $x^2 + \dfrac{2a - b}{3}x = \dfrac{ab}{3}$.

Alfo, if $\sqrt{a^2 + x^2} - 3b = \dfrac{2ab}{x}$:

then $\sqrt{a^2 + x^2} = 3b + \dfrac{2ab}{x} = \dfrac{3bx + 2ab}{x}$,

and $a^2 + x^2 = \dfrac{9b^2x^2 + 12ab^2x + 4a^2b^2}{x^2}$,

and $a^2x^2 + x^4 = 9b^2x^2 + 12ab^2x + 4a^2b^2$,

and $x^4 + \overline{a^2 - 9b^2} \cdot x^2 - 12ab^2x = 4a^2b^2$.

Extracting or finding the Roots of EQUATIONS.

This is finding the value or values of the unknown letter in an Equation, the rules for which are various, according to the degree of the Equation.

1. *For the Root of a Simple* EQUATION.

Having reduced the equation as above, by bringing the unknown terms to one fide, and the known ones to the other, freeing the former from radicals and fractions, by their counter operations, and laftly dividing by the coefficients of the unknown quantity, the value of it is then found: as in the firft and 2d examples of reduction above given.

2. *For the Roots of Quadratic* EQUATIONS.

Thefe are ufually found by what is called completing the fquare; which confifts in fquaring half the coefficient of the 2d term, and adding it to both fides of the equation; for then the unknown fide is a complete fquare of a binomial, and the other fide confifts only of known quantities. Therefore, extract the root on both fides, fo fhall the root of the firft fide be a binomial, one part of which is the unknown letter, and the

other a known or given quantity, and the root of the other fide is taken either + or —, fince the fquare of either of thefe is the fame given quantity: laftly, bringing over the known part of the binomial root to the other fide, with a contrary fign, gives the two roots or values of the unknown letter fought.

Thus, if $x^2 + 2ax = b^2$ be a general quadratic Equation, $2a$ being the coefficient of the 2d term, and b^2 the abfolute known term, both with their figns. Then, a is half that coefficient, and a^2 its fquare; which being added, gives $x^2 + 2ax + a^2 = a^2 + b^2$; and the root extracted gives $x + a = \pm \sqrt{a^2 + b^2}$; then, tranfpofing a, it is $x = -a \pm \sqrt{a^2 + b^2}$, the two roots, or values of x.

And if $x^2 - 4x = 5$: then, adding 2^2 or 4,
it is $x^2 - 4x + 4 = 9$;
the root is $x - 2 = \pm 3$,
then $x = 2 \pm 3 = 5$ or -1,
the two roots, or values of x.

And if $x^2 + 4x = 5$:
then $x^2 + 4x + 4 = 9$;
and $x + 2 = \pm 3$;
or $x = -2 \pm 3 = 1$, or -5, the two roots, which are the fame numbers as before, but with the figns changed.

Alfo, if $x^2 - 4x = -5$:
then $x^2 - 4x + 4 = 4 - 5 = -1$;
and $x^2 - 2 = \pm \sqrt{-1}$;
or $x = 2 \pm \sqrt{-1}$, two imaginary roots.

3. *For the Roots of Cubic* EQUATIONS.

A Cubic EQUATION is that in which the unknown letter afcends to the 3d power; as $x^3 + a^3 = b^3$, or $x^3 + ax^2 + bx = c$.

The 2d term of every Cubic Equation being taken away, thofe equations may all be reduced to this form, $x^3 + ax = b$; and the general value of one root is
$$x = \sqrt[3]{\tfrac{1}{2}b + \sqrt{\tfrac{1}{4}b^2 + \tfrac{1}{27}a^3}} + \sqrt[3]{\tfrac{1}{2}b - \sqrt{\tfrac{1}{4}b^2 + \tfrac{1}{27}a^3}}.$$
This rule is ufually called Cardan's, becaufe firft publifhed by him, but it was invented both by Scipio Ferreus, and Nich. Tartalea, by whom it was communicated to Cardan. See the article ALGEBRA.

When the 2d term is negative, or the equation of this form, $x^3 - ax = \pm b$, the radical $\sqrt{\tfrac{1}{4}b^2 + \tfrac{1}{27}a^3}$ becomes $\sqrt{\tfrac{1}{4}b^2 - \tfrac{1}{27}a^3}$, which will be imaginary or impoffible when $\tfrac{1}{27}a^3$ is greater than $\tfrac{1}{4}b^2$, for $\sqrt{\tfrac{1}{4}b^2 - \tfrac{1}{27}a^3}$ will then be the fquare root of a negative quantity: and yet, in this cafe, the root x is a real quantity; though algebraifts have never been able to find a real finite general expreffion for it. And this is called the Irreducible or Impracticable Cafe.

This cafe may indeed be refolved by the trifection of an arc or angle; or by any of the ufual methods of converging; or by general expreffions in infinite feries. See Saunderfon's Algebra, pa. 713; Philof. Tranf. vol. 18, pa. 136, or Abr. vol. 1, pa. 87; alfo vol. 70, pa. 415. See alfo the article CUBIC *Equations.*

Mr. Cotes obferves, in his Logometria, pa. 29, that the folution of all cubic Equations depends either upon the trifection of a ratio, or of an angle. See this method explained in Saunderfon's Alg. p. 718.

Biquadratic

Biquadratic EQUATIONS, or those that are of 4 dimensions, are resolved after various methods. The first rule was given by Lewis Ferrari, the companion of Cardan, which is one of the best. A 2d method was given by Des Cartes, and another by Mr. Simpson and Dr. Waring. For the explanation of which, see BIQUADRATIC *Equations*.

EQUATIONS of the Higher Degrees or Orders.

There is no general rule to express algebraically the roots of Equations above those of the 4th degree; and therefore methods of approximation are here made use of, which, though not accurately, are yet practically true. Some of these excel in ease and simplicity, and others in quickness of converging. Among these may be reckoned first, Double Position, or Trial-and-Error, both in respect of ease and universality, as it applies in the simplest manner to all sorts of Equations whatever, not excepting even exponential ones, radical expressions of ever so complex a form, expressions of logarithms, of arches by the sines or tangents, of arcs of curves by the abscisses, or any other fluents, or roots of fluxional Equations. For an explanation of this and other methods of converging to the roots of equations, by Halley, Newton, Raphson, &c, &c, see APPROXIMATION, and CONVERGING.

Besides the methods above adverted to, there have been some others, given in the Memoirs of several Academies, and elsewhere. As, by M. Daniel Bernoulli, in the Acta Petropolitana, tom. 3, p. 92; and by M. L. Euler, in the same, vol. 6, New Series, and tom. 5, p. 63 & 82; by Mr. Thos. Simpson, in his Essays, p. 82; in his Dissertations, p. 102; in his Algebra, p. 158; and in his Select Exercises, p. 215.

Absolute EQUATION. See ABSOLUTE.

Adfected, or *Affected* EQUATION. See AFFECTED.

Differential EQUATION, is the Equation of Differences or Fluxions.

Eminential EQUATION. See EMINENTIAL.

Exponential EQUATION, one in which the exponents of the powers are variable or unknown quantities. See EXPONENTIAL.

Fluential EQUATION, is the Equation of the fluents.

Fluxional EQUATION, is the Equation of the fluxions.

Literal EQUATION, is a general Equation expressed in letters, as contradistinguished from a

Numeral Equation, one expressed in numbers.

Transcendental EQUATION. See TRANSCENDENTAL.

EQUATION, in Astronomy, as *Annual* EQUATION, is either of the mean motion of the sun and moon, or of the moon's apogee and nodes.

The Annual Equation of the sun's centre being given, the other three corresponding annual Equations, will be also given, and therefore a table of the first will serve for all of them. Thus, if the annual Equation of the sun's centre, taken from such a table for any time, be called s; and if $\frac{1}{10}s = A$, and $\frac{1}{6}s = B$; then shall the other annual Equations for that time be thus,

$A + \frac{1}{30}A = m$, that of the moon's mean motion;

and $B + \frac{1}{30}B = a$, that of the moon's apogee;

and $\frac{1}{2}B - \frac{1}{170}B = n$, that of her nodes.

And here note, that when s, or the Equation of the sun's centre, is additive; then m is negative, a is positive, and n is negative. But on the contrary, when s is negative or subductive; then m is positive, a negative, and n positive.

There is also an *Equation of the moon's mean motion*, depending on the situation of her apogee in respect of the sun; which is greatest when the moon's apogee is in an octant with the sun, and is nothing at all when it is in the quadratures or syzygies. This Equation when greatest, and the sun in perigee, is 3′ 56″. But it is never above 3′ 34″ when the sun is in apogee. At other distances of the sun from the earth, this Equation when greatest, is reciprocally as the cube of that distance. But when the moon's apogee is any where out of the octants, this Equation grows less, and is mostly, at the same distance between the earth and sun, as the sine of double the distance of the moon's apogee from the next quadrature or syzygy, is to radius. This is to be added to the moon's motion while her apogee passes from a quadrature with the sun to a syzygy; but is to be subtracted from it, while the apogee moves from the syzygy to the quadrature.

There is moreover another *Equation of the moon's motion*, which depends on the aspect of the nodes of the moon's orbit with respect to the sun: and this is greatest when her nodes are in octants to the sun, and quite vanishes when they come to their quadratures or syzygies. This Equation is proportional to the sine of double the distance of the node from the next syzygy or quadrature; and at the greatest is only 47″. This must be added to the moon's mean motion while the nodes are passing from the syzygies with the sun to their quadratures; but subtracted while they pass from the quadratures to the syzygies.

From the sun's true place subtract the equated mean motion of the lunar apogee, as was shewn above, the remainder will be the annual argument of the said apogee; from whence the eccentricity of the moon and the 2d Equation of her apogee may be compared. See *Theory of the* MOON's *motions*, &c.

EQUATION *of the Centre*, called also *Prosthaphæresis*, and *Total Prosthaphæresis*, is the difference between the true and mean place of a planet, or the angle made by the lines of the true and mean place; or, which amounts to the same, between the mean and equated anomaly.

The greatest Equation of the Centre may be obtained by finding the sun's longitude at the times when he is near his mean distances; for then the difference will give the true motion for that interval of time: next find the sun's mean motion for the same interval of time; then half the difference between the true and mean motions will shew the greatest Equation of the Centre.

For Example, by observations made at the Royal Observatory at Greenwich, it appears that at the following mean times the sun's longitudes were as annexed; viz,

Mean times.	Sun's longitudes.
1769 Oct. 1 at 23^h 49^m 12^s - -	6^s 9° 32′ 0·6″
1770 Mar. 29 at 0 4 50 · - -	0 8 50 27·5

dif. of time 178^d 0 15 38; true dif. lon. 5 29 18 27

tropical year = 365^d 5^h 48^m 42^s = 365·2421527;

observed interval = 178 0 15 38 = 178·0108565:

Then

Then

365·2421527 : 178·01085648 :: 360° : 175·455948

or 175° 27′ 21″ the mean motion.

Therefore - 175° 27′ 21″ of mean motion,
anſwers to · 179 18 27 of true motion;
their difference is 3 51 6
and its half - 1 55 33
is the greateſt Equation of the Centre according to
theſe obſervations.

To find the Equation of the Centre, or to reſolve Kepler's problem, is a very troubleſome operation, eſpecially in the more eccentric orbits. How this is to be
done, has been ſhewn by Newton, Gregory, Keil,
Machin, La Caille, and others, by methods little differing from one another; which conſiſt chiefly in finding a certain intermediate angle, called the eccentric
anomaly; having known the mean anamoly, and the
dimenſions of the ſun's orbit. The mean anomaly is
eaſily found, by determining the exact time when the
ſun is in the aphelion, and uſing the following proportion, viz,

As the time of a tropical revolution, or ſolar year,

Is to the interval between the aphelion and given time,

So is 360 degrees, to the degrees of the mean anomaly.

Or it may be found by taking the ſun's mean motion
at the given time out of tables.

To find the Eccentric Anomaly, ſay,

As the aphelion diſtance,

Is to the perihelion diſtance;

So is the tangent of half the mean anomaly,

To the tangent of an arc.

Which arc added to half the mean anomaly, gives
the eccentric anomaly. Then,

To find the True Anomaly, ſay,

As the ſquare root of the aphelion diſtance,

Is to the ſquare root of the perihelion diſtance;

So is the tangent of half the eccentric anomaly,

To the tangent of half the true anomaly.

Then, the difference between the true and mean
anomaly, gives the Equation of the Centre, ſought.
Which is ſubtractive, from the aphelion to the perihelion, or in the firſt 6 ſigns of anomaly; and additive,
from the perihelion to the aphelion, or in the laſt 6
ſigns of anomaly; and hence called *Proſthaphereſis*.

By this problem a table may eaſily be formed.
When the Equations of the Centre for every degree
of the firſt 6 ſigns of mean anomaly are found, they
will ſerve alſo for the degrees of the laſt 6 ſigns, becauſe equal anomalies are at equal diſtances on both
ſides of their apſes. Then ſet theſe equations orderly
to their ſigns and degrees of anomaly; the firſt 6 being
reckoned from the top of the table downwards, and
ſigned *ſubtract*; the laſt 6, for which the ſame Equations ſerve, in a contrary order, being reckoned from
the bottom upwards, and marked *add*. Let alſo the
difference between every adjacent two Equations, called
Tabular Differences, be ſet in another column. Hence,
from theſe Equations of the Centre, augmented or diminiſhed by the proportional parts of their reſpective
tabular differences, for any given minutes and ſeconds,
may eaſily be deduced Equations of the Centre to any
mean anomaly propoſed. Robertſon's Elem. of Navig.
book·5, p. 286, 290, 295, and 308, where ſuch a table
of Equations is given.

5

The late excellent Mr. Euler has particularly conſidered this ſubject, in the Mem. de l'Acad. de Berlin,
tom. 2, p. 225 & ſeq. where he reſolves the following
problems:

1. To find the true and mean anomaly correſponding to the planet's mean diſtance from the ſun; that
is, where the planet is in the extremity of the conjugate axis of its orbit.

2. The eccentricity of a planet being given, to find
the eccentric anomaly correſponding to the greateſt
Equation.

3. The eccentricity being given, to find the mean
anomaly correſponding to the greateſt Equation.

4. From the ſame data, to find the true anomaly
correſponding to this Equation.

5. From the ſame data, to find the greateſt Equation.

6. The greateſt Equation being given, to find the
eccentricity.

Mr. Euler obſerves, that this problem is very difficult, and that it can only be reſolved by approximation
and tentatively, in the manner he mentions: but if the
eccentricity be not great, it may then be found directly
from the greateſt Equation. Thus, if the greateſt
Equation be $= m$, and the eccentricity $= n$;

then is

$$m = 2n + \frac{11}{2^{\cdot}3}n^3 + \frac{599}{2 \cdot 3 \cdot 5 \cdot 7}n^5 + \&c.$$

Whence by reverſion

$$n = \tfrac{1}{2}m - \frac{11}{2^{\cdot}3}m^3 - \frac{587}{2^1 \cdot 3 \cdot 5}m^5 - \&c.$$

Where the greateſt equation m muſt be expreſſed in
parts of the radius, which may be done by reducing
the angle m into ſeconds, and adding 4·6855749 to the
log. of the reſulting number, which will be the log. of
the number m.

The mean anomaly to which this greateſt equation
correſponds, will be

$$x = 90° + \tfrac{5}{8}m - \frac{5}{2^{\cdot}3}m^3 - \frac{1}{2^{\cdot}5}m^5 - \&c. \quad \text{Whence,}$$

if 90° be added to ⅝ of the greateſt Equation, the ſum
will be the mean anomaly ſufficiently exact.

Mr. Euler ſubjoins a table, by which may be found
the greateſt Equations, with the mean and eccentric
anomalies correſponding to theſe greateſt Equations
for every 100th part of unity, which he ſuppoſes equal
to the greateſt eccentricity, or when the tranſverſe and
diſtance of the foci become infinite. The laſt column
of the table gives alſo the logarithm of that diſtance of
the planet from the ſun where its Equation is greateſt.
By means of this table, any eccentricity being given,
by interpolation will be found the correſponding greateſt
Equation. But the chief uſe of the table is to determine the eccentricity when the greateſt Equation is
known; and without this help Mr. Euler thinks the
problem cannot be reſolved.

EQUATION *of Time*, denotes the difference between
mean and apparent time, or the reduction of the apparent unequal time, or motion of the ſun or a planet, to
equal and mean time, or motion; or the Equation of
time is the difference between the ſun's mean motion,
and his right aſcenſion. Apparent time is that which
takes its beginning from the paſſage of the ſun's centre
over the meridian of any place; and had the ſun no

motion

motion in the ecliptic, or was his motion reduced to the equator or in right ascension uniform, he would always return to the meridian after equal intervals of time. But his apparent motion in the ecliptic being continually varying, and his motion in right ascension being rendered farther unequal on account of the obliquity of the ecliptic to the equator, from these causes it arises that the intervals of his return to the meridian become unequal, and the sun will gradually come too slow or too soon to the meridian for an equable motion, such as that of clocks and watches ought to be; and this retardation or acceleration of the sun's coming to the meridian, is called the equation of time.

Now, computing the celestial motions according to equal time, it is necessary to turn that time back again into apparent time, that they may correspond to observation: on the contrary, any phenomenon being observed, the apparent time of it must be converted into equal time, to have it correspond with the times marked in the astronomical tables.

The Equation of time is nothing at four different times in the year, when the whole mean and unequal motions exactly agree; viz, about the 15th of April, the 15th of June, the 31st of August, and the 24th of December: at all other times the sun is either too fast or too slow for mean, equal, or clock time, by a certain number of minutes and seconds, which at the greatest is 16′ 14″, and happens about the 1st of November; every other day throughout the year having a certain quantity of this difference belonging to it; which however is not exactly the same every year, but only every 4th year; for which reason it is necessary to have 4 tables of this Equation, viz, one for each of the four years in the period of leap years. Instead of these, may be here inserted, as follows, one general equation of time, according to the place of the sun, in every point of the ecliptic: where it is to be observed, that the sign of the ecliptic is placed at the tops of the columns, and the particular degree of the sun's place, in each sign, in the first and last columns; and in the angle of meeting in all the other columns, is the equation of time, in minutes and seconds, when the sun has any particular longitude: supposing the obliquity of the ecliptic 23° 28′, and the sun's apogee in 9° of ♋.

Deg.	♈ 0 m s	♉ 1 m s	♊ 2 m s	♋ 3 m s	♌ 4 m s	♍ 5 m s	♎ 6 m s	♏ 7 m s	♐ 8 m s	♑ 9 m s	♒ 10 m s	♓ 11 m s	Deg.
0	7 +36	1 − 9	3 −51	1 +13	5 +57	2 +20	7 −38	15 −31	13 −33	1 −11	11 +28	14 +19	0
1	7 17	1 23	3 47	1 26	5 59	2 4	7 58	15 39	13 17	0 42	11 45	14 13	1
2	6 58	1 36	3 41	1 40	6 0	1 48	8 19	15 46	13 0	0 12	12 1	14 6	2
3	6 39	1 48	3 37	1 53	6 1	1 31	8 40	15 52	12 42	0 +17	12 17	13 59	3
4	6 20	2 0	3 32	2 1	6 1	1 14	9 1	15 57	12 23	0 46	12 32	13 51	4
5	6 1	2 11	3 26	2 20	6 0	0 56	9 21	16 2	12 4	1 16	12 46	13 43	5
6	5 42	2 22	3 19	2 33	5 59	0 38	9 41	16 6	11 44	1 45	12 59	13 34	6
7	5 24	2 32	3 12	2 45	5 57	0 20	10 1	16 9	11 23	2 14	13 12	13 24	7
8	5 5	2 42	3 4	2 58	5 54	0 1	10 20	16 11	11 1	2 43	13 24	13 14	8
9	4 47	2 51	2 56	3 11	5 51	0 −18	10 39	16 13	10 39	3 11	13 35	13 3	9
10	4 28	3 0	2 47	3 23	5 47	0 37	10 57	16 13	10 16	3 39	13 45	12 51	10
11	4 9	3 8	2 38	3 35	5 42	0 57	11 15	16 13	9 53	4 7	13 54	12 39	11
12	3 50	3 16	2 29	3 46	5 37	1 17	11 33	16 12	9 29	4 35	14 2	12 27	12
13	3 32	3 23	2 19	3 58	5 31	1 38	11 51	16 10	9 5	5 2	14 9	12 14	13
14	3 13	3 30	2 8	4 9	5 24	1 58	12 8	16 7	8 40	5 29	14 16	12 0	14
15	2 55	3 36	1 57	4 19	5 17	2 19	12 25	16 4	8 14	5 56	14 22	11 46	15
16	2 37	3 41	1 46	4 29	5 9	2 40	12 41	16 0	7 48	6 22	14 27	11 31	16
17	2 19	3 46	1 35	4 39	5 1	3 1	12 57	15 55	7 22	6 48	14 31	11 16	17
18	2 1	3 50	1 23	4 48	4 52	3 22	13 12	15 49	6 55	7 13	14 35	11 1	18
19	1 43	3 53	1 11	4 57	4 43	3 44	13 27	15 42	6 28	7 37	14 38	10 46	19
20	1 26	3 56	0 59	5 4	4 33	4 5	13 42	15 35	6 0	8 1	14 40	10 30	20
21	1 9	3 58	0 46	5 13	4 22	4 26	13 56	15 26	5 32	8 24	14 41	10 14	21
22	0 52	4 0	0 34	5 20	4 11	4 47	14 9	15 17	5 4	8 47	14 42	9 58	22
23	0 36	4 1	0 21	5 27	3 59	5 9	14 21	15 7	4 36	9 9	14 41	9 41	23
24	0 20	4 1	0 8	5 33	3 46	5 30	14 33	14 56	4 8	9 31	14 40	9 24	24
25	0 4	4 1	0 +5	5 39	3 33	5 52	14 44	14 44	3 39	9 53	14 39	9 6	25
26	0 −11	4 0	0 19	5 44	3 19	6 13	14 53	14 31	3 10	10 14	14 37	8 48	26
27	0 26	3 59	0 31	5 48	3 4	6 35	15 5	14 17	2 41	10 34	14 34	8 30	27
28	0 40	3 57	0 46	5 52	2 50	6 56	15 14	14 3	2 11	10 53	14 30	8 12	28
29	0 53	3 54	0 59	5 55	2 35	7 17	15 23	13 48	1 41	11 11	14 25	7 54	29
30	1 9	3 51	1 13	5 57	2 20	7 28	15 31	13 33	1 11	11 28	14 19	7 36	30

A Table of the Equation of Time, for every Degree of the Sun's Longitude.

The

The Equations with +, are to be added to the apparent time, to have the mean time; those with —, are to be subtracted from apparent time, to give the mean time.

The preceding mark, whether + or —, at the top of any column, belongs to all the numbers or equations in that column till the sign changes; after which, the remainder of the column belongs to the contrary sign.

The Equation answering to any point of longitude between one degree and another, or any number of minutes or parts of a degree, is to be found by proportion in the usual way, viz, as 1° or 60', to that number of minutes, so is the whole difference in the Equation from the given whole degree of longitude to the next degree, to the proportional part of it answering to the given number of minutes.

See Tables of the Equation of Time computed for every year, in the Nautical Almanac, by a method proposed and illustrated by Dr. Maskelyne, the astronomer royal, viz, by taking the difference between the sun's true right ascension and his mean longitude, corrected by the Equation of the equinoxes in right ascension, and turning it into time at the rate of 1 minute of time to 15' of right ascension. Philos. Transf. vol. 54, p. 336.

EQUATION *of a Curve*, is an Equation shewing the nature of a curve by expressing the relation between any abscis and its corresponding ordinate, or else the relation of their fluxions, &c. Thus, the Equation to the circle, is $ax - x^2 = y^2$, where a is its diameter, x any abscis, or part of that diameter, and y the ordinate at that point of the diameter; the meaning being, that whatever abscis is denoted by x, then the square of its corresponding ordinate will be $ax - x^2$. In like manner the Equation

of the ellipse is $\frac{p}{a} \cdot \overline{ax - x^2} = y^2$,

of the hyperbola is $\frac{p}{a} \cdot \overline{ax + x^2} = y^2$,

of the parabola is - - - $px = y^2$.

Where a is an axis, and p the parameter.

And in like manner for any other curves.

This method of expressing the nature of curves by algebraical equations, was first introduced by Des Cartes, who, by thus connecting together the two sciences of algebra and geometry, made them mutually assisting to each other, and so laid the foundation of the greatest improvements that have been made in every branch of them since that time. See Des Cartes's Geometry; also Newton's Lines of the 3d Order, and many other similar works on curve lines, by several authors.

EQUATOR, in Geography, a great circle of the earth, equally distant from its two poles, and dividing it into two equal parts, or hemispheres, the northern and southern. The Equator is sometimes simply called the Line.

The circle in the heavens conceived directly over the Equator, is the Equinoctial. See EQUINOCTIAL.

The greatest height of the Equator above the horizon, is equal to the latitude of the place.

TABLE *for turning Degrees and Minutes into Time, and the Contrary.*

D	H	M	D	H	M	D	H	M
M	M	S		M	S		M	S
1	0	4	61	4	4	121	8	4
2	0	8	62	4	8	122	8	8
3	0	12	63	4	12	123	8	12
4	0	16	64	4	16	124	8	16
5	0	20	65	4	20	125	8	20
6	0	24	66	4	24	126	8	24
7	0	28	67	4	28	127	8	28
8	0	32	68	4	32	128	8	32
9	0	36	69	4	36	129	8	36
10	0	40	70	4	40	130	8	40
11	0	44	71	4	44	131	8	44
12	0	48	72	4	48	132	8	48
13	0	52	73	4	52	133	8	52
14	0	56	74	4	56	134	8	56
15	1	0	75	5	0	135	9	0
16	1	4	76	5	4	136	9	4
17	1	8	77	5	8	137	9	8
18	1	12	78	5	12	138	9	12
19	1	16	79	5	16	139	9	16
20	1	20	80	5	20	140	9	20
21	1	24	81	5	24	141	9	24
22	1	28	82	5	28	142	9	28
23	1	32	83	5	32	143	9	32
24	1	36	84	5	36	144	9	36
25	1	40	85	5	40	145	9	40
26	1	44	86	5	44	146	9	44
27	1	48	87	5	48	147	9	48
28	1	52	88	5	52	148	9	52
29	1	56	89	5	56	149	9	56
30	2	0	90	6	0	150	10	0
31	2	4	91	6	4	151	10	4
32	2	8	92	6	8	152	10	8
33	2	12	93	6	12	153	10	12
34	2	16	94	6	16	154	10	16
35	2	20	95	6	20	155	10	20
36	2	24	96	6	24	156	10	24
37	2	28	97	6	28	157	10	28
38	2	32	98	6	32	158	10	32
39	2	36	99	6	36	159	10	36
40	2	40	100	6	40	160	10	40
41	2	44	101	6	44	161	10	44
42	2	48	102	6	48	162	10	48
43	2	52	103	6	52	163	10	52
44	2	56	104	6	56	164	10	56
45	3	0	105	7	0	165	11	0
46	3	4	106	7	4	166	11	4
47	3	8	107	7	8	167	11	8
48	3	12	108	7	12	168	11	12
49	3	16	109	7	16	169	11	16
50	3	20	110	7	20	170	11	20
51	3	24	111	7	24	171	11	24
52	3	28	112	7	28	172	11	28
53	3	32	113	7	32	173	11	32
54	3	36	114	7	36	174	11	36
55	3	40	115	7	40	175	11	40
56	3	44	116	7	44	176	11	44
57	3	48	117	7	48	177	11	48
58	3	52	118	7	52	178	11	52
59	3	56	119	7	56	179	11	56
60	4	0	120	8	0	180	12	0

As one whole revolution of the earth, or of the 360° of the Equator, is performed in 24 hours, which is at the rate of 15° to the hour; hence the number of degrees of the Equator, answering to any other given time, or the time answering to any given degrees of the Equator, will be easily found by proportion, viz,

as 1hr. : 15° :: any time : its degrees,
or as 15° : 1hr. :: any degrees : their time.

And thus is computed the foregoing Table for turning time into degrees of the Equator, and the contrary.

EQUATORIAL, *Universal*, or PORTABLE OBSERVATORY, is an instrument intended to answer a number of useful purposes in practical astronomy, independent of any particular observatory. It may be employed in any steady room or place, and it performs most of the useful problems in the science of astronomy. The following is the description of one lately invented, and named the *Universal Equatorial*.

The principal parts of this instrument (fig. 2, plate viii.) are, 1st, The azimuth or horizontal circle A, which represents the horizon of the place, and moves on a long axis B, called the vertical axis. 2d, The Equatorial or hour-circle C, representing the Equator, placed at right angles to the polar axis D, or the axis of the earth, upon which it moves. 3d, The semicircle of declination E, on which the telescope is placed, and moving on the axis of declination, or the axis of motion of the line of collimation F. Which circles are measured and divided as in the following Table:

Measures of the several circles and divisions on them.	Radius Inches	Limb divided to	Non. of 30 gives seconds	Divid. on limb into pts. of Inc.	Divid. by Non. into pts. of Inc.
Azimuth or horizontal circle	5·1	15′	30″	45th	1350th
Equatorial or hour circle	5·1	15′ or 1 m. in time	30″ 2″	45th	1350th
Vertical semicircle for declination or latitude	5·	15′	30″	42d	1260th

4th, The telescope, which is an achromatic refractor with a triple object-glass, whose focal distance is 17 inches, and its aperture 2·45 inc., and it is furnished with 6 different eye tubes; so that its magnifying powers extend from 44 to 168. The telescope in this Equatorial may be brought parallel to the polar axis, as in the figure, so as to point to the pole-star in any part of its diurnal revolution; and thus it has been observed near noon, when the sun has shone very bright. 5th, The apparatus for correcting the error in altitude occasioned by refraction, which is applied to the eye-end of the telescope, and consists of a slide G moving in a groove or dove-tail, and carrying the several eye-tubes of the telescope, on which slide there is an index corresponding

to five small divisions engraved on the dove-tail; a very small circle, called the refraction circle H, moveable by a finger-screw at the extremity of the eye-end of the telescope; which circle is divided into half minutes, one whole revolution of it being equal to 3′ 18″, and by its motion it raises the centre of the cross-hairs on a circle of altitude; and also a quadrant I of 1½ inc. radius, with divisions on each side, one expressing the degree of altitude of the object viewed, and the other expressing the minutes and seconds of error occasioned by refraction, corresponding to that degree of altitude. To this quadrant is joined a small round level K, which is adjusted partly by the pinion that turns the whole of this apparatus, and partly by the index of the quadrant; for which purpose the refraction circle is set to the same minute &c, which the index points to on the limb of the quadrant; and if the minute &c, given by the quadrant, exceed the 3′ 18″ contained in one entire revolution of the refraction circle, this must be set to the excess above one or more of its entire revolutions; then the centre of the cross-hairs will appear to be raised on a circle of altitude to the additional height which the error of refraction will occasion at that altitude.

The principal adjustment in this instrument, is that of making the line of collimation to describe a portion of an hour circle in the heavens: in order to which, the azimuth circle must be truly level; the line of collimation, or some corresponding line represented by the small brass rod M parallel to it, must be perpendicular to the axis of its own proper motion; and this last axis must be perpendicular to the polar axis. On the brass rod M there is occasionally placed a hanging level N, the use of which will appear in the following adjustments:

The azimuth circle may be made level by turning the instrument till one of the levels be parallel to an imaginary line joining two of the feet screws; then adjust that level with these two feet screws; turn the circle 180°, or half round; and if the bubble be not then right, correct half the error by the screw belonging to the level, and the other half error by the two foot screws, repeating this operation till the bubble come right; then turn the circle 90° from the two former positions, and set the bubble right, if it be wrong, by the foot screw at the end of the level; when this is done, adjust the other level by its own screw, and the azimuth circle will be truly level. The hanging level must then be fixed to the brass rod by two hooks of equal length, and made truly parallel to it: for this purpose, make the polar axis perpendicular or nearly perpendicular to the horizon; then adjust the level by the pinion of the declination semicircle: reverse the level, and if it be wrong, correct half the error by a small steel screw that lies under one end of the level, and the other half error by the pinion of the declination-semicircle, repeating the operation till the bubble be right in both positions. To make the brass rod, on which the level is suspended, at right angles to the axis of motion of the telescope, or line of collimation, make the polar axis horizontal, or nearly so; set the declination semicircle to 0°, and turn the hour-circle till the bubble come right; then turn the declination-circle

5

Plate IX.

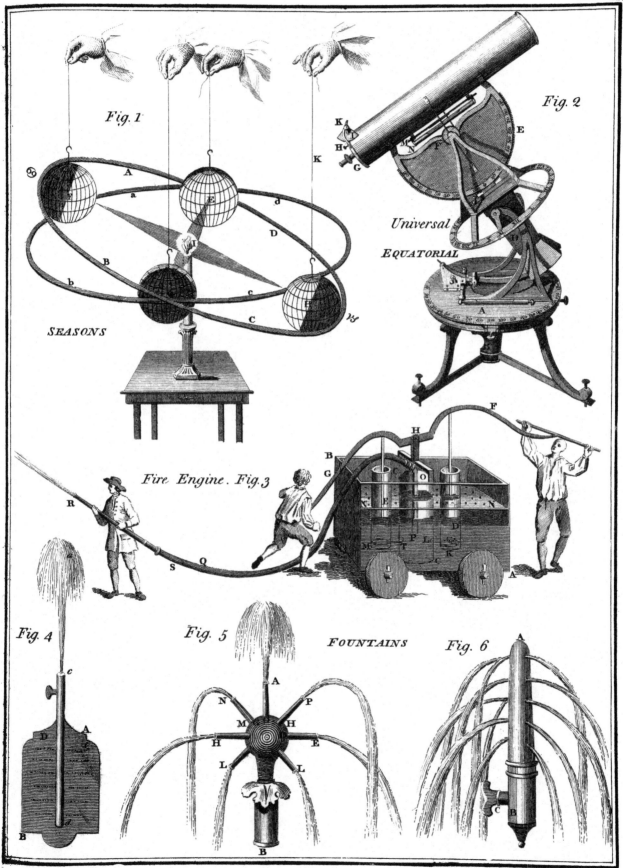

Fig. 1

SEASONS

Fig. 2

Universal
EQUATORIAL

Fire Engine. Fig. 3

Fig. 4

Fig. 5

FOUNTAINS

Fig. 6

circle to 90°; adjuſt the bubble by raiſing or depreſ-ſing the polar axis (firſt by hand till it be nearly right, afterwards tighten with an ivory key the ſocket which runs on the arch with the polar axis, and then apply the ſame ivory key to the adjuſting ſcrew at the end of the ſaid arch till the bubble come quite right); then turn the declination-circle to the oppoſite 90°; if the level be not then right, correct half the error by the aforeſaid adjuſting ſcrew at the end of the arch, and the other half error by the two ſcrews that raiſe or de-preſs the end of the braſs rod. The polar axis remain-ing nearly horizontal as before, and the declination-ſemicircle at 0°, adjuſt the bubble by the hour-circle; then turn the declination-ſemicircle to 90°, and adjuſt the bubble by raiſing or depreſſing the polar axis; then turn the hour-circle 12 hours; and if the bubble be wrong, correct half the error by the polar axis, and the other half error by the two pair of capſtan ſcrews at the feet of the two ſupports on one ſide of the axis of motion of the teleſcope; and thus this axis will be at right angles to the polar axis. The next adjuſt-ment, is to make the centre of the croſs hairs remain on the ſame object, while the eye-tube is turned quite round by the pinion of the refraction apparatus: for this adjuſtment, ſet the index on the ſlide to the firſt diviſion on the dove-tail; and ſet the diviſion marked 18″ on the refraction-circle to its index; then look through the teleſcope, and with the pinion turn the eye-tube quite round; then if the centre of the hairs does not remain on the ſame ſpot during that revolu-tion, it muſt be corrected by the four ſmall ſcrews, 2 and 2 at a time, which will be found upon unſcrewing the neareſt end of the eye-tube that contains the firſt eye-glaſs; repeating this correction till the centre of the hairs remain on the ſpot looked at during a whole revolution. To make the line of collimation parallel to the braſs rod on which the level hangs, ſet the polar axis horizontal, and the declination-circle to 90°, adjuſt the level by the polar axis; look through the teleſcope on ſome diſtant horizontal object, covered by the cen-tre of the croſs hairs: then invert the teleſcope, which is done by turning the hour-circle half round; and if the centre of the croſs hairs does not cover the ſame object as before, correct half the error by the upper-moſt and lowermoſt of the 4 ſmall ſcrews at the eye-end of the large tube of the teleſcope; this correction will give a ſecond object now covered by the centre of the hairs, which muſt be adopted inſtead of the firſt object; then invert the teleſcope as before; and if the ſecond object be not covered by the centre of the hairs, correct half the error by the ſame two ſcrews as were uſed before: this correction will give a third object, now covered by the centre of the hairs, which muſt be adopted inſtead of the ſecond object; repeat this ope-ration till no error remain; then ſet the hour-circle ex-actly to 12 hours, the declination-circle remaining a 90° as before; and if the centre of the croſs hairs do not cover the laſt object fixed on, ſet it to that object by the two remaining ſmall ſcrews at the eye-end of the large tube, and then the line of collimation will be parallel to the braſs rod. For rectifying the nonius of the declination and Equatorial circles, lower the tele-ſcope as many degrees &c below 0° or Æ on the decli-

nation-ſemicircle as are equal to the complement of the latitude; then elevate the polar axis till the bubble be horizontal; and thus the Equatorial circle will be ele-vated to the co-latitude of the place: ſet this circle to 6 hours; adjuſt the level by the pinion of the declina-tion-circle; then turn the Equatorial circle exactly 12 hours from the laſt poſition; and if the level be not right, correct one half of the error by the Equatorial circle, and the other half by the declination-circle: then turn the Equatorial circle back again exactly 12 hours from the laſt poſition; and if the level be ſtill wrong, repeat the correction as before, till it be right, when turned to either poſition: that being done, ſet the nonius of the Equatorial circle exactly to 6 hours, and the nonius of the declination-circle exactly to 0°.

The chief uſes of this Equatorial are,

1ſt, To find the meridian by one obſervation only: for this purpoſe, elevate the Equatorial circle to the co-latitude of the place, and ſet the declination-ſemi-circle to the ſun's declination for the day and hour of the day required; then move the azimuth and hour-circles both at the ſame time, either in the ſame or con-trary direction, till you bring the centre of the croſs hairs in the teleſcope exactly to cover the centre of the ſun; when that is done, the index of the hour-circle will give the apparent or ſolar time at the inſtant of obſervation; and thus the time is gained, though the ſun be at a diſtance from the meridian; then turn the hour-circle till the index points preciſely at 12 o'clock, and lower the teleſcope to the horizon, in order to ob-ſerve ſome point there in the centre of the glaſs; and that point is the meridian mark, found by one obſer-vation only. The beſt time for this operation is 3 hours before, or 3 hours after 12 at noon.

2d, To point the teleſcope on a ſtar, though not on the meridian, in full day-light. Having elevated the equatorial circle to the co-latitude of the place, and ſet the declination-ſemicircle to the ſtar's declination, move the index of the hour-circle till it ſhall point to the pre-ciſe time at which the ſtar is then diſtant from the me-ridian, found in the tables of the right aſcenſion of the ſtars, and the ſtar will then appear in the glaſs.

Beſides theſe uſes, peculiar to this inſtrument, it may alſo be applied to all the purpoſes to which the princi-pal aſtronomical inſtruments are applied; ſuch as a tranſit inſtrument, a quadrant, and an equal-altitude inſtrument.

See the deſcription and drawing of an Equatorial te-leſcope, or portable obſervatory, invented by Mr. Short, in the Philoſ. Tranſ. number 493, or vol. 46, p. 242; and another by Mr. Nairne, vol. 61, p. 107.

EQUIANGULAR *Figure*, is one that has all its angles equal among themſelves; as the ſquare, and all the regular figures.

An equilateral figure inſcribed in a circle, is always Equiangular. But an Equiangular figure inſcribed in a circle, is not always equilateral, except when it has an odd number of ſides: If the ſides be of an even number, then they may either be all equal, or elſe half of them will always be equal to each other, and the other half to each other, the equals being placed alter-nately. See the demonſtration in my Mathematical Miſcellany, pa. 272.

3 L 2

EQUIANGULAR

EQUIANGULAR, is also said of any two figures of the same kind, when each angle of the one is equal to a corresponding angle in the other, whether each figure, separately considered in itself, be an equiangular figure or not, that is, having all its angles equal to each other. Thus, two triangles are Equiangular to each other, if, ex. gr. one angle in each be of 30°, a second angle in each of 50°, and the third angle of each equal to 100 degrees.

Equiangular triangles have not their like sides necessarily equal, but only proportional to each other; and such triangles are always similar to each other.

EQUICRURAL *Triangle*, is one that has two of its sides equal to each other; and is more usually called an Isosceles triangle.

EQUICULUS, EQUULEUS, or EQUUS *Minor*, a constellation of the northern hemisphere. See EQUULEUS.

EQUIDIFFERENT, are such things as have equal differences, or arithmetically proportional. If the terms have all the same difference, viz, the 1st and 2d, the 2d and 3d, the 3d and 4th, &c, they are said to be Continually Equidifferent; as the numbers 3, 6, 9, 12, &c, where the common difference is 3. But if the several different couplets only have the same difference, as the 1st and 2d, the 3d and 4th, the 5th and 6th, &c, they are said to be Discretely Equidifferent; as the terms 3 and 6, 7 and 10, 9 and 12, &c. See ARITHMETICAL *Progression* and *Proportion*.

EQUILATERAL *Figure*, is one that has all its sides equal to each other. Such as the square, and all the regular figures or polygons, or a triangle that has all its sides equal. See EQUIANGULAR.

EQUILATERAL *Hyperbola*, is that which has the two axes equal to each other, and every pair of conjugate diameters also equal to each other. The asymptotes also are at right angles to each other, and make each half a right angle with either axis. Also, such an hyperbola is equal to its opposite hyperbola, and likewise to its conjugate hyperbola; so that all the four conjugate hyperbolas are mutually equal to each other.

Moreover, as the 3d proportional to the two axes is the parameter, therefore in such a figure, the parameter and two axes are all three equal to one another. Hence, as the general equation to hyperbolas is

$$y^2 = \frac{p}{t} . \overline{tx + x^2} \text{ or } = \frac{c^2}{t^2} . \overline{tx + x^2},$$ where t is the

transverse axis, c the conjugate, p the parameter, x the absciss, and y the ordinate; then making t, c, and p all equal, the equation, for the Equilateral Hyperbola, becomes $y^2 = tx + x^2$; differing from the equation of the circle only in the sign of the term x^2, which in the circle is $-$.

EQUILIBRIUM, is an equality between two equal forces acting in opposite directions; so that they mutually balance each other; like the two equal arms, or scales, of a balance, &c.

EQUILIBRIUM, in solid bodies, forms a considerable part of the science of Statics. And Equilibrium of fluids, a considerable part of the doctrine of Hydrostatics.

EQUIMULTIPLES, the products of quantities equally multiplied.

Thus, $3a$ and $3b$ are Equimultiples of a and b; and ma and mb are Equimultiples of a and b.

Equimultiples of any quantities, have the same ratio as the quantities themselves.

Thus $a : b : : 3a : 3b : : ma : mb$.

EQUINOCTIAL, a great circle in the heavens under which the equator moves in its diurnal motion. The poles of this circle are the poles of the world. It divides the sphere into two equal parts, the northern and southern. It cuts the horizon, of any place, in the east and west points; and at the meridian its elevation above the horizon is equal to the co-latitude of the place. The Equinoctial has also various other properties; as,

1. Whenever the sun comes to this circle, he makes equal days and nights all round the globe; because he then rises due east, and sets due west. Hence it has the name Equinoctial. All stars which are under this circle, or have no declination, do also rise due east, and set due west.

2. All people living under this circle, or upon the equator, or line, have their days and nights at all times equal to each other.

3. From this circle, on the globe, is counted, upon the meridian, the declination in the heavens, and the latitude on the earth.

4. Upon the Equinoctial, or equator, is counted the longitude, making in all 360° quite round, or else 180° east, and 180° west.

5. And as the time of one whole revolution is divided into 24 hours; therefore 1 hour answers to 15°, or the 24th part of 360°. Hence,

1° of longitude answers to 4 min. of time,
15' - - - - - - - - - to 1 min. of time,
1' - - - - - - - - - to 4 sec. of time, &c.

6. The shadows of those people who live under this circle are cast to the southward of them for one half of the year, and to the northward of them during the other half; and twice in a year, viz, at the time of the equinoxes, the sun at noon casts no shadow, being exactly in their zenith.

EQUINOCTIAL *Colure*, is the great circle passing through the poles of the world and the Equinoctial points, or first points of Aries and Libra.

EQUINOCTIAL *Dial*, is one whose plane is parallel to the Equinoctial. The properties or principles of this dial are,

1. The hour lines are all equally distant from one another, quite round the circumference of a circle; and the style is a straight pin, or wire, set up in the centre of the circle, perpendicular to the plane of the dial.

2. The sun shines upon the upper part of this dial-plane from the 21st of March to the 22d of September, and upon the under part of the plane the other half of the year.

Some of these dials are made of brass, &c; and set up in a frame, to be elevated to any given latitude.

EQUINOCTIAL *Points*, are the two opposite points where the Ecliptic and Equinoctial cross each other;
the

3

the one point being in the beginning of Aries, and called the Vernal point, or Vernal Equinox; and the other in the beginning of Libra and called the Autumnal point, or Autumnal Equinox.

It is found by observation, that the Equinoctial points, and all the other points of the ecliptic, are continually moving backwards, or in antecedentia, i. e. westwards. This retrograde motion of the Equinoctial points, is that phenomenon called the Precession of the Equinoxes, and is made at the rate of 50″ every year nearly.

EQUINOXES, the times when the sun enters the Equinoctial points, or about the 21st of March and 22d of September: the former being the Vernal or Spring Equinox, and the latter time the Autumnal Equinox.

As the sun's motion is unequal, being sometimes quicker and sometimes slower, it hence happens that there are about 8 days more from the vernal to the autumnal Equinox, or while the sun is on the northern side of the equator, than while he is in moving through the southern signs from the autumnal to the vernal Equinox, or on the southern side of the equator. According to the observations of M. Cassini, the

sun is 186^d 14^h 53^m in the northern signs,
and only 178 14 56 in the southern signs,
so that 7 23 57 is the difference of them, or nearly 8 days.

EQUINUS *Barbatus*, a kind of comet. See HIPPEUS.

EQUULEUS, EQUICULUS, and EQUUS *Minor*, *Equi sectio*, the *Horse's Head*, one of the 48 old constellations, in the northern hemisphere. Its stars, in Ptolomy's catalogue, are 4, in Tycho's 4, in Hevelius's 6, and in Flamsteed's 10

ERECT *Vision*. See VISION.

ERECT *Dials*, such as stand perpendicular to the horizon, and are of various kinds; as Erect Direct, when they face exactly one of the four cardinal points, east, west, north, south; and Erect Declining, when declined from the cardinal points. See DIAL.

To ERECT *a Perpendicular*, is a popular problem in practical geometry, and denotes to raise a perpendicular from a given line &c, as distinguished from Demitting or letting a perpendicular fall upon a line &c, from some point out of it. See PERPENDICULAR.

ERIDANUS, the *River*, a constellation of the southern hemisphere, and one of the 48 old asterisms: The stars in this constellation, in Ptolomy's catalogue, are 34, in Tycho's 19, and in the British catalogue 84.

ERRATIC, an epithet applied to the planets, which are called Erratic or wandering stars, in contradistinction to the fixed stars.

ESCALADE, or SCALADE, a furious attack of a wall or a rampart; carried on with ladders, to pass the ditch, or mount the rampart; without proceeding in form, breaking ground, or carrying on regular works to secure the men.

ESPAULE, or EPAULE. See EPAULE.

ESPLANADE, in Fortification, called also Glacis, a part which serves as a parapet to the counterscarp, or covert way; being a declivity or slope of earth, commencing from the top of the counterscarp, and losing itself insensibly in the level of the champaign.

Esplanade also means the ground which has been levelled from the glacis of the counterscarp, to the first houses; or the vacant space between the works and the houses of the town.

The term is also applied, in the general, to any piece of ground that is made flat or level, and which before had some eminence that incommoded the place.

ESTIVAL *Occident*, *Orient*, or *Solstice*. See OCCIDENT, ORIENT, SOLSTICE.

EVAPORATION, the act of dissipating the humidity of a body in fumes or vapour; differing from exhalation, which is properly a dispersion of dry particles issuing from a body.

Evaporation is usually produced by heat, and by the change of air: thus, common salt is formed by evaporating all the humidity in the brine or salt water; which evaporation is either performed by the heat of the sun, as in the salt-works on the sea-coast, &c; or by means of fire, as at the salt-springs, &c: and it is well known how useful a brisk wind is in drying wet clothes, or the surface of the ground; while in a calm, still atmosphere, they dry extreme slowly.

But, though Evaporation be generally considered as an effect of the heat and motion of the air, yet M. Gauteron, in the Memoires de l'Acad. des Scienc. an. 1705, shews, that a quite opposite cause may have the same effect, and that fluids lose more of their parts in the severest frost than when the air is moderately warm: thus, in the great frost of the year 1708, he found that the greater the cold, the more considerable the evaporation; and that ice itself lost full as much as the warmer liquors that did not freeze.

There are indeed few subjects of philosophical investigation that have occasioned a greater variety of opinion than the theory of Evaporation, or of the ascent of water in such a fluid as air, between 8 and 9 hundred times lighter than itself, to different heights according to the different densities of the air; in which case it must be specifically lighter than the air through which it ascends. The Cartesians account for it by supposing, that by the action of the sun upon the water, small particles of the water are formed into hollow spheres and filled with the *materia subtilis*, which renders them specifically lighter than the ambient air, so that they are buoyed up by it.

Dr. Nieuwentyt, in his Religious Philosopher, cont. 19, and several others, have alleged, that the sun emits particles of fire which adhere to those of water, and form moleculæ, or small bodies, lighter than an equal bulk of air, which consequently ascend till they come to a height where the air is of the same specific gravity with themselves; and that these particles being separated from the fire with which they are incorporated, coalesce and descend in dew or rain.

Dr. Halley has advanced another hypothesis, which has been more generally received: he imagined, that by the action of the sun on the surface of the water, the aqueous particles are formed into hollow spherules, that are filled with a finer air highly rarefied, so as to become specifically lighter than the external air. Philos. Trans. number 192, or Abr. vol. 2, p. 126.

Dr. Desaguliers, dissatisfied with these two hypotheses, proposes another in the Philos. Trans. number 407, or Abr. vol. 7, pa. 61. See also his Course of Experimental

Experimental Philosophy, vol. 2, p. 336. He supposes that heat acts more powerfully on water than on common air; that the same degree of heat which rarefies air two-thirds, will rarefy water near 14,000 times; and that a very small degree of heat will raise a steam or vapour from water, even in winter, whilst it condenses the air; and thus the particles of water are converted into vapour by being made to repel each other strongly, and, deriving electricity from the particles of air to which they are contiguous, are repelled by them and by each other, so as to form a fluid which, being lighter than the air, rises in it, according to their relative gravities. The particles of this vapour retain their repellent force for a considerable time, till, by some diminution of the density of the air in which they float, they are precipitated downwards, and brought within the sphere of each other's attraction of cohesion, and so join again into drops of water.

Many objections have been urged against this opinion, by Mr. Clare in his Treatise of the Motion of Fluids, pa. 294, and by Mr. Rowning in his System of Philosophy, part 2, diss. 6; to which Dr. Hamilton has added the two following, viz, that if heat were the only cause of evaporation, water would evaporate faster in a warm close room, than when exposed in a colder place, where there is a constant current of air; which is contrary to experience; and that the evaporation of water is so far from depending on its being rarefied by heat, that it is carried on even whilst water is condensed by the coldness of the air, till it freezes; and since it evaporates even when frozen into hard ice, it must also evaporate in all the lesser degrees of cold. And therefore heat does not seem to be the principal, much less the only cause of Evaporation.

Others have more successfully accounted for the phenomena of Evaporation on another principle, viz that of solution; and shewn, from a variety of experiments, that what we call Evaporation, is nothing more than a gradual solution of water in air, produced and supported by the same means, viz, attraction, heat, and motion, by which other solutions are effected.

It seems the Abbé Nollet first started this opinion, though without much pursuing it, in his Leçons de Physique Experimentale, first published in 1743: he offers it as a conjecture, that the air of the atmosphere serves as a solvent or sponge, with regard to the bodies that encompass it, and receives into its pores the vapours and exhalations that are detached from the masses to which they belong in a fluid state; and he accounts for their ascent on the same principles with the ascent of liquors in capillary tubes. On his hypothesis, the condensation of the air contributes, like the squeezing of a sponge, to their descent.

Dr. Franklin, in a paper of Philosophical and Meteorological Observations, Conjectures and Suppositions, delivered to the Royal Society about the year 1747, and read in 1756, suggested a similar hypothesis: he observes, that air and water mutually attract each other; and hence he concludes, that water will dissolve in air, as salt in water; every particle of air assuming one or more particles of water; and when too much is added, it precipitates in rain. But as there is not the same contiguity between the particles of air as of water, the solution of water in air is not carried on without

a motion of the air, so as to cause a fresh accession of dry particles. A small degree of heat so weakens the cohesion of the particles of water, that those on the surface easily quit it, and adhere to the particles of air: a greater degree of heat is necessary to break the cohesion between water and air; for its particles being by heat repelled to a greater distance from each other, thereby more easily keep the particles of water that are annexed to them from running into cohesions that would obstruct, refract, or reflect the heat: and hence it happens that when we breathe in warm air, though the same quantity of moisture may be taken up from the lungs as when we breathe in cold air, yet that moisture is not so visible. On these principles he accounts for the production and different appearances of fogs, mists, and clouds. He adds, that if the particles of water bring electrical fire when they attach themselves to air, the repulsion between the particles of water electrified, joins with the natural repulsion of the air to force its particles to a greater distance, so that the air being more dilated, rises and carries up with it the water; which mutual repulsion of the particles of air is increased by a mixture of common fire in the particles of water. When air, loaded with surrounding particles of water, is compressed by adverse winds, or by being driven against mountains, &c, or condensed by taking away the fire that assisted it in expanding, the particles will approach one another, and the air with its water will descend as a dew; or if the water surrounding one particle of air come in contact with the water surrounding another, they coalesce and form a drop, producing rain; and since it is a well-known fact, that vapour is a good conductor of electricity, as well as of common fire, it is reasonable to conclude with Mr. Henley, that Evaporation is one great cause of the clouds becoming at times surcharged with this fluid. Philof. Tranf. vol. 67, pa. 134. See also vol. 55, p. 182, or Franklin's Letters and Papers on Philosophical Subjects, p. 42 &c, and pa. 182, ed. 1769.

M. le Roi, of the Acad. of Sciences at Paris, has also advanced the same opinion, and supported it by a variety of facts and observations in the Memoirs for the year 1751. He shews, that water does undergo in the air a real dissolution, forming with it a transparent mixture, and possessing the same properties with the solutions of most salts in water; and that the two principal causes which promote the solution of water in the air, are heat and wind; that the hotter the air is, within a certain limit, the more water it will dissolve; and that at a certain degree of heat the air will be saturated with water; and by determining at different times the degree of the air's saturation, he estimates the influence of those causes on which the quantity depends that is suspended in the air in a state of solution. Accordingly, the air, heated by evaporating substances to which it is contiguous, becomes more rare and light, rises and gives way to a denser air; and, by being thus removed, contributes to accelerate the Evaporation. The fixed air contained in the internal parts of evaporating bodies, put into action by heat, seems also to increase their Evaporation. The wind is another cause of the increase of Evaporation, chiefly by changing and renewing the air which immediately encompasses the evaporating substances; and from the consideration

of

of these two causes combined, it appears why the quantity of vapour raised in the night is less than that of the day, since the air is then both less heated and less agitated. To the objection urged against this hypothesis, on account of the Evaporation of water in a vacuum, this ingenious writer replies, that the water itself contains a great quantity of air, which gradually disengages itself, and causes the Evaporation; and that it is impossible that a space containing water which evaporates should remain perfectly free from air. To this objection a late writer, Dr. Dobson of Liverpool, replies, that though air appears, by unquestionable experiments, to be a chemical solvent of water, and as such, is to be considered as one cause of its Evaporation, heat is another cause, acting without the intervention of air, and producing a copious Evaporation in an exhausted receiver; agreeably to an experiment of Dr. Irving, who says, that in an exhausted receiver water rises in vapour more copiously at 180° of Fahrenheit's thermometer, than in the open air at 212°, its boiling point. Dr. Dobson farther adds, that water may exist in air in three different states; in a state of perfect solution, when the air will be clear, dry, and heavy, and its powers of solution still active; in a state of beginning precipitation, when it becomes moist and foggy, its powers of solution are diminished, and it becomes lighter in proportion as its water is deposited; and also, when it is completely precipitated, which may happen either by a slower process, when the dissolved water falls in a drizzling rain, or by a more sudden process, when it descends in brisk showers. Philos. Transf. vol. 67, p. 257, and Phipps's Voyage towards the North Pole, p. 211.

Dr. Hamilton, professor of philosophy in the university of Dublin, transmitted to the Royal Society in 1765, a long Dissertation on the nature of Evaporation, in which he proposes and establishes this theory of solution; and though other writers had been prior in their conjectures, and even in their reasoning on this subject, Dr. Hamilton assures us, that he has not represented any thing as new which he was conscious had ever been proposed by any one before him, even as a conjecture. Dr. Hamilton having evinced the agreement between Solution and Evaporation, concludes, that Evaporation is nothing more than a gradual solution of water in air, produced and promoted by attraction, heat, and motion, just as other solutions are effected.

To account for the ascent of aqueous vapours into the atmosphere, this ingenious writer observes, that the lowest part of the air being pressed by the weight of the upper against the surface of the water, and continually rubbing upon it by its motion, attracts and dissolves those particles with which it is in contact, and separates them from the rest of the water. And since the cause of solution in this case is the stronger attraction of the particles of water towards the air, than towards each other, those that are already dissolved and taken up, will be still farther raised by the attraction of the dry air that lies over them, and thus will diffuse themselves, rising gradually higher and higher, and so leave the lowest air not so much saturated but that it will still be able to dissolve and take up fresh particles of water; which process is greatly promoted by

the motion of the wind. When the vapours are thus raised and carried by the winds into the higher and colder parts of the atmosphere, some of them will coalesce into small particles, which slightly attracting each other, and being intermixed with air, will form clouds; and these clouds will float at different heights, according to the quantity of vapour borne up, and the degree of heat in the upper parts of the atmosphere: and thus clouds are generally higher in summer than in winter. When the clouds are much increased by a continual addition of vapours, and their particles are driven close together by the force of the winds, they will run into drops heavy enough to fall down in rain. If the clouds be frozen before their particles are gathered into drops, small pieces of them being condensed and made heavier by the cold, fall down in thin flakes of snow. When the particles are formed into drops before they are frozen, they become hailstones. When the air is replete with vapours, and a cold breeze springs up, which checks the solution of them, clouds are formed in the lower parts of the atmosphere, and compose a mist or fog, which usually happens in a cold morning, and is dispersed when the sun has warmed the air, and made it capable of dissolving these watry particles. Southerly winds commonly bring rain, because, being warm and replete with aqueous vapours, they are cooled by coming into a colder climate; and therefore they part with some of them, and suffer them to precipitate in rain: whereas northerly winds, being cold, and acquiring additional heat by coming into a warmer climate, are ready to dissolve and receive more vapour than they before contained; and therefore, by long continuance, they are dry and parching, and commonly attended with fair weather.

Changes of the air, with respect to its density and rarity, as well as its heat and cold, will produce contrary effects in the solution of water, and the consequent ascent or fall of vapours. Several experiments prove that air, when rarefied, cannot keep so much water dissolved as it does in a more condensed state; and therefore when the atmosphere is saturated with water, and changes from a denser to a rarer state, the high and colder parts of it will let go some of the water before dissolved, forming new clouds, and disposing them to fall down in rain: but a change from a rarer to a denser state will stop the precipitation of the water, and enable the air to dissolve, either in whole or in part, some of those clouds that were formed before, and render their particles less apt to run into drops and fall down in rain: on this account, we generally find that the rarefied and condensed states of the atmosphere are respectively attended with rain or fair weather. See more on this subject in the Philos. Transf. vol. 55, pa. 146, or Hamilton's Philosophical Essays, p. 33.

Dr. Halley, before mentioned, has furnished some experiments on the Evaporation of water; the result of which is contained in the following articles: 1. That water salted to about the same degree as sea-water, and exposed to a heat equal to that of a summer's day, did, from a circular surface of about 8 inches diameter, evaporate at the rate of 6 ounces in 24 hours: whence by a calculus he finds that, in such circumstances, the water evaporates 1-10th of an inch deep in 12 hours:

which

which quantity, he observes, will be found abundantly sufficient to furnish all the rains, springs, dews, &c. By this experiment, every 10 square inches of surface of the water yield in vapour *per diem* a cubic inch of water: and each square foot half a wine pint; every space of 4 feet square, a gallon; a mile square, 6914 tuns; and a square degree, of 69 English miles, will evaporate 33 millions of tuns a day; and the whole Mediterranean, computed to contain 160 square degrees, at least 5280 millions of tuns each day. Philof. Tranf. number 189, or Abridg. vol. 2, pa. 108.——2. A surface of 8 square inches, evaporated purely by the natural warmth of the weather, without either wind or fun, in the course of a whole year, 16292 grains of water, or 64 cubic inches; consequently, the depth of water thus evaporated in one year, amounts to 8 inches. But this being too little to answer the experiments of the French, who found that it rained 19 inches of water in one year at Paris; or those of Mr. Townley, who found the annual quantity of rain in Lancashire above 40 inches; he concludes, that the sun and wind contribute more to Evaporation than any internal heat or agitation of the water. In effect, Dr. Halley fixes the annual Evaporation of London at 48 inches; and Dr. Dobson states the same for Liverpool at 36¾ inches. Philof. Tranf. vol. 67, p. 252.

3. The effect of the wind is very considerable, on a double account; for the same observations shew a very odd quality in the vapours of water, viz, that of adhering and hanging to the surface that exhaled them, which they clothe as it were with a fleece of vapourous air; which once investing the vapour, it afterwards rises in much less quantity. Whence, the quantity of water lost in 24 hours, when the air is very still, was very small, in proportion to what went off when there was a strong gale of wind abroad to dissipate the fleece, and make room for the emission of vapour; and this, even though the experiment was made in a place as close from the wind as could be contrived. Add, that this fleece of water, hanging to the surface of waters in still weather, is the occasion of very strange appearances, by the refraction of the vapours differing from and exceeding that of common air: whence every thing appears raised, as houses like steeples, ships as on land above the water, the land raised, and as it were lifted from the sea, &c.

4. The same experiments shew that the Evaporation in May, June, July, and August, which are nearly equal, are about three times as great as those in the months of November, December, January, and February. Philof. Tranf. numb. 212, or Abr. vol. 2, pa. 110.

Dr. Brownrigg, in his Art of making common salt, pa. 189, fixes the Evaporation of some parts of England at 73·8 inches during the months of May, June, July, and August; and the Evaporation of the whole year at more than 140 inches. But the Evaporation of the four summer months at Liverpool, on a medium of 4 years, was found to be only 18·88 inches. Also Dr. Hales calculates the greatest annual Evaporation from the surface of the earth in England at 6·66 inches; and therefore the annual Evaporation from a surface of water, is to the annual Evaporation from the surface of the earth at Liverpool, nearly as 6 to 1. Philof. Tranf. vol. 67, ubi supra.

In the Transactions of the American Philosophical Society, vol. 3, pa. 125, there is an ingenious paper on Evaporation, by Dr. Wistar. It is there shewn, that evaporation arises when the moist body is warmer than the medium it is inclosed in. And, on the contrary, it acquires moisture from the air, when the body is the colder. This carrying off, and acquiring of moisture, it is shewn, is by the passage of heat out of the body, or into it.

EUCLID, of Megara, a celebrated philosopher and logician; he was a disciple of Socrates, and flourished about 400 years before Christ. The Athenians having prohibited the Megarians from entering their city on pain of death, this philosopher disguised himself in women's clothes to attend the lectures of Socrates. After the death of Socrates, Plato and other philosophers went to Euclid at Megara, to shelter themselves from the tyrants who governed Athens.—This philosopher admitted but one chief good; which he at different times called *God*, or the *Spirit*, or *Providence*.

EUCLID, the celebrated mathematician, according to the account of Pappus and Proclus, was born at Alexandria, in Egypt, where he flourished and taught mathematics, with great applause, under the reign of Ptolomy Lagos, about 280 years before Christ. And here, from his time, till the conquest of Alexandria by the Saracens, all the eminent mathematicians were either born, or studied; and it is to Euclid, and his scholars, we are beholden for Eratosthenes, Archimedes, Apollonius, Ptolomy, Theon, &c, &c. He reduced into regularity and order all the fundamental principles of pure mathematics, which had been delivered down by Thales, Pythagoras, Eudoxus, and other mathematicians before him, and added many others of his own discovering: on which account it is said he was the first who reduced arithmetic and geometry into the form of a science. He likewise applied himself to the study of mixed mathematics, particularly to astronomy and optics.

His works, as we learn from Pappus and Proclus, are the *Elements*, *Data*, *Introduction to Harmony*, *Phenomena*, *Optics*, *Catoprics*, a *Treatise of the Division of Superficies*, *Porisms*, *Loci ad Superficiem*, *Fallacies*, and *Four books of Conics*. The most celebrated of these, is the first work, the Elements of Geometry; of which there have been numberless editions, in all languages; and a fine edition of all his works, now extant, was printed in 1703, by David Gregory, Savilian professor of astronomy at Oxford.

The Elements, as commonly published, consist of 15 books, of which the two last it is suspected are not Euclid's, but a comment of Hypsicles of Alexandria, who lived 200 years after Euclid. They are divided into three parts, viz, the Contemplation of Superficies, Numbers, and Solids: The first 4 books treat of planes only; the 5th of the proportions of magnitudes in general; the 6th of the proportion of plane figures; the 7th, 8th, and 9th give us the fundamental properties of numbers; the 10th contains the theory of commensurable and incommensurable lines and spaces; the 11th, 12th, 13th, 14th, and 15th, treat of the doctrine of solids.

There is no doubt but, before Euclid, Elements of Geometry were compiled by Hippocrates of Chius, Eudoxus,

Eudoxus, Leon, and many others, mentioned by Proclus, in the beginning of his second book; for he affirms that Euclid new ordered many things in the Elements of Eudoxus, completed many things in those of Theatetus, and besides strengthened such propositions as before were too slightly, or but superficially established, with the most firm and convincing demonstrations.

History is silent as to the time of Euclid's death, or his age. He is represented as a person of a courteous and agreeable behaviour, and in great esteem and familiarity with king Ptolomy; who once asking him, whether there was any shorter way of coming at geometry than by his Elements, Euclid, as Proclus testifies, made answer, that there was no royal way or path to geometry.

EUDIOMETER, an instrument for determining the purity of the air, or the quantity of pure and dephlogisticated or vital air contained in it, chiefly by means of its diminution on a mixture with nitrous air.

Instruments of this kind have been but lately made, and that in consequence of the experiments and discoveries of Dr. Priestley, for determining the salubrity of different kinds of air. That writer having discovered, that when nitrous air is mixed with any other air, their original bulk is diminished; and that the diminution is nearly, if not exactly, in proportion to its salubrity; he was hence led to adopt nitrous air as a true test of the purity of respirable air; and nothing more seemed to be necessary but an easy, expeditious, and accurate method of estimating the degree of diminution in different cases; and for this purpose, the Eudiometer was contrived; of which several kinds have been invented, the principal of which are the following.

I. The Eudiometer originally used by Dr. Priestley is a divided glass tube, into which, after having filled it with common water, and inverted it into the same, one measure or more of common air, and an equal quantity of nitrous air, are introduced by means of a small phial, which is called the measure; and thus the diminution of the volume of the mixture, which is seen at once by means of the graduations of the tube, instantly discovers the purity of the air required.

II. The discovery of Dr. Priestley was announced to the public in the year 1772; and several persons, both at home and abroad, presently availed themselves of it, by framing other more accurate instruments. The first of these was contrived by M. Landriani; an account of which is published in the 6th volume of Rosier's Journal, for the year 1775. It consists of a glass tube, fitted by grinding to a cylindrical vessel, to which are joined two glass cocks and a small bason; the whole being fitted to a wooden frame. In this instrument quicksilver is used instead of water; though that is attended with an inconvenience, because the nitrous air acts upon the metal, and renders the experiment ambiguous.

III. In 1777, Mr. Magellan published an account of three Eudiometers invented by himself, consisting of glass vessels of rather difficult construction, and troublesome use. Mr. Cavallo observes, that the construction of all the three is founded on a supposition, that the mixture of nitrous and atmospherical air, having continued for some time to diminish, afterwards increases

again; which it seems is a mistake: neither do they give accurate or uniform results in any two experiments made with nitrous and common air of precisely the same quality.

IV. A preferable method of discovering the purity of the air by means of an Eudiometer, is recommended by M. Fontana, of very great accuracy. The instrument is originally nothing more than a divided glass tube, though the inventor afterwards added to it a complicated apparatus, perhaps of little or no use. The first simple Eudiometer consisted only of a glass tube, uniformly cylindrical, about 18 inches long, and 3-4ths of an inch diameter within side, the outside being marked with a diamond at such distances as are exactly filled by equal measures of elastic fluids: and when any parts of these divisions are required, the edge of a ruler, divided into inches and smaller parts, is held against the tube, so as that the first division of the ruler may coincide with one of the marks on the tube. The nitrous and atmospherical air are introduced into this tube, in order to be diminished, and thence the purity of the atmospheric air ascertained.

V. M. Saussure of Geneva has also invented an Eudiometer, which he thinks is more exact than any of those before described; the apparatus of which is as follows: 1. A cylindrical glass bottle, with a ground stopple, containing about 5½ ounces, which serves as a receiver for mixing the two airs.—2. A small glass phial, to serve as a measure, and is about one-third the size of the receiver.—3. A small pair of scales that may weigh very exactly.—4. Several glass bottles, for containing the nitrous or other air to be used, and which may supply the place of the recipient when broken. The method of using it is as follows: The receiver is to be filled with water, closed exactly with its glass stopper, wiped dry on the outside, and then weighed very nicely. Being then immersed in a vessel of water, and held with the mouth downwards, the stopple is removed, and, by means of a funnel, two measures of common and one of nitrous air are introduced into it, one after another: these diminish as soon as they come into contact; in consequence of which the water enters the recipient in proportionable quantity. After being stopped and well shaken, to promote the diminution, the receiver is to be opened again under water; then stopped and shaken again, and so on for three times successively, after which the bottle is stopped for the last time under water, then taken out, wiped very clean and dry, and exactly weighed as before. It is plain that now, the bottle being filled partly with elastic fluid and partly with water, it must be lighter than when quite full of water; and the difference between those two weights, shows nearly what quantity of water would fill the space occupied by the diminished elastic fluid. Now, in making experiments with airs of different degrees of purity, the said difference will be greater when the diminution is less, or when the air is less pure, and vice versa; by which means the comparative purity between two different kinds of air is determined.

VI. But as this method, notwithstanding the encomiums bestowed on it by the inventor, is subject to several errors and inconveniences; to remedy all these, another instrument was invented by Mr. Cavallo; the description

tion of which, being long, may be seen in his Treatise on the Nature and Properties of Air, pa. 344.

Other conftructions of the Eudiometer have alfo been given by Mr. Cavendifh and Mr. Scheele. For farther information, fee Magellan's Letter to Dr. Prieftley, containing the Defcription of a Glafs Apparatus, &c, and of New Eudiometers &c, 1777, pa. 15 &c; Prieft-ley's Exp. and Obf. on Air, vol. 3, preface and appen-dix; the methods of Dr. Ingenhoufz in Philof. Tranf. vol. 66, art. 15; fee alfo the Philof. Tranf. vol. 73; and Cavallo's Treatife on Air, pa. 274, 315, 316, 317, 328, 340, 344, and 834.

EUDOXUS, of Cnidus, a city of Caria in Afia Minor, flourifhed about 370 years before Chrift. He learned geometry from Archytas, and afterwards tra-velled into Egypt to learn aftronomy and other fcien-ces. There he and Plato ftudied together, as Laertius informs us, for the fpace of 13 years; and afterwards came to Athens, fraught with all forts of knowledge, which they had imbibed from the mouths of the priefts. Here Eudoxus opened a fchool; which he fupported with fo much glory and renown, that even Plato, though his friend, is faid to have envied him. Eudoxus com-pofed Elements of Geometry, from whence Euclid li-berally borrowed, as mentioned by Proclus. Cicero calls Eudoxus the greateft aftronomer that had ever lived: and Petronius fays, he fpent the latter part of his life upon the top of a very high mountain, that he might contem-plate the ftars and the heavens with more convenience and lefs interruption: and we learn from Strabo, that there were fome remains of his obfervatory at Cnidus, to be feen even in his time. He died in the 53d year of his age.

EVECTION, is ufed by fome aftronomers for the Libration of the moon; being an inequality in her mo-tion, by which, at or near the quadratures, fhe is not in a line drawn through the centre of the earth to the fun, as fhe is at the fyzygies, or conjunction and oppo-fition, but makes an angle with that line of about 2° 51′. The motion of the moon about her axis only is equable, which rotation is performed exactly in the fame time as fhe revolves about the earth; for which reafon it is that fhe turns always the fame face towards the earth nearly, and would do fo exactly were it not that her menftrual motion about the earth, in an elliptic or-bit, is not equable; on which account the moon, feen from the earth, appears to librate a little upon her axis, fometimes from eaft to weft, and fometimes from weft to eaft; or fome parts in the eaftern limb of the moon go backwards and forwards a fmall fpace, and fome that were confpicuous, are hid, and then appear again.

The term EVECTION is ufed by fome aftronomers to denote that equation of the moon's motion, which is proportional to the fine of double the diftance of the moon from the fun, diminifhed by the moon's ano-maly: this equation is not yet accurately determined; fome ftate it at 1° 30′, others at 1° 16′, &c. It is the greateft of all the moon's equations, except the Equation of the Centre.

EVEN Number, is that which can be divided into two equal whole numbers; fuch as the feries of alter-nate numbers 2, 4, 6, 8, 10, &c.

EVENLY Even Number, is that which an even num-ber meafures by an even number; as 16, which the even number 8 meafures by the even number 2.

EVENLY Odd Number, is that which an even number meafures by an odd one; as 30, which the even num-ber 6 meafures by the odd number 5.

EVERARD's Sliding Rule, a particular fort of one invented by Mr. Thomas Everard, for the purpofe of gauging. See SLIDING RULE.

EULER (LEONARD), one of the moft extraordi-nary, and even prodigious, mathematical geniufes, that the world ever produced. He was a native of Bafil, and was born April 15, 1707. The years of his in-fancy were paffed at Richen, where his father was mi-nifter. He was afterwards fent to the univerfity of Bafil; and as his memory was prodigious, and his ap-plication regular, he performed his academical tafks with great rapidity; and all the time that he faved by this, was confecrated to the ftudy of mathematics, which foon became his favourite fcience. The early progrefs he made in this ftudy, added frefh ardour to his appli-cation; by which too he obtained a diftinguifhed place in the attention and efteem of profeffor John Bernoulli, who was then one of the chief mathematicians in Europe.

In 1723, M. Euler took his degree as mafter of arts; and delivered on that occafion a Latin difcourfe, in which he drew a comparifon between the philofophy of Newton and the Cartefian fyftem, which was received with the greateft applaufe. At his father's defire, he next applied himfelf to the ftudy of theology and the oriental languages: and though thefe ftudies were fo-reign to his predominant propenfity, his fuccefs was confiderable even in this line: however, with his father's confent, he afterward returned to mathematics as his principal object. In continuing to avail himfelf of the counfels and inftructions of M. Bernoulli, he contracted an intimate friendfhip with his two fons Nicholas and Daniel; and it was chiefly in confequence of thefe con-nections that he afterwards became the principal orna-ment of the philofophical world.

The project of erecting an academy at Peterfburg, which had been formed by Peter the Great, was execut-ed by Catharine the 1ft; and the two young Bernoul-lis being invited to Peterfburg in 1725, promifed Euler, who was defirous of following them, that they would ufe their endeavours to procure for him an advanta-geous fettlement in that city. In the mean time, by their advice, he made clofe application to the ftudy of philofophy, to which he made happy applications of his mathematical knowledge, in a differtation on the na-ture and propagation of found, and an anfwer to a prize queftion concerning the mafting of fhips; to which the Academy of Sciences adjudged the accefjit, or fecond rank, in the year 1727. From this latter difcourfe, and other circumftances, it appears that Euler had very early embarked in the curious and ufeful ftudy of naval architecture, which he afterward enriched with fo many valuable difcoveries. The ftudy of mathematics and philofophy however did not folely engage his attention, as he in the mean time attended the medical and bota-nical lectures of the profeffors at Bafil.

Euler's merit would have given him an eafy admif-fion to honourable preferment either in the magiftracy or univerfity of his native city, if both civil and acade-mical

mical honours had not been there diſtributed by lot. The lot being againſt him in a certain promotion, he left his country, ſet out for Peterſburg, and was made joint profeſſor with his countrymen Hermann and Daniel Bernoulli in the univerſity of that city.

At his firſt ſetting out in his new career, he enriched the academical collection with many memoirs, which excited a noble emulation between him and the Bernoullis; an emulation that always continued, without either degenerating into a ſelfiſh jealouſy, or producing the leaſt alteration in their friendſhip. It was at this time that he carried to new degrees of perfection the integral calculus, invented the calculation by ſines, reduced analytical operations to a greater ſimplicity, and thus was enabled to throw new light on all the parts of mathematical ſcience.

In 1730, M. Euler was promoted to the profeſſorſhip of natural philoſophy; and in 1733 he ſucceeded his friend D. Bernoulli in the mathematical chair. In 1735, a problem was propoſed by the academy, which required expedition, and for the calculation of which ſome eminent mathematicians had demanded the ſpace of ſome months. The problem was undertaken by Euler, who completed the calculation in three days, to the great aſtoniſhment of the academy: but the violent and laborious efforts it coſt him threw him into a fever, which endangered his life, and deprived him of the uſe of his right eye, which afterward brought on a total blindneſs.

The Academy of Sciences at Paris, which in 1738 had adjudged the prize to his memoir Concerning the Nature and Properties of Fire, propoſed for the year 1740 the important ſubject of the Tides of the Sea; a problem whoſe ſolution comprehended the theory of the ſolar ſyſtem, and required the moſt arduous calculations. Euler's ſolution of this queſtion was adjudged a maſter-piece of analyſis and geometry; and it was more honourable for him to ſhare the academical prize with ſuch illuſtrious competitors as Colin Maclaurin and Daniel Bernoulli, than to have carried it away from rivals of leſs magnitude. Seldom, if ever, did ſuch a brilliant competition adorn the annals of the academy; and perhaps no ſubject, propoſed by that learned body, was ever treated with ſuch force of genius and accuracy of inveſtigation, as that which here diſplayed the philoſophical powers of this extraordinary triumvirate.

In the year 1741, M. Euler was invited to Berlin to direct and aſſiſt the academy that was there riſing into fame. On this occaſion he enriched the laſt volume of the Miſcellanies (Melanges) of Berlin with five memoirs, which form an eminent, perhaps the principal, figure in that collection. Theſe were followed, with amazing rapidity, by a great number of important reſearches, which are diſperſed through the memoirs of the Pruſſian academy; a volume of which has been regularly publiſhed every year ſince its eſtabliſhment in 1744. The labours of Euler will appear more eſpecially aſtoniſhing, when it is conſidered, that while he was enriching the academy of Berlin with a profuſion of memoirs, on the deepeſt parts of mathematical ſcience, containing always ſome new points of view, often ſublime truths, and ſometimes diſcoveries of great importance; he ſtill continued his philoſophical contributions to the Peterſburg academy, whoſe memoirs diſplay the

marvellous fecundity of his genius, and which granted him a penſion in 1742.

It was with great difficulty that this extraordinary man; in 1766, obtained permiſſion from the king of Pruſſia to return to Peterſburg, where he wiſhed to paſs the remainder of his days. Soon after his return, which was graciouſly rewarded by the munificence of Catharine the 2d, he was ſeized with a violent diſorder, which ended in the total loſs of his ſight. A cataract, formed in his left eye, which had been eſſentially damaged by the loſs of the other eye, and a too cloſe application to ſtudy, deprived him entirely of the uſe of that organ. It was in this diſtreſſing ſituation that he dictated to his ſervant, a taylor's apprentice, who was abſolutely devoid of mathematical knowledge, his Elements of Algebra; which by their intrinſic merit in point of perſpicuity and method, and the unhappy circumſtances in which they were compoſed, have equally excited wonder and applauſe. This work, though purely elementary, plainly diſcovers the proofs of an inventive genius; and it is perhaps here alone that we meet with a complete theory of the analyſis of Diophantus.

About this time M. Euler was honoured by the Academy of Sciences at Paris with the place of one of the foreign members of that learned body; after which, the academical prize was adjudged to three of his memoirs, Concerning the Inequalities in the Motions of the Planets. The two prize queſtions propoſed by the ſame Academy for 1770 and 1772 were deſigned to obtain from the labours of aſtronomers a more perfect Theory of the Moon. M. Euler, aſſiſted by his eldeſt ſon, was a competitor for theſe prizes, and obtained them both. In this laſt memoir, he reſerved for farther conſideration ſeveral inequalities of the moon's motion, which he could not determine in his firſt theory, on account of the complicated calculations in which the method he then employed had engaged him. He afterward reviſed his whole theory, with the aſſiſtance of his ſon and Meſſrs Krafft and Lexell, and purſued his reſearches till he had conſtructed the new tables, which appeared, together with the great work, in 1772. Inſtead of confining himſelf, as before, to the fruitleſs integration of three differential equations of the ſecond degree, which are furniſhed by mathematical principles, he reduced them to the three ordinates, which determine the place of the moon: he divided into claſſes all the inequalities of that planet, as far as they depend either on the elongation of the ſun and moon, or upon the eccentricity, or the parallax, or the inclination of the lunar orbit. All theſe means of inveſtigation, employed with ſuch art and dexterity as could only be expected from a genius of the firſt order, were attended with the greateſt ſucceſs; and it is impoſſible to obſerve without admiration, ſuch immenſe calculations on the one hand, and on the other the ingenious methods employed by this great man to abridge them, and to facilitate their application to the real motion of the moon. But this admiration will become aſtoniſhment, when we conſider at what period and in what circumſtances all this was effectuated. It was when our author was totally blind, and conſequently obliged to arrange all his computations by the ſole powers of his memory and his genius: it was when he was embarraſſed in his domeſtic affairs by a dreadful fire,

that

that had confumed great part of his fubftance, and forced him to quit a ruined houfe, every corner of which was known to him by habit, which in fome meafure fupplied the want of fight. It was in thefe circumftances that Euler compofed a work which alone was fufficient to render his name immortal.

Some time after this, the famous oculift Wentzell, by couching the cataract, reftored fight to our author; but the joy produced by this operation was of fhort duration. Some inftances of negligence on the part of his furgeons, and his own impatience to ufe an organ, whofe cure was not completely finifhed, deprived him a fecond time and for ever of his fight: a relapfe which was alfo accompanied with tormenting pain. With the affiftance of his fons, however, and of Meffrs Krafft and Lexell, he continued his labours: neither the infirmities of old age, nor the lofs of his fight, could quell the ardour of his genius. He had engaged to furnifh the academy of Peterfburg with as many memoirs as would be fufficient to complete its acts for 20 years after his death. In the fpace of 7 years he tranfmitted to the Academy above 70 memoirs, and above 200 more, left behind him, were revifed and completed by a friend. Such of thefe memoirs as were of ancient date were feparated from the reft, and form a collection that was publifhed in the year 1783, under the title of *Analytical Works*.

The general knowledge of our author was more extenfive than could well be expected in one who had purfued, with fuch unremitting ardour, mathematics and aftronomy as his favourite ftudies. He had made a very confiderable progrefs in medical, botanical, and chemical fcience. What was ftill more extraordinary, he was an excellent fcholar, and poffeffed in a high degree what is generally called *erudition*. He had attentively read the moft eminent writers of ancient Rome; the civil and literary hiftory of all ages and all nations was familiar to him; and foreigners, who were only acquainted with his works, were aftonifhed to find in the converfation of a man, whofe long life feemed folely occupied in mathematical and phyfical refearches and difcoveries, fuch an extenfive acquaintance with the moft interefting branches of literature. In this refpect, no doubt, he was much indebted to a very uncommon memory, which feemed to retain every idea that was conveyed to it, either from reading or from meditation. He could repeat the Æneid of Virgil, from the beginning to the end, without hefitation, and indicate the firft and laft line of every page of the edition he ufed.

Several attacks of a vertigo, in the beginning of September 1783, which did not prevent his computing the motions of the aeroftatic globes, were however the forerunners of his mild paffage out of this life. While he was amufing himfelf at tea with one of his grandchildren, he was ftruck with an apoplexy, which terminated his illuftrious career at 76 years of age.

M. Euler's conftitution was uncommonly ftrong and vigorous. His health was good; and the evening of his long life was calm and ferene, fweetened by the fame that follows genius, the public efteem and refpect that are never withheld from exemplary virtue, and feveral domeftic comforts which he was capable of feeling, and therefore deferved to enjoy.

The catalogue of his works has been printed in 50 pages, 14 of which contain the manufcript works.—The printed ones confift of works publifhed feparately, and works to be found in the memoirs of feveral Academies, viz, in 38 volumes of the Peterfburg Acts, (from 6 to 10 papers in each volume);—in feveral volumes of the Paris Acts;—in 26 volumes of the Berlin Acts, (about 5 papers to each volume);—in the Acta Eruditorum, in 2 volumes;—in the Mifcellanea Taurinenfia;—in vol. 9 of the Society of Ulyffingue;—in the Ephemerides of Berlin;—and in the Memoires de la Société Oeconomique for 1766.

EVOLVENT, in the Higher Geometry, a term ufed by fome writers for the Involute, or curve refulting from the evolution of a curve, in contradiftinction to that evolute, or curve fuppofed to be opened or evolved. See EVOLUTE, and INVOLUTE.

EVOLUTE, in the Higher Geometry, a curve firft propofed by M. Huygens, and fince much ftudied by the later mathematicians. It is any curve fuppofed to be evolved or opened, by having a thread wrapped clofe upon it, faftened at one end, and beginning to evolve or unwind the thread from the other end, keeping the part evolved, or wound off, tight ftretched; then this end of the thread will defcribe another curve called the Involute. Or the fame involute is defcribed the contrary way, by wrapping the thread upon the Evolute, keeping it always ftretched.

Thus, if EFGH be any curve, and AE either a part of the curve, or a right line; then if a thread be wound clofe upon the curve from A to H, where it is fixed, and then be unwound from A; the curve AEFGH, from which it is evolved, is called the Evolute; and the other curve ABCD defcribed by the end of the thread, as it evolves or unwinds, is the Involute. Or, if the thread HD, fixed at H, be wound or wrapped upon the Evolute HGFEA, keeping it always tight, as at the feveral pofitions of it HD, GC, FB, EA, the extremity will defcribe the Involute curve DCBA.

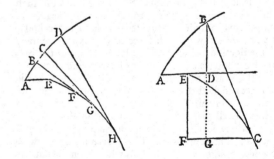

From this defcription it appears, 1. That the parts of the thread at any pofitions, as EA, FB, GC, HD, &c, are radii of curvature, or ofculatory radii, of the involute curve, at the points A, B, C, D.

2. The fame parts of the thread are alfo equal to the correfponding lengths AE, AEF, AEFG, &c, of the Evolute; that is,

AE = AE is the rad. of curvature to the point A,
BF = AF - - - - - B,
CG = AG - - - - - C,
DH = AH - - - - - D.

3. Any

3. Any radius of curvature BF, is perpendicular to the involute at the point B, and is a tangent to the Evolute curve at the point F.

4. The Evolute is the locus of the centre of curvature of the involute curve.

The finding the radii of Evolutes, is a matter of great importance in the higher speculations of geometry; and is even sometimes useful in practice; as is shewn by Huygens, the inventor of this theory, in applying it to the pendulum. Horol. Oscil. part 3. The doctrine of the Oscula of Evolutes is owing to M. Leibnitz, who first shewed the use of Evolutes in the measuring of curvatures.

To find the Evolute *and Involute Curves, the one from the other.*

For this purpose, put

$x = $ AD the abscifs of the involute,
$y = $ DB its ordinate,
$z = $ AB the involute curve,
$r = $ BC its radius of curvature,
$v = $ EF the abscifs of the Evolute,
$u = $ FC its ordinate, and
$a = $ AE a given line, (fig. 2 above).

Then, by the nature of the radius of curvature, it is

$$r = \frac{z^3}{\dot{y}\ddot{x} - \dot{x}\ddot{y}} = \text{BC} = \text{AE} + \text{EC};$$ also by sim. triangles,

$$\dot{z} : \dot{x} :: r : \text{GB} = \frac{r\dot{x}}{z} = \frac{\dot{x}z^2}{\dot{y}\ddot{x} - \dot{x}\ddot{y}},$$

$$\dot{z} : \dot{y} :: r : \text{CC} = \frac{r\dot{y}}{z} = \frac{\dot{y}z^2}{\dot{y}\ddot{x} - \dot{x}\ddot{y}}.$$

Hence EF $= $ GB $-$ DB $= \dfrac{\dot{x}z^2}{\dot{y}\ddot{x} - \dot{x}\ddot{y}} - y = v$; and

FC $= $ AD $-$ AE $+$ GC $= x - a + \dfrac{\dot{y}z^2}{\dot{y}\ddot{x} - \dot{x}\ddot{y}} = u$;

which are the values of the abscifs and ordinate of the Evolute curve EC; and therefore these may be found when the involute is given.

On the other hand, if v and u, or the Evolute be given: then, putting the given curve EC $= s$; since CB $= $ AE $+$ EC, or $r = a + s$, this gives r the radius of curvature. Also, by similar triangles, there result these proportions, viz,

$$\dot{s} : \dot{v} :: r : \frac{rv}{s} = \frac{a+s}{s}\dot{v} = \text{GB},$$

$$\dot{s} : \dot{u} :: r : \frac{ru}{s} = \frac{a+s}{s}\dot{u} = \text{GC}; \text{ theref.}$$

AD $= $ AE $+$ FC $-$ CC $= a + u - \dfrac{a+s}{s}\dot{u} = x$,

and DB $= $ GB $-$ EF $= \dfrac{a+s}{s}\dot{v} - v = y$;

which are the abscifs and ordinate of the involute curve, and which may therefore be found when the Evolute is given. Where it may be noted that $\dot{s}^2 = \dot{v}^2 + \dot{u}^2$, and $\dot{z}^2 = \dot{x}^2 + \dot{y}^2$. Also either of the quantities x, y, may be supposed to flow equably, in which case the respective second fluxion $\ddot{x}$ or $\ddot{y}$ will be nothing, and the corresponding term in the denominator $\dot{y}\ddot{x} - \dot{x}\ddot{y}$ will vanish, leaving only the other term in it; which

will have the effect of rendering the whole operation simpler.

For Ex. Suppose it were required to find the Evolution EC when the given involute AB is the common parabola, whose equation is $px = y^2$, the parameter being p.

Here $y = \sqrt{px}$, $\dot{y} = \frac{1}{2}p^{\frac{1}{2}}x^{-\frac{1}{2}}\dot{x}$, and $\ddot{y} = -\frac{1}{4}p^{\frac{1}{2}}x^{-\frac{3}{2}}\dot{x}$, making $\ddot{x} = 0$. Then, to find first AE the radius of curvature of the parabola AB at the vertex, when $\ddot{x} = 0$, the general value of the radius of curvature above given becomes $r = \dfrac{\dot{z}^3}{-\dot{x}\ddot{y}} = \dfrac{\overline{\dot{x}^2 + \dot{y}^2}^{\frac{3}{2}}}{-\dot{x}\ddot{y}} = $ (by substituting the value of y and $\dot{y}$ &c) $\dfrac{\overline{p + 4x}^{\frac{3}{2}}}{2\sqrt{p}}$ which is the general value of r or BC, the radius of curvature, for any value of x or AD; and when x or AD is $= 0$ or nothing, the value of r, or AE, becomes then

$$a = \frac{p^{\frac{3}{2}}}{2\sqrt{p}} = \tfrac{1}{2}p$$ only; that is half the parameter of the axis is the radius of curvature at the vertex of the parabola.

Again, in the general values of v and u above given by substituting the values of $\dot{y}$, $\ddot{y}$, and z, also 0 for $\ddot{x}$, and $\frac{1}{2}p$ for a: those quantities become.

$$v = \frac{\dot{z}^2}{-\dot{y}} - y = \frac{\dot{x}^2 + \dot{y}^2}{-\dot{y}} - y = \frac{4x^{\frac{3}{2}}}{p^{\frac{1}{2}}} = 4x\sqrt{\frac{x}{p}}; \text{ and}$$

$$u = x - a + \frac{\dot{y}z^2}{-\dot{x}\ddot{y}} = 3x + \tfrac{1}{2}p - a = 3x.$$

Hence then, comparing the values of v and u, there is found $3p^{\frac{1}{2}}v = 4x^{\frac{1}{2}}u$, and $27pv^2 = 16u^3$; which is the equation between the abscifs and ordinate of the Evolute curve EC, shewing it to be the semicubical parabola.

In like manner the Evolute to any other curve is found.—The Evolute to the common cycloid, is an equal cycloid; a property first demonstrated by Huygens. and which he used as a contrivance to make a pendulum vibrate in the curve of a cycloid. See his Horolog. Oscil. See also, on the subject of Evolute and Involute Curves, the Fluxions of Newton, Maclaurin, Simpson, De l'Hôpital, &c, Wolf. Elem. Math. tom. 1, &c, &c.

M. Varignon has applied the doctrine of the radius of the Evolute, to that of central forces; so that having the radius of the Evolute of any curve, there may be found the value of the central force of a body; which, moving in that curve, is found in the same point where that radius terminates; or reciprocally, having the central force given, the radius of the Evolute may be determined. Hist. de l'Acad. an. 1706.

The variation of curvature of the line described by the Evolution of a curve, is measured by the ratio of the radius of curvature of the Evolute, to the radius of curvature of the line described by the Evolution. See Maclaurin's Flux. art. 402, prop. 36.

Imperfect Evolute, a name given by M. Reaumur to a new kind of Evolute. The mathematicians had hitherto only considered the perpendiculars let fall from the Involute on the convex side of the Evolute: but if

other

other lines not perpendicular be drawn upon the same points, provided they be all drawn under the same angle, the effect will still be the same; that is, the oblique lines will all intersect in the curve, and by their intersections form the infinitely small sides of a new curve, to which they would be so many tangents.— Such a curve is a kind of Evolute, and has its radii; but it is an Imperfect one, since the radii are not perpendicular to the first curve, or Involute. Hist. de l'Acad. &c, an. 1709.

EVOLUTION, in Arithmetic and Algebra, denotes the Extraction of the roots out of powers. In which sense it stands opposed to Involution, which is the raising of powers. The note or character that has been used by some Algebraists, to denote Evolution, is _vv_; as the sign of involution is ℺: characters I think first used by Dr. Pell.

EVOLUTION, in Geometry, the opening, or unfolding of a curve, and making it describe an Evolvent.

The equable Evolution of the periphery of a circle, or other curve, is such a gradual approach of the circumference to rectitude, as that its parts do all concur, and equally evolve or unbend; so that the same line becomes successively a less arc of a reciprocally greater circle; till at last they change into a straight line.— In the Philos. Transf. N° 260, a new quadratix to the circle is found by this means, being the curve described by the equable Evolution of its periphery.

EURYTHMY, in Architecture, Painting, and Sculpture, is a kind of majesty, elegance, and easiness appearing in the composition of certain members or parts of a body, building, or painting, and resulting from the fine and exact proportions of them.

EUSTYLE, is the best manner of placing columns, with regard to their distance; which, according to Vitruvius, should be four modules, or two diameters and a quarter.

EXAGON. See HEXAGON.

EXALTATION, in Astrology, is a dignity which a planet acquires in certain signs of the zodiac; which dignity, it is supposed, gives the planet an extraordinary virtue, efficacy, and influence. The opposite side of the zodiac is called the Dejection of the planet.—— Thus, the 15th degree of Cancer is the Exaltation of Jupiter, according to Albumazar, because it was the ascendant of that planet at the time of the creation; that of the sun is in the 19th degree of Aries; and its dejection in Libra; that of the moon is in Taurus, &c. Ptolomy gives the reason of this in his first book De Quadripartita.

EXCENTRIC, is applied to such figures, circles, spheres, &c, as have not the same centre; as opposed to Concentric, which have the same centre.

EXCENTRIC, or _Excentric Circle_, in the ancient Ptolomaic astronomy, was the very orbit of the planet itself, which it was supposed to describe about the earth, and which was conceived Excentric with it; called also the Deferent.

Instead of these Excentric Circles round the earth, the moderns make the planets describe elliptic orbits about the sun; which accounts for all the irregularities of their motions, and their various distances from the earth, &c, more justly and naturally.

EXCENTRIC, or _Excentric Circle_, in the New Astronomy, is the circle described from the centre of the orbit of a planet, with half the greatest axis as a radius; or it is the circle that circumscribes the elliptic orbit of the planet; as the circle AQB.

EXCENTRIC _Anomaly_, or _Anomaly of the Centre_, is an arc AQ of the Excentric circle, intercepted between the aphelion A, and the right line QH, drawn through the centre P of the planet perpendicular to the line of the apses AB.

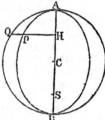

EXCENTRIC _Equation_, in the Old Astronomy, is an angle made by a line drawn from the centre of the earth, with another line drawn from the centre of the Excentric, to the body or place of any planet. This is the same with the prosthapheresis; and is equal to the difference, accounted in an arch of the ecliptic, between the real and apparent place of the sun or planet. See EQUATION _of the Centre_.

EXCENTRIC _Place of a planet_, in its orbit, is the Heliocentric place, or that in which it appears as seen from the sun.

EXCENTRIC _Place in the ecliptic_, is the point of the ecliptic to which the planet is referred as viewed from the sun; and which coincides with the heliocentric longitude.

EXCENTRICITY, is the distance between the centres of two circles, or spheres, which have not the same centre.

EXCENTRICITY, _in the Old Astronomy_, is the distance between the centre of a planet and the centre of the earth.——That the planets have such an Excentricity, is allowed on all sides, and may be evinced from various circumstances; and especially this, that the planets at some times appear larger, and at others less: which can only proceed from hence, that their orbits being Excentric to the earth, in some parts of those orbits the planets are nearer to us, and in others more remote. And as to the Excentricities of the sun and moon, it is thought they are sufficiently proved, both from eclipses, from the moon's greater and less parallax at the same distance from the zenith, and from the sun's continuing longer by 8 days in the northern hemisphere than in the southern one.

EXCENTRICITY, _in the New Astronomy_, is the distance CS between the sun S and the centre C of a planet's orbit; or the distance of the centre from the focus of the elliptic orbit; called also the _Simple_ or _Single Excentricity_.

When the greatest Equation of the centre is given, the Excentricity of the earth's orbit may be found by the following proportion; viz,

As the diameter of a circle in degrees,
Is to the diameter in equal parts;
So the greatest equat. of the centre in degrees,
To the Excentricity in equal parts. Thus,
Greatest equat. of the cent. 1° 55′ 33″ = 1°·9258333 &c. The diam. of a circ. being 1, its circumf. is 3·1415926. Then 3·1415926 : 1 :: 360° : 114°·5915609 diam. in deg. And 114·5915609 : 1 :: 1·9258333 : 0·016806, the Ex

Hence,

Hence, by adding this to 1, and subtracting it from 1, gives 1·016806 = AS the aphelion distance, and 0·983194 = BS the perihelion distance. See Robertson's Elem. of Navig. book 5, pa. 286.

Otherwise, thus: Since it is found that the sun's greatest apparent semi-diameter is to his least, as 32′ 43″ to 31′ 38″, or as 1963″ to 1898″; the sun's greatest distance from the earth will be to his least, or AS to SB, as 1963 to 1898; of which,

the half dif. is $32\frac{1}{2}$ = CS,

and half sum $1930\frac{1}{2}$ = CB; wherefore,

as $1930\frac{1}{2} : 32\frac{1}{2} :: 1 : \cdot016835$ = CS the Excentricity, to the mean distance or semi-axis 1; which is nearly the same as before.

The Excentricities of the orbits of the several planets, in parts of their own mean distances 1000, and also in English miles, are as below, viz, the Excentricity of the orbit of

	Parts.	Miles.
Mercury - -	210 - -	7,730,000
Venus - -	7 - -	482,000
Earth - -	17 - -	1,618,000
Mars - -	93 - -	13,486,000
Jupiter - -	48 - -	23,760,000
Saturn - -	55 - -	49,940,000
Georgian - -	$47\frac{1}{2}$ - -	86,000,000

Double EXCENTRICITY, is the distance between the two foci of the elliptic orbit, and is equal to double the Single Excentricity above given.

EXCHANGE, in Arithmetic, is the bartering or exchanging the money of one place for that of another; or the finding what quantity of the money of one place is equal to a given sum of another, according to a given course of exchange.

The several operations in this case are only different applications of the Rule of Three. See most books of Arithmetic.

Arbitration of EXCHANGE, is the method of remitting to, and drawing upon, foreign places, in such a manner as shall turn out the most profitable.

Arbitration is either Simple or Compound.

Simple Arbitration respects three places only. Here, by comparing the par of arbitration between a first and second place, and between the 1st and a 3d, the rate between the 2d and 3d is discovered; from whence a person can judge how to remit or draw to the most advantage, and to determine what that advantage is.

Compound Arbitration respects the cases in which the exchanges among three, four, or more places are concerned. A person who knows at what rate he can draw or remit directly, and also has advice of the course of exchange in foreign parts, may trace out a path for circulating his money, through more or fewer of such places, and also in such order, as to make a benefit of his skill and credit: and in this lies the great art of such negociations. See my Arithmetic, pa. 105, &c.

EXCURSION, in Astronomy. See ELONGATION.

Circles of EXCURSION. See CIRCLES.

EXEGESIS, or EXEGETICA, in Algebra, is the finding, either in numbers or lines, the roots of the equation of a problem, according as the problem is either numeral or geometrical.

EXHALATION, a fume or steam Exhaling, or issuing, from a body, and diffusing itself in the atmosphere.

The terms Exhalation and Vapour are often used indifferently; but the more accurate writers distinguish them, appropriating the term Vapour to the moist fumes raised from water and other liquid bodies; and the term Exhalation to the dry ones emitted from solid bodies; as earth, fire, minerals, &c. In this sense, Exhalations are dry and subtle corpuscles, or effluvia, loosened from hard terrestrial bodies, either by the heat of the sun, or the action of the air, or some other cause: being emitted upwards to a certain height in the atmosphere, where, mixing with the vapours, they help to constitute clouds, and return back in dews, mists, rains, &c.

Sir Isaac Newton thinks, that true and permanent air is formed from the Exhalations raised from the hardest and most compact bodies.

EXHAUSTED *Receiver*, is a glass, or other vessel, applied on the plate of an air-pump, to have the air extracted out of it by the working of the pump.—Things placed in such an Exhausted Receiver, are said to be *in vacuo*.

EXHAUSTIONS, or the *Method of* EXHAUSTIONS, a method of demonstration founded upon a kind of Exhausting a quantity by continually taking away certain parts of it.

The method of Exhaustions was of frequent use among the ancient mathematicians; as Euclid, Archimedes, &c. It is founded on what Euclid says in the 10th book of his Elements; viz, that those quantities are equal, whose difference is less than any assignable quantity. Or thus, two quantities A and B are equal, when, if to or from one of them as A, any other quantity as *d* be subtracted, however small it be, then the sum or difference is respectively greater or less than the other quantity B: viz, *d* being an indefinitely small quantity,

if A + *d* be greater than B,

and A – *d* less than B,

then is A equal to B.

This principle is used in the 1st prop. of the 10th book, which imports, that if from the greater of two quantities be taken more than its half, and from the remainder more than its half, and so on; there will at length remain a quantity less than either of those proposed. On this foundation it is demonstrated, that if a regular polygon of infinite sides be inscribed in a circle, or circumscribed about it; then the space, which is the difference between the circle and the polygon, will by degrees be quite exhausted, and the circle become ultimately equal to the polygon. And in this way it is that Archimedes demonstrates, that a circle is equal to a right-angled triangle, whose two sides about the right angle, are equal, the one to the semidiameter, and the other to the perimeter of the circle. Prop. 1 De Dimensione Circuli.

Upon the Method of Exhaustions depends the Method of Indivisibles introduced by Cavalerius, which is but a shorter way of expressing the method of Exhaustions; as also Wallis's Arithmetic of Infinites, which is a farther improvement of the Method of Indivisibles; and hence also the Methods of Increments, Differentials, Fluxions,

and

and Infinite Series. See some account of the Method of Exhauſtions in Wallis's Algebra, chap. 73, and in Ronayne's Algebra, part 3, pa. 395.

EXPANSION, is the dilating, ſtretching, or ſpreading out of a body; whether from any external cauſe, as the cauſe of rarefaction, or from an internal cauſe, as elaſticity. Bodies naturally expand by heat beyond their dimenſions when cold; and hence it happens that their dimenſions and ſpecific gravities are different in different temperatures and ſeaſons of the year. Air compreſſed or condenſed, as ſoon as the compreſſing or condenſing force is removed, expands itſelf by its elaſtic power to its former dimenſions.

In ſome few caſes indeed bodies ſeem to expand as they grow cold, as water in the act of freezing: but it ſeems this is owing to the extrication of a number of air bubbles from the fluid at a certain time; and is not at all a regular and gradual expanſion like that of metals, &c, by means of heat. Mr. Boyle, in his Hiſtory of Cold, ſays that ice takes up one 12th part more ſpace than water; but by Major Williams's experiments on the force of freezing water, I have found it occupies but about the 17th or 18th part more ſpace. Tranſac. of the R. Soc. of Edinb. vol. 2, pa. 28. In certain metals alſo, an Expanſion takes place when they paſs from a fluid to a ſolid ſtate: but this too is not to be accounted any proper effect of cold, but of the arrangement of the parts of the metal in a certain manner; and is therefore to be accounted a kind of cryſtallization, rather than any thing elſe.

The Expanſion of different bodies by heat is very various; and many experiments upon it are to be met with in the volumes of the Philoſ. Tranſ. and elſewhere. In the 48th vol. in particular, Mr. Smeaton has given a table of the Expanſion of many different ſubſtances, as determined by experiment, from which the following particulars are extracted. Where it is to be noted, that the quantities of Expanſion which anſwer to 180 degrees of Fahrenheit's thermometer, are expreſſed in ten-thouſandth parts of an Engliſh inch, each ſubſtance being 1 foot or 12 inches in length.

White glaſs barometer tube - - -	100
Martial regulus of antimony - - -	130
Bliſtered ſteel - - - -	138
Hard ſteel - - - -	147
Iron - - - - -	151
Biſmuth - - - -	167
Copper hammered - - - -	204
Copper 8 parts, mixed with 1 of tin - -	218
Caſt braſs - - - - -	225
Braſs 16 parts, with tin 1 - - -	229
Braſs wire - - - -	232
Speculum metal - - - -	232
Spelter ſolder, viz. braſs 2 parts, zink 1 -	247
Fine pewter - - - -	274
Grain tin - - - -	298
Soft ſolder, viz. lead 2, tin 1 - -	301
Zink 8 parts, tin 1, a little hammered -	323
Lead - - - - -	344
Zink or Spelter - - - -	353
Zink hammer'd ½ an inch per foot - -	373

By other experiments too it has been found that, for each degree of heat of the thermometer, mercury, wa-

ter, and air, expand by the following parts of their own bulk, viz,

Mercury the 9600th ⎫
Water the 6666th ⎬ part of its bulk.
Air the 435th ⎭

From the foregoing table it appears, that there is no general rule for the degree of Expanſion to which bodies are ſubject by the ſame degree of heat, either from their ſpecific gravity or otherwiſe. Zink, which is much lighter than lead, expands more with heat; while glaſs, which is lighter than either, expands much leſs; and copper, which is heavier than a mixture of braſs and tin, expands leſs.

It ſeems too that metals obſerve a proportion of Expanſion in a fluid ſtate, quite different from what they do in a ſolid one: For regulus of antimony ſeemed to ſhrink in fixing, after being melted, conſiderably more than zink.

But of all known ſubſtances, thoſe of the aërial kind expand moſt by an equal degree of heat; and in general the greater quantity of latent heat that any ſubſtance contains, the more eaſily is it expanded; though even here no general rule can be formed. It is indeed certain that the denſeſt fluids, ſuch as mercury, oil of vitriol, &c, are leſs expanſible than water, ſpirit of wine, or ether. Which laſt is ſo eaſily expanded, that were it not for the preſſure of the atmoſphere it would be in a continual ſtate of vapour. And indeed this is the caſe, in ſome meaſure, with perhaps all fluids; as it has been found, by experiments with the beſt air-pumps, that water, and other fluids, aſcend in vapours the more as the exhauſtion is the more perfect; from which it would ſeem that water would wholly riſe in vapour, in any temperature, if the preſſure of the atmoſphere was entirely taken off.

After bodies are reduced to a vaporous ſtate, their Expanſion ſeems to go on without any limitation, in proportion to the degree of heat applied; though it may be impoſſible to ſay what would be the ultimate effects of that principle upon them in this way. The force with which theſe vapours expand on the application of high degrees, is very great; nor does it appear that any obſtacle whatever is inſuperable by them.

On this principle depend the ſteam engines, ſo much uſed in various mechanical operations; likewiſe ſome hydraulic machines; and the inſtruments called manometers, which ſhew the variation of gravity in the external atmoſphere, by the expanſion or condenſation of a ſmall quantity of air confined in a proper veſſel. On this principle alſo, perpetual movements might be conſtructed ſimilar to thoſe invented by Mr. Coxe, on the principle of the barometer. And a variety of other curious machines may be conſtructed on the principle of aërial expanſion; an account of ſome of which is given under HYDROSTATICS and PNEUMATICS.

On the principle of the Expanſion of fluids are conſtructed Thermometers. And for the effects of the different Expanſions of metals in correcting the errors of machines for meaſuring time, ſee the article PENDULUM.

The Expanſion of ſolid bodies is meaſured by an inſtrument called the Pyrometer; and the force with which they expand is ſtill greater than that of aërial vapours; the flame of a farthing candle produces an

Expanſion

Expanſion in a bar of iron capable of counteracting a weight of 500 pounds. The quantity of expanſion however is ſo ſmall, that it has never been applied to the movement of any mechanical engine.

EXPECTATION, in the Doctrine of Chances, is applied to any contingent event, upon the happening of which ſome benefit &c is expected. This is capable of being reduced to the rules of computation : for a ſum of money in Expectation when a particular event happens, has a determinate value before that event happens. Thus, if a perſon is to receive any ſum, as 10l. when an event takes place which has an equal chance or probability of happening and failing, the value of the Expectation is half that ſum or 5l. : but if there are 3 chances for failing, and only 1 for its happening, or one chance only in its favour out of all the 4 chances ; then the probability of its happening is only 1 out of 4, or ¼, and the value of the Expectation is but ¼ of 10l. which is only 2l. 10s. or half the former ſum. And in all caſes, the value of the Expectation of any ſum is found by multiplying that ſum by the fraction expreſſing the probability of obtaining it. So the value of the Expectation on 100l. when there are 3 chances out of 5 for obtaining it, or when the probability of obtaining it is ⅗, is ⅗ of 100l. which is 60l. And if s be any ſum expected on the happening of an event, h the chances for that event happening, and f the chances for its failing ; then, there being h chances out of f + h for its happening, the probability will be $\frac{h}{f+h}$, and the value of the expectation is $\frac{h}{f+h} \times s$. See Simpſon's or De Moivre's Doctrine of Chances.

EXPECTATION of Life, in the Doctrine of Life Annuities, is the ſhare, or number of years of life, which a perſon of a given age may, upon an equality of chance, expect to enjoy.

By the Expectation or ſhare of life, ſays Mr. Simpſon (Select Exerciſes pa. 273), is not here to be underſtood that particular period which a perſon hath an equal chance of ſurviving ; this laſt being a different, and more ſimple conſideration. The Expectation of a life, to put it in the moſt familiar light, may be taken as the number of years at which the purchaſe of an annuity, granted upon it, without diſcount of money, ought to be valued. Which number of years will differ more or leſs from the period above-mentioned, according to the different degrees of mortality to which the ſeveral ſtages of life are incident. Thus it is much more than an equal chance, according to the table of the probability of the duration of life (p. 254 ut ſupra), that an infant, juſt come into the world, arrives not to the age of 10 years ; yet the Expectation or ſhare of life due to it, upon an average, is near 20 years. The reaſon of which wide difference, is the great exceſs of the probability of mortality in the firſt tender years of life, above that reſpecting the more mature and ſtronger ages. Indeed if the numbers that die at every age were to be the ſame, the two quantities above ſpecified would alſo be equal ; but when the ſaid numbers become continually leſs and leſs, the Expectation muſt of conſequence be the greater of the two.

Mr. Simpſon has given a table and rule for finding this Expectation, pa. 255 and 273 as above. Thus,

A Table of the EXPECTATIONS of Life in London.					
Age	Expectation	Age	Expectation	Age	Expectation
1	27·0	28	24·6	55	14·2
2	32·0	29	24·1	56	13·8
3	34·0	30	23·6	57	13·4
4	35·6	31	23·1	58	13·1
5	36·0	32	22·7	59	12·7
6	36·0	33	22·3	60	12·4
7	35·8	34	21·9	61	12·0
8	35·6	35	21·5	62	11·6
9	35·2	36	21·1	63	11·2
10	34·8	37	20·7	64	10·8
11	34·3	38	20·3	65	10·5
12	33·7	39	19·9	66	10·1
13	33·1	40	19·6	67	9·8
14	32·5	41	19·2	68	9·4
15	31·9	42	18·8	69	9·1
16	31·3	43	18·5	70	8·8
17	30·7	44	18·1	71	8·4
18	30·1	45	17·8	72	8·1
19	29·5	46	17·4	73	7·8
20	28·9	47	17·0	74	7·5
21	28·3	48	16·7	75	7·2
22	27·7	49	16·3	76	6·8
23	27·2	50	16·0	77	6·4
24	26·6	51	15·6	78	6·0
25	26·1	52	15·2	79	5·5
26	25·6	53	14·9	80	5·0
27	25·1	54	14·5		

For Example, if it be required to find the Expectation or ſhare of life, due to a perſon of 30 years old. Oppoſite the given age in the firſt column of the table, ſtands 23·6 in the ſecond col. for the years in the Expectation ſought.

See De Moivre's Doctrine of Chances applied to the Valuation of Annuities, p. 288 ; or Dr. Price's Obſervations on Reverſionary Payments, p. 168, 364, 374, &c ; or Philoſ. Tranſ. vol. 59, p. 89.

EXPERIMENT, in Philoſophy, a trial of the effect or reſult of certain applications and motions of natural bodies, in order to diſcover ſomething of their laws and relations, &c.

The making of experiments is grown into a kind of art ; and there are now many collections of them, moſtly under the denomination of Courſes of Experimental Philoſophy. Sturmius made a curious collection of the principal Diſcoveries and Experiments of the laſt age, under the title of Collegium Experimentale. Other Courſes of Experiments have been publiſhed by Graveſande, Deſaguliers, Helſham, Cotes, and others.

EXPERIMENTAL Philoſophy, is that which proceeds on Experiments, or which deduces the laws of nature and the properties and powers of bodies, and their actions upon each other, from ſenſible Experiments and obſervations.

Experiments are of the utmoſt importance in philoſophy ; and the great advantages the modern phyſics have

have over the ancient, is chiefly owing to this, viz, that we abound much more in Experiments, and that we make more use of the Experiments we have. The method of the ancients, was chiefly to begin with the causes of things, and thence argue to the phenomena and effects; on the contrary, that of the moderns proceeds from Experiments and Observations, from whence the properties and laws of natural things are deduced, and general theories are formed.

Several of the ancients indeed thought as highly of Experiments as the moderns, and practised them also. Plato omits no occasion of speaking of the advantages of them; and Aristotle's history of animals bears ample testimony for *him*. Democritus's great employment was to make experiments; and even Epicurus himself owes part of his glory to the same cause.

Among the moderns, the making of Experiments was chiefly begun by Friar Bacon, in the 13th century, who it seems spent a great deal of money and labour in this way. After him, the lord chancellor Bacon is looked upon as the founder of the present mode of philosophising by Experiments. And his method has been prosecuted with laudable emulation by the Academy del Cimento, the Royal Society, the Royal Academy at Paris; by Mr. Boyle, and, over all, by Sir Isaac Newton, with many other illustrious names.

Indeed, Experiments, within the last century, are come so much into vogue, that nothing will pass in philosophy, but what is either founded on Experiments, or confirmed by them; so that the new philosophy is almost wholly Experimental.

Yet there are some, even among the learned, who speak of Experiments in a different manner, or perhaps rather of the abuse of them, and in derision of the pretenders to this practice. Thus, though Dr. Keil allows that philosophy has received very considerable advantages from the makers of Experiments; yet he complains of their disingenuity, in too often wresting and distorting their Experiments and Observations to favour some darling theories they had espoused. Nay more, M. Hartsoeker, in his Recueil de plusieurs Pieces de Physique, undertakes to shew, that such as employ themselves in the making of Experiments, are not properly philosophers, but as it were the labourers or operators of philosophers, that work under them, and for them, furnishing them with materials to build their systems and hypotheses upon. And the learned M. Dacier, in the beginning of his discourse on Plato, at the head of his translation of the works of that philosopher, deals still more severely with the makers of Experiments. He breaks out with a kind of indignation at a tribe of idly curious people, whose sole employment consists in making Experiments on the gravity of the air, the equilibrium of fluids, the loadstone, &c, and yet arrogate to themselves the noble title of philosophers. But his honest indignation would have exceeded all bounds, had he lived to see the contemptible fall of one of the principal societies above-mentioned; while its members first amuse themselves with magnetical conundrums, spinning electrical wheels, torturing the unseen and unknown phlogistic particles; and finally polluting the source of science, and the streams of wisdom, with the folly of hunting after cockle-shells, caterpillars, and butterflies!

I

EXPLOSION, a sudden and violent expansion of an elastic fluid, by which it instantly throws off any obstacle that happens to be in the way, sometimes with astonishing force and rapidity, as the Explosion of fired gun-powder, &c.

Explosion differs from expansion, in that the latter is a gradual and continued power, acting uniformly for some certain time; whereas the former is always sudden, and only of momentary or immensurably short duration. The expansions of solid substances do not terminate in violent explosions, on account of their slowness, and the small space through which the expanding substance moves; though their strength may be equally great with that of the most active aerial fluids. Thus we find that though wedges of wood, when wetted, will cleave solid blocks of stone, they never throw them to any distance, as is the case with gunpowder. On the other hand, it is seldom that the expansion of any elastic fluid bursts a solid substance without throwing the fragments of it to a considerable distance, with effects that are often very terrible.

The most part of explosive substances are either aerial, or convertible into such, and raised into an elastic fluid. Thus gun-powder, whose essence seems to consist in common air fixed in the nitre, or at least an air of similar elasticity, where it is condensed into a bulk many hundred times less than the natural state of the atmosphere; which air being suddenly disengaged by the firing of the gun-powder, and the decomposition of its parts, it rapidly expands itself again with a force proportioned to the degree of its condensation when fixed in the gun-powder, and so explodes, and produces all those terrible effects that attend the explosion. The elastic fluid generated by the fired gun-powder expands itself with a velocity of about 10,000 feet per second, and with a force more than 1000 times greater than the pressure of the atmosphere on the same base.

The Electric Explosions seem to be still much more strong and astonishing; as in the cases of lightning, earthquakes, and volcanoes; and even in the artificial electricity produced by the ordinary machines. The astonishing strength of electric explosions, which is beyond all possible means of measuring it, manifests itself by the many tremendous effects we hear of fire-balls and lightning.

In cases where the electric matter acts like common fire, the force of the explosions, though very great, is capable of measurement, by comparing the distances to which bodies are thrown, with their weight. This is most evident in volcanoes, where the projections of the burning rocks and lava manifest the greatness of the power, at the same time that they afford a method of measuring it: and these explosions are owing to the extrication of aerial vapours, and their rarefaction by intense heat.

Next in strength to the aerial vapours, are those of aqueous and other liquids. Very remarkable effects of these are observed in steam-engines; and there is one case from which it has been inferred that aqueous steam is even vastly stronger than fired gun-powder. This is when water is thrown upon melted copper: for here the explosion is so strong as almost to exceed imagination; and the most terrible accidents have happened, even from so slight a cause as one of the workmen spit-

ting

ting in the furnace where copper was melting; arising probably from a sudden decomposition of the water. Explosions happen also from the application of water to other melted metals, though in a lower degree, when the fluid is applied in small quantities, and even to common fire itself, as every person's own experience must have informed him; and this seems to be occasioned by the sudden rarefaction of the water into steam. Examples of this kind often occur when workmen are fastening cramps of iron into stones; where, if there happen to be a little water in the hole into which the lead is poured, this will fly out in such a manner as sometimes to burn them severely. Terrible accidents of this kind have sometimes happened in founderies, when large quantities of melted metal have been poured into wet or damp moulds. In these cases, the sudden expansion of the aqueous steam has thrown out the metal with great violence; and if any decomposition has taken place at the same time, so as to convert the aqueous vapour into an aerial one, the explosion must be still greater.

To this last kind of explosion must be referred that which takes place on pouring cold water into boiling or burning oil or tallow, or in pouring the latter upon the former; the water however being always used in a small quantity.

Another remarkable kind of Explosion is that produced by inflammable and dephlogisticated air, when mixed together, and set on fire; a kind of explosion that often happens in coal mines, &c. This differs from any of the cases before mentioned; for here is an absolute condensation rather than an expansion throughout the whole of the operation; and could the airs be made to take fire throughout their whole substance absolutely at the same instant, there would be no Explosion, but only a sudden production of heat.

Though Explosions be sometimes very destructive, they are likewise of considerable use in life, as in removing obstacles that could scarcely be overcome by any mechanical power whatever. The principal of these are the blowing up of rocks, the separating of stones in quarries, and other purposes of that kind. The destruction occasioned by them in times of war, and the machines formed upon the principle of Explosion for the destruction of the human race, are well known; and if we cannot call these useful, they must be allowed at least to be necessary evils.

The effects of Explosions, when violent, are felt at a considerable distance, by reason of the concussions they give to the atmosphere. Sir Wm. Hamilton relates, that at the explosions of Vesuvius, in 1767, the doors and windows of the houses at Naples flew open if unbolted, and one door was burst open that had been locked, though at the distance of 6 miles: and the explosion of a powder-magazine, or a powder-mill, it is well known, spreads destruction for many miles round; and even kills people by the mere concussion of the air. A curious effect of them too is, that they electrify the air, and even glass windows, at a considerable distance. This is always observable in firing the guns at the Tower of London: and some years ago, after an Explosion of some powder-mills near that city, many people were alarmed by a rattling and breaking of their china-ware. In this respect however, the effects of electrical Explo-

sions are the most remarkable, though not in the uncommon way just mentioned; but it is certain that the influence of a flash of lightning is diffused for a great way round the place where the Explosion happens, producing very perceptible changes both on the animal and vegetable creation.

EXPONENT *of a Power*, in Arithmetic and Algebra, denotes the number or quantity expressing the degree or elevation of the power, or which shews how often a given power is to be divided by its root before it be brought down to unity or 1. Thus, the Exponent or index of a square number, or the 2d power, is 2; of a cube 3; and so on; the square being a power of the 2d degree; the cube, of a 3d, &c. It is otherwise called the Index.

Exponents, as now used, are rather of modern invention. Diophantus, with the Arabian and the first European authors, denoted the powers of quantities by subjoining an abbreviation of the name of the power; though with some variation, and difference from one another. The names of the powers, and the marks for denoting them, according to Diophantus, are as follow: viz.

Names, μονας, αριθμος, δυναμις, κυβος, δυναμοδυναμις,
Marks, μ° , ς , δ^υ , κ^υ , δ^υ δ^υ ,
δυναμοκυβος , κυβοκυβος, &c.
δ^υ κ^υ   κ^υ κ^υ

which we now denote by

1, a, a^2, a^3, a^4, a^5, a^6, &c.

F. Lucas Paciolus, or De Burgo, for the root, square, cube, &c. uses the terms *cosa, censo, cubo, relato (primo, secundo, tertio, &c)*, or the abbreviations *co. ce. cu.*; and ℞ for root or radicality.

Cardan used the Latin contractions of the names of the powers; and other contemporary, as well as succeeding, authors, especially the Germans, as Stifelius, Scheubelius, Pelitarius, &c, used the like contractions, but somewhat varied, as thus:

8, 4, 3, cf, 33, √3, 3 cf, &c.
or 1, 4, 3, cf, 33, √s, 3 cf, &c.
or 1, ℞, q, cf, qq, √s, q cf, &c.
Exp. 0, 1, 2, 3, 4, 5, 6, &c.

But besides that way, the same authors also made use of the numbers as in the last line here above, and it was Stifelius who first called them by the name *Exponent*.

Bombelli, whose Algebra was published in 1579, denotes the *res*, or unknown quantity, by this mark ⌣, and the powers by numeral Exponents set over it, thus: ⌣, ⌣, ⌣, &c. And

Stevinus, who published his Arithmetic in 1585, and his Algebra soon afterwards, has such another method, but instead of ⌣ he uses a small circle ○, within which he places the numeral Exponent of the power; thus ①, ①, ①, ①, &c: and in this way he extends his notation to fractional Exponents, and even to radical ones; thus ①, ①, ①, ①, &c.

Vieta after this used words again to denote the powers. Afterwards Harriot denoted the powers by a repetition of the root; as a, aa, aaa, for the 1st, 2d, and 3d powers. Instead of which, Des Cartes again restored the numeral Exponents, placing them after the root, when the power is high, to avoid a too frequent repetition of the letter of the root; as a^3 a^4, &c, as at

present.

present. Also Albert Girard, in 1629, used the Exponents to roots, thus; $\sqrt{}$, $\sqrt[2]{}$, $\sqrt[3]{}$, &c.

The notation of powers and roots by the present way of Exponents, has introduced a new and general arithmetic of Exponents or powers; for hence powers are multiplied by only adding their Exponents, divided by subtracting the Exponents, raised to other powers, or roots of them extracted, by multiplying or dividing the Exponent by the index of the power or root.—

So $a^2 \times a^3 = a^5$, and $a^{\frac{1}{2}} \times a^{\frac{1}{4}} = a^{\frac{3}{4}}$;

$a^5 \div a^3 = a^2$, and $a^{\frac{3}{4}} \div a^{\frac{1}{2}} = a^{\frac{1}{4}}$;

the 2d power of a^3 is a^6,

and the 3d root of a^6 is a^2.

This algorithm of powers led the way to the invention of logarithms, which are only the indices or Exponents of powers: and hence the addition and subtraction of logarithms, answer to the multiplication and division of numbers; while the raising of powers, and extracting of roots, is effected by multiplying the logarithm by the index of the power, or dividing the logarithm by the index of the root.

EXPONENT *of a Ratio*, is, by some, understood as the quotient arising from the division of the antecedent of the ratio by the consequent: in which sense, the Exponent of the ratio of 3 to 2 is $\frac{3}{2}$; and that of the ratio of 2 to 3 is $\frac{2}{3}$.

But others, and those among the best mathematicians, understand logarithms as the Exponents of ratios; in which sense they coincide with the idea of measures of ratios, as delivered by Kepler, Mercator, Halley, Cotes, &c.

EXPONENTIAL *Calculus*, the method of differencing, or finding the fluxions of, Exponential quantities, and of summing up those differences, or finding their fluents. See CALCULUS, FLUXIONS, and FLUENTS.

EXPONENTIAL *Curve*, is that whose nature is defined or expressed by an Exponential equation; as the curve denoted by $a^x = y$, or by $x^x = y$.

EXPONENTIAL *Equation*, is one in which is contained an exponential quantity: as the equation $a^x = b$, or $x^x = ab$, &c.

Exponential Equations are commonly best resolved by means of logarithms, viz, first taking the log. of the given equation: thus, taking the log. of the equation $a^x = b$, it is $x \times$ log. of $a =$ log. of b; and hence

$$x = \frac{\log. \, b}{\log. \, a}$$

Also, the log. of the equation $x^x = ab$, is $x \times$ log. $x =$ log. ab; and then x is easily found by trial-and-error, or the double rule of position.

EXPONENTIAL *Quantity*, is that whose power is a variable quantity; as the expression a^x, or x^x.

Exponential quantities are of several degrees, and orders, according to the number of exponents or powers, one over another. Thus,

a^x is an Exponential of the 1st order,

a^{x^y}, is one of the 2d order,

$a^{x^{y^z}}$ is one of the 3d order, and so on.

See Bernoulli Oper. tom. 1, pa. 182, &c.

8

EXPRESSION, in Algebra, is any algebraical quantity, simple or compound: as the expression, $3a$, or $2ab$, or $\sqrt{a^2 + c^2}$.

EXTENSION, one of the common and essential properties of body; or that by which it possesses or takes up some part of universal space, called the place of that body.

The extension of a body, is properly in every direction whatever; but it is usual to consider it as extended only in length, breadth, and thickness.

EXTERIOR *Polygon*, or *Talus*, is the outer or circumscribing one. See POLYGON and TALUS.

EXTERMINATION, or EXTERMINATING, in Algebra, is the taking away, or expelling of something from an expression, or from an equation: as to Exterminate surds, fractions, or any particular letter or quantity out of equations.

Thus, to take away the fractional form from this equation $\frac{a^2 + x^2}{2c} = \frac{ab}{d}$; multiply each numerator by the other's denominator, and the equation becomes $a^2d + dx^2 = 2abc$, out of fractions.

Also, to take away the radicality from the equation $3\sqrt{a^2 - x^2} = 2c$, raise each to the 2d power, and it becomes $9a^2 - 9x^2 = 4c^2$.

For Exterminating any quantity out of equations, there are various rules and methods, according to the form of the equations; of which many excellent specimens may be seen in Newton's Algebra, pa. 60, ed. 1738; or in Maclaurin's Algebra, part 1, chap. 12. For example, to Exterminate y out of these two equations, $a + x = b + y$,

and $3b = 2x + y$;

subtract the upper equation from the under, so shall there arise $3b - a - x = 2x - b$; then, by the known methods of transposition &c, there is obtained $4b - a = 3x$,

and hence $x = \dfrac{4b - a}{3}$

EXTERNAL *Angles*, are the angles formed without-side of a figure, by producing its sides out.

In a triangle, any External angle is equal to the sum of both the two internal opposite angles taken together: and, in any right-lined figure, the sum of all the external angles, is always equal to 4 right angles.

EXTRA-*Constellary Stars*, such as are not properly included in any constellation.

EXTRA-*Mundane Space*, is the infinite, empty, void space, which is by some supposed to be extended beyond the bounds of the universe, and consequently in which there is really nothing at all.

EXTRACTION *of Roots*, is the finding the roots of given numbers, or quantities, or equations.

The roots of quantities are denominated from their powers; as the square or 2d root, the cubic or 3d root, the biquadratic or 4th root, the 5th root, &c; which are the roots of the 2d, 3d, 4th, 5th, &c powers. The Extraction of roots has always made a part of arithmetical calculation, at least as far back as the composition of powers has been known: for the composition of powers always led to their resolution, or Extraction of roots, which is performed by the rules exactly reverse of the former. Thus, if any root be considered

fidered

fidered as confifting of two parts $a + x$, of which the former a is known, and the latter x unknown, then the fquare of this root being $a^2 + 2ax + x^2$, which is its compofition, this indicated the method of refolution, fo as to find out the unknown part x; for having fub-tracted the nearest fquare a^2 from the given quantity, there remains $2ax + x^2$ or $\overline{2a + x} \times x$; therefore di-vide this remainder by $2a$, the double of the firft mem-ber of the root, the quotient will be nearly x the other member; then to $2a$ add this quotient x, and multiply the fum $2a + x$ by x, and the product will make up the remaining part $2ax + x^2$ of the given power.

The compofition of the cubic or 3d power next pre-fented itfelf, which confifts of thefe four terms $a^3 + 3a^2x + 3ax^2 + x^3$; by means of which the cubic roots of numbers have been extracted; viz. by fubtracting the neareft cube a^3 from the given power, dividing the remainder by $3a^2$, which gives x nearly for the quo-tient; then completing the divifor up to $3a^2 + 3ax + x^2$, multiply it by x for the other part of the power to be fubtracted. And this was the extent of the Ex-traction of roots in the time of Lucas de Burgo, who, from 1470 to 1500, wrote feveral pieces on arithmetic and algebra, which were the firft works of this kind printed in Europe.

It was not long however before the nature and com-pofition of all the higher powers became known, and general tables of coefficients formed for raifing them; the firft of which is contained in Stifelius's arithmetic, printed at Norimberg in 1543, where he fully explains their ufe in Extracting the roots of all powers what-ever, by methods fimilar to thofe for the fquare and cubic roots, as above defcribed; and thus completed the Extraction of all forts of roots of numbers, at leaft fo far as refpects that method of refolution. Since that time, however, many new methods of Extraction have been devifed, as well as improvements made in the old way.

The Extraction of roots of equations followed clofe-ly that of known numbers. In De Burgo's time they extracted the roots of quadratic equations, the fame way as at prefent. Ferreus, Tartalea, and Cardan ex-tracted the roots of cubic equations, by general rules. Soon afterwards the roots of higher equations were ex-tracted, at leaft in numbers, by approximation. And the late improvements in analytics have furnifhed gene-ral rules for Extracting the roots, in infinite feries, of all equations whatever. All which methods may be feen in moft books of arithmetic and algebra. Of which it may fuffice to give here a fhort fpecimen of fome of the eafieft rules for Extracting the roots of quantities and equations, as they here follow.

I. *To Extract the Square Root of any Number.*— Point off, or divide the number, from the place of units, into portions of two figures each, as here of the number 99856, fetting a point or mark over the fpace between each portion of two figures. Then, beginning at the left hand, take the greateft root 3, of the firft part 9, placing it on the right hand for the firft figure of the root, and fubtracting its fquare 9 from the faid firft part; to the re-

$$
\begin{array}{r}
\text{root} \\
99856\ (\ 316 \\
9 \\
\hline
\end{array}
$$

$$
\begin{array}{r|r}
61 & 98 \\
1 & 61 \\
\hline
626 & 3756 \\
6 & 3756 \\
\end{array}
$$

mainder, which here is 0, bring down the 2d part 98, and on the left hand of it place 6 the double of the firft figure 3, for a divifor; conceive a cipher add-ed to this, making it 60, and then divide the 98 by the 60, the quotient is 1 for the fecond figure of the root, which is accordingly placed there, after the 3, alfo in the divifor after the 6 and below the fame; then multiply thefe as they ftand, the 61 by the 1, and the product 61 fet below the 98, and fubtract it from the fame, which leaves 37 for the next remainder; to this bring down the 3d period 56, making 3756 for the next refolvend: then form its divifor as before, viz, doubling the root 31, or adding, as they ftand in the divifor, the 1 to the 61, either way making 62, which with a cipher makes 620, by which divide the refolvend 3756; the quotient of this divifion is 6, to be placed, as before, both as the next figure of the root, and at the end of the divifor 62, and below itfelf there; then multiply as they ftand the whole divifor 626 by the 6, the product 3756 is exactly the fame as the refolvend, and therefore the number 316 is accurately the fquare root of the given number 99856, as required.

When the root is to be carried into deccimals, couplets of ci-phers are to be added, inftead of figures, as far as may be wanted. In which cafe too, a good abbre-viation is made, after the work has been carried on to half the number of figures, by continuing it to the other half only by the contracted way of divifion; as here in the annexed example for the fquare root of 2 to eight de-cimals, or nine places of figures in all.

$$
\begin{array}{r}
2\ (\ 1{\cdot}41421356 \\
1 \\
\hline
\end{array}
$$

$$
\begin{array}{r|r}
24 & 100 \\
4 & 96 \\
\hline
281 & 400 \\
1 & 281 \\
\hline
2824 & 11900 \\
4 & 11296 \\
\hline
28282 & 60400 \\
2 & 56564 \\
\hline
28284 &)\ 3836\,(1356 \\
\cdots & 2828 \\
\hline
& 1008 \\
& 848 \\
\hline
& 160 \\
& 141 \\
\hline
& 19 \\
& 17 \\
\hline
& 2
\end{array}
$$

II. *To extract the cubic root, or any other root whatever.* This is eafieft done by one general rule, which I have invented, and pub-lifhed in my Tracts, vol. 1, pa. 49, which is to this effect: Let N be any number or power, whofe nth root is to be extracted; and let R be the neareft rational root of N, of the fame kind, or R^n the neareft rational power to N, either greater or lefs than it; then fhall the true root be very nearly equal to

$$
\frac{\overline{n+1}.N + \overline{n-1}.R^n}{\overline{n-1}.N + \overline{n+1}.R^n} \times R;
$$

which rule is general for any root whofe index is denoted by n. And by ex-pounding n fucceffively by all the numbers 2, 3, 4, 5, &c, this theorem will give the following particular rules for the feveral roots, viz, the

2d or fqu. root, $\dfrac{3N + R^2}{N + 3R^2} \times R$;

3d or cube root, $\dfrac{4N + 2R^3}{2N + 4R^3} \times R$, or $\dfrac{2N + R^3}{N + 2R^3} \times R$;

4th root $\dfrac{5N + 3R^4}{3N + 5R^4} \times R$;

5th

5th root $\dfrac{6N+4R^5}{4N+6R^5} \times R$, or $\dfrac{3N+2R^5}{2N+3R^5} \times R$;

6th root $\dfrac{7N+5R^6}{5N+7R^6} \times R$;

7th root $\dfrac{8N+6R^7}{6N+8R^7} \times R$, or $\dfrac{4N+3R^7}{3N+4R^7} \times R$;

&c. &c.

Or the theorem may be stated in the form of a proportion, thus:

as $\overline{n-1}.N+\overline{n+1}.R^n : \overline{n+1}.N+\overline{n-1}.R^n ::$ R : the root sought very nearly.

For ex. suppose the problem proposed, of doubling the cube, or to find the cube root of the number 2. Here N = 2, n = 3, and the nearest power, and root too, is 1:

Hence $2N+R^3 = 4+1 = 5$,

and $N+2R^3 = 2+2 = 4$;

then $4 : 5 :: 1 : \frac{5}{4} = 1\cdot25$ the first approximation.

Again, taking $R = \frac{5}{4}$, and conseq. $R^3 = \frac{125}{64}$:

Hence $2N+R^3 = 4 + \frac{125}{64} = \frac{381}{64}$,

and $N+2R^3 = 2 + \frac{250}{64} = \frac{378}{64}$;

then $378 : 381 :: \frac{5}{4} : \frac{635}{504} = 1\cdot259921$, for the cube root of 2, which is exact in the very last figure.

And again by taking $\frac{635}{504}$ for the value of R, a great many more figures may be found.

III. *To Extract the Roots of Algebraic Quantities.*— This is done by the same rules, and in the same manner as for the roots of numbers in arithmetic, as above taught. Thus, to Extract the square root of $4a^2 + 12ax + 9x^2$.

$$4a^2 + 12ax + 9x^2 \,(2a + 3x \text{ the root}$$
$$4a^2$$

| $4a + 3x$ | $12ax + 9x^2$ |
| $3x$ | $12ax + 9x^2$ |

So also the root is carried out in an infinite series, in imitation of the like Extraction of numbers in infinite decimals: thus, for the square root of $a^2 + x^2$.

$$a^2 + x^2 \,(a + \dfrac{x^2}{2a} - \dfrac{x^4}{8a^3} + \dfrac{x^6}{16a^5} \&c.$$
$$a^2$$

$$2a + \dfrac{x^2}{2a} \,\Big|\, x^2$$
$$\dfrac{x^2}{2a} \,\Big|\, x^2 + \dfrac{x^4}{4a^2}$$
$$2a + \dfrac{x^2}{a} - \dfrac{x^4}{8a^3} \,\Big|\, -\dfrac{x^4}{4a^2}$$
$$-\dfrac{x^4}{8a^3} \,\Big|\, -\dfrac{x^4}{4a^2} - \dfrac{x^6}{8a^4} + \dfrac{x^8}{4a^6}$$
$$2a + \dfrac{x^2}{a} - \dfrac{x^4}{4a^3} + \dfrac{x^6}{16a^5} \,\Big|\, \dfrac{x^6}{8a^4} + \dfrac{x^8}{4a^6}$$
$$\dfrac{x^6}{16a^5} \,\Big|\, \dfrac{x^6}{8a^4} + \dfrac{x^8}{16a^6} \&c.$$
$$\&c.$$

To extract the cube root of $a^3 - x^3$ by the general rule in the 2d article.—Here $N = a^3 - x^3$, $R = a$, or $R^3 = a^3$; hence $N + 2R^3 = 3a^3 - x^3$, and $2N + R^3 = 3a^3 - 2x^3$; therefore, by the rule,

$3a^3 - x^3 : 3a^3 - 2x^3 :: a : a - \dfrac{x^3}{3a^2} - \dfrac{x^6}{9a^5} - \dfrac{x^9}{27a^7}$ &c, which is the cube root of $a^3 - x^3$ very nearly.

But these sorts of roots are best extracted by the Binomial Theorem; which see.

IV. *To Extract the Roots of Equations.*—This is the same thing as to find the value of the unknown quantity in an equation; which is effected by various means, depending on the form of the equation, and the height of the highest power of the unknown quantity in it: for which, see the respective terms, EQUATION, ROOT, QUADRATIC, CUBIC, &c.

The most general, as well as the most easy, method of Extracting the roots of all equations, is by Double Position, or Trial-and-Error; as it easily applies to all sorts of equations whatever, be they ever so complex, even logarithmic and exponential ones. There are also several other good methods of approximating to the roots of equations, given by Newton, Halley, Raphson, De Moivre, &c; of which the most general is a rule for Extracting the root of the following indefinite equation,

viz. $az + bz^2 + cz^3 + dz^4 + ez^5$ &c

$= gy + hy^2 + iy^3 + ky^4 + ly^5$ &c,

given by M. De Moivre in the Philof. Tranf. vol. 20. p. 190, or Abr. vol. 1, pa. 101.

EXTRADOS, the outside of an arch of a bridge, vault, &c.

EXTREME-*and-Mean Proportion*, is when a line, or any quantity is so divided, as that the whole line is to the greater part, as that greater part is to the less part. Hence, in any line so divided, the rectangle of the whole line and the less segment, is equal to the square of the greater segment.

Euclid shews how to divide a line in Extreme-and-mean ratio, in his Elements, book 2, prop. 11, to this effect: Let AB be the given line; to which draw AE perpendicular and equal to half AB; in EA produced take EF = EB, so shall AF be equal to the greater part; consequently if AG be taken equal to AF, the line AB will be divided in G as required.

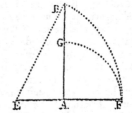 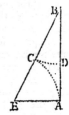

The same may be done otherwise thus:

As before, make AE (fig. 2.) perpendicular and $=$ $\frac{1}{2}$AB; join EB, on which take EC = EA, and then take BD = BC, so shall the line be divided in D as required.

No number can be divided into extreme and mean proportion, so that its two parts shall be rational; as is well demonstrated by Clavius, in his Commentary upon the 9th book of Euclid's Elements; and the same thing will also appear from the following algebraical solution of the same problem: Let a denote the whole line, and x the greater part; then shall $a - x$ be the less part, and the

the rectangle of the whole and lefs part being put equal to the fquare of the greater part, gives this equation, $x^2 = a \times \overline{a-x} = a^2 - ax$; hence $x^2 + ax = a^2$ and by completing the fquare, and extracting the root, &c, there is at laft $x = \dfrac{\sqrt{5}-1}{2}a$ the greater part; confequently $a - x = \dfrac{3-\sqrt{5}}{2}a$ is the lefs part. And as the fquare root of 5, which cannot be exactly extracted, makes a portion of both thefe parts, it is manifeft that neither of them can be obtained in rational numbers.

Euclid makes great ufe of this problem, viz, in feveral parts of the 13th book of the Elements; and by means of it he conftructs that notable propofition, viz the 10th of the 4th book, which is to conftruct an ifofceles triangle having each angle at the bafe double the angle at the vertex.

EXTREMES *Conjunct*, and EXTREMES Disjunct, in Spherical Trigonometry, are the former the two circular parts that lie next the affumed middle part, and the latter are the two that lie remote from the middle part. Thefe were terms applied by lord Napier, in his univerfal theorem for refolving all right-angled and quadrantal fpherical triangles, and publifhed in his Logarithmorum Canonis Defcriptio, an. 1614. In this theorem, Napier condenfes into one rule, in two parts, the rules for all the cafes of right-angled fpherical triangles, which had been feparately demonftrated by Pitifcus, Lanfbergius, Copernicus, Regiomontanus, and others. In this theorem, neglecting the right angle, Napier calls the other five parts, circular parts, which are, the two legs about the right angle, and the complements of the other three, viz of the hypothenufe, and the two oblique angles. Then, taking any three of thefe five parts, one of them will be in the middle between the other two, and thefe two are the Extremes Conjunct when they are immediately adjacent to that middle part, or they are the Extremes Disjunct when they are each feparated from the middle one by another part. Thus, the five parts being AB, AC, and the complements of BC and of the two angles B and C: then if the three parts be AB, and the complements of the angle B and hypothenufe BC be taken, thefe three are contiguous to each other, the angle B lying in the middle between the other two; therefore the comp. of B is middle part, and AB with the comp. of BC the Extremes Conjunct. But if the three fides be taken; BC is equally feparated from the two legs AB and AC, by two angles B and C; and therefore thefe two legs AB and AC are Extremes Disjunct, and the comp. of BC the middle part.

Napier's rule for refolving each cafe is in two parts, as below:

The rectangle contained by radius and the fine of the middle part, is equal to the rectangle of the tangents of the Extremes conjunct, or equal to the rectangle of the fines of the Extremes disjunct. Which rule comprehends all the cafes that can happen in right-angled fpherical triangles; in the application of which rule, the equal rectangles are divided into a proportion or analogy, in fuch manner that the term fought may be the laft of the four terms that are concerned, and confequently its corresponding term in the fame rectangle muft be the firft of thofe terms.

EYE, the organ of fight, confifting of feveral parts, and of fuch forms as beft to anfwer the purpofe for which it was formed.

As vifion or fight is effected by a refraction of light through the humours of the eye to the bottom or farther internal part of it, where the images of external objects are formed on a fine expanfion of the optic nerve, called the *Retina*, and therefore the forepart of the eye muft be of a convex figure, and of fuch a precife degree of convexity as the particular refractive power of the feveral humours require for forming the image of an object at a given focal diftance, viz, the diameter of the eye. Hence we find,

1ft; The external part of the eye-ball CD (Plate 2, fig. 8.) is a ftrong pellucid fubftance, properly convex, and which, when dried, has fome refemblance to a piece of tranfparent horn, for which reafon it is called the *Cornea*, or horny coat of the eye.

2dly; Immediately behind this coat there is a fine clear humour which, from its likenefs to water, is called the *Aqueous* or watery humour, and is contained in the fpace between CD and GFE.

3dly; In this fpace there is a membrane or diaphragm, called the *Uvea*, with a hole in the middle as at F, called the *Pupil*, of a mufcular contexture for altering the dimenfions of that hole, for the adjufting or admitting a due quantity of light.

4thly; Juft behind this diaphragm is placed a lenticular-formed fubftance GE, of a confiderable confiftence, called from its tranfparency the *Cryftalline* humour. This is contained in a fine tunic called the *Choroides*, and is fufpended in the middle of the eye by a ring of mufcular fibres called the *Ligamentum Ciliare*, as at G and E; by which means it is moved a little nearer to, or farther from, the bottom of the eye, to alter the focal diftance.

5thly; All the remaining interior part of the eye, conftituting the great body of it, from GHE to IMK, is made up of a large quantity of a jelly-like fubftance, called the *Vitreous* or glaffy humour; though it refembles glafs in nothing except its tranfparency; it being moft like the white of an egg of any thing.

6thly; On one fide of the hinder part of the Eye, as at K, the optic nerve enters it from the brain, and is expanded over all the interior part of the eye to G and E quite around, the expanfion being named the *Retina*. On this delicate membrane, the image IM of every external object OB, is formed according to the optic laws of nature, in the following manner.

Let OB be any very diftant object. Then a pencil of rays proceeding from any point L, will fall on the cornea CD, and be refracted by the aqueous humour under it, to a point in the axis of that pencil continued out. Then the radius of convexity of the cornea being nearly $\frac{1}{2}$ of an inch; and the fine of incidence in air to that of refraction in the aqueous humour, being nearly as 4 to 3, fuppofing the rays parallel, or the object very far diftant, the focal diftance after the firft refraction, by the proper theorem $\dfrac{mr}{m-n}$, will be found $1\frac{1}{2}$ inch from the cornea: r being the radius $\frac{1}{2}$, and m to n as 4 to 3.

The

The rays thus refracted by the cornea, fall converging on the cryſtalline humour, and tend to a point 1·228 inch behind it; alſo the radii of convexity in the ſaid humour are $\frac{1}{2}$ and $\frac{1}{4}$ reſpectively; and the ſine of incidence to that of refraction of the aqueous into the cryſtalline humour, being as 13 to 12; therefore, by this theorem $\dfrac{mdr}{md - nd + nr}$, the focal diſtance after refraction in the cryſtalline, will be 1·02 inch from the fore part of it: where $m = 13$, $n = 12$, $r = \frac{1}{2}$, and $d = 1\cdot228$.

The rays now paſs from the cryſtalline to the vitreous humour ſtill in a converging ſtate, and the ſines of incidence and refraction being here as 12 to 13, as found by experiment; and ſince the ſurface of the vitreous humour is concave which receives the rays, and is the ſame with the convexity of the hinder ſurface of the cryſtalline, the radius will be the ſame, viz $\frac{1}{4}$ of an inch. Therefore the focal diſtance after this third refraction will be found, by this theorem, $\dfrac{mdr}{nd - md + nr}$, to be 6 tenths of an inch nearly from the hinder part of the cornea: where $m = 12$, $n = 13$, $r = \frac{1}{2}$, and $d = \cdot82$; the thickneſs of the lens of the cornea being nearly $\frac{1}{2}$ of an inch.

Now experience ſhews that the diſtance of the retina in the back part of the eye, behind the cornea, is nearly equal to that focal diſtance; and therefore it follows that all objects at a great diſtance have their images formed on the retina in the bottom or hinder part of the eye, and thus diſtinct viſion is produced by this wonderful organ of optic ſenſation.

When the diſtance of objects is not very great, the focal diſtance, after the laſt refraction in the vitreous humour, will be a little increaſed; and to do this we can move the cryſtalline a little nearer the cornea by means of the ligamentum ciliare, and thus on all occaſions it may be adjuſted for a due focal diſtance for every diſtance of objects, excepting that which is leſs than 6 or 7 inches, in good eyes. Many are of opinion, however, that this is effected by a power in the eye to alter the convexity of the cryſtalline humour as occaſion requires; though this is rather doubtful.

By what has been ſaid it appears, that rays of light flowing from every part of an object OB, placed at a proper diſtance from the eye, will have an image IM thereby formed on the retina in the bottom of the eye; and ſince the rays OM, BI, which come from the extreme parts of the object, croſs each other in the middle of the pupil, the poſition of the image IM will be contrary to that of the object, or inverted, as in the caſe of a lens.

The apparent place of any part of an object is in the axis, and conjugate focus of that pencil of rays by which that part or point is formed in the image. Thus, OM is the axis, and O the focus proper to the rays by which the point M in the image is made; therefore the ſenſation of the place of that part will be conceived in the mind to be at O; in like manner the idea of place belonging to the point I, will be referred in the axis IB, to the proper focus B; therefore the apparent place of the whole image IM, will be conceived in the mind to occupy all the ſpace between O and B, and at the diſtance AL from the eye.

Hence likewiſe appears the reaſon, why we ſee an object upright by means of an inverted image; for ſince the apparent place of every point M will be in the axis MO at O; and this axis croſſing the axis of the eye HL in the pupil, it follows, that the ſenſible place O of that point will lie, without the eye, on the contrary ſide of the axis of the eye, to that of the point in the eye; and ſince this is true of all other parts or points in the image, it is evident that the poſition of every part of the object will be on the contrary ſide of the axis to every correſponding part in the image, and therefore the whole object OB will have a contrary poſition to that of the image IM, or will appear upright.

If the convexity of the cornea CD happens not exactly to correſpond to the diameter of the eye, conſidered as the natural focal diſtance, then the image will not be formed on the retina, and conſequently no diſtinct viſion can be effected in ſuch an eye.

If the cornea be too convex, the focal diſtance in the eye will be leſs than its diameter, and the image will be formed ſhort of the retina. Hence the reaſon why people having ſuch eyes are obliged to hold things very near them, to lengthen the focal diſtances; and alſo why they uſe concave glaſſes to counteract or remedy the exceſs of convexity, in order to view diſtant objects diſtinctly.

When the eye has leſs than a juſt degree of convexity, or is too flat, as is generally the caſe with old eyes, by a natural deficiency of the aqueous humour, then the rays tend to a point or focus beyond the retina or bottom of the eye; and to ſupply this want of convexity in the cornea, we uſe convex lenſes in thoſe frames called ſpectacles, or viſual glaſſes.

Since the rays of light OA, BA, which conſtitute the viſual angle OAB, will, when they are intercepted by a lens, be refracted ſooner to the axis; the ſaid angle will thereby be enlarged, and the object of courſe become magnified; which is the reaſon why thoſe lenſes are called magnifiers, or reading-glaſſes.

The dimenſions, or magnitude, of an object OB, are judged of by the quantity of the angle OAB which it ſubtends at the eye. For if the ſame object be placed at two different diſtances L and N, the angles OAB, oAb, which in theſe two places it ſubtends at the eye, will be of different magnitude; and the lineal dimenſions, viz length and breadth, will be at N and at L, as the angle oAb is to the angle OAB. But the ſurfaces of the objects will be as the ſquares of thoſe angles, and the ſolidities as the cubes of them.

It is found by experience, that two points O, L, in any object, will not be diſtinctly ſeen by the Eye, till they are near enough to ſubtend an angle OAL of one minute. And hence when objects, however large they may be, are ſo remote as not to be ſeen under an angle of one minute, they cannot properly be ſaid to have any apparent dimenſions or magnitude at all; ſuch as is the caſe of the large bodies of the planets, comets, and fixed ſtars. But the optic ſcience has ſupplied means of enlarging this natural ſmall angle under which moſt diſtant objects appear, and thereby increaſing their apparent magnitudes to a very ſurpriſing degree, in the inſtance of that noble inſtrument the teleſcope.

On the other hand, there are in creation an infinity of

of objects, of such small dimensions, that they will not subtend the requisite angle, if brought to the nearest limits of distinct vision, viz 6, 7, or 8 inches from the Eye, as found by experience; and therefore to render them visible at a very near distance, we have a variety of glasses, and instruments of different constructions, usually called microscopes, by which those minute objects appear many thousand times larger than to the naked Eye; and thereby enrich the mind with discoveries of the sublimest nature, in regard to creative power, wisdom, and œconomy.

EYE-*glass*, in Optical Instruments, is that which is next the Eye in using the machine. This is usually a lens convex on both sides; but Eustachia Divini long since invented a microscope of this kind, the power of which he places very greatly above that of the common sort; and this chiefly depending on the Eye-glass, which was double, consisting of two plano-convex glasses, so placed as to touch one another in the middle of their convex surface. This instrument is well spoken of by Fabri in his Optics, and as possessing this peculiar excellence, that it shews all the objects flat, and not crooked, and takes in a large area, though it magnifies very much.

Bull's EYE, a star of the first magnitude, in the Eye of the constellation Taurus, the bull, and by the Arabs called *Aldebaran*.

F.

FACE, or FAÇADE, in Architecture, is sometimes used for the front or outward part of a building, which immediately presents itself to the eye; or the side where the chief entrance is, or next the street, &c.

FACE, FACIA, or FASCIA, also denotes a flat member, having a considerable breadth, and but a small projecture. Such are the bands of an architrave, larmier, &c.

FACE, in Astrology, is used for the 3d part of a sign. —Each sign is supposed to be divided into three faces, of 10 degrees each : the first 10 degrees compose the first face; the next 10 degrees, the 2d face; and the last 10, the 3d face.—Venus is in the 3d face of Taurus; that is, in the last 10 degrees of it.

FACES *of a Bastion*, in Fortification, are the two foremost sides, reaching from the flanks to the outermost point of the bastion, where they meet, and form the saliant angle of the bastion. These are usually the first parts that are undermined, or beaten down; because they reach the farthest out, are the least flanked, and are therefore the weakest.

FACE *of a Place*, is the extent between the outermost points of two adjacent bastions; containing the curtain, the two flanks, and the two faces of those bastions that look towards each other. This is otherwise called the Tenaille of the place.

FACE *Prolonged*, is that part of a line of defence rasant, which is between the angle of the epaule or shoulder of a bastion and the curtain; or the line of a defence rasant diminished by the face of the bastion.

FACIA, in Architecture. See FACE, and FASCIA.

FACTORS, in Multiplication, in Arithmetic, a name given to the two numbers that are multiplied together, viz, the multiplicand and multiplier; so called because they are to facere productum, make or constitute the factum or product.

FACTUM, the product of two quantities multiplied together. As, the factum of 3 and 4 is 12; and the factum of $2a$ and $5b$ is $10ab$.

FACULÆ, in Astronomy, a name given by Scheiner, and others after him, to certain bright spots on the sun's disc, that appear more bright and lucid than the rest of his body.

Hevelius assures us that, on July 20, 1634, he observed a facula whose breadth was equal to a 3d part of the sun's diameter. He says too that the maculæ often change into Faculæ; but these seldom or never into maculæ. And some authors even contend that all the maculæ degenerate into Faculæ before they quite disappear. Many authors, after Kircher and Scheiner, have represented the sun's body full of bright, fiery spots, which they conceive to be a sort of volcanos in the body of the sun : but Huygens, and others of the latest and best observers, finding that the best telescopes discover nothing of the matter, agree entirely to explode the phenomena of Faculæ. All the foundation he could see for the notion of Faculæ, he says, was, that in the darkish clouds which frequently surround the maculæ, there are sometimes seen little points or sparks brighter

brighter than the reft. Their caufe is attributed by thefe authors to the tremulous agitation of the vapours near our earth; the fame as fometimes fhews a little un-evennefs in the circumference of the fun's difc when viewed through a telefcope. Strictly then, the Faculæ are not eructations of fire, and flame, but refractions of the fun's rays in the rarer exhalations, which, being condenfed near that fhade, feem to exhibit a light great-er than that of the fun.

FACULTY, denotes the feveral parts of an univer-fity, divided according to the arts or fciences taught or profeffed there. In moft univerfities there are four Faculties; that of arts, which includes philofophy and the humanities or languages, and is the moft ancient and extenfive; the 2d is that of theology; the 3d, that of medicine; and the 4th, jurifprudence, or laws.

FALCATED, one of the phafes of the planets, vulgarly called horned. The aftronomers fay, the moon, or any planet, is Falcated, when the enlightened part appears in form of a crefcent, like a fickle, or reap-ing-hook, which by the Latins is called falx. The moon is Falcated while fhe moves from the 3d quarter to the conjunction, and fo on from hence to the firft quarter; the bright part appearing then like a crefcent, viz during the firft and laft quarters. But during the 2d and 3d quarters, the light part appears gibbous, and the dark part Falcated.

FALCON or FAUCON, and FALCONET or FAUCO-NET, certain old fpecies of cannon, now long difufed.

FALL, the defcent or natural motion of bodies to-wards the centre of the earth, &c. Galileo firft dif-covered the ratio of the acceleration of falling bodies; viz, that the fpaces defcended from reft are as the fquares of the times of defcent; or, which comes to the fame thing, that if the whole time of Falling be divided into any number of equal parts, whatever fpace it falls through in the firft part of the time, it will Fall 3 times as far in the 2d part of time, and 5 times as far in the 3d portion of time, and fo on, according to the uneven numbers 1, 3, 5, 7, &c. See ACCELERA-TION, DESCENT, GRAVITY, &c.

FALSE-BRAYE, in Fortification. See FAUSSE-BRAYE.

FALSE *Pofition*, in Arithmetic. See POSITION.

FALSE *Root*, a name given by Cardan, to the nega-tive roots of equations, and numbers. So the root of 9 may be either 3 or − 3, the former he calls the true, and the latter the falfe or fictitious root; alfo of this equation $x^2 - x = 6$, the two roots are 3 and −2, the former true, and the latter Falfe.

FASCIA, in Architecture. See FACIA and FACE.

FASCIÆ, in Aftronomy, are certain ftripes or rows of bright parts, obferved on the bodies of fome of the planets, like fwathes, bands, or belts; efpecially on the planet Jupiter.

The Fafciæ, or belts of Jupiter, are more lucid than the reft of the difc, and are terminated by parallel lines. They are fometimes broader and fometimes narrower; nor do they always poffefs the fame part of the difc.

M. Huygens alfo obferved a very large kind of Fafcia in Mars, in the year 1656; but it was darker than the reft of the difc, and occupied the middle part of it.

FASCINES, in Fortification, are faggots made of the twigs and fmall branches of trees and brufh wood, bound up in bundles; thefe, being mixed with earth, ferve to fill up ditches, to make the parapets of trenches, batteries, &c.

FATHOM, an Englifh meafure of the length of 6 feet or 2 yards; and is taken from the utmoft extent of both arms when ftretched into a right line.

FATUUS *Ignis*. See IGNIS *Fatuus*.

FAUCON, and FAUCONET, the fame as Falcon and Falconet; the old names of certain fpecies of ordnance; which, as well as many other names, are now no longer in ufe, as it has been for fome time the practice to de-nominate the feveral fizes of cannon from the weight of their ball, inftead of calling them by thofe fanciful and unmeaning names.

FAUSSE-BRAYE, in Fortification, an elevation of earth, about three feet above the level ground; round the foot of the rampart on the outfide, defended by a parapet about four or five fathoms diftant from the up-per parapet, which parts it from the berme, and the edge of the ditch. The Fauffe-braye is the fame with what is otherwife called Chemin des rondes, and Baffe enceinte; and its ufe is for the defence of the ditch.

FEATHER-EDGED, is a term ufed by workmen, for fuch boards as are thicker on one edge, or fide, than on the other.

FEBRUARY, the 2d month of the year, contain-ing 28 days for three years, and every fourth year 29 days.—In the firft ages of Rome, February was the laft month of the year, and preceded January, till the De-cemviri made an order that February fhould be the 2d month of the year, and come after January.

FELLOWSHIP, COMPANY, or PARTNERSHIP, is a rule in arithmetic, of great ufe in balancing accounts among merchants, and partners in trade, teaching how to affign to every one of them his due fhare of the gain or lofs, in proportion to the ftock he has contributed, and the time it has been employed, or according to any other conditions. Or, more generally, it is a method of dividing a given number, or quantity, into any num-ber of parts, that fhall have any affigned ratios to one another. And hence comes this general rule: Having added into one fum the feveral numbers that exprefs the proportions of the parts, it will be,

As that fum of the proportional numbers :
Is to the given quantity that is to be divided : :
So is each proportional number :
To the correfponding fhare of the given quantity.

For Ex. Suppofe it be required to divide the number 120 into three parts that fhall be in proportion to each other as the numbers 1, 2, 3.—Here 120 is the quan-tity to be divided, and 6 is the fum of the numbers 1, 2, and 3, which exprefs the proportions of the parts; therefore as

$$6 : 120 :: \begin{cases} 1 : 20 \text{ the 1ft part,} \\ 2 : 40 \text{ the 2d part,} \\ 3 : 60 \text{ the 3d part.} \end{cases}$$

This rule is ufually diftinguifhed into two cafes, one in which time is concerned, or in which the ftocks of partners are continued for different times; and the other in which time is not confidered; this latter being called Single Fellowfhip, and the former Double Fel-lowfhip.

Single

Single FELLOWSHIP, or FELLOWSHIP *without Time*, is the cafe in which the times of continuance of the fhares of partners are not confidered, becaufe they are all the fame ; and in this cafe, the rule will be as above, viz,

　As the whole ftock of the partners :
　Is to the whole gain or lofs : :
　So is each one's particular ftock :
　To his fhare of the gain or lofs.

Ex. Two partners, A and B, form a joint ftock, of which A contributed 75l, and B 45l ; with which they gain 30l : how much of it muft each one have ?

$$\text{As } 120 : 30 :: \begin{cases} 75 : 18l.\ 15s. = \text{A's fhare,} \\ 45 : 11l.\ 5s. = \text{B's fhare.} \end{cases}$$

Double FELLOWSHIP, or FELLOWSHIP *with Time*, is the cafe in which the times of the ftocks continuing are confidered, becaufe they are not all the fame.

In this cafe, the fhares of the gain or lofs muft be proportional, both to the feveral fhares of the ftock, and to the times of their continuance, and therefore proportional to the products of the two. Hence this Rule : Multiply each particular fhare of the ftock by the time of its continuance, and add all the products together into one fum ; then fay,

　As that fum of the products :
　Is to the whole gain or lofs : :
　So is each feveral product :
　To the correfponding fhare of the gain or lofs.

For Ex. A had in company 50l. for 4 months, and B 60l. for 5 months ; and their gain was 24l : how muft it be divided between them ?

$$\begin{array}{cc} 50 & 60 \\ 4 & 5 \\ \hline 200 & 300 \\ 300 & \end{array}$$

$$\text{As } 500 : 24 :: \begin{cases} 200 : 9l.\ 12s. = \text{A's fhare,} \\ 300 : 14l.\ 8s. = \text{B's fhare.} \end{cases}$$

FERGUSON (JAMES), an eminent experimental philofopher, mechanift, and aftronomer, was born in Bamffshire, in Scotland, 1710, of very poor parents. At the very earlieft age his extraordinary genius began to unfold itfelf. He firft learned to read, by overhearing his father teach his elder brother : and he had made this acquifition before any one fufpected it. He foon difcovered a peculiar tafte for mechanics, which firft arofe on feeing his father ufe a lever. He purfued this ftudy a confiderable length, while he was yet very young ; and made a watch in wood-work, from having once feen one. As he had at firft no inftructor, nor any help from books, every thing he learned had all the merit of an original difcovery ; and fuch, with inexpreffible joy, he believed it to be.

As foon as his age would permit, he went to fervice ; in which he met with hardfhips, which rendered his conftitution feeble through life. While he was fervant to a farmer (whofe goodnefs he acknowledges in the modeft and humble account of himfelf which he prefixed to one of his publications), he contemplated and learned to know the ftars, while he tended the fheep ; and began the ftudy of aftronomy, by laying down, from his own obfervations only, a celeftial globe. His kind mafter, obferving thefe marks of his ingenuity, procured him the countenance and affiftance of fome neighbouring gentlemen. By their help and inftructions he went on gaining farther knowledge, having by their means been taught arithmetic, with fome algebra, and practical geometry. He had got fome notion of drawing, and being fent to Edinburgh, he there began to take portraits in miniature, at a fmall price ; an employment by which he fupported himfelf and family for feveral years, both in Scotland and England, while he was purfuing more ferious ftudies. In London he firft publifhed fome curious aftronomical tables and calculations ; and afterwards gave public lectures in experimental philofophy, both in London and moft of the country towns in England, with the higheft marks of general approbation. He was elected a fellow of the Royal Society, and was excufed the payment of the admiffion fee and the ufual annual contributions. He enjoyed from the king a penfion of 50 pounds a year, befides other occafional prefents, which he privately accepted and received from different quarters, till the time of his death ; by which, and the fruits of his own labours, he left behind him a fum to the amount of about fix thoufand pounds, inftead of which all his friends had always entertained an idea of his great poverty. He died in 1776, at 66 years of age, though he had the appearance of many more years.

Mr. Fergufon muft be allowed to have been a very uncommon genius, efpecially in mechanical contrivances and executions, for he executed many machines himfelf in a very neat manner. He had alfo a good tafte in aftronomy, with natural and experimental philofophy, and was poffeffed of a happy manner of explaining himfelf in an eafy, clear, and familiar way. His general mathematical knowledge, however, was little or nothing. Of algebra he underftood but little more than the notation ; and he has often told me he could never demonftrate one propofition in Euclid's Elements ; his conftant method being to fatisfy himfelf, as to the truth of any problem, with a meafurement by fcale and compaffes. He was a man of a very clear judgment in any thing that he profeffed, and of unwearied application to ftudy : benevolent, meek, and innocent in his manners as a child : humble, courteous, and communicative : inftead of pedantry, philofophy feemed to produce in him only diffidence and urbanity.

The lift of Mr. Ferguson's public works, is as follows :

1. Aftronomical Tables and Precepts, for calculating the true times of New and Full Moons, &c ; 1763. —2. Tables and Tracts, relative to feveral arts and fciences ; 1767.—3. An Eafy Introduction to Aftronomy, for Young Gentlemen and Ladies ; 2d edit. 1769.—4. Aftronomy explained upon Sir Ifaac Newton's Principles ; 5th edit. 1772.—5. Lectures on Select Subjects in Mechanics, Hydroftatics, Pneumatics, and Optics ; 4th edit. 1772.—6. Select Mechanical Exercifes ; with a fhort Account of the life of the author, by himfelf ; 1773.—7. The Art of Drawing in Perfpective made eafy ; 1775.—8. An Introduction to Electricity ; 1775.—9. Two Letters to the Rev. Mr. John Kennedy ; 1775.—10. A Third Letter to the Rev. Mr. John Kennedy ; 1775.

FERMAT

FERMAT (PETER) counfellor of the parliament of Touloufe, in France, who flourifhed in the 17th century, and died in 1663. He was a general fcholar, and a univerfal genius, cultivating jurifprudence, poetry, and mathematics, but efpecially the latter, for his amufement. He was contemporary and intimately connected with Merfenne, Torricelli, Des Cartes, Pafcal, Roberval, Huygens, Frenicle, and Carcavi, and feveral others the moft celebrated philofophers of their time. He was a firft-rate mathematician, and poffeffed the fineft tafte for pure and genuine geometry, which he contributed greatly to improve, as well as algebra.

Fermat was author of 1. A Method for the Quadrature of all forts of Parabolas.—2. Another on Maximums and Minimums: which ferves not only for the determination of plane and folid problems; but alfo for drawing tangents to curve lines, finding the centres of gravity in folids, and the refolution of queftions concerning numbers: in fhort a method very fimilar to the Fluxions of Newton.—3. An Introduction to Geometric Loci, plane and folid.—4. A Treatife on Spherical Tangencies: where he demonftrates in the folids, the fame things as Vieta demonftrated in planes.—5. A Reftoration of Apollonius's two books on Plane Loci. —6. A General Method for the dimenfion of Curve Lines. Befides a number of other fmaller pieces, and many letters to learned men; feveral of which are to be found in his *Opera Varia Mathematica*, printed at Touloufe, in folio, 1679.

FERMENTATION, an inteftine motion, arifing fpontaneoufly among the fmall and infenfible particles of a mixed body, thereby producing a new difpofition, and a different combination of thofe parts. Fermentation differs from diffolution, as the caufe from its effect, the latter being only a refult or effect of the former.

FESTOON, in Architecture &c, a decoration in form of a garland or clufter of flowers.

FICHANT FLANK. See FLANK.

FICHANT *Line of Defence.* See FIXED *Line of Defence.*

FIELD-FORT. See FORTINE.

FIELD-*Pieces*, are fmall cannon, ufually carried along with an army in the field: fuch as, one pounders, one and a half, two, three, four, fix, nine, and 12 pounders; which, being light and fmall, are eafily carried,

FIELD-*Staff*, is a ftaff carried by the gunners, in which they fcrew lighted matches, when they are on fervice; which is called arming the Field-ftaffs. See LINSTOCK.

FIELD *of View*, or *of Vifion*, is the whole fpace or extent within which objects can be feen through an optical machine, or at one view of the eye without turning it.

The precife limits of this fpace are not eafily afcertained, for the natural view of the eye. In looking at a fmall diftance, we have an imperfect glimpfe of objects through almoft the extent of a hemifphere, or at leaft for above 60 degrees each way from the optic axis; but towards the extremity of this fpace, objects are very imperfectly feen; and the diameter of the field of diftinct vifion does not fubtend an angle of more than 5 degrees at moft, fo that the diameter of a diftinct image on the retina is lefs than $\frac{1}{180}$ of an inch; but it is probably much lefs.

3

FIELD-*Book*, in Surveying, a book ufed for fetting down angles, diftances, and other things, remarkable in taking furveys.

The pages of the Field-book may be conveniently divided into three columns. In the middle column are to be entered the angles taken at the feveral ftations by the theodolite, with the diftances meafured from ftation to ftation. And the offsets, taken with the offset-ftaff, on either fide of the ftation line, are to be entered in the columns on either fide of the middle column, according to their pofition, on the right or left, with refpect to that line: alfo on the right or left of thefe are to be fet down the names and characters of the objects, with proper remarks, &c. See a fpecimen in my Treatife on Menfuration, pa. 517, ed. 2d.

FIFTH, in Mufic, one of the harmonical intervals or concords; called by the ancients Diapente.

The Fifth is the 3d in order of the concords, and the ratio of the chords that produce it, is that of 3 to 2. It is called Fifth, becaufe it contains five terms, or founds, between its extremes, and four degrees; fo that in the natural fcale of mufic it comes in the 5th place, or order, from the fundamental.—The imperfect, or defective Fifth, by the ancients called Semidiapente, is lefs than the Fifth by a mean femitone.

FIGURAL, the fame as FIGURATE numbers; which fee.

FIGURATE *Numbers*, fuch as do or may reprefent fome geometrical figure, in relation to which they are always confidered; as triangular, pentagonal, pyramidal, &c, numbers.

Figurate numbers are diftinguifhed into orders, according to their place in the fcale of their generation, being all produced one from another, viz, by adding continually the terms of any one, the fucceffive fums are the terms of the next order, beginning from the firft order which is that of equal units 1, 1, 1, 1, &c; then the 2d order confifts of the fucceffive fums of thofe of the 1ft order, forming the arithmetical progreffion 1, 2, 3, 4, &c; thofe of the 3d order are the fucceffive fums of thofe of the 2d, and are the triangular numbers 1, 3, 6, 10, 15, &c; thofe of the 4th order are the fucceffive fums of thofe of the 3d, and are the pyramidal numbers 1, 4, 10, 20, 35, &c; and fo on, as below:

Order.	Name.		Numbers.			
1.	Equals.	1, 1,	1,	1,	1,	&c.
2.	Arithmeticals,	1, 2,	3,	4,	5,	&c.
3.	Triangulars,	1, 3,	6,	10,	15,	&c.
4.	Pyramidals,	1, 4,	10,	20,	35,	&c.
5.	2d Pyramidals,	1, 5,	15,	35,	70,	&c.
6.	3d Pyramidals,	1, 6,	21,	56,	126,	&c.
7.	4th Pyramidals,	1, 7,	28,	84,	210,	&c.

The above are all confidered as different forts of triangular numbers, being formed from an arithmetical progreffion whofe common difference is 1. But if that common difference be 2, the fucceffive fums will be the feries of fquare numbers: if it be 3, the feries will be pentagonal numbers, or pentagons; if it be 4, the feries will be hexagonal numbers, or hexagons; and fo on. Thus:

Arith.

FIG [469] FIG

Arithme-ticals.	1st *Sums,* or *Polygons.*				2d *Sums,* or 2d *Polygons.*			
1, 2, 3, 4,	Tri.	1, 3,	6,	10	1, 4,	10,	20	
1, 3, 5, 7,	Sqrs.	1, 4,	9,	16	1, 5,	14,	30	
1, 4, 7, 10,	Pent.	1, 5,	12,	22	1, 6,	18,	40	
1, 5, 9, 13,	Hex.	1, 6,	15,	28	1, 7,	22,	50	
&c.								

And the reason of the names triangles, squares, pentagons, hexagons, &c, is, that those numbers may be placed in the form of these regular figures or polygons, as here below:

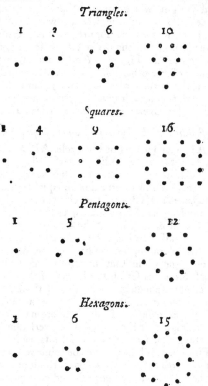

Triangles.

1 3 6 10

Squares.

1 4 9 16

Pentagons.

1 5 12

Hexagons.

1 6 15

But the Figurate numbers of any order may also be found without computing those of the preceding orders; which is done by taking the successive products of as many of the terms of the arithmeticals 1, 2, 3, 4, 5, &c, in their natural order, as there are units in the number which denominates the order of Figurates required, and dividing those products always by the first product: thus, the triangular numbers are found by dividing the products 1 × 2, 2 × 3, 3 × 4, 4 × 5, &c, each by the 1st pr. 1 × 2; the first pyramids by dividing the products 1 × 2 × 3, 2 × 3 × 4, 3 × 4 × 5, &c, by the first 1 × 2 × 3. And, in general, the figurate numbers of any order *n*, are found by substituting successively 1, 2, 3, 4, 5, &c, instead of *x* in this general expression $\frac{x . \overline{x + 1} . \overline{x + 2} . \overline{x + 3} . \&c}{1 . 2 . 3 . 4 . \&c}$; where the factors in the numerator and denominator are supposed to be multiplied together, and to be continued till the number in each be less by 1 than that which expresses the

order of the Figurates required. See Maclaurin's Fluxions, art. 351, in the notes; also Simpson's Algebra, pa. 213; or Malcolm's Arithmetic, pa. 396, where the subject of Figurates is treated in a very extensive and perspicuous manner.

FIGURE, in general, denotes the surface or terminating extremes of a body.——All finite bodies have some figure, form, or shape; whence, figurability is reckoned among the essential properties of body, or matter: a body without Figure, would be an infinite body.

FIGURES, in Architecture and Sculpture, denote representations of things made in solid matter; such as statues, &c.

FIGURES, in Arithmetic, are the numeral characters, by which numbers are expressed or written, as the ten digits, 1, 2, 3, 4, 5, 6, 7, 8, 9, 0. These are usually called the Arabic, and Indian figures, from which people it is supposed they have been derived. They were brought into Europe by the Moors of Spain, and into England about 1130, as Dr. Wallis apprehends: see his Algebra, pa. 9. However, from some ancient dates, supposed to. consist wholly or in part of Arabian figures, some have concluded that these Figures, originally Indian, were known and used in this country at least as early as the 10th century. The oldest date discovered by Dr. Wallis, was on a chimney piece, at Helmdon, in Northamptonshire, thus M133, that is 1133. Other dates discovered since, are 1090, at Colchester, in Essex; M16 or 1016, at Widgel-hall, near Buntingford, in Hertfordshire; 1011 on the north front of the parish church of Rumsey in Hampshire; and 975 over a gate-way at Worcester.

Dr. Ward, however, has urged several objections against the antiquity of these dates. As no example occurs of the use of these figures in any ancient manuscript, earlier than some copies of Johannes de Sacro Bosco, who died in. 1256, he thinks it strange that these Figures should have been used by artificers so long before they appear in the writings of the learned; and he also disputes the fact. The Helmdon date, according to him, should be 1233; the Colchester date 1490; that at Widgel-hall has in it no Arabic Figures, the 1 and 6 being I and G, the initial letters of a name; and the date at Worcester consists, he supposes, of Roman numerals, being really MXV. Martyn's Abridg. Philos. Transf. vol. 9, pa. 420.

Mr. Gibbon observes (in his History of the Decline and Fall of the Roman Empire, vol. v. pa. 321). that " under the reign of the caliph Waled, the Greek language and characters were excluded from the accounts of the public revenue. If this change was productive of the invention or familiar use of our present numerals, the Arabic characters or cyphers, as they are commonly styled, a regulation of office has promoted the most important discoveries of arithmetic, algebra, and the mathematical sciences."

On. the other hand it may be observed that, " according to a new, though probable notion, maintained by M. de Villaison) Anecdota Graeca, tom. ii. p. 152, 157), our cyphers are not of Indian or Arabic invention. They were used by the Greek and Latin arithmeticians long before the age of Boethius. After the extinction of science in the west, they were adopted in the Arabic versions from the original manuscript

and

and reſtored to the Latins about the eleventh century.

FIGURE, in Aſtrology, a deſcription, draught, or conſtruction of the ſtate and diſpoſition of the heavens, at a certain point of time; containing the places of the planets and ſtars, marked down in a Figure of 12 triangles, called houſes. This is alſo called a Horoſcope, and Theme.

FIGURE, of an Eclipſe, in Aſtronomy, denotes a repreſentation upon paper &c, of the path or orbit of the ſun or moon, during the time of the eclipſe; with the different phaſes, the digits eclipſed, and the beginning, middle, and end of darkneſs, &c.

FIGURE, or Delineation, of the full moon, ſuch as, viewed through a teleſcope with two convex glaſſes, is of conſiderable uſe in obſervations of eclipſes, and conjunctions of the moon with other luminaries. In this Figure are uſually repreſented the maculæ or ſpots of the moon, marked by numbers; beginning with the ſpots that uſually enter firſt within the ſhade at the time of the eclipſes, and alſo emerge the firſt.

FIGURE, in Conic Sections, according to Apollonius, is the rectangle contained under the latus-rectum and the tranſverſe axis, in the ellipſe and hyperbola.

FIGURE, in Fortification, is the plan of any fortified place; or the interior polygon, &c. When the ſides, and the angles, are all equal, it is called a regular Figure; but when unequal, an irregular one.

FIGURE, in Geomancy, is applied to the extremes of points, lines, or numbers, thrown or caſt at random: on the combinations or variations of which, the ſages of this art found their divinations.

FIGURE, in Geometry, denotes a ſurface or ſpace incloſed on all ſides; and is either ſuperficial or ſolid; ſuperficial when it is incloſed by lines, and ſolid when it is incloſed or bounded by ſurfaces.

Figures are either ſtraight, curved, or mixed, according as their bounds are ſtraight, or curved, or both.— The exterior bounds of a Figure, are called its ſides; the loweſt ſide, its baſe; and the angular point oppoſite the baſe, the vertex of the Figure; alſo its height, is the diſtance of the vertex from the baſe, or the perpendicular let fall upon it from the vertex.

For Figures, equal, equiangular, equilateral, circumſcribed, inſcribed, plane, regular, irregular, ſimilar, &c; ſee the reſpective adjectives.

Apparent FIGURE, in Optics, that Figure, or ſhape, under which an object appears, when viewed at a diſtance. This is often very different from the true figure; for a ſtraight line viewed at a diſtance may appear but as a point; a ſurface as a line; a ſolid as a ſurface; and a crooked figure as a ſtraight one. Alſo, each of theſe may appear of different magnitudes, and ſome of them of different ſhapes, according to their ſituation with regard to the eye. Thus an arch of a circle may appear a ſtraight line; a ſquare or parallelogram, a trapezium, or even a triangle; a circle, an ellipſis; angular magnitudes, round; a ſphere, a circle; &c.

Alſo any ſmall light, as a candle, ſeen at a diſtance in the dark, will appear magnified, and farther off than it really is. Add to this, that when ſeveral objects are ſeen at a diſtance, under angles that are ſo ſmall as to be inſenſible, as well as each of the angles ſubtended by

any one of them, and that next to it; then all theſe objects appear not only as contiguous, but as conſtituting and ſeeming but one continued magnitude.

FIGURE of the Sines, Coſines, Verſed-ſines, Tangents, or Secants, &c, are Figures made by conceiving the circumference of a circle extended out in a right line, upon every point of which are erected perpendicular ordinates equal to the Sines, Coſines, &c, of the correſponding arcs; and then drawing the curve line through the extremity of all theſe ordinates; which is then the Figure of the Sines, Coſines, &c.

It would ſeem that theſe Figures took their riſe from the circumſtance of the extenſion of the meridian line by Edward Wright, who computed that line by collecting the ſucceſſive ſums of the ſecants, which is the ſame thing as the area of the Figure of the ſecants, this being made up of all the ordinates, or ſecants, by the conſtruction of the Figure. And in imitation of this, the Figures of the other lines have been invented. By means of the Figure of the ſecants, James Gregory ſhewed how the logarithmic tangents may be conſtructed, in his Exercitationes Geometricæ, 4to, 1668.

Conſtruction of the Figures of Sines, Coſines, &c.

Let ADB &c (fig. 1) be the circle, AD an arc, DE its ſine, CE its coſine, AE the verſed ſine, AF the tangent, GH the cotangent, CF the ſecant, and CH the coſecant. Draw a right line aa equal to the whole circumference ADGBA of the circle, upon which lay off alſo the lengths of ſeveral arcs, as the arcs at every 10°, from 0 at a, to 360° at the other end at a; upon theſe points raiſe perpendicular ordinates, upwards or downwards, according as the ſine, coſine, &c, is affirmative or negative in that part of the circle; laſtly, upon theſe ordinates ſet off the length of the ſines, coſines, &c, correſponding to the arcs at thoſe points of the line or circumference aa, drawing a curve line through the extremities of all theſe ordinates; which will be the Figure of the ſines, coſines, verſedſines, tangents, cotangents, ſecants, and coſecants, as in the annexed Figures. Where it may be obſerved, that the following curves are the ſame, viz, thoſe of the ſines and coſines, thoſe of the tangents and cotangents, and thoſe of the ſecants and coſecants; only ſome of their parts a little differently placed.

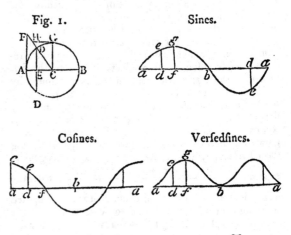

Fig. 1.   Sines.

Coſines.   Verſedſines.

Tangents.

FIG [471] FIG

Tangents.

Cotangents.

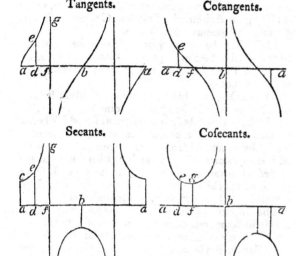

Secants.

Cofecants.

It may be known when any of thefe lines, viz, the fines, cofines, &c, are affirmative or negative, i. e. to be fet upwards or downwards, by obferving the following general rules for thofe lines in the 1ft, 2d, 3d, and 4th quadrants of the circle.

The fines   in the 1ft and 2d   are affirmative,
                 in the 3d and 4th   negative :
The cofines   in the 1ft and 4th are affirmative,
                 in the 2d and 3d   negative :
The tangents   in the 1ft and 3d   are affirmative,
                 in the 2d and 4th   negative :
The cotangents in the 1ft and 3d   are affirmative,
                 in the 2d and 4th   negative :
The fecants   in the 1ft and 4th are affirmative,
                 in the 2d and 3d   negative :
The cofecants   in the 1ft and 2d   are affirmative,
                 in the 3d and 4th   negative :
And all the verfedfines are affirmative.

To find the Equation and Area, &c, to each of thefe Curves.

Draw any ordinate *de ;* putting r = the radius AC of the given circle, x = *ad* or AD any abfcifs or arc, and y = *de* its ordinate, which will be either the fine DE = s, cofine CE = c, verfedfine AE = v, tangent AF = t, cotangent GH = τ, fecant CF = f, or cofecant CH = σ, according to the nature of the particular conftruction. Now, from the nature of the circle, are obtained thefe following general equations, expreffing the relations between the fluxions of a circular arc and its fine, or cofine, &c.

$$\dot{x} = \frac{r\dot{s}}{\sqrt{r^2 - s^2}} = \frac{-r\dot{c}}{\sqrt{r^2 - c^2}} = \frac{r\dot{v}}{\sqrt{2rv - vv}} = \frac{r^2\dot{t}}{r^2 + t^2}$$

$$= \frac{-r^2\dot{\tau}}{r^2 + \tau^2} = \frac{r^2\dot{f}}{f\sqrt{f^2 - r^2}} = \frac{-r^2\dot{\sigma}}{\sigma\sqrt{\sigma^2 - r^2}}.$$

And thefe alfo exprefs the relation between the abfcifs and ordinate of the curves in queftion, each in the order in which it ftands ; where x is the common abfcifs to all of them, and the refpective ordinates are s, c, v, t, τ, f, and σ. And hence the area &c, of any of thefe curves may be found, as follows :

1. *In the* FIGURE *of Sines.*—Here $x = ad$, and $s =$ the ordinate *de ;* and the equation of the curve, as above, is $\dot{x} = \dfrac{r\dot{s}}{\sqrt{r^2 - s^2}}$. Hence the fluxion of the area, or $s\dot{x}$ is $\dfrac{r\dot{s}s}{\sqrt{r^2 - s^2}}$; the correct fluent of which is $r^2 \mp r\sqrt{r^2 - s^2} = r^2 - rc = rv$ the rectangle of radius and verf. i. e. — or + as s is increafing or decreafing ; which is a general expreffion for the area *ade* in the Figure of fines. When $s = o$, as at a or b, this expreffion becomes o or $2r^2$; that is o at a, and $2r^2$ = the area *aeb* ; or r^2 = the area of *afg* when *ad* becomes a quadrant *af.*

2. *In the* FIGURE *of Cofines.*—Here $x = ad$ and $c = de ;$ and the equation of the curve is $\dot{x} = \dfrac{-r\dot{c}}{\sqrt{r^2 - c^2}}$. Hence the fluxion of the area is $c\dot{x} = \dfrac{-r\dot{c}c}{\sqrt{r^2 - c^2}}$; and the fluent of this is $r\sqrt{r^2 - c^2} = rs$, the rectangle of radius and fine, for the general area *adec.* When $s = r$, or $c = o$, this becomes r^2 = the area *afc*, whofe abfcifs *af* is equal to a quadrant of the circumference ; the fame as in the Figure of the fines, upon an equal abfcifs.

3. *In the* FIGURE *of Verfedfines.*—Here $x = ad$, and $v = de ;$ and the equation of the curve is $\dot{x} = \dfrac{r\dot{v}}{\sqrt{2rv - vv}}$. Hence the fluxion of the area is $v\dot{x} = \dfrac{r\dot{v}v}{\sqrt{2rv - vv}} = \dfrac{rv\sqrt{\dot{v}}}{\sqrt{2r - v}}$; and the fluent of this is $rx - rs = r \times \overline{AD - DE}$ for the area *ade* in the Figure of verfed fines. When AD or *ad* is a quadrant AG or *af*, this becomes $\dfrac{3\cdot1416}{2}r^2 - r^2 =$ the area *afg.* And when AD or *ad* is a femicircle *ab*, it becomes $3\cdot1416r^2$ = the area *abg* in the Figure of verfedfines.

4. *In the* FIGURE *of Tangents.*—Here $x = ad$, and $t = de ;$ and the equation of the curve is $\dot{x} = \dfrac{r^2\dot{t}}{r^2 + t^2}$. Hence the fluxion of the area is $t\dot{x} = \dfrac{r^2\dot{t}t}{r^2 + t^2}$; and the correct fluent of this is $\frac{1}{2}r^2 \times$ hyp. log. of $\dfrac{r^2 + t^2}{r^2} = r^2 \times$ hyp. log. of $\dfrac{\sqrt{r^2 + t^2}}{r} = r^2 \times$ hyp. log. of $\dfrac{f}{r}$. And hence the Figure of the tangents may be ufed for conftructing the logarithmic fecants ; a property that was remarked by Gregory at the end of his Exercit. Geomet.

When *ad* becomes a quadrant *af*, t being then infinite, this becomes infinite for the area *afg.* And the fame for the Figure of cotangents, beginning at f inftead of *a.*

5. *For*

5. *For the* FIGURE *of the Secants.*—Here $x = ad$, and $f = de$; and the equation of the curve is $\dot{x} = \dfrac{r^2 f}{f \sqrt{f^2 - r^2}}$.

Hence the fluxion of the area is $f\dot{x} = \dfrac{r^2 f}{\sqrt{f^2 - r^2}}$; the fluent of which is $r^2 \times$ hyp. log. of $\dfrac{f + \sqrt{f^2 - r^2}}{r}$ for the general area *ade*. And when *ad* becomes the quadrant *af*, this expression becomes infinite for the area *afg*.

The same procefs will ferve for the Figure of cofecants, beginning at *f* inftead of *a*.

From hence the meridional parts in Mercator's chart may be calculated for any latitude AD or *ad :* For the merid. parts : are to the arc of latitude AD : : as the fum of the fecants : to the fum of as many radii or : : as the area *ade* : to *ad* × radius *ac* or AD × AC in the firft figure.

FILLET, in Architecture, any little fquare member or ornament ufed in crowning a larger moulding.

FINÆUS (ORONTIUS), in French *Fine*, profeffor of mathematics in the Royal-college of Paris, was the fon of a phyfician, and was born at Briançon in Dauphiné in 1494. He went young to Paris, where his friends procured him a place in the college of Navarre. He applied himfelf there to philofophy and polite literature; but more efpecially to mathematics, in which, having a natural propenfity, he made a confiderable proficiency. Particularly he made a good progrefs in mechanics; in which, having both a genius to invent inftruments, and a fkilful hand to make them, he gained much reputation by the fpecimens he gave of his ingenuity.

Finæus firft made himfelf publicly known by correcting and publifhing Siliceus's *Arithmetic*, and the *Margarita Philofophica.* He afterwards read private lectures in mathematics, and then taught that fcience publicly in the college of Gervais: from the reputation of which, he was recommended to Francis the 1ft, as the propereft perfon to teach mathematics in the new college which that prince had founded at Paris. And here, though he fpared no pains to improve his pupils, he yet found time to publifh a great many books upon moft parts of the mathematics. But neither his genius, his labours, his inventions, and the efteem which numberlefs perfons fhewed him, could fecure him from that fate which fo often befalls men of letters. He was obliged to ftruggle all his life time with poverty; and when he died, left a numerous family deeply in debt. However, as merit muft always be efteemed in fecret, though it feldom has the luck to be rewarded openly; fo Finæus's children found Mecænafes, who for their father's fake affifted his family.—He died in 1555, at 61 years of age.

Like all other mathematicians and aftronomers of thofe times, he was greatly addicted to aftrology; and had the misfortune to be a long time imprifoned for having predicted fome things, that were not acceptable to the court of France. He was alfo one of thofe, who vainly boafted of having found out the quadrature of the circle. An edition of his works, tranflated into the Italian language, was publifhed in 4to, at Venice, 1587; confifting of Arithmetic, Practical Geometry, Cofmography, Aftronomy, and Dialling.

FINITE, the property of any thing-that is bounded or limited, either in its power, or extent, or duration, &c; as diftinguifhed from the property of infinite, or without bounds.

FINITOR, the horizon; being fo called, becaufe it finifhes or bounds the fight or profpect.

FIRE, is that fubtile invifible caufe by which bodies are made hot to the touch, and expanded or enlarged in bulk; by which fluids are rarefied into vapour; or folid bodies become fluid, and at laft either diffipated and carried off in vapour, or elfe melted into glafs. It feems alfo to be the chief agent in nature on which animal and vegetable life have an immediate dependence.

The difputes concerning fire, which long exifted among philofophers, have now in a great meafure fubfided. Thofe celebrated philofophers of the laft century, Bacon, Boyle, and Newton, were of opinion, that Fire was not a fubftance diftinct from other bodies, but that it entirely confifted in the violent motion of the parts of any body, which was produced by the mechanical force of impulfion, or of attrition. So Boyle fays, when a piece of iron becomes hot by hammering, "there is nothing to make it fo, except the forcible motion of the hammer impreffing a vehement and varioufly determined agitation on the fmall parts of the iron." And Bacon defines heat, which he makes fynonymous with Fire, an "expanfive undulatory motion in the minute particles of a body, whereby they tend with fome rapidity from a centre towards a circumference, and at the fame time a little upwards." And according to Newton, Fire is a body heated fo hot as to emit light copioufly; for what elfe, fays he, is red-hot iron, but Fire? and what elfe is a fiery coal than red-hod wood? by which he fuggefts, that bodies which are not Fire, may be changed and converted into it.

On the other hand, the chemifts ftrenuoufly contended that Fire was a fluid of a certain kind, diftinct from all others, and univerfally prefent throughout the whole globe. Boerhaave particularly maintained this doctrine; and in fupport of it brought this argument, that flint and fteel would ftrike fire, and produce the fame degree of heat in Nova Zembla as they would do under the equator. Other arguments were drawn from the increafed weight of metallic calces, which they thought proceeded from the fixing of the element of Fire in the fubftance whofe weight was thus increafed. For a long time however, the matter was moft violently difputed; but the mechanical philofophers at laft prevailed through the deference paid to the principles of Newton, though he himfelf had fcarcely taken any active part in the conteft.

The experiments of Dr. Black however feemed to bring the difpute to a decifion, and that in favour of the chemifts, concerning what he called latent-heat. From thefe difcoveries it appears, that Fire may exift in bodies in fuch a manner, as not to difcover itfelf in any other way than by its action on the minute parts of the body; but that fuddenly this action may be changed in fuch a manner, as no longer to be directed upon the particles of the body itfelf, but upon external objects; in which cafe we then perceive its action by our fenfe of feeling,

feeling, or difcover it by the thermometer, and call it heat, or fenfible Fire.

From this difcovery, and others in electricity, it is now pretty generally allowed, that Fire is a diftinct fluid, capable of being transferred from one body to another. But when this was difcovered, another queftion no lefs perplexing arofe, viz, what kind of a fluid it was; or whether it bears any analogy to thofe with which we are better acquainted. Now there are found two fluids, viz, the folar light, and the electric matter, both of which occafionally act as fire, and which therefore feem likely to be all the fame at the bottom; and popularly the matter has been long fince determined; the folar rays and the electric fluid having been indifferently accounted elementary Fire. Some indeed have imagined both thefe fluids to be mere phlogifton itfelf, or at leaft containing a large portion of it; and Mr. Scheele went fo far in this way as to 'form an hypothefis, which he endeavoured to fupport by experiments, that Fire is compofed of phlogifton and dephlogifticated air. But it is now afcertained beyond doubt, that the refult of fuch a combination is not Fire, but fixed air.

It was long fince obferved by Newton, that heat was certainly conveyed by a medium more fubtile than the common air; for two thermometers, one included in the vacuum of an air-pump, the other placed in the open air, at an equal diftance from the fire, would grow equally hot in nearly the fame time. This and other experiments fhew, that Fire exifts and acts where there is no other matter, and of confequence it is a fluid per fe, independent of every terreftrial fubftance, without being generated or compounded of any thing we are yet acquainted with. To determine the nature of the fluid, we have only to confider whether any other can be difcovered which will pafs through the perfect vacuum juft mentioned, and act there as Fire. Such a fluid is found in the folar light, which is well known to act even in vacuo as the moft violent Fire. The folar light will likewife act in the very fame manner in the moft intenfe cold; for M. de Sauffure has found, that on the cold mountain top the fun-beams are equally powerful as on the plain below, if not more fo. It appears therefore, that the folar light will produce heat independent of any other fubftance whatever; that is, where no other body is prefent, at leaft as far as we can judge, except the light itfelf, and the body to be acted upon. We cannot therefore avoid concluding, that a certain modification of the folar light is the caufe of heat, expanfion, vapour, &c, and anfwers to the reft of the characters given in the foregoing definition of Fire, and that independent of any other fubftance whatever.

It is very probable too, that the electric matter is no other than the folar light abforbed by the earth, and thus becoming fubject to new laws, and affuming many properties apparently different from what it has when it acts as light. Even in this cafe it manifefts its identity with Fire or light, viz, by producing a moft intenfe heat where a large quantity of it paffes through a fmall fpace. So that at any rate, the experiments which have already been made, and the proofs drawn from the phenomena of nature, fhew fuch a ftrong affinity between the elements of Fire, light and electricity, that we may not only affert their identity upon the

moft probable grounds, but lay it down as a pofition againft which at prefent no argument of any weight has an exiftence.

Fire-*Arrow*, is a fmall iron dart, furnifhed with fprings and bars, and alfo a match impregnated with fulphur and powder, which is wound about its fhaft. It is chiefly ufed by privateers and pirates to fire the fails of the enemy's fhip, and for this purpofe it is difcharged from a mufketoon, or a fwivel gun. The match being kindled by the explofion, it communicates the flame to the fail againft which it is directed, where the arrow faftens itfelf by means of its bars and fprings. This weapon is peculiar to hot climates, particularly the Weft Indies; the fails being very dry, are quickly fet on Fire, and the Fire is foon conveyed to the mafts, rigging, and finally to the veffel itfelf.

Fire-*Balls*, in Artillery, are certain balls compofed of combuftible matters, fuch as fine or mealed powder, fulphur, faltpetre, rofin, pitch, &c. Thefe are thrown into the enemy's works in the night time, to difcover where they are; or to fet Fire to houfes, galleries, or blinds of the befiegers, &c. They are fometimes armed with fpikes or hooks of iron, that they may not roll off, but ftick or hang where they are to take effect.

Fire-*Balls*, or *Fiery-Meteors*, in Meteorology, a kind of luminous bodies ufually appearing at a great height above the earth, with a fplendor furpaffing that of the moon, and fometimes apparently as large. They have not been found to obferve any regular courfe or motion, but, on the contrary, moving in all directions, and with very different degrees of celerity; frequently breaking into feveral fmaller ones; fometimes making a ftrong hiffing found, fometimes burfting or vanifhing with a loud report, and fometimes not.

Thefe luminous appearances doubtlefs conftitute one part of the ancient prodigies, blazing ftars, or comets, which laft they fometimes refemble in being attended with a train; but more often they appear round. The firft of thefe of which we have any accurate account, was obferved by Dr. Halley, and fome other philofophers at different places, in the year 1719, the height of which above the furface of the earth was computed at more than 70 miles. Many others have been pretty accurately obferved fince that time, and defcribed by philofophers, as in the French Memoirs, and in the Philof. Tranf. vols. 30, 41, 42, 43, 46, 47, 48, 51, 53, 54, 63, 74, &c. The velocities, directions, appearances, and heights of all thefe were found to be very various; though the height of all of them was fuppofed above the limits affigned to our atmofphere, or where it lofes its refractive power. The moft remarkable of thofe on record, appeared on the 18th of Auguft 1783, about 9 o'clock in the evening. It was feen to the northward of Shetland, and took a foutheafterly direction for an immenfe fpace, being obferved as far as the fouthern provinces of France, and by fome it was faid to have been feen at Rome, paffing over a fpace of 1000 miles in about half a minute of time, and at a very great height. During its courfe it appeared feveral times to change its fhape; fometimes appearing in the form of one ball, fometimes of two or more; fometimes with a train, and fometimes without one.

There are divers opinions concerning the nature and

origin of these Fire-balls. The first thing that occurred to philosophers on this subject was, that the meteors in question were burning bodies rising from the surface of the earth, and flying through the atmosphere with great rapidity. But this hypothesis was soon rejected, on considering that there was no power known by which such bodies could either be raised to a sufficient height, or projected with the velocity of the meteors. The next hypothesis was, that, instead of one single body, they consist of a train of sulphureous vapours, extending a vast way through the atmosphere, and being kindled at one end, display the luminous appearances in question by the fire running from one end of the train to the other. But it is not easy to conceive how such matters can exist and be disposed in such lines in so rare a part of the atmosphere, and even to burn there, in an almost perfect vacuum. For which reason this hypothesis was abandoned, for another, which was, that those meteors are permanent solid bodies, not rising from the earth, but revolving round it in very excentric orbits, and thus in their perigeon moving with vast rapidity. But as the various appearances of one and the same meteor, to observers at different places, are not compatible with the idea of a single body so revolving, this hypothesis has also been given up in its turn.. Another hypothesis that has sometimes been advanced is this; viz, that these meteors are a kind of bodies that take. fire as soon as they come within the atmosphere of the earth. But this cannot be supposed without implying a previous knowledge of these bodies, which it is impossible we can have. The only opportunity we can have of seeing them, is when they are on fire. Before that time they are in an invisible and unknown state; and it is surely improper to argue concerning them in this state, or pretend to determine any one of their properties, when it is not in our power at all to see or investigate them. As these meteors therefore never manifest themselves to our senses but when they are on fire, the only rational conclusion we can draw from thence is, that they have no existence in any other state ; and consequently that their substance must be composed of that fluid which, when acting after a certain manner, becomes luminous and shews itself as fire; remaining invisible and eluding our researches in every other case. On this hypothesis, it is now pretty generally concluded, that the Fire-balls are great bodies of electric matter, moving from one part of the heavens where to our conception, it is superabundant, to another where it is deficient : a conclusion attended with much probability, from the analogy observed between electricity and the phenomena of these meteors : and hence these Fire-balls appear to be of the same family with shooting-stars, lightning, the aurora-borealis, &c, being all referred to the same origin, viz, the electricity of the atmosphere.

Fire-*Engine*, is a machine for extinguishing accidental Fires by means of a stream or jet of water. The common squirting Fire-engine consists of a lifting pump placed in a vessel of water, and wrought by two levers that act always together. During the stroke, the water raised by the piston of the pump spouts forcibly through a pipe joined to the pump-barrel, and made capable of any degree of elevation by means of a yield-ing leather pipe, or by a ball and socket turning every way, screwed on the top of the pump. The vessel containing the water is covered with a strainer, to prevent the mud, &c, which is poured into it with the water, from choking the pump-work. Between the strokes of this engine the stream is discontinued, for want of an air vessel. However, in some cases, Engines of this construction have their use, because the stream, though interrupted, is much smarter than when the engine is made to throw water in a continued stream. See these Engines particularly described in Desaguliers's Exper. Philos. vol. 2, pa. 505 ; or Martin's Philos. Britan. vol. 2, pa. 69. See also the figure of them, plate viii. fig. 3.

Fire-*Engine*, is also sometimes used for the machine employed in raising water by steam, and more properly called Steam-*Engine* ; which see.

Fire-*Lock*, or *Fusil*, a small gun or musket, which fires with a flint and steel ; as distinguished from the old musket, or match-lock, which was fired with a match. The Fire-lock is now in common use with the European armies, and carried by the foot-soldiers. It is usually about 3 feet 8 inches in the barrel ; and, including the stock, 4 feet 8 inches, carrying a leaden bullet, of which 29 make 2lb. The diameter of the ball is ·55, and that of the barrel ·56 parts of an inch. The time of the invention of Fire-locks is uncertain ; but they were used in 1690.

Fire-*Places*, are contrivances for communicating heat to rooms, and also for answering various purposes of art and manufacture.

The general properties of air and fire, on which their construction chiefly depends, are the following, viz, that air is rarefied by heat, and condensed by cold; i. e. the same quantity of air takes up more space when warm than when cold. Air rarefied and expanded by heat, is specifically lighter than it was before, and will rise in other air of greater density so that a fire being made in any chimney, the air about and over the fire is rarefied by the heat, thence becomes lighter, and so rises in the funnel, and goes out at the top of the chimney : the other air in the room, flowing towards the chimney, supplies its place, is then rarefied in its turn, and rises likewise ; and the place of the air thus carried out of the room, is supplied by fresh air coming in through doors and windows, or, if they be shut, through every crevice with violence ; or if the avenues to the room be so closed up, that little or no fresh supply of air can be obtained, the current up the funnel must flag ; and the smoke, no longer driven up, float about in the room.

Upon these principles, various contrivances and kinds of Fire-grates and stoves have been devised, from the old very open and wide chimney places, down to the present modish ones, which are much narrowed in the front, opening, by side and back jambs, and a low breast or mantle, besides the convenience of a flap, called a register, that covers the top of the Fire-stove, but opening to any degree with a small winch, which lifts the back part sloping upwards, and so throws the smoke freely up the funnel, and yet admitting as little air to pass as you please ; by which simple means the warm air is kept very much in the room, while the very narrow and sloping orifice promotes the brisk ascent of
the

the smoke, and yet prevents its return down again, for the same reason.

Another very ingenious, but more complex, apparatus, called the Pensylvania Fire-place, was invented by Dr. Franklin, by which a room is kept very warm by a constant supply of fresh hot air, that passes into it through the stove itself. See the description in his Letters and Papers on Philosophical Subjects.

FIRE-*Pot*, in the Military Art, is a small earthen pot, into which is put a grenade, filled with fine powder till the grenade be covered; the pot is then covered with a piece of parchment, and two pieces of match laid across and lighted. This pot being thrown where it is designed to do execution, breaks and fires the powder, and this again fires the powder in the grenade, which ought to have no fuze, that its operation may be the quicker.

Rafant, or *Razant* FIRE, is a fire from the artillery and small arms, directed parallel to the horizon, or to those parts of the works of a place that are defended.

Running FIRE, is when ranks of men fire one after another; or when the lines of an army are drawn out to fire on account of a victory; in which case each squadron or battalion takes the fire from that on its right, from the right of the first line to the left, and from the left to the right of the second line, &c.

FIRE-*Ships*, in the Navy, are vessels charged with combustible materials or artificial Fire-works; which having the wind of an enemy's ship, grapple her, and set her on fire.

Anderson, in his History of Commerce, vol. 1, pa. 432, ascribes the invention to the English, in this instance, viz. some vessels being filled with combustible matter, and sent among the Spanish ships composing the Invincible Armada in 1588; and hence arose it is said the terrible invention of Fire-ships.

But Livy informs us, that the Rhodians had invented a kind of Fire-ships, which were used in junction with the Roman fleet in their engagement with the Syrians, in the year 190 before Christ: cauldrons of combustible and burning materials were hung out at their prows, so that none of the enemies' ships durst approach them: for these fell on the enemies' gallies, struck their beaks into them, and at the same time set them on fire. Livy, lib. 37, cap. 30.

Wild-FIRE, is a kind of artificial or factitious fire, that burns even under water, and that with greater violence than out of it. It is composed of sulphur, naphtha, pitch, gum, and bitumen, and it is only extinguishable by vinegar, mixed with sand and urine, or by covering it with raw hides. It is said its motion is contrary to that of natural fire, always following the direction in which it is thrown, whether it be downwards, sideways, or otherwise.

The French call it Greek Fire, or Feu Gregeois, because first used by the Greeks about the year 660, as is observed by the Jesuit Petavius, on the authority of Nicetas, Theophanes, Cedrenus, &c. The inventor, according to the same author, was an engineer of Heliopolis, in Syria, named Callinicus, who first applied it in the sea-fight commanded by Constantine Pogonates, against the Saracens, near Cyzicus, in the Hellespont; and with such effect, that he burnt the whole fleet, which contained 30,000 men. But others refer it to a much older date, and ascribe the invention to Marcus Gracchus; an opinion which is supported by several passages, both in the Greek and Roman writers; which shew that it was anciently used by both these nations in their wars. See Scaliger against Cardan.

The successors of Constantine used it on several occasions, with great advantage: and it is remarkable that they were able to keep the secret of the composition to themselves; so that no other nation knew it in the year 960.

It is recorded by Chorier, in his Hist. de Dauph that Hugh, king of Burgundy, demanding ships of the emperor Leo for the siege of Fresne, desired also the Greek Fire.

And F. Daniel gives a good description of the Greek Fire, in his account of the siege of Damietta, under St. Louis. Every body, says he, was astonished with the Greek Fire, which the Turks then prepared; and the secret of which is now lost. They threw it out of a kind of mortar, and sometimes shot it with an odd sort of cross-bow, which was strongly bent by means of a handle, or winch, of much greater force than the bare arm. That which was thrown from the mortar sometimes appeared in the air of the size of a tun, with a long tail, and a noise like that of thunder. The French, by degrees, got the secret of extinguishing it; in which they succeeded several times.

After all, perhaps the invention of the Wild-Fire is to be ascribed to other nations, and to a still older date, and that it was the same as that used among the Indians in Alexander's invasion, when it was said they fought with thunder and lightning, or shot Fire with a terrible noise.

FIRE-*Works*, otherwise called Pyrotechnia, are artificial Fires, or preparations made of gunpowder, sulphur, and other inflammable and combustible ingredients, used on occasion of public rejoicings, and other solemnities. The principal of these are rockets, serpents, stars, hail, mines, bombs, garlands, letters, and other devices.

The invention of Fire-works is attributed, by M. Mahudel, to the Florentines and people of Sienna; who found out likewise the method of adding to them decorations of statues, with fire issuing from their eyes and mouths.

FIRKIN, an English measure of capacity; being the 4th part of a barrel; and containing 8 gallons of ale, soap, butter, or herrings; or 9 gallons of beer.

FIRLOT, a dry measure used in Scotland. The oat-firlot contains 21¼ pints of that country, or about 85 English pints; and the barley-firlot, 31 standard pints. The wheat firlot contains about 2211 cubic inches; and therefore exceeds the English bushel by 60 cubic inches, or almost an English quart.

FIRMAMENT, by some old astronomers, is the orb of the fixed stars, or the highest of all the heavens But in scripture and common language it is used for the middle regions, or the space or expanse appearing like an arch quite around or above us in the heavens. Many ancients and moderns also accounted the Firmament a fluid matter; but those who gave it the name of Firmament must have taken it for a solid one.

FIRMNESS, is the consistence of a body; or that state when its sensible parts cohere, or are united together, so that the motion of one part induces a motion

tion

tion of the reft. In which fenfe firmnefs ftands oppo-fed to fluidity.

The firmnefs of bodies then depends on the connexion or cohefion of their particles; and the caufe of cohefion the Newtonians hold to be an attractive force, inherent in bodies, which binds their fmall particles together; exerting itfelf only at the points of contact, or extreme-ly near them, and vanifhing at greater diftances.

FIRST *Mover*, in the old Aftronomy, is the Pri-mum Mobile, or that which gives motion to the other parts of the univerfe.

FISSURES, in the Hiftory of the Earth, are cer-tain interruptions, moftly parallel to each other, that di-vide or feparate the ftrata of it from one another, in nearly horizontal directions; and the parts of the fame ftratum in nearly vertical directions.

FIXED *Line of Defence*, a line drawn along the face of the baftion, and terminating in the curtain.

FIXED *Signs of the Zodiac*, according to fome, are the four figns Taurus, Leo, Scorpio, Aquarius. They are fo called becaufe the fun paffes them refpectively in the middle of each quarter, when that feafon is more fettled and Fixed than under the figns which begin and end it.

FIXED *Stars*, are fuch as conftantly retain the fame pofition and diftance with refpect to each other; by which they are contradiftinguifhed from erratic or wan-dering ftars, which are continually varying their fitua-tion and diftance.—The Fixed Stars only are properly and abfolutely called ftars; the reft having their pecu-liar denomination of planet or comet.

FIXITY, or FIXEDNESS, the quality of a body which determines it fixed; or a property which enables it to endure the fire and other violent agents.

A body may be faid to be fixed in two refpects: 1ft, When on being expofed to the fire, or a corrofive men-ftruum, its particles are indeed feparated, and the body rendered fluid, but without being refolved into its firft elements. The ad, when the body fuftains the active force of the fire or menftruums whilft its integral parts are not carried off in fumes. Each kind of Fixity is the refult of a ftrong or intimate cohefion between the particles.

FLAME, the fubtleft and brighteft part of the fuel, afcending above it in a pyramidal or conical figure, and heated red-hot. Sir Ifaac Newton defines Flame as only red-hot fmoke, or the vapour of any fubftance raifed from it by fire, and heated to fuch a degree as to emit light copioufly. Is not Flame, fays he, a va-pour, fume, or exhalation, heated red-hot; that is, fo hot as to fhine? For bodies do not Flame without emitting a copious fume; and this fume burns in the Flame. The ignis fatuus is a vapour fhining without heat; and is there not the fame difference between this vapour and Flame, as between rotten wood fhining with-out heat, and burning coals of fire? In diftilling hot fpirits, if the head of the ftill be taken off, the vapour which afcends will take fire at the Flame of a candle, and turn into Flame. Some bodies, heated by motion or fermentation, if the heat grow intenfe, fume copi-oufly; and if the heat be great enough, the fumes will fhine, and become Flame. Metals in fufion do not Flame, for want of a copious fume. All flaming bo-dies, as oil, tallow, wax, wood, foffil coal, pitch, ful-

8

phur, &c, by burning, wafte in fmoke, which at firft is lucid; but at a little diftance from the body ceafes to be fo, and only continues hot. When the Flame is put out, the fmoke is thick, and frequently fmells ftrongly: but in the Flame it lofes its fmell; and, ac-cording to the nature of the fuel, the Flame is of divers colours. That of fulphur and fpirit of wine is blue; that of copper opened with fublimate, green; that of tallow, yellow; of camphire, white; &c. Newton's Op-tics, p. 318.

FLAMSTEED (JOHN), an eminent Englifh aftro-nomer, being indeed the firft aftronomer royal, for whofe ufe the royal obfervatory was built at Greenwich, thence called Flamfteed Houfe. He was born at Denby in Derbyfhire the 19th of Auguft 1646. He was edu-cated at the free fchool of Derby, where his father lived; and at 14 years of age was afflicted with a fevere illnefs, which rendered his conftitution tender ever after, and prevented him then from going to the univerfity, for which he was intended. He neverthelefs profecuted his fchool education with the beft effect; and then, in 1662, on quitting the grammar fchool, he purfued the natural bent of his genius, which led him to the ftudy of aftronomy, and clofely perufed Sacrobofco's book *De Sphæra*, which fell in his way, and which laid the ground-work of all that mathematical and aftronomical know-ledge, for which he became afterward fo juftly famous. He next procured other more modern books of the fame kind, and among them Streete's *Aftronomia Carolinæ*, then lately publifhed, from which he learned to calcu-late eclipfes and the planets' places. Some of thefe being fhewn to a Mr. Halton, a confiderable mathema-tician, he lent him Riccioli's *Almageftum Novum*, and Kepler's *Tabulæ Rudolphinæ*, which he profited much by. In 1669, having calculated fome remarkable eclipfes of the moon, he fent them to lord Brouncker, prefi-dent of the Royal Society, which were greatly approv-ed by that learned body, and procured him a letter of thanks from Mr. Oldenburg their fecretary, and another from Mr. John Collins, with whom, and other learned men, Mr. Flamfteed for a long time afterwards kept up a correfpondence by letters on literary fubjects. In 1670, his father, obferving he held correfpondence with thefe ingenious gentlemen, advifed him to take a jour-ney to London, to make himfelf perfonally acquainted with them; an offer which he gladly embraced, and vifited Mr. Oldenburg and Mr. Collins, who introduced him to Sir Jonas Moore, which proved the means of his greateft honour and preferment. He here got the knowledge and practice of aftronomical inftruments, as telefcopes, micrometers, &c. On his return, he called at Cambridge, and vifited Dr. Barrow, Mr. Ifaac New-ton, and other learned men there, and entered himfelf a ftudent of Jefus College. In 1672 he extracted feve-ral obfervations from Mr. Gafcoigne's and Mr. Crab-tree's letters, which improved him greatly in dioptrics. In this year he made many celeftial obfervations, which, with calculations of appulfes of the moon and planets to fixed ftars for the year following, he fent to Mr. Olden-burg, who publifhed them in the Philof. Tranf.

In 1673, Mr. Flamfteed wrote a fmall tract concern-ing the true diameters of all the planets, when at their greateft and leaft diftances from the earth; which he lent to Mr. Newton in 1685, who made fome ufe of it it in the 4th book of his Principia.—In 1674 he wrote an ephemeris, to shew the falsity of Astrology, and the ignorance of those who pretended to it: with calculations of the moon's rising and setting; also occultations and appulses of the moon and planets to the fixed stars. To which, at Sir Jonas Moore's request, he added a table of the moon's southings for that year; from which, and from Philips's theory of the tides, the high-waters being computed, he found the times come very near. In 1674 too, he drew up an account of the tides, for the use of the king. Sir Jonas also shewed the king, and the duke of York, some barometers and thermometers that Mr. Flamsteed had given him, with the necessary rules for judging of the weather; and otherwise took every opportunity of speaking favourably of Flamsteed to them, till at length he brought him a warrant to be the king's astronomer, with a salary of 100l. per annum, to be paid out of the office of Ordnance, because Sir Jonas was then Surveyor General of the Ordnance. This however did not abate our author's propensity for holy orders, and he was accordingly ordained at Ely by bishop Gunning.

On the 10th of August, 1675, the foundation of the Royal Observatory at Greenwich was laid; and during the building of it, Mr. Flamsteed's temporary observatory was in the queen's house, where he made his observations of the appulses of the moon and planets to the fixed stars, and wrote his Doctrine of the Sphere, which was afterward published by Sir Jonas, in his System of the Mathematics.

About the year 1684 he was presented to the living of Burstow in Surry, which he held as long as he lived. Mr. Flamsteed was equally respected by the great men his contemporaries, and by those who have succeeded since his death. Dr. Wotton, in his Reflections upon Ancient and Modern Learning, styles our author one of the most accurate Observers of the Planets and Stars, and says he calculated tables of the eclipses of the several satellites, which proved very useful to the astronomers. And Mr. Molyneux, in his *Dioptrica Nova*, gives him a high character; and, in the admonition to the reader prefixed to the work, observes, that the geometrical method of calculating a ray's progress is quite new, and never before published; and for the first hint of it, says he, I must acknowledge myself obliged to my worthy friend Mr. Flamsteed. He wrote several small tracts, and had many papers inserted in the Philosophical Transactions, viz, several in almost every volume, from the 4th to the 29th, too numerous to be mentioned in this place particularly.

But his great work, and that which contained the main operations of his life, was the *Historia Cælestis Britannica*, published in 1725, in 3 large folio volumes. The first of which contains the observations of Mr. William Gascoigne, the first inventor of the method of measuring angles in a telescope by means of screws, and the first who applied telescopical sights to astronomical instruments, taken at Middleton, near Leeds in Yorkshire, between the years 1638 and 1643; extracted from his letters by Mr. Crabtree; with some of Mr. Crabtree's observations about the same time; and also those of Mr. Flamsteed himself, made at Derby between the years 1670 and 1675; besides a multitude of curious observations, and necessary tables to be used with

them, made at the Royal Observatory, between the years 1675 and 1689.—The 2d volume contains his observations, made with a mural arch of near 7 feet radius, and 140 degrees on the limb, of the meridional zenith distances of the fixed stars, sun, moon, and planets, with their transits over the meridian; also observations of the diameters of the sun and moon, with their eclipses, and those of Jupiter's satellites, and variations of the compass, from 1689 to 1719: with tables shewing how to render the calculation of the places of the stars and planets easy and expeditious. To which are added, the moon's place at her oppositions, quadratures, &c; also the planets' places, derived from the observations.—The 3d volume contains a catalogue of the right-ascensions, polar-distances, longitudes, and magnitudes of near 3000 fixed stars, with the corresponding variations of the same. To this volume is prefixed a large preface, containing an account of all the astronomical observations made before his time, with a description of the instruments employed; as also of his own observations and instruments; with a new Latin version of Ptolomy's catalogue of 1026 fixed stars; and Ulegh-beig's places annexed on the Latin page, with the corrections; a small catalogue of the Arabs: Tycho Brahe's of about 780 fixed stars: the Landgrave of Hesse's of 386: Hevelius's of 1534: and a catalogue of some of the southern fixed stars not visible in our hemisphere, calculated from the observations made by Dr. Halley at St. Helena, adapted to the year 1726.

This work he prepared in a great measure for the press, with much care and accuracy: but through a natural weakness of constitution, and the declines of age, he died of a strangury before he had finished it, December the 19th, 1719, at 73 years of age; leaving the care of finishing and publishing his work to his friend Mr. Hodgson.—A less perfect edition of the *Historia Cælestis* had before been published, without his consent, viz, in 1712, in one volume folio, containing his observations to the year 1705.

Thus then, as Dr. Keil observed, our author, with indefatigable pains, for more than 40 years watched the motions of the stars, and has given us innumerable observations of the sun, moon, and planets, which he made with very large instruments, accurately divided, and fitted with telescopic sights; whence we may rely much more on the observations he has made, than on former astronomers, who made their observations with the naked eye, and without the like assistance of telescopes.

FLANK, in Fortification, is that part of the bastion which reaches from the curtain to the face; and it defends the curtain, with the opposite face and flank.

Oblique or *Second* FLANK, or FLANK *of the Curtain*, is that part of the curtain from whence the face of the opposite bastion can be seen, being contained between the lines rasant and fichant, or the greater and less lines of defence; or the part of the curtain between the Flank and the point where the fichant line of defence terminates.

Covered, *Low*, or *Retired* FLANK, is the platform of the casemate, which lies hid in the bastion; and is otherwise called the Orillon.

Fichant FLANK, is that from whence a cannon playing,

playing, fires directly on the face of the opposite baſtion.

Raſant or *Razant* FLANK, is the point from whence the line of defence begins, from the conjunction of which with the curtain, the ſhot only raſeth the face of the next baſtion, which happens when the face cannot be diſcovered but from the Flank alone.

Simple FLANKS, are lines going from the angle of the ſhoulder to the curtain; the chief office of which is for the defence of the moat and place.

FLANK, is alſo a term of war, uſed by way of analogy for the ſide of a battalion, army, &c, in contradiſtinction to the front and rear. So, to attack the enemy in Flank, is to diſcover and fire upon them on one ſide.

FLANKED, is uſed of ſomething that has Flanks, or may be approached on the Flank. As

FLANKED *Angle*, which is that formed by the two faces of the baſtion, and forming its point or angle. Alſo,

FLANKED *Line of Defence*, FLANKED *Tenaille*, &c.

FLANKING, in general, is the diſcovering and firing upon the ſide of a place, body, battalion, &c. To Flank a place, or other work, is to diſpoſe it in ſuch a manner as that every part of it may be played upon both in front and rear.

FLANKING means alſo defending. Any fortification that has no defence, but juſt right forwards, is faulty; and to render it complete, one part ought to be made to Flank the other. Hence the curtain is always the ſtrongeſt part of any place, becauſe it is flanked at each end.

Battalions alſo are ſaid to be Flanked by the wings of the cavalry. And a houſe is ſometimes ſaid to be Flanked with two pavilions, or two galleries; meaning it has a gallery, &c, on each ſide.——There are alſo Flanking Angle, Flanking Line of Defence, &c.

FLAT *Baſtion*, is that which is built on a right line, as on the middle of the curtain, &c.

FLEXIBLE, is the property or quality of a body that may be bent.

FLEXION, the ſame as Flexure.

FLEXURE, or FLEXION, is the bending or curving of a line or figure.

When a line firſt bends one way, and then gradually changes to a bend the contrary way, the point where the two parts join, or where the bending changes to the other ſide, is called the point of inflexion, or of contrary Flexure.

FLIE, or FLY, that part of the mariner's compaſs, on which the 32 points of the wind are drawn, and over which the needle is placed, and faſtened underneath.

FLOAT-*Boards*, the boards fixed to the outer rim of underſhot water wheels, ſerving to receive the impulſe of the ſtream, by which the wheel is carried round.——There may be too many of theſe boards on a wheel. It is thought to be the beſt rule, to have their diſtance aſunder ſuch, that each of them may come out of the water as ſoon as poſſible, after it has received and acted with its full impulſe; or, which comes to the ſame thing, when the ſucceeding one is in a direction perpendicular to the ſurface of the water.

FLOATING *Bridge*, is a bridge of boats, caſks, &c, covered with planks, firmly bound together for the paſſage of men, horſes, or carriages, &c.

FLOOD, a Deluge or inundation of water.

FLOOD, is alſo uſed in ſpeaking of the tide, when it is riſing or flowing up; in contradiſtinction to the Ebb, which is when it is decreaſing or running out.

FLOORING, in Carpentry, is commonly underſtood of the boarding of the Floors. The meaſurement of Flooring is eſtimated in ſquares, of 100 ſquare feet each, or of 10 feet on each ſide every way; for 10 times 10 are 100. Hence the length of the floor being multiplied by its breath, in feet, and two figures cut off on the right-hand, gives the ſquares, and feet, or decimals cut off. Thus, a Floor being 22 feet long, and 16 wide;

then   22 length
     16 breadth
    ———
    132
    22
    ———
3,52 the content is therefore 3 ſquares, and 52 feet, or decimals.

FLUENT, or *Flowing Quantity*, in the Doctrine of Fluxions, is the variable quantity which is conſidered as increaſing or decreaſing; or the Fluent of a given fluxion, is that quantity whoſe fluxion being taken, according to the rules of that doctrine, ſhall be the ſame with the given fluxion. See FLUXIONS.

Contemporary FLUENTS, are ſuch as flow together or for the ſame time. And the ſame is to be underſtood of Contemporary Fluxions.——When Contemporary Fluents are always equal, or in any conſtant ratio; then alſo are their fluxions reſpectively either equal, or in that ſame conſtant ratio. That is, if $x = y$, then is $\dot{x} = \dot{y}$; or if $x : y :: n : 1$, then is $\dot{x} : \dot{y} :: n : 1$; or if $x = ny$, then is $\dot{x} = n\dot{y}$.

It is eaſy to find the fluxions to all the given forms of Fluents; but, on the contrary, it is difficult to find the Fluents of many given fluxions; and indeed there are numberleſs caſes in which this cannot at all be done, excepting by the quadrature and rectification of curve lines, or by logarithms, or infinite ſeries.

This doctrine, as it was firſt invented by Sir Iſaac Newton, ſo it was carried by him to a conſiderable degree of perfection, at leaſt as to the moſt frequent, and moſt uſeful forms of fluents; as may be ſeen in his Fluxions, and in his Quadrature of Curves. Maclaurin, in his Treatiſe of Fluxions, has made ſeveral inquiries into Fluents, reducible to the rectification of the ellipſe and hyperbola: and D'Alembert has purſued the ſame ſubject, and carried it farther, in the Memoires de l'Acad. de Berlin, tom. 2, p. 200. To the celebrated Mr. Euler this doctrine is greatly indebted, in many parts of his various writings, as well as in the Inſtitutio Calculi Integralis, in 3 vols 4to, Petr. 1768. The ingenious Mr. Cotes contributed very much to this doctrine, in his Harmonia Menſurarum, concerning the meaſures of ratios and angles, in a large collection of different forms of fluxions, with their correſponding Fluents. And this ſubject was farther proſecuted in the ſame way by Walmeſley, in his Analyſe des Meſures des Rapportes et des Angles, a large vol. in 4to, 1749. Beſides many other Authors who, by their ingenious labours, have greatly contributed to facilitate and extend the doctrine of Fluents; as Emerſon, Simpſon, Landen, Waring, &c, in this country; with l'Hôpital,

pital, and many other learned foreigners. Lastly, in 1785 was published at Vienna, by M. Paccaffi, a German nobleman, Udhandlung uber eine neue Methode zu Integriren, being a method of integrating, or finding the Fluents of given fluxions, by the rules of the direct method, or by taking again the fluxion of the given fluxion, or the 2d fluxion of the fluent fought; and then making every flowing quantity its fluxion, and 2d fluxion, in geometrical progreffion; a method however, which, it feems, only holds true in the eafieft cafes or forms, whofe fluents are eafily had by the moft common methods. See this method farther explained in the rules following.

As it is only in certain particular forms and cafes that the Fluents of given fluxions can be found; there being no method of performing this univerfally a priori, by a direct inveftigation; like finding the fluxion of a given fluent quantity; we can do little more than lay down a few rules for fuch forms of fluxions as are known, from the direct method, to belong to fuch and fuch kinds of Fluents or flowing quantities: and thefe rules, it is evident, muft chiefly confift in performing fuch operations as are the reverfe of thofe by which the fluxions are found to given flowing quantities. The principal cafes of which are as follow:

I. *To find the Fluent of a fimple fluxion;* or that in which there is no variable quantity, and only one fluxional quantity. This is done by barely fubftituting the variable or flowing quantity inftead of its fluxion, and is the refult or reverfe of the notation only. Thus,

The Fluent of $a\dot{x}$ is ax.

The Fluent of $a\dot{y} + 2\dot{y}$ is $ay + 2y$.

The Fluent of $\sqrt{a^2 + x^2}$ is $\sqrt{a^2 + x^2}$.

II. *When any power of a flowing quantity is multiplied by the fluxion of the root.* Then, having fubftituted, as before, the flowing quantity for its fluxion, divide the refult by the new index of the power. Or, which is the fame thing, take out, or divide by, the fluxion of the root; add 1 to the index of the power; and divide by the index fo increafed.

So if the fluxion propofed be - $3x^5\dot{x}$;
Strike out $\dot{x}$, then it is - - $3x^5$;
add 1 to the index, and it is - - $3x^6$;
divide by the index 6, and it is - $\frac{3}{6}x^6$ or $\frac{1}{2}x^6$;
which is the Fluent of the propofed fluxion $3x^5\dot{x}$.

In like manner, the Fluent

of $4ax\dot{x}$ is $2ax^2$;

of $3x^{\frac{1}{2}}\dot{x}$ is $2x^{\frac{3}{2}}$;

of $ax^n\dot{x}$ is $\frac{a}{n+1}x^{n+1}$;

of $\frac{\dot{z}}{z^2}$ or $z^{-2}\dot{z}$ is $-z^{-1}$ or $\frac{-1}{z}$.

of $(a^3 + z^3)^4z^2\dot{z}$ is $\frac{1}{15}(a^3 + z^3)^5$.

III. *When the root under a vinculum is a compound quantity; and the index of the part or factor without the vinculum increafed by 1, is fome multiple of that under the vinculum:* Put a fingle variable letter for the compound root; and fubftitute its powers and fluxion inftead of thofe, of the fame value, in the given quantity; fo will it be reduced to a fimpler form, to which the preceding rule can then be applied.

So, if the given fluxion be $\dot{F} = (a^2 + x^2)^{\frac{2}{3}}x^3\dot{x}$; where 3, the index of the quantity without the vinculum, increafed by 1, makes 4, which is double of 2, the exponent of x^2 within the fame; therefore putting $z = a^2 + x^2$, thence $x^2 = z - a^2$, and its fluxion is $2x\dot{x} = \dot{z}$; hence then $x^3\dot{x} = \frac{1}{2}x^2\dot{z} = \frac{1}{2}z(z - a^2)$, and the given quantity F or $(a^2 + x^2)^{\frac{2}{3}}x^3\dot{x}$ is $= \frac{1}{2}z^{\frac{2}{3}}\dot{z}(z - a^2)$ or $= \frac{1}{2}z^{\frac{5}{3}}\dot{z} - \frac{1}{2}a^2z^{\frac{2}{3}}\dot{z}$; and the Fluent of each term gives

$F = \frac{3}{16}z^{\frac{8}{3}} - \frac{3}{10}a^2z^{\frac{5}{3}} = 3z^{\frac{5}{3}}(\frac{1}{16}z - \frac{1}{10}a)$; or, by fubftituting the value of z inftead of it, the fame Fluent is $3(a^2 + x^2)^{\frac{5}{3}} \times (\frac{1}{16}x^2 - \frac{3}{10}a^2)$, or $\frac{3}{16}.\overline{a^2 + x^2}^{\frac{5}{3}}.\overline{x^2 - \frac{3}{5}a^2}$.

IV. *When there are feveral terms involving two or more variable quantities, having the fluxion of each multiplied by the other quantity or quantities:* Take the Fluent of each term, as if there was only one variable quantity in it, namely that whofe fluxion is contained in it, fuppofing all the others to be conftant in that term; then if the Fluents of all the terms fo found, be the very fame quantity, that quantity will be the Fluent of the whole.

Thus, if the given fluxion be $\dot{x}y + x\dot{y}$. Then, the Fluent of $\dot{x}y$ is xy, fuppofing y conftant; and the Fluent of $x\dot{y}$ is alfo xy, when x is conftant; therefore the common refulting quantity xy is the required Fluent of the given fluxion $\dot{x}y + x\dot{y}$.

And, in like manner, the Fluent of $\dot{x}yz + x\dot{y}z + xy\dot{z}$ is xyz.

V. *When the given fluxional expreffion is in this form* $\frac{\dot{x}y - x\dot{y}}{y^2}$, *viz, a fraction including two quantities, being the fluxion of the former drawn into the latter, minus the fluxion of the latter drawn into the former, and divided by the fquare of the latter:* then the Fluent is the fraction $\frac{x}{y}$, or of the former quantity divided by the latter. That is,

The Fluent of $\frac{\dot{x}y - x\dot{y}}{y^2}$ is $\frac{x}{y}$;

and the Fluent of $\frac{2x\dot{x}y^2 - 2x^2y\dot{y}}{y^4}$ is $\frac{x^2}{y^2}$.

Though the examples of this cafe may be performed by the foregoing one. Thus the given fluxion $\frac{\dot{x}y - x\dot{y}}{y^2}$ reduces to $\frac{\dot{x}}{y} - \frac{x\dot{y}}{y^2}$ or $\frac{\dot{x}}{y} - x\dot{y}y^{-2}$; of which the Fluent of $\frac{\dot{x}}{y}$ is $\frac{x}{y}$ when y is conftant; and the Fluent of $x\dot{y}y^{-2}$ is $+ xy^{-1}$ or $\frac{x}{y}$ when x is conftant; and therefore, by that cafe, $\frac{x}{y}$ is the Fluent of the whole $\frac{\dot{x}y - x\dot{y}}{y^2}$.

VI. *When the fluxion of a quantity is divided by the quantity itfelf:* Then the Fluent is equal to the hyperbolic logarithm of that quantity; or, which is the fame thing, the Fluent is equal to 2·30258509 &c, multiplied by the common log. of the fame quantity.

So,

3

So, the Fluent

of $\dfrac{\dot{x}}{x}$ or $x^{-1}\dot{x}$ is the hyp. log. of x.

of $\dfrac{2\dot{x}}{x}$ is $2 \times$ hyp. log. of x, or $=$ h. l. of x^2;

of $\dfrac{a\dot{x}}{x}$ is $a \times$ h. l. of x, or h. l. of x^a;

of $\dfrac{\dot{x}}{a+x}$ is the h. l. of $a+x$;

of $\dfrac{3x^2\dot{x}}{a+x^3}$ is the h. l. of $a+x^3$.

VII. Many Fluents may be found by the direct method of fluxions, thus : Take the fluxion again of the given fluxional expreſſion, or the 2d fluxion of the Fluent ſought ; into which ſubſtitute $\dfrac{\dot{x}^2}{x}$ for $\ddot{x}$, and $\dfrac{\dot{y}^2}{y}$ for $\ddot{y}$, &c, that is, make $x, \dot{x}, \ddot{x}$, as alſo $y, \dot{y}, \ddot{y}$, &c, in continual proportion, or $x : \dot{x} :: \dot{x} : \ddot{x}$, and $y : \dot{y} :: \dot{y} : \ddot{y}$, &c : then divide the ſquare of the given fluxional expreſſion by the 2d fluxion, juſt found, and the quotient will be the Fluent ſought in many caſes.

Or the ſame rule may be otherwiſe delivered thus : In the given fluxion $\dot{F}$, write x for $\dot{x}$, y for $\dot{y}$, &c, and call the reſult G, taking alſo the fluxion of this quantity, $\dot{G}$; then make $\dot{G} : \dot{F} :: G : F$, ſo ſhall the 4th proportional F be the Fluent as before. And this is the rule of M. Paccaſſi.

It may be proved if this be the true Fluent, by taking the fluxion of it again, which, if it agree with the propoſed fluxion, will ſhew that the Fluent is right ; otherwiſe, it is wrong.

Thus, if it be propoſed to find the Fluent of $nx^{n-1}\dot{x}$. Here $\dot{F} = nx^{n-1}\dot{x}$; write firſt x for $\dot{x}$, and it is $nx^{n-1}x$ or $nx^n = G$; the fluxion of this is $\dot{G} = n^2x^{n-1}\dot{x}$; therefore $\dot{G} : \dot{F} :: G : F$ becomes $n^2x^{n-1}\dot{x} : nx^{n-1}\dot{x} :: nx^n : x^n = F$, the Fluent ſought.

For a 2d ex. ſuppoſe it be propoſed to find the Fluent of $\dot{x}y + x\dot{y}$. Here $F = \dot{x}y + x\dot{y}$; then, writing x for $\dot{x}$, and y for $\dot{y}$, it is $xy + xy$ or $2xy = G$; the fluxion of which is $2\dot{x}y + 2x\dot{y} = \dot{G}$; then $\dot{G} : \dot{F} :: G : F$ becomes $2\dot{x}y + 2x\dot{y} : \dot{x}y + x\dot{y} :: 2xy : xy = F$, the Fluent ſought.

VIII. *To find Fluents by means of a table of forms of Fluxions and Fluents.*

In the following table are contained the moſt uſual forms of fluxions that occur in the practical ſolution of problems, with their correſponding Fluents ſet oppoſite to them ; by means of which, viz, comparing any propoſed fluxion with the correſponding form here, the Fluent of it will be found.

Where it is to be noted, that the logarithms in the ſaid forms, are the hyperbolic ones, which are found by multiplying the common logs. by 2.3025850929940 &c. Alſo the arcs whoſe ſine, or tangent, &c, are mentioned, have the radius 1, and are thoſe in the common tables of ſines, tangents, &c.—And the numbers m, n, &c. are to be ſome quantities, as the forms fail when $n = 0$, or $m = 0$, &c.

Forms	Fluxions.	Fluents.
I	$x^{n-1}\dot{x}$	$\dfrac{1}{n}x^n$
II	$(a \pm x^n)^{m-1}\dot{x}x^{n-1}\dot{x}$	$\pm\dfrac{1}{mn}(a \pm x^n)^m$
III	$\dfrac{x^{mn-1}\dot{x}}{(a \pm x^n)^{m+1}}$	$\dfrac{1}{mna} \times \dfrac{x^{mn}}{(a \pm x^n)^m}$
IV	$\dfrac{(a \pm x^n)^{m-1}\dot{x}}{x^{mn+1}}$	$\dfrac{-1}{mna} \times \dfrac{(a \pm x^n)^m}{x^{mn}}$
V	$(m\dot{x}y + nx\dot{y})x^{m-1}y^{n-1}$, or $\left(\dfrac{m\dot{x}}{x} + \dfrac{n\dot{y}}{y}\right)x^m y^n$	$x^m y^n$
VI	$mx^{m-1}\dot{x}y^nz^r + nx^my^{n-1}\dot{y}z^r + rx^my^nz^{r-1}\dot{z}$, or $(m\dot{x}yz + nx\dot{y}z + rxy\dot{z})x^{m-1}y^{n-1}z^{r-1}$, or $\left(\dfrac{m\dot{x}}{x} + \dfrac{n\dot{y}}{y} + \dfrac{r\dot{z}}{z}\right)x^my^nz^r$	$x^m y^n z^r$
	$\dfrac{\dot{x}}{x}$ or $x^{-1}\dot{x}$	log. of x

Forms.	Fluxions.	Fluents.
VIII	$\dfrac{x^{n-1}\dot{x}}{a \pm x^n}$	$\pm \dfrac{1}{n} \log.$ of $a \pm x^n$
IX	$\dfrac{x^{-1}\dot{x}}{a \pm x^n}$	$\dfrac{1}{na} \log.$ of $\dfrac{x^n}{a \pm x^n}$
X	$\dfrac{x^{\frac{1}{2}n-1}\dot{x}}{a - x^n}$	$\dfrac{1}{n\sqrt{a}} \log.$ of $\dfrac{\sqrt{a}+\sqrt{x^n}}{\sqrt{a}-\sqrt{x^n}}$
XI	$\dfrac{x^{\frac{1}{2}n-1}\dot{x}}{a + x^n}$	$\dfrac{2}{n\sqrt{a}} \times$ arc to tang. $\sqrt{\dfrac{x^n}{a}}$, or $\dfrac{1}{n\sqrt{a}} \times$ arc to cosine $\dfrac{a-x^n}{a+x^n}$
XII	$\dfrac{x^{\frac{1}{2}n-1}\dot{x}}{\sqrt{\pm a + x^n}}$	$\dfrac{2}{n} \log.$ of $\sqrt{x^n} + \sqrt{\pm a + x^n}$
XIII	$\dfrac{x^{\frac{1}{2}n-1}\dot{x}}{\sqrt{a-x^n}}$	$\dfrac{2}{n} \times$ arc to fin. $\sqrt{\dfrac{x^n}{a}}$, or $\dfrac{1}{n} \times$ arc to verf. $\dfrac{2x^n}{a}$
XIV	$\dfrac{x^{-1}\dot{x}}{\sqrt{a \pm x^n}}$	$\dfrac{1}{n\sqrt{a}} \log.$ of $\dfrac{\pm\sqrt{a \mp x^n} \pm \sqrt{a}}{\sqrt{a \pm x^n}+\sqrt{a}}$
XV	$\dfrac{x^{-1}\dot{x}}{\sqrt{-a+x^n}}$	$\dfrac{2}{n\sqrt{a}} \times$ arc to fecant $\sqrt{\dfrac{x^n}{a}}$, or $\dfrac{1}{n\sqrt{a}} \times$ arc to cofin. $\dfrac{2a-x^n}{x^n}$
XVI	$\dot{x} \sqrt{dx - x^2}$	$\dfrac{1}{2}$ circ. feg. to diam. d and verf. x
XVII	$c^{nx}\dot{x}$	$\dfrac{c^{nx}}{n \log. c}$
XVIII	$\dot{x}y^x \log. y + xy^{x-1}\dot{y}$	y
XIX	$(a + x^n)^m x^{rn-1}\dot{x}$	$\dfrac{(a+x^n)^{m+1} x^{rn-n}}{ns} \times \left(\dfrac{1}{1} - \dfrac{r-1.a}{s-1.x^n} + \dfrac{r-1.r-2.a^2}{s-1.s-2.x^{2n}} - \dfrac{r-1.r-2.r-3.a^3}{s-1.s-2.s-3.x^{3n}} \&c \right)$
	Putting $s = m + r$	or $\dfrac{(a+x^n)^{m+1} x^{rn}}{rna} \times \left(\dfrac{1}{1} - \dfrac{s+1.x^n}{r+1.a} + \dfrac{s+1.s+2.x^{2n}}{r+1.r+2.a^2} - \dfrac{s+1.s+2.s+3.x^{3n}}{r+1.r+2.r+3.a^3} \&c \right)$

The Ufe of the foregoing Table of Forms of Fluxions and Fluents.—In the ufe of this table, it is to be obferved, that the firft column ferves only to fhew the number of the form, as a mark of reference ; in the 2d column are the feveral forms of fluxions, which are of different kinds or claffes ; and in the 3d or laft column are the correfponding Fluents.

The method of ufing the table is this. Having any fluxion given, whofe Fluent it is propofed to find : Firft, compare the given fluxion with the feveral forms of fluxions in the 2d column of the table, till one of the forms be found that agrees with it ; which is done by comparing the terms of the given fluxion with the like parts of the tabular fluxion, viz, the radical quantity of the one, with that of the other ; and the exponents of the variable quantities of each, both within and without the vinculum ; all which, being found to agree or correfpond, will give the particular values of the general quantities in the tabular form. Then fubftitute thefe particular values, for the fame quantities in the general or tabular form of the Fluent, and the result will be the particular Fluent fought; after it is multiplied by any coefficient the propofed fluxion may have.

For Ex. To find the Fluent of the given fluxional expreffion $3 x^{\frac{3}{3}} \dot{x}$. This agrees with the firft form; and by comparing the fluxions, it appears that $x = x$, and $n - 1 = \frac{3}{3}$, or $n = \frac{8}{3}$; which being fubftituted in the tabular Fluent, or $\frac{1}{n} x^n$, gives, after multiplying by 3 the coefficient, $3 \times \frac{3}{8} x^{\frac{8}{3}}$ or $\frac{9}{8} x^{\frac{8}{3}}$ for the Fluent fought.

Again, To find the Fluent of $5 x^2 \dot{x} \sqrt{c^3 - x^3}$, or $5 x^2 \dot{x} \cdot \overline{c^3 - x^3}|^{\frac{1}{2}}$. This belongs to the 2d form; for $a = c^3 - x^n = - x^3$, $n = 3$ under the vinculum, $m - 1 = \frac{1}{2}$, or $m = \frac{3}{2}$, and the exponent^{n-1} of x^{n-1} without the vinculum, by ufing 3 for n, is $n - 1 = 2$, which agrees with x^2 in the fluxion given; and therefore all the parts of the form are found to anfwer. Then, fubftituting thefe values into the general Fluent, $-\frac{1}{mn} (a - x^n)^m$, it becomes $-\frac{5 \times 2}{3 \times 3} (c^3 - x^3)^{\frac{3}{2}} = -\frac{10}{9} (c^3 - x^3)^{\frac{3}{2}}$.

Thirdly, To find the Fluent of $\frac{x^2 \dot{x}}{1 + x^3}$. This agrees with the 8th form; where $\pm x^n = + x^3$ in the denominator, or $n = 3$; and the numerator x^{n-1} then becomes x^2, which agrees with the numerator in the given fluxion; alfo $a = 1$. Hence then, by fubftituting in the general form of the Fluent $\frac{1}{n}$ logarithm of $a + x^n$, it becomes $\frac{1}{3}$ logarithm of $1 + x^3$.

IX. *To find Fluents by means of Infinite Series.*— When a finite form cannot be found to agree with a propofed fluxion, it is then ufual to throw it into an infinite feries, either by divifion, or extraction of roots, or by the binomial theorem, &c; after which, the Fluents of all the terms are taken feparately.

For Ex. To find the Fluent of $\frac{1 - x}{1 + x - x^2} \dot{x}$. Here, by dividing the numerator by the denominator, this becomes $\dot{x} - 2 x \dot{x} + 3 x^2 \dot{x} - 5 x^3 \dot{x} + 8 x^4 \dot{x}$ &c; and, the Fluents of all the terms being taken, give $x - x^2 + x^3 - \frac{5}{4} x^4 + \frac{8}{5} x^5$ &c, for the Fluent fought.

To Correct a FLUENT.—The Fluent of a given fluxion, found as above, fometimes wants a correction, to make it contemporary with that required by the problem under confideration, &c: for the Fluent of any given fluxion, as $\dot{x}$, may be either x (which is found by the rule) or it may be $x \pm c$, that is x plus or minus fome conftant quantity c; becaufe both x and $x \pm c$ have the fame fluxion $\dot{x}$: and the finding of this conftant quantity, is called correcting the Fluent. Now this correction is to be determined from the nature of the problem in hand, by which we come to know the relation which the Fluent quantities have to each other at fome certain point or time. Reduce therefore the general Fluential equation, found by the rules above, to that point or time; then if the equation be true at that point, it is correct; but if not, it wants a correction, and the quantity of that correction is the dif-

ference between the two general fides of the equation when reduced to that particular ftate. Hence the general rule for the correction is this:

Connect the conftant, but indeterminate, quantity c with one fide of the Fluential equation, as determined by the foregoing rules; then, in this equation, fubftitute for the variable quantities fuch values as they are known to have at any particular ftate, place, or time; and then from that particular ftate of the equation find the value of c, the conftant quantity of the correction.

Ex. To find the Correct Fluent of $\dot{z} = a x^3 \dot{x}$. Firft the general Fluent of this is $z = a x^4$, or $z = a x^4 + c$, taking in the correction c.

Now if it be known that z and x begin together, or that $z = 0$, when $x = 0$; then writing 0 both for x and z, the general equation becomes $0 = 0 + c$, or $c = 0$; fo that, the value of c being 0, the Correct Fluents are $z = a x^4$.

But if z be $= 0$, when x is $= b$, any known quantity; then fubftituting 0 for z, and b for x, in the general equation, it becomes $0 = a b^4 + c$, from which is found $c = - a b^4$; and this being written for it in the general equation, this becomes $z = a x^4 - a b^4$, for the correct, or contemporary Fluents.

Or laftly, if it be known that z is $=$ fome quantity d, when x is equal fome other quantity, as b; then fubftituting d for z, and b for x, in the general Fluential equation $z = a x^4 + c$, it becomes $d = a b^4 + c$; and hence is deduced the value of the correction, viz, $c = d - a b^4$; confequently, writing this value for c in the general equation, it becomes $z = a x^4 - a b^4 + d$, for the Correct equation of the Fluents in this cafe.

And hence arifes another eafy and general way of correcting the Fluents, which is this: In the general equation of the Fluents, write the particular values of the quantities which they are known to have at any certain time; then fubtract the fides of the refulting particular equation, from the correfponding fides of the general one, and the remainders will give the Correct equation of the Fluents fought. So, as above, the general equation being $z = a x^4$; write d for z, and b for x, then $d = a b^4$; hence by fubtraction $z - d = a x^4 - a b^4$, or $z = a x^4 - a b^4 + d$, the Correct Fluents as before.

FLUID, or FLUID *Body*, according to Newton, is that whofe parts yield to the fmalleft force impreffed, and by yielding are eafily moved among each other; in which fenfe it ftands oppofed to a folid. This is the definition of a perfect fluid: if the Fluid require fome fenfible force to move its parts, it is imperfect in proportion to that force; fuch as perhaps all the fluids we know of in nature.

That Fluids have vacuities in their fubftance is evident, becaufe certain bodies may be diffolved in them without increafing their bulk. Thus, water will diffolve a certain quantity of falt; after which it will receive a little fugar, and after that a little alum; and all this without increafing its firft dimenfions. Which fhews that the particles of thefe folids are fo far feparated as to become fmaller than thofe of the Fluid, and to be received and contained in the interftices between them.

Fluids are either elaftic, fuch as air; or non-elaftic,

as water, mercury, &c. Thefe latter occupy the fame fpace, or are of the fame bulk, under all preffures or forces; but the former dilate and expand themfelves continually by taking off the external preffure from them; for which reafon it is that the denfity and elafticity of fuch fluids, are proportional to the force or weight that compreffes them. The doctrine and laws of Fluids are of the greateft extent in philofophy: the properties of elaftic Fluids conftituting the doctrine of Pneumatics; thofe of the non-elaftic ones, that of Hydroftatics; and their motions, Hydraulics. For which fee thefe refpective articles. Alfo,

For the laws of the preffure and gravitation in Fluids fpecifically heavier or lighter than the bodies immerged, fee SPECIFIC *Gravity*.

For the laws of the refiftance of Fluids, or the retardation of folid bodies moving in Fluids, fee RESISTANCE. And

For the afcent of Fluids in capillary tubes, or between glafs planes, &c, fee ASCENT, and CAPILLARY *Tubes*.

FLUTES, or FLUTINGS, are certain channels or cavities cut along the fhaft of a column, or pilafter.

FLUIDITY, that ftate or affection of bodies, which denominates or renders them Fluid; or that property by which they yield to the fmalleft force impreffed: in contradiftinction to Solidity or Firmnefs.

Fluidity is to be carefully diftinguifhed from Liquidity or Humidity, which latter implies wetting or adhering. Thus, air, ether, mercury, and other melted metals, and even fmoke and flame itfelf, are Fluid bodies, but not Liquid ones; whilft water, beer, milk, urine, &c, are both Fluids and Liquids at the fame time.

The nature and caufes of Fluidity have been varioufly affigned. The Gaffendifts, and ancient corpufcularians, require only three conditions as neceffary to it; viz, a fmallnefs and fmoothnefs of the particles of the body, vacuities interfperfed between them, and a fpherical figure. The Cartefians, and after them Dr. Hook, Mr. Boyle, &c, befide thefe circumftances, require alfo a certain internal or inteftine motion of the particles as chiefly contributing to Fluidity Thus, Mr. Boyle, in his Hiftory of Fluidity, argues from various experiments: for example, a little dry powder of alabafter, or plaifter of Paris, finely fifted, being put into a veffel over the fire, foon begins to boil like water; exhibiting all the motions and phenomena of a boiling liquor: it will tumble varioufly in great waves like that; will bear ftirring with a ftick or ladle like that, without refifting; and if ftrongly ftirred near the fide of the veffel, its waves will apparently dafh againft it: yet it is all the while a dry parched powder.

The like is obferved in fand; a difh of which being fet on a drum-head, brifkly beaten by the fticks, or on the upper ftone of a mill, it in all refpects emulates the properties of a Fluid body. A heavy body, ex. gr. will immediately fink in it to the bottom, and a light one emerge to the top: each grain of fand has a conftant vibratory and dancing motion; and if a hole be made in the fide of the difh, the fand will fpin out like water.

The Cartefians bring divers confiderations to prove that the parts of Fluids are in continual motion: as 1ft,

The change of folids into Fluids, ex. gr. ice into water, and vice verfa; the chief difference between the body in thofe two ftates confifting in this, that the parts, being fixed and at reft in the one, refift the touch; whereas in the other, being already in motion, they give way to the flighteft impulfe. 2dly, The effects of Fluids, which commonly proceed from motion: fuch are the infinuation of Fluids among the pores of bodies; the foftening and diffolving hard bodies; the actions of corrofive menftruums; &c: Add, that no folid can be brought to a ftate of Fluidity, without the intervention of fome moving or moveable body, as fire, air, or water. Air, the fame gentlemen hold to be the firft fpring of thefe caufes of Fluidity, it being this that gives motion to fire and water, though itfelf receives its motion and action from the ether, or fubtle medium.

But Boerhaave pleads ftrenuoufly that fire is the firft mover, and the caufe of all Fluidity in other bodies, as air. water, &c: without this, he fhews that the atmofphere itfelf would fix into one folid mafs. And in like manner, Dr. Black, of Edinburgh, mentions Fluidity as an effect of heat. The different degrees of heat which are required to bring different bodies into a ftate of Fluidity, he fuppofes may depend on fome particulars in the mixture and compofition of the bodies themfelves: which is rendered farther probable from confidering that the natural ftate of bodies in this refpect is changed by certain mixtures; thus, when two metals are compounded, the mixture is commonly more fufible than either of them feparately.

Newton's idea of the caufe of Fluidity is different: he makes it to be the great principle of attraction. The various inteftine motion and agitation among the particles of Fluid bodies, he thinks is naturally accounted for, by fuppofing it a primary law of nature, that as all the particles of matter attract each other when within a certain diftance; fo at all greater diftances, they avoid and fly from one another. For then, though their common gravity, together with the preffure of other bodies upon them, may keep them together in a mafs, yet their continual endeavour to avoid one another fingly, and the adventitious impulfes of heat and light, or other external caufes, may make the particles of Fluids continually move round about one another, and fo produce this quality.

As therefore the caufe of cohefion of the parts of folid bodies appears to be their mutual attraction; fo, on this principle, the chief caufe of Fluidity feems to be a contrary motion, impreffed on the particles of Fluids; by which they avoid and fly from one another, as foon as they come at, and as long as they keep at, fuch a diftance from each other.

It is obferved alfo in all Fluids, that the direction of their preffure againft the veffels which contain them, is in lines perpendicular to the fides of fuch veffels; which property, being the neceffary refult of the fpherical figure of the particles of any Fluid, fhews that the parts of all Fluids are fo, or of a figure very nearly approaching to it.

FLUX, in Hydrography, a regular and periodical motion of the fea, happening twice in 24 hours and 48 minutes, nearly; in which time the water is raifed, and driven violently againft the fhores. The Flux, or Flow, is one of the motions of the tide: the other, by which the water sinks and retires, being called the Reflux, or Ebb. See Tide.

Between the Flux and Reflux there is a kind of rest or cessation, of about half an hour; during which time the water is at its greatest height, called High-water.

The Flux of the sea follows chiefly the course of the moon; and is always highest and greatest at new and full moons, particularly near the time of the equinoxes. In some parts, as at Mount St. Michael, it rises 80 or 90 feet, though in the open sea it never rises above a foot or two; and in some places, as about the Morea, there is no flux at all. It runs up some rivers above 120 miles: though up the river Thames it goes only about 80, viz, near to Kingston in Surry. Above London-bridge, the water flows 4 hours, and ebbs 8; and below the bridge, it flows 5 hours, and ebbs 7.

FLUXION, in the Newtonian analysis, denotes the rate or proportion at which a flowing or varying quantity increases its magnitude or quantity; and it is proportional to the magnitude by which the flowing quantity would be uniformly increased, in a given time, by the generating quantity continuing of the invariable magnitude it has at the moment of time for which the Fluxion is taken: by which it stands contradistinguished from fluent, or the flowing quantity, which is gradually, and indefinitely increasing, after the manner of a space which a body in motion describes.

Mr. Simpson observes, that there is an advantage in thus considering Fluxions, not as mere velocities of increase at a certain point, but as the magnitudes which would be uniformly generated in a given finite time: the imagination is not here confined to a single point, and the higher orders of Fluxions are rendered much more easy and intelligible. And though Sir Isaac Newton defines Fluxions to be the velocities of motions, yet he has recourse to the moments or increments, generated in equal particles of time, to determine those velocities, which he afterwards directs to expound by finite magnitudes of other kinds.

As to the illustration of this definition, and the rules for finding the Fluxions of all sorts of fluent quantities, see the following article, or the Method of Fluxions.

Method of Fluxions, is the algorithm and analysis of Fluxions, and fluents or flowing quantities.

Most foreigners define this as the method of differences or differentials, being the analysis of indefinitely small quantities. But Newton, and other English authors, call these infinitely small quantities, moments; considering them as the momentary increments of variable quantities; as of a line considered as generated by the flux or motion of a point, or of a surface generated by the flux of a line. Accordingly, the variable quantities are called Fluents, or flowing quantities; and the method of finding either the Fluxion, or the fluent, the method of Fluxions.

M. Leibnitz considers the same infinitely small quantities as the differences, or differentials, of quantities; and the method of finding those differences, he calls the Differential Calculus.

Besides this difference in the name, there is another in the notation. Newton expresses the Fluxion of a quantity, as of x, by a dot placed over it, thus $\dot{x}$;

while Leibnitz expresses his differential of the same x, by prefixing the initial letter d, as dx. But, setting aside these circumstances, the two methods are just alike.

The Method of Fluxions is one of the greatest, most subtle, and sublime discoveries of perhaps any age: it opens a new world to our view, and extends our knowledge, as it were, to infinity; carrying us beyond the bounds that seemed to have been prescribed to the human mind, at least infinitely beyond those to which the ancient geometry was confined.

The history of this important discovery, recent as it is, is a little dark, and embroiled. Two of the greatest men of the last age have both of them claimed the invention, Sir I. Newton, and M. Leibnitz; and nothing can be more glorious for the method itself, than the zeal with which the partisans of either side have asserted their title.

To exhibit a just view of this dispute; and of the pretensions of each party, we may here advert to the origin of the discovery, and mark where each claim commenced, and how it was supported.

The principles upon which the Method of Fluxions is founded, or which conducted to it, had been laying, and gradually developing, from the beginning of the last century, by Fermat, Napier, Barrow, Wallis, Slusius, &c, who had methods of drawing tangents, of maxima and minima, of quadratures, &c, in certain particular cases, as of rational quantities, upon nearly the same principles. And it was not wonderful that such a genius as Newton should soon after raise those faint beginnings into a regular and general system of science, which he did about the year 1665, or sooner.

The first time however that the method appeared in print, was in 1684, when M. Leibnitz gave the rules of it in the Leipsic Acts of that year; but without the demonstrations. The two brothers however, John and James Bernoulli, being greatly struck with this new method, applied themselves diligently to it, found out the demonstrations, and applied the calculus with great success.

But before this, M. Leibnitz had proposed his Differential Method, viz, in a letter, dated Jan. 21, 1677, in which he exactly pursues Dr. Barrow's method of tangents, which had been published in 1670: and Newton communicated his method of drawing tangents to Mr. Collins, in a letter dated Dec. 10, 1672; which letter, together with another dated June 13, 1676, were sent to Mr. Leibnitz by Mr. Oldenburgh, in 1676. So that there is a strong presumption that he might avail himself of the information contained in these letters, and other papers transmitted with them, and also in 1675, before the publication of his own letter, containing the first hint of his differential method. Indeed it sufficiently appears that Newton had invented his method before the year 1669, and that he actually made use of it in his Compendium of Analysis and Quadrature of Curves before that time. His attention seems to have been directed this way, even before the time of the plague which happened in London in 1665 and 1666, when he was about 28 years of age.

This is all that is heard of the method, till the year 1687, when Newton's admirable Principia came out, which

which is almoſt wholly built on the ſame calculus. The common opinion then was, that Newton and Leibnitz had each invented it about the ſame time: and what ſeemed to confirm it was, that neither of them made any mention of the other; and that, though they agreed in the ſubſtance of the thing, yet they differed in their ways of conceiving it, calling it by different names, and uſing different characters. However, foreigners having firſt learned the method through the medium of Leibnitz's publication, which ſpread the method through Europe, thoſe geometricians were inſenſibly accuſtomed to look upon him as the ſole, or principal inventor, and became ever after ſtrongly prejudiced in favour of his notation and mode of conceiving it.

The two great authors themſelves, without any ſeeming concern, or diſpute, as to the property of the invention, enjoyed the glorious proſpect of the progreſſes continually making under their auſpices, till the year 1699, when the peace began to be diſturbed.

M. Facio, in a treatiſe on the Line of Swifteſt Deſcent, declared, that he was obliged to own Newton as the firſt inventor of the differential calculus, and the firſt by many years; and that he left the world to judge, whether Leibnitz, the ſecond inventor, had taken any thing from him. This preciſe diſtinction between firſt and 2d inventor, with the ſuſpicion it inſinuated, raiſed a controverſy between M. Leibnitz, ſupported by the editors of the Leipſic Acts, and the Engliſh mathematicians, who declared for Newton. Sir Iſaac himſelf never appeared on the ſcene; his glory was become that of the nation; and his adherents, warm in the cauſe of their country, needed not his aſſiſtance to animate them.

Writings ſucceeded each other but ſlowly, on either ſide; probably on account of the diſtance of places; but the controverſy grew ſtill hotter and hotter: till at length M. Leibnitz, in the year 1711, complained to the Royal Society, that Dr. Keil had accuſed him of publiſhing the Method of Fluxions invented by Sir I. Newton, under other names and characters. He inſiſted that nobody knew better than Sir Iſaac himſelf, that he had ſtolen nothing from him; and required that Dr. Keil ſhould diſavow the ill conſtruction which might be put upon his words.

The Society; thus appealed to as a judge, appointed a committee to examine all the old letters, papers, and documents, that had paſſed among the ſeveral mathematicians, relating to the point; who, after a ſtrict examination of all the evidence that could be procured, gave in their report as follows: " That Mr. Leibnitz " was in London in 1673, and kept a correſpondence " with Mr. Collins by means of Mr. Oldenburgh, till " Sept. 1676, when he returned from Paris to Hano- " ver, by way of London and Amſterdam: that it did " not appear that Mr. Leibnitz knew any thing of the " differential calculus before his letter of the 21ſt of " June, 1677, which was a year after a copy of a let- " ter, written by Newton in the year 1672, had been " ſent to Paris to be communicated to him, and above " 4 years after Mr. Collins began to communicate that " letter to his correſpondents; in which the Method of " Fluxions was ſufficiently explained, to let a man of his " ſagacity into the whole matter: and that Sir I. New-

" ton had even invented his method before the year " 1669, and conſequently 15 years before M. Leibnitz " had given any thing on the ſubject in the Leipſic " Acts." From which they concluded that Dr. Keil had not at all injured M. Leibnitz in what he had ſaid.

The Society printed this their determination, together with all the pieces and materials relating to it, under the title of Commercium Epiſtolicum de Analyſi Promota, 8vo, Lond. 1712. This book was carefully diſtributed through Europe, to vindicate the title of the Engliſh nation to the diſcovery; for Newton himſelf, as already hinted, never appeared in the affair: whether it was that he truſted his honour with his compatriots, who were zealous enough in the cauſe; or whether he felt himſelf even ſuperior to the glory of it.

M. Leibnitz and his friends however could not ſhew the ſame indifference: he was accuſed of a theft; and the whole Commercium Epiſtolicum either expreſſes it in terms, or inſinuates it. Soon after the publication therefore, a looſe ſheet was printed at Paris, in behalf of M. Leibnitz, then at Vienna. It is written with great zeal and ſpirit; and it boldly maintains that the Method of Fluxions had not preceded the Method of Differences; and even inſinuates that it might have ariſen from it. The detail of the proofs however, on each ſide, would be too long, and could not be underſtood without a large comment, which muſt enter into the deepeſt geometry.

M. Leibnitz had begun to work upon a Commercium Epiſtolicum, in oppoſition to that of the Royal Society; but he died before it was completed.

A ſecond edition of the Commercium Epiſtolicum was printed at London in 1722; when Newton, in the preface, account, and annotations, which were added to that edition, particularly anſwered all the objections which Leibnitz and Bernoulli were able to make ſince the Commercium firſt appeared in 1712; and from the laſt edition of the Commercium, with the various original papers contained in it, it evidently appears that Newton had diſcovered his Method of Fluxions many years before the pretenſions of Leibnitz. See alſo Raphſon's Hiſtory of Fluxions.

There are however, according to the opinion of ſome, ſtrong preſumptions in favour of Leibnitz; i. e. that he was no plagiary. for that Newton was at leaſt the firſt inventor, is paſt all diſpute; his glory is ſecure; the reaſonable part, even among the foreigners, allow it: and the queſtion is only, whether Leibnitz took it from him, or fell upon the ſame thing with him; for, in his theory of abſtract notions, which he dedicated to the Royal Academy in 1671, before he had ſeen any thing of Newton's, he already ſuppoſed infinitely ſmall quantities, ſome greater than others; which is one of the great principles of his ſyſtem.

Before proſecuting farther the hiſtory and improvements of this ſcience, it will be proper to premiſe ſomewhat of the principles and practice of it, according to the ideas of the inventor.

Principles of the Method of FLUXIONS.

1. In the doctrine of Fluxions, magnitudes or quantities, of all kinds, are conſidered, not as made up of a number

number of fmall parts, but as generated by continued motion, by means of which they increafe or decreafe: as a line by the motion of a point; a furface by the motion of a line; and a folid by the motion of a fur-face: which is no new principle in geometry; having been ufed by Euclid and Archimedes. So likewife, time may be confidered as reprefented by a line, increaf-ing uniformly by the motion of a point. And quan-tities of all kinds whatever, which are capable of in-creafe and decreafe, may in like manner be reprefented by, lines, furfaces, or folids, confidered as generated by motion.

2. Any quantity, thus generated, and variable, is called a Fluent, or a flowing quantity. And the rate or proportion according to which any flowing quantity increafes, at any pofition or inftant, is the Fluxion of the faid quantity, at that pofition or inftant: and it is proportional to the magnitude by which the flowing quantity would be uniformly increafed, in a given time, with the generating celerity uniformly continued du-ring that time.

3. The fmall quantities that are actually generated or defcribed, in any fmall given time, and by any con-tinued motion, either uniform or variable, are called In-crements.

4. Hence, if the motion of increafe be uniform, by which increments are generated, the increments will in that cafe be proportional, or equal, to the meafures of the Fluxions: but if the motion of increafe be accele-rated, the increments fo generated, in a given finite time, will exceed the Fluxion; and if it be a decreafing motion, the increment fo generated, will be lefs than the Fluxion. But if the time be indefinitely fmall, fo that the motion be confidered as uniform for that in-ftant; then thefe nafcent increments will always be pro-portional or equal to the Fluxions, and may be fubfti-tuted for them, in any calculation.

5. To illuftrate thefe definitions: Suppofe a point *m* be conceived to move from the pofition A, and to

generate a line AP, with a motion any-how regulated; and fuppofe the celerity of the point *m*, at any pofition P, to be fuch, as would, if from thence it fhould become, or continue, uniform, be fufficient to defcribe, or pafs uniformly over, the diftance P*p*, in the given time allowed for the Fluxion: then will the faid line P*p* reprefent the Fluxion of the faid fluent or flowing line AP, at that pofition.

6. Again, fuppofe the right line *mn* to move, from the pofition AB, continually parallel to itfelf,

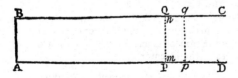

with any continued motion, fo as to generate the fluent, or flowing rectangle ABQP, whilft the point *m* de-fcribes the line AP; alfo let the diftance P*p* be taken,

as above, to exprefs the Fluxion of the line or bafe AP; and complete the rectangle PQ *qp*. Then, like as P*p* is the Fluxion of the line AP, fo is the fmall pa-rallelogram P*q* the Fluxion of the flowing parallelo-gram, AQ; both thefe Fluxions or increments being uniformly defcribed in the fame time.

7. In like manner, if the folid AERP be con-ceived as generated by the plane PQR moving,

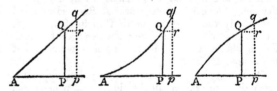

from the pofition ABE, always parallel to itfelf, along the line AD; and if P*p* denote the Fluxion of the line AP. Then, like as the parallelogram P*q*, or P*p* × PQ, exprefles the Fluxion of the flowing rec-tangle AQ, fo likewife fhall the Fluxion of the varia-ble folid or prifm AR be exprefled by the prifm P*r*, or P*p* × the plane PR. And in both thefe laft two cafes, it appears that the Fluxion of the generated rec-tangle, or prifm, is equal to the product of the gene-rant, whether line or plane, drawn into the Fluxion of the line along which it moves.

8. Hitherto the generant, or generating line or plane, has been confidered as of a conftant or invaria-ble magnitude; in which cafe the fluent, or quantity generated, is a parallelogram, or a prifm, the former being defcribed by the motion of a line, and the latter by the motion of a plane. In like manner are other figures, whether plane or folid, conceived to be de-fcribed, by the motion of a variable magnitude, whether it be a line or a plane. Thus, let a variable line PQ be carried with a parallel motion along AP, or whilft a point P is carried along, and defcribes, the line AP,

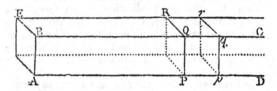

fuppofe another point Q to be carried by a motion perpendicular to the former, and to defcribe the line PQ: let *pq* be another pofition of PQ, indefinitely near to the former; and draw Q*r* parallel to AP. Now in this cafe there are feveral fluents or flowing quantities, with their refpective Fluxions: viz, the line or fluent AP, the Fluxion of which is P*p*, or Q*r*; the line or fluent PQ, the Fluxion of which is *qr*; the curve, or oblique line AQ, defcribed by the oblique motion of the point, the Fluxion of which is Q*q*; and laftly the furface APQ, defcribed by the variable line PQ, and the Fluxion of which is the rectangle PQ*rp*, or PQ × P*p*. And in the fame manner may any folid be conceived to be defcribed, by the motion of a vari-able plane parallel to itfelf, fubftituting the variable plane for the variable line; in which cafe, the Fluxion of the folid, at any pofition, is reprefented by the va-

riable

riable plane, at that pofition, drawn into the Fluxion of the line along which it is carried.

9. Hence then it follows generally, that the Fluxion of any figure, whether plane or folid, at any pofition, is equal to the fection of it, at that pofition, drawn into the Fluxion of the axis, or line along which the variable fection is fuppofed to be perpendicularly carried; i. e. the Fluxion of the figure AQP, is equal the plane PQ × Pp when that figure is a folid, or to the ordinate PQ × Pp when the figure is a furface.

10. It alfo follows, from the fame premifes, that, in any curve, or oblique line, AQ, whofe abfcifs is AP, and ordinate is PQ, the Fluxions of thefe three form a fmall right-angled plane triangle Q qr; for Qr = Pp is the Fluxion of the abfcifs AP, qr the Fluxion of the ordinate PQ, and Qq the Fluxion of the curve or right line AQ. And confequently that, in any curve, the fquare of the Fluxion of the curve, is equal to the fum of the fquares of the Fluxions of the abfcifs and ordinate, when thefe two lines are at right angles to each other.

11. From the premifes it alfo appears, that contemporaneous fluents, or quantities that flow or increafe together, which are always in a conftant ratio to each other, have their Fluxions alfo in the fame conftant ratio at every pofition. For, let AP and BQ be two

A   P p B   Q q

contemporaneous fluents, defcribed in the fame time by the motion of the points P and Q, the contemporaneous pofitions being P, Q, and p, q; and let AP be to BQ, or Ap to Bq, in the conftant ratio of n to 1.

Then is - - - AP = n × BQ,
and - - - Ap = n × Bq;
therefore by fubtraction, - Pp = n × Qq;
that is, the Fluxion Pp : Fluxion Qq :: n : 1,
the fame as Fluent AP : Fluent BQ :: n : 1;
or the Fluxions and Fluents are in the fame conftant ratio.

But if the ratio of the fluents be variable, fo will that of the Fluxions be alfo, though not in the fame variable ratio with the former, at every pofition.

The Notation, &c, in FLUXIONS.

12. To apply the foregoing principles to the determination of the Fluxions of algebraic quantities, by means of which thofe of all other kinds are determined, it will be neceffary firft to premife the notation ufed in this fcience, with fome obfervations. As, firft, that the final letters of the alphabet z, y, x, w, &c, are ufed to denote variable or flowing quantities; and the initial letters a, b, c, d, &c, conftant or invariable ones: Thus, the variable bafe AP of the flowing rectangular figure ABQP, at art. 6, may be reprefented by x; and the invariable altitude PQ, by a: alfo the variable bafe or abfcifs AP, of the figures in art. 8, may be reprefented by x; the variable ordinate PQ, by y; and the variable curve or line AQ, by z.

Secondly, that the Fluxion of a quantity denoted by a fingle letter, is reprefented by the fame letter with a point over it: Thus the Fluxion of x is expreffed

by $\dot{x}$, that of y by $\dot{y}$, and that of z by $\dot{z}$. As to the Fluxions of conftant or invariable quantities, as of a, b, c, &c, they are equal to o or nothing, becaufe they do not flow, or change their magnitude.

Thirdly, that the increments of variable or flowing quantities, are alfo denoted by the fame letters with a fmall (′) over them: So the increments of x, y, z, are $\dot{x}$, $\dot{y}$, $\dot{z}$.

13. From thefe notations, and the foregoing principles, the quantities and their Fluxions, there confidered, will be denoted as below.

In all the foregoing figures, put
the variable or flowing line - - - AP = x,
in art. 6, the conftant line - - - PQ = a,
in art. 8, the variable ordinate - - PQ = y,
the variable curve or right line - - AQ = z;
Then fhall the feveral Fluxions be thus reprefented, viz,
$\dot{x}$ = Pp the Fluxion of the line AP,
$a\dot{x}$ = PQ qp the Fluxion of ABQP in art. 6,
$y\dot{x}$ = PQrp the Fluxion of APQ in art. 8,
$\dot{z}$ = Qq = $\sqrt{\dot{x}^2 + \dot{y}^2}$ the Fluxion of AQ,
and $a\dot{x}$ = Pr the Flux. of the folid in art. 7, if
a denote the conftant generating plane PQR.
Alfo $n x$ = BQ in the figure to art. 11,
and $n\dot{x}$ = Qq the Fluxion of the fame.

14. The principles and notation being now laid down, we may proceed to the practice and rules of this doctrine, which confifts of two principal parts, called the direct and inverfe method of Fluxions; viz, the direct method, which confifts in finding the Fluxion of any propofed fluent or flowing quantity; and the inverfe method, which confifts in finding the fluent of any propofed Fluxion. As to the former of thefe two problems, it can always be determined, and that in finite algebraic terms; but the latter, or finding of fluents, only in fome certain cafes, except by means of infinite feries.—Firft then, of

The Direct Method of FLUXIONS.

15. To find the Fluxion of the product or rectangle of two variable quantities;
let ARQP = xy be the flowing or variable rectangle, generated by two lines RQ and PQ moving always perpendicular to each other, from the pofitions AP and AR; denoting the one by x, and the other by y; and

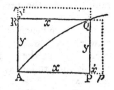

fuppofe x and y to be fo related, that the curve AQ always paffes through their interfection Q, or the oppofite angle of the rectangle.

Now this rectangle confifts of the two trilineal fpaces APQ, ARQ, of which the Fluxion of the former is PQ × Pp or $\dot{x}y$, and that of the latter is RQ × Rr or $x\dot{y}$, by art. 8; therefore the fum of the two, $\dot{x}y + x\dot{y}$, is the Fluxion of the whole rectangle xy or ARQP.

The fame otherwife.—Let the fides of the rectangle, x and y, by flowing, become $x + \dot{x}$ and $y + \dot{y}$: then the product of the two, or $xy + \dot{x}y + x\dot{y} + \dot{y}\dot{y}$ will be the new or contemporaneous value of the flowing rectangle PR or xy; fubtract the one value from the other,

and

and the remainder $x\dot{y} + \dot{x}y + \dot{x}\dot{y}$, will be the increment generated in the same time as $\dot{x}$ or $\dot{y}$; of which the last term $\dot{x}\dot{y}$ is nothing, or indefinitely small in respect of the other two terms, because $\dot{x}$ and $\dot{y}$ are indefinitely small in respect of x and y; which term being therefore omitted, there remains $x\dot{y} + \dot{x}y$ for the value of that increment: and hence, by substituting $\dot{x}$ and $\dot{y}$ for $\dot{x}$ and $\dot{y}$, to which they are proportional, there arises $x\dot{y} + \dot{x}y$ for the value of the Fluxion of xy; the same as before.

17. Hence may be derived the Fluxions of all powers and products, and of all other forms of algebraic quantities whatever. And first for the continual products of any number of quantities, as xyz, or $wxyz$, or $vwxyz$, &c. For xyz put q or pz, so that $p = xy$, and $xyz = pz = q$. Now, taking the Fluxion of $q = pz$, by the last article, it is $\dot{q} = \dot{p}z + p\dot{z}$; but $p = xy$, and so $\dot{p} = \dot{x}y + x\dot{y}$ by the same article; substituting therefore these values of p and $\dot{p}$ instead of them, in the value of $\dot{q}$, this becomes $\dot{q} = \dot{x}yz + x\dot{y}z + xy\dot{z}$, the Fluxion of xyz required; which is therefore equal to the sum of the products arising from the Fluxion of each letter or quantity multiplied by the product of the other two.

Again, to determine the Fluxion of $wxyz$, the continual product of four variable quantities; put this product, viz. $wxyz$ or $qw = r$, where $q = xyz$ as above; then, taking the Fluxion by the last article, $\dot{r} = \dot{q}w + q\dot{w}$; and this, by substituting for q and $\dot{q}$ their values as above, becomes $\dot{r} = \dot{w}xyz + w\dot{x}yz + wx\dot{y}z + wxy\dot{z}$, the Fluxion of $wxyz$ as required; consisting of the Fluxion of each quantity drawn into the products of the other three.

In the very same manner it is found that the Fluxion of $vwxyz$ is $\dot{v}wxyz + v\dot{w}xyz + vw\dot{x}yz + vwx\dot{y}z + vwxy\dot{z}$; and so on, for any number of quantities whatever; in which it is always found that there are as many terms as there are variable quantities in the proposed fluent, and that these terms consist of the Fluxion of each variable quantity multiplied by the product of all the rest of the quantities.

18. From hence is easily derived the Fluxion of any power of a variable quantity, as of x^2, or x^3, or x^4, &c. For, in the rectangle or product xy, if $x = y$, then is the product $xy = xx$ or x^2, and also its Fluxion $\dot{x}y + x\dot{y} = \dot{x}x + x\dot{x}$ or $2x\dot{x}$, the Fluxion of x^2.

Again, if all the three x, y, z be equal; then is the product of the three $xyz = xxx$ or x^3; and its Fluxion $\dot{x}yz + x\dot{y}z + xy\dot{z} = \dot{x}xx + x\dot{x}x + xx\dot{x}$ or $3x^2\dot{x}$, the Fluxion of x^3.

And in the same manner it will appear that the Fluxion of x^4 is $= 4x^3\dot{x}$,

 that of x^5 is $= 5x^4\dot{x}$,

  that of x^n is $= nx^{n-1}\dot{x}$;

where n is any positive whole number. That is, the Fluxion of any positive integral power, is equal to the exponent of the power (n), multiplied by the next less power of the same quantity (x^{n-1}), and by the Fluxion of the root $(\dot{x})$.

19. Next, for the Fluxion of any fraction, as $\frac{\dot{x}}{y}$ of one variable quantity divided by another; put the proposed fraction $\frac{x}{y} = q$; then multiplying by the denominator, $x = qy$; and, taking the Fluxions, $\dot{x} = \dot{q}y + q\dot{y}$, or $\dot{q}y = \dot{x} - q\dot{y}$; and, by division, $\dot{q} = \frac{\dot{x}}{y} - \frac{q\dot{y}}{y} =$ (by substituting the value of q, or $\frac{x}{y}$) $\frac{\dot{x}}{y} - \frac{x\dot{y}}{y^2} = \frac{\dot{x}y - x\dot{y}}{y^2}$ the Fluxion of $\frac{x}{y}$, as required. That is, the Fluxion of any fraction, is equal to the Fluxion of the numerator drawn into the denominator, minus the Fluxion of the denominator drawn into the numerator, and the remainder divided by the square of the denominator.

20. Hence too is easily derived the Fluxion of any negative integer power of a variable quantity, as of x^{-n} or $\frac{1}{x^n}$, which is the same thing. For here the numerator of the fraction is 1, whose Fluxion is nothing; and therefore, by the last article, the Fluxion of such a fraction, or negative power, is barely equal to minus the Fluxion of the denominator, divided by the square of the said denominator. That is, the Fluxion of x^{-n}, or $\frac{1}{x^n}$, is $-\frac{nx^{n-1}\dot{x}}{x^{2n}}$ or $-\frac{n\dot{x}}{x\cdot n^1}$ or $-nx^{-n-1}\dot{x}$; which is the same rule as before for integral powers.

Or, the same thing is otherwise derived immediately from the Fluxion of a rectangle or product, thus: put the proposed fraction, or quotient, $\frac{1}{x^n} = q$; then is $qx^n = 1$; and, taking the Fluxions, $\dot{q}x^n + qnx^{n-1}\dot{x} = 0$; hence $\dot{q}x^n = -qnx^{n-1}\dot{x}$, and (dividing by x^n), $\dot{q} = -\frac{qn\dot{x}}{x} =$ (by substituting $\frac{1}{x^n}$ for q), $\frac{-n\dot{x}}{x^{n+1}}$ or $-nx^{-n-1}\dot{x}$; the same as before.

21. Much in the same manner is obtained the Fluxion of any surd, or fractional power of a fluent quantity, as of $x^{\frac{m}{n}}$ or $\sqrt[n]{x^m}$. For, putting the proposed quantity $x^{\frac{m}{n}} = q$, then, raising each to the n power, $x^m = q^n$; take the Fluxions, $mx^{m-1}\dot{x} = nq^{n-1}\dot{q}$; divide by nq^{n-1}, $\dot{q} = \frac{mx^{m-1}\dot{x}}{nq^{n-1}} = \frac{mx^{m-1}\dot{x}}{nx^{m-\frac{m}{n}}} = \frac{m}{n}x^{\frac{m}{n}-1}\dot{x}$: which is still the same rule as before, for finding the Fluxion of any power of a fluent quantity, and which is therefore general, whether the exponent be positive or negative, or integral or fractional.

22. For

22. For the Fluxions of Logarithms: Let A be the principal vertex of an hyperbola, having its afymptotes CD, CP, with the ordinates DA, BA, PQ, &c, parallel to them. Then, from the nature of the hyperbola, and of logarithms, it is known that any fpace ABPQ is the log. of the ratio of CB to CP, to the modulus ABCD. Now put $1 =$ CB or BA the fide of the fquare or rhombus DB; $m =$ the modulus, or area of DB, or fine of the angle C to the radius 1; alfo the abfcifs CP $= x$, and the ordinate PQ $= y$. Then, by the nature of the hyperbola, CP $\times$ PQ is always equal to DB, that is, $xy = m$; hence $y = \dfrac{m}{x}$, and the fluxion of the fpace,

or $\dot{x}y$ is $\dfrac{m\dot{x}}{x} =$ PQ qp the fluxion of the log. of x, to the modulus m. And in the ordinary hyp. logarithms the modulus m being 1, therefore $\dfrac{\dot{x}}{x}$ is the fluxion of the hyp. log. of x; which is therefore equal to the Fluxion of the quantity, divided by the quantity itfelf. And the fame might be brought out in feveral other ways, independent of the figure of the hyperbola.

23. By means of the Fluxions of logarithms, are determined thofe of exponential quantities, i. e. quantities which have their exponent alfo a flowing or variable quantity. Thefe exponentials are of two kinds, viz, when the root is a conftant quantity, as e^x; and when the root is variable, as y^x.

In the former cafe, put the propofed exponential $e^x = z$, a fingle variable quantity; then take the logarithm of each, fo fhall log. of $z = x \times$ log. of e; take the fluxions of thefe, fo fhall $\dfrac{\dot{z}}{z} = \dot{x} \times$ log. of e; hence $\dot{z} = z\dot{x} \times$ log. of $e = e^x \dot{x} \times$ log. of e, the fluxion of the propofed exponential e^x; and which therefore is equal to the faid propofed quantity, drawn into the fluxion of the exponent, and alfo into the log. of the root.

24. Alfo in the 2d cafe, put the exponential $y^x = z$; then the logarithms give log. $z = x \times$ log. y, and the fluxions give $\dfrac{\dot{z}}{z} = \dot{x} \times$ log. $y + x \times \dfrac{\dot{y}}{y}$; hence $\dot{z} =$ $z\dot{x} \times$ log. $y + \dfrac{zx\dot{y}}{y} =$ (by fubftituting y^x for z) $y^x\dot{x} \times$ log. $y + xy^{x-1}\dot{y}$, is the fluxion of the propofed exponential y^x; which therefore confifts of two terms, of which the one is the fluxion of the propofed quantity confidering the exponent only as conftant, and the other is the fluxion of the fame quantity confidering the root as conftant.

Of Second, Third, &c FLUXIONS.—Having explained the manner of confidering and determining the firft fluxions of flowing or variable quantities; it remains now to confider thofe of the higher orders, as 2d, 3d, 4th, &c, fluxions.

25. If the rate or celerity with which any flowing quantity changes its magnitude, be conftant, or the fame, at every pofition; then is the fluxion of it alfo

conftantly the fame. But if the variation of magnitude be continually changing, either increafing or decreafing; then will there be a certain degree of fluxion peculiar to every point or pofition; and the rate of variation or change in the fluxion, is called the *Fluxion of the Fluxion*, or the *fecond Fluxion* of the given fluent quantity. In like manner, the variation or fluxion of this 2d fluxion is called the *third Fluxion* of the firft propofed fluent quantity; and fo on.

And thefe orders of fluxions are denoted by the fluent letter or quantity, with the corresponding number of points over it; viz, 2 points for the 2d fluxion, 3 for the 3d fluxion, 4 for the 4th fluxion, and fo on. So the different orders of the fluxions of x, are $\dot{x}, \ddot{x}, \dddot{x}, \ddddot{x}$, &c; where each is the fluxion of the one next before it.

26. This defcription of the higher orders of fluxions may be illuftrated by the three figures at the 8th article; where, if x denote the abfcifs AP, and y the ordinate PQ; and if the ordinate PQ or y flow along the abfcifs AP or x, with an uniform motion; then the fluxion of x, viz $\dot{x} =$ Pp or Qr is a conftant quantity, or $\ddot{x} = 0$, in all the figures. Alfo, in fig. 1, in which AQ is a right line, $\dot{y}$ is $= rq$, or the fluxion of PQ, is a conftant quantity, or $\ddot{y} = 0$; for, the angle Q, $=$ the angle A, being conftant, Qr is to rq, or $\dot{x}$ to $\dot{y}$, in a conftant ratio. But in the 2d figure, rq, or the fluxion of PQ, continually increafes more and more; and in fig. 3 it continually decreafes more and more; and therefore in both thefe cafes y has a 2d fluxion, being pofitive in fig. 2, but negative in fig. 3: and fo on for the other orders of fluxions.

27. Thus, if for inftance, the nature of the curve be fuch, that x^3 is everywhere equal to a^2y; then, taking the fluxions, it is $a^2\dot{y} = 3x^2\dot{x}$; and, confidering $\dot{x}$ always as a conftant quantity, and taking always the fluxions, the equations of the feveral orders of fluxions will be as below; viz,

the 1ft fluxions $a^2\dot{y} = 3x^2\dot{x}$,
the 2d fluxions $a^2\ddot{y} = 6x\dot{x}^2$,
the 3d fluxions $a^2\dddot{y} = 6\dot{x}^3$,
the 4th fluxions $a^2\ddddot{y} = 0$,

and all the higher fluxions $= 0$ or nothing.

Alfo the higher orders of fluxions are found in the fame manner as the lower ones. Thus,

The 1ft flux. of y^3 is - - $3y^2\dot{y}$;

its 2d flux. or flux. of $3y^2\dot{y}$, confidered as the rectangle of $3y^2$ and $\dot{y}$, is - - $\Big\}$ $3y^2\ddot{y} + 6y\dot{y}^2$;

and the flux. of this again, or the 3d fluxion of y^3, is - $\Big\}$ $3y^2\dddot{y} + 18y\dot{y}\ddot{y} + 6\dot{y}^3$.

28. In the foregoing articles, it has been fuppofed that the fluents increafe; or that their fluxions are pofitive; but it often happens that fome fluents decreafe, and that therefore their fluxions are negative: and whenever this is the cafe, the fign of the fluxion muft be changed, or made contrary to that of the fluent. So, of the rectangle xy, when both x and y increafe together, the fluxion is $\dot{x}y + x\dot{y}$: but if one of them, as y, decreafe, while the other, x, increafes; then the fluxion of y being $- \dot{y}$, the fluxion of xy will in that cafe be $\dot{x}y - x\dot{y}$. This may be illuftrated by the annexed rec-

tangle

tangle APQR $= xy$, supposed to be generated by the motion of the line PQ from A towards C, and by the motion of the line RQ from B towards A : For, by the motion of PQ, from A towards C, the rectangle is increased, and its fluxion is $+ \dot{x}y$; but by the motion of RQ, from B towards A, the rectangle is decreased, and the fluxion of the decrease is $x\dot{y}$; therefore, taking the fluxion of the decrease from that of the increase, the fluxion of the rectangle xy, when x increases and y decreases, is $\dot{x}y - x\dot{y}$.

For the Inverse Method, or the finding of fluents, see FLUENT. And for the several applications of this science to MAXIMA and MINIMA, the drawing of TANGENTS, &c, see the respective articles.

An idea of the principles of Fluxions being now delivered, as above, we may next consider somewhat of the chief writings and improvements that have been made by divers authors, since the first discovery of them : indeed some of the chief improvements may be learned by consulting the preface to Dr. Waring's Meditationes Analyticæ.

The inventor himself brought the doctrine of Fluxions to a considerable degree of perfection; as may be seen by many specimens of this science, given by him; particularly in his Principia, in his Tract on Quadratures, and in his Treatise on Fluxions, published by Mr. Colson; from all which it will appear, that he not only laid down the whole theory of this method, both direct and inverse; but also applied it in practice, to the solution of many of the most useful and important problems in mathematics and philosophy.

Various improvements however have been made by many illustrious authors on this science; particularly by John Bernoulli, who treated of the fluents belonging to the fluxions of exponential expressions; James Bernoulli, Craig, Cheyne, Cotes, Manfredi, Riccati, Taylor, Fagnanus, Clairaut, D'Alembert, Euler, Condorcet, Walmesley, Le Grange, Emerson, Simpson, Landen, Waring, &c. There are several other treatises also on the principles of Fluxions, by Hayes, Newyentyt, L'Hôpital, Hodson, Rowe, &c, &c, delivering the elements of this science in an easy and familiar way.

FLY, in Mechanics, a heavy weight applied to certain machines, to regulate their motions, as in a jack, or to increase their effect, as in the coining engine, &c; by means of which the force of the power is not only preserved, but equally distributed in all the parts of the revolution.

The Fly is either like a cross, with heavy weights at the ends of its arms; or like a heavy wheel at right-angles to the axis of motion. It may be applied to various sorts of engines, whether moved by men, horses, wind, or water; and is of great use in those parts of an engine having a quick circular motion, and where the power or resistance acts unequally in the different parts of a revolution. In this case the Fly becomes a moderator, making the motion of revolution almost everywhere equal.

FLYERS, in Architecture, such stairs as go straight, and do not wind round, nor have the steps made taper-ing, but equally broad at both ends. Hence, if one flight do not rise to the top of the story &c, there is a broad half pace, and then commonly another set of flyers.

FLYING, the progressive motion of a bird, or other winged animal, through the air.

The parts of birds chiefly concerned in Flying, are the wings and tail : by the former, the bird sustains and wafts himself along; and by the latter he is assisted in ascending and descending, to keep his body poised and upright, and steady. : The wings are extended or stretched quite out, and then struck forcibly downwards against the air, which by its resistance raises the bird upwards; then to make another stroke, the wing, by means of its joints, readily closes in some degree, presenting the sharp edge of the pinion foremost to cut the air, and drawing the collapsed feathers after it like a flag, to diminish the resistance to the ascent as much as possible; the wing and feathers are then stretched out horizontally again, and another downward stroke made, which raises the bird still higher; and so on as far as he pleases, or as the density of the air will sustain him; performing those motions of the wings very rapidly, that the flight may be the quicker.

Artificial FLYING, is that attempted by men, &c, by the assistance of mechanics.

The art of flying has been attempted by several persons in all ages. Friar Bacon, about 500 years ago, not only asserts the possibility of flying, but affirms that he himself knew how to make a machine with which a man might be able to convey himself through the air like a bird; and further adds, that it had been tried with success. Though the fact is to be doubted, if, as it was said, it consisted in the following method, viz, in a couple of large thin hollow copper globes, exhausted of air; which being much lighter than air, would sustain a chair on which a person might sit. Father Francisco Lana, in his Prodromo, proposes the same thing, as his own thought. He computes, that a round vessel of plate-brass, 14 feet in diameter, weighing 3 ounces the square foot, will only weigh 1848 ounces; whereas a quantity of air of the same bulk will weigh near 2156 ounces; so that the globe will not only be sustained in the air, but will carry with it a weight of 304 ounces; and by increasing the size of the globe, the thickness of the metal remaining the same, he adds, a vessel might be made to carry a much greater weight. But the fallacy is obvious : a globe of the dimensions he describes, as Dr. Hook observes, would not sustain the pressure of the air, but he crushed inwards. Indeed it is not probable that such a globe can be made of a thinness sufficient to float in the atmosphere after it is exhausted of air, and yet be strong enough to sustain the compressing force of the atmosphere. But for this purpose it seems that the globe should be filled with an air as elastic or strong as the atmosphere, and yet be very much lighter; such as has lately been used in the Mongolfiers and Balloons; the former of which is filled with common air heated, so as to be more elastic, and less heavy; and the latter with inflammable air, which is as elastic as the common air, with only about one tenth of its weight. And thus the idea of flying, or rather floating, in the air, has been lately realized by the moderns, using however a different sort of air. See AEROSTATION.

The

The same author deſcribes a machine for Flying, invented by the Sieur Beſnier, a ſmith of Sable, in the county of Main. See Philoſ. Collec. numb. 1.

By the foregoing method however, at beſt, only a method of floating can be obtained, like a log floating in a current ; but not of Flying, which conſiſts in moving through the air, independent of any current ; and which muſt be effected by ſomething in the nature of wings. Attempts of this latter kind alſo have indeed been made by ſeveral perſons of late years ; but it does not appear that any of them have been attended with ſuch ſucceſs as to induce the authors of thoſe attempts to make them public. The philoſophers of king Charles the ſecond's reign were much buſied about this art ; and the celebrated biſhop Wilkins was ſo confident of ſucceſs in it, that he ſays, he does not queſtion but in future ages it will be as uſual to hear a man call for his wings, when he is going a journey, as it is now to call for his boots.

The ſtory of the flight of Dædalus is well known.

FLYING *Pinion*, is part of a clock having a fly or fan with which to gather the air, and ſo bridle the rapidity of the clock's motion, when the weight deſcends in the ſtriking part.

FOCAL *Diſtance*, the Diſtance of the Focus, which is ſometimes underſtood as its diſtance from the vertex, as in the parabola ; and ſometimes its diſtance from the centre, as in the ellipſe or hyperbola.

FOCUS, in Geometry and the Conic Sections, is applied to certain points in the Ellipſe, Hyperbola, and Parabola, where the rays reflected from all parts of theſe curves do concur or meet ; that is, rays iſſuing from a luminous point in the one focus, and falling on all points of the curves, are reflected into the other Focus, or into the line directed to the other Focus, viz. into the other Focus in the ellipſe and parabola, and directly from it in the hyperbola. Which is the reaſon of the name Focus, or Burning-point. Hence, as the one Focus of the parabola is at an infinite diſtance ; and conſequently all rays drawn from it, to any finite part of the curve about the vertex, are parallel to one another ; therefore if rays from the ſun, or any other object ſo diſtant as that thoſe rays may be accounted parallel, fall upon the curve of a parabola or concave ſurface of a paraboloidal figure, thoſe rays will all be reflected into its Focus.

Thus, the rays Pf, from the Focus f, are reflected in the direction PF, into the other Focus F, in the ellipſe and parabola, and form the Focus F, into FQ, in the hyperbola.

In all the three curves, the double ordinate CD drawn through the Focus F, is the parameter of the axis, or a 3d proportional to AB and ab, the tranſverſe and conjugate axes.

In the ellipſe and parabola, the tranſverſe axis is equal to the ſum of the two lines $PF + Pf$, drawn from the Foci to any point P in the curve ; but in the hyperbola, the tranſverſe is equal to the difference of thoſe two lines. That is,

$AB = PF + Pf$ in the ellipſe and parabola;
$AB = PF - Pf$ in the hyperbola.

In the ellipſe and parabola, the ſquare of the diſtance between the Foci, is equal to the difference of the ſquares of the two axes ; and in the hyperbola, it is equal to the ſum of their ſquares : that is

$Ff^2 = AB^2 - ab^2$ in the ellipſe and parabola.
$Ff^2 = AB^2 + ab^2$ in the hyperbola.

Therefore the two ſemi-axes, with the diſtance of the Focus from the centre, form always a right-angled triangle FaE, or AaE.

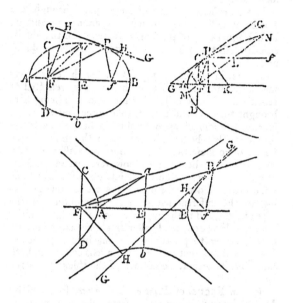

In all the curves, the conjugate ſemi-axis is a mean proportional between the diſtances of either Focus from either end of the tranſverſe axis : that is,

AF : Ea :: Ea : FB,
or Ea^2 = AF . FB.

If there be any tangent to theſe curves, and two lines drawn from the Foci to the point of contact ; theſe two lines will make equal angles with that tangent.

So, if GPG touch the curve at P ;
then is the angle FPG = ∠ fPG.

If a line be drawn from either Focus, perpendicularly upon a tangent ; the diſtance of their interſection from the centre will be equal to the ſemi tranſverſe axis. So, if FH or fH be perpendicular to the tangent PH ; then is EH = EA or EB. Conſequently, the circle deſcribed on the diameter AB, will paſs through all the points H.

The foregoing are the chief properties that are common to the Foci of all the three conic ſections. To which may be added the following properties which are peculiar to the parabola : viz.

In the parabola, the diſtance from the Focus to the vertex, is equal to $\frac{1}{4}$ of the parameter, or half the ordinate at the Focus : viz, AF = $\frac{1}{2}$ FC.

Alſo, a line drawn from the Focus to any point in the curve, is equal to the ſum of the Focal diſtance from the vertex and the abſciſs of the ordinate to that point : i. e. FP = AF + Al.

If from any point of a parabola there be drawn a tangent, and a perpendicular to it PK, both to meet the axis produced ; then the focus will be equally diſtant from the two interſections and the point of contact : i. e. FG = FP = FK.

3 R 2　　　　　　　Hence

Hence also the subnormal IK is $= 2AF$ or $= FC$ the semi-parameter.

The line drawn from the Focus to any point of the curve, is equal to $\frac{1}{4}$ the parameter of the diameter to that point: i. e. $FP = \frac{1}{4}$ the parameter of the diameter $P f$.

If an ordinate to any diameter pass through the Focus, it will be equal to half its parameter; and its absciss equal to $\frac{1}{4}$ of the same parameter; or the absciss equal to half the ordinate: i. e. $PL = \frac{1}{4}MN = \frac{1}{2}LM$ or $\frac{1}{2}LN$.

Focus, in Optics, is a point in which several rays meet, and are collected, after being either reflected or refracted. It is so called, because the rays being here brought together and united, their force and effect are increased, insomuch as to be able to burn; and therefore it is that bodies are placed in this point to be burnt, or to shew the effect of burning glasses, or mirrors.—It is to be observed however, that in practice, the Focus is not an absolute point, but a space of some small breadth, over which the rays are scattered; owing to the different nature and refrangibility of the rays of light, and to the imperfections in the figure of the lens, &c. However, the smaller this space is, the better, or the nearer to perfection the machine approaches. Huygens shews that the Focus of a lens convex on both sides, has its breadth equal to $\frac{5}{6}$ of the thickness of the lens.

Virtual Focus, or *Point of Divergence*, so called by Mr. Molyneux, is the point from whence rays tend, after refraction or reflection; being in this respect opposed to the ordinary Focus, or Point of Concurrence, where rays are made to meet after refraction or reflection. Thus, the Foci of an hyperbola are mutually Virtual Foci to each other; but, in an ellipse, they are common Foci to each other: for the rays are reflected *from* the other Focus in the hyperbola, but *towards* it in the ellipse; as appears by the figures at the beginning of this article.

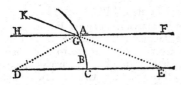

And, in Dioptrics, let ABC be the concavity of a glass, whose centre is D, and axis DE: Let FG be a ray of light falling on the glass, parallel to the axis DE; this ray FG, after it has passed through the glass, at its emersion at G will not proceed directly to H, but be refracted from the perpendicular DG, and will become the ray GK, which being produced to meet the axis in E, this point E is the Virtual Focus, as the ray is refracted directly *from* this point.

Rules for the Foci *of Lenses and Mirrors.*

I. *In Catoptrics, or Lenses.*

1. The Focus of a convex glass, i. e. the point where parallel rays transmitted through a convex glass,

whose surface is the segment of a sphere, do unite, is distant from the pole or vertex of the glass, almost a diameter and half of the convexity.—2. In a Plano-Convex glass, the Focus of parallel rays is distant from the pole of the glass a diameter of the convexity, if the segment do not exceed 30 degrees. Or the rule in Plano-Convex glasses is, As $107 : 193 : :$ so is the radius of convexity: to the refracted ray taken to its concourse with the axis; which in glasses of larger spheres is almost equal to the distance of the Focus taken in the axis.—3. In Double Convex glasses of the same sphere, the Focus is distant from the pole of the glass about the radius of the convexity, if the segment be but 30 degrees. But when the two convexities are unequal, or segments of different spheres, then the rule is, As the sum of the radii of both convexities: to the radius of either convexity alone: : so is double the radius of the other convexity: to the distance of the Focus. —Here observe, that the rays which fall nearer the axis of any glass, are not united with it so near the pole of the glass as those farther off: nor will the Focal distance be so great in a plano-convex glass, when the convex side is towards the object, as when the plane side is towards it. And hence it is truly concluded, that, in viewing any object by a plano-convex glass, the convex side should always be turned outward; as also in burning by such a glass.

II. *For the Virtual Focus,* observe

1. That in Concave glasses, when a ray falls from air parallel to the axis, the Virtual Focus, by its first refraction, becomes at the distance of a diameter and a half of the concavity.—2. In Plano-Concave glasses, when the rays fall parallel to the axis, the Virtual Focus is distant from the glass, the diameter of the concavity.—3. In Plano-Concave glasses, as $107 : 193 : :$ so is the radius of the concavity: to the distance of the Virtual Focus.—4. In Double Concaves of the same sphere, the Virtual Focus of parallel rays is at the distance of the radius of the concavity. But, whether the concavities be equal or unequal, the Virtual Focus, or point of divergency of the parallel rays, is determined by this rule; As the sum of the radii of both concavities: is to the radius of either concavity: : so is double the radius of the other concavity: to the distance of the Virtual Focus.—5. In Concave glasses, exposed to converging rays, if the point to which the incident ray converges, be farther distant from the glass than the Virtual Focus of parallel rays, the rule for finding the Virtual Focus of this ray, is this; As the difference between the distance of this point from the glass, and the distance of the Virtual Focus from the glass: is to the distance of the Virtual Focus: : so is the distance of this point of convergence from the glass: to the distance of the Virtual Focus of this converging ray.—6. In Concave glasses, if the point to which the incident ray converges, be nearer to the glass than the Virtual Focus of parallel rays, the rule to find where it crosses the axis, is this; As the excess of the Virtual Focus, more than this point of convergency: is to the Virtual Focus: : so is the distance of this point of convergency from the glass: to the distance of the point where this ray crosses the axis.

III.

III. *Practical Rules for finding the Foci of Glasses.*

1. To find, by experiment, the Focus of a convex spherical glass, being of a small sphere; apply it to the end of a scale of inches and decimal parts, and expose it before the sun; upon the scale may be seen the bright intersection of the rays measured out: or, expose it in the hole of a dark chamber; and where a white paper receives the distinct representation of distant objects, there is the Focus of the glass.—2. For a glass of a pretty long Focus, observe some distant object through it, and recede from the glass till the eye perceives all in confusion, or the object begins to appear inverted; then the eye is in the Focus.—3. For a Plano-Convex glass: make it reflect the sun against the wall; on the wall will then be seen two sorts of light, a brighter within another more obscure: withdraw the glass from the wall, till the bright image be in its least dimensions; then is the glass distant from the wall about a fourth part of its Focal length.—4. For a Double Convex: expose each side to the sun in like manner; and observe both the distances of the glass from the wall: then is the first distance about half the radius of the convexity turned from the sun; and the second is about half the radius of the other convexity. The radii of the two convexities being thus known, the Focus is then found by this rule; As the sum of the radii of both convexities : is to the radius of either convexity : : so is double the radius of the other convexity : to the distance of the Focus.

IV. *To find the Foci of all Glasses Geometrically.*

Dr. Halley has given a general method for finding the Foci of spherical glasses of all kinds, both concave and convex; exposed to any kind of rays, either parallel, converging, or diverging; as follows: To find the Focus of any parcel of rays diverging from, or converging to, a given point in the axis of a spherical lens, and making the same angle with it; the ratio of the sines of refraction being given.

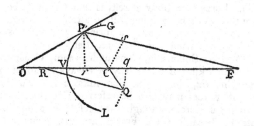

Suppose GL a lens; P a point in its surface; V its pole; C the centre of the spherical segment; O the object, or point in the axis, to or from which the rays proceed; and OP a given ray: and suppose the ratio of refraction to be as *r* to *s*. Then making CR to CO, as *s* to *r* for the immersion of a ray, or as *r* to *s* for the emersion (i. e. as the sines of the angles in the medium which the ray enters, to the corresponding sines in the medium out of which it comes); and laying CR from C towards O, the point R will be the same for all the rays of the point O. Lastly, drawing the radius PC, continued if necessary; with the centre R, and distance OP, describe an arc intersecting PC in

Q. The line QR, being drawn, shall be parallel to the reflected ray; and PF, being made parallel to it, shall intersect the axis in the point F, the Focus sought. —Or, make as CQ : CP : : CR : CF, which will be the distance of the Focus from the centre of the sphere. —And from this general construction, he adverts to a number of particular simple cases.

Dr. Halley gave also an universal algebraical theorem to find the Focus of all sorts of optic glasses, or lenses. See the Philos. Transf. N° 205, or Abr. vol. 1, pa. 191.

V. *In Catoptrics, or Foci by Reflection.*

These are easily found for any known curve, from this principle, that the angle of reflection is always equal to the angle of incidence.

The same are also easily found by experiment, being exposed to any object.

The increase of heat from collecting the sun's rays into a Focus, has been found in many cases of burning glasses, to be astonishingly great; the effect being increased as the square of the diameter of the glass exceeds that of the Focus. If, for instance, there be a burning glass of 12 inches diameter; this will collect or crowd together all the rays of the sun which fall upon the glass into the compass of about ⅛ part of an inch : then, the areas of the two spaces being as the square of 12 to the square of ⅛, or as the square of 96 to the square of 1, that is, as 9216 to 1; it follows, that the heat in the Focus will be 9216 times greater than the sun's common heat. And this will have an effect as great as the direct rays of the sun would have on a body placed at the 96th part of the earth's distance from the sun; or the same as on a planet that should move round the sun at but a very little more than a diameter of the sun's distance from him, or that would never appear farther from him than about 36 minutes.

Besides Dr. Halley in his method for finding the Foci, several other authors have written upon this subject; as Mr. Ditton, in his Fluxions; Dr. Gregory, in his Elements of Dioptrics; M. Carré, and Guisnee, in the Memoires de l'Acad.: Dr. Barrow and Sir I. Newton have also neat and elegant ways of finding geometrically the Foci of spherical glasses; which may be seen in Barrow's Optical Lectures.

FOLIATE, a name given by some to a curve of the 2d order, expressed by the equation $x^3 + y^3 = axy$, being one species of defective hyperbolas, with one asymptote, and consisting of two infinite legs crossing each other, forming a sort of leaf. It is the 42d species of Newton's Lines of the 3d Order.

FOLKES (MARTIN), an English mathematician, philosopher, and antiquary, was born at Westminster about 1690; and was greatly distinguished as a member of the Royal Society in London, and of the Academy of Sciences at Paris. He was admitted into the former at 24 years of age; made one of their council two years after; named vice-president by Sir Isaac Newton himself; and, after Sir Hans Sloane, became president. Coins, ancient and modern, were a great object with him; and his last production was a book upon the English Silver Coin, from the Conquest to his own times. He died at London in 1754. Dr. Birch had drawn up materials for a life of Mr. Folkes, which

arc

are preferved at large in the Anecdotes of Bowyer, p. 562. There are many memoirs of Mr. Folkes's in the Philof. Tranf. from vol. 30 to vol. 46, both inclufive; viz, 1. Account of an Aurora Borealis, vol. 30.—2. Of Lieuwenhoek's curious Microfcope, vol. 32.—3. On the Standard Meafures in the Capitol at Rome, vol. 39.—4. Obfervations of three Mock-funs, vol. 40.—5. On the frefh water Polypus, vol. 42.—6. On human bones petrified, vol. 43.—7. On a paffage in Pliny's Natural Hiftory, vol. 44.—8. On an Earthquake at London, vol. 46.—9. Ditto at Kenfington, vol. 46.—10. Ditto at Newton, vol. 46.

FOMAHAUT, or FOMALHAUT, a ftar of the firft magnitude in the water of the conftellation Aquarius, or in the mouth of the fouthern fifh. Its latitude is 21° 6′ 28″ fouth, and mean longitude to the beginning of 1760, 11ˢ 0° 28′ 55″.

FONTENELLE (BERNARD DE), a celebrated French author, was born at Rouen in 1657, and died in 1756, when he was near 100 years old. He was a univerfal genius: at a very early age he wrote feveral comedies and tragedies of confiderable merit; and he did the fame at a very advanced age. Voltaire declares him the moft univerfal genius the age of Lewis the 14th produced; and compares him to lands fituated in fo happy a climate, as to produce all forts of fruits. His laft comedies, though they fhewed the elegance of Fontenelle, were however, little fitted for the ftage; he then alfo produced an Apology for the Vortices of Des Cartes; upon which Voltaire fays, " We muft excufe his comedies, on account of his great age; and his Cartefian opinions, as they were thofe of his youth, when they were univerfally received all over Europe."

In his poetical performances and Dialogues of the Dead, the fpirit of Voiture was difcerned, though more extended and more philofophical. His Plurality of Worlds is a work fingular in its kind: his defign in this was, to prefent that part of philofophy to view in a gay and pleafing drefs.

Fontenelle applied himfelf alfo to mathematics and natural philofophy; in which he proved not lefs fuccefsful than he had been in polite literature. Having been appointed perpetual fecretary to the Academy of Sciences, he difcharged that truft above 40 years with univerfal applaufe: his Hiftory of the Academy often throws great light upon their memoirs, which are fometimes obfcure; and it has been faid, he was the firft who introduced elegance into the fciences. The Eloges, which he pronounced on the deceafed members of the Academy, have this peculiar merit, that they excite a refpect for the fciences as well as for the authors.

Upon the whole, Fontenelle muft be looked upon as the great mafter of the new art of treating abftract fciences, in a manner that make their ftudy at once eafy and agreeable: nor are any of his works of other kinds void of merit. All thofe talents which he poffeffed from nature, were affifted by a good knowledge of hiftory and languages: and he perhaps furpaffes all men of learning who have not had the gift of invention.

Befide his poetical and theatrical works, with thofe of Belles lettres, &c, he publifhed *Elemens de Geometrie de l'Infini*, in 4to, 1727; alfo the *Theorie des Tourbillons Cartefiens*; and *Difcours moraux & philofophiques*.

All his different works were collected in eleven velumes, 12mo, under the title of *Oeuvres Diverfes*.

FOOT, a meafure of length, divided into 12 inches, and each inch fuppofed to contain 3 barley-corns in length. Geometricians divide the Foot into 10 digits, and the digit into 10 lines, &c. The French divide their Foot, as we do, into 12 inches; but their inch they divide into 12 lines.

It feems this meafure has been taken from the length of the human Foot; but it is of different lengths in different countries. The Paris royal Foot is to the Englifh Foot, as 4263 to 4000, and exceeds the Englifh by 9½ lines; the ancient Roman Foot of the Capitol confifted of 4 palms; equal to 11 inches and 7/10 Englifh; the Rhinland, or Leyden Foot, ufed by the northern nations, is to the Roman Foot, as 19 to 20. For the proportions of the Foot of feveral nations, compared with the Englifh, fee the article MEASURE.

Square FOOT, is a fquare whofe fide is 1 foot, or 12 inches, and confequently its area is 144 fquare inches.

Cubic FOOT, is a cube whofe fide is one Foot, or 12 inches, and confequently it contains 12³ or 1728 cubic inches.

FOOT-*bank*, or FOOT-*ftep*, in Fortification. See BANQUETTE.

FORCE, *Vis*, or *Power*, in Mechanics, Philofophy, &c, denotes the caufe of the change in the ftate of a body, with refpect to motion, reft, preffure, &c; as well as its endeavour to oppofe or refift any change made in fuch its ftate. Thus, whenever a body, which was at reft, begins to move; or when its motion is either not uniform, or not direct; the caufe of this change in the ftate of the body, is what is called *Force*, and is an external Force. Or, while a body remains in the fame ftate, either of reft, or of uniform and rectilinear motion, the caufe of its remaining in fuch ftate, is in the nature of the body, being an innate internal Force, and is called its *Inertia*.

Mechanical Forces may be reduced to two forts; one of a body at reft, the other of a body in motion.

The Force of a body at reft, is that which we conceive to be in a body lying ftill on a table, or hanging by a rope, or fupported by a fpring, &c; and this is called by the names of Preffure, Tenfion, Force, or Vis Mortua, Solicitatio, Conatus Movendi, Conamen, &c; which kind of Force may be always meafured by a weight, viz, the weight that fuftains it. To this clafs of Forces may alfo be referred Centripetal and Centrifugal Forces, though they refide in a body in motion; becaufe thefe Forces are homogeneous to weights, preffures, or tenfions of any kind. The preffure, or Force of gravity in any body, is proportional to the quantity of matter in it.

The Force of a body in motion, is a power refiding in that body, fo long as it continues its motion; by means of which, it is able to remove obftacles lying in its way; to leffen, deftroy, or overcome the Force of any other moving body, which meets it in an oppofite direction; or to furmount any the largeft dead preffure or refiftance, as tenfion, gravity, friction, &c, for fome time; but which will be leffened or deftroyed by fuch refiftance as leffens or deftroys the motion of the body. This is called Vis Motrix, Moving Force, or Motive Force, and by fome late writers Vis Viva, to diftinguifh it it from the Vis Mortua spoken of before; and by these appellations, however different, the same thing is understood by all mathematicians; namely, that power of displacing, of withstanding opposite moving Forces, or of overcoming any dead resistance, which resides in a moving body, and which, in whole or in part, continues to accompany it, as long as the body moves; and may be otherwise called Percussive Force, or Momentum.

But concerning the measure of this kind of Force, mathematicians have been divided into two parties. It is allowed by both sides, that the measure of this Force depends partly upon the mass of matter in the body, or its weight, and partly upon the velocity of its motion; so that upon any increase of either weight or velocity, the moving Force will become greater. It is also agreed, that the velocity being given, or being the same in two moving bodies, their Forces will be in proportion to their masses or weights. But, when two bodies are equal, and the velocities with which they move are different, the two parties no longer agree about the measure of the moving Force.

The Cartesians and Newtonians maintain, that, in this case, the moving Force is in proportion simply as the velocity with which a body moves; so that with a double velocity it has a double Force, &c: But the Leibnitians assert, that the moving Force is proportional to the square of the velocity; so as, with a double velocity to have a quadruple Force, &c. Or, when the bodies are different, the former hold, that the momentum or moving Force of bodies, is in the compound ratio of their weights and velocities: But the Leibnitians hold, that it is in the compound ratio of the weights and squares of the velocities.

Though Leibnitz was the first who expressly asserted, that the Force of a body in motion is as the square of its velocity, which was in a paper inserted in the Leipsic Acts for the year 1686, yet it is thought that Huygens led him into that notion, by some demonstrations in the 4th part of his book De Horologio Oscillatorio, relating to the centre of oscillation, and by his dissertations, in answer to the objections of the abbot Catalan, one of which was published in 1684. This eminent mathematician had demonstrated, that in the collision of two bodies that are perfectly elastic, the sum of the products of each body multiplied by the square of its velocity, was the same after the shock as before; (though the same thing is true of the sums of the products of the bodies multiplied simply by their velocities). Now that proposition is so far general as to obtain in all collisions of bodies that are perfectly elastic: and it is also true, when bodies of a perfect elasticity strike any immoveable obstacle, as well as when they strike one another; or when they are constrained by any power or resistance to move in directions different from those in which they impel one another. These considerations might have induced Huygens to lay it down as a general rule, that bodies constantly preserve their Ascensional Force, i. e. the product of their mass by the height to which their centre of gravity can ascend, which is as the square of the velocity; and therefore, in a given system of bodies, the sum of the squares of their velocities will remain the same, and not be altered by the action of the bodies among

themselves, nor against immoveable obstacles. Leibnitz's metaphysical system led him to think that the same quantity of action or Force subsisted in the universe; and finding this impossible, if Force were estimated by the quantity of motion, he adopted Huygens's principle of the preservation of the Ascensional Force, and made it the measure of moving Forces. But it is to be observed that Huygens's principle, above-mentioned, is general only when bodies are perfectly elastic; and in some other cases which Maclaurin has endeavoured to distinguish: shewing at the same time that no useful conclusion in mechanics is affected by the disputes concerning the measure of the Force of bodies in motion, which have been objected to mathematicians. Analyst, Query 9. See Maclaurin's Fluxions, vol. 2, art. 533; Huygens Oper. tom. 1, pa. 248; &c.

Leibnitz's principle was adopted by several persons; as Wolfius, the Bernoullis, &c. Mr. Dan. Bernoulli, in his Treatise, has assumed the preservation of the Vis Ascendens of Huygens, or, as others express it, the Conservatio Virium Vivarum; and, in Bernoulli's own expression, *æqualitas inter descensum actualem ascensumque potentialem*, as an hypothesis of wonderful use in mechanics. But a late author contends, that the conclusions drawn from this principle are oftener false than true. See De Conservat. Virium Vivarum Differt. Lond. 1744.

Catalan and Papin answered Leibnitz's paper published in 1686; and from that time the controversy became more general, and was carried on for several years by Leibnitz, John and Daniel Bernoulli, Poleni, Wolfius, s'Gravesande, Camus, Muschenbroek, &c, on one side; and Pemberton, Eames, Desaguliers, Dr. S. Clark, M. de Mairan, Jurin, Maclaurin, Robins, &c, on the other. See Act. Erud. 1686, 1690, 1691, 1695; Nouv. de la Rep. des Let. Sept. 1686, 1687, art. 2; Comm. Epist. inter Leibn. et Bern. Ep. 24, p. 143; Discourses sur les Loix de la Comm. du Mouvement, Oper. tom. 3, & Diss. de vera Notione Virium Vivarum, ib.; Act. Petropol. tom. 1, p. 131, &c; Hydronamica, sect. 1; Herman, in Act. Petrop. tom. 1, p. 2, &c; Polen. de Castellis; Wolf. in Act. Petrop. tom. 1, p. 217, &c, and in Cosmol. Gener.; Gravef. in Journ. Lit. and Phys. Elem. Math. 1742, lib. 2, cap. 2 and 3; Memoir. de l'Acad. des Sciences 1728; Muschenbr. Int. ad Phil. Nat. 1762, vol. 1, p. 83 &c; Pemb. &c, in Phil. Transf. number 371, 375, 376, 396, 400, 401, or Abridg. vol. 6, p. 216 &c. Mairan in Memo. de l'Acad. des Sc. 1728, Phil. Transf. numb. 459, or Abridg. vol. 8, p. 236, Philof. Transf. vol. 43, p. 423 &c; Maclaurin's Acc. of Newton's Discoveries, p. 117 &c, Flux. ubi supr. & Recueil des Pieces qui ont emporte le Prix &c. tom. 1: Desagul. Course Exp. Philof. vol. 1, p. 393 &c, vol. 2, p. 49 &c; and Robins's Tracts, vol. 2, p. 135.

The nature and limits of this work will not admit of a full account of the arguments and experiments that have been urged on both sides of this question; but they may be found chiefly in the preceding references. A few of them however may be considered, as follows.

The defenders of Leibnitz's principle, beside the arguments above-mentioned, refer to the spaces that bodies ascend to, when thrown upwards, or the penetrations of bodies let fall into soft wax, tallow, clay, snow, and

and other foft fubftances, which fpaces are always as the fquares of the velocities of the bodies. On the other hand, their opponents retort, that fuch fpaces are not the meafures of the force in queftion, which is rather percuffive and momentary, as thofe above are paffed over in unequal times, and are indeed the joint effect of the Forces and times.

Defaguliers brings an argument from the familiar experiment of the balance, and the other fimple mechanic powers, fhewing that the effect is in proportion to the velocity multiplied by the weight; for example, 4 pounds being placed at the diftance of 6 inches from the centre of motion of a balance, and 2 pounds at the diftance of 12 inches; thefe will have a Vis Viva if the balance be put into a fwinging motion. Now it appears that thefe Forces are equal, becaufe, with contrary directions, they foon deftroy each other; and they are to each other in the fimple ratio of the velocity multiplied by the mafs, viz $4 \times 6 = 24$, and $2 \times 12 = 24$ alfo.

Mr. Robins, in his remarks on J. Bernoulli's treatife, entitled, Difcours fur les Loix de la Communication du Mouvement, informs us, that Leibnitz adopted this opinion through miftake; for though he maintained that the quantity of Force is always the fame in the univerfe, he endeavours to expofe the error of Des Cartes, who alfo afferted, that the quantity of motion is always the fame; and in his difcourfe on this fubject in the Acta Eruditorum for 1686, he fays that it is agreed on by the Cartefians, and all other philofophers and mathematicians, that there is the fame Force requifite to raife a body of 1 pound to the height of 4 yards, as to raife a body of 4 pounds to the height of 1 yard; but being fhewn how much he was miftaken in taking that for the common opinion, which would, if allowed, prove the force of the body to be as the fquare of the velocity it moved with, he afterwards, rather than own himfelf capable of fuch a miftake, endeavoured to defend it as true; fince he found it was the neceffary confequence of what he had once afferted; and maintained, that the force of a body in motion was proportional to the height from which it muft fall, to acquire that velocity; and the heights being as the fquares of the velocities, the Forces would be as the maffes multiplied by them; whereas, when a body defcends by its gravity, or is projected perpendicularly upwards, its motion may be confidered as the fum of the uniform and continual impulfes of the power of gravity, during its falling in the former cafe, and till they extinguifh it in the latter. Thus when a body is projected upwards with a double velocity, thefe uniform impulfes muft be continued for a double time, in order to deftroy the motion of the body; and hence it follows, that the body, by fetting out with a double velocity, and afcending for a double time, muft arife to a quadruple height, before its motion is exhaufted. But this proves that a body with a double velocity moves with a double Force, fince it is produced or deftroyed by the fame uniform power continued for a double time, and not with a quadruple force, though it rifes to a quadruple height; fo that the error of Leibnitz confifted in his not confidering the time, fince the velocities alone are not the caufes of the fpaces defcribed, but the times and the velocities together; yet this is the fallacious argument on which he firft built his new doctrine;

and thofe which have been fince much infifted on, and derived from the indentings or hollows produced in foft bodies by others falling into them, are much of the fame kind. Robins's Tracts, vol. 2, p. 178.

But many of the experiments and reafonings, that have been urged on both fides in this controverfy, have been founded in the different fenfes applied to the term Force. The Englifh and French philofophers, by the word Force, mean the fame thing as they do by momentum, motion, quantity of motion, percuffion, or inftantaneous preffure, which is meafured by the mafs drawn into the velocity, and may be known by its effect; and when they confider bodies as moving through a certain fpace, they allow for the time in which that fpace is defcribed: whereas the Dutch, Italian, and German philofophers, who have efpoufed the new opinion, mean by the word Force, or Force inherent in a body in motion, that which it is able to produce; or, in other words, the Force is always meafured by the whole effect produced by the body in motion; until its whole Force be entirely communicated or deftroyed, without any regard to the time employed in producing this total effect. Thus, fay they, if a point runs through a determinate fpace, and preffes with a certain given Force, or intenfity of preffure, it will perform the fame action whether it move faft or flow, and therefore the time of the action in this cafe ought not to be regarded. s'Gravefande, Phyf. El. Math. § 723—728.

Mr. Euler obferves, with regard to this difpute concerning the meafure of vivid Force, or living Force, as it is fometimes called, that we cannot abfolutely afcribe any Force to a body in motion, whether we fuppofe this Force proportional to the velocity, or to the fquare of the velocity: for the Force exerted by a body, ftriking another at reft, is different from that which it exerts in ftriking the fame body in motion; fo that this Force cannot be afcribed to any body confidered in itfelf, but only relatively to the other bodies it meets with. There is no force in a body abfolutely confidered, but its inertia, which is always the fame, whether the body be in motion or at reft. But if this body be forced by others to change its ftate, its inertia then exerts itfelf as a Force, properly fo called, which is not abfolutely determinable; becaufe it depends on the changes that happen in the ftate of the body.

A fecond obfervation which has been made by fome eminent writers, is, that the effect of a fhock of two or more bodies, is not produced in an inftant, but requires a certain interval of time. If this be fo, the heterogeneity between the Vis Viva and Mortua, or Living and Dead Force, will vanifh; fince a preffure may always be affigned, which in the fame time, however little, fhall produce the fame effect. If then the Vis Viva be homogeneous to the Vis Mortua, and having a perfect meafure and knowledge of the latter, we need require no other meafure of the former than that which is derived from the Vis Mortua equivalent to it.

Now that the change in the ftate of two bodies, by their fhock, does not happen in an inftant, appears evidently from the experiments made on foft bodies: in thefe, percuffion forms a fmall cavity, vifible after the fhock, if the bodies have no elafticity. Such a cavity cannot certainly be made in an inftant. And if the

fhock

shock of soft bodies require a determinate time, we must certainly say as much of the hardest, though this time may be so small as to be beyond all our ideas. Neither can an instantaneous shock agree with that constant law of nature, by virtue of which nothing is performed per saltum. But it is needless to insist farther upon this, since the duration of any shock may be determined from the most certain principles.

There can be no shock or collision of bodies, without their making mutual impressions on each other: these impressions will be greater or less, according as the bodies are more or less soft, other circumstances being the same. In bodies called hard, the impressions are small; but a perfect hardness, which admits of no impression, seems inconsistent with the laws of nature; so that while the collision lasts, the action of bodies is the result of their mutually pressing each other. This pressure changes their state; and the Forces exerted in percussion are really pressures, and truly Vires Mortuæ, if we will use this expression, which is no longer proper, since the pretended infinite difference between the Vires Vivæ and Mortuæ ceases.

The Force of percussion, resulting from the pressures that bodies exert on each other, while the collision lasts, may be perfectly known, if these pressures be determined for every instant of the shock. The mutual action of the bodies begins the first moment of their contact; and is then least; after which this action increases, and becomes greatest when the reciprocal impressions are strongest. If the bodies have no elasticity, and the impressions they have received remain, the Forces will then cease. But if the bodies be elastic, and the parts compressed restore themselves to their former state, then will the bodies continue to press each other till they separate. To comprehend therefore perfectly the Force of percussion, it is requisite first to define the time the shock lasts, and then to assign the pressure corresponding to each instant of this time; and as the effect of pressures in changing the state of any body may be known, we may thence come at the true cause of the change of motion arising from collision. The Force of percussion therefore is no more than the operation of a variable pressure during a given time; and to measure this Force, we must have regard to the time, and to the variations according to which the pressure increases and decreases.

Mr. Euler has given some calculations relative to these particulars; and he illustrates their tendency by this instance: Suppose that the hardness of the two bodies, A and B, is equal; and such, that being pressed together with the force of 100lb, the impression made on each is of the depth of $\frac{1}{1000}$th part of a foot. Suppose also that B is fixed, and that A strikes it with the velocity of 100 feet in a second; according to Mr. Euler, the greatest Force of compression will be equivalent to 400lb, and this Force will produce in each of these bodies an impression equal to $\frac{1}{25}$ of a foot; and the duration of the collision, that is, till the bodies arrive at their greatest compression, will be about $\frac{1}{800}$ of a second. Mr. Euler, in his calculations, supposes the hardness of a body to be proportional to the Force or pressure requisite to make a given impression on it; so that the Force, by which a given impression is made on a body, is in a compound ratio of the hardness of

the body and of the quantity of the impression. But he observes, that regard must be had to the magnitude of the bodies, as the same impression cannot be made on the least bodies as on the greatest, from the defect of space through which their component particles must be driven: he considers therefore only the least impressions, and supposes the bodies of such magnitudes, that with respect to them the impressions may be looked upon as nothing. What he supposes concerning the hardness of bodies, neither implies elasticity nor the want of it, as elasticity only produces a restitution of figure and impression when the pressing Force ceases; but this restitution need not be here considered. It is also supposed, that the bodies which strike each other, have plane and equal bases, by which they touch each other in the collision; so that the impression hereby made diminishes the length of each body. It is farther to be observed, that in Mr. Euler's calculations, bodies are supposed so constituted, that they may not only receive impressions from the Forces pressing them, but that a greater Force is requisite to make a greater impression. This excludes all bodies, fluid or solid, in which the same Force may penetrate farther and farther, provided it have time, without ever being in equilibrio with the resistance: thus a body may continually penetrate farther into soft wax, although the force impelling it be not increased: in these, and the like cases, nothing is required but to surmount the first obstacles; which being once done, and the connection of parts broken, the penetrating body always advances, meeting with the same obstacles as before, and destroying them by an equal Force. But Mr. Euler only considers the first obstacles which exist before any separation of parts, and which are doubtless such, that a greater impression requires a greater Force. Indeed this chiefly takes place in elastic bodies; but it seems likewise to obtain in all bodies when the impressions made on them are small, and the contexture of their parts is not altered.

These things being premised, let the mass or weight of the body A be expressed in general by A, and let its velocity before the shock be that which it might acquire by falling from the height a. Farther, let the hardness of A be expressed by M, and that of B by N, and let the area of the base, on which the impression is made, be cc; then will the greatest compression be

made with the Force $\sqrt{\dfrac{MNcc}{M \div N}} \times Aa$. Therefore

if the hardness of the two bodies, and the plane of their contact during the whole time of their collision be the same, this Force will be as $\sqrt{Aa}$, that is, as the square root of the Vis Viva of the striking body A. And as $\sqrt{a}$ is proportional to the velocity of the body A, the Force of percussion will be in a compound ratio of the velocity and of the subduplicate ratio of the mass of the body striking; so that in this case neither the Leibnitian nor the Cartesian propositions take place. But this Force of percussion depends chiefly on the hardness of the bodies; the greater this is, the greater will the Force of percussion be. If M = N, this Force will be as $\sqrt{Mcc} \times Aa$, that is, in a compound subduplicate ratio of the Vis Viva of the striking body of the hardness, and of the plane of contact. But if

M, the hardneſs of one of the bodies, be infinite, the Force of percuſſion will be as $\sqrt{Ncc \times Aa}$; at the ſame time, if M = N, this Force will be as $\sqrt{\frac{1}{2}Ncc \times Aa}$. Therefore, all other things being equal, the Force of percuſſion, if the ſtriking body be infinitely hard, will be to the Force of percuſſion when both the bodies are equally hard, as $\sqrt{2}$ to 1.

Mr. Euler farther deduces from his calculation, that the impreſſion received by the bodies A and B will be as follows ; viz, as

$$\sqrt{\frac{N \times Aa}{M+N \times Mcc}} \text{ and } \sqrt{\frac{M \times Aa}{M+N \times Ncc}} \text{ reſpectively.}$$

If therefore the hardneſs of A, that is M, be infinite, it will ſuffer no impreſſion ; whereas that on B will extend to the depth of $\sqrt{\frac{Aa}{Ncc}}$. But if the hardneſs of the two bodies be the ſame, or M = N, they will each receive equal impreſſions of the depth $\sqrt{\frac{Aa}{2Ncc}}$. So that the impreſſion received by the body B in this caſe, will be to the impreſſion it receives in the former, as 1 to $\sqrt{2}$.

Mr. Euler has likewiſe conſidered and computed the caſe where the ſtriking body has its anterior ſurface convex, with which it ſtrikes an immoveable body whoſe ſurface is plane. He has alſo examined the caſe when both bodies are ſuppoſed immoveable ; and from his formulæ he deduces the known laws of the colliſion of elaſtic and non-elaſtic bodies. He has alſo determined the greateſt preſſures the bodies receive in theſe caſes ; and likewiſe the impreſſions made on them. In particular he ſhews, that the impreſſions received by the body ſtruck, or B, if moveable, is to the impreſſion received by the ſame body when immoveable, as $\sqrt{B}$ to $\sqrt{A+B}$.

There are ſeveral curious as well as uſeful obſervations in Deſaguliers's Experimental Philoſophy, concerning the comparative Forces of men and horſes, and the beſt way of applying them. A horſe draws with the greateſt advantage when the line of direction is level with his breaſt ; in ſuch a ſituation, he is able to draw 200lb for 8 hours a day, walking about $2\frac{1}{2}$ miles an hour. But if the ſame horſe be made to draw 240lb, he can work only 6 hours a day, and cannot go quite ſo faſt. On a carriage, indeed, where friction alone is to be overcome, a middling horſe will draw 1000lb. But the beſt way to try the Force of a horſe, is to make him draw up out of a well, over a ſingle pulley or roller; and in that caſe, an ordinary horſe will draw about 200lb, as before obſerved.

It is found that 5 men are of equal Force with one horſe, and can with equal eaſe puſh round the horizontal beam of a mill, in a walk 40 feet wide; whereas 3 men will do it in a walk only 19 feet wide.

The worſt way of applying the Force of a horſe is to make him carry or draw up hill: for if the hill be ſteep, 3 men will do more than a horſe, each man climbing up faſter with a burden of 100lb weight, than a horſe that is loaded with 300lb: a difference which is owing to the poſition of the parts of the human body being better adapted to climb, than thoſe of a horſe.

On the other hand, the beſt way of applying the Force of a horſe, is the horizontal direction, in which a man can exert the leaſt Force : thus, a man that weighs 140lb, when drawing a boat along by means of a rope coming over his ſhoulders, cannot draw above 27lb, or exert above 1-7th part of the Force of a horſe employed to the ſame purpoſe ; ſo that in this way the Force of a horſe is equal to that of 7 men.

The beſt and moſt effectual poſture in a man, is that of rowing ; when he not only acts with more muſcles at once for overcoming the reſiſtance, than in any other poſition ; but alſo as he pulls backwards, the weight of his body aſſiſts by way of lever. See Deſaguliers's Exp. Philoſ. vol. 1, p. 241, where ſeveral other obſervations are made relative to Force acquired by certain poſitions of the body ; from which that author accounts for moſt feats of ſtrength and activity. See alſo a Memoir on this ſubject by M. De la Hire, in the Mem. Roy. Acad. 1729 ; or in Deſaguliers's Exp. &c. p. 267 &c, who has publiſhed a tranſlation of part of it with remarks.

Force is diſtinguiſhed into Motive and Accelerative or Retardive.

Motive Force, otherwiſe called Momentum, or Force of Percuſſion, is the abſolute Force of a body in motion, &c; and is expreſſed by the product of the weight or maſs of matter in the body multiplied by the velocity with which it moves. But

Accelerative Force, or *Retardive* Force, is that which reſpects the velocity of the motion only, accelerating or retarding it ; and it is denoted by the quotient of the motive Force divided by the maſs or weight of the body. So,

if m denote the motive Force,
and b the body, or its weight,
and f the accelerating or retarding Force,

then is f as $\frac{m}{b}$.

Again, Forces are either Conſtant or Variable.

Conſtant Forces are ſuch as remain and act continually the ſame for ſome determinate time. Such, for example, is the Force of gravity, which acts conſtantly the ſame upon a body while it continues at the ſame diſtance from the centre of the earth, or from the centre of Force, wherever that may be. In the caſe of a conſtant Force F acting upon a body b, for any time t, we have theſe following theorems ; putting

f = the conſtant accelerating Force = F ÷ b,
v = the velocity at the end of the time t,
s = the ſpace paſſed over in that time, by the conſtant action of that Force on the body :

and g = $16\frac{1}{12}$ feet, the ſpace generated by gravity in 1 ſecond, and calling the accelerating Force of gravity τ ; then is

$$s = \tfrac{1}{2}tv = gft^2 = \frac{v^2}{4gf} ;$$

$$v = 2gft = \frac{2s}{t} = \sqrt{4gfs} ;$$

$$t = \frac{v}{2gf} = \frac{2s}{v} = \sqrt{\frac{s}{gf}} ;$$

$$f = \frac{v}{2gt} = \frac{s}{gt^2} = \frac{v^2}{4gs}.$$

Variable Forces are ſuch as are continually changing

ing

ing in their effect and intensity; such as the Force of gravity at different distances from the centre of the earth, which decreases in proportion as the square of the distance increases. In variable Forces, theorems similar to those above may be exhibited by using the fluxions of quantities, and afterwards taking the fluents of the given fluxional equations. And herein consists one of the great excellencies of the Newtonian or modern analysis, by which we are enabled to manage, and compute the effects of all kinds of variable Forces, whether accelerating or retarding. Thus, using the same notation as above for constant forces, viz, f the accelerating Force at any instant, t the time a body has been in motion by the action of the variable Force, v the velocity generated in that time, s the space run over in that time, and $g = 16\frac{1}{12}$ feet; then is

$$\dot{s} = \frac{v\dot{v}}{2gf} = vt;$$

$$\dot{v} = \frac{2gf\dot{s}}{v} = 2gf\dot{t};$$

$$\dot{t} = \frac{\dot{s}}{v} = \frac{\dot{v}}{2gf};$$

$$f = \frac{v\dot{v}}{2gs} = \frac{\dot{v}}{2gt}.$$

In these four theorems, the Force f, though variable, is supposed to be constant for the indefinitely small time t; and they are to be used in all cases of variable Forces, as the former ones in constant Forces; viz, from the circumstances of the problem under consideration, deduce a general expression for the value of the force f, at any indefinite time t; then substitute it in one of these theorems, which shall be proper to the case in hand; and the equation thence resulting will determine the corresponding values of the other quantities in the problem.

It is also to be observed, that the foregoing theorems equally hold good for the destruction of motion and velocity, by means of retarding or resisting Forces, as for the generation of the same by means of accelerating Forces.

Many applications of these theorems may be seen in my Select Exercises, p. 172 &c.

There are many other denominations and kinds of Forces; such as attractive, central, centrifugal, &c, &c; for which see the respective words.

FORCER, in Mechanics, is properly a piston without a valve. For, by drawing up such a piston, the air is drawn up, and the water follows; then pushing the piston down again, the water, being prevented from descending by the lower valve, is forced up to any height above, by means of a side branch between the two.

See the ways of making these in Desaguliers's Exper. Philos. vol. 2, p. 161 &c. See also Clare's Motion of Fluids, p. 60.

FORCING *Pump*, one that acts, or raises water, by a Forcing piston. See above.

FORELAND, or FORENESS, in Navigation, a point of land jutting out into the sea.

FORELAND, in Fortification, is a small piece of ground between the wall of a place and the moat; called also Berme and Liziere.

FORE-STAFF, an instrument used at sea, for taking the altitudes of the heavenly bodies; being so called, because the observer, in using it, turns his face forward or towards the object, in contradistinction to the Back-staff, with which he turns his back to the object. It is also called the Cross-staff, because it consists of several pieces set across a staff.

The Fore-staff is formed of a straight square staff AB, of about 3 feet long, having each of its four sides graduated like a line of tangents, and four crosses, or vanes, FF, EE, DD, CC, sliding upon it, of unequal lengths, the halves of which represent the radii to the lines of tangents on the different sides of the staff. The first or shortest of these vanes, FF, is called the Ten Cross, or Ten Vane, and belongs to the 10 scale, or that side of the instrument on which divisions begin at 3 degrees, and end at 10. The next longer cross, EE, is called the 30 Cross, belonging to that side of the staff where the divisions begin at 10 degrees, and end at 30, called the 30 scale. The third vane DD, is called the 60 Cross, and belongs to that side where the divisions begin at 20 degrees, and end at 60. The last or longest vane CC, called the 90 Cross, belongs to the side where the divisions begin at 30 degrees, and end at 90.

The chief use of this instrument, is to take the height of the sun, and stars, or the distance between two stars: and the 10, 30, 60, or 90 Cross is to be used, according as the altitude is more or less; that is, if the altitude be less than 10 degrees, the 10 Cross is to be used; if above 10, but less than 30, the 30 Cross is to be used; and so on.

To observe an Altitude with the Fore-staff. Apply the flat end of the staff to the eye, and slide one of the crosses backwards and forwards upon it, till over the upper end of the Cross be just seen the centre of the sun or star, and over the under end the extreme horizon; then the degrees and minutes cut by the cross on the side of the staff proper to the vane in use, gives the altitude above the horizon.

In like manner, for the Distance between two luminaries; the Staff being set to the eye, bring the cross just to subtend or cover that distance, by having the one luminary just at the one end of it, and the other luminary at the other end of it; and the degrees and minutes, in the distance, will be cut on the proper side of the staff, as before.

FORMULA, a theorem or general rule, or expression, for resolving certain particular cases of some problem, &c. So $\frac{1}{2}s + \frac{1}{4}d$ is a general Formula for

the

the greater of two quantities whose sum is s and difference d; and $\frac{1}{2}s - \frac{1}{2}d$ is the Formula, or general value, for the less quantity. Also $\sqrt{dx - x^2}$ is the Formula, or general value, of the ordinate to a circle, whose diameter is d, and abscifs x.

FORT, a little castle or fortress; or a place of small extent, fortified either by nature or art.

The Fort is usually encompassed with a moat, rampart, and parapet, to secure some high ground, or passage of a river; to make good or strengthen an advantageous post; or to fortify the lines and quarters of a siege.

Field FORT, otherwise called *Fortin*, or *Fortlet*, and sometimes *Sconce*, is a small Fort, built in haste, for the defence of a pass or post; but particularly constructed for the defence of a camp in the time of a siege, where the principal quarters are usually joined, or made to communicate with each other, by lines defended by fortins and redoubts. Their figure and size are various, according to the nature of the situation, and the importance of the service for which they are intended; but they are most commonly made square, each side about 100 toises, the perpendicular 10, and the faces 25; the ditch about this Fort may be 10 or 12 toises wide; the parapet is made of turf, and fraised, and the ditch pallisadoed when dry. There may be made a covert-way about this Fort, or else a row of pallisades might be placed on the outside of the ditch. Some of these are fortified with bastions, and some with demi-bastions.——A Fort differs from a citadel, as this last is erected to command and guard some town; and from a redoubt, as it is closed on all sides, while the redoubt is open on one side.

Royal FORT, is one whose line of defence is at least 26 fathoms long.

Star FORT, is a sconce or redoubt, constituted by re-entering and saliant angles, having commonly from five to eight points, and the sides flanking each other.

Forts are sometimes made triangular, only with half bastions; or of various other figures, regular or irregular, and sometimes in the form of a semicircle, especially when they are situated near a river, or the sea, as at the entrance of a harbour, for the convenience of firing at ships quite around them on that side. In the construction of all Forts, it should be remembered, that the figure of fewest sides and bastions, that can probably answer the proposed defence, is always to be preferred; as works on such a plan are sooner executed, and with less expence; besides, fewer troops will serve, and they are more readily brought together in case of necessity.

FORTIFICATION, called also Military Architecture, is the art of fortifying or strengthening a town or other place, by making certain works around it, to secure it and defend it from the attacks of enemies.

Fortification has been undoubtedly practised by all nations, and in all ages; being at first doubtless very rude and simple, and varying in its nature and manner, according to the mode of attack, and the weapons made use of. Thus, when villages and towns were first formed, it was found necessary, for the common safety, to encompass them with walls and ditches, to prevent all violence and sudden surprises from their neighbours. When offensive and missive weapons came to be used, walls were made as a defence against the assailants, and look holes or loop holes made through the walls to annoy the enemy, by shooting arrows &c through them. But finding that as soon as the enemy got close to the walls, they could no longer be seen nor annoyed by the besieged, these added square towers along the wall, at proper distances from each other, so that all the intervening parts of the wall might be seen and defended from the adjacent sides of the towers. However, this manner of inclosing towns was found to be rather imperfect, because there remained still the outer face of the towers which fronted the field, that could not be seen and defended from any other part. To remedy this imperfection, they next made the towers round instead of square, as seeming better adapted both for strength to resist the battering engines, and for being defended from the other parts of the wall. Nevertheless, a small part of these towers still remained unseen, and incapable of being defended, for which reason they were again changed for square ones, as before, but with this difference, that now they presented an angle of the square outwards to the field, instead of a face or side; and thus such a disposition of the works was obtained, as that no part could be approached by the enemy without being seen and attacked.

Since the use of gun-powder, it has been found necessary to add thick ramparts of earth to the walls, and the towers have been enlarged into bastions, as well as many other things added, that have given a new appearance to the whole art of defence, and the name of Fortification, on account of the strength afforded by it, which was about the year 1500, when the round towers were changed into bastions.

But notwithstanding all the improvements made in this art since the invention of gun-powder, that of attacking is still superior to it: the superiority of the besieger's fire, together with the greater number of men, sooner or later compels the besieged to submit. A special advantage was added to the art of attacking by M. Vauban, at the siege of Ath, in the year 1697, viz in the use of ricochet firing, or at a low elevation of the gun, by which the shot was made to run and roll a great way along the inside of the works, to the great annoyance of the besieged.

The chief authors who have treated of Fortification, since it has been considered as a particular art, are the following, and mostly in the order of time: viz, La Treille, Alghisi, Marchi, Pasino, Ramelli, Cataneo, and Speckle, who, as Mr. Robins says, was one of the greatest geniuses that has applied to this art: he was architect to the city of Strasburgh, and died in the year 1589: he published a treatise on Fortification in the German language, which was reprinted at Leipsic in 1736. Afterwards, Errard, who was engineer to Henry the Great of France; Stevinus, engineer to the prince of Orange; with Marolois, the chevalier de Ville, Lorini, Cochorn, the count de Pagan, and the marshal de Vauban; which last two noble authors have contributed greatly to the perfection of the art; besides Scheiter, Mallet, Belidor, Blond l, Muller, Montalambert, &c. Also a list of several works on the art of Fortification may be added, as follows: viz,

Melder's

Melder's Praxis Fortificatoria : Les Fortifications du Comte de Pagan : L'Ingénieur Parfait du Sieur de Ville : Sturmy's Architectura Militaris Hypothetical. : Blondel's Nouvelle Maniere de Fortifier les Places : The Abbé de Fay's Veritable Maniere de Bien Fortifier : Vauban's Ingénieur François : Coehorn's Nouvelle Fortification tant pour un Terrain bas & humide, que fec & elevé : Alexander de Grotte's Fortification : Donatus Roselli's Fortification : Medrano's Ingénieur François : The chevalier de St. Julien's Architecture Militaire : Lanfberg's Nouvelle Maniere de Fortifier les Places : An anonymous treatife in French, called Nouvelle Maniere de Fortifier les Places, tirée des Methodes du chevalier de Ville, &c : Ozanam's Traité de Fortification : Memoires de l'Artillerie de Surirey de St. Remy : Muller's treatifes of Elementary and Practical Fortification : and Montalambert's Fortification Perpendiculaire.

Maxims in FORTIFICATION. From the nature and circumstances of this art, certain general rules, or maxims, have been drawn, and laid down. Thefe may indeed be multiplied to any extent, but the principal of them, are the following : viz,

1. That the manner of fortifying fhould be accommodated to that of attacking. So that no one manner can be affured always to hold, unlefs it be affured that the manner of befieging is incapable of being altered. Alfo, to judge of the perfection of a Fortification, the method of befieging at the time when it was built muft be confidered.

2. All the parts of a Fortification fhould be equally ftrong on all fides, where there is equal danger ; and they fhould be able to refift the moft powerful machines ufed in befieging.

3. A Fortification fhould be fo contrived, as to be defended with the feweft men poffible : which confideration, when well attended to, faves a great deal of expence.

4. That the defendants may be in the better condition, they muft not be expofed to the enemies' artillery ; but the aggreffors muft be expofed to theirs. Hence,

5 All the parts of a Fortification fhould be fo difpofed, as that they may defend each other. In order to this, every part ought to be flanked, i. e. feen fideways, capable of being feen and defended from fome other part ; fo that there be no place where an enemy can lodge himfelf, either unfeen, or under fhelter.

6. All the campaign around muft lie open to the defendants ; fo that no hill or eminence muft be allowed, behind which the enemy might fhelter himfelf from the guns of the Fortification ; or from which he might annoy them with his own. Hence, the fortrefs is to command all the place round about ; and confequently the outworks muft all be lower than the body of the place.

7. No line of defence muft exceed the point-blank mufket-fhot, which is from 120 to 150 fathoms.

8. The more acute the angle at the centre, the ftronger is the place ; as confifting of the more fides, and confequently more defenfible.

9. All the defences fhould be as nearly direct as poffible.

10. The works that are moft remote from the centre

of the place, ought always to be open to thofe that are more near.

Thefe are the general laws and views of Fortification. As to the particular ones, or fuch as refpect the feveral members or parts of the work, they are given under thofe articles refpectively.

Fortification is either theoretical or practical.

Theoretical FORTIFICATION, confifts in tracing the plans and profiles of a work on paper, with fcales and compaffes ; and in examining the fyftems propofed by different authors, to difcover their advantages and defects. And

Practical FORTIFICATION, confifts in forming a project of a work according to the nature of the ground, and other neceffary circumftances, tracing it on the ground, and executing the project, together with all the military buildings, fuch as magazines, ftorehoufes, bridges, &c.

Again, Fortification is either Defenfive or Offenfive.

Defenfive FORTIFICATION is the art of defending a town that is befieged, with all the advantages the Fortification of it will admit. And

Offenfive FORTIFICATION is the fame with the attack of a place, being the art of making and conducting all the different works in a fiege, in order to gain poffeffion of the place

FORTIFICATION is alfo ufed for the place fortified ; or the feveral works raifed to defend and flank it, and keep off the enemy.

All Fortifications confift of lines and angles, which have names according to their various offices. The principal lines are thofe of circumvallation, of contravallation, of the capital, &c. The principal angles are thofe of the centre, the flanking angle, flanked angle, angle of the epaule, &c.

Fortifications are either durable or temporary.

Durable FORTIFICATION, is that which is built and intended to remain a long time. Such are the ufual Fortifications of cities, frontier places, &c. And a

Temporary FORTIFICATION, is that which is erected on fome emergent occafion, and only for a fhort time. Such are field-works, thrown up for the feizing and maintaining a poft, or paffage ; thofe about camps, or in fieges ; as circumvallations, contravallations, redoubts, trenches, batteries, &c.

Again, Fortifications are either regular or irregular. A

Regular FORTIFICATION, is that in which the baftions are all equal ; or which is built in a regular polygon, the fides and angles of which are ufually about a mufket fhot from each other. A Regular Fortification, having the parts all equal, has the advantage of being equally defenfible ; fo that there are no weak places. And an

Irregular FORTIFICATION, is that in which the baftions are unequal, and unlike ; or the fides and angles not all equal, and equidiftant.

In an Irregular Fortification, the defence and ftrength being unequal, it is neceffary to reduce the irregular fhape of the ground, as near as may be, to a regular figure : i. e. by infcribing it in an oval, inftead of a circle ; fo that one half may be fimilar and equal to the other half.

Marine

Marine FORTIFICATION, is sometimes used for the art of raising works on the sea coast &c, to defend harbours against the attacks of shipping.——See a neat treatise on Marine Fortification, at the end of Robertson's Elements of Navigation.

There are many modes of Fortification that have been much esteemed and used; a small specimen of a comparative view of the principal of these, is represented in plate x, viz, those of Count Pagan, and Mess. Vauban, Coehorn, Belidor, and Blondel; the explanation of which is as follows:

1. *Pagan's System.*

A Half Bastions.
B Ravelin and Counterguard.
C Counterguards before the bastions.
D The Ditch.
E The Glacis.
G The place of Arms.
H Retired Flanks.
a Line of Defence.

2. *Vauban's System.*

b Angle of the Bastion, or Flanked angle.
c Angle of the Shoulder.
d Angle of the Flank.
e Saliant Angle.
f Face of the bastion.
g The Flank.
h The Curtain.
i Tenailles.
k Traverses in the covert way.

3. *Coehorn's System.*

1 Concave Flanks.
2 The Curtains.
3 Redoubts in the re-entering angles.
4 Traverses.
5 Stone lodgments.
6 Round Flanks.
7 Redoubt.
8 Coffers planked on the sides, and above covered overhead with a foot of earth.

4 *Belidor's System.*

I Cavaliers.
K Rams-horns, or Tenailles.
L Retrenchments within the detached bastions.
M Circular Curtain.
N The Ravelin.
P Lunettes with retired batteries.
Q Redoubt.
R Detached Redoubt.
S An Arrow.
P Small Traverses.

5. *Blondel's System.*

I Retired Battery.
m Lunettes.
n Ravelin, with Retired Bastion.
o Orillons.

Another, or new method of Fortification has lately been proposed by M. Montalambert, called *Fortification Perpendiculaire*, because the faces of the works are made by a series of lines running zigzag perpendicular to one another.

Profile of a FORTIFICATION, is a representation of a vertical section of a work; serving to shew those dimensions which cannot be represented in plans, and are necessary in the building of a Fortification. The names and dimensions of the principal parts are as follow (see fig. 8, pl. vii), where the numbers or dimensions are all expressed in feet

AB The level of the ground plane,
AC $= 27$,
CD $= 18$, and CW $= 16\frac{1}{2}$; also DN is paral. to AB.
DE $= 30$,
EF $= 2$, FG $= 3$, GH $= 3$, HI $= 4\frac{1}{2}$, IL $= 1\frac{1}{2}$,
LK $= 18$, KM $= 2\frac{1}{2}$, NP $= 36$, NO $= 5$, PR $= 7$,
RS $= 1$, ST $= 12$ or.18, OV $= 9$, P$n = 120$,
$mz = 3$, $nu = 3$, $mc = 30$, $cd = 2$,
$de = 3$, $ef = 3$, $fl = 4\frac{1}{2}$, $rg = 120$, $lh = 1$.
AW the interior talus or slope of the rampart,
WE, or DE the terre-plein of ditto,
FG the talus, and GH the upper part, of the banquette,
HL the interior side of the parapet,
LM the upper part of ditto,
N the cordon of 1 foot radius,
NP the depth, and Pn the breadth of the ditch,
OQ interior side of revetement,
NR the scarp or exterior side of ditto,
ST the depth of the foundation,
YZ revetement of the parapet,
nu the counterscarp,
mc the covert-way,
ce talus of the banquette,
ef the upper part of ditto,
fh parapet of the covert-way,
hg the glacis.

Other sections are at fig. 2, pl. 12.

FORTIFIED *Place*, a Fortress, or Fortification, i. e. a place well flanked, and sheltered with works.

FORTIN, or *Fortlet*, a diminitive of the word Fort, meaning a little fort, or sconce, called also Field Fort.

Star FORTIN, is that whose sides flan keach other, &c.

FORTRESS, the same as Fort, or a Fortification.

FOSTER (SAMUEL), an English mathematician, and astronomy professor of Gresham-college, was born in Northamptonshire, and admitted a Sizer at Emanuel-college Cambridge in 1616. He took the degree of bachelor of arts in 1619, and of master in 1623. He applied early to the mathematics, and attained a great proficiency in it, of which he gave the first specimen in 1624, in a treatise on *The Use of the Quadrant*.

On the death of Mr. Gellibrand, he was chosen to succeed him, in 1636, as astronomy professor in Gresham-college, London. He quitted it again however the same year, though for what reason does not appear, and was succeeded by Mr. Mungo Murray, professor of philosophy at St. Andrews in Scotland. But this gentleman marrying, the professorship again became vacant, and Mr. Foster was re-elected in 1641.

Mr. Foster was one of those gentlemen who held private meetings for cultivating philosophy and useful knowledge,

Plate X.

General View of FORTIFICATION.

Pagan's System

Blondel's System

Vauban's System

Belidor's System

Coehorn's System

knowledge, which afterwards gave rife to the Royal Society. In 1646, Dr. Wallis, who was one of thofe affociating gentlemen, received from Mr. Fofter a theorem *de triangulo fphærico*, which he publifhed in his *Mechanics*. Neither was it only in this branch of fcience that he excelled, but he was likewife well verfed in the ancient languages ; as appears from his revifing and correcting the *Lemmata* of Archimedes, which had been tranflated into Latin from an Arabic manufcript by Mr. John Greaves. He made alfo feveral curious obfervations upon eclipfes of the fun and moon, in various places ; and was particularly noted for inventing, as well as improving, aftronomical and other mathematical inftruments. After a long declining ftate of health, he died at Grefham-college in 1652.

His printed works are as follow ; of which the firft two articles were publifhed before his death, and the reft of them after it.

1. *The Defcription and Ufe of a fmall Portable Quadrant, for the eafy finding the hour of Azimuth*; 4to, 1624. Originally publifhed at the end of Gunter's Defcription of the Crofs Staffe, as an appendix to it.

2. *The Art of Dialling*; 4to, 1638. Reprinted, with additions, in 1675.

3. *Pofthuma Fofteri*; by Wingate; 4to, 1652.

4. *Four Treatifes of Dialling*; 4to, 1654.

5. *Mifcellanies, or Mathematical Lucubrations*. Publifhed by John Twyfden, with additions of his own ; and an appendix by Leybourne; folio, 1659.

6. *The Sector altered, and other Scales added, &c.* Publifhed by Leybourne in 1661, in 4to.

There have been two other perfons of the fame name, who have publifhed fome mathematical pieces. The firft was,

WILLIAM FOSTER, who was a difciple of Mr. Oughtred, and afterward a teacher of the Mathematics in London. He diftinguifhed himfelf by a book, which he dedicated to Sir Kenelm Digby, entitled, *The Circles of Proportion, and the Horizontal Inftrument*, &c ; 4to, 1633.—The other was

MARK FOSTER, who lived later in point of time than either of the other two, and publifhed a treatife entitled, *Arithmetical Trigonometry being the folution of all the ufual Cafes in Plain Trigonometry by Common Arithmetic, without any tables whatfoever* 12mo, 1690.

FOUNDATION, that part of a building which is underground : or the mafs which fupports a building, and upon which it ftands : or it is the coffer or bed dug below the level of the ground, to raife a building upon.

FOUNTAIN, in Philofophy, a fpring or fource of water rifing out of the Ground. See SPRING.

FOUNTAIN, or *Artificial* FOUNTAIN, in Hydraulics, a machine or contrivance by which water is violently fpouted or darted up ; called alfo a Jet d'eau.

There are various kinds of Artificial Fountains, but all formed by a preffure of one fort or another upon the water &c, viz, either the preffure or weight of a head of water, or the preffure arifing from the fpring and elafticity of the air, &c. When thefe are formed by the preffure of a head of water, or any other fluid of the fame kind with the Fountain, or jet, then will this fpout up nearly to the fame height as that head, abating only a little for the refiftance of the air, with that of

the adjutage &c, in the fluid's rufhing through ; but, when the Fountain is produced by any other force than the preffure of a column of the fame fluid with itfelf, it will rife to fuch a height as may be nearly equal to the altitude of a column of the fame fluid whofe preffure is equal to the given force that produces the fountain.

To Conftruct an Artificial Fountain, playing by the preffure of the water. This is to be effected by making a clofe connection between a head or elevated piece of water, and the lower place, where the Fountain is to play ; which may be done in this manner : Having a head of water, naturally, or, for want of fuch, make an artificial one, raifing the water by pumps, or other machinery : from this head convey the water in clofe pipes, in any direction, down to the place where the Fountain is to play ; and there let it iffue through an adjutage, or fmall hole, turned upwards by which means it will fpout up nearly as high as the head of the water comes from, as above mentioned.

To Conftruct an Artificial Fountain, playing by the fpring or elafticity of the air. A veffel proper for a refervoir, as AB, fig. 4, plate viii, is provided either of metal, or glafs, or the like, ending in a fmall neck *c*, at the top : through this neck is put a tube *cd*, till the lower end come near the bottom of the veffel, this being about half full of water. The neck is fo contrived, as that a fyringe, or condenfing pipe, may be fcrewed upon the tube ; by means of which a large quantity of air may be intruded through the tube into the water, out of which it will difengage itfelf, and emerge into the vacant part of the veffel, and lie over the furface of the water CD.

Now, the water in the veffel being thus preffed by the air, which is, for ex. double the denfity of the external air ; and the elaftic force of air being proportional to its denfity, or to its gravitating force, the effect will be the fame, as if the weight of the column of air, over the furface of the water, were double that of the column preffing in the tube ; fo that the water muft be forced to fpout up through, when the fyringe is removed, with a force equal to the excefs of preffure of the included air, above that of the external, that is, in this inftance, with a force equal to the preffure of an entire column of the atmofphere ; which being equal to the preffure of a column of 33 or 34 feet of water, it follows that the Fountain will play to nearly 33 or 34 feet high.

Thefe aereal or aquatic Fountains may be applied in different ways, fo as to exhibit various appearances ; and from thefe alone arifes the greateft part of our artificial water-works : even the engine for extinguifhing fire, is a Fountain playing by the force of confined air.

A Fountain fpouting the water in various directions. Suppofe AB the vertical tube, or fpout, in which the water rifes, (fig. 5, pl. viii) : into this let feveral other tubes be fitted ; fome horizontal, others oblique, or inclining, or reclining, &c, as at E, H, L, N, P, &c. Then, as all water retains the direction of the aperture through which it comes, that iffuing through A will rife perpendicularly ; and the reft will tend different ways, defcribing arches of different magnitudes.

Or thus: Suppofe the vertical tube AB (fig. 6) through which the water rifes, to be ftopped at the
top,

top, as in A; and, inftead of pipes or cocks, let it be only perforated with little holes all around or only half round its furface : then will the water fpin out, in all directions, through the little holes, to different diftances.

And hence, if the tube AB be about the height of a man, and having a turn-cock at C; upon opening this cock, the fpectators will be fprinkled unexpectedly with a fhower.

FOUNTAIN *playing by drawing the breath.* Suppofe AB (fig. 8, plate vi), a globe of glafs, or metal, in which is fitted a tube CD, having a fmall orifice in C, and reaching almoft to D, the bottom of the globe. Then if the air be fucked, or drawn with the mouth, out of the tube CD, and the orifice C be immediately immerged under cold water, the water will afcend through the tube into the fphere. Thus proceeding, by repeated exfuctions, till the veffel be above half full of water ; then applying the mouth to C, and blowing air into the tube, upon removing the mouth, the water will fpout forth.

Or, if the globe be put into hot water, the air being thus rarefied, will make the water fpout as before.

And this kind of Fountain is called Pila Heronis, or Hero's Ball, from the name of its inventor.

FOUNTAIN, *whofe ftream raifes and plays a brafs ball.* Provide a hollow brafs ball A (fig. 9, pl. vi), made very thin, that its weight may not be too great for the force of the water ; and let the tube BC, through which the water rifes, be exactly perpendicular to the horizon. Then the ball, being laid in the bottom of the cup or bafon B, will be taken up in the ftream, and fuftained at a confiderable height, playing a little up and down.

FOUNTAIN *which fpouts water in form of a fhower.* To the tube in which the water is to rife, fit a fpherical, or lenticular head AB (fig. 1, pl. xi) made of a plate of metal, and perforated at the top with a great number of little holes. The water rifing violently towards AB, will be there divided into innumerable little threads, and afterwards broken, and difperfed into the fineft drops.

FOUNTAIN *which fpreads the water in form of a table-cloth.* To the tube AB (fig. 2, pl. xi) folder two fpherical fegments, C and D, almoft touching each other, with a fcrew E, to contract or amplify the interftice or chink, at pleafure. Some choofe to make a fmooth and even cleft, in a fpherical or lenticular head, fitted upon the tube. The water fpouting through this chink or cleft will expand itfelf like a cloth.—And thus, the fountain may be made to fpout out in the figure of men, or other animals.

FOUNTAIN, *which, when it has done fpouting, may be turned like an hour-glafs.* Provide two glaffes, A and B, (fig. 3, pl. xi) to be fo much the larger as the fountain is to play the longer, and placed at fo much the greater diftance from each other as the water is defired to fpout the higher. Let CDE be a crooked tube, furnifhed in E with a jet ; and GHI another bent tube, furnifhed with a jet in I: GF and KL are to be other leffer tubes, open at both ends, and reaching near the

bottom of the veffels A and B, to which the tubes CD and GH are likewife to reach.

Now if the veffel A be filled with water, it will defcend through the tube CD ; and it will fpout up through the jet E, by the preffure of the column of water CD. But unlefs the pipe GF were open at G, to let the air run up to F, and prefs at the top of the furface of the water in the cavity A, the water could not run down and fpout at E. After its fall again, it will fink through the little tube KL, into the veffel B, and expel the air through the tube GI. At length, when all the water is emptied out of the veffel A, by turning the machine upfide down, the veffel B will be the refervoir, and make the water fpout up through the jet I, the pipe KL fupplying B with air to let the water defcend in the direction GHI.

Hence, if the veffels A and B contain juft as much water as will be fpouted up in an hour's time, we fhall have a fpouting clipfydra, or water-clock ; which may be graduated or divided into quarters, minutes, &c.

FOUNTAIN *of Command.* This depends on the fame principles with thofe of the former : CAE (fig. 4, pl. xi) is a veffel of water fecured againft the entrance of the air, except through the pipe GF, when the cock C, by which it is filled, is fhut. There is another pipe EDHB, going from the bottom of the water to the jet B in the bafon DB ; but this is ftopped by the cock H. At the loweft part of the bafon DB, there is a fmall hole at I, to let the water of the bafon DB run into the bafon GH under it ; there is alfo a fmall triangular hole or notch, in the bottom of the pipe FG, at G. Turn the cock H, and the fountain will play for fome time, then ftop, then play again alternately for feveral times together. When thofe times of playing and ftopping are known before hand, you may command the Fountain to play or ftop ; whence its name. The caufe of this phenomenon is as follows : the water coming down the pipe EDHB, would not come out at B, if the air Ss, above the water, were not fupplied as it dilated : now it is fupplied by the pipe GF, which takes it in at the notch G, and delivers it out at F ; but after fome time the water, which was fpouted out at B, falling down into the bafon DB, rifes high enough to come above the notch G, which ftops the paffage of the air ; fo that the air Ss, above the water in the veffel CAE, wanting a fupply, cannot fufficiently prefs, and the Fountain ceafes playing : But when the water of DB has run down into the lower bafon GH, through the hole I, till it falls below the top of the notch G, the air runs up into the upper receptacle, and fupplies that at Ss, and the Fountain plays again. This is feen a little before hand, by a fkin of water on the notch G, before the air finds a paffage, and then you may command the Fountain to play. It is evident that the hole I muft be lefs than the hole of the jet, or elfe all the water would run out into the lower bafon, without rifing high enough to ftop the notch G.

FOUNTAIN *that begins to play upon the lighting of candles, and ceafes as they go out.* Provide two cylindrical veffels, AB and CD, (fig. 7, pl. xi.) connect them by tubes, open at both ends, KL, BE, &c, fo that the air may defcend out of the higher into the lower :

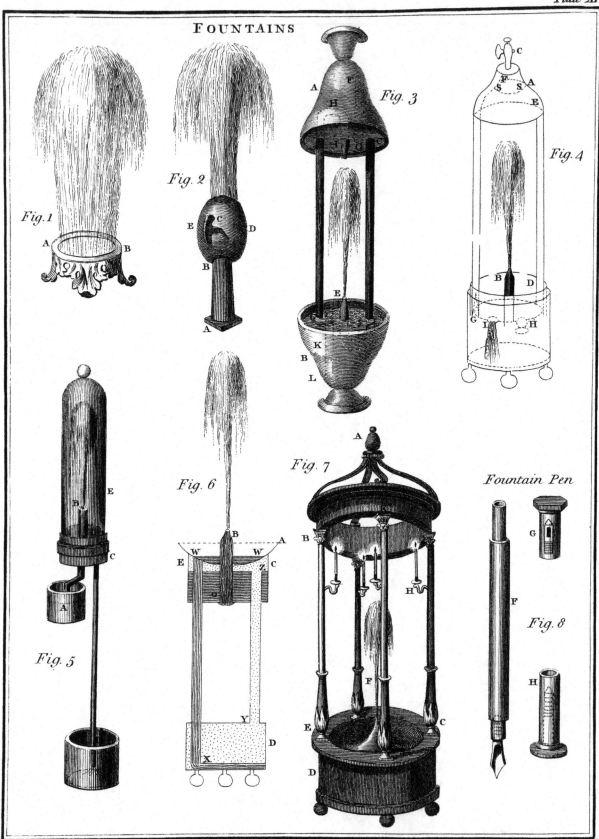

Plate XI.

FOUNTAINS

Fig. 1

Fig. 2

Fig. 3

Fig. 4

Fig. 5

Fig. 6

Fig. 7

Fountain Pen

Fig. 8

to the tubes folder candlesticks, H, &c, and to the hollow cover of the lower vessel CE, fit a little tube or jet, FG, furnished with a cock G, and reaching almost to the bottom of the vessel. In G let there be an aperture, furnished with a screw, by which water may be poured into CD. Then, upon lighting the candles H, &c, the air in the contiguous tubes becoming rarefied by the heat, the water will begin to spout through GF.

By the same contrivance may a statue be made to shed tears upon the presence of the sun, or on the lighting of a candle, &c: all that is here required, being only to lay tubes from the cavity where the air is rarefied, to some other cavities placed near the eyes, full of water.

A FOUNTAIN *by the Rarefaction of the air*, may be made in the following manner: Let AB and CD, fig. 5, pl. xi, be two pipes fixed to a brass head C, made to screw into a glass vessel E, which having a little water in it, is inverted till the pipes are screwed on; then reverting it suddenly, so as to put A, the lower end of the spouting pipe AB, into a jar of water A, and the lower end of the descending pipe CD, into a receiving vessel D, the water will spout up from the jar A into the tall glass vessel E, from which it will go down at the mouth C, through CD, into the vessel D, till the water is wholly emptied out of A, making a Fountain in E, into D. The reason of the play of the Fountain is this: the pipe CD, being 2 feet 9 inches long, lets down a column of water, which rarefies the air 1-12th part in the vessel E, where it presses against the water spouting at B with 1-12th of the force by which the water is pushed up at the hole A, by the pressure of the common air on the water in the vessel A; so that the water spouts up into E, when the air is rarefied 1-12th, with the difference of the pressure of the atmosphere, and the forementioned rarefied air; i. e. of 33 to 2¾, or of 12 to 1. This would raise the water 2 feet 9 inches; but the length of the pipe A, of 9 inches, being deducted, the jet will only rise 2 feet. This, says Defaguliers, may be called a syphon Fountain, where AB is the driving leg, and CD the issuing leg.

FOUNTAIN of *Hero of Alexandria*, so called, because it was contrived by him. In the second Fountain above described, the air is compressed by a syringe; in this, (see fig. 6, pl. xi) the air, being only compressed by the concealed fall of water, makes a jet, which, after some continuance, is considered by the ignorant as a perpetual motion; because they imagine that the same water which fell from the jet rises again. The boxes CE and DYX, being close, we see only the bason ABW, with a hole at W, into which the water spouting at B falls; but that water does not come up again; for it runs down through the pipe WX into the box DYX, from whence it drives out the air, through the ascending pipe YZ, into the cavity of the box CE, where, pressing upon the water that is in it, it forces it out through the spouting pipe OB, as long as there is any water in CE; so that this whole play is only whilst the water contained in CE, having spouted out, falls down through the pipe WX, or of the boxes CE and DY above one another: the height of the water, measured from the bason ABW to the surface of the water in the lower box DYX, is always equal to the

height measured from the top of the jet to the surface of the water in the middle cavity at CE. Now, since the surface CE is always falling, and the water in DY always rising, the height of the jet must continually decrease, till it is shorter by the depth of the cavity CE, which is emptying, added to the depth of the cavity DY, which is always filling; and when the jet is fallen so low, it immediately ceases. The air is represented by the points in this figure.

To prepare this Fountain for playing, which should be done unobserved, pour in water at W, till the cavity DXY is filled; then invert the Fountain, and the water will run from the cavity DXY into the cavity CE, which may be known to be full, when the water runs out at B held down. Set the Fountain up again, and, to make it play, pour in about a pint of water into the bason ABW; and as soon as it has filled the pipe WX it will begin to play, and continue as long as there is any water in CE. You may then pour back the water left in the bason ABW, into any vessel, and invert the Fountain, which, being set upright again, will be made to play, by putting back the water poured out into ABW; and so on as often as you please.

Spouting FOUNTAIN, or Jet d'Eau, is any Fountain whose water is darted forth impetuously through jets, or ajutages, and returns in form of rains, nets, folds, or the like.

FOUNTAIN-*Pen*, is a pen contrived to contain a quantity of ink, and let it flow very gently, so as to supply the writer a long time without the necessity of taking fresh ink.

The Fountain-pen, represented fig. 8, pl. xi, consists of divers pieces of metal, F, G, H, the middle piece F carrying the pen, which is screwed into the inside of a little pipe; and this again, is soldered into another pipe of the same size as the lid G; in which lid is soldered a male screw, for screwing on the cover; as also for stopping a little hole at the place, and hindering the ink from passing through it: at the other end of the piece F is a little pipe, on the outside of which may be screwed the top cover H. A porte-craion goes in the cover, to be screwed into the last mentioned pipe, to stop the end of the pipe into which the ink is to be poured by a funnel.

To use the Pen, the cover G must be taken off, and the pen a little shaken, to make the ink run more freely.

FOURTH, in Music, one of the harmonic intervals, or concords. It consists in the mixture of two sounds, which are in the ratio of 4 to 3; i. e. of two sounds produced by chords, whose lengths are to each other as 4 to 3.

FRACTION, or Broken Number, in Arithmetic and Algebra, is a part, or some parts, of another number or quantity considered as a whole, but divided into a certain number of parts; as 3-4ths, which denotes 3 parts out of 4, of any quantity.

Fractions are usually divided into Vulgar, Decimal, Duodecimal, and Sexagesimal. For the last three sorts, see the respective words.

Vulgar FRACTIONS, called also simple *Fractions*, are usually denoted by two numbers, the one set under the other, with a small line between them: thus ¾ denotes the Fraction three-fourths, or 3 parts out of 4, of some whole quantity confidered as divided into 4 equal parts.

The lower number 4, is called the Denominator of the Fraction, fhewing into how many parts the whole or integer is divided; and the upper number 3, is called the Numerator, and fhews how many of thofe equal parts are contained in the Fraction. Hence it follows, that as the numerator is to the denominator, fo is the Fraction itfelf, to the whole of which it is a Fraction; or as the denominator is to the numerator, fo is the whole or integer, to the Fraction: thus, the integer being denoted by 1, as $4 : 3 :: 1 : \frac{3}{4}$ the Fraction.— And hence there may be innumerable Fractions all of the fame value, as there may be innumerable quantities all in the fame ratio, viz, of 4 to 3; fuch as 8 to 6, or 12 to 9, &c. So that if the two terms of any Fraction i. e. the numerator and denominator, be either both multiplied or both divided by any number, the refulting Fraction will ftill be of the fame value: thus, $\frac{3}{4}$ or $\frac{6}{8}$ or $\frac{9}{12}$ or $\frac{12}{16}$ &c, are all of the fame value with each other.

Fractional expreffions are ufually diftinguifhed into Proper and Improper, Simple and Compound, and Mixt Numbers.

A Proper FRACTION, is that whofe numerator is lefs than the denominator; and confequently the Fraction is lefs than the whole or integer; as $\frac{3}{4}$.

Improper FRACTION, is when the numerator is either equal to, or greater than, the denominator; and confequently the Fraction either equal to, or greater than, the whole integer, as $\frac{4}{4}$, which is equal to the whole; or $\frac{5}{4}$, which is greater than the whole.

Simple FRACTIONS, or *Single* FRACTIONS, are fuch as confift of only one numerator, and one denominator; as $\frac{3}{4}$, or $\frac{5}{4}$, or $12\frac{2}{3}$.

Compound FRACTIONS are Fractions of Fractions, and confift of feveral Fractions, connected together by the word *of*: as $\frac{3}{4}$ of $\frac{3}{4}$, or $\frac{1}{2}$ of $\frac{2}{3}$ of $\frac{3}{4}$.

A Mixt Number confifts of an integer and a Fraction joined together: as $1\frac{3}{4}$, or $12\frac{2}{3}$.

The arithmetic of Fractions confifts in the Reduction, Addition, Subtraction, Multiplication, and Divifion of them.

Reduction of FRACTIONS is of feveral forts; as 1. *To reduce a given whole number into a Fraction of any given denominator.* Multiply the given integer by the propofed denominator, and the product will be the numerator. Thus, it is found that $3 = \frac{6}{2}$, and $5 = \frac{20}{4}$, or $7 = \frac{35}{5}$.

If no denominator be given, or it be only propofed to exprefs the integer Fraction-wife, or like a Fraction; fet 1 beneath it, for its denominator. So $3 = \frac{3}{1}$, and $5 = \frac{5}{1}$, and $7 = \frac{7}{1}$.

2. *To reduce a given Fraction to another Fraction equal to it, that fhall have a given denominator.* Multiply the numerator by the propofed denominator, and divide the product by the former denominator, then the quotient fet over the propofed denominator will form the Fraction required. Thus, if it be propofed to reduce $\frac{3}{4}$ to an equal Fraction whofe denominator fhall be 8; then $3 \times 8 = 24$, and $24 \div 4 = 6$ the numerator, fo that $\frac{6}{8}$ is the Fraction fought, being $= \frac{3}{4}$, and having 8 for its denominator.

3 *To Abbreviate, or reduce Fractions to lower terms.*

Divide their terms, i. e. numerator and denominator, by any number that will divide them both without a remainder, fo fhall the quotients be the correfponding terms of a new Fraction, equal to the former, but in fmaller numbers. In like manner abbreviate thefe new terms again, and fo on till there be no number greater than 1 that will divide them without a remainder, and then the Fraction is faid to be in its leaft terms. Thus, to abbreviate $\frac{15}{60}$; firft divide both terms by 5, and the Fraction becomes $\frac{3}{12}$; next divide thefe by 3, and it becomes $\frac{1}{4}$: fo that $\frac{15}{60} = \frac{3}{12} = \frac{1}{4}$, which is in its leaft terms.

4. *To reduce Fractions to other equivalent ones of the fame denominator.* Multiply each numerator, feparately taken, by all the denominators except its own, and the products will be the new numerators; then multiply all the denominators continually together, for the common denominator, to thefe numerators. Thus, $\frac{2}{3}$ and $\frac{4}{5}$ reduce to $\frac{10}{15}$ and $\frac{12}{15}$; and $\frac{2}{3}$, $\frac{3}{4}$, and $\frac{4}{5}$ reduce to $\frac{40}{60}$, $\frac{45}{60}$, and $\frac{48}{60}$.

5. *To find the value of a Fraction, in the known parts of its integer.* Multiply always the numerator by the number of parts of the next inferior denomination, and divide the products by the denominator. So, to find the value of $\frac{9}{16}$ of a pound fterling; multiply 9 by 20 for fhillings, and dividing by 16, gives 11 for the fhillings; then multiply the remainder 4 by 12 pence, and dividing by 16 gives 3 for pence: fo that 11s. 3d. is the value of $\frac{9}{16}$l. as required.

$$\begin{array}{r} 9 \\ 20 \\ \hline 16)\overline{180}(11s \\ 16 \\ \hline 20 \\ 16 \\ \hline 4 \\ 12 \\ \hline 16)\overline{48}(3d \\ 48 \end{array}$$

6. *To reduce a mixt number to an equivalent improper Fraction.* Multiply the integer by the denominator, and to the product add the numerator, for the new numerator, to be fet over the fame denominator as before. Thus $3\frac{5}{8}$ becomes $\frac{29}{8}$.

7. *To reduce an improper Fraction to its equivalent whole or mixt number.* Divide the numerator by the denominator; fo fhall the quotient be the integral part, and the remainder fet over the denominator will form the fractional part of the equivalent mixt number. Thus $\frac{29}{8}$ reduces to $3\frac{5}{8}$, and $\frac{32}{4} = 8$.

8. *To reduce a compound Fraction to a fimple one.* Multiply all the numerators together for the numerator, and all the denominators together for the denominator, of the fimple Fraction fought. Thus, $\frac{1}{2}$ of $\frac{3}{4} = \frac{3}{8}$, and $\frac{2}{3}$ of $\frac{3}{5}$ of $\frac{7}{9} = \frac{56}{135}$.

To reduce a Vulgar Fraction to a decimal. See DECIMALS. And for feveral other particulars concerning Reduction, as well as the other operations in Fractions; fee my Arithmetic.

Addition of FRACTIONS. Firft reduce the Fractions to their fimpleft form, and reduce them alfo to a common denominator, if their denominators are different; then add all the numerators together, and fet the fum over the common denominator, for the fum of all the Fractions as required.

Thus, $\frac{2}{5} + \frac{3}{5} = \frac{5}{5} = 1$;
And $\frac{2}{7} + \frac{3}{4} = \frac{8}{28} + \frac{21}{28} = \frac{29}{28} = 1\frac{1}{28}$.

Subtraction of FRACTIONS. Reduce the Fractions the fame as for addition; then fubtract the one numerator from

from the other, and set the difference over the common denominator.

So $\frac{7}{3} - \frac{2}{5} = \frac{4}{5}$;

And $\frac{5}{9} - \frac{4}{7} = \frac{42}{63} - \frac{36}{63} = \frac{11}{21}$.

To Multiply FRACTIONS *together.* Reduce them all to the form of simple Fractions, if they are not so; then multiply all the numerators together for the numerator, and all the denominators together for the denominator of the product sought.

Thus $\frac{2}{3} \times \frac{4}{5} = \frac{8}{15}$;

And $\frac{3}{4} \times \frac{5}{8} \times \frac{7}{10} = \frac{105}{400} = \frac{21}{80} = \frac{1}{2}$.

To Divide FRACTIONS. Divide the numerator by the numerator, and the denominator by the denominator, if they will exactly divide. Thus, $\frac{8}{15} \div \frac{4}{5} = \frac{2}{3}$.

But if they will not divide without a remainder, then multiply the dividend by the reciprocal of the divisor, that is, by the Fraction obtained by inverting or changing its terms. Thus, $\frac{8}{15} \div \frac{3}{4} = \frac{8}{15} \times \frac{4}{3} = \frac{32}{45}$.

Algebraic FRACTIONS, or FRACTIONS *in Species*, are exactly similar to vulgar Fractions, in numbers, and all the operations are performed exactly in the same way; therefore the rules need not be repeated, and it may be sufficient here to set down a few examples to the foregoing rules. Thus,

1. The Fraction $\frac{aab}{bc}$ abbreviates to $\frac{aa}{c}$.

2. $\frac{6a^3 - 9ax^2}{6a^2 + 3ax} = \frac{2a^2 - 3x^2}{2a + x}$, by dividing by $3a$.

3. $\frac{a^3 - a^2x + ax^2 - x^3}{a^2 - ax} = \frac{a^2 + x^2}{a}$, by dividing by $a - x$. See COMMON *Measure.*

4. $\frac{a}{b}$ and $\frac{c}{d}$ become $\frac{ad}{bd}$ and $\frac{bc}{bd}$, when reduced to a common denominator.

5. $\frac{a}{b} + \frac{c}{d} = \frac{ad + bc}{bd}$.

6. $\frac{a}{x} - \frac{b}{z} = \frac{az - bx}{xz}$.

7. $\frac{a}{b} \times \frac{c}{d} = \frac{ac}{bd}$.

8. $\frac{a}{x} \div \frac{b}{z} = \frac{a}{x} \times \frac{z}{b} = \frac{az}{bx}$.

Continued FRACTION, is used for a Fraction whose denominator is an integer with a Fraction, which latter Fraction has for its denominator an integer and a Fraction, and the same for this last Fraction again, and so on, to any extent, whether supposed to be infinitely continued, or broken off after any number of terms. Euler, Analys. Inf. vol. 1, p. 295.

As $\frac{1}{2} + \frac{1}{3}$, or $\frac{1}{2} + \frac{1}{3} + \frac{1}{4}$,

or $\frac{1}{2} + \frac{2}{3} + \frac{3}{4} + \frac{4}{5} + $ &c.

Or, using letters instead of numbers,

$$\frac{1}{a} + \frac{1}{b + \dfrac{1}{c + \dfrac{1}{d + \dfrac{1}{e}}}} \text{ &c.}$$

or $\dfrac{A}{a} + \dfrac{B}{b} + \dfrac{C}{c} + \dfrac{D}{d} + \dfrac{E}{e} + $ &c.

When these series are not far extended, it is not difficult to collect them by common arithmetic.

Lord Brounker, it seems, was the first who considered Continued Fractions, or at least, who applied them to the quadrature of curves, in Wallis's Arith. Infin. prop. 191, vol. 1, p. 469 &c, where this author explains the manner of forming them, giving several numeral examples, in approximating ratios, as well as the general series

$$\frac{a}{a} \frac{b}{\beta} \frac{c}{\gamma} \frac{d}{\delta} \text{ &c, as he denotes it.}$$

Huygens also used it for the like purpose, viz, to approximate the ratios of large numbers, in his Descrip. Autom. Planet. in Oper. Relig. p. 173 &c, edit. Amst. 1728. And a special treatise on Continued Fractions was given by Euler, in his Analys. Infin. vol. 1, pa. 295 &c.

This subject is perhaps capable of much improvement, though it has been rather neglected, as very little use has been made of it, except, by those authors, in approximating to the value of Fractions, and ratios, that are expressed in large numbers; besides a method of Goniometry by De Lagny, explained in the Introduction to my Logarithms, pa. 78; as also some use I have made of it in summing very slowly converging series, in my Tracts, p. 38 & seq.

As to the reducing of common Fractions, and ratios, that are expressed in large numbers, to Continued Fractions, it is no more than the common method of finding the greatest common measure of those two numbers, by dividing the greater by the less, and the last divisor always by the last remainder; for then the several quotients are the denominators of the Fractions, the numerators being always 1 or unity. Thus, to find approximating values of the Fraction $\frac{31415926535}{10000000000}$, or to the ratio of 31415926535 to 10000000000, being the ratio of the circumference of a circle to its diameter, by means of a Continued Fraction; or, to change the said Common Fraction to a Continued Fraction: Dividing the greater term always by the less, the same as to find the greatest common measure of the said numbers or terms, the several quotients will be 3, 7, 15, 1, 292, 1, 1, &c, which, after the first, will be the denominators, to the common numerator 1; and therefore the said Fraction will be changed into this Continued Fraction,

$3 + \dfrac{1}{7} + \dfrac{1}{15} + \dfrac{1}{1} + \dfrac{1}{292} + \dfrac{1}{1} + \dfrac{1}{1} + $ &c.

Hence, stopping at any part of these single Fractions, one after another, will give several values of the proposed ratio, all successively nearer and nearer the truth, but alternately too great and too little. So, stopping at $\frac{1}{7}$, it is $3\frac{1}{7} = \frac{22}{7} = 3\cdot142857$ too great, or 22 to 7, the ratio of the circumference to the diameter as given by Archimedes. Again, stopping at $\frac{1}{15}$, it is

$$3\frac{1}{7\frac{1}{15}} = 3\frac{15}{106} = \frac{333}{106} = 3\cdot141509 \text{ &c, too little.}$$

But stopping at $\frac{1}{1}$,

it is $3\frac{1}{7\frac{1}{15}+\frac{1}{1}} = 3\frac{1}{7\frac{1}{15}} = 3\frac{16}{113} = \frac{355}{113}$ (the ratio of

Metius) $= 3\cdot1415929$ &c, which is rather too great. And so on, always nearer and nearer, but alternately too great and too little.

And, in like manner is any algebraic Fraction thrown into a Continued Fraction. As the Fraction

$$\frac{a\beta\gamma\delta + a\beta d + ac\delta}{a\beta\gamma\delta + a\beta d + ac\delta + b\gamma\delta + bd}, \text{ which being in like}$$

manner divided, the quotients are $\frac{\alpha}{a}, \frac{\beta}{b}, \frac{\gamma}{c}, \frac{\delta}{d}$;

which single Fractions being considered as denominators to other Fractions whose common numerator is 1, these will be the reciprocals of the former, and so will become $\frac{a}{\alpha}, \frac{b}{\beta}, \frac{c}{\gamma}, \frac{d}{\delta}$; and hence the proposed common Fraction is equal to this terminate Continued Fraction,

$$\frac{a}{\alpha} + \frac{b}{\beta} + \frac{c}{\gamma} + \frac{d}{\delta}.$$

On the other hand, any Continued Fraction being given, its equivalent common Fraction will be found, by beginning at the last denominator, or lowest end of the given Continued Fraction, and gradually collecting the Fractions backwards, till we arrive at the first, when the whole will thus be collected together into one common Fraction; as was done above in collecting the Fractions

$$\frac{1}{7} + \frac{1}{15} + \frac{1}{1} + \frac{1}{292} + \text{&c.}$$

And in like manner the Continued Fraction

$$\frac{a}{\alpha} + \frac{b}{\beta} + \frac{c}{\gamma} + \frac{d}{\delta} \text{ collects into the Fraction}$$

$$\frac{a\beta\gamma\delta + a\beta d + ac\delta}{a\beta\gamma\delta + a\beta d + ac\delta + b\gamma\delta + bd}.$$

When the given Continued Fraction is an infinite one, collect it successively, first one term, then two together, three together, &c, till the sum is sufficiently exact. Or, if these collected sums converge too slowly to the true value, having collected a few of the terms into successive sums, these being alternately too great and too little, the true value will be found as near as you please by the method of arithmetical means, explained in my Tracts, vol. 1, Tract 2, pa. 11.

Vanishing FRACTIONS. Such Fractions as have both their numerator and denominator vanish, or equal to 0, at the same time, may be called *Vanishing Fractions*. We are not to conclude that such Fractions are equal to nothing, or have no value; for that they have a certain determinate value, has been shewn by the best ma-

thematicians. The idea of such Fractions as these, first originated in a very severe contest among some French mathematicians, in which Varignon and Rolle were the two chief opposite combatants, concerning the then new or differential calculus, of which the latter gentleman was a strenuous opponent. Among other arguments against it, he proposed an example of drawing a tangent to certain curves at the point where the two parts cross each other; and as the fractional expression for the subtangent, by that method, had both its numerator and denominator equal to 0 at the point proposed, Rolle looked upon it as an absurd expression, and as an argument against the method of solution itself. The seeming mystery however was soon explained, and first of all by John Bernoulli. See an account of this affair in Montucla, Hist. Math. vol. 2, pa. 366.

Since that time, such kind of fractions have often been contemplated by mathematicians. As, by Maclaurin, in his Fluxions, vol. 2, pa. 698: Saunderson, in his Algebra, vol. 2, art. 469: De Moivre, in Miscel. Anal. pa. 165: Emerson, in his Algebra, pa. 212: and by many others. The same fractions have also proved a stumbling-block to more mathematicians than one, and the cause of more violent controversies: witness that between Powell and Waring, when they were competitors for the professorship at Cambridge. In the specimen of a work published on occasion of that competition, by Waring, was the fraction $\frac{p-p^5}{1-p}$, which he said became 4 when p was $= 1$. This was struck at by Powell, as absurd, because when $p = 1$, then the fraction $\frac{p-p^5}{1-p} = \frac{1-1}{1-1} = \frac{0}{0}$, which was one chief cause of his not succeeding to the professorship. Waring replied that $\frac{p-p^5}{1-p}$ is $= p + p^2 + p^3 + p^4$ (by common division) $= 1 + 1 + 1 + 1 = 4$, when p is $= 1$. See the controversial pamphlets that passed between those two gentlemen at that time.

There are two modes of finding the value of such fractions, that have been given by the gentlemen above quoted. The one is by considering the terms of the fraction as two variable quantities, continually decreasing, till they both vanish together; or finding the ultimate value of the ratio denoted by the fraction. In this way of considering the matter, it appears that, as the terms of the fraction are supposed to decrease till they vanish, or become only equal to their fluxions or their increments, the value of the fraction at that state, will be equal to the fluxion or increment of the numerator divided by that of the denominator. Hence then, taking the example $\frac{x-x^5}{1-x}$ when $x = 1$; the fluxion of the numerator is $\dot{x} - 5x^4\dot{x}$, and of the denominator $-\dot{x}$; therefore

$$\frac{\dot{x}-5x^4\dot{x}}{-\dot{x}} = \frac{1-5x^4}{-1} = \frac{5x^4-1}{1} = 5x^4-1 = 5-1 = 4,$$

the value of the fraction $\frac{x-x^5}{1-x}$ when $x = 1$.——Or, thus, because $x = 1$, therefore $\frac{x-x^5}{1-x} = \frac{1-x^4}{1-x}$; then the fluxion of the numerator, $-4x^3\dot{x}$, divided by the

the fluxion of the denominator, or $-x$, gives $4x^3$ or 4, the fame as before.

The other method is by reducing the given expreffion to another, or fimple form, and then fubftituting the values of the letters. So in the above example $\frac{x-x^5}{1-x}$, or $\frac{1-x^4}{1-x}$, when $x=1$; divide the numerator by the denominator, and it becomes $1+x+x^2+x^3$, which when $x=1$, becomes 4, for the given fraction, the fame as before.—Again, to find the value of $\frac{a\sqrt{ax-xx}}{a-\sqrt{ax}}$ when x is $=a$, in which cafe both the numerator and denominator become $=0$. Divide the numerator by the denominator, and the quotient is

$\sqrt{ax}+x+x\sqrt{\frac{x}{a}}$; which when $x=a$, becomes

$a+a+a=3a$, for the value of the fraction in that ftate of it.

FRAISE, in Fortification, a kind of defence, confifting of pointed ftakes, driven almoft parallel to the horizon into the retrenchments of a camp, &c, to ward off and prevent any approach or fcalade.

FRANKLIN (Dr. Benjamin), one of the moft celebrated philofophers and politicians of the 18th century, was born in Bofton in North America in the year 1706, being the youngeft of 13 children. His father was a tallow chandler in Bofton, and young Franklin was taken from fchool at 10 years of age, to affift him in his bufinefs. In this fituation he continued two years; but difliking that occupation, he was bound apprentice to an elder brother, who was then a printer in Bofton, but had learned that bufinefs in London, and who in the year 1721 began to print a newfpaper, being the fecond ever publifhed in America; the copies of which our author was fent to diftribute, after having affifted in compofing and printing it. Upon this occafion, our young philofopher enjoyed the fecret and fingular pleafure of being the much admired author of many effays in this paper; a circumftance which he had the addrefs to keep a fecret even from his brother himfelf; and this when he was only 15 years of age. The frequent ill ufage from his brother, induced young Franklin to quit his fervice, which he did, at the age of 17, and went to New York. But not meeting employment here, he went forward to Philadelphia, where he worked with a printer a fhort time; after which, at the inftance of Sir William Keith, governor of the province, he returned to Bofton to folicit pecuniary affiftance from his father to fet up a printing-houfe for himfelf at Philadelphia, upon the promife of great encouragement from Sir William, &c. His father however thought fit to refufe fuch aid, alleging that he was yet too young (18 years old) to be entrufted with fuch a concern; fo our author again returned to Philadelphia without it. Upon this, Sir William faid he would advance the fum that might be neceffary, and our young philofopher fhould go to England, and purchafe all the types and materials, for which purpofe he would give him letters of credit. He could never however get thefe letters, yet by dint of fair promifes of their being fent on board the fhip after him, he failed for England, expecting thefe letters of credit were in the governor's packet, which he was to receive upon its

being opened. In this however he was cruelly deceived, and thus he was fent to London without either money, friends, or credit, at 18 years of age.

He foon found employment, however, as a journeyman printer, firft at a Mr. Palmer's, and afterward with Mr. Watts, with whom he worked a confiderable time, and by whom he was greatly efteemed, being alfo treated with fuch kindnefs, that it was always moft gratefully remembered by our philofopher.

After a ftay of 18 months in London, he returned to Philadelphia, viz in 1726, along with a merchant of that town, as his clerk, on a falary of 50 pounds a year. But his mafter dying the year after, he again engaged to direct the printing bufinefs of the fame perfon with whom he had worked before. After continuing with him the beft part of a year, our philofopher, in partnerfhip with another young man, at length fet up a printing-houfe himfelf.

A little before this time, young Franklin had gradually affociated a number of perfons, like himfelf, of a rational and philofophical turn of mind, and formed them into a club or fociety, to hold meetings, to converfe and communicate their fentiments together, for their mutual improvement in all kinds of ufeful knowledge, which was in high repute for many years after. Among many other ufeful regulations, they agreed to bring fuch books as they had into one place, to form a common library. This refource being found defective, at Franklin's perfuafion they refolved to contribute a fmall fum monthly towards the purchafe of books for their ufe from London. Thus their ftock began to increafe rapidly; and the inhabitants of Philadelphia, being defirous of having a fhare in their literary knowledge, propofed that the books fhould be lent out on paying a fmall fum for the indulgence. Thus in a few years the fociety became rich, and poffeffed more books than were perhaps to be found in all the other colonies: the collection was advanced into a public library; and the other colonies, fenfible of its advantages, began to form fimilar plans; from whence originated the libraries at Bofton, New York, Charleftown, &c; that of Philadelphia being now not inferior to any in Europe.

About 1728 or 1729, young Franklin fet up a newfpaper, the fecond in Philadelphia, which proved very profitable, and otherwife ufeful, as affording an opportunity of making himfelf known as a political writer, by inferting feveral of his writings of that kind into it. In addition to his printing-houfe, he fet up a fhop to fell books and ftationary; and in 1730 he married his wife, who proved very ufeful in affifting to manage the fhop, &c. He afterward began to have fome leifure, both for reading books, and writing them, of which he gave many fpecimens from time to time. In 1732 he began to publifh Poor Richard's Almanac, which was continued for many years. It was always remarkable for the numerous and valuable concife maxims which it contained, for the œconomy of human life, all tending to exhort to induftry and frugality: and in the almanac for the laft year, all the maxims were collected in an addrefs to the reader, entitled, The Way to Wealth. This has been tranflated into various languages, and inferted in different publications. It has alfo been printed on a large fheet, proper to be framed, and hung up in confpicuous places in all houfes, as it

very

5

very well deserves to be. Mr. Franklin became gradually more known for his political talents, and in the year 1736, he was appointed clerk to the General Assembly of Pennsylvania; and was re-elected by succeeding assemblies for several years, till he was chosen a representative for the city of Philadelphia; and in 1737 he was appointed post-master of that city. In 1738, he formed the first fire-company there, to extinguish and prevent fires and the burning of houses: an example which was soon followed by other persons, and other places. And soon after, he suggested the plan of an association for insuring houses and ships from losses by fire, which was adopted; and the association continues to this day. In the year 1744, during a war between France and Great Britain, some French and Indians made inroads upon the frontier inhabitants of the province, who were unprovided for such an attack: the situation of the province was at this time truly alarming, being destitute of every means of defence. At this crisis Franklin stepped forth, and proposed to a meeting of the citizens of Philadelphia, a plan of a voluntary association for the defence of the province. This was approved of, and signed by 1200 persons immediately. Copies of it were circulated through the province; and in a short time the number of signers amounted to 10,000. Franklin was chosen colonel of the Philadelphia regiment; but he did not think proper to accept of the honour.

Pursuits of a different nature now occupied the greatest part of his attention for some years. Being always much addicted to the study of natural philosophy; and the discovery of the Leyden experiment in electricity having rendered that science an object of general curiosity; Mr. Franklin applied himself to it, and soon began to distinguish himself eminently in that way. He engaged in a course of electrical experiments with all the ardour and thirst for discovery which characterized the philosophers of that day. By these, he was enabled to make a number of important discoveries, and to propose theories to account for various phenomena; which have been generally adopted, and which will probably endure for ages. His observations he communicated, in a series of letters, to his friend Mr. Collinson; the first of which is dated March 28, 1747. In these he makes known the power of points in drawing and throwing off the electric matter, which had hitherto escaped the notice of electricians. He also made the discovery of a *plus* and *minus*, or of a *positive* and *negative* state of electricity; from whence in a satisfactory manner he explained the phenomena of the Leyden phial, first observed by Cuneus or Muschenbroeck, which had much perplexed philosophers. He shewed that the bottle, when charged, contained no more electricity than before, but that as much was taken from one side as was thrown on the other; and that, to discharge it, it was only necessary to make a communication between the two sides, by which the equilibrium might be restored, and that then no signs of electricity would remain. He afterwards demonstrated by experiments, that the electricity did not reside in the coating, as had been supposed, but in the pores of the glass itself. After a phial was charged, he removed the coating, and found that upon applying a new coating the shock might still be received. In the year 1749, he first suggested his idea of explaining the phenomena

of thunder-gusts, and of the aurora borealis, upon electrical principles. He points out many particulars in which lightning and electricity agree; and he adduces many facts, and reasoning from facts, in support of his positions. In the same year he conceived the bold and grand idea of ascertaining the truth of his doctrine, by actually drawing down the forked lightning, by means of sharp-pointed iron rods raised into the region of the clouds; from whence he derived his method of securing buildings and ships from being damaged by lightning. It was not until the summer of 1752 that he was enabled to complete his grand discovery the experiment of the electrical kite, which being raised up into the clouds, brought thence the electricity or lightning down to the earth; and M. D'Alibard made the experiment about the same time in France, by following the track which Franklin had before pointed out. The letters which he sent to Mr. Collinson, it is said, were refused a place among the papers of the Royal Society of London; and Mr. Collinson published them in a separate volume, under the title of *New Experiments and Observations on Electricity, made at Philadelphia, in America*; which were read with avidity, and soon translated into different languages. His theories were at first opposed by several philosophers, and by the members of the Royal Society of London; but in 1755, when he returned to that city, they voted him the gold medal which is annually given to the person who presents the best paper on some interesting subject. He was also admitted a member of the Society, and had the degree of doctor of laws conferred upon him by different universities: but at this time, by reason of the war which broke out between Britain and France, he returned to America, and interested himself in the public affairs of that country. Indeed he had done this long before; for although philosophy was a principal object of Franklin's pursuit for several years, he did not confine himself to it alone. In the year 1747 he became a member of the General Assembly of Pennsylvania, as a burgess for the city of Philadelphia. Being a friend to the rights of man from his infancy, he soon distinguished himself as a steady opponent of the unjust schemes of the proprietaries. He was soon looked up to as the head of the opposition; and to him have been attributed many of the spirited replies of the assembly, to the messages of the governors. His influence in the body was very great. This arose not from any superior powers of eloquence; he spoke but seldom and he never was known to make any thing like an elaborate harangue. His speeches often consisted of a single sentence, or of a well-told story, the moral of which was always obviously to the point. He never attempted the flowery fields of oratory. His manner was plain and mild. His style in speaking was, like that of his writings, simple, unadorned, and remarkably concise. With this plain manner, and his penetrating and solid judgment, he was able to confound the most eloquent and subtle of his adversaries, to confirm the opinions of his friends, and to make converts of the unprejudiced who had opposed him. With a single observation, he has rendered of no avail a long and elegant discourse, and determined the fate of a question of importance.

In the year 1749, he proposed a plan of an academy,

to

to be erected in the city of Philadelphia, as a foundation for posterity to erect a seminary of learning, more extensive and suitable to future circumstances; and in the beginning of 1750, three of the schools were opened, namely, the Latin and Greek school, the Mathematical, and the English schools. This foundation soon after gave rise to another more extensive college, incorporated by charter May 27, 1755, which still subsists, and in a very flourishing condition. In 1752, he was instrumental in the establishment of the Pennsylvania Hospital, for the cure and relief of indigent invalids, which has proved of the greatest use to that class of persons. Having conducted himself so well as Postmaster of Philadelphia, he was, in 1753, appointed Deputy Postmaster-general for the whole British colonies.

The colonies being much exposed to depredations in their frontier by the Indians and the French, at a meeting of commissioners from several of the provinces, Mr. Franklin proposed a plan for the general defence, to establish in the colonies a general government, to be administered by a president-general, appointed by the crown, and by a grand council, consisting of members chosen by the representatives of the different colonies; a plan which was unanimously agreed to by the commissioners present. The plan however had a singular fate: It was disapproved of by the ministry of Great Britain, because it gave too much power to the representatives of the people; and it was rejected by every assembly, as giving to the president general, who was to be the representative of the crown, an influence greater than appeared to them proper, in a plan of government intended for freemen. Perhaps this rejection, on both sides, is the strongest proof that could be adduced of the excellence of it, as suited to the situation of Great Britain and America at that time. It appears to have steered exactly in the middle, between the opposite interests of both. Whether the adoption of this plan would have prevented the separation of America from Great Britain, is a question which might afford much room for speculation.

In the year 1755, General Braddock, with some regiments of regular troops, and provincial levies, was sent to dispossess the French of the posts upon which they had seized in the back settlements. After the men were all ready, a difficulty occurred, which had nearly prevented the expedition. This was the want of waggons. Franklin now stepped forward, and, with the assistance of his son, in a little time procured 150. After the defeat of Braddock, Franklin introduced into the assembly a bill for organizing a militia, and had the dexterity to get it passed. In consequence of this act a very respectable militia was formed; and Franklin was appointed colonel of a regiment in Philadelphia, which consisted of 1200 men; in which capacity he acquitted himself with much propriety, and was of singular service; though this militia was soon after disbanded by order of the English ministry.

In 1757, he was sent to England, with a petition to the king and council, against the proprietaries, who refused to bear any share in the public expences and assessments; which he got settled to the satisfaction of the state. After the completion of this business, Franklin remained at the court of Great Britain for some time, as agent for the province of Pennsylvania;

and also for those of Massachusetts, Maryland, and Georgia. Soon after this, he published his Canada pamphlet, in which he pointed out, in a very forcible manner, the advantages that would result from the conquest of this province from the French. An expedition was accordingly planned, and the command given to General Wolfe; the success of which is well known. He now divided his time indeed between philosophy and politics, rendering many services to both. Whilst here, he invented the elegant musical instrument called the *Armonica*, formed of glasses played on by the fingers. In the summer of 1762 he returned to America; on the passage to which he observed the singular effect produced by the agitation of a vessel, containing oil floating on water: the upper surface of the oil remained smooth and undisturbed, whilst the water was agitated with the utmost commotion. On his return he received the thanks of the Assembly of Pennsylvania, which having annually elected him a member in his absence, he again took his seat in this body, and continued a steady defender of the liberties of the people.

In 1764, by the intrigues of the proprietaries, Franklin lost his seat in the assembly, which he had possessed for 14 years; but was immediately appointed provincial agent to England, for which country he presently set out. In 1766 he was examined before the parliament relative to the stamp-act; which was soon after repealed. The same year he made a journey into Holland and Germany; and another into France; being everywhere received with the greatest respect by the literati of all nations. In 1773 he attracted the public attention by a letter on the duel between Mr. Whately and Mr. Temple, concerning the publication of Governor Hutchinson's letters, declaring that he was the person who had discovered those letters. On the 29th of January next year, he was examined before the privy-council on a petition he had presented long before as agent for Massachusetts Bay against Mr. Hutchinson: but this petition being disagreeable to ministry, it was precipitately rejected, and Dr. Franklin was soon after removed from his office of Postmaster-general for America. Finding now all efforts to restore harmony between Great Britain and her colonies useless, he returned to America in 1775; just after the commencement of hostilities. Being named one of the delegates to the Continental Congress, he had a principal share in bringing about the revolution and declaration of independency on the part of the colonies. In 1776 he was deputed by Congress to Canada, to negociate with the people of that country, and to persuade them to throw off the British yoke; but the Canadians had been so much disgusted with the hot-headed zeal of the New Englanders, who had burnt some of their chapels, that they refused to listen to the proposals, though enforced by all the arguments Dr. Franklin could make use of. On his return to Philadelphia, Congress, sensible how much he was esteemed in France, sent him thither to put a finishing hand to the private negociations of Mr. Silas Deane; and this important commission was readily accepted by the Doctor, though then in the 71st year of his age. The event is well known; a treaty of alliance and commerce was signed between France and America; and M. Le Roi asserts, that the Doctor had a great share in the transaction, by strongly advising M. Maurepas not

to

to lose a single moment, if he wished to secure the friendship of America, and to detach it from the mother-country. In 1777 he was regularly appointed plenipotentiary from Congress to the French court; but obtained leave of dismission in 1780. Having at length seen the full accomplishment of his wishes by the conclusion of the peace in 1783, which gave independency to America, he became desirous of revisiting his native country. He therefore requested to be recalled; and, after repeated solicitations, Mr. Jefferson was appointed in his stead. On the arrival of his successor, he repaired to Havre de Grace, and crossing the channel, landed at Newport in the Isle of Wight; from whence, after a favourable passage, he arrived safe at Philadelphia in September 1785. He was received amidst the acclamations of a vast multitude who flocked from all parts to see him, and who conducted him in triumph to his own house; where in a few days he was visited by the members of the Congress and the principal inhabitants of Philadelphia. He was afterward twice chosen president of the Assembly of Philadelphia; but his increasing infirmities obliged him to ask permission to retire, and to spend the remainder of his life in tranquillity; which was granted, in 1788. After this, the infirmities of age increased fast upon him; he became more and more afflicted with the gout and the stone, till the time of his death, which happened the 17th of April 1790, about 11 at night, at 84 years of age; leaving one son, governor William Franklin, a zealous loyalist, who now resides in London; and a daughter, married to Mr. William Bache, merchant in Philadelphia.

Doctor Franklin was author of many tracts on electricity, and other branches of natural philosophy, as well as on politics, and miscellaneous subjects. He had also many papers inserted in the Philosophical Transactions, from the year 1757 to 1774.

FREEZE, or Frize, in Architecture, a large flat member, being that part of the entablature of columns that separates the architrave from the cornice.

FREEZING, or Congelation, the fixing of a fluid body into a firm or solid mass by the action of cold: in which sense the term is applied to water when it freezes into ice; to metals when they resume their solid form after being melted by heat; or to glass, wax, pitch, tallow, &c, when they harden again after having been rendered fluid by heat. But it differs from crystallization, which is rather a separation of the particles of a solid from a fluid in which it had been dissolved more by the moisture than the action of heat.

The process of congelation is always attended with the emission of heat, as is found by experiments on the freezing of water, wax, spermaceti, &c; for in such cases it is always found that a thermometer dipt into the fluid mass, keeps continually descending as this cools, till it arrive at a certain point, being the point of freezing, which is peculiar to each fluid, where it is rather stationary, and then rises for a little, while the congelation goes on. But by what means it is that fluid bodies should thus be rendered solid by cold, or fluid by heat, or what is introduced into the bodies by either of those principles, are matters the learned have never yet been able to discover, or to satisfy themselves upon. The following phenomena however are usually observed.

Water, and some other fluids, suddenly dilate and expand in the act of Freezing, so as to occupy a greater space in the form of ice than before, in consequence of which it is that ice is specifically lighter than the same fluid, and floats in it. And the degree of expansion of water, in the state of ice, is by some authors computed at about $\frac{1}{10}$ of its volume. Oil however is an exception to this property, and quicksilver too, which shrinks and contracts still more after Freezing. Mr. Boyle relates several experiments of vessels made of metal, very thick and strong; in which, when filled with water, close stopped, and exposed to the cold, the water being expanded in Freezing, and not finding either room or vent, burst the vessels. A strong barrel of a gun, with water in it close stopped and frozen, was rent the whole length. Huygens, to try the force with which it expands, filled a cannon with it, whose sides were an inch thick, and then closed up the mouth and vent, so that none could escape; the whole being exposed to a strong Freezing air, the water froze in about 12 hours, and burst the piece in two places. Mathematicians have computed the force of the ice upon this occasion; and they say, that such a force would raise a weight of 27720 pounds. Lastly, Major Edward Williams, of the Royal Artillery, made many experiments on the force of it, at Quebec, in the years 1784 and 1785. He filled all sizes of iron bomb shells with water, then plugged the fuze hole close up, and exposed them to the strong Freezing air of the winter in that climate; sometimes driving in the iron plugs as hard as possible with a sledge hammer; and yet they were always thrown out by the sudden expansion of the water in the act of Freezing, like a ball shot by gunpowder, sometimes to the distance of between 400 and 500 feet, though they weighed near 3 pounds; and when the plugs were screwed in, or furnished with hooks or barbs, to lay hold of the inside of the shell by, so that they could not possibly be forced out, in this case the shell was always split in two, though the thickness of the metal of the shell was about an inch and three quarters. It is farther remarkable, that through the circular crack, round about the shells, where they burst, there stood out a thin film or sheet of ice, like a fin; and in the cases when the plugs were projected by Freezing water, there suddenly issued out from the fuze-hole, a bolt of ice, of the same diameter, and stood over it to the height sometimes of 8 inches and a half. And hence we need not be surprised at the effects of ice in destroying the substance of vegetables and trees, and even splitting rocks, when the frost is carried to excess.

It is also observed that water loses of its weight by Freezing, being found lighter after thawing again, than before it was frozen. And indeed it evaporates almost as fast when frozen, as when it is fluid.

It is said too that water does not freeze in vacuo; requiring for that purpose the presence and contiguity of the air. But this circumstance is liable to some doubt, and it may be suspected that the degree of cold has not been carried far enough in these instances; as it is found that mercury in thermometers has even been frozen

frozen, though it requires a vaſtly greater degree of cold to freeze mercury, than water.

That water which has been boiled freezes more readily than that which has not been boiled; and that a ſlight diſturbance of the fluid diſpoſes it to freeze more ſpeedily; having ſometimes been cooled ſeveral degrees below the Freezing point, without congealing when kept quite ſtill, but ſuddenly freezing into ice on the leaſt motion or diſturbance.

That the water, being covered over with a ſurface of oil of olives, does not freeze ſo readily as without it; and that nut oil-abſolutely preſerves it under a ſtrong froſt, when olive oil would not.

That rectified ſpirit of wine, nut oil, and oil of turpentine, ſeldom freeze.

That the ſurface of the water, in Freezing, appears all wrinkled; the wrinkles being ſometimes in parallel lines, and ſometimes like rays, proceeding from a centre to the circumference.

FREEZING *Mixture*, a preparation for the artificial congelation of water, and other fluids.

According to Mr. Boyle, all kinds of ſalts, whether alkaline or acid; and even all ſpirits, as ſpirit of wine, &c; as alſo ſugar, and ſaccharum ſaturni, mixed with ſnow, are capable of Freezing moſt fluids; and the ſame effect is produced, in a very high degree, by a mixture of oil of vitriol, or ſpirit of nitre, with ſnow.

M. Homberg remarks the ſame of equal quantities of corroſive ſublimate, and ſal ammoniac, with four times the quantity of diſtilled vinegar.

Boerhaave gives a method of producing artificial froſt without either ſnow or ice: we muſt have for this purpoſe, at any ſeaſon of the year, the coldeſt water that can be procured; this is to be mixed with a proper quantity of any ſalt (ſal ammoniac will anſwer the intention beſt), at the rate of about 3 ounces to a quart of water. Another quart of water muſt be prepared in the ſame manner with the firſt; the ſalt, by being diſſolved in each, will make the water much colder than it was before. The two quarts are then to be mixed together, and this will make them colder ſtill. Two quarts more of water prepared and mixed in the ſame manner are to be mixed with theſe, which will increaſe the cold to a much higher degree in all. The whole of this operation is to be carried on in a cold cellar; and a glaſs of common water is then to be placed in the veſſel of the fluid thus artificially cooled, and it will be turned into ice in the ſpace of 12 hours.

There is alſo a method of making artificial ice by means of ſnow, without any kind of ſalt. For this purpoſe fill a ſmall pewter diſh with water, and upon that ſet a common pewter plate filled, but not heaped, with ſnow, Bring this ſimple apparatus near the fire, and ſtir the ſnow in the plate: the ſnow will diſſolve, and the ice will be formed on the back of the plate, which was ſet in the diſh of water.

Mr. Reaumur tried the effect of ſeveral ſalts, and examined the various degrees of cold by an ice thermometer, which being placed in the fluid to be frozen, ſhewed very exactly the degree of cold by the deſcent of the ſpirit.

Nitre, or ſaltpetre, uſually paſſes for a ſalt that may be very ſerviceable in theſe artificial congelations; but the experiments of this gentleman prove that this opinion is erroneous. The moſt perfectly refined ſaltpetre employed in the operation ſunk the ſpirit in the thermometer only 3 degrees and a half below the fixed point. Leſs refined nitre ſunk the thermometer lower, and gave a greater degree of cold; owing to the common or ſea-ſalt that it contains when leſs pure, which has a greater effect than the pure ſaltpetre itſelf.

Two parts of common ſalt being mixed with three parts of powdered ice in very hot weather, the ſpirit in the thermometer immediately deſcended 15 degrees, which is half a degree lower than it would have deſcended in the ſevereſt cold of our winters. Mr. Reaumur then tried the ſalts all round, determining with great regularity and exactneſs, what was the degree of cold occaſioned by each in a given doſe. Among the neutral ſalts, none produced a greater degree of cold than the common ſea ſalt. Among the alkalies, ſal ammoniac ſunk the thermometer only to 13 degrees. Pot-aſhes ſunk it juſt as low as well refined ſaltpetre.

For the common uſes of the table, the ice is not required to be very hard, or ſuch as is produced by long continuance of violent cold: it is rather deſired to be like ſnow. Saltpetre, which is no very powerful freezer, is therefore more fit for the purpoſe than a more potent ſalt. It is not neceſſary that the congelation ſhould be very ſuddenly made; but that it may retain its form as long as may be, when made, is of great importance.

If it be deſired to have ices very hard and firm, and very ſuddenly prepared, then ſea ſalt is of all others moſt to be choſen for the operation. The ices thus made will be very hard, but they will ſoon run. Pot-aſhes afford an ice of about the hardneſs that is uſually required. This forms indeed very ſlowly, but then it will preſerve a long time. And common wood aſhes will perform the buſineſs very nearly in the ſame manner as the pot-aſhes; but for this purpoſe, the wood which is burnt, ought to be freſh.

The ſtrong acid ſpirits of the neutral ſalts act much more powerfully in theſe congelations than the ſalts themſelves, or indeed than any ſimple ſalt can do. Thus, ſpirit of nitre mixed with twice its quantity of powdered ice, immediately ſinks the ſpirit in the thermometer to 19 degrees, or 4 degrees more than that obtained by means of ſea ſalt, the moſt powerful of all the ſalts in making artificial cold. A much greater degree of cold may be given to this mixture, by piling it round with more ice mixed with ſea ſalt. This gives a redoubled cold, and ſinks the thermometer to 24 degrees. If this whole matter be covered with a freſh mixture of ſpirit of nitre and ice, a ſtill greater degree of cold is produced, and ſo on; the cold being by this method of freſh additions to be increaſed almoſt without bounds: but it is to be obſerved, that every addition gives a ſmaller increaſe than the former.

It is very remarkable in the acid ſpirits, that though ſea ſalt is ſo much more powerful than nitre in ſubſtance in producing cold, yet the ſpirit of nitre is much ſtronger than that of ſea ſalt; and another not leſs wonderful phenomenon is, that ſpirit of wine, which is little elſe than liquid fire, has as powerful an effect in congelations, or very nearly ſo, as the ſpirit of nitre itſelf.

The several liquid substances which produce cold, in the same manner as the dry salts on being mixed with ice, are much more speedy in their action than the salts: because they immediately and much more intimately come into contact with the particles of the ice, than the salts can. Of this nature are spirit of nitre, spirit of wine, &c. To produce the expected degree of cold, it is always necessary that the ice and the added matter, whatever it be, should both run together, and, intimately uniting, form one clear fluid. It is hence that no new cold is produced with oil, which, though it melts the ice, yet cannot mix itself into a homogeneous liquid with it, but must always remain floating on the surface of the water that is produced by the melting of the ice. Mem. Acad. Scienc. Par. 1734.

It has been discovered, that fluids standing in a current of air, grow by this means much colder than before. Fahrenheit had long since observed, that a pond, which stands quite calm, often acquires a degree of cold much beyond what is sufficient for Freezing, and yet no congelation ensued: but if a slight breath of air happens in such a case to brush over the surface of the water, it stiffens the whole in an instant. It has also been discovered, that all substances grow colder by the evaporation of the fluids which they contain, or with which they are mixed. If both these methods therefore be practised upon the same body at the same time, they will increase the cold to almost any degree of intenseness we please.

But the most extraordinary instances of artificial Freezing, have since been made in Russia, at Hudson's bay, and other parts, by which quicksilver was frozen into a solid mass of metal. And the same thing had before happened from the natural cold of the atmosphere alone, in Siberia. In the winter of 1733, Professor Gmelin, with two other gentlemen of the Russian Academy, were sent by Anne Ivanouna, the new empress, to explore and describe the different parts of her Asiatic dominions, with the communication of Asia and America. In the winter of 1734 5, Mr. Gmelin being at Yeneseisk in 58° 30' north lat. and 92° long. east from Greenwich, first observed such a descent of the mercury, as must have been attended with congelation, being far below its Freezing point, now fixed at —40 of Fahrenheit's thermometer. " Here, says he, we first experienced the truth of what various travellers have related with respect to the extreme cold of Siberia; for, about the middle of December, such severe weather set in, as we were sure had never been known in our time at Petersburg. The air seemed as if it were frozen, with the appearance of a fog, which did not suffer the smoke to ascend as it issued from the chimneys. Birds fell down out of the air as dead, and froze immediately, unless they were brought into a warm room. Whenever the door was opened, a fog suddenly formed round it. During the day, short as it was, parhelia and haloes round the sun were frequently seen; and in the night mock moons, and haloes about the moon. Finally, our thermometer, not subject to the same deception as the senses, left us no doubt of the excessive cold; for the quicksilver in it was reduced, on the 5th of January, old style, to — 120° of Fahrenheit's scale, lower than it had ever been observed in nature."

The next instance of congelation happened at Yakutsk, in 62° north lat. and 150° east longitude. The weather here was unusually mild for the climate, yet the thermometer fell to —72°; and one person informed the professor by a note, that the mercury in his barometer was frozen. He hastened immediately to his house to behold such a surprising phenomenon; but though he was witness to the fact, observing that the mercury did not continue in one column, but was divided in different places as into little cylinders, which appeared frozen, yet the prejudice he had entertained against the possibility of the congelation, would not allow him to believe it.

Another set of observations, in the course of which the mercury must frequently have been congealed, were made by professor Gmelin at Kirenga fort, in 57½ north lat., and 108 east long.; his thermometer, at different times, standing at —108, —86, —100, —113, and many other intermediate degrees; in the course of the winter of 1737-8. On the 27th of November, after the thermometer had been standing for two days at —46°, he found it sunk at noon to —108. Suspecting some mistake, after he had noted down the observation, he instantly ran back, and found it at —102; but ascending with such rapidity, that in the space of half an hour it had risen to 19°. This phenomenon, which appeared so surprising, doubtless depended on the expansion of the mercury frozen in the bulb of the thermometer, and which now melting, forced upwards the small thread in the stem. And similar appearances were observed on other days afterwards, when the thread of quicksilver in the thermometer was separated about 6 degrees.

A second instance where a natural congelation of mercury has certainly been observed, is recorded in the transactions of the Royal Academy of Sciences at Stockholm, as made by Mr. Andrew Hellant. The weather, in January 1760, was remarkably cold in Lapland; so that on the 5th of that month, the thermometers fell to —76, —128, or lower; on the 23d and following days they fell to —58, —79, —92, and below —238 entirely into the ball. This was observed at four different places in Lapland, situated between the 65th and 78th degrees of north lat. and the 21st and 28th degree of east longitude.

But the congelation of quicksilver, by an artificial Freezing mixture, was first observed, and put beyond doubt, by Mr. Joseph Adam Braun, professor of philosophy at Petersburg. This gentleman wishing to try how many degrees of cold he could produce, availed himself of a good opportunity which offered for that purpose on the 14th of December 1759, when the mercury in the thermometer stood in the natural cold at —34, which it is now known is only 5 or 6 degrees above its point of congelation. Assisting this natural cold therefore with a mixture prepared of aquafortis and pounded ice, his thermometer was sunk to —69. Part of the quicksilver must now have been really congealed, but unexpected by him, and he only thought of pursuing his object of producing still greater degrees of cold; and having expended all his pounded ice, he was obliged to use snow instead of it. With this fresh mixture the mercury sunk to —100, —240, and —350°. Taking the thermometer out, he found it whole,

whole, but the quickfilver fixed, and it continued fo for 12 minutes. On repeating the experiment, with another thermometer which had been graduated no lower than —220, all the mercury funk into the ball, and became folid as before, and did not re-afcend till after a ftill longer interval of time. Mr. Braun now fufpected that the quickfilver was really frozen, and prepared for making a decifive experiment. This was accomplifhed on the 25th of the fame month, and the bulb of the thermometer broken as foon as the metal was congealed; when it appeared that the mercury was changed into a folid and fhining metallic mafs, which flatted and extended under the ftrokes of a peftle, being rather lefs hard than lead, and yielding a dull found like that metal. Mr. Æpinus made fimilar experiments at the fame time, employing as well thermometers as tubes of a larger bore; in which laft he remarked, that the quickfilver fell fenfibly on being frozen, affuming a concave furface, and likewife that the congealed pieces funk in fluid mercury: alfo, in their farther experiments, they invariably found that the mercury funk lower when the whole of it was congealed, than if any part of it remained fluid: all fhewing that, contrary to water, mercury contracted in Freezing. It was farther obferved, that the mercury when congealed looked like the moft polifhed filver, and when beaten flat, it was eafily cut with a penknife, like foft thin fheet lead.

The fact being thus eftablifhed, and fluidity no longer to be confidered as an effential property of quickfilver, Mr. Braun communicated an account of his experiments to the Peterfburg academy, on the 6th of September 1760; of which a large extract was inferted in the Philof. Tranf. vol. 52, pa. 156. He afterwards declared that he never fuffered a winter to pafs without repeating the experiment of Freezing quickfilver, and never failed of fuccefs when the natural cold was of a fufficient ftrength for the purpofe; and this degree of natural cold he fuppofes at —10 of Fahrenheit; though fome commencement of the congelation might be perceived when the temperature of the air was as high as + 2.

The refults of all his experiments were, that with the abovementioned frigorific mixtures, and once with rectified fpirits and fnow, when the natural cold was at —28°, he congealed the quickfilver, and difcovered that it is a real metal that melts with a very fmall degree of heat. However, not perceiving the neceffary confequence of its great contraction in Freezing, he always confounded its point of congelation with that of its greateft contraction in Freezing, and thus marked the former a great deal too low.

In the procefs of his obfervations, Mr. Braun found that double aquafortis was more effectual than fpirit of nitre; but with this fimple fpirit, which feldom brings the mercury lower than —148, this metal may be frozen in the following manner: Six glaffes being filled with fnow as ufual, and the thermometer put in one of them, the fpirit of nitre was poured upon it; when the mercury would fall no lower in this, the thermometer was removed to the fecond, and fo on to the third and fourth, in which fourth immerfion the mercury was congealed.

Mr. Æpinus gives the following direction for ufing the fuming fpirit of nitre: Take fome of this fpirit, cooled as much as poffible, and put it into a wine glafs till it be about half full, filling it up with fnow, and ftirring them till the mixture become of the confiftence of pap, by which means you obtain, almoft in an inftant, the neceffary degree of cold for the Freezing of quickfilver.

It is remarked by Mr. Braun, that by the mixture of fnow and fpirit of nitre, which froze the mercury, he never was able to bring thermometers, filled with the moft highly rectified fpirit of wine, lower than —148: fo that the cold which will freeze mercury, will not freeze fpirit of wine; and therefore fpirit thermometers are the moft fit to determine the degree of coldnefs in frigorific mixtures, till we can conftruct folid metallic thermometers with fufficient accuracy. Mr. Braun tried the effects of different Fluids in his frigorific mixtures: he always found that Glauber's fpirit of nitre and double aquafortis were the moft powerful; and from a number of experiments made when the temperature of the air was between 21 and 28 of Fahrenheit, he concludes, that fpirit of falt pounded upon fnow increafed the natural cold 36°; fpirit of fal ammoniac, 12; oil of vitriol, 42; Glauber's fpirit of nitre, 70; aquafortis, 48; fimple fpirit of nitre, 36; dulcified fpirit of vitriol, 24; Hoffman's anodyne liquor, 38; fpirit of hartfhorn, 12; fpirit of fulphur, 12; fpirit of wine rectified, 24; camphorated fpirit, 18; French brandy, 14; and feveral kinds of wine increafed the natural cold to 7, 8, or 9 degrees.

The moft remarkable congelation of mercury, by natural cold, that has ever been obferved, was that related by Dr. Peter Simon Pallas, who had been fent by the emprefs of Ruffia, with fome other gentlemen, on an expedition fimilar to that of Mr. Gmelin. Being at Krafnoyarfk in the year 1772, in north lat. 56° 30′, and eaft long. 93°, he had an opportunity of obferving the phenomenon we fpeak of. On the 6th and 7th of December that year, fays he, there happened the greateft cold I have ever experienced in Siberia: the air was calm at the time, and feemingly thickened; fo that, though the fky was in other refpects clear, the fun appeared as through a fog. I had only one fmall thermometer left, in which the fcale went no lower than —7°; and on the 6th in the morning, I remarked that the quickfilver in it funk into the ball, except fome fhort columns which ftuck faft in the tube. When the ball of the thermometer, as it hung in the open air, was touched with the finger, the quickfilver rofe; and it could plainly be feen that the folid columns ftuck and refifted a good while, and were at length pufhed upward with a fort of violence. He alfo placed upon the gallery, on the north fide of his houfe, fome quickfilver in an open bowl. Within an hour he found the edges and furface of it frozen folid; and fome minutes afterward the whole was condenfed by the natural cold into a foft mafs very much like tin. While the inner part was ftill fluid, the frozen furface exhibited a great variety of branched wrinkles; but in general it remained pretty fmooth in Freezing. The congealed mercury was more flexible than lead; but on being bent fhort, it was found more brittle than tin; and when hammered out thin, it feemed fomewhat granulated. When the hammer was not perfectly cooled, the

3 U 2   quickfilver

quickfilver melted away under it in drops; and the same thing happened when the metal was touched with the finger, by which alfo the finger was immediately benumbed. When the frozen mafs was broken to pieces in the cold, the fragments adhered to each other and to the bowl in which they lay. In the warm room it thawed on its furface gradually, by drops, like wax on the fire, and did not melt all at once. Although the froft feemed to abate a little towards night, yet the congealed quickfilver remained unaltered, and the experiment with the thermometer could ftill be repeated. On the 7th of December he had an opportunity of making the fame obfervations all day; but fome hours after funfet, a northweft wind fprung up, which raifed the thermometer to —46°, when the mafs of quickfilver began to melt.

The experiments of Mr. Braun were fuccefsfully repeated at Gottingen, in 1774, by Mr. John Frederick Blumenbach; being encouraged to this attempt by the exceffive cold of the winter that year, efpecially the night of January the 11th, when he made the experiment, the thermometer ftanding at —10 in the open air. Mr. Blumenbach at 5 in the evening, put 3 drachms of quickfilver into a fmall fugar glafs, and covered it with a mixture of fnow and Egyptian fal ammoniac, fetting the glafs out in the air upon a mixture alfo of fal ammoniac. At one the next morning, the mercury was found frozen quite folid, and hard to the glafs; and did not melt again till 7 or 8 the next morning. The colour of the frozen mercury was a dull pale white with a blueifh caft, like zinc, very different from the natural appearance of quickfilver.

In the year 1775, by fimilar means, quickfilver was twice frozen by Mr. Hutchins, governor of Albany fort, in Hudfon's bay, viz, in the months of January and February of that year. And the fame was done on the 28th of January 1776, by Dr. Lambert Bicker, fecretary of Rotterdam. The temperature of the atmofphere was then at + 2°; and the loweft it could reduce the thermometer by artificial cold, was —94; when, on breaking the glafs, the mercury was found frozen.

In the beginning of the year 1780 M. Von Elterlein of Vytegra, a town of Ruffia, in lat. 61° north, and long. 36° eaft, froze quickfilver by natural cold. On the 4th of January 1780, the cold being increafed to --34 that evening at Vytegra, he expofed to the open air 3 ounces of very pure quickfilver in a china teacup, covered with paper pierced full of holes. Next day, at 8 in the morning, he found it folid, and looking like a piece of caft lead, with a confiderable depreffion in the middle. On attempting to loofen it in the cup, his knife raifed fhavings from it as if it had been lead, which remained fticking up; and at length the metal feparated from the bottom of the cup in one mafs. He then took it in his hand to try if it would bend: it was ftiff like glue, and broke into two pieces; but his fingers immediately loft all feeling, and could fcarcely be reftored in an hour and a half by rubbing with fnow. At 8 o'clock the thermometer ftood at —57; but half after 9 it was rifen to —40; and then the two pieces of mercury which lay in the cup had loft fo much of their hardnefs, that they could no longer be broken, or cut into fhavings, but refembled

a thick amalgam, which, though it became fluid when preffed by the fingers, immediately afterwards refumed the confiftence of pap. With the thermometer at —39, the quickfilver became fluid. The cold was never lefs on the 5th than —28, and by 9 in the evening it had increafed again to —33. This experiment feems to fix the Freezing point of mercury at —40 of Fahrenheit's thermometer, or 40 below 0; which is 72° below the Freezing point of water.

In the winter of 1781 and 82, Mr. Hutchins refumed the fubject of Freezing quickfilver by artificial cold, with fuch fuccefs, that from his experiments and thofe of M. Von Elterlein, laft mentioned, the Freezing point of mercury is now almoft as well fettled, viz at —40, as that of water is at + 32. Other philofophers indeed had not been altogether inattentive to this fubject. Profeffor Braun himfelf had taken great pains to inveftigate it; but for want of a proper attention to the difference between the contraction of the fluid mercury by cold, and that of the congealing metal by Freezing, he could not determine any thing certain concerning it.

An inftance of the natural congelation of quickfilver alfo occurred in Jemptland, one of the provinces of Sweden, on the 1ft of January 1782; and laftly, on the 26th of the fame month, Mr. Hutchins obferved the fame effect of the cold at Hudfon's bay; when he found that at the point of its Freezing a mercurial thermometer ftood at —40, and a fpirit thermometer at —30.

On this fubject, fee the Philof. Tranf. vol. 51, pa. 672; vol. 52, pa. 156; vol. 66, pa. 174; vol. 73, pa. 303 and 325; vol. 76, pa. 241; vol. 77, pa. 285; vol. 78, pa. 43; and feveral others, particularly vol. 79, pa. 199, &c, being experiments on the congelation of quickfilver in England, by Mr. Richard Walker, where he proves that mercury may be frozen not only in England in fummer, but even in the hotteft climate, at any feafon of the year, and without the ufe of ice or fnow.

FREEZING *Point*, denotes the point or degree of cold, fhewn by a mercurial thermometer, at which certain fluids begin to freeze, or, when frozen, at which they begin to thaw again. On Fahrenheit's thermometer, this point is at + 32 for water, and at —40 for quickfilver, thefe fluids freezing at thofe two points refpectively. It would alfo be well if the Freezing points for other fluids were afcertained, and the whole arranged in a table.

FREEZING *Rain*, or *Raining Ice*, a very uncommon kind of fhower which fometimes falls, particularly one in December 1672, in the weft of England: of which fome accounts are given in the Philof. Tranf. number 90.

This rain, as foon as it touched any thing above ground, as a bough, or the like, immediately fettled into ice; and by enlarging and multiplying the icicles, it broke all down with its weight. The rain that fell on the fnow, immediately froze into ice, without finking in the fnow at all.

It made an amazing deftruction of trees, beyond any thing in all hiftory. " Had it concluded with fome guft of wind, fays a gentleman on the fpot, it might have been of terrible confequence. Having weighed the

the fprig of an afh tree, the wood of which was juft three quarters of a pound, the ice upon it amounted to 16 pounds. Some were frighted with the noife in the air; till they difcerned it was the clatter of icy boughs dafhed againft each other." Dr. Beale obferves, that there was no confiderable froft perceived on the ground during the whole; from which he concludes, that a froft may be very fierce and dangerous on the tops of fome hills, while in other places it keeps at fome feet above the ground; and may wander about very furious in fome places, and be remifs in others not far off. This rain was followed by glowing heats, and a wonderful forwardnefs of vegetation.

FRENICLE (Bernard), a celebrated French mathematician of the 17th century. He was the contemporary and companion of Des Cartes, Fermat, and the other learned mathematicians of their time. He was admitted Geometrician of the French Academy in 1666; and died in 1675.

He had many papers inferted in the Ancient Memoirs of the Academy, of 1666, particularly in vol. 5, of that collection, viz, 1. A method of refolving problems by Exclufions.—2. Treatife of right-angled Triangles in Numbers.—3. Short tract on Combinations. —4. Tables of Magic Squares.—5. General method of making Tables of Magic Squares.

FRESCO, is a fort of painting which is made upon the plaftering of walls before it is dry.

FRIABILITY, the property of a body that is Friable.

FRIABLE, a quality of bodies by which they are rendered tender and brittle, eafily crumbled or reduced to powder between the fingers; their force of cohefion being fuch as eafily expofes them to fuch folution. Such are pumice, and all calcined ftones, burnt allum, &c.

It is fuppofed that Friability arifes from hence, that the body confifts wholly of dry parts irregularly combined, and which are readily feparated, as having nothing unctuous or glutinous to bind them together.

FRICTION, the act of rubbing or grating the furfaces of bodies againft or over each other, called alfo Attrition.

The phenomena arifing from the Friction of divers bodies, under different circumftances, are very numerous and confiderable. Mr. Hawkfbee gives a number of experiments of this kind; particularly of the attrition or Friction of glafs, under various circumftances; the refult of which was, that it yielded light, and became electrical. Indeed all bodies by Friction are brought to conceive heat; many of them to emit light; particularly a cat's back, fugar, beaten fulphur, mercury, fea water, gold, copper, &c, but above all diamonds, which when brifkly rubbed againft glafs, gold, or the like, yield a light equal to that of a live coal when blowed by the bellows.

FRICTION, in Mechanics, denotes the refiftance a moving body meets with from the furface on which it moves.

Friction arifes from the roughnefs or afperity of the furface of the body moved on, and that of the body moving: for fuch furfaces confifting alternately of eminences and cavities, either the eminences of the one muft be raifed over thofe of the other, or they muft be both broken and worn off: but neither can happen without motion, nor can motion be produced without a force impreffed. Hence the force applied to move the body is either wholly or partly fpent on this effect; and confequently there arifes a refiftance, or Friction, which will be greater as the eminences are greater, and the fubftance the harder; and as the body, by continual Friction, becomes more and more polifhed, the Friction diminifhes.

As the Friction is lefs in a body that rolls, than when it flides, hence in machines, left the Friction fhould employ a great part of the power, care is to be taken that no part of the machine flide along another, if it can be avoided; but rather that they roll, or turn upon each other. With this view it may be proper to lay the axes of cylinders, not in a groove or concave matrix, as ufual, but between little wheels, called Friction wheels, moveable on their respective axes: for by this contrivance, the Friction is transferred from the circumference of thofe wheels to their pivots. And in like manner the Friction may be ftill farther diminifhed, by making the axis of thofe wheels reft upon other Friction wheels that turn round with them. This was long fince recommended by P. Cafabus; and experience confirms the truth of it. Hence alfo it is, that a pulley moveable on its axis refifts lefs than if it were fixed, and the cord fliding over the circumference. And the fame may be obferved of the wheels of coaches, and other carriages. Indeed about 20 years ago Friction balls or rollers were placed within the naves of carriage wheels by fome perfons, particularly a Mr. Varlo; and lately Mr. Garnett had a patent for an improved manner of applying Friction wheels to any axis, as of carriages, blocks or pulleys, fcale beams, &c, in which the inclofed wheels or rollers are kept always at the fame diftance by connecting rods or bars.

From thefe principles, with the affiftance of the higher geometry, Olaus Roemer determined the figure of the teeth of wheels, that fhould make the leaft refiftance poffible, which he found fhould be epicycloids. And the fame was afterwards demonftrated by De la Hire, and Camus.

M. Amontons, by experiment, attempted to fettle a foundation for the precife calculation of the quantity of Friction; which M. Parent endeavoured to confirm from reafoning and geometry. M. Amontons' principle is, that the Friction of two bodies depends only on the weight or force with which they prefs each other, being always more or lefs in proportion to that preffure; efteeming it a vulgar error, that the quantity of Friction has any dependence on the extent of the furface that is rubbed, or that the Friction increafes with the furface: arguing that it will require the fame weight to draw along a plane, a piece of wood on its narrow edge, as on its broad and flat fide; becaufe, though on the broad fide there be 4 times the number of touching particles, yet each particle is preffed with but $\frac{1}{4}$ of the weight bearing on thofe of the narrow fide; and fince 4 times the number multiplied by $\frac{1}{4}$ of the weight is equal to $\frac{1}{4}$ of the number multiplied by 4 times the weight, it is plain that the effect, that is, the refiftance, is equal in both cafes, and therefore requires the fame force to overcome it.

On the firft propofal of this paradox, M. De la Hire very

very properly had recourse to experiments, as the best test, had they been judiciously performed : such as they were however, they succeeded in favour of this system. He laid several pieces of rough wood on a rough table ; their sizes were unequal ; but he laid weights on them, so as to render them all equally heavy : and he found that the same precise force, or weight, applied to them by a little pulley, was required to put each in motion, notwithstanding all the inequality of the surfaces. The experiment succeeded in the same manner with pieces of marble, laid on a marble table. After this, by reasoning, M. de la Hire gave a physical solution of the effect. And M. Amontons settled a calculus of the value of Friction, with the loss sustained by it in machines, on the foundation of this new principle. In wood, iron, lead, and brass, which are the chief materials used in machines, he makes the resistance caused by Friction to be nearly the same in all, when those materials are anointed with oil or fat : and the quantity of this resistance, independent of the magnitude of the surface, he makes nearly equal to a third part of the weight of the body moved, or of the force with which the two bodies are pressed together. Others have observed, that if the surfaces be hard and well polished, the Friction will be less than a third part of the weight ; but if the parts be soft or rugged, it will be much greater. It was farther observed, that in a cylinder moved on two small gudgeons, or on a small axis, the Friction would be diminished in the same proportion as the diameter of these gudgeons is less than the diameter of the cylinder ; because in this case, the parts on which the cylinder moves and rubs, will have less velocity than the power which moves it in the same proportion, which is in effect making the Friction to be proportional to the velocity. So that, from the whole of their observations, this general proposition is deduced, viz, That the resistances arising from Friction, are to one another in a ratio compounded of the pressures of the rubbing parts, and the velocities of their motions. Principles which, it is now known from better experiments, are both erroneous ; notwithstanding the hypothesis of M. Amontons has been adopted, and attempted to be confirmed by Camus, Desaguliers, and others.

M. Muschenbroek and the abbé Nollet, however, on the other hand, have concluded from experiments, that the Friction of bodies depends on the magnitude of their surface, as well as on their weight. Though the former says, that in small velocities the Friction varies very nearly as the velocity, but that in great velocities the proportion increases faster : he has also attempted to prove, that by increasing the weight of a body, the Friction does not always increase exactly in the same ratio. Introd. ad Phil. Nat. vol. 1, c. 9, and Lect. Phys. Exp. tom. 1, p. 241. Helsham and Ferguson, from the same kind of experiments, have endeavoured to prove, that the Friction does not vary by changing the quantity of surface on which the body moves ; and the latter of these asserts, that the Friction increases very nearly as the velocity ; and that by increasing the weight, the Friction is increased in the same ratio. Indeed there is scarce any subject of experiment, with regard to which, different persons have formed such various conclusions. Of those who have written on the theory, no one has established it altogether on true principles, till the experiments

lately made by Mr. Vince of Cambridge : Euler, whose theory is extremely elegant, and would have been quite satisfactory had his principles been founded on good experiments, supposes the Friction to vary in proportion to the velocity of the body, and its pressure upon the plane ; neither of which is true : and others, though they have justly imagined that Friction is a uniformly retarding force, have yet retained the other supposition, and so rendered their solutions not at all applicable to the cases for which they were intended.

For these reasons a new and ingenious set of experiments was successfully instituted by the rev. Samuel Vince, A. M. of Cambridge, which are published in the 75th vol. of the Philos. Transf. p. 165. The object of these experiments was to determine,

1st, Whether Friction be a uniformly retarding force.

2d, The quantity of Friction.

3d, Whether Friction varies in proportion to the pressure or weight.

4th, Whether the Friction be the same on whichever of its surfaces a body moves.

Mr. Vince says, " the experiments were made with the utmost care and attention, and the several results agreed so very exactly with each other, that I do not scruple to pronounce them to be conclusive."—" A plane was adjusted parallel to the horizon, at the extremity of which was placed a pulley, which could be elevated or depressed in order to render the string which connected the body and the moving force parallel to the plane or horizon. A scale accurately divided was placed by the side of the pulley perpendicular to the horizon, by the side of which the moving force descended ; upon the scale was placed a moveable stage, which could be adjusted to the space through which the moving force descended in any given time, which time was measured by a well regulated pendulum clock vibrating seconds. Every thing being thus prepared, the following experiments were made to ascertain the law of Friction. But let me first observe, that if Friction be a uniform force, the difference between it and the given force of the moving power must be also uniform, and therefore the moving body must descend with a uniformly accelerated velocity, and consequently the spaces described from the beginning of the motion must be as the squares of the times, just as when there was no Friction, only they will be diminished on account of the Friction." Accordingly the experiments are then related, which are performed agreeably to these ingenious and philosophical ideas, and from them are deduced these general conclusions, which may be considered as established and certain facts or maxims : viz,

1st, That Friction is a uniformly retarding force in hard bodies, not subject to alteration by the velocity ; except when the body is covered with cloth, woollen, &c, and in this case the Friction increases a little with the velocity.

2dly, Friction increases in a less ratio than the quantity of matter, or weight of the body. This increase however is different for the different bodies, more or less ; nor is it yet sufficiently known, for any one body,

body, what proportion the increase of Friction bears to the increase of weight.

3dly, The smallest surface has the least Friction; the weight being the same. But the ratio of the Friction to the surface is not yet accurately known.

Mr. Vince's experiments consisted in determining how far the sliding bodies would be drawn, in given times, by a weight hanging freely over a pulley. This method would both shew him if the Friction were a constant retarding force, and the other conclusions above stated. For as the spaces described by any constant force, in given times, are as the squares of the times; and as the weight drawing the body is a constant force, if the Friction, which acts in opposition to the weight, should also be a constant force, then their difference, or the force by which the body is urged, will also be constant, in which case the spaces described ought to be as the squares of the times; which happened accordingly in the experiments.

Mr. Vince adds some remarks on the nature of the experiments which have been made by others. These, he observes, the authors "have instituted, To find what moving force would *just* put a body at rest in motion: and they concluded from thence, that the accelerative force was then equal to the Friction; but it is manifest, that any force which will put a body in motion must be greater than the force which opposes its motion, otherwise it could not overcome it; and hence, if there were no other objection than this, it is evident, that the Friction could not be very accurately obtained; but there is another objection, which totally destroys the experiment, so far as it tends to shew the quantity of Friction, which is the strong cohesion of the body to the plane when it lies at rest." This he confirms by several experiments, and then adds, "From these experiments therefore it appears, how very considerable the cohesion was in proportion to the Friction when the body was in motion; it being, in one case almost $\frac{1}{2}$, and in another it was found to be very nearly equal to the whole Friction. All the conclusions therefore deduced from the experiments, which have been instituted to determine the Friction from the force necessary to *put* a body in motion (and I have never seen any described but upon such a principle) have manifestly been totally false; as such experiments only shew the resistance which arises from the cohesion and Friction conjointly." Philos. Trans. vol. 75, pa. 165.

Mr. Emerson, in his Principles of Mechanics, deduces from experiments the following remarks relating to the quantity of Friction: When a cubic piece of soft wood of 8 pounds weight, moves upon a smooth plane of soft wood, at the rate of 3 feet per second, its Friction is about $\frac{1}{3}$ of the weight; but if it be rough, the Friction is little less than half the weight: on the same supposition, when both the pieces of wood are very smooth, the Friction is about $\frac{1}{4}$ of the weight: the Friction of soft wood on hard, or of hard wood upon soft, is $\frac{1}{5}$ or $\frac{1}{6}$ of the weight; of hard wood upon hard wood, $\frac{1}{7}$ or $\frac{1}{8}$; of polished steel moving on steel or pewter, $\frac{1}{4}$; moving on copper or lead, $\frac{1}{5}$ of the weight. He observes in general, that metals of the same sort have more Friction than those of different sorts; that lead makes much resistance; that iron or steel running in brass makes the least Friction of any; and that metals oiled make the Friction less than when polished, and twice as little as when unpolished. Desaguliers observes that, in M. Camus's experiments on small models of sledges in actual motion, there are more cases in which the Friction is less than where it is more than $\frac{1}{3}$ of the weight. See a table, exhibiting the Friction between various substances, formed from his experiments in Desag. Exp. Philos. vol. 1, p. 193 &c. also p. 133 to 138, and p. 182 to 254, and p. 458 to 460. On the subject of Friction, see several vols. of the Philos. Trans. as vol. 1, p. 206; vol. 34, p. 77; vol. 37, p. 394; vol. 53, p. 139, &c.

FRIDAY, the 6th day of the week, so called from Friga, or Friya, a goddess worshipped by the Saxons on this day. It is a fast-day in the church of England, in memory of our Saviour's crucifixion, unless Christmas-day happen to fall on Friday, which is always a festival.

Good FRIDAY, the Friday next before Easter, representing the day of our Saviour's crucifixion.

FRIGID *Zone*, the space about either pole of the earth to which the sun never rises for one whole day at least in their winter. These two zones extend to about $23\frac{1}{2}$ degrees every way from the pole, as their centre.

FRIGORIFIC, in Physics, something belonging to, or that occasions cold.—Some philosophers, as Gassendus, and other corpuscularians, denying cold to be a mere privation, or absence of heat, contend that there are actual Frigorific corpuscles or particles, as well as fiery ones: whence proceed cold and heat. But later philosophers allow of no other Frigorific particles beside those nitrous salts &c, which float in the air in cold weather, and occasion freezing.

FRIZE, FRIEZE, or FREEZE, in Architecture, a part of the entablature of columns, between the architrave and cornice.

FRONT, in Architecture, denotes the principal face or side of a building; or that presented to their chief aspect and view.

FRONT, in Perspective, a projection or representation of the face, or forepart of an object, or of that part directly opposite to the eye, called also more usually *orthography*.

FRONTISPIECE, in Architecture, the portale, or principal face of a fine building.

FRONTON, in Architecture, an ornament among us more usually called Pediment.

FROST, such a state of the atmosphere as causes the congelation or freezing of water or other fluids into ice.

The nature and effects of Frost in different countries, are mentioned under the articles Congelation, and Freezing. In the more northern parts of the world, even solid bodies are affected by Frost, though this is only or chiefly in consequence of the moisture they contain, which being frozen into ice, and so expanding as water is known to do when frozen, it bursts and rends any thing in which it is contained, as plants, trees, stones, and large rocks. Some fluids expand by Frost, as water, which expands about $\frac{1}{10}$th part, for which reason ice floats in water; but others again contract, as quicksilver, and hence frozen quicksilver sinks in the fluid metal.

Frost, being derived from the atmosphere, naturally proceeds from the upper parts of bodies downwards, as the water and the earth: so, the longer a Frost is continued,

continued, the thicker the ice becomes upon the water in ponds, and the deeper into the earth the ground is frozen. In about 16 or 17 days Froſt, Mr. Boyle found it had penetrated 14 inches into the ground. At Moſcow, in a hard ſeaſon, the Froſt will penetrate 2 feet deep in the ground; and Capt. James found it penetrated 10 feet deep in Charlton iſland, and the water in the ſame iſland was frozen to the depth of 6 feet. Scheffer aſſures us, that in Sweden the Froſt pierces 2 cubits, or Swediſh ells into the earth, and turns what moiſture is found there into a whitiſh ſubſtance, like ice; and ſtanding waters to 3 ells, or more. The ſame author alſo mentions ſudden cracks or rifts in the ice of the lakes of Sweden, 9 or 10 feet deep, and many leagues long; the rupture being made with a noiſe not leſs loud than if many guns were diſcharged together. By ſuch means however the fiſhes are furniſhed with air; ſo that they are rarely found dead.

The natural hiſtories of Froſts furniſh very extraordinary effects of them. The trees are often ſcorched, and burnt up, as with the moſt exceſſive heat; and ſplit or ſhattered. In the great Froſt in 1683, the trunks of oak, aſh, walnut, &c, were miſerably ſplit and cleft, ſo that they might be ſeen through, and the cracks often attended with dreadful noiſes like the exploſion of fire-arms. Philoſ. Tranſ. number 165.

The cloſe of the year 1708, and the beginning of 1709, were remarkable, throughout the greateſt part of Europe, for a ſevere Froſt. Dr. Derham ſays, it was the greateſt in degree, if not the moſt univerſal, in the memory of man; extending through moſt parts of Europe; though ſcarcely felt in Scotland or Ireland.

In very cold countries, meat may be preſerved by the Froſt 6 or 7 months, and prove tolerable good eating. See Capt. Middleton's obſervations made in Hudſon's bay, in the Philoſ. Tranſ. no. 465, ſect. 2. In that climate the Froſt ſeems never out of the ground, it having been found hard frozen in the two ſummer months. Brandy and ſpirit of wine, ſet out in the open air, freeze to ſolid ice in 3 or 4 hours. Lakes and ſtanding waters, not above 10 or 12 feet deep, are frozen to the ground in winter, and all their fiſh periſh. But in rivers, where the current of the tide is ſtrong, the ice does not reach ſo deep, and the fiſh are preſerved. Id. ib.

Some remarkable inſtances of Froſt in Europe, and chiefly in England, are recorded as below: In the year

220, Froſt in Britain that laſted 5 months.
250, The Thames frozen 9 weeks.
291, Moſt rivers in Britain frozen 6 weeks.
359, Severe Froſt in Scotland for 14 weeks.
508, The rivers in Britain frozen for 2 months.
558, The Danube quite frozen over.
695, Thames frozen 6 weeks; booths built on it.
759, Froſt from Oct. 1 till Feb. 26, 760.
827, Froſt in England for 9 weeks.
859, Carriages uſed on the Adriatic ſea.
908, Moſt rivers in England frozen 2 months.
923, The Thames frozen 13 weeks.
987, Froſt laſted 120 days; began Dec. 22.
998, The Thames frozen 5 weeks.

1035, Severe Froſt on June 24: the corn and fruits deſtroyed.
1063, The Thames frozen 14 weeks.
1076, Froſt in England from Nov. till April.
1114, Several wooden bridges carried away by ice.
1205, Froſt from Jan. 14 till March 22.
1407, Froſt that laſted 15 weeks.
1434, From Nov. 24 till Feb. 10. Thames frozen down to Graveſend.
1683, Froſt for 13 weeks.
170$\frac{8}{9}$, Severe Froſt for many weeks.
1715, The ſame for many weeks.
1739, One for 9 weeks. Began Dec. 24.
1742, Severe Froſt for many weeks.
1747, Severe Froſt in Ruſſia.
1754, Severe one in England.
1760, The ſame in Germany.
1776, The ſame in England.
1788, Thames frozen below bridge; booths on it.

Hoar FROST, is the dew frozen or congealed, early in cold mornings; chiefly in autumn. Though many Carteſians will have it formed of a cloud; and either congealed in the cloud, and ſo let fall; or ready to be congealed as ſoon as it arrives at the earth.

Hoar Froſt, M. Regis obſerves, conſiſts of an aſſemblage of little parcels of ice cryſtals; which are of various figures, according to the different diſpoſition of the vapours, when met and condenſed by the cold.

FRUSTUM, in Geometry, is the part of a ſolid next the baſe, left by cutting off the top, or ſegment, by a plane parallel to the baſe: as the Fruſtum of a pyramid, of a cone, of a conoid, of a ſpheroid, or of a ſphere, which is any part compriſed between two parallel circular ſections; and the Middle Fruſtum of a ſphere, is that whoſe ends are equal circles, having the centre of the ſphere in the middle of it, and equally diſtant from both ends.

For the Solid Content of the Fruſtum of a cone, or of any pyramid, whatever figure the baſe may have. Add into one ſum, the areas of the two ends and the mean proportional between them; then $\frac{1}{3}$ of that ſum will be a mean area, or the area of an equal priſm, of the ſame altitude with the Fruſtum; and conſequently that mean area being multiplied by the height of the Fruſtum, the product will be the ſolid content of it. That is, if A denote the area of the greater end, *a* that of the leſs, and *h* the height;

then $\overline{A + a + \sqrt{Aa}} \times \frac{1}{3}h$ is the ſolidity.

Other rules for pyramidal or conic Fruſtums may be ſeen in my Menſuration, p. 189, 2d edit. 1788.

The curve Surface of the Zone or Fruſtum of a ſphere, is had by multiplying the circumference of the ſphere by the height of the Fruſtum. Menſur. p. 197.

And the Solidity of the ſame Fruſtum is found, by adding together the ſquares of the radii of the two ends, and $\frac{1}{3}$ of the ſquare of the height of the Fruſtum, then multiplying the ſum by the ſaid height and by the number 1·5708. That is, $\overline{R^2 + r^2 + \frac{1}{3}h^2} \times \frac{1}{2}ph$ is the ſolid content of the ſpheric Fruſtum, whoſe height is *h*, and the radii of its ends R and *r*, *p* being = 3·1416. Menſur. p. 209.

For the Fruſtums of ſpheroids, and conoids, either parabolic or hyperbolic, ſee Menſur. p. 326, 328, 332, 382, 435. And in p. 486 &c, are general theorems concerning the Fruſtum of a ſphere, cone, ſpheroid, or conoid,

conoid, terminated by parallel planes, when compared with a cylinder of the same altitude, on a base equal to the middle section of the Frustum made by a parallel plane. The difference between the Frustum and the cylinder is always the same quantity, in different parts of the same, or of similar solids, or whatever the magnitude of the two parallel ends may be; the inclination of those ends to the axis, and the altitude of the Frustum being given; and the said constant difference is $\frac{1}{4}$ part of a cone of the same altitude with the Frustum, and the radius of its base is to that altitude, as the fixed axis is to the revolving axis of the Frustum. Thus, if BEC be any conic section, or a right line, or a circle, whose axis, or a part of it, is AD; AB and CD the extreme ordinates, FE the middle ordinate, AF being = FD; then taking, as AD to DK, so is the whole fixed axis, of which AD is a part, to its conjugate axis; and completing the parallelogram AGHD: then if the whole figure revolve about the axis AD, the line BEC will generate the Frustum of the cone or conoid, according as it is a right line or a conic section, or it will generate the whole solid when AB vanishes, or A and B meet in the same point; likewise AGHD will generate a cylinder, and ADK a cone: then is the 4th part of this cone always equal to the difference between the said cylinder generated by AGHD and the solid or Frustum generated by ABECD; having all the same altitude or axis AD.

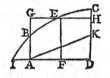

In the parabolic conoid, this difference and the cone vanish, and the Frustum, or whole conoid ABECD, is always equal to the cylinder AGHD, of the same altitude.

In the sphere, or spheroid, the Frustum ABECD is *less* than the cylinder AGHD, by $\frac{1}{4}$ of the cone AKD. And

In the cone or hyperboloid, that Frustum is *greater* than the cylinder, by $\frac{1}{4}$ of the said cone AKD, which is similar to the other cone IBCD.

It may be observed, that the same relations are true, whether the ends of the Frustum are perpendicular or oblique to the axis. And the same will hold for the Frustum of any pyramid, whether right or oblique; and such a Frustum of a pyramid will exceed the prism, of the same altitude, and upon the middle section of the Frustum, by $\frac{1}{4}$ of the same cone.

It has been observed, that the difference, or $\frac{1}{4}$ of the cone AKD, is the same, or constant, when the altitude and inclination of the ends of the Frustum remain the same. But when the inclination of the ends varies, the altitude being constant; then the said difference varies so as to be always reciprocally as the cube of the conjugate to the diameter AD. And when both the altitude and inclination of the ends vary, the differential cone is as the cube of the altitude directly, and the cube of the said conjugate diameter reciprocally: but if they vary so, as that the altitude is always reciprocally as that diameter, then the difference is a constant quantity.

Another general theorem for Frustums, is this. In the Frustum of any solid, generated by the revolution of any conic section about its axis, if to the sum of the

two ends be added 4 times the middle section, $\frac{1}{6}$ of the last sum will be a mean area, and being drawn into the altitude of the solid, will produce the content. That is, $\overline{AE + DC + 4FE} \times \frac{1}{6}AD$ is the content of ABCD.

And this theorem is general for all Frustums, as well as the complete solids, whether right or oblique to the axis, and not only of the solids generated from the circle or conic sections, but also of all pyramids, cones, and in short of any solid whose parallel sections are similar figures.

The same theorem also holds good for any parabolic area ABECD, and is very *nearly true* for the area of any other curve whatever, or for the content of any other solid than those above mentioned.

FUGUE, in Music, is when the different parts of a musical composition follow each other, each repeating in order what the first had performed.

FULCRUM, or *Prop*, in Mechanics, is the fixed point about which a lever &c turns and moves.

FULGURATING *Phosphorus*, a term used by some English writers, to express a substance of the phosphorus kind. It was prepared both in a dry and liquid state, but the preparation it seems was not well known to any but the inventor of it. This matter not only shone in the dark in both states, but communicated its light to any thing it was rubbed on. When inclosed in a glass vessel well stopped, it sometimes would Fulgurate, or throw out little flashes of light, and sometimes fill the whole phial with waves of flame. It does not need recruiting its light at the fire, or in the sunshine, like the phosphorus of the Bolognian stone, but of itself continues in a state of shining for several years together, and is seen as soon as exposed in the dark; the solid or dry matter always resembling a burning coal of fire, though not consuming itself. Philos. Transf. N° 134.

FULIGINOUS, an epithet applied to thick smoke or vapour replete with soot or other crass matter.

In the first fusion of lead, there exhales a great deal of Fuliginous vapour, which being retained and collected, makes what is called Litharge. And Lampblack is what is gathered from the Fuliginous vapours of pines, and other resinous wood, when burnt.

FULMINANT, Fulminans, or Fulminating, an epithet applied to something that thunders, or makes a noise like thunder.

Aurum Fulminans. See Aurum.

Pulvis Fulminans, is a composition of 3 parts of nitre, 2 parts of salt of tartar, and 1 of sulphur.—Both the Aurum and Pulvis Fulminans produce their effect chiefly downwards; in which they differ from gunpowder, which acts in orbem, or all around, but principally upwards. When the composition is laid in brass ladles, and so set on fire, after fulmination, the ladles are often found perforated. It differs also from gunpowder in this, that it does not require to be confined, in order to fulminate, and it must be slowly and gradually heated. Some instants before explosion, a light blue flame appears on its surface, proceeding from the vapours beginning to kindle. No more fire or flame is perceived during the fulmination, being suffocated and extinguished by the quickness and violence of the commotion. Nor does the Fulminating powder generally kindle the combustible bodies in contact with it, because the time of its inflammation is too short.

FULMINATING *Damp.* See DAMP.

FULMINATION, or FULGURATION, a vehement noise or shock resembling thunder, caused by the sudden explosion and inflammation of divers preparations ; as aurum fulminans, &c, when set on fire.

FUNCTION, a term used in analytics, for an algebraical expression any how compounded of a certain letter or quantity with other quantities or numbers : and the expression is said to be a Function of that letter or quantity. Thus $a - 4x$, or $ax + 3x^2$, or $2x - a \sqrt{a^2 - x^2}$, or x^c, or c^x, is each of them a Function of the quantity x.

On the subject of Functions, their divisions, transformations, explication by infinite series, &c, see Euler's Analyf. Infinitorum, c. 1, where the subject is fully treated.

FURLONG, an English long measure, containing 660 feet, or 220 yards, or 40 poles or perches, or the 8th part of a mile.

FURNITURE, in Dialling, certain additional points and lines drawn on a dial, by way of ornament. Such as the signs of the zodiac, length of days, parallels of declination, azimuths, points of the compass, meridians of chief cities, Babylonic, Jewish, or Italian hours, &c.

FUSAROLE, in Architecture, a small round member cut in form of a collar, with oval beads, under the echinus, or quarter-round, in the Doric, Ionic, and Composite capitals.

FUSEE, or FUSY, in Watch-work, is that part resembling a low cone with its sides a little sunk or concave, which is drawn by the spring, and about which the chain or string is wound.

The spring of a watch is the first mover. It is rolled up in a cylindrical box, against which it acts, and which it turns round in unbending itself. The chain, which at one end is wound about the Fusee, and at the other fastened to the spring-box, disengages itself from the Fusee in proportion as the box is turned. And hence the motion of all the other parts of the spring-watch. Now the effort or action of the spring is continually diminishing from first to last ; and unless that inequality was rectified, it would draw the chain with more force, and wind a greater quantity of it upon the box, at one time than another ; so that the movement would never keep equal time.

To correct this irregularity of the spring, it was very happily contrived to have the spring applied to the arms of levers, which are continually longer as the force of the spring is weaker : this foreign assistance, always increasing as it is most needed, maintains the action and effect of the spring in an equality.

It is for this reason then that the Fusee is made tapering somewhat conical, its radius at every point of the axis answering to the corresponding strength of the spring.

Now if the action of the spring diminished equally, as the parallels to the base of a triangle do ; the cone, which is generated of a triangle, would be the precise figure required for the Fusee ; but it is certain that the weakening of the spring is not in that proportion ; and therefore the Fusee should not be exactly conical ; and in fact experience shews that it should be a little hollowed about the middle, because the action of the spring is not there sufficiently diminished of itself.

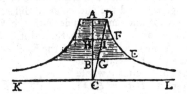

Mr. Varignon has investigated the figure of the Fusee, or the nature of the curve by whose revolution about its axis, shall be produced the solid whose figure the Fusee is to have. This curve it may easily be shewn is an hyperbola whose asymptote is the axis of the Fusee. Thus, let DFE be the curve of the Fusee, its axis being ABC : let AD express the greatest strength of the spring when the watch is quite wound up, or when the spring acts at D, and BG the least strength when the watch is down, or when the spring acts at E ; so as that BE : AD :: AD : BG, or BE × BG = AD²; join DG, producing it to meet the axis produced in C ; then shall HI denote the strength of the spring acting at the corresponding point F of the Fusee ; and the nature of it must be such that the rectangle HI × HF be equal to a constant quantity, or HF must be reciprocally as HI,
or HI : BG :: BE : HF ;
but because   AD, HI, BG,
are directly proportional to   CA, CH, CB,
theref. these are reciprocally propor. to AD, HF, BE ; and consequently the curve DFE is an hyperbola, whose centre is C, and asymptotes AC and KL : so that the figure of the Fusee is the solid generated by an equilateral hyperbola revolved about its asymptote. See also Martin's Mathem. Instit. vol. 2, p. 364.

FUSEE, FUSE, or FUZE, in Artillery, is a wooden tap or tube used to set fire to the powder in a bombshell. The bore of this tube is filled with a composition, of sulphur one part, saltpetre 3 parts, and mealed powder 3, 4, or 5 parts. The tube is driven hard into the hole in the shell, having first cut it to the exact length answering to the time of the intended flight of the shell, so that, the composition in the Fuse catching fire by the discharge of the shell from the mortar, it just burns down to its lower end, and so sets fire to the powder in the shell, and thereby bursts it, at the moment when it arrives at the end of its range or flight.

FUST, in Architecture, the shaft of a column, or the part comprehended between the base and the capital, called also the Naked.

FUZE, or FUZEE. See FUSEE.

G.

GABIONS, in Fortification, are large cylindrical baskets, open at both ends, made of ozier twigs, of 3 or 4 feet in diameter, and from 3 to 6 feet high. These, being filled with earth, are sometimes used as merlons for the batteries, and sometimes as a parapet for the lines of approach, when the attacks are carried on through a stony or rocky ground, and to advance them with extraordinary vigour. They serve also to make lodgments in some posts, and to secure other places from the shot of the enemy; who, on their part, endeavour to burn and destroy the Gabions, by throwing pitched faggots among them.

GABLE, or GABLE-*end,* of a house, is the upright triangular end, from the cornice or eaves to the top of its roof.

GAGE, in Hydrostatics, Pneumatics, &c, is an instrument for ascertaining measures of various kinds. As

GAGE *of the Air-pump,* is adapted for shewing the degree to which the air is rarefied, or the receiver is exhausted, at any time by the air-pump. This is either the common barometer-gage, both long and short, or the pear gage, which at first was thought a great improvement, but afterwards it was discovered that its seeming accuracy was founded on a fallacy, which gave an erroneous indication of exhaustion. See AIR-*pump.*

GAGE *of the Barometer,* is a contrivance for estimating the exact degree of the rise and fall of the mercury in the tube of that instrument. It is well known that whilst the mercury rises in the tube, it sinks in the cistern, and vice versa; and consequently the divisions on the scale fixed near the top of the tube had their distance from the surface of the mercury in the cistern always various; from which there must often happen errors in determining the height of the mercury in the tube. To remedy this inconvenience, a line is cut upon a round piece of ivory, which is fixed near the cistern: this line is accurately placed at a given distance from the scale; for example at 27 inches; and a small float of cork, with a cylindrical piece of ivory fixed to its upper surface, on which a line is cut at the exact distance of 2 inches from the under side of the cork, is left to play freely on the quicksilver, and the cylinder works in a groove made in the other piece. From this construction it appears, that if these marks are made to coincide, by raising or lowering the screw which acts on the quicksilver, then the divisions on the scale will express the true measure of the distance from the surface.

GAGE *of the Condenser,* is a glass tube of a particular construction, adapted to the condensing engine, and designed to shew the exact density and quantity of the air contained at any time in the condenser. See Desaguliers's Exper. Philos. vol. 2, p. 394.

Sea GAGE, an instrument for finding the depth of the sea. Several sorts of these have been invented by Dr. Hales, Dr. Desaguliers, and others. Formerly, the machines for this purpose consisted of two bodies, the one specifically lighter, and the other specifically heavier than the water, so joined together, that as soon as the heavy one came to the bottom, the lighter should get loose from it, and emerge; and the depth was to be estimated by the time the compound was in falling from the top to the bottom of the water, together with the time the lighter body was in rising, reckoned from the disappearing of the machine, till the emergent body was seen again; but no certain conclusion could be drawn from so precarious and incomplete an experiment.

But that invented by Drs. Hales and Desaguliers was of a more exact nature, depending on the pressure of the fluid only. For as the pressure of fluids in all directions is the same at the same depth, a Gage which discovers what the pressure is at the bottom of the sea, will shew what the true depth of the sea is in that place, whether the time of the machine's descent be longer or shorter.

Dr. Hales, in his Vegetable Statics, describes his Gage for estimating the pressures made in opaque vessels; where honey being poured over the surface of mercury in an open vessel, rises upon the surface of the mercury as it is pressed up into a tube whose lower orifice is immersed into the honey and mercury, and whose top is hermetically sealed. Now as by the pressure, the air in the tube is condensed, and the mercury rises, so the mercury comes down again when the pressure is taken off, and would leave no mark of the height to which it had risen; but the honey (or treacle, which does better) which is upon the mercury, sticking to the inside of the tube, leaves a mark, which shews the height to which it had risen, and consequently gives the quantity of pressure, and the height of the surface of the fluid.

Desaguliers's addition to this machine, consisted in a contrivance to carry it down to the bottom of the sea by means of a heavy weight, which was immediately disengaged by striking the bottom, and the Gage, made very light for the purpose, re-ascended to the top.

Dr. Hales afterwards made more experiments of this sort, and proposed another Sea Gage for vast depths, which is described in the Philos. Trans. N° 405, and is to this effect. Suppose a pretty long tube of copper or iron, close at the upper end, to be let down into the

sea,

sea, to any depth, the water will rise in the tube to a height bearing a certain proportion to the depth of the sea to which the machine is sunk. And this proportion is as follows: 33 feet of sea water being nearly equal to the mean pressure of the atmosphere, therefore at 33 feet deep, the air in the tube will be compressed into half the length of the tube, or the water will rise and fill half way up the tube; in like manner at 66 feet deep, the water will occupy ⅔ of the tube; at 99 feet deep it will fill ¾ of the tube; at 132 feet deep it will fill ⅘ of the tube; and so on. Hence therefore, by knowing the height to which the water rises in the tube, there will be known the consequent depth of the sea.

But, in very great depths, the scale near the top of the tube would be so small, and the divisions so close, that there would be no accuracy in the experiment, unless the tube were of a very great length, and this again would render it both liable to be broken, and quite impracticable.

To remedy this inconvenience, he made the following contrivance: To the bottom of the tube he screwed a large hollow globe of copper, with a small orifice, or a short pipe at bottom of the globe, to let in the water; by which means he had a very great quantity of air, and the scale enlarged. See also Desagul. Exp. Phil. vol. 2, p. 224 and 241.

Bucket Sea GAGE, is an instrument contrived by Dr. Hales to find the different degrees of coolness and saltness of the sea at different depths. This Gage consists of a common pale or bucket, with two heads: these heads have each a round hole in the middle, about 4 inches in diameter, covered with square valves opening upward; and that they may both open and shut together, there is a small iron rod, having one end fixed to the upper side of the lower valve, and the other end to the lower side of the upper valve. So that as the bucket descends with its sinking weight into the sea, both the valves may open by the force of the water, which by that means has a free passage through the bucket. But when the bucket is drawn up, then both the valves shut by the force of the water at the upper end of the bucket; so that the bucket is drawn up full of the lowest sea water to which it has descended, and immediately the mercurial thermometer, fixed within it, is examined, to see the degree of temperature; and the degree of saltness is afterwards examined at leisure. Philos. Transf. numb. 9, p. 149, and numb. 24, p. 447, or Abridg. vol. 2, p. 260.

Lord Charles Cavendish adapted a thermometer for the temperature of the sea water, at different depths. See Philos. Transf. vol. 50, p. 300, and Phipps's Voyage towards the North Pole, p. 142. &c.

Aqueo-mercurial GAGE is the name of an apparatus contrived by Dr. Hales, and applied, in various forms, to the branches of trees, to determine the force with which they imbibe moisture. Vegetable Statics, vol. I, ch. 2, p. 84.

Sliding GAGE, a tool used by mathematical instrument makers, for measuring and setting off distances; consisting of a beam, tooth, sliding socket, and the shoulder of the socket.

Tide GAGE, an instrument used for determining the height of the tides by Mr. Bayly, in the course of a

voyage towards the south pole &c, in the Resolution and Adventure, in the years 1772, 1773, 1774, and 1775. This instrument consists of a glass tube, whose internal diameter was 7-10ths of an inch, lashed fast to a 10 foot fir rod, divided into feet, inches, and parts; the rod being fastened to a strong post fixed firm and upright in the water. At the lower end of the tube was an exceeding small aperture, through which the water was admitted. In consequence of this construction, the surface of the water in the tube was so little affected by the agitation of the sea, that its height was not altered the 10th part of an inch when the swell of the sea was 2 feet; and Mr. Bayly was certain, that with this instrument he could discern a difference of the 10th of an inch in the height of the tide.

Water GAGE. See ALTITUDE, and HYDROMETER.

Wind GAGE, an instrument for measuring the force of the wind upon any given surface. Several have been invented formerly, and one was lately invented by Dr. Lind, which is described in the Philos. Transf. vol. 65. See several also under the article ANEMOMETER.

GAGER, see GAUGER.

GAGING, see GAUGING.

GALAXY, or *Milky-Way*, or *Via Lactea*, in Astronomy, that long, whitish, luminous track, which seems to encompass the heavens like a swath, scarf, or girdle; and which is easily seen in a clear night, especially when the moon is not up. It is of a considerable, though unequal breadth; being also in some parts double, but in others single.

The Galaxy passes through many of the constellations in its circuit round the heavens, and keeps its exact place or position with respect to them.

There have been various strange and fabulous stories and opinions concerning the Galaxy.

The ancient poets, and even some of the philosophers, speak of it as the road or way by which the heroes went to heaven. But the Egyptians called it the Way of Straw, from the story of its rising from burning straw, thrown behind the goddess Isis in her flight from the giant Typhon. While the Greeks, who affect to derive every thing in the heavens from some of their own fables, have two origins for it; the one, that Juno, without perceiving it, accidentally gave suck to Mercury when an infant, but that as soon as she turned her eyes upon him, she threw him from her, and as the nipple was drawn from his mouth, the milk ran about for a moment; and the other, that the infant Hercules being laid by the side of Juno when asleep, on waking she gave him the breast; but soon perceiving who it was, she threw him from her, and the heavens were marked by the wasted milk.

Some other philosophers however gave it a different turn, and different origin: these esteemed it to be a tract of liquid fire, spread in this manner along the skies: and others again, supposing a celestial region beyond all that was visible, and imagining that fire, at some time let loose from thence, was to consume the world, made this a part of that celestial fire, and appealed to it as a presage of what would surely happen. This diffused brightness they considered as a crack in the vault or wall of heaven, and fancied this a glimmering of the celestial fire through it, and that there required nothing more than the undoing of this crack erack by some accident in nature, or by the will of the Gods, to make the whole frame start, and let out the fire of destruction.

Aristotle makes the Galaxy a kind of meteor, formed of a crowd of vapours, drawn into that part by certain large stars disposed in the region of the heavens answering to it. Others, finding that the Galaxy was seen all over the globe, that it always corresponded to the same fixed stars, and that it was far above the highest planets, set Aristotle's opinion aside, and placed the Galaxy in the firmament or region of the fixed stars; and concluded that it was nothing else but an assemblage of an infinite number of minute stars. And since the invention of telescopes, this opinion has been abundantly confirmed. For, by directing a good telescope to any part of the milky way, we perceive an innumerable multitude of very small stars, where before we only observed a confused whiteness, arising from the assemblage and union of their joint light; like as any thing powdered with fine white powder, at a distance we only observe the confused whiteness, but on examining it very near we perceive all the small particles of the powder separately; as Milton finely expresses it,

> A broad and ample road, whose dust is gold,
> And pavement stars, as stars to thee appear,
> Seen in the Galaxy, that milky way,
> Which nightly, as a circling zone thou seest
> Powder'd with stars.

There are other such marks in the heavens; as the nebulæ, or, nebulous stars, and certain whitish parts about the south pole, called Magellanic clouds, which are all of the same nature, appearing to be vast clusters of small stars when viewed through a telescope, which are too faint to affect the eye singly.

M. le Monnier however, not being able to discover more stars in this space than in other parts of the heavens, disputes the opinion above recited as to the reason of the whiteness, and supposes that this and the nebulous stars are occasioned by some other kind of matter. Inst. Ast. p. 60.

GALILEI (GALILEO,) a most excellent philosopher, mathematician and astronomer, was born at Pisa in Italy, in 1564. From his infancy he had a strong propensity to philosophy and mathematics, and soon made a great progress in these sciences. So that in 1592 he was chosen professor of mathematics at Padua. While he was professor there, visiting Venice, then famous for the art of glass-making, he heard that in Holland a glass had been invented, through which very distant objects were seen distinctly as if near at hand. This was sufficient for Galileo; his curiosity was raised, and put him upon considering what must be the form of such a glass, and the manner of making it. The result of his enquiry was the invention of the telescope, produced from this hint, without having seen the Dutch glass. All the discoveries he made in astronomy were easy and natural consequences of this invention, which opening a way, till then unknown, into the heavens, thence brought the finest discoveries. One of the first of these, was that of 4 of Jupiter's satellites, which he called the Medicean stars or planets, in honour of Cosmo the 2d, grand-duke of Tuscany, who was of

that family. Cosmo sent for our astronomer from Padua, and made him professor of mathematics at Pisa in 1611; and soon after inviting him to Florence, gave him the office and title of *principal philosopher and mathematician to his highness.*

He had been but a few years at Florence, before the Inquisition began to be very busy with him. Having observed some solar spots in 1612, he printed that discovery the following year at Rome; in which, and in some other pieces, he ventured to assert the truth of the Copernican system, and brought several new arguments to confirm it. For these he was cited before the Inquisition at Rome, in 1615: after some months imprisonment, he was released, and sentence pronounced against him that he should renounce his heretical opinions, and not defend them by word or writing, or insinuate them into the minds of any persons. But having afterwards, in 1632, published at Florence his Dialogues of the two Great Systems of the World, the Ptolomaic and Copernican, he was again cited before the holy-office, and committed to the prison of that ecclesiastical court at Rome. The inquisitors convened in June that year; and in his presence pronounced sentence against him and his books, obliging him to abjure his errors in the most solemn manner; committed him to the prison of their office during pleasure; and enjoined him, as a saving penance, for three years to come, to repeat once a week the seven penitential psalms: referring to themselves, however, the power of moderating, changing, or taking away altogether or in part, the said punishment and penance. On this sentence, he was detained in prison till 1634; and his Dialogues of the System of the World were burnt at Rome.

Galileo lived ten years after this; seven of which were employed in making still further discoveries with his telescope. But by the continual application to that instrument, added to the damage his sight received from the nocturnal air, his eyes grew gradually weaker, till he became totally blind in 1639. He bore this calamity with patience and resignation, worthy of a great philosopher. The loss neither broke his spirit, nor stopped the course of his studies. He supplied the defect by constant meditation; by which means he prepared a large quantity of materials, and began to arrange them by dictating his ideas; when, by a distemper of three months continuance, wasting away by degrees, he expired at Arcetri near Florence, in January 1642, being the 78th year of his age.

Galileo was in his person of small stature, though of a venerable aspect, and vigorous constitution. His conversation was affable and free, and full of pleasantry. He took great delight in architecture and painting, and designed extremely well. He played exquisitely on the lute; and whenever he spent any time in the country, he took great pleasure in husbandry. His learning was very extensive; and he possessed in a high degree a clearness and acuteness of wit. From the time of Archimedes, nothing had been done in mechanical geometry, till Galileo, who being possessed of an excellent judgment, and great skill in the most abstruse points of geometry, first extended the boundaries of that science, and began to reduce the resistance of solid bodies

bodies to its laws. Besides applying geometry to the doctrine of motion, by which philosophy became established on a sure foundation, he made surprising discoveries in the heavens by means of his telescope. He made the evidence of the Copernican system more sensible, when he shewed from the phases of Venus, like to those of the moon, that Venus actually revolves about the sun. He proved the rotation of the sun on his axis, from his spots; and thence the diurnal rotation of the earth became more credible. The satellites that attend Jupiter in his revolution about the sun, represented, in Jupiter's smaller system, a just image of the great solar system; and rendered it more easy to conceive how the moon might attend the earth, as a satellite, in her annual revolution. By discovering hills and cavities in the moon, and spots in the sun constantly varying, he shewed that there was not so great a difference between the celestial bodies and the earth as had been vainly imagined.

He rendered no less service to science by treating, in a clear and geometrical manner, the doctrine of motion, which has justly been called the key of nature. The rational part of mechanics had been so much neglected, that hardly any improvement was made in it for almost 2000 years. But Galileo has given us fully the theory of equable motions, and of such as are uniformly accelerated or retarded, and of these two compounded together. He, first of any, demonstrated that the spaces described by heavy bodies, from the beginning of their descent, are as the squares of the times; and that a body, projected in any direction not perpendicular to the horizon, describes a parabola. These were the beginnings of the doctrine of the motion of heavy bodies, which has been since carried to so great a height by Newton. In geometry, he invented the cycloid, or trochoid; though the properties of it were afterwards chiefly demonstrated by his pupil Torricelli. He invented the simple pendulum, and made use of it in his astronomical experiments: he had also thoughts of applying it to clocks; but did not execute that design: the glory of that invention was reserved for his son Vicenzio, who made the experiment at Venice in 1649; and Huygens afterward carried this invention to perfection. Of Galileo's invention also, was the machine, with which the Venetians render their Laguna fluid and navigable. He also discovered the gravity of the air, and endeavoured to compare it with that of water; besides, opening up several other enquiries in natural philosophy. In short, he was not esteemed and followed by philosophers only, but was honoured by persons of the greatest distinction of all nations.

Galileo had scholars too that were worthy of so great a master, by whom the gravitation of the atmosphere was fully established, and its varying pressure accurately and conveniently measured, by the column of quicksilver of equal weight sustained by it in the barometrical tube. The elasticity of the air, by which it perpetually endeavours to expand itself, and, while it admits of condensation, resists in proportion to its density, was a phenomenon of a new kind (the common fluids having no such property), and was of the utmost importance to philosophy. These principles opened a vast field of new and useful knowledge, and explained a great variety of phenomena, which had been accounted for before that time in a very absurd manner. It seemed as if the air, the fluid in which men lived from the beginning, had been then but first discovered. Philosophers were every where busy enquiring into its various properties and their effects: and valuable discoveries rewarded their industry. Of the great number who distinguished themselves on this occasion, may be mentioned Torricelli and Viviani in Italy, Pascal in France, Otto Guerick in Germany, and Boyle in England.

Galileo wrote a number of treatises, many of which were published in his life-time. Most of them were also collected after his death, and published by Mendessi in 2 vols 4to, under the title of *L'Opere di Galileo Galilei Lynceo*, in 1656. Some of these, with others of his pieces, were translated into English and published by Thomas Salisbury, in his Mathematical Collections, in 2 vols folio. A volume also of his letters to several learned men, and solutions of several problems, were printed at Bologna in 4to. His last disciple, Vincenzo Viviani, who proved a very eminent mathematician, methodized a piece of his master's, and published it under this title, *Quinto libro de gli Elementi d'Euclidi, &c*; at Florence in 1674, 4to. Viviani published some more of Galileo's things, being Extracts from his letters to a learned Frenchman, where he gives an account of the works which he intended to have published, and a passage from a letter of Galileo dated at Arcetri, Oct. 30, 1635, to John Camillo, a mathematician of Naples, concerning the angle of contact. Besides all these, he wrote many other pieces, which were unfortunately lost through his wife's devotion; who, solicited by her confessor, gave him leave to peruse her husband's manuscripts; of which he tore and took away as many as he said were not fit to be published.

GALLERY, in Architecture, a covered place in a building, much longer than broad; which is usually placed in the wings of the building, and serving to walk in, and to place pictures in. It denotes a little aisle, or walk, serving as a common passage to several rooms placed in a line, or row.

GALLERY, in Fortification, a covered walk, or passage, made across the ditch of a besieged town, with timbers fastened in the ground and covered over.

GALLON, an English measure of capacity, for things both liquid and dry, containing 2 pottles, or 4 quarts, or 8 pints. But those pints and quarts, and consequently the Gallon itself, are different, according to the quality of the things measured: the wine Gallon, for instance, contains 231 cubic inches, and holds 8lb 5⅔ oz, avoirdupois, of pure water; the beer and ale Gallon contains 282 cubic inches, and holds 10lb 3¼ oz of water; and the Gallon dry measure, for grain, meal, &c, contains 268⅘ cubic inches, and holds 9lb 11¼ oz of water.

GALLOPER, in Artillery, the name of a carriage serving for the very small guns, and having shafts so as to be drawn without a limber.

GAMING. See CHANCES, and LAWS *of* CHANCE.

GARDECAUT, or GUARD DU CORD, in a watch, is that which stops the fusee, when wound up, and for that end is driven up by the spring. Some call it Guard-cock; others Guard du Gut.

GAR-

GARRISON Guns, such as are mounted and used in a Garrison, consisting of the following weights, viz the 42, 32, 24, 18, 12, 9, and 6 pounders; being made either of brass or iron.

Table of the Weight and Dimensions of Garrison Guns.					
Brass Garrison Guns.			Iron Garrison Guns.		
Shot	Length	Weight.	Shot	Length	Weight
lb.	f. in.	Cw. qr. lb.	lb.	f. in.	Cw. qr. lb.
42	10 0	64 0 0	32	9 8	56 0 0
32	9 2	49 2 18	24	9 8	48 0 0
24	8 4	37 0 0	18	9 0	36 0 0
18	7 6	27 3 0	12	7 8	24 0 0
12	6 7	18 2 0	9	7 0	18 0 0
9	6 0	13 3 0	6	6 1	12 0 0
6	5 3	9 1 0	4	5 4	8 0 0

GASSENDI (Peter), one of the most celebrated philosophers France has produced, was born at Chantersier, about 3 miles from Digne in Provence, in the year 1592. When a child, he took great delight in gazing at the moon and stars whenever they appeared. This pleasure often drew him into bye-places, that he might feast his eyes freely and undisturbed; by which means his parents had him often to seek, not without many anxious fears and apprehensions. In consequence of this promising disposition, he was sent to the best schools, to cultivate it with the instructions of the first masters. He profited so well of these aids, that he was invited to be professor of rhetoric at Digne, before he was quite 16 years of age. After filling this office three years, upon the death of his master at Aix, he was appointed to succeed him as professor of philosophy. After a few years residence here, he composed his Paradoxical Exercitations; which coming to the hands of Nicholas Peiresc, that great patron of learning joined with Joseph Walter, prior of Valette, in promoting him; and, having entered into holy orders, he was first made canon of the church of Digne and doctor of divinity, and then warden or rector of the same church.

Gassendi's fondness for astronomy grew up with his years; and his reputation daily increasing, he was appointed the king's professor of mathematics at Paris in 1645. This institution being chiefly intended for astronomy, our author read lectures on that science to a crowded audience. However, he did not long enjoy this situation; for a dangerous cough and inflammation of the lungs obliged him, in 1647, to return to Digne for the benefit of his native air. Having thus, and by the intermission of his studies, recovered his health, he again returned to Paris in 1653; where, after first writing and publishing the lives of Tycho Brahe, Copernicus, Purbach, and Regiomontanus, in 1654 he again renewed his astronomical labours, with the design of completing the system of the heavens.

But while he was thus employed, too intensely for the feeble state of his health, he relapsed into his former disorder, under which, with the aid of too copious and numerous bleedings, by order of three physicians, he sunk in the year 1655, at 63 years of age.

Gassendi wrote against the metaphysical meditations of Des Cartes; and divided with that great man the philosophers of his time, almost all of whom were either Cartesians or Gassendists. To his knowledge in philosophy and mathematics, he joined profound erudition and deep skill in the languages. He wrote, 1. Three volumes on Epicurus's philosophy; and six others, which contain his own philosophy.—2. Astronomical Works.—3. The lives of Nicholas de Peiresc, Epicurus, Copernicus, Tycho Brahe, Purbach, and Regiomontanus.—4. Epistles, and other treatises. All his works were collected together, and printed at Lyons in 1658, in 6 volumes folio.

Gassendi was the first person that saw the transit of Mercury over the sun, viz. Nov. 7, 1631; as Horrox first predicted and shewed the transit of Venus.—His library was large and valuable: to which he added an astronomical and philosophical apparatus, which, for their accuracy and magnitude, were purchased by the emperor Ferdinand the 3d.—It appears by his letters, printed in the 6th volume of his works, that he was often consulted by the most celebrated astronomers of his time, as Kepler, Longomontanus, Snell, Hevelius, Galileo, Kircher, Bulliald, and others: and he has generally been esteemed one of the founders of the reformed philosophy, in opposition to the groundless hypotheses and empty subtleties of Aristotle and the schoolmen.

GAUGE-*Line*, a line on the common. Gauging rod, used for the purpose of gauging liquids. See GAUGING-*Rod*.

GAUGE-*Point*, of a solid measure, is the diameter of a circle, whose area is expressed by the same number as the solid content of that measure. Or it is the diameter of a cylinder, whose altitude is 1, and its content the same as of that measure.

Thus, the solid content of a wine gallon being 231 cubic inches; if a circle be conceived to contain so many square inches, its diameter will be 17.15; which is therefore the Gauge-point for wine measure. And an ale gallon containing 282 cubic inches; by the same rule, the Gauge-point for ale measure will be found to be 18.95. And after the same manner may the Gauge-point for any other measure be determined.

Hence it follows, that when the diameter of a cylinder in inches is equal to the Gauge-point in any measure, given likewise in inches, every inch in its length will contain an integer of the same measure. So in a cylinder whose diameter is 17.15 inches, every inch in height contains one entire gallon in wine measure; and in another, whose diameter is 18.95, every inch in length contains one ale gallon.

GAUGER, an officer appointed by the commissioners of excise, to Gauge, measure, or examine, all casks, tuns, pipes, barrels, hogsheads, of beer, wine, oil, &c.

GAUGING, the art or act of measuring the capacities or contents of all kinds of vessels, and determining the quantity of fluids, or other matters contained in them. These are principally pipes, tuns, barrels, rundlets, and other casks; also backs, coolers, vats, &c.

As to the solid contents of all prifmatical veffels, as cubes, parallelopipedons, cylinders, &c, they are found by multiplying the area of the. bafe by their altitude. And the contents of all pyramidal bodies, and cones, are equal to 1 3d of the fame.

In Gauging, it has been ufual to divide cafks into four varieties or forms, denominated as follows, from the fuppofed refemblance they bear to the fruftums of folids of the fame names : viz,

1. The middle fruftum of a fpheroid,
2. The middle fruftum of a parabolic fpindle,
3. The two equal fruftums of a paraboloid,
4. The two equal fruftums of a cone.

And particular rules, adapted to each of thefe forms, may be found in moft books of Gauging, and in my Menfuration, p. 575 &c. But as the form is imaginary, and only gueffed at, it hardly ever happens that a true folution is brought out in this way ; befide which, it is very troublefome and inconvenient to have fo many rules to put in practice. I fhall therefore give here one rule only, from p. 592 of that book, which is not only general for all cafks that are commonly met with, but quite eafy, and very accurate, as having been often verified and proved by filling the cafks with a true gallon measure.

General Rule. Add into one fum,
39 times the fquare of the bung diameter,
25 times the fquare of the head diameter, and
26 times the product of thofe diameters ;
multiply the fum by the length of the cafk, and the product by the number ·00034; then this laft product divided by 9 will give the wine gallons, and divided by 11 will give the ale gallons.

Or, $\overline{39B^2 + 25H^2 + 26BH} \times \dfrac{L}{114}$ is the content in inches; which being divided by 231 for wine gallons, or by 282 for ale gallons, will be the content.

For Ex. If the length of a cafk be 40 inches, the bung diameter 32, and the head diameter 24.

Here	-	$32^2 \times 39$	=	39936
and	-	$24^2 \times 25$	=	14400
and	-	$32 \times 24 \times 26$	=	19968

| the fum | - | 74304 |
| multiplied by | - | 40 |

and divid. by 114)2972160
gives 26071 cubic inches;
this divided by 231 gives 112 wine gallons,
or divided by 282 gives 92 ale gallons.

But the common practice of Gauging is performed mechanically, by means of the Gauging or Diagonal Rod, or the Gauging Sliding Rule, the defcription and ufe of which here follow.

GAUGING, or *Diagonal, Rod,* is a rod or rule adapted for determining the contents of cafks, by meafuring the diagonal only, viz the diagonal from the bung to the extremity of the oppofite ftave next the head. It is a fquare rule, having 4 fides or faces, being ufually 4 feet long, and folding together by means of joints.

Upon one face of the rule is a fcale of inches, for taking the meafure of the diagonal ; to thefe are adapted the areas, in ale gallons, of circles to the correfponding diameters, like the lines on the under fides

of the three flides in the fliding rule, defcribed below. And upon the oppofite face are two fcales, of ale and wine gallons, expreffing the contents of cafks having the corresponding diagonals; and thefe are the lines which chiefly conftitute the difference between this inftrument and the fliding rule ; for all the other lines upon it are the fame with thofe in that inftrument, and are to be ufed in the fame manner.

To ufe the Diagonal Rod. Unfold the rod ftraight out, and put it in at the bung hole of the cafk to be gauged, till its end arrive at the interfection of the head and oppofite ftave, or to the fartheft poffible diftance from the bung-hole, and note the inches and parts cut by the middle of the bung; then draw out the rod, and look for the fame inches and parts on the oppofite face of it, and annexed to them are found the contents of the cafk, both in ale and wine gallons.

For Ex. Let it be required to find, by this rod, the content of a cafk whofe diagonal meafures 34·4 inches ; which anfwers to the cafk in the foregoing example, whofe head and bung diameters are 32 and 24, and length 40 inches ; for if to the fquare of 20, half the length, be added the fquare of 28, half the fum of the diameters, the fquare root of the fum will be 34·4 nearly.

Now, to this diagonal 34·4, correfponds, upon the rule, the content 91 ale gallons, or 111 wine gallons ; which are but 1 lefs than the content brought out by the former general rule above given.

GAUGING *Rule,* or *Sliding Rule,* is a fliding rule particularly adapted to the purpofes of Gauging. It is a fquare rule, of four faces or fides, three of which are furnifhed with fliding pieces running in grooves. The lines upon them are moftly logarithmic ones, or diftances which are proportional to the logarithms of the numbers placed at the ends of them ; which kind of lines was placed upon rulers, by Mr. Edmund Gunter, for expeditioufly performing arithmetical operations, ufing a pair of compaffes for taking off and applying the feveral logarithmic diftances : but inftead of the compaffes, fliding pieces were added, by Mr. Thomas Everard, as more certain and convenient in practice, from whom this fliding rule is often called Everard's Rule. For the more particular defcription and ufes of this rule, fee my Menfuration, p. 564, 2d edition.

The writers on Gauging are, Beyer, Kepler, Dechales, Hunt, Everard, Dougherty, Shettleworth, Shirtcliffe, Leadbetter, &c.

GAZONS, in Fortification, turfs, or pieces of frefh earth covered with grafs, cut in form of a wedge, about a foot long, and half a foot thick, to line or face the outfide of works made of earth, to keep them up, and prevent their mouldering.

GELLIBRAND (HENRY), profeffor of aftronomy at Grefham-college, was born in London the 27th of Nov. 1597. He was fent to Trinity-college, Oxford, in 1615, and took his degree in arts 1619. He then entered into orders, and became curate of Chiddingftone in Kent. Afterwards, taking a great fancy to mathematics, by happening to hear one of Sir Henry Saville's lectures in that fcience, he immediately fet himfelf to the clofe ftudy of that noble fcience, and relinquifhed his fair profpects in the church. Contenting himfelf therefore

fore with his private patrimony, which was now come into his hands by the death of his father, the same year he entered again a student at Oxford, making mathematics his sole employment. He made such proficiency in this science before he proceeded A. M., which was in 1623, that he drew the attention and intimate friendship of Mr. Henry Briggs, then lately removed from the geometry professorship in Gresham-college to that of Savilian professor of geometry at Oxford, by the founder Sir Henry Savile, and who, upon the death of Mr. Gunter, procured for our author the professorship of astronomy in Gresham-college, to which he was elected in the beginning of the year 1627. His friend Mr. Briggs dying in 1630, before he had finished the introduction to his Trigonometria Britannica, he recommended the completing and publishing of that work to our author. Gellibrand accordingly added a preface, and the application of the logarithms to plane and spherical trigonometry, &c, and the whole was printed at Gouda, under the care of Adrian Vlacq, in 1633.

While Mr. Gellibrand was preparing that work, he was brought into trouble in the high-commission court, by Dr. Laud, then bishop of London, on account of an almanac, published by William Beale, servant to Mr. Gellibrand, for the year 1631, with the approbation of his master. In this almanac, the popish saints, then usually put into calendars, were omitted, and the names of other saints and martyrs, mentioned in the Book of Martyrs, were placed in their stead, as they stand in Fox's calendars. This it seems gave offence to the bishop, and occasioned the prosecution. But when the cause came to be heard, it appeared that other almanacs of the same kind had formerly been printed; upon which, both master and man were acquitted by Abp. Abbot and the whole court, Laud only excepted; which was afterward made one of the articles against him on his own trial.

It seems Gellibrand was strongly attached to the old Ptolomaic system. For when he went over to Holland, about the printing of Briggs's book abovementioned, he had some discourse with Lansberg, an eminent brother astronomer in Zealand, who affirming that he was fully persuaded of the truth of the Copernican system; our author observes, " that this so styled a truth he should " receive as an hypothesis; and so be easily led on to " the consideration of the imbecility of man's appre- " hension, as not able rightly to conceive of this ad- " mirable opifice of God, or frame of the world, with- " out falling foul of so great an absurdity:" so firmly was he fixed in his adherence to the Ptolomaic system. Gellibrand wrote several things after this, chiefly tending to the improvement of navigation, which would probably have been further advanced by him, had his life been continued longer; but he was untimely carried off by a fever, in 1636, at 39 years of age.

The character of Mr. Gellibrand is that of a plain, plodding, industrious, well-intentioned man, with little invention or genius. His writings are chiefly as below:

1. *Trigonometria Britannica*; or the Doctrine of Triangles, being the 2d part of Briggs's work above-mentioned.

2. A small Tract concerning the longitude.

3. A Discourse on the *Variation of the Magnetic Needle*; annexed to Wright's Errors in Navigation detected.

4. *Institution Trigonometrical*, with its application to astronomy and navigation; 8vo, 1635.

5. *Epitome of Navigation*, with the necessary tables; 8vo.

6. Several manuscripts never published; as, The Doctrine of Eclipses.—A Treatise of Lunar Astronomy.—A Treatise of Ship-building, &c.

GEMINI, a constellation of the northern hemisphere, one of the 48 old constellations, and the 3d in order of the zodiacal signs, Aries, Taurus, Gemini, &c. This constellation consists of two children, twins, called Castor and Pollux, and denoted by the mark, ♊, being a rude drawing of the same.

This constellation was, more anciently, depicted by a couple of young kids, by the Egyptians and eastern nations, as denoting that part of the spring when these animals appear; but the Greeks altered them to two children, which some of them make to be Castor and Pollux, some of them again Hercules and Apollo, and others Triptolemus and Jason; but the Arabians afterwards changed the figures into two peacocks, their religion not allowing them to paint or draw any human figure. Sir Isaac Newton thinks the figures had some reference to the Argonautic expedition.

The ancients attributed to every sign of the zodiac one of the principal deities for its tutelary power. Phœbus had the care of Gemini, and hence all the jargon of astrologers about the agreement of the sun and this constellation.

The stars in the sign Gemini are, in Ptolomy's catalogue 25, in Tycho's 25, in Hevelius's 38, and in the Britannic catalogue 85.

GENERATED, is used by some mathematical writers for whatever is produced by arithmetical operation, or in geometry by the motion of other magnitudes. Thus 20 is the product Generated of 4 and 5; *ab* that of *a* and *b*, 4, 8, 16, &c, the powers generated of or from the root 2, and a^2, a^3, a^4, &c, those from the root *a*. So also, a circle is Generated by the revolution of a line about one of its extremities, a cone by the rotation of a right-angled triangle about its perpendicular, a cylinder by the rotation of a rectangle about one of its sides, or, otherwise, by the motion of a circle in the direction of a right line, and keeping always parallel to itself.

GENERATING Line or Figure, in Geometry, is that which, by any kind of supposed motion, may generate, or produce, any other figure, plane, or solid.

Thus a line according to Euclid, generates a circle; or a right-angled triangle, a cone; &c; and thus also Archimedes supposes his spirals to be generated by the motions of Generating points and lines; the figure thus generated, is called the *Generant*.

It is a general theorem in geometry, that the measure of any generant, or figure produced by any kind of motion of any other figure, or Generating quantity, is equal to the product of this Generating quantity drawn into the length of the path described by its centre of gravity, whatever the kind of motion may be, whether rotatory, or direct, &c.

Vol. I.

3 Y

In

In the modern analyfis, or nuxions, all forts of quantities are confidered as Generated by fome fuch motion, and the quantity hereby generated is called a Fluent.

GENERATION, in Mathematics, is ufed for the formation or production of any geometrical figure, or other quantities. Such as of the figures mentioned in the foregoing articles, or the Generation of equations, curves, folids, &c.

GENESIS, in Geometry, means much the fame as Generation mentioned above, being the formation of a line, furface, or folid, by the motion or flux of a point, line, or furface; as of a globe by the rotation of a femi-circle about its diameter, &c.

In the Genefis of figures, the line or furface which moves, is called the *Defcribent*; and the line round which, or according to which, the revolution or other motion is made, the *Dirigent*.

GENETHLIACI, in Aftrology, are perfons who erect horofcopes, or pretend to foretell what fhall befall a perfon, by means of the ftars which prefided at his nativity.

The ancients called them Chaldæi, and by the general name Mathematici: accordingly, the feveral civil and canon laws, which we find made againft the mathematicians, only refpect the Genethliaci, or aftrologers.

They were expelled Rome, by a formal decree of the fenate; and yet found fo much protection from the credulity of the people, that they remained in the city unmolefted.

Antipater and Archinapolus have fhewn that Genethliology fhould rather be founded on the time of the conception than on that of the birth.

GEOCENTRIC, is faid of a planet or its orbit, to denote its having the earth for its centre. The moon alone is properly geocentric. And yet the motions of all the planets may be confidered in refpect of the earth, or as they appear from the earth, and thence called their Geocentric motions.—Hence alfo the terms Geocentric place, or latitude, or longitude, &c, as explained below.

GEOCENTRIC *Place*, of a planet, is the place where it appears to us, from the earth; or it is a point in the ecliptic, to which a planet, feen from the earth, is referred.

GEOCENTRIC *Latitude*, of a planet, is its latitude as feen from the earth; or the inclination of a line, connecting the planet and the earth, to the plane of the earth's (or true) ecliptic. Or it is the angle which the faid line (connecting the planet and the earth) makes with a line drawn to meet a perpendicular let fall from the planet to the plane of the ecliptic.

GEOCENTRIC *Longitude*, of a planet, is the diftance meafured on the ecliptic, in the order of the figns, between the Geocentric place and the firft point of Aries.

GEODESIA, is properly that part of practical geometry that teaches how to divide or lay out lands and fields, among feveral owners.

GEODESIA is alfo applied, by fome writers, to all meafurements in the field, and as fynonymous with furveying.

GEODESIA is defined by Vitalis, as the art of mea-furing furfaces and folids, not by imaginary lines, as is done in geometry, but by fenfible and vifible things; or by the fun's rays, &c.

GEOGRAPHER, a perfon fkilled in Geography.

GEOGRAPHICAL, fomething relating to Geography, as

GEOGRAPHICAL *Mile*, which is the fea-mile or minute, being the 60th part of a degree of a great circle.

GEOGRAPHICAL *Table*. See MAP.

GEOGRAPHY, the fcience that teaches and explains the nature and properties of the earth, as to its figure, place, magnitude, motions, celeftial appearances, &c, with the various lines, real or imaginary, on its furface.

Geography is diftinguifhed from Cofmography, as a part from the whole; this latter confidering the whole vifible world, both heaven and earth. And from Topography and Chorography, it is diftinguifhed, as the whole from a part.

Golnitz confiders Geography as either exterior or interior: but Varenius more juftly divides it into General and Special; or Univerfal and Particular.

General or *Univerfal* GEOGRAPHY, is that which confiders the earth in general, without any regard to particular countries, or the affections common to the whole globe: as its figure, magnitude, motion, land, fea, &c.

Special or *Particular* GEOGRAPHY, is that which contemplates the conftitution of the feveral particular regions, or countries; their bounds, figure, climate, feafons, weather, inhabitants, arts, cuftoms, language, &c.

Hiftory of GEOGRAPHY. The ftudy and practice of Geography muft have commenced at very early ages of the world. By the accounts we have remaining, it feems this fcience was in ufe among the Babylonians and Egyptians, from whom it paffed to the Greeks firft of any Europeans, and from thefe fucceffively to the Romans, the Arabians, and the weftern nations of Europe. Herodotus fays the Greeks firft learned the pole, the gnomon, and the 12 divifions of the day, from the Babylonians. But Pliny and Diogenes Laertius affert, that Thales of Miletus, in the 6th century before Chrift, firft found out the paffage of the fun from tropic to tropic, and it is faid was the author of two books, the one on the tropic, and the other on the equinox; both probably determined by means of the gnomon; whence he was led to the difcovery of the four feafons of the year, which are determined by the equinoxes and folftices; all which however it is likely he learned of the Egyptians, as well as his divifion of the year into 365 days. This it is faid was invented by the fecond Mercury, furnamed Trifmegiftus, who, according to Eufebius, lived about 50 years after the Exodus. Pliny exprefsly fays that this difcovery was made by obferving when the fhadow returned to its marks; a clear proof that it was done by the gnomon. It is farther faid that Thales conftructed a globe, and reprefented the land and fea upon a table of brafs. Farther that Anaximander, a difciple of Thales, firft drew the figure of the earth upon a globe; and that Hecate, Democritus, Eudoxus, and others, formed Geographical

phical maps, and brought them into common use in Greece.

Meton and Euctemon observed the summer solstice at Athens, on the 27th of June 432 years before Christ, by watching narrowly the shadow of the gnomon, with the design of fixing the beginning of their cycle of 19 years.

Timocharis and Aristillus, who began their observations about 295 B.C., it seems first attempted to fix the latitudes and longitudes of the fixed stars, by considering their distances from the equator, &c. One of their observations gave rise to the discovery of the precession of the equinoxes, which was first remarked by Hipparchus about 150 years after; who also made use of their method, for delineating the parallels of latitude and the meridians, on the surface of the earth; thus laying the foundation of this science as it now appears.

The latitudes and longitudes, thus introduced by Hipparchus, were not however much attended to till Ptolomy's time. Strabo, Vitruvius, and Pliny, have all of them entered into a minute geographical description of the situation of places, according to the length of the shadows of the gnomon, without noticing the longitudes and latitudes.

Maps at first were little more than rude outlines, and topographical sketches of different countries. The earliest on record were those of Sesostris, mentioned by Eustathius; who says, that " this Egyptian king, having traversed great part of the earth, recorded his march in maps, and gave copies of them not only to the Egyptians, but to the Scythians, to their great astonishment." Some have imagined with much probability, that the Jews made a map of the Holy Land, when they gave the different portions to the nine tribes at Shiloh: for Joshua tells us that they were sent to walk through the land, and that they *described it in seven parts in a book;* and Josephus relates that when Joshua sent out people from the different tribes to measure the land, he gave them, as companions, persons well skilled in geometry, who could not be mistaken in the truth.

The first Grecian map on record, was that of Anaximander, mentioned by Strabo, lib. 1, p. 7, supposed to be the one referred to by Hipparchus under the designation of the *ancient map.* Herodotus minutely describes a map made by Aristagoras tyrant of Miletus, which will serve to give some idea of the maps of those times. He relates, that Aristagoras shewed it to Cleomenes king of Sparta, to induce him to attack the king of Persia at Susa, in order to restore the Ionians to their ancient liberty. It was traced upon brass or copper, and seems to have been a mere itinerary, containing the route through the intermediate countries which were to be traversed in that march, with the rivers Halys, the Euphrates, and Tigris, which Herodotus mentions as necessary to be crossed in that expedition. It contained one straight line called the *Royal Road* or *Highway,* which took in all the stations or places of encampment from Sardis to Susa; being 111 in the whole journey, and containing 13,500 stadia, or 1687¼ Roman miles of 5000 feet each.

These itinerary maps of the places of encampment were indispensably necessary in all armies and marches;

and indeed war and navigation seem to be the two grand causes of the improvements both in Geography and astronomy. Athenæus quotes Bæton as author of a work intitled, *The encampments of Alexander's march;* and likewise Amyntas to the same purpose. Pliny observes that Diognetus and Bæton were the surveyors of Alexander's marches, and then quotes the exact number of miles according to their mensuration; which he afterwards confirms by the letters of Alexander himself. The same author also remarks that a copy of this great monarch's surveys was given by Xenocles his treasurer to Patrocles the geographer, who was admiral of the fleets of Seleucus and Antiochus. His book on geography is often quoted both by Strabo and Pliny; and it seems that this author furnished Eratosthenes with the principal materials for constructing his map of the oriental part of the world.

Eratosthenes first attempted to reduce Geography to a regular system, and introduced a regular parallel of latitude, which began at the straits of Gibraltar, passed eastwards through the isle of Rhodes, and so on to the mountains of India, noting all the intermediate places through which it passed. In drawing this line, he was not regulated by the same latitude, but by observing where the longest day was 14 hours and a half, which Hipparchus afterwards determined was the latitude of 36 degrees.

This first parallel through Rhodes was ever after considered with a degree of preference, in constructing all the ancient maps; and the longitude of the then known world was often attempted to be measured in stadia and miles, according to the extent of that line, by many succeeding geographers

Eratosthenes soon after attempted not only to draw other parallels of latitude, but also to trace a meridian at right angles to these, passing through Rhodes and Alexandria, down to Syene and Meroe; and at length he undertook the arduous task of determining the circumference of the globe, by an actual measurement of a segment of one of its great circles. To find the magnitude of the earth, is indeed a problem which has engaged the attention of astronomers and geographers ever since the spherical figure of it was known. It seems Anaximander was the first among the Greeks who wrote upon this subject. Archytas of Tarentum, a Pythagorean, famous for his skill in mathematics and mechanics, also made some attempts in this way; and Dr. Long conjectures that these are the authors of the most ancient opinion that the circumference of the earth is 400,000 stadia: and Archimedes makes mention of the ancients who estimated the circumference of the earth at only 30,000 stadia.

As to the methods of measuring the circumference of the earth, it would seem, from what Aristotle says in his treatise De Cœlo, that they were much the same as those used by the moderns, deficient only in the accuracy of the instruments. That philosopher there says, that different stars pass through our zenith, according as our situation is more or less northerly; and that in the southern parts of the earth stars come above our horizon, which are no longer visible if we go northward. Hence it appears that there are two ways of measuring the circumference of the earth; one by observing stars which pass through the zenith of one place, and do

not

not pafs through that of another; the other, by obferving fome ftars which come above the horizon of one place, and are obferved at the fame time to be in the horizon of another. The former of thefe methods, which is the beft, was followed by Eratofthenes at Alexandria in Egypt, 256 years before Chrift. He knew that at the fummer folftice, the fun was vertical to the inhabitants of Syene, a town on the confines of Ethiopia, under the tropic of cancer, where they had a well made to obferve it, at the bottom of which the rays of the fun fell perpendicularly the day of the fummer folftice: he obferved by the fhadow of a wire fet perpendicularly in an hemifpherical bafon, how far the fun was on that day at noon diftant from the zenith of Alexandria; when he found that diftance was equal to the 50th part of a great circle in the heavens. Then fuppofing Syene and Alexandria under the fame meridian, he inferred that the diftance between them was the 50th part of a great circle upon the earth; and this diftance being by meafure 5000 ftadia, he concluded that the whole circumference of the earth was 250,000 ftadia. But as this number divided by 360 would give 694⅘ ftadia to a degree, either Eratofthenes himfelf or fome of his followers affigned the round number 700 ftadia to a degree; which multiplied by 360, makes the circumference of the earth 252,000 ftadia; whence both thefe meafures are given by different authors, as that of Eratofthenes.

In the time of Pompey the Great, Pofidonius determined the meafure of the circumference of the earth by the 2d method above hinted by Ariftotle, viz, the horizontal obfervations. Knowing that the ftar called Canopus was but juft vifible in the horizon of Rhodes, and at Alexandria finding its meridian height was the 48th part of a great circle in the heavens, or 7½ deg., anfwering to the like quantity of a circle on the earth: Then, fuppofing thefe two places under the fame meridian, and the diftance between them 5000 ftadia, the circumference of the earth will be 240,000 ftadia; which is the firft meafure of Pofidonius. But according to Strabo, Pofidonius made the meafure of the earth to be 180,000 ftadia, at the rate of 500 ftadia to a degree. The reafon of this difference is thought to be, that Eratofthenes meafured the diftance between Rhodes and Alexandria, and found it only 3750 ftadia: taking this for a 48th part of the earth's circumference, which is the meafure of Pofidonius, the whole circumference will be 180,000 ftadia. This meafure was received by Marinus of Tyre, and is ufually afcribed to Ptolomy. But this meafurement is fubject to great uncertainty, both on account of the great refraction of the ftars near the horizon, the difficulty of meafuring the diftance at fea between Rhodes and Alexandria, and by fuppofing thofe places under the fame meridian, when they are really very different.

Several geographers afterwards made ufe of the different heights of the pole in diftant places under the fame meridian, to find the dimenfions of the earth. About the year 800, the khalif Almamun had the diftance meafured between two places that were 2 degrees afunder, and under the fame meridian in the plains of Sinjar near the Red Sea. And the refult was, that the degree at one time was found equal to 56 miles, and at another 56⅓ or 56⅔ miles.

The next attempt to find out the circumference of the earth, was in 1525, by Fernelius, a learned philofopher of France. For this purpofe, he took the height of the pole at Paris, going from thence directly northwards, till he came to the place where the height of the pole was one degree more than at that city. The length of the way was meafured by the number of revolutions made by one of the wheels of his carriage; and after proper allowances for the declivities and turnings of the road, he concluded that 68 Italian miles were equal to a degree on the earth.

According to thefe methods many other meafurements of the earth's circumference have fince that time been made, with much greater accuracy: a particular account of which is given under the article Degree.

Though the maps of Eratofthenes were the beft of his time, they were yet very imperfect and inaccurate. They contained little more than the ftates of Greece, and the dominions of the fucceffors of Alexander, digefted according to the furveys abovementioned. He had indeed feen, and has quoted, the voyages of Pythias into the great Atlantic ocean, which gave him fome faint idea of the weftern parts of Europe; but fo imperfect, that they could not be realized into the outlines of a chart. Strabo fays he was very ignorant of Gaul Spain, Germany, and Britain; and he was equally ignorant of Italy, the coafts of the Adriatic, Pontus, and all the countries towards the north.

Such was the ftate of Geography, and the nature of the maps, before the time of Hipparchus. He made a clofer connection between Geography and aftronomy, by determining the latitudes and longitudes from celeftial obfervations.

War has ufually been the occafion of making or improving the maps of countries; and accordingly Geography made great advances from the progrefs of the Roman arms. In all the provinces occupied by that people, camps were every where conftructed at proper intervals, and good roads made for communication between them; and thus civilization and furveying were carried on according to fyftem, through the whole extent of that large empire. Every new war produced a new furvey and itinerary of the countries where the fcenes of action paffed; fo that the materials of Geography were accumulated by every additional conqueft. Polybius fays, that at the beginning of the fecond Punic war, when Hannibal was preparing his expedition againft Rome, the countries through which he was to pafs were carefully meafured by the Romans. And Julius Cæfar caufed a general furvey of the Roman Empire to be made, by a decree of the fenate. Three furveyors had this tafk affigned them, which they completed in 25 years. The Roman itineraries that are ftill extant, alfo fhew what care and pains they had been at in making furveys in all the different provinces of their empire; and Pliny has filled the 3d, 4th, and 5th books of his Natural Hiftory with the Geographical diftances that were thus meafured. Other maps are alfo ftill preferved, known by the name of the Pentingerian Tables, publifhed by Welfer and Bertius, which give a good fpecimen of what Vegetius calls the *Itinera Picta*, for the better direction of their armies in their march.

The Roman empire had been enlarged to its greateft extent, and all its provinces well known and furveyed, when

when Ptolomy, about 150 years after Chrift, compofed his fyftem of Geography. The chief materials he employed in compofing this work, were the proportions of the gnomon to its fhadow, taken by different aftronomers at the times of the equinoxes and folftices; calculations founded on the length of the longeft days; the meafured or computed diftances of the principal roads contained in their furveys and itineraries; and the various reports of travellers and navigators. All thefe were compared together, and digefted into one uniform body or fyftem; and afterwards were tranflated by him into a new mathematical language, expreffing the different degrees of latitude and longitude, after the invention of Hipparchus, which had been neglected for 250 years.

Ptolomy's fyftem of Geography, notwithftanding it was ftill very imperfect, continued in vogue till the laft three or four centuries, within which time the great improvements in aftronomy, the many difcoveries of new countries by voyagers, and the progrefs of war and arms, have contributed to bring it to a very confiderable degree of perfection; the particulars of which will be found treated under their refpective articles in this work.

Among the moderns, the chief authors on the fubject of Geography are Johannes de Sacrobofco, or John Hallifax, who wrote a treatife on the fphere; Sebaftian Munfter, in his Cofmographia Univerfalis, in 1559; Clavius, on the fphere of Sacrobofco; Piccieli's Geographia et Hydrographia Reformata; Weigelius's Speculum Terræ; De Chales's Geography, in his Mundus Mathematicus; Cellarius's Geography; Cluverius's Introductio in Univerfam Geographiam; Leibnecht's Elementa Geographiæ generalis; Stevenius's Compendium Geographicum; Wolfius's Geographia, in his Elementa Matheseos; Bufching's New Syftem of Geography; Gordon's, Salmon's, and Guthrie's Grammars; and above all, Varenius's Geographia generalis, with Jurin's additions, the moft fcientific and fyftematical of any.

GEOMETER, more properly GEOMETRICIAN; which fee.

GEOMETRICAL, fomething that has a relation to Geometry, or done after the manner, or by the means of Geometry. As, a Geometrical conftruction, a Geometrical curve, a Geometrical demonftration, genius, line, method, Geometrical ftrictnefs, &c.

GEOMETRICAL *Conftruction*, of an equation, is the drawing of lines and figures, fo as to exprefs by them the fame general property and relation, as are denoted by the algebraical equation. See CONSTRUCTION *of Equations*.

GEOMETRICAL *Curve* or *Line*, called alfo an *Algebraical one*, is that in which the relations between the abfciffes and ordinates may be expreffed by a finite algebraical equation. See ALGEBRAICAL *Curves*.

GEOMETRICAL *Lines*, as obferved by Newton, are diftinguifhed into claffes, orders, or genera, according to the number of the dimenfions of the equation that expreffes the relation between the ordinates and abfciffes; or, which comes to the fame thing, according to the number of points in which they may be cut by a right line.

Thus, a line of the firft order, is a right line, fince it

can be only once cut by another right line, and is expreffed by the fimple equation $y + ax + b = 0$: thofe of the 2d, or quadratic order, will be the circle, and the conic fections, fince all of thefe may be cut in two points by a right line, and expreffed by the equation $y^2 + \overline{ax + b} \cdot y + cx^2 + dx + e = 0$: thofe of the 3d or cubic order, will be fuch as may be cut in 3 points by a right line, whofe moft general equation is $y^3 + \overline{ax + b} \cdot y^2 + \overline{cx^2 + dx} + e \cdot y + fx^3 + gx^2 + hx + i = 0$; as the cubical and Neilian parabola, the ciffoid, &c. And a line of an infinite order, is that which a right line may cut in infinite points; as the fpiral, the cycloid, the quadratrix, and every line that is generated by the infinite revolutions of a radius, or circle, or wheel, &c.

In each of thofe equations, x is the abfcifs, y its correfponding ordinate, making any given angle with it; and a, b, c, &c, are given or conftant quantities, affected with their figns $+$ and $-$, of which one or more may vanifh, be wanting or equal to nothing, provided that by fuch defect the line or equation does not become one of an inferior order.

It is to be noted that a curve of any kind is denominated by a number next lefs than the line of the fame kind: thus, a curve of the 1ft order, (becaufe the right line cannot be reckoned among curves) is the fame with a line of the 2d order; and a curve of the 2d kind, the fame with a line of the 3d order; &c.

It is to be obferved alfo, that it is not fo much the equation, as the conftruction or defcription, that makes any curve, geometrical, or not. Thus, the circle is a geometrical line, not becaufe it may be expreffed by an equation, but becaufe its defcription is a poftulate: and it is not the fimplicity of the equation, but the eafinefs of the defcription, that is to determine the choice of the lines for the conftruction of a problem. The equation that expreffes a parabola, is more fimple than that which expreffes a circle; and yet the circle, by reafon of its more fimple conftruction, is admitted before it. Again, the circle and the conic fections, with refpect to the dimenfions of the equations, are of the fame order; and yet the circle is not numbered with them in the conftruction of problems, but by reafon of its fimple defcription is depreffed to a lower order, viz, that of a right line; fo that it is not improper to exprefs that by a circle, which may be expreffed by a right line; but it is a fault to conftruct that by the conic fections, which may be conftructed by a circle.

Either, therefore, the law muft be taken from the dimenfions of equations, as obferved in a circle, and fo the diftinction be taken away between plane and folid problems: or the law muft be allowed not to be ftrictly obferved in lines of fuperior kinds, but that fome, by reafon of their more fimple defcription, may be preferred to others of the fame order, and be numbered with lines of inferior orders.

In conftructions that are equally Geometrical, the moft fimple are always to be preferred: and this law is fo univerfal as to be without exception. But algebraical expreffions add nothing to the fimplicity of the conftruction; the bare defcriptions of the lines here are only to be confidered; and thefe alone were confidered by thofe geometricians who joined a circle with a right line.

line. And as these are easy or hard, the construction becomes easy or hard: and therefore it is foreign to the nature of the thing, from any other circumstance to establish laws relating to constructions.

Either, therefore, with the ancients, we must exclude all lines beside the circle, and perhaps the conic sections, out of geometry; or admit all, according to the simplicity of the description. If the trochoid were admitted into geometry, by means of it we might divide an angle in any given ratio; would it be right therefore to blame those who would make use of this line to divide an angle in the ratio of one number to another; and contend, that you must make use only of such lines as are defined by equations, and therefore not of this line, which is not so defined? If, when an angle is proposed to be divided, for instance, into 10001 parts, we should be obliged to bring a curve defined by an equation of more than 100 dimensions to do the business; which nobody could describe, much less understand; and should prefer this to the trochoid, which is a line well known, and easily described by the motion of a wheel, or circle: who would not see the absurdity?

Either therefore the trochoid is not to be admitted at all in geometry; or else, in the construction of problems, it is to be preferred to all lines of a more difficult description; and the reason is the same for other curves. Hence the trisections of an angle by a conchoid, which Archimedes in his Lemmas, and Pappus in his Collections, have preferred to the inventions of all others in this case, must be allowed as good; because we must either exclude all lines, beside the circle and right line, out of geometry, or admit them according to the simplicity of their descriptions; in which case the conchoid yields to none, except the circle. Equations are expressions of arithmetical computation, and properly have no place in geometry, excepting so far as quantities truly Geometrical (that is, lines, surfaces, solids, and proportions) may be said to be some equal to others. Multiplications, divisions, and such like computations, are newly received into Geometry, and that unwarily, and contrary to the first design of this science. For whoever considers the construction of problems by a right line and a circle, found out by the first geometricians, will easily perceive that geometry was introduced, that by drawing lines, we might easily avoid the tediousness of computation. For which reason the two sciences ought not to be confounded together: the ancients so carefully distinguished between them, that they never introduced arithmetical terms into geometry; and the moderns, by confounding them, have lost the simplicity in which all the elegance of geometry consists. In short, that is arithmetically more simple, which is determined by the more simple equations; but that is Geometrically more simple, which is determined by the more simple drawing of lines; and in geometry that ought to be reckoned best, which is geometrically most simple. Newton's Arith. Univers. appendix. See Curves.

Geometrical *Locus*, or *Place*, called also simply *Locus*, is the path or track of some certain Geometrical determination, in which it always falls. See Locus.

Geometrical *Medium*. See Medium.

Geometrical *Method of the Ancients*. The ancients established the higher parts of their geometry on the same principles as the elements of that science, by demonstrations of the same kind: and they were careful not to suppose any thing done, till by a previous problem they had shewn that it could be done by actually performing it. Much less did they suppose any thing to be done that cannot be conceived; such as a line or series to be actually continued to infinity, or a magnitude diminished till it become infinitely less than what it is. The elements into which they resolved magnitudes were finite, and such as might be conceived to be real. Unbounded liberties have of late been introduced; by which geometry, which ought to be perfectly clear, is filled with mysteries. Maclaurin's Fluxions, Introd. p. 39.

Geometrical *Pace*, is a measure of 5 feet long.

Geometrical *Plan*, in Architecture. See Plan.

Geometrical *Plane*. See Plane.

Geometrical *Progression*, a progression in which the terms have all successively the same ratio: as 1, 2, 4, 8, 16, &c, where the common ratio is 2.

The general and common property of a Geometrical progression is, that the product of any two terms, or the square of any one single term, is equal to the product of every other two terms that are taken at an equal distance on both sides from the former. So of these terms,

$$1, \ 2, \ 4, \ 8, \ 16, \ 32, \ 64, \ \&c,$$
$$1 \times 64 = 2 \times 32 = 4 \times 16 = 8 \times 8 = 64.$$

In any Geometrical Progression, if

a denote the least term,

z the greatest term,

r the common ratio,

n the number of the terms,

s the sum of the series, or all the terms;

then any of these quantities may be found from the others, by means of these general values, or equations, viz.

$$r = \sqrt[n-1]{\frac{z}{a}},$$
$$z = a \times r^{n-1},$$
$$a = z \div r^{n-1},$$
$$n = \frac{\log.\frac{rz}{a}}{\log. r} = \frac{\log. r + \log. z - \log. a}{\log. r},$$
$$s = \frac{r^n - 1}{r - 1} \times a = \frac{r^n - 1}{r - 1} \times \frac{z}{r^{n-1}} = \frac{rz - a}{r - 1} =$$
$$\frac{z^{\frac{n}{n-1}} - a^{\frac{n}{n-1}}}{z^{\frac{1}{n-1}} - a^{\frac{1}{n-1}}}.$$

When the series is infinite, then the least term a is nothing, and the sum $s = \dfrac{rz}{r-1}$. See also Progression.

Geometrical *Proportion*, called also simply *Proportion*, is the similitude or equality of ratios.

Thus, if $a : b :: c : d$, or $a : b = c : d$, the terms a, b, c, d are in Geometrical Proportion; also 6, 3, 14, 7, are in Geometrical Proportion, because $6 : 3 :: 14 : 7$, or $6 : 3 = 14 : 7$.

In

In a Geometrical Proportion, the product of the extremes, or 1st and 4th terms, is equal to the product of the means, or the 2d and 3d terms : so $ad = bc$, and $6 \times 7 = 3 \times 14 = 42$. See Proportion.

Geometrical *Solution*, of a problem, is when the problem is directly resolved according to the strict rules and principles of geometry, and by lines that are truly Geometrical. This expression is used in contradistinction to an arithmetical, or a mechanical, or instrumental solution; the problem being resolved only by a ruler and compasses.

The same term is likewise used in opposition to all indirect and inadequate kinds of solutions, as by approximation, infinite series, &c. So, we have no Geometrical way of finding the quadrature of the circle, the duplicature of the cube, or two mean proportionals; though there are mechanical ways, and others, by infinite series, &c.

Pappus informs us, that the ancients endeavoured in vain to trisect an angle, and to find out two mean proportionals, by means of the right line and circle. Afterwards they began to consider the properties of several other lines; as the conchoid, the cissoid, and the conic sections; and by some of these they endeavoured to resolve some of those problems. At length, having more thoroughly examined the matter, and the conic sections being received into geometry, they distinguished Geometrical problems and solutions into three kinds; viz,

1. *Plane* ones, which, deriving their origin from lines on a plane, may be properly resolved by a right line and a circle.—2. *Solid* ones, which are resolved by lines deriving their original from the consideration of a solid; that is, of a cone.—3. *Linear* ones, to the solution of which are required lines more compounded.

According to this distinction, we are not to resolve solid problems by other lines than the conic sections; especially if no other lines beside the right line, circle, and the conic sections, must be received into geometry.

But the moderns, advancing much farther, have received into geometry all lines that can be expressed by equations; and have distinguished, according to the dimensions of the equations, those lines into classes or orders; and have laid it down as a law, not to construct a problem by a line of a higher order, that may be constructed by one of a lower.

GEOMETRICIAN, a person skilled in Geometry.

GEOMETRY, the science or doctrine of local extension, as of lines, surfaces, and solids, with that of ratios, &c.

The name Geometry literally signifies measuring of the earth, as it was the necessity of measuring the land that first gave occasion to contemplate the principles and rules of this art; which has since been extended to numberless other speculations; insomuch that together with arithmetic, Geometry forms now the chief foundation of all the mathematics.

Herodotus (lib. 2), Diodorus (lib 1), Strabo (lib. 17), and Proclus, ascribe the invention of Geometry to the Egyptians, and assert that the annual inundations of the Nile gave occasion to it; for those waters bearing away the bounds and land-marks of estates and farms, covering the face of the ground uniformly with mud, the people, say they, were obliged every year to distinguish and lay out their lands by the consideration of their figure and quantity; and thus by experience and habit they formed a method or art, which was the origin of Geometry. A farther contemplation of the draughts of figures of fields thus laid down, and plotted in proportion, might naturally lead them to the discovery of some of their excellent and wonderful properties; which speculation continually improving, the art continually gained ground, and made advances more and more towards perfection.

Josephus however seems to ascribe the invention to the Hebrews: and others of the ancients make Mercury the inventor. Polyd. Virgil, de Invent. Rer. lib. 1, cap. 18.

From Egypt, this science passed into Greece, being carried thither by Thales; where it was much cultivated and improved by himself, as also by Pythagoras, Anaxagoras of Clazomene, Hippocrates of Chios, and Plato, who testified his conviction of the necessity and importance of Geometry to the successful study of philosophy, by this inscription over the door of his academy, *Let no one ignorant of Geometry enter here.* Plato thought the word Geometry too mean a name for this science, and substituted instead of it the more extensive name of *Mensuration*; and after him others gave it the title of *Pantometry*. But even these are now become too scanty in their import, fully to comprehend its extent; for it not only inquires into, and demonstrates the quantities of magnitudes, but also their qualities, as the species, figures, ratios, positions, transformations, descriptions, divisions, the finding of their centres, diameters, tangents, asymptotes, curvature, &c. Some again define it as the science of inquiring, inventing, and demonstrating all the affections of magnitude. And Proclus calls it the knowledge of magnitudes and figures, with their limitations; as also of their ratios, affections, positions, and motions of every kind.

About 50 years after Plato, lived Euclid, who collected together all those theorems which had been invented by his predecessors in Egypt and Greece, and digested them into 15 books, called the Elements of Geometry; demonstrating and arranging the whole in a very accurate and perfect manner. The next to Euclid, of those ancient writers whose works are extant, is Apollonius Pergæus, who flourished in the time of Ptolomy Euergetes, about 230 years before Christ, and about 100 years after Euclid. He was author of the first and principal work on Conic Sections; on account of which, and his other accurate and ingenious geometrical writings, he acquired from his patron the emphatical appellation of *The Great Geometrician.* Contemporary with Apollonius, or perhaps a few years before him, flourished Archimedes, celebrated for his mechanical inventions at the siege of Syracuse, and not less so for his very many ingenious Geometrical compositions.

We can only mention Eudoxus of Cnidus, Archytas of Tarentum, Philolaus, Eratosthenes, Aristarchus of Samos, Dinostratus, the inventor of the quadratrix, Menechmus, his brother and the disciple of Plato, the two Aristeus's, Conon, Thracidius, Nicoteles, Leon, Theudius, Hermotimus, Hero, and Nicomedes the inventor

I

ventor of the conchoid : befides whom, there are many other ancient geometricians, to whom this science has been indebted.

The Greeks continued their attention to it, even after they were fubdued by the Romans. Whereas the Romans themfelves were fo little acquainted with it, even in the moft flourifhing time of their republic, that Tacitus informs us they gave the name of mathematicians to thofe who purfued the chimeras of divination and judicial aftrology. Nor does it appear they were more difpofed to cultivate Geometry during the decline, and after the fall of the Roman empire. But the cafe was different with the Greeks; among whom are found many excellent Geometricians fince the commencement of the Chriftian era, and after the tranflation of the Roman empire. Ptolomy lived under Marcus Aurelius; and we have ftill extant the works of Pappus of Alexandria, who lived in the time of Theodofius; the commentary of Eutocius, the Afcalonite, who lived about the year of Chrift 540, on Archimedes's menfuration of the circle; and the commentary on Euclid, by Proclus, who lived under the empire of Anaftafius.

The confequent inundation of ignorance and barbarifm was unfavourable to Geometry, as well as to the other fciences; and the few who applied themfelves to this fcience, were calumniated as magicians. However, in thofe times of European darknefs, the Arabians were diftinguifhed as the guardians and promoters of fcience; and from the 9th to the 14th century, they produced many aftronomers, geometricians, geographers, &c; from whom the mathematical fciences were again received into Spain, Italy, and the reft of Europe, fomewhat before the year 1400. Some of the earlieft writers after this period, are Leonardus Pifanus, Lucas Paciolus or De Burgo, and others between 1400 and 1500. And after this appeared many editions of Euclid, or commentaries upon him: thus, Orontius Finæus, in 1530, publifhed a commentary on the firft 6 books; as did James Peletarius, in 1557; and about the fame time Nicholas Tartaglia publifhed a commentary on the whole 15 books. There have been alfo the editions, or commentaries, of Commandine, Clavius, Billingfly, Scheubelius, Herlinus, Dafypodius, Ramus, Herigon, Stevinus, Saville, Barrow, Taquet, Dechales, Furnier, Scarborough, Keill, Stone, and many others; but the completeft edition of all the works of Euclid, is that of Dr. Gregory, printed at Oxford 1703, in Greek and Latin: the edition of Euclid, by Dr. Robert Simfon of Glafgow, containing the firft 6 books, with the 11th and 12th, is much efteemed for its correctnefs. The principal other elementary writers, befides the editors of Euclid, are Borelli, Pardies, Marchetti, Wolfius, Simfon, &c. And among thofe who have gone beyond Euclid in the nature of the Elementary parts of Geometry, may be chiefly reckoned, Apollonius, in his Conics, his Loci Plani, De Sectione Determinata, his Tangencies, Inclinations, Section of a Ratio, Section of a Space, &c; Archimedes, in his treatifes of the Sphere and Cylinder, the Dimenfion of the Circle, of Conoids and Spheroids, of Spirals, and the Quadrature of the Parabola; Theodofius, in his Spherics; Serenius, in his Sections of the Cone and Cylinder; Kepler's Nova Stereometria; Cavalerius's

7

Geometria Indivifibilium; Torricelli's Opera Geometrica; Viviani, in his Divinationes Geometricæ, Exercitatio Mathematica, De Locis Solidis, De Maximis & Minimis, &c; Vieta, in his Effectio Geometrica, Supplement. Geometriæ, Sectiones Angulares, Refponfum ad Problema, Apollonius Gallus, &c; Gregory St. Vincent's Quadratura Circuli; Fermat's Varia Opera Mathematica; Dr. Barrow's Lectiones Geometricæ; Bulliald de Lineis Spiralibus; Cavalerius; Schooten and Gregory's Exercitationes Geometricæ, and Gregory's Pars Univerfalis, &c; De Billy's treatife De Proportione Harmonica; La Lovera's Geometria veterum promota; Slufius's Mefolabium, Problemata Solida, &c; Wallis, in his treatifes De Cycloide, Ciffoide, &c; De Proportionibus, De Sectionibus Conicis, Arithmetica Infinitorum, De Centro Gravitatis, De Sectionibus Angularibus, De Angulo Contactus, Cuno-Cuneus, &c, &c; Hugo De Omerique, in his Analyfis Geometrica; Pafcal on the Cycloid; Step. Angeli's Problemata Geometrica; Alex. Anderfon's Suppl. Apollonii Redivivi, Variorum Problematum Practice, &c; Baronius's Geomet. Prob. &c; Guido Grandi Geometr. Demonftr. &c; Ghetaldi Apollonius Redivivus, &c; Ludolph van Colen or a Ceulen, de Circulo et Adfcriptis, &c; Snell's Apollonius Batavus, Cyclometricus, &c; Herberftein's Diotome Circulorum; Palma's Exercit. in Geometriam; Guldini Centro-Baryca; with feveral others equally eminent, of more modern date, as Dr. Rob. Simfon, Dr. Mat. Stewart, Mr. Tho. Simpfon, &c. Since the introduction of the new Geometry, or the Geometry of Curve Lines, as expreffed by algebraical equations, in this part of Geometry, the following names, among many others, are more efpecially to be refpected, viz, Des Cartes, Schooten, Newton, Maclaurin, Brackenridge, Cramer, Cotes, Waring, &c, &c.

As to the fubject of Practical Geometry, the chief writers are Beyer, Kepler, Ramus, Clavius, Mallet, Tacquet, Ozanam, Wolfius, Gregory, with innumerable others.

Geometry is diftinguifhed into Theoretical or Speculative, and Practical.

Theoretical or *Speculative* GEOMETRY, treats of the various properties and relations in magnitudes, demonftrating the theorems, &c. And

Practical GEOMETRY, is that which applies thofe fpeculations and theorems to particular ufes in the folution of problems, and in the meafurements in the ordinary concerns of life.

Speculative Geometry again may be divided into Elementary and Sublime.

Elementary or *Common* GEOMETRY, is that which is employed in the confideration of right lines and plane furfaces, with the folids generated from them. And the

Higher or *Sublime* GEOMETRY, is that which is employed in the confideration of curve lines, conic fections, and the bodies formed of them. This part has been chiefly cultivated by the moderns, by help of the improved ftate of Algebra, and the modern analyfis or Fluxions.

GIBBOUS, is ufed for the fhape of one ftate of the enlightened part of the moon, being that in which fhe appears

appears more than half full or enlightened, which is the time between the firſt quarter and the full moon, and from the full moon to the laſt quarter; appearing then Gibbous, that is, bunched out, or convex on both ſides of the enlightened part; as contradiſtinguiſhed from the ſtate when ſhe is leſs than half full, when ſhe is ſaid to be horned, or a creſcent.

GIMBOLS, are the braſs rings by which a ſea-compaſs is ſuſpended in its box that uſually ſtands in the binacle.

GIN, in Artillery and Mechanics, is a machine for raiſing great weights, uſually compoſed of three long legs, &c.

GIRDERS, in Architecture, are the largeſt beams or pieces of timber ſupporting the floors. Their ends are uſually faſtened into the ſummers, or breaſt-ſummers; and the joiſts are framed in at one end to the Girders. By the ſtatute for rebuilding London, no Girder is to lie leſs than 10 inches into the wall, and their ends to be always laid in loam, &c. The ſhorter bearings a Girder has, and the oftener it is ſupported by the internal or partition walls, ſo much the better. The eſtabliſhed breadth and depth of a Girder, according to its length of bearing, are as in the following tablet:

Girders and Summers in length	From	to	must be in	
			Breadth	Depth
	10 feet	15 ft.	11 inc.	8 inc.
	15	18	13	9
	18	21	14	10
	21	24	16	12
	24	26	17	14

GIRT, in Timber-meaſuring, is the circumference of a tree; though ſome uſe this word for the quarter or 4th part of the circumference only, on account of the great uſe that is made of it; for the ſquare of this 4th part is eſteemed and uſed as equal to the area of the ſection of the tree; which ſquare therefore multiplied by the length of the tree, is accounted the ſolid content. This content however is always about one-fourth part leſs than the true quantity; being nearly equal to what this will be after the tree is hewed ſquare in the uſual way: ſo that it ſeems intended to make an allowance for the ſquaring of the tree.

GIRT-*Line*, is a line on the common or carpenter's ſliding rule, employed in caſting up the contents of trees by means of their Girt.

GIVEN, *Datum*, a term very often uſed in mathematics, and ſignifies ſomething that is ſuppoſed to be known.

Thus, if a magnitude be known, or if we can find another equal to it, it is ſaid to be Given in magnitude. Or when the poſition of any thing is known, it is ſaid to be Given in poſition. And when the diameter of a circle is known, the circle is Given in magnitude. Or the circle is Given in poſition when its centre is Given in poſition. When the kind or ſpecies of a figure is known, or remains the ſame, it is Given in ſpecie. And ſo on.

Vol. I.

Euclid wrote a book of *Data*, or concerning things Given, in 95 propoſitions, uſually accompanying his Elements, in the beſt editions, and which Pappus reckons as one of the beſt ſpecimens of the analytical works of the ancients.

GLACIS, in Fortification, a ſloping bank reaching from the parapet of the counterſcarp, or covered-way, to the level ſide of the field, commonly at the diſtance of about 40 yards.

GLOBE, a round or ſpherical body, more uſually called a ſphere, bounded by one uniform convex ſurface, every point of which is equally diſtant from a point within called its centre. Euclid defines the Globe, or ſphere, to be a ſolid figure deſcribed by the revolution of a ſemi-circle about its diameter, which remains unmoved. Alſo, its axis is the fixed line or diameter about which the ſemi-circle revolves; and its centre is the ſame with that of the revolving ſemi-circle, a diameter of it being any right line that paſſes through the centre, and terminated both ways by the ſuperficies of the ſphere. Elem. 11. def. 14, 15, 16, 17.

Euclid, at the end of the 12th book, ſhews that ſpheres are to one another in the triplicate ratio of their diameters, that is, their ſolidities are to one another as the cubes of their diameters. And Archimedes determines the real magnitudes and meaſures of the ſurfaces and ſolidities of ſpheres and their ſegments, in his treatiſe de Sphæra et Cylindro: viz, 1, That the ſuperficies of any Globe is equal to 4 times a great circle of it.—2, That any ſphere is equal to ⅔ of its circumſcribing cylinder, or of the cylinder of the ſame diameter and altitude.—3, That the curve ſurface of the ſegment of a globe, is equal to the circle whoſe radius is the line drawn from the vertex of the ſegment to the circumference of the baſe.—4, That the content of a ſolid ſector of the Globe, is equal to a cone whoſe altitude is the radius of the Globe, and its baſe equal to the curve ſuperficies or baſe of the ſector. With many other properties. And from hence are eaſily deduced theſe practical rules for the ſurfaces and ſolidities of Globes and their ſegments; viz,

1. *For the Surface of a Globe*, multiply the ſquare of the diameter by 3·1416; or multiply the diameter by the circumference.

2. *For the Solidity of a Globe*, multiply the cube of the diameter by ·5236 (viz ⅙ of 3·1416); or multiply the ſurface by ⅙ of the diameter.

3. *For the Surface of a Segment*, multiply the diameter of the Globe by the altitude of the ſegment and the product again by 3·1416.

4. *For the Solidity of a Segment*, multiply the ſquare of the diameter of the Globe by the difference between 3 times that diameter and 2 times the altitude of the ſegment, and the product again by ·5236, or ⅙ of 3·1416.

Hence, if d denote the diameter of the Globe, c the circumference, a the altitude of any ſegment, and $p = 3\cdot1416$; then

	The ſurface.	The ſolidity
In the Globe	$pd^2 = cd$	$\frac{1}{6} pd^3$
In the Segt.	pad	$\frac{1}{6} pa^2 \times \overline{3d - 2a}$

See the art. Sphere, and my Menſuration, p. 197 &c, 2d edit.

The

The GLOBE, or *Terraqueous* GLOBE, is the body or mass of the earth and water together, which is nearly globular.

GLOBE, or *Artificial* GLOBE, is more particularly used for a Globe of metal, plaister, paper, pasteboard, &c, on the surface of which is drawn a map, or representation of either the heavens or the earth, with the several circles conceived upon them. And hence

GLOBES are of two kinds, Terrestrial, and Celestial; which are of considerable use in geography and astronomy, by serving to give a lively representation of their principal objects, and for performing and illustrating many of their operations in a manner easy to be perceived by the senses, and so as to be conceived even without any knowledge of the mathematical grounds of those sciences.

Description of the Globes.

The fundamental parts that are common to both Globes, are an axis, representing the axis of the world, passing through the two poles of a spherical shell, representing those of the world, which shell makes the body of the Globe, upon the external surface of which is drawn the representation of the whole surface of the earth, sea, rivers, islands, &c, for the Terrestrial Globe, and the stars and constellations of the heavens, for the Celestial one; besides the equinoctial and ecliptic lines, the zodiac, the two tropics and polar circles, and a number of meridian lines. There is next a brazen meridian, being a strong circle of brass, circumscribing the Globe, at a small distance from it quite round, in which the globe is hung by its two poles, upon which it turns round within this circle, which is divided into 4 times 90 degrees, beginning at the equator on both sides, and ending with 90 at the two poles. There are also two small hour circles, of brass, divided into twice 12 hours, and fitted on the meridian round the poles, which carry an index pointing to the hour. The whole is set in a wooden ring, placed parallel to, and representing the horizon, in which the Globe slides by the brass meridian, elevating or depressing the pole according to any proposed latitude. There is also a thin slip of brass, called a Quadrant of Altitude, made to fit on occasionally upon the brass meridian, at the highest or vertical-point, to measure the altitude of any thing above the horizon. A magnetic compass is sometimes set underneath. See the figure of the Globes so mounted, at fig. 1, plate xii.

Such is the plain and simple construction of the artificial Globe, whether celestial or terrestrial, as adapted to the time only for which it is made. But as the angle formed by the equator and ecliptic, as well as their point of intersection, is always changing; to remedy these inconveniences, several contrivances have been made, so as to adapt the same Globes to any other time, either past or to come; as well as other contrivances to answer particular purposes.

Thus, Mr. Senex, a celebrated maker of Globes, had a contrivance which, by means of a nut and screw, caused the pole of the equator to revolve about the pole of the ecliptic, by any quantity answering to the precession of the equinoxes, since the time for which the Globe was made. Philos. Transf. number 447, or Abr. vol. 8, p. 217, also Philos. Transf. vol. 46, p. 290.

Mr. Joseph Harris, late assay-master of the Mint, made some contrivances to shew the effects of the earth's motions. He fixed two horary circles under the brass meridian, to the axis, one at each pole, so as to turn round with the Globe, and that meridian served as an index to cut the horary divisions. The Globe in this state serves equally for resolving problems in both north and south latitudes, as also in places near the equator; whereas, in the common construction, the axis and horary circle prevent the brass meridian from being moveable quite round in the horizon. This Globe is also adapted for shewing how the vicissitudes of day and night, and the alteration of their lengths, are really occasioned by the motion of the earth: for this purpose, he divides the brass meridian, at one of the poles, into months and days, according to the sun's declination, reckoning from the pole. Therefore, by bringing the day of the month to the horizon, and rectifying the Globe according to the time of the day, the horizon will represent the circle separating light and darkness, and the upper half of the Globe the illuminated hemisphere, the sun being in the zenith. Mr. Harris also gives an account of a cheap machine for shewing how the annual motion of the earth in its orbit causes the change of the sun's declination, without the great expence of an orrery. Philos. Transf. number 456, or Abr. vol. 8, p. 352.

The late Mr. George Adams made also some useful improvements in the construction of the Globes. Besides what is usual, his Globes have a thin brass semicircle moveable about the poles, with a small thin sliding circle upon it. On the terrestrial Globe, the former of these is a moveable meridian, and the latter is the visible horizon of any particular place to which it is set. But on the celestial Globe, the semi-circle is a moveable circle of declination, and its small annexed circle an artificial sun or planet. Each Globe has a brass wire circle, placed at the limits of the twilight. The terrestrial Globe has many additional circles, as well as the rhumb lines, for resolving all the necessary geographical and nautical problems: and on the celestial Globe are drawn, on each side of the ecliptic, 8 parallel circles, at the distance of one degree from each other, including the zodiac; which are crossed at right-angles by segments of great circles at every 5th degree of the ecliptic, for the more readily noting the place of the moon or of any planet upon the Globe. On the strong brass circle of the terrestrial Globe, and about 23½ degrees on each side of the north pole, the days of each month are laid down according to the sun's declination: and this brass circle is so contrived, that the Globe may be placed with the north and south poles in the plane of the horizon, and with the south pole elevated above it. The equator, on the surface of either Globe, serves the purpose of the horary circle, by means of a semi-circular wire placed in the plane of the equator, carrying two indices, one of which is occasionally to be used to point out the time. For a farther account of these Globes, with the method of using them, see Mr. Adams's treatise on their construction and use.

There are also what are called Patent Globes, made by Mr. Neale; by means of which he resolves several astronomical problems, which do not admit of solution by the common Globes.

Mr.

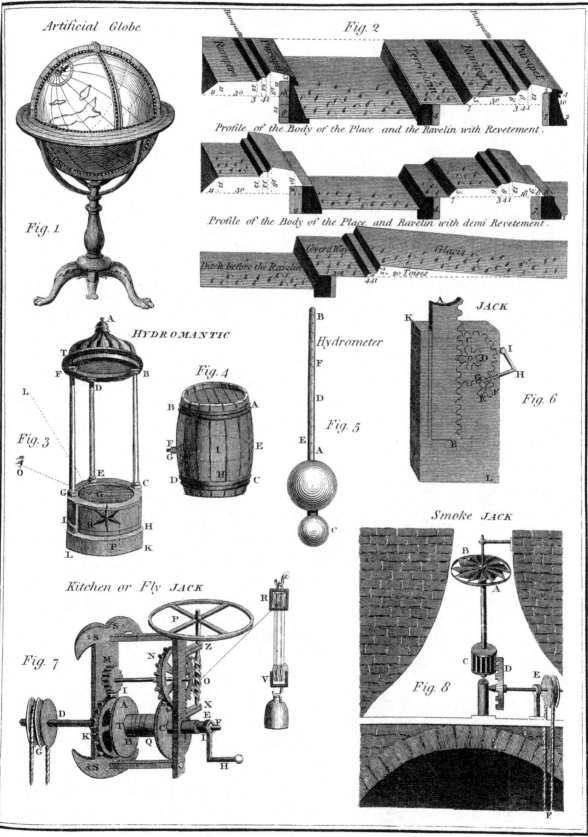

Plate XII.

Artificial Globe

Fig. 2

Banquette

Banquette

Rempart

Terreplin

Rempart

Parapet

Profile of the Body of the Place and the Ravelin with Revetement.

Fig. 1

Profile of the Body of the Place and Ravelin with demi Revetement.

Coverd Way

Glacis

Ditch before the Ravelin

20 Toises

HYDROMANTIC

B

JACK

K

A

Fig. 4

F

D

Hydrometer

I

H

Fig. 3

B

A

Fig. 5

Fig. 6

O

E

E

C

E

C

D

H

C

L

Smoke JACK

B

A

Kitchen or Fly JACK

R

C

D

E

P

Fig. 8

Fig. 7

N

Z

V

M

O

I

D

X

F

K

C

H

G

S

F

Mr. Ferguson likewife made feveral improvements of the Globes, particularly one for conftructing dials, and another called a planetary Globe. See Philof. Tranf. vol. 44, p. 535, and Ferguson's Aftron. p. 291, and 292.

Laftly, in the Philof. Tranf. for 1789, vol. 79, p. 1, Mr. Smeaton has propofed fome improvements of the celeftial Globe, efpecially with refpect to the quadrant of altitude, for the refolution of problems relating to the azimuth and altitude. The difficulty, he obferves, that has occurred in fixing a femicircle, fo as to have a centre in the zenith and nadir points of the Globe, at the fame time that the meridian is left at liberty to raife the pole to its defired elevation, I fuppofe, has induced the Globe-makers to be contented with the ftrip of thin flexible brafs, called the quadrant of altitude; and it is well known how imperfectly it performs its office. The improvement I have attempted, is in the application of a quadrant of altitude of a more folid conftruction; which being affixed to a brafs focket of fome length, and this ground, and made to turn upon an upright fteel fpindle, fixed in the zenith, fteadily directs the quadrant, or rather arc, of altitude to its true azimuth, without being at liberty to deviate from a vertical circle to the right hand or left: by which means the azimuth and altitude are given with the fame exactnefs as the meafure of any other of the great circles. For a more particular defcription of this improvement, illuftrated with figures, fee the place above quoted.

The ufe of the Terreftrial GLOBE.

PROB. I. *To find the latitude and longitude of any place.* —Bring the place to the graduated fide of the firft meridian: then the degree of the meridian it cuts is the latitude fought; and the degree of the equator then under the meridian is the longitude.

II. *To find a place, having a given latitude and longitude.*—Find the degree of longitude on the equator, and bring it to the brafs meridian; then find the degree of latitude on the meridian, either north or fouth of the equator, as the given latitude is north or fouth; then the point of the Globe juft under that degree of latitude is the place required.

III. *To find all the places on the Globe that have the fame latitude, and the fame longitude, or hour, with a given place, as fuppofe London.*—Bring the given place London to the meridian, and obferve what places are juft under the edge of it, from north to fouth; and all thofe places have the fame longitude and hour with it. Then turn the Globe quite round; and all thofe places which pafs juft under the given degree of latitude on the meridian, have the fame latitude with the given place.

IV. *To find the Antœci, Periœci and Antipodes, of any given place, fuppofe London.*—Bring the given place London to the meridian, then count 51½ the fame degree of latitude fouthward, or towards the other pole, and the point thus arrived at will be the Antœci, or where the hour of the day or night is always the fame at both places at the fame time, and where the feafons and lengths of days and nights are alfo equal, but at half a year diftance from each other, becaufe their fea-

fons are oppofite or contrary. London being ftill under the meridian, fet the hour index to 12 at noon, or pointing towards London; then turn the Globe juft half round, or till the index point to the oppofite hour, or 12 at night; and the place that comes under the fame degree of the meridian where London was, fhews where the Periœci dwell, or thofe people that have the fame feafons and at the fame time as London, as alfo the fame length of days and nights &c at that time, but only their time or hour is juft oppofite, or 12 hours diftant, being day with one when night with the other, &c. Laftly, as the Globe ftands, count down by the meridian the fame degree of latitude fouth, and that will give the place of the Antipodes of London, being diametrically under or oppofite to it; and fo having all its times, both hours and feafons oppofite, being day with the one when night with the other, and fummer with the one when winter with the other.

V. *To find the Diftance of two places on the Globe.*— If the two places be either both on the equator, or both on the fame meridian, the number of degrees in the diftance between them, reduced into miles, at the rate of 70 Englifh miles to the degree, (or more exact 69½), will give the diftance nearly. But in any other fituations of the two places, lay the quadrant of altitude over them, and the degrees counted upon it, from the one place to the other, and turned into miles as above, will give the diftance in this cafe.

VI. *To find the Difference in the Time of the day at any two given places, and thence the Difference of Longitude.*—Bring one of the places to the meridian, and fet the hour index to 12 at noon; then turn the Globe till the other place comes to the meridian, and the index will point out the difference of time; then by allowing 15° to every hour, or 1° to 4 minutes of time, the difference of longitude will be known.—Or the difference of longitude may be found without the time, thus:

Firft bring the one place to the meridian, and note the degree of longitude on the equator cut by it; then do the fame by the other place; which gives the longitudes of the two places; then fubtracting the one number of degrees from the other, gives the difference of longitude fought.

VII. *The time being known at any given place, as fuppofe London, to find what hour it is in any other part of the world.*—Bring the given place, London, to the meridian, and fet the index to the given hour; then turn the Globe till the other place come to the meridian, and look at what hour the index points, which will be the time fought.

VIII. *To find the Sun's place in the ecliptic, and alfo on the Globe, at any given time.*—Look into the calendar on the wooden horizon for the month and day of the month propofed, and immediately oppofite ftands the fign and degree which the fun is in on that day. Then in the ecliptic drawn upon the Globe, look for the fame fign and degree, and that will be the place of the fun required.

IX. *To find at what place on the earth the fun is vertical, at a given moment of time at another place, as fuppofe London.*—Find the fun's place on the Globe by the laft problem, and turn the Globe about till

that

that place come to the meridian, and note the degree of the meridian just over it. Then turn the Globe till the given place, London, come to the meridian, and set the index to the given moment of time. Lastly, turn the Globe till the index points to 12 at noon; then the place of the earth, or Globe, which stands under the before noted degree, has the sun at that moment in the zenith.

X. *To find how long the sun shines without setting, in any given place in the frigid zones.*——Subtract the degrees of latitude of the given place from 90, which gives the complement of the latitude, and count the number of this complement upon the meridian from the equator towards the pole, marking that point of the meridian; then turn the Globe round, and carefully observe what two degrees of the ecliptic pass exactly under the point marked on the meridian. Then look for the same degrees of the ecliptic on the wooden horizon, and just opposite to them stand the months and days of the months corresponding, and between which two days the sun never sets in that latitude.

If the beginning and end of the longest night be required, or the period of time in which the sun never rises at that place; count the same complement of latitude towards the south or farthest pole, and then the rest of the work will be the same in all respects as above.

Note, that this solution is independent of the horizontal refraction of the sun, which raises him rather more than half a degree higher, by that means making the day so much longer, and the night the shorter; therefore in this case, set the mark on the meridian half a degree higher up towards the north pole, than what the complement of latitude gives; then proceed with it as before, and the more exact time and length of the longest day and night will be found.

XI. *A place being given in the torrid zone, to find on what two days of the year the sun is vertical at that place.*——Turn the Globe about till the given place come to the meridian, and note the degree of the meridian it comes under. Next turn the Globe round again, and note the two points of the ecliptic passing under that degree of the meridian. Lastly, by the wooden horizon, find on what days the sun is in those two points of the ecliptic; and on these days he will be vertical to the given place.

XII. *To find those places in the torrid zone to which the sun is vertical on a given day.*——Having found the sun's place in the ecliptic, as in the 8th problem, turn the Globe to bring the same point of the ecliptic on the Globe to the meridian; then again turn the Globe round, and note all the places which pass under that point of the meridian; which will be the places sought.

After the same manner may be found what people are Ascii for any given day. And also to what place of the earth, the moon, or any other planet, is vertical on a given day; finding the place of the planet on the globe by means of its right ascension and declination, like finding a place from its longitude and latitude given.

XIII. *To rectify the Globe for the latitude of any place.*——By sliding the brass meridian in its groove, elevate

the pole as far above the horizon as is equal to the latitude of the place; so for London, raise the north pole $51\frac{1}{2}$ degrees above the wooden horizon: then turn the Globe on its axis till the place, as London, come to the meridian, and there set the index to 12 at noon. Then is the place exactly on the vertex, or top point of the Globe, at 90° every way round from the wooden horizon, which represents the horizon of the place. And if the frame of the Globe be turned about till the compass needle point to $22\frac{1}{2}$ degrees, or two points west of the north point (because the variation of the magnetic needle is nearly $22\frac{1}{2}$ degrees west), so shall the Globe then stand in the exact position of the earth, with its axis pointing to the north pole.

XIV. *To find the length of the day or night, or the sun's rising or setting, in any latitude; having the day of the month given.*——Rectify the Globe for the latitude of the place; then bring the sun's place on the globe to the meridian, and set the index to 12 at noon, or the upper 12, and then the Globe is in the proper position for noon day. Next turn the Globe about towards the east till the sun's place come just to the wooden horizon, and the index will then point to the hour of sunrise; also turn the Globe as far to the west side, or till the sun's place come just to the horizon on the west side, and then the index will point to the hour of sunset. These being now known, double the hour of setting will be the length of the day, and double the rising will be the length of the night.——And thus also may the length of the longest day, or the shortest day, be found for any latitude.

XV. *To find the beginning and end of Twilight on any day of the year, for any latitude.*——It is twilight all the time from sunset till the sun is 18° below the horizon, and the same in the morning from the time the sun is 18° below the horizon till the moment of his rise. Therefore, rectify the Globe for the latitude of the place, and for noon by setting the index to 12, and screw on the quadrant of altitude. Then take the point of the ecliptic opposite the sun's place, and turn the Globe on its axis westward, as also the quadrant of altitude, till that point cut this quadrant in the 18th degree below the horizon, then the index will shew the time of dawning in the morning; next turn the Globe and quadrant of altitude towards the east, till the said point opposite the sun's place meet this quadrant in the same 18th degree, and then the index will shew the time when twilight ends in the evening.

XVI. *At any given day, and hour of the day, to find all those places on the Globe where the sun then rises, or sets, as also where it is noon day, where it is day light, and where it is in darkness.*——Find what place the sun is vertical to, at that time; and elevate the Globe according to the latitude of that place, and bring the place also to the meridian; in which state it will also be in the zenith of the Globe. Then is all the upper hemisphere, above the wooden horizon, enlightened, or in day light; while all the lower one, below the horizon, is in darkness, or night: those places by the edge of the meridian, in the upper hemisphere, have noon day, or 12 o'clock; and those by the meridian below, have it midnight: lastly, all those places by the eastern side of the horizon, have the sun just setting,

ting,

ting, and thofe by the weſtern horizon have him juſt rifing.

Hence, as in the middle of a lunar eclipfe the moon is in that degree of the ecliptic oppoſite to the ſun's place; by the preſent problem it may be ſhewn what places of the earth then ſee the middle of the eclipfe, and what the beginning or ending; by uſing the moon's place inſtead of the ſun's place in the problem.

XVII. *To find the bearing of one place from another, and their angle of poſition.*——Bring the one place to the zenith, by rectifying the Globe for its latitude, and turning the Globe till that place come to the meridian; then ſcrew the quadrant of altitude upon the meridian at the zenith, and make it revolve till it come to the other place on the Globe; then look on the wooden horizon for the point of the compaſs, or number of degrees from the ſouth, where the quadrant of altitude cuts it, and that will be the bearing of the latter place from the former, or the angle of poſition ſought.

The Uſe of the Celeſtial GLOBE.

The Celeſtial Globe differs from the terreſtrial only in this; inſtead of the ſeveral parts of the earth, the images of the ſtars and conſtellations are defigned. The meridian circle drawn through the two poles and through the point Cancer, repreſents the ſolſtitial colure; but that through the point Aries, repreſents the equinoctial colure.

PROB. XVIII. *To exhibit the true repreſentation of the face of the heavens at any given time and place.*——Rectify for the lat. of the place, by prob. 13, ſetting the Globe with its pole pointing to the pole of the world, by means of a compaſs. Find the ſun's place in the ecliptic, and turn the Globe to bring it to the meridian, and there ſet the index to 12 at noon. Again revolve the Globe on its axis, till the index point to the given hour of the day or night: ſo ſhall the Globe in this poſition exactly repreſent the face of the heavens as it appears at that time, every conſtellation and ſtar, in the heavens, anſwering in poſition to thoſe on the Globe; ſo that, by examining the Globe, it will immediately appear which ſtars are above or below the horizon, which on the eaſt or weſtern parts of the heavens, which lately rifen, and which going to ſet, &c. And thus the poſitions of the ſeveral planets, or comets, may alſo be exhibited; having marked the places of the Globe where they are, by means of their declination and right aſcenſion.

XIX. *To find the Declination and Right-aſcenſion of any ſtar upon the Globe.*——Turn the Globe till the ſtar come to the meridian: then the number of degrees on the meridian, between the equator and the ſtar, is its declination; and the degree of the equator cut by the meridian, is the right-aſcenſion of the ſtar.——In like manner are found the declination and right-aſcenſion of the ſun, or any other point.

XX. *To find the Latitude and Longitude of any ſtar drawn upon the Globe.*——Bring the ſolſtitial colure to the meridian, and there fix the quadrant of altitude over the pole of the ecliptic in the ſame hemiſphere with the ſtar, and bring its graduated edge to the ſtar: then the degree on the quadrant cut by the ſtar is its latitude, counted from the ecliptic; and the degree of the ecliptic cut by the quadrant its longitude.

XXI. *To find the place of a ſtar, planet, comet, &c. on the Globe; its declination and right-aſcenſion being known.*——Find the given point of right-aſcenſion on the equinoctial, and bring it to the meridian; then count the degrees of declination upon the meridian from the equinoctial, and there make a mark on the Globe, which will be the place of the planet, &c, ſought.

XXII. *To find the place of a ſtar, planet, comet, or other object on the Globe; its latitude and longitude being given.*——Bring the pole of the ecliptic to the meridian, and there fix the quadrant of altitude, which turn round till its edge cut the given longitude on the ecliptic; then count the given latitude, from the ecliptic, upon the quadrant of altitude, and there make a mark on the Globe, which will be the place of the planet, &c, ſought.——The place on the Globe, of any ſuch planet, &c, being found by this or the foregoing problem, its riſing, or ſetting, or any other circumſtance concerning it, may then be found, the ſame as the ſun, by the proper problems.

XXIII. *To find the riſing, ſetting, and culminating of a ſtar, planet, ſun, &c; with its continuance above the Lorizon, for any place and day; as alſo its oblique aſcenſion and deſcenſion, with its eaſtern and weſtern amplitude and azimuth.*——Adjuſt the Globe to the ſtate of the heavens at 12 o'clock that day. Bring the ſtar, &c, to the eaſtern ſide of the horizon: which will give its eaſtern amplitude and azimuth, and the time of riſing, as for the ſun. Again, turn the Globe to bring the ſame ſtar to the weſtern ſide of the horizon: ſo will the weſtern amplitude and azimuth, with the time of ſetting, be found. Then, the time of riſing, ſubtracted from that of ſetting, leaves the continuance of the ſtar above the horizon: this continuance above the horizon taken from 24 hours, leaves the time it is below the horizon. Laſtly, bring the ſtar to the meridian, and the hour to which the index then points is the time of its culmination, or ſouthing.

XXIV. *To find the altitude of the ſun, or ſtar, &c, for any given hour of the day or night.*——Adjuſt the Globe to the poſition of the heavens, and turn it till the index point at the given hour. Then fix on the quadrant of altitude, at 90 degrees from the horizon, and turn it to the place of the ſun or ſtar: ſo ſhall the degrees of the quadrant, intercepted between the horizon and the ſun or ſtar, be the altitude ſought.

XXV. *Given the altitude of the ſun by day, or of a ſtar by night, to find the hour of the day or night.*——Rectify the Globe as in the foregoing problem; and turn the Globe and quadrant, till ſuch time as the ſtar or degree of the ecliptic the ſun is in, cut the quadrant in the given degree of altitude; then will the index point at the hour required.

XXVI. *Given the azimuth of the ſun or a ſtar, to find the time of the day or night.*——Rectify the Globe, and bring the quadrant to the given azimuth in the horizon; then turn the Globe till the ſun or ſtar come to the quadrant, and the index will then ſhew the time of the day or night.

GLOBULAR, relating to, or partaking the property or ſhape of, the Globe. As Globular chart, Globular projection, or Globular ſailing, &c.

GLOBULAR *Chart*, is a repreſentation of the ſurface,

or

or part of the furface, of the terraqueous Globe upon a plane ; in which the parallels of latitude are circles nearly concentric ; and the meridians are curves bending towards the poles ; the rhumb-lines being curves alſo.

The merits of this chart conſiſt in theſe particulars, viz, that the diſtances between places on the ſame rhumb are all meaſured by the ſame ſcale of equal parts ; and the diſtance of any two places in the arch of a great circle, is nearly repreſented in this chart by a ſtraight line.

Land maps alſo made according to this projection would have great advantages over thoſe made in any other way. But for ſea charts for the uſe of navigation, Mercator's are preferable, as both the meridians and parallels, as alſo the rhumbs, are all ſtraight lines.

This projection is not new, though not much noticed till of late. It is mentioned by Ptolomy, in his Geography ; and alſo by Blundeville, in his Exerciſes.

For Globular projection of maps or charts, ſee MAP.

GLOBULAR *Sailing*, is the method of reſolving the caſes of ſailing upon principles deduced from the ſpherical figure of the earth. Such as Mercator's ſailing, or Great-circle ſailing ; which ſee.

GLOSSOCOMON, in Mechanics, is a name given by Heron to a machine compoſed of divers dented wheels with pinions, ſerving to raiſe huge weights.

GNOMON, in Aſtronomy, is an inſtrument or apparatus for meaſuring the altitudes, declinations, &c, of the ſun and ſtars. The Gnomon is uſually a pillar, or column, or pyramid, erected upon level ground, or a pavement. For making the more conſiderable obſervations, both the ancients and moderns have made great uſe of it, eſpecially the former ; and many have preferred it to the ſmaller quadrants, both as more accurate, eaſier made, and more eaſily applied.

The moſt ancient obſervation of this kind extant, is that made by Pytheas, in the time of Alexander the Great, at Marſeilles, where he found the height of the Gnomon was in proportion to the meridian ſhadow at the ſummer ſolſtice, as 213¼ to 600 ; juſt the ſame as Gaſſendi found it to be, by an obſervation made at the ſame place, almoſt 2000 years after, viz, in the year 1636. Ricciol. Almag. vol. 1, lib. 3, cap. 14.

Ulugh Beigh, king of Parthia, &c, uſed a Gnomon in the year 1437, which was 180 Roman feet high. That erected by Ignatius Dante, in the church of St. Petronius, at Bologna, in the year 1576, was 67 feet high. M. Caſſini erected another of 20 feet high, in the ſame church, in the year 1655.

The Egyptian obeliſks were alſo uſed as Gnomons ; and it is thought by ſome modern travellers that this was the very uſe they were deſigned and built for ; it has alſo been found that their four ſides ſtand exactly facing the four cardinal points of the compaſs. It may be added, that the Spaniards in their conqueſt of Peru, found pillars of curious and coſtly workmanſhip, ſet up in ſeveral places, by the meridian ſhadows of which their amatas or philoſophers had, by long experience and repeated obſervations, learned to determine the times of the equinoxes ; which ſeaſons of the year were celebrated with great feſtivity and rich offerings, in honour of the ſun. Garcillaſſo de la Vega, Hiſt. Peru. lib. 2, cap. 22.

USE *of the* GNOMON, *in taking the meridian altitude of the Sun, and thence finding the Latitude of the place.*——
A meridian line being drawn through the centre of the Gnomon, note the point where the ſhadow of the Gnomon terminates when projected along the meridian line, and meaſure the diſtance of that extreme point from the centre of the Gnomon, which will be the length of its ſhadow. Then having the height of the Gnomon, and the length of the ſhadow, the ſun's altitude is thence eaſily found.

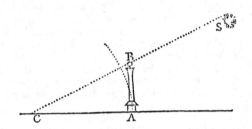

Suppoſe, ex. gr. AB the Gnomon, and AC the length of the ſhadow. Here in the right-angled triangle ABC, are given the baſe AC, and the perpendicular AB, to find the angle C, or the ſun's altitude, which will be found by this analogy, as CA : AB :: radius : the tang. of ∠ C, that is, as the length of the ſhadow is to the height of the Gnomon, ſo is radius to the tangent of the ſun's altitude above the horizon.

The following example will ſerve to illuſtrate this propoſition : Pliny ſays, Nat. Hiſt. lib. 2, cap. 72, that at Rome, at the time of the equinoxes, the ſhadow is to the Gnomon as 8 to 9 ; therefore as

$$8 : 9 :: 1 \text{ or radius} : \frac{9}{8} = 1\cdot125 \text{ a tangent, to which}$$

anſwers the angle 48° 22', which is the height of the equator at Rome, and its complement 41° 38' is therefore the height of the pole, or the latitude of the place.

Riccioli remarks the following defects in the obſervations of the ſun's height, made with the Gnomon by the ancients, and ſome of the moderns : viz, that they neglected the ſun's parallax, which makes his apparent altitude leſs, by the quantity of the parallax, than it would be, if the Gnomon were placed at the centre of the earth : 2d, they neglected alſo the refraction, by which the apparent height of the ſun is a little increaſed : and 3dly, they made the calculations from the length of the ſhadow, as if it were terminated by a ray coming from the centre of the ſun's diſc, whereas the ſhadow is really terminated by a ray coming from the upper edge of the ſun's diſc ; ſo that, inſtead of the height of the ſun's centre, their calculations gave the height of the upper edge of his diſc. And therefore, to the altitude of the ſun found by the Gnomon, the ſun's parallax muſt be added, and from the ſum muſt be ſubtracted the ſun's ſemidiameter, and refraction, which is different at different altitudes ; which being done, the correct height of the equator at Rome will be 48° 4' 13", the complement of which, or 41° 55' 46", is the latitude. Ricciol. Geogr. Refor. lib. 7, cap. 4.

The

The preceding problem may be refolved more accurately by means of a ray of light let in through a fmall hole, than by a fhadow, thus: Make a circular perforation in a brafs plate, to tranfmit enough of the fun's rays to exhibit his image on the floor, or a ftage; fix the plate parallel to the horizon in a high place, proper for obfervation, the height of which above the floor let be accurately meafured with a plummet. Let the floor, or ftage, be perfectly plane and horizontal, and coloured over with fome white fubftance, to fhew the fun more diftinctly. Upon this horizontal plane draw a meridian line paffing through the foot or centre of the Gnomon, i. e. the point upon which the plummet falls from the centre of the hole; and upon this line note the extreme points I and K of the fun's image or diameter, and from each end fubtract the image of half the diameter of the aperture, viz KH and LI: then will HL be the image of the fun's diameter, which, when bifected in B, gives the point on which the rays fall from the centre of the fun.

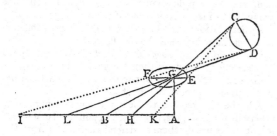

Now having given the line AB, and the altitude of the Gnomon AG, befide the right angle A, the angle B, or the apparent altitude of the fun's centre, is eafily found, thus: as AB : AG :: radius : tang. angle B.

GNOMON, in Dialling, is the ftyle, pin, or cock of a dial, the fhadow of which points out the hours. This is always fuppofed to reprefent the axis of the world, to which it is therefore parallel, or coincident, the two ends of it pointing ftraight to the north and fouth poles of the world.

GNOMON, in Geometry, is a figure formed of the two complements, in a parallelogram, together with either of the parallelograms about the diameter. Thus the parallelogram AC being divided into four parallelograms by the two lines DG, EF parallel to the fides, forming the two complements AB and BC, with the two DE, FG about the diameter HI: then the two Gnomons are AB + BC + DE, and AB + BC + FG.

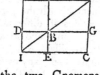

GNOMONIC *Projection of the Sphere*, is the reprefentation of the circles of an hemifphere upon a plane touching it in the vertex, by the eye in the centre, or by lines or rays iffuing from the centre of the hemifphere, to all the points in the furface.

In this projection of the fphere, all the great circles are projected into right lines, on the plane, of an indefinite length; and all leffer circles that are parallel to

5

the plane, into circles; but if oblique to the plane, then are they projected either into ellipfes or hyperbolas, according to their different obliquity. It has its name from Gnomonics, or Dialling, becaufe the lines on the face of every dial are from a projection of this fort: for if the fphere be projected on any plane, and upon that fide of it on which the fun is to fhine; alfo the projected pole be made the centre of the dial, and the axis of the globe the ftyle or Gnomon, and the radius of projection its height; you will have a dial drawn with all its furniture. See Emerfon's Projection of the Sphere.

GNOMONICS, the fame as DIALLING; or the art of drawing fun and moon dials, on any given plane; being fo called, becaufe it fhews how to find the hour of the day or night by the fhadow of a Gnomon or ftyle.

GOLDEN *Number*, is the particular year of the Metonic or Lunar Cycle. See *Lunar* CYCLE.

To find the Golden Number: Add 1 to the given year, and divide the fum by 19, and what remains will be the Golden Number; unlefs 0 remain, for then 19 is the Golden Number.	1791 1 19)1792(94 171 —— 82 76 ——
Thus, the Golden Number for the year 1791, is 6; as by the operation in the margin.	Golden No. 6

GOLDEN *Rule*, a rule fo called on account of its excellent ufe, in arithmetic, and efpecially in ordinary calculations, by which numbers are found in certain proportions, viz, having three numbers given, to find a 4th number, that fhall have the fame proportion to the 3d as the 2d hath to the 1ft. On this account, it is otherwife called *The Rule of Three*, and *The Rule of Proportion*. See RULE *of Three*.

Having ftated, or fet down in a line, the three terms, in the order in which they are proportional, multiply the 2d and 3d together, and divide the product by the 1ft, fo fhall the quotient be the anfwer, or the 4th term fought.

Thus, if 3 yards of cloth coft a guinea or 21 fhillings, what will 20 yards coft. Here the two prices or values muft bear the fame proportion to each other as the two quantities, or number of yards of cloth, i. e. 3 muft bear the fame proportion to 20, as 21s, the value of the former, muft bear to the value of the latter: and therefore the ftating and operation of the numbers will be thus,

Then multiplying the 2d and 3d together, and dividing the product by the 1ft, it gives 140s. or 7l. for the anfwer, being the coft of 20 yards.

3 : 20 :: 21 : 140s. 20 ——— 3)420 Quot. 140s. or 7£.

GONIOMETRICAL *Lines*, are lines ufed for meafuring or determining the quantity of angles: fuch as fines, tangents, fecants, verfed fines, &c.

Mr. Jones, in the Philof. Tranf. number 483, fect. 26, gave a paper, containing a commodious difpofition of equations for exhibiting the relations of Goniometrical Lines; from which a multitude of curious theorems.

rems may be derived. See also Robertson's Elem. of Navigation, vol. 1, p 181, Edit. 4.

GONIOMETRY, a method of measuring angles, so called by M. de Lagny, who gave several papers, on this method, in the Memoirs of the Royal Acad. an. 1724, 1725, 1729. M. de Lagny's method of Goniometry consists in measuring the angles with a pair of compasses, and that without any scale whatever, except an undivided semicircle. Thus, having any angle drawn upon paper, to be measured; produce one of the sides of the angle backwards behind the angular point; then with a pair of fine compasses describe a pretty large semicircle from the angular point as a centre, cutting the sides of the proposed angle, which will intercept a part of the semicircle. Take then this intercepted part very exactly between the points of the compasses, and turn them successively over upon the arc of the semicircle, to find how often it is contained in it, after which there is commonly some remainder: then take this remainder in the compasses, and in like manner find how often it is contained in the last of the integral parts of the 1st arc, with again some remainder: find in like manner how often this last remainder is contained in the former; and so on continually, till the remainder become too small to be taken and applied as a measure. By this means he obtains a series of quotients, or fractional parts, one of another, which being properly reduced into one fraction, give the ratio of the first arc to the semicircle, or of the proposed angle to two right angles, or 180 degrees, and consequently that angle itself in degrees and minutes.

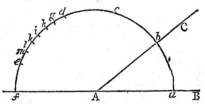

Thus, suppose the angle BAC be proposed to be measured. Produce BA out towards f; and from the centre A describe the semicircle $abcf$, in which ab is the measure of the proposed angle. Take ab in the compasses, and apply it 4 times on the semicircle, as at b, c, d, and e; then take the remainder fe, and apply it back upon ed, which is but once, viz at g; again take the remainder gd, and apply it 5 times on ge, as at h, i, k, l, and m; lastly, take the remainder me, and it is contained just 2 times in ml. Hence the series of quotients is 4, 1, 5, 2; consequently the 4th or last arc em is $\frac{1}{2}$ the third ml or gd, and therefore the 3d arc gd is $\frac{1}{5\frac{1}{2}}$ or $\frac{2}{11}$ of the 2d arc ef; and therefore again this 2d arc ef is $\frac{1}{1\frac{2}{11}}$ or $\frac{11}{13}$ of the 1st arc ab; and consequently this 1st arc ab is $\frac{1}{4\frac{11}{13}}$ or $\frac{13}{63}$ of the whole semicircle af. But $\frac{13}{63}$ of 180° are 37¼ degrees, or 37° 8' 34''⁴⁄₇, which therefore is the measure of the angle sought. When the operation is nicely performed, this angle may be within 2 or 3 minutes of the truth; though M. de Lagny pretends to measure much nearer than that.

It may be added, that the series of fractions forms what is called a continued fraction. Thus, in the example above, the continued fraction, and its reduction, will be as follow:

$$\frac{1}{4} + \frac{1}{1} + \frac{1}{5\frac{1}{2}} = \frac{1}{4} + \frac{1}{1\frac{2}{11}} = \frac{1}{4\frac{11}{13}} = \frac{13}{63};$$

the quotients being the successive denominators, and 1 always for each numerator.

GORGE, or *Neck*, in Architecture, is the narrowest part of the Tuscan or Doric capitals, lying above the shaft of the pillar, between the astragal and the annulets.

It is also a kind of concave moulding, serving for compartments &c, larger than a scotia, but not so deep.

GORGE, in Fortification, is the entrance into a bastion, or a ravelin, or other out-work.

The GORGE *of a Bastion*, is what remains of the sides of the polygon of a place, after cutting off the curtains; in which case it makes an angle in the centre of the bastion, viz, the angle made by two adjacent curtains produced to meet within the bastion.

In flat bastions, the Gorge is a right line on the curtain, reaching between the two flanks.

GORGE *of a Half-moon*, or *of a Ravelin*, is the space between the two ends of their faces next the place.

GORGE of the other out-works, is the interval between their sides next the great ditch.

All the Gorges are to be made without parapets: otherwise the besiegers, having taken possession of a work, might make use of them to defend themselves from the shot of the place. So that they are only fortified with pallisadoes, to prevent a surprize.

The *Demi-*GORGE, or *Half* the GORGE, is that part of the polygon between the flank and the centre of the bastion.

GOTHIC *Architecture*, is that which deviates from the manner, character, proportions, &c, of the antique; having its ornaments wild and chimerical, and its profiles incorrect. This manner of building came originally from the North, whence it was brought, in the 5th century, by the Goths into Germany, and has since been introduced into other countries. The first or most ancient style of Gothic building was very solid, heavy, massive and simple, with semicircular arches, &c: but the more modern style of the Gothic is exceedingly rich, light, and delicate; having an abundance of little whimsical ornaments, with sharp-pointed arches formed by the intersections of different circular segments; also lofty and light spires and steeples, large ramified windows, clustered pillars, &c. Of this kind are our English cathedrals, and many other old buildings.

GRADUATION, is used for the act of Graduating, or dividing any thing into degrees.

For an account of the various methods of Graduating mathematical and astronomical instruments, by straight and circular diagonals, and by concentric arcs, &c; see *Plain* SCALE, *Nonius*, and *Vernier*. And for an account of Mr. Bird's improved method of dividing astronomical instruments, see MURAL *Arch*.

Mr. Ramsden, an ingenious mathematical instrument-maker of London, has lately published, by encouragement of the commissioners of longitude, an explanation

and

and defcription of an engine contrived by him for dividing mathematical inftruments, accompanied with proper drawings; in confideration of which, the faid commiffioners have granted to him the fum of 615*l.* See his book, 4to, 1777.

On the fubject of dividing a foot into many thoufand parts, for mathematical purpofes, fee Philof. Tranf. vol. 2, p. 457, 459, 541, or Abr. vol. 1, pa. 218, 220, &c. And for an account of various other methods and Graduations, fee a paper of Mr. Smeaton's in the Philof. Tranf. vol. 76, for the year 1786, p. 1; being " Obfervations on the Graduation of aftronomical inftruments; with an explanation of the method invented by the late Mr. Henry Hindley, of York, clockmaker, to divide circles into any given number of parts."

GRAHAM (George), clock and watch maker, the moft ingenious and accurate artift in his time, was born at Gratwick, a village in the north of Cumberland, in 1675. In 1688 he came up to London, and was put apprentice to a perfon in that profeffion; but after being fome time with his mafter, he was received, purely on account of his merit, into the family of the celebrated Mr. Tompion, who treated him with a kind of parental affection as long as he lived. That Mr. Graham was, without competition, the moft eminent of his profeffion, is but a fmall part of his character: he was the beft general mechanic of his time, and had a complete knowledge of practical aftronomy; fo that he not only gave to various movements for meafuring time a degree of perfection which had never before been attained, but invented feveral aftronomical inftruments, by which confiderable advances have been made in that fcience: he made great improvements in thofe which had before been in ufe; and, by a wonderful manual dexterity, conftructed them with greater precifion and accuracy than any other perfon in the world.

A great mural arch in the obfervatory at Greenwich was made for Dr. Halley, under Mr. Graham's immediate infpection, and divided by his own hand: and from this incomparable original, the beft foreign inftruments of the kind are copies made by Englifh artifts. The fector by which Dr. Bradley firft difcovered two new motions in the fixed ftars, was of his invention and fabric. He comprifed the whole planetary fyftem within the compafs of a fmall cabinet; from which, as a model, all the modern orreries have been conftructed. And when the French Academicians were fent to the north, to make obfervations for afcertaining the figure of the earth, Mr. Graham was thought the fitteft perfon in Europe to fupply them with inftruments; by which means they finifhed their operations in one year; while thofe who went to the fouth, not being fo well furnifhed, were very much embarraffed and retarded in their operations.

Mr. Graham was many years a member of the Royal Society, to which he communicated feveral ingenious and important difcoveries, viz, from the 31ft to the 42d volume of the Philof. Tranfactions, chiefly on aftronomical and philofophical fubjects; particularly a kind of horary alteration of the magnetic needle; a quickfilver pendulum, and many curious particulars relating to the true length of the fimple pendulum, upon

which he continued to make experiments till almoft the year of his death, which happened in 1751, at 76 years of age.

His temper was not lefs communicative than his genius was penetrating; and his principal view was the advancement of fcience, and the benefit of mankind. As he was perfectly fincere, he was above fufpicion; and as he was above envy, he was candid.

GRANADO, in Artillery, is a little fhell or hollow globe of iron, or other matter, which, being filled with powder, is fired by means of a fmall fufee, and thrown either by the hand, or a piece of ordnance. As foon as it is kindled, the cafe flies in pieces, to the great danger of all that ftand near it. Granadoes ferve to fet fire to clofe and narrow paffages, and are often thrown with the hand among the foldiers, to diforder their ranks; more efpecially in thofe pofts where they ftand thickeft, as in trenches, redoubts, lodgments, &c.

GRAVESANDE (William James), a very celebrated Dutch mathematician and philofopher, was born at Bois-le-duc, Sept. 27, 1688. He ftudied the civil law at Leyden, but mathematical learning was his favourite amufement. When he had taken his doctor's degree, in 1707, he went and fettled at the Hague, where he practifed at the bar, and cultivated an acquaintance with learned men; with a Society of whom he publifhed a periodical Review, entitled *Le Journal Litteraire*, which was continued without interruption from the year 1713 to the year 1722. The parts of it written or extracted by Gravefande were chiefly thofe relating to geometry and phyfics. But he enriched it alfo with feveral original pieces entirely of his own compofition; viz, Remarks on the Conftruction of Pneumatical Engines: A Moral Effay on Lying: And a celebrated Effay on the Collifion of Bodies; which, as it oppofed the Newtonian philofophy, was attacked by Dr. Clarke, and many other learned men.

In 1715, when the States fent to congratulate George the 1ft, on his acceffion to the throne, Dr. Gravefande was appointed fecretary to the embaffy. During his ftay in England he was admitted a member of the Royal Society, and became intimately acquainted with Sir Ifaac Newton. On his return to Holland, he was chofen profeffor of mathematics and aftronomy at Leyden; where he had the honour of firft teaching the Newtonian philofophy, which was then in its infancy. He died in 1742, at 54 years of age.

Gravefande was a man amiable in his private character, and refpectable in his public one; for few men of letters have rendered more eminent fervices to their country. The minifters of the republic confulted him on all occafions when his talents were requifite to affift them, which his fkill in calculation often enabled him to do in money matters. He was of great fervice as a decypherer, in detecting the fecret correfpondence of their enemies. And in his own profeffion none ever applied the powers of nature with more fuccefs, or to more ufeful purpofes.

Of his publications, the principal are,

1. An Introduction to the Newtonian Philofophy; or, a Treatife on the Elements of Phyfics, confirmed by Experiments. This performance, being only a more

perfect copy of his public lectures, was first printed in 1720; and hath since gone through many editions, with considerable improvements : the 6th edit. is in English, in 2 large vols. 4to, by Dr. Defaguliers, in 1747, under the title of Mathematical Elements of Natural Philosophy, confirmed by Experiments.

2. A treatise on the Elements of Algebra, for the use of Young Students ; to which is added a Specimen of a Commentary on Newton's Universal Arithmetic ; as also, A New Rule for determining the Form of an Assumed Infinite Series.

3. An Essay on Perspective. This was written at 19 years of age.

4. A New Theory of the Collision of Bodies.

5. A Course of Logic and Metaphysics.
With several smaller pieces.

His whole mathematical and philosophical works, except the first article above, were collected and pub· lished at Amsterdam, in 2 vols. 4to, to which is prefix· ed a critical account of his life and writings, by Professor Allamand.

GRAVITATION, the exercise of gravity, or the pressure a body exerts on another body beneath it by the power of gravity.

This is sometimes distinguished from gravity. Thus M. Maupertuis, in his *Figure de la Terre*, takes gravity for that force by which a body would fall to the earth supposed at rest ; and Gravitation for the same, but diminished by the centrifugal force. It is only Gravitation, or gravity thus blended with the centrifugal force, that we can usually measure by our experiments. Methods however have been found to distinguish what remains of the primitive gravity, and what has been destroyed by the centrifugal force.

It is one of the laws of nature, discovered by Newton, and now received by all philosophers, that every particle of matter in nature gravitates towards every other particle ; which law is the main principle in the Newtonian philosophy. But what is called Gravitation with respect to the gravitating body, is usually called attraction with respect to the body gravitated to. The planets, both primary and secondary, as also the comets, do all gravitate towards the sun, and towards each other ; as well as the sun towards them ; and that in proportion to the quantity of matter in each of them.

The Peripatetics &c hold, that bodies only gravitate or weigh when out of their natural places, and that Gravitation ceases when they are restored to the same, the purpose of nature being then fulfilled : and they maintain that the final cause of this faculty is only to bring elementary bodies to their proper place, where they may rest. But the moderns shew that bodies exercise gravity even when at rest, and in their proper places. This is particularly shewn of fluids ; and it is one of the laws of hydrostatics, demonstrated by Boyle and others, that fluids gravitate in proprio loco, the upper parts pressing on the lower, &c.

For the laws of Gravitation of bodies in fluids specifically lighter or heavier than themselves, see SPECIFIC GRAVITY. Also for the centre or line or plane of Gravitation, see CENTRE, LINE, or PLANE.

GRAVITY, in Physics, the natural tendency or inclination of bodies towards the centre. And in this sense Gravity agrees with Centripetal force.

Gravity however is, by some, defined more generally as the natural tendency of one body towards another ; and again by others still more generally as the mutual tendency of each body, and each particle of a body, towards all others : in which sense the word answers to what is more usually called Attraction. Indeed the terms Gravity, weight, centripetal force, and attraction, denote in effect all the same thing, only in different views and relations ; all which however it is very common to confound, and use promiscuously. But, in propriety, when a body is considered as tending towards the earth, the force with which it so tends is called Gravity, Force of Gravity, or Gravitating Force ; when the body is considered as immediately tending to the centre of the earth, it is called Centripetal Force ; but when we consider the earth, or mass to which the body tends, it is called Attraction, or Attractive Force ; and when it is considered in respect of an obstacle or another body in the way of its tendency, upon which it acts, it is called Weight.

Philosophers think differently on the subject of Gravity. Some consider it as an inactive property or innate power in bodies, by which they endeavour to join their centre. Others hold Gravity in this sense to be an occult quality, and to be exploded as such out of all sound philosophy. Newton, though he often calls it a vis, power, or property in bodies, yet explains himself, that he means nothing more by the word but the effect or phenomenon : he does not consider the principle, the cause by which bodies tend downwards, but the tendency itself, which is no occult quality, but a sensible phenomenon, be its causes what they may ; whether a property essential to body, as some make it, or superadded to it, as others ; or even an impulse of some body from without, as others.

It is a law of nature long observed, that all bodies near the earth have a Gravity or weight, or a tendency towards its centre, or at least perpendicular to its surface ; which law the moderns, and especially Sir I. Newton, from certain observations have found to be much more extensive, and holding universally with respect to all known bodies and matter in nature. It is therefore at present acknowledged as a principle or law of nature, that all bodies, and all the particles of all bodies, mutually gravitate towards each other: from which single principle it is that Newton has happily deduced all the great phenomena of nature.

Hence Gravity may be distinguished into Particular and General.

Particular GRAVITY, is that which respects the earth, or by which bodies descend, or tend towards the centre of the earth ; the phenomena or properties of which are as follow :

1. All circumterrestrial bodies do hereby tend towards a point, which is either accurately or very nearly the centre of magnitude of the terraqueous globe. Not that it is meant that there is really any virtue or charm in the point called the centre, by which it attracts bodies ; but because this is the result of the gravitation of bodies towards all the parts of which the earth consists.

2. This point or centre is fixed within the earth, or at least has been so far as any authentic history reaches. For a consequence of its shifting, though ever so little, would

5

would be the overflowing of the low lands on that side of the globe towards which it should approach. Dr. Halley suggests, that it would well account for the universal deluge, to have the centre of gravitation removed for a time towards the middle of the then inhabited world; for the change of its place but the 2000th part of the radius of the earth, or about 2 miles, would be sufficient to lay the tops of the highest hills under water.

3. In all places equidistant from the centre of the earth, the force of Gravity is nearly equal. Indeed all parts of the earth's surface are not at equal distances from the centre, because the equatorial parts are higher than the polar parts by about 17 miles; as has been proved by the necessity of making the pendulum shorter in those places, before it will swing seconds. In the new Petersburg Transactions, vol. 6 and 7, M. Krafft gives a formula for the proportion of Gravity in different latitudes on the earth's surface, which is this:

$$y = (1 + 0\cdot0052848 \; \text{sine} \; {}^2\lambda) \, g \; ;$$

where g denotes the Gravity at the equator, and y the Gravity under any other latitude λ. On this subject, see also the articles DEGREE, and EARTH.

4. Gravity equally affects all bodies, without regard either to their bulk, figure, or matter: so that, abstracting from the resistance of the medium, the most compact and loose, the greatest and smallest bodies would all descend through an equal space in the same time; as appears from the quick descent of very light bodies in an exhausted receiver. The space which bodies do actually fall, in vacuo, is $16\frac{1}{12}$ feet in the first second of time, in the latitude of London; and for other times, either greater or less than that, the spaces descended from rest are directly proportional to the squares of the times, while the falling body is not far from the earth's surface.

5. This power is the greatest at the earth's surface, from whence it decreases both upwards and downwards, but not both ways in the same proportion; for upwards the force of Gravity is less, or decreases, as the square of the distance from the centre increases, so that at a double distance from the centre, above the surface, the force would be only 1-4th of what it is at the surface; but below the surface, the power decreases in such sort that its intensity is in the direct ratio of the distance from the centre; so that at the distance of half a semidiameter from the centre, the force would be but half what it is at the surface; at ⅓ of a semidiameter the force would be ⅓, and so on.

6. As all bodies gravitate towards the earth, so does the earth equally gravitate towards all bodies; as well as all bodies towards particular parts of the earth, as hills, &c, which has been proved by the attraction a hill has upon a plumb line, insensibly drawing it aside.— Hence the gravitating force of entire bodies consists of those of all their parts: for by adding or taking away any part of the matter of a body, its Gravity is increased or decreased in the proportion of the quantity of such portion to the whole mass. Hence also the gravitating powers of bodies, at the same distance from the centre, are proportional to the quantities of matter in the bodies.

General or Universal GRAVITY, is that by which all the planets tend to one another, and indeed by which all the bodies and particles of matter in the universe tend towards one another.

The existence of the same principle of Gravitation in the superior regions of the heavens, as on the earth, is one of the great discoveries of Newton, who made the proof of it as easy as that on the earth. At first it would seem this was only conjecture with him: he observed that all bodies near the earth, and in its atmosphere, had the property of tending directly towards it; he soon conjectured that it probably extended much higher than any distance to which we could reach, or make experiments; and so on, from one distance to another, till he at length saw no reason why it might not extend as far as to the moon, by means of which she might be retained in her orbit as a stone in a sling is retained by the hand; and if so, he next inferred why might not a similar principle exist in the other great bodies in the universe, the sun and all the other planets, both primary and secondary, which might all be retained in their orbits, and perform their revolutions, by means of the same universal principle of gravitation.

These conjectures he soon realized and verified by mathematical proofs. Kepler had found out, by contemplating the motions of the planets about the sun, that the area described by a line connecting the sun and planet, as this revolved in its orbit, was always proportional to the time of its description, or that it described equal areas in equal times, in whatever part of its orbit the planet might be, moving always so much the quicker as its distance from the sun was less. And it is also found that the satellites, or secondary planets, respect the same law in revolving about their primaries. But it was soon proved by Newton, that all bodies moving in any curve line described on a plane, and which, by radii drawn to any certain point, describe areas about the point proportional to the times, are impelled or acted on by some power tending towards that point. Consequently the power by which all these planets revolve, and are retained in their orbits, is directed to the centre about which they move, viz, the primary planets to the sun, and the satellites to their several primaries.

Again, Newton demonstrated, that if several bodies revolve with an equable motion in several circles about the same centre, and that if the squares of their periodical times be in the same proportion as the cubes of their distances from the common centre, then the centripetal forces of the revolving bodies, by which they tend to their central body, will be in the reciprocal or inverse ratio of the squares of the distances. Or if bodies revolve in orbits approaching to circles, and the apses of those orbits be at rest, then also the centripetal forces of the revolving bodies will be reciprocally proportional to the squares of the distances. But it had been agreed on by the astronomers, and particularly Kepler, that both these cases obtain in all the planets. And therefore he inferred that the centripetal forces of all the planets are reciprocally proportional to the squares of the distances from the centres of their orbits.

Upon the whole it appears, that the planets are retained in their orbits by some power which is continually acting upon them: that this power is directed towards the centre of their orbits: that the intensity or efficacy of this power increases upon an approach

towards

towards the centre, and diminishes on receding from the same, and that in the reciprocal duplicate ratio of the distances: and that, by comparing this centripetal force of the planets with the force of gravity on the earth, they are found to be perfectly alike, as may easily be shewn in various instances. For example, in the case of the moon, the nearest of all the planets. The rectilinear spaces described in any given time by a falling body, urged by any powers, reckoning from the beginning of its descent, are proportional to those powers. Consequently the centripetal force of the moon revolving in her orbit, will be to the force of Gravity on the surface of the earth, as the space which the moon would describe in falling during any small time, by her centripetal force towards the earth, if she had no circular motion at all, to the space a body near the earth would describe in falling by its Gravity towards the same.

Now by an easy calculation of those two spaces, it appears that the former force is to the latter, as the square of the semi-diameter of the earth is to the square of that of the moon's orbit. The moon's centripetal force therefore is equal to the force of Gravity; and consequently these forces are not different, but they are one and the same: for if they were different, bodies acted on by the two powers conjointly would fall towards the earth with a velocity double to that arising from the sole power of Gravity.

It is evident therefore that the moon's centripetal force, by which she is retained in her orbit, and prevented from running off in tangents, is the very power of Gravity of the earth extended thither. See Newton's Princip. lib. 1, prop. 45, cor. 2, and lib. 3, prop. 3; where the numeral calculation may be seen at full length.

The moon therefore gravitates towards the earth, and reciprocally the earth towards the moon. And this is also farther confirmed by the phenomena of the tides.

The like reasoning may also be applied to the other planets. For as the revolutions of the primary planets round the sun, and those of the satellites of Jupiter and Saturn round their primaries, are phenomena of the same kind with the revolution of the moon about the earth; and as the centripetal powers of the primary are directed towards the centre of the sun, and those of the satellites towards the centres of their primaries; and lastly as all these powers are reciprocally as the squares of the distances from the centres, it may safely be concluded that the power and cause are the same in all.

Therefore, as the moon gravitates towards the earth, and the earth towards the moon; so do all the secondaries to their primaries, and these to their secondaries; and so also do the primaries to the sun, and the sun to the primaries. Newton's Princip. lib. 3, prop. 4, 5, 6; Greg. Astron. lib. 1, sect. 7, prop. 46 and 47.

The laws of Universal Gravity are the same as those of bodies gravitating towards the earth, before laid down.

Cause of GRAVITY. Various theories have been advanced by the philosophers of different ages to account for this grand principle of Gravitation. The ancients, who were only acquainted with particular Gravity, or the tendency of sublunar bodies towards the earth, aimed no farther than a system that might answer the

I

more obvious phenomena of it. However, some hints are found concerning the Gravitation of celestial bodies in the account given of the doctrine of Thales and his successors; and it would seem that Pythagoras was still better acquainted with it, to which it is supposed he had a view in what he taught concerning the Harmony of the Spheres.

Aristotle and the Peripatetics content themselves with referring Gravity or weight to a native inclination in heavy bodies to be in their proper place or sphere, the centre of the earth. And Copernicus ascribes it to an innate principle in all parts of matter, by which, when separated from their wholes, they endeavour to return to them again the nearest way. In answer to Aristotle and his followers, who considered the centre of the earth as the centre of the universe, he observed that it was reasonable to think there was nothing peculiar to the earth in this principle of Gravity; that the parts of the sun, moon, and stars tended likewise to each other, and that their spherical figure was preserved in their various motions by this power. Copern. Revol. lib. 1, cap. 9. But neither of these systems assigns any physical cause of this great effect: they only amount to this, that bodies descend because they are inclined to descend.

Kepler, in his preface to the commentaries concerning the planet Mars, speaks of Gravity as of a power that was mutual between bodies, and says that the earth and moon tend towards each other, and would meet in a point so many times nearer to the earth than to the moon, as the earth is greater than the moon, if their motions did not hinder it. He adds, that the tides arise from the Gravity of the waters towards the moon. To him we also owe the important discovery of the analogy between the distances of the several planets from the sun, and the periods in which they complete their revolutions, viz, that the squares of their periodic times are always in the same proportion as the cubes of their mean distances from the sun. However, Kepler, Gassendi, Gilbert, and others, ascribe Gravity to a certain magnetic attraction of the earth; conceiving the earth to be one great magnet continually emitting effluvia, which take hold of all bodies, and draw them towards the earth. But this is inconsistent with the several phenomena.

Des Cartes and his followers, Rohault &c, attribute Gravity to an external impulse or trusion of some subtle matter. By the rotation of the earth, say they, all the parts and appendages of it necessarily endeavour to recede from the centre of rotation; but whence they cannot all actually recede, as there is no vacuum or space to receive them. But this hypothesis, founded on the supposition of a plenum, is overthrown by what has been since proved of the existence of a vacuum.

Dr. Hook inclines to an opinion much like that of Des Cartes. Gravity he thinks deducible from the action of a most subtle medium, which easily pervades and penetrates the most solid bodies; and which, by some motion it has, detrudes all earthly bodies from it, towards the centre of the earth. Vossius too, and many others, give partly into the Cartesian notion, and suppose Gravity to arise from the diurnal rotation of the earth round its axis.

Dr. Halley, despairing of any satisfactory theory, chooses

chooses to have immediate recourse to the agency of the Deity. So Dr. Clarke, from a view of several properties of Gravity, concludes that it is no adventitious effect of any motion, or subtle matter, but an original and general law impressed by God on all matter, and preserved in it by some efficient power penetrating the very solid and intimate substance of it; being found always proportional, not to the surfaces of bodies or corpuscles, but to their solid quantity and contents. It should therefore be no more inquired why bodies gravitate, than how they came to be first put in motion. Annot. in Rohault. Phys. part 1, cap. 11.

Gravesande, in his Introduct. ad Philos. Newton. contends that the cause of Gravity is utterly unknown; and that we are to consider it no otherwise than as a law of nature originally and immediately impressed by the Creator, without any dependence on any second law or cause at all. Of this he thinks the three following considerations sufficient proof. 1. That Gravity requires the presence of the gravitating or attracting body: so the satellites of Jupiter, for ex. gravitate towards Jupiter, wherever he may be. 2. That the distance being supposed the same, the velocity with which bodies are moved by the force of Gravity, depends on the quantity of matter in the attracting body: and the velocity is not changed, whatever the mass of the gravitating body may be. 3. That if Gravity do depend on any known law of motion, it must be some impulse from an extraneous body; so that as Gravity is continual, a continual stroke must also be required. Now if there be any such matter continually striking on bodies, it must be fluid, and subtle enough to penetrate the substance of all bodies: but how shall a body subtle enough to penetrate the substance of the hardest bodies, and so rare as not sensibly to hinder the motion of bodies, be able to impel vast masses towards each other with such force? how does this force increase the ratio of the mass of the body, towards which the other body is moved? whence is it that all bodies move with the same velocity, the distance and body gravitated to being the same? can a fluid which only acts on the surface either of the bodies themselves, or their internal particles, communicate such a quantity of motion to bodies, which in all bodies shall exactly follow the proportion of the quantity of matter in them?

Mr. Cotes goes yet farther. Giving a view of Newton's philosophy, he asserts that Gravity is to be ranked among the primary qualities of all bodies; and deemed equally essential to matter as extension, mobility, or impenetrability. Præfat. ad Newt. Princip. But Newton himself disclaims this notion; and to shew that he does not take Gravity to be essential to bodies, he declares his opinion of the cause; choosing to propose it by way of query, not being yet sufficiently satisfied about its experiments. Thus, after having shewn that there is a medium in nature vastly more subtle than air, by whose vibrations sound is propagated, by which light communicates heat to bodies, and by the different densities of which the refraction and reflection of light are performed; he proceeds to inquire: " Is not this medium much rarer within the dense bodies of the sun, stars, planets, and comets, than

in the empty celestial spaces between them? and in passing from them to greater distances, doth it not grow denser and denser perpetually, and thereby cause the Gravity of those great bodies towards one another, and of their parts towards the bodies; every body endeavouring to recede from the denser parts of the medium towards the rarer?

For if this medium be supposed rarer within the sun's body than at its surface, and rarer there than at the hundredth part of an inch from his body, and rarer there than at the fiftieth part of an inch from his body, and rarer there than at the orb of Saturn; I see no reason why the increase of density should stop any where, and not rather be continued through all distances from the Sun to Saturn, and beyond.

And though this increase of density may at great distances be exceeding slow; yet if the elastic force of this medium be exceeding great, it may suffice to impel bodies from the denser parts of the medium towards the rarer with all that power which we call Gravity.

And that the elastic force of this medium is exceeding great, may be gathered from the swiftness of its vibrations. Sounds move about 1140 English feet in a second of time, and in seven or eight minutes of time, they move about one hundred English miles: light moves from the Sun to us in about seven or eight minutes of time, which distance is about 70000000 English miles, supposing the horizontal parallax of the Sun to be about twelve seconds; and the vibrations, or pulses of this medium, that they may cause the alternate fits of easy transmission, and easy reflection, must be swifter than light, and by consequence above 700000 times swifter than sounds; and therefore the elastic force of this medium, in proportion to its density, must be above 700000×700000 (that is, above 490000000000) times greater than the elastic force of the air is in proportion to its density: for the velocities of the pulses of elastic mediums are in a subduplicate ratio of the elasticities and the rarities of the mediums taken together.

As Magnetism is stronger in small loadstones than in great ones, in proportion to their bulk; and Gravity is stronger on the surface of small planets, than those of great ones, in proportion to their bulk; and small bodies are agitated much more by electric attraction than great ones: so the smallness of the rays of light may contribute very much to the power of the agent by which they are refracted; and if any one should suppose, that æther (like our air) may contain particles which endeavour to recede from one another (for I do not know what this æther is), and that its particles are exceedingly smaller than those of air, or even than those of light; the exceeding smallness of such particles may contribute to the greatness of the force, by which they recede from one another, and thereby make that medium exceedingly more rare and elastic than air, and of consequence, exceedingly less able to resist the motions of projectiles, and exceedingly more able to press upon gross bodies by endeavouring to expand itself." Optics, p. 325 &c.

GRAVITY, in Mechanics, denotes the conatus or tendency of bodies towards the centre of the earth.—That part of mechanics which considers the equilibrium

or

or motion of bodies arifing from Gravity or weight, is particularly called ftatics.

Gravity in this view is diftinguifhed into Abfolute and Relative.

Abfolute GRAVITY is that with which a body defcends freely and perpendicularly through an unrefifting medium. The laws of which fee under DESCENT OF BODIES, ACCELERATION, MOTION, &c.

Relative GRAVITY is that with which a body defcends on an inclined plane, or through a refifting medium, or as oppofed by fome other refiftance. The laws of which fee under the articles INCLINED PLANE, DESCENT, FLUID, RESISTANCE, &c.

GRAVITY, in Hydroftatics. The laws of bodies gravitating in Fluids make the bufinefs of Hydroftatics.

Gravity is here divided into Abfolute and Specific.

Abfolute or *True* GRAVITY, is the whole force with which the body tends downwards.

Specific GRAVITY, is the relative, comparative, or apparent Gravity in any body, in refpect of that of an equal bulk or magnitude of another body; denoting that Gravity or weight which is peculiar to each fpecies or kind of body, and by which it is diftinguifhed from all other kinds.

In this fenfe a body is faid to be Specifically Heavier than another, when under the fame bulk it contains a greater weight than that other; and reciprocally the latter is faid to be Specifically Lighter than the former. Thus, if there be two equal fpheres, each one foot in diameter; the one of lead, and the other of wood: fince the leaden one is found heavier than the wooden one, it is faid to be Specifically, or in Specie, Heavier; and the wooden one Specifically Lighter.

This kind of Gravity is by fome called Relative; in oppofition to Abfolute Gravity, which increafes in proportion to the quantity or mafs of the body.

Laws of the SPECIFIC GRAVITY of bodies.

I. If two bodies be equal in bulk, their fpecific gravities are to each other as their weights, or as their denfities.

II. If two bodies be of the fame fpecific gravity or denfity, their abfolute weights will be as their magnitudes or bulks.

III. In bodies of the fame weight, the fpecific gravities are reciprocally as their bulks.

IV. The fpecific gravities of all bodies are in a ratio compounded of the direct ratio of their weights, and the reciprocal ratio of their magnitudes. And hence again the fpecific gravities are as the denfities.

V. The abfolute gravities or weights of bodies are in the compound ratio of their fpecific gravities and magnitudes or bulks.

VI. The magnitudes of bodies are directly as their weights, and reciprocally as their fpecific gravities.

VII. A body fpecifically heavier than a fluid, lofes as much of its weight when immerfed in it, as is equal to the weight of a quantity of the fluid of the fame bulk or magnitude.

Hence, fince the Specific Gravities are as the abfo-

lute gravities under the fame bulk; the Specific Gravity of the fluid, will be to that of the body immerged, as the part of the weight loft by the folid, is to the whole weight.

And hence the Specific Gravities of fluids are as the weights loft by the fame folid immerged in them.

VIII. *To find the Specific Gravity of a Fluid, or of a Solid.*—On one arm of a balance fufpend a globe of lead by a fine thread, and to the other faften an equal weight, which may juft balance it in the open air. Immerge the globe into the fluid, and obferve what weight balances it then, and confequently what weight is loft, which is proportional to the Specific Gravity as above. And thus the proportion of the Specific Gravity of one fluid to another is determined by immerfing the globe fucceffively in all the fluids, and obferving the weights loft in each, which will be the proportions of the Specific Gravities of the fluids fought.

This fame operation determines alfo the Specific Gravity of the folid immerged, whether it be a globe or of any other fhape or bulk, fuppofing that of the fluid known. For the Specific Gravity of the fluid is to that of the folid, as the weight loft is to the whole weight.

Hence alfo may be found the Specific Gravity of a body that is lighter than the fluid, as follows:

IX. *To find the Specific Gravity of a Solid that is lighter than the fluid, as water, in which it is put.*—Annex to the lighter body another that is much heavier than the fluid, fo as the compound mafs may fink in the fluid. Weigh the heavier body and the compound mafs feparately, both in water and out of it; then find how much each lofes in water, by fubtracting its weight in water from its weight in air; and fubtract the lefs of thefe remainders from the greater.

Then, As this laft remainder,
  Is to the weight of the light body in air,
  So is the Specific Gravity of the fluid,
  To the Specific Gravity of that body.

X. The Specific Gravities of bodies of equal weight, are reciprocally proportional to the quantities of weight loft in the fame fluid. And hence is found the ratio of the Specific Gravities of folids, by weighing in the fame fluids, maffes of them that weigh equally in air, and noting the weights loft by each.

The Specific Gravities of many kinds of bodies, both folid and fluid, have been determined by various authors. Marinus Ghetaldus particularly tried the Specific Gravities of various bodies, efpecially metals; which were taken from thence by Oughtred. In the Philof. Tranf. are feveral ample tables of them, by various authors, particularly thofe of Mr. Davis, vol. 45, p. 416, or Abr. vol. 10, p. 206. Some tables of them were alfo publifhed by P. Merfenne, Mufchenbroeck, Ward, Cotes, Emerfon, Martin, &c.

It will be fufficient here to give thofe of fome of the moft ufual bodies, that have been determined with the greater certainty. The numbers in this table exprefs the number of Avoirdupois ounces in a cubic foot of each body, that of common water being juft 1000 ounces, or 62¼ lb.

TABLE

Table of Specific Gravities.

I. Solids.

Platina, pure	23000
Fine gold	19640
Standard gold	18888
Lead	11325
Fine Silver	11091
Standard Silver	10535
Copper	9000
Copper halfpence	8915
Gun metal	8784
Fine brafs	8350
Caft brafs	8000
Steel	7850
Iron	7645
Pewter	7471
Caft Iron	7425
Tin	7320
Lapis calaminaris	5000
Loadftone	4930
Mean of the whole Earth	4500
Crude Antimony	4000
Diamond	3517
Granite	3500
White lead	3160
Ifland cryftal	2720
Marble	2705
Pebble ftone	2700
Jafper	2666
Rock cryftal	2650
Pearl	2630
Green glafs	2600
Flint	2570
Onyx ftone	2510
Common ftone	2500
Cryftal	2210
Clay	2160
Oyfter fhells	2092
Brick	2000
Common earth	1984
Nitre	1900
Vitriol	1880
Alabafter	1874
Horn	1840
Ivory	1825
Sulphur	1810
Chalk	1793
Solid gunpowder	1745
Alum	1714
Dry bone	1660
Human calculus	1542
Sand	1520
Lignum vitæ	1327
Coal	1250
Jet	1238
Ebony	1177
Pitch	1150
Rofin	1100
Mahogany	1063
Amber	1040
Brazil wood	1031
Boxwood	1030
Common water	1000

Bees wax	955
Butter	940
Oak	925
Gunpowder, fhaken	922
Logwood	913
Ice	908
Afh	800
Maple	755
Beech	700
Elm	600
Fir	550
Saffafras wood	482
Charcoal	
Cork	240
New fallen fnow	86

II. Fluids.

Quickfilver	13600
Oil of Vitriol	1700
Oil of Tartar	1550
Honey	1450
Spirit of Nitre	1315
Aqua Foitis	1300
Treacle	1290
Aqua Regia	1234
Human blood	1054
Urine	1032
Cow's milk	1031
Sea Water	1030
Ale	1028
Vinegar	1026
Tar	1015
Water	1000
Diftilled Water	993
Red Wine	990
Proof Spirits	931
Olive Oil	913
Pure Spirits of Wine	866
Oil of Turpentine	800
Æther	726
Common Air	1·232
or very nearly	$1\frac{7}{30}$.

Thefe numbers being the weight of a cubic foot, or 1728 cubic inches, of each of the bodies, in Avoirdupois ounces, by proportion the quantity in any other weight, or the weight of any other quantity, may be readily known.

For ex. Required the content of an irregular block of common ftone which weighs 1 cwt, or 112lb, or 1792 ounces. Here, as 2500 : 1792 :: 1728 : 1228$\frac{4}{5}$ cubic inches the content.

Ex. 2. To find the weight of a block of granite, whofe length is 63 feet, and breadth and thicknefs each 12 feet; being the dimenfions of one of the ftones, of granite, in the walls of Balbeck. Here, 63 × 12 × 12 = 9072 feet is the content of the ftone; therefore as 1 : 9072 :: 3500 oz : 31752000 oz or 885 tons 18 cwt. 3 qrs. the weight of the ftone.

XI. A body defcends in a fluid fpecifically lighter, or afcends in a fluid fpecifically heavier, with a force equal to the difference between its weight and that of an equal bulk of the fluid.

XII. A body finks in a fluid fpecifically heavier, fo

far

far as that the weight of the body is equal to the weight of a quantity of the fluid of the same bulk as the part immersed. Hence, as the Specific Gravity of the fluid is to that of the body, so is the whole magnitude of the body, to the magnitude of the part immersed.

XIII. The Specific Gravities of equal solids are as their parts immerged in the same fluid.

The several theorems here delivered, are both demonstrable from the principles of mechanics, and are also equally conformable to experiment, which answers exactly to the calculation; as is abundantly evident from the courses of philosophical experiments, so frequently exhibited; where the laws of specific gravitation are well illustrated.

GREAT BEAR, one of the constellations in the northern hemisphere. See Ursa Major.

GREAT Circles, of the Globe or Sphere, are those whose planes pass through the centre, dividing it into two equal parts or hemispheres, and therefore having the same centre and diameter with the sphere itself. The principal of these are, the equator, the ecliptic, the horizon, the meridians, and the two colures.

Great-Circle Sailing, is the art of conducting a ship along the arc of a great circle. And it is also that part of the theory of navigation which treats of sailing in the arc of a great circle. See Navigation.

GREAVES (John), an eminent astronomer, antiquary and linguist, was born in 1602, being the eldest son of John Greaves rector of Colemore, near Alresford in Hampshire, and master of a grammar school, where his son of course was well grounded in the primary rules of literature. He then went to Baliol college in Oxford, in 1617; but afterward, on account of his skill in philosophy and polite literature, he was the first of five that were elected into Merton college. Having read over all the ancient Greek and Latin writers, he applied to the study of natural philosophy and mathematics; and having contracted an intimacy with Mr. Briggs, Savilian professor of geometry at Oxford, and Dr. Bainbridge, Savilian professor of astronomy there, he was animated by their examples to prosecute that study with the greatest industry; and not content with reading the writings of Purbach, Regiomontanus, Copernicus, Tycho Brahe, Kepler, and other celebrated astronomers of that and the preceding age, he made the ancient Greek, Arabian, and Persian authors familiar to him, having before gained an accurate skill in the oriental languages. These accomplishments procured him the professorship of geometry in Gresham-college London, in 1630; and at the same time he held his fellowship of Merton-college.

In a journey to the Continent, in 1635, he visited the celebrated Golius, professor of Arabic at Leyden, and Claud Hardy at Paris, to converse about the Persian language. Hence he passed through Italy, and accurately surveyed the venerable remains of antiquity at Rome, visiting and corresponding everywhere with the most learned men of every nation. After visiting Padua, Florence, and Leghorn, he hence embarked for Constantinople, where he arrived in 1638. From thence he passed over to Rhodes, and Alexandria in Egypt, where he staid four or five months, and made a great number of curious observations. He next went to

Grand Cairo, measured the pyramids; and while there he adjusted the measure of the foot, observed by all nations. From hence he returned again through Italy, and arrived in England in the year 1640, after storing his mind with a variety of curious knowledge, and collecting many valuable oriental manuscripts and ancient curiosities; and while at Rome he made a particular inquiry into the true state of the ancient weights and measures.

On the death of Dr. John Bainbridge, in 1643, he was chosen Savilian professor of astronomy at Oxford, and principal reader of Linacre's lecture in Merton college; an appointment for which he was eminently qualified, from his critical acquaintance with the works of the ancient and modern astronomers. In 1645 he proposed a method of reforming the calendar, by omitting the intercalary day for 40 years to come: the paper relating to which, was published by Dr. Thomas Smith, in the Philos. Transf. for 1699. In 1646, he published his Pyramidographia, or a Description of the Pyramids of Egypt; and, in 1647, his Discourse on the Roman Foot and Denarius; from which, as from two principles, the measures and weights used by the Ancients may be deduced. He also published several other curious works concerning antiquities, &c.

Soon after publishing the last mentioned book, he was ejected, by the parliament visitors, from the professorship of astronomy and fellowship of Merton-college; and the soldiers committed many outrages, breaking open his chests, and destroying many of his manuscripts; which greatly affected him. On this occasion he retired to London, where he afterwards married, and prosecuted his studies with great vigour, as appears from several of his philosophical and theological writings. This however proved but a transient happiness to him; for he died at London, the 8th of October 1652, before he was quite 50 years of age; and left his astronomical instruments to the Savilian library in Oxford, where they are deposited.

GREEK Orders, in Architecture, are the Doric, Ionic and Corinthian; in contradistinction to the two Latin orders, viz the Tuscan and Composite.

GREEN, One of the original colours of the rays of light, or of the prismatic colours exhibited by the refraction of the rays of light.

Green is the pleasantest of all the colours to the sight. And hence it has been inferred as a proof of the wisdom and goodness of the Deity, that almost all vegetables, cloathing the surface of the earth, are green; which they are when growing in the open air; though those in subterraneous places, or places inaccessible to fresh air, are white or yellow. See Chromatics, and Colours.

GREGORIAN Calendar, so called from Pope Gregory the 13th, is the new or reformed Calendar, shewing the new and full moons, with the time of Easter, and the other moveable feasts depending upon it, by means of epacts disposed through the several months of the Gregorian year.

Gregorian *Epoch*, is the epoch or time, from which the Gregorian calendar, or computation took place. This began in the year 1582; so that the year 1800 is the 218th of this epoch.

GREGORIAN *Telescope*, a particular sort of telescope, invented by Mr. James Gregory. See TELESCOPE.

GREGORIAN *Year*, the new account, or new style, introduced upon the reformation of the calendar, by Pope Gregory the 13th, in the year 1582, and from whom it took its name. This was introduced to reform the old, or Julian year, established by Julius Cæsar, which consisted of 365 days 6 hours, or 365 days and a quarter, that is three years of 365 days each, and the fourth year of 366 days. But as the mean tropical year consists only of 365ds 5hrs 48m 57sec. the former lost 11min. 3sec. every year, which in the time of Pope Gregory had amounted to 10 days, and who, by adding these 10 days, brought the account of time to its proper day again, and at the same time appointed that every century after, a day more should be added, thereby making the years of the complete centuries, viz 1600, 1700, 1800, &c, to be common years of 365 days each, instead of leap years of 366 days, which makes the mean Gregorian year equal to 365ds 5hrs 45m. 36sec.

This computation was not introduced into the account of time in England, till the year 1752, when the Julian account had lost 11 days, and therefore the 3d of September was in that year, by act of parliament, accounted the 14th, thereby restoring the 11 days which had thus been omitted. See YEAR.

GREGORY (JAMES), professor of mathematics, first in the university of St. Andrews, and afterwards in that of Edinburgh, was one of the most eminent mathematicians of the 17th century. He was a son of the Rev. Mr. John Gregory minister of Drumoak in the county of Aberdeen, and was born at Aberdeen in November 1638. His mother was a daughter of Mr. David Anderson of Finzaugh, or Finshaugh, a gentleman who possessed a singular turn for mathematical and mechanical knowledge. This mathematical genius was hereditary in the family of the Andersons, and from them it seems to have been transmitted to their descendants of the names of Gregory, Reid, &c. Alexander Anderson, cousin-german of the said David, was professor of mathematics at Paris in the beginning of the 17th century, and published there several valuable and ingenious works; as may be seen in the memoirs of his life and writings, under the article ANDERSON. The mother of James Gregory inherited the genius of her family; and observing in her son, while yet a child, a strong propensity to mathematics, she instructed him herself in the elements of that science. His education in the languages he received at the grammar school of Aberdeen, and went through the usual course of academical studies in the Marischal college; but he was chiefly delighted with philosophical researches, into which a new door had lately been opened by the key of the mathematics. Galileo, Kepler, Des Cartes, &c, were the great masters of this new method: their works therefore became the principal study of young Gregory, who soon began to make improvements upon their discoveries in Optics. The first of these improvements was the invention of the reflecting telescope; the construction of which instrument he published in his *Optica Promota*, in 1663, at 24 years of age. This discovery soon attracted the attention of the mathematicians, both of our own and of foreign countries, who

immediately perceived its great importance to the sciences of optics and astronomy. But the manner of placing the two specula upon the same axis appearing to Newton to be attended with the disadvantage of losing the central rays of the larger speculum, he proposed an improvement on the instrument, by giving an oblique position to the smaller speculum, and placing the eye-glass in the side of the tube. It is observable however, that the Newtonian construction of that instrument was long abandoned for the original or Gregorian, which is now always used when the instrument is of a moderate size; though Herschel has preferred the Newtonian form for the construction of those immense telescopes, which he has of late so successfully employed in observing the heavens.

About the year 1664 or 1665, coming to London, he became acquainted with Mr. John Collins, who recommended him to the best optic glass-grinders there, to have his telescope executed. But as this could not be done for want of skill in the artists to grind a plate of metal for the object speculum into a true parabolic concave, which the design required, he was much discouraged with the disappointment; and after a few imperfect trials made with an ill-polished spherical one, which did not succeed to his wish, he dropped the pursuit, and resolved to make the tour of Italy, then the mart of mathematical learning, that he might prosecute his favourite study with greater advantage. And the university of Padua being at that time in high reputation for mathematical studies, Mr. Gregory fixed his residence there for some years. Here it was that he published, in 1667, *Vera Circuli et Hyperbolæ Quadratura*; in which he propounded another discovery of his own, the invention of an infinitely converging series for the areas of the circle and hyperbola. He sent home a copy of this work to his friend Mr. Collins, who communicated it to the Royal Society, where it met with the commendations of lord Brounker and Dr. Wallis. He reprinted it at Venice the year following, to which he added a new work, entitled *Geometriæ Pars Universalis, inserviens Quantitatum Curvarum Transmutationi et Mensuræ*; in which he is allowed to have shewn, for the first time, a method for the transmutation of curves. These works engaged the notice, and procured the author the correspondence of the greatest mathematicians of the age, Newton, Huygens, Wallis, and others. An account of this piece was also read by Mr. Collins before the Royal Society, of which Mr. Gregory, being returned from his travels, was chosen a member the same year, and communicated to them an account of a controversy in Italy about the motion of the earth, which was denied by Riccioli and his followers.— Through this channel, in particular, he carried on a dispute with Mr. Huygens on the occasion of his treatise on the quadrature of the circle and hyperbola, to which that great man had started some objections; in the course of which our author produced some improvements of his series. But in this dispute it happened, as it generally does on such occasions, that the antagonists, though setting out with temper enough, yet grew too warm in the combat. This was the case here, especially on the side of Gregory, whose defence was, at his own request, inserted in the Philosophical Transactions. It is unnecessary to enter into particulars:

suffice it therefore to say that, in the opinion of Leibnitz, who allows Mr. Gregory the highest merit for his genius and discoveries, M. Huygens has pointed out, though not errors, some considerable deficiencies in the treatise above mentioned, and shewn a much simpler method of attaining the same end.

In 1668, our author published at London another work, entitled, *Exercitationes Geometricæ*, which contributed still much farther to extend his reputation. About this time he was elected professor of mathematics in the university of St. Andrew's, an office which he held for six years. During his residence there, he married, in 1669, Mary, the daughter of George Jameson, the celebrated painter, whom Mr. Walpole has termed the Vandyke of Scotland, and who was fellow disciple with that great artist in the school of Rubens at Antwerp.

In 1672, he published " The Great and New Art of Weighing Vanity : or a Discovery of the Ignorance and Arrogance of the Great and New Artist, in his Pseudo philosophical Writings. By M. Patrick Mathers, Arch-bedal to the University of St. Andrews. To which are annexed some Tentamina de Motu Penduli & Projectorum." Under this asumed name, our author wrote this little piece to expose the ignorance of Mr. Sinclare, professor at Glasgow, in his hydrostatical writings, and in return for some ill usage of that author to a colleague of Mr. Gregory's. The same year, Newton, on his wonderful discoveries in the nature of light, having contrived a new reflecting telescope, and made several objections to Mr. Gregory's, this gave birth to a dispute between those two philosophers, which was carried on during this and the following year, in the most amicable manner on both sides; Mr. Gregory defending his own construction, so far, as to give his antagonist the whole honour of having made the catoptric telescopes preferable to the dioptric; and shewing, that the imperfections in these instruments were not so much owing to a defect in the object speculum, as to the different refrangibility of the rays of light. In the course of this dispute, our author described a burning concave mirror, which was approved by Newton, and is still in good esteem. Several letters that passed in this dispute, are printed by Dr. Desaguliers, in an Appendix to the English edition of Dr. David Gregory's Elements of Catoptrics and Dioptrics.

In 1674, Mr. Gregory was called to Edinburgh, to fill the chair of mathematics in that university. This place he had held but little more than a year, when, in October 1675, being employed in shewing the satellites of Jupiter through a telescope to some of his pupils, he was suddenly struck with total blindness, and died a few days after, to the great loss of the mathematical world, at only 37 years of age.

As to his character, Mr. James Gregory was a man of a very acute and penetrating genius. His temper seems to have been warm, as appears from his conduct in the dispute with Huygens ; and, conscious perhaps of his own merits as a discoverer, he seems to have been jealous of losing any portion of his reputation by the improvements of others upon his inventions. He possessed one of the most amiable characters of a true philosopher, that of being content with his fortune in his situation. But the most brilliant part of his character is that of his mathematical genius as an inventor, which was of the first order ; as will appear by the following list of his inventions and discoveries. Among many others may be reckoned, his Reflecting Telescope ;—Burning Concave Mirror ;—Quadrature of the Circle and Hyperbola, by an infinite converging series ;—his method for the Transformation of Curves ;—a Geometrical Demonstration of lord Brounker's series for Squaring the Hyperbola—his Demonstration that the Meridian Line is analogous to a scale of Logarithmic Tangents of the Half Complements of the Latitude ;—he also invented and demonstrated geometrically, by help of the hyperbola, a very simple converging series for making the logarithms ;—he sent to Mr. Collins the solution of the famous Keplerian problem by an infinite series ;—he discovered a method of drawing Tangents to Curves geometrically, without any previous calculations ;—a rule for the Direct and Inverse method of Tangents, which stands upon the same principle (of exhaustions) with that of fluxions, and differs not much from it in the manner of application ; a Series for the length of the Arc of a Circle from the Tangent, and vice versa ; as also for the Secant and Logarithmic Tangent and Secant, and vice versa :—These, with others, for measuring the length of the elliptic and hyperbolic curves, were sent to Mr. Collins, in return for some received from him of Newton's, in which he followed the elegant example of this author, in delivering his series in simple terms, independent of each other. These and other writings of our author are mostly contained in the following works, viz,

1. *Optica Promota* ; 4to, London 1663.

2. *Vera Circuli et Hyperbolæ Quadratura* ; 4to, Padua 1667 and 1668.

3. *Geometriæ Pars Universalis* ; 4to, Padua 1668.

4. *Exercitationes Geometricæ* ; 4to, London 1668.

5. *The Great and New Art of Weighing Vanity*, &c. 8vo, Glasgow 1672.

The rest of his inventions make the subject of several letters and papers, printed either in the *Philos. Trans.* vol. 3 ; the *Commerc. Epistol. Joh. Collins et Aliorum*, 8vo, 1715 ; in the Appendix to the English edition of Dr. David Gregory's *Elements of Optics*, 8vo, 1735, by Dr. Desaguliers ; and some series in the *Exercitatio Geometrica* of the same author, 4to, 1684, Edinburgh ; as well as in his little piece on Practical Geometry.

GREGORY (*Dr. David*), Savilian professor of astronomy at Oxford, was nephew of the above-mentioned Mr. James Gregory, being the eldest son of his brother Mr. David Gregory of Kinardie, a gentleman who had the singular fortune to see three of his sons all professors of mathematics, at the same time, in three of the British Universities, viz, our author David at Oxford, the second son James at Edinburgh, and the third son Charles at St. Andrews. Our author David, the eldest son was born at Aberdeen in 1661, where he received the early parts of his education, but completed his studies at Edinburgh ; and, being possessed of the mathematical papers of his uncle, soon distinguished himself likewise as the heir of his genius. In the 23d year of his age, he was elected professor of mathematics in the university of Edinburgh ; and, in the same year, he published *Exercitatio Geometrica de Dimensione Figurarum*,

Figurarum, five Specimen Methodi generalis Dimetiendi quafvis Figuras, Edinb. 1684, 4to. He very foon perceived the excellence of the Newtonian philofophy; and had the merit of being the firft that introduced it into the fchools, by his public lectures at Edinburgh. "He had (fays Mr. Wifton, in the Memoirs of his own Life, i. 32) already caufed feveral of his fcholars to keep acts, as we call them, upon feveral branches of the Newtonian philofophy; while we at Cambridge, poor wretches, were ignominioufly ftudying the fictitious hypothefis of the Cartefian."

In 1691, on the report of Dr. Bernard's intention of refigning the Savilian profefforfhip of aftronomy at Oxford, our author went to London; and being patronifed by Newton, and warmly befriended by Mr. Flamfteed the aftronomer royal, he obtained the vacant profefforfhip, though Dr. Halley was a competitor. This rivalfhip, however, inftead of animofity, laid the foundation of friendfhip between thefe eminent men; and Halley foon after became the colleague of Gregory, by obtaining the profefforfhip of geometry in the fame univerfity. Soon after his arrival in London, Mr. Gregory had been elected a fellow of the Royal Society; and, previoufly to his election into the Savilian profefforfhip, had the degree of doctor of phyfic conferred on him by the univerfity of Oxford.

In 1693, he publifhed in the Philof. Tranf. a refolution of the Florentine problem *de Teftudine veliformi quadrabili*; and he continued to communicate to the public, from time to time, many ingenious mathematical papers by the fame channel.

In 1695, he printed at Oxford, *Catoptricæ et Dioptricæ Sphæricæ Elementa*; a work which, we are informed in the preface, contains the fubftance of fome of his public lectures read at Edinburgh, eleven years before. This valuable treatife was republifhed in Englifh, firft with additions by Dr. William Brown, with the recommendation of Mr. Jones and Dr. Defaguliers; and afterwards by the latter of thefe gentlemen, with an appendix containing an account of the Gregorian and Newtonian telefcopes, together with Mr. Hadley's tables for the conftruction of both thofe inftruments. It is not unworthy of remark, that, in the conclufion of this treatife, there is an obfervation which fhews, that the conftruction of achromatic telefcopes, which Mr. Dollond has carried to fuch great perfection, had occurred to the mind of David Gregory, from reflecting on the admirable contrivance of nature in combining the different humours of the eye. The paffage is as follows: "Perhaps it would be of fervice to make the object lens of a different medium, as we fee done in the fabric of the eye; where the cryftalline humour (whofe power of refracting the rays of light differs very little from that of glafs) is by nature, who never does any thing in vain, joined with the aqueous and vitreous humours (not differing from water as to their power of refraction) in order that the image may be painted as diftinct as poffible upon the bottom of the eye."

In 1702 our author publifhed at Oxford, in folio, *Aftronomiæ Phyficæ et Geometricæ Elementa*; a work which is accounted his mafter-piece. It is founded on the Newtonian doctrines, and was efteemed by Newton himfelf as a moft excellent explanation and defence of his philofophy. In the following year he gave to the world an edition, in folio, of the works of Euclid, in Greek and Latin; being done in profecution of a defign of his predeceffor Dr. Bernard, of printing the works of all the ancient mathematicians. In this work, which contains all the treatifes that have been attributed to Euclid, Dr. Gregory has been careful to point out fuch as he found reafon, from internal evidence, to believe to be the productions of fome inferior geometrician. In profecution of the fame plan, Dr. Gregory engaged foon after, with his colleague Dr. Halley, in the publication of the Conics of Apollonius; but he had proceeded only a little way in this undertaking, when he died at Maidenhead in Berkfhire, in 1710, being the 49th year of his age only.

Befides thofe works publifhed in our author's life time, as mentioned above, he had feveral papers inferted in the Philof. Tranf. vol. 18, 19, 21, 24, and 25, particularly a paper on the catenarian curve, firft confidered by our author. He left alfo in manufcript, *A Short Treatife of the Nature and Arithmetic of Logarithms*, which is printed at the end of Keill's tranflation of Commandine's Euclid; and a *Treatife of Practical Geometry*, which was afterwards tranflated, and publifhed in 1745, by Mr. Maclaurin.

Dr. David Gregory married, in 1695, Elizabeth, the daughter of Mr. Oliphant of Langtown in Scotland. By this lady he had four fons, of whom, the eldeft, David, was appointed regius profeffor of modern hiftory at Oxford by king George the 1ft, and died at an advanced age in 1767, after enjoying for many years the dignity of dean of Chriftchurch in that univerfity.

When David Gregory quitted Edinburgh, he was fucceeded in the profefforfhip at that univerfity by his brother *James*, likewife an eminent mathematician; who held that office for 33 years, and, retiring in 1725, was fucceeded by the celebrated Maclaurin. A daughter of this profeffor James Gregory, a young lady of great beauty and accomplifhments, was the victim of an unfortunate attachment, that furnifhed the fubject of Mallet's well known ballad of *William and Margaret*.

Another brother, Charles, was created profeffor of mathematics at St. Andrews by Queen Anne, in 1707. This office he held with reputation and ability for 32 years; and, refigning in 1739, was fucceeded by his fon, who eminently inherited the talents of his family, and died in 1763.

Some farther Particulars of the Family of the Gregorys and Anderfons, communicated by Dr. Thomas Reid, Profeffor of Moral Philofophy in the Univerfity of Glafgow, a Nephew of the late Dr. David Gregory Savilian Profeffor at Oxford.

Some account of the family of the Gregorys at Aberdeen, is given in the Life of the late Dr. John Gregory prefixed to his works, printed at Edinburgh for A. Strahan and T. Cadell, London, and W. Creech, Edinburgh, 1788, in four fmall 8vo volumes.

Who was the author of that Life, or whence he had his information, I do not know. I have heard it afcribed to Mr. Tytler the younger, whofe father was appointed

one

one of the guardians of Dr. John Gregory's children. Some additions to what is contained in it, and remarks upon it, is all I can furnish upon this subject.

Page 3. I know nothing of the education of David Anderson of Finzaugh. He seems to have been a self-taught Engineer. Every public work which surpassed the skill of common artists, was committed to the management of David. Such a reputation he acquired by his success in works of this kind, that with the vulgar he got the by-name of *Davie do a' thing*, that is in the Scottish dialect, *David who could do every thing*. By this appellation he is better known than by his proper name. He raised the great bells into the steeple of the principal church: he cut a passage for ships of burden through a ridge of rock under water, which crossed the entrance into the harbour of Aberdeen. In a long picture gallery at Cullen House, the seat of the earl of Findlater, the wooden ceiling is painted with several of the fables of Ovid's Metamorphosis. The colours are still bright, and the representation lively. The present earl's grandfather told me that this painting was the work of David Anderson my ancestor, whom he acknowledged as a friend and relation of his family.

Such works, while they gave reputation to David, suited ill with his proper business, which was that of a merchant in Aberdeen. In that he succeeded ill; and having given up mercantile business, from a small remainder of his fortune began a trade of making malt; and having instructed his wife in the management of it, left it to her care, and went into England to try his fortune as an engineer; an employment which in his own country he had practised gratuitously. Having in that way made a fortune which satisfied him, he returned to Aberdeen, where his wife had also made money by her malting business.

After making such provision for their children as they thought reasonable, they agreed that the longest liver of the two should enjoy the remainder, and at death should bequeath it to certain purposes in the management of the magistrates of Aberdeen.

The wife happened to live longest, and fulfilled what had been concerted with her husband. Her legacies, well known in Aberdeen, are called after her name *Jane Guld's Mortifications*, a mortification in Scots law signifying a bequeathment for some charitable purpose. They consist of sums for different purposes. For orphans, for the education of boys and girls, for unmarried gentlewomen, and for widows; and they still continue to be useful to many in indigent circumstances. She was the daughter of Dr. Guild a minister of Aberdeen. Besides her money, she bequeathed a piece of tapestry, wrought by her own hand, and representing the history of queen Esther, from a drawing made by her husband. The tapestry continues to ornament the wall of the principal church.

In the same page it is said that Alexander Anderson, professor of mathematics at Paris, was the cousingerman of David above-mentioned. I know not the writer's authority for this: I have always heard that they were brothers; but for this I have only family tradition.

P. 4. It is here said that James Gregory was in-

structed in the Elements of Euclid by his mother, the daughter of David Anderson.

The account I have heard differs from this. It is, that his brother David, being ten or eleven years older, had the direction of his education after their father's death, and, when James had finished his course of philosophy, was at a loss to what literary profession he should direct him. After some unsuccessful trials, he put Euclid's Elements into his hand, and finding that he applied to Euclid with great avidity and success, he encouraged and assisted him in his mathematical studies.

This tradition agrees with what James Gregory says in the preface to his *Optica Promota*; where after mentioning his advance to the 26th proposition, he adds, *Ubi diu hæsi omne spe progrediendi orbatus, sed continuis hortatibus et auxiliis fratris mei Davidis Gregorii, in Mathematicis non parum versati (cui si quid in hisce Scientiis præstitero, me illud debere non inficias ibo) animatus, tandem incidi &c*. Whether David had been instructed in mathematics by his mother, or had any living instructor, I know not.

P. 5, 6. In these two pages I think the merit of Gregory compared with that of Newton in the invention of the catoptric telescope, is put in a light more unfavourable to Newton than is just. Gregory believing that the imperfection of the dioptric telescope arose solely from the spherical figure of the glasses, invented his telescope to remedy that imperfection. Being less conversant in the practice of mechanics, he did not attempt to make any model. The specula of his telescope required a degree of polish and a figure which the best opticians of that age were unable to execute. Newton demonstrated that the imperfection of the dioptric telescope arose chiefly from the different refrangibility of the rays of light; he demonstrated also that the catoptric telescope required a degree of polish far beyond what was necessary for the dioptric. He made a model of his telescope; and finding that the best polish which the opticians could give, was insufficient, he improved the polish with his own hand, so as to make it answer the purpose, and has described most accurately the manner in which he did this. And, had he not given this example of the practicability of making a reflecting telescope, it is probable that it would have passed as an impracticable idea to this day.

P. 11. To what is said of this James Gregory might have been added, that he was led by analogy to the true law of Refraction, not knowing that it was discovered by Des Cartes before (see Preface to *Optica Promota*); and that in 1670 having received in a letter from Collins, a Series for the Area of the Zone of a Circle, and as Newton had invented an universal method by which he could square all Curves Geometrical and Mechanical by Infinite Series of that kind; Gregory after much thought discovered this universal method, or an equivalent one. Of this he perfectly satisfied Newton and the other mathematicians of that time, by a letter to Collins in Feb. 1671. He was strongly solicited by his brother David to publish his Universal Method of Series without delay, but excused himself upon a point of honour; that as Newton was the first inventor, and as he had

been

been led to it by an account of Newton's having such a method, he thought himself bound to wait till Newton should publish his method. I have seen the letters that passed between the brothers on this subject.

With regard to the controversy between James Gregory and Huygens, I take the subject of that controversy to have been, not whether J. Gregory's Quadrature of the Circle by a converging series was just, but whether he had demonstrated, as in one of his propositions he pretended to do, That it is impossible to express perfectly the Area of a Circle in any known Algebraical form, besides that of an infinite converging series. Huygens excepted to the demonstration of this proposition, and Gregory defended it; neither of them convinced his antagonist, nor do I know that Leibnitz improved upon what Gregory had done.

P. 12. David Gregory of Kinardie deserved a more particular account than is here given.

It is true that he served an apprenticeship to a mercantile house in Holland, but he followed that profession no longer than he was under authority, having a stronger passion for knowledge than for money. He returned to his own country in 1655, being about 28 years of age, and from that time led the life of a philosopher. Having succeeded to the estate of Kinardie by the death of an elder brother, he lived there to the end of that century. There all his children were born, of whom he had thirty-two by two wives.

Kinardie is above 40 English miles north from Aberdeen, and a few miles from Bamf, upon the river Diveron. He was a jest among the neighbouring gentlemen for his ignorance of what was doing about his own farm, but an oracle in matters of learning and philosophy, and particularly in medicine, which he had studied for his amusement, and begun to practise among his poor neighbours. He acquired such a reputation in that science, that he was employed by the nobility and gentlemen of that county, but took no fees. His hours of study were singular. Being much occupied through the day with those who applied to him as a physician, he went early to bed, rose about two or three in the morning, and, after applying to his studies for some hours, went to bed again and slept an hour or two before breakfast.

He was the first man in that country who had a barometer; and by some old letters which I have seen, it appeared, that he had corresponded with some philosophers on the continent about the changes in the barometer and in the weather, particularly with Mariotte the French philosopher. He was once in danger of being prosecuted as a conjurer by the Presbytery on account of his barometer. A deputation of that body having waited upon him to enquire into the ground of certain reports that had come to their ears, he satisfied them so far as to prevent the prosecution of a man known to be so extensively useful by his knowledge of medicine.——About the beginning of this century he removed with his family to Aberdeen, and in the time of queen Anne's war employed his thoughts upon an improvement in artillery, in order to make the shot of great guns more destructive to the enemy, and executed a model of the engine he had conceived. I have conversed with a clock-maker in Aberdeen who was employed in making this model; but having made many different pieces by direction without knowing their intention, or how they were to be put together, he could give no account of the whole. After making some experiments with this model, which satisfied him, the old gentleman was so sanguine in the hope of being useful to the allies in the war against France, that he set about preparing a field equipage with a view to make a campaign in Flanders, and in the mean time sent his model to his son the Savilian professor, that he might have his and Sir Isaac Newton's opinion of it. His son shewed it to Newton, without letting him know that his own father was the inventor. Sir Isaac was much displeased with it, saying, that if it tended as much to the preservation of mankind as to their destruction, the inventor would have deserved a great reward; but as it was contrived solely for destruction, and would soon be known by the enemy, he rather deserved to be punished, and urged the professor very strongly to destroy it, and if possible to suppress the invention. It is probable the professor followed this advice. He died soon after, and the model was never found.

When the rebellion broke out in 1715, the old gentleman went a second time to Holland, and returned when it was over to Aberdeen, where he died about 1720, aged 93.

He left an historical manuscript of the Transactions of his own Time and Country, which my father told me he had read.

I was well acquainted with two of this gentleman's sons, and with several of his daughters, besides my own mother. The facts abovementioned are taken from what I have occasionally heard from them, and from other persons of his acquaintance.

P. 14. In confirmation of what is said in this page, that the two brothers David and James were the first who taught the Newtonian philosophy in the Scotch Universities; I have by me a *Thesis*, printed at Edinburgh in 1690, by James Gregory, who was at that time a professor of philosophy at St. Andrews, and succeeded his brother David in the profession of mathematics at Edinburgh. In this *Thesis*, after a dedication to Viscount Tarbet, follow the names of twenty-one of his scholars who were candidates for the degree of A. M. then twenty-five positions or *Theses*. The first three relate to logic, and the abuse of it in the Aristotelian and Cartesian philosophy. He defines logic to be the art of making a proper use of things granted, in order to find what is sought, and therefore admits only two *Categories* in logic, viz, *Data* and *Quæsita*. The remaining twenty-two positions are a compend of Newton's Principia. This *Thesis*, as was the custom at that time in the Scotch universities, was to be defended in a public disputation, by the candidates, previous to their taking their degree.

The famous Dr. Pitcairn was a fellow student and intimate companion of these two Gregories, and during the vacation of the college was wont to go north with them to Kinardie, their father's house.

David Gregory was appointed a preceptor to the duke of Gloucester, queen Anne's son; but his entering upon that office was prevented by the death of that prince in the eleventh year of his age.

P. 19. D. Gregory's Euclid is said to have been wrote in

in profecution of a defign of his predeceffor Dr. Bernard, of printing the works of all the antient mathematicians. This defign ought to have been afcribed to Savile, who left in charge to the two profeffors of his foundation, to print the mathematical works of the antients, and I think left a fund for defraying the expence. Wallis did fomething in confequence of this charge; Gregory and Halley did a great deal; but I think nothing has been done in this defign by the Savilian profeffors fince their time.

P. 20. Befides what is mentioned, Dr. Gregory left in manufcript a Commentary on Newton's Principia, which Newton valued, and kept by him for many years after the author's death. It is probable that in what relates to aftronomy, this commentary may coincide in a great meafure with the author's aftronomy, which indeed is an excellent Commentary upon that part of the Principia.

P. 24. This David Gregory publifhed in Latin, a very good compend of arithmetic and algebra, with the title *Arithmeticæ et Algebræ Compendium, in Ufum Juventutis Academicæ*. Edinb. 1736. He had a defign of publifhing his uncle's Commentary on the Principia, with extracts from the papers left by James Gregory his grand uncle; but the expence being too great for his fortune, and he too gentle a folicitor of the affiftance of others, the defign was dropped. His fon David, yet alive, was mafter of an Eaft India fhip.

P. 40. To the projectors of the fociety at Aberdeen, ought to have been added John Stewart profeffor of mathematics in the Marifchal college at Aberdeen. He publifhed an explanation of two treatifes of Sir Ifaac Newton, viz. his Quadrature of Curves, and his Analyfis by Equations of an infinite number of terms. He was an intimate friend of Dr. Reid's.

Another of the firft members of that fociety was Dr. David Skene, who, befides his eminence in the practice of medicine, had applied much to all parts of natural hiftory, particularly to botany, and was a correfpondent of the celebrated Linnæus.

Dr. John Gregory and Dr. David Skene were the firft who attempted a college of medicine at Aberdeen. The firft gave lectures to his pupils in the theory and practice of medicine, and in chemiftry; the laft, in anatomy, materia medica, and midwifery, in order to prepare them for attending the medical college at Edinburgh.

T. R.

The following additional lines by Mr. James Millar, Profeffor of Mathematics, Glafgow.

Another inftance of the prevalence of mathematical genius in the family of Gregory or Anderson, whether produced by an original and inexplicable determination of the mind, or communicated by the force of example, and the confcioufnefs of an intimate connection with a reputation already acquired in a particular line, is the celebrated Dr. Reid, profeffor of moral philofophy in the univerfity of Glafgow; a nephew, by his mother, of the late Dr. David Gregory, Savilian profeffor at Oxford.

This gentleman, well known to the public by his moral and metaphyfical writings, and remarkable for that liberality, and that ardent fpirit of enquiry, which neither overlooks nor undervalues any branch of fcience, is

peculiarly diftinguifhed by his abilities and proficiency in mathematical learning. The objects of literary purfuit are often directed by accidental occurrences. And apprehenfion of the bad confequences which might refult from the philofophy of the late Mr. Hume, induced Dr. Reid to combat the doctrines of that eminent author; and produced a work, which has excited univerfal attention, and feems to have given a new turn to fpeculations upon that fubject. But it is well known to Dr. Reid's literary acquaintance, that thefe exertions have not diminifhed the original bent of his genius, nor blunted the edge of his inclination for mathematical refearches; which, at a very advanced age, he ftill continues to profecute with a youthful attachment, and with unremitting affiduity.

It may farther be obferved, of the extraordinary family above mentioned, that Dr. James Gregory, the prefent learned profeffor of phyfic and medicine in the univerfity of Edinburgh, is the fon of the late Dr. John Gregory, upon the memoirs of whofe life the above remarks have been written by Dr. Reid; the faid James has lately publifhed a moft ingenious work, intitled, *Philofophical and Literary Effays*, in 2 volumes 8vo, Edinb. 1792; and he feems to be another worthy inheritant of the fingular genius of his family.

GREGORY (*St. Vincent*), a very refpectable Flemifh geometrician, was born at Bruges in 1584, and became a Jefuit at Rome at 20 years of age. He ftudied mathematics under the learned Jefuit Clavius. He afterward became a reputable profeffor of thofe fciences himfelf, and his inftructions were folicited by feveral princes: he was called to Prague by the emperor Ferdinand the 2d; and Philip the 4th, king of Spain, was defirous of having him to teach mathematics to his fon the young prince John of Auftria. He was not lefs eftimable for his virtues than his fkill in the fciences. His well-meant endeavours were very commendable, when his holy zeal, though for a falfe religion, led him to follow the army in Flanders one campaign, to confefs the wounded and dying foldiers, in which he received feveral wounds himfelf. He died of an apoplexy at Ghent, in 1667, at 83 years of age.

As a writer, Gregory St. Vincent was very diffufe and voluminous, but he was an excellent geometrician. He publifhed, in Latin, three mathematical works, the principal of which was his *Opus Geometricum Quadraturæ Circuli, et Sectionum Coni*, Antwerp, 1647, 2 vol. folio. Although he has not demonftrated, in this work, the Quadrature of the circle, as he pretends to have done, the book neverthelefs contains a great number of truths and important difcoveries; one of which is this, viz, that if one afymptote of an hyperbola be divided into parts in geometrical progreffion, and from the points of divifion ordinates be drawn parallel to the other afymptote, they will divide the fpace between the afymptote and curve into equal portions; from whence it was fhewn by Merfenne, that, by taking the continual fums of thofe parts, there would be obtained areas in arithmetical progreffion, adapted to abfciffes in geometrical progreffion, and which therefore were analogous to a fyftem of logarithms.

GRENADE, or GRENADO. See GRANADE.

GRUS, the *Crane*, one of the new conftellations, in
the

the fouthern hemifphere; containing, according to Mr. Sharp's catalogue, 13 ftars.

GRUS is alfo one of the Arabian conftellations, and anfwers to our Ophiucus, to which they changed this conftellation, their religion prohibiting them from drawing any human figures.

GRY, a meafure containing one-tenth of a line. A line is one-tenth of a digit, and a digit is one-tenth of a foot, and a philofophical foot, one-third of a pendulum, whofe diadromes, or vibrations, in the latitude of 45 degrees, are each equal to one fecond of time, or one-fixtieth of a minute.

GUARDS, a name that has been fometimes applied to the two ftars neareft the north pole; being in the hind part of the chariot, at the tail of Urfa Minor or little bear; one of them being alfo called the pole ftar.

GUERICKE (OTTO or OTHO), counfellor to the elector of Brandenbourg and burgomafter of Magdebourg, was born in 1602, and died in 1686 at Hambourg. He was one of the greateft philofophers of his time. It was Guericke that invented the air-pump; the two brafs hemifpheres, which being applied to each other, and the air exhaufted, 16 horfes were not able to draw them afunder; the marmoufet of glafs which defcended in a tube in rainy weather, and rofe again on the return of ferene weather. This laft machine fell into difufe on the invention of the barometer, efpecially after Huygens and Amontons gave theirs to the world. Guericke made ufe of his marmoufet to foretell ftorms; from whence he was looked upon as a forcerer by the people; fo that the thunder having one day fallen upon his houfe, and fhivered to pieces feveral machines which he had employed in his experiments, they failed not to fay it was a punifhment from heaven that was angry with him.——Guericke was author of feveral works in natural philofophy, the principal of which was his *Experimenta Magdeburgica*, in folio, which contains his experiments on a vacuum.

GUERITE, in Fortification, a centry-box; being a fmall tower of wood, or ftone, ufually placed on the point of a baftion, or on the angles of the fhoulder, to hold a centinel, who is to take care of the ditch, and watch againft a furprife.

GUEULE, in Architecture. See GULA.

GUINEA, a gold coin ftruck in England.——The value or rate of the guinea has varied. It was at firft equal to 20 fhillings; but by the fcarcity of gold it was afterwards advanced to 21 *s*. 6*d*.; though it is now funk to 21*s*.

The pound weight troy of gold is cut into 44 parts and a half, and each part makes a guinea, which is therefore equal to $\frac{2}{89}$lb, or $\frac{4}{89}$oz, or 5dwts $9\frac{39}{89}$. gr.

This coin took its name, Guinea, from the circumftance of the gold of which it was firft ftruck being brought from that part of Africa called Guinea, for which reafon alfo it bore the impreffion of an elephant.

GULA, GUEULE, or GOLA, in Architecture, a wavy member whofe contour refembles the letter S, commonly called an Ogee.

GULBE, in Architecture, the fame as Gorge.

GULF, or GULPH, in Geography, a part of the ocean running up into the land through a narrow paffage, or ftrait, and forming a bay within. As, the Gulf of Venice, or Adriatic fea; the Gulf of Arabia, or of Perfia, which is the Red Sea; the Gulf of Conftantinople, or the Black Sea; the Gulf of Mexico; &c.

GUN, a fire-arm, or weapon of offence, which forcibly difcharges a ball or other matter through a cylindrical tube, by means of inflamed gun-powder.

The word Gun now includes moft of the fpecies of fire-arms; mortars and piftols being almoft the only ones excepted from this denomination. They are divided into great and fmall guns: the former including all that are ufually called cannon, ordnance, or artillery; and the latter includes mufquets, firelocks, carabines, mufquetoons, blunderbuffes, fowling-pieces, &c.

It is not certainly known at what time thefe weapons were firft invented. And though the introduction of guns into the weftern part of the world is but of modern date, comparatively fpeaking; yet it is certain that in fome parts of Afia they have been ufed for many ages, though in a very rude and imperfect manner. Philoftratus fpeaks of a city near the river Hyphafis in the Indies, which was faid to be impregnable, and that its inhabitants were relations of the gods, becaufe they threw thunder and lightning upon their enemies; and other Greek authors, as alfo Quintus Curtius, fpeak of the fame thing having happened to Alexander the Great. Hence fome have imagined that guns were ufed by the eaftern nations in his time, while others fuppofe the thunder and lightning alluded to by thofe authors, were only certain artificial fire-works, or rockets, fuch as we know are ufed in the wars by the Indians even in the prefent day againft the Europeans. Be this however as it may, it is afferted by many modern travellers, that Guns were ufed in China as far back as the year of Chrift 85, and have continued in ufe ever fince.

The firft hint of the invention of Guns in Europe, is in the works of Roger Bacon, who flourifhed in the 13th century. In a treatife written by him about the year 1280, he propofes to apply the violent explofive force of gun-powder for the deftruction of armies. And though it is certainly known that the compofition of gun-powder is defcribed by Bacon in the faid work, yet the invention has ufually, though improperly, been afcribed to Bartholdus Schwartz, a German monk, who it is faid difcovered it only in the year 1320; and the invention is related in the following manner. Schwartz having, for fome purpofe, pounded nitre, fulphur, and charcoal together, in a mortar, which he afterwards covered imperfectly with a ftone, a fpark of fire accidentally fell into the mortar, which fetting the mixture on fire, the explofion blew the ftone to a confiderable diftance. Hence it is probable that Schwartz might be taught the fimpleft method of applying it in war; for it rather feems that Bacon conceived the manner of ufing it to be by the violent effort of the flame unconfined, and which is indeed capable of producing aftonifhing effects. (See GUNPOWDER.) And the figure and name of *mortars* given to a fpecies of old artillery, and their employment, in throwing large ftone bullets at an elevation, very much favour this conjecture.

Soon

Soon after the time of Schwartz, we find Guns commonly ufed as inftruments of war. Great guns were firft ufed. Thefe were originally made of iron-bars foldered together, and fortified with ftrong iron hoops or rings; feveral of which are ftill to be feen in the Tower of London, and in the Warren at Woolwich. Others were made of thin fheets of iron rolled up together and hooped: and on particular emergencies fome have been made of leather, and of lead, with plates of iron or copper. Thefe firft pieces were executed in a rude and imperfect manner, like the firft effays of moft new inventions. Stone balls were thrown out of them, and a fmall quantity of powder ufed on account of their weaknefs. They were of a cylindrical form, without ornaments, and were placed on their carriages by rings.

When, or by whom they were firft made, is uncertain. It is known however that the Venetians ufed cannon at the fiege of Claudia Jeffa, now called Chioggia, in 1366, which were brought thither by two Germans, with fome powder and leaden balls; as likewife in their wars with the Genoefe in 1379. But before that, king Edward the 3d made ufe of cannon at the battle of Creffy in 1346, and at the fiege of Calais in 1347. Cannon were employed by the Turks at the fiege of Conftantinople, then in poffeffion of the Chriftians, in 1394, and in that of 1452, which threw a weight of 100lb; but they commonly burft at the 1ft, 2d, or 3d firing. Louis the 12th had one caft at Tours, of the fame fize, which threw a ball from the Baftile to Charenton: one of thefe extraordinary cannon was taken at the fiege of Dieu in 1546, by Don John de Caftro, and is now in the caftle of St. Julian da Barra, 10 miles from Lifbon: the length of is 20 feet 7 inches, its diameter at the middle 6 feet 3 inches, and it threw a ball of 100lb weight. It has neither dolphins, rings, nor button; is of an unufual kind of metal; and it has a large Indoftan infcription upon it, which fays it was caft in 1400.

Formerly, cannon were dignified with uncommon names. Thus, Lewis the 12th, in 1503, had 12 brafs cannon caft, of an extraordinary fize, called after the names of the 12 peers of France. The Spanifh and Portuguefe called them after their faints. The emperor Charles the 5th, when he marched againft Tunis, founded the 12 apoftles. At Milan there is a 70 pounder, called the Pimontelle; and one at Bois-le-duc, called the Devil. A 60 pounder at Dover-caftle, called Queen Elizabeth's pocket-piftol. An 80 pounder in the Tower of London, brought there from Edinburgh-caftle, called Mounts-meg. An 80 pounder in the royal arfenal at Berlin, called the Thunderer. An 80 pounder at Malaga, called the Terrible. Two curious 60 pounders in the arfenal at Bremen, called the Meffenger of bad news. And laftly an uncommon 70 pounder in the caftle of St. Angelo at Rome, made of the nails that faftened the copper-plates which covered the ancient Pantheon, with this infcription upon it, Ex clavis trabalibus porticus Agrippæ.

In the beginning of the 15th century thefe uncommon names were generally abolifhed, and the following more univerfal ones took place, viz,

Names.		Wt. of ball, Pounders.		Wt. of piece in cwts. about Cwt.
Cannon royal, or carthoun,		48	-	90
Baftard cannon, or ¾ carthoun,		36	-	79
Demi-carthoun,	-	24	-	60
Whole culverins,	-	18	-	50
Demi-culverins,	-	9	-	30
Falcon,	-	6	-	25
Sacker { largeft fize		8	-	18
ordinary		6	-	15
loweft fort		5	-	13
Bafilifk	-	48	-	85
Serpentine	-	4	-	8
Afpic	-	2	-	7
Dragon	-	6	-	12
Syren	-	60	-	81
Falconet	-	3, 2, and 1		15,10,5
Rabinet	-	1		
Moyens	-	10 or 12 oz.		

Thefe curious names of beafts and birds of prey were adopted on account of their fwiftnefs in motion, or of their cruelty; as the falconet, falcon, facker, and culverin, &c, for their fwiftnefs in flying; the bafilifk, ferpentine, afpic, dragon, fyren, &c, for their cruelty.

But, at prefent, cannon take their names from the weight of their proper ball. Thus, a piece that difcharges a caft-iron ball of 24 pounds, is called a 24 pounder; one that carries a ball of 12 pounds, is called a 12 pounder; and fo of the reft, divided into the following forts, viz,

Ship-guns, confifting in 42, 36, 32, 24, 18, 12, 9, 6, and 3 pounders.

Garrifon-guns, in 42, 32, 24, 18, 12, 9, and 6 pounders.

Battering-guns, in 24, 18, and 12 pounders.

Field-pieces, in 12, 9, 6, 3,2, 1½, 1, and ½ pounders.

Mortars, it is thought, have been at leaft as ancient as cannon. They were employed in the wars of Italy, to throw balls of red-hot iron, ftones, &c, long before the invention of fhells. Thefe laft, it is fuppofed, were of German invention, and the ufe of them in war fhewn by the following accident; viz, a citizen of Venlo, at a feftival celebrated in honour of the duke of Cleves, throwing a number of fhells, one of them fell on a houfe and fet it on fire, by which misfortune the greateft part of the town was reduced to afhes. The firft account of fhells ufed for military purpofes, is in 1435, when Naples was befieged by Charles the 8th. Hiftory informs us, with more certainty, that fhells were thrown out of mortars at the fiege of Wachtendonk, in Guelderland, in 1588, by the earl of Mansfield; and Cyprian Lucar wrote upon the method of filling and throwing fuch fhells in his Appendix to the Colloquies of Tartaglia, printed at London in 1588; where alfo the compounding and throwing of carcaffes and various forts of fire-works are fhewn.

Mr. Malter, an Englifh engineer, firft taught the French the art of throwing fhells, which they practifed at the fiege of Motte in 1634. The method of throwing red-hot balls out of mortars was firft certainly put in

5

in practice at the siege of Stralsund in 1675 by the elector of Brandenburgh : though some say in 1653 at the siege of Bremen.

Another species of ordnance has been long in use, by the name of Howitzer, which is a kind of medium as to its length, between the cannon and the mortar, and is a very useful piece, for discharging either shells or large balls, which is done either at point-blanc, or at a small elevation.

A new species of ordnance has lately been introduced by the Carron company, and thence called a Carronade, which is only a very short howitzer, and which possesses the advantage of being very light and easy to work.

The species of Guns before mentioned, are now made chiefly of cast iron ; except the howitzer, which is of brass, as well as some cannon and mortars.

Muskets were first used at the siege of Rhege in the year 1521. The Spaniards were the first who armed part of their foot with these weapons. At first they were very heavy, and could not be used without a rest. They had match-locks, and did execution at a great distance. On their march the soldiers carried only the rests and ammunition, having boys to bear their muskets after them. They were very slow in loading, not only by reason of the unwieldiness of their pieces, and because they carried the powder and ball separate, but from the time it took to prepare and adjust the match ; so that their fire was not near so brisk as ours is now. Afterwards a lighter match-lock musket came in use : and they carried their ammunition in bandeliers, to which were hung several little cases of wood covered with leather, each containing a charge of powder. The muskets with rests were used as late as the beginning of the civil wars in the time of Charles the 1st. The lighter kind succeeded them, and continued till the beginning of the present century, when they also were disused, and the troops throughout Europe armed with firelocks. These are usually made of hammered iron. For the dimensions, construction, and practice of every species of Gun, &c, see the several articles CANNON, MORTAR, &c. See also GUNNERY.

GUNNERY, the art of charging, directing, and exploding fire-arms, as cannon, mortars, muskets, &c, to the best advantage.

Gunnery is sometimes considered as a part of the military art, and sometimes as a part of pyrotechny. To the art of Gunnery too belongs the knowledge of the force and effect of gunpowder, the dimensions of the pieces, and the proportions of the powder and ball they carry, with the methods of managing, charging, pointing, spunging, &c. Also some parts of Gunnery are brought under mathematical consideration, which among mathematicians are called absolutely by the name Gunnery, viz, the rules and method of computing the range, elevation, quantity of powder, &c, so as to hit a mark or object proposed, and is more particularly called PROJECTILES ; which see.

HISTORY of GUNNERY.

Long before the invention of gunpowder, and of Gunnery, properly so called, the art of artillery, or projectiles, was actually in practice. For, not to mention the use of spears, javelins, or stones thrown with the hand, or of bows and arrows, all which are found among the most barbarous and ignorant people, ac-

counts of the larger machines for throwing stones, darts, &c, are recorded by the most ancient writers. Thus, one of the kings of Judah, 800 years before the christian æra, erected engines of war on the towers and bulwarks of Jerusalem, for shooting arrows and great stones for the defence of that city. 2 Chron. xxvi. 15. Such machines were afterwards known among the Greeks and Romans by the names of Ballista, Catapulta, &c, which produced effects by the action of a spring of a strongly twisted cordage, formed of tough and elastic animal substances, no less terrible than the artillery of the moderns. Such warlike instruments continued in use down to the 12th and 13th centuries, and the use of bows still longer ; nor is it probable that they were totally laid aside till they were superseded by gunpowder and the modern ordnance.

The first application of gunpowder to military affairs, it seems, was made soon after the year 1300, for which the proposal of friar Bacon, about the year 1280, for applying its enormous explosion to the destruction of armies, might give the first hint ; and Schwartz, to whom the invention of gunpowder has been erroneously ascribed, on account of the accident abovementioned under the article GUN, might have been the first who actually applied it in this way, that is in Europe ; for as to Asia, it is probable that the Chinese and Indians had something of the kind many ages before. Thus, only to mention the prohibition of fire-arms in the code of Gentoo laws, printed by the East India Company in 1776, which seems to confirm the suspicion suggested by a passage in Quintus Curtius, that Alexander the Great found some weapons of that kind in India : Cannon in the Shanscrit idiom is called shet-aghnee, or the weapon that kills a hundred men at once.

However, the first pieces of artillery, which were charged with gunpowder and stone bullets of a prodigious size, were of very clumsy and inconvenient structure and weight. Thus, when Mahomet the 2d besieged Constantinople in 1453, he battered the walls with stones of this kind, and with pieces of the calibre of 1200 pounds ; which could not be fired more than four times a day. It was however soon discovered that iron bullets, of much less weight than stone ones, would be more efficacious if impelled by greater quantities of stronger powder. This occasioned an alteration in the matter and form of the cannon, which were now cast of brass. These were lighter and more manageable than the former, at the same time that they were stronger in proportion to their bore. This change took place about the close of the 15th century.

By this means came first into use such powder as is now employed over all Europe, by varying the proportion of the materials. But this change of the proportion was not the only improvement it received. The practice of graining it is doubtless of considerable advantage. At first the powder had been always used in the form of fine meal, such as it was reduced to by grinding the materials together. And it is doubtful whether the first graining of powder was intended to increase its strength, or only to render it more convenient for filling into small charges and the charging of small arms, to which alone it was applied for many years, whilst meal-powder was still used for cannon.

But at laſt the additional ſtrength which the grained powder was found to poſſeſs, doubtleſs from the free paſſage of the air between the grains, occaſioned the meal-powder to be entirely laid aſide.

For the laſt 200 years, the formation of cannon has been very little improved; the beſt pieces of modern artillery differing little in their proportions from thoſe uſed in the time of Charles the 5th. Indeed lighter and ſhorter pieces have been often propoſed and tried; but though they have their advantages in particular caſes, it is agreed they are not ſufficient for general ſervice. Yet the ſize of the pieces has been much decreaſed; the ſame purpoſes being now accompliſhed, by ſmaller pieces than what were formerly thought neceſſary. Thus the battering cannon now approved, are thoſe that formerly were called demi cannon, carrying a ball of 24 pounds weight; this weight having been found fully ſufficient. The method alſo of making a breach, by firſt cutting off the whole wall as low as poſſible before its upper part is attempted to be beaten down, ſeems to be a conſiderable modern improvement in the practical part of gunnery. But the moſt conſiderable improvement in the practice, is the method of firing with ſmall quantities of powder, and elevating the piece but a little, ſo that the bullet may juſt go clear of the parapet of the enemy, and drop into their works, called ricochet firing: for by this means the ball, coming to the ground at a ſmall angle, and with a ſmall velocity, does not bury itſelf, but bounds or rolls along a great way, deſtroying all before it. This method was firſt practiſed by M. Vauban at the ſiege of Aeth, in the year 1692. A practice of this kind was ſucceſsfully uſed by the king of Pruſſia at the battle of Roſbach in 1757. He had ſeveral ſix-inch mortars, made with trunnions, and mounted on travelling carriages, which were fired obliquely on the enemy's lines, and among their horſe. Theſe being charged with only 8 ounces of powder, and elevated at one degree and a quarter, did great execution: for the ſhells rolling along the lines with burning fuſes made the ſtouteſt of the enemy not wait for their burſting.

The uſe of fire-arms was however long known before any theory of projectiles was formed. The Italians were the firſt people that made any attempts at the theory, which they did about the beginning of the 16th century, and amongſt them it ſeems the firſt who wrote profeſſedly on the flight of cannon ſhot, was Nicholas Tartalia, of Breſcia, the ſame author who had ſo great a ſhare in the invention of the rules for cubic equations. In 1537 he publiſhed, at Venice, his *Nova Scientia*, and in 1546 his *Queſiti & Inventioni diverſi*, in both which he treats profeſſedly on theſe motions, as well as in another work, tranſlated into Engliſh with additions by Cyprian Lucar, under the title of Colloquies concerning the Art of Shooting in great and ſmall Pieces of Artillery, and publiſhed at London in 1588. He determined, that the greateſt range of a ſhot was when diſcharged at an elevation of 45°: and he aſſerted, contrary to the opinion of his contemporaries, that no part of the path deſcribed by a ball is a right line; although the curvature in the firſt part of it is ſo ſmall, that it need not be attended to. He compared it to the ſurface of the ſea; which, though it appears to be a plane, is yet doubtleſs incurvated round the centre of the earth. He ſays he invented the gunner's quadrant, for laying a piece of ordnance at any point or degree of elevation; and though he had but little opportunity of acquiring any practical knowledge by experiments, he yet gave ſhrewd gueſſes at the event of ſome untried methods.

The philoſophers of thoſe times alſo took part in the queſtions ariſing upon this ſubject; and many diſputes on motion were held, eſpecially in Italy, which continued till the time of Galileo, and probably gave riſe to his celebrated Dialogues on Motion. Theſe were not publiſhed till the year 1638; and in the interval there were publiſhed many theories of the motion of military projectiles, as well as many tables of their comparative ranges; though for the moſt part very fallacious, and inconſiſtent with the motions of theſe bodies.

It is remarkable however that, during theſe conteſts, ſo few of thoſe who were intruſted with the care of artillery, thought it worth while to bring their theories to the teſt of experiment. Mr. Robins informs us, in the preface to his New Principles of Gunnery, that he had met with no more than four authors who had treated experimentally on this ſubject. The firſt of theſe is Collado, in 1642, who has given the ranges of a falconet, carrying a three-pound ſhot, to every point of the gunner's quadrant, each point being the 12th part, or 7° and a half. But from his numbers it is manifeſt that the piece was not charged with its uſual allotment of powder. The reſult of his trials ſhews the ranges at the point-blanc, and the ſeveral points of elevation, as below.

Collado's Experiments.

Elevation at Points.		Deg.	Range in paces.
0	or	0	268
1	-	7½	594
2	-	15	794
3	-	22½	954
4	-	30	1010
5	-	37½	1040
6	-	45	1053
7	-	52½	between the 3d and 4th
8	-	60	between the 2d and 3d
9	-	67½	between the 1ſt and 2d
10	-	75	between the 0 and 1ſt
11	-	82½	fell very near the piece.

The next was by Wm. Bourne, in 1643, in his *Art of Shooting in Great Ordnance.* His elevations were not regulated by the points of the Gunner's quadrant, but by degrees; and he gives the proportions between the ranges at different elevations and the extent of the point-blanc ſhot, thus: if the extent of the point-blanc ſhot be repreſented by 1, then the proportions of the ranges at ſeveral elevations will be as below, viz.

Bourne's Proportion of Ranges.

Elevation.		Range.
0°	-	1
5	-	$2\frac{2}{5}$
10	-	$3\frac{1}{3}$
15	-	$4\frac{1}{3}$
20	-	$4\frac{5}{7}$

and the greateſt random $5\frac{1}{2}$;

which

which greateſt random, he ſays, in a calm day is at 42° elevation ; but according to the ſtrength of the wind, and as it favours or oppoſes the flight of the ſhot, the elevation may be from 45° to 36°.—He does not ſay with what piece he made his trials ; though from his proportions it ſeems to have been a ſmall one. This however ought to have been mentioned, as the relation between the extent of different ranges varies extremely according to the velocity and denſity of the bullet.

After him, Eldred and Anderſon, both Engliſhmen, alſo publiſhed treatiſes on this ſubject. The former of theſe was many years gunner of Dover Caſtle, where moſt of his experiments were made, the earlieſt of which are dated in 1611, though his book was not publiſhed till 1646, and was intitled *The Gunner's Glaſs.* His principles were ſufficiently ſimple, and within certain limits very near the truth, though they were not rigorouſly ſo. He has given the actual ranges of different pieces of artillery at ſmall elevations, all under 10 degrees. His experiments are numerous, and appear to be made with great care and caution ; and he has honeſtly ſet down ſome, which were not reconcilable to his method : upon the whole he ſeems to have taken more pains, and to have had a juſter knowledge of his buſineſs, than is to be found in moſt of his practical brethren.

Galileo printed his Dialogues on Motion in the year 1646. In theſe he pointed out the general laws obſerved by nature in the production and compoſition of motion, and was the firſt who deſcribed the action and effects of gravity on falling bodies : on theſe principles he determined, that the flight of a cannon-ſhot, or of any other projectile, would be in the curve of a parabola, unleſs ſo far as it ſhould be diverted from that track by the reſiſtance of the air. He alſo propoſed the means of examining the inequalities which ariſe from thence, and of diſcovering what ſenſible effects that reſiſtance would produce in the motion of a bullet at ſome given diſtance from the piece.

Notwithſtanding theſe determinations and hints of Galileo, it ſeems that thoſe who came after him never imagined that it was neceſſary to conſider how far the operations of Gunnery were affected by this reſiſtance. Inſtead of this, they boldly aſſerted, without making the experiment, that no great variation could ariſe from the reſiſtance of the air in the flight of ſhells or cannon ſhot. In this perſuaſion they ſupported themſelves chiefly by conſidering the extreme rarity of the air, compared with thoſe denſe and ponderous bodies ; and at laſt it became an almoſt generally eſtabliſhed maxim, that the flight of theſe bodies was nearly in the curve of a parabola.

Thus, Robert Anderſon, in his *Genuine Uſe and Effects of the Gunne,* publiſhed in 1674, and again in his book, *To hit a Mark,* in 1690, relates a great many experiments ; but proceeding on the principles of Galileo, he ſtrenuouſly aſſerts that the flight of all bullets is in the curve of a parabola ; undertaking to anſwer all objections that could be brought to the contrary. The ſame thing was alſo undertaken by Blondel, in his *Art de jetter les Bombes,* publiſhed in 1683 ; where, after long diſcuſſion, he concludes, that the variations from the air's reſiſtance are ſo ſlight as not to deſerve any notice. The ſame ſubject is treated of in the Philoſ.

Tranſ. N° 216, p. 68, by Dr. Halley ; who alſo, ſwayed by the very great diſproportion between the denſity of the air and that of iron or lead, thought it reaſonable to believe that the oppoſition of the air to large metal-ſhot is ſcarcely diſcernible ; although in ſmall and light ſhot he owns that it muſt be accounted for.

But though this hypotheſis went on ſmoothly in ſpeculation ; yet Anderſon, who made a great number of trials, found it impoſſible to ſupport it without ſome new modification. For though it does not appear that he ever examined the comparative ranges of either cannon or muſket ſhot when fired with their uſual velocities, yet his experiments on the ranges of ſhells thrown with velocities that were but ſmall, in compariſon of thoſe above mentioned, convinced him that their whole track was not parabolical. But inſtead of making the proper inferences from hence, and concluding that the reſiſtance of the air was of conſiderable efficacy, he framed a new hypotheſis ; which was, that the ſhell or bullet at its firſt diſcharge flew to a certain diſtance in a right line, from the end of which line only it began to deſcribe a parabola : and this right line, which he calls the line of the impulſe of the fire, he ſuppoſes is the ſame for all elevations. So that, by aſſigning a proper length to this line of impulſe, it was always in his power to reconcile any two ſhots made at any two different angles ; though the ſame method could not ſucceed with three ſhots ; nor indeed does he ever inform us of the event of his experiments when three ranges were tried at one time.

But after the publication of Newton's Principia, it might have been expected, that the defects of the theory would be aſcribed to their true cauſe, which is the great reſiſtance of the air to ſuch ſwift motions ; as in that work he particularly conſidered the ſubject of ſuch motions, and related the reſult of experiments, made on ſlow motions at leaſt ; by which it appeared, that in ſuch motions the reſiſtance increaſes as the ſquare of the velocities, and he even hints a ſuſpicion that it will increaſe above that law in ſwifter motions, as is now known to be the caſe. So far however were thoſe who treated this ſubject ſcientifically, from making a proper allowance for the reſiſtance of the atmoſphere, that they ſtill neglected it, or rather oppoſed it, and their theories ſtill differed moſt egregiouſly from the truth. Huygens alone ſeems to have attended to this principle : for in the year 1690 he publiſhed a treatiſe on gravity, in which he gave an account of ſome experiments tending to prove that the track of all projectiles, moving with very ſwift motions, was widely different from that of a parabola. The reſt of the learned generally acquieſced in the juſtneſs and ſufficiency of Galileo's doctrine, and accordingly very erroneous calculations concerning the ranges of cannon were given. Nor was any farther notice taken of theſe errors till the year 1716, at which time Mr. Reſſons, a French officer of artillery, of great merit and experience, gave in a memoir to the Royal Academy, importing that, " although it was agreed that theory joined with practice did conſtitute the perfection of every art ; yet experience had taught him that theory was of very little ſervice in the uſe of mortars : That the works of M. Blondel had juſtly enough deſcribed the ſeveral parabolic lines, according to the different degrees of the elevation of the piece; but that practice had convinced him there was no theory in the effect of gunpowder; for having endeavoured, with the greatest precision, to point a mortar according to these calculations, he had never been able to establish any solid foundation upon them."—— One instance only occurs in which D. Bernoulli applies the doctrine of Newton to the motions of projectiles, in the Com. Acad. Petrop. tom. 2, pa. 338 &c. Besides which nothing farther was done in this business till the time of Mr. Benjamin Robins, who published a treatise in 1742, intitled *New Principles of Gunnery*, in which he treated particularly, not only of the resistance of the atmosphere, but also of the force of gunpowder, the nature and effects of different guns, and almost every thing else relating to the flight of military projectiles; and indeed he carried the theory of gunnery nearly to its utmost perfection.

The first thing considered by Mr. Robins, and which is indeed the foundation of all other particulars relating to Gunnery, is the explosive force of gunpowder. M. De la Hire, in the Hist. of the Acad. of Sciences for the year 1702, supposed that this force may be owing to the increased elasticity of the air contained in, and between the grains, in consequence of the heat and fire produced at the time of the explosion: a cause not adequate to the 200th part of the effect. On the other hand, Mr. Robins determined, by irrefragable experiments, that this force was owing to an elastic fluid, similar to our atmosphere, existing in the powder in an extremely condensed state, which being suddenly freed from the powder by the combustion, expanded with an amazing force, and violently impelled the bullet, or whatever may oppose its expansion.

The intensity of this force of exploded gunpowder Mr. Robins ascertained in different ways, after the example of Mr. Hawksbee, related in the Philos. Transf. N° 295, and his Physico-Mechan. Exper. pa. 81. One of these is by firing the powder in the air thus: A small quantity of the powder is placed in the upper part of a glass tube, and the lower part of the tube is immerged in water, the water being made to rise so near the top, that only a small portion of air is left in that part where the powder is placed: then in this situation the communication between the upper part of the tube and the external air being closed, the powder is fired by means of a burning glass, or otherwise; the water descends upon the explosion, and stands lower in the tube than before, by a space proportioned to the quantity of powder fired.

Another way was by firing the powder in vacuo, viz, in an exhausted receiver, by dropping the grains of powder upon a hot iron included in the receiver. By this means a permanent elastic fluid was generated from the fired gunpowder, and the quantity of it was always in proportion to the quantity of powder that was used, as was found by the proportional sinking of the mercurial gage annexed to the air pump. The result of these experiments was, that the weight of the elastic air thus generated, was equal to $\frac{3}{10}$ of the compound mass of the gunpowder which yielded it; and that its bulk, when cold and expanded to the rarity of common atmospheric air, was about 240 times the bulk of the powder; and consequently in the same proportion

would such fluid at first, if it were cold, exceed the force or elasticity of the atmosphere. But as Mr. Robins found, by another ingenious experiment, that air heated to the extreme degree of the white heat of iron, has its elasticity quadrupled, or is 4 times as strong; he thence inferred that the force of the elastic air generated as above, at the moment of the explosion, is at least 4 times 240, or 960, or in round numbers about 1000 times as strong as the elasticity or pressure of the atmosphere, on the same space.

Having thus determined the force of the gunpowder, or intensity of the agent by which the projectile is to be urged, Mr. Robins next proceeds to determine the effects it will produce, or the velocity with which it will impel a shot of a given weight from a piece of ordnance of given dimensions; which is a problem strictly limited, and perfectly soluble by mathematical rules, and is in general this: Given the first force, and the law of its variation, to determine the velocity with which it will impel a given body in passing through a given space, which is the length of the bore of the gun.

In the solution of this problem, Mr. Robins assumes these two postulates, viz, 1, That the action of the powder on the bullet ceases as soon as the bullet is out of the piece; and 2d, That all the powder of the charge is fired and converted into elastic fluid before the bullet is sensibly moved from its place: assumptions which for good reasons are found to be in many cases very near the truth. It is to be noted also, that the law by which the force of the elastic fluid varies, is this, viz, that its intensity is directly as its density, or reciprocally proportional to the space it occupies, being so much the stronger as the space is less: a principle well known, and common to all elastic fluids. Upon these principles then Mr. Robins resolves this problem, by means of the 39th prop. of Newton's Principia in a direct way, and the result is equivalent to this theorem, when the quantities are expressed by algebraic symbols; viz, the velocity of the ball

$$v = 27130 \sqrt{\frac{10a}{cd} \times \log. \frac{b}{a}}$$

$$\text{or} = 100 \sqrt{\frac{223ad^2}{w} \times \log. \frac{b}{a}};$$

where v is the velocity of the ball,
  a the length of the charge of powder,
  b the whole length of the bore,
  c the spec. grav. of the ball, or wt. of a cubic foot of the same matter in ounces,
  d the diam. of the bore,
  w the wt. of the ball in ounces.

For example, Suppose $a = 2\frac{1}{3}$ inc., $b = 45$ inches, $c = 11345$ oz, for a ball of lead, and $d = \frac{3}{4}$ inches; then $v = 27130 \sqrt{\frac{7}{2269} \times \log. \frac{120}{7}} = 1674$ feet per second, the velocity of the ball.

Or, if the wt. of the bullet be $w = 1\frac{9}{20}$ oz $= \frac{29}{20}$ oz. Then $v = 100 \sqrt{\frac{1115 \times 189}{29 \times 32} \times \log. \frac{120}{7}} = 1674$ feet, as before.

" Having in this proposition, says Mr. Robins, shewn how the velocity, which any bullet acquires from the

the force of powder, may be computed upon the principles of the theory laid down in the preceding propositions; we shall next shew, that the actual velocities, with which bullets of different magnitudes are impelled from different pieces, with different quantities of powder, are really the same with the velocities assigned by these computations; and consequently that this theory of the force of powder, here delivered, does unquestionably ascertain the true action and modification of this enormous power.

" But in order to compare the velocities communicated to bullets by the explosion with the velocities resulting from the theory by computation; it is necessary that the actual velocities with which bullets move, should be capable of being discovered, which yet is impossible to be done by any methods hitherto made public. The only means hitherto practised by others for that purpose, have been either by observing the time of the flight of the shot through a given space, or by measuring the range of the shot at a given elevation; and thence computing, on the parabolic hypothesis, what velocity would produce this range. The first method labours under this insurmountable difficulty, that the velocities of these bodies are often so swift, and consequently the time observed is so short, that an imperceptible error in that time may occasion an error in the velocity thus found, of 2, 3, 4, 5, or 600 feet in a second. The other method is so fallacious, by reason of the resistance of the air (to which inequality the first is also liable), that the velocities thus assigned may not be perhaps the 10th part of the actual velocities sought.

" To remedy then these inconveniences, I have invented a new method of finding the real velocities of bullets of all kinds; and this to such a degree of exactness (which may be augmented too at pleasure), that in a bullet moving with a velocity of 1700 feet in 1″, the error in the estimation of it need never amount to its 500th part; and this without any extraordinary nicety in the construction of the machine."

Mr. Robins then gives an account of the machine by which he measures the velocities of the balls, which machine is simply this, viz, a pendulous block of wood suspended freely by a horizontal axis, against which block are to be fired the balls whose velocities are to be determined.

" This instrument thus fitted, if the weight of the pendulum be known, and likewise the respective distances of its centre of gravity, and of its centre of oscillation, from its axis of suspension, it will thence be known what motion will be communicated to this pendulum by the percussion of a body of a known weight moving with a known degree of celerity, and striking it in a given point; that is, if the pendulum be supposed at rest before the percussion, it will be known what vibration it ought to make in consequence of such a determined blow; and, on the contrary, if the pendulum, being at rest, is struck by a body of a known weight, and the vibration, which the pendulum makes after the blow, is known, the velocity of the striking body may from thence be determined.

" Hence then, if a bullet of a known weight strikes the pendulum, and the vibration, which the pendulum makes in consequence of the stroke, be ascertained;

the velocity with which the ball moved, is thence to be known."

Mr. Robins then explains his method of computing velocities from experiments with this machine; which method is rather troublesome and perplexed, as well as the rules of Euler and Antoni, who followed him in this business, but a much simpler rule is given in my Tracts, vol. 1, p. 119, where such experiments are explained at full length, and this rule is expressed by either of the two following formulas,

$$v = 5 \cdot 6727 cg \times \frac{p+b}{bir} \quad \sqrt{o} = 614 \cdot 58 cg \times \frac{p+b}{birn}, \text{ the}$$

velocity; where v denotes the velocity of the ball when it strikes the pendulum, p the weight of the pendulum, b the weight of the ball, c the chord of the arc described by the vibration to the radius r, g the distance below the axis of motion to the centre of gravity, o the distance to the centre of oscillation, i the distance to the point of impact, and n the number of oscillations the pendulum will perform in one minute, when made to oscillate in small arcs. The latter of these two theorems is much the easiest, both because it is free of radicals, and because the value of the radical $\sqrt{o}$, in the former, is to be first computed from the number n, or number of oscillations the pendulum is observed to make.

With such machines Mr. Robins made a great number of experiments, with musket barrels of different lengths, with balls of various weights, and with different charges or quantities of powder. He has set down the results of 61 of these experiments, which nearly agree with the corresponding velocities as computed by his theory of the force of powder, and which therefore establish that theory on a sure foundation.

From these experiments, as well as from the preceding theory, many important conclusions were deduced by Mr. Robins; and indeed by means of these it is obvious that every thing may be determined relative both to the true theory of projectiles, and to the practice of artillery. For, by firing a piece of ordnance, charged in a similar manner, against such a ballistic pendulum from different distances, the velocity lost by passing through such spaces of air will be found, and consequently the resistance of the air, the only circumstance that was wanting to complete the theory of Gunnery, or military projectiles; and of this kind I have since made a great number of experiments with cannon balls, and have thereby obtained the whole series of resistances to such a ball when moving with every degree of velocity, from o up to 2000 feet per second of time. In the structure of artillery, they may likewise be of the greatest use: For hence may be determined the best lengths of guns; the proportions of the shot and powder to the several lengths; the thickness of a piece, so as it may be able to confine, without bursting, any given charge of powder; as also the effect of wads, chambers, placing of the vent, ramming the powder, &c. For the many other curious circumstances relating to this subject, and the various other improvements in the theory and practice of Gunnery, made by Mr. Robins, consult the first vol. of his Tracts, collected and published by Dr. Wilson, in the year 1761, where ample information may be found.

Soon after the first publication of Robins's New Principles of Gunnery, in 1742, the learned in several other nations, treading in his steps, repeated and farther extended the same subject, sometimes varying and enlarging the machinery; particularly Euler in Germany, D'Antoni in Italy, and Meffrs. D'Arcy and Le Roy in France. But most of these, like Mr. Robins, with small fire-arms, such as muskets, and fufils.

But in the year 1775, in conjunction with several able officers of the Royal Artillery, and other ingenious gentlemen, I undertook a course of experiments with the ballistic pendulum, in which we ventured to extend the machinery to cannon shot of 1, 2, and 3 pounds weight. An account of these experiments was published in the Philof. Tranf. for 1778, and for which the Royal Society honoured me with the prize of the gold medal. " These were the only experiments that I know of which had been made with cannon balls for this purpose, although the conclusions to be deduced from such, are of the greatest importance to those parts of natural philosophy which are dependent on the effects of fired gunpowder; nor do I know of any other practical method of ascertaining the initial velocities within any tolerable degree of the truth. The knowledge of this velocity is of the utmost consequence in Gunnery: by means of it, together with the law of the resistance of the medium, every thing is determinable relative to that business; for, besides its being an excellent method of trying the strength of different forts of powder, it gives us the law relative to the different quantities of powder, to the different weights of shot, and to the different lengths and sizes of guns. Besides these, there does not seem to be any thing wanting to answer any inquiry that can be made concerning the flight and ranges of shot, except the effects arising from the resistance of the medium. In these experiments the weights of the pendulums employed were from 300 to near 600 pounds. In that paper is described the method of constructing the machinery, of finding the centres of gravity and oscillation of the pendulum, and of making the experiments, which are all set down in the form of a journal, with all the minute and concomitant circumstances; as also the investigation of the new and easy rule, set down just above, for computing the velocity of the ball from the experiments. The charges of powder were varied from 2 to 8 ounces, and the shot from 1 to near 3 pounds. And from the whole were clearly deduced these principal inferences, viz,

" 1. First, That gunpowder fires almost instantaneously.—2. That the velocities communicated to balls or shot, of the same weight, by different quantities of powder, are nearly in the subduplicate ratio of those quantities: a small variation, in defect, taking place when the quantities of powder became great.—3. And when shot of different weights are employed, with the same quantity of powder, the velocities communicated to them, are nearly in the reciprocal subduplicate ratio of their weights.—4. So that, universally, shot which are of different weights, and impelled by the firing of different quantities of powder, acquire velocities which are directly as the square roots of the quantities of powder, and inversely as the square roots of the weights of the shot, nearly.—5. It would therefore be a great improvement in artillery, to make use of shot of a long form, or of heavier matter; for thus the momentum of a shot, when fired with the same weight of powder, would be increased in the ratio of the square root of the weight of the shot.—6. It would also be an improvement to diminish the windage; for by so doing, one-third or more of the quantity of powder might be saved.—7. When the improvements mentioned in the last two articles are considered as both taking place, it is evident that about half the quantity of powder might be saved, which is a very considerable object. But important as this saving may be, it seems to be still exceeded by that of the article of the guns; for thus a small gun may be made to have the effect and execution of another of two or three times its size in the present mode, by discharging a shot of two or three times the weight of its natural ball or round shot. And thus a small ship might discharge shot as heavy as those of the greatest now made use of.

" Finally, as the above experiments exhibit the regulations with regard to the weights of powder and balls, when fired from the same piece of ordnance, &c; so by making similar experiments with a gun, varied in its length, by cutting off from it a certain part before each course of experiments, the effects and general rules for the different lengths of guns may be certainly determined by them. In short, the principles on which these experiments were made, are so fruitful in consequences, that, in conjunction with the effects resulting from the resistance of the medium, they seem to be sufficient for answering all the enquiries of the speculative philosopher, as well as those of the practical artillerist.

In the year 1786 was published the first volume of my Tracts, in which is detailed, at great length, another very extensive course of experiments which were carried on at Woolwich in the years 1783, 1784, and 1785, by order of the Duke of Richmond, Master General of the Ordnance. The objects of this course were very numerous, but the principal of them were the following:

" 1. The velocities with which balls are projected by equal charges of powder, from pieces of the same weight and calibre, but of different lengths.

" 2. The velocities with different charges of powder, the weight and length of the gun being the same.

" 3. The greatest velocity due to the different lengths of guns, to be obtained by increasing the charge as far as the resistance of the piece is capable of sustaining.

" 4. The effect of varying the weight of the piece; every thing else being the same.

" 5. The penetration of balls into blocks of wood.

" 6. The ranges and times of flight of balls; to compare them with their initial velocities for determining the resistance of the medium.

" 7. The effect of wads;
of different degrees of ramming;
of different degrees of windage;
of different positions of the vent;
of chambers, and trunnions, and every other circumstance necessary to be known for the improvement of artillery."

All these objects were obtained in a very perfect and accurate manner; excepting only the article of ranges, which were not quite so regular and uniform as might be be wifhed. The balls too were moft of them of one pound weight; but the powder was increafed from 1 ounce, up till the bore was quite full; and the pendulum was from 600 to 800lb. weight. The conclufions from the whole were as follow:

" 1. That the former law, between the charge and velocity of ball, is again confirmed, viz, that the velocity is directly as the fquare root of the weight of powder, as far as to about the charge of 8 ounces: and fo it would continue for all charges, were the guns of an indefinite length. But as the length of the charge is increafed, and bears a more confiderable proportion to the length of the bore; the velocity falls the more fhort of that proportion.

" 2. That the velocity of the ball increafes with the charge to a certain point, which is peculiar to each gun, where it is greateft; and that by farther increafing the charge, the velocity gradually diminifhes, till the bore is quite full of powder. That this charge for the greateft velocity is greater as the gun is longer, but not greater however in fo high a proportion as the length of the gun is, fo that the part of the bore filled with powder bears a lefs proportion to the whole in the long guns, than it does in the fhort ones; the part of the whole which is filled being indeed nearly in the reciprocal fubduplicate ratio of the length of the empty part. And the other circumftances are as in this table.

Table of Charges producing the Greateft Velocity.				
Gun Num.	Length of the bore.	Length filled.	Part of the whole.	Wt. of the powder.
	inches.	inches.		oz.
1	28·2	8·2	$\frac{3}{10}$	12
2	38·1	9·5	$\frac{1}{12}$	14
3	57·4	10·7	$\frac{3}{16}$	16
4	79·9	12·1	$\frac{3}{20}$	18

" 3. It appears that the velocity continually increafes as the gun is longer, though the increafe in velocity is but very fmall in refpect of the increafe in length, the velocities being in a ratio fomewhat lefs than that of the fquare roots of the length of the bore, but fomewhat greater than that of the cube roots of the length, and is indeed nearly in the middle ratio between the two.

" 4. The range increafes in a much lefs ratio than the velocity, and indeed is nearly as the fquare root of the velocity, the gun and elevation being the fame. And when this is compared with the property of the velocity and length of gun in the foregoing paragraph, we perceive that very little is gained in the range by a great increafe in the length of the gun, the charge being the fame. And indeed the range is nearly as the 5th root of the length of the bore; which is fo fmall an increafe, as to amount only to about ⅓th part more range for a double length of gun.

" 5. It alfo appears that the time of the ball's flight is nearly as the range; the gun and elevation being the fame.

" 6. It appears that there is no fenfible difference caufed in the velocity or range, by varying the weight of the gun, nor by the ufe of wads, nor by different degrees of ramming, nor by firing the charge of powder in different parts of it.

" 7. But a great difference in the velocity arifes from a fmall degree of windage. Indeed with the ufual eftablifhed windage only, namely, about $\frac{1}{20}$th of the caliber, no lefs than between ⅓ and ¼ of the powder efcapes and is loft. And as the balls are often fmaller than that fize, it frequently happens that half the powder is loft by unneceffary windage.

" 8. It appears that the refifting force of wood, to balls fired into it, is not conftant. And that the depths penetrated by different velocities or charges, are nearly as the logarithms of the charges, inftead of being as the charges themfelves, or, which is the fame thing, as the fquare of the velocity.

" 9. Thefe, and moft other experiments, fhew that balls are greatly deflected from the direction they are projected in; and that fo much as 300 or 400 yards in a range of a mile, or almoft ¼th of the range, which is nearly a deflection of an angle of 15 degrees.

" 10. Finally, thefe experiments furnifh us with the following concomitant data, to a tolerable degree of accuracy, namely, the dimenfions and elevation of the gun, the weight and dimenfions of the powder and fhot, with the range and time of flight, and the firft velocity of the ball. From which it is to be hoped that the meafure of the refiftance of the air to projectiles, may be determined, and thereby lay the foundation for a true and practical fyftem of Gunnery, which may be as well ufeful in fervice as in theory."

Since the publication of thofe Tracts, we have profecuted the experiments ftill farther, from year to year, gradually extending our aim to more objects, and enlarging the guns and machinery, till we have arrived at experiments with the 6 pounder guns, and pendulums of 1800 pounds weight. One of the new objects of enquiry, was the refiftance the atmofphere makes to military projectiles; to obtain which, the guns have been placed at many different diftances from the pendulum, againft which they are fired, to get the velocity loft in paffing through thofe fpaces of air; by which, and the ufe of the whirling machine, defcribed near the end of the 1ft vol. of Robins's Tracts, for the flower motions, I have inveftigated the refiftance of the air to given balls moving with all degrees of velocity, from 0 up to 2000 feet per fecond; as well as the refiftance for many degrees of velocity, to planes and figures of other fhapes, and inclined to their path in all varieties of angles; from which I have deduced general laws and formulas for all fuch motions.

Mr. Robins made alfo fimilar experiments on the refiftance of the air; but being only with mufket bullets, on account of their fmallnefs, and of then change of figure by the explofion of the powder, I find they are very inaccurate, and confiderably different from thofe above mentioned, which were accurately made with pretty confiderable cannon balls, of iron. For which reafon we may omit here the rules and theory deduced from them by Mr. Robins, till others more correct fhall

have

have been established. All these experiments indeed
agree in evincing the very enormous resistance the air
makes to the swift motions of military projectiles,
amounting in some cases to 20 or 30 times the weight
of the ball itself; on which account the common rules
for projectiles, deduced from the parabolic theory, are
of little or no use in real practice; for, from these experiments it is clearly proved, that the track described
by the flight even of the heaviest shot, is neither a parabola, nor yet approaching any thing near it, except
when they are projected with very small velocities; in
so much that some balls, which in the air range only
to the distance of one mile, would in vacuo, when
projected with the same velocity, range above 10 or 20
times as far. For the common rules of the parabolic
theory, see PROJECTILES. And for a small specimen
of my experiments on resistances, see the 2d vol. of the
Edinburgh Philos. Trans.; as also my Conic Sections
and Select Exercises, at the end, also the articles
FORCE, and RESISTANCE, in this Dictionary.

Mr. Benjamin Thompson instituted a very considerable course of experiments of the same kind as those of
Mr. Robins, with musket barrels, which was published
in the Philos. Trans. vol. 71, for the year 1781. In
these experiments, the conclusions of Mr. Robins are
generally confirmed, and several other curious circumstances in this business are remarked by Mr. Thompson. This gentleman also pursues a hint thrown out
by Mr. Robins relative to the determining the velocity
of a ball from the recoil of the pendulous gun itself.
Mr. Robins, in prop. 11, remarks that the effect of the
exploded powder upon the recoil of the gun, is the
same, whether the gun is charged with a ball, or without one; and that the chord, or velocity, of recoil with
the powder alone, being subtracted from that of the recoil when charged with both powder and ball, leaves the
velocity which is due to the ball alone. From whence
Mr. Thompson observes, that the inference is obvious,
viz, that the momentum thus communicated to the
gun by the ball alone, being equa. to the momentum
of the ball, this becomes known; and therefore being
divided by the known weight of the ball, the quotient
will be its velocity. Mr. Thompson sets a great value on this new rule, the velocities by means of which,
he found to agree nearly with several of those deduced
from the motion of the pendulum; and in the other cases
in which they differed greatly from these, he very incon-
sistently supposes that these latter ones are erroneous.
In the experiments however contained in my Tracts, a
great multitude of those cases are compared together,
and the inaccuracy of that new rule is fully proved.

Having in the 9th prop compared together a number of computed and experimented velocities of balls, to
verify his theory: in the 10th prop. Mr. Robins assigns
the changes in the force of powder, which arise from
the different state of the atmosphere, as to heat and
moisture, both which he finds have some effect on it,
but especially the latter. In prop. 11 he investigates
the velocity which the flame of gunpowder acquires
by expanding itself, supposing it fired in a given piece
of artillery, without either a bullet or any other body
before it. This velocity he finds is upwards of 7000 feet
per second. But the celebrated Euler, in his commentary on this part of Mr. Robins's book, thinks it may be

still much greater. And in this prop. too it is that
Mr. Robins declares his opinion, above alluded to, viz,
that the effect of the powder upon the recoil of the gun
is the same, in all cases, whether fired with a ball, or
without one.—In prop. 12 he ascertains the manner in
which the flame of powder impels a ball which is laid
at a considerable distance from the charge; shewing
here that the sudden accumulation and density of the
fluid against the ball, is the reason that the barrel is
so often burst in those cases.—In prop. 13 he enumerates the various kinds of powder, and describes the
properest methods of examining its goodness. He
here shews that the best proportion of the ingredients,
is when the saltpetre is ¾ of the whole compound mass
of the powder, and the sulphur and charcoal the other
¼ between them, in equal quantities. In this prop.
Mr. Robins takes occasion to remark upon the use of
eprouvettes, or methods of trying powder; condemning
the practice of the English in using what is called the
vertical eprouvette; as well as that of the French, in
using a small mortar, with a very large ball, and a small
charge of powder: and instead of these, he strongly
recommends the use of his ballistic pendulum, for its
great accuracy: But for still more dispatch, he says
he should use another method, which however he re-
serves to himself, without giving any particular description of it. From what has been done by Mr.
Robins upon this head, several persons have introduced
his method of suspending the gun as a pendulum, and
noting the quantity of its oscillating recoil when fired
with a certain quantity of powder; and of this kind I
have contrived a machine, which possesses several advantages over all others, being extremely simple, accurate, and expeditious; so much so indeed, that the
weighing out of the powder is the chief part of the
trouble. See GUNPOWDER, and EPROUVETTE.

The other or 2d chapter of Mr. Robins's work, in
8 propositions, treats " of the resistance of the air,
and of the track described by the flight of shot and
shells." And of these, prop. 1 describes the general
principles of the resistance of fluids to solid bodies
moving in them. Here Mr. Robins discriminates between continued and compressed fluids, which immediately rush into the space quitted by a body moving in
them, and whose parts yield to the impulse of the body
without condensing and accumulating before it; and
such fluids as are imperfectly compressed, rushing into
a void space with a limited velocity, as in the case of
our atmosphere, which condenses more and more before the ball as this moves quicker, and also presses
the less behind it, by following it always with only a
given velocity: hence it happens that the former fluid
will resist moving bodies in proportion to the square
of the velocity, while the latter resists in a higher
proportion. Proposition 2 is " to determine the resistance of the air to projectiles by experiments." One
of the methods for this purpose, is by the ballistic
pendulum, placing the gun at different distances from
it, by which he finds the velocity lost in passing through
certain spaces of air, and consequently the force of resistance to such velocities as the body moves with in the
several parts of its path. And another way was by
firing balls, with a known given velocity, over a large
piece of water, in which the fall and plunge of the ball
could

could be seen, and consequently the space it passed over in a given time. By these means Mr. Robins determined the resistances of the air to several different velocities, all which shewed that there was a gradual increase of the resistance, over the law of the square of the velocity, as the body moved quicker.—In the remaining propositions of this chapter, he proceeds a little farther in this subject of the resistance of the air; in which he lays down a rule for the proportion of the resistance between two assigned velocities; and he shews that when a 24 pound ball, fired with its full charge of powder, first issues from the piece, the resistance it meets with from the air is more than 20 times its weight. He farther shews that "the track described by the flight of shot or shells is neither a parabola, nor nearly a parabola, unless they are projected with small velocities;" and that "bullets in their flight are not only depressed beneath their original direction by the action of gravity, but are also frequently driven to the right or left of that direction by the action of some other force: and in the 8th or last proposition, he pretends to shew that the depths of penetration of balls into firm substances, are as the squares of the velocities. But this is a mistake; for neither does it appear that his trials were sufficiently numerous or various, nor were his small leaden balls fit for this purpose; and I have found, from a number of trials with iron cannon balls, that the penetrations are in a much lower proportion, and that the resisting force of wood is not uniform. See my TRACTS.

In the following small tracts, added to the principles, in this volume, Mr. Robins prosecutes the subject of the resistance of the air much farther, and lays down rules for computing ranges made in the air. But these must be far from accurate, as they are founded on the two following principles, which I know, from numerous experiments, are erroneous: viz, 1st, "That till the velocity of the projectile surpasses that of 1100 feet in a second, the resistance may be esteemed to be in the duplicate proportion of the velocity. 2d, That if the velocity be greater than that of 11 or 1200 feet in a second, then the absolute quantity of that resistance in these greater velocities will be near 3 times as great, as it should be by a comparison with the smaller velocities." For, instead of leaping at once from the law of the square of the velocities, and ever after being about 3 times as much, my experiments prove that the increase of the resistance above the law of the square of the velocity, takes place at first in the smallest motions, and increases gradually more and more, to a certain point, but never rises so high as to be 3 times that quantity, after which it decreases again. To render this evident, I have inserted the following table of the actual quantities of resistances, which are deduced from accurate experiments, and which shew also the nature of the law of the variations, by means of the columns of differences annexed; reserving the detail of the experiments themselves to another occasion. These resistances are, upon a ball of 1·965 inc. diameter, in avoirdupois ounces, and are for all velocities, from 0, up to that of 2000 feet per second of time.

VOL. I.

The quantity of the resistance of the air to a ball of 1·965 inc. diameter.

Veloc. in feet	Resist. in ounces	1st Differences	2d Differences
0	0·000		
5	0·006		
10	0·025		
15	0·054		
20	0·100		
25	0·155		
30	0·23		
40	0·42		
50	0·67		
100	2¾		
200	11	8¾	5¾
300	25	14	6
400	45	20	7
500	72	27	8
600	107	35	9
700	151	44	10
800	205	54	12
900	271	66	13
1000	350	79	13
1100	442	92	12
1200	546	104	11
1300	661	115	9
1400	785	124	7
1500	916	131	4
1600	1051	135	0
1700	1186	135	2
1800	1319	133	5
1900	1447	128	6
2000	1569	122	

The additional tracts of Mr. Robins, in the latter part of this volume, which contain many useful and important matters, are numbered and titled as follows, viz, Number 1, "Of the resistance of the air. Number 2, Of the resistance of the air; together with the method of computing the motions of bodies projected in that medium. Number 3, An account of the experiments, relating to the resistance of the air, exhibited at different times before the Royal Society, in the year 1746. Number 4, Of the force of fired gunpowder, together with the computation of the velocities thereby communicated to military projectiles. Number 5, A comparison of the experimental ranges of cannon and mortars with the theory contained in the preceding papers.—Practical Maxims relating to the effects and management of artillery, and the flight of shells and shot.—A proposal for increasing the strength of the British navy, by changing all the guns, from the 18 pounders downwards, into others of equal weight, but of a greater bore." With several letters, and other papers, "On pointing, or the directing of cannon to strike distant objects; Of the nature and advantage of rifled barrel pieces," &c.

4 D

I have

I have dwelt thus long on Mr. Robins's New Principles of Gunnery, becauſe it is the firſt work that can be conſidered as attempting to eſtabliſh a practical ſyſtem of gunnery, and projectiles, on good experiments, on the force of gunpowder, on the reſiſtance of the air, and on the effects of different pieces of artillery. Thoſe experiments are however not ſufficiently perfect, both on account of the ſmallneſs of the bullets, and for want of good ranges, to form a proper theory upon. I have ſupplied ſome of the neceſſary deſiderata for this purpoſe, viz, the reſiſtance of the air to cannon balls moving with all degrees of velocity, and the velocities communicated by given charges of powder to different balls, and from different pieces of artillery. But there are ſtill wanting good experiments with different pieces of ordnance, giving the ranges and times of flight, with all varieties of charges, and at all different angles of elevation. A few however of thoſe I have obtained, as in the following ſmall table, which are derived from experiments made with a medium one-pounder gun, the iron ball being nearly 2 inches in diameter.

Powder	Elev. of gun	Veloc. of ball	Range	Time of flight
oz	o	feet	feet	''
2	15	860	4100	9
4	15	1230	5100	12
8	15	1640	6000	14½
12	15	1680	6700	15½
2	45	860	5100	21

The celebrated Mr. Euler added many excellent diſſertations on the ſubject of Gunnery, in his tranſlation of Robins's Gunnery into the German language; which were again farther improved in Brown's tranſlation of the ſame into Engliſh in the year 1777. See alſo Antoni's *Examen de la Poudre*; the experiments of MM. D'Arcy and Le Roy, in the Memoirs of the Academy in 1751; and D'Arcy's *Eſſai d'une theorie d'artillerie* in 1760: my Tracts; and paper on the force of fired gunpowder in the Philoſ. Tranſ. for 1778: and Thompſon's paper on the ſame ſubject in 1781. Of the common or parabolic theory of Gunnery, Mr. Simpſon gave a very neat and conciſe treatiſe in his Select Exerciſes. And other authors on this part, are Starrat, Gray, Williams, Glenie, &c.

GUNPOWDER, a compoſition of nitre, ſulphur, and charcoal, mixed together, and uſually granulated. This eaſily takes fire; and when fired, it rarefies or expands with great vehemence, by means of its elaſtic force.—It is to this powder that we owe all the effect and action of guns, and ordnance of all ſorts. So that fortification, with the modern military art, &c, in a great meaſure depends upon it. The above definition however is not general, for inſtead of the nitre, it has lately been diſcovered that the marine acid anſwers much better.

The invention of Gunpowder is aſcribed, by Polydore Virgil, to a chemiſt; who having accidentally put ſome of this compoſition in a mortar, and covered it with a ſtone, it happened to take fire, and blew up the ſtone. Thevet ſays that the perſon here ſpoken of was a monk of Fribourg, named Conſtantine Anelzen; but Belleforet and other authors, with more probability, hold it to be Bartholdus Schwartz, or the black, who diſcovered it, as ſome ſay, about the year 1320; and the firſt uſe of it is aſcribed to the Venetians, in the year 1380, during the war with the Genoeſe. But there are earlier accounts of its uſe, after the accident of Schwartz, as well as before it. For Peter Mexia, in his Various Readings, mentions that the Moors being beſieged in 1343, by Alphonſus the 11th, king of Caſtile, diſcharged a kind of iron mortars upon them, which made a noiſe like thunder; and this is ſeconded by what is related by Don Pedro, biſhop of Leon, in his chronicle of king Alphonſus, who reduced Toledo, viz, that in a ſea-combat between the king of Tunis and the Mooriſh king of Seville, about that time, thoſe of Tunis had certain iron tubs or barrels, with which they threw thunderbolts of fire.

Du-Cange adds, that there is mention made of gunpowder in the regiſters of the chambers of accounts in France as early as the year 1338.

But it appears that Roger Bacon knew of Gunpowder near 100 years before Schwartz was born. He tells us, in his Treatiſe De Secretis Operibus Artis & Naturæ, & de Nullitate Magiæ, cap. 6, (which is ſuppoſed by ſome to have been publiſhed at Oxford in 1216, and which was undoubtedly written before his Opus Majus, in 1267), " that from ſaltpetre, and other ingredients, we are able to make a fire that ſhall burn at what diſtance we pleaſe." And Dr. Plott, in his Hiſtory of Oxfordſhire, pa. 236, aſſures us that theſe " other ingredients were explained in a MS. copy of the ſame treatiſe, in the hands of Dr. G. Langbain, and ſeen by Dr. Wallis, to be ſulphur and wood coal." Farther, in the life of Friar Bacon in Biographia Britannica, vol. 1, we are told that Bacon himſelf has divulged the ſecret of this compoſition in a cypher, by tranſpoſing the letters of the two words in chap. xi. of the ſaid treatiſe; where it is thus expreſſed: *ſed tamen ſalis petræ* LURA MOPE CAN UBRE (i. e. *carbonum pulvere*) *et ſulphuris; et ſic facies tonitrum & coruſcationem, ſi ſcias artificium:* and from hence the biographer apprehends the words *carbonum pulvere* were transferred to the 6th chapter of Langbain's MS. In this ſame chapter Bacon expreſsly ſays that ſounds like thunder, and coruſcations, may be formed in the air, much more horrible than thoſe that happen naturally. And farther adds, that there are many ways of doing this, by which a city or an army might be deſtroyed: and he ſuppoſes that by an artifice of this kind Gideon defeated the Midianites with only 300 men: Judges, chap. 7. There is alſo another paſſage to the ſame purpoſe, in the treatiſe De Scientia Experimentali. See Dr. Jebb's edition of the Opus Majus, p. 474.

Mr. Robins, in the preface to his Gunnery, apprehends that Bacon deſcribes Gunpowder not as a new compoſition firſt propoſed by himſelf, but as the application of an old one to military purpoſes, and that it was known long before his time.

But Mr. Dutens carries the antiquity of Gunpowder

der still much higher, and refers to the writings of the ancients themselves for the proof of it. " Virgil, says he, and his Commentator Servius (Æneid, lib. 6, v. 585), Hyginus (Fabul. 61 and 650), Eustathius (ad Odyss, λ 234, pa. 1682, lib. 1), La Cerda (in Virgil. loc. cit.), Valerius Flaccus (lib. i. 662), and many other authors (as Raphael Volatarran. in Commentar. Cornelius Agrippa poster. Oper. de Verbo Dei, c. 100, p. 237.—Gruteri Fax Artium Liberal tom. 2, p. 1236), speak in such a manner of Salmoneus's attempts to imitate thunder, as suggest to us that this prince used for that purpose a composition of the nature of gunpowder. Eustathius in particular speaks of him on this occasion, as being so very expert in mechanics, that he formed machines, which imitated the noise of thunder; and the writers of fable, whose surprise in this respect may be compared to that of the Mexicans when they first beheld the fire-arms of the Spaniards, give out that Jupiter, incensed at the audacity of this prince, slew him with lightning, as he was employing himself in launching his thunder. But it is much more natural to suppose that this unfortunate prince, the inventor of Gunpowder, gave rise to these fables, by having accidentally fallen a victim to his own experiments. Dion (Hist. Rom. in Caligula, p. 662) and Joannes Antiochenus (in Chronico, &c. a Valesio edita, Paris 1634, p. 804), report the very same thing of Caligula, assuring us that this emperor imitated thunder and lightning by means of certain machines, which at the same time emitted stones. Themistius informs us that the Brachmans encountered one another with thunder and lightning, which they had the art of launching from on high at a considerable distance; (Themist. Oratio 27, p. 337). And in another place he relates, that Hercules and Bacchus, attempting to assail them in a fort where they were entrenched, were so roughly received by reiterated strokes of thunder and lightning, launched upon them from on high by the besieged, that they were obliged to retire, leaving behind them an everlasting monument of the rashness of their enterprise. Agathias the historian reports of Anthemius Traliensis, that having fallen out with his neighbour Zeno the rhetorician, he set fire to his house with thunder and lightning. It appears from all these passages, that the effects ascribed to these engines of war, especially those of Caligula, Anthemius, and the Indians, could be only brought about by Gunpowder. And what is still more, we find in Julius Africanus a receipt for an ingenious composition to be thrown upon an enemy, which very nearly resembles that powder. But what places this beyond all doubt, is a clear and positive passage of an author called Marcus Græcus, whose work in manuscript is in the royal library at Paris, intitled Liber Ignium. Dr. Mead had the same also in manuscript, and a copy of that is now in my hands. (See above). The author describes several ways of encountering an enemy, by launching fire upon him; and among others gives the following. Mix together one pound of live sulphur, two of charcoal of willow, and 6 of saltpetre; reducing them to a very fine powder in a marble mortar. He adds, that a certain quantity of this is to be put into a long, narrow, and well compacted cover, and so discharged into the air. Here we have the description

of a rocket. The cover with which thunder is imitated, he represents as short, thick, but half-filled, and strongly bound with packthread; which is exactly the form of a cracker. He then treats of different methods of preparing the match, and how one squib may set fire to another in the air, by having it inclosed within it. In short, he speaks as clearly of the composition and effects of Gunpowder, as any person in our times could do. I own I have not yet been able precisely to determine when this author lived, but probably it was before the time of the Arabian physician Mesue, who speaks of him, and who flourished in the beginning of the 9th century. Nay, there is reason to believe that he is the same of whom Galen speaks; in which case he will be of antiquity sufficient to support what I advance." It appears too, from many authors, and many circumstances, that this composition has been known to the Chinese and Indians for thousands of years. See what is said on this head under the article GUN.

To this history of Gunpowder it may be added, that it has lately been discovered that saltpetre or nitre is not essential to this composition, but that its place may be supplied by other substances; for new Gunpowder, of double the strength of the old, has lately been made in France, by the chemists in that country, without any nitre at all; and in the year 1790 I tried some of this new powder, that was made at Woolwich, with my eprouvette, when I found it about double the strength of the ordinary sort. This is effected by substituting, instead of the nitre, the like quantity of the marine acid.

But perhaps this new composition may not come into common and general use; both because of the great expence in procuring or making the acid, and of the trouble and danger of preventing it from taking fire by the heat of making it; for it is found to catch fire and explode from a very small degree of heat, and without the aid of a spark.

As to the Preparation of GUNPOWDER; there are divers compositions of it, with respect to the proportions of the three ingredients, to be met with in pyrotechnical writings; but the process of making it up is much the same in all.

For some time after the invention of artillery, Gunpowder was of a much weaker composition than that now in use, or that described by Marcus Græcus; which was chiefly owing to the weakness of their first pieces. See GUN and CANNON. Of 23 different compositions, used at different times, and mentioned by Tartaglia in his Ques. and Inv. lib. 3, quest. 5, the first, which was the oldest, contained equal parts of the three ingredients. But when guns of modern structure were introduced, Gunpowder of the same composition as the present came also into use. In the time of Tartaglia the cannon powder was made of 4 parts of nitre, one of sulphur, and one of charcoal; and the musket powder of 48 parts of nitre, 7 parts of sulphur, and 8 parts of charcoal; or of 18 parts of nitre, 2 parts of sulphur, and 3 parts of charcoal. But the modern composition is 6 parts of nitre, to one of each of the other two ingredients. Though Mr. Napier says, he finds the strength commonly to be greatest when the proportions are, nitre 3lb, charcoal about 9oz, and sulphur about 3oz. See his paper on Gunpowder in the Transactions of the Royal Irish Academy, vol. 2. The can-

non

non powder was in meal, and the musket powder grained. And it is certain that the graining of powder, which is a very considerable advantage, is a modern improvement. See the preface to Robins's Math. Tracts, pa. 32.

To make Gunpowder duly, regard is to be had to the purity or goodness of the ingredients, as well as the proportions of them ; for the strength of the powder depends much on that circumstance, and also on the due working or mixing of them together.

To purify the nitre, by taking away the fixt or common salt, and earthy part. Dissolve it in a quantity of hot water over the fire ; then filtrate it through a flannel bag, into an open vessel, and set it aside to cool, and to crystallize. These crystals may in like manner be dissolved and crystallized again ; and so on, till they become quite pure and white. Then put the crystals into a dry kettle over a moderate fire, which gradually increase till it begins to smoke, evaporate, lose its humidity, and grow very white : it must be kept continually stirring with a ladle, lest it should return to its former figure, by which its greasiness would be taken away : after that, so much water is to be poured into the kettle as will cover the nitre ; and when it is dissolved, and reduced to the consistency of a thick liquor, it must be continually stirred with a ladle till all the moisture is again evaporated, and it be reduced to a dry and white meal.

The like regard is to be had to the sulphur ; choosing that which is in large lumps, clear and perfectly yellow ; not very hard, nor compact, but porous ; nor yet too much shining ; and if, when set on fire, it freely burns all away, it is a sign of its goodness : so likewise, if it be pressed between two iron plates that are hot enough to make it run, and in the running appear yellow, and that which remains of a reddish colour, it is then fit for the purpose. But in case it be foul, it may be purified in this manner : melt the sulphur in a large iron ladle, or pot, over a very gentle coal fire, well kindled, but not flaming ; then scum off all that rises on the top, and swims upon the sulphur ; take it presently after from the fire, and strain it through a double linen cloth, letting it pass leisurely ; so will it be pure, the gross matter remaining behind in the cloth.

For the charcoal, the third ingredient, such should be chosen as is large, clear, and free from knots, well burnt, and cleaving. The charcoal of light woods is mostly preferred, as of willow, and that of the branches or twigs of a moderate thickness, as of an inch or two in diameter. Dogwood is now much esteemed for this purpose. And a method of charring the wood in a large iron cylinder has lately been recommended, and indeed proved, as yielding better charcoal than formerly.

The charcoal not only concurs with the sulphur in supplying the inflammable matter, which causes the detonation of the nitre, but also greatly adds to the explosive power of it by the quantity of elastic vapour expelled during its combustion.

These three ingredients, in their purest state, being procured, long experience has shewn that they are then to be mixed together in the proportion before mentioned, to have the best effect, viz, three-quarters of the composition to be nitre, and the other quarter made up

of equal parts of the other two ingredients ; or, which is the same thing, 6 parts nitre, 1 part sulphur, and 1 part charcoal.

But it is not the due proportion of the materials only, which is necessary to the making of good powder; another circumstance, not less essential, is the mixing them well together : if this be not effectually done, some parts of the composition will have too much nitre in them, and others too little ; and in either case there will be a defect of strength in the powder. Robins, pa. 119.

After the materials have been reduced to fine dust, they are mixed together, and moistened with water, or vinegar, or urine, or spirit of wine, &c, and then beaten together with wooden pestles for 24 hours, either by hand, or by mills, and afterwards pressed into a hard, firm, and solid cake. When dry, it is grained or corned ; which is done by breaking the cake of powder into small pieces, and so running it through a sieve ; by which means the grains may have any size given them, according to the nature of the sieve employed, either finer or coarser ; and thus also the dust is separated from the grains, and again mixed with other manufacturing powder, or worked up into cakes again.

Powder is smoothed, or glazed, as it is called, for small arms, by the following operation : a hollow cylinder or cask is mounted on an axis, turned by a wheel ; this cask is half filled with powder, and turned for 6 hours ; and thus by the mutual friction of the grains of powder it is smoothed, or glazed. The fine mealy part, thus separated or worn off from the rest, is again granulated.

The Nature, Effects, &c, of Powder.—When the powder is prepared as above, if the least spark be struck upon it from a steel and flint, the whole will immediately inflame, and burst out with extreme violence.—The effect is not hard to account for : the charcoal part of the grain upon which the spark falls, catching fire like tinder, the sulphur and nitre are readily melted, and the former also breaks into flame ; the contiguous grains at the same time undergoing the same fate.

Sir Isaac Newton reasons thus upon the point : The charcoal and sulphur in Gunpowder easily take fire, and kindle the nitre ; and the spirit of the nitre, being thereby rarefied into vapour, rushes out with an explosion much after the manner that the vapour of water rushes out of an eolipile ; the sulphur also, being volatile, is converted into vapour, and augments the explosion : add, that the acid vapour of the sulphur, namely that which distils under a bell into oil of sulphur, entering violently into the fixt body of the nitre, lets loose the spirit of the nitre, and excites a greater fermentation, by which the heat is farther augmented, and the fixt body of the nitre is also rarefied into fume ; and the explosion is thereby made more vehement and quick.

For if salt of tartar be mixed with Gunpowder, and that mixture be warmed till it takes fire, the explosion will be far more violent and quick than that of Gunpowder alone ; which cannot proceed from any other cause, than the action of the vapour of the Gunpowder upon the salt of tartar, by which that salt is rarefied.

The explosion of Gunpowder therefore arises from the violent action, by which all the mixture being quickly and

and vehemently heated, is rarefied and converted into fume and vapour; which vapour, by the violence of that action becoming so hot as to shine, appears in the form of a flame.

M. De la Hire, in the History of the French Academy for 1702, ascribes all the force and effect of Gunpowder to the spring or elasticity of the air inclosed in the several grains of it, and in the intervals or spaces between the grains: the powder being kindled, sets the springs of so many little parcels of air a-playing, and dilates them all at once, whence the effect; the powder itself only serving to light a fire which may put the air in action; after which the whole is done by the air alone.

But it appears from the experiments and observations of Mr. Robins, that if this air be in its natural state at the time when the powder is fired, the greatest addition its elasticity could acquire from the flame of the explosion, would not amount to five times its usual quantity, and therefore could not suffice for the 200th part of the effort which is exerted by fired powder.

To understand the force of Gunpowder, it must be considered that, whether it be fired in a vacuum or in air, it produces by its explosion a permanent elastic fluid. See Philos. Transf. number 295; also Haukſbee's Physf. Mechan. Exp. p. 81. It also appears from experiment, that the elasticity or pressure of the fluid produced by the firing of Gunpowder, is, cæteris paribus, directly as its density.

To determine the elasticity and quantity of this elastic fluid, produced from the explosion of a given quantity of Gunpowder, Mr. Robins premises, that the elasticity of this fluid increases by heat, and diminishes by cold, in the same manner as that of the air; and that the density of this fluid, and consequently its weight, is the same with the weight of an equal bulk of air, having the same elasticity and the same temperature. From these principles, and from the experiments by which they are established (for a detail of which we must refer to the book itself, so often cited in these articles), he concludes that the fluid produced by the firing of Gunpowder is nearly $\frac{3}{10}$ of the weight of the generating powder itself; and that the volume or bulk of this air or fluid, when expanded to the rarity of common atmospheric air, is about 244 times the bulk of the said generating powder.—Count Saluce, in his Miscel. Phil. Mathem. Soc. Priv. Taurin. p. 125, makes the proportion as 222 to 1; which he says agrees with the computation of Mess. Haukſbee, Amontons, and Belidor.

Hence it appears, that any quantity of powder fired in any confined space, which it adequately fills, exerts at the instant of its explosion against the sides of the vessel containing it, and the bodies it impels before it, a force at least 244 times greater than the elasticity of common air, or, which is the same thing, than the pressure of the atmosphere; and this without considering the great addition arising from the violent degree of heat with which it is endued at that time; the quantity of which augmentation is the next head of Mr. Robins's enquiry. He determines that the elasticity of the air is augmented in a proportion somewhat greater than that of 4 to 1, when heated to the extremest heat of red hot iron; and supposing that the

flame of fired Gunpowder is not of a less degree of heat, increasing the former number a little more than 4 times, makes nearly 1000; which shews that the elasticity of the flame, at the moment of explosion, is about 1000 times stronger than the elasticity of common air, or than the pressure of the atmosphere. But, from the height of the barometer, it is known that the pressure of the atmosphere upon every square inch, is on a medium 14¾ lb; and therefore 1000 times this, or 14750lb, is the force or pressure of the flame of Gunpowder, at the moment of explosion, upon a square inch, which is very nearly equivalent to 6 tons and a half.

This great force however diminishes as the fluid dilates itself, and in that proportion, viz, in proportion to the space it occupies, it being only half the strength when it occupies a double space, one third the strength when triple the space, and so on.

Mr. Robins farther supposes the degree of heat above mentioned to be a kind of medium heat; but that in the case of large quantities of powder the heat will be higher, and in very small quantities lower; and that therefore in the former case the force will be somewhat more, and in the latter somewhat less, than 1000 times the force of the atmosphere.

He farther found that the strength of powder is the same in all variations in the density of the atmosphere. But that the moisture of the air has a great effect upon it; for the same quantity which in a dry season would discharge a bullet with a velocity of 1700 feet in one second, will not in damp weather give it a velocity of more than 12 or 1300 feet in a second, or even less, if the powder be bad, and negligently kept. Robins's Tracts, vol. 1, p. 101, &c. Farther, as there is a certain quantity of water which, when mixed with powder, will prevent its firing at all, it cannot be doubted but every degree of moisture must abate the violence of the explosion; and hence the effects of damp powder are not difficult to account for.

It is to be observed, that the moisture imbibed by powder does not render it less active when dried again. Indeed, if powder be exposed to very great damps without any caution, or when common salt abounds in it, as often happens through negligence in refining the nitre, in such cases the moisture it imbibes may perhaps be sufficient to dissolve some part of the nitre: which is a permanent damage that no drying can retrieve. But when tolerable care is taken in preserving powder, and the nitre it is composed of has been well purged from common salt, it will retain its force for a long time; and it is said that powder has been known to have been preserved for 50 years without any apparent damage from its age.

The velocity of expansion of the flame of Gunpowder, when fired in a piece of artillery, without either bullet or other body before it, is prodigiously great, viz, 7000 feet per second, or upwards, as appears from the experiments of Mr. Robins. But Mr. Bernoulli and Mr. Euler suspect it is still much greater. And I suspect it may not be less, at the moment of explosion, than 4 times as much.

It is this prodigious celerity of expansion of the flame of fired Gunpowder, which is its peculiar excellence, and the circumstance in which it so eminently surpasses

surpasses all other inventions, either ancient or modern : for as to the momentum of these projectiles only, many of the warlike machines of the ancients produced this in a degree far surpassing that of our heaviest cannon shot or shells ; but the great celerity given to these bodies cannot be in the least approached by any other means but the flame of powder.

To prove Gunpowder. There are several ways of doing this. 1, By sight : thus, if it be too black, it is a sign that it is moist, or else that it has too much charcoal in it ; so also, if rubbed upon white paper, it blackens it more than good powder does : but if it be of a kind of azure colour, somewhat inclining to red, it is a sign of good powder. 2, By touching : for if in crushing it with the fingers ends, the grains break easily, and turn into dust, without feeling hard, it has too much coal in it ; or if, in pressing it under the fingers upon a smooth hard board, some grains feel harder than the rest, it is a sign the sulphur is not well mixed with the nitre. Also by thrusting the hand into the parcel of powder, and grasping it, as if to take out a handful, you will feel if it is dry and equal grained, by its evading the grasp, and running mostly out of the hand. 3, By burning ; and here the method most commonly followed for this purpose with us, says Mr. Robins, is to fire a small heap of it on a clean board, and to attend nicely to the flame and smoke it produces, and to the marks it leaves behind on the board : but besides this uncertain method, there are other contrivances made use of, such as powder triers acting by a spring, commonly sold at the shops, and others again that move a great weight, throwing it upwards, which is a very bad sort of eprouvette. But these machines, says Mr. Robins, though more perfect than the common powder-triers, are yet liable to great irregularities ; for as they are all moved by the instantaneous stroke of the flame, and not by its continued pressure, they do not determine the force of the fired powder with sufficient certainty and uniformity. Another method is to judge from the range given to a large solid ball, thrown from a very short mortar, charged with a small quantity of powder ; which is also an uncertain way, both on account of the great disproportion between the weight of the ball and powder, and the unequal resistance of the air ; not to mention that it is too tedious to prove large quantities of powder in this way ; for, " if each barrel of powder was to be proved in this manner, the trouble of charging the mortar, and bringing back the ball each time, would be intolerable, and the delay so great, that no business of this kind could ever be finished ; and if a number of barrels are received on the merit of a few, it is great odds, but some bad ones would be amongst them, which may prove a great disappointment in time of service." These exceptions do " noways hold, continues Mr. Robins, against the method by which I have tried the comparative strength of different kinds of powder, which has been by the actual velocity given to a bullet, by such a quantity of powder as is usually esteemed a proper charge for the piece : and as this velocity, however great, is easily discovered by the motion which the pendulum acquires from the stroke of the bullet, it might seem a good amendment to the method used by the French (viz, that of the small mortar above

mentioned) to introduce this trial by the pendulum instead of it. But though I am satisfied, that this would be much more accurate, less laborious, and readier than the other, yet, as there is some little attention and caution required in this practice, which might render it of less dispatch than might be convenient, when a great number of barrels were to be separately tried, I should myself choose to practise another method not less certain, but prodigiously more expeditious ; so that I could engage, that the weighing out of a small parcel of powder from each barrel should be the greatest part of the labour ; and, doubtless, three or four hands could, by this means, examine 500 barrels in a morning : besides, the machines for this purpose, as they might be made of cast iron, would be so very cheap, that they might be multiplied at pleasure." Robins, page 123. It is not certainly known what might be the particular construction of the eprouvette here hinted at, but it was probably a piece of ordnance suspended like a pendulum, as he had made several experiments with a barrel in that manner. Be this however as it may, several persons, from those ideas and experiments of Mr. Robins, have made eprouvettes on this principle, which seems to be the best of any ; and on this idea also I have lately made a machine for this purpose, which has several peculiar contrivances, and advantages over all others, both in the nature of its motion, and the divisions on its arc, &c. It is a small cannon, the bore of which is about one inch in diameter, and is usually charged with 2 ounces of powder, and with powder only, as a ball is not necessary, and the strength of the powder is accurately shewn by the arc of the gun's recoil. The whole machine is so simple, easy, and expeditious, that, as Mr. Robins observed above, the weighing of the powder is the chief part of the trouble ; and so accurate and uniform, that the successive repetitions or firings with the same quantity of the same sort of powder, hardly ever yield a difference in the recoil of the 100th part of itself.

To recover damaged Powder. The method of the powder merchants is this ; they put part of the powder on a sail-cloth, to which they add an equal weight of what is really good ; then with a shovel they mingle it well together, dry it in the sun, and barrel it up, keeping it in a dry and proper place.

Others again, if it be very bad, restore it by moistening it with vinegar, water, urine, or brandy ; then they beat it fine, sift it, and to every pound of powder add an ounce, or an ounce and a half, or two ounces (according as it is decayed), of melted nitre ; and afterwards these ingredients are to be moistened and well mixed, so that nothing may be discerned in the composition ; which may be known by cutting the mass, and then they granulate it as useful.

In case the powder be quite spoiled, the only way is to extract the saltpetre with water, in the usual way, by boiling, filtrating, evaporating, and crystallizing ; and then, with fresh sulphur and charcoal, to make it up afresh.

On the subject of Gunpowder, see also Euler on Robins's Gunnery, Antoni Examen de la Poudre, Baumé's Chemistry, and Thompson's Experiments in the Philos. Transf. for 1781.

GUNTER (EDMUND), an excellent English mathematician,

mematician, was born in Hertfordshire in 1581. He was educated at Westminster school under Dr. Busby, and from thence was elected to Christ-church College, Oxford, in 1599, where he took the degree of master of arts in 1606, and afterwards entered into holy orders; and in 1615 he took the degree of bachelor of divinity. But being particularly distinguished for his mathematical talents, when Mr. Williams resigned the professorship of astronomy in Gresham College, London, Mr. Gunter was chosen to succeed him, the 6th of March, 1619; where he greatly distinguished himself by his lectures and writings, and where he died in 1626, at only 45 years of age, to the great loss of the mathematical world.

Mr. Gunter was the author of many useful inventions and works. About the year 1606, he merited the title of an inventor, by the new projection of his Sector, which he then described in a Latin treatise, not printed however till some time afterwards.—In 1618 he had invented a small portable quadrant, for the more easy finding the hour and azimuth, and other useful purposes in astronomy.—And in 1620 or 1623, he published his *Canon Triangulorum*, or Table of Artificial Sines and Tangents, to the radius, 10,000,000 parts, to every minute of the quadrant, being the first tables of this kind published; together with the first 1000 of Briggs's logarithms of common numbers, which were in later editions extended to 10,000 numbers.—In 1622, he discovered, by experiment made at Deptford, the variation or changeable declination of the magnetic needle; his experiment shewing that the declination had changed by 5 degrees in the space of 42 years; and the same was confirmed and established by his successor Mr. Gellibrand.—He applied the logarithms of numbers, and of sines and tangents, to straight lines drawn on a scale or rule; with which proportions in common numbers and trigonometry were resolved by the mere application of a pair of compasses; a method founded on this property, that the logarithms of the terms of equal ratios are equidifferent. This was called Gunter's Proportion, and Gunter's Line; and the instrument, in the form of a two-foot scale, is now in common use for navigation and other purposes, and is commonly called the Gunter. He also greatly improved the Sector and other instruments for the same uses; the description of all which he published in 1624.—He introduced the common measuring chain, now constantly used in land-surveying, which is thence called Gunter's Chain.—Mr. Gunter drew the lines on the dials in Whitehall-garden, and wrote the description and use of them, by the direction of prince Charles, in a small tract; which he afterwards printed at the desire of king James, in 1624.—He was the first who used the word *co-sine*, for the sine of the complement of an arc. He also introduced the use of Arithmetical Complements into the logarithmical arithmetic, as is witnessed by Briggs, cap. 15, Arith. Log. And it has been said that he first started the idea of the Logarithmic Curve, which was so called, because the segments of its axis are the logarithms of the corresponding ordinates.

His works have been collected, and various editions of them have been published; the 5th is by Mr. William Leybourn, in 1673, containing the Description and Use of the Sector, Cross-staff, Bow, Quadrant, and other instruments; with several pieces added by Samuel Foster, Henry Bond, and William Leybourn.

GUNTER's CHAIN, the chain in common use for measuring land, according to true or statute measure; so called from Mr. Gunter its reputed inventor.

The length of the chain is 66 feet, or 22 yards, or 4 poles of $5\frac{1}{2}$ yards each; and it is divided into 100 links, of 7.92 inches each.

This Chain is the most convenient of any thing for measuring land, because the contents thence computed are so easily turned into acres. The reason of which is, that an acre of land is just equal to 10 square chains, or 10 chains in length and 1 in breadth, or equal to 100000 square links. Hence, the dimensions being taken in chains, and multiplied together, it gives the content in square chains; which therefore being divided by 10, or a figure cut off for decimals, brings the content to acres; after which the decimals are reduced to roods and perches, by multiplying by 4 and 40. But the better way is to set the dimensions down in links as integers, considering each chain as 100 links; then, having multiplied the dimensions together, producing square links, divide these by 100000, that is, cut off five places for decimals, the rest are acres, and the decimals are reduced to roods and perches, as before.

Ex. Suppose in measuring a rectangular piece of ground, its length be 795 links, and its breadth 480 links.

$$
\begin{array}{r}
63600 \\
3180 \\
\hline
\text{Ac. } 3.81600 \\
4 \\
\hline
\text{Ro. } 3.264 \\
40 \\
\hline
\text{Per. } 10.560
\end{array}
$$

So the content is 3 ac. 3 roods 10 perches.

GUNTER's LINE, a Logarithmic line, usually graduated upon scales, sectors, &c; and so called from its inventor Mr. Gunter.

This is otherwise called the *line of lines*, or *line of numbers*, and consists of the logarithms transferred upon a ruler, &c, from the tables, by means of a scale of equal parts, which therefore serves to resolve problems instrumentally, in the same manner as logarithms do arithmetically. For, whereas logarithms resolve proportions, or perform multiplication and division, by only addition and subtraction, the same are performed on this line by turning a pair of compasses over this way or that, or by sliding one slip of wood by the side of another, &c.

This line has been contrived various ways, for the advantage of having it as long as possible. As, first, on the two feet ruler or scale, by Gunter. Then, in 1627 the logarithms were drawn by Wingate, on two separate rulers, sliding against each other, to save the use of compasses in resolving proportions. They were also in 1627 applied to concentric circles by Oughtred. Then in a spiral form by Mr. Milburne of Yorkshire, about the year 1650. Also, in 1657, on the present common sliding rule, by Seth Partridge.

Lastly, Mr. William Nicholson has proposed another disposition of them, on concentric circles, in the Philos. Transf. an. 1787, pa. 251. His instrument is equivalent to a ftraight rule of 28½ inches long. It confifts of three concentric circles, engraved and graduated on a plate of about 1½ inch in diameter. From the centre proceed two legs, having right-lined edges in the direction of radii; which are moveable either fingly, or together. To ufe this inftrument; place the edge of one leg at the antecedent of any proportion, and the other at the confequent, and fix them to that angle: the two legs being then moved together, and the antecedent leg placed at any other number, the other leg gives its confequent in the like pofition or fituation on the lines.

The whole length of the line is divided into two equal intervals, or radii, of 9 larger divifions in each radius, which are numbered from 1 to 10, the 1 ftanding at the beginning of the line, becaufe the logarithm of 1 is 0, and the 10 at the end of each radius; alfo each of thefe 9 fpaces is fubdivided into 10 other parts, unequal according to the logarithms of numbers; the fmaller divifions being always 10ths of the larger; thus, if the large divifions be units or ones, the fmaller are tenth-parts; if the larger be tens, the fmaller are ones; and if the larger be 100's, the fmaller are 10's; &c.

Ufe of Gunter's Line. 1. *To find the product of two numbers.* Extend the compaffes from 1 to either of the numbers, and that extent will reach the fame way from the other number to the product. Thus, to multiply 7 and 5 together; extend the compaffes from 1 to 5, and that extent will reach from 7 to 35, which is the product.

2. *To divide one number by another.* Extend the compaffes from the divifor to 1, and that extent will reach the fame way from the dividend to the quotient. Thus, to divide 35 by 5; extend the compaffes from 5 to 1, and that extent will reach from 35 to 7, which is the quotient.

3. *To find a 4th Proportional to three given Numbers;* as fuppofe to 6, 9, and 10. Extend from 6 to 9, and that extent will reach from 10 to 15, which is the 4th proportional fought. And the fame way a 3d proportional is found to two given terms, extending from the 1ft to the 2d, and then from the 2d to the 3d.

4. *To find a Mean Proportional between two given numbers,* as fuppofe between 7 and 28. Extend from 7 to 28, and bifect that extent; then its half will reach from 7 forward, or from 28 backward, to 14, the mean proportional between them.—Alfo, to extract the fquare root, as of 25, which is only finding a mean proportional between 1 and the given fquare 25, bifect the diftance between 1 and 25, and the half will reach from 1 to 5, the root fought.—In like manner the cubic or 3d root, or the 4th, 5th, or any higher root, is found, by taking the extent between 1 and the given power; then take fuch part of it as is denoted by the index of the root, viz, the 3d part for the cube root, the 4th part for the 4th root, and fo on, and that part will reach from 1 to the root fought.

If the Line on the Scale or Ruler have a flider, this is to be ufed inftead of the compaffes.

GUNTER's QUADRANT, is a quadrant made of wood, brafs, or fome other fubftance; being a kind of ftereographic projection on the plane of the equinoctial, the eye being fuppofed in one of the poles: fo that the tropic, ecliptic, and horizon, form the arches of circles,

but the hour circles other curves, drawn by means of feveral altitudes of the fun, for fome particular latitude every day in the year.

The ufe of this inftrument, is to find the hour of the day, the fun's azimuth, &c, and other common problems of the fphere or globe; as alfo to take the altitude of an object in degrees. See QUADRANT.

GUNTER's SCALE, ufually called by feamen *the Gunter,* is a large plain fcale, having various lines upon it, of great ufe in working the cafes or queftions in Navigation.

This Scale is ufually 2 feet long, and about an inch and a half broad, with various lines upon it, both natural and logarithmic, relating to trigonometry, navigation, &c.

On the one fide are the natural lines, and on the other the artificial or logarithmic ones. The former fide is firft divided into inches and tenths, and numbered from 1 to 24 inches, running the whole length near one edge. One half the length of this fide confifts of two plane diagonal fcales, for taking off dimenfions to three places of figures. On the other half or foot of this fide, are contained various lines relating to trigonometry, in the natural numbers, and marked thus, viz,

Rumb, the rumbs or points of the compafs,
Chord, the line of chords,
Sine, the line of fines,
Tang. the tangents,
S. T. the femitangents,
and at the other end of this half are
Leag. leagues, or equal parts,
Rumb. another line of rumbs,
M. L. miles of longitude,
Chor. another line of chords.

Alfo in the middle of this foot are *L.* and *P.* two other lines of equal parts. And all thefe lines on this fide of the fcale ferve for drawing or laying down the figures to the cafes in trigonometry and navigation.

On the other fide of the fcale are the following artificial or logarithmic lines, which ferve for working or refolving thofe cafes; viz,

S. R. the fine rumbs,
T. R. the tangent rumbs,
Numb. line of numbers,
Sine, Sines,
V. S. the verfed fines,
Tang. the tangents,
Meri. Meridional parts.
E. P. Equal parts.

The late Mr. John Robertfon, librarian to the Royal Society, greatly improved this fcale, both as to fize and accuracy, for the ufe of mariners. He extended it to 30 inches long, 2 inches broad, and half an inch thick; upon which the feveral lines are very accurately laid down by Meffrs. Nairne and Blunt, ingenious inftrument makers. Mr. Robertfon died before his improved fcales were publifhed; but the account and defcription of them were fupplied and drawn up by his friend Mr. William Mountaine, and publifhed in 1778.

GUTTÆ, or *Drops,* in Architecture, are ornaments in form of little bells or cones, ufed in the Doric order, on the architrave, below the tryglyphs. There are ufually fix of them.

HADLEY's

H.

HADLEY's *Quadrant, Sextant*, &c, an excellent instrument so called from its inventor John Hadley, Esq. See its description and use under the article QUADRANT.

HAIL, or HAILSTONES, an aqueous concretion, usually in form of white or pellucid spherules, descending out of the atmosphere.

Hailstones assume various shapes, being sometimes round, at other times pyramidal, crenated, angular, thin, and flat, and sometimes stellated, with six radii like the small crystals of snow.

It is very difficult to account for the phenomena of hail in a satisfactory manner; and there are various opinions upon this head. It is usually conceived that hail is formed of drops of rain, frozen in their passage through the middle region. Others, as the Cartesians, take it for the fragments of a frozen cloud, half melted, and thus precipitated and congealed again. Signior Beccaria supposes, that it is formed in the higher regions of the air, where the cold is intense, and where the electric matter is very copious. In these circumstances, a great number of particles of water are brought near together, where they are frozen, and in their descent they collect other particles; so that the density of the substance of the Hailstone grows less and less from the centre; this being formed first in the higher regions, and the surface being collected in the lower. Accordingly, in mountains, Hailstones as well as drops of rain, are very small; and both agree in this circumstance, that the more intense is the electricity that forms them, the larger they are.

It is frequently observed that Hail attends thunder and lightning; and hence Beccaria observes, that as motion promotes freezing, so the rapid motion of the electrified clouds may promote that effect in the air.

Natural histories furnish us with a great variety of curious instances of extraordinary showers of Hail. See the Philos. Transf. number 203, 229; and Hist. de France, tom. 2, pa. 339.

HALF-MOON, in Fortification, is an outwork having only two faces, forming together a saliant angle, which is flanked by some part of the place, and of the other bastions. See DEMILUNE and RAVELIN.

HALF-TANGENTS, are the tangents of the half arcs. See SCALE and SEMITANGENTS.

HALLEY (Dr. EDMUND), a most eminent English mathematician, philosopher and astronomer, was born in the parish of St. Leonard, Shoreditch, near

London, Oct. 29, 1656. His father, a wealthy citizen and soap-boiler, resolving to improve the promising disposition observed in his son, put him first to St. Paul's school, where he soon excelled in all parts of classical learning, and made besides a considerable advance in the mathematics; so that, as Wood observes, he had perfectly learnt the use of the celestial globe, and could make a complete dial; and we are informed by Halley himself, that he observed the change of the variation of the magnetic needle at London in 1672, one year before he left school. In 1673 he was sent to Oxford, where he chiefly applied himself to mathematics and astronomy, in which he was greatly assisted by a curious apparatus of instruments, which his father, willing to encourage his son's genius, had purchased for him. At 19 years of age he began to oblige the world with new observations and discoveries (which he continued to do to the end of a very long life), by publishing " A Direct and Geometrical method of finding the Aphelia and Excentricity of the planets." Besides various particular observations, made from time to time upon the celestial phenomena; he had, from his first admission into college, pursued a general scheme for ascertaining the true places of the fixed stars, and so to correct the errors of Tycho Brahe. His original view in this was, to carry on the design of that first restorer of astronomy, by completing the catalogue of those stars from his own observations; but upon farther enquiry, finding this province taken up by Hevelius and Flamsteed, he dropped that pursuit, and formed another; which was, to perfect the whole scheme of the heavens, by the addition of the stars which lie so near the south pole, that they could not be observed by those astronomers, as never rising above the horizon either at Dantzick or at Greenwich. With this view he left the University, before he had taken any degree, and embarked for the island of St. Helena in Nov. 1676, when he was only 20 years of age, and arrived there after a voyage of three months. He immediately set about his task with such diligence, that he completed his catalogue, and, returning home, landed in England in Nov. 1678, after an absence of two years only. The university of Oxford immediately conferred upon him the degree of A. M. and the Royal Society of London elected him one of their members.

In 1679 he was pitched upon by the Royal Society to go to Dantzick, to endeavour to adjust a dispute between Hevelius and Mr. Hooke, concerning the preference as to plain and glafs fights in aftrofcopical inftruments. He arrived at Dantzick the 26th of May, when he immediately, in conjunction with Hevelius, fct about their aftronomical obfervations, which they clofely continued till the 18th of July, when Halley left Dantzick, and returned to England.

In the year 1680 he undertook what is called the grand tour, accompanied by his friend the celebrated Mr. Nelfon. In the way from Calais to Paris, Mr. Halley had a fight of a remarkable comet, as it then appeared a fecond time that year, in its return from the fun. He had the November before feen it in its defcent; and he now haftened to complete his obfervations upon it, by viewing it from the royal obfervatory of France. His defign in this part of his tour was, to fettle a friendly correfpondence between the two royal aftronomers of Greenwich and Paris; and in the mean time to improve himfelf under fo great a mafter as Caffini. From thence he went to Italy, where he fpent great part of the year 1681; but his affairs calling him home, he then returned to England.

Soon after his return, he married the daughter of Mr. Tooke, auditor of the exchequer, and took up his refidence at Iflington, where he fet up his tube and fextant, and eagerly purfued his favourite ftudy: in the fociety of this amiable lady he lived happily for five-and-fifty years. In 1683 he publifhed his *Theory of the Variation of the Magnetical Compafs*; in which he fuppofes the whole globe of the earth to be one great magnet, having four magnetical poles or points of attraction, &c. The fame year alfo he entered upon a new method of finding out the longitude, by an accurate obfervation of the moon's motion. His purfuits it feems were now a little interrupted by the death of his father, who having fuffered greatly by the fire of London, as well as by a fecond marriage, into which he had imprudently entered, was found to have wafted his fortunes. Our author foon refumed his purfuits however; for in the beginning of 1684 he turned his thoughts to the fubject of Kepler's fefqui-alterate proportion; when, after fome meditation, he concluded from it, that the centripetal force muft decreafe in proportion to the fquare of the diftance reciprocally. He found himfelf unable to make it out in any geometrical way; and therefore, after applying in vain for affiftance to Mr. Hooke and Sir Chriftopher Wren, he went to Cambridge to Mr. Newton, who fully fupplied him with what he fo ardently fought. But Halley having now found an immenfe treafure in Newton, could not reft, till he had prevailed with the owner to enrich the public with it; and to this interview the world is in fome meafure indebted for the *Principia Mathematica Philofophiæ Naturalis*. That great work was publifhed in 1686; and Halley, who had the whole care of the impreffion, prefixed to it a difcourfe of his own, giving a general account of the aftronomical part of the book; and alfo a very elegant copy of verfes in Latin.

In 1687 he undertook to explain the caufe of a natural phenomenon, which had till then baffled the refearches of the ableft geographers. It is obferved that the Mediterranean fea never fwells in the leaft, although there is no vifible difcharge of the prodigious quantity of water that runs into it from nine large rivers, befides feveral fmall ones, and the conftant fetting in of the current at the mouth of the Streights. His folution of this difficulty gave fo much fatisfaction to the Society, that he was requefted to profecute thefe enquiries. He did fo; and having fhewn, by accurate experiments, how that vaft acceffion of water was actually carried off in vapours raifed by the action of the fun and wind upon its furface, he proceeded with the like fuccefs to point out the method ufed by nature to return the faid vapours into the fea. This circulation he fuppofes to be carried on by the winds driving thefe vapours to the mountains; where being collected, they form fprings, which uniting become rivulets or brooks, and many of thefe again meeting in the valleys, grow into large rivers, emptying themfelves at laft into the fea: thus demonftrating, in the moft beautiful manner, the way in which the equilibrium of receipt and expence is continually preferved in the univerfal ocean.

He next ranged in the field of fpeculative geometry, where, obferving fome imperfections in the methods before laid down for conftructing folid problems, or equations of the 3d and 4th powers, he furnifhed new rules, which were both more eafy and more elegant than any of the former; together with a new method of finding the number of roots of fuch equations, and the limits of the fame.

Mr. Halley next undertook to publifh a more correct Ephemeris for the year 1688, there being then great want of proper ephemerides of any tolerable exactnefs, the common ones being juftly complained of by Mr. Flamfteed.—In 1691 he publifhed exact tables of the conjunctions of Venus and Mercury; and he afterwards fhewed one extraordinary ufe to be made of thofe tables, viz. for difcovering the fun's parallax, and thence the true diftance of the earth from the fun.—In 1692, our author produced his tables for fhewing the value of annuities on lives, calculated from bills of mortality; and his univerfal theorem for finding the foci of optic glaffes.

But it would be endlefs to enumerate all his valuable difcoveries now communicated to the Royal Society, and publifhed in the Philof. Tranf. of which, for many years, his pieces were the chief ornament and fupport. Their various merit is thrown into one view by the writer of his eloge in the Paris Memoirs; who, having mentioned his Hiftory of the Trade-winds and Monfoons, proceeds in thefe terms: " This was immediately followed by his eftimation of the quantity of vapours which the fun raifes from the fea; the circulation of vapours; the origin of fountains; queftions on the nature of light and tranfparent bodies; a determination of the degrees of mortality, in order to adjuft the valuation of annuities on lives; and many other works, all the fciences relating to aftronomy, geometry, and algebra, optics and dioptrics, baliftics and artillery, fpeculative and experimental philofophy, natural hiftory, antiquities, philology, and criticifm; being about 25 or 30 differtations, which he produced during the 9 or 10 years of his refidence at London; and all abounding with ideas new, fingular, and ufeful."

In 1691, the Savilian profefforfhip of aftronomy at Oxford being vacant, he applied for that office, but without fuccefs. Whifton, in the Memoirs of his own Life, tells us from Dr. Bentley, that Halley " being thought of for fucceffor to the mathematical chair at Oxford,

5

Oxford, bishop Stillingfleet was desired to recommend him at court; but hearing that he was a sceptic and a banterer of religion, the bishop scrupled to be concerned, till his chaplain Bentley should talk with him about it, which he did. But Halley was so sincere in his infidelity, that he would not so much as pretend to believe the christian religion, though he thereby was likely to lose a professorship; which he did accordingly, and it was then given to Dr. Gregory."

Halley had published his Theory of the Variation of the Magnetical Compass, as has been already observed, in 1683; which, though it was well received both at home and abroad, he found, upon a review, liable to great and insuperable objections. Yet the phenomena of the variation of the needle, upon which it is raised, being so many certain and indisputed facts, he spared no pains to possess himself of all the observations relating to it he could possibly come at. To this end he procured an application to be made to king William, who appointed him commander of the Paramour pink, with orders to search out by observations the discovery of the rule of variations, and to lay down the longitudes and latitudes of the English settlements in America.— He set out on this attempt on the 24th of November, 1698: but having crossed the line, his men grew sickly; and his first lieutenant mutinying, he returned home in June 1699. Having got the lieutenant tried and cashiered, he set sail a second time in September following, with the same ship, and another of less bulk, of which he had also the command. He now traversed the vast Atlantic ocean from one hemisphere to the other, as far as the ice would permit him to go; and having made his observations at St. Helena, Brazil, Cape Verde, Barbadoes, the Madeiras, the Canaries, the coast of Barbary, and many other latitudes, he arrived in England in September 1700; and the next year published a general chart, shewing at one view the variation of the compass in all those places.

Captain Halley, as he was now called, had been at home little more than half a year, when he was sent by the king, to observe the course of the tides, with the longitude and latitude of the principal head-lands in the British channel; which having executed with his usual expedition and accuracy, he published a large map of the channel.

Soon after, the emperor of Germany resolving to make a convenient harbour for shipping in the Adriatic, captain Halley was sent by queen Anne to view the two ports on the coast of Dalmatia. He embarked on the 22d of November 1702; passed over to Holland; and going through Germany to Vienna, he proceeded to Istria: but the Dutch opposing the design, it was laid aside; yet the emperor made him a present of a rich diamond ring from his finger, and honoured him with a letter of recommendation, written with his own hand, to queen Anne. Presently after his return, he was sent again on the same business; when passing through Hanover, he supped with the electoral prince, who was afterward king George the 1st, and his sister the queen of Prussia. On his arrival at Vienna, he was the same evening presented to the emperor, who sent his chief engineer to attend him to Istria, where they repaired the fortifications of Trieste, and added new ones.

Mr. Halley returned to England in Nov. 1703; and the same year he was made professor of geometry in the university of Oxford, instead of Dr. Wallis then just deceased, and he was at the same time honoured by the university with the degree of doctor of laws. He was scarcely settled in Oxford, when he began to translate into Latin, from the Arabic, *Apollonius de Sectione Rationis*; and to restore the two books *De Sectione Spatii* of the same author, which are lost, from the account given of them by Pappus; and he published the whole work in 1706. He afterwards had a share in preparing for the press Apollonius's Conics; and ventured to supply the whole 8th book, the original of which is also lost. To this work he added Serenus on the Section of the Cylinder and Cone, printed from the original Greek, with a Latin translation, and published the whole in folio 1710. Beside these, the *Miscellanea Curiosa*, in 3 volumes 8vo, had come out under his direction in 1708.

In 1713, he succeeded Doctor, afterwards Sir, Hans Sloane, in the office of Secretary to the Royal Society. And, upon the death of Mr. Flamsteed in 1719, he was appointed to succeed him at Greenwich as Astronomer Royal; upon which occasion, that he might be more at leisure to attend the duties of this office, he resigned that of secretary to the Royal Society in 1721. Although he was 63 or 64 years of age when he entered upon his office at Greenwich, for the space of 18 years he watched the heavens with the closest attention, hardly ever missing an observation during all that time, and, without any assistant, performed the whole business of the observatory himself.

Upon the accession of the late king, his consort queen Caroline made a visit at the Royal Observatory; and being pleased with every thing she saw, took notice that Dr. Halley had formerly served the crown as a captain in the navy: and she soon after obtained a grant of his half-pay for that commission, which he accordingly enjoyed from that time during his life. An offer was also made him of being appointed mathematical preceptor to the duke of Cumberland; but he declined that honour, on account of his advanced age, and the duties of his office. In 1729 he was chosen a foreign member of the Academy of Sciences at Paris.

About 1737 he was seized with a paralytic disorder in his right hand, which, it is said, was the first attack he ever felt upon his constitution: however, he came as usual once a week, till within a very short time of his death, to meet his friends in town on Thursdays, before the meeting of the Royal Society, at what is yet called Dr. Halley's club. His paralytic disorder increasing, his strength gradually wore away, till he expired Jan. 14, 1742, in the 86th year of his age; and his corps was interred in the church-yard of Lee near Blackheath.——Beside the works before mentioned, his principal publications are, 1. *Catalogus Stellarum Australium.* 2. *Tabulæ Astronomicæ.* 3. The Astronomy of Comets. With a great multitude of Papers in the Philos. Transf. from vol. 11 to vol. 60.

HALIFAX (John). See Sacrobosco.

HALO, or Corona, a coloured circle appearing round the body of the sun, moon, or any of the larger stars.

Naturalists conceive the Halo to arise from a refrac-

tion

tion of the rays of light in paffing through the fine rare veficulæ of a thin vapour towards the top of the atmo-fphere.

Des Cartes obferves, that the Halo never appears when it rains; whence he concludes that this phenome-non is occafioned by the refraction of light in the round particles of ice, which are then floating in the atmo-fphere; and to the different protuberance of thefe parti-cles he afcribes the variation in the diameter of the Halo. Gaffendi fuppofes, that a Halo is occafioned in the fame manner as the rainbow; the rays of light being, in both cafes, twice refracted and once reflected within each drop of rain or vapour, and that the difference between them is wholly owing to their different fitua-tion with refpect to the obferver. Dechales alfo endea-vours to fhew that the generation of the Halo is fimilar to that of the rainbow; and that the reafon why the colours of the Halo are more dilute than thofe of the rainbow, is owing chiefly to their being formed, not in large drops of rain, but in very fmall vapour. But the moft confiderable and generally received theory, re-lating to the generation of Halos, is that of Mr. Huy-gens. This celebrated author fuppofes Halos, or cir-cles round the fun, to be formed by fmall round grains of hail, compofed of two different parts, the one of which is tranfparent, inclofing the other, which is opaque; which is the general ftructure actually obferved in hail. He farther fuppofes that the grains or glo-bules, that form thefe Halos, confifted at firft of foft fnow, and that they have been rounded by a continual agitation in the air, and thawed on their outfide by the heat of the fun, &c. And he illuftrates his ideas of their formation by geometrical figures.

Mr. Weidler endeavours to refute Huygens's manner of accounting for Halos, by a vaft number of fmall vapours, each with a fnowy nucleus, coated round with a tranf-parent covering. He fays, that when the fun paints its image in the atmofphere, and by the force of its rays puts the vapours in motion, and drives them toward the furface, till they are collected in fuch a quantity, and at fuch a diftance from the fun on each fide, that its rays are twice refracted, and twice reflected, when they reach the eye they exhibit the appearance of a Halo, adorned with the colours of the rainbow; which may happen in globular pellucid vapours without fnowy nuclei, as appears by the experiment of hollow glafs fpheres filled with water: therefore, whenever thofe fpherical vapours are fituated as before mentioned, the refractions and reflections will happen every where alike, and the figure of a circular crown, with the ufual order of colours, will be the confequence. Philof. Tranf. number 458.

Newton's theory of Halos may be feen in his Optics, p. 155. And this curious theory was confirmed by actual obfervation in June 1692, when the author faw by reflection, in a veffel of ftagnated water, three Halos, crowns, or rings of colours, about the fun, like three little rainbows concentric to his body. Thefe crowns inclofed one another immediately, fo that their colours proceeded in this continual order from the fun outward: blue, white, red; purple, blue, green, pale yellow, and red; pale blue, pale red. The like crowns fometimes appear about the moon. The more equal the globules of water or ice are to one another, the

more crowns of colours will appear, and the colours will be the more lively. Optics, p. 288.

There are feveral ways of exhibiting phenomena fimilar to thefe. The flame of a candle, placed in the midft of a fteam in cold weather, or placed at the dif-tance of fome feet from a glafs window that has been breathed upon, while the fpectator is alfo at the dif-tance of fome feet from another part of the window, or placed behind a glafs receiver, when air is admitted into the vacuum within it to a certain denfity, in each of thefe circumftances will appear to be encompaffed by a coloured Halo. Alfo, a quantity of water being thrown up againft the fun, as it breaks and difperfes into drops, forms a kind of Halo or iris, exhibiting the colours of the natural rainbow. Muffchenbroek obferved, that when the glafs windows of his room were covered with a thin plate of ice on the infide, the moon feen through it was furrounded with a large and varioufly coloured Halo; which, upon opening the window, he found arofe entirely from that thin plate of ice, becaufe none was feen except through this plate. Muffchenbroek concludes his account of coronas with obferving, that fome denfity of vapour, or fome thicknefs of the plates of ice, divides the light in its tranfmiffion either through the fmall globules or their interftices, into its fepa-rate colours; but what that denfity is, or what the fize of the particles which compofe the vapour, he does not pretend to determine. Introd. ad Phil. Nat. p. 1037.

It has often been obferved that a Halo about the fun or moon, does not appear circular and concentric to the luminary, but oval and excentric, with its longeft dia-meter perpendicular to the horizon, and extended from the moon farther downward than upward. Dr. Smith afcribes this phenomenon to the apparent concave of the fky being lefs than a hemifphere. When the angle which the diameter of a Halo fubtends at the eye is 45° or 46°, and the bottom of the Halo is near the ho-rizon, and confequently its apparent figure is moft oval, the apparent vertical diameter is divided by the moon in the proportion of about 2 to 3 or 4, and is to the ho-rizontal diameter drawn through the moon, as 4 to 3, nearly.—See farther on the fubject of this article, Prieftley's Hift. of Difcoveries relating to Vifion, p. 596—613; and Smith's Optics, art. 167, 513, 526, 527, &c.

HAMEL (JOHN BAPTISTE DU), a very learned French philofopher and writer, in the 17th century, was born in lower Normandy in 1614. At 18 years of age he publifhed a treatife, in which he explained, in a very fimple manner, and by one or two figures, Theo-dofius's 3 books upon Spherics; to which he added a tract upon trigonometry extremely perfpicuous, and de-figned as an introduction to aftronomy. He publifhed afterwards various other works on aftronomy and phi-lofophy. Natural philofophy, as it was then taught, was only a collection of vague, knotty, and barren queftions; when our author undertook to eftablifh it upon right principles, and publifhed his *Aftronomia Phyfica*.

In 1666 M. Colbert propofed to Louis the 14th a fcheme, which was approved of, for eftablifhing a royal academy of fciences; and appointed our author fecre-tary of it. In 1678, his *Philofophia Vetus & Nova* was

was printed at Paris in 4 vols, 12mo; and in 1681 it was enlarged and printed there in 6 vols. He wrote several other pieces; and his works in this way were collected and published at Nuremberg 1681, in 4 volumes 4to, under the title of *Opera Philosophica & Astronomica*. These were highly valued then, though the improvements in philosophy since that time have rendered them of little or no use now.

In 1697 he resigned his place of secretary of the Royal Academy of Sciences; in which he was succeeded by M. Fontenelle. However, he published, in 1698, *Regiæ Scientiarum Academiæ Historia*, 4to, in four books; which being much liked, he afterwards augmented with two books more. This work contains an account of the foundation of the Royal Academy of Sciences, and its transactions, from 1666 to 1700, and is now the most useful of all his works. He was Régius Professor of Philosophy, in which office he was succeeded by M. Varignon, at his death, which happened Aug. 6, 1706, in the 93d year of his age.

HANCES, Hanches, or Hanses, in architecture, are certain small intermediate parts of arches between the key or crown and the spring at the bottom, being perhaps about one-third of the arch, and situated nearer the bottom than the top or crown; and are otherwise called the *spandrels*.

HANDSPIKE, or Handspec, a lever or piece of ash, elm, or other strong wood, for raising by the hand great weights, &c. It is 5 or 6 feet long, cut thin and crooked at the lower end, that it may get the easier between things that are to be separated, or under any thing that is to be raised. It is better than a crow of iron, because its length allows a better poise.

HARD *Bodies*, are such as are absolutely inflexible to any pressure or percussion whatever; differing from soft bodies, whose parts yield and are easily moved amongst one another, without restoring themselves again; and from elastic bodies, the parts of which also yield and give way, but presently restore themselves again to their former state and situation. Hence, hard bodies do not bend, or indent, but break. It is probable however there are no bodies in nature that are absolutely or perfectly either hard, soft, or elastic; but all possessing these qualities, more or less, in some degree. M. Bernoulli goes so far as to say that Hardness, in the common sense, is absolutely impossible, being contrary to the law of continuity.

The laws of motion for hard bodies are the same as for soft ones, both being supposed to adhere together on their impact. And these two sorts of bodies might be comprized under the common name of Unelastic.

HARDENING, the giving a greater degree of hardness to bodies than they had before.

There are several ways of Hardening iron and steel; as by hammering them, quenching them in cold water, &c.

Case-Hardening, is a superficial conversion of iron into steel, as if it were casing it, or covering it with a thin coat of harder matter. It is thus performed: Take cow horn or hoof, dry it well in an oven, and beat it to powder; put equal quantities of this powder and of bay salt into stale urine, or white wine vinegar, and mix them well together; cover the iron or steel all

3

over with this mixture, and wrap it up in loam, or plate iron, so as the mixture touch every part of the work; then put it in the fire, and blow the coals to it, till the whole lump have a blood red heat, but no higher; lastly, take it out, and quench it.—See Steel, under which article are described other processes for this purpose.

HARDNESS, or Rigidity, that quality in bodies by which their parts so cohere as not to yield inward, or give way to an external impulse, without instantly going beyond the distance of their mutual attraction; and therefore are not subject to any motion in respect of each other, without breaking the body.

There were many fanciful opinions among the ancients concerning the cause of hardness; such as, heat, cold, dryness, the hooked figure of the particles of matter. The Cartesians make the Hardness of bodies to consist in rest, as that of soft and fluid ones in the motion of their particles.

Newton shews that the primary particles of all bodies, whether solid or fluid, are perfectly hard; and are not capable of being broken or divided by any power in nature. These particles, he maintains, are connected together by an attractive power; and according to the circumstances of this attraction, the body is either hard, or soft, or even fluid. If the particles be so disposed or fitted for each other, as to touch in large surfaces, the body will be hard; and the more so as those surfaces are the larger. If, on the contrary, they only touch in small surfaces, the body, by the weakness of the attraction, will remain soft.

At present, many philosophers think that Hardness consists in the absence or want of the action of the universal fluid, or elementary fire, among the particles of the body, or a deficiency of what is called latent heat; while on the contrary, fluidity, according to them, consists in the motion of the particles, in consequence of the action of that elementary fire.

Hardness appears to diminish the cohesion of bodies, in some degree, though their frangibility or brittleness does not by any means keep pace with their hardness. Thus, though glass be very hard and very brittle; yet flint is still harder, though less brittle. Among the metals, these two properties seem to be more connected, though even here the connection is by no means complete: for though steel be both the hardest and most brittle of all the metals; yet lead, which is the softest, is not the most ductile. Neither is Hardness connected with the specific gravity of bodies; for a diamond, the hardest substance in nature, is little more than half the weight of the lightest metal. And as little is it connected with the coldness, or electrical properties, or any other quality with which we are acquainted. Some bodies are rendered hard by cold, and others by different degrees of heat.

Mr. Quist and others have constructed tables of the Hardness of different substances. And the manner of constructing these tables, was by observing the order in which they were able to cut or make any impression upon one another. The following table, extracted from Magellan's edition of Cronstedt's Mineralogy, was taken from Quist, Bergman, and Kirwan. The first column shews the Hardness, and the second the specific gravity.

Diamond

	Hardness.	Spec. Grav.
Diamond from Ormus	20	3·7
Pink diamond	19	3·4
Blueish diamond	19	3·3
Yellowish diamond	19	3·3
Cubic diamond	18	3·2
Ruby	17	4·2
Pale blue sapphire	17	3·8
Pale ruby from Brazil	16	3·5
Deep blue sapphire	16	3·8
Topaz	15	4·2
Whitish ditto	14	3·5
Ruby spinell	13	3·4
Emerald	12	2·8
Garnet	12	4·4
Agate	12	2·6
Onyx	12	2·6
Sardonyx	12	2·6
Bohemian topaz	11	2·8
Occid. amethyst	11	2·7
Crystal	11	2·6
Carnelian	11	2·7
Green jasper	11	2·7
Schoerl	10	3·6
Tourmaline	10	3·0
Quartz	10	2·7
Opal	10	2·6
Chryfolite	10	3·7
Reddish yellow jasper	9	2·6
Zeolyte	8	2·1
Fluor	7	3·5
Calcareous spar	6	2·7
Gypsum	5	2·3
Chalk	3	2·7

HARMONICA, Harmonics, a branch or division of the ancient music; being that part which considers the differences and proportions of sounds, with respect to acute and grave; as distinguished from Rhythmica, and Metrica.

Mr. Malcolm has made a very industrious and learned enquiry into the Harmonica, or harmonic principles, of the ancients.

HARMONICA, the name of a musical instrument invented by Dr. Franklin, consisting of the glasses, called musical glasses.

It is said that the first hint of musical glasses is to be found in an old English book, in which a number of various amusements were described. That author directs his pupil to choose half a dozen drinking-glasses; to fill each of them with water in proportion to the gravity or acuteness of the sound which he intended it to give; and having thus adjusted them one to another, he might entertain the company with a church tune. These were perhaps the rude hints which Mr. Puckeridge, an Irish gentleman, afterwards improved, and after him, Mr. E. Delaval, an ingenious member of the Royal Society; and finally brought to perfection by the celebrated Franklin. See the history and description in his Letters; particularly in that to Beccaria.

HARMONICAL, or Harmonic, something relating to Harmony. Thus,

HARMONICAL *Arithmetic*, is so much of the theory and doctrine of numbers, as relates to making the comparisons and reductions of musical intervals, which are expressed by numbers, for finding their mutual relations, compositions, and resolutions.

HARMONICAL *Composition*, in its general sense, includes the composition both of harmony and melody; i. e. of music, or song, both in a single part, and in several parts.

HARMONICAL *Interval*, the difference between two sounds, in respect of acute and grave: or that imaginary space terminated with two sounds differing in acuteness or gravity.

HARMONICAL *Proportion*, or *Musical Proportion*, is that in which the first term is to the third, as the difference of the first and second is to the difference of the 2d and 3d; or when the first, the third, and the said two differences, are in geometrical proportion. Or, four terms are in Harmonical proportion, when the 1st is to the 4th, as the difference of the 1st and 2d is to the difference of the 3d and 4th. Thus, 2, 3, 6, are in harmonical proportion, because $2 : 6 :: 1 : 3$. And the four terms 9, 12, 16, 24 are in harmonical proportion, because $9 : 24 :: 3 : 8$.—If the proportional terms be continued in the former case, they will form an harmonical progression, or series.

1. The reciprocals of an arithmetical progression are in Harmonical progression; and, conversely, the reciprocals of Harmonicals are arithmeticals. Thus, the reciprocals of the Harmonicals 2, 3, 6, are $\frac{1}{2}, \frac{1}{3}, \frac{1}{6}$, which are arithmeticals; for $\frac{1}{2} - \frac{1}{3} = \frac{1}{6}$, and $\frac{1}{3} - \frac{1}{6} = \frac{1}{6}$ also: and the reciprocals of the arithmeticals 1, 2, 3, 4, &c, are $\frac{1}{1}, \frac{1}{2}, \frac{1}{3}, \frac{1}{4}$, &c, which are Harmonicals; for $\frac{1}{1} : \frac{1}{3} :: \frac{1}{1} - \frac{1}{2} : \frac{1}{2} - \frac{1}{3}$; and so on. And, in general, the reciprocals of the arithmeticals $a, a + d, a + 2d, a + 3d$, &c, viz, $\frac{1}{a}, \frac{1}{a+d}, \frac{1}{a+2d}, \frac{1}{a+3d}$, &c, are Harmonicals; et e contra.

2. If three or four numbers in Harmonical proportion be either multiplied or divided by some number, the products, or the quotients, will still be in Harmonical proportion. Thus,

the Harmonicals   6, 8, 12,
multiplied by 2 give   12, 16, 24,
or divided by 2 give   3, 4, 6,

which are also Harmonicals.

3. To find a Harmonical mean proportional between two terms: Divide double their product by their sum.

4. To find a 3d term in Harmonical proportion to two given terms: Divide their product by the difference between double the 1st term and the 2d term.

5. To find a 4th term in Harmonical proportion to three terms given: Divide the product of the 1st and 3d by the difference between double the 1st and the 2d term.

Hence,

Hence, of the two terms a and b;

the Harmonical mean is $\dfrac{2ab}{a+b}$,

the 3d Harmonical propor. is $\dfrac{ab}{2a-b}$,

also to a, b, c, the 4th Harm. is $\dfrac{ac}{2a-b}$.

6. If there be taken an arithmetical mean, and a Harmonical mean, between any two terms, the four terms will be in geometrical proportion. Thus, between 2 and 6,

the arithmetical mean is 4, and
the Harmonical mean is 3;
and hence 2 : 3 :: 4 : 6.

Also, between a and b,

the arithmetical mean is $\dfrac{a+b}{2}$, and

the Harmonical mean is $\dfrac{2ab}{a+b}$;

but $a : \dfrac{2ab}{a+b} :: \dfrac{a+b}{2} : b$.

HARMONY, in Music, the agreeable result of an union of several musical sounds, heard at one and the same time; or the mixture of divers sounds, which together have an agreeable effect on the ear.

As a continued succession of musical sounds produces melody, so a continued combination of them produces Harmony.

Among the ancients however, as sometimes also among the moderns, Harmony is used in the strict sense of consonance; and so is equivalent to the symphony.

The words *concord* and Harmony do really signify the same thing; though custom has made a little difference between them. Concord is the agreeable effect of two sounds in consonance; and Harmony the effect of any greater number of agreeable sounds in consonance.

Again, Harmony always implies consonance; but concord is also applied to sounds in succession; though never but where the terms can stand agreeably in consonance. The effect of an agreeable succession of several sounds, is called *melody*; as that of an agreeable consonance is called Harmony.

Harmony is well defined, the sum or result of the combination of two or more concords; that is, of three or more simple sounds striking the ear all together; and different compositions of concords make different Harmony.

The ancients seem to have been entirely unacquainted with Harmony, the soul of the modern music. In all their explications of the melopœia, they say not one word of the concert or Harmony of parts. We have instances, indeed, of their joining several voices, or instruments, in consonance; but then these were not so joined, as that each had a distinct and proper melody, so making a succession of various concords; but they were either unisons, or octaves, in every note; and so all performed the same individual melody, and constituted one song.

When the parts differ, not in the tension of the whole, but in the different relations of the successive notes, it is this that constitutes the modern art of Harmony.

HARMONY *of the Spheres*, or *Celestial Harmony*, a kind of music much spoken of by many of the ancient philosophers and fathers, supposed to be produced by the sweetly tuned motions of the stars and planets. This Harmony they attributed to the various proportionate impressions of the heavenly bodies upon one another, acting at proper intervals. They think it impossible that such prodigious bodies, moving with such rapidity, should be silent: on the contrary, the atmosphere, continually impelled by them, must yield a set of sounds proportionate to the impression it receives; and that consequently, as they run all in different circuits, and with various degrees of velocity, the different tones arising from the diversity of motions, directed by the hand of the Almighty, must form an agreeable symphony or concert.

They therefore supposed, that the moon, as being the lowest of the planets, corresponded to *mi*; Mercury, to *fa*; Venus, to *sol*; the Sun, to *la*; Mars; to *si*; Jupiter, to *ut*; Saturn, to *re*; and the orb of the fixed stars, as being the highest of all, to *mi*, or the octave.

It is thought that Pythagoras had a view to the gravitation of celestial bodies, in what he taught concerning the Harmony of the spheres.

A musical chord gives the same note as one double in length, when the tension or force with which the latter is stretched is quadruple; and the gravity of a planet is quadruple of the gravity of a planet at a double distance. In general, that any musical chord may become unison to a lesser chord of the same kind, its tension must be increased in the same proportion as the square of its length is greater; and that the gravity of a planet may become equal to the gravity of another planet nearer the sun, it must be increased in proportion as the square of its distance from the sun is greater. If therefore we should suppose musical chords extended from the sun to each planet, that all these chords might become unison, it would be requisite to increase or diminish their tensions in the same proportions as would be sufficient to render the gravities of the planets equal; and from the similitude of those proportions, the celebrated doctrine of the Harmony of the spheres is supposed to have been derived.

Kepler wrote a large work, in folio, on the Harmonies of the world, and particularly of that of the celestial bodies. He first endeavoured to find out some relation between the dimensions of the five regular solids and the intervals of the planetary spheres; and imagining that a cube, inscribed in the sphere of Saturn, would touch by its six planes the sphere of Jupiter, and that the other four regular solids in like manner fitted the intervals that are between the spheres of the other planets, he became persuaded that this was the true reason why the primary planets were precisely six in number, and that the author of the world had determined their distances from the sun, the centre of the system, from a regard to this analogy. But afterwards finding that the disposition of the five regular solids amongst the planetary spheres, was not agreeable to the intervals between their orbits, he endeavoured to discover other schemes of Harmony. For this purpose he compared the motions of the same planet at its greatest and least distances, and of the different planets in their several orbits, as they would appear viewed

from

from the sun; and here he fancied that he found a similitude to the divisions of the octave in music. Lastly, he imagined that if lines were drawn from the earth, to each of the planets, and the planets appended to them, or stretched by weights proportional to the planets, these lines would then sound all the notes in the octave of a musical chord.

See his Harmonics; also Plin. lib. 2, cap. 22; Macrob. in Somn. Scip. lib. 2, cap. 1; Plutarch de Animal. Procreatione, è Timæo; and Maclaurin's View of Newton's Discov. book 1, chap. 2.

HARQUEBUSS, a hand-gun, or a fire-arm of a proper length and weight to be borne in the arm. Hanzelet prescribes its proper length to be 40 calibres, or diameters of its bore; and the weight of its ball 1 oz. and ⅞; its charge of powder as much.

HARRIOT (THOMAS), a very eminent English mathematician and astronomer, was born at Oxford in 1560, and died at London July 2, 1621, in the 61st year of his age. Harriot has hitherto been known to the world only as an algebraist, though a very eminent one; but from his manuscript papers, that have been but lately discovered by Dr. Zach, astronomer to the duke of Saxe-Gotha, it appears that he was not less eminent as an astronomer and geometrician. Dr. Zach has printed an account of those papers, in the Astronomical Ephemeris of the Royal Academy of Sciences at Berlin, for the year 1788; of which, as it is very curious, and contains a great deal of information, I shall here give a translation, to serve as memoirs concerning the life and writings of this eminent man; afterwards adding only some necessary remarks of my own.

" I here present to the world (says Dr. Zach), a short account of some valuable and curious manuscripts, which I found in the year 1784, at the seat of the earl of Egremont, at Petworth in Sussex, in hopes that this learned and inquisitive age will either think my endeavours about them worthy of its assistance, or else will be thereby induced to attempt some other means of publishing them. The only undeniable proof I can now produce of the usefulness of such an undertaking, is by giving a succinct report of the contents of these materials, and briefly shewing what may be effected by them. And although I come to the performance of such an enterprize with much less abilities than the different parts of it require, yet I trust that my love for truth, my design and zeal to vindicate the honour due to an Englishman, the author of these manuscripts, which are the chief reasons that have influenced me in this undertaking, will serve as my excuse.

" A predecessor of the family of lord Egremont, viz, that noble and generous earl of Northumberland, named Henry Percy, was not only a generous favourer of all good learning, but also a patron and Mæcenas of the learned men of his age. Thomas Harriot, the author of the said manuscripts, Robert Hues (well known by his Treatise upon the Globes), and Walter Warner, all three eminent mathematicians, who were known to the earl, received from him yearly pensions; so that when the earl was committed prisoner to the Tower of London in the year 1606, our author, with Hues and Warner, were his constant companions; and were usually called the earl of Northumberland's three Magi.

" Thomas Harriot is a known and celebrated mathematician among the learned of all nations, by his excellent work, *Artis Analyticæ Praxis, ad æquationes algebraicas nova expedita & generali methodo, resolvendas, Tractatus posthumus*; Lond. 1631: dedicated to Henry earl of Northumberland; published after his death by Walter Warner. It is remarkable, that the fame and the honour of this truly great man were constantly attacked by the French mathematicians; who could not endure that Harriot should in any way diminish the fame of their Vieta and Des Cartes, especially the latter, who was openly accused of plagiarism from our author. [*See Montucla's Histoire des Mathematiques, part 3, p. 485 & seq.—Lettres de M. Des Cartes, tom. 3, pa. 457, edit. Paris 1667, in 4to.—Dictionnaire de Moreri, word Harriot.—Encyclopedie, word Algebra.—Lettres de M. de Voltaire, sur la nation Angloise, lettre 14.—Memoire de l' Abbé de Gua dans les Mem. de l' Acad. des Sciences de Paris pour 1741.—Jer. Collier's great Historical Dictionary, word Harriot.—Dr. Wallis's preface to his Algebra.—To which may be added the article Algebra, in this dictionary.*]

" Des Cartes published his Geometry 6 years after Harriot's work appeared, viz, in the year 1637. Sir Charles Cavendish, then ambassador at the French court at Paris, when Des Cartes's Geometry made its first appearance in public, observed to the famous geometrician Roberval, that these improvements in Analysis had been already made these 6 years in England, and shewed him afterwards Harriot's *Artis Analyticæ Praxis*, which as Roberval was looking over, at every page he cried out, *Oui! oui! il l'a vu! Yes! yes! he has seen it!* Des Cartes had also been in England before Harriot's death, and had heard of his new improvements and inventions in Analysis. A critical life of this man, which his papers would enable me to publish, will shew more clearly what to think upon this matter, which I hope may be discussed to the due honour of our author.

" Now all this relates to Harriot the celebrated analyst; but it has not hitherto been known that Harriot was an eminent astronomer, both theoretical and practical, which first appears by these manuscripts; among which, the most remarkable are 199 observations of the Sun's Spots, with their drawings, calculations and determinations of the sun's rotation about his axis. There is the greatest probability that Harriot was the first discoverer of these spots, even before either Galileo or Scheiner. The earliest intelligence we have of the first discovered solar spots, is of one Joh. Fabricius Phrysius, who in the year 1611 published at Wittemberg a small treatise, intitled, *De Maculis in Sole observatis & apparente eorum cum Sole conversione narratio.* Galileo, who is commonly accounted the first discoverer of the Solar Spots, published his book, *Istoria e Dimonstrazioni intorne alle Macchie Solare e loro accidenti*, at Rome, in the year 1613. His first observation in this work, is dated June 2, 1612. Angelo de Filiis, the editor of Galileo's work, who wrote the dedication and preface to it, mentions, pa. 3, that Galileo had not only discovered these spots in the month of April in the year 1611, at Rome, in the Quirinal Garden, but had shewn them several months before *(molti mesi innanzi)* to his friends in Florence. And that the observations of the disguised Apelles (the Jesuit Scheiner, a pretender

tender to this firft difcovery) were not later than the month of October in the fame year; by which the epoch of this difcovery was fixed to the beginning of the year 1611. But a paffage in the firft letter of Galileo's works, pa. 11, gives a more precife term to this difcovery. Galileo there fays in plain terms, that he had obferved the Spots in the Sun 18 months before. The date of this letter is May 24, 1612; which brings the true epoch of this difcovery to the month of November, 1610. However, Galileo's firft produced obfervations are only from June 2, 1612, and thofe of father Scheiner of the month of October, in the fame year. But now it appears from Harriot's manufcripts, that his firft obfervations of thefe Spots are of Dec. 8, 1610. It is not likely that Harriot could have this notice from Galileo, for I do not find this mathematician's name ever quoted in Harriot's papers. But I find him mentioning Jofephus a Cofta's book 1, chap. 2, of his *Natural and Moral Hiftory of the Weft Indies*, in which he relates that in Peru there are Spots to be feen in the Sun which are not to be feen in Europe. It rather feems that Harriot had taken the hint from thence. Befides, it is very likely that Harriot, who lived with fo generous a patron to all good learning and improvements, had got the new invention of telefcopes in Holland much fooner in England, than they could reach Galileo, who at that time lived at Venice. Harriot's very careful and exact obfervations of thefe Spots, fhew alfo that he was in poffeffion of the beft and moft improved telefcopes of that time; for it appears he had fome with magnifying powers of 10, 20, and 30 times. At leaft there are no earlier obfervations of the Solar Spots extant than his: they run from December 8, 1610, till January 18, 1613. I compared the correfponding ones with thefe obferved by Galileo, between which I found an exact agreement. Had Harriot had any notion about Galileo's difcoveries, he certainly would have alfo known fomething about the phafes of Venus and Mercury, and efpecially about the fingular fhape of Saturn, firft difcovered by Galileo; but I find not a word in all his papers concerning the particular figure of that planet.

"*Of Jupiter's Satellites.* I found among his papers a large fet of obfervations, with their drawing, pofition, and calculations of their revolutions and periods. His firft obfervation of thofe difcovered Satellites, I find to be of January 16, 1610; and they go till February 26, 1612. Galileo pretends to have difcovered them January 7, 1610; fo that it is not improbable that Harriot was likewife the firft difcoverer of thefe attendants of Jupiter.

" Among his other obfervations of the Moon, of fome Eclipfes, of the planet Mars, of Solftices, of Refraction, of the Declination of the Needle; there are moft remarkable ones of the noted Comets of 1607, and of 1618, the latter, for there were two this year (*fee Kepler de Cometis, pa. 49*). They were all obferved with a crofs-ftaff, by meafuring their diftances from fixed ftars; whence thefe obfervations are the more valuable, as comets had before been but grofsly obferved: Kepler himfelf obferved the comet of 1607 only with the naked eye, pointing out its place by a coarfe eftimation, without the aid of an inftrument; and the elements of their orbits could, in defect of better obferva-

tions, be only calculated by them. The obfervations of the comet of the year 1607, are of the more importance, even now for modern aftronomy, as this is the fame comet that fulfilled Dr. Halley's prediction of its return in the year 1759. That prediction was only grounded upon the elements afforded him by thefe coarfe obfervations; for which reafon he only affigned the term of its return to the fpace of a year. The very intricate calculations of the perturbations of this comet, afterwards made by M. Clairaut, reduced the limits to a month's fpace. But a greater light may now be thrown upon this matter by the more accurate obfervations on this comet by Mr. Harriot. In the month of October 1785, when I converfed upon the fubject of Harriot's papers, and efpecially on this comet, with the very celebrated mathematician M. de la Grange, director of the Royal Academy of Sciences at Berlin; he then fuggefted to me an idea, which, if brought into execution, will clear up an important point in aftronomy. It is well known to aftronomers how difficult a matter it is, to determine the mafs, or quantity of matter, in the planet Saturn; and how little fatisfactory the notions of it are, that have hitherto been formed. The whole theory of the perturbations of comets depending upon this uncertain datum, feveral attempts and trials have been made towards a more exact determination of it by the moft eminent geometricians of this age, and particularly by la Grange himfelf; but never having been fatisfied with the few and uncertain data heretofore obtained for the refolution of this problem, he thought that Harriot's obfervations on the comet of 1607, and the modern ones of the fame comet in 1759, would fuggeft a way of refolving the problem *a pofteriori*; that of determining by them the elements of its ellipfis. The retardation of the comet compared to its period, may clearly be laid to the account of the attraction and perturbation it has fuffered in the region of Jupiter and Saturn; and as the part of it belonging to Jupiter is very well known, the remainder muft be the fhare which is due to Saturn; from whence the mafs of the latter may be inferred. In confequence of this confideration I have already begun to reduce moft of Harriot's obfervations of this comet, in order to calculate by them the true elements of its orbit on an elliptical hypothefis, to complete M. la Grange's idea upon this matter.

" I forbear to mention here any more of Harriot's analytical papers, which I found in a very great number. They contain feveral elegant folutions of quadratic, cubic, and biquadratic equations; with fome other folutions and *loca geometrica*, that fhew his eminent qualifications, and will ferve to vindicate them againft the attacks of feveral French writers, who refufe him the juftice due to his fkill and accomplifhments, merely to fave Des Cartes's honour, who yet, by fome impartial men of his own nation, was accufed of public plagiarifm.

" Thomas Harriot was born at Oxford, in the year 1560. After being inftructed in the rudiments of languages, he became a commoner of St. Mary's-Hall, where he took the degree of bachelor of arts in 1579. He had then fo diftinguifhed himfelf by his uncommon fkill in mathematics, as to be recommended foon after to Sir Walter Raleigh, as a proper preceptor to him in that fcience. Accordingly that noble knight became his firft patron, took him into his family, and allowed him a handsome pension. In 1584 he went over with Sir Walter's first colony to Virginia; where he was employed in discovering and surveying the country, &c; maps of which I have found (says Dr. Zach) very neatly done among his papers. After his return he published *A Brief and True Report of the Newfoundland of Virginia, of the Commodities there found to be raised*, &c; Lond. 1588. This was reprinted in the 3d volume of Hakluyt's Voyages: it was also translated into Latin, and printed at Frankfort in the year 1590. Sir Walter introduced him to the acquaintance of the earl of Northumberland, who allowed him a yearly pension; Wood says, of 120l. only; but by some of his receipts, which I have found among his papers, it appears that he had 300l, which indeed was a very large sum at that time. Wood, in his *Athen. Oxon.* mentions nothing of Harriot's papers, except a manuscript in the library at Sion College, London, entitled *Ephemeris Chyrometrica*. I got access to this library and manuscripts, and was indeed in hopes of finding something more of Harriot's; for most of his observations are dated from Sion College; but I could not find any thing from Harriot himself. I found indeed some other papers of his friends: he mentions, in his observations, one Mr. Standish, at Oxford, and Nicol. Torperly, who also was of the acquaintance of the earl of Northumberland, and had a yearly pension: from the former I found two observations of the same comet of 1618, made in Oxford, which he communicated to Mr. Harriot.

"Thomas Harriot died July 2, 1621. His disease was a cancerous ulcer in the lip, which some pretend he got by a custom he had of holding the mathematical brass instruments, when working, in his mouth. I found several of his letters, and answers to them, from his physician Dr. Alexander Rhead, who in his treatise mentions Harriot's disease. His body was conveyed to St. Christopher's church, in London. Over his grave was soon after erected a monument, with a large inscription, which was destroyed with the church itself by the dreadful fire of September 1666. He was but 60 years of age."

The peculiar nature and merits of Harriot's Algebra, we have spoken largely and particularly of, under the art. *Algebra*, page 89. As to his manuscripts lately discovered by Dr. Zach, as above mentioned, it is with pleasure I can announce, that they are in a fair train to be published: they have been presented to the university of Oxford, on condition of their printing them; with a view to which, they have been lately put into the hands of an ingenious member of that learned body, to arrange and prepare them for the press.

HARRISON (JOHN), a most accurate mechanic, the celebrated inventor of the famous *time-keeper* for ascertaining the longitude at sea, and also of the compound or *gridiron-pendulum*; was born at Foulby, near Pontefract in Yorkshire, in 1693. His father was a carpenter, in which profession the son assisted; occasionally also, according to the miscellaneous practice of country artists, surveying land, and repairing clocks and watches; and young Harrison always was, from his early childhood, greatly attached to any machinery moving by wheels. In 1700 he removed with his father to Barrow in Lincolnshire; where, though his opportunities of acquiring knowledge were very few, he eagerly improved every incident from which he might

collect information; frequently employing all or great part of his nights in writing or drawing: and he always acknowledged his obligations to a clergyman who came every Sunday to officiate in the neighbourhood, who lent him a MS. copy of professor Sanderson's lectures; which he carefully and neatly transcribed, with all the diagrams. His native genius exerted itself superior to these solitary disadvantages; for in the year 1726, he had constructed two clocks, mostly of wood, in which he applied the escapement and compound pendulum of his own invention: these surpassed every thing then made, scarcely erring a second in a month. In 1728, he came up to London with the drawings of a machine for determining the longitude at sea, in expectation of being enabled to execute one by the Board of Longitude. Upon application to Dr. Halley, the astronomer royal, he referred him to Mr. George Graham; who advised him to make his machine, before applying to that Board. He accordingly returned home to perform his task; and in 1735 came to London again with his first machine; with which he was sent to Lisbon the next year, to make trial of it. In this short voyage, he corrected the dead reckoning about a degree and a half; a success which procured him both public and private encouragement. About the year 1739 he completed his second machine, of a construction much more simple than the former, and which answered much better: this, though not sent to sea, recommended Mr. Harrison yet stronger to the patronage of his friends and the public. His third machine, which he produced in 1749, was still less complicated than the second, and more accurate, as erring only 3 or 4 seconds in a week. This he conceived to be the *ne plus ultra* of his attempts; but by endeavouring to improve pocket-watches, he found the principles he applied to surpass his expectations, so much, as to encourage him to make his fourth time-keeper, which is in the form of a pocket-watch, about 6 inches diameter. With this time-keeper his son made two voyages, the one to Jamaica, and the other to Barbadoes; in which experiments it corrected the longitude within the nearest limits required by the act of the 12th of queen Anne; and the inventor had therefore, at different times, more than the proposed reward, receiving from the Board of Longitude at different times almost 24,000l. besides a few hundreds from the East India Company, &c. These four machines were given up to the Board of Longitude. The three former were not of any use, as all the advantages gained by making them, were comprehended in the last: being worthy however of preservation, as mechanical curiosities, they are deposited in the Royal Observatory at Greenwich. The fourth machine, emphatically distinguished by the name of *The time-keeper*, was copied by the ingenious Mr. Kendal; and that duplicate, during a three years circumnavigation of the globe in the southern hemisphere by captain Cook, answered as well as the original.

The latter part of Mr. Harrison's life was employed in making a fifth improved time-keeper, on the same principles with the preceding one; which, after a ten-weeks trial, in 1772, at the king's private observatory at Richmond, erred only 4½ seconds. Within a few years of his death, his constitution visibly declined; and he had frequent fits of the gout, a disorder that

never

never attacked him before his 77th year. His constitution at last yielding to the infirmities of old age, he died at his house in Red-Lion Square, in 1776, at 83 years of age.

Like many other mere mechanics, Mr. Harrison found a difficulty in delivering his sentiments in writing (at least in the latter periods of his life, when his faculties were much impaired) in which he adhered to a peculiar and uncouth phraseology. This was but too evident in his *Description concerning such Mechanism as will afford a nice or true Mensuration of Time*, &c. 8vo, 1775. This small work includes also an account of his new musical scale; being a mechanical division of the octave, according to the proportion which the radius and diameter of the circle have respectively to the circumference. He had in his youth been the leader of a band of church-singers; had a very delicate ear for music; and his experiments on sound, with a curious monochord of his own improvement, it has been said were not less accurate than those he was engaged in for the mensuration of time.

HAUTEFEUILLE (JOHN), an ingenious mechanic, born at Orleans in 1674. He made a great progress in mechanics in general, but had a particular taste for clock-work, and made several discoveries in it that were of singular use. It was he it seems who found out the secret of moderating the vibration of the balance by means of a small steel-spring, which has since been made use of. This discovery he laid before the members of the Academy of Sciences in 1694; and these watches are, by way of eminence, called *pendulum-watches*; not that they have real pendulums, but because they nearly approach to the justness of pendulums. M. Huygens perfected this happy invention; but having declared himself the inventor and obtained a patent for making watches with spiral springs, the abbé Feuille opposed the registering of it, and published a piece on the subject against Huygens. He died in 1724, at 50 years of age. Besides the above,

He wrote a great many other pieces, most of which are small pamphlets, but very curious: as, 1. His Perpetual Pendulum. 2. New Inventions. 3. The Art of Breathing under Water, and the means of preserving a Flame shut up in a Small Place. 4. Reflections on machines for Raising Water. 5. His Opinion on the different Sentiments of Mallebranche and Regis, relating to the Appearance of the Moon when seen in the Horizon. 6. The Magnetic Balance. 7. A Placet to the king on the Longitude. 8. Letter on the Secret of the Longitude. 9. A New System on the Flux and Reflux of the Sea. 10. The means of making Sensible Experiments that prove the Motion of the Earth: and many other pieces.

HAYES (CHARLES, Esq.), a very singular person, whose great erudition was so concealed by his modesty, that his name is known to very few, though his publications are many. He was born in 1678, and died in 1760, at 82 years of age. He became distinguished in 1704 by a Treatise of Fluxions, in folio, being we believe the first treatise on that science ever published in the English language; and the only work to which he ever set his name. In 1710 came out a small 4to pamphlet, in 19 pages, intitled, A New and easy Method to find out the Longitude from observing the Altitudes of the Celestial

bodies. Also, in 1723, he published, The Moon, a Philosophical Dialogue; tending to shew, that the moon is not an opaque body, but has native light of her own.

To a skill in the Greek and Latin, as well as the modern languages, he added the knowledge of the Hebrew: and he published several pieces relating to the translation and chronology of the scriptures. During a long course of years he had the chief management of the late African company, being annually elected sub-governor. But on the dissolution of that company in 1752, he retired to Down in Kent, where he gave himself up to study; from whence however he returned in 1758, to chambers in Gray's inn, London, where he died in 1760, as mentioned above.

He left a posthumous work, that was published in 8vo, under the title of *Chronographia Asiatica et Ægyptiaca &c.*

HEAT, the opposite to cold, being a relative term denoting the property of fire, or of those bodies we denominate hot; being in us a sensation excited by the action of fire.

Heat, as it exists in the hot body, or that which constitutes and denominates a body hot, and enables it to produce such effects on our organs, is variously considered by the philosophers: some making it a quality, others a substance, and others only a mechanical affection. The former principle is laid down by Aristotle and the Peripatetics. While the Epicureans, and other corpuscularians, define Heat not as an accident of fire, but as an essential power or property of it, the same in reality with it, and only distinguished from it in the manner of our conception. So that Heat, on their principles, is no other than the volatile substance of fire itself, reduced into atoms, and emitted in a continual stream from ignited bodies; so as not only to warm the objects within its reach, but also, if they be inflammable, to kindle them, turn them into fire, and conspire with them to make flame. In effect, these corpuscles, say they, flying off from the ignited body, constitute fire while yet contained within the sphere of its flame; but when fled, or got beyond the same, and dispersed every way, so as to escape the apprehension of the eye, and only to be perceived by the feeling, they take the denomination of Heat, inasmuch as they excite in us that sensation.

The Cartesians, improving on this doctrine, assert that Heat consists in a certain motion of the insensible particles of a body, resembling the motion by which the several parts of our body are agitated by the motion of the heart and blood.

Our latest and best writers of mechanical, experimental, and chemical philosophy, differ very considerably about Heat. The chief difference is, whether it be a peculiar property of one certain immutable body, called fire, or phlogiston, or electricity; or whether it may be produced mechanically in other bodies, by inducing an alteration in their particles. The former tenet, which is as ancient as Democritus, and the system of atoms, had given way to that of the Cartesians, and other mechanists; but is now with great address retrieved, and improved on, by some of the latest writers, particularly Homberg, the younger Lemery, Gravesande, Boerhaave in his lectures on fire, Black, Crawford, and other chemical philosophers.

4 F 2

The

The thing called fire, according to Boerhaave, is a body sui generis, created such ab origine, unalterable in its nature and properties, and not either producible de novo from any other body, nor capable of being reduced into any other body, or of ceasing to be fire. This fire, he contends, is diffused equably every where, and exists alike, or in equal quantity, in all the parts of space, whether void, or possessed by bodies; but that naturally, and in itself, it is perfectly latent and imperceptible; being only discovered by certain effects which it produces, and which are cognizable by our senses. These effects are Heat, light, colour, rarefaction, and burning, which are all indications of fire, as being none of them producible by any other cause: so that wherever we observe any of these, we may safely infer the action and presence of fire. But though the effect cannot be without the cause, yet the fire may remain without any of these effects; any, we mean, gross enough to affect our senses, or become objects of them: and this, he adds, is the ordinary case; there being a concurrence of other circumstances, which are often wanting, necessary to the production of such sensible effects.

The mechanical philosophers, particularly Bacon, Boyle, and Newton, conceive otherwise of Heat; considering it not as an original inherent property of any particular sort of body; but as mechanically producible in any body. The former, in an express treatise De Forma Calidi, from a particular enumeration of the several phenomena and effects of Heat, deduces several general properties of it; and hence he defines Heat, an expansive undulatory motion in the minute particles of the body; by which they tend, with some rapidity, towards the circumference, and at the same time incline a little upwards.

Mr. Boyle, in a Treatise on the Mechanical Origin of Heat and Cold, strongly supports the doctrine of the producibility of Heat, with new observations and experiments; as in the instance of a smith briskly hammering a small piece of iron, which, though cold before, soon becomes exceedingly hot.

This system is also farther supported by Newton, who does not conceive fire as any particular species of body, originally endued with such and such properties. Fire, according to him, is only a body much ignited, that is heated hot, so as to emit light copiously: what else, says he, is red-hot iron but fire? and what else is a burning charcoal but red-hot wood? or flame itself, but red-hot smoke? It is certain that flame is only the volatile part of the fuel heated red-hot, i. e. so hot as to shine; and hence only such bodies as are volatile, that is, such as emit a copious fume, will flame; nor will they flame longer than they have fume to burn. In distilling hot spirits, if the head of the still be taken off, the ascending vapours will catch fire from a candle, and turn into a flame. And in the same manner several bodies, much heated by motion, attrition, fermentation, or the like, will emit lucid fumes, which, if they be copious enough, and the heat sufficiently great, will be flame; and the reason why fused metals do not flame, is the smallness of their fume; this is evident, because spilter, which fumes most copiously, does likewise flame. Add, that all flaming bodies, as oil, tallow, wax, wood, pitch, sulphur, &c, by flaming, waste and vanish into burning smoke. And do not all fixed bodies, when heated beyond a certain degree, emit light,

and shine? and is not this emission performed by the vibrating motion of their parts? and do not all bodies, which abound with terrestrial and sulphureous parts, emit light as often as those parts are sufficiently agitated, whether that agitation be made by external fire, or by friction, or percussion, or putrefaction, or by any other cause? Thus, sea water, in a storm; quicksilver agitated in vacuo; the back of a cat, or the neck of a horse, obliquely rubbed in a dark place; wood, flesh, and fish, while they putrefy; vapours from putrefying waters, usually called ignes fatui; stacks of moist hay or corn; glow-worms; amber and diamonds by rubbing; fragments of steel struck off with a flint, &c, all emit light. Are not gross bodies and light convertible into one another? and may not bodies receive much of their activity from the particles of light which enter their composition? I know no body less apt to shine than water; and yet water, by frequent distillations, changes into fixed earth, which, by a sufficient Heat, may be brought to shine like other bodies.

Add, that the sun and stars, according to Newton's conjecture, are no other than great earths vehemently heated: for large bodies, he observes, preserve their Heat the longest, their parts heating one another; and why may not great, dense, and fixed bodies, when heated beyond a certain degree, emit light so copiously, as by the emission and reaction of it, and the reflections and refractions of the rays within the pores, to grow still hotter, till they arrive at such a period of Heat as is that of the sun? Their parts also may be farther preserved from fuming away, not only by their fixity, but by the vast weight and density of their atmospheres incumbent on them, thus strongly compressing them, and condensing the vapours and exhalations arising from them. Hence we see warm water, in an exhausted receiver, shall boil as vehemently as the hottest water open to the air; the weight of the incumbent atmosphere, in this latter case, keeping down the vapours, and hindering the ebullition, till it has conceived its utmost degree of Heat. So also a mixture of tin and lead, put on a red-hot iron in vacuo, emits a fume and flame; but the same mixture in the open air, by reason of the incumbent atmosphere, does not emit the least sensible flame.

Thus much for the system of the producibility of Heat.

On the other hand, M. Homberg, in his Essai du Soufre Principe, holds, that the chemical principle or element, sulphur, which is supposed one of the simple, primary, pre-existent ingredients of all natural bodies, is real fire, and consequently that fire is co-eval with body. Mem. de l'Acad. an. 1705.

Dr. Gravesande goes upon much the same principle. According to him, fire enters the composition of all bodies, is contained in all bodies, and may be separated or procured from all bodies, by rubbing them against each other, and thus putting their fire in motion. But fire, he adds, is by no means generated by such motion. Elem. Phys. tom. 2, cap. 1. Heat, in the hot body, he says, is an agitation of the parts of the body, made by means of the fire contained in it; by such agitation a motion is produced in our bodies, which excites the idea of Heat in our minds: so that Heat, in respect of us, is nothing but that idea, and in the hot body nothing but motion. If such motion expel

the

the fire in right lines, it may give us the idea of light; if in a various and irregular motion, only of Heat.

Lemery, the younger, agrees with these two authors, in asserting this absolute and ingenerable nature of fire; but he extends it farther. Not contented with confining it as an element to bodies, he endeavours to shew, that it is equably diffused through all space; that it is present in all places, even in the void spaces between the bodies, as well as in the insensible interstices between their parts. And this last sentiment falls in with that of Boerhaave above delivered. Mem. de l'Acad. an. 1713.

Philosophers have lately distinguished Heat into Absolute, and Sensible. By Absolute Heat, or fire, they mean that power or element which, when it is in a certain degree, excites in animals the sensation of Heat; and by Sensible Heat, the same power considered in its relation to the effects which it produces: thus, two bodies are said to have equal quantities of sensible Heat, when they produce equal effects upon the mercury in the thermometer; but as bodies of different kinds have different capacities for containing Heat, the absolute Heat in such bodies will be different, though the sensible Heat be the same. Thus, a pound of water and a pound of antimony, of the same temperature, have equal sensible Heat; but the former contains a much greater quantity of absolute Heat than the latter.

M. De Luc has evinced, by a variety of experiments, that the expansions of mercury between the freezing and boiling points of water, correspond precisely to the quantities of absolute Heat applied, and that its contractions are proportionable to the diminution of this element within these limits. And from hence it may be inferred, that if the mercury were to retain its fluid form, its contractions would be proportionable to the decrements of the absolute Heat, though the diminution were continued to the point of total privation. But the comparative quantities of absolute Heat, which are communicated to different bodies, or separated from them, cannot be determined in a direct manner by the thermometer.

Some philosophers have apprehended that the quantities of absolute Heat in bodies, are in proportion to their densities. While others, as Boerhaave, imagined that Heat is equally diffused through all bodies, the densest as well as the rarest, and therefore that the quantities of Heat in bodies are in proportion to their bulk or magnitude: and, at his desire, Fahrenheit attempted to determine the fact by experiment. For this purpose, he took equal quantities of the same fluid, and gave them different degrees of Heat, then upon mixing them intimately together, he found that the temperature of the mixture was a just medium, or arithmetical mean, between the two. But if this experiment be made with water and mercury, in the same circumstances, viz in equal bulks, the result will be different, as the temperature of the mixture will not be a mean between the two, but always nearer to that of the water than to the quicksilver; so that, when the water is the hotter, the temperature of the mixture is above the mean, and below it when the water is the colder. And from experiments of this kind it has been inferred, that the comparative quantities of the absolute Heats of these fluids, are reciprocally proportional to the changes which

are produced in their sensible Heats, when they are mixed together at different temperatures: and this fact has been publicly taught, for several years, by Dr. Black, and Dr. Irvine, in the universities of Edinburgh and Glasgow. This rule however does not apply to those substances which, in mixture, excite sensible Heat by chemical action.

From the experiments and reasoning employed by Dr. Crawford, it more fully appears, that the quantities of absolute Heat in different bodies, are not as their densities; or that equal weights of heterogeneous substances, as air and water, having the same temperature, may contain unequal quantities of absolute Heat: he also shews, that if phlogiston be added to a body, a quantity of the absolute Heat of that body will be extricated; and if the phlogiston be separated again, an equal quantity of Heat will be absorbed. So that Heat and phlogiston appear to be two opposite principles in nature. But this ingenious writer has not presumed absolutely to decide the question that has been long agitated, whether Heat be a substance or a quality.— He inclines to the former opinion however, and observes, that if we adopt the opinion, that Heat is a distinct substance, or an element sui generis, the phenomena will be found to admit of a simple and obvious interpretation, and to be perfectly agreeable to the analogy of nature. See Crawford's Experiments and Observations on Animal Heat and the Inflammation of Combustible Bodies.

Animal HEAT. The Heat of animals is very various, both according to the variety of their kinds, and the difference of the seasons: accordingly, zoologists have divided them into hot and cold blooded, reckoning those to be hot that are near or above our own temperature, and all others cold whose Heat is below ours, and consequently affect us with the sense of cold; thus making the human species a medium between the hot and cold blooded animals, or at least the lowest order of the hot blooded.

The Heat of the human body, in its natural state, according to Dr. Boerhaave, is such as to raise the mercury in the thermometer to 92° or at most to 94°; and Dr. Pitcairn makes the heat of the human skin the same. Indeed it is evident that different parts of the human body, and its different states, as well as the different seasons, will make it shew of different temperatures. Thus, by various experiments at different times, the Heat of the human body is made various by the following authors:

Boerhaave and Pitcairn	92°
Amontons	91, 92, or 93
Sir Isaac Newton	95½
Fahrenheit and Musschenbroek, the blood,	96
Dr. Martine, the skin	97 or 98
————, the urine	99
Dr. Hales, the skin	97
————, the urine	103
Mr. John Hunter, under his tongue,	97
————, in his rectum	98½
————, his urethra at 1 inch,	92
at 2 inches,	93
at 4 inches,	94
the ball of the thermom. at the bulb of the urethra	97

For the powers of animals to bear various degrees of Heat, see the Philof. Tranf. vol. 65, 68, &c.

There is hardly any fubject of philofophical inveftigation that has afforded a greater variety of hypothefes, conjectures, and experiments, than the caufe of animal Heat. The firft opinion which has very generally obtained, is, that the Heat of animal bodies is owing to the attrition between the arteries and the blood. All the obfervations and reafoning brought in favour of this opinion however, only fhew that the Heat and the motion of the arteries are generally proportional to each other; without fhewing which is the caufe, and which the effect; or indeed that either is the caufe or effect of the other, fince both may be the effects of fome other caufe.

Dr. Douglas, in his Effay on the Generation of Heat in animals, afcribes it folely to the friction of the globules of blood in their circulation through the capillary veffels.

Another opinion is, that the lungs are the fountain of Heat in the human body: and this opinion is fupported by much the fame fort of arguments as the former, and feemingly to little better purpofe.

A third opinion is, that the caufe of animal Heat is owing to the action of the folid parts upon one another. And as the heart and arteries move moft, it has been thought natural to expect that the Heat fhould be owing to this motion. But even this does not feem very plaufible, from the following confiderations: 1ft, The moving parts, however we term them folid, are neither hard nor dry; which two conditions are abfolutely requifite to make them fit to generate Heat by attrition. 2d, None of their motions are fwift enough to promife Heat in this way. 3d, They have but little change of furface in their attritions. And 4thly, The moveable fibres have fat, mucilage, or liquors everyway furrounding them, to prevent their being deftroyed, or heated by attrition.

A fourth caufe affigned for the Heat of our bodies, is that procefs by which our aliment and fluids are perpetually undergoing fome alteration. And this opinion is chiefly fupported by Dr. Stevenfon, in the Edinburgh Medical Effays, vol. 5, art. 77.

The late ingenious Dr. Franklin inclines to this opinion, when he fays, that the fluid fire, as well as the fluid air, is attracted by plants in their growth, and becomes confolidated with the other materials of which they are formed, and makes a great part of their fubftance; that when they come to be digefted, and to undergo a kind of fermentation in the veffels, part of the fire, as well as part of the air, recovers its fluid active ftate again, and diffufes itfelf on the body digefting and feparating it; &c. Exper. and Obf. on Electricity, p. 346.

Dr. Mortimer thinks the Heat of animals explicable from the phofphorus and air they contain. Phofphorus exifts, at leaft in a dormant ftate, in animal fluids; and it is alfo known that they all contain air; it is therefore only neceffary to bring the phofphoreal and aereal particles into contact, and Heat muft of confequence be generated; and were it not for the quantity of aqueous humours in animals, fatal accenfions would frequently happen. Philof. Tranf. number 476.

Dr. Black fuppofes, that animal Heat is generated altogether in the lungs, by the action of the air on the principle of inflammability, and is thence diffufed over the reft of the body by means of the circulation. But Dr. Leflie urges feveral arguments againft this hypothefis, tending to fhew that it is repugnant to the known laws of the animal machine; and he advances another hypothefis inftead of it, viz. that the fubtle principle by chemifts termed phlogifton, which enters into the compofition of natural bodies, is in confequence of the action of the vafcular fyftem gradually evolved through every part of the animal machine, and that during this evolution Heat is generated. This opinion, he candidly acknowledges, was firft delivered by Dr. Duncan of Edinburgh; and that fomething fimilar to it is to be found in Dr. Franklin's works, and in a paper of Dr. Mortimer's in the Philof. Tranf.

The laft hypothefis we fhall mention, is the very plaufible one of Dr. Crawford, lately publifhed in his Experiments and Obfervations on Animal Heat. This ingenious gentleman has inferred, from a variety of experiments, that Heat and phlogifton, fo far from being connected, as moft philofophers have imagined, act in fome meafure in oppofition to each other. By the action of Heat on bodies, the force of their attraction of phlogifton is diminifhed, and by the action of phlogifton, a part of their abfolute Heat is expelled. He has alfo demonftrated, that atmofpherical air contains a greater quantity of abfolute Heat than the air which is expired from the lungs of animals: he makes the proportion of the abfolute Heat of atmofpherical air to that of fixed air, as 67 to 1; and the Heat of dephlogifticated air to that of atmofpherical air as 4.6 to 1; and obferving that Dr. Prieftley has proved, that the power of this dephlogifticated air in fupporting animal life, is 5 times as great as that of atmofpherical air, he concludes that the quantity of abfolute Heat contained in any kind of air fit for refpiration, is very nearly in proportion to its purity, or to its power of fupporting animal life; and fince the air exhaled by refpiration, is found to contain only the 67th part of the Heat which was contained in the atmofpherical air, previous to infpiration, it is very reafonably inferred, that the latter muft neceffarily depofit a very great proportion of its abfolute Heat in the lungs. Dr. Crawford has alfo fhewn, that the blood which paffes from the lungs to the heart, by the pulmonary vein, contains more abfolute Heat than that which paffes from the heart to the lungs, by the pulmonary artery; the abfolute Heat of florid arterial blood being to that of venous blood, as $11\frac{1}{2}$ to 10: therefore, fince the blood which is returned by the pulmonary vein to the heart has the quantity of its abfolute Heat increafed, it muft have acquired this Heat in its paffage through the lungs; fo that in the procefs of refpiration a quantity of abfolute Heat is feparated from the air, and abforbed by the blood. Dr. Prieftley has alfo proved, that in refpiration, phlogifton is feparated from the blood, and combined with air.

This theory however has been contefted and difputed, and, it has been faid, Dr. Crawford's experiments repeated, with contrary refults. Though no regular and fyftematical theory has yet been formed in its ftead.

HEAT of Combuftible and Inflammable Bodies. Dr. Crawford's theory with refpect to the inflammation of com-

combuftible bodies, is founded on the fame principles as his doctrine concerning the Heat of animals. According to him, the Heat which is produced by combuftion, is derived from the air, and not from the inflammable body. Inflammable bodies, he fays, abound with phlogifton, and contain little abfolute Heat : the atmofphere, on the contrary, abounds with abfolute Heat, and contains little phlogifton. In the procefs of inflammation, the phlogifton is feparated from the inflammable body, and combined with the air; the air is phlogifticated, and gives off a great proportion of its abfolute Heat, which, when extricated fuddenly, burfts forth into flame, and produces an intenfe degree of fenfible Heat. And fince it appears by calculation, that the Heat produced by converting atmofpherical into fixed air, is fuch, if it were not diffipated, as would be fufficient to raife the air fo changed, to more than 12 times the Heat of red-hot iron, it follows, that in the procefs of inflammation a very great quantity of Heat is derived from the air. But, on the contrary, no part of the Heat can be derived from the combuftible body; becaufe this body, during the inflammation, being deprived of its phlogifton, undergoes a change fimilar to that of the blood by the procefs of refpiration, in confequence of which its capacity of containing Heat is increafed; and therefore it will not give off any part of its abfolute Heat, but, like the blood in its paffage through the lungs, it will abforb Heat.

A fimilar theory of Heat has lately been publifhed by Mr. Elliot. See his Philofophical Obfervations on the fenfes of Vifion and Hearing; to which is added an Effay on Combuftion and Animal Heat. 8vo, 1780.

HEAT, in Geography, is that which relates to the earth. There is a great variety in the Heat of different places and feafons. Naturalifts have commonly laid it down, that the nearer any place is to the centre of the earth, the hotter it is found; but this does not hold ftrictly true. And if it were, the effect might be otherwife accounted for, and more fatisfactorily, than from their imagined central fire.

Mr. Boyle, who had been at the bottom of fome mines himfelf, with more probability fufpects that this degree of Heat, at leaft in fome of them, may arife from the peculiar nature of the minerals there produced. And he inftances a mineral of the vitriolic kind, dug up in large quantities, in feveral parts of England, which, by the bare effufion of common water, will grow fo hot as almoft to take fire. To which may be added, that fuch places, in the bowels of the earth, ufually feel hot, from the confined and ftagnant ftate of the air in them, in which the heat is retained, through the want of a current or change of air to carry the Heat off.

On the other hand, on afcending high mountains, the air grows more and more cold and piercing. Thus, the tops of the Pike of Teneriffe, the Alps, and feveral other mountains, even in the moft fultry countries, are found always invefted with fnow and ice, which the Heat is never fufficient to thaw. In fome of the mountains of Peru there is no fuch thing as running water, but all ice : plants vegetate a little about the bottom of the mountains, but near the top no vegetable can live, for the intenfenefs of the cold. This effect is attributed to the thinnefs of the air, and the little furface of the earth there is to reflect the rays, as well as the great diftance of the general furface of the earth which reflects the rays back into the atmofphere.

As to the diverfity in the Heat of different climes and feafons, it arifes from the different angles under which the fun's rays ftrike upon the furface of the earth. In the Philof. Tranf. Abr. vol. 2, p. 165, Dr. Halley has given a computation of this Heat, on the principle, that the fimple action of the fun's rays, like other impulfes or ftrokes, is more or lefs forcible, according to the fines of the angles of incidence, or to the fines of the fun's altitudes, at different times or places.

Hence it follows, that, the time of continuance, or the fun's fhining on any place, being taken for a bafis, and the fines of the fun's altitudes perpendicularly erected upon it, and a curve line drawn through the extremities of thofe perpendiculars, the area thus comprehended will be proportional to the collection of all the Heat of the fun's beams in that fpace of time.

Hence it will likewife follow, that at the pole, the collection of all the Heat of a tropical day, is proportional to the rectangle or product of the fine of $23\frac{1}{2}$ degrees into 24 hours, or the circumference of a circle, or as $\frac{8}{10}$ into 12 hours, the fine of $23\frac{1}{2}$ degrees being nearly $\frac{4}{10}$ of radius. Or the polar Heat will be equal to that of the fun continuing 12 hours above the horizon at 53 degrees height; and the fun is not 5 hours more elevated than this under the equinoctial.

But as it is the nature of Heat to remain in the fubject, after the luminary is removed, and particularly in the air, under the equinoctial the 12 hours abfence of the fun abates but little from the effect of his Heat in the day; but under the pole, the long abfence of the fun for 6 months has fo chilled the air, that it is in a manner frozen, and after the fun has rifen upon the pole again, it is long before his beams can make any impreffion, being obftructed by thick clouds and fogs.

From the foregoing principle Dr. Halley computes the following table, exhibiting the Heat to every 10th degree of latitude, for the equinoctial and tropical fun, and from which an eftimate may eafily be made for the intermediate degrees.

Lat.	Sign that the Sun is in.		
	♈ or ♎	♋	♑
0	20000	18341	18341
10	19696	20290	15834
20	18794	21737	13166
30	17321	22651	10124
40	15321	23048	6944
50	12855	22991	3798
60	10000	22773	1075
70	6840	23543	0
80	3473	24073	0
90	0	25055	0

From the fame principles, and table, alfo are deduced the following corollaries, viz,

That

1, That the equatorial Heat, when the sun becomes vertical, is as twice the square of the radius.—2, That at the equator the Heat is as the sine of the sun's declination.—3, That in the frigid zones, when the sun sets not, the Heat is as the circumference of a circle into the sine of the altitude at 6: And consequently that in the same latitude, these aggregates of Heat are as the sines of the sun's declination; and at the same declination of the sun, they are as the sines of the latitudes; and generally they are as the sines of the latitudes into the sines of declination.—4, That the equatorial day's Heat is everywhere as the cosine of the latitude.—5, In all places where the sun sets, the difference between the summer and winter Heats, when the declinations are contrary, is equal to a circle into the sine of the altitude at 6, in the summer parallel; and consequently those differences are as the rectangles of the sines of the latitude and declination.—6, The tropical sun has the least force of any at the equator; and at the pole it is greatest of all.

Many objections have been urged against this theory of Dr. Halley. Some have objected, that the effect of the sun's Heat is not in the simple, but in the duplicate ratio of the sines of the angles of incidence; like the law of the impulse of fluids. And indeed, the quantity of the sun's direct rays received at any place, being evidently as the sine of the angle of incidence, or of the sun's altitude, *if* the Heat be also proportional to the force with which a ray strikes, like the mechanical action or impulse of any body, then it will follow that the Heat must be in the compound ratio of both, that is, as the square of the sine of the sun's altitude. But this last principle is here only assumed gratis, as we do not know a priori that the Heat is proportional to the force of a striking body; and it is only experiment that can determine this point.

It is certain that Heat communicated by the sun to bodies on the earth, depends also much upon other circumstances beside the direct force of his rays. These must be modified by our atmosphere, and variously reflected and combined by the action of the earth's surface itself, to produce any remarkable effects of Heat. So that if it were not for these additional circumstances, it is probable the naked Heat of the sun would not be very sensible.

Dr. Halley himself was well apprised, that many other circumstances, besides the direct force of the sun's rays, contributed to augment or diminish the effect of this, and the Heat resulting from it, in different climates; and therefore no calculation, formed on the preceding theory, can be supposed to correspond exactly with observation and experiment. It has also been objected that, according to the foregoing theory, the greatest Heat in the same place should be at the summer solstice, and the most extreme cold at the winter solstice; which is contrary to experience. To this objection it may be replied, that Heat is not produced in bodies by the sun instantaneously, nor do the effects of his Heat cease immediately when his rays are withdrawn; and therefore those parts which are once heated, retain the Heat for some time; which, with the additional Heat daily imparted, makes it continue to increase, though the sun declines from us: and this is the reason why July is hotter than June, although the sun has

withdrawn from the summer tropic: as we also find it is generally hotter at one, two, or three in the afternoon, when the sun has declined towards the west, than at noon, when he is on the meridian. As long as the heating particles, which are constantly received, are more numerous than those which fly away or lose their force, the Heat of bodies must continually increase. So, after the sun has left the tropic, the number of particles, which Heat our atmosphere and earth, constantly increases, because we receive more in the day than we lose at night, and therefore our Heat must also increase. But as the days decrease again, and the action of the sun becomes weaker, more particles will fly off in the night time than are received in the day, by which means the earth and air will gradually cool. Farther, those places which are well cooled, require time to be heated again; and therefore January is mostly colder than December, although the sun has withdrawn from the winter tropic, and begun to emit his rays more perpendicularly upon us.

But the chief cause of the difference between the Heat of summer and winter is, that in summer the rays fall more perpendicularly, and pass through a less dense part of the atmosphere; and therefore with greater force, or at least in greater number in the same place: and besides, by their long continuance, a much greater degree of Heat is imparted by day than can fly off by night.

For the calculations and opinions of several other philosophers on this head, see Keill's Astron. Lect. 8; Ferguson's Astron. chap. 10; Long's Astron. § 777; Memo. Acad. Scienc. 1719.

As to the temperature or Heat of our atmosphere, it may be observed that the mercury seldom falls under 16° in Fahrenheit's thermometer; but we are apt to reckon it very cold at 24°, and it continues coldish to 40° and a little above. However, such colds have been often known as bring it down to 0°, the beginning of the scale, or nearly the cold produced by a mixture of snow and salt, often near it, and in some places below it. Thus, the degree of the thermometer has been observed at various times and places as follows:

Places	Latit.	Year	Thermom.
Pensylvania	40° 0′	1732	5°
Paris	48 50	1709 & 1710	8
Leyden	52 10	1729	5
Utrecht	52 8		4
London	51 31	1709 & 1710	0
Copenhagen	55 43	1709	0
Upsal	59 56	1732	—1
Petersburg	59 56	——	—28
Torneo	65 51	1736-7	—33
Hudson's Bay	52 24	1775	—37

The middle temperature of our atmosphere is about 48°, being nearly a medium of all the seasons. The French make it somewhat higher, reckoning it equal to the cave of their royal observatory, or 53°. In cold countries, the air is found agreeable enough to the inhabitants while it is between 40 and 50°. In our climate we are best pleased with the heat of the air from 50 to 60°; while in the hot countries the air is generally at a medium about 70°. With us, the air is not reckoned

reckoned warm till it arrives at about 64°, and it is very warm and fultry at 80°. It is to be noted that the foregoing obfervations are to be underftood of the ftate of the air in the fhade; for as to the Heat of bodies acted upon by the direct rays of the fun, it is much greater: thus, Dr. Martine found dry earth heated to above 120°; but Dr. Hales found a very hot fun-fhine Heat in 1727 to be about 140°; and Muff-chenbroek once obferved it fo high as 150°; but at Montpelier the fun was fo very hot, on one day in the year 1705, as to raife M. Amontons's thermometer to the mark of boiling water itfelf, which is our 212°.

It appears from the regifter of the thermometer kept at London by Dr. Heberden for 9 years, viz, from the end of 1763 to the end of 1772, that the mean Heat at 8 in the morning was 47°·4; and by another regifter kept at Hawkhill, near Edinburgh, that the mean Heat in that place, during the fame period of time, was 46°. Alfo by regifters kept in London and at Hawkhill, for the three years 1772, 1773, 1774, it appears, that the mean Heat of thefe three years in London, at 8 in the morning, was 48°·5, and at 2 in the afternoon 56°, but the mean of both morning and afternoon 52°·2; while the mean Heat at Hawkhill for the fame time,

at 8 in the morning was   45°·4
and at 2 in the afternoon   50·1
and the mean of both   47·7·

The mean Heat of fprings near Edinburgh feems to be 47°, and at London 51°. Philof. Tranf. vol. 65, art. 44.

Laftly, from the meteorological journals of the Royal Society, publifhed in the Philof. Tranf. it appears that the mean heights of the thermometer, for the whole years, kept without and within the houfe, are as below:

	Therm. Without	Therm. Within
For 1775	51·5	52·7
1776	51·1	52·9
1777	51·0	·53·0
1778	52·0	53·1
mean of all	51·4	52·9

HEAVEN, an azure tranfparent orb invefting our earth, where the celeftial bodies perform their motions. It is of divers denominations, as the higheft or empyrean Heaven, the etherial or ftarry Heaven, the planetary Heaven, &c.

Formerly the Heavens were confidered as folid fubftances, or elfe as fpaces full of folid matter; but Newton has abundantly fhewn that the Heavens are void of almoft all refiftance, and confequently of almoft all matter: this he proves from the phenomena of the celeftial bodies; from the planets perfifting in their motions, without any fenfible diminution of their velocity; and the comets freely paffing in all directions towards all parts of the Heavens.

Heaven, taken in this general fenfe, or the whole expanfe between our earth and the remoteft regions of the fixed ftars, may be divided into two very unequal parts, according to the matter occupying them; viz, the atmofphere or aereal Heaven, poffeffed by air; and the ethereal Heaven, poffeffed by a thin and unrefifting medium, called ether.

HEAVEN is more particularly ufed, in Aftronomy, for an orb, or circular region, of the ethereal Heaven.

The ancient aftronomers affumed as many different Heavens as they obferved different celeftial motions. All thefe they made folid, thinking they could not otherwife fuftain the bodies fixed in them; and of a fpherical form, as being the moft proper for motion. Thus they had feven Heavens for the feven planets; viz, the Heavens of the Moon, Mercury, Venus, the Sun, Mars, Jupiter, and Saturn. The 8th was for the fixed ftars, which they particularly called the firmament. Ptolomy added a 9th Heaven, which he called the primum mobile. After him two cryftalline Heavens were added by king Alphonfus, &c, to account for fome irregularities in the motions of the other Heavens. And laftly an empyrean Heaven was drawn over the whole for the refidence of the Deity; which made the number 12.

But others admitted many other Heavens, according as their different views and hypothefes required. Eudoxus fuppofed 23, Calippus 30, Regiomontanus 33, Ariftotle 47, and Fracaftor no lefs than 70.

The aftronomers however did not much concern themfelves whether the Heavens they thus allowed, were real or not; provided they ferved a purpofe in accounting for any of the celeftial motions, and agreed with the phenomena.

HEAVINESS, the fame as Gravity, which fee.

Heavy bodies do not tend precifely to the very centre of the earth, except at the poles, and the equator, on account of the fpheroidal figure of the earth; their direction being every where perpendicular to the furface of the fpheroid.

HEGIRA, a term in Chronology, fignifying the epoch, or account of time, ufed by the Mahomedans, who begin their computation from the day that Mahomet was forced to make his efcape from the city of Mecca, which happened on Friday the 16th of July 622.

The years of the Hegira are lunar ones, confifting only of 354 days. Hence, to reduce thefe years to the Julian calendar, that is, to find what Julian year a given year of the Hegira anfwers to: reduce the year of the Hegira into days, by multiplping by 354, divide the product by 365¼, and to the quotient add 622, the year the Hegira commenced.

HEIGHT, the third dimenfion of a body, confidered with regard to its elevation above the ground.

HEIGHT, of a figure, the fame as its altitude, being the perpendicular from its vertex to the bafe.

HEIGHT of the Pole, &c. See Altitude of the Pole, &c.

HELIACAL, fomething relating to the fun. Thus,

HELIACAL *Rifing*, of a ftar or planet, is when it rifes with, or at the fame time, as the fun. And Heliacal fetting, the fame as the fetting with the fun.

Or, a ftar rifes Heliacally, when, after it has been in conjunction with the fun, and fo invifible, it gets at fuch a diftance from him as to be feen in the morning before the fun's rifing. And it is faid to fet Heliacally, when it approaches fo near the fun as to be hid by his beams. So that, in ftrictnefs, the Heliacal rifing and fetting are only an apparition and occultation.

HELICE *Major* and *Minor*; the fame as Urfa Major and Minor.

HELICOID *Parabola*, or the *Parabolic Spiral*, is a curve arising from the supposition that the common or Apollonian parabola is bent or twisted, till the axis come into the periphery of a circle, the ordinates still retaining their places and perpendicular positions with respect to the circle, all these lines still remaining in the same plane. Thus, the axis of a parabola being bent into the circumference BCDM, and the ordinates CF, DG, &c, still perpendicular to it, then the parabola itself, passing through the extremities of the ordinates, is twisted into the curve BFG, &c, called the Helicoid, or Parabolic Spiral.

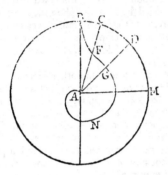

Hence all the ordinates CF, DG, &c, tend to the centre of the circle, being perpendicular to the circumference

Also, the equation of the curve remains the same as when it was a parabola; viz, putting $x =$ any circular abscifs BC, and $y =$ CF the corresponding ordinate, then is $px = y^2$, where p is the parameter of the parabola.

HELIOCENTRIC *Place of a Planet*, is the place in which a planet would appear to be when viewed from the sun; or the point of the ecliptic, in which a planet viewed from the sun would appear to be. And therefore the Heliocentric place coincides with the longitude of a planet viewed from the sun.

HELIOCENTRIC *Latitude of a Planet*, is the inclination of the line drawn between the centre of the sun and the centre of a planet, to the plane of the ecliptic. The greatest Heliocentric Latitude is equal to the inclination of the planet's orbit to the plane of the ecliptic.

HELIOCOMETES, Comet of the Sun, a phenomenon sometimes observed at the setting of the sun; thus denominated by Sturmius and Pylen, who had seen it, because it seems to make a comet of the sun, being a large tail, or column of light, fixed or hung to that luminary, and dragging after it, at its setting, like the tail of a comet.

HELIOMETER, or ASTROMETER; an instrument for measuring, with particular exactness, the diameters of the sun, moon, and stars.

This instrument was invented by M. Bouguer in 1747, and is a kind of telescope, consisting of two object glasses of equal focal distance, placed by the side of each other, so that the same eye-glass serves for both. The tube of this instrument is of a conical form, larger at the upper end, which receives the two object-glasses, than at the lower, which is furnished with an eye-glass and micrometer. By the construction of this instrument,

8

two distinct images of an object are formed in the focus of the eye-glass, whose distance, depending on that of the two object-glasses from one another, may be measured with great accuracy. Mem. Acad. Sci. 1748.

Mr. Servington Savery discovered a similar method of improving the micrometer, which was communicated to the Royal Society in 1743.

HELIOSCOPE, a kind of telescope peculiarly adapted for viewing and observing the sun without hurting the eye.

There are various kinds of this instrument, usually made by employing coloured glass for the object or eye-glass, or both; and sometimes only using an eye-glass blacked by holding it over the smoke or flame of a lamp or candle, which is Huygens's way.—See Dr Hooke's treatise on Helioscopes.

HELIOSTATA, an instrument invented by Dr. Gravesande, and so called from its property of fixing the sun-beam in one position, viz, in a horizontal direction across the dark chamber while it is used. See Gravesande's Physices Element. Mathematica, tom. 2, p. 715 ed. 3tia 1742, for an account of the principles, construction and use of this instrument.

HELISPHERICAL *Line*, is the Rhumb-line in Navigation; being so called, because on the globe it winds round the pole helically or spirally, coming still nearer and nearer to it.

HELIX, a Spiral line. See SPIRAL.

HEMISPHERE, the half of a sphere or globe, when divided in two by a plane passing through its centre.

Hemisphere is also used for a map or projection of half the terrestrial globe, or of half the celestial sphere, on a plane; being more frequently called a planisphere.

The centre of gravity of a Hemisphere, is five-eighths of the radius distant from the vertex.

A glass Hemisphere unites the parallel rays at the distance of four-thirds of a diameter from the pole of the glass.

HEMITONE, in Music, a half note.

HENDECAGON, a figure of eleven sides, or the Endecagon; which see.

HENIOCHAS, or HENIOCHUS, a northern constellation, the same as Auriga, which see.

HEPTAGON, in Geometry, a figure of seven sides and seven angles.—When those sides and angles are all equal, the Heptagon is said to be regular, otherwise it is irregular.

In a regular Heptagon, the angle C at the centre is $= 51°\frac{3}{7}$, the angle DAB of the polygon is $= 128°\frac{4}{7}$, and its half CAB $= 64°\frac{2}{7}$. Also the area is $=$ the square of the side $AB^2 \times 3.6339124$ or $= AB^2 \times \frac{7}{4}t$, where t is the tangent of the angle CAB of $64°\frac{2}{7}$ to the radius 1; or t is the root of the equation;

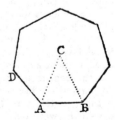

$t^{12} - 26t^{10} + 143t^8 - 245t^6 + 143t^4 - 26t^2 + 1 = 0$; or

$$t = \sqrt{\frac{1+x}{1-x}} = \frac{1 + \sqrt{1-y^2}}{y} = \frac{y}{1 - \sqrt{1-y^2}},$$

where

where the value of x and y are the roots of the equations

$$x^6 - \tfrac{5}{4}x^4 + \tfrac{3}{8}x^2 - \tfrac{1}{16} = 0,$$
$$y^6 = \tfrac{7}{4}y^4 + \tfrac{5}{8}y^2 - \tfrac{7}{8}f = 0.$$

See my Mensuration, p. 21, 114, and 116, 2d edition.

HEPTAGON, in Fortification, a place fortified or strengthened with seven baftions for its defence.

HEPTAGONAL *Numbers*, are a kind of polygonal numbers in which the difference of the terms of the corresponding arithmetical progreffion is 5. Thus,

Arithmeticals, 1, 6, 11, 16, 21, 26, &c.
Heptagonals, 1, 7, 18, 34, 55, 81, &c.

where the Heptagonals are formed by adding continually the terms of the arithmeticals, above them, whofe common difference is 5.

One property, among many others, of thefe Heptagonal numbers is, that if any one of them be multiplied by 40, and to the product add 9, the fum will be a fquare number.

Thus $1 \times 40 + 9 = 49 = 7^2$;
and $7 \times 40 + 9 = 289 = 17^2$;
and $18 \times 40 + 9 = 729 = 27^2$;
and $34 \times 40 + 9 = 1369 = 37^2$; &c.

Where it is remarkable that the feries of fquares fo formed is 7^2, 17^2, 27^2, 37^2, &c, the common difference of whofe roots is 10, the double of the common difference of the arithmetical feries from which the Heptagonals are formed.—See POLYGONALS.

HEPTANGULAR *Figure*, in Geometry, is one that has feven angles; and therefore alfo feven fides.

HERCULES, in Aftronomy, a conftellation of the northern hemifphere, and one of the 48 old conftellations mentioned by ancient writers.

It is not known by what name it was diftinguifhed by the Egyptians and others before the Greeks. Thefe latter, perhaps not knowing its real name, firft called it fimply the kneeling man, becaufe he is drawn in that pofture; but they afterwards fuccefsively afcribed it to, and called it by the names of, Cetheus, Thefeus, and laftly Hercules, which it ftill retains.

The ftars in this conftellation, in Ptolomy's catalogue, are 29; in Tycho's 28; and in the Britannic catalogue, 113.

HERISSON, in Fortification, a beam armed with iron fpikes, having their points turned outward. It is fupported in the middle by a ftake, having a pivot on which it turns; and ferves as a barrier to block up a paffage.—Heriffons are frequently placed before gates, efpecially the pofterns of a town or fortrefs, to fecure thofe paffages which muft of neceffity be often opened.

HERMANN (JAMES), a learned mathematician of the Academy of Berlin, and member of the Academy of Sciences at Paris, was born at Bafil in 1678. He was a great traveller; and for 6 years was profeffor of mathematics at Padua. He afterwards went to Ruffia, being invited thither by the Czar in 1724, as well as his compatriot Daniel Bernoulli. On his return to his native country, he was appointed profeffor of morality and natural law at Bafil; where he died in 1733, at 55 years of age.

He wrote feveral mathematical and philofophical pieces, in the Memoirs of different Academies, and elfewhere; but his principal work, is the *Phoronomia*, or two Books on the Forces and Motions of both Solid and Fluid bodies; 4to, 1716: a very learned work on the new mathematical phyfics.

HERMETIC, or HERMETICAL *Art*, a name given to chemiftry, on a fuppofition that Hermes Trifmegiftus was its inventor.

HERMETICAL *Philofophy*, is that which undertakes to folve and explain all the phenomena of nature from the three chemical principles, falt, fulphur, and mercury.——A confiderable addition was made to the ancient Hermetical Philofophy, by the modern doctrine of alcali and acid.

HERMETICAL *Seal*, or *Hermetical Sealing*, a manner of ftopping or clofing glafs veffels, for chemical and other operations, fo very clofely, that no fubftance can poffibly exhale or efcape. This is ufually done by heating the neck of the veffel in the flame of a lamp, with a blow-pipe, till it be ready to melt, and then with a pair of hot pincers twifting it clofe together.

HERSCHEL, the name by which the French, and moft other European nations, call the new planet, difcovered by Dr. Herfchel in the year 1781. Its mark or character is ♅. The Italians call it Ouranos, or Urania, but the Englifh, the GEORGIAN *Planet*, which fee.

HERSE, in Fortification, a lattice or portcullice, in the form of a harrow, befet with iron fpikes, to block up a gate way, &c.

HERSILLON, or little Herfe, in Fortification, is a plank armed with iron fpikes, for the fame ufe as the Herfe, and alfo to impede the march of the infantry or cavalry.

HESSE (WILLIAM Prince of), rendered his name immortal by his encouragement of learning, by his ftudies, and by his obfervations, for many years, of the celeftial bodies. For this purpofe, he erected an obfervatory at Caffel, and furnifhed it with good inftruments, well adapted to that defign; calling alfo to his affiftance two eminent artifts, Chriftopher Rothmann and Jufte Byrge. His obfervations, which are of a very curious nature, were publifhed at Leyden, in the year 1618, by Willebrord Snell; and are in part mentioned by Tycho Brahe, as well in his epiftles as in the 2d volume of his *Progymnafmata*; a fignal example to all princely and heroic minds, to undertake the promoting the arts of peace, and advancing this truly noble and celeftial fcience. This prince died in the year 1597.

HETERODROMUS *Vectis*, or *Lever*, in Mechanics, a lever in which the fulcrum, or point of fufpenfion, is between the weight and the power; being the fame as what is otherwife called a lever of the firft kind.

HETEROGENEAL, the fame as HETEROGENEOUS; which fee.

HETEROGENEOUS, literally imports things of different natures, or fomething that confifts of parts of different or diffimilar kinds; in oppofition to Homogeneous. Thus,

HETEROGENEOUS *Bodies*, are fuch as have their parts of unequal denfity.

HETEROGENEOUS *Line*, is that which confifts of parts or rays of different refrangibility, reflexibility, and colour.

HETEROGENEOUS *Numbers*, are mixed numbers, confifting of integers and fractions.

HETEROGENEOUS *Particles*, are fuch as are of different kinds, natures, and qualities; of which generally all bodies confift.

HETEROGENEOUS *Quantities*, in Mathematics, are

thofe

thofe which cannot have proportion, or be compared together as to greater and lefs; being of fuch different kind and confideration, as that one of them taken any number of times, never equals or exceeds the other. As lines, furfaces, and folids in geometry.

HETEROGENEOUS *Surds*, are fuch as have different radical figns; as $\sqrt{a}$ and $\sqrt[3]{b^2}$; or $\sqrt[5]{10}$ and $\sqrt[7]{20}$.

HETEROSCII, in Geography, are fuch inhabitants of the earth as have their fhadows at noon projected always the fame way with regard to themfelves, or always contrary ways with refpect to each other. Thus, all the inhabitants without the torrid zone are Heterofcii, with regard to themfelves, fince any one fuch inhabitant has his fhadow at noon always the fame way, viz, always north of him in north latitude, and always fouth of him in fouth latitude; or thefe two fituations are Heterofcii to each other, having fuch fhadows projected contrary ways at all times of the year.

HEVELIUS (John), a very celebrated aftronomer, and a burgomafter of Dantzick, was born in that city in 1611. He ftudied mathematics under Peter Cruger, in which he made a wonderful progrefs. He afterwards fpent feveral years on his travels through Holland, England, Germany, and France, for his improvement in the fciences. On his return, he conftructed excellent telefcopes himfelf, and began diligently to obferve the heavens, an employment he clofely followed during the courfe of a long life, which was terminated only in 1687, at 76 years of age. Hevelius was author of feveral notable difcoveries in the heavens. He was the firft that obferved the phenomenon called the libration of the moon, and made feveral other important obfervations on the other planets. He alfo difcovered feveral fixed ftars, which he named the *firmament of Sobiefki*, in honour of John the 3d, king of Poland. He framed a large catalogue of the ftars, and collected multitudes of the unformed ones into new conftellations of his own framing. His wife was alfo well fkilled in aftronomy, and made a part of the obfervations that were publifhed by her hufband. His principal publications are, his *Selenographia*, or an exact defcription of the moon; in which he has engraved all her phafes, and remarkable parts, diftinguifhed by names, and afcertained their refpective bounds by the help of telefcopes; containing alfo a delineation of the feveral vifible fpots, with the various motions, changes, and appearances, difcovered by the telefcope, as alfo in the fun and other planets, 1647.——In 1654, two epiftles; one to the celebrated aftronomer Riccioli, concerning the Libration of the Moon; and the other to Bulliald, on the Eclipfes of both luminaries.—In 1656, a Differtation *De Natura Saturni faciei*, &c.—In 1668, his *Cometographia*, reprefenting the whole nature of comets, their fituation parallaxes, diftances, diverfe appearances, and furprifing motions, with a hiftory of all the comets from the beginning of the world down to the prefent time; being enriched with curious fculpture of his own execution: to which he added a treatife on the planet Mercury, feen in the fun at Dantzick, May 3, 1661; with the hiftory of a new ftar appearing in the neck of Cetus, and another in the beak of Cygnus; befides an Illuftration of fome aftronomical difcoveries of the late Mr. Horrox, in his treatife on Venus feen in the fun, Nov. 24, 1639; with a difcourfe of fome curious Parafelena and Parhelia obferved at Dantzick. He fent copies of this work to feveral members of the Royal Society at London, and among them to Mr. Hooke, in return for which, this gentleman fent to Hevelius a defcription of the Dioptric Telefcope, with an account of the manner of ufing it; and recommending it to him, as much preferable to telefcopes with plain fights. This gave rife to a difpute between them, viz, "whether diftances and altitudes could be taken with plain fights any nearer than to a minute." Hooke afferted that they could not; but that, with an inftrument of a fpan radius, by the help of a telefcope, they might be determined to the exactnefs of a fecond. Hevelius, on the other hand, infifted that, by the advantage of a good eye and long practice, he was able with his inftruments to come up even to that exactnefs; and, appealing to experience and facts, fent by way of challenge 8 diftances, each between two different ftars, to be examined by Hooke. Thus the affair refted for fome time with outward decency, but not without fome inward grudge between the parties.

In 1673, Hevelius publifhed the firft part of his *Machina Coelestis*, as a fpecimen of the exactnefs both of his inftruments and obfervations; and fent feveral copies as prefents to his friends in England, but omitting Mr. Hooke. This, it is fuppofed, occafioned that gentleman to print, in 1674, *Animadverfions on the Firft Part of the Machina Coelestis*; in which he treated Hevelius with a very magifterial air, and threw out feveral unhandfome reflections, which were greatly refented; and the difpute grew afterwards to fuch a height, and became fo notorious, that in 1679 Dr. Halley went, at the requeft of the Royal Society, to examine both the inftruments and the obfervations made with them. Of both thefe, Halley gave a favourable account, in a letter to Hevelius; and Hooke managed the controverfy fo ill, that he was univerfally condemned, though the preference has fince been given to telefcopic fights. However, Hevelius could not be prevailed on to make ufe of them: whether he thought himfelf too experienced to be informed by a young aftronomer, as he confidered Hooke; or whether, having made fo many obfervations with plain fights, he was unwilling to alter his method, left he might bring their exactnefs into queftion; or whether, being by long practice accuftomed to the ufe of them, and not thoroughly apprehending the ufe of the other, nor well underftanding the difference, is uncertain. Befides Halley's letter, Hevelius received many others in his favour, which he took the opportunity of inferting among the aftronomical obfervations in his *Annus Climactericus*, printed in 1685. In a long preface to this work, he fpeaks with more confidence and greater indignation than he had done before, and particularly exclaimed againft Hooke's dogmatical and magifterial manner of affuming a kind of dictatorfhip over him. This revived the conteft, and occafioned feveral learned men to engage in it. The book itfelf being fent to the Royal Society, at their requeft an account of it was given by Dr. Wallis; who, among other things, took notice, that " Hevelius's obfervations had been mifreprefented, fince it appeared from this book, that he could diftinguifh by plain fights to a fmall part of a minute." About the fame time Mr. Molyneux alfo wrote a letter to the fociety, in vindication of Hevelius, againft Hooke's animadverfions. To all which, Hooke drew up a letter in anfwer, which was read before the fociety, containing many qualifying

and

and accommodating expreſſions, but ſtill at leaſt expreſſing the ſuperiority of teleſcopic ſights over plain
ones, excellent as the obſervations were that had been
made with theſe.

In 1679, Hevelius had publiſhed the ſecond part of
his *Machina Cœleſtis* ; but the ſame year, while he was
at a ſeat in the country, he had the misfortune to have
his houſe at Dantzic burnt down. By this calamity it
is ſaid he ſuſtained ſeveral thouſand pounds damage ;
having not only his obſervatory and all his valuable inſtruments and aſtronomical apparatus deſtroyed, but alſo a great many copies of his *Machina Cœleſtis*, an accident which has made this ſecond part very ſcarce, and
conſequently very dear.

In 1690, were publiſhed a deſcription of the heavens, called, *Firmamentum Sobieſcianum*, in honour of
John the 3d, king of Poland, as above mentioned ; and
alſo *Prodromus Aſtronomiæ, & Novæ Tabulæ Solares, una
cum Catalogo Fixarum* ; in which he lays down the neceſſary preliminaries for taking an exact catalogue of
the ſtars.

But both theſe works were poſthumous ; for Hevelius died the 28th of January 1687, exactly 76 years of
age, as above ſaid, and univerſally admired and re-
ſpected ; abundant evidence of which appears in a collection of letters between him and many other perſons,
that was printed at Dantzic in 1683.

HEXACHORD, a certain interval or muſical concord, uſually called a ſixth.

HEXAEDRON, or HEXAHEDRON, one of the
five regular or Platonic bodies ; being indeed the ſame
as the cube ; and is ſo called from its having 6 faces.—
The ſquare of the ſide or edge of a Hexahedron, is one-
third of the ſquare of the diameter of the circumſcribing ſphere : and hence the diameter of a ſphere is to
the ſide of its inſcribed Hexahedron, as $\sqrt{3}$ to 1.

In general, if A, B, and C be put to denote reſpectively the linear ſide, the ſurface, and the ſolidity of
a Hexahedron or cube, alſo *r* the radius of the inſcribed ſphere, and R the radius of the circumſcribed one ; then we have theſe general equations or
relations :

1. $A = 2r \quad = \frac{2}{3}R\sqrt{3} = \sqrt{\frac{1}{6}B} = \sqrt[3]{C}.$
2. $B = 24r^2 \quad = 8R^2 \quad = 6A^2 \quad = 6\sqrt[3]{C^2}.$
3. $C = 8r^3 \quad = \frac{8}{9}R^3\sqrt{3} = A^3 \quad = \frac{1}{6}B\sqrt{\frac{1}{6}B}.$
4. $R = r\sqrt{3} = \frac{1}{2}A\sqrt{3} = \frac{1}{2}\sqrt{\frac{1}{2}B} = \frac{1}{2}\sqrt{3} \times \sqrt[3]{C}.$
5. $r = \frac{1}{3}R\sqrt{3} = \frac{1}{2}A \quad = \frac{1}{2}\sqrt{\frac{1}{6}B} = \frac{1}{2}\sqrt[3]{C}.$

HEXAGON, in Geometry, a figure of ſix ſides,
and conſequently of as many angles. When theſe are
equal, it is a regular Hexagon.—The angles of a Hexagon are each equal to 120°, and its ſides are each equal
to the radius of its circumſcribing circle. Hence a regular Hexagon is inſcribed in a circle, by ſetting the
radius off 6 times upon the periphery. And hence alſo, to deſcribe a regular Hexagon upon a given line,
deſcribe an equilateral triangle upon it, the vertex of
which will be the centre of the circumſcribing circle.

The ſide of a Hexagon being *s*, its area will be
$2\cdot5980762s^2 = \frac{3s^2}{2} \times \text{tang. } 60° = \frac{3}{2}s^2\sqrt{3}.$

HEXASTYLE, in the Ancient Architecture, a
building with 6 columns in front.

HIERO's *Crown*, in Hydroſtatics. The hiſtory
of this crown, and of the important hydroſtatical propoſition which it gave occaſion to, is as follows : Hiero,
king of Syracuſe, having furniſhed a workman with a
quantity of gold for making a crown, ſuſpected that he
had been cheated, by the workman uſing a greater alloy of ſilver than was neceſſary in making it ; and he
applied to Archimedes to diſcover the fraud, without
defacing the crown.

This celebrated mathematician was led by chance to
a method of detecting the impoſture, and of determining preciſely the quantities of gold and ſilver compoſing
the crown : for he obſerved, when bathing in a tub of
water, that the water ran over as his body entered it,
and he preſently concluded that the quantity ſo running
over was equal to the bulk of his body that was im-
merſed. He was ſo pleaſed with the diſcovery, that it
is ſaid he ran about naked crying out, ευρηκα, ευρηκα,
I have found it ; and ſome affirm that he offered a heta-
comb to Jupiter for having inſpired him with the
thought.

On this principle he procured a ball or maſs of gold,
and another of ſilver, exactly of the ſame weight with
the crown ; conſidering, that, if the crown were of pure
gold, it would be of equal bulk and expel an equal
quantity of water as the golden ball ; and if it were of
ſilver, then it would be of equal bulk and expel an
equal quantity of water with the ball of ſilver ; but of
intermediate quantity, if it conſiſted of a mixture of the
two, gold and ſilver ; which, upon trial, he found to
be the caſe ; and hence, by a compariſon of the quantities of water diſplaced by the three maſſes, he diſcovered the exact portions of gold and ſilver in the crown.

Now, ſuppoſe, for example, that each of the hree
maſſes weighed 100 ounces ; and that on immerſing
them ſeverally in water, there were diſplaced 5 ounces
of water by the golden ball, 9 ounces by the ſilver,
and 6 ounces by the compound, or crown ; that is,
their reſpective or comparative bulks are as 5, 9, and
6, the ſum of which is 20.

Then the method of operation is this :

From 9   6
Take 6   5
     ―――   ―――
rem.   3   1, whoſe ſum is 4.

Therefore 4 : 100 :: 3 : 75 oz. of gold,
     and 4 : 100 :: 1 : 25 oz. of ſilver.

That is, the crown conſiſted of 75 ounces of gold, and
25 ounces of ſilver.

See Cotes Hydroſ. Lect. p. 81 ; or Martin's Philoſ.
Britan, vol. 1, p. 305, &c. See alſo SPECIFIC *Gravity.*

HIGH-*Water*, that ſtate of the tides when they
have flowed to their greateſt height, or have ceaſed
to flow or riſe. At High-water the motion commonly
ceaſes for a quarter or half an hour, before it begin to
ebb again. The times of High-water of every day of
the moon's age, is uſually computed from that which is
obſerved on the day of the full or change ; viz, by
taking 4-5ths of the moon's age on any day of the
month, and adding it to the time of High-water on
the day of the full or change ; then is the ſum nearly
equal to the time of High-water on the day of the
month propoſed. And as to the times of High-water,
on the day of the full and change of the moon, at
many different places ; they have been obſerved as
they are ſet down in the followiug table.

Table

TABLE of the Times of High-water on the Days of the New and Full Moons, at many different Places.

Names of Places.	Countries.	High-w.	Names of Places.	Countries.	High-w.
Aberdeen.	Scotland	0h 45m	Dort	Holland	3h 0m
Aldborough	England	9 45	Dover	England	11 30
Alderney I.	England	12 0	Downs	England	1 15
Amazons River	South America	6 0	Dublin	Ireland	9 15
Amsterdam	Holland	3 0	Dunbar	Scotland	2 30
Amsterdam I. of	South Seas	8 30	Dundee	Scotland	2 15
Andrew's St.	Scotland	2 15	Dungarvan	Ireland	4 30
Anholt I.	Denmark	0 0	Dungeness	England	9 45
Antwerp	Flanders	6 0	Dunkirk	France	0 0
Archangel	Ruffia	6 0	Dunnose	I. of Wight	9 45
Arran I.	Ireland	11 0	Dufky Bay	N. Zealand	10 57
Afhley Riv.	Carolina	0 45	Eafter Ifle	Chili	2 0
Auguftine St.	Florida	7 30	Edyftone	Englifh Channel	5 30
Bajador Ca.	Negroland	0 0	Elbe R.	Germany	0 0
Baltimore	Ireland	4 30	Embden	Germany	0 0
Barfleur Ca.	France	7 30	Eftaples	France	11 0
Bayonne	France	3 30	Falmouth	England	5 30
Beachy-head	England	0 0	Flamborough H.	England	4 0
N. and S. Bear	Labradore	12 0	C. Florida	Florida	7 30
Belfaft	Ireland	10 0	Flufhing	Holland	0 45
Bellifle	France	3 30	N. Foreland	England	9 45
Bermudas I.	Bahama I.	7 0	Foulnefs	England	6 45
Berwick	England	2 30	Fowey	England	5 15
Blackney	England	6 0	Fayal Ifl.	Azores	2 20
Blanco Cap.	Negroland	9 45	Garonne R.	France	3 0
Blavet	France	3 0	Gibraltar	Spain	0 0
Bourdeaux	France	3 0	C. Good Hope	Caffers	3 0
Boulogne	France	10 30	Goree (Ifle)	Atlantic Ocean	1 30
Bremen	Germany	6 0	Granville	France	7 0
Breft	France	3 45	Gravelines	Flanders	0 0
Bridlington B.	England	3 45	Gravefend	England	1 30
Brill	Holland	1 30	Groin	Spain	3 3
Briftol	England	6 45	Guernfey I.	Englifh Channel	1 30
Buchanefs	Scotland	3 0	Hague	Holland	8 15
Button's Ifles	New Brit.	6 50	Halifax	Nova Scotia	7 30
Cadiz	Spain	4 30	Hamburgh	Germany	6 0
Caen	France	9 0	Hare Ifle	Canada	3 30
Calais	France	11 30	Harlem	Holland	9 0
Canaria I.	Canaries	3 0	Hartlepool	England	3 0
C. Cantin	Barbary	0 0	Harwich	England	11 15
Cape Town	Caffers	2 30	Havre de Grace	France	9 0
Cafkets	Guernfey	8 15	Holy Head	Wales	1 30
Cathnefs Po.	Scotland	9 0	Honfleur	France	9 0
Charles Town	Carolina	3 0	Hull	England	6 0
Q. Charlotte's S.	New Zealand	9 0	Humber R.	England	5 13
Cherbourg	France	7 30	St. John's	Newfoundland	6 0
Churchill R.	Hudfon's Bay	7 20	St. Julian (Port)	Patagonia	4 45
Ca. Cleare	Ireland	4 30	Kentifhnock	Englifh coast	0 0
Concarneau	France	3 0	Kinfale	Ireland	5 15
Conquet	France	2 15	Land's End	England	7 30
Coquet Ifle	England	3 0	Leith	Scotland	4 30
Corke	Ireland	6 30	Leoftoff	England	9 45
C. Corfe	Guinea	3 30	Lifbon	Portugal	2 15
Cromer	England	7 0	Liverpool	England	11 15
Dartmouth	England	6 30	Lizard	England	7 30
St. David's H.	Wales	6 0	Loire (Riv.)	France	3 0
Dieppe	France	10 30	London	England	3 0

Names of Places.	Countries.	High-w.	Names of Places.	Countries.	High-w.
Lundy (Isle)	England	5 h 15 m	Sandwich	England	11h 30m
Madeira	Atl. Ocean	12 4	Scarborough H.	England	3 45
St. Maloes	France	6 0	Scilly Isles	England	3 45
Isle of Man	England	9 0	Senegal	Negroland	10 30
Margate	England	11 15	Severn, (Mouth.)	England	6 0
St. Mary's (Isle)	Scilly Isles	3 45	Sheerness	England	0 0
Milford	Wales	5 15	Sierra Leona	Guinea	8 15
Mount's Bay	England	4 30	Shetland I.	Scotland	3 0
Nantes	France	3 0	Isle of Sky	Scotland	5 30
Naze	Norway	11 15	Spurn	England	5 15
Needles	England	10 15	Start Point	England	6 45
Newcastle	England	3 15	Stockton	England	5 15
Nieuport	Flanders	12 0	Sunderland	England	3 20
Nore	England	0 0	Tanna	Pacific Ocean	3 0
North Cape	Lapland	3 0	Teneriff	Canaries	3 0
Orfordness	England	9 45	Texel (Isle)	Holland	7 30
Orkneys	Scotland	3 0	Thames Mouth	England	1 30
Ostend	Flanders	12 0	Tinmouth	England	3 0
Placentia	Newfoundland	9 0	Torbay	England	5 15
Plymouth	England	6 0	St. Valery	France	10 30
Portland	England	8 15	Vannes	France	3 45
Porto Praya	Cape Verdes	11 0	Ushant	France	4 30
Portsmouth	England	11 15	Waterford	Ireland	6 30
Quebec	Canada	7 30	Wells	England	6 0
Rhée (Isle)	France	3 0	Weymouth	England	7 20
Resolution (Bay)	Ohitahoo	2 30	Whitby	England	3 0
Robin Hood's B.	England	3 0	Isle of Wight	England	0 0
Rochefort	France	4 15	Winchelsea	England	0 45
Rochelle	France	3 45	Wintertoness	England	9 0
Rochester	England	0 45	Yarmouth	England	9 45
Rotterdam	Holland	3 0	New York	America	3 0
Rouen	France	1 15	Youghall	Ireland	4 30
Rye	England	11 15	Zuric Sea	Holland	3 0

HIPS, in Architecture, are those pieces of timber placed at the corners of a roof. These are much longer than the rafters, because of their oblique position.

HIP means also the angle formed by two parts of the roof, when it rises outwards.

HIP-Roof, called also Italian Roof, is one in which two parts of the roof meet in an angle, rising outwards: the same angle being called a valley when it sinks inwards.

HIPPARCHUS, a celebrated astronomer among the ancients, was born at Nice in Bithynia, and flourished between the 154th and the 163d olympiads; that is, between 60 and 135 years before Christ; for in this space of time it is that his observations are dated. He is accounted the first, who from vague and scattered observations, reduced astronomy into a science, and prosecuted the study of it systematically. Pliny often mentions him, and always with great commendation. He was the first, he tells us, who attempted to count the number of the fixed stars; and his catalogue is preserved in Ptolomy's Almagest, where they are all noted according to their longitudes and apparent magnitudes. Pliny places him among those men of a sublime genius, who, by foretelling the eclipses, taught mankind, that they ought not to be frightened at these phénomena. Thales was the first among the Greeks, who could discover when there was to be an eclipse. Sulpitius Gallus among the Romans began to succeed in this kind of prediction; and he gave an essay of his skill very seasonably, the day before a battle was fought. After these two, Hipparchus improved that science very much; making ephemerides, or catalogues of eclipses, for 600 years. He admires him for making a review of all the stars, acquainting us with their situations and magnitudes; for by these means, says he, posterity will be able to discover, not only whether they are born and die, but also whether they change their places, and whether they increase or decrease. He mentioned a new star which was produced in his days; and by its motion, at its first appearance, he began to doubt whether this did not frequently happen, and whether those stars, which we call fixed, do not likewise move. Hipparchus is also memorable for being the first who discovered the precession of the equinoxes, or a very slow apparent motion of the fixed stars from east to west, by which in a great number of years they will seem to have performed a complete revolution. He endeavoured also to reduce to rule the many discoveries he made, and invented new instruments, by which he marked their magnitudes and places in the heavens; so that by means of them it might be easily observed, not only whether they appear and disappear, but likewise whether they pass by one another, or move, and whether they increase or decrease.

The first observations he made, were in the isle of Rhodes;

Rhodes; whence he got the name Rhodius; but afterwards he cultivated this science in Bithynia and Alexandria only. One of his works is still extant, viz, his Commentary upon Aratus's Phenomena. He composed several other works; and upon the whole it is agreed, that astronomy is greatly indebted to him, for laying that rational and solid foundation, upon which all succeeding astronomers have since built their superstructure.

HIRCUS, in Astronomy, a fixed star of the first magnitude, the same with Capella.

HIRCUS is also used by some writers for a comet, encompassed as it were with a mane, seemingly rough and hairy.

HIRE (PHILIP DE LA), an eminent French mathematician and astronomer, was born at Paris in 1640. His father who was painter to the king, intending him for the same occupation, taught him drawing and such branches of mathematics as relate to it: but died when the son was only 17 years of age. Three years after this, he travelled into Italy for improvement in that art, where he spent 4 years. He applied himself also to mathematics, which gradually engrossed all his attention. On his return to Paris, he continued his mathematical studies with great eagerness, and he afterwards published some works, which gained him so much reputation, that he was named a member of the Academy of Sciences in 1678.

The minister Colbert having formed a design for a better chart or map of France than any former ones, De la Hire was appointed, with Picard, to make the necessary observations for that purpose. This occupied him some years in several of the provinces; and, beside the main object of his peregrinations, he was not unmindful of other branches of knowledge, but philosophized upon every thing that occurred, and particularly upon the variations of the magnetic needle, upon refractions, and upon the height of mountains, as determined by the barometer.

In 1683, de la Hire was employed in continuing the meridian line, which Picard had begun in 1669. He continued it from Paris northward, while Cassini carried it on to the south: but Colbert dying the same year, the work was dropped before it was finished. De la Hire was next employed, with other members of the academy, in taking the necessary levels for the grand aqueducts, which Louis the 14th was about to make.

The great number of works published by our author, together with his continual employments, as professor of the Royal College and of the Academy of Architecture, give us some idea of the great labours he underwent. His days were always spent in study; his nights very often in astronomical observations; seldom seeking any other relief from his labours, than a change of one for another. In his manner, he had the exterior politeness, circumspection, and prudence of Italy; on which account he appeared too reserved in the eyes of his countrymen; though he was always esteemed as a very honest disinterested man. He died in 1718, at 78 years of age.

Of the numerous works which he published, the principal are, 1. Traité de Mechanique; 1665.—2. Nouvelle Methode en Geometrie pour les Sections des Superficies Coniques & Cylindriques; 1673, 4to.

—3. De Cycloide; 1677, 12mo.—4. Nouveaux Elemens des Sectiones Coniques: les Lieux Geometriques: la Construction, ou Effection des Equations; 1678, 12mo.—5. La Gnomonique, &c; 1682. 12mo.—6. Traite du Nivellement de M. Picard, avec des additions; 1684. —7. Sectiones Conicæ in novem libros distributæ; 1685, folio. This was considered as an original work, and gained the author great reputation all over Europe —8. Traité du Mouvement des Eaux, &c; 1686.— 9. Tabulæ Astronomicæ; 1687 and 1702, 4to.—10. Ecole des Arpenteurs; 1689.—11. Veterum Mathematicorum Opera, Græcè & Latinè, pleraque nunc primum edita; 1693, folio. This edition had been begun by Thevenot; who dying, the care of finishing it was committed to de la Hire. It shews that our author's strong application to mathematical and astronomical studies had not hindered him from acquiring a very competent knowledge of the Greek tongue. Beside these, and other smaller works, there are a vast number of his pieces scattered up and down in Journals, and particularly in the Memoirs of the Academy of Sciences, viz, from 1666 till the year 1718.

HOBBES (THOMAS), a famous writer and philosopher, was born at Malmsbury in Wiltshire, in 1588, being the son of a clergyman of that place. He completed his studies at Oxford, and was afterwards governor to the eldest son of William Cavendish earl of Devonshire, with whom he travelled through France and Italy, applying himself closely to the study of polite literature. In 1626 his patron the earl of Devonshire died; and 1628 his son also; the same year Mr. Hobbes published his translation of Thucydides in English. He soon after went abroad a second time as governor to the son of Sir Gervase Clifton; but shortly after returned, to resume his concern for the hopes of the Devonshire family, to whom he had so early attached himself; the countess dowager having desired to put the young earl under his care, then about 13 years of age. This charge was very agreeable to Mr. Hobbes's inclinations, and he discharged the trust with great diligence and fidelity. In 1634 he accompanied his young pupil to Paris, where he employed his own vacant hours in the study of natural philosophy, frequently conversing with Father Mersenne, Gassendi, and other eminent philosophers there. From Paris he attended his pupil into Italy, where he became acquainted with the celebrated Galileo, who freely communicated his notions to him; and from hence he returned with his ward into England. But afterwards, foreseeing the civil wars, he went to seek a retreat at Paris; where he soon made acquainted with Des Cartes and the other learned philosophers there, with whom he afterwards held a correspondence upon several mathematical subjects, as appears from the letters of Mr. Hobbes published in the works of Des Cartes.

In 1642, Mr. Hobbs printed his famous book *De Cive*, which raised him many adversaries, who charged him with instilling principles of a dangerous tendency. Among many illustrious persons who, from the troubles in England, retired to France for safety, was Sir Charles Cavendish, brother to the Duke of Newcastle: and this gentleman, being well skilled in the mathematics, proved a constant friend and patron to Mr. Hobbes; who, by embarking in 1645 in a controversy

about

about fquaring the circle, was grown fo famous by it, that in 1647 he was recommended to inftruct Charles prince of Wales, afterwards king Charles the 2d, in mathematical learning. During this he employed his vacant time in compofing his Leviathan, which was publifhed in England in 1651. After the publication of this work, he returned to England, and paffed the remainder of his long life in a very retired and ftudious manner, in the houfe of the Earl of Devonfhire, moftly at his feat in Derbyfhire, but accompanying the earl always to London, fearing to be left out of his immediate protection, left he fhould be feized by officers from the parliament or government, on account of the freedom of his opinions in politics and religion. He received great marks of refpect from king Charles the 2d at the reftoration in 1660, with a penfion of 100l. a year. From that time, till his death, he applied himfelf to his ftudies, and in oppofing the attacks of his adverfaries, who were very numerous : in mathematical fubjects difputes rofe to a great height between him and Dr. Wallis, on account of his pretended Quadrature of the Circle, Cubature of the Sphere, and Duplication of the Cube, which he obftinately defended without ever acknowledging his error.

His long life was that of a perfectly honeft man ; a lover of his country, a good friend, charitable and obliging. He accuftomed himfelf much more to thinking, than reading ; and was fond of a well-felected, rather than a large library. He had a hatred to the clergy, having been perfecuted by them, on account of the freedom of his doctrine, and having a very indifferent opinion of their knowledge and their principles. In his laft ficknefs he was very anxious to know whether his difeafe was curable ; and when intimations were given, that he might have eafe, but no remedy, he faid, ' I fhall be glad to find a hole to creep out of the world at.' He died the 4th of Dec. 1679, at 91 years of age.

His chief publications were,

1. An Englifh tranflation of Thucydides's Hiftory of the Grecian war.

2. De Mirabilibus Pecci, and Memoirs of his own Life, both in Latin verfe.

3. Elements of Philofophy.

4. Anfwer to Sir William Davenant's Epiftle, or Preface to Gondibert.

5. Human Nature, or the Fundamental Elements of Policy.

6. Elements of Law.

7. Leviathan ; or the Matter, Form, and Power of a Commonwealth.

8. A Compendium of Ariftotle's Rhetoric.

9. A Letter on Liberty and Neceffity.

10. The Queftions, concerning Neceffity and Chance, ftated.

11. Six Leffons to the Profeffors of Mathematics, of the Inftitution of Sir Henry Saville.

12. The marks of Abfurd Geometry, &c.

13. Dialogues of Natural Philofophy.

Befides many other pieces on Polity, Theology, Mathematics, and other mifcellaneous fubjects, to the number of 41.

HOBITS, in Gunnery. See HOWITZ.

HOGSHEAD, a meafure, or veffel, of wine or

Vol. I.

oil ; containing the 4th part of a tun, the half of a pipe, or 63 gallons.

HOLDER (WILLIAM), a learned and philofophical Englifhman, was born in Nottinghamfhire, educated at Cambridge, and in 1642 became rector of Blechingdon in Oxfordfhire. In 1660 he proceeded D. D. he became afterwards canon of Ely, Fellow of the Royal Society, canon of St. Paul's, fub-dean of the royal chapel, and fub-almoner to the king. He was a general fcholar, a very accomplifhed perfon, and a great virtuofo.

Dr. Holder greatly diftinguifhed himfelf, by giving fpeech to a young gentleman of the name of Popham, who was born deaf. This was effected at his own houfe at Blechingdon in 1659 ; but the young man lofing what he had been taught by Holder after he was called home to his friends, he was fent to Dr. Wallis, who brought him to his fpeech again. Holder publifhed a book, intitled " the Elements of Speech ; an effay or inquiry into the natural Production of Letters : with an appendix, concerning perfons that are deaf and dumb, 1669," 8vo. In the appendix he relates how foon, and by what methods, he brought young Popham to fpeak. In the Philof. Tranf. for July 1670, was inferted a letter from Dr. Wallis, in which he claims to himfelf the honour of bringing that gentleman to fpeak. By way of anfwer to which, in 1678, Dr. Holder publifhed in 4to, " A Supplement to the Philof. Tranf. of July 1670, with fome reflections on Dr. Wallis's letter there inferted." Upon which the latter foon after publifhed " A Defence of the Royal Society, and the Philofophical Tranfactions, particularly thofe of July 1670, in anfwer to the cavils of Dr. William Holder, 1678," 4to.

Dr. Holder's accomplifhments were very general. He was fkilled in the theory and practice of mufic, and wrote " A Treatife of the Natural Grounds and Principles of Harmony, 1694," 8vo. He wrote alfo " A Treatife concerning Time, with applications of the Natural Day, Lunar Month, and Solar Year, &c, 1694," 8vo. He died at Amen Corner in London, Jan. 24, 1697, and was buried in St. Paul's.

HOLLOW, in Architecture, a concave moulding, about a quarter of a circle, by fome called a Cafement, by others an Abacus.

HOLLOW-*Tower*, in Fortification, is a rounding made of the remainder of two brifures, to join the curtin to the crillon, where the fmall fhot are played, that they may not be fo much expofed to the view of the enemy.

HOLY *Thurfday*, otherwife called Afcenfion day, being the 39th day after Eafter Sunday, and kept in commemoration of Chrift's afcenfion up into heaven.

HOLY *Rood*, or *Holy Crofs*, a feftival kept on the 14th of September, in memory of the exaltation of our Saviour's crofs.

HOLY *Week*, is the laft week of Lent, called alfo Paffion Week,

HOLYWOOD (JOHN), or HALIFAX, or *Sacrobofco*, was, according to Leland, Bale, and Pitts, born at Halifax in Yorkfhire : according to Stainhurft, at Holywood near Dublin ; and according to Dempfter and Mackenzie, in Nithfdale in Scotland. Though

4 H there

there may perhaps have been more than one of the name. Mackenzie informs us, that having finished his studies, he entered into orders, and became a canon regular of the order of St. Augustin in the famous monastery of Holywood in Nithsdale. The English biographers, on the contrary, tell us that he was educated at Oxford. They all agree however in asserting, that he spent most of his life at Paris; where, says Mackenzie, he was admitted a member of the university, June 5, 1221, under the syndics of the Scotch nation; and soon after was elected professor of mathematics, which he taught with applause for many years. According to the same author, he died in 1256, as appears from the inscription on his monument in the cloisters of the convent of St. Maturine at Paris.

Holywood was contemporary with Roger Bacon, but probably older by about 20 years. He was certainly the first mathematician of his time; and he wrote, 1. *De Sphæra Mundi;* a work often reprinted, and illustrated by various commentators.—2. *De Anni Ratione, seu de Computo Ecclesiastico.*—3. *De Algorismo,* printed with *Comm. Petri Cirvilli Hisp*: Paris, 1498.

HOMOCENTRIC, the same as Concentric.

HOMODROMUS *Vectis,* or *Lever,* in Mechanics, is a lever in which the weight and power are both on the same side of the fulcrum, as in the lever of the 2d and 3d kind; being so called because here the weight and power move both in the same direction, whereas in the Heterodromus they move in opposite directions.

HOMOGENEAL, or Homogeneous, consisting of similar parts, or of the same kind and nature, in contradistinction from heterogeneous, where the parts are of different kinds.——Natural bodies are usually composed of Homogeneous parts, as a diamond, a metal, &c. But artificial bodies, on the contrary, are assemblages of heterogeneous parts, or parts of different kinds; as a building, of stone, wood, &c.

Homogeneal *Light,* is that whose rays are all of one and the same colour, refrangibility, &c.

Homogeneal *Numbers,* are those of the same kind and nature.

Homogeneal *Surds,* are such as have one common radical sign; as $\sqrt{27}$ and $\sqrt{30}$, or $\sqrt{a}$ and $\sqrt{b}$, or $2\frac{3}{c}$ and $\frac{3}{d}$.

HOMOGENEUM *Adfectionis,* a name given by Vieta to the second term of a compound or affected equation, being that which makes it adfected.

Homogeneum *Comparationis,* in Algebra, a name given by Vieta to the absolute known number or term in a compound or affected equation. This he places on the right-hand side of the equation, and all the other terms on the left.

HOMOLOGOUS, in Geometry, is applied to the corresponding sides of similar figures, or those that are opposite to equal or corresponding angles, and are so called because they are proportional to each other. For all similar figures have their like sides Homologus, or proportional to one another, also their areas or surfaces are Homologous or proportional to the squares of the like sides, and their solid contents Homologous or proportional to the cubes of the same.

HOOKE (Robert), a very eminent mathematician and philosopher, was born, 1635, at Freshwater in the Isle of Wight, where his father was minister. He was intended for the church; but being of a weakly constitution, and very subject to the head-ach, all thoughts of that nature were laid aside. Thus left to himself, the boy followed the bent of his genius, which was turned to mechanics; and employed his time in making little toys, which he did with wonderful art and dexterity. He had also a good turn for drawing; for which reason, after his father's death, which happened in 1648, he was placed with Sir Peter Lely; but the smell of the oil-colours increasing his head-ach, he quitted painting in a very short time. He was afterwards kindly taken by Dr. Busby into his house, and supported there, while he attended Westminster-school; where he not only acquired a competent share of Greek and Latin, together with an insight into Hebrew and some other Oriental languages, but also made himself master of a good part of Euclid's Elements; and, as Wood asserts, invented 30 different ways of flying.

About the year 1653 he went to Christ-church in Oxford; and in 1655 was introduced to the Philosophical Society there; where, discovering his mechanic genius, he was first employed to assist Dr. Willis in his chemical operations, and was afterwards recommended to Mr. Robert Boyle, whom he served several years in the same capacity. He was also instructed in astronomy about this time by Dr. Seth Ward, Savilian Professor of that science; and from henceforward distinguished himself by many noble inventions and improvements of the mechanic kind. He also invented several astronomical instruments, for making observations both at sea and land, and was particularly serviceable to Mr. Boyle in completing the invention of the air-pump. In 1662 he was appointed Curator of Experiments to the Royal Society; and when that body was established by royal charter, he was in the list of those, who were first named by the council in May 20, 1663; and he was admitted accordingly June 3, with a peculiar exemption from all payments. Sept. 28 of the same year, he was named by lord Clarendon, chancellor of Oxford, for the degree of M. A.; and Oct. 19 it was ordered that the repository of the Royal Society should be committed to his care: the white gallery in Gresham-college being appropriated to that use. In May 1664, he began to read the astronomy lecture at Gresham college for the professor Dr. Pope, then in Italy; and the same year he was made Professor of Mechanics to the Royal Society by Sir John Cutler, with a salary of 50l. per annum, which that gentleman, the founder, settled upon him for life. Jan. 11, 1665, that society granted a salary also of 30l. a year, for his office of Curator of Experiments for life; and the month of March the same year he was elected professor of geometry in Gresham-college.

In 1665 too, he published in folio, his " Micrographia, or some Philosophical Descriptions of Minute Bodies, made by Magnifying Glasses, with Observations and Enquiries thereupon." And the same year, during the recess of the Royal Society on account of the plague, he attended Dr. Wilkins and other ingenious gentlemen into Surry, where they made several experiments. In 1666 he produced to the Royal Society eiety a model for rebuilding the city of London, then deftroyed by the great fire, with which the Society was well pleafed; and the Lord Mayor and Aldermen preferred it to that of the city furveyor, though it happened not to be carried into execution. The rebuilding of the city according to the act of parliament requiring able perfons to fet out the ground for the proprietors, Mr. Hooke was appointed one of the furveyors; an employment in which he got moft part of his eftate, as appeared from a large iron cheft of money found after his death, locked down with a key in it, and a date of the time, which fhewed it to have been fo fhut up above 30 years. From 1668 he was engaged for many years in a warm conteft with Hevelius, concerning the difference in accuracy between obferving with aftronomical inftruments with plain and telefcopic fights; in which difpute many learned men afterwards engaged, and in which Hooke managed fo ill, as to be univerfally condemned, though it has fince been agreed that he had the better fide of the queftion.—In 1771 he attacked Newton's " New Theory of Light and Colours;" where, though he was obliged to fubmit in refpect to the argument, it is faid he came off with more credit. The Royal Society having commenced their meetings at Grefham-college, November 1674, the Committee in December allowed him 40l. to erect a turret over part of his lodgings, for trying his inftruments, and making aftronomical obfervations: and the year following he publifhed " A Defcription of Telefcopes, and fome other inftruments made by R. H. with a Poftfcript," complaining of fome injuftice done him by their fecretary Mr. Oldenburg, who publifhed the Philofophical Tranfactions, in regard to his invention of pendulum watches. This charge drew him into a difpute with that gentleman, which ended in a declaration of the Royal Society in their fecretary's favour.—Mr. Oldenburg dying in 1677, Mr. Hooke was appointed to fupply his place, and began to take minutes at the meeting in October, but did not publifh the Tranfactions.—Soon after this, he grew more referved than formerly; and though he read his Cutlerian Lectures, often made experiments, and fhewed new inventions before the Royal Society, yet he feldom left any account of them to be entered in their regifters; defigning, as he faid, to publifh them himfelf, which however he never performed.—In 1686, when Newton's work the Principia was publifhed, Hooke laid claim to his difcovery concerning the force and action of gravity, which was warmly refented by that great philofopher. Hooke, though a great inventor and difcoverer himfelf, was yet fo envious and ambitious, that he would fain have been thought the only man who could invent and difcover. This made him often lay claim to the inventions and difcoveries of other perfons; on which occafions however, as well as in the prefent cafe, the thing was generally carried againft him.

In the beginning of the year 1687, his brother's daughter, Mrs. Grace Hooke, who had lived with him feveral years, died: and he was fo affected with grief at her death, that he hardly ever recovered it, but was obferved from that time to become lefs active, more melancholy, and more cynical than ever. At the fame time, a chancery fuit in which he was concerned with Sir John Cutler, on account of his falary for reading

the Cutlerian Lectures, made him uneafy, and increafed his diforder.—In 1691, he was employed in forming the plan of the hofpital near Hoxton, founded by Robert Afk, alderman of London, who appointed archbifhop Tillotfon one of his executors; and in December the fame year, Hooke was created M. D. by a warrant from that prelate. In July 1696, the chancery fuit with Sir John Cutler was determined in his favour, to his inexpreffible fatisfaction. His joy on that occafion was found in his diary thus expreffed; DOMSHLGISSA; that is, *Deo, Optimo, Maximo, fit honor, laus, gloria, in fæcula fæculorum, Amen.* " I was born on this day of July 1635, and God hath given me a new birth: may I never forget his mercies to me! while he gives me breath may I praife him!"—In the fame year 1696, an order was granted to him for repeating moft of his experiments at the expence of the Royal Society, upon a promife of his finifhing the accounts, obfervations, and deductions from them, and of perfecting the defcription of all the inftruments contrived by him: but his increafing illnefs and general decay rendered him unable to perform it. He continued fome years in this wafting condition; and thus languifhing till he was quite emaciated, he died March 3, 1702, in his 67th year, at his lodgings in Grefham college, and was buried in St. Helen's church, Bifhopfgate ftreet; his corps being attended by all the members of the Royal Society then in London.

As to Mr. Hooke's character, it is not in all refpects one of the moft amiable. In his perfon he made rather a defpicable figure, being but of a fhort ftature, very crooked, pale, lean, and of a meagre afpect, with dark-brown hair, very long, and hanging over his face lank and uncut. Suitable to his perfon, his temper was penurious, melancholy, and miftruftful: and, though poffeffed of great philofophical knowledge, he had fo much ambition, that he would be thought the only man who could invent or difcover; and hence he often laid claim to the inventions and difcoveries of others, while he boafted of many of his own which he never communicated. In the religious part of his character, he was fo exemplary, that he always expreffed a great veneration for the Deity; and feldom received any remarkable benefit in life, or made any confiderable difcovery in nature, or invented any ufeful contrivance, or found out any difficult problem, without fetting down his acknowledgment to God, as many places in his diary plainly fhew.—His chief publications are,

1. *Lectiones Cutlerianæ,* or the Cutlerian Lectures.

2. *Micrographia,* or Defcriptions of Minute Bodies made by Magnifying Glaffes.

3. A Defcription of Heliofcopes.

4. A Defcription of fome Mechanical Improvements of Lamps and Water-poifes.

5. Philofophical Collections.

6. Pofthumous Works, collected from his papers by Richard Waller fecretary to the Royal Society. Befides a number of papers in the Philof. Tranf. volumes 1, 2, 3, 5, 6, 9, 16, 17, 22.

HORARY, fomething relating to Hours. As,

HORARY *Circles,* hour lines or circles, marking the hours, or drawn at the diftance of hours from one another.

HORARY *Motion,* is the motion or fpace moved in an hour.

hour. Thus, the Horary motion of the earth on her axis, is 15°; for, completing her revolution of 360°, in 24 hours, therefore the motion in one hour will be the 24th part of 360°, which is 15 degrees.

HORIZON, in Aftronomy, a great circle of the fphere, dividing the world into two parts, or hemifpheres; the one upper, and vifible; the other lower, and hid.

The Horizon is either Rational or Senfible.

HORIZON, *Rational*, *True*, or *Aftronomical*, called alfo fimply and abfolutely the Horizon, is a great circle having its plane paffing through the centre of the earth, and its poles are the zenith and nadir. Hence all the points of the Horizon, quite around, are at a quadrant diftance from the zenith and nadir. Alfo the meridian and vertical circles cut the Horizon at right angles, and into two equal parts.

HORIZON, *Apparent*, *Senfible*, or *Vifible*, is a leffer circle of the fphere, parallel to the rational Horizon, dividing the vifible part of the fphere from the invifible, and whofe plane touches the fpherical furface of the earth.

The fenfible Horizon is divided into Eaftern and Weftern; the Eaftern or Ortive being that in which the heavenly bodies rife; and the Weftern, or Occidual, being that in which they fet.

HORIZON, in Geography, is a circle dividing the vifible part of the earth and heavens from that which is invifible. This is peculiarly called Senfible or Apparent Horizon, to diftinguifh it from the Rational or True, which paffes through the centre of the earth; as already obferved. Thefe two Horizons, though diftant from one another by the femidiameter of the earth, will appear to coincide when continued to the fphere of the fixed ftars; becaufe the earth compared with this fphere is but a point.

By Senfible Horizon is alfo often meant a circle which determines the fegment of the furface of the earth, over which the eye can reach; called alfo the Phyfical Horizon. And in this fenfe we fay, a fpacious Horizon, a narrow or fcanty Horizon, &c; depending chiefly on the height the eye is elevated above the earth,

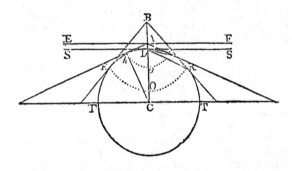

For, it is evident that the higher the eye is placed, the farther is the vifible Horizon extended. Thus, if the eye be at A, at the height AD above the earth; draw the two tangents A*h*, A*r*; and let one of thefe lines A*h*, be moved round the point A, and in its revolution always touch the furface of the earth; then

the other point *h* will defcribe the vifible Horizon *h o r*, &c. But if the eye be placed higher as at B, the tangents BH and BR will reach farther, and the vifible Horizon HOR will be larger.

The vifible Horizon is moft accurately obferved at fea, and is therefore fometimes called the Horizon of the fea. In obferving this Horizon, the vifual rays A*h* and A*r* will, on account of the curve furface of the fea, always point a little below the true fenfible Horizon SS or EF, and confequently below the rational Horizon TT, which is parallel to it.

To find the Depreffion of the Horizon of the fea below the true Horizon, which varies with the height of the eye, and in a fmall degree with the variation of the refractive power of the atmofphere, fee DEPRESSION.

As to the right-lined diftance, or tangent E*h*, it may be found thus; as radius : fin. $\angle$ C : : CA : A*h*,

or thus; as radius : tan. $\angle$ C : : C*h* : A*h*, either of which will be nearly the fame as the arc or curved diftance D*h*. Or, without finding the angle C, thus; the fquare of A*h* is equal to the difference of the fquares of CA and C*h*, i. e. $Ah^2 = CA^2 - Ch^2 = \overline{CA + Ch} \times \overline{CA - Ch} = \overline{CA + Ch} \times AD$, and hence $Ah = \sqrt{\overline{CA + Ch} \times AD}$, which is alfo equal to A*h* nearly.

The diftance on a perfect globe, if the vifual rays came to the eye in a ftraight line, would be as above ftated: but by means of the refraction of the atmofphere, diftant objects on the Horizon appear higher than they really are, or appear lefs depreffed below the true Horizon SS, and may be feen at a greater diftance, efpecially on the fea. M. Legendre, in his Memoir on Meafurements of the Earth, in the Mem. Acad. Sci. for the year 1787, fays that, from feveral experiments, he is induced to allow for refraction a 14th part of the diftance of the place obferved, expreffed in degrees and minutes of a great circle. Thus, if the diftance be 14000 toifes, the refraction will be 1000 toifes, equal to the 57th part of a degree, or 1′ 3′.

HORIZON *of the Globe*, a broad wooden circle. See GLOBE.

HORIZONTAL, fomething that relates to the Horizon, or that is taken in the Horizon, or on a level with or parallel to it. Thus, we fay, a Horizontal plane, Horizontal line, Horizontal diftance, &c.

HORIZONTAL *Dial*, is one drawn on a plane parallel to the horizon; having its gnomon or ftyle elevated according to the altitude of the pole of the place it is defigned for.

HORIZONTAL *Diftance*, is that eftimated in the direction of the horizon.

HORIZONTAL *Line*, in Perfpective, is a right line drawn through the principal point, parallel to the horizon; or it is the interfection of the Horizontal and perfpective planes.

HORIZONTAL *Line*, or bafe of a hill, in Surveying, a line drawn on the Horizontal plane of the hill, or that on which it ftands.

HORIZONTAL *Moon*. See *Apparent* MAGNITUDE.

HORIZONTAL *Parallax*. See PARALLAX.

HORIZONTAL *Plane*, is that which is parallel to the horizon of the place, or not inclined to it.

HORIZONTAL

HORIZONTAL *Plane*, in Perspective. See PLANE.

HORIZONTAL *Projection*. See PROJECTION, and MAP.

HORIZONTAL *Range*, of a piece of ordnance, is the distance at which it falls on, or strikes the horizon, or on a Horizontal plane, whatever be the angle of elevation or direction of the piece. When the piece is pointed parallel to the horizon, the range is then called the point-blank or point-blanc range.

The greatest Horizontal range, in the parabolic theory, or in a vacuum, is that made with the piece elevated to 45 degrees, and is equal to double the height from which a heavy body must freely fall to acquire the velocity with which the shot is discharged. Thus, a shot being discharged with the velocity of v feet per second ; because gravity generates the velocity $2g$ or $32\frac{1}{6}$ feet in the first second of time, by falling $16\frac{1}{12}$ or g feet, and because the spaces descended are as the squares of the velocities, therefore as $4g^2 : v^2 :: g : \frac{v^2}{4g}$ the space a body must descend to acquire the velocity v of the shot or the space due to the velocity v; consequently the double of this, or $\frac{v^2}{2g} = \frac{v^2}{32\frac{1}{6}}$ is the greatest Horizontal range with the velocity v, or at an elevation of 45 degrees ; which is nearly half the square of a quarter of the velocity.

In other elevations, the Horizontal range is as the sine of double the angle of elevation ; so that, any other elevation being e, it will be,

as radius 1 : sin. $2e$ :: $\frac{v^2}{32\frac{1}{6}} : \frac{v^2}{32\frac{1}{6}} \times$ sin. $2e$, the range at the elevation e, with the velocity v.

But in a resisting medium, like the atmosphere, the actual ranges fall far short of the above theorems, in so much that with the great velocities, the actual or real ranges may be less than the 10th part of the potential ranges ; so that some balls, which actually range but a mile or two, would in vacuo range 20 or 30 miles. And hence also it happens that the elevation of the piece, to shoot farthest in the resisting medium, is always below 45°, and gradually the more below it as the velocity is greater, so that the greater velocities with which balls are discharged from cannon with gunpowder, require an elevation of the gun equal to but about 30°, or even less. And the less the size of the balls is too, the less must this angle of elevation be, to shoot the farthest with a given velocity. See PROJECTILE, and GUNNERY.

HORIZONTAL *Refraction*. See REFRACTION.

HORIZONTAL *Speculum*, one to find a horizon at Sea, &c, when the atmosphere is hazy near the horizon, by which the sight of it is prevented.

A speculum of this kind was invented by a Mr. Serson, on the principle of a top spinning, which always keeps its upright position, notwithstanding the motion of the substance it spins upon. This curious instrument, as it has since been improved by Mr. Smeaton, consists of a well polished metal speculum, of about 3 inches and a half in diameter, inclosed within a circular rim of brass ; so fitted that the centre of gravity of the whole shall fall near the point on which it spins. This is the end of a steel axis running through the

centre of the speculum, above which it finishes in a square, for the conveniency of fitting a roller on it, which sets it in motion by means of a piece of tape wound about the roller.

Various other contrivances to form artificial horizons have been invented by different persons, as glass planes floating on mercury, &c. See HADLEY'S QUADRANT, and several inventions of this sort in the Philos. Transf. by Elton, Halley, Leigh, &c. vol. xxxvii, p. 273, vol. xxxviii, p. 167, vol. xl, p. 413, 417, &c.

HORN-WORK, in Fortification, a sort of out-work, advancing toward the field, to cover and defend a curtin, bastion, or other place, suspected to be weaker than the rest ; as also to possess a height ; carrying in the fore-part, or head, two demi-bastions, resembling horns : these horns, epaulments, or shoulderings, being joined by a curtin, shut up on the side by two wings, parallel to one another, are terminated at the gorge of the work, and so present themselves to the enemy.

HOROGRAPHY, the art of making or constructing dials ; called also Dialling, Horologiography, Gnomonica, Sciatherica, Photosciatherica, &c.

HOROLOGIUM, a common name, among ancient writers, for any instrument or machine for measuring the hours. See CLOCK, WATCH, SUN-DIAL, CHRONOMETER, CLEPSYDRA, &c.

HOROMETRY, the art of measuring or dividing time by hours, and keeping the account of time.

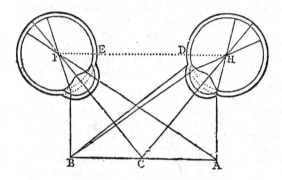

HOROPTER, in Optics, is a right line drawn through the point where the two optic axes meet, parallel to that which joins the centres of the two eyes, or the two pupils. As the line AB drawn through C the point of concourse of the optic axes of the eyes, and parallel to HI joining the centres of the eyes.— This line is called the Horopter, because it is found to be the limit of distinct vision. It has several properties in Optics, which are described at large in Aguillonius, Opt. lib. 2, diss. 10.

HOROSCOPE, in Astrology, is the ascendant or first house, being that part of the zodiac which is just rising in the eastern side of the horizon at any proposed time, when a scheme is to be set or calculated, or a prediction made of any event. See ASCENDANT.

HOROSCOPE is also used for a scheme or figure of the 12 celestial houses ; i. e. the 12 signs of the zodiac, in which is marked the disposition of the heavens for any given time. Thus it is said, To draw or construct a Horoscope or scheme, &c. And it is more peculiarly called,

called, Calculating a nativity, when the life and fortune of a person are the subject of prediction.

Lunar HOROSCOPE, is the point the moon issues out of, when the sun is in the ascending point of the east; and is also called the Part of Fortune.

HOROSCOPE was also a mathematical instrument, in manner of a planisphere; but now disused. It was invented by J. Paduanus, who wrote a special treatise upon it.

HORROR *of a Vacuum*, an imaginary principle among the more ancient philosophers, to which they ascribed the ascent of water in pumps, and other similar phenomena, which are now known to be occasioned by the weight of the air.

HORROX (JEREMIAH), an eminent English astronomer, was born at Toxteth in Lancashire, about the year 1619. From a grammar school in the country, he was sent to Cambridge, where he spent some time in academical studies. About 1633 he began to apply himself to the study of astronomy: but living at that time with his father at Toxteth, in very moderate circumstances, and being destitute of books and other assistances for such studies, he could not make any considerable progress in it. About the year 1636, he formed an acquaintance with Mr. William Crabtree, of Broughton near Manchester, who was engaged in the same studies, with whom a mutual correspondence was carried on till his death; sometimes communicating their improvements to Mr. Samuel Foster, professor of geometry at Gresham College in London. Having now obtained a companion in his studies, Mr. Horrox assumed new vigour, procured other instruments and books, and was pursuing his studies and observations with great assiduity, when he was suddenly cut off by death, the 3d of January 1640, in the 22d year of his age.

What we have of his writings is sufficient to shew how great a loss the world had by his death. He had just finished his *Venus in Sole visa*, 1639, a little before, as appears by some of the letters to his friend Mr. Crabtree, by which also it appears that he made his observations on that phenomenon at Hool near Liverpool. This tract, of Venus seen in the Sun, was published at Dantzick in 1668, by Hevelius, together with his own *Mercurius in Sole visus* May 3, 1661. His other posthumous works, or rather his imperfect papers, were published by Dr. Wallis, in 1673, 4to, with some account of his life; in which we find he first asserts and promotes the Keplerian astronomy against the hypothesis of Lansberg; which he proves to be inconsistent with itself, and neither agreeing with observations nor theory. He likewise reasons very justly concerning the celestial bodies and their motions, vindicates Tycho Brahe from some objections made to his hypothesis, and gives a new theory of the moon: to which are added the Lunar Numbers of Mr. Flamsteed. There are also extracts from several letters between him and Mr. Crabtree, upon various astronomical subjects; with a catalogue of astronomical observations.

There are two things particularly which will perpetuate the memory of this very extraordinary young man. The one is, that he was the first that ever predicted or saw the planet Venus in the sun; for we do not find

that any persons, besides himself and Mr. Crabtree, ever beheld such a phenomenon. Though he was not apprised of the great use that was to be made of it, in discovering the parallax and distance of the sun and planets, yet he made from it many useful observations, corrections, and improvements in the theory of the motions of Venus.—Secondly, his New Theory of Lunar Motions, which Newton himself made the ground work of all his astronomy, relative to the moon, who always spoke of our author as a genius of the first rank.

HORSE-SHOE, in Fortification, is a work sometimes of a round, sometimes of an oval figure, inclosed with a parapet, raised in the ditch of a marshy place, or in low grounds; sometimes also to cover a gate; or to serve as a lodgment for soldiers, to prevent surprises, or relieve an over-tedious defence.

HOSPITAL, *or* HOPITAL (WILLIAM-FRANCIS-ANTHONY, *marquis of*), a celebrated French mathematician, was born of an ancient family in 1661. He was a mathematician almost from his infancy; for being one day at the duke of Rohan's, where some able mathematicians were speaking of a problem of Pascal's, which appeared to them very difficult, he ventured to say, that he believed he could resolve it. They were surprised at such presumption in a boy of 15, for he was then no more; however, in a few days he sent them the solution.

M. l'Hospital entered early into the army, and was a captain of horse; but being very short-sighted, and on that account exposed to perpetual inconveniences and errors, he at length quitted the army, and applied himself entirely to his favourite amusement.—He contracted a friendship with Malbranche, and took his opinion upon all occasions.—In 1699 he was received an honorary member of the Academy of Sciences at Paris.

He was the first person in France who wrote upon Newton's analysis, and on this account was regarded almost as a prodigy. His work was entitled *l'Analyse des Infinimens Petits*, 1696. He engaged afterwards in another mathematical work, in which he included *Les Sections Coniques, les Lieux Géometriques, la Construction des Equations, et une Théorie des Courbes Méchaniques:* but, a little before he had finished it he was seized with a fever, which carried him off, the 2d of February 1704, at 43 years of age. The work was published after his death, viz, in 1707. There are also six of his pieces inserted in different volumes of the Memoirs of the Academy of Sciences.

HOUR, in Chronology, an aliquot part of a natural day, usually the 24th, but sometimes a 12th part. With us, it is the 24th part of the earth's diurnal rotation, or the time from noon to noon, and therefore it answers to 15 degrees of the whole circle of longitude, or of 360°. The hour is divided by 60ths, viz, first into 60 minutes, then each minute into 60 seconds, &c.

The division of the day into Hours is very ancient; as is shewn by Kircher, Oedip. Ægypt. tom. 2, par. 2, class 7, cap. 8. The most ancient Hour is that of the 12th part of the day. Herodotus, lib. 2, observes, that the Greeks learnt from the Egyptians, among other things, the method of dividing the day into 12 parts. And the astronomers of Cathaya, &c, still retain this division.

2   The

The division of the day into 24 hours, was not known to the Romans before the Punic war. Till that time they only regulated their days by the rising and setting of the sun. They divided the 12 hours of their day into four; viz, Prime, which commenced at 6 o'clock, Third at 9, Sixth at 12, and None at 3. They also divided the night into four watches, each containing 3 Hours.

There are various kinds of Hours, used by chronologers, astronomers, dialists, &c. Sometimes too, Hours are divided into Equal and Unequal.

Equal Hours, are the 24th parts of a day and night precisely; that is, the time in which the 15 degrees of the equator pass the meridian. These are also called Equinoctial Hours, because measured on the equinoctial; and Astronomical, because used by astronomers.

Astronomical Hours, are equal Hours, reckoned from noon to noon, in a continued series of 24.

Babylonish Hours, are equal Hours, reckoned from sun-rise in a continued series of 24.

European Hours, used in civil computation, are equal Hours, reckoned from midnight; 12 from thence till noon, and 12 more from noon till midnight.

Jewish, or *Planetary*, or *Ancient* Hours, are 12th parts of the artificial day and night. They are called Ancient or Jewish Hours, because used by the ancients, and still among the Jews. They are called Planetary Hours, because the astrologers pretend, that a new planet comes to predominate every Hour; and that the day takes its denomination from that which predominates the first Hour of it; as Monday from the moon, &c.

Italian Hours, are equal Hours, reckoned from sunset, in a continued series of 24.

Unequal or *Temporary* Hours, are 12th parts of the artificial day and night. The obliquity of the sphere renders these more or less unequal at different times; so that they only agree with the equal Hours at the times of the equinoxes.

Hour-*Circles*, or Horary-*Circles*, are great Circles, meeting in the poles of the globe or world, and crossing the equinoctial or equator at right angles; the same as meridians. They are supposed to be drawn through every 15th degree of the equinoctial and equator, each answering to an hour, and dividing them into 24 equal parts; and on both globes they are supplied by the meridian Hour-circle and index.

Hour-*Glass*, a popular kind of chronometer or clepsydra, serving to measure time by the descent or running of sand, water, &c, out of one glass vessel into another. —The best, it is said, are such as, instead of sand, have egg-shells, well dried in the oven, then beaten fine and sifted.

Hour-*Lines*, on a Dial, are lines which arise from the intersections of the plane of the Dial, with the several planes of the Hour-circles of the sphere; and therefore must be all right lines on a plane Dial.

Hour-*Scale*, a divided line on the edge of Collins's quadrant, being only two lines of tangents of 45 degrees each, set together in the middle. Its use, together with the lines of latitude, is to draw the Hour-lines of Dials that have centres, by means of an equilateral triangle, drawn on the dial-planes.

HOWITZ, or Howitzer, in Artillery, a kind of mortar, or something between a cannon and mortar, partaking of the nature of both, being either a very short gun or a long mortar. It is of German invention, and is mounted upon a carriage like a travelling gun-carriage, with its trunnions placed nearly in the middle. The Howitz is one of the most useful kinds of ordnance, as it can be employed occasionally either as a cannon or mortar, discharging either shells or grape shot, as well as balls, and so doing great execution. They are also very easily travelled about from place to place.

HUMIDITY, or moisture, the power or quality of wetting or moistening other bodies, and adhering to them.

Fluids are moist to some bodies, and not to others. Thus, quicksilver is not moist in respect to our hands or clothes, and other things, which it will not stick to; but it may be called Humid in reference to gold, tin, or lead, to the surfaces of which it will presently adhere, and render them soft and moist. Even water itself, which wets almost every thing, and is the great standard of moisture and Humidity, is not capable of wetting all things; for it stands or runs off in globular drops from any thing greased or oiled, or the leaves of cabbages, and many other planets; and it will not wet the feathers of ducks, geese, swans, and other water-fowl.

HUNDRED, the number of ten times ten, or the square of 10. The place of Hundreds makes the third in order in the Arabic or modern numeration, being denoted thus 100. In the Roman notation it is denoted by the letter C, being the initial of its name, Centum.

Hundred *Weight*, or the great Hundred, contains 112 pounds weight. It is subdivided into 4 quarters, and each quarter into 28 lbs.

HURTERS, in Fortification, denote pieces of timber, about 6 inches square, placed at the lower end of the platform, next to the parapet, to prevent the wheels of the gun-carriages from damaging the parapet.

HUYGENS (Christian), a very eminent astronomer and mathematician, was born at the Hague in Holland, in 1629, being the son of Constantine Huygens, lord of Zuylichem, who had served three successive princes of Orange in the quality of secretary. He spent his whole life in cultivating the mathematics; and not in the speculative way only, but also in making them subservient to the uses of life. From his infancy he discovered an extraordinary fondness for the mathematics; in a short time made a great progress in them; and perfected himself in those studies under professor Schooten, at Leyden. In 1649 he went to Holstein and Denmark, in the retinue of Henry count of Nassau; and was extremely desirous of going to Sweden, to visit Des Cartes who was then in that country with the queen Christina, but the count's short stay in Denmark would not permit him.—In 1651 he gave the world a specimen of his genius for mathematics, in a treatise intitled, *Theoremata de Quadratura Hyperboles, Ellipsis, & Circuli, ex dato Portionum Gravitatis Centro*; in which he clearly shewed what might be expected from him afterwards.——In 1655 he travelled into France, and took the degree of LL.D. at Angers.—— In 1658 he published his *Horologium Oscillatorium, sive de Motu Pendulorum*, &c, at the Hague. He had exhibited

hibited in a former work, intitled, *Brevis Inftitutio de Ufu Horologiorum ad inveniendas Longitudines*, a model of a new invented pendulum; but as fome perfons, envious of his reputation, were labouring to deprive him of the honour of the invention, he wrote this book to explain the conftruction of it; and to fhew that it was very different from the pendulum of aftronomers invented by Galileo.—In 1659 he publifhed his *Syftema Saturninum*, &c; in which he firft of any one explained the ring of Saturn, and difcovered alfo one of the fatellites belonging to that planet, which had hitherto efcaped the eyes of aftronomers: new difcoveries, made with glaffes of his own forming, which gained him a high rank among the aftronomers of his time.

In 1660, he took a fecond journey into France, and the year after paffed over into England, where he communicated his art of polifhing glaffes for telefcopes, and was made Fellow of the Royal Society. About this time the air-pump was invented, which received confiderable improvements from him. This year alfo he difcovered the laws of the collifion of elaftic bodies; as did alfo about this time Wallis and Wren, with whom he had a difpute about the honour of this difcovery. Upon his return to France, in 1663, the minifter Colbert, being informed of his great merit, fettled a confiderable penfion upon him, to engage him to fix at Paris; to which Mr. Huygens confented, and ftaid there from the year 1666 to 1681, where he was admitted a member of the Academy of Sciences. All this time he fpent in mathematical purfuits, wrote feveral books, which were publifhed from time to time, and invented and perfected feveral ufeful inftruments and machines: particularly he had a difpute, about the year 1668, with Mr. James Gregory, concerning the Quadrature of the Circle and Hyperbola of the latter, then juft publifhed, in which Huygens it feems had the better fide of the queftion. But continual application gradually impaired his health; and though he had vifited his native country twice, viz, in 1670 and 1675, for the recovery of it, he was now obliged to betake himfelf to it altogether. Accordingly he left Paris in 1681, and retired to his own country, where he fpent the remainder of his life in the fame purfuits and employments. He died at the Hague, June 8, 1695, in the 67th year of his age, while his *Cofmotheoros*, or treatife concerning a plurality of worlds, was printing; fo that this work did not appear till 1698.

Mr. Huygens loved a quiet and ftudious manner of life, and frequently retired into the country to avoid interruption, but did not contract that morofenefs which is fo commonly the effect of folitude and retirement. He was one of the pureft and moft ingenious mathematicians of his age, and indeed of any other; and made many valuable difcoveries. He was the firft who difcovered Saturn's ring, and a third fatellite of that planet, as mentioned above. He invented the means of rendering clocks exact, by applying the pendulum, and of rendering all its vibrations equal, by the cycloid. He brought telefcopes to perfection, and made many other ufeful difcoveries.

He was the author of many excellent works. The principal of thefe are now contained in two collections, of 2 volumes each, printed in 4to, under the care of profeffor Gravefande. The firft was at Leyden

in 1682, under the title of *Opera Varia*; and the fecond at Amfterdam, in 1728, entitled *Opera Reliqua*.

HYADES, a clufter of 5 ftars in the face of the conftellation Taurus, or the Bull.

HYALOIDES, the vitreous humour of the eye contained between the tunica-retina and the uvea.

HYBERNAL *Occident*. See *Occident*.

HYBERNAL *Orient*. See *Orient*.

HYDATOIDES, the watery humour of the eye contained between the cornea and the uvea.

HYDRA, a fouthern conftellation, confifting of a number of ftars, imagined to reprefent a water ferpent. The ftars in the conftellation Hydra, in Ptolomy's catalogue, are 27; in Tycho's, 19; in Hevelius's, 31; and in the Britannic catalogue, 60.

HYDRAULICS, the fcience of the motion of water and other fluids, with its application in artificial water-works of all forts.—As to what refpects merely the equilibrium of fluids, or their gravitation or action at reft, belongs to Hydroftatics. Upon removing or deftroying that equilibrium, motion enfues; and here Hydraulics commence. Hydraulics therefore fuppofe Hydroftatics; and many writers, from the near relation between them, like mechanics and ftatics, join the two together, and treat of them conjointly as one fcience.

The laws of Hydraulics are given under the word Fluid. And the art of raifing water, with the feveral machines employed for that purpofe, are defcribed under their feveral names, Fountain, Hydrocanifterium Pump, Siphon, Syringe, &c.

The principal writers who have cultivated and improved Hydraulics and Hydroftatics, are Archimedes, in his Libris de Infidentibus Humido; Hero of Alexandria, in his Liber Spiritualium; Marinus Ghetaldus, in his Archimedes promotus; Mr. Oughtred; Jo. Ceva, in his Geometria Motus; Jo. Bap. Balianus, De Motu Naturali Gravium, Solidorum et Liquidorum; Mariotte, in his treatife of the Motion of Water and other fluids; Boyle, in his Hydroftatical Paradoxes; Fran. Tertius de Lanis, in his Magifterium Naturæ et Artis; Lamy, in his Traite de l'Equilibre des Liqueurs; Rohault; Dr. Wallis, in his Mechanics; Dechales; Newton, in his Principia; Gulielmeni, in his Menfura Aquarum Fluentium; Herman; Wolfius; Gravefande; Mufchenbroek; Leopold; Schottus, in his Mechanica Hydraulico-Pneumatica; Geo. Andr. Bockler, in his Architectura Curiofa Germanica; Auguft. Rammilleis; Lucas Antonius Portius; Sturmy, in his treatife on the Conftruction of Mills; Switzer's Hydroftatics; Varignon, in the Mem. Acad. Sci.; Jurin; Belidor; Bernoulli; Defaguliers; Clare; Emerfon; Ferguson; Ximenes; Boffu; D'Alembert; Buat; &c, &c.

HYDRAULICO-PNEUMATICAL, a term applied by fome authors to fuch engines as raife water by means of the weight or fpring of the air.

HYDROGRAPHICAL *Charts* or *Maps*, more ufually called fea-charts, are projections of fome part of the fea, or coaft, for the ufe of navigation.

In thefe are laid down all the rhumbs or points of the compafs, the meridians, parallels, &c, with the coafts, capes, iflands, rocks, fhoals, fhallows, &c, in their proper places, and proportions.

The

The making and felling thefe charts was for fome time the employment of Columbus, the firft difcoverer of America. The ftory goes, that happening to be heir to the memoirs and journals of one Alonzo Sanchez de Huelva, a noted pilot and captain of a fhip, who by chance had been driven by a ftorm to the ifland of St. Domingo, and dying at Columbus's houfe foon after his return, this gave Columbus the firft hint to attempt a difcovery of the Weft Indies.

For the conftruction and ufe of the feveral kinds of Hydrographical Maps, fee CHART, and SAILING.

HYDROLOGY, is that part of natural hiftory which examines and explains the nature and properties of water in general.

HYDROMANCY, the act or art of divining or foretelling future events by means of water.

This is one of the four general kinds of divination: the other three refpecting the other elements, fire, air, and earth, are denominated refpectively pyromancy, aeromancy, and geomancy.

Varro mentions the Perfians as the firft inventors of Hydromancy, adding, that Numa Pompilius and Pythagoras made ufe of it.

The writers on optics furnifh us with divers Hydromantic machines, veffels, &c. For example,

To conftruct an Hydromantic machine, by means of which an image or object fhall be removed out of the fight of the fpectator, and reftored again at pleafure, without altering the pofition, either of the one or the other. Provide two veffels, ABF, CGLK (Plate 12, fig. 3), the uppermoft filled with water, and fupported by three little pillars, one of which BC is hollow, and furnifhed with a cock B. Let the lower veffel CL be divided by a partition HI into two parts, the lower of which may be opened or clofed by means of a cock at P. Upon the partition place an object, or image, which the fpectator at O cannot fee by a direct ray GL.

If now the cock B be opened, the water defcending into the cavity CI, the ray GL will be refracted from the perpendicular GR to O; fo that the fpectator will now fee the object by the refracted ray OG. And again, fhutting the cock B, and opening the other P, the water will defcend into the lower cavity HL; where, the refraction ceafing, no rays will now come from the object to the eye: but upon fhutting the cock P, and opening the other B, the water will fill the cavity again, and bring the object in fight of O afrefh.

To make an Hydromantic veffel, which fhall exhibit the images of external objects as if fwimming in water. Provide a cylindrical veffel ABCD (fig. 4, pl. 12.) divided into two cavities by a glafs EF, not perfectly polifhed: in G apply a lens convex on both fides; and in H incline a plane mirror, of an elliptic figure, to an angle of 45 degrees; and let IH and HG be fomething lefs than the diftance of the focus of the lens G, fo that the place of the images of an object radiating through the fame may fall within the cavity of the upper veffel: let the inner cavity be blacked, and the upper filled with clear water.—If now the veffel be difpofed in a dark place, fo as the lens be turned towards an object illuminated by the fun, its image will be feen as fwimming in the water.

HYDROMETER, an inftrument for meafuring

the properties and effects of water, as its denfity, gravity, force, velocity, &c.

That with which the fpecific gravity of water is determined, is often called an aerometer, or water-poife.

The general principle on which the conftruction and ufe of the Hydrometer depends, has been illuftrated under the article Specific GRAVITY; where it is fhewn that a body fpecifically lighter than feveral fluids, will ferve to find out their fpecific gravities; becaufe it will fink deepeft in the fluids whofe fpecific gravity is the leaft. So if AB (fig. 5, pl. 12) be a fmall even glafs tube, hermetically fealed, having a fcale of equal divifions marked upon it, with a hollow ball of about an inch in diameter at bottom, and a fmaller ball C under it, communicating with the firft; into the little ball is put mercury or fmall fhot, before the tube is fealed, fo that it may fink in water below the ball, and float or ftand upright, the divifions on the ftem fhewing how far it finks.—If this inftrument be dipped in common water, and fink to D, it will fink only to fome lower point E in falt water; but in port wine it will fink to fome higher point F, and in brandy perhaps to B.

It is evident that an Hydrometer of this kind will only fhew that one liquid is fpecifically heavier than another; but the true fpecific weight of any liquid cannot be determined without a calculation for this particular inftrument, the tube of which fhould be truly cylindrical. Befides, thefe inftruments will not ferve for fluids whofe denfities are much different.

Mr. Clarke conftructed a new Hydrometer, fhewing whether any fpirits be proof, or above or below proof, and in what degree. This inftrument was made of a ball of copper (becaufe ivory imbibes fpirituous liquors, and glafs is apt to break), to which is foldered a brafs wire about a quarter of an inch thick; upon this wire is marked the point to which it exactly finks in proof fpirits; as alfo two other marks, one above and one below the former, exactly anfwering to one-tenth above proof and one-tenth below proof. There are alfo a number of fmall weights made to add to it, fo as to anfwer to the other degrees of ftrength befides thofe above, and for determining the fpecific gravities of different fluids. Philof. Tranf. Abr. vol. vi, p. 326.

Dr. Defaguliers contrived an Hydrometer for determining the fpecific gravities of different waters, to fuch a degree of nicety, that it would fhew when one kind of water was but the 40,000th part heavier than another. It confifts of a hollow glafs ball of about 3 inches in diameter, charged with fhot to a proper degree, and having fixed in it a long and very flender wire, of only the 40th part of an inch in diameter, and divided into tenths of inches, each tenth anfwering to the 40,000th part, as above. See his Exper. Philof. vol. 2, p. 234.

Mr. Quin and other perfons have alfo conftructed Hydrometers, with other and various contrivances, and with different degrees of accuracy; but all nearly on the fame general principles.

But there is one circumftance which deferves particular attention in the conftruction and graduation of Hydrometers, for determining the precife ftrength of different brandies, and other fpirituous liquors. Mr. Reaumur difcovered, in making his fpirit thermometers, that when rectified fpirit and water, or phlegm,

the

the other conftituent part of brandy, are mixed together, there appears to be a mutual penetration of the
two liquors, and not merely juxtapofition of parts; fo
that a part of the one fluid feems to be received into
the pores of the other; by which it happens, that if
a pint of rectified fpirit be added to a pint of water,
the mixture will be fenfibly lefs than a quart. The variations hence produced in the bulk of the mixed fluid
render the Hydrometer, when graduated in the ufual
way by equal divifions, an erroneous meafure of its
ftrength; becaufe the fpecific gravity of the compound
is found not to correfpond to the mean gravity of the
two ingredients. M. Montigny conftructed a fcale
for this inftrument in the manner before fuggefted by
Dr. Lewis, on actual obfervation of the finking or
rifing of the Hydrometer in various mixtures or alcohol and water, made in certain known proportions.
Hift. de l'Acad. Roy. des Sci. 1768; alfo Neumann's
Chem. by Lewis, p. 450, note r.

M. De Luc has lately publifhed a fcheme for the
conftruction of a comparable Hydrometer, fo that a
workman, after having conftructed one upon his principles, may make all others fimilar to each other, and
capable of indicating the fame degree on the fcale, when
immerfed in the fame liquor of the fame temperature.
This inftrument is propofed to be conftructed of a ball
of flint glafs, communicating with a fmall hollow cylinder, containing fuch a quantity of quickfilver for a
ballaft, that the inftrument may fink nearly to the top,
in the moft fpirituous liquor, made as hot as poffible;
to which is alfo attached a thin filvered tube, for
a fcale, &c. The whole defcription may be feen at
large in the Philof. Tranf. vol. 68, p. 500.

M. Le Roi alfo publifhed a propofal for conftructing
comparable Hydrometers. See Hift. de l'Acad. des
Scien. for 1770, Mem. 7.

HYDROMETRIA, HYDROMETRY, the menfura-
tion of water and other fluid bodies, their gravity,
force, velocity, quantity, &c; including both hydroftatics and hydraulics.

HYDROSCOPE, an inftrument anciently ufed for
the meafure of time. It was a kind of water-clock,
confifting of a cylindrical tube, conical at bottom: the
cylinder was graduated with divifions, to which the
top of the water becoming fucceffively contiguous, as
it trickled out of the vertex of the cone, pointed out
the hour.

HYDROSTATICAL *Balance,* a kind of balance
contrived for the exact and eafy finding the fpecific
gravities of bodies, both folid and fluid, and thereby
of eftimating the degree of purity of bodies of all
kinds, with the quality and richnefs of metals, ores,
minerals, &c, and the proportions in any mixture, adul-
teration, or the like.

This is effected by weighing the body both in water, or other fluid, and out of it; and for this purpofe
one of the fcales has ufually a hook at the bottom, for
fufpending the body by fome very fine thread. And
the ufe of the inftrument is founded on this theorem
of Archimedes, that any body weighed in water, lofes
as much of its weight as is equal to the weight of the
fame bulk of the water. Thus then is known the proportion of the fpecific gravities of the folid and fluid, or
the proportion of their weights under the fame bulk,

viz, the proportion of the weight of the body weighed
out of water, to the difference between the fame and its
weight in water. Hence alfo, by doing the fame thing
for feveral different folids, with the fame fluid, or different fluids with the fame folid, all their fpecific gravities become known.

The inftrument needs but little defcription. AB is

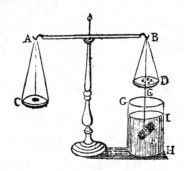

a nice balance beam, with its fcales C and D, turning
with the fmall part of a grain, the one of them, D,
having a hook in the bottom, to receive the loop of a
horfe hair &c, E, by which the body F is fufpended.
GH is a jar of water, in which the body is immerfed
when weighing.

The pieces in the fcale C denote the weight of the
body out of water; then, upon immerging it, put weights
in the fcale D to reftore the balance again, and they
will fhew the fpecific gravity of the body.

There have been various kinds of the Hydroftatical
balance, and improvements made on it, by different
perfons. Thus, Dr. Defaguliers fet three fcrews in the
foot of the ftand, to move any fide higher or lower,
till the ftem be quite upright, which is known by a
plummet hanging over a fixed point in the pedeftal.
Defag. Exp. Philof. vol. 2, p. 196. And for fundry
other conftructions of this inftrument, defigned for
greater accuracy than the common fort, fee Martin's
Phil. Britan. or Gravefande's Phyfices Elem. Math.
tom. 1, lib. 3, cap. 3, &c.

The fpecific gravities of fmall weights may be determined by fufpending them in loops of horfe hair, or
fine filken threads, to the hook at the bottom of the
fcale. Thus, if a guinea fufpended in air weigh 129
grains, and upon being immerfed in water require $7\frac{1}{5}$
grains to be put in the fcale over it, to reftore the
equilibrium; we thus find that a quantity of water of
equal bulk with the guinea, weighs $7\frac{1}{5}$ grains, or 7·2;
therefore dividing the 129 by the 7·2, the quotient
17·88 fhews that the guinea is fo many times heavier
than its bulk of water. Whence, if any piece of gold
be tried, by weighing it firft in air, then in water,
and if, upon dividing the weight in air by the lofs in
water, the quotient be 17·88, the gold is good; if the
quotient be 18 or more, the gold is more fine; but if it
be lefs than 17·88, the gold is too much alloyed with
other metal. If filver be tried in the fame manner,
and found to be 11 times heavier than water, it is very
fine; if it be $10\frac{1}{2}$ times heavier, it is ftandard; but if
lefs, it is mixed with fome lighter metal, fuch as tin.

When the body, whofe fpecific gravity is fought, is
lighter than water, fo that it will not quite fink; annex

to

to it a piece of another body heavier than water, fo that the mafs compounded of the two may fink together. Weigh the denfer body, and the compound mafs, feparately, both in water and out of it, thereby finding how much each lofes in water; and fubtract the lefs of thefe two loffes from the greater; then fay,

As the remainder
is to the weight of the light body in air,
fo is the fpecific gravity of water
to the fpecific gravity of the light body.

HYDROSTATICAL *Bellows*, a machine for fhewing the upward preffure of fluids and the Hydroftatical paradox. It confifts of two thick boards, A, D, each about 16 or 18 inches diameter, more or lefs, covered or connected firmly with leather round the edges, to open and fhut like a common bellows, but without valves; only a pipe, B, about 3 feet high is fixed into the bellows at *e*, Now let water be poured into the pipe at C, and it will run into the bellows, gradually feparating the boards, by raifing the upper one. Then if feveral weights, as three hundred weights, be laid upon the upper board, by pouring the water in at the pipe till it be full, it will fuftain all the weights, though the water in the pipe fhould not weigh a quarter of a pound; for the pipe or tube may be as fmall as we pleafe, provided it be but long enough, the whole effect depending upon the height, and not at all on the width of the pipe: for the proportion is always this,

As the area of the orifice of the pipe
is to the area of the bellows board,
fo is the weight of water in the pipe
to the weight it will fuftain on the board.

Hence if a man ftand upon the upper board, and blow into the pipe B, he will raife himfelf upon the board; and the fmaller the pipe, the eafier he will be able to raife himfelf; and then by putting his finger upon the top of the pipe, he can fupport himfelf as long as he pleafes, provided the bellows be air-tight.

Mr. Ferguson has defcribed another machine, which may be fubftituted inftead of this common Hydroftatical bellows. It is however on the fame principle of the Hydroftatical paradox; and may be feen in the Supplement to his Lectures, p. 19.

HYDROSTATICAL *Paradox*, is a principle in Hydroftatics, fo called becaufe it has a paradoxical appearance at firft view, and it is this; that any quantity of water, or other fluid, how fmall foever, may be made to balance and fupport any quantity, or any weight, how great foever. This is partly illuftrated in the laft article, on the Hydroftatical bellows, where it appears that any weight whatever may be blown up and fupported by the breath from a perfon's mouth. And the principle may be explained as follows: It is well-known that water in a pipe or canal, open at both ends, always rifes to the fame height at both ends, whether thofe ends be wide or narrow, equal or unequal. Thus, the fmall pipe GH being clofe joined to another open veffel AI, of any fize whatever; then

pouring water into the one of thefe, it will rife up in the other, and ftand at the fame height, or horizontal line DF in both of them, and that whether they are upright, or inclined in any pofition. So that all the water that is in the large veffel from A to I, is fupported by that which is in the fmall veffel from D to I only. And as there is no limit to this latter one, but that it may be made as fine even as a hair, it hence evidently appears that any quantity of water may be thus fupported by any other the fmalleft quantity.

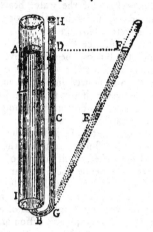

Since then the preffure of fluids is directly as their perpendicular heights, without any regard to their quantities, it appears that whatever the figure or fize of the veffels may be, if they are but of equal heights, and the areas of their bottoms equal, the preffures of equal heights of water are equal upon the bottoms of thefe veffels; even though the one fhould contain a thoufand or ten thoufand times as much as the other.

Mr. Fergufon confirms and illuftrates this paradox by the following experiment.

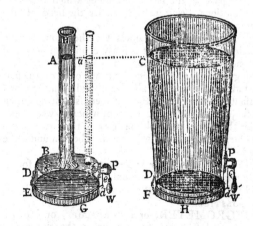

Let two veffels be prepared of equal heights, but very unequal contents, fuch as AB and CD; each veffel being open at both ends, and their bottoms E and F of equal widths. Let a brafs bottom G and H be exactly fitted to each veffel, not to go into it, but for it to ftand upon; and let a piece of wet leather be put between each vessel and its brass bottom, for the sake of closeness. Join each bottom to its vessel by a hinge D, so that it may open like the lid of a box; and let each bottom be kept up to its vessel by equal weights W, hung to lines which go over the pulleys P, whose blocks are fixed to the sides of the vessel at *f*, and the lines tied to hooks at *d*, fixed in the brass bottoms opposite to the hinges D. Things being thus prepared and fitted, hold one vessel upright in the hands over a bason on a table, and cause water to be poured slowly into it, till the pressure of the water bears down its bottom at the side *d*, and raises the weight E; and then part of the water will run out at *d*. Mark the height at which the surface H of the water stood in the vessel, when the bottom began to give way at *d*; and then, holding up the other vessel in the same manner, cause water to be poured into it; and it will be seen that when the water rises in this vessel just as high as it did in the former, its bottom will also give way at *d*, and it will lose part of the water.

The natural reason of this surprising phenomenon is, that since all parts of a fluid at equal depths below the surface, are equally pressed in all manner of directions, the water immediately below the fixed part B*f* will be pressed as much upward against its lower surface within the vessel, by the action of the column A*g*, as it would be by a column of the same height, and of any diameter whatever; and therefore, since action and reaction are equal and contrary to each other, the water immediately below the surface B*f* will be pressed as much downward by it, as if it were immediately touched and pressed by a column of the height A*g*, and of the diameter B*f*; and therefore the water in the cavity BD*df* will be pressed as much downward upon its bottom G, as the bottom of the other vessel is pressed by all the water above it. Lectures, p. 105.

HYDROSTATICS, is the science which treats of the nature, gravity, pressure, and equilibrium of fluids; and of the weighing of solids in them.

That part of the science of fluids which treats of their motions, being included under the head of Hydraulics.

Hydrostatics and hydraulics together constitute a branch of philosophy that is justly considered as one of the most curious, ingenious, and useful of any; affording theorems and phenomena not only of the first use and importance, but also surprisingly amusing and pleasant; as appears in the numberless writings upon the subject; the principal points of which may be found under the several particular articles of this work; and the chief writings on this science may be seen under the article Hydraulics.

HYDRUS, or *Water Serpent*, one of the new southern constellations, including only ten stars.

HYEMAL *Solstice*, the same with Winter Solstice. See Solstice.

HYGROMETER, or Hygroscope, or Notiometer, an instrument for measuring the degrees of moisture in the air.

There are various kinds of Hygrometers; for whatever body either swells by moisture, or shrinks by dryness, is capable of being formed into an Hygrometer. Such are woods of most kinds, particularly deal, ash, poplar, &c. Such also is catgut, the beard of a wild

8

oat, and twisted cord, &c. The best and most usual contrivances for this purpose are as follow.

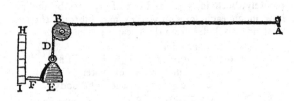

1. Stretch a common cord, or a fiddle-string, ABD along a wall, passing it over a pulley B; fixing it at one end A, and to the other end hanging a weight E, carrying a style or index F. Against the same wall fit a plate of metal HI, graduated, or divided into any number of equal parts; and the Hygrometer is complete.

For it is matter of constant observation, that moisture sensibly shortens cords and strings; and that, as the moisture evaporates, they return to their former length again. The like may be said of a fiddle-string: and from hence it happens that such strings are apt to break in damp weather, if they are not slackened by the screws of the violin. Hence it follows, that the weight E will ascend when the air is more moist, and descend again when it becomes drier. By which means the index F will be carried up and down, and, by pointing to the several divisions on the scale, will shew the degrees of moisture or dryness.

2. Or thus, for a more sensible and accurate Hygro-

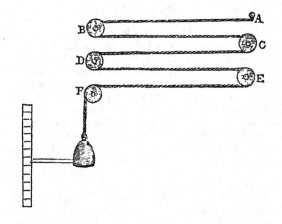

meter: strain a whipcord, or catgut, over several pulleys B, C, D, E, F; and proceed as before for the rest of the construction. Nor does it matter whether the several parts of the cord be parallel to the horizon, as expressed in the annexed figure, or perpendicular to the same, or in any other position; the advantage of this, over the former method, being merely the having a greater length of cord in the same compass; for the longer the cord, the greater is the contraction and dilatation, and consequently the degrees of variation of the index over the scale, for any given change of moisture in the air.

3. Or

3. Or thus: Faſten a twiſted cord, or fiddle-ſtring, AB, by one end at A, ſuſtaining a weight at B, carrying an index C round a circular ſcale DE deſcribed on a horizontal board or table.——For a cord or catgut twiſts itſelf as it moiſtens, and untwiſts again as it dries. Hence, upon an increaſe or decreaſe of the humidity of the air, the index will ſhew the quantity of twiſting or untwiſting, and conſequently the increaſe or decreaſe of moiſture or dryneſs.

4. Thoſe Dutch toys, called weather houſes, where a ſmall image of a man, and one of a woman, are fixed upon the ends of an index, are conſtructed upon this principle. For the index, being ſuſtained by a cord or twiſted catgut, turns backwards and forwards, bringing out the man in wet weather, and the woman in dry.

5. Or thus: Faſten one end of a cord, or catgut, AB, to a hook at A; and to the other end a ball D of about one pound weight; upon which draw two concentric circles, and divide them into any number of equal parts, for a ſcale; then fit a ſtyle or index EC into a proper ſupport at E, ſo as the extremity C may almoſt touch the diviſions of the ball.——Here the cord twiſting or untwiſting, as in the former caſe, will indicate the change of moiſture, by the ſucceſſive application of the diviſions of the circular ſcale, as the ball turns round, to the index C.

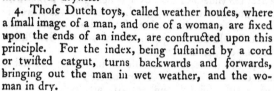

6. Or an Hygrometer may be made of the thin boards of aſh or fir, by their ſwelling or contracting. But this, and all the other kinds of this inſtrument, above deſcribed, become in time ſenſibly leſs and leſs accurate; till at laſt they loſe their effect entirely, and ſuffer no alteration from the weather. But the following ſort is much more durable, ſerving for many years with tolerable accuracy.

very clean; and, when dry again, in water or vinegar in which there has been diſſolved ſal ammoniac, or ſalt of tartar; after which let it dry again.——Now, if the air become moiſt, the ſponge will imbibe it and grow heavier, and conſequently will preponderate, and turn the index towards C; on the contrary, when the air becomes drier, the ſponge becomes lighter, and the index turns towards A; and thus ſhewing the ſtate of the air.

8. In the laſt mentioned Hygrometer, Mr. Gould, in the Philoſ. Tranſ. inſtead of a ſponge, recommends oil of vitriol, which grows ſenſibly lighter or heavier from the degrees of moiſture in the air; ſo that being ſaturated in the moiſteſt weather, it afterwards retains or loſes its acquired weight, as the air proves more or leſs moiſt. The alteration in this liquor is ſo great, that in the ſpace of 57 days it has been known to change its weight from 3 drachms to 9; and has ſhifted a tongue or index of a balance 30 degrees. So that in this way a pair of ſcales may afford a very nice Hygrometer. The ſame author ſuggeſts, that oil of ſulphur or campanam, or oil of tartar per deliquium, or the liquor of fixed nitre, might be uſed inſtead of the oil of vitriol.

9. This balance may be contrived in two ways; by either having the pin in the middle of the beam, with a ſlender tongue a foot and a half long, pointing to the diviſions on an arched plate, as repreſented in the laſt figure above. Or the ſcale with the liquor may be hung to the point of the beam near the pin, and the other

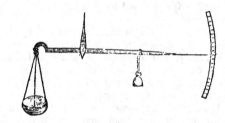

extremity made ſo long, as to deſcribe a large arch on a board placed for the purpoſe; as in the figure here annexed.

10. Mr. Arderon has propoſed ſome improvement in the ſponge Hygrometer. He directs the ſponge A to

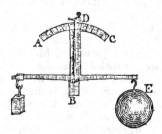

7. Take the Manoſcope, deſcribed under that article, and inſtead of the exhauſted ball E, ſubſtitute a ſponge, or other body, that eaſily imbibes moiſture. To prepare the ſponge, it may be proper firſt to waſh it in water

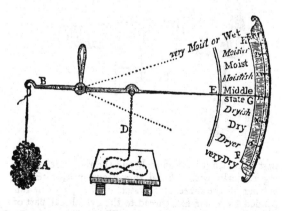

be ſo cut, as to contain as large a ſuperficies as poſſible, and to hang by a fine thread of ſilk upon the beam of a balance

balance B, and exactly balanced on the other side by another thread of silk at D, strung with the smallest lead shot, at equal distances, so adjusted as to cause an index E to point at G, the middle of a graduated arch FGH, when the air is in a middle state between the greatest moisture and the greatest dryness. Under this silk so strung with shot, is placed a little table or shelf I, for that part of the silk or shot to rest upon which is not suspended. When the moisture imbibed by the sponge increases its weight, it will raise the index, with part of the shot, from the table, and vice versa when the air is dry. Philof. Tranf. vol. 44, p. 96.

11. From a series of Hygroscopical observations, made with an apparatus of deal wood, described in the Philof. Tranf. number 480, Mr. Coniers concludes, 1st, That the wood shrinks most in summer, and swells most in winter, but is most liable to change in the spring and fall. 2d, That this motion happens chiefly in the day time, there being scarce any variation in the night. 3d, That there is a motion even in dry weather, the wood swelling in the morning, and shrinking in the afternoon. 4th, That the wood, by night as well as by day, usually shrinks when the wind is in the north, north-east, and east, both in summer and winter. 5th, That by constant observation of the motion and rest of the wood, with the help of a thermometer, the direction of the wind may be told nearly without a weather-cock. He adds, that even the time of the year may be known by it; for in spring it moves more and quicker than in winter; in summer it is more shrunk than in spring; and has less motion in autumn than in summer.

See an account of a method of constructing these and other Hygrometers, in Phil. Tranf. Abr. vol. 2, p. 30, &c, and plate 1 annexed. See also Philof. Tranf. vol. 11. p. 647 and 715, vol. 15, p. 1032, vol. 43, p. 6, vol. 44, p. 95, 169 and 184, vol. 54, p. 259, vol. 61, p. 198, vol. 63, p. 404, &c.

12. Dr. Hook's Hygrometer was made of the beard of a wild oat, set in a small box, with a dial plate and an index. See his Micrographia, p. 150.

13. The Doctors Hales and Defaguliers both contrived another form of sponge Hygrometer, on this principle. They made an horizontal axis, having a small

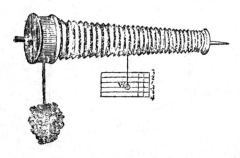

part of its length cylindrical, and the remainder tapering conically with a spiral thread cut in it, after the manner of the fuzee of a watch. The sponge is suspended by a fine silk thread to the cylindrical part of the axis, upon which it winds. This is balanced by a small weight W, suspended also by a thread, which

winds upon the spiral fuzee. Then when the sponge grows heavier, in moist weather; it descends and turns the axis, and so draws up the weight, which coming to a thicker part of the axis it becomes a balance to the sponge, and its motion is shewn by an attached scale. And vice versa when the air becomes drier.——Salt of tartar, or any other salt, or pot ashes, may be put into the scale of a balance, and used instead of the sponge. Defag. Exper. Philof. vol. 2, p. 300.

14. Mr. Ferguson made an Hygrometer of a thin deal pannel; and to enlarge the scale, and so render its variations more sensible, he employed a wheel and axle, making one cord pass over the axle, which turned a wheel ten times as large, over which passed a line with a weight at the end of it, whose motion was therefore ten times as much as that of the deal pannel. The board should be changed in 3 or 4 years. See Philof. Tranf. vol. 54, art. 47.

15. Mr. Smeaton gave also an ingenious and elaborate construction of an Hygrometer; which may be seen in the Philof. Tranf. vol. 61, art. 24.

16. Mr. De Luc's contrivance for an Hygrometer is very ingenious, and on this principle. Finding that even ivory swells with moisture, and contracts with dryness, he made a small and very thin hollow cylinder of ivory, open only at the upper end, into which is fitted the under or open end of a very fine long glass tube, like that of a thermometer. Into these is introduced some quicksilver, filling the ivory cylinder, and a small part of the length up the glass tube. The consequence is this: when moisture swells the ivory cylinder, its bore or capacity grows larger, and consequently the mercury sinks in the fine glass tube; and vice versa, when the air is drier, the ivory contracts, and forces the mercury higher up the tube of glass. It is evident that an instrument thus constructed is in fact also a thermometer, and must necessarily be affected by the vicissitudes of heat and cold, as well as by those of dryness and moisture; or that it must act as a thermometer as well as an Hygrometer. The ingenious contrivances in the structure and mounting of this instrument may be seen in the Philof. Tranf. vol. 63, art. 38; where it may be seen how the above imperfection is corrected by some simple and ingenious expedients, employed in the original construction and subsequent use of the instrument; in consequence of which, the variations in the temperature of the air, though they produce their full effects on the instrument, as a thermometer, do not interfere with or embarrass its indications as an Hygrometer.

17. In the Philof. Tranf. for 1791, Mr. De Luc has given a second paper on Hygrometry. This has been chiefly occasioned by a Memoir of M. de Saussure on the same subject, entitled Essais sur l'Hygrometrie, in 4to, 1783. In this work M. de S. describes a new Hygrometer of his construction, on the following principle. It is a known fact that a hair will stretch when it is moistened, and contract when dried: and M. de Saussure found, by repeated experiments, that the difference between the greatest extension and contraction, when the hair is properly prepared, and has a weight of about 3 grains suspended by it, is nearly one 40th of its whole length, or one inch in 40. This circumstance suggested the idea of a new Hygrometer. To

render

render thefe fmall variations of the length of the hair perceptible, an apparatus was contrived, in which one of the extremities of the hair is fixed, and the other, bearing the counterpoife abovementioned, furrounds the circumference of a cylinder, which turns upon an axis to which a hand is adapted, marking upon a dial in large divifions the almoft infenfible motion of this axis. About 12 inches high is recommended as the moft convenient and ufeful : and to render them portable, a contrivance is added, by which the hand and the counterpoife can be occafionally fixed.

But M. de Luc, in his Idees fur la Meteorologie, vol. 1, anno 1786, fhews that hairs, and all the other animal or vegetable hygrofcopic fubftances, taken lengthwife, or in the direction of their fibres, undergo contrary changes from different variations of humidity ; that when immerfed in water, they lengthen at firft, and afterwards fhorten ; that when they are near the greateft degree of humidity, if the moifture be increafed, they fhorten themfelves ; if it be diminifhed, they lengthen themfelves firft before they contract again. Thefe irregularities, which render them incapable of being true meafures of humidity, he fhews to be the neceffary confequence of their organic reticular ftructure De Sauffure takes his point of extreme moifture from the vapours of water under a glafs bell, keeping the fides of the bell continually moiftened ; and affirms, that the humidity is, there, conftantly the fame in all temperatures ; the vapours even of boiling water having no other effect than thofe of cold. De Luc, on the contrary, fhews that the differences in humidity under the bell are very great, though De Sauffure's Hygrometer was not capable of difcovering them ; and that the real undecompofed vapour of boiling water has the directly oppofite effect to that of cold, the effect of extreme drynefs ; and on this point he mentions an interefting fact, communicated to him by Mr. Watt, viz, that wood cannot be employed in the fteam engine, for any of thofe parts where the vapour of the boiling water is confined, becaufe it dries fo as to crack as if expofed to the fire.

To thefe charges of M. De Luc, a reply is made by M. De Sauffure, in his Defence of the Hair Hygrometer, in 1788 ; where he attributes the general difagreement between the two inftruments, to irregularities of M. De Luc's ; and affigns fome aberrations of his own Hygrometer, which could not have proceeded from the above caufe, but to its having been out of order ; &c.

This has drawn from M. De Luc a fecond paper on Hygrometry, publifhed in the Philof. Tranf. for 1791, p. 1, and 389. This author here refumes the four fundamental principles which he had fketched out in the former paper, viz, 1ft, That fire is a fure, and the only fure means of obtaining extreme drynefs. 2d, That water, in its liquid ftate, is a fure, and the only fure means of determining the point of extreme moifture. 3d, There is no reafon, a priori, to expect, from any hygrofcopic fubftance, that the meafurable effects, produced in it by moifture, are proportional to the intenfities of that caufe.—But, 4th, perhaps the comparative changes of the dimenfions of a fubftance, and of the weight of the fame or other fubftances, by the fame variations of moifture, may lead to fome difcovery in that refpect. On thefe heads M. De Luc expatiates

at large in this paper, fhewing the imperfections of M. De Sauffure's principles of Hygrometry, and particularly as to a hair, or any fuch fubftance when extended lengthwife, being properly ufed as an Hygrometer. On the other hand, he fhews that the expanfion of fubftances acrofs the fibres, or grain, renders them, in that refpect, by far the moft proper for this purpofe. He choofes fuch as can be made very thin, as ivory, or deal fhavings, but over all he finds whalebone to be far the beft of any. But, for all the reafonings of thefe ingenious philofophers on this interefting fubject, and complete information, fee the publications above quoted, as alfo the Monthly Review, vol. 51, p. 224, vol. 71, p. 213, vol. 76, p. 316, vol. 78, p. 236, and vol. 6, of the new feries for the year 1791, p. 133.

HYGROMETRY, the fcience of the meafurement of the moifture of the atmofphere. The chief writings on this fcience are thofe of M. De Luc and M. De Sauffure, for which fee the laft article on Hygrometers.

HYGROSCOPE, is commonly ufed in the fame fenfe with Hygrometer. Wolfius, however, regarding the etymology of the word, makes fome difference. According to him, the Hygrofcope only fhews the alterations of the air in refpect of humidity and drynefs, but the hygrometer meafures them. A Hygrofcope therefore is only an indefinite or lefs accurate hygrometer.

HYPATIA, a very learned and beautiful lady, was born at Alexandria, about the end of the 4th century, as fhe flourifhed about the year of Chrift 430. She was the daughter of Theon, a celebrated philofopher and mathematician, and prefident of the famous Alexandrian fchool. Her father, encouraged by her extraordinary genius, had her not only educated in all the ordinary qualifications of her fex, but inftructed in the moft abftrufe fciences. She made fuch great progrefs in philofophy, geometry, aftronomy, and the mathematics in general, that fhe paffed for the moft learned perfon of her time. She publifhed commentaries on Apollonius's Conics, on Diophantus's Arithmetic, and other works. At length fhe was thought worthy to fucceed her father in that diftinguifhed and important employment, the government of the fchool of Alexandria ; and to deliver inftructions out of that chair where Ammonius, Hierocles, and many other great men, had taught before ; and this at a time too when men of great learning abounded both at Alexandria and in many other parts of the Roman empire. Her fame being fo extenfive, and her worth fo univerfally acknowledged, it was no wonder that fhe had a crowded auditory. " She explained to her hearers (fays Socrates, an ecclefiaftical hiftorian of the 5th century, born at Conftantinople) the feveral fciences that go under the general name of philofophy ; for which reafon there was a confluence to her, from all parts, of thofe who made philofophy their delight and ftudy."

Her fcholars were not lefs eminent than they were numerous. One of them was the celebrated Synefius, who was afterwards bifhop of Ptolomais. This ancient Chriftian Platonift always expreffes the ftrongeft, as well as the moft grateful, teftimony of the virtue of his tutorefs ; never mentioning her without the moft profound refpect, and fometimes in terms of affection but

little

little fhort of adoration. But it was not Synefius only, and the difciples of the Alexandrian fchool, who admired Hypatia for her virtue and learning: never was woman more careffed by the public, and yet never had woman a more unfpotted character. She was held as an oracle for her wifdom, for which fhe was confulted by the magiftrates in all important cafes; a circumftance which often drew her among the greateft concourfe of men, without the leaft cenfure of her manners. In fhort, when Nicephorus intended to pafs the higheft compliment on the princefs Eudocia, he thought he could not do it better than by calling her another Hypatia.

While Hypatia thus reigned the brighteft ornament of Alexandria, Oreftes was governor of that place for the emperor Theodofius, and Cyril was bifhop or patriarch. Oreftes, having had a liberal education, could not but admire Hypatia and as a wife governor often confulted her. This, together with an averfion which Cyril had againft Oreftes, proved fatal to the lady. About 500 monks affembling, attacked the governor one day, and would have killed him, had he not been refcued by the townfmen; and the refpect which Oreftes had for Hypatia caufing her to be traduced among the Chriftian multitude, they dragged her from her chair, tore her in pieces, and burnt her limbs.

Cyril is ftrongly fufpected of having fomented this tragedy. Cave indeed endeavours to remove the imputation of fo horrid an action from the patriarch; and lays it upon the Alexandrian mob in general, whom he calls " a very trifling inconftant people." But though Cyril fhould be allowed neither to have been the perpetrator, nor even the contriver, of it, yet it is much to be fufpected that he did not difcountenance it in the manner he ought to have done: a fufpicion which muft needs be greatly confirmed by reflecting, that he was fo far from blaming the outrage committed by the monks upon Oreftes, that he afterwards received the dead body of Ammonius, one of the moft forward in that outrage, who had grievoufly wounded the governor, and who was juftly punifhed with death. Upon this riotous ruffian Cyril made a panegyric in the church where he was laid, in which he extolled his courage and conftancy, as one that had contended for the truth; and changing his name to Thaumafius, or the Admirable, ordered him to be confidered as a martyr. " However (continues Socrates), the wifeft part of Chriftians did not approve the zeal which Cyril fhewed on this man's behalf, being convinced that Ammonius had juftly fuffered for his defperate attempt."

HYPERBOLA, one of the conic fections, being that which is made by a plane cutting a cone fo, that, entering one fide of the cone, and not being parallel to the oppofite fide, it may cut the circular bafe when the oppofite fide is ever fo far produced below the vertex, or fhall cut the oppofite fide of the cone produced above the vertex, or fhall make a greater angle with the bafe than the oppofite fide of the cone makes; all thefe three circumftances amounting to the fame thing, but in other words.

1. Thus, the figure DAE is an Hyperbola, made by a plane entering the fide VQ of a cone PVQ at A, and either cutting the bafe PEQ when the plane is not parallel to VP, and this is ever fo far produced; or when

the angle ARQ is greater than the angle VPQ; or when the plane cuts the oppofite fide in B above the vertex.

2. By the Hyperbola is fometimes meant the whole plane of the fection, and fometimes only the curve line of the fection.

3. Hence, the cutting plane meets the oppofite cone in B, and there forms another Hyperbola d B e, equal to the former one, and having the fame tranfverfe axis AB; and the fame vertices A and B Alfo the two are called Oppofite Hyperbolas.

4. The centre C is the middle point of the tranverfe axis.

5. The femi-conjugate axis is CL, a mean proportional between CI and CK, the diftances to the fides of the oppofite cone, when CI is drawn parallel to the diameter PQ of the bafe of the cone. Or the whole conjugate axis is a mean proportional between AF and BH, which are drawn parallel to the bafe of the cone.

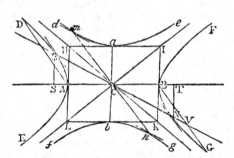

6. If DAE and FBG be two oppofite Hyperbolas, having the fame tranfverfe and conjugate axes AB and a b, perpendicularly bifecting each other; and if d a e and f b g be two other oppofite Hyperbolas, having the fame axes with the two former, but in the contrary order, viz, having a b for their firft or tranfverfe axis, and AB for their fecond or conjugate axis: then any two adjacent curves are called Conjugate Hyperbolas, and the whole figure formed by all the four curves, the Figure of the Conjugate Hyperbolas. And if the rectangle HIKL be infcribed within the four conjugate Hyperbolas, touching the vertices A, B, a, b, and having their fides parallel and equal to the two axes; and if then the two diagonals HCK, ICL, of the parallelogram be drawn, thefe diagonals are the afymptotes of the curves, being lines that continually approach nearer and nearer to the curves, without meeting them, except at an infinite diftance, where each afymptote and the two adjacent fides of the two conjugate Hyperbolas may be fuppofed all to meet; the afymptote being there a common tangent to them both, viz, at that infinite diftance.

7. Hence the four Hyperbolas, meeting and running into each other at the infinite diftance, may be confidered as the four parts of one entire curve, having the fame axes, tangents, and other properties.

8. A Diameter in general, is any line, as MN, drawn through the centre C, and meeting, or terminate

nated by the opposite legs of the opposite Hyperbolas. And if parallel to this diameter there be drawn two tangents, at *m* and *n*, to the opposite legs of the other two opposite Hyperbolas, the line *mCn* joining the points of contact, is the conjugate diameter to MN, and the two mutually conjugates to each other. Or, if to the points M or N there be drawn a tangent, and through the centre C the line *mn* parallel to it, that line will be the conjugate to MN. The points where each of these meet the curves, as M, N, *m*, *n*, are the vertices of the diameters; and the tangents to the curves at the two vertices of any diameter, are parallel to each other, and also to the other or conjugate diameter.

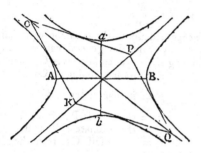

9. Moreover, if those tangents to the four Hyperbolas, at the vertices of two conjugate diameters, be produced till they meet, they will form a parallelogram OPQR; and the diagonals OQ and PR of the parallelogram will be the asymptotes of the curves; which therefore pass through the opposite angles of all the parallelograms so inscribed between the curves. Also it is a property of these parallelograms, that they are all equal to each other, and therefore equal to the rectangle of the two axes; as will be farther noticed below. Farther, if these diagonals or asymptotes make a right angle between them, or if the inscribed parallelogram be a square, or if the two axes be equal to each other, then the Hyperbola is called a right-angled or an equilateral one.

10. An Ordinate to any diameter, is a line drawn parallel to its conjugate, or to the tangent at its vertex, and terminated by the diameter produced and the curve. So MS and TN are ordinates to the axis AB; also AD and BG are ordinates to the diameter MN (last fig. but one). Hence the ordinates to the axis are perpendicular to it; but ordinates to the other diameters are oblique to them.

11. An abscis is a part of any diameter, contained between its vertex and an ordinate to it; and every ordinate has two abscisses: as AT and BT, or MV and NV.

12. The Parameter of any diameter, is a third proportional to the diameter and its conjugate.—The Parameter of the axis is also equal to the line AG or B*g* (fig. 1), if FG be drawn to make the angle AFG = the angle BAV, or the line H*g* to make the angle BH*g* = the angle ABV.

13. The Focus is the point in the axis where the ordinate is equal to half the parameter of the axis; as

S and T (fig. 2) if MS and TN be half the parameter, or the 3d proportional to CA and C*a*. Hence there are two Foci, one on each side the vertex, or one for each of the opposite Hyperbolas. These two points in the axis are called Foci, or burning points, because it is found by opticians that rays of light issuing from one of them, and falling upon the curve of the Hyperbola, are reflected into lines that verge towards the other point or Focus.

To describe an Hyperbola, in various ways.

14. (1st *Way by points.*)—In the transverse axis AF produced, take the foci F and *f*, by making CF and C*f* = A*a* or B*a*, assume any point I : Then with the

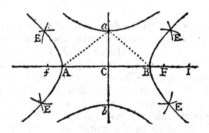

radii AI, BI, and centres F, *f*, describe arcs intersecting in E, which will give four points in the curve. In like manner, assuming other points I, as many other points will be found in the curve. Then, with a steady hand, draw the curve line through all the points of intersection E.—In the same manner are to be constructed the other pair of opposite Hyperbolas, using the axis *ab* instead of AB.

15. (2d *Way by points, for a Right-angled Hyperbola only.*)—On, the axis produced if necessary, take any

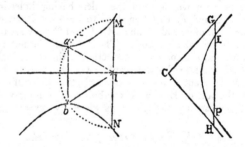

point I, through which draw a perpendicular line, upon which set off IM and IN equal to the distance I*a* or I*b* from I to the extremities of the other axis; and M and N will be points in the curve.

16. (3d *Way by points, to describe the curve through a given point.*)—CG and CH being the asymptotes, and P the given point of the curve; through the point P draw any line GPH between the asymptotes, upon which take GI = PH, so shall I be another point of the curve. And in this manner may any number of points be found, drawing as many lines through the given point P.

17. (4th·

17. (4th *Way by a continued Motion.*)—If one end

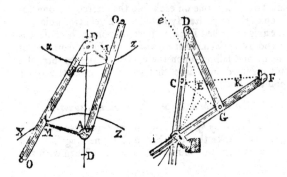

of a long ruler *f*MO be faftened at the point *f*, by a pin on a plane, fo as to turn freely about that point as a centre. Then take a thread FMO, fhorter than the ruler, and fix one end of it in F, and the other to the end O of the ruler. Then if the ruler *f*MO be turned about the fixed point *f*, at the fame time keeping the thread OMF always tight, and its part MO clofe to the fide of the ruler, by means of the pin M; the curve line AX defcribed by the motion of the pin M is one part of an Hyperbola. And if the ruler be turned, and move on the other fide of the fixed point F, the other part AZ of the fame Hyperbola may be defcribed after the fame manner.—But if the end of the ruler be fixed in F, and that of the thread in *f*, the oppofite Hyperbola *xaz* may be defcribed.

18. (5th *Way, by a continued Motion.*)—Let C and F be the two foci, and E and K the two vertices of the Hyperbola. (See the laft fig. above.) Take three rulers CD, DG, GF, fo that CD = GF = EK, and DG = CF; the rulers CD and GF being of an indefinite length beyond C and G, and having flits in them for a pin to move in; and the rulers having holes in them at C and F, to faften them to the foci C and F by means of pins, and at the points D and G they are to be joined by the ruler DG. Then, if a pin be put in the flits, viz, the common interfection of the rulers CD and GF, and moved along, caufing the two rulers GF, CD, to turn about the foci C and F, that pin will defcribe the portion E*e* of an Hyperbola.—The foregoing are a few among various ways given by feveral authors.

Some of the chief Properties of the Hyperbola.

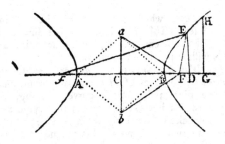

19. (1ft) The fquares of the ordinates, of any diameter, are to each other, as the rectangles of their

abfciffes; i. e. DE² : GH² :: AD . BD : AG . BG.

20. As the fquare of any diameter, is to the fquare of its conjugate; fo is the rectangle of two abfciffes, to the fquare of their ordinate. That is, AB² : *ab*² :: AD . BD : DE².

Or, becaufe the rectangle AD . BD is = the difference of the fquares CD² — CB², the fame property is,

As AB² : *ab*² :: CD² — CB² : DE²,

Or AB : $\frac{ab^2}{AB}$ :: CD² — CB² : DE²,

That is AB : *p* :: CD² — CB² : DE², where *p* is the parameter of the diameter AB, or the 3d proportional $\frac{ab^2}{AB}$.

And hence is deduced the common equation of the Hyperbola, by which its general nature is expreffed. Thus, putting *d* = the femidiameter CA or CB,

c = its femiconjugate C*a* or C*b*,

p = its parameter or $\frac{2d^2}{c}$,

x = the abfcifs BD from the vertex,

y = the ordinate DE, and

v = the abfcifs CD from the centre ;

Then is $d^2 : c^2 :: \overline{2d + x} . x : y^2$,

or $d^2 : c^2 :: \overline{d + v} . x : y^2$,

or $d^2 : c^2 :: v^2 - d^2 : y^2$,

or $d : p :: v^2 - d^2 : y^2$; fo that

$$ y^2 = \frac{2dx + x^2}{d^2} . c^2 = \frac{v^2 - d^2}{d^2} . c^2 = \frac{v^2 - d^2}{d} . p ; $$

any of which equations or proportions exprefs the nature of the curve. And hence arifes the name Hyperbola, fignifying to exceed, becaufe the ratio of d^2 to c^2, or of d to p, exceeds that of $2dx$ to y^2; that ratio being equal in the parabola, and defective in the ellipfe, from which circumftances alfo thefe take their names.

21. The diftance between the centre and the focus, is equal to the diftance between the extremities of the tranfverfe and conjugate axes. That is, CF = A*a* or A*b*, where F is the focus.

22. The conjugate femi-axis is a mean proportional between the diftances of the focus from both vertices of the tranfverfe. That is, C*a* is a mean between AF and BF, or AF : C*a* :: C*a* : BF, or AF . BF = C*a*².

23. The difference of two lines drawn from the foci, to meet in any point of the curve, is equal to the tranfverfe axis. That is, *f*E — FE = AB, where F and *f* are the two foci.

24. All the parallelograms infcribed between the four conjugate Hyperbolas are equal to one another, and each equal to the rectangle of the two axes. That is, the parallelogram OPQR = AB . *ab* (fig. to art. 9).

25. The difference of the fquares of every pair of conjugate diameters, is equal to the fame conftant quantity, viz, the difference of the fquares of the two axes. That is, MN² — *mn*² = AB² — *ab*², (fig. to art. 6) where MN and *mn* are any two conjugate diameters.

4

26. The rectangles of the parts of two parallel lines, terminated by the curve, are to one another, as the rectangles of the parts of any other two parallel lines, any where cutting the former. Or the rectangles of the parts of two interfecting lines, are as the fquares of their parallel diameters, or fquares of their parallel tangents.

27. All the rectangles are equal which are made of the fegments of any parallel lines, cut by the curve, and limited by the afymptotes, and each equal to the fquare of their parallel diameter. That is, HE . EK or He . eK = CQ² or CP².

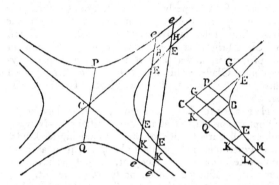

28. All the parallelograms are equal, which are formed between the afymptotes and curve, by lines parallel to the afymptotes. That is, the paral. CGEK = CPBQ.—Hence is obtained another method of expreffing the nature of the curve by an equation, involving the abfcifs taken on one afymptote, and ordinate parallel to the other afymptote. Thus, if $x =$ CK, $y =$ KE, $a =$ CQ, and $b =$ BQ the ordinate at the vertex B of the curve; then, by the property in this article, $ab = xy$, or $a : x :: y : b$; that is, the rectangle of the abfcifs and ordinate is every where of the fame magnitude, or any ordinate is reciprocally as its abfcifs.

29. If the abfciffes CQ, CK, CL, &c, taken on the one afymptote, be in geometrical progreffion increafing; then fhall the ordinates QB, KE, LM, &c, parallel to the other afymptote, be a like geometrical progreffion in the fame ratio, but decreafing; and all the rectangles are equal, under every abfcifs and its ordinate, viz, CQ. QB = CK . KE = CL . LM, &c.

30. The abfciffes CQ, CK, CL, &c, being taken in geometrical progreffion; the fpaces or afymptotic areas BQKE, EKLM, &c, will be all equal; or, the fpaces BQKE, BQLM, &c, will be in arithmetical progreffion; and therefore thefe fpaces are the hyperbolic logarithms of thofe abfciffes.

Thefe, and many other curious properties of the Hyperbola, may be feen demonftrated in my Treatife on Conic Sections, and feveral others. See alfo Conic Sections.

Acute Hyperbola, one whofe afymptotes make an acute angle.

Ambigenal Hyperbola, is that which has one of its infinite legs falling within an angle formed by the afymptotes, and the other falling without that angle. This is one of Newton's triple Hyperbolas of the 2d order. See his Enumeratio Lin. tert. Ord. See alfo Ambigenal.

Common, or *Conic* Hyperbola, is that which arifes from the fection of a cone by a plane; called alfo the Apollonian Hyperbola, being that kind treated on by the firft and chief author Apollonius.

Conjugate Hyperbolas, are thofe formed or lying together, and having the fame axes, but in a contrary order, viz, the tranfverfe of each equal the conjugate of the other; as the two Conjugate Hyperbolas *Pee* and EEE in the laft figure but one.

Equilateral, or *Rectanglar* Hyperbola, is that whofe two axes are equal to each other, or whofe afymptotes make a right angle.—Hence, the property or equation of the equilateral Hyperbola, is $y^2 = ax + x^2$, where a is the axis, x the abfcifs, and y its ordinate; which is fimilar to the equation of the circle, viz, $y^2 = ax - x^2$, differing only in the fign of the fecond term, and where a is the diameter of the circle.

Infinite Hyperbolas, or Hyperbolas *of the higher kinds*, are expreffed or defined by general equations fimilar to that of the conic or common Hyperbola, but having general exponents, inftead of the particular numeral ones, but fo as that the fum of thofe on one fide of the equation, is equal to the fum of thofe on the other fide. Such as, $ay^{m+n} = bx^m(d + x)^n$, where x and y are the abfcifs and ordinate to the axis or diameter of the curve; or $x^m y^n = a^{m+n}$, where the abfcifs x is taken on one afymptote, and the ordinate y parallel to the other.

As the Hyperbola of the firft kind, or order, viz the conic Hyperbola, has two afymptotes; that of the 2d kind or order has three; that of the 3d kind, four; and fo on.

Obtufe Hyperbola, is that whofe afymptotes form an obtufe angle.

Rectangular Hyperbola, the fame as Equilateral Hyperbola.

Hyperbolic *Arc*, is the arc of an Hyperbola. Put $a =$ CA the femitranfverfe axe, $c =$ Ca the femiconjugate, $y =$ an ordinate PQ to the axe drawn from the end Q of the arc AQ, beginning at the vertex A: then putting $q = \dfrac{aa + cc}{c^4}$,

$A =$ the hyp. log. of $\dfrac{y + \sqrt{cc + yy}}{c}$

$= 2\cdot302585093 \times$ common log. of $\dfrac{y + \sqrt{cc + yy}}{c}$,

$B = \dfrac{y\sqrt{cc + yy} - ccA}{2}$, $C = \dfrac{y^3\sqrt{cc + yy} - 3ccB}{4}$,

$D = \dfrac{y^5\sqrt{cc + yy} - 5ccC}{6}$, &c;

then is the length of the arc AQ expreffed by

$c \times (A + \dfrac{q}{2} B - \dfrac{q^2}{2\cdot4} C + \dfrac{1\cdot3q^3}{2.4.6} D - \dfrac{1\cdot3\cdot5q^4}{2.4.6.8} E$ &c;

or

or by $\dfrac{15c^2t + (21c^2 + 19t^2)\,x}{15c^2t + (21c^2 + 9t^2)\,x} \times y$, nearly; where t is the whole transverse axe $2CA$, $c = 2Ca$ the conjugate, $x = AP$ the abscifs, and $y = PQ$ the ordinate.

These and other rules may be seen demonstrated in my Mensuration, p. 408, &c, 2d edit.

HYPERBOLIC *Area*, or *Space*, the area or space included by the Hyperbolic curve and other lines.

Putting $a = CA$ the semitransverse, $c = Ca$ the semiconjugate, $y = PQ$ the ordinate, and $v = CP$ its distance from the centre; then is the

area $APQ = \frac{1}{2}vy - \frac{1}{2}ac \times$ hyp. log. of $\dfrac{ay + cv}{ac}$;

fector $CAQ = \frac{1}{2}ac \times$ hyp. log. of $\dfrac{ay + cv}{ac}$;

area $APQ = 2xy \times (\frac{1}{3} - \dfrac{z^1}{1\cdot3\cdot5} - \dfrac{z^2}{3\cdot5\cdot7} - \dfrac{z^3}{5\cdot7\cdot9} - \dfrac{z^4}{7\cdot9\cdot11}$ &c),

where $x = AP$, and $z = \dfrac{x}{2a + x}$; or

$APQ = \dfrac{2cx}{15a} \times (\sqrt{2ax} + 4\sqrt{2ax + \frac{3}{4}xx})$ nearly.

Let CT and CE be the two asymptotes, and the ordinates DA, EF parallel to the other asymptote CT; then the asymptotic space ADEF or fector CAF is

$= CD \times DA \times$ hyp. log. of $\dfrac{CE}{CD}$ or

$= CD \times DA \times$ hyp. log. $\dfrac{DA}{EF}$ or

$= CD \times DA \times \dfrac{DE}{CD} - \dfrac{DE^2}{2CD^2} + \dfrac{DE^3}{3CD^3} - \dfrac{DE^4}{4CD^4}$ &c;

and this last series was first given by Mercator in his Logarithmotechnia.

See my Mensuration, p. 413, &c, 2d edit.

Generally, if $x^m y^n = a^{m+n}$ be an equation expressing an Hyperbola of any order; then its asymptotic

area will be $\dfrac{n}{n - m}\,xy$; which space therefore is always quadrable, in all the orders of Hyperbolas, except the first or common Hyperbola only, in which m and n being each 1, the denominator $n - m$ becomes 0 or nothing.

HYPERBOLIC *Conoid*, a solid formed by the revolution of an Hyperbola about its axis, otherwise called an Hyperboloid.

To find the Solid Content of an Hyperboloid.

Let AC be the semitransverse of the generating Hyperbola, and AH the height of the solid; then as $2AC + AH$ is to $3AC + AH$, so is the cone of the same base and altitude, to the content of the Conoid.

To find the Curve Surface of an Hyperboloid.

Let AC be the semitransverse, and AB perpendicular to it, and equal to the semiconjugate of ADE the generating Hyperbola, or section through the axis of the solid. Join CB; make $CF = CA$, and on CA let fall the perpendicular FG; then with the semitransverse CG, and semiconjugate $GH = AB$, describe the Hyperbola GIK; then as the diameter of a circle is to its circumference, so is the Hyperbolic frustum ILAMK to the curve surface of the Conoid generated by DAE. See my Mensur. p. 429, &c, 2d edit.

HYPERBOLIC *Cylindroid*, a solid formed by the revolution of an Hyperbola about its conjugate axis, or line through the centre perpendicular to the transverse axis. This solid is treated of in the Philof. Transf. by Sir Christopher Wren, where he shews some of its properties, and applies it to the grinding of Hyperbolical Glasses; affirming that they must be formed this way, or not at all. See Philof. Transf. vol. 4, pa. 961.

HYPERBOLIC *Leg*, of a curve, is that having an asymptote, or tangent at an infinite distance.—Newton reduces all curves, both of the first and higher kinds, into Hyperbolic and parabolic legs, i. e. such as have asymptotes, and such as have not, or such as have tangents at an infinite distance, and such as have not.

HYPERBOLIC *Line*, is used by some authors for what is more commonly called the Hyperbola itself, being the curve line of that figure; in which sense the surface terminated by it is called the Hyperbola.

HYPERBOLIC *Logarithm*, a logarithm so called as being similar to the asymptotic spaces of the Hyperbola. The Hyperbolic logarithm of a number, is to the common logarithm, as 2·30258509929940457 to 1, or as 1 to ·43429448190325518. The first invented logarithms, by Napier, are of the Hyperbolic kind; and so are Kepler's. See LOGARITHM.

HYPERBOLIC *Mirror*, is one ground into that shape.

HYPERBOLIC *Space*, that contained by the curve of the Hyperbola, and certain other lines. See HYPERBOLIC AREA.

HYPERBOLICUM *Acutum*, a solid made by the revolution of the infinite area or space contained between the curve of the Hyperbola, and its asymptote. This produces a solid, which though infinitely long and generated by an infinite area, is nevertheless equal to a finite solid body; as is demonstrated by Torricelli, who gave it this name.

HYPERBOLIFORM *Figures*, are such curves as approach, in their properties, to the nature of the Hyperbola; called also Hyperboloides.

HYPERBOLOIDS, are Hyperbolas of the higher kind, whose nature is expressed by this equation, $ay^{m+n} = bx^m(a+x)^n$. See HYPERBOLA. It also means the Hyperbolic Conoid. See that article.

HYPERBOREANS, the most northern nations, or regions, as dwelling beyond or about the wind Boreas: as the Siberians, Samoieds, &c.

HYPERTHYRON, in Architecture, a sort of table, usually placed over gates or doors of the Doric order, above the chambranle, in form of a frize.

HYPETHRE, in Ancient Architecture, two rows

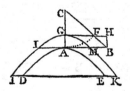

of pillars furrounding, and ten at each face of a temple, &c, with a periftyle within of fix columns.

HYPOGEUM, in the ancient Architecture, a name common to all the parts of a building that are under ground; as the cellars, butteries, &c.

HYPOGEUM, in Aftrology, a name given to the ce-leftial houfes that are below the horizon; and efpe-cially the imum coeli, or bottom point of the heavens.

HYPOMOCHLION, the fulcrum or prop of a lever; or the point which fuftains its preffure, when employed either in raifing or lowering bodies. The Hypomochlion is frequently a roller fet under the lever; or under ftones or pieces of timber, &c, that they may be the more eafily lifted up, or removed.

HYPOTENUSE, or HYPOTHENUSE, in a right-angled triangle, is the fide which fubtends, or is oppo-fite to the right angle, and is always the longeft of the three fides; as the fide AC, oppofite to the right angle B.

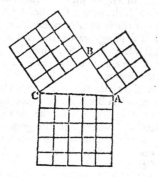

It is a celebrated theorem in Plane Geometry, being the 47th prop. of the 1ft book of Euclid, that in every right-angled triangle ABC, the fquare formed upon the Hypothenufe AC, is equal to both the two fquares formed upon the other two fides AB and BC; or that $AC^2 = AB^2 + BC^2$. This is particularly called the Pythagorean theorem, from its reputed in-ventor Pythagoras, who it is faid facrificed a whole hecatomb to the mufes, in gratitude for the difcovery. But the fame thing is true of circles or any other fimi-lar figures, viz, that any figure defcribed on the Hypo-tenufe, is equal to the fum of the two fimilar figures defcribed on both the other two fides.

HYPOTHENUSE. See HYPOTENUSE.

HYPOTHESIS, in Geometry, or Mathematics, means much the fame thing with fuppofition, being a fuppofition or an affumption of fomething as a condi-tion, upon which to raife a demonftration, or from which to draw an inference.

Dr. Barrow fays, Hypothefes, or poftulatums, are propofitions affuming or affirming fome evidently pof-fible mode, action, or motion of a thing, and that there is the fame affinity between hypothefes and pro-blems, as between axioms and theorems: a problem fhewing the manner, and demonftrating the poffibility of fome ftructure, and an Hypothefis affuming fome conftruction which is manifeftly poffible.

HYPOTHESIS, in Philofophy, denotes a kind of fyftem laid down from our own imagination, by which to account for fome phenomenon or appearance of na-ture. Thus there are Hypothefes to account for the tides, for gravity, for magnetifm, for the deluge, &c.

The real and fcientific caufes of natural things ge-nerally lie very deep: obfervation and experiment, the proper means of arriving at them, are in moft cafes ex-tremely flow; and the human mind is very impatient: hence we are often induced to feign or invent fomething that may feem like the caufe, and which is calculated to anfwer the feveral phenomena, fo that it may poffi-bly be the true caufe.

Philofophers are divided as to the ufe of fuch fictions or Hypothefes, which are much lefs current now than they were formerly. The lateft and beft writers are for excluding Hypothefes, and ftanding intirely on ob-fervation and experiment. Whatever is not deduced from phenomena, fays Newton, is an Hypothefis; and Hypothefes, whether metaphyfical, or phyfical, or mechanical, or of occult qualities, have no place in ex-perimental philofophy. Phil. Nat. Prin. Math. in Calce.

HYPOTHESIS is more particularly applied, in Aftro-nomy, to the feveral fyftems of the heavens; or the divers manners in which different aftronomers have fuppofed the heavenly bodies to be ranged, or moved. The principal Hypothefes are the Ptolomaic, the Tychonic, and the Copernican. This laft is now fo generally received, and fo well eftablifhed and war-ranted by obfervation, that it is thought derogatory to it to call it an Hypothefis.

HYPOTRACHELION, in Architecture, is ufed for a little frize in the Tufcan and Doric capital, be-tween the aftragal and annulets; called alfo the colerin and gorgerin.

The word is applied by fome authors in a more ge-neral fenfe, to the neck of any column, or that part of its capital below the aftragal.

I. J.

JACK, in Mechanics, is an inftrument in common ufe for raifing heavy timber, or very great weights of any kind; being a certain very powerful combination of teeth and pinions, and the whole inclofed in a ftrong wooden ftock or frame BC, and moved by a winch or handle HP; the outfide appearing as in fig. 1, here annexed.

In fig. 6, pl. 12, the wheel or rack-work is fhewn, being the view of the infide when the ftock is removed. Though it is not drawn in the juft proportions and dimenfions, for the rack AB muft be fuppofed at leaft four times as long in proportion to the wheel Q, as the figure reprefents it; and the teeth, which will be then four times more in number to have about 3 in the inch. Now if the handle HP be 7 inches long, the circumference of this radius will be 44 inches, which is the diftance or fpace the power moves through in one revolution of the handle: but as the pinion of the handle has but 4 leaves, and the wheel Q fuppofe 20 teeth, or 5 times the number, therefore to make one revolution of the wheel Q, it requires 5 turns of the handle, in which cafe it paffes through 5 times 44 or 220 inches: but the wheel having a pinion R of 3 leaves, thefe will raife the rack 3 teeth, or one inch, in the fame fpace. Hence then, the handle or power moving 220 times as faft as the weight, will raife or balance a weight of 220 times its own power. And if this be the hand of a man, who can fuftain 100 pounds weight, he will, by help of this Jack, be able to raife, or fuftain a weight or force of 22000lb, or about 10 tons weight.

This machine is fometimes open behind from the bottom almoft up to the wheel Q, to let the lower claw, which in that cafe is turned up as at B, draw up any weight. When the weight is drawn or pufhed fufficiently high, it is kept from going back by hanging the end of the hook S, fixed to a ftaple, over the curved part of the handle at *h*.

JACK is alfo the name of a well-known engine in the kitchen, ufed for turning a fpit. Here the weight is the power applied, acting by a fet of pulleys; the friction of the parts, and the weight with which the

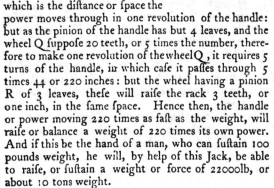

fpit is charged, make the force to be overcome; and a fteady uniform motion is maintained by means of the fly.

See the fig. of this machine, pl. 12, fig. 7.

Smoke JACK, is an engine ufed for the fame purpofe with the common Jack, and is fo called from its being moved by means of the fmoke, or rarefied air, afcending the chimney, and ftriking againft the fails of the horizontal wheel AB (plate 12, fig. 8), which being inclined to the horizon, is moved about the axis of the wheel, together with the pinion C, which carries the wheels D and E; and E carries the chain F, which turns the fpit. The wheel AB fhould be placed in the narrow part of the chimney, where the motion of the fmoke is fwifteft, and where alfo the greateft part of it muft ftrike upon the fails.—The force of this machine depends upon the draught of the chimney, and the ftrength of the fire.

JACK-*arch*, in Architecture, is an arch of one brick thicknefs.

JACK-*head*, in Hydraulics, a part fometimes annexed to the forcing pump.

JACOB's-*Staff*, a mathematical inftrument for taking heights and diftances; the fame with the Crofs-ftaff; which fee.

JACOBUS, a gold coin, worth 25 fhillings; fo called from king James the firft of England, in whofe reign it was ftruck. They diftinguifhed two kinds of the Jacobus, the old and the new; the former valued at 25 fhillings, weighing 6 dwts 10 grs; the latter, called alfo Carolus, valued at 23 fhillings, and weighing 5 dwts 20 grains.

JAMBS, or JAUMS, in Architecture, are the upright fides of chimneys, from the hearth to the mantle-tree. Alfo door pofts, or the upright pofts at the ends of the window frames.

St JAMES's *Day*, a feftival in the calendar, obferved on the 25th of July, in honour of St. James the apoftle.

JANUARY, the firft month of the year, according to the computation now ufed in the Weft, and containing 31 days; fo called by the Romans from Janus, one of their divinities, to whom they gave two faces; becaufe on the one fide, the firft day of this month looked towards the new year, and on the other towards the old one. The name may alfo be derived from Janua, a gate; this month, being the firft of the year, may be confidered as the gate or entrance of it.

January

January and February were introduced into the year by Numa Pompilius; Romulus's year beginning with the month of March.

JAUMS. See JAMBS.

ICE, a brittle transparent body, formed of some fluid, frozen or fixed by cold. The specific gravity of Ice to water, is various, according to the nature and circumstances of the water, degree of cold, &c. Dr. Irving (Phipps's Voyage towards the North Pole) found the densest Ice he could meet with about a 14th part lighter than water. M. de Mairan found it, at different trials, 1-14th, 18th, or 19th lighter than water; and when the water was previously purged of air, only a 22d part.

The rarefaction of Ice has been supposed owing to the air-bubbles produced in Ice while freezing; these, being considerably large in proportion to the water frozen, render the Ice so much specifically lighter. It is well known that a considerable quantity of air is lodged in the interstices of water, though it has there little or no elastic property, on account of the disunion of its particles; but upon these particles coming closer together, and uniting as the water freezes, light, expansive, and elastic air-bubbles are thus generated, and increase in bulk as the cold grows stronger, and by their elastic force bursts to pieces any vessel in which the water is closely contained. But snow-water, or any water long boiled over the fire, affords an Ice more solid, and with fewer bubbles. Pure water long kept in vacuo and frozen afterwards there, freezes much sooner, on being exposed to the same degree of cold, than water unpurged of its air and set in the open atmosphere. And the Ice made of water thus divested of its air, is much harder, more solid and transparent, and heavier than common Ice.

But M. de Mairan, in a dissertation on Ice, attributes the increase of the bulk of the water under this form, chiefly to a different arrangement of its parts: the icy skin on water being composed of filaments which are found to be joined constantly and regularly at an angle of 60°, and which, by this disposition, occupy a greater volume than if they were parallel. Besides, after Ice is formed, he found it continue to expand by cold; a piece of Ice, which was at first only a 14th part specifically lighter than water, on being exposed some days to the frost, became a 12th part lighter; and thus he accounts for the bursting of Ice in ponds.

It appears from an experiment of Dr. Hooke, in 1663, that Ice refracts the light less than water; whence he infers, that the lightness of Ice, which causes it to swim in water, is not produced merely by the small bubbles which are visible in it, but that it arises from the uniform constitution or general texture of the whole mass: a fact which was afterward confirmed by M. de la Hire. See Hooke's Exper. by Derham, p. 26, Acad. Per. 1693, Mem. p. 25.

Sir Robert Barker thus describes the process of making Ice in the East Indies, in a country where he never saw any natural Ice. On a large plain they dig three or four pits, each about 30 feet square, and 2 feet deep; the bottoms of which are covered, about 8 or 12 inches thick, with sugar-cane, or the stems of the large Indian corn, dried. On this bed are placed in rows a number of small shallow unglazed earthen pans, formed of a very porous earth, a quarter of an inch thick, and about an inch and a quarter deep; which, at the dusk of the evening, they fill with soft water that has been boiled. In the morning before sunrise the Ice-makers attend at the pits, and collect what has been frozen in baskets, which they convey to the place of preservation. This is usually prepared in some high and dry situation, by sinking a pit 14 or 15 feet deep, which they line first with straw, and then with a coarse kind of blanketing. The Ice is deposited in this pit, and beaten down with rammers, till at length its own accumulated cold again freezes it, and it forms one solid mass. The mouth of the pit is well secured from the exterior air with straw and blankets, and a thatched roof is thrown over the whole. Philos. Transf. vol. 65, p. 252.

ICHNOGRAPHY, in Architecture, is a transverse or horizontal section of a building, exhibiting the plot of the whole edifice, and of the several rooms and apartments in any story; together with the thickness of the walls and partitions; the dimensions of the doors, windows, and chimneys; the projectures of the columns and piers, with every thing visible in such a section.

ICHNOGRAPHY, in Fortification, is the plan or representation of the length and breadth of a fortress; the distinct parts of which are marked out, either on the ground itself, or upon paper.

ICHNOGRAPHY, in Perspective, the view of any thing cut off by a plane parallel to the horizon, just by the base or bottom of it; being the same with what is otherwise called the plan, geometrical plan, or ground-plot, of any thing, and is opposed to Orthography or Elevation.

ICOSAEDRON, or ICOSAHEDRON, one of the five regular bodies or solids, terminated by twenty equilateral and equal triangles. It may be considered as consisting of 20 equal and similar triangular pyramids, whose vertices meet in the centre of a sphere conceived to circumscribe it, and therefore having all their heights and bases equal; therefore the solidity of one of those pyramids multiplied by 20, the number of them, gives the solid content of the Icosaedron.

To form or make the Icosaedron.—Describe upon a card paper, or some other such like substance, 20 equilateral triangles, as in the figure at the article *Regular Body*. Cut it out by the extreme edges, and cut all the other lines half through, then fold the sides up by these edges half cut through, and the solid will be formed.

The linear edge or side of the Icosaedron being A, then will the surface be $5 A^2 \sqrt{3} = 8 \cdot 6602540 A^2$, and the solidity =

$$\tfrac{5}{6} A^3 \sqrt{\frac{7 + 3\sqrt{5}}{2}} = 2 \cdot 1816950 A^3.$$

More generally, put A = the linear edge or side, B the surface, and C the solid content of the Icosaedron, also *r* the radius of the inscribed, and R the radius of the circumscribing sphere, then we have these general equations, viz,

1st,

$$1st,\ A = r\sqrt{42 - 18\sqrt{5}} = R\sqrt{\frac{10 - 2\sqrt{5}}{5}} =$$

$$\sqrt{\frac{B\sqrt{3}}{15}} = \sqrt[3]{\tfrac{6}{5}}\,C\sqrt{\frac{7 - 3\sqrt{5}}{2}}.$$

$$2d,\ B = 3r^2 \times \sqrt{7\sqrt{3} - 3\sqrt{15}} = 2R^2 \times \sqrt{5\sqrt{3} - \sqrt{15}}$$

$$= 5A^2\sqrt{3} = \sqrt[3]{70\sqrt{3} - 30\sqrt{15}}.$$

$$3d,\ C = 10r^3 \times \sqrt{7\sqrt{3} - 3\sqrt{15}} = \tfrac{2}{5}R^3\sqrt{10 + 2\sqrt{5}}$$

$$= \tfrac{1}{5}A^3\sqrt{\frac{7 + 3\sqrt{5}}{2}} = \frac{B}{18}\sqrt{\frac{7\sqrt{3} + 3\sqrt{15}}{10}}B.$$

$$4th,\ R = r\sqrt{15 - 6\sqrt{5}} = \tfrac{1}{2}A\sqrt{\frac{5 + \sqrt{5}}{2}}$$

$$= \tfrac{1}{2}\sqrt{\frac{5\sqrt{3} + \sqrt{15}}{30}}B = \sqrt[3]{\tfrac{1}{4}}\,C\sqrt{\frac{5 - \sqrt{5}}{10}}.$$

$$5th,\ r = R\sqrt{\frac{5 + 2\sqrt{5}}{15}} = \tfrac{1}{2}A\sqrt{\frac{7 + 3\sqrt{5}}{6}}$$

$$= \tfrac{1}{6}\sqrt{\frac{7\sqrt{3} + 3\sqrt{15}}{10}}B = \tfrac{1}{2}\sqrt[3]{\frac{7\sqrt{3} + 3\sqrt{15}}{30}}C.$$

See my Menfuration, p. 258, 2d edit.

IDES, in the Roman Calendar, a name given to a feries of 8 days in each month; which, in the full months, March, May, July, and October, commenced on the 15th day; and in the other months, on the 13th day; from thence reckoned backward, fo as in thofe four months to terminate on the 8th day, and in the reft on the 6th. Thefe came between the calends and the nones. And this way of counting is ftill ufed in the Roman Chancery, and in the Calendar of the Breviary.

The Ides of May were confecrated to Mercury; the Ides of March were always efteemed unhappy, after the death of Cæfar; the time after the Ides of June was reckoned fortunate for thofe who entered into matrimony; the Ides of Auguft were confecrated to Diana, and were obferved as a feaft by the flaves; on the Ides of September, auguries were taken for appointing the magiftrates, who formerly entered into their offices on the Ides of May, and afterwards on thofe of March.

JET D'EAU, a French word, fignifying a fountain that throws up water to fome height in the air.

A Jet of water is thrown up by the weight of the column of water above its ajutage, or orifice, up to its fource or refervoir; and therefore it would rife to the fame height as the head or refervoir, if certain caufes did not prevent it from rifing quite fo high. For firft, the velocity of the lower particles of the Jet being greater than that of the upper, the lower water ftrikes that which is next above it; and as fluids prefs every way, by its impulfe it widens, and confequently fhortens the column. Secondly, the water at the top of the Jet does not immediately fall off, but forms a kind of ball or head, the weight of which depreffes the Jet; but if the Jet be a little inclined, or not quite upright, it will play higher, though it will not be quite fo beautiful. Thirdly, the friction againft the fides of the pipe and hole of the ajutage, will prevent the Jet from rifing quite fo high, and a fmall one will be more impeded than a large one. And the fourth caufe is the refiftance of the air, which

is proportional to the fquare of the velocity of the water nearly; and therefore the defect in the height will be nearly in the fame proportion, which is alfo the fame as the proportion of the heights of the refervoirs above the ajutage. Hence, and from experience, it is found that a Jet, properly conftructed, will rife to different heights according to the height of the refervoir, as in the following table of the heights of refervoirs and the heights of their corresponding Jets; the former in feet, and the latter in feet and tenths of a foot.

Heights of Reservoirs and their Jets.							
Ref.	Jet.	Ref.	Jet.	Ref.	Jet.	Ref.	Jet.
5	4·9	31	28·3	57	49·0	82	67·0
6	5·9	32	29·2	58	49·7	83	67·7
7	6·8	33	30·0	59	50·5	84	68·4
8	7·8	34	30·8	60	51·2	85	69·1
9	8·7	35	31·6	61	52·0	86	69·8
10	9·7	36	32·5	62	52·7	87	70·5
11	10·6	37	33·3	63	53·5	88	71·1
12	11·6	38	34·1	64	54·2	89	71·8
13	12·5	39	34·9	65	54·9	90	72·5
14	13·4	40	35·7	66	55·7	91	73·2
15	14·3	41	36·6	67	56·4	92	73·8
16	15·2	42	37·4	68	57·1	93	74·5
17	16·1	43	38·1	69	57·8	94	75·2
18	17·0	44	38·9	70	58·6	95	75·8
19	17·9	45	39·8	71	59·3	96	76·5
20	18·8	46	40·5	72	60·0	97	77·2
21	19·7	47	41·3	73	60·7	98	77·8
22	20·6	48	42·1	74	61·4	99	78·5
23	21·5	49	42·9	75	62·1	100	79·1
24	22·3	50	43·7	76	62·8	110	85·6
25	23·2	51	44·4	77	63·5	120	91·9
26	24·1	52	45·2	78	64·2	130	98·0
27	24·9	53	46·0	79	64·9	140	104
28	25·8	54	46·7	80	65·6	150	110
29	26·6	55	47·5	81	66·3	160	116
30	27·5	56	48·2				

By various experiments that have been made by Mariotte, Defaguliers, and others, it has been found, that if the refervoir be 5 feet high, a conduct pipe $1\frac{3}{4}$ inch diameter will admit a hole in the ajutage from $\frac{1}{4}$ to $\frac{3}{8}$ of an inch; and fo on as in the following table:

Height Reservoir.	Diam. of the Ajutage.	Diam. of the Conduct Pipe.
5 feet	$\frac{1}{4}$ to $\frac{3}{8}$ inch	$1\frac{3}{4}$ inch
10	$\frac{1}{4}$ to $\frac{1}{2}$	2
15	$\frac{1}{2}$	$2\frac{1}{4}$
20	$\frac{1}{2}$	$2\frac{1}{2}$
25	$\frac{1}{2}$	$2\frac{3}{4}$
30	$\frac{1}{2}$ to $\frac{3}{4}$	$3\frac{1}{4}$
40	$\frac{3}{4}$	$4\frac{1}{4}$
50	$\frac{3}{4}$	$5\frac{1}{4}$
60	1	$5\frac{1}{4}$ or 6
80	$1\frac{1}{4}$	$6\frac{1}{4}$ or 7
100	$1\frac{1}{4}$ to $1\frac{1}{2}$	7 or 8

But

But the fize of the pipe will be more or lefs with the diftance.

If it be required to keep any number of Jets of given dimenfions playing, by one common conduct-pipe; the diameter of an ajutage muft be found that fhall be equal to all the fmall ones that are given, and from this its proper conduct pipe. Thus, if there be 4 ajutages, each ¼ of an inch diameter; then the fquare of ¼ is $\frac{1}{16}$, which multiplied by 4, the number of them, makes $\frac{4}{16}$, the fquare root of which is $\frac{2}{4}$ or 1½, the diameter of an ajutage equal to all the other four; to which in the table anfwers a pipe of 8 inches diameter. In general, the diameter of the conduct-pipe fhould be about 6 times that of the ajutage.

See Mariotte's Mouvement des Eaux; Defaguliers's Exper. Philof. vol. 2, p. 127, &c; Clare's Motion of Fluids, p. 109; &c.

JETTE, the border made round the ftilts under a pier, in certain old bridges, being the fame with ftarling, confifting of a ftrong framing of timber filled with ftones, chalk, &c; to preferve the foundations of the piers from injury.

IGNIS FATUUS, a common meteor, chiefly feen in dark nights about meadows, marfhes, and other moift places, as alfo in burying grounds, and near dung-hills. It is known among the people by the appellations, Will with a Wifp, and Jack with a Lantern.

Dr. Shaw defcribes a remarkable Ignis Fatuus, which he faw in the Holy Land, that was fometimes globular, or in the form of the flame of a candle; and prefently afterward it fpread itfelf fo much as to involve the whole company in a pale harmlefs light, and then contract itfelf again, and fuddenly difappear. But in lefs than a minute it would become vifible as before; or, running along from one place to another, with a fwift progreffive motion, would expand itfelf at certain intervals over more than 2 or 3 acres of the adjacent mountains. The atmofphere had been thick and hazy, and the dew on the horfes' bridles was uncommonly clammy and unctuous. In the fame weather he obferved thofe luminous appearances, which fkip about the mafts and yards of fhips at fea, and which the failors call corpufanfe, by a corruption of the Spanifh cuerpofanto. Shaw's Travels, p. 363.

Newton calls it a vapour fhining without heat; and fuppofed it to be of the fame nature with the light iffuing from putrefcent fubftances. Willughby and Ray were of opinion that it is occafioned by fhining infects: but all the appearances of it obferved by Derham, Beccaria, and others, fufficiently evince that it muft be an ignited vapour. Inflammable air has been found to be the moft common of all the factitious airs in nature; and that it is the ufual product of the putrefaction and decompofition of vegetable fubftances in water. Signor Volta writes to Dr. Prieftley, that he fires inflammable air by the electric fpark, even when the electricity is very moderate: and he fuppofes that this experiment explains the inflammation of the Ignes Fatui, provided they confift of inflammable air, iffuing from marfhy ground by help of the electricity of fogs, and by falling ftars, which have probably an electrical origin. See Prieftley's Obf. on Air, vol. 3, p. 382; the Philof. Tranf. Abr. vol. 7, p. 147 &c.

Vol. I.

ILLUMINATION, the act or effect of a luminous body, or a body that emits light; fometimes alfo the ftate of another body that receives it.

Circle of ILLUMINATION. See CIRCLE.

ILLUMINATIVE *Lunar Month*, the fpace of time that the moon is vifible, between one conjunction and another.

IMAGE, in Optics, is the fpectre or appearance of an object, made either by reflection or refraction.

In all plane mirrors, the Image is of the fame magnitude as the object; and it appears as far behind the mirror as the object is before it. In convex mirrors, the Image appears lefs than the object; and farther diftant from the centre of the convexity, than from the point of reflection. Mr. Molyneux gives the following rule for finding the diameter of an Image, projected in the diftinct bafe of a convex mirror, viz, As the diftance of the object from the mirror, is to the diftance from the Image to the glafs; fo is the diameter of the object, to the diameter of the Image. See LENS, MIRROR, REFLECTION, and REFRACTION.

IMAGINARY *Quantities*, or Impoffible Quantities, in Algebra, are the even roots of negative quantities; which expreffions are Imaginary, or impoffible, or oppofed to real quantities; as $\sqrt{-aa}$, or $\sqrt[4]{-a^4}$, &c. For, as every even power of any quantity whatever, whether pofitive or negative, is neceffarily pofitive, or having the fign +, becaufe + by +, or — by — give equally +; from hence it follows that every even power, as the fquare for inftance, which is negative, or having the fign —, has no poffible root; and therefore the even roots of fuch powers or quantities are faid to be impoffible or Imaginary. The mixt expreffions arifing from Imaginary quantities joined to real ones, are alfo Imaginary; as $a - \sqrt{-aa}$, or $b + \sqrt{-aa}$.

The roots of negative quantities were, perhaps, firft treated of in Cardan's Algebra. As to the uneven roots of fuch quantities, he fhews that they are negative, and he affigns them: but the even roots of them he rejects, obferving that they are nothing as to common ufe, being neither one thing nor another; that is, they are merely Imaginary or impoffible. And fince his time, it has gradually become a part of Algebra to treat of the roots of negative quantities. Albert Girard, in his *Invention Nouvelle en l'Algebre*, p. 42, gives names to the three forts of roots of equations, calling them, greater than nothing, lefs than nothing, and *envelopee*, as $\sqrt{-3}$: but this was foon after called Imaginary or impoffible, as appears by Wallis's Algebra, p. 264, &c; where he obferves that the fquare root of a negative quantity, is a mean proportional between a pofitive and a negative quantity; as $\sqrt{-bc}$ is the mean proportional between $+b$ and $-c$, or between $-b$ and $+c$; and this he exemplifies by geometrical conftructions. See alfo p. 313.

The arithmetic of thefe Imaginary quantities has not yet been generally agreed upon; viz, as to the operations of multiplication, divifion, and involution; fome authors giving the refults with +, and others on the contrary with the negative fign —. Thus, Euler, in his Algebra, p. 106 &c, makes the fquare of $\sqrt{-3}$ to be —3, of $\sqrt{-1}$ to be —1, &c; and yet he makes the product of two impoffibles, when they are unequal, to be poffible and real: as $\sqrt{-2} \times \sqrt{-3} = \sqrt{6}$;

4 L

$= \sqrt{6}$; and $\sqrt{-1} \times \sqrt{-4} = \sqrt{4}$ or 2. But how can the equality or inequality of the factors cause any difference in the signs of the products?

If $\sqrt{-2} \times \sqrt{-3}$ be $= \sqrt{+6}$, how can $\sqrt{-3} \times \sqrt{-3}$, which is the square of $\sqrt{-3}$, be -3? Again, he makes $\sqrt{-3} \times \sqrt{+5} = \sqrt{-15}$. Also in division, he makes $\sqrt{-4} \div \sqrt{-1}$ to be $= \sqrt{+4}$ or 2; and $\sqrt{+3} \div \sqrt{-3} = \sqrt{-1}$; also that 1 or $\sqrt{+1} \div \sqrt{-1} = \sqrt{\frac{+1}{-1}} = \sqrt{-1}$;

consequently, multiplying the quotient root $\sqrt{-1}$ by the divisor $\sqrt{-1}$, must give the dividend $\sqrt{+1}$; and yet, by squaring, he makes the square of $\sqrt{-1}$, or the product $\sqrt{-1} \times \sqrt{-1}$, equal to -1.

But Emerson makes the product of Imaginaries to be Imaginary; and for this reason, that "otherwise a real product would be raised from impossible factors, which is absurd. Thus, $\sqrt{-a} \times \sqrt{-b} = \sqrt{-ab}$, and $\sqrt{-a} \times -\sqrt{-b} = -\sqrt{-ab}$, &c.

Also $\sqrt{-a} \times \sqrt{-a} = -a$, and $\sqrt{-a} \times -\sqrt{-a} = +a$, &c."

And thus most of the writers on this part of Algebra, are pretty equally divided, some making the product of impossibles real, and others Imaginary.

In the Philof. Tranf. for 1778, p. 318 &c, Mr. Playfair has given an ingenious differtation "On the Arithmetic of Impossible Quantities." But this relates chiefly to the applications and uses of them, and not to the algorithm of them, or rules for their products, quotients, squares, &c. From some operations however here performed, we learn that he makes the product of $\sqrt{-1}$ by $\sqrt{-1}$, or the square of $\sqrt{-1}$, to be -1; and yet in another place he makes the product of $\sqrt{-1}$ and $\sqrt{1-z^2}$ to be $\sqrt{-1+z^2}$. Mr. Playfair concludes, " that Imaginary expressions are never of use in investigations but when the subject is a property common to the measures both of ratios and of angles; but they never lead to any consequence which might not be drawn from the affinity between those measures; and that they are indeed no more than a particular method of tracing that affinity. The deductions into which they enter are thus reduced to an argument from analogy, but the force of them is not diminished on that account. The laws to which this analogy is subject; the cases in which it is perfect, in which it suffers certain alterations, and in which it is wholly interrupted, are capable of being precisely afcertained. Supported on so sure a foundation, the arithmetic of impossible quantities will always remain an useful inftrument in the discovery of truth, and may be of service when a more rigid analysis can hardly be applied. For this reason, many researches concerning it, which in themselves might be deemed absurd, are neve thelefs not destitute of utility. M. Bernoulli has found, for example, that if r be the radius of a circle, the circumference is $= \frac{4 \log. \sqrt{-1}}{\sqrt{-1}} r$. Confidered as a quadrature of the circle, this Imaginary theorem is wholly insignificant, and would defervedly pass for an abuse of calculation; at the same time we learn from it, that if in any equation the quantity $\frac{\log. \sqrt{-1}}{\sqrt{-1}}$

should occur, it may be made to disappear, by the substitution of a circular arch, and a property, common to both the circle and hyperbola, may be obtained. The same is to be observed of the rules which have been invented for the transformation and reduction of impossible quantities*; they facilitate the operations of this imaginary arithmetic, and thereby lead to the knowledge of the most beautiful and extensive analogy which the doctrine of quantity has yet exhibited.

* The rules chiefly referred to, are those for reducing the impossible roots of an equation to the form $A + B\sqrt{-1}$."

IMAGINARY *Roots*, of an equation, are those roots or values of the unknown quantity in an equation, which contain some Imaginary quantity. So the roots of the equation $xx + aa = 0$, are the two Imaginary quantities $+\sqrt{-aa}$ and $-\sqrt{-aa}$, or $+a\sqrt{-1}$ and $-a\sqrt{-1}$; also the two roots of the equation $xx + ax + aa = 0$, are the Imaginary quantities $\pm \frac{1}{2}a - \frac{1}{2}a\sqrt{-3}$; and the three roots of the equation $x^3 - 1 = 0$, or $x^3 = 1$, are 1 and $\frac{-1 + \sqrt{-3}}{2}$ and $\frac{-1 - \sqrt{-3}}{2}$, the first real, and the two latter Imaginary. Sometimes too the real root of an equation may be expressed by Imaginary quantities; as in the irreducible case of cubic equations, when the root is expressed by Cardan's rule; and that happens whenever the equation has no Imaginary roots at all; but when it has two Imaginary roots, then the only real root is expressed by that rule in an Imaginary form. See my paper on Cubic Equations, in the Philof. Tranf. for 1780, p. 406 &c.

Albert Girard first treated expressly on the impossible or Imaginary roots of equations, and shewed that every equation has as many roots, either real or Imaginary, as the index of the highest power denotes. Thus, the roots of the biquadratic equation $x^4 = 4x - 3$, he shews are two real and two Imaginary, viz, 1, 1, $-1 + \sqrt{-2}$, and $-1 - \sqrt{-2}$; and he renders the relation general, between all the roots and the coefficients of the terms of the equation. See his *Invention Nouvelle en l Algebre*, anno 1629, theor. 2, pa. 40 &c.

M. D'Alembert demonftrated, that every Imaginary root of any equation can always be reduced to the form $e + f\sqrt{-1}$, where e and f are real quantities. And hence it was also shewn, that if one root of an equation be - $e + f\sqrt{-1}$, another root of it will always be - $e - f\sqrt{-1}$: and hence it appears that the number of the Imaginary roots in any equation is always even, if any; i. e. either none, or else two, or four, or six, &c. Memoirs of the Academy of Berlin, 1746.

To discover how many impossible roots are contained in any proposed equation, Newton gave this rule, in his Algebra, viz, Constitute a series of fractions, whose denominators are the series of natural numbers 1, 2, 3, 4, 5, &c, continued to the number shewing the index or exponent of the highest term of the equations, and their numerators the same series of numbers in the contrary order: and divide each of these fractions by that next before it, and place the resulting quotients over the intermediate terms of the equation; then under each of the intermediate terms, if its square multiplied

by

by the fraction over it, be greater than the product of the terms on each side of it, place the sign $+$; but if not, the sign $-$; and under the first and last term place the sign $+$. Then will the equation have as many Imaginary roots as there are changes of the underwritten signs from $+$ to $-$, and from $-$ to $+$. So for the equation $x^3 - 4x^2 + 4x - 6 = 0$, the series of fractions is $\frac{3}{1}, \frac{2}{2}, \frac{1}{3}$; then the second divided by the first gives $\frac{2}{6}$ or $\frac{1}{3}$, and the third divided by the second gives $\frac{1}{4}$ also; hence these quotients placed over the intermediate terms, the whole will stand

thus, $x^3 - \overset{\frac{1}{3}}{4x^2} + \overset{\frac{1}{3}}{4x} - 6$.
$\quad\quad + \quad + \quad - \quad +$

Now because the square of the 2d term multiplied by its superscribed fraction, is $\frac{16}{3}x^4$, which is greater than $4x^4$ the product of the two adjacent terms, therefore the sign $+$ is set below the 2d term; and because the square of the 3d term multiplied by its overwritten fraction, is $\frac{16}{3}x^2$, which is less than $24x^2$ the product of the terms on each side of it, therefore the sign $-$ is placed under that term; also the sign $+$ is set under the first and last terms. Hence the two changes of the underwritten signs $+ + - +$, the one from $+$ to $-$, and the other from $-$ to $+$, shew that the given equation has two impossible roots.

When two or more terms are wanting together, under the place of the 1st of the deficient terms write the sign $-$, under the 2d the sign $+$, under the 3d $-$, and so on, always varying the signs, except that under the last of the deficient terms must always be set the sign $+$, when the adjacent terms on both sides of the deficient terms have contrary signs. As in the equation
$x^5 + ax^4 * * * + a^5 = 0$,
$\quad + \quad + - + - \quad +$
which has four Imaginary roots.

The author remarks, that this rule will sometimes fail of discovering all the impossible roots of an equation, for some equations may have more of such roots than can be found by this rule, tho' this seldom happens.

Mr. Maclaurin has given a demonstration of this rule of Newton's, together with one of his own, that will never fail. And the same has also been done by Mr. Campbell. See Philos. Trans. vol. 34, p. 104, and vol. 35, p. 515.

The real and imaginary roots of equations may be found from the method of fluxions, applied to the doctrine of maxima and minima, that is, to find such a value of x in an equation, expressing the nature of a curve, made equal to y, an abscissa which corresponds to the greatest and least ordinate. But when the equation is above 3 dimensions, the computation is very laborious. See Stirling's treatise on the lines of the 3d order, Schol. pr. 8, pa. 59, &c.

IMBIBE, is commonly used in the same sense as absorb, viz, where a dry porous body takes up another that is moist.

IMMENSE, that whose amplitude or extension cannot be equalled by any measure whatsoever, or how often soever repeated.

IMMERSION, the act of plunging into water, or some other fluid.

IMMERSION, in Astronomy, is when a star, or planet comes so near the sun, that it cannot be seen; being as it were enveloped, and hid in the rays of that luminary.

IMMERSION also denotes the beginning of an eclipse, or of an occultation, when the body, or any part of it just begins to disappear, either behind the edge of another body, or in its shadow. As, in an eclipse of the moon, when she begins to be darkened by entering into the shadow of the earth: or the beginning of an eclipse of the sun, when the moon's disc just begins to cover him: or the beginning of the eclipses of any of the satellites, as those of Jupiter, by entering into his shadow: or, lastly, the beginning of an occultation of any star or planet, by passing behind the body of the moon or another planet. In all these cases, the darkened body is said to immerge, or to be immerged, or begin to be hid, by dipping as it were into the shade. In like manner, when the darkened body begins to appear again, it is said to emerge, or come out of darkness again.

IMPACT, the simple or single action of one body upon another to put it in motion. Point of Impact, is the place or point where a body acts.

IMPENETRABILITY, a quality by which a thing cannot be pierced or penetrated; or a property of body by which it fills up certain spaces, so that there is no room in them for any other body.

IMPENETRABLE, that cannot be penetrated.

IMPERFECT *Number*, is that whose aliquot parts, taken all together, do not make a sum that is equal to the number itself, but either exceed it, or fall short of it; being an abundant number in the former case, and a defective number in the latter. Thus, 12 is an abundant Imperfect number, because the sum of all its aliquot parts, 1, 2, 3, 4, 6, makes 16, which exceeds the number 12. And 10 is a defective Imperfect number, because its aliquot parts, 1, 2, 5, taken all together, make only 8, which is less than the number 10 itself.

IMPERIAL *Table*, is an instrument made of brass, with a box and needle, and staff, &c, used for measuring of land.

IMPERVIOUS, not to be pervaded or entered either because of the closeness of the pores, or the particular configuration of its parts.

IMPETUS, in Mechanics, force, momentum, motion, &c.

IMPOSSIBLE *Quantity*, or *Root*, the same as IMAGINARY ones; which see.

IMPOST, in Architecture, a capital or plinth, to a pillar, or pilaster, or pier, that supports an arch, &c.

IMPROPER *Fraction*, is a fraction whose numerator is either equal to, or greater than, its denominator. As $\frac{5}{5}$ or $\frac{7}{3}$ or $\frac{19}{7}$. An Improper fraction is reduced to a whole or mixt number, by dividing the numerator by the denominator; the quotient is the integer, and the remainder set over the divisor makes the fractional part of the value of the original Improper fraction. Thus $\frac{4}{4} = 1$, and $\frac{5}{3} = 1\frac{2}{3}$, and $\frac{19}{7} = 3\frac{1}{7}$. So that when the numerator is just equal to the denominator, the Improper fraction is exactly equal to unity or 1; but when the numerator is the greater, the fraction is greater than 1.

4 L 2

IMPULSE,

IMPULSE, the single or momentary action or force by which a body is impelled; in contradistinction to continued forces; like the blow of a hammer, &c.

IMPULSIVE, a term applied to actions by impulse.

INACCESSIBLE *Height* or *Distance*, is that which cannot be approached, or measured by actual measurement, by reason of some impediment in the way; as water, &c.

See HEIGHTS and DISTANCES,

INCEPTIVE, of *Magnitude*, a term used by Dr. Wallis, to express such moments, or first principles, as, though of no magnitude themselves, are yet capable of producing such as are. See INFINITE, and INDIVISIBLE. Thus, a point has no magnitude itself, but is inceptive of a line, which it produces by its motion. Also a line, though it has no breadth, is yet Inceptive of breadth; that is, it is capable, by its motion, of producing a surface, which has breadth.

INCH, a common English measure, being the 12th part of a foot, or 3 barley corns in length.

INCIDENCE, or *line of* INCIDENCE, in Mechanics, implies the direction or inclination in which one body strikes or acts on another.——In the incursions of two moving bodies, their Incidence is said to be Direct or Oblique, as the directions of their motion make a straight line, or an angle at the point of Impact.

Angle of INCIDENCE, by some writers, denotes the angle comprehended between the line of Incidence, and a perpendicular to the body acted on at the point of Incidence. Thus, suppose AB an Incident line, and BF a perpendicular to the plane CB at the incident point B; then ABF is the angle of Incidence, or of inclination.

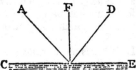

But, according to Dr. Barrow, and some other writers, the Angle of Incidence is the complement of the former, or the angle made between the incident line, and the plane acted on, or a tangent at the point of Incidence; as the angle ABC.

It is demonstrated by optical writers, 1st, That the Angle of Incidence, of the rays of light, is always equal to the angle of reflection; and that they lie in the same plane. And the same is proved by the writers on Mechanics, concerning the reflection of elastic bodies. That is, the ∠ABF = the ∠FBD, or the ∠ABC = the ∠DBE.—2d, That the sines of the Angles of Incidence and refraction are to each other, either accurately, or very nearly, in a given or constant ratio.—3dly, That from air into glass, the sine of the Angle of Incidence, is to the sine of the angle of refraction, as 300 to 193, or nearly as 14 to 9: and, on the other hand, that out of glass into air, the sign of the Angle of Incidence, is to the sine of the angle of refraction, as 193 to 300, or as 9 to 14 nearly.

INCIDENCE *of Eclipse.* See ECLIPSE and IMMERSION.

Axis of INCIDENCE, is the line FB perpendicular to the reflecting plane at the point of Incidence B.

Cathetus of INCIDENCE. See CATHETUS, and REFLECTION.

Line of INCIDENCE, in Catoptrics, denotes a right line, as AB, in which light is propagated from a radiant point A, to a point B, in the surface of a speculum. The same line is also called an Incident ray.

Line of INCIDENCE, in Dioptrics, is a right line, as AB, in which light is propagated unrefracted, in the same medium, from the radiant point to the surface of the refracting body, CBE.

Point of INCIDENCE, is the point B on the surface of the reflecting or refracting medium, on which the Incident ray falls.

Scruples of INCIDENCE. See SCRUPLES.

INCIDENT *Ray*, is the line or ray AB, falling on the surface of any body, at B.

INCLINATION, in Geometry, Mechanics, or Physics, denotes the mutual tendency of two lines, planes, or bodies, towards one another; so that their directions make at the point of concourse some certain angle.

INCLINATION *of the Axis of the Earth*, is the angle it makes with the plane of the ecliptic; or the angle between the planes of the equator and ecliptic.

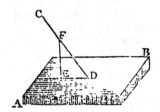

INCLINATION *of a Line to a plane*, is the acute angle, as CDE, which the line CD makes with another line DE drawn in the plane through the incident point D and the foot of a perpendicular E from any point of the line upon the plane.

INCLINATION *of an Incident ray*, is the angle of inclination, or angle of incidence.

INCLINATION *of the Magnetical needle.* See DIPPING *Needle.*

INCLINATION *of Meridians*, in Dialling, is the angle that the hour-line on the globe, which is perpendicular to the dial-plane, makes with the meridian.

INCLINATION *of the Orbit of a planet*, is the angle formed by the planes of the ecliptic and of the orbit of the planet. The quantity of this Inclination for the several planets, is as follows, viz.

Mercury	-	-	6°	54'
Venus	-	-	3	20
Earth	-	-	0	0
Moon	-	-	5	18
Mars	-	-	1	52
Jupiter	-	-	1	20
Saturn	-	-	2	30
Herschel	-	-	0	48

INCLINATION *of a Plane*, in Dialling, is the arch of a vertical, a vertical circle, perpendicular both to the plane and the horizon, and intercepted between them.

INCLINATION *of a Planet*, is the arch or angle comprehended between the ecliptic and the place of the planet in its orbit. The greateſt Inclination, or declination, is the ſame as the Inclination of the orbit; which ſee above.

INCLINATION *of a Reflected ray*, is the angle which a ray after reflection makes with the axis of Inclination; as the angle FBD, in the laſt fig. but one.

INCLINATION *of Two Planes*, is the angle made by two lines drawn in thoſe planes perpendicular to their common interſection, and meeting in any point of that interſection.

Angle of INCLINATION, is the ſame as what is otherwiſe called the angle of incidence.

Argument of INCLINATION. See ARGUMENT.

INCLINED *Plane*, in Mechanics, is a plane inclined to the horizon, or making an angle with it. It is one of the ſimple mechanic powers, and the double inclined plane makes the wedge.

1. The power gained by the Inclined plane, is in proportion as the length of the plane is to its height, or as radius to the ſine of its inclination; that is, a given weight hanging freely, will balance upon the plane another weight, that ſhall be greater in that proportion. So, when the greater weight W on the plane, is balanced by the leſs weight *w* hanging perpendicularly, then is *w* : W :: BC : AC :: ſin. ∠A : radius. Or, in other words, the relative gravity of a body upon the plane, or its force in deſcending down the plane, is to its abſolute gravity or weight, in the ſame proportion of the height of the plane to its length, or of the ſine of inclination to radius.

2. Hence therefore the relative gravities of the ſame body on different inclined planes, or their forces to deſcend down the planes, are to each other, as the ſines of the angles of inclination, to radius 1, or directly as the heights of the planes, and inverſely as their lengths.

3. Hence, if the planes have the ſame height, and abſolute weights of the bodies be directly proportional to the lengths of the planes, then the forces to deſcend will be equal. Conſequently, if the bodies be then connected by a ſtring acting parallel to the planes, they will exactly balance each other; as in the annexed figure.

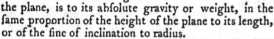

4. The relative force of gravity upon the plane being in a conſtant ratio to the abſolute weight of the body, viz, as ſine of inclination to radius; therefore all the laws relating to the perpendicular free deſcents of bodies by gravity, hold equally true for the deſcents on inclined planes; ſuch as, that the motion is a uniformly accelerated one; that the velocities are directly as the times, and the ſpaces as the ſquare of either of them; uſing only the relative force upon the plane for the abſolute weight of the body, or inſtead of $32\frac{1}{4}$ feet, the velocity generated by gravity in the firſt ſecond of time, uſing $32\frac{s}{4}s$, where *s* is the ſine of the inclination to the radius 1.

5. The velocity acquired by a body in deſcending down an Inclined plane AC, when the body arrives at A, is the ſame as the velocity acquired by deſcending freely down the perpendicular altitude BC, when it arrives at B. But the times are very different; for the time of deſcending down the Inclined plane, is greater than down the perpendicular, in the

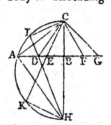

ſame proportion as the length of the plane AC, is to the height CB: and ſo the time of deſcending from any point C to a horizontal line or plane ABG &c, down any oblique line, or Inclined plane, is directly proportional to the length of that plane, CA, or CD, or CE, or CB, or CF, &c.

6. Hence, if there be drawn AH perpendicular to AC, meeting CB produced in H; then the time of deſcending down any plane CA, is equal to the time of deſcending down the perpendicular CH. So that, if upon CH as a diameter a circle be deſcribed, the times of deſcent will be exactly equal, down every chord in the circle, beginning at C, and terminating any where in the circumference, as CI, CA, CK, CH, &c, or beginning any where in the circumference, and terminating at the loweſt point of the circle, as CH, IH, AH, KH, &c.

7. When bodies aſcend up Inclined planes, their motion is uniformly retarded; and all the former laws for deſcents, or the generation of motion, hold equally true for aſcents, or the deſtruction of as much motion.

INCLINED *Towers*, are towers inclined, or leaning out of the perpendicular. See TOWERS.

INCLINERS, in Dialling, are inclined dials. See DIAL.

INCOMMENSURABLE, Lines, or Numbers, or Quantities in general, are ſuch as have no common meaſure, or no line, number, or quantity of the ſame kind, that will meaſure or divide them both without a remainder. Thus, the numbers 15 and 16 are Incommenſurable, becauſe, though 15 can be meaſured by 3 and 5, and 16 by 2, 4, and 8, there is yet no ſingle number that will divide or meaſure them both.

Euclid demonſtrates (prop. 117, lib. 10) that the ſide of a ſquare and its diagonal are Incommenſurable to each other. And Pappus, prop. 17, lib. 4, ſpeaks of Incommenſurable angles.

INCOMMENSURABLE *in Power*, is ſaid of quantities whoſe 2d powers, or ſquares, are Incommenſurable. As $\sqrt{2}$ and $\sqrt{3}$, whoſe ſquares are 2 and 3, which are Incommenſurable. It is commonly ſuppoſed that the diameter and circumference of a circle are Incommenſurable to each other; at leaſt their commenſurability has never been proved. And Dr. Barrow ſurmiſes even that they are infinitely Incommenſurable, or that all poſſible powers of them are Incommenſurable.

INCOMPOSITE *Numbers*, are the ſame with thoſe called by Euclid prime numbers, being ſuch as are not compoſed by the multiplication together of other numbers. As 3, 5, 7, 11, &c.

INCREMENT,

INCREMENTS, is the small increase of a variable quantity. Newton, in his Treatise on Fluxions, calls these by the name Moments, and observes that they are proportional to the velocity or rate of increase of the flowing or variable quantities, in an indefinitely small time; he denotes them by subjoining a cipher o, to the flowing quantity whose moment or Increment it is; thus o the moment of x. In the doctrine of Increments, by Dr. Brooke Taylor and Mr. Emerson, they are denoted by points below the variable quantities; as x. Some have also denoted them by accents underneath the letter, as : but it is now more usual to express them by accents over the same letter; as .

INCREMENTS, *Method of*, a branch of Analytics, in which a calculus is founded on the properties of the successive values of variable quantities, and their differences, or Increments.

The inventor of the Method of Increments was the learned Dr. Taylor, who, in the year 1715, published a treatise upon it; and afterwards gave some farther account and explication of it in the Philos. Transf. as applied to the finding the sums of series. And another ingenious and easy treatise on the same, was published by Mr. Emerson, in the year 1763. The method is nearly allied to Newton's Doctrine of Fluxions, and arises out of it. Also the Differential method of Mr. Stirling, which he applies to the summation and interpolation of series, is of the same nature as the Method of Increments, but not so general and extensive.

From the Method of Increments, Mr. Emerson observes, " The principal foundation of the Method of Fluxions may be easily derived. For as in the Method of Increments, the Increment may be of any magnitude, so in the Method of Fluxions, it must be supposed infinitely small; whence all preceding and successive values of the variable quantity will be equal, from which equality the rules for performing the principal operations of fluxions are immediately deduced. That I may give the reader, continues he, a more perfect idea of the nature of this method: suppose the abscissa of a curve be divided into any number of equal parts, each part of which is called the Increment of the abscissa; and imagine so many parallelograms to be erected thereon; either circumscribing the curvilineal figure, or inscribed in it; then the finding the sum of all these parallelograms is the business of the Method of Increments. But if the parts of the abscissa be taken infinitely small, then these parallelograms degenerate into the curve; and then it is the business of the Method of Fluxions, to find the sum of all, or the area of the curve. So that the Method of Increments finds the sum of any number of finite quantities; and the Method of Fluxions the sum of any infinite number of infinitely small ones: and this is the essential difference between these two methods." Again, " There is such a near relation between the Method of Fluxions, and that of Increments, that many of the rules for the one, with little variation, serve also for the other. And here, as in the Method of Fluxions, some questions may be solved, and the integrals found, in finite terms; whilst in others we are forced to have recourse to infinite series for a solution. And the like difficulties will occur in the Method of Increments, as usually happen in Fluxions. For whilst some fluxionary quantities have no fluents, but what are expressed by series; so some Increments have no integrals, but what infinite series afford; which will often, as in fluxions, diverge and become useless."

By means of the Method of Increments, many curious and useful problems are easily resolved, which scarcely admit of a solution in any other way. As, suppose several series of quantities be given, whose terms are all formed according to some certain law, which is given; the Method of Increments will find out a general series, which comprehends all particular cases, and from which all of that kind may be found.

The Method of Increments is also of great use in finding any term of a series proposed: for the law being given by which the terms are formed; by means of this general law, the Method of Increments will help us to this term, either expressed in finite quantities, or by an infinite series.

Another use of the Method of Increments, is to find the sums of series; which it will often do in finite terms. And when the sum of a series cannot be had in finite terms, we must have recourse to infinite series; for the integral being expressed by such a series, the sum of a competent number of its terms will give the sum of the series required. This is equivalent to transforming one series into another, converging quicker: and sometimes a very few terms of this series will give the sum of the series sought.

Definitions in the Method of Increments.

1. When a quantity is considered as increasing, or decreasing, by certain steps or degrees, it is called an Integral.

2. The increase of any quantity from its present value, to the next succeeding value, is called an Increment: or, if it decreases, a Decrement.

3. The increase of any Increment, is the Second Increment; and the increase of the 2d Increment, is the 3d Increment; and so on.

4. Succeeding Values, are the several values of the integral, succeeding one another in regular order, from the present value; and Preceding Values, are such as arise before the present value. All these are called by the general term Factors.

5. A Perfect quantity is such as contains any number of successive values without intermission; and a Defective quantity, is that which wants some of the successive values. Thus $x\, x\, x\, x$ is a Perfect quantity; and $x\, x\, x$, an Imperfect or defective one.

Notation. This, according to Mr. Emerson's method, is as follows:

1. Simple Integral quantities are denoted by any letters whatever, as z, y, x, u, &c.

2. The several values of a simple integral, are denoted by the same letter with small figures under them: so if z be an integral, then z, z, z, z, &c are the present value, and the 1st, 2d, 3d, &c, successive values of it; and the preceding values are denoted by figures with negative signs, thus z, z, z, z, are the 1st, 2d,

3d,

3d, 4th preceding values; and the figure denoting any value, is the characteristic.

3. The Increments are denoted with the same letters, and points under them: thus, $\dot{x}$ is the Increment of x, and $\dot{z}$ is the increment of z. Also $\overset{\shortmid}{x}$ is the Increment of $\overset{\shortmid}{x}$; and $\overset{\shortparallel}{x}$ of $\overset{\shortparallel}{x}$, &c.

4. The 2d, 3d, and other Increments, are denoted with two, three, or more points: so $\ddot{z}$ is the 2d Increment of z, and $\dddot{z}$ is the 3d Increment of z, and so on. And these are denominated Increments of such an order, according to the number of points.

5. If $\dot{x}$ be any Increment, then $[\dot{x}]$ is the integral of it; also $^2[\dot{x}]$ denotes the integral of $[\dot{x}]$, or the 2d integral of $\dot{x}$; and $^3[\dot{x}]$ is the 3d integral of $\dot{x}$, or an integral of the 3d order, &c.

6. Quantities written thus, $x \ldots x$ mean the same as $x\,x\,x\,x$, or signify that the quantities are continued from the first to the last, without break or interruption.

To find the Increment of any Integral, or variable quantity.

Rule 1. If the proposed quantity be not fractional, and be a perfect integral, consisting of the successive values of the variable quantity which increases uniformly: Multiply the proposed integral by the number of factors, and change the lowest factor for an Increment. So the Increment of $a - 3x + 6z$ is $-3\dot{x} + 6\dot{z}$; for the Increment of the constant quantity a is 0 or nothing. So likewise,

The Increment of $c\,x\,x\,x\,x$, is $4c\dot{x}\,x\,x\,x$.

The Increment of $ax\,x\,x$, is $3ax\,x\,\dot{x}$.

The Increment of $x \ldots x$, is $\overline{m+n+1}.\dot{x}\,x \ldots x$.

Rule 2. In fractional quantities, where the denominator is perfect, and the variable quantity increases uniformly: Multiply the proposed integral by the number of factors, and by the constant Increment with a negative sign, and take the next succeeding value into the denominator. Thus,

The Increment of $\dfrac{a}{x \ldots x}$, is $\dfrac{-5a\dot{x}}{x \ldots x}$.

The Increment of $\dfrac{c}{x \ldots x}$, is $\dfrac{-9c\dot{x}}{x \ldots x}$.

Rule 3. The Increment of any power, as x^n is $\overline{x+\dot{x}}|^n - x^n$; that is, the difference between the present value x^n and the next succeeding value $\overline{x+\dot{x}}|^n$. And generally, the Increment of any quantity whatever, is found by subtracting the present value, or the given quantity, from its next succeeding value. Also by expanding the compound quantity in a series, and subtracting x^n from it, the Increment will be either

$$= nx^{n-1}\dot{x} + \frac{n.n-1}{2}x^{n-2}\dot{x}^2 + \frac{n.n-1.n-2}{2.3}x^{n-3}\dot{x}^3 \text{ &c; or}$$

$$= x^n \times : \frac{n\dot{x}}{x} + \frac{n.n-1}{2}\cdot\frac{\dot{x}^2}{x^2} + \frac{n.n-1.n-2}{2.3}\cdot\frac{\dot{x}^3}{x^3} \text{ &c.}$$

So the Increment of x^4, is $\overline{x+\dot{x}}|^4 - x^4 = 4x^3\dot{x} + 6x^2\dot{x}^2 + 4x\dot{x}^3 + \dot{x}^4$.

The Increment of $\dfrac{1}{x^3}$ or x^{-3}

is $\dfrac{1}{\overline{x+\dot{x}}|^3} - \dfrac{1}{x^3} = \overline{x+\dot{x}}|^{-3} - x^{-3}$

$= -3x^{-4}\dot{x} + 6x^{-5}\dot{x}^2 - 10x^{-6}\dot{x}^3$ &c

$= -\dfrac{3\dot{x}}{x^4} + \dfrac{6\dot{x}^2}{x^5} - \dfrac{10\dot{x}^3}{x^6}$ &c.

The Increment of a^x, a being constant, is $a^{\overset{\shortmid}{x}} - a^x = a^{x+\dot{x}} - a^x = a^x.(a^{\dot{x}} - 1)$.

The Increment of $\dfrac{1}{a^x}$ is $\dfrac{1}{a^{\overset{\shortmid}{x}}} - \dfrac{1}{a^x}$

$= \dfrac{a^x - a^{\overset{\shortmid}{x}}}{a^x a^{\overset{\shortmid}{x}}} = \dfrac{a^x - a^{x+\dot{x}}}{a^x a^{x+\dot{x}}} = \dfrac{1 - a^{\dot{x}}}{a^{x+\dot{x}}}$.

The Increment of xz is $\overset{\shortmid}{x}\overset{\shortmid}{z} - xz = \overline{x+\dot{x}}.\overline{z+\dot{z}} - xz = z\dot{x} + x\dot{z} + \dot{x}\dot{z}$.

And so on for any form of Integral whatever, subtracting the given quantity from its next succeeding value. So,

The Increment of the log. of x is log. $\overset{\shortmid}{x}$ − log. x $= $ log. $\overline{x+\dot{x}}$ − log. $x = $ log. $\dfrac{x+\dot{x}}{x}$, which, by the nature of logarithms, is

$\dfrac{\dot{x}}{x} - \dfrac{\dot{x}^2}{2x^2} + \dfrac{\dot{x}^3}{3x^3} - \dfrac{\dot{x}^4}{4x^4}$ &c.

Schol. From hence may be deduced the principles and rules of fluxions; for the method of fluxions is only a particular case of the method of Increments, fluxions being infinitely small Increments; therefore if in any form of Increments the Increment be taken infinitely small, the form or expression will be changed into a fluxional one.

Thus, in $z\dot{x} + x\dot{z} + \dot{x}\dot{z}$, which is the Increment of the rectangle xz, if $\dot{x}$ and $\dot{z}$ be changed for $\dot{x}$ and $\dot{z}$, the expression will become $z\dot{x} + x\dot{z} + \dot{x}\dot{z}$ for the fluxion of xz, or only $z\dot{x} + x\dot{z}$, because $\dot{x}\dot{z}$ is infinitely less than the rest.

So likewise, if $\dot{x}$ be changed for $\dot{x}$ in this

$$nx^{n-1}\dot{x} + \frac{n.n-1}{2}x^{n-2}\dot{x}^2 + \frac{n.n-1.n-2}{2.3}x^{n-3}\dot{x}^3$$

&c, which is the Increment of x^n, it becomes

$$nx^{n-1}\dot{x} + \frac{n.n-1}{2}x^{n-}\dot{x}^2 + \frac{n.n-1.n-2}{2.3}x^{n-3}\dot{x}^3$$

&c, or only $nx^{n-1}\dot{x}$, for the fluxion of the power x^n, as all the terms after the first will be nothing, because $\dot{x}^2$ and ^{3} &c are infinitely less than $\dot{x}$.

And thus may all the other forms of fluxions be derived from the corresponding Increments. And in like manner, the finding of the integrals, is only a more general way of finding fluents, as appears in what follows.

To find out the Integral of any given Increment.

Rule 1. When the variable quantity increases uniformly, and the proposed integral consists of the successive values of it multiplied together, or is a perfect Increment not fractional: Multiply the given Increment

ment by the next preceding value of the variable quantity, then divide by the new number of factors, and by the constant Increment.

Ex. Thus, the integral of $4cyxxx$ is $cxxxx$.
123   123

The integral of $3ax \; x \; x$ is $ax \; x \; x$.

Rule 2. In a fractional-expression, where the variable quantity increases uniformly, and the denominator is perfect, containing the successive values of the variable quantity: Throw out the greatest value of the variable letter, then divide by the new number of factors, and by the constant Increment with a negative sign. So,

The integral of $\dfrac{ax}{x \ldots x}$ is $\dfrac{-a}{5x \ldots x}$.

The integral of $\dfrac{-cx}{x \ldots \ldots x}$ is $\dfrac{c}{x \ldots \ldots x}$.

Rule 3. Various other particular rules are given, but these and the two foregoing are all best included in the following general table of the most useful forms of Increments and integrals, to be used in the same way as the similar table of fluxions and fluents, to which these correspond.

Forms	Increments	Integrals
	A Table of Increments and their Integrals.	
1	x when constant not constant	x, or x, or x, or x &c. x only.
2	$\dfrac{x \ldots \ldots xx}{m}$ x constant	$\dfrac{x \ldots \ldots x}{m+n}$
3	$\dfrac{ax}{x \ldots \ldots x}$ $\overline{m}$ x constant	$\dfrac{-a}{m+n-2.x \ldots \ldots x}$
4	$zx + xz$	xz
5	$\dfrac{zx - xz}{zz}$	$\dfrac{x}{z}$
6	a^x x given	$\dfrac{a^x}{a^x - 1}$

Integrals, when found from given Increments, are corrected in the very same way as fluents when found from given fluxions, viz. instead of every several variable quantity in the integral, substituting such a determinate value of them as they are known to have in some particular case; and then subtracting each side of the resulting equation from the corresponding side of the integral, the remaining equation will be the correct form of the integrals.

For an example of the use of the Method of Increments, suppose it were required to find the sum of any number of terms of the series $1.2 + 2.3 + 3.4 + 4.5$ &c.

Let x be the number of the terms, and z the sum of them;

Then, by the progression of the series, the last or the x term is xx, and the next term after that will be xx, that is $z = xx$, where $x = 1$. Hence the integral is $z = \frac{1}{3}x \, xx = \frac{1}{3}x \cdot x + 1 \cdot x + 2$, which is the sum of x terms of the given series. So if the number of terms x be 10, this becomes $\frac{1}{3}.10.11.12 = 440$, which is the sum of 10 terms of the given series $1.2 + 2.3 + 3.4$ &c. Or, when $x = 100$, the sum of 100 terms of the same series is $\frac{1}{3}.100.101.102 = 100.101.34 = 343400$.

Again, to find the sum z of n terms of the series $\dfrac{1}{1.3.5} + \dfrac{1}{3.5.7} + \dfrac{1}{5.7.9}$ &c.

Here the nth term is $\dfrac{1}{2n-1.2n+1.2n+3}$. Put $x = 2n-1$; then is $x = 2n = 2$, and the nth term is $\dfrac{1}{x.x + 2.x + 4} = \dfrac{1}{xxx}$; and the $n + 1$th term or z is $\dfrac{1}{xxx}$; the general integral of which is $z = \dfrac{-1}{2xxx} = \dfrac{-1}{4xx}$.

But this wants a correction; for when $n = 0$ or no terms, then $x = -1$, and the sum $z = 0$, and the integral becomes z or $0 = \dfrac{-1}{4.1.3} = \dfrac{-1}{12}$; that is $-\dfrac{1}{12}$ is the correction, and being subtracted, the correct state of the integrals becomes

$z = \dfrac{1}{12} - \dfrac{1}{4xx} = \dfrac{1}{12} - \dfrac{1}{4.2n + 1.2n + 3}$, which is the sum of n terms of the proposed series. And when n is infinite, the latter fraction is nothing, and the sum of the infinite series, or the infinite number of the terms, is accurately $\dfrac{1}{12}$.

When $n = 100$, the sum of 100 terms of the series becomes $z = \dfrac{1}{12} - \dfrac{1}{4.201.203} = \dfrac{1}{12} - \dfrac{1}{163212}$.

For more ample information and application on this science, see Emerson's Increments, Taylor's Methodus Incrementorum, and Stirling's Summatio & Interpolatio Serierum.

INCURVATION *of the rays of Light.* See LIGHT, and REFRACTION.

INDEFINITE, *Indeterminate*, that which has no certain bounds, or to which the human mind cannot affix any. Des Cartes uses the word, in his Philosophy, instead of infinite, both in numbers and quantities, to signify an inconceivable number, or number so great, that an unit cannot be added to it; and a quantity so great, as not to be capable of any addition. Thus, he says, the stars, visible and invisible, are in number Indefinite, and not, as the Ancients held, infinite; and that

that quantity may be divided into an Indefinite number of parts, not an infinite number.

Indefinite is now commonly ufed for indeterminate, number or quantity, that is, a number or quantity in general, in contradiftinction from fome particular known and given one.

INDETERMINED, or INDETERMINATE, in Geometry, is underftood of a quantity, which has no certain or definite bounds.

INDETERMINATE *Problem*, is that which admits of innumerable different folutions, and fometimes perhaps only of a great many different anfwers; otherwife called an unlimited problem.

In problems of this kind the number of unknown quantities concerned, is greater than the number of the conditions or equations by which they are to be found; from which it happens that generally fome other conditions or quantities are affumed, to fupply the defect, which being taken at pleafure, give the fame number of anfwers as varieties in thofe affumptions.

As, if it were required to find two fquare numbers whofe difference fhall be a given quantity d. Here, if x^2 and y^2 denote the two fquares, then will $x^2 - y^2 = d$, by the queftion, which is only one equation, for finding two quantities. Now by affuming a third quantity z fo that $z = x + y$ the fum of the two roots; then is $x = \dfrac{z^2 + d}{2z}$, and $y = \dfrac{z^2 - d}{2z}$, which are the two roots having the difference of their fquares equal to the given quantity d, and are expreffed by means of an affumed quantity z; fo that there will be as many anfwers to the queftion, as there can be taken values of the Indeterminate quantity z, that is, innumerable.

Diophantus was the firft writer on Indeterminate problems, viz, in his Arithmetic or Algebra, which was firft publifhed in 1575 by Xilander, and afterwards in 1621 by Bachet, with a large commentary, and many additions to it. His book is wholly upon this fubject; whence it has happened, that fuch kind of queftions have been called by the name of Diophantine problems. Fermat, Des Cartes, Frenicle, in France, and Wallis and others in England, particularly cultivated this branch of Algebra, on which they held a correfpondence, propofing difficult queftions to each other; an inftance of which are thofe two curious ones, propofed by M. Fermat, as a challenge to all the mathematicians of Europe, viz 1ft, To find a cube number which added to all its aliquot parts fhall make a fquare number; and 2d, To find a fquare number which added to all its aliquot parts fhall make a cubic number; which problems were anfwered after feveral ways by Dr. Wallis, as well as fome others of a different nature. See the Letters that paffed between Dr. Wallis, the lord Brounker, Sir Kenelm Digby, &c, in the Doctor's Works; and the Works of Fermat, which were collected and publifhed by his fon. Moft authors on Algebra have alfo treated more or lefs on this part of it, but more efpecially Kerfey, Preftet, Ozanam, Kirkby, &c. But afterwards, mathematicians feemed to have forgot fuch queftions, if they did not even defpife them as ufelefs, when Euler drew their attention by fome excellent compofitions, demonftrating fome general theorems, which had only been known by induction. M. la Grange has alfo taken up the fubject, having refolved very difficult problems in a general way, and difcovered more direct methods than heretofore. The 2d volume of the French tranflation of Euler's Algebra contains an elementary treatife on this branch, and, with la Grange's additions, an excellent theory of it; treating very generally of Indeterminate problems, of the firft and fecond degree, of folutions in whole numbers, of the method of Indeterminate coefficients, &c.

Finally, Mr. John Leflie has given, in the 2d volume of the Edinburgh Philof. Tranfactions, an ingenious paper on the refolution of Indeterminate problems, refolving them by a new and general principle. "The doctrine of Indeterminate equations," fays Mr. Leflie, "has been feldom treated in a form equally fyftematic with the other parts of Algebra. The folutions commonly given are devoid of uniformity, and often require a variety of affumptions. The object of this paper is to refolve the complicated expreffions which we obtain in the folution of Indeterminate problems, into fimple equations, and to do fo, without framing a number of affumptions, by help of a fingle principle, which though extremely fimple, admits of a very extenfive application."

"Let A × B be any compound quantity equal to another, C × D, and let m be any rational number affumed at pleafure; it is manifeft that, taking equimultiples, A × mB = C × mD. If, therefore, we fuppofe that A = mD, it muft follow that mB = C, or B = $\dfrac{C}{m}$. Thus two equations of a lower dimenfion are obtained. If thefe be capable of farther decompofition, we may affume the multiples n and p, and form four equations ftill more fimple. By the repeated application of this principle, an higher equation admitting of divifors, will be refolved into thofe of the firft order, the number of which will be one greater than that of the multiples affumed."

For example, refuming the problem at firft given, viz, to find two rational numbers, the difference of the fquares of which fhall be a given number. Let the given number be the product of a and b; then by hypothefis, $x^2 - y^2 = ab$; but thefe compound quantities admit of an eafy refolution, for $\overline{x + y} \times \overline{x - y} = a \times b$. If therefore we fuppofe $x + y = ma$, we fhall obtain $x - y = \dfrac{b}{m}$; where m is arbitrary, and if rational, x and y muft alfo be rational. Hence the refolution of thefe two equations gives the values of x and y, the numbers fought, in terms of m; viz $x = \dfrac{m^2 a + b}{2m}$, and $y = \dfrac{m^2 a - b}{2m}$.

INDEX, in Arithmetic, is the fame with what is otherwife called the characteriftic or the exponent of a logarithm; being that which fhews of how many places the abfolute or natural number belonging to the logarithm confifts, and of what nature it is, whether an integer or a fraction; the Index being lefs by 1 than the number of integer figures in the natural number, and is pofitive for integer or whole numbers, but negative in fractions, or in the denominator of a fraction; and in decimals, the negative index is 1 more than the number of ciphers in the decimal, after the point, and before the first significant figure; or, still more generally, the Index shews how far the first figure of the natural number is distant from the place of units, either towards the left hand, as in whole numbers, or towards the right, as in decimals; these opposite cases being marked by the correspondent signs + and −, of opposite affections, the sign − being set over the Index, and not before it, because it is this Index only which is understood as negative, and not the decimal part of the logarithm. Thus, in this logarithm 2·4234097, the figures of whose natural number are 2651, the 2 is the Index, and being positive, it shews that the first figure of the number must be two places removed from the units place, or that there will be three places of integers, the number of these places being always 1 more than the Index; so that the natural number will be 265·1. But if the same Index be negative, thus 2̄·4234097, it shews that the natural number is a decimal, and that the first significant figure of it is in the 2d place from units, or that there is one cipher at the beginning of the decimal, being 1 less than the negative Index; and consequently that the natural number of the logarithm in this case is ·02651. Hence, by varying the natural number, with respect to the decimal places in it as in the former of the two columns here annexed, the Index of their logarithm will vary as in the 2d column.

Number.	Logarithm.
2651	3·4234097
265·1	2·4234097
26·51	1·4234097
2·651	0·4234097
·2651	1̄·4234097
·02651	2̄·4234097
·002651	3̄·4234097

Mr. Townly introduced a peculiar way of noting these Indices, when they become negative, or express decimal figures, which is now much in use, especially in the log. sines and tangents, &c, viz, by taking, instead of the true Index, its arithmetical complement to 10; so that, in this way, the logarithm 2̄·4234097 is written 8·4234097.

For the addition and subtraction of Indices, see LOGARITHM.

INDEX of a Globe, is a little style fitted on to the north-pole, and turning round with it, pointing out the divisions of the hour-circle.

INDEX of a Quantity, in Arithmetic and Algebra, otherwise called the exponent, is the number that shews to what power it is understood to be raised: as in 10³, or a³, the figure 3 is the Index or exponent of the power, signifying that the root or quantity, 10 or a, is raised to the 3d power. See this fully treated under EXPONENT.

INDICTION, or Roman INDICTION, a kind of epoch, or manner of counting time, among the Romans; containing a cycle or revolution of 15 years.

The popes have dated their acts by the year of the Indiction, which was fixed to the 1st of January anno Domini 313, ever since Charlemagne made them sovereign; before that time, they dated them by the years of the Emperors.

At the time of reforming the calendar, the year 1582 was reckoned the 10th year of the Indiction; so that beginning to reckon from hence, and dividing the number of years elapsed between that time and this, by 15, the remainder, with the addition of 10, rejecting 15 if the sum be more, will be the year of the Indiction.

But the Indiction will be easier found thus: Add 3 to the given year of Christ; divide the sum by 15, and the remainder after the division, will be the year of the Indiction: if there be no remainder, the Indiction is 15. In either of these ways, the Indiction for the year 1795 is 13.

INDIVISIBLES, are those indefinitely small elements, or principles, into which any body or figure may ultimately be divided.

A line is said to consist of points, a surface of parallel lines, and a solid of parallel surfaces: and because each of these elements is supposed Indivisible, if in any figure a line be drawn perpendicularly through all the elements, the number of points in that line, will be the same as the number of the elements.

Whence it appears, that a parallelogram, or a prism, or a cylinder, is resolvable into elements, as Indivisibles, all equal to each other, parallel, and like or similar to the base; for which reason, one of these elements multiplied by the number of them, that is the base of the figure multiplied by its height, gives the area or content. And a triangle is resolvable into lines parallel to the base, but decreasing in arithmetical progression; so also do the circles, which constitute the parabolic conoid, as well as those which constitute the plane of a circle, or the surface of a cone. In all which cases, as the last or least term of the arithmetic progression is 0, and the length of the figure the same thing as the number of the terms, therefore the greatest term, or base, being multiplied by the length of the figure, half the product is the sum of the whole, or the content of the figure.

And in any other figure or solid, if the law of the decrease of the elements be known, and thence the relation of the sum to the greatest term, which is the base, the whole number of them being the altitude of the figure, then the said sum of the elements is always the content of the figure.

A cylinder may also be resolved into cylindrical curve surfaces, having all the same height, and continually decreasing inwards, as the circles of the base do, on which they insist.

This way of considering magnitudes, is called the Method of Indivisibles, which is only the ancient method of exhaustions, a little disguised and contracted. And it is found of good use, both in computing the contents of figures in a very short and easy way, as above instanced, and in shortening other demonstrations in mathematics; an instance of which may here be given in that celebrated proposition of Archimedes, that a sphere is two-thirds of its circumscribed cylinder. Thus,

Suppose a cylinder, a hemisphere, and an inverted cone, having all the same base and altitude, and cut by an infinite number of planes all parallel to the base, of which EFGH is one; it is evident that the square of EI, the radius of the cylinder, is every where equal

to

to the fquare of SF, the radius of the fphere; and alfo that the fquare of EI, or of SF, is equal to the fum of the fquares of IF and IS, or of IF and IK, becaufe IK = IS; that is, IE² = IF² + IK², in every pofition; but IE is the radius of the cylinder, IF the correfponding radius of the fphere, and IK that of the cone; and the circular fections of thefe bodies, are as the fquares of their radii; therefore the fection of the cylinder is every where equal to the fum of the fections of the hemifphere and cone; and, as the number of all thofe fections, which is the common height of the figures, is the fame, therefore all the fections, or elements, of the cylinder, will be equal to the fum of all thofe of the hemifphere and cone taken together; that is, the cylinder is equal to both the hemifphere and cone: but as the cone itfelf is equal to one-third part of the cylinder; therefore the hemifphere is equal to the other two-thirds of it.

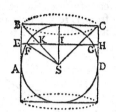

The Method of Indivifibles was introduced by Càvalerius, in 1635, in his Geometria Indivifibilium. The fame was alfo purfued by Torricelli in his works, printed 1644: and again by Cavalerius himfelf in another treatife, publifhed in 1647.

INERTIÆ *Vis.* See Vis *Inertiæ.*

INFINITE, is applied to quantities which are either greater or lefs than any affignable ones. In which fenfe it differs but little from the terms Indefinite and Indeterminate. Thus, an

INFINITE, or *Infinitely great* line, denotes only an indefinite or indeterminate line; or a line to which no certain bounds or limits are prefcribed.

INFINITE *Quantities.* Though the idea of magnitude infinitely great, or fuch as exceeds any affignable quantity, does include a negation of limits, yet all fuch magnitudes are not equal among themfelves; but befides Infinite length, and Infinite area, there are no lefs than three feveral forts of Infinite folidity; all of which are quantities *fui generis*; and thofe of each fpecies are in given proportions.

Infinite length, or a line infinitely long, may be confidered, either as beginning at a point, and fo infinitely extended one way; or elfe both ways from the fame point.

As to Infinite furface or area, any right line infinitely extended both ways on a plane infinitely extended every way, divides that plane into two equal parts, one on each fide of the line. But if from any point in fuch a plane, two right lines be infinitely extended, making an angle between them; the Infinite area, intercepted between thefe Infinite right lines, is to the whole Infinite plane, as that angle is to 4 right angles. And if two Infinite and parallel lines be drawn at a given diftance on fuch an Infinite plane, the area intercepted between them will be likewife Infinite; but yet it will be infinitely lefs than the whole plane; and even infinitely lefs than the angular or fectoral fpace, intercepted between two Infinite lines, that are inclined, though at never fo fmall an angle; becaufe in the one cafe, the given finite diftance of the parallel lines diminifhes the Infinity in one of the dimenfions; whereas

in a fector, there is Infinity in both dimenfions. And thus there are two fpecies of Infinity in furfaces, the one infinitely greater than the other.

In like manner there are fpecies of Infinites in folids, according as only one, or two, or as all their three dimenfions, are Infinite; which, though they be all infinitely greater than a finite folid, yet are they in fucceffion infinitely greater than each other.

Some farther properties of Infinite quantities are as follow:

The ratio between a finite and an Infinite quantity, is an Infinite ratio.

If a finite quantity be multiplied by an infinitely fmall one, the product will be an infinitely fmall one; but if the former be divided by the latter, the quotient will be infinitely great.

On the contrary, a finite quantity being multiplied by an infinitely great one, the product is infinitely great; but the former divided by the latter, the quotient will be infinitely little.

The product or quotient of an infinitely great or an infinitely little quantity, by a finite one, is refpectively infinitely great, or infinitely little.

An infinitely great multiplied by an infinitely little, is a finite quantity; but the former divided by the latter, the quotient is infinitely Infinite.

The mean proportional between infinitely great, and infinitely little, is finite.

Arithmetic of INFINITES. See ARITHMETIC. Alfo Wallis's treatife of this fubject; and another by Emerfon, at the beginning of his Conic Sections; alfo Bulliald's treatife *Arithmetica Infinitorum.*

INFINITE *Decimals,* fuch as do not terminate, but go on without end; as ·333 &c = $\frac{1}{3}$, or ·142857 &c = $\frac{1}{7}$. See REPETEND.

INFINITELY *Infinite Fractions,* or all the powers of the fractions whofe numerator is 1; which are all together equal to unity, as is demonftrated by Dr. Wood, in Hook's Philof. Coll. Nº 3, p. 45; where fome curious properties are deduced from the fame.

INFINITE *Series,* a feries confidered as infinitely continued as to the number of its terms. See SERIES.

INFINITESIMALS, are certain infinitely or indefinitely fmall parts; as alfo the method of computing by them.

In the method of Infinitefimals, the element by which any quantity increafes or decreafes, is fuppofed to be infinitely fmall, and is generally expreffed by two or more terms, fome of which are infinitely lefs than the reft, which being neglected as of no importance, the remaining terms form what is called the *difference* of the propofed quantity. The terms that are neglected in this manner, as infinitely lefs than the other terms of the element, are the very fame which arife in confequence of the acceleration, or retardation, of the generating motion, during the infinitely fmall time in which the element is generated; fo that the remaining terms exprefs the element that would have been produced in that time, if the generating motion had continued uniform. Therefore, thofe *differences* are accurately in the fame ratio to each other, as the generating motions or fluxions. And hence, though in this method, Infinitefimal parts of the elements are neglected, the conclufions are accurately true, without

even

even an infinitely small error, and agree precisely with those that are deduced by the method of fluxions.

But however safe and convenient this method may be, some will always scruple to admit infinitely little quantities, and infinite orders of Infinitesimals, into a science that boasts of the most evident and accurate principles, as well as of the most rigid demonstrations. In order to avoid such suppositions, Newton considers the simultaneous increments of the flowing quantities as finite, and then investigates the ratio which is the limit of the various proportions which those increments bear to each other, while he supposes them to decrease together till they vanish; which ratio is the same with the ratio of the fluxions. See Maclaurin's Treatise of Fluxions, in the Introduc. p. 39 &c, also art. 495 to 502.

INFLAMMABILITY, that property of bodies by which they kindle, or catch fire.

INFLECTION, in Optics, called also Diffraction, and Deflection of the rays of light, is a property of them, by reason of which, when they come within a certain distance of any body, they will either be bent from it, or towards it; being a kind of imperfect reflection or refraction.

Some writers ascribe the first discovery of this property to Grimaldi, who first published an account of it, in his Treatise De Lumine, Coloribus, & Iride, printed in 1666. But Dr. Hook also claims the discovery of it, and communicated his observations on this subject to the Royal Society, in 1672. He shews that this property differs both from reflection and refraction; and that it seems to depend on the unequal density of the constituent parts of the ray, by which the light is dispersed from the place of condensation, and rarefied or gradually diverged into a quadrant; and this deflection, he says, is made towards the superficies of the opaque body perpendicularly.

Newton discovered, by experiments, this Inflection of the rays of light; which may be seen in his Optics.

M. De la Hire observed, that when we look at a candle, or any luminous body, with our eyes nearly shut, rays of light are extended from it, in several directions, to a considerable distance, like the tails of comets. The true cause of this phenomenon, which has exercised the sagacity of Des Cartes, Rohault, and others, seems to be, that the light passing among the eyelashes, in this situation of the eye, is inflected by its near approach to them, and therefore enters the eye in a great variety of directions. He also observes, that he found that the beams of the stars being observed, in a deep valley, to pass near the brow of a hill, are always more refracted than if there were no such hill, or the observation was made on the top of it; as if the rays of light were bent down into a curve, by passing near the surface of the mountain.

Point of INFLECTION, or of *contrary flexure*, in a curve, is the point or place in the curve where it begins to bend or turn a contrary way; or which separates the concave part from the convex part, and lying between the two; or where the curve changes from concave to convex, or from convex to concave, on the same side of the curve: such as the point E in the annexed figures; where the former of the two is con-

cave towards the axis AD from A to E, and convex from E to F; but, on the contrary, the latter figure is convex from A to E, and concave from E to F.

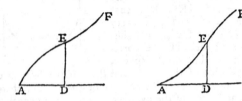

There are various ways of finding the point of Inflexion; but the following, which is new, seems to be the simplest and easiest of all. From the nature of curvature it is evident that, while a curve is concave towards an axis, the fluxion of the ordinate decreases, or is in a decreasing ratio, with regard to the fluxion of the abscis; but, on the contrary, that the said fluxion increases, or is in an increasing ratio to the fluxion of the abscis, where the curve is convex towards the axis; and hence it follows that those two fluxions are in a constant ratio at the point of Inflection, where the curve is neither concave nor convex. That is, if $x = $ AD the abscis, and $y = $ DE the ordinate, then $\dot{x}$ is to $\dot{y}$ in a constant ratio, or $\dfrac{\dot{x}}{\dot{y}}$ or $\dfrac{\dot{y}}{\dot{x}}$ is a constant quantity. But constant quantities have no fluxion, or their fluxion is equal to nothing; so that in this case the fluxion of $\dfrac{\dot{x}}{\dot{y}}$ or of $\dfrac{\dot{y}}{\dot{x}}$ is equal to nothing. And hence we have this general rule: viz,

Put the given equation of the curve into fluxions; from which equation of the fluxions find either $\dfrac{\dot{x}}{\dot{y}}$ or $\dfrac{\dot{y}}{\dot{x}}$; then take the fluxion of this ratio or fraction, and put it equal to 0 or nothing; and from this last equation find also the value of the same $\dfrac{\dot{x}}{\dot{y}}$ or $\dfrac{\dot{y}}{\dot{x}}$: then put this latter value equal to the former, which will be an equation from whence, and the first given equation of the curve, x and y will be determined, being the abscis or ordinate answering to the point of Inflection in the curve.

Or, putting the fluxion of $\dfrac{\dot{x}}{\dot{y}}$ equal to 0, that is, $\dfrac{\ddot{x}\dot{y} - \dot{x}\ddot{y}}{\dot{y}^2} = 0$, or $\ddot{x}\dot{y} - \dot{x}\ddot{y} = 0$, or $\ddot{x}\dot{y} = \dot{x}\ddot{y}$, or $\dot{x} : \dot{y} :: \ddot{x} : \ddot{y}$, that is, the 2d fluxions have the same ratio as the 1st fluxions, which is a constant ratio; and therefore if $\dot{x}$ be constant, or $\ddot{x} = 0$, then shall $\ddot{y}$ be $= 0$ also; which gives another rule, viz; Take both the 1st and 2d fluxions of the given equation of the curve, in which make both $\ddot{x}$ and $\ddot{y} = 0$, and the resulting equations will determine the values of x and y, or abscis and ordinate answering to the point of Inflection.

For example, if it be required to find the point of Inflection in the curve whose equation is

ax

$ax^2 = a^2y + x^2y$. Now the fluxion of this is $2ax\dot{x} = a^2\dot{y} + 2xy\dot{x} + x^2\dot{y}$, which gives $\dfrac{\dot{x}}{\dot{y}} = \dfrac{a^2 + x^2}{2ax - 2xy}$. Then the fluxion of this again made $= 0$, gives $2x\dot{x} \cdot ax - xy = \overline{a^2 + x^2} \cdot \overline{a\dot{x} - y\dot{x} - x\dot{y}}$; and this gives again. $\dfrac{\dot{x}}{\dot{y}} = \dfrac{a^2 + x^2}{a^2 - x^2} \times \dfrac{x}{a - y}$. Lastly, this value of $\dfrac{\dot{x}}{\dot{y}}$ put $=$ the former, gives $\dfrac{a^2 + x^2}{a^2 - x^2} \cdot \dfrac{x}{a - y} = \dfrac{a^2 + x^2}{2x} \cdot \dfrac{1}{a - y}$; and hence $2x^2 = a^2 - x^2$, or $3x^2 = a^2$, and $x = a\sqrt{\frac{1}{3}}$, the abscifs.

Hence also, from the original equation, $y = \dfrac{ax^2}{a^2 + x^2} = \dfrac{\frac{1}{3}a^3}{\frac{4}{3}a^2} = \frac{1}{4}a$, the ordinate to the point of Inflection sought.

When the curve has but one point of Inflection, it will be determined by a simple equation, as above; but when there are several points of Inflection, by the curve bending several times from the one side to the other, the resulting equation will be of a degree corresponding to them, and its roots will determine the abscisses or ordinates to the same.

Other methods of determining the points of Inflection in curves, may be seen in most books on the doctrine of fluxions.

To know whether a curve be concave or convex towards any point assigned in the axis; find the value of $\ddot{y}$ at that point; then if this value be positive, the curve will be convex towards the axis, but if it be negative, it will be concave.

INFORMED *Stars*, or INFORMES *Stellæ*, are such stars as have not been reduced into any constellation; otherwise called Sporades.—There was a great number of this kind left by the ancient astronomers; but Hevelius and some others of the moderns have provided for the greater part of them, by making new constellations.

INGINEER. See ENGINEER.

INGRESS, in Astronomy, the sun's entrance into one of the signs, especially Aries.

INNOCENTS *Day*, a feast celebrated on the 28th day of December, in commemoration of the infants murdered by Herod.

INORDINATE *Proportion*, is where the order of the terms compared, is disturbed or irregular. As, for example, in two ranks of numbers, three in each rank, viz, in one rank,   -   -   2, 3, 9, and in the other rank,   -   -   8, 24, 36, which are proportional, the former to the latter, but in a different order, viz,   -   2 : 3 :: 24 : 36, and   -   3 : 9 :: 8 : 24. then, casting out the mean terms in each rank, it is concluded that   2 : 9 :: 8 : 36, that is, the first is to the 3d in the first rank, as the first is to the 3d in the 2d rank.

INSCRIBED *Figure*, is one that has all its angular points touching the sides of another figure in which the former is said to be inscribed.

INSCRIBED *Hyperbola*, is one that lies wholly within the angle of its asymptotes; as the common or conical hyperbola doth.

INSTANT, otherwise called a Moment, an infinitely small part of duration, or in which we perceive no succession, or which takes up the time of only one idea in our mind.

It is a maxim in mechanics, that no natural effect can be produced in an Instant, or without some definite time; also that the greater the time, the greater the effect. And hence may appear the reason, why a burthen seems lighter to a person, the faster he carries it; and why, the faster a person slides or scates on the ice, the less liable it is to break, or bend.

INSULATE, or INSULATED, a term applied to a column or other edifice, which stands alone, or free and detached from any adjacent wall, &c, like an island in the sea.

INSULATED, in Electricity, is a term applied to bodies that are supported by electrics, or non-conductors; so that their communication with the earth, by conducting substances, is interrupted.

INTACTÆ, are right lines to which curves do continually approach, and yet can never meet them more usually called Asymptotes.

INTEGERS, denote whole numbers: as contradistinguished from fractions.—Integers may be considered as numbers which refer to unity, as a whole to a part.

INTEGRAL *Number*, an integer; not a fraction.

INTEGRAL *Calculus*, in the New Analysis, is the reverse of the differential calculus, and is the finding the Integral from a given differential; being similar to the inverse method of fluxions, or the finding the fluent to a given fluxion.

INTEGRANT *Parts*, in *Philosophy*, are the similar parts of a body, or parts of the same nature with the whole; as filings of iron are the Integrant parts of iron, having the same nature and properties with the bar or mass they were filed off from.

INTENSITY, or INTENSION, in Physics, is the degree or rate of the power or energy of any quality; as heat, cold, &c. The Intensity of qualities, as gravity, light, heat, &c, vary in the reciprocal ratio of the squares of the distances from the centre of the radiating quality.

INTERCALARY *Day*, denotes the odd day inserted in the leap-year. See BISSEXTILE.

INTERCEPTED *Axis*, in Conic Sections, the same with what is otherwise called the abscifs or abscissa.

INTERCOLUMNATION, or INTERCOLUMNIATION, is the space between column and column.

INTEREST, is a sum reckoned for the loan or forbearance of another sum, or principal, lent for, or due at, a certain time, according to some certain rate or proportion; being estimated usually at so much per cent. or by the 100. This forms a particular rule in Arithmetic. The highest legal Interest now allowed in England, is after the rate of 5 per cent. per annum, or the 20th part of the principal for the space of a year, and so in proportion for other times, either greater or less. Except in the case of pawn-brokers, to whom it has lately been made legal to take a higher interest, for one of the worst and most destructive purposes that can be suffered in any state.

Interest

Interest is either Simple or Compound.

Simple INTEREST, is that which is connted and allowed upon the principal only, for the whole time of forbearance.

The sum of the Principal and Interest is called the Amount.

As the Interest of any sum, for any time, is directly proportional to the principal sum and time; therefore the Interest of 1 pound for one year being multiplied by any proposed principal sum, and by the time of its forbearance, in years and parts, will be its Interest for that time. That is, if

$r =$ the rate of Interest of $1l.$ per annum,
$p =$ any principal sum lent,
$t =$ the time it is lent for, and
$a =$ the amount, or sum of principal and Interest;

then is $prt =$ the Interest of the sum p, for the time t, at the rate r; and consequently $p + prt = p \times \overline{1 + rt} = a$, the amount of the same for that time. And from this general theorem, other theorems can easily be deduced for finding any of the quantities above mentioned; which collected all together, will be as follow:

1st, $a = p + prt$ the amount,

2d, $p = \dfrac{a}{1 + rt}$ the principal,

3d, $r = \dfrac{a - p}{pt}$ the rate,

4th, $t = \dfrac{a - p}{pr}$ the t

For example, let it be required to find in what time any principal sum will double itself, at any rate of Simple Interest. In this case we must use the 1st theorem $a = p + prt$, in which the amount a must be $= 2p$ or double the principal, i. e. $p + prt = 2p$; and hence $t = \dfrac{1}{r}$; where r being the interest of $1l.$ for one year, it follows that the time of doubling at Simple Interest, is equal to the quotient of any sum divided by its Interest for one year. So that, if the rate of Interest be 5 per cent. then $\dfrac{100}{5} = 20$ is the time of doubling.

Or the 4th theorem immediately gives

$$t = \dfrac{a - p}{pr} = \dfrac{2p - p}{pr} = \dfrac{2 - 1}{r} = \dfrac{1}{r}.$$

For more readily computing the Interest on money, various Tables of numbers are calculated and formed; such as a Table of Interest of $1l.$ for any number of years, and for any number of months, or weeks, or days, &c, and at various rates of Interest.

Another Table is the following, by which may be readily found the Interest of any sum of money, from 1 to a million of pounds, for any number of days, at any rate of Interest.

Numb.	l.	s.	d.	q.	No.	l.	s.	d.	q.
1000000	2739	14	6	0·99	100	0	5	5	3·01
900000	2465	15	0	3·29	90	0	4	11	0·71
800000	2191	15	7	1·59	80	0	4	4	2·41
700000	1917	16	1	3·89	70	0	3	10	0·11
600000	1643	16	8	2·19	60	0	3	3	1·81
500000	1369	17	3	0·49	50	0	2	8	3·51
400000	1095	17	9	2·79	40	0	2	2	1·21
300000	821	18	4	1·10	30	0	1	7	2·90
200000	547	18	10	3·40	20	0	1	1	0·60
100000	273	19	5	1·70	10	0	0	6	2·30
90000	246	11	6	0·33	9	0	0	5	3·67
80000	219	3	6	2·96	8	0	0	5	1·04
70000	191	15	7	1·59	7	0	0	4	2·41
60000	164	7	8	0·22	6	0	0	3	3·78
50000	136	19	8	2·85	5	0	0	3	1·15
40000	109	11	9	1·48	4	0	0	2	2·55
30000	82	3	10	0·11	3	0	0	1	3·89
20000	54	15	10	2 74	2	0	0	1	1·26
10000	27	7	11	1·37	1	0	0	0	2·63
9000	24	13	1	3·23	0·9	0	0	0	2·37
8000	21	18	4	1·10	0·8	0	0	0	2·10
7000	19	3	6	2·96	0·7	0	0	0	1·84
6000	16	8	9	0·82	0·6	0	0	0	1·58
5000	13	13	11	2·68	0·5	0	0	0	1·32
4000	10	19	2	0·55	0·4	0	0	0	1·05
3000	8	4	4	2·41	0·3	0	0	0	0·79
2000	5	9	7	0·27	0·2	0	0	0	0·53
1000	2	14	9	2·14	0·1	0	0	0	0·26
900	2	9	3	3·12	0·09	0	0	0	0 24
800	2	3	10	0·11	0·08	0	0	0	0·21
700	1	18	4	1·10	0·07	0	0	0	0·18
600	1	12	10	2·08	0·06	0	0	0	0·16
500	1	7	4	3·07	0·05	0	0	0	0·13
400	1	1	11	0 05	0·04	0	0	0	0·11
300	0	16	5	1·04	0·03	0	0	0	0·08
200	0	10	11	2·03	0·02	0	0	0	0·05
100	0	5	5	3·01	0·01	0	0	0	0 03

The Rule for using the Table is this:

Multiply the principal by the rate, both in pounds; multiply the product by the number of days, and divide this last product by 100; then take from the Table the several sums which stand opposite the several parts of the quotient, and adding them together will give the interest required.

Ex. What is the interest of $225l. 10s.$ for 23 days, at $4\frac{1}{2}$ per cent. per annum?

princ. 225·5
rate 4·5

1014·74
75 days 23

100)23339·25

233·3925

Then in the Table

			l.	s.	d.	q.
against	200	is	0	10	11	2·03
	30	—	0	1	7	2·90
	3	—	0	0	1	3·89
	0·3	—	0	0	0	0·79
	0·09	—	0	0	0	0·24

Ans. 0 12 9 1·85 true in the last place of decimals.

Another ingenious and general method of computing

puting Intereſt, is by the following ſmall but comprehenſive Table.

A General Intereſt Table,
By which the Intereſt of any Sum, at any Rate, and for any Time, may be readily found.

Days.	3 per Cent.				3½ per Cent.				4 per Cent.				4½ per Cent.				5 per Cent.			
	l.	s.	d.	q.	l.	s.	d.	q.	l.	s.	d.	q.	l.	s.	d.	q.	l.	s.	d.	q.
1			1	3			2	1			2	2			3	0			3	1
2			3	3			4	2			5	1			6	0			6	2
3			5	3			6	3			7	3			8	3			9	3
4			7	3			9	0			10	2			11	3		1	1	0
5			9	3			11	2		1	1	1		1	2	3		1	4	1
6			11	3		1	1	3		1	3	3		1	5	3		1	7	2
7		1	1	3		1	4	0		1	6	1		1	8	3		1	11	0
8		1	3	3		1	6	1		1	9	0		1	11	3		2	2	1
9		1	5	3		1	8	2		1	11	2		2	2	2		2	5	2
10		1	7	2		1	11	0		2	2	1		2	5	2		2	8	3
20		3	3	1		3	10	0		4	4	2		4	11	1		5	5	3
30		4	11	0		5	9	0		6	6	3		7	4	2		8	2	2
40		6	6	3		7	8	0		8	9	0		9	10	1		10	11	2
50		8	2	2		9	7	0		10	11	2		12	3	3		13	8	1
60		9	10	1		11	6	0		13	1	3		14	9	2		16	5	1
70		11	6	0		13	5	0		15	4	0		17	3	1		19	2	0
80		13	1	3		15	4	0		17	6	1		19	8	3	1	1	11	0
90		14	9	2		17	3	0		19	8	2	1	2	2	1	1	4	7	3
100		16	5	1		19	2	0	1	1	11	0	1	4	8	0	1	7	4	3
200	1	12	10	2	1	18	4	1	2	3	10	0	2	9	3	3	2	14	9	2
300	2	9	3	3	2	17	6	1	3	5	9	0	3	13	11	3	4	2	2	1

N. B. This Table contains the intereſt of 100l. for all the ſeveral days in the 1ſt column, and at the ſeveral rates of 3, 3½, 4, 4½, and 5 per cent. in the other 5 columns.

To find the Intereſt of 100l. *for any other time*, as 1 year and 278 days, at 4½ per cent. Take the ſums for the ſeveral days as here below.

The Int. for 1 year		4	10	0	0
Againſt 200 ds.	is	2	9	3	3
—— 70 ds.	-	0	17	3	1
—— 8 ds.	-	0	1	11	0
Intereſt required	-	7	18	6	0

For any other Sum than 100l. Firſt find for 100l. as above, and take it ſo many times or parts as the ſum is of 100l. Thus, to find for 355l. at 4½, for 1 year and 278 days.

Firſt, 3 times the above ſum, (for 300l.)	is	-	23	15	8 1
¼ (for 50l.)	is	-	3	19	3 1
⅒ of this (for 5l.)			0	7	11 0
So for 355 it is		-	28	2	10 2

When the intereſt is required for any other rate than thoſe in the Table, it may be eaſily made out from them. So ½ of 5 is 2½, ½ of 4 is 2, ½ of 3 is 1½, ⅓ of 3 is 1, 1-6th of 3 is ½, and 1-12th of 3 is ¼. And ſo, by parts, or by adding or ſubtracting, any rate may be made out.

Compound INTEREST, called alſo *Intereſt-upon-Intereſt*, is that which is counted not only upon the principal ſum lent, but alſo for its Intereſt, as it becomes due, at the end of each ſtated time of payment.

Although it be not lawful to lend money at Compound Intereſt, yet in purchaſing annuities, penſions, &c, and taking leaſes in reverſion, it is uſual to allow Compound Intereſt to the purchaſer for his ready money; and therefore it is very neceſſary to underſtand this ſubject.

Beſides the quantities concerned in Simple Intereſt, viz, the principal p, the rate or Intereſt of 1l. for 1 year r, the amount a, and the time t, there is another quantity employed in Compound Intereſt, viz, the ratio of the rate of Intereſt, which is the amount of 1l. for 1 time of payment, and which here let be denoted by R, viz, $R = 1 + r$. Then, the particular amounts for the ſeveral times may be thus computed, viz, As 1 pound is to its amount for any time, ſo is any propoſed principal ſum to its amount for the ſame time; i. e.

1l. : R :: p : pR the 1ſt year's amount,
1l. : R :: pR : pR^2 the 2d year's amount,
1l. : R :: pR^2 : pR^3 the 3d year's amount,
and ſo on.

Therefore in general, $pR^t = a$ is the amount for the year, or t time of payment. From whence the following general theorems are deduced:

1ſt, $a = pR^t$ the amount,

2d, $p = \frac{a}{R^t}$ the principal,

3d, $R = \sqrt[t]{\frac{a}{p}}$ the ratio,

4th, $t = \frac{\text{log. of } a - \text{log. of } p}{\text{log. of R}}$ the time.

From which any one of the quantities may be found, when the reſt are given.

For example, ſuppoſe it were required to find in how many years any principal ſum will double itſelf, at any rate of Intereſt. In this caſe we muſt employ the 4th theorem, where a will be $= 2p$, and then it is $t = \frac{\text{l. } a - \text{l. } p}{\text{log. R}} = \frac{\text{l. } 2p - \text{l. } p}{\text{log. R}} = \frac{\text{log. } 2}{\text{log. R}}$.

So, if the rate of Intereſt be 5 per cent. per annum; then $R = 1 + \cdot05 = 1\cdot05$, and hence

$$t = \frac{\text{log. } 2}{\text{log. } 1\cdot05} = \frac{\cdot3010300}{\cdot0211893} = 14\cdot2067 \text{ nearly};$$

that is, any ſum doubles in 14⅕ years nearly, at the rate of 5 per cent. per annum Compound Intereſt.

Hence, and from the like queſtion in Simple Intereſt, above given, are deduced the times in which any ſum doubles itſelf, at ſeveral rates of Intereſt, both ſimple and compound: viz,

At		At Simp. Int. Years.	At Comp. Int. Years.
2		50	35·0028
2½		40	28·0701
3		33⅓	23·4498
3½		28⁴⁄₇	20·1488
4	per cent. per an.	25	17·6730
4½	Intereſt, 1l. or	22²⁄₉	15·7473
5	any other ſum	20	14·2067
6	will double in	16⅔	11·8957
7		14²⁄₇	10·2448
8		12½	9·0065
9		11⅑	8·0432
10		10	7·2725

 The

The following Table will facilitate the calculation of Compound Intereſt for any ſum, and any number of years, at various rates of Intereſt.

The Amount of 1l. in any Number of Years.						
Yrs.	3	3½	4	4½	5	6
1	1·0300	1·0350	1·0400	1·0450	1·0500	1·0600
2	1·0609	1·0712	1·0816	1·0920	1·1025	1·1236
3	1·0927	1·1087	1·1249	1·1412	1·1576	1·1910
4	1·1255	1·1475	1·1699	1·1925	1·2155	1·2625
5	1·1593	1·1877	1·2167	1·2462	1·2763	1·3382
6	1·1941	1·2293	1·2653	1·3023	1·3401	1·4185
7	1·2299	1·2723	1·3159	1·3609	1·4071	1·5036
8	1·2668	1·3168	1·3686	1·4221	1·4775	1·5939
9	1·3048	1·3629	1·4233	1·4861	1·5513	1·6895
10	1·3439	1·4106	1·4802	1·5530	1·6289	1·7909
11	1·3842	1·4600	1·5395	1·6229	1·7103	1·8983
12	1·4258	1·5111	1·6010	1·6959	1·7959	2·0122
13	1·4685	1·5640	1·6651	1·7722	1·8856	2·1329
14	1·5126	1·6187	1·7317	1·8519	1·9799	2·2609
15	1·5580	1·6753	1·8009	1·9353	2·0789	2·3966
16	1·6047	1·7340	1·8730	2·0224	2·1829	2·5404
17	1·6528	1·7947	1·9479	2·1134	2·2920	2·6928
18	1·7024	1·8575	2·0258	2·2085	2·4066	2·8543
19	1·7535	1·9225	2·1068	2·3079	2·5270	2·0256
20	1·8061	1·9898	2·1911	2·4117	2·6533	2·2071

The uſe of this Table, which contains all the powers R', to the 20th power, or the amounts of 1l. is chiefly to calculate the Intereſt, or the amount, of any principal ſum, for any time, not more than 20 years. For example, required to find to how much 523l. will amount in 15 years, at the rate of 5l. per cent. per annum Compound Intereſt.

In the Table, on the line 15 and column 5 per cent, is the amount of 1l. viz.   2·0789,
this multiplied by the principal   523,

gives the amount   1087·2647
or   -   -   -   -   1087l. 5s. 3¼d.
and therefore the Intereſt is   564l. 5s. 3¼d.

See Annuities; Diſcount; Reverſion; Smart's Tables of Intereſt; the Philoſ. Tranſ. vol. 6, p. 508; and moſt books on Arithmetic.

INTERIOR Figure, Angle of. See ANGLE.
INTERIOR Polygon. See POLYGON.
INTERIOR Talus. See TALUS.
INTERNAL Angles, are all angles made within any figure, by the ſides of it. In a triangle ABC, the two

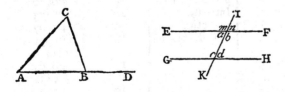

angles A and C are peculiarly called Internal and oppo-

ſite, in reſpect of the external angle CBD, which is equal to them both together.

INTERNAL Angle is alſo applied to the two angles formed between two parallels, by a line interſecting thoſe parallels, on each ſide of the interſecting line. Such are the angles a, b, c, d, formed between the parallels EF and GH, on each ſide of the interſecting line.——The two adjacent Internal angles a and b, or c and d, are together equal to two right angles.

INTERNAL and Oppoſite Angles, is alſo applied to the two angles a and b, which are reſpectively equal to the two n and m, called the external and oppoſite angles.

Alſo the alternate Internal angles are equal to one another; viz, $a = d$, and $b = c$.

INTERPOLATION, in the modern Algebra, is uſed for finding an intermediate term of a ſeries, its place in the ſeries being given.

The Method of Interpolation was firſt invented by Mr. Briggs, and applied by him to the calculation of logarithms, &c, in his Arithmetica Logarithmica, and his Trigonometria Britannica; where he explains, and fully applies the method of Interpolation by differences. His principles were followed by Reginal and Mouton in France, and by Cotes and others in England. Wallis made uſe of the method of Interpolation in various parts of his works; as his Arithmetic of Infinites, and his Algebra, for quadratures, &c. The ſame was alſo happily applied by Newton in various ways: by it he inveſtigated his binomial theorem, and quadratures of the circle, ellipſe, and hyperbola: ſee Wallis's Algebra, chap. 85, &c. Newton alſo, in lemma 5, lib. 3 Princip. gave a moſt elegant ſolution of the problem for drawing a curve line through the extremities of any number of given ordinates; and in the ſubſequent propoſition, applied the ſolution of this problem to that of finding from certain obſerved places of a comet, its place at any given intermediate time. And Dr. Waring, who adds, that a ſolution ſtill more elegant, on ſome accounts, has been ſince diſcovered by Meſſ. Nichol and Stirling, has alſo reſolved the ſame problem, and rendered it more general, without having recourſe to finding the ſucceſſive differences. Philoſ. Tranſ. vol. 69, part 1, art. 7.

Mr. Stirling indeed purſued this branch as a diſtinct ſcience, in a ſeparate treatiſe, viz, Tractatus de Summatione et Interpolatione Serierum Infinitarum, in the year 1730.

When the 1ſt, 2d, or other ſucceſſive differences of the terms of a ſeries become at laſt equal, the Interpolation of any term of ſuch a ſeries may be found by Newton's Differential Method.

When the Algebraic equation of a ſeries is given, the term required, whether it be a primary or intermediate one, may be found by the reſolution of affected equations; but when that equation is not given, as it often happens, the value of the term ſought muſt be exhibited by a converging ſeries, or by the quadrature of curves. See Stirling, ut ſupra, p. 86. Meyer, in Act. Petr. tom. 2, p. 180.

A general theorem for Interpolating any term is as follows: Let A denote any term of an equidiſtant ſeries of terms, and a, b, c, &c, the firſt of the 1ſt, 2d, 3d, &c orders of differences; then the term z, whoſe
diſtance

distance from A is expressed by x, will be this, viz,
Theorem 1,

$$z = A + xa + x \cdot \frac{x-1}{2} b + x \cdot \frac{x-1}{2} \cdot \frac{x-2}{3} c \, \&c.$$

Hence, if any of the orders of differences become equal to one another, or $= 0$, this series for the interpolated term will break off, and terminate, otherwise it will run out in an infinite series.

Ex. To find the 20th term of the series of cubes 1, 8, 27, 64, 125, &c. or 1^3, 2^3, 3^3, 4^3, 5^3, &c.

Set down the series in a column, and take their continual differences as here annexed, where the 4th differences, and all after it become $= 0$, also $A = 1$, $a = 7$, $b = 12$, $c = 6$, and $x = 19$; therefore the 20th term sought is barely

A			
	a		
1	b		
	7	c	
8	19	12	d
	37	18	0
27			6
	61	24	
64			
125			

$$z = 1 + 19 \times 7 + 19 \times \frac{18}{2} \times 12 + 19 \cdot \frac{18}{2} \cdot \frac{17}{3} \cdot 6$$

$$= 1 + 133 + 2052 + 5814 = 8000.$$

Theor. 2. In any series of equidistant terms, a, b, c, d, &c, whose first differences are small; to find any term wanting in that series, having any number of terms given. Take the equation which stands against the number of given terms, in the following Table; and by reducing the equation, that term will be found.

No.	Equations.
1	$a - b = 0$
2	$a - 2b + c = 0$
3	$a - 3b + 3c - d = 0$
4	$a - 4b + 6c - 4d + e = 0$
5	$a - 5b + 10c - 10d + 5e - f = 0$
6	$a - 6b + 15c - 20d + 15e - 6f + g = 0$
&c.	&c.
n	$a - nb + n \cdot \frac{n-1}{2} c - n \cdot \frac{n-1}{2} \cdot \frac{n-2}{3} d \, \&c = 0.$

where it is evident that the coefficients in any equation, are the unciæ of a binomial $1 + 1$ raised to the power denoted by the number of the equation.

Ex. Given the logarithms of 101, 102, 104, and 105; to find the log of 103.

Here are 4 quantities given; therefore we must take the 4th equation $a - 4b + 6c - 4d + e = 0$, in which it is the middle quantity or term c that is to be found, because 103 is in the middle among the numbers 101, 102, 104, 105; then that equation gives the value of c as follows, viz $c = \dfrac{4 \cdot \overline{b+d} - \overline{a+e}}{6}$.

Now the logs. of the given numbers will be thus:

$$\begin{aligned}
2 \cdot 0043214 &= a \\
2 \cdot 0086002 &= b \\
2 \cdot 0170333 &= d \\
2 \cdot 0211893 &= e \\
\hline
4 \cdot 0256335 &= b + d \\
4 & \\
\hline
16 \cdot 1025340 &= 4 \cdot \overline{b+d} \\
\text{subtr.} \quad 4 \cdot 0255107 &= a + e \\
\hline
6)12 \cdot 0770233 & \\
2 \cdot 0128372 \ \text{the} &\ \text{log. of 103.}
\end{aligned}$$

Theor. 3. When the terms a, b, c, d, &c, are at unequal distances from each other; to find any intermediate one of these terms, the rest being given.

Let p, q, r, s, &c, be the several distances of those terms from each other; then let

$$B = \frac{b-a}{p}, \quad C = \frac{B_1 - B}{p+q}, \quad D = \frac{C_1 - C}{p+q+r}, \quad E = \frac{D_1 - D}{p+q+r+s}.$$

$$B_1 = \frac{c-b}{q}, \quad C_1 = \frac{B_2 - B_1}{q+r}, \quad D_1 = \frac{C_2 - C_1}{q+r+s},$$

$$B_2 = \frac{d-c}{r}, \quad C_2 = \frac{B_3 - B_2}{r+s},$$

$$B_3 = \frac{e-d}{s},$$

&c. &c. &c.

Then the term z, whose distance from the beginning is x, will be

$$z = a + Bx + Cx \cdot \overline{x-p} + Dx \cdot \overline{x-p} \cdot \overline{x-p-q}$$
$$+ Ex \cdot \overline{x-p} \cdot \overline{x-p-q} \cdot \overline{x-p-q-r} + \&c;$$

to be continued to as many terms as there are terms in the given series.

By this series may be found the place of a comet, or the sun, or any other object at a given time; by knowing the places of the same for several other given times.

Other methods of Interpolation may be found in the Philos. Transf. number 362; or Stirling's Summation and Interpolation of Series.

INTERSCENDENT, in Algebra, is applied to quantities, when the exponents of their powers are radical quantities. Thus $x^{\sqrt{2}}$, $x^{\sqrt{a}}$, &c, are interscendent quantities. See FUNCTION.

INTERSECTION, the cutting of one line, or plane, by another; or the point or line in which two lines or two planes cut each other.——The mutual intersection of two planes is a right line. The centre of a circle, or conic section, &c, is in the intersection of two diameters; and the central point of a quadrangle is the Intersection of two diagonals.

INTERSTELLAR, a word used by some authors, to express those parts of the universe, that are without and beyond the limits of our solar system.

In the Interstellar regions, it is supposed there are several other systems of planets moving round the fixed stars, as the centres of their respective motions. And if it be true, as it is not improbable, that each fixed star is

thus

thus a fun to fome habitable orbs, or earths, that move round it, the Interftellar world will be infinitely the greateft part of the univerfe.

INTERTIES, or INTERDUCES, in Architecture, thofe fmall pieces of timber which lie horizontally between the fummers, or between them and the cell or raifing plate.

INTERVAL, in Mufic, the difference between two founds, in refpect of acute and grave. Authors diftinguifh feveral divifions of an Interval, as firft into Simple and Compound. The

Simple INTERVAL is that without parts, or divifion; fuch are the octave, and all that are within it; as the 2d, 3d, 4th, 5th, 6th, and 7th, with their varieties.

Compound INTERVAL confifts of feveral leffer Intervals; as the 9th, 10th, 11th, 12th, &c, with their varieties.

This Simple Interval was by the ancients called a Diaftem, and the Compound they called a Syftem.

An Interval is alfo divided into Juft or True, and into Falfe. The

Juft or *True* INTERVALS, are fuch as all thofe above mentioned, with their varieties, whether major or minor. And the

Falfe INTERVALS, are the diminutive or fuperfluous ones.

An Interval is alfo divided into the Confonance and Diffonance; which fee.

INTESTINE *Motion of the parts of fluids*, that which is among its corpufcles or component parts.

When the attracting corpufcles of any fluid are elaftic, they muft neceffarily produce an inteftine motion; and that, greater or lefs, according to the degrees of their elafticity and attractive force. For, two elaftic particles, after meeting, will fly from each other, with the fame degree of velocity with which they met; abftracting from the refiftance of the medium. But when, in leaping back from each other, they approach other particles, their velocity will be increafed.

INTRADOS, the interior and lower fide, or curve, of the arch of a bridge, &c. In contradiftinction from the extrados, or exterior curve, or line on the upper fide of the arch. See my Treatife on Bridges &c. prop. 5.

INVERSE, is applied to a manner of working the rule of three, or proportion, which feems to go backward, i. e. reverfe or contrary to the order of the common and direct rule: So that, whereas, in the direct rule, more requires more, or lefs requires lefs; in the Inverfe rule, on the contrary, more requires lefs, or lefs requires more.

For inftance, in the direct rule it is faid, If 3 yards of cloth coft 20 fhillings, how much will 6 yards coft? the anfwer is 40 fhillings: where more yards require more money, and lefs yards require lefs money. But in the Inverfe rule it is faid, If 20 men perform a piece of work in 4 days, in how many days will 40 men perform as much? where the anfwer is 2 days; and here the more men require the lefs time, and the fewer men the more time.

INVERSE *Method of Fluxions*, is the method of finding fluents, from the fluxions being given; and is fimilar to what the foreign mathematicians call the Calculus Integralis. See FLUENTS.

INVERSE *Method of Tangents*, is the method of finding the curve belonging to a given tangent; as oppofed to the direct method, or the finding the tangent to a given curve.

As, to find a curve whofe *fubtangent* is a third proportional to $r - y$ and y, or whofe *fubtangent* is equal to the femiordinate, or whofe fubnormal is a conftant quantity.——The folution of this problem depends chiefly on the Inverfe method of fluxions. See TANGENT.

INVERSE *Proportion*, or INVERSE *Ratio*, is that in which more requires lefs, or lefs requires more. As for inftance, in the cafe of light, or heat from a luminous object, the light received is lefs at a greater diftance, and greater at a lefs diftance; fo that here more, as to diftance, gives lefs, as to light, and lefs diftance gives more light. This is ufually expreffed by the term Inverfely, or Reciprocally; as in the cafe above, where the light is Inverfely, or Reciprocally as the fquare of the diftance; or in the Inverfe or Reciprocal duplicate ratio of the diftance.

INVERSION, *Invertendo*, or *by Inverfion*, according to the 14th definition of Euclid, lib. 5, is Inverting the terms of a proportion, by changing the antecedents into confequents, and the confequents into antecedents. As in thefe, $a : b :: c : d$, then by Inverfion $b : a :: d : c$.

INVESTIGATION, the fearching or finding any thing out, by means of certain fteps, traces, or ways.

INVOLUTE *Figure* or *Curve*, is that which is traced out by the outer extremity of a ftring as it is folded or wrapped upon another figure, or as it is unwound from off it.——The Involute of a cycloid, is alfo a cycloid equal to the former, which was firft difcovered by Huygens, and by means of which he contrived to make a pendulum vibrate in the curve of a cycloid, and fo theoretically at leaft vibrate always in equal times whether the arch of vibration were great or fmall, which is a property of that curve.——For the doctrine and nature of Involutes and Evolutes, fee EVOLUTE.

INVOLUTION, in Arithmetic and Algebra, is the raifing of powers from a given root; as oppofed to Evolution, which is the extracting, or developing of roots from given powers. So the Involution of the number 3, or its powers, are thus raifed:

3 - or 3^1 or 3 is the 1ft power, or root,
3 × 3 or 3^2 or 9 is the 2d power, or fquare,
3 × 3 × 3 or 3^3 or 27 is the 3d power, or cube,
and fo on.

And hence, to find any power of a given root, or quantity, let the root be multiplied by itfelf a number of times which is one lefs than the number of the index; i. e. once multiplied for the 2d root, twice for the 3d root, thrice for the 4th root, &c.

Thus, to Involve ·12 to the 3d power.

·12

·0144 fquare, or 2d power.
·12

·001728 cube, or 3d power.

So also, in Algebra, to Involve the binomial $a + b$, or raise its powers.

$$
\begin{array}{ll}
a + b & \text{1st power, or root} \\
a + b &
\end{array}
$$

$$
\begin{array}{l}
a^2 + ab \\
\quad + ab + b^2
\end{array}
$$

$$
\begin{array}{ll}
a^2 + 2ab + b^2 & \text{2d power} \\
a + b &
\end{array}
$$

$$
\begin{array}{l}
a^3 + 2a^2b + 2ab^2 \\
\quad + a^2b + ab^2 + b^3
\end{array}
$$

$$
\begin{array}{ll}
a^3 + 3a^2b + 3ab^2 + b^3 & \text{3d power} \\
a + b &
\end{array}
$$

$$
\begin{array}{l}
a^4 + 3a^3b + 3a^2b^2 + ab^3 \\
\quad + a^2b + 3a^2b^2 + 3ab^3 + b^4
\end{array}
$$

$$a^4 + 4a^3b + 6a^2b^2 + 4ab^3 + b^4 \quad \text{4th power}$$

And in like manner for any other quantities, whatever the number of their terms may be. But compound algebraic quantities are best involved by the BINOMIAL *Theorem*; which see.

Simple quantities are Involved, by raising the numeral coefficients to the given power, and the literal quantities are raised by multiplying their indices by that of the root; that is, the raising of powers is performed by the multiplication of indices, the same as the multiplication of logarithms. Thus,

The 2d power of a is a^2.

The 2d power of $2a^2$ is $2^2a^2 \times 2$ or $4a^4$

The 3d power of $3a^2b^3$ is $27a^6b^9$.

The 3d power of $a^{\frac{1}{2}}b^{\frac{2}{3}}$ is $a^{\frac{3}{2}}b^2$.

The nth power of $a^m c^p$ is $a^{mn}c^{pn}$ or $a^{m}c^{p^{n}}$

INWARD *Flanking Angle*, in Fortification, is that made by the curtin and the razant flanking line of defence.

JOINTS, in Architecture, are the separations between the stones or bricks; which may be filled with mortar, plaster, or cement.

JOINT, in Carpentry, &c, is applied to several manners of assembling, setting, or fixing pieces of wood together. As by a mortise Joint, a dove-tail Joint, &c.

Universal JOINT, in Mechanics, an excellent invention of Dr. Hook, adapted to all kinds of motions and flexures; of which he has given a large account in his Cutlerian Lectures, printed in 1678. This seems to have given occasion to the gimbols used in suspending the sea compasses; the mechanism of which is the same with that of Desaguliers's rolling lamp.

JOINT-*Lives*, are such as continue during the same time, or that exist together. See LIFE-*Annuities*.

JOISTS, or JOYSTS, those pieces of timber framed into the girders and summers, and on which the boarding of floors is laid.

" JONES (WILLIAM), F. R. S. a very eminent mathematician, was born at the foot of Bodavon mountain [Mynydd Bodafon] in the parish of Llanfihangel tre'r Bard, in the Isle of Anglesy, North Wales, in the year 1675. His father's name was John George,

his surname being the proper name of his father. For it is a custom in several parts of Wales for the proper name of the father to become the surname of his children. John George the father was commonly called Sion Siors of Llanbabo, to which place he moved, and where his children were brought up. Accordingly our author, whose proper name was William, took the surname of Jones from the proper name of his father, who was a farmer, and of a good family, being descended from Hwfa ap Cynddelw, one of the 15 tribes of North Wales. He gave his two sons the common school education of the country, reading, writing, and accounts, in English, and the Latin Grammar. Harry his second son took to the farming business; but William the eldest, having an extraordinary turn for mathematical studies, determined to try his fortune abroad from a place where the same was but of little service to him. He accordingly came to London, accompanied by a young man, Rowland Williams, afterwards an eminent perfumer in Wych-street. The report in the country is, that Mr. Jones soon got into a merchant's counting house, and so gained the esteem of his master, that he gave him the command of a ship for a West India voyage; and that upon his return he set up a mathematical school, and published his book of Navigation; and that upon the death of the merchant he married his widow: that, lord Macclesfield's son being his pupil, he was made secretary to the chancellor, and one of the deputy tellers of the exchequer:—and they have a story of an Italian wedding which caused great disturbance in lord Macclesfield's family, but was compromised by Mr. Jones; which gave rise to a saying, " that Macclesfield was the making of Jones, and Jones the making of Macclesfield." The foregoing account of Mr. Jones, I found among the papers of the late Mr. John Robertson, librarian and clerk to the Royal Society, who had been well known to Mr. Jones, and possessed many of his papers.

Mr. Jones having by his industry acquired a competent fortune, lived upon it as a private gentleman for many years, in the latter part of his life, in habits of intimacy with Sir Isaac Newton and others the most eminent mathematicians and philosophers of his time; and died July 3, 1749, at 74 years of age, being one of the vice-presidents of the Royal Society; leaving at his death one daughter, and his widow with child, which proved a son, who is the present Sir William Jones, now one of the judges in India, and highly esteemed for his great abilities, extensive learning, and eminent patriotism.——Mr. Jones's publications are,

1. *A new Compendium of the Whole Art of Navigation*, &c; in small 8vo, London, 1702. This is a neat little piece, and dedicated to the Rev. Mr. John Harris, the same I believe who was author of the *Lexicon Technicum*, or Universal Dictionary of Arts and Sciences, in whose house Mr. Jones says he composed his book.

2. *Synopsis Palmariorum Matheseos*: Or a New Introduction to the Mathematics, &c; 8vo, London, 1706. Being a very neat and useful compendium of all the mathematical sciences, in about 300 pages.

His papers in the Philos. Trans. are the following:

3. A Compendious Disposition of Equations for exhibiting the relations of Goniometrical Lines; vol. 44, p. 560.

4. A Tract on Logarithms; vol. 61, pa. 455.

5. Account of the person killed by lightning in Tottenham Court-Chapel, and its effects on the building; vol. 62, pa. 131.

6. Properties of the Conic Sections, deduced by a compendious method; vol. 63, pa. 340.

In all these works of Mr. Jones, a remarkable neatness, brevity, and accuracy, every where prevails. He seemed to delight in a very short and comprehensive mode of expression and arrangement; in so much that sometimes what he has contrived to express in two or three pages, would occupy a little volume in the ordinary style of writing.

Mr. Jones it is said possessed the best mathematical library in England; scarcely any book of that kind but what was there to be found. He had collected also a great quantity of manuscript papers and letters of former mathematicians, which have often proved useful to writers of their lives, &c. After his death, these were dispersed, and fell into different persons hands; many of them, as well as of Mr. Jones's own papers, were possessed by the late Mr. John Robertson, before mentioned, at whose death I purchased a considerable quantity of them. From such collections of these it was that Mr. Jones was enabled to give that first and elegant edition, in 4to, 1711, of several of Newton's papers, that might otherwise have been lost, intitled, *Analysis per quantitatum Series, Fluxiones, ac Differentias : cum Enumeratione Linearum Tertii Ordinis.*

IONIC *Column,* or *Order,* the 3d of the five orders, or columns, of Architecture. The first idea of this order was given by the people of Ionia; who, according to Vitruvius, formed it on the model of a young woman, dressed in her hair, and of an easy elegant shape, as the Doric had been formed on the model of a strong robust man.

This column is a medium between the massive and the more delicate orders, the simple and the rich. It is distinguished from the Composite, by having none of the leaves of acanthus in its capital; and from the Tuscan, Doric, and Corinthian, by the volutes, or rams horns, which adorn its capital; and from the Tuscan and Doric too, by the channels, or fluting, in its shaft.

The height of this column is 18 modules, or 9 diameters of the column taken at the bottom: indeed at first its height was but 16 modules; but, to render it more beautiful than the Doric, its height was augmented by adding a base to it, which was unknown in the Doric. M. le Clerc makes its entablature to be 4 modules and 10 minutes, and its pedestal 6 modules; so that the whole order makes 28 modules 10 minutes.

JOURNAL, in Merchants Accounts, is a book into which every particular article is posted out of the Waste-book, according to the order of time, specifying the debtor and creditor in each account and transaction.

JOURNAL, in Maritime Affairs, is a register kept by the pilot. and others, noticing every thing that happens to the ship, from day to day, and from hour to hour, with regard to the winds, the rhumbs or courses, the knots or rate of running, the rake, soundings, astronomical observations, for the latitudes and longitudes, &c; to enable them to adjust the reckoning, and determine the place where the ship is.

In all sea Journals, the day, or what is called the 24 hours, is divided into twice 12 hours, those before noon marked A. M. for ante meridiem, and those from noon to midnight marked P. M. post meridiem, or afternoon.

There are various ways of keeping a sea Journal, according to the different notions of mariners concerning the articles that are to be entered. Some writers direct the keeping such a kind of Journal as is only an abstract of each day's transactions, specifying the weather, what ships or lands were seen, accidents on board, the latitude, longitude, meridional distance, course, and run : these particulars are to be drawn from the ship's log-book, or from that kept by the person himself. Other authors recommend the keeping only of one account, including the log-book, and all the work of each day, with the deductions drawn from it.

JOURNAL is also used for the title of several books which come out at stated times; and give accounts and abstracts of the new books that are published, with the new improvements daily made in arts and sciences.

The first Journal of this kind was, the Journal des Sçavans, printed at Paris : the design was set on foot for the ease of such as are too busy, or too lazy, to read the entire books themselves. It seems an excellent way of satisfying a man's curiosity, and becoming learned upon easy terms : and so useful has it been found, that it has been executed in most other countries, though under a great variety of titles.

Of this kind are the Acta Eruditorum of Leipsic; the Nouvelles de la Republique des lettres of Mr. Bayle, &c; the Bibliotheque Universelle, Choisie, et Ancienne et Moderne, of M. le Clerc; the Memoirs de Trevoux, &c. In 1692, Juncker printed in Latin, An Historical Treatise of the Journals of the Learned, published in the several parts of Europe; and Wolfius, Struvius, Morhoff, Fabricius, &c, have done something of the same kind.

The Philosophical Transactions of London; the Memoirs of the Royal Academy of Sciences; those of the Academy of Belles Lettres; the Miscellanea Naturæ Curiosorum; the Experiments of the Academy del Cimento, the Acta Philo-exoticorum Naturæ et Artis, which appeared from March 1686 to April 1687, and which are a history of the Academy of Bresse; the Miscellanea Berolinensia, or Memoirs of the Academy of Berlin; the Commentaries of the Academy of Petersburgh; the Memoirs of the Institute at Bologna; the Acta Literaria Sueciæ; the Memoirs of the Royal Academy of Stockholm, begun in 1740; the Commentarii Societatis Regiæ Gottingensis, begun in 1750, &c, &c, are not so properly Journals, though they are frequently ranked in the number.

Juncker and Wolfius give the honour of the first invention of Journals to Photius. His Bibliotheca, however, is not altogether of the same nature with the modern Journals; nor was his design the same. It consists of abridgments, and extracts of books which he had read during his embassy in Persia. M. Salo first began the Journal des Sçavans at Paris, in 1665, under the name of the Sieur de Hedonville; but his death soon after interrupted the work. The abbé Gallois then took it up, and he, in the year 1674, gave way to the abbé de la Roque, who continued it nine years, and was succeeded by M. Cousin, who carried it on till the year 1702, when the abbé Bignon

instituted

inftituted a new Society, and committed the care of continuing the Journal to them, who improved and publifhed it under a new form. This Society is ftill continued, and M. de Loyer has had the infpection of the Journal; which is no longer the work of any fingle author, but of a great number.

The other French Journals are the Memoirs and Conferences of Arts and Sciences, by M. Dennis, during the years 1672, 1673, and 1674; New Difcoveries in all the parts of Phyfic, by M. de Blegny; the Journal of Phyfic, begun in 1684, and fome others, difcontinued almoft as foon as begun.

Rozier's Journal de Phyfique, begun in July 1771, and continued, till in the year 1780, there were 19 vols. quarto.

The Nouvelles de la Republique des Lettres, News from the Republic of Letters, were begun by M. Bayle in 1684, and carried on by him till the year 1687, when M. Bayle being difabled by ficknefs, his friends, M. Bernard and M. de la Roque, took them up, and continued them till 1699. After an interruption of nine years, M. Bernard refumed the work, and continued it till the year 1710. The Hiftory of the Works of the Learned, by M. Bafnage, was begun in the year 1686, and ended in 1710. The Univerfal Hiftorical Library, by M. le Clerc, was continued to the year 1693, and contained twenty-five volumes. The Bibliotheque Choifie of the fame author, began in 1703. The Mercury of France is one of the moft ancient Journals of that country, and is continued by different hands: the Memoirs of a Hiftory of Sciences and Arts, ufually called Memoires des Trevoux, from the place where they are printed, began in 1701. The Effays of Literature reached but to a twelfth volume in 1702, 1703, and 1704; thefe only take notice of ancient authors. The Journal Literaire, by Father Hugo, began and ended in 1705. At Hamburgh they have made two attempts for a French Journal, but the defign failed: an Ephemerides Sçavantes has alfo been undertaken, but that foon difappeared. A Journal des Sçavans, by M. Dartis, appeared in 1694, and was dropt the year following. That of M. Chauvin, begun at Berlin in 1696, held out three years; and an effay of the fame kind was made at Geneva. To thefe may be added, the Journal Literaire begun at the Hague 1715, and that of Verdun, and the Memoires Literaires de la Grande Bretagne by M. de la Roche; the Bibliotheque Angloife, and Journal Britannnique, which are confined to Englifh books alone. The Italian Journals are, that of abbot Nazari, which lafted from 1668 to 1681, and was printed at Rome. That of Venice began in 1671, and ended at the fame time with the other: the authors were Peter Moretti, and Francis Miletti. The Journal of Parma, by Roberti and Father Bacchini, was dropped in 1690, and refumed again in 1692. The Journal of Ferrara, by the abbé de la Torre, began and ended in 1691. La Galerio di Minerva, begun in 1696, is the work of a Society of men of letters. Seignior Apoftolo Zeno, Secretary to that Society, began another Journal in 1710, under the protection of the Grand Duke: it is printed at Venice, and feveral perfons of diftinction have a hand in it.

The Fafti Euriditi della Bibliotheca Volante, were publifhed at Parma. There has appeared fince, in Italy, the Giornale dei Letterati.

The principal among the Latin Journals, is that of Leipfic, under the title of Acta Eruditorum, begun in 1682: P. P. Manzani began another at Parma. The Nova Literaria Maris Balthici lafted from 1698 to 1708. The Nova Literaria Germaniæ, collected at Hamburgh, began in 1703. The Acta Literaria ex Manufcriptis, and the Bibliotheca curiofa, begun in 1705, and ended in 1707, are the work of Struvius. Meff. Kufter and Sike, in 1697, began a Bibliotheca Novorum Librorum, and continued it for two years. Since that time, there have been many Latin Journals; fuch, befides others, is the Commentarii de Rebus in Scientia Naturali et Medicina geftis, by M. Ludwig. The Swifs Journal, called Nova Literaria Helvetiæ, was begun in 1702, by M. Scheuchzer; and the Acta Medica Hafnenfia, publifhed by T. Bartholin, make five volumes from the year 1671 to 1679. There are two Low-Dutch Journals; the one under the title of Boockzal van Europe; it was begun at Rotterdam in 1692, by Peter Rabbus; and continued from 1702 to to 1708, by Sewel and Gavern: the other was done by a phyfician, called Ruiter, who began it in 1710. The German Journals of beft note are, the Monathlichen Unterredungen, which continued from 1689 to 1698. The Bibliotheca Curiofa, began in 1704, and ended in 1707, both by M. Tenzel. The Magazin d'Hambourg, begun in 1748: the Phyficalifche Beluftigunzen, or Philofophical Amufements, begun at Berlin in 1751. The Journal of Hanover began in 1700, and continued for two years by M. Eccard, under the direction of M. Leibnitz, and afterwards carried on by others. The Theological Journal, publifhed by M. Loefcher, under the title of Altes und Neues, that is, Old and New. A third at Leipfic and Francfort, the authors Meff. Walterck, Kraufe, and Grofchuffius; and a fourth at Hall, by M. Turk.

The Englifh Journals are, The Hiftory of the Works of the Learned, begun at London in 1699. Cenfura Temporum, in 1708. About the fame time there appeared two new ones, the one under the title of Memoirs of Literature, containing little more than an Englifh tranflation of fome articles in the foreign Journals, by M. de la Roche; the other a collection of loofe tracts, entitled, Bibliotheca curiofa, or a Mifcellany. Thefe, however, with fome others, are now no more, but are fucceeded by the Annual Regifter, which began in 1758; the New Annual Regifter, begun in 1780; the Monthly Review, which began in the year 1749, and gives a character of all Englifh literary publications, with the moft confiderable of the foreign ones: the Critical Review, which began in 1756, and is nearly on the fame plan: as alfo the London Review, by Dr. Kenrick, from 1775 to 1780; Maty's Review, from Feb. 1782 to Aug. 1786; the Englifh Review begun in Jan. 1783; and the Analytical Review, begun in May 1788, and ftill continues with much reputation. Befides thefe, we have feveral monthly pamphlets, called Magazines, which, together with a chronological Series of occurrences, contain letters from correfpondents, communicating extraordinary difcoveries in nature and art, with controverfial pieces on all fubjects. Of thefe, the principal are thofe called, the Gentleman's Magazine, *whicb* which began with the year 1731; the London Magazine, which began a few months after, and has lately been difcontinued; the Univerfal Magazine, which is nearly of as old a date.

IRIS, another name for the RAINBOW; which fee.

IRIS alfo denotes the ftriped variegated circle round the pupil of the eye, formed of a duplicature of the uvea.

In different fubjects, the Iris is of feveral very different colours; whence the eye is called grey, or black &c. In its middle is a perforation, through which appears a fmall black fpeck, called the fight, pupil, or apple of the eye, round which the Iris forms a ring.

IRIS is alfo applied to thofe changeable colours, which fometimes appear in the glaffes of telefcopes, microfcopes, &c; fo called from their fimilitude to a rainbow.

The fame appellation is alfo given to that coloured fpectrum, which a triangular prifmatic glafs will project on a wall, when placed at a proper angle in the funbeams.

IRIS *Marina*, the *Sea-Rainbow*. This elegant appearance is generally feen after a violent ftorm, in which the fea water has been in vaft emotions. The celeftial rainbow however has great advantage over the marine one, in the brightnefs and variety of the colours, and in their diftinctnefs one from the other; for in the fea-rainbow, there are fcarce any other colours than a dufky yellow on the part next the fun, and a pale green on the oppofite fide. The other colours are not fo bright or fo diftinct as to be well determined; but the fea-rainbows are more frequent and more numerous than the others: it is not uncommon to fee 20 or 30 of them at a time at noon-day.

IRRATIONAL *Numbers*, or *Quantities*, are the fame as furds, or fuch roots as cannot be accurately extracted, being incommenfurable to unity. See SURDS.

IRREDUCIBLE *Cafe*, in Algebra, is ufed for that cafe of cubic equations where the root, according to Cardan's rule, appears under an impoffible or imaginary form, and yet is real. Thus, in the equation $x^3 - 15x = 4$, the root, according to Cardan's rule, will be $x = \sqrt[3]{2 + \sqrt{-121}} + \sqrt[3]{2 - \sqrt{-121}}$, which is in the form of an impoffible expreffion, and yet it is equal to the quantity 4:

for $\sqrt[3]{2 + \sqrt{-121}} = 2 + \sqrt{-1}$,

and $\sqrt[3]{2 - \sqrt{-121}} = 2 - \sqrt{-1}$,

therefore there fum is $x = 4$.

The other two roots of the equation are alfo real.

Algebraifts, for almoft three centuries, have in vain endeavoured to refolve this cafe, and to bring it under a real form; and the problem is not lefs celebrated among them, than the fquaring of the circle is among geometricians.

It is to be obferved, that, as in fome other cafes of cubic equations, the value of the root, though rational, is found under an irrational or furd form; becaufe the root in this cafe is compounded of two equal furds with contrary figns, which deftroy each other; as if $x = 2 + \sqrt{3} + 2 - \sqrt{3}$, then $x = 4$. In like manner, in the Irreducible cafe, where the root is rational, there are two equal imaginary quantities, with contrary figns, joined to real quantities; fo that the ima-

ginary quantities deftroy each other; as in the cafe above of the root of the equation $x^3 - 15x = 4$, which was found to be $2 + \sqrt{-1} + 2 - \sqrt{-1} = 4$.

It is remarkable that this cafe always happens, viz one root, by Cardan's rule, in an impoffible form, whenever the equation has three real roots, and no impoffible ones, but at no time elfe.

If we were poffeffed of a general rule for accurately extracting the cube root of a binomial radical quantity, it is evident we might refolve the Irreducible cafe generally, which confifts of two of fuch cubic binomial roots. But the labours of the algebraifts, from Cardan's down to the prefent time, have not been able to remove this difficulty. Dr. Wallis thought that he had difcovered fuch a rule; but, like moft others, it is merely tentative, and can only fucceed in certain particular circumftances.

Mr. Maferes, curfitor baron of the exchequer, has lately deduced, by a long train of algebraical reafoning, from Newton's celebrated binomial theorem, an infinite feries, which will refolve this cafe, without any mention of either impoffible or negative quantities. And I have alfo difcovered feveral other feries which will do the fame thing, in all cafes whatever; both inferted in the Phil. Tranf. See Cardan's Algebra; the articles Algebra, Cubic Equations; Wallis's Algebra, chap. 48; De Moivre in the Appendix to Sanderfon's Algebra, p. 744; Philof. Tranf. vol. 68, part 1, art. 42, and vol. 70, p. 387.

IRREGULAR, fomething that deviates from the common forms or rules. Thus, we fay an Irregular fortification, an Irregular building, &c.

IRREGULAR *Figure*, in Geometry, whether plane or folid, is that whofe fides, as well as angles, are not all equal and fimilar among themfelves.

IRREGULARITIES *in the Moon's motion*. See MOON.

ISAGONE, in Geometry, is fometimes ufed for a figure confifting of equal angles.

ISLAND, or ISLE, a tract of dry land encompaffed by water; whether by the fea, a river, or lake, &c. In which fenfe Ifland ftands contradiftinguifhed from Continent, or terra firma; like Great Britain, Ireland, Jerfey, Sicily, Minorca, &c.

Some naturalifts imagine that Iflands were formed at the deluge: others think they have been rent and feparated from the continent by violent ftorms, inundations, and earthquakes; while others are thrown up by volcanoes, or otherwife grow or emerge from the fea.

Varenius thinks moft of thefe opinions true in fome inftances, and believes that there have been Iflands produced each of thefe ways. St. Helena, Afcenfion, and other fteep rocky Iflands, he fuppofes have become fo, by the fea's overflowing their neighbouring champaigns. By the heaping up huge quantities of fand &c, he thinks the Iflands of Zealand, Japan, &c, were formed: Sumatra and Ceylon, and moft of the Eaft India Iflands, he rather thinks were rent off from the main land, as England probably was from France. It is alfo certain that fome have emerged from the bottom of the fea; as Santorini formerly, and three other Ifles near it lately; the laft in 1707, which rofe from the bottom of the fea, after an earthquake: the ancients had a tradition that Delos rofe from the

bottom

bottom of the sea.; and Seneca obferves that the Ifland Therafia rofe out of the Ægean fea in his time, of which the mariners were eye witneffes; as they have been within thefe 10 years, in the fea between Norway and Iceland, where an Ifland has juft emerged. The late circumnavigators too have made it probable, that many of the South-fea Iflands have had their foundations, of coral rock, gradually increafing, and growing out of the fea.

ISLES, or rather Ailes, in Architecture, the fides or wings of a building.

ISOCHRONAL, or ISOCHRONOUS, is applied to fuch vibrations of a pendulum as are performed in equal times. Of which kind are all the vibrations of the fame pendulum in a cycloidal curve, and in a circle nearly, whether the arcs it defcribes be longer or fhorter; for when it defcribes a fhorter arc, it moves fo much the flower; and when a long one, proportionably fafter.

ISOCHRONAL *Line*, is that in which a heavy body is fuppofed to defcend with a uniform velocity, or without any acceleration.

Leibnitz, in the Act. Erud. Lipf. for April 1689, has a difcourfe on the Linea Ifochrona, in which he fhews, that a heavy body, with the velocity acquired by its defcent from any height, may defcend from the fame point by an infinite number of Ifochronal curves, which are all of the fame fpecies, differing from one another only in the magnitude of their parameters-(fuch as are all the quadratocubical paraboloids), and confequently fimilar to one another. He fhews alfo, how to find a line, in which a heavy body defcending, fhall recede uniformly from a given point, or approach uniformly to it.

ISOMERIA, in Algebra, a term of Vieta, denoting the freeing an equation from fractions; which is done by reducing all the fractions to one common denominator, and then multiplying each member of the equation by that common denominator, that is rejecting it out of them all.

ISOPERIMETRICAL *Figures*, are fuch as have equal perimeters, or circumferences.

It is demonftrated in geometry, that among Ifoperimetrical figures, that is always the greateft which contains the moft fides or angles. From whence it follows, that the circle is the moft capacious of all figures which have the fame perimeter with it.

That of two Ifoperimetrical triangles, which have the fame bafe, and one of them two fides equal, and the other unequal; that is the greater whofe fides are equal.

That of Ifoperimetrical figures, whofe fides are equal in number, that is the greateft which is equilateral, and equiangular. And hence arifes the folution of that popular problem, To make the hedging or walling, which will fence in a certain given quantity of land, alfo to fence in any other greater quantity of the fame. For, let x be one fide of a rectangle that will contain the quantity aa of acres; then will $\frac{aa}{x}$ be its other fide,

and double their fum, viz, $2x + \frac{2aa}{x}$, will be the

perimeter of the rectangle: let alfo bb be any greater number of acres, in the form of a fquare, then is b one fide of it, and $4b$ its perimeter, which muft be equal to that of the rectangle; and hence the equation $2x + \frac{2aa}{x} = 4b$, or $x^2 + a^2 = 2bx$, in which quadratic equation the two roots are $x = b \pm \sqrt{b^2 - a^2}$, which are the lengths of the two dimenfions of the rectangle, viz, whofe area b^2 is in any proportion lefs than the fquare a^2, of the fame perimeter. As, for example, if one fide of a fquare be 10, and one fide of a rectangle be 19, but the other only 1; fuch fquare and parallelogram will be Ifoperimetrical, viz, each perimeter 40; yet the area of the fquare is 100, and of the parallelogram only 19.

Ifoperimetrical lines and figures have greatly engaged the attention of mathematicians at all times. The 5th book of Pappus's Collections is chiefly upon this fubject; where a great variety of curious and important properties are demonftrated, both of planes and folids, fome of which were old in his time, and many new ones of his own. Indeed it feems he has here brought together into this book all the properties relating to Ifoperimetrical figures then known, and their different degrees of capacity.

The analyfis of the general problem concerning figures that, among all thofe of the fame perimeter, produce maxima and minima, was given by Mr. James Bernoulli, from computations that involve 2d and 3d fluxions. And feveral enquiries of this nature have been fince profecuted in like manner, but not always with equal fuccefs. Mr. Maclaurin, to vindicate the doctrine of fluxions from the imputation of uncertainty, or obfcurity, has illuftrated this fubject, which is confidered as one of the moft abftrufe parts of this doctrine, by giving the refolution and compofition of thefe problems by firft fluxions only; and in a manner that fuggefts a fynthetic demonftration, ferving to verify the folution. See Maclaurin's Fluxions, p. 486; Analyfis Magni problematis Ifoperimetrici Act. Erud. Lipf. 1701, p. 213; Mem. Acad. Scienc. 1705, 1706, 1718; and the works of John Bernoulli, tom. 1, p. 202, 208, 424, and tom. 2, p. 235; where is contained what he and his brother James publifhed on this problem. Mr. John Bernoulli, in his firft paper, confidered only two fmall fucceffive fides of the curve; whereas the true method of refolving this problem in general, requires the confidering three fuch fmall fides, as may be perceived by examining the two folutions.

M. Euler has alfo publifhed, on this fubject, many profound refearches, in the Peterfburg commentaries; and there was printed at Laufanne, in 1744, a pretty large work upon it, intitled, Methodus inveniendi lineas curvas, maximi minimive proprietate gaudentes: five Solutio problematis Ifoperimetrici in latiffimo fenfu accepti.

M. Cramer too, in the Berlin Memoirs for 1752, has given a paper, in which he propofes to demonftrate in general, what can be demonftrated only of regular figures in the elements of geometry, viz, that the circle is the greateft of all Ifoperimetrical figures, regular or irregular.

Oa

On this head, see also Simpson's Tracts, p. 98; and the Philos. Transf. vol. 49 and 50.

ISOSCELES *Triangle*, is a triangle that has two sides equal. In the 5th prop. of Euclid's 1st book, which prop. is usually called the Pons Asinorum, or Asses bridge, it is demonstrated, that the angles, *a* and *b*, at the base of the Isosceles triangle, are equal to each other; and that if the equal sides be produced, the two angles, *c* and *d*, below the base, will also be equal. It is also inferred, that every equilateral triangle is also equiangular.

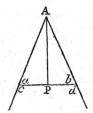

Other properties of this figure are, that the perpendicular AP, from the vertex to the base, bisects the base, the vertical angle, and also the whole triangle. And that if the vertical angles of two Isosceles triangles be equal, the two triangles will be equiangular.

ISTHMUS, in Geography, a narrow neck or slip of land, that joins two other large tracts together, and separating two seas, or two parts of the same sea.

The most remarkable Isthmuses are, that of Panama or Straits of Darien, joining north and south America; that of Suez, which connects Asia and Africa; that of Corinth, or Peloponnesus, in the Morea; that of Crim Tartary, otherwise called Taurica Chersonesus; that of the peninsula Romania and Erisso, or the Isthmus of the Thracian Chersonesus, 12 furlongs broad, and which Xerxes undertook to cut through. The Ancients had several designs of cutting the Isthmus of Corinth, which is a rocky hillock, about 10 miles over; but without effect, the invention of sluices being not then known. There have also been attempts for cutting the Isthmus of Suez, to make a communication between the Mediterranean and the Red-sea.

JUDICIAL, or JUDICIARY *Astrology*; that relating to the forming of judgments, and making prognostications. See ASTROLOGY.

JULIAN *Calendar*, is that depending on, and connected with the Julian year and account of time; so called from Julius Cæsar, by whom it was established. See CALENDAR.

JULIAN *Epoch*, is that of the institution of the Julian reformation of the calendar, which began the 46th year before Christ.

JULIAN *Period*, is a cycle of 7980 consecutive years, invented by Julius Scaliger, from whom it was named; though some say his name was Joseph Scaliger, and that it was called the Julian Period, because he made use of Julian years. This period is formed by multiplying continually together the three following cycles, viz, that of the sun of 28 years, that of the moon of 19 years, and that of the indiction of 15 years; so that this epoch, although but artificial or feigned, is yet of good use; in that every year within the period is distinguishable by a certain peculiar character; for the year of the sun, moon, and indiction, will not be the same again till the whole 7980 years have revolved. Scaliger fixed the beginning of this period 764 years before the creation, or rather the period naturally reduces to that year, taking the numbers of the three given cycles as he then found them; and accounting

3950 years from the creation to the birth of Christ, this makes the 1st year of the Christian era answer to the 4714th year of the Julian period; therefore, to find the year of this period, answering to any proposed year of Christ, to the constant number 4713, add the given year of Christ, and the sum will be the year of the Julian period: thus, to 4713 adding 1791, the sum 6504 is the year of this period for the year of Christ 1791. Hence the first revolution of the Julian period will not be completed till the year of Christ 3267, after which a new revolution of this period will commence.

But the year of this period may be found for any time, from the numbers of the three cycles that compose it, without making use of the given year of Christ, thus: multiply the

numbers $\begin{Bmatrix} 4845 \\ 4200 \\ 6916 \end{Bmatrix}$ respectively by the year of the $\begin{Bmatrix} \text{sun,} \\ \text{moon,} \\ \text{indiction;} \end{Bmatrix}$

then add the three products together, and divide the sum by 7980, so shall the remainder after division be the year of the Julian period corresponding to the given years of the other three cycles. Thus, for the year 1791, the years of the solar, lunar, and indiction cycles, are 8, 6, and 9; therefore multiplying by these, &c, according to the rule, thus

$$
\begin{array}{ccc}
4845 & 4200 & 6916 \\
8 & 6 & 9 \\
\hline
38760 & 25200 & 62244 \\
 & 38760 & \\
 & 62244 & \\
\end{array}
$$

$$7980\,)\,126204\,(\,15$$
$$7980$$
$$\overline{46404}$$
$$39900$$

remains   6504 the year of the Jul. period.

JULIAN *Year*, is the old account of the year, established by Julius Cæsar, and consisted of 365¼ days. This year continued in use in all Europe till it was superseded in most parts by the new or Gregorian account, in the year 1582. In England however it continued to be used till the year 1752, when it was abolished by act of parliament, and eleven days added to the account, to bring it up to the new style. In Russia, the old or Julian year and style are still in use.

JULY, the 7th month of the year, consisting of 31 days; about the 21st of which the sun usually enters the sign ♌ leo. This month was so named by Mark Antony, from Julius Cæsar, who was born in this month.

JUPITER, ♃, one of the superior planets, remarkable for its brightness, being the brightest of all, except sometimes the planet Venus, and is much the largest of all the planets.

Jupiter is situated between Mars and Saturn, being the 5th in order of the primary planets from the sun. His diameter is more than 10 times the diameter of the earth, and therefore his magnitude more than 1000 times.

I

times. His annual revolution about the fun, is performed in 11 years 314 days 12 hours 20 minutes 9 feconds, going at the rate of more than 25 thoufand miles per hour; and he revolves about his own axis in the fhort fpace of 9 hours 56 minutes, by which his equatorial parts are carried round at the amazing rate of 26 thoufand miles per hour, which is about 25 times fafter than the like parts of our earth revolve.

Jupiter is furrounded by faint fubftances, called zones or belts, in which fo many changes appear, that they are generally afcribed to clouds: for fome of them have been firft interrupted and broken, and then have vanifhed entirely. They have fometimes been obferved of different breadths, and afterwards have all become nearly of the fame breadth. Large fpots have been feen in thefe belts; and when a belt vanifhes, the contiguous fpots difappear with it. The broken ends of fome belts have often been obferved to revolve in the fame time with the fpots: only thofe nearer the equator in fomewhat lefs time than thofe nearer the poles; perhaps on account of the fun's greater heat near the equator, which is parallel to the belts and courfe of the fpots. Several large fpots, which appear round at one time, grow oblong by degrees, and then divide into two or three round fpots. The periodical time of the fpots near the equator is 9 hours 50 minutes, but of thofe near the poles 9 hours 56 minutes. See Dr. Smith's Optics, § 1105 and 1109.

The axis of Jupiter is fo nearly perpendicular to his orbit, that he has no fenfible change of feafons; which is a great advantage, and wifely ordered by the Author of Nature. For, if the axis of this planet were inclined any confiderable number of degrees, juft fo many degrees round each pole would in their turn be almoft 6 years together in darknefs. And, as each degree of a great circle on Jupiter contains about 706 miles, it is eafy to judge what vaft tracts of land would be rendered uninhabitable by any confiderable inclination of his axis.

The difference between the equatorial and polar diameters of Jupiter, is upwards of 6000 miles; the former being to the latter as 13 to 12: fo that his poles are more than 3000 miles nearer his centre than the equator is. This happens from his quick motion round his axis; for the fluids, together with the light particles, which they can carry or wafh away with them, recede from the poles which are at reft, towards the equator where the motion is quickeft, until there be a fufficient number accumulated to make up the deficiency of gravity loft by the centrifugal force, which always arifes from a quick motion round an axis: and when the deficiency of weight or gravity of the particles is made up by a fufficient accumulation, there is then an equilibrium, and the equatorial parts rife no higher.

Jupiter's orbit is 1° 20′ inclined to the ecliptic. The place of his aphelion 9° 10′ of ♎, the place of his afcending node 7° 29′ of ♋, and that of his fouth or defcending node 7° 29′ of ♑. The excentricity of his orbit is $\frac{1}{20}$ of his mean diftance from the fun.

The fun appears to Jupiter but the 48th part fo large as to us; and his light and heat are in the fame

fmall proportion, but compenfated by the quick returns of them, and by 4 moons, fome of them larger than our earth, which revolve about him; fo that there is fcarce any part of this huge planet but what is, during the whole night, enlightened by one or more of thefe moons, except his poles, whence only the fartheft moons can be feen, and where their light is not wanted, becaufe the fun conftantly circulates in or near the horizon, and is very probably kept in view of both poles by the refraction of Jupiter's atmofphere, which, if it be like ours, has certainly refractive power enough for that purpofe. This planet feen from its neareft moon, appears 1000 times as large as our moon does to us; increafing and waneing in all her monthly fhapes, every 42½ hours. The periods, diftances, in femidiameters of Jupiter, and angles of the orbits of thefe moons, feen from the earth are as follow:

No.	Periods round Jupiter.	Diftances.	Angles of orbits.
1	1ᵈ 18ʰ 36ᵐ	5⅔	3′ 55″
2	3 13 15	9	6 14
3	7 3 59	14⅓	9 58
4	16 18 30	25⅓	17 30

The three neareft moons of Jupiter fall into his fhadow, and are eclipfed in every revolution: but the orbit of the 4th fatellite is fo much inclined, that it paffeth by its oppofition to Jupiter, without falling into his fhadow, two years in every fix. By thefe eclipfes, aftronomers have not only difcovered that the fun's light takes up 8 minutes of time in coming to us; but have alfo by them determined the longitudes of places on this earth, with greater certainty and facility, than by any other method yet known. The outermoft of thefe fatellites will appear nearly as large as the moon does to us. See M. De la Place's Theory of Jupiter's Satellites, in the Memoires de l'Acad. and in the Connoiffance des Temps for 1792, pa. 273.

Though there be 4 primary planets below Jupiter, yet an eye placed on his furface would never perceive any of them; unlefs perhaps as fpots paffing over the fun's difc, when they happen to come between the eye and the fun.—The parallax of the fun, viewed from Jupiter, will fcarce be fenfible, being not much above 20 feconds; and the fun's apparent diameter in Jupiter, but about 6 minutes.—Dr. Gregory adds, that an aftronomer in Jupiter would eafily diftinguifh two kinds of planets, four nearer him, viz his fatellites, and two more remote, viz the fun and Saturn: the former however will fall vaftly fhort of the fun in brightnefs, notwithftanding the great difproportion in the diftances and apparent magnitude.

JURIN (Dr. JAMES), a very diftinguifhed perfon in feveral walks of literature, particularly medicine, mathematics, and philofophy, which he cultivated with equal fuccefs. He was fecretary of the Royal Society in London, as well as prefident of the College of Phyficians there, at the time of his death, which happened March 22, 1750.

Doctor Jurin was author of feveral ingenious compofitions; particularly "an Effay upon Diftinct and Indiftinct

diftinct Vifion;" printed at the end of the 2d volume of Dr. Smith's Syftem of Optics; alfo feveral controverfial papers; againft Michellotti, upon the momentum of running waters; againft Robins, upon diftinct vifion; and againft the partifans of Leibnitz, upon the forces of moving bodies; &c. His papers inferted in the Philof. Tranf. are the following:

1. On the Sufpenfion of Water in Capillary Tubes: vol. 30, p. 739.

2. Obfervations on the Motion of Running Water: p. 748.

3. On an old Roman Infcription: p. 813.

4. A Difcourfe on the Power of the Heart: p. 863 and 929.

5. On the Specific Gravity of Human Blood: p. 1000.

6. Defence of his Doctrine of the Power of the Heart againft the Objections of Dr. Keill: p. 1039.

7. On the Action of Glafs-Tubes upon Water and Quickfilver: p. 1083.

8. On the Specific Gravity of Solids when weighed in Water: vol. 31, p. 223.

9. On the Motion of Running Water, againft Michellotti: vol. 32, p. 179.

10. Remarkable Inftance of the Small-pox: vol. 32, p. 191.

11. Inoculated and Natural Small-pox compared: vol. 32, pa. 213.

12. On Meteorological Diaries: vol. 32, p. 422.

13. On the Meafure and Motion of Running Water: vol. 41, p. 5 and 65.

14. Meteorological Obfervations in Charles Town: vol. 42, p. 491.

15. On the Action of Springs: vol. 43, p. 46.

16. On the Force of Bodies in Motion: p. 423.

17. Dynamic Principles, or Meteorological Principles of Mechanics: vol. 66, p. 103.

END OF VOLUME I.